Comprehensive Natural Products Chemistry

Comprehensive Natural Products Chemistry

Editors-in-Chief

Sir Derek Barton†
Texas A&M University, USA

Koji Nakanishi
Columbia University, USA

Executive Editor

Otto Meth-Cohn
University of Sunderland, UK

Volume 3

CARBOHYDRATES AND THEIR DERIVATIVES INCLUDING
TANNINS, CELLULOSE, AND RELATED LIGNINS

Volume Editor

B. Mario Pinto
Simon Fraser University, Canada

1999

ELSEVIER

AMSTERDAM – LAUSANNE – NEW YORK – OXFORD – SHANNON – SINGAPORE – TOKYO

Elsevier Science Ltd., The Boulevard, Langford Lane, Kidlington, Oxford,
OX5 1GB, UK

First edition 1999

Library of Congress Cataloging-in-Publication Data
Comprehensive natural products chemistry / editors-in-chief, Sir Derek
Barton, Koji Nakanishi ; executive editor, Otto Meth-Cohn. -- 1st ed.
    p. cm.
  Includes index.
  Contents: v. 3. Carbohydrates and their derivatives including tannins,
cellulose, and related lignins / volume editor B. Mario Pinto
    1. Natural products. I. Barton, Derek, Sir, 1918-1998. II. Nakanishi,
Koji, 1925- . III. Meth-Cohn, Otto.
QD415.C63 1999
547.7--dc21   98-15249

British Library Cataloguing in Publication Data
Comprehensive natural products chemistry
    1. Organic compounds
    I. Barton, Sir Derek, 1918-1998 II. Nakanishi Koji III. Meth-Cohn Otto
    572.5

ISBN 0-08-042709-X (set : alk. paper)
ISBN 0-08-043155-0 (Volume 3 : alk. paper)

∞™   The paper used in this publication meets the minimum requirements of the American National Standard for
Information Sciences—Permanence of Paper for Printed Library Materials, ANSI Z39.48–1984.

Typeset by BPC Digital Data Ltd., Glasgow, UK.
Printed and bound in Great Britain by BPC Wheatons Ltd., Exeter, UK.

Contents

Introduction

For many decades, Natural Products Chemistry has been the principal driving force for progress in Organic Chemistry.

In the past, the determination of structure was arduous and difficult. As soon as computing became easy, the application of X-ray crystallography to structural determination quickly surpassed all other methods. Supplemented by the equally remarkable progress made more recently by Nuclear Magnetic Resonance techniques, determination of structure has become a routine exercise. This is even true for enzymes and other molecules of a similar size. Not to be forgotten remains the progress in mass spectrometry which permits another approach to structure and, in particular, to the precise determination of molecular weight.

There have not been such revolutionary changes in the partial or total synthesis of Natural Products. This still requires effort, imagination and time. But remarkable syntheses have been accomplished and great progress has been made in stereoselective synthesis. However, the one hundred percent yield problem is only solved in certain steps in certain industrial processes. Thus there remains a great divide between the reactions carried out in living organisms and those that synthetic chemists attain in the laboratory. Of course Nature edits the accuracy of DNA, RNA, and protein synthesis in a way that does not apply to a multi-step Organic Synthesis.

Organic Synthesis has already a significant component that uses enzymes to carry out specific reactions. This applies particularly to lipases and to oxidation enzymes. We have therefore, given serious attention to enzymatic reactions.

No longer standing in the wings, but already on-stage, are the wonderful tools of Molecular Biology. It is now clear that multi-step syntheses can be carried out in one vessel using multiple cloned enzymes. Thus, Molecular Biology and Organic Synthesis will come together to make economically important Natural Products.

From these preliminary comments it is clear that Natural Products Chemistry continues to evolve in different directions interacting with physical methods, Biochemistry, and Molecular Biology all at the same time.

This new Comprehensive Series has been conceived with the common theme of "How does Nature make all these molecules of life?" The principal idea was to organize the multitude of facts in terms of Biosynthesis rather than structure. The work is not intended to be a comprehensive listing of natural products, nor is it intended that there should be any detail about biological activity. These kinds of information can be found elsewhere.

The work has been planned for eight volumes with one more volume for Indexes. As far as we are aware, a broad treatment of the whole of Natural Products Chemistry has never been attempted before. We trust that our efforts will be useful and informative to all scientific disciplines where Natural Products play a role.

D. H. R. Barton†   K. Nakanishi   O. Meth-Cohn

Preface

It is surprising indeed that this work is the first attempt to produce a "comprehensive" overview of Natural Products beyond the student text level. However, the awe-inspiring breadth of the topic, which in many respects is still only developing, is such as to make the job daunting to anyone in the field. Fools rush in where angels fear to tread and the particular fool in this case was myself, a lifelong enthusiast and reader of the subject but with no research base whatever in the field!

Having been involved in several of the *Comprehensive* works produced by Pergamon Press, this omission intrigued me and over a period of gestation I put together a rough outline of how such a work could be written and presented it to Pergamon. To my delight they agreed that the project was worthwhile and in short measure Derek Barton was approached and took on the challenge of fleshing out this framework with alacrity. He also brought his long-standing friend and outstanding contributor to the field, Koji Nakanishi, into the team. With Derek's knowledge of the whole field, the subject was broken down into eight volumes and an outstanding team of internationally recognised Volume Editors was appointed.

We used Derek's 80th birthday as a target for finalising the work. Sadly he died just a few months before reaching this milestone. This work therefore is dedicated to the memory of Sir Derek Barton, Natural Products being the area which he loved best of all.

OTTO METH-COHN
Executive Editor

SIR DEREK BARTON

Sir Derek Barton, who was Distinguished Professor of Chemistry at Texas A&M University and holder of the Dow Chair of Chemical Invention died on March 16, 1998 in College Station, Texas of heart failure. He was 79 years old and had been Chairman of the Executive Board of Editors for Tetrahedron Publications since 1979.

Barton was considered to be one of the greatest organic chemists of the twentieth century whose work continues to have a major influence on contemporary science and will continue to do so for future generations of chemists.

Derek Harold Richard Barton was born on September 8, 1918 in Gravesend, Kent, UK and graduated from Imperial College, London with the degrees of B.Sc. (1940) and Ph.D. (1942). He carried out work on military intelligence during World War II and after a brief period in industry, joined the faculty at Imperial College. It was an early indication of the breadth and depth of his chemical knowledge that his lectureship was in physical chemistry. This research led him into the mechanism of elimination reactions and to the concept of molecular rotation difference to correlate the configurations of steroid isomers. During a sabbatical leave at Harvard in 1949–1950 he published a paper on the "Conformation of the Steroid Nucleus" (*Experientia*, 1950, **6**, 316) which was to bring him the Nobel Prize in Chemistry in 1969, shared with the Norwegian chemist, Odd Hassel. This key paper (only four pages long) altered the way in which chemists thought about the shape and reactivity of molecules, since it showed how the reactivity of functional groups in steroids depends on their axial or equatorial positions in a given conformation. Returning to the UK he held Chairs of Chemistry at Birkbeck College and Glasgow University before returning to Imperial College in 1957, where he developed a remarkable synthesis of the steroid hormone, aldosterone, by a photochemical reaction known as the Barton Reaction (nitrite photolysis). In 1978 he retired from Imperial College and became Director of the Natural Products Institute (CNRS) at Gif-sur-Yvette in France where he studied the invention of new chemical reactions, especially the chemistry of radicals, which opened up a whole new area of organic synthesis involving Gif chemistry. In 1986 he moved to a third career at Texas A&M University as Distinguished Professor of Chemistry and continued to work on novel reactions involving radical chemistry and the oxidation of hydrocarbons, which has become of great industrial importance. In a research career spanning more than five decades, Barton's contributions to organic chemistry included major discoveries which have profoundly altered our way of thinking about chemical structure and reactivity. His chemistry has provided models for the biochemical synthesis of natural products including alkaloids, antibiotics, carbohydrates, and DNA. Most recently his discoveries led to models for enzymes which oxidize hydrocarbons, including methane monooxygenase.

The following are selected highlights from his published work:

The 1950 paper which launched Conformational Analysis was recognized by the Nobel Prize Committee as the key contribution whereby the third dimension was added to chemistry. This work alone transformed our thinking about the connection between stereochemistry and reactivity, and was later adapted from small molecules to macromolecules e.g., DNA, and to inorganic complexes.

Barton's breadth and influence is illustrated in "Biogenetic Aspects of Phenol Oxidation" (*Festschr. Arthur Stoll*, 1957, 117). This theoretical work led to many later experiments on alkaloid biosynthesis and to a set of rules for *ortho-para*-phenolic oxidative coupling which allowed the predication of new natural product systems before they were actually discovered and to the correction of several erroneous structures.

In 1960, his paper on the remarkably short synthesis of the steroid hormone aldosterone (*J. Am. Chem. Soc.*, 1960, **82**, 2641) disclosed the first of many inventions of new reactions—in this case nitrite photolysis—to achieve short, high yielding processes, many of which have been patented and are used worldwide in the pharmaceutical industry.

Moving to 1975, by which time some 500 papers had been published, yet another "Barton reaction" was born—"The Deoxygenation of Secondary Alcohols" (*J. Chem. Soc. Perkin Trans. 1*, 1975, 1574), which has been very widely applied due to its tolerance of quite hostile and complex local environments in carbohydrate and nucleoside chemistry. This reaction is the chemical counterpart to ribonucleotide→ deoxyribonucleotide reductase in biochemistry and, until the arrival of the Barton reaction, was virtually impossible to achieve.

In 1985, "Invention of a New Radical Chain Reaction" involved the generation of carbon radicals from carboxylic acids (*Tetrahedron*, 1985, **41**, 3901). The method is of great synthetic utility and has been used many times by others in the burgeoning area of radicals in organic synthesis.

These recent advances in synthetic methodology were remarkable since his chemistry had virtually no precedent in the work of others. The radical methodology was especially timely in light of the significant recent increase in applications for fine chemical syntheses, and Barton gave the organic community an entrée into what will prove to be one of the most important methods of the twenty-first century. He often said how proud he was, at age 71, to receive the ACS Award for Creativity in Organic Synthesis for work published in the preceding five years.

Much of Barton's more recent work is summarized in the articles "The Invention of Chemical Reactions—The Last 5 Years" (*Tetrahedron*, 1992, **48**, 2529) and "Recent Developments in Gif Chemistry" (*Pure Appl. Chem.*, 1997, **69**, 1941).

Working 12 hours a day, Barton's stamina and creativity remained undiminished to the day of his death. The author of more than 1000 papers in chemical journals, Barton also held many successful patents. In addition to the Nobel Prize he received many honors and awards including the Davy, Copley, and Royal medals of the Royal Society of London, and the Roger Adams and Priestley Medals of the American Chemical Society. He held honorary degrees from 34 universities. He was a Fellow of the Royal Societies of London and Edinburgh, Foreign Associate of the National Academy of Sciences (USA), and Foreign Member of the Russian and Chinese Academies of Sciences. He was knighted by Queen Elizabeth in 1972, received the Légion d'Honneur (Chevalier 1972; Officier 1985) from France, and the Order of the Rising Sun from the Emperor of Japan. In his long career, Sir Derek trained over 300 students and postdoctoral fellows, many of whom now hold major positions throughout the world and include some of today's most distinguished organic chemists.

For those of us who were fortunate to know Sir Derek personally there is no doubt that his genius and work ethic were unique. He gave generously of his time to students and colleagues wherever he traveled and engendered such great respect and loyalty in his students and co-workers, that major symposia accompanied his birthdays every five years beginning with the 60th, ending this year with two celebrations just before his 80th birthday.

With the death of Sir Derek Barton, the world of science has lost a major figure, who together with Sir Robert Robinson and Robert B. Woodward, the cofounders of *Tetrahedron*, changed the face of organic chemistry in the twentieth century.

Professor Barton is survived by his wife, Judy, and by a son, William from his first marriage, and three grandchildren.

A. I. SCOTT
Texas A&M University

Reprinted from *Tetrahedron*, 1998, **54**, 8847
Photograph courtesy of Library and Information Centre, Royal Society of Chemistry. © The Nobel Foundation

Contributors to Volume 3

Dr. R. H. Atalla
University of Wisconsin, Forest Products Laboratory, 1 Gifford Pinchot Drive, Madison, WI 53705-2398, USA

Dr. M. Basu
Department of Chemistry and Biochemistry, University of Notre Dame, Notre Dame, IN 46556, USA

Professor S. Basu
Department of Chemistry and Biochemistry, University of Notre Dame, Notre Dame, IN 46556, USA

Dr. R. Bekker
Department of Chemistry, University of the Orange Free State, PO Box 339, Bloemfontein 9300, South Africa

Dr. G. P. Bolwell
Department of Biological Sciences, Royal Holloway and Bedford New College, University of London, Egham Hill, Egham, Surrey TW20 0EX, UK

Dr. I. Brockhausen
Department of Biochemistry, Hospital for Sick Children, 555 University Avenue, Toronto, ON, M5G 1X8, Canada

Dr. T. D. H. Bugg
Department of Chemistry, University of Southampton, Highfield, Southampton, SO17 1BJ, UK

Dr. S. Dastgheib
Department of Chemistry and Biochemistry, University of Notre Dame, Notre Dame, IN 46556, USA

Dr. L. B. Davin
Institute of Biological Chemistry, Washington State University, PO Box 646340, Pullman, WA 99164-6340, USA

Dr. V. Eckert
Medizinisches Zentrum für Hygiene und Medizinische Mikrobiologie, Arbeitsgruppe Parasitologie, Phillips-Universität, Robert-Koch-Strasse 17, D-35037 Marburg, Germany

Dr. A. D. Elbein
Department of Biochemistry and Molecular Biology, University of Arkansas for Medical Sciences, College of Medicine, 4301 West Markham, Slot 516, Little Rock, AR 72205, USA

Dr. D. Ferreira
National Centre for the Development of Natural Products, School of Pharmacy, University of Mississippi, University, MS 38677, USA

Dr. P. Gerold
Medizinisches Zentrum für Hygiene und Medizinische Mikrobiologie, Arbeitsgruppe Parasitologie, Phillips-Universität, Robert-Koch-Strasse 17, D-35037 Marburg, Germany

Dr. A. Gregory
Current Biology Ltd, Middlesex House, 34-42 Cleveland Street, London, W1P 6LB, UK

Dr. D. Grimmecke
Forschungszentrum Borstel, Zentrum für Medizin und Biowissenschaften, Abteilung Immunchemie und Biochemische Mikrobiologie, Parkallee 22, D-23845 Borstel, Germany

Professor G. G. Gross
Universität Ulm, Abteilung Allgemeine Botanik, D-89069 Ulm, Germany

Dr. J. W. Hawes
Biochemistry and Biotechnology Faculty, Indiana University School of Medicine, Indianapolis, IN 46202, USA

Dr. D. P. Henderson
Department of Chemistry, Duke University, Box 90346, Durham, NC 27708-0346, USA

Dr. A. Herscovics
McGill Cancer Centre, McIntyre Medical Sciences Building, 3655 Drummond Street, Montreal, PQ, H3G 1Y6, Canada

Dr. O. Holst
Forschungszentrum Borstel, Zentrum für Medizin und Biowissenschaften, Abteilung Immunchemie und Biochemische Mikrobiologie, Parkallee 22, D-23845 Borstel, Germany

Dr. D. A. Johnson
Department of Chemistry, University of Minnesota, 207 Pleasant Street SE, 139 Smith Hall, Minneapolis, MN 55455-0431, USA

Professor N. G. Lewis
Director, Institute of Biological Chemistry, Washington State University, PO Box 646340, Pullman, WA 99164-6340, USA

Dr. H.-W. Liu
Department of Chemistry, University of Minnesota, 207 Pleasant Street SE, 139 Smith Hall, Minneapolis, MN 55455-0431, USA

Dr. U. Mamat
Forschungszentrum Borstel, Zentrum für Medizin und Biowissenschaften, Abteilung Immunchemie und Biochemische Mikrobiologie, Parkallee 22, D-23845 Borstel, Germany

Dr. D. Mohnen
Complex Carbohydrate Research Center and Department of Biochemistry and Molecular Biology, University of Georgia, 220 Riverbend Road, Athens, GA 30602, USA

Dr. R. J. Molyneux
Western Regional Research Center, Agricultural Research Service, US Department of Agriculture, 800 Buchanan Street, Albany, CA 94710, USA

Dr. R. J. J. Nel
Department of Chemistry, University of the Orange Free State, PO Box 339, Bloemfontein 9300, South Africa

Professor B. M. Pinto
Department of Chemistry, Simon Fraser University, Burnaby, BC, V5A 1S6, Canada

Dr. J. Preiss
Department of Biochemistry, Michigan State University, East Lansing, MI 48824-1319, USA

Professor E. Th. Rietschel
Forschungszentrum Borstel, Zentrum für Medizin und Biowissenschaften, Abteilung Immunchemie und Biochemische Mikrobiologie, Parkallee 22, D-23845 Borstel, Germany

Professor K. Sandhoff
Institut für Organische Chemie und Biochemie der Universität Bonn, Gerhard-Domagk Strasse 1, D-53121 Bonn, Germany

Dr. S. Sarkanen
Department of Wood and Paper Science, University of Minnesota, Kaufert Laboratory, Room 106, 2004 Folwell Avenue, St. Paul, MN 55108, USA

Dr. H. Schachter
Department of Biochemistry, Hospital for Sick Children, 555 University Avenue, Toronto, ON, M5G 1X8, Canada

Dr. R. T. Schwarz
Medizinisches Zentrum für Hygiene und Medizinische Mikrobiologie, Arbeitsgruppe Parasitologie,
Phillips-Universität, Robert-Koch-Strasse 17, D-35037 Marburg, Germany

Professor U. Seydel
Forschungszentrum Borstel, Zentrum für Medizin und Biowissenschaften, Abteilung Immunchemie
and Biochemische Mikrobiologie, Parkallee 22, D-23845 Borstel, Germany

Dr. M. Sivak
Department of Biochemistry, Michigan State University, East Lansing, MI 48824-1319, USA

Dr. E. J. Toone
Department of Chemistry, Duke University, Box 90346, Durham, NC 27708-0346, USA

Dr. G. van Echten-Deckert
Institut für Organische Chemie und Biochemie der Universität Bonn, Gerhard-Domagk Strasse 1,
D-53121 Bonn, Germany

Dr. T. N. Wight
Department of Pathology, University of Washington, Seattle, WA 98195-7470, USA

Abbreviations

The most commonly used abbreviations in *Comprehensive Natural Products Chemistry* are listed below. Please note that in some instances these may differ from those used in other branches of chemistry

A	adenine
ABA	abscisic acid
Ac	acetyl
ACAC	acetylacetonate
ACTH	adrenocorticotropic hormone
ADP	adenosine 5'-diphosphate
AIBN	2,2'-azobisisobutyronitrile
Ala	alanine
AMP	adenosine 5'-monophosphate
APS	adenosine 5'-phosphosulfate
Ar	aryl
Arg	arginine
ATP	adenosine 5'-triphosphate
B	nucleoside base (adenine, cylosine, guanine, thymine or uracil)
9-BBN	9-borabicyclo[3.3.1]nonane
BOC	*t*-butoxycarbonyl (or carbo-*t*-butoxy)
BSA	*N,O*-bis(trimethylsilyl)acetamide
BSTFA	*N,O*-bis(trimethylsilyl)trifluoroacetamide
Bu	butyl
Bu^n	*n*-butyl
Bu^i	isobutyl
Bu^s	*s*-butyl
Bu^t	*t*-butyl
Bz	benzoyl
CAN	ceric ammonium nitrate
CD	cyclodextrin
CDP	cytidine 5'-diphosphate
CMP	cytidine 5'-monophosphate
CoA	coenzyme A
COD	cyclooctadiene
COT	cyclooctatetraene
Cp	η^5-cyclopentadiene
Cp*	pentamethylcyclopentadiene
12-Crown-4	1,4,7,10-tetraoxacyclododecane
15-Crown-5	1,4,7,10,13-pentaoxacyclopentadecane
18-Crown-6	1,4,7,10,13,16-hexaoxacyclooctadecane
CSA	camphorsulfonic acid
CSI	chlorosulfonyl isocyanate
CTP	cytidine 5'-triphosphate
cyclic AMP	adenosine 3',5'-cyclic monophosphoric acid
CySH	cysteine
DABCO	1,4-diazabicyclo[2.2.2]octane
DBA	dibenz[*a*,*h*]anthracene
DBN	1,5-diazabicyclo[4.3.0]non-5-ene

DBU	1,8-diazabicyclo[5.4.0]undec-7-ene
DCC	dicyclohexylcarbodiimide
DEAC	diethylaluminum chloride
DEAD	diethyl azodicarboxylate
DET	diethyl tartrate (+ or -)
DHET	dihydroergotoxine
DIBAH	diisobutylaluminum hydride
Diglyme	diethylene glycol dimethyl ether (or bis(2-methoxyethyl)ether)
DiHPhe	2,5-dihydroxyphenylalanine
Dimsyl Na	sodium methylsulfinylmethide
DIOP	2,3-*O*-isopropylidene-2,3-dihydroxy-1,4-bis(diphenylphosphino)butane
dipt	diisopropyl tartrate (+ or -)
DMA	dimethylacetamide
DMAD	dimethyl acetylenedicarboxylate
DMAP	4-dimethylaminopyridine
DME	1,2-dimethoxyethane (glyme)
DMF	dimethylformamide
DMF-DMA	dimethylformamide dimethyl acetal
DMI	1,3-dimethyl-2-imidazalidinone
DMSO	dimethyl sulfoxide
DMTSF	dimethyl(methylthio)sulfonium fluoroborate
DNA	deoxyribonucleic acid
DOCA	deoxycorticosterone acetate
EADC	ethylaluminum dichloride
EDTA	ethylenediaminetetraacetic acid
EEDQ	*N*-ethoxycarbonyl-2-ethoxy-1,2-dihydroquinoline
Et	ethyl
EVK	ethyl vinyl ketone
FAD	flavin adenine dinucleotide
Fl	flavin
FMN	flavin mononucleotide
G	guanine
GABA	4-aminobutyric acid
GDP	guanosine 5'-diphosphate
GLDH	glutamate dehydrogenase
gln	glutamine
Glu	glutamic acid
Gly	glycine
GMP	guanosine 5'-monophosphate
GOD	glucose oxidase
G-6-P	glucose-6-phosphate
GTP	guanosine 5'-triphosphate
Hb	hemoglobin
His	histidine
HMPA	hexamethylphosphoramide (or hexamethylphosphorous triamide)
Ile	isoleucine
INAH	isonicotinic acid hydrazide
IpcBH	isopinocampheylborane
Ipc$_2$BH	diisopinocampheylborane
KAPA	potassium 3-aminopropylamide
K-Slectride	potassium tri-*s*-butylborohydride

LAH	lithium aluminum hydride
LAP	leucine aminopeptidase
LDA	lithium diisopropylamide
LDH	lactic dehydrogenase
Leu	leucine
LICA	lithium isopropylcyclohexylamide
L-Selectride	lithium tri-*s*-butylborohydride
LTA	lead tetraacetate
Lys	lysine
MCPBA	*m*-chloroperoxybenzoic acid
Me	methyl
MEM	methoxyethoxymethyl
MEM-Cl	ß-methoxyethoxymethyl chloride
Met	methionine
MMA	methyl methacrylate
MMC	methyl magnesium carbonate
MOM	methoxymethyl
Ms	mesyl (or methanesulfonyl)
MSA	methanesulfonic acid
MsCl	methanesulfonyl chloride
MVK	methyl vinyl ketone
NAAD	nicotinic acid adenine dinucleotide
NAD	nicotinamide adenine dinucleotide
NADH	nicotinamide adenine dinucleotide phosphate, reduced
NBS	*N*-bromosuccinimider
NMO	*N*-methylmorpholine *N*-oxide monohydrate
NMP	*N*-methylpyrrolidone
PCBA	*p*-chlorobenzoic acid
PCBC	*p*-chlorobenzyl chloride
PCBN	*p*-chlorobenzonitrile
PCBTF	*p*-chlorobenzotrifluoride
PCC	pyridinium chlorochromate
PDC	pyridinium dichromate
PG	prostaglandin
Ph	phenyl
Phe	phenylalanine
Phth	phthaloyl
PPA	polyphosphoric acid
PPE	polyphosphate ester (or ethyl *m*-phosphate)
Pr	propyl
Pr^i	isopropyl
Pro	proline
Py	pyridine
RNA	ribonucleic acid
Rnase	ribonuclease
Ser	serine
Sia$_2$BH	disiamylborane
TAS	tris(diethylamino)sulfonium
TBAF	tetra-*n*-butylammonium fluoroborate
TBDMS	*t*-butyldimethylsilyl
TBDMS-Cl	*t*-butyldimethylsilyl chloride
TBDPS	*t*-butyldiphenylsilyl
TCNE	tetracyanoethene

TES	triethylsilyl
TFA	trifluoracetic acid
TFAA	trifluoroacetic anhydride
THF	tetrahydrofuran
THF	tetrahydrofolic acid
THP	tetrahydropyran (or tetrahydropyranyl)
Thr	threonine
TMEDA	N,N,N',N',tetramethylethylenediamine[1,2-bis(dimethylamino)ethane]
TMS	trimethylsilyl
TMS-Cl	trimethylsilyl chloride
TMS-CN	trimethylsilyl cyanide
Tol	toluene
TosMIC	tosylmethyl isocyanide
TPP	tetraphenylporphyrin
Tr	trityl (or triphenylmethyl)
Trp	tryptophan
Ts	tosyl (or p-toluenesulfonyl)
TTFA	thallium trifluoroacetate
TTN	thallium(III) nitrate
Tyr	tyrosine
Tyr-OMe	tyrosine methyl ester
U	uridine
UDP	uridine 5'-diphosphate
UMP	uridine 5'-monophosphate

Contents of All Volumes

An Historical Perspective of Natural Products Chemistry

KOJI NAKANISHI

Columbia University, New York, USA

To give an account of the rich history of natural products chemistry in a short essay is a daunting task. This brief outline begins with a description of ancient folk medicine and continues with an outline of some of the major conceptual and experimental advances that have been made from the early nineteenth century through to about 1960, the start of the modern era of natural products chemistry. Achievements of living chemists are noted only minimally, usually in the context of related topics within the text. More recent developments are reviewed within the individual chapters of the present volumes, written by experts in each field. The subheadings follow, in part, the sequence of topics presented in Volumes 1–8.

1. ETHNOBOTANY AND "NATURAL PRODUCTS CHEMISTRY"

Except for minerals and synthetic materials our surroundings consist entirely of organic natural products, either of prebiotic organic origins or from microbial, plant, or animal sources. These materials include polyketides, terpenoids, amino acids, proteins, carbohydrates, lipids, nucleic acid bases, RNA and DNA, etc. Natural products chemistry can be thought of as originating from mankind's curiosity about odor, taste, color, and cures for diseases. Folk interest in treatments for pain, for food-poisoning and other maladies, and in hallucinogens appears to go back to the dawn of humanity

For centuries China has led the world in the use of natural products for healing. One of the earliest health science anthologies in China is the Nei Ching, whose authorship is attributed to the legendary Yellow Emperor (thirtieth century BC), although it is said that the dates were backdated from the third century by compilers. Excavation of a Han Dynasty (206 BC -AD 220) tomb in Hunan Province in 1974 unearthed decayed books, written on silk, bamboo, and wood, which filled a critical gap between the dawn of medicine up to the classic Nei Ching; Book 5 of these excavated documents lists 151 medical materials of plant origin. Generally regarded as the oldest compilation of Chinese herbs is Shen Nung Pen Ts'ao Ching (Catalog of Herbs by Shen Nung), which is believed to have been revised during the Han Dynasty; it lists 365 materials. Numerous revisions and enlargements of Pen Ts'ao were undertaken by physicians in subsequent dynasties, the ultimate being the Pen Ts'ao Kang Mu (General Catalog of Herbs) written by Li Shih-Chen over a period of 27 years during the Ming Dynasty (1573–1620), which records 1898 herbal drugs and 8160 prescriptions. This was circulated in Japan around 1620 and translated, and has made a huge impact on subsequent herbal studies in Japan; however, it has not been translated into English. The number of medicinal herbs used in 1979 in China numbered 5267. One of the most famous of the Chinese folk herbs is the ginseng root *Panax ginseng*, used for health maintenance and treatment of various diseases. The active principles were thought to be the saponins called ginsenosides but this is now doubtful; the effects could well be synergistic between saponins, flavonoids, etc. Another popular folk drug, the extract of the Ginkgo tree, *Ginkgo biloba* L., the only surviving species of the Paleozoic era (250 million years ago) family which became extinct during the last few million years, is mentioned in the Chinese Materia Medica to have an effect in improving memory and sharpening mental alertness. The main constituents responsible for this are now understood to be ginkgolides and flavonoids, but again not much else is known. Clarifying the active constituents and mode of (synergistic) bioactivity of Chinese herbs is a challenging task that has yet to be fully addressed.

The Assyrians left 660 clay tablets describing 1000 medicinal plants used around 1900–400 BC, but the best insight into ancient pharmacy is provided by the two scripts left by the ancient Egyptians, who were masters of human anatomy and surgery because of their extensive mummification practices. The Edwin Smith Surgical Papyrus purchased by Smith in 1862 in Luxor (now in the New York Academy of Sciences collection), is one of the most important medicinal documents of the ancient Nile Valley, and describes the healer's involvement in surgery, prescription, and healing practices using plants, animals, and minerals. The Ebers Papyrus, also purchased by Edwin Smith in 1862, and then acquired by Egyptologist George Ebers in 1872, describes 800 remedies using plants, animals, minerals, and magic. Indian medicine also has a long history, possibly dating back to the second millennium BC. The Indian materia medica consisted mainly of vegetable drugs prepared from plants but also used animals, bones, and minerals such as sulfur, arsenic, lead, copper sulfate, and gold. Ancient Greece inherited much from Egypt, India, and China, and underwent a gradual transition from magic to science. Pythagoras (580–500 BC) influenced the medical thinkers of his time, including Aristotle (384–322 BC), who in turn affected the medical practices of another influential Greek physician Galen (129–216). The Iranian physician Avicenna (980–1037) is noted for his contributions to Aristotelian philosophy and medicine, while the German-Swiss physician and alchemist Paracelsus (1493–1541) was an early champion who established the role of chemistry in medicine.

The rainforests in Central and South America and Africa are known to be particularly abundant in various organisms of interest to our lives because of their rich biodiversity, intense competition, and the necessity for self-defense. However, since folk-treatments are transmitted verbally to the next generation via shamans who naturally have a tendency to keep their plant and animal sources confidential, the recipes tend to get lost, particularly with destruction of rainforests and the encroachment of "civilization." Studies on folk medicine, hallucinogens, and shamanism of the Central and South American Indians conducted by Richard Schultes (Harvard Botanical Museum, emeritus) have led to renewed activity by ethnobotanists, recording the knowledge of shamans, assembling herbaria, and transmitting the record of learning to the village.

Extracts of toxic plants and animals have been used throughout the world for thousands of years for hunting and murder. These include the various arrow poisons used all over the world. *Strychnos* and *Chondrodendron* (containing strychnine, etc.) were used in South America and called "curare," *Strophanthus* (strophantidine, etc.) was used in Africa, the latex of the upas tree *Antiaris toxicaria* (cardiac glycosides) was used in Java, while *Aconitum napellus*, which appears in Greek mythology (aconitine) was used in medieval Europe and Hokkaido (by the Ainus). The Colombian arrow poison is from frogs (batrachotoxins; 200 toxins have been isolated from frogs by B. Witkop and J. Daly at NIH). Extracts of *Hyoscyamus niger* and *Atropa belladonna* contain the toxic tropane alkaloids, for example hyoscyamine, belladonnine, and atropine. The belladonna berry juice (atropine) which dilates the eye pupils was used during the Renaissance by ladies to produce doe-like eyes (belladona means beautiful woman). The Efik people in Calabar, southeastern Nigeria, used extracts of the calabar bean known as esere (physostigmine) for unmasking witches. The ancient Egyptians and Chinese knew of the toxic effect of the puffer fish, fugu, which contains the neurotoxin tetrodotoxin (Y. Hirata, K. Tsuda, R. B. Woodward).

When rye is infected by the fungus *Claviceps purpurea*, the toxin ergotamine and a number of ergot alkaloids are produced. These cause ergotism or the "devil's curse," "St. Anthony's fire," which leads to convulsions, miscarriages, loss of arms and legs, dry gangrene, and death. Epidemics of ergotism occurred in medieval times in villages throughout Europe, killing tens of thousands of people and livestock; Julius Caesar's legions were destroyed by ergotism during a campaign in Gaul, while in AD 994 an estimated 50,000 people died in an epidemic in France. As recently as 1926, a total of 11,000 cases of ergotism were reported in a region close to the Urals. It has been suggested that the witch hysteria that occurred in Salem, Massachusetts, might have been due to a mild outbreak of ergotism. Lysergic acid diethylamide (LSD) was first prepared by A. Hofmann, Sandoz Laboratories, Basel, in 1943 during efforts to improve the physiological effects of the ergot alkaloids when he accidentally inhaled it. "On Friday afternoon, April 16, 1943," he wrote, "I was seized by a sensation of restlessness... ." He went home from the laboratory and "perceived an uninterrupted stream of fantastic dreams ... ." (*Helvetica Chimica Acta*).

Numerous psychedelic plants have been used since ancient times, producing visions, mystical fantasies (cats and tigers also seem to have fantasies?, see nepetalactone below), sensations of flying, glorious feelings in warriors before battle, etc. The ethnobotanists Wasson and Schultes identified "ololiqui," an important Aztec concoction, as the seeds of the morning glory *Rivea corymbosa* and gave the seeds to Hofmann who found that they contained lysergic acid amides similar to but less potent than LSD. Iboga, a powerful hallucinogen from the root of the African shrub *Tabernanthe iboga*, is used by the Bwiti cult in Central Africa who chew the roots to obtain relief from fatigue and hunger; it contains the alkaloid ibogamine. The powerful hallucinogen used for thousands of years by the American Indians, the peyote cactus, contains mescaline and other alkaloids. The Indian hemp plant, *Cannabis sativa*, has been used for making rope since 3000 BC, but when it is used for its pleasure-giving effects it is called cannabis and has been known in central Asia, China, India, and the Near East since ancient times. Marijuana, hashish (named after the Persian founder of the Assassins of the eleventh century, Hasan-e Sabbah), charas, ghanja, bhang, kef, and dagga are names given to various preparations of the hemp plant. The constituent responsible for the mind-altering effect is 1-tetrahydrocannabinol (also referred to as 9-THC) contained in 1%. R. Mechoulam (1930–, Hebrew University) has been the principal worker in the cannabinoids, including structure determination and synthesis of 9-THC (1964 to present); the Israeli police have also made a contribution by providing Mechoulam with a constant supply of marijuana. Opium (morphine) is another ancient drug used for a variety of pain-relievers and it is documented that the Sumerians used poppy as early as 4000 BC; the narcotic effect is present only in seeds before they are fully formed. The irritating secretion of the blister beetles, for example *Mylabris* and the European species *Lytta vesicatoria*, commonly called Spanish fly, was used medically as a topical skin irritant to remove warts but was also a major ingredient in so-called love potions (constituent is cantharidin, stereospecific synthesis in 1951, G. Stork, 1921–; prep. scale high-pressure Diels–Alder synthesis in 1985, W. G. Dauben, 1919–1996).

Plants have been used for centuries for the treatment of heart problems, the most important being the foxgloves *Digitalis purpurea* and *D. lanata* (digitalin, diginin) and *Strophanthus gratus* (ouabain). The bark of cinchona *Cinchona officinalis* (called quina-quina by the Indians) has been used widely among the Indians in the Andes against malaria, which is still one of the major infectious diseases; its most important alkaloid is quinine. The British protected themselves against malaria during the occupation of India through gin and tonic (quinine!). The stimulant coca, used by the Incas around the tenth century, was introduced into Europe by the conquistadors; coca beans are also commonly chewed in West Africa. Wine making was already practiced in the Middle East 6000–8000 years ago; Moors made date wines, the Japanese rice wine, the Vikings honey mead, the Incas maize chicha. It is said that the Babylonians made beer using yeast 5000–6000 years ago. As shown above in parentheses, alkaloids are the major constituents of the herbal plants and extracts used for centuries, but it was not until the early nineteenth century that the active principles were isolated in pure form, for example morphine (1816), strychnine (1817), atropine (1819), quinine (1820), and colchicine (1820). It was a century later that the structures of these compounds were finally elucidated.

2. DAWN OF ORGANIC CHEMISTRY, EARLY STRUCTURAL STUDIES, MODERN METHODOLOGY

The term "organic compound" to define compounds made by and isolated from living organisms was coined in 1807 by the Swedish chemist Jons Jacob Berzelius (1779–1848), a founder of today's chemistry, who developed the modern system of symbols and formulas in chemistry, made a remarkably accurate table of atomic weights and analyzed many chemicals. At that time it was considered that organic compounds could not be synthesized from inorganic materials *in vitro*. However, Friedrich Wöhler (1800–1882), a medical doctor from Heidelberg who was starting his chemical career at a technical school in Berlin, attempted in 1828 to make "ammonium cyanate," which had been assigned a wrong structure, by heating the two inorganic salts potassium cyanate and ammonium sulfate; this led to the unexpected isolation of white crystals which were identical to the urea from urine, a typical organic compound. This well-known incident marked the beginning of organic chemistry. With the preparation of acetic acid from inorganic material in 1845 by Hermann Kolbe (1818–1884) at Leipzig, the myth surrounding organic compounds, in which they were associated with some vitalism was brought to an end and organic chemistry became the chemistry of carbon compounds. The same Kolbe was involved in the development of aspirin, one of the earliest and most important success stories in natural products chemistry. Salicylic acid from the leaf of the wintergreen plant had long been used as a pain reliever, especially in treating arthritis and gout. The inexpensive synthesis of salicylic acid from sodium phenolate and carbon dioxide by Kolbe in 1859 led to the industrial production in 1893 by the Bayer Company of acetylsalicylic acid "aspirin," still one of the most popular drugs. Aspirin is less acidic than salicylic acid and therefore causes less irritation in the mouth, throat, and stomach. The remarkable mechanism of the anti-inflammatory effect of aspirin was clarified in 1974 by John Vane (1927–) who showed that it inhibits the biosynthesis of prostaglandins by irreversibly acetylating a serine residue in prostaglandin synthase. Vane shared the 1982 Nobel Prize with Bergström and Samuelsson who determined the structure of prostaglandins (see below).

In the early days, natural products chemistry was focused on isolating the more readily available plant and animal constituents and determining their structures. The course of structure determination in the 1940s was a complex, indirect process, combining evidence from many types of experiments. The first effort was to crystallize the unknown compound or make derivatives such as esters or 2,4-dinitrophenylhydrazones, and to repeat recrystallization until the highest and sharp melting point was reached, since prior to the advent of isolation and purification methods now taken for granted, there was no simple criterion for purity. The only chromatography was through special grade alumina (first used by M. Tswett in 1906, then reintroduced by R. Willstätter). Molecular weight estimation by the Rast method which depended on melting point depression of a sample/camphor mixture, coupled with Pregl elemental microanalysis (see below) gave the molecular formula. Functionalities such as hydroxyl, amino, and carbonyl groups were recognized on the basis of specific derivatization and crystallization, followed by redetermination of molecular formula; the change in molecular composition led to identification of the functionality. Thus, sterically hindered carbonyls, for example the 11-keto group of cortisone, or tertiary hydroxyls, were very difficult to pinpoint, and often had to depend on more searching experiments. Therefore, an entire paper describing the recognition of a single hydroxyl group in a complex natural product would occasionally appear in the literature. An oxygen function suggested from the molecular formula but left unaccounted for would usually be assigned to an ether.

Determination of C-methyl groups depended on Kuhn–Roth oxidation which is performed by drastic oxidation with chromic acid/sulfuric acid, reduction of excess oxidant with hydrazine, neutralization with alkali, addition of phosphoric acid, distillation of the acetic acid originating from the C-methyls, and finally its titration with alkali. However, the results were only approximate, since *gem*-dimethyl groups only yield one equivalent of acetic acid, while primary, secondary, and tertiary methyl groups all give different yields of acetic acid. The skeletal structure of polycyclic compounds were frequently deduced on the basis of dehydrogenation reactions. It is therefore not surprising that the original steroid skeleton put forth by Wieland and Windaus in 1928, which depended a great deal on the production of chrysene upon Pd/C dehydrogenation, had to be revised in 1932 after several discrepancies were found (they received the Nobel prizes in 1927 and 1928 for this "extraordinarily difficult structure determination," see below).

In the following are listed some of the Nobel prizes awarded for the development of methodologies which have contributed critically to the progress in isolation protocols and structure determination. The year in which each prize was awarded is preceded by "Np."

Fritz Pregl, 1869–1930, Graz University, Np 1923. Invention of carbon and hydrogen microanalysis. Improvement of Kuhlmann's microbalance enabled weighing at an accuracy of 1 µg over a 20 g range, and refinement of carbon and hydrogen analytical methods made it possible to perform analysis with 3–4 mg of sample. His microbalance and the monograph *Quantitative Organic Microanalysis* (1916) profoundly influenced subsequent developments in practically all fields of chemistry and medicine.

The Svedberg, 1884–1971, Uppsala, Np 1926. Uppsala was a center for quantitative work on colloids for which the prize was awarded. His extensive study on ultracentrifugation, the first paper of which was published in the year of the award, evolved from a spring visit in 1922 to the University of Wisconsin. The ultracentrifuge together with the electrophoresis technique developed by his student Tiselius, have profoundly influenced subsequent progress in molecular biology and biochemistry.

Arne Tiselius, 1902–1971, Ph.D. Uppsala (T. Svedberg), Uppsala, Np 1948. Assisted by a grant from the Rockefeller Foundation, Tiselius was able to use his early electrophoresis instrument to show four bands in horse blood serum, alpha, beta and gamma globulins in addition to albumin; the first paper published in 1937 brought immediate positive responses.

Archer Martin, 1910–, Ph.D. Cambridge; Medical Research Council, Mill Hill, and Richard Synge, 1914–1994, Ph.D. Cambridge; Rowett Research Institute, Food Research Institute, Np 1952. They developed chromatography using two immiscible phases, gas–liquid, liquid–liquid, and paper chromatography, all of which have profoundly influenced all phases of chemistry.

Frederick Sanger, 1918–, Ph.D. Cambridge (A. Neuberger), Medical Research Council, Cambridge, Np 1958 and 1980. His confrontation with challenging structural problems in proteins and nucleic acids led to the development of two general analytical methods, 1,2,4-fluorodinitrobenzene (DNP) for tagging free amino groups (1945) in connection with insulin sequencing studies, and the dideoxynucleotide method for sequencing DNA (1977) in connection with recombinant DNA. For the latter he received his second Np in chemistry in 1980, which was shared with Paul Berg (1926–, Stanford University) and Walter Gilbert (1932–, Harvard University) for their contributions, respectively, in recombinant DNA and chemical sequencing of DNA. The studies of insulin involved usage of DNP for tagging disulfide bonds as cysteic acid residues (1949), and paper chromatography introduced by Martin and Synge 1944. That it was the first elucidation of any protein structure lowered the barrier for future structure studies of proteins.

Stanford Moore, 1913–1982, Ph.D. Wisconsin (K. P. Link), Rockefeller, Np 1972; and William Stein, 1911–1980, Ph.D. Columbia (E. G. Miller); Rockefeller, Np 1972. Moore and Stein cooperatively developed methods for the rapid quantification of protein hydrolysates by combining partition chromatography, ninhydrin coloration, and drop-counting fraction collector, i.e., the basis for commercial amino acid analyzers, and applied them to analysis of the ribonuclease structure.

Bruce Merrifield, 1921–, Ph.D. UCLA (M. Dunn), Rockefeller, Np 1984. The concept of solid-phase peptide synthesis using porous beads, chromatographic columns, and sequential elongation of peptides and other chains revolutionized the synthesis of biopolymers.

High-performance liquid chromatography (HPLC), introduced around the mid-1960s and now coupled on-line to many analytical instruments, for example UV, FTIR, and MS, is an indispensable daily tool found in all natural products chemistry laboratories.

3. STRUCTURES OF ORGANIC COMPOUNDS, NINETEENTH CENTURY

The discoveries made from 1848 to 1874 by Pasteur, Kekulé, van't Hoff, Le Bel, and others led to a revolution in structural organic chemistry. Louis Pasteur (1822–1895) was puzzled about why the potassium salt of tartaric acid (deposited on wine casks during fermentation) was dextrorotatory while the sodium ammonium salt of racemic acid (also deposited on wine casks) was optically inactive although both tartaric acid and "racemic" acid had identical chemical compositions. In 1848, the 25 year old Pasteur examined the racemic acid salt under the microscope and found two kinds of crystals exhibiting a left- and right-hand relation. Upon separation of the left-handed and right-handed crystals, he found that they rotated the plane of polarized light in opposite directions. He had thus performed his famous resolution of a racemic mixture, and had demonstrated the phenomenon of chirality. Pasteur went on to show that the racemic acid formed two kinds of salts with optically active bases such as quinine; this was the first demonstration of diastereomeric resolution. From this work Pasteur concluded that tartaric acid must have an element of asymmetry within the molecule itself. However, a three-dimensional understanding of the enantiomeric pair was only solved 25 years later (see below). Pasteur's own interest shifted to microbiology where he made the crucial discovery of the involvement of "germs" or microorganisms in various processes and proved that yeast induces alcoholic fermentation, while other microorganisms lead to diseases; he thus saved the wine industries of France, originated the process known as "pasteurization," and later developed vaccines for rabies. He was a genius who made many fundamental discoveries in chemistry and in microbiology.

The structures of organic compounds were still totally mysterious. Although Wöhler had synthesized urea, an isomer of ammonium cyanate, in 1828, the structural difference between these isomers was not known. In 1858 August Kekulé (1829–1896; studied with André Dumas and C. A. Wurtz in Paris, taught at Ghent, Heidelberg, and Bonn) published his famous paper in Liebig's *Annalen der Chemie* on the structure of carbon, in which he proposed that carbon atoms could form C–C bonds with hydrogen and other atoms linked to them; his dream on the top deck of a London bus led him to this concept. It was Butlerov who introduced the term "structure theory" in 1861. Further, in 1865 Kekulé conceived the cyclo-hexa-1:3:5-triene structure for benzene (C_6H_6) from a dream of a snake biting its own tail. In 1874, two young chemists, van't Hoff (1852–1911, Np 1901) in Utrecht, and Le Bel (1847–1930) in Paris, who had met in 1874 as students of C. A. Wurtz, published the revolutionary three-dimensional (3D) structure of the tetrahedral carbon Cabcd to explain the enantiomeric behavior of Pasteur's salts. The model was welcomed by J. Wislicenus (1835–1902, Zürich, Würzburg, Leipzig) who in 1863 had demonstrated the enantiomeric nature of the two lactic acids found by Scheele in sour milk (1780) and by Berzelius in muscle tissue (1807). This model, however, was criticized by Hermann Kolbe (1818–1884, Leipzig) as an "ingenious but in reality trivial and senseless natural philosophy." After 10 years of heated controversy, the idea of tetrahedral carbon was fully accepted, Kolbe had died and Wislicenus succeeded him in Leipzig.

Emil Fischer (1852–1919, Np 1902) was the next to make a critical contribution to stereochemistry. From the work of van't Hoff and Le Bel he reasoned that glucose should have 16 stereoisomers. Fischer's doctorate work on hydrazines under Baeyer (1835–1917, Np 1905) at Strasbourg had led to studies of osazones which culminated in the brilliant establishment, including configurations, of the Fischer sugar tree starting from D-(+)-glyceraldehyde all the way up to the aldohexoses, allose, altrose, glucose, mannose, gulose, idose, galactose, and talose (from 1884 to 1890). Unfortunately Fischer suffered from the toxic effects of phenylhydrazine for 12 years. The arbitrarily but luckily chosen absolute configuration of D-(+)-glyceraldehyde was shown to be correct sixty years later in 1951 (Johannes-Martin Bijvoet, 1892–1980). Fischer's brilliant correlation of the sugars comprising the Fischer sugar tree was performed using the Kiliani (1855–1945)–Fischer method via cyanohydrin intermediates for elongating sugars. Fischer also made remarkable contributions to the chemistry of amino acids and to nucleic acid bases (see below).

4. STRUCTURES OF ORGANIC COMPOUNDS, TWENTIETH CENTURY

The early concept of covalent bonds was provided with a sound theoretical basis by Linus Pauling (1901–1994, Np 1954), one of the greatest intellects of the twentieth century. Pauling's totally interdisciplinary research interests, including proteins and DNA is responsible for our present understanding of molecular structures. His books *Introduction to Quantum Mechanics* (with graduate student E. B. Wilson, 1935) and *The Nature of the Chemical Bond* (1939) have had a profound effect on our understanding of all of chemistry.

The actual 3D shapes of organic molecules which were still unclear in the late 1940s were then brilliantly clarified by Odd Hassel (1897–1981, Oslo University, Np 1969) and Derek Barton (1918–1998, Np 1969). Hassel, an X-ray crystallographer and physical chemist, demonstrated by electron diffraction that cyclohexane adopted the chair form in the gas phase and that it had two kinds of bonds, "standing (axial)" and "reclining (equatorial)" (1943). Because of the German occupation of Norway in 1940, instead of publishing the result in German journals, he published it in a Norwegian journal which was not abstracted in English until 1945. During his 1949 stay at Harvard, Barton attended a seminar by Louis Fieser on steric effects in steroids and showed Fieser that interpretations could be simplified if the shapes ("conformations") of cyclohexane rings were taken into consideration; Barton made these comments because he was familiar with Hassel's study on *cis*- and *trans*-decalins. Following Fieser's suggestion Barton published these ideas in a four-page *Experientia* paper (1950). This led to the joint Nobel prize with Hassel (1969), and established the concept of conformational analysis, which has exerted a profound effect in every field involving organic molecules.

Using conformational analysis, Barton determined the structures of many key terpenoids such as ß-amyrin, cycloartenone, and cycloartenol (Birkbeck College). At Glasgow University (from 1955) he collaborated in a number of cases with Monteath Robertson (1900–1989) and established many challenging structures: limonin, glauconic acid, byssochlamic acid, and nonadrides. Barton was also associated with the Research Institute for Medicine and Chemistry (RIMAC), Cambridge, USA founded by the Schering company, where with J. M. Beaton, he produced 60 g of aldosterone at a time when the world supply of this important hormone was in mg quantities. Aldosterone synthesis ("a good problem") was achieved in 1961 by Beaton ("a good experimentalist") through a nitrite photolysis, which came to be known as the Barton reaction ("a good idea") (quotes from his 1991 autobiography published by the American Chemical Society). From Glasgow, Barton went on to Imperial College, and a year before retirement, in 1977 he moved to France to direct the research at ICSN at Gif-sur-Yvette where he explored the oxidation reaction selectivity for unactivated C–H. After retiring from ICSN he made a further move to Texas A&M University in 1986, and continued his energetic activities, including chairman of the *Tetrahedron* publications. He felt weak during work one evening and died soon after, on March 16, 1998. He was fond of the phrase "gap jumping" by which he meant seeking generalizations between facts that do not seem to be related: "In the conformational analysis story, one had to jump the gap between steroids and chemical physics" (from his autobiography). According to Barton, the three most important qualities for a scientist are "intelligence, motivation, and honesty." His routine at Texas A&M was to wake around 4 a.m., read the literature, go to the office at 7 a.m. and stay there until 7 p.m.; when asked in 1997 whether this was still the routine, his response was that he wanted to wake up earlier because sleep was a waste of time—a remark which characterized this active scientist approaching 80!

Robert B. Woodward (1917–1979, Np 1965), who died prematurely, is regarded by many as the preeminent organic chemist of the twentieth century. He made landmark achievements in spectroscopy, synthesis, structure determination, biogenesis, as well as in theory. His solo papers published in 1941–1942 on empirical rules for estimating the absorption maxima of enones and dienes made the general organic chemical community realize that UV could be used for structural studies, thus launching the beginning of the spectroscopic revolution which soon brought on the applications of IR, NMR, MS, etc. He determined the structures of the following compounds: penicillin in 1945 (through joint UK–USA collaboration, see Hodgkin), strychnine in 1948, patulin in 1949, terramycin, aureomycin, and ferrocene (with G. Wilkinson, Np 1973—shared with E. O. Fischer for sandwich compounds) in 1952, cevine in 1954 (with Barton Np 1966, Jeger and Prelog, Np 1975), magnamycin in 1956, gliotoxin in 1958, oleandomycin in 1960, streptonigrin in 1963, and tetrodotoxin in 1964. He synthesized patulin in 1950, cortisone and cholesterol in 1951, lanosterol, lysergic acid (with Eli Lilly), and strychnine in 1954, reserpine in 1956, chlorophyll in 1960, a tetracycline (with Pfizer) in 1962, cephalosporin in 1965, and vitamin B_{12} in 1972 (with A. Eschenmoser, 1925–, ETH Zürich). He derived biogenetic schemes for steroids in 1953 (with K. Bloch, see below), and for macrolides in 1956, while the Woodward–Hoffmann orbital symmetry rules in 1965 brought order to a large class of seemingly random cyclization reactions.

Another central figure in stereochemistry is Vladimir Prelog (1906–1998, Np 1975), who succeeded Leopold Ruzicka at the ETH Zürich, and continued to build this institution into one of the most active and lively research and discussion centers in the world. The core group of intellectual leaders consisted of P. Plattner (1904–1975), O. Jeger, A. Eschenmoser, J. Dunitz, D. Arigoni, and A. Dreiding (from Zürich University). After completing extensive research on alkaloids, Prelog determined the structures of nonactin, boromycin, ferrioxamins, and rifamycins. His seminal studies in the synthesis and properties of 8–12 membered rings led him into unexplored areas of stereochemisty and chirality. Together with Robert Cahn (1899–1981, London Chemical Society) and Christopher Ingold (1893–1970, University College, London; pioneering mechanistic interpretation of organic reactions), he developed the Cahn–Ingold–Prelog (CIP) sequence rules for the unambiguous specification of stereoisomers. Prelog was an excellent story teller, always had jokes to tell, and was respected and loved by all who knew him.

4.1 Polyketides and Fatty Acids

Arthur Birch (1915–1995) from Sydney University, Ph.D. with Robert Robinson (Oxford University), then professor at Manchester University and Australian National University, was one of the earliest chemists to perform biosynthetic studies using radiolabels; starting with polyketides he studied the biosynthesis of a variety of natural products such as the C_6-C_3-C_6 backbone of plant phenolics, polyene macrolides, terpenoids, and alkaloids. He is especially known for the Birch reduction of aromatic rings, metal-ammonia reductions leading to 19-norsteroid hormones and other important products (1942–) which were of industrial importance. Feodor Lynen (1911–1979, Np 1964) performed studies on the intermediary metabolism of the living cell that led him to the demonstration of the first step in a chain of reactions resulting in the biosynthesis of sterols and fatty acids.

Prostaglandins, a family of 20-carbon, lipid-derived acids discovered in seminal fluids and accessory genital glands of man and sheep by von Euler (1934), have attracted great interest because of their extremely diverse biological activities. They were isolated and their structures elucidated from 1963 by S. Bergström (1916–, Np 1982) and B. Samuelsson (1934–, Np 1982) at the Karolinska Institute, Stockholm. Many syntheses of the natural prostaglandins and their nonnatural analogues have been published.

Tetsuo Nozoe (1902–1996) who studied at Tohoku University, Sendai, with Riko Majima (1874–1962, see below) went to Taiwan where he stayed until 1948 before returning to Tohoku University. At National Taiwan University he isolated hinokitiol from the essential oil of *taiwanhinoki*. Remembering the resonance concept put forward by Pauling just before World War II, he arrived at the seven-membered nonbenzenoid aromatic structure for hinokitiol in 1941, the first of the troponoids. This highly original work remained unknown to the rest of the world until 1951. In the meantime, during 1945–1948, nonbenzenoid aromatic structures had been assigned to stipitatic acid (isolated by H. Raistrick) by Michael J. S. Dewar (1918–) and to the thujaplicins by Holger Erdtman (1902–1989); the term tropolones was coined by Dewar in 1945. Nozoe continued to work on and discuss troponoids, up to the night before his death, without knowing that he had cancer. He was a remarkably focused and warm scientist, working unremittingly. Erdtman (Royal Institute of Technology, Stockholm) was the central figure in Swedish natural products chemistry who, with his wife Gunhild Aulin Erdtman (dynamic General Secretary of the Swedish Chemistry Society), worked in the area of plant phenolics.

As mentioned in the following and in the concluding sections, classical biosynthetic studies using radioactive isotopes for determining the distribution of isotopes has now largely been replaced by the use of various stable isotopes coupled with NMR and MS. The main effort has now shifted to the identification and cloning of genes, or where possible the gene clusters, involved in the biosynthesis of the natural product. In the case of polyketides (acyclic, cyclic, and aromatic), the focus is on the polyketide synthases.

4.2 Isoprenoids, Steroids, and Carotenoids

During his time as an assistant to Kekulé at Bonn, Otto Wallach (1847–1931, Np 1910) had to familiarize himself with the essential oils from plants; many of the components of these oils were compounds for which no structure was known. In 1891 he clarified the relations between 12 different monoterpenes related to pinene. This was summarized together with other terpene chemistry in book form in 1909, and led him to propose the "isoprene rule." These achievements laid the foundation for the future development of terpenoid chemistry and brought order from chaos.

The next period up to around 1950 saw phenomenal advances in natural products chemistry centered on isoprenoids. Many of the best natural products chemists in Europe, including Wieland, Windaus, Karrer, Kuhn, Butenandt, and Ruzicka contributed to this breathtaking pace. Heinrich Wieland (1877–1957) worked on the bile acid structure, which had been studied over a period of 100 years and considered to be one of the most difficult to attack; he received the Nobel Prize in 1927 for these studies. His friend Adolph Windaus (1876–1959) worked on the structure of cholesterol for which he also received the Nobel Prize in 1928. Unfortunately, there were chemical discrepancies in the proposed steroidal skeletal structure, which had a five-membered ring B attached to C-7 and C-9. J. D. Bernal, Mineralogical Museums, Cambridge University, who was examining the X-ray patterns of ergosterol (1932) noted that the dimensions were inconsistent with the Wieland–Windaus formula. A reinterpretation of the production of chrysene from sterols by Pd/C dehydrogenation reported by Diels (see below) in 1927 eventually led Rosenheim and King and Wieland and Dane to deduce the correct structure in 1932. Wieland also worked on the structures of morphine/strychnine alkaloids, phalloidin/amanitin cyclopeptides of toxic mushroom *Amanita phalloides*, and pteridines, the important fluorescent pigments of butterfly wings. Windaus determined the structure of ergosterol and continued structural studies of its irradiation product which exhibited antirachitic activity "vitamin D." The mechanistically complex photochemistry of ergosterol leading to the vitamin D group has been investigated in detail by Egbert Havinga (1927–1988, Leiden University), a leading photochemist and excellent tennis player.

Paul Karrer (1889–1971, Np 1937), established the foundations of carotenoid chemistry through structural determinations of lycopene, carotene, vitamin A, etc. and the synthesis of squalene, carotenoids, and others. George Wald (1906–1997, Np 1967) showed that vitamin A was the key compound in vision during his stay in Karrer's laboratory. Vitamin K (K from "Koagulation"), discovered by Henrik Dam (1895–1976, Polytechnic Institute, Copenhagen, Np 1943) and structurally studied by Edward Doisy (1893–1986, St. Louis University, Np 1943), was also synthesized by Karrer. In addition, Karrer synthesized riboflavin (vitamin B$_2$) and determined the structure and role of nicotinamide adenine dinucleotide phosphate (NADP$^+$) with Otto Warburg. The research on carotenoids and vitamins of Karrer who was at Zürich University overlapped with that of Richard Kuhn (1900–1967, Np 1938) at the ETH Zürich, and the two were frequently rivals. Richard Kuhn, one of the pioneers in using UV-vis spectroscopy for structural studies, introduced the concept of "atropisomerism" in diphenyls, and studied the spectra of a series of diphenyl polyenes. He determined the structures of many natural carotenoids, proved the structure of riboflavin-5-phosphate (flavin-adenine-dinucleotide-5-phosphate) and showed that the combination of NAD-5-phosphate with the carrier protein yielded the yellow oxidation enzyme, thus providing an understanding of the role of a prosthetic group. He also determined the structures of vitamin B complexes, i.e., pyridoxine, *p*-aminobenzoic acid, pantothenic acid. After World War II he went on to structural studies of nitrogen-containing oligosaccharides in human milk that provide immunity for infants, and brain gangliosides. Carotenoid studies in Switzerland were later taken up by Otto Isler (1910–1993), a Ruzicka student at Hoffmann-La Roche, and Conrad Hans Eugster (1921–), a Karrer student at Zürich University.

Adolf Butenandt (1903–1998, Np 1939) initiated and essentially completed isolation and structural studies of the human sex hormones, the insect molting hormone (ecdysone), and the first pheromone, bombykol. With help from industry he was able to obtain large supplies of urine from pregnant women for estrone, sow ovaries for progesterone, and 4,000 gallons of male urine for androsterone (50 mg, crystals). He isolated and determined the structures of two female sex hormones, estrone and progesterone, and the male hormone androsterone all during the period 1934–1939 (!) and was awarded the Nobel prize in 1939. Keen intuition and use of UV data and Pregl's microanalysis all played important roles. He was appointed to a professorship in Danzig at the age of 30. With Peter Karlson he isolated from 500 kg of silkworm larvae 25 mg of α-ecdysone, the prohormone of insect and crustacean molting hormone, and determined its structure as a polyhydroxysteroid (1965); 20-hydroxylation gives the insect and crustacean molting hormone or ß-ecdysone (20-hydroxyecdysteroid). He was also the first to isolate an insect pheromone, bombykol, from female silkworm moths (with E. Hecker). As president of the Max Planck Foundation, he strongly influenced the postwar rebuilding of German science.

The successor to Kuhn, who left ETH Zürich for Heidelberg, was Leopold Ruzicka (1887–1967, Np 1939) who established a close relationship with the Swiss pharmaceutical industry. His synthesis of the 17- and 15-membered macrocyclic ketones, civetone and muscone (the constituents of musk) showed that contrary to Baeyer's prediction, large alicyclic rings could be strainless. He reintroduced and refined the isoprene rule proposed by Wallach (1887) and determined the basic structures of many sesqui-, di-, and triterpenes, as well as the structure of lanosterol, the key intermediate in cholesterol biosynthesis. The "biogenetic isoprene rule" of the ETH group, Albert Eschenmoser, Leopold Ruzicka, Oskar Jeger, and Duilio Arigoni, contributed to a concept of terpenoid cyclization (1955), which was consistent with the mechanistic considerations put forward by Stork as early as 1950. Besides making the ETH group into a center of natural products chemistry, Ruzicka bought many seventeenth century Dutch paintings with royalties accumulated during the war from his Swiss and American patents, and donated them to the Zürich Kunsthaus.

Studies in the isolation, structures, and activities of the antiarthritic hormone, cortisone and related compounds from the adrenal cortex were performed in the mid- to late 1940s during World War II by Edward Kendall (1886–1972, Mayo Clinic, Rochester, Np 1950), Tadeus Reichstein (1897–1996, Basel University, Np 1950), Philip Hench (1896–1965, Mayo Clinic, Rochester, Np 1950), Oskar Wintersteiner (1898–1971, Columbia University, Squibb) and others initiated interest as an adjunct to military medicine as well as to supplement the meager supply from beef adrenal glands by synthesis. Lewis Sarett (1917–, Merck & Co., later president) and co-workers completed the cortisone synthesis in 28 steps, one of the first two totally stereocontrolled syntheses of a natural product; the other was cantharidin (Stork 1951) (see above). The multistep cortisone synthesis was put on the production line by Max Tishler (1906–1989, Merck & Co., later president) who made contributions to the synthesis of a number of drugs, including riboflavin. Besides working on steroid reactions/synthesis and antimalarial agents, Louis F. Fieser (1899–1977) and Mary Fieser (1909–1997) of Harvard University made huge contributions to the chemical community through their outstanding books *Natural Products related to Phenanthrene* (1949), *Steroids* (1959), *Advanced Organic Chemistry* (1961), and *Topics in Organic Chemistry* (1963), as well as their textbooks and an important series of books on Organic Reagents. Carl Djerassi (1923–, Stanford University), a prolific chemist, industrialist, and more recently a novelist, started to work at the Syntex laboratories in Mexico City where he directed the work leading to the first oral contraceptive ("the pill") for women.

Takashi Kubota (1909–, Osaka City University), with Teruo Matsuura (1924–, Kyoto University), determined the structure of the furanoid sesquiterpene, ipomeamarone, from the black rotted portion of spoiled sweet potatoes; this research constitutes the first characterization of a phytoallexin, defense substances produced by plants in response to attack by fungi or physical damage. Damaging a plant and characterizing the defense substances produced may lead to new bioactive compounds. The mechanism of induced biosynthesis of phytoallexins, which is not fully understood, is an interesting biological mechanistic topic that deserves further investigation. Another center of high activity in terpenoids and nucleic acids was headed by Frantisek Sorm (1913–1980, Institute of Organic and Biochemistry, Prague), who determined the structures of many sesquiterpenoids and other natural products; he was not only active scientifically but also was a central figure who helped to guide the careers of many Czech chemists.

The key compound in terpenoid biosynthesis is mevalonic acid (MVA) derived from acetyl-CoA, which was discovered fortuitously in 1957 by the Merck team in Rahway, NJ headed by Karl Folkers (1906–1998). They soon realized and proved that this C_6 acid was the precursor of the C_5 isoprenoid unit isopentenyl diphosphate (IPP) that ultimately leads to the biosynthesis of cholesterol. In 1952 Konrad Bloch (1912–, Harvard, Np 1964) with R. B. Woodward published a paper suggesting a mechanism of the cyclization of squalene to lanosterol and the subsequent steps to cholesterol, which turned out to be essentially correct. This biosynthetic path from MVA to cholesterol was experimentally clarified in stereochemical detail by John Cornforth (1917–, Np 1975) and George Popják. In 1932, Harold Urey (1893–1981, Np 1934) of Columbia University discovered heavy hydrogen. Urey showed, contrary to common expectation, that isotope separation could be achieved with deuterium in the form of deuterium oxide by fractional electrolysis of water. Urey's separation of the stable isotope deuterium led to the isotopic tracer methodology that revolutionized the protocols for elucidating biosynthetic processes and reaction mechanisms, as exemplified beautifully by the cholesterol studies. Using MVA labeled chirally with isotopes, including chiral methyl, i.e., -CHDT, Cornforth and Popják clarified the key steps in the intricate biosynthetic conversion of mevalonate to cholesterol in stereochemical detail. The chiral methyl group was also prepared independently by Duilio Arigoni (1928–, ETH, Zürich). Cornforth has had great difficulty in hearing and speech since childhood but has been helped expertly by his chemist wife Rita; he is an excellent tennis and chess player, and is renowned for his speed in composing occasional witty limericks.

Although MVA has long been assumed to be the only natural precursor for IPP, a non-MVA pathway in which IPP is formed via the glyceraldehyde phosphate-pyruvate pathway has been discovered (1995–1996) in the ancient bacteriohopanoids by Michel Rohmer, who started working on them with Guy Ourisson (1926–, University of Strasbourg, terpenoid studies, including prebiotic), and by Duilio Arigoni in the ginkgolides, which are present in the ancient *Ginkgo biloba* tree. It is possible that many other terpenoids are biosynthesized via the non-MVA route. In classical biosynthetic experiments, ^{14}C-labeled acetic acid was incorporated into the microbial or plant product, and location or distribution of the ^{14}C label was deduced by oxidation or degradation to specific fragments including acetic acid; therefore, it was not possible or extremely difficult to map the distribution of all radioactive carbons. The progress in ^{13}C NMR made it possible to incorporate ^{13}C-labeled acetic acid and locate all labeled carbons. This led to the discovery of the nonmevalonate pathway leading to the IPP units. Similarly, NMR and MS have made it possible to use the stable isotopes, e.g., ^{18}O, ^{2}H, ^{15}N, etc., in biosynthetic studies. The current trend of biosynthesis has now shifted to genomic approaches for cloning the genes of various enzyme synthases involved in the biosynthesis.

4.3 Carbohydrates and Cellulose

The most important advance in carbohydrate structures following those made by Emil Fischer was the change from acyclic to the current cyclic structure introduced by Walter Haworth (1883–1937). He noticed the presence of α- and ß-anomers, and determined the structures of important disaccharides including cellobiose, maltose, and lactose. He also determined the basic structural aspects of starch, cellulose, inulin, and other polysaccharides, and accomplished the structure determination and synthesis of vitamin C, a sample of which he had received from Albert von Szent-Györgyi (1893–1986, Np 1937). This first synthesis of a vitamin was significant since it showed that a vitamin could be synthesized in the same way as any other organic compound. There was strong belief among leading scientists in the 1910s that cellulose, starch, protein, and rubber were colloidal aggregates of small molecules. However, Hermann Staudinger (1881–1965, Np 1953) who succeeded R. Willstätter and H. Wieland at the ETH Zürich and Freiburg, respectively, showed through viscosity measurements and various molecular weight measurements that macromolecules do exist, and developed the principles of macromolecular chemistry.

In more modern times, Raymond Lemieux (1920–, Universities of Ottawa and Alberta) has been a leader in carbohydrate research. He introduced the concept of *endo-* and *exo-*anomeric effects, accomplished the challenging synthesis of sucrose (1953), pioneered in the use of NMR coupling constants in configuration studies, and most importantly, starting with syntheses of oligosaccharides responsible for human blood group determinants, he prepared antibodies and clarified fundamental aspects of the binding of oligosaccharides by lectins and antibodies. The periodate–potassium permanganate cleavage of double bonds at room temperature (1955) is called the Lemieux reaction.

4.4 Amino Acids, Peptides, Porphyrins, and Alkaloids

It is fortunate that we have China's record and practice of herbal medicine over the centuries, which is providing us with an indispensable source of knowledge. China is rapidly catching up in terms of infrastructure and equipment in organic and bioorganic chemistry, and work on isolation, structure determination, and synthesis stemming from these valuable sources has picked up momentum. However, as mentioned above, clarification of the active principles and mode of action of these plant extracts will be quite a challenge since in many cases synergistic action is expected. Wang Yu (1910–1997) who headed the well-equipped Shanghai Institute of Organic Chemistry surprised the world with the total synthesis of bovine insulin performed by his group in 1965; the human insulin was synthesized around the same time by P. G. Katsoyannis, A. Tometsko, and C. Zaut of the Brookhaven National Laboratory (1966).

One of the giants in natural products chemistry during the first half of this century was Robert Robinson (1886–1975, Np 1947) at Oxford University. His synthesis of tropinone, a bicyclic amino ketone related to cocaine, from succindialdehyde, methylamine, and acetone dicarboxylic acid under Mannich reaction conditions was the first biomimetic synthesis (1917). It reduced Willstätter's 1903 13-step synthesis starting with suberone into a single step. This achievement demonstrated Robinson's analytical prowess. He was able to dissect complex molecular structures into simple biosynthetic building blocks, which allowed him to propose the biogenesis of all types of alkaloids and other natural products. His laboratory at Oxford, where he developed the well-known Robinson annulation reaction (1937) in connection with his work on the synthesis of steroids became a world center for natural products study. Robinson was a pioneer in the so-called electronic theory of organic reactions, and introduced the use of curly arrows to show the movements of electrons. His analytical power is exemplified in the structural studies of strychnine and brucine around 1946–1952. Barton clarified the biosynthetic route to the morphine alkaloids, which he saw as an extension of his biomimetic synthesis of usnic acid through a one-electron oxidation; this was later extended to a general phenolate coupling scheme. Morphine total synthesis was brilliantly achieved by Marshall Gates (1915–, University of Rochester) in 1952.

The yield of the Robinson tropinone synthesis was low but Clemens Schöpf (1899–1970) , Ph.D. Munich (Wieland), Universität Darmstadt, improved it to 90% by carrying out the reaction in buffer; he also worked on the stereochemistry of morphine and determined the structure of the steroidal alkaloid salamandarine (1961), the toxin secreted from glands behind the eyes of the salamander.

Roger Adams (1889–1971, University of Illinois), was the central figure in organic chemistry in the USA and is credited with contributing to the rapid development of its chemistry in the late 1930s and 1940s, including training of graduate students for both academe and industry. After earning a Ph.D. in 1912 at Harvard University he did postdoctoral studies with Otto Diels (see below) and Richard Willstätter (see below) in 1913; he once said that around those years in Germany he could cover all *Journal of the American Chemical Society* papers published in a year in a single night. His important work include determination of the structures of tetrahydrocannabinol in marijuana, the toxic gossypol in cottonseed oil, chaulmoogric acid used in treatment of leprosy, and the Senecio alkaloids with Nelson Leonard (1916–, University of Illinois, now at Caltech). He also contributed to many fundamental organic reactions and syntheses. The famous Adams platinum catalyst is not only important for reducing double bonds in industry and in the laboratory, but was central for determining the number of double bonds in a structure. He was also one of the founders of the *Organic Synthesis* (started in 1921) and the *Organic Reactions* series. Nelson Leonard switched interests to bioorganic chemistry and biochemistry, where he has worked with nucleic acid bases and nucleotides, coenzymes, dimensional probes, and fluorescent modifications such as ethenoguanine.

The complicated structures of the medieval plant poisons aconitine (from *Aconitum*) and delphinine (from *Delphinium*) were finally characterized in 1959–1960 by Karel Wiesner (1919–1986, University of New Brunswick), Leo Marion (1899–1979, National Research Council, Ottawa), George Büchi (1921–, mycotoxins, aflatoxin/DNA adduct, synthesis of terpenoids and nitrogen-containing bioactive compounds, photochemistry), and Maria Przybylska (1923–, X-ray).

The complex chlorophyll structure was elucidated by Richard Willstätter (1872–1942, Np 1915). Although he could not join Baeyer's group at Munich because the latter had ceased taking students, a close relation developed between the two. During his chlorophyll studies, Willstätter reintroduced the important technique of column chromatography published in Russian by Michael Tswett (1906). Willstätter further demonstrated that magnesium was an integral part of chlorophyll, clarified the relation between chlorophyll and the blood pigment hemin, and found the wide distribution of carotenoids in tomato, egg yolk, and bovine corpus luteum. Willstätter also synthesized cyclooctatetraene and showed its properties to be wholly unlike benzene but close to those of acyclic polyenes (around 1913). He succeeded Baeyer at Munich in 1915, synthesized the anesthetic cocaine, retired early in protest of anti-Semitism, but remained active until the Hitler era, and in 1938 emigrated to Switzerland.

The hemin structure was determined by another German chemist of the same era, Hans Fischer (1881–1945, Np 1930), who succeeded Windaus at Innsbruck and at Munich. He worked on the structure of hemin from the blood pigment hemoglobin, and completed its synthesis in 1929. He continued Willstätter's structural studies of chlorophyll, and further synthesized bilirubin in 1944. Destruction of his institute at Technische Hochschule München, during World War II led him to take his life in March 1945. The biosynthesis of hemin was elucidated largely by David Shemin (1911–1991).

In the mid 1930s the Department of Biochemistry at Columbia Medical School, which had accepted many refugees from the Third Reich, including Erwin Chargaff, Rudolf Schoenheimer, and others on the faculty, and Konrad Bloch (see above) and David Shemin as graduate students, was a great center of research activity. In 1940, Shemin ingested 66 g of ^{15}N-labeled glycine over a period of 66 hours in order to determine the half-life of erythrocytes. David Rittenberg's analysis of the heme moiety with his home-made mass spectrometer showed all four pyrrole nitrogens came from glycine. Using ^{14}C (that had just become available) as a second isotope (see next paragraph), doubly labeled glycine ^{15}NH$_2$^{14}CH$_2$COOH and other precursors, Shemin showed that glycine and succinic acid condensed to yield δ-aminolevulinate, thus elegantly demonstrating the novel biosynthesis of the porphyrin ring (around 1950). At this time, Bloch was working on the other side of the bench.

Melvin Calvin (1911–1997, Np 1961) at University of California, Berkeley, elucidated the complex photosynthetic pathway in which plants reduce carbon dioxide to carbohydrates. The critical ^{14}CO$_2$ had just been made available at Berkeley Lawrence Radiation Laboratory as a result of the pioneering research of Martin Kamen (1913–), while paper chromatography also played crucial roles. Kamen produced ^{14}C with Sam Ruben (1940), used ^{18}O to show that oxygen in photosynthesis comes from water and not from carbon dioxide, participated in the *Manhattan* project, testified before the House UnAmerican Activities Committee (1947), won compensatory damages from the US Department of State, and helped build the University of California, La Jolla (1957). The entire structure of the photosynthetic reaction center (>10 000 atoms) from the purple bacterium *Rhodopseudomonas viridis* has been established by X-ray crystallography in the landmark studies performed by Johann Deisenhofer (1943–), Robert Huber (1937–), and Hartmut Michel (1948–) in 1989; this was the first membrane protein structure determined by X-ray, for which they shared the 1988 Nobel prize. The information gained from the full structure of this first membrane protein has been especially rewarding.

The studies on vitamin B_{12}, the structure of which was established by crystallographic studies performed by Dorothy Hodgkin (1910–1994, Np 1964), are fascinating. Hodgkin also determined the structure of penicillin (in a joint effort between UK and US scientists during World War II) and insulin. The formidable total synthesis of vitamin B_{12} was completed in 1972 through collaborative efforts between Woodward and Eschenmoser, involving 100 postdoctoral fellows and extending over 10 years. The biosynthesis of fascinating complexity is almost completely solved through studies performed by Alan Battersby (1925–, Cambridge University), Duilio Arigoni, and Ian Scott (1928–, Texas A&M University) and collaborators where advanced NMR techniques and synthesis of labeled precursors is elegantly combined with cloning of enzymes controlling each biosynthetic step. This work provides a beautiful demonstration of the power of the combination of bioorganic chemistry, spectroscopy and molecular biology, a future direction which will become increasingly important for the creation of new "unnatural" natural products.

4.5 Enzymes and Proteins

In the early days of natural products chemistry, enzymes and viruses were very poorly understood. Thus, the 1926 paper by James Sumner (1887–1955) at Cornell University on crystalline urease was received with ignorance or skepticism, especially by Willstätter who believed that enzymes were small molecules and not proteins. John Northrop (1891–1987) and co-workers at the Rockefeller Institute went on to crystallize pepsin, trypsin, chymotrypsin, ribonuclease, deoyribonuclease, carboxypeptidase, and other enzymes between 1930 and 1935. Despite this, for many years biochemists did not recognize the significance of these findings, and considered enzymes as being low molecular weight compounds adsorbed onto proteins or colloids. Using Northrop's method for crystalline enzyme preparations, Wendell Stanley (1904–1971) at Princeton obtained tobacco mosaic virus as needles from one ton of tobacco leaves (1935). Sumner, Northrop, and Stanley shared the 1946 Nobel prize in chemistry. All these studies opened a new era for biochemistry.

Meanwhile, Linus Pauling, who in mid-1930 became interested in the magnetic properties of hemoglobin, investigated the configurations of proteins and the effects of hydrogen bonds. In 1949 he showed that sickle cell anemia was due to a mutation of a single amino acid in the hemoglobin molecule, the first correlation of a change in molecular structure with a genetic disease. Starting in 1951 he and colleagues published a series of papers describing the alpha helix structure of proteins; a paper published in the early 1950s with R. B. Corey on the structure of DNA played an important role in leading Francis Crick and James Watson to the double helix structure (Np 1962).

A further important achievement in the peptide field was that of Vincent Du Vigneaud (1901–1978, Np 1955), Cornell Medical School, who isolated and determined the structure of oxytocin, a posterior pituitary gland hormone, for which a structure involving a disulfide bond was proposed. He synthesized oxytocin in 1953, thereby completing the first synthesis of a natural peptide hormone.

Progress in isolation, purification, crystallization methods, computers, and instrumentation, including cyclotrons, have made X-ray crystallography the major tool in structural. Numerous structures including those of ligand/receptor complexes are being published at an extremely rapid rate. Some of the past major achievements in protein structures are the following. Max Perutz (1914, Np 1962) and John Kendrew (1914–1997, Np 1962), both at the Laboratory of Molecular Biology, Cambridge University, determined the structures of hemoglobin and myoglobin, respectively. William Lipscomb (1919–, Np 1976), Harvard University, who has trained many of the world's leaders in protein X-ray crystallography has been involved in the structure determination of many enzymes including carboxypeptidase A (1967); in 1965 he determined the structure of the anticancer bisindole alkaloid, vinblastine. Folding of proteins, an important but still enigmatic phenomenon, is attracting increasing attention. Christian Anfinsen (1916–1995, Np 1972), NIH, one of the pioneers in this area, showed that the amino acid residues in ribonuclease interact in an energetically most favorable manner to produce the unique 3D structure of the protein.

4.6 Nucleic Acid Bases, RNA, and DNA

The "Fischer indole synthesis" was first performed in 1886 by Emil Fischer. During the period 1881–1914, he determined the structures of and synthesized uric acid, caffeine, theobromine, xanthine, guanine, hypoxanthine, adenine, guanine, and made theophylline-D-glucoside phosphoric acid, the first synthetic nucleotide. In 1903, he made 5,5-diethylbarbituric acid or Barbital, Dorminal, Veronal, etc. (sedative), and in 1912, phenobarbital or Barbipil, Luminal, Phenobal, etc. (sedative). Many of his syntheses formed the basis of German industrial production of purine bases. In 1912 he showed that tannins are gallates of sugars such as maltose and glucose. Starting in 1899, he synthesized many of the 13 α-amino acids known at that time, including the L- and D-forms, which were separated through fractional crystallization of their salts with optically active bases. He also developed a method for synthesizing fragments of proteins, namely peptides, and made an 18-amino acid peptide. He lost his two sons in World War I, lost his wealth due to postwar inflation, believed he had terminal cancer (a misdiagnosis), and killed himself in July 1919. Fischer was a skilled experimentalist, so that even today, many of the reactions performed by him and his students are so delicately controlled that they are not easy to reproduce. As a result of his suffering by inhaling diethylmercury, and of the poisonous effect of phenylhydrazine, he was one of the first to design fume hoods. He was a superb teacher and was also influential in establishing the Kaiser Wilhelm Institute, which later became the Max Planck Institute. The number and quality of his accomplishments and contributions are hard to believe; he was truly a genius.

Alexander Todd (1907–1997, Np 1957) made critical contributions to the basic chemistry and synthesis of nucleotides. His early experience consisted of an extremely fruitful stay at Oxford in the Robinson group, where he completed the syntheses of many representative anthocyanins, and then at Edinburgh where he worked on the synthesis of vitamin B_1. He also prepared the hexacarboxylate of vitamin B_{12} (1954), which was used by D. Hodgkin's group for their X-ray elucidation of this vitamin (1956). M. Wiewiorowski (1918–), Institute for Bioorganic Chemistry, in Poznan, has headed a famous group in nucleic acid chemistry, and his colleagues are now distributed worldwide.

4.7 Antibiotics, Pigments, and Marine Natural Products

The concept of one microorganism killing another was introduced by Pasteur who coined the term antibiosis in 1877, but it was much later that this concept was realized in the form of an actual antibiotic. The bacteriologist Alexander Fleming (1881–1955, University of London, Np 1945) noticed that an airborne mold, a *Penicillium* strain, contaminated cultures of *Staphylococci* left on the open bench and formed a transparent circle around its colony due to lysis of *Staphylococci*. He published these results in 1929. The discovery did not attract much interest but the work was continued by Fleming until it was taken up further at Oxford University by pathologist Howard Florey (1898–1968, Np 1945) and biochemist Ernst Chain (1906–1979, Np 1945). The bioactivities of purified "penicillin," the first antibiotic, attracted serious interest in the early 1940s in the midst of World War II. A UK/USA team was formed during the war between academe and industry with Oxford University, Harvard University, ICI, Glaxo, Burroughs Wellcome, Merck, Shell, Squibb, and Pfizer as members. This project resulted in the large scale production of penicillin and determination of its structure (finally by X-ray, D. Hodgkin). John Sheehan (1915–1992) at MIT synthesized 6-aminopenicillanic acid in 1959, which opened the route for the synthesis of a number of analogues. Besides being the first antibiotic to be discovered, penicillin is also the first member of a large number of important antibiotics containing the ß-lactam ring, for example cephalosporins, carbapenems, monobactams, and nocardicins. The strained ß-lactam ring of these antibiotics inactivates the transpeptidase by acylating its serine residue at the active site, thus preventing the enzyme from forming the link between the pentaglycine chain and the D-Ala-D-Ala peptide, the essential link in bacterial cell walls. The overuse of ß-lactam antibiotics, which has given rise to the disturbing appearance of microbial resistant strains, is leading to active research in the design of synthetic ß-lactam analogues to counteract these strains. The complex nature of the important penicillin biosynthesis is being elucidated through efforts combining genetic engineering, expression of biosynthetic genes as well as feeding of synthetic precursors, etc. by Jack Baldwin (1938–, Oxford University), José Luengo (Universidad de León, Spain) and many other groups from industry and academe.

Shortly after the penicillin discovery, Selman Waksman (1888–1973, Rutgers University, Np 1952) discovered streptomycin, the second antibiotic and the first active against the dreaded disease tuberculosis. The discovery and development of new antibiotics continued throughout the world at pharmaceutical companies in Europe, Japan, and the USA from soil and various odd sources: cephalosporin from sewage in Sardinia, cyclosporin from Wisconsin and Norway soil which was carried back to Switzerland, avermectin from the soil near a golf course in Shizuoka Prefecture. People involved in antibiotic discovery used to collect soil samples from various sources during their trips but this has now become severely restricted to protect a country's right to its soil. M. M. Shemyakin (1908–1970, Institute of Chemistry of Natural Products, Moscow) was a grand master of Russian natural products who worked on antibiotics, especially of the tetracycline class; he also worked on cyclic antibiotics composed of alternating sequences of amides and esters and coined the term depsipeptide for these in 1953. He died in 1970 of a sudden heart attack in the midst of the 7th IUPAC Natural Products

Symposium held in Riga, Latvia, which he had organized. The Institute he headed was renamed the Shemyakin Institute.

Indigo, an important vat dye known in ancient Asia, Egypt, Greece, Rome, Britain, and Peru, is probably the oldest known coloring material of plant origin, Indigofera and Isatis. The structure was determined in 1883 and a commercially feasible synthesis was performed in 1883 by Adolf von Baeyer (see above, 1835–1917, Np 1905), who founded the German Chemical Society in 1867 following the precedent of the Chemistry Society of London. In 1872 Baeyer was appointed a professor at Strasbourg where E. Fischer was his student, and in 1875 he succeeded J. Liebig in Munich. Tyrian (or Phoenician) purple, the dibromo derivative of indigo which is obtained from the purple snail Murex bundaris, was used as a royal emblem in connection with religious ceremonies because of its rarity; because of the availability of other cheaper dyes with similar color, it has no commercial value today. K. Venkataraman (1901–1981, University of Bombay then National Chemical Laboratory) who worked with R. Robinson on the synthesis of chromones in his early career, continued to study natural and synthetic coloring matters, including synthetic anthraquinone vat dyes, natural quinonoid pigments, etc. T. R. Seshadri (1900–1975) is another Indian natural products chemist who worked mainly in natural pigments, dyes, drugs, insecticides, and especially in polyphenols. He also studied with Robinson, and with Pregl at Graz, and taught at Delhi University. Seshadri and Venkataraman had a huge impact on Indian chemistry. After a 40 year involvement, Toshio Goto (1929–1990) finally succeeded in solving the mysterious identity of commelinin, the deep-blue flower petal pigment of the Commelina communis isolated by Kozo Hayashi (1958) and protocyanin, isolated from the blue cornflower Centaurea cyanus by E. Bayer (1957). His group elucidated the remarkable structure in its entirety which consisted of six unstable anthocyanins, six flavones and two metals, the molecular weight approaching 10 000; complex stacking and hydrogen bonds were also involved. Thus the pigmentation of petals turned out to be far more complex than the theories put forth by Willstätter (1913) and Robinson (1931). Goto suffered a fatal heart attack while inspecting the first X-ray structure of commelinin; commelinin represents a pinnacle of current natural products isolation and structure determination in terms of subtlety in isolation and complexity of structure.

The study of marine natural products is understandably far behind that of compounds of terrestrial origin due to the difficulty in collection and identification of marine organisms. However, it is an area which has great potentialities for new discoveries from every conceivable source. One pioneer in modern marine chemistry is Paul Scheuer (1915–, University of Hawaii) who started his work with quinones of marine origin and has since characterized a very large number of bioactive compounds from mollusks and other sources. Luigi Minale (1936–1997, Napoli) started a strong group working on marine natural products, concentrating mainly on complex saponins. He was a leading natural products chemist who died prematurely. A. Gonzalez Gonzalez (1917–) who headed the Organic Natural Products Institute at the University of La Laguna, Tenerife, was the first to isolate and study polyhalogenated sesquiterpenoids from marine sources. His group has also carried out extensive studies on terrestrial terpenoids from the Canary Islands and South America. Carotenoids are widely distributed in nature and are of importance as food coloring material and as antioxidants (the detailed mechanisms of which still have to be worked out); new carotenoids continue to be discovered from marine sources, for example by the group of Synnove Liaaen-Jensen, Norwegian Institute of Technology). Yoshimasa Hirata (1915–), who started research at Nagoya University, is a champion in the isolation of nontrivial natural products. He characterized the bioluminescent luciferin from the marine ostracod *Cypridina hilgendorfii* in 1966 (with his students, Toshio Goto, Yoshito Kishi, and Osamu Shimomura); tetrodotoxin from the fugu fish in 1964 (with Goto and Kishi and co-workers), the structure of which was announced simultaneously by the group of Kyosuke Tsuda (1907–, tetrodotoxin, matrine) and Woodward; and the very complex palytoxin, $C_{129}H_{223}N_3O_{54}$ in 1981–1987 (with Daisuke Uemura and Kishi). Richard E. Moore, University of Hawaii, also announced the structure of palytoxin independently. Jon Clardy (1943–, Cornell University) has determined the X-ray structures of many unique marine natural products, including brevetoxin B (1981), the first of the group of toxins with contiguous *trans*-fused ether rings constituting a stiff ladder-like skeleton. Maitotoxin, $C_{164}H_{256}O_{68}S_2Na_2$, MW 3422, produced by the dinoflagellate *Gambierdiscus toxicus* is the largest and most toxic of the nonbiopolymeric toxins known; it has 32 alicyclic 6- to 8-membered ethereal rings and acyclic chains. Its isolation (1994) and complete structure determination was accomplished jointly by the groups of Takeshi Yasumoto (Tohoku University), Kazuo Tachibana and Michio Murata (Tokyo University) in 1996. Kishi, Harvard University, also deduced the full structure in 1996.

The well-known excitatory agent for the cat family contained in the volatile oil of catnip, *Nepeta cataria*, is the monoterpene nepetalactone, isolated by S. M. McElvain (1943) and structure determined by Jerrold Meinwald (1954); cats, tigers, and lions start purring and roll on their backs in response to this lactone. Takeo Sakan (1912–1993) investigated the series of monoterpenes neomatatabiols, etc.

from Actinidia, some of which are male lacewing attractants. As little as 1 fg of neomatatabiol attracts lacewings.

The first insect pheromone to be isolated and characterized was bombykol, the sex attractant for the male silkworm, *Bombyx mori* (by Butenandt and co-workers, see above). Numerous pheromones have been isolated, characterized, synthesized, and are playing central roles in insect control and in chemical ecology. The group at Cornell University have long been active in this field: Tom Eisner (1929–, behavior), Jerrold Meinwald (1927–, chemistry), Wendell Roeloff (1939–, electrophysiology, chemistry). Since the available sample is usually minuscule, full structure determination of a pheromone often requires total synthesis; Kenji Mori (1935–, Tokyo University) has been particularly active in this field. Progress in the techniques for handling volatile compounds, including collection, isolation, GC/MS, etc., has started to disclose the extreme complexity of chemical ecology which plays an important role in the lives of all living organisms. In this context, natural products chemistry will be play an increasingly important role in our grasp of the significance of biodiversity.

5. SYNTHESIS

Synthesis has been mentioned often in the preceding sections of this essay. In the following, synthetic methods of more general nature are described. The Grignard reaction of Victor Grignard (1871–1935, Np 1912) and then the Diels–Alder reaction by Otto Diels (1876–1954, Np 1950) and Kurt Alder (1902–1956, Np 1950) are extremely versatile reactions. The Diels–Alder reaction can account for the biosynthesis of several natural products with complex structures, and now an enzyme, a Diels–Alderase involved in biosynthesis has been isolated by Akitami Ichihara, Hokkaido University (1997).

The hydroboration reactions of Herbert Brown (1912–, Purdue University, Np 1979) and the Wittig reactions of Georg Wittig (1897–1987, Np 1979) are extremely versatile synthetic reactions. William S. Johnson (1913–1995, University of Wisconsin, Stanford University) developed efficient methods for the cyclization of acyclic polyolefinic compounds for the synthesis of corticoid and other steroids, while Gilbert Stork (1921–, Columbia University) introduced enamine alkylation, regiospecific enolate formation from enones and their kinetic trapping (called "three component coupling" in some cases), and radical cyclization in regio- and stereospecific constructions. Elias J. Corey (1928–, Harvard University, Np 1990) introduced the concept of retrosynthetic analysis and developed many key synthetic reactions and reagents during his synthesis of bioactive compounds, including prostaglandins and gingkolides. A recent development is the ever-expanding supramolecular chemistry stemming from 1967 studies on crown ethers by Charles Pedersen (1904–1989), 1968 studies on cryptates by Jean-Marie Lehn (1939–), and 1973 studies on host–guest chemistry by Donald Cram (1919–); they shared the chemistry Nobel prize in 1987.

6. NATURAL PRODUCTS STUDIES IN JAPAN

Since the background of natural products study in Japan is quite different from that in other countries, a brief history is given here. Natural products is one of the strongest areas of chemical research in Japan with probably the world's largest number of chemists pursuing structural studies; these are joined by a healthy number of synthetic and bioorganic chemists. An important Symposium on Natural Products was held in 1957 in Nagoya as a joint event between the faculties of science, pharmacy, and agriculture. This was the beginning of a series of annual symposia held in various cities, which has grown into a three-day event with about 50 talks and numerous papers; practically all achievements in this area are presented at this symposium. Japan adopted the early twentieth century German or European academic system where continuity of research can be assured through a permanent staff in addition to the professor, a system which is suited for natural products research which involves isolation and assay, as well as structure determination, all steps requiring delicate skills and much expertise.

The history of Japanese chemistry is short because the country was closed to the outside world up to 1868. This is when the Tokugawa shogunate which had ruled Japan for 264 years was overthrown and the Meiji era (1868–1912) began. Two of the first Japanese organic chemists sent abroad were Shokei Shibata and Nagayoshi Nagai, who joined the laboratory of A. W. von Hoffmann in Berlin. Upon return to Japan, Shibata (Chinese herbs) started a line of distinguished chemists, Keita and Yuji Shibata (flavones) and Shoji Shibata (1915–, lichens, fungal bisanthraquinonoid pigments, ginsenosides); Nagai returned to Tokyo Science University in 1884, studied ephedrine, and left a big mark in the embryonic era of organic chemistry. Modern natural products chemistry really began when three extraordinary organic chemists returned from Europe in the 1910s and started teaching and research at their respective faculties:

Riko Majima, 1874–1962, C. D. Harries (Kiel University); R. Willstätter (Zürich): Faculty of Science, Tohoku University; studied urushiol, the catecholic mixture of poison ivy irritant.

Yasuhiko Asahina, 1881–1975, R. Willstätter: Faculty of pharmacy, Tokyo University; lichens and Chinese herb.

Umetaro Suzuki, 1874–1943, E. Fischer: Faculty of agriculture, Tokyo University; vitamin B_1(thiamine).

Because these three pioneers started research in three different faculties (i.e., science, pharmacy, and agriculture), and because little interfaculty personnel exchange occurred in subsequent years, natural products chemistry in Japan was pursued independently within these three academic domains; the situation has changed now. The three pioneers started lines of first-class successors, but the establishment of a strong infrastructure takes many years, and it was only after the mid-1960s that the general level of science became comparable to that in the rest of the world; the 3rd IUPAC Symposium on the Chemistry of Natural Products, presided over by Munio Kotake (1894–1976, bufotoxins, see below), held in 1964 in Kyoto, was a clear turning point in Japan's role in this area.

Some of the outstanding Japanese chemists not already quoted are the following. Shibasaburo Kitazato (1852–1931), worked with Robert Koch (Np 1905, tuberculosis) and von Behring, antitoxins of diphtheria and tetanus which opened the new field of serology, isolation of microorganism causing dysentery, founder of Kitazato Institute; Chika Kuroda (1884–1968), first female Ph.D., structure of the complex carthamin, important dye in safflower (1930) which was revised in 1979 by Obara *et al.*, although the absolute configuration is still unknown (1998); Munio Kotake (1894–1976), bufotoxins, tryptophan metabolites, nupharidine; Harusada Suginome (1892–1972), aconite alkaloids; Teijiro Yabuta (1888–1977), kojic acid, gibberrelins; Eiji Ochiai (1898–1974), aconite alkaloids; Toshio Hoshino (1899–1979), abrine and other alkaloids; Yusuke Sumiki (1901–1974), gibberrelins; Sankichi Takei (1896–1982), rotenone; Shiro Akabori (1900–1992), peptides, C-terminal hydrazinolysis of amino acid ; Hamao Umezawa (1914–1986), kanamycin, bleomycin, numerous antibiotics; Shojiro Uyeo (1909–1988), lycorine; Tsunematsu Takemoto (1913–1989), inokosterone, kainic acid, domoic acid, quisqualic acid; Tomihide Shimizu (1889–1958), bile acids; Kenichi Takeda (1907–1991), Chinese herbs, sesquiterpenes; Yoshio Ban (1921–1994), alkaloid synthesis; Wataru Nagata (1922–1993), stereocontrolled hydrocyanation.

7. CURRENT AND FUTURE TRENDS IN NATURAL PRODUCTS CHEMISTRY

Spectroscopy and X-ray crystallography has totally changed the process of structure determination, which used to generate the excitement of solving a mystery. The first introduction of spectroscopy to the general organic community was Woodward's 1942–1943 empirical rules for estimating the UV maxima of dienes, trienes, and enones, which were extended by Fieser (1959). However, Butenandt had used UV for correctly determining the structures of the sex hormones as early as the early 1930s, while Karrer and Kuhn also used UV very early in their structural studies of the carotenoids. The Beckman DU instruments were an important factor which made UV spectroscopy a common tool for organic chemists and biochemists. With the availability of commercial instruments in 1950, IR spectroscopy became the next physical tool, making the 1950 Colthup IR correlation chart and the 1954 Bellamy monograph indispensable. The IR fingerprint region was analyzed in detail in attempts to gain as much structural information as possible from the molecular stretching and bending vibrations. Introduction of NMR spectroscopy into organic chemistry, first for protons and then for carbons, has totally changed the picture of structure determination, so that now IR is used much less frequently; however, in biopolymer studies, the techniques of difference FTIR and resonance Raman spectroscopy are indispensable.

The dramatic and rapid advancements in mass spectrometry are now drastically changing the protocol of biomacromolecular structural studies performed in biochemistry and molecular biology. Herbert Hauptman (mathematician, 1917–, Medical Foundation, Buffalo, Np 1985) and Jerome Karle (1918–, US Naval Research Laboratory, Washington, DC, Np 1985) developed direct methods for the determination of crystal structures devoid of disproportionately heavy atoms. The direct method together with modern computers revolutionized the X-ray analysis of molecular structures, which has become routine for crystalline compounds, large as well as small. Fred McLafferty (1923–, Cornell University) and Klaus Biemann (1926–, MIT) have made important contributions in the development of organic and bioorganic mass spectrometry. The development of cyclotron-based facilities for crystallographic biology studies has led to further dramatic advances enabling some protein structures to be determined in a single day, while cryoscopic electron micrography developed in 1975 by Richard Henderson and Nigel Unwin has also become a powerful tool for 3D structural determinations of membrane proteins such as bacteriorhodopsin (25 kd) and the nicotinic acetylcholine receptor (270 kd).

Circular dichroism (c.d.), which was used by French scientists Jean B. Biot (1774–1862) and Aimé Cotton during the nineteenth century "deteriorated" into monochromatic measurements at 589 nm after R.W. Bunsen (1811–1899, Heidelberg) introduced the Bunsen burner into the laboratory which readily emitted a 589 nm light characteristic of sodium. The 589 nm $[\alpha]_D$ values, remote from most chromophoric maxima, simply represent the summation of the low-intensity readings of the decreasing end of multiple Cotton effects. It is therefore very difficult or impossible to deduce structural information from $[\alpha]_D$ readings. Chiroptical spectroscopy was reintroduced to organic chemistry in the 1950s by C. Djerassi at Wayne State University (and later at Stanford University) as optical rotatory dispersion (ORD) and by L. Velluz and M. Legrand at Roussel-Uclaf as c.d. Günther Snatzke (1928–1992, Bonn then Ruhr University Bochum) was a major force in developing the theory and application of organic chiroptical spectroscopy. He investigated the chiroptical properties of a wide variety of natural products, including constituents of indigenous plants collected throughout the world, and established semiempirical sector rules for absolute configurational studies. He also established close collaborations with scientists of the former Eastern bloc countries and had a major impact in increasing the interest in c.d. there.

Chiroptical spectroscopy, nevertheless, remains one of the most underutilized physical measurements. Most organic chemists regard c.d. (more popular than ORD because interpretation is usually less ambiguous) simply as a tool for assigning absolute configurations, and since there are only two possibilities in absolute configurations, c.d. is apparently regarded as not as crucial compared to other spectroscopic methods. Moreover, many of the c.d. correlations with absolute configuration are empirical. For such reasons, chiroptical spectroscopy, with its immense potentialities, is grossly underused. However, c.d. curves can now be calculated nonempirically. Moreover, through-space coupling between the electric transition moments of two or more chromophores gives rise to intense Cotton effects split into opposite signs, exciton-coupled c.d.; fluorescence-detected c.d. further enhances the sensitivity by 50- to 100-fold. This leads to a highly versatile nonempirical microscale solution method for determining absolute configurations, etc.

With the rapid advances in spectroscopy and isolation techniques, most structure determinations in natural products chemistry have become quite routine, shifting the trend gradually towards activity-monitored isolation and structural studies of biologically active principles available only in microgram or submicrogram quantities. This in turn has made it possible for organic chemists to direct their attention towards clarifying the mechanistic and structural aspects of the ligand/biopolymeric receptor interactions on a more well-defined molecular structural basis. Until the 1990s, it was inconceivable and impossible to perform such studies.

Why does sugar taste sweet? This is an extremely challenging problem which at present cannot be answered even with major multidisciplinary efforts. Structural characterization of sweet compounds and elucidation of the amino acid sequences in the receptors are only the starting point. We are confronted with a long list of problems such as cloning of the receptors to produce them in sufficient quantities to investigate the physical fit between the active factor (sugar) and receptor by biophysical methods, and the time-resolved change in this physical contact and subsequent activation of G-protein and enzymes. This would then be followed by neurophysiological and ultimately physiological and psychological studies of sensation. How do the hundreds of taste receptors differ in their structures and their physical contact with molecules, and how do we differentiate the various taste sensations? The same applies to vision and to olfactory processes. What are the functions of the numerous glutamate receptor subtypes in our brain? We are at the starting point of a new field which is filled with exciting possibilities.

Familiarity with molecular biology is becoming essential for natural products chemists to plan research directed towards an understanding of natural products biosynthesis, mechanisms of bioactivity triggered by ligand–receptor interactions, etc. Numerous genes encoding enzymes have been cloned and expressed by the cDNA and/or genomic DNA-polymerase chain reaction protocols. This then leads to the possible production of new molecules by gene shuffling and recombinant biosynthetic techniques. Monoclonal catalytic antibodies using haptens possessing a structure similar to a high-energy intermediate of a proposed reaction are also contributing to the elucidation of biochemical mechanisms and the design of efficient syntheses. The technique of photoaffinity labeling, brilliantly invented by Frank Westheimer (1912–, Harvard University), assisted especially by advances in mass spectrometry, will clearly be playing an increasingly important role in studies of ligand–receptor interactions including enzyme–substrate reactions. The combined and sophisticated use of various spectroscopic means, including difference spectroscopy and fast time-resolved spectroscopy, will also become increasingly central in future studies of ligand–receptor studies.

Organic chemists, especially those involved in structural studies have the techniques, imagination, and knowledge to use these approaches. But it is difficult for organic chemists to identify an exciting and worthwhile topic. In contrast, the biochemists, biologists, and medical doctors are daily facing exciting life-related phenomena, frequently without realizing that the phenomena could be understood or at least clarified on a chemical basis. Broad individual expertise and knowledge coupled with multidisciplinary research collaboration thus becomes essential to investigate many of the more important future targets successfully. This approach may be termed "dynamic," as opposed to a "static" approach, exemplified by isolation and structure determination of a single natural product. Fortunately for scientists, nature is extremely complex and hence all the more challenging. Natural products chemistry will be playing an absolutely indispensable role for the future. Conservation of the alarming number of disappearing species, utilization of biodiversity, and understanding of the intricacies of biodiversity are further difficult, but urgent, problems confronting us.

That natural medicines are attracting renewed attention is encouraging from both practical and scientific viewpoints; their efficacy has often been proven over the centuries. However, to understand the mode of action of folk herbs and related products from nature is even more complex than mechanistic clarification of a single bioactive factor. This is because unfractionated or partly fractionated extracts are used, often containing mixtures of materials, and in many cases synergism is most likely playing an important role. Clarification of the active constituents and their modes of action will be difficult. This is nevertheless a worthwhile subject for serious investigations.

Dedicated to Sir Derek Barton whose amazing insight helped tremendously in the planning of this series, but who passed away just before its completion. It is a pity that he was unable to write this introduction as originally envisaged, since he would have had a masterful overview of the content he wanted, based on his vast experience. I have tried to fulfill his task, but this introduction cannot do justice to his original intention.

ACKNOWLEDGMENT

I am grateful to current research group members for letting me take quite a time off in order to undertake this difficult writing assignment with hardly any preparation. I am grateful to Drs. Nina Berova, Reimar Bruening, Jerrold Meinwald, Yoko Naya, and Tetsuo Shiba for their many suggestions.

8. BIBLIOGRAPHY

"A 100 Year History of Japanese Chemistry," Chemical Society of Japan, Tokyo Kagaku Dojin, 1978.
K. Bloch, *FASEB J.*, 1996, **10**, 802.
"Britannica Online," 1994–1998.
Bull. Oriental Healing Arts Inst. USA, 1980, **5**(7).
L. F. Fieser and M. Fieser, "Advanced Organic Chemistry," Reinhold, New York, 1961.
L. F. Fieser and M. Fieser, "Natural Products Related to Phenanthrene," Reinhold, New York, 1949.
M. Goodman and F. Morehouse, "Organic Molecules in Action," Gordon & Breach, New York, 1973.
L. K. James (ed.), "Nobel Laureates in Chemistry," American Chemical Society and Chemistry Heritage Foundation, 1994.
J. Mann, "Murder, Magic and Medicine," Oxford University Press, New York, 1992.
R. M. Roberts, "Serendipity, Accidental Discoveries in Science," Wiley, New York, 1989.
D. S. Tarbell and T. Tarbell, "The History of Organic Chemistry in the United States, 1875–1955," Folio, Nashville, TN, 1986.

3.01
The World of Carbohydrates and Associated Natural Products

B. MARIO PINTO

Simon Fraser University, Burnaby, BC, Canada

3.01.1 INTRODUCTION

The fields of glycochemistry and glycobiology now feature prominently as mature disciplines. Indeed, the world of carbohydrates and associated natural products appears to have come into its own.[1-4] The diversity of structures made possible by Nature's carbohydrate building set is greater than that of oligonucleotides or oligopeptides,[5] and has given carbohydrates pivotal roles in different areas of biology and chemistry. These range from interacting systems in embryonic development and the control of cell adhesion and cell activation to the provision of energy sources and structural platforms. The rapid development of more sensitive physical methods and analytical techniques[6,7] has led to significant advances in the understanding of the structure, dynamics, and biological functions of carbohydrates. Thus, NMR spectroscopic techniques have evolved to the point that subtle events can now be probed, e.g. the role of structure and dynamics in the binding of oligo-saccharides to complementary receptors,[8-10] or the changes in pK_as of catalytic groups during enzyme action.[11] Significantly, the advent of nanoprobe techniques has opened up new frontiers for the analysis of microgram quantities of complex carbohydrates.[12] Mass spectrometric analysis of carbohydrate-containing macromolecules has undergone a revolution with matrix-assisted laser desorption/time of flight and electrospray ionization techniques,[6,7] and high-performance capillary electrophoresis techniques are now used to probe cellular glycosylation events, with the ultimate goal of single-cell analysis.[13] Structural information derived from X-ray crystallography is now used to infer molecular mechanism, as in the translocation of sugars across a membrane by a transport protein[14] or the formation of a distorted sugar ring or covalent intermediate in a retaining glycosidase reaction.[11] Chemical and enzymatic[15] synthetic methodology now provides key compounds with which to probe the role of oligosaccharide-mediated or oligosaccharide-triggered biological events. A noteworthy contribution is the synthesis of a pentasaccharide related to heparin that stimulates the antithrombin III-mediated inhibition of blood coagulation factor Xa more effectively than the natural pentasaccharide ligand.[16] The emergence of several modern textbooks in the area of carbohydrate chemistry attests to the rapid advances in the synthetic field.[17-24]

The advances described above have been matched by impressive developments in the field of molecular biology and the application of molecular biological techniques to problems in structural biology. The tools have been exploited very effectively in biosynthetic studies of carbohydrates and their derivatives. Indeed, the combination of classical and modern biosynthetic probes has opened up new vistas in the fields of glycobiology and glycochemistry. A logical, unifying theme that links the diverse types of carbohydrates is one of biosynthesis. Knowledge of biosynthetic pathways can be used to advantage in the treatment of disease, the engineering of desirable properties in carbohydrate-processing enzymes or carbohydrate polymers, and the design of carbohydrate-based therapeutics, immunodiagnostics, and vaccines. Accordingly, this volume presents the different aspects of carbohydrates and associated natural products along biosynthetic principles. Although the biological activities of the compounds are an important aspect of their chemical interest, priority has not been given to this subject. Similarly, aspects of isolation, structure elucidation, and synthesis have not been surveyed, although synthesis and structural aspects that relate to biosynthesis have been included in certain cases. The reader is referred to books on carbohydrate chemistry[3,4,17-24] for leading references in the synthetic areas not surveyed in this volume.

3.01.2 OVERVIEW

This volume, containing 20 chapters, focuses on the biosynthesis of the different classes of carbohydrates, their derivatives, and associated compounds. The first part of this volume concentrates on the three classes of glycan-bearing molecules that mediate biorecognition events in eukaryotic systems, namely glycoproteins, glycolipids, and proteoglycans. Thus, Chapters 3.02–3.04 highlight the key role played by glycosidases and glycosyltransferases in the processing of glycan chains in glycoproteins. The next two chapters feature the biosynthesis of glycosphingolipids, sphingolipid transport, and degradation, and the regulation of glycolipid biosynthesis in developing tissues and tumor cells. Chapter 3.07 then deals with the naturally occurring glycosidase inhibitors of the alkaloid class. Chapter 3.08 is devoted to the biosynthesis of proteoglycans such as heparin and heparan sulfate.

In the second part of the volume, some molecules of importance in protozoans and bacteria are surveyed. Thus, Chapter 3.09 describes the biosynthesis of lipopolysaccharides, while Chapters 3.10 and 3.11 feature aspects of bacterial peptidoglycan biosynthesis and its inhibition, and the biosynthesis of glycosylphosphatidylinositol anchors, respectively.

Part three of the volume deals with two topics, namely the occurrence, genetics, and biosynthesis of deoxysugars, and a comprehensive survey of the action of different aldolases.

Part four of this volume highlights the chemistry of carbohydrates, their derivatives, and associated molecules of importance in energy storage, structure, and protection. Topics covered in this section include the biosynthesis of starch and glycogen, and the biosynthesis of pectins, galactomannans, celluloses, and hemicelluloses in plant cell walls. A new perspective on lignin assembly *in vivo* and the synthesis of condensed and hydrolyzable tannins are also presented in this section.

An important topic that has not been surveyed in this volume is the biosynthesis of lipochitin oligosaccharides (LCOs) that act as signaling and regulator molecules to cause root nodule formation in leguminous plants. These molecules are produced by *Rhizobium* bacteria which live in symbiosis with the plants. LCOs contain a core that is comprised of a polymer of β-1,4-linked N-acetylglucosamine with a fatty acyl moiety replacing the acetyl group on the terminal nonreducing unit. The reader is referred to Chapter 13 in Volume 1 of this series and Ref. 25 of this chapter for a discussion of the biosynthesis of LCOs and their role in nodulation.

3.01.3 PART 1: GLYCOSIDASES, GLYCOSYLTRANSFERASES, *N*- AND *O*-LINKED GLYCOPROTEINS, GLYCOSPHINGOLIPIDS, GLYCOSIDASE INHIBITORS, AND PROTEOGLYCANS

The glycan chains linked to proteins and lipids are implicated in a variety of biological processes such as intercellular interactions in embryonic development, differentiation and maturation, cell signaling events, intracellular targeting of enzymes, and cell adhesion. Changes in the composition of glycan chains have also been linked to the metastatic cancer state. In addition, the glycan chains can modulate the physical and biological properties of their carrier molecules, which may themselves be signaling molecules. The protein-bound glycans occur as oligosaccharides linked through asparagine (Asn) to give *N*-linked glycans or through serine (Ser) or threonine (Thr) to give *O*-linked glycans. The assembly or processing of the *N*-linked oligosaccharide units is complex and employs both glycosidase and glycosyltransferase enzymes. In Chapter 3.02, Herscovics describes the biosynthesis of *N*-glycans in eukaryotes with particular emphasis on the role and importance of glycosidases in glycoprotein processing. She reviews the different types of *N*-glycans and then traces the biosynthesis of the initially formed lipid-linked oligosaccharide precursor from dolichol phosphate and monosaccharides in the membranes of the endoplasmic reticulum (ER). The oligosaccharide precursor $Glc_3Man_9GlcNAc_2$ is then transferred to specific asparagine residues on polypeptide chains in the lumen of the ER by an oligosaccharyltransferase complex. Following transfer, the oligosaccharide precursor is modified or processed by glycosidases and glycosyltransferases to give mature *N*-glycans. The processing pathways of the ER and subsequently the Golgi in mammals and yeast are surveyed. The author stresses that unlike protein and nucleic acid biosynthesis, which is template based, *N*-glycan biosynthesis depends on the specificity of the glycosidases and glycosyltransferases and their proper subcellular localization. She concludes with insights into the role of the processing glucosidases as a quality control mechanism to ensure proper folding of proteins, proper membrane localization, and transport of glycoproteins out of the ER, and into the influence of mannose trimming by mannosidases that leads to proteolytic degradation of malfolded proteins. Chapter 3.03 by Schachter focuses on the role of glycosyltransferases in *N*-glycan synthesis and highlights the advances in molecular biology that have revealed new glycosyltransferase genes. He begins with an overview of the structure and biosynthesis of *N*-glycans and then reviews the structural features of the glycosyltransferases. In particular, genetic studies that link the various domains of these enzymes to catalytic functions or targeting and membrane anchoring functions are surveyed. Control of gene transcription, the organization of the transferase genes, and the variation of gene expression in normal and disease states are focal points of this chapter. In conclusion, the author indicates that recent discoveries have arisen from screening DNA databases for homologous genes or expressed sequence tags, and hints that this approach might lead to the identification and cloning of many more glycosyltransferase genes, thus permitting definition of the functional roles of glycan chains in development and differentiation.

In Chapter 3.04, Brockhausen describes the different structures of *O*-glycans on mucins, the main class of *O*-glycosylated proteins, and surveys the action of glycosyltransferases involved in their assembly. Unlike *N*-glycan processing, glycosidases do not appear to be involved here and glycosyltransferases are the main players. The terminal structures of *O*-glycans resemble those on *N*-glycans; however, the core structures and the early-acting enzymes of the *O*-glycan pathways appear to be specific to *O*-glycan processing. Although several different sugars are linked to Ser and Thr via *O*-glycosidic linkages, the author focuses on *O*-glycans that are based on core structures which contain Gal*N*Ac α-linked to Ser or Thr of glycoproteins. The synthesis of the *O*-glycan core structures and their subsequent elongation, branching, and modification to give mature structures are surveyed. The nature of the *O*-glycan structures involved in the display or masking of blood group and tissue antigens and cancer-associated antigens, and their roles in modulating cell-surface and cell-adhesion properties and in growth, differentiation, and disease states are also presented.

In Chapter 3.05, Van Echten-Deckert and Sandhoff present the organization and topology of glycosphingolipid (GSL) biosynthesis. Although most of the GSLs are integral parts of the plasma membranes of vertebrate cells, their biosynthesis takes place at the membranes of the ER and continues on the Golgi membranes. Degradation occurs in the lysosomal compartment following endocytosis. GSL degradation products such as sphingosine and ceramide have been implicated in signal transduction. The authors present a detailed account of the localization of GSL biosynthesis and of intracellular movement and transport of molecules as GSLs move to the cell surface from the Golgi, and the topology of endocytosis and lysosomal degradation as a portion of these molecules moves from the plasma membrane to the intracellular organelles. The recycling and salvage of catabolic products in GSL metabolism, particularly of "high-energy" molecules such as sialic acid and sphingosine, are also discussed. The assistance of activator proteins required for the hydrolytic degradation of GSLs is also described; the authors propose that the role of the simultaneous action of a hydrolase and an activator protein might be to protect the plasma membrane from unwanted degradation. The authors indicate that GSL species are detected in almost all intracellular membranes. Remaining to be clarified is the mechanism of movement of glycolipids to these sites and their functional roles. Further challenges are elucidation of the details of endocytosis of GSLs from the plasma membrane and the roles and mechanism of action of the activator proteins. In Chapter 3.06, Basu *et al.* present a comprehensive survey of the many glycosyltransferases involved in GSL biosynthesis that have been isolated or cloned and characterized. Particular note is made of the acceptor specificity of the enzymes and their requirement for other motifs such as the hydrophobic domain provided by the ceramide moiety. The authors conclude with a discussion of the regulation of GSL biosynthesis and suggest that further understanding of the regulation of expression of specific GSLs on the plasma membrane or intracellular membranes during development or metastasis will require an understanding of their gene structure and transcriptional regulation.

A great deal of activity in the synthetic field has focused on the design and synthesis of novel glycosidase inhibitors. These compounds are intended to block specific steps in the trimming of *N*-linked glycans in order to produce aberrant oligosaccharide structures for probing molecular recognition events mediated by the carbohydrate structures on glycoproteins. In some cases, the compounds are candidate therapeutic agents for the treatment of disease states, e.g. metastatic cancer or retroviral infection. It is beyond the scope of this volume to deal with the different classes of natural and synthetic glycosidase inhibitors. Rather, Elbein and Molyneux, in Chapter 3.07, focus their discussion on naturally-occurring glycosidase inhibitors of alkaloid origin. These compounds are polyhydroxy alkaloids from plant sources and microorganisms which possess glycosidase inhibitory properties. Their occurrence in microorganisms may suggest an endophytic relationship with plants and is likely to be of ecological significance. The authors present a useful compilation of the natural sources of the various alkaloids together with their enzyme targets and then proceed to discuss the biological activities of the compounds, which range from insecticidal and antimicrobial activity and plant growth inhibition to mammalian toxicity and therapeutic activity. Sections on *N*-linked glycoprotein processing (as in Chapters 3.02 and 3.03) and the consequences of the inhibition of particular glycosidase enzymes along the processing pathway by selected candidate inhibitors are then presented.

Proteoglycans on mammalian cell surfaces differ from glycoproteins and glycolipids in that their carbohydrate chains can be *N*- or *O*-linked oligosaccharides or glycosaminoglycans (GAGs). The GAGs contain a polyanionic backbone consisting of *N*-sulfonyl or *N*-acetyl hexosamine residues alternating with hexuronic acid or galactose residues. The GAGs chondroitin/chondroitin sulfate, dermatan/dermatan sulfate, and heparin/heparan sulfate are all bound *O*-glycosidically to a serine residue of the core protein via a tetrasaccharide linker, whereas keratan/keratan sulfate is linked *N*-glycosidically to an asparagine residue of the core protein. Hyaluronan, another GAG, consists of a polymer of glucuronic acid and *N*-acetylglucosamine and is not covalently attached to protein. In addition to modulating cell adhesion and proliferation, these molecules are important players in wound repair, coagulation, and lipolysis. In Chapter 3.08, Wight discusses core protein biosynthesis and GAG assembly on specific serine residues within core proteins. A useful grouping of proteoglycans according to location in tissues and according to similarities in core protein and gene structure is presented. The author offers insight into the genetic regulation of core protein biosynthesis and the action of cytokines and growth factors in activation of specific genes. The synthesis of a particular GAG appears to be regulated by the amino acid sequences around the glycosylation sites coupled with the genetic background of the cells. Core protein biosynthesis and initial glycosylation take place in the ER, and further glycosylation and GAG assembly then take place in the Golgi by alternate addition of hexosamines and glucuronic acid or galactose to a linkage tetrasaccharide, GlcUA-Gal-Gal-Xyl, attached to serine (in the case of hyaluronan, biosynthesis occurs at the plasma membrane). Thus, hexosaminyltransferase enzymes specific for the transfer of α- or β-Glc*N*Ac or β-Gal*N*Ac likely recognize the linkage tetrasaccharide, the amino acid sequences flanking the attachment sites and/or the local protein conformation to initiate the synthesis of different GAG chains. The growing chains are modified by epimerases (to give iduronic acid), deacetylases, and sulfotransferases. The sulfation appears to occur as the GAG chains are being assembled, not subsequent to their assembly, and is closely linked to the epimerase activity.

3.01.4 PART 2: LIPOPOLYSACCHARIDES, PEPTIDOGLYCAN, GLYCOSYL-PHOSPHATIDYLINOSITOLS, AMINOGLYCOSIDE AND AMINOCYCLITOL ANTIBIOTICS

Gram-negative bacteria display at their surface different macromolecules of which lipopolysaccharides (LPS) form an important class that is essential for bacterial viability. Located at the outer membrane, LPS are the main surface antigens of Gram-negative bacteria and aid in their elimination by the immune system. In contrast, LPS can play a role in bacterial virulence by interfering with complement activation and phagocytosis and can also lead to a wide range of toxic effects, hence their alternate name endotoxin. The diverse functions of LPS associated with increased bacterial virulence have led to its identification as a potential target for new therapeutic agents. Consequently, a great deal of work has been aimed at the elucidation of the structures, biosynthesis, biological activities, and functions of LPS. In Chapter 3.09, Mamat *et al.* present a comprehensive account of all these aspects. The biological activities of LPS that are described offer insights into the intricacies of its interaction with humoral and cellular targets in host systems and suggest points of attack to interfere with endotoxic properties. A detailed analysis of the chemical structures of the LPS of the major pathogenic bacteria then follows. An interesting section on physicochemical properties is then presented in which the effect of conformation, aggregate structure, and phase state on the interaction with host cell membranes is probed. Here it is suggested that endotoxicity is dependent on the conformation of the lipid A component and that a shift in aggregate–monomer equilibrium leads to intercalation of the monomer units in the phospholipid membrane. The remainder of the chapter provides the reader with a fascinating account of the biosynthesis of the sugars present on the *O*-chain, their assembly into the *O*-polysaccharide repeating units, the biosynthesis of lipid A and the inner core, and finally the outer core. Throughout this chapter, emphasis is placed on both biochemical and genetic data to account for structural polymorphism, and the participation of various gene products in LPS biosynthesis is highlighted. The mechanisms of translocation of carrier lipid-linked *O*-repeating units across the inner membrane and their polymerization into *O*-polysaccharides are also outlined. An intriguing aspect of outer core assembly in nonenteric bacteria is presented, namely the ability to synthesize LPS forms in response to changes in the micro-environment of the host. Thus, variable oligosaccharides expressed on the LPS of human mucosal pathogens such as *Neisseria meningitidis* and *Haemophilus influenzae* can mimic those on human glycosphingolipids, thereby escaping immune detection and also leading to functional mimicry of host molecules. The authors conclude by posing future challenges which include the understanding and characterization of the regulatory steps in LPS biosynthesis, the response of LPS biosynthesis to extra- and intracellular signals, and the mechanisms of polysaccharide transport across the cytoplasmic membrane, through the periplasm, and its incorporation into the outer membrane.

The peptidoglycan layer is an important constituent of both Gram-positive and -negative bacterial cell walls, the former containing a thick outer layer of peptidoglycan and the latter having a thin layer that forms part of the outer membrane. The main function of the peptidoglycan layer is structural in that it prevents lysis of the bacterial cell, but this layer is closely associated with other molecules. In Gram-positive organisms, the association is with teichoic acids, strongly anionic polyol phosphates, whereas in Gram-negative organisms, the association is with lipoproteins of the outer membrane. In Chapter 3.10, Bugg presents an exhaustive account of peptidoglycan biosynthesis and its inhibition. The bacterial cell-wall peptidoglycan is composed of glycan chains, a pentapeptide side chain containing D-amino acids, and interstrand peptide cross-links. Following a discussion of the structures of the different components, the author focuses on their biosynthesis and assembly to give peptidoglycan. Of particular interest is the synthesis of lipid intermediate I containing conjugates of *N*-acetylmuramic acid with the pentapeptide and its elaboration to the lipid-linked disaccharide (Mur*N*Ac–Glc*N*Ac)-pentapeptide conjugate (lipid II) on the cytoplasmic face of the membrane. After peptide cross-linking, the modified lipid intermediate II is translocated across the cytoplasmic membrane, possibly assisted by protein. This intermediate is then polymerized by transglycosylation with concomitant release of the lipid carrier. The author draws an analogy with the mechanism of *N*-glycan biosynthesis (see Chapter 3.02), which also takes place initially on a lipid carrier and is subsequently transferred to the protein to give the *N*-linked glycoprotein, and of LPS biosynthesis (see Chapter 3.09) in which lipid-linked repeating units are flipped across the cytoplasmic membrane and subsequently transglycosylated. The final transformation into structurally rigid peptidoglycan occurs by cross-linking of the peptide chains by a transpeptidase enzyme. The transglycosylase and transpeptidase enzymes belong to the family of penicillin-binding proteins. Once formed, the peptidoglycan framework is not static and changes occur in response to changes in shape during growth and cell division. Thus, the breakdown of peptidoglycan in localized areas is of some importance and takes place through the action of glycosidase and peptidase enzymes. In this chapter, the author highlights molecular aspects of the hydrolysis reactions and discusses a pathway for recycling of peptidoglycan fragments. An interesting section on peptidoglycan assembly in antibiotic-resistant bacteria is also presented. Increased bacterial resistance to antibiotics that interfere with peptidoglycan assembly is a serious concern.

Vancomycin-resistant *Enterococci* and methicillin-resistant *Staphylococcus aureus* have developed alternative pathways for peptidoglycan assembly. Vancomycin inhibits peptidoglycan assembly by preventing transglycosylation and transpeptidation. In resistant strains, it is proposed that vancomycin induces the expression of enzymes that implement a modified pathway for peptido-glycan synthesis which involves the incorporation of D-lactate or D-2-hydroxybutyrate (X) into the peptidoglycan. An initially formed D-Ala–D-X ester instead of the D-Ala–D-Ala amide is further elaborated to give a tetrapeptide–D-X unit linked to MurNAc. Translocation occurs as before and cross-linking between strands then occurs by aminolysis of the D-X ester linkage. In addition, resistant strains appear to contain a peptidase that is specific for the regular D-Ala-D-Ala dipeptide precursors, and whose hydrolytic action aids in the switch from the normal to the modified biosyn-thetic pathway. The methicillin-resistant strains of *Staphylococcus aureus* appear to express a novel penicillin binding protein that has a very low affinity for penicillins but are still capable of peptidoglycan assembly. Since peptidoglycan biosynthesis has no direct counterpart in eukaryotic systems and is necessary for bacterial survival, its inhibition has been the subject of intensive investigation. The biosynthesis occurs in the cytoplasm, the membrane, and extracellularly, and a detailed account of approaches to the inhibition of all these stages is presented.

In Chapter 3.11, Eckert *et al.* describe advances in the biosynthetic studies of glycosylphos-phatidylinositols (GPIs) in parasitic protozoa and higher eukaryotes. The authors trace how an exotic motif for anchoring surface antigens in protozoa was later shown to be a general feature in eukaryotic systems. Thus, the main surface proteins in protozoa were shown to be linked to the membrane-associated phosphatidylinositol via ethanolamine and a carbohydrate bridge. Sub-sequently, GPI anchors were shown to be widely distributed in eukaryotes (except for plant cells), linking glycoproteins to cell surfaces, in contrast to their usual anchoring via transmembrane domains. Whereas GPI anchoring appears to be a general phenomenon among protozoans, it is associated mainly with proteins of specialized function in eukaryotes. The dense packing of GPI-anchored proteins coupled with antigenic variation is thought to protect parasites from the host immune system. The authors present the structural details of GPI anchors that have been conserved through eukaryotic evolution and then go on to describe the biosynthesis of the different components with particular attention to genetic aspects. The core glycan structure and its assembly are conserved, while the modifications of the carbohydrate backbone and the hydrophobic moiety and the order of modification vary markedly in different organisms. Thus, modifications may precede or follow elaboration of the core glycan, e.g. the addition of an ester-linked fatty acid to the inositol ring. Similarly, modifications may occur before or after transfer of the GPI unit to the protein. In some organisms, the protein is absent and protein-free GPIs occur as metabolic end products. The topology of GPI anchor biosynthesis is another interesting aspect. Biosynthesis of GPI anchor precursors occurs on the cytoplasmic leaflet of the ER and these are translocated across the ER membrane to the luminal side where transfer to protein, itself previously translocated into the ER, takes place. This occurs via a transamidase reaction in which the original carboxy terminus is cleaved and replaced by the GPI anchor via the amino function of the terminal ethanolamine phosphate. The roles of the different domains of the protein in signaling its translocation and anchoring in the ER membrane, and the recognition sites for transamidase action are described. The authors propose several functions for GPI anchors which include anchoring membrane proteins, transmembrane signaling, second messenger activity, and increasing parasite pathogenicity. Differ-ences in properties, e.g. tyrosine kinase activation, between transmembrane and GPI-anchored forms of the membrane proteins have led to the suggestion that GPI anchors are involved in signal transduction processes. A section describing various GPI biosynthesis inhibitors that have aided the elucidation of GPI structures and biosynthesis is presented. The individual aspects of GPI anchor biosynthesis in protozoa, yeast, and mammalian cells are also surveyed. The authors conclude with a section on lipophosphoglycans (LPGs) and glycosylinositolphospholipids (GIPLs) which contain the common structural motif present in GPI anchors but are highly modified and form a protective coat of nonprotein-bound glycolipids.

The aminoglycosides and aminocyclitols constitute an important class of antibiotics although their use has been overshadowed by other antibiotics such as the third- and fourth-generation cephalosporins. Investigations, especially of the well-known streptomycin, have led to an increased understanding of the biosynthesis and genetic regulation of this class of molecules. Resistance to the aminoglycoside antibiotics is a significant clinical problem and the molecular biology of resist-ance has been a primary focus of several investigations. Regarding the mechanism of bactericidal action of aminoglycoside antibiotics, it is generally accepted that streptomycin interacts with riboso-mal protein S12, causing inaccurate protein synthesis.[26] The incorporation of mismade proteins into the membrane increases the permeability of the cell, allowing more streptomycin to enter, and at high concentrations the antibiotic completely blocks protein synthesis, causing cell death. The other aminoglycoside antibiotics act in a fundamentally similar fashion, varying in the specificity of their interactions with the ribosome and their susceptibility to inactivation by resistance enzymes.

The aminoglycoside and aminocyclitol antibiotics have not been treated explicitly in this volume. For more details of the biosynthesis and regulation of this class of compounds, and molecular biological aspects of resistance, the reader is referred to two excellent articles[27,28] and to Section 3.12.5.3.4 of this volume.

3.01.5 PART 3: DEOXYSUGARS, ALDOLASES

In Chapter 3.12, Johnson and Liu present a cogent account of the genetics and mechanisms of biosynthesis of deoxysugars from a variety of plant, bacterial, and mammalian sources. These sugars are found in LPS, glycoproteins, and glycolipids, and are involved in mediating cellular and molecular recognition events. Deoxysugars also occur as components of bacterial antibiotics, where they play a crucial role in targeting and binding. The chapter deals with deoxysugars not containing amino functionality. The authors begin with a useful compilation of the sources of naturally-occurring deoxysugars and then proceed to describe their biological activities as components of LPS, different types of bacterial antibiotics, and cardiac glycosides. The main portions of the chapter are dedicated to the mechanisms of deoxysugar biosynthesis with exquisite details of the biosynthetic enzymes involved in the transformations. Insights gained from molecular biological studies are highlighted. Thus, the location and sequencing of genes has permitted disruption of the individual genes and examination of the resulting metabolites. Comparisons of the deduced protein sequences with those of well-characterized enzymes have led to the proposal of several biosynthetic pathways for which little or no biochemical evidence existed previously. This section describes the state-of-the-art in the genetics of deoxysugars in *O*-chains of LPS and antibiotics, and in class I and II reductases. Of particular interest will be the exploitation of this knowledge for engineering recombinant strains to produce novel antibiotics.

Enzyme-catalyzed aldol reactions are invaluable in forming carbon–carbon bonds. The aldolase enzymes that accomplish this task are varied and employ several different mechanisms. In Chapter 3.13, Henderson and Toone present an incisive account of the action of aldolases. They focus their attention on aldolases that use either a Schiff base or divalent zinc for nucleophilic activation, enzymes that catalyze aldol or benzoin-type reactions with the nucleophilic assistance of pyridoxal or thiamine cofactors, transaldolase or transketolase enzymes of glycolysis, and the aldol reactions of phosphoenolpyruvate, a preformed enolate nucleophile. Several examples of reactions proceeding in the retro-aldol direction are also described. Such reactions are of importance in catabolism. The authors have presented a comprehensive account of the different types of aldolases that might find use in synthesis. They include descriptions of the sources of the enzymes, their *in vivo* roles, optimal reaction conditions, X-ray structural and mechanistic information, substrate specificities, and synthetic applications.

3.01.6 PART 4: STARCH AND GLYCOGEN, PECTINS AND GALACTOMANNANS, CELLULOSES, HEMICELLULOSES, LIGNIN, CONDENSED AND HYDROLYZABLE TANNINS

In Chapter 3.14, Preiss and Sivak discuss the biochemistry and molecular biology of starch biosynthesis in plants and of glycogen biosynthesis in bacteria and mammals. These 1,4-linked glucans containing 1,6-linked branches serve as storage reserves whereby degradative enzymes, the amylases and phosphorylases, process the different branches simultaneously and quickly to give glucose. The discovery of sugar nucleotides by Leloir *et al.*[29–32] led quickly to the realization that glycosyl transfers from sugar nucleotides to acceptors would lead to the elaboration of oligo- and polysaccharides. In the case of the synthesis of bacterial glycogen and starch, the glucose nucleotide, derived from glucose 1-phosphate and nucleoside triphosphates, is transferred to a primer glucan to give a linear polysaccharide that can undergo rearrangement to give branched polysaccharides. The authors describe the roles of glycogen in bacteria and mammals and of starch in plants and then proceed to present details of their biosynthesis. Whereas the glucose donor for glycogen synthesis in mammals is UDP-glucose, ADP-glucose serves as the donor for starch synthesis in plants and algae and for glycogen synthesis in bacteria. In bacteria, there are alternative pathways leading to the formation of glucans, namely from sucrose or maltose or from glucose 1-phosphate via the phosphorylase reaction. A detailed account of the structural and mechanistic aspects of the different enzymes involved in starch and glycogen biosynthesis, together with insights gained fom molecular biological studies, form a significant portion of the chapter. An interesting hypothesis is advanced to account for the biosynthesis of amylose and amylopectin regarding the specific roles of the starch synthases and the branching and debranching enzymes. The questions are of relevance to the formation of starch granules and the variety in their number and size per cell in different plant species. A comprehensive treatment of the modes of regulation of plant and algal starch synthesis, bacterial glycogen synthesis, and mammalian glycogen synthesis is also presented. Whereas in bacteria and plants ADP-glucose is used only for (1-4)-glucan synthesis, and is regulated at the level of ADP-glucose formation, in mammals, UDP-glucose is used as a substrate for the synthesis of several other cellular constituents and is regulated at the glycogen synthase step. The mechanism of control in the latter case is by allosteric regulation and also by covalent modification (phosphorylation/dephosphorylation). The effects of phosphorylation at individual sites and the synergistic inactivation of the enzyme by phosphorylation at different sites is highlighted. The insulin-induced dephosphorylation at specific sites provides a mechanism for stimulation of glycogen synthesis. Biochemical and genetic evidence is also presented to suggest that glycogen synthesis in mammals requires the associated protein glycogenin and that glucosylation of this protein gives a primer for glucan synthesis.

The next section of the volume deals with the structural components of plant cell walls. Primary cell walls are composed mainly of polysaccharides and are found in different parts of the plant, e.g. around growing cells, cells in leaves, and in the junctions between cells. New wall is continually laid down at the plasma membrane and the older wall is pushed outward. Secondary walls often have altered polysaccharide composition and morphology and may be associated with lignin. The polysaccharides in the primary wall are celluloses (β-(1-4)-linked glucans), hemicelluloses (xyloglucans), and pectin (polysaccharides containing D-galacturonic acid that are partially methyl esterified), while those in the secondary wall include the galactomannans. The latter are referred to as gums or mucilages and are of importance as food reserves for the plant and as food additives for man. In Chapter 3.15, Mohnen focuses mainly on three pectic substances of the primary wall, namely homogalacturonan, rhamnogalacturonan I (RG-I), and rhamnogalacturonan II (RG-II). RG-I contains a backbone of D-galacturonic acid α-(1-2)-linked to L-rhamnopyranose to which are attached oligosaccharide chains containing arabinosyl and/or galactosyl residues. The former units exist in the furanose form. RG-II has a complex structure and contains up to 11 different types of glycosyl residues including the unusual sugars L-aceric acid and D-apiose. Interestingly, RG-II in the wall exists as a dimer that is cross-linked by a borate diester. The author describes the synthesis of the different nucleoside diphosphate donors via the nucleotide interconversion pathway or the salvage pathway, and their translocation from the cytosolic side of the Golgi membranes to the lumen of the ER and Golgi. The action of glycosyltransferases then yields the pectic polysaccharides which are then transported to the plasma membrane. The issue of whether partial esterification takes place prior to insertion into the cell wall is addressed. The action of the nonglycosyltransferase enzymes such as the methyltransferases or acetyltransferases for modification of the glycosyl units is then described. The coordinated regulation of the multiple enzymes required for pectin bio-synthesis, e.g. in the transition from primary to secondary wall synthesis, is particularly critical and the limited available information is summarized. Finally, the biosynthesis of galactomannans of the secondary wall in endosperm cell walls and seeds in legumes is reviewed. The author makes the point here that the careful biochemical analysis of galactomannan galactosyltransferase activity in cell extracts has led to models for the mechanism of coordination of the multiple enzymes involved in the production of species-specific polysaccharides, prior to the cloning of genes or purification of the enzymes. The value of examining the secondary gene products (polysaccharides) in addition to the primary gene products is stressed.

Cellulose is the key structural component of wood, cotton, and paper and, as such, is one of the most important natural resources known to man. In Chapter 3.16, Atalla presents an overview of this important polysaccharide with emphasis on its states of aggregration and the manner in which this affects its properties. Native celluloses occur in diverse forms and are produced by plants, bacteria, and marine organisms. Although these structures are generally highly ordered, the cor-relation with the state of aggregation is not obvious. The author addresses the issues of aggregation, tertiary structure, and morphology of cellulose in order to provide a better understanding of its biosynthesis, biological function, and biodegradation. These aspects are relevant to the industrial utilization of cellulosic materials. The author presents a fascinating account of the elucidation of the structures of the two forms of native cellulose. The application of X-ray diffractometry, X-ray crystallography, Raman and infrared spectroscopy, and electron miscroscopy is described and new insights obtained by CP/MAS solid-state NMR spectroscopy are highlighted. The extension of the methods to explore differences between native celluloses of different biological origins is also described. Such information derived at the nanoscale level is used to interpret the organization of cellulose into supramolecular assemblies at the microscale level. The author then focuses on the biogenesis of celluloses at two levels. The first summarizes biosynthetic studies of a bacterial cellulose that have been facilitated by isolation of the cellulose synthase complex. The assembly of cellulose chains proceeds via a direct displacement on a UDP-glucose unit by another glucosyl moiety, without the formation of covalent enzyme intermediates or lipid-linked oligosaccharide intermediates. A critical activator and regulator of cellulose synthase, cyclic diguanylic acid, and the enzymatic control of its levels are described. The identification of the bacterial synthase genes has permitted extension of studies to similar genes coding for binding of UDP-glucose and β-(1-4)-glycosidic bond formation in plants. At the second level of biogenesis, the author addresses issues of ultrastructure, that is, the control and directionality of glucan chain polymerization and its crystallization to form cellulose microfibrils.

The organization in native states of celluloses has implications for the action of cellulolytic enzymes and chemical agents, and the use of solvent systems for solubilization and swelling. These properties are closely related to states of aggregation. For example, differences in lattice order brought about by penetration of semicrystalline domains by solvent can result in swelling of cellulose without dissolution. The author describes microfibril structure and states of aggregation in native celluloses and co-aggregation with other components such as the hemicelluloses. In the final section, the action of hydrolytic enzymes in the biodegradation of celluloses is described. The knowledge is of particular interest for modification of the properties of cellulosic fibers and for the de-inking process in the recycling of paper.

Hemicelluloses are best defined and distinguished from pectin and cellulosic polysaccharides in terms of their chemical structures. The primary wall hemicelluloses contain different types of highly branched xyloglucans. Some consist of β-(1-4)-linked glucose units with attached α-D-xylopyranose (Xyl*p*) units which may have, in turn, galactopyranose or fucose substituents. Others are glucuronarabinoxylans, comprised of backbones of (1-4)-linked β-D-Xyl*p* units, substituted with arabinosyl, galactosyl, and glucuronyl residues, or glucans with alternating β-(1-3) and β-(1-4) units. Hemicelluloses of the secondary wall consist of (1-4)-linked β-D-Xyl*p* units, substituted with 4-*O*-methyl-D-glucuronic acid, D-glucuronic acid, arabinose, and acetate. Glucomannans also form part of the hemicellulosic component in secondary walls. In Chapter 3.17, Gregory and Bolwell present a detailed account of the biosynthesis of hemicelluloses in relation to its interaction with other cell wall components. Immunolocalization of the various glycans in different portions of the cell wall has led to a more precise definition of the order of deposition of the glycans in the newly formed wall and to the localization of the assembly of hemicellulosic glycans to the *trans*-Golgi. It is now clear that xyloglucan is the main hemicellulose of the primary wall and xylan and its variants are the main components of the secondary wall. The authors have effectively summarized the associations between hemicellulose glycan assembly and the other cell-wall components. The current state of knowledge of the glycosyltransferase activity associated with hemicellulose glycan biosynthesis is summarized, together with the genetic studies to date. The regulation of hemicellulose biosynthesis during growth leads to changes in the types and quantities of glycans synthesized. The control mechanisms involved in synthesis of the nascent hemicelluloses and their subsequent modification are discussed. An interesting section on the regulation of development and differentiation by extracellular signals supplied by molecules such as auxin, cytokinin, sucrose, ethene, and the more complex oligosaccharins is presented. The turnover and modification of polysaccharides by extracellular enzymes such as the glycanases, transglycosylases, and endoglycan transferases after initial synthesis and deposition is also described. The authors conclude with the suggestion that modification of the hemicellulose content in the cell wall by genetic manipulation might lead to improved fiber quality in paper and might, therefore, be an attractive target for plant biotechnology.

Lignins, suberins, and covalently-bound hydroxycinnamic acids form an integral part of plant cell walls. Their assembly is intimately linked to the biosynthesis of other constituents such as the hemicelluloses, and its regulation. Lignins provide mechanical support but have also been associated with control of diffusion and defense against microorganisms. Phenylpropanoid metabolites in the cytoplasm are converted into monolignols which are transported into the cell wall where lignin deposition occurs. The suberins are lignin-like substances that contain both phenylpropanoid and esterified fatty acid and alcohol components and are deposited in an analogous manner. In Chapter 3.18, Lewis *et al.* treat the complex issues of lignin biogenesis and biodegradation. Significantly, progress regarding structure elucidation *in vivo* has led to a new hypothesis for lignin assembly. The point is of importance because previous hypotheses were based on the isolation of lignin preparations that were altered by chemical treatment or on studies of model compounds aimed at probing biodegradation mechanisms. Attention is focused on the intracellular transport of monolignols to the cell wall and the possible involvement of monolignol UDPGlc glucosyltransferase and β-glucosidases in this process. The current understanding of the roles of the oxidative enzymes responsible for generating the free-radical intermediates that lead to dehydrogenative coupling of monolignols to give lignin is presented. The involvement of dirigent proteins in controlling the assembly of lignins is discussed. The orchestration of lignification with the deposition of the other cell-wall components is stressed.

In addition to the primary metabolites described thus far, plants also synthesize secondary metabolites such as alkaloids, terpenes, and tannins. The tannins are an important class of compounds that serve to protect the plant from microbial infection and insect attack. Their biosynthesis consumes a significant amount of stored photosynthetic energy and is closely linked to nitrogen cycling. Historically, hydrolyzable tannins were classified according to their susceptibility to acid hydrolysis and were shown to be compounds based on gallic acid and hexahydroxydiphenic acid. The condensed tannins were classified as those derived from flavan-3,4-diol, although subsequently they too were shown sometimes to contain gallic acid substituents. Today, hydrolyzable tannins are known as gallo- or ellagitannins and condensed tannins as proanthocyanidins or polyflavonoids. In Chapter 3.19, Ferreira *et al.* present a chemical approach to trace the conversion of the monomeric flavan alcohols and flavans to oligomeric proanthocyanidins. Thus, chemical transformations and semisynthesis of oligoflavonoids are used to infer the principles and pathways of their biosynthesis. The chemistry of oligomeric proanthocyanidins also provides an understanding ranging of the processes involved in the *post mortem* aging of the corresponding polymers in wood and bark and of the requirements for certain compounds in leather tanning and in the production of adhesives. An interesting section on the astringent taste produced on the palate by oligomeric proanthocyanidins is included. The astringency of beverages such as red wine are attributed to the precipitation in mucous secretions of proline-rich salivary proteins (PRPs) and mucopolysaccharides by the proanthocyanidins. The PRPs have been implicated in a defense role against polyphenols in the digestive tract. The authors conclude with comments on the beneficial effects of tannins in red wine, for example, in preventing diseases such as coronary heart disease through their action as antioxidants and free radical scavengers.

In Chapter 3.20, Gross begins with a historical perspective of the use of polyphenols for a variety of uses including the tanning of hides to leather, and then presents a biosynthetic perspective of the assembly of hydrolyzable tannins. The author traces the biosynthesis of gallic acid and its further elaboration to β-glucogallin and then to penta-*O*-galloyl-β-D-glucopyranose, the key precursor of the two subclasses of hydrolyzable tannins, namely the gallotannins and ellagitannins. The former class of compounds derives from addition of additional galloyl residues to the pentagalloylglucose core via depside bond formation (an esterification). In contrast, ellagitannins result from oxidative processes that lead to C—C bond formation between adjacent galloyl groups. In addition, dimer and oligomer formation results from C—C or C—O bond formation via intermolecular oxidative reactions. The details of the formation of the depside bonds by galloyltransferase to form gallotannins are then presented, followed by an account of the frustrating attempts to unravel the enzymes responsible for the oxidative transformation of pentagalloylglucose to give ellagitannins. An interesting section on the catabolism of hydrolyzable tannins is presented. Degradation is performed not only by the plants but also by fungi, yeasts, and bacteria. The enzyme is a tannase (tannin acyl hydrolase) that cleaves the ester bonds between aryl groups and the central glucose moiety to give gallic acid and the aryl-*O*-aryl moieties in the case of ellagitannins. The depsidic ester linkages of the gallotannins are also hydrolyzed by tannase. The author proposes that the degradation of tannins could play a role in reducing astringency in ripening fruits. He concludes by posing several challenges which include an understanding of the regulation of the synthesis and degradation of tannins, the cellular localization of the synthetic enzymes and their products, and the ability of the enzymes to process plant polyphenols which normally lead to precipitation of proteins, as in the tanning process.

3.01.7 CONCLUSIONS

The discussion presented in the foregoing sections attests to the diversity of function of carbo-hydrates and their associated molecules. From such diversity comes strength. It is striking that one class of molecules could be responsible for such a wide array of phenomena ranging from molecular recognition events that mediate cellular communication to providing energy sources or mechanical strength. It is clear that the network of interactions within the different carbohydrate components and between these components and other biomolecules is critical to many of the processes described above. Regulation of biosynthetic pathways and cell signaling are therefore of prime importance. Rapid advances in biochemical and molecular biological techniques promise to help unravel some of the factors associated with cell signaling and coordination of enzyme action. Such advances can also be predicted to lead to the engineering of higher quality materials based on carbohydrates. Increasingly sensitive physical and analytical techniques are certain to provide new insights into the intermolecular interactions of oligosaccharides with their complementary receptors, the translocation of carbohydrates across membranes, and the morphology of polysaccharides. The deciphering of the complex chemical code of carbohydrates to reveal the information content of oligo- and polysaccharides promises to be a challenging yet attainable objective.

This chapter is dedicated, with respect, to the memories of three of my mentors, Mr. F. C. Pinto, Professor J. K. N. Jones, and Professor Sir D. H. R. Barton.

ACKNOWLEDGMENTS

I am grateful to M. A. Johnson for a critical reading of this chapter and to K. D. Randell for checking the references in this volume.

3.01.8 REFERENCES

1. T. Feizi and D. R. Bundle, *Curr. Opin. Struct. Biol.*, 1996, **6**, 659.
2. J. Rini and K. Drickamer, *Curr. Opin. Struct. Biol.*, 1997, **7**, 615.
3. H.-J. Gabius and S. Gabius, eds., "Glycosciences: Status and Perspectives," Chapman & Hall, Weinheim, 1997.
4. M. Fukuda and O. Hindsgaul, eds., "Molecular Glycobiology," Oxford University Press, Oxford, 1994.
5. R. A. Laine, in "Glycosciences: Status and Perspectives," eds. H.-J. Gabius and S. Gabius, Chapman & Hall, Weinheim, 1997, p. 1.
6. R. R. Townsend and A. T. Hotchkiss, Jr., eds., "Techniques in Glycobiology," Marcel Dekker, New York, 1997.
7. P. Jackson and J. T. Gallagher, eds., "A Laboratory Guide to Glycoconjugate Analysis," Birkhauser, Berlin, 1997.
8. T. Peters and B. M. Pinto, *Curr. Opin. Struct. Biol.*, 1996, **6**, 710.
9. E. F. Hounsell and M. Ragazzi, *Carbohydr. Res.*, 1997, **300**, 1.
10. A. D. French and I. G. Csizmadia, *J. Mol. Struct. Theochem.*, 1997, **395 396**, 1.
11. A. White and D. R. Rose, *Curr. Opin. Struct. Biol.*, 1997, **7**, 645.
12. A. E. Manzi and P. A. Keifer, in "Techniques in Glycobiology," eds. R. R. Townsend and A. T. Hotchkiss, Jr., Marcel Dekker, New York, 1997, p. 1.
13. X. C. Le, W. Tan, C. Scaman, A. Szpacenko, E. Arriaga, Y. Zhang, N. J. Dovichi, O. Hindsgaul, and M. M. Palcic, *Glycobiology*, in press.
14. R. Dutzler, Y.-F. Wang, P. Rizkallah, J. P. Rosenbusch, and T. Schirmer, *Structure*, 1996, **4**, 127.
15. G. M. Watt, P. A. S. Lowden, and S. L. Flitsch, *Curr. Opin. Struct. Biol.*, 1997, **7**, 652.
16. P. Westerduin, J. E. M. Basten, M. A. Broekhoven, V. De Kimpe, W. H. A. Kuijpers, and C. A. A. Van Boeckel, *Angew. Chem. Int. Ed. Engl.*, 1996, **35**, 331.
17. H. Ogura, A. Hasegawa, and T. Suami, eds., "Carbohydrates: Synthetic Methods and Applications in Medicinal Chemistry," VCH, Weinheim, 1992.
18. P. M. Collins and R. J. Ferrier, "Monosaccharides," Wiley, New York, 1995.
19. S. H. Khan and R. A. O'Neill, eds., "Modern Methods in Carbohydrate Synthesis," Harwood Academic, Amsterdam, 1996.
20. J. F. Robyt, "Essentials of Carbohydrate Chemistry," Springer, New York, 1998.
21. S. David, "The Molecular and Supramolecular Chemistry of Carbohydrates," Oxford University Press, Oxford, 1997.
22. S. Hanessian, ed., "Preparative Carbohydrate Chemistry," Marcel Dekker, New York, 1997.
23. G.-J. Boons, ed., "Carbohydrate Chemistry," Blackie Academic & Professional, London, 1998.
24. H. Driguez and J. Thiem, eds., "Glycoscience," *Topics in Current Chemistry*, Springer, Berlin, 1997, vols. 186, 187.
25. J.-C. Promé, *Curr. Opin. Struct. Biol.*, 1996, **6**, 671.
26. B. D. Davis, *Microbiol. Rev.*, 1987, **51**, 341.
27. W. Piepersberg, in "Genetics and Biochemistry of Antibiotic Production," eds. L. C. Vining and C. Stuttard, Butterworth-Heinemann, Boston, 1995, Chapter 19, p. 531.
28. K. Hotta, J. Davies, and M. Yagisawa, in "Genetics and Biochemistry of Antibiotic Production," eds. L. C. Vining and C. Stuttard, Butterworth-Heinemann, Boston, 1995, Chapter 19, p. 571.
29. R. Caputto, L. F. Leloir, R. E. Trucco, C. E. Cardini, and A. C. Paladini, *J. Biol. Chem.*, 1949, **179**, 479.
30. C. E. Cardini, A. C. Paladini, R. Caputto, and L. F. Leloir, *Nature*, 1950, **165**, 191.
31. R. Caputto, L. F. Leloir, C. E. Cardini, and A. C. Paladini, *J. Biol. Chem.*, 1950, **184**, 333.
32. L. F. Leloir and C. E. Cardini, in "The Enzymes," eds. P. D. Boyer, H. Lardy, and K. Myrback, Academic Press, New York, 1960, vol. 2, p. 39.

3.02
Glycosidases of the Asparagine-linked Oligosaccharide Processing Pathway

ANNETTE HERSCOVICS

McGill University, Montréal, PQ, Canada

3.02.1   INTRODUCTION

3.02.1.1   Importance of Protein Glycosylation in Biological Systems

Glycosylation is a major post-translational modification of membrane and secreted proteins in eucaryotic cells. The carbohydrate groups on glycoproteins play a structural role and influence the conformation, solubility, and stability of proteins. In addition, specific carbohydrate structures serve as molecular recognition signals in both intracellular and extracellular interactions. Specific glycan recognition mediates certain types of intracellular transport, such as targeting of lysosomal enzymes, and the interaction of cells with their environment. Carbohydrate-mediated interactions are particularly important in multicellular organisms, and are essential to normal embryonic development,[1,2] and to a variety of physiological processes. N-Glycosylation of proteins is necessary for cellular viability, and alterations in N-glycan structures are associated with a variety of diseases[3] including metastatic cancer[4] and a group of heterogeneous genetic disorders named carbohydrate-deficient glycoprotein syndrome (CDGS).[5] A bibliography of the many functions ascribed to protein-bound glycans has been compiled by Varki.[6]

3.02.1.2   Glycosidases

Glycosidases are ubiquitous intracellular and extracellular enzymes responsible for the hydrolysis of glycosidic linkages. A large number of glycosidases have been studied, and a compilation of these enzymes into more than 50 families based on related amino acid sequences is available on the World Wide Web in the SWISS-PROT database. Within each family, it is highly likely that the enzymes have a similar three-dimensional structure and catalytic mechanism, but this knowledge is available for only some of the glycosidases.[7] There are two types of glycosidases: (i) *exo*-glycosidases that release a single monosaccharide from the nonreducing terminus of an oligosaccharide; and (ii) *endo*-glycosidases that cleave internal glycosidic bonds. These enzymes are specific for the anomeric configuration, and hydrolyze glycosidic bonds with either retention or inversion of the anomeric configuration. Based on mechanistic studies with a variety of glycosidases,[8–11] hydrolysis of glycosidic bonds with inversion of the anomeric configuration most likely occurs by a single displacement mechanism and involves the concerted action of two ionizable amino acids, usually aspartic and glutamic acid, whereby one of these acts as a general acid catalyst by protonating the glycosidic oxygen atom, while the other acts as a general base extracting a proton from nucleophilic water. In contrast, retaining glycosidases function through a double displacement mechanism in which a glycosyl enzyme intermediate is formed and hydrolyzed by acid/base catalysis mediated by the carboxylic side chains of aspartic or glutamic acid.

Glycosidases have different functions, with the vast majority being required as degradative enzymes for the digestion of extracellular carbohydrates to monosaccharides. Similarly, glycosidases perform important degradative intracellular functions. They are required for the catabolism of polysaccharides, in response to physiological requirements (e.g., glycogen degradation as a source of energy) and for the turnover of complex carbohydrates (e.g., lysosomal degradation of cellular components). In contrast to these degradative enzymes, there is a group of intracellular glycosidases that are exceptional since they participate in a biosynthetic pathway. These enzymes are low-abundance α-glucosidases, α-mannosidases, and N-acetylglucosaminidases that are required for the maturation of asparagine-linked oligosaccharides (N-glycans or N-linked oligosaccharides) on secretory and membrane glycoproteins. The major focus of this chapter is to review the current state of knowledge of the specific glycosidases involved in N-glycan processing with relevant comparisons to glycosidases that are required for glycoprotein degradation. Emphasis is placed on processing glycosidases of mammalian cells and of the yeast *Saccharomyces cerevisiae*, for which there is the most information, but processing glycosidases most likely occur in all eukaryotes, including plants, fungi, insects, etc. Reviews covering some aspects of this topic have been published,[12,13] and may be referred to for additional details and for a more complete bibliography of articles published prior to 1994.

3.02.2   N-GLYCANS

3.02.2.1   Structure of N-Glycans

The asparagine-linked oligosaccharides represent only one class of covalently bound carbohydrates found on glycoproteins, but within this class there are many structural variations.[14,15] They have been classified into three general types: the high-mannose, the complex, and the hybrid N-glycans, as depicted in Figure 1. All N-glycans have the same basic pentasaccharide core structure of $Man_3GlcNAc_2$ (GlcNAc denotes N-acetyl-D-glucosamine) covalently attached through the GlcNAc at the reducing end to the amide group of asparagine within the tripeptide sequon Asn–X–Ser–(Thr), where X is any amino acid except proline. The invariant core may be substituted with a monosaccharide such as fucose or N-acetylglucosamine, and it may be further decorated with various carbohydrate structures in a species-specific and cell-specific manner. Furthermore, the structure of each oligosaccharide at different sites within the same glycoprotein depends on its location within each polypeptide chain. The variability in the structure of N-glycans determines the specificity of interactions in carbohydrate mediated biological recognition.

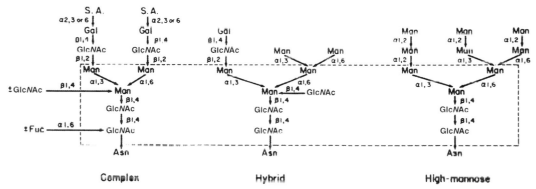

Figure 1   Structure of N-glycans The major types of asparagine-linked oligosaccharides are shown: (i) complex bi-antennary oligosaccharide; (ii) hybrid oligosaccharide; and (iii) high-mannose oligosaccharide. The invariant pentasaccharide core structure present in all N-linked oligosaccharides is surrounded by the rectangle. (Reproduced by permission of Annual Reviews Inc. from *Annu. Rev. Biochem.*, 1985, **54**, 633.)

3.02.2.2   Biosynthesis of N-Glycans

3.02.2.2.1   Formation of N-glycan precursor

In spite of the heterogeneity found in mature oligosaccharides, the early steps in the biosynthesis of N-glycans have been remarkably well conserved through eucaryotic evolution. The different types of N-glycans arise from the same oligosaccharide precursor. In most species this precursor is $Glc_3Man_9GlcNAc_2$ attached through a pyrophosphate linkage to a lipid called dolichol. Dolichols are polyprenols containing a variable number of isoprene units (19 ± 2 in mammalian cells) with a characteristic saturated α-isoprene unit. The dolichol-linked oligosaccharide precursor is formed in the membranes of the endoplasmic reticulum (ER) by the stepwise addition of each monosaccharide to dolichol phosphate in a series of enzymatic reactions that are similar in all eukaryotes.[16] Once formed, the oligosaccharide precursor is transferred from dolichol to newly synthesized polypeptide chains emerging from membrane-bound ribosomes in the lumen of the ER. The transfer of the oligosaccharide precursor to specific asparagine residues in the growing polypeptide chain is catalyzed by an oligosaccharyltransferase complex consisting of several distinct polypeptides.[17] Studies on various yeast mutants blocked in the early stages of assembly of the dolichol-linked

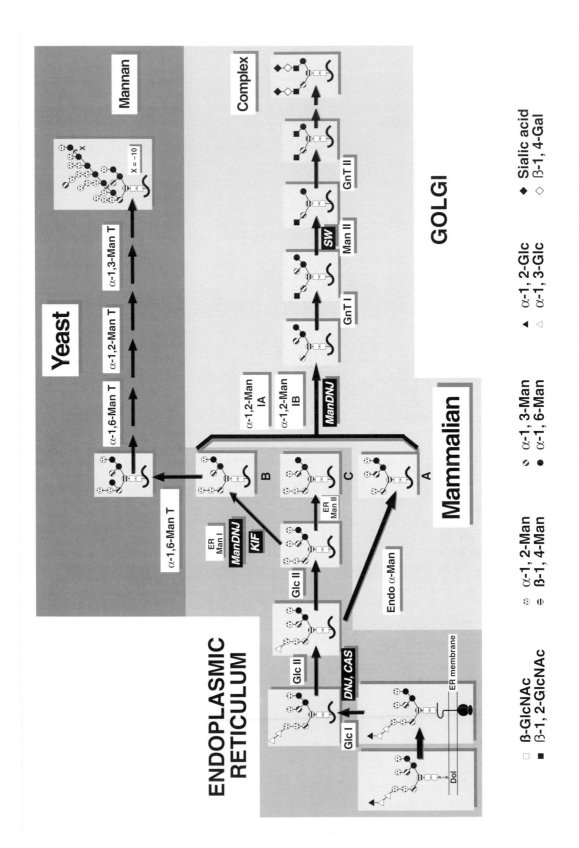

oligosaccharide precursor have shown that *N*-linked glycosylation is an essential function, and that the $Man_3GlcNAc_2$ pentasaccharide core is a minimum *N*-glycan structure compatible with cellular viability.[16]

3.02.2.2.2   *Maturation of* N-*glycans*

Following its attachment to protein, the $Glc_3Man_9GlcNAc_2$ oligosaccharide precursor undergoes an elaborate process of maturation, also called processing, to yield the final structures found on glycoproteins. Processing of *N*-glycans is tightly coupled with the intracellular secretory pathway, and consists of consecutive enzymatic reactions catalyzed by specific glycosidases and glycosyltransferases in the lumen of the ER and of the Golgi, as depicted in Figure 2. This maturation is not essential to cell viability, but it is important for cellular interactions, particularly in multicellular organisms. Reviews describing the earlier work on this pathway in mammalian cells and in yeast have been published.[12,14,16,18]

The initial stages of *N*-glycan processing are similar in most eukaryotes, and begin immediately following transfer of $Glc_3Man_9GlcNAc_2$ to protein with the removal of the glucose residues. As seen in Figure 2, two enzymes are required for glucose trimming within the lumen of the ER. α-Glucosidase I cleaves the terminal α-1,2-linked glucose, and α-glucosidase II removes the two α-1,3-linked glucose residues. In contrast, subsequent trimming of mannose residues is much more variable. In *S. cerevisiae*, only one α-1,2-linked mannose is removed in the ER with the formation of a single $Man_8GlcNAc_2$ isomer (isomer B). The yeast glycoproteins are then modified by a variety of mannosyltransferases in the Golgi. In mammalian cells, there are several different α-mannosidases, and the extent of mannose trimming depends on the cell type and the specific site within each glycoprotein. As will be discussed in detail in this chapter, there is evidence for the existence of distinct ER α-1,2-mannosidases that have different properties and form different $Man_8GlcNAc_2$ isomers (isomer B and isomer C). In addition, there is a Golgi *endo*-α-mannosidase that produces another $Man_8GlcNAc_2$ isomer (isomer A). There are additional α1,2-mannosidases in the ER and in the Golgi that collectively remove up to four α-1,2-mannose residues to form $Man_5GlcNAc_2$. This oligosaccharide is essential for the biosynthesis of complex *N*-glycans in most mammalian cells since it is a substrate for GlcNAc transferase I, the first Golgi glycosyltransferase that initiates the branches of complex *N*-glycans with the formation of $GlcNAcMan_5GlcNAc_2$. Following the action of GlcNAc transferase I, Golgi α-mannosidase II removes the terminal α-1,3- and α-1,6-linked mannose residues to yield $GlcNAcMan_3GlcNAc_2$. The $GlcNAcMan_3GlcNAc_2$ oligosaccharide thus formed can be extended by other GlcNAc transferases (II–V) to initiate the branches of bi-, tri-, and tetra-antennary complex oligosaccharides that are subsequently modified by Golgi glycosyltransferases specific for the addition of Gal, GalNAc (*N*-acetyl-D-galactosamine), GlcNAc, Fuc, and sialic acid residues (see Chapter 3.03).

3.02.2.2.3   *Importance of glycosidases in glycoprotein biosynthesis*

From the pathway depicted in Figure 2, it is evident that the structures of the oligosaccharides on glycoproteins depend entirely on the specificity of the glycosidases and glycosyltransferases and on their proper localization within the secretory apparatus, and not on a template as is the case for protein and nucleic acid biosynthesis. The expression, intracellular localization, and specificity of

---

Figure 2   Processing of yeast and mammalian *N*-linked oligosaccharides. Schematic representation of the processing pathway leading to complex *N*-linked oligosaccharides in mammalian cells and to high mannose oligosaccharides in the yeast *Saccharomyces cerevisiae*. The glycosidases and a few glycosyltransferases involved in this pathway are shown. Also, specific inhibitors of the processing glycosidases are indicated at the appropriate steps. The abbreviations for the enzymes are as follows: Glc I, α-glucosidase I; Glc II, α-glucosidase II; Endo α-Man, *endo* α-mannosidase; ER Man I, endoplasmic reticulum α-mannosidase I; ER Man II, endoplasmic reticulum α-mannosidase II; α-1,2-Man IA, α-1,2-mannosidase IA; α-1,2-Man IB, α-1,2-mannosidase IB; Man II, α-mannosidase II; GnT I, *N*-acetylglucosaminyltransferase I; GnT II, *N*-acetylglucosaminyltransferase II; α-1,6-Man T, yeast α-1,6-mannosyltransferase; α-1,2-Man T, yeast α-1,2-mannosyltransferase; α-1,3-Man T, yeast α-1,3-mannosyltransferase. The abbreviations for the inhibitors are: *DNJ*, 1-deoxynojirimycin; *CAS*, castanospermine; *KIF*, kifunensine; *ManDNJ*, 1-deoxymannojirimycin; *SW*, swainsonine.

the glycosidases and glycosyltransferases therefore determine the various forms of mature *N*-glycans found on glycoproteins.

The existence of processing glycosidases was first demonstrated in the mid-1970s. The approach used in earlier work was to study the time course of incorporation of radioactively labeled sugar precursors into glycoproteins of cells in culture, and then to characterize the properties of gly- cosidases in crude membrane preparations obtained from different sources. Specific glycosidase inhibitors and cell mutants were extremely useful to establish the sequence of reactions occurring in the processing pathway. In this way the biochemical properties of glycosidases involved in *N*-glycan processing were determined. Eventually, some of the enzymes were purified sufficiently for microsequencing, and partial amino acid sequences were used to design degenerate oligo- nucleotides for DNA amplification using polymerase chain reaction (PCR). The amplified DNA fragments were then utilized as probes to screen genomic or cDNA libraries in order to isolate the genes and cDNAs encoding processing glycosidases. More recently, as more glycosidases are being cloned, advantage has been taken of sequence similarities between members of the same family to design oligonucleotide primers for PCR. It is thus becoming possible to clone additional members of each family from different tissues or species without having to purify the enzymes. The availability of cloned DNAs encoding processing glycosidases is beginning to provide recombinant glycosidases for detailed studies of their structure, enzymatic mechanism, and function in cell biology.

3.02.3 α-GLUCOSIDASES

3.02.3.1 α-Glucosidase I

The first step in *N*-glycan processing is catalyzed by α-glycosidase I, which only removes the terminal α-1,2-linked glucose from $Glc_3Man_9GlcNAc_2$ immediately following its transfer to newly formed polypeptide chains. The enzyme is present in the membranes of the rough ER, most likely close to sites of translocation since removal of a glucose residue can occur cotranslationally.[19] α-Glucosidase I has been characterized and purified from several mammalian tissues,[20–33] from plants,[34] and from *S. cerevisiae*.[35,36] It has an optimum pH of 6.2–6.8, no apparent requirement for divalent cation, and does not act on aryl α-D-glucosides. The efficiency of the mammalian enzyme,[25,27] but not of the yeast enzyme,[37] decreases considerably as an increasing number of mannose residues are removed from the α-1,6 branch of $Glc_3Man_9GlcNAc_2$. Oligosaccharides with three glucose residues but only 5–7 mannose residues are therefore poor substrates of mammalian α-glucosidase I. The enzyme can also remove glucose from the lipid-linked oligosaccharide, and thus may regulate the amount of $Glc_3Man_9GlcNAc_2$ precursor available for transfer to protein.[38] The mammalian enzymes are tetramers consisting of subunits of about 85 kDa on sodium dodecyl sulfate–polyacrylamide gel electrophoresis (SDS–PAGE), while the yeast enzyme has a subunit size of about 95 kDa.

The mammary gland enzyme was shown to be greatly inhibited by sulfhydryl reagents, an inhibition that could be prevented by the glucose analog 1-deoxynojirimycin, suggesting that cysteine is important for enzyme activity.[39] Arginine and tryptophan residues have also been implicated in substrate binding by demonstrating inactivation of these residues with group-selective chemical modifiers.[40] The exact catalytic mechanism of α-glucosidase I has not yet been established.

A cDNA encoding α-glucosidase I has been isolated from a human cDNA library.[41] This cDNA encodes a type II transmembrane protein of 92 kDa with an *N*-terminus cytoplasmic domain of about 37 amino acids followed by a transmembrane region of about 20 amino acids and a large catalytic domain with a single *N*-linked glycosylation site facing the lumen of the ER. Following transfection of mammalian cells in culture, this cDNA caused overexpression of α-glucosidase I activity, and the resulting protein was immunolocalized to the nuclear envelope and the ER. The α-glucosidase I gene has been localized to human chromosome 2p12-p13 by fluorescence *in situ* hybridization and analysis of somatic cell hybrids.[42]

A Chinese hamster ovary cell mutant, Lec 23, was shown to lack α-glucosidase I. This recessive mutant was selected for its resistance to the leukoagglutinin from *Phaseolus vulgaris* (L-PHA), a lectin that recognizes specific tri- and tetraantennary complex oligosaccharides.[43] In this mutant there is an accumulation of high-mannose oligosaccharides containing three glucose residues, but some complex oligosaccharides are still formed by an alternative pathway that has not yet been identified.

Two *S. cerevisiae* mutants lacking α-glucosidase I have been isolated independently. In earlier studies, the recessive *gls1* mutant was described, but the gene could not be cloned due to the lack of an easily detectable phenotype.[44] More recently, the *cwh41* mutant was isolated by selection for hypersensitivity to calcofluor white, an index of altered cell surface properties. This mutant exhibits increased K1 killer toxin resistance, and a reduction in cell wall β1,6-glucans that are important as killer toxin receptors.[45] The *CWH41* gene was cloned by functional complementation of the calcofluor white sensitivity phenotype and was shown to encode an ER type II transmembrane protein.[45] The *CWH41* gene was found to have significant similarity in amino acid sequence to human α-glucosidase I.[46] Disruption of the *CWH41* gene causes loss of α-glucosidase I activity both *in vivo* and *in vitro*, thereby demonstrating that the gene encodes yeast α-glucosidase I.[46] These observations indicate that proper glucose trimming of *N*-glycans is important for normal cell wall β-1,6-glucan biosynthesis for reasons that are not presently understood.

3.02.3.2 α-Glucosidase II

α-Glucosidase II removes the two α-1,3-linked glucose residues from the oligosaccharide precursor in the ER. Pulse labeling studies of cells in culture showed that these reactions occur more slowly than removal of the terminal α-1,2-linked glucose, particularly so for the innermost glucose residue.[47,48] α-Glucosidase II is even more sensitive than α-glucosidase I to a reduced number of mannose residues on the α-1,6 branch of the oligosaccharide, so that even $Glc_{1-2}Man_5GlcNAc$ oligosaccharides are poor substrates of both the yeast and mammalian enzymes.[27,37] This property of α-glucosidase II may partly account for the relatively slower rate of removal of the α-1,3-linked glucose residues, since trimming of up to two mannose residues can occur on glucosylated oligosaccharides in the ER.

α-Glucosidase II can cleave aryl α-D-glucosides, has no divalent cation requirement, and has a pH optimum ranging from 5.8 to 7.5, depending on the source. It has been immunolocalized on the nuclear envelope and on both rough and smooth ER of pig hepatocytes.[49] However, in kidney tubular cells, α-glucosidase II immunoreactivity was found predominantly in endocytotic structures beneath the plasma membrane and in purified brush border preparations.[50] The functional significance of this localization in the kidney is not known.

α-Glucosidase II is more readily solubilized from the ER than α-glucosidase I. It has been characterized and in some cases purified from several mammalian tissues,[24,26,28,51–54] from plants,[55] and from yeast.[37] Earlier studies suggested that mammalian α-glucosidase II is a tetramer consisting of 123 kDa subunits,[54] although during purification various enzymatically active proteolytically released fragments as small as 62 kDa were obtained. However, it has been reported that purified rat liver α-glucosidase II is a dimer of two different subunits, α and β, that could not be separated without loss of enzyme activity.[56] Comparisons of partial peptide sequences obtained from the rat α subunit with sequences in the database identified a corresponding human partial cDNA and the homologous *S. cerevisiae* gene. Disruption of the gene in *S. cerevisiae* causes disappearance of α-glucosidase II activity without affecting α-glucosidase I activity or growth. Partial peptide sequences of the β subunit are identical to a human cDNA in the database encoding a 58 kDa soluble protein with a hydrophobic signal sequence, two putative EF hand calcium-binding motifs, stretches of consecutive glutamic acids, and the putative ER retention sequence HDEL. It was concluded that the α subunit is the soluble catalytic domain of α-glucosidase II and that the β subunit may be responsible for α-glucosidase II localization to the ER. However, this possibility remains controversial.[57] The human α-glucosidase II gene, previously designated neutral α-glucosidase AB, has been localized to the long arm of human chromosome 11.[58,59]

A mouse lymphoma cell mutant, BW5147PHA^{R}2.7, that is resistant to the *P. vulgaris* leuko-agglutinin and lacks α-glucosidase II activity has been isolated. In this mutant there is an accumulation of $Glc_2Man_{8-9}GlcNAc_2$ oligosaccharides, but the cells are still able to form some complex *N*-glycans through an alternative pathway utilizing the Golgi *endo*-α-mannosidase described below.[60,61]

3.02.3.3   Glucosidase Inhibitors

There are a number of glucosidase inhibitors that have been extremely useful to determine the role of glucose trimming in the formation of glycoproteins. Several detailed reviews on processing glycosidase inhibitors have been published.[62–64] Both α-glucosidases are inhibited by micromolar concentrations of the glucose analogues 1-deoxynojirimycin, its synthetic derivatives *N*-methyl-1-

deoxynojirimycin and *N*-butyl-1-deoxynojirimycin, and by the plant alkaloid castanospermine.[62–64] Another plant alkaloid, australine, was shown to be a more specific inhibitor of α-glucosidase I, but only relatively high concentrations of this compound are effective.[65] *In vivo*, these compounds interfere with the maturation of *N*-glycans to complex oligosaccharides,[37] and usually cause the accumulation of $Glc_3Man_{7-9}GlcNAc_2$ on glycoproteins.[66–68] However, 1-deoxynojirimycin itself is not quite as effective as its derivatives in inhibiting α-glucosidase I, and a mixture of oligosaccharides with a variable number of glucose and mannose residues, $Glc_{1-3}Man_{7-9}GlcNAc_2$, are found on glycoproteins of treated cells.[67] It has also been shown that 1-deoxynojirimycin under some conditions may inhibit the addition of glucose to $Man_9GlcNAc_2$-PP-dolichol.[69] In addition, bromoconduritol, another glucose analogue, was shown to inhibit α-glucosidase II and to cause the accumulation of $Glc_1Man_{7-9}GlcNAc_2$ on glycoproteins *in vivo*.[70,71] Kinetic studies of α-glucosidase II from rat liver have shown that the enzyme has two binding sites with different affinities. There is a high-affinity binding site that is not affected by bromoconduritol and a low-affinity binding site that is sensitive to this inhibitor. It was suggested that removal of the terminal glucose from $Glc_2Man_9GlcNAc_2$ occurs at the high-affinity site whereas the bromoconduritol-sensitive removal of glucose from $Glc_1Man_9GlcNAc_2$ most likely takes place at the other site.[72,73] This kinetic analysis may explain the differential effect of bromoconduritol for the two steps catalyzed by the same enzyme. Furthermore, the different affinities of the two sites may also contribute to the different half-lives of the two glucose residues observed *in vivo*.

There are many studies indicating that even with the complete elimination of α-glucosidase activity by α-glucosidase inhibitors, or in α-glucosidase-deficient mutant cells, the formation of complex *N*-glycans cannot be entirely prevented. The extent of complex *N*-glycan synthesis in the presence of α-glucosidase inhibitors varies in different cells, and is due to the occurrence of alternative pathways that are independent of trimming by the α-glucosidases. The major alternative pathway relies on the action of the Golgi *endo*-α-mannosidase that can form the $Man_8GlcNAc_2$ isomer A from $Glc_{1-3}Man_9GlcNAc_2$, as described earlier. Another alternative pathway involves the transfer of nonglucosylated oligosaccharides directly from dolichol intermediates to protein. This transfer is a major pathway in trypanosomes,[74] in some yeast mutants that do not synthesize glucosylated dolichol-linked oligosaccharides,[16] and also occurs as an alternative pathway in mouse teratocarcinoma cells.[75]

3.02.4   *endo*-α-MANNOSIDASE

A specific Golgi *endo*-α-mannosidase provides an alternative processing pathway to eliminate the glucose residues[33,76–78] when glucose removal by the α-glucosidases in the ER has not occurred. This situation prevails in cells treated with α-glucosidase inhibitors[78,79] and in mutants lacking α-glucosidase activity.[61] The *endo*-α-mannosidase therefore allows some degree of maturation to complex oligosaccharides in the absence of α-glucosidase activity. It is the only processing glycosidase that cleaves an internal glycosidic linkage. Although *in vitro* the *endo*-α-mannosidase prefers monoglucosylated oligosaccharides as substrates with the release of the disaccharide $Glc\alpha1,3Man$,[77] it is capable of cleaving $Glc_{1-3}Man_{4-9}GlcNAc$ to yield $Man_{3-8}GlcNAc$ and $Glc_{1-3}Man$. In contrast to α-glucosidase I and II, the activity of the *endo*-α-mannosidase is enhanced when mannose residues are removed from the other branches. The $Man_8GlcNAc_2$ isomer resulting from *endo*-α-mannosidase action on $Glc_{1-3}Man_9GlcNAc_2$ is isomer A, as shown in Figure 2. Some evidence has been obtained to indicate that *endo*-α-mannosidase action may occur *in vivo* even in the absence of an α-glucosidase blockade.[80] The *endo*-α-mannosidase has no divalent ion requirement, has a pH optimum around 7, and is specifically inhibited by the disaccharides $Glc\alpha1,3$-(1-deoxymannojirimycin) and $Glc\alpha1,3$-(1,2-dideoxy)mannose.[81] It has been purified from rat liver Golgi membranes by affinity chromatography on a column of $Glc\alpha1,3$-Man-*O*-$(CH_2)_8CO$-NH-Affi-Gel.[82] Two protein bands of 56 and 60 kDa were seen on SDS–PAGE. It was subsequently shown that the 60 kDa protein is the chaperone calreticulin that copurified with the *endo*-α-mannosidase.[83] Both *endo*-α-mannosidase and calreticulin recognize the disaccharide affinity ligand, but additional work is required to determine whether they are both present in the same intracellular compartment. The *endo*-α-mannosidase is widely distributed in mammalian cells, with the exception of chinese hamster ovary cells, but there is no evidence for its presence in *S. cerevisiae*. It seems that the appearance of the *endo*-α-mannosidase is a relatively late event in eucaryotic evolution, being mostly limited in its occurrence to chordates.[84]

3.02.5   *exo*-α-MANNOSIDASES

3.02.5.1   Classification of Eukaryotic α-Mannosidases

Processing α-mannosidases cleave mannose residues from the oligosaccharide precursor within the lumen of the ER and of the Golgi. These enzymes differ from the lysosomal α-mannosidases involved in glycoprotein catabolism by their higher pH optimum, which ranges from 5.6 to 6.5. Although earlier classifications of α-mannosidases were based on their biochemical characteristics such as substrate specificity, cation requirement, and on their role as either biosynthetic or catabolic enzymes, more recent cloning studies have revealed that there are two distinct classes of α-mannosidases based on their amino acid sequences, irrespective of other criteria. These have been termed class 1 and class 2 α-mannosidases.[12] The class 1 enzymes only cleave α-1,2-mannose residues whereas the class 2 enzymes are capable of cleaving α-1,2-, α-1,3-, and α-1,6-linked mannose residues. In addition to their different amino acid sequences and specificities, the two classes of α-mannosidases also exhibit different susceptibilities to inhibitors, and differences in their cation requirements. The class 1 α-mannosidases are inhibited by pyranose monosaccharide analogues whereas the class 2 enzymes are affected by furanose analogues.[13] Although little is known regarding the catalytic mechanisms of processing glycosidases, these observations suggest that the two classes of α-mannosidases have different structures and enzymatic mechanisms.

3.02.5.2   Class 1 α-Mannosidases (α-1,2-Mannosidases)

α-1,2-Mannosidases participate in the early stages of *N*-glycan processing immediately following, or concurrently with, glucose trimming in the ER.[19,85] In many species, trimming by α-1,2-mannosidases continues in the Golgi but the number of processing α-1,2-mannosidases involved in *N*-glycan maturation is variable. In the budding yeast *S. cerevisiae* there is only one processing α-1,2-mannosidase in the ER, whereas the fission yeast *Schizosaccharomyces pombe* does not appear to have this activity.[86] In mammalian cells, the number of α-1,2-mannosidases involved in *N*-glycan maturation in the ER and in the Golgi is still unclear, but it is evident that the expression of different α-1,2-mannosidases is species-specific and is also cell-specific in multicellular organisms.

3.02.5.2.1   Yeast ER α-1,2-mannosidase

In the yeast *S. cerevisiae* there is only one highly specific processing α-1,2-mannosidase that removes a single mannose residue from Man$_9$GlcNAc$_2$ to form the Man$_8$GlcNAc$_2$ isomer B, as shown in Figure 2.[87] This enzyme is distinct from the nonspecific vacuolar α-mannosidase involved in glycoprotein catabolism.[88] It was purified to homogeneity as a proteolytically released soluble form of about 60 kDa[89] and as the intact glycoprotein of about 67 kDa.[90] Treatment with *endo*-β-*N*-acetylglucosaminidase H decreases the molecular size of the enzyme by about 4 kDa,[88–90] consistent with the presence of three *N*-glycan core structures. The yeast α-1,2-mannosidase has a pH optimum of 6.5–6.8 and does not utilize aryl α-D-mannopyranoside as substrate. It is inhibited by EDTA, an effect that is completely reversed by Ca^{2+} ions and partially by Mg^{2+}, but not by any other divalent cation. The yeast enzyme is inhibited by the mannose analogue 1-deoxymannojirimycin, and it is even more sensitive to kifunensine (F. Lipari and A. Herscovics, unpublished findings).

The yeast α-1,2-mannosidase *MNS1* gene was the first class 1 α-mannosidase to be cloned using partial amino acid sequences obtained from the purified enzyme to design degerate oligonucleotides for PCR.[91] It encodes a type II integral membrane protein of 63 kDa with a single hydrophobic region of about 18–20 amino acids and no significant cytoplasmic tail. The transmembrane domain is followed by a large catalytic domain facing the lumen of the ER[92] and containing three *N*-linked glycosylation sites and a putative EF hand Ca^{2+}-binding consensus sequence. The yeast α-1,2-mannosidase was shown to be a resident protein of the ER by immunocytochemistry by light and electron microscopy.[93]

Overexpression of the *MNS1* gene on a high-copy plasmid causes an 8–10-fold increase in α-1,2-mannosidase activity,[91] and disruption of this gene completely eliminates trimming of Man$_9$GlcNAc$_2$ to Man$_8$GlcNAc$_2$ both *in vivo* and *in vitro*.[91,94] These results indicate that it is the only functional processing α-1,2-mannosidase in *S. cerevisiae*. The yeast mutant lacking the processing α-1,2-mannosidase exhibits normal growth and is capable of outer chain biosynthesis. However, invertase formed in the null mutant has a slightly different mobility on SDS–PAGE, suggestive of a small change in *N*-glycan structure.[94]

The catalytic domain of the yeast enzyme has been expressed in milligram quantities as a glycoprotein secreted from *S. cerevisiae*[95] and from the methylotropic yeast *Pichia pastoris*.[96] This recombinant α-1,2-mannosidase has the same enzymatic properties as the purified endogenous enzyme, and provides large amounts of protein for structure–function studies, thus serving as a model for class 1 α-1,2-mannosidases. Using proton nuclear magnetic resonance spectroscopy the yeast α-1,2-mannosidase was shown to be an inverting glycosidase.[97] The yeast α-1,2-mannosidase catalytic domain contains two disulfide bonds (Cys340–Cys385 and Cys468–Cys471) and one free sulfhydryl group (Cys485).[96] Using the recombinant α-1,2-mannosidase it was observed that enzyme activity is lost in the presence of dithiothreitol with first-order kinetics, suggesting that at least one of the disulfide bonds is essential, but there was no effect of sulfhydryl-reactive reagents. Mutagenesis of each of the cysteine residues to serine demonstrated that Cys340 and Cys385 are necessary for production of the recombinant enzyme whereas none of the other cysteine residues are required for synthesis of active enzyme.[96] The disulfide bond between Cys340 and Cys385 is therefore essential for proper structure of the yeast α-1,2-mannosidase. Since these two cysteine residues are highly conserved they may also be important for structural integrity of the other class 1 α-1,2-mannosidases. Mutagenesis that eliminates each of the *N*-glycosylation sites individually had no effect on the production or activity of the recombinant enzyme, but deletion of all three glycosylation sites completely prevented recombinant protein production by *P. pastoris* (F. Lipari and A. Herscovics, unpublished findings). The yeast recombinant α1,2-mannosidase has been crystallized, and preliminary X-ray crystallographic data have been collected.[98]

3.02.5.2.2 *Mammalian ER α-1,2-mannosidases*

Although the Golgi was initially believed to be the major site of mannose trimming, there is a great deal of evidence to indicate that several α-1,2-mannosidase activities are functional in the ER of mammalian cells.[85,99–102] Experiments *in vivo* indicate the presence of a 1-deoxymannojirimycin-sensitive ER activity that trims Man$_9$GlcNAc to Man$_{5,6}$GlcNAc,[103] and for both 1-deoxymannojirimycin-sensitive and -insensitive ER activities that can produce isomer B of Man$_8$GlcNAc.[102,103] Moreover, an ER α-1,2-mannosidase activity that produces the C isomer of Man$_8$GlcNAc has also been described.[104] This enzyme activity is insensitive to kifunensine at concentrations that completely inhibit isomer B formation. The ER α-1,2-mannosidase producing isomer B has properties similar to the yeast processing enzyme, and is expected to be a class 1 α-1,2-mannosidase, but an ER mammalian enzyme with this specificity has not yet been purified or cloned. On the other hand, the kifunensine-resistant ER α-1,2-mannosidase that produces isomer C has properties similar to the class 2 ER/cytosolic rat liver enzyme described below that also produces the C isomer of Man$_8$GlcNAc.[105,106]

3.02.5.2.3 *Mammalian Golgi α-1,2-mannosidases*

Several mammalian α-1,2-mannosidases have been purified from different sources. In most cases, the purified enzymes are truncated smaller soluble forms lacking their transmembrane domains due to partial proteolysis. Rat liver Golgi α-mannosidase I was the first of these enzymes to be characterized. It could be separated into two forms, IA and IB, by ion exchange chromatography and gel filtration. Both enzyme fractions were shown to remove the α-1,2-linked mannose residues from Man$_9$GlcNAc, to form Man$_5$GlcNAc.[107–109] The two enzymes are inhibited by 1-deoxymannojirimycin and EDTA, have an optimum pH of 6.0, and share some antigenic determinants. α-Mannosidase IA was highly purified and shown to be a tetramer consisting of 57 kDa subunits.[109] It was not established in these studies whether rat liver Golgi α-mannosidases IA and IB were alternative forms derived from the same gene or whether they arose from different genes, but the cloning of two distinct mouse α-1,2-mannosidase cDNAs derived from different genes, described below, favors the latter possibility.

Using antibodies to the purified rat liver α-mannosidase IA, immunoreactivity was observed primarily in the medial and/or trans-cisternae of the Golgi of different cells in the rat, but this distribution within the Golgi is variable.[110] In some cell types, all cisternae of the Golgi were reactive, and the enzyme was also found within secretory granules and at the cell surface. Although the rat liver Golgi enzymes have not been cloned and the effect of Ca^{2+} ions was not examined, their enzymatic properties indicate that they are class 1 α-mannosidases.

A Ca^{2+}-dependent α-1,2-mannosidase was purified from rabbit liver.[111-113] This enzyme has properties similar to the rat liver Golgi α-1,2-mannosidase in that it can trim $Man_9GlcNAc$ to $Man_5GlcNAc$ and is sensitive to 1-deoxymannojirimycin. Its activity is stimulated by specific phospholipids.[114] The enzyme that migrated as a 52 kDa band on SDS–PAGE was purified to homogeneity, and partial amino acid sequence information was used to design degenerate oligonucleotides for PCR. The resulting amplified fragment was used as a probe to screen cDNA libraries.[115] The full-length cDNA isolated from murine cells encodes a type II transmembrane protein of 73 kDa, indicating that partial proteolysis of the *N*-terminus occurred during purification of the enzyme. The *C*-terminal catalytic domain contains two potential *N*-glycosylation sites and a putative EF hand Ca^{2+}-binding consensus sequence. Transfection of this cDNA into murine cells in culture caused a 20-fold increase in α-1,2-mannosidase activity. Immunofluorescence using the antibodies raised against rat liver Golgi α-1,2-mannosidase IA localized the enzyme to the Golgi of transfected mammalian cells in culture.[115] Because of this cross-reactivity, the enzyme was designated murine Golgi α-1,2-mannosidase IA.

It was observed that the catalytic domain of the murine Golgi α-1,2-mannosidase IA is about 35% identical in amino acid sequence to the yeast processing α-1,2-mannosidase, indicating that the α-1,2-mannosidase enzyme family has been conserved through eucaryotic evolution.[116] Taking advantage of the highly conserved regions to design primers for PCR, another cDNA was isolated from a murine 3T3 cDNA library. This cDNA encodes a 73 kDa type II transmembrane protein that was called α-1,2-mannosidase IB.[116,117] It has a *C* terminal catalytic domain that is about 65% identical in amino acid sequence to the murine α-1,2-mannosidase IA. The catalytic domain contains an EF-hand consensus sequence and a single *N*-glycosylation site. The enzyme is also localized to the Golgi of transfected mammalian cells in culture,[116] cleaves $Man_9GlcNAc$ to $Man_5GlcNAc$, is inhibited by 1-deoxymannojirimycin, and requires Ca^{2+} ions for activity.[117] A variant cDNA with three point mutations (U to C) that inactivate the enzyme and may be the result of RNA editing was also isolated.[116,118,119] The transmembrane domains of murine α1,2-mannosidases IA and IB are nearly identical, but their cytoplasmic regions of about 35 amino acids are very different. Southern blot analysis indicated that the two enzymes are products of distinct genes, and Northern blotting showed that they exhibit very different patterns of tissue-specific expression in adult mouse tissues and during mouse embryonic development (A. Herscovics, unpublished data).[116,120]

The murine α-1,2-mannosidase IB gene has been isolated and characterized.[119] It spans at least 80 kb of the genome, and consists of 13 exons. Fluorescence *in situ* hybridization localized the gene to mouse chromosome 3F2. Another α-1,2-mannosidase IB-related gene, or pseudogene, was also found on mouse chromosome 4A13.[119] The corresponding human cDNA has also been isolated.[120] It encodes a protein with about 94% amino acid identity to the mouse α-1,2-mannosidase IB, but there is evidence for alternatively spliced transcripts in human placenta. The human gene has been localized by fluorescence *in situ* hybridization to chromosome 1p13, a region syntenic with mouse chromosome 3F2. The human gene has a very similar intron–exon structure as the mouse gene.[120]

Ca^{2+}-dependent α-1,2-mannosidases with similar properties to the murine and rabbit enzymes were purified from calf[121] and pig[122] liver. These enzymes (called Man_9-mannosidases) were 56 and 49 kDa on SDS–PAGE, respectively, and 65 kDa on Western blots of crude pig liver microsomes. The protein was immunolocalized in the ER, but not in the Golgi, of pig hepatocytes.[123] The pig liver enzyme removes three mannose residues from $Man_9GlcNAc_2$, but does not as readily cleave the fourth mannose residue.[124] Although initially it was believed that the Man_9-mannosidases have a different specificity from the other mammalian α-1,2-mannosidases,[121,122] their specificity was shown to be similar to the rabbit liver enzyme when the same oligosaccharides were used as substrates.[124] Amino acid sequence information derived from the purified pig liver enzyme was used for PCR to isolate a human kidney cDNA encoding a 73 kDa type II membrane protein[125] that is about 35% identical in amino acid sequence to the yeast processing α-1,2-mannosidase[91] and about 88% identical to murine α-1,2-mannosidase IA.[115] The size of the cDNA indicates that the α-1,2-mannosidases purified from calf and pig livers had undergone some partial proteolysis while retaining their catalytic activity. The enzyme has the highly conserved regions characteristic of class 1 α-1,2-mannosidases, and catalytic properties that resemble the murine and rabbit enzymes. Furthermore, when transfected into mammalian cells in culture, the derived protein is immunolocalized to the Golgi,[126] in contrast to the ER localization observed in pig hepatocytes.[123] The reasons for the different intracellular localization of the pig and human enzymes are not known. However, the similarities in their catalytic domains and in their enzymatic properties indicate that the Man_9-mannosidases are species variants of α-1,2-mannosidase IA.

3.02.5.2.4 *Insect α-1,2-mannosidases*

Class 1 α-1,2-mannosidases have been cloned from *Drosophila melanogaster*[127] and from Sf9 cells derived from the lepidopteran *Spodoptera frugiperda*.[128] The *D. melanogaster mas-1* gene was shown to encode two proteins from the use of alternative promoters and alternative exons 1a and 1b. The *mas-1a* and *mas-1b* genes encode type II transmembrane proteins of 72.5 and 75 kDa, respectively, that have significant sequence similarity with the other class 1 α-1,2-mannosidases including the EF-hand consensus sequence. The tissue-specific expression of these two genes is developmentally regulated, and a null mutant of *mas-1a* exhibits developmental abnormalities, particularly in the peripheral nervous system and sensory organs, suggesting a role of the enzyme in the development of the fly nervous system.[127] The specific enzymatic properties of the derived proteins have not yet been reported.

Lepidopteran insect cells such as Sf9 cells are widely used as hosts for heterologous glycoprotein production by recombinant baculoviruses.[129] *N*-Glycans from these insect cells contain Man_{5-9}-$GlcNAc_2$ high-mannose oligosaccharides, as well as smaller $Man_{2-3}GlcNAc_2$ oligosaccharides (so-called pauci-mannose oligosaccharides) that can be terminated with $β$-1,2-GlcNAc residues and may also have α-1,3- an α-1,6-fucose residues on the core GlcNAc.[130] α-1,2-Mannosidase activity acting on $Man_9GlcNAc_2$ has been described in insect cells. It was shown that $Man_6GlcNAc_2$ accumulates in extracts of uninfected cells, whereas trimming to $Man_5GlcNAc_2$ and even to Man_3-$GlcNAc_2$ occurred following baculovirus infection.[131] It was shown that baculovirus infection causes an increase in enzyme activity that degrades $Man_6GlcNAc_2$. An enzyme of 63 kDa with the properties of class 1 α-1,2-mannosidases has been purified from membranes of recombinant baculovirus infected *Spodoptera frugiperda* cells. This enzyme is Ca^{2+}-dependent, does not utilize *p*-nitrophenyl-α-D-mannopyranoside, and is inhibited by 1-deoxymannojirimycin but not by swainsonine. The purified enzyme utilizes reduced $Man_{6-9}GlcNAc_2$-ol as substrates but exhibits some preference for trimming reduced $Man_6GlcNAc_2$-ol to $Man_5GlcNAc_2$-ol.[132]

An α-1,2-mannosidase cDNA that encodes a 75 kDa type II membrane protein with about 57% identity to the *Drosophila mas* encoded protein and significant sequence similarity to the other class 1 enzymes has been cloned from Sf9 cells.[128] The exact specificity of this enzyme has not yet been studied, but its expression is not a function of baculovirus infection. Its relationship with the enzyme activity that is increased following baculovirus infection[131] has not been established, but since its expression is not affected by baculovirus infection, it is not likely to encode the same enzyme. Southern blotting, however, indicates that there may be two α-1,2-mannosidase-related genes in the Sf9 genome.[128]

3.02.5.2.5 *Miscellaneous α-1,2-mannosidases*

Two α-1,2-mannosidases have been cloned from *Aspergillus saitoi*[133] and *Penicillium citrinum*[134] that are 70% identical in amino acid sequence to each other. They have sufficient similarity to other eucaryotic enzymes to be classified as members of the class 1 α-1,2-mannosidases. However, the role of these two enzymes in *N*-glycan processing is not established since they are both secreted enzymes with cleavable *N*-terminal signal sequences and a relatively acidic pH optimum of 5.0. They are both inhibited by 1-deoxymannojirimycin, do not utilize *p*-nitrophenyl-α-D-mannopyranoside as substrate, but unlike other class 1 enzymes they do not require Ca^{2+} for activity. The EF-hand consensus sequence includes an extra amino acid that may disrupt EF-hand helix formation, thereby eliminating its potential for Ca^{2+} binding.[134] The *Penicillium* enzyme was inactivated by chemical modification of acidic residues with a water-soluble carbodiimide, and a specific aspartic acid residue (Asp375) was shown to be protected from this inactivation by the inhibitor 1-deoxymannojirimycin.[135] A conclusive role of this residue in catalysis has not yet been established. It will be interesting to determine whether the catalytic function and the structure of these enzymes differ significantly from the other class 1 α-1,2-mannosidases. Enzymes with properties of the class 1 α-1,2-mannosidases have also been purified from plants[136] and from hen oviduct,[137] but these have not yet been cloned.

3.02.5.2.6 *Comparison between α-1,2-mannosidases*

The dendogram presented in Figure 3 depicts the amino acid sequence relationships between the currently known class 1 α-mannosidases, and illustrates that this enzyme family has been highly conserved through eucaryotic evolution, its members occurring in unicellular and multicellular organisms from yeast to mammals. The amino acid sequence similarity is observed throughout their catalytic domains. There are three conserved cysteine residues and several highly conserved peptide sequences, but no preserved *N*-linked glycosylation sites. Within this family, the mammalian enzymes form two distinct groups with catalytic domains that are about 65% identical in amino acid sequence and are about 35% identical to the yeast enzyme. Within each group the members from the different mammalian species are over 90% identical in amino acid sequence. These structurally related enzymes are all Ca^{2+}-dependent and contain a putative EF-hand Ca^{2+}-binding consensus sequence; they are all inhibited by 1-deoxymannojirimycin, but not by swainsonine, and they do not utilize *p*-nitrophenylmannoside as a substrate. As mentioned above, the *Aspergillus* and *Penicillium* enzymes have significant amino acid similarity to the other class 1 enzymes, but it is possible that their role and catalytic mechanisms are distinct since they were reported to be secreted enzymes that do not require Ca^{2+} for activity.

The mannose analogue 1-deoxymannojirimycin is an inhibitor of all class 1 α-1,2-mannosidases. When mammalian cells in culture are treated with this compound, maturation to complex *N*-glycan synthesis is completely prevented and $Man_{8-9}GlcNAc_2$ oligosaccharides accumulate on glycoproteins. As discussed earlier, 1-deoxymannojirimycin-sensitive α-1,2-mannosidases are present in both the ER and the Golgi. Kifunensine, a cyclic oxamide derivative of 1-aminomannojirimycin, is an even more effective inhibitor of the yeast (F. Lipari and A. Herscovics, unpublished findings), plant, and mammalian α-1,2-mannosidases,[138] and much lower concentrations of kifunensine than of 1-deoxymannojirimycin are required for inhibition both *in vivo* and *in vitro*.

3.02.5.3   Class 2 α-Mannosidases

Class 2 α-mannosidases are a heterogeneous group of enzymes with some similarities in amino acid sequence and enzymatic properties, but different intracellular localization and functions. Some of the class 2 enzymes are involved in the *N*-glycan-processing pathway, while others are required for *N*-glycan catabolism. They are less specific than the class 1 enzymes since they can cleave α-1,2-, α-1,3-, and α-1,6-linked mannose residues and aryl α-D-mannosides. These enzymes are usually inhibited by different concentrations of swainsonine, and are not usually affected by 1-deoxymannojirimycin.

3.02.5.3.1   *Golgi-α-mannosidase II*

α-Mannosidase II was the first Golgi processing glycosidase described, and shown to be distinct from lysosomal and cytosolic α-mannosidases.[139–141] Although Golgi α-mannosidase II utilizes *p*-nitrophenyl-α-D-mannopyranoside as a substrate, it has a restricted specificity toward its oligosaccharide substrate. It cleaves the terminal α-1,3- and α-1,6-linked mannose residues from $GlcNAcMan_5GlcNAc_2$ of the oligosaccharide-processing pathway, but it will not utilize $Man_5GlcNAc_2$ as a substrate. Its function therefore depends upon the prior action of GlcNAc transferase I.[108,141,142] It is a membrane-bound enzyme that has no divalent cation requirement and a pH optimum of 5.6–6.5, depending on the source. It is activated by sulfhydryl reagents and is inhibited by thiol alkylating agents.[140] The enzyme has been purified from rat liver and shown to form dimers consisting of subunits with an apparent molecular size of 124 kDa on SDS–PAGE.[140,143] Using partial amino acid sequence information obtained from the rat liver enzyme to design primers for PCR, an α-mannosidase II cDNA has been isolated from a murine 3T3 cDNA library.[144] It encodes a type II transmembrane protein of 132 kDa with a short cytoplasmic tail of five amino acids and a lumenal *C*-terminal catalytic domain that can be purified following mild chymotrypsin digestion.[143] Transfection of mammalian cells in culture with the cDNA results in 10–12-fold overexpression of enzyme activity and Golgi localization of the resulting protein. It has been immunolocalized to the medial and trans-Golgi of a variety of cells in the rat, but with significant cell-dependent variation in its intra-Golgi localization similar to that described above for α-mannosidase IA.[110]

A human α-mannosidase II cDNA, encoding a protein exhibiting about 80% amino acid identity with the murine enzyme, was also isolated and used to screen a genomic library. A related, but distinct, human gene, termed α-mannosidase II^X, was found. The α-mannosidase II^X and α-mannosidase II genes map to chromosomes 15q25 and 5q21–22, respectively.[144,145] Evidence was

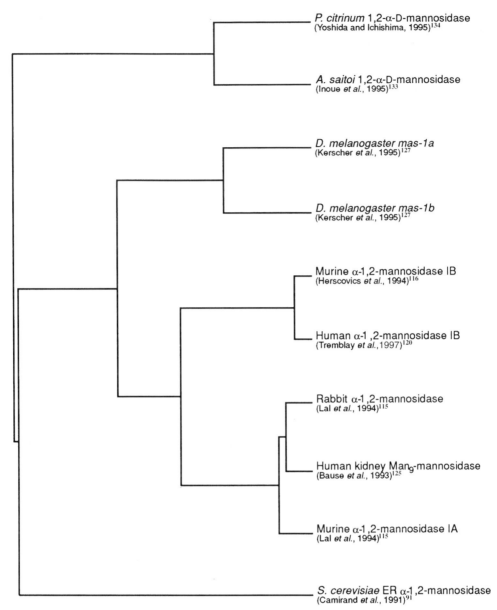

Figure 3 Class 1 α-mannosidases. Dendogram prepared using the Pileup program of the University of Wisconsin Genetics Computer Group (version 8), depicting the amino acid sequence relationships between members of the class 1 α-mannosidases. (Reprinted with permission from the thesis at McGill University "Genomic organization and chromosomal mapping of the murine α1,2-mannosidase IB gene involved in *N*-glycan maturation," p. 30. Copyright © 1996 Nathalie V. Campbell Dyke.)

obtained for the existence of alternatively spliced variants of α-mannosidase II[X] and for significant differences in tissue-specific expression of the respective α-mannosidase II and II[X] transcripts. The exact specificity of α-mannosidase II[X] has not yet been reported as it was assayed with *p*-nitrophenyl-α-D-mannopyranoside.

A membrane-bound α-mannosidase II with properties similar to the rat liver enzyme has been purified from mung bean seedlings,[146] and from insect cell lines.[147,148] The enzyme has a strict requirement for the oligosaccharide Glc*N*AcMan$_5$Glc*N*Ac$_2$ as a substrate with a pH optimum of 6.0–6.5, does not require a divalent cation for activity, and is highly sensitive to swainsonine. The insect, mung bean, and *Xenopus* liver class II α-mannosidases were shown to exhibit some branch specificity whereby the α-1,6-linked mannose residue is removed from Glc*N*AcMan$_5$Glc*N*Ac$_2$ before the α-1,3-linked mannose unit.[147,148] Using degenerate primers to conserved regions of class II α-mannosidases for PCR, a cDNA was isolated from the lepidopteran insect Sf9 cells that encodes an α-mannosidase that is about 37% identical in amino acid sequence to human Golgi α-mannosidase II. Northern blot analysis indicates that the expression of this enzyme varies in different lepidopteran cell lines, suggesting that it may be a limiting factor for the elaboration of complex *N*-glycans in the baculovirus expression system.[149] An α-mannosidase II cDNA homologue called *GmII* has also been isolated from *D. melanogaster* using the murine Golgi α-mannosidase II cDNA for low-stringency screening of a *Drosophila* library.[150] It encodes a protein of about 127 kDa with significant similarity to other members of the class 2 α-mannosidase family. The gene was localized to the right arm of chromosome 3 by *in situ* hybridization.

α-Mannosidase II is strongly inhibited by the plant alkaloid swainsonine,[62] and to different extents by mannostatin A,[151] but not by 1-deoxymannojirimycin. Swainsonine causes the accumulation of hybrid *N*-glycans in place of complex oligosaccharides. Similarly, the ricin-resistant mammalian cell lines Ric[15] and Ric[19], which have decreased α-mannosidase II activity, form mostly hybrid *N*-glycans.[152] The lack of α-mannosidase II is also responsible for some cases of the human genetic disease HEMPAS, in which erythrocyte glycoproteins are deficient in complex-type *N*-glycans called polylactosaminoglycans.[153,154] The fact that the glycosylation defect is restricted to a few specific cell types in these patients suggests that there may be more than one gene controlling α-mannosidase II-like activity, but it remains to be determined whether the newly discovered α-mannosidase II[X] gene accounts for the restricted localization of the *N*-glycan defect in these HEMPAS patients. Another possibility is that there may be a Golgi α-mannosidase in unaffected tissues of these patients that does not require the prior action of GlcNAc transferase I, as some of the enzymes described below.

3.02.5.3.2 *ER/cytosolic α-mannosidase*

There is evidence for α-mannosidase activity in the cytosol of mammalian cells.[155–158] This activity is distinguished from lysosomal α-mannosidase activity by its higher pH optimum, different order of removal of mannose from Man$_9$GlcNAc, and different cation requirements. The cytosolic α-mannosidase is involved in the catabolism of free oligosaccharides that originate in the ER either from lipid-linked oligosaccharides[38,159] or from glycoproteins.[160–163] To account for the origin of cytosolic oligosaccharides, there exists an ATP-dependent transport of free oligosaccharides from the lumen of the ER into the cytosol that has been demonstrated in permeabilized mammalian cells.[164] The cytosolic α-mannosidase is selective for oligosaccharides with a single GlcNAc residue at the reducing end.[163,165] The cytosolic enzyme produces a truncated Man$_5$GlcNAc with the same structure as the Man$_5$GlcNAc intermediate formed during the biosynthesis of dolichol pyrophosphate oligosaccharide.[163,166] This Man$_5$GlcNAc is then translocated from the cytosol to the lysosomes, where it can be further degraded.[167] The cytosolic enzyme was purified from rat[168,170] and bovine[158] liver. The rat liver enzyme was shown to be a tetramer consisting of 110 kDa subunits with a pH optimum around 6, using *p*-nitrophenyl-α-D-mannopyranoside as a substrate. The enzyme is stabilized by Co^{2+}, Mn^{2+}, and Fe^{2+} ions, by dithiothreitol, and by the inhibitor mannosylamine. It is not affected by 1-deoxymannojirimycin, and is inhibited by swainsonine, albeit at higher concentrations than Golgi α-mannosidase II and lysosomal α-mannosidase. It is also inhibited by 1,4-dideoxy-1,4-imino-D-mannitol.[106] The purified cytosolic enzyme can utilize Man$_{5-9}$GlcNAc oligosaccharides as substrates, and is therefore not specific for the mannose linkage.[169] In contrast to lysosomal α-mannosidase, the cytosolic enzyme has a preference for branched rather than linear high-mannose oligosaccharides.[105,156] The bovine liver cytosolic enzyme was found to have a molecular size of 500 kDa on gel filtration.[158] Its properties are similar to the rat liver cytosolic enzyme and to a neutral α-mannosidase purified from quail oviduct[165] in that it utilizes *p*-nitrophenyl-α-D-mannopyranoside as a substrate with a pH optimum of 6.0–6.5 and is greatly stimulated by Co^{2+} and slightly by Mn^{2+}. It has a marked preference for oligosaccharides with one GlcNAc at the reducing end, consistent with a role in oligosaccharide catabolism.

There are two studies suggesting that the mammalian cytosolic enzyme is related to an ER α-mannosidase.[106,169] In the first report, antibodies to the soluble purified α-mannosidase were found to cross-react with an ER α-mannosidase, suggesting the possibility that the purified cytosolic enzyme was derived from a membrane-bound ER α-mannosidase by partial proteolysis.[169] In the later report, antibodies raised against a synthetic peptide derived from the *C*-terminal region of the purified cytosolic enzyme were found to cross-react with the kifunensine-insensitive ER α-mannosidase,[106] but in this instance the ER protein was smaller (82 kDa) than the cytosolic enzyme (105 kDa).[106] The reasons for the difference between these two reports are not clear, and the exact molecular relationship between the cytosolic and ER α-mannosidases needs to be investigated further.

The cDNA encoding the rat liver ER/cytosolic enzyme was isolated from a cDNA library using degenerate oligonucleotides derived from the amino acid sequence of the purified soluble enzyme for PCR.[171] It was shown to encode a 116 kDa protein with no hydrophobic region that could serve as a signal sequence or a transmembrane domain. For this reason, its role in *N*-glycan processing is difficult to envisage without invoking some unknown mechanism that would allow its catalytic function on the lumenal side of the ER. The cDNA encodes an enzyme that exhibits 33% amino acid identity and 58% similarity to the yeast vacuolar α-mannosidase, a nonspecific enzyme associated with the inner surface of the vacuolar membrane.[172] The yeast vacuolar enzyme also lacks a signal sequence or transmembrane domain, and is transported to the vacuoles by an unknown mechanism that is independent of the secretory pathway.[173] The role of the cloned rat liver ER/cytosolic enzyme remains to be clarified.

3.02.5.3.3 *Lysosomal and vacuolar α-mannosidases*

There are α-mannosidases in lysosomes and in yeast vacuoles that have significant amino acid similarity with the class 2 α-mannosidases described above. The detailed properties and function of lysosomal enzymes are described in a review.[13] Unlike most of the α-mannosidases involved in glycoprotein maturation, the lysosomal and vacuolar enzymes function at an acidic pH optimum of 4.0–4.5. They cleave α-1,2-, α-1,3-, and α-1,6-linked mannose residues during glycoprotein catabolism. The major lysosomal α-mannosidase of mammalian cells has a broad specificity, and removes mannose residues from Man$_9$GlcNAc in a nonrandom manner that is different from either cytosolic or Golgi α-mannosidases. It utilizes aryl α-D-mannopyranosides as substrates, is activated by Zn^{2+}, and is inhibited by swainsonine. The lysosomal enzyme has been purified from different sources, and the *Dictyostelium discoideum*, human, and murine enzymes have been cloned, taking advantage of regions that are conserved in class 2 α-mannosidases to design primers for PCR.[174–176] The human gene encoding the major lysosomal α-mannosidase has been localized to chromosome 19. It is deficient in patients with mannosidosis, a disease that causes mental retardation and is characterized by the excretion of large amounts of high-mannose oligosaccharides in the urine. Lysosomes also have an α-1,6-mannosidase that specifically cleaves the α-1,6-mannose residue from the pentasaccharide core of *N*-glycans,[177] but this enzyme has not yet been cloned.

3.02.5.3.4 *Miscellaneous α-mannosidases*

There are several mammalian α-mannosidases that cleave α-1,2-, α-1,3-, and α-1,6-linked mannose residues with enzymatic properties similar to members of the class 2 α-mannosidase family, but these have not yet been cloned; definitive classification of these enzymes awaits their structural determination. One of these enzymes was partially purified from rat brain microsomes and was shown to cleave Man$_{4-9}$GlcNAc to Man$_3$GlcNAc with similar efficiencies.[178] The rat brain α-mannosidase has a pH optimum of 6.0, does not require the prior action of GlcNAc transferase I, has minimal activity with *p*-nitrophenyl-α-D-mannopyranoside, and is unaffected by either swainsonine or 1-deoxymannojirimycin at concentrations that greatly inhibit Golgi α-mannosidase II and α-1,2-mannosidases, respectively. The rat brain microsomal α-mannosidase may therefore produce Man$_3$GlcNAc$_2$ for the initiation of complex *N*-glycan formation by GlcNAc transferase I, thereby replacing α-mannosidase II that was not detectable in rat brain. Also, its presence may partly account for the observation that, unlike other species, rats do not accumulate hybrid *N*-glycans in the brain following the administration of swainsonine. The rat brain microsomal α-mannosidase was shown to cross-react with antibodies raised against the rat liver cytosolic α-mannosidase, but rat brain cytosolic α-mannosidase activity was greatly inhibited by Zn^{2+} at concentrations that did not affect the microsomal activity.

A broad-specificity α-mannosidase has been purified from rat liver microsomes.[179–181] The enzyme catalyzes the ordered removal of α-1,2-, α-1,3-, and α-1,6-linked mannose residues from Man$_{4-9}$-GlcNAc oligosaccharide substrates, with the limit digestion product being Man$_3$GlcNAc. The purified enzyme is a dimer of 110 kDa subunits, with a pH optimum of 6.1–6.5. The enzyme is stabilized by Co^{2+}, but is inhibited by Zn^{2+}, Cu^{2+}, and Fe^{2+}. The enzyme is inhibited by swainsonine and 1-deoxymannojirimycin at 50–500-fold higher concentrations than those required to inhibit Golgi α-mannosidase II and α-mannosidase I, respectively. It differs from the rat liver cytosolic α-mannosidase by its inability to utilize *p*-nitrophenyl-α-D-mannopyranoside as a substrate and from the rat brain microsomal α-mannosidase described above by its lack of action on GlcNAc-Man₅GlcNAc. This α-mannosidase was immunolocalized in the ER, the Golgi, and endosomes of rat liver, with a lumenal orientation.[181] From its properties and localization, it is conceivable that this enzyme participates in *N*-glycan processing, but this role has not yet been demonstrated.

An enzyme that can cleave α-1,2-, α-1,3-, and α-1,6-linked mannose residues from Man₈GlcNAc with the formation of Man₃GlcNAc was identified as an ecto-enzyme on rat sperm plasma membranes, where it may play a role in sperm–egg interactions.[182] A soluble enzyme activity, most likely derived from the sperm plasma membrane, was also found in rat epididymal fluid and shown to have similar properties.[183] It consists of four subunits of 115 kDa. This α-mannosidase has an optimum pH of 6.2–6.5 and does not utilize *p*-nitrophenyl-α-D-mannopyranoside, and so it is clearly distinct from a sperm acrosomal α-mannosidase that has an acid pH optimum. It is activated by Co^{2+} and Mn^{2+}, and is inhibited by Cu^{2+} and Zn^{2+}. It is not sensitive to either swainsonine or 1-deoxymannojirimycin, and does not cross-react with any of the antibodies toward other rat α-mannosidases. It therefore appears to be a unique enzyme that may be specific to the testis.

3.02.5.3.5   *Comparison of class 2 α-mannosidases*

The dendogram in Figure 4 indicates that the class 2 enzymes can be further subdivided into three subgroups according to their amino acid sequence homology. The first group consists of the ER/cytosolic mammalian enzyme and the yeast vacuolar α-mannosidase that do not have a signal sequence. The yeast vacuolar enzyme is involved in glycoprotein catabolism, but, as discussed above, the role of the ER/cytosolic α-mannosidase is still unclear. The second group includes the *N*-glycan-processing α-mannosidase II from different species that are Golgi type II transmembrane proteins. The third category consists of soluble lysosomal α-mannosidases required for glycoprotein catabolism in different species.

3.02.6   *N*-ACETYLGLUCOSAMINIDASES

3.02.6.1   *N*-Acetylglucosamine-1-phosphodiester α-*N*-Acetylglucosaminidase

The major pathway for targeting of soluble enzymes to the lysosomes is mediated by two mannose 6-phosphate receptors in the Golgi that specifically recognize mannose 6-phosphate on high-mannose oligosaccharides and transport the lysosomal enzymes to their destination.[184] Processing of the *N*-glycans on lysosomal enzymes includes a highly specific step catalyzed by GlcNAc-phosphotransferase, an enzyme that recognizes high-mannose oligosaccharides specifically on lysosomal enzymes by binding to a unique conformation-dependent protein recognition domain. This enzyme transfers *N*-acetylglucosamine 1-phosphate to the C-6 position α-1,2-linked mannose residues on high-mannose oligosaccharides in an early Golgi compartment. Following this modification, a highly specific *N*-acetylglucosamine-1-phosphodiester α-*N*-acetylglucosaminidase removes the terminal *N*-acetylglucosamine residue, to expose mannose 6-phosphate for binding to the receptors. This uncovering α-*N*-acetylglucosaminidase has been purified from different sources.[185–191] It has a pH optimum of 6.7–7.0, and does not require any divalent cation for activity. Unlike the phosphotransferase, the α-*N*-acetylglucosaminidase does not specifically recognize a protein determinant on lysosomal enzymes. It has a preference for substrates containing GlcNAc-α-P-Man-α1,2-Man. Acetaminodeoxycastanospermine is a potent inhibitor of the enzyme.[191] The bovine liver enzyme was shown to migrate as a 129 kDa protein and a proteolytically released 121 kDa species on SDS–PAGE.[189] A smaller soluble form (118 kDa) of the enzyme was found in human serum.[189,190] The bovine liver enzyme was shown to be a dimer and to be present in an early Golgi compartment by immunofluorescence of mammalian cells in culture.[192]

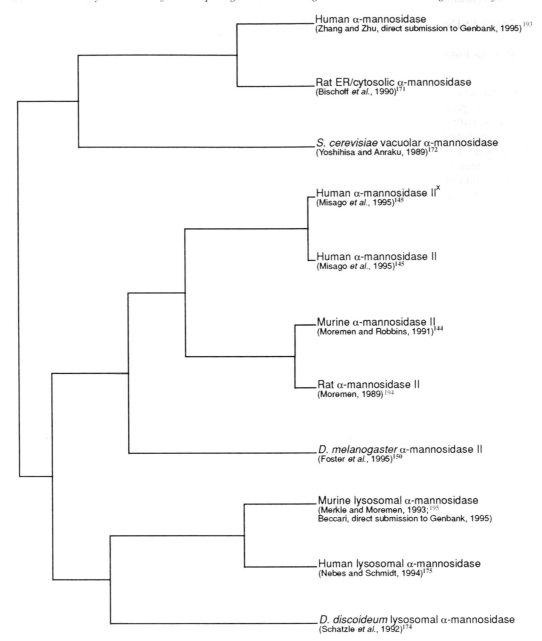

Figure 4   Class 2 α-mannosidases. Dendogram prepared using the Pileup program of the University of Wisconsin Genetics Computer Group (version 8), depicting the amino acid sequence relationships between members of the class 2 α-mannosidases. (Reprinted with permission from the thesis at McGill University "Genomic organization and chromosomal mapping of the murine α1,2-mannosidase IB gene involved in *N*-glycan maturation," p. 35. Copyright © 1996 Nathalie V. Campbell Dyke.)

3.02.6.2   Insect *β-N*-Acetylglucosaminidase

An unusual *β-N*-acetylglucosaminidase has been described in insect cells that utilize 4-nitrophenyl *β-N*-acetylglucosaminide, tri-*N*-acetylchitotriose, and an *N*-linked complex biantennary oligosaccharide, from which it only removes the terminal *β*-1,2-*N*-acetylglucosamine residue from the α-1,3 arm.[196] This enzyme, which was found in both soluble and membrane-bound forms, does not utilize Glc*N*AcMan$_5$Glc*N*Ac$_2$ as a substrate and has no divalent cation requirement. It was suggested that the *β-N*-acetylglucosaminidase participates in *N*-glycan processing following Glc*N*Ac transferase II to produce the truncated complex *N*-glycans with a single *β*-1,2-Glc*N*Ac residue that have been characterized in insect cells.[130,197]

3.02.7   PROCESSING GLYCOSIDASES AND QUALITY CONTROL

3.02.7.1   Protein Folding in the ER

It has been known for some time that the retention of glucose on *N*-linked glycans in cells treated with α-glucosidase inhibitors interferes with the secretion and proper membrane localization of some, but not all, glycoproteins. Such incompletely trimmed glycoproteins accumulate in the ER, and in some cases are rapidly degraded.[62,198–201] Glucose trimming is therefore required for transport of certain glycoproteins out of the ER, and inhibiting this process has a variety of biological consequences that include inhibition of myoblast fusion,[71,202] inhibition of virus assembly and infectivity,[203,204] and reversal of the transformed phenotype induced by some oncogenes.[199] More recent evidence has shown that glucose trimming contributes to an elaborate system of quality control in the ER, whose function is to ensure proper folding of newly formed polypeptide chains and to recognize and degrade misfolded proteins. The ability of polypeptides to fold into functional three-dimensional structures is a function of their amino acid sequence, but proper protein folding is greatly facilitated by interaction with specific proteins known as molecular chaperones.[205] Incompletely folded *N*-glycosylated proteins have been shown to associate transiently in the ER with two chaperones, the membrane-bound calnexin, and its soluble homologue calreticulin, that bind glycoproteins containing high-mannose oligosaccharides with a single glucose residue.[83,206–211] The interaction of newly formed glycoproteins with calnexin and calreticulin promotes folding, prevents degradation and premature oligomerization, and also controls the rate of glycoprotein transport out of the ER.[212,213] Once released from the calnexin or calreticulin complex, completely folded glycoproteins can leave the ER and be targeted to their proper destination. However, if the released glycoproteins have not acquired their proper conformation, they are substrates for UDP-Glc:glycoprotein glucosyltransferase, a soluble lumenal ER enzyme that specifically adds a single glucose residue to the protein-bound high-mannose oligosaccharides. This glucosyltransferase acts as a sensor for malfolded glycoproteins since it only functions with denatured glycoproteins,[214] A model presented in Figure 5[215] has been proposed to explain the role of the α-glucosidases and of the glucosyltransferase in glycoprotein folding mediated by calnexin and calreticulin.[207] In this model, the presence of monoglucosylated high-mannose oligosaccharides on partially folded glycoproteins may arise either from removal of two glucose residues by the sequential action of α-glucosidases I and II, or from reglucosylation by the glucosyltransferase. Repeated cycles of deglucosylation by α-glucosidase II and reglucosylation by the glucosyltransferase can take place until complete folding has occurred. Although for some glycoproteins such as ribonuclease B[216,217] binding to calnexin was shown to be independent of protein conformation, for other glycoproteins the initial recognition is also carbohydrate-dependent but there is evidence for subsequent protein-mediated interaction with calnexin.[211,218] The role of the α-glucosidases in this process was demonstrated in cells treated with the α-glucosidase inhibitors 1-deoxynojirimycin and castanospermine, and in the Lec 23 and PHA[R 2.7] mutant cells that lack α-glucosidases I and II, respectively.[209] Interaction of newly formed glycoproteins with calnexin and calreticulin does not occur in these mutants that cannot form monoglucosylated oligosaccharides. Both α-glucosidases are required to form the monoglucosylated ligand. α-Glucosidase II plays a dual role since it promotes association with the chaperones by producing the monoglucosylated oligosaccharide, and also prevents further association by removing the last glucose residue.

3.02.7.2   Degradation of Malfolded Proteins

Malfolded glycoproteins can be rapidly degraded in the ER or in a post-ER compartment by mechanisms that are still poorly understood. There is some evidence indicating that the extent of mannose trimming of malfolded glycoproteins can influence their proteolytic degradation. It was shown that when the yeast prepro-α-factor is expressed in mammalian cells it is efficiently translocated into the lumen of the ER, but is then rapidly degraded.[219] This degradation was prevented by inhibiting α1,2-mannosidase activity with 1-deoxymannojirimycin, whereas the α-glucosidase inhibitor 1-deoxynojirimycin promoted proteolysis. Similarly, in *S. cerevisiae* is was shown that a malfolded carboxypeptidase Y mutant was rapidly degraded in the ER, and that its degradation was significantly reduced in yeast cells lacking the processing α-1,2-mannosidase.[220] These results suggest that the extent of trimming by α-1,2-mannosidases plays a role in some aspects of recognition leading to proteolytic degradation. The carbohydrate may play a direct role in this process or it may be involved indirectly by affecting protein conformation. The latter hypothesis is supported by

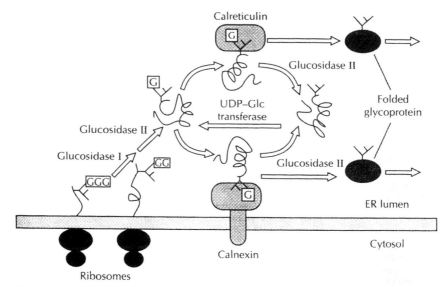

Figure 5 Role of glucosidases in the interaction of newly synthesized glycoproteins with the chaperones calnexin and calreticulin to promote glycoprotein folding in the endoplasmic reticulum. (Reproduced by permission of Current Opinion in Cell Biology from *Curr. Opin. Cell Biol.*, 1995, **7**, 525.)

a study[221] that shows differential stability of the α and β subunits of the T-cell receptor as a function of their N-glycan structures. The TCRα subunit was found to be more readily degraded in cells that synthesize truncated N-glycans such as $Glc_3Man_{5-7}GlcNAc_2$ than in cells that make Glc_3Man_9-$GlcNAc_2$, whereas the stability of the TCRβ subunit was independent of the N-glycan structure.

ACKNOWLEDGMENTS

The work in the author's laboratory is supported by National Institute of Health Grant GM31265 and by an operating grant from the Medical Research Council of Canada.

3.02.8   REFERENCES

1. M. Metzler, A. Gertz, M. Sarkar, H. Schachter, J. W. Schrader, and J. D. Marth, *EMBO J.*, 1994, **13**, 2056.
2. E. Ioffe and P. Stanley, *Proc. Natl. Acad. Sci. USA*, 1994, **91**, 728.
3. I. Brockhausen, *Crit. Rev. Clin. Lab. Sci.*, 1993, **30**, 65.
4. P. E. Goss, M. A. Baker, J. P. Carver, and J. W. Dennis, *Clin. Cancer Res.*, 1995, **1**, 935.
5. J. Jaeken and H. Carchon, *J. Inherit. Metab. Dis.*, 1993, **16**, 813.
6. A. Varki, *Glycobiology*, 1993, **3**, 97.
7. B. Henrissat and A. Romeu, *Biochem. J.*, 1995, **311**, 350.
8. P. Lalégerie, G. Legler, and J. M. Yon, *Biochimie*, 1982, **64**, 977.
9. G. Legler, *Adv. Carbohydrate Chem. Biochem.*, 1990, **48**, 319.
10. M. L. Sinnott, *Chem. Rev.*, 1990, **90**, 1171.
11. S. G. Withers, *Pure Appl. Chem.*, 1995, **67**, 1673.
12. K. W. Moremen, R. B. Trimble, and A. Herscovics, *Glycobiology*, 1994, **4**, 113.
13. P. F. Daniel, B. Winchester, and C. D. Warren, *Glycobiology*, 1994, **4**, 551.
14. R. Kornfeld and S. Kornfeld, *Annu. Rev. Biochem.*, 1985, **54**, 631.
15. H. Schachter, *Biochem. Cell Biol.*, 1986, **64**, 163.
16. A. Herscovics and P. Orlean, *FASEB J.*, 1993, **7**, 540.
17. S. Silberstein and R. Gilmore, *FASEB J.*, 1996, **10**, 849.
18. S. C. Hubbard and R. J. Ivatt, *Annu. Rev. Biochem.*, 1981, **50**, 555.
19. P. H. Atkinson and J. T. Lee, *J. Cell Biol.*, 1984, **98**, 2245.
20. R. A. Ugalde, R. J. Staneloni, and L. F. Leloir, *FEBS Lett.*, 1978, **91**, 209.
21. W. W. Chen and W. J. Lennarz, *J. Biol. Chem.*, 1978, **253**, 5780.
22. M. G. Scher and C. J. Waechter, *J. Biol. Chem.*, 1979, **254**, 2630.
23. R. A. Ugalde, R. J. Staneloni, and L. F. Leloir, *Biochem. Biophys. Res. Commun.*, 1979, **91**, 1174.
24. L. S. Grinna and P. W. Robbins, *J. Biol. Chem.*, 1979, **254**, 8814.
25. R. G. Spiro, M. J. Spiro, and V. D. Bhoyroo, *J. Biol. Chem.*, 1979, **254**, 7659.
26. R. A. Ugalde, R. J. Staneloni, and L. F. Leloir, *Eur. J. Biochem.*, 1980, **113**, 97.
27. L. S. Grinna and P. W. Robbins, *J. Biol. Chem.*, 1980, **255**, 2255.

28. J. M. Michael and S. Kornfeld, *Arch. Biochem. Biophys.*, 1980, **199**, 249.
29. J. J. Elting, W. W. Chen, and W. J. Lennarz, *J. Biol. Chem.*, 1980, **255**, 2325.
30. H. Hettkamp, G. Legler, and E. Bause, *Eur. J. Biochem.*, 1984, **142**, 85.
31. E. Bause, J. Schweden, A. Gross, and B. Orthen, *Eur. J. Biochem.*, 1989, **183**, 661.
32. K. Shailubhai, M. A. Pratta, and I. K. Vijay, *Biochem. J.*, 1987, **247**, 555.
33. D. R. P. Tulsiani, V. D. Coleman, and O. Touster, *Arch. Biochem. Biophys.*, 1990, **277**, 114.
34. T. Szumilo, G. P. Kaushal, and A. D. Elbein, *Arch. Biochem. Biophys.*, 1986, **247**, 261.
35. R. D. Kilker, Jr., B. Saunier, J. S. Tkacz, and A. Herscovics, *J. Biol. Chem.*, 1981, **256**, 5299.
36. E. Bause, R. Erkens, J. Schweden, and L. Jaenicke, *FEBS Lett.*, 1986, **206**, 208.
37. B. Saunier, R. D. Kilker, Jr., J. S. Tkacz, A. Quaroni, and A. Herscovics, *J. Biol. Chem.*, 1982, **257**, 14155.
38. M. J. Spiro and R. G. Spiro, *J. Biol. Chem.*, 1991, **266**, 5311.
39. B. S. Pukazhenthi, N. Muniappa, and I. K. Vijay, *J. Biol. Chem.*, 1993, **268**, 6445.
40. A. Romaniouk and I. K. Vijay, *Glycobiology*, 1997, **7**, 399.
41. B. Kalz-Füller, E. Bieberich, and E. Bause, *Eur. J. Biochem.*, 1995, **231**, 344.
42. B. Kalz-Füller, C. Heidrich-Kaul, M. Nöthen, E. Bause, and G. Schwanitz, *Genomics*, 1996, **34**, 442.
43. M. K. Ray, J. Yang, S. Sundaram, and P. Stanley, *J. Biol. Chem.*, 1991, **266**, 22818.
44. B. Esmon, P. C. Esmon, and R. Schekman, *J. Biol. Chem.*, 1984, **259**, 10322.
45. B. Jiang, J. Sheraton, A. F. J. Ram, G. J. P. Dijkgraaf, F. M. Klis, and H. Bussey, *J. Bacteriol.*, 1996, **178**, 1162.
46. P. A. Romero, G. J. P. Dijkgraaf, S. Shahinian, A. Herscovics, and H. Bussey, *Glycobiology*, 1997, **7**, 997.
47. S. Kornfeld, E. Li, and I. Tabas, *J. Biol. Chem.*, 1978, **253**, 7771.
48. S. C. Hubbard and P. W. Robbins, *J. Biol. Chem.*, 1979, **254**, 4568.
49. J. M. Lucocq, D. Brada, and J. Roth, *J. Cell Biol.*, 1986, **102**, 2137.
50. D. Brada, D. Kerjaschki, and J. Roth, *J. Cell Biol.*, 1990, **110**, 309.
51. D. M. Burns and O. Touster, *J. Biol. Chem.*, 1982, **257**, 9991.
52. D. Brada and U. C. Dubach, *Eur. J. Biochem.*, 1984, **141**, 149.
53. S. Saxena, K. Shailubhai, B. Dong-Yu, and I. K. Vijay, *Biochem. J.*, 1987, **247**, 563.
54. Y. Hino and J. E. Rothman, *Biochemistry*, 1985, **24**, 800.
55. G. P. Kaushal, I. Pastuszak, K.-i. Hatanaka, and A. D. Elbein, *J. Biol. Chem.*, 1990, **265**, 16271.
56. E. S. Trombetta, J. F. Simons, and A. Helenius, *J. Biol. Chem.*, 1996, **271**, 27509.
57. T. Flura, D. Brada, M. Ziak, and J. Roth, *Glycobiology*, 1997, **7**, 617.
58. F. Martiniuk, M. Smith, A. Ellenbogen, R. J. Desnick, K. Astrin, J. Mitra, and R. Hirschhorn, *Cytogenet. Cell Genet.*, 1983, **35**, 110.
59. F. Martiniuk, A. Ellenbogen, and R. Hirschhorn, *J. Biol. Chem.*, 1985, **260**, 1238.
60. M. L. Reitman, I. S. Trowbridge, and S. Kornfeld, *J. Biol. Chem.*, 1982, **257**, 10357.
61. K. Fujimoto and R. Kornfeld, *J. Biol. Chem.*, 1991, **266**, 3571.
62. A. D. Elbein, *FASEB J.*, 1991, **5**, 3055.
63. R. Datema, S. Olofsson, and P. A. Romero, *Pharmacol. Ther.*, 1987, **33**, 221.
64. B. Winchester and G. W. J. Fleet, *Glycobiology*, 1992, **2**, 199.
65. J. E. Tropea, R. J. Molyneux, G. P. Kaushal, Y. T. Pan, M. Mitchell, and A. D. Elbein, *Biochemistry*, 1989, **28**, 2027.
66. Y. T. Pan, H. Hori, R. Saul, B. A. Sanford, R. J. Molyneux, and A. D. Elbein, *Biochemistry*, 1983, **22**, 3975.
67. P. A. Romero, B. Saunier, and A. Herscovics, *Biochem. J.*, 1985, **226**, 733.
68. G. B. Karlsson, T. D. Butters, R. A. Dwek, and F. M. Platt, *J. Biol. Chem.*, 1993, **268**, 570.
69. P. A. Romero, P. Friedlander, and A. Herscovics, *FEBS Lett.*, 1985, **183**, 29.
70. R. Datema, P. A. Romero, G. Legler, and R. T. Schwartz, *Proc. Natl. Acad. Sci. USA*, 1982, **79**, 6787.
71. G. C. Trudel, A. Herscovics, and P. C. Holland, *Biochem. Cell Biol.*, 1988, **66**, 1119.
72. J. M. Alonso, A. Santa-Cecilia, and P. Calvo, *Biochem. J.*, 1991, **278**, 721.
73. J. M. Alonso, A. Santa-Cecilia, and P. Calvo, *Eur. J. Biochem.*, 1993, **215**, 37.
74. A. J. Parodi, *Glycobiology*, 1993, **3**, 193.
75. P. A. Romero and A. Herscovics, *J. Biol. Chem.*, 1986, **261**, 15936.
76. W. A. Lubas and R. G. Spiro, *J. Biol. Chem.*, 1987, **262**, 3775.
77. W. A. Lubas and R. G. Spiro, *J. Biol. Chem.*, 1988, **263**, 3990.
78. S. E. H. Moore and R. G. Spiro, *J. Biol. Chem.*, 1990, **265**, 13104.
79. C. Rabouille and R. G. Spiro, *J. Biol. Chem.*, 1992, **267**, 11573.
80. S. Weng and R. G. Spiro, *Glycobiology*, 1996, **6**, 861.
81. S. Hiraizumi, U. Spohr, and R. G. Spiro, *J. Biol. Chem.*, 1993, **268**, 9927.
82. S. Hiraizumi, U. Spohr, and R. G. Spiro, *J. Biol. Chem.*, 1994, **269**, 4697.
83. R. G. Spiro, Q. Zhu, V. Bhoyroo, and H.-D. Söling, *J. Biol. Chem.*, 1996, **271**, 11588.
84. K. Dairaku and R. G. Spiro, *Glycobiology*, 1997, **7**, 579.
85. D. Godelaine, M. J. Spiro, and R. G. Spiro, *J. Biol. Chem.*, 1981, **256**, 10161.
86. F. D. Ziegler, T. R. Gemmill, and R. B. Trimble, *J. Biol. Chem.*, 1994, **269**, 12527.
87. J. C. Byrd, A. L. Tarentino, F. Maley, P. H. Atkinson, and R. B. Trimble, *J. Biol. Chem.*, 1982, **257**, 14657.
88. S. Jelinek-Kelly, A. Akiyama, B. Saunier, J. S. Tkacz, and A. Herscovics, *J. Biol. Chem.*, 1985, **260**, 2253.
89. S. Jelinek-Kelly and A. Herscovics, *J. Biol. Chem.*, 1988, **263**, 14757.
90. F. D. Ziegler and R. B. Trimble, *Glycobiology*, 1991, **1**, 605.
91. A. Camirand, A. Heysen, B. Grondin, and A. Herscovics, *J. Biol. Chem.*, 1991, **266**, 15120.
92. B. Grondin and A. Herscovics, *Glycobiology*, 1992, **2**, 369.
93. J. Burke, F. Lipari, S. Igdoura, and A. Herscovics, *Eur. J. Cell Biol.*, 1996, **70**, 298.
94. R. Puccia, B. Grondin, and A. Herscovics, *Biochem. J.*, 1993, **290**, 21.
95. F. Lipari and A. Herscovics, *Glycobiology*, 1994, **4**, 697.
96. F. Lipari and A. Herscovics, *J. Biol. Chem.*, 1996, **271**, 27615.
97. F. Lipari, B. J. Gour-Salin, and A. Herscovics, *Biochem. Biophys. Res. Commun.*, 1995, **209**, 322.
98. K. Dole, F. Lipari, A. Herscovics, and P. L. Howell, *J. Struct. Biol.*, 1997, **120**, 69.
99. S. Hickman, J. L. Theodorakis, J. M. Greco, and P. H. Brown, *J. Cell Biol.*, 1984, **98**, 407.

100. M. G. Rosenfeld, E. E. Marcantonio, J. Hakimi, V. M. Ort, P. H. Atkinson, D. Sabatini, and G. Kreibich, *J. Cell Biol.*, 1984, **99**, 1076.
101. J. Bischoff and R. Kornfeld, *J. Biol. Chem.*, 1983, **258**, 7907.
102. L. J. Rizzolo and R. Kornfeld, *J. Biol.. Chem.*, 1988, **263**, 9520.
103. J. Bischoff, L. Liscum, and R. Kornfeld, *J. Biol. Chem.*, 1986, **261**, 4766.
104. S. Weng and R. G. Spiro, *J. Biol. Chem.*, 1993, **268**, 25 656.
105. J.-F. Haeuw, G. Strecker, J.-M. Wieruszeski, J. Montreuil, and J.-C. Michalski, *Eur. J. Biochem.*, 1991, **202**, 1257.
106. S. Weng and R. G. Spiro, *Arch. Biochem. Biophys.*, 1996, **325**, 113.
107. I. Tabas and S. Kornfeld, *J. Biol. Chem.*, 1979, **254**, 11 655.
108. D. R. P. Tulsiani, S. C. Hubbard, P. W. Robbins, and O. Touster, *J. Biol. Chem.*, 1982, **257**, 3660.
109. D. R. P. Tulsiani and O. Touster, *J. Biol. Chem.*, 1988, **263**, 5408.
110. A. Velasco, L. Hendricks, K. W. Moremen, D. R. P. Tulsiani, O. Touster, and M. G. Farquhar, *J. Cell Biol.*, 1993, **122**, 39.
111. W. T. Forsee and J. S. Schutzbach, *J. Biol. Chem.*, 1981, **256**, 6577.
112. W. T. Forsee, C. F. Palmer, and J. S. Schutzbach, *J. Biol. Chem.*, 1989, **264**, 3869.
113. J. S. Schutzbach and W. T. Forsee, *J. Biol. Chem.*, 1990, **265**, 2546.
114. W. T. Forsee, J. D. Springfield, and J. S. Schutzbach, *J. Biol. Chem.*, 1982, **257**, 9963.
115. A. Lal, J. S. Schutzbach, W. T. Forsee, P. J. Neame, and K. W. Moremen, *J. Biol. Chem.*, 1994, **269**, 9872.
116. A. Herscovics, J. Schneikert, A. Athanassiadis, and K. W. Moremen, *J. Biol. Chem.*, 1994, **269**, 9864.
117. J. Schneikert and A. Herscovics, *Glycobiology*, 1994, **4**, 445.
118. J. Schneikert and A. Herscovics, *J. Biol. Chem.*, 1995, **270**, 17 736.
119. N. Campbell Dyke, A. Athanassiadis, and A. Herscovics, *Genomics*, 1997, **41**, 155.
120. L. Tremblay, N. Campbell Dyke, and A. Herscovics, *Glycobiology*, 1997, **6**, 757.
121. J. Schweden, G. Legler, and E. Bause, *Eur. J. Biochem.*, 1986, **157**, 563.
122. J. Schweden and E. Bause, *Biochem. J.*, 1989, **264**, 347.
123. J. Roth, D. Brada, P. M. Lackie, J. Schweden, and E. Bause, *Eur. J. Cell Biol.*, 1990, **53**, 131.
124. E. Bause, W. Breuer, J. Schweden, R. Roeser, and R. Geyer, *Eur. J. Biochem.*, 1992, **208**, 451.
125. E. Bause, E. Bieberich, A. Rolfs, C. Völker, and B. Schmidt, *Eur. J. Biochem.*, 1993, **217**, 535.
126. E. Bieberich and E. Bause, *Eur. J. Biochem.*, 1995, **233**, 644.
127. S. Kerscher, S. Albert, D. Wucherpfennig, M. Heisenberg, and S. Schneuwly, *Dev. Biol.*, 1995, **168**, 613.
128. Z. Kawar, A. Herscovics, and D. L. Jarvis, *Glycobiology*, 1997, **7**, 433.
129. D. L. Jarvis and L. A. Guarino, *Methods Mol. Biol.*, 1995, **39**, 187.
130. V. Kubelka, F. Altmann, G. Kornfeld, and L. März, *Arch. Biochem. Biophys.*, 1994, **308**, 148.
131. D. J. Davidson, R. K. Bretthauer, and F. J. Castellino, *Biochemistry*, 1991, **30**, 9811.
132. J. Ren, R. K. Bretthauer, and F. J. Castellino, *Biochemistry*, 1995, **34**, 2489.
133. T. Inoue, T. Yoshida, and E. Ichishima, *Biochim. Biophys. Acta*, 1995, **1253**, 141.
134. T. Yoshida and E. Ichishima, *Biochim. Biophys. Acta*, 1995, **1263**, 159.
135. T. Yoshida, K. Maeda, M. Kobayashi, and E. Ichishima, *Biochem. J.*, 1994, **303**, 97.
136. T. Szumilo, G. P. Kaushal, H. Hori, and A. D. Elbein, *Plant Physiol.*, 1986, **81**, 383.
137. N. Hamagashira, H. Oku, T. Mega, and S. Hase, *J. Biochem.*, 1996, **119**, 998.
138. A. D. Elbein, J. E. Tropea, M. Mitchell, and G. P. Kaushal, *J. Biol. Chem.*, 1990, **265**, 15 599.
139. B. Dewald and O. Touster, *J. Biol. Chem.*, 1973, **248**, 7223.
140. D. R. P. Tulsiani, D. J. Opheim, and O. Touster, *J. Biol. Chem.*, 1977, **252**, 3227.
141. I. Tabas and S. Kornfeld, *J. Biol. Chem.*, 1978, **253**, 7779.
142. N. Harpaz and H. Schachter, *J. Biol. Chem.*, 1980, **255**, 4894.
143. K. W. Moremen, O. Touster, and P. W. Robbins, *J. Biol. Chem.*, 1991, **266**, 16 876.
144. K. W. Moremen and P. W. Robbins, *J. Cell Biol.*, 1991, **115**, 1521.
145. M. Misago, Y.-F. Liao, S. Kudo, S. Eto, M.-G. Mattei, K. W. Moremen, and M. N. Fukuda, *Proc. Natl. Acad. Sci. USA*, 1995, **92**, 11 766.
146. G. P. Kaushal, T. Szumilo, I. Pastuszak, and A. D. Elbein, *Biochemistry*, 1990, **29**, 2168.
147. F. Altmann and L. März, *Glycoconjugate J.*, 1995, **12**, 150.
148. J. Ren, F. J. Castellino, and R. K. Bretthauer, *Biochem. J.*, 1997, **324**, 951.
149. D. L. Jarvis, D. A. L. Bohlmeyer, Y.-F. Liao, K. K. Lomax, R. K. Merkle, C. Weinkauf, and K. W. Moremen, *Glycobiology*, 1997, **7**, 113.
150. J. M. Foster, B. Yudkin, A. E. Lockyer, and D. B. Roberts, *Gene*, 1995, **154**, 183.
151. J. E. Tropea, G. P. Kaushal, I. Pastuszak, M. Mitchell, T. Aoyagi, R. J. Molyneux, and A. D. Elbein, *Biochemistry*, 1990, **29**, 10 062.
152. R. C. Hughes and J. Feeney, *Eur. J. Biochem.*, 1986, **158**, 227.
153. M. N. Fukuda, K. A. Masri, A. Dell, L. Luzzatto, and K. W. Moremen, *Proc. Natl. Acad. Sci. USA*, 1990, **87**, 7443.
154. M. N. Fukuda, *Glycobiology*, 1990, **1**, 9.
155. C. A. Marsh and G. C. Gourlay, *Biochim. Biophys. Acta*, 1971, **235**, 142.
156. D. R. P. Tulsiani and O. Touster, *J. Biol. Chem.*, 1987, **262**, 6506.
157. H. Oku, S. Hase, and T. Ikenaka, *J. Biochem.*, 1991, **110**, 29.
158. M. Kumano, K. Omichi, and S. Hase, *J. Biochem.*, 1996, **119**, 991.
159. R. Cacan, R. Cecchelli, and A. Verbert, *Eur. J. Biochem.*, 1987, **166**, 469.
160. K. R. Anumula and R. G. Spiro, *J. Biol. Chem.*, 1983, **258**, 15 274.
161. C. Villers, R. Cacan, A.-M. Mir, O. Labiau, and A. Verbert, *Biochem. J.*, 1994, **298**, 135.
162. S. E. H. Moore and R. G. Spiro, *J. Biol. Chem.*, 1994, **269**, 12 715.
163. T. Grard, V. Herman, A. Saint-Pol, D. Kmiecik, O. Labiau, A.-M. Mir, C. Alonso, A. Verbert, R. Cacan, and J.-C. Michalski, *Biochem. J.*, 1996, **316**, 787.
164. S. E. H. Moore, C. Bauvy, and P. Codogno, *EMBO J.*, 1995, **14**, 6034.
165. H. Oku and S. Hase, *J. Biochem.*, 1991, **110**, 982.
166. D. Kmiécik, V. Herman, C. J. M. Stroop, J.-C. Michalski, A.-M. Mir, O. Labiau, A. Verbert, and R. Cacan, *Glycobiology*, 1995, **5**, 483.

167. A. Saint-Pol, C. Bauvy, P. Codogno, and S. E. H. Moore, *J. Cell Biol.*, 1997, **136**, 45.
168. V. A. Shoup and O. Touster, *J. Biol. Chem.*, 1976, **251**, 3845.
169. J. Bischoff and R. Kornfeld, *J. Biol. Chem.*, 1986, **261**, 4758.
170. T. Grard, A. Saint-Pol, J.-F. Haeuw, C. Alonso, J.-M. Wieruszeski, G. Strecker, and J.-C. Michalski, *Eur. J. Biochem.*, 1994, **223**, 99.
171. J. Bischoff, K. Moremen, and H. F. Lodish, *J. Biol. Chem.*, 1990, **265**, 17110.
172. T. Yoshihisa and Y. Anraku, *Biochem. Biophys. Res. Commun.*, 1989, **163**, 908.
173. T. Yoshihisa and H. Anraku, *J. Biol. Chem.*, 1990, **265**, 22418.
174. J. Schatzle, J. Bush, and J. Cardelli, *J. Biol. Chem.*, 1992, **267**, 4000.
175. V. L. Nebes and M. C. Schmidt, *Biochem. Biophys. Res. Commun.*, 1994, **200**, 239.
176. Y.-F. Liao, A. Lal, and K. W. Moremen, *J. Biol. Chem.*, 1996, **271**, 28348.
177. R. De Gasperi, P. F. Daniel, and C. D. Warren, *J. Biol. Chem.*, 1992, **267**, 9706.
178. D. R. P. Tulsiani and O. Touster, *J. Biol. Chem.*, 1985, **260**, 13081.
179. E. Monis, P. Bonay, and R. C. Hughes, *Eur. J. Biochem.*, 1987, **168**, 287.
180. P. Bonay and R. C. Hughes, *Eur. J. Biochem.*, 1991, **197**, 229.
181. P. Bonay, J. Roth, and R. C. Hughes, *Eur. J. Biochem.*, 1992, **205**, 399.
182. D. R. P. Tulsiani, M. D. Skudlarek, and M.-C. Orgebin-Crist, *J. Cell Biol.*, 1989, **109**, 1257.
183. D. R. P. Tulsiani, M. D. Skudlarek, S. K. Nagdas, and M. C. Orgebin-Crist, *Biochem. J.*, 1993, **290**, 427.
184. S. Kornfeld and I. Mellman, *Annu. Rev. Cell Biol.*, 1989, **5**, 483.
185. A. Varki and S. Kornfeld, *J. Biol. Chem.*, 1981, **256**, 9937.
186. A. Varki, W. Sherman, and S. Kornfeld, *Arch. Biochem. Biophys.*, 1983, **222**, 145.
187. A. Waheed, A. Hasilik, and K. von Figura, *J. Biol. Chem.*, 1981, **256**, 5717.
188. R. Couso, L. Lang, R. M. Roberts, and S. Kornfeld, *J. Biol. Chem.*, 1986, **261**, 6326.
189. K. G. Mullis, M. Huynh, and R. H. Kornfeld, *J. Biol. Chem.*, 1994, **269**, 1718.
190. J.K. Lee and M. Pierce, *Arch. Biochem. Biophys.*, 1995, **319**, 413.
191. T. Page, K.-W. Zhao, L. Tao, and A. L. Miller, *Glycobiology*, 1996, **6**, 619.
192. K. G. Mullis and R. H. Kornfeld, *J. Biol. Chem.*, 1994, **269**, 1727.
193. L. X. Zhang, L. P. Zhu, W. Shi, and F. R. Ma, *Chung Hua Wei Sheng Wu Hsueh Ho Mieh I Hsueh Tsa Chih*, 1997, **17**, 34.
194. K. W. Moremen, *Proc. Natl. Acad. Sci. USA*, 1989, **86**, 5276.
195. R. K. Merkle and K. W. Moremen, *Glycoconjugate J.*, 1996, **10**, 240.
196. F. Altmann, H. Schwihla, E. Staudacher, J. Glössl, and L. März, *J. Biol. Chem.*, 1995, **270**, 17344.
197. F. Altmann, *Trends Glycosci. Glycotechnol.*, 1996, **8**, 101.
198. H. F. Lodish and N. Kong, *J. Cell Biol.*, 1984, **98**, 1720.
199. E. J. Nichols, R. Manger, S. Hakomori, A. Herscovics, and L. R. Rohrschneider, *Mol. Cell. Biol.*, 1985, **5**, 3467.
200. K.-T. Yeo, T.-K. Yeo, and K. Olden, *Biochem. Biophys. Res. Commun.*, 1989, **161**, 1013.
201. S. E. H. Moore and R. G. Spiro, *J. Biol. Chem.*, 1993, **268**, 3809.
202. P. C. Holland and A. Herscovics, *Biochem. J.*, 1986, **238**, 335.
203. A. Karpas, G. W. Fleet, R. A. Dwek, S. Petursson, S. K. Namgoong, N. G. Ramsden, G. S. Jacob, and T. W. Rademacher, *Proc. Natl. Acad. Sci. USA*, 1988, **85**, 9229.
204. T. M. Block, X. Lu, F. M. Platt, G. R. Foster, W. H. Gerlich, B. S. Blumberg, and R. A. Dwek, *Proc. Natl. Acad. Sci. USA*, 1994, **91**, 2235.
205. J. P. Hendrick and F.-U. Hartl, *Annu. Rev. Biochem.*, 1993, **62**, 349.
206. E. Degen and D. B. Williams, *J. Cell Biol.*, 1991, **112**, 1099.
207. C. Hammond, I. Braakman, and A. Helenius, *Proc. Natl. Acad. Sci. USA*, 1994, **91**, 913.
208. D. N. Hebert, B. Foellmer, and A. Helenius, *Cell*, 1995, **81**, 425.
209. A. Ora and A. Helenius, *J. Biol. Chem.*, 1995, **270**, 26060.
210. J. R. Peterson, A. Ora, P. N. Van, and A. Helenius, *Mol. Biol. Cell.*, 1995, **6**, 1173.
211. F. E. Ware, A. Vassilakos, P. A. Peterson, M. R. Jackson, M. A. Lehrman, and D. B. Williams, *J. Biol. Chem.*, 1995, **270**, 4697.
212. W.-J. Ou, P. H. Cameron, D. Y. Thomas, and J. J. M. Bergeron, *Nature*, 1993, **364**, 771.
213. J. J. M. Bergeron, M. B. Brenner, D. Y. Thomas, and D. B. Williams, *Trends Biochem. Sci.*, 1994, **19**, 124.
214. M. Sousa and A. J. Parodi, *EMBO J.*, 1995, **14**, 4196.
215. C. Hammond and A. Helenius, *Curr. Opin. Cell Biol.*, 1995, **7**, 523.
216. A. R. Rodan, J. F. Simons, E. S. Trombetta, and A. Helenius, *EMBO J.*, 1996, **15**, 6921.
217. A. Zapun, S. M. Petrescu, P. M. Rudd, R. A. Dwek, D. Y. Thomas, and J. J. M. Bergeron, *Cell*, 1997, **88**, 29.
218. D. B. Williams, *Biochem. Cell. Biol.*, 1995, **73**, 123.
219. K. Su, T. Stoller, J. Rocco, J. Zemsky, and R. Green, *J. Biol. Chem.*, 1993, **268**, 14301.
220. M. Knop, N. Hauser, and D. H. Wolf, *Yeast*, 1996, **12**, 1229.
221. J. E. M. Van Leeuwen and K. P. Kearse, *J. Biol. Chem.*, 1997, **272**, 4179.

3.03
Glycosyltransferases Involved in *N*-Glycan Synthesis

HARRY SCHACHTER

University of Toronto and Hospital for Sick Children, ON, Canada

---

3.03.1   INTRODUCTION

The major biological macromolecules are proteins, nucleic acids, and glycoconjugates (glyco-proteins and glycolipids). Nucleic acids and proteins are linear polymers in which the building blocks (nucleotides and amino acids) are respectively joined together by identical 3′,5′-phosphodiester and -CONH- amide bonds. In contrast, glycan polymers are often branched and the monosaccharide building blocks may be joined to one another in either alpha or beta glycosidic linkages, connecting the anomeric carbon of one sugar to one of several different carbon positions on the adjoining sugar. The number of all possible linear and branched isomers of a hexasaccharide has been calculated to be over 1×10^{12}.[1] This fact makes it very difficult to determine complete oligosaccharide structures with the same assurance that can be achieved for peptide or DNA sequencing and also explains the difficulties encountered by organic chemists who attempt to synthesize relatively large oligosaccharides.

A living cell manages the complex task of oligosaccharide biosynthesis by using an assembly-line approach. Whereas protein and nucleic acid polymers can be assembled by copying every new molecule from a pre-existing molecule using a template approach, it is clear that this method cannot readily be applied to a branched molecule with a variety of different linkages between component monomers. The cell's glycan assembly line is the endomembrane system—the endoplasmic reticulum and Golgi apparatus.[2] The "workers" are glycosidases, glycosyltransferases, and enzymes which modify oligosaccharides (sulfation, phosphorylation, acetylation, etc.). These enzymes are chained to the assembly line with their catalytic domains within the lumen. They act sequentially on the growing oligosaccharide as it moves along the lumen of the endomembrane system.

There are many factors which control and integrate this biosynthetic assembly line:[2–5] (i) synthesis of nucleotide-sugars, dolichol-linked sugars, and other precursors in the cytoplasm; (ii) transport of these precursors into the lumen of the endomembrane assembly line; (iii) transcription and translation of oligosaccharyltransferases, glycosyltransferases, glycosidases, and other enzymes; (iv) targeting of these enzymes to their correct locations on the endomembrane assembly line; (v) "substrate-level" control factors such as competition between assembly line enzymes for common substrates, substrate specificities of these enzymes, and factors such as cations and pH which control enzyme activity; (vi) rate of movement of nascent polypeptide from the ribosome into the endomembrane lumen and along the lumen to the cell surface, since these factors affect the residence time within the lumen; (vii) conformation of both the polypeptide and the protein-bound oligosaccharide and their accessibility to modifying enzymes; (viii) binding of nascent glycoproteins to chaperones such as calnexin and calreticulin; and (ix) tissue-specific and time-specific expression of assembly-line enzymes. This chapter will be limited to a discussion of the glycosyltransferases which act on *N*-glycans.

Over 100 different glycosidic linkages have been reported and for every linkage there is usually at least one specific glycosyltransferase; there are a few exceptions to this rule in which a single enzyme makes more than one linkage, for example, the Lewis blood group-dependent α1,3/4-fucosyltransferase. The World Wide Web has a guide maintained by Iain Wilson which lists all the glycosyltransferases whose genes have been cloned.[6] There were 35 cloned vertebrate glycosyltransferases reported in early 1997 and the genes encoding many of these enzymes have been cloned from several species.

This chapter will discuss the major glycosyltransferases involved in the biosynthesis of the peripheral antennae of protein-bound *N*-glycans (Asn-Glc*N*Ac *N*-glycosidic linkage, Figure 1). The reaction catalyzed by these enzymes is:

$$\text{Donor}\,(X\!-\!\text{sugar}) + \text{acceptor}\,(R\!-\!OH) \rightarrow \text{Sugar}\!-\!O\!-\!R + XH \qquad (1)$$

where "sugar" is a monosaccharide, X is a nucleotide, and the physiological substrate R-OH is a growing protein-bound *N*-glycan. When these glycosyltransferases are assayed *in vitro*, R-OH is usually a glycopeptide, a free saccharide, or a saccharide linked to an aglycone such as a methyl, benzyl, or octyl group.

The glycosyltransferases are Type II intrinsic membrane proteins firmly bound to the endo-membrane assembly line with their C-terminal catalytic domains within the lumen. Detergent treatment is required for solubilization and full expression of enzymatic activity *in vitro*. Most of these enzymes require a divalent cation, which probably serves to bind the negatively charged nucleotide-sugar to the protein. Exogenous cation is not required for the assay of sialyltransferases and several *β*1,6-*N*-acetylglucosaminyltransferases. Glycosyltransferases usually show precise substrate specificities. However, under nonphysiological conditions *in vitro*, some enzymes can catalyze interesting promiscuous reactions at relatively low rates. For example, the human blood group B *α*1,3-galactosyltransferase can transfer Gal*N*Ac instead of Gal to its acceptor substrate *in vitro* and thereby make the human blood group A epitope.[7] If this reaction were to occur *in vivo* a severe blood incompatibility would result.

(a)

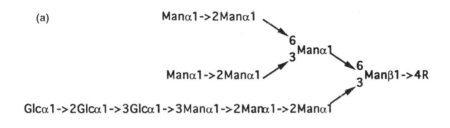

(b)

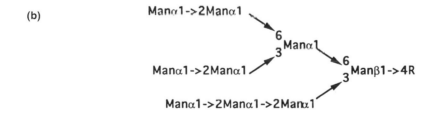

(c)

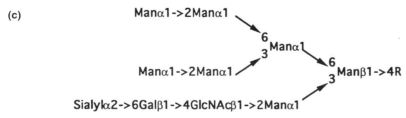

Figure 1 (a)–(e) Typical *N*-linked oligosaccharides. R = *β*1–4Glc*N*Ac*β*1–4Glc*N*Ac-Asn-X. (a) Precursor oligosaccharide derived from Glc₃Man₉Glc*N*Ac₂-PP-Dol. (b) Oligomannose type *N*-glycan. (c) A hybrid *N*-glycan. (d) A bisected hybrid *N*-glycan. (e) A bisialylated complex *N*-glycan. (f) GnT I to VI incorporate Glc*N*Ac residues into the Man*α*1–6[Man*α*1–3]Man*β*-R *N*-glycan core.

(d)

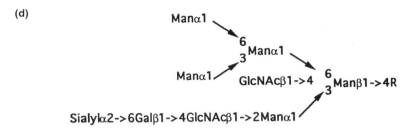

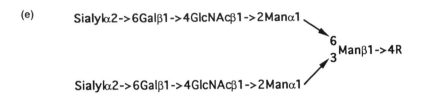

(e)

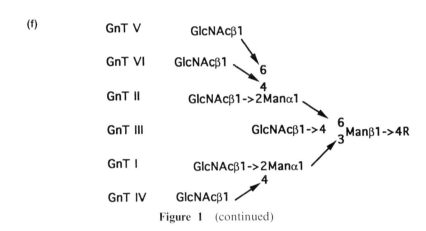

(f)

Figure 1   (continued)

The purification, kinetic properties, substrate specificity, cell biology, and molecular biology of the N-glycan glycosyltransferases have been reviewed previously by many authors.[2,8–16] This chapter will deal primarily with data from the 1990s.

3.03.2   AN OVERVIEW OF THE STRUCTURE AND BIOSYNTHESIS OF N-GLYCANS

Many reviews have been published on the structure[8,17–19] and biosynthesis[2,4,13,16,20–23] of N-glycans and only a relatively brief overview will be presented here.

3.03.2.1   Structure of N-Glycans

The carbohydrate structure data bank CarBank[17] contains over 15 000 unique structures and the list is growing rapidly. Since the 1970s, there have been enormous improvements in the methods for the isolation and structural characterization of glycans; novel methods have been developed for the chemical and enzymatic release of glycans from protein, for the fractionation of the released glycans by high pressure liquid chromatography, lectin affinity chromatography, gel electrophoresis and capillary electrophoresis, and for structural determination by methylation analysis, high performance nuclear magnetic resonance spectroscopy, and mass spectrometry. It is beyond the scope of this chapter to review this literature.

All N-glycans share the same pentasaccharide core structure Manα1–6(Manα1–3)Manβ1–4GlcNAcβ1–4GlcNAc-Asn-X. The oldest structures from an evolutionary point of view are called oligomannose *N*-glycans (Figure 1(b)), in which from one to six additional Man residues are attached to the core. Complex *N*-glycans (Figure 1(e)) carry from two to five "antennae" attached to the core structure. Every antenna is initiated by a specific *N*-acetylglucosaminyltransferase (GnT, Figure 1(f)), which adds a Glc*N*Ac in $\beta1$–2, $\beta1$–4, or $\beta1$–6 linkage to one or other of the terminal Man residues of the core. The most common antenna is sialylα2–6Galβ1–4Glc*N*Ac- but many variations are found, e.g., (i) sialic acid may be linked α2–3 to Gal; (ii) sialic acid may be replaced by Fucα1–2 or Galα1–3 linked to Gal; (iii) Galβ1–4 may be replaced by Galβ1–3 or Gal*N*Acβ1–4; (iv) Fuc may be added in an α1–3 or α1–4 linkage to Glc*N*Ac; (v) antennae may be sulfated at various positions; (vi) antennae may be terminated with various antigenic epitopes such as the human blood group ABO, H and Lewis structures, [24] and the poly-*N*-acetyllactosamine-containing i and I epitopes;[25,] and (vii) antennae may be truncated to Glc*N*Ac or Gal-Glc*N*Ac. The glycosyltransferases which synthesize Galα1–3Gal, Galβ1–3Glc*N*Ac, Gal*N*Acβ1–4Glc*N*Ac, the human blood group A and B epitopes, and poly-*N*-acetyllactosamine-containing structures will not be discussed in this chapter. Structures which carry one or two antennae on the Manα1–3 arm of the core and only Man residues on the Manα1–6 arm are termed hybrid *N*-glycans (Figure 1(c)). Both hybrid and complex *N*-glycans may be "bisected" by a Glc*N*Ac residue linked β1–4 to the β-linked Man residue of the core (Figure 1(d)); this residue is incorporated by GnT III (Figure 1(f)) and is not further substituted.

It is clear from the above discussion that a large variety of *N*-glycans can be made by the cell. It has been found that an Asn residue at a specific position in a specific polypeptide chain can carry more than one type of *N*-glycan. This leads to the phenomenon of "microheterogeneity" in which a glycoprotein preparation that is "pure" in the sense of containing only a single peptide chain may in fact contain a large number of different "glycoforms,"[8] molecules which differ from one another in glycan structure. Microheterogeneity creates a difficult problem for the analytical chemist and its role in cellular function is not clear.

3.03.2.2 Synthesis of Dolichol-pyrophosphate Oligosaccharide

The synthesis of *N*-glycans begins with the transfer within the lumen of the endoplasmic reticulum of Glc$_3$Man$_9$Glc*N*Ac$_2$ (Figure 1(a)) from Glc$_3$Man$_9$Glc*N*Ac$_2$-pyrophosphate-dolichol to an asparagine residue of the nascent polypeptide chain.[22,23] Dolichol pyrophosphate (Dol-PP) is a phosphorylated polyisoprenoid alcohol upon which the Glc$_3$Man$_9$Glc*N*Ac$_2$ glycan is assembled in a stepwise manner in the rough endoplasmic reticulum.[22] The first committed step is catalyzed by *N*-acetylglucosaminyl-1-phosphate transferase, which transfers GlcNAc-1-phosphate from UDP-GlcNAc to Dol-P to form dolichol pyrophosphate *N*-acetylglucosamine (Dol-PP-GlcNAc). This step is inhibited by tunicamycin, a potent inhibitor of *N*-glycan synthesis.[26] The further stepwise addition of one GlcNAc, nine Man, and three Glc residues to Dol-PP-GlcNAc leads to Glc$_3$Man$_9$GlcNAc$_2$-PP-Dol. The glycosyltransferases involved in this process will not be discussed in this chapter. The initial steps in the synthesis of Glc$_3$Man$_9$Glc*N*Ac$_2$-PP-Dol occur on the cytoplasmic side of the endoplasmic reticulum. There is a "flip-flop" during the synthesis, probably at the Man$_5$GlcNAc$_2$-PP-Dol stage, from the cytoplasmic to the lumenal face of the membrane. This allows utilization of the final Glc$_3$Man$_9$Glc*N*Ac$_2$-PP-Dol product for synthesis of nascent glycoprotein on the lumenal side.

3.03.2.3 Oligosaccharyltransferase

Oligosaccharyltransferase (OST) transfers Glc$_3$Man$_9$Glc*N*Ac$_2$ (Figure 1(a)) "en bloc" from Glc$_3$Man$_9$Glc*N*Ac$_2$-PP-Dol to an Asn residue of the nascent polypeptide chain.[23,27] Although OST prefers glucosylated donors, it can also use truncated nonglucosylated Dol-PP-oligosaccharides, including Dol-PP-GlcNAc$_2$, but at an appreciably slower rate. The acceptor Asn residue must occur in an Asn-X-Ser/Thr "sequon," where X can be any amino acid except Pro. However, only 16% of the potential "sequons" are glycosylated, primarily due to the effects of the protein environment on enzyme activity. Denaturation of protein acceptors favours OST action.

Vertebrate and yeast OST have been purified and shown to be protein complexes consisting of three to six nonidentical polypeptide subunits. Two of the subunits of vertebrate OST are ribophorins I and II, abundant integral membrane protein components of the rough endoplasmic reticulum, suggesting that OST is located near the protein translocation channel. Yeast OST has at least six subunits, three of which are homologous to subunits of the vertebrate OST. Genes encoding five of the yeast subunits (OST1, *Z46719*; WBP1, *X61388*; OST3, *U25052*; SWP1, *X67705*; OST2, *U32307*) and three of the vertebrate subunits (Ribophorin I, *Y00281*, *X05300*; Ribophorin II, *Y00282*, *X55298*; OST48, *A44654*, *A44561*) have been cloned.[27]

3.03.2.4   Processing of Protein-bound *N*-Glycans

Oligosaccharide processing by glucosidases I and II, endoplasmic reticulum mannosidase, and Golgi mannosidase I forms Man$_5$GlcNAc$_2$-Asn-X, which is the entry point for the formation of hybrid and complex *N*-glycans due to the action of UDP-GlcNAc:Manα1–3R [GlcNAc to Manα1–3] β1,2-*N*-acetylglucosaminyltransferase I (GnT I, Figure 1(f), see Table 3). This enzyme step is essential for subsequent action of several enzymes in the processing pathway, i.e., α3/6-mannosidase II (Figure 2), GnT II (Figure 2), GnT III and GnT IV, and the α1,6-fucosyltransferase, which adds fucose in an α1–6 linkage to the Asn-linked GlcNAc.[4,5] It is of interest that GnT I action is also a prerequisite for the activities of GnT II and β1,2-xylosyltransferase in plants[28] and in snail.[29]

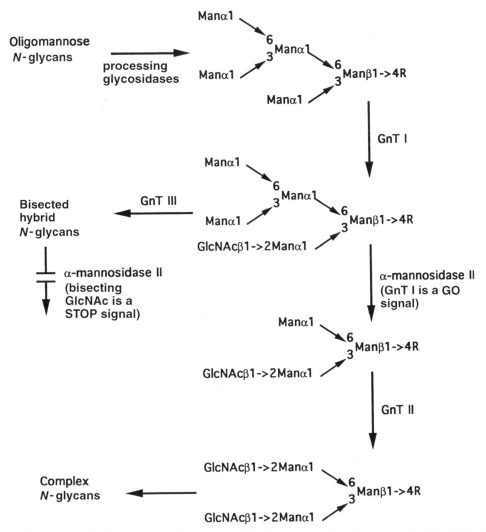

Figure 2   Conversion of oligomannose *N*-glycans to hybrid and complex *N*-glycans. GnT I, UDP-GlcNAc: Manα1–3R [GlcNAc to Manα1–3] β1,2-*N*-acetylglucosaminyltransferase I. GnT II, UDP-GlcNAc:Manα1–6R [GlcNAc to Manα1–6] β1,2-*N*-acetylglucosaminyltransferase II.

GnT II to VI (Figure 1(f)) act on the product of α-mannosidase II to initiate the various complex *N*-glycan "antennae." The most common antenna is sialylα2–6Galβ1–4GlcNAc-. Sialic acid can also be incorporated in an α2,3 linkage to Gal; the Galβ1–4 residue may be replaced by a Galβ1–3 residue or by a GalNAcβ1–4 residue; the terminal sialic acid residue may be replaced by a Galα1–3 residue; Gal and GlcNAc residues may be fucosylated, poly-N-acetyllactosamine, sulfate or phosphate groups may be present, and other modifications have also been reported.[3]

If GnT III acts on the product of GnT I before α-mannosidase II to form the bisected hybrid structure (Figures 1(d) and 2), the pathway is committed to hybrid structures because α-mannosidase II cannot act on bisected oligosaccharides.[30] The reverse order of action leads to complex N-glycans. The relative abundance of GnT III and α-mannosidase II in a particular tissue therefore controls the pathway towards hybrid or complex N-glycans. The route taken at such a divergent branch point is dictated primarily by the relative activities of glycosyltransferases, which compete for a common substrate. The insertion of a bisecting GlcNAc by GnT III prevents the actions of GnT II, IV and V, α-mannosidase II, and core α1,6-fucosyltransferase, and is an example of a glycosyl residue acting as a STOP signal, whereas the action of GnT I is a GO signal. Competition and STOP and GO signals are "substrate-level controls" of the biosynthetic pathways as opposed to control at the transcriptional or translational levels.

3.03.3   UDP-Gal:GlcNAc-R β1,4-GALACTOSYLTRANSFERASE (E.C. 2.4.1.38/90; 2.4.1.22)

UDP-Gal:GlcNAc-R β1,4-galactosyltransferase (β4GalT)[14,31,32] is ubiquitous and acts on N-glycans, O-glycans, and glycolipids. The enzyme alters its substrate specificity when it complexes with α-lactalbumin to form lactose synthetase, which transfers Gal to Glc to make the milk sugar lactose. Although β4GalT can transfer GalNAc from UDP-GalNAc to GlcNAc at a relatively low rate,[33] the addition of α-lactalbumin greatly stimulates this activity,[34] which may explain the presence of the GalNAcβ1–4GlcNAc moiety on bovine milk glycoproteins.

Bovine β4GalT cDNA was isolated in 1986, the first glycosyltransferase gene to be cloned.[35–37] Human, mouse, rat, and chicken genes have also been cloned (see Table 1 for accession numbers). The human β4GalT gene is located on chromosome 9p13–21.[51] Recombinant human β4GalT has been expressed in *Saccharomyces cerevisiae* and purified. Analysis by isoelectric focusing (IEF) revealed considerable heterogeneity. Removal of the single N-glycosylation consensus sequence by site-directed mutagenesis and of O-glycans by jack bean α-mannosidase treatment resulted in a homogeneous enzyme by IEF with kinetic parameters and physical properties similar to the native enzyme.[52]

Table 1   UDP-Gal:GlcNAcR β1,4-galactosyltransferase involved in N-glycan synthesis.

Abbreviations	Enzyme product	Tissue	E.C. no.	Comments	Acc. no.[a]	Ref.
β4GalT	Gal(β1–4)GlcNAcR	Cow	2.4.1.38/90	Makes lactose in the presence of α-lactalbumin	X14558	37
			2.4.1.22		M13569	35
					M25398	38
					J05217	39
		Human			M22921	40
					X55415	41
					X13223	42
					M13701	43
					M70421-33	44
		Mouse			D00314	45
					J03880	46
					D37790	47
					M27917-23	48
		Rat			P80225	49
		Chicken			L12565	50

[a] EMBL/GenBank database accession number.

3.03.3.1   The Glycosyltransferase Domain Structure

All glycosyltransferases cloned up until 1997 are type II integral membrane proteins (N_{in}/C_{out} orientation) with a typical domain structure (Figure 3).[11,13,53] For example, β4GalT has a short amino-terminal cytoplasmic domain (11–24 residues), a 20-residue noncleavable signal/anchor transmembrane domain and a long intralumenal carboxy-terminal catalytic domain (386–402 residues).

The amino-terminal, transmembrane, and stem domains are not required for catalytic activity but are essential for accurate targeting and anchoring of the enzyme to a specific region of the Golgi membrane. Comparison of the amino acid sequences of bovine, human, and murine β4GalT shows over 90% sequence similarly in the transmembrane anchor and C-terminal catalytic domains, but higher variability occurs in the stem region. Site-directed mutagenesis was utilized to identify a tetrapeptide region in the catalytic domain as a binding site for UDP-galactose.[54] A β4GalT probe has been used to clone the gene for a snail *N*-acetylglucosaminyltransferase,[55] but no similarities to other proteins have been detected.

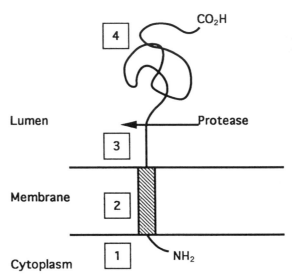

Figure 3 Domain structure of the glycosyltransferases. All glycosyltransferases cloned to date (1997) have an amino acid sequence compatible with a type II transmembrane glycoprotein (N_{in}/C_{out}). (1) Short amino-terminal cytoplasmic domain. (2) Trans-membrane noncleavable signal/anchor domain. (3) Stem or neck region which acts as a tether to hold the catalytic domain in the lumen and can be cleaved by proteases to release a catalytically active soluble enzyme. (4) Carboxy-terminal catalytic domain.

3.03.3.2   Control of β4GalT Gene Transcription

The bovine, human, and murine β4GalT genes have two in-frame ATG codons at the 5'-end of the coding region. Northern analysis showed the presence of two sets of transcripts (3.9 and 4.1 kb).[46,48] The first transcription initiation site was upstream of the first ATG codon and the second site was between the two ATG codons, indicating that at least two promoters control the gene. Translation from the two in-frame ATG codons predicts proteins differing in length by 13 amino acids. *In vitro* translation experiments showed that only a single protein product of the predicted size was obtained from both the short and long mRNAs, respectively, proving that the long transcript initiated translation only at the upstream Met codon. When *in vitro* translation was carried out in the presence of dog pancreas microsomes, both the long and short proteins increased in size by about 3 kDa. Endoglycosidase H treatment removed most of this extra material, indicating that both proteins had been glycosylated. It was concluded that both proteins are oriented as type II integral membrane proteins (Figure 3).[39]

The long and short forms of β4GalT are expressed in a tissue-specific manner in the mouse and provide a mechanism for regulation of enzyme levels.[56] Tissues which express relatively low levels of β4GalT (such as brain) are under the control of the upstream promoter and contain only the long transcript. Most somatic mouse tissues express intermediate levels of enzyme and are under the control of both promoters. Tissues which express β4GalT at very high levels (lactating mammary gland) have 10 times more short transcript than long. The upstream promoter appears to be a typical constitutive "house-keeping" promoter. It lacks classical CAAT and TATA boxes and has six upstream GC boxes (*cis*-acting positive regulatory elements) under the control of the Sp1 transcription factor. Between the two promoters are three more GC boxes, at least two mammary-gland-specific positive regulatory elements, and a putative negative *cis*-acting regulatory element. It was suggested that lactation turns on the synthesis of mammary-gland-specific transcription factors which bind to the positive regulatory elements and deactivate the negative control, thereby activating the tissue-specific downstream promoter and causing a marked increase in production of the short form of β4GalT. DNase I footprinting and electrophoretic mobility shift assays have confirmed this hypothesis.[57] It was shown that the region immediately upstream of the 4.1 kb start site is occupied mainly by the ubiquitous transcription factor Sp1. In contrast, the region adjacent to the 3.9 kb start site is bound by multiple proteins which include the tissue-restricted factor AP2, a mammary-gland-specific form of CTF/NF1, Sp1, as well as a candidate negative regulatory factor that represses transcription from the 3.9 kb start site. These data indicate that the 3.9 kb start site has been introduced into the mammalian β4GalT gene to effect lactose biosynthesis in the lactating mammary gland.

The presence of a germ-cell-specific promoter which regulates expression of β4GalT in haploid round spermatids has been reported.[58–60] Shur's group has provided evidence that β4GalT on the mouse sperm surface mediates fertilization by binding to terminal GlcNAc residues on the egg zona pellucida glycoprotein ZP3.[31,61]

Two groups have reported the preparation of mice with targeted mutations in the β4GalT gene.[62,63] Although both groups found that mice homozygous for the β4GalT mutant gene progressed normally through embryonic development, one study[63] experienced 90% lethality in the first few weeks of life, whereas 50% of the mice from the other study[62] survived. The surviving mice were fertile, showing that β4GalT is not essential for sperm–egg interaction. Mutant mice showed stunted growth, thin skin, sparse hair, dehydration, and evidence of pituitary insufficiency. This may be due to the formation of underglycosylated hormones which are known hormone antagonists.

3.03.3.3 Targeting to the Golgi Apparatus

The domains responsible for targeting β4GalT to the *trans*-cisternae of the Golgi apparatus[32] have been investigated with chimeric cDNAs encoding hybrid proteins in which various domains of β4GalT are connected to reporter proteins not normally retained in the Golgi apparatus.[64 67] Following either stable or transient expression of these hybrid constructs in mammalian cells, the intracellular destination of the reporter protein is determined by immunofluorescence and immunoelectron microscopy. The transmembrane domain is essential for accurate Golgi retention, while sequences outside the transmembrane domain play accessory roles for β4GalT[68] and for other glycosyltransferases.[69–72]

There is a divergence of opinion on the role of the long and short transcripts of β4GalT in targeting. It has been suggested that the long form may be preferentially targeted to the cell surface, where it plays a role in cell–cell recognition,[73–76] while other workers have concluded that both the long and short forms are targeted to the Golgi apparatus.[14,56,65–67,77,78] Escape of the long form, but not the short form, from the Golgi apparatus to the plasma membrane may be due to the abnormally high β4GalT expression levels in transient expression system.[79]

A possible mechanism for the specific retention of proteins within the Golgi apparatus is the presence of a unique retention signal on every protein and a unique Golgi membrane receptor for every such signal. However, since Golgi retention is not saturable even at the very high levels of β4GalT expression of transient transfection experiments, a more likely mechanism is retention due to homo-oligomerization of the glycosyltransferase or hetero-oligomer formation with other Golgi proteins. The large aggregate may be unable to enter budding vesicles either because of its size or due to interaction with the Golgi membrane lipid bilayer.[80] Evidence for such hetero-oligomers or "kin oligomers" has been obtained.[80–83] Kin recognition seems to be mediated primarily by the lumenal domain rather than the transmembrane domain[68,83,84] but it has been reported that oligomer formation and Golgi retention of β4GalT are both abolished by mutations in the transmembrane domain.[85] The transmembrane domains of plasma membrane-targeted proteins are broader and more hydrophobic than those of Golgi-targeted proteins and the length of the transmembrane domain has been shown to be critical for accurate Golgi localization,[77,84,86] suggesting that sorting may be mediated by interaction with lipid microdomains of different thicknesses.

A hybrid protein containing the membrane anchor region of yeast α1,2-mannosyltransferase (Mnt1p) fused to human β4GalT was constructed and expressed in the yeast *S. cerevisiae*. The hybrid β4GalT localized to the yeast Golgi apparatus and was functional in the molecular environment of the yeast Golgi, indicating conservation between yeast and human cells.[87]

3.03.4   SIALYLTRANSFERASES

The "one linkage–one glycosyltransferase" rule suggests that there are at least 12 different sialyltransferases required for the synthesis of all the known sialylated structures present in *N*- and *O*-glycans and glycolipids. *N*-glycan synthesis involves primarily two of these enzymes, CMP-sialic acid:Galβ1–4GlcNAcR α2,6-sialyltransferase (E.C. 2.4.99.1) and CMP-sialic acid:Galβ1–3/4GlcNAcR α2,3-sialyltransferase (E.C. 2.4.99.6). Abbreviations for the α2,6- and α2,3-sialyltransferases are, respectively, ST6Gal I and ST3Gal III (Table 2).[117] Extensive reviews of the sialyltransferases are available.[118,119]

3.03.4.1   CMP-sialic acid:Galβ1–4Glc*N*AcR α2,6-Sialyltransferase (ST6Gal I, E.C. 2.4.99.1)

CMP-sialic acid:Galβ1–4Glc*N*AcR α2,6-sialyltransferase (ST6Gal I) catalyzes the incorporation of sialic acid in an α2–6 linkage to the Galβ1–4Glc*N*Ac termini of *N*-glycans, with a marked preference for the Manα1–3 arm in biantennary substrates.[105,120] Studies with modified Galβ1–4Glc*N*AcR substrates have shown that the 2-amido group of Glc*N*Ac and the 6-hydroxy group of Gal are essential for activity, but Gal can be replaced by sugars such as Man, Glc, Glc*N*Ac, and Gal*N*Ac.[121–125] Modification of either the C-3 or C-4 hydroxys of the Gal residue reduced activity of ST6Gal I to various degrees and some of these substrate analogues inhibited the enzyme.[126]

Expression of ST6Gal I varies dramatically between different tissues and within the same tissue during development and cell differentiation.[99,100,103,127–130] Cell surface sialic acid variations have also been shown to occur with cell transformation and tumor progression.[131] In general, cells with reduced sialylation show attenuated metastatic potential although overexpression of ST6Gal I also causes reduction in metastatic phenotype.[132] However, colon cancer cells with a high expression of α2,6-sialylated glycans are more tumorigenic and metastatic in mice.[133,134] Transformation of rat fibroblasts with the oncogene c-Ha-*ras* induced a 10-fold increase in ST6Gal I due to elevated mRNA levels and protein expression,[135,136] whereas other oncogenes did not have this effect; the ST6Gal I levels of the *ras*-transformed fibroblasts correlated with increased invasiveness.[137] Human cancer has also been associated with high ST6Gal I levels.[133,138,139]

3.03.4.1.1   *Control of* α2,6-*sialyltransferase gene transcription*

The tissue-specific expression of ST6Gal I in rats and humans is due to the action of multiple promoters acting on a single gene to produce transcripts differing primarily in their 5′-untranslated regions. Analysis of rat[102] and human[94,95,140] cDNAs encoding ST6Gal I indicate that the enzyme has a typical glycosyltransferase domain structure (Figure 3). The rat ST6Gal I gene spans at least 80 kb of genomic DNA and contains at least 11 exons. There are at least three promoters responsible for the production of five or more messages from a single gene sequence.[99,100,103,127] Promoter P_L produces a liver-specific 4.3 kb mRNA, promoter P_C is constitutive in several tissues and produces a 4.7 kb mRNA, and promoter P_K is restricted to the kidney and produces several 3.6 kb mRNAs missing the 5′-half of the coding sequence. The proteins encoded by the 3.6 kb messages are not catalytically active.[128] The human gene is localized on chromosome 3q21–q28[93] and also has at least three promoters,[93,97] one of which is hepatic-specific.[98]

Cell type-specific transcription of genes is controlled by transcription factors which themselves have a restricted pattern of expression. Promoter P_L was demonstrated to be about 50-fold more active in the hepatoma cell (HepG2) known to express ST6Gal I than in a cell line (Chinese hamster ovary) which does not express this enzyme.[101] The P_L promoter sequence contains consensus binding sites for liver-restricted transcription factors,[141] thereby accounting for the high level expression of ST6Gal I in rat liver.

In mature B-lymphocytes, cell surface α2,6-sialyl structures serve as recognition ligands for adhesion molecules CD22β on B-cells and CD45 on T-cells.[142,143] B-lymphocyte maturation is accompanied by increased expression of ST6Gal I[93] due to the appearance of a B-cell specific isoform that is probably responsible for the synthesis of the sialyl ligands for CD22β and CD45. The B-cell specific mRNA is under the control of a promoter that is distinct from the promoter which produces the hepatic-specific transcript.[144,145]

Table 2 Sialyltransferases involved in *N*-glycan synthesis.

Abbreviations	Enzyme product	Tissue	E.C. no.	Comments	Acc. no.[a]	Ref.
ST3Gal III (ST3N I)	Sia(α2–3)-Gal(β1–3/4)GlcNAc-	Human	2.4.99.6	Gal(β1–3)GlcNAc > Gal(β1–4)GlcNAc	L23768	88
		Mouse			D28941	89
		Rat			M97754, M98462	90
ST3Gal III (ST3N II)	Sia(α2–3)-Gal(β1–3/4)GlcNAc-	Human		Gal(β1–4)GlcNAc > Gal(β1–3)GlcNAc	X74570	91
ST3 Gal IV	Sia(α2–3)-Gal(β1–3)GalNAcSia(α2–3)-Gal(β1–4)GlcNAc-	Human		Does not act on Gal(β1–3)GlcNAc-; acts on *N*- and *O*-glycans, glycolipids	L23767, L29553	92
ST6Gal I	Sia(α2–6)-Gal(β1–4)GlcNAc-	Human	2.4.99.1		S55689, S55693, S55697-9	93
					X17247	94
					X62822	95
					X54363	96
					L11720	97
					Z35760	98
		Rat			M73985-7	99, 100
					M54999	101
					M18769	102
					M83142-4	103
					D16106	104
ST8Sia II	[Sia(α2–8)]ₙSia(α2–3)-Gal(β1–3/4)GlcNAc-	Mouse		Fetal and neonatal brain; acts on Sia(α2–3)-; is a polysialic acid synthase; no action on glycolipids	X75558	105
		Chick embryo			L29556	106
		Human			U33551	107
		Mouse			U82762	108
					X83562	109
					X83562, X99645-99651	110
ST8Sia III	Sia(α2–8)Sia(α2–3)-Gal(β1–4)GlcNAc-	Rat		*N*-glycans > glycolipids; in brain and testis	L13445	111
ST8Sia IV	[Sia(α2–8)]ₙSia(α2–3)-Gal(β1–3/4)GlcNAc-	Mouse		Acts on Sia(α2–3)-; is a polysialic acid synthase	X80502	112, 113
		Chinese hamster			Z46801	114
		Mouse			X86000	115
		Human			L41680	116

[a] EMBL/GenBank database accession number.

3.03.4.1.2   Targeting to the Golgi apparatus

Rat ST6Gal I has been localized to the *trans*-Golgi cisternae and the *trans*-Golgi network of hepatocytes, hepatoma cells, and intestinal goblet cells but may be more diffusely distributed in the Golgi apparatus of other cell types.[146] The cytoplasmic, transmembrane, and stem domains of ST6Gal I are not required for catalytic activity. The transmembrane domain is, however, essential for proper Golgi targeting and both the stem and cytoplasmic domains may also play a role which is as yet (1997) not understood.[69,70,147]

In bovine and rat liver Golgi membranes, 30% of ST6Gal I exists as a disulfide-bonded 100 kDa homodimer that can be converted to the 50 kDa monomer form of the enzyme upon reduction.[148] The dimer form of the enzyme possesses no significant catalytic activity but binds strongly to galactose and galactose-terminated substrates, suggesting that the dimer may act as a galactose-specific lectin in the Golgi. Rat liver expresses two forms of ST6Gal I that differ by a single nucleotide;[149] both forms are functional enzymes but whereas one form is retained within the cell, the other is rapidly cleaved and secreted.

Recombinant full-length human ST6Gal I expressed in *S. cerevisiae* shows kinetic properties similar to the native rat enzyme and is retained in the yeast endoplasmic reticulum as a fully active enzyme.[150] The transmembrane domain of rat ST6Gal I is required for targeting this enzyme to the Golgi apparatus of *S. cerevisiae*.[151]

3.03.4.1.3   The sialylmotif

Comparison of the amino acid sequences of ST6Gal I, CMP-sialic acid:Galβ1–3GalNAcR α2,3-sialyltransferase (ST3Gal I), and CMP-sialic acid:Galβ1–3/4GlcNAcR α2,3-sialyltransferase I (ST3Gal III) reveals a region of 55 amino acids with extensive homology (40% identity; 58% conservation) in the middle of the catalytic domain.[90] There is a second 23 residue region of similarity near the CO_2H-terminus.[152] This so-called "sialylmotif" has been used to clone many other sialyltransferases (Table 2) using the polymerase chain reaction and degenerate oligonucleotide primers. Site-directed mutagenesis experiments have suggested that the sialylmotif participates in the binding of CMP-sialic acid to the enzyme.[153]

3.03.4.2   CMP-sialic acid:Galβ1–3/4GlcNAcR α2,3-Sialyltransferase I (ST3Gal III, ST3N I, E.C. 2.4.99.6)

Recombinant rat liver[90] and human placental[88] CMP-sialic acid:Galβ1–3/4GlcNAcR α2,3-sialyl-transferase I (ST3 Gal III, ST3N I) catalyze the incorporation of sialic acid in an α2–3 linkage to both the Galβ1–3GlcNAc and Galβ1–4GlcNAc termini of *N*-glycans with a marked preference for the type 1 chain, Galβ1–3GlcNAc-, thereby indicating that this enzyme is different from the previously reported human placental enzyme, which showed preferential activity towards type 2 chains, Galβ1–4GlcNAc-.[154] The amino acid sequences deduced from rat[90] and human[88] cDNAs encoding ST3N I indicated a domain structure typical of the glycosyltransferases (Figure 3). Expression of a truncated form of ST3N I lacking the cytoplasmic and transmembrane domains showed that these domains are not required for enzyme activity but are necessary for targeting to the Golgi apparatus. Modification of either the C-3 or C-4 hydroxy residues of the Gal residue of the substrate Galβ1–4GlcNAcR resulted in an inactive or almost inactive substrate and these compounds were good inhibitors of the enzyme.[126]

3.03.4.3   CMP-sialic acid:Galβ1–3/4GlcNAcR α2,3-Sialyltransferase II (ST3N II)

Selection with a cytotoxic lectin and an expression cloning approach were used to clone a human melanoma cDNA encoding CMP-sialic acid:Galβ1–3/4GlcNAcR α2,3-sialyltransferase II (ST3N II), which differs from the previously cloned ST3N I (above) in showing a 3 : 1 preferential activity towards type 2 chains, Galβ1–4GlcNAc, relative to type 1 chains, Galβ1–3GlcNAc.[91] Human ST3N I and II have only 34% homology but both enzymes have the sialylmotif typical of all sialyltransferases cloned up until the 1990s. The relationship of ST3N II to a similar human placental enzyme[154] is as yet unknown.

3.03.4.4 CMP-sialic acid:Galβ1–4Glc$N$AcR/Galβ1–3GalNAcR α2,3-Sialyltransferase (ST3Gal IV)

The sialylmotif was used to clone a novel CMP-sialic acid:Galβ1–4Glc$N$AcR/Galβ1–3GalNAcR α2,3-sialyltransferase (ST3Gal IV) (Table 2), which transfers sialic acid in an α2–3-linkage to the terminal Gal of Galβ1–4Glc$N$Ac or Galβ1–3GalNAc of oligosaccharide, glycoprotein, and glycolipid acceptors.[92] The human enzyme could not act on Galβ1–3Glc$N$Ac termini[92] and was therefore different from ST3N I and II described above. The enzyme may be the glycolipid α2,3-sialyltransferase SAT-3.[155] The gene is localized to human chromosome 11(q23–q24).[156] It spans more than 25 kb of human genomic DNA and is distributed over 14 exons that range in size from 61 to 679 base pairs. The gene produces at least five transcripts in human placenta, which code for identical protein sequences except at the 5' ends.[92,156] These transcripts are produced by a combination of alternative splicing and alternative promoter utilization. Northern analysis indicated that one of them is specifically expressed in placenta, testis, and ovary, indicating that its expression is independently regulated.

3.03.4.5 Brain-Specific CMP-sialic acid:[sialylα2–8]$_n$sialylα2–3Galβ1–3/4Glc$N$AcR α2-8-Sialyltransferase (ST8Sia II)

There are at least three α2–8-sialyltransferases (Table 2), which attach sialic acid in an α2–8 linkage to α2–3 or α2–8-linked sialic acid residues of N-glycans (ST8Sia II, III, and IV). The sialylmotif was used to clone cDNAs encoding a novel CMP-sialic acid:[sialylα2–8]$_n$sialylα2–3Galβ1–3/4Glc$N$AcR α2–8-sialyltransferase (STX, ST8Sia II), which is expressed in the brains of fetal rats,[111,157] mice,[109,158] and humans[108] but is poorly expressed in adult brain and in other adult and fetal tissues. The expression of ST8Sia II is therefore both tissue-specific and developmentally regulated. The human enzyme is located on chromosome 15q26.[108] The mouse ST8Sia II gene spans about 80 kb and is composed of 6 exons.[110] The promoter region is GC-rich and lacks both TATA and CCAAT boxes.

The enzyme was inactive towards nonsialylated substrates and gangliosides but transferred sialic acid in an α2–8 linkage to the terminal sialic acid of sialylα2–3Galβ1–3/4Glc$N$Ac-glycoproteins[109] and was also able to extend these chains with polysialic acid [sialylα2–8]$_n$ both *in vitro* and *in vivo*.[107,158] The enzyme can therefore both initiate and extend polysialic acid synthesis. ST8Sia II can effect the specific polysialylation of neural cell adhesion molecule (N-CAM) *in vivo*;[159] this activity requires the presence of a core α1,6-fucose residue (attached to the Asn-linked Glc$N$Ac).

3.03.4.6 CMP-sialic acid:sialylα2–3Galβ1–4Glc$N$AcR α2–8-Sialyltransferase (ST8Sia III)

The sialylmotif was used to clone two other mouse genes encoding α2–8-sialyltransferases which act on N-glycan antennae (ST8Sia III and IV, Table 2).[112,115] The predicted amino acid sequence of mouse brain ST8Sia III[112] showed 27.6% and 34.4% identity with mouse ST8Sia I and ST8Sia II, respectively. ST8Sia III attaches sialic acid in an α2–8 linkage to the terminal sialic acid of both protein- and lipid-linked sialylα2–3Galβ1–4Glc$N$Ac, whereas ST8Sia I acts only on sialylated glycolipids, and ST8Sia II and IV both act only on sialylated glycoproteins. However, the kinetic properties of ST8Sia III revealed that it is much more specific to N-linked oligosaccharides of glycoproteins than glycosphingolipids. The ST8Sia III gene was expressed only in brain and testis and it appeared first in postcoitum embryonal brain and then decreased. In contrast, ST8Sia IV is strongly expressed in lung, heart and spleen but only weakly in brain.[115]

3.03.4.7 CMP-sialic acid:[sialylα2–8]$_n$sialylα2–3Galβ1–3/4Glc$N$AcR α2–8-Sialyltransferase (ST8Sia IV)

Mouse ST8Sia IV[115] exhibits relatively low amino acid sequence identities with ST8Sia I (15%), II (56%), and III (26%) but shows 99% identity with cloned hamster[114] and human fetal brain[116] polysialyltransferases. ST8Sia IV can initiate polysialic acid synthesis by attaching an α2–8-linked sialic acid to an α2–3-linked sialic acid and can also effect polymerization of polysialic acid *in vitro* and *in vivo*. The enzyme is expressed in fetal brain, lung and kidney and in adult brain (weakly), heart, spleen, thymus, peripheral blood leukocytes, and other tissues.[108] The human enzyme is located on chromosome 5p21.[108]

Polysialic acid (PSA) is a developmentally regulated product of post-translational modification of N-CAM and may act as a regulator of N-CAM mediated cell–cell adhesion.[160] N-CAM with PSA is abundant in embryonic brain but adult N-CAM lacks this glycan structure. The expression of PSA in N-CAM facilitates neurite outgrowth.[114,116] HeLa cells doubly transfected with ST8Sia II and ST8Sia IV supported neurite outgrowth much better than HeLa cells expressing N-CAM alone.[108] In the adult, PSA becomes restricted to regions of permanent neural plasticity and regenerating neural and muscle tissues. Hamster ST8Sia IV induced PSA synthesis in all N-CAM-expressing cell lines tested.[114] The human enzyme[116] can attach PSA to N-CAM and is strongly expressed in fetal brain but weakly in adult brain. Recombinant ST8Sia IV can add PSA not only to N-CAM *in vitro* but also to other glycoproteins with $\alpha2,3$-terminating N-glycans ($\alpha1$-acid glycoprotein, fetuin).[161] The enzyme can also perform autopolysialylation.[162]

Both ST8Sia II and IV can effect the synthesis of polysialylated N-CAM and the functions of these two enzymes in brain development are not clear. The expression of PSA, ST8Sia II, and ST8Sia IV was therefore studied during *in vitro* neuronal differentiation of mouse embryonal carcinoma P19 cells.[163] It was found that during neuronal differentiation only ST8Sia II was upregulated in parallel with the expression of PSA.

3.03.5   *N*-ACETYLGLUCOSAMINYLTRANSFERASES

Six *N*-acetylglucosaminyltransferases (GnT I to VI) are involved in the synthesis of complex N-glycans (Figure 1(f), Table 3).[4,5,13] The molecular biology of the *N*-acetylglucosaminyltransferases has been reviewed.[187]

3.03.5.1   UDP-GlcNAc:Manα1–3R [GlcNAc to Manα1–3] β1,2-*N*-acetylglucosaminyltransferase I (GnT I, EC 2.4.1.101)

GnT I controls the synthesis of hybrid and complex N-glycans (Figure 2) by initiating the synthesis of the first antenna. GnT I has been purified to homogeneity from rabbit liver.[188] Detailed kinetic analysis of the rabbit enzyme has shown that catalysis is by an ordered sequential Bi–Bi mechanism in which UDP-GlcNAc binds first and UDP leaves last. Mn^{2+} is essential for activity. Although the physiological substrate is $Man_5GlcNAc_2Asn-X$ (Figure 2), the minimum substrate requirement is Manα1–3Manβ1R where R can be a 4GlcNAc residue or a hydrophobic octyl group.[188,189] Essential substrate groups are an unsubstituted equatorial hydroxy at C-4 and an unsubstituted axial hydroxy at C-2 of the Manβ residue; modifications at C-6 of the Manβ residue caused variations in K_M but no major alterations in enzyme activity. Manα1–6(2-deoxyManα1–3)Manβ1R is not a competitive inhibitor and removal of the hydroxy groups at C-3, 4, or 6 of the Manα1–3 residue leads either to a poor or inactive substrate.[190] Manα1–6(6-*O*-methylManα1–3)Manβ1-octyl is a competitive inhibitor ($K_i = 0.76$ mM).

The genes for rabbit, human, mouse, rat, Chinese hamster, chicken, and frog GnT I have been cloned (Table 3) and there are at least three *Caenorhabditis elegans* genes listed in the EMBL/GenBank database which are homologous to mammalian GnT I. The amino acid sequences of the mammalian enzymes are over 90% identical but there is no sequence similarity to any other known glycosyltransferase. GnT I has the type II integral membrane protein domain structure typical of all cloned glycosyltransferases (Figure 3). The transmembrane segment of GnT I is essential for retention in medial-Golgi cisternae but the other domains also play a role.[71,72] The protein is not N-glycosylated because there are no Asn-X-Ser(Thr) sequons but it is *O*-glycosylated.[81,191] Recombinant GnT I produced in the baculovirus/Sf9 insect cell system[192] has been used to convert various derivatives of Manα1–6(Manα1–3)Manβ1-octyl to Manα1–6(GlcNAcβ1–2Manα1–3)Manβ1-octyl and these compounds have been used to study the substrate specificity of GnT II.[193,194]

3.03.5.1.1   *Organization of the GnT I gene*

The human GnT I gene *MGAT1* has been localized to chromosome 5q35.[175] Part of the 5'-untranslated region and all of the coding and 3'-untranslated regions of the human and mouse GnT I genes are on a single 2.5 kb exon (exon 2).[165,168,169] The remaining 5'-untranslated sequence of the human GnT I gene is on an exon (exon 1) which is between 5.6 and 15 kb upstream.[195] The mouse

Table 3 *N*-acetylglucosaminyltransferases involved in *N*-glycan synthesis.

Abbreviations[a]	Enzyme product	Tissue	E.C. no.	Comments	Acc no.[b]	Ref.
GnT I	GlcNAc(β1–2)-Man(α1–3)Manβ	Rabbit	2.4.1.101	First step towards hybrid and complex *N*-glycans	M57301	164
		Human			M61829	165
		Mouse			M55621	166
					X77487-8	167
					M73491	168
					L07037	169
		Rat			D16302	170
		Chicken				171
		Frog				172
		Chinese hamster				173
		Caenorhabditis elegans			U65791-2	174
GnT II	GlcNAc(β1–2)-Man(α1–6)Manβ	Human	2.4.1.143	Synthesis of b antennary *N*-glycans	Z46381, U23516, U28735	175
		Rat			U15128, L36537	176
		Frog			U21662	172
GnT III	GlcNAc(β1–4)-Manβ	Human	2.4.1.144	Synthesis of bisected *N*-glycans	D13789	177
		Rat			D10852	178
		Mouse			L39373	179
					U66844	180
GnT IV	GlcNAc(β1–4)-[GlcNAc(β1–2)]Man(α1–3)Manβ	Hen oviduct	2.4.1.145	Initiation of antenna on Man(α1–3) arm		181
						182
GnT V	GlcNAc(β1–6)-[GlcNAc(β1–2)]Man(α1–6)Manβ	Human	2.4.1.155	Initiation of antenna on Man(α1–6) arm	D17716	183
		Rat			L14284	184
		Chinese hamster			U62587-8	185
GnT VI	GlcNAc(β1–4)-[GlcNAc(β1–2)][GlcNAc-(β1–6)]Man(α1–6)Manβ	Chicken		Initiation of antenna on Man(α1–6) arm	Not cloned	186

[a]GnT, *N*-acetylglucosaminyltransferase. [b]EMBL/GenBank database accession number.

GnT I gene has a similar organization except that there are at least two upstream noncoding exons.[167] There is only a single copy of the gene in the haploid human and mouse genomic DNA.

There are multiple transcription start sites for exon 1 compatible with the expression by several human cell lines and tissues of two transcripts, a broad band ranging in size from 2.7 kb to 2.95 kb and a sharper band at 3.1 kb.[195] The 5′-flanking region of exon 1 has a GC content of 81% and has no canonical TATA or CCAAT boxes but contains potential binding sites for transcription factor Sp1. CAT expression was observed on transient transfection into HeLa cells of a fusion construct containing the CAT gene and a genomic DNA fragment from the 5′-flanking region of exon 1. It is concluded that *MGAT1* is a typical housekeeping gene.

3.03.5.2   UDP-Glc*N*Ac:Manα1–6R [Glc*N*Ac to Manα1–6] β1,2-*N*-acetylglucosaminyltransferase II (GnT II, E.C. 2.4.1.143)

GnT II initiates the first antenna on the Manα1–6 arm of the *N*-glycan core and is therefore essential for normal complex *N*-glycan formation (Figure 2). GnT II has been purified to homogeneity from rat liver.[196] Detailed kinetic analysis of the rat enzyme[197] has shown that catalysis is by an ordered sequential Bi–Bi mechanism in which UDP-Glc*N*Ac binds first and UDP leaves last. Mn^{2+} is essential for activity. The minimal substrate is Manα1–6(Glc*N*Acβ1–2Manα1–3)Manβ1R where R can be a hydrophobic aglycone. The 2-deoxyManα1–6(Glc*N*Acβ1–2Manα1–3)Manβ1R analogue is a competitive inhibitor ($K_i = 0.13$ mM) but the other hydroxy groups of the Manα1–6 residue are not essential for activity.[194] Substitution of the C-4 hydroxy of the Manβ residue but not its removal leads to an inactive substrate. Glc*N*Acβ1–2Manα1–3Manβ1-octyl is a good inhibitor of the enzyme ($K_i = 0.9$ mM), indicating that this trisaccharide moiety is required for substrate binding to the enzyme.

The human and rat GnT II genes have been cloned (Table 3). The enzyme has a typical glycosyltransferase domain structure (Figure 3). There is no sequence homology to any previously cloned glycosyltransferase including GnT I. Recombinant GnT II produced in the baculovirus/Sf9 insect cell system[175] has been used to convert various derivatives of Manα1–6(Glc*N*Acβ1–2Manα1–3)Manβ1-octyl to Glc*N*Acβ1–2Manα1–6(Glc*N*Acβ1–2Manα1–3)Manβ1-octyl; these compounds can be used to study the substrate specificity of GnT III, IV, and V.[198]

Fibroblasts from two children with carbohydrate-deficient glycoprotein syndrome (CDGS) Type II have been shown to lack GnT II activity[199,200] due to point mutations in the GnT II gene.[201,202] CDGS is a group of autosomal recessive diseases with multisystemic abnormalities including a severe disturbance of nervous system development.[203] Mouse embryos lacking a functional GnT I gene die prenatally at about 10 days of gestation[204,205] with multisystemic abnormalities and no obvious cause of death.[206] Among the anatomical defects noted was a failure of normal neural tube development. These findings indicate that complex *N*-glycans are essential for normal post-implantation embryogenesis and development, particularly of the nervous system. Since carbohydrates are also believed to be important in pre-implantation development, it was surprising that the GnT I-null embryos survived for 9.5 days. Evidence has been obtained which suggests that the mother provides either GnT I or complex *N*-glycans to the pre-implantation GnT I-null embryo.[206]

3.03.5.2.1   *Organization of the GnT II gene*

The human GnT II gene (*MGAT2*) is on chromosome 14q21.[175] The open reading frame and 3′-untranslated region of the human and rat genes were shown to be on a single exon.[175,176] Work using 5′-RACE (rapid amplification of cDNA ends) and RNase protection analyses of the human gene showed multiple transcription initiation sites at −440 to −489 bp relative to the ATG translation start codon (+1), proving that the entire GnT II gene is on a single exon.[207,208] The gene has three AATAAA polyadenylation sites downstream of the translation stop codon. 3′-RACE using RNA from the human cell line LS-180 indicated that all three sites were utilized for transcription termination, yielding transcripts at 2.0, 2.7, and 2.9 kb. The gene has a CCAAT box at −587 bp but lacks a TATA box and the 5′-untranslated region is GC-rich and contains consensus sequences suggestive of multiple binding sites for Sp1; these properties are typical for a housekeeping gene. A series of chimeric constructs containing different lengths of the 5′-untranslated region fused to the chloramphenicol acetyltransferase (CAT) reporter gene were tested in transient transfection experiments using HeLa cells. The CAT activity of the construct containing the longest insert (−1076 bp relative to the ATG start codon) showed an ∼38-fold increase compared to that of the control. Removal of the region between −636 and −553 bp caused a dramatic decrease in CAT activity, indicating this to be the main promoter region of the gene.

3.03.5.3 UDP-GlcNAc:R₁-Manα1–6[GlcNAcβ1–2Manα1–3]Manβ1–4R₂ [GlcNAc to Manβ1–4] β1,4-N-acetylglucosaminyltransferase III (GnT III, E.C. 2.4.1.144)

GnT III incorporates a bisecting GlcNAc residue into the *N*-glycan core (Figure 1(f)). It is of interest because it causes a STOP signal, that is, several enzymes involved in *N*-glycan synthesis cannot act on bisected substrates (GnT II, IV, and V and mannosidase II, Figures 1(f), 2). Increased levels of GnT III are therefore expected to inhibit the synthesis of highly branched sialylated complex *N*-glycans. Whereas the genes encoding GnT I and II are typical housekeeping genes and are expressed in all tissues tested, GnT III has a distinct tissue distribution, for example, it is poorly expressed in liver and strongly expressed in kidney.

3.03.5.3.1 *Organization of the GnT III gene*

The genes for human, mouse and rat GnT III have been cloned (Table 3). The enzyme has a typical glycosyltransferase domain structure (Figure 3). There is no apparent sequence similarity to any known glycosyltransferase. Analysis of the human GnT III gene shows that the entire coding region is on a single exon (exon 1) on chromosome 22q.13.1.[177] The human GnT III promoter region has been analyzed.[209,210] There are at least three upstream noncoding exons designated H15 exon-1, H15 exon-2, and H20 exon-1 and at least three different mRNA transcripts formed by alternative splicing of these exons:[210] H15 (H15 exons −1 and −2 and exon 1), H20 (H20 exon −1 and exon 1), and H204 (exon 1). Assays using luciferase as a reporter protein in a human hepatoblastoma cell line demonstrated promoter activity upstream of transcripts H15 and H204 but not H20. None of the promoter regions contained either a TATA or CCAAT box but consensus sequences for many putative transcription factor binding sites (Ets, Myb, Myc, etc.) were present.

3.03.5.3.2 *Variation of GnT III expression*

The enzyme has been shown to be elevated in various types of rat hepatoma[211–214] and human leukemia.[215,216] Transfection of the GnT III gene into various cell lines has produced interesting biological effects: (i) suppression of hepatitis B virus gene expression by a GnT III-transfected hepatoma cell line;[217] (ii) suppression of natural killer (NK) cell toxicity with resultant increased spleen colonization of GnT III-transfected NK-sensitive K562 cells in nude mice;[218] (iii) suppression of the metastatic potential of GnT III-transfected B16-hm mouse melanoma cells[219] and elevated expression of E-cadherin by the latter cells;[220] (iv) GnT III transection of a human glioma cell line which expresses the epidermal growth factor (EGF) receptor on its cell surface blocked EGF binding and EGF receptor autophosphorylation;[221] (v) overexpression of GnT III in rat pheochromocytoma PC12 cells resulted in inhibition of growth response and of tyrosine phosphorylation of the Trk/nerve growth factor (NGF) receptor following addition of NGF;[222] and (vi) transfection of the GnT III gene into a swine endothelial cell line reduced the antigenicity to human natural antibodies, presumably by reducing the synthesis of complex *N*-glycans which carry the Galα1–3Gal epitope that causes rejection of pig xenotransplants.[223] These effects are presumably mediated by the effect of the bisecting GlcNAc on the conformation of complex *N*-glycans[224–226] and by the inhibition of complex *N*-glycan synthesis due to over-expression of GnT III.

Treatment of hepatoma cells with forskolin, an adenylyl cyclase activator, causes a dramatic increase of GnT III levels, probably due to enhanced transcription,[227] with a resultant increase in E-PHA-reactive glycoproteins (indicative of an increase in bisected *N*-glycans) and a decrease in L-PHA-reactive glycoproteins (indicative of a decrease in tri- and tetra-antennary *N*-glycans, probably due to inhibition of GnT IV and V activities). This increase in bisected *N*-glycans was observed on intracellular glycoproteins, whereas cell surface glycoproteins showed a decrease, suggesting that the bisecting GlcNAc residue may play a role in intracellular glycoprotein sorting.[227]

Surprisingly, mice in which the GnT III gene has been "knocked out" show either no phenotype[180] or a relatively mild phenotype.[228] Overexpression of GnT III in transgenic mice reduced the antigenicity of some organs to natural human antibodies, presumably by suppression of xenoantigens.[229]

3.03.5.4 UDP-GlcNAc:R₁Manα1–3R₂ [GlcNAc to Manα1–3] β1,4-N-acetylglucosaminyltransferase IV (GnT IV, E.C. 2.4.1.145)

GnT IV adds a GlcNAc in a β1–4 linkage to the Manα1–3Manβ arm of the N-glycan core (Figure 1(f)).[181] When the biantennary substrate GlcNAcβ1–2Manα1–6[GlcNAcβ1–2Manα1–3]Manβ1–4GlcNAcβ1–4GlcNAc-Asn-X is presented to a hen oviduct extract, GnT IV acts only on the Manα1–3Manβ arm. However, hen oviduct extracts will add a GlcNAc in a β1–4 linkage to the Manα residue in the linear oligosaccharides GlcNAcβ1–2Manα1–6Manβ1R (R = methyl, methoxycarbonyloctyl, *o*-nitrophenyl), GlcNAcβ1–2Manα1–6Glcβ1R (R = *o*-nitrophenyl, allyl), GlcNAcβ1–2Manα1–3Manβ1-methoxycarbonyloctyl, and even GlcNAcβ1–2Manα1-methyl, that is, GlcNAcβ1–2ManαR is converted to GlcNAcβ1–4[GlcNAcβ1–2]ManαR.[186,230] Thus, the branch specificity displayed by the biantennary substrate is lost when a linear substrate is used. It is possible that more than one enzyme may be involved in these various activities. GlcNAcβ1–6ManαR compounds do not serve as substrates.

GnT IV has recently been purified from bovine small intestine[231] and the bovine and human genes have been cloned.[182] The open reading frame of the bovine enzyme has 1605 bp and encodes a typical Type II membrane protein. There is no homology to known glycosyltransferases. It is of great interest that there are two distinct human GnT IV genes which show only a 62% homology to one another at the amino acid sequence level. The bovine gene and both human genes have been expressed and all three proteins show GnT IV activity.

3.03.5.5 UDP-GlcNAc:R₁Manα1–6R₂ [GlcNAc to Manα1–6] β1,6-N-acetylglucosaminyltransferase V (GnT V, E.C. 2.4.1.155)

GnT V adds a GlcNAc in a β1–6 linkage to the Manα1–6Manβ arm of the N-glycan core (Figure 1(f)). Kinetic analysis has shown that, like GnT I and II, GnT V follows an ordered sequential Bi–Bi mechanism.[232] Several studies have been published which map the substrate requirements of this enzyme.[233–247] The minimal substrate is GlcNAcβ1–2Manα1–6Manβ1R, where R can be a hydrophobic aglycone although compounds with a biantennary structure are significantly better substrates than those with linear structures. The β-linked Man residue can be replaced by Glc. The enzyme is not inhibited by EDTA.[248] GnT V, like GnT II and IV, cannot act on substrates with a bisecting GlcNAc residue attached to the C-4 hydroxy of the β-linked Man residue, although methyl substitution of this hydroxy increases enzyme activity and its removal has minimal effects on enzyme activity. Galactosylation or removal of the C-4 hydroxy of the GlcNAc residue linked to the Manα1–6 residue results in loss of enzyme activity. A substrate analogue, GlcNAcβ1–2(6-deoxy)Manα1–6Glcβ1-octyl, was shown to be an excellent competitive inhibitor (K_i = 0.07 mM).[233,236]

One of the most common alterations in transformed or metastatic malignant cells is the presence of larger N-glycans due primarily to a combination of increased GlcNAc branching, sialylation, and poly-N-acetyllactosamine content.[131,249–259] GnT V plays a major role in these effects. Cells transformed with polyoma virus, Rous sarcoma virus or T24 H-*ras*[233,239,257,260–264] treated with transforming growth factor-β or phorbol ester[265] were shown to have significantly increased GnT V activity. Poly-N-acetyllactosamine chains have been shown to carry cancer-associated antigens and the initiation of these chains is favored on the antenna initiated by GnT V.[131,266,267] Transfection of the GnT III gene into a highly metastatic mouse melanoma cell line resulted in decreased β1–6 branching of N-glycans without altering GnT V enzyme levels[219] due to the fact that GnT V cannot act on bisected N-glycans;[4] the GnT III-transfected cells showed a marked reduction in metastatic potential. Transfection of the GnT V gene into premalignant epithelial cells resulted in relaxation of growth controls and reduced substratum adhesion.[255] Evidence has been obtained for post-translational activation of GnT V by phosphorylation.[268] A peptide encoded by an intron sequence of the GnT V gene has been shown to be a human melanoma-specific antigen recognized by human cytolytic T lymphocytes.[269]

3.03.5.5.1   *Organization of the GnT V gene*

Rat[248] and human[237] GnT V have been purified and the rat, human, and Chinese hamster ovary genes have been cloned (Table 3) and expressed.[184,185,270] The human gene has been mapped to chromosome 2q21 and contains 17 exons spanning over 155 kb.[271] It is of interest that the promoter regions of both GnT II[208] and GnT V[271] have putative binding sites for the products of the *Ets* and c-*Myb* families of proto-oncogenes since increased GnT V activity is associated with metastatic activity and GnT II action is a prerequisite for GnT V action (see above). Ets is a nuclear phosphoprotein transcription factor that binds to purine-rich DNA sequences and is associated with transformation properties.[272] Deletion analysis of the GnT V gene promoter region using luciferase as a reporter protein identified two regions as positive regulatory elements.[273] Both regions contained an Ets consensus sequence. Gel mobility shift experiments showed that both regions bound the Ets protein. Co-transfection of luciferase constructs controlled by either of the two putative promoter regions and an Ets expression plasmid showed stimulation of luciferase by the Ets product. These experiments suggest that the induction of GnT V expression in metastatic cells may be due, at least in part, to the Ets protein.

3.03.5.6   UDP-GlcNAc:R$_1$(R$_2$)Manα1–6R$_3$ [GlcNAc to Manα1–6] β1,4-*N*-acetylglucosaminyltransferase VI (GnT VI)

GnT VI adds a GlcNAc residue in a β1–4 linkage to the Manα1–6Manβ arm of the *N*-glycan core (Figure 1(f)).[186] The minimal substrate for GnT VI is the trisaccharide GlcNAcβ1–6[GlcNAcβ1–2]Manα1R (R = methyl, 6Manβ1-methyl, or 6Manβ1-methoxycarbonyloctyl). The enzyme therefore requires the prior actions of GnT I, II, and V. Unlike *N*-glycan GnT I to V, GnT VI can act on both bisected and nonbisected substrates. The enzyme has been demonstrated in birds[186] and fish[274] but not in mammalian tissues. The enzyme has not been purified nor has the gene been cloned.

3.03.6   FUCOSYLTRANSFERASES

Fucose is found in *N*- and *O*-glycans and glycolipids attached to Gal, Glc, or GlcNAc residues in α-linkage to carbons 2, 3, 4, or 6. Several fucosylated structures form human blood group antigenic epitopes (A, B, H, Le^a, Le^b, Le^x, and Le^y). Fucosylated oligosaccharides are often expressed in a regulated manner in development, differentiation, and progression of metastasis.[275–279]

3.03.6.1   Human Blood Group H and Se GDP-Fuc:Galβ-R α1,2-Fucosyltransferases

There are at least two human and three rabbit GDP-Fuc:Galβ-R α1,2-fucosyltransferases (α2FucT), encoded respectively by the H (E.C. 2.4.1.69) and Se (secretory) loci.[280–284] The H locus is expressed mainly in tissues derived from mesoderm (hematopoietic tissues, plasma) or ectoderm, while Se locus expression is restricted to tissues derived from endoderm (secretory fluids and epithelial cells lining salivary glands, stomach, and intestine).[285] The H and Se loci are closely linked on human chromosome 19q13.3 separated by only about 35 kb.[286–288] The genes for both enzymes have been cloned (Table 4) and encode proteins with the typical glycosyltransferase domain structure (Figure 3). The nonsecretor phenotype occurs in about 20% of individuals and at least some of these are due to homozygosity for a nonsense allele at the Se locus.[311,314–317] Neither the H nor Se gene is essential for normal human survival. However, inactivating point mutations in the coding regions of both alleles of the H gene (such as occur in the rare Bombay and para-Bombay phenotypes) can cause serious problems if blood transfusion is required.[280] The Bombay phenotype may also be associated with leukocyte adhesion deficiency type II, a congenital disease which exhibits severe mental retardation, due to defective synthesis of GDP-fucose.[318–322]

Transgenic mice were bred which expressed a fusion gene containing cDNA encoding the human H α1,2FucT under the control of the murine whey acidic protein promoter to direct gene expression primarily to the lactating mammary gland.[323] Milk samples from these animals contained soluble active α1,2FucT and large quantities of 2′-fucosyllactose and modified glycoproteins containing the H antigen.

Table 4 Fucosyltransferases involved in N-glycan synthesis.

Abbreviations[a]	Enzyme product	E.C. no.	Comments	Tissue	Acc no.[b]	Ref.
α3/4FucT III	Fuc(α1-3/4)[Gal(β1-4/3)]GlcNAc-	2.4.1.65	Broad specificity, makes Le^a, Le^b, Le^x, Le^y, $SiaLe^a$, $SiaLe^x$	Human	X53578	289
					U27326-8	290
					S52874, S52967-9	291
					D89324-5	292, 293
α3FucT	Fuc(α1-3)[Gal(β1-4)]GlcNAc-		Lewis type	Cow	X87810	294
				Chicken	U73678	295
α3FucT IV	Fuc(α1-3)[Gal(β1-4)]GlcNAc-		Narrow specificity, makes only Le^x; myeloid type	Caenorhabditis elegans	Z66497, U40028, U80846	174
				Human	M65030	296
					S65161	297
					M58596-7	298
					U33457-8	299
					D63379-80	300
				Mouse	U58860	301
α3FucT V	Fuc(α1-3)[Gal(β1-4)]GlcNAc-		Makes Le^x, $SiaLe^x$	Rat	M81485	302
				Human	U27329-30	290
α3FucT VI	Fuc(α1-3)[Gal(β1-4)]GlcNAc-		Makes Le^x, $SiaLe^x$; Plasma type	Human	L01698	303
					M98825	304
					U27331-7	290
					S52874, S52967-9	291
α3FucT VII	Fuc(α1-3)[Gal(β1-4)]GlcNAc-		Makes $SiaLe^x$ but not Le^x; candidate for making E-selectin ligand on leukocyte cell surface	Human	X78031	305
					U08112, U11282	306
				Mouse	U45980	307
Hz2FucT	Fuc(α1-2)Gal/βR	2.4.1.69	Blood group H epitope	Human	M35531	308
					S79196	288
				Rat	L26009-10	309
				Rabbit	X80226	283
				Pig	L50534	310
Sec2FucT	Fuc(α1-2)Gal/βR		Secretory gene	Human	U17894-5	311
				Rabbit	X80225	283
					X91269	284
core α6FucT	Fuc(α1-6)[R]GlcNAc-Asn-X	2.4.1.68	Makes core Fuc(α1-6)GlcNAc-Asn-X	Pig	D86723	312
core α3FucT	Fuc(α1-3)[R]GlcNAc-Asn-X		Makes core Fuc(α1-3)GlcNAc-Asn-X	Plants	Not cloned	28
				Insects		313

[a] FucT, fucosyltransferase; Le, Lewis; Sia, sialic acid.   [b] EMBL GenBank database accession number.

3.03.6.2   Human Blood Group Lewis GDP-Fuc:Galβ1–4/3GlcNAc (Fuc to GlcNAc) α1,3/4-Fucosyltransferase III (FucT III, E.C. 2.4.1.65)

There are at least five distinct human GDP-Fuc:Galβ1–4GlcNAc (Fuc to GlcNAc) α1,3-fucosyltransferases (α3FucT)[324–328] named FucT III to VII (Table 4). FucT III (the Lewis type enzyme) has the broadest α3FucT substrate specificity since it can incorporate Fuc in either an α1–3 linkage to Galβ1–4GlcNAc (to make Le[x]) or an α1–4 linkage to Galβ1–3GlcNAc (to make Le[a]) even when the Gal residue in these structures is substituted by a fucose (to make Le[y] or Le[b], respectively) or sialic acid (to make sialyl-Le[x] or sialyl-Le[a], respectively). FucT III, IV, and V have been shown to act on sulfated oligosaccharide substrates.[329]

The gene for human FucT III has been cloned (Table 4) and contains an open reading frame on a single exon. The human FucT III, V, and VI genes form a cluster on chromosome 19p13.3.[288,290,330] The protein encoded by the FucT III gene shows the domain structure typical of all glycosyltransferases cloned to date (Figure 3). The FucT III gene is inactivated by a single amino acid substitution in Lewis histo-blood type negative (le/le) individuals.[292,293,331,332]

FucT III, V, and VI share ~85–90% amino acid sequence identity while FucT IV shares only about 60% identity with the other α3FucTs. The α3FucT family shows no sequence similarities to any other glycosyltransferases. Domain swapping experiments between FucT III, V, and VI have shown that these enzymes discriminate between different oligosaccharide acceptor substrates through a discrete 11 amino acid peptide fragment in a "hypervariable" region of the catalytic domain.[333] Sixty one and 75 amino acids could be eliminated from the N-terminus of FucT III and V, respectively, without a significant loss of enzyme activity.[334] In contrast, the truncation of one or more amino acids from the C-terminus of FucT V resulted in a dramatic or total loss of enzyme activity.

3.03.6.3   Human GDP-Fuc:Galβ1–4GlcNAc (Fuc to GlcNAc) α1,3-Fucosyltransferase IV (FucT IV, Myeloid Type)

Human FucT IV, V, VI, and VII differ from FucT III in that they either cannot form the Fucα1–4GlcNAc linkage or do so relatively poorly. FucT IV, found in myeloid tissues, has the narrowest acceptor specificity in this group since it is not effective in the synthesis of the sialylα2–3Galβ1–4[Fucα1–3]GlcNAc moiety (sialyl-Le[x]). The origin of the sialyl-Le[x] structure on neutrophils has aroused a great deal of interest because sialyl-Le[x] is the most likely ligand for an endothelial cell receptor called endothelial-leukocyte adhesion molecule-1 (ELAM-1; E-selectin). E-selectin is essential for a normal inflammatory response involving inflammation-activated homing of leukocytes to endothelial cells.[296,297,328,335–340] Although FucT IV is a myeloid enzyme, it is not primarily responsible for synthesis of the myeloid cell E-selectin ligand since it is relatively ineffective in sialyl-Le[x] synthesis.[341] FucT VII is the main enzyme involved in the synthesis of neutrophil sialyl-Le[x], while FucT IV plays an accessory role.[306,342,343] The gene for FucT IV has been cloned (Table 4) and maps to chromosome 11q21.[344] The open reading frame is on a single exon and encodes a protein with a domain structure typical for the glycosyltransferases (Figure 3).

3.03.6.4   Human GDP-Fuc:Galβ1–4GlcNAc (Fuc to GlcNAc) α1,3-Fucosyltransferases V and VI (FucT V, VI)

A probe prepared from the cDNA encoding human FucT III was used to screen a human genomic DNA library under conditions of low stringency and two genes were isolated (Table 4), each with an open reading frame on a single exon, encoding two proteins (FucT V and VI) with domain structures typical of glycosyltransferases (Figure 3).[302,303] Although FucT V and VI differ quantitatively in substrate specificity, both can synthesize Le[x] and sialyl-Le[x], but not Le[a] or sialyl-Le[a]. Comparison of FucT V and VI substrate specificities with those of the plasma-type[345] and lung carcinoma-type[328] enzymes indicates that FucT VI is probably the plasma-type enzyme but that neither enzyme type corresponds to FucT V. Study of an Indonesian population which lacks plasma α3FucT activity proves that FucT VI is the plasma-type enzyme.[346] FucT VI (but not FucT III or V) is responsible for the α3-fucosylation of serum glycoproteins produced by the liver.[347] The genes for FucT III and VI are closely linked on human chromosome 19.[291] Replacement of the stem region and the transmembrane domain of FucT V by protein A results in an enzyme with GDP-fucose hydrolyzing activity.[348]

3.03.6.5   Human GDP-Fuc:Galβ1–4GlcNAc (Fuc to GlcNAc) α1,3-Fucosyltransferase VII (FucT VII)

The gene for a novel human leukocyte α3FucT (FucT VII) has been cloned (Table 4).[306,349] FucT VII differs from the other α3FucTs in that it can make sialyl-Le[x] but not Le[x], Le[a], nor sialyl-Le[a]. The gene for FucT VII is localized to chromosome 9 and shares about 39% amino acid sequence identity with FucT III (a prototype of chromosome 19-localized FucT III, V, and VI) and about 38% with chromosome 11-localized FucT IV.

FucT VII is responsible for synthesis of the selectin ligands required for inflammation-activated homing of leukocytes to endothelial cells.[307,342,343,350,351] Mice deficient in FucT VII exhibit a leukocyte adhesion deficiency characterized by absent leukocyte E- and P-selectin ligand activity and deficient high endothelial venule (HEV) L-selectin ligand activity.[350] Selectin ligand deficiency in these mice is indicated by blood leukocytosis, impaired leukocyte extravasation in inflammation and faulty lymphocyte homing. These observations demonstrate an essential role for FucT VII in E-, P-, and L-selectin ligand biosynthesis and imply that this locus can control leukocyte trafficking in health and disease.

Leukemia cells in patients with adult T-cell leukemia (ATL) strongly express sialyl-Le[x]. Human T-cell leukemia virus type 1 (HTLV-1), the etiological agent of ATL, produces Tax protein, which is implicated in pathogenesis. It has been shown that HTLV-1 Tax protein can transactivate the FucT VII gene, suggesting that this may be the mechanism for sialyl-Le[x] synthesis in ATL cells.[352]

3.03.6.6   GDP-Fuc:β-N-acetylglucosaminide (Fuc to Asn-linked GlcNAc) α1,3- and α1,6-Fucosyltransferases

Fucose can be transferred to the asparagine-linked N-acetylglucosamine of the N-glycan core either in an α1–6 linkage (mammals, insects) or α1–3 linkage (plants, insects) by GDP-Fuc:β-N-acetylglucosaminide (Fuc to Asn-linked GlcNAc) α1,6-fucosyltransferase (core α6FucT, E.C. 2.4.1.68) or α1,3-fucosyltransferase (core α3FucT), respectively (Table 4). GlcNAcβ1–2Manα1–3Manβ- on the N-glycan core, due to prior action of GnT I, is essential for both α6FucT[353,354] and α3FucT.[28,313,355] Human[356] and porcine[312] α6FucT and mung bean α3FucT[357] have been purified. The porcine[312] and human[358] genes have been cloned. Insects have been shown to contain both α3FucT and α6FucT, which can act in concert to place two fucose residues on the same Asn-linked GlcNAc residue.[313,359–362]

3.03.7   CONCLUDING REMARKS

Since the late 1980s there has been an explosion in our knowledge of the structure and biosynthesis of protein- and lipid-bound glycans, particularly of N-glycans. The sophisticated analytical techniques mentioned earlier in this chapter are adding new structures to our already large structural library at an ever increasing rate. There has been a somewhat slower advance in the characterization of the hundreds of glycosyltransferases that must be required to make these diverse glycan structures. However, new glycosyltransferase genes are being added to the database at an accelerating pace due to advances in expression cloning[363] and homology screening. The latter can be carried out not only by the traditional experimental approach, which uses degenerate oligonucleotides to probe libraries or to carry out polymerase chain reaction screens, but also by screening the DNA databases for homologous genes or expressed sequence tags. For example, several glycosyltransferases have been discovered by homology screening of the *C. elegans* genomic DNA database, for example, GnT I,[364] core 2 β6GnT and GnT V,[365] and UDP-GalNAc:polypeptide N-acetylgalactosaminyltransferase.[366] Since the purification of glycosyltransferases is a tedious and often difficult procedure, it is hoped that these newer approaches will lead to the cloning of many more glycosyltransferases in the near future. However, homology screening has its pitfalls. For example, a homologous gene may not encode the enzyme used for screening. When a probe based on the bovine β4GalT sequence was used to screen a pond snail (*Lymnaea stagnalis*) genomic DNA library, a gene was obtained which encoded a type II membrane protein with considerable sequence similarity to mammalian β4GalT.[55] However, this protein had no β4GalT activity and proved to be a novel β4GnT which synthesizes GlcNAcβ1–4GlcNAc. The importance of this enzyme to the snail is not understood. The finding indicates that it may be difficult to find an enzyme activity for a protein obtained by homology screening.

Although recent advances in glycobiology have been impressive, many problems remain to be solved. Several questions relate to the genomic organization of glycosyltransferases. Why do mammalian glycosyltransferase genes fall into two categories,[10] single-exonic and multi-exonic? The entire coding regions for FucT III to VI, GnT I, GnT II, and O-glycan core 2β6GnT are within a single exon whereas the coding regions for ST6Gal I, β4GalT, and α3GalT are distributed over five or more exons. The genes for the latter enzymes span 35 kb or more of genomic DNA and there are one or more noncoding exons several kilobases upstream of the coding region. Many of these genes have a single long (2–3 kb) 3'-terminal exon which carries the entire relatively long 3'-untranslated region, the translational stop signal, and part of the 3'-terminal coding region.

Another interesting question relates to the tissue- and time-specific expression of some glycosyltransferase genes, a phenomenon that is believed to be involved in differentiation and development. The expression mechanism relies at least in part on differential activation of multiple promoters, for example, β4GalT,[56,58–60] ST6Gal I,[93,100,101,103,127] GnT I,[167] GnT III,[367] and GnT V.[271,368]

The evolution of glycosyltransferases is another aspect which will be greatly advanced by a continually enlarging sequence database. There are at least five glycosyltransferase gene families:[10,15,16,55] (i) α3FucT (Table 4), (ii) α3GalT/α3GalNAcT, (iii) O-glycan core 2 β6GnT, (iv) the sialyltransferases (Table 2), and (v) the β4GalT/β4GnT family. The α3GalT/α3GalNAcT family includes the human blood group B α3GalT and human blood group A α1,3-N-acetylgalacto-saminyltransferase. The O-glycan core 2 β6GnT family has at least two members (core 2 β6GnT and the blood group antigen I β6GnT) localized to human chromosome 9q21. The members of a family share varying degrees of sequence similarity. The fact that homologous regions for the β6GnT and sialyltransferase families are distributed over more than one exon suggests that evolution occurred through gene duplication followed by intron insertion rather than by exon shuffling.[369]

Perhaps the most frustrating problem in glycobiology has been and remains the functional significance of the thousands of glycan structures present in living cells.[370] Especially difficult to understand is the biological significance of microheterogeneity (described earlier in this chapter). Much evidence has been obtained for the role of cell surface glycans in the interaction of cells with receptors in the environment, for example, the role of selectins in the inflammatory process.[338,371] However, if cell surface glycans are a language of communication, this language will clearly be difficult to decipher. A very effective approach to elucidate glycan functions is to create mutant cell lines and animals. Indeed, many cell line mutants have been created which either lack or overexpress a glycosyltransferase.[372,373] However, this technique, while useful for a variety of experimental applications (elucidation of biosynthetic pathways, production of engineered glycoproteins, expression cloning, Golgi targeting, study of cell–cell adhesion, etc.), has not provided information on the role of glycans in development and differentiation. It is of great interest that cell lines lacking a functional GnT I gene show normal growth and viability whereas mouse embryos with a null mutation of the GnT I gene die at about 9–10 days after fertilization.[204,205] The study of mice which either overexpress a particular glycosyltransferase transgene or have a null mutation in such a gene may provide a Rosetta stone for defining the functional role of glycans *in vivo*.[374]

Congenital diseases involving defects in glycan synthesis are providing similar information. Several diseases with genetic defects in N-glycan synthesis have been reported, that is, inclusion cell disease (I-cell disease), a lysosomal storage disease with a defective N-acetylglucosaminyl-1-phosphotransferase;[375] carbohydrate-deficient glycoprotein syndrome type I, in which most patients show a defect in phosphomannomutase required for normal N-glycan assembly;[203,376] carbohydrate-deficient glycoprotein syndrome type II, two cases of which have been attributed to a defect in GnT II;[199,201,202] hereditary erythroblastic multinuclearity with a positive acidified serum lysis test (HEMPAS, congenital dyserythropoietic anemia type II), one case of which has been shown to be due to a defect in α-mannosidase II;[377,378] and leukocyte adhesion deficiency type II (LADII), an immunodeficiency disease due to defective synthesis of GDP-fucose and consequently of the ligands for selectins.[318–321]

The fact that patients with carbohydrate-deficient glycoprotein syndrome as well as mouse embryos with a null mutation in the GnT I gene show severe developmental abnormalities, particularly of the nervous system, indicates the importance of complex N-glycans in development. Indeed, various studies have shown that carbohydrates play important roles in vertebrate development.[206] Although work will undoubtedly continue on mutant mice, other development models should also be considered. *C. elegans* is a particularly attractive model because of the detailed information available on its development, its relatively simple architecture and the rapid progress being made in the sequencing of its genome. Several glycosyltransferases are already being studied in this nematode[364–366] and others are sure to follow.

The future looks bright for glycobiology. Nucleic acids and proteins have dominated the bio-chemical literature for many years but the glycans, like these molecules, are also a class of biological polymer with an enormous capacity for carrying information. The role of glycans is probably more subtle than that of nucleic acids and proteins in that their information transfer role is restricted to complex biological systems and therefore they cannot be readily studied in tissue culture models. As genetic tools for studying complex organisms become more sophisticated, the glycans will almost certainly take their proper place in the biological firmament.

ACKNOWLEDGMENTS

This work was supported by the Medical Research Council of Canada, the Protein Engineering Network of Centres of Excellence (PENCE), and the Mizutani Foundation for Glycosciences.

3.03.8 REFERENCES

1. R. A. Laine, *Glycobiology*, 1994, **4**, 759.
2. R. Kornfeld and S. Kornfeld, *Annu. Rev. Biochem.*, 1985, **54**, 631.
3. R. D. Cummings, in "Glycoconjugates. Composition, Structure and Function," eds. H. J. Allen and E. C. Kisailus, Marcel Dekker, New York, 1992, p. 333.
4. H. Schachter, *Biochem. Cell Biol.*, 1986, **64**, 163.
5. H. Schachter, *Glycobiology*, 1991, **1**, 453.
6. I. B. H. Wilson, 1996, http://bellatrix.pel.ox.ac.uk/people/iain/glycosyltransferase.html.
7. P. Greenwell, A. D. Yates, and W. M. Watkins, *Carbohydr. Res.*, 1986, **149**, 149.
8. T. W. Rademacher, R. B. Parekh, and R. A. Dwek, *Annu. Rev. Biochem.*, 1988, **57**, 785.
9. G. W. Hart, *Curr. Opin. Cell. Biol.*, 1992, **4**, 1017.
10. D. H. Joziasse, *Glycobiology*, 1992, **2**, 271.
11. J. C. Paulson and K. J. Colley, *J. Biol. Chem.*, 1989, **264**, 17 615.
12. H. Schachter, *Curr. Opin. Struct. Biol.* 1991, **1**, 755.
13. H. Schachter, in "Glycoproteins," eds. J. Montreuil, J. F. G. Vliegenthart, and H. Schachter, Elsevier, Amsterdam, 1995, p. 153.
14. J. H. Shaper and N. L. Shaper, *Curr. Opin. Struct. Biol.*, 1992, **2**, 701.
15. D. H. Van den Eijnden and D. H. Joziasse, *Curr. Opin. Struct. Biol.* 1993, **3**, 711.
16. R. Kleene and E. G. Berger, *Biochim. Biophys. Acta*, 1993, **1154**, 283.
17. J. F. G. Vliegenthart and J. Montreuil, in "Glycoproteins," eds. J. Montreuil, J. F. G. Vliegenthart, and H. Schachter, Elsevier, Amsterdam, 1995, p. 13.
18. A. Kobata, *Eur. J. Biochem.*, 1992, **209**, 483.
19. H. Lis and N. Sharon, *Eur. J. Biochem.*, 1993, **218**, 1.
20. T. A. Beyer, J. E. Sadler, J. I. Rearick, J. C. Paulson, and R. L. Hill, in "Advances in Enzymology," ed. A. Meister, Wiley, New York, 1981, vol. 52, p. 23.
21. M. D. Snider, in "Biology of Carbohydrates," eds. V. Ginsburg and P. W. Robbins, Wiley, New York, 1984, Vol 2, p. 163.
22. F. W. Hemming, in "Glycoproteins," eds. J. Montreuil, J. F. G. Vliegenthart and H. Schachter, Elsevier, Amsterdam, 1995, p. 127.
23. A. Verbert, in "Glycoproteins," eds. J. Montreuil, J. F. G. Vliegenthart, and H. Schachter, Elsevier, Amsterdam, 1995, p. 145.
24. W. M. Watkins, in "Glycoproteins," eds. J. Montreuil, J. F. G. Vliegenthart, and H. Schachter, Elsevier, Amsterdam, 1995, p. 313.
25. M. Fukuda, in "Molecular Glycobiology," eds. M. Fukuda and O. Hindsgaul, Oxford University Press, Oxford, 1994, p. 1.
26. Y. T. Pan and A. D. Elbein, in "Glycoproteins," eds. J. Montreuil, J. F. G. Vliegenthart, and H. Schachter, Elsevier, Amsterdam, 1995, p. 415.
27. S. Silberstein and R. Gilmore, *FASEB J.*, 1996, **10**, 849.
28. K. D. Johnson and M. J. Chrispeels, *Plant Physiol.*, 1987, **84**, 1301.
29. H. Mulder, F. Dideberg, H. Schachter, B. A. Spronk, M. De Jong-Brink, J. P. Kamerling, and J. F. G. Vliegenthart, *Eur. J. Biochem.*, 1995, **232**, 272.
30. N. Harpaz and H. Schachter, *J. Biol. Chem.*, 1980, **255**, 4894.
31. B. D. Shur, *Glycobiology*, 1991, **1**, 563.
32. G. J. Strous, *CRC Crit. Rev. Biochem.*, 1986, **21**, 119.
33. M. M. Palcic and O. Hindsgaul, *Glycobiology*, 1991, **1**, 205.
34. K. Y. Do, S. I. Do, and R. D. Cummings, *J. Biol. Chem.*, 1995, **270**, 18 447.
35. H. Narimatsu, S. Sinha, K. Brew, H. Okayama, and P. K. Qasba, *Proc. Natl. Acad. Sci. USA*, 1986, **83**, 4720.
36. N. L. Shaper, J. H. Shaper, J. L. Meuth, J. L. Fox, H. Chang, I. R. Kirsch, and G. F. Hollis, *Proc. Natl. Acad. Sci. USA*, 1986, **83**, 1573.
37. G. D'Agostaro, B. Bendiak, and M. Tropak, *Eur. J. Biochem.*, 1989, **183**, 211.
38. A. S. Masibay and P. K. Qasba, *Proc. Natl. Acad. Sci. USA*, 1989, **86**, 5733.
39. R. N. Russo, N. L. Shaper, and J. H. Shaper, *J. Biol. Chem.*, 1990, **265**, 3324.
40. K. A. Masri, H. E. Appert, and M. N. Fukuda, *Biochem. Biophys. Res. Commun.*, 1988, **157**, 657.

41. G. Watzele and E. G. Berger, *Nucleic Acids Res.*, 1990, **18**, 7174.
42. T. Uejima, M. Uemura, S. Nozawa, and H. Narimatsu, *Cancer Res.*, 1992, **52**, 6158.
43. H. E. Appert, T. J. Rutherford, G. E. Tarr, J. S. Wiest, N. R. Thomford, and D. J. McCorquodale, *Biochem. Biophys. Res. Commun.*, 1986, **139**, 163.
44. L. Mengle-Gaw, M. F. McCoy-Haman, and D. C. Tiemeier, *Biochem. Biophys. Res. Commun.*, 1991, **176**, 1269.
45. K. Nakazawa, T. Ando, T. Kimura, and H. Narimatsu, *J. Biochem. (Tokyo)*, 1988, **104**, 165.
46. N. L. Shaper, G. F. Hollis, J. G. Douglas, I. R. Kirsch, and J. H. Shaper, *J. Biol. Chem.*, 1988, **263**, 10 420.
47. F. Uehara and T. Muramatsu, *EMBL/GeneBank Listing*, 1996, AN: D37790.
48. G. F. Hollis, J. G. Douglas, N. L. Shaper, J. H. Shaper, J. M. Stafford-Hollis, R. J. Evans, and I. R. Kirsch, *Biochem. Biophys. Res. Commun.*, 1989, **162**, 1069.
49. B. Bendiak, L. D. Ward, and R. J. Simpson, *Eur. J. Biochem.*, 1993, **216**, 405.
50. S. Ghosh, B. S. Sankar, and S. Basu, *Biochem. Biophys. Res. Commun.*, 1992, **189**, 1215.
51. A. M. V. Duncan, M. M. McCorquodale, C. Morgan, T. J. Rutherford, H. E. Appert, and D. J. McCorquodale, *Biochem. Biophys. Res. Commun.*, 1986, **141**, 1185.
52. M. Malissard, L. Borsig, S. Di Marco, M. G. Grutter, U. Kragl, C. Wandrey, and E. G. Berger, *Eur. J. Biochem.*, 1996, **239**, 340.
53. H. Schachter, in "Molecular Glycobiology," eds. M. Fukuda and O. Hindsgaul, Oxford University Press, Oxford, 1994, p. 88.
54. H. Y. Zu, M. N. Fukuda, S. S. Wong, Y. Wang, Z. D. Liu, Q. S. Tang, and H. E. Appert, *Biochem. Biophys. Res. Commun.*, 1995, **206**, 362.
55. H. Bakker, M. Agterberg, A. Van Tetering, C. A. M. Koeleman, D. H. Van den Eijnden, and I. Van Die, *J. Biol. Chem.*, 1994, **269**, 30 326.
56. A. Harduin-Lepers, J. H. Shaper, and N. L. Shaper, *J. Biol. Chem.*, 1993, **268**, 14 348.
57. B. Rajput, N. L. Shaper, and J. H. Shaper, *J. Biol. Chem.*, 1996, **271**, 5131.
58. N. L. Shaper, W. W. Wright, and J. H. Shaper, *Proc. Natl. Acad. Sci. USA*, 1990, **87**, 791.
59. A. Harduin-Lepers, N. L. Shaper, J. A. Mahoney, and J. H. Shaper, *Glycobiology*, 1992, **2**, 361.
60. N. L. Shaper, A. Harduin-Lepers, and J. H. Shaper, *J. Biol. Chem.*, 1994, **269**, 25 165.
61. D. J. Miller, M. B. Macek, and B. D. Shur, *Nature*, 1992, **357**, 589.
62. M. Asano, K. Furukawa, M. Kido, S. Matsumoto, Y. Umesaki, N. Kochibe, and Y. Iwakura, *EMBO J.*, 1997, **16**, 1850.
63. Q. X. Lu, P. Hasty, and B. D. Shur, *Dev. Biol.*, 1997, **181**, 257.
64. P. A. Gleeson, R. D. Teasdale, and J. Burke, *Glycoconjugate J.*, 1994, **11**, 381.
65. T. Nilsson, J. M. Lucocq, D. Mackay, and G. Warren, *EMBO J.*, 1991, **10**, 3567.
66. R. D. Teasdale, G. D'Agostaro, and P. A. Gleeson, *J. Biol. Chem.*, 1992, **267**, 4084.
67. D. Aoki, N. Lee, N. Yamaguchi, C. Dubois, and M. N. Fukuda, *Proc. Natl. Acad. Sci. USA*, 1992, **89**, 4319.
68. R. D. Teasdale, F. Matheson, and P. A. Gleeson, *Glycobiology*, 1994, **4**, 917.
69. S. Munro, *EMBO J.*, 1991, **10**, 3577.
70. R. Y. Dahdal and K. J. Colley, *J. Biol. Chem.*, 1993, **268**, 26 310.
71. J. Burke, J. M. Pettitt, D. Humphris, and P. A. Gleeson, *J. Biol. Chem.*, 1994, **269**, 12 049.
72. J. Burke, J. M. Pettitt, H. Schachter, M. Sarkar, and P. A. Gleeson, *J. Biol. Chem.*, 1992, **267**, 24 433.
73. L. C. Lopez and B. D. Shur, *Biochem. Biophys. Res. Commun.*, 1988, **156**, 1223.
74. L. C. Lopez, C. M. Maillet, K. Oleszkowicz, and B. D. Shur, *Mol. Cell. Biol.*, 1989, **9**, 2370.
75. L. C. Lopez, A. Youakim, S. C. Evans, and B. D. Shur, *J. Biol. Chem.*, 1991, **266**, 15 984.
76. S. C. Evans, L. C. Lopez, and B. D. Shur, *J. Cell. Biol.*, 1993, **120**, 1045.
77. A. S. Maslbay, P. V. Balaji, E. E. Doeggeman, and P. K. Qasba, *J. Biol. Chem.*, 1993, **268**, 9908.
78. R. N. Russo, N. L. Shaper, D. J. Taatjes, and J. H. Shaper, *J. Biol. Chem.*, 1992, **267**, 9241.
79. A. Dinter and E. G. Berger, in "International Symposium on Glycoconjugates, Seattle, WA, 1995," Abstr. in *Glycoconjugate J.*, 1995, **12**, 448.
80. T. Nilsson, P. Slusarewicz, M. H. Hoe, and G. Warren, *FEBS Lett.*, 1993, **330**, 1.
81. T. Nilsson, M. H. Hoe, P. Slusarewicz, C. Rabouille, R. Watson, F. Hunte, G. Watzele, E. G. Berger, and G. Warren, *EMBO J.*, 1994, **13**, 562.
82. P. Slusarewicz, T. Nilsson, N. Hui, R. Watson, and G. Warren, *J. Cell. Biol.*, 1994, **124**, 405.
83. T. Nilsson, C. Rabouille, N. Hui, R. Watson, and G. Warren, *J. Cell Sci.*, 1996, **109**, 1975.
84. S. Munro, *EMBO J.*, 1995, **14**, 4695.
85. N. Yamaguchi and M. N. Fukuda, *J. Biol. Chem.*, 1995, **270**, 12 170.
86. S. Munro, *Biochem. Soc. Trans.*, 1995, **23**, 527.
87. T. Schwientek, H. Narimatsu, and J. F. Ernst, *J. Biol. Chem.*, 1996, **271**, 3398.
88. H. Kitagawa and J. C. Paulson, *Biochem. Biophys. Res. Commun.*, 1993, **194**, 375.
89. K. Sasaki, E. Watanabe, K. Kawashima, N. Hanai, T. Nishi, and M. Hasegawa, *EMBL/GenBank listing*, 1994, AN: D28941.
90. D. X. Wen, B. D. Livingston, K. F. Medzihradszky, S. Kelm, A. L. Burlingame, and J. C. Paulson, *J. Biol. Chem.*, 1992, **267**, 21 011.
91. K. Sasaki, E. Watanabe, K. Kawashima, S. Sekine, T. Dohi, M. Oshima, N. Hanai, T. Nishi, and M. Hasegawa, *J. Biol. Chem.*, 1993, **268**, 22 782.
92. H. Kitagawa and J. C. Paulson, *J. Biol. Chem.*, 1994, **269**, 1394.
93. X. C. Wang, A. Vertino, R. L. Eddy, M. G. Byers, S. N. Jani-Sait, T. B. Shows, and J. T. Y. Lau, *J. Biol. Chem.*, 1993, **268**, 4355.
94. U. Grundmann, C. Nerlich, T. Rein, and G. Zettlmeissl, *Nucleic Acids Res.*, 1990, **18**, 667.
95. B. J. E. G. Bast, L. J. Zhou, G. J. Freeman, K. J. Colley, T. J. Ernst, J. M. Munro, and T. F. Tedder, *J. Cell. Biol.*, 1992, **116**, 423.
96. I. Stamenkovic, H. C. Asheim, A. Deggerdal, H. K. Blomhoff, E. B. Smeland, and S. Funderud, *J. Exp. Med.*, 1990, **172**, 641.
97. H. C. Aasheim, D. A. Aas-Eng, A. Deggerdal, H. K. Blomhoff, S. Funderud, and E. B. Smeland, *Eur. J. Biochem.*, 1993, **213**, 467.

98. D. A. Aas-Eng, H. C. Aasheim, A. Deggerdal, E. Smeland, and S. Funderud, *Biochem. Biophys. Acta*, 1995, **1261**, 166.
99. T. P. O'Hanlon, K. M. Lau, X. C. Wang, and J. T. Y. Lau, *J. Biol. Chem.*, 1989, **264**, 17 389.
100. X.-C. Wang, T. P. O'Hanlon, R. F. Young, and J. T. Y. Lau, *Glycobiology*, 1990, **1**, 25.
101. E. C. Svensson, B. Soreghan, and J. C. Paulson, *J. Biol. Chem.*, 1990, **265**, 20 863.
102. J. Weinstein, E. U. Lee, K. McEntee, P.-H. Lai, and J. C. Paulson, *J. Biol. Chem.*, 1987, **262**, 17 735.
103. D. X. Wen, E. C. Svensson, and J. C. Paulson, *J. Biol. Chem.*, 1992, **267**, 2512.
104. T. Hamamoto, M. Kawasaki, N. Kurosawa, T. Nakaoka, Y.-C. Lee and S. Tsuji, *Bioorg. Med. Chem.*, 1993, **1**, 141.
105. N. Kurosawa, M. Kawasaki, T. Hamamoto, T. Nakaoka, Y. C. Lee, M. Arita, and S. Tsuji, *Eur. J. Biochem.*, 1994, **219**, 375.
106. H. Kitagawa and J. C. Paulson, *J. Biol. Chem.*, 1994, **269**, 17 872.
107. E. P. Scheidegger, L. R. Sternberg, J. Roth and J. B. Lowe, *J. Biol. Chem.*, 1995, **270**, 22 685.
108. K. Angata, J. Nakayama, B. Fredette, K. Chong, B. Ranscht, and M. Fukuda, *J. Biol. Chem.*, 1997, **272**, 7182.
109. N. Kojima, Y. Yoshida, N. Kurosawa, Y. C. Lee, and S. Tsuji, *FEBS Lett.*, 1995, **360**, 1.
110. Y. Yoshida, N. Kurosawa, T. Kanematsu, N. Kojima, and S. Tsuji, *J. Biol. Chem.*, 1996, **271**, 30 167.
111. B. D. Livingston and J. C. Paulson, *J. Biol. Chem.*, 1993, **268**, 11 504.
112. Y. Yoshida, N. Kojima, N. Kurosawa, T. Hamamoto, and S. Tsuji, *J. Biol. Chem.*, 1995, **270**, 14 628.
113. Y. Yoshida, N. Kurosawa, T. Kanematsu, A. Taguchi, M. Arita, N. Kojima, and S. Tsuji, *Glycobiology*, 1996, **6**, 573.
114. M. Eckhardt, M. Mühlenhoff, A. Bethe, J. Koopman, M. Frosch, and R. Gerardy-Schahn, *Nature*, 1995, **373**, 715.
115. Y. Yoshida, N. Kojima, and S. Tsuji, *J. Biochem. (Tokyo)*, 1995, **118**, 658.
116. J. Nakayama, M. N. Fukuda, B. Fredette, B. Ranscht, and M. Fukuda, *Proc. Natl. Acad. Sci. USA*, 1995, **92**, 7031.
117. S. Tsuji, A. K. Datta, and J. C. Paulson, *Glycobiology*, 1996, **6**, R5.
118. A. Harduin-Lepers, M.-A. Recchi, and P. Delannoy, *Glycobiology*, 1995, **5**, 741.
119. S. Tsuji, *J. Biochem. (Tokyo)*, 1996, **120**, 1.
120. D. H. Joziasse, W. E. C. M. Schiphorst, D. H. Van den Eijnden, J. A. Van Kuik, H. Van Halbeek, and J. F. G. Vliegenthart, *J. Biol. Chem.*, 1987, **262**, 2025.
121. C. H. Hokke, J. G. M. Van der Ven, J. P. Kamerling, and J. F. G. Vliegenthart, *Glycoconjugate J.*, 1993, **10**, 82.
122. J. Van Pelt, L. Dorland, M. Duran, C. H. Hokke, J. P. Kamerling, and J. F. G. Vliegenthart, *FEBS Lett.*, 1989, **256**, 179.
123. K. B. Wlasichuk, M. A. Kashem, P. V. Nikrad, P. Bird, C. Jiang, and A. P. Venot, *J. Biol. Chem.*, 1993, **268**, 13 971.
124. J. Van Pelt, C. H. Hokke, L. Dorland, M. Duran, J. P. Kamerling, and J. F. G. Vliegenthart, *Clin. Chim. Acta*, 1990, **187**, 55.
125. J. Van Pelt, L. Dorland, M. Duran, C. H. Hokke, J. P. Kamerling, and J. F. G. Vliegenthart, *J. Biol. Chem.*, 1990, **265**, 19 685.
126. J. A. L. M. Van Dorst, J. M. Tikkanen, C. H. Krezdorn, M. B. Streiff, E. G. Berger, J. A. Van Kuik, J. P. Kamerling, and J. F. G. Vliegenthart, *Eur. J. Biochem.*, 1996, **242**, 674.
127. J. C. Paulson, J. Weinstein, and A. Schauer, *J. Biol. Chem.*, 1989, **264**, 10 931.
128. T. P. O'Hanlon and J. T. Y. Lau, *Glycobiology*, 1992, **2**, 257.
129. X. C. Wang, T. J. Smith and J. Lau, *J. Biol. Chem.*, 1990, **265**, 17 849.
130. A. Vertino-Bell, J. Ren, J. D. Black, and J. Y. T. Lau, *Dev. Biol.*, 1994, **165**, 126.
131. J. W. Dennis, in "Cell Surface Carbohydrates and Cell Development," ed. M. Fukuda, CRC Press, Boca Raton, FL, 1992, p. 161.
132. R. Takano, E. Muchmore, and J. W. Dennis, *Glycobiology*, 1994, **4**, 665.
133. F. Dall'Olio, N. Malagolini, and F. Serafini-Cessi, *Int. J. Cancer*, 1992, **50**, 325.
134. J. Morgenthaler, W. Kemmner, and R. Brossmer, *Biochem. Biophys. Res. Commun.*, 1990, **171**, 860.
135. N. Le Marer, V. Laudet, E. C. Svensson, H. Cazlaris, B. Van Hille, C. Lagrou, D. Stehelin, J. Montreuil, A. Verbert, and P. Delannoy, *Glycobiology*, 1992, **2**, 49.
136. P. Delannoy, H. Pelczar, V. Vandamme, and A. Verbert, *Glycoconjugate J.*, 1993, **10**, 91.
137. N. Le Marer and D. Stéhelin, *Glycobiology*, 1995, **5**, 219.
138. F. Dall'Olio, N. Malagolini, G. Di Stefano, F. Minni, D. Marrano, and F. Serafini-Cessi, *Int. J. Cancer*, 1989, **44**, 434.
139. F. Dall'Olio, N. Malagolini, S. Guerrini, J. T. Y. Lau, and F. Serafini-Cessi, *Glycoconjugate J.*, 1996, **13**, 115.
140. P. Lance, K. M. Lau, and J. T. Lau, *Biochem. Biophys. Res. Commun.*, 1989, **164**, 225.
141. E. C. Svensson, P. B. Conley, and J. C. Paulson, *J. Biol. Chem.* 1992, **267**, 3466.
142. I. Stamenkovic, D. Sgroi, A. Aruffo, M. S. Sy, and T. Anderson, *Cell*, 1991, **66**, 1133.
143. I. Stamenkovic, D. Sgroi, and A. Aruffo, *Cell*, 1992, **68**, 1003.
144. N.-W. Lo and J. T. Y. Lau, *Glycobiology*, 1996, **6**, 271.
145. N.-W. Lo and J. T. Y. Lau, *Biochem. Biophys. Res. Commun.*, 1996, **228**, 380.
146. K. J. Colley, E. U. Lee, B. Adler, J. K. Browne, and J. C. Paulson, *J. Biol. Chem.*, 1989, **264**, 17 619.
147. S. H. Wong, S. H. Low and W. Hong, *J. Cell Biol.*, 1992, **117**, 245.
148. J. Y. Ma and K. J. Colley, *J. Biol. Chem.*, 1996, **271**, 7758.
149. J. Y. Ma, R. Qian, F. M. Rausa III, and K. J. Colley, *J. Biol. Chem.*, 1997, **272**, 672.
150. L. Borsig, S. X. Ivanov, G. F. Herrmann, U. Kragl, C. Wandrey, and E. G. Berger, *Biochem. Biophys. Res. Commun.*, 1995, **210**, 14.
151. T. Schwientek, C. Lorenz, and J. F. Ernst, *J. Biol. Chem.*, 1995, **270**, 5483.
152. K. Drickamer, *Glycobiology*, 1993, **3**, 2.
153. A. K. Datta, and J. C. Paulson, *J. Biol. Chem.* 1995, **270**, 1497.
154. M. Nemansky, and D. H. Van den Eijnden, *Glycoconjugate J.*, 1993, **10**, 99.
155. S. C. Basu, *Glycobiology*, 1991, **1**, 469.
156. H. Kitagawa, M. G. Mattei, and J. C. Paulson, *J. Biol. Chem.*, 1996, **271**, 931.
157. B. Livingston, H. Kitagawa, D. Wen, K. Medzihradsky, A. Burlingame, and J. C. Paulson, in "21st Annual Meeting of the Society for Complex Carbohydrates, Nashville, Tennessee, 1992." Abstr. in *Glycobiology*, 1992, **2**, 489.
158. N. Kojima, Y. Yoshida, and S. Tsuji, *FEBS Lett.*, 1995, **373**, 119.
159. N. Kojima, Y. Tachida, Y. Yoshida, and S. Tsuji, *J. Biol. Chem.*, 1996, **271**, 19 457.
160. F. A. Troy II, *Glycobiology*, 1992, **2**, 5.

161. J. Nakayama and M. Fukuda, *J. Biol. Chem.*, 1996, **271**, 1829.
162. M. Muhlenhoff, M. Eckhardt, A. Bethe, M. Frosch, and R. Gerardy-Schahn, *EMBO J.*, 1996, **15**, 6943.
163. N. Kojima, M. Kono, Y. Yoshida, Y. Tachida, M. Nakafuku, and S. Tsuji, *J. Biol. Chem.*, 1996, **271**, 22058.
164. M. Sarkar, E. Hull, Y. Nishikawa, R. J. Simpson, R. L. Moritz, R. Dunn, and H. Schachter, *Proc. Natl. Acad. Sci. USA*, 1991, **88**, 234.
165. E. Hull, M. Sarkar, M. P. N. Spruijt, J. W. M. Höppener, R. Dunn, and H. Schachter, *Biochem. Biophys. Res. Commun.*, 1991, **176**, 608.
166. R. Kumar, J. Yang, R. D. Larsen and P. Stanley, *Proc. Natl. Acad. Sci. USA*, 1990, **87**, 9948.
167. J. Yang, M. Bhaumik, Y. Liu, and P. Stanley, *Glycobiology*, 1994, **4**, 703.
168. S. Pownall, C. A. Kozak, K. Schappert, M. Sarkar, E. Hull, H. Schachter, and J. D. Marth, *Genomics*, 1992, **12**, 699.
169. R. Kumar, J. Yang, R. L. Eddy, M. G. Byers, T. B. Shows, and P. Stanley, *Glycobiology*, 1992, **2**, 383.
170. T. Fukada, K. Iida, N. Kioka, H. Sakai, and T. Komano, *Biosci. Biotechnol. Biochem.*, 1994, **58**, 200.
171. S. Narasimhan, R. Yuen, C. J. Fode, J. Ali, S. Rajalakshmi, K. Schappert, J. W. Dennis, and H. Schachter, in "22nd Annual Meeting of the Society for Complex Carbohydrates, Puerto Rico, 1993." Abstr. in *Glycobiology*, 1993, **3**, 531.
172. J. Mucha, S. Kappel, H. Schachter, W. Hane, and J. Glössl, in "XIIIth International Symposium on Glycoconjugates, Seattle, WA, 1995." Abstr. in *Glycoconjugate J.*, 1995, **12**, 473.
173. H. Puthalakath, J. Burke, and P. A. Gleeson, *J. Biol. Chem.*, 1996, **271**, 27818.
174. R. Wilson, R. Ainscough, K. Anderson, C. Baynes, M. Berks, J. Bonfield, J. Burton, M. Connell, T. Copsey, J. Cooper, A. Coulson, M. Craxton, S. Dear, Z. Du, R. Durbin, A. Favello, L. Fulton, A. Gardner, P. Green, T. Hawkins, L. Hillier, M. Jier, L. Johnston, M. Jones, J. Kershaw, J. Kirsten, P. Laister, P. Latreille, J. Lightning, C. Lloyd, A. McMurray, B. Mortimore, M. O'Callaghan, J. Parsons, C. Percy, L. Rifken, A. Roopra, D. Saunders, R. Shownkeen, N. Smaldon, A. Smith, E. Sonnhammer, R. Staden, J. Sulston, J. Thierry-Mieg, K. Thomas, M. Vaudin, K. Vaughan, R. Waterston, A. Watson, L. Weinstock, J. Wilkinson-Sproat, and P. Wohldman, *Nature*, 1994, **368**, 32.
175. J. Tan, C. A. F. D'Agostaro, B. Bendiak, F. Reck, M. Sarkar, J. A. Squire, P. Leong, and H. Schachter, *Eur. J. Biochem.*, 1995, **231**, 317.
176. G. A. F. D'Agostaro, A. Zingoni, R. L. Moritz, R. J. Simpson, H. Schachter and B. Bendiak, *J. Biol. Chem.* 1995, **270**, 15211.
177. Y. Ihara, A. Nishikawa, T. Tohma, H. Soejima, N. Niikawa, and N. Taniguchi, *J. Biochem. (Tokyo)*, 1993, **113**, 692.
178. A. Nishikawa, Y. Ihara, M. Hatakeyama, K. Kangawa, and N. Taniguchi, *J. Biol. Chem.*, 1992, **267**, 18199.
179. M. Bhaumik, M. F. Seldin, and P. Stanley, *Gene*, 1995, **164**, 295.
180. J. J. Priatel, M. Sarkar, H. Schachter, and J. D. Marth, *Glycobiology*, 1997, **7**, 45.
181. P. A. Gleeson and H. Schachter, *J. Biol. Chem.*, 1983, **258**, 6162.
182. M. T. Minowa, A. Yoshida, T. Hara, A. Iwamatsu, S. Oguri, H. Ikenaga, and M. Takeuchi, in "International Symposium on Glycosyltransferases and Cellular Communications," Ministry of Education, Science, Sports and Culture, Osaka, Japan, 1997, p. 125.
183. H. Saito, A. Nishikawa, J. G. Gu, Y. Ihara, H. Soejima, Y. Wada, C. Sekiya, N. Niikawa, and N. Taniguchi, *Biochem. Biophys. Res. Commun.*, 1994, **198**, 318.
184. M. Shoreibah, G. S. Perng, B. Adler, J. Weinstein, R. Basu, R. Cupples, D. Wen, J. K. Browne, P. Buckhaults, N. Fregien, and M. Pierce, *J. Biol. Chem.*, 1993, **268**, 15381.
185. J. Weinstein, S. Sundaram, X. H. Wang, D. Delgado, R. Basu and P. Stanley, *J. Biol. Chem.*, 1996, **271**, 27462.
186. I. Brockhausen, E. Hull, O. Hindsgaul, H. Schachter, R. N. Shah, S. W. Michnick, and J. P. Carver, *J. Biol. Chem.*, 1989, **264**, 11211.
187. N. Taniguchi and Y. Ihara, *Glycoconjugate J.*, 1995, **12**, 733.
188. Y. Nishikawa, W. Pegg, H. Paulsen, and H. Schachter, *J. Biol. Chem.*, 1988, **263**, 8270.
189. G. Möller, F. Reck, H. Paulsen, K. J. Kaur, M. Sarkar, H. Schachter, and I. Brockhausen, *Glycoconjugate J.*, 1992, **9**, 180.
190. F. Reck, M. Springer, E. Meinjohanns, H. Paulsen, I. Brockhausen, and H. Schachter, *Glycoconjugate J.*, 1995, **12**, 747.
191. M. H. Hoe, P. Slusarewicz, T. Misteli, R. Watson and G. Warren, *J. Biol. Chem.*, 1995, **270**, 25057.
192. M. Sarkar, *Glycoconjugate J.*, 1994, **11**, 204.
193. F. Reck, M. Springer, H. Paulsen, I. Brockhausen, M. Sarkar, and H. Schachter, *Carbohydr. Res.*, 1994, **259**, 93.
194. F. Reck, E. Meinjohanns, M. Springer, R. Wilkens, J. A. L. M. Van Dorst, H. Paulsen, G. Möller, I. Brockhausen, and H. Schachter, *Glycoconjugate J.*, 1994, **11**, 210.
195. B. Yip, S. H. Chen, H. Mulder, J. W. M. Hoppener, and H. Schachter, *Biochem. J.*, 1997, **321**, 465.
196. B. Bendiak and H. Schachter, *J. Biol. Chem.*, 1987, **262**, 5775.
197. B. Bendiak and H. Schachter, *J. Biol. Chem.*, 1987, **262**, 5784.
198. F. Reck, E. Meinjohanns, J. Tan, A. A. Grey, H. Paulsen, and H. Schachter, *Carbohydr. Res.*, 1995, **275**, 221.
199. J. Jaeken, H. Schachter, H. Carchon, P. DeCock, B. Coddeville, and G. Spik, *Arch. Dis. Child.*, 1994, **71**, 123.
200. J. Jaeken, G. Spik, and H. Schachter, in "Glycoproteins and Disease," eds. J. Montreuil, J. F. G. Vliegenthart, and H. Schachter, Elsevier, Amsterdam, 1996, p. 457.
201. J. H. M. Charuk, J. Tan, M. Bernardini, S. Haddad, R. A. F. Reithmeier, J. Jaeken, and H. Schachter, *Eur. J. Biochem.*, 1995, **230**, 797.
202. J. Tan, J. Dunn, J. Jaeken, and H. Schachter, *Am. J. Hum. Genet.*, 1996, **59**, 810.
203. J. Jaeken, H. Carchon and H. Stibler, *Glycobiology*, 1993, **3**, 423.
204. M. Metzler, A. Gertz, M. Sarkar, H. Schachter, J. W. Schrader, and J. D. Marth, *EMBO J.*, 1994, **13**, 2056.
205. E. Ioffe and P. Stanley, *Proc. Natl. Acad. Sci. USA*, 1994, **91**, 728.
206. R. M. Campbell, M. Metzler, M. Granovsky, J. W. Dennis, and J. D. Marth, *Glycobiology*, 1995, **5**, 535.
207. S. Chen, J. Tan and H. Schachter, in "24th Meeting of the Society for Glycobiology, Boston, MA, November, 1996." Abstr. in *Glycobiology*, 1996, **6**, 760.
208. S. Chen, S. Zhou, J. Tan, and H. Schachter, *Glycoconjugate J.*, 1998, in press.
209. Y. J. Kim, J. H. Park, K. S. Kim, J. E. Chang, J. H. Ko, M. H. Kim, D. H. Chung, T. W. Chung, I. S. Choe, Y. C. Lee, and C. H. Kim, *Gene*, 1996, **170**, 281.
210. N. Koyama, E. Miyoshi, Y. Ihara, R. J. Kang, A. Nishikawa, and N. Taniguchi, *Eur. J. Biochem.*, 1996, **238**, 853.

211. S. Narasimhan, H. Schachter and S. Rajalakshmi, *J. Biol. Chem.*, 1988, **263**, 1273.
212. K. Ishibashi, A. Nishikawa, N. Hayashi, A. Kasahara, N. Sato, S. Fujii, T. Kamada, and N. Taniguchi, *Clin. Chim. Acta*, 1989, **185**, 325.
213. A. Nishikawa, J. Gu, S. Fujii and N. Taniguchi, *Biochim. Biophys. Acta*, 1990, **1035**, 313.
214. E. Miyoshi, A. Nishikawa, Y. Ihara, J. Gu, T. Sugiyama, N. Hayashi, H. Fusamoto, T. Kamada, and N. Taniguchi, *Cancer Res.*, 1993, **53**, 3899.
215. M. Yoshimura, A. Nishikawa, Y. Ihara, T. Nishiura, H. Nakao, Y. Kanayama, Y. Matuzawa, and N. Taniguchi, *Int. J. Cancer*, 1995, **60**, 443.
216. M. Yoshimura, Y. Ihara, and N. Taniguchi, *Glycoconjugate J.*, 1995, **12**, 234.
217. E. Miyoshi, Y. Ihara, N. Hayashi, H. Fusamoto, T. Kamada, and N. Taniguchi, *J. Biol. Chem.*, 1995, **270**, 28 311.
218. M. Yoshimura, Y. Ihara, A. Ohnishi, N. Ijuhin, T. Nishiura, Y. Kanakura, Y. Matsuzawa, and N. Taniguchi, *Cancer Res.*, 1996, **56**, 412.
219. M. Yoshimura, A. Nishikawa, Y. Ihara, S. Taniguchi, and N. Taniguchi, *Proc. Natl. Acad. Sci. USA*, 1995, **92**, 8754.
220. M. Yoshimura, Y. Ihara, Y. Matsuzawa and N. Taniguchi, *J. Biol. Chem.*, 1996, **271**, 13 811.
221. A. Rebbaa, H. Yamamoto, T. Saito, E. Meuillet, P. Kim, D. S. Kersey, E. G. Bremer, N. Taniguchi, and J. R. Moskal, *J. Biol. Chem.*, 1997, **272**, 9275.
222. Y. Ihara, Y. Sakamoto, M. Mihara, K. Shimizu, and N. Taniguchi, *J. Biol. Chem.*, 1997, **272**, 9629.
223. M. Tanemura, S. Miyagawa, Y. Ihara, S. Mikata, H. Matsuda, R. Shirakura, and N. Taniguchi, *Transplant. Proc.*, 1997, **29**, 891.
224. J. P. Carver and J.-R. Brisson, in "Biology of Carbohydrates," eds. V. Ginsburg and P. W. Robbins, Wiley, New York, 1984, vol. 2, p. 289.
225. J. P. Carver, *Biochem. Soc. Trans.*, 1984, **12**, 517.
226. J. P. Carver and D. A. Cumming, *Pure Appl. Chem.*, 1987, **59**, 1465.
227. A. S. Sultan, E. Miyoshi, Y. Ihara, A. Nishikawa, Y. Tsukada, and N. Taniguchi, *J. Biol. Chem.*, 1997, **272**, 2866.
228. M. Bhaumik, S. Sundaram, and P. Stanley, in "24th Meeting of the Society for Glycobiology, Boston, MA, November, 1996." Abstr. in *Glycobiology*, 1996, **6**, 720.
229. M. Tanemura, S. Miyagawa, Y. Ihara, A. Nishikawa, M. Suzuki, K. Yamamura, H. Matsuda, R. Shirakura, and N. Taniguchi, *Transplant. Proc.*, 1997, **29**, 895.
230. I. Brockhausen, G. Möller, J. M. Yang, S. H. Khan, K. L. Matta, H. Paulsen, A. A. Grey, R N. Shah, and H. Schachter, *Carbohydr. Res.*, 1992, **236**, 281.
231. S. Oguri, M. T. Minowa, Y. Ihara, N. Taniguchi, H. Ikenaga, and M. Takeuchi, in "International Symposium on Glycosyltransferases and Cellular Communications," Ministry of Education, Science, Sports and Culture, Osaka, Japan, 1997, p. 124.
232. N. Zhang, K. C. Peng, L. Chen, D. Puett, and M. Pierce, *J. Biol. Chem.*, 1997, **272**, 4225.
233. M. M. Palcic, J. Ripka, K. J. Kaur, M. Shoreibah, O. Hindsgaul, and M. Pierce, *J. Biol. Chem.*, 1990, **265**, 6759.
234. R. D. Cummings, I. S. Trowbridge, and S. Kornfeld, *J. Biol. Chem.*, 1982, **257**, 13 421.
235. O. Hindsgaul, S. H. Tahir, O. P. Srivastava, and M. Pierce, *Carbohydr. Res.*, 1988, **173**, 263.
236. O. Hindsgaul, K. J. Kaur, G. Srivastava, M. Blaszczyk-Thurin, S. C. Crawley, L. D. Heerze, and M. M. Palcic, *J. Biol. Chem.*, 1991, **266**, 17 858.
237. J. Gu, A. Nishikawa, N. Tsuruoka, M. Ohno, N. Yamaguchi, K. Kanagawa, and N. Taniguchi, *J. Biochem. (Tokyo)*, 1993, **113**, 614.
238. O. P. Srivastava, O. Hindsgaul, M. Shoreibah, and M. Pierce, *Carbohydr. Res.*, 1988, **179**, 137.
239. S. C. Crawley, O. Hindsgaul, G. Alton, M. Pierce, and M. M. Palcic, *Anal. Biochem.*, 1990, **185**, 112.
240. I. Brockhausen, J. P. Carver and H. Schachter, *Biochem. Cell Biol.*, 1988, **66**, 1134.
241. I. Brockhausen, F. Reck, W. Kuhns, S. Khan, K. L. Matta, E. Meinjohanns, H. Paulsen, R. N. Shah, M. A. Baker, and H. Schachter, *Glycoconjugate J.*, 1995, **12**, 371.
242. O. Kanie, S. C. Crawley, M. M. Palcic, and O. Hindsgaul, *Carbohydr. Res.*, 1993, **243**, 139.
243. S. H. Khan, S. C. Crawley, O. Kanie and O. Hindsgaul, *J. Biol. Chem.*, 1993, **268**, 2468.
244. S. H. Khan and K. L. Matta, *Carbohydr. Res.* 1993, **243**, 29.
245. S. H. Khan and K. L. Matta, *J. Carbohydr. Chem.*, 1993, **12**, 335.
246. S. H. Khan and K. L. Matta, *Carbohydr. Res.*, 1995, **278**, 351.
247. T. Linker, S. C. Crawley, and O. Hindsgaul, *Carbohydr. Res.*, 1993, **245**, 323.
248. M. G. Shoreibah, O. Hindsgaul, and M. Pierce, *J. Biol. Chem.*, 1992, **267**, 2920.
249. J. W. Dennis, S. Laferte, C. Waghorne, M. L. Breitman, and R. S. Kerbel, *Science*, 1987, **236**, 582.
250. J. W. Dennis and S. Laferte, in "UCLA Symposia on Molecular and Cellular Biology, New Series, Altered Glycosylation in Tumor Cells," eds. C. L. Reading, S.-I. Hakomori, and D. M. Marcus, Alan R. Liss, New York, 1988, p. 257.
251. L. A. Smets and W. P. Van Beek, *Biochim. Biophys. Acta*, 1984, **738**, 237.
252. S. Yagel, R. Feinmesser, C. Waghorne, P. K. Lala, M. L. Breitman, and J. W. Dennis, *Int. J. Cancer*, 1989, **44**, 685.
253. S. Yousefi, E. Higgins, Z. Daoling, A. Pollex-Krüger, O. Hindsgaul, and J. W. Dennis, *J. Biol. Chem.*, 1991, **266**, 1772.
254. B. Fernandes, U. Sagman, M. Auger, M. Demetrio, and J. W. Dennis, *Cancer Res.*, 1991, **51**, 718.
255. M. Demetriou, I. R. Nabi, M. Coppolino, S. Dedhar, and J. W. Dennis, *J. Cell Biol.*, 1995, **130**, 383.
256. M. Asada, K. Furukawa, K. Segawa, T. Endo, and A. Kobata, *Cancer Res.*, 1997, **57**, 1073.
257. J. W. Dennis, K. Kosh, D.-M. Bryce, and M. L. Breitman, *Oncogene*, 1989, **4**, 853.
258. J. W. Dennis and S. Laferte, *Cancer Res.*, 1989, **49**, 945.
259. E. W. Easton, I. Blokland, A. A. Geldof, B. R. Rao, and D. H. Van den Eijnden, *FEBS Lett.*, 1992, **308**, 46.
260. K. Yamashita, Y. Tachibana, T. Ohkura, and A. Kobata, *J. Biol. Chem.*, 1985, **260**, 3963.
261. J. Arango and M. Pierce, *J. Cell. Biochem.*, 1988, **37**, 225.
262. Y. Lu and W. Chaney, *Mol. Cell Biochem.*, 1993, **122**, 85.
263. M. Pierce and J. Arango, *J. Biol. Chem.*, 1986, **261**, 10 772.
264. E. W. Easton, J. G. M. Bolscher, and D. H. Van den Eijnden, *J. Biol. Chem.*, 1991, **266**, 21 674.
265. E. Miyoshi, A. Nishikawa, Y. Ihara, H. Saito, N. Uozumi, N. Hayashi, H. Fusamoto, T. Kamada, and N. Taniguchi, *J. Biol. Chem.*, 1995, **270**, 6216.
266. R. D. Cummings and S. Kornfeld, *J. Biol. Chem.*, 1984, **259**, 6253.

267. D. H. van den Eijnden, A. H. L. Koenderman, and W. E. C. M. Schiphorst, *J. Biol. Chem.*, 1988, **263**, 12461.
268. T.-Z. Ju, H.-L. Chen, J.-X. Gu, and H. Qin, *Glycoconjugate J.*, 1995, **12**, 767.
269. Y. Guilloux, S. Lucas, V. G. Brichard, A. Van Pel, C. Viret, E. DePlaen, F. Brasseur, B. Lethe, F. Jotereau, and T. Boon, *J. Exp. Med.*, 1996, **183**, 1173.
270. L. Chen, N. Zhang, B. Adler, J. Browne, N. Freigen, and M. Pierce, *Glycoconjugate J.*, 1995, **12**, 813.
271. H. Saito, J. G. Gu, A. Nishikawa, Y. Ihara, J. Fujii, Y. Kohgo, and N. Taniguchi, *Eur. J. Biochem.*, 1995, **233**, 18.
272. B. Wasylyk, S. L. Hahn, and A. Giovane, *Eur. J. Biochem.*, 1993, **211**, 7.
273. R. Kang, H. Saito, Y. Ihara, E. Miyoshi, N. Koyama, Y. Sheng, and N. Taniguchi, *J. Biol. Chem.*, 1996, **271**, 26706.
274. T. Taguchi, K. Kitajima, S. Inoue, Y. Inoue, J. M. Yang, H. Schachter, and I. Brockhausen, *Biochem. Biophys. Res. Commun.*, 1997, **230**, 533.
275. T. Feizi, *Biochem. Soc. Trans.*, 1988, **16**, 930.
276. T. Feizi, *Environ. Health Perspect.*, 1990, **88**, 231.
277. H. C. Gooi, T. Feizi, A. Kapadia, B. B. Knowles, D. Solter, and M. J. Evans, *Nature*, 1981, **292**, 156.
278. S.-I. Hakomori, *Ann. Rev. Biochem.*, 1981, **50**, 733.
279. S.-I. Hakomori, *Adv. Cancer Res.*, 1989, **52**, 257.
280. R. J. Kelly, L. K. Ernst, R. D. Larsen, J. G. Bryant, J. S. Robinson, and J. B. Lowe, *Proc. Natl. Acad. Sci. USA*, 1994, **91**, 5843.
281. J. Le Pendu, J. P. Cartron, R. U. Lemieux, and R. Oriol, *Am. J. Hum. Genet.*, 1985, **37**, 749.
282. A. Sarnesto, T. Köhlin, O. Hindsgaul, J. Thurin, and M. Blaszczyk-Thurin, *J. Biol. Chem.*, 1992, **267**, 2737.
283. S. Hitoshi, S. Kusunoki, I. Kanazawa, and S. Tsuji, *J. Biol. Chem.*, 1995, **270**, 8844.
284. S. Hitoshi, S. Kusunoki, I. Kanazawa, and S. Tsuji, *J. Biol. Chem.*, 1996, **271**, 16975.
285. R. Oriol, *J. Immunogenet.*, 1990, **17**, 235.
286. R. Mollicone, A. M. Dalix, A. Jacobsson, B. E. Samuelsson, G. Gerard, K. Crainic, T. Caillard, J. Le Pendu, and R. Oriol, *Glycoconjugate J.*, 1988, **5**, 499.
287. S. Rouquier, J. B. Lowe, R. J. Kelly, A. L. Fertitta, G. C. Lennon, and D. Giorgi, *J. Biol. Chem.*, 1995, **270**, 4632.
288. I. Reguigne Arnould, P. Couillin, R. Mollicone, S. Faure, A. Fletcher, R. J. Kelly, J. B. Lowe, and R. Oriol, *Cytogenet. Cell Genet.*, 1995, **71**, 158.
289. J. F. Kukowska Latallo, R. D. Larsen, R. P. Nair, and J. B. Lowe, *Genes Dev.*, 1990, **4**, 1288.
290. H. S. Cameron, D. Szczepaniak, and B. W. Weston, *J. Biol. Chem.*, 1995, **270**, 20112.
291. S. Nishihara, M. Nakazato, T. Kudo, H. Kimura, T. Ando, and H. Narimatsu, *Biochem. Biophys. Res. Commun.*, 1993, **190**, 42.
292. S. Nishihara, H. Narimatsu, H. Iwasaki, S. Yazawa, S. Akamatsu, T. Ando, T. Seno, and I. Narimatsu, *J. Biol. Chem.*, 1994, **269**, 29271.
293. S. Nishihara, S. Yazawa, H. Iwasaki, M. Nakazato, T. Kudo, T. Ando, and H. Narimatsu, *Biochem. Biophys. Res. Commun.*, 1993, **196**, 624.
294. A. Oulmouden, A. Wierinckx, J. M. Petit, and R. Julien, *EMBL/GenBank listing*, 1995, AN: X87810.
295. K. P. Lee, L. M. Carlson, J. B. Woodcock, N. Ramachandra, T. L. Schultz, T. A. Davis, J. B. Lowe, C. B. Thompson, and R. D. Larsen, *J. Biol. Chem.*, 1996, **271**, 32960.
296. J. B. Lowe, J. F. Kukowska-Latallo, R. P. Nair, R. D. Larsen, R. M. Marks, B. A. Macher, R. J. Kelly, and L. K. Ernst, *J. Biol. Chem.*, 1991, **266**, 17467.
297. R. Kumar, B. Potvin, W. A. Muller, and P. Stanley, *J. Biol. Chem.*, 1991, **266**, 21777.
298. S. E. Goelz, C. Hession, D. Goff, B. Griffiths, R. Tizard, B. Newman, G. Chi-Rosso, and R. Lobb, *Cell*, 1990, **63**, 1349.
299. K. M. Gersten, S. Natsuka, M. Trinchera, B. Petryniak, R. J. Kelly, N. Hiraiwa, N. A. Jenkins, D. J. Gilbert, N. G. Copeland, and J. B. Lowe, *J. Biol. Chem.*, 1995, **270**, 25047.
300. M. Ozawa and T. Muramatsu, *J. Biochem. (Tokyo)*, 1996, **119**, 302.
301. E. M. Sajdel-Sulkowska, F. I. Smith, G. Wiederchain, and R. H. McCluer, *EMBL/GenBank listing*, 1996, AN: K58860.
302. B. W. Weston, R. P. Nair, R. D. Larsen, and J. B. Lowe, *J. Biol. Chem.*, 1992, **267**, 4152.
303. B. W. Weston, P. I. Smith, R. J. Kelly, and J. B. Lowe, *J. Biol. Chem.*, 1992, **267**, 24575.
304. K. L. Koszdin, and B. R. Bowen, *Biochem. Biophys. Res. Commun.*, 1992, **187**, 152.
305. K. Sasaki, K. Kurata, K. Funayama, M. Nagata, E. Watanabe, S. Ohta, N. Hanai, and T. Nishi, *J. Biol. Chem.*, 1994, **269**, 14730.
306. S. Natsuka, K. M. Gersten, K. Zenita, R. Kannagi, and J. B. Lowe, *J. Biol. Chem.*, 1994, **269**, 16789.
307. P. L. Smith, K. M. Gersten, B. Petryniak, R. J. Kelly, C. Rogers, Y. Natsuka, J. A. Alford III, E. P. Scheidegger, S. Natsuka, and J. B. Lowe, *J. Biol. Chem.*, 1996, **271**, 8250.
308. R. D. Larsen, L. K. Ernst, R. P. Nair, and J. B. Lowe, *Proc. Natl. Acad. Sci. USA*, 1990, **87**, 6674.
309. J. P. Piau, N. Labarriere, G. Dabouis, and M. G. Denis, *Biochem. J.*, 1994, **300**, 623.
310. S. Cohney, E. Mouhtouris, I. F. C. McKenzie, and M. S. Sandrin, *Immunogenetics*, 1996, **44**, 76.
311. R. J. Kelly, S. Rouquier, D. Giorgi, G. G. Lennon, and J. B. Lowe, *J. Biol. Chem.*, 1995, **270**, 4640.
312. N. Uozumi, S. Yanagidani, E. Miyoshi, Y. Ihara, T. Sakuma, C.-X. Gao, T. Teshima, S. Fujii, T. Shiba, and N. Taniguchi, *J. Biol. Chem.*, 1996, **271**, 27810.
313. E. Staudacher, F. Altmann, J. Glössl, L. März, H. Schachter, J. P. Kamerling, K. Hård, and J. F. G. Vliegenthart, *Eur. J. Biochem.*, 1991, **199**, 745.
314. T. Kudo, H. Iwasaki, S. Nishihara, N. Shinya, T. Ando, I. Narimatsu, and H. Narimatsu, *J. Biol. Chem.*, 1996, **271**, 9830.
315. S. Henry, R. Mollicone, J. B. Lowe, B. Samuelsson, and G. Larson, *Vox Sang.*, 1996, **70**, 21.
316. S. Henry, R. Mollicone, P. Fernandez, B. Samuelsson, R. Oriol, and G. Larson, *Glycoconjugate J.*, 1996, **13**, 985.
317. L. C. Yu, Y. H. Yang, R. E. Broadberry, Y. H. Chen, Y. S. Chan, and M. Lin, *Biochem. J.*, 1995, **312**, 329.
318. A. Etzioni, M. Frydman, S. Pollack, I. Avidor, M. L. Phillips, J. C. Paulson, and R. Gershoni-Baruch, *N. Engl. J. Med.*, 1992, **327**, 1789.
319. U. H. Von Andrian, E. M. Berger, L. Ramezani, J. D. Chambers, H. D. Ochs, J. M. Harlan, J. C. Paulson, A. Etzioni, and K. E. Arfors, *J. Clin. Invest.*, 1993, **91**, 2893.
320. M. L. Phillips, B. R. Schwartz, A. Etzioni, R. Bayer, H. D. Ochs, J. C. Paulson, and J. M. Harlan, *J. Clin. Invest.*, 1995, **96**, 2898.

321. A. Etzioni, L. M. Phillips, J. C. Paulson, and J. M. Harlan, in "Cell Adhesion and Human Disease," eds. J. Marsh and J. A. Goode, Wiley, Chichester, 1995, p. 51.
322. Y. Shechter, A. Etzioni, C. Levene, and P. Greenwell, *Transfusion*, 1995, **35**, 773.
323. P. A. Prieto, P. Mukerji, B. Kelder, R. Erney, D. Gonzalez, J. S. Yun, D. F. Smith, K. W. Moremen, C. Nardelli, M. Pierce, Y. S. Li, X. Chen, T. E. Wagner, R. D. Cummings, and J. J. Kopchick, *J. Biol. Chem.*, 1995, **270**, 29 515.
324. R. Mollicone, J. J. Candelier, B. Mennesson, P. Couillin, A. P. Venot, and R. Oriol, *Carbohydr. Res.*, 1992, **228**, 265.
325. R. Mollicone, A. Gibaud, A. François, M. Ratcliffe, and R. Oriol, *Eur. J. Biochem.*, 1990, **191**, 169.
326. P. A. T. Tetteroo, H. T. de Heij, D. H. van den Eijnden, F. J. Visser, E. Schoenmarker, and A. H. M. G. van Kessel, *J. Biol. Chem.*, 1987, **262**, 15 984.
327. P. Couillin, R. Mollicone, M. C. Grisard, A. Gibaud, N. Ravisé, J. Feingold, and R. Oriol, *Cytogenet. Cell Genet.*, 1991, **56**, 108.
328. B. A. Macher, E. H. Holmes, S. J. Swiedler, C. L. M. Stults, and C. A. Srnka, *Glycobiology*, 1991, **1**, 577.
329. E. V. Chandrasekaran, R. K. Jain, R. D. Larsen, K. Wlasichuk, R. A. Di Cioccio, and K. L. Matta, *Biochemistry*, 1996, **35**, 8925.
330. R. S. McCurley, A. Recinos III, A. S. Olsen, J. C. Gingrich, D. Szczepaniak, H. S. Cameron, R. Krauss, and B. W. Weston, *Genomics*, 1995, **26**, 142.
331. R. Mollicone, I. Reguigne, R. J. Kelly, A. Fletcher, J. Watt, S. Chatfield, A. Aziz, H. S. Cameron, B. W. Weston, J. B. Lowe, and R. Oriol, *J. Biol. Chem.*, 1994, **269**, 20 987.
332. T. F. Orntoft, E. M. Vestergaard, E. Holmes, J. S. Jakobsen, N. Grunnet, M. Mortensen, P. Johnson, P. Bross, N. Gregersen, K. Skorstengaard, U. B. Jensen, L. Bolund, and H. Wolf, *J. Biol. Chem.*, 1996, **271**, 32 260.
333. D. J. Legault, R. J. Kelly, Y. Natsuka, and J. B. Lowe, *J. Biol. Chem.*, 1995, **270**, 20 987.
334. X. H. Xu, L. Vo and B. A. Macher, *J. Biol. Chem.*, 1996, **271**, 8818.
335. G. Walz, A. Aruffo, W. Kolanus, M. Bevilacqua, and B. Seed, *Science*, 1990, **250**, 1132.
336. J. B. Lowe, L. M. Stoolman, R. P. Nair, R. D. Larsen, T. L. Behrend, and R. M. Marks, *Cell*, 1990, **63**, 475.
337. J. B. Lowe, L. M. Stoolman, R. P. Nair, R. D. Larsen, T. L. Behrend, and R. M. Marks, *Biochem. Soc. Trans.*, 1991, **19**, 649.
338. M. L. Phillips, E. Nudelman, F. C. Gaeta, M. Perez, A. K. Singhal, S. Hakomori, and J. C. Paulson, *Science*, 1990, **250**, 1130.
339. M. J. Polley, M. L. Phillips, E. Wayner, E. Nudelman, A. K. Singhal, S. Hakomori, and J. C. Paulson, *Proc. Natl. Acad. Sci. USA*, 1991, **88**, 6224.
340. B. K. Brandley, S. J. Swiedler, and P. W. Robbins, *Cell*, 1990, **63**, 861.
341. S. Sueyoshi, S. Tsuboi, R. Sawada-Itirai, U. N. Dang, J. B. Lowe, and M. Fukuda, *J. Biol. Chem.*, 1994, **269**, 32 342.
342. J. L. Clarke and W. M. Watkins, *J. Biol. Chem.*, 1996, **271**, 10 317.
343. R. N. Knibbs, R. A. Craig, S. Natsuka, A. Chang, M. Cameron, J. B. Lowe, and L. M. Stoolman, *J. Cell Biol.*, 1996, **133**, 911.
344. I. Reguigne, M. R. James, C. W. Richard III, R. Mollicone, A Seawright, J. B. Lowe, R. Oriol, and P. Couillin, *Cytogenet. Cell Genet.*, 1994, **66**, 104.
345. A. Sarnesto, T. Köhlin, O. Hindsgaul, K. Vogele, M. Blaszczyk-Thurin, and J. Thurin, *J. Biol. Chem.*, 1992, **267**, 2745.
346. R. Mollicone, I. Reguigne, A. Fletcher, A. Aziz, M. Rustam, B. W. Weston, R. J. Kelly, J. B. Lowe, and R. Oriol, *J. Biol. Chem.*, 1994, **269**, 12 662.
347. E. C. M. Brinkman-Van der Linden, R. Mollicone, R. Oriol, G. Larson, D. H. Van den Eijnden, and W. Van Dijk, *J. Biol. Chem.*, 1996, **271**, 14 492.
348. T. De Vries, C. A. Srnka, M. M. Palcic, S. J. Swiedler, D. H. Van den Eijnden, and B. A. Macher, *J. Biol. Chem.*, 1995, **270**, 8712.
349. S. Natsuka, K. M. Gersten, K. Zenita, R. Kannagi, and J. B. Lowe, *J. Biol. Chem.*, 1994, **269**, 20 806.
350. P. Maly, A. D. Thall, B. Petryniak, C. E. Rogers, P. L. Smith, R. M. Marks, R. J. Kelly, K. M. Gersten, G. Y. Cheng, T. L. Saunders, S. A. Camper, R. T. Camphausen, F. X. Sullivan, Y. Isogai, O. Hindsgaul, U. H. Von Andrian, and J. B. Lowe, *Cell*, 1996, **86**, 643.
351. A. J. Wagers, J. B. Lowe, and G. S. Kansas, *Blood*, 1996, **88**, 2125.
352. N. Hiraiwa, M. Hiraiwa, and R. Kannagi, *Biochem. Biophys. Res. Commun.*, 1997, **231**, 183.
353. G. D. Longmore and H. Schachter, *Carbohydr. Res.*, 1982, **100**, 365.
354. M. C. Shao, C. W. Sokolik, and F. Wold, *Carbohydr. Res.*, 1994, **251**, 163.
355. F. Altmann, G. Kornfeld, T. Dalik, E. Staudacher, and J. Glossl, *Glycobiology*, 1993, **3**, 619.
356. J. A. Voynow, R. S. Kaiser, T. F. Scanlin, and M. C. Glick, *J. Biol. Chem.*, 1991, **266**, 21 572.
357. E. Staudacher, T. Dalik, P. Wawra, F. Altmann, and L. März, *Glycoconjugate J.*, 1995, **12**, 780.
358. S. Yanagidani, N. Uozumi, Y. Ihara, E. Miyoshi, N. Yamaguchi, and N. Taniguchi, *J. Biochem. (Tokyo)*, 1997, **121**, 626.
359. E. Staudacher, F. Altmann, L. März, K. Hård, J. P. Kamerling, and J. F. G. Vliegenthart, *Glycoconjugate J.*, 1992, **9**, 82.
360. V. Kubelka, F. Altmann, E. Staudacher, V. Tretter, L. März, K. Hård, J. P. Kamerling, and J. F. G. Vliegenthart, *Eur. J. Biochem.*, 1993, **213**, 1193.
361. V. Kubelka, F. Altmann, and L. Marz, *Glycoconjugate J.*, 1995, **12**, 77.
362. L. März, F. Altmann, E. Staudacher, and V. Kubelka, in "Glycoproteins," eds. J. Montreuil, J. F. G. Vliegenthart, and H. Schachter, Elsevier, Amsterdam, 1995, p. 543.
363. M. Fukuda, M. F. A. Bierhuizen, and J. Nakayama, *Glycobiology*, 1996, **6**, 683.
364. M. Sarkar, S. Zhou, S. Chen, A. Spence, and H. Schachter, in "International Symposium on Glycosyltransferases and Cellular Communications," Ministry of Education, Science, Sports and Culture, Osaka, Japan, 1997, p. 105.
365. C. E. Warren, P. J. Roy, J. G. Culotti, and J. W. Dennis, in "24th Meeting of the Society for Glycobiology, Boston, MA, November, 1996." Abstr. in *Glycobiology*, 1996, **6**, 759.
366. F. K. Hagen, C. A. Gregoire and L. A. Tabak, *Glycoconjugate J.*, 1995, **12**, 901.
367. N. Koyama, E. Miyoshi, Y. Ihara, A. Nishikawa, and N. Taniguchi, in "XIIIth International Symposium on Glyco-conjugates, Seattle, WA, August, 1995." Abstr. in *Glycoconjugate J.*, 1995, **12**, 475.
368. G.-S. Perng, M. Shoreibah, I. Margitich, M. Pierce, and N. Fregien, *Glycobiology*, 1994, **4**, 867.

369. M. F. A. Bierhuizen, K. Maemura, S. Kudo, and M. Fukuda, *Glycobiology*, 1995, **5**, 417.
370. A. Varki, *Glycobiology*, 1993, **3**, 97.
371. A. Varki, *J. Clin. Invest.*, 1997, **99**, 158.
372. P. Stanley, *Glycobiology*, 1992, **2**, 99.
373. P. Stanley and E. Ioffe, *FASEB J.*, 1995, **9**, 1436.
374. J. D. Marth, *Glycoconjugate J.*, 1994, **11**, 3.
375. S. Kornfeld, *Biochem. Soc. Trans.*, 1990, **18**, 367.
376. E. Van Schaftingen and J. Jaeken, *FEBS Lett.*, 1995, **377**, 318.
377. M. N. Fukuda, K. A. Masri, A. Dell, L. Luzzatto, and K. W. Moremen, *Proc. Natl. Acad. Sci. USA*, 1990, **87**, 7443.
378. M. N. Fukuda, *Glycobiology*, 1990, **1**, 9.

3.04
Glycosyltransferases Involved in the Synthesis of Ser/Thr-GalNAc O-Glycans

INKA BROCKHAUSEN

University of Toronto and Hospital for Sick Children, ON, Canada

3.04.1 INTRODUCTION

An extraordinary variety of complex *O*-glycan chains occur on glycoproteins. Many functions have been proposed for these *O*-glycans, including roles in cell adhesion, in the immune system, fertilization, protection, and lubrication. Glycosyltransferases assembling these sugar chains appear to exist as families of homologous enzymes. The early-acting glycosyltransferases are usually specific for *O*-glycans, while later-acting enzymes may also be involved in the synthesis of *N*-glycans and glycolipids with similar structures. These enzyme activities are expressed in a tissue-specific and developmentally regulated fashion, and often change in disease. Although several glycosyltransferases have been extensively studied, many more enzymes involved in the biosynthesis of the core structures and extended backbone of *O*-glycans remain to be characterized.

Carbohydrate chains may be linked to the Ser or Thr residues of proteins via the *O*-glycosidic linkages. Several different types of these *O*-linkages exist, including Manα-*O*-Ser/Thr in yeast and mammalian glycoproteins, GlcNAcβ-*O*-Ser/Thr (GlcNAc denotes *N*-acetylglucosamine) in nuclear and cytoplasmic proteins, and Glcβ-*O*-Ser/Thr and Fucα-*O*-Ser/Thr in blood clotting factors. In addition, Xylβ-*O*-Ser is found in proteoglycans and Galβ-*O*-hydroxy-Lys in collagens. This chapter deals with *O*-glycans that are based on GalNAc α-linked to Ser or Thr of glycoproteins. *O*-Glycans are found in secreted and membrane-bound glycoproteins of mammals, birds, fish, insects, frogs, snails, and other species.[1]

The factors controlling the biosynthesis of *O*-glycans have been extensively reviewed,[1-7] and will be summarized here. Mucins are the main class of glycoproteins carrying *O*-glycans, with about 50–80% by weight carbohydrate due to *O*-glycans. A series of glycosyltransferases and sulfotransferases is involved in the synthesis of the hundreds of *O*-glycan chains with different structures found on glycoproteins. The functions of these *O*-glycan chains, however, still remain to be defined more precisely;[8] they may include control of protein folding and conformation, protection from degradation of the underlying carbohydrate and peptide, lubrication of the mucus-secreting internal ducts and intestines, control of antigen expression, provision of ligands for cell adhesion, regulation of immune functions, and control of sperm–egg binding during fertilization. Drastic changes of *O*-glycan structures and the extent of *O*-glycosylation can occur in disease as well as during growth, development, and cellular differentiation. These changes are often associated with different biological behavior of cells. It is thus of foremost importance to understand the control of biosynthesis in normal and diseased cells.

Each of the glycosyltransferases assembling *O*-glycans synthesizes a specific linkage, either between GalNAc and Thr or Ser of the acceptor substrate (α-*O*-linkage) or between two sugars (either α or β linkages). For most glycosyltransferases, the linkages synthesized are specific; however, a number of different acceptor structures may serve as substrates, *in vivo* and *in vitro*, and the nucleotide sugar donor substrate sometimes may be replaced *in vitro* with nucleotide sugars of related structures. The general reaction catalyzed by these enzymes is

nucleotide-(P)-sugar donor + acceptor (HO-R) → sugar-O-R product + nucleotide-(P)

In contrast to Asn-linked *N*-glycans, *O*-glycans are not preassembled on a dolichol derivative in the endoplasmic reticulum[9] but sugars are added individually from nucleotide sugar donors. Glycosidases do not appear to be involved in the processing of *O*-glycans, and the addition of the first sugar, GalNAc (*N*-acetylgalactosamine), to the peptide occurs mainly in the cis-Golgi without a requirement for a specific amino acid sequon. However, the terminal structures of *O*-glycans often resemble those of *N*-glycans and certain glycolipids, although they may be functionally distinct, probably due to characteristic differences in the presentation to their biological environments. These terminal structures may be assembled by the same transferases that act on similar substrate structures of *N*-glycans and glycolipids, while the core structures of *O*-glycans and the early-acting enzymes of the *O*-glycan pathways are usually *O*-glycan-specific.

All *O*-glycan glycosyltransferases cloned thus far are type II membrane glycoproteins, localized in the Golgi. They have a common domain structure, with a short amino terminus directed toward the cytoplasm, a membrane anchor region, a stem region, often rich in Pro and Thr, and a catalytic region at the carboxy terminus that is directed toward the lumen of the Golgi. Enzymes that catalyze similar reactions often occur as families with certain region(s) of homology among them. A small proportion of these membrane-bound transferases may be cleaved, releasing catalytically active soluble enzymes. Sometimes, glycosyltransferases may be found on cell surfaces where they conceivably act as cell adhesion molecules by binding to cell surface glycoprotein substrates.[10,11]

The enzymes synthesizing *O*-glycans (and other complex glycans) are thought to be localized in the Golgi in an assembly line where an enzyme-produced product may serve as a substrate for the subsequent enzymatic step. The organization of this assembly line is not well understood; it must function with high efficiency, since often an almost complete glycosylation is seen on glycoproteins; this may not be reproducible in *in vitro* assays where the Golgi organization is severely disturbed by homogenization or detergent treatment. While the initial *O*-glycosylation takes place mainly in the cis-Golgi compartment, the synthesis of core structures appears to occur in intermediate compartments while terminal reactions can occur in the medial and trans-Golgi. Very few *O*-glycan-specific enzymes have been carefully studied by immunoelectron microscopy; much of the evidence for enzyme localization comes from cellular fractionation or less direct studies.

3.04.2   *O*-GLYCAN STRUCTURES AND FUNCTIONS

The sugars commonly found in *O*-glycans are GalNAc, Gal, GlcNAc, sialic acid, and fucose. Gal and GlcNAc, and possibly other sugars, may be sulfated. Sulfate esters and sialic acids contribute to the acidic properties of *O*-glycans. Glycoproteins from non-mammalian species such as frogs may display a variety of *O*-glycan structures, sometimes containing unusual acidic sugars, and these structures may differ from those of *O*-glycans found in mammalian glycoproteins.

The only sugar common to all *O*-glycans is GalNAc. GalNAcα-Ser/Thr in an unsubstituted form is rare on normal glycoproteins, but is often found in cancer, and has been named the Tn antigen. The sialylated Tn antigen has the structure sialylα2-6-GalNAcα-Ser/Thr, and is found in sub-maxillary mucins and other glycoproteins; it appears in some cancers and is associated with a poor prognosis.[12]

Eight different core structures of *O*-glycans have been described in mammalian mucins:

Core 1:   Galβ1,3 GalNAcα-Ser/Thr-R
Core 2:   GlcNAcβ1,6 (Galβ1,3) GalNAcα-Ser/Thr-R
Core 3:   GlcNAcβ1,3 GalNAcα-Ser/Thr-R
Core 4:   GlcNAcβ1,6 (GlcNAcβ1,3) GalNAcα-Ser/Thr-R
Core 5:   GalNAcα1,3 GalNAcα-Ser/Thr-R
Core 6:   GlcNAcβ1,6 GalNAcα-Ser/Thr-R
Core 7:   GalNAcα1,6 GalNAcα-Ser/Thr-R
Core 8:   Galα1,3 GalNAcα-Ser/Thr-R

The core structures may be substituted by monosaccharides such as sialic acid; they may be elongated by linear or branched Gal-GlcNAc sequences (*N*-acetyllactosamines), or may be gal-actosylated, fucosylated, sialylated, sulfated, and contain blood group or tissue antigens. The ABO and Lewis types of blood group antigens as well as the i and I antigens are commonly found on *O*-glycans, especially on mucins (Table 1). Many of these antigens appear in a developmentally controlled fashion. At the nonreducing end of *O*-glycan chains, sugars are often present in α anomeric linkages.[4,6]

Core 1 is the most common core structure in mucins as well as in other secreted and cell surface-bound glycoproteins. Core 1, or T antigen, is not usually exposed in glycoproteins, but sialylated forms of core 1 are found. *O*-Glycans with core 2 structures are also frequently present, although core 2 is expressed in a cell type-specific fashion. It has been found to change during activation of lymphocytes and differentiation of cells. *O*-Glycans with core 3 to 8 structures have only been found on mucins and not on other glycoproteins. Core 3 and 4 structures are predominantly found in the gastrointestinal tract, but are also present in human lung and salivary mucins.[13] Core 5 occurs in glycoproteins from several species, and has been found in human adenocarcinoma[14] and meconium.[15] To date, oligosaccharides with core 6 have only been reported on human glycoproteins, including meconium and ovarian cyst mucins.[4] Core 7 has recently been described in bovine submaxillary mucin,[16] while core 8 occurs in human bronchial mucin and in certain types of frog mucins.[17,18]

Little is known about the biological significance of these *O*-glycan structures. Terminal sugars often serve as carbohydrate antigens and ligands,[8] and may be attached to different types of core structures. These core structures may sterically present antigens in unique ways, thus resulting in different biological activities. The role of *O*-glycans has been investigated using GalNAcα-benzyl or -phenyl derivatives which diffuse through membranes and compete with the synthesis of *O*-glycan core structures.[19] As a consequence of GalNAc-benzyl treatment, mucins express more unprocessed GalNAc residues and show greatly reduced sialylation.[20] The inhibition of *O*-glycan synthesis in human cancer cells led to a decreased binding to E-selectin and attachment to endothelial cells.[21] It has thus been postulated that E-selectin binding to *O*-glycans plays a major role in the metastatic behavior of cancer cells. However, this may depend on the system under study. GalNAcα-benzyl treatment exposes more core 1 on the cell adhesion molecule CD44, and this enhances the metastatic

Table 1   Structures of blood group and tissue antigens found on
O-glycans.

Tn	GalNAcα-Thr/Ser
Sialyl-Tn	SAα2,6GalNAcα-Thr/Ser
T	Galβ1,3GalNAcα-Thr/Ser
Sialyl-T	SAα2,3Galβ1,3GalNAcα-Thr/Ser
A	GalNAcα1,3Galβ-
	|α1,2
	Fuc
B	Galα1,3Galβ-
	|α1,2
	Fuc
H or O	Galβ
	|α1,2
	Fuc
i	Gal1,4GlcNAcβ1,3Galβ-
I	Galβ1,4GlcNAc
	|β1,6
	Galβ1,4GlcNAcβ1,3Galβ-
Sd^d (Cad)	GalNAcβ1,4Galβ-
	|α2,3
	SA
Le^a	Galβ1,3GlcNAcβ1,3Gal-
	|α1,4
	Fuc
Le^b	Galβ1,3GlcNAcβ1,3Gal-
	|α1,2 |α1,4
	Fuc Fuc
Le^x	Galβ1,4GlcNAcβ1,3Gal-
(SSEA-1)	|α1,3
	Fuc
Le^y	Galβ1,4GlcNAcβ1,3Gal-
	|α1,2 |α1,3
	Fuc Fuc
Sialyl-Le^a	SAα2,3Galβ1,3GlcNAcβ1,3Gal-
	|α1,4
	Fuc
Sialyl-Le^x	SAα2,3Galβ1,4GlcNAcβ1,3Gal-
	|α1,3
	Fuc

capacity of cancer cells.[22] Normal *O*-glycans of CD44 appear to regulate the adhesion of lower molecular weight species of CD44 to hyaluronate in colon cancer cells.[23]

The recognition by P-selectin of the ligand PSGL-1 appears to require the *O*-glycan core 2 structure.[24] *O*-Glycans are recognized by a number of intercellular adhesion molecules, antibodies, lectins, and other carbohydrate-binding molecules, bacteria, and viruses.[1,25] In addition, glycoproteins containing *O*-glycans have been shown to be involved in the process of fertilization, in the control of the immune system, in infectious diseases, and in the spread of cancer cells.

Individual mucin molecules may be differentially glycosylated and sulfated. The changing expression of specific mucins in cancer may therefore lead to altered glycosylation. Newly expressed carbohydrate and peptide antigens found on cancer cell mucins due to altered glycosylation have been exploited in developing diagnostics and immunotherapeutic agents for cancer.[26,27]

3.04.3   FACTORS CONTROLLING *O*-GLYCAN BIOSYNTHESIS

O-Glycans are assembled in the Golgi apparatus through the sequential addition of individual sugars transferred from nucleotide sugars by glycosyltransferases, some of which are specific for *O*-glycans (Table 2). It is likely that *O*-glycan synthesis is controlled in a species-, tissue-, and growth-specific manner; many of the possible controlling factors differ between cell types and change during growth and differentiation, as well as in disease.

Table 2 Glycosyltransferases acting specifically on *O*-glycans.

1.	UDP-GalNAc: polypeptide α-GalNAc-transferase (polypeptide GalNAc-T)
2.	UDP-Gal: GalNAcβ1,3-Gal-transferase (core 1 β3-Gal-T)
3.	UDP-GlcNAc: GalNAcβ1,3-GlcNAc-transferase (core 3 β3-GlcNAc-T)
4.	UDP-GlcNAc: Galβ1,3GalNAc (GlcNAc to GalNAc) β1,6-GlcNAc-transferase (core 2 β6-GlcNAc–T)
5.	UDP-GlcNAc: GlcNAcβ1,3GalNAc (GlcNAc to GalNAc) β1,6-GlcNAc-transferase (core 4 β6-GlcNAc-T)
6.	UDP-GlcNAc: Galβ1,3 (R-)GalNAc (GlcNAc to Gal) β1,3-GlcNAc-transferase (elongation β3-GlcNAc-T)
7.	CMP-sialic acid: GalNAc α2,6-sialyltransferase (α6-sialyl-T)
8.	CMP-sialic acid: Galβ1,3 (R-)GalNAc α2,3-sialyltransferase (α3-sialyl-T)

The microenvironment within the membrane and the lumen of the Golgi may control enzyme activities and specificities. The activities of membrane-bound glycosyltransferases may be influenced by the membrane composition and the association with other proteins. For example, the interaction with other transferases may stimulate activities.[28] Glycosyltransferases have to be localized in their respective subcompartments in the Golgi in order to act properly in the assembly line. The mechanisms of Golgi localization appear to involve retention through interaction between Golgi membranes and the membrane anchor region of the glycosyltransferases, as well as the adjacent regions and possibly other recognition sites in the protein.[29]

The numbers of individual oligosaccharide structures of glycoproteins are determined by the relative activities of competing glycosyltransferases and their intracellular distribution, as well as the transport rates of substrates through the Golgi compartments. The substrate specificities of glycosyltransferases control the biosynthetic pathways and limit the possible number of *O*-glycan structures. The specificities of some glycosyltransferases result in STOP and GO signals for subsequent reactions. For example, the attachment of sialic acid in an α2,6 linkage to GalNAc is a STOP signal, since sialylα2,6GalNAc is not a substrate for any other glycosyltransferase. Conversely, the addition of Gal in a β1,3 linkage to GalNAc is a necessary GO signal before the branching reaction by core 2 β6-GlcNAc-T in the synthesis of core 2 can occur. The peptide portion of substrates also has an effect on *O*-glycan processing. In particular, enzymes that act early in the *O*-glycosylation pathway[30,31] are influenced by the structure as well as the glycosylation of the peptide substrate. Due to this site directed processing, the *O*-glycan structures at various sites of a glycoprotein may be different.

Glycosyltransferase and sulfotransferase activities are also influenced by divalent metal ions.[6] Most UDP-sugar-binding enzymes require an unphysiologically high concentration of a divalent cation (usually manganese ions) for maximum activity *in vitro*. Fucosyltransferases (Fuc-T), sialyltransferases (sialyl-T), and β1,6-*N*-acetylglucosaminyltransferases (GlcNAc-T), however, can function well without divalent metal ions. Certain metal ions may have a strong inhibitory effect.

To date, physiological binding proteins have not been identified for *O*-glycan glycosyltransferases, with the exception of lactalbumin that binds to and has the ability to change the kinetic behavior of β4-galactosyltransferase (Gal-T).[4]

Glycosyltransferases are often glycoproteins themselves, and enzyme glycosylation may influence stability, transport, and activity, and could be a possible factor in feedback regulation or compensatory mechanisms. Other types of possible regulation include phosphorylation by associated protein kinases, for example for β4-Gal-T.[32]

Now that a large number of glycosyltransferase genes have been cloned, the mechanisms of gene regulation may be elucidated. Glycosyltransferases have various numbers of exons, and often several promoters appear to function in a cell type-specific fashion. In addition, binding sites for transcription factors have been identified.[7,33–35] All of these factors contribute to the complex control mechanisms of *O*-glycan biosynthesis.

3.04.4 INITIATION OF *O*-GLYCAN BIOSYNTHESIS. UDP-GalNAc: POLYPEPTIDE α-*N*-ACETYLGALACTOSAMINYLTRANSFERASE (POLYPEPTIDE GalNAc-T; EC 2.4.1.41)

The first step in the synthesis of all *O*-glycans is catalyzed by polypeptide GalNAc-T (Figure 1). A number of studies have shown that the initiation of *O*-glycan synthesis occurs mainly in the Golgi, and possibly also in earlier or later compartments.[36–39] The enzyme has been localized by immunoelectron microscopy studies to the cis-Golgi,[40] although recent data suggest a more wide-spread intracellular distribution (T. Nilsson, personal communication), possibly depending on the cell type and the differentiation status.

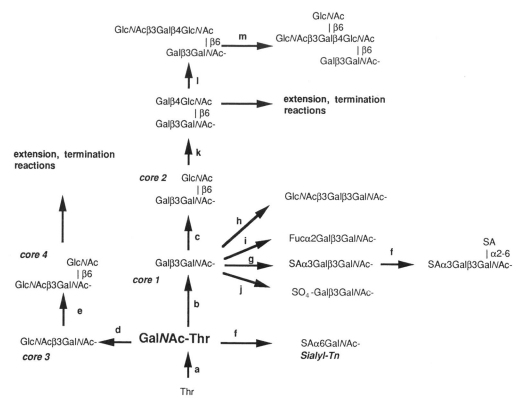

Figure 1   Composite known biosynthetic pathways specific for *O*-glycans. The biosynthesis of the common *O*-glycan core structures 1 to 4 and some of the elongation and sialylation reactions are shown. Letters indicate the enzymes catalyzing the reactions. Path a, polypeptide α-Gal*N*Ac-T; path b, core 1 *β*3-Gal-T; path c, core 2 *β*6-Glc*N*Ac-T; path d, core 3 *β*3-Glc*N*Ac-T; path e, core 4 *β*6-Glc*N*Ac-T; path f, α6-sialyl-T; path g, α3-sialyl-T; path h, elongation *β*3-Glc*N*Ac-T; path i, α2-Fuc-T; path j, core 1 3-sulfotransferase; path k, *β*4-Gal-T; path l, i *β*3-Gal-T; path m, I *β*6-Glc*N*Ac-T.

Polypeptide Gal*N*Ac-T has been purified from several tissues as well as in recombinant forms.[4,41] The specificities of the enzymes isolated from different sources, and of the various recombinant enzymes, are often characteristically different, and it appears that there is a preference of each enzyme for certain glycoprotein acceptor structures.[42] No single amino acid sequon is required in the acceptor substrate for the transfer of Gal*N*Ac, but the acceptor peptide structure does determine the enzyme activity.[41] Several attempts have been made to define the peptide sequences of glyco-proteins that are likely to be *O*-glycosylated,[31,43–45] but the presence of a family of enzymes with various specificities and tissue distribution makes this task very difficult. Pro is usually found near *O*-glycosylation sites, and may serve to expose Ser/Thr residues. Thus, the three-dimensional structure of the substrate, its charge and hydrophobicity, as well as the accessibility of the hydroxy group of Thr/Ser may be important factors in determining *O*-glycosylation. In addition, the existing glycosylation of peptide substrates near the glycosylation site is an important regulator of the enzyme activity.[31] *In vitro*, the transfer of Gal*N*Ac by bovine colostrum polypeptide Gal*N*Ac-T to multiple glycosylation sites is generally more difficult. However, the *in vivo* transfer to mucin substrates with multiple sites appears to be extremely efficient. It is clear that polypeptide Gal*N*Ac-T can act on many different substrates but that the selectivity of this process depends on the spectrum of enzymes and substrates expressed in each cell.

For reasons that are not yet clear, crude and purified polypeptide Gal*N*Ac-T that have been studied *in vitro* efficiently transfer Gal*N*Ac only to Thr, but not to Ser of the peptide substrate.[46–48] This is in contrast to the fact that many glycoproteins and mucins have glycosylated Ser residues.

The cDNAs encoding several members of this glycosyltransferase family with distinct substrate specificities have been cloned from bovine, human, and other sources.[42,46–50] Although there is homology between polypeptide GalNAc-T proteins (GalNAc-T motif) within the catalytic region that was useful in cloning several of the enzyme species,[49] no significant homology to other types of glycosyltransferases has been identified.

In humans, the genes encoding four distinct polypeptide GalNAc-T (T1 to T4)[41] and a pseudogene[51] have been identified; in addition, there are other members of this gene family that remain to be investigated. Three of the cloned polypeptide GalNAc-T (T1, T2, and T3) are encoded in multiple exons while polypeptide GalNAc-T4 is encoded by only one exon and may represent an evolutionary older member of this glycosyltransferase family.[41] The human genes and the pseudogene are all located on different chromosomes. All mammalian cells appear to express at least one of the members of this enzyme family.

Polypeptide GalNAc-T may have O-glycosylation sites themselves, and the enzyme proteins have at least one N-glycosylation site, although it is not known if occupation of glycosylation sites is necessary for enzyme activity. The bovine colostrum enzyme is a glycoprotein containing two N-glycans, mostly of the complex type.[52] Many of the Cys residues in the enzyme appear to be conserved, and may be important for the enzyme activity,[46,53,54] as is the case for other glycosyltransferases such as Gal-T,[55] sialyl-T,[56,57] Fuc-T,[58] and GlcNAc-T (D. Toki, J. Schutzbach, and I. Brockhausen, unpublished findings).

O-Glycosylation by polypeptide GalNAc-T is not mediated by a dolichol or lipid intermediate, in contrast to the O-glycosylation of yeast mannoproteins where mannose is transferred from dolichol-phosphomannose. O-Mannosylation in yeast is catalyzed by a family of polypeptide mannosyltransferases.[59] A total inhibition of mannosylation could be achieved by the gene knock-out of a combination of these enzymes while single knock-outs were not effective. Similarly, a single gene knock-out of polypeptide GalNAc-T in mice did not lead to a change in phenotype,[60] presumably because several enzymes are expressed. In fact, no mammalian cell lacking polypeptide GalNAc-T activity has yet been found, although the expression levels may differ, and changes in various conditions and in disease may occur.[1] This suggests that the enzyme, and O-glycosylation, is essential for survival of cells and animals, and therefore, multiple apparently redundant enzyme species have evolved.

3.04.5 SYNTHESIS OF *O*-GLYCAN CORE STRUCTURES

3.04.5.1 UDP-Gal: GalNAc-R β1,3-Gal-Transferase (Core 1 β3-Gal-T, EC 2.4.1.122)

O-Glycans based on core 1, Galβ1,3 GalNAc-, and the enzyme synthesizing core 1, core 1 β3-Gal-T, are present in most mammalian cells (Figure 1). The specificity of the enzyme toward GalNAcα-peptides and other GalNAc derivatives has been extensively studied.[61,62] The activity requires all the substituents of the GalNAc ring with the exception of the 6-hydroxy group, which may be substituted by GlcNAc but not by sialic acid. This means that the enzyme can act on core 6, GlcNAcβ1,6 GalNAc- to synthesize core 2, GlcNAcβ1,6 (Galβ1,3) GalNAc-, although this pathway remains to be demonstrated *in vivo*.

The activity is strongly influenced by the composition, length, and sequence of the peptide backbone of the substrate and the attachment position and number of sugar residues present.[30,61] These studies, and those of other enzymes processing O-glycans, suggest that the synthesis of core structures at individual O-glycosylation sites is regulated by the peptide structure and glycosylation near these sites. This site-directed processing of O-glycans may in part explain the differences in O-glycan structures between glycoproteins, and at various sites of the same glycoprotein. It may also explain why in disease states, where different relative levels of proteins are expressed, O-glycan structures change.

A deficiency of core 1 β3-Gal-T is present in erythrocytes (Tn erythrocytes) and a small proportion of lymphocytes from patients with permanent mixed-field polyagglutinability.[63] The activity can be reactivated with certain differentiation agents. The human T-lymphoblastoid cell line Jurkat[64] and a human colon cancer cell line LSC[65] also show core 1 β3-Gal-T deficiency associated with a lack of O-glycan processing which does not appear to be reversible. The activity of this enzyme changes during differentiation of human colonic adenocarcinoma Caco-2 cells.[66]

3.04.5.2   UDP-GlcNAc: GalNAc-R β1,3-GlcNAc-Transferase (Core 3 β3-GlcNAc-T, EC 2.4.1.147)

The *O*-glycan core 3 structure occurs in colonic, bronchial, lung, and other mucins, and is synthesized by core 3 β3-GlcNAc-T (Figure 1).[67] The enzyme activity is expressed in a tissue-specific fashion, and is reduced in colon cancer tissue.[68,69] Although colonic tissue is a rich source of the enzyme, the activity is undetectable in many cultured cells.[70,71]

The enzyme has been characterized in regard to substrate specificity[67,71] but has not yet been purified or cloned. The specificity of the enzyme requires all the substituents of the GalNAc ring. However, core 6 structures are substrates for a β3-GlcNAc-T activity in ovarian tissue.[72]

3.04.5.3   UDP-GlcNAc: Galβ1,3GalNAc-R (GlcNAc to GalNAc) β1,6-GlcNAc-Transferase (Core 2 β6-GlcNAc-T, EC 2.4.1.102) and UDP-GlcNAc: GlcNAcβ1,3GalNAc-R (GlcNAc to GalNAc) β1,6-GlcNAc-Transferase (Core 4 β6-GlcNAc-T, EC 2.4.1.148)

O-Glycan core 2, GlcNAcβ1,6 (Galβ1,3) GalNAc-, is found on many glycoproteins and mucins, while the similar branched structure core 4, GlcNAcβ1,6 (GlcNAcβ1,3) GalNAc-, appears to be restricted to mucins. The branch of the core 2 structure allows for the multiple attachment of various antigenic determinants. PSGL-1 is a high-affinity ligand for P- and E-selectin from neutrophils and other myeloid cells.[24] Transfection of core 2 β6-GlcNAc-T into CHO cells expressing PSGL-1 resulted in greatly enhanced binding ability to P-selectin.[73] The Lewis type epitopes recognized by selectins are therefore likely to be attached to the GlcNAcβ1,6 branch of the core 2 structure.

The activity synthesizing core 2 from core 1, core 2 β6-GlcNAc-T (Figure 1), is relatively ubiquitous. It appears to be regulated during cellular differentiation and activation, in cancer cells, and in other diseased cells.[66,74–77] The activity that occurs in leukocytes has been named the L enzyme;[78] it only synthesizes core 2 and is not accompanied by other β6-GlcNAc-T activities.

The gene encoding the human L enzyme has been cloned and was localized to chromosome 9q21,[79,80] near the locus for the blood group A and B transferases and the I β6-GlcNAc-T. The gene contains potential binding sites for transcription factors, such as those that are active in activated T lymphocytes.[35] Core 2 β6-GlcNAc-T L and, possibly, other branching β6-GlcNAc-T appear to be regulated during activation of lymphocytes.[74]

Homologous regions of the β6-GlcNAc-T gene are distributed over more than one exon; this indicates that evolution occurred through gene duplication followed by intron insertion rather than by exon shuffling.[35] A pseudogene resembling that of core 2 β6-GlcNAc-T has also been identified.[81]

The recombinant L enzyme, lacking the amino terminus and the membrane anchor region, expressed in Sf9 insect cells, is *N*-glycosylated and requires two *N*-linked glycans at the amino terminus for full activity.[82] The L enzyme has a relatively restricted substrate specificity and an absolute requirement for the 4- and 6-hydroxyl groups and the 2-acetamido group of GalNAc, and the 6-hydroxyl group of the Gal residue of Galβ1,3 GalNAc-R substrate.[78] The substrate derivative lacking the 6-hydroxyl group of GalNAc forms a poor competitive inhibitor, while the *p*-nitrophenyl derivative of core 1 upon UV irradiation is a powerful irreversible inhibitor.[83] Any substitution of core 1 inhibits core 2 formation.[84,85] Similarly, core 4 cannot be formed after galactosylation of core 3. Thus, branching to form core 2 or core 4 has to occur before extension of these chains (I. Brockhausen, unpublished findings).

Normal T lymphocytes and K562 cells are very low in core 2 β6-GlcNAc-T L activity and have negligible amounts of core 2 structures.[77] However, high activities are found in various leukemias such as chronic myelogenous leukemia (CML) and acute myelogenous leukemia (AML)[76] as well as in T-lymphocytic leukemia cells from patients with acute lymphocytic leukemia (ALL) and chronic lymphocytic leukemia (CLL).[77] These studies suggest that cellular maturation is associated with changes of core 2 expression, and the increase of core 2 β6-GlcNAc-T activity in leukemia may be due to the relative immaturity of cells.

Breast cancer cells produce mucin with less core 2 than normal cells.[86,87] The biosynthetic mechanisms underlying these changes involve increased α3-sialyl-T and decreased core 2 β6-GlcNAc-T L expression.[88]

A high expression level of core 2 β6-GlcNAc-T is found in human postnatal but not embryonal cortical thymocytes.[89] In the mouse, the enzyme is widely expressed at early developmental stages and then becomes more restricted to mucin- and cartilage-producing tissues.[90] Core 2 β6-GlcNAc-T activity is significantly increased in the heart tissue of diabetic rats. Differential display of mRNA of normal and streptozotocin-induced diabetic rats revealed an increased expression of core 2 β6-GlcNAc-T specifically in the heart tissue.[91]

The enzyme synthesizing core 4, core 4 β6-GlcNAc-T (Figure 1), has been purified from bovine tracheal tissue.[92] This enzyme also has the ability to synthesize core 2 as well as the GlcNAc1β1,6 Gal- branch of the I antigen. It was named the M enzyme, since it occurs in mucin-secreting cells.[78] Different ratios of β6-GlcNAc-T activities differing in substrate specificities occur in various cell types.[6,71] In addition, core 2 β6-GlcNAc-T (L enzyme) and the β6-GlcNAc-T that synthesizes the I antigen have regions of homology.[80] These enzymes may therefore be members of a large β6-GlcNAc-T family.

The activity synthesizing core 4 is reduced relative to core 2 β6-GlcNAc-T L in a number of model cancer cells.[66,70,71] In colon cancer tissue, core 2 β6-GlcNAc-T activity, and especially the accompanying core 4 and I β6-GlcNAc-T activities, are reduced.[68] This suggests that the L enzyme, and in particular the M enzyme, are down-regulated in cancer.

3.04.5.4   Synthesis of Core 5 to Core 8

Core 5, GalNAcα1,3GalNAcα-, has been found in mucins from several mammalian and non-mammalian species, in human meconium,[15] and glycoproteins from colonic adenocarcinoma.[14] UDP-GalNAc: GalNAc-mucin α3-GalNAc-T, core 5 α3-GalNAc-T[93] was found in detergent extracts of one of six human intestinal cancerous tissues using asialo-bovine submaxillary mucin as a substrate. There is no report to date on the formation of core 7 by a core 7 α6-GalNAc-T or of core 8 by core 8 α3-Gal-T. An activity has been described in human ovarian tissue[72] that catalyzes the synthesis of core 6. The relationship of this β6-GlcNAc-T to other β6-GlcNAc-T synthesizing branches or linear structures of *O*-glycan chains remains to be shown.

3.04.6   ELONGATION AND BRANCHING OF *O*-GLYCANS

3.04.6.1   UDP-GlcNAc: Galβ1,3 (R1–6) GalNAc-R (GlcNAc to Gal) β1,3-GlcNAc-Transferase (Elongation β3-GlcNAc-T, EC 2.4.1.146)

Many different cell types are capable of elongating *O*-glycans.[6] Core structures 1 to 4 are elongated by repeating GlcNAcβ1,3Galβ1,4 units, synthesized by i β3-GlcNAc-T and β4-Gal-T. In addition, the Gal moiety of core 1 and core 2 structures may be elongated by an elongation β3-GlcNAc-T (Figure 1).[94] This elongation reaction forms the basis for the attachment of poly-*N*-acetyllactosamine chains at the Gal moiety. The substrate specificity and tissue distribution of this elongation enzyme differ from those of the i β3 GlcNAc T and core 3 β3-GlcNAc-T.[67,95]

The elongation enzyme is relatively restricted in its distribution,[6] and is present in human intestinal tissue and colon cancer cell lines.[66,68,71] The activity appears to be turned off in leukemic leukocytes[76] and in human breast cancer cells T47D.[88]

3.04.6.2   UDP-GlcNAc: Galβ1,4 GlcNAc (GlcNAc to Gal) β1,3-GlcNAc-Transferase (i β3-GlcNAc-T; EC 2.4.1.149)

The poly-*N*-acetyl-lactosamine chains of *O*-glycans are probably synthesized by the repeated and concerted actions of β4-Gal-T (described in Chapter 3.03) and i β3-GlcNAc-T (Figure 1).[95,96] i β3-GlcNAc-T may extend the glycan chains of *N*- and *O*-glycans and glycolipids.

The i blood group antigen of poly-*N*-acetyllactosamine structure is a developmentally regulated determinant on erythrocytes that is made in early fetal life.[97] Poly-*N*-acetyllactosamines linked to *O*-glycans occur on many secreted and membrane-bound glycoproteins, and often carry terminal carbohydrate recognition sequences.

i β3-GlcNAc-T has been described in many different cell types.[6,66,70,71,76,95,98] The activity is relatively high in human serum, suggesting that i β3-GlcNAc-T is one of the glycosyltransferases that is released from Golgi membranes by proteolytic cleavage. i β3-GlcNAc-T activity appears to be increased in acute myeloid leukemia cells.[76]

Another i β3-GlcNAc-T activity synthesizing the GlcNAcβ1-3Gal linkage in lactotriaosylceramide can be distinguished from the i β3-GlcNAc-T synthesizing poly-*N*-acetyllactosamine chains by its metal ion activation, pH optimum, and kinetics, and is found in the myeloid but not lymphoid cell lineages.[99] This suggests that there is also a family of β3-GlcNAc-T.

3.04.6.3   UDP-GlcNAc: GlcNAcβ1,3 Gal (GlcNAc to Gal) β1,6-GlcNAc-Transferase (I β6-GlcNAc-T)

A number of different β6-GlcNAc-T activities can be distinguished by their substrate specificities and tissue distribution, suggesting that a large family of β6-GlcNAc-T acting on Gal or GalNAc residues of *O*-glycans exists.[6,80,85,96,100–102] I β6-GlcNAc transferases are involved in the synthesis of the I antigen by adding a GlcNAc β1,6 branch either to terminal Gal or internal Gal residues (Figure 1). Various cell types may contain different ratios of these enzyme activities. Activities synthesizing linear GlcNAcβ1,6 Gal and GlcNAcβ1,6 GalNAc structures have been described in Novikoff ascites tumor cells and ovarian tissue, and may be characteristic of human cells.[4,72,98]

Both the i and the I antigens are found on many glycoproteins and mucins. The expression of the i antigen and the branched I antigen, Galβ1,4 GlcNAcβ1-6 (Galβ1,4 GlcNAcβ1,3) Gal-, on erythrocytes is regulated during development.[97] In the mouse, the I antigen as well as the I β6-GlcNAc-T is expressed in many tissues in epithelial and dividing cells.[103]

Recombinant human I β6-GlcNAc-T[80] transfected into CHO cells lacking the I antigen caused an increased appearance of I branched structures on cell surfaces, relative to the i antigen. Glycoprotein carrying newly appearing I antigen was detected with an antibody, although I β6-GlcNAc-T activity was not detectable. The cDNA sequence of the enzyme revealed homology in the putative catalytic domain to core 2 β6-GlcNAc-T but not to other glycosyltransferases. Both I β6-GlcNAc-T and core 2 β6-GlcNAc-T genes were localized to chromosome 9, q21.

3.04.6.4   UDP-Gal: GlcNAc-R β1,4-Galactosyltransferase (β4-Gal-T)

UDP-Gal:GlcNAc-R β1,4-Gal-T (EC 2.4.1.38; EC 2.4.1.90), β4-Gal-T, is ubiquitous and has been purified and cloned from many sources and has been dealt with in detail in Chapter 3.03. The specificity of the enzyme is regulated in mammary glands by binding to α-lactalbumin, which changes the kinetics of β4-Gal-T to favor the synthesis of lactose in milk. The enzyme has been localized to the trans-Golgi compartment, but is also found on cell surfaces.[10,104] Cell surface β4-Gal-T can bind to terminal GlcNAc residues on other cells, and thereby may play a role in cell adhesion, as has been suggested for sperm–egg binding in the mouse.[11]

3.04.6.5   UDP-Gal: GlcNAc-R β1,3-Galactosyltransferase (β3-Gal-T)

Another elongating enzyme, β3-Gal-T, has been characterized and purified from pig trachea[105] and the activity has been found in a number of species and cell types.[106–109] This enzyme is clearly different from the β4-Gal-T. The activity is not influenced by α-lactalbumin. It adds Gal in a β1,3-linkage to GlcNAc residues of *O*-glycan core 3 and other structures with terminal β-GlcNAc residues, to synthesize type 1 chains.[110]

A β3-Gal-T activity is present in the mammary gland of the Tammar wallaby,[107] and acts on the Gal residue of lactose. Both the latter β3-Gal-T and β4-Gal-T appear to be subject to developmental regulation in the mammary tissue of the Tammar wallaby. Another β3-Gal-T has been characterized in human adenocarcinoma cells. This enzyme can be separated from the β4-Gal-T by α-lactalbumin affinity chromatography, and synthesizes the type 1 chains in glycolipids.[111] The activity in normal colonic tissue and colonic adenocarcinoma also acts on *N*-glycans.[112] It appears that the β3- but not the β4-Gal-T activity is decreased in adenocarcinoma. The relationship between these 3β-Gal-T activities remains to be established.

3.04.7   TERMINATION REACTIONS IN THE SYNTHESIS OF *O*-GLYCANS

Many of the terminal *O*-glycan structures are recognized as blood group or tissue antigens, for example the Lewis, ABO, Cad, and S^d blood groups. The addition of these α-linked sugars to nonreducing terminal or internal residues of *O*-glycan chains may mask underlying antigens, or produce new antigens, and may terminate chain growth. Some of the rare epitopes occurring in mammalian glycoproteins include α1,4-linked GlcNAc, which is recognized by a monoclonal antibody[113,114] and appears in gastrointestinal mucins. These terminal determinants are usually common to several types of glycoconjugates. Studies of the glycosyltransferases synthesizing these antigens have been reviewed.[6,115]

Terminal sialic acid residues on *O*-glycans appear in a developmentally regulated and tissue-specific fashion. Sialylated *O*-glycans may also be part of several cancer-associated antigens such as sialyl-T, sialyl-Tn, and sialyl-Lewis antigens, and play an important role in the cell surface and cell adhesion properties of cells.

A few terminal structures are found exclusively on *O*-glycans. A number of enzymes exist therefore that require *O*-glycan types of substrates (Table 2). *O*-Glycan-specific enzymes include α6-sialyltransferase (ST6 GalNAc) I and II, synthesizing the sialyl-Tn and sialyl-T antigens, the α3-sialyltransferase (ST3 Gal) that sialylates *O*-glycan core 1 and 2, and ST6 GalNAc III that requires sialylα2,3 Galβ1,3 GalNAc-R as a substrate. In addition, there are several glycosyltransferases that act on *O*-glycans although they can synthesize similar structures on other types of glycan chains.

Sialylation is not always a chain-terminating event. Thus, α6-sialylation of core 1 prevents formation of core 2 by core 2 β6-GlcNAc-T and inhibits elongation of core 1 by elongation β3-GlcNAc-T.[94] However, in the pathways to sialylated blood group A or B structures, α6-sialyl-T I must act first, followed by α2-Fuc-T. The synthesis of the Cad or Sd[a] determinant also requires sialylation before the addition of β4-linked GalNAc.

Several sialyl-T activities acting on asialofetuin, and especially the activity of α6-sialyl-T measured by the incorporation of sialic acid into native fetuin (containing sialylα2,3Galβ1,3 GalNAc-), change during development and are high in embryonic rat brain but low in the adult brain.[116] Cell surface hypersialylation is a common observation in transformed and metastatic cells as well as in leukemia cells.[1,5]

A comparison of the amino acid sequences of various sialyl-T indicates that there are regions with extensive homology within the catalytic domain.[117] This "sialyl motif" region has been used to clone many sialyltransferases. Site-directed mutagenesis experiments have suggested that the sialyl motif participates in the binding of CMP-sialic acid to the enzyme.[118]

3.04.7.1 CMP-sialic acid: Galβ1,3 GalNAc-R α3-Sialyltransferase (ST3 Gal I, α3-sialyl-T, EC 2.4.99.4)

α3-Sialyl-T transfers sialic acid in an α2,3 linkage to the Gal residue of *O*-glycan core 1 and 2 substrates (Figure 1).[78] Several α3-sialyl-T have been cloned, based on regions of high homology.[34,119–125] Two species of the enzyme, ST3O I and II, are probably responsible for the synthesis of sialylated *O*-glycans, although ST3O II preferably acts on glycolipids.[122]

Substrate specificity studies indicate that the enzyme from placenta and from AML cells[78] has an absolute requirement only for the 3-hydroxyl group of the Gal residue of core 1 substrates and can be competitively inhibited by 3-deoxy-Galβ1,3 GalNAcα-benzyl. Reboul *et al.*[126] postulated that the α3-sialyl-T may be regulated by a phosphorylation/dephosphorylation mechanism. The enzyme from C6 glioma cells appears to depend on *N*-glycosylation for full activity. Activity was inhibited by treating cells with tunicamycin, and the partially purified enzyme was inactivated by peptide-*N*-glycosidase.[127]

The product of α3-sialyl-T interacts with a number of sialic acid-binding molecules. Sendai virus infection of bovine kidney cells can be prevented by removing sialic acid with neuraminidase; the virus receptor is restored with *O*-glycan α3-sialyl-T while an α6-sialyl-T is ineffective.[128] Sialoadhesins are present on macrophages and are known to function in removing erythrocytes and lymphocytes from the circulation. Sialylα2,3Galβ1,3GalNAc- has been identified as a ligand for a mouse sialoadhesin which is involved in interactions between different cells of the hemotopoietic system.[129]

O-Glycans containing the sialylα2,3Gal- moiety vary between cells at different stages of myeloid differentiation.[130] The *O*-glycan α3-sialyl-T also appears to be very active in leukemia-derived cell lines during differentiation.[131] The expression of the α3-sialyl-T is increased in AML and CML, which show abnormal growth characteristics.[132] It is also increased in breast cancer cells and colon cancer tissues.[68,88] This increase in α3-sialylation is usually associated with the occurrence of smaller, truncated, and sialylated *O*-glycans.

The enzyme is regulated during the maturation of thymocytes.[133] Saitoh *et al.*[134] investigated *O*-glycan biosynthesis in HL60 cell lines resistant to retinoic acid- and 6-thioguanidine-induced differentiation. The sialylα2,3 Galβ1,3 GalNAc- structure on leukosialin as well as the *O*-glycan α3-sialyl-T activity were much more prevalent in the wild type cell line than in the altered cells, suggesting a role for *O*-glycan α3-sialyl-T in cellular differentiation. Ha-ras oncogene transfection into FR3T3 rat fibroblasts caused a decreased expression of the *O*-glycan α3-sialyl-T and an increased expression of α6-sialyl-T acting on *N*-glycans.[135]

The DNA encoding α3-sialyl-T containing a myc tag has been transfected into normal and cancerous human mammary cells that express significant levels of the enzyme. Using an anti-myc-tag antibody and immunoelectron microscopy, the enzyme has been localized mainly to the medial and trans-Golgi compartments in both normal and cancerous mammary cells.[136] In addition, the transfection resulted in an increase in sialylated core 1 structures and decreased GlcNAc content, suggesting that overexpression of the α3-sialyl-T resulted in an effective reduction of the synthesis of core 2. It is therefore likely that the sialyl-T competes with chain branching and extension enzymes that are thought to be localized in intermediate Golgi compartments. Although sialylation is a chain termination event, its increase may thus cause a premature termination of O-glycan processing and result in shorter, more sialylated O-glycans.

3.04.7.2   CMP-sialic acid: R$_1$-GalNAc-R α6-Sialyltransferases (ST6 GalNAc I, EC 2.4.99.3; ST6 GalNAc II and III, α6-Sialyl-T)

GalNAc can be processed to form O-glycan core structures or, alternatively, to form sialylα2-6 GalNAc- which cannot be converted to core structures (Figure 1). The cDNA encoding three species of the α6-sialyl-T family acting on GalNAc have been cloned based on the sialyl motif.[137,138] ST6 GalNAc I uses GalNAcα-peptide as a substrate and terminates O-glycan chain growth since sialylα2-6 GalNAc is not a substrate for any of the known glycosyltransferases. ST6 GalNAc II preferably acts on Galβ1,3 GalNAc-peptide. ST6 GalNAc III does not require peptide in the substrate but can also act on substrates where peptide is replaced by a hydrophobic group such as a nitrophenyl group. It is, however, specific for the carbohydrate moiety of substrates and can only act on sialic acid α2,3 Galβ1,3 GalNAc-R.[139] It is unknown which α6-sialyl-T are responsible for sialylation of GalNAc of cores 3 and 5.

Inhibition studies with N-ethylmaleimide showed that the α6-sialyl-T acting on asialofetuin contains sulfhydryl groups that are important for activity. Since the inhibition could be prevented by CMP-sialic acid, these sulfhydryl groups may be near the CMP-sialic acid-binding site.[57]

Sialylα2-6 GalNAc-chains are found in mucins as the sialyl-Tn antigen. Sialyl-Tn antigen is rarely present in normal cells but is found in cancer tissues and is associated with a poor prognosis.[12,140,141] Although the mechanism of sialyl-Tn increase is not known, it is possible that this structure arises due to high α6-sialyl-T activity or relatively low activity of core-synthesizing enzymes, or by intracellular rearrangements, allowing α6-sialyl-T I to act first. The expression of sialyl-Tn antigen may also be due to decreased masking of sialic acid residues by O-acetylation. A lack of O-glycan processing concomitant with sialyl-Tn expression has been found in human colon cancer LSC cells.[65] These cells lack core 1 β3-Gal-T, which normally synthesizes O-glycan core 1; this deficiency allows GalNAc residues to become sialylated instead of being converted to core 1 and 2 structures.

3.04.7.3   *O*-Glycan Sulfation

Sulfotransferases[142–144] that synthesize sulfate esters at the 3- and 6-positions of Gal, the 6-position of GlcNAc, and possibly other linkages, regulate the acidic properties of glycan chains. The same sugar residue may be substituted with sulfate as well as sialic acid, for example, in mucin from the human colon cancer line CL.16E.[145] The roles of these sulfated O-glycan chains may range from protection of O-glycans from degradation to recognition by cell adhesion molecules. Sulfated sialyl-Lewis^x epitopes of the cell adhesion molecule GLYCAM-1 are recognized by L-selectins.[146]

The number of sulfated chains in mucins may vary in cancer and other diseases. Increased secretion of sulfated mucins is observed in cystic fibrosis,[1,147] while decreased sulfation is found in mucins from patients with intestinal cancer and inflammatory bowel disease.[148,149]

It is not known how many different sulfotransferases act on O-glycans. Several sulfotransferases acting on O-glycans have been described but have not been purified or cloned. A sulfotransferase activity that synthesizes the 3-SO$_4$-Gal linkage of O-glycan core 1 (Figure 1) from rat colon has been characterized.[150] Another sulfotransferase activity that acts on the 6-position of GlcNAc has been described.[142] Sulfotransferases are present in a number of cell types including human lung,[151] thyroid,[152] and other tissues. The core 1 3-sulfotransferase appears to be turned off in cancer; the activity is reduced in human colon cancer tissue[68] and in breast cancer cell lines compared to normal mammary cell lines.[88]

Sulfate esters of *O*-glycans play an important role in the regulation of glycosylation. Sulfation may block further branching and extension of *O*-glycan chains. For example, sulfation of the Gal residue of core 1 structures blocks the action of core 2 *β*6-GlcNAc-T that synthesizes core 2, as well as other core 1-processing reactions. Thus, branching has to occur before sulfation. The attachment of repeating Gal*β*1,4 GlcNAc*β*1,3 sequences is dependent on the position of sulfate esters. Sulfation at the 3- or 4-position of Gal prevents the growth of poly-*N*-acetyllactosamine chains.[143,153] However, 6-sulfated GlcNAc but not 3- or 4-sulfated GlcNAc is a substrate for the *β*4-Gal-T. Depending on the position of the sulfate esters, the addition of *α*1,2 or *α*1,3 linked Fuc to *N*-acetyllactosamine chains may also be blocked.[154,155]

3.04.7.4 Fucosylation of *O*-Glycans

Fucosylated structures may designate part of tissue-specific blood group antigens or ligands for selectins. Certain mucins are particularly rich in these fucosylated determinants. These antigens may change during growth and differentiation, and in some diseases. Lewis antigens have been found to change especially in cancer.[1,156,157] Lewis^y structures are associated with apoptosis.[158] The *α*2-, *α*3-, and *α*4-Fuc-T synthesizing these structures are described in Chapter 3.03. It is likely that all of these enzymes act on *O*-glycans as well as on other glycoconjugates.

3.04.7.5 Blood Group A-dependent *α*3-GalNAc-transferase and Blood Group B-dependent *α*3-Gal-transferase

Blood group A and B determinants (Table 1) occur on *O*-glycans and are synthesized from the H-determinant by A-dependent *α*3-GalNAc-transferase and B-dependent *α*3-Gal-transferase, respectively. The expression of their genes depends on the blood group status of individuals. The blood group A and B enzymes act on the same acceptor substrates but differ in their binding to nucleotide sugar donors. The similarity of the two enzyme proteins is explained by the finding that they differ only by four amino acids and the genes by a few base pairs.[115,159,160] In fact, the difference in kinetics and specificity is determined by only one of these amino acids, as shown by expressing a number of mutants in *Escherichia coli*.[161]

3.04.7.6 UDP-Gal: Gal-R *α*1,3-Gal-transferase (*α*3-Gal-T)

The Gal*α*1,3 Gal*β*1,4 GlcNAc- sequence (linear B antigen) is a terminal structure found on *O*-glycans in most mammals with the exception of humans, Old World monkeys, and apes.[162,163] This structure has been implicated in sperm–egg binding in the mouse,[33,164,165] although recent knock-out experiments showed that the structure is not essential for fertilization.[166] Humans do not display the linear B antigen without Fuc *α*1,2-linked to Gal. Human serum therefore contains antibody to the linear B antigen, which has been proposed to be an important factor in xenograft rejection.[163]

The gene encoding *α*3-Gal-T synthesizing the linear B antigen is nonfunctional in humans; instead, a pseudogene is present which has significant homology to that of the blood group B *α*3-Gal-T.[167] The pseudogene is localized to human chromosome 9q34 near the blood group A and B transferase genes, suggesting that the blood group genes and the human *α*3-Gal-T pseudogene are derived from the same ancestral gene by gene duplication and subsequent divergence.

In animal cells, the expression of Gal*α*1,3Gal may mask other antigens or may compete with the expression of other terminal carbohydrate epitopes such as the Lewis antigens. Differentiation appears to influence the expression of these epitopes. Retinoic acid differentiation of mouse teratocarcinoma cells increases the expression of the Gal*α*1,3Gal epitope as well as the mRNA levels for *α*3-Gal-T.[168] The expression of Gal*α*1,3Gal epitopes on cell surfaces can be suppressed by expression of *α*2-Fuc-T in COS cells, suggesting that the *α*2-Fuc-T is localized in the same or an earlier compartment. However, if chimeric enzymes are produced where the *α*3-Gal-T contains the cytoplasmic tail of the *α*2-Fuc-T, and vice versa, the Gal-T appears to act before the Fuc-T, and Gal*α*1,3Gal is expressed.[169] These studies suggest that the overexpression of *α*2-Fuc-T in mammalian donor organs may suppress the Gal*α*1,3Gal epitope and this may overcome a serious problem in xenotransplantation.

Differentiation of mouse teratocarcinoma cells increases the proportion of secreted α3-Gal-T into the cell culture medium.[168] Interestingly, a recombinant soluble and secreted form of the α3-Gal-T expressed in human cells is capable of galactosylating glycoproteins. This may indicate that not only the membrane-bound form but also soluble forms of glycosyltransferases are active in the Golgi.[170]

3.04.7.7   Blood Group Cad or Sd^a-dependent β1,4-GalNAc-transferases

Several β1,4-GalNAc-T exist that differ in substrate specificity and synthesize the Cad and Sd^a blood group antigens found on *N*- and *O*-glycans. The presence of the sialyl residue α3 linked to Gal of the substrate is essential for β4-GalNAc-T activities.[171] The Cad determinant and the β4-GalNAc-T is lacking in a mutant mouse cytotoxic cell line resistant to the GalNAc binding *Vicia villosa* lectin.[172]

The expression of the Sd^a antigen and the β4-GalNAc-T activity are regulated during differentiation of cells, during development, and differ among functionally distinct T-cell clones in the mouse. The expression is often drastically reduced in cancer tissue and cells.[173-175] The murine cDNA encoding the β4-GalNAc-T has been isolated[176] and appears to be homologous to the human counterpart. The expression of β4-GalNAc-T mRNA correlates with the expression of the Sd^a epitope in human gastric mucosa, but is lacking in most specimens of gastric cancer.[173]

3.04.8   REFERENCES

1. I. Brockhausen and W. Kuhns, in "Glycoproteins and Human Disease," Medical Intelligence Unit, CRC Press and Mosby-Year Book, Chapman & Hall, New York, 1997.
2. T. A. Beyer, J. E. Sadler, J. I. Rearick, J. C. Paulson, and R. L. Hill, *Adv. Enzymol.*, 1981, **52**, 23.
3. J. E. Sadler, in "Biology of Carbohydrates," eds. V. Ginsburg and P. W. Robbins, Wiley, New York, 1984, vol. 2, p. 199.
4. H. Schachter and I. Brockhausen, in "Glycoconjugates. Composition, Structure and Function," eds. H. J. Allen and E. C. Kisailus, Dekker, New York, 1992, p. 263.
5. I. Brockhausen, *Crit. Rev. Clin. Lab. Sci.*, 1993, **30**, 65.
6. I. Brockhausen, in "Glycoproteins," eds. J. Montreuil, J. F. G. Vliegenthart, and H. Schachter, Elsevier, Amsterdam, 1995, vol. 29a, p. 201.
7. I. Brockhausen and H. Schachter, in "Glycosciences: Status and Perspectives," eds. H.-J. Gabius and S. Gabius, Chapman & Hall, Weinheim, 1997, p. 79.
8. A. Varki, *Glycobiology*, 1993, **3**, 97.
9. P. Babczinski, *FEBS Lett.*, 1980, **117**, 207.
10. A. Passaniti and G. W. Hart, *Cancer Res.*, 1990, **50**, 7261.
11. B. D. Shur, *Biochim. Biophys. Acta*, 1989, **988**, 389.
12. S. H. Itzkowitz, E. J. Bloom, W. A. Kokal, G. Modin, S.-I. Hakomori, and Y. S. Kim, *Cancer*, 1990, **66**, 1960.
13. E. F. Hounsell, M. J. Davies, and D. V. Renouf, *Glycoconjugate J.*, 1996, **13**, 19.
14. A. Kurosaka, H. Nakajima, J. Funakoshi, M. Matsuyama, T. Nagayo, and I. Yamashina, *J. Biol. Chem.*, 1983, **258**, 11 594.
15. E. F. Hounsell, A. M. Lawson, J. Feeney, H. C. Gooi, N. J. Pickering, M. S. Stoll, S. C. Lui, and T. Feizi, *Eur. J. Biochem.*, 1985, **148**, 367.
16. W. G. Chai, E. F. Hounsell, G. C. Cashmore, J. R. Rosankiewicz, C. J. Bauer, J. Feeney, T. Feizi, and A. M. Lawson, *Eur. J. Biochem.*, 1992, **203**, 257.
17. H. van Halbeek, A.-M. Strang, M. Lhermitte, H. Rahmoune, G. Lamblin, and P. Roussel, *Glycobiology*, 1994, **4**, 203.
18. E. Maes, D. Florea, F. Delplace, J. Lemoine, Y. Plancke, and G. Strecker, *Glycoconjugate J.*, 1997, **14**, 127.
19. S. F. Kuan, J. C. Byrd, C. Basbaum, and Y. S. Kim, *J. Biol. Chem.*, 1989, **264**, 19 271.
20. P. Delannoy, I. Kim, N. Emery, C. De Bolos, A. Verbert, P. Degand, and G. Huet, *Glycoconjugate J.*, 1996, **13**, 717.
21. N. Kojima, K. Handa, W. Newman, and S.-I. Hakomori, *Biochem. Biophys. Res. Commun.*, 1992, **182**, 1288.
22. T. Nakano, T. Matsui, and T. Ota, *Anticancer Res.*, 1996, **16**, 3577.
23. A. Dasgupta, K. Takahashi, M. Cutler, and K. K. Tanabe, *Biochem. Biophys. Res. Commun.*, 1996, **227**, 110.
24. F. Li, P. P. Wilkins, S. Crawley, J. Weinstein, R. D. Cummings, and R. P. McEver, *J. Biol. Chem.*, 1996, **271**, 3255.
25. L. H. Regimbald, L. M. Pilarski, B. M. Longenecker, M. A. Reddish, G. Zimmermann, and J. C. Hugh, *Cancer Res.*, 1996, **56**, 4244.
26. R. A. Graham, J. M. Burchell, and J. Taylor-Papadimitriou, *Cancer Immunol. Immunother.*, 1996, **42**, 71.
27. J. Taylor-Papadimitriou, B. D'Souza, J. Burchell, N. Kyprianon, and F. Berdichersky, *Ann. N.Y. Acad. Sci*, 1993, **698**, 31.
28. M. Moscarello, M. M. Mitranic, and G. Vella, *Biochim. Biophys. Acta*, 1985, **831**, 192.
29. K. J. Colley, *Trends in Glycosci. Glycotech.*, 1997, **9**, 267.
30. M. Granovsky, T. Bielfeldt, S. Peters, H. Paulsen, M. Meldal, J. Brockhausen, and I. Brockhausen, *Eur. J. Biochem.*, 1994, **221**, 1039.
31. I. Brockhausen, D. Toki, J. Brockhausen, S. Peters, T. Bielfeldt, A. Kleen, H. Paulsen, M. Meldal, F. Hagen, and L. A. Tabak, *Glycoconjugate J.*, 1996, **13**, 849.

32. B. A. Bunnell, D. E. Adams, and V. J. Kidd, *Biochem. Biophys. Res. Commun.*, 1990, **171**, 196.
33. D. H. Joziasse, *Glycobiology*, 1992, **2**, 271.
34. A. Harduin-Lepers, M.-A. Recchi, and P. Delannoy, *Glycobiology*, 1995, **5**, 741.
35. M. F. A. Bierhuizen, K. Maemura, S. Kudo, and M. Fukuda, *Glycobiology*, 1995, **5**, 417.
36. J. Roth, *J. Cell Biol.*, 1984, **98**, 399.
37. J. Perez-Vilar, J. Hidalgo, and A. Velasco, *J. Biol. Chem.*, 1991, **266**, 23 967.
38. M. Deschuyteneer, A. E. Eckhardt, J. Roth, and R. L. Hill, *J. Biol. Chem.*, 1988, **263**, 2452.
39. S. Röttger, E. P. Bennett, T. White, H. Clausen, and T. Nilsson, "4th International Workshop on Carcinoma-Associated Mucins," Cambridge, 1996.
40. J. Roth, Y. Wang, A. E. Eckhardt, and R. L. Hill, *Proc. Natl. Acad. Sci. USA*, 1994, **91**, 8935.
41. H. Clausen and E. P. Bennett, *Glycobiology*, 1996, **6**, 635.
42. T. Sorensen, T. White, H. H. Wandall, A. K. Kristensen, P. Roepstorff, and H. Clausen, *J. Biol. Chem.*, 1995, **270**, 24 166.
43. B. C. O'Connell, F. K. Hagen, and L. A. Tabak, *J. Biol. Chem.*, 1992, **267**, 25 010.
44. A. P. Elhammer, R. A. Poorman, E. Brown, L. L. Maggiora, J. G. Hoogerheide, and F. J. Kezdy, *J. Biol. Chem.*, 1993, **268**, 10 029.
45. A. A. Gooley and K. L. Williams, *Glycobiology*, 1994, **4**, 413.
46. T. White, E. P. Bennett, K. Takio, T. Sorensen, N. Bonding, and H. Clausen, *J. Biol. Chem.*, 1995, **270**, 24 156.
47. F. L. Homa, T. Hollander, D. J. Lehman, D. R. Thomsen, and A. P. Elhammer, *J. Biol. Chem.*, 1993, **268**, 12 609.
48. F. K. Hagen, B. van Wuyckhuyse, and L. A. Tabak, *J. Biol. Chem.*, 1993, **268**, 18 960.
49. E. P. Bennett, H. Hassan, and H. Clausen, *J. Biol. Chem.*, 1996, **271**, 17 006.
50. F. K. Hagen, K. G. T. Hagen, T. M. Beres, M. M. Balys, B. C. van Wuyckhuyse, and L. A. Tabak, *J. Biol. Chem.*, 1997, **272**, 13 843.
51. J. A. Meurer, R. F. Drong, F. L. Horna, J. L. Slightom, and A. P. Elhammer, *Glycobiology*, 1996, **6**, 231.
52. Y. Wang, J. L. Abernethy, A. E. Eckhardt, and R. L. Hill, *J. Biol. Chem.*, 1992, **267**, 12 709.
53. F. K. Hagen, C. A. Gregoire, and L. A. Tabak, *Glycoconjugate J.*, 1995, **12**, 901.
54. F. K. Hagen, C. A. Gregoire, and L. A. Tabak, *Glycobiology*, 1994, **4**, 739.
55. Y. Wang, S. S. Wong, M. N. Fukuda, H. Zu, Z. Lio, Q. Tang, and H. E. Appert, *Biochem. Biophys. Res. Commun.*, 1994, **204**, 701.
56. K. Drickamer, *Glycobiology*, 1993, **3**, 2.
57. H. Baubichon-Cortay, P. Broquet, P. George, and P. Louisot, *Glycoconjugate J.*, 1989, **6**, 115.
58. F. H. Holmes, Z. Xu, A. L. Sherwood, and B. A. Macher, *J. Biol. Chem.*, 1995, **270**, 8145.
59. M. Gentzsch and W. Tanner, *EMBO J.*, 1996, **15**, 5752.
60. T. Hennet, F. K. Hagen, L. A. Tabak, and J. D. Marth, *Proc. Natl. Acad. Sci. USA*, 1995, **92**, 12 070.
61. I. Brockhausen, G. Möller, G. Merz, K. Adermann, and H. Paulsen, *Biochemistry*, 1990, **29**, 10 206.
62. I. Brockhausen, G. Möller, A. Pollex-Krüger, V. Rutz, H. Paulsen, and K. L. Matta, *Biochem. Cell Biol.*, 1992, **70**, 99.
63. J. P. Cartron, J. Andreu, J. Cartron, G. W. Bird, C. Salmon, and A. Gerbal, *Eur. J. Biochem.*, 1978, **92**, 111.
64. M. Thurnher, S. Rusconi, and E. G. Berger, *J. Clin. Invest.*, 1993, **91**, 2103.
65. I. Brockhausen, N. Dickinson, S. Ogata, and S. Itzkowitz, *Glycoconjugate J.*, 1995, **12**, 566.
66. I. Brockhausen, P. Romero, and A. Herscovics, *Cancer Res.*, 1991, **5**, 3136.
67. I. Brockhausen, K. L. Matta, J. Orr, and H. Schachter, *Biochemistry*, 1985, **24**, 1866.
68. J. M. Yang, J. C. Byrd, B. B. Siddiki, Y. S. Chung, M. Okuno, M. Sowa, Y. S. Kim, K. L. Matta, and I. Brockhausen, *Glycobiology*, 1994, **4**, 873.
69. M. J. King, A. Chan, R. Roe, B. F. Warren, A. Dell, H. R. Morris, D. C. Bartolo, P. Dudley, and A. P. Corfield, *Glycobiology*, 1994, **4**, 267.
70. F. Vavasseur, K. Dole, J. Yang, K. L. Matta, N. Myerscough, A. Corfield, C. Paraskeva, and I. Brockhausen, *Eur. J. Biochem.*, 1994, **222**, 415.
71. F. Vavasseur, J. M. Yang, K. Dole, H. Paulsen, and I. Brockhausen, *Glycobiology*, 1995, **5**, 351.
72. S. Yazawa, S. A. Abbas, R. Madiyalakan, J. J. Barlow, and K. L. Matta, *Carbohydr. Res.*, 1986, **149**, 241.
73. R. Kumar, R. T. Camphausen, F. X. Sullivan, and D. A. Cumming, *Blood*, 1996, **88**, 3872.
74. F. Piller, V. Piller, R. I. Fox, and M. Fukuda, *J. Biol. Chem.*, 1988, **263**, 15 146.
75. M. Hefferman, R. Lotan, B. Amos, M. Palcic, R. Takano, and J. W. Dennis, *J. Biol. Chem.*, 1993, **268**, 1242.
76. I. Brockhausen, W. Kuhns, H. Schachter, K. L. Matter, D. R. Sutherland, and M. A. Baker, *Cancer Res.*, 1991, **51**, 1257.
77. O. Saitoh, F. Piller, R. I. Fox, and M. Fukuda, *Blood*, 1991, **77**, 1491.
78. W. Kuhns, V. Rutz, H. Paulsen, K. L. Matta, M. A. Baker, M. Barner, M. Granovsky, and I. Brockhausen, *Glycoconjugate J.*, 1993, **10**, 381.
79. M. F. Bierhuizen and M. Fukuda, *Proc. Natl. Acad. Sci. USA*, 1992, **89**, 9326.
80. M. F. Bierhuizen, M. G. Mattei, and M. Fukuda, *Genes Dev.*, 1993, **7**, 468.
81. M. F. Bierhuizen, K. Maemura, and M. Fukuda, *Glycoconjugate J.*, 1995, **12**, 857.
82. D. Toki, M. Sarker, B. Yip, F. Reck, D. Joziasse, M. Fukuda, H. Schachter, and I. Brockhausen, *Biochem. J.*, 1997, **329**, 63.
83. D. Toki, M. A. Granovsky, F. Reck, W. Kuhns, M. A. Baker, K. L. Matta, and I. Brockhausen, *Biochem. Biophys. Res. Commun.*, 1994, **198**, 417.
84. D. Williams and H. Schachter, *J. Biol. Chem.*, 1980, **255**, 11 247.
85. I. Brockhausen, K. L. Matta, J. Orr, H. Schachter, A. H. L. Koenderman, and D. H. van den Eijnden, *Eur. J. Biochem.*, 1986, **157**, 463.
86. S. R. Hull, A. Bright, K. L. Carraway, M. Abe, D. F. Hayes, and D. W. Kufe, *Cancer Commun.*, 1989, **1**, 261.
87. K. O. Lloyd, J. Burchell, V. Kudryashov, B. W. T. Yin, and J. Taylor-Papadimitriou, *J. Biol. Chem.*, 1996, **271**, 33 325.
88. I. Brockhausen, J. M. Yang, J. Burchell, C. Whitehouse, and J. Taylor-Papadimitriou, *Eur. J. Biochem.*, 1995, **233**, 607.
89. L. G. Baum, M. Pang, N. L. Perillo, T. Wu, A. Delegeane, C. H. Uittenbogaart, M. Fukuda, and J. J. Seilhammer, *J. Exp. Med.*, 1995, **181**, 877.

90. M. Granovsky, C. Fode, C. E. Warren, R. M. Campbell, J. D. Marth, M. Pierce, N. Fregien, and J. W. Dennis, *Glycobiology*, 1995, **5**, 797.
91. Y. Nishio, C. E. Warren, J. A. Buczek-Thomas, J. Rulfs, D. Koya, L. P. Aiello, E. P. Feener, T. B. Miller, Jr., J. W. Dennis, and G. L. King, *J. Clin. Invest.*, 1995, **96**, 1759.
92. P. A. Ropp, M. R. Little, and P. W. Cheng, *J. Biol. Chem.*, 1991, **266**, 23 863.
93. A. Kurosaka, I. Funakoshi, M. Matsuyama, T. Nagayo, and I. Yamashina, *FEBS Lett.*, 1985, **190**, 259.
94. I. Brockhausen, D. Williams, K. L. Matta, J. Orr, and H. Schachter, *Can. J. Biochem. Cell Biol.*, 1983, **61**, 1322.
95. D. H. van den Eijnden, A. H. L. Koenderman, and W. E. C. M. Schiphorst, *J. Biol. Chem.*, 1988, **263**, 12 461.
96. A. H. Koenderman, P. L. Koppen, and D. H. van den Eijnden, *Eur. J. Biochem.*, 1987, **166**, 199.
97. T. Feizi, *Nature*, 1985, **314**, 53.
98. M. Basu and S. Basu, *J. Biol. Chem.*, 1984, **259**, 12 557.
99. C. L. M. Stults and B. A. Macher, *Arch. Biochem. Biophys.*, 1993, **303**, 125.
100. F. Piller, J.-P. Cartron, A. Maranduba, A. Veyrieres, Y. Leroy, and B. Fournet, *J. Biol. Chem.*, 1984, **259**, 13 385.
101. A. Leppänen, L. Penttilä, R. Niemelä, J. Helin, A. Seppo, S. Lusa, and O. Renkonen, *Biochemistry*, 1991, **30**, 9287.
102. J. Gu, A. Nishikawa, S. Fujii, S. Gasa, and N. Taniguchi, *J. Biol. Chem.*, 1992, **267**, 2994.
103. A. D. Magnet and M. Fukuda, *Glycobiology*, 1997, **7**, 285.
104. B. D. Shur and C. A. Neely, *J. Biol. Chem.*, 1988, **263**, 17 706.
105. B. T. Sheares, J. T. Lau, and D. M. Carlson, *J. Biol. Chem.*, 1982, **257**, 599.
106. P. Bailly, F. Piller, J.-P. Cartron, Y. Leroy, and B. Fournet, *Biochem. Biophys. Res. Commun.*, 1986, **141**, 84.
107. M. Messer and K. R. Nicholas, *Biochim. Biophys. Acta*, 1991, **1077**, 79.
108. D. H. Joziasse, H. C. M. Damen, M. de Jong-Brink, H. T. Edzes, and D. H. van den Eijnden, *FEBS Lett.*, 1987, **221**, 139.
109. H. Mulder, H. Schachter, M. De Jong-Brink, J. G. Van der Ven, J. P. Kamerling, and J. F. Vliegenthart, *Eur. J. Biochem.*, 1991, **201**, 459.
110. B. T. Sheares and D. M. Carlson, *J. Biol. Chem.*, 1983, **258**, 9893.
111. E. H. Holmes, *Arch. Biochem. Biophys.*, 1989, **270**, 630.
112. A. Seko, T. Ohkura, H. Kitamura, S. Yonezawa, E. Sato, and K. Yamashita, *Cancer Res.*, 1996, **56**, 3468.
113. H. van Halbeek, G. J. Gerwig, J. F. G. Vliegenthart, H. L. Smits, P. J. M. van Kerkhof, and M. F. Kramer, *Biochim. Biophys. Acta*, 1983, **747**, 107.
114 K. Ishihara, M. Kurihara, Y. Goso, T.Urata, H. Ota, T. Katsuyama, and K. Hotta, *Biochem. J.*, 1996, **318**, 409.
115. W. M. Watkins, in "Glycoproteins," eds. J. Montreuil, J. F. G. Vliegenthart, and H. Schachter, Elsevier, Amsterdam, 1995, vol. 29a, p. 313.
116. F. Dall'Olio, N. Malagolini, G. Di Stefano, M. Ciambella, and F. Serafini-Cessi, *Biochem. J.*, 1990, **270**, 519.
117. D. X. Wen, B. D. Livingston, K. F. Medzihradszky, S. Kelm, A. L. Burlingame, and J. C. Paulson, *J. Biol. Chem.*, 1992, **267**, 21 011.
118. A. K. Datta and J. C. Paulson, *J. Biol. Chem.*, 1995, **270**, 1497.
119. W. Gillespie, S. Kelm, and J. C. Paulson, *J. Biol. Chem.*, 1992, **267**, 21 004.
120. H. Kitagawa and J. C. Paulson, *J. Biol. Chem.*, 1994, **269**, 1394.
121. Y. C. Lee, N. Kurosawa, T. Hamamoto, T. Nakoaka, and S. Tsuji, *Eur. J. Biochem.*, 1993, **216**, 377.
122. Y. C. Lee, C. S. Park, and C. A. Strott, *J. Biol. Chem.*, 1994, **269**, 15 838.
123. M. L. Chang, R. L. Eddy, T. B. Shows, and J. T. Y. Lau, *Glycobiology*, 1995, **5**, 319.
124. N. Kurosawa, T. Hamamoto, M. Inoue, and S. Tsuji, *Biochim. Biophys. Acta*, 1995, **1244**, 216.
125. Y. J. Kim, K. S. Kim, S. H. Kim, C. H. Kim, J. H. Ko, I. S. Choe, S. Tsuji, and Y. C. Lee, *Biochem. Biophys. Res. Commun.*, 1996, **228**, 324.
126. P. Reboul, P. George, J. Geoffroy, P. Louisot, and P. Broquet, *Biochem. Biophys. Res. Commun.*, 1992, **186**, 1575.
127. P. Broquet, P. George, J. Geoffroy, P. Reboul, and P. Louisot, *Biochem. Biophys. Res. Commun.*, 1991, **178**, 1437.
128. M. A. K. Markwell and J. C. Paulson, *Proc. Natl. Acad. Sci. USA*, 1980, **77**, 5693.
129. P. R. Crocker, S. Kelm, C. Dubois, B. Martin, A. S. McWilliam, D. M. Shotton, J. C. Paulson, and S. Gordon, *EMBO J.*, 1991, **10**, 1661.
130. S. R. Carlsson, H. Sasaki, and M. Fukuda, *J. Biol. Chem.*, 1986, **261**, 12 787.
131. A. Kanani, D. R. Sutherland, E. Fibach, K. L. Matta, A. Hindenburg, I. Brockhausen, W. Kuhns, R. N. Taub, D. H. van den Eijnden, and M. A. Baker, *Cancer Res.*, 1990, **50**, 5003.
132. M. A. Baker, A. Kanani, I. Brockhausen, H. Schachter, A. Hindenburg, and R. N. Taub, *Cancer Res.*, 1987, **47**, 2763.
133. W. Gillespie, J. C. Paulson, S. Kelm, M. Pang, and L. G. Baum, *J. Biol. Chem.*, 1993, **268**, 3801.
134. O. Saitoh, R. E. Gallagher, and M. Fukuda, *Cancer Res.*, 1991, **51**, 2854.
135. P. Delannoy, H. Pelczar, V. Vandamme, and A. Verbert, *Glycoconjugate J.*, 1993, **10**, 91.
136. C. Whitehouse, J. Burchell, S. Gschmeissner, I. Brockhausen, K. O. Lloyd, and J. Taylor-Papadimitriou, *J. Cell Biol.*, 1997, **137**, 1229.
137. N. Kurosawa, N. Kojima, M. Inoue, T. Hamamoto, and S. Tsuji, *J. Biol. Chem.*, 1994, **269**, 19 048.
138. N. Kurosawa, T. Hamamoto, Y. C. Lee, T. Nakaoka, N. Kojima, and S. Tsuji, *J. Biol. Chem.*, 1994, **269**, 1402.
139. M. L. Bergh, G. J. M. Hooghwinkel, and D. H. van den Eijnden, *J. Biol. Chem.*, 1983, **258**, 7430.
140. H. Nakasaki, T. Mitomi, T. Noto, K. Ogoshi, H. Hanaue, Y. Tanaka, H. Makuuchi, H. Clausen, and S.-I. Hakomori, *Cancer Res.*, 1989, **49**, 3662.
141. S. Yonezawa, T. Tachikawa, S. Shin, and E. Sato, *Am. J. Clin. Pathol.*, 1992, **98**, 167.
142. S. R. Carter, A. Slomiany, K. Gwozdzinski, Y. H. Liau, and B. L. Slomiany, *J. Biol. Chem.*, 1988, **263**, 11 977.
143. Y. Goso and K. Hotta, *Glycoconjugate J.*, 1993, **10**, 226.
144. W. J. Kuhns, M. M. Burger, and G. Misevic, *Biol. Bull.*, 1995, **189**, 223.
145. C. Capon, C. L. Laboisse, J.-M. Wieruszeski, J.-J. Maoret, C. Augeron, and B. Fournet, *J. Biol. Chem.*, 1992, **267**, 19 248.
146. S. Hemmerich, C. Bertozzi, H. Leffler, and S. D. Rosen, *Biochemistry*, 1994, **33**, 4820.
147. K. V. Chace, D. S. Leahy, R. Martin, R. Carubelli, M. Flux, and G. P. Sachdev, *Clin. Chim. Acta*, 1983, **132**, 143.
148. T. Yamori, H. Kimura, K. Stewart, D. M. Ota, K. R. Cleary, and T. Irimura, *Cancer Res.*, 1987, **47**, 2741.
149. A. H. Raouf, H. H. Tsai, N. Parker, J. Hoffman, R. J. Walker, and J. M. Rhodes, *Clinical Science*, 1992, **83**, 623.

150. W. J. Kuhns, R. K. Jain, K. L. Matta, H. Paulsen, M. A. Baker, R. Geyer, and I. Brockhausen, *Glycobiology*, 1995, **5**, 689.
151. J. M. Lo-Guidice, J. M. Perini, J. J. Lafitte, M. P. Ducourouble, P. Roussel, and G. Lamblin, *J. Biol. Chem.*, 1995, **270**, 27 544.
152. Y. Kato and R. Spiro, *J. Biol. Chem.*, 1989, **264**, 3364.
153. W. J. Kuhns, O. Popescu, M. M. Burger, and G. Misevic, *J. Cell Biochem.*, 1995, **58**, 1.
154. E. V. Chandrasekaran, R. K. Jain, R. D. Larsen, K. Wlasichuk, R. A. DiCioccio, and K. L. Matta, *Biochemistry*, 1996, **35**, 8925.
155. E. V. Chandrasekaran, R. K. Jain, R. D. Larsen, K. Wlasichuk, and K. L. Matta, *Biochemistry*, 1996, **35**, 8914.
156. T. Feizi, *Biochem. Soc. Trans.*, 1988, **16**, 930.
157. S.-I. Hakomori, *Adv. Cancer Res.*, 1989, **52**, 257.
158. K. Hiraishi, K. Suzuki, and S.-I. Hakomori, *Glycobiology*, 1993, **3**, 381.
159. F.-I. Yamamoto and S.-I. Hakomori, *J. Biol. Chem.*, 1990, **265**, 19 257.
160. F.-I. Yamamoto, H. Clausen, T. White, J. Marken, and S.-I. Hakomori, *Nature*, 1990, **345**, 229.
161. N. O. L. Seto, M. M. Palcic, C. A. Compston, H. Li, D. R. Bundle, and S. A. Narang, *J. Biol. Chem.*, 1997, **272**, 14 133.
162. U. Galili, *Lancet*, 1989, **2**, 358.
163. U. Galili, *Immunol. Today*, 1993, **14**, 480.
164. J. D. Bleil and P. M. Wassarman, *Proc. Natl. Acad. Sci. USA*, 1988, **85**, 6778.
165. A. Cheng, T. Le, M. Palacios, L. H. Bookbinder, P. M. Wassarman, F. Suzuki, and J. D. Bleil, *J. Cell Biol.*, 1994, **125**, 867.
166. A. D. Thall, P. Maly, and J. B. Lowe, *J. Biol. Chem.*, 1995, **270**, 21 437.
167. U. Galili and K. Swanson, *Proc. Natl. Acad. Sci. USA*, 1991, **88**, 7401.
168. S. K. Cho, J. Yeh, M. Cho, and R. D. Cummings, *J. Biol. Chem.*, 1996, **271**, 3238.
169. N. Osman, I. F. C. McKenzie, E. Mouhtouris, and M. S. Sandrin, *J. Biol. Chem.*, 1996, **271**, 33 105.
170. S. K. Cho and R. D. Cummings, *J. Biol. Chem.*, 1997, **272**, 13 622.
171. A. Takeya, O. Hosomi, and T. Kogure, *J. Biochemistry (Tokyo)*, 1987, **101**, 251.
172. A. Conzelmann and C. Bron, *Biochem. J.*, 1987, **242**, 817.
173. T. Dohi, Y. Yuyama, Y. Natori, P. L. Smith, J. B. Lowe, and M. Oshima, *Int. J. Cancer*, 1996, **67**, 626.
174. N. Malagolini, F. Dall'Olio, G. Di Stefano, F. Minni, D. Marrano, and F. Serafini-Cessi, *Cancer Res.*, 1989, **49**, 6466.
175. N. Malagolini, F. Dall'Olio, and F. Serafini-Cessi, *Biochem. Biophys. Res. Commun.*, 1991, **180**, 681.
176. P. L. Smith and J. B. Lowe, *J. Biol. Chem.*, 1994, **269**, 15 162.

3.05
Organization and Topology of Sphingolipid Metabolism

GERHILD VAN ECHTEN-DECKERT and KONRAD SANDHOFF
Universität Bonn, Germany

3.05.1   INTRODUCTION

Glycosphingolipids (GSLs) are amphiphilic components of plasma membranes of all vertrebrate cells.[1,2] They also occur in intracellular membranes of the secretory and endocytotic pathways, e.g., in the Golgi, (trans Golgi network) TGN, endosomal, and lysosomal membranes.

Their hydrophobic ceramide moieties anchor complex GSLs in the outer leaflet of the plasma membrane, so that their hydrophilic oligosaccharide residues face the extracellular space. Today more than 300 different GSL structures are known which can be classified in a few families as illustrated in Figure 1. Gangliosides are a group of sialic acid containing GSLs (Figure 2) which show a great molecular diversity with numerous novel minor components.[4,5] Gangliosides are enriched in the brain prevailing in neuronal, and particularly synaptic membranes, as well as in growth cones.[6,7] It is well known that the cell-surface carbohydrate profile is cell and species specific and characteristically changes in development, differentiation, organ regeneration, and oncogenic transformation (Figure 3), suggesting its significance for cell–cell interactions and cell adhesion.[4,8]

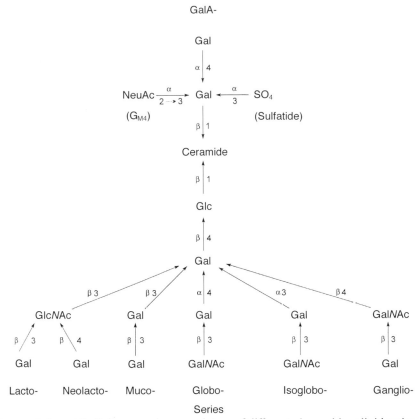

Figure 1   Scheme of glycosidic linkages and nomenclature of different glycosphingolipid series. Ceramide, *N*-acylsphingosine; Gal, D-galactose; GalNAc, *N*-acetyl-D-galactosamine; Glc, D-glucose; GlcNAc, *N*-acetyl-glucosamine; NeuAc, *N*-acetylneuraminic acid.

The oligosaccharide chains of GSLs are binding sites for lectins, specific carbohydrate recognizing proteins such as bacterial toxins, binding proteins of viruses and antibodies, which by means of their binding to cell surfaces, might influence on cellular activity.[5]

Sphingolipid metabolites have been identified as endogenous signal-transducing molecules. Sphingomyelin (SM), the major membrane sphingolipid can be hydrolyzed by sphingomyelinases to form ceramide, which stimulates differentiation, inhibits proliferation and has also been associated with apoptosis.[9–11] Moreover, sphingosine and sphingosine-1-phosphate (SPP), originally proposed as negative regulators of protein kinase C (PKC),[12] were shown to play alternative signaling roles as mitogenic second messengers.[11]

Although most of the GSLs are concentrated on the plasma membrane, their biosynthesis and degradation are localized intracellularly. Whereas GSL biosynthesis starts at the membranes of the endoplasmic reticulum (ER) and continues on the Golgi membranes, catabolism occurs after endocytosis in the lysosomal compartment.

3.05.2   GLYCOSPHINGOLIPID BIOSYNTHESIS IN THE ER–GOLGI COMPLEX

Our current knowledge on localization of GSL biosynthesis essentially emerged from subcellular fractionation studies or from cell culture experiments in which inhibitors of cellular transport through the ER–Golgi complex, like the fungal macrolide brefeldin A (BFA) or the antibiotic monensin, have been used. It is generally accepted that GSL formation, like the formation of glycoproteins, is coupled to a vesicular membrane flow, from the ER through the cisternae of the Golgi complex to the plasma membrane.[13] However, the involvement of glycolipid binding and/or

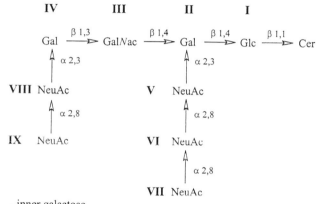

(a) G_{D1a}

(b) G_{P1c} (NeuAcα2 → 8NeuAcα2 → 3Galβ1 → 3Gal*N*Acβ1 → 4(NeuAcα2 → 8NeuAcα2 → 8NeuAcα2 → 3)Galβ1 → 4Glcβ1 → 1Cer):

		IV		III		II		I		

$$\text{Gal} \xrightarrow{\beta\,1,3} \text{Gal}N\text{ac} \xrightarrow{\beta\,1,4} \text{Gal} \xrightarrow{\beta\,1,4} \text{Glc} \xrightarrow{\beta\,1,1} \text{Cer}$$

     ↑ α 2,3   ↑ α 2,3

VIII NeuAc   **V** NeuAc

     ↑ α 2,8   ↑ α 2,8

IX NeuAc   **VI** NeuAc

                    ↑ α 2,8

                  VII NeuAc

II   - inner galactose
IV   - outer galactose

The following structures of GSLs are part of the GP1c structure and contain in addition to the hydrophobic ceramide (Cer) backbone the following sugar residues marked by Roman numbers:

GlcCer	- glucosylceramide, **I**		G_{D1b}	- **I** - **VI**
LacCer	- lactosylceramide, **I, II**		G_{T1c}	- **I** - **VII**
G_{M3}	- **I, II, V**		G_{M1b}	- **I** - **IV, VIII**
G_{D3}	- **I, II, V, VI**		G_{D1a}	- **I** - **V, VIII**
G_{T3}	- **I, II, V** - **VII**		G_{T1b}	- **I** - **VI, VIII**
G_{A2}	- **I** - **III**		G_{Q1c}	- **I** - **VIII**
G_{M2}	- **I** - **III, V**		G_{D1c}	- **I** - **IV, VIII, IX**
G_{D2}	- **I** - **III, V, VI**		G_{T1a}	- **I** - **V, VIII, IX**
G_{T2}	- **I** - **III, V** - **VII**		G_{Q1b}	- **I** - **VI, VIII, IX**
G_{A1}	- **I** - **IV**		G_{P1c}	- **I** - **IX**
G_{M1A}	- **I** - **V**			

Figure 2   Structure of (a) ganglioside G_{D1a} and (b) of ganglioside G_{P1c} and of related glycolipids. The terminology used for gangliosides is that of Svennerholm.[3] For abbreviations, see the caption to Figure 1.

transfer proteins in the transport of sphingolipids, especially during initial steps of biosynthesis cannot be excluded.

3.05.2.1   Dihydroceramide Formation at the ER

The first steps of sphingolipid biosynthesis leading to the formation of dihydroceramide (DHCer) (Figure 4) are catalyzed by membrane-bound enzymes active at the cytosolic face of the ER.[14,15]

The pathway of *de novo* sphingolipid biosynthesis starts with the condensation of serine and palmitoyl-CoA to 3-dehydrosphinganine, a reaction catalyzed by serine palmitoyltransferase.[16,17] 3-Dehydrosphinganine is immediately reduced to D-*erythro*-sphinganine by a D-3-dehydro-sphinganine-NADH-oxidoreductase.[17]

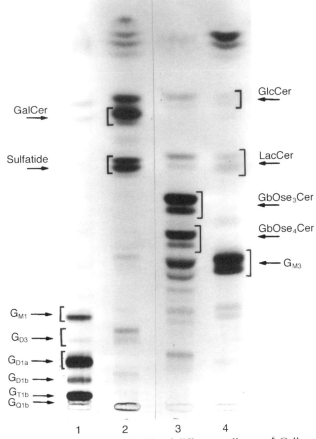

Figure 3 Biosynthetic labeling of glycosphingolipids of different cell types.[6] Cells were incubated for 48 h in the presence of [^{14}C]galactose (2 μCi ml^{-1}) and then harvested. Glycolipids were extracted, desalted, separated by thin layer chromatography and visualized by fluorography. Lane 1, primary cultured cerebellar neurons; lane 2, oligodendrocytes; lane 3, fibroblasts; lane 4, neuroblastoma cells (B104). The mobility of standard lipids is indicated. GbOse$_3$Cer, Galα1 → 4Galβ1 → 4Glcβ1 → 1Cer; GbOse$_4$Cer, GalNAcβ1 → 3 Galα1 → 4Galβ1 → 4Glcβ1 → 1Cer.

Acylation of the amino group of sphinganine by acyl-CoA yields dihydroceramide.[13] Its conversion to ceramide by introducing a *trans*-4,5 double bond seems to be physiologically significant for the cells. Both ceramide and sphingosine were shown to be involved in signal-transduction processes, triggering apoptosis and mitosis, respectively, while their saturated analogues dihydroceramide and sphinganine (dihydrosphingosine) were much less effective or failed to show these effects at all.[18–20] Therefore, it is interesting to know at which molecular level the *trans*-4 double bond of sphingolipids is introduced. As first suggested by Ong and Brady[21] and Stoffel and Bister,[22] later shown by Merrill and Wang[23] and finally demonstrated by Rother *et al.*,[24] the introduction of the *trans*-4 double bond occurs by desaturation of dihydroceramide and not at the level of sphinganine. Therefore, sphingosine is not an intermediate of GSL biosynthesis, but rather a catabolic product of ceramide, which may originate either as a degradation product of sphingomyelin (SM) or GSLs or as a biosynthetic intermediate of GSL formation. An *in vitro* assay of dihydroceramide desaturase activity has been established using rat liver microsomes as an enzyme source.[25] Relative to *N*-octanoyl derivatives of dihydroceramide, dihydrosphingomyelin was desaturated only by about 20%, while dihydro-GlcCer and sphinganine were not desaturated at all. On the other hand, in cultured cells C6-NBD-DHCer (*N*-6-(7-nitro-2,1,3-benzoxadiazol-4-yl)aminohexanoyl-dihydroceramide), was mainly converted to various saturated dihydro(glyco)sphingolipids. Introduction of the double bond apparently occurred slowly, presumably during various cycles of degradation and resynthesis, at the level of the C6-NBD-DHCer analogue.[26] Also, newly synthesized sphingolipids of a murine macrophage-like cell line (J774A.1) contain mainly DHCer as hydrophobic backbone, which is later on slowly replaced by Cer.[27] Apparently, only a small fraction of native DHCer is converted to Cer during initial *de novo* sphingolipid biosynthesis. Sphinganine, released after catabolism of saturated GSLs from the lysosomal compartment can be reacylated to DHCer, which is

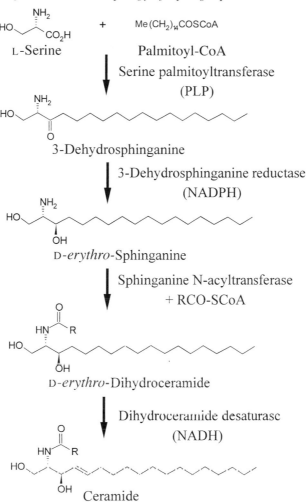

Figure 4 Scheme of ceramide biosynthesis. All enzymatic steps are located to the cytosolic leaflet of the endoplasmic reticulum.

then desaturated to Cer and reused for sphingolipid biosynthesis in a salvage pathway (see Figure 5).

From preliminary experiments it seems very likely that dihydroceramide desaturase activity is associated with the cytosolic leaflet of the ER.[28] Interestingly, it has been reported that not only ceramide derived from hydrolysis of plasma membrane associated sphingomyelin but also *de novo* biosynthesized ceramide might trigger cellular responses to induce apoptosis[29] and differentiation.[30]

3.05.2.2 Early Sphingolipid Glycosylation

3.05.2.2.1 *Intracellular movement of ceramide from the ER to the Golgi*

We now know that the biosynthesis of DHCer and possibly also that of ceramide occurs at the cytosolic surface of the ER. But, the main site of Cer consumption is the Golgi, where transfer of head groups onto Cer by SM synthase to give SM, by glucosyltransferase to give GlcCer, and by galactosyltransferase to form GalCer occurs (see below). The rate and mechanism of ceramide transport from the ER to the Golgi are, however, not yet clear.

Although it has been suggested that ceramide is transported from ER to the Golgi by vesicles,[31] a direct demonstration of this type of ceramide movement is noticeably absent from the literature.[32]

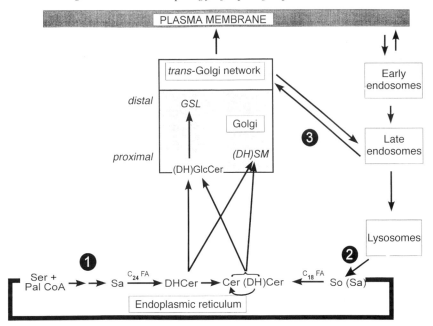

Figure 5 Pathways for the formation of sphingolipids. Route 1 represents *de novo* glycosphingolipid (GSL) biosynthesis. Route 2 illustrates the sphingosine (So) salvage pathway. Route 3 shows recycling of native or partially hydrolysed sphingolipids. Ser, serine; PalCoA, palmitoyl-coenzyme A; Sa, sphinganine; DHCer, dihydroceramide; Cer, ceramide; (DH)GlcCer, (dihydro)glucosylceramide; (DH)SM, (dihydro)sphingomyelin.

There is, on the other hand, evidence for a protein facilitated transport of Cer between the ER and the Golgi apparatus. In mitotic HeLa cells, in which all vesicular transport pathways are inhibited, SM and GlcCer are still synthesized, but not higher GSLs, suggesting that Cer transport between ER and Golgi apparatus occurs even in the absence of vesicular traffic.[33] Further evidence to support the facilitated transport of (DH)Cer from the ER to the Golgi apparatus comes from studies using stereoisomers. All four stereoisomers of sphinganine are acylated *in vivo*, but only the D-*erythro* and to some extent the L-*threo*-isomers are subsequently converted to SM and GlcCer[34] perhaps implying that D-*erythro*-DHCer is transported out of the ER by a protein facilitated mechanism that is stereospecific towards the long-chain base. Observations in rat liver Golgi that SM synthase is able to metabolize all four Cer stereoisomers and GlcCer synthase is able to metabolize both D-*erythro* and L-*threo*-Cer[35] support this idea.

An experiment we have performed with primary cultured neurons also argues in favor of a protein-facilitated ceramide transport. The lysosomotropic amine primaquine has previously been shown to block intracellular transport by inhibiting the formation of functional transport vesicles by irreversibly inactivating the membrane that forms transport vesicles (donor), but not the membranes that are the destination of those vesicles (acceptor).[36]

Pretreatment of cultured neurons with primaquine (1 mM) followed by biosynthetic labeling of cellular sphingolipids with [^{14}C]serine almost completely inhibited *de novo* sphingolipid biosynthesis (Figure 6). Only SM and GlcCer, the immediate derivatives of Cer, which at least in part are synthesized in an early Golgi compartment (see below), were still labeled, suggesting a nonvesicular transport of Cer from the ER to the early Golgi compartment.

Once (DH)Cer reached the Golgi complex, where most enzymes involved in GSL biosynthesis are localized, it has access to both membrane leaflets by rapid transbilayer movement, presumably, as it lacks a polar headgroup.

3.05.2.2.2 *Galactosylation of ceramide*

Biosynthesis of GalCer (Figure 1) is mainly associated with myelination and appears to be a characteristic property of oligodendrocytes. Besides the myelinating tissue of the nervous system

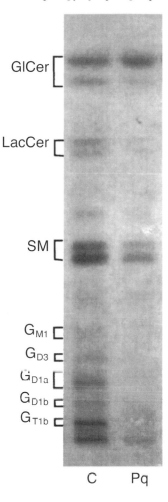

GlCer

LacCer

SM

G_{M1}
G_{D3}
G_{D1a}
G_{D1b}
G_{T1b}

C   Pq

Figure 6 The effect of primaquine on glycosphingolipid biosynthesis of primary cultured neurons. Cells were incubated for 30 min in the absence (C) or presence of 100 mM primaquine (Pq). Then [^{14}C]serine (2 μCi ml^{-1}) was added to the medium and the incubation was continued. After 6 h, cells were harvested and lipids were analyzed as described in the caption to Figure 3.

in humans, GalCer is also abundant in the epithelia of the intestine and kidney. Conflicting results have been reported on the localization of the enzyme responsible for GalCer synthesis, UDP-Gal:ceramide:galactosyltransferase. The ER[37] and/or the Golgi[38] have been suggested as possible sites of GalCer biosynthesis. Using short-chain ceramides with either 2-hydroxy fatty acid (HFA) or a normal fatty acid (NFA), Burger *et al.*[39] determined two different sites with a different topography of GalCer biosynthesis, in dog kidney MDCK II cells and also in the rat Schwann cell line D6P2T, depending on the nature of the ceramide fatty acid. Whereas the HFA-GalCer synthesizing activity colocalized with an ER marker, the NFA-GalCer synthesizing activity fractionated at the Golgi density of a sucrose gradient. In oligodendrocytes, where NFA-Cer is the precursor for GalCer, NFA-GalCer was found to be a *cis*-Golgi marker and its formation requires the transport of NFA-Cer to the Golgi complex.[40]

In cell homogenates and permeabilized cells short-chain NFA-GalCer was immediately accessible to serum albumin, suggesting a cytosolic topology for the Golgi associated Gal-transferase activity, whereas short-chain HFA-GalCer was protected against serum albumin, suggesting an opposite, anticytosolic topology for its formation. These topological results imply that NFA-GalCer must translocate from the cytosolic to the lumenal leaflet of the Golgi membrane, to become available to the next biosynthetic enzymes.[39] On the other hand, HFA-GalCer formed at the lumenal face of the ER may follow the exocytotic pathway through the Golgi compartment by vesicular membrane flow.

3.05.2.2.3   *Formation of GlcCer and LacCer*

Apart from the fact that GlcCer is the most abundant GSL in most cells, it is the common precursor for virtually all GSLs (except the gala series), including gangliosides (except G_{M4}) (Figure 1). In contrast to complex GSLs, GlcCer is located in both the outer (anticytosolic) and the inner (cytosolic) leaflet of the plasma membrane. In the literature GlcCer synthase activity has been assigned to the cytosolic leaflet of the Golgi apparatus.[41–44] However, the exact site of GlcCer synthesis has not unambiguously been proven. Reports analyzing subcellular fractions from porcine submaxillary glands suggested that GlcCer synthase is primarily associated with the Golgi apparatus,[41] whereas Futerman and Pagano[42] showed that GlcCer synthase is more widely distributed among microsomal subfractions of rat liver. Although GlcCer synthesis in rat liver mainly occurred in a *cis/medial*-Golgi subfraction, significant synthesis was also detected in two microsomal fractions, one of which could be attributed to an intermediate compartment between ER and *cis/medial*-Golgi. In studies on GSL metabolism in primary cultured neurons using a 15 °C temperature block, in which vesicular transport is arrested between the ER and *cis*-Golgi an accumulation of GlcCer occurred, suggesting either a preGolgi compartment for the glucosylation of Cer[6] or alternatively, a nonvesicular transport of Cer to the Golgi complex. GlcCer was also found to accumulate when the vesicular transport between the *medial*- and *trans*-cisternae of the Golgi apparatus was disrupted by monensin, strongly suggesting that GlcCer synthesis occurs in the "early" proximal Golgi.[6,45] In contrast to these reports, Jeckel *et al.*,[43] who used Golgi fractions from rat and rabbit liver as well as from CHO and HepG2 cells, found GlcCer synthase activity in two different Golgi fractions: one containing a proximal (early) Golgi marker and the other a distal (late) Golgi marker. The "late" synthesized GlcCer may be beyond the site of LacCer synthesis and could explain why a considerable amount of GlcCer is never used for GSL biosynthesis. It seems likely that late Golgi GlcCer is translocated to the lumenal side and directly transported to the cell surface. Synthesis of GlcCer late in the Golgi could also explain the very rapid kinetics of transport of truncated GlcCer to the plasma membrane. Thus, the time for delivery of newly synthesized short-chain GlcCer and ganglioside G_{M3} to the surface of CHO cells was 5 min, while newly synthesized sphingomyelin reached the cell surface only after 14 min.[46] However, in the same cell type newly synthesized long-chain GlcCer reached the plasma membrane after 7.2 min, whereas the respective long-chain G_{M3} required 21.5 min to reach the plasma membrane.[47] Furthermore, the transport of G_{M3} and of sphingomyelin was prevented at 15 °C, as well as in the presence of BFA, while that of GlcCer was not affected.[47] From these results the authors conclude that a major fraction of newly synthesized GlcCer is rapidly transported to the plasma membrane by a non-Golgi pathway, independent of the vesicular pathways used for the transport of proteins and complex GSL. It cannot be excluded that GlcCer exposed to the cytosol may interact with cytosolic transfer proteins and thus be delivered to target membranes. A GSL transfer protein has been described which is capable of transferring GlcCer from donor to acceptor membranes *in vitro*.[48] It is, however, not clear whether glycolipid transfer proteins have a similar function *in vivo*. On the other hand, kinetic data suggest that most of the GlcCer, like SM, is transported to the cell surface by a vesicular mechanism.[49]

Galactosylation of GlcCer to form LacCer is the next biosynthetic step of most GSL series (Figure 1). From various cell culture studies in which inhibitors of the vesicular traffic through the Golgi compartment were used,[50] as well as from subcellular fractionation experiments performed in rat liver Golgi,[51] LacCer synthase (Gal-transferase I) activity was assigned mainly to the early (*cis*) Golgi cisternae. In contrast to these results, Lannert *et al.*[52] found that in rat liver Golgi LacCer biosynthesis resides in the late (*trans*) Golgi compartments. The discrepancy between the results obtained by the latter two groups could be explained, at least in part, by the different methodological approaches used. In contrast to the previous study by Trinchera *et al.*,[51] Lannert *et al.*[52] used a truncated analogue of ceramide and GlcCer, respectively, with only eight carbon atoms in both its sphingosine and fatty acid moieties for the determination of transferase activities in different Golgi subfractions. As the truncated substrates readily permeate membranes, enzyme assays could be performed in the absence of detergents in intact membranes. Although most LacCer synthase activity was found in the late Golgi fraction, there was significant activity of this enzyme also in the early Golgi fraction, suggesting an overlapping rather than a clearcut compartmentalization of the enzyme in the two Golgi regions.

Conflicting results have also been reported on the topology of LacCer synthase. Trinchera *et al.*[44] found that besides GlcCer synthesis the subsequent transfer of galactose to form LacCer also occurs on the cytosolic leaflet of the Golgi apparatus. This conclusion is based on a comparison of intact and detergent-permeabilized rat liver Golgi, and on the assumption that Golgi permeabilization should increase the apparent activity of lumenal enzymes by increasing the availability of their substrates. The reliability of this conclusion is, however, questionable since inhibitory effects of detergent on LacCer synthase activity, using C6-NBD-GlcCer as a substrate, have been reported.[39] The finding that cells lacking the UDP-Gal carrier, which translocates UDP-Gal into the Golgi lumen, can only synthesize GlcCer[53,54] strongly argues in favor of a lumenal topology for Gal-transferase I. These results have been confirmed by Lannert *et al.*[55] and Burger *et al.*,[39] who performed sidedness experiments, indicating that GSL biosynthesis depends on translocation of GlcCer to the lumenal leaflet of the Golgi membrane. At present, however, it is unknown how GlcCer is translocated from the cytosolic to the lumenal leaflet of the Golgi membrane. Studies with a truncated epoxy-GlcCer ((2*S*,3*R*,4*E*)-1-[4,7-anhydro-4-*C*-(hydroxymethyl)-β-D-gluco-pyranosyloxy]-2-(dodecanoylamino)-4-dodecen-3-ol), a compound which strongly reduces LacCer synthase activity in cultured neurons, led to the assumption that LacCer synthase itself or a protein complex containing the enzyme might facilitate the translocation of GlcCer from the cytosolic to the lumenal surface of Golgi membranes.[56]

3.05.2.3   Sialylation Steps: Ganglioside Biosynthesis

3.05.2.3.1   *Substrate specificity and topology of glycosyltransferases*

Transfer of subsequent sugars to LacCer to form various trihexosylceramides, thus defining the core structure and diversity of the respective GSL series (Figure 1), as well as the sequential addition of further monosaccharide or sialic acid residues to the growing oligosaccharide chain, is catalyzed by membrane bound glycosyltransferases, which have been shown to be restricted to the lumenal leaflet of the Golgi apparatus.[57–60] In other words, higher GSLs are synthesized in the Golgi lumen and they cannot translocate towards the cytosolic leaflet.[55] Their transport follows the kinetics of vesicular traffic.[61,62] As demonstrated for rat liver Golgi, sequential glycosylation of analogous precursors, which differ only in the number of neuraminic acid residues bound to the inner galactose residue of the oligosaccharide chain, is catalyzed by a set of rather unspecific glycosyltransferases (Figure 7).[50] As illustrated by Figure 7, the number of sialic acid residues bound to the inner galactose of the carbohydrate chain (0, 1, 2 or 3) determines to which series (asialo, a, b or c, respectively) a certain ganglioside belongs. It has been shown, in rat liver Golgi, that only one GalNAc-transferase catalyzes the reaction from LacCer, G_{M3}, G_{D3}, and G_{T3} to G_{A2}, G_{M2}, G_{D2}, and G_{T2}, respectively,[64,65] and accordingly, only one galactosyltransferase is responsible for the formation of G_{A1}, G_{M1a}, and G_{D1b} from G_{A2}, G_{M2} and G_{D2}, respectively. Similar results were obtained for sialyltransferase IV[64] and sialyltransferase V.[66] Kinetic studies in rat liver Golgi came to the conclusion that sialyltransferases I and II which catalyze the initial sialylation steps are more specific for their substrate.[67] On the other hand, Nara *et al.*[68] reported that cloned sialyltransferase II (G_{D3} synthase) isolated from human melanoma cells[69] also has sialyltransferase V activity, catalyzing the formation of G_{D3} and of G_{D1c}, G_{T1a} and G_{Q1b} *in vitro*. Furthermore, by transfection of the cloned human α 2,8-sialyltransferase cDNA, transient and stable expression of G_{T1a} and G_{Q1b} was also observed in COS-7 cells as well as in Swiss 3T3 cells, both of which originally lacked sialyltransferase II and sialyltransferase V activities.

Several other sialyltransferases have been cloned and analyzed.[70] The overlapping substrate specificities of different sialyltransferase families observed *in vitro* must not necessarily be relevant for the much more complex *in vivo* situation.[70] A species-specific substrate specificity of different sialyltransferases, as well as the existence of isoenzymes in different cell types can also not be excluded. Thus, a ganglioside-specific sialyltransferase, catalyzing the formation of both G_{D3} and G_{T3} is specifically expressed in neural tissues.[71] Among human tissues the expression of its mRNA is highly restricted to fetal and adult brain.[71]

3.05.2.3.2   *Sub-Golgi localization of glycosyltransferases*

Although many glycosyltransferases involved in ganglioside biosynthesis have been cloned and characterized (see Chapter 3.06), it is not yet clear where exactly in the Golgi stack individual glycosylation reactions take place. If different steps of ganglioside biosynthesis are localized in different Golgi compartments and the biosynthetic process is coupled to a vesicle bound membrane flow of the growing molecules through different Golgi compartments, inhibitors of exocytotic membrane flow should attenuate formation of complex gangliosides. In a system in which cultured cells are fed with radioactive precursors of GSL biosynthesis, the biosynthetic labeling of intermediates before the respective block should be increased and consequently labeling of more complex GSLs beyond the drug-induced transport block should be decreased.

One of the first drugs used for this purpose was the cationic ionophore monensin, shown to impede primarily vesicular membrane flow between the proximal and distal Golgi cisternae.[72] When primary cultured neurons were incubated in the presence of this drug incorporation of radioactivity from [^{14}C]galactose decreased remarkably for the gangliosides G_{M1a}, G_{D1a}, G_{D1b}, G_{T1b}, and G_{Q1b}, whereas relative labeling of GlcCer, LacCer, G_{M3}, G_{D3}, and G_{M2} increased significantly, suggesting that complex gangliosides, e.g., G_{T1b} and G_{Q1b}, are synthesized distal to the monensin block, in the TGN (*trans*-Golgi network) or in the *trans*-Golgi cisternae, while the less glycosylated gangliosides like G_{M3} and G_{D3} as well as GlcCer and LacCer are formed prior to the monensin block, most probably in *cis/medial*-Golgi elements.[6] Similar results of monensin on GSL biosynthesis were reported previously for other cells. In cultured neurotumor cells,[73] as well as in cultured human fibroblasts,[45,74] a monensin-induced increase of radiolabel incorporation into GlcCer and LacCer with a concomitant reduction of label incorporation into gangliosides, as well as globosides, were observed. These results again argue for the biosynthesis of the more highly glycosylated neutral GSLs and gangliosides in a Golgi site beyond the monensin-induced transport block. Another drug used to obtain further evidence for the localization of different glycosylation steps is the antibiotic brefeldin A (BFA). BFA is a macrocyclic lactone synthesized from palmitate by different fungi.[75] Initially, BFA was shown to inhibit protein secretion at an early step in the secretory pathway. Although retained in the ER of BFA treated cells, proteins were further processed by Golgi enzymes. Different studies of Golgi markers finally revealed that within 1 h of BFA treatment the Golgi apparatus is disassembled and redistributed to the ER. In contrast, compartments of the *trans*-Golgi network (TGN) did not redistribute into the ER. Independent of the labeled precursor (galactose, serine, sphingosine, or palmitic acid) used, BFA caused a dramatic reduction of label incorporation into gangliosides G_{M1a}, G_{D1a}, G_{T1b}, and G_{Q1b} of primary cultured cerebellar neurons with simultaneous accumulation of label in GlcCer, LacCer, G_{M3}, and G_{D3}.[76] These results suggest that biosynthesis of complex gangliosides is localized to the TGN. Using similar methodological approaches, Rosales Fritz and Maccioni[77] found that BFA blocked the synthesis of complex gangliosides without affecting that of G_{M3}, G_{D3}, and also G_{T3}, in chick embryo retina cells, and concluded that, similar to G_{M3} and G_{D3}, synthesis of G_{T3} is localized to the proximal Golgi. However, more detailed studies with monensin revealed that, although localized in the proximal Golgi, the site of G_{T3} synthesis is different from those for LacCer, G_{M3} and G_{D3} synthesis.[78] Their results point to the *cis/medial*-Golgi as the main compartment for coupled synthesis of LacCer, G_{M3}, and G_{D3} and to the *trans*-Golgi as the main compartment of G_{T3} synthesis in cultured retina cells from chick embryos. Thus, a main conclusion from studies with BFA is that the synthesis of the precursors of a, b, and c series gangliosides occurs in the *cis/medial/trans*-Golgi. These intermediates are then transported via vesicular membrane flow beyond the *trans*-Golgi, most probably to the TGN for further glycosylation (Figure 7).

Rat liver Golgi subfractionation studies,[79] as well as subcellular fractionation of primary cultured neurons,[80] indicated that ganglioside sialyltransferases distribute among Golgi subfractions in the order in which they act. Thus, transferases involved in the synthesis of less- and more-glycosylated gangliosides are enriched in early (proximal) and late (distal) Golgi compartments, respectively. Maxzúd *et al.*,[81] who studied the *in vitro* labeling of the endogenous gangliosides of Golgi preparations from chick embryo retina cells, also concluded that glycosylation steps catalyzed by GalNAc-transferase, Gal-transferase II, and sialyltransferase IV colocalize and are functionally coupled in the TGN while proximal Golgi Gal-transferase I, sialyltransferase I, and sialyltransferase II as well as their corresponding GSL acceptors extend their presence to the TGN. Taken together all these results suggest an overlapping rather than a clearcut compartmentalization of glycosyltransferases in the Golgi complex, with a *cis*-Golgi towards *trans*-Golgi/TGN decreasing gradient distribution of the "early" glycosyltransferases (e.g., sialyltransferase I) and a *cis*-Golgi towards *trans*-Golgi/TGN increasing gradient distribution of the "late" glycosyltransferases (e.g., sialyltransferase IV). In contrast to these findings, Lannert *et al.*[52] conclude from their rat liver Golgi subfractionation studies (see also 3.05.2.2.3.), that all the reactions involved in GSL biosynthesis starting with the formation of LacCer reside in the lumen of the *trans*-Golgi and TGN. Like previous subfractionation studies their results also indicate overlapping rather than clearcut compartmentalization of sialyltransferase activities within the Golgi stack. Cell-type-specific enzyme distribution within the Golgi apparatus could also account for some of the conflicting results obtained in different studies.

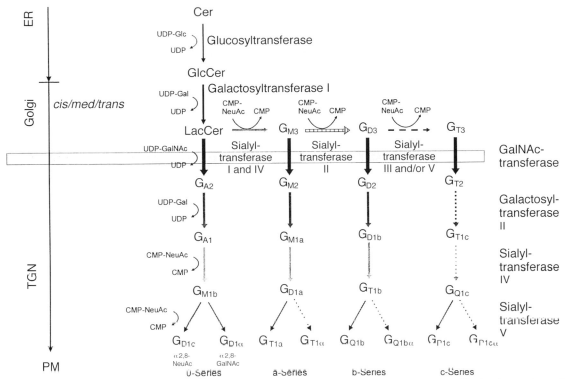

Figure 7 General scheme for ganglioside biosynthesis, as established in rat liver Golgi.[50] The formation of G_{D1z} from G_{A1} via G_{M1b} has been described by Hidari *et al.*[63] All the glycosyltransferases given are associated with the Golgi membrane. See the caption to Figure 2 for the terminology of glycosphingolipids. Cer, ceramide; ER, endoplasmic reticulum; TGN, *trans*-Golgi network; PM, plasma membrane; BFA, brefeldin A.

3.05.2.4 Formation of Sphingomyelin, a Phosphosphingolipid

Since the discovery of the "sphingomyelin cycle" as an ubiquitous, evolutionarily conserved signaling system, analogous to well-known second messenger systems such as the cAMP and phosphoinositide pathway, sphingomyelin has emerged to the focus of interest in many research laboratories.[81,82] Sphingomyelin accounts for about 10% of the cellular lipids and is preferentially concentrated in the outer leaflet of the plasma membrane of mammalian cells. Originally it was considered to be a structural element of the plasma membrane, providing a rigid barrier to the environment. It is synthesized by the transfer of phosphocholine (the water soluble phospholipid specific head group) from phosphatidylcholine onto ceramide, the hydrophobic sphingolipid specific backbone. There is almost no cellular organelle that has not been proposed to be the site of SM biosynthesis. In earlier studies, SM synthase activity has been attributed to the mitochondria, the ER, the plasma membrane, the Golgi apparatus, and, in more recent reports, also to the endosomes.[83] Subcellular fractionation studies have now assigned the bulk of SM synthase activity to the lumen of the *cis*-Golgi, while a minor part has been localized to the cell surface.[84–86] Making use of 3,3-diaminobenzidine cytochemistry, van Helvoort *et al.*[87] could discriminate between Golgi vesicles and endosomes and came to the conclusion that little if any SM synthase localizes to the endocytic pathway of HepG2 and BHK-21 cells. Instead, similar to the report by Futerman *et al.*,[84] most SM synthase activity was found at the Golgi and less than 10% on the cell surface. Possibly, the SM synthase activities localized in the Golgi and in the plasma membrane represent two isoforms of the same enzyme. As SM synthase is also able to catalyze the reverse reaction, namely the formation of phosphatidylcholine from SM and diacylglycerol (DAG), it has been suggested that the plasma membrane associated enzyme may regulate the concentration of two lipid second messengers, DAG and ceramide, in this compartment during signal transduction processes.[86]

Synthesis of SM at the lumenal face of the *cis*-Golgi is consistent with the topological orientation of SM on the anticytosolic face of the plasma membrane and implies that SM is not translocated to the cytosolic surface of the Golgi, but transported to the plasma membrane via the exocytotic vesicular pathway.[88] Both BFA and mitoses, known to block the vesicular traffic between the Golgi complex and the plasma membrane, inhibited transport of native SM to the cell surface in fibroblastic cells like BHK-21 cells and CHO cells,[89,90] indicating that the mechanism for SM transport is one mediated by a vesicular bulk flow process which is also known to transport proteins. In contrast to these findings, BFA did not inhibit transport to the cell surface neither of SM in rat hepatocytes[91] nor of the truncated, fluorescent analogue C6-NBD-SM in HepG2 cells,[92] suggesting that in these cells transport of SM to the plasma membrane occurs independently of protein secretion and most likely does not require the passage through the Golgi, even though SM is normally synthesized in that organelle. However, one cannot exclude that the apparent discrepancy between different studies is due not only to the different cell types used but also to differences in methodology.

The effect of BFA on SM biosynthesis also appears to be cell-type specific, as well as dependent on the methodological approaches used.[93] In primary cultured neurons, BFA reduced SM formation by about 50%, irrespective of the biosynthetic precursor supplied, suggesting that SM formation also occurs distal to the BFA block, in the TGN or in the plasma membrane. Alternatively, taking into account that in these cells, in contrast to GSL biosynthesis, SM formation is much more sensitive (by one order of magnitude) to the Cer level available,[94] an increased employment of administered C6-NBD-Cer, as well as of native ceramide, for the biosynthesis of GlcCer, LacCer, G_{M3} and G_{D3}, which were shown to strongly accumulate in these cells after BFA treatment, could also explain the reduced SM formation.[75] In contrast to neurons which are enriched in complex GSLs, fibroblastic cells (BHK-21, CHO) are rather poor in GSLs. Therefore, as expected, in fibroblastic cells BFA greatly increased SM synthesis,[89,95] confirming its localization in the early Golgi compartment. Furthermore, formation of GlcCer and cholesterol ester was also increased by BFA, whereas that of cholesterol, triacylglycerol, and phosphatidylcholine was diminished.[89] Although quite contradictory the BFA studies demonstrate the close interdependence of SM metabolism and transport with that of GSLs, cholesterol, triacylglycerol, and phosphatidylcholine.

3.05.3 PATHWAYS OF SPHINGOLIPID TRANSPORT

To complete the picture of sphingolipid metabolism, we give a brief summary of the actual concept concerning intracellular GSL flow. Detailed reviews on the mechanism and selectivity of intracellular lipid traffic are avilable in the literature.[31,32,49,88,96,97]

Two major pathways of intracellular GSL flow in eukaryotic cells can be discriminated. First, the exocytic flow of native GSLs from the Golgi to the plasma membrane and second, the endocytic flow of GSLs in the opposite direction from the plasma membrane to intracellular organelles, mainly endosomes and lysosomes, where they are degraded.[98]

3.05.3.1 Exocytosis: Antero- and Retrograde Transport in the ER–Golgi–Plasma Membrane System

Although vesicular pathways connect the Golgi to both the plasma membrane and the ER, GSLs and sphingomyelin, as well as cholesterol, are enriched only in the former. In view of the retrograde transport through the Golgi to the ER,[99] a unidirectional pathway of sphingolipids to the plasma membrane seems quite unlikely. In the mid-1980s Matyas and Morre[100] provided data that may support lipid recycling between the Golgi apparatus and the ER. They observed that newly synthesized radiolabeled gangliosides from rat liver appeared first in the Golgi, mitochondria, and supernatant fractions and only later in the ER, plasma membrane, and nuclear membrane fractions. Similarly, fluorescent lipids delivered to the Golgi by liposome fusion redistributed to the ER within 30 min.[101] Van Meer[31] suggested that sphingolipids and cholesterol aggregate in the lumenal leaflet of Golgi membranes and that these aggregates are preferentially included into anterograde vesicles. Moreover, it has been suggested that at the site of these domains the bilayer will be thicker and therefore select for membrane proteins with larger transmembrane domains.[102] Conversely, these proteins could be responsible for the incorporation of these domains into the anterograde vesicles.[88]

3.05.3.2 Topology of Endocytosis and of Lysosomal Degradation

Having reached the cell surface GSLs display dynamic nature. A part of them is continuously internalized via endocytosis. The significance of this event is largely unknown. One possible explanation could be the importance of different GSL degradation products, like sphingosine, SPP, and ceramide in cellular signal transduction.[11]

In the conventional model components of the plasma membrane reach the lysosome by endocytic membrane flow via the early and late endosomes[103] (Figure 8(a)). In the endosomal compartment, which appears to be crucial for the final destination of membrane molecules, a sorting process occurs directing molecules either to the lysosome, or to the Golgi compartment or even back to the plasma membrane.[104] According to the conventional model, membrane fragments, after a series of vesicle budding and fusion events, are finally incorporated into the lysosomal membrane. The question arises as to how these former plasma membrane fragments are then degraded without destroying the lysosomal membrane. Taking into account that the inner leaflet of the lysosomal membrane is covered with a thick glycocalix composed of limps (lysosomal integral membrane proteins) and lamps (lysosomal associated membrane proteins),[105] one can hardly imagine how molecules of the lysosomal membrane could be rapidly digested.

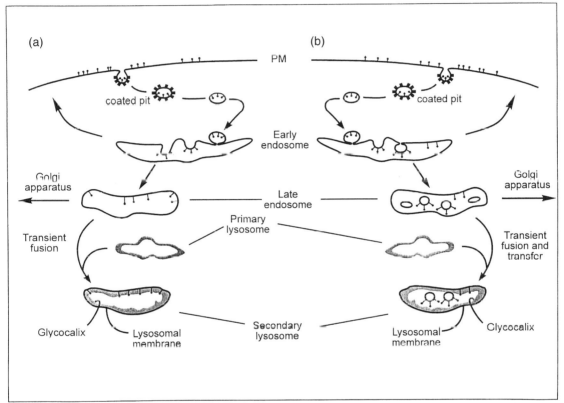

Figure 8 Two models for the topology of endocytosis and lysosomal digestion of glycosphingolipids (GSLs) derived from the plasma membrane (PM).[104] (a) Conventional model: degradation of plasma membrane derived GSLs occurs selectively within the lysosomal membrane. (b) Alternative model: during endocytosis GSLs of the PM become incorporated into the lipid bilayer of intraendosomal vesicles or into other membrane fragments (multivesicular bodies). These vesicles and membrane fragments are transferred into the lysosome by fusion of the late endosome with the primary lysosome. ꙮ, glycosphingolipid.

In an alternative model for the topology of endocytosis, proposed by Fürst and Sandhoff,[106] components of the plasma membrane pass through the endosomal compartment as intraendosomal vesicles that finally become intralysosomal vesicles or membrane fragments (Figure 8(b)). The vesicles are initially formed in the early endosomes by selective budding-in (invagination) of endosomal membrane fragments enriched with components of the plasma membrane. While the surrounding endosomal membranes pass along the endocytic pathway being subjected to successsive events of membrane fission and fusion, the intraendosomal vesicles are carried along as passengers and normally do not undergo fusion and fission. When the vesicles finally reach the lysosome, GSLs originating from the outer leaflet of the plasma membrane face the lysosol as components of the outer leaflet of intralysosomal vesicles or membrane structures. In this model the glycocalix barrier is elegantly bypassed. This yet hypothetical model is supported by the following observations. (i) Multivesicular bodies occur at the level of early and late endosomal reticulum.[107-110] (ii) The epidermal growth factor receptor derived from the plasma membrane and internalized into lysosomes of hepatocytes is not integrated into the lysosomal membrane.[111] It has been shown that multivesicular endosomes containing internalized EGF-EGF receptor complexes fuse directly with lysosomes by a single heterotypic fusion step.[112] (iii) Spherical multivesicular bodies are the predominant endocytic compartment in HEp-2 cells and these multivesicular bodies mature within 60-90 min into lysosomes that still contain internal vesicles.[113] (iv) Multivesicular storage bodies accumulate in cells from patients with sphingolipid storage diseases. They were observed in cerebral cells of G_{M1} gangliosidosis patients in the 1960s.[114] More recently, multivesicular storage bodies were observed in fibroblasts and Kupffer cells of a patient with a sphingolipid storage disease caused by a combined activator protein deficiency.[115,116] Analysis of cultured fibroblasts of this patient localized the storage vesicles to the late endosomal or lysosomal compartments, which proved to be still functionally active, except that they failed to degrade sphingolipids with short oligosaccharide head groups. When the missing precursor of the sphingolipid activator proteins was added to the culture medium of the mutant cells, not only the degradation block was abolished, but also the number and size of storage bodies was reduced.[118]

Although budding-in of vesicles at the level of early and/or late endosomes has not been directly proven yet, it appears very likely, since it explains both the occurence of intralysosomal vesicles and the selective degradation of GSLs derived from the plasma membrane, in contrast to those of the lysosomal membrane.

As mentioned above, the lysosome is not necessarily the ultimate target of endocytosed sphingolipids. Especially in lipid uptake studies in which fluorescent, short-chain acyl analogues were used, the lysosomal compartment was not the final destination of the internalized sphingolipids.[96] C6-NBD-Ceramide has been shown to label the Golgi apparatus of cultured cells.[118] Pütz and Schwarzmann[119] have reported that this Golgi labeling in cultured fibroblasts requires ceramide metabolism. Thus, with increasing Golgi labeling formation of C6-NBD-GlcCer and C6-NBD-sphingomyelin was observed. No Golgi labeling could be observed with the metabolically inert 1-O-methyl-C6-NBD-ceramide.

Most fluorescent NBD-sphingomyelin, on the other hand, has been found to be internalized but then returned to the plasma membrane from "early" endosomes, with only a small percentage being delivered to the lysosomes.[120] Comparable results were obtained with NBD-GlcCer.[121] The fluorescent ganglioside C6-NBD-G_{M1} after internalization was transported to a pool of recycling endosomes in the cell body of hippocampal neurons, with little transport to lysosomes, as indicated by lack of degradation and the reappearance of intact C6-NBD-G_{M1} at the cell surface after recycling.[122] In contrast to the fluorescent short-chain acyl NBD-analogues, radiolabeled gangliosides with naturally occuring acyl chain length (C_{16}–C_{18}) were found to be directed after endocytosis mainly to the lysosome, with a minor fraction being directed to the Golgi compartment and glycosylated to more complex gangliosides. Direct glycosylation of exogenously applied gangliosides has been first demonstrated in various types of cultured human fibroblasts deficient in G_{M2} catabolism, in which labeled G_{D1a} was formed by successive glycosylation of the administered labeled G_{M2}.[123] Furthermore, when fibroblasts from a patient suffering from Tay-Sachs disease were incubated with G_{M1} ^{3}H-labeled in the terminal galactose and ^{14}C-labeled in the fatty acid residue, the ^{3}H/^{14}C ratios of the cell-bound G_{M1} and G_{D1a} indicate that the doubly labeled G_{D1a} was formed by direct sialylation of the incorporated G_{M1}.[124] These results strongly suggest that a part of the endocytosed G_{M1} bypassed the lysosome, being directly translocated to the Golgi compartment, most likely the TGN.

3.05.3.3 Salvage of Catabolic Products of GSL Metabolism: Site of Recycling

Little is known about the coordination between the endocytic and the exocytic sphingolipid flux. Also, little is known about the extent of metabolic recycling of the individual components of gangliosides produced during degradation and the contribution of these salvage processes to the overall GSL turnover.

Figure 5 depicts three different routes of GSL biosynthesis. The contribution of each route to the overall GSL content has been studied using different methodological approaches. Labeling cellular GSLs with either [^{3}H]serine or with ^{14}C-labeled sugars and inhibiting *de novo* sphinganine biosynthesis with β-chloroalanine Gillard *et al.*[125] found in SW 13 cells (ATCC CC1 105), derived from a human small cell carcinoma of the adrenal cortex, that *de novo* GSL biosynthesis constituted only 20–40% of total GSL formation while 60–80% of the GSLs were formed in the recycling pathway. Similar results were obtained in human fibroblasts and C6 glioma cells. The observation that GlcCer was not formed by recycling of ceramide from endosomes to the Golgi (Figure 5, pathway 3), could be due either to the fact that ceramide is not recycled by this pathway or, more likely, that ceramide-containing recycled endosomal vesicles fuse with a Golgi compartment distal (downstream) to GlcCer synthase site of action.

There are great differences in the degree of recycling of different degradation products formed when cultured cells are fed with labeled gangliosides. Cultured fibroblasts were fed with ganglioside G_{M3} carrying a radioactive tag either in the sialic acid or in the sphingosine moiety or finally in the stearoyl chain.[126] After 15 h pulse time and 72 h chase period, the radioactive products from both ganglioside catabolism and salvage processes of catabolic fragments were analyzed. Labeled sialic acid was mostly recycled for the biosynthesis of gangliosides and sialoglycoproteins. Salvage of sphingosine was also high and labeled sphingosine was found to be recycled for the biosynthesis of gangliosides, neutral GSLs, and sphingomyelin. Free radioactive sphingosine was hardly detectable but its catabolism to tritiated water was quite marked. Salvage of stearic acid for ganglioside biosynthesis was found to be of minor if any importance, most of the labeled stearic acid being used for the biosynthesis of glycerolipids. Similar experiments were conducted with cultured cerebellar granule cells which were subjected to pulse chase with ganglioside G_{M1} labeled either at the level of NeuAc, Gal or sphingosine.[127] Then the formation of ^{3}H-labeled catabolites, including tritiated water and the formation of tritium-labeled biosynthetic products obtained by recycling of [^{3}H]NeuAc, [^{3}H]Gal, and [^{3}H]sphingosine released during intralysosomal ganglioside degradation (salvage) was determined (Figure 5, route 2). While [^{3}H]NeuAc was almost quantitatively recycled in polysialogangliosides (G_{D1a}, G_{D1b}, G_{T1b}, *O*-acetylated G_{T1b}) and sialylated proteins, most of [^{3}H]Gal was degraded to tritiated water. Interestingly, [^{3}H]sphingosine was found to be substantially recycled for the biosynthesis of sphingomyelin. The capacity to recycle [^{3}H]sphingosine, liberated by the catabolism of exogenous [^{3}H]G_{M1}, labeled in the sphingosine chain to produce [^{3}H]sphingomyelin, was also observed in astrocytes, skin fibroblasts, neuroblastoma cells, and HeLa cells, albeit to a cell-type-specific extent.[127]

At the present time, no information on the exit of sphingosine from the lysosome is available. We do know, however, that sphingosine must leave the lysosome to reach the endoplasmic reticulum to be processed to ceramide. Interestingly, different spectra of fatty acid residues are found in *de novo* formed ceramide, by acylation of sphinganine (Figure 5, pathway 1) and salvage formed ceramide by acylation of sphingosine (Figure 5, pathway 2). Using suboptimal concentrations of fumonisin B1, a specific inhibitor of ceramide synthase,[94] Gillard *et al*[125] observed that in SW 13 cells (see above), ceramide synthesized from sphinganine contains very-long-chain fatty acids (C_{22}–C_{24}), while ceramide synthesized from sphingosine contains shorter-chain fatty acids (C_{16}–C_{18}). Similar conclusions were drawn by the group of Sonnino, who studied recycling of sphinganine and sphingosine, after feeding cultured human fibroblasts with either sphinganine or sphingosine containing G_{M3} species, both ^{3}H-labeled at carbon atom 3 of the long-chain base (V. Chigorno and S. Sonnino, personal communication). Both long-chain bases were found to be recycled in the biosynthesis of G_{D3}, albeit in two different molecular species of this ganglioside. Recycling of sphinganine yielded a radioactive G_{D3} species containing very-long-chain fatty acids, thus supporting the idea that sphinganine is acylated mainly with C_{22}–C_{24} fatty acids. Recycling of sphingosine yielded mainly G_{D3} containing shorter-chain fatty acids (C_{16}–C_{18}). Endogenous cellular gangliosides are often split into two spots or components by thin layer chromatography (TLC).[6,128] The upper spots should correspond to the respective species containing very-long-chain fatty acids and result virtually from *de novo* biosynthesis, while the lower spots should contain shorter-chain fatty acids and result from salvage pathways. It is not clear, at the present time, whether there exist different sites for these two pathways or whether isoenzymes with different substrate specificities are involved. There is evidence that the fatty acid moiety of ceramide may influence the utilization of GSLs as substrates by glycosyltransferases.[129] Thus, individuals, whose erythrocytes display the recessive p phenotype, are unable to synthesize larger compounds of the globo series. Their erythrocytes contain an abnormally large quantity of LacCer, the precursor of globosides (Figure 1), and virtually all of this excess LacCer contains C_{22}–C_{24} fatty acids, in contrast to the LacCer of normal erythrocytes. This observation suggests that only a subset of LacCer molecules that contains fatty acids of a certain length is selectively used for globoside biosynthesis.

A major conclusion emerging from all these studies is that salvage processes particularly for sialic acid and sphingosine, the biosynthesis of which is more demanding from the energetic point of view, represent a major route of GSL metabolism.

3.05.4 FUNCTIONAL IMPORTANCE OF ACTIVATOR PROTEINS IN LYSOSOMAL DEGRADATION OF SPHINGOLIPIDS

Within the lysosome degradation of GSLs occurs by the stepwise action of specific acid exohydrolases, starting at the hydrophilic end of the molecule (Figure 9).[104] More than ten different exohydrolases are involved in GSL degradation. The deficiency of any one of the degrading hydrolases causes an accumulation of the corresponding lipid substrate in the lysosomal compartment, leading to the so-called sphingolipid storage diseases. In these inherited diseases accumulation of lipids occurs mainly in those cell types and organs in which the lipids are predominantly synthesized, although acid hydrolases are present in all the cells of the organism with the exception of human erythrocytes. Therefore, different lipid storage diseases are rather heterogeneous from the biochemical, as well as from the clinical point of view.[130]

In contrast to membrane bound glycosyltransferases catalyzing GSL biosynthesis, exohydrolases involved in GSL degradation are water soluble. This fact obviously has a series of consequences. It has been shown that enzymatic reactions, in which both enzyme and substrate are membrane bound, are independent of the incubation volume indicating that the reaction proceeds mainly at the surface or within individual membranes and not through the aqueous phase.[131] It is assumed that glycosyltransferases and their respective lipid substrate meet via lateral diffusion within the membrane. In contrast, water-soluble exohydrolases can act only at the surface of membranes containing their lipid substrate. Thus hydrolases recognize and cleave only those oligosaccharide chains that extrude far enough into the aqueous space.[132] To degrade membrane-bound GSLs with short-chain oligosaccharides, these enzymes require the assistance of small glycoprotein cofactors, the so-called sphingolipid activator proteins (SAPs or saposins and the G_{M2} activator).[106] Thus, the simultaneous use of two components for GSL degradation might protect the plasma membrane from inappropriate degradation. Cell damage by missorted hydrolases ending up at the cell surface, albeit at a low concentration, is prevented by both: the neutral pH (hydrolases being fully active at a low pH), as well as the low concentration of sphingolipid activator proteins at the cell surface.

Since the discovery of the sulfatide activator protein (SAP-B) in 1964,[133] several other factors required for hydrolytic degradation of GSLs have been described. When sequence data became available, it turned out that only two genes encode for the five known SAPs.[106] One gene encodes for the G_{M2} activator protein and the second for the SAP-precursor protein (prosaposin) which is processed to four homologous proteins: SAP-A, -B, -C and -D. Most of our current knowledge of SAPs has emerged from studies of patients with atypical lipid storage diseases. A deficiency of SAP-C (Gaucher factor) causes a juvenile variant of Gaucher disease, characterized by an accumulation of GlcCer despite normal glucosylceramidase levels (Figure 9).[135,136] Complete deficiency of the SAP-precursor protein caused simultaneous accumulation of Cer, GlcCer, LacCer, ganglioside G_{M3}, and also of GalCer, sulfatides, digalactosylceramides, as well as of globotriaosylceramide in the patients' cells.[136] Treatment of the mutant cells with different activator proteins specifically prevents the accumulation of one or more of the stored sphingolipids, indicating the *in vivo* function of these molecules. While the SAP-precursor protein abolished at nanomolar concentrations the storage of all mentioned sphingolipids,[117] SAP-B, known to have a broad GSL specificity,[137] only stimulated at somewhat higher concentrations the degradation of accumulated LacCer and G_{M3} but not that of Cer and GlcCer.[128] SAP-D, on the other hand, clearly stimulated Cer degradation.[128] The physiological function as well as the mechanism of action of SAPs is however far from clear at the present time. Some of them act as sphingolipid binding proteins or liftases, forming a water soluble complex with the lipid, thus raising it out of the membrane.[106] Other mechanisms of action are also described. SAP-C can directly activate glucosylceramide β-glucosidase[138] while the G_{M2} activator binds GSLs, such as ganglioside G_{M2}, and interacts specifically with β-hexosaminidase A.[139]

The physiological function of SAPs is even more complex than generally believed. Thus, the SAP-precursor protein has been shown to rescue hippocampal CAT neurons from lethal ischemic damage and to promote peripheral nerve regeneration *in vivo*.[140,141] Furthermore the SAP-precursor protein and SAP-C stimulated neurite outgrowth in different neuroblastoma cell lines, as well as in primary cultured cerebellar neurons.[141–143]

3.05.5 CONCLUDING REMARKS

Although it is generally accepted that GSLs are present on the outer leaflet of the plasma membrane and on the lumenal side of the Golgi membrane, some GSL species were detected in

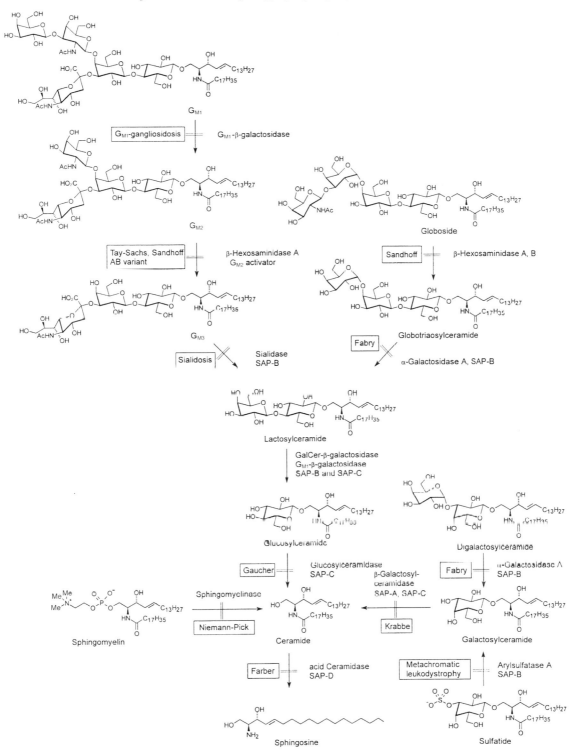

Figure 9 Lysosomal degradation of sphingolipids.[104] Sphingolipid activator proteins (SAPs), exohydrolases and the eponyms of known storage diseases are shown. AB variant, AB variant of G_{M2} gangliosidosis (deficiency of G_{M2} activator protein). Heterogeneity within the ceramide moiety due to varying degrees of saturation, hydroxylation, and chain length are not considered.

almost all intracellular membranes including the nuclear envelope.[144,145] It is not clear how glycolipids are moved to these sites and what their functional roles are.

The use of the short-chain (C_6) fluorescent (NBD) analogues has helped to define potential pathways of sphingolipid metabolism and transport. However, while the majority of the internalized fluorescent lipid returns to the plasma membrane, in compartments identical to those of the "recycling" transferrin receptor,[97] in vivo ocurring GSLs with longer chain length (C_{16}–C_{24}) end up to a large extent in the lysosome or are subjected, to a much lesser extent, to direct glycosylation in the Golgi apparatus. To explain different results obtained with analogues of different chain length, the following points should be considered. (i) The metabolism of short-chain analogues in cultured cells is quantitatively different from that of their endogenous counterparts (unpublished results from the authors' laboratory). (ii) The short-chain fatty acid renders the analogues more water-soluble than their endogenous counterparts. As a result they can undergo spontaneous transfer between donor and acceptor membranes, including protein-facilitated diffusion through the aqueous phase. Thus, they can be readily integrated into cellular membranes but also reextracted into serum protein containing media.[97] (iii) In contrast to endogenous GSLs, the C6-NBD-analogues are able to undergo spontaneous transbilayer movement ("flip-flop").[97] Furthermore, the multidrug resistance P-glycoprotein is capable to translocate short-chain lipid analogues across the plasma membrane.[146] Therefore, some caution is required in the interpretation of data obtained with truncated analogues which significantly differ from native sphingolipids in their metabolism, as well as in their intracellular transport and topology.

Biochemical pathways of sphingolipid metabolism are quite well established. It appears that the sequential glycosylation of GSLs by different glycosyltransferases occurs as the maturing GSL moves vectorially from the ER through the Golgi compartment to the plasma membrane. Precise intracellular location of some biosynthetic enzymes is, however, not yet clear and awaits isolation of the enzyme, followed by immunolocalization. Furthermore, the mechanism that regulates GSL biosynthesis in different cells and in cells at different periods of development is still a fertile field of research. In addition to the transcriptional level, the compartmental organization of the biosynthetic machinery, as well as the availability of sugar nucleotides, which might be limited by a highly active UDP-sugar pyrophosphatase[147] could be relevant in regulating the expression of GSLs by cells.

Details of the endocytic pathway of GSLs derived from the plasma membrane are still to be elucidated. How does the cell decide whether GSLs of the plasma membrane follow the conventional endocytic pathway or are incorporated into intraendosomal vesicles? Furthermore the mechanism of action and the complex physiological role of SAPs has to be defined.

ACKNOWLEDGMENTS

We thank Dr. Anthony H. Futerman for critical reading the manuscript and Judith Weisgerber, Ute Schepers, Dr. Bernd Ließem and Dr. Thomas Kolter for their help in preparing the figures.

3.05.6 REFERENCES

1. R. W. Ledeen and R. K. Yu, Methods Enzymol., 1982, 83, 139.
2. H. Wiegandt, New Compr. Biochem., 1985, 10, 199.
3. L. Svennerholm, J. Neurochem., 1963, 10, 613.
4. L. Svennerholm, A. K. Asbury, R. A. Reisfeld, K. Sandhoff, K. Suzuki, G. Tettamanti, and G. Toffano (eds.), "Biological Function of Gangliosides," Progress in Brain Research, vol. 101, Elsevier, Amsterdam, 1994.
5. Y. Nagai, Behav. Brain Res., 1995, 66, 99.
6. G. van Echten and K. Sandhoff, J. Neurochem., 1989, 52, 207.
7. A. Schwarz and A. H. Futerman, Biochim. Biophys. Acta, 1996, 1286, 247.
8. S.-L. Hakomori, J. Biol. Chem., 1990, 265, 18 713.
9. C. C. Kan and R. Kolesnick, Trends Glycosci. Glycotechnol., 1993, 5, 99.
10. Y. A. Hannun and L. M. Obeid, Trends Biochem. Sci., 1995, 20, 73.
11. S. Spiegel, D. Foster, and R. Kolesnick, Curr. Opin. Cell Biol., 1996, 8, 159.
12. Y. A. Hannun and R. M. Bell, Science, 1989, 243, 500.
13. K. Sandhoff and G. van Echten, Adv. Lipid Res., 1993, 26, 119.
14. E. C. Mandon, I. Ehses, J. Rother, G. van Echten, and K. Sandhoff, J. Biol. Chem., 1992, 267, 11 144.
15. K. Hirschberg, J. Rodger, and A. H. Futerman, Biochem. J., 1993, 290, 751.
16. P. E. Braun and E. E. Snell, J. Biol. Chem., 1968, 243, 3775.
17. W. Stoffel, D. LeKim, and G. Sticht, Hoppe-Seyler's Z. Physiol. Chem., 1968, 349, 664.
18. Y. A. Hannun, J. Biol. Chem., 1994, 269, 3125.
19. T. Okazaki, N. Domae, R. M. Bell, and Y. A. Hannun, Trends Glycosci. Glycotechnol., 1994, 6, 278.
20. A. Olivera, H. Zhang, R. O. Carlson, M. E. Matte, R. R. Schmidt, and S. Spiegel, J. Biol. Chem., 1994, 269, 17 924.
21. D. E. Ong and R. N. Brady, J. Biol. Chem., 1973, 248, 3884.
22. W. Stoffel and K. Bister, Hoppe-Seyler's Z. Physiol Chem., 1974, 355, 911.
23. A. H. Merrill, Jr. and E. Wang, J. Biol. Chem., 1986, 261, 3764.
24. J. Rother, G. van Echten, G. Schwarzmann, and K. Sandhoff, Biochem. Biophys. Res. Commun., 1992, 189, 14.

25. C. Michel, G. van Echten-Deckert, J. Rother, K. Sandhoff, E. Wang, and A. H. Merrill, Jr., *J. Biol. Chem.*, 1997, **272**, 22432.
26. J. W. Kok, M. Nikolova-Karakashian, K. Klappe, C. Alexander, and A. H. Merrill, Jr., *J. Biol. Chem.*, 1997, **272**, 21128.
27. E. R. Smith and A. H. Merrill, Jr., *J. Biol. Chem.*, 1995, **270**, 18749.
28. C. Michel and G. van Echten-Deckert, *FEBS Lett.*, 1997, **416**, 153.
29. R. Bose, M. Verheij, A. Haimovitz-Friedman, K. Scotto, Z. Fuks, and R. Kolesnick, *Cell*, 1995, **82**, 405.
30. K. Yokoyama, H. Nojiri, M. Suzuki, M. Setaka, A. Suzuki, and S. Nojima, *FEBS Lett.*, 1995, **368**, 477.
31. G. van Meer, *Annu. Rev. Cell Biol.*, 1989, **5**, 247.
32. R. E. Pagano, *Curr. Opin. Cell Biol.*, 1990, **2**, 652.
33. R. N. Collins and G. Warren, *J. Biol. Chem.*, 1992, **267**, 24906.
34. W. Stoffel and K. Bister, *Hoppe-Seyler's Z. Physiol. Chem.*, 1973, **354**, 169.
35. R. E. Pagano and O. C. Martin, *Biochemistry*, 1988, **27**, 4439.
36. R. R. Hiebsch, T. J. Raub, and B. W. Wattenberg, *J. Biol. Chem.*, 1991, **266**, 20323.
37. A. Carruthers and E. M. Carey, *J. Neurochem.*, 1983, **41**, 22.
38. H. P. Siegrist, T. Burkart, U. N. Wiesmann, N. N. Herschkowitz, and M. A. Spycher, *J. Neurochem.*, 1979, **33**, 497.
39. K. N. J. Burger, P. van der Bijl, and G. van Meer, *J. Cell Biol.*, 1996, **133**, 15.
40. A. Kendler and G. Dawson, *J. Neurosci. Res.*, 1992, **31**, 205.
41. H. Coste, M. B. Martel, and R. Got, *Biochim. Biophys. Acta*, 1989, **858**, 6.
42. A. H. Futerman and R. E. Pagano, *Biochem. J.*, 1991, **280**, 295.
43. D. Jeckel, A. Karrenbauer, K. N. J. Burger, G. van Meer, and F. Wieland, *J. Cell Biol.*, 1992, **117**, 259.
44. M. Trinchera, M. Fabbri, and R. Ghidoni, *J. Biol. Chem.*, 1991, **266**, 20907.
45. M. Saito, M. Saito, and A. Rosenberg, *Biochemistry*, 1984, **23**, 1043.
46. A. Karrenbauer, D. Jeckel, W. Just, R. Birk, R. R. Schmidt, J. E. Rothmann, and F. T. Wieland, *Cell*, 1990, **63**, 259.
47. D. E. Warnrock, M. S. Lutz, W. A. Blackburn., W. W. Young, Jr., and J. U. Baenziger, *Proc. Natl. Acad. Sci. USA*, 1994, **91**, 2708.
48. T. Sasaki, *Experimentia*, 1990, **46**, 611.
49. G. van Meer and K. N. J. Burger, *Trends Cell Biol.*, 1992, **2**, 332.
50. G. van Echten and K. Sandhoff, *J. Biol. Chem.*, 1993, **268**, 5341.
51. M. Trinchera, A. Fiorilli, and R. Ghidoni, *Biochemistry*, 1991, **30**, 2719.
52. H. Lannert, K. Gorgas, I. Meißner, F. T. Wieland, and D. Jeckel, *J. Biol. Chem.*, 1998, **273**, 2939.
53. S. L. Deutscher and C. B. Hirschberg, *J. Biol. Chem.*, 1986, **261**, 96.
54. A. W. Brandli, G. C. Hanoon, E. Rodriguez Boulan, and K. Simons, *J. Biol. Chem.*, 1988, **263**, 16283.
55. H. Lannert, C. Bünning, D. Jeckel, and F. T. Wieland, *FEBS Lett.*, 1994, **342**, 91.
56. C. Zacharias, G. van Echten-Deckert, M. Plewe, R. R. Schmidt, and K. Sandhoff, *J. Biol. Chem.*, 1994, **269**, 13313.
57. D. J. Carey and C. W. Hirschberg, *J. Biol. Chem.*, 1981, **256**, 989.
58. K. E. Creek and D. J. Morré, *Biochim. Biophys. Acta*, 1981, **643**, 292.
59. B. Fleischer, *J. Cell Biol.*, 1981, **89**, 246.
60. H. K. M. Yusuf, G. Pohlentz, and K. Sandhoff, *Proc. Natl. Acad. Sci. USA*, 1983, **80**, 7075.
61. H. Miller-Podraza and P. H. Fishman, *Biochemistry*, 1982, **21**, 3265.
62. W. W. Young, Jr., M. S. Lutz, and W. A. Blackburn, *J. Biol. Chem.*, 1992, **267**, 12011.
63. K. I.-P. J. Hidari, I. Kawashima, T. Tai, F. Inagaki, Y. Nagai, and Y. Sanai, *Eur. J. Biochem.*, 1994, **221**, 603.
64. G. Pohlentz, D. Klein, G. Schwarzmann, D. Schmitz, and K. Sandhoff, *Proc. Natl. Acad. Sci. USA*, 1988, **85**, 7044.
65. H. Iber, C. Zacharias, and K. Sandhoff, *Glycobiology*, 1992, **2**, 137.
66. H. Iber and K. Sandhoff, *FEBS Lett.*, 1989, **254**, 124.
67. H. Iber, G. van Echten, and K. Sandhoff, *Eur. J. Biochem.*, 1991, **195**, 115.
68. K. Nara, Y. Watanabe, I. Kawashima, T. Tai, Y. Nagai, and Y. Sanai, *Eur. J. Biochem.*, 1996, **238**, 647.
69. K. Nara, Y. Watanabe, K. Maruyama, K. Kasahara, Y. Nagai, and Y. Sanai, *Proc. Natl. Acad. Sci. USA*, 1994, **91**, 7952.
70. S. Tsuji, *J. Biochem.*, 1996, **120**, 1.
71. J. Nakayama, M. N. Fukuda, Y. Hirabayashi, A. Kanamori, K. Sasaki, T. Nishi, and M. Fukuda, *J. Biol. Chem.*, 1996, **271**, 3684.
72. A. M. Tartakoff, *Cell*, 1983, **32**, 1026.
73. H. Miller-Prodraza and P. H. Fishman, *Biochim. Biophys. Acta*, 1984, **804**, 44.
74. M. Saito and A. Rosenberg, *Biochemistry*, 1985, **24**, 3054.
75. R. D. Klausner, J. G. Donaldson, and J. Lippincott-Schwartz, *J. Cell Biol.*, 1992, **116**, 1071.
76. G. van Echten, H. Iber, H. Stotz, A. Takatsuki, and K. Sandhoff, *Eur. J. Cell Biol.*, 1990, **51**, 135.
77. V. M. Rosales Fritz and H. J. F. Maccioni, *J. Neurochem.*, 1995, **65**, 1859.
78. V. M. Rosales Fritz, M. K. Maxzúd, and H. J. F. Maccioni, *J. Neurochem.*, 1996, **67**, 1393.
79. M. Trinchera and R. Ghidoni, *J. Biol. Chem.*, 1989, **264**, 15766.
80. H. Iber, G. van Echten, and K. Sandhoff, *J. Neurochem.*, 1992, **58**, 1533.
81. M. K. Maxzúd, J. L. Daniotti, and H. J. F. Maccioni, *J. Biol. Chem.*, 1995, **270**, 20207.
82. R. Kolesnick and D. W. Golde, *Cell*, 1994, **77**, 325.
83. D. Allan and K.-J. Kallen, *Trends Cell Biol.* 1994, **4**, 350.
84. A. H. Futerman, B. Stieger, A. L. Hubbard, and R. E. Pagano, *J. Biol. Chem.*, 1990, **265**, 8650.
85. D. Jeckel, A. Karrenbauer, R. Birk, R. R. Schmidt, and F. Wieland, *FEBS Lett.*, 1990, **261**, 155.
86. A. van Helvoort, W. van't Hof, T. Ritsema, A. Sandra, and G. van Meer, *J. Biol. Chem.*, 1994, **269**, 1763.
87. A. van Helvoort, W. Stoorvogel, G. van Meer, and K. N. J. Burger, *J. Cell Sci.*, 1997, **110**, 781.
88. A. van Helvoort and G. van Meer, *FEBS Lett.*, 1995, **369**, 18.
89. K.-J. Kallen, P. Quinn, and D. Allan, *Biochem. J.*, 1993, **289**, 307.
90. A. van Helvoort, M. L. Giudici, M. Thielemans, and G. van Meer, *J. Cell Sci.*, 1997, **110**, 75.
91. Y.-J. Shiao and J. E. Vance, *J. Biol. Chem.*, 1993, **268**, 26085.
92. G. van Meer and W. van't Hof, *J. Cell. Sci.*, 1993, **104**, 833.

93. K. Sandhoff and G. van Echten, in "Progress in Brain Research," eds. L. Svennerholm, A. K. Asbury, R. A. Reisfeld, K. Sandhoff, G. Tettamanti, and G. Toffano, Elsevier, Amsterdam, 1994, vol. 101, p. 17.
94. A. H. Merrill, Jr., G. van Echten, E. Wang, and K. Sandhoff, *J. Biol. Chem.*, 1993, **268**, 27 299.
95. A. Brüning, A. Karrenbauer, E. Schnabel, and F. T. Wieland, *J. Biol. Chem.*, 1992, **267**, 5052.
96. D. Hoekstra and J. W. Kok, *Biochim. Biophys. Acta*, 1992, **1113**, 277.
97. A. G. Rosenwald and R. E. Pagano, *Adv. Lipid Res.*, 1993, **26**, 101.
98. G. Schwarzmann and K. Sandhoff, *Biochemistry*, 1990, **29**, 10 865.
99. F. Letourneur, E. C. Gaynor, S. Hennecke, C. Démollière, R. Duden, S. D. Emr, H. Riezman, and P. Cosson, *Cell*, 1994, **79**, 1199.
100. G. R. Matyas and D. J. Morré, *Biochim. Biophys. Acta*, 1987, **921**, 599.
101. P. M. Hoffmann and R. E. Pagano, *Eur. J. Cell Biol.*, 1993, **60**, 371.
102. M. S. Bretscher and S. Munro, *Science*, 1993, **261**, 1280.
103. G. Griffiths, B. Hoflack, K. Simons, I. Mellman, and S. Kornfeld, *Cell*, 1988, **52**, 329.
104. K. Sandhoff and T. Kolter, *Trends Cell Biol.*, 1996, **6**, 98.
105. S. R. Carlsson, J. Roth, F. Piller, and M. Fukuda, *J. Biol. Chem.*, 1988, **263**, 18 911.
106. W. Fürst and K. Sandhoff, *Biochim. Biophys. Acta*, 1992, **1126**, 1.
107. J. W. Kok, T. Babia, and D. Hoekstra, *J. Cell Biol.*, 1991, **114**, 231.
108. S. Zachgo, B. Dobberstein, and G. Griffiths, *J. Cell Sci.*, 1992, **103**, 811.
109. C. R. Hopkins, A. Gibson, M. Shipman, and K. Miller, *Nature*, 1990, **346**, 335.
110. J. A. McKanna, H. T. Haigler, and S. Cohen, *Proc. Natl. Acad. Sci. USA*, 1979, **76**, 5689.
111. C. A. Renfrew and A. L. Hubbard, *J. Biol. Chem.*, 1991, **266**, 21 265.
112. C. E. Futter, A. Pearse, L. J. Hewlett, and C. R. Hopkins, *J. Cell Biol.*, 1996, **132**, 1011.
113. B. van Deurs, P. K. Holm, L. Kayser, K. Sandvig, and S. H. Hansen, *Eur. J. Cell Biol.*, 1993, **61**, 208.
114. K. Suzuki, K. Suzuki, and G. C. Chen, *Pathol. Eur.*, 1968, **3**, 389.
115. K. Harzer, B. C. Paton, A. Poulos, B. Kustermann-Kuhn, W. Roggendorf, T. Grisar, and M. Popp, *Eur. J. Pediatr.*, 1989, **149**, 31.
116. D. Schnabel, M. Schröder, W. Fürst, A. Klein, R. Hurwitz, T. Zenk, J. Weber, K. Harzer, B. C. Paton, A. Poulos, K. Suzuki, and K. Sandhoff, *J. Biol. Chem.*, 1992, **267**, 3312.
117. J. K. Burkhardt, S. Hüttler, A. Klein, W. Möbius, A. Habermann, G. Griffiths, and K. Sandhoff, *Eur. J. Cell Biol.*, 1997, **73**, 10.
118. N. G. Lipsky and R. E. Pagano, *Science*, 1985, **228**, 745.
119. U. Pütz and G. Schwarzmann, *Eur. J. Cell Biol.*, 1995, **68**, 113.
120. M. Koval and R. E. Pagano, *J. Cell Biol.*, 1989, **108**, 2169
121. J. W. Kok, S. Eskelinen, K. Hoekstra, and D. Hoekstra, *Proc. Natl. Acad. Sci. USA*, 1989, **86**, 9896.
122. A. Sofer, G. Schwarzmann, and A. H. Futerman, *J. Cell Sci.*, 1996, **109**, 2111.
123. S. Sonderfeld, E. Conzelmann, G. Schwarzmann, J. Burg, U. Hinrichs, and K. Sandhoff, *Eur. J. Biochem.*, 1985, **149**, 247.
124. D. Klein, P. Leinekugel, G. Pohlentz, G. Schwarzmann, and K. Sandhoff, in "New Trends in Ganglioside Research: Neurochemical and Neuroregenerative Aspects," *FIDIA Research Series*, eds. R. W. Ledeen, E. L. Hogan, G. Tettamanti, A. J. Yates, and R. K. Yu, Springer-Verlag, Berlin, 1988, vol. 14, p. 247.
125. B. K. Gillard, R. G. Harrell and D. M. Marcus, *Glycobiology*, 1996, **6**, 33.
126. V. Chigorno, G. Tettamanti, and S. Sonkino, *J. Biol. Chem.*, 1996, **271**, 21 738.
127. L. Riboni, R. Bassi, A. Prinetti, and G. Tettamanti, *FEBS Lett.*, 1996, **391**, 336.
128. A. Klein, M. Henseler, C. Klein, K. Suzuki, K. Harzer, and K. Sandhoff, *Biochem. Biophys. Res. Commun.*, 1994, **200**, 1440.
129. D. M. Marcus, M. Naiki, and S. K. Kundu, *Proc. Natl. Acad. Sci. USA*, 1976, **73**, 3263
130. C. Scriver, A. L. Beaudet, W. S. Sly, and D. Valle (eds.) "The Metabolic and Molecular Basis of Inherited Disease," McGraw-Hill, New York, 7th edn., vol. 2, 1995.
131. G. Scheel, E. Acevedo, E. Conzelmann, H. Nehrkorn, and K. Sandhoff, *Eur. J. Biochem.*, 1982, **127**, 245.
132. E. M. Meier, G. Schwarzmann, W. Fürst, and K. Sandhoff, *J. Biol. Chem.*, 1991, **266**, 1879.
133. E. Mehl and H. Jatzkewitz, *Hoppe-Seyler's Z. Physiol Chem.*, 1964, **339**, 260.
134. H. Christomanou, A. Aignesberger, and R. P. Linke, *Biol. Chem. Hoppe-Seyler*, 1986, **367**, 879.
135. D. Schnabel, M. Schröder, and K. Sandhoff, *FEBS Lett.*, 1991, **284**, 57.
136. V. Bradova, F. Smid, B. Ulrich-Bott, W. Roggendorf, B. C. Paton, and K. Harzer, *Hum. Genet.*, 1993, **92**, 143.
137. S.-C. Li, S. Sonnino, G. Tettamanti, and Y.-T. Li, *J. Biol. Chem.*, 1988, **263**, 6588.
138. M. W. Ho and J. S. O'Brien, *Proc. Natl. Acad. Sci. USA*, 1971, **68**, 2810.
139. H.-J. Kytzia and K. Sandhoff, *J. Biol. Chem.*, 1985, **260**, 7568.
140. Y. Kotani, S. Matsuda, M. Sakanaka, K. Kondoh, S. Ueno, and A. Sano, *J. Neurochem.* 1996, **66**, 2019.
141. Y. Kotani, S. Matsuda, T. C. Wen, M. Sakanaka, J. Tanaka, N. Maeda, K. Kondoh, S. Ueno, and A. Sano, *J. Neurochem.* 1996, **66**, 2197.
142. J. S. O'Brien, G. S. Carson, H. C. Seo, M. Hiraiwa, and Y. Kishimoto, *Proc. Natl. Acad. Sci. USA*, 1994, **91**, 9593.
143. J. S. O'Brien, G. S. Carson, H. C. Seo, M. Hiraiwa, S. Weiler, J. M. Tomich, J. A. Barranger, M. Kahn, N. Azuma, and Y. Kishimoto, *FASEB J.*, 1995, **9**, 681.
144. G. Wu, Z.-H. Lu, and R. W. Ledeen, *J. Neurochem.*, 1995, **65**, 1419.
145. A. Sofer and A. H. Futerman, *J. Neurochem.*, 1996, **67**, 2134.
146. A. van Helvoort, A. J. Smith, H. Sprong, I. Fritzsche, A. H. Schinkel, P. Borst, and G. van Meer, *Cell*, 1996, **87**, 507.
147. J. A. Martina, J. L. Daniotti, and H. J. F. Maccioni, *J. Neurochem.*, 1995, **64**, 1274

3.06
Biosynthesis and Regulation of Glycosphingolipids

SUBHASH BASU, MANJU BASU, and SARA DASTGHEIB
University of Notre Dame, IN, USA

and

JOHN W. HAWES
Indiana University School of Medicine, Indianapolis, IN, USA

3.06.1   INTRODUCTION

The glycosphingolipid (GSL) field encompasses the two fields dealing with glycoconjugates and lipids. Several reviews on their biosynthesis focusing on glycoconjugate moieties have been published.[1-5] The purpose of this chapter is to discuss progress made in this field in the biosynthesis and regulation of the glycose moieties of several GSLs. Structures of sialic- and/or fucose-containing GSLs are shown in Figure 1.

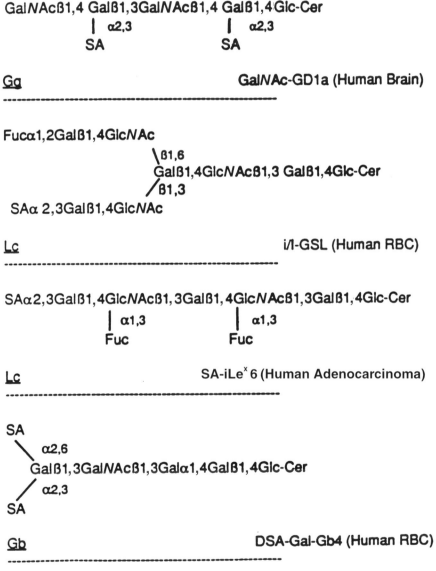

Figure 1   Structures of glycosphingolipids containing tetraglycosyl-core structures of ganglio (top panel) lacto (middle two panels), and globo (bottom panel) families.

During the 1970s–1990s, 18–20 glycolipid glycosyltransferases (GSL:GLTs) have been characterized from eukaryotic (normal or developing) tissues and cells (normal, rapid growing, or tumors), which catalyze *in vitro* biosynthesis of over 50 physiologically relevant and well-characterized GSLs. Several different glycoplipid:galactosyltransferases (GalTs),[4,5,6–15] fucosyltransferases (FucTs),[4,5,16–30] sialyltransferases (SATs),[4,5,31–50] N-acetylgalactosaminyltransferases (GalNAcTs),[4,5,51–74] N-acetylglucosaminyltransferases (GlcNAcTs),[4,5,75–81] glucosyltransferases (GlcT),[4,5,82–88] glucuronyltransferases (GlcATs),[4,5,89–94] and cerebroside synthase (GalT-1)[4,95–99] have been characterized from different eukaryotic systems. Each glycosyltransferase is specific for its donor sugar nucleotide substrate and catalyzes the transfer of specific monosaccharide residues from the sugar nucleotide donor to the nonreducing terminus of the growing oligosaccharide chain of an acceptor glycosphingolipid molecule. The specificity of an acceptor molecule may reside on the sugar chain or on both the ceramide and sugar moieties. The specificity of a glycosyltransferase reaction is characterized by the enzymes of high affinity with respect to both the "donor" and the acceptor structure. It is believed that the glycosyltransferase specificity for the nucleotide sugar donor substrate is absolute. Specificities concerning the sugar acceptor substrate can vary. Some GSL:GLTs exhibit a broad acceptor specificity, catalyzing the transfer of sugars at comparative rates to glycolipids, glycoproteins, or oligosaccharides.[3–5] Other glycosyltransferases exhibit a most stringent requirement regarding the nature of the acceptor and/or the oligosaccharide moiety of the acceptor, and will be discussed in later sections. The substrate specificities of a transferase catalyzing this same reaction *in vitro* may also vary depending on its tissue of origin.[1,3,5] Most of these specificity studies have been carried out with highly purified GLTs of tissue or cell origin.[2,5,12–15,28,39,44,45,59,61,65,99] However, studies completed with the recombinant, overexpressed proteins will be mentioned in later sections.

3.06.2 GLYCOLIPID GALACTOSYLTRANSFERASES (GalTs)

The GSL:GalTs which form the largest family (Table 1) catalyze reactions that utilize UDP-α-galactose as the sugar donor and are described in the literature using different designations. However, one unified nomenclature (Table 1)[4,5] will be helpful for researchers in the field and also for the discussion presented here.

Table 1 Different glycolipid galactosyltransferases isolated from animal cells (donor: UDP-galactose).

Abbreviation	Alternative names	Acceptor	Linkage	Ref.
GalT-1 or CGalT	Cerebroside synthase (IUBEN-2.4.1.62)[a]	HFACer	$\beta 1,1$	4,8,96
GalT-2	Lactosylceramide synthase	Glc-Cer	$\beta 1,4$	4,5,7,13
GalT-3	GM1 synthase (IUBEN-2.4.1.62)	GM2	$\beta 1,3$	4–6,14,31,100
GalT-4	GT or $\beta 1,4$-Galtransferase; (IUBEN-2,4,1,86)	LcOse3Cer	$\beta 1,4$	4,5,9,15
GalT-5	$\alpha 1,3$-Galtransferase (IUBEN-2.4.1.87)	nLcOse4Cer	$\alpha 1,3$	4,10,101,102
GalT-6	$\alpha 1,4$-Galtransferase	LcOse2Cer	$\alpha 1,4$	103

[a] IUBEN, International Union of Biochemistry Enzyme Nomenclature.

The galactosyltransferases that catalyze different positional and anomeric linkage formation in various GSLs comprise a family of more than seven individual gene products (Table 1; Figures 2(a) and (b)). Most of these activities have been characterized and purified from animal cells.[13–15,98,99] The putative amino acid sequences from cDNA sequences of at least three GSL:GalTs have been reported and will be discussed in the following sections.[104,105]

3.06.2.1 UDP-galactose: HFA-ceramide $\beta 1,1$-Galactosyltransferase (GalT-1 or CGalT)

GalT-1 or CGalT catalyzes the transfer of galactose from UDP-galactose to ceramide containing an α-hydroxy-fatty acid to form galactosylcerebroside (Gal-Cer$_{hfa}$).

The enzyme was first detected in embryonic chicken brain[8,95] and rat brain.[96] It has been partially purified[97,98] from rat brain and its kinetic properties have been studied in the presence of detergents and phospholipids.[99] The same GalT-1 has been shown to catalyze the synthesis of nonhydroxy-galactosylceramide and galactosyldiglyceride.[106] GalT-1 has been cloned from a mouse gene,[107] and chromosomal mapping has also been completed.[103] The human gene of GalT-1 has been mapped in chromosome 4, band g 26.[108]

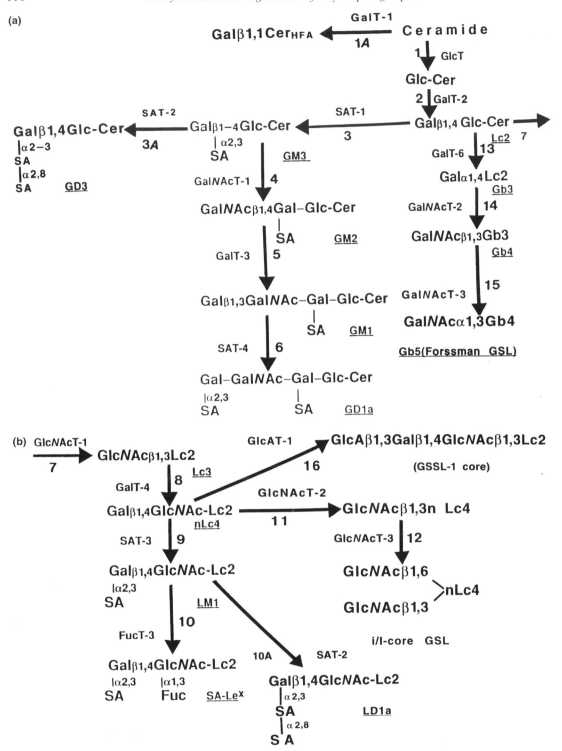

Figure 2 Proposed biosynthetic pathways: (a) for ganglio and globo family glycolipids and (b) for glucurono and lacto family glycolipids.

3.06.2.2 UDP-galactose: Glc-Cer β1,4-Galactosyltransferase (GalT-2)

The biosynthesis of lactosylceramide from glucosylceramide was first achieved in an *in vitro* embryonic chicken brain membrane system enriched in Golgi bodies.[7] It has been shown to be ubiquitous in other animal tissues.

The activity has been solubilized from 13-day-old embryonic chicken brain Golgi-rich membranes.[4,7] The Mg^{2+} requirement of the ECB GalT-2 distinguishes this enzyme from other known galactosyltransferases which require Mn^{2+} for optimal activity when tested in *in vitro* assay systems. GalT-2 has been purified from normal human kidney[13] and a high titer polyclonal antiserum specific to the peptide sequence of the enzyme has been produced to aid its future cloning work. The transfer of galactose from UDP-galactose to glucosylceramide is catalyzed by a specific β1,4-galactosyltransferase (GalT-2) that was first characterized in embryonic chicken brain[5,8] and rat spleen.[109] The product of the GalT-2-catalyzed reaction, lactosylceramide, has been found to be the intermediate in the synthesis of more complex GSLs of the ganglio, globo, and lacto families. The level of lactosylceramide has been found to be elevated in human tumors and other animal tissues.[110]

After the discovery of GSL:GLTs, the physiological role of lactosylceramide has been assigned as a precursor of the other complex glycoplipids.

3.06.2.3   UDP-galactose: GM2 β1,3-Galactosyltransferase (GalT-3)

Biosynthesis of GM1 ganglioside (Galβ1,3GalNAcβ1,4(NeuAcα2,3)Galβ1,4Glc-ceramide) from GM2 ganglioside (GalNAcβ1,4(NeuAcα2,3)Galβ1,4Glc-ceramide) is catalyzed by a β1,3-galactosyltransferase (GalT-3) (Figure 2(a), step 5), first characterized in embryonic chicken brains (13–19 days old).[4-6] The enzyme activity was subsequently detected in frog and rat brains,[111] chick neural retinal cells,[112] chick embryo liver,[113] and rat liver.[114] Compared with β1,4GalT, the GalT-3 activity level is much lower in cultured cells (TSD) derived from Tay–Sachs diseased brains.[115] GalT-3 is developmentally regulated in the embryonic chicken brain (ECB)[14,116] and is especially enriched in 19-day-old ECB. During solubilization from the ECB Golgi-rich membranes, a β1,4galactosyltransferase, GalT-4 (Figure 2(b), step 8) was cosolubilized and separated.[117] A high-titer polyclonal antibody against highly purified ECB GalT-3 has been produced[14,116] for further cloning purposes.[118-120] Kinetic studies and immunological characterization of purified GalT-3 from 19-day-old ECB have been reported.[14,121]

Purified GalT-3 demonstrates a low K_m for both donor (UDP-galactose) and acceptor (GM2) substrates.[14,116,121,122] Substrate specificity studies with the purified enzyme clearly show the absence of any other glycosyltransferase activities. It is also noteworthy that GalT-3 exhibits a very stringent specificity for acceptor structure; thus, the enzyme is not active on free *N*-acetylgalactosamine (GalNAc) or *p*-nitrophenyl-GalNAc. This property is in marked contrast to that found for GalT-4 discussed below.[123] The potential glycoprotein substrate, asialo ovine submaxillary mucin, also is not an acceptor for GalT-3 under the published experimental conditions, but inhibits the transfer of galactose to GM2 ganglioside in a concentration-dependent manner.

The specificity of GalT-3 for GM2 has been analyzed for the contribution of acceptor substrate structure on GalT-3 and GalT-4 activities. Modified glycosphingolipids, either completely deacylated (lyso-GM2, lyso-Lc3) or with the long-chain (C_{16}–C_{18}) fatty acid in the ceramide moiety replaced by an acetyl (C_2) group (acetyl-GM2, acetyl-Lc3), were chemically synthesized. These compounds, which differ from one another with respect to both structure and hydrophobicity, were tested as potential substrates for GalT-3 and GalT-4. A comparison of the kinetic parameters of lyso- and acetyl-GM2 and natural GM2 indicated that both the K_m and V_{max} values with GalT-3 are unfavorably changed (10-fold increase in K_m, 4–8-fold decrease in V_{max}) with the modified substrates.[116,122] This clearly shows that, in addition to its specificity for the *N*-acetylgalactosaminyl acceptor terminus, GalT-3 also recognizes a hydrophobic domain on the acceptor (GM2) structure provided by the ceramide moiety. Whether this requirement is specific for a particular structure or arrangement, or is more of a general requirement for hydrophobicity (where the effects of the ceramide fatty acid may be replaced by hydrophobic amino acid residues of glycoproteins around an *O*-glycosylation site), remains to be seen.

Purified GalT-4 (see below), on the other hand, does not exhibit such a stringent requirement for a hydrophobic domain[124] in its substrate structure, implying that the composition of the carbohydrate component is the sole determinant of enzyme activity. This result was anticipated based on the observation that GalT-4 is capable of transferring galactose to acceptor GlcNAc termini on glyco-lipids, glycoproteins, and even free saccharides. However, there appears to be a strict requirement for the *N*-acetylglucosaminyl group, since lyso-Lc3 does not behave as an active substrate in the range of concentrations tested.[124] This leads one to speculate that GalT-4 contains a specific recognition site (at or near the catalytic domain) for the acetamido group of the *N*-acetylglucosamine residue, in the absence of which binding of the substrate to the enzyme is not favorable enough to sustain catalysis. Based on the above observation, GalT-3 has been classified as an HY-CAR enzyme (dual recognition for hydrophobic and carbohydrate domains on the substrate (Figure 3 and Table 2) and GalT-4 is categorized as a CAR enzyme (recognition for carbohydrate domains alone).[5] Another enzyme of the ganglioside biosynthesis pathway, SAT-1 (CMP-sialic acid: Lc2 sialyltransferase (see later)), has also been shown to belong to the category of an HY-CAR enzyme.[5] It appears that both GalT-3 and SAT-1 are probably unique regulatory enzymes in the ganglioside biosynthetic pathway and, therefore, exhibit the distinguishing specificity for carbohydrate–lipid dual recognition, whereas GalT-4 belongs to the class of transferases whose specificities are restricted only to the oligosaccharide acceptors.

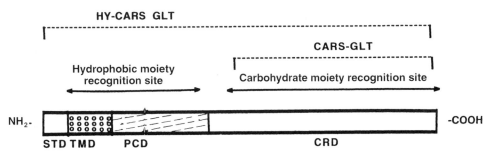

Figure 3 Classification of GSL glycosyltransferases from recognition of binding sites for hydrophobic and carbohydrate moieties of the acceptors: proposed domain structure of a type-2 membrane-bound GLT. STD, short terminal domain; TMD, transmembrane domain; PCD, proteolytic cleavage domain; CRD, carbohydrate recognition domain; GLT, GSL glycosyltransferase.

Table 2 Classification of 15 glycolipid glycosyltransferases according to CARS- and HY-CARS-specific properties.

GLT abbreviation	GLT reaction	Hydrophobicity classification	Ref.
GlcT-1	Cer → Glc-Cer	HY-CARS	4,5,7,82,88,125
GalT-1	HFACer → Gal-Cer	HY-CARS	4,5,8,96
GalT-2	Gal-Cer → Lac-Cer	HY-CARS	4,5,7,13
GalT-3	GM2 → GM1	HY-CARS	4–6,14,31,100
GalT-4	Lc3Cer → nLc4Cer	CARS	4,5,9,15
GalT-5	nLc4Cer → nlc5Cer	CARS	4,10,126
GlcNAcT-1	Lc2Cer → Lc3Cer	HY-CARS	4,5,76,79
FucT-2	nLc4Cer → H-Type	CARS	4,16,17,19
FucT-3	LM1 → SA-Le^x	HY-CARS	4,18,20–22
GalNAcT-1	GM3 → GM2	HY-CARS	4,5,31,51,63,65,68,70,73,127
GalNAcT-2	Gb3Cer → Gb4Cer	CARS	54,61,63,65,68
SAT-1	Lc2Cer → GM3	HY-CARS	4,5,31,32,36
SAT-2	nLc4Cer → LM1	CARS	4,5,34,36,37,47,50
SAT-3	LM1 → LD1A	CARS	4,33,35,36,46,49
SAT-4	GM1 → GD1a	HY-CARS	4,5,31,43–45,48

3.06.2.4   UDP-galactose: LcOse3Cer β1,4-Galactosyltransferase (GalT-4)

Among the galactosyltransferases, a milk β1,4-GalT (lactose synthase A protein, EC 2.4.1.22) is capable of using glucose as an acceptor in the presence of α-lactalbumin (lactose synthase B protein) leading to the biosynthesis of lactose.[123,128,129] During the 1980s and 1990s, the homologous protein A, catalyzing transfer of galactose to glycolipid (without modulation by α-lactalbumin), has been characterized and isolated from rabbit bone marrow,[9,101] sera,[130–132] embryonic chicken brain,[113,114,133,134] rat prostate tumor,[135,136] neuroblastoma,[137–139] human colon carcinoma,[20,24,27] adenocarcinoma,[140,141] and mouse T-lymphoma.[142] In addition to milk A protein,[123,128,143] the GalT-4 from mouse T-lymphomas (104 000-fold)[15,144] and embryonic chicken brain (49 000-fold)[14,116,121] have been purified to homogeneity and polyclonal antibodies against ECB-GalT-4 have been produced[116] to study distributions in subcellular fractionations and for cloning work. Kinetic properties of both

GalT-4 (104 000-fold) and GalT-3 (5000-fold) have been studied in detail with purified enzymes from T-lymphoma[15] and ECB,[14,116,121] respectively. The ability of the purified (40 000-fold) GalT-4 from ECB[116] and T-lymphomas to transfer galactose to glycosphingolipids (GlcNAcβ1,3Galβ1,4Glc-ceramide (LcOse3Cer) or GlcNAcβ1,3Galβ1,4GlcNAcβ1,3Galβ1,4Glc-ceramide (GlcNAc-nLcOse4Cer), glycoproteins (S$\bar{\text{A}}$, G$\bar{\text{al}}$,α_1 AGP), and free oligosaccharides have been tested.

A 1.3 kb cDNA clone coding for bovine galactosyltransferases was first isolated from a λgt 11 expression library by immunological screening with monospecific polyclonal antisera to the affinity-purified bovine milk GalT.[145] The nucleotide sequence predicted an open reading frame (ORF) which codes for 334 amino acids with an M_r of 37 645. Based on the $M_r = 57 000$ for the membrane-bound enzyme, the clone accounted for 61% of the coding sequences from the NH$_2$-terminal end of the protein. A 1.7 kb insert was used from λgt 10 human liver cDNA library for human β1,4-GalT, the sequencing of which revealed a 783 bp galactosyltransferase coding sequence, with the remainder of the sequence corresponding to the 31 regions of the mRNA downstream from the termination codon. The homology at the amino acid level was 80%, and 91% to the partial sequences of bovine and human milk β1,4-GalT, respectively.[146] A full-length murine β1,4-GalT was cloned from a murine cDNA library using a bovine β1,4-GalT probe. The sequence of a full-length murine cDNA β1,4-GalT clone, an ORF of 399 amino acids, revealed two sets of start sites for transcriptional initiation.[126,147] No apparent sequence similarity was detected between bovine β1,4-GalT and α1,3-GalT (GalT-5; described below) which was cloned by immunological screening of a bovine λgt 11 library.[102] However, both of the proteins were found to contain a cysteine residue (Cys[298] in α1,3-GalT and Cys[339] in β1,4-GalT), followed at a distance of 5–6 amino acid residues by a hexapeptide with the sequence B-Asp-Lys-Lys-Asn A (A = Glu/Asp; B = Arg/Lys). The hexapeptide is also conserved in murine and human GalTs. However, a corresponding sequence is absent in rat α2,6-SAT(ST6N or ST6GalI).[148]

An $\sim$600 bp ECB cDNA fragment (truncated GalT-4), a homologue[116,118,119] of mammalian β1,4 GalT, has been expressed in *E. coli* as a GST fusion protein.[149–153]

The expressed GST-GalT-4 (48 kDa) has been found to be catalytically active with similar substrate specificity to that of native ECB. Purified GalT-4 catalyzes the *in vitro* biosynthesis of nLcOse4Cer (Galβ1,4GlcNAcβ1,3Galβ1,4Glc-ceramide) from LcOse3 Cer(GlcNAcβ1,3-Galβ1,4Glc-ceramide).[149–153] The fusion protein has been found to be immunoreactive, on Western blot analysis, with antibody raised against the purified ECB GalT-4.[150,153]

According to a reported differential trace acetylation experiment,[154] the binding of α-lactalbumin to β1,-GalT results in a decrease in the reactivities of lysine 93 and 181 and an increase in the reactivities of one or more of lysines 230, 237, and 241. On the basis of this observation it was proposed that the NH$_2$-terminal region of bovine β1,4-GalT between residues 79 and $\sim$250 is involved in the interaction with α-lactalbumin and, by implication, with acceptor Glc or GlcNAc. In studies[149–153] with ECB truncated GalT-4, it has been observed that amino acid residues corresponding to 220–250 of the bovine sequence present in the GST-GalT-4 (truncated fusion protein construct) appears sufficient for interaction with α-lactalbumin and to confer lactose synthase activity.

3.06.2.5 UDP-galactose: Galβ1,4GlcNAc-R α1,3-Galactosyltransferase (GalT-3)

In vitro synthesis of a fucose-free blood group B-active pentaglycosylceramide, nLcOse5Cer (Galα1,3Galβ1,4GlcNAcβ1,3Galβ1,4Glc-ceramide), from nLcOse4Cer (Galβ1,4GlcNAcβ, 1,3Galβ1,4Glc-ceramide) was first reported using a solubilized enzyme fraction from rabbit bone marrow.[4,10,101] Anti-Gal antibody, which is present in large amounts in humans and which interacts specifically with α-galactosyl epitopes, has been described.[155] GalT-5 has been cloned[102] and Galα1,3Galβ1,4-GlcNAcR epitopes on human red cells have been synthesized using recombinant primate GalT-5 expressed in *E. coli*.[155]

3.06.3 GLYCOLIPID: GLUCOSYLTRANSFERASE (GlcT)

3.06.3.1 UDP-glucose: nFA-ceramide β1,1-Glucosyltransferase (GlcT-1)

The synthesis of cerebrosides (glucosyl- and galactosyl-) was postulated to be catalyzed by the two different gene products (glucosyl- and galactosyltransferases) expressed in developing chicken[82] and mouse brains.[83] However, the sequences determined from the cDNA sequences of GlcT-1 (Figure 2, step 1)[88] and GalT-1 (Figure 2, step 1A)[107] show little structural homology. Glucosylceramide is proved to be the precursor of most of the longer-chain glycosphingolipids of all three families of glycolipids (ganglio-, globo-, and lacto-) (Figures 2(a) and 2(b)).[2,5] During the myelination process of the neurons, GalT-1 (Figure 2, step 1a) is expressed and galactosylceramide is sulfated by sulfotransferase[107] to produce the sulfatide[156] as an end product of the pathway. Sulfation of intermediate GSLs of the globo- and glucuronylβ1–3nLcOse4Cer is also under study. Both L-PDMP (1-phenyl-2-decanoylamino-3-morpholino-1-propanol·HCl) and L-PPMP (1-phenyl-2-hexadecanoylamino-3-morpholino-1-propanol·HCl) inhibit GlcT-1,[85] perhaps by binding to the hydrophobic sites of the enzyme (Figure 3). However, mixed inhibition kinetics have been observed with the enzyme isolated from tissues.[85] Substrate competition studies with L-PDMP and L-PPMP with the cloned and expressed protein (GlcT-1) will provide better insight into the hydrophobic domain of this GlcT-1 (UDP-glucose: nFA-ceramide β1,1-glucosyltransferase). The *de novo* enzyme has been reported to be heat sensitive.[82] The gene regulation of expression of GlcT-1 by hormones would be an interesting field of research to be explored after the sequence of this enzyme is known.[88]

3.06.4 GLYCOLIPID: FUCOSYLTRANSFERASES (FucT OR FT)

The biosynthesis of fucosylglycolipids (Figures 1 and 2) has become an important area of research in recent years for the regulated expression of fuco- (Figure 1; Le^x [110]) and sialo-fucosyl glycosphingolipids on developing cells and metastatic cancer cells, respectively (Figure 1). Of course, ABO blood group glycoconjugates on human red blood cells[157] had been identified as glycolipids containing a ceramide moiety in 1963.[157,158] The characterization of many fucosyltransferases has been reported from a variety of tissues and cultured tumor and cancer cells.[110,157–160] A wide variety of these enzymes, which catalyze the synthesis of H, Le^x or SA-Le^x epitopes attached to cell surfaces, have also been cloned.[25,26,159] However, enzymes which catalyze the syntheses of epitopes attached to a ceramide moiety will be reviewed in the following section.

Table 3 Different glycolipid fucosyltransferases of animal cells (donor: GDP-fucose).

Abbreviation	Alternative names	Acceptor	Linkage	Ref.
FucT-2	α1,2-Fucosyltransferase (IUBEN-2.4.1.89)^a or FUT-1/α1,2FT	nLcOse4Cer	α1,2	4,16,17,19
FucT-3	α1,3-Fucosyltransferase (FUT3–6/FucT-III–VI)	SA-nLcOse4Cer	α1,3	4,5,18,20,21,22,25,160,161

^{a}IUBEN, International Union of Biochemistry Enzyme Nomenclature.

FucTs, like other glycosyltransferases, are type II transmembrane proteins sharing a common domain structure.[159,161] They have a short NH$_2$-terminal cytoplasmic chain, a 16–20 amino acid transmembrane or signal anchor domain, and an extended stem region of larger globular protein, the COOH-terminal domain or catalytic domain.[162–164] The catalytic domain is in the Golgi lumen and catalyzes fucosylation of oligosaccharides at the terminal galactose moiety (by α1,2-FucT) or an internal *N*-acetylglucosamine moiety (by α1,3-FucT).[4,18,20,160] The α-fucosyltransferases catalyze the transfer of a fucose residue from GDP-β-fucose, with inversion of configuration at the anomeric center, to form α-linked fucosyl bonds with glycolipid and/or glycoprotein acceptors. Their specificity most likely resides in the carbohydrate recognition site of the COOH globular domain (Figure 3). Seven nonhomologous (or paralogous) human FucT genes (named FUT-1 to FUT-7 or FucT-I to FucT-VII) have been cloned.[25,159] Their nomenclature has been compared along with their sequence alignment and fold recognition.[161]

3.06.4.1 GDP-fucose: nLcOse4Ceramide α1,2-Fucosyltransferase (FucT-2)

All α1,2-fucosyltransferases catalyze the biosynthesis of α-fucosyl linkages on β-linked galactose residues by the general reaction shown in Equation (1).

$$GDP\text{-}\beta\text{-}Fuc + Gal\beta\text{-}R \longrightarrow Fuc\alpha1,2Gal\beta1,R + GDP \tag{1}$$

However, the α1,2-fucosyltransferases reported to date can be classified into three types according to the structures of their preferred fucosyl acceptor substrates. The first type, and perhaps the most thoroughly studied, is specific only for the presence of a terminal β-linked galactose, with little specificity for its positional linkage (β1,3 vs. β1,4) to the penultimate sugar residue, GlcNAc.[165] In fact, the activity of this type of enzyme is commonly measured using a synthetic compound, phenyl β-D-galactopyranoside, as the fucosyl acceptor substrate which cannot be utilized by other types of α1,2-fucosyltransferases.[166] This type of enzyme is also very active with lactose (Galβ1–4Glc), *N*-acetyllactosamine (Galβ1,4GlcNAc), and lacto-*N*-biose (Galβ1,3GlcNAc) as acceptor substrates. Using nLcOse4Cer, the H-active glycosphingolipid biosynthesis *in vitro* was achieved with Golgi-rich membranes from bovine spleen[16] and 9–19-day-old embryonic chicken brains.[20,22] The second class of α1,2-fucosyltransferases consists of enzymes that are completely specific for the type 2 core structure (*N*-acetyllactosamine), but do not transfer fucose to lactose or to the type 1 core structure (Galβ1,3GlcNAc).[20–22] The third type of α1,2-fucosyltransferase is specific for acceptor substrates containing the "ganglio" core structure (Galβ1–3GalNAc) such as mucins and gangliosides.[23]

3.06.4.1.1 *Soluble α1,2-FucTs of human serum*

The best known of the first type of fucosyltransferases (less specific) is the H gene-specified α1,2-fucosyltransferase of human serum, which has been studied for many years by a number of different laboratories.[16,22,165,168] Partial characterization of this enzyme from the serum of type O individuals using primarily glycolipid substrates has been reported.[17,166] The enzyme is most active with nLcOse4Cer (type 2 chain) and less so with LcOse4Cer (type 1 chain) and GgOse4Cer (Galβ1,3 GalNAcβ1,4Galβ1,4Glc-Cer). The product formed with nLcOse4Cer cross-reacted with the α1,2-fucose-specific lectin from *Ulex europeus*. Unfortunately, some characteristics of the *in vitro* reaction, such as reactivity with phenyl β-D-galactopyranoside or the detergent specificity, were not examined. However, in this study perhaps the most important observation was that H glycolipid was not synthesized from nLcOse4Cer by serum from individuals of the rare Bombay phenotype who lack the H antigen on their erythrocytes. A more thorough study of the saccharide specificity of this enzyme was reported for two different α1,2-fucosyltransferases from type O serum which were separated by ion-exchange chromatography on DEAE-Sephadex.[167] The enzyme believed to be the H gene product displayed the lowest K_m value for phenyl β-D-galactopyranoside (4.3 mM), followed by lactose (5.5 mM). Although the enzyme was purified only 65-fold by successive ion-exchange chromatography, it was also shown to be a glycoprotein by its ability to bind to the mannose-specific lectin ConA.

The other α1,2-fucosyltransferase discovered in serum is now believed to be the product of the Se or secretor gene because it is nearly identical with the α1,2-fucosyltransferase in human milk.[168] This enzyme is also active with phenyl β-D-galactopyranoside and lactose but differs from the H gene product by having a lower K_m for Galβ1,3GlcNAc (lacto-*N*-biose) and a higher K_m for *N*-acetyllactosamine (25–30 mM). The two enzymes were separated[166] based on their difference in charge, by passage through S-Sepharose, a very strong cation exchanger to which the Se enzyme bound but the H enzyme did not. The enzymes thus separated displayed kinetic properties nearly identical to those previously reported. The H gene-specified fucosyltransferase (unretained on S-Sepharose) was further purified to apparent homogeneity by affinity chromatography on GDP-hexanolamine-Sepharose and gel filtration–HPLC. The purified enzyme obtained in this way had apparent molecular weights of 200 kDa and 50 kDa as determined by sodium dodecyl sulfate–polacrylamide gel electrophoresis (SDS–PAGE) under nonreducing and reducing conditions. The enzyme purified in this study was also found to contain *N*-linked oligosaccharide moieties based on its ability to bind to several lectin-Sepharose resins, including lentil (LCH), ConA (specific for α-mannose), abrin, and RCA-I (specific for terminal galactose residues). Another aspect of the fucosyltransferases from serum and milk that is poorly understood is their unique existence as soluble forms of α1,2-FucTs. The majority of FucTs exist as membrane-bound enzymes and relationships between membrane-bound and soluble forms of fucosyltransferases are unknown.

3.06.4.1.2 *Membrane-bound α1,2-FucT of spleen and brain*

The second class of α1,2-fucosyltransferases differs from the enzymes reported above in their high specificity for terminal Galβ1,4GlcNAc structures on glycoproteins and glycolipids. This type of enzyme, unlike the serum or submaxillary gland enzymes, is not active with either phenyl β-D-galactopyranoside or mucin-type substrates (Galβ1,3GalNAc) as fucosyl-acceptor substrates. Nor are they highly active with lactose or lacto-N-biose (Galβ1,3GlcNAc). Perhaps the best known activity of this type is that from bovine spleen.[16,19] The membrane-bound α1,2-FucT first reported from the bovine spleen system[16,19] was highly specific for Galβ1,4GlcNAc-containing substrates such as nLcOse$_4$Cer[16] and nLcOse$_5$Cer.[19] The product formed from nLcOse$_4$Cer migrated as a pentaglycosylceramide on TLC and also cross-reacted with the *Ulex europeus* lectin. The activity was highest in the presence of the cationic detergent G-3634-A, a mixture of quaternary alkylamines. Sodium taurocholate was 30% and all other detergents tested were $\leqslant 5\%$ as effective as G-3634-A. The reason for this detergent specificity is unknown.

3.06.4.1.3 GM1: α1,2-FucTs of rat bone marrow and hepatomas

The third class of α1,2-fucosyltransferases, and perhaps the least understood, is specific for structures containing terminal Galβ1,3GalNAc sequences. This enzyme may be strictly dedicated to the synthesis of fucogangliosides such as fucosyl-GM$_1$. Unfortunately, a comparison of gangliosides and mucins (with similar carbohydrate structures) as fucosyl-acceptor substrates for any one enzyme has never been provided. A GM$_1$:α1,2-fucosyltransferase activity is highly expressed in rat hepatomas and in the precancerous livers of rats fed the chemical carcinogen N-2-acetyl-aminofluorene.[23,169] The enzyme transferred fucose to GM$_1$ and also to asialo-GM$_1$, and the product formed with GM$_1$ comigrated with standard fucosyl GM$_1$ on TLC analysis. However, the activity was not examined with free oligosaccharides or with any mucin-type glycoprotein. One very unusual property of the rat hepatoma activity was extreme inhibition by nearly every detergent tested. Unlike any other glycosyltransferase previously reported, this enzyme is active only in the absence of detergent. An enzyme that transfers fucose specifically to GM$_1$ and to asialo-GM$_1$ was also reported in rat bone marrow homogenate.[170] This enzyme has not been thoroughly characterized. However, it has been shown to be most active in the presence of Triton-series detergents and only 10% as active in the presence of deoxycholate. It should be noted that GM$_1$ α1,2-fucosyltransferase activity has not been characterized, or even reported, from brain, the tissue from which fucosyl-GM$_1$ was first isolated.

It appears that α1,2-fucosyltransferases are a diverse and not fully understood family of enzymes. Further research with cloned α1,2-FucTs will be required to elucidate the structural and functional relationships among the numerous activities reported and the mechanisms involved in the regulation of their expression.

3.06.4.2 GDP-fucose: Gal-GlcNAc-R α1,3-Fucosyltransferases (FucT-3)

The other general class of fucosyltransferase activities consists of those enzymes which transfer fucose to the penultimate GlcNAc residues of type 1 or type 2 core structures to synthesize the Lewis blood group antigens (Le^a and Le^x). Many of these enzymes catalyze only the synthesis of α1,3-fucosyl linkages on type 2, neolacto, core structures by the reactions shown in Equations (2) and (3).

$$\text{GDP-}\beta\text{-Fuc} + \text{Gal}\beta1,4\text{GlcNAc-R} \longrightarrow \text{Gal}\beta1,4\text{GlcNAc-R} + \text{GDP} \qquad (2)$$
$$\begin{array}{cc} & |\alpha1,3 \\ (\text{Le}^x) & \text{Fuc} \end{array}$$

$$\text{GDP-}\beta\text{-Fuc} + \text{Gal}\beta1,4\text{GlcNAc-R} \longrightarrow \text{Gal}\beta1,4\text{GlcNAc-R} \qquad (3)$$
$$\begin{array}{ccc} |\alpha2,3 & |\alpha2,3 \quad |\alpha1,3 \\ \text{SA} & (\text{SA-Le}^x) \quad \text{SA} \quad \text{Fuc} \end{array}$$

Other enzymes are known that catalyze the reactions shown in Equations (2) and (3) and also the synthesis of α1,4-fucosyl linkages on the penultimate GlcNAc residues of type 1 core structures (Equations (4) and (5)).

$$\text{GDP-}\beta\text{-Fuc} + \text{Gal}\beta1,3\text{Glc}N\text{Ac-R} \longrightarrow \text{Gal}\beta1,3\text{Glc}N\text{Ac-R} + \text{GDP} \qquad (4)$$
$$\hspace{6.5cm} |\alpha1,4$$
$$\hspace{3.5cm} (\text{Le}^a) \qquad \text{Fuc}$$

$$\text{GDP-}\beta\text{-Fuc} + \text{Gal}\beta1,3\text{Glc}N\text{Ac-R} \longrightarrow \text{Gal}\beta1,3\text{Glc}N\text{Ac-R} + \text{GDP} \qquad (5)$$
$$\hspace{1.5cm}|\alpha2,3 \hspace{4cm} |\alpha2,3 \quad |\alpha1,4$$
$$\hspace{1.5cm}\text{SA} \hspace{4.5cm} \text{SA} \quad \text{Fuc}$$

Such dual specificity is highly unusual among glycosyltransferases.

3.06.4.2.1 α1,3-FucTs of cultured neuroblastomas

As with adenocarcinomas, cell culture techniques are also useful for studies of glycoconjugate metabolism in cells of neuronal origin. Cell culture systems of great utility for studies of neuronal differentiation are the numerous mouse and human neuroblastomas. Both α1,2- and α1,3-FucT activities have been reported in a membrane preparation isolated from cultured human neuro-blastoma IMR-32 cells.[18,139] In this study, the incorporation of radiolabeled fucose into endogenous glycoproteins and glycolipids of growing cells was measured before and after neuronal differentiation chemically induced by 6-mercaptoguanosine (sGuo). It was found that after 72 h of treatment with sGuo, incorporation of fucose into both the glycolipid and glycoprotein fractions was doubled. However, the levels of *in vitro* FucT activities with exogenous acceptor substrates were exactly the same in both treated and untreated cells. Although the differentiating IMR-32 cells underwent significant morphological changes such as neurite outgrowth, neither the mechanism nor the biological role of these observed changes in fucosylation of cellular glycolipids and glycoproteins has been elucidated.

3.06.4.2.2 α1,3-FucTs (FucT-3) of brain tissue

Although the presence of various fucose-containing glycolipids in neuronal tissues was established only in the 1990s, the *in vivo* incorporation of radiolabeled fucose into brain glycoproteins has been known for many years.[171] It is surprising, then, that very little is known about the expression of fucosyltransferases in the brain. It is also somewhat ironic that one of the very first reports of *in vitro* fucosyltransferase activity was that from mouse brain.[171] In this study, the incorporation of [14C]fucose from GDP-[14C]fucose into both endogenous and exogenous glycoprotein substrates was measured in a microsomal membrane preparation of whole adult mouse brain. The enzyme was active only with mucin-type glycoproteins containing typical O-linked oligosaccharides, and thus may be similar to the GM1-FucT or to the submaxillary gland enzyme. The mouse brain enzyme was not active with β galactosidase-treated glycoproteins and was, therefore, presumed to catalyze the synthesis of α1,2-fucosyl linkages. The activity was not stimulated by the addition of divalent metal cations. However, no other properties, such as detergent specificity, were examined. An α1,3-FucT from ECB has also been solubilized and characterized[20,22] which catalyzes transfer of fucose (α1,3) to the internal GlcNAc of nLcOse4Cer.

3.06.4.2.3 The oncofetal nature of α1,3-FucTs

The importance of these studies using cultured carcinoma cell lines lies in the fact that the corresponding normal adult tissues of colon, lung, and liver do not contain α1,3-fucosyltransferase activities,[172] nor do they express the SSEA-1 and related antigens at their cell surfaces.[173] Therefore, it appears that the change from a normal to a cancerous epithelial cell may occur with the expression of previously silent fucosyltransferase genes. The role of this change of gene expression in cancer cell biology is not exactly known. However, it most likely represents a retrogenetic expression of an embryonic cell type. It has become well established that the tumor-specific, fucose-containing antigens not expressed in normal adult tissues are present in many embryonic tissues.[173] This oncofetal nature of fucosyltransferase expression might be best exemplified by the α1,3-FucT of EC (endoderm cells) cells described above. It has been found that the expression of the enzyme stops when EC cells are chemically induced to differentiate into parietal endoderm cells by treatment with retinoic acid or dibutyryl-cAMP.[174] The FucT activity is only associated with the embryonic, undifferentiated form of this cell line. Recent experiments, also involving EC cells, suggest that the SSEA-1 epitope may play a role in cell recognition and adhesion during embryonic development.[173] In this study, homotypic aggregation of F9 EC cells was specifically inhibited by the presence of lacto-*N*-fucopentaose III (Le^x oligosaccharide). The F9 cells were also shown to aggregate specifically with liposomes containing Le^x glycolipid. Surprisingly, though, further observations suggested that the cell surface receptor that recognizes the Le^x carbohydrate structure may be Le^x itself. Detergent-solubilized, radiolabeled F9 cell surface components that bound to an Le^x-octyl-Sepharose column were shown to be glycoproteins reactive with anti-Le^x antibodies.[174] Furthermore, liposomes containing Le^x glycolipid were found to self-aggregate, whereas liposomes containing nLcOse4Cer or LM1 did not. These results led to the novel hypothesis that carbohydrate–carbohydrate interactions between Le^x structures on interacting cells are perhaps part of the mechanism for cell adhesion in embryonic and tumorigenic stages. Support for this hypothesis has come from studies of the carbohydrate binding properties of the LEC adhesion proteins.[175,176] All of the proteins in this diverse family of adhesion proteins have been found to have the ability to bind specifically to sialo-Le^x and Le^x-type structures, although the nature of this interaction has not been fully characterized.[176] These proteins also contain other peptide domains homologous to Ca^{2+}-dependent animal lectins[177] and the EGF receptor.[178] Thus, the proposed interactions between opposing Le^x structures may be part of a complex mechanism for regulating cell adhesion in embryonic and tumorigenic cells involving the oncofetal expression of α1,3-fucosyltransferases.

3.06.4.2.4   *α1,3-FucTs of cultured human adenocarcinomas*

Among the most important systems for the study of α1,3-fucosyltransferase expression have been the various cultured human adenocarcinoma cell lines. These cell lines are of great interest since the many human adenocarcinomas so far examined are known to express structures such as Le^x and sialo-Le^x as tumor-specific antigens. Colo-205 is a human cell line derived from an HJT-90 malignant colon carcinoma.[179] These cells express an α1,3-FucT that is highly active with both nLcOse4Cer and LM1 and appear similar to the LEC11[180] and HAF[181] FucTs.[20,22] The Colo-205 FucT has been found to display an important kinetic property with respect to fucosyl-acceptor substrate specificity.[20,22] Although the V_{max} was nearly identical with those for nLcOse4Cer and LM1, the K_m was fivefold less for the sialylated substrate.[20] Apparently, the active site of this enzyme has the ability to recognize specifically the terminal sialic acid residue in such a way as to provide a higher affinity for the sialylated *N*-acetyllactosamine structure. This kinetic phenomenon has not been examined in other systems such as LEC11 or EC cells, and raises an important question concerning the biosynthesis of tumor-associated antigens. Are Le^x and sialo-Le^x synthesized by the same or by different α1,3-fucosyltransferases? An enzyme similar to the α1,3-FucT of Colo-205 cells has also been reported from the human lung adenocarcinoma line NCl-H69.[182] This JHT-93 αFucT activity, like Colo-205 FucT, was not affected by divalent metal ions and was optimally active at neutral pH. The detergent-solubilized enzyme from NCl-H69 was also highly active with both sialylated and nonsialylated glycolipids, although kinetic parameters were not determined.

3.06.4.2.5   *α1,3-FucT of human amniotic fluid*

The majority of α1,3-FucTs reported thus far appear to be active with both terminally sialylated and nonsialylated substrates. A soluble α1,3-FucT from human amniotic fluid (HAF) has been purified to near homogeneity by chromatography on fetuin-agarose.[182] The purified HAF fraction contained a major band with a molecular weight of 62 kDa upon native SDS–PAGE, and was equally active with nLcOse4Cer and LM1. The purified HAF enzyme was also equally active with both sialylated and desialylated glycoproteins, and thus appears to be similar in substrate specificity to the LEC11 enzyme. Other potential fucosyl-acceptor substrates, such as lacto-*N*-biose, were not tested, nor is the physiological substrate of this enzyme known.

3.06.4.2.6   *α1,3-FucTs of embryonal carcinoma cells*

A membrane-bound α1,3-fucosyltransferase has been well characterized in mouse embryonal carcinoma F9 (EC-F9) cells,[182] the cell line from which the SSEA-I antigen was first discovered.

This activity was highly specific for *N*-acetyllactosamine and could utilize no other substrate. The EC-F9 cell enzyme was solubilized with Triton X-100 and purified to apparent homogeneity by ion-exchange and affinity chromotography. The enzyme displayed a native molecular weight of 65 kDa on SDS–PAGE but was obtained in only a low quantity (30 μg). EC cells express a large number of glycoproteins containing polylactosamine sequences.[173] The functional role of these glycoproteins in embryonal carcinoma cell biology is not known. It is believed that the above enzyme functions mainly to catalyze the synthesis of α1,3-fucosyl linkages on polylactosamine-containing glyco-proteins.

3.06.4.2.7 α1,3-FucTs in Chinese hamster ovary cells

Another cell culture system useful for the study of fucosyltransferases is the Chinese hamster ovary (CHO) cell and its lectin-resistant glycosylation mutants. The CHO mutants LEC11 and LEC12 were selected for their resistance to wheat germ agglutinin, and were also found to have an increased sensitivity to the β-galactose-binding toxin ricin, compared with the parent cell lines.[180,183] Although the exact nature of the mutations is not known, LEC11 and LEC12 were both found to express α1,3-fucosyltransferase activities not detected in the parent cell line.[180,181] Furthermore, both LEC11 and LEC12 produce cell-surface components that bind anti-SSEA-1 monoclonal antibodies.[180] However, detergent extracts of these two cell lines displayed a fundamental difference in fucosyl-acceptor substrate specificity when various glycoproteins and glycolipids were used in the *in vitro* assay system.[184,185] The α1,3-FucT activity of LEC11 was equally active with terminally sialylated substrates and their desialylated derivatives, such as fetuin and asialofetuin, or LM1 (NeuAcα2,3nLcOse4Cer) and nLcOse4Cer. The LEC12 enzyme, on the other hand, was only active with nonsialylated substrates, but did synthesize an authentic Le^x product able to bind to anti-SSEA-1 antibodies. Thus, it was proposed that LEC11 and LEC12 express two distinct α1,3-FucT genes that are not expressed in wild-type CHO cells. Whether or not both of these types of activities are expressed in other cells or tissues is not completely clear, but is an important question since α1,3-fucosyl linkages are found on both sialylated and nonsialylated structures in developing and tumorigenic systems.

3.06.5 GLYCOLIPID: SIALYLTRANSFERASES

In eukaryotic cells, sialic acids (Neu5Ac or Neu5Gc) occur essentially as terminal sugars in α2,3- and α2,6-linkages to galactose or *N*-acetylgalactosaminyl residues of oligosaccharides which are attached to proteins[1,186–188] or to sphingolipids.[4,5,36] Sialyltransferases (SATs) are a family of glycosyl-transferases that transfer sialic acid from the donor substrate CMP-NeuAc to the acceptor for oligosaccharide or are bound to glycoproteins or glycolipids. Several comprehensive reviews have been published on sialyltransferases.[4,36,189–192] At least 13 distinct sialyltransferase activities have been characterized and cDNA have been cloned.[36,192] However, in this section we will review some publications on the sialyltransferases that specifically catalyze glycolipid biosynthesis. The nomenclature used in this chapter for glycolipid sialyltransferases has been published previously,[4,36,192] and in Table 4 a comparison is made with the suggested new nomenclature.

Table 4 Different glycolipid sialyltransferases of animal cells (donor: CMP-NeuAc).

Abbreviation	Alternative names	Acceptor	Linkage	Ref.
SAT-1	GM3 synthase	LcOse2Cer	α2,3	4,31,32,36
SAT-2	GD3 synthase	GM3/LM1	α2,8	4,33,35,36,47,49
SAT-3	STZ or ST3Gal IV	nLcOse4Cer	α2,3	4,5,34,36,37,46
SAT-4	GDIa synthase	GM1/GgOse4Cer	α2,3	4,5,31,43–45,48
SAT-6	ST6GalI (ST6N, SialT-1)	nLcOse4Cer	α2,6	148,162,193

3.06.5.1   CMP-NeuAc: Lactosylceramide α2,3-Sialyltransferase (SAT-1)

The sialyltransferase activity that catalyzes the transfer of sialic acid from CMP-NeuAc to lactosylceramide to form GM3 ganglioside was first characterized in 9-day-old embryonic chicken brains[4,5,31,32,34,52] and has subsequently been detected in rat liver[11] and other animal tissues.[39,40] It has been purified[194,195] but the production of an antibody against the polypeptide chain for immuno-screening work of any cDNA library has not been reported. From the kinetic studies it appears that this glycolipid α2,3-sialyltransferase (SAT-1) is different from that of rat mammary SAT which catalyzes the transfer of sialic acid to lactose disaccharide.[190]

3.06.5.2   CMP-NeuAc: GM3 or LM1 α2,8-Sialyltransferase (SAT-2)

The α2,8-sialyltransferases which catalyze the transfer of sialic acid from CMP-NeuAc to GM3 and LM1 (Figure 2(a), step 3A and Figure 2(b), step 10A) have been proposed to be the same gene product based on the kinetic studies in the presence of a detergent-solubilized Golgi membrane preparation isolated from embryonic chicken brains.[4,5,33] The *in vitro* biosynthesis of LD1a (NeuAc-α2,8NeuAcα2,3nLcOse4Cer) (Figure 2(b), step 10A) has been reported in solubilized SAT-2[35] from embryonic chicken-brain Golgi membranes. This SAT-2 activity has been cloned from human melanoma WM266-4,[196] human SK-Mel-28,[47] rat fetal brain,[50] and KF 3027-Hyg5[197] cells. The sequence of α2,8-sialyltransferase (SAT-2 or GD3 synthase) showed a high level of similarity with other SATs at two conserved regions. The cloned and expressed SAT-2 catalyzed the formation of 2,8-linkages in GD3 and GQ1b gangliosides with relative rates of 100 and 16, respectively. As predicted, mRNA (2.6 kb) of this SAT-2 gene is strongly expressed in human melanoma lines.

3.06.5.3   CMP-NeuAc: nLcOse4Cer α2–3Sialyltransferase (SAT-3) and CMP-NeuAc: GM1 α2,3-Sialyltransferase (SAT-4)

Sialoglycoconjugates (glycoproteins and glycolipids) are ubiquitous in animals, either as components of cellular membranes or as extracellular fluids such as serum, cerebrospinal fluid, and nacrotic fluid. The sialoglycoconjugates (sialoglycoproteins and sialoglycolipids) enriched on the cancer cell surfaces are important determinants in the social behavior of the cells[110] and are believed to be regulated during cell proliferation and metastasis.[157] Reports on the role of Lewis X (Le^x), sialyl-Lewis X (SA-Le^x), sialyl-diLewis X(SA-diLe^x)(NeuAcα2,3Galβ1,4(Fucα1,3)GlcNAcβ1,3 Galβ1,4GlcNAc Gal-Glc-Cer), or sulfated sialyl-Lewis X as ligands for the selectins (L-, E-, and P-)[198] suggest a hypothesis for the role of these glycoconjugates on tumor cell surfaces during metastasis and cancer cell proliferation.[199]

A stepwise biosynthetic pathway of Le^x and SA-Le^x (Figure 2(b)) has been reported from embryonic chicken brain[5,20] and human colon carcinoma Colo-205 cells.[20,24,27,200] However, little is known about the expression of Lewis X and sialyl-Lewis X blood-group glycoconjugates with multi-lactosamine (or polylactosamine) chain-bound to ceramide.

A novel sialyltransferase, SAT-3 (CMP-NeuAc:nLcOse4Cer α2,3-sialyltransferase), has been characterized in bovine spleen,[36,201,202] embryonic chicken brain,[4,5,34,36,37] human colon carcinoma Colo-205,[20,24,27,36,37] and melanoma WM266-4[203] cells, and the reaction is given in Equation (6).

Activities of two glycolipid sialyltransferases, SAT-3 and SAT-4, have been solubilized from Colo-205 membrane preparation[20,27,38,150] using sodium taurocholate, which catalyzes the synthesis of LM1 and GD1a gangliosides, respectively (Equations (6) and (7)).

SAT-3-catalyzed reaction (Figure 2(b), step 9):

CMP-[^{14}C]-NeuAc + **Galβ1,4GlcNAc**-Gal-Glc-Cer ⟶ NeuAcα2–3Galβ1,4GlcNAc-Gal-Glc-Cer   (6)
                          (nLc4Cer)   (LM1)

SAT-4-catalyzed reaction (Figure 2(a), step 6):

CMP-[^{14}C]-NeuAc + **Galβ1,3GalNAc**-Gal-Glc-Cer ⟶ NeuAcα2–3Galβ1,3GalNAc-Gal-Glc-Cer   (7)
                          | |
                    (GM1a) NeuAc   (GD1a) NeuAc

Both sialyltransferase activities, SAT-3 (CMP-NeuAc: nLcOse4Cer α2,3-sialyltransferase) and SAT-4 (CMP-NeuAc: GgOse4Cer α2,3-sialyltransferase), in Colo 205 cells catalyze the transfer of sialic acid to the terminal galactose of GlcNAc- and GalNAc-containing glycolipid substrates, respectively. Competition kinetic studies with nLcOse4Cer and GM1 as substrates in a sialyltransferase assay showed that these two activities are catalyzed by two different catalytic entities.[37] The two enzymes were co-solubilized with taurocholate and resolved by DEAE-Cibacron Blue-Sepharose column chromatography into two elution peaks. The column eluent with SAT-3 activity failed to transfer sialic acid to asialo-α1-acid glycoproteins, indicating that this enzyme is different from the sialyltransferase (ST3N) that synthesizes the NeuAcα2,3Gal linkage in asparagine-linked oligosaccharides of glycoproteins. However, Colo-205 SAT-3 activity can be immunoprecipitated with a polyclonal antibody produced against a truncated protein expressed in *E. coli* as a GST-fusion protein from an ECB cDNA homologue[38] of an α2,3-sialyltransferase (SAT-3 or STZ) that has been cloned from human placenta[46] and human melanoma cell.[47] A concentration-dependent decrease in the residual SAT-3 activity relative to SAT-4 activity was observed in Colo-205 supernatant after precipitation of the immune complex. Expression of SAT-3 (STZ) cDNA was also detected in Colo 205 cells by RT–PCR, followed by sequence analysis of the RT–PCR product. Characterization of the catalytic reaction products of SAT-3 and SAT-4 by TLC, sialidase treatment, and binding to specific antibodies indicated that both SAT-3 and SAT-4 catalyze the formation of an α2,3-linkage between sialic acid and a terminal galactose unit of glycolipid substrates.[37]

The existence of two different gene products was also suggested by two different pH optima (SAT-3, pH 6.8; SAT-4, pH 6.4). Among the seven detergents tested,[34,36,121,150] sodium taurocholate was found to be the most efficient in solubilizing SAT-3 and SAT-4 activities from Colo-205 cells, at a protein to detergent ratio of 2:1. The major portion of SAT 3 activities can be recovered in the detergent-solubilized supernatant (DSS) prepared with taurocholate and Zwittergent; all other detergents tested failed to reach complete solubilization of the SAT-3 activity. About five times more activity was recovered in the case of SAT-4 when taurocholate was used for solubilization. Triton X-100, Cutscum, and Zwittergent are also effective in solubilizing SAT-4 activity from Colo-205 homogenate. G3634A, a positively charged detergent, inhibited the activities of both the enzymes, but to different extents.

The apparent K_m values for SAT-3 and SAT-4 activities with nLcOse4Cer and GgOse4Cer are 0.19 mM and 0.93 mM, respectively. Competition experiments with nLc4 and Gg4 indicated that SAT-3 and SAT-4 are possibly two different catalytic entities as the experimental curve matched with the curve obtained from the theoretical calculation for a two-enzyme-catalyzed reaction.[37,38] Results from similar experiments with GM1 and GgOse4Cer as substrates suggested that perhaps the same SAT-4 catalyzes the transfer of sialic acid to these acceptors to form GD1a and GM1b, respectively.[37,121,150] Newcastle disease virus α2,3-sialidase cleaved both of the radioactive products ([14C]NeuAc-nLcOse4Cer, 90%; [14C]NeuAc-GgOse4Cer, 79%), indicating that the sialic acid is linked α2,3 in the products of both enzyme-catalyzed reactions.

A SAT-3 gene from human placenta has been cloned and expressed.[46] Expression of a SAT-3 clone-containing plasmid pSTZ in COS-1 cells produced an active α2,3-sialyltransferase, which used oligosaccharide, glycoproteins, and glycolipid acceptor substrates with terminal galactose in the Galβ1,3GalNAc and Galβ1,4GlcNAc but not in the Galβ1,3GlcNAc sequences. The cloning results confirm that this human placenta sialyltransferaces gene is perhaps the enzyme previously described as SAT-3 (CMP-NeuAc:nLcOse4Cer α2,3-sialyltransferase) from embryonic chicken brain.[34] However, two different glycolipid sialyltransferases have been characterized from human colon carcinoma Colo-205 cells,[37] SAT-3 and SAT-4, which catalyze the transfer of sialic acid to nLcOse4Cer (Galβ1,4GlcNAcβ1,3Galβ1,4Glc-Cer) and GgOse4Cer (Galβ1,3GalNAcβ1,4Galβ1,4Glc-Cer), respectively.[37,38,150]

Based on the similarity in enzymatic properties of SAT-3, a developmentally regulated α2,3-sialyltransferase from embryonic chicken brain (ECB),[34] and an α2,3-sialyltransferase overexpressed in a eukaryotic system from cloned human placenta cDNA[46] and human melanoma cell line WM 266-4,[203] we postulated a genetic similarity among these three sources. A partial homology is expected with gene structure from human colon carcinoma Colo-205 also. Using an RT–PCR-based approach, the partial cDNAs for both ECB and Colo-205 SAT-3 have been isolated and sequenced.[37,38,150]

3.06.5.4   CMP-NeuAc: nLcOse4Cer α2,6-Sialyltransferase: ST6GalI or ST6N or SiaT-1 (SAT-6)

The α2,6-sialyltransferase (ST6N or ST6Gal-I or SiaT-1) was purified and cloned from rat liver.[148,204] This ST6GalI-GalI has been proved to catalyze the transfer of sialic acid from CMP-sialic acid to free disaccharide (Galβ1,4GlcNAc-) or when it is bound to glycoproteins. The cloned enzyme is also believed to transfer both to glycosphingolipids and glycoproteins containing the same sugar terminal sequences Galβ1,4GlcNAc.[50] Kinetic analysis of the S-sialylmotif of ST6GalI showed a change of K_m values for both the donor and the acceptor substrates.[205]

3.06.6　GLYCOLIPID: GLUCURONYLTRANSFERASE (GlcAT)

3.06.6.1　UDP-GlcA: nLcOse4Cer β1,3-Glucuronyltransferases

HNK-carbohydrate (or epitope) is now considered to be a hallmark of several cell adhesion molecules. HNK-1 epitope has been implicated in the migration of neural crest cells, the adhesion of astrocytes and neurons to laminin, the outgrowth of neuritic and astrocytic processes, the preferential outgrowth of neurites from motor neurons, and the homophilic binding of neural cell to cell adhesion molecules.[93] In cerebellar granule neurones it promotes differentiation and neurite growth.[206] HNK-epitope is highly immunogenic. Since the early 1980s, HNK-1 antigens (glycolipids, glycoproteins, or proteoglycans) have also been found in tumor cells.[207] The major glycolipid containing the HNK-1 epitope SGGL-1 (3-O-sulfate GlcAβ1,3Galβ1,4GlcANAcβ1,3Galβ1,4 Glc1,1Cer) and its highest homologue (3-O-sulfate GlcAβ1,3Galβ1,4GlcNAcβ1,3nLcOse4-Cer) were isolated from human peripheral nerve.[208] The transfer of glucuronic acid to nLcOse4Cer to form the core glycolipid of HNK-1 epitope (GlcAβ1,3Galβ1,4GlcNAcβ1,3Galβ1–4Glc-Cer) is catalyzed by a β1,3glucuronyltransferase (GlcAT-1) (Table 5), characterized from 19-day-old embryonic chicken brain.[89,90] This activity is inhibited by 10–100 μM sphingosine, a negative modulator in the signal transduction pathway.[89] Subsequently, using rat brain extract as the enzyme source, it has been suggested that the glucuronyltransferase catalyzing transfer to glycolipids is different from that involved in transfer to the glycoprotein.[94] The unambiguous proof will perhaps come from the sequence comparison[209] of the two different GlcAT catalyzing glycolipid and glyco-protein HNK-1 molecules. Partial purification and solubilization of the glycolipid glucuronyl-transferase (GlcAT-1) has been reported.[89]

Table 5 Two different glycosyltransferases involved in the biosynthesis of glucuronylglycosphingolipids.

Abbreviation	Alternative names	Acceptor	Linkage	Ref.
GlcT-1	Glucucerebroside synthase (IUBEN: 2.41.80)[a]	Ceramide (nonhydroxy-fatty acid)	β1,1	4,5,7,82,88,125
GlcAT-1	SGGI-1 core synthase	nLcOse4Cer	β1,3	89,90,93

[a] IUBEN, International Union of Biochemistry Enzyme Nomenclature.

3.06.7　GLYCOLIPID N-ACETYLGLUCOSAMINYLTRANSFERASES (GlcNAcTs)

Lacto-series glycolipids are characterized by having an N-acetylglucosamine residue as the third sugar linked to the terminal galactosyl group of lactosylceramide, LcOse2Cer(Galβ1,4-Glc-Cer), in their core structure. The triglycosylceramide, LcOse3Cer(GlcNAcβ1,3Galβ1,4Glc-Cer), is the core structure for both Type-1, Type-2 chain-associated antigenic glycoconjugates of ABH and Lewis (Le[a] and Le[x]) families.[157] The presence of polylactosaminylglycolipids has been characterized by the presence of Galβ1,4GlcNAc units either joint, tandemly, or in a branched manner. The branched structure (GlcNAcβ1,3(GlcNAcβ1,6)Galβ1,4Glc-Ceramide) present in the li-antigen has been of particular interest: it is described in the following section with the emphasis on the synthetic route of this glycosphingolipid.

Different glycolipid N-acetylglucosaminyltransferases isolated from animal cells are listed in Table 6. A β1,3-N-acetylglucosaminyltransferase (GlcNAcT-1) which catalyzes the biosynthesis of LcOse3Cer from UDP-GlcNAc and lactosylceramide was reported from rabbit bone marrow,[75] mouse lymphoma, and human colon carcinoma cells.[78,79] The biosynthesis of two pentaglycosyl-ceramides containing a terminal GlcNAc moiety attached to nLcOse4Cer by β1,3GlcNAc transferase (GlcNAcT-1) and β1,6GlcNAc transferase (GlcNAcT-3) have been reported from human Colo-205[79] and mouse P-1798,[76] respectively. However, both GlcNAcT-1 and GlcNAcT-2 are present in human colon carcinoma Colo-205 cells.[78,79] These two transferases showed different pH optima, different cation and anion effects, and a differential heat-inactivation pattern at 55 °C. Permethylation studies of the radioactive products isolated from both of the enzyme-catalyzed reactions indicated the presence of a 1,3-linked β-D-GlcNAc group at the nonreducing end in both cases.

Table 6 Different glycolipid *N*-acetylglucosaminyltransferases isolated from animal cells.

Abbreviation	Alternative names	Acceptor	Linkage	Ref.
GlcNAcT-1	Paragloboside core synthase	LcOse2Cer	$\beta1,3$	4,5,75,76,78,79,81,101
GlcNAcT-2	i.core synthase	nLcOse4Cer	$\beta1,3$	4,76,79
GlcNAcT-3	i/I core synthase	GlcNAcβ1,3nLcOse4Cer	$\beta1,6$	76,80

Purification of these enzymes to homogeneity is under investigation in several laboratories, and cloning sequences are not available. Whether these glycolipid GlcNAcTs have any sequence homology to any of the six GnTs or GlcNAcTs (I–VI) in the pathway of *N*-linked oligosaccharide biosynthesis will be of immediate interest.[210,211] Many of these glycoprotein GlcNAc transferases have been cloned[212] and their sequences show little homology.[210–213] It is proposed that perhaps GlcNAcT-1 regulates the expression of the sulfoglucuronyl glycolipids (GlcAβ1,3(SO$_3$-*O*-3) Galβ1,4GlcNAcβ1,3Galβ1,4Glc-Cer) in specific cell types in the cerebellum during development.[214]

3.06.8   GLYCOLIPID *N*-ACETYLGALACTOSAMINYLTRANSFERASES (GalNAcTs)

The gangliosides constitute an important class of glycosphingolipids, and are characterized by the fact that they contain at least one sialic acid moiety in the oligosaccharide chain attached to ceramide.[215,216] The gangliosides with a triglycosylganglio core structure are identified by the *N*-acetylgalactosaminyl group as a third sugar.[12] However, in globo or Forssman families, the *N*-acetylgalactosaminyl group exists as fourth and fifth sugars, respectively. Three different glycolipid *N*-acetylgalactosaminyltransferases have been characterized from animal tissues.[4,5] (Table 7).

Table 7 Different glycolipid *N*-acetylgalactosaminyltransferases isolated from animal cells.

Abbreviation	Alternative names	Acceptor	Linkage	Ref.
GalNAcT-1	GM2 synthase	GM3/GD3	$\beta1,4$	4,5,31,61,63,65,68,70,73,127
GalNAcT-2	Globoside synthase	GbOse3Cer	$\beta1,3$	54,61,63,65,68,73
GalNAcT-3	αGalNAc transferase	GbOse4Cer	$\alpha1,3$	59,64,72

3.06.8.1   UDP-GalNAc: GM3 β1,4*N*-Acetylgalactosaminyltransferase (GalNAcT-1)

Ganglioside GM2 (GalNAcβ1,4(NeuAcα2,3)Galβ1,4Glc-Cer) is widely distributed in animal brains,[4,5,31,215,216] and other animal tissues. The *in vitro* biosynthesis of GM2 ganglioside from UDP-GalNAc and ganglioside GM3 was first achieved using a Golgi-rich membrane preparation isolated from embryonic chicken brain (ECB).[31,51,52,58,65] The GalNAcT-1 activity has been solubilized and separated from GalNAcT-2 activity (see next section) of 19-day-old ECB.[63,65] In addition to ECB, using GM3 as an acceptor, the GalNAcT-1 activity has been characterized in rat liver,[11] rat brain,[53,217] mouse liver,[61,62] fetal pig brain,[31,52,58] NIL hamster cells,[218] mouse neuroblastoma cells, and 3T3 cells.[83] The guinea pig RBC glycolipid GgOse3Cer (GalNAcβ1,4Galβ1,4Glc-Cer) has been synthesized using guinea pig bone marrow GalNAcT-1.[55] The GalNAcT-1 has been purified to homogeneity[68] and expression cloned.[70,73] The cloned enzyme is also specific for GM3 or GD3 substrates and is not active with glycolipids without sialic acid (e.g., lactosylceramide) or a disaccharide (lactose) as observed with the purified enzyme from ECB.[63,65]

Based on kinetic studies, it has been concluded that both GA2 (GgOse3Cer) and GD2 synthase activities and GM1b, GD1a, and GT1b synthase activities are catalyzed by a single enzyme (GalNAcT-1) present in Golgi vesicles from rat liver.[43] However, these results need to be revisited using available cloned enzymes. Mice with disrupted GM2/GD2 synthase genes lack complex gangliosides and exhibit only subtle defects in their nervous system.[219] Genomic organization and chromosomal assignment of the human GalNAcT-1 gene with multiple transcription units has been reported.[219]

3.06.8.2   UDP-Gal*N*Ac: GbOse3Cer *β*1,3*N*-Acetylgalactosaminyltransferase (Gal*N*AcT-2)

The *β*1,3*N*-acetylgalactosaminyltransferase which catalyzes the transfer of Gal*N*Ac from UDP-Gal*N*Ac to GbcOse3Cer (Gal*α*1,4Gal*β*1,4Glc-Cer) was initially characterized in embryonic chicken brains.[54] Subsequently, the activity has been solubilized from Golgi-rich membranes of ECB[63,65] and chemically transformed guinea pig tumor 104 cells,[64] and purified for studies of its substrate specificity. This activity has also been purified to homogeneity from canine spleen.[61]

3.06.8.3   UDP-Gal*N*Ac: GbOse4Cer *α*1,3-*N*-Acetylgalactosaminyltransferase (Gal*N*AcT-3)

Biosynthesis of the Forssman hapten (Gal*N*Ac*α*1,3-Gal*N*Ac*β*1,3Gal*α*1,4Gal*β*1,4Glc-Cer) from globoside (Gal*N*Ac*β*1,3Gal*α*1,4Gal*β*1,4Glc-Cer) is catalyzed by a specific *α*1,3-*N*-acetyl-galactosaminyltransferase characterized from guinea pig kidney[57] and mouse adrenol Y-1 tumor cells.[56] The activity has been solubilized and partially purified from dog spleen.[59] Complete purification and characterization of an *α*1,3Gal*N*Ac-transferase encoded by the human blood group A gene has been reported.[72]

3.06.9   REGULATION OF GLYCOLIPID BIOSYNTHESIS

Glycosphingolipids containing fucose and sialic acid residues of Gg, Lc, or Gb families are expressed on the eukaryotic cell surfaces in a developmentally regulated,[5,12,110,157,220] tissue-specific, or tumor-specific manner.[157,199] Such structures function as cell–cell adhesion or as receptors for specific proteins (antibodies or lectins) and are also implicated as tumor-specific markers.[220,221] The glycosphingolipids of the ganglio family (gangliosides) are ubiquitous constituents of eukaryotic tissues, and their oligosaccharide structures undergo alterations[215] during cellular development, differentiation, and aging in keeping their less known physiological functions. Regulation of SAT-1 by protein kinase-C catalyzed phosphorylation has been reported.[222] Increased endogenous GM3 may play an important role in regulating cellular differentiation as evidenced by antisense oligo-nucleotides against Gal*N*AcT-1 and SAT-2 genes in HL-60 cells.[223] Several reviews[157,199,215,224–226] are available on probable functions of glycosphingolipids and this topic is not reviewed here. However, since the mid-1980s several glycolipid glycosyltransferases have been cloned. An understanding of their gene structure and transcriptional regulation would provide us with more detailed knowledge of the regulation of expression of specific glycosphingolipids on the cell surface or intracellular membranes during tissue development or cancer cell metastasis.

It has been shown[227,228] for GT (or GalT-4) that the distal promoter region in the immediate upstream of the 4.1 kb start site is bound primarily by the ubiquitous transcription factor Sp-1. In contrast, the proximal region adjacent to the 3.9 kb start site is a target for binding by multiple proteins which include a candidate negative regulatory factor, Sp-1, a mammary gland-specific form of CTF/NF1, and the tissue-restriction factor, AP2. By mutations in the conserved sialyl motif region of SAT-6 (ST-6 Gal$_l$), binding of this region to the CMP-NeuAc donor has been suggested.[204,205] However, mutations in the sialyl motif-conserved region and their effect on the K_m values of CMP-NeuAc or glycolipids with ECB or Colo-205 SAT-6 or SAT-3 are not known. The upstream 5′-sequence of the SAT-6 promoter region has also been recognized and appears to bind with several transcription factors.[229] Regulation of binding of these transcription factors to glycosyltransferase genes during development and oncogenic processes will be of great importance.

ACKNOWLEDGMENTS

We thank Mrs. Dorisanne Nielsen and Mrs. Rosemary Patti for their help during the preparation of the manuscript. The work reviewed in this chapter was partly supported by United States Public Health Services Grant NS-18005 (The Jacob Javits Award) to S.B.

3.06.10   REFERENCES

1. H. Schachter and S. Roseman, in "The Biochemistry of Glycoproteins and Proteoglycans," ed. W. J. Lennarz, Plenum, New York, 1980, pp. 85–160.

2. S. Basu, M. Basu, J.-L. Chien, and K. A. Presper, in "Structure and Function of Gangliosides," ed. L. Svennerholm, H. Dreyfus, and P.-F. Urban, *Adv. Exp. Med. Biol.*, Plenum, New York, 1980, vol. 125, pp. 213–226.

3. Y. Kishimoto, in "The Enzymes," ed. P. D. Boyer, Academic Press, New York, 1983, vol. 16, pp. 357–407.

4. M. Basu, T. De, K. K. Das, J. W. Kyle, H. C. Chon, R. J. Schaeper, and S. Basu, *Methods Enzymol.*, 1987, **138**, 575.

5. S. Basu, *Glycobiology*, 1991, **1**, 469.

6. S. Basu, B. T. Kaufman, and S. Roseman, *J. Biol. Chem.*, 1965, **240**, 4115.

7. S. Basu, B. Kaufman, and S. Roseman, *J. Biol. Chem.*, 1968, **243**, 5802.

8. S. Basu, A. Schultz, M. Basu, and S. Roseman, *J. Biol. Chem.*, 1971, **246**, 4272.

9. M. Basu and S. Basu, *J. Biol. Chem.*, 1972, **247**, 1489.

10. M. Basu and S. Basu, *J. Biol. Chem.*, 1973, **248**, 1700.

11. T. W. Keenan, D. J. Morre, and S. Basu, *J. Biol. Chem.*, 1974, **249**, 310.

12. S. Basu and M. Basu, in "The Glycoconjugates," ed. M. Horowitz, Academic Press, New York, 1982, vol. 3, pp. 265–285.

13. S. Chatterjee, N. Ghosh, and S. Khurana, *J. Biol. Chem.*, 1992, **267**, 7148.

14. S. Ghosh, J. W. Kyle, S. Dastgheib, F. Daussin, Z. Li, and S. Basu, *Glycoconjugate J.*, 1995, **12**, 838.

15. M. Basu, S.-A. Weng, H. Tang, F. Khan, F. Rossig, and S. Basu, *Glycoconjugate J.*, 1996, **13**, 423.

16. S. Basu, M. Basu, and J.-L. Chien, *J. Biol. Chem.*, 1975, **250**, 2956.

17. T. Pacuszka and J. Koscielak, *Eur. J. Biochem.*, 1976, **64**, 499.

18. K. A. Presper, M. Basu, and S. Basu, *Proc. Natl. Acad. Sci. USA*, 1978, **75**, 289.

19. K. A. Presper, M. Basu, and S. Basu, *J. Biol. Chem.*, 1982, **257**, 169.

20. M. Basu, J. W. Hawes, Z. Li, S. Ghosh, F. A. Khan, B.-J. Zhang, and S. Basu, *Glycobiology*, 1991, **1**, 527.

21. E. Holmes, G. Ostrander, and S. Hakomori, *J. Biol. Chem.*, 1985, **260**, 7619.

22. J. W. Hawes, Ph.D. Thesis, University of Notre Dame, 1991.

23. E. H. Holmes and S.-I. Hakomori, *J. Biochem.*, 1987, **101**, 1095.

24. M. Basu, K. K. Das, B. Zhang, F. A. Khan, and S. Basu, *Indian J. Biochem. Biophys.*, 1988, **25**, 112.

25. B. W. Weston, R. P. Nair, R. D. Larsen, and J. B. Lowe, *J. Biol. Chem.*, 1992, **267**, 4152.

26. B. W. Weston, P. L. Smith, R. J. Kelly, and J. B. Lowe, *J. Biol. Chem.*, 1992, **267**, 24575.

27. M. Basu, S. S. Basu, Z. Li, H. Tang, and S. Basu, *Indian J. Biochem. Biophys.*, 1993, **30**, 324.

28. P. H. Johnson, A. S. R. Donald, and W. M. Watkins, *Glycoconjugate J.*, 1993, **10**, 152.

29. E. H. Holmes and B. A. Macher, *Arch. Biochem. Biophys.*, 1993, **301**, 190.

30. H. Kojima, K. Nakamura, R. Mineta-Kitajima, Y. Sone, and Y. Tamai, *Glycoconjugate J.*, 1996, **13**, 445.

31. S. Basu, 1996, Ph.D. Thesis, University of Michigan, Ann Arbor, M.I., 1966.

32. B. Kaufman and S. Basu, *Methods Enzymol.*, 1966, **8**, 365.

33. B. Kaufman, S. Basu, and S. Roseman, *J. Biol. Chem.*, 1968, **243**, 5804.

34. M. Basu, S. Basu, A. Stoffyn, and P. Stoffyn, *J. Biol. Chem.*, 1982, **257**, 12765.

35. H. Higashi, M. Basu, and S. Basu, *J. Biol. Chem.*, 1985, **260**, 824.

36. S. Basu, M. Basu, and S. S. Basu, in "Biology of Sialic Acids," ed. A. Rosenberg, Plenum, New York, 1995, chap. 3, pp. 69–94.

37. S. S. Basu, M. Basu, Z. Li, and S. Basu, *Biochemistry*, 1996, **35**, 5166.

38. S. S. Basu, M. Basu, S. Dastgheib, S. Ghosh, and S. Basu, *Indian J. Biochem. Biophys.*, 1997, **34**, 97.

39. R. K. Yu and S. H. Lee, *J. Biol. Chem.*, 1976, **251**, 198.

40. S. S. Ng and J. A. Dain, *J. Neurochem.*, 1977, **29**, 1075.

41. P. Stoffyn and A. Stoffyn, *Carbohydr. Res.*, 1979, **74**, 279.

42. V. Liepkans, A. Jolif, and G. Larson, *Biochemistry*, 1988, **27**, 8683.

43. G. Pohlentz, D. Klein, G. Schwarzmann, D. Schmitz, and K. Sandhoff, *Proc. Natl. Acad. Sci. USA*, 1988, **85**, 7044.

44. S. Dasgupta, J.-L. Chien, and E. L. Hogan, *Biochim. Biophys. Acta*, 1990, **1036**, 11.

45. T.-J. Gu, X.-B. Gu, T. Ariga, and R. K. Yu, *FEBS Lett.*, 1990, **275**, 83.

46. H. Kitagawa and J. C. Paulson, *J. Biol. Chem.*, 1994, **269**, 1394.

47. K. Nara, Y. Watanabe, K. Maruyama, K. Kasahara, Y. Nagai, and Y. Sanai, *Proc. Natl. Acad. Sci. USA*, 1994, **91**, 7952.

48. M. Pitto, P. Palestini, and M. Masserini, *FEBS Lett.*, 1996, **383**, 223.

49. M. Kono, Y. Yoshida, N. Kojima, and S. Tsuji, *J. Biol. Chem.*, 1966, **271**, 29366.

50. G. Zeng, L. Gao, and R. K. Yu, *Gene*, 1997, **187**, 131.

51. J. C. Steigerwald, B. Kaufman, S. Basu, and S. Roseman, *Fed. Proc., Fed. Am. Soc. Exp. Biol.*, 1996, **25**, 587.

52. B. Kaufman, S. Basu, and S. Roseman, in "Inborn Disorders of Sphingolipid Metabolism," *Proc. Int. Symp. Cerebral Sphingolipidoses*, ed. S. M. Aronson and B. W. Volk, Pergamon, New York, 1967, pp. 193–213.

53. S. Handa and R. M. Burton, *Lipids*, 1969, **4**, 589.

54. J.-L. Chien, T. Williams, and S. Basu, *J. Biol. Chem.*, 1973, **248**, 1778.

55. M. Basu, J.-L. Chien, and S. Basu, *Biochem. Biophys. Res. Commun.*, 1974, **60**, 1097.

56. K.-K. Yeung, J. R. Moskal, J. L. Chien, D. A. Gardner, and S. Basu, *Biochem. Biophys. Res. Commun.*, 1974, **59**, 252.

57. T. Ishibashi, S. Kijimoto, and A. Makita, *Biochim. Biophys. Acta*, 1974, **337**, 92.

58. J. C. Steigerwald, S. Basu, B. Kaufman, and S. Roseman, *J. Biol. Chem.*, 1975, **250**, 6727.

59. N. Taniguchi, N. Yokosawa, S. Gasa, and A. Makita, *J. Biol. Chem.*, 1982, **257**, 10631.

60. T. Taki, R. Kamada, and M. Matsumoto, *J. Biochem.*, 1982, **91**, 665.

61. N. Taniguchi and A. Makita, *J. Biol. Chem.*, 1984, **259**, 5637.

62. Y. Hashimoto, M. Abe, Y. Kiuchi, A. Suzuki, and T. Yamakawa, *J. Biochem.*, 1984, **95**, 1543.

63. R. J. Schaeper, Ph.D. Thesis, University of Notre Dame, 1987.

64. K. K. Das, M. Basu, S. Basu and C. H. Evans, *Carbohydr. Res.*, 1986, **149**, 119.

65. R. J. Schaeper, K. K. Das, Z. Li, and S. Basu, *Carbohydr. Res.*, 1992, **236**, 227.

66. T. Dohi, A. Nishikawa, I. Ishizuka, M. Totani, K. Yamaguchi, K. Nakagawa, O. Saitoh, S. Ohshiba, and M. Oshima, *Biochem. J.*, 1992, **288**, 161.

67. A. Takeya, O. Hosomi, N. Shimoda, and S. Yazawa, *J. Biochem.*, 1992, **112**, 389.

68. Y. Hashimoto, M. Sekine, K. Iwasaki, and A. Suzuki, *J. Biol. Chem.*, 1993, **268**, 25857.

69. N. Malagolini, F. Dall'olio, S. Guerrini, and F. Serafini-Cessi, *Glycoconjugate J.*, 1994, **11**, 89.
70. K. Sango, O. N. Johnson, C. A. Kozak, and R. L. Proia, *Genomics*, 1995, **27**, 362.
71. A. Yoshida, T. Hara, H. Ikenaga, and M. Takeuchi, *Glycoconjugate J.*, 1995, **12**, 824.
72. D. B. Haslam and J. U. Baenziger, *Proc. Natl. Acad. Sci. USA*, 1996, **93**, 10697.
73. Y. Nagata, S. Yamashiro, J. Yodoi, K. O. Lloyd, H. Shiku, and K. Furukawa, *J. Biol. Chem.*, 1992, **267**, 12082.
74. F.-i. Yamamoto, J. Marken, T. Tsuji, T. White, H. Clausen, and S. Hakomori, *J. Biol. Chem.*, 1990, **265**, 1146.
75. S. Basu, M. Basu, H. Den, and S. Roseman, *Fed. Proc.*, *Fed. Am. Soc. Exp. Biol.*, 1970, **29**, 410.
76. M. Basu and S. Basu, *J. Biol. Chem.*, 1984, **259**, 12557.
77. F. Piller, J. P. Cartson, A. Maranduba, A. Veyrières, Y. Leroy, and B. Fournet, *J. Biol. Chem.*, 1984, **259**, 13385.
78. E. H. Holmes, S.-i. Hakomori, and G. K. Ostrander, *J. Biol. Chem.*, 1977, **262**, 15649.
79. M. Basu, F. A. Khan, K. K. Das, and B.-J. Zhang, *Carbohydr. Res.*, 1991, **209**, 261.
80. S. Yousefi, E. Higgins, D. Zhuang, A. Pollex-Krueger, O. Hindsgaul, and J. W. Dennis, *J. Biol. Chem.*, 1991, **266**, 1772.
81. D. K. H. Chou and F. B. Jungalwala, *J. Biol. Chem.*, 1993, **268**, 21727.
82. S. Basu, B. Kaufman, and S. Roseman, *J. Biol. Chem.*, 1973, **248**, 1388.
83. N. S. Radin, L. Hof, R. M. Bradley, and R. O. Brady, *Brain Res.*, 1969, **14**, 497.
84. G. S. Shukla, A. Shukla, and N. S. Radin, *J. Neurochem.*, 1991, **56**, 2125.
85. N. S. Radin, J. A. Shayman, and J. I. Inokuchi, *Adv. Lipid Res.*, 1993, **26**, 183.
86. Y. Lavie, H.-T. Cao, S. L. Bursten, A. E. Giuliano, and M. C. Cabot, *J. Biol. Chem.*, 1996, **271**, 19530.
87. Y. Lavie, H.-T. Cao, A. Volner, A. Lucci, T.-Y. Han, V. Geffen, A. E. Giuliano, and M. C. Cabot, *J. Biol. Chem.*, 1997, **272**, 1682.
88. S. Ichikawa, H. Sakiyama, G. Suzuki, K. I.-P. J. Hidari, and Y. Hirabayashi, *Proc. Natl. Acad. Sci. USA*, 1996, **93**, 4638.
89. K. K. Das, M. Basu, Z. Li, S. Basu, and F. Jungalwala, *Indian J. Biochem. Biophys.*, 1990, **27**, 396.
90. K. K. Das, M. Basu, S. Basu, D. K. H. Chou, and F. B. Jungalwala, *J. Biol. Chem.*, 1991, **266**, 5238.
91. H. Yokoto, A. Yuasa, and R. Sato, *J. Biochem.*, 1992, **112**, 182.
92. C. Kawashima, K. Terayama, M. Ii, S. Oka, and T. Kawasaki, *Glycoconjugate J.*, 1993, **9**, 307.
93. F. B. Jungalwala, *Neurochem. Res.*, 1994, **19**, 945.
94. S. Oka, K. Terayama, C. Kawashima, and T. Kawasaki, *J. Biol. Chem.*, 1992, **267**, 22711.
95. S. Basu, A. Schultz, and M. Basu, *Fed. Proc.*, *Fed. Am. Soc. Exp. Biol.*, 1969, **27**, 346.
96. P. Morell and N. S. Radin, *Biochemistry*, 1969, **8**, 506.
97. E. Costantino-Ceccarini and K. Suzuki, *Arch. Biochem. Biophys.*, 1975, **167**, 646.
98. N. M. Neskovic, L. L. Sarlieve, and P. Mandel, *Biochim. Biophys. Acta*, 1976, **429**, 342.
99. N. M. Neskovic, P. Mandel, and S. Galt, in "Enzymes of Lipid Metabolism," eds. S. Galt, L. Freysz, and P. Mandel, *Adv. Exp. Med. Biol.*, Plenum, New York, 1977, vol. 101, pp. 613–630.
100. H. Miyazaki, S. Fukumoto, K. Okada, T. Hasegawa, K. Furukawa, and K. Furukawa, *J. Biol. Chem.*, 1997, **272**, 24794.
101. M. Basu, D.Sc. Thesis, University of Calcutta, 1972.
102. D. H. Joziasse, J. H. Shaper, E. H. van den Eignden, A. J. van Tunen, and N. L. Shaper, *J. Biol. Chem.*, 1989, **264**, 14290.
103. T. Coetzee, X. Li, N. Fujita, J. Marcus, K. Suzuki, U. Francke, and B. Popko, *Genomics*, 1966, **35**, 215.
104. R. Barker, K. W. Olsen, J. H. Shaper, I. P. Trayer, and R. L. Hill, *J. Biol. Chem.*, 1972, **247**, 7135.
105. N. Navaratnam, S. Ward, C. Fisher, N. J. Kuhn, and J. B. C. Findlay, *Eur. J. Biochem.*, 1988, **171**, 623.
106. P. Van der Bijl, G. J. Strous, M. Lopes-Cardozo, J. Thomas-Oates, and G. Van Meer, *Biochem. J.*, 1996, **317**, 589.
107. A. Bosio, E. Binczcek, and W. Stoffel, *Genomics*, 1996, **35**, 223.
108. A. Bosio, E. Binczek, M. M. Le Beau, A. A. Fernald, and W. Stoffel, *Genomics*, 1996, **34**, 69.
109. P. Stoffyn, A. Stoffyn, and G. Hauser, *J. Biol. Chem.*, 1973, **248**, 1920.
110. S. Hakomori, *Annu. Rev. Immunol.*, 1984, **2**, 103.
111. G. B. Yip and J. A. Dain, *Biochim. Biophys. Acta*, 1970, **206**, 252.
112. M. Pierce, *J. Cell. Biol.*, 1982, **93**, 76.
113. K. Furukawa and S. Roth, *Biochem. J.*, 1985, **227**, 573.
114. F. Kaplan and P. Hechtman, *J. Biol. Chem.*, 1983, **258**, 770.
115. M. Basu, K. A. Presper, S. Basu, L. M. Hoffman, and S. E. Brooks, *Proc. Natl. Acad. Sci. USA*, 1979, **76**, 4270.
116. S. Ghosh, Ph.D. Thesis, University of Notre Dame, 1992.
117. J. Kyle, Ph.D. Thesis, University of Notre Dame, 1985.
118. S. Gosh, S. S. Basu, and S. Basu, *Biochem. Biophys. Res. Commun.*, 1992, **189**, 1215.
119. S. Basu, S. Ghosh, S. S. Basu, J. W. Kyle, T. Li, and M. Basu, *Indian J. Biochem. Biophys.*, 1993, **30**, 315.
120. S. Dastgheib, M. Basu, and S. Basu, in "Proceedings of Glyco XIV, Zurich, September 6–12, 1997," in press.
121. S. Basu, M. Basu, K. K. Das, F. Daussin, R. J. Schaeper, P. Banerjee, F. A. Khan, and I. Suzuki, *Biochimie*, 1988, **70**, 1551.
122. S. Ghosh, K. K. Das, F. Daussin, and S. Basu, *Indian J. Biochem. Biophys.*, 1990, **27**, 379.
123. U. Brodbeck, W. L. Denton, N. Tanahashi, and K. E. Ebner, *J. Biol. Chem.*, 1967, **242**, 1391.
124. S. Basu, G. Ghosh, M. Basu, J. W. Hawes, K. K. Das, B.-J. Zhang, Z. Li, S.-A. Weng, and C. Westervelt, *Indian J. Biochem. Biophys.*, 1990, **27**, 386.
125. R. Watanabe, K. Wu, P. Paul, D. L. Marks, T. Kobayashi, M. R. Pittelkow, and R. E. Pagano, *J. Biol. Chem.*, 1998, **273**, 9651.
126. K. Nakazawa, T. Ando, T. Kimura, and H. Narimatsu, *J. Biochem. (Tokyo)*, 1988, **104**, 165.
127. K. Furukawa, H. Soejima, N. Niikawa, H. Shiku, and K. Furukawa, *J. Biol. Chem.*, 1996, **271**, 20836.
128. I. P. Trayer and R. L. Hill, *J. Biol. Chem.*, 1971, **246**, 6666.
129. Y. Fujita-Yamaguchi and A. Yoshida, *J. Biol. Chem.*, 1981, **256**, 2701.
130. J. Moskal, Ph.D. Thesis, University of Notre Dame, 1977.
131. S. Basu, M. Basu, J. R. Moskal, J.-L. Chien, and D. A. Gardner, in "Glycolipid Methodology," ed. L. A. Witting, American Oil Chemists' Society Press, Champaign, IL, 1976, pp. 123–139.

132. J. R. Moskal, J.-L. Chien, M. Basu, and S. Basu, *Fed. Proc., Fed. Am. Soc. Exp. Biol.*, 1975, **34**, 645.
133. S. Basu, M. Basu, J. W. Kyle, T. De, K. Das, and R. J. Schaper, in "Enzymes of Lipid Metabolism," eds. L. Freysz and S. Gatt, Plenum Press, New York, 1986, pp. 232–245.
134. S. Basu, K. Das, R. J. Schaeper, P. Banerjee, F. Daussin, M. Basu, F. A. Khan, and B. J. Zhang, in "Ganglioside Research: Neurochemical and Neuroregenerative Aspects," ed. R. Ledeen, Fidia Research Series, Italy, 1988, pp. 259–273.
135. D. M. Jenis, Ph.D. Thesis, University of Notre Dame, 1981.
136. D. M. Jenis, S. Basu, and M. Pollard, *Cancer Biochem. Biophys.*, 1982, **6**, 37.
137. J. R. Moskal, D. A. Gardner, and S. Basu, *Biochem. Biophys. Res. Commun.*, 1974, **61**, 701.
138. S. Basu, J. R. Moskal, and D. A. Gardner, in "Biochemical and Pharmacological Implications of Ganglioside Function," ed. G. Porcellati, *Adv. Exp. Med. Biol.*, Plenum Press, New York, 1976, vol. 71, pp. 45–64.
139. M. Basu, F. Rossi, N. M. Young, and S. Basu, *Glycoconjugate J.*, 1996, **6**, 723.
140. E. H. Holmes, *Arch. Biochem. Biophys.*, 1989, **270**, 630.
141. E. H. Holmes, *J. Biol. Chem.*, 1992, **267**, 25328.
142. M. Basu, S. Basu, and M. Potter, in "Cell Surface Glycolipids," ed. C. C. Sweeley, *Amer. Chem. Soc. Symp. Ser.*, American Chemical Society, Washington, DC, 1980, vol. 128, pp. 187–212.
143. T. A. Beyer and R. L. Hill, in "The Glycoconjugates," ed. M. Horowitz, Academic Press, New York, 1982, vol. 3, part A, pp. 25–44.
144. F. A. Khan, M. Basu, K. K. Das, B. J. Zhang, K. D. Graber, I. Suzuki, and S. Basu, *J. Cell Biol.*, 1988, **107**, 209.
145. N. L. Shaper, J. H. Shaper, J. L. Meuth, J. L. Fox, H. Chang, I. R. Kirsch, and G. F. Hollis, *Proc. Natl. Acad. Sci. USA*, 1986, **83**, 1573.
146. H. E. Appert, T. J. Rutherford, G. E. Tarr, F. S. Wiest, N. R. Thomford, and D. J. McCorquodale, *Biochem. Biophys. Res. Commun.*, 1986, **139**, 163.
147. N. L. Shaper, G. F. Hollis, J. G. Douglas, I. R. Kirsch, and J. H. Shaper, *J. Biol. Chem.*, 1988, **263**, 10420.
148. J. Weinstein, E. U. Lee, K. McEntee, P. H. Lai, and J. C. Paulson, *J. Biol. Chem.*, 1987, **262**, 17735.
149. S. S. Basu, M. Basu, S. Ghosh, R. Chaudhuri, T. Welborn, and S. Basu, *FASEB J.*, 1994, **8**, A-1425.
150. S. S. Basu, Ph.D. Thesis, University of Notre Dame, 1996.
151. S. S. Basu, S. Ghosh, and S. Basu, *Glycobiology*, 1994, **4**, 738.
152. S. Ghosh, S. S. Basu, J. C. Strum, S. Basu, and R. M. Bell, *Anal. Biochem.*, 1995, **225**, 376.
153. S. S. Basu, S. Dastgheib, S. Ghosh, M. Basu, P. Kelly, and S. Basu, *Acta Biochim. Pol.*, 1998, **45**, 451.
154. K. Brew and S. K. Sinha, *J. Biol. Chem.*, 1981, **256**, 4193.
155. U. Galili and F. Anarati, *Glycobiology*, 1995, **5**, 775.
156. K. Honke, M. Yamane, A. Ishii, T. Kobayashi, and A. Makita, *J. Biochem.*, 1996, **119**, 421.
157. S. Hakomori, *Annu. Rev. Biochem.*, 1981, **50**, 733.
158. S. Handa, *Jpn. J. Exp. Med.*, 1963, **33**, 347.
159. J. B. Lowe, *Semin. Cell Biol.*, 1981, **2**, 289, 307.
160. W. M. Watkins, in "Glycoproteins," ed. J. Montreuil, H. Schachter, and J. F. G. Hart, Elsevier, Amsterdam, 1995, chap. 5.
161. C. Brenton, R. Oriol, and A. Imberty, *Glycobiology*, 1996, **6**, vii.
162. J. C. Paulson and K. J. Colley, *J. Biol. Chem.*, 1989, **264**, 17615.
163. D. H. Joziasse, *Glycobiology*, 1992, **2**, 271.
164. R. Kleen and E. G. Berger, *Biochim. Biophys. Acta*, 1993, **1154**, 283.
165. A. Betteridge and W. M. Watkins, *Biochem. Soc. Trans.*, 1985, **13**, 1126.
166. M. A. Chester, A. D. Yates, and W. M. Watkins, *Eur. J. Biochem.*, 1976, **69**, 583.
167. T. Kumazaki and A. Yoshida, *Proc. Natl. Acad. Sci. USA*, 1984, **81**, 4193.
168. L. Shen, E. F. Grollman, and V. Ginsburg, *Proc. Natl. Acad. Sci. USA*, 1968, **59**, 224.
169. E. H. Holmes and S.-i. Hakomori, *J. Biol. Chem.*, 1982, **258**, 3706.
170. T. Taki, R. Kamada, and M. Matsumoto, *J. Biochem.*, 1982, **91**, 665.
171. M. Zatz and S. H. Barondes, *J. Neurochem.*, 1971, **18**, 1625.
172. E. H. Holmes, G. K. Ostrander, H. Clausen, and N. Graem, *J. Biol. Chem.*, 1987, **262**, 11331.
173. H. Muramatsu and T. Muramatsu, *FEBS Lett.*, 1983, **163**, 181.
174. I. Eggans, B. Fenderson, T. Toyokuni, B. Dean, M. Stroud, and S.-i. Hakomori, *J. Biol. Chem.*, 1989, **264**, 9476.
175. B. K. Brandley, S. J. Swiedler, and P. W. Robbins, *Cell*, 1990, **63**, 861.
176. K. Drickamer, *J. Biol. Chem.*, 1988, **263**, 9557.
177. M. P. Bevilacqua, J. S. Pober, D. L. Mendrick, R. S. Cotran, and M. A. Gimbrone, Jr., *Proc. Natl. Acad. Sci. USA*, 1987, **84**, 9238.
178. A. Mitsakos and F. G. Hanisch, *Biol. Chem. Hoppe Seyler*, 1989, **370**, 239.
179. T. U. Semple, L. A. Quinn, L. K. Woods, and G. E. Moore, *Cancer Res.*, 1978, **38**, 1345.
180. C. Campbell and P. Stanley, *Cell*, 1983, **35**, 303.
181. T. Muramatsu, G. Gachelin, M. Damonneville, C. Delarbre, and F. Jacob, *Cell*, 1979, **18**, 183.
182. H. Muramatsu, Y. Kamada, and T. Muramatsu, *Eur. J. Biochem.*, 1989, **157**, 71.
183. P. Stanley, *Somatic Cell Genet.*, 1983, **9**, 593.
184. C. Campbell and P. Stanley, *J. Biol. Chem.*, 1984, **259**, 11208.
185. D. R. Howard, M. Fukuda, M. N. Fukuda, and P. Stanley, *J. Biol. Chem.*, 1987, **262**, 16830.
186. S. Roseman, *Chem. Phys. Lipids*, 1970, **5**, 270.
187. J. I. Rearick, J. E. Sadler, J. C. Paulson, and R. L. Hill, *J. Biol. Chem.*, 1979, **254**, 4444.
188. H. Kitagawa and J. C. Paulson, *J. Biol. Chem.*, 1994, **269**, 1394.
189. P. Broquet, H. Baubichon-Cortay, P. George, and P. Louisot, *Int. J. Biochem.*, 1991, **23**, 385.
190. T. Miyag, *Tanpakushitsu Kakusan Koso*, 1985, **30**, 461.
191. S. Tsuji, *J. Biochem.*, 1996, **120**, 1.
192. S. Tsuji, A. K. Datta, and J. C. Paulson, *Glycobiology*, 1996, **6**, v.
193. M. Nakamura, A. Tsunoda, K. Yanagisawa, Y. Furukawa, T. Sakai, G. Larson, and M. Saito, in press.
194. L. J. Melkerson-Watson and C. C. Sweeley, *J. Biol. Chem.*, 1991, **266**, 4448.

195. U. Preuss, X. Gu, T. Gu, and R. K. Yu, *J. Biol. Chem.*, 1996, **268**, 26 273.
196. K. Sasaki, K. Kurata, N. Kojima, A. Kurosawa, S. Ohta, N. Hanai, S. Tsuji, and T. Nishi, *J. Biol. Chem.*, 1994, **269**, 15 950.
197. M. Haraguchi, S. Yamashiro, A. Yamamoto, K. Furukawa, K. Takamiya, K. O. Lloyd, H. Shiku, and K. Furukawa, *Proc. Natl. Acad. Sci. USA*, 1994, **91**, 10 455.
198. M. L. Phillips, E. Nudelman, F. C. A. Gaeta, M. Perez, A. K. Singhal, S.-i. Hakomori, and J. C. Paulson, *Science*, 1990, **250**, 1130.
199. S.-i. Hakomori and Y. Igarashi, *J. Biochem.*, 1995, **118**, 1091.
200. E. H. Holmes and S. B. Levery, *Arch. Biochem. Biophys.*, 1989, **274**, 633.
201. J. L. Chien, Ph.D. Thesis, University of Notre Dame, 1975.
202. P. Banerjee, M. Basu, K. K. Das, and S. Basu, *Fed. Proc., Fed. Am. Soc. Exp. Biol.*, 1987, **46**, 1927.
203. K. Sasaki, E. Watanabe, T. Nishi, and M. Hasegawa, *J. Biol. Chem.*, 1993, **268**, 22 782.
204. A. K. Datta and J. C. Paulson, *J. Biol. Chem.*, 1994, **270**, 1497.
205. A. K. Datta, A. Sinha, and J. C. Paulson, *J. Biol. Chem.*, 1998, **273**, 9608.
206. D. K. H. Chou, S. A. Tobel, and F. B. Jungalwala, *J. Biol. Chem.*, 1998, **273**, 8508.
207. D. K. H. Chou, A. A. Ilyas, J. E. Evans, C. Costello, R. H. Quarles, and F. B. Jungalwala, *J. Biol. Chem.*, 1986, **261**, 11 717.
208. T. Ariga, T. Kohriyama, L. Freddo, N. Latov, M. Saito, K. Kon, S. Ando, M. Suzuki, M. E. Hemling, K. L. Rinehart, Jr., S. Kusunoki, and R. K. Yu, *J. Biol. Chem.*, 1987, **262**, 848.
209. K. Terayama, S. Oka, T. Seiki, Y. Miki, A. Nakamura, Y. Kozutsumi, K. Takio, and T. Kawasaki, *Proc. Natl. Acad. Sci. USA*, 1997, **94**, 6093.
210. R. Kumar, J. Yang, R. D. Larsen, and P. Stanley, *Proc. Natl. Acad. Sci. USA*, 1990, **87**, 9948.
211. M. Sarkar, E. Hull, Y. Nishikawa, R. J. Simpson, R. L. Moritz, R. Dunn, and H. Schachter, *Proc. Natl. Acad. Sci. USA*, 1991, **88**, 234.
212. H. Schachter, *Glycobiology*, 1991, **1**, 453.
213. N. Tamiguchi, M. Yoshimura, E. Miyoshi, Y. Ihara, A. Nishikawa, and S. Fujii, *Glycobiology*, 1996, **6**, 691.
214. D. K. H. Chou and F. B. Jungalwala, *J. Biol. Chem.*, 1996, **271**, 28 868.
215. H. Wiegandt, *Adv. Neurochem.*, 1982, **4**, 149.
216. L. Svennerholm, *Biochem. Biophys. Res. Commun*, 1962, **9**, 436.
217. J. L. Dicesare and J. A. Dain, *Biochim. Biophys. Acta*, 1971, **231**, 385.
218. M. W. Lockney and C. Sweeley, *Biochim. Biophys. Acta*, 1982, **712**, 234.
219. K. Takamiya, A. Yamamoto, K. Furukawa, S. Yamashiro, M. Shin, M. Okada, S. Fukumoto, M. Haraguchi, N. Takeda, K. Fujimura, M. Sakae, M. Kishikawa, H. Shiku, K. Furukawa, and S. Aizawa, *Proc. Natl. Acad. Sci. USA*, 1996, **93**, 10 662.
220. S. Hokomori and N. Kajima, *Glycobiology*, 1991, **1**, 613.
221. W. M. Watkins, P. O. Skacel, and J. L. Clarke, in "Glycoimmunology," eds. A. Alavi and J. S. Anford, *Adv. Exp. Med. Biol.*, Plenum Press, New York, 1995, vol. 376, pp. 83–93.
222. X. Gu, U. Preub, T. Gu, and R. K. Yu, *J. Neurochem.*, 1995, **64**, 2295.
223. G. Zeng, T. Ariga, X. Gu, and R. K. Yu, *Proc. Natl. Acad. Sci. USA*, 1995, **92**, 8670.
224. Y. Nagai and M. Iwamari, in "Biology of Sialic Acids," ed. A. Rosenberg, Plenum Press, New York, 1995, pp. 197–241.
225. S.-I. Hakomori, *J. Biol. Chem.*, 1990, **265**, 18 713.
226. G. van Echten and K. Sandhoff, *J. Biol. Chem.*, 1993, **268**, 5341.
227. A. Harduin-Lepers, J. H. Shaper, and N. L. Shaper, *J. Biol. Chem.*, 1993, **268**, 14 348.
228. B. Rajput, N. L. Shaper, and J. H. Shaper, *J. Biol. Chem.*, 1996, **271**, 5131.
229. E. C. Svensson, B. Soregham, and J. C. Paulson, *J. Biol. Chem.*, 1990, **265**, 20863.

3.07
Alkaloid Glycosidase Inhibitors

ALAN D. ELBEIN

University of Arkansas for Medical Sciences, Little Rock, AR, USA

and

RUSSELL J. MOLYNEUX

US Department of Agriculture, Albany, CA, USA

Alkaloid Glycosidase Inhibitors

3.07.1 INTRODUCTION

Polyhydroxy alkaloids with glycosidase inhibitory properties have been isolated and identified in the 1980s and 1990s, with few exceptions. Discovery of the indolizidine alkaloids swainsonine[1] and castanospermine,[2] with their potent and specific inhibitory activities towards α-mannosidase and α- and β-glucosidase, respectively, created a recognition that additional nitrogen-containing analogues of simple sugars might have similar properties and stimulated the search for new members of the class. As a result, more than 50 naturally occurring members of the group have been discovered, almost doubling the number discussed in a previous review.[3] Another review has discussed these alkaloids, with particular reference to their ecological significance.[4] Numerous synthetic analogues have been prepared, but the scope of this chapter will be restricted to the chemistry and bioactivity of those alkaloids isolated from natural sources, their glycosidase-inhibitory properties and consequent effects on glycoprotein processing.

3.07.2 CHEMISTRY OF ALKALOID GLYCOSIDASE INHIBITORS

3.07.2.1 Structural Classes

The alkaloid glucosidase inhibitors discovered up until 1998 do not conform to a single structural class but do have several features in common, including two or more hydroxyl groups and a nitrogen atom, generally heterocyclic in character. A small group of glycosidase inhibitors isolated from microorganisms also exists, which are structurally more closely related to amino sugars. However, it is possible to integrate the major class of heterocyclic compounds into structural groups based upon five- and six-membered rings, which may also be fused into bicyclic ring systems. Five different subclasses can be defined, from the simple monocyclic examples to the more complex bicyclic rings, as follows. (Commonly used alternative or abbreviated names for individual alkaloids are shown in parentheses.)

3.07.2.1.1 Pyrrolidines

Alkaloids of the pyrrolidine class, with five-membered rings, are exemplified by 2,5-dihydroxy-methyl-3,4-dihydroxypyrrolidine (DMDP), (1, R = OH), which is fully (tetra-)substituted at all carbon atom ring positions.[5] The trisubstituted representatives are 6-deoxy-DMDP (1, R = H),[6] 1,4-dideoxy-1,4-imino-D-arabinitol (D-AB1) (2, R = β-OH),[7] 1,4-dideoxy-1,4-imino-D-ribitol (2, R = α-OH),[8] 3,4-dihydroxy-5-hydroxymethyl-1-pyrroline (nectrisine) (2, R = β-OH; 1,5-double bond),[9] and N-hydroxyethyl-2-hydroxymethyl-3-hydroxypyrrolidine (3),[10] the only alkaloid in this group bearing a substituent on the nitrogen atom. Only a single disubstituted member of the group is known, namely 2-hydroxymethyl-3-hydroxypyrrolidine (CYB3) (2, R = H).[11]

A pentahydroxy alkaloid, 2,5-dideoxy-2,5-imino-DL-*glycero*-D-*manno*-heptitol (homoDMDP) (1, R = CH$_2$OH) and its 7-apioside (1, R = CH$_2$O-apiose) have been isolated and structurally identified.[12] HomoDMDP is thus the most highly hydroxylated representative of the pyrrolidine class.

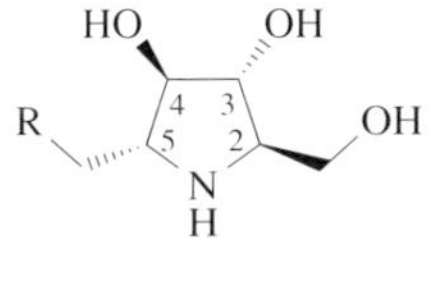

(1) 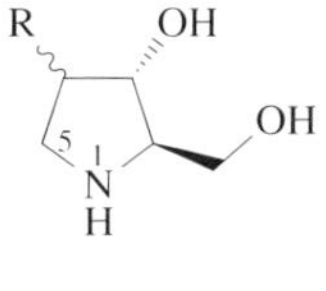   (2) 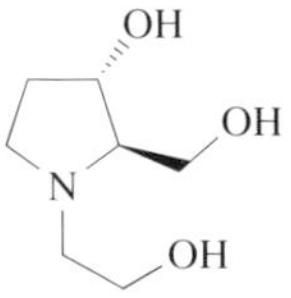   (3)

3.07.2.1.2 Piperidines

Alkaloids with six-membered rings of the piperidine class encompass nine members, one of which, 6-deoxyfagomine (4, R^1 = H, R^2 = β-OH),[13] is disubstituted, while two, namely fagomine

(**4**, R^1 = OH, R^2 = β-OH)[14] and 3-*epi*-fagomine (**4**, R^1 = OH, R^2 = α-OH),[8] are trisubstituted. An additional three alkaloids, 1-deoxynojirimycin (DNJ) (**5**, R = α-OH)[15] and its *N*-methyl derivative,[8] and 1-deoxymannojirimycin (DMJ) (**5**, R = β-OH),[16] are tetrasubstituted. The latter has also been found to occur as a series of glycosides, namely: 2-*O*, 3-*O*, and 4-*O*-α-D-glucopyranosides; 2-*O*, 3-*O*, 4-*O*, and 6-*O*-β-D-glucopyranosides; and, 2-*O* and 6-*O*-α-D-galactopyranosides.[8] The remaining four alkaloids are characterized by complete substitution at all carbon atoms, and include the glucose analogue, nojirimycin (**6**, R^1 = α-OH, R^2 = α-OH),[17] the mannose analogue, nojirimycin B (mannojirimycin) (**6**, R^1 = β-OH, R^2 = α-OH),[18] and the galactose analogue, galactostatin (**6**, R^1 = α-OH, R^2 = β-OH).[19] α-Homonojirimycin (HNJ, (**7**)) has a hydroxymethyl group at the 1-position, in place of the hydroxy group found at that position in nojirimycin, and the alkaloid has also been isolated as its 7-*O*-β-D-glucopyranoside.[20]

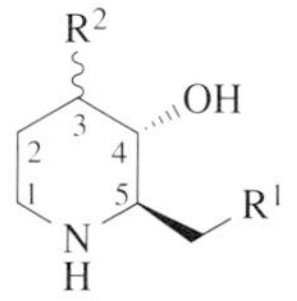

(**4**) 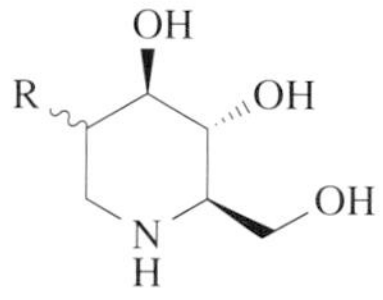   (**5**) 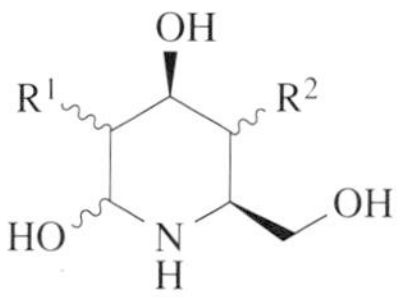   (**6**) 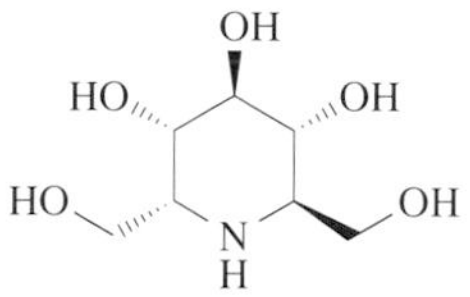   (**7**)

3.07.2.1.3 Pyrrolizidines

The pyrrolizidine alkaloids that are inhibitors of glycosidases may be regarded in a formal structural sense as the result of fusion of two pyrrolidine ring systems, with the common nitrogen atom at the bridgehead. The tetrasubstituted pyrrolizidines, australine (**8**)[21] and alexine (**9**)[22] differ only in the stereochemistry at the bridgehead carbon atom (C-7a), all other substituents having identical configurations. A certain amount of confusion has arisen in the naming of epimers of these compounds because those having a bridgehead configuration identical to that of australine have been classified as 7a-*epi*-alexines. In fact, alexine itself is the only member of this group isolated to date which has an α bridgehead proton.

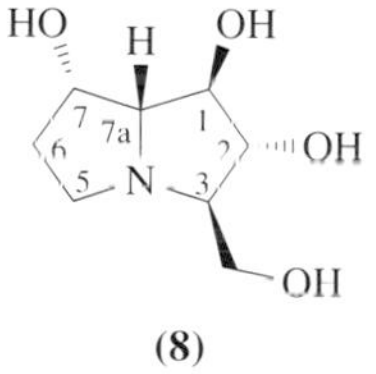

(**8**) 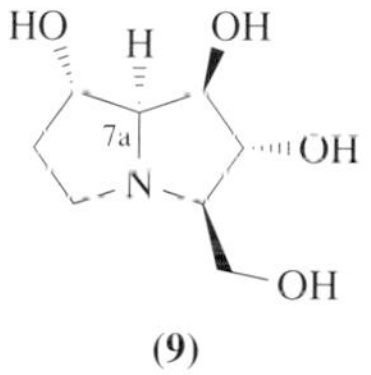   (**9**)

Harris *et al.*[23] have proposed that all alkaloids having the *R* stereochemistry be named as australines and those with the *S* stereochemistry as alexines. Adopting this convention, the three known naturally occurring epimers would therefore be named as follows, (with the alternate name in parentheses): 1-*epi*-australine (1,7a-di-*epi*-alexine) (**8**, 1-OH, α)[24] 3-*epi*-australine (3,7a-di-*epi*-alexine) (**8**, 3-CH$_2$OH, α),[23] and 7-*epi*-australine (7,7a-di-*epi*-alexine) (**8**, 7-OH, β).[24]

A unique tetrasubstituted pyrrolizidine alkaloid is 7a-*epi*-alexaflorine,[25] which also has a 7a-(*R*) bridgehead configuration, consistent with all the other alkaloids except alexine, and may be regarded as an oxidized form of australine. On the basis of its physical properties, including resistance to melting and insolubility in all solvents except for water, together with evidence of a carboxylate ion in its infrared spectrum, this alkaloid was shown to exist in the zwitterionic form (**10**).

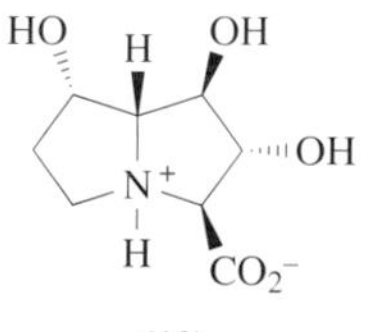

(**10**)

An interesting addition to the class has been casuarine (**11**, R = H),[26] a highly oxygenated penta-substituted pyrrolidine. This alkaloid has also been found as the 6-glucoside (**11**, R = α-D-glucosyl).[26] The occurrence of several australine/alexine epimers suggests that epimeric forms of casuarine will ultimately be discovered.

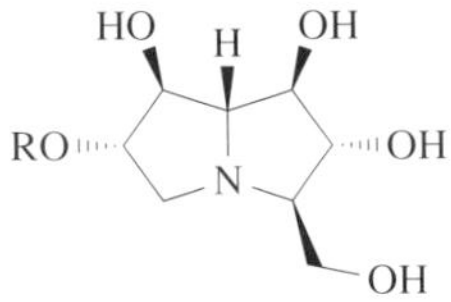

(**11**)

3.07.2.1.4 Indolizidines

In an analogous manner to the pyrrolizidine alkaloids, the indolizidine group may be visualized as a pyrrolidine ring fused with a piperidine ring, yielding a bicyclic 5/6 ring system. Seven naturally occurring members have been discovered, the simplest of which are the dihydroxylated alkaloids, lentiginosine (**12**, R = β-OH) and 2-*epi*-lentiginosine (**12**, R = α-OH).[27]

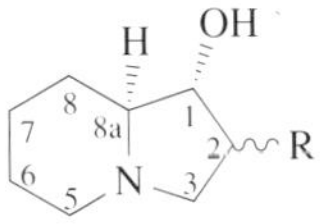

(**12**)

The familiar trihydroxylated alkaloid swainsonine (**13**)[1,28–30] is unique within the indolizidine class as the only member with an 8a-(*R*) bridgehead configuration. A second trihydroxyindolizidine, 7-deoxy-6-*epi*-castanospermine (**14**, R^1 = β-OH, R^2 = H)[31] has the 8a-(*S*) configuration characteristic of the tetrahydroxy alkaloid, castanospermine (**15**),[2] and its epimers 6-*epi*-castanospermine (**14**, R^1 = β-OH, R^2 = β-OH)[32] and 6,7-di-*epi*-castanospermine (**14**, R^1 = β-OH, R^2 = α-OH).[10]

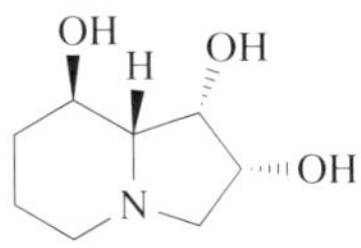

(**13**) 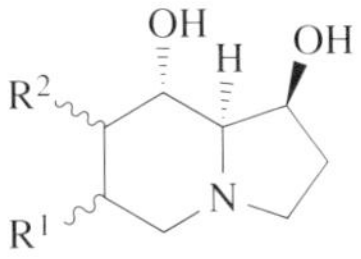   (**14**) 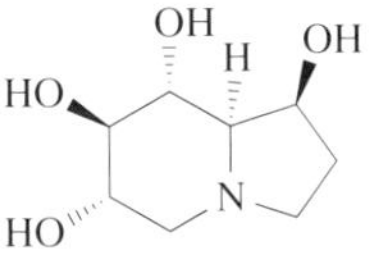   (**15**)

Theoretically, pentahydroxylated indolizidines, corresponding to casuarine, could occur but none have yet been isolated from natural sources.

3.07.2.1.5 Nortropanes

The polyhydroxy pyrrolidine, piperidine, pyrrolizidine, and indolizidine groups have been established for some time but the nortropane group is a relatively new addition to the catalog of alkaloid classes with glycosidase-inhibitory properties. Whereas tropane alkaloids are well-known in nature, nortropanes (i.e., compounds in which the nitrogen atom is not methylated) are relatively rare. The nortropane ring system can be conceptualized as a result of fusion of a five-membered pyrrolidine ring with a six-membered piperidine ring, but in contrast to the indolizidines the fusion points are α to the nitrogen atom of each monocyclic system.

The polyhydroxy nortropane group now consists of more individual alkaloids than any of the other classes, and the chemistry of these compounds has been the subject of a review.[33] The alkaloids have been named calystegines after the source of the first member to be isolated, the bindweed *Calystegia sepium*.[34,35] A consistent feature of all calystegines, in addition to the absence of *N*-methylation, is the presence of an α-OH group at the bridgehead junction (C-1) of the bicyclic ring system (i.e., an aminoketal functionality). Three subclasses have been defined, namely calystegines A, B, and C, each of which corresponds to tri-, tetra- and pentahydroxylation, respectively.

Four trihydroxylated alkaloids, calystegines A_3 (**16**, $R^1 = \alpha$-OH, $R^2 =$ H),[35] A_5 (**16**, $R_1 =$ H, $R^2 = \alpha$-OH),[36] A_6 (**17**),[37] and A_7 (**18**)[13] are known. Although the majority of calystegines bear an equatorial hydroxyl group at the C-3 position, the latter two alkaloids lack this substituent, while calystegine A_6 is unique within the A subgroup in possessing a secondary hydroxyl group on the five-membered ring moiety.

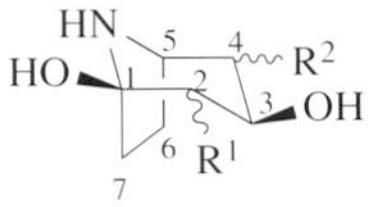

(**16**) 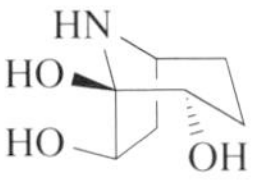 (**17**) 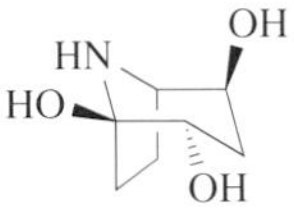 (**18**)

The calystegine B alkaloids consist of five tetrahydroxylated compounds, namely B_1 (**19**),[35] B_2 (**16**, $R^1 = \alpha$-OH, $R^2 = \alpha$-OH),[35] B_3 (**16**, $R^1 = \beta$-OH, $R^2 = \alpha$-OH),[36] B_4 (**16**, $R^1 = \alpha$-OH, $R^2 = \beta$-OH),[38] and B_5 (**20**).[13] Calystegine B_5 is the only alkaloid within this subgroup that does not have a 3-OH substituent. Although both calystegines B_1 and B_5 have secondary hydroxy groups on the five-membered ring moiety, these occur at different positions, namely C-6 and C-7, respectively. The remaining three members, calystegines B_2, B_3 and B_4, differ only in the stereochemistry of the hydroxy groups located at C-2 and C-4 on the six-membered ring; the C-3 hydroxy substituent is β in all three alkaloids. An alkaloid named calystegine N_1,[37] corresponding to calystegine B_2 but with an amino group, rather than a hydroxy group, at the bridgehead C-1 position has also been obtained. However, reactions of nojirimycin derivatives with ammonia-saturated methanol, resulting in replacement of the 2-OH group by an NH_2 substituent,[39] suggest that calystegine N_1 is an artifact of the isolation procedure, which involves elution from an ion-exchange column with dilute ammonium hydroxide.

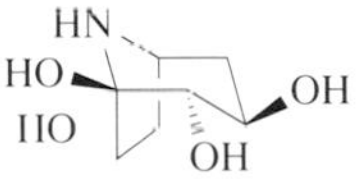

(**19**) 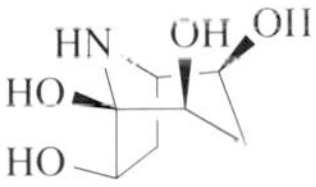 (**20**)

Two pentahydroxylated calystegine C alkaloids are known, having identical substitution patterns, including a hydroxy group at C-6 analogous to calystegine B_1. These alkaloids, calystegines C_1 (**21**, $R = \alpha$-OH)[8] and C_2 (**21**, $R = \beta$-OH),[40] differ only in the stereochemistry of the C-2 hydroxy substituent.

(**21**)

Two additional alkaloids, bearing axially oriented methyl groups on the nitrogen atom, have been isolated and structurally characterized.[13] These compounds, *N*-methylcalystegine B_2 (**22**, $R =$ H) and *N*-methylcalystegine C_1 (**22**, $R =$ OH) should strictly be classified as tropane alkaloids but the preponderance of polyhydroxy nortropanes isolated to date suggests that these new alkaloids are the result of *N*-methylation of the latter rather than products of the normal biosynthetic route to tropane alkaloids. For the purposes of this chapter they are therefore classified within the nortropane group.

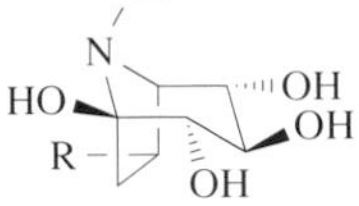

(**22**)

3.07.2.1.6 *Miscellaneous glycosidase inhibitors*

A few nitrogen-containing glycosidase inhibitors, although they are polyhydroxylated, do not fall readily within the above structural classifications. These include the aminocyclopentanes, mannostatin A (**23**),[41] and the much more complex glycosylated cyclic urea derivative, trehazolin (**24**),[42] and kifunensine (**25**)[43] and nagstatin (**26**),[44] which may be regarded as highly modified piperidines. All of these compounds are metabolites isolated from various microorganisms.

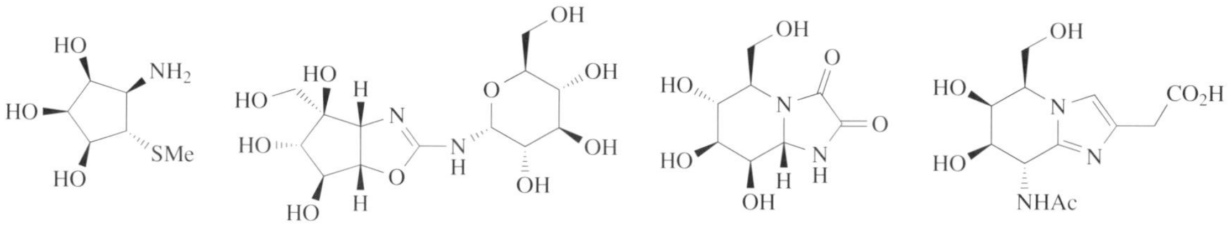

(**23**)   (**24**)   (**25**)   (**26**)

3.07.2.2   Occurrence and Isolation from Natural Sources

3.07.2.2.1   *Occurrence*

The polyhydroxy alkaloid glycosidase inhibitors have been isolated primarily from plant sources, but also occur in microorganisms and have occasionally been found in insects.[4] The sources of the individual alkaloids are listed in Table 1. Many of the earliest polyhydroxy alkaloids to be discovered, particularly the bicyclic pyrrolizidines and indolizidines, were found in the plant family Leguminosae. This apparent taxonomic relationship has now become far less secure with the isolation of casuarine (**11**) from the Casuarinaceae and Myrtaceae.[26] Moreover, swainsonine (**13**) has been identified as a constituent of several *Ipomoea* species (Convolvulaceae), co-occurring with calystegines.[45] Similarly, the initial isolation of calystegines ((**15**)–(**18**)) from the Convolvulaceae[34,35,46] has now been overshadowed by a much more widespread occurrence in the Solanaceae,[13,36–38,40,47,48] and a limited presence in *Morus* species (Moraceae).[49]

Certain individual alkaloids, predominantly DMDP and swainsonine, have a particularly widespread pattern of occurrence. Thus, DMDP (**1**, R = OH) has been isolated from plants in the families Araceae, Campanulaceae, Euphorbiaceae, Hyacinthaceae, and Leguminosae,[5,19,20,50] as well as from the body of a lepidopteran (*Urania fulgens*),[20] and from a *Streptomyces* species.[51] Similarly, swainsonine (**13**) has also been discovered in two unrelated microorganisms, *Rhizoctonia leguminicola* and *Metarhizium anisopliae*,[29,30] in addition to its quite widespread occurrence in plants.[52] It has been shown that the biosynthetic pathways to swainsonine in the Diablo locoweed, *Astragalus oxyphysus*, and *R. leguminicola* are identical, implying either a direct or indirect relationship between plant and microorganism.[53] Thus, the genetic ability to produce this alkaloid could have been transferred from one to the other in the course of evolution. Alternatively, microorganisms capable of producing the alkaloid may have an endophytic association with the plants. The presence of a calystegine-catabolizing *Rhizobium meliloti* strain in roots of *Calystegia sepium* but not within plants that do not produce calystegines emphasizes the complexity of such interactions.[54] In contrast to the previous examples, castanospermine (**15**) and its epimers (**14**)[2,10,31,32] and the australine/alexine ((**8**)–(**11**))[21–24] alkaloids have so far been restricted to the monotypic *Castanospermum australe* and species of *Alexa*, which are closely related genera in the Leguminosae.

It is apparent from these examples that no consistent conclusions can be drawn regarding the distribution of polyhydroxy alkaloids at the present time. It may be that these natural products are quite widely distributed and many new sources will be discovered in the future. The comparative newness of their discovery relative to many other classes of alkaloids is probably a consequence of their cryptic nature, due to exceptional water solubility and relative insolubility in non-hydroxylic organic solvents.[55] The increasing number, regio- and stereochemical potential for structural variation and significant biological properties of these glycosidase inhibitors will no doubt result in discovery of new members of the known classes. The identification of the nortropane group is also an indicator that new structural groups may yet remain to be discovered.

3.07.2.2.2   *Isolation*

The hydrophilicity of the polyhydroxy alkaloids renders them incapable of being isolated by conventional extraction and purification methods which involve extraction into nonpolar organic

Table 1 Natural source and enzyme inhibition properties of polyhydroxy alkaloids.

Alkaloid	Natural source	Enzyme inhibited	Ref.
Pyrrolidines			
CYB-3 (**2**, R = H)	*Castanospermum australe* (Leguminosae)	α-Glucosidase (weak)	**56**
6-Deoxy-DMDP (**1**, R = H)	*Angylocalyx pynaertii* (Leguminosae)	β-Mannosidase	**6**
D-AB1 (**2**, R = β-OH)	*Hyacinthoides non-scripta* (Hyacinthaceae) *Angylocalyx* spp. (Leguminosae) *Morus bombycis* (Moraceae) *Arachniodes standishii* (Polypodiaceae)	α-Glucosidase α-D-Arabinosidase	**56,57** **58**
1,4-Dideoxy-1,4-imino-D-ribitol (**2**, R = α-OH)	*Morus alba* (Moraceae)	α-Glucosidase (weak)	**59**
Nectrisine (**2**, R = β-OH; 1,5-double bond)	*Nectria lucida* F-4490 (Ascomycetes)	α-Glucosidase α-Mannosidase	**9** **60**
N-Hydroxyethyl-2-hydroxymethyl-3-hydroxypyrrolidine (**3**)	*Castanospermum australe* (Leguminosae)	Undetermined	
DMDP (**1**, R = OH)	*Aglaonema* spp., *Nephthytis poissoni* (Araceae). *Omphalea diandra*; *Endospermum* spp. (Euphorbiceae). *Hyacinthoides non-scripta* (Hyacinthaceae). *Derris elliptica*; *Lonchocarpus* spp. (Leguminosae); *Urania fulgens* (Lepidoptera); *Streptomyces* sp. KSC-5791	α- and β-Glucosidase β-Mannosidase Invertase Trehalase	**12,61,62** **63** **50** **51**
HomoDMDP (**1**, R = CH₂OH)	*Hyacinthoides non-scripta* (Hyacinthaceae)	α- and β-Glucosidase	**12**
Piperidines			
6-Deoxyfagomine (**4**, R¹ = H, R² = β-OH)	*Lycium chinense* (Solanaceae)	Undetermined	
Fagomine (**4**, R¹ = OH, R² = β-OH)	*Fagopyrum esculentum* (Fagaceae), *Xanthocercis zambesiaca* (Leguminosae); *Morus* spp. (Moraceae)	β-Galactosidase α-Glucosidase (weak)	**64** **56**
3-*epi*-Fagomine (**4**, R¹ = OH, R² = α-OH)	*Morus alba* (Moraceae)	β-Galactosidase	**64**
1-Deoxynojirimycin (DNJ) (**5**, R = α-OH)	*Morus* spp. (Moraceae); *Bacillus* spp.; *Streptomyces lavandulae*	α- and β-Glucosidase Invertase Trehalase	**15**
N-Methyl-DNJ	*Morus alba* (Moraceae)	α-Glucosidase	**65**
1-Deoxymannojirimycin (DMJ) (**5**, R = β-OH)	*Omphalea diandra* (Euphorbiaceae); *Lonchocarpus* spp. (Leguminosae); *Streptomyces lavandulae*	α-Mannosidase α-Fucosidase	**66,67** **66**
Nojirimycin (**6**, R¹ = α-OH, R² = α-OH)	*Streptomyces* spp.	α- and β-Glucosidase	**68**

Table 1 (continued)

Alkaloid	Natural source	Enzyme inhibited	Ref.
Mannojirimycin (**6**, $R^1 = \beta$-OH, $R^2 = \alpha$-OH)	*Streptomyces lavandulae*	α-Mannosidase	**18**
Galactostatin (**6**, $R^1 = \alpha$-OH, $R^2 = \beta$-OH)	*Streptomyces lydicus*	β-Galactosidase	**69**
α-Homonojirimycin	*Omphalea diandra* (Euphorbiaceae); *Hyacinthoides non-scripta* (Hyacinthaceae); *Urania fulgens* (Lepidoptera)	α-Glucosidase	**56,61,62**

Pyrrolizidines

Alkaloid	Natural source	Enzyme inhibited	Ref.
Australine (**8**)	*Castanospermum australe* (Leguminosae)	Amyloglucosidase	**21,70**
Alexine (**9**)	*Alexa* spp. (Leguminosae)	Amyloglucosidase Trehalase	**24**
1-*epi*-Australine (**8**, 1-OH, α)	*C. australe*; *Alexa* spp. (Leguminosae)	Amyloglucosidase α-Glucosidase	**24**
3-*epi*-Australine (**8**, 3-CH$_2$OH, α)	*Castanospermum australe* (Leguminosae)	Amyloglucosidase	**23**
7-*epi*-Australine (**8**, 7-OH, β)	*C. australe*; *Alexa* spp. (Leguminosae)	Amyloglucosidase α-Glucosidase	**24**
7a-*epi*-Alexaflorine (**10**)	*Alexa grandiflora* (Leguminosae)	Amyloglucosidase	**25**
Casuarine (**11**, R = H)	*Casuarina equisetifolia* (Casuarinaceae)	Undetermined	

Indolizidines

Alkaloid	Natural source	Enzyme inhibited	Ref.
Lentiginosine (**12**, R = β-OH)	*Astragalus lentiginosus* (Leguminosae)	Amyloglucosidase	**27**
2-*epi*-Lentiginosine (**12**, R = α-OH)	*Astragalus lentiginosus* (Leguminosae)	None	**27**
Swainsonine (**13**)	*Swainsona* spp.; *Astragalus* spp.; *Oxytropis* spp. (Leguminosae). *Ipomoea* spp. (Convolvulaceae). *Rhizoctonia leguminicola*; *Metarhizium anisopliae*	α-Mannosidase	**71**
7-Deoxy-6-*epi*-castanospermine (**14**, $R^1 = \beta$-OH, $R^2 = H$)	*Castnospermum australe* (Leguminosae)	Amyloglucosidase	**31**
Castanospermine (**15**)	*C. australe*; *Alexa* spp. (Leguminosae)	α- and β-Glucosidase	**72**
6-*epi*-Castanospermine (**14**, $R^1 = \beta$-OH, $R^2 = \beta$-OH)	*Castanospermum australe* (Leguminosae)	Amyloglucosidase	**32**
6,7-Di-*epi*-castanospermine (**14**, $R^1 = \beta$-OH, $R^2 = \alpha$-OH)	*Castanospermum australe* (Leguminosae)	Amyloglucosidase β-Glucosidase	**10**

Nortropanes

Alkaloid	Natural source	Enzyme inhibited	Ref.
Calystegine A$_3$ (**16**, $R^1 = \alpha$-OH, $R^2 = H$)	*Calystegia* spp.; *Convolvulus arvensis*; *Ipomoea* spp. (Convolvulaceae). *Atropa belladonna*; *Datura wrightii*; *Hyoscyamus niger*; *Lycium chinense*; *Mandragora officianarum*; *Physalis alkekengi* var. *francheti*; *Scopolia japonica*; *Solanum* spp. (Solanaceae)	β-Glucosidase Trehalase	**36,46**

Table 1 (continued)

Alkaloid	Natural source	Enzyme inhibited	Ref.
Calystegine A₅ (**16**, R¹ = H, R² = α-OH)	*Hyoscyamus niger*; *Lycium chinense*; *Physalis alkekengi* var. *francheti*; *Scopolia japonica* (Solanaceae)	None	**36**
Calystegine A₆ (**17**)	*Hyoscyamus niger*; *Lycium chinense* (Solanaceae)	Undetermined	
Calystegine A₇ (**18**)	*Lycium chinense* (Solanaceae)	Trehalase	**13**
Calystegine B₁ (**19**)	*Calystegia sepium*; *Convolvulus arvensis* (Convolvulaceae). *Duboisia leichhardtii*; *Hyoscyamus niger*; *Lycium chinense*; *Mandragora officianarum*; *Physalis alkekengi* var. *francheti*; *Scopolia japonica* (Solanaceae)	β-Galactosidase β-Glucosidase	**36,46**
Calystegine B₂ (**16**, R¹ = α-OH, R² = α-OH)	*Calystegia* spp.; *Convolvulus arvensis*; *Ipomoea* spp. (Convolvulaceae) *Atropa belladonna*; *Datura wrightii*; *Duboisia leichhardtii*; *Hyoscyamus niger*; *Lycium chinense*; *Mandragora officianarum*; *Physalis alkekengi* var. *francheti*; *Scopolia japonica*; *Solanum* spp. Solanaceae)	α-Galactosidase β-Glucosidase Trehalase	**13** **36,48** **13**
Calystegine B₃ (**16**, R¹ = β-OH, R² = α-OH)	*Lycium chinense*; *Physalis alkekengi* var. *francheti*; *Scopolia japonica* (Solanaceae)	β-Glucosidase (weak) Trehalase	**13**
Calystegine B₄ (**16**, R¹ = α-OH, R² = β-OH)	*Duboisia leichhardtii*; *Scopolia japonica* (Solanaceae)	β-Glucosidase Trehalase	**38**
Calystegine B₅ (**20**)	*Lycium chinense* (Solanaceae)	Undetermined	
N-Methylcalystegine B₂ (**22**, R = H)	*Lycium chinense* (Solanaceae)	α-Galactosidase Trehalase	**13,54**
Calystegine C₁ (**21**, R = α-OH)	*Morus alba* (Moraceae) *Duboisia leichhardtii*; *Lycium chinense*; *Scopolia japonica* (Solanaceae)	α-Galactosidase β-Galactosidase β-Glucosidase Trehalase	**13** **8,13,36** **8,13,36** **13**
Calystegine C₂ (**21**, R = β-OH)	*Duboisia leichhardtii*; *Lycium chinense* (Solanaceae)	α-Mannosidase	**40**
N-Methylcalystegine C₁ (**22**, R = OH)	*Lycium chinense* (Solanaceae)	α-Galactosidase	**13**

	Miscellaneous		
Mannostatin A (**23**)	*Streptoverticillium verticillus*	α-Mannosidase	**73**
Trehazolin (**24**)	*Micromonospora* sp.	Trehalase	**42**
Kifunensine (**25**)	*Kitasatosporia kifunense*	α-Mannosidase	**74**
Nagstatin (**26**)	*Streptomyces amakusaensis*	β-*N*-Acetyl-glucosaminidase	**44**

solvents and partitioning between aqueous acid and base. Ion-exchange chromatography is therefore generally employed for purification, following extraction from the natural source by water, methanol or ethanol, either alone or in various mixtures. Subsequent separation can be achieved by paper, column, or thin-layer chromatography. The alkaloids are particularly amenable to detection by thin-layer chromatography in association with specific spray reagents, gas chromatography with flame ionization or mass spectrometric detection, and by their glycosidase inhibitory properties. All of these techniques have been reviewed in detail.[75]

Structural determination places a particular reliance on nuclear magnetic resonance spectroscopy which generally permits establishment of the specific ring system present, the substitution pattern, and relative stereochemistry of the hydroxy groups. Mass spectrometry provides similar information, with the exception of stereochemistry. The isolation of increasing numbers of these alkaloids has furnished a spectroscopic database which renders the determination of structures increasingly facile. Determination of the absolute stereochemistry is dependent upon X-ray crystallography, which can be used whenever well-refined crystal data can be obtained, either from the alkaloid itself or a crystalline derivative such as the hydrochloride salt. Alternatively, circular dichroism techniques may be applied, especially the benzoate chirality method.[16] Although this technique may have the most general utility, being independent of the physical state of the alkaloid, it has so far had only very limited application.

3.07.3 GLYCOSIDASE INHIBITION

3.07.3.1 Glycosidase Inhibitory Activity

The inhibitory activity of individual alkaloids may be remarkably specific, as with swainsonine, which inhibits only α-mannosidase and Golgi mannosidase II, or can be more general, showing a spectrum of activity against a series of glycosidases. Additionally, the potency may vary with the source of a particular enzyme, its purity, and the conditions, such as pH, under which the assay is performed. For these reasons the inhibitory properties of individual alkaloids are presented here only in a summary form (Table 1). The inhibition of N-linked glycoprotein processing by the most potent and specific of the alkaloids is discussed in detail in Section 3.07.6; particulars regarding other alkaloids should be obtained from the publications referenced in Table 1.

3.07.3.2 Structure–Activity Relationships

Early approaches to correlation of structure of the polyhydroxy alkaloids with their glycosidase inhibitory properties appeared to indicate a rather straightforward relationship.[3] Swainsonine (**13**) was perceived as an aza-analogue of D-mannopyranose, lacking the hydroxymethine group at C-4, but otherwise having the same relative disposition of the remaining hydroxyl groups, which therefore accounted for its ability to inhibit α-mannosidase.[71] The structures of 1-deoxynojirimycin (**5**, R = α-OH) and castanospermine (**15**) correlated even more closely, as monocyclic and bicyclic "aza sugars", with that of glucose, and they inhibited glucosidases as expected. This naive approach had to be reconsidered with the isolation of 6-*epi*-castanospermine (**14**, R^1 = β-OH, R^2 = β-OH) which, in spite of its stereochemical similarity to mannose, failed to inhibit either α- or β-mannosidase but instead proved to be an effective inhibitor of α-glucosidase, with a level of activity only slightly less than that of castanospermine.[32] Numerous additional examples of inhibitory specificities due to both naturally occurring alkaloids and synthetic analogues have further undermined this empirical approach and it is obvious that structure–activity correlations can only be developed with the aid of sophisticated molecular modeling techniques.

Molecular orbital calculations and molecular modeling have been applied to a series of known mannosidase inhibitors and others which were expected to inhibit but failed to do so. The results showed that good inhibitors fit closely with a single low-energy conformer of the mannosyl cation and demonstrated that 6-*epi*-castanospermine did not comply with the structural requirements.[76,77] The electronegative binding groups present in the inhibitor necessary for specificity and activity were established, as were those which were of little significance. Additional studies of this type should provide valuable information regarding the receptor sites on the various enzymes but the inhibition data available is compromised by the variability in enzymes and the conditions under which measurements have been made. A comprehensive screening program using standardized conditions would provide much more useful information for structure–activity correlations and consequently the design of specific and potent inhibitors.

The crystal structures of glucoamylase and its complex with the inhibitor 1-deoxynojirimycin (**5**, R = α-OH) have recently been reported.[78] This structural data has now been used in a molecular modeling study, using 1-deoxynojirimycin and other deoxynojirimycin derivatives, DMDP (**1**, R = OH), australine (**8**), and castanospermine (**15**), to probe the active site of the enzyme.[79] Preliminary results indicated that binding to specific residues within the active site were essential for inhibitory activity and that the inhibitory potency was dependent upon the number of hydrogen bonds involved in such binding. However, although castanospermine is an excellent inhibitor of the enzyme it lacked these requirements and therefore did not conform to the model. Nevertheless, this approach illustrates the potential value of such methods for understanding enzyme–inhibitor interactions, which should prove useful with increasing refinements in the models and available structural data.

In the absence of more comprehensive molecular modeling studies, the inhibition results obtained have been rationalized on the basis of generally accepted models for glycosidase inhibition. This approach has been developed most effectively for the calystegines, which provide a comprehensive series of structurally related natural polyhydroxy alkaloids. For β-glucosidase inhibition, the model involves the presence of two carboxylic acid groups at the active site of the enzyme, one responsible for generation and the other for stabilization of the glycosyl cation intermediate.[36] It has been speculated that for calystegines B_1 (**19**) and C_1 (**21**, R = α-OH), the *exo* hydroxy group at the 6-position is protonated by the acidic group responsible for catalytic activity within the active site, in an analogous manner to the inhibitor conduritol B epoxide. In contrast, calystegine B_2 (**16**, R^1 = α-OH, R^2 = α-OH), which shows a similar level of inhibitory activity towards β-glucosidase, is supposed to be bound to the glucosyl cation binding site through the hydroxyl group at the 4-position. The essential requirement of equatorial hydroxyl groups at the 2- and 3-positions is in accord with earlier studies of interaction of other inhibitors with β-glucosidase. Thus, the interaction of inhibitory calystegines with glycosidases can be envisioned as binding to the sites determining specificity and to the catalytic center, through specific hydroxyl groups and through the imino group.

The mechanism of galactosidase inhibitory activity is less apparent. Calystegines B_1, B_2 and C_1 are potent inhibitors of either α or β-galactosidase, yet calystegine B_3 (**16**, R^1 = β-OH, R^2 = α-OH), with a much closer configurational similarity to D-galactose than any of the former, has no inhibitory activity against these enzymes, an observation which is reminiscent of the situation with 6-*epi*-castanospermine in the indolizidine alkaloid series. Obviously, a much larger set of natural or synthetic epimers, enantiomers and structural analogues is needed before a complete understanding of structure–activity relationships can be applied to prediction of inhibitory activity. Some progress in this direction has been made through a comparison of glycosidase inhibition by synthetic analogues and derivatives of (+)-calystegine B_2. The nonnatural (−)-enantiomer showed no glycosidase inhibitory properties, whereas N-methylation of natural B_2 suppressed inhibition of β-glucosidase while activity towards α-galactosidase was retained.[54]

3.07.3.3 Synthetic Polyhydroxy Alkaloids

In addition to the synthesis of known naturally occurring alkaloids for the purpose of structural confirmation, many epimers, enantiomers, and structural analogues have been prepared. The number of these synthetic alkaloids, particularly those related to swainsonine, castanospermine, and australine, now approaches or perhaps exceeds those isolated from natural sources. The natural product focus of this review does not permit a comprehensive survey of these compounds. Various aspects of the synthetic approaches, either *a priori* syntheses or those routes commencing from carbohydrate-based templates, have been summarized in a number of publications.[80–82] Nonnatural epimers have been prepared by modification of natural alkaloids which are available in large quantities, such as castanospermine,[83] and ring-expanded analogues of pyrrolizidine and indolizidine alkaloids have also been synthesized.[84,85]

It is probable that at least some of the synthetic compounds, especially epimers of known naturally occurring alkaloids, will subsequently be found to occur in nature. In addition, new structural classes have already been generated which might reasonably be expected to be biosynthesized by plants. Predominant among these are polyhydroxy quinolizidine alkaloids, consisting of two six-membered rings fused into a bicyclic system, which are ring-expanded homologues of the indolizidine alkaloids.[84,86] Although quinolizidine alkaloids are a well-established class of natural products, none have yet been isolated that bear more than two hydroxyl groups. This is probably a consequence of the high water solubility of polyhydroxylated alkaloids which renders them unextractable into the nonhydroxylic solvents normally used for alkaloid purification. The combination of novel natural polyhydroxy alkaloids, together with synthetic analogues tailored to have specific structural features, will ultimately lead to a full comprehension of the interaction of these alkaloids with receptor sites on the enzyme which results in their glycosidase inhibitory properties.

3.07.4 BIOLOGICAL ACTIVITY OF GLYCOSIDASE INHIBITORS

3.07.4.1 Mammalian Toxicity

As might be expected from a class of compounds that inhibits glycosidases and consequently the fundamental cellular function of glycoprotein processing, the polyhydroxy alkaloids exhibit an exceptional diversity of biological activities. Discovery and isolation of many of the alkaloids has been a result of observations of the ultimate clinical effects which result from the consumption by animals of plants containing these bioactive compounds. Predominant among such examples is the occurrence of swainsonine (**13**) in *Swainsona* species (poison peas) of Australia[1] and *Astragalus* and *Oxytropis* species (locoweeds) of North America.[52] The potent α-mannosidase inhibitory activity of swainsonine disrupts glycoprotein processing by mannosidase II in the Golgi, resulting in neuronal vacuolation due to abnormal storage of mannose-rich oligosaccharides, leading to the neurological damage so characteristic of the locoism syndrome. However, the clinical effects are not limited to the nervous system since emaciation, reproductive failure in both males and females, and congestive right-heart failure are also observed. Since the discovery of swainsonine as the causative agent, locoweed poisoning has now been established as a widespread phenomenon, with additional occurrences being reported from South America and many parts of China and Tibet.[87]

Swainsonine has been reported to co-occur with calystegines B$_2$ (**16**, R^1 = α-OH, R^2 = α-OH) and C$_1$ (**21**, R = OH) in *Ipomoea* species of Australia which cause poisoning of sheep and cattle,[45] and in *I. carnea*, resulting in toxicity to goats in Mozambique. The clinical signs of poisoning are characterized by the expected neurological damage resulting from swainsonine ingestion but these are exacerbated by muscle-twitching, tremors and epileptiform seizures. Histological examination of tissues showed vacuolation of Purkinje cells in addition to swainsonine-induced cytoplasmic vacuolation of neurons and axonal dystrophy. The calystegines inhibit β-glucosidase and α-galactosidase which would produce phenocopies of the genetic lysosomal storage defects, Gaucher's disease and Fabry's disease, respectively, and the additional syndromes are significant indicators of the latter.

In contrast to the above examples which exhibit a complexity of effects, the alkaloids concentrated in the chestnut-like seeds of *Castanospermum australe* (Black Bean), primarily castanospermine (**15**) and australine (**8**), together with several less potent epimers of both, produce gastrointestinal disturbances in livestock and humans but no discernable neurological damage.[88] This is consistent with the ability of the alkaloids to inhibit α- and β-glucosidase, resulting in a syndrome phenotypic of the genetic defect, Pompe's disease. Although this relationship has not been directly established in field cases of poisoning, rodent feeding experiments with castanospermine resulted in vacuolation of hepatocytes and skeletal myocytes, and glycogen accumulation, consistent with Pompe's disease or type II glycogenesis.[89] Gastrointestinal problems and lethargy have also been observed in livestock grazing bluebells (*Hyacinthoides non-scripta*) in the UK, and the demonstration of the presence of DMDP and homoDMDP in this plant may account for the syndrome.[12]

All of the above poisoning syndromes are relatively obvious once signs develop, although this may take several weeks of consumption of the plant because the alkaloids implicated often are present at very low levels. Nevertheless, they are potent inhibitors and it has been estimated that a swainsonine content of 0.001% of the dry weight of the plant may be sufficient to induce locoism.[87] For those alkaloids which are less active or which are present at extremely low levels, it seems probable that the signs of poisoning would be subclinical, with no overt changes being apparent. In such cases, toxicity may only be manifested as minor digestive disturbances, failure to gain weight and other deviations from optimal health which could be attributed to stress or infectious diseases. The occurrence of various calystegines in human food plants from the family Solanaceae, such as potatoes, eggplant and peppers, could account for a variety of complaints, primarily gastrointestinal, reported in certain individuals consuming these vegetables.[90]

3.07.4.2 Insecticidal Activity

It should be anticipated that compounds capable of inhibiting glycosidases would have an inhibitory effect on digestive enzymes, and defense against herbivorous insects may be one of the roles played by the polyhydroxy alkaloids in plants which contain them. Conversely, it is well established that insects co-evolve with their host plants to circumvent such defenses and utilize the active constituents for their own defense. Such strategies involving specific alkaloids have been demonstrated for several plant–insect relationships.

Castanospermine (**15**) added to an artificial diet is highly inhibitory to feeding by the pea aphid, *Acyrthosiphon pisum*, with a 50% deterrency level of 20 ppm, and a consequent very low survival rate.[91] Although the alkaloid does not inhibit aphid trehalase, it has been shown differentially to inhibit a number of disaccharidases from a wide taxonomic distribution of insects.[92] Castanospermine has also been shown to be an antifeedant compound to the Egyptian cotton leafworm, *Spodoptera littoralis*, as are D-AB1 (**2**, R = β-OH), DMDP (**1**, R = OH), and swainsonine (**13**).[93] DMDP also appears to be a particularly effective feeding deterrent to nymphs of the locusts *Schistocerca gregaria* and *Locusta migratoria*, at levels as low as 0.001% of the body weight.[94] Since these alkaloids inhibit different enzymes, it is difficult to correlate antifeedant activity with inhibition of digestive enzymes alone. It is possible that deterrency may also be a consequence of blocking of the sensory response to glucose.[93]

Insect resistance to the effects of the alkaloids has been observed. Thus, the bruchid beetle *Callosobruchus maculatus*, a feeder on legumes that do not produce DMDP, has a gut α-glucosidase which is 100 times more sensitive to the alkaloid than that of *Ctenocolum tuberculatum*, which has adapted to feed exclusively on DMDP-containing species of the legume subtribe Lonchocarpinae.[4] Among the Lepidoptera, the aposematically-colored moth, *Urania fulgens*, accumulates DMDP and α-homonojirimycin (**7**) from its food plant, the vine *Omphalea diandra*, but does not sequester the other alkaloid present, 1-deoxymannojirimycin (**5**, R = β-OH),[20] while the Death's-Head hawkmoth procures calystegines from its Solanaceous hosts.[47] The mechanism of resistance to the effects of the alkaloids is not understood but it appears likely that those which are accumulated serve a protective role in the insect. In contrast, alkaloids which may be harmful can be specifically excreted. For example, pea aphids feeding upon the spotted locoweed, *Astragalus lentiginosus*, excrete in their honeydew swainsonine (**13**) acquired from the phloem of the plant, while showing no feeding deterrency.[91] Since this plant was colonized opportunistically in the laboratory and is not a normal host for the pea aphid, the implication is that certain insects my have a general ability to compartmentalize and eliminate polyhydroxy alkaloids that might otherwise be harmful.

3.07.4.3 Plant Growth Inhibition

Polyhydroxy alkaloids from several of the structural classes have been shown to be inhibitory to the growth of plants. Particularly noteworthy in this respect is castanospermine (**15**) which has been demonstrated to be a potent root elongation inhibitor of lettuce, *Lactuca sativa*, alfalfa, *Medicago sativa*, barnyard grass, *Echinochloa crusgalli*, and red millet, *Panicum miliaceum*.[95] The alkaloid was much more effective against the dicots, showing 50% inhibition of root length growth at 300 ppb, while the monocots were 1000 times less sensitive. The structurally related indolizidine alkaloid, swainsonine (**13**), failed to exhibit any phytotoxic activity against these species, indicating that the bioactivity is a consequence of α- or β-glucosidase inhibition but not of α-mannosidase inhibition. Nojirimycin (**6**, $R^1 = \alpha$-OH, $R^2 = \alpha$-OH) is inhibitory to cell extension of *Pisum sativum* stem segments and of coleoptiles of *Avena* and *Triticum*, induced by auxins. There is considerable evidence that elongation is a consequence of cell-wall loosening due to degradation or depolymerization of xyloglucans by *exo-β*-glucanases and inhibition of these enzymes by the alkaloid could therefore account for the failure of the cells to elongate.[96]

The phytotoxic effects of the polyhydroxy alkaloids may confer a major competitive advantage upon plants which biosynthesize them through the phenomenon of allelopathy. The alkaloids are highly water soluble so that excretion into the surrounding soil or leaching from various parts of the plant can suppress the growth of encroaching species through creation of a zone of inhibition. At the same time, movement of water would transport the compounds through the soil so that concentrations in the vicinity of the secreting plant itself do not attain levels high enough to induce self-inhibition. However, there is some evidence that *Castanospermum australe* seeds may be inhibited from germination by the presence of castanospermine (**15**). Considerable irrigation is required before the seeds commence to sprout and this may be a valuable strategy in the native environment where rainfall is highly seasonal, enabling germination and rooting to take place only when the rainy season is well-established.

Natural calystegine B_2, that is the (+)-enantiomer (**16**, R^1 = α-OH, R^2 = α-OH), showed significant inhibition of alfalfa seed germination, and growth and lateral production of roots transformed by *Agrobacterium rhizogenes*, but corresponding effects were not observed with the unnatural (−)-enantiomer.[54] Root length was reduced by 40% after treatment for 43 h with 10 mM (+)-calystegine B_2, while under the same conditions the unnatural (synthetic) alkaloid caused an 18% increase in root length. Such results demonstrate the dependency of bioactivity upon specific structural conformations and stereochemistry.

3.07.4.4   Antimicrobial Activity

There has been little information reported in regard to the effect of polyhydroxy alkaloid inhibitors on growth or function of microorganisms, although nojirimycin (**6**, R^1 = α-OH, R^2 = α-OH) was discovered as a result of the antimicrobial activity of *Streptomyces nojiriencis*, *S. roseochromogenes*, and *S. lavandulae* against a drug-resistant strain of *Shigella flexneri*.[66] The antibiotic activity of the same alkaloid towards *Xanthomonas oryzae* renders it capable of preventing the bacterial leaf blight of rice.[66]

The calystegines were first isolated from roots of the bindweed, *Calystegia sepium*.[34] Although these alkaloids have now been detected in other plant parts, there appears to be a relatively high abundance in subterranean organs of the Convolvulaceae and Solanaceae and they are therefore believed to be nutritional mediators between such plants and associated rhizosphere bacteria. Over 20% of the bacteria isolated from the rhizospheres of calystegine-producing plants were capable of catabolizing the alkaloids, whereas no bacteria with this ability were obtained from plants which did not elaborate calystegines.[54] In addition, wild-type *Rhizobium meliloti* 41 was capable of using natural (+)-calystegine B_2 (**16**, R^1 = α-OH, R^2 = α-OH) as an exclusive source of carbon and nitrogen, whereas a catabolism-deficient strain of *R. meliloti* was not. Furthermore, neither organism could utilize the unnatural, synthetic enantiomer, (−)-calystegine B_2. The ability to catabolize such compounds, which at the same time may have antibiotic properties towards other microorganisms, has an obvious competitive advantage for those specific bacteria capable of utilizing them.

3.07.4.5   Therapeutic Activity

The capability of polyhydroxy alkaloids to disrupt the general cellular function of glycoprotein processing leads to the expectation that these compounds should have therapeutic potential for the treatment of various disease states. The significant mammalian toxicity of certain of the alkaloids is an obvious hindrance to their utility. However, this is frequently true of many drug candidates and it is not unreasonable to assume that an appropriate dose–response relationship could be achieved. Moreover, adverse effects, such as the neurological damage caused by swainsonine, often develop quite slowly and appear to be reversible if ingestion of the alkaloid is terminated, as would be the situation with most drug regimens. Investigation of the alkaloids for therapeutic potential has so far concentrated on three major disease states, namely for treatment of cancer and inhibition of metastasis, as antidiabetic drugs, and for antiviral activity.

Swainsonine (**13**) has received particular attention as an antimetastatic agent. *In vivo* experiments with mice have shown that pulmonary colonization is reduced by over 80% if the animals are provided with drinking water containing 3 µg mL^{-1} of swainsonine for 24 h prior to injection with B16-F10 murine melanoma cells.[97] This effect has been shown to be due to enhancement of natural killer T-cells and increased susceptibility of cancerous cells to their effect.[98] The pharmacokinetics of swainsonine in such experiments indicate that the levels of alkaloid and period of administration would not be sufficient to produce neurological damage.[99] It has been suggested that post-operative metastasis of tumor cells in humans could be suppressed by intravenous administration of the alkaloid prior to and following the surgery. Clinical trials in humans with very advanced malignancies showed that lysosomal α-mannosidases and Golgi mannosidase II were inhibited and some improvement in clinical status occurred.[100] Castanospermine has also been reported to suppress metastasis in mice[101] but experiments with this alkaloid have not been as extensive as those with swainsonine.

Castanospermine (**15**) and 1-deoxynojirimycin (**5**, R = α-OH) have been shown to be capable of suppressing the infectivity of a number of retro viruses, including the human immunodeficiency virus (HIV) responsible for AIDS.[102–105] This effect is a consequence of inhibition of glycoprotein processing which results in changes in the structure of the glycoprotein coat of the virus. Cellular recognition of the host is thus prevented and syncytium formation is suppressed. In spite of this significant effect, both of these alkaloids suffer from the disadvantage that they are highly water-soluble and therefore excreted very rapidly. This defect has been overcome by derivatization to give 6-*O*-butyryl-castanospermine and *N*-butyl-deoxynojirimycin,[106,107] and both of these compounds have undergone clinical trials against AIDS in humans, either alone or in combination with AZT. As might be expected, gastrointestinal disturbances have been reported as a significant side effect.

Another structural modification of 1-deoxynojirimycin, the *N*-hydroxyethyl derivative, miglitol, an inhibitor of α-glucosidase, has been clinically evaluated and released as an antidiabetic drug in insulin- and noninsulin-dependent diabetes. The alkaloid was shown potently to inhibit glucose-induced insulin release and also suppressed islet α-glucoside hydrolase activity, thus controlling postprandial glycemia.[108] The structurally related alkaloids, 2-*O*-α-D-galactopyranosyl-DNJ and fagomine, have also been shown to have antihypoglycemic activity in streptozocin-induced diabetic mice but have not been tested in humans.[109]

The ability of polyhydroxy alkaloid glycosidase inhibitors to prevent cellular recognition has resulted in their evaluation for clinical situations where suppression of an immune response would be desirable, or for use against parasitic diseases. Thus, *in vivo* experiments have shown that castanospermine can be used as an immunosuppressive drug, promoting heart and renal allograft survival in rats.[110] Parasitic diseases may also be controlled by altering cellular recognition processes. Castanospermine provides protection against cerebral malaria by preventing adhesion of *Plasmodium falciparum* to infected erythrocytes,[111] while swainsonine inhibits the association of *Trypanosoma cruzi*, the causative agent of Chagas' disease, with host cells by formation of defective mannose-rich oligosaccharides on the cell surface.[112]

There is no doubt that the polyhydroxy alkaloids have considerable potential for treatment of a variety of disease states in humans and animals. The primary challenge in introducing them as commercial drugs is to minimize their toxicity and enhance the specificity of their beneficial effects. Improvement of their pharmacokinetic properties should result in much lower dose rates being necessary so that undesirable side-effects are limited. Increased specificity of action can be achieved by preparation of synthetic derivatives and a comprehensive understanding of structure–activity relationships.

3.07.5 PROCESSING OF *N*-LINKED OLIGOSACCHARIDES

3.07.5.1 Introduction

Glycoproteins are widespread in nature, being found in all eucaryotic cells.[113] They have also been shown to be present in various archaebacteria as well as in some lower bacteria.[114,115] In addition, it has become eminently clear that carbohydrate sequences on glycoproteins, glycolipids, and proteoglycans are critically important as ligands in molecular recognition.[116] At least with regard to the *N*-linked glycoproteins, on which this review focuses, these molecules have been implicated in a number of important physiological functions, especially cell–cell recognition reactions involving such critical phenomena as inflammation,[117] pathogenesis,[118] parasitism,[119] development,[120] cell adhesion,[121] and symbiosis,[122] to mention only a few.

N-linked oligosaccharides are also involved in lysosomal enzyme targeting,[123] in the uptake or removal of glycoproteins from the blood,[124] in protein folding in the endoplasmic reticulum,[125] and in many other physiological phenomena of potential significance.[126,127] Although the carbohydrate portion of the glycoprotein has not been shown to participate in every case of recognition, specific oligosaccharide structures are clearly central to many of these cases. Thus, inhibitors that block specific steps in the assembly of the various *N*-linked oligosaccharides and cause the formation of altered or immature oligosaccharide structures should be valuable tools for probing the role of carbohydrates in glycoprotein function.[128]

Figure 1 shows three representative structures of the *N*-linked oligosaccharides. All of these oligosaccharides have the same core structure shown within the box, and are composed of a branched trimannose structure linked to a disaccharide of GlcNAc (i.e., *N*,*N'*-diacetylchitobiose). The immature or initially synthesized oligosaccharide is a high mannose structure shown in (A), and this oligosaccharide is the biosynthetic precursor that gives rise to all of the other *N*-linked oligo-saccharides. High-mannose (or oligomannose-type) oligosaccharides are most commonly found in glycoproteins from lower eucaryotes such as fungi and yeast, although a small percentage of the *N*-linked oligosaccharides of animal cell surface proteins are of the high-mannose type.

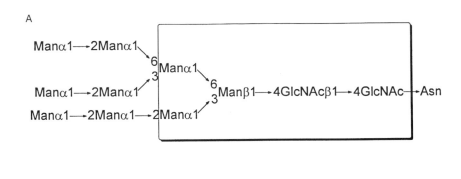

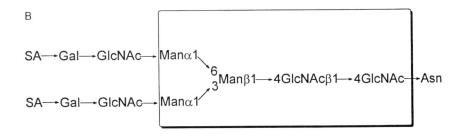

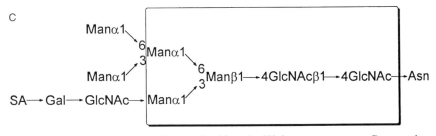

Figure 1   Structural classes of *N*-linked oligosaccharides. A, High-mannose type; B, complex type; and C, hybrid type.

The lower structure (C) of Figure 1 is a hybrid type of oligosaccharide that is produced by partial processing down to the Glc*N*Ac transferase I step, and then addition of various sugars to the 3-linked mannose branch. However, hybrid structures are apparently the result of an absence of mannosidase II action or activity. It is not clear whether hybrid structures are formed normally, but they are found in glycoproteins produced in individuals with HEMPAS disease, a condition where individuals lack mannosidase II activity. Hybrid structures can also be induced by treating cultured cells with swainsonine (**13**). The middle structure (B) in Figure 1 is an example of one type of complex oligosaccharide that is frequently found in cell surface glycoproteins of higher eucaryotes, such as the low density lipoprotein receptor and many other membrane receptors. This particular structure is referred to as a biantennary complex chain, but other complex oligosaccharides may have three of the sialic acid-galactose-Glc*N*Ac chains (triantennary chains), or four of these trisaccharide sequences (tetraantennary chains).

3.07.5.2 Biosynthesis of *N*-Linked Oligosaccharides

The biosynthesis of the *N*-linked oligosaccharide chains involves two rather distinct series of reactions. The first of these pathways gives rise to the precursor, or immature oligosaccharide, which is then transferred cotranslationally to the protein chain while it is being synthesized on membrane-bound polysomes.[129] In contrast, the second series of reactions involves the modification of this precursor oligosaccharide by the removal of some sugars and the addition of others, to give a large number of different oligosaccharide structures.[130] This first pathway requires the participation of a lipid carrier and the involvement of lipid-linked saccharide intermediates. The reactions leading to the production of the final lipid-linked oligosaccharide precursor are presented in Figure 2.

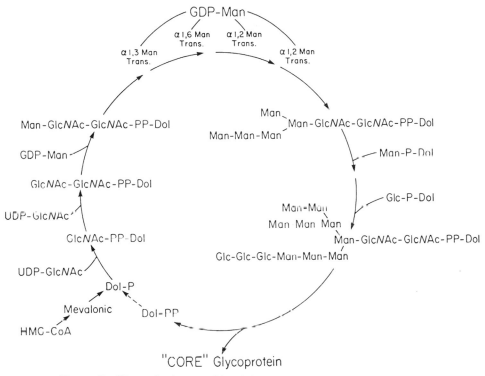

Figure 2 Biosynthetic assembly of the core *N*-linked oligosaccharides.

As shown in Figure 2, the assembly of the *N*-linked oligosaccharide chain is initiated in the endoplasmic reticulum (ER) by the transfer of a GlcNAc-1-P from UDP-GlcNAc to dolichyl-P to form GlcNAc-PP-dolichol.[131] A second GlcNAc is then added, also from UDP-GlcNAc, to produce GlcNAcβ1,4GlcNAc-PP-dolichol.[132] Then, five mannose residues are added, the first in a β1,4 linkage to the terminal GlcNAc, and the next four in α linkages to form the important intermediate, Man$_5$GlcNAc$_2$-PP-dolichol.[133] These first seven reactions are believed to occur on the cytosolic side of the ER membrane, since they involve nucleoside diphosphate sugars as the sugar donors, and these activated sugar donors are biosynthesized in the cytoplasm by soluble sugar nucleotide pyrophosphorylases. It seems likely, therefore, that the sugar acceptor, dolichyl-P, is initially oriented in the ER membrane in such a way that the phosphate group is exposed to the cytoplasm, and is therefore able to accept sugars from the cytosol. After the addition of the first seven sugars to give Man$_5$GlcNAc$_2$-PP-dolichol, this lipid-linked oligosaccharide is believed to undergo a "flip-flop" in the membrane so that the oligosaccharide chain now becomes oriented towards the lumen of the ER.[134]

The assembly of the oligosaccharide is completed by the addition of four more mannose residues and then three glucose units to give a Glc$_3$Man$_9$GlcNAc$_2$-PP-dolichol.[135] These last seven sugars (i.e., four mannose and three glucose units) are all added in the lumen of the ER, and are donated by the activated lipid precursors, mannosyl-P-dolichol and glucosyl-P-dolichol.[136,137] These two sugar donors are synthesized using the sugar nucleotides, GDP-mannose and UDP-glucose, by transfer of the respective sugar to dolichyl-P.[138] The reactions for the synthesis of the activated lipid-linked monosaccharides are proposed to occur on the cytosolic side of the ER membrane and are catalyzed by the enzymes, dol-P-man synthase and dol-P-glc synthase.[139,140]

The final step in this pathway is the transfer of the $Glc_3Man_9GlcNAc_2$ from its lipid carrier to specific asparagine residues on the polysome-bound protein, catalyzed by the enzyme oligosaccharyltransferase.[141,142] The asparagine residue that acts as the acceptor of this oligosaccharide chain must be in the tripeptide consensus sequence, Asn-X-Ser(Thr), where X can be any amino acid except proline, but certain amino acids are favored over others.[143] In addition, the tripeptide sequence must be in a specific conformation or orientation, such as a β-turn of the protein, in order to be glycosylated.[144] In spite of the fact that all of the reactions in this pathway are well known, it is still not clear how the pathway is regulated, nor where the control points are located.

3.07.5.3 Processing of *N*-Linked Oligosaccharides

After the oligosaccharide is transferred to protein and while the protein chain is still being synthesized in the ER, the oligosaccharide begins to undergo a number of processing or trimming reactions. The initial reactions in this second pathway encompass the removal of three glucose residues and up to six mannose residues, but later processing reactions involve the addition of a number of other sugars, principally GlcNAc, galactose, neuraminic acid, L-fucose, and possibly GalNAc.[145] The processing pathway is outlined in Figure 3.

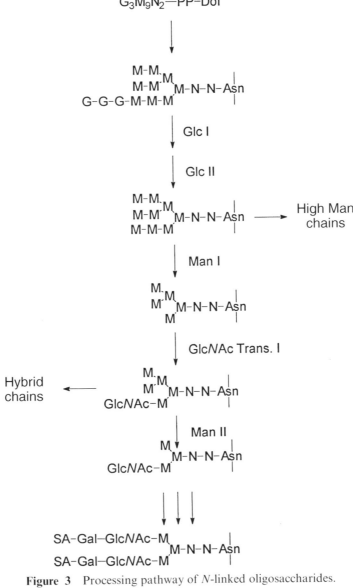

Figure 3 Processing pathway of *N*-linked oligosaccharides.

The first processing step involves a membrane-bound glucosidase, called glucosidase I, which removes the outermost α1,2-linked glucose.[146] This enzyme is quite distinct from the common glycosidases, such as the lysosomal enzymes that are involved in the degradation of polysaccharides, glycolipids and other complex carbohydrates, since those enzymes usually have a pH optimum of around 5, whereas glucosidase I has a pH optimum of about 6.4 to 6.8.[147] In addition, the common glycosidases are only specific for the sugar at the nonreducing terminus and the anomeric configuration of the glycosidic bond, but do not have strong specificity for the group to which this sugar is attached, nor the specific glycosidic linkage if that group is another sugar. Glucosidase I, on the other hand, will only cleave a terminal glucose that is attached in α1,2-linkage to another glucose. Thus, glucosidase I will not work with p-nitrophenyl-α-D-glucopyranoside.[148] Finally, these kinds of enzyme can be distinguished by their location; the processing glucosidases are in the ER, while the other hydrolytic α-glucosidases are usually in the lysosomes.

Glucosidase I is the enzyme that initiates the trimming or maturation of the N-linked oligosaccharide chains and therefore may play a key role in controlling the rate of transport or exit of newly formed glycoproteins from the ER to the Golgi apparatus. This enzyme has been purified from a number of sources, including calf[149] and porcine[150] liver, and bovine mammary glands,[151] as well as plants (mung bean seedlings)[148] and yeast (*Saccharomyces cerevisiae*).[152] The pig liver glucosidase I was cloned from a human hippocampus cDNA library and expressed in COS 1 cells. The expressed enzyme had a molecular mass of 95 kDa and was degraded by endoglucosaminidase H (Endo H) to a 93 kDa form, indicating that the enzyme has a high-mannose oligosaccharide at the asparagine 655 glycosylation site.[153] The hydrophobicity profile of the enzyme and the fact that trypsin treatment of microsomes released a 4 kDa fragment, support the view that the glucosidase I is a transmembrane glycoprotein containing a short cytoplasmic domain of about 37 amino acids, followed by a transmembrane domain and a large C-terminal catalytic domain on the luminal side of the ER membrane.[153]

A yeast mutant gls1, has been isolated that is lacking glucosidase I and produces glycoproteins with Glc₃Man₉GlcNAc₂ structures.[154] This alteration in the normal structure of the oligosaccharides on these yeast proteins has no effect on their secretion. A CHO mutant that is missing glucosidase I was also isolated by virtue of its resistance to the toxic effects of the lectin L-PHA. The mutation in these cells, called Lec 23, has profound effects on the secretion or targeting of glycoproteins.[155]

A second glucosidase, located in the lumen of the ER and called glucosidase II, removes the other two α1,3-linked glucoses to give a Man₉GlcNAc₂-protein. Interestingly, this enzyme removes the outermost α1,3-linked glucose quite rapidly ($t_{1/2}$ = 5 min), whereas removal of the innermost α1,3-linked glucose is considerably slower ($t_{1/2}$ = 20–30 min).[156] Those earlier observations on the activity of this enzyme correlate well with the more recently described role of this enzyme in protein folding. That is, a single α1,3-linked glucose on the high-mannose chain functions as a recognition site to bind a chaperone to those proteins that are improperly folded or denatured, and that chaperone expedites or assists their proper folding.

Thus, it has been shown that the ER contains a protein called calnexin that functions to help newly synthesized membrane proteins fold into their proper conformation, a step that is apparently necessary for many of these proteins to be transported to the Golgi apparatus at the proper rate.[125] Calnexin is a lectin that recognizes a single α1,3-linked glucose on the high mannose chains of unfolded or denatured proteins.[157] Since glucosidase II acts fairly slowly on the final α1,3-linked glucose, there must be a time period when the glycoprotein has only a single glucose on its oligosaccharide. This glucose on the high-mannose chains of unfolded proteins is the recognition site for calnexin to bind to those proteins that have not yet assumed their proper conformation.[158–160]

The ER also contains a safety mechanism to assure that unfolded or improperly folded glycoproteins can interact with this chaperone to obtain the conformation that is required for exit from the ER into the Golgi apparatus. Thus, an unusual glucosyltransferase that is localized in the ER functions to transfer a glucose from UDP-glucose to high mannose chains on denatured, but not on native, glycoproteins.[161] Once this glucose has been added, calnexin can recognize and assist this protein in its proper folding and transfer to the Golgi.[162] As a result, a glycoprotein that has had all of its three glucose residues removed by glucosidase I and II but has failed to fold into the proper conformation can be reglucosylated by this novel enzyme, and this signal then allows the protein another opportunity to interact with calnexin and fold properly. This mechanism, involving the removal of glucoses by the glucosidases and reglucosylation by the glucosyltransferase, is postulated to be part of a unique "glycoprotein-specific folding and quality control mechanism" in the ER that allows this organelle to control and pass properly folded glycoproteins on to the next step in transport and processing.

Glucosidase II has a fairly high pH optimum of about 6.5 to 7.0, but also hydrolyzes *p*-nitrophenyl-α-D-glucoside.[163] On the other hand, the enzyme does appear to be fairly specific for the α1,3-linked glucose since hydrolysis of $Glc_2Man_9GlcNAc_2$ is inhibited by nigerose, an α1,3-linked disaccharide of glucose, but not by the corresponding α1,2-, α1,4-, or α1,6-linked disaccharides of glucose.[164] The enzyme from pig kidney was shown to have a subunit molecular mass of 100 kDa and to contain a high-mannose oligosaccharide,[165] while the enzyme from mung bean seedlings had two 110 kDa subunits as well as high-mannose oligosaccharides,[166] although in some other animal systems, glucosidase II subunits were reported to have molecular masses of 65 kDa.[167,168] This enzyme has been reported to be located in the rough and smooth ER of pig hepatocytes[169] but has also been located in post-Golgi structures in tubular cells of pig kidney.[170] The cDNA for glucosidase II was cloned using degenerate oligonucleotides based on the amino acid sequences derived from a purified pig liver glucosidase II. A 3.9 kb cDNA was isolated with an open reading frame of 2.9 kb. The amino acid sequence did not contain any known ER retention signals or any hydrophobic regions that might represent transmembrane domains, but it did contain a single *N*-linked oligosaccharide consensus site near the amino terminus.[171]

The processing glucosidases can best be assayed, *in vitro*, using the radiolabeled oligosaccharide substrates, $[^3H]Glc_3Man_9GlcNAc$ and $[^3H]Glc_2Man_9GlcNAc$. These substrates are readily prepared in cultured animal cells infected with an enveloped virus, such as influenza virus, that has an *N*-linked glycoprotein coat. Thus, MDCK cells are infected with influenza virus, and progeny virus are produced in these cells in the presence of a glucosidase or mannosidase processing inhibitor to prevent the removal of those specific sugars.[172] For example, if the virus is grown in the presence of castanospermine (**15**), the oligosaccharide chains on its envelope glycoproteins will be mostly of the $Glc_3Man_9GlcNAc_2$ structure, whereas if the virus is grown in the presence of deoxymannojirimycin (**5**, R − β-OH) or kifunensine (**25**), it would have mostly $Man_9GlcNAc_2$ structures.[173] The oligosaccharides are radiolabeled by growing the virus in the presence of either $[^3H]$galactose to label the three glucose residues of the oligosaccharides, or in $[2-^3H]$mannose to label the nine mannose units. The virus-infected MDCK cells are incubated for 40 h to allow the virus to replicate and lyse the cells and the virus particles are isolated from the culture medium by ultracentrifugation. The viral pellet is then treated exhaustively with pronase to digest the proteins and the resulting glycopeptides are isolated by gel filtration. These glycopeptides are then incubated with Endo H (i.e., endo-glucosaminidase H) to cleave the high-mannose and glucose-containing high-mannose glyco-peptides, and the resulting oligosaccharides, having a single GlcNAc at the reducing end, are isolated by gel filtration on columns of Biogel P-4.[174]

Once the two glucosidases have removed all three glucoses from the *N*-linked oligosaccharide as shown in Figure 3, a number of α-mannosidases can remove one or more of the four α1,2-linked mannose residues to ultimately give a $Man_5GlcNAc_2$-protein (i.e., Manα1,3(Manα1,6)Manα1,6 [Manα1,3]Manβ1,4GlcNAcβ1,4GlcNAc-protein).[175] There are believed to be at least three different α1,2-mannosidases involved in the conversion of $Man_9GlcNAc_2$ to $Man_5GlcNAc_2$; an ER α-mannosidase, a Golgi Man₉-mannosidase, and a Golgi mannosidase I.[176] These enzymes differ in a number of properties including their substrate specificity, their sensitivity to various mannosidase inhibitors, and their intracellular location. The ER mannosidase presumably removes only a single mannose to generate a unique and specific $Man_8GlcNAc_2$ structure. This enzyme is reported to cleave the α1,2-mannosidic linkage in $Man_9GlcNAc_2$ that is normally resistant to hydrolysis by the Golgi Man₉-mannosidase.[177] However, a soluble form of the ER α-mannosidase has been shown to exhibit rather low specificity, in that it can release several different α1,2-linked mannose residues from the $Man_9GlcNAc$ substrate. These mannoses are removed in a random fashion so that three different $Man_8GlcNAc$ structures are produced, as well as a number of $Man_7GlcNAc$ isomers.[178] The discrepancy in specificity between the ER mannosidase and the soluble mannosidase reported in these two studies may be due to the effects of the protein itself on substrate specificity (i.e., the ER α-mannosidase may act differently in its specificity on the free oligosaccharide) compared with the protein-bound oligosaccharide.

The Man₉-mannosidase, at least the enzyme from pig liver, cleaves both free and peptide-bound $Man_9GlcNAc_2$ to give a specific $Man_6GlcNAc_2$ isomer.[179] Thus, the ER mannosidase and the Man₉-mannosidase may be complementary to each other. Another α1,2-mannosidase, isolated from rat liver Golgi and requiring Ca^{2+}, apparently cleaves each of the four α1,2-mannoses in the $Man_9GlcNAc_2$ at a comparable rate, indicating that it alone could produce the $Man_5(GlcNAc)_2$ that is involved in the formation of complex types of oligosaccharides.[180,181] The exact function of these different α-mannosidases is not currently known. The fact that each of these enzymes removes α1,2-linkages, and that there is considerable redundancy in their action, indicates that each has a specific role in the processing, and perhaps the targeting pathway, and that they may function to produce oligosaccharides with specific signals for particular roles in the cell.

In addition to these *exo*-α1,2-mannosidases, some animal cells and tissues contain an *endo*-α1,2-mannosidase that cleaves the glucose branch of the $Glc_{3-1}Man_9GlcNAc_2$ between the two terminal mannoses to release a Glc_3Man, Glc_2Man or Glc_1Man from the oligosaccharide and leave a $Man_8GlcNAc_2$-protein.[182] This enzyme presumably prefers oligosaccharides with a single glucose on the high-mannose chain and may represent an alternate route to that utilizing glucosidase I and glucosidase II. Nevertheless, the specific role of this interesting enzyme in the processing pathway is still not clear; it may represent a new targeting route in some cells.

The cDNA encoding an endoplasmic reticulum α-mannosidase was isolated from a rat liver gt11 library. Two degenerate oligonucleotides were prepared based on the amino acid sequences obtained from the purified enzyme. These oligonucleotides were used as primers in PCR with liver cDNA as the template to generate an unambiguous cDNA probe. The 524 base-pair cDNA fragment was then used to isolate cDNA clones by hybridization. Two overlapping clones were used to construct a full length cDNA of 3392 bases which encoded an open reading frame of 1040 amino acids and a 116 kDa protein that contained six of the known peptide sequences. No signal sequence or membrane spanning domains were found in the amino acid sequence. Northern blots of various animal tissues using the cDNA as a probe revealed that a 3.5 kb mRNA was present in all tissues examined, but was enriched in adrenal glands and testis and was less abundant in spleen, intestine, and muscle. The rat liver ER α-mannosidase bears striking homology to the vacuolar α-mannosidase from *Saccharomyces cerevisiae*.[183]

The Man$_9$ mannosidase was also cloned in gt10, using a mixed pig liver cDNA library. Three isolated clones allowed the construction of a 2731 base-pair full length cDNA. This cDNA construct contained an open reading frame of 1977 bp and encoded a 73 kDa protein of 659 amino acids. The 73 kDa active enzyme expressed in COS cells had the same substrate specificity, sensitivity to inhibitors and metal ion requirements as a previously isolated 49 kDa active fragment. Structural and hydrophobicity analysis of the coding region as well as other studies indicated that this enzyme is a nonglycosylated, type II transmembrane protein with a 48 residue cytosolic tail, followed by a 22 amino acid membrane anchor, a luminal 100 residue stem and a 49 kDa C-terminal catalytic domain.[184] Immunofluorescence studies indicted that the pig liver enzyme expressed in COS cells resides in the ER. On the other hand, the human kidney enzyme expressed in COS cells was localized in the Golgi apparatus.[185] The authors speculate that localization is likely to be sequence dependent.

After removal of the four α1,2-linked mannose units, the $Man_5GlcNAc_2$-protein is a substrate for GlcNAc transferase I, a glycosyltransferase in the medial Golgi stacks, that transfers a GlcNAc from UDP-GlcNAc to the mannose on the α1,3-branch to give $GlcNAc$-$Man_5GlcNAc_2$ protein.[186,187] This enzyme was purified to homogeneity from various sources and shown to be a type II integral membrane protein. The enzyme is specific for the $Man\alpha1,3,Man\beta1,4GlcNAc$ arm of the *N*-glycan core, and transfers a GlcNAc in β1-2-linkage to the terminal 1,3-linked mannose.[188,189] This reaction is necessary before mannosidase II can remove the α1,3 and α1,6 mannoses from the $Man\alpha1,6$ arm to give the trimannose structure. The gene for this enzyme was disrupted by homologous recombination in embryonic stem cells and transmitted to the germ line. Mice lacking GlcNAc transferase I activity did not survive to term, and biochemical and morphological analysis of embryos showed that they were developmentally retarded especially in regard to neural tissue.[190]

Once the GlcNAc has been added to the 3-linked mannose, mannosidase II can remove the two mannoses that are linked to the α1,6-linked mannose branch. The result of this reaction is a $GlcNAc\beta1,2Man\alpha1,3(Man\alpha1,6)Man\beta1,4GlcNAc\beta1,4GlcNAc$-protein.[191] Mannosidase II has been purified to homogeneity from rat liver[192] and mung bean seedlings.[193] The animal enzyme and the plant enzyme had apparent molecular masses of about 125 kDa on SDS gels, and both enzymes appeared to be glycoproteins.[192,193] However, the primary sequence of the murine mannosidase II derived from cloning studies predicted a molecular mass of 132 kDa for the deglycosylated enzyme.[194] This discrepancy may be explained by anomalous migration on SDS-PAGE (sodium dodecyl sulfate-polyacrylamide gel electrophoresis) by the deglycosylated or glycosylated protein, since the glycosylated enzyme migrates as a 124 kDa protein.

The full length mannosidase II cDNA has been isolated from a 3T3 cDNA library. The murine enzyme is a type II transmembrane glycoprotein with a cytoplasmic tail of five amino acids, a single transmembrane domain, and a luminally oriented catalytic domain.[194] The cDNA was overexpressed in COS cells, resulting in the appearance of immunoreactive material in a perinuclear membrane array indicating Golgi localization. The human α-mannosidase cDNA has also been isolated and this gene was mapped to chromosome 5.[194] Although the enzyme has been located in the Golgi apparatus, its "subGolgi" location depends on the cell type.[195] Thus, in exocrine pancreatic cells, hepatocytes, and intestinal goblet cells, the enzyme is found in the medial to trans Golgi. But in CHO cells, it was restricted to the medial Golgi.[195] Thus, in some cells, mannosidase I and mannosidase II appear to colocalize in the same region of the Golgi.

α-Mannosidase II activity has been demonstrated in all mammalian tissues that have been examined. However, the level of the enzyme is very low in brain.[196] Interestingly enough, this tissue has been found to have an alternate hydrolytic enzyme that has α1,2, α1,3, and α1,6-mannosidase activity and can cleave $Man_9GlcNAc_2$ down to $Man_3GlcNAc_2$.[197] This enzyme is clearly distinct from mannosidase II in terms of its substrate specificity and its reaction to various mannosidase inhibitors (see Section 3.07.6.3). Its specific role in glycoprotein processing is still to be determined.

A lack of mannosidase II has also been observed in HEMPAS disease, a hereditary affliction that is characterized by altered expression of one or several of the glycoprotein processing enzymes.[198] One form of the disease results from a deficiency in mRNA expression of α-mannosidase II. Lymphocytes derived from patients having this defect contain less than 10% of control mannosidase II levels, and their glycoproteins contain mostly hybrid types of oligosaccharides.[199]

The catalytic domain of the murine mannosidase II cDNA shows a considerable amount of similarity in sequence to the lysomal α-mannosidase cloned from the slime mold, *Dictyostelium discoideum*.[200] Nevertheless, these two enzymes have considerable differences in pH optimum, substrate specificity, and localization within the cell. Based on the sequence similarity, it has been proposed that the two enzymes were derived from the duplication and divergence of a primordial α-mannosidase gene with later acquisition of localization information and substrate specificity. A lesser degree of sequence similarity was observed between murine α-mannosidase II and the endoplasmic reticulum α-mannosidase or its cytoplasmic homologue, or the yeast vacuolar α-mannosidase.[201]

Following the action of the various glycosidases in the trimming part of the pathway, a number of glycosyltransferases act on the $GlcNAcMan_3GlcNAc_2$-protein to produce the complex types of N-linked oligosaccharides. Thus, in the trans-Golgi apparatus, there are a number of GlcNAc transferases, galactosyltransferases, fucosyltransferases, and sialyltransferases, that can add these sugars to the N-linked chains to give a great diversity of complex chains, having biantennary, triantennary, or tetraantennary structures. Many of these enzymes have been well characterized and a number of the genes for these important proteins have now been cloned.[202] Although there are not any good inhibitors of these enzymes currently available, the search for, or the chemical synthesis of, such compounds should be a rewarding future goal.

3.07.6 INHIBITORS OF N-LINKED GLYCOPROTEIN PROCESSING

3.07.6.1 Introduction

A number of low molecular mass compounds have been isolated from natural sources, or synthesized chemically, that specifically inhibit the glycosidases in the trimming pathway. These inhibitors have become valuable tools to use in biological systems to determine the role of N-linked oligosaccharide processing on the function of various membrane or secretory glycoproteins. The inhibitors are of special interest since they are small molecules which are able to permeate most cells and therefore can be used with intact cells and tissues to study "*in vivo*" situations. In addition, these inhibitors have been very useful in distinguishing the various processing enzymes from each other. The best example is shown in Table 2 where it is clear that the many different α-mannosidases have very different sensitivities to the various mannosidase inhibitors.[128,173] The remaining sections of this chapter describe the chemistry and biological activities of the various classes of alkaloidal and alkaloidal-like compounds that function as inhibitors of N-linked oligosaccharide processing.

A number of naturally occurring, sugar-like compounds, in which the ring oxygen is replaced by a nitrogen, have been isolated and are described in Section 3.07.2. Many of these alkaloids have been shown to be potent inhibitors of various glycosidases. The nitrogen in the ring apparently mimics the catalytic intermediate in the reaction (i.e., an oxycarbanion intermediate) but these compounds are still specifically recognized and bound to the active site of a particular glycosidase because of the resemblance in chirality to specific sugars like D-glucose and D-mannose. Thus, they function as valuable inhibitors of glycosidases, such as those that are involved in glycoprotein processing.

Table 2 Effect of processing inhibitors on various α-mannosidases.

Enzyme	Alkaloid				
	Swainsonine (μM)	Deoxymannojirimycin (μM)	Kifunensine (μM)	Mannostatin (μM)	Mannoamidrazone (μM)
ER-Man-ase				?	0.5–1
M$_9$N-Man-ase (ER)		5–7	?	?	?
Man-ase IA (Golgi)		1–2	?	?	?
Man-ase I (Mung bean)		40–50	0.02–0.05		4
Man-ase II (Rat liver)	0.2			?	?
Man-ase II (Mung bean)	0.09			0.09	0.1

3.07.6.2 Glucosidase Inhibitors

Castanospermine (**15**), as indicated earlier, is an indolizidine alkaloid that was first isolated from the seeds of the Australian tree, *Castanospermum australe*.[2] The initial studies on the effect of this compound in biological systems demonstrated that it was a reasonably potent inhibitor of β-glucosidase.[72] Later studies also showed that castanospermine inhibited a number of isolated α-glucosidases, including the glycoprotein processing enzymes, glucosidase I and glucosidase II, sucrase, maltase and lysosomal α-glucosidase.[203] Since this compound is such a potent inhibitor of intestinal maltase and sucrose, it prevents the degradation of the disaccharides sucrose and maltose, and therefore blocks the normal digestion of starch and sucrose. As a result, the seeds of *Castanospermum australe* are toxic to animals and cause severe diarrhea and other gastrointestinal upsets.[89] In addition, when castanospermine is fed to mice over a four or five day period, it inhibits the lysosomal α-glucosidase and causes the accumulation of partially degraded glycogen particles within the lysosomes (i.e., a situation similar to that which occurs in Pompe's disease, a genetic disease where afflicted individuals are lacking the lysosomal α-glucosidase).[204]

When various cultured animal cells are grown in the presence of castanospermine, the processing of the N-linked oligosaccharides is blocked at the first step (i.e., glucosidase I), and the asparagine linked glycoproteins have mostly oligosaccharides with Glc$_3$Man$_{9-7}$GlcNAc$_2$ structures.[205] However, in some cells there is an endomannosidase in the Golgi that can release a Glc$_{1-3}\alpha$1,3Man from glucose-containing N-linked oligosaccharides.[206] Although this enzyme prefers to act on the monoglucosylated oligosaccharide and release the disaccharide Glcα1,3Man, it can apparently also cleave the oligosaccharide containing three glucose residues. Thus, cells that contain this enzyme may be able to get around a castanospermine block. As mentioned above, the role of the endomannosidase in glycoprotein processing is not yet understood.

There are other glucosidase inhibitors that act at the level of glucosidase I and have similar effects to that of castanospermine but may have somewhat different levels of activity, or different specificities. These include 1-deoxynojirimycin (**5**, R = α-OH), which is a polyhydroxylated piperidine analogue that corresponds to D-glucopyranose, but has a nitrogen in the ring. This compound also inhibits α- and β-glucosidases.[207] Another inhibitor is the pyrrolidine alkaloid, 2,5-dihydroxymethyl-3,4-dihydroxypyrrolidine (DMDP) (**1**, R = OH).[208] The latter compound is much less effective than the above two inhibitors, which suggests that a six-membered ring structure is preferred for inhibitory activity. Nevertheless, DMDP does inhibit α- and β-glucosidase.[209]

The effect of preventing the removal of the glucose residues from the N-linked oligosaccharides on the targeting of the glycoproteins can be quite dramatic. Thus, when the hepatocyte cell line, Hep-G2, was incubated for various times in the presence of 1-deoxynojirimycin, the rate of secretion of the serum protein, α_1-antitrypsin, was greatly diminished, while the rate of secretion of other serum N-linked glycoproteins, such as ceruloplasmin and the C-3 component of the complement, were only marginally affected.[210] Cell fractionation studies indicated that the antitrypsin had accumulated or was held up in the ER–Golgi compartment, suggesting that the presence of glucose on the oligosaccharides might retard the movement of those proteins from the ER to, or through, the Golgi apparatus. Similar results were obtained when the biosynthesis and targeting of the low density lipoprotein receptor of fibroblasts and smooth muscle cells were examined, in the absence and presence of castanospermine. In these studies, it could be shown that cells grown in the presence of the inhibitor had only about one-half the number of receptor molecules at their cell surface, and therefore bound much less ^{125}I-LDL. However, these inhibited cells still had the same total number of LDL receptor molecules in the cells. The missing receptor molecules were found to be located in the ER or Golgi, based on cell fractionation studies.[211]

An interesting study was done in IM-9 lymphocytes where castanospermine was used to examine the role of oligosaccharide processing in the biosynthesis and targeting of the insulin receptor. Cells treated with castanospermine had a 50% decrease in the number of insulin receptors at the cell surface, as demonstrated by the binding of ^{125}I-insulin. The studies showed that removal of glucose residues from the N-linked glycoprotein was not necessary for the cleavage of the insulin proreceptor, that is for the maturation of the receptor. However, as shown in other systems, the presence of glucose apparently slowed the transport of this glycoprotein out of the ER to the Golgi, resulting in a decrease in the number of receptor molecules at the cell surface.[212]

In the case of the E_2 glycoprotein of coronavirus, both castanospermine and deoxynojirimycin caused a significant drop by log2 in the formation of virus, and also a dramatic inhibition in the appearance of E_2 glycoprotein at the cell surface. Significantly, the E_2 that was formed in the presence of the glucosidase inhibitors was still acylated with fatty acids as was the control viral E_2. However, the drug-induced E_2 accumulated in an intracellular compartment that was not definitively identified, but was probably the ER.[213]

Another study dealing with the sodium channel of rat brain neurons also showed that addition of palmitic acid to this protein was not prevented by the processing inhibitors.[214] The sodium channel is composed of α- and β-subunits that form a complex during maturation of the channel. The α-subunit undergoes post-translational modification by the addition of a palmitate, and the incorporation of this fatty acid into the glycoproteins was prevented by tunicamycin, a glycosylation inhibitor that completely prevents formation of N-linked oligosaccharides. On the other hand, castanospermine prevented processing of the oligosaccharide chains and the addition of sialic acids, but had no effect on the addition of palmitic acid. This alkaloid also did not affect the covalent assembly of the α- and β-subunits or the biological function of the channel.[214] Thus, the oligosaccharide is apparently necessary for palmitate addition, but the specific structure of the oligosaccharide (i.e., high-mannose or complex) is presumably not critical for the addition of palmitate groups.

GP120 is the envelope protein of HIV, the AIDS associated virus, and this protein is a glycoprotein with many oligosaccharide chains. These oligosaccharides are involved in the recognition and mechanism of attachment of HIV to the CD4 receptor on T lymphocytes and other susceptible cells. GP120 interacts with target molecules on the susceptible cells to cause the fusion of the cells with the formation of syncytia, which are necessary for viral formation and infectivity. The glucosidase inhibitors, 1-deoxynojirimycin (DNJ) and castanospermine, caused a significant decrease in the formation of new virus and in syncytium formation.[102,104,215] As a result of these interesting results, these inhibitors have been tested in human clinical trials as potential antiAIDS drugs. Although the results have not been published, one reported side effect in humans was the occurrence of diarrhea and other gastrointestinal problems in individuals taking these compounds. As shown in Table 1, there are a number of other compounds in addition to castanospermine and (DNJ) that are also inhibitors of glucosidases and glycoprotein processing. One such compound is the pyrrolidine alkaloid DMDP (**1**, R = OH), which occurs in several different plant families. When placed in a medium of cultured animal cells, DMDP inhibits the same step and gives the same oligosaccharide structure (i.e., $Glc_3Man_{9-7}GlcNAc_2$) as do castanospermine and DNJ.[208] However, DMDP is much less effective than these other inhibitors and therefore considerably higher concentrations are necessary in the medium. The fact that a five-membered ring structure can show glycosidase activity against enzymes that act on hexopyranosides is significant and would certainly warrant modeling studies of this structure in comparison to the indolizidine and piperidine alkaloids. Several other unusual structures that show increased selectivity towards the two processing glucosidases (i.e., glucosidase I and glucosidase II) are discussed below.

Australine (**8**) is a tetrahydroxypyrrolizidine alkaloid that was found in the same seeds that contain castanospermine, namely *Castanospermum australe*.[21] However, australine is present in the seeds in much lower amounts than is castanospermine. This compound is a good inhibitor of fungal amyloglucosidase, but it also inhibits the processing glucosidase I. However, in contrast to the other glucosidase I inhibitors discussed above which are also fairly effective against glucosidase II, australine is a very poor inhibitor of glucosidase II.[70] Thus, australine is the first glucosidase inhibitor to distinguish between these two processing enzymes. Nevertheless, the key affect of australine in cell culture is to block glucosidase I and cause the accumulation of glycoproteins having Glc_3Man_9 $GlcNAc_2$ structures. Additional compounds such as australine, and especially ones with more potent activity, will be useful tools to help understand the differences between glucosidase I and glucosidase II inhibitors.

Another interesting glucosidase inhibitor is 2,6-diamino-2,6-imino-7-O-(β-D-glucopyranosyl)-D-*glycero*-L-guloheptitol (MDL 25 637). This compound, referred to in the following discussion as MDL, was synthesized chemically to resemble a disaccharide that would function as a transition state analogue of the intestinal enzyme, sucrase.[216] As anticipated, MDL did inhibit rat intestinal maltase, sucrase, isomaltase, glucoamylase, and trehalase when present in micromolar amounts. Most interesting was the observation that MDL also showed specificity for the glucosidases but in the opposite manner to that of australine. Thus, MDL was much more effective against glucosidase II than it was against glucosidase I.[217] In cell culture, MDL was quite different from the other glucosidase inhibitors in that it caused the accumulation of glycoproteins having Glc_2Man_9 $(GlcNAc)_2$ structures. However, the overall effects of MDL on glycoprotein function in cell culture are likely to be similar to those observed with castanospermine and other inhibitors of glucosidase I.

A compound named trehazolin (24) was isolated as a trehalase inhibitor and has also been tested as an inhibitor of the processing glucosidases.[218] This compound inhibited glucosidase I quite well, but was a very poor inhibitor of glucosidase II.[219] The isolation and demonstration that structures like australine, MDL or trehazolin do exist, and that these compounds have selective actions against the processing glucosidases should stimulate the search for more and better inhibitors. Such inhibitors will be useful tools for additional studies on the role of carbohydrate and especially of the glucose residues in the function and localization of N-linked glycoproteins.

In the last few years, it has become clear why and how inhibitors of glucosidase I cause many N-linked glycoproteins to accumulate in the ER. Helenius[220] as well as other investigators have elegantly shown that the ER has a "protein correction and folding system" that helps newly synthesized ER proteins fold into the proper conformation that is necessary for transport to the Golgi apparatus. This system involves the action of a chaperone (i.e., a protein that helps other proteins fold). The chaperone, named calnexin, is also a lectin that recognizes a monoglucosylated high-mannose oligosaccharide on the unfolded glycoprotein. In the presence of castanospermine or other glucosidase I inhibitors, the first glucose cannot be removed, and therefore the unfolded protein cannot be recognized by calnexin and cannot be helped to fold. Most proteins will fold on their own given enough time, but the folding of some may be very slow and interaction with calnexin can help speed up this process. Thus, proteins like the LDL (low density lipoprotein) receptor, or the insulin receptor, or α_1-antitrypsin, are transported to the Golgi at a much slower rate in the presence of glucosidase inhibitors because of the inability of calnexin to bind to the protein.

3.07.6.3 Mannosidase Inhibitors

A number of α-mannosidase inhibitors have been identified from natural sources or synthesized chemically. In addition to their use as tools to examine the role of mannose oligosaccharides in the function of N-linked glycoproteins, they have also been valuable in distinguishing the various α-mannosidase activities from each other.

The first glycoprotein processing inhibitor to be reported was the indolizidine alkaloid, swainsonine (13),[1] an inhibitor of mannosidase II.[221] This compound was initially shown to be an inhibitor of the lysosomal α-mannosidase and to cause symptoms of the lysosomal storage disease α-mannosidosis when administered to animals.[71] Thus, swainsonine was essentially the prototype which chemists could use to design other glycosidase inhibitors. That is, based on the structures of swainsonine, castanospermine and 1-deoxynojirimycin, it appeared evident that a useful glycosidase inhibitor should have the following characteristics:

(i) a ring structure, probably of the pyranose type, with nitrogen replacing the heterocyclic oxygen;
(ii) a number (unknown at the time and still not certain) of hydroxyl groups; and
(iii) stereochemistry of the hydroxyl groups matching that of the sugar for which the glycosidase to be inhibited is specific.

In this section on mannosidase inhibitors, they will be discussed in the order in which they act in the glycoprotein processing pathway (Figure 2), rather than in order of their historical identification.

Based on the fact that 1-deoxynojirimycin (DNJ) (**5**, R = α-OH) was a good inhibitor of α-glucosidases, it was reasonable to assume that a related structure, but with mannose chirality, would be an inhibitor of α-mannosidases. The 2-epimer of DNJ, namely 1-deoxymannojirimycin (DMJ) (**5**, R = β-OH) was synthesized chemically and was indeed found to be a potent inhibitor of the glycoprotein processing mannosidase I.[222,223] Most interestingly, DMJ did not inhibit jack bean or lysosomal α-mannosidase, nor did it inhibit mannosidase II. Those observations on the selective specificity of DMJ demonstrate that it is dangerous to screen for new glycosidase inhibitors by using the commonly occurring aryl-glycosidases (i.e., α- and β-glucosidase, galactosidase, or mannosidase) to test for the inhibitory activity. That is, if the goal is to find a new glycoprotein processing inhibitor, such as an inhibitor of ER α-mannosidase, then one would desire a specific inhibitor that does not work on Golgi mannosidase I or mannosidase II, or jack bean or lysosomal α-mannosidase. Thus, if one used the enzymes that hydrolyze aryl-mannosides (such as *p*-nitrophenyl–D-manno-pyranoside) to screen for such a compound, the screens would obviously be negative and any potential inhibitor would be discarded.

In the period since deoxymannojirimycin was synthesized and shown to be a specific inhibitor of Golgi mannosidase I, a number of other neutral α-mannosidase activities have been reported in animal cells. These enzymes have all been discussed in Section 3.07.5.3 on glycoprotein processing, although it is still not clear what role, if any, some of them play in the trimming of *N*-linked oligosaccharides. As also indicated earlier, these enzymes have different substrate specificities from mannosidase I, and thus many of them are resistant to inhibition by DMJ. As these new mannosidases are purified and separated from each other, and from other competing activities, and as rapid assays for measuring their activities become available, it will be easier to identify or synthesize specific new inhibitors for each of these enzymes. Nevertheless, at this time, a number of α-mannosidase inhibitors have been identified and the activities of these various compounds on different α-mannosidases are presented in Table 2.

In animal cells, DMJ inhibited the Golgi mannosidase IA/B and caused the accumulation of glycoproteins having a high mannose oligosaccharide, mostly of the $Man_9GlcNAc_2$ structure.[224] In contrast to the effect of the glucose analogue DNJ, which prevented the secretion of IgD and IgM by cells in culture, DMJ had no effect.[225] As suggested above, this effect of DNJ is due to the function of calnexin on protein folding and its interaction with glucose. However, once the protein has folded and the glucoses are removed, the protein is treated normally with respect to targeting, regardless of whether it has a high mannose or modified chain.

In one interesting study, DMJ was used as a tool to determine whether glycoproteins were recycled through the Golgi during the endocytic process. In this experiment, membrane glycoproteins were synthesized in CHO cells in the presence of DMJ to inhibit mannose trimming, together with [2-³H]mannose to label the *N*-linked glycoproteins. After an appropriate incubation, the medium was changed to remove inhibitor and label and the cells were incubated for additional times. During this second period, the oligosaccharide structure of the transferrin receptor was determined under conditions where it would undergo endocytosis. Before the chase, the oligosaccharide structure of the transferrin receptor was of the high mannose type, but during the chase period, a small percentage of the recycled receptor molecules underwent processing and gave complex types of structure. These studies indicated that some endocytosed glycoproteins do recycle through the Golgi compartments and may undergo oligosaccharide processing.[226] However, the amount of glycoprotein molecules that were actually modified in this experiment was small, indicating that recycling through the Golgi is probably not a major route.

UT-1 cells were used to examine the role of the ER α-mannosidase in glycoprotein targeting and function. UT-1 cells are cells that overexpress HMG CoA reductase, a glycoprotein enzyme that resides in the ER of the cell. The oligosaccharide chains of this protein are of the high mannose type and mostly $Man_8GlcNAc_2$ and $Man_6GlcNAc_2$ structures. Since previous studies had shown that the ER mannosidase is not inhibited by DMJ, this inhibitor was used to determine whether the initial trimming of mannoses involved the ER mannosidase. In these studies, the HMG CoA reductase produced in the presence of DMJ had mostly $Man_8GlcNAc_2$ structures and the smaller oligosaccharides were not found, indicating that the ER enzyme was involved in the removal of the first mannose, but other mannoses were trimmed by DMJ-sensitive mannosidase(s).[227]

DIM (1,4-dideoxy-1,4-imino-D-mannitol) is another inhibitor that was synthesized from benzyl-α-D-mannopyranose and shown to be a good inhibitor of jack bean α-mannosidase.[228] It also inhibited glycoprotein processing in cultured MDCK cells, and gave rise to glycoproteins having mostly $Man_9GlcNAc_2$ structures suggesting that it inhibited the Golgi α-mannosidase I.[229] In keeping with these observations, *in vitro* studies with a partially purified preparation of mannosidase I showed that DIM did inhibit release of [^{3}H]mannose from [^{3}H]Man$_9$GlcNAc.[229] However, DIM is not nearly as effective an inhibitor of α-mannosidases as is either swainsonine or kifunensine (see below). On the other hand, DIM is of considerable interest as an inhibitor since:

(i) it has a furanose rather than a pyranose ring structure, and
(ii) it is synthesized chemically and therefore can be produced in large amounts and readily modified to produce various structural analogues.

It is not clear whether this compound also inhibits the ER mannosidase since this activity may not be present in MDCK cells.

Kifunensine (**25**) is an alkaloid produced by the actinomycete, *Kitasatosporia kifunense*, and it corresponds in structure to the cyclic oxamide derivative of 1-amino-DMJ.[43] This alkaloid is a very weak inhibitor of jack bean α-mannosidase, as is DMJ, but is a strong inhibitor of the Golgi mannosidase I ($IC_{50} = 2$ to 5×10^{-8} M). This inhibition is almost 100 times higher than the inhibition of mannosidase I by DMJ. Interestingly, kifunensine had no effect on either the ER mannosidase or on mannosidase II.[74] Influenza virus-infected MDCK cells incubated in the presence of kifunensine produced influenza virus particles in which the envelope glycoproteins had *N*-linked oligosaccharides mostly having Man$_9$GlcNAc$_2$ structures. This is the same effect as that seen in the presence of DMJ. However, kifunensine was much more effective in causing this change in structure and only 1/50 as much of this inhibitor was needed compared with DMJ.[74]

A compound that mimics the mannopyranosyl cation, the intermediate proposed as being involved in the enzymatic hydrolysis of α-mannopyranosides, was synthesized chemically and named mannonolactam amidrazone.[230] This compound not only inhibited Golgi mannosidase I with an IC_{50} of 4 μM, and mannosidase II with an IC_{50} of 100 nM, but was also a potent inhibitor of ER α-mannosidase (IC_{50} of 1 μM).[231] Furthermore, the compound also inhibited the aryl-α-mannosidase (IC_{50} of 400 nM) and the aryl-β-mannosidase (IC_{50} of 150 μM), although it clearly preferred α-linkages. In cell culture studies, mannonolactam amidrazone gave rise to glycoproteins with the same type of high mannose oligosaccharide as seen with DMJ and kifunensine. Thus inhibition of Golgi mannosidase I (and/or ER mannosidase) appears to prevent trimming of most if not all mannose residues.[231] The designers of this compound[230] hypothesize that the reason that it is so effective as a general mannosidase inhibitor is that it is the first analogue of mannose that mimics the true half-chair conformation of the cationic intermediate that is believed to be involved in catalysis of the α mannosides. Mannonolactam should serve as a model for the synthesis of more specific mannosidase inhibitors.

As mentioned earlier, the first processing inhibitor to be described was the indolizidine alkaloid, swainsonine (**13**).[1] In early studies, swainsonine was added to the culture media of MDCK cells infected with influenza virus, and these cultures were labeled by the addition of [2-^{3}H]mannose. This inhibitor caused a significant inhibition in the amount of mannose-labeled, Endo H-resistant oligosaccharides (i.e., complex oligosaccharides) and a great increase in the amount of mannose-labeled Endo H-sensitive structures. These latter oligosaccharides were shown to be hybrid types of oligosaccharides.[221,232] However, the change in the structure of the viral oligosaccharides from complex to hybrid types did not affect the production, maturation or release of the influenza virus particles.

These early studies did not identify the specific site of swainsonine inhibition, but later *in vitro* studies with the purified α-mannosidases demonstrated that swainsonine specifically inhibited mannosidase II, and was inactive towards mannosidase I.[233] In keeping with this site of action, swainsonine caused the formation of hybrid structures when it was added to the medium of cultured animal cells producing VSV glycoproteins (i.e., G protein),[234] fibronectin,[235] and BHK cell surface glycoproteins.[236] In most studies where swainsonine was used to determine the effect of changes in oligosaccharide structure on glycoprotein function, this inhibitor had little effect on functional aspects of the proteins in question, although it did cause alterations in structure to hybrid chains. The inhibitor did prevent the receptor-mediated uptake of mannose-terminated glycoproteins by macrophages. This inhibition was probably due to the formation of hybrid structures on the macrophage surface which could then react with and bind the mannose receptors.[237]

Swainsonine proved to be a valuable tool in determining the sequence of addition of certain sugars during the assembly of the *N*-linked oligosaccharides. Thus, the addition of L-fucose or sulfate to the influenza viral protein was studied in the presence of various processing inhibitors. When the glycoproteins were produced in the presence of castanospermine or DMJ, there was no [^{3}H]fucose[238] or [^{35}S]sulfate[239] associated with the glycoproteins, suggesting that fucose and sulfate were added after the mannosidase I step in processing. However, in the presence of swainsonine, the glycoproteins contained both L-fucose and sulfate indicating that the transferases that added these groups worked after the GlcNAc transferase I processing step. These results agree with the reported acceptor oligosaccharide specificity (i.e., GlcNAc-Man$_5$GlcNAc$_2$ of the fucosyltransferase and the sulfotransferase.

In some studies, swainsonine did cause a loss in the function of specific proteins. Thus, glucocorticoid stimulation of resorptive cells, involving the attachment of osteoblasts to bone, is inhibited by swainsonine.[240] Treatment of either the parasite, *Trypanosoma cruzi*, or the macrophages with swainsonine inhibits the interaction of these cells with each other.[112] This alkaloid also caused a dramatic decline in the ability of B16 melanoma cells to colonize the lungs of experimental animals.[241] As a result of these and similar studies, swainsonine has been undergoing tests and consideration as a drug to treat certain types of cancers. These are only a few of the many studies that have been done with this interesting compound. Many of these other studies are summarized in a review.[209]

Another inhibitor of mannosidase II, named mannostatin (**23**), was isolated from the fungus, *Streptoverticillium verticillus*.[41] This compound is of special interest because it has a very unusual structure with an exocyclic nitrogen, a five-membered ring, and a thiomethyl group, but is still a glycosidase inhibitor. Mannostatin was found to be a potent inhibitor of jack bean α-mannosidase as well as mannosidase II ($IC_{50} = 100$ nM). In cell culture studies, mannostatin caused the formation of the same types of hybrid oligosaccharides as are formed in the presence of swainsonine.[73] Interestingly, acetylation of the amino group of mannostatin resulted in loss of mannosidase activity. While this compound does not have any functional advantage over swainsonine as an inhibitor, it is of considerable interest, since it adds a great deal of additional structural information to our understanding of the requirements necessary for a compound to be a glycosidase inhibitor.

3.07.7 REFERENCES

1. S. M. Colegate, P. R. Dorling, and C. R. Huxtable, *Aust. J. Chem.*, 1979, **32**, 2257.
2. L. D. Hohenschutz, E. A. Bell, P. J. Jewess, D. P. Leworthy, R. J. Pryce, E. Arnold, and J. Clardy, *Phytochemistry*, 1981, **20**, 811.
3. A. D. Elbein and R. J. Molyneux, in "Alkaloids: Chemical and Biological Perspectives," ed. S. W. Pelletier, Wiley-Interscience, New York, 1987, vol. 5, p. 1.
4. R. J. Nash, A. A. Watson, and N. Asano, in "Alkaloids: Chemical and Biological Perspectives," ed. S. W. Pelletier, Pergamon, Oxford, UK, 1996, vol. 11, p. 345.
5. A. Welter, J. Jadot, G. Dardenne, M. Marlier, and J. Casimir, *Phytochemistry*, 1976, **15**, 747.
6. R. J. Molyneux, Y. T. Pan, J. E. Tropea, A. D. Elbein, C. H. Lawyer, D. J. Hughes, and G. W. J. Fleet, *J. Nat. Prod.*, 1993, **56**, 1356.
7. D. W. C. Jones, R. J. Nash, E. A. Bell, and J. M. Williams, *Tetrahedron Lett.*, 1985, **26**, 3125.
8. N. Asano, K. Oseki, E. Tomioka, H. Kizu, and K. Matsui, *Carbohydr. Res.*, 1994, **259**, 243.
9. T. Shibata, O. Nakayama, Y. Tsurumi, M. Okuhara, H. Terano, and M. Kohsaka, *J. Antibiot. (Tokyo)*, 1988, **41**, 296.
10. R. J. Molyneux, Y. T. Pan, J. E. Tropea, M. Benson, G. P. Kaushal, and A. D. Elbein, *Biochemistry*, 1991, **30**, 9981.
11. R. J. Nash, E. A. Bell, G. W. J. Fleet, R. H. Jones, and J. M. Williams, *J. Chem. Soc., Chem. Commun.*, 1985, 738.
12. A. A. Watson, R. J. Nash, M. R. Wormald, D. J. Harvey, S. Dealler, E. Lees, N. Asano, H. Kizu, A. Kato, R. C. Griffiths, A. J. Cairns, and G. W. J. Fleet, *Phytochemistry*, 1997, **46**, 255.
13. N. Asano, A. Kato, M. Miyauchi, H. Kizu, T. Tomimori, K. Matsui, R. J. Nash, and R. J. Molyneux, *Eur. J. Biochem.*, 1997, **248**, 296.
14. M. Koyama and S. Sakamura, *Agric. Biol. Chem.*, 1974, **38**, 1111.
15. S. Murao and S. Miyata, *Agric. Biol. Chem.*, 1980, **44**, 219.
16. L. E. Fellows, E. A. Bell, D. G. Lynn, F. Pilkiewicz, I. Miura, and K. Nakanishi, *J. Chem. Soc., Chem. Commun.*, 1979, 977.
17. S. Inouye, T. Tsuruoka, T. Ito, and T. Niida, *Tetrahedron*, 1968, **24**, 2125.
18. T. Niwa, T. Tsuruoka, H. Goi, Y. Kodama, J. Itoh, S. Inouye, Y. Yamada, T. Niida, M. Nobe, and Y. Ogawa, *J. Antibiot. (Tokyo)*, 1984, **37**, 1579.
19. J. V. Dring, G. C. Kite, R. J. Nash, and T. Reynolds, *Bot. J. Linn. Soc.*, 1995, **117**, 1.
20. G. C. Kite, J. M. Horn, J. T. Romeo, L. E. Fellows, D. C. Lees, A. M. Scofield, and N. G. Smith, *Phytochemistry*, 1990, **29**, 103.
21. R. J. Molyneux, M. Benson, R. Y. Wong, J. E. Tropea, and A. D. Elbein, *J. Nat. Prod.*, 1988, **51**, 1198.
22. R. J. Nash, L. E. Fellows, J. V. Dring, G. W. J. Fleet, A. E. Derome, T. A. Hamor, A. M. Scofield, and D. J. Watkin, *Tetrahedron Lett.*, 1988, **29**, 2487.
23. C. M. Harris, T. M. Harris, R. J. Molyneux, J. E. Tropea, and A. D. Elbein, *Tetrahedron Lett.*, 1989, **30**, 5685.
24. R. J. Nash, L. E. Fellows, J. V. Dring, G. W. J. Fleet, A. Girdhar, N. G. Ramsden, J. M. Peach, M. P. Hegarty, and A. M. Scofield, *Phytochemistry*, 1990, **29**, 111.
25. A. C. de S. Pereira, M. A. C. Kaplan, J. G. S. Maia, O. R. Gottlieb, R. J. Nash, G. W. J. Fleet, L. Pearce, D. J. Watkin, and A. M. Scofield, *Tetrahedron*, 1991, **47**, 5637.
26. R. J. Nash, P. I. Thomas, R. D. Waigh, G. W. J. Fleet, M. R. Wormald, P. M. de Q. Lilley, and D. J. Watkin, *Tetrahedron Lett.*, 1994, **35**, 7849.
27. I. Pastuszak, R. J. Molyneux, L. F. James, and A. D. Elbein, *Biochemistry*, 1990, **29**, 1886.

28. R. J. Molyneux and L. F. James, *Science*, 1982, **216**, 190.
29. M. J. Schneider, F. S. Ungemach, H. P. Broquist, and T. M. Harris, *Tetrahedron*, 1983, **39**, 29.
30. M. Hino, O. Nakayama, Y. Tsurumi, K. Adachi, T. Shibata, H. Terano, M. Kohsaka, H. Aoki, and H. Imanaka, *J. Antibiot. (Tokyo)*, 1985, **38**, 926.
31. R. J. Molyneux, J. E. Tropea, and A. D. Elbein, *J. Nat. Prod.*, 1990, **53**, 609.
32. R. J. Molyneux, J. N. Roitman, G. Dunnheim, T. Szumilo, and A. D. Elbein, *Arch. Biochem. Biophys.*, 1986, **251**, 450.
33. R. J. Molyneux, R. J. Nash, and N. Asano, in "Alkaloids: Chemical and Biological Perspectives," ed. S. W. Pelletier, Pergamon, Oxford, UK, 1996, vol. 11, p. 303.
34. D. Tepfer, A. Goldmann, N. Pamboukdjian, M. Maille, A. Lepingle, D. Chevalier, J. Dénarié, and C. Rosenberg, *J. Bacteriol.*, 1988, **170**, 1153.
35. A. Goldmann, M.-L. Milat, P.-H. Ducrot, J.-Y. Lallemand, M. Maille, A. Lepingle, I. Charpin, and D. Tepfer, *Phytochemistry*, 1990, **29**, 2125.
36. N. Asano, A. Kato, K. Oseki, H. Kizu, and K. Matsui, *Eur. J. Biochem.*, 1995, **229**, 369.
37. N. Asano, A. Kato, Y. Yokoyama, M. Miyauchi, M. Yamamoto, H. Kizu, and K. Matsui, *Carbohydr. Res.*, 1996, **284**, 169.
38. N. Asano, A. Kato, H. Kizu, K. Matsui, A. A. Watson, and R. J. Nash, *Carbohydr. Res.*, 1996, **293**, 195.
39. H. Yoon, S. B. King, and B. Ganem, *Tetrahedron Lett.*, 1991, **32**, 7199.
40. A. Kato, N. Asano, H. Kizu, K. Matsui, S. Suzuki, and M. Arisawa, *Phytochemistry*, 1997, **45**, 425.
41. T. Aoyagi, T. Yamamoto, K. Kojiri, H. Morishima, M. Nagai, M. Hamada, T. Takeuchi, and H. Umezawa, *J. Antibiot. (Tokyo)*, 1989, **42**, 883.
42. O. Ando, H. Satake, K. Itoi, A. Sato, M. Nakajima, S. Takahashi, H. Haruyama, Y. Ohkuma, T. Kinoshita, and R. Enokita, *J. Antibiot. (Tokyo)*, 1991, **44**, 1165.
43. H. Kayakiri, S. Takase, T. Shibata, M. Okamoto, H. Terano, M. Hashimoto, T. Tada, and S. Koda, *J. Org. Chem.*, 1989, **54**, 4015.
44. T. Aoyagi, T. Suda, K. Uotani, F. Kojima, T. Aoyama, K. Horiguchi, M. Hamada, and T. Takeuchi, *J. Antibiot. (Tokyo)*, 1992, **45**, 1404.
45. R. J. Molyneux, R. A. McKenzie, B. M. O'Sullivan, and A. D. Elbein, *J. Nat. Prod.*, 1995, **58**, 878.
46. R. J. Molyneux, Y. T. Pan, A. Goldmann, D. A. Tepfer, and A. D. Elbein, *Arch. Biochem. Biophys.*, 1993, **304**, 81.
47. R. J. Nash, M. Rothschild, E. A. Porter, A. A. Watson, R. D. Waigh, and P. G. Waterman, *Phytochemistry*, 1993, **34**, 1281.
48. B. Drager, A. van Almsick, and G. Mruchatz, *Planta Med.*, 1995, **61**, 577.
49. N. Asano, E. Tomioka, H. Kizu, and K. Matsui, *Carbohydr. Res.*, 1994, **253**, 235.
50. S. V. Evans, L. E. Fellows, T. K. M. Shing, and G. W. J. Fleet, *Phytochemistry*, 1985, **24**, 1953.
51. S. Watanabe, H. Kato, K. Nagayama, and H. Abe, *Biosci. Biotech. Biochem.*, 1995, **59**, 936.
52. R. J. Molyneux, L. F. James, K. E. Panter, and M. H. Ralphs, *Phytochem. Anal.*, 1991, **2**, 125.
53. C. M. Harris, B. C. Campbell, R. J. Molyneux, and T. M. Harris, *Tetrahedron Lett.*, 1988, **29**, 4815.
54. A. Goldmann, B. Message, D. Tepfer, R. J. Molyneux, O. Duclos, F.-D. Boyer, Y. T. Pan, and A. D. Elbein, *J. Nat. Prod.*, 1996, **59**, 1137.
55. R. J. Molyneux, in "Bioactive Natural Products: Detection, Isolation and Structural Identification," eds. S. M. Colegate and R. J. Molyneux, CRC Press, Boca Raton, FL, 1993, p. 59.
56. A. M. Scofield, L. E. Fellows, R. J. Nash, and G. W. J. Fleet, *Life Sci.*, 1986, **39**, 645.
57. G. W. J. Fleet, S. J. Nicholas, P. W. Smith, S. V. Evans, L. E. Fellows, and R. J. Nash, *Tetrahedron Lett.*, 1985, **26**, 3127.
58. M. T. H. Axamawaty, G. W. J. Fleet, K. A. Hannah, S. K. Namgoong, and M. L. Sinnott, *Biochem. J.*, 1990, **266**, 245.
59. B. Winchester and G. W. J. Fleet, *Glycobiology*, 1992, **2**, 199.
60. H. Kayakiri, K. Nakamura, S. Takase, H. Setoi, I. Uchida, H. Terano, M. Hashimoto, T. Tada, and S. Koda, *Chem. Pharm. Bull. (Tokyo)*, 1991, **39**, 2807.
61. A. M. Scofield, P. Witham, R. J. Nash, G. C. Kite, and L. E. Fellows, *Comp. Biochem. Physiol.*, 1995, **112A**, 187.
62. A. M. Scofield, P. Witham, R. J. Nash, G. C. Kite, and L. E. Fellows, *Comp. Biochem. Physiol.*, 1995, **112A**, 197.
63. B. Winchester, *Biochem. Soc. Trans.*, 1992, **20**, 699.
64. N. Asano, K. Oseki, H. Kizu, and K. Matsui, *J. Med. Chem.*, 1994, **37**, 3701.
65. H. Hettkamp, G. Legler, and E. Bause, *Eur. J. Biochem.*, 1984, **142**, 85.
66. N. Ishida, K. Kumagai, T. Niida, T. Tsuruoka, and H. Yumoto, *J. Antibiot. (Tokyo)*, 1967, **20**, 66.
67. J. Bischoff and R. Kornfeld, *Biochem. Biophys. Res. Commun.*, 1984, **125**, 324.
68. T. Niwa, S. Inouye, T. Tsuruoka, Y. Koaze, and T. Niida, *Agric. Biol. Chem.*, 1970, **34**, 966.
69. Y. Miyake and M. Ebata, *Agric. Biol. Chem.*, 1988, **52**, 661.
70. J. E. Tropea, R. J. Molyneux, G. P. Kaushal, Y. T. Pan, M. Mitchell, and A. D. Elbein, *Biochemistry*, 1989, **28**, 2027.
71. P. R. Dorling, C. R. Huxtable, and S. M. Colegate, *Biochem. J.*, 1980, **191**, 649.
72. R. Saul, J. P. Chambers, R. J. Molyneux, and A. D. Elbein, *Arch. Biochem. Biophys.*, 1983, **221**, 593.
73. J. E. Tropea, G. P. Kaushal, I. Pastuszak, M. Mitchell, T. Aoyagi, R. J. Molyneux, and A. D. Elbein, *Biochemistry*, 1990, **29**, 10062.
74. A. D. Elbein, J. E. Tropea, M. Mitchell, and G. P. Kaushal, *J. Biol. Chem.*, 1990, **265**, 15599.
75. R. J. Molyneux, in "Methods in Plant Biochemistry: Volume 8—Alkaloids and Sulphur Compounds," ed. P. G. Waterman, Academic Press, London, 1993, p. 511.
76. D. A. Winkler and G. Holan, *J. Med. Chem.*, 1989, **32**, 2084.
77. D. A. Winkler, *J. Med. Chem.*, 1996, **39**, 4332.
78. E. M. S. Harris, A. E. Aleshin, L. M. Firsov, and R. B. Honzatko, *Biochemistry*, 1993, **32**, 1618.
79. M. Ebner, C. E. Ekhart, and A. E. Stütz, in "Electronic Conference on Heterocyclic Chemistry '96," eds. H. S. Rzepa, J. Snyder, and C. Leach, Royal Society of Chemistry, London, 1997 (see also: http://www.ch.ic.ac.uk/ectoc/echet96/).
80. G. W. J. Fleet, in "Swainsonine and Related Glycosidase Inhibitors," eds. L. F. James, A. D. Elbein, R. J. Molyneux, and C. D. Warren, Iowa State University Press, Ames, 1980, p. 382.
81. B. Ganem, *Acc. Chem. Res.*, 1996, **29**, 340.

82. K. Burgess and I. Henderson, *Tetrahedron*, 1992, **48**, 4045.
83. R. H. Furneaux, G. J. Gainsford, J. M. Mason, P. C. Tyler, O. Hartley, and B. G. Winchester, *Tetrahedron*, 1997, **53**, 245.
84. W. H. Pearson and E. J. Hembre, *J. Org. Chem.*, 1996, **61**, 5537.
85. W. H. Pearson and E. J. Hembre, *J. Org. Chem.*, 1996, **61**, 5546.
86. G. Gradnig, A. Berger, V. Grassberger, A. E. Stütz, and G. Legler, *Tetrahedron Lett.*, 1991, **32**, 4889.
87. R. J. Molyneux, L. F. James, M. H. Ralphs, J. A. Pfister, K. E. Panter, and R. J. Nash, in "Poisonous Plants of the World: Agricultural, Phytochemical and Ecological Aspects," eds. S. M. Colegate and P. R. Dorling, CAB International, Wallingford, UK, 1994, p. 107.
88. S. L. Everist, "Poisonous Plants of Australia," Angus and Robertson, Sydney, 1994, p. 403.
89. R. Saul, J. J. Ghidoni, R. J. Molyneux, and A. D. Elbein, *Proc. Natl. Acad. Sci. USA*, 1985, **82**, 93.
90. N. Asano, A. Kato, K. Matsui, A. A. Watson, R. J. Nash, R. J. Molyneux, L. Hackett, J. Topping, and B. Winchester, *Glycobiology*, 1997, **7**, 1085.
91. D. L. Dreyer, K. C. Jones, and R. J. Molyneux, *J. Chem. Ecol.*, 1985, **11**, 1045.
92. B. C. Campbell, R. J. Molyneux, and K. C. Jones, *J. Chem. Ecol.*, 1987, **13**, 1759.
93. M. S. J. Simmonds, W. M. Blaney, and L. E. Fellows, *J. Chem. Ecol.*, 1990, **16**, 3167.
94. W. M. Blaney, M. S. J. Simmonds, S. V. Evans, and L. E. Fellows, *Entomol. Exp. Appl.*, 1984, **36**, 209.
95. K. L. Stevens and R. J. Molyneux, *J. Chem. Ecol.*, 1988, **14**, 1467.
96. C.-L. Rosenfield and P. Matile, *Plant Cell Physiol.*, Tokyo, 1979, **20**, 605.
97. M. J. Humphries, K. Matsumoto, S. L. White, R. J. Molyneux, and K. Olden, *Clin. Exp. Metastasis*, 1990, **8**, 89.
98. M. J. Humphries, K. Matsumoto, S. L. White, R. J. Molyneux, and K. Olden, *Cancer Res.*, 1988, **48**, 1410.
99. D. Bowden, J. Adir, S. L. White, C. D. Bowen, K. Matsumoto, and K. Olden, *Anticancer Res.*, 1993, **13**, 841.
100. P. E. Goss, J. Baptiste, B. Fernandes, M. Baker, and J. W. Dennis, *Cancer Res.*, 1994, **54**, 1450.
101. G. K. Ostrander, N. K. Scribner, and L. R. Rohrschneider, *Cancer Res.*, 1988, **48**, 1091.
102. R. A. Gruters, J. J. Neefjes, M. Tersmette, R. E. Y. de Goede, A. Tulp, H. G. Huisman, F. Miedema, and H. L. Ploegh, *Nature (London)*, 1987, **330**, 74.
103. A. S. Tyms, E. M. Berrie, T. A. Ryder, R. J. Nash, M. P. Hegarty, M. A. Mobberley, J. M. Davis, E. A. Bell, D. J. Jeffries, D. Taylor-Robinson, and L. E. Fellows, *Lancet*, 1987, **2**, 1025.
104. B. D. Walker, M. Kowalski, W. C. Goh, K. Kozarsky, M. Krieger, C. Rosen, L. Rohrschneider, W. A. Haseltine, and J. Sodroski, *Proc. Natl. Acad. Sci. USA*, 1987, **84**, 8120.
105. D. L. Taylor, L. E. Fellows, G. H. Farrar, R. J. Nash, D. Taylor-Robinson, M. A. Mobberley, T. A. Ryder, D. J. Jeffries, and A. S. Tyms, *Antiviral Res.*, 1988, **10**, 11.
106. C. G. Bridges, S. P. Ahmed, M. S. Kang, R. J. Nash, E. A. Porter, and A. S. Tyms, *Glycobiology*, 1995, **5**, 243.
107. G. W. J. Fleet, A. Karpas, R. A. Dwek, L. E. Fellows, A. S. Tyms, S. Petursson, S. K. Namgoong, N. G. Ramsden, P. W. Smith, J. C. Son, F. Wilson, D. R. Witty, G. S. Jacob, and T. W. Rademacher, *FEBS Lett.*, 1988, **237**, 128.
108. R. H. Taylor, H. M. Barker, E. A. Bowey, and J. E. Canfield, *Gut*, 1986, **27**, 1471.
109. F. Chen, N. Nakashima, I. Kimura, M. Kimura, N. Asano, and S. Koya, *Biol. Pharm. Bull.*, 1995, **18**, 1676.
110. P. M. Grochowicz, A. D. Hibberd, Y. C. Smart, K. M. Bowen, D. A. Clark, W. B. Cowden, and D. O. Willenborg, *Transpl. Immunol.*, 1996, **4**, 275.
111. P. S. Wright, D. E. Cross-Doersen, K. K. Schroeder, T. L. Bowlin, P. P. McCann, and A. J. Bitonti, *Biochem. Pharmacol.*, 1991, **41**, 1855.
112. F. Villalta and F. Kierszenbaum, *Mol. Biochem. Parasitol.*, 1985, **16**, 1.
113. A. Gottschalk, "Glycoproteins: Their Composition, Structure and Function," Elsevier, New York, 1972.
114. M. F. Mescher, U. Hansen, and J. L. Strominger, *J. Biol. Chem.*, 1976, **251**, 7289.
115. F. A. Troy, II, *Glycobiology*, 1992, **2**, 5.
116. A. Varki, *Glycobiology*, 1993, **3**, 97.
117. T. A. Yednock and S. D. Rosen, *Adv. Immunol.*, 1989, **44**, 313.
118. T. Feizi, *Nature (London)*, 1985, **314**, 53.
119. P. A. Haynes, M. A. J. Ferguson, and G. A. M. Cross, *Glycobiology*, 1996, **6**, 869.
120. H. Leffler, F. R. Masiarz, and S. H. Barondes, *Biochemistry*, 1989, **28**, 9222.
121. Q. Zhou and R. D. Cummings, in "Cell Surface Carbohydrates and Cell Development," ed. M. Fukuda, CRC Press, Boca Raton, FL, 1992, p. 99.
122. A. Kobata, *Eur. J. Biochem.*, 1992, **209**, 483.
123. S. Kornfeld, *J. Clin. Invest.*, 1986, **77**, 1.
124. G. Ashwell and J. Harford, *Annu. Rev. Biochem.*, 1982, **51**, 531.
125. C. Hammond, I. Braakman, and A. Helenius, *Proc. Natl. Acad. Sci. USA*, 1994, **91**, 913.
126. K. Olden, J. B. Parent, and S. L. White, *Biochim. Biophys. Acta*, 1982, **650**, 209.
127. R. U. Margolis, and R. K. Margolis, "Neurobiology of Glycoconjugates," Plenum Press, New York, 1989.
128. A. D. Elbein, *FASEB J.*, 1991, **5**, 3055.
129. R. J. Staneloni and L. F. Leloir, *CRC Crit. Rev. Biochem.*, 1982, **12**, 289.
130. R. Kornfeld and S. Kornfeld, *Ann. Rev. Biochem.*, 1985, **54**, 631.
131. G. P. Kaushal and A. D. Elbein, *J. Biol. Chem.*, 1985, **260**, 16 303.
132. L. F. Leloir, R. J. Staneloni, H. Carminatti, and N. H. Behrens, *Biochem. Biophys. Res. Commun.*, 1973, **52**, 1285.
133. C. B. Sharma, G. P. Kaushal, Y. T. Pan, and A. D. Elbein, *Biochemistry*, 1990, **29**, 8901.
134. C. B. Hirschberg and M. D. Snider, *Annu. Rev. Biochem.*, 1987, **56**, 63.
135. C. M. D'Souza, C. B. Sharma, and A. D. Elbein, *Anal. Biochem.*, 1992, **203**, 211.
136. J. S. Rush, J. G. Shelling, N. S. Zingg, P. H. Ray, and C. J. Waechter, *J. Biol. Chem.*, 1993, **268**, 13 110.
137. W. Jankowski, T. Mankowski, and T. Chojnacki, *Biochim. Biophys. Acta*, 1974, **337**, 153.
138. J. P. Spencer and A. D. Elbein, *Proc. Natl. Acad. Sci. USA*, 1980, **77**, 2524.
139. A. Haselbeck and W. Tanner, *Proc. Natl. Acad. Sci. USA*, 1982, **79**, 1520.
140. P. Gold and M. Green, *J. Biol. Chem.*, 1983, **258**, 12 967.
141. D. J. Kelleher, G. Kreibich, and R. Gilmore, *Cell*, 1992, **69**, 55.
142. V. Kumar, F. S. Heinemann, and J. Ozols, *Biochem. Mol. Biol. Int.*, 1995, **36**, 817.

143. E. Bause, W. Breuer, and S. Peters, *Biochem. J.*, 1995, **312**, 979.
144. D. K. Struck and W. J. Lennarz, in "The Biochemistry of Glycoproteins and Glycolipids," ed. W. J. Lennarz, Plenum Press, New York, 1980, p. 35.
145. M. D. Snider, in "Biology of Carbohydrates," eds. V. Ginsburg and P. W. Robbins, Wiley and Sons, New York, 1984, vol. 2, p. 163.
146. S. J. Turco, B. Stetson, and P. W. Robbins, *Proc. Natl. Acad. Sci. USA*, 1977, **74**, 4411.
147. J. Schweden, C. Borgmann, G. Legler, and E. Bause, *Arch. Biochem. Biophys.*, 1986, **248**, 335.
148. T. Szumilo, G. P. Kaushal, and A. D. Elbein, *Arch. Biochem. Biophys.*, 1986, **247**, 261.
149. H. Hettkamp, G. Legler, and E. Bause, *Eur. J. Biochem.*, 1984, **142**, 85.
150. E. Bause, J. Schweden, A. Gross, and B. Orthen, *Eur. J. Biochem.*, 1989, **183**, 661.
151. K. Shailubhai, M. A. Pratta, and I. K. Vijay, *Biochem. J.*, 1987, **247**, 555.
152. E. Bause, R. Erkens, J. Schweden, and L. Jaenicke, *FEBS Lett.*, 1986, **206**, 208.
153. B. Kalz-Fuller, E. Bieberich, and E. Bause, *Eur. J. Biochem.*, 1995, **231**, 344.
154. B. Esmon, P. C. Esmon, and R. Scheckman, *J. Biol. Chem.*, 1984, **259**, 10 322.
155. P. Stanley, S. Salustio, S. S. Krag, and B. Dunn, *Somat. Cell Mol. Genet.*, 1990, **16**, 211.
156. S. C. Hubbard and P. W. Robbins, *J. Biol. Chem.*, 1979, **254**, 4568.
157. I. Braakman, K. Hoover-Litty, K. R. Wagner, and A. Helenius, *J. Cell Biol.*, 1991, **114**, 401.
158. F. E. Ware, A. Vassilakos, P. A. Peterson, M. R. Jackson, M. A. Lehrman, and D. B. Williams, *J. Biol. Chem.*, 1995, **270**, 4697.
159. R. G. Spiro, Q. Zhu, V. Bhoyroo, and H. D. Soling, *J. Biol. Chem.*, 1996, **271**, 11 588.
160. A DeSilva, I. Braakman, and A. Helenius, *J. Cell Biol.*, 1993, **120**, 647.
161. M. Sousa, M. A. Ferrero-Garcia, and A. J. Parodi, *Biochemistry*, 1992, **31**, 97.
162. P. Choudhury, Y. Liu, R. J. Bick, and R. N. Sifers, *J. Biol. Chem.*, 1997, **272**, 13 446.
163. D. M. Burns and O. Touster, *J. Biol. Chem.*, 1982, **257**, 9991.
164. G. P. Kaushal, I. Pastuszak, K. Hatanaka, and A. D. Elbein, *J. Biol. Chem.*, 1990, **265**, 16 271.
165. D. Brada and U. C. Dubach, *Eur. J. Biochem.*, 1984, **141**, 149.
166. G. P. Kaushal, Y. Zheng, and A. D. Elbein, *J. Biol. Chem.*, 1993, **268**, 14 536.
167. S. Saxena, K. Shailubhai, B. Dong-Yu, and I. K. Vijay, *Biochem. J.*, 1987, **247**, 563.
168. F. Martiniuk, A. Ellenbogen, and R. Hirschhorn, *J. Biol. Chem.*, 1985, **260**, 1238.
169. J. M. Lucocq, D. Brada, and J. Roth, *J. Cell. Biol.*, 1986, **102**, 2137.
170. D. Brada, D. Kerjaschki, and J. Roth, *J. Cell. Biol.*, 1990, **110**, 309.
171. G. J. Strous, P. Van Kerkhof, R. Brok, J. Roth, and D. Brada, *J. Biol. Chem.*, 1987, **262**, 3620.
172. T. Szumilo and A. D. Elbein, *Anal. Biochem.*, 1985, **151**, 32.
173. A. D. Elbein, *Annu. Rev. Biochem.*, 1987, **56**, 497.
174. F. Maley, R. B. Trimble, A. L. Tarentino, and T. H. Plummer, Jr., *Anal. Biochem.*, 1989, **180**, 195.
175. R. B. Trimble, K. W. Moremen, and A. Herscovics, in "Guidebook to the Secretory Pathway," ed. T. Stevens, P. Novick, and J. A. Kornblatt, Scientific Publishers, Dallas, TX, 1992, p. 1.
176. P. F. Daniel, B. Winchester, and C. D. Warren, *Glycobiology*, 1994, **4**, 551.
177. F. Lipari and A. Herscovics, *J. Biol. Chem.*, 1996, **271**, 27 615.
178. J. Bischoff and R. Kornfeld, *J. Biol. Chem.*, 1983, **258**, 7907.
179. J. Schweden, G. Legler, and E. Bause, *Eur. J. Biochem.*, 1986, **157**, 563.
180. I. Tabas and S. Kornfeld, *J. Biol. Chem.*, 1979, **254**, 11 655.
181. W. T. Forsee and J. S. Schutzbach, *J. Biol. Chem.*, 1981, **256**, 6577.
182. W. A. Lubas and R. G. Spiro, *J. Biol. Chem.*, 1987, **262**, 3775.
183. J. Bischoff, K. Moremen, and H. F. Lodish, *J. Biol. Chem.*, 1990, **265**, 17 110.
184. E. Bieberich, K. Treml, C. Völker, A. Rolfs, B. Kalz Füller, and E. Bause, *Eur. J. Biochem.*, 1997, **246**, 681.
185. E. Bause, E. Bieberich, A. Rolfs, C. Volker, and B. Schmidt, *Eur. J. Biochem.*, 1993, **217**, 535.
186. H. Schachter, *Biochem. Cell Biol.*, 1986, **64**, 163.
187. M. Metzler, A. Gertz, M. Sarkar, H. Schachter, J. W. Schrader, and J. D. Marth, *EMBO J.*, 1994, **13**, 2056.
188. N. Harpaz and H. Schachter, *J. Biol. Chem.*, 1980, **255**, 4885.
189. B. Yip, S.-H. Chen, H. Mulder, J. W. M. Hoppener, and H. Schachter, *Biochem. J.*, 1997, **321**, 465.
190. E. Ioffe and P. Stanley, *Proc. Natl. Acad. Sci. USA*, 1994, **91**, 728.
191. D. P. R. Tulsiani, S. C. Hubbard, P. W. Robbins, and O. Touster, *J. Biol. Chem.*, 1982, **257**, 3660.
192. K. W. Moremen and O. Touster, *J. Biol. Chem.*, 1986, **261**, 10 945.
193. G. P. Kaushal, T. Szumilo, I. Pastuszak, and A. D. Elbein, *Biochemistry*, 1990, **29**, 2168.
194. K. W. Moremen and P. W. Robbins, *J. Cell Biol.*, 1991, **115**, 1521.
195. A. Velasco, L. Hendricks, K. W. Moremen, D. R. P. Tulsiani, O. Touster, and M. G. Farquhar, *J. Cell Biol.*, 1993, **122**, 39.
196. L. S. Wilkerson and O. Touster, *Arch. Biochem. Biophys.*, 1993, **303**, 238.
197. D. R. P. Tulsiani, and O. Touster, *J. Biol. Chem.*, 1985, **260**, 13 081.
198. M. N. Fukuda, K. A. Masri, A. Dell, L. Luzzatto, and K. W. Moremen, *Proc. Natl. Acad. Sci. USA*, 1990, **87**, 7443.
199. M. N. Fukuda, *Glycobiology*, 1990, **1**, 9.
200. J. Schatzle, J. Bush, and J. Cardelli, *J. Biol. Chem.*, 1992, **267**, 4000.
201. K. W. Moremen, R. B. Trimble, and A. Herscovics, *Glycobiology*, 1994, **4**, 113.
202. R. Kleene and E. G. Berger, *Biochim. Biophys. Acta*, 1993, **1154**, 283.
203. Y. T. Pan, J. Ghidoni, and A. D. Elbein, *Arch. Biochem. Biophys.*, 1992, **303**, 134.
204. P. Baudhuin, H. G. Hers, and H. Loeb, *Lab. Invest.*, 1964, **13**, 1139.
205. Y. T. Pan, H. Hori, R. Saul, B. A. Sanford, R. J. Molyneux, and A. D. Elbein, *Biochemistry*, 1983, **22**, 3975.
206. S. E. Moore and R. G. Spiro, *J. Biol. Chem.*, 1990, **265**, 13 104.
207. D. D. Schmidt, W. Frommer, L. Müller, and E. Truscheit, *Naturwissenschaften*, 1979, **66**, 584.
208. A. D. Elbein, M. Mitchell, B. A. Sanford, L. E. Fellows, and S. V. Evans, *J. Biol. Chem.*, 1984, **259**, 12 409.
209. A. D. Elbein, in "Cell Surface and Extracellular Glycoconjugates," eds. R. P. Meecham and D. D. Roberts, Academic Press, NY, 1993, p. 119.

210. H. F. Lodish and N. Kong, *J. Cell Biol.*, 1984, **98**, 1720.
211. E. H. Edwards, E. A. Sprague, J. L. Kelley, J. J. Kerbacher, C. J. Schwartz, and A. D. Elbein, *Biochemistry*, 1989, **28**, 7679.
212. R. F. Arakaki, J. A. Hedo, E. Collier, and P. Gorden, *J. Biol. Chem.*, 1987, **262**, 11 886.
213. R. Repp, T. Tamura, C. B. Boschek, H. Wege, R. T. Schwarz, and H. Niemann, *J. Biol. Chem.*, 1985, **260**, 15 873.
214. J. W. Schmidt and W. A. Catterall, *J. Biol. Chem.*, 1987, **262**, 13 713.
215. D. C. Montefiori, W. E. Robinson, Jr., and W. M. Mitchell, *Proc. Natl. Acad. Sci. USA*, 1988, **85**, 9248.
216. P. S. Liu, *J. Org. Chem.*, 1987, **52**, 4717.
217. G. P. Kaushal, Y. T. Pan, J. E. Tropea, M. Mitchell, P. Liu, and A. D. Elbein, *J. Biol. Chem.*, 1988, **263**, 17 278.
218. C. Uchida, H. Kitahashi, S. Watanabe, and S. Ogawa, *J. Chem. Soc., Perkin Trans. I.*, 1995, 1707.
219. Y.-C. Zeng and A. D. Elbein, *Eur. J. Biochem.*, submitted for publication.
220. A. Helenius, *Mol. Biol. Cell*, 1994, **5**, 253.
221. A. D. Elbein, R. Solf, P. R. Dorling, and K. Vosbeck, *Proc. Natl. Acad. Sci. USA*, 1981, **78**, 7393.
222. G. Legler and E. Julich, *Carbohydr. Res.*, 1984, **128**, 61.
223. U. Fuhrmann, E. Bause, G. Legler, and H. Ploegh, *Nature (London)*, 1984, **307**, 755.
224. A. D. Elbein, G. Legler, A. Tlusty, W. McDowell, and R. Schwarz, *Arch. Biochem. Biophys.*, 1984, **235**, 579.
225. V. Gross, K. Steube, T. A. Tran-Thi, W. McDowell, R. T. Schwarz, K. Decker, W. Gerok, and P. C. Heinrich, *Eur. J. Biochem.*, 1985, **150**, 41.
226. M. D. Snider and O. C. Rogers, *J. Cell Biol.*, 1986, **103**, 265.
227. J. Bischoff, L. Liscum, and R. Kornfeld, *J. Biol. Chem.*, 1986, **261**, 4766.
228. G. W. J. Fleet, P. W. Smith, S. V. Evans, and L. E. Fellows, *J. Chem. Soc., Chem. Commun.*, 1984, 1240.
229. G. Palamarczyk, M. Mitchell, P. W. Smith, G. W. J. Fleet, and A. D. Elbein, *Arch. Biophys. Biochem.*, 1985, **243**, 35.
230. B. Ganem and G. Papandreou, *J. Am. Chem. Soc.*, 1991, **113**, 8984.
231. Y. T. Pan, G. P. Kaushal, G. Papandreou, B. Ganem, and A. D. Elbein, *J. Biol. Chem.*, 1992, **267**, 8313.
232. A. D. Elbein, P. R. Dorling, K. Vosbeck, and M. Horisberger, *J. Biol. Chem.*, 1982, **257**, 1573.
233. D. P. Tulsiani, H. P. Broquist, L. F. James, and O. Touster, *Arch. Biochem. Biophys.*, 1984, **232**, 76.
234. M. S. Kang and A. D. Elbein, *J. Virol.*, 1983, **46**, 60.
235. R. G. Arumugham and M. L. Tanzer, *J. Biol. Chem.*, 1983, **258**, 11 883.
236. L. Foddy, J. Feeny, and R. C. Hughes, *Biochem. J.*, 1986, **233**, 697.
237. K.-N. Chung, V. L. Sheperd, and P. D. Stahl, *J. Biol. Chem.*, 1984, **259**, 14 637.
238. P. M. Schwartz and A. D. Elbein, *J. Biol. Chem.*, 1985, **260**, 14 452.
239. R. Merkle, A. D. Elbein, and A. Heifetz, *J. Biol. Chem.*, 1985, **260**, 1083.
240. Z. Bar-Shavit, A. J. Kahn, L. E. Pegg, K. R. Stone, and S. L. Teitelbaum, *J. Clin. Invest.*, 1984, **73**, 1277.
241. M. J. Humphries, K. Matsumoto, S. L. White, and K. Olden, *Proc. Natl. Acad. Sci. USA*, 1986, **83**, 1752.

3.08
Biosynthesis of Proteoglycans

THOMAS N. WIGHT

University of Washington, Seattle, WA, USA

3.08.1 INTRODUCTION

Proteoglycans are glycoconjugates that consist of a protein backbone to which three types of carbohydrate chains can be covalently attached. These three types of carbohydrates are: (i) N-linked oligosaccharides; (ii) O-linked oligosaccharides; and (iii) glycosaminoglycans (GAGs). The presence of GAGs attached to core protein separates this family from other glycoconjugates. A GAG is a linear polysaccharide consisting of N-sulfonylglucosamine (GlcNSO$_3$) or N-acetylglucosamine (GlcNAc) or N-acetylgalactosamine (GalNAc) residues alternating in glycosidic linkages with glucuronic acid (GlcUA), iduronic acid (IdUA), or galactose (Gal) residues. Specific disaccharide repeat patterns give rise to different types of unbranched GAGs and sizes of 20–40 kDa. The types of GAGs are chondroitin/chondroitin sulfate (CS), dermatan/dermatan sulfate (DS), heparin/heparan sulfate (HS), and keratan/keratan sulfate (KS). These GAGs are substituted to varying degrees with sulfate linked to the free amino and/or hydroxyl group of the hexosamine or to galactose for KS and to a lesser extent to hydroxyl groups in the uronic acids (Figure 1).[1-6] The GAGs exist covalently attached to a core protein through O-glycosidic linkage to the amino acid serine. Another GAG, hyaluronan (HA) is not covalently attached to protein. Usually one type of GAG predominates on a given core protein but hybrid proteoglycans do exist. All the GAGs are negatively charged by virtue of their sulfate and carboxyl groups with heparin being the most anionic substance found in living tissues. This feature of high negative charge distinguishes proteoglycans from other glycoconjugates.

Proteoglycans are present throughout the animal and plant kingdom. Their roles vary from forming critical structural elements and shaping tissues to influencing cell behaviors and regulating activities of enzymes, cytokines, and growth factors.[7-10] Thus, they possess both structural and metabolic roles in maintaining tissue homeostasis. These molecules are found in the extracellular

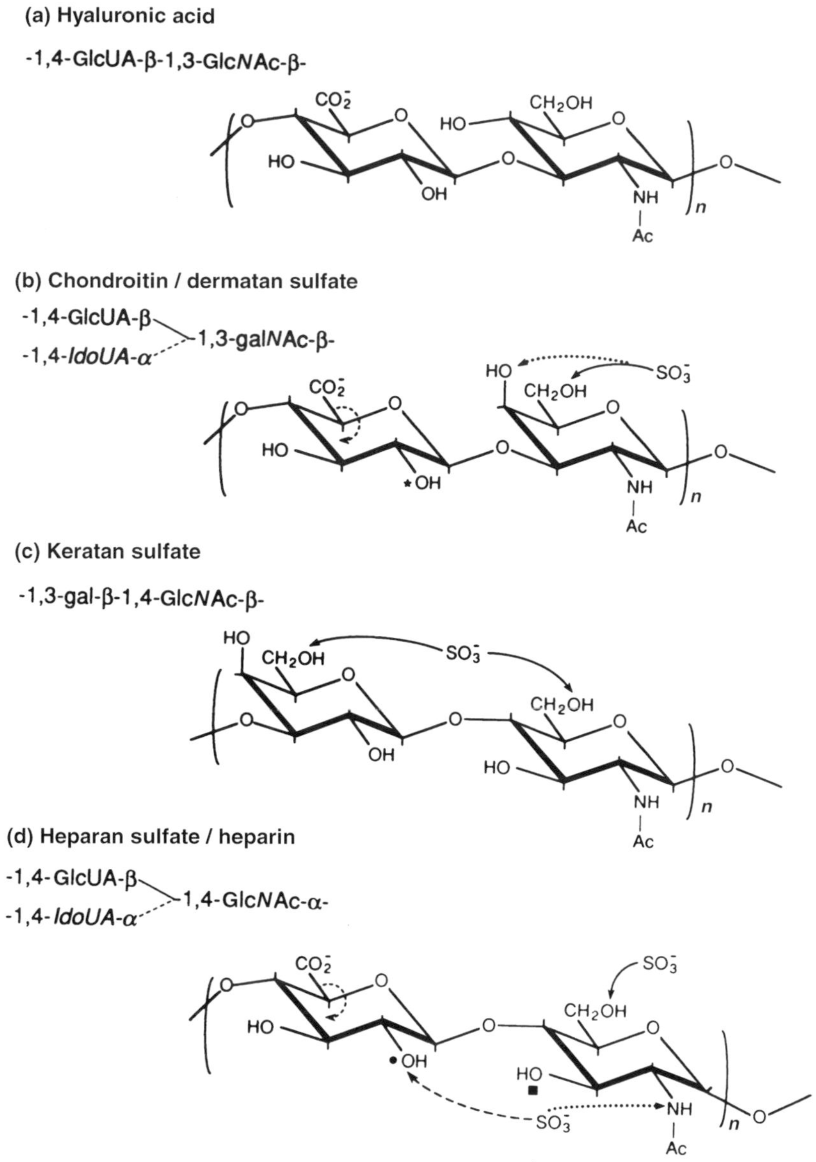

(a) Hyaluronic acid

-1,4-GlcUA-β-1,3-GlcNAc-β-

(b) Chondroitin / dermatan sulfate

-1,4-GlcUA-β
-1,4-*IdoUA-α*   -1,3-galNAc-β-

(c) Keratan sulfate

-1,3-gal-β-1,4-GlcNAc-β-

(d) Heparan sulfate / heparin

-1,4-GlcUA-β
-1,4-*IdoUA-α*   -1,4-GlcNAc-α-

Figure 1 Diagram showing the chemical structure of the different types of GAGs and the sites within the disaccharide repeat patterns that undergo modification by epimerases and sulfotransferases (reproduced by permission of Plenum Press from "Cell Biology of Extracellular Matrix," 1991).

matrix (ECM) and associated with specialized structures of the ECM such as basement membranes. They are also found as part of plasma membranes where they function as receptors and coreceptors influencing macromolecular uptake[10] as well as cellular cytokine and growth factor responses.[7-9] These molecules also occur intracellularly in cells of the immune system where they can act as "chaperones" for components in immune reactions (see reviews[5,6]). Abnormalities in the biosynthesis and turnover of these molecules lie at the basis of many diseases such as chondrodystrophies,[11,12] atherosclerosis,[13] different types of cancer,[14] Alzheimer's disease,[15] and macular corneal dystrophy[16] to cite a few.

Proteoglycan biosynthesis involves: (i) synthesis of core protein; (ii) xylosylation of specific serine residues in the core protein; (iii) addition of Gal to Xyl-Ser; (iv) addition of second Gal to Gal-Xyl-Ser; (v) addition of UA to Gal-Gal-Xyl-Ser to form the linkage tetrasaccharide of GlcUA-Gal-Gal-Xyl-Ser; (vi) repeat addition of hexosamines alternating with GlcUA or Gal; and (vii) modification of the growing chain by deacetylases, epimerases, and sulfotransferases. These reactions occur in both temporal and spatial patterns within the cell. Each step in their biosynthesis is controlled by particular genes which specify sets of proteins organized into complex multifunctional catalytic domains within membranes of the secretory apparatus of the cell.[1] This organization allows the sequential assembly of the molecule and involves the events of transcription, post-transcription, and post-translational processing. Regulation of proteoglycan biosynthesis can occur at any one or more of these steps.

The pathways for biosynthesis involve different compartments of the cell. Core protein synthesis and initial glycosylation take place in the endoplasmic reticulum (ER) with subsequent completion of glycosylation and GAG chain synthesis occurring in different parts of the Golgi. The completed proteoglycans are packaged into secretory vesicles to be either inserted into the plasma membrane via fusion of the secretory vesicle with this organelle or released into the ECM compartment[1] (Figure 2). A proportion of these proteoglycans inserted into the plasma membrane may be proteolytically cleaved and released[2,3] or recycled by endocytosis and degraded[3] (Figure 3). The *N*-linked and *O*-linked oligosaccharides are synthesized and added to the core protein as they are for glycoproteins and this aspect of oligosaccharide synthesis will be covered in Chapters 3.02, 3.03, and 3.04. This chapter will consider proteoglycan core protein and GAG biosynthesis and highlight more recent findings since excellent reviews have been previously written on this topic.[4,6,17,18]

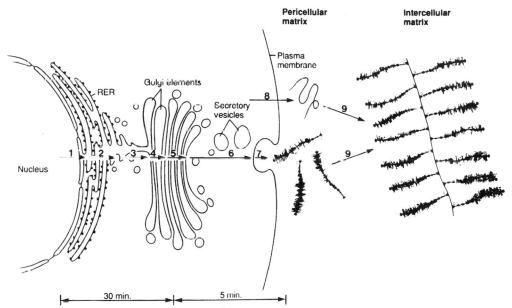

Figure 2 Diagram illustrating the synthesis of the proteoglycan aggrecan and the associated molecules, link protein and hyaluronan by a chondrocyte. Aggrecan and link protein are synthesized by routes common to secreted glycoproteins and involve (1) transcription of the aggrecan and link protein to form specific mRNAs, (2) translation of the mRNAs in the rough endoplasmic reticulum (RER) and intital xylosylation of the core protein, (3) movement of the core and link proteins from the RER to (4) the *cis* and (5) medial-*trans* Golgi compartments where the GAGs are synthesized and added to the core protein and modified by sulfotransferases, (6) packaged into secretory vesicles and transported to the plasma membrane, and (7) released into the extracellular matrix. (8) Hyaluronan which is a GAG that is not attached to a core glycoprotein is synthesized separately at the plasma membrane. Once outside of the cell, aggrecan, link protein (○), and hyaluronan interact to form a high-molecular weight aggregate that imparts compressive resilience to the tissue (reproduced by permission of Harwood Academic Publishers from "Extracellular Matrix," 1996).

3.08.2 NOMENCLATURE AND CLASSIFICATION

The nomenclature of these molecules has changed over the years.[19] First identified in the late nineteenth century in cartilage, they were known as "chondrogen" or "mucoprotein." In the 1930s to 1950s, the GAG nature of the molecules was defined and they were termed "acid mucopolysaccharides." Subsequently, it was recognized that the GAGs were attached to specific proteins and hence the name "proteoglycans" was adopted. Today, more than 30 different proteoglycans are known, each a product of separate genes and each containing a distinct complement of GAGs.[20–22] Before the cloning and sequencing of core proteins, these molecules were identified by the predominant type of GAG associated with the core protein and specified as CSPGs, DSPGs, KSPGs,

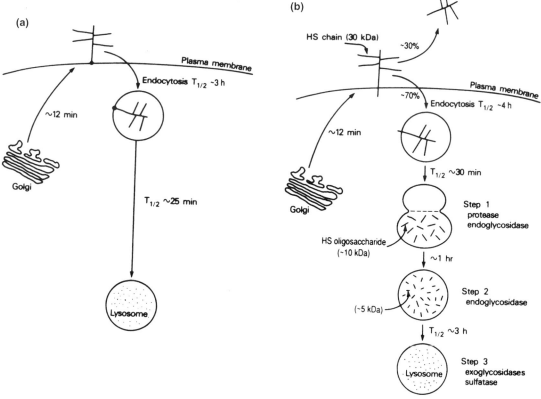

Figure 3 Model of the synthetic secretory pathway of the family of proteoglycans that associates with the plasma membrane of the cell. (a) Represents fate of GPI anchored proteoglycans while (b) depicts fate of those proteoglycans intercalated within the plasma membrane. The diagram shows the kinetic parameters of those proteoglycans involving transport from the Golgi to the cell surface, shedding to the extracellular space or endocytosis and degradation through a series of steps involved in the stepwise degradation of the proteoglycan (reproduced by permission of the American Society for Biochemistry and Molecular Biology from *J. Biol. Chem.*, 1992, **267**, 9451).

and HSPGs. However, with the identification of core protein sequence, it became apparent that common structural elements existed among the various core protein genes so that it has been possible to group the proteoglycans and their genes into distinct families. In addition, trivial names have been given to different proteoglycans based either on some aspect of their structure or some function they fulfill. Table 1 lists some of the proteoglycans whose core proteins have been cloned, their common names and some of their characteristics. In Table 1, they have been grouped as to where they are found in tissue. Table 2 lists some of the known proteoglycan gene families, separated according to similarities in gene and core protein structure.[20–22] Figure 4 depicts the structure of four different proteoglycans to illustrate the heterogeneity of this family of molecule.

3.08.3 CORE PROTEIN BIOSYNTHESIS

The structural characteristics of many proteoglycan genes have been elucidated and genomic organization of most of these genes indicates that they are modular and have utilized exon shuffling and duplication during evolution (see the review by Iozzo[21]). An interesting feature of different proteoglycans with similar properties, such as those that bind HA (aggrecan, versican, neurocan, and brevican) is that there is high conservation of exon/intron junctions in the two regions of these molecules that exhibit protein homology (i.e., the HA binding regions and the selectin-like domains). The hyaluronan binding region in these genes is encoded by four exons with identical conservation of exon size. In addition, introns flanking individual modules are capable of undergoing alternative splicing, thereby generating different isoforms of the core protein such as observed for the splicing of versican. This splicing takes place in the central coding region and involves two large exons that code for the GAG binding domains of the core protein. These two exons have been designated α

Table 1 Some common proteoglycans and their locations.

	Core protein	Type of GAG (no. of chains)	Location
Extracellular matrix			
aggrecan	220	CS (100)	cartilage
versican/PGM	265–370	CS (10–30)	most soft tissues
decorin	40	DS (1)	all connective tissues (collagen)
biglycan	40	DS (2)	all connective tissues
perlecan	467	HS/CS (3)	basement membrane
lumican	38	KS (3–4)	all connective tissues (collagen)
fibromodulin	42	KS (2–3)	all connective tissues (collagen)
Cell surface			
syndecans (1–4)	31–45	HS/CS (1–4)	epithelia and most soft tissues
glypicans (1–4)	60	HS (3)	epithelia and most soft tissues
NG-2	251	CS (2–3)	neural cells, embryonic SMC
betaglycan	110	CS/HS (1–2)	fibroblasts
Intracellular			
serglycin	10–19	HS/CS (10–15)	myeloid cells

PGM, proteoglycan (mesenchyme); SMC, smooth muscle cell.

Table 2 Some gene families of proteoglycans.

SLRPs (Small Leucine Rich Proteoglycans)
Decorin (DSPG)
Biglycan (DSPG)
Fibromodulin (KSPG)
Lumican (KSPG)
Epiphycan (DSPG)
Keratocan (KSPG)

Hyalectans/lecticans (hyaluronan binding proteoglycan)
Versican/PGM (CSPG)
Aggrecan (CSPG)
Neurocan (CSPG)
Brevican (CSPG)

Membrane spanning (integral membrane proteoglycans)
Syndecan 1 (syndecan) (HS/CS PG)
Syndecan 2 (fibroglycan) (HSPG)
Syndecan 3 (*N*-syndecan) (HSPG)
Syndecan 4 (ryudocan/amphiglycan) (HSPG)
NG-2 (melanoma associated protein) (CSPG)
Betaglycan (CS/HS PG)

Membrane Associated (GPI Anchored)
Glypican-1 (glypican) (HSPG)
Glypican-2 (cerebroglycan) (HSPG)
Glypican-3 (OCI-5) (HSPG)
Glypican-4 (K-glypican) (HSPG)

Basement Membrane
Perlecan (HSPG)
Bamacan (CSPG)

and β.[23] Such alternative splicing produces four variants of versican: V0 which contains both α and β GAG domains, V1 which contains the β GAG domain, V2 which contains the α GAG domain, and V3 which contains neither the α or β domains[23–25] (Figure 5). Whether the different isoforms of versican are differentially regulated awaits further study but evidence exists regarding tissue specific occurrence. For example, the V2 isoform of versican appears to be confined to nervous tissue,[26] whereas V1 predominates in blood vessels and other "soft tissues" and arterial smooth muscle cells.[27,28]

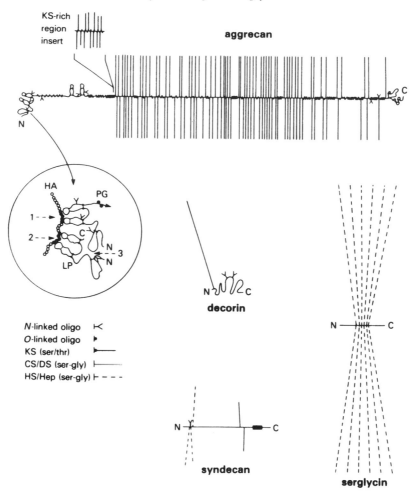

Figure 4 Diagram of the structure of four different proteoglycans demonstrating differences in the sizes of the core protein and in the number and sizes of the GAGs attached to the core protein (reproduced by permission of Plenum Press from "Cell Biology of Extracellular Matrix," 1991).

The genomic structures of other proteoglycan genes such as those of the SLRP gene family (Table 2) have also shown modular design with four domains which supports a gene duplicating theory of molecular evolution (see review by Iozzo[21]). Decorin and biglycan are two prototypic members of the SLRP gene family whose proteins exhibit similar modular design.[29,30] The proteins each contain a signal peptide and a propeptide which is unique to these proteoglycans.[31] The propeptide may play a role in enzyme recognition since transfection of mammalian cells with constructs containing deletions of the propeptide secrete proteoglycans with short GAG chains.[32] Furthermore, a recombinant biglycan that lacks the propeptide results in a core protein devoid of GAG chains.[33] The second domain of both decorin and biglycan is occupied by evenly spaced cysteine and GAG attachment sites. The third domain contains the leucine rich repeats while the fourth domain is characterized by a loop structure with two cysteine residues. This similarity in protein structure is repeated at the genomic level. Both proteoglycans are encoded by eight exons with similar intron/exon boundaries (see review, by Iozzo[21]).

The synthesis of core proteins of the proteoglycans is controlled by specific promoter elements in the mRNA. These elements in part regulate tissue-specific expression of specific types of proteoglycans as well as control the biosynthesis of core proteins in response to activation by cytokines and growth factors. All the proteoglycan promoters sequenced to date contain specific motifs that regulate transcription factor binding that either activates or "silences" proteoglycan gene transcription and core protein synthesis. Thus the human perlecan promoter has several AP2 binding sites, NFκB sites, TGF β responsive elements, all of which have been shown to be functionally active.[34,35] In fact, TGF-β1 induces both mRNA and protein levels for perlecan.[36] Furthermore, the

(a)

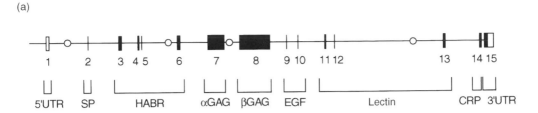

(b)

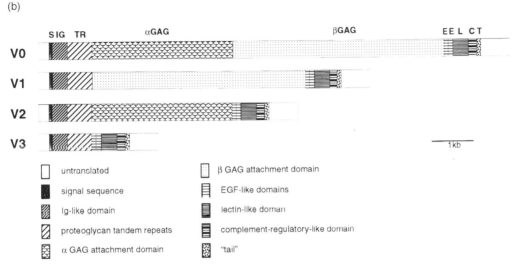

Figure 5 Domain structure of human versican isoforms. Alternative splicing generates four different mRNAs (V0, V1, V2, V3) leading to proteoglycan forms that differ in number of GAG chains. The genomic structure is shown in (a) (after Naso *et al.*[23]) and the isoforms are shown in (b) (after Lemire *et al.*[28]). Thus far, protein products have been demonstrated for the V0, V1, and V2 forms only.

expression of perlecan core protein is developmentally regulated. For example, the pattern of perlecan mRNA and protein expression correlates inversely with the degree of smooth muscle cell replication during rat aortic development.[37] In fact, perlecan expression correlates universally with the maturation of many tissues.[38]

Promoter sequences for the hyaluranan binding proteoglycans (hyalectans) are known. The versican and neurocan promoters both contain TATA boxes whereas the aggrecan and brevican promoters do not. Like the aggrecan promoter, the versican promoter contains multiple AP2 sites. However, the 150 bp promoter region immediately upstream of the transcriptional start site of the versican gene lacks SP1 or AP2 sites. Such differences suggest different mechanisms for transcriptional initiation of the aggrecan and versican genes.[23,39,40] However, both genes respond to similar activation in some cases. For example, transforming growth factor (TGF) β1 increases aggrecan mRNA in chondrocytes and versican mRNA in smooth muscle cells[41] while decreasing decorin mRNA in both cell types. On the other hand, both families of proteoglycan mRNAs are influenced oppositely by interleukin-1 (IL-1) in both cell types.[40,42,43] It may be that the regulation of core protein synthesis by the same cytokine in these two cell types involves differences between transcription and post-transcription processing.

The genes controlling synthesis of the major DS-containing proteoglycans, decorin and biglycan, have rather complex promoter regions.[21,30,44] For example, the decorin gene has two promoter regions that exhibit different activities, a proximal promoter region that contains tumor necrosis factor responsive elements and an AP1 site that functions both as a repressor of decorin gene expression in response to TNFα and as an inducer of decorin gene expression by IL-1.[21,44] The distal region of the promoter contains a number of *cis*-acting elements such as AP1, AP5, and NFκB and several repeats of TGF β negative elements. Although decorin and biglycan genes are structurally similar, their promoter elements are quite different. The biglycan gene does not contain TATA or

CAAT boxes but is enriched in G-C content. This gene appears more highly conserved than the decorin gene. It also contains numerous motifs that show binding to members of the Ets family of oncogenes. Such differences in gene structure suggest that the synthesis of decorin and biglycan core protein are regulated differently. For example, TGF-β1 downregulates decorin mRNA while upregulating biglycan mRNA[43] in arterial smooth muscle cells and in several different cells, indicating situations in which the two genes are oppositely regulated.[21] Interestingly, experiments examining the regulation of the human biglycan gene failed to identify transcriptional regulation by TGF-β1 in tumor cell lines suggesting post-transcriptional regulation for this gene in response to this cytokine.[45] Another feature of biglycan gene expression is the observation that it does not follow the rules of X chromosomal inactivation even though it is located on the X chromosome.[46] Moreover, additional Y chromosomes increase biglycan expression even though the biglycan gene is not present on the Y chromosome. Such results indicate that biglycan behaves as a psuedo-autosomal gene and that Y chromosomal factors may regulate transcriptional activity of this gene.

What is clear is that different cytokines regulate the level of mRNA for different proteoglycan core proteins in very specific ways and this regulation appears to be very cell type specific.[47,48] For example, in arterial smooth muscle cells, PDGF increases mRNA transcripts for versican and induces the synthesis of this proteoglycan but does not affect biglycan or decorin synthesis.[41,43] On the other hand, TGF-β1 increases both versican and biglycan core protein synthesis while decreasing decorin mRNA. IL-1 appears to have effects opposite to TGF-β1 by decreasing versican[49] and increasing decorin mRNA expression.[42] Interestingly, both PDGF and TGF-β1 cause elongation of GAG chains on versican, biglycan, and decorin in arterial smooth muscle cells[41,43] indicating that the selective effect that these cytokines have on core protein synthesis in these cells does not exist at the post-translational level. These results also indicate that regulation of core protein and GAG synthesis may be under separate signaling control mechanisms.

The manner in which cytokines and growth factors influence the cell's ability to synthesize specific molecules can be through the generation of specific secondary signals which lead to the activation of specific transcription factors and thus the activation of specific genes. For example, PDGF generates a series of secondary signals by binding to its receptor. Binding activates the receptor tyrosine kinase which phosphorylates the receptor and initiates a cascade of signals resulting in the division of the cell. Interestingly, some of the same signals control proteoglycan synthesis as well. For example, inhibiting tyrosine kinase activity by genestein inhibits versican synthesis induced by PDGF in arterial smooth muscle cells but does not affect the post-translational processing of the GAG chains.[50] Further downstream signals also appear to be critical since inhibition of protein kinase C and mitogen associated protein kinase kinase (MAPKK) also inhibits PDGF induction of versican synthesis.[51] However, blockage of MAPKK inhibits versican mRNA content but does not affect CS GAG chain elongation on versican induced by PDGF. Transfection of osteosarcoma cells with the biglycan promoter receptor genes following forskolin treatment shows increased transcriptional activity which could be blocked by an inhibitor of cAMP-dependent protein kinase.[52] Thus, for these cells, protein kinase A (PKA) appears to be involved in the regulation of biglycan expression. Inhibition of PKA in arterial smooth muscle cells had no effect on PDGF stimulated expression of versican.[51] Such results indicate that different signaling events control different aspects in the biosynthesis of these molecules and indicate a complex process of biosynthetic regulation.

The synthesis of the core protein takes part in the cisternae of the RER. It is clear that organelles involved in the biosynthesis and processing of proteoglycans such as the ER and Golgi are compartmentalized and may carry out very different aspects of proteoglycan biosynthesis. For example, the core protein precursor of aggrecan is found within a subcompartment of the ER in chondrocytes and separated from other secretory proteins such as collagen.[53,54] The lethal chicken mutation, nanomelia, is due to a mutation in the aggrecan gene that leads to a truncated core protein precursor that accumulates in the ER and is not secreted.[54] Interestingly, cell-free translation studies show that the mutant precursor was modified by *N*-linked oligosaccharide addition and by addition of xylose but was not further processed. Such studies indicate the need for translocation to the appropriate membrane compartment for completion of biosynthesis. Evidence for distinct subcompartments of secretory organelles involved in specific proteoglycan synthesis comes from studies using agents such as brefeldin A that disrupt translocation of the proteoglycan core proteins from the ER to the Golgi compartments.[55] Interestingly, chondrocytes treated with brefeldin A were unable to elongate GAG chains on aggrecan but could do so on decorin. This drug fuses the ER to part of the Golgi and separates reactions taking place in different parts of the Golgi. Such results suggest that different core proteins are segregated into discrete ER-Golgi compartments.

3.08.4 GLYCOSAMINOGLYCAN BIOSYNTHESIS

Chemical synthesis of GAGs is difficult owing to the large number of substitution sites on the monosaccharide building blocks and the problem of substrate selectivity and the formation of glycosidic bonds. Thus, the bulk of information regarding the biosynthesis of GAGs comes from studies using natural cell systems.

3.08.4.1 Chain Initiation

The synthesis of GAGs attached to the core protein takes place in the Golgi and the sugars that form the backbone of the GAGs come from the cell's cytoplasm. These sugars are converted to high energy intermediates as sugar nucleotides in the cytoplasm and require specialized transporter proteins to translocate them into the lumen of the ER and Golgi.[56,57] This transport takes place in such a fashion that a nucleotide monophosphate precursor moves out of the Golgi as a nucleotide diphosphate precursor enters, efficiently recycling the nucleotide precursors and providing sufficient amounts of activated sugars to meet the biosynthetic demand. The sugars are added to the core protein by a series of glycosyltransferases and modified by different sulfotransferases and epimerases (Figure 6). The specificity of these enzymes in large part regulates the structure of the GAG. It is also presumed that differential expression of these enzymes is key for the control of the biosynthesis of these molecules. However, at the time of writing few studies address the degree to which these specific glycosyltransferases are differentially expressed. However, the activities of these enzymes can be measured using appropriate substrates and a few studies have shown that glycosyltransferases and sulfotransferases are developmentally regulated.[58-60]

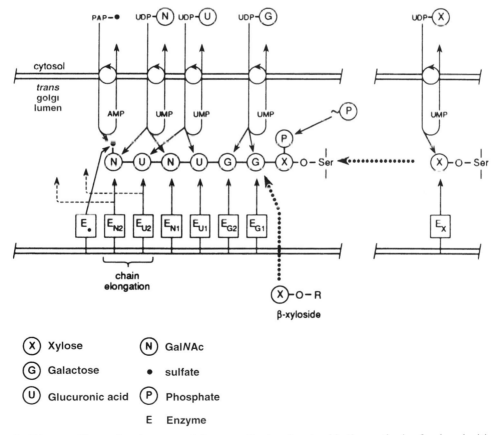

Figure 6 Diagram illustrating the sequential enzymatic steps involved in the synthesis of a chondroitin sulfate GAG (reproduced by permission of Plenum Press from "Cell Biology of Extracellular Matrix," 1991).

The first step in the biosynthesis of the GAG chain is to link xylose to serine residues in the core protein by xylosyltransferase using UDP-xylose as donor. While a preferred consensus sequence in the core protein does not exist for xylose addition, there are particular characteristics of amino acid sequence that govern xylose addition.[6,22] For example, the attachment site almost always has a Gly following Ser and at least two acidic amino acids as part of the amino acid clusters surrounding the Ser-Gly site. Xylosylation is an incomplete process and not all Ser-Gly repeat patterns with the above characteristics get xylosylated. Furthermore, site-directed mutagenesis studies show that converting glycine adjacent to serine to alanine has little effect on GAG assembly,[61] suggesting that xylosyltransferase may recognize serine residues in a particular conformation rather then in a specific sequence pattern. This enzyme is key in the biosynthetic initiation of all proteoglycans except KS. Mutant Chinese hamster ovary (CHO) cells defective in xylosyltransferase fail to synthesize CS, DS, and HS suggesting a common enzymatic pathway involved in the initiation of GAG synthesis for proteoglycans containing these GAGs (see review by Esko[22]). Thus, there may exist only one xylosyltransferase for CSPG, DSPG, and HSPG synthesis. Most evidence suggests that xylosylation begins in the RER and continues in the Golgi.[52,53,62,63] The xylose residue may be phosphorylated in a transient manner[64] and this appears to be a Golgi event and may play a role in directing the core protein to specific locations for GAG synthesis. However, there does not appear to be at this point any level of specificity since this xylose in CSPG, DSPG, and HSPG has been shown to be phosphorylated.[1]

The next step in the biosynthesis of the linkage region is the stepwise addition of two Gal residues from UDP-Gal by two specific membrane bound galactosyltransferases, appearing to lie in different locations in the *cis* and *trans* Golgi.[1,65,66] Like xylosyltranferase, CHO mutants defective in galactosyl transferase I do not synthesize CS, DS, HS indicating a common enzyme involvement in the synthesis of this portion of the linkage region.[22] The Gal residues in the linkage region may be sulfated to varying degrees at the 4- or 6- position. It is of interest that 6-sulfation of Gal is found predominantly as part of the linkage region of chondroitin-6-sulfate. Since the enzyme for 6-*O*-sulfation of Gal in KS is the same enzyme used for 6-sulfation of GalNAc in chondroitin (see below), it is likely that 6-sulfation of Gal in the linkage region may be catalyzed by the same enzyme.[60] The functional significance of this modification is not yet known.

The last step in the synthesis of the common tetrasaccharide linkage region in CS, DS, HS is addition of GlcUA from UDP-GlcUA by GlcUA transferase I. Substrate competition experiments have indicated that this enzyme is distinct from the enzyme GlcUA transferase II which is involved in the synthesis of CS polymer,[1,67,68] although both of these enzymes occur close together with core protein in distinct membrane subfractions distinct from Gal transferases I and II.[1,67–69] Success has been achieved in the cloning of glucuronyltransferase I and reveals that this enzyme is specific for Galβ1-3Galβ1-4Xylβ1-*O* Ser of the linkage region and separate from the polymerizing enzymes.[70] Such studies further support the notion that distinct enzyme complexes are organized in discrete subcompartments within secretory organelles of the cell. This step completes the linkage region which is common to CS, DS, and HS but not KS.

3.08.4.2 Chain Elongation

From here, the biosynthetic pathway diverges giving rise to different GAG structures, involving different transferase enzymes. These reactions occur by the addition of β-GalNAc to initiate CS synthesis, and the addition of αGlcNAc to initiate HS synthesis. Again, available evidence suggests that the enzyme that adds the first GalNAc or GlcNAc to the growing chain is distinct and different from the GalNAc and GlcNAc transferases involved in CS and HS polymerization.[1,22] What exactly controls whether the precursor core protein plus linkage region becomes CS or HS is not known but may involve specific enzyme recognition signals found within specific amino acid sequences in the core protein. For example, the addition of αGlcNAc (HS) usually involves core proteins that contain clusters of acidic amino acids near the HS attachment site.[22,71–73] Interestingly, recombinant perlecan/aggrecan chimeras add HS chains when domain I of perlecan is coupled to domain II and III of perlecan and expressed in COS cells, but if domain I of perlecan is coupled to G3 domain of aggrecan, CS chains are attached.[74] Thus utilization of the attachment sites for HS and CS can be influenced by non GAG bearing domains of the protein as well as conformation of the protein around the attachment site.[75] In fact, there are examples where the same core protein can be substituted with either HS chains or CS chains and this exists for serglycin in those cells of the hemopoietic lineage.[76] Such examples argue against core specific sequences in determining type of GAG chain attached on some proteoglycans in particular cell types.

3.08.4.2.1 *Chondroitin/dermatan sulfate*

Once the linkage region is complete, chain elongation requires two glycosyltransferases to add alternating residues of GlcUA and GalNAc.[1,77] These enzymes are membrane bound and there is some evidence that these enzymes may be located in different parts of the Golgi giving rise to different GSPGs. The chondroitin polymer is modified by sulfation at the 4- or 6-position of the GalNAc residues by the transfer of sulfate from adenosine 3′-phosphate 5′-phosphosulfate (PAPS). Like the activated sugars, PAPS is synthesized in the cytosol requiring ATP and transported into the Golgi by an antiport mechanism which exchanges one PAPS for one AMP. Two enzymes, GalNAc 4-*O*-sulfotransferase and GalNAc 6-*O*-sulfotransferase, add sulfate to the hexosamines. Available data suggest that sulfation occurs as the GAGs are actively growing rather than after completion of the chains.[1,60,78] The sulfotransferases appear to operate in an efficient manner and sulfation tends to occur in an all or none manner, probably because of the proximity of membrane embedded growing nascent proteoglycan and membrane embedded sulfotransferases. Success has been achieved in the purification and cloning of the GalNAc 6-*O*-sulfotransferase.[79] This enzyme is capable of transferring sulfate to chondroitin and keratan sulfate and contains a transmembrane domain similar to other glycosyltransferases and heparin/heparan sulfate *N*-sulfotransferase/*N*-deacetylases.[80,81] The polymerization process results in CS chains usually between 20–40 kDa although longer chains have been observed. Factors that determine the length of the growing chains are not understood. It has been suggested that chain length is, in part, controlled by the level of the sulfation reaction.[1,4,60] For example, sulfation appears to start prior to completion of the growing chain but proceeds faster than chain elongation. As sulfation reaches the end of the growing chain, the nonreducing terminal galactosamine will become sulfated. *In vitro* studies have shown that a 4-sulfated nonreducing terminal galactosamine functions poorly as an acceptor for GlcUA. The presence of 4-sulfate on a preterminal GalNAc with a terminal GlcUA has a similar effect in abolishing incorporation of GalNAc onto GlcUA. Thus, the incorporation of a 4-*O*-sulfate in these positions could be considered a mechanism by which to limit chain elongation.[4,60] Other factors such as steric events and association constants of the growing chain with the glycosyltransferases and sulfotransferases within the membrane may also be a consideration in dictating chain size.

It should be noted that chondroitin sulfates may be sulfated in positions other than the 4- or 6-position of hexosamine. For example, sulfate esters may exist on the C-2 of GlcUA. In addition, GalNAc may contain both 4-*O*- and 6-*O*-sulfates and these disulfated disaccharides are prominent in the secretory granules of mast cells but can be synthesized in other locations as well. Factors that regulate these altered sulfation patterns are poorly understood.[60]

Dermatan sulfate is synthesized by modifying the CS backbone by epimerizing some of the GlcUA to IdUA.[02,84] This epimerase, which appears specific for CS modification is tightly coupled to sulfation such that prevention of sulfation shifts the equilibrium in favor of GlcUA instead of IdUA. These results argue for close proximity of the epimerase with the sulfotransferases in the Golgi membranes. The C-5 epimerization and *O*-sulfation reactions are typically incomplete leaving unmodified regions giving rise to microheterogeneity in the DS polymer.[85–88] The degree of epimerization may vary between 1% and 90%. For example, in skin up to 90% of the hexuronic acids are converted to IdUA but only 35% in tendon. In addition, a variable proportion of the IdUA may become sulfated on C-2 whereas analogous sulfation of GlcUA only very occasionally occurs. Different contents of IdUA may have functional consequences as well. For example, DS chains high in IdUA self-associate to a greater extent than those low in IdUA and these chains also have a greater inhibitory activity on cell proliferation.[88] The precise subcellular location of the enzymes involved in converting CS to DS is not known. Treatment of fibroblasts with monensin which blocks transport through the medial part of the Golgi results in separation of polymerization and 6-*O*-sulfation from epimerization and 4-*O*-sulfation[60] indicating possible coupling of the activities of the epimerase and GalNAc 4-*O*-sulfotransferase, as mentioned above. Studies of decorin synthesis using brefeldin A have identified intermediates that may be products of different multienzyme complexes located in different parts of the synthetic machinery and catalyze the building of a limited section of the chain at a time.[86] Most of these data are taken from experiments which disrupt the normal secretory pathways, thus creating artefactual mixing of different cellular compartments and therefore should be interpreted with caution. However, such studies have been useful in the identification of biosynthetic intermediates and in determining the existence of core protein specific pathways regulating post-translational processing.

The two major DS-containing proteoglycans, decorin and biglycan, when produced by the same cell have GAGs whose chain composition is similar and may even be identical.[89,90] Such similarities in structure suggest that these two proteoglycans may travel similar routes in any given tissue during post-translational processing and future studies will be needed to determine whether compositional differences exist in the GAG chains among different DSPGs synthesized by different cells. The glucuronosyl and iduronosyl residues in DS exist in different conformations and promote distinct secondary structural differences. Such differences may lead to differences in affinities of the different DS chains for other ligands. The conversion of decorin core protein to decorin proteoglycan bearing one dermatan sulfate chain occurs with a half life of 12 min in fibroblasts indicating a rapid and coordinated action of the glycosyltransferases, glucuronsyl epimerase, and the sulfotransferases.[1,60,87] Clearly, the uronosyl C-5 epimerase acts as a key regulator in the synthesis of DS chains. Since epimerization of GlcUA to IdUA is followed by 4-*O*-sulfation of the adjacent GalNAc, both of these enzymes are tightly coupled. This is not true of the 6-*O*-sulfotransferases so GAGs that contain predominantly GalNac 6-*O*-sulfate lack IdUA although there are some exceptions. Interestingly, monensin, which blocks the secretory route to the medial part of the Golgi has only minute effects on 6-*O*-sulfation of decorin but dramatically reduces epimerization and 4-*O*-sulfation suggesting that 6-*O*-sulfation occurs in an earlier Golgi compartment.[1,60,87]

3.08.4.2.2 *Heparin/heparan sulfate*

Heparin and heparan sulfate differ in content of sulfated residues and percent composition of IdUA so their biosynthesis should involve the same steps with differences of degree in the modifications of the growing GAG chains. Heparin is produced exclusively by mast cells and the heparin chains are attached to a Ser-Gly enriched protein termed serglycin (Tables 1, 2 and Figure 4). Heparan sulfate proteoglycans are produced by virtually all cells and are present on different core proteins and constitute four major families of proteoglycans: perlecan, syndecans, betaglycan, and glypicans.[91]

Polymerization and modification of heparin/HS chains follow the same pattern as that of chondrotin/dermatan but with more modification steps and the transfer of GlcNAc instead of GalNAc to the tetrasaccharide linkage region.[1,4,17,18,91,92] Thus, the heparin/HS GAG chain is formed by alternating transfer of GlcUA and GlcNAc monosaccharide units from the corresponding UDP-sugar nucleotides to the nonreducing termini of the nascent chains by a copolymerase (i.e., a 70 kDa enzyme that can promote the two transferase reactions[93]). There are a number of modifications that occur in the growing chain and these modifications have spatial and temporal patterns (Figure 7). The sequence in which the modifications occur has been elegantly described.[18] The first modification is *N*-deacetylation and *N*-sulfation of GlcNAc catalyzed by a single 110 kDa enzyme.[94–98] This enzyme appears to occur in two forms, one form associated with the synthesis of heparin and found in mast cells while the other form is associated with the synthesis of HS. Both enzymes have been cloned[94–96] and are produced by separate and distinct genes. Thus, transfection of an HS producing cell line with cDNA encoding the mast cell enzyme produces highly *N*-sulfated heparin like HS.[97] This enzyme reacts with some GlcNAc residues clustered along the chain to remove acetyl groups and replace them with *N*-sulfate groups in such a way as to spread in both directions to create blocks of modified GlcNAc separated by large blocks of unmodified GlcNAc residues.[91] An epimerase then acts on GlcUA residues adjacent to the GlcNS to create IdUA.[99] The growing chain is sulfated by a series of reactions catalyzed by different sulfotransferases.[100–103] Following epimerization, IdUA is sulfated at the 2-position and then sulfate is added to the 6-OH of the GlcN residues adjacent to the uronic acid. These sulfotransferase reactions are believed to be catalyzed by different enzymes and may be differentially regulated during HS biosynthesis. Next, certain sulfated substituted sugars and uronic acids are sulfated at the C-3 by one of the 3-*O*-sulfotransferase isozymes.[99] For example 3-OST-1 sulfates GlcNS residues when the upstream uronic acid is GlcUA whereas 3-OST-2 and 3-OST-3 sulfate GlcNS residues when the upstream uronic acid is IdUA2S or GlcUA2S. Such enzyme specificities give rise to polymers with defined structures within the chains and these specific blocks appear to have selected affinities for particular ligands.[91,92]

All of these reactions appear to occur in the same region of the Golgi so that the modifying enzymes may all exist as part of a supramolecular membrane Golgi complex. Since HS chains differ more between cell type than between core proteins expressed by the same cell type, each cell may possess a specific pattern of biosynthetic enzymes that determine specific HS sequence and structure.

Once HS biosynthesis is complete, the proteoglycan and/or the heparin chains may have several fates. In the case of heparin proteoglycan in the mast cell, an endo-β-D-glucuronidase cleaves the heparin chains attached to the core protein into heparin fragments of 5–25 kDa and these fragments are packaged into secretory granules to be released with other mast cell granule constituents such

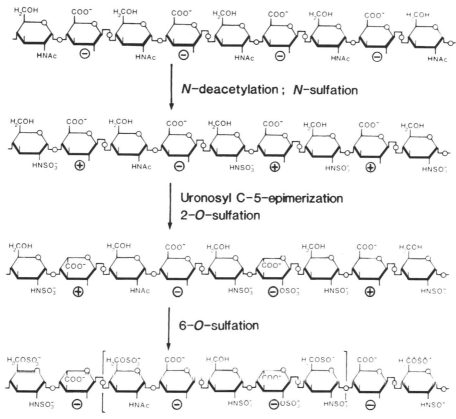

Figure 7 Diagram illustrating the stepwise modification involved in the biosynthesis of heparin/HS. The substrate specificity of the uronosyl-C-5-epimerase is illustrated by those uronic acid residues that act as substrates for the enzyme (+) and those that do not (−). This scheme does not show the 3-*O*-sulfate reaction. However, with the success of the cloning of the 3-*O*-sulfotransferases, the substrate specificities for these enzymes should be forthcoming.[91] Interestingly, it is the 3-*O*-sulfate on the glucosamine that provides the pentasaccharide sequence shown in brackets with anti-thrombin binding activity (see review by Rosenberg *et al.*[91]) (reproduced by permission of Academic Press from "Biology of Extracellular Matrix," 1987).

as histamine.[104,105] In the case of HSPGs, these macromolecules are transferred to the cell surface to be either inserted into the plasma membrane as is the case for syndecan or released to become incorporated into specialized structures such as basement membrane, as is the case for perlecan. Membrane HSPGs such as the syndecans may be shed by proteolytic cleavage of dibasic residues near the membrane spanning domain of the core protein or rerouted back into the cytoplasm by endocytosis to be catabolized and recycled.[2,3]

3.08.4.2.3 *Keratan sulfate*

Keratan sulfate is a sulfated polylactosamine of the type found in glycoproteins and mucins.[106] However, it resembles the other GAGs by the backbone structure of alternating β1,3 and β1,4 bonds. The position of the hexuronic acid in the other GAGs is occupied by GlcNAc in KS and that of the hexosamine by galactose (see Figure 1). The nonsulfated backbone of KS which is present on a number of glycoproteins is referred to as lactosaminoglycans. In KSPGs, the lactosaminoglycans are normally sulfated on the 6-position of either or both sugars. Two types of KSPGs exist depending upon their linkage to the core protein. KSI which is prominent in the cornea is linked through *N*-glycosylamine bonds to asparagine residues in the core protein in a mannose-containing linkage oligosaccharide. This oligosaccharide is identical to biantennary oligosaccharides found in complex-type glycoproteins, with KS extending one branch of the oligosaccharide and sialic acid terminating the second branch. Thus, synthesis of this type of linkage occurs through the same pathways involving dolichol (lipid intermediate precursor) as outlined for *N*-linked sugar addition in the synthesis of glycoproteins. This is the main KS in the cornea and is part of at least three KSPGs, fibromodulin, lumican, and keratocan.[21,107] These are members of the small leucine-rich proteoglycans (SLRPs). Chain elongation involves the glycosyl transfer of galactose and GlcNAc. UDP galactose is an essential precursor. For example, cell mutants that have deficiency in the transport of UDP-galactose into Golgi vesicles fail to synthesize KS but do synthesize CS and HS.[57] UDP-galactose is formed from UDP-glucose and requires UDP-glucose-4-epimerase. UDP-glucose is used for either the formation of UDP-glucuronic acid for CS synthesis or UDP-galactose for KS synthesis. The enzyme UDP-glucose dehydrogenase which permits CS synthesis is inhibited by NADH so that NAD/NADH levels are also important regulators of KS biosynthesis.[6]

The mode of biosynthesis of corneal KSPG is similar to that of serum glycoproteins, especially in the linkage region since tunicamycin inhibits KS synthesis.[4-6] The Golgi membrane fractions of corneal cells contain a KS acetylglucosaminyl transferase and galactosyltransferase activity and sulfation is catalyzed by at least two sulfotransferases, GlcNAc 6-*O*-sulfotransferase and Gal 6-*O*-sulfotransferase.[108,109] The Gal 6-*O*-sulfotransferase from cornea is specific for KS and does not act on chondroitin. In fact, the activity of these polymerizing enzymes decreases with increasing molecular mass and sulfation degree, suggesting a feedback in which degree of sulfation may dictate chain termination. Thus, these enzymes are key regulators of KS biosynthesis.

The second type of KS is linked to core protein via a linkage to Ser/Thr as in the mucins and follows rules outlined for the synthesis of mucins. This form, found in skeletal muscle is frequently referred to as KS II.

3.08.4.2.4 *Hyaluronan*

Hyaluronan is a unique GAG since it has the characteristics of a repeat disaccharide pattern of a hexosamine (GlcNAc) and a hexuronic acid (GlcUA) but it does not involve any type of linkage to a glycoprotein but interacts with a number of different molecules in the ECM.[110-112] However, chain elongation (i.e., polymerization) occurs like the other GAGs and involves the formation of GlcNAc β1–3 GlcUAβ1–4 repeated several times, catalyzed by the enzyme HA synthase (HAS). What is not clear is what primes this reaction since core proteins for HA do not exist. The HA synthases have been cloned and represent a family of enzymes.[113-115] What is very unique is that the biosynthesis of HA does not take place in the Golgi where other GAGs are synthesized but occurs in the plasma membrane.[116,117] Such a location would allow unconstrained polymer growth and newly synthesized HA can reach molecular weights of 10^6 and several microns in length. Another unique feature of HA biosynthesis is that chain elongation occurs at the reducing end of the molecule rather than the nonreducing end, as for other GAGs. Thus, HA is elongated at its reducing end by adding UDP-*N*-acetylglucosamine with displacement of UDP from UDP-glucuronosyl. The elongating molecule appears to be extruded directly into the ECM and may associate with the surface of the cell through HA receptors such as RHAMM and/or CD44. Such associations create a pericellular coat enriched in HA which may facilitate cell proliferation and migration.[111,112,118] Thus HA synthesis is upregulated when cells are stimulated to migrate and/or proliferate.[111,112]

Considerable success has been achieved in the identification and cloning of different HAS. For example, three mammalian HAS genes have been identified along with a related *Xenopus laevis* gene (DG42) and a gene sequence in a virus that infects chlorella-like green algae[113-115] that code for the HA synthetic enzyme. The three mammalian gene products display homology to bacterial HAS from *Streptococcus pyogenes* which is the enzyme responsible for the synthesis of the capsular coat of HA in bacteria. There is high identity between the bacterial and eukaryotic HAS genes suggesting a common ancestral gene. Interestingly the different HAS genes are differentially expressed and have related but distinct enzymatic properties. Thus, HAS1 is expressed early in development while HAS2 and 3 are more prominent in later stages of development. HAS2 and HAS3 appear to synthesize HA of different sizes, HAS2 producing a much longer HA than HAS3 when transfected in COS-1 cells.[113] The common predicted structural features shared by all the HAS proteins indicate several membrane spanning domains, suggesting that the enzyme is embedded within the membrane of several sites.

A number of growth factors and cytokines such as PDGF, bFGF, and TGFβ stimulate HA synthesis in a variety of cells but the mechanism(s) responsible for this stimulation have not been determined. With the successful cloning of the HAS genes, it should now be possible to determine whether selective activation of specific elements in the different HAS genes control HA biosynthesis in these cells. The well documented importance of HA in cellular proliferation and migration and in tissue development makes this approach an exciting one.

3.08.5 CONCLUDING REMARKS

Defining the biosynthesis of proteoglycans has been a long and arduous task involving protein and carbohydrate biochemists, enzymologists, and molecular biologists. The research began by building an understanding of the complexities of GAG structure and synthesis and, with the advent of molecular biology, moved into defining mechanisms of core protein synthesis. Investigators of these marvelous molecules are now trying to piece these two aspects of biosynthetic pathways together to better understand the factors that regulate the production of these molecules. Proteoglycan biosynthesis involves thousands of reactions that occur in a stepwise manner, each step dependent upon the previous one. What is remarkable is that the cell can carry out these reactions with making but a few mistakes. Rapid progress is expected because of advances in protein and carbohydrate sequencing and successful cloning and identification of the enzymes involved in their biosynthesis.

ACKNOWLEDGMENTS

This work was supported by NIH Grants 18645 and 11251. The author would like to thank Ms. Alyssa Stephenson for the typing of the manuscript.

3.08.6 REFERENCES

1. J. E. Silbert and G. Sugumaran, *Biochim. Biophys. Acta*, 1995, **1241**, 371.
2. M. Bernfield, R. Kokenyesi, M. Kato, M. T. Hinkes, J. Spring, R. L. Gallo, and E. J. Lose, *Annu. Rev. Cell Biol.*, 1992, **8**, 365.
3. M. Yanagishita and V. C. Hascall, *J. Biol. Chem.*, 1992, **267**, 9451.
4. D. Heinegård and M. Paulsson, in "Extracellular Matrix Biochemistry," eds. K. A. Piez and A. Reddi, Elsevier Science, New York, 1984, p. 277.
5. T. N. Wight, D. K. Heinegård, and V. C. Hascall, in "Cell Biology of Extracellular Matrix," ed. E. D. Hay, Plenum Press, New York, 1991, p. 45.
6. V. C. Hascall, D. K. Heinegård, and T. N. Wight, in "Cell Biology of Extracellular Matrix," ed. E. D. Hay, Plenum Press, New York, 1991, p. 149.
7. J. T. Gallagher, *Curr. Opin. Cell Biol.*, 1989, **1**, 1201.
8. E. Ruoslahti and Y. Yamaguchi, *Cell*, 1991, **64**, 867.
9. T. N. Wight, M. G. Kinsella, and E. E. Qwarnstrom, *Curr. Opin. Cell Biol.*, 1992, **4**, 793.
10. K. J. Williams and I. V. Fuki, *Curr. Opin. Lipidol.*, 1997, **8**, 253.
11. S. A. Bingel, R. D. Sande, and T. N. Wight, *Lab. Invest.*, 1985, **53**, 479.
12. A. Superti-Furga, J. Hastbacka, A. Rossi, J. J. van der Harten, W. R. Wilcox, D. H. Cohn, D. L. Rimoin, B. Steinmann, E. S. Lander, and R. Gitzelmann, *Ann. N.Y. Acad. Sci.*, 1996, **785**, 195.
13. T. N. Wight, in "Atherosclerosis and Coronary Artery Disease," eds. V. Fuster, R. Ross, and E. J. Topol, Raven Publishers, Philadelphia, PA, 1996, p. 421.
14. R. V. Iozzo, *Cancer Metastasis Rev.*, 1988, **7**, 39.
15. A. D. Snow and T. N. Wight, *Neurobiol. Aging*, 1989, **10**, 481.
16. J. R. Hassell, N. Sundar Raj, C. Cintron, R. Midura, and V. C. Hascall, in "Keratan Sulfate—Chemistry, Biology, Chemical Pathology," eds. H. Greiling and J. E. Scott, The Biochemical Society, London, 1989, p. 215.
17. L. Rodèn, in "The Biochemistry of Glycoproteins and Proteoglycans," ed. W. J. Lennarz, Plenum Press, New York, 1980, p. 267.
18. U. Lindahl and L. Kjellen, in "Biology of Proteoglycans," eds. T. N. Wight and R. P. Mecham, Academic Press, Orlando, FL, 1987, p. 59.
19. M. Yanagishita, *Experientia*, 1993, **49**, 366.
20. R. V. Iozzo and A. D. Murdoch, *FASEB J.*, 1996, **10**, 598.
21. R. V. Iozzo, *Annu. Rev. Biochem.*, 1998, **67**, 609.
22. J. D. Esko, *Curr. Opin. Cell Biol.*, 1991, **3**, 805.
23. M. F. Naso, D. R. Zimmermann, and R. V. Iozzo, *J. Biol. Chem.*, 1994, **269**, 32 999.
24. M. T. Dours-Zimmermann and D. R. Zimmermann, *J. Biol. Chem.*, 1994, **269**, 32 992.
25. M. Zako, T. Shinomura, M. Ujita, K. Ito, and K. Kimata, *J. Biol. Chem.*, 1995, **270**, 3914.
26. W. Paulus, I. Baur, M. T. Dours-Zimmermann, and D. R. Zimmermann, *J. Neuropathol. Exp. Neurol.*, 1996, **55**, 528.
27. L. Y. Yao, C. Moody, E. Schönherr, T. N. Wight, and L. J. Sandell, *Matrix Biol.*, 1994, **14**, 213.
28. J. Lemire, K. R. Braun, P. Maurel, E. Kaplan, S. M. Schwartz, and T. N. Wight, *Arterioscler. Thromb. Vasc. Biol.*, in press.
29. K. G. Danielson, A. Fazzio, I. Cohen, L. A. Cannizzaro, I. Eichstetter, and R. V. Iozzo, *Genomics*, 1993, **15**, 146.
30. L. W. Fisher, A. M. Heegaard, U. Vetter, W. Vogel, W. Just, J. D. Termine, and M. F. Young, *J. Biol. Chem.*, 1991, **266**, 14 371.
31. R. S. Sawhney, T. M. Hering, and L. J. Sandell, *J. Biol. Chem.*, 1991, **266**, 9231.
32. Å. Oldberg, P. Antonsson, J. Moses, and L. Å. Fransson, *FEBS Lett.*, 1996, **386**, 29.
33. A. M. Hocking and D. J. McQuillan, *Glycobiology*, 1996, **6**, 717.
34. I. R. Cohen, S. Grassel, A. D. Murdoch, and R. V. Iozzo, *Proc. Natl. Acad. Sci. USA*, 1993, **90**, 10 404.

35. R. V. Iozzo, J. Pillarisetti, B. Sharma, A. D. Murdoch, K. G. Danielson, J. Uitto, and A. Mauviel, *J. Biol. Chem.*, 1997, **272**, 5219.
36. G. R. Dodge, I. Kovalszky, J. R. Hassell, and R. V. Iozzo, *J. Biol. Chem.*, 1990, **265**, 18 023.
37. M. C. Weiser, J. K. Belknap, S. S. Grieshaber, M. G. Kinsella, and R. A. Majack, *Matrix Biol.*, 1996, **15**, 331.
38. M. Handler, P. D. Yurchenco, and R. V. Iozzo, *Dev. Dyn.*, 1997, **210**, 130.
39. H. Li and N. B. Schwartz, *J. Mol. Evol.*, 1995, **41**, 878.
40. W. B. Valhmu, G. D. Palmer, J. Dobson, S. G. Fischer, and A. Ratcliffe, *J. Biol. Chem.*, 1998, **273**, 6196.
41. E. Schönherr, H. T. Jarvelainen, L. J. Sandell, and T. N. Wight, *J. Biol. Chem.*, 1991, **266**, 17 640.
42. J. Lemire, C. Tsoi, E. Qwarnström, and T. N. Wight, *J. Biol. Chem.*, submitted for publication, 1998.
43. E. Schönherr, H. T. Jarvelainen, M. G. Kinsella, L. J. Sandell, and T. N. Wight, *Arterioscler. Thromb.*, 1993, **13**, 1026.
44. M. Santra, K. G. Danielson, and R. V. Iozzo, *J. Biol. Chem.*, 1994, **269**, 579.
45. H. Ungefroren and N. B. Krull, *J. Biol. Chem.*, 1996, **271**, 15 787.
46. C. Geerkens, U. Vetter, W. Just *et al.*, *Hum. Genet.*, 1995, **96**, 44.
47. E. J. Kovacs and L. A. DiPietro, *FASEB J.*, 1994, **8**, 854.
48. J. J. Nietfeld, O. Huber-Bruning, and J. W. J. Bÿlsma, in "Proteoglycans," ed. P. Jolles, Birkhauser Verlag, Basel, 1994, p. 215.
49. E. E. Qwarnström, H. T. Jarvelainen, M. G. Kinsella, C. O. Ostberg, L. J. Sandell, R. C. Page, and T. N. Wight, *Biochem J.*, 1993, **294**, 613.
50. E. Schönherr, M. G. Kinsella, and T. N. Wight, *Arch. Biochem. Biophys.*, 1997, **339**, 353.
51. L. Cardoso, K. Bornfeldt, E. Qwarnström, and T. N. Wight, *J. Biol. Chem.*, submitted for publication 1998.
52. H. Ungefroren, T. Cikos, N. B. Krull, and H. Kalthoff, *Biochem. Biophys. Res. Commun.*, 1997, **235**, 413.
53. B. M. Vertel, A. Velasco, S. LaFrance, L. Walters, and K. Kaczman-Daniel, *J. Cell Biol.*, 1989, **109**, 1827.
54. B. M. Vertel, B. L. Grier, H. Li, and N. B. Schwartz, *Biochem. J.*, 1994, **301**, 211.
55. S. Wong-Palms and A. H. Plaas, *Arch. Biochem. Biophys.*, 1995, **319**, 383.
56. C. B. Hirschberg and M. D. Snider, *Annu. Rev. Biochem.*, 1987, **56**, 63.
57. L. Toma, M. A. Pinhal, C. P. Dietrich, H. B. Nader, and C. B. Hirschberg, *J. Biol. Chem.*, 1996, **271**, 3897.
58. H. Kitagawa, M. Ujikawa, and K. Sugahara, *J. Biol. Chem.*, 1996, **271**, 6583.
59. H. Kitagawa, K. Tsutsumi, Y. Tone, and K. Sugahara, *J. Biol. Chem.*, 1997, **272**, 31 377.
60. J. E. Silbert, *Glycoconjugate J.*, 1996, **13**, 907.
61. D. M. Mann, Y. Yamaguchi, M. A. Bourdon, and E. Ruoslahti, *J. Biol. Chem.*, 1990, **265**, 5317.
62. H. P. Hoffmann, N. B. Schwartz, L. Rodèn, and D. J. Prockop, *Connect. Tissue Res.*, 1984, **12**, 151.
63. B. M. Vertel, L. M. Walters, N. Flay, A. E. Kearns, and N. B. Schwartz, *J. Biol. Chem.*, 1993, **268**, 11 105.
64. J. Moses, Å. Oldberg, F. Cheng, and L. Å. Fransson, *Eur. J. Biochem.*, 1997, **248**, 521.
65. G. Sugumaran and J. E. Silbert, *J. Biol. Chem.*, 1991, **266**, 9565.
66. G. Sugumaran, M. Katsman, and J. E. Silbert, *J. Biol. Chem.*, 1992, **267**, 8802.
67. T. Helting and L. Rodèn. *J. Biol. Chem.*, 1969, **244**, 2799.
68. A. Brandt, J. Distler, and G. W. Jourdian, *Proc. Natl. Acad. Sci. USA*, 1969, **64**, 374.
69. G. Sugumaran, M. Katsman, and J. E. Silbert, *Biochem J*, 1998, **329**, 203.
70. H. Kitagawa, Y. Tone, J. Tamura, K. W. Neumann, T. Ogawa, S. Oka, T. Kawasaki, and K. Sugahara, *J. Biol. Chem.*, 1998, **273**, 6615.
71. L. Zhang and J. D. Esko, *J. Biol. Chem.*, 1994, **269**, 19 295.
72. J. D. Esko and L. Zhang, *Curr. Opin. Struct. Biol.*, 1996, **6**, 663.
73. M. Dolan, T. Horchar, B. Rigatti, and J. R. Hassell, *J. Biol. Chem.*, 1997, **272**, 4316.
74. K. Doege, X. Chen, P. K. Cornuet, and J. Hassell, *Matrix Biol.*, 1997, **16**, 211.
75. M. Costell, K. Mann, Y. Yamada, and R. Timpl, *Eur. J. Biochem.*, 1997, **243**, 115.
76. R. V. Tantravahi, R. L. Stevens, K. F. Austen, and J. H. Weiss, *Proc. Natl. Acad. Sci. USA*, 1986, **83**, 9207.
77. M. E. Richmond, S. DeLuca, and J. E. Silbert, *Biochemistry*, 1973, **12**, 3904.
78. G. Sugumaran and J. E. Silbert, *J. Biol. Chem.*, 1990, **265**, 18 284.
79. M. Fukuda, K. Uchimura, K Nakashima, M. Kato, K. Kimata, T. Shinomura, and O. Habuchi, *J. Biol. Chem.*, 1995, **270**, 18 575.
80. O. Habuchi, Y. Matsui, Y. Kotoya, Y. Aoyama, Y. Yasuda, and M. Noda, *J. Biol. Chem.*, 1993, **268**, 21 968.
81. G. Sugumaran, M. Katsman, and R. R. Drake, *J. Biol. Chem.*, 1995, **270**, 22 483.
82. Å. Malmstrom and L. Å. Fransson, *J. Biol. Chem.*, 1975, **250**, 3419.
83. Å. Malmstrom, *Biochem J.*, 1981, **198**, 669.
84. J. E. Silbert, M. E. Palmer, D. E. Humphries, and C. K. Silbert, *J. Biol. Chem.*, 1986, **261**, 13 397.
85. R. J. Linhardt and R. E. Hileman, *Gen. Pharmacol.*, 1995, **26**, 443.
86. J. Moses, A. Oldberg, E. Eklund, and L. Å. Fransson, *Eur. J. Biochem.*, 1997, **248**, 767.
87. H. Kresse, H. Hausser, E. Schönherr, and K. Bittner, *Eur. J. Clin. Chem. Clin. Biochem.*, 1994, **32**, 259.
88. H. Kresse, H. Hausser, and E. Schönherr, *Experientia*, 1993, **49**, 403.
89. H. U. Choi, T. L. Johnson, S. Pal, L. H. Tang, L. Rosenberg, and P. J. Neame, *J. Biol. Chem.*, 1989, **264**, 2876.
90. L. Å. Fransson, A. Schmidtchen, L. Coster, and Å. Malmstrom, *Glycoconjugate J.*, 1991, **8**, 108.
91. R. D. Rosenberg, N. W. Shworak, J. Liu, J. J. Schwartz, and L. Zhang, *J. Clin. Invest.*, 1997, **99**, 2062.
92. M. Salmivirta, K. Lidholt, and U. Lindahl, *FASEB J.*, 1996, **10**, 1270.
93. T. Lind, U. Lindahl, and K. Lidholt, *J. Biol. Chem.*, 1993, **268**, 20 705.
94. I. Eriksson, D. Sandback, B. Ek, U. Lindahl, and L. Kjellen, *J. Biol. Chem.*, 1994, **269**, 10 438.
95. Y. Hashimoto, A. Orellana, G. Gil, and C. B. Hirschberg, *J. Biol. Chem.*, 1992, **267**, 15 744.
96. A. Orellana, C. B. Hirschberg, Z. Wei, S. J. Swiedler, and M. Ishihara, *J. Biol. Chem.*, 1994, **269**, 2270.
97. W. F. Cheung, I. Eriksson, M. Kusche-Gullberg, U. Lindahl, and L. Kjellen, *Biochemistry*, 1996, **35**, 5250.
98. Z. Wei, S. J. Swiedler, M. Ishihara, A. Orellana, and C. B. Hirschberg, *Proc. Natl. Acad. Sci. USA*, 1993, **90**, 3885.
99. P. Campbell, H. H. Hannesson, D. Sandback, L. Rodèn, U. Lindahl, and J. P. Li, *J. Biol. Chem.*, 1994, **269**, 26 953.
100. M. Kobayashi, H. Habuchi, O. Habuchi, M. Saito, and K. Kimata, *J. Biol. Chem.*, 1996, **271**, 7645.
101. H. Habuchi, O. Habuchi, and K. Kimata, *J. Biol. Chem.*, 1995, **270**, 4172.
102. J. Liu, N. W. Shworak, L. M. S. Fritze, J. M. Edelberg, and R. D. Rosenberg, *J. Biol. Chem.*, 1996, **271**, 27 072.

103. H. Wlad, M. Maccarana, I. Eriksson, L. Kjellen, and U. Lindahl, *J. Biol. Chem.*, 1994, **269**, 24 538.

104. Å. Oldberg, C. H. Heldin, A. Wasteson, C. Busch, and M. Hook, *Biochemistry*, 1980, **19**, 5755.

105. L. Thunberg, G. Backstrom, A. Wasteson, H. C. Robinson, S. Ogren, and U. Lindahl, *J. Biol. Chem.*, 1982, **257**, 10 278.

106. H. Greiling, in "Proteoglycans," ed. P. Jolles, Birkhauser Verlag, Basel, 1994, p. 101.

107. J. Funderburgh, in "Proteoglycans: Structure, Biology and Molecular Interactions," ed. R. V. Iozzo, Marcel-Dekker, New York, 1998, in press.

108. M. Fukuta, J. Inazawa, T. Torii, K. Tsuzuki, E. Shimada, and O. Habuchi, *J. Biol. Chem.*, 1997, **272**, 32 321.

109. S. Degroote, J. M. Lo-Guidice, G. Strecker, M. P. Ducourouble, P. Roussel, and G. Lamblin, *J. Biol. Chem.*, 1997, **272**, 29 493.

110. J. R. E. Fraser and T. C. Laurent, in "Extracellular Matrix 2: Molecular Components and Interactions," ed. W. D. Comper, Harwood Academic Publications, Amsterdam, 1996, p. 141.

111. B. P. Toole, in "Cell Biology of Extracellular Matrix," ed. E. D. Hay, Plenum, New York, 1991, p. 305.

112. C. B. Knudson and W. Knudson, *FASEB J.*, 1993, **7**, 1233.

113. A. P. Spicer and J. A. McDonald, *J. Biol. Chem.*, 1998, **273**, 1923.

114. P. H. Weigel, V. C. Hascall, and M. Tammi, *J. Biol. Chem.*, 1997, **272**, 13 997.

115. P. L. DeAngelis, W. Jing, M. V. Graves, D. E. Burbank, and J. L. Van Etten, *Science*, 1997, **278**, 1800.

116. P. Prehm, *Biochem J.*, 1983, **211**, 181.

117. P. Prehm, *Biochem J.*, 1983, **211**, 191.

118. S. Evanko, J. Angelo, and T. N. Wight, *Atheroscler. Thromb. Vasc. Biol.*, 1998, in press.

3.09
Lipopolysaccharides

UWE MAMAT, ULRICH SEYDEL, DIETER GRIMMECKE,
OTTO HOLST, and ERNST TH. RIETSCHEL
Forschungszentrum Borstel, Germany

3.09.1 INTRODUCTION

Gram-negative bacteria, which include many human pathogens such as *Escherichia coli*, *Haemophilus influenzae*, *Vibrio cholerae*, *Pseudomonas aeruginosa*, and *Chlamydia pneumoniae*, express at their surface various amphiphilic structures including the capsular antigens, lipoproteins, the enterobacterial common antigen, and the lipopolysaccharides (LPS). Of these macromolecules, LPS are of particular microbiological, immunological, and medical significance. Bacterial mutants with defects in early steps of LPS biosynthesis are not viable. Therefore, it appears that LPS are essential for bacterial survival, their vital role being based on their participation in the proper organization and function of the bacterial outer membrane. Within this outer membrane, LPS are assembled in the outer leaflet and are, therefore, located at the interface between the bacterial cell and its aqueous environment. Thus, LPS represent the main surface antigens (O-antigens) of Gram-negative bacteria. In their membrane-associated and exposed location they are targets for bacteriophages, they harbor binding sites for antibodies and nonimmunoglobulin serum factors, and, thus, are involved in the specific recognition and elimination of bacteria by the host organism's defense systems. On the other hand, LPS may function to prevent the activation of complement and uptake of bacteria by phagocytic cells and, by shielding pathogens from cellular host defenses, they play an important role in bacterial virulence.

Administration of isolated LPS or LPS released from dividing or blebbing bacteria expresses in higher organisms a broad spectrum of toxic activities such as pyrogenicity, tachypnoe, tachycardia, hypotension, and irreversible shock. To emphasize these activities, LPS have been termed endotoxins. Because of their endotoxic properties, LPS contribute to the pathogenic potential of Gram-negative bacteria, and they are known as major factors involved in the disastrous manifestations and clinical consequences of severe Gram-negative infections and generalized inflammation. Finally, LPS activate B- and T-lymphocytes, granulocytes, and mononuclear cells and, hence, are potent immunostimulators. By virtue of this property they also seem to be involved in certain physiological host–parasite interactions.

In view of their fascinating and diverse spectrum of pathological and physiological activities, as well as their important function in bacteria, LPS have been studied in many laboratories using genetic, biological, immunological, chemical, and physical approaches. In fact, considerable progress has been made in the understanding of the pathogenesis of local and generalized inflammatory reactions induced by Gram-negative bacteria and, in particular, of the role of endotoxins in such processes. Thus, the endotoxic principle of LPS was identified, its primary structure was elucidated, it was chemically synthesized, and quantitative relationships between its constitution and bioactivity were established. In addition, the *in vivo* mechanisms of endotoxin bioactivity involving mononuclear cells and monocyte-derived bioactive peptide and lipid mediators as well as activated oxygen and nitroxide species and the intracellular regulation of mediator production have been studied in some detail. Also, the molecular basis of the primary steps of the interaction between endotoxin and the host cell, which involves humoral and cellular endotoxin binding proteins and receptors, and a particular conformation of endotoxins and which initiates the cascade of events leading to endotoxic manifestations, have been characterized at the molecular level. Finally, the genetic determination and biosynthetic pathways of LPS have been largely elucidated, as have the parameters determining the capacity of LPS to form membranes.

Despite the enormous progress made, many questions have not been answered. In this chapter, advances in understanding of the biosynthesis, primary structure, three-dimensional conformation, and biological activity of endotoxins are discussed. Of the considerable literature available, original publications and summarizing overviews mainly from the 1980s and 1990s are cited.

3.09.2 BIOLOGY OF ENDOTOXIN (LIPOPOLYSACCHARIDE)

The mechanisms involved in the biological activity of endotoxin are now largely understood.[1] When Gram-negative bacteria multiply, shed membrane fragments, elongate under the influence of antibiotics, or die on exposure to antibiotics or complement, endotoxin is released in the form of free LPS or complexed to the outer membrane protein A (ompA).[2] Circulating endotoxins constitute a particular class of toxins, which induce in host organisms the production of bioactive mediators ultimately responsible for the effects observed during endotoxemia. For these effects, the lipid A component constitutes the essential LPS component. LPS (or lipid A), after associating with certain serum factors, is either neutralized or interacts with receptors expressed by endotoxin target cells such as granulocytes, lymphocytes, vascular cells and, in particular, monocytes/macrophages. In response to LPS, these cells form and secrete endogenous mediators, which are endowed with potent intrinsic bioactivities and which ultimately induce the typical endotoxin effects potentially leading to the clinical picture of septic shock.

3.09.2.1 Humoral LPS Targets and Cellular Binding Molecules

Prominent among the humoral factors interacting with LPS is high density lipoprotein (HDL), which is able, like low density lipoprotein, to attenuate LPS effects.[3] The LPS-neutralizing properties are shared by bactericidal permeability-increasing protein (BPI) and sCD14, a soluble form of the membrane-linked LPS receptor mCD14.[4,5] With regard to the expression of endotoxicity the most important serum protein appears to be the LPS binding protein (LBP), which dramatically augments LPS and lipid A activity, rendering femto- to picogram amounts of LPS bioactive.[6]

LBP is synthesized in hepatocytes as a glycosylated 58 kDa protein, which is constitutively secreted into the blood stream. The concentration of LBP in normal serum is ~ 14–$22\ \mu g\ mL^{-1}$ with a maximum of up to $200\ \mu g\ mL^{-1}$ in acute-phase serum 24 h after induction. LBP was first purified from rabbit and human plasma and later from murine sources.[7,8] Rabbit and human LBP have been sequenced and 69% similarity was revealed between the two species. The domain responsible for LPS binding is located in the N-terminal region. It is characterized by an accumulation of positively charged amino acids and expresses hydrophobic properties. LBP shares 44% sequence similarity with human BPI, the lipid A binding region of which is also located in the N-terminal region.[9] A putative LPS-binding domain, characterized by the presence of positively charged amino acids and located in the proximity of a hydrophobic domain is also present in the structurally well-characterized endotoxin-neutralizing protein (ENP) naturally present in the hemolymph of the horseshoe crab *Limulus polyphemus*.[10] ENP, like BPI does not (as LBP) augment but rather inhibits LPS bioactivity. These different properties of the three LPS-binding proteins may depend on structural features of their carboxy terminus.

The biological activity of LBP is based on its capacity to transport lipids carrying negative charges such as phosphatidylserine and LPS. LBP (like BPI) interacts with LPS aggregates to release single molecules or small aggregates and to transport these to membrane systems of target cells.[11,12] LPB may thus be regarded as a biological amplifier, which enables host organisms to detect small amounts of LPS that signal the invasion of Gram-negative bacteria, i.e., infection. The host organism therefore uses LPS/LBP complexes to activate its defense system to deal appropriately with the invading microorganism. It has been demonstrated that LBP catalytically transfers LPS not only to responsive cells, i.e., the CD14 receptor (see below), but also to HDL, thus catalyzing endotoxin neutralization.[13] Since gene-deficient mice, which do not express LBP, are relatively LPS-resistant, it appears that the augmenting rather than the detoxifying LBP-mediated mechanisms prevail.[14]

The particular significance of CD14 as a primary LPS receptor was recognized in 1990.[15] It was then observed that CD14 serves as an LPS-binding site provided LPS has been allowed to complex to LBP. The importance of CD14 is underlined by the observation that certain anti-CD14 mAbs inhibit LPS/LBP complex-induced TNF (tumor necrosis factor) production and protein tyrosine phosphorylation and that certain cell types transfected with CD14 become highly reactive to LPS/LBP complexes.[16] LPS binding takes place via the lipid A component and is of high affinity ($K_p = 3 \times 10^{-8}$ M).[17] The binding domain of CD14 has been initially defined to reside between amino acids 57 and 64, but later studies point to a region comprising amino acids 39 to 44.[18]

CD14 is not a transmembrane molecule, but is linked to the cell surface via a GPI anchor.[19] As this type of membrane anchor does not allow direct signal transduction, the mechanism of LPS/LBP-induced and CD14-mediated cell activation is not yet understood. One possible model suggests that individual LPS-molecules are guided to cells by LBP, where they first associate with CD14, which subsequently mediates the interaction of LPS with a second, so far unknown receptor, which can transduce a signal to the cellular interior.

CD11c/CD18 has been demonstrated to function as a transmembrane signaling receptor for endotoxin, which, however, appears to be operative in the absence of CD14.[20] In looking for a CD14-dependent functional LPS receptor a ligand blotting assay was used to investigate the binding of LPS to membrane proteins of the human monocytic cell line Mono-Mac-6. Among membrane proteins, an 80 kDa protein was identified, which binds LPS or free lipid A only in the presence of serum.[21] Subsequent experiments identified the serum factors mediating binding of lipid A to the 80 kDa membrane protein as sCD14 and LBP. Thus, the 80 kDa protein fulfills an important prerequisite of a putative signal transducing molecule: it recognizes LPS or lipid A only in the context of LBP/CD14. The 80 kDa membrane protein is also present in membrane preparations of human peripheral blood monocytes and endothelial cells.

CD14 also exists in a soluble form (sCD14). As such, it is present in the circulation at a concentration of 2–6 μg mL^{-1}. In normal serum, several types of sCD14 (48 kDa, 53 kDa, 55 kDa) are present which either result from shedding of membrane-bound CD14 or from cellular production of GPI-free CD14 forms. It has been demonstrated that sCD14 is capable of interacting directly with LPS.[22] The sCD14/LPS complex is capable of binding to CD14-negative cells such as endothelial cells and of activating these to produce cytokines.[23,24]

Studies show that activation of monocytes by certain other bacterial immunomodulators such as peptidoglycan, mannuronan, and lipoarabinomannan also proceeds via mCD14.[25–30] The exact structural requirements of the glycoconjugate for binding to mCD14, cell activation, and cytokine secretion are not presently understood. It appears, however, that D-*gluco* or D-*manno* configurated glycosyl residues are involved indicating that CD14 may represent a lectin.[31,32] Monocyte activation by glycosphingolipids of *Sphingomonas paucimobilis*, which share chemical and physical features with LPS, is not dependent on mCD14.

3.09.2.2 Cellular Targets of LPS

The most important target cells of endotoxin are components of the cellular immune system. Thus, defense cells of almost all species have the ability to recognize minute amounts of LPS, thereby sensing invading microorganisms. Four cell types can be distinguished which recognize LPS but respond to it in different ways, i.e., by phagocytosis, differentiation, proliferation, and mediator secretion.

Polymorphonuclear leukocytes (PMN) take up bacteria and bacterial membrane fragments including LPS, and their phagocytic capacity is greatly enhanced by LPS.[33] As phagocytosis may be regarded as a first and direct measure of the host to eliminate threatening microorganisms, the phagocytosis-augmenting property of LPS emphasizes its important role in early nonspecific steps of the host defense against microbial invasion. On the other hand, PMN have enzymes which degrade (i.e., de-*O*-acylate and dephosphorylate) LPS and lipid A to nontoxic partial structures.[34] Furthermore, PMN contain polycationic proteins, which, like BPI, are capable of interacting with LPS. Finally, LPS-activated PMN may attach to endothelial cells and, by causing damage to the endothelial lining and by penetrating through the vessel wall into the tissue, may contribute substantially to endotoxin-induced inflammatory reactions.

B-lymphocytes of murine origin are stimulated by LPS to proliferation, differentiation, and secretion of antibodies.[35] This polyclonal activation may also be regarded as an early defense mechanism of the host against pathogenic microorganisms as it yields antibodies of various antibacterial specificities. Also T-lymphocytes (TH1-type) of human origin are, in a monocyte-dependent fashion, activated by LPS to proliferate and to secrete lymphokines, in particular interferon (IFN) γ.[36] In addition, LPS acts on a subset of murine T-lymphocytes (CD8$^+$ CD4$^-$) which are capable of suppressing the humoral immune response to bacterial polysaccharides such as pneumococcal type III polysaccharide. Lipid A has been shown to downregulate these T-suppressor cells, thereby augmenting antipolysaccharide antibody formation.[37]

Monocytes and tissue macrophages are activated by LPS to produce a large variety of bioactive protein mediators, which include interleukin (IL)-1, IL-6, IL-8, IL-10, IL-12, macrophage migration-inhibitory factor, and, in particular, tumor necrosis factor α (TNF).[38,39] Many host cells carry receptors for these mediators and are capable of responding to them, for example by enhanced activity, chemotaxis, or by apoptosis. If produced in small amounts, these monokines help to eliminate and inactivate invading microorganisms, for example by causing moderate fever, inducing leukocytosis, attracting defense cells to the infectious focus, activating intracellular microbicidal mechanisms, and initiating an acute-phase response.[40] If overproduced, however, these hormone-like mediators become a threat to the host organism by causing damage to cells and organs.[41] Thus, multiorgan failure and irreversible shock may be the result of severe Gram-negative infection resulting from overwhelming mediator production. On exposure to endotoxin, macrophages produce, in addition to monokines, reduced oxygen species (superoxide anion, hydrogen peroxide, hydroxy radical, and nitric oxide), bioactive metabolites of arachidonic acid (prostaglandins (PG), thromboxane, and leukotrienes (LT)) and of linoleic acid (e.g., (*S*)-13-hydroxylinoleic acid), and platelet-activating factor (PAF).

Finally, vascular cells such as endothelial or smooth muscle cells have the capacity to produce, upon stimulation by LPS, proteinaceous mediators such as the cytokines IL-1, IL-6, and IL-8. These cells also form a variety of other mediators, including prostacyclin, nitric oxide, PAF, interferons, and colony-stimulating factors.[42,43] In addition to mediator release, the expression of adhesion molecules is of great importance in the regulation of inflammatory responses. The expression of adhesion molecules, leading to adherence of PMNs, is induced by LPS as well as by IL-1 and TNF.[44]

The molecular events initiated *in vivo* after injection of endotoxin or after its release from bacteria can be summarized in the following way (Figure 1). That portion of LPS which is not detoxified by humoral (e.g., BPI, sCD14, or HDL) or cellular (e.g., PMN) host components interacts with LBP, and/or sCD14 and subsequently activates target cells including monocytes/macrophages, endothelial cells, granulocytes, and lymphocytes to produce and release endogenous mediators or to express adhesion molecules. Prominent mediators such as TNF, IL-1, IL-6, IL-8, IL-12, IFNγ, and MIF are formed which are capable of activating susceptible (i.e., cytokine receptor carrying) cells to produce secondary mediators such as PAF, LT, and PG, reduced oxygen species, nitric oxide, and proteases such as elastase and collagenase to induce cellular adherence and to call defense cells to the site of infection. If this inflammatory reaction cascade is limited and if it is confined to a local area, it is highly beneficial as it helps to activate the defense system and to destroy invading microorganisms. If, however, these mediators are overproduced and released into the circulation, systemic inflammatory reactions and clinically relevant situations such as septic shock may result. Finally, it should be emphasized that the host is capable of counterbalancing this hyperinflammatory response by the formation of antiphlogistic mediators such as IL-10, TGFb, and PGE2. Hyper-inflammation may, therefore, be followed by a phase of hyporesponsiveness of the immune system during which bacteria, due to the absence of immunostimulatory cytokines, are not prevented from growth and multiplication and maintain their capacity for inducing toxic reactions. Septic shock, therefore, may be regarded as the result of a vicious cycle of alternating over- and underproduction of mediators rather than the detrimental consequence of a unidirectional mediator cascade.

3.09.3 CHEMISTRY OF LIPOPOLYSACCHARIDE

3.09.3.1 General Architecture of Lipopolysaccharide

LPS of various Gram-negative bacteria are made up according to the same architectural principle (Figure 2). They consist of a polysaccharide portion and a covalently bound lipid component, termed lipid A. In the classical case of enterobacterial wild-type LPS, the polysaccharide component consists of two regions represented by the O-specific polysaccharide chain and the core oligo-saccharide, which can be further divided into an inner and an outer core. These regions differ in their chemical structure, degree of structural conservation, principles of biosynthesis, and genetic determination.

The lipid A component of enterobacteria and many other Gram-negative bacteria consists of a disaccharide, 2-amino-2-deoxy-6-O-(2-amino-2-deoxy-4-phospho-β-D-glucopyranosyl)-α-D-gluco-pyranose 1-phosphate [4-P-GlcN-(1→6)-GlcN-1→P] substituted by N- and O-bound fatty acids. The presence of fatty acids in lipid A appears to be essential for anchoring the molecule in the outer leaflet of the outer membrane, where LPS replaces the phospholipids usually found in biological membranes.

The core oligosaccharide and O-specific chain together represent the polysaccharide domain of LPS. Varying numbers of repeating units in the O-polysaccharide are one of the reasons for molecular variability of LPS, which may be expressed in a ladder-like banding pattern revealed in polyacrylamide gel electrophoresis (PAGE) of sodium dodecyl sulfate (SDS)- or deoxycholate-containing gels, often visualized using staining with silver nitrate. Distances between bands charac-terize the size (i.e., number of glycosyl residues) of repeating units. The size of a repeating unit may vary from one monosaccharide (e.g., *Legionella pneumophila* O:1) to eight (*Hafnia alvei*),[45,46] and up to 60 different monosaccharides were identified as constituents.[47–49] Furthermore, modification by amidation, acylation, esterification, phosphorylation, and etherification contributes to structural variation of the O-specific chains.[47,49]

Certain groups of Gram-negative bacteria have lost the ability to produce parts of the poly-saccharide chain of LPS. Bacterial colonies of, e.g., *Salmonella enterica* and *E. coli* strains, on agar plates may visually change from smooth to rough phenotype due to loss of the O-specific polysaccharide. LPS of smooth colonies, possessing a complete core and an extended polysaccharide

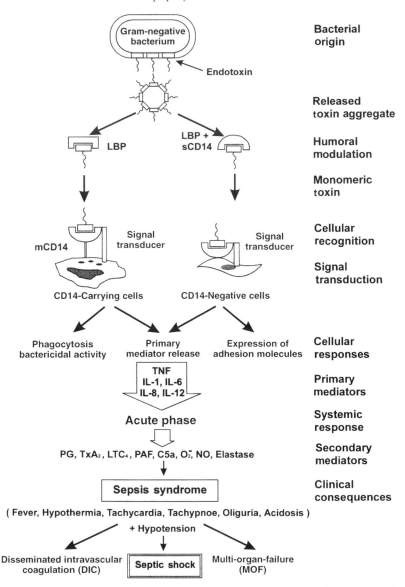

Figure 1 Humoral and cellular pathways involved in endotoxin-induced septic shock.

chain, is termed smooth (S)-form LPS and those with one repeating unit semirough (SR)-LPS. Those LPS lacking the O-specific chain and having a complete or truncated core oligosaccharides are termed rough (R)-form LPS.

Polydispersity of the O-specific chain is controlled genetically and may result in populations of molecules of LPS having one repeating unit, e.g., in the *S. enterica* serovar Typhimurium mutant SH777 (SR-form),[50,51] *E. coli* O:111,[52] *Shigella sonnei* phase I PhI2,[53,54] and *V. cholerae* O:139.[55,56] In S-form LPS, the number of repeats varies considerably. For *E. coli* O:111, the degree of polymerization of the repeating units has been determined by chemical analysis and by counting bands on SDS–PAGE analysis to be 1–40.[52] In *S. enterica* O-specific chains, the number of repeating units was found to range from 1 to more than 30,[57,58] whereas the average degree of polymerization varies in the range from 3 to 8.[59,60] However, enterobacterial LPS may also contain populations of molecules not being substituted by O-specific chains, and the regulatory mechanism controlling the R-/S-form ratio is not understood at present. Hence, enterobacteria and a large number of other Gram-negative bacteria may have both, R-form LPS built up from lipid A and core oligosaccharide, and S-form LPS consisting of lipid A, core oligosaccharide, and O-specific polysaccharide.[61] As shown for *E. coli* and *S. enterica*, the fatty acid content of lipid A of R-form LPS is higher than that of S-form LPS.[61]

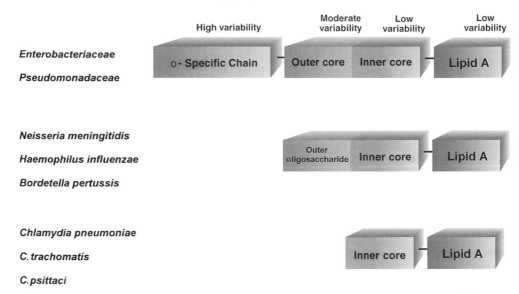

Figure 2 Schematic architecture of LPS of various Gram-negative bacteria. High variability in mono-saccharide constituents, nature of glycosidic bonds, size of repeating units, and chain length has been observed for the O-specific chain. The outer core exhibits a moderate variability, while the inner core region and the lipid A domain represent the structurally most conserved part of the LPS molecule. A variety of non-enterobacterial wild-type strains form LPS which lack the O-specific chain resembling LPS of enterobacterial rough-(R)-mutants.

As shown in Figure 2, certain bacteria, comprising human mucosal pathogens, express LPS lacking an O-specific polysaccharide. Among these bacteria are important pathogens like *Neisseria meningitidis, Neisseria gonorrhoeae, H. influenzae, Bordetella pertussis*, and *Chlamydia* spp., the core of the latter LPS consisting only of lipid A and an oligosaccharide of 3-deoxy-D-*manno*-octulopyranosonic acid (Kdo).

Among the numerous genera of Gram-negative bacteria, the Enterobacteriaceae and some pathogenic species have been most intensively investigated. Common classification schemes differentiate on a serological basis according to O-, K- and H-antigen specificity, resulting from different structures of the O-specific side chain of LPS, capsular polysaccharides (CPS), and flagellar proteins, respectively. Comprehensive schemes like that of *E. coli*,[62,63] the Kauffmann–White scheme for *Salmonella*,[64] and those for *Pseudomonas* (Habs, Fischer, Lanyi, and IATS) cover all pathogenic species of the respective genera and have been permanently extended by new sero-types.[65–67] Furthermore, progress made by microbiological, biochemical, and genetic research, as well as structural data from chemical and immunochemical investigation improves the schemes continuously.

3.09.3.2 The O-Specific Chains and Core Oligosaccharides

3.09.3.2.1 Escherichia coli

Diseases associated with *E. coli* cover a wide spectrum of clinical manifestations, pathogenic and epidemiological characteristics. In developing countries, *E. coli* is the major cause of intestinal infections associated with high childhood morbidity and mortality. In contrast, in developed countries *E. coli* more often is the causative agent of community-acquired and nosocomial extraintestinal infections. Currently, intestinal *E. coli* infections are classified into four distinct groups: entero-toxigenic (ETEC), enteroinvasive (EIEC), enteropathogenic (EPEC), and enterohemorrhagic (EHEC) *E. coli* strains. Among extraintestinal infections, *E. coli* is the most frequent etiologic agent of urinary tract infections, neonatal menigitis, and bacteremia.

In order to differentiate intestinal and extraintestinal pathogenic *E. coli*, serotyping based on the somatic O, capsular K, and flagellar H antigens is used. The O-specific chain represents a heteropolymeric polysaccharide chain and usually acts as a strong immunogen inducing high titers of O-specific antibodies. The species *E. coli* exhibits more than 150 different serotypes expressing different structures of O-specific chains, some of which (e.g., O:1, O:9, O:18, O:20) are divided into subgroups.[63] Differentiation by serotyping of bacteria having S-form LPS is based on structural differences in the carbohydrate structures of the repeating unit of the O-side chain and on covalently linked noncarbohydrate substituents such as *O*-alkyl, *O*-acyl, or *N*-acyl residues and phosphodiester.[49] Most of the O-specific polysaccharides of *E. coli* and those of *Shigella* (see below) are hexosaminoglycans, e.g., *E. coli* O:7, O:16, and O:111 (Table 1).

Table 1   Structure of repeating units of *Escherichia coli* O-specific chains.

Serotype	Structure[a]	Ref.
O:7	→ 3)-α-D-GlcpNAc-(1 → 3)-β-D-Quip4NAc-(1 → 2)-α-D-Manp-(1 → 4)-β-D-Galp-(1 →	68
O:8	3-*O*-Me-α-D-Manp-(1 → [3)-β-D-Manp-(1 → 2)-α-D-Manp-(1 → 2)-α-D-Manp-(1 →]$_n$	69
O:9	→ 3)-α-D-Manp-(1 → 3)-α-D-Manp-(1 → [2)-α-D-Manp-(1 →]$_n$ $n = 2$ for *E. coli* O:9a $n = 3$ for *E. coli* O:9	70–72
O:16	→ 2)-β-D-Galf-(1 → 6)-α-D-Glcp-(1 → 3)-α-L-Rhap-(1 → 3)-α-D-GlcpNAc-(1 → 2 | OAc	73, 74
O:16 (*E. coli* K-12)	→ 2)-β-D-Galf-(1 → 6)-α-D-Glcp-(1 → 3)-α-L-Rhap-(1 → 3)-α-D-GlcpNAc-(1 → 2 6 | ↑ OAc 1 α-D-Glcp	74
O:111	α-Colp 1 ↓ 6 → 3)-β-D-GalpNAc-(1 → 4)-α-D-Glcp-(1 → 4)-α-D-Galp-(1 → 3 ↑ 1 α-Colp	75

[a]In all tables, schematic structures for O-antigenic polysaccharides core oligosaccharides are given in the extended system according to IUPAC–IUBMB recommendations.[76] In addition, substitutions with phosphate groups are not shown.

There are also exceptions such as *E. coli* O:8 and O:9, which constitute unbranched homopolymers built up of D-Man containing only α-(1→2)- and α-(1→3)-linkages differing in the sequence of the linkages (Table 1).[77] It is important to note that the nonreducing terminus of the O:8 chain was identified as 3-*O*-methyl-α-D-mannose.[69] The biological importance of the *O*-methylation at the distal terminus of *E. coli* O:8 remains unclear. The O-chains O:8 and O:9 in combination with defined CPS are frequently found in ETEC strains causing endemic diarrhea.[63]

E. coli O:7 and O:16 strains belong to those microorganisms which cause neonatal meningitis. Furthermore, *E. coli* O:7 causes urinary tract infections and bacteremia. However, no structural or functional relationships of the O:7 and O:16 O-side chains are evident which would explain the frequent appearance during extraintestinal infections.

In addition to the widespread 2-acetamido-2-deoxyhexosamines, many unusual monosaccharide constituents are found in O-side chains of *E. coli* LPS (and other enterobacteria). Thus, the *E. coli* O:7 polysaccharide (Table 1) contains the rare sugar 4-acetamido-4,6-dideoxy-D-*gluco*-hexose (D-Qui4NAc, viosamine),[68] and in *E. coli* O:10 and O:157 (EHEC strains), its D-*galacto*- and D-*manno*-isomers (tomosamine and perosamine, respectively) were identified.[49] If colitose (3,6-dideoxy-L-*xylo*-hexose, Col) is present, forming immunodominant assemblies, it always covers the O-side chains as a terminal sugar, e.g., in the O:111 O-chain by twofold substitution of the glucose residue in the main chain,[75] in O:55, in *S. enterica* serovar Z, and *Yersinia pseudotuberculosis* serotype VI.[49,77] For each group of intestinal and extraintestinal *E. coli* strains, restricted sets of O-serotypes were identified. The O-specific chains may determine the organisms' immunogenicity and serum sensitivity, which together with CPS, adhesins, and toxins determine their pathogenicity.[78,79]

E. coli constitutes one of the best studied bacterial species in terms of structural chemistry of polysaccharides and proteins, biochemistry, microbiology, and genetics. Most genetic work was done on strain K-12.[80] Phenotypically, *E. coli* K-12 is rough and possesses a complete core. Investigations, however, showed that two independent mutational events occurred in different lines of *E. coli* K-12 strains, both resulting in the loss of O-side chains.[81] The complementation of an *rfb*-50 mutation in strain EMG2 by the corresponding functional gene of strain WG1 resulted in a strain which produced an O-specific polysaccharide serologically typed as O:16 with cross-reactivity to O:17. Structural analysis demonstrated that the O-chain of *E. coli* O:16 and *E. coli* O:16 (K-12) has the same carbohydrate backbone.[73,74] Surprisingly, also the *O*-acetylation of the L-Rha residue, as found in the wild type O:16 strain, has been conserved in the construct. However, the *E. coli* O:16 (K-12) repeating unit was additionally substituted by D-Glc at position 6 of the D-GlcNAc residue (Table 1) resulting in serological cross-reaction with *E. coli* O:17, as shown recently by structural analysis of the *E. coli* O:17 O-specific chain,[82] which, in fact, possesses the identical epitope α-D-Glc*p*-(1→6)-α-D-Glc*p*NAc.

Five different LPS core types exist in *E. coil*, i.e., the K-12 and R1 to R4 cores (Table 2).[83] Investigations on the core type distribution among clinical isolates demonstrated R1 to be the most frequent.[84] Common to all core types is the same carbohydrate backbone of the inner core region built up of Kdo and L-*glycero*-D-*manno*-heptose (L,D-Hep). *E. coli* K-12 strain W3100 exhibits a partial substitution at Kdo II by a third Kdo residue at O-4 or by an L-Rha residue at O-5, which may be lacking in other strains.[85-87] In the case of strain AB1133, which possesses an LPS free of L-Rha, the K-12 core consists of three oligosaccharides (OS1, OS2, and OS3, see Table 2), which were structurally investigated after chromatographic separation from the LPS of strain AB1133.[88] OS1 represents the smallest oligomer (40%), OS2 (40%) and OS3 (20%) have the OS1 basal structure substituted by the disaccharide L-α-D-Hep*p*-(1→6)-α-D-Glc*p*-(1→, or the trisaccharide β-D-Glc*p*NAc-(1→7)-L-α-D-Hep*p*-(1→6)-α-D-Glc*p*-(1→, respectively.[88] These results are in good agreement with earlier reports on LPS of AB1133 demonstrating the presence of three distinct bands in SDS–PAGE analyses.[87,89] In the R2 core, D-Gal (at O-7) instead of L-Rha may terminate the Kdo side chain.[83,90] The outer core, also termed the hexose region, is composed of four or five residues of D-Glc and D-Gal arranged in a characteristic sequence for each core type. The hexose regions of R2 and R3 are additionally substituted by D-GlcNAc,[83] and the R1 and R3 core oligosaccharides have characteristic D-GlcNAc substitutions at L,D-Hep III of the inner core region.[91,92]

3.09.3.2.2 Salmonella enterica

Salmonellae represent a diverse group of primarily intestinal microorganisms of vertebrates. This genus comprises more than 2000 serotypes, of which 46 are of pathological importance for man.[64] Acute enteric diseases (salmonellosis) like typhus (*S. enterica* serovar Typhi) appear annually worldwide, and in North America and Western Europe salmonellosis is associated with the highest lethality among infectious diseases. Control of these diseases is difficult because *Salmonella* are able to persist intracellularly and on mucosal surfaces for a long time without causing acute disease. The colonized host may set free bacteria, which, usually unrecognized, are distributed in the environment, e.g., via faeces or food products.

S. enterica serovars A (Paratyphi), B (Typhimurium), and D1 (Enteritidis and Typhi) possess in their O-specific chain a common main glycosyl sequence (Table 3), but differ in substituents of this backbone. Serovar A expresses 3,6-dideoxy-α-D-*ribo*-hexose (paratose, Par), serovar B 3,6-dideoxy-α-D-*xylo*-hexose (abequose, Abe), and 2-*O*-acetyl-3,6-dideoxy-α-D-*xylo*-hexose (2-*O*-Ac-Abe), and serovar D1 3,6-dideoxy-α-D-*arabino*-hexose (tyvelose, Tyv), all 3,6-dideoxyhexoses (D-ddHex) being linked via a (1→3) bond to the D-Man residue of the backbone. In the case of serovars A and B, the main chain D-Gal residue is substituted at position 4, and 4 or 6, respectively, by D-Glc.

Whereas most of *S. enterica* O-side chains are genetically encoded by the chromosomal *rfb* gene cluster, the O:54 polysaccharide of *S. enterica* serovar Borreze was reported to require functions located on a plasmid (see below). The O-chain of this serotype was identified to consist of a D-ManNAc homopolymer with alternating (1→3) and (1→4) linkages (Table 3).[105]

S. enterica expresses two core oligosaccharides (Table 2). The inner core regions of *S. enterica* and *E. coli* (Kdo and Hep assembly) are built up of the same monosaccharides, but the outer core regions differ in the arrangement of the hexoses. As in the *E. coli* R3 core, the *S. enterica* core contains the sequence α-D-Glc*p*-(1→2)-α-D-Gal*p*-(1→3)-α-D-Glc*p*-(1→, whereas the lateral D-Gal and D-GlcNAc residues are absent from the *E. coli* R3, but present in the R2 core (Table 2). Using a monoclonal antibody (T6), the terminal D-GlcNAc residue was shown to be absent from almost one fifth of the strains tested.[110] Structural analysis of *S. enterica* serovars IV and Arizonae proved

Table 2 Structure of *Escherichia coli* and *Salmonella enterica* core oligosaccharides.

Core type	Species, strain, serovar, and structure	Ref.
K-12	*E. coli* W3100, W3110, AB1133	88, 86, 93

R → 2)-α-D-Glc*p*-(1 → 3)-α-D-Glc*p*-(1 → 3)-L-α-D-Hep*p*-(1 → 3)-L-α-D-Hep*p*-(1 → 5)-α-Kdo

```
                  6                    7                            4
                  ↑                    ↑                            ↑
                  1                    1                            2
              α-D-Gal p          L-α-D-Hep p            α-Kdo-(2 → 4)ᵃ-α-Kdo
                                                                    5
                                                                    ↑
                                                                    1
                                                                   Rᴸᵃᵈ
```

OS1   R = H
OS2   R = L-α-D-Hep*p*-(1 → 6)-α-D-Glc*p*-(1 →
OS3   R = β-D-Glc*p*NAc-(1 → 7)-L-α-D-Hep*p*-(1 → 6)-α-D-Glc*p*-(1 →

| R1 | | 94, 95 |

α-D-Gal*p*-(1 → 2)-α-D-Gal*p*-(1 → 2)-α-D-Glc*p*-(1 → 3)-L-α-D-Hep*p*-(1 → 3)-L-α-D-Hep*p*-(1 → 5)-α-Kdo

```
                  3                    7                            4
                  ↑                    ↑                            ↑
                  1                    1                            2
              β-D-Glc p          L-α-D-Hep p            α-Kdo-(2 → 4)ᵃ-α-Kdo
                                       7
                                       ↑
                                  α-D-Glc p Nᵃ
```

| R2 | *E. coli* EH100, F576 | 90, 94–96 |

α-D-Glc*p*-(1 → 2)-α-D-Glc*p*-(1 → 3)-α-D-Glc*p*-(1 → 3)-L-α-D-Hep*p*-(1 → 3)-L-α-D-Hep*p*-(1 → 5)-α-Kdo

```
  2                    6                    7                            4
  ↑                    ↑                    ↑                            ↑
  1                    1                    1                            2
α-D-Glc p NAc      α-D-Gal p          L-α-D-Hep p            α-Kdo-(2 → 4)ᵃ-α-Kdo
                                                                         7
                                                                         ↑
                                                                         1
                                                                       R²·ᵃᵉ
```

| R3 | | 92, 94 |

α-D-Glc*p*-(1 → 2)-α-D-Glc*p*-(1 → 2)-α-D-Gal*p*-(1 → 3)-α-D-Glc*p*-(1 → 3)-L-α-D-Hep*p*-(1 → 3)-L-α-D-Hep*p*-(1 → 5)-α-Kdo

```
                       3                              7                        4
                       ↑                              ↑                        ↑
                       1                              1                        2
                  α-D-Glc p NAc              L-α-D-Hep p          α-Kdo-(2 → 4)ᵃ-α-Kdo
                                                     7
                                                     ↑
                                                     1
                                              α-D-Glc p NAcᵃ
```

| R4 | | 97 |

α-D-Gal*p*-(1 → 2)-α-D-Gal*p*-(1 → 2)-α-D-Glc*p*-(1 → 3)-α-D-Glc*p*-(1 → Inner Coreᵇ

```
                                            4
                                            ↑
                                            1
                                        β-D-Gal p
```

| Ra | *S. enterica* sv. Minnesota, sv. Arizonae | 98–104 |

Oagᶜ → 4)-α-D-Glc*p*-(1 → 2)-α-D-Gal*p*-(1 → 3)-α-D-Glc*p*-(1 → 3)-L-α-D-Hep*p*-(1 → 3)-L-α-D-Hep*p*-(1 → 5)-α-Kdo

```
       2                              6                    7                        4
       ↑                              ↑                    ↑                        ↑
       1                              1                    1                        2
      R³ᶠ                         α-D-Gal p          L-α-D-Hep p          α-Kdo-(2 → 4)ᵃ-α-Kdo
```

ᵃ Substitutions in nonstoichiometric amounts. ᵇ A detailed structure has not yet been determined. ᶜ Oag = O-specific chain. ᵈ R¹ = α-L-Rha*p* in strains W3100 and W3110. ᵉ R² = α-D-Gal*p* in strain EH100. ᶠ R³ = α-D-Glc*p*NAc in sv. Minnesota and α-D-Glc*p* in sv. Arizonae.

that the terminal D-GlcNAc residue was changed to a terminal D-Glc residue.[103,104] However, the T6 epitope was detected in all *S. enterica* strains of serovars A to E, but only in 71% of the serovars F to 67. The attachment site of the O-specific chain to the core has been identified as position 4 of the distal D-Glc residue (Table 2).[108]

3.09.3.2.3 Klebsiella pneumoniae

Klebsiella pneumoniae represents an opportunistic pathogen causing septicemia, pneumonia, and urinary tract infections in humans. Its LPS constitutes an important virulence determinant and

Table 3 Structure of repeating units of *Salmonella enterica* O-specific chains.

Serovar	Structure			Ref.

$$[\rightarrow 2)\text{-}\alpha\text{-}D\text{-}Manp\text{-}(1 \rightarrow 4)\text{-}\alpha\text{-}L\text{-}Rhap\text{-}(1 \rightarrow 3)\text{-}\alpha\text{-}D\text{-}Galp\text{-}(1 \rightarrow]_n$$

```
                    3                              X
                    ↑                              ↑
                    1                              1
               α-D-ddHex                      α-D-Glcp
```

Serovar	α-D-*ddHex*[a]	X	O-Factor	Ref.
Paratyphi (A)	Par	4	1, 2, 12	47, 106
Typhimurium (B)	Abe	4 or 6	1, 4, 5, 12	47, 107
Typhimurium (B)	Abe, 2-*O*-Ac-Abe	4	4, 5, 12	47, 107
Typhimurium (B)	Abe or 2-*O*-Ac-Abe	6	4, 5, 12	47, 106, 107
Enteritidis (D1)	Tyv	no Glc	9, 12, 27$_D$	47, 108
Typhi (D1)	Tyv	4	9, 12	77, 106, 109
Borreze (O:54)	$\rightarrow$ 3)-β-D-Man*p*NAc-(1 $\rightarrow$ 4)-β-D-Man*p*NAc-(1 $\rightarrow$			105

[a]α-D-ddHex – dideoxyhexose.

belongs to a toxic complex additionally containing CPS and proteins released from the cell surface during infection and causing characteristic lung tissue damage.[111] Chemically, the O-side chain of the most important serotypes O:1, O:2, and O:8 exhibits very simple repeating units built up of D-Gal*p* and D-Gal*f* (O:1, O:2(2a,2b), O:2(2a,2e,2h), and O:8), as well as D-Gal*f* and D-Glc*p*NAc (O.2(2a,2c)) (Table 4). Galactan I represents a regular polysaccharide made up from D-Gal*f* and D-Gal*p* and occurs in the serotypes O:1, O:2, and O:8. In contrast to galactan I, galactan II consists exclusively of (1→3)-linked D-Gal*p* residues. In serotypes O:8, O:2(2a,2e), and O:2(2a,2e,2h), galactan I is partially *O*-acetylated, which in serotype O:8 concerns the hydroxy groups in positions 2 and 6 of the D-Gal*f* residue. Characteristic for differentiation of serotypes O:2(2a,2b) and O:2(2a,2c) is the presence of a second polymer in the latter serotype, which is built up from repeating units of →5)-β-D-Gal*f*-(1→3)-β-D-Glc*p*NAc-(1→ (O:2(2c)-determinant). Serogroups O:2(2a,2e) and O:9 share identical branched galactans (one branch per two repeating units of galactan I), whereas the galactan in serotype O:2(2a,2c,2h) possesses one branch per repeating unit of galactan I (Table 4). These examples show how bacteria can vary their O-side chain structure by minor modifications leading to new serotypes.

Table 4 Structure of repeating units of *Klebsiella pneumoniae* O-specific chains.

Serotype	Structure[a]	Ref.
O.1	D-Galactan I — $\rightarrow$ 3)-β-D-Gal*f*-(1 $\rightarrow$ 3) α-D-Gal*p*-(1 $\rightarrow$ and	112, 113, 117
O:8	D-Galactan II — $\rightarrow$ 3)-β-D-Gal*p*-(1 $\rightarrow$ 3)-α-D-Gal*p*-(1 $\rightarrow$	112, 113, 117
O:2a, 2b	D-Galactan I — $\rightarrow$ 3)-β-D-Gal*f*-(1 $\rightarrow$ 3)-α-D-Gal*p*-(1 $\rightarrow$ and an additional unknown constituent responsible for 2b reactivity	114
O:2a, 2c	D-Galactan I — $\rightarrow$ 3)-β-D-Gal*f*-(1 $\rightarrow$ 3)-α-D-Gal*p*-(1 $\rightarrow$ and $\rightarrow$ 5)-β-D-Gal*f*-(1 $\rightarrow$ 3)-β-D-Glc*p*NAc-(1 $\rightarrow$	114

O:2 (2a, 2e) O:9

$$\rightarrow 3)\text{-}\beta\text{-}D\text{-}Galf\text{-}(1 \rightarrow 3)\text{-}\alpha\text{-}D\text{-}Galp\text{-}(1 \rightarrow 3)\text{-}\beta\text{-}D\text{-}Galf\text{-}(1 \rightarrow 3)\text{-}\alpha\text{-}D\text{-}Galp\text{-}(1 \rightarrow$$

```
                                                                        2
                                                                        ↑
                                                                        1
                                                                   α-D-Galp
```
115, 116

O:2 (2a, 2e, 2h)

$$\rightarrow 3)\text{-}\beta\text{-}D\text{-}Galf\text{-}(1 \rightarrow 3)\text{-}\alpha\text{-}D\text{-}Galp\text{-}(1 \rightarrow$$

```
                2
                ↑
                1
         α-D-Galp (67%)
```
116

[a]For *O*-acetyl substitutions see text.

The core oligosaccharide of *K. pneumoniae* contains an inner region, which like other Enterobacteriaceae is made up of two Kdo and three L-D-Hep residues. However, L-D-Hep I is substituted at position O-4 by the disaccharide β-D-Gal*p*A-(1→6)-β-D-Glc*p* modifying the inner core region. L-D-Hep II is substituted at O-3 by the outer core region, formed by α-D-Glc*p*-(1→4)-α-D-Gal*p*A in a *K. pneumoniae* mutant deficient in O:8 specificity,[118] or by D-α-D-Hep*p*-(1 → 2)-D-α-D-Hep*p*-(1 → 2)-D-α-D-Hep*p*-(1 → 2)-D-α-D-Hep*p*-(1 → 6)-α-D-Glc*p*N-(1 → 4)-α-D-Gal*p*A in an O:1 deficient mutant (Table 5).[119,120] Remarkably, the *K. pneumoniae* cores lack phosphate groups, and it is tempting to speculate that uronic acid residues possibly substitute for phosphate groups, providing the negative charges required for certain biological LPS functions such as accumulation of cations at the bacterial surface.

Table 5   Structure of *Klebsiella pneumoniae* core oligosaccharides.

Strain (serotype)	Structure	Ref.

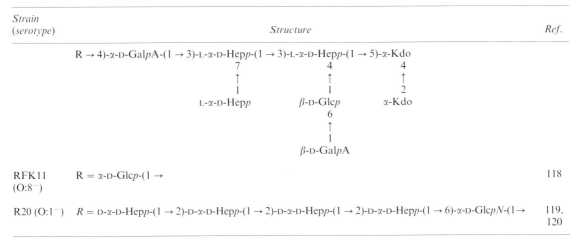

| RFK11 (O:8⁻) | R = α-D-Glc*p*-(1 → | 118 |
| R20 (O:1⁻) | R = D-α-D-Hep*p*-(1 → 2)-D-α-D-Hep*p*-(1 → 2)-D-α-D-Hep*p*-(1 → 2)-D-α-D-Hep*p*-(1 → 6)-α-D-Glc*p*N-(1→ | 119, 120 |

3.09.3.2.4   Serratia marcescens

The primary structures of the O-side chains of 29 serotypes were elucidated.[121-124] The sizes of repeating units were found to vary between 1 and 7 glycosyl residues. The monosaccharide constituents represent usual pentoses, hexoses, 6-deoxyhexoses, *N*-acetylaminosugars, and uronic acids. As noncarbohydrate substituents, *O*-acetyl groups (serotypes O:4, O:5, O:6, O:7, O:14), and pyruvic and lactic acid (O:4 and O:18, respectively) were found. Both neutral (e.g., O:1, O:4 to O:10, O:16, O:27, and O:28) and acidic (e.g., O:2 to O:6, O:13, and O:23) polysaccharides were isolated from LPS extracts and characterized. Several *S. marcescens* serotypes (e.g., O:2 to O:4, O:13, O:15) express both neutral and acidic O-side chains. Further investigations have shown that the neutral polymers correspond to O-antigenic polysaccharides, while the acidic polymers are derived from K-antigenic capsules.[125,126] *S. marcescens* O:16 expresses two neutral polymers, the D-galactan O-side chain (Table 6), identical to D-galactan I of *K. pneumoniae* O:1 (Table 4), and a homopolymer of D-Rib*f*.[127] Variation of O-specificity by expression of two types of O-side chains is rather unusual and has so far been identified in LPS of *S. marcescens*, *Burkholderia*,[128,129] and *Acinetobacter*.[130]

3.09.3.2.5   Shigella spp.

Within the genus of *Shigella*, more than 42 serotypes based on O-antigen classification have been recognized: 14 for *Shigella flexneri*, 12 for *Shigella dysenteriae*, 15 for *Shigella boydii*, and 1 for *S. sonnei*. The LPS core oligosaccharide of *Shigella* spp. appears to be structurally conserved. The *S. sonnei* Phase I PhI 3 and *S. flexneri* core oligosaccharides are structurally identical (Table 7). Like the *E. coli* R1 and R3 cores, they carry a D-GlcN substitution at position 7 of L,D-Hep III of the inner core region. Furthermore, in the *S. sonnei* phase I PhI 3 LPS, the complete *E. coli* R1 core was found, and the first repeating unit of the O-specific chain was found to be linked to the β-D-Glc residue (Table 7).

All *Shigella* spp. O-specific chains characterized so far harbor at least one aminosugar in the repeating unit and also contain usual hexoses and uronic acids. The O-chains of *S. flexneri* possess a characteristic L-Rha-trisaccharide-D-GlcNAc main chain (Table 6). Serotype variation is caused

Table 6 Structure of repeating units of various enterobacterial O-specific chains.

Species, serotype	Structure	Ref.
Serratia marcescens O:16	→ 3)-β-D-Gal$f$-(1 → 3)-α-D-Gal$p$-(1 → and → 2)-β-D-Rib$f$-(1 →	127
Shigella dysenteriae type 1	→ 2)-α-D-Gal$p$-(1 → 3)-α-D-Glc$p$NAc-(1 → 3)-α-L-Rha$p$-(1 → 3)-α-L-Rha$p$-(1 →	131
S. flexneri Y	→ 3)-β-D-Glc$p$NAc-(1 → 2)-α-L-Rha$p$-(1 → 2)-α-L-Rha$p$-(1 → 3)-α-L-Rha$p$-(1 → 4 ↑ H	132–135
S. flexneri O:2a	→ 3)-β-D-Glc$p$NAc-(1 → 2)-α-L-Rha$p$-(1 → 2)-α-L-Rha$p$-(1 → 3)-α-L-Rha$p$-(1 → 4 ↑ 1 α-D-Glc$p$	132
Yersinia enterocolitica O:1, 2a, 3	→ 2)-β-L-6d-Alt$p$-(1 → 2)-β-L-6d-Alt$p$-(1 → 3)-β-L-6d-Alt$p$-(1 →[a]	136
Y. enterocolitica O:3	→ 2)-β-L-6d-Alt$p$-(1 →[a]	136
Y. pseudotuberculosis IA	→ 3)-β-D-Glc$p$NAc-(1 → 3)-α-D-Gal$p$-(1 → 4 ↑ 1 α-D-Par$p$-(1 → 3)-β-D-6d-Hep$p$[b]	49
Y. pseudotuberculosis IB	→ 3)-β-D-Glc$p$NAc-(1 → 2)-α-D-Man$p$-(1 → 4)-α-D-Man$p$-(1 → 3)-α-L-Fuc$p$ (1 → 3 ↑ 1 α-D-Par$f$	49
Y. pseudotuberculosis IIa	→ 3)-α-D-Gal$p$-(1 → 4 ↑ 1 α-D-Abe$p$-(1 → 3)-α-D-6d-Hep$p$	137, 138

[a] β-L-6d-Alt$p$ = 6-deoxy-L-altropyranose.   [b] β-D-6d-Hep$p$ = 6-deoxy-D-*manno*-heptopyranose.

Table 7 Structure of *Shigella* core oligosaccharides.

R^2 → 3)-α-D-Glc$p$-(1 → 3)-α-D-Glc$p$-(1 → 3)-L-α-D-Hep$p$-(1 → 3)-L-α-D-Hep$p$-(1 → 5)-α-Kdo
2   7
↑   ↑
R^1 1
α-D-Glc$p$N-(1 → 7)-L-α-D-Hep$p$

Species, core type	Structure	Ref.
S. sonnei R-type	R^1 = R^2 = H	53, 54
S. sonnei Phase I PhI1	R^1 = α-D-Gal$p$-(1 → 2)-α-D-Gal$p$-(1 → R^2 = [→ 4)-α-L-Alt$p$NAcA-(1 → 3)-β-D-Fuc$p$NAc4N-(1 →]$_4$ → 3)-β-D-Glc$p$-(1 →	53, 54
S. sonnei Phase I PhI2	R^1 = α-D-Gal$p$-(1 → 2)-α-D-Gal$p$-(1 → R^2 = α-L-Alt$p$NAcA-(1 → 3)-β-D-Fuc$p$NAc4N-(1 → 3)-β-D-Glc$p$-(1 →	53, 54
S. sonnei Phase I PhI3	R^1 = α-D-Gal$p$-(1 → 2)-α-D-Gal$p$-(1 → R^2 = β-D-Glc$p$-(1 →	53, 54
S. flexneri	R^1 = α-D-Gal$p$-(1 → 2)-α-D-Gal$p$-(1 → R^2 = β-D-Glc$p$-(1 →	54

by different branches with one or two units of D-Glc and by *O*-acetylation. In contrast, the *S. dysenteriae* O-side chains, e.g., type 1 (Table 6), do not follow such simple rules (except for the general make-up) indicating a greater genetic diversity of this genus.[49] Unusual for *Shigella* spp. is the presence of pseudaminic acid in *S. boydii* type 7 and 2,4-diamino-2,4,6-trideoxy-D-galactose (D-Fuc*N*Ac4*N*) in *S. sonnei* O-side chains, respectively (Table 7).[49] In general, however, accessibility to diverse, biosynthetic easily available sugar precursors is an important basis for the large and epidemiologically important group of *S. enterica*, *E. coli*, and *Shigella* spp. in order to create structural and, thus, serological diversity in the O-side chain region.

3.09.3.2.6   Yersinia *spp.*

The genus *Yersinia* comprises important human pathogens such as *Yersinia enterocolitica*, the causative agent of intestinal yersiniosis, *Y. pestis*, the etiologic agent of plague, and *Y. pseudotuberculosis*, which causes acute mesenteric lymphadenitis.

Serologically, *Y. enterocolitica* has been subdivided into 34 O-antigenic variants. Serotype O:3 is characterized by a homopolymeric O-specific chain of (1→2)-linked 6-deoxy-β-L-altropyranose (L-6d-Alt) (Table 6).[136] Strains with the serological formulas O:1,2a,3 and O:2a,2b,3 have identical trisaccharide repeating units in the O-specific polysaccharides consisting of L-6d-Alt residues linked (1→2), (1→2), and (1→3). Serological differences between O:2a,2b,3 result from partial *O-acetylation of the (1→2)-linked monomeric units at the hydroxy group at position* 3. *Y. enterocolitica* serotype O:9 expresses a homopolymer with a monosaccharide repeating unit of 4,6-dideoxy-4-formamido-α-D-mannose (4-*N*-formyl-D-perosamine),[139] a structure which is also found as O-side chain in *V. cholerae* (see below). Common to other *Y. enterocolitica* serotypes (O:4,32, O:5, O:5,27, O:8, O:10), which have branched structures, is the presence of one or more deoxysugars indicating an important role of hydrophilic and hydrophobic arrangements in the O-specific polysaccharide.[49]

The O-specific chains of *Y. pseudotuberculosis* consist of a neutral, branched hexosaminoglycan with repeating units of four and five monomers, and in each case branches represent or contain deoxysugars such as 6-deoxy-D-*manno*-heptose (serotypes IA, IIA), paratose (IA, IB, III), abequose (IIC), tyvelose (IVA), ascarylose (VA), 6-deoxy-altrose (VB), colitose (VI, VII), and yersiniose A (VI).[49,137] Since most of these lateral glycosyl residues are also found in other bacteria, serological cross-reactions, e.g., with *S. enterica* and *E. coli* are to be expected.

3.09.3.2.7   Acetobacter methanolicus

Acetobacter methanolicus strains belong to facultatively methylotrophic acetic acid bacteria, which are not pathogenic for humans, but possess some importance for biotechnological applications such as acetic acid production and food fermentation. Although little is known about their LPS, some *A. methanolicus* strains have been structurally characterized with regard to their O-specific chains (Table 8). The type strain of this genus expresses an O-side chain with alternating D-Gal*f* and D-Gal*p* units, a structural feature which is known from *K. pneumoniae* O:1 (Table 4) and *S. marcescens* O:16 (Table 6). Furthermore, *A. methanolicus* 135 (Table 8) expresses an O-side chain structurally similar to that of *E. coli* O:8 and O:9 (Table 1) and of *K. pneumoniae* O:3 and O:5.[77]

3.09.3.2.8   Legionella pneumophila

The genus *Legionella* now comprises more than 40 species which are typed into 61 serogroups. Most species are nonpathogenic for man or rarely cause disease. However, *L. pneumophila*, the etiologic agent of legionellosis (Legionnaires' disease) causes severe pneumonia. *L. pneumophila* expresses a unique S-form LPS, which exhibits very narrow bands in SDS–PAGE analysis indicating a polydisperse nature and small repeating units.[151,152] The O-specific side chain of serotype 1 was identified as a homopolymer (Table 8) composed of 5-acetamidino-7-acetamido-8-*O*-acetyl-3,5,7,9-tetradeoxy-L-*glycero*-D-*galacto*-nonulosonic acid, the monomers being bridged by (2→4)-linkages.[45] There is no free hydroxy group in the repeating unit as position 18 is *O*-acetylated, positions 5 and 7 are substituted by acetamidoyl and *N*-acetyl groups, respectively, and position 4 is involved in the ketosidic interresidue linkages. The lack of free hydroxy groups, the presence of deoxy functions in position 3 and 9, and *N*-acetyl groups render the polymer highly hydrophobic. Interestingly, the

Table 8 Structure of repeating units of various nonenterobacterial O-specific chains.

Species, serotype	Structure	Ref.
Acetobacter methanolicus 58/4	→ 2)-β-D-Gal*f*-(1 → 3)-β-D-Gal*p*-(1 →	140
A. methanolicus 58	→ 6)-α-D-Glc*p*-(1 → 2)-α-D-Gal*p*-(1 → 6)-α-D-Gal*p*-(1 →	141
A. methanolicus 70	→ 2)-β-D-Gal*f*-(1 → 3)-β-D-Gal*p*-(1 → and → 2)-α-D-Glc*p*-(1 → 6)-α-D-Glc*p*-(1 →	142
A. methanolicus 135	→ 4)-α-D-Man*p*-(1 → 2)-α-D-Man*p*-(1 → 2)-α-D-Man*p*-(1 → 2)-α-D-Man*p*-(1 → 2)-α-D-Man*p*-(1 →	143
Legionella pneumophila O:1	→ 4)-α-Sug*p*-(2 → [a] 8 ↑ Ac	45, 144
Pseudomonas aeruginosa (Common A-band LPS)	→ 3)-α-D-Rha*p*-(1 → 3)-α-D-Rha*p*-(1 → 2)-α-D-Rha*p*-(1 →	145, 146
P. aeruginosa O:5	→ 4)-β-D-Man*p*NAc3NA-(1 → 4)-β-D-Man*p*NAc3NAcA-(1 → 3)-α-D-Fuc*p*NAc-(1 → | Me-C=NH	147
Vibrio cholerae O:1 Ogawa	3-O-Me-α-D-Rha*p*4NR-(1 → [2)-α-D-Rha*p*4NR-(1]$_n$ → R = CH$_2$OH-CH$_2$-CH$_2$-CH$_2$OH(S)-CO-	148, 149
V. cholerae O:1 Inaba	α-D-Rha*p*4NR-(1 → [2)-α-D-Rha*p*4NR-(1]$_n$ → R = CH$_2$OH-CH$_2$-CH$_2$-CH$_2$OH(S)-CO-	148, 150

[a]Sug = 5-acetamidino-7-acetamido-3,5,7,9-tetradeoxy-L-*glycero*-D-*galacto*-nonulosonic acid

hydrophobic surface has been shown to be associated with the pathogenic potential of *L. pneumo phila* as it supports the adherence to alveolar macrophages in the initial phase of infection.[153] It is noteworthy that the loss of the 8-*O*-acetyl group results in the loss of serological reactivity with specific poly- and monoclonal antibodies.[153]

The core oligosaccharide of *L. pneumophila* serotype 1 was isolated from hydrolysates obtained after treatment of LPS at pH 4.4 containing long chain (~45 residues), medium chain (~30 residues), and short chain (~12 residues of the nonulosonic acid) polysaccharide, as well as small amounts of the unsubstituted core.[151] Kdo was identified as the reducing end glycosyl residue. Heptose is not present, but the core contains an extended hexose region (6 hexoses) lacking phosphate residues. The distal part of the core oligosaccharide contains three units of deoxysugars, i.e., two residues of L-Rha and D-Qui*N*Ac, each of which is *O*-acetylated at positions 2, 2, and 4, respectively (Table 9). Hence, also the outer core region exhibits considerable hydrophobicity.

3.09.3.2.9 Pseudomonas aeruginosa

Pseudomonas aeruginosa is a major causative factor of postoperative and posttraumatic septic complications. It causes pneumonia in patients with cystic fibrosis and is associated with high lethality and corneal infection. Among the large and heterogeneous family of pseudomonads, *P. aeruginosa* and its LPS have been studied most intensively. The architecture of the O-specific chain of the numerous serotypes of *P. aeruginosa* follows some general rules. Typical components of O-side chains of *P. aeruginosa* LPS are aminosugars (2-amino-2-deoxyhexoses, 2-amino-2,6-dideoxy-hexoses, D-Qui*N*4*N*), acidic mono- and diamino-sugars (D-Gal*N*A, L-Gal*N*A, L-Alt*N*A, D-Glc*N*3*N*A, D-Man*N*3*N*A, D-Gal*N*3*N*A, L-Gal*N*3*N*A), and 5,7-diamino-3,5,7,9-tetradeoxy-nonulosonic acids (L-*glycero*-L-*manno*- (pseudaminic acid), and D-*glycero*-L-*galacto*-configuration)). The repeating units comprise two to four sugar residues. With one exception (serogroup O:13a,13c), monosaccharides are arranged in a linear order. Frequently, O-side chains contain structural modifications by *O*- and *N*-linked noncarbohydrate substituents, which include acetyl, formyl, (*R*)- and (*S*)-2-hydroxybutyryl-, and acetamidoyl residues.[154]

Table 9 Structure of some nonenterobacterial core oligosaccharides.

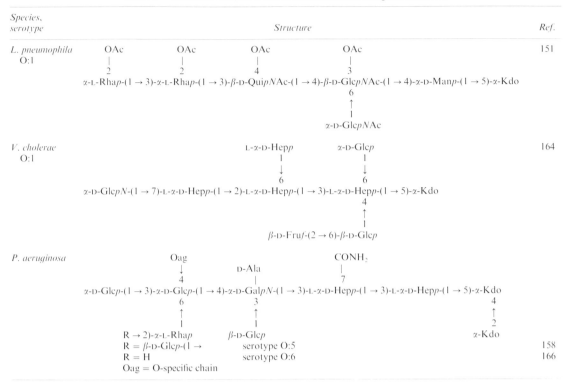

Species, serotype	*Structure*	*Ref.*
L. pneumophila O:1	OAc OAc OAc OAc | | | | 2 2 4 3 α-L-Rha*p*-(1 → 3)-α-L-Rha*p*-(1 → 3)-β-D-Qui*p*NAc-(1 → 4)-β-D-Glc*p*NAc-(1 → 4)-α-D-Man*p*-(1 → 5)-α-Kdo 6 ↑ 1 α-D-Glc*p*NAc	151
V. cholerae O:1	L-α-D-Hep*p* α-D-Glc*p* 1 1 ↓ ↓ 6 6 α-D-Glc*p*N-(1 → 7)-L-α-D-Hep*p*-(1 → 2)-L-α-D-Hep*p*-(1 → 3)-L-α-D-Hep*p*-(1 → 5)-α-Kdo 4 ↑ 1 β-D-Fru*f*-(2 → 6)-β-D-Glc*p*	164
P. aeruginosa	Oag CONH$_2$ ↓ D-Ala | 4 | 7 α-D-Glc*p*-(1 → 3)-α-D-Glc*p*-(1 → 4)-α-D-Gal*p*N-(1 → 3)-L-α-D-Hep*p*-(1 → 3)-L-α-D-Hep*p*-(1 → 5)-α-Kdo 6 3 4 ↑ ↑ ↑ 1 1 2 R → 2)-α-L-Rha*p* β-D-Glc*p* α-Kdo	
	R = β-D-Glc*p*-(1 → serotype O:5	158
	R = H serotype O:6	166
	Oag = O-specific chain	

Several strains of *P. aeruginosa* coexpress two distinct forms of LPS, designated A-band and B-band LPS, according to the appearance of different banding patterns in SDS–PAGE analysis.[155] B-band LPS contains the serotype-specific heteropolymeric O-side chain as described above, whereas A-band LPS carries a homopolymer composed of D-Rha arranged as a trisaccharide repeating unit linked (1→2), (1→3), and (1→3) (Table 8).[154] Because of its frequent appearance in various *P. aeruginosa* strains, A-band LPS is considered to be a common antigen[135]. However, phytopathogenic pseudomonads also have O-chain structures similar to that of A-band LPS.[156] As determined by SDS–PAGE analyses and Western immunoblot of LPS, A-band LPS exhibits shorter O-chain length and is less variable than the type-specific LPS, which may exhibit up to four different ladder-like patterns. Long-chain B-band LPS sterically shields A-band LPS. Therefore, A-band specific monoclonal antibodies are not able to agglutinate bacteria. However, strong agglutination was observed in the case of strains deficient in the type-specific O-antigen and in mucoid (alginate producing) phenotypes isolated from cystic fibrosis patients.[146,157]

P. aeruginosa PAO1 (serotype IATS O:5), a wild type strain causing corneal infection, expresses an O-side chain consisting of 2-acetamido-3-acetimidoyl- and 2,3-diacetamido-2,3-dideoxy-β-D-mannuronic acid, as well as of 2-acetamido-2,6-dideoxy-α-D-galactose (D-Fuc*N*Ac) (Table 8).[49,147] The O-side chain was shown to be responsible for bacterial adherence to epithelial cells. For efficient cell ingestion, however, the terminal D-Glc residue of the outer core (Table 9) is necessary. Adherence and ingestion could be inhibited by purified O-side chain and, in particular, core oligosaccharide preparations.[158] This example demonstrates that, in addition to pili and fimbriae, LPS may also be involved in the association and ingestion of bacteria to the epithelial cells and, thus, contribute to the initial process of bacterial infection.

P. aeruginosa harbors characteristic structural features in the core oligosaccharide, which are absent from enterobacterial and most nonenterobacterial LPS. In all *P. aeruginosa* strains expressing a complete or deficient core structure such as strain PAC 605, the D-Gal*N* residue is *N*-acylated by D-Ala.[159,160] The *P. aeruginosa* PAO1 core is characterized by a high concentration of phosphate residues,[161] of which so far only two have been localized at position 2 and 4 of the L-D-Hep I residue.[158]

Also common to all *P. aeruginosa* serotypes is a unique modification of L-D-Hep II represented by a carbamoyl substitution at position 7 (Table 9).[158] This structure was found for all strains of

RNA group 1, which, in addition to *P. aeruginosa*, includes *Pseudomonas fluorescens*, *Pseudomonas putida*, and *Pseudomonas syringae*. In other RNA groups, the carbamoyl substitution appears to be absent.

3.09.3.2.10   Vibrio cholerae

Vibrio cholerae is an important human pathogen causing severe diarrhea, which is still associated with high mortality, particularly in third world countries. On the basis of differences in O-chain structures, *V. cholerae* is, on a serological basis, currently classified into O:1 and non-O:1. The former group is further subdivided into two major O-forms, Ogawa and Inaba, and the latter into the O-forms O:2 to O:155. It was serologically established that *V. cholerae* O:1 expresses three antigenic factors, i.e., the group antigenic factor A, the Ogawa antigen factor B, and the Inaba antigen factor C. Hence, Ogawa possesses the antigenic formula AB and Inaba the formula AC.[150,162] The O-specific polysaccharides of LPS from both the Ogawa and the Inaba strains are linear homopolymers of $(1\rightarrow2)$-D-perosamine, the amino functions of which are acylated with 3-deoxy-L-*glycero*-tetronic acid (Table 8).[148] 2-*O*-Methyl-*N*-(3-deoxy-L-*glycero*-tetronyl)-D-perosamine was identified at the nonreducing terminus of the Ogawa O-specific chain, whereas that of the Inaba was *N*-(3-deoxy-L-*glycero*-tetronyl)-D-perosamine.[163] The structural basis for their cross-reactivity is the identity of the O-side chains (factor A).

The core oligosaccharides of the *V. cholerae* O:1 Ogawa and the Inaba strains were found to be identical.[164] Interestingly, the O:1 core harbors only one Kdo residue, which is phosphorylated at position 4, and the inner core is formed by four residues of L-D-Hep. Hep I is substituted by D-Glc and the disaccharide β-D-Fru*f*-$(2\rightarrow6)$-β-D-Glc*p*. The fructose residue seems to have biological importance for the serological properties of LPS.[165]

3.09.3.2.11   Chlamydia *spp.*

Chlamydiaceae constitute a monogeneric family of pathogenic, obligatory intracellular bacteria, which cause acute and chronic diseases in animals and humans. *Chlamydia psittaci* finds its natural reservoir among animals, but human infections are known from avian strains causing severe pneumonia. *Chlamydia trachomatis*, serovar A through C, is the causative agent of chronic eye infections leading to blindness in the late stage of infection. Serovars D to K cause sexually transmitted diseases in men (urethritis and prostatitis) and women (urethritis, cervicitis, and salpingitis often resulting in infertility). *C. pneumoniae* is emerging as a causative agent of atypical pneumoniae and has recently been associated with the pathogenesis and clinical consequences of atherosclerosis. As *Chlamydia* spp. are obligatory intracellular bacteria, they cannot be grown in large amounts. Thus, structural analysis of their LPS was limited. However, using recombinant bacteria, in which the cloned gene for the Kdo-transferase could be expressed,[167] larger amounts of LPS became available. It was established that chlamydial LPS contains a chemically and antigenically unique structure of a trisaccharide of the sequence α-Kdo-$(2\rightarrow8)$-α-Kdo-$(2\rightarrow4)$-α-Kdo-$(2\rightarrow$.[168–171] Monoclonal antibodies directed against this trisaccharide epitope are genus-specific and recognize only bacteria of the genus *Chlamydia*. Serological properties of the native LPS were identical to both the de-*O*- and de-*N*-acylated recombinant carbohydrate backbone and the respective pentasaccharide obtained by total chemical synthesis.[172–174]

3.09.3.3   The Lipid A Component

Lipid A represents the covalently bound lipid content of LPS.[1] It can be separated from the polysaccharide region by treatment of LPS with mild acid, which preferentially cleaves the linkage between the Kdo I residue of the inner core and lipid A. Lipid A was discovered during studies on *S. enterica* and *E. coli* LPS and characterized as a peculiar phosphoglycolipid possessing an architecture which is unique in nature. Figure 3 shows the lipid A structure of four different types of Gram-negative bacteria, which all express biologically highly active LPS (*E. coli*, *H. influenzae*, *Chromobacterium violaceum*, and *N. meningitidis*).[175] Structurally, these lipid A share a 1,4′-bisphosphorylated β-$(1\rightarrow6)$-linked D-GlcN disaccharide (lipid A backbone), which carries free hydroxy groups in position 4 (GlcN I) and 6′ (GlcN II), the latter serving as the attachment site of

Kdo, i.e., the inner core in LPS. The lipid A backbone is acylated by four (*R*)-3-hydroxy fatty acids at positions 2, 3, 2′, and 3′. In each case, the acyl group at position 2′ of GlcN II carries at its 3-hydroxy group a further (fifth) fatty acid. The four structures, however, differ in the location of a sixth acyl group (R^1 or R^2) and the chain length of fatty acids (symbols m, n, and o). As Figure 3 shows, lipid A of *E. coli* and *H. influenzae* carries this sixth fatty acid at GlcN II and thus possesses an asymmetric distribution of acyl groups over GlcN I and GlcN II (4 + 2), whereas a symmetric acyl arrangement (3 + 3) is present in lipid A of *C. violaceum* and *N. meningitidis*. Importantly, the average length of acyl chains is smaller in the latter group (mainly 12 carbon atoms) than in the former (mainly 14 carbon atoms).

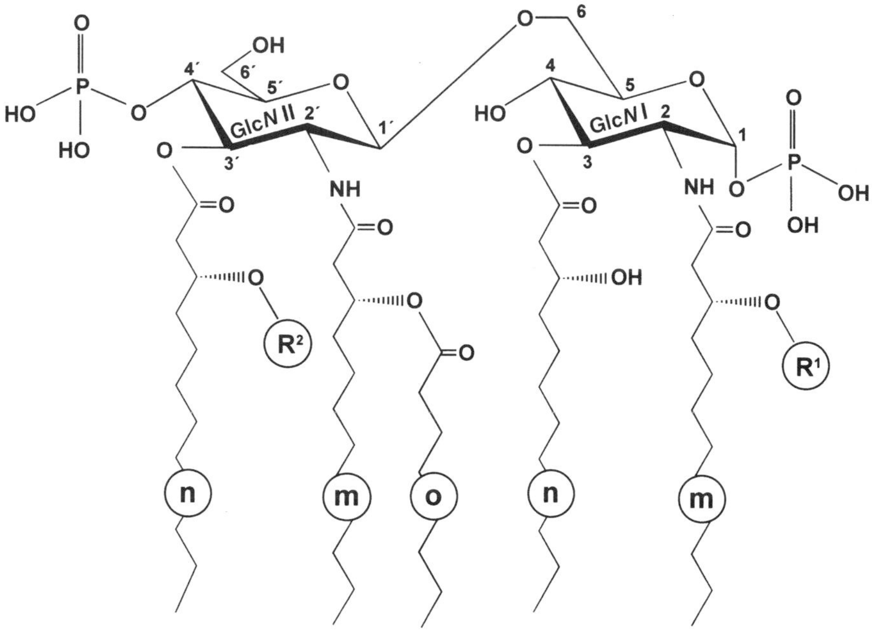

	Nature of		Number of carbon atoms		
Bacterial species	R^1	R^2	*m*	*n*	*o*
Escherichia coli	H	14:0[a]	14	14	12
Haemophilus influenzae	H	14:0	14	14	14
Neisseria meningitidis	12:0[b]	H	14	12	12
Chromobacterium violaceum	12:0	H	12	10	12

[a] 14:0 = myristic acid.   [b] 12:0 = lauric acid.

Figure 3   Primary chemical structure of the lipid A component of various Gram-negative bacteria. The fully protonated form is shown.

These structural examples demonstrate that lipid A of various origins exhibit a similar architecture, but variations do exist concerning the hexosamine backbone, the nature of acyl residues, and the substitution of phosphate groups. Thus, GlcN may be replaced by 2,3-diamino-2,3-dideoxy-D-glucose (D-GlcN3N) as in the case in *Campylobacter jejuni* and, in particular, in *L. pneumophila*, which contains a β-(1 → 6)-linked GlcN3N–GlcN3N disaccharide.[176,177] The *Legionella* lipid A backbone carries, like several bacterial species of the α-2-subclass, long-chain *n*-2 hydroxylated fatty acids. This type of unusual hydroxy fatty acid harbors the functional hydroxy group at the penultimate position of the hydrocarbon chain and not at position 3. The most prominent representative, 27-hydroxyoctacosanoic acid (28:0(27-OH)), possesses the double length of the usual 14:0(3-OH) acid and may stretch through the entire outer membrane. In addition, unusual acyl groups such as 28:0(27oxo), 18:0(3-OH), 20:0(3-OH), 22:0(3-OH), and *i*14:0(2,3-diOH) are present, and the chemical structure of *L. pneumophila* lipid A is shown in Figure 4.

In *Rhodobacter sphaeroides* and all species of the α-3 branch of the phylogenetic tree, the amide-linked fatty acids of the disaccharide backbone were identified as 3-oxotetradecanoic acid. Some of

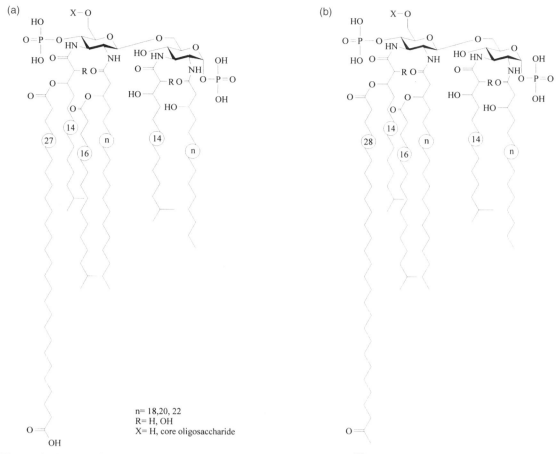

n= 18,20, 22
R= H, OH
X= H, core oligosaccharide

Figure 4 Proposed chemical structure of *L. pneumophila* O·1 lipid A.[177] Two populations of lipid A molecules have been identified, which differ in the *N*-acyloxylacyl group at position 3′ of the carbohydrate backbone, carrying either heptacosandioic (27:0-dioic) (a) or 27-oxooctacosanoic acid (28:0(27-oxo)) (b). The fully protonated form is shown.

these species contain, in addition, unsaturated fatty acids as lipid A constituents, e.g., Δ-7-tetradecenoic acid in *R. sphaeroides*. These lipid A are of particular biological interest, as they not only lack endotoxic activity but possess LPS-antagonizing properties.[1]

Substitution of the backbone phosphate groups constitutes a common feature of lipid A, and in Table 10 such substituting groups are summarized. In general, they are charged and not present in stoichiometric amounts. Substituents at either the glycosidic or the ester-bound (position 4) backbone phosphate group include phosphate (*E. coli*), aminoethanol (*S. enterica*), aminoethanolphosphate (*N. meningitidis*),[178] and L-Ara4N (*Proteus mirabilis, Y. pestis, K. pneumoniae*).[179]

Finally, phosphate groups may be partly or completely absent from lipid A. This is the case, for example, in *Bacteroides fragilis* lipid A, which lacks the nonglycosidic phosphate group at position 4.[184] A most remarkable phosphate-free structural variant of lipid A has been identified in *Rhizobium leguminosarum* bv. Phaseoli.[185,186] Here, during biosynthesis, the 4′-phosphate group is enzymatically removed and replaced by an α-linked D-GalA residue. Also, the glycosidic phosphate is cleaved off, followed by oxidation of GlcN I at C-1, yielding 2-amino-2-deoxygluconic acid carrying (long chain) acyl residues in amide and ester linkage. Thus, lipid A of *R. leguminosarum* carries a total of two negative charges at the nonreducing and the reducing backbone units, however, not in the form of phosphate residues, but as carboxylate groups.

Lipid A has been shown to constitute the endotoxic principle of LPS. It further represents, together with the Kdo-containing inner core, the structurally most conserved region of LPS. Based on the results of chemical analyses, *E. coli* type lipid A has been chemically synthesized.[187] The demonstration of identity of bacterial and synthetic *E. coli* lipid A in all chemical, physicochemical, physical, and in particular biological parameters, unequivocally verified the previously deduced and proposed structure to be correct.

Table 10 Phosphate-linked substituents of the lipid A backbone [4′-P-β-D-GlcpN-(1 → 6)-α-D-GlcpN-1-P].

	Substituents		
	C-4′	*C-1*	*Ref.*
Escherichia coli	—	P[a]	180
Klebsiella pneumoniae	L-Arap4N[bc]	L-Arap4N[c]	120, 181
Legionella pneumophila			153
Neisseria meningitidis	P-EtN[d]	P-EtN	178
Pseudomonas aeruginosa			182
Salmonella enterica sv. Minnesota	L-Arap4N	EtN	180
Vibrio cholerae	—	P-EtN	183
Yersinia pestis	L-Arap4N	D-Araf[e]	179

[a]P = nonstoichiometric phosphate substitution. [b]L-Arap4N = 4-amino-4-deoxy-L-*arabino*-pentopyranose. [c]In lipid A of serotype O:3, but not in lipid A of serotype O:1. [d]EtN = 2-aminoethanol. [e]D-Araf = D-*arabino*-pentofuranose.

3.09.4 PHYSICOCHEMISTRY OF LIPOPOLYSACCHARIDE

3.09.4.1 Aggregate Structure, Molecular Conformation, and Phase State

Lipids and glycolipids, like LPS, are amphiphilic molecules consisting of a hydrophilic, polar headgroup and hydrophobic apolar hydrocarbon chains and are main constituents of cell membranes. A prerequisite for normal cell functioning is the maintenance of a particular composition of the lipid matrix at given ambient conditions.[188] Disturbances of this composition, for example, by uptake of extraneous lipids which differ in their chemical structure (e.g., acylation pattern, headgroup conformation, net electrical charge) from that of the normal constituents of the cell matrix, may lead to: (i) alterations of membrane fluidity and/or permeability, (ii) phase separation and domain formation, (iii) disturbance of the lamellar membrane architecture, and (iv) internalization of the extraneous lipids. In many cases, the cell may be able to compensate for the changes by altering the composition of the lipid matrix, a phenomenon termed "homoviscous adaptation."[189] If this is not possible, certainly not on a timescale of minutes, anyone of these membrane alterations may cause severe dysfunctions of the cell. These may manifest themselves, for example, in transient or permanent alterations in the functioning of transmembrane proteins, which might be involved in signal transduction. The membrane alterations and their influence on cell functioning will be the more severe the more the chemical structures of the constituents and the interacting lipids differ.

The interaction of endotoxins or free lipid A with host cells of the immune system (monocytes/macrophages) may be discussed in this context, because the intercalation of LPS molecules is considered to be an important step in the activation cascade.[190,191] Various types of interaction between endotoxin and host cells, receptor-dependent and receptor-independent, which may finally lead to an intercalation have been described in the literature and have been briefly reviewed above. At low LPS concentrations, the receptor-dependent and at high concentrations, the receptor-independent pathway seem to be effective. The prevalent LPS concentration, however, influences strongly the availability of endotoxin molecules in the monomeric or aggregated form.

Amphiphilic molecules, in general, tend to aggregate and to form multimeric clusters in an aqueous environment above a critical aggregate concentration (CAC). The absolute values of the CAC for LPS of various chemical structures will be likely to depend on the acylation pattern, the length of the sugar moiety, and the charges. Therefore, of all endotoxin structures, they should be lowest for lipid A, but have not yet been determined due to extreme experimental difficulties in the low concentration range. From a comparison of the limited number of available data for other lipids, only a rough estimate can be made. Thus, from the change in CAC from 5×10^{-10} M for dipalmitoylphosphatidylcholine to 7×10^{-6} M for lysopalmitoylphosphatidylcholine, which have the same headgroup, but differ in the number of fatty acid residues by a factor of two, for hexaacyl lipid A, a value of well below 10^{-10} M must be assumed on the basis of a value ($< 10^{-7}$ M) published for the lipid A precursor lipid Ia, a tetraacyl lipid A.[192,193] Under normal experimental *in vitro* conditions, such concentrations seem realistic, and endotoxins should therefore form aggregates. The structure of these supramolecular aggregates depends in its geometry on the primary chemical structure of the contributing LPS molecules and on their secondary structure, which corresponds to the molecular shape or conformation of the individual molecules. Aggregation is promoted by hydrophobic interaction or, more precisely, the minimization of the Gibbs free energy of the water–amphiphile system, which is accomplished by the tendency of free water to increase its entropy.[192] The molecular conformation of a given LPS is not constant, but rather depends on the phase state of its hydrocarbon moiety. Thus, the same LPS molecule may exist in a highly ordered gel state below and in a less ordered liquid crystalline (fluid) state above a phase transition temperature T_c which depends, among other parameters, on the length and the degree of saturation of the acyl chains. Changes in the phase state may, therefore, also provoke a change in the supramolecular structure. This interrelationship is further complicated by the influence of other ambient conditions such as water concentration and the presence of ions, negative or positive, which interact with charged groups of the endotoxin molecules.

As outlined above, the conformation and the state of order of the acyl chains are characteristics of the individual LPS molecules. Only theoretical methods like molecular modeling using energy minimization calculations allow the determination of these characteristics for an individual molecule; however, they are normally performed for a molecule in vacuum without consideration of ambient conditions.[194] An experimental determination of these characteristics has to be performed on the aggregate. The relation between the molecular shape of an individual molecule and the three-dimensional supramolecular structure of aggregates formed above the CAC can be described in a simple geometric model by a shape factor $S - a_h/a_0$, which relates the final structure to the ratio of the effective cross-sectional areas of the hydrophilic polar, a_0, and the hydrophobic apolar, a_h, regions of the endotoxin molecules.[192,195] A further factor influencing the final supramolecular structure is the presence or absence of a prominent axis of the molecules (anisotropy) originating, for example, from headgroup charges or from an asymmetric distribution of the acyl chains. In Figure 5, the most relevant supramolecular structures for lipid A—lamellar (L), cubic (Q), hexagonal (H_{II})[196]—are depicted, together with synchrotron small angle X-ray diffraction patterns obtained from these structures. This figure points out, at the same time, the possibility of obtaining information on the conformation of individual lipid A molecules from the X-ray patterns.

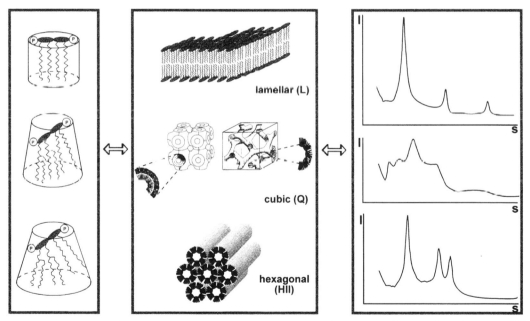

Figure 5 Molecular conformation, supramolecular structures, and corresponding small-angle X-ray diffraction spectra of endotoxins.

It is obvious that the conformation is influenced by the phase state of the hydrocarbon chains via an increase or decrease of a_h due to the presence (liquid crystalline state) or absence (gel state) of *gauche*-conformers or via an increase or decrease of a_0 by the interaction of positive or negative ions with headgroup charges due to bridging effects, and by variations in the hydration state, in particular of the sugar moiety. The phase behavior of LPS from Enterobacteriaceae shows a characteristic dependence on the length of the sugar moiety of the various LPS chemotypes and of lipid A. Thus, the T_c values as determined by Fourier-transform infrared or fluorescence spectroscopy are highest for lipid A (around 45 C), lowest for deep rough (e.g., Re) mutant LPS (around 30 C), and increase with increasing sugar length toward completion of the O-chain (wild-type LPS) up to around 37 °C. These values have been determined under near physiological water contents at neutral pH and in the absence of divalent cations.[197–200] Reducing the water content leads to a gradual vanishing of the phase transition, which might be explained by the strong lyotropic behavior of endotoxins. The presence of divalent cations leads to a concentration-dependent shift of T_c to higher values.[198,201]

In previous investigations it has been found that a peculiar chemical structure of lipid A is a prerequisite for the expression of full biological activity.[1,175,202,203] The lipid A backbone composed of a β-(1→6)-linked D-GlcN disaccharide substituted with two phosphate groups (positions 1 and 4′) and six fatty acid residues of defined chain length (10–16 carbon atoms) and in a defined linked distribution to the reducing and nonreducing GlcN residues constitute the minimal structure for the expression of various biological effects. Concomitant with changes in the chemical structure, however, are changes in the physical conformation. The concept of a unique endotoxic lipid A conformation was, therefore, extended by correlating the three-dimensional structures of supra-molecular aggregates of various lipid A isolates from entero- and nonenterobacterial sources, differing in their primary chemical structure, with their biological activities. Figure 6 shows an example of X-ray small-angle diffraction patterns of a biologically active lipid A (isolated from LPS of the *S. enterica* serovar Minnesota Re mutant R595) at near physiological water content for two different molar ratios of lipid A and Mg^{2+} and for various temperatures. Clearly, for both Mg^{2+}-concentrations, a triphasic behavior is observed with a structural sequence $Q_1 \leftrightarrow Q_2 \leftrightarrow H_{II}$ for the Mg^{2+}-free (a) and $L \leftrightarrow Q \leftrightarrow H_{II}$ for the Mg^{2+}-containing sample (b). Q, Q_1, and Q_2 are cubic structures with different symmetries. Detailed studies have investigated lipid A samples which comprised those of bis- and monophosphoryl (lacking the 1-phosphate group) structures of *E. coli* and *S. enterica* serovar Minnesota and the lipid A from *C. violaceum*, *Rhodocyclus gelatinosus*, *Rhodobacter capsulatus*, *Rhodopseudomonas viridis*, *Rhodospirillum fulvum*, and *C. jejuni*.[204–206] In Figure 7, the three-dimensional supramolecular structure and molecular conformation, the state of order at 37 °C, which is roughly correlated with the phase transition temperature T_c, and the biological activities are listed for the various lipid A samples. For the latter, only relative statements can be made as endotoxic activities like TNF-induction may, for example, vary between different laboratories and may be dependent on the isolation and purification procedure of lipid A. Therefore, differences in biological activity are assumed to be significant only if they comprise at least one order of magnitude (see below).

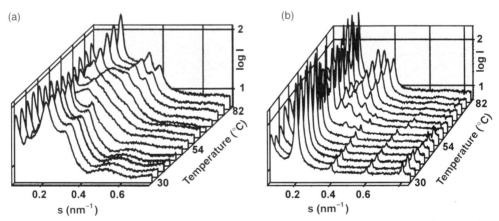

Figure 6 Small-angle X-ray diffraction patterns of free lipid A from *S. enterica* serovar Minnesota R595 LPS in dependence on temperature at 85% water content in the absence (a) and the presence (b) of an equimolar content of Mg^{2+}.

It was found that the lipid A of *C. violaceum*, *R. capsulatus*, *R. viridis*, and *R. fulvum* assume a lamellar structure, the bisphosphorylated lipid A from *E. coli* and *S. enterica* serovar Minnesota a cubic structure, and the lipid A from *R. gelatinosus* an inverted hexagonal structure, respectively, whereas the three-dimensional structures of the monophosphoryl lipid A of *S. enterica* serovar Minnesota and lipid A of *C. jejuni* are a mixture of lamellar and cubic phases.[204,205]

Literature data for biological activities are not available for all lipid A samples. In these cases, data for the parent LPS were used which appear justified in light of the fact that lipid A represents the endotoxic principle of LPS. It was found that in comparison to LPS and lipid A from *S. enterica* serovar Minnesota, lethality (LD_{50} in mice) and pyrogenicity (MPD-3 in rabbits) are three to four orders of magnitude lower for LPS and lipid A of *R. capsulatus* and *R. viridis*, but are similar for

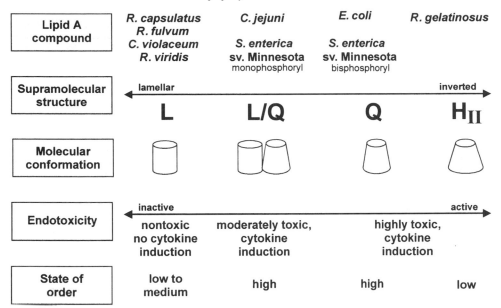

Lipid A compound	*R. capsulatus* *R. fulvum* *C. violaceum* *R. viridis*	*C. jejuni* *S. enterica* sv. Minnesota monophosphoryl	*E. coli* *S. enterica* sv. Minnesota bisphosphoryl	*R. gelatinosus*
Supramolecular structure	lamellar ← **L**	**L/Q**	**Q**	inverted → **H$_{II}$**
Molecular conformation				
Endotoxicity	inactive ← nontoxic no cytokine induction	moderately toxic, cytokine induction	highly toxic, cytokine induction	active →
State of order	low to medium	high	high	low

Figure 7 Supramolecular structure and molecular conformation, endotoxicity, and state of order of the acyl chains of various lipid A from entero- and nonenterobacterial strains.

LPS from *R. gelatinosus*.[207] Similarly, LPS of *R. viridis* lacks TNF-induction capacity (murine peritoneal macrophages). The lethality and pyrogenicity doses of monophosphoryl lipid A from *S. enterica* serovar Minnesota were one to two orders of magnitude lower than observed with the bisphosphoryl compound or the parent LPS,[208,209] whereas cytokines such as IL-1 and TNF were induced to a comparable degree.[210,211] The biological activity of *C. jejuni* LPS as estimated from pyrogenicity and TNF-inducing capacity was found to be 50- to 100-fold lower than that of *Salmonella* LPS,[212] whereas that of lipid A from *R. fulvum* and *C. violaceum* (IL-6 inducing capacity) is three to four orders of magnitude lower than that of *Salmonella* LPS (authors' own unpublished results). The tendency of lipid A to adopt a nonlamellar (cubic or H$_{II}$) structure is, thus, directly related to the ability to express biological activity or, vice versa, lipid A samples preferring a lamellar organization exhibit no or only low activity.

From these results it is concluded that the basic determinant for endotoxicity is the conformation of the lipid A component, both in its free form and as a constituent of LPS. Addition of glycosyl residues such as Kdo and Hep does not change the conformation of the lipid A component. Such substituents rather modify the CAC and the "solubility" (aggregate size) within aqueous media, and the fluidity of the hydrocarbon chains (at 37 °C) and, therefore, may modify endotoxin activity. From the results for lipid A of *R. capsulatus* and *R. gelatinosus* (see Figure 7) it becomes obvious that the fluidity *per se* is not a determinant of biological activity. In both cases, the state of order is very low (high fluidity), but only lipid A from *R. gelatinosus* with its preference for the H$_{II}$ structure is biologically active.

However, data obtained with lipid A of *C. jejuni* suggest a modulating influence of fluidity. Even though the main fraction of lipid A of *C. jejuni* adopts a cubic structure, it exhibits only relatively low activity. The main difference of this lipid A as compared to all lipid A investigated so far is its very low acyl chain mobility (highest order parameter).[212] Therefore, due to the limited amount of data available, at present it is not possible to define the significance of fluidity more precisely.

3.09.4.2 LPS–Cell Membrane Interactions

In previous studies, a nonspecific intercalation of LPS into liposomes and mammalian cells by mere hydrophobic interaction was described and discussed as a possible mechanism of cell activation.[11,213,214] However, an intercalation of endotoxin aggregates (free lipid A and R-LPS of *S. enterica* serovar Minnesota) on a short timescale (up to 0.5 h) with a phospholipid membrane resembling the composition of the macrophage membrane could be excluded. On a long timescale, however, clear evidence for a nonspecific uptake of endotoxin molecules in phospholipid membranes was found.[11] This can be understood on the basis that even above the CAC, LPS is also present in the monomeric form—the monomers being in equilibrium with the aggregates—and that these monomers may be able to intercalate into the phospholipid membrane, thus shifting the equilibrium and releasing more and more monomers from the aggregates. An LBP-mediated, CD14-independent intercalation of LPS into phospholipid membranes has been described.[12] Here, LBP obviously acts as a lipid transfer protein, disaggregating the LPS assemblies and inserting the monomers or oligomers into the phospholipid bilayer.[215]

Such a monomer or oligomer driven process would provide an explanation for the previous observation of a more pronounced biological activity of LPS in a highly disaggregated as compared with an aggregated form,[216] would be in agreement with results describing an enhancing effect of LBP due to its disaggregating capacity,[217,218] and would contribute to an understanding of the observation of cell activation at higher endotoxin concentrations (0.5–1 µg mL^{-1}) also in the absence of CD14.[219] However, if endotoxin monomers or small oligomers were the biologically active units, the importance of a consideration of the headgroup influence on the physicochemical behavior of an individual endotoxin molecule in a phospholipid environment with very different physicochemical properties than an environment of identical endotoxin molecules, had to be judged. However, this represents an unsolved problem.

It seems likely that, independent of the kind of aggregation, an important prerequisite for biological activity is the conical shape of the individual lipid A molecule with a slightly higher cross-section of the hydrophobic than of the hydrophilic moiety leading, in aggregated form, to nonlamellar inverted structures. This molecular shape deviating from a cylindrical geometry may cause a strong disturbance in the target cell membrane—the latter being necessarily in a lamellar state—and may provide a trigger signal to a specific membrane protein for cell activation. Whether the action of endotoxin in the target cell membrane results from monomeric units or from domains with a larger number of single molecules cannot be decided.

Another interesting aspect of cell activation by endotoxin arises from the fact that a membrane being composed only on one side of phospholipids but on the other side of LPS has a completely different inner membrane potential than a phospholipid bilayer.[220] Thus, the presence of LPS molecules only on one side of the host cell membrane, the outer leaflet, in the vicinity of a signalling protein—possibly a voltage-dependent ion channel—may cause changes in membrane potential finally leading to channel gating.[221,222]

3.09.5 BIOSYNTHESIS OF LIPOPOLYSACCHARIDE

Significant progress has been made in the elucidation of various important steps of LPS biosynthesis during the 1980s and 1990s. The construction of specific LPS mutants by traditional genetic approaches in combination with molecular genetics and recombinant DNA technology not only greatly facilitated the understanding of individual enzymatic steps of LPS biosynthesis but also provided insights into the complex assembly processes of its constituent parts to give the complete LPS molecule in the outer membrane. LPS biosynthesis of a growing list of various Gram-negative bacteria is being investigated, but is still best understood for *E. coli* and *S. enterica* strains. Biosynthesis of LPS requires a series of coordinated individual steps including the synthesis of activated precursors in the cytoplasm of the bacterial cell, formation of polysaccharide repeating units, polymerization of the repeating units, translocation across the cytoplasmic membrane, transport to and integration into the outer membrane of the assembled molecules, and regulation of both the individual steps and the entire biosynthetic pathway. In addition, gene products of temperate bacteriophages such as transferases or polymerases are able to replace or modify bacterial enzymes for further modifications of the LPS structure. Although each of these steps has been studied to a varying extent, it became evident from structural, biochemical, or genetic data that chemical heterogeneity of the surface-exposed LPS molecule and polymorphism of its determining genes emphasize the enormous flexibility of Gram-negative bacteria to respond to changing environmental conditions. However, despite the structural variability of LPS and its genetic polymorphism, LPS biosynthesis was shown to be based on general principles. Obviously, to facilitate the generation of a complex molecule such as LPS, pathways for synthesizing the constituent parts of the LPS molecule as separate blocks have been evolved. Accordingly, LPS is generated from two separate components, the O-specific polysaccharide side chain and the lipid A-core oligosaccharide portion, followed by their ligation and modification. In fact, lipid A, the core oligosaccharide, and the O-specific polysaccharide can be differentiated not only by their different degree of structural conservation, but also by their biosynthesis and genetic determination. The genes for biosynthesis of the constituent parts of the LPS molecule are primarily organized into clusters of contiguous genes found in different regions on the bacterial chromosome. Lipid A biosynthesis is mainly determined by the *lpx* genes. Genes for the synthesis of the core oligosaccharide are located in the *rfa* gene cluster, whereas the genes involved in the synthesis of the O-specific polysaccharide are clustered in the *rfb* region of the chromosome. In accordance with the individual steps required for LPS assembly, the *rfa* and *rfb* gene clusters carry information for formation of the polysaccharide repeating units, their polymerization, translocation, and synthesis of unique precursors for LPS-specific sugars. Some of the LPS precursors are not exclusively used for LPS biosynthesis but represent common intermediates in the housekeeping metabolism of the bacterial cell. Their synthesis is usually not directed by the LPS gene clusters, but is encoded on general housekeeping genes. The interplay of manifold biosynthetic reactions in LPS assembly requires complex regulatory mechanisms. Therefore, it is not surprising that knowledge regarding regulation of LPS biosynthesis is limited. Moreover, multiple sensing mechanisms of the bacterial cell and correlations in the rate of synthesis of cell surface components as a whole make the characterization of the regulatory systems difficult.

Numerous excellent reviews have been published, which focus on general or specific aspects of LPS and polysaccharide biosynthesis in different Gram-negative bacteria.[107,223–228] Since an overview of main features of LPS biosynthesis is presented, the reader is referred to these publications for detailed information on specific topics or various bacteria, the LPS biosynthesis of which is being studied. In accordance with a proposal for a new Bacterial Polysaccharide Gene Nomenclature (BPGN) scheme,[229] an attempt is made to keep, in addition to former gene designations, the recommended new BPGN names, which have been applied particularly to known whole polysaccharide gene clusters of *E. coli*, *S. enterica*, *K. pneumoniae*, *Y. enterocolitica*, *Y. pseudotuberculosis*, *P. aeruginosa*, and *B. pertussis*. As a gene/protein designation will appear for the first time, the proposed new name will be given in square parentheses and, instead of the old gene/protein symbol, will be further used throughout the text. To avoid confusion, however, the traditional names *rfb* and *rfa* will be used to specify the polymorphic loci involved as a whole in the biosynthesis of O-polysaccharide and core oligosaccharide, respectively.

3.09.5.1 Biosynthesis of Activated Monosaccharide Precursors

3.09.5.1.1 Link between central metabolism and LPS biosynthesis

Initial steps in LPS biosynthesis involve the synthesis of the constituent monosaccharide residues as high energy nucleotide sugars. In particular, the central metabolic intermediates UDP-Glc, UDP-Gal, and UDP-GlcNAc can serve as glycosyl donors for both the core oligosaccharide and the O-polysaccharide. They are synthesized in several successive reactions usually encoded by various housekeeping genes (Figure 8). Up to five genes are involved in the catabolism of exogenous galactose via the LeLoir pathway in enterobacteria.[230] Galactokinase (*galK*) catalyzes the phosphorylation of Gal to Gal1P, galactose-1-phosphate-uridylyltransferase (*galT*) the transfer of the UDP residue from UDP-Glc to Gal1P, UDP-galactose-4-epimerase (*galE*) the reversible conversions of UDP-Gal and UDP-Glc, UTP-glucose-1-phosphate-uridylyltransferase (*galU*) the formation of Glc1P, and phosphoglucomutase (*pgm*) the reversible transformation of Glc1P into Glc6P. In *E. coli*, *S. enterica*, and *Klebsiella* species, the structural genes *galK*, *galT*, and *galE* are organized in an operon inducible with galactose,[231–233] whereas the arrangement of these genes in other Gram-negative bacteria was found to be highly variable. Remarkably, it is not uncommon that particularly *galE* lacks a linkage with *galT–galK* or that homologues of *galE* have been found in bacteria unable to utilize galactose by the LeLoir pathway. The *galE* genes of *Erwinia amylovora* or *Erwinia stewartii* were shown to be separated from the remaining *gal* operon and were located near the genes for exopolysaccharide (EPS) synthesis.[234,235] Furthermore, independently of the presence of Gal in the culture medium, *galE* was constitutively expressed, and mutations in *galE* affected the synthesis of both the EPS and the O-specific polysaccharide. The observation that *galE* in *E. amylovora* is flanked by repeated sequences suggested a transposition event during evolutionary separation of the gene from the *gal* operon. In *H. influenzae* type b, *galE* was shown to be one of the genes of the *lic3* locus for LPS biosynthesis.[236] Deletions of the metabolic genes *galE* and *galK* of the remaining *gal* locus resulted in a loss of the Gal-containing structure in the LPS molecule.[237,238] In *Y. enterocolitica* serotype O:3, *galE* probably belongs to the *trs* [*wbc*] operon which, in addition to the *rfa* gene cluster, appears to be required for LPS core assembly.[239] In contrast, although *galE* has been shown to be linked to the other genes of the galactose utilization pathway in *Y. enterocolitica* serotype O:8, an additional gene (*lse*) with high degree of similarity to *galE* could be identified in the locus for O-antigen biosynthesis.[240] However, studies on Lse revealed that it was indeed involved in the completion of the LPS molecule, although it did not exhibit UDP-galactose-4-epimerase but, presumably, UDP-*N*-acetylglucosamine-4-epimerase activity.[241] In all of the *N. meningitidis* group A and group B strains tested, two copies of *galE* have been found in the chromosome, from which only one was a functional *galE* gene located in the *cps* locus for biosynthesis of CPS.[242] The second copy lacked part of the coding sequence and, therefore, did not encode a functional protein. The essential role of *galE* for the incorporation of Gal into LPS could be shown once more by mutagenesis of *galE*, which led to an apparent reduction in LPS molecular weight and a loss of reactivity of monoclonal antibodies recognizing Gal-containing structures. In addition, the correlation of lack of *galE* activity with the absence of Gal in LPS could be confirmed by compositional analysis of LPS from the transposon *Tn*916 insertion-derived *N. meningitidis* mutant NMB-SS3.[243] Apart from the apparent absence of a second *galE* copy in *N. gonorrhoeae*, *galE* was located in the gonococcal homologue of the meningococcal B capsule region D, and mutagenesis of the gene resulted in a deep-rough LPS phenotype.[244] Interestingly, the gonococcus is unable to synthesize a capsule probably due to a loss of the other CPS-encoding regions except the *galE*-containing region D, emphasizing the importance of *galE* for LPS biosynthesis. Finally, the predicted amino acid sequence of *orf6* within the *rfb* region of *V. cholerae* O:139 strain MO45 displayed a high degree of similarity to the *galE* gene products of *N. gonorrhoeae* and *E. amylovora*.[245] Taken together, the metabolic gene *galE* is closely associated with LPS biosynthesis responsible for maintaining the balance between UDP-Glc and UDP-Gal levels. Since Gal is often found in high amounts in macromolecular EPS, CPS, or LPS, their synthesis depends on the availability of UDP-Gal. Therefore, the splitting of the *gal* locus appears to enable the bacterial cell to regulate the *gal* genes differentially and to coordinate the expression of *galE* with the synthesis of Gal-containing cell surface molecules.

Definitive proof has been presented that the *galU* gene of *E. coli* codes for the enzyme UTP-glucose-1-phosphate-uridylyltransferase catalyzing the reversible conversion of Glc1*P* and UTP into UDP-Glc and inorganic pyrophosphate.[246] Mutations in *galU* interfere with the synthesis of UDP-Glc, the reduced supply of which probably could have direct influence (or via UDP-Gal) on the biosynthesis of cell surface carbohydrate structures. For example, *galU* mutants of *E. coli*, *K. pneumoniae*, or *S. flexneri* were unable to incorporate the first Glc into the core oligosaccharide, which resulted in the synthesis of a truncated LPS molecule (compare Tables 2, 5, and 7).[247-251] However, the phenotypes of *galU* mutants appeared rather pleiotropic as shown for *E. coli*, which additionally lost the ability to export the minor outer membrane protein TolC or to produce flagella.[252,253] The reduced amounts of flagellin were attributed to lowered transcription rates of flagellar genes. This is of particular interest, because UDP-Glc was proposed recently as a potential intracellular signal molecule that controls in *E. coli* the expression of the σ^s subunit of RNA polymerase and σ^s-dependent genes.[254] Notably, similar effects on gene expression could be observed for mutants of phosphoglucomutase (*pgm*) or phosphoglucose isomerase (*pgi*), the latter being responsible for the reversible isomerization of Glc6*P* and Fru6*P*,[255] but not for mutants defective in *galE*. In any case, it remains to be elucidated, whether LPS biosynthesis directly depends on the supply of UDP-Glc as a glucosyl donor and/or whether transcription of the respective genes is regulated by a cellular system sensing the levels of UDP-Glc. The derived amino acid sequence of *galU* displayed high similarity to that of *galF* (*rfb*2.8) within or immediately adjacent to the *rfb* gene cluster for O-polysaccharide biosynthesis in *S. enterica*,[256,257] *E. coli* strains,[258,259] or *S. flexneri*.[260] Therefore, it was previously suggested that *galF* probably codes for a second UTP-glucose-1-phosphate-uridylyltransferase not essential for O-polysaccharide formation, but presumably necessary for a sufficient supply of glycosyl donors for LPS and CPS biosynthesis.[246,256] However, the originally proposed function of GalF to be involved in the modification of GalU was supported by genetic and biochemical data.[261,262] Evidence was provided that the GalU and GalF proteins of *E. coli* serotype O:7 are not functionally interchangeable but, indeed, interact physically to form an oligomeric enzyme.[263] Obviously, GalF represents a novel regulatory component of GalU, conferring a higher thermal stability on the enzyme and increasing the intracellular amount of UDP-Glc by suppression of pyrophosphorolysis, the reversible synthesis reaction from UDP-Glc to Glc1*P* in the presence of inorganic pyrophosphate.[263]

Mutations in the *pgm* gene of several *N. meningitidis* and *N. gonorrheae* strains caused a complete loss of phosphoglucomutase activity, which blocked the formation of Glc1*P* from Glc6*P*.[264,265] With this defect, neither UDP-Glc nor UDP-Gal could be generated preventing the attachment of Glc to the growing LPS chain. Analogous results were obtained for *E. coli* K-12 *pgm* deletion mutants that produced a rough-type LPS lacking Glc and Gal.[266,267] The Glc1*P* metabolism was not completely

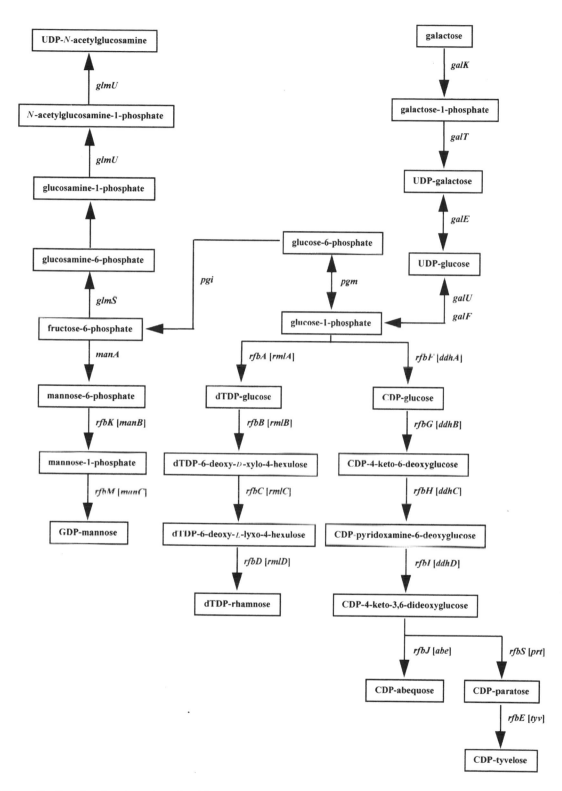

Figure 8 Supply of activated nucleotide sugar precursors by-products of housekeeping and polysaccharide biosynthesis genes for the O-specific chain in *S. enterica*. The new BPGN names for LPS genes are given in square parentheses.[229]

abolished in these strains, but there was no indication of the presence of other enzymes which exhibit phosphoglucomutase activity. It is intriguing to note that the *pgm* mutation in *E. coli* could be complemented by the *algC* gene of *P. aeruginosa* previously shown to encode a protein with phosphomannomutase activity.[268,269] Indeed, it was demonstrated that AlgC is bifunctional, involved in both the synthesis of Glc1*P* and Man1*P*, suggesting an important physiological role in the formation of EPS and LPS. Studies on *algC* mutants revealed that the loss of alginate production was accompanied by the inability of bacteria to attach Glc residues to Gal*N* in the LPS core and to incorporate Rha into the O-specific polysaccharide (see Tables 8 and 9). Therefore, the lack of the antigenically conserved A-band LPS was not merely a result of the inability to bind the O-specific polysaccharide to a truncated LPS core, but indicated the need for AlgC activity for both O-side chain and core synthesis.[268,270,271]

Apart from being a main precursor for peptidoglycan and the enterobacterial common antigen (ECA), UDP-Glc*N*Ac is utilized for synthesis of the Glc*N* domain of lipid A, the indispensable function in the outer membrane of which has hindered the construction of viable mutants defective in precursor formation. It is, therefore, assumed that UDP-Glc*N*Ac is synthesized from Fru6*P* by a four-step reaction (Figure 8). In *E. coli*, the first step is mediated by the gene *glmS*, which codes for the enzyme L-glutamine-D-fructose-6-phosphate-amidotransferase previously shown to be essential for the bacterial cell, if Glc*N* or Glc*N*Ac were not supplied as a carbon source in the culture medium.[272–274] The coding gene *glmM* for the second reaction was identified in 1996,[275] although phosphoglucosamine mutase activity could be detected previously in *E. coli* crude extracts suggesting the conversion of Glc*N*6*P* to Glc*N*1*P* in the pathway leading to UDP-Glc*N*Ac.[276] Notably, evidence was provided that GlmM undergoes phosphorylation resulting in an enzymatically active form of the protein. Thus, the results obtained appeared indicative of a specific regulatory mechanism in form of the phosphorylation extent of GlmM, which could adjust the synthesis of Glc*N*1*P* to specific requirements of LPS and peptidoglycan biosynthesis.[275] The third and fourth steps in the formation of UDP-Glc*N*Ac, catalyzed by glucosamine-1-phosphate-acyltransferase and *N*-acetylglucosamine-1-phosphate-uridylytransferase, are carried out by the bifunctional enzyme GlmU.[276,277] The encoding gene could be identified immediately upstream of *glmS*, and mutational studies on the essential *glmU*, only possible in the presence of the wild-type gene on a plasmid with a thermosensitive replicon, revealed under nonpermissive temperatures that the resulting inhibition of peptidoglycan and LPS biosynthesis induced a progressive loss of the rod shape of bacterial cells before their lysis occurred, indicating the location of GlmU at a branch point in the synthesis of essential components of the cell envelope.[278] It is interesting to note that the separable active sites for acetyl- and uridyl transfer could be assigned to discrete N-terminal and C-terminal fragments of GlmU, respectively, catalyzing first the formation of Glc*N*Ac1*P* and subsequently the synthesis of UDP-Glc*N*Ac.[276,277] Furthermore, the identification of an active site for acetyl transfer within GlmU was consistent with the finding that the N-terminus of GlmU displayed significant similarities to prokaryotic acyl- or acetyltransferases including the gene product of *firA/lpxD*, UDP-3-*O*-(*R*-3-hydroxymyristoyl)-glucosamine-*N*-acyltransferase, which mediates the third step of lipid A biosynthesis (see below).[276]

The biosynthetic steps leading to the provision of the central metabolic intermediates were suggested as suitable points for regulation of polymer synthesis.[227] One interesting example for the modulation of the intracellular carbon flux in *E. coli* has been provided by the discovery of the *csrA* gene product as part of a potential adaptive response pathway, which regulates several essential enzymes of the central carbohydrate metabolism such as phosphoglucomutase or phosphoglucose-isomerase being negatively or positively regulated by *csrA*, respectively.[279]

3.09.5.1.2 *Synthesis of unique precursors of O-specific sugars*

As summarized above, gene products of the housekeeping metabolism are not primarily confined to LPS biosynthesis, but play an essential role in providing glycosyl donors in the form of activated nucleotide sugars. The nucleotide sugars are either directly incorporated into the growing LPS chain or, to complete the process of precursor formation, may be used by gene products of the LPS gene clusters for further synthesis of unique precursors for LPS-specific sugars. For example, correlations between the products of housekeeping and LPS-specific genes in the synthesis of precursors of O-specific polysaccharides in *S. enterica* are outlined in Figure 8. The activated nucleotide sugars are all, with the exception of GDP-Man, derived from Glc1*P* and a nucleoside triphosphate. The coordination of biosynthesis of each activated precursor appears to be achieved by clustering of the corresponding genes for each biosynthetic pathway and, therefore, by organization of the *rfb* cluster into three blocks (Figure 9).

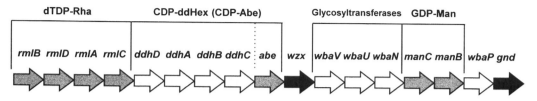

Figure 9 Schematic presentation of the *rfb* gene cluster of *S. enterica* serovar Typhimurium (group B).[107,229]

The formation of the activated precursor GDP-Man requires a three-step reaction. The isomerization of Fru6P to Man6P is mediated by the action of phosphomannose isomerase encoded by *manA* outside the *rfb* gene cluster.[280–282] Mutations in the *manA* gene, which is also involved in Man catabolism, resulted in the inability of the bacterial cells to form Man-containing O-specific polysaccharides.[283,284] The synthesis of GDP-Man is completed via Man1P by the enzymes phosphomannomutase and GDP-mannose-pyrophosphorylase, which are specific to the GDP-Man pathway, in enterobacteria usually encoded by the genes *rfbK* [*manB*] and *rfbM* [*manC*], respectively. The genes were located adjacent to one another on various chromosomal *rfb* regions for the synthesis of O-specific polysaccharides in *S. enterica* serovars Typhimurium,[256] Montevideo,[285] and Anatum,[257] as well as *E. coli* strains VW187 (O:7, K:1),[286] F719 (O:9, K⁻),[287] and E69 (O:9, K:30).[288] Furthermore, *manB*$_{EcO111}$ and *manC*$_{EcO111}$ were identified as part of the GDP-Col pathway via GDP-Man in *E. coli* O:111,[289] and homologues of the genes presumably might be involved in the formation of GDP-Man in *V. cholerae* O.139.[245] However, the GDP-Man pathway genes were highly variable even within the serovars of *S. enterica*. The close structural relationship of *S. enterica* serovars A, B, D1, D2, and E1, sharing a Man–Rha–Gal trisaccharide backbone (see Table 3), correlated with a conservation among their *manC-manB* genes, while in serovars C1 and C2 these genes were found to be different not only from those in the other serovars but also from each other.[256,257,290–294] These variations at the gene level were shown to find their expression in different repeating unit backbones in group C2 and C1 strains, which are composed of Rha–(Man)$_2$-Gal and (Man)$_4$-GlcNAc, respectively. Based on comparisons of the deduced amino acid sequences, it was shown that *manB* of group C1 appeared to be more closely related to the *cpsG* [*manB*$_{CA}$] gene of *S. enterica* group B strains than to the corresponding *manB*$_{Oag}$ gene.[297] It is intriguing that a duplicated *manC*²–*manB*² region could be identified in *E. coli* E69 (O9:K30), which was located downstream of *manC*¹–*manB*¹ near the remaining genes of the *rfb* cluster.[288] Despite the observation of an extensive polymorphism within the *manC* and *manB* regions among a series of *E. coli* strains of serotypes O.8 and O.9, the *rfb* genes were demonstrated to be involved in the synthesis of group I K:30 CPS. ManB$_{EcO9}$ and ManC$_{EcO9}$ were closely related to ManB$_{CA}$ and CpsB [ManC$_{CA}$], respectively, that were identified as isozymes of phosphomannomutase and GDP-mannose-pyrophosphorylase in *E. coli* K-12 and *S. enterica* serovar Typhimurium, thought to participate in the synthesis of GDP-Man on the pathway to GDP-fucose (GDP-Fuc).[295,296] GDP-Fuc may serve as the precursor for Fuc, which constitutes a component of the slime polysaccharide colanic acid.[297] It is interesting that enterobacteria obviously utilize separate phosphomannomutase and GDP-mannose-pyrophosphorylase genes for biosynthetic pathways leading to the formation of LPS or CPS, while *Xanthomonas campestris* and *P. aeruginosa* exploit the bifunctional enzymes *xanA* and *algC* with phosphoglucomutase/phosphomannomutase and *xanB* and *algA* with phosphomannose-isomerase/GDP-mannose-pyrophosphorylase activities for the generation of GDP-Man for both LPS and CPS synthesis.[270,298,299] In *V. cholerae* O:1, the *rfbA* [*manAC*$_{VcO1}$] gene codes for a protein, which has been predicted to act as a bifunctional enzyme displaying both phosphomannose isomerase and GDP-mannose-pyrophosphorylase activities, while *rfbB* [*manB*$_{VcO1}$] presumably codes for a putative monofunctional phosphomannomutase.[300] The two genes were assigned to an operon within the *rfb* region obviously involved in the synthesis of the O-antigenic polymer 4-amino-4,6-dideoxy-mannose (Rha4N, perosamine, see Table 8). In any case, on the basis of available sequence data, a number of genes encoding GDP-mannose-pyrophosphorylase or phosphomannomutase isozymes, respectively, were found to be related. In particular, the regions specifying the potential critical functional domains, such as the serine residue of the active site, the metal-binding pocket, and the sugar-binding site, could be assigned to bacterial phosphomannomutase/phosphoglucomutase enzymes by comparison with the crystal structure of the phosphoglucomutase from rabbit muscle.[270,301]

The GDP-Man pathway may be extended for further synthesis of unique precursors. Bacteria such as *E. coli* O:111, *Y. pseudotuberculosis*, *Y. enterocolitica* O:8, or various *S. enterica* serovars, having Fuc or Col in their O-specific polysaccharides, are expected to utilize phospho-mannomutase/GDP-mannose-pyrophosphorylase activities for the synthesis of GDP-Fuc or GDP-Col via the intermediate GDP-4-keto-6-deoxy-D-mannose.[241,289,294,302,303] The assumption that GDP-Man could be used in *P. aeruginosa* for biosynthesis of A-band common antigen LPS, mostly composed of Rha trisaccharide repeating units,[304] was initially based on observations of the inability of *algC* mutants to form A-band LPS (see above). A GDP-Man conversion protein for A-band LPS biosynthesis (*gca*) has been cloned and partially characterized.[305,306] Several lines of evidence suggested *gca* to code either for a dehydratase or a bifunctional enzyme facilitating the conversion of GDP-Man to GDP-Rha, for which a GDP-D-mannose-dehydrase and a GDP-4-keto-D-rhamnose-reductase are thought to be required.[307] Interestingly, although *gca* was not universally present in all *Pseudomonas* species tested, hybridization data for all 20 O-serotypes of *P. aeruginosa* indicated that the *gca* gene is well conserved.

The first of four successive reactions in the formation of dTDP-Rha, the immediate precursor for Rha residues of LPS, is the synthesis of dTDP-Glc, which is catalyzed by glucose-1-phosphate-thymidylyltransferase (Figure 8).[308,309] The subsequent conversion of dTDP-Glc to dTDP-6-deoxy-D-*xylo*-4-hexulose is mediated by dTDP-D-glucose-4,6-dehydratase, followed by the formation of dTDP-Rha, which is catalyzed by the two enzymes dTDP-6-deoxy-D-*xylo*-4-hexulose-3,5-epimerase and NADPH:dTDP-6-deoxy-L-*lyxo*-4-hexulose 4-reductase.[310,311] The enzymes for the dTDP-Rha pathway were shown to be encoded at the 5′-end of the *rfb* cluster by the corresponding genes *rfbA* [*rmlA*], *rfbB* [*rmlB*], *rfbC* [*rmlC*], and *rfbD* [*rmlD*] (Figure 9). Apart from varying degrees of similarity, comparative sequence analysis and hybridization data revealed a strong conservation of the dTDP-Rha pathway genes among enteric bacteria such as of the different serovars of *S. enterica*,[294], *S. dysenteriae* type 1,[312] *S. flexneri*,[260,313] *Y. enterocolitica* serotype O:3,[314] *E. coli* serotypes O:1, O:2, O:4, O:7, O:75, or O:141,[259] and *E. coli* K-12 W3110.[258] Interestingly, the functional order of the *rmlABCD* genes did not correlate with their map order within the *rfb* regions. The arrangement into the *rmlBDAC* gene block was found to be quite consistent among enterobacteria, while in *X. campestris* or *Leptospira interrogans* serovar Copenhageni, despite all similarities, the genes were suggested to be organized in the order *rmlBADC* and *rmlCDBA*, respectively.[315,316] Sequence similarities of *rmlB* and *rmlA* from *S. flexneri* and *E. coli* K-12 to *E. coli* K-12 genes *o355* and *o292* indicated that functional homologues of the dTDP-Rha pathway genes have been evolved within the *rfe-rff* [*wec*] gene cluster for synthesis of ECA, comparable with the functional duplications of the GDP-Man biosynthesis genes.[74,260] Based on biochemical and genetic studies on the dTDP-Rha biosynthesis region of *E. coli* VW187 (O7:K1), the *rffG/o355* [*rmlB*$_{ECA}$] and *rffH/o292* [*rmlA*$_{ECA}$] genes have been proven to be the functional homologues of *rmlB*$_{EcO7}$ and *rmlA*$_{EcO7}$ coding for dTDP-glucose-dehydratase and glucose-1-phosphate-thymidylyltransferase activities, respectively.[259] Since *N. gonorrhoeae* synthesizes a rough-type LPS lacking repeating units, it is intriguing that homologues for *rmlB*, *rmlA*, and *rmlD* were found downstream of the *galE* gene in the gonococcal homologue of the meningococcal B capsule region D.[317] However, null mutations in the *rmlBAD*$_{Ng}$ gene block did not lead to changes of the LPS phenotype. Therefore, it remains to be elucidated, whether the absence of a homologue for *rmlC* precludes the existence of a functional dTDP-Rha pathway or whether the homologues exhibit so far unrecognized functions.

3,6-Dideoxyhexoses (ddHex) have been identified as important antigenic determinants, which contribute to the serological specificity of immunologically active polysaccharides.[318] They were found almost exclusively within O-specific polysaccharides of LPS, e.g., Abe, Par, Tyv, and Col were shown to be present in *S. enterica* and, additionally ascarylose (Asc), in *Y. pseudotuberculosis* strains, whereas in *E. coli* the formation of ddHex appeared to be limited to Col. In *S. enterica* serogroups A, B, C2, D1, and D2, the formation of the cytidyl 5′-diphosphate sugars CDP-Abe, CDP-Par, and CDP-Tyv as precursors for ddHex was demonstrated to share an initial common pathway proceeding from Glc1P to CDP-4-keto-3,6-dideoxyglucose (Figure 8). The four genes *rfbF* [*ddhA*] (coding for glucose-1-phosphate-cytidylyltransferase), *rfbG* [*ddhB*] (coding for CDP-glucose-4,6-dehydratase), *rfbH* [*ddhC*] (coding for CDP-6-deoxy-D-*xylo*-4-hexulose-3-dehydratase), and *rfbI* [*ddhD*] (coding for CDP-6-deoxy-Δ3,4-glucoseen-reductase) of this pathway were located adjacent to the genes for synthesis of dTDP-Rha and were found in several ddHex-containing *S. enterica* serovars to be conserved with regard to their arrangement and relative position in the central section of the *rfb* clusters (Figure 9).[256,290,293] To complete ddHex formation, three additional genes are required contributing to the synthesis of serovar-specific sugars. CDP-abequose-synthase, encoded by *rfbJ* [*abe*], catalyzes the conversion of CDP-4-keto-3,6-dideoxyglucose to CDP-Abe in serovars B and C2.[319] In contrast, serovars A and D contain Par and Tyv in their repeating units, respectively.

In the presence of NAD(P)H as the hydrogen donor, the reduction to CDP-Par is mediated by the action of the *rfbS* [*prt*]-encoded CDP-paratose-synthase, a common biosynthetic step in both serovars. Finally, CDP-tyvelose-2-epimerase, encoded on *rfbE* [*tyv*], catalyzes the formation of CDP-Tyv in serovar D. Although *tyv* has been identified in all serovar A strains examined, the gene was nonfunctional due to a frameshift mutation converting the fourth codon to an amber stop codon and accounting for serovar A strains synthesizing Par instead of Tyv.[320] The ddHex pathway genes in *S. enterica* serovar B and *Y. pseudotuberculosis* were found to be similar, since the complete set of genes specifying the common ddHex pathway (*ddhABCD*) could be identified in almost all *Y. pseudotuberculosis* strains of different serotype containing Abe, Par, Tyv, or Asc in their O-polysaccharides.[303,321–323] However, the observed divergence among the genes for the final steps in ddHex formation suggested alternative genes in generating the final serological specificity in different *Y. pseudotuberculosis* strains. Thus, in *Y. pseudotuberculosis* serogroup VA, the last two reactions in ddHex formation were shown to be carried out by the epimerase AscE and the reductase AscF, converting CDP-4-keto-3,6-dideoxyglucose to CDP-Asc.[323] Although the O-polysaccharide of *Y. enterocolitica* serotype O:8 does not contain any ddHex, it is believed that the organism utilizes the products of the *rfb*-encoded *ddhAB* genes for the first steps of the hypothetical pathway leading to the formation of CDP-6-deoxy-D-glucose.[241]

Results on the molecular characterization of the serotype-specific B-band LPS gene cluster from *P. aeruginosa* serotype O:5 indicated that a clustering of the genes within the *rfb* region, in accordance with the biosynthetic pathway they code for, seems not to be a general rule.[324] Genes thought to be involved in the biosynthetic pathways for the diaminouronic acid residues (*wbpI*, *wbpE*, *wbpA*, *wbpCD*) and *N*-acetylfucosamine (FucNAc) (*wbpM*, *wbpB*, *wbpK*) were found to be scattered along the *rfb* cluster (see Table 8).

Since the O-polysaccharide represents the structurally most variable part of the LPS molecule and contains unusual and rare sugars, it is reasonable to assume that further studies on *rfb* regions of different origin will discover new enzymes for biosynthesis of the corresponding precursors. For example, investigations provided insights into the organization of the *rfb* region and the putative pathways for biosynthesis of Rha4N and 3-deoxy-L-*glycero*-tetronic acid, which form the basic structure of the O-antigenic determinant of *V. cholerae* O:1.[300,325] Finally, despite the fact that various O-polysaccharide chains of different organisms contain galactofuranosyl (Gal*f*) residues (Tables 1 and 4), the *glf* gene of the cryptic *rfb* region from *E. coli* K-12 and *rfbD* [*glf*$_{KpnO1}$] from *K. pneumoniae* O:1 were identified in the mid-1990s to encode UDP-galactopyranose-mutases for the interconversion of UDP Gal*p* and UDP-Gal*f*.[326,327] The similarity between the predicted amino acid sequences of Glf$_{EcK-12}$ and Glf$_{KpnO1}$ was striking, and complementation experiments demonstrated their equivalent functions. However, it is interesting to note that the enzymes differed in their cofactor requirement in that Glf$_{KpnO1}$, but not Glf$_{EcK-12}$, was absolutely dependent on the presence of NADH or NADPH.[327]

3.09.5.2 Assembly and Polymerization of O-Polysaccharide Repeating Units

All of the nucleotide sugars are synthesized by cytosolic enzymes and, therefore, their assembly into O-units must occur at the cytoplasmic face of the cytoplasmic membrane. This assumption is supported by the fact that the glycosyltranferases, catalyzing the formation of glycosidic bonds, in general do not contain significant hydrophobic segments which could act as transmembrane domains or membrane anchors. Since most of the transferases are basic proteins, a direct but loose association with the cytoplasmic face of the cytoplasmic membrane can be assumed. With very few known exceptions (see below), glycosyltransferases appear to represent highly specific enzymes with respect to a correct recognition of the appropriate donor and acceptor molecules and formation of the particular glycosidic bonds. Thus, assembly of the repeating units of O-polysaccharides is realized through a series of sequential reactions, each of which is mediated by specific glycosyltransferases. The sequential transfer of sugar residues occurs on an antigen-carrier lipid (ACL), which has been identified as the C_{55} polyisoprenoid alcohol derivative undecaprenyl pyrophosphate.[328]

Two general mechanisms for polymerization of O-polysaccharide repeating units have been described. One pathway depends on the Rfc [Wzy] enzyme and involves the polymerization of preformed ACL-linked repeating units in a blockwise manner. The second or monomeric pathway is Wzy-independent and was shown to be fundamentally different, since polymerization occurs by sequential transfer of monosaccharide residues from their activated nucleotide sugars to a growing polysaccharide chain which is attached to ACL. A third and new pathway for O-polysaccharide assembly has been proposed for the plasmid-encoded O:54 biosynthesis in *S. enterica* serovar Borreze (see below).

Several lines of evidence indicated that in the Wzy-dependent pathway of *S. enterica*, the formation of O-polysaccharides is a transmembrane process starting with the synthesis of O-units by transfer of sugar residues to ACL at the cytoplasmic side of the cytoplasmic membrane, followed by translocation of the ACL-linked O-units across the inner membrane, and completed by polymerization of the O-antigenic chains at the periplasmic face of the cytoplasmic membrane (Figures 10 and 11).[329] In *S. enterica* serogroups A, B, C2, D, and E1, assembly of the O-units is initiated with the transfer of Gal1*P* to ACL by the action of galactosyl-pyrophosphoryl-undecaprenol-synthetase encoded by the *rfbP* [*wbaP*] gene.[256,330] The transfer of Gal1*P* from the precursor during the initial reaction results in a conservation of the energy of the linkage in the undecaprenol-linked intermediate and, hence, differs from subsequent steps, in which the sugar rather than the sugar phosphate is transferred. Furthermore, WbaP was found to be different in its structure and function in comparison to those transferases of the *rfb* region, which are responsible for transfer of the distal sugars to the intermediates. Besides having five potential transmembrane domains, the protein was suggested to exhibit two functions, which could be assigned to its N- and C-terminal halves, respectively. While the N-terminus appeared to be involved in processing of the O-polysaccharide units, only the C-terminal part of WbaP displayed galactosyl-1-phosphate-transferase activity catalyzing the initial step in formation of the O-units.[331,332] Interestingly, comparison of both the derived amino acid sequences and the profiles of hydrophobic segments revealed that, despite a variable region presumably responsible for sugar specificity, the galactosyl-1-phosphate-transferase domain of WbaP was very similar to those potential sugar transferases of EPS biosynthesis, which catalyze the addition of the first sugar residue to a lipid carrier such as ExoY of *Rhizobium meliloti*,[333] AmsG of *E. amylovora*,[334] or GumD of *X. campestris*.[335] Like the transferases interacting with undecaprenol, WbpL was found to be the most hydrophobic protein of the three putative *rfb*-encoded transferases of *P. aeruginosa* serotype O:5 and, hence, it was assumed that the protein initiates biosynthesis of O-units by addition of Fuc*N*Ac to ACL.[324] In *S. enterica* serogroup B strains, the subsequent reactions in completion of the O-units are determined by the genes *rfbN* [*wbaN*], *rfbU* [*wbaU*], and *rfbV* [*wbaV*], which code for the sequential transfer of Rha, Man, and Abe, respectively (Figure 10). Biochemical studies on cells carrying cloned *rfb* genes for potential transferases enabled the identification and characterization of the rhamnosyl- and mannosyl-transferase genes in *S. enterica* serogroups B, E1, and C2.[336] Although the transferases were shown to exhibit a low degree of similarity, correlations between their relatedness at the sequence level and their specific transfer reactions could be observed.[294] Thus, in serogroups A, B, D1, D2, and E1, the rhamnosyltransferases, which add Rha in α-(→3)-linkage to the galactosyl pyrophosphoryl undecaprenyl intermediate, appeared to be well conserved. In contrast, the rhamnosyltransferase RfbQ [WbaQ] of serogroup C2, adding Rha to Man, was different. A similar situation was found among related mannosyltransferases in serogroups A, B, and D1, catalyzing the formation of α-(1→4)-linkages between Rha and Man, whereas the mannosyltransferase-encoding genes *wbaU* from serogroup B and *rfbO* [*wbaO*] from serogroup E1 strains had little homology, since group E1 strains form a β-(1→4)-linkage to Rha.[257,336] It is not surprising, therefore, that serogroup C2 strains utilize the two different mannosyltransferases RfbZ [WbaZ] and RfbW [WbaW] for transfer of the first and the second Man residue onto the O-unit intermediate, respectively.[336] Despite all the observed variabilities, hydrophobic cluster analysis revealed, however, a global similarity among prokaryotic mannosyltransferases including several invariant and probably catalytic amino acid residues in well conserved regions of the enzymes.[337] With the identification of the ddHex-transferase genes in *S. enterica* serogroups A, B, C2, D1, and D2, the list of the *Salmonella* transferase genes responsible for the assembly of the O-units has been completed.[338] The abequosyl-, tyvelosyl-, and paratosyl-transferases of serogroups B, D1, and A, respectively, all designated WbaV, were shown to be closely related, catalyzing the addition of ddHex to the same Man–Rha–Gal backbone. These WbaV transferases displayed significant similarity to the potential abequosyltransferase Orf8.7 [WbaV$_{\text{YpsIIA}}$] of *Y. pseudotuberculosis* serovar IIA, but exhibited only a low degree of identity to the corresponding RfbR [WbaR] transferase of *S. enterica* serogroup C2, which utilizes different substrates for the Abe transfer. Although a limited identity appears to be a typical feature of glycosyltransferases, sequence comparisons revealed numerous open reading frames within *rfb* clusters, for example, of *E. coli* K-12,[258] *V. cholerae* O:1 and O:139,[245] *S. dysenteriae* type 1,[312] *S. flexneri*,[339] or *P. aeruginosa* serotype O:5,[324] thought to code for potential glycosyltransferases.

The mechanisms for translocation of the ACL-linked O-units across the cytoplasmic membrane are still unclear and await further investigation. However, accumulation of single ACL-linked O-polysaccharide units at the cytoplasmic side of the cytoplasmic membrane in *S. enterica* strains,

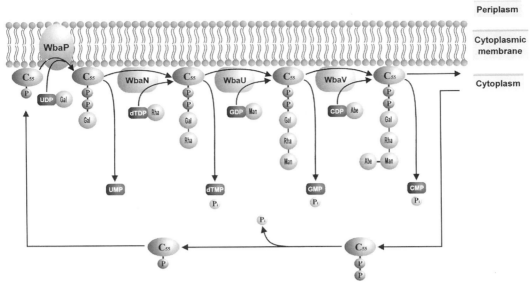

Figure 10 Biosynthesis of ACL-linked O-units in *S. enterica* serovar Typhimurium (group B).[227]

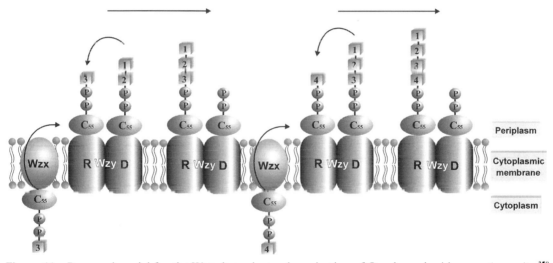

Figure 11 Proposed model for the Wzy-dependent polymerization of O-polysaccharide repeating units.[350]

defective in the *rfbX* [*wzx*] gene, suggested Wzx to have a flippase function in translocation of O-polysaccharide units across the inner membrane.[340] Wzx has been found in the majority of *rfb* gene clusters coding for Wzy-dependent O-antigen biosynthesis. Despite a low degree of sequence similarity among the Wzx proteins, they were shown to exhibit an apparent similarity at the structural level suggesting a close and specific association of these proteins with their particular O-polysaccharides. Wzx proteins are very hydrophobic with 12 potential transmembrane domains, which utilize rare codons within their first 25 amino acid residues and, hence, could be regulated by available isoaccepting tRNA molecules.[241,341]

In accordance with the block model (Figure 11),[342] the polysaccharide grows successively by one repeating unit at the reducing end of the O-polysaccharide. The blockwise transfer of partially polymerized chains from one ACL to a single repeating unit on another ACL enables the elongating polymer to remain attached on the lipid carrier until polymerization into a long-chain O-poly-saccharide is completed. The process of glycosidic linkage formation between the subunits is catalyzed by the O-antigen polymerase, which is encoded by the *wzy* gene. The requirement of Wzy for the polymerization process was primarily demonstrated on *wzy* mutants, which synthesized an incomplete LPS molecule consisting of a single repeating unit on a complete LPS core, known as the semi-rough LPS phenotype.[343] O-antigen polymerase genes have been cloned and sequenced from several bacteria including different *S. enterica* serovars,[293,344] *E. coli* O:4,[345] *S. dysenteriae* type

1,[312] *S. flexneri*,[346] *Y. enterocolitica* serotype O:8,[241] and *P. aeruginosa* serotype O:5.[324,347,348] Cloning and mutational analysis have clearly demonstrated that the chromosomal location of the *wzy* gene is not predictable. The *wzy* genes of *S. enterica* serogroups A, B, and D were shown to map outside the *rfb* gene cluster, whereas the *S. enterica* serogroups C2 and E1, *E. coli* O:4, the two *Shigella* species, and *P. aeruginosa* were found to carry the gene inside the *rfb* region. As expected, *wzy* displayed sequence similarity among those bacterial strains of *S. enterica* serogroups A, B, and D1 or *P. aeruginosa* serotypes O:2, O:5, O:16, and O:20, in which similar O-repeating backbone structures are joined by Wzy.[344,348] However, apart from the fact that the O-antigen polymerases of *S. enterica* serogroup B and D1 strains were able to form glycosidic bonds between O-repeating units of either serotype,[349] it appears more general that the O-antigen polymerases cannot polymerize other than their own repeating units. Nevertheless, all the Wzy proteins of different origin investigated so far have several features in common. Wzy enzymes have been reported as very hydrophobic proteins with 11 to 13 transmembrane domains suggesting their location in the cytoplasmic membrane. Since the polymerization process obviously occurs at the periplasmic face of the cytoplasmic membrane, it was hypothesized that the functional domains of Wzy might be located in the periplasm.[346] The O-antigen polymerase has two substrates and, therefore, it was further proposed that Wzy carries two sites, designated R (receiving) and D (donating), which bind the new ACL-linked O-unit and the ACL-linked O-polymer, respectively.[350] According to this model for polymerization, the ACL-linked O-chain will be donated by transfer to the ACL-linked O-unit at the R site. Each polymerization cycle is completed after the extended ACL-linked chain moves back to the D site without dissociation, replacing the spent ACL and allowing the R site to bind a new ACL-linked O-unit (Figure 11). Although Wzy resembles porin proteins in several aspects, a possible role of Wzy in translocation of the ACL-linked O-repeating units across the cytoplasmic membrane, however, remains highly speculative.[344] It is of particular interest that the *wzy* gene products have not only a high content of the amino acids leucine, isoleucine, and phenylalanine, representing ∼30% of the total amino acid composition, but use, like Wzx, codons for rare tRNA species within the first 25 amino acids, which makes their translation susceptible for the availability of the rare isoaccepting tRNA molecules.[351] Therefore, it is attractive to speculate that the atypical codon usage results in a tight regulation of Wzy by corresponding tRNA levels and emphasizes Wzy as a key enzyme in determining an optimal O-polysaccharide formation under varying environmental conditions.[346]

The Wzy-independent mechanism of O-polysaccharide assembly, proposed to designate ABC-transporter-dependent pathway,[352] has been demonstrated for biosynthesis of mannan- or galactan-containing homopolymeric chains in *E. coli* O:9,[353] *K. pneumoniae* serotype O:1, and *S. marcescens* serotype O:16.[354] This mechanism differs from the Wzy-dependent pathway in that the homopolymeric chains are entirely formed by glycosyltransferases at the cytoplasmic face of the cytoplasmic membrane, transferring processively the monosaccharide residues from their activated precursors to the nonreducing end of the growing O-polysaccharide chain (Figure 12). Therefore, there should be no requirement for an O-antigen polymerase activity, but a requirement for a postpolymerization export system. The formation of the mannan- or galactan-containing homopolymers is catalyzed by several mannosyl- and galactosyl-transferases, respectively. The galactosyltransferase RfbF [WbbO] is involved in the initiation of O-unit formation in *K. pneumoniae* O:1 and *S. marcescens* O:16, and sequence relationships between the enzymes of both microorganisms emphasized their functional identity.[354] It is of interest that WbbO was suggested to contain dual galactopyranosyl- and galactofuranosyl-transferase activity able to add each of Gal*p* and Gal*f* to a GlcNAc-primed lipid intermediate.[355] As expected, there was no obvious similarity between the deduced amino acid sequences of the galactosyltransferases WbbO from *K. pneumoniae* and the C-terminal part of WbaP from *S. enterica*. However, the C-termini of WbbO and the plasmid-encoded galactopyranosyltransferase RfpB [WbbP], involved in O-unit assembly in *S. dysenteriae* type 1, shared significant similarity.[312,356,357] Moreover, WbbO of *Klebsiella* was capable of functionally replacing WbbP in the biosynthesis of the *S. dysenteriae* type 1 O-polysaccharide, which indicated that both galactosyltransferases indeed required a GlcNAc-primed lipid intermediate as an acceptor molecule (see below). Apart from that, however, the pathway of O-polysaccharide biosynthesis of *S. dysenteriae* type 1 resembled the Wzy-dependent block mechanism described for *S. enterica* strains.[312,355] To complete the D-galactan I structure, the *K. pneumoniae* *rfbC* [*wbbM*] and *rfbE* [*wbbN*] genes were proposed to code for additional enzymes with galactosyltransferase activity.[358] In *E. coli* O:9, biosynthesis of the O-polysaccharide needs fewer transferases than sugars are present in the repeating unit. The first mannosyltransferase MtfC [WbdC] was shown to catalyze the initial reaction for growth of the O-polysaccharide chain, which is completed subsequently by the two mannosyltransferases MtfA [WbdA] and MtfB [WbdB] (Figure 12).[353] It should be noted that WbdB exhibited dual enzymatic activity able to transfer Man residues to two different Man-linked acceptor molecules.

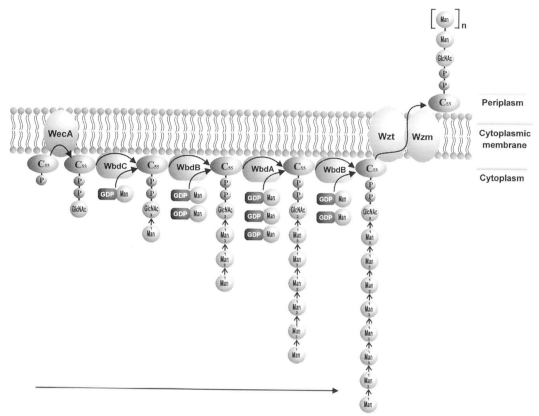

Figure 12   Proposed model for the ABC-transporter-dependent O-polysaccharide assembly in *E. coli* O:9.[353]

An ATP-binding cassette (ABC) transporter, consisting of an ATP-binding and a transmembrane component, was shown to be essential for translocation of the completed O-chains across the cytoplasmic membrane. For example, *K. pneumoniae* serotype O:1 mutants, defective in the ATP-binding and membrane components RfbB [Wzt$_{KpnO1}$] and RfbA [Wzm$_{KpnO1}$], respectively, produced rough-type LPS and accumulated the O-antigenic D-galactan I in the cytoplasm.[359] Furthermore, transposon mutagenesis of the genes *rfbD* [wzm$_{YenO3}$] and *rfbE* [wzt$_{YenO3}$] resulted as well in cytoplasmic accumulation of the O-antigen and, hence, indicated that they code for potential O-antigen exporters in *Y. enterocolitica* serotype O:3 known to synthesize a homopolymeric O-antigen of L-6d-Alt (see Table 6).[314] Similarly, a centrally located ABC-transporter, encoded by the genes *rfbI* [wzt$_{VcO1}$] and *rfbH* [wzm$_{VcO1}$], was proposed for the *rfb* cluster of *V. cholerae* O:1.[360,361] It is thought that the ATP-binding component of ABC-transporters is responsible for coupling the energy derived from ATP hydrolysis to the integral membrane component, which performs the energy-dependent translocation of the polysaccharide. The homopolymeric O-polysaccharide biosynthesis is very reminiscent of group II CPS biosynthesis in *E. coli*, *N. meningitidis*, and *H. influenzae*.[227] Apart from the fact that the O- and CPS-chains are both polymerized at the cytoplasmic face of the cytoplasmic membrane, the O-polysaccharide transporters were found to have their counterparts in cytoplasmic membrane transport systems involved in export of the CPS intermediates, although they do not seem to be functionally exchangeable.[359] Comparison of the deduced amino acid sequences revealed that the O-chain-transporter ATP-binding components Wzt from *K. pneumoniae* serotype O:1 and *S. marcescens* O:16, Wzt$_{YenO3}$ from *Y. enterocolitica* serotype O:3, Wzt$_{VcO1}$ from *V. cholerae*, and Orf431 [Wzt$_{EcO9}$] from *E. coli* O:9 are related to each other and to the corresponding CPS export proteins BexA from *H. influenzae*,[362] CtrD from *N. meningitidis*,[363] or to KpsT [Wzt] from *E. coli* K:1 and K:5.[364,365] They all contain highly conserved regions including the A and B sites of the ATP-binding consensus motif.[366] The sequence similarity of the integral membrane components was not as striking, but the highly hydrophobic proteins shared a common secondary structure with five to six potential transmembrane domains, which implied their involvement in similar functions.[359] A different ABC polysaccharide transport system has been described for *Aeromonas salmonicida*.[367]

The identified ATP-binding protein AbcA exhibited significant similarity to the ATP-binding components of the polysaccharide export systems mentioned above, but was not flanked by a gene for a potential integral membrane component. The inability of AbcA mutants to produce smooth-type LPS in conjunction with a significant decrease in O-polysaccharide biosynthesis suggested AbcA to be involved not only in energizing the polysaccharide export, but also to be required for optimal synthesis of O-polysaccharides.[367]

LPS biosynthesis can be additionally divided into *rfe* [*wecA*]-dependent and *wecA*-independent pathways. This classification is based on the fact that some O-polysaccharides require for their biosynthesis the *wecA* gene coding for the transfer of GlcNAc1P from UDP-GlcNAc to undecaprenyl monophosphate, which results in the formation of *N*-acetylglucosamine-pyrophosphorylundecaprenol during the first step of ECA synthesis.[368] It was proposed that WecA obviously plays a general role in biosynthesis of GlcNAc-containing O-polysaccharides by providing GlcNAc-primed lipid intermediate for subsequent synthesis steps.[368] WecA can be utilized in both the Wzy-dependent and the ABC-transporter-dependent pathway for O-chain assembly. In the Wzy-dependent pathways of *S. enterica* serogroup C1 and *S. dysenteriae* type 1 as well as in *E. coli* K-12 (O:16), O:7, O:18, O:75, and O:111, the formation of each O-polysaccharide unit is initiated by the WecA-catalyzed transfer of GlcNAc residues to ACL.[74,258,312,369] After polymerization of the completed O-repeating units, each unit contains GlcNAc as the first monosaccharide residue. Therefore, WecA appears to be a functional analogue of WbaP, both being able to initiate the synthesis of O-repeating units.[312] The WbaP dependence on the initiation of O-unit synthesis represents an indication for the block model of O-unit polymerization, while such a striking correlation does not exist for the WecA-dependent synthesis. In the ABC-transporter-dependent pathways of homopolymeric O-unit assembly, the repeating unit structures do not contain GlcNAc residues and, hence, WecA was shown to play a different role. As demonstrated for biosynthesis of D-galactan I structures in *K. pneumoniae* serotype O:1 and *S. marcescens* serotype O:16,[354,355] and mannosyl homopolymers in *E. coli* serotypes O:8,[353,370] WecA was responsible for priming the synthesis of the homopolymers by providing acceptor molecules in the form of undecaprenyl pyrophosphoryl-linked intermediates. However, the galactosyltransferases WbbM and WbbN of *K. pneumoniae* serotype O:1 as well as the mannosyltransferases WbdA and WbdB of *E. coli* O:9 are not capable of accepting the GlcNAc-primed ACL as a substrate for completion of the homopolymeric O-chains.[353,358] Therefore, the polymerization process depends on an additional initiation step catalyzed by discrete glycosyltransferases forming acceptor molecules suitable for the glycosyltransferases of the subsequent polymerization reactions. This priming function is carried out by the galactosyltransferases WbbO of *K. pneumoniae* serotype O:1 and *S. marcescens* serotype O:16 as well as the mannosyltransferase WbdC of *E. coli* O:9 (Figure 12), for which a requirement for a WecA-activated lipid intermediate has been demonstrated.

3.09.5.3   Regulation of O-Polysaccharide Polymerization

On silver-stained or radiolabeled SDS–PAGE, each band of the typical ladder-like pattern of smooth LPS from wild-type bacteria corresponds to an O-polysaccharide chain of specific length attached to a lipid A/core oligosaccharide molecule. It is known that the distribution of O-chain length is not uniform but bimodal with characteristic strain-dependent preferences for the lengths of O-chains. Thus, O-polysaccharides with chain lengths above one to five repeating units become progressively less abundant forming the molecules with intermediate O-antigen sizes. The abundance increases again with molecules of about 15 to 19 repeating units and reaches a plateau of high-molecular-mass molecules with chain lengths from about 20 to 35 O-units, which represent the most abundant portion of total O-polysaccharide. Based on a mathematical model for synthesis of O-polysaccharides and their distribution in LPS, it was proposed previously that the bimodal distribution of LPS cannot be explained without an additional mechanism having specificity for preferred O-polysaccharide chain lengths.[371] The elucidation of the *rol/cld* [*wzz*] gene, coding for the regulator of O-length (or chain length determinant), provided evidence of an enzymatic specificity for control of the extent of O-chain polymerization.[350,372] In mutants which lacked the *wzz* gene, the bimodal distribution of O-chain length was eliminated and was replaced by LPS with an unregulated unimodal banding pattern on SDS–PAGE. The *wzz* or *wzz*-like genes have been identified in close proximity to their respective *rfb* gene clusters in different *E. coli* serotypes,[350,372–374] *S. enterica* serogroups B and C2,[290,350,373] *Shigella* strains,[312,375] *Y. pseudotuberculosis* IIA,[376] or *P. aeruginosa* serotype O:5 and O:16.[324,377] Hydropathy profiles of the deduced amino acid sequences indicated a structural conservation among all Wzz proteins so far investigated. They contain two potential membrane-spanning regions near their amino- and carboxy-terminal ends separated by a large hydrophilic central region. Experimental data confirmed the predicted anchoring of the N- and C-terminal ends of Wzz within the cytoplasmic membrane and suggested the majority of the protein to be located in the periplasm.[350,375] The Wzz-dependent regulation of O-chain length has been described so far exclusively for those microorganisms, in which LPS biosynthesis proceeds via the Wzy-dependent pathway of O-antigen polymerization. However, the mechanism by which the Wzz protein could influence the extent of O-polysaccharide polymerization remains unknown. Unfortunately, there is no direct experimental proof for either hypothesis that: (i) Wzz could form a complex with the O-antigen polymerase Wzy to facilitate a time-dependent transition between the process of O-chain extension and the termination of this process by transfer of the grown O-antigen to the O-antigen-core-lipid-A-ligase RfaL [WaaL]; or (ii) Wzz could interact as a molecular chaperone to control the kinetics of the ligation of the O-chain to the core by modulating interactions between Wzy and WaaL.[350,375] Investigations on the mechanism for generating the observed modal distribution with a preferred O-chain length in *E. coli* O:8 and O:9 revealed *wzz* homologues in those bacterial strains, which coexpressed group IB capsular K antigens with the Wzz-independent homopolymeric O-polysaccharides.[374,378] The results obtained allowed the conclusion that the chain length of the capsule-related form of LPS, K_{LPS}, is regulated by Wzz, whereas Wzz did not interact with the synthesis of O:8 or O:9 polysaccharides. Since there is obviously no gene product for determination of the O-polysaccharide chain length in bacteria with the ABC-transporter-dependent pathway, it is assumed that the modality of chain length distribution might be mediated by components of the transport machinery.[352]

In several strains of *P. aeruginosa*,[379,380] *S. enterica* serovar Anatum,[381] *S. marcescens*,[382] or *Y. enterocolitica* serotypes O:3 and O:8,[383,384] the banding pattern of LPS on SDS–PAGE has been demonstrated to be influenced by growth temperature suggesting an additional or different mechanism for regulation of O-chain length. Indeed, the temperature-dependent decrease of long O-polysaccharide chains in *Y. enterocolitica* serotype O:3 was probably mediated by a temperature-inducible repressor able to down-regulate the transcription of genes within the *rfb* cluster.[383] Whether the encoding gene for the putative repressor is similar to a novel locus downstream of the *rfb* cluster of *Y. enterocolitica* serotype O:8 shown to be involved in the reduction of O-antigen expression at 37 C, remains to be investigated.[384]

3.09.5.4 Modification of the O-Polysaccharide Chain

Numerous O-polysaccharides can undergo modifications in the form of O-acetylation or glucosylation of individual sugar residues within the O-units or changes of the Wzy generated linkages between the O-units. The usually nonstoichiometric modifications can be a permanent and typical feature of the respective LPS structure and, hence, account for structural and antigenic variations between serotypes within one O-group. Furthermore, O-polysaccharides may be modified at different stages of their biosynthesis such as at the cytoplasmic face of the cytoplasmic membrane prior to the translocation of the ACL-linked O-unit across the membrane or after polymerization of the O-units at the periplasmic face of the inner membrane. Almost all of the genes responsible for O-polysaccharide modifications were found to map outside the *rfb* clusters either chromosomal or, in most of the cases, on prophages. The only exceptions known to date are the putative O-acyltransferases WbpC/D of *P. aeruginosa* serotype O:5 as well as the O-acetyltransferases Orf9 [WbbJ] of *E. coli* K-12 (O:16) and Orf14.9/RfbL [WbaL] of *S. enterica* serogroup C2, which are encoded within their respective *rfb* gene clusters.[74,258,324,338,385] It is of interest that WbaL presumably transferred the acetyl group to the Rha residue only of the completed ACL-linked Rha–(Man)₂–Gal backbone and not to a truncated substrate. As demonstrated by sequence comparisons, WbbJ and WbaL were related to each other and to the protein family of cytoplasmic transacetylases, which include, for example, the thiogalactoside transacetylase LacA of *E. coli* or various chloramphenicol acetyltransferases.[386] However, the acetyltransferase OafA from *S. enterica* serovar Typhimurium, shown to confer the O:5 serotype (O-factor 5) by acetylation of the 2-hydroxy group of the Abe residue within the O-unit,[387,388] was not related to the enzymes mentioned above, but could be assigned by similarity to a family of integral membrane transferases, which perform the acetylation of carbohydrate structures in various prokaryotes such as the O-acetyltransferase Oac of the *S. flexneri* bacteriophage Sf6.[389] Since even the semi-rough LPS phenotype with one repeating unit displayed O:5 specificity and, presumably, acetyl coenzyme A serves as the acetyl donor, it is likely that the OafA-mediated reaction is carried out on the completed ACL-linked O-unit at the cytoplasmic face of the inner membrane.[343,389] On the other hand, prophage ε^{15} prevented the acetylation of the Gal residue (factor-10-antigen) within the O-units of *S. enterica* converting the serogroup E1 into E2.[390] In addition to the phage-encoded disappearance of the factor-10-antigen, the linkage between the repeating units was changed from the α- to the β-configuration, the stereochemical effect of which resulted immunochemically in the appearance of the new factor-15-specificity.[391] Although the mechanism for the ε^{15}-directed synthesis of β-linked O-units is not clear, it seems logical to propose that the specificity of the O-antigen polymerase Wzy has been altered by a prophage-encoded polymerase able to inhibit, replace, or modify the bacterial enzyme. Similarly, the temperate bacteriophage D3 of *P. aeruginosa* PAO1 converted the linkage between the tri-saccharide repeating units from α-$(1\rightarrow4)$ to β-$(1\rightarrow4)$ and, additionally, introduced an acetyl group into position 4 of the Fuc*N* residue.[392] Another example of lysogenic conversion of the phage receptor is the prophage Sf6-mediated change from serotype Y (group antigen 3,4) to the 3b serotype (group antigen 6,3,4) by *O*-acetylation of the O-polysaccharide chain in *S. flexneri*. The *O*-acetyl-transferase-encoding gene *oac* of Sf6, meanwhile cloned and sequenced, was demonstrated to be sufficient for the conversion of the serotype.[393,394]

In *S. enterica* serogroups B and D, the Gal residue can be glucosylated at positions C-4 or C-6, representing the two different antigenic factors 12_2 and 1, respectively. The C-4 glucosylation of the repeating unit is determined by the chromosomal *oafR-oafE* locus,[395] while the antigenic factor 1 was shown to be encoded on genes of bacteriophage P22.[396] In contrast to the OafA-mediated modification immediately after completion of the ACL-linked O-unit (see above), the single O-units of semi-rough LPS molecules were not glucosylated and, therefore, it was suggested that addition of the Glc residue, which is derived from UDP-Glc via the intermediate glucosyl phosphoryl undecaprenol, requires a partially or completely polymerized O-chain as a substrate.[107,223]

A completely different phenotypic appearance has been demonstrated for strains of *S. enterica* serovar Choleraesuis lysogenized with bacteriophage 14. Data on an increase of the average length of O-polysaccharide chains in the lysogens implied that phage 14-encoded gene product(s) to enhance the efficiency of O-chain biosynthesis by replacing or complementing the bacterial O-antigen polymerase with a more efficient phage-encoded enzyme.[397] Investigations on an unusual lysogenic conversion in *A. methanolicus* strain 58/4, stimulated by the finding that the host strain completely lacked the O-polysaccharide after lysogenization with phage Acm1 (see Table 8), pro-vided evidence for the action of a phage-encoded antisense RNA, also involved in the suppression of O-chain biosynthesis in various *E. coli* strains.[398] Taken together, it is reasonable to assume, as biochemical and genetic data will provide a more complete picture of the biosynthetic pathways of LPS in various bacterial species, that more examples will be found in which accessory genetic elements, like temperate bacteriophages, either take direct effect on different steps in LPS assembly or certain pathways derived from evolutionary events in which accessory genetic elements were involved.

3.09.5.5 Plasmid-Encoded Determinants of O-Polysaccharide Biosynthesis

As outlined above, most of the genetic determinants for O-polysaccharide biosynthesis are located on the bacterial chromosome. However, there are a few examples known for plasmid-encoded functions involved in O-chain synthesis. In some strains of *S. flexneri* serotype 2a, the plasmid pHS-2, obviously associated with reactive arthritis, was found to carry an additional O-chain length determinant, which probably acts independently of its counterpart located on the chromosome.[376] The plasmid-borne Wzz$_{pHS-2}$ protein was structurally related to the known Wzz proteins and was shown to be responsible for the generation of an O-polysaccharide portion of extreme chain length. It remains to be proven whether Wzz$_{pHS-2}$ is associated with the development of reactive arthritis as suggested by the authors. Several other reports described the involvement of plasmids in O-antigen biosynthesis in different *Shigella* species. A large 120 MDa plasmid was demonstrated to carry at least four gene clusters responsible for the expression of the form I O-antigen in *S. sonnei*.[399–401] Both the plasmid pHW400-encoded galactosyltransferase WbbP$_{pHW400}$ and the chro-mosomal *rfb* genes were required for synthesis of a complete O-polysaccharide in *S. dysenteriae* serotype 1.[357,402–404] WbbP$_{pHW400}$ initiated the O-unit synthesis by adding the first Gal residue to a GlcNAc-primed ACL, and the gene products of the *rfb* region mediated all the subsequent steps in completion of the repeating unit. It is intriguing that the expression of the O:111 polysaccharide in *E. coli* strain M92 was determined by the chromosomal *rfb* gene cluster,[289,405] while it was encoded on the 54 MDa plasmid pYR111 in the *E. coli* serotype O:111 strain B171.[406] In *S. enterica* serovar Dublin, a *wzy*-like activity was suggested to be located on a large plasmid, since a cured strain with a semi-rough LPS phenotype obviously was defective in polymerization of the O-antigen.[407]

Studies have reported on a correlation between the presence of a plasmid and the synthesis of the O:54 antigen in *S. enterica* serovar Borreze.[408] In the mid-1990s, definitive proof was provided that pWQ799 of serovar Borreze is the only plasmid in *S. enterica* described so far, which contains a complete *rfb* gene cluster directing the synthesis of the only known homopolymeric O-polysaccharide in *Salmonella* (Table 3).[105,409] Interestingly, biosynthesis of the O:54 polysaccharide was *wecA*-dependent and additionally required the activity of the *rffE* [*wecC*$_{ECA}$] gene coding for UDP-*N*-acetylglucosamine-2-epimerase. Thus, the WecA-requirement and the *wecC*-directed provision of the O:54 precursor UDP-Man*N*Ac represent further examples of functional interactions between LPS and ECA biosynthesis. Results of the genetic organization of the *rfb*$_{O:54}$ cluster and functional characterization of its gene products deserve particular attention as a third and new pathway for O-polysaccharide assembly has been demonstrated—the synthase-dependent pathway.[352,410] The *rfb*$_{O:54}$ cluster was shown to contain the three open reading frames *rfbA* [*wbbE*], *rfbB* [*wbbF*], and *rfbC* [*wecC*$_{pWQ799}$]. While WecC$_{pWQ799}$ appeared to be a functional homologue of WecC$_{ECA}$ and, therefore, was not required for O:54 biosynthesis in ECA-synthesizing enteric bacteria, the *N*-acetylmannosaminyl-transferase WbbE and the processive *N*-acetylmannosaminyl-transferase WbbF were sufficient to direct the assembly of the O:54-polysaccharide catalyzing the transfer of the first Man*N*Ac residue to the ACL-linked Glc*N*Ac and the completion of the O:54-polysaccharide, respectively. Thus, O:54 biosynthesis resembled the ABC-transporter-dependent pathway of homopolymeric O-unit assembly. However, there was no indication for the utilization of an ABC-transporter or a Wzx O-unit-transport protein in the synthesis of the O:54-polysaccharide. Instead, the predicted topology of WbbF suggested the protein to contain, in addition to the glycosyltransferase activity, a putative transport function in the form of a transmembrane channel at its C-terminus presumably able to link the polymerization reaction with the transport of the growing chain.[410]

3.09.5.6 Genetic Polymorphism and Structural Variations Among O-Polysaccharides

The enormous structural variability among O-polysaccharides has been suggested to result from adaptations of Gram-negative bacteria to selective pressures of particular environmental niches leading to the development of clonal structures within bacterial populations.[411] Evolutionary mechanisms obviously enabled changes of the most exposed surface structure of the clones, i.e., the O-specific polysaccharide chain, either to evade niche-specific selection pressures and/or maintain advantageous adaptations to a particular niche. The structural diversity of O-polysaccharides reflects an extensive genetic polymorphism among the *rfb* gene clusters. It has been proposed that genetic exchange events between different microorganisms as well as recombination processes account for the genetic divergence among the O-chain-specifying gene sequences.[412] It is of special interest that the *rfb* clusters investigated so far usually possess a G+C content at least 10% below the characteristic species average and include a number of discrete G+C contents.[256] These observations provide strong evidence for the assumption that the *rfb* genes originated from different progenitors with A+T-rich DNAs. The low G+C content of the *rfb*-encoded genes for the dTDP-Rha pathway in *L. interrogans* serovar Copenhageni, almost identical to the genomic G+C content of the species, together with the assumption that spirochetes diverged from eubacterial groups at an early stage of evolution, gave reason to speculate about an acquisition of *rfb* genes by Gram-negative bacteria from a spirochete ancestor.[316]

Further indications for an intergeneric lateral transfer and recombination of *rfb* genes were provided by sequence analysis of *gnd* genes from various enterobacterial strains.[413,414] In enteric bacteria, the *gnd* gene is part of the pentose phosphate pathway and codes for 6-phosphogluconate-dehydrogenase, a key enzyme of intermediary carbohydrate metabolism, which, hence, is expected to be well conserved among the species. However, *gnd* sequence variations were shown to be surprisingly high, which suggested recombination events at *gnd* with high frequency. The important metabolic function of *gnd*, however, makes the high recombination rate on the gene itself unlikely, but appears to be influenced by the adjacent *rfb* region, which is subject to diversifying selection.[414] Taken together, it is most likely that the close physical association of *gnd* and the *rfb* gene cluster enabled the lateral cotransfer of both loci. The questions about common ancestors or the mechanisms for dissemination of the O-polysaccharides among strains remain open. The distinctive feature of bacteriophages or plasmids to mediate transduction or conjugation, respectively, make the accessory genetic elements attractive candidates for carriers of genetic information. The first direct evidence for such a mechanism of lateral transfer of O-polysaccharide-encoding genes has been demonstrated for the ColE1-related plasmid pWQ799 of *S. enterica* serovar Borreze, a derivative of which could be mobilized in the presence of a conjugative helper plasmid conferring an O:54 serotype on the *E. coli* K-12 recipient strain.[409] The sequence similarities between pHS-2 plasmid DNA of *S. flexneri* serotype 2a and the replication/transfer regions of *E. coli* plasmids ColE1 and F may indicate that pHS-2 represents another potential candidate for providing further evidence for the proposed mechanism of dissemination of O-polysaccharide determinants.[376]

Several additional mechanisms were suggested to explain the polymorphism among the *rfb* gene clusters. In *S. enterica* serogroup D2, it was proposed that the *rfb* region has been evolved by intraspecific recombination possibly mediated by Hinc-repeats resembling insertion sequences with short flanking inverted repeat sequences, which were found to be embedded in the *rfb* region.[415] Interestingly, Hinc-repeats were originally discovered as components of *Rhs* elements that have been classified as accessory genetic elements comprising a family of large and complex genetic repetitions on chromosomes of various, but not all, *E. coli* strains.[416-418] As mentioned above, a frameshift mutation within the *tyv* gene of *S. enterica* serogroup A strains was shown to be responsible for the O-antigenic distinction between strains of serogroups A and D.[320] For the O-antigenic conversion of serotypes Ogawa to Inaba of *V. cholerae* serogroup O:1 (see Table 8) a similar mechanism has been proposed. As demonstrated, a single gene, designated *rfbT* and present in both serotypes, undergoes frameshift mutations in correlation with the host immune response during a cholera infection, causing a premature transcription termination and expression of truncated *rfbT* gene products in Inaba strains.[360,419-421] Multiple mechanisms appear to be responsible for the genesis of the new *V. cholerae* serotype O:139 Bengal strains. The strains presumably originated from a *V. cholerae* O:1, biotype El Tor strain by the acquisition of a novel DNA fragment, which contained seven open reading frames (*otnA-H*) and replaced almost the complete O:1 *rfb* region retaining *rfaD* and a slightly modified *rfbQRS* locus, and acquired the ability to produce an O-antigenic capsule.[245,422-425] Thus, *V. cholerae* O:139 strains acquired the unusual ability to code for a distinct O-polysaccharide and CPS by the same genetic locus.[245] Several interesting features have been observed within the new generated region composed of a mosaic of genes of both the O:1 and the O:139 strain.[425] The *rfbQRS* region was demonstrated to be closely related to the Hinc-repeats of *E. coli*. Furthermore, small 7-bp sequence repeats were identified to flank the *otnEFG* region, for which homologous genes could be found between repetitive extragenic palindrome sequences in *E. coli* thought to be involved in recombination processes. Based on the genetic organization of the *rfaD-otn-rfbQRS* region from *V. cholerae* O:139, it was suggested that the repeat motif sequences should be able to facilitate gene transfer processes between *V. cholerae* strains. The identification of *his* genes within the O-antigen cluster, usually located adjacent to *rfb*, and the novel insertion sequence, designated *IS1209* and dividing the *rfb* cluster into an O:5-serogroup-specific and nonspecific region in *P. aeruginosa* serotype O:5, may support the hypothesis that transfer and recombination processes account for the mosaic-like organization of the *rfb* region.[324]

In contrast, several examples are known, which demonstrate that an identical or very similar chemical composition and structure of an O-specific polysaccharide chain does not necessarily correlate with similarities at the genetic level, in spite of possible functional conservations. This situation has been found in strains of *S. boydii* type 12 and *E. coli* O:7, which possess O-polysaccharides of very similar chemical structure but no significant similarities between their *rfb* regions.[426] Apart from a gene arrangement in the same order and a high degree of similarity between the ABC-transporter-encoding genes *wzm* and *wzt*, the homology at the nucleotide sequence level was shown to be quite low for the remaining *rfb* genes coding for the D-galactan I structure in *S. marcescens* serotype O:16 and *K. pneumoniae* serotype O:1.[354] It was proposed that stringent requirements for the function of the ABC-transporter permitted only a limited sequence drift in *wzm* and *wzt*, whereas the galactosyltransferases could accommodate many more sequence variations without affecting their activity and specificity.[358] Moreover, clonal diversity based on significant variations in the *rfb* regions has been reported within a given bacterial species, in strains synthesizing very similar O-polysaccharides such as *E. coli* O:101,[427] *S. boydii* type 12,[426] or within *Klebsiella* species.[113,358] The obtained results of identical carbohydrate backbone structures of the O-polysaccharides and divergence of the respective *rfb* clusters from *K. pneumoniae* serotypes O:1 and O:8 are unique to date as they indicate that there should be no obvious biosynthetic reason for extensive DNA variations on the genetic level.

3.09.5.7 Biosynthesis of Lipid A and Inner Core

3.09.5.7.1 *Formation of the lipid A precursor Ia*

Biochemical and genetic data demonstrated unequivocally that the biosynthetic pathways leading to the formation of the lipid A component (Figure 3) and the inner core of LPS are closely linked to each other by a series of sequential reactions, in which the intermediates are derived from the precursors UDP-GlcNAc, *R*-3-hydroxymyristoyl-acyl carrier protein (-ACP), ATP, CMP-Kdo, myristoyl-ACP, and lauroyl-ACP.[224] The two key precursors, UDP-GlcNAc and *R*-3-hydroxy-myristoyl-ACP, are situated at branch points in the *E. coli* metabolism, as UDP-GlcNAc is also the GlcN donor for peptidoglycan biosynthesis, and *R*-3-hydroxymyristoyl-ACP can be used addition-ally in synthesis of palmitate residues for membrane glycerophospholipids.[225]

The transfer of the acyl group from *R*-3-hydroxymyristoyl-ACP to position 3 of UDP-GlcNAc was shown to be the initial reaction in the synthesis of lipid A.[428,429] This reaction is catalyzed by UDP-*N*-acetylglucosamine-*O*-acyltransferase encoded by the *lpxA* gene.[430,431] It is intriguing to note that the characteristic *R*-3-hydroxymyristate found at position 3 of lipid A reflects an extraordinary specificity of the acyltransferase for *R*-3-hydroxymyristoyl-ACP as the acyl donor,[428] proven not only for *E. coli* but also for various enterobacterial strains of *P. mirabilis*, *Citrobacter freundii*, *Klebsiella oxytoca*, or *S. marcescens*.[432] A remarkable feature of the *O*-acylation reaction is its thermodynamically unfavorable equilibrium constant, in which the formation of the thioester substrate *R*-3-hydroxymyristoyl-ACP appeared to be greatly favored over that of the oxygen ester UDP-3-*O*-(*R*-3-hydroxymyristoyl)-GlcNAc.[429,433] Therefore, the subsequent second reaction, the *N*-deacetylation of the *O*-acylated UDP-GlcNAc residue, is the first irreversible step in lipid A formation.[434] The UDP-3-*O*-(*R*-3-hydroxymyristoyl)-*N*-acetylglucosamine-deacetylase is encoded by the *envA*/*lpxC* gene, which was located at the 3′-end of a large cluster of genes for cell division and peptidoglycan biosynthesis.[435,436] With respect to a global regulatory network sensing the rate of lipid A biosynthesis and/or the lipid A content in the outer membrane, the *N*-deacetylation has been implicated as a step of regulation considering that the *O*-acylation of UDP-GlcNAc is revers-ible. Indeed, analysis of mutants defective in either the acyltransferases *lpxA* and *lpxD* (see below) as well as certain conditions leading to a reduced lipid A content in bacterial cells, revealed a significant increase in the specific activity of the deacetylase.[429,437,438] However, the upregulation of the deacetylase activity was obviously not a result of alterations in the transcription rate of the gene but might be, as hypothesized, associated with a translation factor or specific protease, which reacts to changes in lipid A levels.[438] Moreover, with regard to an adjustment of lipid A biosynthesis to the bacterial growth rate, it is attractive to speculate that the location of *lpxC* within a gene cluster for cell envelope components and cell division processes could make their coordinated regulation possible.[435] The removal of the acetyl moiety from UDP-3-*O*-(*R*-3-hydroxymyristoyl)-GlcNAc enables the third step in lipid A assembly, the *N*-acylation with another *R*-3-hydroxymyristate from *R*-3-hydroxymyristoyl-ACP to generate UDP-2,3-diacyl-GlcN, a reaction as specific as that for the *O*-acylation reaction and consistent with the characteristic composition of lipid A in various enterobacterial strains.[432,434] The gene encoding UDP-3-*O*-(*R*-3-hydroxymyristoyl)-glucosamine-*N*-acyltransferase has been identified as *firA*/*lpxD*.[437,439] Mutations in *lpxD* affected LPS biosynthesis pleiotropically in *E. coli* and *S. enterica* serovar Typhimurium strains, going along with a reduction of lipid A 4′-kinase activity, the enzyme of the sixth step in lipid A synthesis (see below), or attachment of hexadecanoic acid residues to lipid A molecules as partial substituents not found in wild-type strains.[439] However, there is still uncertainty regarding additional functions of LpxD such as a physical association with the RNA polymerase holoenzyme and, hence, playing a regulatory role through the transcriptional machinery.[440] In *E. coli*, the *lpxA* and *lpxD* genes are part of a remarkable complex operon of 11 genes encoding enzymes of DNA replication,[441] biosynthesis of glycerophospholipids,[442,443] and lipid A. The genes encoding the acyltransferases are separated by the *fabZ* gene, the predicted amino acid sequence of which displayed a high degree of similarity to FabA,[444] an *R*-3-hydroxydecanoyl-ACP-dehydrase involved in unsaturated fatty acid synthesis.[445] Based on these similarities and the finding that mutations in *fabZ* presumably could suppress an *lpxA* mutation by the resulting availability of an increased *R*-3-hydroxymyristoyl-ACP pool, FabZ was suggested to function as an *R*-3-hydroxymyristoyl-ACP-dehydrase located at a branch point, at which *R*-3-hydroxymyristoyl-ACP is being used for lipid A or for palmitate biosynthesis.[444] Taken together, LpxA and LpxD are not only functionally related by performing acylation reactions with the same substrate *R*-3-hydroxymyristoyl-ACP, but were shown to be essentially homologous. Furthermore, the proteins contain an unusual structure of repeating hexapeptides, which appear conserved among LpxA and LpxD proteins of *E. coli*, *S. enterica* serovar Typhimurium, *Y. enterocolitica*,[446] and *Rickettsia rickettsii*.[447]

Part of the direct precursor of the nonreducing sugar of lipid A, UDP-2,3-diacyl-GlcN, was demonstrated to undergo pyrophosphatase(s)-catalyzed reaction(s) with the release of UMP to generate 2,3-diacyl-GlcN1P, also known as lipid X and representing the immediate precursor of the reducing sugar of the lipid A molecule.[448–450] The subsequent condensation of one molecule of UDP-2,3-diacyl-GlcN with one molecule of 2,3-diacyl-GlcN1P, mediated by the product of the *lpxB* gene, the lipid A disaccharide synthase, forms the characteristic β-(1→6)-linked D-GlcN disaccharide backbone of the lipid A molecule.[449–451] The *lpxB* gene has been identified immediately downstream of *lpxA* within the complex gene cluster noted above, and the genetic organization suggests that both genes are cotranscribed and translationally coupled.[430,451] It is noteworthy that the enzymes involved in the pathway leading to the synthesis of the β-(1→6)-linked D-GlcN disaccharide backbone of the lipid A molecule were found predominantly in the cytosolic fraction of bacteria cells, although a transient association with the cytoplasmic membrane cannot be excluded at least for the latter reactions, since the biosynthetic intermediates become hydrophobic and, as shown for lipid X, primarily integrated in the membrane.[449] The demonstrated physical association of LpxB and the aerobic glycerol-3-phosphate dehydrogenase GlpD suggested that the latter enzyme acts as an adaptor for binding LpxB to the inner membrane.[452] Whole-genome sequencing of *H. influenzae* strain Rd revealed several *lpx* homologues, which indicates a high degree of conservation among the genes specifying the structurally very similar lipid A structures in *H. influenzae* and *E. coli*.[453] Indeed, cloned *lpxA* and *lpxB* genes from *H. influenzae* type b were able functionally to replace the corresponding genes in *E. coli* and additionally displayed the same gene order of *fabZ-lpxA-lpxB-rnhB*.[454]

In the sixth step of lipid A assembly, a membrane-bound 4′-kinase, presumably highly specific for disaccharides, was found to add a monophosphate residue to position 4′ of the tetraacyl-disaccharide 1-phosphate intermediate, which results in the formation of tetraacyldisaccharide 1,4′-bisphosphate, the immediate lipid A precursor Ia composed of a β-(1→6)-linked disaccharide of D-GlcN that is acylated at positions 2, 3, 2′, and 3′ as well as phosphorylated at positions 1 and 4′.[455] Despite the observations that mature *E. coli* lipid A may contain very small amounts of palmitate, a membrane-bound transacylase has been identified that is able to catalyze the addition of a palmitoyl residue to the 3-hydroxy group of the *N*-linked *R*-3-hydroxymyristoyl residue of both the tetraacyldisaccharide 1,4′-bisphosphate and the lipid X, obviously using any common diacylglycerophospholipid with a palmitoyl moiety at the *sn*-1 position as the acyl donor and converting the lipid X to lipid Y.[456] The biological significance of this reaction is uncertain, but it could be a mechanism for lipid A modification in response to varying environmental conditions.

3.09.5.7.2 *Attachment of Kdo and late acylation reactions*

Mature lipid A molecules of *E. coli* wild-type strains contain two additional acyl chains, primarily laurate and myristate, attached to the *R*-3-hydroxymyristoyl group of the nonreducing glucosamine to form the characteristic acyloxyacyl units of lipid A (Figure 3). However, previous studies on *E. coli* and *S. enterica* serovar Typhimurium mutants, defective in Kdo biosynthesis, revealed a predominant accumulation of the underacylated lipid Ia precursor in the cells under nonpermissive conditions.[457–459] This phenomenon can be explained by the discovery of a lauroyl- and a myristoyl-transferase exhibiting extremely high specificity not only for their acyl donors, lauroyl- or myristoyl-ACP, but also for their substrate, (Kdo)$_2$-lipid Ia.[460] Thus, the inability of the late acyltransferases to transfer laurate and myristate to precursor Ia suggests that the sequential addition of the Kdo residues onto the lipid Ia molecule must precede the incorporation of the acyl chains.

Biosynthesis of Kdo occurs by two enzymatically catalyzed reactions. Kdo-8-phosphate-synthase catalyzes the aldol condensation of D-Ara5P and phosphoenolpyruvate to yield Kdo8P,[461–463] followed by removal of the phosphate by a specific phosphatase to give Kdo.[464] Prior to incorporation into lipid A, Kdo requires a CTP-activation in a reaction carried out by CMP-Kdo-synthetase.[465,466] The genes *kdsA* and *kdsB*, coding for Kdo-8-phosphate-synthase and CMP-Kdo-synthetase, respectively, have been cloned, sequenced, and functionally characterized from *E. coli* and *Chlamydia* spp.,[466–472] and homologues of the enterobacterial genes have been identified by sequence comparisons in the genome of *H. influenzae* strain Rd.[453,473] Despite the fact that the *kdsA* and *kdsB* genes were derived from distantly related bacteria, their deduced amino acid sequences showed a high degree of similarity, respectively, emphasizing a conservation of the pathway for synthesis and activation of Kdo at least for the Gram-negative bacteria studied. Moreover, the presence of conserved sequence motifs within the Kdo-8-phosphate-synthase and 3-deoxy-D-*arabino*-heptulosonate-7-phosphate-synthases of different origin suggested the proteins to be members of a family of enzymes, which catalyze the formation of a phosphorylated 3-deoxy-α-keto sugar acid by aldol condensation of phosphoenolpyruvate with a phosphorylated sugar.[462,471,474,475] Notably, *kdsA* and *kdsB* were found separated from the LPS gene clusters at different positions on the chromosomes of *E. coli* and *H. influenzae* Rd.[466,473,476] Since their gene products are involved in the same biosynthetic pathway, it should be reasonable to assume a coordinated regulation of both genes, while a second CMP-Kdo-synthetase, encoded by the *kpsU* gene, was shown to be temperature regulated and associated with capsule expression of group II capsule synthesizing *E. coli* strains.[477–479] In order to investigate regulatory aspects of *kdsA* and *kdsB* from *E. coli*, it was demonstrated that both genes undergo independently of the growth rate a growth-phase dependent regulation at the transcriptional level with a rapid repression of mRNA expression as bacterial cells enter the stationary growth phase.[480] Furthermore, *kdsA* of *E. coli* was part of an operon and was cotranscribed with two downstream located open reading frames of unknown function.[480] A close linkage to at least two open reading frames has been proposed as well for the *kdsA* gene of *C. psittaci* 6BC.[471] It is intriguing that *kdsB*, coding for the CTP-utilizing CMP-Kdo-synthetase, was shown to form a transcriptional unit with the gene for the CTP-generating CTP-synthetase of *C. trachomatis*, indicative of a potential role of the enzyme in LPS biosynthesis.[470,481] However, further investigations are necessary to elucidate whether the obviously developmentally regulated expression of the CTP-synthetase correlates with varying demands for CTP in chlamydial LPS biosynthesis

In the seventh step of lipid A assembly in *E. coli*, two CMP-Kdo molecules are finally used for the sequential transfer onto the lipid A precursor lipid Ia, leading to an α-(2→6)-linkage between the GlcN backbone and a first Kdo residue, and an α-(2→4)-linkage between a second Kdo residue and the first one. The formation of two different glycosidic bonds is catalyzed by the unusual bifunctional Kdo transferase KdtA [WaaA] tolerating those acceptor molecules, which can be acylated to various extents, but exhibiting an absolute requirement for the 4′*P* on the tetra-acyldisaccharide 1,4′-bisphosphate intermediate.[482–484] Moreover, biochemical studies on the assembly of the genus-specific epitope in *Chlamydia* spp., composed of the trisaccharide α-Kdo-(2→8)-α-Kdo-(2→4)-α-Kdo, provided definite proof that the single Kdo-transferases were capable of recognizing lipid Ia as a substrate and attaching three Kdo residues to the lipid A precursor.[485–487] Kdo-transferases have been cloned and sequenced from various Gram-negative bacteria such as *E. coli*,[483] different *C. trachomatis* serotypes,[167,485,488] *C. psittaci* 6BC,[489] *C. pneumoniae* TW183,[486] *S. marcescens* N28b,[490] *B. pertussis* BP536,[491] as well as *H. influenzae* type b and strain Rd.[453,492,493] The enterobacterial WaaA proteins were nearly identical, while comparison of the deduced amino acid sequences of all Kdo transferases displayed a low degree of similarity with merely very short stretches of identical amino acid residues along the whole sequences. The high degree of sequence variations among the chlamydial Kdo-transferases within the same genus was unexpected, since they have the same function, i.e., the assembly of the very same chlamydial genus-specific LPS epitope.[486,489] From structural analysis of the LPS of *H. influenzae* and *B. pertussis* it is known that their LPS molecules contain only one Kdo residue,[228] and it is still not clear why the single Kdo-transferase of *E. coli* is able to add two, the chlamydial WaaA proteins three, and those of *H. influenzae* and *B. pertussis* apparently only one Kdo residue to the lipid A precursor Ia. Furthermore, there is also uncertainty about the role of an open reading frame in LPS biosynthesis found downstream of *waaA* in different bacteria. By sequence comparison, the gene product of the *kdtX* gene from *S. marcescens* appeared related to the protein H10653 (KdtB) of *H. influenzae*, but not to the product of *kdtB* from *E. coli*.[453,483,490]

In *E. coli*, the Kdo-dependent lauroyltransferase was shown to be encoded by the *htrB* [*waaM*] gene, since a significant reduction of laurate amounts in LPS correlated with a lack of lauroyl-transferase activity in *waaM* deficient mutants.[494,495] WaaM was first described as a membrane-associated protein required for viability of the bacterial cells in rich media above 33 °C.[496] The unique temperature requirement of WaaM did not correlate with a heat-inducible transcription of its message and, hence, it was proposed that *waaM* is not a typical heat shock gene, but rather belongs to a new class of genes, the products of which appear to be essential for bacterial growth at high temperatures.[497] Possibly, WaaM could be involved as a regulatory factor in LPS biosynthesis under conditions of rapid growth.[495] It is of interest to note that a homologue of *waaM* which was identified upstream of *rfaE* [*waaE*] for ADP-heptose-synthase in the nontypable *H. influenzae* strain 2019 was able functionally to complement a *waaM* mutation in *E. coli*.[498,499] WaaM of *H. influenzae* has been suggested as a multifunctional protein, since the phenotypes of the mutant strains appeared rather diverse and complex. The strains consisted of a reduced number of acyl chain substituents in the lipid A component and a modified core structure resulting from an alteration in the overall hexose/phosphoethanolamine ratio.[499] Unlike the temperature-sensitive *waaM* mutants of *E. coli*, the *waaM* deficient *H. influenzae* strains restored the ability to grow at temperatures of the wild-type strain after a few passages at 30 °C, which suggested the development of factor(s) suppressing the *waaM* mutation.[499] Suppressors of *waaM* mutations have been identified in *E. coli*, capable of suppressing both the temperature-sensitive growth phenotype and the morphological changes of *waaM*⁻ strains. Spontaneously arising mutations in the *accBC* operon, encoding two subunits of the acetyl coenzyme A carboxylase enzyme complex and mediating the first step in fatty acid biosynthesis in *E. coli*, might be responsible for alterations of functions that are affected by null mutations in the *waaM* gene.[445,495] Another suppressor of *waaM* null mutations was shown to be encoded by the *msbA* gene, only having suppressing function when present as multiple copies on cloning vectors.[500] Apart from being highly similar to proteins comprising the family of ABC-transporters, MsbA is obviously the only ATP-dependent translocator known to date, which seems to be essential for bacterial growth or viability.[500,501] Studies on the function of MsbA provided evidence that the protein presumably plays an important role in the translocation process of the lipid A-core-portion of LPS or its precursor(s) across the cytoplasmic membrane.[501] Although *msbA* codes for an ABC-transporter with an integral membrane component on the same polypeptide, the product of the downstream located *orfE* gene appeared to be necessary as well for the transport across the membrane. Taken together, it was demonstrated that in *E. coli*, deficient in the lauroyltransferase WaaM, a large quantity of defective underacylated lipid A-core-precursors accumulate at the cytoplasmic membrane at nonpermissive temperatures, while only a sufficient amount of MsbA, expressed from a cloning vector, is able to restore the translocation process and, hence, suppress the temperature-sensitive phenotype of *waaM*⁻ bacteria.[501] Finally, *msbB* [*waaN*], previously found to code for another multicopy suppressor of *waaM*⁻-related phenotypic appearances in *E. coli*,[502] has been suggested as the structural gene for the second potential Kdo-dependent late acyltransferase adding myristate to (Kdo)$_2$-lipid Ia.[503,504] This suggestion was supported by the findings that the predicted amino acid sequences of WaaM and WaaN displayed significant similarity, and a mutation within the *waaN* gene resulted in LPS lacking the myristoyl, but obviously not the lauroyl, moiety.[502,504] Thus, WaaM and WaaN seem indeed to be functionally related, but are clearly distinguishable by the synthesis steps they mediate in lipid A assembly.

Biochemical data on the presence of all the enzymes for the formation of (Kdo)$_2$-lipid Ia from UDP-GlcNAc in two biovars of *R. leguminosarum* provided strong evidence for a high conservation of the early steps in lipid A synthesis even in genetically diverse bacteria like *E. coli* and *R. leguminosarum*.[505] However, the lipid A structure of *R. leguminosarum* differs from that of *E. coli* as it is acylated with a unique 27-hydroxyoctacosanoic acid and completely lacks phosphate groups containing galacturonic acid and aminogluconic acid residues in place of the 4'- and the 1-phosphate group, respectively.[185,506] A set of novel (Kdo)$_2$-lipid Ia recognizing enzymes for the further processing of this key intermediate has been identified in rhizobial protein extracts including 1- and 4'-phosphatases as well as a long-chain acyltransferase.[507,508] Thus, it became evident that *E. coli* and *R. leguminosarum* utilize different pathways to generate their species-specific lipid A molecules only after completion of (Kdo)$_2$-lipid Ia by the same seven synthesis steps summarized above. In contrast, attachment of Kdo to the lipid Ia precursor does not seem to be an absolute prerequisite for late acylation reactions in all Gram-negative bacteria. Based on previous results of *P. aeruginosa* cells, accumulating completely acylated lipid A molecules when CMP-Kdo-synthetase activity was suppressed,[509] it could be shown that *Pseudomonas* cell extracts possess a distinct lauroyltransferase able to transfer laurate from lauroyl-ACP to lipid Ia in the absence of the Kdo disaccharide.[510]

3.09.5.7.3 Heptosylation and completion of the inner core

Prior to transfer to the inner core, ADP-L-*glycero*-D-*manno*-heptose (ADP-L,D-Hep) is thought to be synthesized from sedoheptulose 7-phosphate (Sed7P) in a four-step reaction. First, Sed7P is converted by a phosphoheptose-isomerase to yield D-*glycero*-D-*manno*-heptose 7P (D,D-Hep7P), followed by synthesis of D,D-Hep1P in a phosphoheptose-mutase catalyzed reaction. In the third step, an ADP-heptose-synthase is responsible for the generation of ADP-D,D-Hep, which is finally converted to ADP-L,D-Hep by an epimerase.[511,512] Bacterial strains defective in the enzymes of the pathway proceeding from Sed7P to ADP-D,D-Hep produce heptose-deficient LPS, whereas mutations in the epimerase gene result in a more complex phenotype including the incorporation of some D,D-Hep into the LPS molecule. The phosphoheptose-isomerases LpcA [GmhA] of enterobacteria and *H. influenzae* were shown to be highly conserved and, together with known aldose/ketose isomerases, may comprise a novel family of phosphosugar isomerases.[513-515] The genes *waaE* and *rfaD* [*gmhD*] have been first identified in enterobacterial strains coding for ADP-heptose-synthase and ADP-D-*glycero*-D-*manno*-heptose-6-epimerase in the synthesis of ADP-L,D-Hep, respectively.[516-518] Meanwhile, a number of structural and functional homologues of WaaE and GmhD were found in various Gram-negative bacteria indicating a conservation of the heptose pathway even in distantly related microorganisms. Thus, the predicted amino acid sequence of *gmhD* from *P. aeruginosa* PAO1 was almost identical to that of *E. coli*, and the cloned $gmhD_{Pa}$ was capable of complementing an *E. coli gmhD* mutation.[519] As mentioned above, *waaE* was located immediately downstream of the *waaM* gene in *H. influenzae* 2019, and both genes seem to be transcribed from the same promoter into opposite directions making the coordinated regulation of their expression possible.[498] The WaaE protein of *N. gonorrhoeae*, like its counterpart from *H. influenzae*, and the gonococcal epimerase GmhD were able as well to complement the corresponding defects in *S. enterica* serovar Typhimurium.[520,521] The gonococcal *waaE* and *gmhD* genes were found adjacent to each other and presumably might form a transcriptional unit.[521] This is of particular interest, since a *gmhD* homologue was closely linked to the *rfb* region for O-antigen biosynthesis in *V. cholerae* O:1 strains of the Inaba and Ogawa serotypes,[522] while in enteric bacteria the gene was located near the left end of the *rfa* cluster for core oligosaccharide biosynthesis (Figure 13).[226,523,524] It deserves attention that a screening for heat shock proteins in *E. coli* identified the *htrM* gene, which has been finally proven to be identical to the *gmhD* gene.[525] The transcription of the *htrM/gmhD* gene from *E. coli* requires at least two promoters, one of which resembled the consensus sequence of an $E\sigma^{32}$-dependent heat shock promoter.[525] In *S. enterica* serovar Typhimurium, however, the *gmhD* gene lacks the heat shock promoter and presumably is only transcribed from a typical $E\sigma^{70}$ promoter used by housekeeping genes.[524] The physiological importance of this observation is not clear but may indicate differences in expression of the gene in *E. coli* and *S. enterica* serovar Typhimurium. The transfer of ADP-L,D-Hep to the first Kdo (Kdo I) of $(Kdo)_2$-lipid Ia is catalyzed by heptosyltransferase I, the product of the *rfaC* [*waaC*] gene in Enterobacteriaceae.[518,526] However, there are still gaps in our knowledge, whether the transfer of the first Hep proceeds before or after acylation of $(Kdo)_2$-lipid Ia, or the acyltransferases and heptosyltransferase I may act concurrently on the lipid A precursor in *S. enterica* serovar Typhimurium and *E. coli*, as proposed on the basis of biochemical data.[518] Mutational analysis, complementation tests, molecular cloning, and sequence analysis revealed functional and structural WaaC counterparts in *N. gonorrhoeae*, *H. influenzae* strain Rd, and *B. pertussis*.[473,491,527] A membrane-associated mannosyltransferase, proposed as a functional analogue of WaaC, was shown to be involved in biosynthesis of the unusual heptoseless inner core region of *R. leguminosarum*, able to transfer Man from GDP-Man to $(Kdo)_2$-lipid Ia in an *in vitro* assay, followed by addition of a Gal residue in the presence of UDP-Gal by another novel rhizobial glycosyltransferase.[508,528] In *R. leguminosarum*, galactose positionally replaces the second heptose found in other Gram-negative bacteria, being incorporated by the heptosyltransferase II. Its encoding gene *rfaF* [*waaF*] has been identified in a variety of genetically diverse bacteria such as *E. coli* and *S. enterica* serovar Typhimurium,[226] *N. gonorrhoeae*,[529-531] *N. meningitidis*,[532] *H. influenzae* strain Rd,[473] or the non-typeable *H. influenzae* strain 2019.[533] Although a functional assignment of a gene or gene product to the reaction of Hep III attachment to the inner core is presently lacking, the RfaQ [WaaQ] protein has been suggested as a potential candidate for carrying out this *rfaP* [*waaP*]-dependent reaction in enteric bacteria.[226,534] This assumption is based on the finding that the heptosyltransferases WaaC/WaaF, the Kdo-transferase, and WaaQ share regions of similarity and, therefore, may comprise a family of related transferases performing glycosylations in the inner LPS core region.[226,483,534,535] Among the proposed 25 putative LPS-related genes in *H. influenzae* Rd, the products of the genes *opsX* [$waaC_{Hi}$], *rfaF* [$waaF_{Hi}$], and *orfH* [$waaQ_{Hi}$], presumably required for the sequential transfer of the three Hep residues to Kdo, appeared to be the structural and functional homologues of the enterobacterial heptosyltransferases, since mutant strains defective in $waaC_{Hi}$, $waaF_{Hi}$, or $waaQ_{Hi}$ produced LPS apparently without, with one, or two Hep residues, respectively.[473] Investigations of genetic determinants for biosynthesis of the inner core region of *N. meningitidis* provided strong evidence that the inner core structure has to be extended prior to completion of the α-chain oligosaccharide of LPS. An inner core extension operon, consisting of the genes tRNA-Ile-*icsB/lgtF-icsA/rfaK* [$waaK_{Nm}$] and separated from the remaining operon for α-chain biosynthesis, was found to encode glycosyltransferases for two subsequent synthesis steps.[536-539] Apparently, the α-chain will be extended from Hep I of the inner core, only if the first Hep has been substituted with Glc in a reaction catalyzed by IcsB/LgtF and dependent on the presence of the γ-chain, a GlcNAc residue at the second Hep, which is added by IcsA/$WaaK_{Nm}$ before the glucosylation can take place.[536–539] While IcsA/WaaK$_{Nm}$ appeared to be a homologue of the *N*-acetylglucosamine-trans-ferase WaaK from *S. enterica* serovar Typhimurium, the deduced amino acid sequence of *icsB/lgtF* displayed similarity not only to β-glycosyltransferases, but also to both the KdtX and KdtB protein of *S. marcescens* and *H. influenzae*, respectively. It is worthy of note that the inner core extension genes presumably form a transcriptional unit, being also transcribed from the probably growth rate-dependent tRNA promoter and, hence, might be under the control of factors that regulate tRNA expression.[536]

Despite an inner core backbone with a relatively stable stoichiometry, several substituents in variable or nonstoichiometric amounts have been found in enterobacterial strains (Table 10). As outlined, the relatively fixed carbohydrate backbone of the inner core is primarily formed by the β-(1→6)-linked disaccharide of D-GlcN, the two Kdo residues, and by three L,D-Hep units, while the heterogeneity of the lipid A/inner core region is caused by partial substitutions with Kdo, L-Ara4N, α-L-Rha, and phosphate or phosphoethanolamine residues at different sites of the molecule.[83] It must be emphasized that the biosynthetic steps for these partial substitutions are mostly unknown, keeping in mind that the substituents may vary among the strains even within a species, probably depending on the specific genetic background of a given strain, and their nature may dramatically change in response to varying environmental conditions. In addition, much of the information available about the variable or nonstoichiometric substituents is frequently derived from the isolation of different partial structures from the core regions of mutant strains, which may not represent the actual structures in the wild-type strains. The best characterized example for a correlation of the level of lipid A/inner core substituents and changes in properties of the cell envelope, presumably depending on physiological demands, is the regulation of the negative net charge on the LPS molecule by genes of the *pmr* locus. Mutations in the *pmrA* gene, conferring a polymyxin B resistant phenotype on the bacterial cell, were demonstrated to be associated with an esterification of the lipid A 4′-monophosphate by L-Ara4N, thought to reduce the net charge on LPS and leading, therefore, to a decreased binding capacity of the cells for polymyxin B.[540–542] It is noteworthy that PmrA of *S. enterica* serovar Typhimurium appeared to be a DNA-binding response-regulator typical for two-component regulatory systems, which consist of sensory kinases for monitoring environmental parameters and response-regulators for mediating changes in the expression of various genes in response to environmental stimuli.[543,544] The *pmrA* gene was located adjacent to the *pmrB* gene for the putative kinase-sensor protein, and both gene products were highly similar to those of the *basRS* locus of *E. coli*, which suggested that an analogous regulatory system might be affected in polymyxin B resistant *E. coli* strains.[543,545] Finally, homologues of BasRS could also be identified in *H. influenzae* and *N. gonorrhoeae*.[228]

Studies on strain-specific LPS banding patterns on high-resolution SDS–PAGE in combination with genetic analysis of the various *E. coli* K-12 strains indicated a functional relationship between genes of the *rfa* cluster for core oligosaccharide assembly and genes of the *rfb* cluster for O-polysaccharide synthesis.[87,89] The partial substitution of Kdo II with α-L-Rha in the inner core region of some *E. coli* K-12 strains (see Table 2), the only one of its kind presently known for enterobacteria,[83,85] appeared to be directed by Rha biosynthesis genes of the *rfb* region and may represent part of a distinct pathway leading to the synthesis of an alternative LPS core population not used for O-antigen attachment.[87]

3.09.5.8   Biosynthesis of the Outer Core

3.09.5.8.1   *Outer core assembly in enteric bacteria*

The very similar *rfa* regions of *E. coli* K-12 and *S. enterica* serovar Typhimurium are complex gene clusters in that they not only contain the genes for all the glycosyltransferases required for the sequential transfer of the hexoses to the nonreducing terminus of the growing outer core oligosaccharide chain, but also include genetic determinants for core modifications, O-polysaccharide attachment, ADP-D-*glycero*-D-*manno*-heptose-6-epimerase, as well as for Kdo- and Hep-transfer (Figure 13).[226,546] Apart from probably being regulated by internal promoters, the LPS core genes of the *rfa* region were shown to be clustered into three operons.[546–548] The operon near the left end contains the genes *gmhD*, *waaFC*, and *waaL* for ADP-D-*glycero*-D-*manno*-heptose-6-epimerase, the heptosyltransferases I and II, and the putative ligase, respectively, while the operon at the right end comprises the genes *waaA* and *kdtB*. The largest operon includes the genes necessary for outer core biosynthesis. Most knowledge of outer core structure and biosynthesis is based on previous investigations of mutant strains (Figure 13). The assignment of the resulting rough phenotype to a gene function was greatly facilitated by the fact that mutations in a particular gene, coding for the attachment of a specific sugar to the growing oligosaccharide chain, could prevent the further addition of all the distal residues (Figure 13).[223,226]

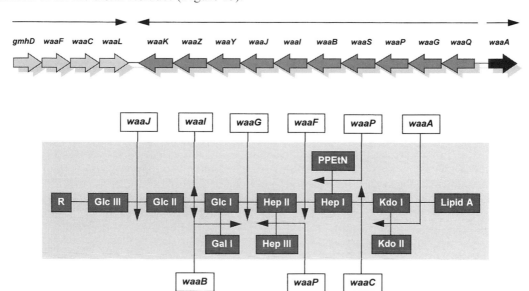

Figure 13 Schematic presentation of the genetic organization of the *rfa* gene cluster (top), of the chemical structure of the core oligosaccharide, and of functions of some *rfa* genes in core oligosaccharide assembly in *E. coli* K-12 (bottom).[226]

In both *E. coli* K-12 and *S. enterica* serovar Typhimurium, the *rfaG* [*waaG*] gene codes for the glucosyltransferase, which adds Glc as the first sugar of the outer core from UDP-Glc to Hep II of the inner core.[535,547,549,550] Mutations in *waaG* were demonstrated to confer a highly pleiotropic phenotype on the strains, including the loss of flagella, pili, K99 fimbriae, or a suppression of the expression of outer-membrane proteins.[549,551] As expected, the remaining outer-core structures with their different sugar compositions are determined by different biosynthetic pathways in *E. coli* K-12 and *S. enterica* serovar Typhimurium. The *rfaI* [*waaI*] gene is thought to be involved in the addition of the second backbone hexose, which is Gal I in *S. enterica* and Glc II in *E. coli* K-12.[552,553] The attachment of the third hexose to the backbone chain, shown to be Glc II in *S. enterica* and Glc III in *E. coli* K-12, requires the *rfaJ* [*waaJ*] gene, the function of which might be dependent on the activity of the *rfaY* [*waaY*] gene.[552,553] Although the WaaI and WaaJ proteins of both microorganisms were shown to be highly similar, they differed in their complementation efficiency and, hence, probably must differ in their substrate specificity.[553] Furthermore, the *waaJ*-mediated completion of the core backbone structure was demonstrated to go along with the phenotypic appearance of multiple forms of the LPS core, which suggested that the basic core structure is modified by the attachment of additional residues as the synthesis of the main hexose chain progresses.[226,553] This suggestion was supported by the finding that *E. coli* K-12 strains, defective in the *rfaB* [*waaB*] gene and unable to add the lateral Gal residue to Glc I, exhibited an Rc chemotype, in which the addition of Glc II to the growing main chain was also disrupted.[553] Thus, WaaI presumably required the lateral Gal residue for an efficient recognition of the acceptor molecule so that the main hexose chain can be extended. However, the assumed interactions between the WaaB and WaaI proteins in outer core biosynthesis, proposed on the basis of the distinct phenotypic appearance of some *waaB* mutants of *S. enterica* serovar Typhimurium, remain to be reevaluated as defined mutants are constructed.[226,553,554]

Although the definitive functions of all the remaining genes of the *rfa* cluster are not clear to date, several lines of evidence suggested the genes *waaP*, *rfaS* [*waaS*], *rfaZ* [*waaZ*], and *waaK* to be involved in the synthesis of other branch residues of the core oligosaccharide. The kinase-like activity of WaaP is thought to be responsible for the attachment of phosphoryl or pyrophosphorylethanolamine substituents to Hep I of the inner core.[549] Mutations in *waaP* result in a rather complex Rc phenotype, including the lack of phosphatidyl ethanolamine on Hep I, the lateral Gal residue on Glc I, and the Hep III residue on Hep II.[535,549] Moreover, a single mutation in *waaP* was obviously sufficient to induce the production of the colanic acid polysaccharide possibly by increasing the expression of RcsC, the membrane sensor of the two-component regulatory system RcsC/RcsB.[549,555] Finally, the results of investigations on hemolysin synthesis in *E. coli* indicated that lesions in the *waaP* gene may cause the secretion of cytolytically inactive hemolysin from the bacterial cells.[556] An interesting feature of *S. enterica* waaP mutants is their apparent leakiness, as they can express some O-polysaccharide with a complete core in addition to the main LPS core population of the RcP⁻ chemotype.[557] However, further studies are needed to elucidate whether the galactosyltransferase WaaI$_{Se}$ can act on a nonphosphorylated Rc core oligosaccharide, though with decreased efficiency. The role of WaaS and WaaZ in core assembly of *E. coli* strains is not clear, but the proteins appear to be required for a branch pathway of core biosynthesis to generate alternative LPS core structures, which cannot serve as substrates for O-polysaccharide attachment (see above).[87,89] The *waaS* and *waaZ* genes were proposed to code for functions that are involved in the partial substitution of Kdo II with α-L-Rha and, hence, mediate the production of an LPS core form to which O-antigen is not attached.[87,89,226] It is noteworthy that *S. enterica* serovar Typhimurium apparently does not produce an abundant alternative core structure, which is consistent with the finding that the bacterium lacks the WaaS-coding sequence and, presumably, carries a nonfunctional *waaZ* gene.[226] The *waaS* gene has an exceptionally low G+C content and may, together with the *waaK* and *waaL* genes, comprise those discrete blocks within the *rfa* clusters of *E. coli* K-12 and *S. enterica* serovar Typhimurium, which have, like the gene blocks of discrete G+C contents within the *rfb* regions, distinct origins and histories. It is not surprising, therefore, that the products of *waaK* and *waaL* are the only *rfa*-encoded proteins of the two organisms, which are poorly conserved at the primary amino acid level.[558,559] Furthermore, apart from being structurally similar, WaaK and WaaL from *E. coli* K-12 and *S. enterica* serovar Typhimurium were shown to be functionally different. Analysis of the phenotypic appearances of *waaK* mutants on high-resolution PAGE suggested that WaaK$_{Ec}$ probably acts at an earlier stage of core assembly than WaaK$_{Se}$.[89] It is believed that *waaK* is the structural gene for a transferase adding GlcNAc to the completed core of *S. enterica* serovar Typhimurium, while in *E. coli* K-12 it is transferred to an inner core constituent, a process which is thought to play a significant role in core completion.[89,226,559] The role of WaaL seems to be quite complex. First, the predicted structure of the protein differs from those of the other Rfa proteins as it contains at least 10 potential membrane-spanning domains.[558,559] Second, WaaL has been proposed as a component of the O-antigen ligase complex required for transfer of *O*-polysaccharide from ACL to the LPS core.[226] Third, WaaL appeared to be involved in both the core modification and the activity of the *waaK* gene.[89] Taken together, the integral membrane protein WaaL is thought to be an important part of a complex between proteins and carbohydrates, which modulates LPS core completion and O-polysaccharide attachment. Based on the general assumption that LPS assembly is a transmembrane process including the final stages of O-polysaccharide ligation to core-lipid A at the periplasmic face of the cytoplasmic membrane, the participation of various gene products in the ligation pathway has to be considered. At this point it should be remembered that several gene products or LPS synthesis steps have been suggested as being associated with this pathway such as MsbA playing an important role in the translocation of the core-lipid A-portion across the cytoplasmic membrane,[501] the N-terminal sequences of WbaP involved in processing of the O-polysaccharide units,[331,332] Wzx having flippase function in translocation of O-polysaccharide units across the inner membrane,[340] Wzy being required for polymerization of the O-antigen,[350] and Wzz interacting with WaaL and/or Wzy to control the kinetics or commit the O-polysaccharide chain to the ligation reaction.[350,375]

Presently, there are no experimental data or models which can explain convincingly the coordinated regulation of the operons within the *rfa* gene cluster. However, results of previous investigations indicated that the RfaH protein may act as a positive transcriptional regulator of the large central outer core operon in *E. coli* K-12 and *S. enterica* serovar Typhimurium, and that the site of interaction probably lies within the intergenic region between the divergently transcribed *waaA* and *waaQ* genes.[534,560–563] Studies on *rfaH*, which is located outside the *rfa* cluster, were greatly stimulated by findings that the gene obviously codes for a novel regulatory protein. It appeared to be required for transcription of several operons, which determine synthesis, export, and cell-surface assembly of outer-membrane-associated molecules, including the *waaQ-K* operon, the *traY-Z* operon for F pilus synthesis,[564,565] at least one *kps* gene cluster for group II capsule production,[566] and the *hlyCABD* operon for synthesis of the secreted toxic form of hemolysin.[567–570] Interestingly, a conserved sequence of eight base pairs was identified upstream of the operons as the only common feature among the RfaH-regulated gene clusters.[569] Moreover, the so-called *ops* element was noticed as part of the conserved 39-bp JUMPstart sequence, which was found previously to be located in the noncoding region upstream of several bacterial gene clusters coding for the synthesis of the outer core, O-polysaccharide components, or group II capsules.[566,568,569,571] The proximity to putative promoters suggested the JUMPstart sequences to be involved in the regulation of transcription, enabling the bacterial cell to coordinate the expression of polysaccharide structures.[571] While experimental data for this assumption are presently missing, it has been shown that the RfaH-dependent suppression of transcriptional polarity of the distal genes of the *hly* operon required the presence of the 5′*cis*-acting *ops* element.[569,570] Therefore, it is reasonable to expect that the distribution of the *ops* elements in all presently known RfaH-regulated operons presumably defines the specificity of their regulation by transcription elongation.[568–570]

The biosynthesis of the core in *Y. enterocolitica* serotype O:3 was found to have some distinctive features. The core structure is probably encoded by at least two different genetic loci, the *rfa* region and the *wbc* gene cluster for synthesis of the inner and outer core, respectively.[239,383] Sequence analysis of the *wbc* operon revealed similarities of the predicted gene products to proteins of other enterobacterial *rfb* gene clusters, the most remarkable of which exists between the TrsA [Wzx$_{YeO3}$] protein and the potential flippase Wzx for translocation of ACL-linked O-polysaccharide units across the inner membrane. It was hypothesized that the *wbc* operon had lost the ability to code for O-antigen biosynthesis during evolution, but remained functional because of its indispensable role in outer core completion.[239] The hypothesis of being an *rfb* relict was supported by the finding that the *wbc* operon has been found immediately downstream of the *adk*/*hemH* genes at a similar location on the chromosome as demonstrated for the *rfb* clusters of *Y. pseudotuberculosis* and *Y. enterocolitica* serotype O:8.[239,241,303]

3.09.5.8.2 *Outer core assembly in nonenteric bacteria*

Significant knowledge has been accumulated on the molecular mechanisms of LPS biosynthesis of important human mucosal pathogens such as *N. gonorrhoeae*, *N. meningitidis*, *H. influenzae*, or *B. pertussis*. A remarkable feature of these organisms is their ability to adjust their LPS biosynthesis to microenvironmental changes in the host and, hence, to establish strategies for infecting human cells and evading host immune defenses. One of the strategies evolved is that the LPS outer core undergoes structural variation, being subject to both antigenic variation by changing the carbohydrate composition and phase variation by reversible on/off-switching of distinct outer core constituents.[228,572,573] This indicates that each bacterial strain can synthesize a set of LPS molecules simultaneously and regardless of the LPS forms produced, the variable oligosaccharides exhibit the extraordinary feature of mimicking those of human glycosphingolipids, which confers a poor immunogenicity on to the bacterial structures and enables the structures to imitate functions of host molecules.[573,574]

In *N. gonorrhoeae*, the *lgt* cluster of five open reading frames has been identified to code for the glycosyltransferases required for both the sequential transfer of the sugar residues comprising the lacto-*N*-neotetraose α-chain (LgtEABD) and the addition of an α-linked Gal residue in the synthesis of an alternative α-chain structure (LgtC).[575,576] Three of the genes, namely *lgtA*, *lgtC*, and *lgtD*, contain poly-G tracts, which are subject to a slippage mechanism leading to frameshifts and premature terminations of the genes by addition or deletion of single guanosines during DNA replication and accounting for the high frequency variation within the α-chain structure.[577] Thus, "turn-off positions" of the *lgtD* and *lgtA* genes would prevent the addition of the terminal GalNAc or the attachment of GlcNAc to the growing α-chain, respectively, and these phase variations of the outer core appear very common among gonococcal strains.[575,578] Accordingly, the variable synthesis of the alternative α-chain structure would be dependent on the expression of the *lgtC* gene.[575] The *lgt* gene cluster was found to be conserved to a considerable degree among several *N. gonorrhoeae* strains.[576,579] However, the previously suggested novel glycosylation pathway for the outer core of the *N. gonorrhoeae* strain 15253, synthesizing a lactosyl group instead of a complete α-chain and a second one linked to Hep II as the β-chain, was consistent with the finding that the strain contained only the genes *lgtA* and *lgtE*. LgtA was shown to be involved in the phase-variable transfer of GlcNAc to the truncated α-chain and the probably bifunctional LgtE in the transfer of a Gal residue to both the α- and the β-chain.[576,580] The terminal trisaccharide of the lacto-*N*-neotetraose epitope in the parent *N. meningitidis* immunotype L3 strain MC58 is encoded by the *lgtABE* locus, the genes of which were demonstrated to be highly similar to those of the gonococcal locus including the poly-G tract in the coding sequence of *lgtA*. The lack of *lgtD* and *lgtC* in the meningococcal strain was consistent with the absence of both the GalNAc modification of lacto-*N*-neotetraose and the terminal Gal residue of the α-Gal-(1→4)-β-Gal epitope, respectively.[581,582] It is intriguing that the structural variation in the outer core of neisserial strains can be additionally extended *in vivo* by an exogenous modification using host-provided cytidine 5′-monophosphate *N*-acetylneuraminic acid (CMP-Neu5*N*Ac) as the sugar nucleotide donor.[578,583-585] The major acceptor site for sialylation is a free terminal Gal on lacto-*N*-neotetraose, but any variation that terminates with a free Gal residue may obviously serve as an acceptor for sialic acid.[586,587] This was confirmed by cloning and functional characterization of the *lst* genes from *N. meningitidis* and *N. gonorrhoeae* encoding the α-2,3-sialyltransferases, which displayed a strong preference for the terminal Gal on lacto-*N*-neotetraose as the natural acceptor.[588] Sialylation of LPS by exogenously synthesized CMP-Neu5*N*Ac is performed by sialyltransferases that have been detected at the bacterial cell surface of various pathogenic neisserial strains, but not in their nonpathogenic counterparts.[586,589] In *N. meningitidis* serogroup B and C strains, synthesizing a sialic acid-containing capsule, modification of LPS with sialic acid additionally depends on the availability of endogenously produced CMP-Neu5*N*Ac, a process which is assumed to occur in the cytoplasm or at the cytoplasmic membrane, as the membrane is impermeable to CMP-Neu5*N*Ac.[590] Despite the fact that the strains utilize an α-2,8-sialyltransferase for CPS synthesis,[591] both the capsule expression and the endogenous sialylation of LPS require the activity of the *siaA* gene essentially involved in CMP-Neu5*N*Ac biosynthesis. Investigations of capsule-negative strains provided evidence for a novel mechanism of genetic variation in meningococci that enables a concurrent on/off-switching of capsular biosynthesis and endogenous LPS sialylation as a result of a reversible insertional inactivation of *siaA* by the new insertion sequence element *IS*1301.[592] All these findings indicate that the microorganisms have evolved several mechanisms for a differential sialylation of the variable oligosaccharide part of LPS to facilitate entry into mucosal epithelial cells and to resist the human immune defence. Bacterial cells that have LPS with low amounts of sialic acid can enter the host cells very efficiently, but are susceptible to killing by complement, whereas after transition to highly sialylated LPS the bacteria become entry deficient, but are resistant to complement- and antibody-mediated killing.[578]

In *H. influenzae* serotype b strains, the *lic1ABCD*, *lic2AB*, and *lic3ABCD* loci have been identified to code for different phase-variable LPS epitopes, which are apparently responsible for an enhancement of the invasive capacity of the microorganisms.[236,593-598] The invasion capacity of various *H. influenzae* strains has been correlated with the phase-variable incorporation of phosphorylcholine into their LPS structures.[599] It was demonstrated that the microorganisms are capable of choline uptake from the culture medium, which appeared to be transported, phosphorylated, converted into an activated nucleoside triphosphate, and transferred onto LPS by the action of the *lic1BACD* gene products, respectively, resembling the eucaryotic pathway for choline incorporation into lipids. The phase-variable expression of the phosphorylcholine structure has been attributed to a translational switching within the *lic1A* gene, due to varying numbers of intragenic tetramer repeats.[599] Instead of having poly-G tracts, the *lic1A* as well as the *lic2A* and *lic3A* genes contain at their 5′-ends multiple tandem repeats of the tetramer 5′-CAAT-3′. Multiple repeats of another tetramer, 5′-GCAA-3′, have been found within the 5′-end of ORF1 in the *lex-2* locus for synthesis of a phase-variable LPS epitope of *H. influenzae* type b strain DL42.[600] A direct correlation between the number of 5′-CAAT-3′ repeats and the phase variation of the α-Gal-(1→4)-β-Gal structure has been shown for the *lic2A* gene, which appeared to play a significant role in the formation of at least three distinct phase-variable LPS epitopes.[597] The *lex-1* gene is involved in the expression of the α-Gal-(1→4)-β-Gal epitope in *H. influenzae* type b strain DL42 and contained, in spite of being almost identical to *lic2A*, 19 in place of 16 copies of the 5′-CAAT-3′ tetramer.[597,601] Further emphasis on the variability of the two genes from strains of the same clonal group was given by the finding that they are located in different chromosomal regions.[597] Based on a high degree of similarity of the predicted amino acid sequences, Lic2A, Lic2B, LgtE, and LgtB from *Neisseria*, as well as LpsA, described in 1995 as a protein involved in LPS biosynthesis of *Pasteurella haemolytica* A1,[602] may comprise a Lic2A-like family of putative glycosyltransferases, which are required for analogous functions in phase-variable LPS biosynthesis.[575,581,598] The main distinctive feature of the Lic2A-like protein family is that *lic2A* has been proven as the only gene containing the repetitive 5′-CAAT-3′ tetramer. Deletion of the multiple tandem repeats did not abolish the incorporation of the α-Gal-(1→4)-β-Gal epitope into the LPS, which suggested that the 5′-CAAT-3′ tetramers in *lic2A* are not needed for the biological activity of the enzyme, but probably only for the generation of phase variations. However, it was observed that a deleted *lic2A* gene could not prevent a phase variation of the α-Gal-(1→4)-β-Gal structure and, additionally, that *lic2A* genes with 5′-CAAT-3′ repeats could not always direct the synthesis of this epitope.[598] Thus, a novel gene of *H. influenzae*, highly similar to *mrp* for a "*metG*-related protein" in *E. coli*,[603] may encode one of the assumed additional factors necessary for the ultimate incorporation of α-Gal-(1→4)-β-Gal into LPS.[604] The interesting features of Mrp$_{Hi}$ of being similar to ATPases and affecting the copy number of the 5′-CAAT-3′ repeat in *lic2A* renders the protein a potential candidate as an energy-supplier in LPS assembly and regulator of the final α-Gal-$(1\rightarrow4)$-β-Gal amount incorporated into LPS.[604] By searching the genome sequence of *H. influenzae* strain Rd for multiple tandem repeats, a neisserial *lgtC* homologue with repetitive 5'-GACA-3' tetramers was identified as a novel *Haemophilus* LPS biosynthesis gene. Mutational analysis of *lgtC*$_{Hi}$ provided strong evidence that the copy number of the repeat correlated with the phase-variable expression of the α-Gal-$(1\rightarrow4)$-β-Gal epitope.[605] Finally, a total of 25 additional candidate LPS genes, potentially involved in precursor supply, sugar transfer, translocation, or regulation, were found to be scattered around the genome of *H. influenzae* strain Rd, creating the basis for future studies on the complex mechanisms involved in phase-variable LPS biosynthesis.[473]

The *B. pertussis* locus for LPS biosynthesis exhibits a unique arrangement of 14 genes specifying the assembly of the core oligosaccharide and a core-distal terminal trisaccharide, which resembles a single O-polysaccharide unit typical for semi-rough LPS of some enterobacterial mutants. The similarity of BplG$_{Bp}$ [WlbG] to several proteins, catalyzing the addition of a sugar residue to ACL, supports the hypothesis that the trisaccharide of GlcNAc, 2,3-diacetamido-2,3-dideoxy-mannuronic acid, and *N*-acetyl-*N*-methylfucosamine are assembled like an O-polysaccharide repeating unit on a carrier lipid.[491] It is of interest that at least part of the *B. pertussis* locus was shown to be present in *Bordetella parapertussis* and *Bordetella bronchiseptica* known to synthesize an O-polysaccharide-like homopolymer of 2,3-diacetamido-2,3-dideoxy-galacturonic acid.[606] Therefore, it is tempting to speculate that *B. pertussis* either lost the ability to produce an O-specific polysaccharide by mutational events or might be able to produce an O-antigen-like structure under certain, but so far unknown, conditions. A report in 1994 on the synthesis of O-antigen-like structures in bacteria of the genus *Chlamydia*, probably induced by currently unknown host-provided signals, may stimulate the discussion about further mechanisms of LPS phase variation, which are advantageous for a microorganism in order to survive in the host environment.[607]

3.09.6   FINAL REMARKS

The biomedical importance of endotoxin has greatly stimulated efforts to elucidate the structure of LPS of different origin, the mechanisms of its biological action, and to understand LPS biosynthesis at the molecular level. Such studies not only represent a fascinating scientific challenge, but may also be seen in the context of application-oriented research as they could be aimed at the identification of potential targets for novel therapeutic drugs. This concept is certainly attractive as LPS is essential for bacterial viability. In view of its indispensable function in the outer membrane and its structural and genetic conservation, the lipid A–Kdo domain of LPS appears to be the most promising domain to interfere with LPS biosynthesis in order to cause bactericidal effects. However, each step of LPS assembly, in particular, if it is used in a large spectrum of Gram-negative bacteria, is of potential interest, if an interference with it would be bactericidal, eliminate endotoxic properties of LPS, or facilitate the removal of bacteria by the host defense system, i.e., reduce microbial pathogenicity.

Molecular approaches have allowed an improved analysis of LPS structure, biosynthesis, activity, function, and relationships between these areas. Future challenges concern the characterization of regulatory systems acting on both individual synthesis steps and the entire pathway of LPS assembly. Further, the elucidation of extra- and intracellular signals, to which LPS biosynthesis responds, including additional aspects of LPS phase variation and resulting structure–function relationships, will represent a field of active LPS research, as will studies of the role of accessory genetic elements in modification of LPS synthesis and structure, as well as of dissemination of genetic determinants of LPS biosynthesis among Gram-negative bacteria. Finally, clarification of the mechanisms of polysaccharide transport across the cytoplasmic membrane, its transfer through the periplasm, and its incorporation into the outer membrane will continue to challenge research groups worldwide.

ACKNOWLEDGMENTS

The financial support of the DFG [Sonderforschungsbereich 367, projects B2 (EThR, UM) and B8 (US); Sonderforschungsbereich 470, projects B1 (OH), B4 (EThR), and B5 (US); Graduiertenkolleg [grant GRK288/1-92, projects A1 (EThR, UM) and A3 (US)], of the BMBF [grant 01K/9471, project B6 (EThR, US); grant POL-150 (OH)], of the GIF [grant 1 169-207.02/94 (EThR)], of the DAAD [grant 314-vigoni-dr (OH)], and the Fonds der Chemischen Industrie (EThR) is gratefully appreciated.

3.09.7 REFERENCES

1. E. Th. Rietschel, H. Brade, O. Holst, L. Brade, S. Müller-Loennies, U. Mamat, U. Zähringer, F. Beckmann, U. Seydel, K. Brandenberg *et al.*, *Curr. Top. Microbiol. Immunol.*, 1996, **216**, 39.
2. D. C. Morrison and J. L. Ryan, *Annu. Rev. Med.*, 1987, **38**, 417.
3. M. A. Freudenberg, T. C. Bog-Hansen, U. Back, and C. Galanos, *Infect. Immun.*, 1980, **28**, 373.
4. P. Elsbach and J. Weiss, *Immunobiology*, 1993, **187**, 417.
5. C. Schütt, T. Schilling, U. Grünwald, W. Schönfeld, and C. Krüger, *Res. Immunol.*, 1992, **143**, 71.
6. R. R. Schumann, S. R. Leong, G. W. Flaggs, P. W. Gray, S. D. Wright, J. C. Mathison, P. S. Tobias, and R. J. Ulevitch, *Science*, 1990, **249**, 1429.
7. P. S. Tobias, K. Soldau, and R. J. Ulevitch, *J. Exp. Med.*, 1986, **164**, 777.
8. P. Gallay, D. Heumann, D. Le Roy, C. Barras, and M. P. Glauser, *Proc. Natl. Acad. Sci. USA*, 1994, **91**, 7922.
9. R. G. Little, D. N. Kelner, E. Lim, D. J. Burke, and P. J. Conlon, *J. Biol. Chem.*, 1994, **269**, 1865.
10. A. Hoess, S. Watson, G. R. Siber, and R. Liddington, *EMBO J.*, 1993, **12**, 3351.
11. A. B. Schromm, K. Brandenburg, E. Th. Rietschel, and U. Seydel, *J. Endotox. Res.*, 1995, **2**, 313.
12. A. B. Schromm, K. Brandenburg, E. Th. Rietschel, H.-D. Flad, S. F. Carroll, and U. Seydel, *FEBS Lett.*, 1996, **399**, 267.
13. M. M. Wurfel, S. T. Kunitake, H. Lichenstein, J. P. Kane, and S. D. Wright, *J. Exp. Med.*, 1994, **180**, 1025.
14. R. S. Jack, X. Fan, M. Bernheiden, G. Rune, M. Ehlers, A. Weber, G. Kirsch, R. Mentel, B. Fürll, M. Freudenberg, G. Schmitz, F. Stelter, and C. Schütt, *Nature*, 1997, **389**, 742.
15. S. D. Wright, R. A. Ramos, P. S. Tobias, R. J. Ulevitch, and J. C. Mathison, *Science*, 1990, **249**, 1431.
16. J.-D. Lee, K. Kato, P. S. Tobias, T. N. Kirkland, and R. J. Ulevitch, *J. Exp. Med.*, 1992, **175**, 1697.
17. T. N. Kirkland, F. Finley, D. Leturcq, A. Moriarty, J.-D. Lee, R. J. Ulevitch, and P. S. Tobias, *J. Biol. Chem.*, 1993, **268**, 24818.
18. F. Stelter, M. Bernheiden, R. Menzel, R. S. Jack, S. Witt, X. Fan, M. Pfister, and C. Schütt, *Eur. J. Biochem.*, 1997, **243**, 100.
19. A. Haziot, S. Chen, E. Ferrero, M. G. Low, R. Silber, and S. M. Goyert, *J. Immunol.*, 1988, **141**, 547.
20. R. R. Ingalls and D. T. Golenbock, *J. Exp. Med.*, 1995, **181**, 1473.
21. J. Schletter, H. Brade, L. Brade, C. Krüger, H. Loppnow, S. Kusumoto, E. Th. Rietschel, H.-D. Flad, and A. J. Ulmer, *Infect. Immun.*, 1995, **63**, 2576.
22. E. Hailman, H. S. Lichenstein, M. M. Wurfel, D. S. Miller, D. A. Johnson, M. Kelley, L. A. Busse, M. M. Zukowski, and S. D. Wright, *J. Exp. Med.*, 1994, **179**, 269.
23. M. Arditi, J. Zhou, R. Dorio, G. W. Rong, S. M. Goyert, and K. S. Kim, *Infect. Immun.*, 1993, **61**, 3149.
24. H. Loppnow, F. Stelter, U. Schönbeck, C. Schlüter, M. Ernst, C. Schütt, and H.-D. Flad, *Infect. Immun.*, 1995, **63**, 1020.
25. B. Weidemann, H. Brade, E. Th. Rietschel, R. Dziarski, V. Bazil, S. Kusumoto, H.-D. Flad, and A. J. Ulmer, *Infect. Immun.*, 1994, **62**, 4709.
26. B. Weidemann, J. Schletter, R. Dziarski, S. Kusumoto, F. Stelter, E. Th. Rietschel, H.-D. Flad, and A. J. Ulmer, *Infect. Immun.*, 1997, **65**, 858.
27. D. Gupta, T. N. Kirkland, S. Viriyakosol, and R. Dziarski, *J. Biol. Chem.*, 1996, **271**, 23310.
28. T. Espevik, M. Otterlei, G. Skjak-Braek, L. Ryan, S. D. Wright, and A. Sundan, *Eur. J. Immunol.*, 1993, **23**, 255.
29. J. Pugin, I. D. Heumann, A. Tomasz, V. V. Kravchenko, Y. Akamatsu, M. Nishijima, M. P. Glauser, P. S. Tobias, and R. J. Ulevitch, *Immunity*, 1994, **1**, 509.
30. E. Mattsson, L. Verhage, J. Rollof, A. Fleer, J. Verhoef, and H. van Dijk, *FEMS Immunol. Med. Microbiol.*, 1993, **7**, 281.
31. E. Th. Rietschel, J. Schletter, B. Weidemann, V. El-Samouti, T. Mattern, U. Zähringer, U. Seydel, H. Brade, H.-D. Flad, S. Kusumoto, D. Gupta, R. Dziarski, and A. J. Ulmer, *Microb. Drug Res.*, 1998, in press.
32. J.-M. Cavaillon, C. Marie, M. Caroff, A. Ledur, I. Godard, D. Poulain, C. Fitting, and N. Haeffner-Cavaillon, *J. Endotox. Res.*, 1996, **3**, 471.
33. U. F. Schade, I. Burmeister, and R. Engel, *Biochem. Biophys. Res. Commun.*, 1987, **147**, 695.
34. A. L. Erwin and R. S. Munford, in "Bacterial Endotoxic Lipopolysaccharides. Molecular Biochemistry and Cellular Biology," eds. D. C. Morrison and J. L. Ryan, CRC Press, Boca Raton, FL, 1992, vol. 1, p. 405.
35. J. Andersson, F. Melchers, C. Galanos, and O. Lüderitz, *J. Exp. Med.*, 1973, **137**, 943.
36. T. Mattern, A. Thanhäuser, N. Reiling, K.-M. Toellner, M. Duchrow, S. Kusumoto, E. Th. Rietschel, M. Ernst, H. Brade, H.-D. Flad, and A. J. Ulmer, *J. Immunol.*, 1994, **153**, 2996.
37. P. J. Baker, *Immunobiology*, 1993, **187**, 372.
38. C. F. Nathan, *J. Clin. Invest.*, 1987, **79**, 319.
39. F. P. Heinzel, R. M. Rerko, P. Ling, J. Hakimi, and D. S. Schoenhaut, *Infect. Immun.*, 1994, **62**, 4244.
40. B. Echtenacher, W. Falk, D. N. Männel, and P. H. Krammer, *J. Immunol.*, 1990, **145**, 3762.
41. S. N. Vogel and M. M. Hogan, in "Immunopharmacology: The Role of Cells and Cytokines in Immunity and Inflammation," eds. J. J. Oppenheim and E. M. Shevack, Oxford University Press, Oxford, 1990, p. 238.
42. P. Libby, H. Loppnow, J. C. Fleet, H. Palmer, H. M. Li, S. J. C. Warner, R. N. Salomon, and S. K. Clinton, in "Arteriosclerosis: Cellular and Molecular Interactions in the Artery Wall," eds. A. I. Gotlieb, B. L. Langille, and B. Federoff, Plenum Press, New York, 1991, p. 161.
43. U. Schönbeck, E. Brandt, F. Petersen, H.-D. Flad, and H. Loppnow, *J. Immunol.*, 1995, **154**, 2375.
44. D. E. Doherty, L. Zagarella, P. M. Henson, and G. S. Worthen, *J. Immunol.*, 1989, **143**, 3673.

45. Y. A. Knirel, E. Th. Rietschel, R. Marre, and U. Zähringer, *Eur. J. Biochem.*, 1994, **221**, 239.
46. A. Gamian, E. Romanowska, U. Dabrowski, and J. Dabrowski, *Biochemistry*, 1991, **30**, 5032.
47. L. Kenne and B. Lindberg, in "The Polysaccharides," ed. G. O. Aspinall, Academic Press, New York, 1983, vol. 2, p. 287.
48. B. Lindberg, *Adv. Carbohydr. Chem. Biochem.*, 1990, **48**, 279.
49. Y. A. Knirel and N. K. Kochetkov, *Biokhimiya (Moscow)*, 1994, **59**, 1784.
50. C. G. Hellerqvist and A. A. Lindberg, *Carbohydr. Res.*, 1971, **16**, 39.
51. P. J. Hitchcock and T. M. Brown, *J. Bacteriol.*, 1983, **154**, 269.
52. R. E. W. Hancock, D. N. Karunaratne, and C. Bernegger-Egli, in "Bacterial Cell Wall," eds. J.-M. Ghuysen and R. Hakenbeck, Elsevier Science, Amsterdam, 1994, p. 263.
53. A. Gamian and E. Romanowska, *Eur. J. Biochem.*, 1982, **129**, 105.
54. C. Lugowski, M. Kulakowska, and E. Romanowska, *J. Immunol. Methods*, 1986, **95**, 187.
55. A. D. Cox, J. R. Brisson, V. Varma, and M. B. Perry, *Carbohydr. Res.*, 1996, **290**, 43.
56. A. D. Cox and M. B. Perry, *Carbohydr. Res.*, 1996, **290**, 59.
57. E. T. Palva and P. H. Mäkelä, *Eur. J. Biochem.*, 1980, **107**, 137.
58. V. Jiminez-Lucho and J. Foulds, *J. Infect. Dis.*, 1990, **162**, 763.
59. S. Schlecht and I. Fromme, *Zentralbl. Bakteriol. [Orig. A]*, 1975, **233**, 199.
60. S. Schlecht and H. Mayer, *Int. J. Med. Microbiol. Virol. Parasitol. Infect. Dis.*, 1994, **280**, 448.
61. C. Galanos, B. H. Jiao, T. Komuro, M. A. Freudenberg, and O. Lüderitz, *J. Chromatogr.*, 1988, **440**, 397.
62. I. Ørskov, F. Ørskov, B. Jann, and K. Jann, *Bacteriol. Rev.*, 1977, **41**, 667.
63. F. Ørskov and I. Ørskov, *Methods Microbiol.*, 1984, **14**, 43.
64. F. Kauffmann, *Zentralbl. Bakteriol. [Orig. A]*, 1973, **223**, 508.
65. T. F. Muraschi, D. M. Bolles, C. Moczulski, and M. Lindsay, *J. Infect. Dis.*, 1966, **116**, 84.
66. B. Lanyi and T. Bergan, *Methods Microbiol.*, 1978, **10**, 93.
67. P. V. Liu, H. Matsumoto, H. Kusama, and T. Bergan, *Int. J. Syst. Bacteriol.*, 1983, **33**, 256.
68. V. L'vov, A. S. Shashkov, B. A. Dmitriev, N. K. Kochetkov, B. Jann, and K. Jann, *Carbohydr. Res.*, 1984, **126**, 249.
69. P. E. Jansson, J. Lönngren, G. Widmalm, K. Leontein, K. Slettengren, S. B. Svenson, G. Wrangsell, A. Dell, and P. R. Tiller, *Carbohydr. Res.*, 1985, **145**, 59.
70. L. A. Parolis, H. Parolis, and G. G. Dutton, *Carbohydr. Res.*, 1986, **155**, 272.
71. P. Prehm, B. Jann, and K. Jann, *Eur. J. Biochem.*, 1976, **67**, 53.
72. Y. A. Knirel, E. V. Vinogradov, A. S. Shashkov, N. K. Kochetkov, V. L'vov, and B. A. Dmitriev, *Carbohydr. Res.*, 1985, **141**, C1.
73. B. Jann, A. S. Shashkov, H. Kochanowski, and K. Jann, *Carbohydr. Res.*, 1994, **264**, 305.
74. G. Stevenson, B. Neal, D. Liu, M. Hobbs, N. H. Packer, M. Batley, J. W. Redmond, L. Lindquist, and P. Reeves, *J. Bacteriol.*, 1994, **176**, 4144.
75. K. Eklund, P. J. Garegg, L. Kenne, A. A. Lindberg, and B. Lindberg, in "Abstracts of the IXth International Symposium on Carbohydrate Chemistry," London, 1978.
76. IUPAC–IUB Joint Commission on Biochemical Nomenclature (JCBN), *Eur. J. Biochem.*, 1982, **126**, 433.
77. K. Jann and B. Jann, in "Handbook of Endotoxin. Chemistry of Endotoxin," ed. E. Th. Rietschel, Elsevier Science, Amsterdam, 1984, vol. 1, p. 138.
78. H. Smith, *J. Gen. Microbiol.*, 1990, **136**, 377.
79. C. A. Lee, *Infect. Agents Dis.*, 1996, **5**, 1.
80. B. J. Bachmann, in "*Escherichia coli* and *Salmonella typhimurium*: Cellular and Molecular Biology," eds. F. C. Neidhardt, J. L. Ingraham, K. B. Low, B. Magasanik, M. Schaechter, and H. E. Umbarger, American Society for Microbiology, Washington, DC, 1987, vol. 2, p. 1190.
81. D. Liu and P. R. Reeves, *Microbiology*, 1994, **140**, 49.
82. H. Masoud and M. B. Perry, *Biochem. Cell Biol.*, 1996, **74**, 241.
83. O. Holst and H. Brade, in "Bacterial Endotoxic Lipopolysaccharides. Molecular Biochemistry and Cellular Biology," eds. D. C. Morrison and J. L. Ryan, CRC Press, Boca Raton, FL, 1992, vol. 1, p. 135.
84. B. J. Appelmelk, Y.-Q. An, T. A. Hekker, L. G. Thijs, D. M. MacLaren, and J. De Graaf, *Microbiology*, 1994, **140**, 1119.
85. O. Holst and H. Brade, *Carbohydr. Res.*, 1990, **207**, 327.
86. O. Holst, U. Zähringer, H. Brade, and A. Zamojski, *Carbohydr. Res.*, 1991, **215**, 323.
87. J. D. Klena and C. A. Schnaitman, *J. Bacteriol.*, 1994, **176**, 4003.
88. H. D. Grimmecke and E. Th. Rietschel, unpublished.
89. J. D. Klena, R. S. Ashford, II, and C. A. Schnaitman, *J. Bacteriol.*, 1992, **174**, 7297.
90. O. Holst, E. Röhrscheidt-Andrzejewski, H.-P. Cordes, and H. Brade, *Carbohydr. Res.*, 1989, **188**, 212.
91. Y. Haishima, O. Holst, and H. Brade, *Eur. J. Biochem.*, 1992, **203**, 127.
92. S. Müller-Loennies, O. Holst, and H. Brade, *Eur. J. Biochem.*, 1994, **224**, 751.
93. Z. Pakulski, A. Zamojski, O. Holst, and U. Zähringer, *Carbohydr. Res.*, 1991, **215**, 337.
94. P.-E. Jansson, B. Lindberg, A. A. Lindberg, and R. Wollin, *Carbohydr. Res.*, 1979, **68**, 385.
95. E. V. Vinogradov, K. van der Drift, J. E. Thomas-Oates, S. Meshkov, H. Brade, and O. Holst, unpublished.
96. G. Hämmerling, O. Lüderitz, O. Westphal, and H. P. Mäkelä, *Eur. J. Biochem.*, 1971, **22**, 331.
97. U. Feige, B. Jann, K. Jann, G. Schmidt, and S. Stirm, *Biochem. Biophys. Res. Commun.*, 1977, **79**, 88.
98. O. Lüderitz, O. Dröge, P. F. Mühlradt, E. Ruschmann, and O. Westphal, in "Proceedings of the XIth International C.N.R.S. Congress," Paris, 1969, p. 95.
99. N. dalla Venezia, M. Bruneteau, P. Binder, O. Creach, R. Fontanges, and G. Michel, *FEMS Microbiol. Immunol.*, 1988, **1**, 133.
100. O. Holst and H. Brade, *Carbohydr. Res.*, 1991, **219**, 247.
101. O. Holst and H. Brade, *Carbohydr. Res.*, 1993, **245**, 159.
102. C. M. John and B. W. Gibson, *Anal. Biochem.*, 1990, **187**, 281.
103. R. S. Tsang, S. Schlecht, and H. Mayer, *Int. J. Med. Microbiol. Virol. Parasitol. Infect. Dis.*, 1992, **276**, 330.
104. M. M. A. Olsthoorn, K. Bock, S. Schlecht, J. Haverkamp, J. E. Thomas-Oates, and O. Holst, *J. Biol. Chem.*, 1998, **273**, 3817.

105. W. J. Keenleyside, M. Perry, L. Maclean, C. Poppe, and C. Whitfield, *Mol. Microbiol.*, 1994, **11**, 437.
106. O. Lüderitz, O. Westphal, A. M. Staub, and H. Nikaido, in "Bacterial Endotoxins," eds. G. Weinbaum, S. Kadis, and S. J. Ajl, Academic Press, New York, 1971, vol. 4, p. 145.
107. P. R. Reeves, in "Bacterial Cell Wall," eds. J.-M. Ghuysen and R. Hakenbeck, Elsevier Science, Amsterdam, 1994, p. 281.
108. C. Galanos, O. Lüderitz, E. Th. Rietschel, and O. Westphal, in "Biochemistry of Lipids II," ed. T. W. Goodwin, University Park Press, Baltimore, MD, 1977, vol. 14, p. 239.
109. A. Weintraub, A. A. Lindberg, P. Lipniunas, and B. O. Nilsson, *Glycoconjugate J.*, 1988, **5**, 207.
110. R. S. Tsang, S. Schlecht, S. Aleksic, K. H. Chan, and P. Y. Chau, *Res. Microbiol.*, 1991, **142**, 521.
111. D. C. Straus, *Infect. Immun.*, 1987, **55**, 44.
112. O. Kol, J.-M. Wieruszeski, G. Strecker, B. Fournet, R. Zalisz, and P. Smets, *Carbohydr. Res.*, 1992, **236**, 339.
113. R. F. Kelly, W. B. Severn, J. C. Richards, M. B. Perry, L. L. Maclean, J. M. Tomás, S. Merino, and C. Whitfield, *Mol. Microbiol.*, 1993, **10**, 615.
114. C. Whitfield, M. B. Perry, L. L. Maclean, and S. H. Yu, *J. Bacteriol.*, 1992, **174**, 4913.
115. L. L. Maclean, C. Whitfield, and M. B. Perry, *Carbohydr. Res.*, 1993, **239**, 325.
116. R. F. Kelly, M. B. Perry, L. L. Maclean, and C. Whitfield, *J. Endotox. Res.*, 1995, **2**, 131.
117. C. Whitfield, J. C. Richards, M. B. Perry, B. R. Clarke, and L. L. Maclean, *J. Bacteriol.*, 1991, **173**, 1420.
118. W. B. Severn, R. F. Kelly, J. C. Richards, and C. Whitfield, *J. Bacteriol.*, 1996, **178**, 1731.
119. M. Süsskind, S. Müller-Loennies, W. Nimmich, H. Brade, and O. Holst, *Carbohydr. Res.*, 1995, **269**, C1.
120. M. Süsskind, H. Brade, and O. Holst, unpublished.
121. S. G. Wilkinson, in "Cellular and Molecular Aspects of Endotoxin Reactions," eds. A. Nowotny, J. J. Spitzer, and E. J. Ziegler, Excerpta Medica, Amsterdam, 1990, p. 95.
122. H. M. Aucken, M. Merkouroglou, A. W. Miller, L. Galbraith, and S. G. Wilkinson, *FEMS Microbiol. Lett.*, 1995, **130**, 267.
123. H. M. Aucken, S. G. Wilkinson, and T. L. Pitt, *FEMS Microbiol. Lett.*, 1996, **138**, 77.
124. O. Holst, H. M. Aucken, and G. Seltmann, *J. Endotox. Res.*, 1997, **4**, 215.
125. M. A. Gaston and T. L. Pitt, *J. Clin. Microbiol.*, 1989, **27**, 2697.
126. H. M. Aucken, S. G. Wilkinson, and T. L. Pitt, *J. Clin. Microbiol.*, 1997, **35**, 59.
127. D. Oxley and S. G. Wilkinson, *Carbohydr. Res.*, 1989, **193**, 241.
128. A. D. Cox, C. J. Taylor, A. J. Anderson, M. B. Perry, and S. G. Wilkinson, *Eur. J. Biochem.*, 1995, **231**, 784.
129. L. M. Beynon, A. D. Cox, C. J. Taylor, S. G. Wilkinson, and M. B. Perry, *Carbohydr. Res.*, 1995, **272**, 231.
130. S. R. Haseley, W. H. Traub, and S. G. Wilkinson, *Eur. J. Biochem.*, 1997, **244**, 147.
131. B. A. Dmitriev, Y. A. Knirel, and N. K. Kochetkov, *Eur. J. Biochem.*, 1976, **66**, 559.
132. L. Kenne, B. Lindberg, K. Petersson, E. Katzenellenbogen, and E. Romanowska, *Eur. J. Biochem.*, 1978, **91**, 279.
133. N. I. A. Carlin, A. A. Lindberg, K. Bock, and D. R. Bundle, *Eur. J. Biochem.*, 1984, **139**, 189.
134. P.-E. Jansson, L. Kenne, and G. Widmalm, *Carbohydr. Res.*, 1989, **188**, 169.
135. P.-E. Jansson, L. Kenne, and T. Wehler, *Carbohydr. Res.*, 1988, **179**, 359.
136. R. P. Gorshkova, E. N. Kalmykova, V. V. Isakov, and Y. S. Ovodov, *Eur. J. Biochem.*, 1985, **150**, 527.
137. K. Samuelsson, B. Lindberg, and R. R. Brubaker, *J. Bacteriol.*, 1974, **117**, 1010.
138. A. C. Kessler, P. K. Brown, L. K. Romana, and P. R. Reeves, *J. Gen. Microbiol.*, 1991, **137**, 2689.
139. M. Caroff, D. R. Bundle, and M. B. Perry, *Eur. J. Biochem.*, 1984, **139**, 195.
140. H. D. Grimmecke, U. Mamat, W. Lauk, A. S. Shashkov, Y. A. Knirel, E. V. Vinogradov, and N. K. Kochetkov, *Carbohydr. Res.*, 1991, **220**, 165.
141. H. D. Grimmecke, U. Mamat, I. Kühn, A. S. Shashkov, and Y. A. Knirel, *Carbohydr. Res.*, 1994, **252**, 309.
142. H. D. Grimmecke, Y. A. Knirel, A. S. Shashkov, B. Kiesel, W. Lauk, and M. Voges, *Carbohydr. Res.*, 1994, **253**, 277.
143. H. D. Grimmecke, M. Voges, Y. A. Knirel, A. S. Shashkov, W. Lauk, and B. Kiesel, *Carbohydr. Res.*, 1994, **253**, 283.
144. Y. A. Knirel, J. H. Helbig, and U. Zähringer, *Carbohydr. Res.*, 1996, **283**, 129.
145. M. Rivera and E. J. McGroarty, *J. Bacteriol.*, 1989, **171**, 2244.
146. M. Y. C. Lam, E. J. McGroarty, A. M. Kropinski, L. A. McDonald, S. S. Pedersen, N. Hoiby, and J. S. Lam, *J. Clin. Microbiol.*, 1989, **27**, 962.
147. Y. A. Knirel, E. V. Vinogradov, N. A. Kocharova, N. A. Paramonov, N. K. Kochetkov, B. A. Dmitriev, E. S. Stanislavsky, and B. Lanyi, *Acta Microbiol. Hung.*, 1988, **35**, 3.
148. L. Kenne, B. Lindberg, P. Unger, B. Gustafsson, and T. Holme, *Carbohydr. Res.*, 1982, **100**, 341.
149. T. Ito, T. Higuchi, M. Hirobe, K. Hiramatsu, and T. Yokota, *Carbohydr. Res.*, 1994, **256**, 113.
150. Y. Isshiki, Y. Haishima, S. Kondo, and K. Hisatsune, *Eur. J. Biochem.*, 1995, **229**, 583.
151. Y. A. Knirel, H. Moll, and U. Zähringer, *Carbohydr. Res.*, 1996, **293**, 223.
152. S. Otten, S. Iyer, W. Johnson, and R. Montgomery, *J. Bacteriol.*, 1986, **167**, 893.
153. J. H. Helbig, P. C. Lück, Y. A. Knirel, W. Witzleb, and U. Zähringer, *Epidemiol. Infect.*, 1995, **115**, 71.
154. Y. A. Knirel, *Crit. Rev. Microbiol.*, 1990, **17**, 273.
155. S. A. Makin and T. J. Beveridge, *Microbiology*, 1996, **142**, 299.
156. Y. A. Knirel and N. A. Kocharova, *Biokhimiya* (*Moscow*), 1995, **60**, 1964.
157. E. J. McGroarty and M. Rivera, *Infect. Immun.*, 1990, **58**, 1030.
158. F. Beckmann, H. Moll, K.-E. Jäger, and U. Zähringer, *Carbohydr. Res.*, 1995, **267**, C3.
159. D. T. Drewry, K. C. Symes, G. W. Gray, and S. G. Wilkinson, *Biochem. J.*, 1975, **149**, 93.
160. P. S. N. Rowe and P. M. Meadow, *Eur. J. Biochem.*, 1983, **132**, 329.
161. A. M. Kropinski, B. Jewell, J. Kuzio, F. Milazzo, and D. Berry, *Antibiot. Chemother.*, 1985, **36**, 58.
162. S. C. Szu, R. Gupta, P. Kovac, D. N. Taylor, and J. B. Robbins, *Bull. Inst. Pasteur*, 1995, **93**, 269.
163. K. Hisatsune, S. Kondo, Y. Isshiki, T. Iguchi, and Y. Haishima, *Biochem. Biophys. Res. Commun.*, 1993, **190**, 302.
164. E. V. Vinogradov, K. Bock, O. Holst, and H. Brade, *Eur. J. Biochem.*, 1995, **233**, 152.
165. W. Kaca, L. Brade, E. Th. Rietschel, and H. Brade, *Carbohydr. Res.*, 1986, **149**, 293.
166. H. Masoud, I. Sadovskaya, T. De Kievit, E. Altman, J. C. Richards, and J. S. Lam, *J. Bacteriol.*, 1995, **177**, 6718.
167. F. E. Nano and H. D. Caldwell, *Science*, 1985, **228**, 742.
168. H. Brade, L. Brade, and F. E. Nano, *Proc. Natl. Acad. Sci. USA*, 1987, **84**, 2508.

169. O. Holst, L. Brade, P. Kosma, and H. Brade, *J. Bacteriol.*, 1991, **173**, 1862.
170. O. Holst, W. Broer, J. E. Thomas-Oates, U. Mamat, and H. Brade, *Eur. J. Biochem.*, 1993, **214**, 703.
171. H. Brade, W. Brabetz, I. Brade, O. Holst, S. Löbau, M. Lukácová, U. Mamat, A. Rozalski, K. Zych, and P. Kosma, *Pure Appl. Chem.*, 1995, **67**, 1617.
172. L. Brade, S. Schramek, U. F. Schade, and H. Brade, *Infect. Immun.*, 1986, **54**, 568.
173. L. Brade, H. Brunnemann, M. Ernst, Y. Fu, O. Holst, P. Kosma, H. Näher, K. Persson, and H. Brade, *FEMS Immunol. Med. Microbiol.*, 1994, **8**, 27.
174. P. Kosma, M. Strobl, G. Allmaier, E. Schmid, and H. Brade, *Carbohydr. Res.*, 1994, **254**, 105.
175. E. Th. Rietschel, T. Kirikae, U. F. Schade, U. Mamat, G. Schmidt, H. Loppnow, A. J. Ulmer, U. Zähringer, U. Seydel, F. Di Padova, M. Schreier, and H. Brade, *FASEB J.*, 1994, **8**, 217.
176. A. P. Moran, U. Zähringer, U. Seydel, D. Scholz, P. Stütz, and E. Th. Rietschel, *Eur. J. Biochem.*, 1991, **198**, 459.
177. U. Zähringer, Y. A. Knirel, B. Lindner, J. H. Helbig, A. Sonesson, R. Marre, and E. Th. Rietschel, in "Bacterial Endotoxins: Lipopolysaccharides from Genes to Therapy," eds. J. Levin, C. R. Alving, R. S. Munford, and H. Redl, Wiley–Liss, New York, 1995, p. 113.
178. V. A. Kulshin, U. Zähringer, B. Lindner, C. A. Frasch, C.-M. Tsai, B. A. Dmitriev, and E. Th. Rietschel, *J. Bacteriol.*, 1992, **174**, 1793.
179. N. Dalla Venezia, S. Minka, M. Bruneteau, H. Mayer, and G. Michel, *Eur. J. Biochem.*, 1985, **151**, 399.
180. E. Th. Rietschel, C. Galanos, O. Lüderitz, and O. Westphal, in "Immunopharmacology and the Regulation of Leucocyte Function," ed. D. R. Webb, Marcel Dekker, New York, 1982, p. 183.
181. I. M. Helander, Y. Kato, I. Kilpeläinen, R. Kostiainen, B. Lindner, K. Nummila, T. Sugiyama, and T. Yokochi, *Eur. J. Biochem.*, 1996, **237**, 272.
182. V. A. Kulshin, U. Zähringer, B. Lindner, K. E. Jäger, B. A. Dmitriev, and E. Th. Rietschel, *Eur. J. Biochem.*, 1991, **198**, 697.
183. K. W. Broady, E. Th. Rietschel, and O. Lüderitz, *Eur. J. Biochem.*, 1981, **115**, 463.
184. A. Weintraub, U. Zähringer, H.-W. Wollenweber, U. Seydel, and E. Th. Rietschel, *Eur. J. Biochem.*, 1989, **183**, 425.
185. U. R. Bhat, L. S. Forsberg, and R. W. Carlson, *J. Biol. Chem.*, 1994, **269**, 14 402.
186. N. P. Price, T. M. Kelly, C. R. H. Raetz, and R. W. Carlson, *J. Bacteriol.*, 1994, **176**, 4646.
187. S. Kusumoto, in "Bacterial Endotoxic Lipopolysaccharides. Molecular Biochemistry and Cellular Biology," eds. D. C. Morrison and J. L. Ryan, CRC Press, Boca Raton, FL, 1992, vol. 1, p. 81.
188. M. Shinitzky, in "Physiology of Membrane Fluidity," ed. M. Shinitzky, CRC Press, Boca Raton, FL, 1984, vol. 1, p. 1.
189. A. R. Cossins and M. Sinensky, in "Physiology of Membrane Fluidity," ed. M. Shinitzky, CRC Press, Boca Raton, FL, 1984, vol. 2, p. 1.
190. S. K. Jackson, P. E. James, C. C. Rowlands, and B. Mile, *Biochim. Biophys. Acta*, 1992, **1135**, 165.
191. S. Kabir, D. L. Rosenstreich, and S. E. Mergenhagen, in "Bacterial Toxins and Cell Membranes," eds. J. Jeljaszewic and T. Wadström, Academic Press, London, 1978, p. 59.
192. J. N. Israelachvili, "Intermolecular and Surface Forces," Academic Press, London, 1991.
193. M. Hofer, R. Y. Hampton, C. R. H. Raetz, and H. Yu, *Chem. Phys. Lipids*, 1991, **59**, 167.
194. M. Kastowsky, T. Gutberlet, and H. Bradaczek, *J. Bacteriol.*, 1992, **174**, 4798.
195. J. N. Israelachvili, S. Marcelja, and R. G. Horm, *Q. Rev. Biophys.*, 1980, **13**, 121.
196. K. Brandenburg, U. Seydel, A. B. Schromm, H. Loppnow, M. H. J. Koch, and E. Th. Rietschel, *J. Endotox. Res.*, 1996, **3**, 173.
197. K. Brandenburg and U. Seydel, *Biochim. Biophys. Acta*, 1984, **775**, 225.
198. K. Brandenburg and U. Seydel, *Eur. J. Biochem.*, 1990, **191**, 229.
199. D. Naumann, C. Schultz, A. Sabisch, M. Kastowsky, and H. Labischinski, *J. Mol. Struct.*, 1989, **214**, 213.
200. U. Seydel and K. Brandenburg, in "Bacterial Endotoxic Lipopolysaccharides. Molecular Biochemistry and Cellular Biology," eds. D. C. Morrison and J. L. Ryan, CRC Press, Boca Raton, FL, 1992, vol. 1, p. 225.
201. N. Wellinghausen, A. B. Schromm, U. Seydel, K. Brandenburg, J. Luhm, H. Kirchner, and L. Rink, *J. Immunol.*, 1996, **157**, 3139.
202. E. Th. Rietschel, T. Kirikae, W. Feist, H. Loppnow, P. Zabel, L. Brade, A. J. Ulmer, H. Brade, U. Seydel, U. Zähringer, M. Schlaak, H.-D. Flad, and U. F. Schade, in "Molecular Aspects of Inflammation," eds. H. Sies, L. Floh, and G. Zimmer, Springer-Verlag, Berlin, 1991, p. 207.
203. E. Th. Rietschel, U. Seydel, U. Zähringer, U. F. Schade, L. Brade, H. Loppnow, W. Feist, M.-H. Wang, A. F. Ulmer, U.-D. Flad, K. Brandenburg, T. Kirikae, D. Grimmecke, O. Holst, and H. Brade, in "Infectious Disease Clinics of North America. Vol. 5: Gram-Negative Septicemia and Septic Shock," eds. R. C. Moellering, Jr., L. S. Young, and M. P. Glauser, W. B. Saunders, Philadelphia, PA, 1991, p. 753.
204. K. Brandenburg, H. Mayer, M. H. J. Koch, J. Weckesser, E. Th. Rietschel, and U. Seydel, *Eur. J. Biochem.*, 1993, **218**, 555.
205. K. Brandenburg, A. B. Schromm, M. H. J. Koch, and U. Seydel, in "Bacterial Endotoxins: Lipopolysaccharides from Genes to Therapy," eds. J. Levin, C. R. Alving, R. S. Munford, and H. Redl, Wiley-Liss, New York, 1995, p. 167.
206. U. Seydel, K. Brandenburg, and E. Th. Rietschel, *Prog. Clin. Biol. Res.*, 1994, **388**, 17.
207. H. Mayer, J. H. Krauss, A. Yokota, and J. Weckesser, in "Endotoxin," eds. H. Friedman, T. W. Klein, M. Nakano, and A. Nowotny, Plenum Press, New York, 1990, p. 45.
208. K. Takayama, N. Qureshi, E. Ribi, and J. L. Cantrell, *Rev. Infect. Dis.*, 1984, **6**, 439.
209. C. R. Alving, J. N. Verma, M. Rao, U. Krzych, S. Amselem, S. M. Green, and N. M. Wassef, *Res. Immunol.*, 1992, **143**, 197.
210. P. A. Kiener, F. Marek, G. Rodgers, P.-F. Lin, G. Warr, and J. Desiderio, *J. Immunol.*, 1988, **141**, 870.
211. H. Loppnow, P. Libby, M. Freudenberg, J. H. Krauss, J. Weckesser, and H. Mayer, *Infect. Immun.*, 1990, **58**, 3743.
212. A. P. Moran, *FEMS Immunol. Med. Microbiol.*, 1995, **11**, 121.
213. D. M. Jacobs, H. Yeh, and R. M. Price, *Adv. Exp. Med. Biol.*, 1990, **256**, 233.
214. N. E. Larsen, R. I. Enelow, E. R. Simons, and R. Sullivan, *Biochim. Biophys. Acta*, 1985, **815**, 1.
215. E. Hailman, J. J. Albers, G. Wolfbauer, A. Y. Tu, and S. D. Wright, *J. Biol. Chem.*, 1996, **271**, 12 172.
216. K. Takayama, D. H. Mitchell, Z. Z. Din, P. Mukerjee, C. Li, and D. L. Coleman, *J. Biol. Chem.*, 1994, **269**, 2241.

217. J. A. Gegner, R. J. Ulevitch, and P. S. Tobias, *J. Biol. Chem.*, 1995, **270**, 5320.
218. B. Yu and S. D. Wright, *J. Biol. Chem.*, 1996, **271**, 4100.
219. W. A. Lynn, Y. Liu, and D. T. Golenbock, *Infect. Immun.*, 1993, **61**, 4452.
220. A. Wiese, J. O. Reiners, K. Brandenburg, K. Kawahara, U. Zähringer, and U. Seydel, *Biophys. J.*, 1996, **70**, 321.
221. D. J. Nelson, B. Jow, and F. Jow, *J. Membr. Biol.*, 1992, **125**, 207.
222. M. F. Wilkinson, M. L. Earle, C. R. Triggle, and S. Barnes, *FASEB J.*, 1996, **10**, 785.
223. P. H. Mäkelä and B. A. D. Stocker, in "Handbook of Endotoxin. Chemistry of Endotoxin," ed. E. Th. Rietschel, Elsevier Science, Amsterdam, 1984, vol. 1, p. 59.
224. C. R. H. Raetz, *Annu. Rev. Biochem.*, 1990, **59**, 129.
225. C. R. H. Raetz, in "Bacterial Endotoxic Lipopolysaccharides. Molecular Biochemistry and Cellular Biology," eds. D. C. Morrison and J. L. Ryan, CRC Press, Boca Raton, FL, 1992, vol. 1, p. 67.
226. C. A. Schnaitman and J. D. Klena, *Microbiol. Rev.*, 1993, **57**, 655.
227. C. Whitfield and M. A. Valvano, *Adv. Microb. Physiol.*, 1993, **35**, 135.
228. A. Preston, R. E. Mandrell, B. W. Gibson, and M. A. Apicella, *Crit. Rev. Microbiol.*, 1996, **22**, 139.
229. P. R. Reeves, M. Hobbs, M. A. Valvano, M. Skurnik, C. Whitfield, D. Coplin, N. Kido, J. Klena, D. Maskell, C. R. H. Raetz, and P. D. Rick, *Trends Microbiol.*, 1996, **4**, 495.
230. S. Adhya, in "*Escherichia coli* and *Salmonella typhimurium*: Cellular and Molecular Biology," eds. F. C. Neidhardt, J. L. Ingraham, K. B. Low, B. Magasanik, M. Schaechter, and H. E. Umbarger, American Society for Microbiology, Washington, DC, 1987, vol. 2, p. 1503.
231. M. J. Weickert and S. Adhya, *Mol. Microbiol.*, 1993, **10**, 245.
232. H.-S. S. Houng, D. J. Kopecko, and L. S. Baron, *J. Bacteriol.*, 1990, **172**, 4392.
233. H.-L. Peng, T.-F. Fu, S.-F. Liu, and H.-Y. Chang, *J. Biochem. (Tokyo)*, 1992, **112**, 604.
234. M. Metzger, P. Bellemann, P. Bugert, and K. Geider, *J. Bacteriol.*, 1994, **176**, 450.
235. P. J. Dolph, D. R. Majerczak, and D. L. Coplin, *J. Bacteriol.*, 1988, **170**, 865.
236. D. J. Maskell, M. J. Szabo, P. D. Butler, A. E. Williams, and E. R. Moxon, *Mol. Microbiol.*, 1991, **5**, 1013.
237. D. J. Maskell, M. J. Szabo, M. E. Deadman, and E. R. Moxon, *Mol. Microbiol.*, 1992, **6**, 3051.
238. E. K. H. Schweda, P.-E. Jansson, E. R. Moxon, and A. A. Lindberg, *Carbohydr. Res.*, 1995, **272**, 213.
239. M. Skurnik, R. Venho, P. Toivanen, and A. Al-Hendy, *Mol. Microbiol.*, 1995, **17**, 575.
240. D. E. Pierson and S. Carlson, *J. Bacteriol.*, 1996, **178**, 5916.
241. L. Zhang, J. Radziejewska-Lebrecht, D. Krajewska-Pietrasik, P. Toivanen, and M. Skurnik, *Mol. Microbiol.*, 1997, **23**, 63.
242. M. P. Jennings, P. van der Ley, K. E. Wilks, D. J. Maskell, J. T. Poolman, and E. R. Moxon, *Mol. Microbiol.*, 1993, **10**, 361.
243. F. K. N. Lee, D. S. Stephens, B. W. Gibson, J. J. Engstrom, D. Zhou, and M. A. Apicella, *Infect. Immun.*, 1995, **63**, 2508.
244. B. D. Robertson, M. Frosch, and J. P. M. van Putten, *Mol. Microbiol.*, 1993, **8**, 891.
245. L. E. Comstock, J. A. Johnson, J. M. Michalski, J. G. Morris, Jr., and J. B. Kaper, *Mol. Microbiol.*, 1996, **19**, 815.
246. A. C. Weissborn, Q. Liu, M. K. Rumley, and E. P. Kennedy, *J. Bacteriol.*, 1994, **176**, 2611.
247. T. A. Sundararajan, A. M. C. Rapin, and H. M. Kalckar, *Proc. Natl. Acad. Sci. USA*, 1962, **48**, 2187.
248. T. Fukasawa, K. Jokura, and K. Kurahashi, *Biochem. Biophys. Res. Commun.*, 1962, **7**, 121.
249. T. Fukasawa, K. Jokura, and K. Kurahashi, *Biochim. Biophys. Acta*, 1963, **74**, 608.
250. H.-Y. Chang, J.-H. Lee, W.-L. Deng, T.-F. Fu, and H.-L. Peng, *Microb. Pathog.*, 1996, **20**, 255.
251. R. C. Sandlin, K. A. Lampel, S. P. Keasler, M. B. Goldberg, A. L. Stolzer, and A. T. Maurelli, *Infect. Immun.*, 1995, **63**, 229.
252. C. Wandersman and S. Létoffé, *Mol. Microbiol.*, 1993, **7**, 141.
253. Y. Komeda, T. Icho, and T. Iino, *J. Bacteriol.*, 1977, **129**, 908.
254. J. Böhringer, D. Fischer, G. Mosler, and R. Hengge-Aronis, *J. Bacteriol.*, 1995, **177**, 413.
255. B. E. Froman, R. C. Tait, and L. D. Gottlieb, *Mol. Gen. Genet.*, 1989, **217**, 126.
256. X.-M. Jiang, B. Neal, F. Santiago, S. J. Lee, L. K. Romana, and P. R. Reeves, *Mol. Microbiol.*, 1991, **5**, 695.
257. L. Wang, L. K. Romana, and P. R. Reeves, *Genetics*, 1992, **130**, 429.
258. Z. Yao and M. A. Valvano, *J. Bacteriol.*, 1994, **176**, 4133.
259. C. L. Marolda and M. A. Valvano, *J. Bacteriol.*, 1995, **177**, 5539.
260. D. F. Macpherson, P. A. Manning, and R. Morona, *Mol. Microbiol.*, 1994, **11**, 281.
261. T. Nakae and H. Nikaido, *J. Biol. Chem.*, 1971, **246**, 4386.
262. T. Nakae and H. Nikaido, *J. Biol. Chem.*, 1971, **246**, 4397.
263. C. L. Marolda and M. A. Valvano, *Mol. Microbiol.*, 1996, **22**, 827.
264. D. Zhou, D. S. Stephens, B. W. Gibson, J. J. Engstrom, C. F. McAllister, F. K. N. Lee, and M. A. Apicella, *J. Biol. Chem.*, 1994, **269**, 11 162.
265. R. C. Sandlin and D. C. Stein, *J. Bacteriol.*, 1994, **176**, 2930.
266. S. Adhya and M. Schwartz, *J. Bacteriol.*, 1971, **108**, 621.
267. M. Lu and N. Kleckner, *J. Bacteriol.*, 1994, **176**, 5847.
268. J. B. Goldberg, K. Hatano, and G. B. Pier, *J. Bacteriol.*, 1993, **175**, 1605.
269. N. A. Zielinski, A. M. Chakrabarty, and A. Berry, *J. Biol. Chem.*, 1991, **266**, 9754.
270. R. W. Ye, N. A. Zielinski, and A. M. Chakrabarty, *J. Bacteriol.*, 1994, **176**, 4851.
271. M. J. Coyne, Jr., K. S. Russell, C. L. Coyle, and J. B. Goldberg, *J. Bacteriol.*, 1994, **176**, 3500.
272. S. Dutka-Malen, P. Mazodier, and B. Badet, *Biochimie*, 1988, **70**, 287.
273. H. C. Wu and T. C. Wu, *J. Bacteriol.*, 1971, **105**, 455.
274. M. Sarvas, *J. Bacteriol.*, 1971, **105**, 467.
275. D. Mengin-Lecreulx and J. van Heijenoort, *J. Biol. Chem.*, 1996, **271**, 32.
276. D. Mengin-Lecreulx and J. van Heijenoort, *J. Bacteriol.*, 1994, **176**, 5788.
277. A. M. Gehring, W. J. Lees, D. J. Mindiola, C. T. Walsh, and E. D. Brown, *Biochemistry*, 1996, **35**, 579.
278. D. Mengin-Lecreulx and J. van Heijenoort, *J. Bacteriol.*, 1993, **175**, 6150.
279. N. A. Sabnis, H. Yang, and T. Romeo, *J. Biol. Chem.*, 1995, **270**, 29 096.

280. F. J. Stevens, P. W. Stevens, J. G. Hovis, and T. T. Wu, *J. Gen. Microbiol.*, 1981, **124**, 219.
281. J. S. Miles and J. R. Guest, *Gene*, 1984, **32**, 41.
282. L. V. Collins and J. Hackett, *Gene*, 1991, **103**, 135.
283. S. R. Maloy and W. D. Nunn, *J. Bacteriol.*, 1981, **145**, 1110.
284. L. V. Collins, S. Attridge, and J. Hackett, *Infect. Immun.*, 1991, **59**, 1079.
285. S. J. Lee, L. K. Romana, and P. R. Reeves, *J. Gen. Microbiol.*, 1992, **138**, 305.
286. C. L. Marolda and M. A. Valvano, *J. Bacteriol.*, 1993, **175**, 148.
287. T. Sugiyama, N. Kido, T. Komatsu, M. Ohta, A. Jann, B. Jann, A. Saeki, and N. Kato, *Microbiology*, 1994, **140**, 59.
288. P. Jayaratne, D. Bronner, P. R. MacLachlan, C. Dodgson, N. Kido, and C. Whitfield, *J. Bacteriol.*, 1994, **176**, 3126.
289. D. A. Bastin and P. R. Reeves, *Gene*, 1995, **164**, 17.
290. P. K. Brown, L. K. Romana, and P. R. Reeves, *Mol. Microbiol.*, 1991, **5**, 1873.
291. D. Liu, N. K. Verma, L. K. Romana, and P. R. Reeves, *J. Bacteriol.*, 1991, **173**, 4814.
292. S. J. Lee, L. K. Romana, and P. R. Reeves, *J. Gen. Microbiol.*, 1992, **138**, 1843.
293. P. K. Brown, L. K. Romana, and P. R. Reeves, *Mol. Microbiol.*, 1992, **6**, 1385.
294. S.-H. Xiang, A. M. Haase, and P. R. Reeves, *J. Bacteriol.*, 1993, **175**, 4877.
295. G. Stevenson, S. J. Lee, L. K. Romana, and P. R. Reeves, *Mol. Gen. Genet.*, 1991, **227**, 173.
296. S. Gottesman and V. Stout, *Mol. Microbiol.*, 1991, **5**, 1599.
297. A. Markovitz, R. J. Sydiskis, and M. M. Lieberman, *J. Bacteriol.*, 1967, **94**, 1492.
298. R. Köplin, W. Arnold, B. Hötte, R. Simon, G. Wang, and A. Pühler, *J. Bacteriol.*, 1992, **174**, 191.
299. D. Shinabarger, A. Berry, T. B. May, R. Rothmel, A. Fialho, and A. M. Chakrabarty, *J. Biol. Chem.*, 1991, **266**, 2080.
300. U. H. Stroeher, L. E. Karageorgos, M. H. Brown, R. Morona, and P. A. Manning, *Gene*, 1995, **166**, 33.
301. J.-B. Dai, Y. Liu, W. J. Ray, Jr., and M. Konno, *J. Biol. Chem.*, 1992, **267**, 6322.
302. A. D. Elbein and E. C. Heath, *J. Biol. Chem.*, 1965, **240**, 1926.
303. A. C. Kessler, A. M. Haase, and P. R. Reeves, *J. Bacteriol.*, 1993, **175**, 1412.
304. T. L. Arsenault, D. W. Huges, D. B. Maclean, W. A. Szarek, A. M. B. Kropinski, and J. S. Lam, *Can. J. Chem.*, 1991, **69**, 1273.
305. J. Lightfoot and J. S. Lam, *Mol. Microbiol.*, 1993, **8**, 771.
306. H. L. Currie, J. Lightfoot, and J. S. Lam, *Clin. Diagn. Lab. Immun.*, 1995, **2**, 554.
307. A. Markovitz, *J. Biol. Chem.*, 1964, **239**, 2091.
308. J. H. Pazur and E. W. Shuey, *J. Biol. Chem.*, 1961, **236**, 1780.
309. L. Lindqvist, R. Kaiser, P. R. Reeves, and A. A. Lindberg, *Eur. J. Biochem.*, 1993, **211**, 763.
310. A. Melo and L. Glaser, *J. Biol. Chem.*, 1968, **243**, 1475.
311. K. Marumo, L. Lindqvist, N. Verma, A. Weintraub, P. R. Reeves, and A. A. Lindberg, *Eur. J. Biochem.*, 1992, **204**, 539.
312. J. D. Klena and C. A. Schnaitman, *Mol. Microbiol.*, 1993, **9**, 393.
313. K. Rajakumar, B. H. Jost, C. Sasakawa, N. Okada, M. Yoshikawa, and B. Adler, *J. Bacteriol.*, 1994, **176**, 2362.
314. L. Zhang, A. Al-Hendy, P. Toivanen, and M. Skurnik, *Mol. Microbiol.*, 1993, **9**, 309.
315. R. Köplin, G. Wang, B. Hötte, U. B. Priefer, and A. Pühler, *J. Bacteriol.*, 1993, **175**, 7786.
316. M. Mitchison, D. M. Bulach, T. Vinh, K. Rajakumar, S. Faine, and B. Adler, *J. Bacteriol.*, 1997, **179**, 1262.
317. B. D. Robertson, M. Frosch, and J. P. M. van Putten, *J. Bacteriol.*, 1994, **176**, 6915.
318. O. Westphal and O. Lüderitz, *Angew. Chem.*, 1960, **72**, 881.
319. P. Wyk and P. Reeves, *J. Bacteriol.*, 1989, **171**, 5687.
320. N. Verma and P. Reeves, *J. Bacteriol.*, 1989, **171**, 5694.
321. J. S. Thorson, S. F. Lo, and H.-W. Liu, *J. Am. Chem. Soc.*, 1993, **115**, 5827.
322. J. S. Thorson, T. M. Kelly, and H. W. Liu, *J. Bacteriol.*, 1994, **176**, 1840.
323. J. S. Thorson, S. F. Lo, O. Ploux, X. He, and H.-W. Liu, *J. Bacteriol.*, 1994, **176**, 5483.
324. L. L. Burrows, D. F. Charter, and J. S. Lam, *Mol. Microbiol.*, 1996, **22**, 481.
325. R. Morona, U. H. Stroeher, L. E. Karageorgos, M. H. Brown, and P. A. Manning, *Gene*, 1995, **166**, 19.
326. P. M. Nassau, S. L. Martin, R. E. Brown, A. Weston, D. Monsey, M. R. McNeil, and K. Duncan, *J. Bacteriol.*, 1996, **178**, 1047.
327. R. Köplin, J.-R. Brisson, and C. Whitfield, *J. Biol. Chem.*, 1997, **272**, 4121.
328. A. Wright, M. Dankert, P. Fennessey, and P. W. Robbins, *Proc. Natl. Acad. Sci. USA*, 1967, **57**, 1798.
329. B. C. McGrath and M. J. Osborn, *J. Bacteriol.*, 1991, **173**, 649.
330. M. J. Osborn and Y. Tze-Yuen, *J. Biol. Chem.*, 1968, **243**, 5145.
331. L. Wang and P. R. Reeves, *J. Bacteriol.*, 1994, **176**, 4348.
332. L. Wang, D. Liu, and P. R. Reeves, *J. Bacteriol.*, 1996, **178**, 2598.
333. T. L. Reuber and G. C. Walker, *Cell*, 1993, **74**, 269.
334. P. Bugert and K. Geider, *Mol. Microbiol.*, 1995, **15**, 917.
335. L. Ielpi, R. O. Couso, and M. A. Dankert, *J. Bacteriol.*, 1993, **175**, 2490.
336. D. Liu, A. M. Haase, L. Lindqvist, A. A. Lindberg, and P. R. Reeves, *J. Bacteriol.*, 1993, **175**, 3408.
337. R. A. Geremia, E. A. Petroni, L. Ielpi, and B. Henrissat, *Biochem. J.*, 1996, **318**, 133.
338. D. Liu, L. Lindqvist, and P. R. Reeves, *J. Bacteriol.*, 1995, **177**, 4084.
339. R. Morona, D. F. Macpherson, L. van den Bosch, N. I. A. Carlin, and P. A. Manning, *Mol. Microbiol.*, 1995, **18**, 209.
340. D. Liu, R. A. Cole, and P. R. Reeves, *J. Bacteriol.*, 1996, **178**, 2102.
341. D. F. Macpherson, P. A. Manning, and R. Morona, *Gene*, 1995, **155**, 9.
342. P. W. Robbins, D. Bray, B. M. Dankert, and A. Wright, *Science*, 1967, **158**, 1536.
343. Y. Naide, H. Nikaido, P. H. Mäkelä, R. G. Wilkinson, and B. A. D. Stocker, *Proc. Natl. Acad. Sci. USA*, 1965, **53**, 147.
344. L. V. Collins and J. Hackett, *J. Bacteriol.*, 1991, **173**, 2521.
345. S. Lukomski, R. A. Hull, and S. I. Hull, *J. Bacteriol.*, 1996, **178**, 240.
346. R. Morona, M. Mavris, A. Fallarino, and P. A. Manning, *J. Bacteriol.*, 1994, **176**, 733.
347. M. J. Coyne, Jr. and J. B. Goldberg, *Gene*, 1995, **167**, 81.
348. T. R. de Kievit, T. Dasgupta, H. Schweizer, and J. S. Lam, *Mol. Microbiol.*, 1995, **16**, 565.

349. M. Nurminen, C. G. Hellerqvist, V. V. Valtonen, and P. H. Mäkelä, *Eur. J. Biochem.*, 1971, **22**, 500.
350. D. A. Bastin, G. Stevenson, P. K. Brown, A. Haase, and P. R. Reeves, *Mol. Microbiol.*, 1993, **7**, 725.
351. G.-F. T. Chen and M. Inouye, *Nucleic Acids Res.*, 1990, **18**, 1465.
352. C. Whitfield, P. A. Amor, and R. Köplin, *Mol. Microbiol.*, 1997, **23**, 629.
353. N. Kido, V. I. Torgov, T. Sugiyama, K. Uchiya, H. Sugihara, T. Komatsu, N. Kato, and K. Jann, *J. Bacteriol.*, 1995, **177**, 2178.
354. M. Szabo, D. Bronner, and C. Whitfield, *J. Bacteriol.*, 1995, **177**, 1544.
355. B. R. Clarke, D. Bronner, W. J. Keenleyside, W. B. Severn, J. C. Richards, and C. Whitfield, *J. Bacteriol.*, 1995, **177**, 5411.
356. I. C. Fält, E. K. H. Schweda, A. Weintraub, S. Sturm, K. N. Timmis, and A. A. Lindberg, *Eur. J. Biochem.*, 1993, **213**, 573.
357. S. Göhmann, P. A. Manning, C.-A. Alpert, M. J. Walker, and K. N. Timmis, *Microb. Pathog.*, 1994, **16**, 53.
358. R. F. Kelly and C. Whitfield, *J. Bacteriol.*, 1996, **178**, 5205.
359. D. Bronner, B. R. Clarke, and C. Whitfield, *Mol. Microbiol.*, 1994, **14**, 505.
360. U. H. Stroeher, L. E. Karageorgos, R. Morona, and P. A. Manning, *Proc. Natl. Acad. Sci. USA*, 1992, **89**, 2566.
361. P. A. Manning, U. H. Stroeher, L. E. Karageorgos, and R. Morona, *Gene*, 1995, **158**, 1.
362. J. S. Kroll, B. Loynds, L. N. Brophy, and E. R. Moxon, *Mol. Microbiol.*, 1990, **4**, 1853.
363. M. Frosch, U. Edwards, K. Bousset, B. Krausse, and C. Weisgerber, *Mol. Microbiol.*, 1991, **5**, 1251.
364. M. J. Pavelka, Jr., L. F. Wright, and R. P. Silver, *J. Bacteriol.*, 1991, **173**, 4603.
365. A. N. Smith, G. J. Boulnois, and I. S. Roberts, *Mol. Microbiol.*, 1990, **4**, 1863.
366. M. Fath and R. Kolter, *Microbiol. Rev.*, 1993, **57**, 995.
367. S. Chu, B. Noonan, S. Cavaignac, and T. J. Trust, *Proc. Natl. Acad. Sci. USA*, 1995, **92**, 5754.
368. U. Meier-Dieter, K. Barr, R. Starman, L. Hatch, and P. D. Rick, *J. Biol. Chem.*, 1992, **267**, 746.
369. D. C. Alexander and M. A. Valvano, *J. Bacteriol.*, 1994, **176**, 7079.
370. P. D. Rick, G. L. Hubbard, and K. Barr, *J. Bacteriol.*, 1994, **176**, 2877.
371. R. C. Goldman and F. Hunt, *J. Bacteriol.*, 1990, **172**, 5352.
372. R. A. Batchelor, G. E. Haraguchi, R. A. Hull, and S. I. Hull, *J. Bacteriol.*, 1991, **173**, 5699.
373. R. A. Batchelor, P. Alifano, E. Biffali, S. I. Hull, and R. A. Hull, *J. Bacteriol.*, 1992, **174**, 5228.
374. C. Dodgson, P. Amor, and C. Whitfield, *J. Bacteriol.*, 1996, **178**, 1895.
375. R. Morona, L. van den Bosch, and P. A. Manning, *J. Bacteriol.*, 1995, **177**, 1059.
376. G. Stevenson, A. Kessler, and P. R. Reeves, *FEMS Microbiol. Lett.*, 1995, **125**, 23.
377. L. L. Burrows, D. Chow, and J. S. Lam, *J. Bacteriol.*, 1997, **179**, 1482.
378. A. V. Franco, D. Liu, and P. R. Reeves, *J. Bacteriol.*, 1996, **178**, 1903.
379. A. M. B. Kropinski, V. Lewis, and D. Berry, *J. Bacteriol.*, 1987, **169**, 1960.
380. S. A. Makin and T. J. Beveridge, *J. Bacteriol.*, 1996, **178**, 3350.
381. M. McConnell and A. Wright, *J. Bacteriol.*, 1979, **137**, 746.
382. K. Poole and V. Braun, *J. Bacteriol.*, 1988, **170**, 5146.
383. A. Al-Hendy, P. Toivanen, and M. Skurnik, *Microb. Pathog.*, 1991, **10**, 81.
384. L. Zhang, P. Toivanen, and M. Skurnik, in "Yersiniosis: Present and Future. Contributions to Microbiology and Immunology," eds. G. Ravagnan and C. Chiesa, Karger, Basel, 1995, vol. 13, p. 310.
385. Z. Yao, H. Liu, and M. A. Valvano, *J. Bacteriol.*, 1992, **174**, 7500.
386. M. A. Hediger, D. F. Johnson, D. P. Nierlich, and I. Zabin, *Proc. Natl. Acad. Sci. USA*, 1985, **82**, 6414.
387. P. H. Mäkelä, *J. Bacteriol.*, 1966, **91**, 1115.
388. C. G. Hellerqvist, B. Lindberg, S. Svensson, T. Holme, and A. A. Lindberg, *Carbohydr. Res.*, 1968, **8**, 43.
389. J. M. Slauch, A. A. Lee, M. J. Mahan, and J. J. Mekalanos, *J. Bacteriol.*, 1996, **178**, 5904.
390. P. W. Robbins, J. M. Keller, A. Wright, and R. L. Bernstein, *J. Biol. Chem.*, 1965, **240**, 384.
391. L. Barksdale and S. B. Arden, *Annu. Rev. Microbiol.*, 1974, **28**, 265.
392. J. Kuzio and A. M. Kropinski, *J. Bacteriol.*, 1983, **155**, 203.
393. C. A. Clark, J. Beltrame, and P. A. Manning, *Gene*, 1991, **107**, 43.
394. N. K. Verma, J. M. Brandt, D. J. Verma, and A. A. Lindberg, *Mol. Microbiol.*, 1991, **5**, 71.
395. P. H. Mäkelä, *J. Bacteriol.*, 1973, **116**, 847.
396. N. Zinder, *Science*, 1957, **126**, 1237.
397. N. A. Nnalue, S. Newton, and B. A. D. Stocker, *Microb. Pathog.*, 1990, **8**, 393.
398. U. Mamat, E. Th. Rietschel, and G. Schmidt, *Mol. Microbiol.*, 1995, **15**, 1115.
399. D. J. Kopecko, O. Washington, and S. B. Formal, *Infect. Immun.*, 1980, **29**, 207.
400. Y. Yoshida, N. Okamura, J. Kato, and H. Watanabe, *J. Gen. Microbiol.*, 1991, **137**, 867.
401. N. Okamura, T. Chida, M. Kinoshita, Y. Yoshida, S. Kondo, and K. Hisatsune, *Microbiol. Immunol.*, 1993, **37**, 331.
402. H. Watanabe and K. N. Timmis, *Infect. Immun.*, 1984, **43**, 391.
403. H. Watanabe, A. Nakamura, and K. N. Timmis, *Infect. Immun.*, 1984, **46**, 55.
404. T. L. Hale, P. Guerry, R. C. Seid, Jr., C. Kapfer, M. E. Wingfield, C. B. Reaves, L. S. Baron, and S. B. Formal, *Infect. Immun.*, 1984, **46**, 470.
405. D. A. Bastin, L. K. Romana, and P. R. Reeves, *Mol. Microbiol.*, 1991, **5**, 2223.
406. L. W. Riley, L. N. Junio, L. B. Libaek, and G. K. Schoolnik, *Infect. Immun.*, 1987, **55**, 2052.
407. K. Kawahara, T. Hamaoka, S. Suzuki, M. Nakamura, S. Y. Murayama, T. Arai, N. Terakado, and H. Danbara, *Microb. Pathog.*, 1989, **7**, 195.
408. M. Y. Popoff and L. Le Minor, *Ann. Inst. Pasteur/Microbiol.*, 1985, **136B**, 169.
409. W. J. Keenleyside and C. Whitfield, *J. Bacteriol.*, 1995, **177**, 5247.
410. W. J. Keenleyside and C. Whitfield, *J. Biol. Chem.*, 1996, **271**, 28 581.
411. P. R. Reeves, *FEMS Microbiol. Lett.*, 1992, **79**, 509.
412. P. R. Reeves, *Trends Genet.*, 1993, **9**, 17.
413. M. Bisercic, J. Y. Feutrier, and P. R. Reeves, *J. Bacteriol.*, 1991, **173**, 3894.
414. K. Nelson and R. K. Selander, *Proc. Natl. Acad. Sci. USA*, 1994, **91**, 10 227.
415. S.-H. Xiang, M. Hobbs, and P. R. Reeves, *J. Bacteriol.*, 1994, **176**, 4357.

416. A. B. Sadosky, J. A. Gray, and C. W. Hill, *Nucleic Acids Res.*, 1991, **19**, 7177.
417. S. Zhao, C. H. Sandt, G. Feulner, D. A. Vlazny, J. A. Gray, and C. W. Hill, *J. Bacteriol.*, 1993, **175**, 2799.
418. C. W. Hill, C. H. Sandt, and D. A. Vlazny, *Mol. Microbiol.*, 1994, **12**, 865.
419. R. Morona, M. S. Matthews, J. K. Morona, and M. H. Brown, *Mol. Gen. Genet.*, 1990, **224**, 405.
420. T. Ito, Y. Ohshita, K. Hiramatsu, and T. Yokota, *FEBS Lett.*, 1991, **286**, 159.
421. T. Ito, K. Hiramatsu, Y. Ohshita, and T. Yokota, *Microbiol. Immunol.*, 1993, **37**, 281.
422. M. K. Waldor, R. Colwell, and J. J. Mekalanos, *Proc. Natl. Acad. Sci. USA*, 1994, **91**, 11388.
423. L. E. Comstock, D. Maneval, Jr., P. Panigrahi, A. Joseph, M. M. Levine, J. B. Kaper, J. G. Morris, Jr., and J. A. Johnson, *Infect. Immun.*, 1995, **63**, 317.
424. E. M. Bik, A. E. Bunschoten, R. D. Gouw, and F. R. Mooi, *EMBO J.*, 1995, **14**, 209.
425. E. M. Bik, A. E. Bunschoten, R. J. L. Willems, A. C. Y. Chang, and F. R. Mooi, *Mol. Microbiol.*, 1996, **20**, 799.
426. M. A. Valvano and C. L. Marolda, *Infect. Immun.*, 1991, **59**, 3917.
427. M. W. Heuzenroeder, D. W. Beger, C. J. Thomas, and P. A. Manning, *Mol. Microbiol.*, 1989, **3**, 295.
428. M. S. Anderson and C. R. H. Raetz, *J. Biol. Chem.*, 1987, **262**, 5159.
429. M. S. Anderson, H. G. Bull, S. M. Galloway, T. M. Kelly, S. Mohan, K. Radika, and C. R. H. Raetz, *J. Biol. Chem.*, 1993, **268**, 19858.
430. D. N. Crowell, M. S. Anderson, and C. R. H. Raetz, *J. Bacteriol.*, 1986, **168**, 152.
431. J. Coleman and C. R. H. Raetz, *J. Bacteriol.*, 1988, **170**, 1268.
432. J. M. Williamson, M. S. Anderson, and C. R. H. Raetz, *J. Bacteriol.*, 1991, **173**, 3591.
433. C. R. H. Raetz, *J. Bacteriol.*, 1993, **175**, 5745.
434. M. S. Anderson, A. D. Robertson, I. Macher, and C. R. H. Raetz, *Biochemistry*, 1988, **27**, 1908.
435. K. Young, L. L. Silver, D. Bramhill, P. Cameron, S. S. Eveland, C. R. H. Raetz, S. A. Hyland, and M. S. Anderson, *J. Biol. Chem.*, 1995, **270**, 30384.
436. N. F. Sullivan and W. D. Donachie, *J. Bacteriol.*, 1984, **160**, 724.
437. T. M. Kelly, S. A. Stachula, C. R. H. Raetz, and M. S. Anderson, *J. Biol. Chem.*, 1993, **268**, 19866.
438. P. G. Sorensen, J. Lutkenhaus, K. Young, S. S. Eveland, M. S. Anderson, and C. R. H. Raetz, *J. Biol. Chem.*, 1996, **271**, 25898.
439. A. M. Roy and J. Coleman, *J. Bacteriol.*, 1994, **176**, 1639.
440. I. B. Dicker and S. Seetharam, *Mol. Microbiol.*, 1992, **6**, 817.
441. H. G. Tomasiewicz and C. S. McHenry, *J. Bacteriol.*, 1987, **169**, 5735.
442. C. R. H. Raetz and W. Dowhan, *J. Biol. Chem.*, 1990, **265**, 1235.
443. S.-J. Li and J. E. Cronan, Jr., *J. Biol. Chem.*, 1992, **267**, 16841.
444. S. Mohan, T. M. Kelly, S. S. Eveland, C. R. H. Raetz, and M. S. Anderson, *J. Biol. Chem.*, 1994, **269**, 32896.
445. K. Magnuson, S. Jackowski, C. O. Rock, and J. E. Cronan, Jr., *Microbiol. Rev.*, 1993, **57**, 522.
446. R. Vuorio, T. Härkönen, M. Tolvanen, and M. Vaara, *FEBS Lett.*, 1994, **337**, 289.
447. E. I. Shaw and D. O. Wood, *Gene*, 1994, **140**, 109.
448. C. E. Bulawa and C. R. H. Raetz, *J. Biol. Chem.*, 1984, **259**, 4846.
449. B. L. Ray, G. Painter, and C. R. H. Raetz, *J. Biol. Chem.*, 1984, **259**, 4852.
450. K. Radika and C. R. H. Raetz, *J. Biol. Chem.*, 1988, **263**, 14859.
451. D. N. Crowell, W. S. Reznikoff, and C. R. H. Raetz, *J. Bacteriol.*, 1987, **169**, 5727.
452. M. E. Milla and C. R. H. Raetz, *Biochim. Biophys. Acta*, 1996, **1304**, 245.
453. R. D. Fleischmann, M. D. Adams, O. White, R. A. Clayton, E. F. Kirkness, A. R. Kerlavage, C. J. Bult, J.-F. Tomb, B. A. Dougherty, J. M. Merrick *et al.*, *Science*, 1995, **269**, 496.
454. S. Servos, S. Khan, and D. Maskell, *Gene*, 1996, **175**, 137.
455. B. L. Ray and C. R. H. Raetz, *J. Biol. Chem.*, 1987, **262**, 1122.
456. K. A. Brozek, C. E. Bulawa, and C. R. H. Raetz, *J. Biol. Chem.*, 1987, **262**, 5170.
457. P. D. Rick, L. W.-M. Fung, C. Ho, and M. J. Osborn, *J. Biol. Chem.*, 1977, **252**, 4904.
458. C. R. H. Raetz, S. Purcell, M. V. Meyer, N. Qureshi, and K. Takayama, *J. Biol. Chem.*, 1985, **260**, 16080.
459. R. C. Goldman, C. C. Doran, and J. O. Capobianco, *J. Bacteriol.*, 1988, **170**, 2185.
460. K. A. Brozek and C. R. H. Raetz, *J. Biol. Chem.*, 1990, **265**, 15410.
461. P. H. Ray, *J. Bacteriol.*, 1980, **141**, 635.
462. G. D. Dotson, R. K. Dua, J. C. Clemens, E. W. Wooten, and R. W. Woodard, *J. Biol. Chem.*, 1995, **270**, 13698.
463. H. M. Salleh, M. A. Patel, and R. W. Woodard, *Biochemistry*, 1996, **35**, 8942.
464. P. H. Ray and C. D. Benedict, *J. Bacteriol.*, 1980, **142**, 60.
465. P. H. Ray, C. D. Benedict, and H. Grasmuk, *J. Bacteriol.*, 1981, **145**, 1273.
466. R. C. Goldman, T. J. Bolling, W. E. Kohlbrenner, Y. Kim, and J. L. Fox, *J. Biol. Chem.*, 1986, **261**, 15831.
467. M. Woisetschläger and G. Högenauer, *J. Bacteriol.*, 1986, **168**, 437.
468. M. Woisetschläger and G. Högenauer, *Mol. Gen. Genet.*, 1987, **207**, 369.
469. R. C. Goldman and W. E. Kohlbrenner, *J. Bacteriol.*, 1985, **163**, 256.
470. G. Tipples and G. McClarty, *J. Biol. Chem.*, 1995, **270**, 7908.
471. W. Brabetz and H. Brade, *Eur. J. Biochem.*, 1997, **244**, 66.
472. J. L. Wylie, E. R. Iliffe, L. L. Wang, and G. McClarty, *Infect. Immun.*, 1997, **65**, 1527.
473. D. W. Hood, M. E. Deadman, T. Allen, H. Masoud, A. Martin, J. R. Brisson, R. Fleischmann, J. C. Venter, J. C. Richards, and E. R. Moxon, *Mol. Microbiol.*, 1996, **22**, 951.
474. W. D. Tolbert, J. R. Moll, R. Bauerle, and R. H. Kretsinger, *Proteins*, 1996, **24**, 407.
475. I. A. Shumilin, R. H. Kretsinger, and R. Bauerle, *Proteins*, 1996, **24**, 404.
476. M. Woisetschläger, A. Hödl-Neuhofer, and G. Högenauer, *J. Bacteriol.*, 1988, **170**, 5382.
477. C. Pazzani, C. Rosenow, G. J. Boulnois, D. Bronner, K. Jann, and I. S. Roberts, *J. Bacteriol.*, 1993, **175**, 5978.
478. C. Rosenow, I. S. Roberts, and K. Jann, *FEMS Microbiol. Lett.*, 1995, **125**, 159.
479. S. Jelakovic, K. Jann, and G. E. Schulz, *FEBS Lett.*, 1996, **391**, 157.
480. H. Strohmaier, P. Remler, W. Renner, and G. Högenauer, *J. Bacteriol.*, 1995, **177**, 4488.
481. J. L. Wylie, J. D. Berry, and G. McClarty, *Mol. Microbiol.*, 1996, **22**, 631.
482. K. A. Brozek, K. Hosaka, A. D. Robertson, and C. R. H. Raetz, *J. Biol. Chem.*, 1989, **264**, 6956.

483. T. Clementz and C. R. H. Raetz, *J. Biol. Chem.*, 1991, **266**, 9687.
484. C. J. Belunis and C. R. H. Raetz, *J. Biol. Chem.*, 1992, **267**, 9988.
485. C. J. Belunis, K. E. Mdluli, C. R. H. Raetz, and F. E. Nano, *J. Biol. Chem.*, 1992, **267**, 18 702.
486. S. Löbau, U. Mamat, W. Brabetz, and H. Brade, *Mol. Microbiol.*, 1995, **18**, 391.
487. C. J. Belunis, T. Clementz, S. M. Carty, and C. R. H. Raetz, *J. Biol. Chem.*, 1995, **270**, 27 646.
488. U. Mamat, S.Löbau, K. Persson, and H. Brade, *Microb. Pathog.*, 1994, **17**, 87.
489. U. Mamat, M. Baumann, G. Schmidt, and H. Brade, *Mol. Microbiol.*, 1993, **10**, 935.
490. J. F. Guasch, N. Piqué, N. Climent, S. Ferrer, S. Merino, X. Rubires, J. M. Tomás, and M. Regué, *J. Bacteriol.*, 1996, **178**, 5741.
491. A. Allen and D. Maskell, *Mol. Microbiol.*, 1996, **19**, 37.
492. S. M. Spinola, Y. A. Kwaik, A. J. Lesse, A. A. Campagnari, and M. A. Apicella, *Infect. Immun.*, 1990, **58**, 1558.
493. Y. A. Kwaik, R. E. McLaughlin, M. A. Apicella, and S. M. Spinola, *Mol. Microbiol.*, 1991, **5**, 2475.
494. T. Clementz, J. J. Bednarski, and C. R. H. Raetz, *J. Biol. Chem.*, 1996, **271**, 12 095.
495. M. Karow, O. Fayet, and C. Georgopoulos, *J. Bacteriol.*, 1992, **174**, 7407.
496. M. Karow, O. Fayet, A. Cegielska, T. Ziegelhoffer, and C. Georgopoulos, *J. Bacteriol.*, 1991, **173**, 741.
497. M. Karow and C. Georgopoulos, *Mol. Microbiol.*, 1991, **5**, 2285.
498. N.-G. Lee, M. G. Sunshine, and M. A. Apicella, *Infect. Immun.*, 1995, **63**, 818.
499. N.-G. Lee, M. G. Sunshine, J. J. Engstrom, B. W. Gibson, and M. A. Apicella, *J. Biol. Chem.*, 1995, **270**, 27 151.
500. M. Karow and C. Georgopoulos, *Mol. Microbiol.*, 1993, **7**, 69.
501. A. Polissi and C. Georgopoulos, *Mol. Microbiol.*, 1996, **20**, 1221.
502. M. Karow and C. Georgopoulos, *J. Bacteriol.*, 1992, **174**, 702.
503. T. Clementz, J. Bednarski, and C. R. H. Raetz, *FASEB J.*, 1995, **9**, A1311.
504. J. E. Somerville, Jr., L. Cassiano, B. Bainbridge, M. D. Cunningham, and R. P. Darveau, *J. Clin. Invest.*, 1996, **97**, 359.
505. N. P. J. Price, T. M. Kelly, C. R. H. Raetz, and R. W. Carlson, *J. Bacteriol.*, 1994, **176**, 4646.
506. R. I. Hollingsworth and R. W. Carlson, *J. Biol. Chem.*, 1989, **264**, 9300.
507. N. P. J. Price, B. Jeyaretnam, R. W. Carlson, J. L. Kadrmas, C. R. H. Raetz, and K. A. Brozek, *Proc. Natl. Acad. Sci. USA*, 1995, **92**, 7352.
508. K. A. Brozek, J. L. Kadrmas, and C. R. H. Raetz, *J. Biol. Chem.*, 1996, **271**, 32 112.
509. R. C. Goldman, C. C. Doran, S. K. Kadam, and J. O. Capobianco, *J. Biol. Chem.*, 1988, **263**, 5217.
510. S. Mohan and C. R. H. Raetz, *J. Bacteriol.*, 1994, **176**, 6944.
511. L. Eidels and M. J. Osborn, *Proc. Natl. Acad. Sci. USA*, 1971, **68**, 1673.
512. L. Eidels and M. J. Osborn, *J. Biol. Chem.*, 1974, **249**, 5642.
513. A. Preston, D. Maskell, A. Johnson, and E. R. Moxon, *J. Bacteriol.*, 1996, **178**, 396.
514. J. S. Brooke and M. A. Valvano, *J. Biol. Chem.*, 1996, **271**, 3608.
515. J. S. Brooke and M. A. Valvano, *J. Bacteriol.*, 1996, **178**, 3339.
516. W. G. Coleman, Jr., *J. Biol. Chem.*, 1983, **258**, 1985.
517. W. G. Coleman, Jr. and K. S. Deshpande, *J. Bacteriol.*, 1985, **161**, 1209.
518. D. M. Sirisena, K. A. Brozek, P. R. MacLachlan, K. E. Sanderson, and C. R. H. Raetz, *J. Biol. Chem.*, 1992, **267**, 18 874.
519. W. G. Coleman, Jr., L. Chen, and L. Ding, in "Pseudomonas: Molecular Biology and Biotechnology," eds. E. Galli, S. Silver, and B. Witholt, American Society for Microbiology, Washington, DC, 1992, p. 161.
520. E. S. Drazek, D. C. Stein, and C. D. Deal, *J. Bacteriol.*, 1995, **177**, 2321.
521. J. C. Levin and D. C. Stein, *J. Bacteriol.*, 1996, **178**, 4571.
522. U. H. Stroeher, L. E. Karageorgos, R. Morona, and P. A. Manning, *Gene*, 1995, **155**, 67.
523. J. C. Pegues, L. S. Chen, A. W. Gordon, L. Ding, and W. G. Coleman, Jr., *J. Bacteriol.*, 1990, **172**, 4652.
524. D. M. Sirisena, P. R. MacLachlan, S.-L. Liu, A. Hessel, and K. E. Sanderson, *J. Bacteriol.*, 1994, **176**, 2379.
525. S. Raina and C. Georgopoulos, *Nucleic Acids Res.*, 1991, **19**, 3811.
526. L. Chen and W. G. Coleman, Jr., *J. Bacteriol.*, 1993, **175**, 2534.
527. D. Zhou, N.-G. Lee, and M. A. Apicella, *Mol. Microbiol.*, 1994, **14**, 609.
528. J. L. Kadrmas, K. A. Brozek, and C. R. H. Raetz, *J. Biol. Chem.*, 1996, **271**, 32 119.
529. E. Petricoin, III, R. J. Danaher, and D. C. Stein, *J. Bacteriol.*, 1991, **173**, 7896.
530. R. J. Danaher, E. F. Petricoin, III, and D. C. Stein, *J. Bacteriol.*, 1994, **176**, 3428.
531. E. T. Schwan, B. D. Robertson, H. Brade, and J. P. M. van Putten, *Mol. Microbiol.*, 1995, **15**, 267.
532. M. P. Jennings, M. Bisercic, K. L. R. Dunn, M. Virji, A. Martin, K. E. Wilks, J. C. Richards, and E. R. Moxon, *Microb. Pathog.*, 1995, **19**, 391.
533. W. A. Nichols, B. W. Gibson, W. Melaugh, N.-G. Lee, M. Sunshine, and M. A. Apicella, *Infect. Immun.*, 1997, **65**, 1377.
534. T. Clementz, *J. Bacteriol.*, 1992, **174**, 7750.
535. C. T. Parker, E. Pradel, and C. A. Schnaitman, *J. Bacteriol.*, 1992, **174**, 930.
536. C. M. Kahler, R. W. Carlson, M. M. Rahman, L. E. Martin, and D. S. Stephens, *J. Bacteriol.*, 1996, **178**, 6677.
537. C. M. Kahler, R. W. Carlson, M. M. Rahman, L. E. Martin, and D. S. Stephens, *J. Bacteriol.*, 1996, **178**, 1265.
538. P. van der Ley, M. Kramer, L. Steeghs, B. Kuipers, S. R. Andersen, M. P. Jennings, E. R. Moxon, and J. T. Poolman, *Mol. Microbiol.*, 1996, **19**, 1117.
539. P. van der Ley, M. Kramer, A. Martin, J. C. Richards, and J. T. Poolman, *FEMS Microbiol. Lett.*, 1997, **146**, 247.
540. M. Vaara, T. Vaara, M. Jensen, I. M. Helander, M. Nurminen, E. Th. Rietschel, and P. H. Mäkelä, *FEBS Lett.*, 1981, **129**, 145.
541. I. M. Helander, I. Kilpeläinen, and M. Vaara, *Mol. Microbiol.*, 1994, **11**, 481.
542. K. Numilla, I. Kilpeläinen, U. Zähringer, M. Vaara, and I. M. Helander, *Mol. Microbiol.*, 1995, **16**, 271.
543. K. L. Roland, L. E. Martin, C. R. Esther, and J. K. Spitznagel, *J. Bacteriol.*, 1993, **175**, 4154.
544. J. B. Stock, A. J. Ninfa, and A. M. Stock, *Microbiol. Rev.*, 1989, **53**, 450.
545. S. Nagasawa, K. Ishige, and T. Mizuno, *J. Biochem. (Tokyo)*, 1993, **114**, 350.
546. C. A. Schnaitman, C. T. Parker, J. D. Klena, E. L. Pradel, N. B. Pearson, K. E. Sanderson, and P. R. MacLachlan, *J. Bacteriol.*, 1991, **173**, 7410.

547. E. A. Austin, J. F. Graves, L. A. Hite, C. T. Parker, and C. A. Schnaitman, *J. Bacteriol.*, 1990, **172**, 5312.
548. C. Roncero and M. J. Casadaban, *J. Bacteriol.*, 1992, **174**, 3250.
549. C. T. Parker, A. W. Kloser, C. A. Schnaitman, M. A. Stein, S. Gottesman, and B. W. Gibson, *J. Bacteriol.*, 1992, **174**, 2525.
550. S. K. Kadam, A. Rehemtulla, and K. E. Sanderson, *J. Bacteriol.*, 1985, **161**, 277.
551. E. Pilipcinec, T. T. Huisman, P. T. J. Willemsen, B. J. Appelmelk, F. K. de Graaf, and B. Oudega, *FEMS Microbiol. Lett.*, 1994, **123**, 201.
552. P. Carstenius, J.-I. Flock, and A. A. Lindberg, *Nucleic Acids Res.*, 1990, **18**, 6128.
553. E. Pradel, C. T. Parker, and C. A. Schnaitman, *J. Bacteriol.*, 1992, **174**, 4736.
554. R. Wollin, E. S. Creeger, L. I. Rothfield, B. A. D. Stocker, and A. A. Lindberg, *J. Biol. Chem.*, 1983, **258**, 3769.
555. V. Stout and S. Gottesman, *J. Bacteriol.*, 1990, **172**, 659.
556. P. L. D. Stanley, P. Diaz, M. J. A. Bailey, D. Gygi, A. Juarez, and C. Hughes, *Mol. Microbiol.*, 1993, **10**, 781.
557. I. M. Helander, M. Vaara, S. Sukupolvi, M. Rhen, S. Saarela, U. Zähringer, and P. H. Mäkelä, *Eur. J. Biochem.*, 1989, **185**, 541.
558. P. R. MacLachlan, S. K. Kadam, and K. E. Sanderson, *J. Bacteriol.*, 1991, **173**, 7151.
559. J. D. Klena, E. Pradel, and C. A. Schnaitman, *J. Bacteriol.*, 1992, **174**, 4746.
560. E. S. Creeger, T. Schulte, and L. I. Rothfield, *J. Biol. Chem.*, 1984, **259**, 3064.
561. A. Farewell, R. Brazas, E. Davie, J. Mason, and L. I. Rothfield, *J. Bacteriol.*, 1991, **173**, 5188.
562. E. Pradel and C. A. Schnaitman, *J. Bacteriol.*, 1991, **173**, 6428.
563. R. Brazas, E. Davie, A. Farewell, and L. I. Rothfield, *J. Bacteriol.*, 1991, **173**, 6168.
564. L. Beutin and M. Achtman, *J. Bacteriol.*, 1979, **139**, 730.
565. L. Beutin, P. A. Manning, M. Achtman, and N. Willetts, *J. Bacteriol.*, 1981, **145**, 840.
566. M. P. Stevens, P. Hänfling, B. Jann, K. Jann, and I. S. Roberts, *FEMS Microbiol. Lett.*, 1994, **124**, 93.
567. M. J. A. Bailey, V. Koronakis, T. Schmoll, and C. Hughes, *Mol. Microbiol.*, 1992, **6**, 1003.
568. J. A. Leeds and R. A. Welch, *J. Bacteriol.*, 1996, **178**, 1850.
569. J. M. Nieto, M. J. A. Bailey, C. Hughes, and V. Koronakis, *Mol. Microbiol.*, 1996, **19**, 705.
570. M. J. A. Bailey, C. Hughes, and V. Koronakis, *Mol. Microbiol.*, 1996, **22**, 729.
571. M. Hobbs and P. R. Reeves, *Mol. Microbiol.*, 1994, **12**, 855.
572. E. R. Moxon and D. Maskell, in "Molecular Biology of Bacterial Infection: Current Status and Future Perspectives," eds. C. E. Hormaeche, C. W. Penn, and C. J. Smyth, Cambridge University Press, Cambridge, 1992, p. 75.
573. J. P. M. van Putten and B. D. Robertson, *Mol. Microbiol.*, 1995, **16**, 847.
574. H. Schneider, T. L. Hale, W. D. Zollinger, R. C. Seid, Jr., C. A. Hammack, and J. M. Griffiss, *Infect. Immun.*, 1984, **45**, 544.
575. E. C. Gotschlich, *J. Exp. Med.*, 1994, **180**, 2181.
576. A. L. Erwin, P. A. Haynes, P. A. Rice, and E. C. Gotschlich, *J. Exp. Med.*, 1996, **184**, 1233.
577. Q.-L. Yang and E. C. Gotschlich, *J. Exp. Med.*, 1996, **183**, 323.
578. J. P. M. van Putten, *EMBO J.*, 1993, **12**, 4043.
579. R. J. Danaher, J. C. Levin, D. Arking, C. L. Burch, R. Sandlin, and D. C. Stein, *J. Bacteriol.*, 1995, **177**, 7275.
580. R. Yamasaki, D. E. Kerwood, H. Schneider, K. P. Quinn, J. M. Griffiss, and R. E. Mandrell, *J. Biol. Chem.*, 1994, **269**, 30345.
581. M. P. Jennings, D. W. Hood, I. R. A. Peak, M. Virji, and E. R. Moxon, *Mol. Microbiol.*, 1995, **18**, 729.
582. W. Wakarchuk, A. Martin, M. P. Jennings, E. R. Moxon, and J. C. Richards, *J. Biol. Chem.*, 1996, **271**, 19166.
583. H. Smith, J. A. Cole, and N. J. Parsons, *FEMS Microbiol. Lett.*, 1992, **79**, 287.
584. H. Smith, N. J. Parsons, and J. A. Cole, *Microb. Pathog.*, 1995, **19**, 365.
585. M. J. Gill, D. P. McQuillen, J. P. M. van Putten, L. M. Wetzler, J. Bramley, H. Crooke, N. J. Parsons, J. A. Cole, and H. Smith, *Infect. Immun.*, 1996, **64**, 3374.
586. R. E. Mandrell, J. M. Griffiss, H. Smith, and J. A. Cole, *Microb. Pathog.*, 1993, **14**, 315.
587. R. F. Rest and R. E. Mandrell, *Microb. Pathog.*, 1995, **19**, 379.
588. M. Gilbert, D. C. Watson, A.-M. Cunningham, M. P. Jennings, N. M. Young, and W. W. Wakarchuk, *J. Biol. Chem.*, 1996, **271**, 28271.
589. R. E. Mandrell, H. Smith, G. A. Jarvis, J. M. Griffiss, and J. A. Cole, *Microb. Pathog.*, 1993, **14**, 307.
590. R. E. Mandrell, J. J. Kim, C. M. John, B. W. Gibson, J. V. Sugai, M. A. Apicella, J. M. Griffiss, and R. Yamasaki, *J. Bacteriol.*, 1991, **173**, 2823.
591. U. Edwards, A. Müller, S. Hammerschmidt, R. Gerardy-Schahn, and M. Frosch, *Mol. Microbiol.*, 1994, **14**, 141.
592. S. Hammerschmidt, R. Hilse, J. P. M. van Putten, R. Gerardy-Schahn, A. Unkmeier, and M. Frosch, *EMBO J.*, 1996, **15**, 192.
593. J. N. Weiser, A. A. Lindberg, E. J. Manning, E. J. Hansen, and E. R. Moxon, *Infect. Immun.*, 1989, **57**, 3045.
594. J. N. Weiser, J. M. Love, and E. R. Moxon, *Cell*, 1989, **59**, 657.
595. J. N. Weiser, D. J. Maskell, P. D. Butler, A. A. Lindberg, and E. R. Moxon, *J. Bacteriol.*, 1990, **172**, 3304.
596. J. N. Weiser, A. Williams, and E. R. Moxon, *Infect. Immun.*, 1990, **58**, 3455.
597. N. J. High, M. E. Deadman, and E. R. Moxon, *Mol. Microbiol.*, 1993, **9**, 1275.
598. N. J. High, M. P. Jennings, and E. R. Moxon, *Mol. Microbiol.*, 1996, **20**, 165.
599. J. N. Weiser, M. Shchepetov, and S. T. H. Chong, *Infect. Immun.*, 1997, **65**, 943.
600. G. P. Jarosik and E. J. Hansen, *Infect. Immun.*, 1994, **62**, 4861.
601. L. D. Cope, R. Yogev, J. Mertsola, J. L. Latimer, M. S. Hanson, G. H. McCracken, Jr., and E. J. Hansen, *Mol. Microbiol.*, 1991, **5**, 1113.
602. M. D. Potter and R. Y. C. Lo, *FEMS Microbiol. Lett.*, 1995, **129**, 75.
603. F. Dardel, M. Panvert, and G. Fayat, *Mol. Gen. Genet.*, 1990, **223**, 121.
604. N. J. High, M. E. Deadman, D. W. Hood, and E. R. Moxon, *FEMS Microbiol. Lett.*, 1996, **145**, 325.
605. D. W. Hood, M. E. Deadman, M. P. Jennings, M. Bisercic, R. D. Fleischmann, J. C. Venter, and E. R. Moxon, *Proc. Natl. Acad. Sci. USA*, 1996, **93**, 11121.
606. J. L. Di Fabio, M. Caroff, D. Karibian, J. C. Richards, and M. B. Perry, *FEMS Microbiol. Lett.*, 1992, **76**, 275.
607. M. Lukácová, M. Baumann, L. Brade, U. Mamat, and H. Brade, *Infect. Immun.*, 1994, **62**, 2270.

3.10
Bacterial Peptidoglycan Biosynthesis and its Inhibition

TIMOTHY D. H. BUGG
University of Southampton, UK

3.10.1   BACTERIAL CELL WALL PEPTIDOGLYCAN STRUCTURE

The peptidoglycan layer is an important structural component of bacterial cell walls, composed of carbohydrate glycan strands with peptide cross-links. The carbohydrate backbone of peptidoglycan is common to all bacteria, but there are important differences in the molecular composition and polymeric architecture of peptidoglycan between Gram-positive and Gram-negative bacteria. This chapter will begin with a discussion of cell wall composition in Gram-positive and Gram-negative bacteria, then discuss in turn the structures of the carbohydrate backbone, the peptide side chains, and the interstrand cross-links. Specific texts and reviews are available describing cell wall structure.[1–6]

3.10.1.1   Composition of Bacterial Cell Walls

3.10.1.1.1   *Gram-positive and Gram-negative bacterial cell walls*

In 1884, Gram[7] discovered a reagent which could be used to stain bacterial cells for microscopic examination. Bacteria were subsequently divided into two classifications based on whether their cell surfaces could be stained with Gram's reagent: Gram-positive and Gram-negative. The difference in reactivity with the Gram stain between the two classes highlights major differences in the cell surface architecture.[8] Gram-positive bacteria contain a thick outer layer of peptidoglycan which reacts strongly with Gram's reagent. Gram-negative bacteria, however, possess an outer membrane encompassing the peptidoglycan layer.[8]

Electron microscopic studies of Gram-positive bacterial cell walls have revealed a thick, amorphous layer of peptidoglycan between 20 nm and 80 nm in depth, covering a thin cytoplasmic membrane (Figure 1). The peptidoglycan layer can be isolated by sonic or mechanical disruption of the bacteria, followed by centrifugation, to give a rigid hollow sacculus which still holds the shape of the original cell. This sacculus contains peptidoglycan together with cell surface proteins, polysaccharides, and teichoic acids. Treatment with trichloroacetic acid or hot formamide yields the murein or peptidoglycan, which can be analyzed structurally.[4]

In Gram-negative bacteria the peptidoglycan layer is relatively thin, 2–3 nm in depth, and is surrounded by a porous outer membrane layer (Figure 1). Neutron and X-ray diffraction studies have shown that 75–80% of the cell wall of *Escherichia coli* contains a single layer of peptidoglycan, with the remainder consisting of three layers.[9] The peptidoglycan layer is positioned in the periplasmic space between the cytoplasmic membrane and the outer membrane, and is physically associated with the outer membrane. However, the concept of a rigid monolayer of peptidoglycan has been superseded by the idea of a periplasmic gel which confers gel-like mechanical properties on the wall.[10] The outer membrane is a complex layer consisting of lipoprotein, lipopolysaccharide (LPS), phospholipids, and channel-forming proteins called porins. The outer membrane is largely

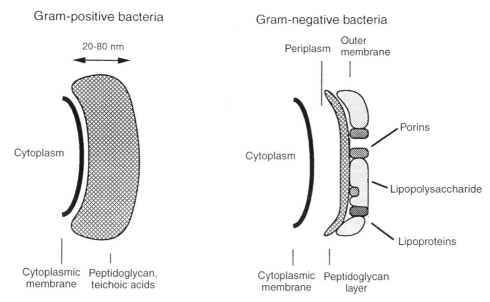

Figure 1 Schematic representation of Gram-positive and Gram-negative bacterial cell walls

responsible for controlling the passage of large molecules into the cell, but there is evidence to suggest that there are also specific transport systems associated with the peptidoglycan layer.[11]

Many Gram-negative and Gram-positive bacteria also contain an outer "capsule" which interacts with the external environment. This layer commonly consists of high molecular weight poly-saccharides, but some polypeptide capsules are also found.[6] The capsule is thought to be important for bacterial virulence, adhesion, and water sequestration.[6]

3.10.1.1.2 *Functions of the peptidoglycan layer*

The primary function of the peptidoglycan layer is to prevent lysis of the bacterial cell through turgor pressure, which arises owing to the higher osmotic pressure inside the cell than in the outside medium. The high internal osmotic pressure would cause water to enter the cell until the turgor pressure is matched by the elastic stretch of the peptidoglycan layer. The internal turgor pressure has been estimated at 0.5 mPa (5 atm) for Gram-negative bacteria and as much as 3 MPa (30 atm) for Gram-positive bacteria.[1] The peptidoglycan layer must therefore be very strong and rigid.

During growth and cell division, changes in cell shape take place, hence the bacterial cell must possess a system for continuously breaking down and rebuilding the peptidoglycan layer. Conse-quently, there are a number of bacteriolytic enzymes which have been discovered which can break down peptidoglycan, which will be discussed in Section 3.10.2.6. The regulation of the peptidoglycan layer is therefore tightly coupled to cell division, cell growth, and the maintenance of cell shape.

Although the primary role of peptidoglycan is structural, in both Gram-positive and Gram-negative bacteria the peptidoglycan layer is associated with other cellular structures. The thick Gram-positive peptidoglycan layer is associated with teichoic acids, strongly anionic polyol phos-phates, which have been implicated in the sequestration of metal cations.[12] In Gram-negative bacteria, the peptidoglycan layer is associated with lipoproteins of the outer membrane.[1] Hence there are also nonstructural roles for the peptidoglycan layer.

3.10.1.2 Carbohydrate Structure of Peptidoglycan

Peptidoglycan is composed of three structural features: the glycan strands, the pentapeptide side chains, and the interstrand peptide cross-links. Figure 2 is a schematic representation of how these components are interlinked. The structures of these components can be determined most easily by enzymatic degradation of the isolated murein by bacteriolytic enzymes, followed by chromatographic analysis of the carbohydrate or peptide fragments.

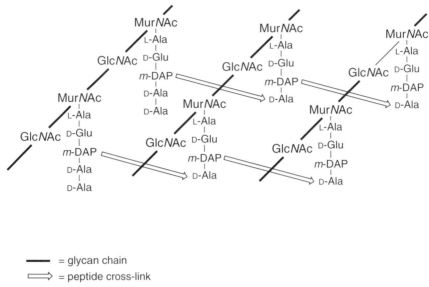

— = glycan chain

⟹ = peptide cross-link

Figure 2   Generalized structure of peptidoglycan.

The glycan component of all peptidoglycans consists of a β-1,4-linked chain of alternating N-acetylglucosamine and N-acetylmuramic acid residues. The monosaccharide D-N-acetylglucosamine (GlcNAc) is a common amino sugar which makes up the polysaccharide chitin found in plants and arthropods. N-Acetylmuramic acid (MurNAc) is found only in bacterial peptidoglycan: its structure is the same as that of GlcNAc except that the C-3 hydroxy group is modified with a lactyl ether appendage. The stereochemistry of the lactyl ether is that of D-lactic acid. The structures of GlcNAc (**1**) and MurNAc (**2**) are shown. The lactyl side chain of muramic acid is attached to the peptide side chain in native peptidoglycan; however, for the purposes of analysis, the peptide side chain can be cleaved using N-acetylmuramyl-L-alanine amidase.

(**1**) D-N-acetylglucosamine

(**2**) N-Acetylmuramic acid

The repeating β-1,4-linked structure shown in Figure 3 is common to all bacterial peptidoglycans examined, with only minor variations.[4] In some strains, e.g., *Staphylococcus aureus*, up to 50% of N-acetylmuramic acid residues contain an O-acetyl group at C-6.[13] In strains such as *Micrococcus lysodeikticus*, up to 40% of the N-acetylmuramic acid residues are not attached to peptide side chains.[14] In strains of *Mycobacteria* the N-acetyl group of MurNAc is hydroxylated, and is found as N-glycolylmuramic acid (**3**).[15] Finally, in the spore cortex of *Bacillus subtilis*, a δ-lactam form (**4**) of muramic acid has been found in which the lactyl ether carboxylate is attached via an amide bond to the 2-amino group of D-glucosamine.[16]

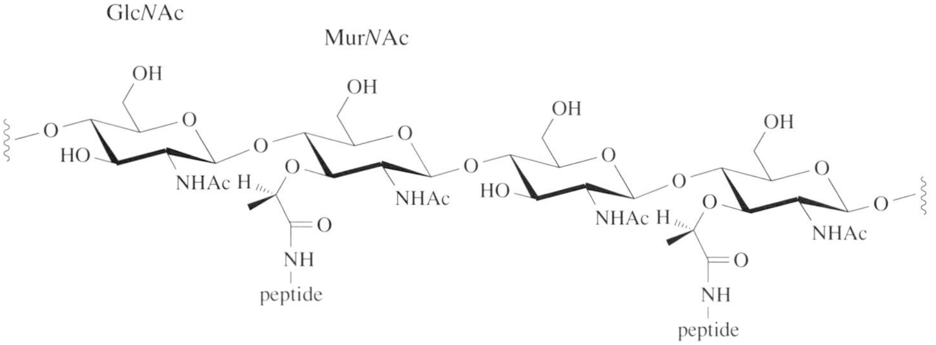

GlcNAc

MurNAc

Figure 3 Structure of glycan chain of peptidoglycan.

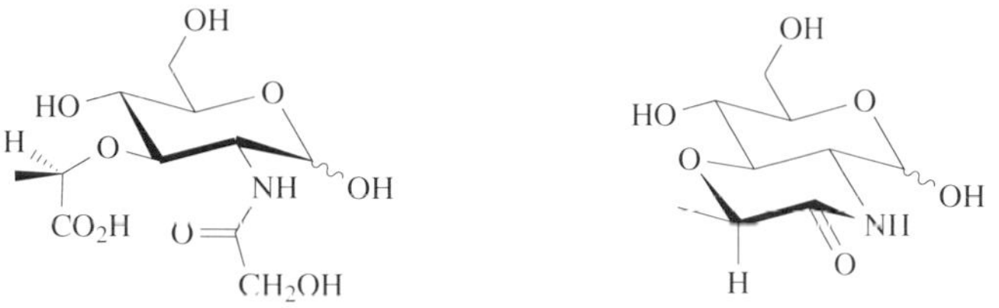

(3) (4)

The length of the glycan chain has been estimated in a number of cases.[4] The *S. aureus* glycan chains are between 12 and 16 disaccharide units in length,[13,17] whereas those of *Lactobacillus casei* and *Bacillus* are about 10 disaccharide units in length.[4] In a range of strains of *E. coli* with impaired murein metabolism, glycan chain lengths of 10–40 disaccharide units were measured, with a higher degree of cross-linking in those strains with shorter chain length.[18] Since the single murein sacculus must enclose the entire bacterial cell, the picture that emerges is therefore one of overlapping glycan strands linked together in an ordered array of chains.

3.10.1.3 The Pentapeptide Side Chain of Peptidoglycan

3.10.1.3.1 Structure of pentapeptide side chain

Attached to the lactyl ether appendage of *N*-acetylmuramic acid is a pentapeptide side chain, which contains D-amino acids that are unique to bacterial peptidoglycan. The common structure of the pentapeptide is L-Ala–γ-D-Glu–X–D-Ala–D-Ala, where X is an L-amino acid containing an amino side chain, commonly L-lysine or *meso*-diaminopimelic acid (*meso*-DAP (*m*-DAP)). The structure of the pentapeptide is shown in Figure 4.[4,5]

The pentapeptide is cross-linked through amino acid X to another peptide strand in mature peptidoglycan. This transpeptidation results in the formation of an amide bond between the amino group of X with the carbonyl group of position 4 of the second peptide chain, and the loss of the terminal D-alanine of the second peptide chain. When murein is treated with *N*-acetylmuramyl-L-alanine amidase, the peptide side chain is released, covalently attached to one or more other peptide chains, as shown in Figure 5. Thus, the pattern of muropeptide fragments obtained by enzymatic treatment of intact murein is quite complex, but can be resolved by HPLC analysis versus authentic standards.[19] In this way, the peptide structures of many bacteria have been analyzed, and certain variations in structure found, as explained below.

Several features of the peptide structure are worthy of note. The presence of D-amino acids is unique, and provides resistance to protease enzymes in the external medium. The presence of a γ-linked D-glutamic acid is also an unusual peptide structure, although γ-linked glutamyl units are also found in folate derivatives[20] and in some bacterial exopolymers.[21] *meso*-Diaminopimelate is also an unusual amino acid, although it is found as an intermediate in the biosynthesis of L-lysine in plants. The non-occurrence of D-amino acids and *meso*-diaminopimelic acid in mammals presents opportunities for the selective inhibition of peptidoglycan biosynthesis, as will be discussed in Section 3.10.3.

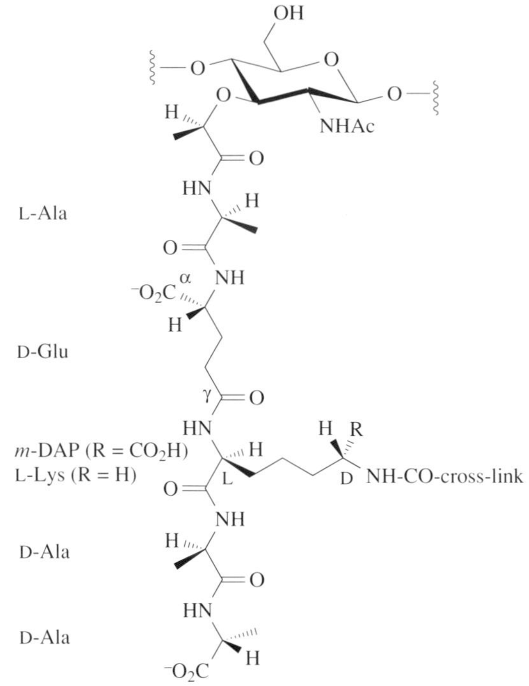

Figure 4   Structure of the pentapeptide side chain.

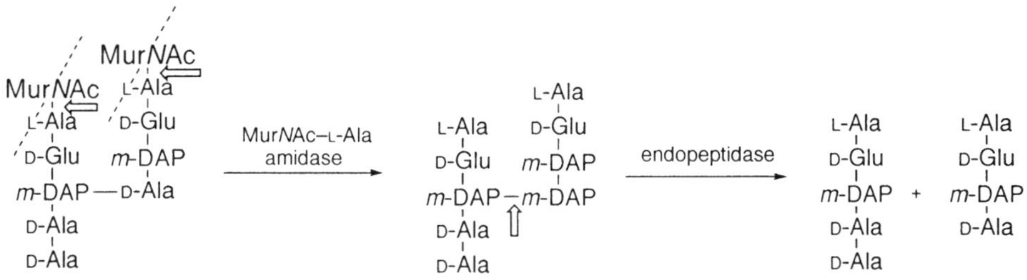

Figure 5   Release of muropeptides by treatment with *N*-acetylmuramyl-L-alanine amidase.

3.10.1.3.2   *Variations in the pentapeptide side chain*

Variations in position 1 are rare, with L-alanine being found in most bacterial peptidoglycans. The exceptions are *Corynebacteria*, which contain glycine at position 1 (and also remarkably contain L-homoserine at position 3), and *Butyribacterium rettgeri*, which contains L-serine at position 1 (and L-ornithine at position 3).[4,5]

D-Glutamic acid is found universally at position 2 of bacterial peptidoglycan, linked through the γ-carboxylate to the remainder of the peptide side chain. However, in a number of strains, including *Streptococcus*, *Enterococcus*, and *Lactobacillus*, the α-carboxylate is amidated (α-$CONH_2$).[4,5] In some strains of *Micrococcus* the α-carboxyl is linked to an additional amino acid, which can be either glycine or D-serine, and in some *Corynebacteria* a peptide cross-link is formed through position 2 (see Section 3.10.1.4.2).[4,5] Finally, *threo*-3-hydroxy-D-glutamic acid has been found in this position in *Microbacterium lacticum*.[22]

Most variation is found in position 3 of the pentapeptide side chain. *meso*-Diaminopimelic acid is found at position 3 in all Gram-negative bacteria (including *E. coli*) and in some strains of *Bacillus*, *Lactobacillus*, *Clostridium*, *Corynebacterium*, and *Propionibacterium*.[4,5] The symmetrical *meso* isomer of 2,6-diaminopimelic acid contains one L center and one D center, and it has been determined in these strains that the L center is found in the pentapeptide chain, with the D center in the side chain, as shown in Figure 6.[23,24] Most Gram-positive bacteria (including *Staphylococcus*, *Enterococcus*, *Lactobacillus*, and *Micrococcus*) contain L-lysine in position 3.[4,5]

A few strains have been found to contain other diamino acids at position 3, as illustrated in Figure 7. L,L-Diaminopimelic acid is found in many Actinomycetes; L-ornithine (containing a δ-

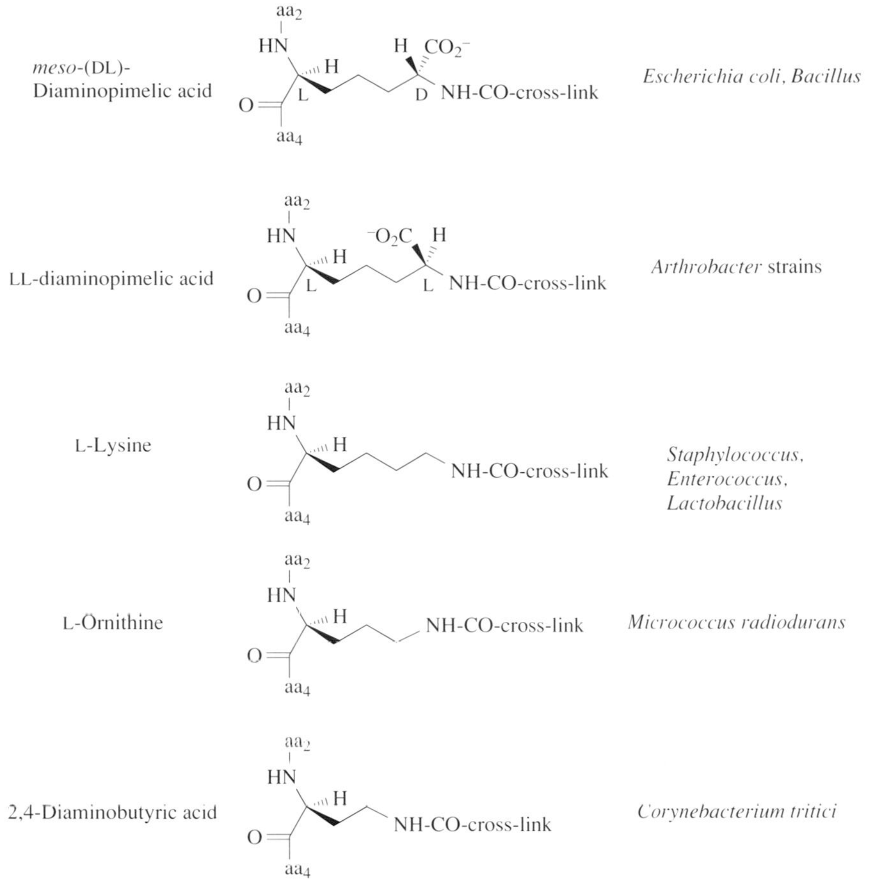

Figure 6 Diamino acids found at position 3 of pentapeptide.

amino group) is found in *Micrococcus radiodurans*, *Lactobacillus bifidus*, *Lactobacillus cellobiosus*, and *Treponema reiteri*; and 2,4-diaminobutyric acid (containing a γ-amino group) is found in *Corynebacterium tritici*.[4,5] In some strains the diamino acid is hydroxylated: hydroxylysine has been found as a minor constituent in *S. faecalis* and *S. pyogenes*, and 2,6-diamino-3-hydroxypimelic acid has been found in several Actinoplannaceae.[4,5] Lanthionine, an analogue of 2,6-diaminopimelic acid containing a sulfur atom in place of C-4, has been found in the peptidoglycan of *Fusobacterium nucleatum*.[25]

Positions 4 and 5 were thought to contain universally the D-Ala–D-Ala peptide. However, the incidence of bacterial resistance to the glycopeptide vancomycin, which is able to recognize specifically the *N*-acyl–D-Ala–D-Ala terminus, has led to the discovery that these vancomycin-resistant strains contain altered substituents at position 5, as shown in Figure 7. Strains of *Enterococcus* expressing high-level vancomycin resistance have been found to contain D-lactate in place of D-alanine at position 5,[26–28] and this hydroxy acid was subsequently found in strains of *Lactobacillus* which are constitutively resistant to vancomycin.[29] The peptidoglycan of a low-level resistant strain of *E. gallinarum* has been found to contain D-serine at position 5.[30] The altered biosynthetic route in these strains will be discussed in Section 3.10.2.7.

3.10.1.4 Interstrand Peptide Cross-links

Peptidoglycan strands are held together by a network of peptide cross-links formed between the pentapeptide side chains. The structures of these cross-links in different bacterial strains have been reviewed in detail.[4,5]

The level of cross-linking is largely responsible for the structural rigidity of the peptidoglycan layer. In Gram-negative bacteria such as *E. coli*, the level of cross-linking is in the range 25–50%, whereas in the multilayered wall of Gram-positive bacteria, the degree of cross-linking is 70–90%.[31]

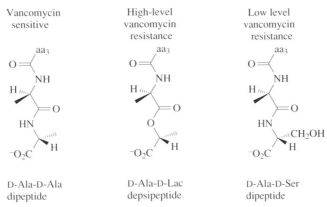

Figure 7 Substituents found at position 5 in vancomycin-resistant strains.

3.10.1.4.1 *Cross-links from* L-*lysine or* **meso-D***AP to* D-*alanine*

The simplest type of linkage is an amide bond formed directly between the ε-amino group of *meso*-DAP and the D-alanine residue at position 4 of a second peptide side chain. This type of linkage is found in *E. coli* and strains of *Bacillus*.[4,5] There are also a few examples of strains which contain a direct link between either L-lysine or L-ornithine and D-alanine.[5]

Bacteria that contain L-lysine at position 3 usually have an intervening amino acid or peptide chain between the ε-amino group of L-lysine and the D-alanine of the second chain. The composition of this peptide cross-link varies greatly between bacteria, and this structural variation has been used for the taxonomic classification of bacteria.[5] The most common structural type amongst Gram-positive bacteria is a peptide bridge composed of L-amino acids or glycine, of 1–5 residues in length.[4,5] The well-studied *S. aureus* Copenhagen strain contains five glycine residues in its peptide cross-link; however, other strains of *Staphylococcus* and *Micrococcus* contain L-alanine, L-serine, or L-threonine.[5] In *Enterococcus faecalis* and strains of *Lactobacillus*, there is an intervening residue of D-aspartate, linked to L-lysine through its β-carboxyl, which in many cases is amidated on its α-carboxyl.[4,5] These cross-links are illustrated in Figure 8.

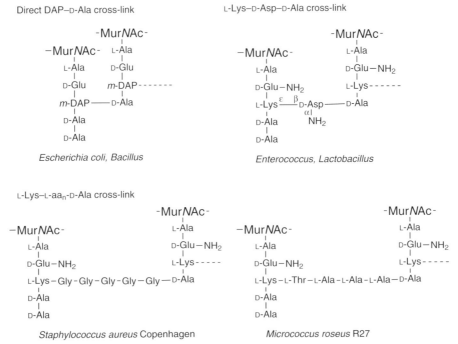

Figure 8 Peptide cross-links through position 3.

3.10.1.4.2 *Other types of interstrand cross-links*

A few Micrococcaceae contain cross-links between L-alanine at position 4 and the α-amino group of L-alanine at position 1 of a second pentapeptide chain, which has become detached from the glycan strand.[4,32] Several such units form a large peptide cross-bridge which eventually terminates at the L-lysine ε-amino group of a glycan-bound chain, as shown in Figure 9.

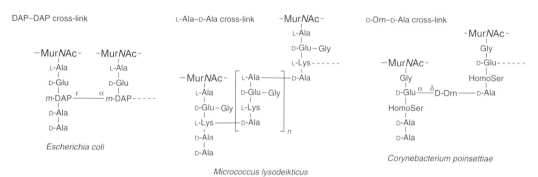

Figure 9 Other types of peptide cross-links.

Several strains of *Corynebacterium* contain L-homoserine in place of a diamino acid at position 3. In these strains a D-ornithine residue is usually attached (via its δ-amino group) to the α position of D-glutamic acid at position 2. The α-amino group of this D-ornithine bridge forms a cross-link with D-alanine at position 4 of a second peptide chain.[4,33] Other coryneform bacteria contain D-lysine or L-lysine in place of D-ornithine.[5]

Finally, the peptidoglycan of *E. coli* has been found to contain linkages between two *meso*-DAP residues of adjacent strands.[31] These linkages were found to consist of an amide bond between the D side chain of *meso*-DAP and the L center on the peptide chain of the second strand. Formation of the DAP–DAP linkage results in the loss of the D-Ala D-Ala dipeptide of the second strand, as shown in Figure 9. These DAP DAP linkages are present at the level of 5 10% of the total linkages in *E. coli* peptidoglycan.[31]

3.10.2 BIOSYNTHESIS OF PEPTIDOGLYCAN

The biosynthesis of bacterial peptidoglycan occurs in several stages, in different parts of the cell. The first stage is the production of the nucleotide-linked monosaccharides for glycan formation. The second stage is the assembly of the uridine-5′-diphospho (UDP)-MurNAc-pentapeptide cytoplasmic precursor. The third stage is the translocation of the phospho-MurNAc-pentapeptide unit across the cytoplasmic membrane by means of an undecaprenyl phosphate lipid carrier. Finally, the mature peptidoglycan is assembled on the cell surface via transglycosylation and transpeptidation reactions. Each of these stages will be reviewed in turn, starting with the cytoplasmic steps shown in Figure 10. There will then be a discussion of enzymes involved in the cellular processing and recycling of peptidoglycan, and a discussion of modifications to peptidoglycan biosynthesis found in antibiotic-resistant bacteria. The intracellular steps of peptidoglycan biosynthesis have been reviewed previously.[34]

3.10.2.1 Biosynthesis of Nucleotide-linked Monosaccharides

3.10.2.1.1 *Biosynthesis of UDP-N-acetylglucosamine*

The activated monosaccharide uridine-5′-diphospho-*N*-acetyl-D-glucosamine (UDPGlcNAc) is used for the assembly of both peptidoglycan and the lipid A component of lipopolysaccharide in bacteria, and is also used for the assembly of chitin in fungi. The assembly of UDPGlcNAc in *E. coli* proceeds from the central metabolite fructose-6-phosphate via the pathway shown in Scheme 1.

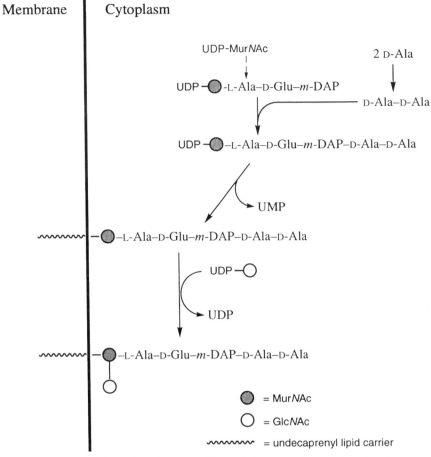

Figure 10 Cytoplasmic steps of bacterial peptidoglycan biosynthesis.

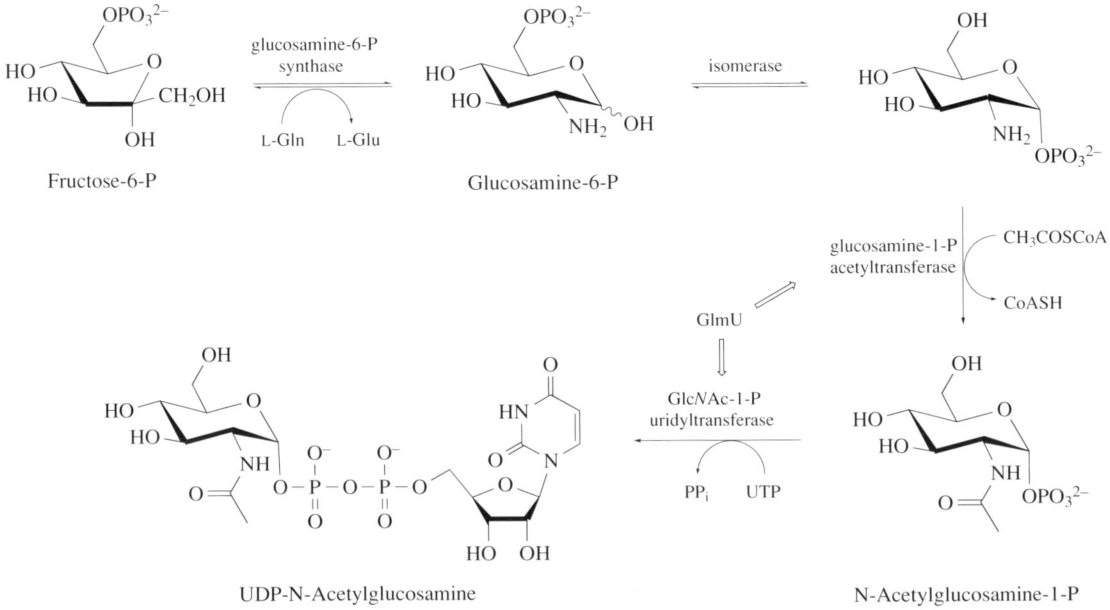

Membrane   Cytoplasm

UDP-MurNAc

2 D-Ala

UDP—⬤-L-Ala–D-Glu–*m*-DAP

D-Ala–D-Ala

UDP—⬤–L-Ala–D-Glu–*m*-DAP–D-Ala–D-Ala

UMP

⬤–L-Ala–D-Glu–*m*-DAP–D-Ala–D-Ala

UDP—◯

UDP

⬤–L-Ala–D-Glu–*m*-DAP–D-Ala–D-Ala

⬤ = MurNAc

◯ = GlcNAc

〜〜〜 = undecaprenyl lipid carrier

Fructose-6-P

glucosamine-6-P
synthase

L-Gln   L-Glu

Glucosamine-6-P

isomerase

glucosamine-1-P
acetyltransferase

CH₃COSCoA

CoASH

GlmU

GlcNAc-1-P
uridyltransferase

PPᵢ   UTP

UDP-N-Acetylglucosamine

N-Acetylglucosamine-1-P

Scheme 1

Fructose-6-phosphate is converted into glucosamine-6-phosphate by the enzyme glucosamine-6-phosphate synthase, which has been overexpressed and purified from *E. coli*.[35] The enzyme requires glutamine as a source of ammonia, and the purified enzyme shows no activity with ammonia itself.[35] The enzyme was found to catalyze the exchange of the C-1 *proR* hydrogen with solvent, and to transfer ~1% of tritium from this position to the C-2 position. Hence a mechanism has been proposed involving the formation of a C-2 imine with ammonia, derived from glutamine, and abstraction of the C-1 *proR* hydrogen, as shown in Scheme 2.[36] Both domains of the *E. coli* enzyme have been crystallized for X-ray crystallographic studies.[37]

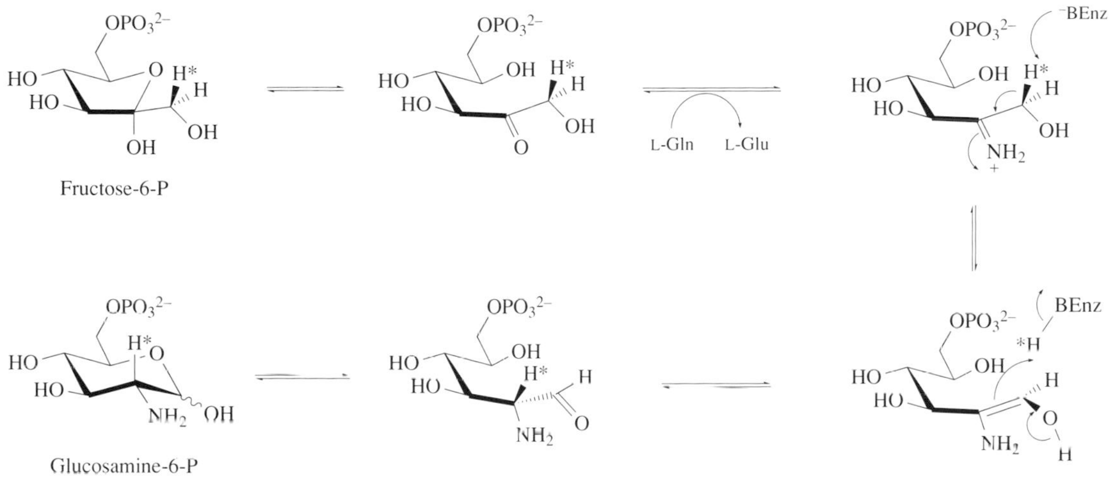

Scheme 2

Glucosamine-6-phosphate is then converted by an isomerase enzyme into glucosamine-1-phosphate, which is converted by a bifunctional enzyme into UDP-*N*-acetylglucosamine. The gene encoding the enzyme responsible for UDPGlc*N*Ac synthesis in *E. coli* was identified as the *glmU* gene, which has been cloned and overexpressed.[38] Using the overexpressed gene construct, the glucosamine-1-phosphate acetyltransferase and Glc*N*Ac-1 phosphate uridyltransferase activities were found to copurify, implying that the GlmU gene product is a bifunctional enzyme.[39]

The amino acid sequence of the 49 kDa GlmU protein contains two distinct domains: the *N*-terminal domain (residues 22–119) shows sequence similarity to several nucleotidyl transferases (pyrophosphorylases) which transfer nucleotide phosphate groups from nucleotide triphosphates to monosaccharide-1-phosphates; the *C*-terminal domain (residues 260–450) shows sequence similarity to a number of acyltransferases which utilize acetyl-CoA.[40] An *N*-terminal proteolytic fragment of GlmU was isolated containing the *N*-terminal 331 amino acid residues. This fragment was found to be catalytically active for uridyl transfer to *N*-acetylglucosamine-1-phosphate, but showed no acetyl transfer activity towards glucosamine-1-phosphate. Subsequently, a fusion protein was constructed which overexpressed the *C*-terminal protein domain, and was found to be catalytically active for transfer of an acetyl group from acetyl-CoA to glucosamine-1-phosphate. Analysis of the reaction kinetics of the bifunctional GlmU protein indicated a pre-steady-state lag in the production of UDP-Glc*N*Ac due to the accumulation of steady-state levels of the intermediate *N*-acetyl-glucosamine-1-phosphate.[40]

Thus, GlmU contains two distinct sites which catalyze first the acetyl transfer from acetyl-CoA to glucosamine-1-phosphate, then the uridyl transfer from UTP to *N*-acetylglucosamine-1-phosphate, as shown in Scheme 3. *In vivo* the equilibrium for this reaction is increased in favor of UDPGlc*N*Ac formation by the enzyme inorganic pyrophosphatase, which catalyzes the hydrolysis of pyrophosphate to inorganic phosphate. Improved methods for the chemical and enzymatic synthesis of UDPGlc*N*Ac have been reported.[41]

3.10.2.1.2 Biosynthesis of UDP-N-acetylmuramic acid

The first committed step of peptidoglycan biosynthesis is the transfer of a 1-carboxyethenyl group from phosphoenolpyruvate (PEP) to UDP-*N*-acetylglucosamine, to give 3-enolpyruvyl-1-UDP-*N*-

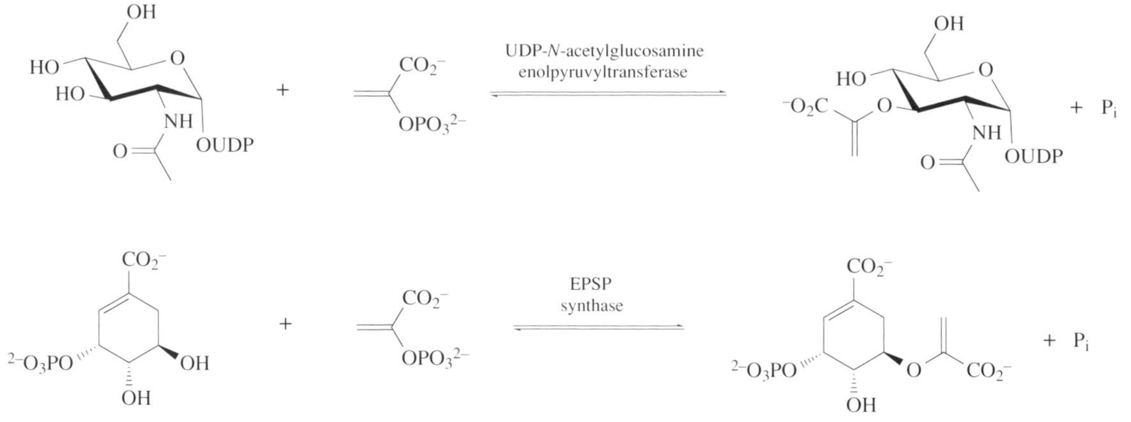

OH

HO
HO
NH
O
OUDP

+

CO_2^-
OPO_3^{2-}

UDP-*N*-acetylglucosamine
enolpyruvyltransferase

OH

HO
^-O_2C
O
NH
O
OUDP

+ P_i

CO_2^-

$^{2-}O_3PO$ OH
OH

+

CO_2^-
OPO_3^{2-}

EPSP
synthase

CO_2^-

$^{2-}O_3PO$
OH
O CO_2^-

+ P_i

Scheme 3

acetylglucosamine, as shown in Scheme 3. This mechanistically unusual transformation is catalyzed by UDPGlc*N*Ac enolpyruvyl transferase, whose corresponding gene has been cloned from *E. coli*,[42] and from *Enterobacter cloacae*.[43] The amino acid sequence of the encoded *E. coli* MurA enzyme was found to share 16–18% sequence identity with amino acid sequences of the enzyme 5-enolpyruvyl-shikimate-3-phosphate (EPSP) synthase, which carries out an analogous reaction on the shikimate pathway, as shown in Scheme 3.

The sequence similarity suggested that these two enzymes might follow mechanistically similar pathways. EPSP synthase has previously been shown, via pre-steady-state kinetics and rapid quench studies, to proceed via a two-step addition–elimination mechanism involving a tetrahedral adduct.[44] Rapid quench studies on the *E. coli* MurA enzyme with the natural substrates UDPGlc*N*Ac and PEP led to the isolation and characterization of such a tetrahedral adduct (**5**).[45] However, studies with the *E. cloacae* enzyme revealed the existence of a covalent adduct formed between PEP and cysteine-115, in the form of a thiohemiketal intermediate (**6**), which was kinetically competent as a reaction intermediate. Cysteine-115 was found to be the site of alkylation of fosfomycin, an antibiotic inhibitor for this enzyme (see Section 3.10.3.1), and mutation of Cys-115 to serine was found to give an inactive mutant protein.[46]

The apparent contradiction between these two lines of evidence was resolved by further site-directed mutagenesis experiments. Sequencing of the *murA* gene from fosfomycin-resistant *Mycobacterium tuberculosis* revealed the presence of an aspartate residue at position 115, and construction of the Cys115Asp mutant of the *E. coli* enzyme gave a highly active enzyme which was resistant to fosfomycin.[47] Thus, cysteine-115 is believed to form a thiohemiketal adduct, which then dissociates prior to reaction with UDPGlc*N*Ac to form the tetrahedral intermediate. Cysteine-115 is believed to act as an active site general acid for protonation of the PEP enol ether, since its function can be fulfilled by replacement with aspartic acid but not serine. The proposed mechanism is shown in Scheme 4. Further evidence for the covalent enzyme adduct has been obtained by pre-steady-state kinetic analysis[48] and by NMR spectroscopic studies.[49]

The enzyme has also been found to process *E* and *Z* isomers of fluoro-PEP to give stable tetrahedral intermediates,[50] allowing a stereochemical analysis of the fluorinated tetrahedral intermediate,[51] a shown in Scheme 5. Processing of the same fluoro-PEP substrates by EPSP synthase gave the tetrahedral intermediate with identical stereochemistry, confirming the similarity in the latter stages of both reaction mechanisms.[52] The enzyme also processes *E* and *Z* isomers of phosphoenolbutyrate stereospecifically.[53]

The conversion of enolpyruvyl-UDP-*N*-acetylglucosamine into UDP-*N*-acetylmuramic acid is catalyzed by an NADPH-dependent reductase enzyme. The *murB* gene encoding this enzyme in *E. coli* has been cloned,[54] allowing overexpression of the enzyme for mechanistic studies. The purified 38 kDa protein contains a stoichiometric amount of tightly bound FAD which is reducible during catalytic turnover.[55] The enzyme catalyzes the enantioselective transfer of the C-4 *proS* hydrogen of NADPH on to flavin, whence the same hydrogen is transferred to the β-position of the substrate. Enzymatic substrate conversions in the presence of ^2H_2O revealed that deuterium was inserted into the α-position.[55] Thus a mechanism involving hydride transfer from flavin to the unsaturated substrate has been proposed, as shown in Scheme 6.

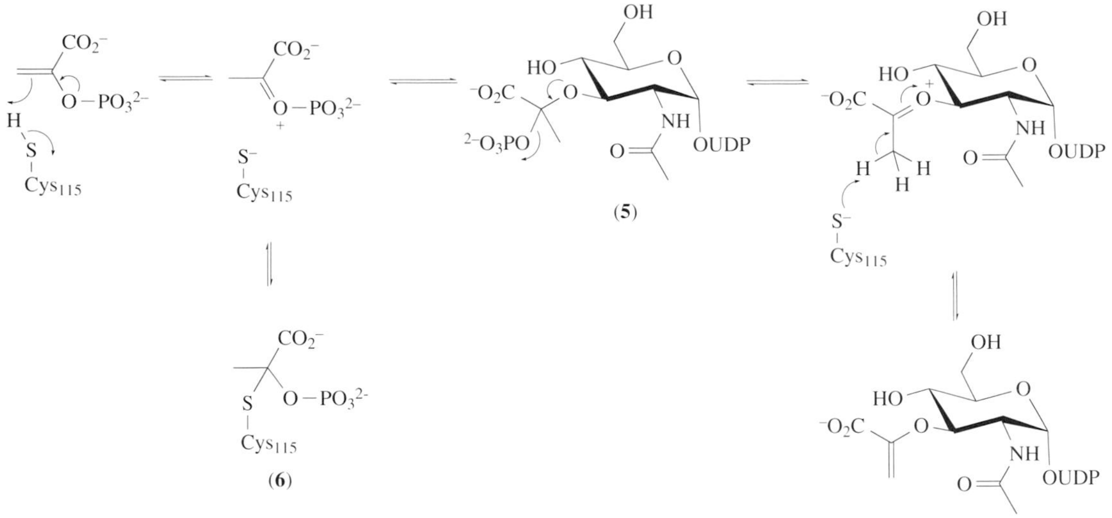

Scheme 4

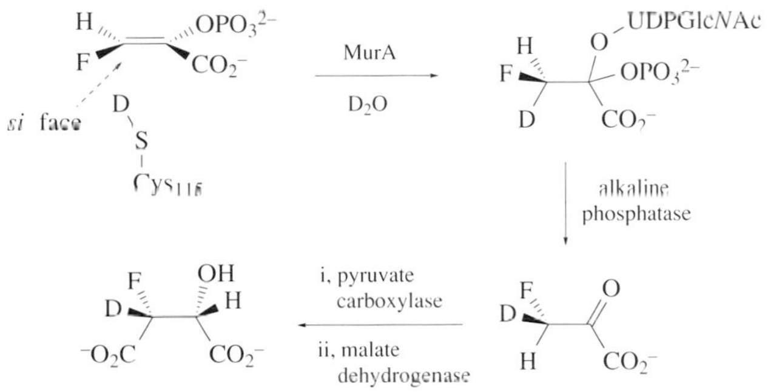

Scheme 5

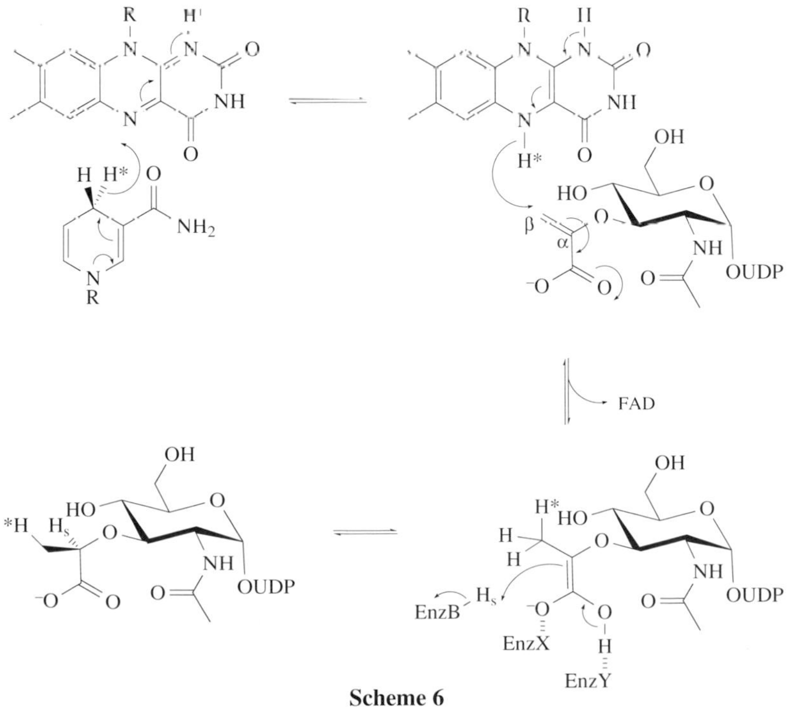

Scheme 6

Enzymatic processing of (*E*)-enolbutyryl-UDP-*N*-acetylglucosamine gave the reduced product, but isomerization to the *Z* isomer was also observed.[56] This implies that the reaction is reversible, and that the carbanion intermediate is sufficiently long-lived to accommodate bond rotation prior to reverse reaction. These results might also be explained by a radical mechanism involving reversible hydrogen atom transfer from flavin, followed by single-electron transfer from a transient flavin semiquinone. Analysis of the steady-state kinetics of the enzymatic reaction have revealed that a ping-pong kinetic mechanism is followed.[57]

The X-ray crystal structure of enolpyruvyl-UDPGlc*N*Ac reductase from *E. coli* has been determined at 2.7 Å resolution,[58] and was subsequently refined at 1.8 Å resolution.[59] The structure revealed that C-3 of the alkene double bond of the substrate was positioned in close proximity to N-5 of bound FAD, consistent with the hydride transfer mechanism proposed.[58] C-2 of the alkene double bond was positioned 3.1 Å away from the hydroxy side chain of Ser-229, suggesting that the latter acts as a proton donor to quench the developing carbanion at C-2. A Ser229Ala mutant enzyme was produced by site-directed mutagenesis, and was found to be catalytically inactive for substrate reduction.[59] The overall structure of the mutant was the same as for the wild-type enzyme, although the mutation caused a reorganization of the hydrogen bond network at the active site.[59]

3.10.2.2   Biosynthesis of D-Amino Acids in Peptidoglycan

One of the most distinctive features of the molecular structure of peptidoglycan is the presence in the pentapeptide side chain of D-amino acids, which are not normally found in higher organisms, except in certain peptide natural products.[60] D-Amino acids are biosynthesized in bacteria either by enzymatic racemization of an L-amino acid[61] or by reductive amination of an α-keto acid by PLP-dependent transaminases or ammonia-linked dehydrogenases. Amino acid racemase enzymes can be further divided into two classes: those that utilize the coenzyme pyridoxal 5′-phosphate (PLP) and those that have no cofactor.[62] Examples of both classes are used in the production of D-alanine, D-glutamic acid, and *meso*-diaminopimelic acid for bacterial peptidoglycan biosynthesis.

3.10.2.2.1   Biosynthesis of D-alanine

D-Alanine is generated in prokaryotes by enzymatic racemization of L-alanine, catalyzed by PLP-dependent alanine racemase. Alanine racemase has been purified from the Gram-negative bacteria *Pseudomonas striata*[63] and *Salmonella typhimurium*;[64,65] and the Gram-positive bacteria *Streptococcus faecalis*,[66] *Staphylococcus aureus*,[67] and *Bacillus stearothermophilus*;[68] the last enzyme has also been crystallized.[69] Two alanine racemase genes have been found in *S. typhimurium*: a constitutively expressed gene *alr* and an inducible gene *dadB*.[70] the *dadB* gene is coexpressed with a gene encoding a D-alanine dehydrogenase in response to the presence of L-alanine in growth media, apparently as a catabolic pathway for utilization of L-alanine for growth.[70] This *dad* operon has also been identified in *E. coli*.[71]

All known alanine racemases utilize pyridoxal 5′-phosphate as a cofactor, which forms an imine linkage with the α-amino group of alanine, hence increasing the acidity of the α-proton. Deprotonation of either alanine enantiomer could proceed via a two-base mechanism, with one base positioned on either side of the active site, or via a single active site base which is able to access both faces of the PLP aldimine.[72] Single turnover experiments with the *P. striata* racemase acting on α-deuterated substrates have shown 0.74–10% internal return of the deuteron into the opposite enantiomer, favoring a one-base mechanism for this enzyme, as shown in Scheme 7.[73] Detailed isotope exchange studies on the *B. stearothermophilus* and *S. typhimurium* racemases have shed light on the kinetic and energetic profiles of these enzymatic reactions, but have not resolved the one-base/two-base issue.[72] Alanine racemase has been the subject of numerous inhibition studies, which will be discussed in Section 3.10.3.2.

The structure of alanine racemase from *B. stearothermophilus* has been solved at 1.9 Å resolution.[74] The tertiary structure of the monomer consists of two domains: an eight-stranded α/β-barrel containing the pyridoxal phosphate cofactor, and a β-stranded C-terminal domain. The cofactor is covalently attached to Lys-39, which upon substrate binding may act as a base for deprotonation of one face of the substrate–PLP adduct. Several active site residues have been identified as binding

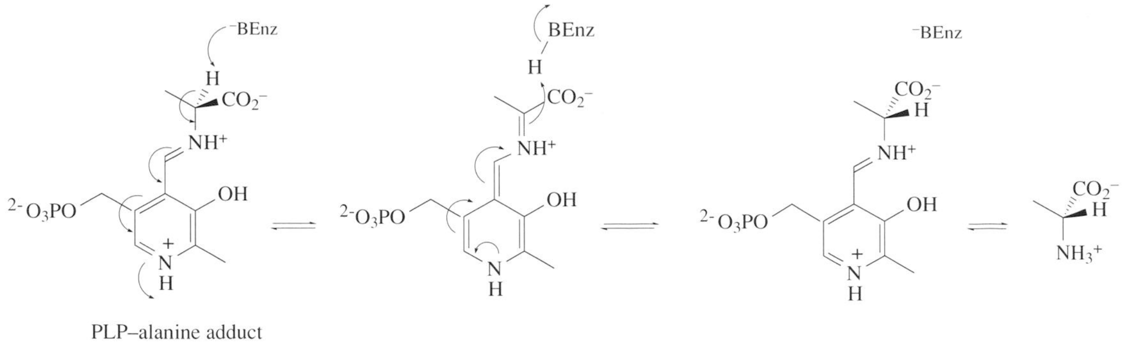

PLP–alanine adduct

Scheme 7

the PLP cofactor, of which Tyr-265F from the second monomer is suitably positioned to act as a second base in a two-base mechanism.[74]

3.10.2.2.2 *Biosynthesis of D-glutamic acid*

Two distinct biosynthetic routes to D-glutamic acid have been found in different bacterial strains: enzymatic racemization of L-glutamic acid by a cofactor-independent racemase and reductive amination of α-ketoglutaric acid by a pyridoxal 5′-phosphate-dependent D-amino acid transaminase.

Glutamate racemase is found in many lactic acid bacteria, where it is believed to supply the D-glutamic acid required for peptidoglycan biosynthesis.[75] The gene encoding glutamate racemase has been cloned from *Pediococcus pentosaceus* and the encoded enzyme purified to homogeneity. The purified 40 kDa enzyme contains no cofactors, shows a strict substrate specificity for glutamic acid, and is inactivated by cysteine-directed reagents.[76] Incubation of one enantiomer of glutamic acid with enzyme in ^3H_2O led to the incorporation of the ^3H label in the opposite enantiomer under single-turnover conditions,[77] consistent with a two-base mechanism, as found previously for proline racemase.[78]

Further mechanistic experiments have been carried out on glutamate racemase from *Lactobacillus fermenti*.[79–81] The enzyme has been overexpressed and was also found to contain no cofactors.[79] Single-turnover ^3H_2O incorporation experiments revealed the incorporation of ^3H label in the product enantiomer, with no incorporation of ^3H label in the starting enantiomer, indicating a two-base mechanism involving monoprotic bases.[80] Observation of the racemization reaction in ^2H_2O by circular dichroism spectroscopy revealed substantial "overshoots," due to the slower processing of the deuteriated product enantiomer. Using this technique, substantial primary kinetic isotope effects were measured for each reaction direction ($^HV_{max}/^DV_{max} = 2.2$ for (*S*)-glutamic acid and 3.1 for (*R*)-glutamic acid).[81]

Two conserved cysteine residues Cys-73 and Cys-184 were investigated as likely active site bases using site-directed mutagenesis. Both C73A and C184A enzymes were inactive as racemases, but were found to catalyze the elimination of HCl from opposite enantiomers of *threo*-3-chloroglutamic acid, a reaction which requires only a single base.[81] These studies indicate that cysteine-73 abstracts the C-2 hydrogen from (*R*)-glutamic acid and cysteine-184 abstracts the C-2 hydrogen from (*S*)-glutamic acid, as shown in Scheme 8.

An essential *murI* gene has been identified at minute 90 of the *E. coli* chromosome which is required for the biosynthesis of D-glutamic acid.[82] The *murI* gene was found to encode a glutamate racemase activity,[83,84] which was overexpressed and purified to homogeneity.[85] The 30 kDa enzyme was found to contain no cofactors, and was found to share 30% amino acid sequence similarity with the *Lactobacillus* enzyme. However, unlike the *Lactobacillus* enzyme, the *E. coli* enzyme was found to be dependent upon the presence of UDPMur*N*Ac–L-Ala for activity. No activity was detected in the absence of UDPMur*N*Ac–L-Ala, for which the enzyme showed a high affinity ($K_d = 4$ μmol L^{-1}) and high specificity. This remarkable enzyme activation appears to be the physiological mechanism by which *E. coli* regulates the amount of D-glutamic acid generated intracellularly, and thus avoids excessive racemization of the intracellular pool of L-glutamic acid.[85] Isotope exchange experiments on the *E. coli* enzyme indicate that it also follows a two-base catalytic mechanism

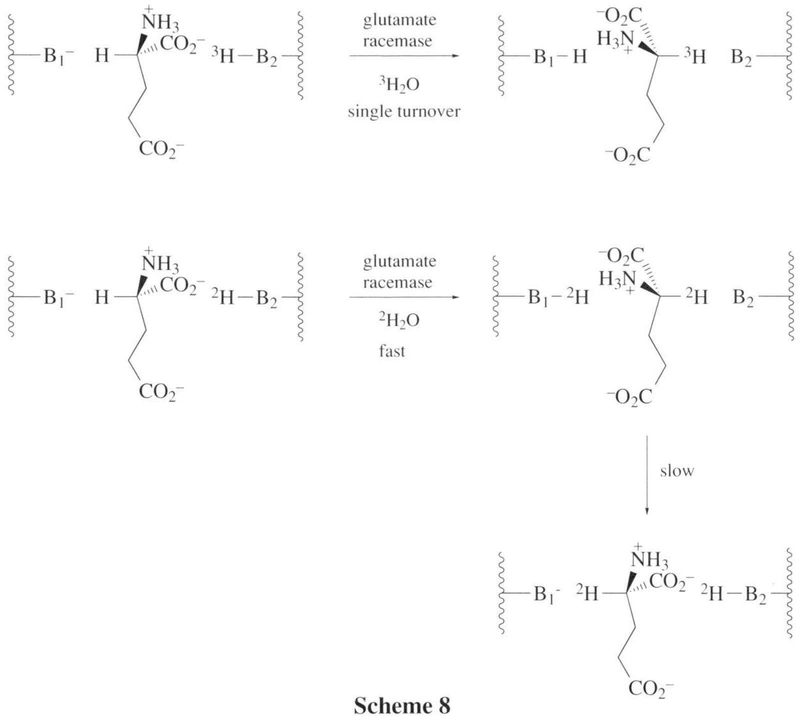

Scheme 8

involving the two cysteine residues implicated in the *Lactobacillus* enzyme, which are conserved in the sequence of the *E. coli* enzyme.[86]

In *Bacillus* D-glutamic acid is generated from α-ketoglutaric acid by the PLP-dependent enzyme D-amino acid aminotransferase.[62,87] This enzyme has been purified from *Bacillus subtilis*,[88] from *Bacillus sphaericus*,[89] and from a thermophilic strain *Bacillus* YM-1.[90] Each enzyme was found to contain 1 mol equiv. of PLP, which acts as a cofactor for transamination.[87] Cloning and sequencing of the gene encoding the *Bacillus* YM-1 enzyme revealed that this enzyme shared 31% sequence identity with *E. coli* branched-chain L-amino acid transaminase, suggesting that a change in stereo-specificity has occurred at some point during the evolution of this enzyme.[91]

PLP-dependent transaminases catalyze the 1,3-shift of a proton from C-4′ of the coenzyme–imine (external aldimine) adduct to the α-position of the amino acid, as shown in Figure 11. Isotope labeling experiments have established that in both the *Bacillus* YM-1 D-amino acid aminotransferase and the *E. coli* branched-chain L-amino acid aminotransferase the C-4′ *proR* hydrogen is transferred on the *re* face, unlike all other transaminases, which utilize the *si* face.[92]

The X-ray crystal structure of the *Bacillus* YM-1 enzyme has been solved to 1.9 Å resolution.[93] The enzyme exists as a dimer of total molecular weight of 65 kDa. The tertiary structure of each monomer is novel, and consists of two discrete domains. The smaller *N*-terminal domain consists of a four-stranded antiparallel β-sheet, whereas the larger *C*-terminal domain consists of two mixed β-sheets. The PMP cofactor is bound at the interface of the two domains, which form a hydrophobic surface. The catalytic lysine-145 extends toward the coenzyme on the *re* face, in the correct alignment to catalyze *pro-R* proton transfer at C-4′.[93]

It is not clear how widely distributed the glutamate racemase and D-amino acid aminotransferase activities are in the bacterial kingdom. However, it has been found that strains such as *S. hemolyticus* possess both enzyme activities.[94]

3.10.2.2.3   *Biosynthesis of* meso-*diaminopimelic acid*

meso-Diaminopimelic acid is found only in plants and microorganisms, which utilize it as an intermediate for L-lysine biosynthesis. The peptidoglycan structure of many bacteria, especially Gram-negative bacteria, contains *meso*-diaminopimelic acid in position 3 of the pentapeptide side chain.[4,5] Mycobacteria have also been found to contain *meso*-diaminopimelate in their peptidoglycan, and the genetic locus for *meso*-DAP biosynthesis has been determined.[95]

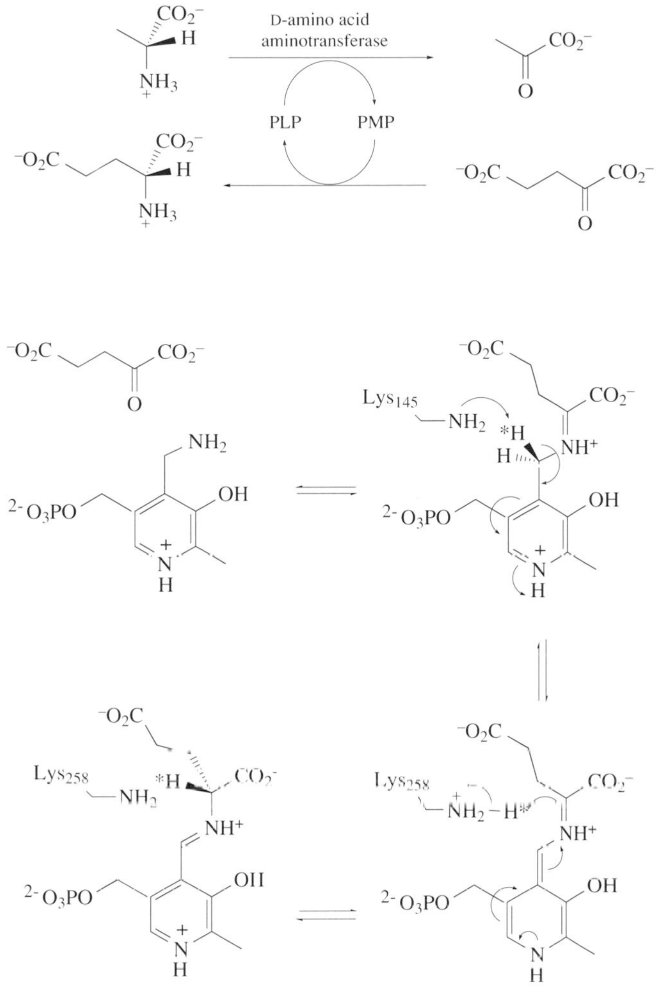

Figure 11 Reactions catalyzed by D-amino acid aminotransferase.

The biosynthetic pathways to *meso*-diaminopimelic acid have been reviewed.[96] In prokaryotes, the major pathway involves the enzymatic epimerization of L,L-diaminopimelate by diaminopimelate epimerase, as shown in Scheme 9. This enzyme has been purified from *E. coli*, and was found to be 34 kDa monomeric protein requiring no cofactors for activity.[97] Single-turnover tritium incorporation experiments have indicated that this enzyme, like glutamate racemase, follows a two-base concerted mechanism.[97] The *dapF* gene encoding this enzyme has been cloned from *E. coli*[98] and the enzyme has been overexpressed.[99] Surprisingly, a *dapF⁻* strain of *E. coli* was found to be viable without supplementation of *meso*-DAP.[99] It is not clear whether this is due to the presence of a second gene for *meso*-DAP biosynthesis or whether the *dapF⁻* strain can incorporate L,L-DAP in place of D,L-DAP.[100]

3.10.2.2.4 *Biosynthesis of other D-amino acids for peptidoglycan biosynthesis*

A wide range of amino acid racemases and epimerases have been identified from bacterial sources.[62] In the context of peptidoglycan biosynthesis, the only other well-characterized amino acid racemase is aspartate racemase, which has been partially purified from *Streptococcus faecalis*[101] and purified to homogeneity from *Streptococcus thermophilus*.[102] The latter enzyme was found to contain no cofactors, but was strongly inhibited by thiol reagents. Racemization in the presence of ^3H_2O led to the preferential incorporation of ^3H into the product enantiomer; hence this enzyme also appears to follow a two-base concerted mechanism similar to glutamate racemase.[102]

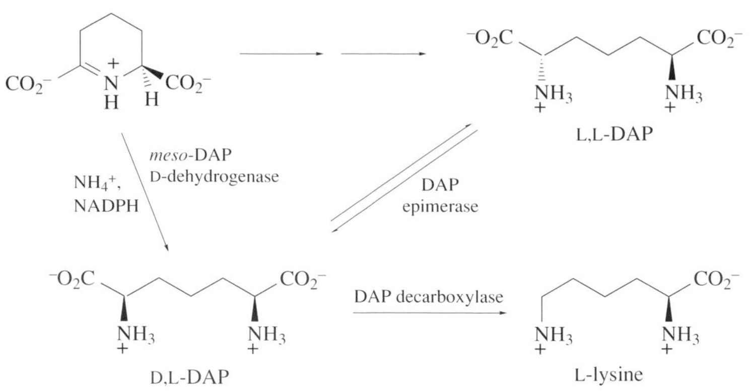

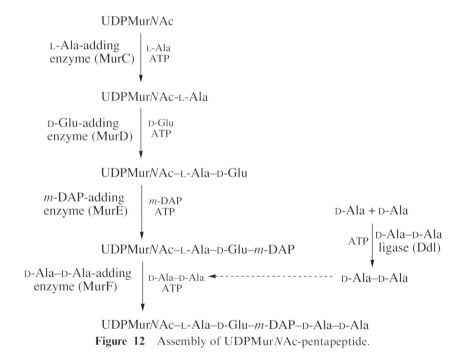

Scheme 9

3.10.2.3   Assembly of UDPMur*N*Ac-pentapeptide

Assembly of the cytoplasmic peptidoglycan precursor UDPMur*N*Ac–L-Ala–D-Glu–*m*-DAP–D-Ala–D-Ala is accomplished by means of a series of ATP-dependent amino acid ligases that add amino acids (hence their usual names are L-Ala adding enzyme, etc.) sequentially on to the lactyl side chain of UDP-*N*-acetylmuramic acid. L-Alanine, D-glutamic acid, and *meso*-diaminopimelic acid are added on by their respective ligases to give UDPMur*N*Ac-tripeptide, then the final two amino acids are added as a D-Ala–D-Ala dipeptide, synthesized by D-Ala–D-Ala ligase (Figure 12).

UDPMur*N*Ac

L-Ala-adding | L-Ala
enzyme (MurC) | ATP

UDPMur*N*Ac-L-Ala

D-Glu-adding | D-Glu
enzyme (MurD) | ATP

UDPMur*N*Ac–L-Ala–D-Glu

m-DAP-adding | *m*-DAP   D-Ala + D-Ala
enzyme (MurE) | ATP

                                        ATP | D-Ala–D-Ala
                                             | ligase (Ddl)

UDPMur*N*Ac–L-Ala–D-Glu–*m*-DAP

D-Ala–D-Ala-adding | D-Ala–D-Ala ◄------------------------- D-Ala–D-Ala
enzyme (MurF) | ATP

UDPMur*N*Ac–L-Ala–D-Glu–*m*-DAP–D-Ala–D-Ala

Figure 12   Assembly of UDPMur*N*Ac-pentapeptide.

3.10.2.3.1   *Assembly of UDPMur*N*Ac-tripeptide*

A cluster of genes involved in cell division and cell wall biosynthesis are situated at minute 2 of the *E. coli* chromosome (Figure 13).[103] Each of the genes in this region has been identified, including all of the *mur* genes encoding the amino acid adding enzymes responsible for assembly of UDPMur-*N*Ac-pentapeptide.[104,105] The nucleotide sequences for the *murC*,[106] *murD*,[107,108] and *murE*[109,110] genes encoding the L-Ala-, D-Glu-, and *meso*-DAP-adding enzymes, respectively, have been determined.

The deduced amino acid sequences show 10–20% sequence identity with one another and with the sequence of MurF, the D-Ala–D-Ala-adding enzyme, suggesting that these enzymes of similar function may be evolutionarily related, and may follow similar mechanisms.[34] The UDPMur*N*Ac–peptide substrates for the respective amino acid-adding enzymes are not commercially available but can be isolated in small quantities from antibiotic-treated bacterial cells, and are separable by HPLC.[111]

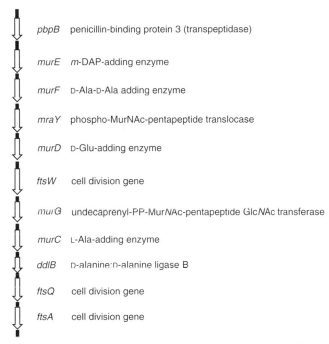

pbpB	penicillin-binding protein 3 (transpeptidase)
murE	*m*-DAP-adding enzyme
murF	D-Ala-D-Ala adding enzyme
mraY	phospho-MurNAc-pentapeptide translocase
murD	D-Glu-adding enzyme
ftsW	cell division gene
murG	undecaprenyl-PP-Mur*N*Ac-pentapeptide Glc*N*Ac transferase
murC	L-Ala-adding enzyme
ddlB	D-alanine:D-alanine ligase B
ftsQ	cell division gene
ftsA	cell division gene

Figure 13 Organization of genes involved in peptidoglycan biosynthesis at minute 2 of the *E. coli* chromosome.

The L-Ala-adding enzyme (or UDPMur*N*Ac: L-alanine ligase) has been purified from *S. aureus*[112] and from *E. coli*,[113] the latter enzyme being overexpressed to high levels.[113] The 50 kDa *E. coli* enzyme shows high specificity for L-alanine, although glycine and L-serine are processed with high K_m values.[113,114] The enzyme shows a 20-fold higher V_{max} for UDPMur*N*Ac rather than the unsaturated precursor 3-enolpyruvyl-UDP-Glc*N*Ac.[113] Enzymatic conversion of UDPMur*N*Ac labeled with ^{18}O in the lactyl carboxylate by the *E. coli* enzyme yielded samples of inorganic phosphate containing one atom of ^{18}O, consistent with the existence of an acyl phosphate intermediate in the reaction mechanism.[115] The purified enzyme was found to exist as a mixture of monomeric and dimeric forms, both of which were catalytically active.[116]

The D-Glu-adding enzyme has been purified 20-fold from *E. coli* and was found to be highly specific for D-glutamate, although it would process phospho-Mur*N*Ac–L-Ala with a 20-fold higher K_m than UDPMur*N*Ac–L-Ala.[117]

Using a construct containing the *murE* gene, the *meso*-DAP-adding enzyme has been overexpressed and purified to near homogeneity.[118] A number of structural analogues of *meso*-diaminopimelate were found to be substrates for this enzyme,[118] including L,L-DAP, which has a 2000-fold higher K_m than D,L-DAP, demonstrating that the enzyme is not completely enantiospecific.[100] The sulfur analogue *meso*-lanthionine has also been found to act as an efficient substrate for this enzyme.[119] Specificity towards the UDPMur*N*Ac–L-Ala–D-Glu site has been shown to be high for the *E. coli* enzyme.[120] Release of [^{14}C]*m*-DAP from UDPMur*N*Ac–L-Ala–D-Glu-[^{14}C]*m*-DAP and exchange of [^{14}C]*m*-DAP into UDPMur*N*Ac–L-Ala–D-Glu–*m*-DAP were both detected in the presence of ADP and inorganic phosphate (P$_i$) with purified enzyme, indicating that the enzyme can also catalyze the reverse reaction.[118]

The L-lysine-adding enzyme of Gram-positive *B. sphaericus* has been purified 980-fold and was found to be activated by up to sevenfold by P$_i$, a product of the enzymic reaction.[121]

3.10.2.3.2   *D-Alanine:D-alanine ligase*

The D-alanine branch of bacterial cell wall biosynthesis has been reviewed previously.[122] It is initiated by alanine racemase, which generates D-alanine by racemization of L-alanine. Two molecules of D-alanine are then ligated by an ATP-dependent ligase enzyme.

Two genes have been identified in *E. coli* which encode D-alanine:D-alanine ligases. The *ddlB* gene maps at minute 2 of the *E. coli* chromosome, immediately following the *murC* gene.[123] The *ddlA* gene, cloned by complementation with a temperature-sensitive mutant,[124] shares 90% sequence identity with a *ddl* gene from *Salmonella typhimurium*,[125] and maps at minute 8.5 of the *E. coli* chromosome. The encoded DdlA and DdlB enzymes, 39 kDa and 32 kDa, respectively, have been overexpressed and purified to homogeneity.[124] Although the two enzymes share only 35% sequence identity, they show very similar kinetic properties. K_m values can be measured for both of the D-alanine binding sites, DdlA yielding values of 5.7 μmol L^{-1} and 0.55 mmol L^{-1}, and DdlB values of 3.3 μmol L^{-1} and 1.2 mmol L^{-1}.[124]

The *E. coli* DdlB enzyme has been crystallized in the presence of an aminoalkylphosphinate transition state analogue and ATP, and the X-ray crystal structure determined to 2.3 Å.[126] The tertiary structure of the enzyme consists of three domains of similar size, with ATP bound at the interface of the central and *C*-terminal domains, similar to the tertiary structure of glutathione synthetase, another ATP-dependent amino acid ligase.[127] Two peptide loops extend into the active site and are involved in substrate recognition: loop 148–153 from the central domain and loop 206–220 from the *C*-terminal domain. Tyrosine-216 and serine-150 from these loops form a triad with glutamate-15 that are positioned to act as catalytic groups in the reaction mechanism, as shown in Scheme 10.[126]

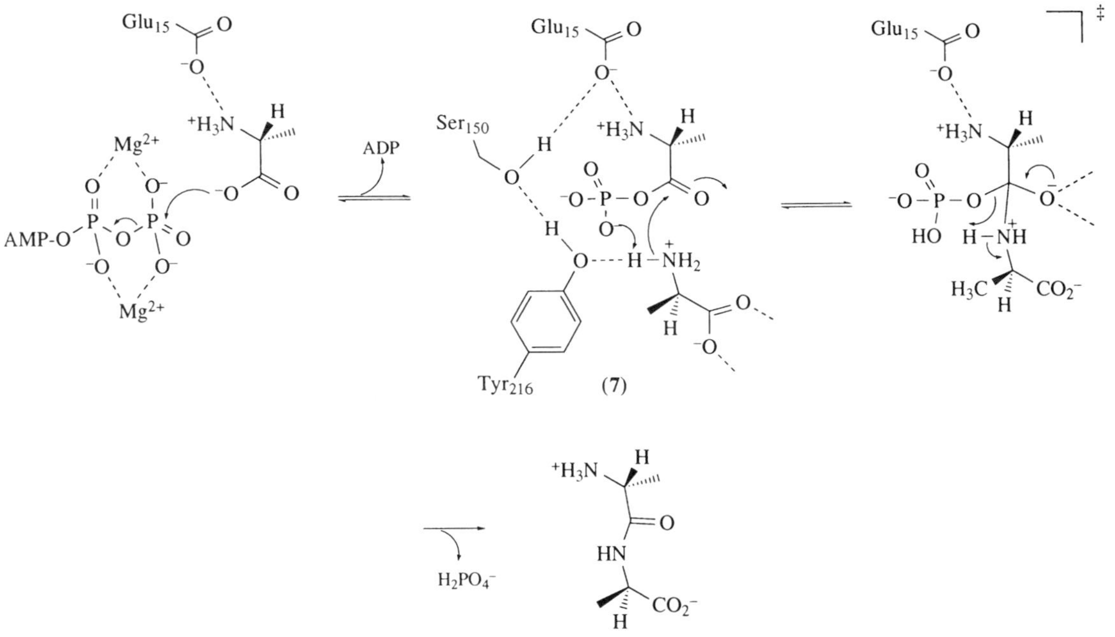

Scheme 10

Evidence has been obtained for an acyl phosphate in the catalytic mechanism of the *S. typhimurium* ligase.[128] Incubation of enzyme with D-Ala–D-Ala and [^{14}C]-D-alanine in the presence of P_i (but not ADP) led to the gradual incorporation of the ^{14}C label into D-Ala–D-Ala, consistent with a D-alanyl phosphate intermediate (**7**). Furthermore, evidence for this intermediate had been obtained from ^{18}O positional isotope exchange.[128] In the crystal structure the first molecule of D-alanine was bound via an electrostatic interaction of its α-NH$_3^+$ group with glutamate-15, and its carboxylate was positioned close to the γ-phosphate of ATP, for formation of the acyl phosphate intermediate.[126]

Inspection of the crystal structure indicated that tyrosine-216 would be well positioned to act as a base to deprotonate the second molecule of D-alanine for attack on the acyl phosphate intermediate (**7**). At this point, transition-state stabilization of the resulting tetrahedral intermediate would be provided by neighboring arginine-255. Loss of P_i then leads to the formation of product D-Ala–D-Ala, as shown in Figure 14.[126]

The active-site architecture also offered an explanation for the altered specificity of a D-alanine: D-lactate ligase VanA involved in high level vancomycin resistance in *Enterococcus faecalis* (see Section 3.10.2.7).[129,130] In VanA, tyrosine-216 is replaced by a lysine residue, which owing to its positive charge would disfavor the binding of a second molecule of D-alanine. Three mutants were constructed that disrupt the catalytic network: Y216F, S150A, and E15Q.[131] All three showed modest catalytic activity for D-Ala–D-Ala synthesis, suggesting that the base used for deprotonation of the attacking α-amino group might be the neighboring phosphate group.[131] However, these mutants were also found to possess novel catalytic activities in the presence of D-lactate.[132] The Y216F and S150A mutants catalyzed the synthesis of the D-Ala–D-Lac depsipeptide, similarly to VanA, implying that disruption of the triad of active-site hydrogen bonds controls the specificity towards the second nucleophile. The E15Q mutant catalyzed the synthesis of the D-Lac–D-Ala amide, which can be explained by the disruption of the electrostatic interaction between Glu-15 and the α-amino group of the *N*-terminal D-alanine molecule,[132] as shown in Figure 14.

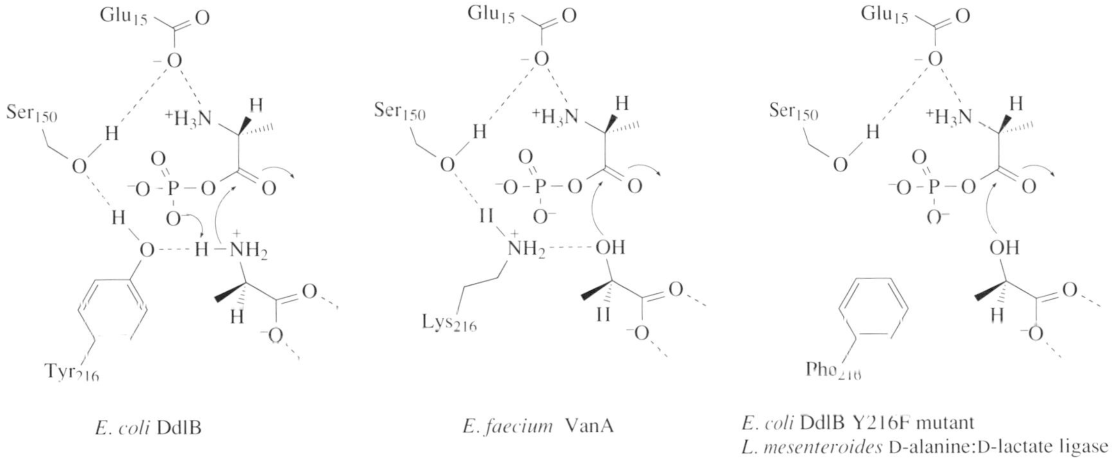

E. coli DdlB   *E. faecium* VanA   *E. coli* DdlB Y216F mutant
L. mesenteroides D-alanine:D-lactate ligase

Figure 14 Reaction specificity of D-alanine:D-alanine and D-alanine:D-lactate ligases.

The control of reaction specificity has been investigated further by examining the corresponding ligase enzyme from *Leuconostoc mesenteroides*, which is intrinsically resistant to vancomycin owing to the presence of D-lactate at position 5 of its pentapeptide sequence. The ligase from this organism was found to catalyze the synthesis of the D-Ala–D-Lac depsipeptide, and to contain phenylalanine in place of tyrosine at position 216.[133] Mutation of Phe-216 back to tyrosine gave a mutant enzyme which catalyzes the synthesis of the D-Ala–D-Ala dipeptide.[133] Hence the residue at position 216 is critical for the control of reaction specificity, as shown in Figure 14.

3.10.2.3.3 *D-Ala–D-Ala-adding enzyme*

D-Ala–D-Ala-adding enzyme catalyzes the ATP-dependent condensation between UDP-Mur*N*Ac–L-Ala–γ-D-Glu–*m*-DAP and D-Ala–D-Ala to give UDP-Mur*N*Ac–L-Ala–γ-D-Glu–*m*-DAP–D-Ala–D-Ala, the final cytoplasmic precursor. The specificity of the *S. faecalis* enzyme was found to be strict for the *C*-terminal D-alanine position, in contrast to D-alanine:D-alanine ligase, which is highly specific for the *N*-terminal position.[134] The combination of the specificity of the two enzymes ensures that D-Ala–D-Ala is the preferred dipeptide incorporated into peptidoglycan.

The *E. coli* enzyme has been purified to homogeneity, and is a 48 kDa monomeric enzyme.[135] The sequence of the corresponding *murF* gene has been cloned and sequenced,[136] and the encoded amino acid sequence shows similarity to the sequences of the earlier adding enzymes. It therefore seems likely that they follow similar mechanistic courses. The purified *E. coli* enzyme has been found to catalyze the reverse reaction, and to catalyze the isotope exchange of [^{14}C]-D-Ala–D-Ala into UDPMur*N*Ac-pentapeptide in the presence of P_i (but not ADP), consistent with an acyl phosphate intermediate.[137] The kinetic mechanism of the *E. coli* D-Ala–D-Ala-adding enzyme has been analyzed: the forward reaction was found to follow a sequential ordered kinetic mechanism in which ATP binds to the free enzyme, followed by UDPMur*N*Ac-tripeptide and D-Ala–D-Ala, respectively.[138]

3.10.2.4 The Intramembrane Cycle of Peptidoglycan Biosynthesis

The next stage of peptidoglycan biosynthesis involves the transfer of the phospho-Mur$N$Ac-pentapeptide portion of UDPMur$N$Ac-pentapeptide across the cytoplasmic membrane, with the assistance of a lipid carrier undecaprenyl phosphate. There is a cycle of intramembrane reactions: transfer of phospho-Mur$N$Ac-pentapeptide on to undecaprenyl phosphate; addition of a second Glc$N$Ac monosaccharide residue and peptide cross-link amino acids; flipping of the lipid-linked intermediate across the membrane; transglycosylation and polymerization of peptidoglycan; and recycling of the lipid carrier. These steps are illustrated in Figure 15.

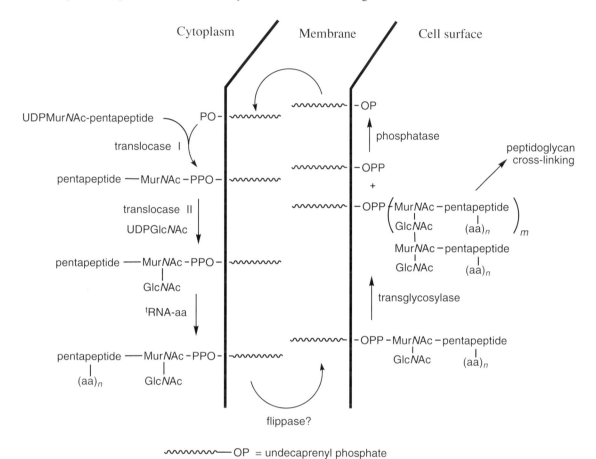

Figure 15 Steps of the intramembrane cycle of peptidoglycan biosynthesis.

3.10.2.4.1 *Biosynthesis of undecaprenyl phosphate*

The lipid carrier was identified by Strominger and co-workers.[139] Undecaprenyl phosphate is a polyprenyl phosphate whose carbon skeleton is biosynthesized by a prenyl transferase enzyme, undecaprenyl pyrophosphate synthetase. This membrane-linked enzyme catalyzes the stepwise addition of isopentenyl pyrophosphate (IPP) units to farnesyl pyrophosphate to give a mixture of C_{50} and C_{55} prenyl pyrophosphates.[140] The stereochemistry of the first three prenyl units derived from farnesyl pyrophosphate is E, whereas the stereochemistry of the subsequent prenyl units is Z. The stereochemistry of carbon–carbon bond formation by this enzyme has been examined, and the addition of IPP was found to occur from the *si* face of IPP.[141]

Undecaprenyl pyrophosphate is then dephosphorylated by a specific phosphatase enzyme, to give undecaprenyl phosphate.[142] Further dephosphorylation to undecaprenol has been observed, and an undecaprenol kinase enzyme has been identified.[143] The dephosphorylation and phosphorylation of undecaprenyl phosphate may be a control mechanism for the flux of Mur$N$Ac-pentapeptide units across the cytoplasmic membrane.[143] The biosynthesis of undecaprenyl phosphate is illustrated in Scheme 11.

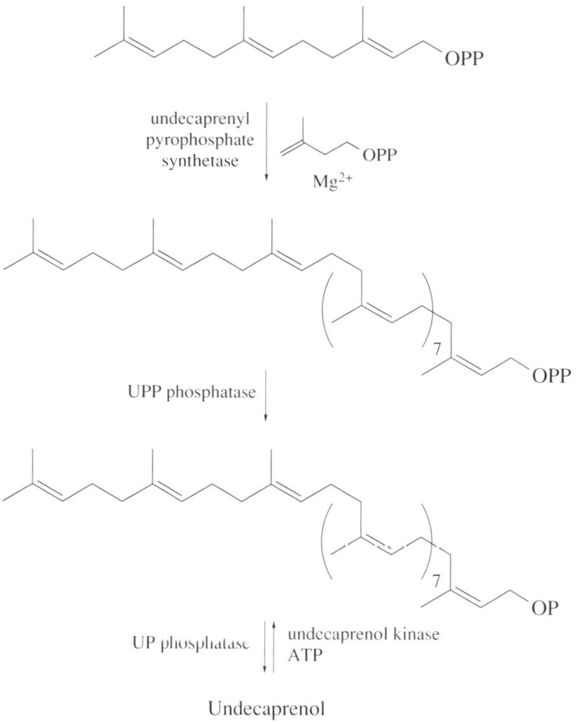

Scheme 11

3.10.2.4.2 Phospho-MurNAc-pentapeptide translocase (translocase I)

The first step in the membrane cycle of reactions is the transfer of phospho-MurNAc-pentapeptide from UDPMurNAc-pentapeptide to undecaprenyl phosphate, catalyzed by phospho-MurNAc-pentapeptide translocase (also called translocase I). This is a phospho-transfer reaction in which the phosphodiester linkage of uridine diphospho-sugar is broken, as shown in Scheme 12.

Translocase I activity was first identified in particulate membrane fragments of *Staphylococcus aureus*, using a radiochemical assay.[144] In addition to catalyzing the transfer of phospho-MurNAc-pentapeptide on to undecaprenyl phosphate, the particulate enzyme was found to catalyze the exchange of [^{14}C]UMP into UDPMurNAc-pentapeptide at rates in excess of the transfer reaction.[145] The enzyme also catalyzed the slow hydrolysis of UDPMurNAc-pentapeptide to UMP and phospho-MurNAc-pentapeptide.[145] These observations are consistent with a two-step mechanism involving a covalent intermediate, as shown in Scheme 12. The particulate enzyme could be solubilized by treatment with detergents such as lauroyl sarcosinate, oleoyl sarcosinate, or Triton X-100.[146]

The substrate specificity of the *S. aureus* particulate enzyme has been investigated. The enzyme was found to process efficiently substrates in which the pentapeptide sequence was modified at position 3 or 5, but UDPMurNAc-tripeptide was processed with 80-fold lower efficiency.[147] The enzyme was found to process efficiently a fluorescent substrate analogue, UDPMurNAc–L-Ala–γ-D-Glu–L-Lys(N^{ε}-dansyl)–D-Ala–D-Ala, leading to a sixfold increase in fluorescence yield in the lipid-linked product.[148]

The gene encoding translocase I in *E. coli* has been identified as the *mraY* gene, located in the cluster of biosynthetic genes at minute 2.[149] Sequence similarity was detected with that of yeast dolichyl phosphate:GlcNAc-1-phosphate transferase, an enzyme which catalyzes the first committed step of eukaryotic glycoprotein biosynthesis.[149] Secondary structural predictions suggest that the MraY gene product is an integral membrane containing at least seven transmembrane helices. Overexpression of the *mraY* gene in *E. coli* was found to give 30–40-fold overproduction of translocase I activity, which could be efficiently solubilized using 1% Triton X-100.[150] The fluorescent substrate analogue UDPMurNAc-L-Ala–γ-D-Glu–*m*-DAP(N^{ε}-dansyl)–D-Ala–D-Ala was used to construct a continuous fluorescence enhancement assay. The solubilized enzyme was found to accept exogenous heptaprenyl phosphate or dodecaprenyl phosphate as lipid substrates, but not farnesyl phosphate. Enzyme activity was activated 5–10-fold by the phospholipid phosphatidylglycerol.[150]

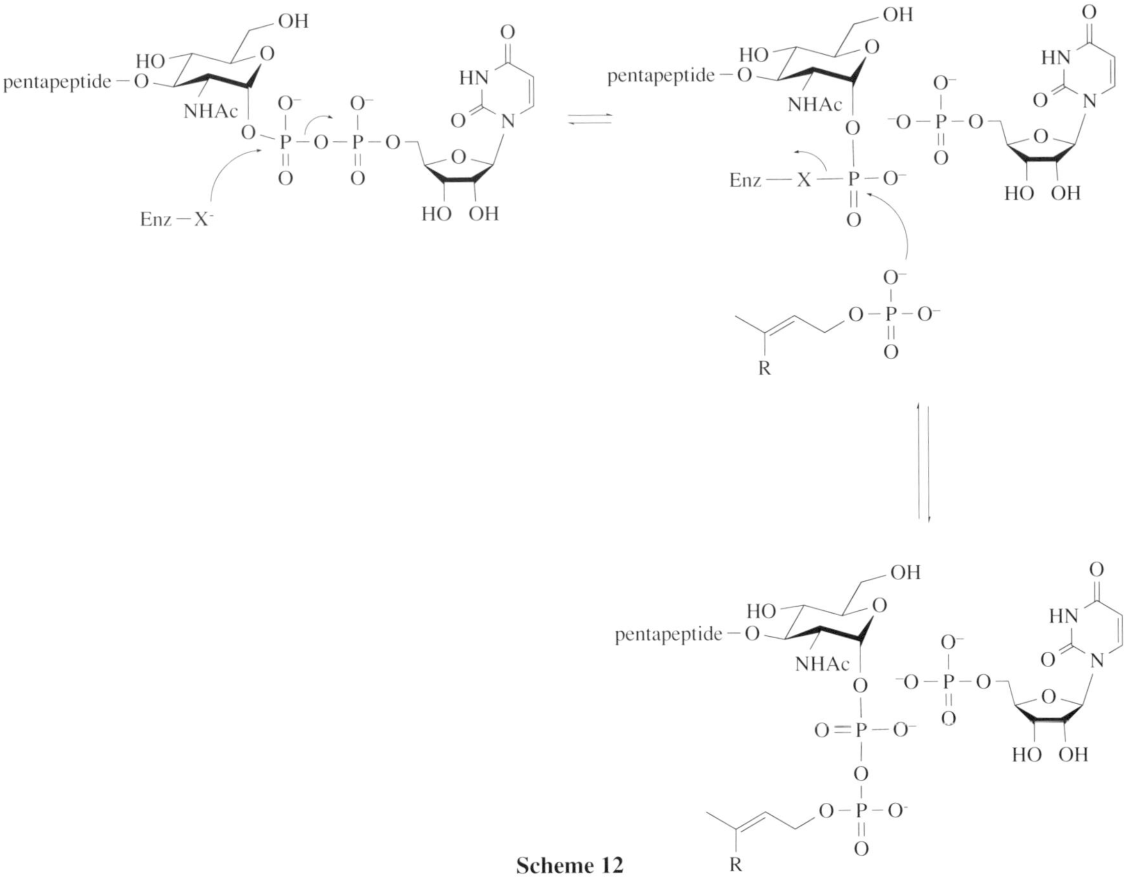

Scheme 12

3.10.2.4.3   *Undecaprenyl-diphospho-MurNAc-pentapeptide:UDPGlcNAc GlcNAc transferase (translocase II)*

To the first lipid-linked intermediate undecaprenyl-diphospho-MurNAc-pentapeptide (also known as lipid intermediate I) is then attached a residue of N-acetylglucosamine from UDPGlcNAc. This reaction is catalyzed by a glycosyl transferase enzyme known as translocase II, yielding lipid intermediate II as the reaction product. The gene encoding translocase II in *E. coli* has been identified as the *murG* gene, situated in the cluster of biosynthetic genes at minute 2.[151] It has been determined that translocase II is associated with the cytoplasmic face of the cytoplasmic membrane, since the enzyme is proteolyzed by trypsin only under conditions where trypsin is able to access the cytoplasm.[152] Procedures have been developed for the generation and isolation of the lipid linked intermediates I and II from *E. coli*. However, very low levels of both intermediates were found, corresponding to copy numbers of 700 and 2000 per cell, respectively.[153]

3.10.2.4.4   *Attachment of the peptide cross-link amino acids*

It has been determined that the pentaglycine peptide cross-link found in *S. aureus* is added at the stage of lipid intermediate II.[154] The glycine donor was found to be glycyl tRNA.[154] Addition of glycine units was found to occur sequentially on to the free amino terminus, starting with the ε-amino terminus of L-lysine, as shown in Figure 16.[155] This direction of elongation is opposite to that which occurs during protein synthesis. An enzyme activity capable of catalyzing the glycine addition reaction was purified 100-fold from *S. aureus*.[155]

The *femA* and *femB* genes of *S. aureus* have been implicated in the formation of the pentaglycine cross-link, since mutations in these genes affect the composition of the bridge in mature peptidoglycan.[156–158] A *femAB*⁻ double mutant was found to contain cross-links containing a single glycine residue.[158] Complementation of this strain with cloned *femA* or *femAB* genes resulted in the extension of the cross-link to a triglycine and a pentaglycine bridge, respectively, implying that the

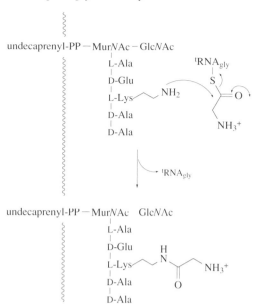

Figure 16 Formation of peptide cross-links in *S. aureus* (first of five steps).

FemA protein is responsible for the formation of glycines 2 and 3 and the FemB protein for the formation of glycines 4 and 5.[158]

The D-aspartate cross-link found in *S. faecalis* and *L. casei* is biosynthesized by activation of D-aspartate by ATP to give β-D-aspartyl phosphate, which can be incorporated into peptidoglycan in a cell-free system.[159] It is not known at what stage in the biosynthesis the D-aspartate is added to the pentapeptide.

3.10.2.4.5 *Transmembrane flipping of the lipid-linked intermediate*

After assembly of lipid intermediate II, complete with peptide cross-link, the intermediate is somehow flipped from the cytoplasmic face of the membrane to the external face. This physical step is remarkable, since the diphospho-disaccharide-pentapeptide appendage is strongly hydrophilic, yet it must somehow be transported across the lipid bilayer. The question of whether this step is protein assisted is still unanswered. Studies of lipid phosphates in phospholipid bilayers using voltammetry and NMR spectroscopy have indicated that the lipid phosphate promotes the formation of non-bilayer structures which may assist flipping across the membrane.[160,161] However, analysis of dansylated undecaprenyl-diphospho-Mur*N*Ac-pentapeptide in *S. aureus* membrane fragments by fluorescence energy transfer has revealed that the rate of non-assisted flipping is much lower than that required to support peptidoglycan synthesis *in vivo*.[162]

The possibility of protein-assisted flipping is reminiscent of the "flippase" protein model for multidrug resistance, which has been reviewed.[163] Gene products have been identified in related lipid-linked pathways for succinoglycan biosynthesis and lipopolysaccharide biosynthesis which are required for the translocation of cytoplasmic precursors to the cell surface.[164,165] Hence it seems likely that the flipping of lipid intermediate II is protein assisted in some way.

The external transglycosylation and transpeptidation steps which will be discussed in the next section generate undecaprenyl pyrophosphate as a by-product. This must be dephosphorylated and flipped back across to the cytoplasmic face of the membrane in order to complete the membrane cycle.

3.10.2.5 Extracellular Steps of Peptidoglycan Assembly

After translocation across the cytoplasmic membrane, the lipid-linked disaccharide-pentapeptide is transformed into peptidoglycan by transglycosylation and transpeptidation. The enzymes that catalyze these reactions are members of the family of penicillin-binding proteins (PBPs). Since the PBPs have been extensively reviewed,[166–168] a fairly brief discussion will be given here, highlighting the molecular aspects of peptidoglycan formation in *E. coli*.

3.10.2.5.1   *Classes of penicillin-binding proteins in* Escherichia coli

The mechanism of action of penicillin, which will be discussed further in Section 3.10.3.5, involves the opening of the β-lactam ring and formation of a stable enzyme–inhibitor adduct. Treatment of membrane preparations with ^{14}C-labeled penicillin, followed by sodium dodecyl sulfate poly-acrylamide gel electrophoresis (SDS–PAGE) and autoradiography, reveals multiple radiolabeled PBPs.[169] Each bacterial strain has a different collection of PBPs, which vary in molecular weight (from 25 kDa to 100 kDa) and in their affinity for penicillin. Genetic knockout of individual PBP genes gives viable mutant strains, suggesting that other PBPs can take the place of mutated enzymes—a strategy which perhaps has been selected in the course of evolution.

In *E. coli*, nine penicillin-binding proteins have been identified, and classified according to their molecular weight. Their properties are summarized in Table 1. The cellular roles of individual PBPs have been elucidated by the study of mutants in the corresponding genes.[170] Mutants of PBP 1A, 1B, 2, and 3 give strains with varied modified morphological properties, which imply that PBP 1A and 1B are involved in cell elongation, PBP 2 is involved in maintenance of cell shape, and PBP 3 is involved in cell division.[170] Mutants of PBP 4, 5, 6, and 7 show no significant morphological changes,[167] although the PBP 7 mutant strain was more susceptible to β-lactams which act on nongrowing cells.[171] Tenfold overproduction of PBP 5 gave a strain with a spherical cell shape.[172]

Table 1   Penicillin-binding proteins of *Escherichia coli*.

Protein	M_r (kDa)	Enzymatic properties	Probable cellular role
PBP 1A	90	Bifunctional transglycosylase/transpeptidase	Cell elongation
PBP 1B	87	Bifunctional transglycosylase/transpeptidase	Cell elongation
PBP 2	66	Transpeptidase/D,D-carboxypeptidase	Cell shape maintenance
PBP 3	60	Transpeptidase/D-D-carboxypeptidase	Cell division/septum formation
PBP 4	49	D,D-carboxypeptidase/*m*-DAP–D-Ala endopeptidase	Murein processing
PBP 5	42	D,D-carboxypeptidase	Murein processing (cell shape)
PBP 6	40	D,D-carboxypeptidase	Murein processing
PBP 7/8	32 (29)	*m*-DAP–D-Ala endopeptidase	Non-growing cell maintenance

PBP 1A and 1B are large, bifunctional proteins which have been found to catalyze both trans-glycosylation and transpeptidation reactions (see below).[166–168] PBP 2 and 3 both exhibit trans-peptidase and associated D,D-carboxypeptidase activities, and are thought to interact with other integral membrane proteins RodA and FtsW *in vivo*.[166–168] PBP 4, 5, and 6 all possess high D,D-carboxypeptidase activity, and this is thought to be their major role *in vivo*,[173] although PBP 4 also shows activity for cleavage of the *m*-DAP–D-Ala amide bond.[174] This endopeptidase activity is the sole catalytic property of PBP 7, which has also been found as a 29 kDa proteolytic fragment of PBP 8.[175]

3.10.2.5.2   *Transglycosylation of lipid intermediate II*

The transglycosylation of lipid intermediate II on the cell surface of *E. coli* is catalyzed by the *N*-terminal domains of PBP 1A and 1B. The reaction catalyzed is a glycosyl transfer reaction involving the displacement of the α-diphospho-undecaprenyl group by the C-4 hydroxyl of GlcNAc, resulting in a β-1,4-linkage as shown in Scheme 13. The displacement therefore proceeds with overall inversion of configuration at the anomeric center.[176]

Purified PBP 1A has been found to catalyze the polymerization and cross-linking of radiolabeled undecaprenyl-diphospho-MurNAc(pentapeptide)-GlcNAc to the level of 8% cross-linking.[177] Purified PBP 1B also catalyzes the polymerization and cross-linking of labeled lipid intermediate II.[178,179] Treatment of PBP 1B with penicillin leads to inactivation of the transpeptidase domain and stimulation of the transglycosylase activity,[178] but no such stimulation was observed by treatment

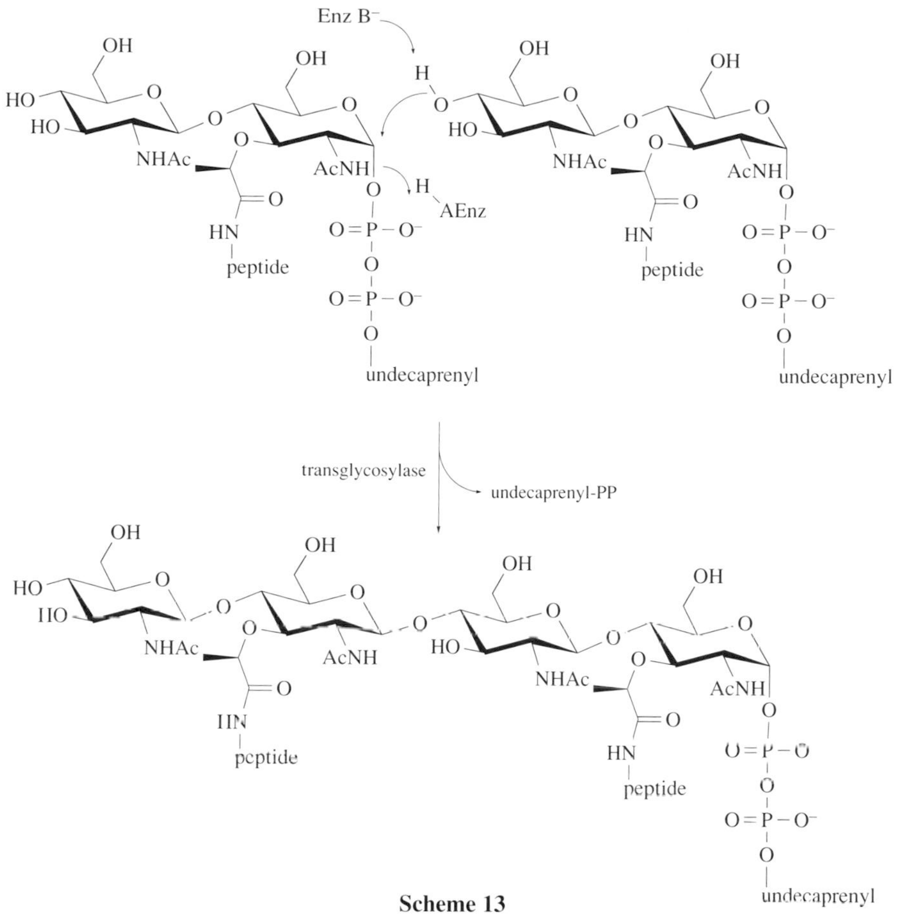

Scheme 13

of PBP 1A with penicillin.[177] The other observed difference is the dependence of PBP 1B on Mg^{2+} ions, not found with PBP 1A.[178] The activities observed for these enzymes *in vitro* are 2–3% of that required *in vivo*, but this may reflect the limitations of the *in vitro* assay.[153]

3.10.2.5.3   D-D-*Transpeptidases*, D,D-*carboxypeptidases, and β-lactamases*

The final step in peptidoglycan biosynthesis is the transpeptidation reaction between the amino terminus of *meso*-diaminopimelic acid (for most Gram-negative bacteria) or a peptide cross-link (for most Gram-positive bacteria) and the carbonyl group of D-alanine at position 4 of a second peptide side chain. This cross-linking provides the structural rigidity of mature peptidoglycan required for maintenance of cell shape and prevention of cell lysis. Transpeptidases follow a similar mechanistic course to the more familiar peptidase enzymes, except that the nucleophilic partner in catalysis is an amino donor rather than water. In the case of peptidoglycan biosynthesis, the D,D-transpeptidases have an active-site serine residue whose function is similar to the catalytic triad of the serine proteases.

The site of acylation by penicillin in PBPs is an active-site serine residue, which acts as a nucleophile in the transpeptidation reaction, as shown in Scheme 14.[166–168] A covalent acyl enzyme intermediate is formed, which in the presence of an exogenous amine forms a new amide bond to complete the transpeptidation reaction. In the absence of an exogenous amine, the acyl enzyme intermediate is hydrolyzed by water, resulting in the hydrolysis of the terminal D-alanine residue. This scheme explains why PBPs exhibit both D,D-transpeptidase and D,D-carboxypeptidase activities, the proportion of these activities being determined by the reactivity of the acyl enzyme intermediate versus amine and water nucleophiles.[167]

The active-site serine has been identified in several PBPs,[180–186] and is found as a sequence motif Ser–X–X–Lys.[167] This sequence motif is also found in *β*-lactamase enzymes, which catalyze the

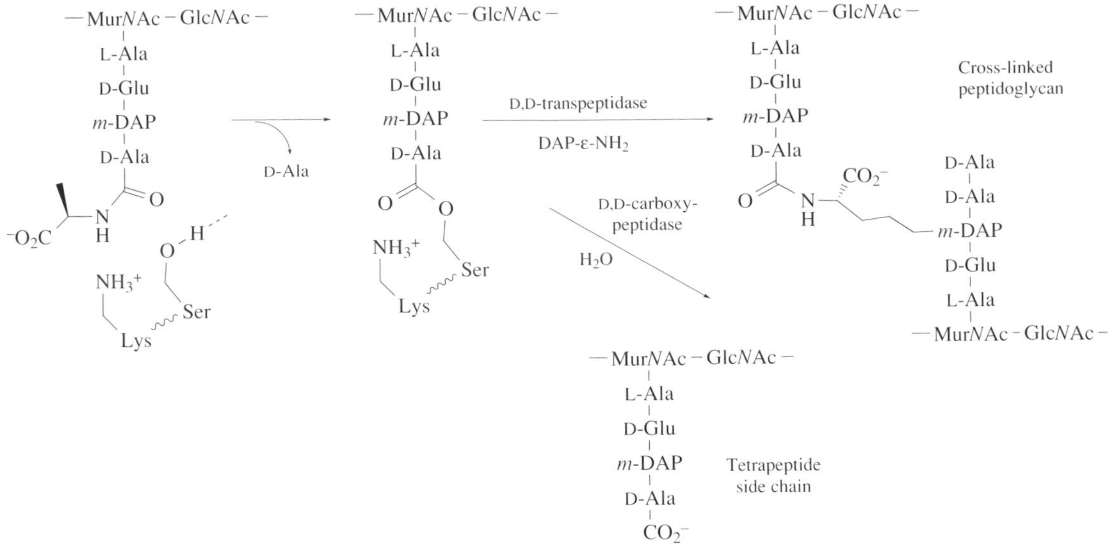

Scheme 14

hydrolysis of penicillin. The reaction mechanism of the β-lactamases proceeds via an acyl enzyme intermediate, which, unlike that formed with PBPs, is hydrolyzed by water (see Scheme 15). The X-ray crystal structures of class A β-lactamases,[187,188] a class of C β-lactamase,[189] and the *Streptomyces* R61 PBP/D,D-peptidase[190] reveal that the lysine residue found in the sequence motif is situated close to the active-site serine residue, which is involved in a complex hydrogen bond network. It has been proposed in the case of the class A β-lactamases that the neighboring lysine acts as an acidic group for protonation of the departing nitrogen atom.[187,188] Deprotonation of the active-site serine in the class A β-lactamases is thought to be carried out by glutamate-166, via an intervening water molecule.[187,188] In the class C β-lactamases deprotonation is thought to be achieved by an active-site tyrosine residue.[189]

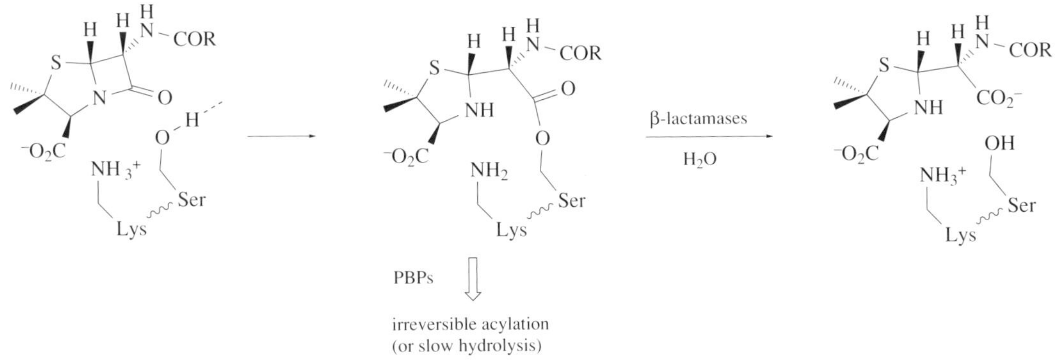

Scheme 15

The functions of the corresponding active site residues in *E. coli* PBP 2 have been analyzed by site-directed mutagenesis, followed by analysis of activity by genetic complementation and penicillin binding.[191] The active-site serine in this enzyme is serine-330. Mutation of active-site lysine-333 gave completely inactive mutant enzymes, indicating that it is essential for acylation. Mutation of aspartate-447 (corresponding to Glu-166 above) gave inactive mutant enzyme, apart from the D447E mutant, which was active. Mutants of the putative active-site tyrosine Tyr393 retained activity, indicating that this is not an essential residue.[191] More detailed analysis of the PBP reaction mechanism awaits further crystallographic data, and preliminary X-ray analysis of *E. coli* PBP 4 indicates that it is related in structure to the class A β-lactamases.[192] The relationship between the PBPs and the β-lactamases with regard to structure and mechanism suggests that the β-lactamases have evolved from an ancestral PBP.[190]

3.10.2.6 Enzymatic Processing and Recycling of Peptidoglycan

Growing bacteria are constantly remodeling and rebuilding their murein sacculus, which requires the presence of enzymes which can break down the peptidoglycan layer in localized areas of the cell wall. Such enzymes are found in all bacterial cells, and are collectively known as the murein hydrolases.[193] These enzymes can be divided into glycosidase enzymes which cleave the glycan strand and peptidase enzymes which cleave the cross-linked peptide chains. The biochemical properties of these enzymes have been reviewed,[193] so this section will highlight the classes of enzyme found in *E. coli* and the molecular aspects of their catalysis. The properties of these enzymes are summarized in Table 2 and their sites of action illustrated in Figure 17.

Table 2 The murein hydrolases of *Escherichia coli*.

Enzyme	M_r (kDa)	Bond broken	Type of substrate
Soluble lytic transglycosylase	65	MurNAc–GlcNAc glycosidic bond	Murein
Membrane lytic transglycosylase	35	MurNAc–GlcNAc glycosidic bond	Murein
β-N-Acetylglucosaminidase	36	GlcNAc–MurNAc glycosidic bond	Muropeptides
N-Acetylmuramyl-L-alanine amidase	39	MurNAc–L-Ala amide bond	Muropeptides
D,D-Endopeptidase (PBP 4)	49	m-DAP(D)–D-Ala cross-link	Murein
Soluble D,D-endopeptidase	30	m-DAP(D)–D-Ala cross-link	Murein
D,D-Carboxypeptidase (PBP 5)	42	D-Ala–D-Ala amide bond	Muropeptides
D,D-Carboxypeptidase (PBP 6)	40	D-Ala–D-Ala amide bond	Muropeptides
L,D-Carboxypeptidase	86	m-DAP(L)–D-Ala amide bond	Muropeptides

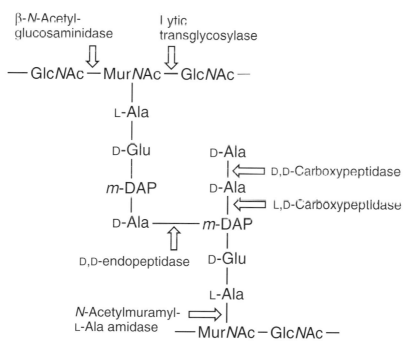

Figure 17 Sites of action of *E. coli* murein hydrolases.

3.10.2.6.1 Lytic glycosidases

The well-known glycosidase enzyme lysozyme, found widely in mammals, catalyzes the hydrolysis of the MurNAc–GlcNAc glycosidic bond. Lysozyme is not found in bacteria, but they do contain a lytic transglycosylase enzyme which catalyzes the cleavage of the MurNAc–GlcNAc glycosidic bond, using the C-6 hydroxyl group of MurNAc as an internal nucleophile.[194] Thus, a 1,6-anhydro-muramic acid is formed with retention of stereochemistry, as shown in Scheme 16. The X-ray crystal structure of the soluble 70 kDa enzyme has been solved, and its tertiary structure forms a doughnut-shaped superhelical ring of α-helices which presumably envelops the glycan strand.[195] The structure of this enzyme complexed with an inhibitor bulgecin A has also been solved.[196] The crystal structure reveals the close proximity of the carboxylate side chain of glutamic acid-478 to the position normally occupied by the scissile glycosidic bond. By analogy with the catalytic mechanism of lysozyme, glutamic acid-478 is thought to act as an acidic catalytic group, protonating the departing hydroxyl group. Stabilization of the oxonium ion may be provided by the neighboring C-2 acetamido group, as shown in Scheme 16.[196]

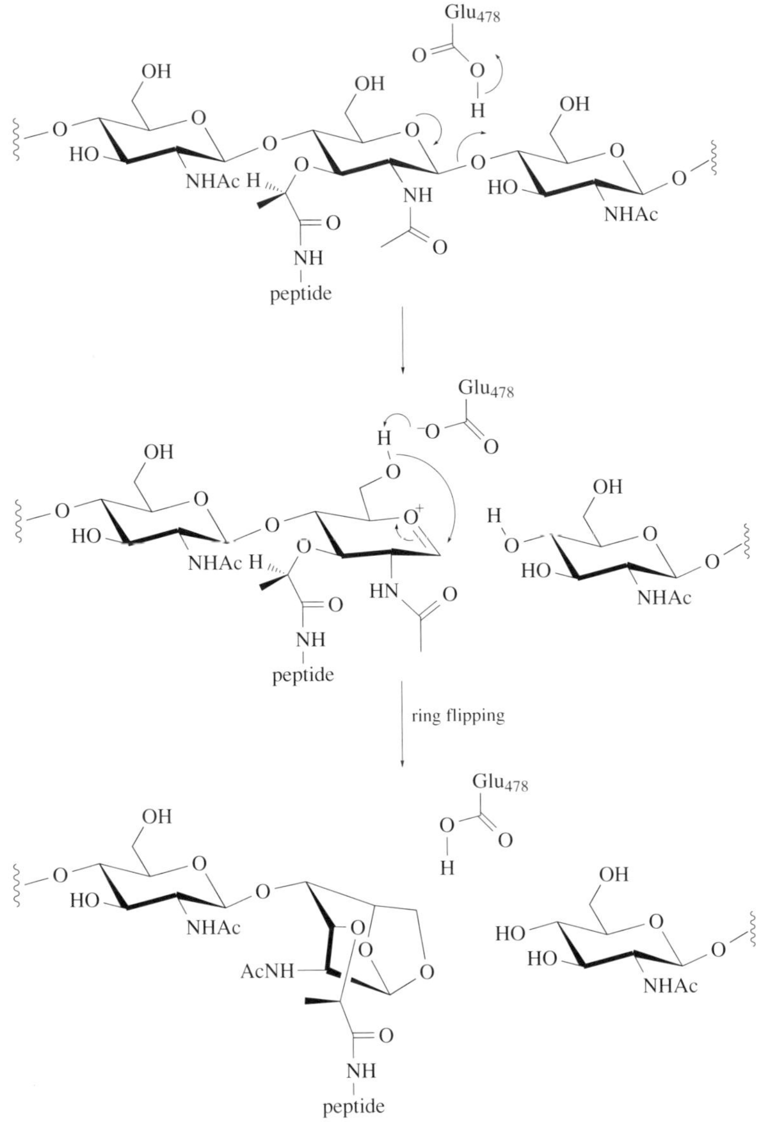

Scheme 16

A second soluble lytic transglycosylase of molecular weight 35 kDa has also been identified in *E. coli*,[197] and a membrane-bound 38 kDa lytic transglycosylase.[198] A β-*N*-acetylglucosaminidase which cleaves the Glc*N*Ac–Mur*N*Ac glycosidic bond of small muropeptides has also been identified.[199]

3.10.2.6.2 *Lytic endopeptidases*

Several enzymes have been identified which can cleave amide bonds found in the cross-linked murein structure. As mentioned above, the penicillin-binding proteins 4, 5, and 6 show high D,D-carboxypeptidase activity.[173] The biosynthetic role of removing terminal D-alanine residues from the pentapeptide side chains is not obvious; however, this process may control the availability of sites for cross-linking. PBP 4 and 7/8 have also been found to catalyze the cleavage of the cross-link formed between the ε-amino group (i.e., α-amino group of the D center) of *meso*-diaminopimelic acid and the D-alanine of position 4 on a second peptide strand.[174,175] This D,D-endopeptidase activity leads to the breakdown of cross-links in mature peptidoglycan. There is also a penicillin-insensitive D,D-endopeptidase of size 30 kDa which acts on intact murein.[200,201]

There are two peptidase enzymes in *E. coli* which can cleave the pentapeptide side chain itself. An *N*-acetylmuramyl-L-alanine amidase has been identified which cleaves the amide bond between the lactyl side chain of MurNAc and L-alanine at position 1 of the pentapeptide.[202] This enzyme is active on muropeptides rather than intact murein; thus, presumably, its role *in vivo* is to release peptide fragments after cleavage of the glycan strand. An L,D-endopeptidase enzyme has also been identified which can cleave the amide bond between the L center of *meso*-DAP and D-alanine at position 4 of the same pentapeptide chain.[203]

3.10.2.6.3 *Recycling of peptidoglycan fragments*

The combined action of the lytic transglycosylase, D,D-endopeptidase, MurNAc–L-Ala amidase, and L,D-endopeptidase breaks down the structure of intact murein in *E. coli* and generates the peptide fragments L-Ala–γ-D-Glu–*m*-DAP–D-Ala and L-Ala–γ-D-Glu–*m*-DAP. *E. coli* cells labeled with [³H]DAP have been found to lose 6–8% of the label per generation in the form of these peptides.[204] However, some of the tripeptide L-Ala–γ-D-Glu–*m*-DAP is taken up into the cell and incorporated into UDPMurNAc-pentapeptide without degradation into *meso*-DAP.[205] A recycling pathway has been proposed to account for these observations, as shown in Figure 18.[205]

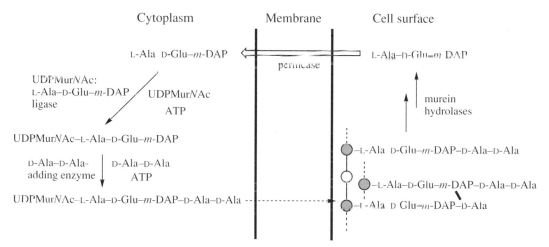

Figure 18 Pathway for recycling of peptidoglycan fragments in *E. coli*.

This recycling pathway involves the uptake of tripeptide fragments by a specific permease,[206] followed by the ligation of L-Ala–γ-D-Glu–*m*-DAP with UDPMurNAc. A ligase activity has been identified which catalyzes this reaction, and is dependent upon ATP and Mg^{2+} for activity.[207] The corresponding gene has been identified as the *mpl* gene, situated at minute 96 of the *E. coli* chromosome, which shows sequence similarity to the *murC* gene encoding the L-Ala-adding enzyme.[207] It is estimated that 30–40% of new cell wall synthesis proceeds via this recycling pathway in *E. coli*. Recycling pathways have not been identified in Gram-positive bacteria, which lose 25–50% of their peptidoglycan as peptide fragments per generation.[208]

3.10.2.7 Peptidoglycan Assembly in Antibiotic-resistant Bacteria

Emerging bacterial resistance to clinically useful antibiotics constitutes a serious threat to the medical treatment of microbial infection.[209,210] There are several classes of antibiotics which act on steps involved in peptidoglycan biosynthesis (to be discussed in Section 3.10.3), and bacterial resistance to these antibiotics has emerged in each case, through a variety of mechanisms.[211,212] Two particular cases are worthy of note at this point, since they involve a significant alteration in the assembly of peptidoglycan in the antibiotic-resistant bacteria. These are the vancomycin-resistant *Enterococci* (VRE) and methicillin-resistant *Staphylococcus aureus* (MRSA).

3.10.2.7.1 An inducible pathway for incorporation of D-lactate into the peptidoglycan of vancomycin-resistant Enterococci

Vancomycin is a member of the glycopeptide family of antibiotics, which inhibit peptidoglycan assembly in Gram-positive bacteria by complexation of peptidyl–D-Ala–D-Ala termini on the cell surface.[213] Vancomycin complexation prevents transglycosylation and transpeptidation of the cell wall.[213] Resistance to vancomycin first became apparent in 1986 in strains of *Enterococcus faecium* and *Enterococcus faecalis*.[214] High-level resistance was accompanied by the appearance of a 38–40 kDa membrane protein, which was found to be encoded by plasmid-borne DNA.[214]

A cluster of five genes was identified in *Enterococcus faecium* BM4147 which was sufficient to confer vancomycin resistance to glycopeptide-susceptible strains.[215] The properties of the five gene products are summarized in Figure 19. The first two genes, *vanS* and *vanR*, encoded a two-component regulatory system for the induction of resistance in response to vancomycin,[216] analogous to other bacterial two-component regulatory systems.[217] The VanS protein is a transmembrane protein whose extracellular domain is able to respond to the presence of vancomycin in the external medium, an event which triggers the autophosphorylation of a histidine residue in the intracellular domain.[218] The histidyl phosphate is then transferred to an aspartate residue on the soluble 27 kDa VanR protein.[218] The phosphorylated VanR protein then binds to a 254 base pair promoter region upstream of the *vanH* gene, inducing the expression of the *vanH*, *vanA*, and *vanX* genes.[219]

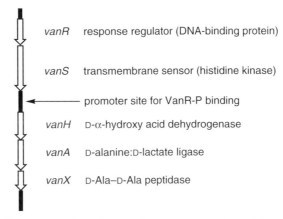

Figure 19 Organization of genes involved in high-level vancomycin resistance in *Enterococcus faecium* BM4147.

The VanH, VanA, and VanX proteins are enzymes which implement a modified pathway for peptidoglycan biosynthesis, illustrated in Figure 20.[220] VanA was found to have catalytic activity as a D-alanine:D-alanine ligase, but with an elevated K_m value for D-alanine (38 mmol L^{-1}) compared with *E. coli* DdlA (0.6 mmol L^{-1}) and DdlB (1.2 mmol L^{-1}), suggesting that its cellular role is not to synthesize D-Ala–D-Ala.[129] The purified enzyme was found to have much broader specificity than DdlA/B, proving capable of synthesizing a range of D-Ala–X dipeptides.[129]

The VanH protein was found to be an NADH-dependent dehydrogenase capable of reducing a range of α-keto acids to the corresponding D-2-hydroxy acids.[130] D-Lactic acid and D-2-hydroxybutyrate were found to be good substrates for ligation by VanA, yielding the depsipeptides D-Ala–D-Lac and D-Ala–D-HBut.[130] As discussed in Section 3.10.2.3.2, the molecular basis of the switch between amide bond formation and ester bond formation in VanA vs. DdlA/B has been elucidated by X-ray crystallography and site-directed mutagenesis.[126,131–133] The depsipeptides D-Ala–D-Lac and D-Ala–D-HBut were found to be good substrates for D-Ala–D-Ala-adding enzyme, yielding UDPMur*N*Ac-tetrapeptide–X esters.[130]

How does the modification at position 5 lead to vancomycin resistance? The molecular recognition of vancomycin for peptidyl–D-Ala–D-Ala involves the formation of several specific hydrogen-bonding interactions, one of which involves the N—H of the terminal D-alanine.[213] The analogues

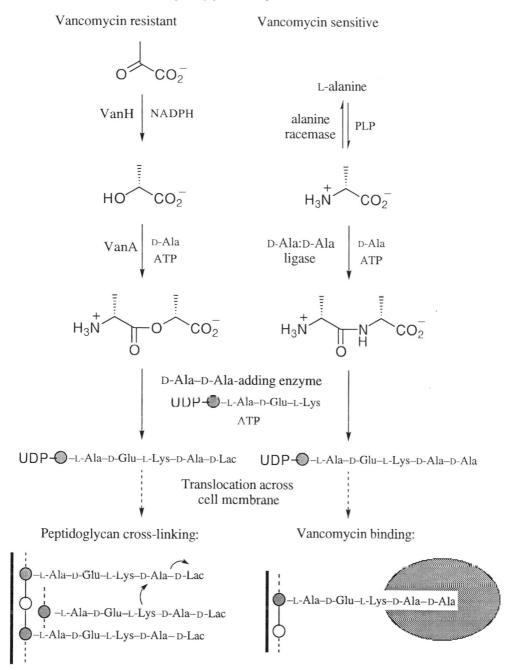

Figure 20 Proposed mechanism for high-level vancomycin resistance.

N-acetyl–D-Ala–D-Lac and N-acetyl–D-Ala–D-HBut, in which this N—H is substituted by an oxygen atom, showed no interaction with vancomycin by UV spectrophotometry.[130] Moreover, it is known that a D-lactate depsipeptide terminus is a good substrate for PBPs (since a hydroxy acid is a better leaving group than an amino acid),[221] so the depsipeptide termini will be able to synthesize intact peptidoglycan.

This proposed pathway was confirmed by analysis of peptidoglycan precursors in vancomycin-resistant bacteria, which revealed the presence of D-lactate at position 5.[26] D-Lactate has also been found in position 5 of strains of *Lactobacillus* which possess intrinsic resistance to vancomycin.[29] D-Serine has been found at position 5 of low-level resistant *Enterococci*;[30] hence it appears that these strains utilize a D-alanine:D-alanine ligase of widened substrate specificity, but do not incorporate a hydroxy acid.

The final resistance protein VanX was found to be a Zn^{2+}-dependent D-Ala–D-Ala peptidase enzyme whose role appears to be the hydrolysis of the regular D-Ala–D-Ala dipeptide precursors.[222] This enzyme assists the takeover of the normal biosynthetic pathway in favor of the modified pathway, and is required for high-level resistance *in vivo*. In common with other zinc-dependent hydrolases, VanX was found to be susceptible to inhibition by sulfur-containing analogues.[223]

3.10.2.7.2   *Peptidoglycan biosynthetic enzymes in methicillin-resistant* Staphylococcus aureus *(MRSA)*

Strains of *S. aureus* resistant to the β-lactam methicillin pose a considerable threat to postoperative and immunocompromised hospital patients. There are several genetic determinants required for the *mec*$^+$ phenotype, denoting resistance to all β-lactam antibiotics. A major determinant is the *mecA* gene, found in all MRSA strains, which encodes a novel penicillin-binding protein PBP 2a.[224] The 78 kDa PBP 2a protein shows a very low affinity for penicillin, but is still capable of assembling peptidoglycan.[225,226] The peptidoglycan composition of MRSA strains has been found to be very similar to that of susceptible *S. aureus* strains, so the catalytic properties of PBP 2a appear to be similar to those of other PBPs.[227] The low affinity of PBP 2a for second- and third-generation β-lactams is a cause for concern in antibiotic therapy.

The *S. aureus fem* genes (*A*, *B*, *C*, *D*, and *X*) are also required for methicillin resistance. As explained in Section 3.10.2.4.4, there is good evidence that the *femA* and *femB* genes are involved in the assembly of the pentaglycine cross-link found in *S. aureus* peptidoglycan.[156–158] The *S. aureus llm* gene has also been identified as determinant in methicillin resistance in a number of clinical isolates.[228] The amino acid sequence of the encoded Llm protein shows sequence similarity to the sequence of *E. coli* phospho-MurNAc-pentapeptide translocase, which catalyzes the first step of the intramembrane cycle.[229] Hence there appear to be a number of factors connected with peptidoglycan biosynthetic enzymes involved in high-level methicillin resistance.

3.10.3   INHIBITION OF PEPTIDOGLYCAN BIOSYNTHESIS

Peptidoglycan biosynthesis is an essential pathway for all bacteria which has no direct counterpart in eukaryotic cells. The consequences of the inhibition of peptidoglycan biosynthesis are that the cell wall can no longer withstand the high internal osmotic pressure, and the cell lyses. The various stages of peptidoglycan biosynthesis therefore offer attractive targets for the development of selective antibacterial agents. This section will describe both naturally occurring and laboratory-made inhibitors of cell wall biosynthetic enzymes. The inhibition of peptidoglycan biosynthesis has been reviewed previously.[3,34,230]

Several steps of peptidoglycan biosynthesis are inhibited by clinically useful antibiotics, as summarized in Table 3. The efficacy of these compounds against a range of bacterial strains (the antibacterial spectrum) is governed by several factors. The ability of the antibiotic to penetrate the cell and access the target enzyme depends on the size and hydrophobicity of the compound, and, in many cases, the ability to be taken up by one of the bacterial transport systems, which differ between bacterial types. In some cases the antibiotic is active against enzymes from some strains but not others. In other cases bacteria are able to efficiently export the antibiotic, modify it into an inactive form, or to degrade the antibiotic.

Table 3 Peptidoglycan biosynthesis as a target for antibiotics.

Target	Antibiotic
UDPGlcNAc enolpyruvyl transferase	Fosfomycin
Alanine racemase	Ala-P, β-chloroalanine
D-alanine:D-alanine ligase	D-Cycloserine
Complexation of lipid carrier	Bacitracin, amphomycin
Translocase I	Tunicamycin, mureidomycin
Transglycosylation	Moenomycin
Transpeptidation	β-Lactams
Complexation of peptidyl–D-Ala–D-Ala	Vancomycin

3.10.3.1   Inhibition of Monosaccharide Biosynthesis

N-Acetylglucosamine is not unique to prokaryotic cells, since it is used in eukaryotic cells for glycoprotein biosynthesis. Nevertheless, there are several inhibitors of glucosamine-6-phosphate synthase which show antibacterial and antifungal activity: A19009 (**8**),[231] Sch37137 (**9**),[232] chlorotetain (**10**),[233] and bacilysin (**11**).[234] The mechanism of inhibition of *E. coli* glucosamine-6-phosphate synthase by (**8**) has been investigated.[235] Compound (**8**) was found to be a time-dependent irreversible inhibitor, and the mechanism of inactivation was found to proceed via Michael addition of the thiol side chain of the *N*-terminal cysteine residue to the fumaroyl moiety of the inhibitor, resulting in covalent modification of the enzyme. Studies using a model peptide based on the *N*-terminal sequence of the protein have shown that following inactivation of a cyclization reaction with the free α-amino group of the *N*-terminal cysteine occurs, resulting in a blocked *N*-terminus.[236]

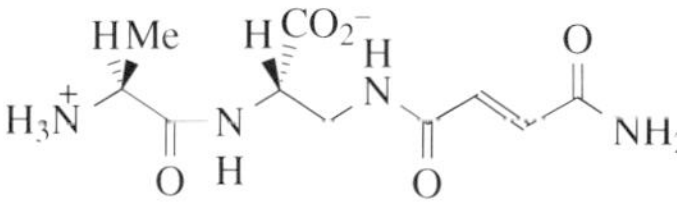

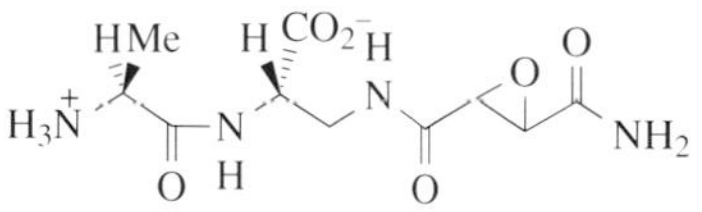

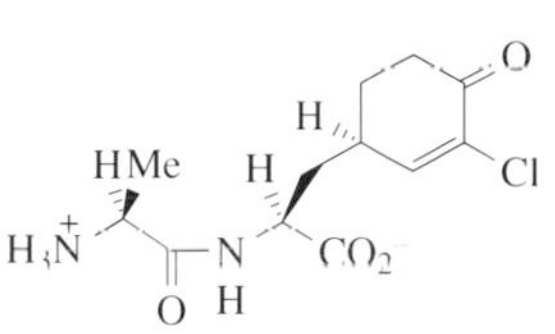

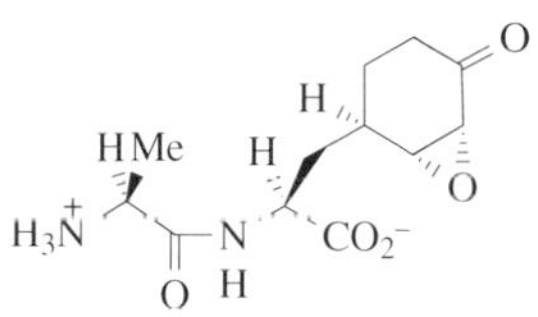

Glucosamine-6-phosphate synthase is also inactivated by glutamate-γ-semialdehyde (**12**), via attack of the active site thiol to form a thiohemiacetal, which acts as a reaction intermediate analogue.[237] The active-site thiol can also be targeted by an analogue (**13**) of GlcNAc-6-phosphate containing an iodoacetyl side chain.[238]

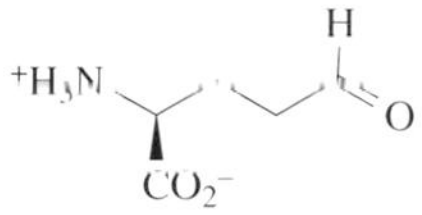

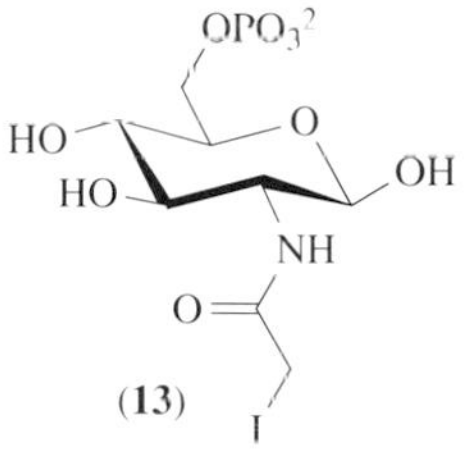

N-Acetylmuramic acid is found only in bacterial cells, and at least one inhibitor of the conversion of UDPGlcNAc to UDPMurNAc is known to have antibacterial activity. UDPGlcNAc enolpyruvyl transferase is irreversibly inactivated by the natural product fosfomycin (or phosphonomycin (**14**)). Using partially purified enzyme from *Micrococcus lysodeikticus*, Kahan *et al.*[239] were able to isolate a cysteine–fosfomycin adduct, implying that an active-site cysteine residue is responsible for inactivation by fosfomycin. The reactive cysteine residue has been identified as Cys-115, which is believed to act as an active-site acidic group in the catalytic mechanism (see Section 3.10.2.1.2).[240]

3.10.3.2   Inhibition of D-Amino Acid Biosynthesis

D-Amino acids are found only in bacterial peptidoglycan and certain peptide natural products. Therefore, inhibitors of D-amino acid biosynthesis will act as selective antibacterial agents. Inhibitors of alanine racemase, which have been reviewed,[122,241] fall into two classes: the first class are mechanism-based "suicide" inhibitors such as β-chloro-alanine (15), which incorporate an electrophilic group X at the β-carbon. β-Chloroalanine inactivates alanine racemase in a time-dependent manner by attachment to the pyridoxal 5′-phosphate cofactor, followed by α-deprotonation and elimination of HCl. The product of elimination is released from the cofactor to give a reactive aminoacrylate species (16) at the active site, which alkylates the pyridoxal–lysine adduct (Scheme 17).[242] β-Halo-D-alanines have also been found to inactivate the PLP-dependent D-amino acid aminotransferase from *B. sphaericus*, presumably via a similar mechanism.[243] β,β,β-Trifluoroalanine also inactivates alanine racemase via elimination of HF, followed by attack on the PLP–aminoacrylate adduct by the ε-amino group of lysine-38.[72]

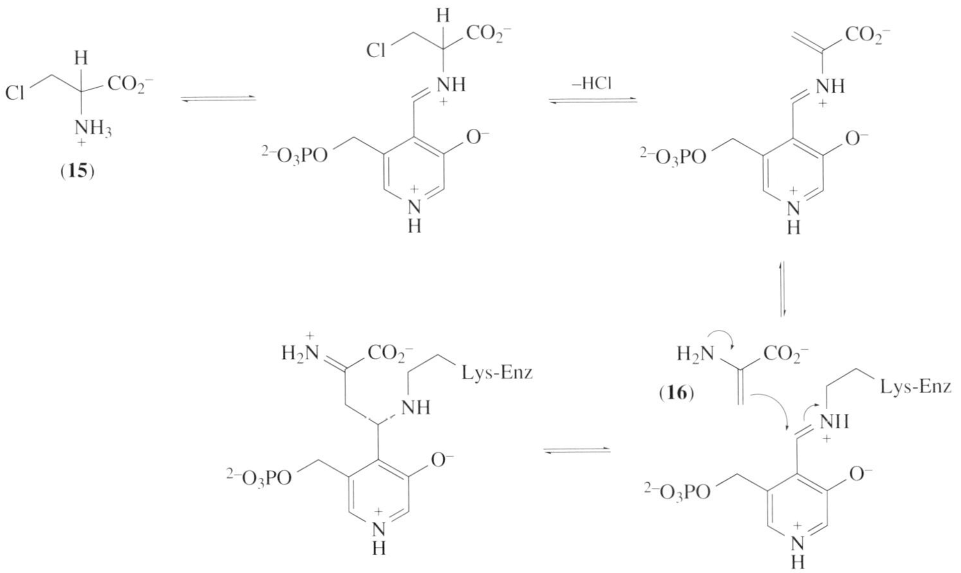

Scheme 17

The second class are slow-binding inhibitors such as alanine phosphonate (17)[244] and alanine boronate (18),[245] which form long-lived complexes at the active site of the racemase. The structure of the complex formed with (17) has been characterized by ^{15}N solid-state NMR spectroscopy, revealing that the imine linkage between pyridoxal phosphate and ^{15}N-labeled alanine phosphonate is present in the inactivated complex.[246]

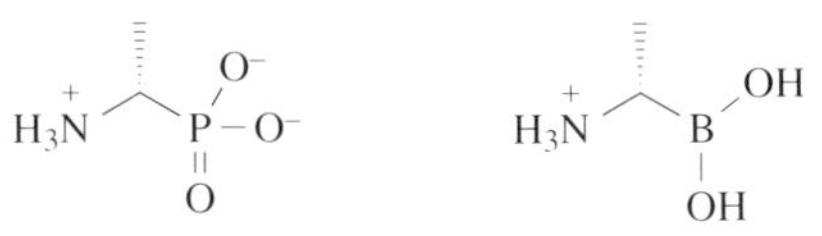

α-Halomethyl-DAPs have been found to act as potent irreversible inhibitors of the cofactor-independent DAP epimerase from *E. coli*.[99] Kinetic analysis has revealed that the enzyme catalyzes the release of halide ion, implying that aziridino-DAP (19) is an intermediate.[247] Peptide mapping studies have established that active site cysteine-73 is specifically alkylated, as shown in Scheme 18.[99] β-Fluoro-substituted *meso*-DAPs are also potent reversible inhibitors of DAP epimerase, inhibition proceeding via elimination of HF after α-deprotonation.[248,249] A series of structural analogues of 2,6-diaminopimelic acid including *meso*-lanthionine (20; $K_i = 0.18$ mmol L^{-1}) and N-hydroxy-DAP (21; $K_i = 5.6$ μmol L^{-1}) were found to be competitive inhibitors for the *E. coli* enzyme.[250] The stereoselective synthesis of phosphonate analogues of 2,6-diaminopimelic acid has been achieved; however, these analogues showed only modest inhibition ($K_i = 4$–20 mmol L^{-1}) of *E. coli* DAP epimerase and DAP dehydrogenase.[251]

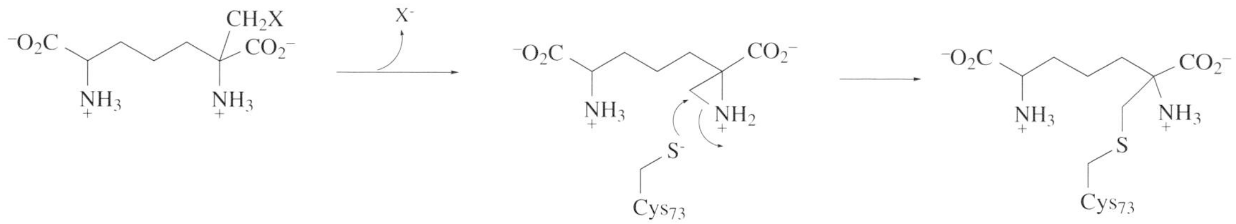

Scheme 18

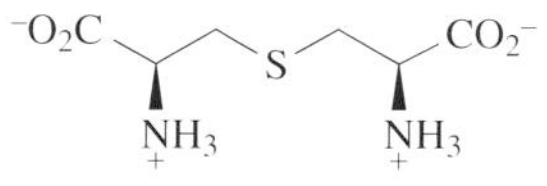

The inhibition of other enzymes in the diaminopimelic acid biosynthetic pathway has been reviewed.[96] Other targets include DAP dehydrogenase, for which the heterocyclic DAP analogue (**22**) has been shown to act as a potent competitive inhibitor ($K_i = 4.2$ μmol L^{-1}),[252] and N-succinyl-L,L-diaminopimelic acid aminotransferase, for which the hydrazine analogue (**23**) has been shown to act as a potent slow-binding inhibitor ($K_i^* = 22$ nmol L^{-1}).[253]

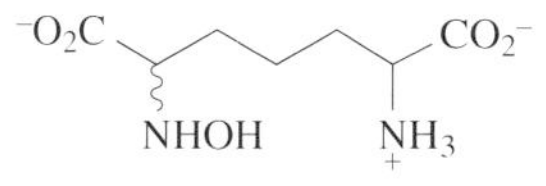

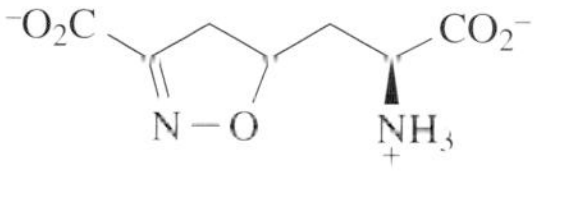

3.10.3.3 Inhibition of ATP-dependent Amino Acid Ligases

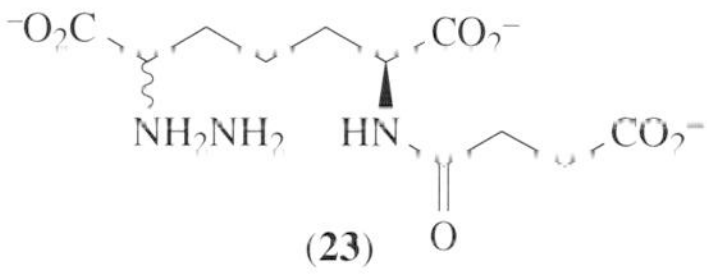

The ATP-dependent amino acid ligases which assemble the cytoplasmic precursor UDPMur*N*Ac–L-Ala–γ-D-Glu–*m*-DAP–D-Ala–D-Ala from UDPMur*N*Ac were discussed in Section 3.10.2.3. It is likely that their respective catalytic mechanisms proceed via carboxylate activation by ATP to give an acyl phosphate intermediate, followed by attack of the donor amino group.

D-Alanine:D-alanine ligase is one site of action of the antibiotic D-cycloserine (**24**),[254] a cyclic analogue of D-alanine which also inhibits alanine racemase. D-Cycloserine is a competitive inhibitor for *E. coli* DdlA and DdlB with K_i values in the range 9–27 μmol L^{-1}, similar in magnitude to the K_m values for the first D-alanine binding site, suggesting that D-cycloserine binds competitively to this site.[124]

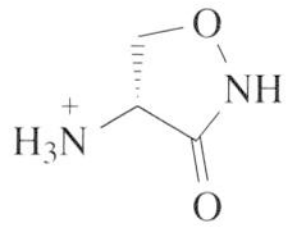

A series of aminoalkylphosphinate analogues of D-Ala–P–D-Ala (**25**) (Scheme 1), which mimic the tetrahedral transition state (**27**) formed upon attack of the second molecule of D-alanine upon the acyl phosphate intermediate, have been synthesized.[255] These phosphinates were found to be potent inhibitors of D-alanine:D-alanine ligase activity from *S. faecalis*.[255] Kinetic analysis of the inhibition of the *S. typhimurium* ligase by phosphinate (**25a**) revealed that the ATP is required, and that the onset of inhibition shows biphasic progress curves, consistent with slow-binding enzyme inhibition.[256] A very stable enzyme–inhibitor EI* complex is formed, which dissociates very slowly from the enzyme active site ($t_{1/2} = 8$ h).[256] The structure of the EI* complex formed by inhibition of *S. typhimurium* D-alanine:D-alanine ligase by phosphinate (**25b**) has been investigated by solid-state rotational resonance ^{31}P NMR spectroscopy.[257] Coupling of 1 kHz was observed between the phosphinate signal and a phosphate ester signal, which identified phosphinophosphate (**26b**) (Scheme 1) as the enzyme-bound species.[257] This structure has been further confirmed by cocrystallization of *E. coli* DdlB with ATP and phosphinate (**26b**).[126]

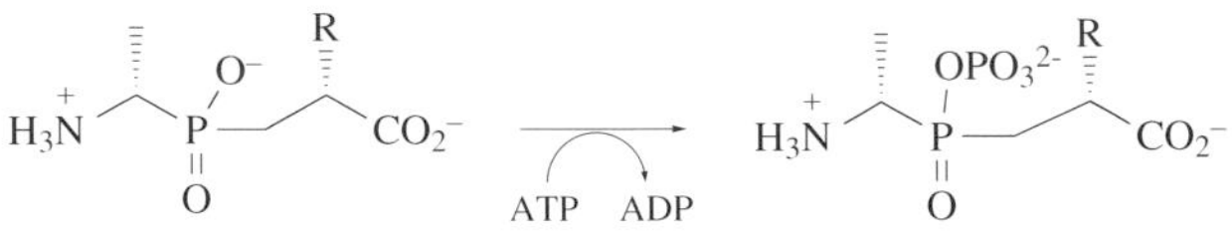

(25)

(**a**) R = C$_7$H$_{15}$
(**b**) R = Me

(26)

(**a**) R = C$_7$H$_{15}$
(**b**) R = Me

Scheme 19

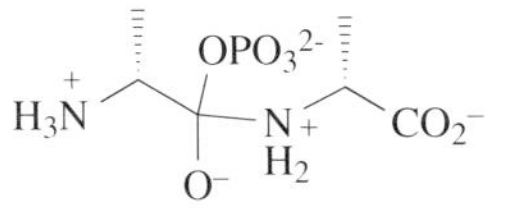

(27)

D-Alanine:D-alanine ligase is also activated by two analogues of D-alanine which act as slow-binding inhibitors. Alanine phosphonate (**17**) and the corresponding boronic acid analogue (**18**) both act as ATP-dependent slow-binding inhibitors of the *S. typhimurium* ligase.[245,256] The EI* complex formed by phosphonate (**17**) is short-lived ($t_{1/2} = 1.7$ min),[256] but the complex formed by (**18**) in the presence of D-alanine is very long-lived ($t_{1/2} = 4.5$ days).[245]

Designing mechanism-based inhibitors for the amino acid-adding enzymes which catalyze the conversion of UDPMur*N*Ac to UDPMur*N*Ac-pentapeptide represents a significant synthetic challenge in view of the structural complexity of the substrates. Aminoalkylphosphinate analogues of L-Ala–P–γ-D-Glu have been synthesized and tested as inhibitors of *E. coli* D-Glu-adding enzyme.[258] An *N*-acetyl–L-Ala–P–γ-D-Glu analogue was found to show only modest inhibition (IC$_{50}$ > 1 mmol L^{-1}), but analogue (**28**), incorporating the uridine diphosphate moiety, was a much more potent inhibitor (IC$_{50}$ = 0.68 μmol L^{-1}), indicating that the UDP group is required for efficient substrate recognition.[258]

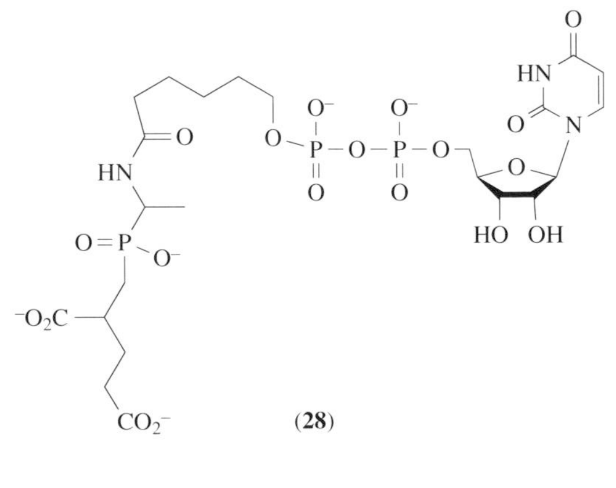

(28)

Aminoalkylphosphinate inhibitors for D-Ala–D-Ala-adding enzyme based on an L-Lys–P–Gly–D-Ala skeleton show a similar pattern for substrate recognition.[259] An *N*-acetyl–L-Lys–P–Gly–D-Ala analogue showed modest inhibition ($K_i = 0.7$ mmol L^{-1}) of the *E. coli* enzyme, whereas a more extended analogue (**29**) mimicking the preceding L-Ala and γ-D-Glu residues showed more effective inhibition ($K_i = 200$ μmol L^{-1}), which was reversible.[259]

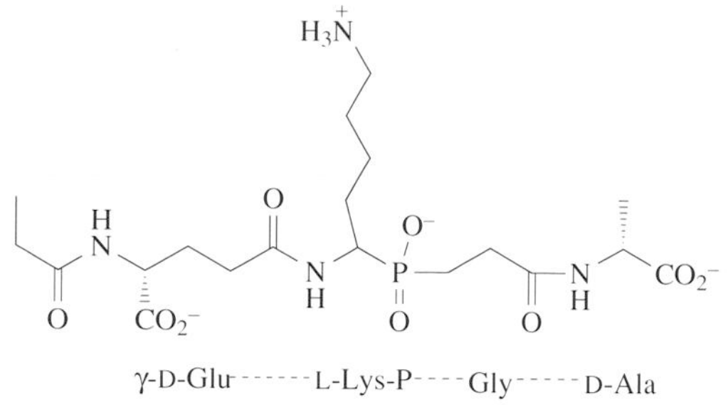

γ-D-Glu ------ L-Lys-P ---- Gly ----- D-Ala

(29)

3.10.3.4   Inhibition of Intramembrane Steps

The intramembrane steps of peptidoglycan assembly involve the assembly of a lipid-linked precursor using an undecaprenyl phosphate lipid carrier, as discussed in Section 3.10.2.4. Naturally occurring inhibitors of both translocase I and translocase II have come to light, in addition to antibacterial agents which function by complexation of the lipid carrier.

3.10.3.4.1   Inhibition of phospho-MurNAc-pentapeptide translocase (translocase I)

The transfer of phospho-MurNAc-pentapeptide to undecaprenyl phosphate, catalyzed by translocase I, is the first step of the intramembrane cycle. Three nucleoside-containing natural products ((30)–(32)) are known to inhibit translocase I. The tunicamycin family of nucleosides contains a uridine moiety attached to a disaccharide and a fatty acid side chain.[260] Tunicamycin (30) has been found to inhibit the formation of lipid intermediate I in cell-free systems.[261] Kinetic analysis of the inhibition of solubilized *E. coli* translocase I by tunicamycin using a continuous fluorescence enhancement assay revealed that the observed inhibition was reversible ($K_i = 0.55$ μmol L^{-1}) and competitive with respect to UDPMurNAc-pentapeptide, but non-competitive with respect to the lipid phosphate substrate.[262] Hence it appears that the uridine nucleoside mimics that of the soluble substrate, but that the fatty acid side chain assists membrane localization rather than mimicking the lipid phosphate substrate. Tunicamycin also inhibits the corresponding phospho-GlcNAc transferase which catalyzes the first step of eukaryotic glycoprotein biosynthesis, and is therefore not a selective antibacterial agent.[263]

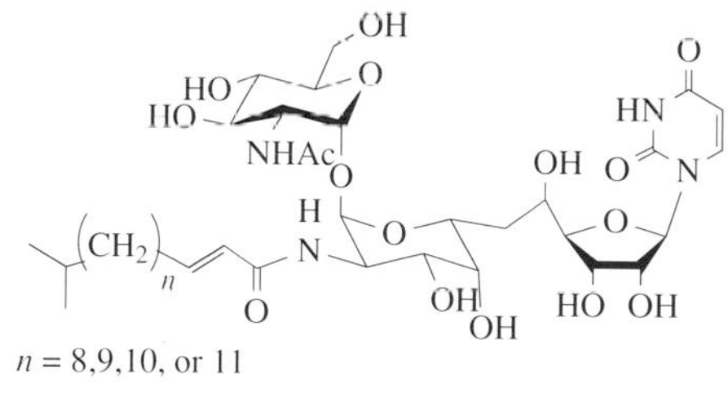

n = 8,9,10, or 11

Tunicamycin (30)

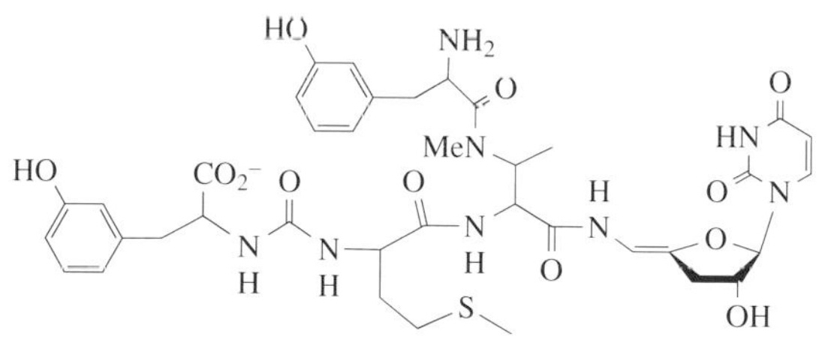

Mureidomycin A (31)

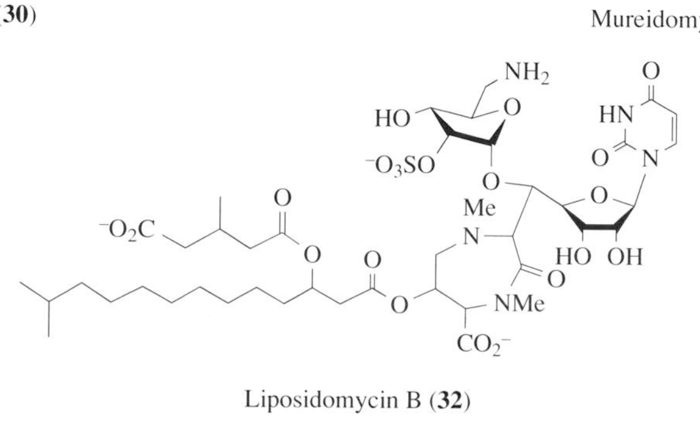

Liposidomycin B (32)

The mureidomycins are a class of peptidylnucleoside antibiotics isolated from *Streptomyces flavidovirens* which show selective antipseudomonal activity.[264] Two other families of natural product antibiotics, the pacidamycins and the napsamycins, have been found to have very similar structures.[265,266] Mureidomycin A (**31**) was shown to inhibit translocase I activity in particulate preparations of *Pseudomonas aeruginosa*, without inhibiting eukaryotic glycoprotein biosynthesis.[267,268] Continuous assays of solubilized *E. coli* translocase I in the presence of mureidomycin A gave a biphasic onset of inhibition, consistent with a slow-binding enzyme inhibition mechanism.[150] Inhibition constants of $K_i = 36$ nmol L^{-1} and $K_i^* = 2$ nmol L^{-1} were measured, and the onset of inhibition was found to be competitive with respect to both UDPMur$N$Ac-pentapeptide and the lipid phosphate substrate.[150]

The liposidomycins are a third family of nucleoside natural products found to inhibit translocase I.[269,270] Their structures contain a uridine nucleoside attached to a sulfated amino sugar, and via a seven-membered heterocyclic ring, to a fatty acid side chain. The inhibition of solubilized *E. coli* translocase I by liposidomycin B (**32**) was also found to exhibit slow-binding inhibition kinetics ($K_i^* = 80$ nmol L^{-1}).[262] The nature of the EI-to-EI* transition in the slow-binding inhibition of translocase I is yet to be determined.

3.10.3.4.2 *Inhibition of translocase II*

A cyclic depsipeptide, ramoplanin (**33**), which contains a fatty acyl side chain, has been found to inhibit the membrane steps of peptidoglycan biosynthesis in *S. aureus* and *Bacillus megaterium*.[271] Using membrane preparations of *B. megaterium*, ramoplanin was found to have no effect on translocase I activity, but was found to inhibit completely the conversion of lipid intermediate I to lipid intermediate II at a 100 μg mL^{-1} concentration.[271] It therefore appears that the primary site of action of ramoplanin is translocase II.

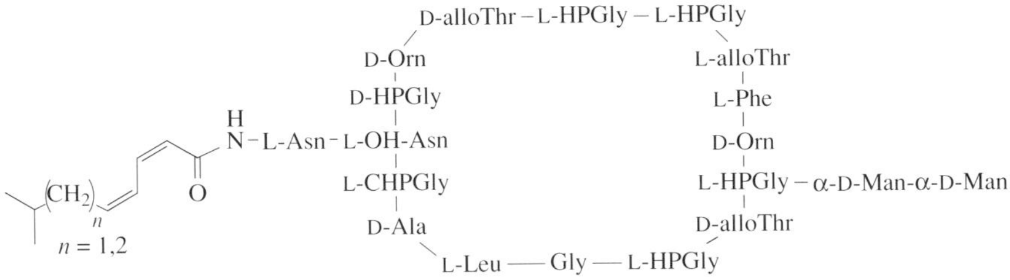

HPGly = *p*-hydroxyphenylglycine; CHPGly = *m*-chloro-*p*-hydroxyphenylglycine; Man = mannose

(**33**)
Ramoplanin

3.10.3.4.3 *Complexation of the lipid carrier*

The bacitracin family of peptide antibiotics produced by *Bacillus licheniformis* inhibit bacterial peptidoglycan biosynthesis and lead to the accumulation of UDPMur$N$Ac-pentapeptide in treated cells.[272] It has been established that bacitracin A (**34**) forms a 1:1 complex with undecaprenyl pyrophosphate in the presence of divalent metal ions such as Mg^{2+}, with an association constant of 10^6 L mol^{-1}.[273,274] The tight sequestration of undecaprenyl pyrophosphate disrupts the recycling of the lipid carrier required for the intramembrane cycle of reactions, leading to cell lysis in Gram-positive bacteria. The structure of the copper(II)–bacitracin complex has been analyzed by EPR spectroscopy.[275]

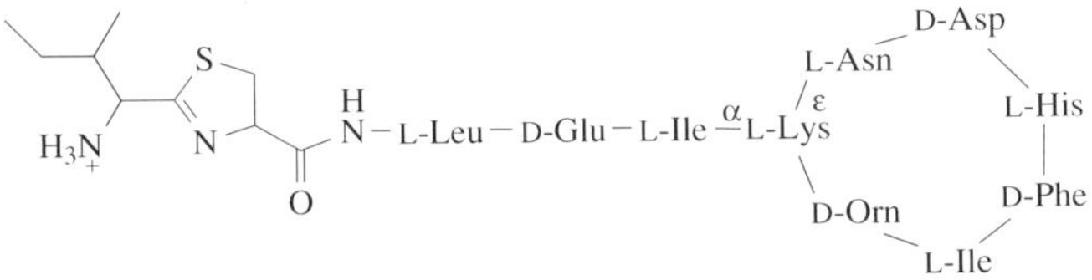

(**34**) Bacitracin A

Strains of *Bacillus subtilis* and *Bacillus thuringiensis* that exhibit bacitracin resistance produce two membrane proteins on treatment with bacitracin.[276] Overexpression of the *E. coli bacA* gene confers bacitracin resistance to susceptible *E. coli* mutants, and it is thought that the BacA gene product may be an undecaprenol kinase which is able to generate elevated levels of the phosphorylated lipid carrier.[277]

Complexation of undecaprenyl phosphate is also precedented: the peptide antibiotic amphomycin (**35**) has been shown to chelate polyprenyl phosphates in the presence of Ca^{2+} ions.[278,279] There are a number of other lipophilic peptide antibiotics, several of which have been shown to inhibit membrane-linked steps of bacterial peptidoglycan biosynthesis,[280] for which complexation of the lipid phosphate or pyrophosphate is a possible mode of action.

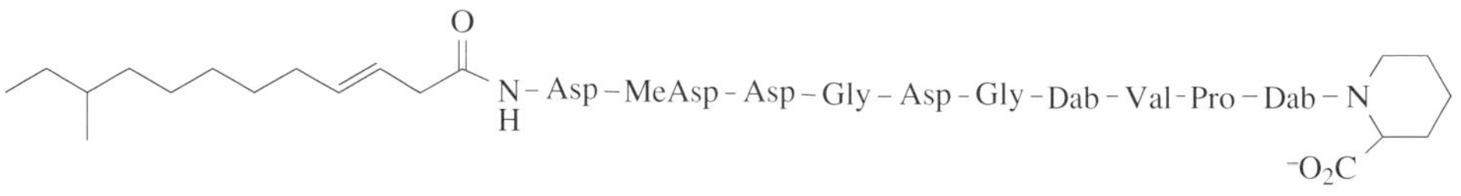

(**35**)
Amphomycin

3.10.3.5 Inhibition of Extracellular Steps of Peptidoglycan Biosynthesis

The biosynthetic reactions occurring on the cell surface are targets for a wide range of antibacterial agents. Since these steps are exposed to the cell medium, access to the target is much more straightforward than for the intracellular enzymes, particularly in Gram-positive bacteria which lack an outer membrane. Since β lactam chemistry is a field in its own right, only a brief summary covering mode of action and structural types will be given here.

3.10.3.5.1 *Inhibition of peptidoglycan transglycosylase*

Moenomycin A (**36**) is a phosphoglycolipid antibiotic isolated from *Streptomyces bambergiensis*, whose structure has been solved by degradation and spectroscopic studies.[281,282] Moenomycin A inhibits the transglycosylase activity of *E. coli* PBP 1A and PBP 2A at concentrations of 10–100 nmol L^{-1}, leading to cell lysis.[283,284] The structure of moenomycin A resembles that of the lipid intermediate II substrate for the transglycosylase enzyme, as shown in Figure 21.

In order to study the interaction of moenomycin A with its membrane target, [^{3}H]decahydro-moenomycin was prepared and incubated with membrane preparations of *E. coli*.[284] Analysis of the incubation by SDS–PAGE followed by autoradiography revealed that no membrane proteins had been labeled, indicating that the inhibition by moenomycin A was not irreversible.[284] Given the structural similarity with the lipid-linked substrate, it seems likely that moenomycin A acts as a reversible competitive inhibitor for transglycosylase. An active disaccharide fragment of moenomycin A has been prepared by degradation of the natural product,[285,286] and a synthetic phosphonate ester analogue of this disaccharide has been synthesized.[287]

The lantibiotic mersacidin, a posttranslationally modified peptide of molecular weight 1825 Da, has also been found to inhibit the incorporation of [^{14}C]UDPGlc*N*Ac into polymeric peptidoglycan at a 100 μg mL^{-1} concentration using an *E. coli* membrane preparation.[288] It is possible that this antibiotic acts via inhibition of the transglycosylase activity, unlike other members of the lantibiotic family, which are thought to act via pore formation in the cytoplasmic membrane.[289]

3.10.3.5.2 *Complexation of peptidyl–D-Ala–D-Ala cell surface intermediates*

The glycopeptide family of antibiotics, typified by vancomycin and teicoplanin, are active against a wide range of Gram-positive bacteria. Their mode of action proceeds via complexation of the peptidyl–D-Ala–D-Ala termini on the cell surface, thus inhibited both transglycosylation and transpeptidation. The complexation of peptidyl–D-Ala–D-Ala by vancomycin involves the formation of a series of specific hydrogen bonds, as shown in Figure 22, and this process of molecular recognition has been reviewed.[213,290]

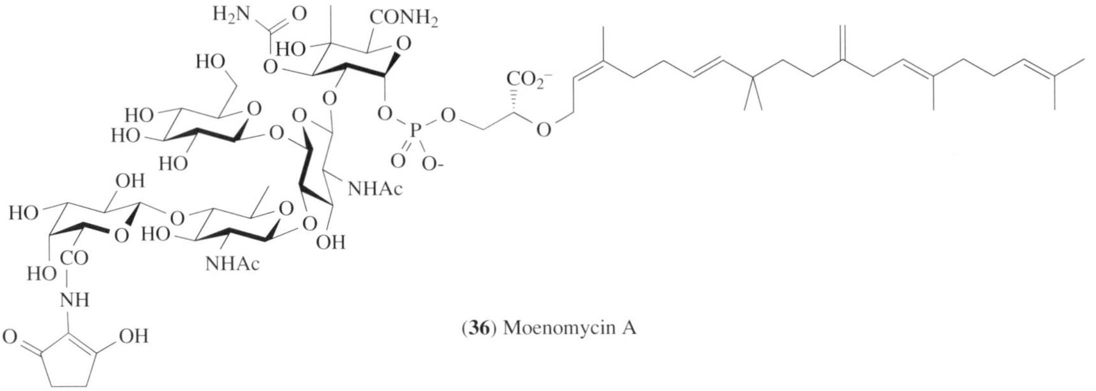

(**36**) Moenomycin A

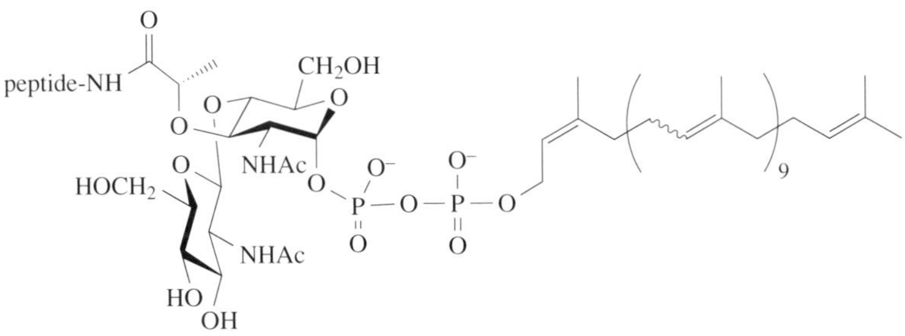

Lipid intermediate II

Figure 21   Structures of moenomycin A and lipid intermediate II.

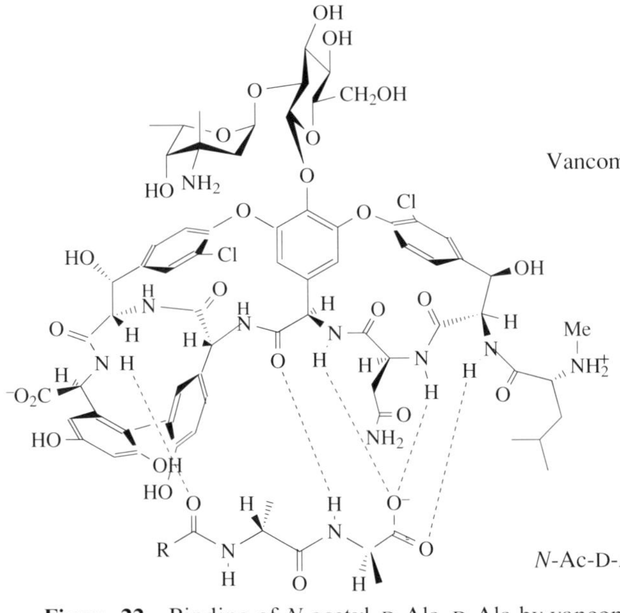

Vancomycin

N-Ac-D-Ala-D-Ala

Figure 22   Binding of *N*-acetyl–D-Ala–D-Ala by vancomycin.

The incidence of vancomycin resistance in the late 1980s in previously susceptible strains of *Enterococcus* prompted a major research effort into the mechanism of resistance. As explained in Section 3.10.2.7, the resistant strains were found to contain altered peptidoglycan precursors, notably the terminal -D-Ala–D-lactate instead of -D-Ala–D-Ala, which showed low affinity toward vancomycin.[34,215,220]

Since 1990, efforts have been directed towards the discovery of modified glycopeptides able to bind the peptidyl–D-Ala–D-lactate termini. Several other glycopeptide natural products have been isolated, and many semisynthetic glycoproteins have been prepared, some of which are active against the vancomycin-resistant *Enterococci*.[291,292] At the same time, it was found that several glycopeptides, notably eremomycin, showed a propensity to form dimers, through surface contacts on the opposite face of the antibiotic to that involved in ligand binding.[293] It was found that the dimerization was enhanced by the presence of cell wall analogues,[293] and furthermore that a consequence of dimerization was to increase the antibacterial potency.[294] This effect can be rationalized as shown in Figure 23: binding of one D-Ala–D-Ala terminus is followed by dimerization of the glycopeptide; binding of the second D-Ala–D-Ala terminus is therefore effectively an intramolecular process.[290,294] This model has been tested by binding studies of decanoyl–L-Lys–D-Ala–D-Ala, which mimics the lipid intermediate II, contained in detergent micelles.[295] This analogue was found to bind to the glycopeptide ristocetin A 100 times more tightly than the soluble analogue N^α-acetyl–L-Lys–D-Ala–D-Ala, owing to the cooperative binding effect upon dimerization.[295]

Other glycopeptides such as teicoplanin do not form dimers, but contain a lipophilic tail which localizes the antibiotic in the cytoplasmic membrane, thus increasing the effective concentration for binding.[293] A semi-synthetic glycopeptide, LY264826 (**37**), containing a *p*-chlorobenzyl side chain, has been found to show activity against vancomycin-resistant strains.[296] Although LY264826 binds peptidyl–D-Ala–D-Lac analogues relatively weakly in solution, it dimerizes strongly and is also membrane associated through the hydrophobic appendage.[297] It is thought that the activity of this glycopeptide is due to a combination of membrane localization, dimerization, and cooperative ligand binding,[290,297] as shown in Figure 23.

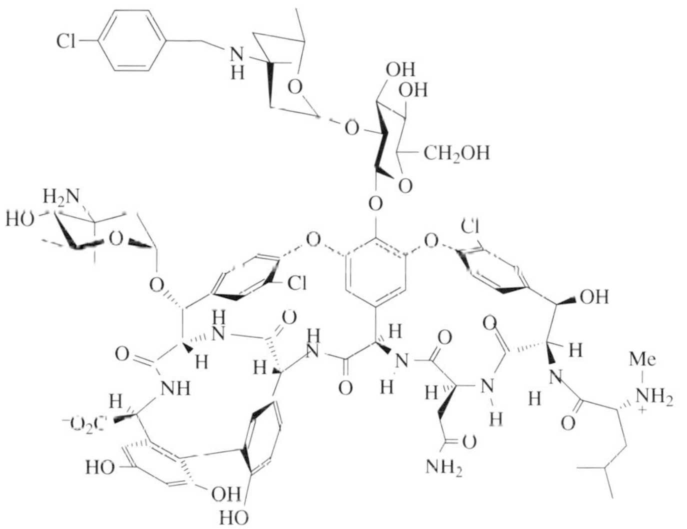

LY264826
(**37**)

3.10.3.5.3 *Inhibition of transpeptidases by β-lactam antibiotics*

The final step of peptidoglycan assembly, the transpeptidation of pentapeptide side chains on parallel glycan strands, is inhibited by the β-lactam class of antibiotics, typified by the penicillins. As explained in Section 3.10.2.5, the multiple transpeptidase enzymes are identified by their ability to bind [^{14}C]penicillin and are known collectively as the penicillin-binding proteins (PBPs).

A rationalization of the inhibitory action of penicillin was proposed in 1965 by Tipper and Strominger,[298] who suggested that the three-dimensional structure of penicillin mimics the peptidyl–D-Ala–D-Ala terminus of peptidoglycan precursors. The mechanism of inactivation involves attack of the catalytic serine residue on the β-lactam carbonyl, leading to the opening of the β-lactam ring and the formation of a stable covalent adduct (see Scheme 15). The bacteriocidal effects of the penicillins are not simply due to inactivation of PBPs: cell lysis is caused by the action of the cell's autolytic enzymes, a process which has been reviewed.[299]

A wide range of natural and semisynthetic β-lactam antibiotics have been prepared, which show differing antibacterial spectra. The chemistry and activity of the β-lactam antibiotics have been extensively reviewed,[300–302] hence the present discussion will mention just a few selected examples.

The most common isolated penicillin is penicillin G (**38**), which contains a phenylacetyl side chain. The side chain can be hydrolyzed using the enzyme penicillin acylase, yielding 6-amino-penicillanic acid (**39**) (Scheme 20). The free amino terminus of (**39**) can be readily acylated to give

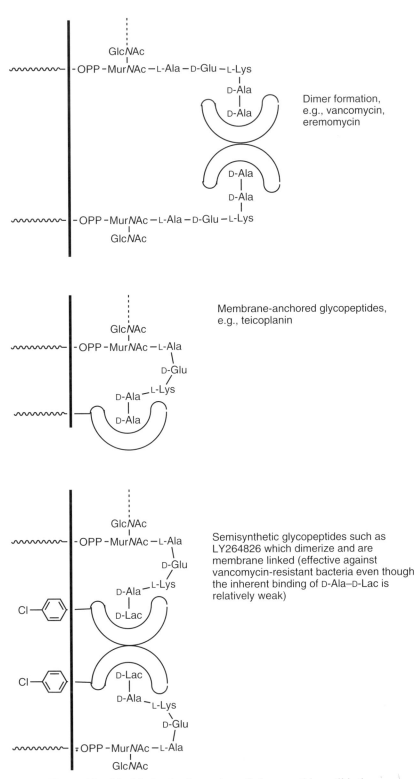

Figure 23   Models for *in vivo* action of glycopeptide antibiotics.

a range of semisynthetic penicillins such as ampicillin (**40**) and methicillin (**41**), which have wider antibacterial spectra than penicillin G.[300,301]

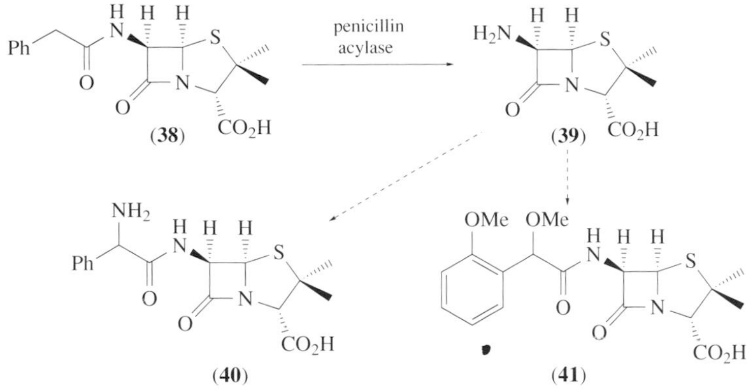

Scheme 20

The cephalosporins also contain a β-lactam ring which is attached to a six- instead of a five-membered ring. The cephalosporins show greater resistance to β-lactamase enzymes which are capable of breaking down the penicillins (see Section 3.10.2.5.3). Examples of cephalosporins are cephalosporin C (**42**) and cephamycin C (**43**).[301] Many semisynthetic cephalosporins have been prepared, several of which show improved β-lactamase stability.[301] Other classes of β-lactam natural products which show potent biological activity have also been isolated, such as thienamycin (**44**).[302]

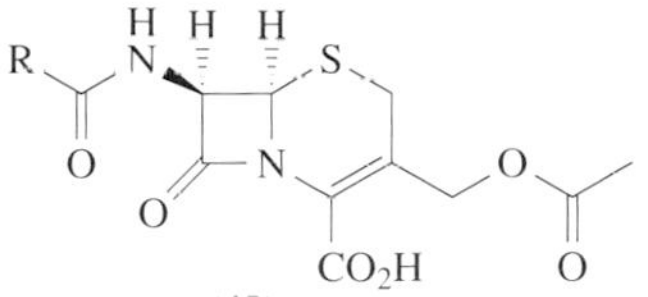

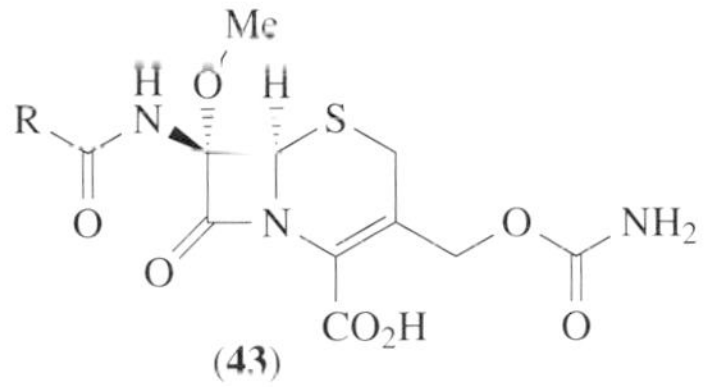

The penicillins and cephalosporins are widely used as clinical antibiotics; however, their use is increasingly undermined by bacterial β-lactam resistance.[303] The major mechanism of resistance is hydrolysis of the β-lactam by β-lactamase enzymes,[304–307] which has led to the development of selective β-lactamase inhibitors such as clavulanic acid (**45**).[302] However, resistance is in many cases due to alterations in the structure of the PBP targets, which can lead to active PBPs with low affinity towards β-lactams, such as the PBP 2a from methicillin-resistant *S. aureus* (MRSA).[308] For a more detailed discussion of this large topic, the reader is directed to the reviews cited above.[300–302]

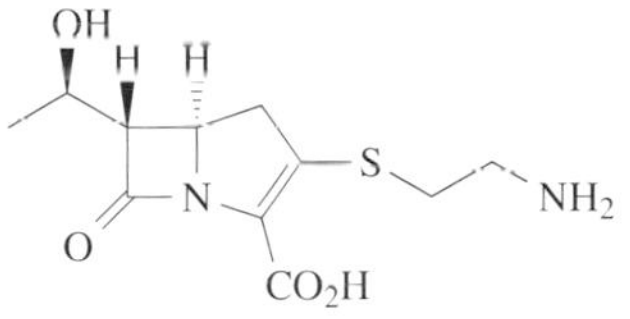

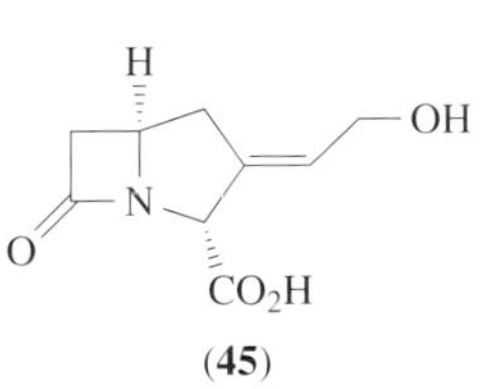

3.10.3.5.4 *Inhibition of autolytic enzymes*

The bulgecins (**46**) are a class of natural products which show an unusual type of biological activity. When bulgecins are applied in combination with a penicillin to enteric bacteria such as *E. coli*, they cause bulges in the bacterial cell which are visible under the microscope, eventually leading to cell lysis.[309,310] These natural products have been found to act as non-competitive inhibitors for the soluble *E. coli* lytic transglycosylase, which catalyzes the cleavage of the glycan strand of peptidoglycan.[311] Inhibition of the autolytic response upsets the balance of cell wall synthesis and breakdown, leading to the observed changes in cell shape. The structure of the bulgecin family shows some similarity to the structure of the Glc$N$Ac–Mur$N$Ac site of action of the lytic transglycosylase, and the cocrystallization of bulgecin A with this enzyme has allowed the elucidation of active-site interactions.[196]

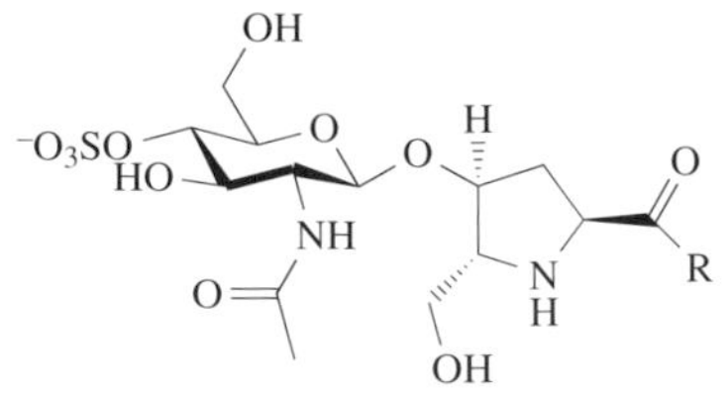

(**46**)

Bulgecin A R = NHCH$_2$CH$_2$SO$_3^-$
Bulgecin B R = NHCH$_2$CH$_2$CO$_2^-$
Bulgecin C R = OH

3.10.4   RELATIONSHIP TO OTHER LIPID-LINKED OLIGOSACCHARIDE BIOSYNTHETIC PATHWAYS

The pathway for bacterial peptidoglycan biosynthesis has many unique features, such as the use of D-amino acids and the final transpeptidation step. However, certain features of peptidoglycan biosynthesis, notably those concerning the membrane cycle of reactions, find parallels in other pathways used for the biosynthesis of oligo- and polysaccharides.[312]

3.10.4.1   Relationship to Fungal Cell Wall Biosynthesis

The major components of fungal cell walls are chitin, β-glucan, and mannoproteins.[313] Chitin is a linear homopolymer composed of β-1,4-linked *N*-acetylglucosamine units. The structure of chitin is therefore related to the glycan structure found in peptidoglycan, but it lacks the lactyl-pentapeptide appendages of the bacterial cell wall.

Fungal cell walls are assembled by membrane-bound chitin synthases, which utilize UDP-Glc*N*Ac as a substrate. Three chitin synthases are found in fungi and yeasts.[314] Chitin synthase is inhibited by two families of antifungal agents: the polyoxins (**47**)[315] and the nikkomycins.[316,317] These nucleoside natural products are competitive inhibitors ($K_i = 0.1–1~\mu$mol L^{-1}) which mimic the uridine-diphospho sugar substrate in a similar fashion to inhibitors of bacterial translocase I such as tunicamycin.

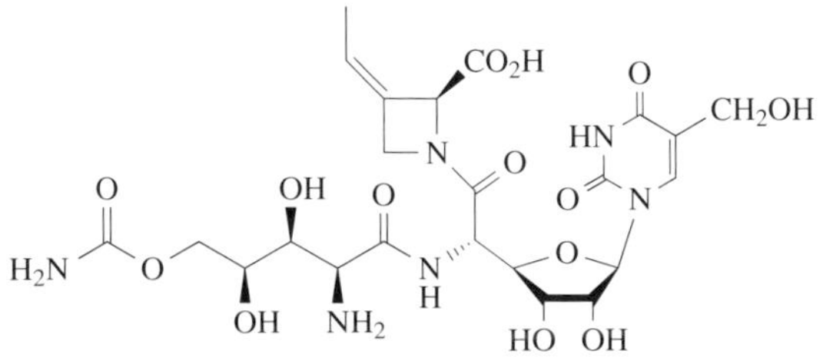

Polyoxin A
(**47**)

3.10.4.2   Relationship to Eukaryotic *N*-Linked Glycoprotein Biosynthesis

The oligosaccharide core of eukaryotic asparagine-linked glycoproteins is assembled via a lipid-linked pathway which bears several similarities to the membrane cycle utilized in bacterial peptidoglycan biosynthesis, as shown in Figure 24.[318,319] The pathway involves a lipid carrier, dolichyl

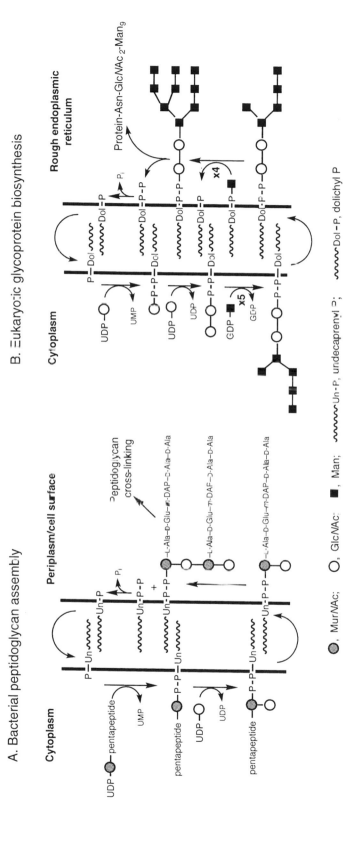

Figure 24 Comparison of lipid-linked cycles for oligosaccharide biosynthesis.

phosphate (**48**), a polyprenyl phosphate related to undecaprenyl phosphate but containing a saturated isoprene unit adjacent to the phosphate terminus.

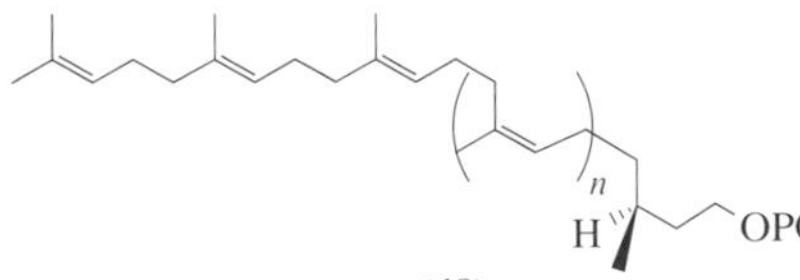

(**48**)
Dolichyl phosphate (n = 13–17)

The eukaryotic cycle commences with the attachment of UDP-GlcNAc to dolichyl phosphate, forming dolichyl-diphospho-GlcNAc and releasing UMP. This phosphotransfer reaction is similar to that catalyzed by bacterial translocase I, and 30–40% sequence identity has been detected between the sequences of the *E. coli* and yeast proteins.[149] The second step in the eukaryotic pathway is the attachment of a second GlcNAc residue, just like the bacterial pathway. Thereafter, five successive mannose residues are added, using GDP-mannose as substrate. The GlcNAc$_2$Man$_5$ unit is then flipped across the membrane of the endoplasmic reticulum, and four further mannose units are added, using dolichyl-phospho-mannose as the donor. The GlcNAc$_2$Man$_9$ core is then transferred to the glycosylation site on the protein, and dolichyl pyrophosphate is recycled by dephosphorylation.[318,319]

Hence there are the following similarities in strategy between the two cycles: (i) use of a phosphorylated polyisoprenoid lipid carrier; (ii) addition of the first sugar unit to the prenyl phosphate by an integral membrane protein, releasing a nucleotide monophosphate; (iii) addition of subsequent sugar units by membrane-associated proteins, releasing a nucleotide diphosphate; and (iv) recycling of the lipid carrier by dephosphorylation and flipping across the membrane.[312]

The antibiotic tunicamycin, which inhibits bacterial translocase I, inhibits the eukaryotic counterpart enzyme 1500 times more tightly than the bacterial enzyme, and consequently tunicamycin is too toxic for clinical use as an antibiotic.[268,320] However, the slow-binding inhibitor mureidomycin A shows 2000-fold selectivity for the bacterial enzyme,[268] hence there appear to be differences in selectivity between the two systems.

3.10.4.3 Other Lipid-linked Oligosaccharide Biosynthetic Pathways

Similar lipid-linked cycles are also used for the assembly of several bacterial exopolysaccharides.[312] The polysaccharide component of outer membrane lipopolysaccharide found in Gram-negative bacteria is also assembled by a lipid-linked pathway, which has been analyzed in detail for the lipid A component of *Salmonella typhimurium*.[321] This pathway utilizes undecaprenyl phosphate as a lipid carrier, to which is first attached UDP-galactose, with loss of UMP. Successive additions of rhamnose, mannose, and abequose residues are followed by flipping across the cytoplasmic membrane and transglycosylation of the lipid-linked tetrasaccharide.[321] Many of the genes involved in this cycle are found in the *rfb* cluster of genes in *S. typhimurium*.[322]

Succinoglycan, an acidic polysaccharide produced by *Rhizobium meliloti*, is also assembled via a similar strategy.[164] UDP-galactose is attached to undecaprenyl phosphate by a transmembrane protein, ExoY.[323] Subsequently, seven UDP-glucose units are attached, to which are added acetyl, succinyl, and pyruvyl modifications, before transglycosylation takes place.[164]

Lipid-linked saccharides are also found in plants, where they are utilized for the biosynthesis of plant glycoproteins, via a cycle very similar to that found in animals.[324,325] There is some evidence that lipid-linked glucose units can be incorporated into plant polysaccharides.[326]

In summary, the lipid-linked cycle used for bacterial peptidoglycan biosynthesis is similar to lipid-linked cycles for oligo- and polysaccharide biosynthesis in prokaryotic and eukaryotic systems. In such systems a lipid carrier is used to solve a topological problem of how to build a glycan chain on the outside of a biological membrane, using components which are biosynthesized on the inside. The complexities of peptidoglycan cross-linking are, however, unique to the bacterial system. The advent of modern genetics is facilitating the detailed study of such pathways at the molecular level, leading to the discovery of new biosynthetic pathways and new possible targets for antibiotic action.

3.10.5 REFERENCES

1. H. J. Rogers, H. R. Perkins, and J. B. Ward, "Microbial Cell Walls and Membranes," Chapman & Hall, London, 1980.
2. M. R. J. Salton, "The Bacterial Cell Wall," Elsevier, New York, 1964.
3. P. Actor, L. Daneo-More, M. L. Higgins, M. R. J. Salton, and G. D. Shockman (eds.), "Antibiotic Inhibition of Bacterial Cell Surface Assembly and Function," American Society for Microbiology, Washington, DC, 1988.
4. J. M. Ghuysen, *Bacteriol. Rev.*, 1968, **32**, 425.
5. K. H. Schleifer and O. Kandler, *Bacteriol. Rev.*, 1972, **36**, 407.
6. S. M. Hammond, P. A. Lambert, and A. Rycroft, "The Bacterial Cell Surface," Croom Helm, Beckenham, 1983.
7. C. Gram, *Fortschr. Med.*, 1884, **2**, 185.
8. M. R. J. Salton, *J. Gen. Microbiol.*, 1963, **30**, 223.
9. H. Labischinski, E. W. Goodell, A. Goodell, and M. L. Hochberg, *J. Bacteriol.*, 1991, **173**, 751.
10. J. A. Hobot, E. Carlemalm, W. Villiger, and E. Kellenberger, *J. Bacteriol.*, 1984, **160**, 143.
11. A. J. Dijkstra and W. Keck, *J. Bacteriol.*, 1996, **178**, 5555.
12. A. H. Hughes, I. C. Hancock, and J. Baddiley, *Biochem. J.*, 1973, **132**, 83.
13. D. J. Tipper, J. L. Strominger, and J. C. Ensign, *Biochemistry*, 1967, **6**, 906.
14. M. Leyh-Bouille, J. M. Ghuysen, D. J. Tipper, and J. L. Strominger, *Biochemistry*, 1966, **5**, 3079.
15. L. Essers and H. J. Schoop, *Biochim. Biophys. Acta*, 1978, **544**, 180.
16. A. D. Warth and J. L. Strominger, *Proc. Natl. Acad. Sci. USA*, 1969, **64**, 528.
17. J.-M. Ghuysen and J. L. Strominger, *Biochemistry*, 1963, **2**, 1119.
18. U. Schwarz and B. Glauner, in "Antibiotic Inhibition of Bacterial Cell Surface Assembly and Function," eds P. Actor, L. Daneo-Moore, M. L. Higgins, M. R. J. Salton, and G. D. Shockman, American Society for Microbiology, Washington, DC, 1988, p. 33.
19. B. Glauner, *Anal. Biochem.*, 1988, **172**, 451.
20. R. L. Kisliuk, *Mol. Cell. Biochem.*, 1981, **39**, 331.
21. F. A. Troy, *J. Biol. Chem.*, 1973, **248**, 305.
22. K. H. Schleifer, R. Plapp, and O. Kandler, *Biochem. Biophys. Res. Commun.*, 1967, **28**, 566.
23. H. Diringer and D. Jusic, *Z. Naturforsch. Teil B*, 1966, **21**, 603.
24. E. Bricas, J. M. Ghuysen, and P. Dezelee, *Biochemistry*, 1967, **6**, 2598.
25. A. Fredriksen, E. N. Vasstrand, and H. B. Jensen, *J. Bacteriol.*, 1991, **173**, 900.
26. S. Handwerger, M. J. Pucci, K. J. Volk, J. Liu, and M. S. Lee, *J. Bacteriol.*, 1992, **174**, 5982.
27. J. Messer and P. E. Reynolds, *FEMS Microbiol. Lett.*, 1992, **73**, 195.
28. N. E. Allen, J. N. Hobbs, J. M. Richardson, and R. M. Riggin, *FEMS Microbiol. Lett.*, 1992, **77**, 109.
29. S. Handwerger, M. J. Pucci, K. J. Volk, J. Liu, and M. S. Lee, *J. Bacteriol.*, 1994, **176**, 260.
30. P. E. Reynolds, H. A. Snaith, A. J. Maguire, S. Dutka-Malen, and P. Courvalin, *Biochem. J.*, 1994, **301**, 5.
31. B. Glauner, J. V. Höltje, and U. Schwarz, *J. Biol. Chem.*, 1988, **263**, 10088.
32. J. M. Ghuysen, E. Bricas, M. Lache, and M. Leyh-Bouille, *Biochemistry*, 1968, **7**, 1450.
33. H. R. Perkins, *Biochem. J.*, 1967, **102**, 29c.
34. T. D. H. Bugg and C. T. Walsh, *Nat. Prod. Rep.*, 1992, **9**, 199.
35. B. Badet, P. Vermoote, P. Y. Haumont, F. Lederer, and F. Le Goffic, *Biochemistry*, 1987, **26**, 1940.
36. B. Golinelli-Pimpaneau, F. Le Goffic, and B. Badet, *J. Am. Chem. Soc.*, 1989, **111**, 3029.
37. G. Obmolova, M. A. Badet-Denisot, B. Badet, and A. Teplyakov, *J. Mol. Biol.*, 1994, **242**, 703.
38. D. Mengin-Lecreulx and J. van Heijenoort, *J. Bacteriol.*, 1993, **175**, 6150.
39. D. Mengin-Lecreulx and J. van Heijenoort, *J. Bacteriol.*, 1994, **176**, 5788.
40. A. M. Gehring, W. J. Lees, D. J. Mindiola, C. T. Walsh, and E. D. Brown, *Biochemistry*, 1996, **35**, 579.
41. J. E. Heidlas, W. J. Lees, P. Pale, and G. M. Whitesides, *J. Org. Chem.*, 1992, **57**, 146.
42. J. L. Marquardt, D. A. Siegele, R. Kolter, and C. T. Walsh, *J. Bacteriol.*, 1992, **174**, 5748.
43. C. Wanke, R. Falchetto, and N. Amrhein, *FEBS Lett.*, 1992, **301**, 271.
44. K. S. Anderson, J. A. Sikorski, A. J. Benesi, and K. A. Johnson, *J. Am. Chem. Soc.*, 1988, **110**, 6577.
45. J. L. Marquardt, E. D. Brown, C. T. Walsh, and K. S. Anderson, *J. Am. Chem. Soc.*, 1993, **115**, 10398.
46. C. Wanke and N. Amrhein, *Eur. J. Biochem.*, 1993, **218**, 861.
47. D. H. Kim, W. J. Lees, K. E. Kempsell, W. S. Lane, K. Duncan, and C. T. Walsh, *Biochemistry*, 1996, **35**, 4923.
48. E. D. Brown, J. L. Marquardt, J. P. Lee, C. T. Walsh, and K. S. Anderson, *Biochemistry*, 1994, **33**, 10638.
49. C. Ramilo, R. J. Appleyard, C. Wanke, F. Krekel, N. Amrhein, and J. N. Evans, *Biochemistry*, 1994, **33**, 15071.
50. D. H. Kim, W. J. Lees, and C. T. Walsh, *J. Am. Chem. Soc.*, 1994, **116**, 6478.
51. D. H. Kim, W. J. Lees, and C. T. Walsh, *J. Am. Chem. Soc.*, 1995, **117**, 6380.
52. D. H. Kim, G. W. Tucker-Kellogg, W. J. Lees, and C. T. Walsh, *Biochemistry*, 1996, **35**, 5435.
53. W. J. Lees and C. T. Walsh, *J. Am. Chem. Soc.*, 1995, **117**, 7329.
54. M. J. Pucci, L. F. Discotto, and T. J. Dougherty, *J. Bacteriol.*, 1992, **174**, 1690.
55. T. E. Benson, J. L. Marquardt, A. C. Marquardt, F. A. Etzkorn, and C. T. Walsh, *Biochemistry*, 1993, **32**, 2024.
56. W. J. Lees, T. E. Benson, J. M. Hogle, and C. T. Walsh, *Biochemistry*, 1996, **35**, 1342.
57. A. M. Dhalla, J. Yanchunas, Jr., H. T. Ho, P. J. Falk, J. J. Villafranca, and J. G. Robertson, *Biochemistry*, 1995, **34**, 5390.
58. T. E. Benson, D. J. Filman, C. T. Walsh, and J. M. Hogle, *Nat. Struct. Biol.*, 1995, **2**, 644.
59. T. E. Benson, C. T. Walsh, and J. M. Hogle, *Biochemistry*, 1997, **36**, 806.
60. H. Kleinkauf and H. von Dohren, *Eur. J. Biochem.*, 1990, **192**, 1.

61. E. Adams, *Adv. Enzymol.*, 1976, **44**, 69.
62. K. Soda and N. Esaki, *Pure Appl. Chem.*, 1994, **66**, 709.
63. G. Rosso, K. Takashima, and E. Adams, *Biochem. Biophys. Res. Commun.*, 1969, **34**, 134.
64. S. A. Wasserman, E. Daub, P. Grisafi, D. Botstein, and C. T. Walsh, *Biochemistry*, 1984, **23**, 5182.
65. N. Esaki and C. T. Walsh, *Biochemistry*, 1986, **25**, 3261.
66. B. Badet and C. T. Walsh, *Biochemistry*, 1985, **24**, 1333.
67. D. Roise, Ph.D. Thesis, Massachusetts Institute of Technology, 1984.
68. K. Inaguki, K. Tanizawa, B. Badet, C. T. Walsh, H. Tanaka, and K. Soda, *Biochemistry*, 1986, **25**, 3268.
69. D. J. Neidhart, M. D. Distefano, K. Tanizawa, K. Soda, C. T. Walsh, and G. A. Petsko, *J. Biol. Chem.*, 1987, **262**, 15 323.
70. S. A. Wasserman, C. T. Walsh, and D. Botstein, *J. Bacteriol.*, 1983, **153**, 1439.
71. M. Lobocka, J. Hennig, J. Wild, and T. Klopotowski, *J. Bacteriol.*, 1994, **176**, 1500.
72. W. S. Faraci and C. T. Walsh, *Biochemistry*, 1988, **27**, 3267.
73. S.-J. Shen, H. G. Floss, H. Kumagai, H. Yamada, N. Esaki, K. Soda, S. A. Wasserman, and C. T. Walsh, *J. Chem. Soc., Chem. Commun.*, 1983, 82.
74. J. P. Shaw, G. A. Petsko, and D. Ringe, *Biochemistry*, 1997, **36**, 1329.
75. N. Nakajima, K. Tanizawa, H. Tanaka, and K. Soda, *Agric. Biol. Chem.*, 1988, **52**, 3099.
76. N. Nakajima, K. Tanizawa, H. Tanaka, and K. Soda, *Agric. Biol. Chem.*, 1986, **50**, 2823.
77. S. Y. Choi, N. Esaki, T. Yoshimura, and K. Soda, *J. Biochem. (Tokyo)*, 1992, **112**, 139.
78. G. J. Cardinale and R. H. Abeles, *Biochemistry*, 1968, **7**, 3970.
79. K. A. Gallo and J. R. Knowles, *Biochemistry*, 1993, **32**, 3981.
80. K. A. Gallo, M. E. Tanner, and J. R. Knowles, *Biochemistry*, 1993, **32**, 3991.
81. M. E. Tanner, K. A. Gallo, and J. R. Knowles, *Biochemistry*, 1993, **32**, 3998.
82. P. Doublet, J. van Heijenoort, and D. Mengin-Lecreulx, *J. Bacteriol.*, 1992, **174**, 5772.
83. P. Doublet, J. van Heijenoort, J. P. Bohin, and D. Mengin-Lecreulx, *J. Bacteriol.*, 1993, **175**, 2970.
84. M. J. Pucci, J. Novotny, L. F. Discotto, and T. J. Dougherty, *J. Bacteriol.*, 1994, **176**, 528.
85. P. Doublet, J. van Heijenoort, and D. Mengin-Lecreulx, *Biochemistry*, 1994, **33**, 5285.
86. H. T. Ho, P. J. Falk, K. M. Ervin, B. J. Krishnan, L. F. Discotto, T. J. Dougherty, and M. J. Pucci, *Biochemistry*, 1995, **34**, 2464.
87. H. Hayashi, H. Wada, T. Yoshimura, N. Esaki, and K. Soda, *Annu. Rev. Biochem.*, 1990, **59**, 87.
88. M. Martinez-Carrion and W. T. Jenkins, *J. Biol. Chem.*, 1965, **240**, 3538.
89. K. Yonaha, H. Misono, T. Yamamoto, and K. Soda, *J. Biol. Chem.*, 1975, **250**, 6983.
90. K. Tanizawa, Y. Masu, S. Asano, H. Tanaka, and K. Soda, *J. Biol. Chem.*, 1989, **264**, 2445.
91. K. Tanizawa, S. Asano, Y. Masu, S. Kuramitsu, H. Kagamiyama, H. Tanaka, and K. Soda, *J. Biol. Chem.*, 1989, **264**, 2450.
92. T. Yoshimura, K. Nishimura, J. Ito, N. Esaki, H. Kagamiyama, J. M. Manning, and K. Soda, *J. Am. Chem. Soc.*, 1993, **115**, 3897.
93. S. Sugio, G. A. Petsko, J. M. Manning, K. Soda, and D. Ringe, *Biochemistry*, 1995, **34**, 9661.
94. M. J. Pucci, J. A. Thanassi, H. T. Ho, P. J. Falk, and T. J. Dougherty, *J. Bacteriol.*, 1995, **177**, 336.
95. J. D. Cirillo, T. R. Weisbrod, A. Banerjee, B. R. Bloom, and W. R. Jacobs, *J. Bacteriol.*, 1994, **176**, 4424.
96. R. J. Cox, *Nat. Prod. Rep.*, 1996, **13**, 29.
97. J. S. Wiseman and J. S. Nichols, *J. Biol. Chem.*, 1984, **259**, 8907.
98. C. Richaud, W. Higgins, D. Mengin-Lecreulx, and P. Stagier, *J. Bacteriol.*, 1987, **169**, 1454.
99. W. Higgins, C. Tardif, C. Richaud, M. A. Krivanek, and A. Cardin, *Eur. J. Biochem.*, 1989, **186**, 137.
100. D. Mengin-Lecreulx, C. Michaud, C. Richaud, D. Blanot, and J. van Heijenoort, *J. Bacteriol.*, 1988, **170**, 2031.
101. H. C. Lamont, W. L. Staudenbauer, and J. L. Strominger, *J. Biol. Chem.*, 1972, **247**, 5103.
102. T. Yamauchi, S. Y. Choi, H. Okada, M. Yohda, H. Kumagai, N. Esaki, and K. Soda, *J. Biol. Chem.*, 1992, **267**, 18 361.
103. T. Miyakawa, H. Matsuzawa, M. Matsuhashi, and Y. Sugino, *J. Bacteriol.*, 1972, **112**, 950.
104. I. N. Maruyama, A. H. Yamamoto, and Y. Hirota, *J. Bacteriol.*, 1988, **170**, 3786.
105. D. Mengin-Lecreulx, C. Parquet, L. R. Desviat, J. Pla, B. Flouret, J. A. Ayala, and J. van Heijenoort, *J. Bacteriol.*, 1989, **171**, 6126.
106. M. Ikeda, M. Wachi, H. K. Jung, F. Ishino, and M. Matsuhashi, *Nucleic Acids Res.*, 1990, **18**, 4014.
107. D. Mengin-Lecreulx and J. van Heijenoort, *Nucleic Acids Res.*, 1990, **18**, 183.
108. M. Ikeda, M. Wachi, F. Ishino, and M. Matsuhashi, *Nucleic Acids Res.*, 1990, **18**, 1058.
109. J.-S. Tao and E. E. Ishiguro, *Can. J. Microbiol.*, 1989, **35**, 1051.
110. C. Michaud, C. Parquet, B. Flouret, D. Blanot, and J. van Heijenoort, *Biochem. J.*, 1990, **269**, 277.
111. B. Flouret, D. Mengin-Lecreulx, and J. van Heijenoort, *Anal. Biochem.*, 1981, **114**, 59.
112. Y. Mizuno, M. Yaegashi, and E. Ito, *J. Biochem. (Tokyo)*, 1973, **74**, 525.
113. D. Liger, A. Masson, D. Blanot, J. van Heijenoort, and C. Parquet, *Eur. J. Biochem.*, 1995, **230**, 80.
114. M. Gubler, Y. Appoldt, and W. Keck, *J. Bacteriol.*, 1996, **178**, 906.
115. P. J. Falk, K. M. Ervin, K. S. Volk, and H. T. Ho, *Biochemistry*, 1996, **35**, 1417.
116. H. Jin, J. J. Emanuele, Jr., R. Fairman, J. G. Robertson, M. E. Hail, H. T. Ho, P. J. Falk, and J. J. Villafranca, *Biochemistry*, 1996, **35**, 1423.
117. C. Michaud, D. Blanot, B. Flouret, and J. van Heijenoort, *Eur. J. Biochem.*, 1987, **166**, 631.
118. C. Michaud, D. Mengin-Lecreulx, J. van Heijenoort, and D. Blanot, *Eur. J. Biochem.*, 1990, **194**, 853.
119. D. Mengin-Lecreulx, D. Blanot, and J. van Heijenoort, *J. Bacteriol.*, 1994, **176**, 4321.
120. M. Abo-Ghalia, C. Michaud, D. Blanot, and J. van Heijenoort, *Eur. J. Biochem.*, 1985, **153**, 81.
121. R. A. Anwar and M. Vlaovic, *Biochem. Cell. Biol.*, 1986, **64**, 297.
122. C. T. Walsh, *J. Biol. Chem.*, 1989, **264**, 2393.
123. A. C. Robinson, D. L. Kenan, J. Sweeney, and W. D. Donachie, *J. Bacteriol.*, 1986, **167**, 809.
124. L. E. Zawadzke, T. D. H. Bugg, and C. T. Walsh, *Biochemistry*, 1991, **30**, 1673.
125. E. Daub, L. E. Zawadzke, D. Botstein, and C. T. Walsh, *Biochemistry*, 1988, **27**, 3701.
126. C. Fan, P. C. Moews, C. T. Walsh, and J. R. Knox, *Science*, 1994, **266**, 439.

127. C. Fan, P. C. Moews, Y. Shi, C. T. Walsh, and J. R. Knox, *Proc. Natl. Acad. Sci. USA*, 1995, **92**, 1172.
128. L. S. Mullins, L. E. Zawadzke, C. T. Walsh, and F. M. Raushel, *J. Biol. Chem.*, 1990, **265**, 8993.
129. T. D. H. Bugg, S. Dutka-Malen, M. Arthur, P. Courvalin, and C. T. Walsh, *Biochemistry*, 1991, **30**, 2017.
130. T. D. H. Bugg, G. D. Wright, S. Dutka-Malen, M. Arthur, P. Courvalin, and C. T. Walsh, *Biochemistry*, 1991, **30**, 10408.
131. Y. Shi and C. T. Walsh, *Biochemistry*, 1995, **34**, 2768.
132. I. S. Park, C. H. Lin, and C. T. Walsh, *Biochemistry*, 1996, **35**, 10464.
133. I. S. Park and C. T. Walsh, *J. Biol. Chem.*, 1997, **272**, 9210.
134. F. C. Neuhaus and W. G. Struve, *Biochemistry*, 1965, **4**, 120.
135. K. Duncan, J. van Heijenoort, and C. T. Walsh, *Biochemistry*, 1990, **29**, 2379.
136. C. Parquet, B. Flouret, D. Mengin-Lecreulx, and J. van Heijenoort, *Nucleic Acids Res.*, 1989, **17**, 5379.
137. T. D. H. Bugg, J. van Heijenoort, and C. T. Walsh, unpublished results.
138. M. S. Anderson, S. S. Eveland, H. R. Onishi, and D. L. Pompliano, *Biochemistry*, 1996, **35**, 16264.
139. Y. Higashi, J. L. Strominger, and C. C. Sweeley, *J. Biol. Chem.*, 1970, **245**, 3697.
140. I. Takahashi and K. Ogura, *J. Biochem. (Tokyo)*, 1982, **92**, 1527.
141. M. Kobayashi, M. Ito, T. Koyama, and K. Ogura, *J. Am. Chem. Soc.*, 1985, **107**, 4588.
142. G. Siewert and J. L. Strominger, *Proc. Natl. Acad. Sci. USA*, 1967, **57**, 767.
143. Y. Higashi, G. Siewert, and J. L. Strominger, *J. Biol. Chem.*, 1970, **245**, 3683.
144. W. G. Struve, R. K. Sinha, and F. C. Neuhaus, *Biochemistry*, 1966, **5**, 82.
145. M. G. Heydanek, Jr., W. G. Struve, and F. C. Neuhaus, *Biochemistry*, 1969, **8**, 1214.
146. M. G. Heydanek, Jr. and F. C. Neuhaus, *Biochemistry*, 1969, **8**, 1474.
147. W. P. Hammes and F. C. Neuhaus, *J. Biol. Chem.*, 1974, **249**, 3140.
148. W. A. Weppner and F. C. Neuhaus, *J. Biol. Chem.*, 1977, **252**, 2296.
149. M. Ikeda, M. Wachi, H. K. Jung, F. Ishino, and M. Matsuhashi, *J. Bacteriol.*, 1991, **173**, 1021
150. P. E. Brandish, M. K. Burnham, J. T. Lonsdale, R. Southgate, M. Inukai, and T. D. H. Bugg, *J. Biol. Chem.*, 1996, **271**, 7609.
151. D. Mengin-Lecreulx, L. Texier, M. Rousseau, and J. van Heijenoort, *J. Bacteriol.*, 1991, **173**, 4625.
152. K. Bupp and J. van Heijenoort, *J. Bacteriol.*, 1993, **175**, 1841.
153. Y. van Heijenoort, M. Gomez, M. Derrien, J. Ayala, and J. van Heijenoort, *J. Bacteriol.*, 1992, **174**, 3549.
154. M. Matsuhashi, C. P. Dietrich, and J. L. Strominger, *J. Biol. Chem.*, 1967, **242**, 3191.
155. T. Kamiryo and M. Matsuhashi, *J. Biol. Chem.*, 1972, **247**, 6306.
156. H. Maidhof, B. Reinicke, P. Blümel, B. Berger-Bächi, and H. Labischinski, *J. Bacteriol.*, 1991, **173**, 3507.
157. U. Henze, T. Sidow, J. Wecke, H. Labischinski, and B. Berger-Bächi, *J. Bacteriol.*, 1993, **175**, 1612.
158. A. M. Stranden, K. Ehlert, H. Labischinski, and B. Berger-Bächi, *J. Bacteriol.*, 1997, **179**, 9.
159. W. Staudenbauer and J. L. Strominger, *J. Biol. Chem.*, 1972, **247**, 5095.
160. T. Janas, J. Kuczera, and T. Chojnacki, *Chem. Phys. Lipids*, 1989, **51**, 227.
161. M. J. Knudsen and F. A. Troy, *Chem. Phys. Lipids*, 1989, **51**, 205.
162. W. A. Weppner and F. C. Neuhaus, *J. Biol. Chem.*, 1978, **253**, 472.
163. M. J. Fath and R. Kolter, *Microbiol. Rev.*, 1993, **57**, 995.
164. M. A. Glucksmann, T. L. Reuber, and G. C. Walker, *J. Bacteriol.*, 1993, **175**, 7045.
165. D. Liu, R. A. Cole, and P. R. Reeves, *J. Bacteriol.*, 1996, **178**, 2102.
166. D. J. Waxman and J. L. Strominger, *Annu. Rev. Biochem.*, 1983, **52**, 825.
167. J. M. Ghuysen, *Annu. Rev. Microbiol.*, 1991, **45**, 37.
168. J. M. Frère and B. Joris, *CRC Crit. Rev. Microbiol.*, 1985, **11**, 299.
169. B. G. Spratt, *Eur. J. Biochem.*, 1977, **72**, 341.
170. B. G. Spratt, *Proc. Natl. Acad. Sci. USA*, 1975, **72**, 2999.
171. E. Tuomanen and J. Schwartz, *J. Bacteriol.*, 1987, **169**, 4912.
172. N. G. Stoker, J. K. Broome-Smith, A. Edelman, and B. G. Spratt, *J. Bacteriol.*, 1983, **155**, 847.
173. H. Amanuma and J. L. Strominger, *J. Biol. Chem.*, 1980, **255**, 11173.
174. T. Tamura, Y. Imae, and J. L. Strominger, *J. Biol. Chem.*, 1976, **251**, 414.
175. T. Romeis and J. V. Höltje, *Eur. J. Biochem.*, 1994, **224**, 597.
176. M. L. Sinnott, *Chem. Rev.*, 1990, **90**, 1171.
177. F. Ishino, K. Mitsui, S. Tamaki, and M. Matsuhashi, *Biochem. Biophys. Res. Commun.*, 1980, **97**, 287.
178. H. Suzuki, Y. van Heijenoort, T. Tamura, J. Mizoguchi, Y. Hirota, and J. van Heijenoort, *FEBS Lett.*, 1980, **110**, 245.
179. J. Nakagawa, S. Tamaki, S. Tamioka, and M. Matsuhashi, *J. Biol. Chem.*, 1984, **259**, 13937.
180. J. M. Frère, C. Duez, J. M. Ghuysen, and J. Vandekerlhove, *FEBS Lett.*, 1976, **70**, 257.
181. R. R. Yocum, D. J. Waxman, J. R. Rasmussen, and J. L. Strominger, *Proc. Natl. Acad. Sci. USA*, 1979, **76**, 2730.
182. R. A. Nicholas, F. Ishino, W. Park, M. Matsuhashi, and J. L. Strominger, *J. Biol. Chem.*, 1985, **260**, 6394.
183. R. A. Nicholas, H. Suzuki, Y. Hirota, and J. L. Strominger, *Biochemistry*, 1985, **24**, 3448.
184. R. A. Nicholas, J. L. Strominger, H. Suzuki, and Y. Hirota, *J. Bacteriol.*, 1985, **164**, 456.
185. W. Keck, B. Glauner, U. Schwarz, J. K. Broome-Smith, and B. G. Spratt, *Proc. Natl. Acad. Sci. USA*, 1985, **82**, 1999.
186. A. Takasuga, H. Adachi, F. Ishino, M. Matsuhashi, T. Ohta, and H. Matsuzawa, *J. Biochem. (Tokyo)*, 1988, **104**, 822.
187. O. Herzberg and J. Moult, *Science*, 1987, **236**, 694.
188. P. C. Moews, J. R. Knox, O. Dideberg, P. Charlier, and J. M. Frère, *Proteins*, 1990, **7**, 156.
189. C. Oefner, A. D-Arcy, J. J. Daly, K. Gubernator, R. L. Charnas, I. Heinze, C. Hubschwerlen, and F. K. Winkler, *Nature (London)*, 1990, **343**, 284.
190. J. A. Kelly, O. Dideberg, P. Charlier, J. P. Wery, M. Libert, P. C. Moews, J. R. Knox, C. Duez, C. Fraipont, B. Joris, J. Dusart, J. M. Frère, and J. M. Ghuysen, *Science*, 1986, **231**, 1429.
191. H. Adachi, M. Ishiguro, S. Imajoh, T. Ohta, and H. Matsuzawa, *Biochemistry*, 1992, **31**, 430.
192. A. M. W. H. Thunnissen, F. Fusetti, B. de Boer, and B. W. Dijkstra, *J. Mol. Biol.*, 1995, **247**, 149.
193. J. V. Höltje and E. I. Tuomanen, *J. Gen. Microbiol.*, 1991, **137**, 441.
194. J. V. Höltje, D. Mirelman, N. Sharon, and U. Schwarz, *J. Bacteriol.*, 1975, **124**, 1067.
195. A. M. W. H. Thunnissen, A. J. Dijkstra, K. Kalk, H. J. Rozeboom, H. Engel, W. Keck, and B. W. Dijkstra, *Nature (London)*, 1994, **367**, 750.

196. A. M. W. H. Thunnissen, H. J. Rozeboom, K. H. Kalk, and B. W. Dijkstra, *Biochemistry*, 1995, **34**, 12 729.
197. H. Engel, A. J. Smink, L. van Wijngaarden, and W. Keck, *J. Bacteriol.*, 1992, **174**, 6394.
198. A. Ursinus and J. V. Höltje, *J. Bacteriol.*, 1994, **176**, 338.
199. D. W. Yem and H. C. Wu, *J. Bacteriol.*, 1976, **125**, 372.
200. S. Tomioka and M. Matsuhashi, *Biochem. Biophys. Res. Commun.*, 1978, **84**, 978.
201. W. Keck and U. Schwarz, *J. Bacteriol.*, 1979, **139**, 770.
202. J. van Heijenoort, C. Parquet, B. Flouret, and Y. van Heijenoort, *Eur. J. Biochem.*, 1975, **58**, 611.
203. B. D. Beck and J. T. Park, *J. Bacteriol.*, 1977, **130**, 1292.
204. E. W. Goodell and U. Schwarz, *J. Bacteriol.*, 1985, **162**, 391.
205. E. W. Goodell, *J. Bacteriol.*, 1985, **163**, 305.
206. J. T. Park, *J. Bacteriol.*, 1993, **175**, 7.
207. D. Mengin-Lecreulx, J. van Heijenoort, and J. T. Park, *J. Bacteriol.*, 1996, **178**, 5347.
208. D. Boothby, L. Daneo-Moore, M. L. Higgins, J. Coyette, and G. D. Shockman, *J. Biol. Chem.*, 1973, **248**, 2161.
209. H. C. Neu, *Science*, 1992, **257**, 1064.
210. M. N. Swartz, *Proc. Natl. Acad. Sci. USA*, 1994, **91**, 2420.
211. J. D. Hayes and C. R. Wolf, *Biochem. J.*, 1990, **272**, 281.
212. G. A. Jacoby and G. L. Archer, *N. Engl. J. Med.*, 1991, **324**, 661.
213. J. C. J. Barna and D. H. Williams, *Annu. Rev. Microbiol.*, 1984, **38**, 339.
214. P. Courvalin, *Antimicrob. Agents Chemother.*, 1990, **34**, 2291.
215. M. Arthur and P. Courvalin, *Antimicrob. Agents Chemother.*, 1993, **37**, 1563.
216. M. Arthur, C. Molinas, and P. Courvalin, *J. Bacteriol.*, 1992, **174**, 2582.
217. R. B. Bourret, K. A. Borkovich, and M. I. Simon, *Annu. Rev. Biochem.*, 1991, **60**, 401.
218. G. D. Wright, T. R. Holman, and C. T. Walsh, *Biochemistry*, 1993, **32**, 5057.
219. T. R. Holman, Z. Wu, B. L. Wanner, and C. T. Walsh, *Biochemistry*, 1994, **33**, 4625.
220. G. D. Wright and C. T. Walsh, *Acc. Chem. Res.*, 1992, **25**, 468.
221. J. R. Rasmussen and J. L. Strominger, *Proc. Natl. Acad. Sci. USA*, 1978, **75**, 84.
222. Z. Wu, G. D. Wright, and C. T. Walsh, *Biochemistry*, 1995, **34**, 2455.
223. Z. Wu and C. T. Walsh, *J. Am. Chem. Soc.*, 1996, **118**, 1785.
224. K. Ubukata, R. Nonoguchi, M. Matsuhashi, and M. Konno, *J. Bacteriol.*, 1989, **171**, 2882.
225. B. J. Hartman and A. Tomasz, *J. Bacteriol.*, 1984, **158**, 513.
226. P. E. Reynolds and C. Fuller, *FEMS Microbiol. Lett.*, 1986, **33**, 251.
227. B. L. M. de Jonge, Y. S. Chang, D. Gage, and A. Tomasz, *J. Biol. Chem.*, 1992, **267**, 11 248.
228. H. Maki, T. Yamaguchi, and K. Murakami, *J. Bacteriol.*, 1994, **176**, 4993.
229. P. E. Brandish, Ph.D. Thesis, University of Southampton, 1995.
230. E. F. Gale, E. Cundliffe, P. E. Reynolds, M. H. Richmond, and M. J. Waring, "The Molecular Basis of Antibiotic Action," 2nd edn., Wiley, Chichester, 1981.
231. B. B. Molloy, D. H. Lively, R. M. Gale, M. Gorman, and L. D. Boeck, *J. Antibiot. (Tokyo)*, 1972, **25**, 137.
232. R. Cooper, A. C. Horan, F. Gentile, V. Gullo, D. Loebenberg, J. Marquez, M. Patel, M. S. Puar, and I. Truumees, *J. Antibiot. (Tokyo)*, 1988, **41**, 13.
233. C. Rapp, G. Jung, W. Katzer, and W. Loeffler, *Angew. Chem., Int. Ed. Engl.*, 1988, **27**, 1733.
234. J. E. Walker and E. P. Abraham, *Biochem. J.*, 1970, **118**, 557.
235. B. Badet, P. Vermoote, and F. Le Goffic, *Biochemistry*, 1988, **27**, 2282.
236. N. Kucharczyk, M. A. Denisot, F. Le Goffic, and B. Badet, *Biochemistry*, 1990, **29**, 3668.
237. S. L. Bearne and R. Wolfenden, *Biochemistry*, 1995, **34**, 11 515.
238. S. L. Bearne, *J. Biol. Chem.*, 1996, **271**, 3052.
239. F. M. Kahan, J. S. Kahan, P. J. Cassidy, and H. Kropp, *Ann. N.Y. Acad. Sci.*, 1974, **235**, 364.
240. J. L. Marquardt, E. D. Brown, W. S. Lane, T. M. Haley, Y. Ichikawa, C. H. Wong, and C. T. Walsh, *Biochemistry*, 1994, **33**, 10 646.
241. F. C. Neuhaus and W. P. Hammes, *Pharmacol. Ther.*, 1981, **14**, 265.
242. B. Badet, D. Roise, and C. T. Walsh, *Biochemistry*, 1984, **23**, 5188.
243. T. S. Soper, W. M. Jones, B. Lerner, M. Trop, and J. M. Manning, *J. Biol. Chem.*, 1977, **252**, 3170.
244. B. Badet, K. Inagaki, K. Soda, and C. T. Walsh, *Biochemistry*, 1986, **25**, 3275.
245. K. Duncan, W. S. Faraci, D. S. Matteson, and C. T. Walsh, *Biochemistry*, 1989, **28**, 3541.
246. V. Copie, W. S. Faraci, C. T. Walsh, and R. G. Griffin, *Biochemistry*, 1988, **27**, 4966.
247. F. Gerhart, W. Higgins, C. Tardif, and J.-B. Ducep, *J. Med. Chem.*, 1990, **33**, 2157.
248. R. J. Baumann, E. H. Bohme, J. S. Wiseman, M. Vaal, and J. S. Nichols, *Antimicrob. Agents Chemother.*, 1988, **32**, 1119.
249. M. H. Gelb, Y. Lin, M. A. Pickard, Y. Song, and J. C. Vederas, *J. Am. Chem. Soc.*, 1990, **112**, 4932.
250. L. K. P. Lam, L. D. Arnold, T. H. Kalantar, J. G. Kelland, P. M. Lane-Bell, M. M. Palcic, M. A. Pickard, and J. C. Vederas, *J. Biol. Chem.*, 1988, **263**, 11 814.
251. Y. Song, D. Niederer, P. M. Lane-Bell, L. K. P. Lam, S. Crawley, M. M. Palcic, M. A. Pickard, D. L. Pruess, and J. C. Vederas, *J. Org. Chem.*, 1994, **59**, 5784.
252. S. D. Abbott, P. M. Lane-Bell, K. P. S. Sidhu, and J. C. Vederas, *J. Am. Chem. Soc.*, 1994, **116**, 6513.
253. R. J. Cox, W. A. Sherwin, L. K. P. Lam, and J. C. Vederas, *J. Am. Chem. Soc.*, 1996, **118**, 7449.
254. F. C. Neuhaus and J. L. Lynch, *Biochemistry*, 1964, **3**, 471.
255. W. H. Parsons, A. A. Patchett, H. G. Bull, W. R. Schoen, D. Taub, J. Davidson, P. L. Combs, J. P. Springer, H. Gadebusch, B. Weissberger, M. E. Valiant, T. N. Mellin, and R. D. Busch, *J. Med. Chem.*, 1988, **31**, 1772.
256. K. Duncan and C. T. Walsh, *Biochemistry*, 1988, **27**, 3709.
257. A. E. McDermott, F. Creuzet, R. G. Griffin, L. E. Zawadzke, Q. Z. Ye, and C. T. Walsh, *Biochemistry*, 1990, **29**, 5767.
258. M. E. Tanner, S. Vaganay, J. van Heijenoort, and D. Blanot, *J. Org. Chem.*, 1996, **61**, 1756.
259. D. J. Miller, S. M. Hammond, D. Anderluzzi, and T. D. H. Bugg, *J. Chem. Soc. Perkin Trans. 1*, in press.
260. G. Tamura, "Tunicamycin," Japan Scientific Press, Tokyo, 1982.
261. G. Tamura, T. Sasaki, M. Matsuhashi, A. Takatsuki, and M. Yamasaki, *Agric. Biol. Chem.*, 1976, **40**, 447.

262. P. E. Brandish, K. I. Kimura, M. Inukai, R. Southgate, J. T. Lonsdale, and T. D. H. Bugg, *Antimicrob. Agents Chemother.*, 1996, **40**, 1640.
263. A. Heifetz, R. W. Keenan, and A. D. Elbein, *Biochemistry*, 1979, **18**, 2186.
264. F. Isono, T. Katayama, M. Inukai, and T. Haneishi, *J. Antibiot. (Tokyo)*, 1989, **42**, 674.
265. J. P. Karwowski, M. Jackson, R. J. Theriault, R. H. Chen, G. J. Barlow, and M. L. Maus, *J. Antibiot. (Tokyo)*, 1989, **42**, 506.
266. S. Chatterjee, S. R. Nadkarni, E. K. S. Vijayakumar, M. V. Patel, B. N. Ganguli, H. W. Fehlhaber, and L. Vertesy, *J. Antibiot. (Tokyo)*, 1994, **47**, 595.
267. F. Isono and M. Inukai, *Antimicrob. Agents Chemother.*, 1991, **35**, 234.
268. M. Inukai, F. Isono, and A. Takatsuki, *Antimicrob. Agents Chemother.*, 1993, **37**, 980.
269. K. Isono, M. Uramato, H. Kusakabe, K. Kimura, K. Isaki, C. C. Nelson, and J. A. McCloskey, *J. Antibiot. (Tokyo)*, 1985, **38**, 1617.
270. M. Ubakata, K.-i. Kimura, K. Isono, C. C. Nelson, J. M. Gregson, and J. A. McCloskey, *J. Org. Chem.*, 1992, **57**, 6392.
271. E. A. Somner and P. E. Reynolds, *Antimicrob. Agents Chemother.*, 1990, **34**, 413.
272. W. A. Toscano, Jr. and D. R. Storm, *Pharmacol. Ther.*, 1982, **16**, 199.
273. K. J. Stone and J. L. Strominger, *Proc. Natl. Acad. Sci. USA*, 1971, **68**, 3223.
274. D. R. Storm and J. L. Strominger, *J. Biol. Chem.*, 1973, **248**, 3940.
275. E. G. Seebauer, E. P. Duliba, D. A. Scogin, R. B. Gennis, and R. L. Belford, *J. Am. Chem. Soc.*, 1983, **105**, 4926.
276. M. Garcia-Patrone, *Antimicrob. Agents Chemother.*, 1990, **34**, 796.
277. B. D. Cain, P. J. Norton, W. Eubanks, H. S. Nick, and C. M. Allen, *J. Bacteriol.*, 1993, **175**, 3784.
278. H. Tanaka, R. Oiwa, S. Matsukura, and S. Omura, *Biochem. Biophys. Res. Commun.*, 1979, **86**, 902.
279. D. K. Banerjee, *J. Biol. Chem.*, 1989, **264**, 2024.
280. P. E. Linnett and J. L. Strominger, *Antimicrob. Agents Chemother.*, 1973, **4**, 231.
281. P. Welzel, F.-J. Witteler, D. Müller, and W. Riemer, *Angew. Chem., Int. Ed. Engl.*, 1981, **20**, 121.
282. H.-W. Fehlhaber, M. Girg, G. Seibert, K. Hobert, P. Welzel, Y. van Heijenoort, and J. van Heijenoort, *Tetrahedron*, 1990, **46**, 1557.
283. Y. van Heijenoort, M. Derrien, and J. van Heijenoort, *FEBS Lett.*, 1978, **89**, 141.
284. Y. van Heijenoort, M. Leduc, H. Singer, and J. van Heijenoort, *J. Gen. Microbiol.*, 1987, **133**, 667.
285. P. Welzel, F. Kunisch, F. Kruggel, H. Stein, J. Scherkenbeck, A. Hiltmann, H. Duddeck, D. Müller, J. E. Maggio, H.-W. Fehlhaber, G. Seibert, Y. van Heijenoort, and J. van Heijenoort, *Tetrahedron*, 1987, **43**, 585.
286. K.-H. Metten, K. Hobert, S. Marzian, U. F. Hackler, U. Heinz, P. Welzel, W. Aretz, D. Böttger, U. Hedtmann, G. Seibert, A. Markus, M. Limbert, Y. van Heijenoort, and J. van Heijenoort, *Tetrahedron*, 1992, **48**, 8401.
287. L. Qiao and J. C. Vederas, *J. Org. Chem.*, 1993, **58**, 3480.
288. H. Brötz, G. Bierbaum, A. Markus, E. Molitor, and H. G. Sahl, *Antimicrob. Agents Chemother.*, 1995, **39**, 714.
289. H. W. van den Hooven, C. A. E. M. Spronk, M. van de Kamp, R. N. H. Konings, C. W. Hilbers, and F. J. M. van de Van, *Eur. J. Biochem.*, 1996, **235**, 394.
290. D. H. Williams, *Nat. Prod. Rep.*, 1996, **13**, 469.
291. R. Nagarajan, *Antimicrob. Agents Chemother.*, 1991, **35**, 605.
292. R. Nagarajan, *J. Antibiot. (Tokyo)*, 1993, **46**, 1181.
293. J. P. Mackay, U. Gerhard, D. A. Beauregard, M. S. Westwell, M. S. Searle, and D. H. Williams, *J. Am. Chem. Soc.*, 1994, **116**, 4581.
294. D. A. Beauregard, D. H. Williams, M. N. Gwynn, and D. J. Knowles, *Antimicrob. Agents Chemother.*, 1995, **39**, 781.
295. M. S. Westwell, B. Bardsley, R. J. Dancer, A. C. Try, and D. H. Williams, *J. Chem. Soc., Chem. Commun.*, 1996, 589.
296. N. E. Allen, J. N. Hobbs, Jr., and T. I. Nicas, *Antimicrob. Agents Chemother.*, 1996, **40**, 2356.
297. N. E. Allen, D. L. LeTourneau, and J. N. Hobbs, Jr., *Antimicrob. Agents Chemother.*, 1997, **41**, 66.
298. D. J. Tipper and J. L. Strominger, *Proc. Natl. Acad. Sci. USA*, 1965, **54**, 1133.
299. A. Tomasz, *Annu. Rev. Microbiol.*, 1979, **33**, 113.
300. M. R. J. Salton and G. D. Shockman (eds.), "β-Lactam Antibiotics: Mode of Action, New Developments and Future Prospects," Academic Press, New York, 1981.
301. C. E. Newall and P. D. Hallam, in "Comprehensive Medicinal Chemistry," ed. P. G. Sammes, Pergamon Press, Oxford, 1990, vol. 2, p. 609.
302. A. G. Brown, M. J. Pearson, and R. Southgate, in "Comprehensive Medicinal Chemistry," ed. P. G. Sammes, Pergamon Press, Oxford, 1990, vol. 2, p. 655.
303. G. A. Jacoby, *Annu. Rev. Med.*, 1996, **47**, 169.
304. K. Bush, *Antimicrob. Agents Chemother.*, 1989, **33**, 264.
305. K. Bush, *Antimicrob. Agents Chemother.*, 1989, **33**, 271.
306. A. Philippon, R. Labia, and G. A. Jacoby, *Antimicrob. Agents Chemother.*, 1989, **33**, 1131.
307. G. A. Jacoby and A. A. Medeiros, *Antimicrob. Agents Chemother.*, 1991, **35**, 1697.
308. N. H. Georgopapadakou, *Antimicrob. Agents Chemother.*, 1993, **37**, 2045.
309. A. Imada, K. Kintaka, M. Nakao, and S. Shinagawa, *J. Antibiot. (Tokyo)*, 1982, **35**, 1400.
310. M. Nakao, K. Yukishige, M. Kondo, and A. Imada, *Antimicrob. Agents Chemother.*, 1986, **30**, 414.
311. M. F. Templin, D. H. Edwards, and J. V. Höltje, *J. Biol. Chem.*, 1992, **267**, 20039.
312. T. D. H. Bugg and P. E. Brandish, *FEMS Microbiol. Lett.*, 1994, **119**, 255.
313. N. H. Georgopapadakou and T. J. Walsh, *Antimicrob. Agents Chemother.*, 1996, **40**, 279.
314. C. E. Bulawa, *Annu. Rev. Microbiol.*, 1993, **47**, 505.
315. K. Isono and S. Suzuki, *Heterocycles*, 1979, **13**, 333.
316. H. Decker, F. Walz, C. Bormann, H. Zahner, H. P. Fiedler, H. Heitsch, and W. A. Konig, *J. Antibiot. (Tokyo)*, 1990, **43**, 43.
317. J. P. Gaughran, M. H. Lai, D. R. Kirsch, and S. J. Silverman, *J. Bacteriol.*, 1994, **176**, 5857.
318. S. C. Hubbard and R. J. Ivatt, *Annu. Rev. Biochem.*, 1981, **50**, 555.
319. C. Abeijon and C. B. Hirschberg, *Trends Biochem. Sci.*, 1992, **17**, 32.
320. A. D. Elbein, *Annu. Rev. Biochem.*, 1987, **56**, 497.

321. C. R. H. Raetz, *Annu. Rev. Biochem.*, 1990, **59**, 129.
322. C. A. Schnaitman and J. P. Klena, *Microbiol. Rev.*, 1993, **57**, 655.
323. P. Muller, M. Keller, W. M. Weng, J. Quandt, W. Arnold, and A. Puhler, *Mol. Plant Microbe Interact.*, 1993, **6**, 55.
324. A. D. Elbein, in "Encyclopedia of Plant Physiology, New Series," eds. W. Tanner and F. A. Loewus, Springer, Berlin, 1981, vol. 13B (Plant Carbohydrates 2), p. 166.
325. R. P. Lezica, *Biochem. Soc. Trans.*, 1979, **7**, 334.
326. H. E. Hopp, P. A. Romero, G. R. Daleo, and R. Pont Lezica, *Eur. J. Biochem.*, 1978, **84**, 561.

3.11

Biosynthesis of Glycosylated Phosphatidylinositol in Parasitic Protozoa, Yeast, and Higher Eukaryotes

VOLKER ECKERT, PETER GEROLD, and RALPH T. SCHWARZ
Philipps-Universität Marburg, Germany

3.11.1 INTRODUCTION

Glycosylated phosphatidylinositols represent a group of glycolipids which are defined by a common structural motif that consists of an inositolphospholipid, linked to glucosamine and followed by a mannose in an α-(1-4) linkage (see Figure 1). The most prominent representatives of this group are the so-called glycosyl-phosphatidylinositol (GPI) membrane anchors which have been found in all eukaryotic systems investigated thus far.[1-9] Since much structural, functional, and genetic data about GPI anchors have been obtained over the past years, the authors will focus predominantly on this type of glycosylated phosphatidylinositol. The other members of this group, the glycosylinositol phospholipids (GIPLs) or the rather complex lipophosphoglycans (LPGs), which are almost exclusively found in Leishmania parasites, share the common structural motif Man-α-(1-4)GlcN-(1-6)-inositol-phosphate with protein-bound GPI anchors but are then highly modified, forming a densely packed glycocalyx of nonprotein bound glycolipids, which is viewed as a protective coat like the variant surface glycoprotein (VSG) in *Trypanosoma brucei brucei*.[3,4,7] Since these rather unusual glycolipids seem to be synthesized predominantly in *Leishmania*, they will be discussed briefly in Section 3.11.3.

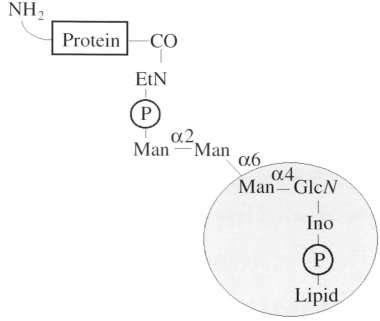

Figure 1   The minimal structure of a GPI anchor. P, phosphodiester bridge; Ino, inositol; GlcN, glucosamine; Man, mannose; EtN, ethanolamine. The shaded area indicates the minimal structure shared by all glycosylated phosphatidylinositols.

GPI membrane anchors, an alternative principle of anchoring proteins in membranes compared with classical transmembrane domains, were first discovered in the parasitic protozoan *T. b. brucei*[10] in which the entire cell surface is covered by a single GPI-anchored surface antigen, the VSG, and in rat brain Thy-1.[11] It was shown that the VSG protein was associated with ethanolamine and a carbohydrate structure linked to a hydrophobic component[12] and could be released from the cell surface as a water-soluble form by a trypanosome phosphatidylinositol-specific phospholipase C (PI-PLC).[1] Together with the data of Low and co-workers showing that PI-PLC was able to selectively release certain membrane proteins such as alkaline phosphatase, acetylcholinesterase, and 5′-nucleotidase,[13] it became clear that these proteins were linked to the membrane via ethanolamine and a carbohydrate bridge that was covalently linked to the membrane-bound phosphatidylinositol and the term glycosyl-phosphatidylinositol anchor was defined.[14] Subsequently, Ferguson and co-workers elucidated the *T. b. brucei* VSG GPI anchor structure,[10] which consists of ethanolaminephosphate-6-Man-α-(1-2)-Man-α-(1-6)-Man-α-(1-4)-GlcN-(1-6)-phosphatidylinositol. This carbohydrate sequence, which is now regarded as the evolutionary conserved core glycan, has been found in all GPI anchor structures spanning the evolutionary distance from protozoans and yeast to mammalian cells.

This evolutionary conserved structure is then modified by a variety of other components such as carbohydrates, ethanolamine phosphate, or fatty acids, and in the case of trypanosomes, the initial fatty acids are exchanged in a process called fatty acid remodeling,[15] giving rise to species-, stage-, and tissue-specific GPI structures.[15]

Initially, GPI anchoring of surface proteins was often regarded as an exotic way of anchoring membrane proteins predominantly used by protozoans in contrast to the "classical" transmembrane domains. Since then, over 100 proteins which are GPI anchored have been described,[1-5] demonstrating that this principle is widely distributed among eukaryotes with GPI anchoring seeming to be a more general principle among protozoans, and higher eukaryotes using this principle predominantly for certain proteins with specialized functions.[2,9]

3.11.2   GLYCOSYLPHOSPHATIDYLINOSITOL MEMBRANE ANCHORS

3.11.2.1   Structure, Genetics, and Functions of GPI Anchor Biosynthesis

As mentioned above, all GPI anchors share a common core structure conserved in all species analyzed thus far. The structural analysis of such structures was pioneered by the work of Masterson,

Menon, and Ferguson, utilizing a combination of radiolabeling techniques, chemical, and enzymatic cleavage reactions and subsequent analysis of the obtained cleavage products by biochemical methods such as size exclusion chromatography, TLC, HPLC, high pH anion exchange chromatography (HPAEC), as well as spectroscopic methods such as NMR and mass spectroscopy.[16,17] This work provided a large amount of structural information and gave first insight into the variety of individual GPI anchor types.

Combinations of *in vivo* and *in vitro* radiolabeling experiments in combination with these analytic techniques also gave first insight into the sequence of reactions in GPI biosynthesis which will be discussed in detail in the following section.

All GPI anchors consist of inositol (mostly myo-inositol) linked to a hydrophobic residue, which provides the membrane-anchoring function. However, there are striking differences among the hydrophobic fragments of GIP anchors from different species. Ester-linked fatty acids, ether-linked hydrophobic groups, and ceramide are described for the hydrophobic fragments of GPIs.[3-5,9] The initial hydrophobic components can then be replaced in an elaborate process called fatty acid remodeling. The inositol ring itself may also be modified by an additional fatty acid, thus rendering the GPI anchor (G)PI-PLC resistant.[5]

Nonacetylated glucosamine, a molecule very rarely found in biological systems, is attached to the inositol ring in an α-(1-6) linkage, followed by three mannoses in a linear array. Phosphoethanolamine is then attached via the phosphate group to the 6-position of the terminal mannose, thus completing the minimal GPI structure that can be transferred to protein in a transamidase-like reaction, involving the amine of the ethanolamine phosphate. This structure is usually, sometimes highly, modified by, e.g., additional ethanolamine phosphates and carbohydrate side chains linked to the trimannosyl core glycan, thus giving rise to a theoretically almost unlimited variety of structures characteristic for a given species as well as for certain stages. Comprehensive descriptions of these structural data can be found in various reviews.[1-9]

Figure 2 gives an overview of some of these modifications. An example for stage-specific modifications is the parasitic protozoan *T. b. brucei*, exhibiting different sugar side chains and lipid moieties on the GPI anchors of the insect stage (procyclic acid repetitive protein (PARP) protein) or the form that lives in the warm-blooded host (VSG protein).[3,4] These different structures may be a reflection of the strikingly different environmental conditions, although a definitive answer is still lacking.

Besides the immediately obvious function of GPI anchors, namely the anchoring of proteins in membranes, it became clear that GPIs play an important role in other cellular mechanisms, some of which will briefly be mentioned. GPI anchoring seems to provide a signal for transport to the cell membrane. In some polarized epithelial cells, GPI-anchored proteins are exclusively transported to the apical surface which makes this anchor function as a sorting and targeting signal.[9] This highly ordered intracellular transport might be explained by the observation that GPI-anchored proteins become sequestered into specialized transport vesicles during their passage through the Golgi apparatus or shortly thereafter.[18] These vesicles are proposed to contain clusters of GPI-anchored proteins, glycosphingolipids, cholesterol, and accessory proteins[19] which are then postulated to form specialized membrane microdomains at the cell surface. This sequestering may also explain the exclusion of these proteins from the clathrin-dependent endocytosis pathway and thus in some cases their usually low turnover rates.[5]

One of the most interesting and controversial aspects of GPI function is their ability to participate in signaling mechanisms or to function directly as second messengers. Most information on GPI-mediated signaling phenomena has been obtained using mammalian lymphocytes[20] where many GPI-anchored proteins are described, showing that the GPI anchor of some T-cell antigens is essential for T-cell mitogenesis. It was further shown that the GPI anchor of such lymphocyte antigens is a prerequisite for their association with tyrosine kinases, predominantly of the src-kinase family.[9]

In addition to the activation of tyrosine kinases, GPIs may serve as second messengers in the plasma membrane. Since they are structurally related to more common second messengers like inositol phosphate, diacyl glycerol, phosphatidic acid, and ceramide, GPIs or their cleavage products (hydrolyzed following receptor ligation) may play a role in cellular signaling and hormone action.[21] Free and protein-released GPIs are reported to mimic the effects of three hormone-like peptides: interleukin-2, nerve-growth-factor, and insulin.[22] Another highly interesting aspect of this involvement of GPI anchors in signal transduction is their involvement in the pathogenicity of protozoan parasites. GPIs are highly abundant in protozoa and some of these parasitic protozoa (e.g., *Plasmodium falciparum*, *T. b. brucei*, *Trypanosoma congolense*) share a striking similarity in their pathology: they exhibit the release of high amounts of tumor necrosis factor α (TNF-α) by their

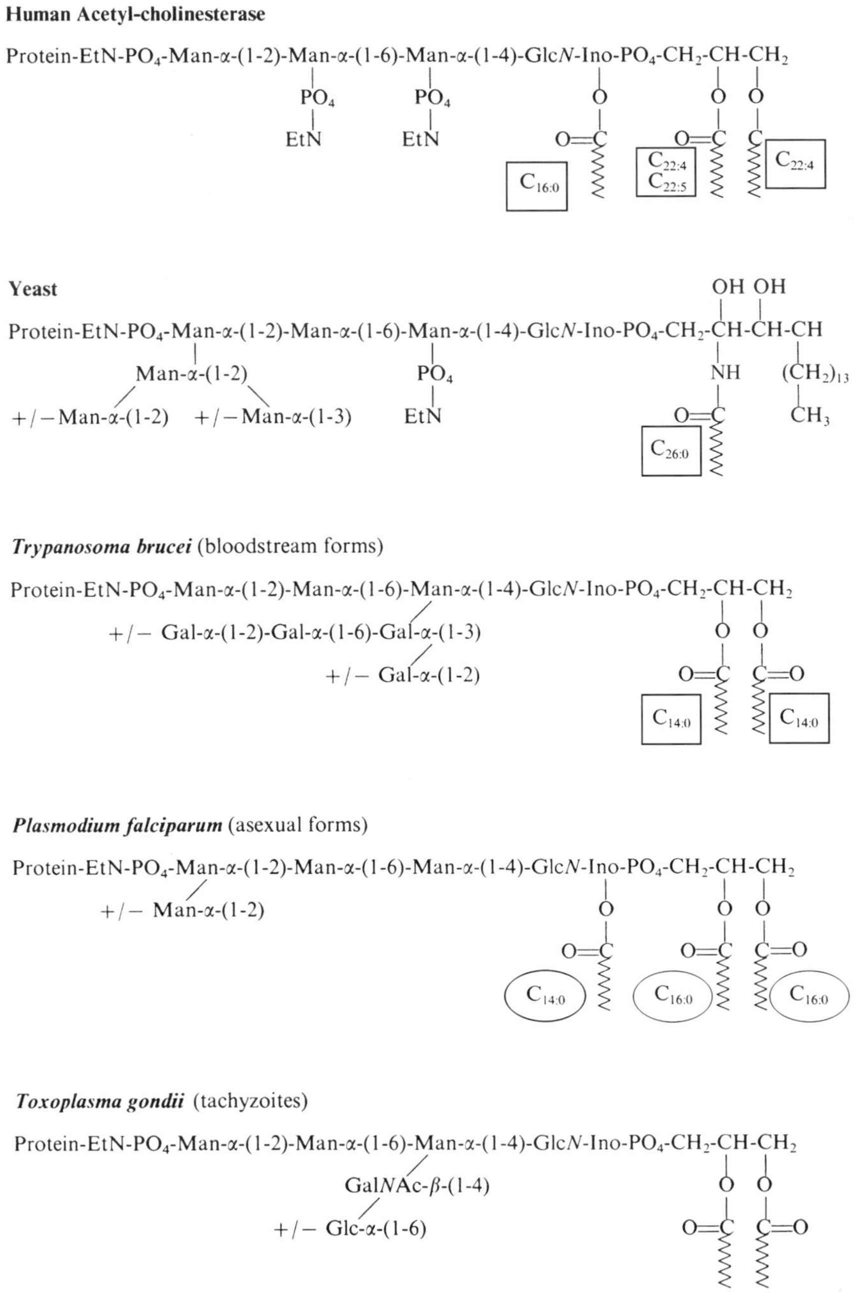

Human Acetyl-cholinesterase

Protein-EtN-PO$_4$-Man-α-(1-2)-Man-α-(1-6)-Man-α-(1-4)-GlcN-Ino-PO$_4$-CH$_2$-CH-CH$_2$
| | | | |
PO$_4$ PO$_4$ O O O
| | | | |
EtN EtN O=C O=C C
| | | |
C$_{16:0}$ C$_{22:4}$ C$_{22:4}$
C$_{22:5}$

Yeast

Protein-EtN-PO$_4$-Man-α-(1-2)-Man-α-(1-6)-Man-α-(1-4)-GlcN-Ino-PO$_4$-CH$_2$-CH-CH-CH
| | OH OH
Man-α-(1-2) PO$_4$ NH (CH$_2$)$_{13}$
/ \ | |
+/−Man-α-(1-2) +/−Man-α-(1-3) EtN O=C CH$_3$
|
C$_{26:0}$

Trypanosoma brucei (bloodstream forms)

Protein-EtN-PO$_4$-Man-α-(1-2)-Man-α-(1-6)-Man-α-(1-4)-GlcN-Ino-PO$_4$-CH$_2$-CH-CH$_2$
/ | |
+/− Gal-α-(1-2)-Gal-α-(1-6)-Gal-α-(1-3) O O
/ | |
+/− Gal-α-(1-2) O=C C=O
| |
C$_{14:0}$ C$_{14:0}$

Plasmodium falciparum (asexual forms)

Protein-EtN-PO$_4$-Man-α-(1-2)-Man-α-(1-6)-Man-α-(1-4)-GlcN-Ino-PO$_4$-CH$_2$-CH-CH$_2$
/ | | |
+/− Man-α-(1-2) O O O
| | |
O=C O=C C=O
| | |
C$_{14:0}$ C$_{16:0}$ C$_{16:0}$

Toxoplasma gondii (tachyzoites)

Protein-EtN-PO$_4$-Man-α-(1-2)-Man-α-(1-6)-Man-α-(1-4)-GlcN-Ino-PO$_4$-CH$_2$-CH-CH$_2$
/ | |
GalNAc-β-(1-4) O O
/ | |
+/− Glc-α-(1-6) O=C C=O

Figure 2 Comparison of five GPI anchor structures demonstrating the structural divergence between different species that is achieved by modifications of the conserved core glycan and through fatty acid exchange processes. Gal, galactose; GalNAc, *N*-acetyl-galactosamine; Man, Mannose; Man-P, mannose-phosphate; GlcNH2, glucosamine, GlcNAc, *N*-acetylglucosamine; Glc, glucose; P, phosphodiester bridge; Ino, inositol; GlcN, glucosamine; EtN-P, ethanolamine phosphate.

hosts. Investigations on human *Malaria tropica* caused by *P. falciparum* and different rodent models of severe malaria confirm the exacerbating and frequently fatal role of high TNF-α levels for malaria pathology.[23] The increase of TNF-α levels during malaria infection is induced by a parasite toxin which was shown to be malarial GPI.[24]

Besides the induction of TNF-α release, malarial GPIs (free and protein bound) were shown to be involved in the upregulation of endothelial cell surface markers,[25] which are receptors for *P. falciparum*-infected erythrocytes, and nitric oxide synthase,[26] which synthesizes nitric oxide, another mediator involved in human cerebral malaria.[27] A similar result was presented for the activation of

TNF-α and interleukin-1 release induced by GPIs of *T. b. brucei*.[28] With the list of functions for GPI anchors growing longer, such functional analysis has gained even more importance by the finding that the human disease paroxysmal nocturnal hemoglobinuria is caused by a defect in GPI anchoring[29] which could be identified as an inactivating somatic mutation in a gene involved in GPI anchor biosynthesis (see below). Furthermore, although several mammalian cell lines with a block in GPI anchor biosynthesis have been established and are perfectly viable in culture, disruption of a gene involved in GPI anchor biosynthesis in mice proved to be embryonically lethal, thus making GPI anchor biosynthesis essential for mammalian embryogenesis.[30] By blocking GPI anchor biosynthesis in keratinocytes via tissue-specific gene targeting in mice, it was also shown that this pathway plays an essential role in the development and maintenance of skin.[31]

With the help of such GPI-negative cell lines, mostly mutant murine thymoma cell lines that were unable to express and synthesize GPI-anchored Thy1 on the cell surface, it was possible to establish a genetic system to clone mammalian genes involved in GPI anchor biosynthesis. These mutant cell lines were grouped into complementation classes called A, B, C, E, F, H, K, and L by somatic cell fusion analysis,[32,33] and their defect was characterized biochemically by analyzing the GPI intermediates which accumulate in these cells. Descriptions of these cell lines are found in reviews by Kinoshita.[32,33] Cloning of genes involved in mammalian GPI biosynthesis was achieved by complementation screens, using the restoration of Thy1 surface expression as a selectable phenotype; the corresponding genes are termed PIG A, B, etc. (see also Figure 3).[32–34]

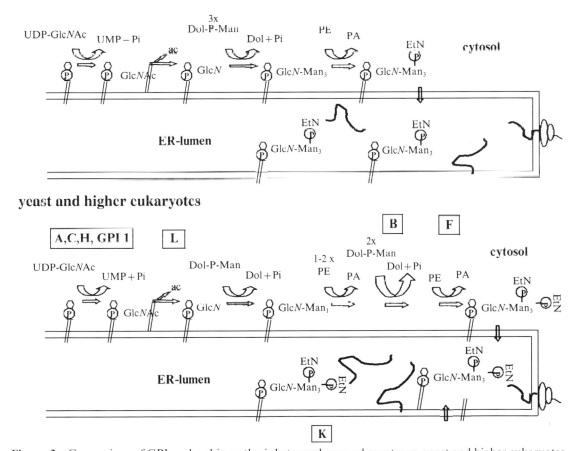

Figure 3 Comparison of GPI anchor biosynthesis between lower eukaryotes vs. yeast and higher eukaryotes. UDP-GlcNAc, UDP-*N*-acetyl-glucosamine; GlcN, glucosamine; Dol-P-Man, Dolichol-Phosphate-Mannose; EtN-P, ethanolamine-phosphate; PE, phosphatidylethanolamine phosphate. The bold letters indicate the complementation groups of mammalian mutant cell lines used in GPI research (Table 1). It may also be noted that the addition of ethanolamine phosphates other than the terminal EtN-P is only transient in yeast (see Section 3.11.2.2.3.(ii)).

Another organism that has meanwhile gained equal importance for the genetic analysis of GPI biosynthesis is the budding yeast *Saccharomyces cerevisiae*. This organism is genetically and biochemically well characterized and it was shown through temperature-sensitive mutants that GPI anchoring of proteins is essential.[35] A block in this pathway leads to a lethal phenotype, thus providing a powerful tool for complementation cloning which led to the identification of several genes for GPI biosynthesis.[9,32–34] As for the protozoa, no such genetic system has been established yet, but since transfection techniques for e.g. *T. b. brucei*, *T. gondii*, and *P. falciparum* have been developed, such screening systems may soon be anticipated.

Those genes that have been identified and cloned thus far as well as the corresponding enzymes will be described in the following section (Section 3.11.2.2) when the individual steps of GPI anchor biosynthesis are being discussed; a compilation of the genes cloned thus far can be found in Table 1.

Table 1   Compilation of cloned yeast and mammalian genes in comparison with mammalian mutant cell lines and their complementation groups.

Step in biosynthesis	Mammalian complementation group	Mammalian gene	Yeast gene
Synthesis of GlcNAc-PI	A	PIG-A	GPI 3/SPT 14
	H	PIG-H	
	C	PIG-C	GPI 2
		GPI 1	GPI 1
Deacetylation	L	PIG-L	
Addition of the third mannose	B	PIG-B	GPI 10
Addition of the terminal EtN-P	F	PIG-F	
		Gaa 1	Gaa 1
Transfer to protein	K	GPI 8	GPI 8
Synthesis of Dol-P-Man	E	DPM 1	DPM 1
	Lec 15	DPM 2	

3.11.2.2 Biosynthesis of GPI Anchors

3.11.2.2.1 *Biosynthesis of the conserved core structure*

The first detailed studies on the pathway for GPI anchor biosynthesis came from the trypanosome system using washed trypanosome membranes or cell lysates and tritiated sugar nucleotides.[3–5,9] Using UDP-*N*-acetyl-[³H]glucosamine and GDP-[³H]mannose, biosynthetic intermediates were readily detectable. Structural analysis of these GPI anchor biosynthesis intermediates led to the identification of nonprotein-bound GPIs which were built by sequential transfer of sugars and ethanolamine-phosphate. Additional information on the GPI anchor biosynthetic pathway of bloodstream forms of *T. b. brucei* came from the use of specific inhibitors which interfere with defined steps of this pathway.[5] Investigations in other systems such as yeast and mammalian cell lines then showed that the GPI anchor biosynthesis of the evolutionary conserved GPI anchor backbone (Figure 1) involves the same overall steps in all eukaryotes investigated thus far and can be divided into three major steps (Figure 3):
 (i) synthesis of glucosamine phosphatidylinositol (GlcN-PI);
 (ii) elongation of GlcN-PI by three mannoses;
(iii) addition of the terminal ethanolamine phosphate and transfer to protein.
Subcellular localization experiments in all systems investigated thus far showed that the complete assembly of the GPI anchor and the subsequent transfer to a protein takes place in the endoplasmic reticulum (ER), whereas subsequent modifications may occur during transport of the protein linked anchor through the secretory pathway.[9]

Differences in this pathway, however, are seen during acylation of the inositol ring, the transfer of additional ethanolamine phosphates and, of course, in the specific modifications which are added after core structure assembly.

(i) The first step: synthesis of GlcN-PI

In the initial reaction that leads to GPI anchor biosynthesis, GlcNAc is transferred from UDP-GlcNAc to phosphatidylinositol to form GlcNAc-PI. This reaction was investigated in protozoan, yeast, and mammalian cell lysates or membrane fractions and it was shown that in all systems the same donor (UDP-GlcNAc) is being used.[3-5,8] There seems to be no clear-cut preference for a specific PI in most systems at this point of biosynthesis. In *T. b. brucei* and *Leishmania*, the lipid part of the molecule is replaced with the correct fatty acids at a later step and is often referred to as fatty acid remodeling (see below).[3,7]

Genetic analysis revealed that at least three genes are involved in this reaction. Three complementation groups were identified in mammalian cells (class A, H, and C) and in yeast (GPI 1, 2, and 3).[32-34] Sequence comparison showed that PIG A is homologous with GPI 3 (also termed SPT14) and PIG C corresponds to GPI 2. PIG H does not show any significant homology to any yeast open reading frame, whereas a human homologue for GPI 1 has been identified and cloned. These findings suggest that yeast only uses three genes for the synthesis of GlcNAc-PI or that this fourth gene differs dramatically from its human counterpart on the sequence level. Biochemical studies using the four cloned human genes revealed that indeed these four proteins form a functional complex in the ER with PIG A being regarded as a catalytic subunit since it contains a region homologous to a bacterial GlcNAc transferase.[35]

After the formation of GlcNAc-PI, the GlcNAc moiety is then deacetylated by a separate enzyme activity since the above-mentioned GlcNAc transferase complex itself was not sufficient to perform this reaction *in vitro*.[35] A rat cDNA that was able to rescue the deacetylase-deficient phenotype of class L cells and the gene was termed PIG L.[36] The deacetylated GlcN-PI could now serve as the substrate for elongation of the three mannoses but in several systems it was shown that acylation of the inositol ring has to precede these elongation steps. For further details about these initial steps of GPI biosyntheses, see references 32–34.

(ii) The second step: elongation of GlcN-PI by three mannoses

In this step the preformed GlcN-PI is elongated by the subsequent addition of three mannoses. The first mannose is added in an α-(1-4) linkage followed by the second mannose linked to the first Man residue in an α-(1-6) configuration. Finally, the third mannose then follows in an α-(1-2) linkage. No yeast mutants or deficient cell lines for the addition of the first and second mannoses have yet been identified and therefore none of the corresponding genes could be cloned up to now. Addition of the third mannose is abolished in class B mutant cell lines and by complementation of this defect the mammalian gene (PIG B) could be cloned.[37] The enzymatic activity of the PIG B protein has not yet been demonstrated but the homology with a yeast mannosyltransferase (Alg9p) makes it probable that this protein is responsible for addition of the third mannose. Subsequently, the corresponding yeast gene (GPI 10) was identified and cloned.[38,39]

The donor for these mannoses in all systems was identified as dolichol-phosphate-mannose (Dol-P-Man),[40] a hydrophobic mannose donor, the formation of which can be inhibited selectively by amphomycin.[5] Biosynthesis of this donor is of particular interest since genetic analysis revealed a rather unusual divergence between different taxa. The enzyme catalyzing the synthesis of Dol-P-Man from GDP-mannose and dolicholphosphate is called Dol-P-Man synthase and the first gene for this protein was cloned from yeast by Orlean and co-workers[41] by complementing a temperature-sensitive yeast strain deficient in Dol-P-Man synthesis. This single polypeptide (DPM1), which shows the usual characteristics of a type I transmembrane protein, was able to restore Dol-P-Man synthesis in yeast and *in vitro* when expressed in *E. coli*, suggesting that this protein is the synthase itself. Using Orlean's yeast strain, Mazhari-Tabrizi and co-workers[42] cloned the Dol-P-Man synthase from *T. b. brucei* by heterologous complementation. Their results also demonstrated that a single polypeptide (DPM1) as in yeast is sufficient for the synthesis of Dol-P-Man.

Through somatic cell fusion experiments with Class E and Lec 15 cell lines it had already been suggested that these cells represent different complementation classes, indicating that two proteins might be necessary for Dol-P-Man synthesis in mammalian cells. Cloning of the human DPM1 gene[43] showed that this enzyme was lacking a C-terminal transmembrane domain. The human DPM1 gene was able to complement mutant Class E cells and is therefore regarded as the catalytic component of the mammalian Dol-P-Man synthase, making it necessary to postulate at least another protein involved in this biosynthesis which would be responsible for the membrane association of human DPM 1. Such a gene (DPM 2) which was able to restore Dol-P-Man synthesis in Lec 15 mutant CHO cells has been cloned recently.[44] It could be shown that this 84 amino acid membrane protein which is localized in the ER forms a complex with DPM 1, which is a prerequisite for the ER localization of DPM 1. In addition, DPM 2 enhances the binding of dolicholphosphate and is responsible for stable expression of DPM 1, making this protein the regulatory subunit of the mammalian Dol-P-Man synthase.

These data leave unanswered questions about the evolution and phylogeny of Dol-P-Man synthesis especially since the two yeasts which were thought to be closely related fall into different classes with respect to the DPM 1 gene.

(iii)   The third step: addition of the terminal ethanolamine phosphate and transfer to protein

After the third mannose is added, ethanolamine phosphate is added to complete the minimal GPI anchor structure which can be transferred to protein. The donor for this reaction has been identified as the phospholipid phosphatidylethanolamine in trypanosomes[45] and yeast.[46] Class F cells which are deficient in this step were used to clone the corresponding gene (PIG F) by complementation.[47] Sequence analysis revealed a highly hydrophobic ER protein as might be expected for a protein having two membrane-bound substrates.

The preassembled GPI anchor can now be transferred to newly synthesized proteins in a transamidase-like reaction on the luminal side of the ER membrane. This reaction proceeds quickly as soon as the protein is appropriately inserted and positioned into the ER membrane.[1,2] Translocation of the nascent protein chain across the ER membrane is a requirement for subsequent GPI anchoring.[2] The original carboxy terminus of such proteins is cleaved off by the transamidase complex and replaced by the GPI anchor via the amino function of the terminal ethanolamine phosphate.[48] The proteins are recognized via a C-terminal stretch of hydrophobic amino acids which is postulated to act as a temporary transmembrane domain and will be cleaved off by the transamidase complex prior to attaching the new carboxy-terminal amino acid to the readily assembled GPI anchor.[48]

Comparison of c-DNA data of various GPI-anchored proteins revealed that the C-terminus of GPI-anchored proteins has no motif-like feature other than the hydrophobicity and the lack of a cytoplasmic tail.[48] However, a stretch of three consecutive small amino acids 10–12 residues aminoterminal to the hydrophobic domain was found in these proteins with the most amino-terminal acid being the residue for attachment of the GPI anchor. Based on those sequence data this cleavage/attachment site has been defined and the so-called $\omega/\omega+1/\omega+2$ rule was postulated with ω being the amino acid the GPI anchor becomes attached to.[48] Each of the three positions is characterized by the amino acids that are tolerated in order to allow efficient recognition by the transamidase: the residues found at the ω site are aspartic acid, asparagine, glycine, alanine, serine, and cysteine. The $\omega+1$ site allows these six amino acids plus glutamic acid and threonine, making this site the least specific. The most restricted position is the $\omega+2$ site allowing only glycine, alanine, and serine with the exception of threonine in the decay accelerating factor.[48] Site-directed mutagenesis or saturation mutagenesis of all three amino acids simultaneously in a model protein confirmed the results obtained by sequence comparison.[49] Similar experiments could show that the positions adjacent to the cleavage/attachment site have no effect on GPI anchoring. This $\omega/\omega+1/\omega+2$ rule is now widely used to predict GPI anchoring and the attachment site from sequence data.

Although this $\omega/\omega+1/\omega+2$ rule applies for all GPI-anchored proteins thus far, individual organisms may exhibit preferences for certain combinations of amino acids in this cleavage/attachment site. This was shown experimentally for yeast[50] and in experiments concerned with the expression of GPI proteins from parasitic protozoa in heterologous systems. Moran and Caras[51] showed that two protozoan surface antigens, the *T. b. brucei* VSG (the prime example for a GPI-anchored protein) and the *P. falciparum* circum-sporozoite protein (where GPI-anchoring has only been deduced from the cDNA sequence), were not GPI anchored when expressed in mammalian (COS) cells.[51] The proteins themselves were expressed in significant amounts but were retained in the ER, thus showing no GPI-anchoring or extremely inefficient GPI anchoring. By exchanging their carboxy termini with a mammalian sequence, correct targeting to the cell membrane could be restored.[51] By further analysis the authors could show that it indeed was the cleavage/attachment site of the protozoal proteins that was responsible for the aberrant processing in COS cells. Similar results have been obtained for expression of the *P. falciparum* circum-sporozoite protein in dictyostelium.[52]

These experiments suggest that although the functional sequences on the protein level fit the established rules, there might be subtle differences in the processing machinery between distantly related taxa such as protozoans and higher eukaryotes which might be useful for an approach to develop new antiparasitic therapeutics via selective inhibitors of GPI anchoring.[53]

Finally, this rather simple recognition signal gives cells the flexibility to express certain proteins in different forms which can be either GPI anchored, secreted, or anchored via a classical trans-membrane domain by simply synthesizing these proteins with different carboxy termini. This is achieved by either differential splicing or expression of similar but distinct genes.[5,48]

As mentioned above, it is believed that more than one protein is involved in the transfer of the GPI anchor to a protein and therefore the term transamidase complex is being used. There are indeed two genes that up to now have been identified as being involved in that reaction. The yeast mutants gaa1 and gpi8 were shown to be unable to transfer the preformed GPI anchor to protein and thus deficient in the transamidase reaction.[54,55] The corresponding genes were cloned and the proteins were shown to be ER membrane proteins. Both proteins possess a luminal domain, which is in agreement with the notion that the transfer to protein takes place in the lumen of the ER. The yeast GPI8 protein may be a catalytic subunit of the transamidase complex since it exhibits homologies to a family of plant endopeptidases.[55] The human GPI8 gene has been cloned and was shown to complement the yeast gpi8 mutant.[55]

These data show that a least two proteins are components of the transmidase complex but since this reaction involves complex recognition and catalytic reactions as well as highly ordered structural requirements, there may be more genes involved that have yet to be identified.

Another interesting aspect is the topology of GPI anchor biosynthesis. Since all donors for GPI anchor precursors such as phosphatidylinositol, UDP-*N*-acetyl-glucosamine, dolichol-phosphate-mannose, and phosphatidylethanolamine are available on both leaflets of the endoplasmic reticulum, it was speculative for a long time where GPI biosynthesis takes place. Experiments provided evidence that GPI anchor precursor biosynthesis occurs at least in part on the cytoplasmic leaflet of the endoplasmic reticulum,[56–58] supported by the fact that the enzymes PIG-A, H, and L are oriented towards the cytoplasmic side of the ER.[36] Dol-P-Man, the mannose donor, is synthesized on the cytoplasmic side of the ER but utilized on the luminal side as was shown for *N*-glucosylation, arguing for a translocation prior to the addition of three mannoses. This would also be supported by the lumenal orientation of PIG B.[37] However, other data on the membrane orientation of mannosylated precursors point towards a cytoplasmic orientation of these precursors,[58] implying that the complete GPI anchor would have to cross the ER membrane. A definitive answer as to when this translocation occurs can therefore not be given at this time. However, these data also imply the existence of an enzyme which helps the preformed GPI anchor precursors to cross the ER membrane to the luminal leaflet where transfer onto the protein will occur. This postulated enzyme, the so-called "flipase," still needs to be identified.

3.11.2.2.2 *Inhibitors of GPI anchor biosynthesis*

GPI biosynthesis inhibitors have been important tools for the elucidation of GPI structures and biosynthetic pathways. We will therefore briefly discuss some of the compounds used in GPI research.

Phenylmethyl sulfonyl fluoride (PMSF): This inhibitor was shown to specifically block inositol acylation and transfer of the terminal ethanolaminephosphate in trypanosomes.[59]

Diisopropyl fluorophosphate (DFP): This compound is also trypanosome-specific and blocks inositol deacylation.[60]

Mannosamine: This mannose analogue efficiently inhibits GPI synthesis in trypanosomes and mammalia.[5] It was shown that Man*N*-Man-Glc*N*-PI accumulates in both systems which demonstrates that elongation with the third mannose is blocked by the 2-amino group of the incorporated mannosamine.[61] Since all three mannoses are added via Dol-P-Man, the formation of Dol-P-Man*N* has to be postulated as an essential intermediate step and could indeed be shown in trypanosomes.

Synthetic GPI-analogues: In such studies phosphatidylinositol was used which was modified by a 2-hydroxyl group on the inositol ring.[62–64] Analysis in cell-free systems from *T. b. brucei*, *Leishmania*, and HeLa cells showed distinct differences between the parasite and mammalian pathways.[62–64] The two kinetoplastid parasites were able to elongate the modified PI by three mannoses. Subsequent addition of the terminal ethanolamine phosphate is then blocked, indicating that inositol acylation is not a prerequisite for mannosylation but is essential for ethanolamine phosphate transfer. In

HeLa membranes, no mannosylation of this synthetic substrate was found, indicating that acylation has to precede mannosylation and that acyl PI is an obligatory substrate for addition of the first mannose.[64]

YW3548: This inhibitor has been identified[65] and was shown to be specific for yeast and mammalia. The protozoan parasites (*P. falciparum, T. gondii, T. b. brucei*, as well as the free living protozoan *Paramecium primaurelia*) were insensitive to this inhibitor even at concentrations 100-fold higher than needed to inhibit yeast and mammalian GPI-biosynthesis.[65] Biochemical analysis indicated that this compound selectively inhibits the addition of ethanolamine phosphate to the first mannose of the Man2-GlcN-PI intermediate.[65] This inhibition shows that addition of extra ethanolamine phosphate(s) is an essential step in mammalia and yeast, whereas this reaction is either not present or not essential in the protozoans investigated.

3.11.2.2.3 *Individual aspects of GPI anchor biosynthesis in different species*

The biosynthetic steps of the GPI anchor core glycan assembly as discussed in the previous section are common to all organisms studied thus far. Since GPI anchor biosynthesis is a phylogenetically ancient pathway that spans the whole evolutionary distance from protozoans to mammals, it is not surprising that differences in the processing of GPI molecules or modification of the core glycan were established in distantly related taxa. These individual aspects will now be discussed in the following sections.

(i) *The protozoa*

Since the protozoa themselves represent a phylogenetically ancient and therefore highly divergent group, the individual taxa exhibit significant differences in their GPI biosynthesis which will therefore be discussed separately.

(*a*) *Trypanosomes.* The protein-bound GPI anchor of *T. b. brucei* is sensitive towards PI-PLC cleavage, demonstrating that the inositol is not acylated. However, GPI precursors have been identified in this parasite that carry an acylated inositol.[60] Only acylated intermediates that have at least one mannose have been identified, making this form the potential substrate for acylation. Although only transient, acylation followed by deacylation seems to be an essential step in trypanosomes since inhibition of the deacylase leads to growth arrest and subsequent cell death.[60] These transiently acylated intermediates are in equilibrium with the nonacylated forms and may serve as a reservoir in the biosynthesis.

The GPIs are initially synthesized carrying stearic acid in the *sn*-1 and a mixture of fatty acids in the *sn*-2 position of the glycerol backbone.[4] The GPI anchor found on the VSG protein, however, has myristic acid in both positions. This exchange reaction in which the original fatty acids are replaced with myristic acid in a stepwise fashion is called fatty acid remodeling and is an elaborate process involving at least 13 intermediates,[15] which themselves may be exchanged or are in equilibrium with each other, suggesting that some of these forms form a backup pool making this reaction highly complex and unique for the bloodstream forms of trypanosomes. This remodeling of the lipid part takes place after the complete anchor is assembled and has to be completed before transfer to protein since only the fully glycerol myristoylated form can be found on the VSG protein.[4] No fatty acid remodeling has been observed in the insect forms of *T. b. brucei*. This stage of the parasite is characterized by an acylated inositol on the protein-bound GPI anchor, a lyso-alkyl group carrying exclusively long-chain fatty acids and different modifications on the glycan part.[5] GPI anchor biosynthesis in the parasite *T. congolense* follows the same pathway as *T. b. brucei* including fatty acid remodeling. The difference from the *T. b. brucei* anchor is limited to modification of the protein-bound GPI anchor by a thus far unique disaccharide.[66]

(*b*) *Leishmania.* Leishmania parasites also use GPIs for the anchoring of surface proteins. The basic structure of these anchors consists of the evolutionary conserved core glycan, which may then be modified by additional carbohydrates (Figure 4).[7] GPI-anchor precursors, like the free glycolipids LPG and GIPL, undergo fatty acid remodeling reactions prior to the transfer to protein.

(*c*) *Plasmodium falciparum.* The malaria parasite *P. falciparum* performs inositol acylation, but in contrast to the trypanosomal system, this reaction has to take place before addition of the first mannose, making GlcN-PI the substrate for acylation and the acylated form an obligatory precursor for mannosylation (Gerold, unpublished data).[67] Data obtained with a cell-free system show that

GPIs in *Leishmania*

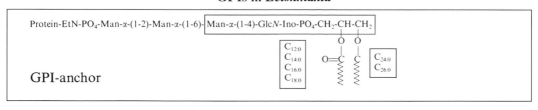

GPI-anchor

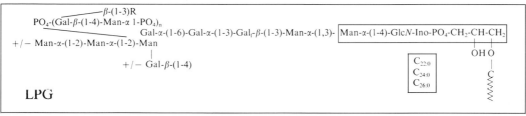

LPG

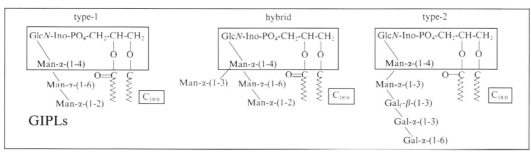

GIPLs

Figure 4 Structures of *Leishmania* glycosylated phosphatidylinositols with the boxed areas indicating the common core structure between these three types of glycolipids. GlcN, glucosamine; Gal, galactose; EtN-PO$_4$, ethanolamine phosphate; Ino, inositol; Man, mannose; PO$_4$, phosphodiester bridge.

myristyl-CoA is the donor for inositol acylation in this parasite. The stages of the intraerythrocytic cycle (blood forms) of *P. falciparum* synthesize two types of mature GPI anchor precursors termed Pfα and Pfβ, with α carrying a Man$_4$ and β carrying a Man$_3$ glycan structure.[68] The synthesis of these two precursors is tightly regulated during this intraerythrocytic cycle in a maturation dependent manner with the relative amounts of these two anchor types varying through the individual steps of this cycle, with Pfβ being preferentially synthesized and transferred to protein in trophozoite stages.[69] Both precursors are used to anchor proteins.[69] Biochemical analysis of the anchor present on the two main merozoite surface proteins MSP1 and MSP2 which are also predominantly synthesized in the trophozoite stage revealed that only Pfα is used to anchor these two proteins, making it necessary to postulate a selection mechanism which very efficiently excludes Pfβ from being used as membrane anchor for MSP1/2.[70]

P. falciparum was also shown to be a marked example for the preferential use of a certain type of protein glycosylation. Hardly any *N*- or *O*-glycosylation can be detected in the blood stages of this parasite, making GPI anchoring the predominant type of protein glycosylation.[71,72]

(*d*) *Toxoplasma gondii*. Inositol acylation has not been found *in vivo* in this protozoan parasite, whereas *in vitro* analysis in a cell-free system provided evidence for acylated and nonacylated Man$_2$GlcN-PI and Man$_3$GlcN-PI precursor structures, but this modification could not be detected in Man1GlcN-PI precursors.[73]

Later intermediates where the core glycan is already modified by GalNAc with or without the terminal ethanolamine phosphate are found as diacyl and monoacyl glycolipids. None of these later forms is inositol acylated which is in agreement with data showing that mature, protein-bound anchors only exhibit a nonacetylated inositol.[74] *T. gondii* synthesizes several potential precursors *in vivo* that could be transferred to protein and are found as free and highly immunogenic GPIs on the parasite surface.[75] These free GPIs were long known as low molecular weight (LMW) antigen until they were identified as free GPIs.[75] Analysis of the GPI anchor present on the major surface antigen SAG1 (P30) demonstrates that this protein can occur in two different glycoforms and carries two distinct glycan types on the membrane anchor as shown in Figure 2 (Zinecker *et al.*).[76]

(ii)   Yeast

In yeast, inositol acylation was found on early intermediates of GPI anchor biosynthesis, whereas the protein-bound forms are deacylated.[5] It is still not clear if only certain precursors carry this modification or if all free GPIs are inositol modified. Deacylation therefore has to take place either during precursor biosynthesis or immediately after transfer to protein. The substrate for acylation is GlcN-PI, with palmitoyl-CoA being the donor for the acyl chains.[5] It is still not known if acylation is a prerequisite for mannose elongation.

In the majority of the mature, protein-bound GPI anchor population, the diacylglycerol backbone is replaced by a ceramide, rendering these mature anchors alkaline resistant.[77] Unlike the fatty acid remodeling in trypanosomes, this process is a replacement of the complete lipid once the anchor is attached to protein.[78] Of particular interest is the use of additional ethanolamine phosphates. It was first believed that yeast like the other lower eukaryotes does not perform this reaction since no such additional ethanolamine phosphates were found on mature, protein-bound GPI anchors. Data, however, showed that *S. cerevisiae* indeed adds an ethanolamine phosphate to the first mannose of Man$_2$GlcN-PI intermediates before elongation with the third mannose can proceed[38,39] since inhibition of ethanolamine transfer by the inhibitor YW3548 leads to accumulation of that intermediate carrying two mannoses.[65] Data by Canivenc-Gansel *et al.* provide evidence for a substantial fraction of mature GPI precursors still carrying this additional modification.[39] This ethanolamine phosphate cannot be found on protein-bound GPIs,[77] making this a transient but essential modification. This puts yeast in the middle between the protozoa and mammalian cells and by the same token may raise the question if this distinctive feature between protozoa and higher eukaryotes indeed is that exclusive or if a transient addition of ethanolamine phosphate in the protozoa simply has not been observed as yet.

(iii)   Mammalia

All mannosylated GPI anchor intermediates were shown to be resistant to cleavage with PI-PLC, suggesting that they are inositol acylated. This acylation has to precede addition of the first mannose, making GlcN-acylPI the substrate for mannosylation. Studies on this acylation reaction *in vitro* showed that this process can be stimulated by either CoA or acyl-CoA.[79,80] In contrast to trypanosomes, this reaction cannot be inhibited by phenylmethyl sulfonyl fluoride (PMSF).[59] Most protein-bound anchors were found to be sensitive towards PI-PLC treatment,[1,5,9] indicating that deacylation has to occur either immediately before or after transfer to protein. A key feature of mammalian cells is the addition of an ethanolamine phosphate to the first mannose being a prerequisite for the addition of the following two mannoses.[81] After addition of the third mannose, the second ethanolamine phosphate, which is needed for the transfer to protein, is added to the 6-position of this mannose.[81] A third ethanolamine phosphate can then be added to the second mannose to form a three mannose–three ethanolamine phosphate core glycan.[81] Both forms have been found on acetylcholinesterase of human erythrocytes but whether they are transferred to protein equally efficiently is still not known.

As shown in these examples, GPI anchor biosynthesis, although being evolutionarily highly conserved, has evolved to a pathway that is characterized by a high degree of divergence when distantly related species are compared. This is also reflected at the molecular level when the individual enzymes are compared. Their degree of conservation on the amino acid sequence level between, e.g., mammalia and yeast, may be as low as 24%,[34] thus also reflecting the marked degree of divergence on the molecular level and even arguing for the need of individually distinct pathways due to selective pressure, especially in the parasitic protozoa.

3.11.3   LIPOPHOSPHOGLYCANS (LPGs) AND GLYCOSYLINOSITOL PHOSPHOLIPIDS (GIPLs)

As mentioned above, these glycosylated phosphatidylinositols which share the common structural motif Man-α-(1-4)-GlcN1-PI with GPI anchors were first found in *Trypanosoma cruzi* (GIPLs) and subsequently in the genus *Leishmania* where they seem to be the major class of free glycolipids.[4,7] They produce a densely packed surface coat (glycocalyx) which might be a key protective feature against the sometimes harsh biochemical environment to which these parasites have adapted.[7] There is evidence that these glycolipids are essential for survival in the mammalian as well as the insect host. These free glycolipids are involved in the attachment to the midgut wall of the insect vector, prevention of complement-mediated lysis in the mammalian host, as well as protection from hydrolases in the macrophage phagolysosome.[7] They were also shown to modulate the host cell immune response by interfering with host cell signal transduction pathways.[7]

3.11.3.1 Lipophosphoglycans (LPGs)

LPG biosynthesis is predominantly seen in the promastigote stage which develops in the insect vector. Only in *Leishmania mayor* is significant LPG biosynthesis seen in other stages of the life cycle.

The basic structure of LPGs consists of a linear backbone that contains the disaccharide repeat P-6-Gal-β-(1-4)-Man-α-1-linked to a unique GPI-like structure containing a hexasaccharide core (see Figure 4) linked to lysoalkyl-PI. The disaccharide repeats may then either be unsubstituted or modified with monosaccharides or oligosaccharide side chains,[4,7] being responsible for the species- and stage-specificity for the LPGs. The terminal galactose of the disaccharide backbone is modified by a so-called cap structure which also exhibits species- and stage-specific carbohydrates. Polymorphism is therefore produced because of developmentally regulated variability in the number of disaccharide repeats, addition of modifications to the repeats, and variation in the capping structure during the parasite's life cycle, predominantly seen as an increase in chain length or capping of the terminal β-Gal residue. Biosynthesis of the complex structures of LPGs involves a set of unique enzymes, some of which have been cloned over the years.[82] These genes are important for parasite virulence.[82]

3.11.3.2 Glycosylinositol Phospholipids (GIPLs)

GIPLs are present at very high levels (5×10^7 molecules per cell) in all developmental stages of the parasite.[4,7] Three distinct series of GIPL structures have been identified, which are synthesized in a species- and stage-specific manner.[7] Since they represent the major cell surface component, their function is predominantly viewed as a protective coat especially in the amastigote stage that lives in the macrophage phagolysosome.[7] The three distinct types or lineages share the Man-GlcN-PI core. Type 1 and 2 GIPLs are then extended in a linear fashion with either α-(1-6)-mannose residues (Type 1) or with α-(1-3)-mannose residues (Type 2). Hybrid-type GIPLs possess branched mannose chains due to the presence of both linkage types.[7] These three mannose core structures can then be modified by additional carbohydrates and ethanolamine phosphate linked directly to the glucosamine (see Figure 4).

Despite similar glycan moieties with respect to GPI or LPG, the lipid moieties of GIPLs are clearly distinct from these other types.[7] The distinct alkyl chain composition of GIPLs compared to protein and LPG anchors is acquired by the selection of specific alkylacyl PI species and not by subsequent remodeling steps, unlike mammalian cells, yeast, and some protozoa where there is no evidence that the early enzymes involved in GPI anchor biosynthesis show selectivity for specific molecular PI species.[4-7] The presence of distinct PI lipid moieties in GIPL and GPI as well as differences in flux of intermediates through these pathways strongly support the notion that GIPLs are the products of a separate pathway instead of being excess precursors for the other two pathways.[83] This is also supported by several lines of evidence pointing to a certain degree of subcellular compartmentalization for GPI and GIPL biosynthetic pathways. Fatty acid remodeling reaction similar to those of the African trypanosomes then occur after assembly of the glycan head group.

Free glycolipids similar to the *Leishmania* LPGs and GIPLs have been identified in other protozoa, mostly kinetoplastid parasites, but no information about their biosynthesis has been obtained to date.[4,7]

In conclusion it can be said that GPI research has evolved from the biochemical/analytical field to a truly interdisciplinary subject involving genetics, cell biology, molecular biology, and medical research, and is now a rapidly growing area with very interesting prospects for the future.

ACKNOWLEDGMENTS

The work of the authors was supported by Deutsche Forschungsgemeinschaft, Fonds der Chemischen Industrie, Stiftung P. E. Kempkes, Hessisches Ministerium für Wissenschaft und Kunst, and DAAD.

3.11.4 REFERENCES

1. G. A. M. Cross, *Annu. Rev. Cell Biol.*, 1990, **6**, 1.
2. P. T. Englund, *Annu. Rev. Biochem.*, 1993, **62**, 121.
3. M. A. J. Ferguson, J. S. Brimacombe, S. Cottaz, R. A. Field, L. S. Güther, S. W. Homans, M. J. McConville, A. Mehlert, K. G. Milne, J. E. Ralton, *et al.*, *Parasitology*, 1994, **108**, 545.
4. M. J. McConville and M. A. J. Ferguson, *Biochem. J.*, 1993, **294**, 305.
5. M. C. Field and A. K. Menon, in "Lipid Modification of Proteins," ed. M. J. Schlesinger, CRC Press, Boca Raton, FL, 1993, p. 83.
6. V. L. Stevens, *Biochem. J.*, 1995, **310**, 361.
7. M. J. McConville, in "Molecular Biology of Parasitic Protozoa," eds. D. F. Smith and M. Parsons, Oxford University Press, Oxford, 1996, p. 205.
8. P. Gerold, V. Eckert, and R. T. Schwarz, *Trends Glycosci. Glycotechnol.*, 1996, **8**, 265.
9. V. Eckert, P. Gerold, and R. T. Schwarz, in "Glycosciences", eds. H.-J. Gabius, and S. Gabius, Chapman & Hall, Weinheim, 1997, p. 223.
10. M. A. J. Ferguson, S. W. Homans, R. A. Dwek, and T. W. Rademacher, *Science*, 1988, **239**, 753.
11. S. W. Homans, M. A. Ferguson, R. A. Dwek, T. W. Rademacher, R. Anand, and A. F. Williams, *Nature*, 1988, **333**, 269.
12. A. A. Holder, *Biochem. J.*, 1983, **209**, 261.
13. M. G. Low and J. B. Finean, *Biochem. J.*, 1977, **167**, 281.
14. M. A. J. Ferguson, M. G. Low, and G. A. M. Cross, *J. Biol. Chem.*, 1985, **260**, 14 547.
15. W. J. Masterson, J. Raper, T. L. Doering, G. W. Hart, and P. T. Englund, *Cell*, 1990, **62**, 73.
16. M. C. Field and A. K. Menon, in "Lipid Modifications of Proteins," eds. N. N. Hooper and A. J. Turner, IRL Press, Oxford, 1992, p. 155.
17. M. A. J. Ferguson, in "Glycobiology: A Practical Approach," eds. M. Fukuda and A. Kobata, IRL Press, Oxford, 1993, p. 349.
18. D. A. Brown and J. K. Rose, *Cell*, 1992, **68**, 533.
19. K. Simons and A. Wandinger-Ness, *Cell*, 1990, **62**, 207.
20. P. J. Robinson, *Immunol. Today*, 1991, **12**, 35.
21. A. R. Saltiel, *J. Bioenerg. Biomembr.*, 1991, **23**, 29.
22. L. Schofield and S. D. Tachado, *Immunol. Cell Biol.*, 1996, **74**, 555.
23. L. H. Miller, M. F. Good, and G. Milon, *Science*, 1994, **264**, 1878.
24. L. Schofield and F. Hackett, *J. Exp. Med.*, 1993, **177**, 145.
25. L. Schofield, S. Novakovic, P. Gerold, R. T. Schwarz, M. J. McConville, and S. D. Tachado, *J. Immunol.*, 1996, **156**, 1886.
26. S. D. Tachado, P. Gerold, M. J. McConville, T. Baldwin, D. Quilici, R. T. Schwarz, and L. Schofield, *J. Immunol.*, 1996, **156**, 1897.
27. I. A. Clark and K. A. Rockett, *Parasitol. Today*, 1994, **10**, 410.
28. S. D. Tachado and L. Schofield, *Biochem. Biophys. Res. Commun.*, 1994, **205**, 984.
29. T. Kinoshita, N. Inoue, and J. Takeda, *Adv. Immunol.*, 1995, **60**, 57.
30. K. Kawagoe, D. Kitamura, M. Okabe, I. Taniuchi, M. Ikawa, T. Watanabe, T. Kinoshita, and J. Takeda, *Blood*, 1996, **87**, 3600.
31. M. Tarutani, S. Itami, M. Okabe, M. Ikawa, T. Tezuka, K. Yoshikawa, T. Kinoshita, and J. Takeda, *Proc. Natl. Acad. Sci., USA*, 1997, **94**, 7400.
32. J. Takeda and T. Kinoshita, *Trends Biochem. Sci.*, 1995, **20**, 367.
33. T. Kinoshita, K. Ohishi, and J. Takeda, *J. Biochem. (Tokyo)*, 1997, **122**, 251.
34. O. Nosjean, A. Briolay, and B. Roux, *Biochem. Biophys. Acta*, 1997, **1331**, 153.
35. R. Watanabe, N. Inoue, B. Westfall, C. H. Taron, P. Orleans, J. Takeda, and T. Kinoshita, *EMBO J.*, 1998, **17**, 877.
36. N. Nakamura, N. Inoue, R. Watanabe, M. Takahashi, J. Takeda, V. L. Stevens, and T. Kinoshita, *J. Biol. Chem.*, 1997, **272**, 15 834.
37. M. Takahashi, N. Inoue, K. Ohishi, Y. Maeda, N. Nakamura, Y. Endo, T. Fujita, J. Takeda, and T. Kinoshita, *EMBO J.*, 1996, **15**, 4254.
38. C. Sütterlin, M. V. Escribano, P. Gerold, Y. Maeda, M. J. Mazon, T. Kinoshita, R. T. Schwarz, and H. Riezman, *Biochem. J.*, 1998, **332**, 153.
39. E. Canivenc-Gansel, I. Imhofa, F. Reggiori, P. Burda, A. Conzelmann, and A. Benachour, *Glycobiology*, 1998, **8**, 761.
40. A. K. Menon, S. Mayor, and R. T. Schwarz, *EMBO J.*, 1990, **9**, 4249.
41. P. Orlean, C. Albright, and P. W. Robbins, *J. Biol. Chem.*, 1988, **263**, 17 499.
42. R. Mazhari-Tabrizi, V. Eckert, M. Blank, R. Müller, D. Mumberg, M. Funk, and R. T. Schwarz, *Biochem. J.*, 1996, **316**, 853.
43. S. Tomita, N. Inoue, Y. Maeda, K. Ohishi, J. Takeda, and T. Kinoshita, *J. Biol. Chem.*, 1998, **273**, 9249.
44. Y. Maeda, S. Tomita, R. Watanabe, K. Ohishi, and T. Kinoshita, *EMBO J.*, 1998, **17**, 4920.
45. A. K. Menon, M. Eppinger, S. Mayor, and R. T. Schwarz, *EMBO J.*, 1993, **12**, 1907.
46. A. K. Menon and V. L. Stevens, *J. Biol. Chem.*, 1992, **267**, 15 277.
47. N. Inoue, T. Kinoshita, T. Orii, and J. Takeda, *J. Biol. Chem.*, 1993, **268**, 6882.

48. S. Udenfriend and S. Kodukula, *Annu. Rev. Biochem.*, 1995, **64**, 563.
49. S. Kodukula, L. D. Gerber, R. Amthauer, L. Brink, and S. Udenfriend, *J. Cell Biol.*, 1993, **120**, 657.
50. C. Nuoffer, A. Horvath, and H. Riezman, *J. Biol. Chem.*, 1993, **268**, 10 558.
51. P. Moran and I. W. Caras, *J. Cell Biol.*, 1993, **125**, 333.
52. C. D. Reymond, C. Beghdadi-Rais, M. Roggero, E. A. Duarte, C. Desponds, M. Bernard, D. Groux, H. Matile, C. Bron, G. Corradin, and N. J. Fasel, *J. Biol. Chem.*, 1995, **270**, 12 941.
53. N. Fasel, H. Stark, C. Bron, and C. Beghdadi-Rais, *Braz. J. Med. Biol. Res.*, 1994, **27**, 189.
54. D. Hamburger, M. Egerton, and H. Riezman, *J. Cell Biol.*, 1995, **129**, 629.
55. M. Benghezal, A. Benachour, S. Rusconi, M. Aebi, and A. Conzelmann, *EMBO J.*, 1996, **15**, 6575.
56. K. Mensa-Wilmot, J. H. LoBowitz, K. P. Chang, A. Al-Qahtani, B. S. McGuire, S. Tucker, and J. C. Morris, *J. Cell Biol.*, 1994, **124**, 935.
57. J. Vidugiriene and A. K. Menon, *J. Cell Biol.*, 1993, **121**, 987.
58. J. Vidugiriene and A. K. Menon, *J. Cell Biol.*, 1994, **127**, 333.
59. M. L. Güther, W. J. Masterson, and M. A. J. Ferguson, *J. Biol. Chem.*, 1994, **269**, 18 694.
60. M. L. Güther and M. A. J. Ferguson, *EMBO J.*, 1995, **14**, 3080.
61. J. E. Ralton, K. G. Milne, M. L. Güther, R. A. Field, and M. A. J. Ferguson, *J. Biol. Chem.*, 1993, **268**, 24 183.
62. T. K. Smith, S. Cottaz, J. S. Brimacombe, and M. A. J. Ferguson, *J. Biol. Chem.*, 1996, **271**, 6476.
63. T. K. Smith, F. C. Milne, D. K. Sharma, A. Crossman, J. S. Brimacombe, and M. A. J. Ferguson, *Biochem. J.*, 1997, **326**, 393.
64. T. K. Smith, D. K. Sharma, A. Crossman, A. Dix, J. S. Brimacombe, and M. A. J. Ferguson, *EMBO J.*, 1997, **16**, 6667.
65. C. Sütterlin, A. Horvath, P. Gerold, R. T. Schwarz, Y. Wang, M. Dreyfuss, and H. Riezman, *EMBO J.*, 1997, **16**, 6374.
66. P. Gerold, B. Striepen, B. Reitter, H. Geyer, R. Geyer, E. Reinwald, H.-J. Risse, and R. T. Schwarz, *J. Mol. Biol.*, 1996, **261**, 181.
67. P. Gerold, unpublished results.
68. P. Gerold, A. Dieckmann-Schuppert, and R. T. Schwarz, *J. Biol. Chem.*, 1994, **269**, 2597.
69. A. Schmidt, R. T. Schwarz, and P. Gerold, *Exp. Parasitol.*, 1998, **88**, 95.
70. P. Gerold, L. Schofield, M. J. Blackman, A. A. Holder, and R. T. Schwarz, *Mol. Biochem. Parasitol.*, 1996, **75**, 131.
71. A. Dieckmann-Schuppert, P. Gerold, and R. T. Schwarz, in "Glycoproteins and Disease," eds. J. Montreuil, H. Schachter, and J. F. G. Vliegenthart, Elsevier Science B.V., Amsterdam, 1996, p. 125.
72. D. C. Gowda, P. Gupta, and E. A. Davidson, *J. Biol. Chem.*, 1997, **272**, 6428.
73. S. Tomavo, J. F. Dubremetz, and R. T. Schwarz, *J. Biol. Chem.*, 1992, **267**, 21 446.
74. S. Tomavo, J. F. Dubremetz, and R. T. Schwarz, *Biol. Cell*, 1993, **78**, 155.
75. B. Stiepen, C. F. Zinecker, J. B. Damm, P. A. Melgers, G. J. Gerwig, M. Koolen, J. F. Vliegenthart, J.-F. Dubremetz, and R. T. Schwarz, *J. Mol. Biol.*, 1997, **266**, 797.
76. C. F. Zinecker, B. Striepen, J. F. Dubremetz, and R. T. Schwarz, manuscript submitted.
77. C. Fankhauser, S. W. Homans, J. E. Thomas-Oates, M. J. McConville, C. Desponds, A. Conzelmann, and M. A. J. Ferguson, *J. Biol. Chem.*, 1993, **268**, 26 365.
78. F. Reggiori, E. Canivenc-Gansel, and A. Conzelmann, *EMBO J.*, 1997, **16**, 3506.
79. V. L. Stevens and H. Zhang, *J. Biol. Chem.*, 1994, **269**, 31 397.
80. W. T. Doerrler, J. Ye, J. R. Falck, and M. A. Lehrman, *J. Biol. Chem.*, 1996, **271**, 27 031.
81. S. Hirose, G. M. Prince, D. Sevlever, L. Ravi, T. L. Rosenberry, E. Ueda, and M. E. Medof, *J. Biol. Chem.*, 1992, **267**, 16 968.
82. S. M. Beverley and S. J. Turco, *Trends Microbiol.*, 1998, **6**, 35.
83. J. E. Ralton and M. J. McConville, *J. Biol. Chem.*, 1998, **273**, 4245.

3.12
Deoxysugars: Occurrence, Genetics, and Mechanisms of Biosynthesis

DAVID A. JOHNSON and HUNG-WEN LIU

University of Minnesota, Minneapolis, MN, USA

3.12.1 INTRODUCTION

Advances in carbohydrate research have enabled investigators to discover and characterize deoxysugars from a variety of plant, bacterial, and mammalian sources. These deoxysugars are derived from common sugars and have at least one hydroxyl group replaced by a hydrogen atom. As a class, deoxysugars exhibit a greater range of activities, due in part to their increased thermodynamic stability and hydrophobicity. A wide variety of deoxysugars are found in lipopolysaccharides (LPSs),[1] glycoproteins,[2,3] and glycolipids.[4] Many biological roles have been assigned to these polymeric units, including purely structural roles, serving as ligands for cell–cell interactions, serving as targets for toxins, antibodies, and microorganisms, and even controlling the half-life of proteins in serum.[5] Similar functions, especially target binding and specificity, have been ascribed to the deoxysugar components of numerous secondary metabolites such as bacterial antibiotics, which are another abundant source of these unusual carbohydrates. Aberrant elaboration of the polysaccharide in glycoconjugates is the cause of many disease states,[6,7] and several antibiotics are used to treat various malignancies as well as bacterial and parasitic infections. Thus, it is medically important to have a complete understanding of the nature of these metabolites so that suitable diagnostic strategies and improved treatment procedures can be developed.

Toward this goal, much effort has been expended to elucidate the structure–activity relationships, biosyntheses, and genetics of the deoxysugars in these natural products. An arsenal of experimental techniques, including chemical synthesis, molecular biology, various spectroscopic methods, and X-ray crystallography, has been used in this effort, and the result is an expansive literature. In this chapter, we will attempt to bring together literature on the occurrence, biosynthesis, and genetics of deoxysugars. Additionally, information on the enzymatic mechanisms of formation and the detailed modes of action for the deoxysugars will be presented where it is available.

While aminosugars are technically deoxysugars, they are components of bacterial LPSs and aminoglycoside antibiotics. Therefore, the scope of this chapter will primarily be limited to deoxysugars in which hydrogen replaces one or more hydroxyl group(s). Branched-chain deoxysugars, and other deoxygenated carbohydrates which are devoid of an amino functionality, will also be presented. The discussion will focus on end products, so transient deoxysugars, such as those occurring in glycolysis, will not be considered. The reader is referred to Chapters 3.06 and 3.07 for a more in-depth discussion of glycolipids.

3.12.2 NATURAL OCCURRENCE OF DEOXYSUGARS

Table 1 provides a partial, but still extensive, listing of deoxysugars with their formal names, trivial names, example source(s), and relevant references. The reader is referred to other reviews for additional listings and information on the synthesis of these compounds.[1,8–13] The system of nomenclature recommended by IUPAC was generally followed in naming the deoxysugars in their acyclic forms. Although trivial names have been historically associated with either the D or L configuration of a deoxysugar, a single trivial name with a suitable prefix is used to denote both configurations in this chapter, unless each enantiomer has an explicit trivial name. Some of the compounds have more than one trivial name based on the natural product source, and we have made an attempt to list these together. Also, methyl ethers of most deoxysugars have different trivial names. The listing is ordered according to the degree of deoxygenation (i.e., mono-, di-, or trideoxy), the carbon position which is deoxygenated, and the configuration of the sugar. Branched-chain sugars are listed separately. With this arrangement, similar sugars should be juxtaposed, thus making it easier to note the different trivial names associated with the various isomers. The structures for the cyclic forms of selected compounds are given on the following pages (Figure 1). The sugars may be referred to using either a formal name or a trivial name throughout this chapter, as deemed appropriate.

3.12.2.1   Deoxygenation at the C-6 Position

As can be seen in Table 1, most of the naturally occurring deoxysugars are 6-deoxyhexoses with or without hydroxyl group replacements at other positions. The most common 6-deoxysugars are L-fucose (13) and L-rhamnose (23), which are often found as components of the polysaccharides of many glycoconjugates.[1,2,25] Other 6-deoxyhexoses and their methyl ether derivatives seem to be found primarily in either LPSs or various secondary metabolites, which are mainly antibiotics. Some noteworthy examples are D- and L-digitalose (15) and (16) from seaweed,[26] and D-antiarose (20),[47] D- and L-acofriose (24) and (25),[52–54] and D-talomethylose (26)[50] from bacterial LPSs. Cardiac glycosides are sources for D-digitalose (16),[14,28,29] D-quinovose (18),[14,46] D-thevetose (19),[28,29] D-antiarose (20),[14] and L-talomethylose (27).[14] The related starfish saponins[44,45] and the antimicrobial viridopentaoses A, B, and C[203] have been shown to include D-quinovose (18) as well. All members of the orthosomycin antibiotics family, which comprises everninomicin, flambamycin, avilamycin, and curamycin, contain D-curacose (17).[39] Two other antibiotics, mycoside G[55] and the calicheamicins,[56,57] are known to bear L-acofriose (25) in their structures. The review by Schaffer[12] includes more information on these 6-deoxysugars. Significantly, sugars of this class serve as precursors for saccharides which have a higher degree of deoxygenation and branching carbons, as discussed below.

3.12.2.2   Deoxygenation at the C-2 Position

Monodeoxygenation at positions other than C-6 in the carbon skeleton are less common, with the exception of 2-deoxy-D-*erythro*-pentose (3), which is the sugar in deoxyribonucleotides and forms the framework of DNA. A few 2-deoxysugars, such as 2-deoxy-D-*xylo*-hexose (1) and 2-deoxy-D-*arabino*-hexose (2), have been isolated from cardiac glycosides.[14] However, most naturally occurring 2-deoxysugars also have their 6-hydroxyl group replaced by hydrogen. A great variety of 2,6-dideoxysugars have been found in antibiotics isolated from *Streptomyces* spp. and in cardiac glycosides isolated from several plant sources. The 3- or 4-hydroxyl group is frequently methylated in these dideoxysugars.

A common member of this class of sugar is 2,6-dideoxy-D-*arabino*-hexose (29), which has several trivial names. For example, it is called 2-deoxy-D-rhamnose in the orthosomycin antibiotics,[39] D-canarose in cardiac glycosides,[59] D-chromose C in chromomycin A$_3$,[60–63] and D-olivose in olivomycin A,[64–66] mithramycin,[66–69] and several other antibiotics (see Table 1). Both enantiomers of oleandrose, (30) and (31), have been isolated from cardiac glycosides,[14,76–80] while only L-oleandrose (31) is found in avermectins.[81,82] Kerriamycin A contains D-kerriose (33), which is an unusual deoxysugar with a keto moiety at the 3-position.[71] Oleandomycin,[84] olivomycin A,[64–66] mithramycin,[66–69] chromocyclomycin,[70] and some cardiac glycosides[14] have 2,6-dideoxy-D-*lyxo*-hexose (D-oliose (34)) in their structures. The enantiomer of this sugar is called 2-deoxy-L-fucose (35), and it is found in esperamicins[86,87] and some anthracycline antibiotics, including aclacinomycins[85] and arugomycin.[88] Cardiac glycosides are also a well-known source for D- and L-diginose (36) and (37),[14,77,78] D-digitoxose (40),[14,77,78,80,96] D-boivinose (46),[14] and both enantiomers of cymarose (42) and (43)[28,29,76–80] and sarmentose (47) and (48).[14,77,78] It should be noted that D-cymarose (42) has also been called D-variose in variamycin.[100–102] Arugomycin[88,89] and decilorubicin[90] both contain L-diginose (37). The antibiotics olivomycin A,[64–66] chromomycin A$_3$,[61,63,91,92] and rhodomycin[93,94] all comprise 2,6-dideoxy-4-*O*-methyl-D-*lyxo*-hexose (38), which is commonly referred to as D-olivomose or D-chromose A. A 4-*O*-acetyl derivative of this sugar known as D-chromose D or D-acetyloliose (39) is also found in chromomycin A$_3$[61–63,92] and olivomycin A.[64–66]

3.12.2.3   Deoxygenation at the C-3 Position

Sugars with the 3-hydroxyl group replaced by a hydrogen atom are more rare than 2-deoxysugars. Two known examples of C-3 monodeoxysugars are 3-deoxy-D-*erythro*-pentose (4) from cordycepin[16] and an *O*-branched form of 3-deoxy-D-*threo*-pentose (5) from agrocin 84.[17] Both of these sugars

Table 1 Partial listing of naturally-occurring deoxysugars.

Formal name	Trivial name	Occurrence	Ref.
Monodeoxysugars			
2-Deoxy-D-xylo-hexose (**1**)	2-Deoxy-D-gulose or 2-deoxy-D-idose	Cardiac glycosides	14
2-Deoxy-D-arabino-hexose (**2**)	2-Deoxy-D-glucose	Cardiac glycosides	14
2-Deoxy-D-erythro-pentose (**3**)	2-Deoxy-D-ribose	Deoxyribonucleic acid	15
3-Deoxy-D-erythro-pentose (**4**)	Cordycepose	Cordycepin	16
3-Deoxy-D-threo-pentose (**5**)		Agrocin 84	17
4-Deoxy-D-arabino-hexose (**6**)	4-Deoxy-D-altrose or 4-deoxy-D-idose	*Citrobacter* species	18, 19
6-Deoxy-D-allose (**7**)	D-Allomethylose	Cardiac glycosides	14
6-Deoxy-2,3-di-O-methyl-D-allose (**8**)	D-Mycinose	Tylosin	20, 21
6-Deoxy-L-altrose (**9**)	L-Altromethylose	Bacterial LPSs	22, 23
6-Deoxy-3-O-methyl-L-altrose (**10**)	L-Vallarose	Cardiac glycosides	14
6-Deoxy-4-O-methyl-D-altrose (**11**)		Sodarin	24
6-Deoxy-D-galactose (**12**)	D-Fucose	Cardiac glycosides	14
6-Deoxy-L-galactose (**13**)	L-Fucose	LPSs	1, 25
		Glycoconjugates	2–4
6-Deoxy-2-O-methyl-D- or L-galactose (**14**)		Seaweed	26
6-Deoxy-2,4-di-O-methyl-D-galactose (**15**)	D-Labilose	Labilomycin	27
6-Deoxy-3-O-methyl-L-galactose (**16**)	L-Digitalose	Seaweed	26
6-Deoxy-3-O-methyl-D-galactose (**16**)	D-Digitalose	Cardiac glycosides	14, 28, 29
		Seaweed	26
6-Deoxy-4-O-methyl-D-galactose (**17**)	D-Curacose	Flambamycin	30–33
		Everninomicins B, C, D, -2	34–39
		Avilamycins	40, 41
		Curamycin	42, 43
6-Deoxy-D-glucose (**18**)	D-Quinovose	Starfish saponins	44, 45
		Cardiac glycosides	14, 46
6-Deoxy-3-O-methyl-D-glucose (**19**)	D-Thevetose	Cardiac glycosides	28, 29
6-Deoxy-D-gulose (**20**)	D-Antiarose	Bacterial LPSs	47
6-Deoxy-D-manno-heptose (**21**)		Cardiac glycosides	14
6-Deoxy-D-mannose (**22**)	D-Rhamnose	Bacterial LPSs	48, 49
6-Deoxy-L-mannose (**23**)	L-Rhamnose	Bacterial LPSs	50
		Various glycosides	1–3, 14, 51
6-Deoxy-3-O-methyl-D-mannose (**24**)	D-Acofriose	*Campylobacter fetus*	52
6-Deoxy-3-O-methyl-L-mannose (**25**)	L-Acofriose	*Klebsiella* K73:O10	53
		Some Gram-negative bacteria	54
6-Deoxy-D-talose (**26**)	D-Talomethylose	Mycoside G	55
		Calicheamicins	56, 57
6-Deoxy-L-talose (**27**)	L-Talomethylose	Bacterial LPSs	1, 50
6-Deoxy-3-O-methyl-D-talose (**28**)	L-Acovenose	Cardiac glycosides	14
		Rhodopseudomonas palustris	58

Dideoxysugars

Sugar	Source	Reference
2,6-Dideoxy-D-*arabino*-hexose (**29**)	Cardiac glycosides	59
D-Canarose	Chromomycin A₃	60–63
D-Chromose C	Olivomycin A	64–66
D-Olivose	Mithramycin	66–69
	Chromocyclomycin	70
	Oxamicetin	71
	Landomycins A–D	72
	Urdamycins A–D	73
	Kerriamycin B	74
	Chlorothricin	75
	Flambamycin	31, 33
	Everninomicins B, C, D, -2	34–39
	Avilamycins	40, 41
	Curamycin	43
2-Deoxy-D-rhamnose	Cardiac glycosides	14, 76–78
	Leptadenia hastata	79
	Aclepias fruticosa	80
2,6-Dideoxy-3-*O*-methyl-D- (**30**) or L-*arabino*-hexose (**31**)	Avermectins	81, 82
L-Oleandrose	Notonesomycin A	83
2,6-Dideoxy-4-*O*-methyl-D-*arabino*-hexose (**32**)	Kerriamycin A	74
2,6-Dideoxy-D-*erythro*-hexos-3-ulose (**33**)	Oleandomycin	84
2,6-Dideoxy-D-*lyxo*-hexose (**34**)	Olivomycin A	64, 65
D-Kerriose	Mithramycin	66–69
D-Oliose	Cardiac glycosides	14
	Chromocyclomycin	70
2,6-Dideoxy-L-*lyxo*-hexose (**35**)	Aclacinomycins	85
	Esperamicins	86, 87
	Arugomycin	88
2,6-Dideoxy-3-*O*-methyl-D- (**36**) or L-*lyxo*-hexose (**37**)	Cardiac glycosides	14, 77, 78
D- or L-Diginose	Arugomycin	88, 89
L-Diginose	Decilorubicin	90
2,6-Dideoxy-4-*O*-methyl-D-*lyxo*-hexose (**38**)	Olivomycin A	64–66
D-Olivomose	Chromomycin A₃	61, 63, 91, 92
D-Chromose A	α-Rhodomycin	93
4-*O*-Acetyl-2,6-dideoxy-D-*lyxo*-hexose (**39**)	β-Rhodomycins S-1b, 2, 3, 4	94
D-Chromose D	Chromomycin A₃	61–63, 92
D-Acetyloliose	Olivomycin A	64–66
2,6-Dideoxy-D-*ribo*-hexose (**40**)	Cardiac glycosides	14, 77, 78
D-Digitoxose	*Aclepias fruticosa*	80
	Lipomycin	95
	Erisimum marschallianum	96
2,6-Dideoxy-L-*ribo*-hexose (**41**)	Kijanimicins	97
L-Digitoxose	Tetrocarcins	98
	Antlerimicin	99
2,6-Dideoxy-3-*O*-methyl-D- (**42**) or L-*ribo*-hexose (**43**)	Cardiac glycosides	14, 28, 29, 76–78
D- or L-Cymarose	*Leptadenia hastata*	79
	Aclepias fruticosa	80

Table 1 (continued)

Formal name	Trivial name	Occurrence	Ref.
2,6-Dideoxy-3-*O*-methyl-D-*ribo*-hexose (**43**)	D-Variose	Variamycin	100–102
		Kijanimicins	97
2,6-Dideoxy-4-*O*-methyl-L-*ribo*-hexose (**44**)		Granaticin A	103, 104
2,6-Dideoxy-D-*threo*-hexos-4-ulose (**45**)			14
2,6-Dideoxy-D-*xylo*-hexose (**46**)	D-Boivinose	Cardiac glycosides	14, 77, 78
2,6-Dideoxy-3-*O*-methyl-D- (**47**) or L-*xylo*-hexose (**48**)	D- or L-Sarmentose	Cardiac glycosides	105
3,6-Dideoxy-D-*arabino*-hexose (**49**)	Tyvelose	Bacterial LPSs	106
		Eubacterium saburreum L32	107
		Trichinella spiralis	
3,6-Dideoxy-L-*arabino*-hexose (**50**)	Ascarylose	Bacterial LPSs	105, 108
		Glycolipid	109, 110
3,6-Dideoxy-L-*glycero*-hexos-4-ulose (**51**)	L-Cinerulose B	Aclacinomycins B1, B2	85
3,6-Dideoxy-D-*ribo*-hexose (**52**)	Paratose	Bacterial LPSs	105, 111, 112
3,6-Dideoxy-D-*xylo*-hexose (**53**)	Abequose	Bacterial LPSs	105
3,6-Dideoxy-L-*xylo*-hexose (**54**)	Colitose	Bacterial LPSs	105, 113–116
4,6-Dideoxy-3-*O*-methyl-D-*ribo*-hexose (**55**)	D-Chalcose	Cardiac glycosides	117
4,6-Dideoxy-3-*O*-methyl-D-*xylo*-hexose (**56**)	D-Lankavose	Chalcomycin	118, 119
		Lankamycin	120, 121
Trideoxysugars			
2,3,6-Trideoxy-D-*erythro*-hexose (**57**)	D-Amicetose	Amicetin	122
2,3,6-Trideoxy-L-*erythro*-hexose (**58**)	L-Amicetose	Landomycin C	72
		Aclacinomycin M1, M2	85
2,3,6-Trideoxy-4-*O*-methyl-D-*erythro*-hexose (**59**)		Tetrocarcins	98
		Dianemycin	123
		Septamycin	124
2,3,6-Trideoxy-D- (**60**) or L-*glycero*-hexos-4-ulose (**61**)	D- or L-Cinerulose A	Aclacinomycins A1, A2, G1, K	85
2,3,6-Trideoxy-L-*glycero*-hexos-2-en-4-ulose (**62**)	L-Aculose	Aclacinomycin Y1	85
		Cardiac glycosides	125
2,3,6-Trideoxy-D-*threo*-hexose (**63**)	D-Rhodinose	Sakyomicins	126, 127
		Rhodomycin	94, 122
2,3,6-Trideoxy-L-*threo*-hexose (**64**)	L-Rhodinose	Streptolydigin	128, 129
		Landomycins A–C	72
		Aclacinomycin N1	85
		Urdamycins A–D	73
		Narbosines	130
		Notonesomycin A	83
Branched-chain deoxysugars			
4-*C*-[1-(*S*)-Methoxyethyl]-2,3-*O*-methylene-L-arabinono-1,5-lactone (**65**)		Everninomicins	39, 131
5-Deoxy-3-*C*-(hydroxymethyl)-L-lyxose (**66**)	L-Dihydrostreptose	Bluensomycin	132, 133
5-Deoxy-3-*C*-formyl-L-lyxose (**67**)	L-Streptose	Streptomycin	134
6-Deoxy-3-*C*-methyl-D-gulose (**68**)	D-Virenose	Virenomycin	135, 136
6-Deoxy-3-*C*-methyl-L-gulose (**69**)	L-Virenose	*Coxiella burnetii*	137, 138
6-Deoxy-5-*C*-methyl-4-*O*-methyl-L-*lyxo*-hexose (**70**)	L-Noviose	Novobiocin	139, 140

Systematic name	Trivial name	Occurrence	References
6-Deoxy-3-C-methyl-D-mannose (71)	D-Evalose	Everninomicins B, C, D, -2	34-39
		Flambamycin	31-33
6-Deoxy-3-C-methyl-2,3,4-O-trimethyl-L-mannose (72)	L-Nogalose	Nogalamycin	141, 142
6-Deoxy-3-C-methyl-2-O-methyl-L-talose (73)	L-Vinelose	*Acetobacter vinelandii*	143, 144
6-Deoxy-4-C-[1-(S)-hydroxyethyl]-2,3-O-methylene-D-galactono-1,5-lactone (74)		Avilamycin C	41, 145, 146
4-C-Acetyl-6-deoxy-2,3-O-methylene-D-galactono-1,5-lactone (75)		Flambamycin	31, 33, 147
		Avilamycin A	40, 41, 145
2-C-Butyl-2,5-dideoxy-3-O-(3-methylbutanoyl)-L-arabinono-1,4-lactone (76)	Blastmycinone	Blastmycin	148
4-O-Acetyl-2,6-dideoxy-3-C-methyl-L-*arabino*-hexose (77)	L-Chromose B	Chromomycin A$_3$	60-63, 92
4-O-Isobutyryl-2,6-dideoxy-3-C-methyl-L-*arabino*-hexose (78)	L-Olivomycose E	Olivomycin A	64-66
	L-Chromose B'	Chromomycin A$_2$	61, 62, 149
2,6-Dideoxy-3-C-methyl-D-*arabino*-hexose (79)	D-Evermicose	Everninomicins C, D, C170, -2	36-39, 150, 151
2,6-Dideoxy-3-C-methyl-3-O-methyl-L-*ribo*-hexose (80)	L-Cladinose	Erythromycin A	152, 153
2,6-Dideoxy-3-C-methyl-D-*ribo*-hexose (81)	D-Mycarose	Mithramycin	66-69
2,6-Dideoxy-3-C-methyl-L-*ribo*-hexose (82)	L-Mycarose	Magnamycin	154, 155
		Carbomycin	156
		Spiramycin	157, 158
		Formacidin	159
		Tylosin	20, 21, 160
		Angolamycin	161
		Leucomycin A$_3$	21, 162
		Erythromycin C	163
		Erythromycin D	164
		Chromocyclomycin	70
		Josamycin	165
		Kitsamycin	165
		Plateomycins	166, 167
		Maridomycin	168, 169
		Kedarcidin	170
2,6-Dideoxy-3-C-methyl-3-O-methyl-L-*xylo*-hexose (83)	L-Arcanose	Lankamycin	171-173
2,6-Dideoxy-3-C-methyl-L-*xylo*-hexose (84)	L-Axenose	Axenomycins A, B, D	174
2,6-Dideoxy-4-C-(1-(S)-hydroxyethyl)-L-*xylo*-hexose (85)	γ-Octose	Quinocycline A	175
		Isoquinocycline B	176-179
4-C-Acetyl-2,6-dideoxy-L-*xylo*-hexose (86)	Trioxacarcinose B	Quinocycline B	175, 176
		Isoquinocycline B	175
2,6-Dideoxy-4-S-methyl-4-thio-D-*ribo*-hexose (87)		Esperamicins	86, 87
3,6-Dideoxy-4-C-[1-(S)-hydroxyethyl]-D-*xylo*-hexose (88)	Yersiniose A	Bacterial LPSs	180-183
3,6-Dideoxy-4-C-[1-(R)-hydroxyethyl]-D-*xylo*-hexose (89)	Yersiniose B	Bacterial LPSs	184-186
3,6-Dideoxy-4-C-(1,3-dimethoxy-4,5,6,7-tetrahydroxyheptyl)-D-*xylo*-hexose (90)		*Mycobacterium gastri*	187, 188
4,6-Dideoxy-3-C-[1-(S)-hydroxyethyl]-D-*ribo*-hexose 3,1'-carbonate (91)	D-Aldgarose	Aldgamycins E	189-191
2,3,6-Trideoxy-3-C-methyl-4-O-methyl-3-nitro-L-*arabino*-hexose (92)	L-Evernitrose	Everninomicins B, C, D	39, 192, 193
2,3,6-Trideoxy-3-C-methyl-3-nitro-L-*ribo*-hexose (93)	L-Decilonitrose	Decilorubicin	90, 194
		Arugomycin	88, 89
2,3,6-Trideoxy-4-C-glycolyl-L-*threo*-hexose (94)	L-Pillarose	Pillaromycin A	195-198
2,3,6-Trideoxy-3-C-methyl-4-O-methyl-3-nitro-L-*xylo*-hexose (95)	L-Rubranitrose	Rubradirin	199-202

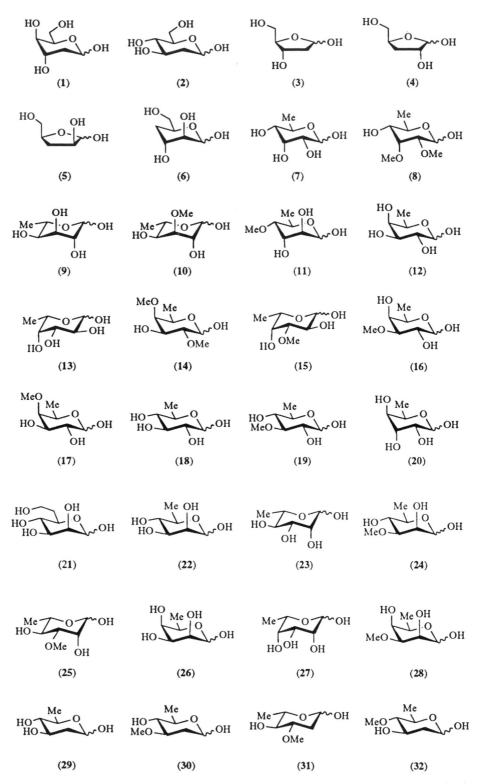

Figure 1   Structures of selected naturally-occurring deoxysugars. The names, occurrences, and references for the compounds are provided in Table 1.

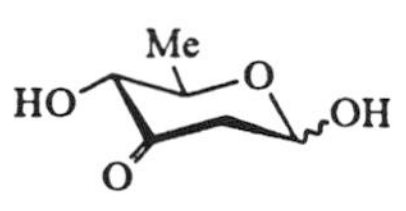

(33) 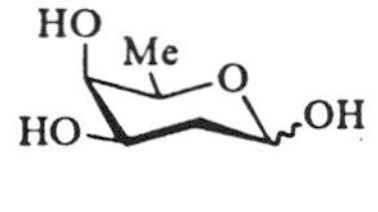 **(34)** 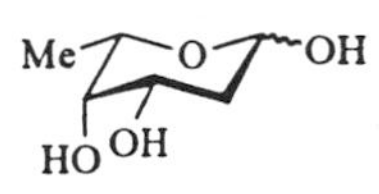 **(35)** 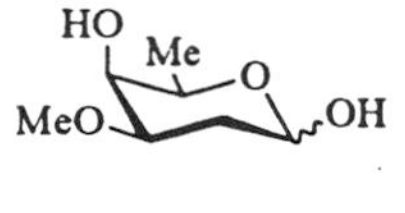 **(36)**

(37) 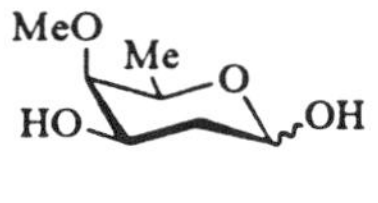 **(38)** 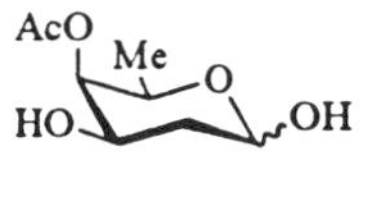 **(39)** 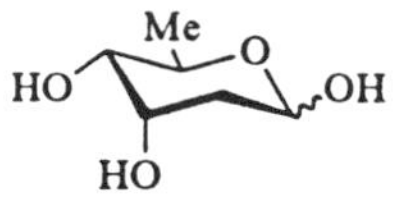 **(40)**

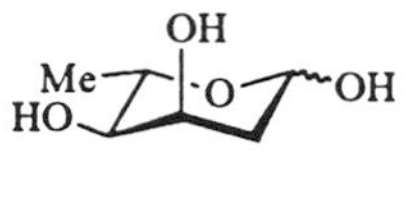

(41) 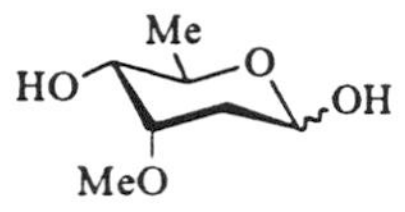 **(42)** 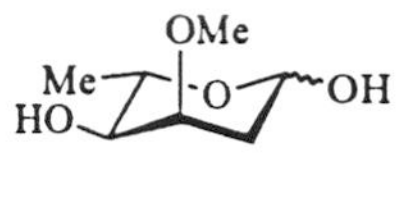 **(43)** 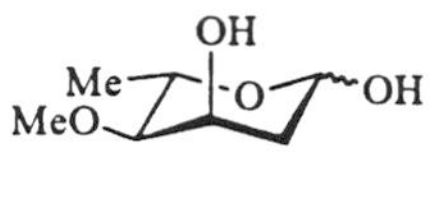 **(44)**

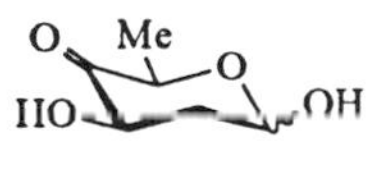

(45) 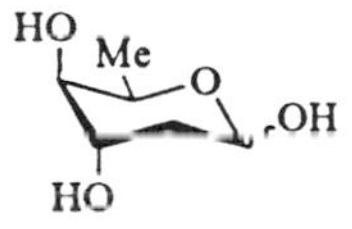 **(46)** 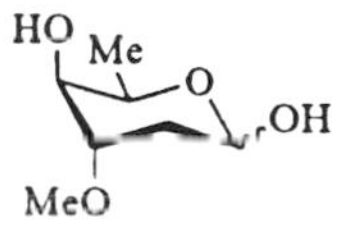 **(47)** 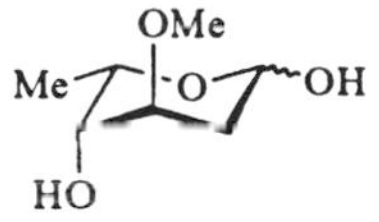 **(48)**

(49) 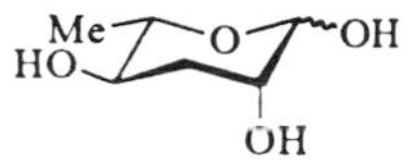 **(50)** 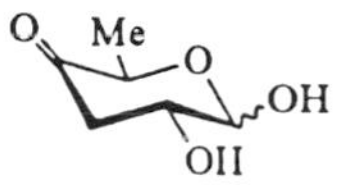 **(51)** 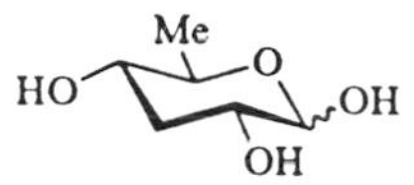 **(52)**

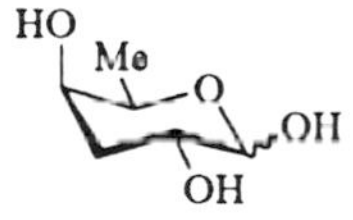

(53) 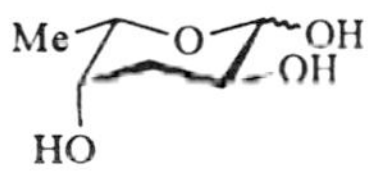 **(54)** 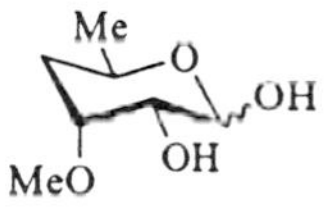 **(55)** 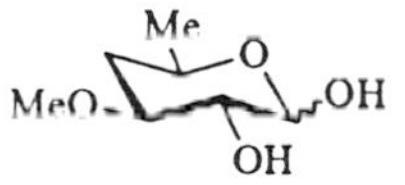 **(56)**

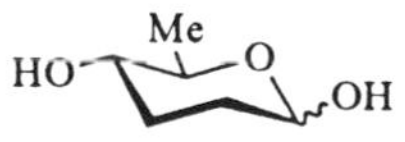

(57) 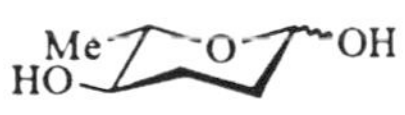 **(58)**  **(59)** 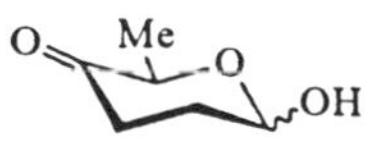 **(60)**

(61)  **(62)** 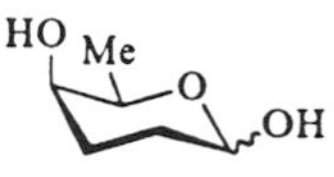 **(63)** 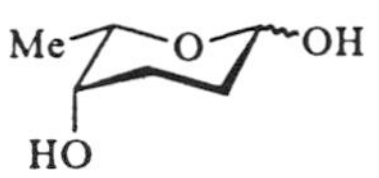 **(64)**

Figure 1 (continued)

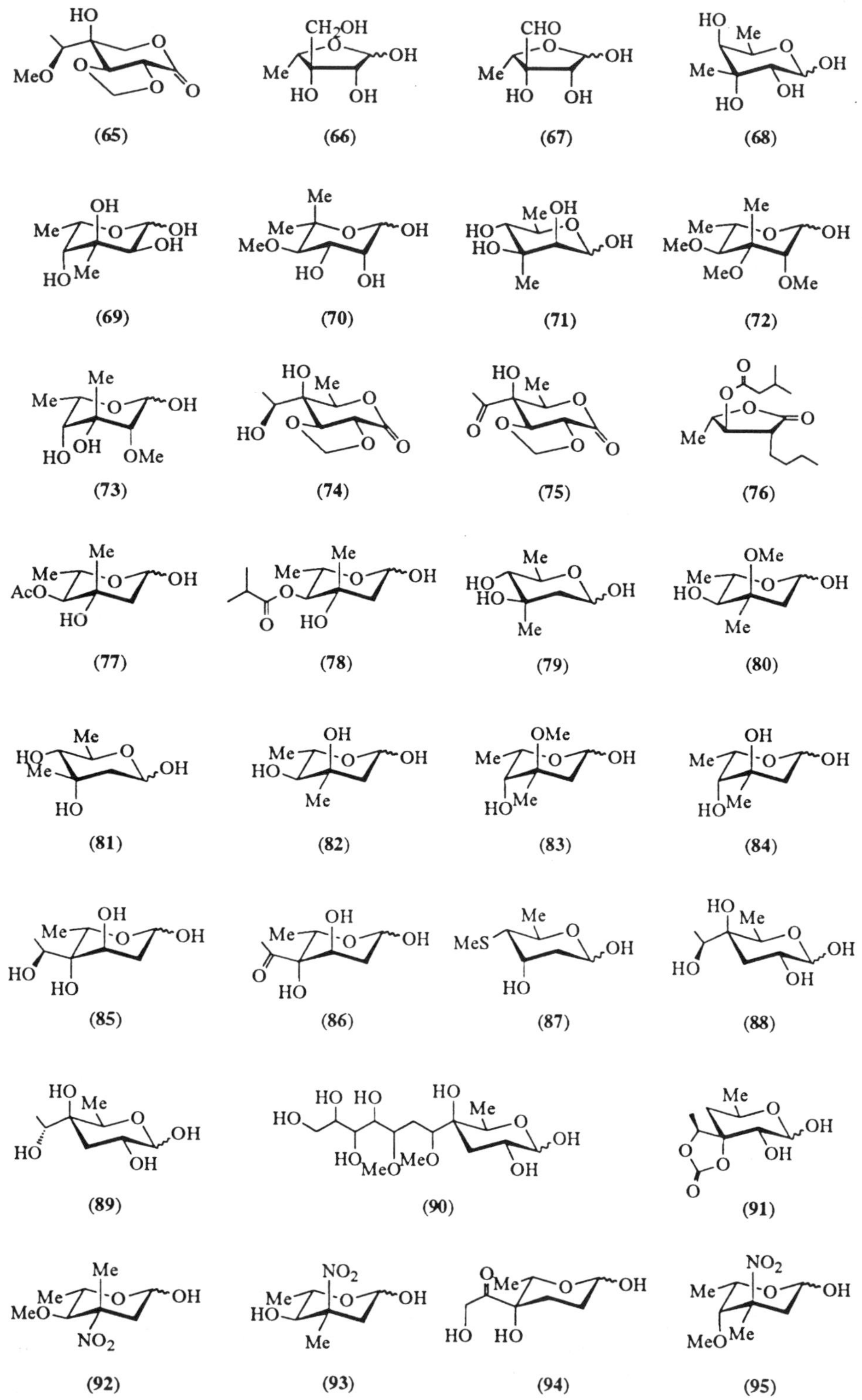

Figure 1 (continued)

are structural components of nucleoside antibiotics.[204] Additional 3-deoxysugars occur as 3,6-dideoxysugars—such as tyvelose (**49**), ascarylose (**50**), paratose (**52**), abequose (**53**), and colitose (**54**)—in the LPSs of Enterobacteriaceae and other pathogenic bacteria (see Table 1). Most of these sugars occur as a terminal α-pyranoside residue, but paratose has been found in the β-furanosyl[111] and β-pyranosyl[112] forms in the LPS of *Yersinia pseudotuberculosis* serogroups IB and III, respectively. In aclacinomycins B1 and B2, a 3,6-dideoxysugar by the name of L-cinerulose B (**51**) has been derivatized to have an unusual diether linkage to another sugar at the C-1 and C-2 hydroxyl groups.[85]

3.12.2.4 Deoxygenation at the C-4 Position

Monosaccharides with a hydrogen substitution at the C-4 position have been discovered in a relatively small number of sources. The LPS of *Citrobacter* spp. is composed of 4-deoxy-D-*arabino*-hexose (**6**),[18,19] and this is the only reported occurrence of a C-4 monodeoxyhexose. However, dideoxysugars lacking the 4-hydroxyl group are known. For example, 4,6-dideoxy-3-*O*-methyl-D-*ribo*-hexose (**55**) has been found in cardiac glycosides,[117] and 4,6-dideoxy-3-*O*-methyl-D-*xylo*-hexose (**56**) has been identified in the antibiotics chalcomycin[118,119] and lankamycin.[120,121] No other non-branched 4-deoxysugars lacking an amino group have been reported thus far, though there are a few aminosugars with a hydrogen replacement at C-4. D-Desosamine (**130**) from erythromycin A[152,153] is one such sugar, and it will be discussed later in this chapter (see Section 3.12.5.3.2).

3.12.2.5 Deoxygenation at the C-2, C-3, and C-6 Positions

Concurrent deoxygenation at C-2, C-3, and C-6 occurs only in a limited number of deoxysugars discovered so far. Bacterial antibiotics are the primary source of these 2,3,6-trideoxysugars, such as amicetose (**57**), cinerulose A (**60**) and (**61**), aculose (**62**), and rhodinose (**63**) and (**64**) (see Table 1). Usually, the antibiotics that contain these highly deoxygenated carbohydrates are members of the anthracycline and angucycline families. Sources other than those listed in Table 1 may be found in the reviews by Lown[205] and Rohr and Thiericke.[206] Some macrolide antibiotics, such as notonesomycin A,[83] have also been shown to contain trideoxysugars. L-Aculose (**62**) has been found in cardiac glycosides.[125] Interestingly, carbohydrate metabolites that are not attached to an aglycone have been isolated from the crude extracts of some soil bacteria.[130] These natural products have been named narbosines, and they contain L-rhodinose (**64**).

3.12.2.6 Branched-chain Deoxysugars

Deoxysugars with a branched carbon skeleton have been characterized from many different sources, most often from antibiotics produced in microorganisms. The appendage at the branching carbon can be methyl, formyl, hydroxymethyl, 1-hydroxyethyl, acetyl, 2-hydroxyacetyl, 1,3-dimethylpropyl, and other side chains. The branching carbon usually has a polar substituent as well, with tertiary alcohols being the most common in an array that includes cyclic carbonates, acetates, amino, and nitro groups. The branched-chain sugars have been divided into two groups by Grisebach[207] based upon their biogenesis. One group includes methyl-branched sugars and sugars with a two-carbon branch. Biosynthesis of these molecules involves the transfer of a C_1 or C_2 unit from appropriate donors. Because the corresponding nonbranched deoxysugar generally occurs elsewhere in nature, it is believed that the branching step must follow the deoxygenation step(s) in the biosynthetic pathways of these sugars.[207] The other group consists of furanose sugars having a hydroxymethyl or formyl branch, which is formed by intramolecular rearrangement of the pyranose ring. Deoxysugars in this group include L-dihydrostreptose (**66**)[132,133] and L-streptose (**67**),[134] both of which are found in antibiotics produced by *Streptomyces* spp.

Among branched-chain sugars, members composed of a one- or two-carbon branched chain are most abundant. All of them are deoxygenated at the C-6 position, and in many cases they also lack a hydroxyl group at C-2. For example, D-virenose (**68**) is a structural component of virenomycin,[135,136] while the L isomer (**69**) is found exclusively in the LPS of *Coxiella burnetii*.[137,138] Studies of novobiocin and nogalamycin have led to the identification of L-noviose (**70**)[139,140] and L-nogalose (**72**),[141,142] respectively. The orthosomycin antibiotics are a source of several branched-chain deoxysugars including D-evalose (**71**), D-evermicose (**79**), and L-evernitrose (**92**) (see Table 1). Blastmycin contains a branched lactone known as blastmycinone (**76**).[148] The 4-ester derivatives of 2,6-dideoxy-3-*C*-methyl-L-*arabino*-hexose have been found in chromomycins[60–63,92] and olivomycin A,[64–66] with the acetyl derivative being called L-chromose B (**77**) and the isobutyryl derivative being called either L-chromose B′ or L-olivomycose E (**78**). Macrolide antibiotics are a recurrent source of L-mycarose (**82**) (see Table 1). By contrast, D-mycarose (**81**) has been found only in mithramycin.[66–69] Other examples of branched-chain sugars include L-cladinose (**80**) in erythromycin A,[152,153] L-arcanose (**83**) in lankamycin,[171–173] and L-axenose (**84**) in axenomycins.[174] The anthracycline antibiotics quinocycline A and B and isoquinocycline A and B are sources of γ-octose (**85**) and trioxacarcinose B (**86**).[175–179]

Two 3,6-dideoxy branched-sugars, yersiniose A and B (**88**) and (**89**), which differ in their side-chain configurations, have been found in *Yersinia* spp.[180–182,184–186] and in *Legionella* spp.[183] A more elaborately branched 3,6-dideoxysugar (**90**) has been found in *Mycobacterium gastri*.[187,188] Reflecting the scarcity of 4-deoxygenated monosaccharides, only one 4,6-dideoxy branched-chain sugar, D-aldgarose (**91**), from aldgamycins E,[189–191] has been isolated so far. Trideoxysugars with a branching carbon have been found in everninomicins,[39] decilorubicin,[90,194] arugomycin,[88,89] pillaromycin A,[195–198] and rubradirin.[199–202] One of these sugars, L-pillarose (**94**), has a two-carbon branch at C-4. The other three, L-evernitrose (**92**), L-decilonitrose (**93**), and L-rubranitrose (**95**), are more unusual in that they also have a nitro group at the branching carbon.

3.12.3 BIOLOGICAL ACTIVITY OF DEOXYSUGARS

The biological functions of sugars are amazingly diverse. As a general observation, the more specific and crucial biological roles assumed by carbohydrates involve the participation of unusual sugars.[5] While some general characterization on the relative importance of these sugars has been reported, only in the 1990s has detailed information on the chemical basis for activity become available. Even so, just a few of the natural products possessing deoxysugars have been studied at the molecular level required for elucidating the chemical mode of action of these intriguing carbohydrates. Controlled acid hydrolysis and antibody engineering studies have proven to be useful for demonstrating the immunogenicity of deoxysugars in bacterial LPSs. Similarly, partial hydrolysis of several classes of bacterial antibiotics has shown that the glycosidic moiety is crucial for binding selectivity and biological activity. Crystallography and NMR techniques are the primary tools used for mapping the hydrogen bonds responsible for the binding interactions and target recognition of these molecules. Structure–activity relationship investigations have clarified the role of the sugar moieties in vancomycin and some cardiac glycosides. Some well-studied examples are summarized below. These studies may serve as excellent models for extending our mechanistic insight to other important glycoconjugates and glycosides.

3.12.3.1 Lipopolysaccharides

Since a more thorough summary of this topic is available in Chapter 3.09 only literature on the biological roles of the deoxysugar components in these structures is summarized here. LPSs, also known as endotoxins, are the immunodominant antigens of most Gram-negative bacteria,[208] and they define many of the properties of host–parasite interactions. Common deoxysugars such as L-rhamnose (**23**) can be found throughout the O antigen of many LPSs, and a few O antigens consist entirely of a single deoxysugar.[19,22,23] However, rare deoxysugars, such as the 3,6-dideoxysugars, are usually located at the nonreducing termini of the O-antigen repeating unit. These dideoxysugars have been found to be the predominant determinants of the immunological reactivity of several pathogenic bacteria, especially Enterobacteriaceae such as *Salmonella* and *Yersinia* spp.[1,209–213] The immunological interactions of one of these dideoxysugars, abequose (**53**), has been studied at the molecular level.

In these studies, a dodecasaccharide derived from the *Salmonella enterica* serogroup B O antigen, which contains abequose (see Section 3.12.5.2), was cocrystallized with an antibody Fab fragment and subsequently analyzed using X-ray diffraction techniques.[214] Consistent with its role in serological specificity, abequose was found to bind tightly to a deep hydrophobic pocket of the Fab fragment.[214] The epitope consists of a α-D-galactose*p*(1→2)[α-D-abequose*p*(1→3)]-α-D-mannose branched trisaccharide unit, which is bound to the Fab fragment with hydrophobic interactions and an extensive hydrogen-bonding network. The most crucial hydrogen bonds in the network are those between O-2, O-4, and O-5 of abequose (53) and amino acid residues in the binding pocket. The importance of the abequose residue in this network is highlighted by the fact that substitution of the epimers tyvelose (49) or paratose (52) for abequose led to abortive complexes. These results were substantiated by subsequent crystallization[215,216] and NMR[217] studies of the trisaccharide epitope bound with a single-chain variable domain (scFv). Interestingly, controlled acid hydrolysis revealed that the antibody preferentially binds one of two abequoses in an O-antigen fragment.[218] Thus, while there are several abequose residues in the LPS O-antigen repeat, it is likely that the immune response is directed to a limited number of these epitopes.

Studies with various *Coxiella burnetii* strains showed that the presence of the branched-chain sugars L-dihydrohydroxystreptose (66) and L-virenose (69) in the LPS changes the morphology of the cell from rough to smooth. More importantly, both neutral sugars are present in the virulent S-LPS-I, but neither sugar is present in the nonvirulent R-LPS-II. Thus, the incorporation of these unusual sugars converts a nonvirulent strain to a virulent strain.[219] Since similar results have been noted in other cases, it appears that the immune system reacts to these unusual carbohydrates, probably in the same specific manner as described above.

3.12.3.2 Bacterial Antibiotics

Antibiotics from microorganisms are a rich source of natural products containing deoxysugars. Those antibiotics belonging to the families of aureolic acids,[220,221] anthracyclines,[205,222,223] and enediynes[224,225] have attracted significant attention because of their potent activity against a variety of human tumors. These compounds are composed of an aglycone and one or more appended saccharide units. Research has established that these drugs exhibit their activity by intercalating into host DNA and thus inhibiting normal cellular functions. In the case of enediynes, further rearrangement to a reactive diradical intermediate leads to DNA damage. Various techniques, including DNA footprinting, crystallography, and NMR, have been used to determine the sequence selectivity, binding orientation, and other pertinent traits of the drug–DNA interactions. Through these efforts, the roles of the deoxysugar moieties in DNA binding have become well characterized. Essentially, the carbohydrates bind in the minor groove of the DNA and at least partially determine the target specificity. The relevant studies are summarized below, and this information should prove valuable in the rational design of DNA binders[226] and/or antitumor agents with decreased side effects. Also, these principles may be applicable to account for the interactions between aminoglycoside antibiotics and RNA.[227,228]

3.12.3.2.1 Aureolic acids

Members of the aureolic acid group of antibiotics include chromomycin A₃, olivomycin A, and mithramycin. The aglycone for chromomycin A₃ (96a) is identical to that of mithramycin (96c), but differs from that of olivomycin A (96b) by a methyl group. Conversely, the sugar components for chromomycin A₃ and olivomycin A are nearly identical, while those in mithyramycin are different ((96a)–(96c) and see Table 1). These drugs require a divalent metal ion, preferably an Mg^{2+} ion,[229,230] and a guanine-containing target for activity.[231,232] They form symmetrical dimer cation complexes and bind in the minor groove of DNA,[233–235] thus inhibiting RNA synthesis.[232] Significantly, dimers of both chromomycin A₃ and mithramycin induce a conformation change in the DNA from the B form to the A form so that they can be accommodated in the wider minor groove. However, mithramycin seems to have a slightly lower affinity for DNA than chromomycin A₃,[235] probably due to the differences in the oligosaccharide side chains. DNA footprinting experiments using (methidiumpropyl-EDTA)iron(II),[236] DNase I,[237] and hydroxyl radicals[238] revealed a minimum of three base pairs with two continuous GC pairs as the preferred binding site. Notably, under high concentrations of these antibiotics, the cleavage patterns were very similar for chromomycin A₃ and olivomycin A, while those for mithramycin were somewhat different. This evidence supports the idea that the carbohydrate moiety plays a role in determining the sequence specificity of the drug, although it is difficult to estimate the magnitude of this role.

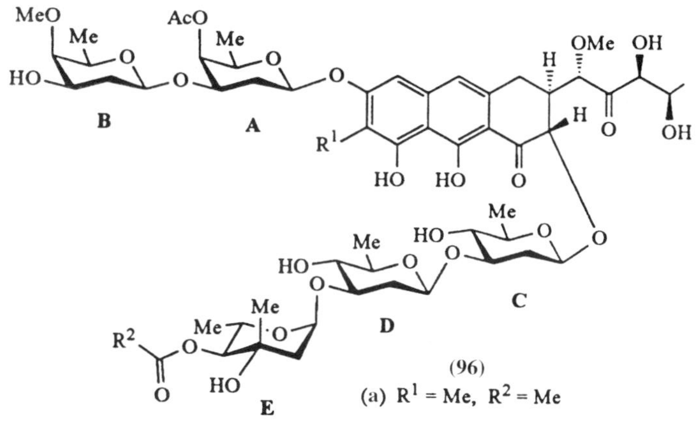

(96)

(a) R^1 = Me, R^2 = Me

(b) R^1 = H, R^2 = Pp^i

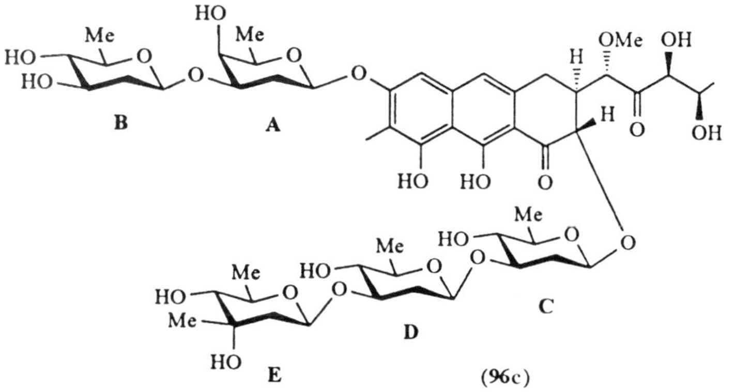

(96c)

The structures of the 2:1 complexes of chromomycin A_3[233,234] and mithramycin[235] with DNA have been resolved by NMR spectroscopy. In these experiments, the hydrogen-bonding interactions of the deoxysugars with both strands of the DNA double helix were unveiled. Three of these sugars (the C-D-E trisaccharide in (96a) and (96c) were found to align with the minor groove, and they are also involved in the stabilization of the 2:1 drug–metal complex before DNA binding occurs.[239,240] Removal of the C-D-E trisaccharide inactivates the drug.[240,241] Of the two other sugars, the A sugar binds weakly in the minor groove, while the B sugar apparently interacts even more weakly, if at all, with the major groove of the DNA helix. Consistent with these observations, partial hydrolysis of the B sugar had a minimal effect on the binding of chromomycin A_3 to DNA.[240,242] Clearly, the deoxysugars in both drugs, and presumably in other aureolic acid antibiotics, are important to facilitate the formation of the cation dimer necessary for DNA binding, and very likely they also play a role in target sequence selectivity.

3.12.3.2.2 *Anthracycline antibiotics*

Anthracycline antibiotics are effective antitumor agents and are widely prescribed in chemotherapeutic treatments for a number of human tumors, with doxorubicin (adriamycin (97a)) and daunorubicin (daunomycin (97b)) being the most popular.[205,222,243] These compounds contain a tetracyclic chromophore to which is attached one or more deoxysugars with or without amino groups. Removal of these sugars results in at least a partial loss of biological activity.[205,222] Several X-ray crystal structures[244–248] and solution NMR structures[249,250] have shown that the tetracyclic chromophore of these drugs intercalates DNA and alters the local topology. In addition, the glycosidic moiety binds in the minor groove (also in the major groove for arugomycin (97c)).[249] Another mode of activity for anthracyclines is direct inhibition of topoisomerase II.[251,252]

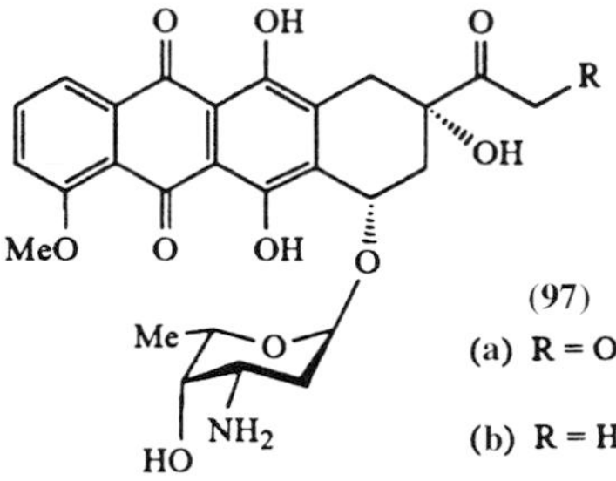

(97)

(a) R = OH

(b) R = H

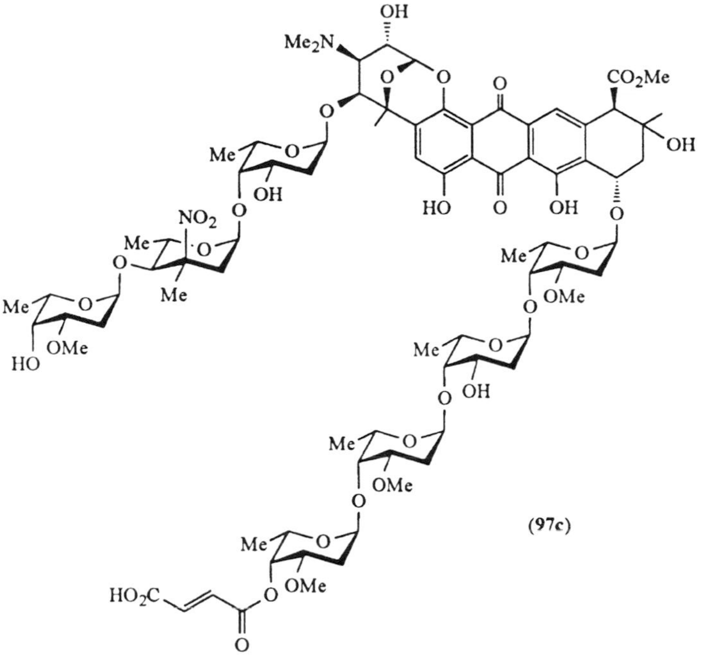

(97c)

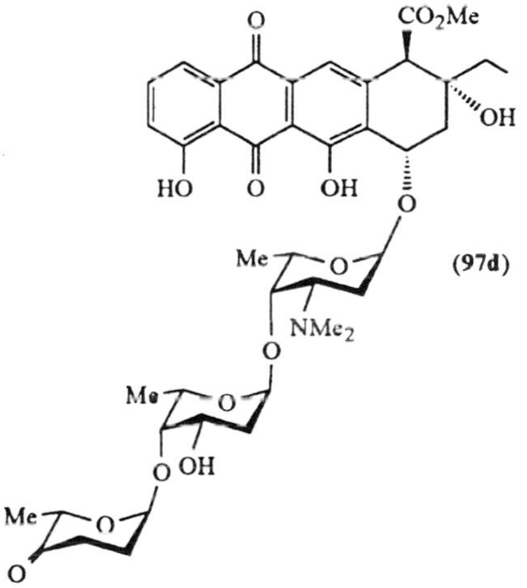

(97d)

DNA footprinting studies have identified GT and GC sequences as preferred binding sites.[253-255] Though longer saccharide chains produced a larger DNA footprint, the base pair preference was largely unaffected when the number and type of sugar components were altered. Thus, it appears that the oligosaccharide chains do not play a significant role in the DNA sequence selectivity. This is consistent with NMR investigations which showed that the sugar chains of arugomycin are flexible and contribute little to the interaction of the antibiotic with DNA.[249] Footprinting with just the glycosidic portions of the drugs showed no evidence for binding to DNA,[255] and similar results were obtained when binding was probed spectroscopically.[256] Therefore, intercalation of the chromophore into DNA seems to be the dominant interaction which both determines preferred binding sequences and directs the binding of the saccharides in the minor groove. Nevertheless, the saccharide side chains have been shown to dramatically enhance the binding affinity of the drugs for DNA.[257] Additionally, in aclacinomycin A (**97d**) the trisaccharide binding forced the DNA to kink toward the major groove with opening of the minor groove.[250] Also, the sugar moieties of aclacinomycin A proved to be essential for the drug to exhibit an inhibitory effect on the chymo-trypsin-like activity of the bovine pituitary 20 S proteasome.[258] Further studies may reveal that the sugars of anthracyclines are similarly important in the inhibition of other proteins, such as topoisomerases and helicases.

3.12.3.2.3   *Enediyne antibiotics*

Since the late 1980s different investigators have discovered several members of a new class of antibiotics which comprise several deoxysugars, a hexasubstituted benzene ring, and an unusual cyclic enediyne chromophore (reviewed by Lee *et al.*,[224] Smith and Nicolaou,[225] and Ellestad and Ding[259]). These enediyne antibiotics, such as calicheamicin γ_1^1 (**98**) and esperamicin A$_1$ (**99**), exhibit phenomenal antitumor activity, with calicheamicin being over 1000 times more potent than doxorubicin (**97a**). These drugs act via a remarkable mode of action in which the (*Z*)-1,5-diyn-3-ene unit is delivered to the minor groove of the DNA helix, whereupon reduction of the chromophore leads to cycloaromatization of the enediyne and 1,4-benzenoid diradical formation. The diradical is positioned to abstract hydrogen atoms from the sugar phosphate backbone of the nearby DNA, thus causing single- or double-strand scission of the helix and forming the molecular basis for the cytotoxic activity of these drugs.

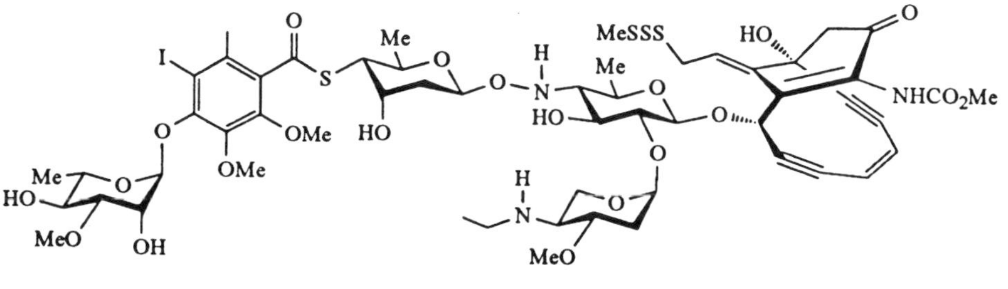

(**98**)

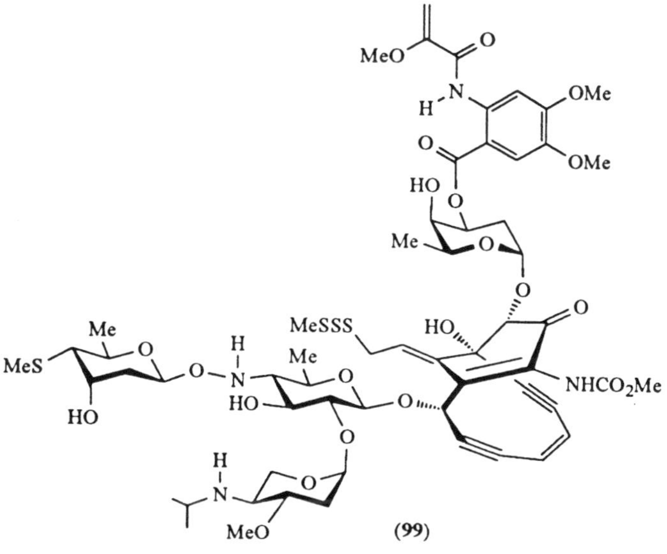

(**99**)

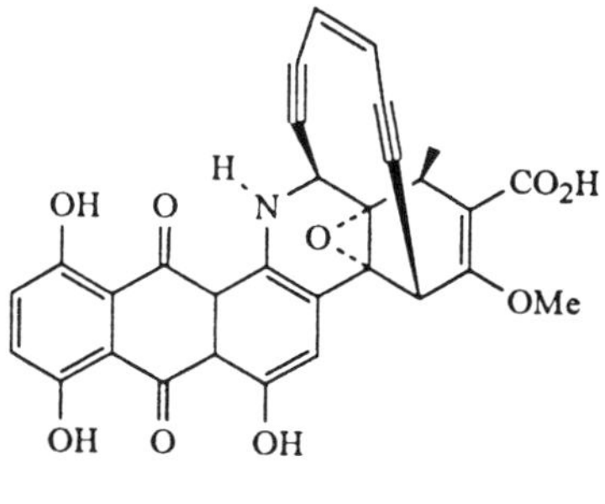

(**100**)

In calicheamicin γ_1^1 (**98**) and presumably in other enediynes, the substituted benzene and sugars form an aryltetrasaccharide unit which, together with the iodine substituent,[260] has been shown to direct the drug to the minor groove of double-helical DNA with a high specificity for sequences such as 5'-TCCT-3' and 5'-TTTT-3'.[225] Binding of the drug distorts the DNA by widening the minor groove, so homo(pyrimidine/purine) sequences may be the preferred binding sites because they are easier to distort relative to other sequences.[261–263] Although the aglycone can bind DNA, it does so with far less affinity and no sequence selectivity.[264] Additionally, the calicheamicinone aglycone is not efficient at double-strand scission,[264] so the sugars are vital to the biological activity of these compounds. Reflecting this importance, the carbohydrate portion without the aglycone is able to bind DNA with nearly the same interactions as those of the carbohydrates in the whole drug,[265] and it can block transcription.[266] The aryltetrasaccharide dimer binds DNA even stronger and more specifically,[267] and its transcription inhibition is also significantly enhanced.[268] These observations are in contrast to the anthracycline antibiotics, whose sequence selectivity is determined primarily by the aglycone and whose carbohydrate moieties are unable to bind DNA in the absence of the aglycone (see Section 3.12.3.2.2). Interestingly, the related dynemycin A (**100**), which is an anthraquinone–enediyne lacking a carbohydrate domain, has little sequence specificity,[269,270] thus adding support for the sequence-selecting role of the deoxysugars in the enediyne family.

3.12.3.2.4 Vancomycin

Vancomycin (**101a**) is a glycopeptide antibiotic that is widely used in the treatment of Gram-positive bacterial infections (reviewed by Williams[271]). It consists of a core heptapeptide with attached saccharide moieties, one of which is the deoxy aminosugar vancosamine. Vancomycin exhibits its antimicrobial activity by binding bacterial cell wall mucopeptide precursors terminating in the sequence L-Lys-D-Ala-D-Ala.[272] Five hydrogen bonds account for this binding specificity, and the disruption of one of these hydrogen bonds by the replacement of the terminal alanine with lactate (D-Ala-D-Lac) in the mucopeptide precursor is the molecular basis for resistance to vancomycin.[271] Because vancomycin and other glycopeptides are the only few drugs effective against several drug resistant bacteria, extensive efforts have been directed toward the discovery and development of vancomycin derivatives with activity against the drug-resistant bacteria.[273]

(101)

(a) $R^1 = OH, R^2 = H, R^3 = H$

(b) $R^1 = H, R^2 = OH, R^3 = $

As a result of these efforts, several derivatives such as LY264826 (**101b**) have been found to be up to 500 times more active than vancomycin (**101a**) against resistant bacteria. The most notable difference between vancomycin and LY264826 is the presence of the aminosugar (R^3) attached to amino acid 6. This extra sugar possibly facilitates the dimerization of the antibiotic and/or the anchoring of the antibiotic to the cell membrane, both of which have been shown to be important for vancomycin activity.[271] Alkylation of the 3-amino group on the disaccharide at residue 4 further enhances the activity, probably by serving as a hydrophobic anchor to the cell membrane.[274] Future experiments using derivatives with different saccharides attached to residues 4 and 6 should clarify the role(s) of the deoxy aminosugars in these important natural products.

3.12.3.3 Cardiac Glycosides

Cardiac glycosides consist of a five-ring cardenolide aglycone, called a genin, with a number of attached monosaccharides that often include deoxysugars. These steroidal compounds are usually isolated from plant sources,[14] but they have been discovered in higher mammals as adrenal cortical hormones (reviewed by Blaustein[275]). Possibly using the same binding site as the natural hormone, cardiac glycosides inhibit Na^+/K^+-ATPases,[276–278] resulting in an inotropic activity that has proven to be useful in the treatment of various heart conditions.[279–281] This activity is enhanced several-fold due to the presence of the sugars in these compounds,[282,283] but the sugars alone cannot bind to the protein.[284–286] Altering the structure of the sugar(s) has a dramatic influence on the activity of the cardiac glycoside.[287–290]

Other studies have revealed that the saccharides lengthen the half-life of activity by preventing the host from modifying the genin and neutralizing its inhibitory activity.[291] Also, the sugar proximate to the genin has the greatest effect on the binding and activity of the drug.[292,293] Though the structure of the cardiac glycoside binding site has been extensively studied,[294–296] its detailed nature remains largely unknown, especially the amino acid residues which interact with the sugar moiety, because a crystal structure is still unavailable. However, a three-step model has been proposed to account for the interaction between a cardiac glycoside and its receptor. The binding event may proceed via reversible binding of the steroid core with the receptor, followed by a conformational change that exposes a sugar-binding motif, which then binds to the saccharides and hinders dissociation.[297,298]

3.12.4 PATHWAYS AND MECHANISMS OF DEOXYSUGAR BIOSYNTHESIS

Though relatively few deoxysugar-containing natural products have been studied closely enough to elucidate the function(s) of the deoxysugar component(s), even fewer have been studied at the biosynthetic level. While feeding experiments using isotopic precursors have increased our understanding of the biogenesis of many important metabolites, biochemical characterization of the enzymes involved in these pathways has been hampered by various difficulties. First, due to their low levels of expression within the cell, the enzymes are very challenging to purify. Compounding this challenge is the scarcity of suitable methods to assay the enzyme activity, due in part to a lack of knowledge about the substrates. In those cases for which an assay was available and a purification protocol was developed, the amount of enzyme obtained was adequate for some insightful biochemical and kinetics studies, but extensive mechanistic investigations were not practical. The advent of genetic techniques has helped to address many of these difficulties as well as expand the tools available to the enzymologist. For instance, genetic probes based on purified native enzymes have enabled the construction and overexpression of recombinant wild-type and mutant enzymes involved in the biogenesis of deoxyribonucleotides, 6-deoxyhexoses, and 3,6-dideoxyhexoses. Studies on both the native and recombinant enzymes have led to a detailed knowledge of the underlying mechanisms in these biosynthetic pathways, and they are summarized below.

3.12.4.1 Mechanism of Deoxyribose Formation

The deoxyribonucleotides required for all DNA replication and repair are derived from ribonucleotides by replacement of the 2′-hydroxyl group of the ribose moiety with a hydrogen atom. Ribonucleotide reductases (RNRs) are the only enzymes capable of catalyzing this reaction. As such, they are ideal targets for antiviral and anticancer drugs. Efforts to elucidate the mechanisms of both catalysis and inactivation of this crucial enzyme have led to at least three classes of RNR with different cofactor requirements and protein structures.[299,300] Despite these differences in the various classes of RNR, their catalyses share a common theme in which a stable protein radical is used to abstract the 3′-hydrogen atom to initiate reduction of the 2′-hydroxyl group on the ribose moiety. While the mechanisms for the generation of this protein radical may be different, once the protein radical is formed, the subsequent catalytic steps appear to be identical for all classes of RNR. Biological reducing agents such as thioredoxin and glutaredoxin provide the electrons for the reduction. Allosteric regulation is an important aspect of the catalysis, as the reaction proceeds very slowly in the absence of the appropriate effector; however, this subject has been reviewed[299,301] and will not be addressed here. Additional information on the structures and mechanisms of RNR enzymes can be found in a number of reviews.[299,302–309] Summarized below are the prototypical enzymes from the three best-understood classes of RNR. The mechanisms of RNR in the following discussion may be applicable to the formation of other deoxynucleotides, such as other 2′-deoxypentoses (Table 1) and the 3′-deoxyadenosine in cordycepin.[310]

3.12.4.1.1 Class I—Escherichia coli RNR

Class I reductases are present in some bacteria and all higher organisms. These enzymes are composed of two nonidentical homodimeric subunits ($\alpha_2\beta_2$), each with distinct functions. The large subunit (α_2 or R1) harbors the binding sites for both substrates and allosteric effectors and carries out the actual reduction, whereas the small subunit (β_2 or R2) contains an oxygen-linked dinuclear iron center and a tyrosyl radical that are essential for activity.[299,304] The *Escherichia coli* class I RNR is the best-studied enzyme in this class, which can be further divided into two subclasses (see Section 3.12.5.1.1). The R1 subunit of *E. coli* RNR (class Ia) is composed of two 86 kDa protomers, while the R2 subunit protomers are each 43 kDa. Because this enzyme catalyzes the aerobic deoxygenation of only ribonucleotide diphosphates, it is often called RDPR. The crystal structures of both subunits have been solved separately,[311,312] and the holoenzyme complex was modeled using the three-dimensional structures of these subunits. This holoenzyme model offers a structural basis for the catalytic mechanism deduced via biochemical and genetic techniques. The combination of these studies has enabled a more thorough understanding of the mechanism of RDPR than of any other ribonucleotide reductase.

Early studies (reviewed by Stubbe[304]) characterized the dinuclear iron center, which was first observed by Reichard and co-workers.[302] The iron center is composed of two high-spin iron atoms antiferromagnetically coupled through a μ-oxo bridge. The ligand sphere and location of the center within R2 have been established by the crystal structure,[311,313] and an illustration of the iron center is shown in Figure 2. The tyrosine radical was identified as Tyr122 by mutagenesis,[314] and quenching of this radical by hydroxyurea inactivates the enzyme.[315] Activity can be restored by removing the Fe^{III} atoms and subsequently adding Fe^{II} and oxygen[316] or by chemically or enzymatically reducing the Fe^{III} atoms to Fe^{II} in the presence of oxygen.[317] Tyr122 is 5 Å away from the closest iron atom,[311] and this distance is close enough to allow magnetic interactions between the iron center and the tyrosyl radical.[318] Such magnetic interactions, as well as a hydrophobic environment provided by residues Phe208, Phe212, and Ile234,[319] are essential to stabilize the tyrosyl radical. The involvement of the iron center in the generation of the Tyr122 radical has been firmly established since the assembly of the diferric cluster/tyrosyl radical cofactor can be directly monitored using rapid kinetics methods.[320] This cofactor assembly process involves several intermediates, including a possible tryptophan radical[321] and a reactive spin-coupled Fe^{III}/Fe^{IV} diiron center with significant delocalization onto the oxygen ligand(s).[322] The oxide that bridges the two ferric ions in the cluster has been shown to be derived from molecular oxygen.[323]

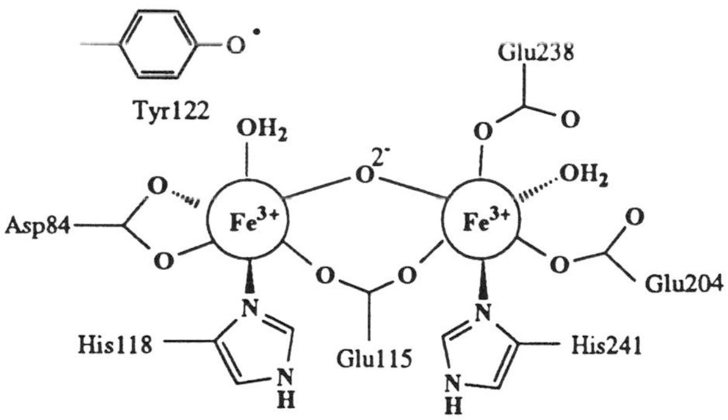

Figure 2 The structure of the μ-oxo dinuclear iron center in *E. coli* RDPR (after Nordlund and Eklund[313]).

Notably, the iron center/tyrosyl radical cofactor is buried in R2 with the Tyr122 radical located 10 Å from the nearest protein surface,[311] which is the presumed binding site of R1. In the modeled holoenzyme complex,[312] there is a total of 35 Å between Tyr122 and the active site of R1, so direct participation of the Tyr122 radical in the reduction of the substrate is highly unlikely. Instead, a long-range electron transfer pathway has been proposed to convey the radical from Tyr122 to an amino acid residue in the active site of R1.[309,312] The residues involved in this proposed relay form a hydrogen-bonded network running through the holoenzyme complex, and they include Tyr122,

Asp84, His118, Asp237, Trp48, and Tyr356 (and possibly Glu350) in the R2 protein and Tyr730, Tyr731, and Cys439 in the R1 protein. Evidence supporting this proposal comes from several studies. One such study with the Y122F mutant has provided indirect evidence indicating that formation of a Trp48 radical intermediate is possible.[324] Participation of Trp48 and Asp237 in the electron transfer pathway is supported by mutagenesis of the corresponding residues in the mouse enzyme (Trp103 and Asp266).[325] Replacement of either Glu350 or Tyr356 caused either partial or total loss of activity, respectively, though the binuclear iron center and the Tyr122 radical remained intact in the mutant proteins.[326] Similarly, the Y730F and Y731F mutants retained the capability of assembling the cofactor but were enzymatically inactive,[327] thus implicating these residues in the formation of an active-site radical. Mutations of Cys439, which is hydrogen bonded to Tyr730 and proposed to be the terminal residue of the electron relay, also caused a total loss of catalytic activity.[328,329] The Cys439 residue is proximal to the substrate and correctly positioned to abstract the 3′-hydrogen atom from the substrate to initiate the reaction.[312]

Scheme 1 depicts the proposed mechanism of ribonucleotide reduction (amino acid numbering for *E. coli* RDPR). After the Cys439 thiyl radical abstracts a 3′-hydrogen atom from the substrate, subsequent protonation of the 2′-hydroxyl group by one of the active-site thiols (assisted perhaps by an active-site carboxylate group)[330] allows loss of the hydroxyl group as a water molecule. The resulting radical intermediate is then reduced by the redox-active thiol pair Cys225 and Cys462, leading to the incorporation of a solvent hydrogen at C-2′ with net retention of configuration. This reduction may occur by stepwise electron transfer (Scheme 1, path a) via a formyl methyl radical (**102a**) and a disulfide radical anion (**102b**) intermediate or by direct hydride transfer to the 2′-carbon (path b). Both routes result in a 3′-carbon radical to which Cys439 transfers the original hydrogen atom removed from this position, thus forming the deoxyribonucleotide product. Regeneration of the active-site thiol pair (Cys225/Cys462) involves another redox-active thiol pair, Cys754 and Cys759, located at the C-terminal end of the R1 subunit. The catalytic cycle is completed when this thiol pair is in turn reduced by NADPH through either the thioredoxin or glutaredoxin systems.

Support for this mechanistic proposal is derived from several lines of evidence, and most of the earlier studies have been summarized in good detail by Stubbe.[303,304] For example, the use of isotopically labeled [3′-³H,U-¹⁴C]- or [3′-²H]NDPs (nucleotide diphosphates) as substrates unambiguously established that the catalysis involves cleavage of the 3′-carbon–hydrogen bond and return of the same hydrogen back to the 3′-position. Although model studies have demonstrated that the observed cleavage could occur via a radical mechanism, a substrate radical intermediate has not yet been detected. However, a possible nucleotide-based radical has been observed during studies with RDPR and the time-dependent inactivators (*E*)- and (*Z*)-2′-fluoromethylene-2′-deoxy-cytidine 5′-diphosphate.[331] The participation of Cys225 and Cys462 as an active-site thiol pair has been implicated by mutagenesis studies,[328,332,333] and it is supported by the crystal structure in which these residues are found to be properly positioned in the active site to interact with the substrate and form a disulfide bond.[312] Interestingly, inactivation studies using the tyrosyl radical scavengers 2′-deoxy-2′-thiouridine 5′-diphosphate[334] and 2′-azido-2′-deoxyuridine 5′-diphosphate[335] revealed a perthiyl radical and a [XN·S$_{Cys}$R1] radical, respectively. These results provide strong evidence that reduction of the substrate radical intermediate may occur via a disulfide radical anion intermediate ((**102b**), Scheme 1, path a). Studies on other mutants revealed that Cys754 and Cys759 serve as a redox shuttle to regenerate the active-site thiols using electrons from thioredoxin and glutaredoxin.[328,332] Presumably, the class Ib RNR (NrdD and NrdG, see Section 3.12.5.1.1) from *E. coli* has similar properties to the class Ia RDPR, but confirmation of this hypothesis will have to await further biochemical and structural studies on this newly discovered subclass.

3.12.4.1.2   Class II—Lactobacillus leichmannii RNR

The 82 kDa monomeric reductase from *Lactobacillus leichmannii* utilizes adenosylcobalamin (AdoCbl) as the cofactor in the aerobic and anaerobic reduction of ribonucleotide triphosphates,[304,306] hence it is often denoted RTPR. This reductase, upon binding with substrate, presumably promotes the homolysis of the intrinsically reactive Co—C bond in the AdoCbl cofactor[336] to form cob(II)alamin and a 5′-deoxyadenosyl radical (5′-dA·).[337] Based on studies using [3′-³H]NTPs (nucleotide triphosphates),[338] the function of 5′-dA· has been proposed to be abstraction of a hydrogen atom from an amino acid residue in the active site rather than direct abstraction of the 3′-hydrogen atom of the ribonucleotide. This protein radical would then abstract the 3′-hydrogen atom from the substrate and initiate the reduction in the same fashion as for the class I reductases

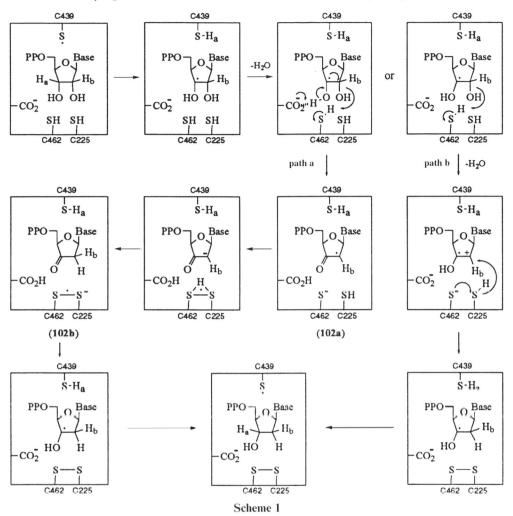

Scheme 1

(Scheme 1). Although the 5'-dA· has not yet been observed, a kinetically competent intermediate has been identified as a thiyl radical coupled to cob(II)alamin.[339] The assignment was supported by the fact that the features of the EPR spectra of this species can be accounted for by model simulations.[340] Mutagenesis studies suggested that this catalytic thiyl radical is Cys408.[341] Additionally, Cys408 is needed for the characteristic dithiol- and effector-dependent isotope exchange between the 5'-methylene hydrogens of AdoCbl and bulk solvent,[341] thus further supporting its proposed role in catalysis.

Interestingly, the local amino acid sequences around Cys408 from *L. leichmannii* and Cys439 from *E. coli* RDPR R1 are highly conserved.[342] Given the homology of this fragment and the identical phenotypes (inactive with respect to nucleotide reduction) of the C408S and C439S mutants,[329,341] the characterization of a kinetically competent thiyl radical in RTPR adds credence to the catalytic role assigned to Cys439 in RDPR (see Section 3.12.4.1.1). The phenotypes of other RTPR mutants enabled the assignments of Cys119 and Cys419 as the active-site cysteines that directly reduce the ribonucleotide substrate, and Cys731 and Cys736 as the redox shuttle thiols that regenerate the active-site cysteines.[341] Other similarities between the class I and II RNRs include the location of the redox shuttle thiols at the C-terminus, the displayed phenotypes of inhibition by 2'-halogenated 2'-deoxynucleotides,[304] and the replacement of the 2'-hydroxyl group by a solvent-exchangeable hydrogen atom with net retention of configuration. Also, the 3'-hydrogen atom is returned to the same position of the ribose at the end of the deoxygenation in both RDPR and RTPR.

These similarities are remarkable considering the differences in the primary and quaternary structures of the class I and II enzymes, and they provide compelling evidence that many aspects of the active sites of these enzymes are very likely the same. Nonetheless, some differences exist since active-site-directed inhibitors are known to selectively inactivate one enzyme over the other. An example of such as class- and/or species-specific inhibitor is 2′-*C*-methyladenosine diphosphate, which is a mechanism-based inactivator of the class II AdoCbl-dependent ribonucleotide diphosphate reductase from *Corynebacterium nephridii* but not of RDPR from *E. coli*.[343] Similar compounds may be helpful in the treatment of viral infections which are dependent upon a self-encoded RNR for sufficient production of DNA precursors. It should be noted that several RNR inhibitors which exploit larger class-dependent differences in allosterism, quaternary structure, and cofactor dependence are currently under investigation.[304]

3.12.4.1.3   *Class III—anaerobic* E. coli *RNR*

As discussed later in Section 3.12.5.1.3, when *E. coli* is grown under anaerobic conditions, a heterodimeric RNR different from the class I RDPR is expressed. Similar to class I RNR, this enzyme consists of a 160 kDa (α_2) and a 35 kDa (β_2) subunit, although it preferentially reduces ribonucleotide triphosphates. Additionally, this RNR (anaRNR) has a strict dependence on AdoMet for activity,[344] and when it is activated with the flavodoxin system a protein radical forms.[345] Only the first 28 amino acid residues of the N-terminus of the large subunit share a limited homology (30%) with the N-terminus from class Ia R1 subunits.[346] No further sequence homology with other classes of RNR is apparent throughout the rest of the molecule. However, a sequence of five amino acids (Arg-Val-Cys-Gly681-Tyr) at the C-terminus demonstrated a compelling homology to a C-terminal pentapeptide (Arg-Val-Ser-Gly734-Tyr) in pyruvate formate lyase (Pfl).[346] Because Gly734 in Pfl has been shown to harbor a catalytically essential protein radical,[347] Gly681 in anaRNR was proposed to assume a similar role and be the important protein radical involved in the reduction reaction.[346] Interestingly, when Pfl is exposed to air, the glycine radical is quenched, and the protein is truncated at Gly734.[347] Likewise, exposure of anaRNR to air also leads to truncation at Gly681.[348] Mutagenesis and EPR experiments confirmed that the protein radical resides at Gly681.[349]

Though this protein radical may participate in ribonucleotide reduction by generating a thiyl radical as in the class I and II RNRs, analysis of the sequence of anaRNR does not reveal any apparent candidates for this catalytic cysteine.[346] Additionally, no obvious redox-active thiol pairs could be identified in this protein, even at the C-terminus. This structural deviation from the class I and II enzymes is accompanied by a significant chemical difference in that the catalysis of anaRNR is thioredoxin- and glutaredoxin-independent. Instead, anaRNR uses formate to regenerate its activity during the reduction of ribonucleotides.[350]

Similar to the class I and II enzymes, when the anaRNR reaction is run in D_2O, deuterium is incorporated into the 2′-position with net retention of configuration.[351] This observation is expected and can be explained by a relatively facile hydrogen exchange between the solvent and the redox-active thiol pair that reduces the 2′-carbon. Notably, there is a 1% deuterium incorporation at the 3′-position of the product under the reaction conditions. This finding is analogous to the ~1% tritium washout detected when [3′-³H]NTPs are reduced by the class I[352] and class II[338] RNRs. This result can be explained by the thiyl radical being able to exchange hydrogen atoms with bulk solvent, but at a rate which is much slower than that of return of the 3′-hydrogen atom back to the 3′-position of the deoxynucleotide. As mentioned above, there is as yet no direct evidence for the involvement of such thiols in anaRNR, but the results of the D_2O experiment is consistent with the general mechanism in Scheme 1. Should further structural and biochemical studies establish the participation of cysteines in the reaction of anaRNR, then the mechanistic similarity of class I, II, and III RNR can be extended to the protein level.

3.12.4.2   Mechanisms of Deoxyhexose Formation

Though there is more mechanistic information available on the biosynthesis of 2-deoxy-D-ribose in deoxyribonucleotides, considerable knowledge has also accumulated on the formation of some 6-deoxyhexoses and 3,6-dideoxyhexoses in bacterial LPSs (Table 1). Mechanistic proposals for deoxygenation at the 2- and 4-positions on the hexose ring have emerged, and they are largely based on insights derived from studies of the biosynthesis of 3,6-dideoxysugars. These proposals are discussed later in Section 3.12.5.2, in conjunction with the limited genetic and biochemical studies on those pathways. Branched-chain deoxysugar synthesis has not been studied at the mechanistic level, although feeding experiments and comparisons to enzymes involved with carbon–carbon bond formations in other metabolic pathways have allowed the formulation of several chemically sound hypotheses. This section summarizes some of these hypotheses and the work responsible for our current understanding of deoxyhexose biosynthetic mechanisms. More information can be found in a few useful reviews on the subject by Grisebach,[207] Glaser,[353] Gabriel,[354] and Liu and Thorson.[355]

3.12.4.2.1 Biosynthesis of 6-deoxyhexoses

Because 6-deoxyhexoses are encountered widely in bacterial LPSs as well as a great variety of plant and mammalian glycoconjugates (see Section 3.12.2), the enzymes responsible for producing these sugars are found throughout nature. As shown in Scheme 2, the general pathway of 6-deoxyhexose biosynthesis is initiated by the formation of a nucleotidyl sugar (**103**) followed by C-4 oxidation and C-6 dehydration (NDP refers to any nucleotide diphosphate). Subsequent stereospecific C-4 reduction, with or without epimerization at C-3 and/or C-5, results in the various isomers of 6-deoxyhexoses. Depending on the specific pathway, more than one nucleotidyl sugar precursor may be associated with a given deoxysugar.[354] For instance, dTDP- or UDP-glucose is the precursor for dTDP- or UDP-rhamnose, respectively, while GDP-mannose is the precursor for GDP-fucose. These nucleotidyl deoxysugar products are recognized by corresponding glycosyltransferases and incorporated into the glycosidic portion of cellular glycoconjugates. In this section, the enzymes involved in 6-deoxyhexose biosynthesis are discussed separately in mechanistic detail.

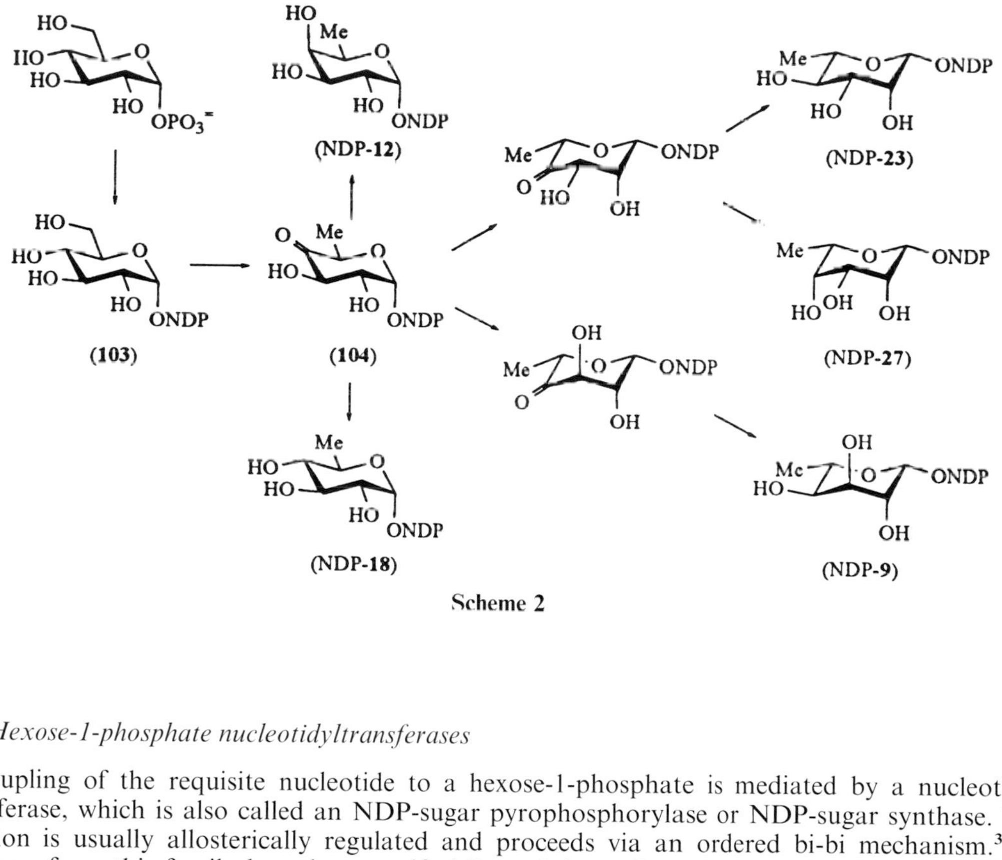

Scheme 2

(i) Hexose-1-phosphate nucleotidyltransferases

Coupling of the requisite nucleotide to a hexose-1-phosphate is mediated by a nucleotidyltransferase, which is also called an NDP-sugar pyrophosphorylase or NDP-sugar synthase. This reaction is usually allosterically regulated and proceeds via an ordered bi-bi mechanism.[356–360] Enzymes from this family have been purified from *Salmonella enterica* serovars Paratyphi[361] and Typhimurium (strain LT2),[359,360] *Azotobacter vinelandii*,[362] *Yersinia* (formerly *Pasteurella*) *pseudotuberculosis*,[363] and a diverse number of other sources.[356,357] A comparison of the deduced amino acid sequences indicates that these enzymes can be classified into two groups: the prokaryotic group with about 300 amino acid residues and the eukaryotic group with about 500 amino acid residues.[364] These enzymes have two highly conserved domains,[364–366] one of which is proposed to be a portion of the effector binding site,[367,368] while the other domain may be involved in substrate binding.[365] Mutagenesis and chemical modification experiments support these assignments and suggest that specific lysine residues are important for the binding of both effectors and substrates.[369–371] All the enzymes studied show an absolute requirement for Mg^{2+} (or another divalent cation) and exhibit optimal activity over a broad pH range between 7.0 and 9.0.[356,357] Although the catalytic mechanism appears to be the same for these enzymes, the substrate and effector specificity depends upon the source of the enzyme and the metabolic pathway in which it is involved.[356,357,365] Additionally, more than one nucleotide or hexose can be utilized as substrates in some cases, although the competency of these compounds is different.

(ii) Nucleotidyl diphosphohexose-4,6-dehydratases

The committing step in almost all deoxyhexose pathways is the irreversible conversion of a nucleotidyl sugar to a 4-keto-6-deoxyhexose derivative by nucleotidyl diphosphohexose-4,6-dehydratases (formerly known as 4,6-oxidoreductases). The product of these enzymes, an NDP-4-keto-6-deoxyhexose (104), is the common branch point for the biosynthesis of most deoxysugars, aminosugars, and branched-chain sugars.[355] Enzymes from this family have been isolated from a number of organisms,[372,373] and the best-studied member is dTDP-D-glucose-4,6-dehydratase (80 kDa homodimer) from *E. coli*. This enzyme, like others in this class,[374] requires NAD^+ as a cofactor for activity. However, it contains one NAD^+ molecule per dimer,[375] even though there are two available binding sites. Studies of the related enzyme CDP-D-glucose-4,6-dehydratase from *Y. pseudotuberculosis*[376] revealed that NADH binds tighter than NAD^+ to the enzyme, and thus the other pyridine nucleotide-binding site may be occupied by NADH in the *E. coli* enzyme. Regardless, these enzymes have a high affinity for NAD^+ and utilize it as a prosthetic group, as opposed to other pyridine nucleotide-dependent enzymes, which use NAD^+ or NADH as a cosubstrate.[374] The catalytic cycle consists of three distinct steps. As depicted in Scheme 3, the first step is oxidation of the nucleotidyl diphosphohexose (103) to the corresponding 4-ketohexose (105), which then undergoes dehydration in the second step to a 4-keto-$\Delta^{5,6}$-glucoseen intermediate (106). The final step involves reduction at C-6 to give the desired product (104). The asterisk refers to isotopic label and NDP is any nucleotide diphosphate.

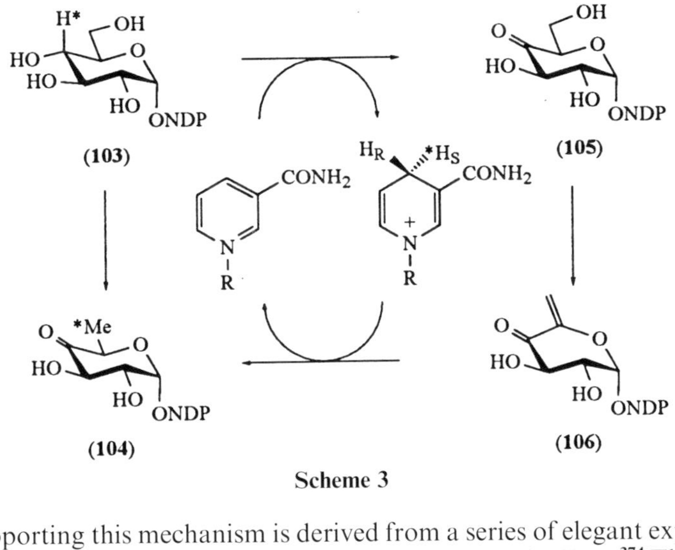

Scheme 3

The evidence supporting this mechanism is derived from a series of elegant experiments by Glaser and Zarkowsky[353] and Gabriel,[354] and has been summarized nicely by Frey.[374] The role of the NAD^+ coenzyme in this intramolecular oxidation–reduction is to mediate the transfer of the C-4 hydrogen atom from the substrate (103) to C-6 of the product (104).[377,378] Studies of both dTDP-D-glucose-4,6-dehydratase[379] and GDP-D-mannose-4,6-dehydratase[380] demonstrated that the displacement of the 6-hydroxyl group by the 4-hydrogen atom occurs with net inversion of configuration. When the reaction is conducted in [³H]H₂O or ²H₂O, a solvent hydrogen atom is incorporated at the C-5 position.[378] Further analysis established that the internal hydrogen transfer to and from NAD^+ proceeds with "*si*-face" stereospecificity.[379–381] Thus, the elimination of water from C-5 and C-6 was concluded to be a *syn* process.

(iii) NDP-6-deoxy-4-hexulose-3,5-epimerases

All D- and L-6-deoxyhexoses studied so far are derived from either D-glucose or D-mannose. Since an epimerization at C-5 is required to convert a D-sugar to an L-sugar, the participation of an epimerase in the biosynthesis of L-6-deoxyhexoses is implicated. In fact, many of the epimerases involved in L-6-deoxyhexose formation are 3,5-epimerases, such as the epimerase responsible for the biosynthesis of GDP-fucose (GDP-(**13**)) in human erythrocytes[382] and in porcine thyroid,[383] of dTDP-6-deoxy-L-talose (dTDP-(**26**)) in *E. coli*,[384] and of dTDP-rhamnose (dTDP-(**23**)) in *S. enterica* LT2[385] and in *Pseudomonas aeruginosa*.[386] Notably, a 3,5-epimerase (55 kDa) has been isolated and characterized from the erythromycin-producing bacterium *S. erythraea*.[387] Also, the gene for a different enzyme (40 kDa) with putative 3,5-epimerase activity based on deduced protein sequence comparisons has been cloned from the same strain.[388] However, Southern analysis indicated that neither epimerase is encoded by a gene within the erythromycin biosynthetic cluster that contains genes for L-cladinose (**80**) and L-desosamine (**130**) formation. Also, their association with other biosynthetic pathways remains to be elucidated.

The epimerases involved in GDP-fucose (GDP-(**13**)) biosynthesis in mammals appear also to have NADPH-dependent 4-ketoreductase activity,[382,383] but this bifunctional nature has not yet been proven. Conversely, all other epimerases found so far are definitely monofunctional. Mixing experiments showed that epimerases and reductases from *P. aeruginosa* and *E. coli* are interchangeable,[384] with the stereospecificity of the reductase dictating the structure of the final product. Illustrated in Scheme 4 is the proposed mechanism for the catalysis of 3,5-epimerases (the asterisk refers to isotopic label and NDP is any nucleotide diphosphate). The intermediacy of the enediols (**107**) and (**108**) is implicated by the loss of tritium at C-3[384] and by the incorporation of solvent hydrogen atoms into the product at the epimerized carbon centers.[389] Currently, there is insufficient evidence to discriminate between an ordered or random process. If the reaction is ordered, it would be interesting to determine which carbon center is epimerized first.

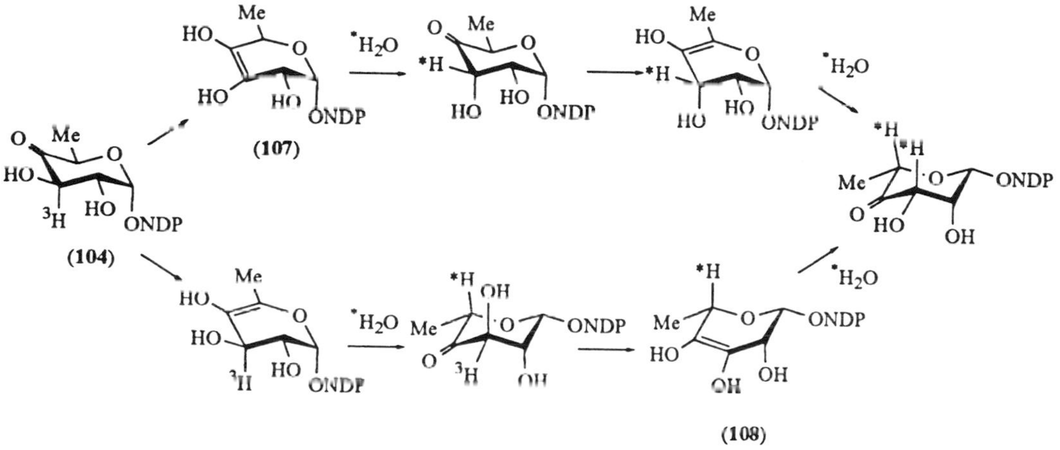

Scheme 4

(iv) NDP-6-deoxy-4-hexulose-4-reductases

The enzyme that catalyzes the stereospecific reduction of the C-4 keto group is the least-studied enzyme in the 6-deoxyhexose pathways. Though a 4-reductase has been partially purified in a fractionation of the crude extracts from *E. coli* and *P. aeruginosa*,[384,386] it was not active in the absence of the 3,5-epimerase. In another demonstration of 4-reductase activity,[390] the enzyme was even less pure. Thus, there are few details available concerning the characteristics of these enzymes. However, it is known that all 4-reductases, including the bifunctional enzymes in the GDP-fucose pathway described above, have a strict dependence on NAD(P)H. Much work remains to be done concerning the stereospecificity of the hydride transfer from the pyridine nucleotide and other kinetic and mechanistic aspects of the reaction.

3.12.4.2.2 3,6-Dideoxyhexoses

The biosynthetic pathways of 3,6-dideoxyhexoses share many common enzymatic steps found in 6-deoxyhexose pathways. Among the five known 3,6-dideoxyhexoses, the formation of ascarylose in *Y. pseudotuberculosis* has been thoroughly studied.[355] Scheme 5 shows the biosynthetic pathway of CDP-ascarylose (**114**) formation. The first step of the pathway is the coupling of α-D-glucose-1-phosphate (**109**) and cytidine triphosphate (CTP) by α-D-glucose-1-phosphate cytidylyltransferase (E$_p$) to yield CDP-D-glucose (**110**). As with the 6-deoxyhexoses, an irreversible dehydration mediated by the NAD$^+$-dependent CDP-D-glucose-4,6-dehydratase (E$_{od}$) commits the nucleotidyl sugar to this particular pathway. The C-3 hydroxyl group of the resulting product, CDP-6-deoxy-L-*threo*-D-*glycero*-4-hexulose (**111**), is removed in a step catalyzed by CDP-6-deoxy-L-*threo*-D-*glycero*-4-hexulose-3-dehydrase (E$_1$) and E$_1$ reductase (E$_3$) to form CDP-3,6-dideoxy-D-*glycero*-D-*glycero*-4-hexulose (**112**). Subsequent epimerization at C-5 by CDP-3,6-dideoxy-D-*glycero*-D-*glycero*-4-hexulose-5-epimerase (E$_{ep}$) gives CDP-3,6-dideoxy-D-*glycero*-L-*glycero*-4-hexulose (**113**). The final step is the stereospecific reduction at C-4 by CDP-3,6-dideoxy-D-*glycero*-L-*glycero*-4-hexulose-4-reductase (E$_{red}$) to yield CDP-L-ascarylose (**114**). The characteristics and mechanisms of these enzymes are discussed in this section.

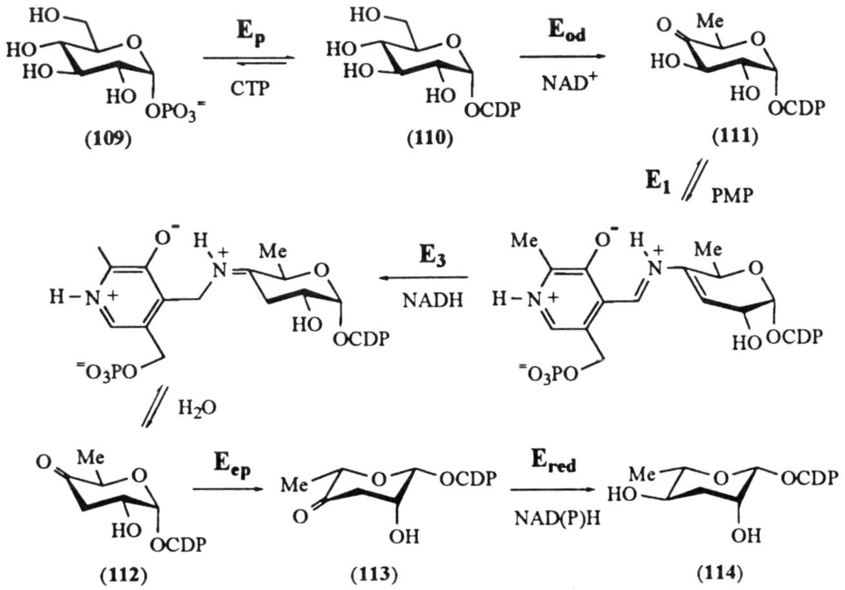

Scheme 5

(i) α-D-Glucose-1-phosphate cytidylyltransferase (E$_p$)

Early efforts to study the enzyme responsible for the coupling of cytidine to glucose-1-phosphate resulted in the isolation of a monomeric protein with an M_r of 110 kDa.[363] However, cloning and overexpression experiments have led to the purification and characterization of the genuine E$_p$, which is a homotetrameric (29 kDa subunit) protein.[366] Like the other enzymes of this class, this genuine E$_p$ has a dependence upon Mg^{2+} for activity and displays a relatively broad pH optimum between pH 7.0 and 9.0.[366] However, E$_p$ was most reactive in the presence of Co^{2+}, while Mg^{2+} and Mn^{2+} provided about 70–75% of this activity. Also, E$_p$ prefers CTP almost exclusively, with UTP being the only other nucleotide that can serve as a substrate, albeit with <5% activity. Protein sequence comparisons showed that E$_p$ exhibits homology with other nucleotidyltransferases (see Section 3.12.4.2.1(i)), and it has the conserved domains for the binding of effectors and substrates.[366] Notably, a diverse array of sugar derivatives, such as CDP-ascarylose (**114**), CDP-abequose (**122**), CDP-4-keto-6-deoxyglucose (**111**), and CDP-fucose, all inhibit this enzyme.[363] It is likely that each plays a regulatory role by binding to the allosteric site. As with the other enzymes in this class, the kinetic mechanism of E$_p$ is presumably ordered bi-bi (see Section 3.12.4.2.1(i)).

(ii) CDP-D-glucose-4,6-dehydratase (E$_{od}$)

The E$_{od}$ enzyme (80 kDa homodimer) from *Y. pseudotuberculosis* VA has been purified from both the native organism[372] and from a heterologous host in which a recombinant form was expressed.[391] Like most nucleotidyl diphosphohexose-4,6-dehydratases, E$_{od}$ appears to bind only one NAD$^+$

molecule per dimer,[376] and its catalytic cycle has been shown to be identical to that described above in Section 3.12.4.2.1(ii) and shown in Scheme 3.[372] However, unlike other dehydratases of this class, which have a tightly bound NAD^+ as a prosthetic group, the purified E_{od} requires exogenous NAD^+ for full catalytic activity of the enzyme, whereas most other enzymes of this class exhibit full activity without exogenous NAD^+.

One reason for this difference may be that E_{od} has a lower affinity for NAD^+ than others in its class. For example, dehydroquinate synthase (DHQase), which is one of the best-studied enzymes of this class, exhibits a K_{NAD} of 2 nM in the absence of substrate,[392] whereas K_{NAD} for E_{od} is much higher ($K_1 = 40$ nM and $K_2 = 540$ nM).[376] Notably, the affinities for NAD^+ in both subunits of E_{od} were resolved, and the binding events exhibit a large degree of anticooperativity. Also, it was found that E_{od} contains sequestered NADH, which binds with a higher affinity ($K_1 = 0.21$ nM and $K_2 = 7.5$ nM) than NAD^+. Both the anticooperativity of NAD^+ binding and the tight NADH binding may explain why the cofactor binding sites of wild-type E_{od}, and possibly of other members in this class, are only half occupied. Interestingly, the sequestered NADH is released upon binding with substrate, CDP-D-glucose (**110**).[376] This substrate-induced release of reduced pyridine nucleotide may be the result of a protein conformational change that establishes a preference for NAD^+ over NADH and fully activates the enzyme. These observations have necessitated the proposal of a slightly different mechanism for E_{od} catalysis. The proposed sequence of binding and catalytic events that occur in the mechanism of E_{od} is shown in Scheme 6,[376] though there is insufficient evidence to conclude whether NAD^+ or the substrate binds first.

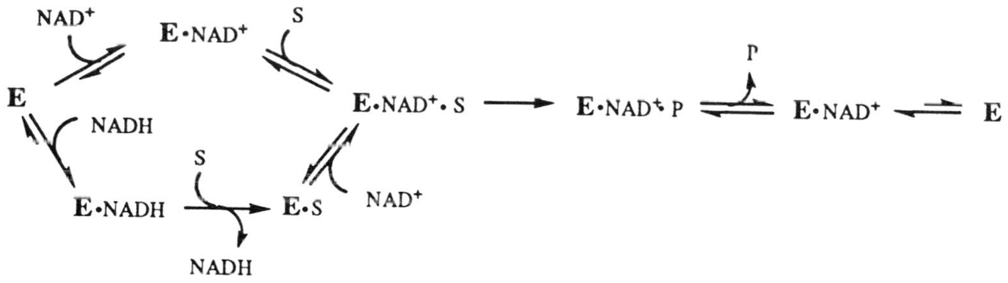

Scheme 6

Concerning the lower NAD^+-binding affinity of E_{od}, analysis of the primary sequence revealed an altered ADP-binding fold (GHTGFKG) relative to the common Rossmann consensus (GXGXXG).[391] This extended cofactor binding fold has a bulky histidine replacing the first glycine and a charged lysine substituting for a hydrophobic residue, thus potentially unfavorable steric and electronic interactions could account for the lowered NAD^+-binding affinity of E_{od}. In support of this hypothesis, mutants in which the histidine was changed to glycine (H17G) and the lysine to isoleucine (K21I) exhibited improved NAD^+ affinities of up to 20-fold.[376] Therefore, the replacement of the histidine seemed to remove the interference with the α-helix dipole,[393] and changing lysine to a hydrophobic residue eliminated adverse charge interactions. It should be noted that dTDP-dihydrostreptose synthase from *Streptomyces griseus*,[394] dTDP-D-glucose-4,6-dehydratase from *S. griseus*,[394] and CDP-D-glucose-4,6-dehydratase from *Salmonella enterica* serovar Typhimurium[385] all contain the same GXXGXXG consensus pattern;[376] thus, they may exhibit similar pyridine nucleotide binding properties to E_{od}.

(iii) CDP-6-deoxy-L-threo-D-glycero-4-hexulose-3-dehydrase (E_1)

The removal of the C-3 hydroxyl group of the 4-keto-6-deoxysugar (**111**) in the third step is mediated by CDP-6-deoxy-L-*threo*-D-*glycero*-4-hexulose-3-dehydrase (E_1) in conjunction with CDP-6-deoxy-L-*threo*-D-*glycero*-4-hexulose-3-dehydrase reductase (E_3). The E_1 enzyme is a pyridoxamine 5'-phosphate (PMP)-dependent iron–sulfur protein that is homodimeric (49 kDa subunit) in its active form. Both radiometric[395] and fluorometric[396] analysis showed that each monomer binds one PMP cofactor. EPR analysis of the protein reduced by either chemical[397] or enzymatic[398] means confirmed that the stoichiometric quantities of iron and sulfur detected in the enzyme form a [2Fe–2S] center that exhibits slightly rhombic symmetry ($g_1 = 2.012$, $g_2 = 1.950$, $g_3 = 1.932$) and resembles that found in adrenodoxin and putidaredoxin. However, examination of the deduced amino acid sequence failed to locate any known [2Fe–2S] center binding motifs, so E_1 may have a novel ligand network for its iron–sulfur center. It should be noted that purified E_1 is a mixture of apo- and holo-enzyme, thus exogenous PMP, iron, and sulfur are required to fully reconstitute the activity of the enzyme.

When E_1 is incubated with substrate (**111**), a Schiff base (**115**) forms between the C-4 keto group of the substrate and the amino group of PMP.[399] As shown in Scheme 7, subsequent abstraction of the *pro-S* 4′-hydrogen atom triggers the elimination of the C-3 hydroxyl group and formation of the $\Delta^{3,4}$-glucoseen intermediate (**116**)[399] (arrows indicating the loss of the C-3 hydroxyl group are for illustrative purposes; the mechanism may not be concerted). The active-site base responsible for the abstraction has been identified as His220.[396] Although the dehydration product (**116**) could not be directly detected, when this incubation was conducted in $[^{18}O]H_2O$, incorporation of ^{18}O into positions C-3 and C-4 of the recovered substrate was noted.[399] These observations clearly demonstrate the reversibility of the E_1-catalyzed dehydration. It has been shown that reduction of the glucoseen intermediate is required to drive the dehydration to completion. This reduction is most effectively mediated by E_3 and NADH, but chemical reductants such as dithionite[400] and other reducing enzyme systems such as diaphorase and methane monooxygenase[399] can also generate the product (**112**) in small amounts.

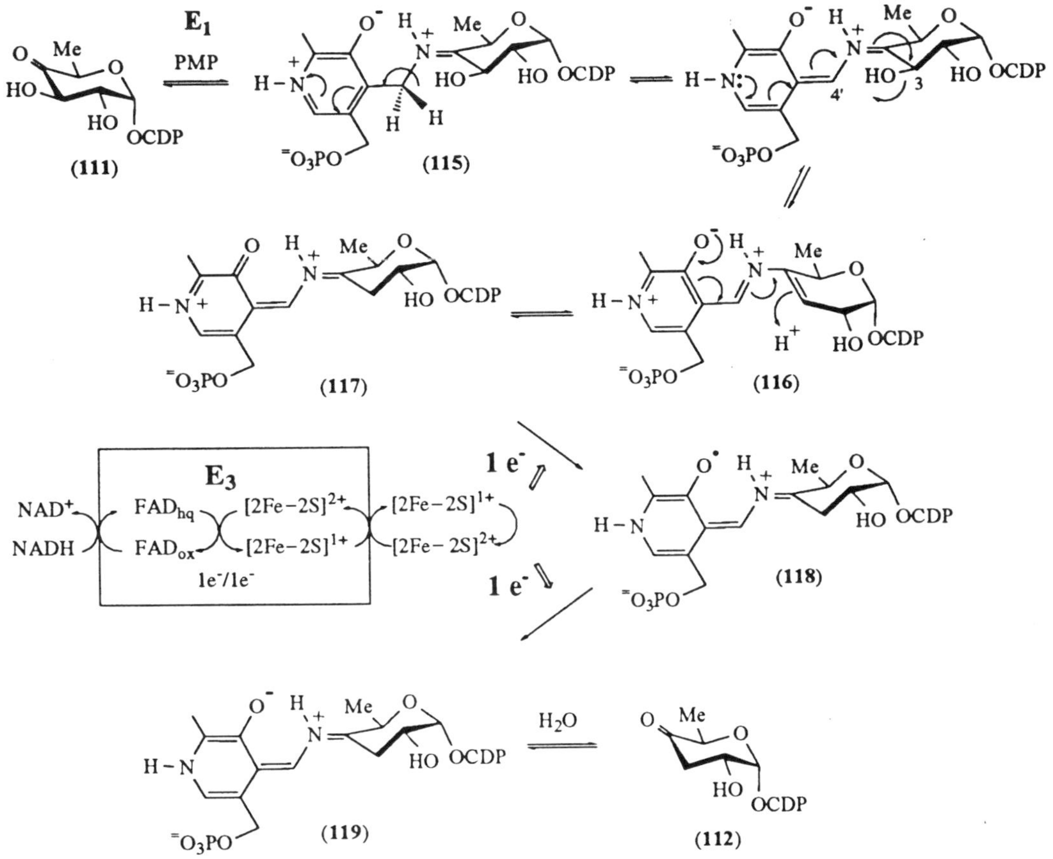

Scheme 7

Reduction of the glucoseen intermediate (**116**) by E_3 leads to the incorporation of a solvent hydrogen atom at C-3, and the net displacement of the 3-hydroxyl group by the solvent hydrogen atom proceeds with retention of configuration.[401] Together with the reversible *pro-S* stereospecificity of the C-4′ hydrogen abstraction, these observations imply that the dehydration is likely a *syn* elimination. Therefore, the overall catalysis in the active site of E_1 may be a suprafacial process occurring on the solvent-exposed *si* face of the PMP–substrate complex. A similar stereochemical course has been found for most coenzyme B_6-dependent enzymes.[402] It should be noted that although E_1 shares low sequence homology with many PLP/PMP enzymes, several invariant amino acid residues that serve as a fingerprint for this class of enzymes are conserved in E_1 (Gly169, Glu191, and Arg403).[396,403] One significant difference between the general class of coenzyme B_6 enzymes and E_1 is the replacement of a highly conserved and catalytically essential lysine with a histidine at position 220. Because His220 has been shown to be the active site base directly involved in E_1

catalysis,[396] it is possible that this single amino acid replacement is responsible for converting a pyridoxal 5′-phosphate (PLP)-dependent transaminase to a PMP-dependent dehydrase.

Though the above results indicated that E_1 behaves similarly to other coenzyme B_6-dependent enzymes, additional experiments have clearly shown that E_1 is in a unique class by itself. Specifically, removal of the [2Fe–2S] center of E_1 had little effect on its ability to abstract the C-4′ hydrogen of PMP, though the apoenzyme lost essentially all ability to catalyze product formation.[397] Since the [2Fe–2S] center is an obligatory one-electron redox coenzyme, its participation in the reaction strongly implicated that reduction of the glucoseen intermediate (**116**) is a stepwise electron transfer process. Thus, this reduction is expected to produce an organic free radical intermediate. Indeed, upon chemical reduction of E_1 in the presence of substrate (**111**), an organic radical with a Lorentzian-type absorption ($g \approx 2$) could be observed using EPR spectroscopy.[400] Subsequently, enzymatic reduction using E_3 and NADH produced a kinetically competent organic radical ($g = 2.003$) exhibiting an EPR lineshape identical to that generated by chemical reduction.[398] The organic radical signal did not display any detectable hyperfine splitting, so the radical seems to couple very weakly, if at all, with neighboring nuclei. The implications of this observation on the detailed mechanism of the C-3 deoxygenation are discussed in the next section on E_3. It is important to point out that although participation of PMP in a dehydration reaction is unprecedented, the involvement of single-electron chemistry truly distinguishes E_1 and establishes a novel subclass of coenzyme B_6-dependent enzymes.

Another member of this subclass may be involved in the conversion of cytosylglucuronic acid (**120**) to blasticidin S (**121**), which is an antifungal antibiotic produced by *Streptomyces griseochromogenes*. As shown in Scheme 8, this biosynthetic pathway includes two deoxygenations at C-2′ and C-3′ and a transamination at C-4′ of a glucose precursor. Isotopic feeding and NMR experiments revealed that the loss of the C-3 hydroxyl group occurs with net retention of configuration,[404] as in the E_1 reaction.[401] Furthermore, inhibition studies demonstrated that the source of the C-4′ amino group is most likely PMP.[404] When PLP and cytosinine are incubated with crude extract from *S. griseochromogenes* in D_2O, deuterium is incorporated at C-2′ and C-4′ of cytosinine,[405] thus corroborating the participation of a coenzyme B_6 cofactor in the reaction. These results are consistent with E_1 being related to an enzyme in the blasticidin S pathway, though the latter enzyme appears to have the capability of transamination as well as deoxygenation.

(iv) CDP-6-deoxy-L-threo-D-glycero-4-hexulose-3-dehydrase reductase (E_3)

As discussed in the previous section, the C-3 deoxygenation in ascarylose biosynthesis is catalyzed by both E_1 and E_3 (formerly CDP-6-deoxy-$\Delta^{3,4}$-glucoseen reductase). Enzyme E_3 is a monomeric protein (36 kDa) containing one equivalent of flavin adenine dinucleotide (FAD) as determined by UV–visible and HPLC analysis.[406] EPR spectroscopy on the reduced enzyme revealed that the stoichiometric iron and sulfur detected in the enzyme formed a plant ferredoxin-type [2Fe–2S] center, which gives rise to a rhombic EPR spectrum with g values of 2.043, 1.960, and 1.877.[406] Single-electron reduction of E_3 produced a flavin semiquinone with a g value of 2.002.[406] Alignments of the deduced amino acid sequence showed that E_3 has a close relationship with members of the ferredoxin-NADP$^+$ reductase family.[407] As is the case with other proteins of this family, E_3 can transfer electrons from NADH to oxygen and a variety of one-electron acceptors with varying levels of efficiency.[408] Thus, E_3 is a NADH oxidase and is capable of acting as a two-electron to one-electron switch. This capability remains after the [2Fe–2S] center is removed, but E_3(apoFeS) is unable to catalyze final product formation in the presence of E_1 and substrate (**111**).[406]

The independent nature of the cofactors in E_3 was confirmed by stopped-flow spectroscopic experiments, which showed that the hydride transfer from NADH to FAD occurred at the same rate in both E_3(apoFeS) and holo-E_3.[409] After the hydride transfer, which exhibits a 10-fold deuterium isotope effect when (4R)-[^{2}H]NADH is substituted for NADH,[409] subsequent intramolecular electron transfer from the reduced FAD hydroquinone to the [2Fe–2S] center was found to be pH-dependent.[409] At pH 7, the equilibrium favored the hydroquinone and oxidized [2Fe–2S]$^{2+}$ state of the two-electron reduced enzyme, while the flavin semiquinone radical and reduced [2Fe–2S]$^+$ state was favored at pH 10. This pH-dependent distribution of electrons may be derived from changes of the redox potentials of both the FAD and the [2Fe–2S] center with respect to pH. Spectroelectrochemical studies have lent credence to this hypothesis, since the midpoint potential of FAD was found to change from -212 mV at pH 7.5 to -273 mV at pH 8.4, while these same conditions produced a less dramatic shift from -257 mV to -279 mV for the [2Fe–2S] center.[410] Results from

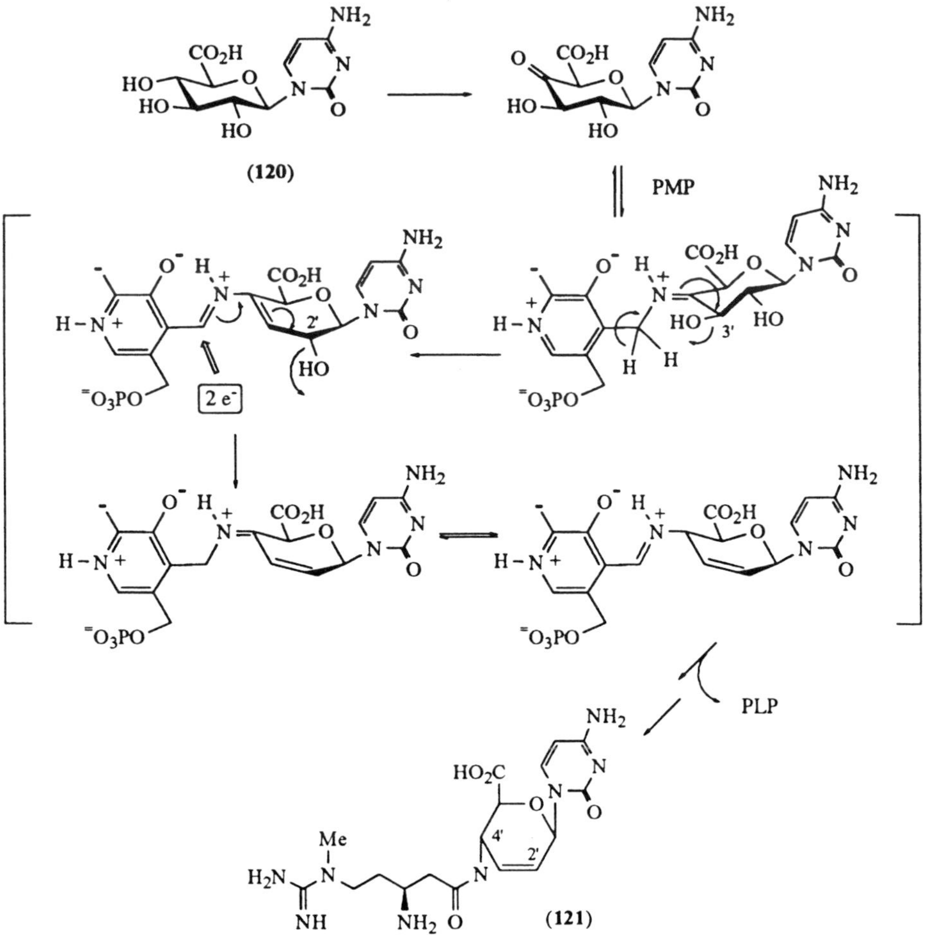

Scheme 8

the stopped-flow experiments suggested that an ionizable group having an estimated pK_a of 7.3 is responsible for the pH regulation.[409] Based on studies with other flavoenzymes,[411] N-1 of the flavin is probably the protic position, and this hypothesis is consistent with the significant pH-dependent change observed in the flavin potential.

The insights gained from the stopped-flow studies of the coupled reaction of E$_1$, E$_3$, NADH, and substrate (**111**) in conjunction with the reactivities of the E$_1$ and E$_3$ apoenzymes described above have allowed the electron transfer pathway from NADH to substrate to be proposed. Scheme 9 shows the proposed electron-transfer pathway from NADH, through the enzyme-bound cofactors, and to the substrate in the E$_1$–E$_3$ coupled reaction.[398] First the binding of NADH to E$_3$ initiates the formation of a charge-transfer complex between the oxidized FAD and reduced pyridine nucleotide (Int-I).[409] Chemical modification studies identified Cys296 as a residue that may participate in the stabilization of this charge-transfer complex.[407] Transfer of a hydride from NADH to FAD reduces the flavin to the hydroquinone (Int-II), which then shuttles electrons one at a time through the [2Fe–2S] centers of both E$_3$ and E$_1$ into the PMP-glucoseen intermediate bound in the E$_1$ active site (Int-III to Int-VI). While one of the organic radicals produced by this one-electron transfer process has been clearly identified as a flavin semiquinone,[409] the other radical detected by rapid freeze–quench EPR[398] is not yet well characterized. However, this species exhibits an EPR signal that lacks hyperfine splitting, as mentioned in the previous section, and a phenoxyl radical is one species that can give rise to such a signal.[412] Thus, the available evidence suggests that a PMP–substrate Schiff base radical species with the unpaired spin localized mainly on the C-3 oxygen of the PMP is a likely candidate for the second radical ((**118**), see Scheme 7).[398] Presumably, the generation of this radical (**118**) requires the tautomerization of the PMP-glucoseen intermediate (**116**) to a reactive quinone methide species (**117**), which is a good one-electron acceptor.[413] Transfer of a second electron results in the two-electron reduced PMP–product Schiff base (**119**), which is subsequently hydrolyzed to release the product (**112**) and end the catalytic cycle (Int-VI to Int-VII). During the first product turnover, two additional NADH molecules are needed to prime the enzyme complex for steady state catalysis (cycling between Int-V, Int-VI, and Int-VII). Such priming is common for redox systems such as P-450 enzymes and dioxygenases.

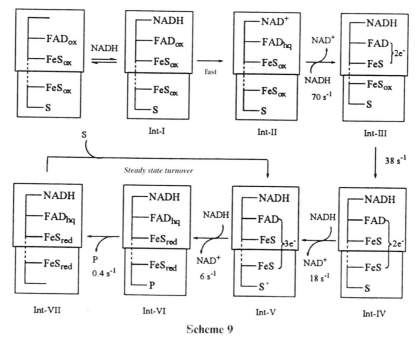

Scheme 9

(v) CDP-3,6-dideoxy-D-glycero-D-glycero-4-hexulose-5-epimerase (E_{ep}) and CDP-3,6-dideoxy-D-glycero-L-glycero-4-hexulose-4-reductase (E_{red})

The two final enzymes catalyzing the epimerization at C-5 (E_{ep}) and the stereospecific reduction at C-4 (E_{red}) have not been fully characterized. An early fractionation of the crude extracts of *Y. pseudotuberculosis* spp. led to a partial separation of a NADH-dependent fraction (labeled E_2) that catalyzes the final two steps of CDP-ascarylose formation.[414] As with the reductases and epimerases responsible for the formation of 6-deoxyhexoses (see Section 3.12.4.2.1), the E_2 fraction from different *Yersinia* strains was interchangeable. The strain-specific E_1 and E_3 enzymes were interchangeable as well, but when different E_1, E_2, and E_3 fractions were mixed, the identity of the final product was solely dependent upon the source of the E_2 fraction (i.e., type II E_2 yields abequose, type V E_2 yields ascarylose, etc.).[414] This observation implies that the pathways of all 3,6-dideoxyhexoses include a common intermediate, CDP-3,6-dideoxy-D-*glycero*-D-*glycero*-4-hexulose (**112**), which is generated by E_1 and E_3 and is processed to the correct 3,6-dideoxyhexose isomer by subsequent enzymes in the pathway (including the E_2 fraction). Scheme 10 gives an overview of biosynthetic routes to CDP-ascarylose (**114**), CDP-abequose (**122**), CDP-paratose (**123**), and CDP-tyvelose (**124**).

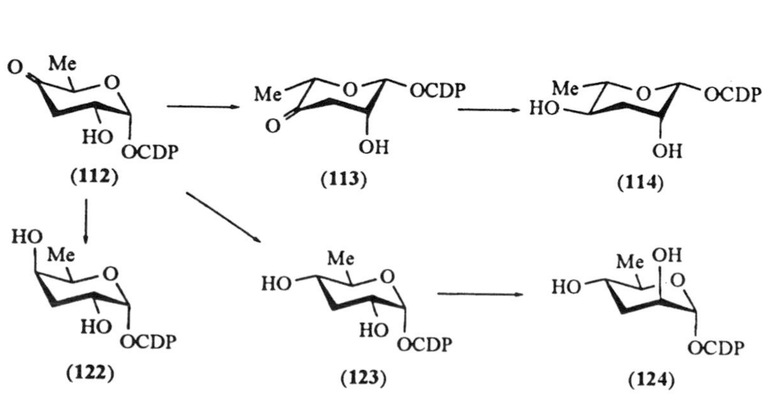

Scheme 10

The genes encoding E_{ep} and E_{red} in the ascarylose pathway were cloned, and crude extracts of the heterologously expressed enzymes were shown to exhibit epimerase and reductase activity by a GC–MS assay.[415,416] Similarly, the reductase activity was detected by an HPLC assay using cell extracts of *E. coli* that contained the reductase gene cloned from the abequose biosynthetic cluster found in *S. enterica* LT2 (there is no epimerase in the abequose pathway).[417] However, in no case has either an epimerase or a reductase from a 3,6-dideoxyhexose pathway been purified to homogeneity, so there are no further biochemical data available on these enzymes. Presumably, E_{ep} and E_{red} utilize mechanisms similar to those discussed for 6-deoxyhexose pathways in Sections 3.12.4.2.1(iii) and 3.12.4.2.1(iv), respectively.

3.12.4.2.3 Branched-chain deoxysugars

While there have been several tracer studies on the biosynthesis of branched-chain sugars,[207] there have been relatively few reports of mechanistic studies on this subject. As with the biosynthesis of other deoxysugars, nucleotide-bound sugars are the substrates for the enzymes in these pathways. Feeding isotopically labeled precursors to branched-chain sugar-producing bacteria showed that L-methionine and pyruvate are the sources of the one- and two-carbon branches, respectively, in L-noviose (**70**), L-vinelose (**73**), L-cladinose (**80**), L-mycarose (**82**), D-aldgarose (**91**), and the quino-cycline sugars.[207] As illustrated in Scheme 11, branching at C-3 is probably initiated by a deprotonation of the 4-keto sugar substrate (**125**) to generate an enediol intermediate (**126**) (dTDP = 2′-deoxythymidine diphosphate). Subsequent C—C bond formation between C-3 and the electrophilic methyl group of *S*-adenosyl-L-methionine (AdoMet or SAM) gives methyl branched sugars such as D-evermicose (**79**). A similar mechanism can be envisioned for the two-carbon branched sugars. Since pyruvate must be activated for this reaction, the coupling is most likely catalyzed by a thiamine pyrophosphate (TPP)-dependent enzyme. Whereas AdoMet has been shown to participate in the methyltransferase reaction, the participation of TPP has not yet been confirmed in the two-carbon transfer reaction.

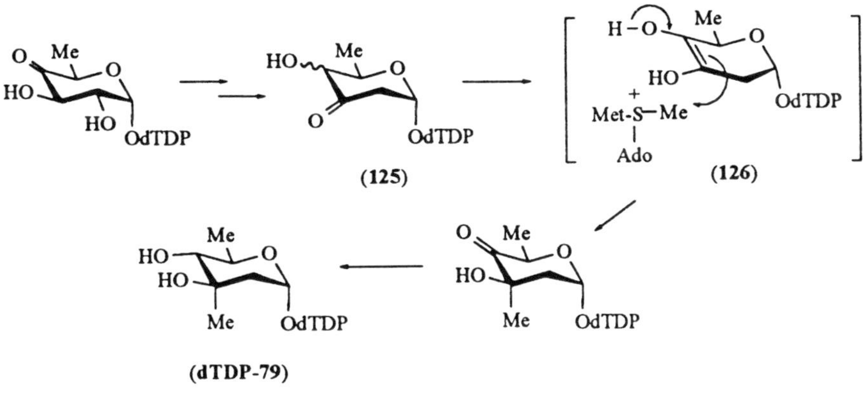

Scheme 11

3.12.5 GENETICS OF DEOXYSUGAR BIOSYNTHESIS

The development of advanced molecular biological techniques and the discovery that most of the genes associated with a bacterial biosynthetic pathway are clustered have significantly facilitated the study of many biosynthetic enzymes. For example, the genes encoding enzymes in 6-deoxysugar and 3,6-deoxysugar biosynthetic pathways have been located, cloned, and heterologously expressed at levels sufficient for thorough biochemical studies. Additionally, site-directed mutagenesis has proven to be an invaluable tool in the elucidation of enzymatic mechanisms, as described in Section 3.12.4. In fact, genetics analysis has allowed the genes encoding pathways for which little direct biochemical evidence is available to be located and sequenced. Study of these uncharacterized pathways commonly includes disruption of individual genes and subsequent analysis of the resulting metabolites. The combination of this information with comparisons of the deduced protein sequences to those of well-characterized enzymes has enabled the postulation of several deoxysugar biosynthetic pathways. These proposed routes along with the genetic basis of the biochemically characterized pathways in Section 3.12.4 are discussed below. Table 2 presents a summary of selected genes and their defined or proposed functions in several deoxysugar biosynthetic pathways.

3.12.5.1 Genetics of RNR

3.12.5.1.1 Class I RNR

The class I RNR can be divided into two subclasses, class Ia and Ib.[424] The large and small subunits of class Ia reductase in *E. coli* are encoded by the *nrdA* and *nrdB* genes,[418,419] respectively, and the isolated proteins represent the best-characterized reductase that serves as a prototype for the class Ia enzymes. The *nrdAB* genes have also been isolated from *S. enterica* serovar Typhimurium and are 87% and 86% homologous, respectively, to the *E. coli* counterparts.[420] Consequently, the translated NrdA and NrdB proteins share 96.5% and 98.5%, respectively, overall amino acid sequence identity with the *E. coli* enzymes.[420] There is a consensus sequence in the promoter regions of the genes from these two species, thus suggesting that the *nrdAB* genes may be under a common control.[420] In fact, expression of these genes, and presumably those of all other RNRs, is regulated by the cell cycle and induced by DNA damage.[458,459] The class Ia genes cloned from *Saccharomyces cerevisiae*[429–431] are named *RNR1*, *RNR2*, and *RNR3*, with the *RNR3* gene product being an alternative large subunit R1 protein that is expressed only in the presence of DNA damage and is not necessary for viability.[430] The RNR1 and RNR3 proteins are 80% identical in the portions of their genes that have been sequenced.[458] Other *nrdAB*-type genes have been cloned and sequenced from mouse,[421] clam,[422] and a number of viral sources.[311,423] While varying, often low, levels of sequence identities were encountered for these reductases, the critical protein residues involved with the dinuclear iron/radical center were highly conserved.[311,460]

The class Ib reductase consisting of the R1E and R2F subunits is encoded by the *nrdE* and *nrdF* genes, which were first isolated from *S. enterica* serovar Typhimurium.[424] The corresponding genes have also been cloned and sequenced from *E. coli*[461] and *Lactococcus lactis*.[425] Sequence comparisons[461] showed that a cloned RNR from *Mycobacterium tuberculosis*[462] is also a member of this class, and its *nrdE* gene product is 71% identical to that from *S. enterica*. Interestingly, *Mycoplasma genitalium* contains the *nrdEF* genes but not the *nrdAB* genes.[476] As in the class Ia reductase, the small subunit of class Ib reductase (R2F) contains the μ-oxo dinuclear iron center and a tyrosyl radical.[463] However, aside from a strict conservation of the ligand residues of the iron center and the catalytically essential tyrosine, there is relatively little (18–28%) sequence identity between the class Ia and Ib enzymes.[420,424,461] Additionally, the class Ib enzymes use a different, redoxin-like electron transporter protein encoded by *nrdH*, in place of thioredoxin or glutaredoxin used by the class Ia RNR, to regenerate the activity.[125] The two classes also differ in their allosteric regulation by dATP.[463]

When a plasmid bearing the *nrdEF* genes from *E. coli* and *S. enterica* are introduced to these strains, deficiencies in the *nrdAB* genes are abolished.[424] However, the chromosomal *nrdEF* genes from these organisms are expressed at levels insufficient to suppress mutations in the *nrdAB* genes,[461,463] so the biological function of the *nrdEF* genes in *E. coli* and *S. enterica* is not yet clear. Nevertheless, the class Ib genes in *E. coli* and *S. enterica* may be under the same control as the class Ia genes in the same species, as they all contain a similar consensus sequence.[461] The role of the *nrdEF* genes is not ambiguous for the other microorganisms mentioned above, since these species have chromosomal *nrdEF* genes that express a functional RNR at levels that can support cellular viability.[252,425,426] Based on these observations, it has been suggested that class Ib proteins may be the prevalent class I reductase in microorganisms.[425]

3.12.5.1.2 Class II RNR

The class II reductases are found in many microorganisms[464] and are represented by the *L. leichmannii* enzyme.[299,304] The gene encoding this protein has been cloned and sequenced from *L. leichmannii*.[342] There is no statistically significant homology between this enzyme and any other enzyme in the protein database, including the class I RNR from *E. coli*. However, there are three fragments that show a local sequence similarity, and these fragments are associated with critical

Table 2   Listing of deoxysugar biosynthetic genes and corresponding products.

Gene(s)	Proposed identity/function of gene product(s)	Ref.
Deoxyribonucleotide biosynthesis (Section 3.12.5.1)		
nrdAB	Both subunits of class Ia RNR (R1 and R2, respectively)	418–423
nrdEF	Both subunits of class Ib RNR (R1E and R2F, respectively)	420, 424–426
nrdDG	Both subunits of class III RNR (large and small subunits, respectively)	345, 346, 427, 428
nrdH	Redoxin-like electron transporter for class Ib RNR	425
RNR1, RNR	Large and small RNR subunits in *Saccharomyces cerevisiae*	429–431
RNR3	Alternative large RNR subunit in *S. cerevisiae* induced by DNA damage	430
dTDP-rhamnose biosynthesis (Section 3.12.5.2.1)		
rfbA	α-D-Glucose-1-phosphate thymidylyltransferase	359, 385, 432–436
rfbB	dTDP-D-glucose-4,6-dehydratase	385, 432–437
rfbC	dTDP-6-deoxy-L-*threo*-D-*glycero*-4-hexulose-3,5-epimerase	385, 432–436
rfbD	dTDP-6-deoxy-D-*erythro*-L-*glycero*-4-hexulose-4-reductase	385, 432–436
dTDP-6-deoxy-L-altrose biosynthesis (Y. enterocolitica *genes) (Section 3.12.5.2.1)*		
rfbA	dTDP-6-deoxy-L-*threo*-D-*glycero*-4-hexulose-5-epimerase	438
rfbF	dTDP-6-deoxy-L-*threo*-L-*glycero*-4-hexulose-4-reductase	438
rfbG	α-D-Glucose-1-phosphate thymidylyltransferase	438
3,6-Dideoxysugar biosynthesis (Section 3.12.5.2.2)		
rfbF/ascA	α-D-Glucose-1-phosphate cytidylyltransferase (E$_p$)	360, 403, 416, 439–441
rfbG/ascB	CDP-D-glucose-4,6-dehydratase (E$_{od}$)	403, 416, 439–441
rfbH/ascC	CDP-6-deoxy-L-*threo*-D-*glycero*-4-hexulose-3-dehydrase (E$_1$)	403, 416, 439–441
rfbI/ascD	CDP-6-deoxy-L-*threo*-D-*glycero*-4-hexulose-3-dehydrase reductase	403, 416, 439–441
rfbJ	CDP-abequose synthase	385, 439, 440–443
rfbK	Phosphomannomutase	444
rfbM	α-D-Mannose 1-phosphate cytidylyltransferase	444
rfbS	CDP-paratose synthetase	433, 440, 445
rfbE	CDP-paratose-2-epimerase (also tyvelose epimerase)	433, 440, 445, 446
ascE	CDP-3,6-dideoxy-D-*glycero*-D-*glycero*-4-hexulose-5-epimerase (E$_{ep}$)	403, 416
ascF	CDP-3,6-dideoxy-D-*glycero*-L-*glycero*-4-hexulose-4-reductase (E$_{red}$)	403, 416
Daunosamine biosynthesis (Section 3.12.5.3.1)		
dnrJ	Aminotransferase	447, 448
dnrL	α-D-Glucose-1-phosphate thymidylyltransferase	449
dnrM	Nonfunctioning dTDP-D-glucose-4,6-dehydratase	449
dnrU	dTDP-4-keto-6-deoxyhexose-3,5-epimerase	355
dnrV	dTDP-4-keto-6-deoxyhexose-4-reductase	355
dnrS	Glycosyltransferase	450
dauH	Glycosyltransferase	451
Mycarose biosynthesis (Section 3.12.5.3.2)		
eryBI	dTDP-4-keto-2,6-dideoxyhexose-3-*C*-methyl transferase	452, 453, 512
eryBII	dTDP-3,4-diketo-2,6-deoxyhexose-3-reductase	452, 512
eryBIV	dTDP-4-keto-3-methyl-2,6-dideoxyhexose-4-reductase	452, 512
eryBV	Glycosyltransferase	452, 512
eryBVI	dTDP-4-keto-6-deoxyhexose-2,3-dehydratase	452, 512
eryBVII	dTDP-4-keto-2,6-dideoxyhexose-5-epimerase	452, 512
eryG	*O*-methyltransferase	453
gdh	dTDP-D-glucose-4,6-dehydratase	388
Desosamine biosynthesis (Section 3.12.5.3.2)		
eryCI	dTDP-3-keto-4,6-dideoxyhexose-3-transaminase	452, 512
eryCII	dTDP-4-keto-6-deoxyhexose 3-isomerase or a reductase	452, 512
eryCIII	Glycosyltransferase	452, 512
eryCIV	dTDP-3-keto-6-deoxyhexose-4-dehydratase	452, 512
eryCV	dTDP-4-keto-6-deoxyhexose 3-isomerase or a reductase	452, 512
eryCVI	*N*-methyltransferase	452, 512
Mycaminose biosynthesis (Section 3.12.5.3.3)		
*tylA*1	α-Glucose-1-phosphate thymidylyltransferase	454
*tylA*2	dTDP-glucose-4,6-dehydratase	454
tylB	dTDP-3-keto-6-deoxyhexose-3-transaminase	403, 454
Streptomycin sugar biosynthesis (Section 3.12.5.3.4)		
Dihydrostreptose		
strD	α-Glucose-1-phosphate thymidylyltransferase	394
strE	dTDP-glucose-4,6-dehydratase	394
strL	dTDP-dihydrostreptose synthase	394
strM	dTDP-6-deoxy-L-*threo*-D-*glycero*-4-hexulose-3,5-epimerase	394

Table 2 (continued).

Gene(s)	Proposed identity/function of gene product(s)	Ref.
Streptidine		
strS	Transaminase	455
Miscellaneous genes from other pathways		
*pur*4	Transaminase (puromycin)	456
graD	α-Glucose-1-phosphate thymidylyltransferase (granaticin, Section 3.12.5.3.5)	457
graE	dTDP-glucose-4,6-dehydratase (granaticin, Section 3.12.5.3.5)	457

thiol residues required for catalysis.[341,342] Despite the vast structural differences, the similarity of these fragments implies a strong resemblance in the catalytic mechanisms of these two classes of RNR, as is discussed above in Section 3.12.4.1.

3.12.5.1.3 Class III RNR

Growing *E. coli* anaerobically induces the expression of a completely different reductase that is the prototype for class III RNR.[306] Early studies on this enzyme presumed that the reductase is a homodimer (α_2) in which Gly681 is activated to a catalytic radical upon incubation with 5′-adenosylmethionine (AdoMet), the flavodoxin system, NADPH, and an unknown activase.[306] The *nrdD* gene encoding the protomers of the dimer has been cloned and sequenced, and the corresponding amino acid sequence has very little homology with those of the other classes of RNR.[346] Subsequent studies revealed that the unknown activase is in fact a homodimer (β_2) whose protomers are conjoined by an oxygen-labile [4Fe–4S] cluster, and this dimer forms a tight $\alpha_2\beta_2$ complex with the NrdD protein.[465] The protomers of the β_2 subunit are encoded by the *nrdG* gene,[345] which shares limited homology with the activase from anaerobic pyruvate formate lyase (Pfl) from *E. coli*. Both genes appear to be under common control in the same operon.[345]

Bacteriophage T4 also produces an anaerobic class III reductase whose large subunit is encoded by an *nrdD* gene formerly called *sunY*, and exhibits 72% sequence similarity to the *E. coli* enzyme.[427] The *nrdG* gene for the small subunit in T4 is located downstream from *nrdD*, and bears limited sequence similarity to the Pfl activase as well.[428] The docking of the large and small subunits in class III RNR appears to be species-specific, as the T4 *nrdD* gene product could not be activated by the *E. coli* NrdG protein.[466] The *nrdDG* genes have also been found in *L. lactis*.[425] Surprisingly, the *nrdD⁻* mutant phenotype of this organism can be suppressed by the *nrdEF* genes from its own genome in the usual anaerobic culture conditions,[425] though stringently anaerobic conditions expectedly eliminate the activity of the aerobic NrdE protein. The genome of *Haemophilus influenzae* has been sequenced, and homologues of *nrdDG* genes are present within this facultative anaerobe,[467] but they have not been proven to encode a reductase.

3.12.5.1.4 Other classes of RNR

No genetic information is currently available for the potential fourth class of RNR. This class IV reductase has been isolated from *Brevibacterium ammoniagenes*, and it is postulated to include a dinuclear oxygen-linked manganese center.[300] This oxo-bridged Mn^{III} complex may serve the same role as the dinuclear iron center in class I reductases to generate and stabilize a catalytic tyrosyl radical.[300] Another possible class of reductase has been isolated from the archaebacterium *Methanobacterium thermoautotrophicum*.[468] However, there are no genetic and little biochemical data reported for this enzyme, aside from observations that it is different from all other known classes of RNR.

3.12.5.2 Genetics of Deoxysugars in O Antigen

This section focuses on the genetics of the deoxysugar components of polymorphic O-specific saccharides, which have been well characterized in a number of bacterial species. Research on various strains of *S. enterica*,[385,446,469] *Y. pseudotuberculosis*,[415,439] *E. coli*,[432,444,470,471] and *Klebsiella pneumoniae*[472] has identified many genes that encode proteins responsible for the formation of the deoxysugar components in the O antigen. The O antigens of the *S. enterica* serovars in serogroups A, B, C, and D (using the naming convention suggested by Le Minor and Popoff)[473] consist of repeating oligosaccharide units with different 3,6-dideoxysugars at the nonreducing termini: paratose (**52**) for serotype A, abequose (**53**) for serogroups B and C2, and tyvelose (**49**) for serogroup D.[474] Colitose (**54**) has been isolated from the LPS of *S. enterica* serovars Greenside and Adelaide.[475] These four dideoxysugars plus ascarylose (**50**) have also been found in the O antigens of *Y. pseudotuberculosis*: abequose (**53**) in serogroups IIA, IIB, and IIC, paratose (**52**) in serogroups IA, IB, and III, tyvelose (**49**) in serogroups IVA and IVB, ascarylose (**50**) in serogroup VA, and colitose (**54**) in serogroups VI and VII.[49,476] As opposed to these species that contain a variety of 3,6-dideoxysugars, *E. coli* and *Citrobacter* spp. appear to produce only colitose and abequose, respectively, in selected serotypes. Most of the specific enzymes involved in the biosynthesis of O antigens are encoded by genes clustered in the *rfb* locus.[477] As shown in Table 3, the organization of deoxysugar gene clusters from different organisms is related, and there is a remarkable homology among several of these genes.[415,416,446] The genetic characterization of the biosynthesis of these sugars in their respective organisms has dramatically influenced views on the evolutionary origin of O antigen variation.[478,479] The letters in Table 3 refer to genes in the *rfb* clusters (e.g., B means *rfbB*). The genes *rfbIFGH* in *Y. pseudotuberculosis* have been called *ascDABC* by Thorson *et al.*[403] The *rfbE* genes with an asterisk denote homologues of *rfbE* that are nonfunctioning due to mutations. Note that the assignments for *rfbBDAC* reflect corrections made in 1994.[436]

Table 3 Organization of the O-antigen deoxysugar gene clusters in *S. enterica* serogroups A, B, C2, and D and in *Y. pseudotuberculosis* serogroups IA, IIA, IVA, and VA.

Organism(s)	3,6-Dideoxysugar gene clusters										Ref.
A	B	D	A	C	I	F	G	H	S	E*	445,446
IA	B	D	A	C	I	F	G	H	S	orf1	440
D, IVA	B	D	A	C	I	F	G	H	S	E	445, 446
B, C2	B	D	A	C	I	F	G	H	J		385, 441, 443, 446
IIA	B	D	A	C	I	F	G	H	J	E*	439
VA	B	D	A	C	I	F	G	H	ascE	ascF	403, 416

3.12.5.2.1 6-Deoxyhexoses

The biosynthesis of dTDP-rhamnose (**127**) in *Salmonella* spp.,[359,385,390,433,437] *Shigella* spp.,[434,435] and *E. coli*[432,436,471] has been well characterized. Scheme 12 shows the biosynthetic pathways for the formation of dTDP-rhamnose (**127**) and dTDP-6-deoxy-L-altrose (**128**). The steps are assigned to genes from *S. enterica* (*S-rfbABCD*) and *Y. enterocolitica* O:3 (*Y-rfbAFG*).[438] The four enzymes involved in the rhamnose pathway are encoded by a cluster composed of four genes, *rfbBDAC*. The first enzyme in the pathway, α-glucose-1-phosphate thymidylyltransferase, is encoded by *rfbA*. The subsequent enzymes, dTDP-D-glucose-4,6-dehydratase, dTDP-6-deoxy-L-*threo*-D-*glycero*-4-hexulose-3,5-epimerase, and dTDP-6-deoxy-D-*erythro*-L-*glycero*-4-hexulose-4-reductase, are encoded by *rfbB*, *rfbC*, and *rfbD*, respectively. It should be noted that various trivial names have often been used to describe these enzymes in the literature, and published corrections to initial open reading frame assignments have been incorporated in this discussion without a detailed explanation.

There is a high degree of sequence homology, ranging from 60% to 92%, among the rhamnose biosynthetic clusters from different enteric bacteria that produce dTDP-rhamnose.[432] This homology extends to several genes in the biosynthetic cluster of dTDP-6-deoxy-L-altrose (**128**), which is the C-3 epimer of dTDP-rhamnose and is the precursor for the primary constituent of the O antigen in several *Yersinia* strains.[22,23] The deduced peptide sequence of *rfbA* (Y-RfbA) from *Y. enterocolitica* O:3 is 63% identical to that of *rfbC* (S-RfbC) from *S. enterica* serovar Typhimurium.[438] Also, Y-RfbF and S-RfbD are 54% identical after adding two gaps, and Y-RfbG and S-RfbA are 30% identical with nine gaps.[438] Based on these sequence similarities, the pathway for 6-deoxy-L-altrose formation in Scheme 12 has been proposed.[438] To achieve the proper configuration in the final product, Y-RfbA is proposed to be a 5-epimerase instead of a 3,5-epimerase (S-RfbC). Future

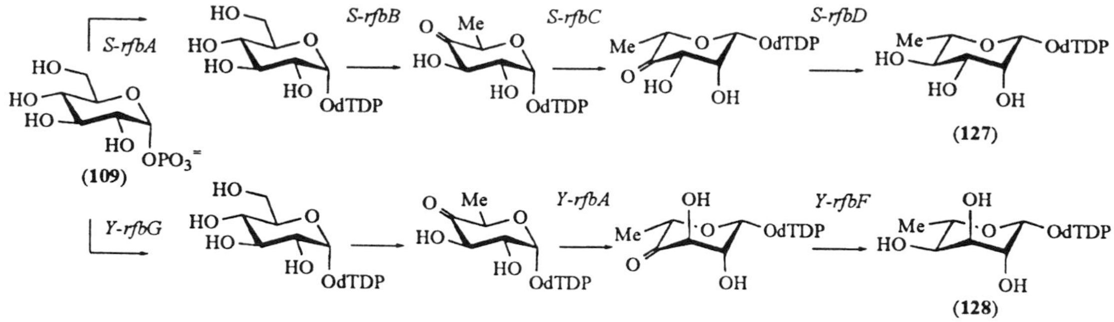

Scheme 12

studies of the expressed gene products using appropriate enzyme assays should confirm and refine this proposal.

The G + C content in many of the aforementioned genes is low relative to the average for the organism.[385,432,434,436-438] This average (about 0.485 for *Yersinia* spp., 0.5 for *E. coli*, and 0.51 for *Salmonella* spp.)[480,481] is the result of a genetic bias in the mutation rates from G · C to A · T and/or from A · T to G · C such that one direction is favored in different organisms.[480] Over time, directional mutation will force the G + C content of acquired genes to match that of the host organism. Based on these theories, the low G + C content of the *rfb* clusters may reflect the acquisition of these genes via interspecific gene transfer.[478] Thus, the genes for O-antigen biosynthesis in *S. enterica*, *Y. pseudotuberculosis*, *S. dysenteriae*, and *Y. enterocolitica*, and possibly in other bacterial species, may be part of the same evolutionary continuum.

3.12.5.2.2 3,6-Dideoxyhexoses

Scheme 13 depicts an integrated pathway for the biosynthesis of CDP-ascarylose (**114**), CDP-abequose (**122**), CDP-paratose (**123**), and CDP-tyvelose (**124**). Each step is labeled with the corresponding gene. Sequence comparison along with enzymatic assays have identified *rfbF* as the gene encoding α-glucose-1-phosphate cytidylyltransferase (E_p); *rfbG* for CDP-D-glucose-4,6-dehydratase (E_{od}); *rfbJ* for CDP-abequose synthase; *rfbS* for CDP-paratose synthetase; and *rfbE* for CDP-paratose-2-epimerase (also tyvelose epimerase).[360,385,439-443,445] The identities of the *rfbH* and *rfbI* genes were unclear until a comparison of the deduced protein sequences with the biochemically characterized E_1 and E_3 enzymes from *Y. pseudotuberculosis* VA (see Section 3.12.4.2.2) revealed significant homologies that supported the mapping of *rfbH* and *rfbI* to E_1 and E_3, respectively (discussed in more detail below). Interestingly, enzymes from the cloned *rfbFGHIJ* genes of *S. enterica* LT2 have been used to successfully synthesize CDP-D-abequose (**122**) *in vitro*.[417]

The *rfbFGHI* genes have a conserved genetic organization (see Table 3), and the individual genes are nearly identical within different strains of the same species.[440,478] Comparison of the deduced amino acid sequences reveals a high identity for RfbF, G, and H (80%, 72%, and 87%, respectively) from *Yersinia* and *Salmonella* strains, but a more modest 51% identity for RfbI.[440] This similarity provides a molecular basis for the interchangeability of these enzymes, as discussed in Section 3.12.4.2.2(v). The remainder of the cluster differs according to the dideoxysugar being produced by the strain. The *rfbJ* gene in *Salmonella* serogroup B is replaced by the *rfbS* gene in serogroups A and D. These two reductase genes are homologous, though they have only a 26% identity at the amino acid level, with several gaps present.[445] The *rfbE* gene, which is needed in the conversion of CDP-paratose (**123**) to CDP-tyvelose (**124**) in serogroup D, is also present in serogroup A, though a mutation has rendered the gene inactive in this strain.[445] Thus, the paratose-producing serogroup A seems to have evolved from the tyvelose-producing strains. Similar gene replacements have also been observed in the corresponding *Yersinia* serogroups that produce the same dideoxysugars,[439,440] though there are some differences. For instance, an inactivated *rfbE* gene is not present in *Y. pseudotuberculosis* serogroup IA,[440] suggesting that the paratose-producing strains in *Yersinia* did not derive from the tyvelose-producing strains. Interestingly, there is a nonfunctioning *rfbE* gene in the abequose-producing *Yersinia* strain (serogroup IIA).[439] The deduced RfbS, E, and J proteins from the *Salmonella* and *Yersinia* pathways are 45%, 68%, and ~24% identical.[439,440] Although the G + C contents and codon usage are similar for the corresponding genes in the *S. enterica* and *Y. pseudotuberculosis* strains, the DNA and amino acid sequences have diverged sufficiently so that the

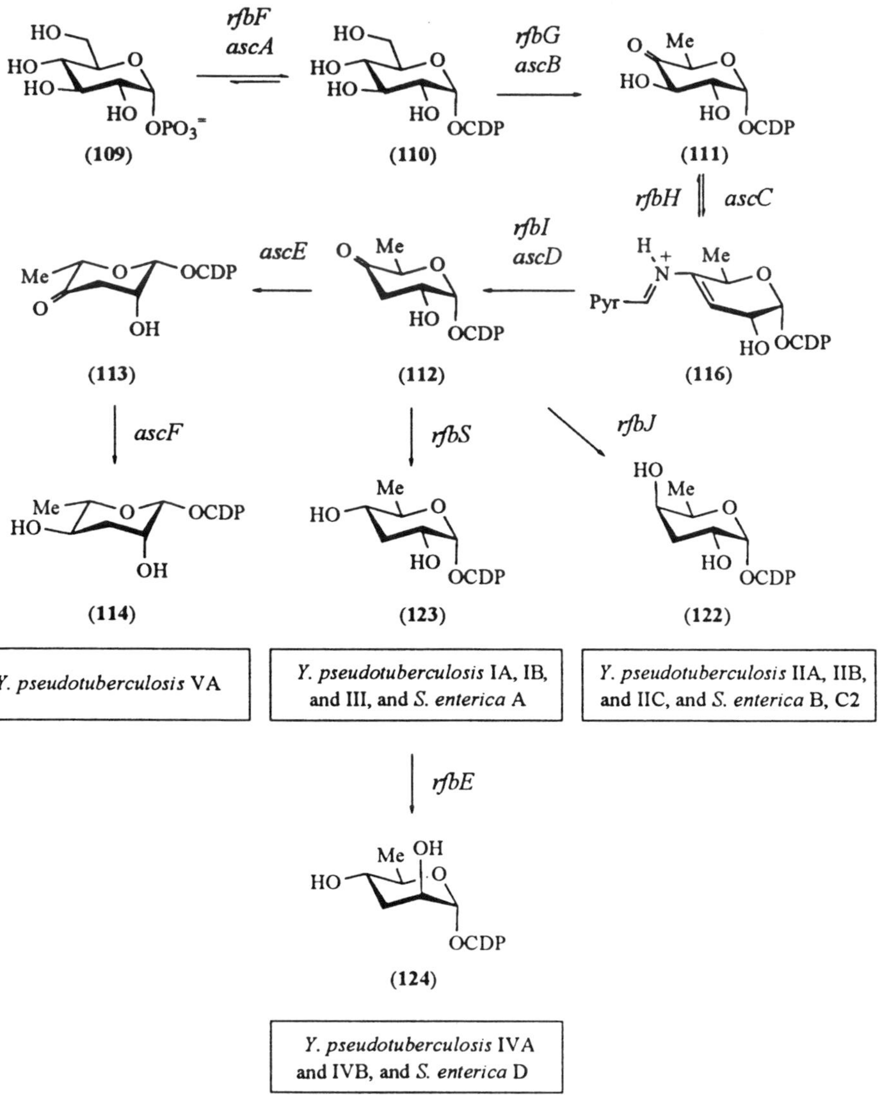

Scheme 13

dideoxyhexose genes may be concluded to have been evolving separately in these two species for quite some time.[439,440,478]

In the ascarylose biosynthetic pathway of *Y. pseudotuberculosis* VA, the gene *ascA* codes for a protein corresponding to that coded by *rfbF*, and *ascB* similarly correlates to *rfbG*.[416] Unlike in the *Salmonella* studies, enzymatic assays allowed unambiguous mapping of *ascC* and *ascD* to E_1 and E_3, respectively. Comparison of these genes from the *asc* cluster with analogous genes from *S. enterica* LT2 revealed an 80% amino acid residue identity for *ascA–rfbF* and 72% for *ascB–rfbG*.[416] Also, there was an 86% residue identity for *ascC–rfbH* and 51% for *ascD–rfbI*, thus supporting the assignment of *rfbH* as the E_1 equivalent and *rfbI* as the E_3 equivalent in the abequose, paratose, and tyvelose biosynthetic pathways in *Salmonella* strains.[416] While an *rfbH* gene was found in *Yersinia* IIA, there was no evidence for an *rfbI* gene.[439] This apparent lack of an E_3 equivalent in the group IIA strains may have interesting implications for the biosynthetic pathway for abequose in *Yersinia* spp. The final two enzymes in the ascarylose pathway, E_{ep} and E_{red}, are encoded by the genes *ascE* and *ascF*, respectively, based on sequence analysis and on evidence produced by a GC–MS assay.[415] As might be expected, no apparent homologues of *ascEF* genes have been found in strains of *S. enterica*.

The biosynthesis of colitose (**54**) is less characterized than that of other 3,6-dideoxysugars, though it is known that GDP-mannose rather than CDP-glucose serves as the precursor.[482] Although most genes in the colitose pathway have not been assigned, genes coding for the synthesis of GDP-mannose have been isolated and sequenced from the colitose-producing *E. coli* O111.[444] These genes are homologous with the GDP-mannose biosynthetic genes (*rfbM* and *rfbK*) in *S. enterica* and in other strains of *E. coli*.[444] Since colitose is the L isomer of abequose, it is expected that the biosynthetic pathways for abequose and colitose should be analogous. Thus, the adjacent open reading frame *orf6.7*, which contains a possible NAD$^+$-binding sequence, may be a dehydrogenase or a reductase in the colitose pathway. Likewise, Orf7.7 could be the E$_1$ equivalent required for colitose formation because it bears some relation to RfbH from *Y. pseudotuberculosis* and *S. enterica*.[444]

Analysis of the *rfb* and *asc* gene clusters revealed that the G + C content is lower than the average for the species,[416,439,444,478] as discussed above for 6-deoxyhexoses (see Section 3.12.5.2.1). Based on these G + C contents, the *rfb* and *asc* genes can be divided into three distinct groups that may have been acquired from different sources and have independent evolutionary histories. The group consisting of *ascABC* and *rfbFGH* has an average G + C content of about 0.43–0.44, and most closely matches that in *Yersinia* and *Salmonella* spp. Assuming the directional mutation is similar for both species, these genes may have been acquired by the organisms first. This is a reasonable assumption given the fact that the *ascAB* and *rfbFG* gene products catalyze the conversion of α-D-glucose-1-phosphate (**109**) to CDP-6-deoxy-L-*threo*-D-*glycero*-4-hexulose (**111**), which is a common precursor for most deoxyhexoses. The inclusion of the *ascC/rfbH* gene, which encodes a PMP-dependent dehydrase (E$_1$) in this group is interesting because Thorson *et al.*[415] have unveiled an evolutionary link between this gene and genes that encode PLP/PMP-dependent transaminases. Therefore, it is possible the *ascC/rfbH* gene product may have been a transaminase involved in aminosugar biosynthesis before it evolved into a dehydrase to catalyze a deoxygenation step in C-3 deoxysugar formation.

The second group of genes (*ascD*, *rfbE*, *rfbKM*, and *rfbI*) that may have been laterally acquired has a G + C content ranging from 0.36 to 0.4. In the third group, which includes *ascEF*, *rfbJ*, and *rfbS*, the ratio is about 0.32, and these genes may be the most recent additions to the genomes of these genera. The unassigned genes from colitose-producing *E. coli* O111 also have a low G + C content ranging from 0.30 to 0.35, but the lack of data prevents the placement of these genes into one of the above groups. Given the homology among several of the genes within these groups, it is likely that interspecific gene transfer accounts for the O-antigen variation. However, not enough information is currently available to adequately elaborate this scenario. The study of additional gene clusters that code for 3,6-dideoxysugar biosynthetic enzymes may provide important details to elucidate the evolutionary relationship between the O antigens of these bacteria. Also, genetic engineering could provide some useful insights into the evolutionary status of 3,6-dideoxysugar biosynthesis, and this technique could potentially produce the three remaining 3,6-dideoxysugar isomers not found in nature.

3.12.5.3 Genetics of Deoxysugars in Antibiotics

Though a large number of antibiotics have been discovered, genetic information is available for only a few of these secondary metabolites. However, this knowledge has become increasingly extensive, and it has shown great potential in the area of hybrid antibiotics development.[483–488] Currently, much of the sequence data pertain to the genes involved in the biosynthesis of the aglycone moiety,[489] though there are several cases in which a partial characterization of the saccharide biosynthetic genes have been reported. These studies have enabled the assignment of some deoxysugar biosynthetic genes and mechanistic speculation on the corresponding gene products. Notably, the proposed biosynthetic pathways of the 2,6- and 4,6-dideoxysugars in antibiotics are based to a large extent on the current knowledge of similar pathways in O-antigen formation (see Section 3.12.5.2). Several of the better-studied examples are described below, and the methodologies utilized should be generally applicable in the genetic investigations of related glycosylated metabolites.

3.12.5.3.1 *Anthracycline antibiotics*

Anthracycline antibiotics such as doxorubicin (**97a**), daunorubicin (**97b**), and nogalamycin each contain a polyketide aglycone and one or more saccharides.[205,222,223] Since the polyketide synthase (PKS) enzymes are highly conserved, a probe derived from the actinorhodin (*act*) PKS genes was designed and used to screen the genome of *Streptomyces peucetius* for the biosynthetic gene cluster of daunorubicin.[447] Subsequent cloning and sequencing efforts combined with selected gene disruption and/or expression experiments have characterized many of the biosynthetic genes.[448,450,490–492] Based on this information, a putative pathway for the formation of the daunosamine moiety has been proposed.

As shown in Figure 3, the first step is catalyzed by a dTDP-glucose pyrophosphorylase possibly encoded by *dnrL*.[449] The adjacent *dnrM* gene encodes a dTDP-glucose-4,6-dehydratase, but expression of this gene yields a nonfunctioning, truncated protein due to a frameshift mutation.[449] Interestingly, there exists another 4,6-dehydratase gene located outside the *dnr* gene cluster,[449,450] and the corresponding protein has been isolated.[373] The *dnrU* and *dnrV* genes may encode a 3,5-epimerase and a 4-ketoreductase, respectively.[355] Study on the *dnrQS* genes[450] showed that *dnrS* appears to encode a glycosyltransferase, and *dnrQ* is needed for daunosamine biosynthesis, though the function of its protein product is currently unknown. Analysis of the *dnrJ* gene revealed that the DnrJ protein is homologous with E₁ (*ascC*) from the ascarylose biosynthetic pathway and is likely to function as a coenzyme B₆-dependent transaminase.[448] While being similar to 3,6-dideoxysugar formation, the pathways for 2,6- and 4,6-dideoxyhexose formation apparently lack an E₃ homologue (*ascD*) in their gene clusters,[403] so the biosyntheses of the various dideoxysugars may be fundamentally distinct. Further enzymatic characterization should provide intriguing insight on this subject.

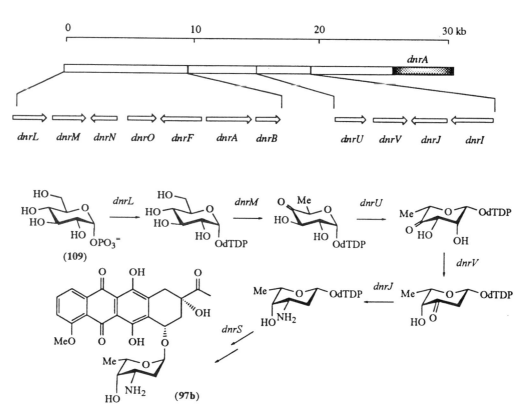

Figure 3 Organization of the *dnr* cluster for daunorubicin (**97b**) biosynthesis in *S. peucetius* and the proposed pathway for the formation of daunosamine. The arrows indicate open reading frames and the direction of transcription within the two enlarged regions.

The daunorubicin biosynthetic gene cluster has also been found in *Streptomyces griseus* JA3933. Sequence analysis of the *lkmB* region revealed that several of the deduced gene products may be homologous to those in the *dnr* cluster.[493] The product of ORF4 resembles the putative 3,5-epimerases from the streptomycin (StrM, see Section 3.12.5.3.4) and rhamnose pathways (RfbC, see Section 3.12.5.2.1),[493] so it may also be related to the *dnrU* gene product. The ORF5 product contains a dinucleotide-binding site at the N-terminus and is distantly homologous to StrE and StrL from the streptomycin pathway,[493] thus it may be involved in daunosamine biosynthesis. Also,

ORF6 encodes a protein which is quite similar to DnrJ from *S. peucetius*.[493] The *dauH* gene from the daunorubicin biosynthetic cluster in *Streptomyces* sp. strain C5 encodes a protein which is 41% identical to the DnrS protein from *S. peucetius*,[451] therefore it has been assigned as a possible glycosyltransferase. Reports have emerged on the partial characterization of the biosynthetic genes of other anthracycline antibiotics such as aclacinomycin,[494] rhodomycin,[495] and nogalamycin,[488] but these have focused on the PKS genes, thus leaving the deoxysugar genes yet to be analyzed.

3.12.5.3.2 Erythromycin

Erythromycin (**129**) is a clinically important macrolide antibiotic with two deoxysugars, L-cladinose (**80**) and D-desosamine (**130**), attached to the aglycone. Genetic studies on the biosynthesis of erythromycin, which is produced by *Saccharopolyspora erythraea* (formerly *Streptomyces erythraeus*), have enabled a significant understanding of the organization and function of the associated genes.[452,496–500,512] As in many antibiotic producers, most of these genes are clustered around the resistance gene (*ermE*). However, the genes associated with deoxysugar biosynthesis are not located consecutively within the *ery* cluster, being separated by the polyketide *eryA* genes (Figure 4). Additionally, in contrast to the macrolide tylosin producer *Streptomyces fradiae* (Figure 5),[454] genes coding for enzymes of the early steps leading to a nucleotide-linked 4-keto-6-deoxyglucose intermediate are not located within the *ery* cluster.[452] Instead, a single copy of the gene coding for a dTDP-glucose-4,6-dehydratase (*gdh*) has been found elsewhere in the genome.[388] The enzyme product, dTDP-4-keto-6-deoxyhexose, is assumed to serve as a common precursor for both L-cladinose and D-desosamine, though this has not yet been confirmed. Presumably, the product of this gene also participates in other biosynthetic pathways required for cellular metabolism, as mutants with a disrupted *gdh* were not viable.[388] The remaining enzymes and their associated genes have been divided into three phenotypic classes based on studies of nonproducing mutants of *S. erythraea*: *eryB* for synthesis or attachment of mycarose (**82**) (the precursor of cladinose), *eryC* for desosamine, and *eryD* for both sugar pathways.[501] Deduced amino acid sequence comparisons and gene disruption studies have permitted pathways for the formation of both mycarose and desosamine to be proposed,[452,453] though the exact order of the steps has not been established. Much of the following discussion is based on the studies by Summers *et al.*[452] Gaisser *et al.* have reported very similar results.[512]

For L-mycarose (**82**), the gene product of *eryBVI* may be a 2,3-dehydratase that converts the 4-keto-6-deoxyglucose to a 3,4-diketo-2,6-dideoxysugar intermediate (**131**). The 3-keto group may be reduced in the next step by the product of *eryBII*, and the eryBVII protein may catalyze the epimerization of the 4-keto 2,6-dideoxyhexose intermediate at the C-5 position. Although its function is still uncertain, *eryBI* might encode the C-methyl transferase that methylates the C-3 position.[453] The reduction at C-4 may be mediated by the *eryBIV* gene product, which shares two conserved sequence motifs with the products of *ascF* from the ascarylose pathway of *Y. pseudotuberculosis*,[416] of *rfbJ* from the abequose pathway in *S. enterica*,[442] and of *strL* from the streptomycin pathway of *S. griseus*.[394] Transfer of the mycarose to the C-3 position of the aglycone may be accomplished by the *eryBV* product, which has a strong sequence identity with the proposed glycosyltransferase encoded by *dnrS* from the daunorubicin pathway.[450] After attachment of mycarose, the product of *eryG* may serve as the O-methyltransferase which converts mycarose to cladinose.[453]

For D-desosamine (**130**), tautomerization of the 4-keto-6-deoxyglucose precursor to a 3-keto sugar may be the first committing step. It has been suggested that this tautomerase may be encoded by *eryCII* or *eryCV*, and one of these genes is also the candidate for a proposed glucoseen reductase gene. The *eryCIV* product may facilitate the C-4 deoxygenation of the 3-keto sugar intermediate (**132**), as this protein is 25% identical to the E_1 dehydratase in the ascarylose pathway of *Y. pseudotuberculosis*, and it is also related to possible transaminases coded by *tylB* from the tylosin pathway and by *dnrJ* from the daunorubicin pathway.[452] Interestingly, the *eryCI* gene product, which was previously suggested to play a regulatory role in the erythromycin pathway,[502] is also related to this group of proteins. One significant difference between EryCI and EryCIV is that EryCI is 61% identical to TylB (see Section 3.12.5.3.3) while EryCIV is only 31% identical.[452] Because of its greater homology to TylB, which has been suggested to be a key component of an aminosugar (mycaminose) biosynthetic pathway, the *eryCI* gene product has been assigned as the C-3 transaminase. Sequence comparison also suggested that N-methylation of the C-3 amino moiety may be catalyzed by the *eryCVI* product, and subsequent transglycosylation of the desosamine to the aglycone may be mediated by the product of *eryCIII*.

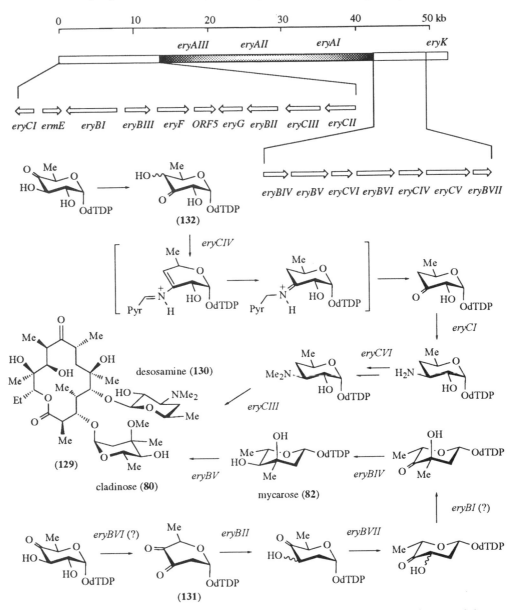

Figure 4 Organization of the *ery* cluster for erythromycin (**129**) biosynthesis in *S. erythraea* and the proposed pathways for the formation of cladinose (**80**) and desosamine (**130**). Open reading frames are denoted by arrows aligned along the direction of transcription.

3.12.5.3.3 *Tylosin*

Tylosin (**133**) is another macrolide antibiotic with medical importance produced by *S. fradiae*. It is composed of a 16-membered branched lactone and three deoxysugars named mycinose (**8**), mycarose (**82**), and mycaminose (**134**). Extensive studies using blocked mutants have mapped and located many of the genes in the tylosin biosynthesis cluster (*tyl*, Figure 5).[500,503] Of these genetic loci, mutations in *tylA* or *tylL* prevented the production or attachment of all three deoxysugars, while defects in either *tylB* or *tylM* only affected mycaminose synthesis or attachment. Mycinose attachment to the hydroxyl anchor group at C-23, which is formed by an oxidase encoded or controlled by *tylH*, occurs after mycaminose is added to the aglycone. Biosynthesis of the mycinose precursor, 6-deoxy-D-allose (**7**), or its attachment to C-23 requires *tylD* and *tylJ*. Enzymes encoded or controlled by *tylE* and *tylF* methylate the hydroxyl groups at positions C-2 and C-3, respectively, of the attached 6-deoxy-D-allose, and convert it to mycinose. Either mycarose synthesis or its linkage to mycaminose is impaired by mutations in *tylC* and *tylK*. Also, mutations in *tylI* result in failure to oxidize *O*-mycaminosyl-tylactone at C-20 of the lactone ring.

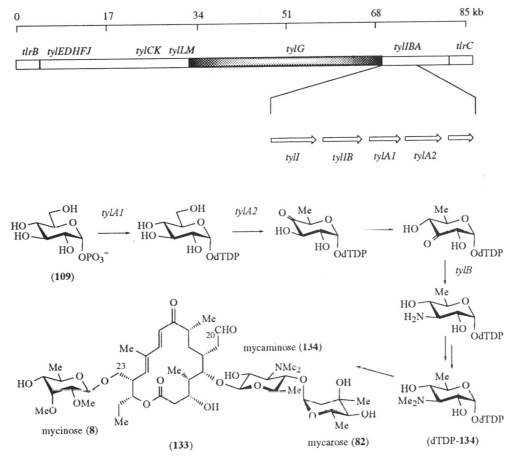

Figure 5 Organization of the *tyl* cluster for tylosin (**133**) biosynthesis in *S. fradiae* and the proposed pathway for the formation of mycaminose (**134**). Open reading frames are denoted by arrows aligned along the direction of transcription. Genetic loci associated with blocked mutants and the various biosynthetic steps are indicated.

Within these genetic loci, the only genes whose sequences have been reported are from the *tylIBA* region.[454] Sequence comparisons of the deduced protein encoded by *tylI* shows that it is a heme-dependent cytochrome P450 enzyme, which explains why *tylI* mutants fail to oxidize the C-20 position of the lactone ring. The *tylB* gene product closely resembles that of *eryCI* (erythromycin),[454] *dnrJ* (daunorubicin),[454] and *pur4* (puromycin),[456] and it has been assigned as a transaminase.[403] A dTDP-glucose pyrophosphorylase is encoded by *tylA1*, as determined by sequence analysis and activity assays of the recombinant protein.[454] The *tylA2* product is a dTDP-glucose-4,6-dehydratase based on strong amino acid identity with the Gdh protein from *S. erythraea* and the StrE protein from *S. griseus*.[454] Since these enzymes are required for the synthesis of deoxysugars, these assignments are consistent with the observed *tylA* mutant phenotype of failure to synthesize or add any of the three deoxysugars to the aglycone.

3.12.5.3.4 Streptomycin

Aminocyclitol antibiotics have also been studied in great detail at the genetic level. In particular, the streptomycin (**135**) biosynthetic clusters (*str*) in both *S. griseus* and *S. glaucescens* have been well characterized.[504] The *str* clusters of these two strains have different genetic organizations, though the individual gene counterparts are significantly homologous. As shown in Figure 6, the genes *strD*, *strE*, *strL*, and *strM* seem to encode proteins involved in the biosynthesis of dTDP-dihydrostreptose (dTDP-(**66**)), which is the precursor of the streptose (**67**) moiety in streptomycin. Specifically, the products of *strD*, *strE*, *strM*, and *strL* may be α-D-glucose-1-phosphate deoxythymidylyltransferase, dTDP-glucose-4,6-dehydratase, dTDP-6-deoxy-L-*threo*-D-*glycero*-4-hexulose-3,5-epimerase, and dTDP-dihydrostreptose synthase, respectively.[394] In the streptidine (**136**) pathway.[455] *strS* is believed to encode a transaminase, due to the similarities between its translated amino acid sequence and that of *ascC* (E₁) in the ascarylose pathway.[403] Interestingly, DNA hybridizing experiments using probes based on genes from the streptomycin pathway have suggested that similar genes may be found in a range of organisms that produce secondary metabolites containing deoxysugars.[505]

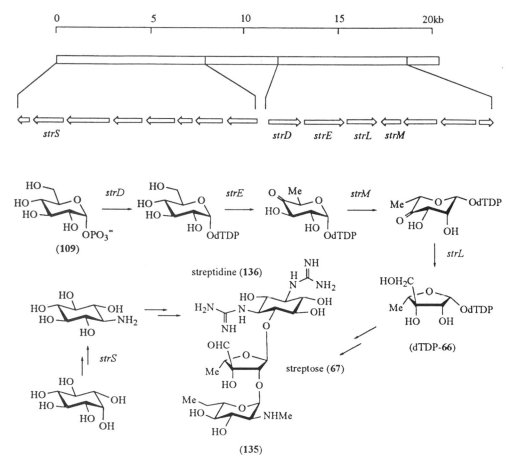

Figure 6 Organization of the *str* cluster for streptomycin (**135**) biosynthesis in *S. griseus* and the proposed pathways for the formation of streptose (**66**) and streptidine (**136**). The arrows indicate open reading frames and the direction of transcription. The putative deoxysugar biosynthetic genes are labeled.

3.12.5.3.5 *Granaticin*

Streptomyces violaceoruber Tü22 produces granaticin (**137**), which is a benzoisochromane quinone antibiotic with a 2,6-dideoxysugar uniquely annealed to the aromatic ring system through two carbon–carbon linkages at C-1 and C-4. Analysis of a gene cluster (*gra*) near the polyketide synthase genes revealed two open reading frames that code for enzymes required for deoxysugar formation.[457] The *graD* gene encodes a dTDP-glucose pyrophosphorylase based on deduced amino acid sequence identity with StrD from *S. griseus*. Similarly, sequence homology with StrE suggested that the *graE* gene product is a dTDP-D-glucose-4,6-dehydratase, and activity assays of the heterologously expressed protein confirmed this hypothesis.[457] Although other enzymes or genes in this pathway have yet to be characterized, a proposed pathway for the formation of the 2,6-dideoxy-D-*threo*-hexos-4-ulose (**138**) precursor for ring fusion can be based upon the C-3 deoxygenation chemistry, as depicted in Scheme 14, path a. This pathway is mechanistically reasonable, and it accounts for the observed loss of hydrogen atoms from C-3 and C-5 of the starting material.[379] An alternative mechanism including a sequence of dehydration and reduction may also yield the dideoxysugar (Scheme 14, path b), but there is currently no evidence to differentiate between the two proposals. Interestingly, loss of the C-2 hydroxyl group in granaticin proceeds with retention of configuration,[379] while C-2 deoxygenation of the D-olivose (**29**) sugar in chlorothricin (**139**) proceeds with inversion of configuration.[506] If the mechanism presented in Scheme 14 is generally applicable to the biosynthesis of other 2,6-dideoxysugars, it appears that enzymes that catalyze C-2 deoxygenations in hexoses may be divided into at least two classes,[104] with each class differing in stereochemistry and possibly in mechanism.

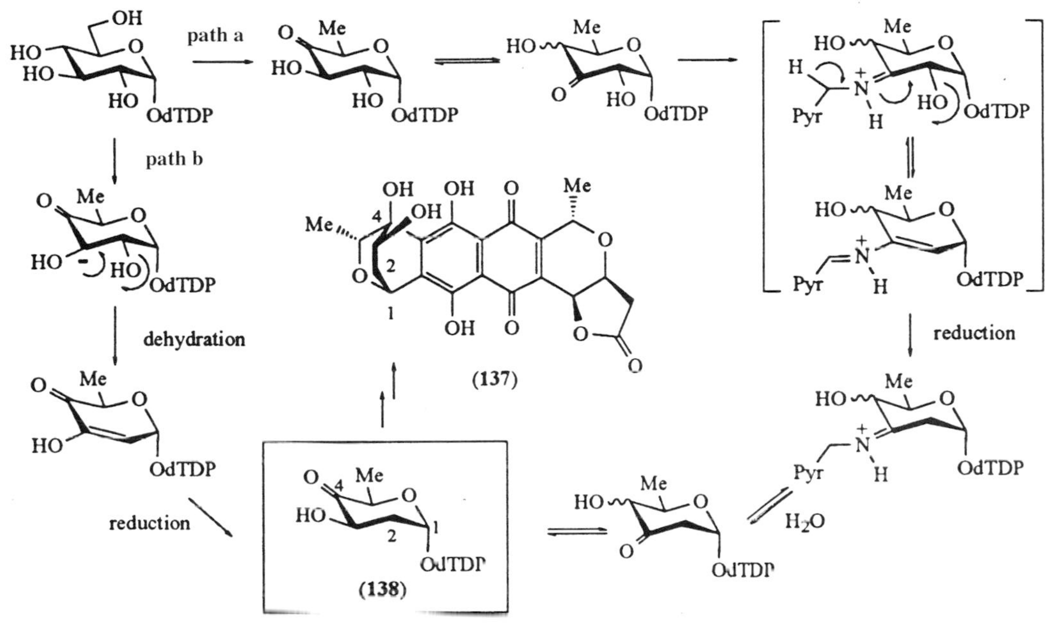

Scheme 14

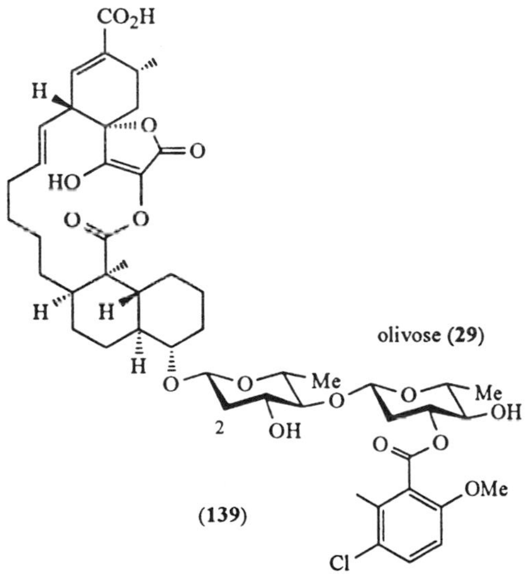

(139)

3.12.5.3.6 Avermectins

Avermectins (**140**) are compounds possessing potent anthelmintic activity and are useful against a broad spectrum of nematodes and arthropods.[82] They exist as a complex of 16-membered macrocyclic lactones containing two L-oleandrose (**31**) units attached to the aglycone.[81] Only a few enzymes involved in the biosynthesis of avermectins have been characterized, so the proposed biosynthetic pathway is mainly based on studies using isotopic precursors and blocked mutants.[507,508] Mutants in the *aveB* phenotypic class are deficient in the synthesis or attachment of the oleandrose units to the aglycone. Functional complementation of a mutant deficient in glycosylation enabled the isolation of a cluster of eight genes (*avrBCDEFGHI*) that are responsible for the biosynthesis and attachment of the oleandrose units.[507,509] In contrast to the multiple loci identified for glycosylation of erythromycin (see Section 3.12.5.3.2) and tylosin (see Section 3.12.5.3.3), all the genes for glycosylation of avermectin aglycones are within a single, contiguous cluster. While the functions of these open reading frames have not been assigned yet, the order of the methylation and glycosyl-transferase steps has been established by biochemical studies. Avermectin derivatives that lacked methyl groups on the oleandrose moiety were not methylated when they were fed to the producing organisms, *Streptomyces avermitilis*.[510] These methyl groups derive from methionine, and they are presumably transferred by an AdoMet-dependent transferase. Also, dTDP-oleandrose was found to be the specific substrate of a glycosyltransferase that catalyzes the stepwise addition of oleandrose to avermectin aglycones.[511] Therefore, methylation of the sugar units in avermectins is not a terminal step, in contrast to the biosynthesis of erythromycin and tylosin (see Sections 3.12.5.3.2 and 3.12.5.3.3, respectively).

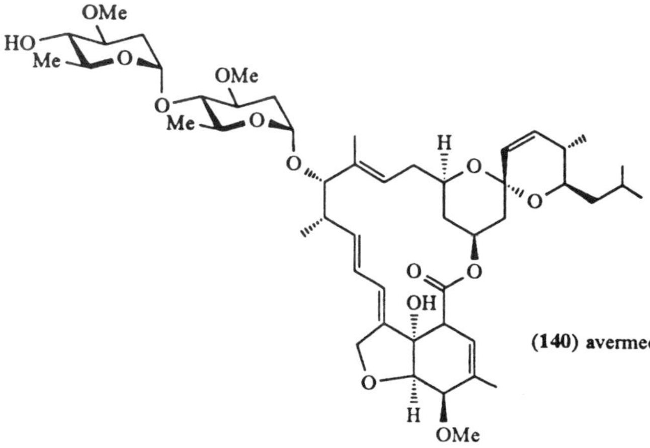

(140) avermectin A1a

3.12.6 CONCLUDING REMARKS

Deoxysugars are important structural components in numerous natural products, including glycoproteins, bacterial endotoxins, and secondary metabolites. Efforts to understand the roles of deoxysugars have unveiled an exciting array of diverse functions for these compounds. For instance, they are involved with intercellular communications, immunogenic responses to pathogenic bacteria, and the biological activity of many antibiotics. Alterations to the number and/or identity of deoxysugars present in these natural products have considerable repercussions on their activity. In fact, many disease states are being attributed to the aberrant elaboration of deoxysugars, and antibiotics with enhanced activity have been produced by modifying the glycosidic moiety of these compounds. Thus, understanding both the genetic and mechanistic aspects of deoxysugar biosynthesis is a critical goal in the design of effective therapeutic strategies and in the development of useful drugs.

The mechanistic studies on the biosynthesis of deoxyribonucleic acids, 6-deoxyhexoses, and 3,6-dideoxyhexoses are a significant stride toward this goal. These efforts have highlighted the elegance and complexities involved with the natural formation of these compounds, and they provide a basis for the continued study of less characterized systems, such as 2,6- and 4,6-dideoxysugar biosynthesis. Genetic studies have been similarly important, and they hold great promise for advancing the pace of the mechanistic studies. Furthermore, genetic engineering techniques are essential for the construction of recombinant strains that overproduce clinically relevant microbial metabolites, including hybrid antibiotics with novel activity or lower toxicity. As mentioned throughout the chapter, much work remains to be done concerning the characterization of deoxysugar function and biosynthesis. Thus, future investigations in the field of deoxysugars have tremendous potential for generating new medical products as well as revealing some intriguing mechanistic information.

3.12.7 REFERENCES

1. B. Lindberg, *Adv. Carbohydr. Chem. Biochem.*, 1990, **48**, 279.
2. H. J. Allen and E. C. Kisailus (eds.), "Glycoconjugates: Composition, Structure, and Function," Dekker, New York, 1992.
3. J. Montreuil, J. F. G. Vleigenthart, and H. Schachter (eds.), "Glycoproteins," Elsevier, Amsterdam, 1995.
4. H. Wiegandt (ed.), "Glycolipids," Elsevier, Amsterdam, 1985.
5. A. Varki, *Glycobiology*, 1993, **3**, 97.
6. A. Alavi and J. S. Axford (eds.), "Glycoimmunology," Plenum Press, New York, 1995.
7. J. Montreuil, J. F. G. Vliegenthart, and H. Schachter (eds.), "Glycoproteins and Disease," Elsevier, Amsterdam, 1996.
8. S. Hanessian, *Adv. Carbohydr. Chem.*, 1966, **21**, 143.
9. S. Hanessian and T. H. Haskell, in "The Carbohydrates: Chemistry and Biochemistry," eds. W. Pigman and D. Horton, Academic Press, New York, 1970, vol. IIA, p. 139.
10. J. E. Courtois and F. Percheron, in "The Carbohydrates: Chemistry and Biochemistry," eds. W. Pigman and D. Horton, Academic Press, New York, 1970, vol. IIA, p. 213.
11. N. R. Williams and J. D. Wander, in "The Carbohydrates: Chemistry and Biochemistry," eds. W. Pigman and D. Horton, Academic Press, New York, 1980, vol. 1B, p. 761.
12. R. Schaffer, in "The Carbohydrates: Chemistry and Biochemistry," eds. W. Pigman and D. Horton, Academic Press, New York, 1972, vol. 1A, p. 69.
13. J. Yoshimura, *Adv. Carbohydr. Chem. Biochem.*, 1984, **42**, 69.
14. T. Reichstein and E. Weiss, *Adv. Carbohydr. Chem.*, 1962, **17**, 65.
15. W. G. Overend and M. Stacey, *Adv. Carbohydr. Chem.*, 1953, **8**, 45.
16. H. R. Bentley, K. G. Cunningham, and F. S. Spring, *J. Chem. Soc.*, 1951, 2301.
17. W. P. Roberts, M. E. Tate, and A. Kerr, *Nature (London)*, 1977, **265**, 379.
18. J. Keleti, H. Mayer, I. Fromme, and O. Lüderitz, *Eur. J. Biochem.*, 1970, **16**, 284.
19. E. Romanowska, A. Romanowska, C. Lugowski, and E. Katzenellenbogen, *Eur. J. Biochem.*, 1981, **121**, 119.
20. H. Achenbach, W. Regel, and W. Karl, *Chem. Ber.*, 1975, **108**, 2481.
21. S. Ōmura, A. Nakagawa, A. Neszmélyi, S. D. Gero, A.-M. Sepulchre, F. Piriou, and G. Lukacs, *J. Am. Chem. Soc.*, 1975, **97**, 4001.
22. J. Hoffman, B. Lindberg, and R. R. Brubaker, *Carbohydr. Res.*, 1980, **78**, 212.
23. R. P. Gorshkova, E. N. Kalmykova, V. V. Isakov, and Y. S. Ovodov, *Eur. J. Biochem.*, 1985, **150**, 527.
24. A. M. Spichtig and A. Vasella, *Helv. Chim. Acta*, 1971, **54**, 1191.
25. E. Percival, *Methods Carbohydr. Chem.*, 1962, **1**, 195.
26. E. Percival and M. Young, *Carbohydr. Res.*, 1974, **32**, 195.
27. E. Akita, K. Maeda, and H. Umezawa, *J. Antibiot. (Tokyo)*, Ser. A, 1964, **17**, 200.
28. T. Yamauchi and F. Abe, *Chem. Pharm. Bull.*, 1990, **38**, 669.
29. T. Yamauchi and F. Abe, *Chem. Pharm. Bull.*, 1990, **38**, 1140.
30. W. D. Ollis, C. Smith, and D. E. Wright, *J. Chem. Soc., Chem. Commun.*, 1974, 881.
31. W. D. Ollis, C. Smith, I. O. Sutherland, and D. E. Wright, *J. Chem. Soc., Chem. Commun.*, 1976, 350.
32. W. D. Ollis, C. Smith, and D. E. Wright, *J. Chem. Soc., Chem. Commun.*, 1976, 348.
33. W. D. Ollis, C. Smith, and D. E. Wright, *Tetrahedron*, 1979, **35**, 105.
34. A. K. Ganguly and A. K. Saksena, *J. Chem. Soc., Chem. Commun.*, 1973, 531.
35. A. K. Ganguly and A. K. Saksena, *J. Antibiot.*, 1975, **28**, 707.
36. A. K. Ganguly and S. Szmulewicz, *J. Antibiot.*, 1975, **28**, 710.
37. A. K. Ganguly, O. Z. Sarre, D. Greeves, and J. Morton, *J. Am. Chem. Soc.*, 1975, **97**, 1982.
38. A. K. Ganguly, S. Szmulewicz, O. Z. Sarre, and V. M. Girijavallabhan, *J. Chem. Soc., Chem. Commun.*, 1976, 609.
39. A. K. Ganguly, in "Topics in Antibiotic Chemistry," ed. P. G. Sammes, Wiley, New York, 1978, vol. 2, p. 59.
40. F. Buzzetti, F. Eisenberg, H. N. Grant, W. Keller-Schierlein, W. Voser, and H. Zähner, *Experientia*, 1968, **24**, 320.
41. J. L. Mertz, J. S. Peloso, B. J. Barker, G. E. Babbitt, J. L. Occolowitz, V. L. Simson, and R. M. Kline, *J. Antibiot.*, 1986, **39**, 877.
42. E. G. Gros, *Carbohydr. Res.*, 1966, **2**, 56.
43. E. G. Gros, V. Deulofeu, O. L. Galmarini, and B. Frydman, *Experientia*, 1968, **24**, 323.
44. E. Finamore, L. Minale, R. Riccio, G. Rinaldo, and F. Zollo, *J. Org. Chem.*, 1991, **56**, 1146.
45. A. J. Roccatagliata, M. S. Maier, A. M. Seldes, M. Iorizzi, and L. Minale, *J. Nat. Prod.*, 1994, **57**, 747.
46. R. Rees, C. R. Gavilanes, W. Meier, A. Fürst, and K. Meyer, *Helv. Chim. Acta*, 1961, **44**, 1607.
47. E. N. Kalmykova, R. P. Gorshkova, V. V. Isakov, and Y. S. Ovodov, *Bioorg. Khim.*, 1988, **14**, 652.
48. H. B. Borén, K. Eklind, P. J. Garegg, B. Lindberg, and Å. Pilotti, *Acta Chem. Scand.*, 1972, **26**, 4143.
49. K. Samuelsson, B. Lindberg, and R. R. Brubaker, *J. Bacteriol.*, 1974, **117**, 1010.
50. A. Markovitz, *J. Biol. Chem.*, 1962, **237**, 1767.
51. Y. A. Knirel, A. S. Shaskhov, B. A. Dimitriev, N. K. Kochetkov, N. V. Kasyanchuk, and I. Y. Zakharova, *Bioorg. Khim.*, 1980, **6**, 1851.
52. S. N. Senchenkova, A. S. Shashkov, Y. A. Knirel, J. J. McGovern, and A. P. Moran, *Eur. J. Biochem.*, 1996, **239**, 434.
53. H. Björndal, B. Lindberg, and W. Nimmich, *Acta Chem. Scand.*, 1970, **24**, 3414.
54. J. Weckesser, H. Mayer, and G. Drews, *Eur. J. Biochem.*, 1970, **16**, 158.
55. C. Villé and M. Gastambide-Odier, *Carbohydr. Res.*, 1970, **12**, 97.
56. M. D. Lee, T. S. Dunne, M. M. Siegel, C. C. Chang, G. O. Morton, and D. B. Borders, *J. Am. Chem. Soc.*, 1987, **109**, 3464.

57. M. D. Lee, T. S. Dunne, C. C. Chang, M. M. Siegel, G. O. Morton, G. A. Ellestad, W. J. McGahren, and D. B. Borders, *J. Am. Chem. Soc.*, 1992, **114**, 985.
58. J. Weckesser, H. Mayer, and I. Fromme, *Biochem. J.*, 1973, **135**, 293.
59. P. Studer, S. K. Pavanaram, C. R. Gavilanes, H. Linde, and K. Meyer, *Helv. Chim. Acta*, 1963, **46**, 23.
60. M. Miyamoto, Y. Kawamatsu, M. Shinohara, K. Nakanishi, Y. Nakadaira, and N. S. Bhacca, *Tetrahedron Lett.*, 1964, 2371.
61. M. Miyamoto, Y. Kawamatsu, K. Kawashima, M. Shinohara, K. Tanaka, S. Tatsuoka, and K. Nakanishi, *Tetrahedron*, 1967, **23**, 421.
62. M. Miyamoto, K. Morita, Y. Kawamatsu, K. Kawashima, and K. Nakanishi, *Tetrahedron*, 1967, **23**, 411.
63. R. Riccio and K. Nakanishi, *J. Org. Chem.*, 1982, **47**, 4589.
64. Y. A. Berlin, S. E. Esipov, M. N. Kolosov, M. M. Shemyakin, and M. G. Brazhnikova, *Tetrahedron Lett.*, 1964, 1323.
65. Y. A. Berlin, S. E. Esipov, M. N. Kolosov, M. M. Shemyakin, and M. G. Brazhnikova, *Tetrahedron Lett.*, 1964, 3513.
66. J. Thiem and B. Meyer, *Tetrahedron*, 1981, **37**, 551.
67. Y. A. Berlin, O. A. Kiseleva, M. N. Kolosov, M. M. Shemyakin, V. S. Soifer, I. V. Vasina, I. V. Yartseva, and V. D. Kuznetow, *Nature (London)*, 1968, **218**, 193.
68. G. P. Bakhaeva, Y. A. Berlin, E. F. Boldyreva, O. A. Chupronova, M. N. Kolosov, V. S. Soifer, T. E. Vasiljeva, and I. V. Yartseva, *Tetrahedron Lett.*, 1968, 3595.
69. J. Thiem and G. Schneider, *Angew. Chem., Int. Ed. Engl.*, 1983, **22**, 58.
70. Y. A. Berlin, M. N. Kosolov, and I. V. Yartseva, *Khim. Prir. Soedin.*, 1973, **9**, 539 (*Chem. Abstr.*, 1974, **80**, 27 439).
71. F. W. Lichtenthaler and T. Kulikowski, *J. Org. Chem.*, 1976, **41**, 600.
72. T. Henkel, J. Rohr, J. M. Beale, and L. Sohwenen, *J. Antibiot.*, 1990, **43**, 492.
73. J. Rohr, J. M. Beale, and H. G. Floss, *J. Antibiot.*, 1989, **42**, 1151.
74. Y. Hayakawa, K. Furihata, H. Seto, and N. Ōtake, *Tetrahedron Lett.*, 1985, **26**, 3475.
75. W. Keller-Schierlein, R. Muntwyler, W. Pache, and H. Zähner, *Helv. Chim. Acta*, 1969, **52**, 127.
76. R. N. Tursunova, V. A. Maslennikova, and N. K. Abubakirov, *Khim. Prir. Soedin.*, 1975, **11**, 171 (*Chem. Abstr.*, 1975, **83**, 114 803).
77. G. F. Pauli, P. Junior, S. Berger, and U. Matthiesen, *J. Nat. Prod.*, 1993, **56**, 67.
78. G. F. Pauli, *J. Nat. Prod.*, 1995, **58**, 483.
79. R. Aquino and C. Pizza, *J. Nat. Prod.*, 1995, **58**, 672.
80. F. Abe, Y. Mori, H. Okabe, and T. Yamauchi, *Chem. Pharm. Bull.*, 1994, **42**, 1777.
81. G. Albers-Schönberg, B. H. Arison, J. C. Chabala, A. W. Douglas, P. Eskola, M. H. Fisher, A. Lusi, H. Mrozik, J. L. Smith, and R. L. Tolman, *J. Am. Chem. Soc.*, 1981, **103**, 4216.
82. H. G. Davies and R. H. Green, *Nat. Prod. Rep.*, 1986, **3**, 87.
83. T. Sasaki, K. Furihata, H. Nakayama, H. Seto, and N. Ōtake, *Tetrahedron Lett.*, 1986, **27**, 1603.
84. H. Els, W. D. Celmer, and K. Murai, *J. Am. Chem. Soc.*, 1958, **80**, 3777.
85. T. Oki, I. Kitamura, Y. Matsuzawa, N. Shibamoto, T. Ogasawara, A. Yoshimoto, T. Inui, H. Naganawa, T. Takeuchi, and H. Umezawa, *J. Antibiot.*, 1979, **32**, 801.
86. J. Golik, J. Clardy, G. Dubay, G. Groenewold, H. Kawaguchi, M. Konishi, B. Krishnan, H. Ohkuma, K.-i. Saitoh, and T. W. Doyle, *J. Am. Chem. Soc.*, 1987, **109**, 3461.
87. J. Golik, G. Dubay, G. Groenewold, H. Kawaguchi, M. Konishi, B. Krishnan, H. Ohkuma, K.-i. Saitoh, and T. W. Doyle, *J. Am. Chem. Soc.*, 1987, **109**, 3462.
88. H. Kawai, Y. Hayakawa, M. Nakagawa, K. Furihata, H. Seto, and N. Ōtake, *Tetrahedron Lett.*, 1984, **25**, 1937.
89. H. Kawai, Y. Hayakawa, M. Nakagawa, K. Furihata, H. Seto, and N. Ōtake, *J. Antibiot.*, 1987, **40**, 1273.
90. K. Ishii, Y. Nishimura, S. Kondo, and H. Umezawa, *J. Antibiot.*, 1983, **36**, 454.
91. M. Miyamoto, Y. Kawamatsu, M. Shinohara, Y. Asahi, Y. Nakadaira, H. Kakisawa, K. Nakanishi, and N. S. Bhacca, *Tetrahedron Lett.*, 1963, 693.
92. J. Thiem and B. Meyer, *J. Chem. Soc., Perkin Trans. 2*, 1979, 1331.
93. H. Brockmann, B. Scheffer, and C. Stein, *Tetrahedron Lett.*, 1973, 3699.
94. H. Brockmann and H. Greve, *Tetrahedron Lett.*, 1975, 831.
95. A. Zeeck, *Justus Liebigs Ann. Chem.*, 1975, 2079.
96. N. P. Maksyutina, *Khim. Prir. Soedin.*, 1975, **11**, 603 (*Chem. Abstr.*, 1976, **84**, 74 570).
97. A. K. Mallams, M. S. Puar, and R. R. Rossman, *J. Am. Chem. Soc.*, 1981, **103**, 3938.
98. F. Tomita, T. Tamaoki, K. Shirahata, M. Kasai, M. Morimoto, S. Ohkubo, K. Mineura, and S. Ishii, *J. Antibiot.*, 1980, **33**, 668.
99. K. Kobinata, M. Uramoto, T. Mizuno, and K. Isono, *J. Antibiot.*, 1980, **33**, 244.
100. H. Takai, H. Yuki, and K. Takiura, *Tetrahedron Lett.*, 1975, 3647.
101. J. S. Brimacombe, A. S. Mengech, and L. C. N. Tucker, *J. Chem. Soc., Perkin Trans. 1*, 1977, 643.
102. A. S. Shashkov, D. V. Iashunskii, I. V. Zhdanovich, and G. B. Lokskin, *Bioorg. Khim.*, 1991, **17**, 410.
103. W. Keller-Schierlein, M. Brufani, and S. Barcza, *Helv. Chim. Acta*, 1968, **51**, 1257.
104. H. G. Floss and J. M. Beale, *Angew. Chem., Int. Ed. Engl.*, 1989, **28**, 146.
105. O. Westphal and O. Lüderitz, *Angew. Chem.*, 1960, **72**, 881.
106. J. Hoffmann, B. Lindberg, T. Hofstad, and N. Skaug, *Carbohydr. Res.*, 1978, **66**, 67.
107. A. J. Reason, L. A. Ellis, J. A. Appleton, N. Wisnewski, R. B. Grieve, M. McNeil, D. L. Wassom, H. R. Morris, and A. Dell, *Glycobiology*, 1994, **4**, 593.
108. T. A. Chowdhury, P. E. Jansson, B. Lindberg, J. Lindberg, B. Gustafsson, and T. Holme, *Carbohydr. Res.*, 1991, **215**, 303.
109. C. Fouquey, J. Polonsky, and E. Lederer, *Bull. Soc. Chim. Biol.*, 1957, **39**, 101.
110. E. Lederer, *Bull. Soc. Chim. Biol.*, 1960, **42**, 1367.
111. V. V. Isakov, R. P. Gorshkova, S. V. Tomshich, Y. S. Ovodov, and A. S. Shashkov, *Bioorg. Khim.*, 1981, **7**, 559.
112. R. P. Gorshkova, N. A. Komandrova, A. Kalinovsky, and Y. S. Ovodov, *Eur. J. Biochem.*, 1980, **107**, 131.
113. B. Lindberg, F. Lindh, J. Lonngren, A. A. Lindberg, and S. B. Svenson, *Carbohydr. Res.*, 1981, **97**, 105.
114. K. Hisatsune, S. Kondo, Y. Isshiki, T. Iguchi, Y. Kawamata, and T. Shimada, *Biochem. Biophys. Res. Commun.*, 1993, **196**, 1309.

115. Y. A. Knirel, L. Paredes, P. E. Jansson, A. Weintraub, G. Widmalm, and M. J. Albert, *Eur. J. Biochem.*, 1995, **232**, 391.
116. Y. A. Knirel, S. N. Senchenkova, P. E. Jansson, A. Weintraub, M. Ansaruzzaman, and M. J. Albert, *Eur. J. Biochem.*, 1996, **238**, 160.
117. F. Abe, T. Yamauchi, T. Fujioka, and K. Mihashi, *Chem. Pharm. Bull.*, 1986, **34**, 2774.
118. P. W. K. Woo, H. W. Dion, and Q. R. Bartz, *J. Am. Chem. Soc.*, 1961, **83**, 3352.
119. P. W. K. Woo, H. W. Dion, and L. F. Johnson, *J. Am. Chem. Soc.*, 1962, **84**, 1066.
120. E. Gäumann, R. Hütter, W. Keller-Schierlein, L. Neipp, V. Prelog, and H. Zähner, *Helv. Chim. Acta*, 1960, **43**, 601.
121. W. Keller-Schierlein and G. Roncari, *Helv. Chim. Acta*, 1962, **45**, 138.
122. C. L. Stevens, K. Nagarajan, and T. H. Haskell, *J. Org. Chem.*, 1962, **27**, 2991.
123. E. J. Czerwinski and L. K. Steinrauf, *Biochem. Biophys. Res. Commun.*, 1971, **45**, 1284.
124. T. J. Petcher and H.-P. Weber, *J. Chem. Soc., Chem. Commun.*, 1974, 697.
125. K. Ohta, E. Mizuta, H. Okazaki, and T. Kishi, *Chem. Pharm. Bull.*, 1984, **32**, 4350.
126. H. Irie, Y. Mizuno, I. Kouno, T. Nagasawa, Y. Tani, H. Yamada, T. Taga, and K. Osaki, *J. Chem. Soc., Chem. Commun.*, 1983, 174.
127. T. Nagasawa, H. Fukao, H. Irie, and H. Yamada, *J. Antibiot.*, 1984, **37**, 693.
128. H. Brockmann and T. Waehneldt, *Naturwissenschaften*, 1963, **50**, 43.
129. K. L. Rinehart, Jr., and D. B. Borders, *J. Am. Chem. Soc.*, 1963, **85**, 4037.
130. T. Henkel, S. Breiding-Mack, A. Zeeck, S. Grabley, P. E. Hammann, K. Hütter, G. Till, R. Thiericke, and J. Wink, *Liebigs Ann. Chem.*, 1971, 575.
131. A. K. Ganguly, O. Z. Sarre, A. T. McPhail, and W. Miller, *J. Chem. Soc., Chem. Commun.*, 1979, 22.
132. B. Bannister and A. D. Argoudelis, *J. Am. Chem. Soc.*, 1963, **85**, 234.
133. T. Miyaki, H. Tsukiura, M. Wakae, and H. Kawaguchi, *J. Antibiot., Ser. A*, 1962, **15**, 15.
134. R. U. Lemieux and M. L. Wolfrom, *Adv. Carbohydr. Chem.*, 1948, **3**, 337.
135. M. G. Brazhnikova, M. K. Kudinova, V. V. Kuliaeva, N. P. Potapova, and V. I. Ponomalenko, *Antibiotiki*, 1977, **22**, 967.
136. V. V. Kuliaeva, M. K. Kudinova, N. P. Potapova, L. M. Rubasheva, M. G. Brazhnikova, B. V. Rosynov, and A. R. Bekker, *Bioorg. Khim.*, 1978, **4**, 1087.
137. S. Schramek, J. Radziejewska-Lebrecht, and H. Mayer, *Eur. J. Biochem.*, 1985, **148**, 455.
138. H. Mayer, J. Radziejewska-Lebrecht, and S. Schramek, *Adv. Exp. Med. Biol.*, 1988, **228**, 577.
139. J. W. Hinman, H. Hoeksema, E. L. Caron, and W. G. Jackson, *J. Am. Chem. Soc.*, 1956, **78**, 1072.
140. E. Walton, J. O. Rodin, C. H. Stammer, F. W. Holly, and K. Folkers, *J. Am. Chem. Soc.*, 1958, **80**, 5168.
141. P. F. Wiley, F. A. MacKeller, E. L. Caron, and R. B. Kelly, *Tetrahedron Lett.*, 1968, 663.
142. P. F. Wiley, R. B. Kelly, E. L. Caron, V. H. Wiley, J. H. Johnson, F. A. MacKeller, and S. A. Mizsak, *J. Am. Chem. Soc.*, 1977, **99**, 542.
143. S. Okuda, N. Suzuki, and S. Suzuki, *J. Biol. Chem.*, 1967, **242**, 958.
144. M. Funabashi, S. Yamazaki, and J. Yoshimura, *Carbohydr. Res.*, 1975, **44**, 275.
145. W. Keller-Schierlein, W. Heilman, W. D. Ollis, and C. Smith, *Helv. Chim. Acta*, 1979, **62**, 7.
146. W. Heilmann, E. Kupfer, W. Keller-Schierlein, H. Zähner, H. Wolf, and H. H. Peter, *Helv. Chim. Acta*, 1979, **62**, 1.
147. L. Ninet, F. Benazet, Y. Charpentie, M. Dubost, J. Florent, J. Lunel, D. Maney, and J. Preud'homme, *Experientia*, 1974, **30**, 1270.
148. H. Yonehara and S. Takeuchi, *J. Antibiot., Ser. A*, 1958, **11**, 254.
149. M. Miyamoto, Y. Kawamatsu, M. Shinohara, Y. Nakadaira, and K. Nakanishi, *Tetrahedron*, 1966, **22**, 2785.
150. A. K. Ganguly and O. Z. Sarre, *J. Chem. Soc., Chem. Commun.*, 1969, 1149.
151. A. K. Ganguly, O. Z. Sarre, and S. Szmulewicz, *J. Chem. Soc., Chem. Commun.*, 1971, 746.
152. P. F. Wiley and O. Weaver, *J. Am. Chem. Soc.*, 1955, **77**, 3422.
153. T. Kaneda, J. C. Butte, Jr., S. B. Taubman, and J. W. Corcoran, *J. Biol. Chem.*, 1962, **237**, 322.
154. P. P. Regna, F. A. Hochstein, R. L. Wagner, Jr., and R. B. Woodward, *J. Am. Chem. Soc.*, 1953, **75**, 4625.
155. F. A. Hochstein and K. Murai, *J. Am. Chem. Soc.*, 1954, **76**, 5080.
156. F. W. Tanner, A. R. English, T. M. Lees, and J. B. Routien, *Antibiot. Chemother.*, 1952, **2**, 441.
157. R. E. Paul and S. Tchelitcheff, *Bull. Soc. Chim. Fr.*, 1957, 443.
158. R. E. Paul and S. Tchelitcheff, *Bull. Soc. Chim. Fr.*, 1960, 150.
159. R. Corbaz, L. Ettlinger, E. Gäumann, W. Keller-Schierlein, F. Kradolfer, F. Kyburz, L. Neipp, V. Prelog, A. Wettstein, and H. Zähner, *Helv. Chim. Acta*, 1956, **39**, 304.
160. R. L. Hamill, M. E. Haney, Jr., M. Stamper, and P. F. Wiley, *Antibiot. Chemother.*, 1961, **11**, 328.
161. R. Hütter, W. Keller-Schierlein, and H. Zähner, *Arch. Mikrobiol.*, 1961, **39**, 158.
162. T. Watanabe, N. Nishida, and K. Satake, *Bull. Chem. Soc. Jpn.*, 1961, **34**, 1285.
163. W. Hofheinz and H. Grisebach, *Z. Naturforsch.*, 1962, **B17**, 852.
164. J. Majer, J. R. Martin, R. S. Egan, and J. W. Corcoran, *J. Am. Chem. Soc.*, 1977, **99**, 1620.
165. K. Kawahara, T. Yoshida, T. Watanabe, K. Miyauchi, B. Nomiya, S. Tada, and S. Kuwahara, *Chemotherapy (Tokyo)*, 1972, **20**, 633.
166. A. Kinumaki, I. Takamori, Y. Sugawara, M. Suzuki, and T. Okuda, *J. Antibiot.*, 1974, **27**, 107.
167. A. Kinumaki, I. Takamori, Y. Sugawara, Y. Seki, and M. Suzuki, *J. Antibiot.*, 1974, **27**, 117.
168. M. Muroi, M. Izawa, and T. Kishi, *Chem. Pharm. Bull.*, 1976, **24**, 450.
169. M. Muroi, M. Izawa, and T. Kishi, *Chem. Pharm. Bull.*, 1976, **24**, 463.
170. J. E. Leet, D. R. Schroeder, S. J. Hofstead, J. Golik, K. L. Colson, S. Huang, S. E. Klohr, T. W. Doyle, and J. A. Matson, *J. Am. Chem. Soc.*, 1992, **114**, 7946.
171. W. Keller-Schierlein and G. Roncari, *Helv. Chim. Acta*, 1962, **45**, 138.
172. W. Keller-Schierlein and G. Roncari, *Helv. Chim. Acta*, 1964, **47**, 78.
173. G. Roncari and W. Keller-Schierlein, *Helv. Chim. Acta*, 1966, **49**, 705.
174. F. Arcamone, G. Barbieri, G. Franceschi, S. Penco, and A. Vigevani, *J. Am. Chem. Soc.*, 1973, **95**, 2008.
175. U. Matern, H. Grisebach, W. Karl, and H. Achenbach, *Eur. J. Biochem.*, 1972, **29**, 1.
176. U. Matern and H. Grisebach, *Eur. J. Biochem.*, 1972, **29**, 5.

177. A. Tulinsky, *J. Am. Chem. Soc.*, 1964, **86**, 5368.
178. J. S. Webb, R. W. Broschard, D. B. Cosulich, J. H. Mowat, and J. E. Lancaster, *J. Am. Chem. Soc.*, 1962, **84**, 3183.
179. D. B. Cosulich, J. H. Mowat, R. W. Broschard, J. B. Patrick, and W. E. Meyer, *Tetrahedron Lett.*, 1963, 453.
180. R. P. Gorshkova, V. A. Zubkov, V. V. Isakov, and Yu. S. Ovodov, *Carbohydr. Res.*, 1984, **126**, 308.
181. R. P. Gorshkova, V. V. Isakov, V. A. Zubkov, and Yu. S. Ovodov, *Bioorg. Khim.*, 1989, **15**, 1627.
182. R. P. Gorshkova, V. V. Isakov, V. A. Zubkov, and I. S. Ovodov, *Bioorg. Khim.*, 1994, **20**, 1231.
183. A. Sonesson, E. Jantzen, K. Bryn, T. Tangen, J. Eng, and U. Zähringer, *Microbiol.*, 1994, **140**, 1261.
184. R. P. Gorshkova, V. A. Zubkov, V. V. Isakov, and I. S. Ovodov, *Bioorg. Khim.*, 1987, **13**, 1146.
185. V. A. Zubkov, R. P. Gorshkova, and I. S. Ovodov, *Bioorg. Khim.*, 1988, **14**, 65.
186. V. A. Zubkov, R. P. Gorshkova, T. I. Burtseva, V. V. Isakov, and I. S. Ovodov, *Bioorg. Khim.*, 1989, **15**, 187.
187. M. Gilleron, J. Vercauteren, and G. Puzo, *J. Biol. Chem.*, 1993, **268**, 3168.
188. M. Gilleron, J. Vercauteren, and G. Puzo, *Biochemistry*, 1994, **33**, 1930.
189. M. P. Kunstmann, L. A. Mitscher, and N. Bohonos, *Tetrahedron Lett.*, 1966, 839.
190. G. A. Ellestad, M. P. Kunstmann, J. E. Lancaster, L. A. Mitscher, and G. Morton, *Tetrahedron*, 1967, **23**, 3893.
191. S. Mizobuchi, J. Mochizuki, H. Soga, H. Tanba, and H. Inoue, *J. Antibiot. (Jpn.)*, 1986, **39**, 1776.
192. A. K. Ganguly, O. Z. Sarre, and H. Reimann, *J. Am. Chem. Soc.*, 1968, **90**, 7129.
193. A. K. Ganguly, O. Z. Sarre, A. T. McPhail, and K. D. Onán, *J. Chem. Soc., Chem. Commun.*, 1977, 313.
194. Y. Nishimura, K. Ishii, and S. Kondo, *J. Antibiot.*, 1990, **43**, 54.
195. M. Asai, E. Mizuta, K. Mizuno, A. Miyake, and S. Tatsuoka, *Chem. Pharm. Bull.*, 1970, **18**, 1720.
196. J. O. Pezzanite, J. Clardy, P. Y. Lau, G. Wood, D. L. Walker, and B. Fraser-Reid, *J. Am. Chem. Soc.*, 1975, **97**, 6250.
197. D. L. Walker and B. Fraser-Reid, *J. Am. Chem. Soc.*, 1975, **97**, 6251.
198. H. Paulsen and W. Koebernick, *Carbohydr. Res.*, 1977, **56**, 53.
199. B. K. Bhuyen, S. P. Owen, and A. Dietz, *Antimicrob. Agents Chemother.*, 1964, 91.
200. F. Ruesser, *Biochemistry*, 1973, **12**, 1136.
201. S. A. Mizsak, H. Hoeksema, and L. M. Pschigoda, *J. Antibiot.*, 1979, **32**, 771.
202. H. Hoeksema, S. A. Mizsak, L. Baczynskyj, and L. M. Pschigoda, *J. Am. Chem. Soc.*, 1982, **104**, 5173.
203. T. Okuda, Y. Ito, T. Yamaguchi, T. Furumai, M. Suzuki, and M. Tsuruoka, *J. Antibiot.*, 1966, **19**, 85.
204. K. Isono, *J. Antibiot.*, 1988, **41**, 1711.
205. J. W. Lown, *Chem. Soc. Rev.*, 1993, **22**, 165.
206. J. Rohr and R. Thiericke, *Nat. Prod. Rep.*, 1992, **9**, 103.
207. H. Grisebach, *Adv. Carbohydr. Chem. Biochem.*, 1978, **35**, 81.
208. H. Mayer, U. R. Bhat, H. Masoud, J. Radziejewska-Lebrecht, C. Widemann, and J. H. Krauss, *Pure Appl. Chem.*, 1989, **61**, 1271.
209. Yu. S. Ovodov, R. P. Gorshkova, S. V. Tomshich, N. A. Komandrova, V. A. Zubkov, E. N. Kalmykova, and V. V. Isakov, *J. Carbohydr. Chem.*, 1992, **11**, 21.
210. O. Lüderitz, A. M. Staub, and O. Westphal, *Bacteriol. Rev.*, 1966, **30**, 192.
211. F. Kauffmann (ed.), "The Bacteriology of Enterobacteriaceae," 2nd edn., Williams and Wilkins, Baltimore, MD, 1966.
212. A. J. Griffiths and D. B. Davies, *Carbohydr. Polymers*, 1991, **14**, 241.
213. A. J. Griffiths and D. B. Davies, *Carbohydr. Polymers*, 1991, **14**, 339.
214. M. Cygler, D. R. Rose, and D. R. Bundle, *Science*, 1991, **253**, 442.
215. A. Zdanov, Y. Li, D. R. Bundle, S.-j. Deng, C. R. MacKenzie, S. A. Narang, N. M. Young, and M. Cygler, *Proc. Natl. Acad. Sci. USA*, 1994, **91**, 6423.
216. D. R. Bundle, E. Eichler, M. A. J. Gidney, M. Meldal, A. Ragauskas, B. W. Sigurskjold, B. Sinnott, D. C. Watson, M. Yaguchi, and N. M. Young, *Biochemistry*, 1994, **33**, 5172.
217. D. R. Bundle, H. Baumann, J.-R. Brisson, S. M. Gagné, A. Zdanov, and M. Cygler, *Biochemistry*, 1994, **33**, 5183.
218. H. Baumann, E. Altman, and D. R. Bundle, *Carbohydr. Res.*, 1993, **247**, 347.
219. K. Amano, J. C. Williams, S. R. Missler, and V. N. Reinhold, *J. Biol. Chem.*, 1987, **262**, 4740.
220. G. F. Gause in "Antibiotics," eds. J. W. Corcoran and F. E. Hahn, Springer-Verlag, New York, 1974, vol. 3, p. 197.
221. J. D. Skarbek and M. K. Speedie, in "Antitumor Compounds of Natural Origin: Chemistry and Biochemistry," ed. A. Aszalos, CRC Press, Boca Raton, FL, 1981, p. 191.
222. J. W. Lown (ed.), "Anthracycline and Anthracenedione-based Anticancer Agents," Elsevier, New York, 1988.
223. C. R. Hutchinson, in "Genetics and Biochemistry of Antibiotic Production," eds. L. C. Vining and C. Stuttard, Butterworth–Heinemann, Boston, MA, 1995, p. 331.
224. M. D. Lee, G. A. Ellestad, and D. B. Borders, *Acc. Chem. Res.*, 1991, **24**, 235.
225. A. L. Smith and K. C. Nicolaou, *J. Med. Chem.*, 1996, **39**, 2103.
226. D. Kahne, *Chem. Biol.*, 1995, **2**, 7.
227. Y. Wang and R. R. Rando, *Chem. Biol.*, 1995, **2**, 281.
228. Y. Wang, J. Killian, K. Hamasaki, and R. R. Rando, *Biochemistry*, 1996, **36**, 12 338.
229. X. L. Gao and D. J. Patel, *Biochemistry*, 1990, **29**, 10 940.
230. X. L. Gao, P. Mirau, and D. J. Patel, *J. Mol. Biol.*, 1992, **223**, 259.
231. D. C. Ward, E. Reich, and I. H. Goldberg, *Science*, 1965, **149**, 1259.
232. B. C. Baguley, *Mol. Cell. Biochem.*, 1982, **43**, 167.
233. X. L. Gao and D. J. Patel, *Biochemistry*, 1989, **28**, 751.
234. D. L. Banville, M. A. Keniry, M. Kam, and R. H. Shafer, *Biochemistry*, 1990, **29**, 6521.
235. D. L. Banville, M. A. Keniry, and R. H. Shafer, *Biochemistry*, 1990, **29**, 9294.
236. M. W. Van Dyke and P. B. Dervan, *Biochemistry*, 1983, **22**, 2373.
237. K. R. Fox and N. R. Howarth, *Nucleic Acids Res.*, 1985, **13**, 8695.
238. B. M. G. Cons and K. R. Fox, *Nucleic Acids Res.*, 1989, **17**, 5447.
239. D. J. Silva, R. Goodnow, Jr., and D. Kahne, *Biochemistry*, 1993, **32**, 463.
240. D. J. Silva and D. E. Kahne, *J. Am. Chem. Soc.*, 1993, **115**, 7962.
241. Y. Kaziro and M. Kamiyama, *J. Biochem.*, 1967, **62**, 424.
242. W. Behr, K. Honikel, and G. Hartmann, *Eur. J. Biochem.*, 1969, **9**, 82.
243. W. Priebe (ed.), "Anthracycline Antibiotics," American Chemical Society, Washington, DC, 1995.

244. A. H.-J. Wang, G. Ughetto, G. J. Quigley, and A. Rich, *Biochemistry*, 1987, **26**, 1152.
245. M. H. Moore, W. N. Hunter, B. L. d'Estaintot, and O. Kennard, *J. Mol. Biol.*, 1989, **206**, 693.
246. C. A. Frederick, L. D. Williams, G. Ughetto, G. A. van der Marel, J. H. van Boom, A. Rich, and A. H.-J. Wang, *Biochemistry*, 1990, **29**, 2538.
247. L. D. Williams, C. A. Frederick, G. Ughetto, and A. Rich, *Nucleic Acids Res.*, 1990, **18**, 5533.
248. B. Gallois, B. L. d'Estaintot, T. Brown, and W. N. Hunter, *Acta Crystallogr.*, 1993, **D49**, 311.
249. M. S. Searle, W. Bicknell, L. P. G. Wakelin, and W. A. Denny, *Nucleic Acids Res.*, 1991, **19**, 2897.
250. D. Yang and A. H.-J. Wang, *Biochemistry*, 1994, **33**, 6595.
251. P. B. Jensen, B. S. Sorensen, M. Sehested, E. J. F. Demant, E. Kjeldsen, E. Friche, and H. H. Hansen, *Biochem. Pharmacol.*, 1993, **45**, 2025.
252. Y. Pommier, in "Anthracycline Antibiotics," ed. W. Priebe, American Chemical Society, Washington, DC, 1995, vol. 574, p. 183.
253. J. B. Chaires, K. R. Fox, J. E. Herrera, M. Britt, and M. J. Waring, *Biochemistry*, 1987, **26**, 8227.
254. J. B. Chaires, in "Anthracycline Antibiotics," ed. W. Priebe, American Chemical Society, Washington, DC, 1995, vol. 574, p. 156.
255. C. J. Shelton, M. M. Harding, and A. S. Prakash, *Biochemistry*, 1996, **35**, 7974.
256. C. J. Shelton and M. M. Harding, *J. Chem. Res. Synop.*, 1995, 158.
257. S. Kunimoto, Y. Takahashi, T. Uchida, T. Takeuchi, and H. Umezawa, *J. Antibiot.*, 1988, **41**, 655.
258. M. E. Figueiredo-Pereira, W. E. Chen, J. Li, and O. Johdo, *J. Biol. Chem.*, 1996, **271**, 16455.
259. G. A. Ellestad and W.-d. Ding, in "Molecular Aspects of Anticancer Drug–DNA Interactions," eds. S. Neidle and M. Waring, Macmillan, London, 1994, vol. 2, p. 130.
260. T. Li, Z. Zeng, V. A. Estevez, K. U. Baldenius, K. C. Nicolaou, and G. F. Joyce, *J. Am. Chem. Soc.*, 1994, **116**, 3709.
261. S. Walker, J. Murnick, and D. Kahne, *J. Am. Chem. Soc.*, 1993, **115**, 7954.
262. M. Uesugi and Y. Sugiura, *Biochemistry*, 1993, **32**, 4622.
263. G. Krishnamurthy, W.-d. Ding, L. O'Brien, and G. A. Ellestad, *Tetrahedron*, 1994, **50**, 1341.
264. J. Drak, N. Iwasawa, S. Danishefsky, and D. M. Crothers, *Proc. Natl. Acad. Sci. USA*, 1991, **88**, 7464.
265. L. Gomez-Paloma, J. A. Smith, W. J. Chazin, and K. C. Nicolaou, *J. Am. Chem. Soc.*, 1994, **116**, 3697.
266. S. N. Ho, S. H. Boyer, S. L. Schreiber, S. J. Danishefsky, and G. R. Crabtree, *Proc. Natl. Acad. Sci. USA*, 1994, **91**, 9203.
267. K. C. Nicolaou, B. M. Smith, K. Ajito, H. Komatsu, L. Gomez-Paloma, and Y. Tor, *J. Am. Chem. Soc.*, 1996, **118**, 2303.
268. C. Liu, B. M. Smith, K. Ajito, H. Komatsu, L. Gomez-Paloma, T. Li, E. A. Theordorakis, K. C. Nicolaou, and P. K. Vogt, *Proc. Natl. Acad. Sci. USA*, 1996, **93**, 940.
269. Y. Sugiura, T. Shiraki, M. Konishi, and T. Oki, *Proc. Natl. Acad. Sci. USA*, 1990, **87**, 3831.
270. T. Shiraki and Y. Sugiura, *Biochemistry*, 1990, **29**, 9795.
271. D. H. Williams, *Nat. Prod. Rep.*, 1996, **13**, 469.
272. H. R. Perkins, *Biochem. J.*, 1969, **111**, 195.
273. R. Nagarajan, *J. Antibiot.*, 1993, **46**, 1181.
274. T. I. Nicas, D. L. Mullen, J. E. Flokowitsch, D. A. Preston, N. J. Snyder, M. J. Zweifel, S. C. Wilkie, M. J. Rodriguez, R. C. Thompson, and R. D. G. Cooper, *Antimicrob. Agents Chemother.*, 1996, **40**, 2194.
275. M. P. Blaustein, *Am. J. Physiol.*, 1993, **264**, C1367.
276. K. Ahmed, D. C. Rohrer, D. S. Fullerton, T. Deffo, E. Kitatsuji, and A. H. L. From, *J. Biol. Chem.*, 1983, **258**, 8092.
277. D. S. Fullerton, M. Kihara, T. Deffo, E. Kitatsuji, K. Ahmed, B. Simat, A. H. L. From, and D. C. Rohrer, *J. Med. Chem.*, 1984, **27**, 256.
278. A. H. L. From, D. S. Fullerton, and K. Ahmed, *Mol. Cell. Biochem.*, 1990, **94**, 157.
279. A. G. Gilman, L. S. Goodman, and A. Gilman (eds.), "The Pharmacological Basis of Therapeutics," 6th edn., Macmillan, New York, 1980.
280. E. Erdmann, *Eur. Heart J., Suppl. F.*, 1995, **16**, 16.
281. S. A. Doi and P. N. Landless, *Br. J. Clin. Prac.*, 1995, **49**, 257.
282. B. Forbush, III, *Curr. Top. Membr. Transp.*, 1983, **19**, 113.
283. D. C. Rohrer, M. Kihara, T. Deffo, H. Rathore, K. Ahmed, A. H. L. From, and D. S. Fullerton, *J. Am. Chem. Soc.*, 1984, **106**, 8269.
284. R. C. Thomas, *Physiol. Rev.*, 1972, **52**, 563.
285. R. Thomas, J. Boutagy, and A. Gelbert, *J. Pharm. Sci.*, 1974, **63**, 1649.
286. F. T. Wallick, B. J. R. Pitts, L. K. Lane, and A. Schwartz, *Arch. Biochem. Biophys.*, 1980, **202**, 442.
287. K. Takiura, H. Yuki, Y. Okamoto, H. Takai, and S. Honda, *Chem. Pharm. Bull.*, 1974, **22**, 2263.
288. M. Takechi and Y. Tanaka, *Phytochemistry*, 1994, **37**, 1421.
289. M. Takechi, C. Uno, and Y. Tanaka, *Phytochemistry*, 1994, **41**, 125.
290. J. Weiland, R. Schon, R. Megges, K. R. H. Repke, and T. R. Watson, *J. Enzyme Inhib.*, 1994, **8**, 197.
291. K. R. H. Repke, W. Schönfeld, J. Weiland, R. Megges, and A. Hache, in "Design of Enzyme Inhibitors as Drugs," eds. M. Sandler and H. J. Smith, Oxford University Press, Oxford, 1989, p. 435.
292. A. Yoda, S. Yoda, and A. M. Sarrif, *Mol. Pharmacol.*, 1973, **9**, 766.
293. J. Beer, R. Kunze, I. Herrmann, H. J. Portius, N. M. Mirsalichova, N. K. Abubakirov, and K. R. H. Repke, *Biochim. Biophys. Acta*, 1988, **937**, 335.
294. J. B. Lingrel and T. Kuntzweiler, *J. Biol. Chem.*, 1994, **269**, 19659.
295. W. J. O'Brien, E. T. Wallick, and J. B. Lingrel, *J. Biol. Chem.*, 1993, **268**, 7707.
296. R. Antolovic, W. Schoner, K. Geering, C. Canessa, B. C. Rossier, and J.-D. Horisberger, *FEBS Lett.*, 1995, **368**, 169.
297. R. J. Adams, A. Schwartz, G. Grupp, I. Grupp, S. W. Lee, E. T. Wallick, T. Powell, V. W. Twist, and P. Gathiram, *Nature (London)*, 1982, **296**, 167.
298. O. Hansen, *Pharmacol. Rev.*, 1984, **36**, 143.
299. P. Reichard, *Science*, 1993, **260**, 1773.
300. A. Willing, H. Follmann, and G. Auling, *Eur. J. Biochem.*, 1988, **170**, 603.
301. S. Eriksson and B.-M. Sjöberg, in "Allosteric Enzymes," ed. G. Hervé, CRC Press, Boca Raton, FL, 1989, p. 189.

302. L. Thelander and P. Reichard, *Annu. Rev. Biochem.*, 1979, **48**, 133.
303. J. Stubbe, *Annu. Rev. Biochem.*, 1989, **58**, 257.
304. J. Stubbe, *Adv. Enzymol. Relat. Areas Mol. Biol.*, 1990, **63**, 349.
305. J. Stubbe, *J. Biol. Chem.*, 1990, **265**, 5329.
306. P. Reichard, *J. Biol. Chem.*, 1993, **268**, 8383.
307. M. Fontecave, P. Nordlund, H. Eklund, and P. Reichard, *Adv. Enzymol. Relat. Areas Mol. Biol.*, 1992, **65**, 147.
308. S. Booker, J. Broderick, and J. Stubbe, *Biochem. Soc. Trans.*, 1993, **21**, 727.
309. B.-M. Sjöberg, *Structure*, 1994, **2**, 793.
310. M. B. Lennon and R. J. Suhadolnik, *Biochim. Biophys. Acta*, 1976, **425**, 532.
311. P. Nordlund, B.-M. Sjöberg, and H. Eklund, *Nature (London)*, 1990, **345**, 593.
312. U. Uhlin and H. Eklund, *Nature (London)*, 1994, **370**, 533.
313. P. Nordlund and H. Eklund, *J. Mol. Biol.*, 1993, **232**, 123.
314. A. Larsson and B.-M. Sjöberg, *EMBO J.*, 1986, **5**, 2037.
315. G. Lassmann, L. Thelander, and A. Gräslund, *Biochem. Biophys. Res. Commun.*, 1992, **188**, 879.
316. C. L. Atkin, L. Thelander, P. Reichard, and G. Lang, *J. Biol. Chem.*, 1973, **248**, 7464.
317. J. Covès, B. Delon, I. Climent, B.-M. Sjöberg, and M. Fontecave, *Eur. J. Biochem.*, 1995, **233**, 357.
318. M. Sahlin, L. Petersson, A. Gräslund, A. Ehrenberg, B.-M. Sjöberg, and L. Thelander, *Biochemistry*, 1987, **26**, 5541.
319. M. Ormö, K. Regnström, Z. Wang, L. Que, Jr., M. Sahlin, and B.-M. Sjöberg, *J. Biol. Chem.*, 1995, **270**, 6570.
320. J. M. Bollinger, Jr., W. H. Tong, N. Ravi, B. H. Huynh, D. E. Edmondson, and J. A. Stubbe, *Methods Enzymol.*, 1995, **258**, 278.
321. J. M. Bollinger, Jr., W. H. Tong, N. Ravi, B. H. Huynh, D. E. Edmondson, and J. A. Stubbe, *J. Am. Chem. Soc.*, 1994, **116**, 8024.
322. B. E. Sturgeon, D. Burdi, S. Chen, B.-H. Huynh, D. E. Edmondson, J. Stubbe, and B. M. Hoffman, *J. Am. Chem. Soc.*, 1996, **118**, 7551.
323. J. Ling, M. Sahlin, B.-M. Sjöberg, T. M. Loehr, and J. Sanders-Loehr, *J. Biol. Chem.*, 1994, **269**, 5595.
324. F. Lendzian, M. Sahlin, F. MacMillan, R. Bittl, R. Fiege, S. Pötsch, B.-M. Sjöberg, A. Gräslund, W. Lubitz, and G. Lassmann, *J. Am. Chem. Soc.*, 1996, **118**, 8111.
325. U. Rova, K. Goodtzova, R. Ingemarson, G. Behravan, A. Gräslund, and L. Thelander, *Biochemistry*, 1995, **34**, 4267.
326. I. Climent, B.-M. Sjöberg, and C. Y. Huang, *Biochemistry*, 1992, **31**, 4801.
327. M. Ekberg, M. Sahlin, M. Eriksson, and B.-M. Sjöberg, *J. Biol. Chem.*, 1996, **271**, 20655.
328. A. Åberg, S. Hahne, M. Karlsson, A. Larsson, M. Ormö, A. Ahgren, and B.-M. Sjöberg, *J. Biol. Chem.*, 1989, **264**, 12249.
329. S. S. Mao, G. X. Yu, D. Chalfoun, and J. Stubbe, *Biochemistry*, 1992, **31**, 9752.
330. J. Stubbe and W. A. van der Donk, *Chem. Biol.*, 1995, **2**, 793.
331. W. A. van der Donk, G. X. Yu, D. J. Silva, J. Stubbe, J. R. McCarthy, E. T. Jarvi, D. P. Matthews, R. J. Resvick, and E. Wagner, *Biochemistry*, 1996, **35**, 8381.
332. S. S. Mao, T. P. Holler, G. X. Yu, J. M. Bollinger, Jr., S. Booker, M. I. Johnston, and J. Stubbe, *Biochemistry*, 1992, **31**, 9733.
333. S. S. Mao, T. P. Holler, J. M. Bollinger, Jr., G. X. Yu, M. I. Johnston, and J. Stubbe, *Biochemistry*, 1992, **31**, 9744.
334. J. Covès, L. Le Hir De Fallois, L. Le Pape, J.-L. Decout, and M. Fontecave, *Biochemistry*, 1996, **35**, 8595.
335. W. A. van der Donk, J. Stubbe, G. J. Gerfen, B. F. Bellew, and R. G. Griffin, *J. Am. Chem. Soc.*, 1995, **117**, 8908.
336. B. P. Hay and R. G. Finke, *J. Am. Chem. Soc.*, 1986, **108**, 4820.
337. M. L. Ludwig, C. L. Drennan, and R. G. Matthews, *Structure*, 1996, **4**, 505.
338. G. W. Ashley, G. Harris, and J. Stubbe, *J. Biol. Chem.*, 1986, **261**, 3958.
339. S. Licht, G. J. Gerfen, and J. Stubbe, *Science*, 1996, **271**, 477.
340. G. J. Gerfen, S. Licht, J.-P. Willems, B. M. Hoffman, and J. Stubbe, *J. Am. Chem. Soc.*, 1996, **118**, 8192.
341. S. Booker, S. Licht, J. Broderick, and J. Stubbe, *Biochemstry*, 1994, **33**, 12676.
342. S. Booker and J. Stubbe, *Proc. Natl. Acad. Sci. USA*, 1993, **90**, 8352.
343. S. C. McFarlan, S. P. Ong, and H. P. C. Hogenkamp, *Biochemistry*, 1996, **35**, 4485.
344. R. Eliasson, M. Fontecave, H. Jörnvall, M. Krook, E. Pontis, and P. Reichard, *Proc. Natl. Acad. Sci. USA*, 1990, **87**, 3314.
345. X. Sun, R. Eliasson, E. Pontis, J. Andersson, G. Buist, B.-M. Sjöberg, and P. Reichard, *J. Biol. Chem.*, 1995, **270**, 2443.
346. X. Sun, J. Harder, M. Krook, H. Jörnvall, B.-M. Sjöberg, and P. Reichard, *Proc. Natl. Acad. Sci. USA*, 1993, **90**, 577.
347. A. F. V. Wagner, M. Frey, F. A. Neugebauer, W. Schäfer, and J. Knappe, *Proc. Natl. Acad. Sci. USA*, 1992, **89**, 996.
348. D. S. King and P. Reichard, *Biochem. Biophys. Res. Commun.*, 1995, **206**, 731.
349. X. Sun, S. Ollagnier, P. P. Schmidt, M. Atta, E. Mulliez, L. Lepape, R. Eliasson, A. Gräslund, M. Fontecave, P. Reichard, and B.-M. Sjöberg, *J. Biol. Chem.*, 1996, **271**, 6827.
350. E. Mulliez, S. Ollagnier, M. Fontecave, R. Eliasson, and P. Reichard, *Proc. Natl. Acad. Sci. USA*, 1995, **92**, 8759.
351. R. Eliasson, P. Reichard, E. Mulliez, S. Ollagnier, M. Fontecave, E. Liepinsh, and G. Otting, *Biochem. Biophys. Res. Commun.*, 1995, **214**, 28.
352. J. Stubbe, M. Ator, and T. Krenitsky, *J. Biol. Chem.*, 1983, **258**, 1625.
353. L. Glaser and H. Zarkowsky, in "The Enzymes," ed. P. D. Boyer, Academic Press, New York, 1971, vol. 5, p. 465.
354. O. Gabriel, in "Carbohydrates in Solution," ed. R. F. Gould, American Chemical Society, Washington, DC, 1973, p. 387.
355. H.-w. Liu and J. S. Thorson, *Annu. Rev. Microbiol.*, 1994, **48**, 223.
356. R. L. Turnquist and R. G. Hansen, in "The Enzymes," ed. P. D. Boyer, Academic Press, New York, 1973, vol. 8A, p. 51.
357. J. Preiss, in "The Enzymes," ed. P. D. Boyer, Academic Press, New York, 1973, vol. 8A, p. 73.
358. T. H. Haugen and J. Preiss, *J. Biol. Chem.*, 1979, **254**, 127.
359. L. Lindqvist, R. Kaiser, P. R. Reeves, and A. A. Lindberg, *Eur. J. Biochem.*, 1993, **211**, 763.
360. L. Lindqvist, R. Kaiser, P. R. Reeves, and A. A. Lindberg, *J. Biol. Chem.*, 1994, **269**, 122.
361. R. M. Mayer and V. Ginsburg, *J. Biol. Chem.*, 1965, **240**, 1900.
362. K. Kimata and S. Suzuki, *J. Biol. Chem.*, 1966, **241**, 1099.

363. P. A. Rubenstein and J. L. Strominger, *J. Biol. Chem.*, 1974, **249**, 3789.
364. S. A. Hossain, K. Tanizawa, Y. Kazuta, and T. Fukui, *J. Biochem.*, 1994, **115**, 965.
365. B. J. Smith-White and J. Preiss, *J. Mol. Evol.*, 1992, **34**, 449.
366. J. S. Thorson, T. M. Kelly, and H.-w. Liu, *J. Bacteriol.*, 1994, **176**, 1840.
367. T. F. Parsons and J. Preiss, *J. Biol. Chem.*, 1978, **253**, 6197.
368. T. F. Parsons and J. Preiss, *J. Biol. Chem.*, 1978, **253**, 7638.
369. T. Katsube, Y. Kazuta, K. Tanizawa, and T. Fukui, *Biochemistry*, 1991, **30**, 8546.
370. J. Sheng, Y.-y. Charng, and J. Preiss, *Biochemistry*, 1996, **35**, 3115.
371. M. A. Hill, K. Kaufmann, J. Otero, and J. Preiss, *J. Biol. Chem.*, 1991, **266**, 12455.
372. Y. Yu, R. N. Russell, J. S. Thorson, L.-d. Liu, and H.-w. Liu, *J. Biol. Chem.*, 1992, **267**, 5868.
373. M. W. Thompson, W. R. Strohl, and H. G. Floss, *J. Gen. Microbiol.*, 1992, **138**, 779.
374. P. A. Frey, in "Pyridine Nucleotide Coenzymes. Chemical, Biochemical, and Medical Aspects. Part B," eds. D. Dolphin, R. Poulson, and O. Avramovic, Wiley Interscience, New York, 1987, p. 461.
375. H. Zarkowsky and L. Glaser, *J. Biol. Chem.*, 1970, **244**, 4750.
376. X. He, J. S. Thorson, and H.-w. Liu, *Biochemistry*, 1996, **35**, 4721.
377. O. Gabriel and L. C. Lindquist, *J. Biol. Chem.*, 1968, **243**, 1479.
378. A. Melo, W. H. Elliott, and L. Glaser, *J. Biol. Chem.*, 1968, **243**, 1467.
379. C. E. Snipes, G.-U. Brillinger, L. Sellers, L. Mascaro, and H. G. Floss, *J. Biol. Chem.*, 1977, **252**, 8113.
380. P. J. Oths, R. M. Mayer, and H. G. Floss, *Carbohydr. Res.*, 1990, **198**, 91.
381. S.-F. Wang and O. Gabriel, *J. Biol. Chem.*, 1970, **245**, 8.
382. M. Tonetti, L. Sturla, A. Bisso, U. Benatti, and A. De Flora, *J. Biol. Chem.*, 1996, **271**, 27274.
383. S. Chang, B. Duerr, and G. Serif, *J. Biol. Chem.*, 1988, **263**, 1693.
384. R. W. Gaugler and O. Gabriel, *J. Biol. Chem.*, 1973, **248**, 6041.
385. X.-M. Jiang, B. Neal, F. Santiago, S. J. Lee, L. K. Romana, and P. R. Reeves, *Mol. Microbiol.*, 1991, **5**, 695.
386. A. Melo and L. Glaser, *J. Biol. Chem.*, 1968, **243**, 1475.
387. B. W. Jarvis and C. R. Hutchinson, *Arch. Biochem. Biophys.*, 1994, **308**, 175.
388. K. J. Linton, B. W. Jarvis, and C. R. Hutchinson, *Gene*, 1995, **153**, 33.
389. V. Ginsburg, *J. Biol. Chem.*, 1960, **235**, 2196.
390. K. Marumo, L. Lindqvist, N. Verma, A. Weintraub, P. R. Reeves, and A. A. Lindberg, *Eur. J. Biochem.*, 1992, **204**, 539.
391. J. S. Thorson, E. Oh, and H.-w. Liu, *J. Am. Chem. Soc.*, 1992, **114**, 6941.
392. S. L. Bender, S. Medhi, and J. R. Knowles, *Biochemistry*, 1989, **28**, 7555.
393. J. Sancho, L. Serrano, and A. R. Fersht, *Biochemistry*, 1992, **31**, 2253.
394. K. Pissowotzki, K. Mansouri, and W. Piepersberg, *Mol. Gen. Genet.*, 1991, **231**, 113.
395. T. M. Weigel, L.-d. Liu, and H.-w. Liu, *Biochemistry*, 1992, **31**, 2129.
396. Y. Lei, O. Ploux, and H.-w. Liu, *Biochemistry*, 1995, **34**, 4643.
397. J. S. Thorson and H.-w. Liu, *J. Am. Chem. Soc.*, 1993, **115**, 7539.
398. D. A. Johnson, G. T. Gassner, V. Bandarian, F. J. Ruzicka, D. P. Ballou, G. H. Reed, and H.-w. Liu, *Biochemistry*, 1996, **35**, 15846.
399. T. M. Weigel, V. P. Miller, and H.-w. Liu, *Biochemistry*, 1992, **31**, 2140.
400. J. S. Thorson and H.-w. Liu, *J. Am. Chem. Soc.*, 1993, **115**, 12177.
401. P. A. Pieper, Z. Guo, and H.-w. Liu, *J. Am. Chem. Soc.*, 1995, **117**, 5158.
402. M. M. Palcic and H. G. Floss, in "Vitamin B_6 Pyridoxal Phosphate. Chemical, Biochemical, and Medical Aspects. Part A," eds. D. Dolphin, R. Poulson, and O. Avramovic, Wiley Interscience, New York, 1986, p. 25.
403. J. S. Thorson, S. F. Lo, H.-w. Liu, and C. R. Hutchinson, *J. Am. Chem. Soc.*, 1993, **115**, 6993.
404. S. J. Gould and J. Guo, *J. Am. Chem. Soc.*, 1992, **114**, 10176.
405. S. J. Gould and Q. Zhang, *J. Antibiot.*, 1995, **48**, 652.
406. V. P. Miller, J. S. Thorson, O. Ploux, S. F. Lo, and H.-w. Liu, *Biochemistry*, 1993, **32**, 11934.
407. O. Ploux, Y. Lei, K. Vatanen, and H.-w. Liu, *Biochemistry*, 1995, **34**, 4159.
408. S. F. Lo, V. P. Miller, Y. Lei, J. S. Thorson, H.-w. Liu, and J. L. Schottel, *J. Bacteriol.*, 1994, **176**, 460.
409. G. T. Gassner, D. A. Johnson, H.-w. Liu, and D. P. Ballou, *Biochemistry*, 1996, **35**, 7752.
410. K. D. Burns, P. A. Pieper, H.-w. Liu, and M. T. Stankovich, *Biochemistry*, 1996, **35**, 7879.
411. F. Müller, J. Vervoort, C. P. M. van Mierlo, S. G. Mayhew, W. J. H. van Berkel, and A. Bacher, in "Flavins and Flavoproteins," eds. D. E. Edmundson and D. B. McCormick, Walter de Gruyter, Berlin, 1987, p. 261.
412. K. Mukai, K. Ueda, K. Ishizu, and S. Yamauchi, *Bull. Chem. Soc. Jpn.*, 1984, **57**, 1151.
413. H.-U. Wagner and R. Gompper, in "The Chemistry of the Quinonoid Compounds," ed. S. Patai, Wiley, New York, 1974, vol. 2, p. 1145.
414. S. Matsuhashi and J. L. Strominger, *J. Biol. Chem.*, 1967, **242**, 3494.
415. J. S. Thorson, S. F. Lo, and H.-w. Liu, *J. Am. Chem. Soc.*, 1993, **115**, 5827.
416. J. S. Thorson, S. F. Lo, O. Ploux, X. He, and H.-w. Liu, *J. Bacteriol.*, 1994, **176**, 5483.
417. L. Lindqvist, K. H. Schweda, P. R. Reeves, and A. A. Lindberg, *Eur. J. Biochem.*, 1994, **225**, 863.
418. J. Carlson, J. A. Fuchs, and J. Messing, *Proc. Natl. Acad. Sci. USA*, 1984, **81**, 4294.
419. O. Nilsson, A. Åberg, T. Lundqvist, and B.-M. Sjöberg, *Nucleic Acids Res.*, 1988, **16**, 4174.
420. A. Jordan, I. Gibert, and J. Barbé, *Gene*, 1995, **167**, 75.
421. L. Thelander and P. Berg, *Mol. Cell. Biol.*, 1986, **6**, 3433.
422. N. M. Standart, S. J. Bray, E. L. George, T. Hunt, and J. V. Ruderman, *J. Cell. Biol.*, 1985, **100**, 1968.
423. M. P. Riggio and D. E. Onions, *Gene*, 1994, **143**, 217.
424. A. Jordan, I. Gibert, and J. Barbé, *J. Bacteriol.*, 1994, **176**, 3420.
425. A. Jordan, E. Pontis, F. Aslund, U. Hellman, I. Gibert, and P. Reichard, *J. Biol. Chem.*, 1996, **271**, 8779.
426. C. M. Fraser, J. D. Gocayne, O. White, M. D. Adams, R. A. Clayton, R. D. Fleischmann, C. J. Bult, A. R. Ketlavage, G. Sutton, J. M. Kelley, *et al.*, *Science*, 1995, **270**, 397.
427. P. Young, M. Öhman, M. Q. Xu, D. A. Shub, and B.-M. Sjöberg, *J. Biol. Chem.*, 1994, **269**, 20229.
428. P. Young, M. Öhman, and B.-M. Sjöberg, *J. Biol. Chem.*, 1994, **269**, 27815.

429. S. J. Elledge and R. W. Davis, *Mol. Cell. Biol.*, 1987, **7**, 2783.
430. S. J. Elledge and R. W. Davis, *Genes Dev.*, 1990, **4**, 740.
431. H. K. Hurd, C. W. Roberts, and J. W. Roberts, *Mol. Cell. Biol.*, 1987, **7**, 3673.
432. C. L. Marolda and M. A. Valvano, *J. Bacteriol.*, 1995, **177**, 5539.
433. S.-H. Xiang, M. Hobbs, and P. R. Reeves, *J. Bacteriol.*, 1994, **176**, 4357.
434. J. D. Klena and C. A. Schnaitman, *Mol. Microbiol.*, 1993, **9**, 393.
435. K. Rajakumar, B. H. Jost, C. Sasakawa, N. Okada, M. Yoshikawa, and B. Adler, *J. Bacteriol.*, 1994, **176**, 2362.
436. G. Stevenson, B. Neal, D. Liu, M. Hobbs, N. H. Packer, M. Batley, J. W. Redmond, L. Lindquist, and P. Reeves, *J. Bacteriol.*, 1994, **176**, 4144.
437. L. K. Romana, F. S. Santiago, and P. R. Reeves, *Biochem. Biophys. Res. Commun.*, 1991, **174**, 846.
438. L. Zhang, A. Al-Hendy, P. Toivanen, and M. Skurnik, *Mol. Microbiol.*, 1993, **9**, 309.
439. A. C. Kessler, A. Haase, and P. R. Reeves, *J. Bacteriol.*, 1993, **175**, 1412.
440. M. Hobbs, and P. R. Reeves, *Biochim. Biophys. Acta*, 1995, **1245**, 273.
441. P. K. Brown, L. K. Romana, and P. R. Reeves, *Mol. Microbiol.*, 1991, **5**, 1873.
442. P. Wyk and P. Reeves, *J. Bacteriol.*, 1989, **171**, 5687.
443. P. K. Brown L. K. Romana, and P. R. Reeves, *Mol. Microbiol.*, 1992, **6**, 1385.
444. D. A. Bastin and P. R. Reeves, *Gene*, 1995, **164**, 17.
445. N. Verma and P. Reeves, *J. Bacteriol.*, 1989, **171**, 5694.
446. D. Liu, N. K. Verma, L. K. Romana, and P. R. Reeves, *J. Bacteriol.*, 1991, **173**, 4814.
447. K. J. Stutzman-Engwall and C. R. Hutchinson, *Proc. Natl. Acad. Sci. USA*, 1989, **86**, 3135.
448. K. Madduri and C. R. Hutchinson, *J. Bacteriol.*, 1995, **177**, 1208.
449. M. A. Gallo, J. Ward, and C. R. Hutchinson, *Microbiology*, 1996, **142**, 269.
450. S. L. Otten, X. Liu, J. Ferguson, and C. R. Hutchinson, *J. Bacteriol.*, 1995, **177**, 6688.
451. M. L. Dickens, J. Ye, and W. R. Strohl, *J. Bacteriol.*, 1996, **178**, 3384.
452. R. G. Summers, S. Donadio, M. J. Staver, E. Wendt-Pienkowski, C. R. Hutchinson, and L. Katz, *Microbiology*, 1997, **143**, 3251.
453. S. F. Haydock, J. A. Dowson, N. Dhillon, G. A. Roberts, J. Cortes, and P. F. Leadlay, *Mol. Gen. Genet.*, 1991, **230**, 120.
454. L. A. Merson-Davies and E. Cundliffe, *Mol. Microbiol.*, 1994, **13**, 349.
455. L. Retzlaff, G. Mayer, S. Beyer, J. Ahlert, S. Verseck, J. Distler, and W. Piepersberg, in "Industrial Microorganisms: Basic and Applied Molecular Genetics," eds. R. H. Baltz, G. D. Hegeman, and P. L. Skatrud, American Society of Microbiology, Washington, DC, 1993, p. 183.
456. J. A. Tercero, J. C. Espinosa, R. A. Lacalle, and A. Jiménez, *J. Biol. Chem.*, 1996, **271**, 1579.
457. A. Bechthold, J. K. Sohng, T. M. Smith, X. Chu, and H. G. Floss, *Mol. Gen. Genet.*, 1995, **248**, 610.
458. S. J. Elledge, Z. Zhou, J. B. Allen, and T. A. Navas, *BioEssays*, 1993, **15**, 333.
459. G. R. Greenberg and J. M. Hilfinger, *Prog. Nucleic Acid Res. Mol. Biol.*, 1996, **53**, 345.
460. G. J. Mann, A. Gräslund, E.-I. Ochiai, R. Ingemarson, and L. Thelander, *Biochemistry*, 1991, **30**, 1939.
461. A. Jordan, E. Aragall, I. Gibert, and J. Barbé, *Mol. Microbiol.*, 1996, **19**, 777.
462. F. Yang, G. Lu, and H. Rubin, *J. Bacteriol.*, 1994, **176**, 6738.
463. A. Jordan, E. Pontis, M. Atta, M. Krook, I. Gibert, J. Barbé, and P. Reichard, *Proc. Natl. Acad. Sci. USA*, 1994, **91**, 12892.
464. R. L. Blakley and H. A. Barker, *Biochem. Biophys. Res. Commun.*, 1964, **16**, 301.
465. S. Ollagnier, E. Mulliez, J. Gaillard, R. Eliasson, M. Fontecave, and P. Reichard, *J. Biol. Chem.*, 1996, **271**, 9410.
466. P. Young, J. Andersson, M. Sahlin, and B.-M. Sjöberg, *J. Biol. Chem.*, 1996, **271**, 20770.
467. R. D. Fleischmann, M. D. Adams, O. White, R. A. Clayton, E. F. Kirkness, A. R. Ketlavage, C. J. Bult, J.-F. Tomb, B. A. Dougherty, J. M. Merrick, *et al.*, *Science*, 1995, **269**, 496.
468. I. S.-Y. Sze, S. C. McFarlan, A. Spormann, H. P. C. Hogenkamp, and H. Follman, *Biochem. Biophys. Res. Commun.*, 1992, **184**, 1101.
469. D. Liu, L. Lindqvist, and P. R. Reeves, *J. Bacteriol.*, 1995, **177**, 4084.
470. D. A. Bastin, L. K. Romana, and P. R. Reeves, *Mol. Microbiol.*, 1991, **5**, 2223.
471. Z. Yao and M. A. Valvano, *J. Bacteriol.*, 1994, **176**, 4133.
472. B. R. Clarke and C. Whitfield, *J. Bacteriol.*, 1992, **174**, 4614.
473. L. Le Minor and M. Y. Popoff, *Int. J. Syst. Bacteriol.*, 1987, **37**, 465.
474. R. J. Roantree, *Annu. Rev. Microbiol.*, 1967, **21**, 443.
475. L. Kenne, B. Lindberg, E. Söderholm, D. R. Bundle, and D. W. Griffith, *Carbohydr. Res.*, 1983, **111**, 289.
476. N. A. Komandrova, R. P. Gorshkova, V. A. Zubkov, and Yu. S. Ovodov, *Bioorg. Khim.*, 1989, **15**, 104.
477. C. A. Schnaitman and J. D. Klena, *Microbiol. Rev.*, 1993, **57**, 655.
478. P. Reeves, *Trends in Genetics*, 1993, **9**, 17.
479. P. Reeves, *Trends Microbiol.*, 1995, **3**, 381.
480. N. Sueoka, *Proc. Natl. Acad. Sci. USA*, 1988, **85**, 2653.
481. D. J. Brenner, J. Ursing, H. Bercovier, A. G. Steigerwalt, G. R. Fanning, J. M. Alonso, and H. H. Mollaret, *Curr. Microbiol.*, 1980, **4**, 195.
482. A. D. Elbein and E. C. Heath, *J. Biol. Chem.*, 1965, **240**, 1926.
483. L. Katz and S. Donadio, *Annu. Rev. Microbiol.*, 1993, **47**, 875.
484. D. A. Hopwood, *Curr. Opin. Biotechnol.*, 1993, **4**, 531.
485. R. McDaniel, S. Ebert-Khosla, D. A. Hopwood, and C. Khosla, *Science*, 1993, **262**, 1546.
486. C. R. Hutchinson, *Bio/Technology* (*New York*), 1994, **12**, 375.
487. J Niemi, K. Ylihonko, J. Hakala, R. Pärssinen, A. Kopio, and P. Mäntsälä, *Microbiol.*, 1994, **140**, 1351.
488. K. Ylihonko, J. Hakala, T. Kunnari, and P. Mäntsälä, *Microbiology*, 1996, **142**, 1965.
489. D. A. Hopwood and D. H. Sherman, *Annu. Rev. Genet.*, 1990, **24**, 37.
490. S. L. Otten, J. Ferguson, and C. R. Hutchinson, *J. Bacteriol.*, 1995, **177**, 1216.
491. S. L. Otten, K. J. Stutzman-Engwall, and C. R. Hutchinson, *J. Bacteriol.*, 1990, **172**, 3427.
492. A. Grimm, K. Madduri, A. Ali, and C. R. Hutchinson, *Gene*, 1994, **151**, 1.

493. H. Krügel, G. Schumann, F. Hänel, and G. Fiedler, *Mol. Gen. Genet.*, 1993, **241**, 193.

494. N. Tsukamoto, I. Fujii, Y. Ebizuka, and U. Sankawa, *J. Antibiot.*, 1992, **45**, 1286.

495. J. Niemi and P. Mäntsälä, *J. Bacteriol.*, 1995, **177**, 2942.

496. J. Vara, M. Lewandowska-Skarbek, Y.-G. Wang, S. Donadio, and C. R. Hutchinson, *J. Bacteriol.*, 1989, **171**, 5872.

497. J. M. Weber, J. O. Leung, G. T. Maine, R. H. B. Potenz, T. J. Paulus, and J. P. DeWitt, *J. Bacteriol.*, 1990, **172**, 2372.

498. S. Donadio, M. J. Staver, J. B. McAlpine, S. J. Swanson, and L. Katz, *Science*, 1991, **252**, 675.

499. S. Donadio, D. Stassi, J. B. McAlpine, M. J. Staver, P. J. Sheldon, M. Jackson, S. J. Swanson, E. Wendt-Pienkowski, Y.-g. Wang, B. Jarvis, C. R. Hutchinson, and L. Katz, in "Industrial Microorganisms: Basic and Applied Molecular Genetics," eds. R. H. Baltz, G. D. Hegeman, and P. L. Skatrud, American Society for Microbiology, Washington, DC, 1993, p. 257.

500. L. Katz and S. Donadio, in "Genetics and Biochemistry of Antibiotic Production," eds. L. C. Vining and C. Stuttard, Butterworth–Heinemann, Boston, MA, 1995, p. 385.

501. J. M. Weber, C. K. Wierman, and C. R. Hutchinson, *J. Bacteriol.*, 1985, **164**, 425.

502. N. Dhillon, R. S. Hale, J. Cortes, and P. F. Leadlay, *Mol. Microbiol.*, 1989, **3**, 1405.

503. R. H. Baltz and E. T. Seno, *Annu. Rev. Microbiol.*, 1988, **42**, 547.

504. W. Piepersberg, in "Genetics and Biochemistry of Antibiotic Production," eds. L. C. Vining and C. Stuttard, Butterworth–Heinamann, Boston, MA, 1995, p. 531.

505. M. Stockmann and W. Piepersberg, *FEMS Microbiol. Lett.*, 1992, **69**, 185.

506. J. J. Lee, J. P. Lee, P. J. Keller, C. E. Cottrell, C.-J. Chang, H. Zähner, and H. G. Floss, *J. Antibiot.*, 1986, **39**, 1123.

507. D. J. MacNeil, in "Genetics and Biochemistry of Antibiotic Production," eds. L. C. Vining and C. Stuttard, Butterworth–Heinemann, Boston, MA, 1995, p. 421.

508. H. Ikeda and S. Omura, *J. Antibiot.*, 1995, **48**, 549.

509. T. MacNeil, K. M. Gewain, and D. J. MacNeil, *J. Bacteriol.*, 1993, **175**, 2552.

510. M. D. Schulman and C. Ruby, *Antimicrob. Agents Chemother.*, 1987, **31**, 964.

511. M. D. Schulman, S. L. Acton, D. L. Valentino, and B. H. Arison, *J. Biol. Chem.*, 1990, **265**, 16965.

512. S. Gaisser, G. A. Böhm, J. Cortés, and P. F. Leadlay, *Mol. Gen. Genet.*, 1997, **256**, 239.

3.13
Aldolases

DARLA P. HENDERSON and ERIC J. TOONE
Duke University, Durham, NC, USA

3.13.1   INTRODUCTION

3.13.1.1   Scope of the Chapter

Enzymes are now widely accepted as catalysts for synthetic organic chemistry. Work during the first fifty years of applied enzymology focused primarily on the use of oxido-reductases and hydrolytic enzymes. The 1990s, however, have seen attention in the field shift to the use of carbon–carbon bond forming catalysts. While facile methods for functional group manipulation and the preparation of enantiomerically pure materials are doubtless of enormous value, the formation of carbon–carbon bonds remains the foundation on which all synthetic organic chemistry is built.

Despite a broad recognition of the importance of carbon–carbon bond forming reactions in biocatalysis, a relatively small number of enzymes have been thoroughly investigated as catalysts for synthetic organic chemistry. The goal of this work is to present a complete description of the state of knowledge of the aldolases. A single accurate definition of the class of enzymes referred to as aldolases is lacking, and as a result the list of enzymes covered here is in some ways arbitrary. Although most aldolases are classified according to the use of either a Schiff base or divalent zinc for nucleophile activation, enzymes that catalyze aldol or benzoin-type reactions through the assistance of pyridoxal or thiamine cofactors are known: these enzymes are included here. Likewise, the nitrile lyases are carbon–carbon bond forming enzymes formally belonging to EC class 2: these enzymes are not considered here. Transaldolase and transketolase (EC class 2 enzymes) are considered here because of their relevance to glycolysis and their formation of aldol products.

Our goal is to provide a ready compilation of those aldolases that might find synthetic utility. For each enzyme considered we have included when available:

   (i) a representation reaction scheme,
   (ii) the *in vivo* role of the enzyme,
   (iii) known sources of the enzyme,
   (iv) optimal reaction conditions including pH, buffer, ionic strength, etc.,
   (v) crystal structures and protein and nucleic acid sequences,
   (vi) mechanistic information, and
   (vii) substrate specificities and synthetic applications.

During the 1990s a plethora of outstanding reviews has appeared on the use of aldolases in organic synthesis.[1-9] These reviews have focused primarily on the dihydroxyacetone phosphate aldolases, most often fructose-1,6-diphosphate (FDP) aldolase, and the pyruvate aldolase, neuraminic acid aldolase. Accordingly, we have limited our discussion of these enzymes and refer the reader to those sources for more extensive discussions.

3.13.1.2   The Mechanisms of Enzymatic Aldol Reactions

Enzymes discussed in this chapter catalyze aldol reactions via five general strategies: (i) Schiff base/enamine formation, (ii) Lewis acid activation of the nucleophile by divalent zinc, (iii) utilization of the preformed enolate nucleophile phosphoenolpyruvate, (iv) formation of a nucleophilic species covalently linked to thiamine, and (v) formation of a nucleophilic species covalently linked to pyridoxal. A brief mechanistic description of each class follows.

3.13.1.2.1 *Nucleophile activation by Schiff base/enamine formation*

Both dihydroxyacetone phosphate and pyruvate can be activated for aldol reaction through formation of the corresponding enamine: such aldolases are known as type I. Formation of a Schiff base (imine) with an active site lysine followed by conversion to the enamine produces the nucleophilic species. The enamine then attacks the electrophilic aldehyde in the usual fashion (Scheme 1). The initially formed Schiff base can be reduced by either sodium borohydride or sodium cyanoborohydride; such inactivation is generally accepted as the definitive test for enamine nucleophile activation.

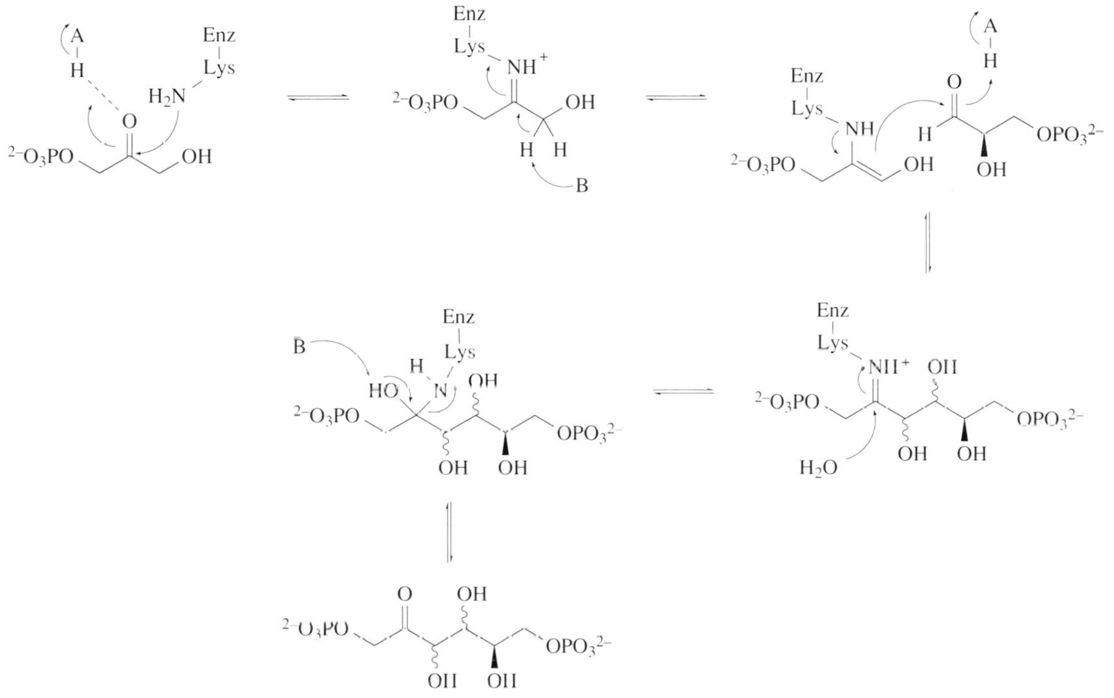

Scheme 1

The mechanism requires the assistance of both a general base and a general acid catalyst for reaction. While some aldolases show bell-shaped pH-activity relationships indicative of the presence of two enzyme-bound residues involved in catalysis, others show a single transition at low pH (< 5). The latter behavior signals either a slow enolization from imine to enamine followed by fast reaction, or activity of solvent or buffer as the acid catalyst during aldol addition.

3.13.1.2.2 *Nucleophile activation by zinc*

So-called type II aldolases activate nucleophiles towards aldol attack through the agency of divalent zinc. The catalytic zinc conceivably functions as a Lewis acid in two roles: catalyzing conversion of the keto form of the nucleophile to the nucleophilic enol and/or stabilizing the incipient negative charge at the transition state on the electrophilic aldehyde oxygen (Scheme 2).

NMR studies suggest that in the yeast FDP aldolase the zinc is too far from the nucleophile for direct coordination. Rather, the metal apparently acts through an intervening amino acid, probably histidine.[10] Type II aldolases are sensitive to EDTA treatment and reversible inactivation by this and other metal chelators is generally accepted as definitive proof of Lewis acid nucleophile activation. Because removal of structural metal ions without catalytic function can abolish enzyme activity, this test should only be accepted as conclusive proof of a type II mechanism in the absence of sensitivity to sodium borohydride.

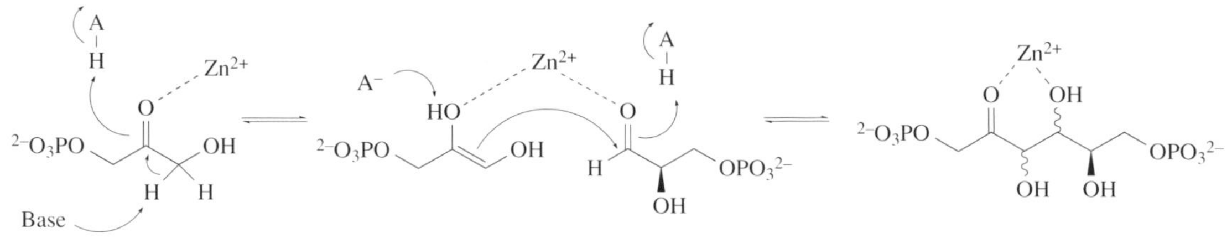

Scheme 2

3.13.1.2.3 Aldol reactions using preformed enolate nucleophiles

A large group of enzymes utilizes the preformed enolate phosphoenolpyruvate as the nucleophilic component in aldol reaction. The mechanism of reaction for these enzymes remains controversial. The only member of this group to receive significant mechanistic study is 3-deoxy-D-manno-2-octulosonate-8-phosphate synthase, a key enzyme in bacterial cell wall biosynthesis. Abeles and co-workers[11] demonstrated that the reaction proceeds with cleavage of the enol phosphate C—O, rather than P—O, bond. Based on this and other observations, a mechanism was proposed that invokes concomitant attack by water at C-2 of pyruvate and nucleophilic attack at the aldehydic carbon followed by loss of phosphate (Scheme 3).

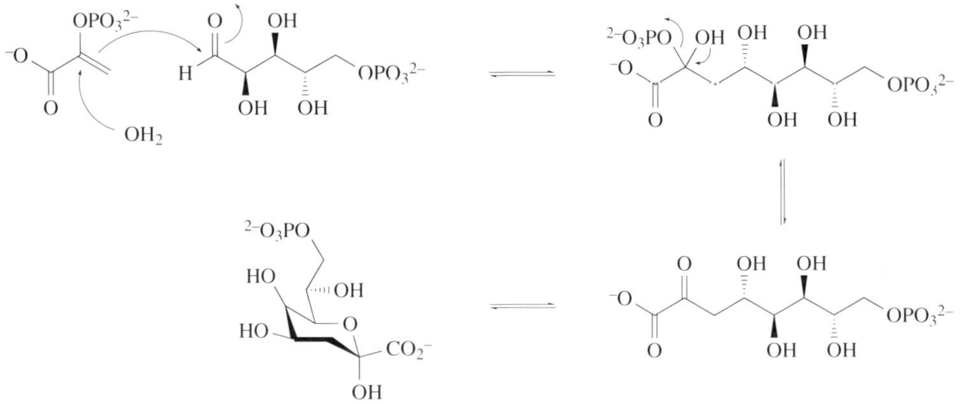

Scheme 3

Based on a series of labeling experiments and the inhibitory activity of a variety of deoxy sugars, Baasov et al.[12] proposed an alternative mechanism (Scheme 4). In this postulate, pyruvate attack on the electrophilic aldehyde proceeds without assistance of water. Rather, nucleophilic attack of pyruvate and reaction of the electrophile C-3 hydroxy with the incipient positive charge on pyruvate produces a glycosyl phosphate intermediate. Reaction of pyruvate with erythrose-4-phosphate may be either concerted or stepwise. Finally, cleavage of the glycosyl phosphate and reaction with water forms the final product. Baasov suggested that the mechanism is concerted and is initiated by nucleophilic attack of the hydroxy of erythrose-4-phosphate to phosphoenolpyruvate with concomitant attack of the double bond on the electrophilic carbonyl.[14]

3.13.1.2.4 Aldol reaction using thiamine pyrophosphate

Appropriately substituted ketones and aldehydes react with thiamine to produce the nucleophilic species 2-hydroxyethylthiamine pyrophosphate (Scheme 5). In some instances, such as pyruvate decarboxylase, the adduct decomposes to yield a decarboxylated aldehyde. Alternatively, the intermediate may be intercepted by an electrophilic carbonyl to produce acyloin products. Several enzymes that utilize thiamine cofactors to form carbon–carbon bonds in this fashion are considered aldolases and are covered here.

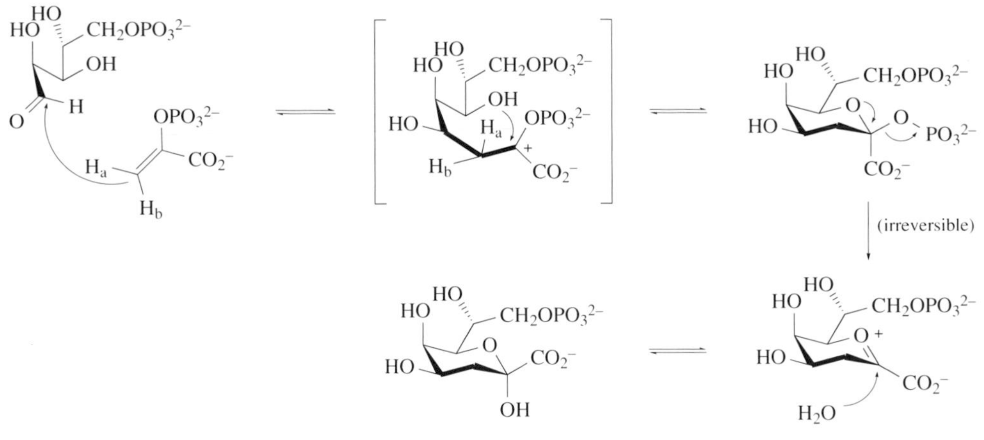

Scheme 4

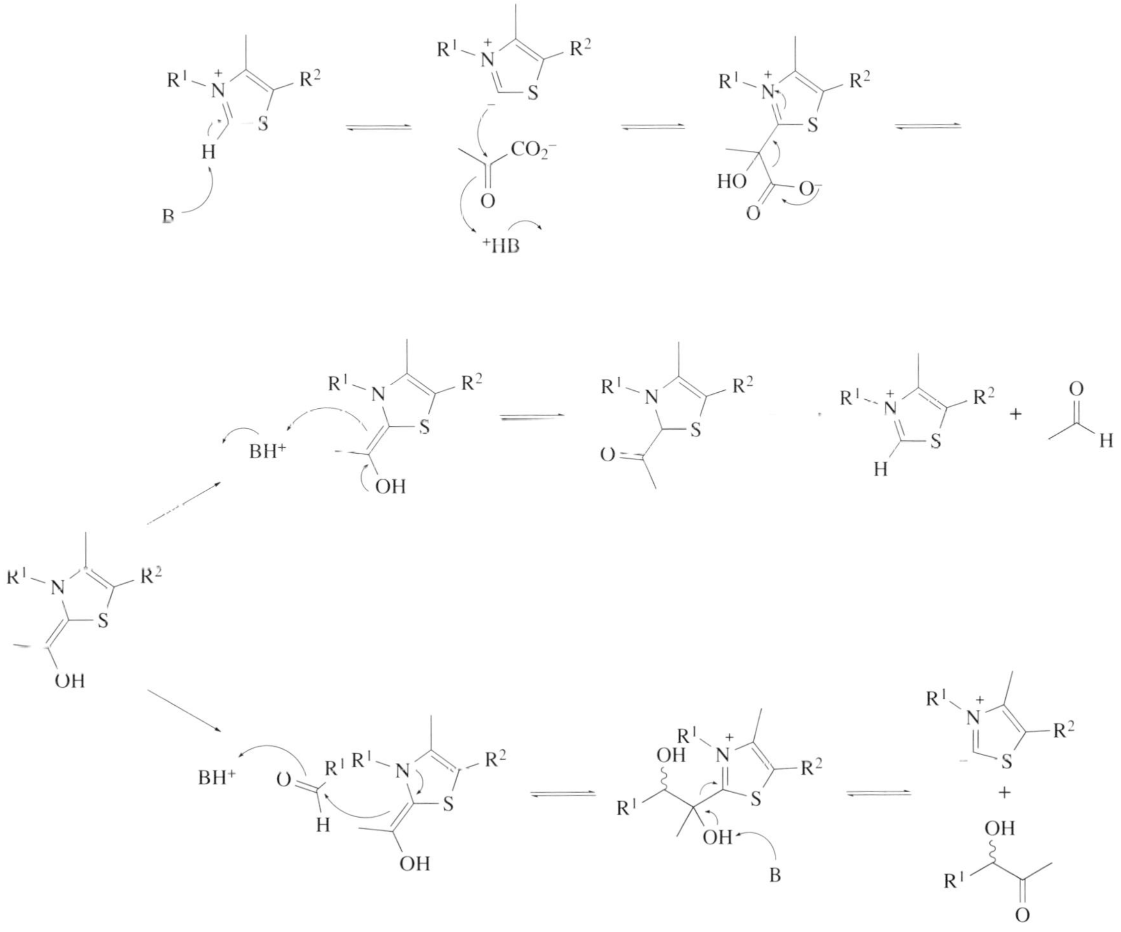

Scheme 5

3.13.1.2.5 *Aldol reaction using pyridoxal*

Pyridoxal in both oxidized and reduced forms produces Schiff base adducts with either an amino or an aldehyde moiety, respectively. In amino acid biosynthesis, this reactivity pattern is employed for the formation of carbon–carbon bonds. Thus, reaction of an amino moiety with oxidized pyridoxal produces an imine which, following rearrangement, proceeds to a nucleophilic enamine

(Scheme 6). Reaction with an electrophilic aldehyde followed by hydrolysis of the imine provides aldol products.

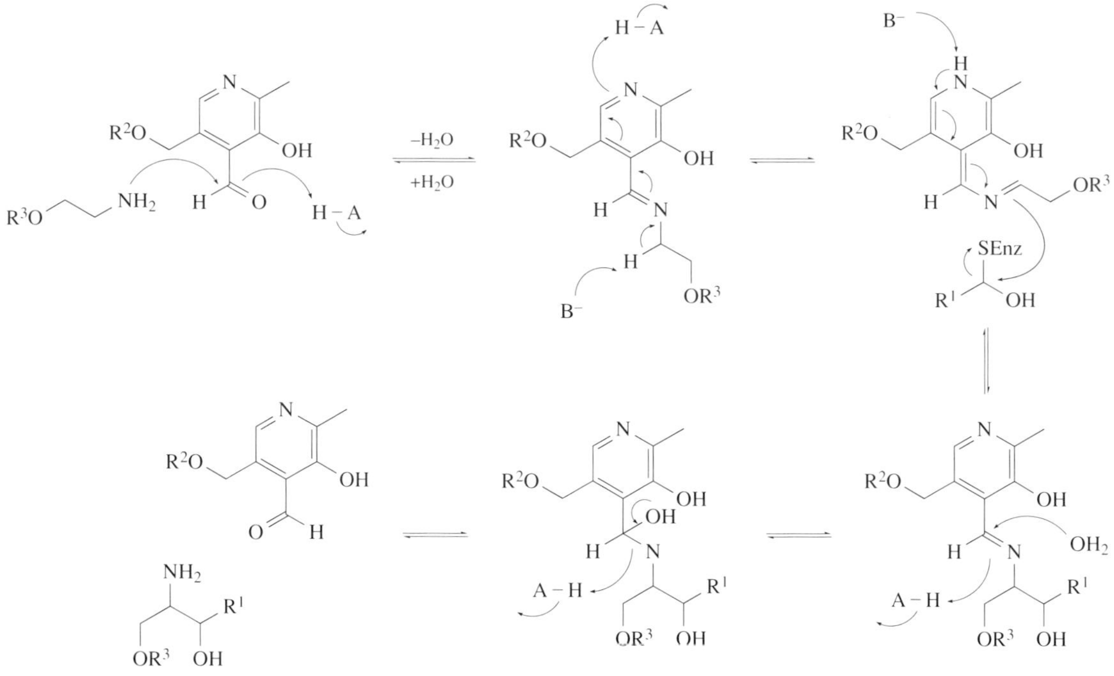

Scheme 6

Below we review known aldolases. In general, aldolases have been grouped according to mechanistic classes discussed above. We have grouped several major enzymes of glycolysis including transketolase, transaldolase, FDP aldolase, 2-keto-3-deoxy-6-phosphogluconate aldolase, fuculose-1-phosphate aldolase, 2-keto-3-deoxy-L-arabonate aldolase, rhamnulose-1-phosphate aldolase, 2-keto-3-deoxy-6-phosphogalactonate aldolase, 2-keto-3-deoxy-D-xylonate aldolase, and tagatose-1,6-diphosphate aldolase together in a separate heading. In several instances these glycolytic pathways represent examples of convergent evolution and several enzymes catalyze identical reactions by differing mechanisms. Accordingly, these enzymes are considered together. Within each classification, enzymes are listed in order of EC number. Those enzymes lacking an EC number and/or CAS registry number have been listed at the end of each mechanistic section.

3.13.2 SCHIFF BASE FORMING ALDOLASES

3.13.2.1 Deoxyribose-5-phosphate Aldolase

Deoxyribose-5-phosphate aldolase (DERA; EC 4.1.2.4; CAS 9026-97-5) is expressed by a wide variety of prokaryotic and eukaryotic organisms and is primarily responsible for the catabolism of deoxyribose from nucleic acids (Scheme 8) although the enzyme appears to be inducible in several bacterial strains by growth on deoxyribose. DERA is the only identified bacterial aldolase that utilizes an aldehydic nucleophile.

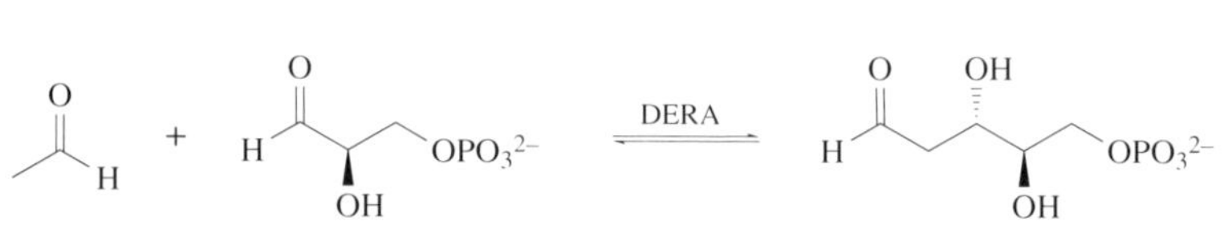

Scheme 7

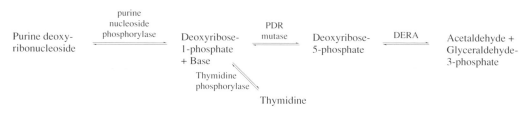

Scheme 8

DERA from mammalian liver, *Escherichia coli* and *Corynebacterium diphtheria* was first reported by Racker[13] in 1951. Since that time, the enzyme has been reported in various animal tissues,[14] *Spiroplasmatacae* species,[15] *Lactobacillus plantarum*,[16] *L. casei*,[17] *L. acidophilus*,[16] *Salmonella typhimurium*,[18] *Bacillus subtilis*,[19] *B. cereus*,[20] *Anaeroplasma intermedium*,[21] *Asteroleplasma anaerobium*,[21] several strains of *Acholeplasma* and *Mycoplasma*,[22-24] and *Haemophilus influenzae*.[25] The enzymes from *M. genitalium*, *M. pneumoniae*, *B. subtilis*, *H. influenzae*, and *E. coli* have been sequenced and/or cloned.[19,22-27] The *E. coli* protein has been overexpressed, and *E. coli* DERA has been crystallized although no structural information has been reported.[28]

Bacterial versions of DERA exist as dimers with subunit molecular weights near 30 kDa (27.7 kDa from *E. coli*,[28] 28.5 kDa from *S. typhimurium*[18]). The rat liver protein is reported to have a total molecular weight of 253 kDa, although no subunit molecular weight data exist.[14] The enzymes exhibit broad pH optima over the range of 6–8. The equilibrium constant for the *in vivo* reaction is ~0.25 mM. Michaelis constants for all three *in vivo* substrates are sub-millimolar at 0.1 mM for 2-deoxyribose-5-phosphate, 0.6 mM for glyceraldehyde-3-phosphate, and 0.3 mM for acetaldehyde.

DERA has been investigated extensively by Wong and co-workers[29] as a catalyst for stereo controlled carbon–carbon bond formation and a large body of data has been compiled regarding the substrate specificity of the enzyme in the direction of synthesis. The enzyme accepts a broad range of electrophilic substrates but shows a preference for 2-hydroxyaldehydes bearing the D-configuration at C-2. The enzyme shows a narrow substrate specificity for the nucleophilic component of the reaction, but accepts propanal, acetone, and fluoroacetone as substrates, albeit at greatly diminished rates relative to acetaldehyde.[30] Sequential two- and three-substrate aldol reactions have been conducted with DERA or a combination of DERA and rabbit muscle FDP aldolase to synthesize a variety of sugar analogues.[31-33] Deoxythiosugars[34] and deoxyazasugars[35,36] are examples of compounds synthesized in a chemo-enzymatic fashion via DERA catalysis.

3.13.2.2 3-Deoxy-D-manno-octulosonic Acid Aldolase

3-Deoxy-D-manno-octulosonic acid (KDO) aldolase (EC 4.1.2.23; CAS 9026-95-3) is an inducible enzyme found in both gram-negative and gram-positive bacteria. KDO occurs as a ketosidic component in all lipopolysaccharides in gram-negative bacteria and has also been identified in acidic exopolysaccharides (K-antigens). The incorporation of KDO appears to be a vital step in lipopolysaccharide biosynthesis.[37] The aldolase catalyzes the reaction between D-arabinose and pyruvate with an *re*-face attack on the aldehydic carbonyl. Early work by Ghalambor and Heath detailed KDO aldolase from KDO-grown *E. coli*[38,39] and *Aerobacter cloacae*.[40] The degradation of KDO is favored ($K_{eq} = 77$ mM); KDO biosynthesis does not proceed by this enzyme but rather via KDO8P synthetase (EC 4.1.2.16, Section 3.13.4.2). Knappmann and Kula[41] have screened a variety of gram-negative microorganisms for KDO aldolase. The enzyme is apparently noncytoplasmic, which presented difficulties in isolation. Wong and co-workers[42] investigated KDO aldolase from *Aureobacterium barkerei* strain KDO-37-2-ATCC 49977, a gram-positive bacterium. The enzyme was purified to a specific activity of 3.7 U mg^{-1} (one unit U is defined as the enzyme required to catalyze conversion of 1 μmol of substrate to product per minute). K_m for the natural electrophile D-arabinose is unusually high at 1.2 M, suggesting that the weakly populated open chain form may be the species bound. In contrast, K_m for KDO is reportedly 6 mM. This value both emphasizes the role of the enzyme in KDO degradation and suggests that the cyclic form of the reaction product is bound by the enzyme.

KDO aldolase exhibits a broad substrate specificity (Table 1) and D-arabinose can be replaced by a variety of aldehydes. In general, the enzyme accepts substrates with the *R* configuration at C-3. With unnatural electrophiles, the reaction is apparently readily reversible: substrates with the

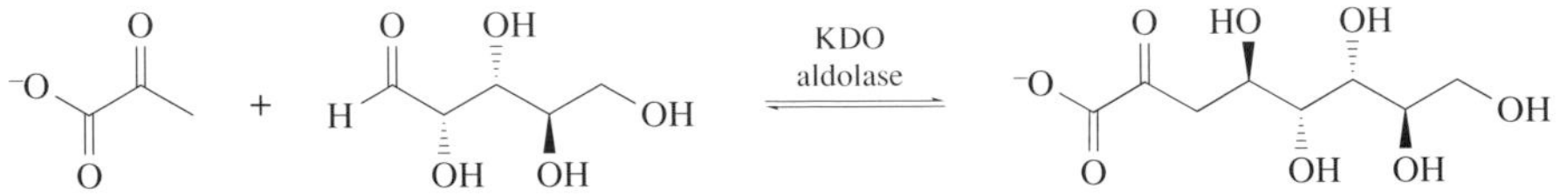

Scheme 9

S configuration at C-2 are favored kinetically while those with the *R* configuration at C-2 are thermodynamically favored.[42] As with other aldolases, substituted pyruvate adducts are not accepted as substrates.

Table 1 Substrate specificity of KDO aldolase from *A. barkerei.*

Substrate	V_{rel}
D-Arabinose	100
D-Threose	128
D-Erythrose	93
D-Ribose	72
2-Deoxy-D-ribose	71
L-Glyceraldehyde	36
D-Glyceraldehyde	23
2-Deoxy-2-fluoro-D-arabinose	46
D-Lyxose	35
5-Azido-2,5-dideoxy-D-Ribose	15
D-Altrose	25
L-Mannose	15

Source: Sugai *et al.*[42]

3.13.2.3 Dihydroneopterin Aldolase

Unlike mammalian cells which import folates via a transport system, most prokaryotes carry out *de novo* biosynthesis of folic acid derivatives. The ultimate progenitor of the pterin nucleus is 3-phosphoglycerate and an early step in the biosynthesis involves aldol cleavage of a glycolaldehyde side chain (Scheme 11).

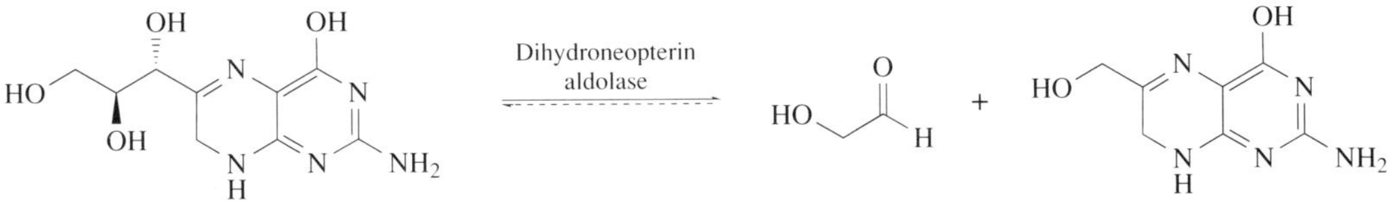

Scheme 10

Dihydroneopterin aldolase (EC 4.1.2.25; CAS 37290-59-8) was first isolated and examined from *E. coli* by Mathis and Brown.[43] The enzyme showed a molecular weight near 100 kDa (SDS–PAGE) and a pH optimum near 9.6. Kinetic examination yielded a Michaelis constant for dihydroneopterin of 9 mM. The authors were unable to observe aldol reaction, although substrate concentrations utilized may have been too low (dihydropterin at 37 mM, glycolaldehyde at 150 mM) for the observation of activity. The enzyme is specific for the L-*threo* configuration of the side chain, and the D-*threo* and L-*erythro* isomers were not cleaved. The authors proposed a mechanism for cleavage that invokes nucleophilic activation, although the mechanism differs from most Schiff base aldolases in that the imine does not involve an enzyme residue (Scheme 12).

More recently, the molecular biology of the folate biosynthesis operon has been examined in several pathogens. In *Streptococcus pneumoniae*, four genes (*sulA*, *-B*, *-C*, and *-D*) make up a 10 kb chromosomal fragment that encodes for folate biosynthesis.[44] The *sulD* gene encodes for a bifunctional protein that catalyzes both retroaldol reaction and subsequent phosphorylation (hydroxymethyldihydropterin pyrophosphokinase). This behavior contrasts with the genetic organization of the dihydroneopterin aldolase and dihydroneopterin pyrophosphokinase domains in *B. subtilis*

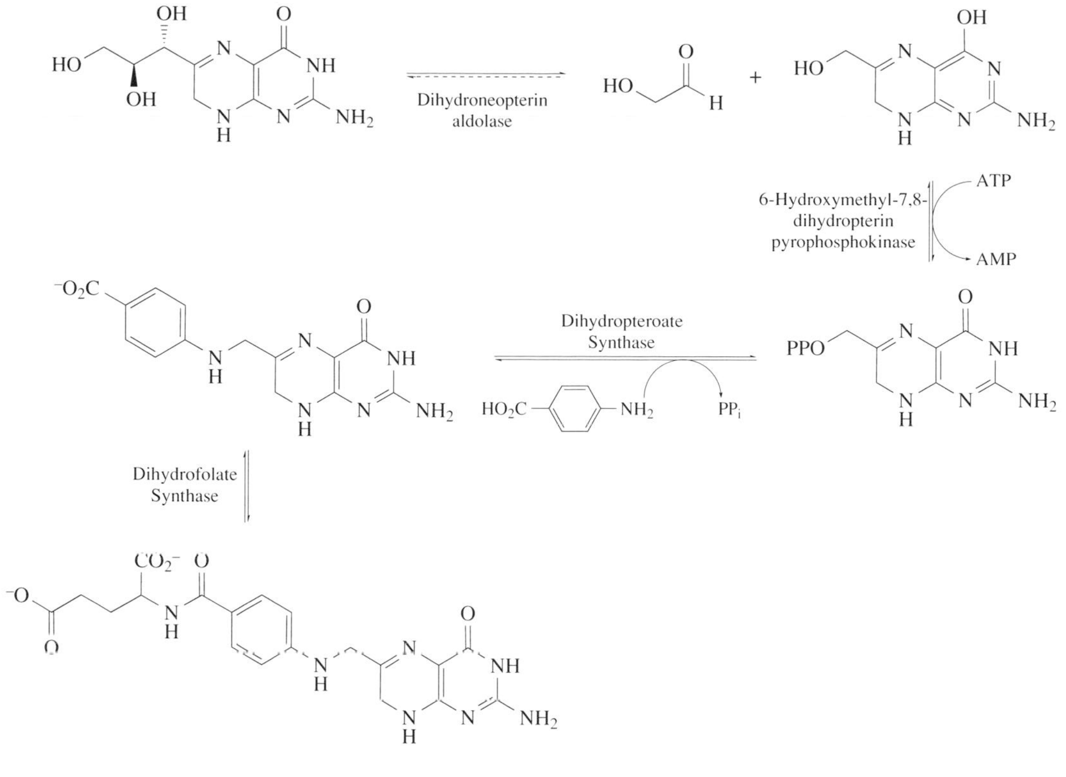

Scheme 11

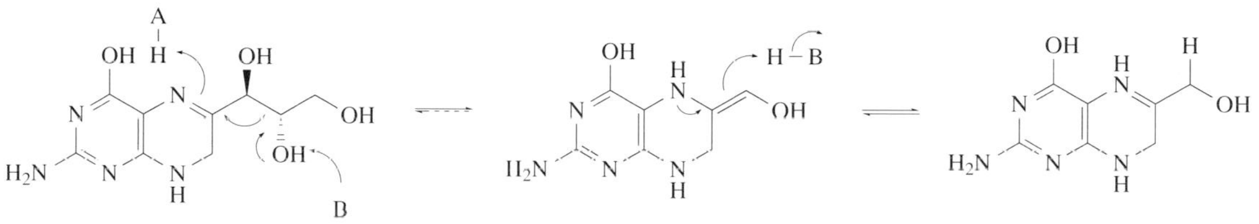

Scheme 12

where although the genes responsible for aldolase and kinase production are linked on a single operon, they encode distinct proteins.[45]

Volpe *et al.*[46,47] reported the genetic organization of the folate synthesis operon from *Pneumocystis carinii*. In this organism, the entire system of enzymes is coded by four genes, designated *fasA*, *-B*, *-C*, and *-D*. Cloning of various segments of this operon utilizing the insect-baculovirus expression system demonstrated that both *fasA* and *fasB* are required for aldolase activity. Finally, Delves and co-workers[48] conducted similar cloning experiments with the folate synthesis system of *Staphylococcus haemolyticus*. In this instance, the *folP* gene coding for the dihydropteroate synthase was cloned and expressed. Examination of the downstream open reading frame, labeled *folQ*, overlaps the *folP* gene and, by homology with the *sulD* gene of *S. pneumoniae*, codes for the dihydroneopterin aldolase. Again, the activity of this gene product appears to be multifunctional.

3.13.2.4 4-(2-Carboxyphenyl)-2-oxobut-3-enoate Aldolase (2′-Hydroxybenzalpyruvate Aldolase)

A large number of microorganisms are capable of growth on phenanthrene, naphthalene, or substituted variants of these nuclei as the sole carbon source.[49–52] Reports from several investigators conclusively demonstrate that phenanthrene/naphthalene degradation is conferred by the presence of catabolic plasmids, the best described of which is NAH7 harbored by numerous strains of *Pseudomonas putida*.[49–54] The molecular biology of this and related plasmids has been studied extensively.[55] In a series of papers, Barnsley and co-workers[56–60] elucidated the aromatic metabolic pathways utilized by a variety of such pseudomonads. Oxidation of the polyaromatic hydrocarbon nucleus apparently proceeds through 1-hydroxy-2-naphthoic acid which is subsequently cleaved to 2'-carboxybenzalpyruvate (Scheme 14). This latter metabolite is subjected to enzyme-catalyzed retroaldol reaction to yield *o*-carboxybenzaldehyde.

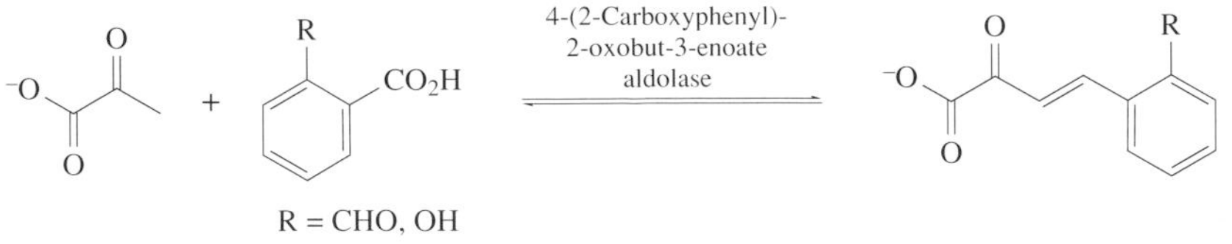

Scheme 13

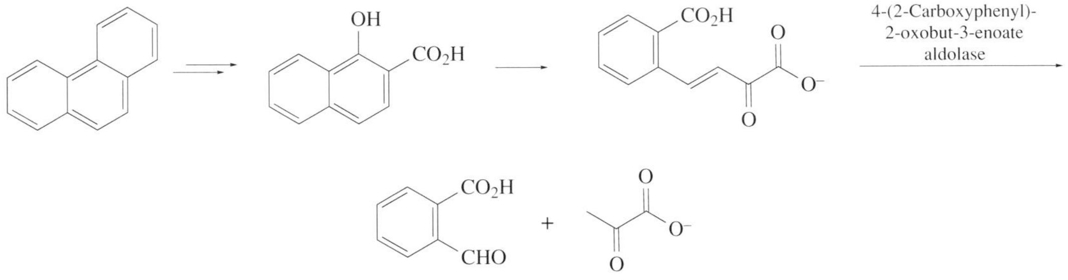

Scheme 14

2'-Hydroxybenzalpyruvate aldolase (EC 4.1.2.34) has been purified to homogeneity from *P. vesicularis* DSM 6383 by Stolz and co-workers.[61] The homogeneous protein shows a specific activity of 24 U mg^{-1} and is a trimer with subunit molecular weight of 38.5 kDa (SDS–PAGE). Catalysis was shown to proceed via Schiff base formation by inactivation with sodium borohydride in the presence of hydroxybenzalpyruvate, salicylaldehyde, or pyruvate. The enzyme is also inactivated by *p*-chloromercuribenzoate. This inhibition is reversible on treatment with dithiothreitol, suggesting that an active-site cysteine is required for activity. The enzyme was insensitive to treatment with EDTA.

The aldolase from *P. putida* G7 (ATCC 17485) has been cloned, expressed and examined by Eaton and Chapman.[62,63] Plasmid pRE701, encoding the aldolase, expresses a single peptide of molecular weight 33 kDa. Eaton later presented a more detailed description of the *nahE* gene encoding the hydroxybenzalpyruvate hydratase/aldolase from *P. putida*.[62] The gene encodes for a 331 amino acid peptide with a molecular weight near 37 kDa and an isoelectric point of 5.43. The gene showed 94% homology to *doxI* from the dox operon of *Pseudomonas* sp. strain C18 that codes for dibenzothiophene oxidation.

Crude cell extracts of several other strains of *P. putida* (NCIB 9816, KT2442, NAH7, NCIB 10535) were also investigated and activities similar to that of the previous *putida* enzyme were identified (Table 2).[63] The *P. vesicularis* protein catalyzes retroaldol addition of 2'-hydroxybenzalpyruvate as well as 2',4'- and 2',6'-dihydroxybenzalpyruvate. Benzalpyruvate was converted slowly while 2-hydroxy-6-oxo-6-phenylhexa-2,4-dienoate, benzylidene acetone, and cinnamic acid were not substrates.[61]

Table 2 V_{rel} and K_m values for 2'-hydroxybenzalpyruvate aldolases.

Strain	BN6		9816		NAH7		10535	
Substrate[a]	V_{rel} (%)[b]	K_m (mM)[b]	V_{rel} (%)	K_m (mM)	V_{rel} (%)	K_m (mM)	V_{rel} (%)	K_m (mM)
2'-HBP	100	17	100	20	100	4	100	ND
2',4'-DHBP	96	15	280	45	145	19	218	ND
2',6'-DHBP	64	6	87	10	118	9	118	ND
BP	1	4000	ND	ND	ND	ND	ND	ND

Source: Eaton[62] and Eaton and Chapman.[63]   [a]BP: Benzalpyruvate; 2'-HBP: 2'-hydroxybenzalpyruvate; 2',4'-DHBP: 2',4'-dihydroxybenzalpyruvate; 2',6'-DHBP: 2',6'-dihydroxybenzalpyruvate.   [b]Values for purified enzyme. Crude cell extracts show similar values.

The aldolase from *P. aeruginosa* PAO1 cleaves several substrates, but requires the presence of an *o*-hydroxy substituent on the aromatic ring (Figure 1).[60] In the synthetic direction, the aldolase accepts substrates other than salicylaldehyde, including substrates lacking the *o*-hydroxy substituent. For example, benzaldehyde is rapidly converted to benzalpyruvate. The authors report that this reaction is irreversible, presumably as a result of dehydration of the initial adduct. No discussion of the stereochemistry of the aldol addition is reported, and is likely made moot by the rapid dehydration of the initial product.

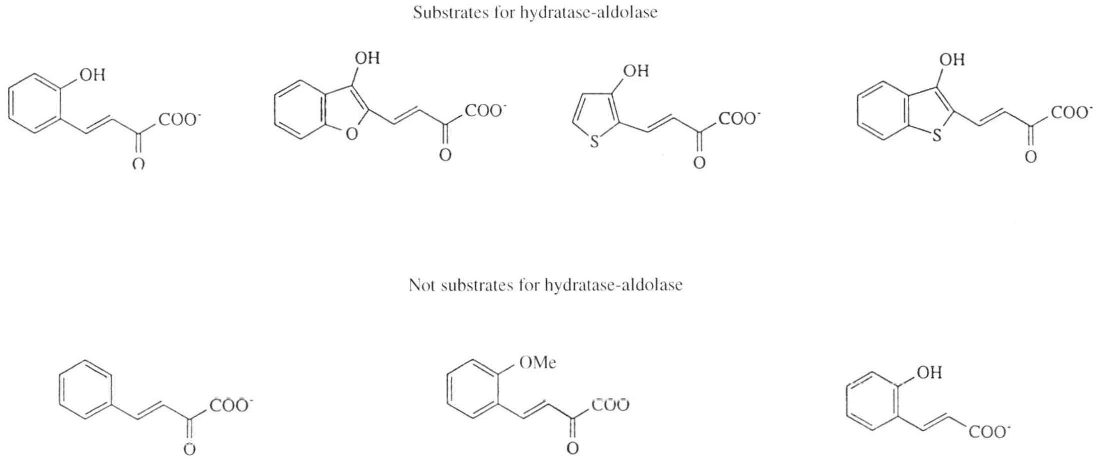

Substrates for hydratase-aldolase

Not substrates for hydratase-aldolase

Figure 1 Substrate spectrum for 2′-hydroxybenzalpyruvate aldolase.

The aldolase reaction shown in Scheme 15 requires both dehydratase and aldolase activity. The question of whether the two reactions are catalyzed by one or two proteins was long debated. 2′-Hydroxybenzalpyruvate is known to be metabolized to salicylaldehyde and pyruvate and requires hydration prior to aldol cleavage. Reports by Eaton and Chapman[63] suggest the two reactions are catalyzed by a single protein, based on the isolation of a DNA fragment expressing both activities. This fragment encodes for a protein of 33 kDa molecular weight, too small to represent two distinct enzymes. These researchers proposed a mechanism for the hydratase activity that invokes participation of a substrate *o*-hydroxy substituent (Scheme 15).

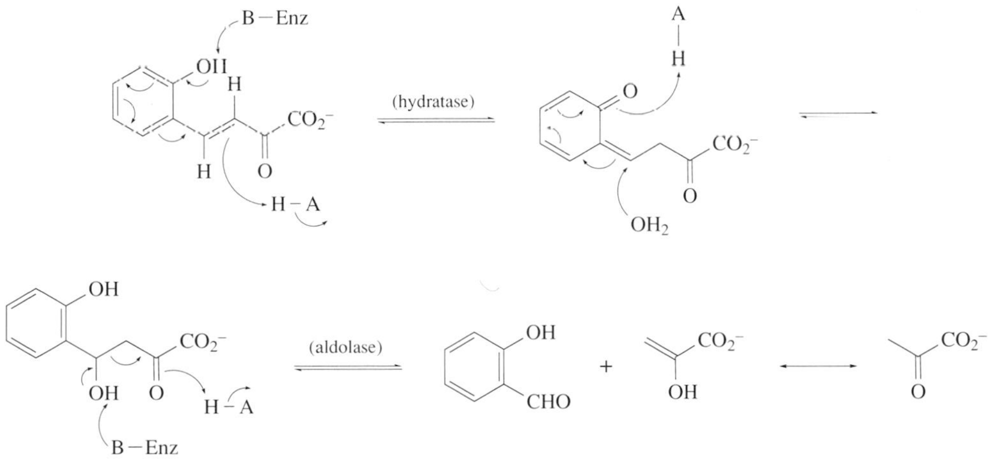

Scheme 15

A variety of reports on the decomposition of substituted phenanthrenes and naphthalenes suggest that aldolases from several species have broad substrate specificities. For example, reports of phenanthrene mineralization, while failing to discuss explicitly an aldolase activity, note the presence of *o*-carboxybenzaldehyde as a metabolite.[64–68] Similarly, catabolic routes for naphthalene sulfonates[69] and the insecticide carbaryl[70] both utilize an aldolase, presumably through oxidation/

hydrolysis to 1-naphthol. Both 1- and 2-methylnaphthalene are utilized by *P. putida* CSV86,[53] while 2,6-dimethylnaphthalene is metabolized by several strains of *Flavobacteria*,[56] suggesting that at least some versions of the aldolase tolerate alkyl substitution of the ring. Finally, 1- and 2-chloronaphthalene are decomposed by several pseudomonads, although at rates too low to support growth. The corresponding chlorosalicylates were isolated in culture media, suggesting that halogen substitution of the aromatic nucleus is also tolerated by the aldolase.

3.13.2.5 *N*-Acetylneuraminate Aldolase

N-Acetylneuraminate (NeuAc) is the best known member of the sialic acids, a 40-membered class of amino sugars. NeuAc and its derivatives are acidic sugars typically located at the terminal positions of glycoproteins and glycolipids. The sialic acids thus play important roles in biochemical recognition processes including viral infection[71] and cellular adhesion.[72,73] Sialic acids also occur at elevated levels in several types of cancer cells.[74,75] In prokaryotic cells, NeuAc has been found as a constituent capsular polysaccharide of some pathogenic bacteria (*E. coli* K1 serotypes and *Neisseria meningitidis* B and C).[76] In this role, NeuAc functions as a pathogenic determinant that protects against host defenses.

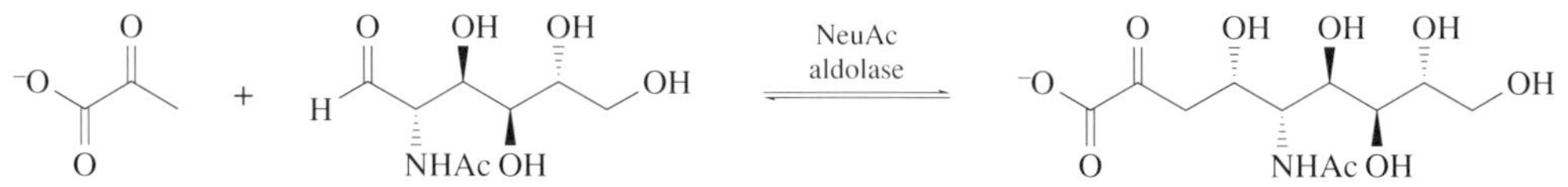

Scheme 16

NeuAc aldolase (EC 4.1.3.3; CAS 9027-60-5) catalyzes the reaction of *N*-acetyl-D-mannosamine and pyruvate to produce NeuAc. NeuAc aldolase is inducible in microorganisms and is produced only in the presence of NeuAc. The enzyme has been detected in a variety of sources including mammalian tissues,[74,75] *Streptococcus oralis*,[77] *Clostridium perfringens*,[78–80] *Corynebacterium diphtheria*,[81–83] *Pasteurella multocida*,[84] *Vibrio cholera*,[85] and *E. coli*.[86]

NeuAc aldolase is the first intracellular enzyme in the pathway of NeuAc catabolism.[87–89] Since the equilibrium for the catalyzed reaction lies in the catabolic direction, NeuAc synthase (EC 4.1.3.19, Section 3.13.4.3) is proposed to be responsible for bacterial NeuAc synthesis, while NeuAc aldolase regulates the pool of intracellular NeuAc by catabolism.

Schauer and Wember[90] isolated and characterized NeuAc aldolase from pig kidney. The enzyme was purified 630-fold by heat treatment, gel filtration, and affinity chromatography on immobilized neuraminic acid β-methyl glycoside. The NeuAc aldolase exhibits a molecular weight of 58 kDa, a pH optimum near 7.2 and a K_m (NeuAc) of 3.7 mM. Substrate specificity studies indicate that the enzyme accepts glycolylneuraminic acid and 9-*O*-acetyl-*N*-acetylneuraminic acid at 55% and 32% the rate of the natural substrate, respectively. *N*-acetyl-4-*O*-acetylneuraminic acid and 2-deoxy-2,3-didehydro-*N*-acetylneuraminic acid are not substrates. Enzyme activity was inhibited by *p*-chloromercuribenzoate, *o*-phenanthroline, cyanide, 5-diazonium-1-*H*-tetrazole, 5,5′-dithiobis(2-nitrobenzoic acid), diethylpyrocarbonate, and Rose Bengal in the presence of light and oxygen. Reduction with sodium borohydride in the presence of NeuAc or pyruvate resulted in irreversible inhibition of enzyme activity, suggesting that the enzyme utilizes a Schiff base. Inhibition experiments with affinity labels further implicate the involvement of histidine, lysine, and thiol residues in enzyme catalysis.

The purification and biochemical characterization of NeuAc aldolase from *E. coli* K1 has also been investigated.[91] This enzyme was purified 312-fold to a specific activity of 15 U mg^{-1} by streptomycin sulfate and ammonium sulfate fractionation, hydrophobic chromatography on butyl-agarose and phenyl Sepharose, gel filtration on Sephacryl, and anion exchange chromatography. The molecular weight of the *E. coli* enzyme is 135 kDa and it exists as a tetramer of 33 kDa subunits. Maximum activity was obtained near pH 7.8 and 37 °C. Michaelis constants were determined for *N*-acetylmannosamine (7.7 mM), pyruvate (8.3 mM), and NeuAc (4.8 mM). The synthetic reaction was activated by Ca^{II} and inhibited by Mn^{II}; NeuAc cleavage was inhibited by Ca^{II}, Mn^{II}, and pyruvate. Of additional interest, Ferrero *et al.*[91] also reported that *E. coli* K1 lacks NeuAc synthase (EC 4.1.3.19, Section 3.13.4.3), in contrast to previous reports; rather the anabolic activity detected is attributable to NeuAc aldolase.

NeuAc aldolase has also been detected in most of the 65 strains of *Pasteurella* and *Pasteurella*-like strains investigated by Müller and Mannheim.[84] No purification or biochemical properties of any of these enzymes were reported.

The metabolism of glycoprotein-derived sialic acid was investigated in nine strains of *S. oralis* isolated from blood cultures of patients with infective endocarditis or from the oral cavity as part of the normal flora. Homer *et al.*[77] detected elevated levels of NeuAc aldolase in cell-free extracts of mucin-grown cultures as compared to markedly repressed activity in cells grown on glucose. Three strains were grown in media supplemented with α_1-acid glycoprotein, a major component of human plasma containing high levels of sialic acid: three strains all expressed high levels of the aldolase. No purification or biochemical properties of any of these enzymes were reported.

Lilley *et al.*[92] cloned the *E. coli* NeuAc aldolase into an inducible expression vector and over-expressed the gene in *E. coli*. The recombinant enzyme was grown on a 110 L scale, producing 4000 mg of enzyme with a purified specific activity of 1.2–2.2 U mg^{-1}.

The three-dimensional structure of NeuAc aldolase from *E. coli* has been determined by X-ray crystallography.[93] In contrast to earlier reports that the enzyme is trimeric, the structure shows a tetrameric protein consisting of an α/β-barrel domain followed by a carboxy terminal extension of three α helices (Figure 2). The active site was identified as a pocket at the carboxy terminal end of the eight-stranded β-barrel. Lys165 lies within this pocket and is likely to be the reactive residue that forms a Schiff base intermediate with the substrate. NeuAc aldolase shows homology to dihydropicolinate synthase and MosA (an enzyme implicated in rhizopine synthesis), suggesting that all three enzymes share similar structures.

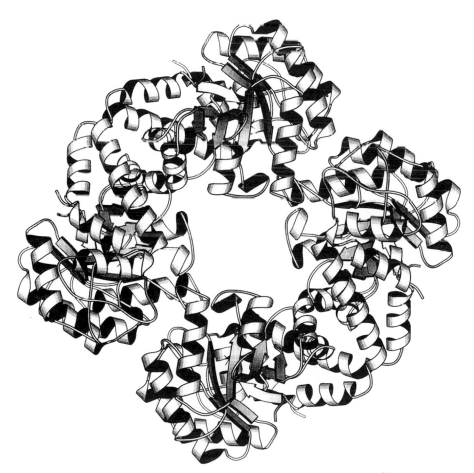

Figure 2 *E. coli* NeuAc aldolase crystal structure at 2.2 Å.

The importance of configuration and the form of the substrate accepted by NeuAc aldolase have been investigated by several groups in an effort to elucidate the mechanism. Deijl and Vliegenthart[94] reported the less stable α-anomer of the NeuAc cyclic pyranose is accepted by the aldolase, while the β-anomer is inert. Friebolin and co-workers[95] confirmed that both substrates and products in NeuAc aldolase-catalyzed synthesis and cleavage are α-pyranoses. Based on these observations, these workers proposed a mechanism in which open chain forms of NeuAc and *N*-acetylmannosamine are not involved in enzyme binding (Scheme 17). Strengthening these proposals, David *et al.*[96] showed that while 4- and 6-*O* methyl ethers of ManNAc (pyranoses) are accepted as substrates, the 5-*O*-methyl ether of ManNAc (a furanose) is inert. Alternatively, C-5-deoxy analogues of ManNAc-substrates limited to furanose forms were synthesized and were shown to be excellent substrates for NeuAc aldolase. Such furanoses are flexible, and modeling studies show that coincidence of the ring substituents for at least some mannofuranose substrates with substituents of mannopyranose is at least possible. Continuing investigations suggest that the determining factors for substrate suitability are an unhindered α-face of the hexose and an anomeric hydroxy group in a hydrophobic environment.

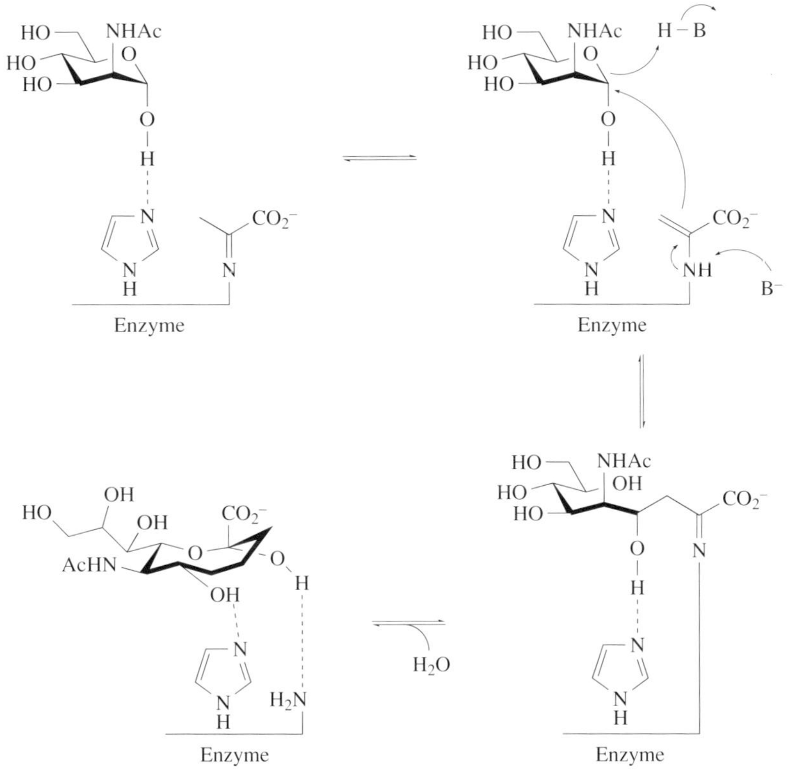

Scheme 17

NeuAc aldolase has been used extensively for the preparation of unnatural sialic acids (Table 3). NeuAc aldolase-catalyzed reactions are readily reversible and the enzyme operates under thermo-dynamic control.[97] Thus, electrophiles that exist in a $^4C_1$ chair conformation with an equatorial hydroxy at C-3 give *si*-face addition of pyruvate. Alternatively, electrophiles that place the C-3 hydroxy in an equatorial position in the $^1C_4$ conformation yield *re*-face attack. Substrates that lack a C-3 substituent or place the C-3 hydroxy in an axial position in the predominant conformation give mixtures of products resulting from both *re* and *si* face addition.

Kragl *et al.*[97] further investigated an enzymatic two-step synthesis of NeuAc from *N*-acetyl-glucosamine and pyruvate. The reaction, which occurs in a membrane reactor, utilizes an epimerase to convert *N*-acetylglucosamine to *N*-acetylmannosamine followed by aldolase-catalyzed reaction with pyruvate. An excess of pyruvate is employed to drive the reaction and NeuAc has been synthesized on a multikilogram scale (Scheme 18). Enzymatic synthesis with an epimerase is also feasible for the preparation of unnatural derivatives; KDO and 4-*epi* KDO were also synthesized in a 5:1 ratio on a preparative scale (8.2 mmol) with 1800 units of enzyme.

Table 3 Substrate specificity of NeuAc aldolase.

Substrate	V_{max} (U mg^{-1})	K_m (mol L^{-1})
L-Xylose	0.44	0.32
D-Xylose	0.84	0.21
L-Arabinose	1.4	0.5
D-Arabinose	1.42	0.84

Source: Kragl *et al.*[97]

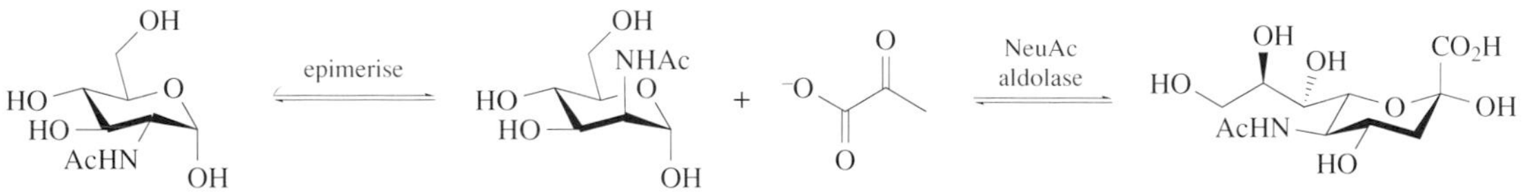

Scheme 18

The substrate specificity of the *E. coli* enzyme in the retroaldol direction was investigated by Aisaka *et al.*[98] N-glycolylneuraminic acid was cleaved at 20% the rate of NeuAc. However, there was no detectable cleavage of colominic acid, the α,α'-2-8 homopolymer of NeuAc, or 2-keto-carboxylic acids such as 2-ketohexanoic acid, 2-ketooctanoic acid, and 2-ketononanoic acid. Thus, NeuAc aldolase shows high specificity for sialic acids. K_m for both N-glycolylneuraminic acid and N-acetylneuraminic acid is 3.3 mM. V_{max} is 14.0 mmol min^{-1} mg^{-1} for N-glycolylneuraminic acid and is 71.4 mmol min^{-1} mg^{-1} for N-acetylneuraminic acid.

NeuAc aldolase has been utilized in chemo-enzymatic syntheses of a variety of NeuAc analogues.[99,100] Zhou *et al.*[101] utilized NeuAc aldolase synthetically to prepare the first 3-substituted analogue of castanospermine via a NeuAc derivative.

3.13.2.6 4-Hydroxy-2-ketoglutarate Aldolase

4-Hydroxy-2-ketoglutarate aldolase (EC 4.1.3.16 (formerly EC 4.1.2.31); CAS 9030-81-3) catalyzes three reactions *in vivo*; cleavage of 4-hydroxy-2-ketoglutarate to produce pyruvate and glyoxylate, cleavage of 4-hydroxy-2-ketobutyrate to produce pyruvate and formaldehyde, and β-decarboxylation of oxaloacetate. This aldolase is the only known enzyme that shows both aldolase and β-decarboxylase activity. β-decarboxylases and aldolases have long been compared structurally; as a bifunctional enzyme, hydroxyketoglutarate aldolase may provide the evolutionary link between the two enzymes.

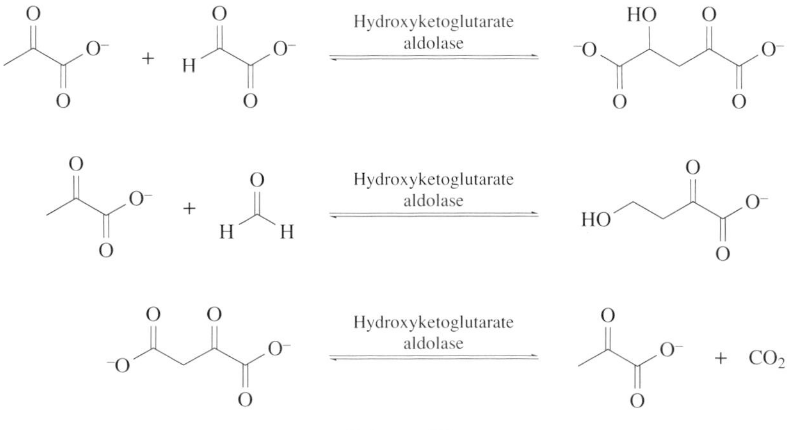

Scheme 19

Both type I and type II versions of the enzyme are known. The aldolase has been isolated from a variety of bacterial[102,103] and mammalian sources.[104-107] In mammalian cells, hydroxyketoglutarate aldolase functions as the terminal step of hydroxyproline metabolism,[108] while in bacterial cells the enzyme is postulated to control intercellular glyoxylate levels.[109] It has also been suggested that hydroxyketoglutarate aldolase may play an important role in the mineralization of glyoxylate to two moles of CO_2, via malate (Scheme 20).[109] In this scheme, glyoxylate is condensed with pyruvate to form hydroxyketoglutarate. Oxidative decarboxylation yields malate, which is in turn converted to oxaloacetate. The latter species is decarboxylated to pyruvate, completing the mineralization of glyoxylate.

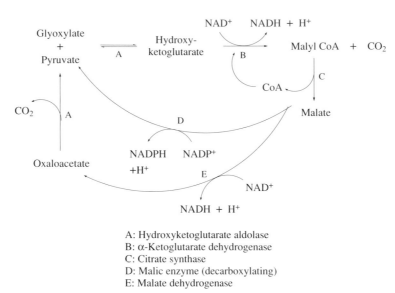

A: Hydroxyketoglutarate aldolase
B: α-Ketoglutarate dehydrogenase
C: Citrate synthase
D: Malic enzyme (decarboxylating)
E: Malate dehydrogenase

Scheme 20

The pH activity range of the enzyme is quite broad, and at least 50% of the maximum activity is observed between pH 5 and 9. The substrate spectrum and stereospecificity of the enzyme varies with the source (Tables 4 and 5). Thus, while the mammalian enzyme produces and cleaves both stereoisomers of hydroxyketoglutarate, the *E. coli* enzyme shows a 10-fold preference for the L(R) isomer. Nishihara and Dekker[110] reported K_m values of 2.3 and 25 mM for L- and D-hydroxy-ketoglutarate, respectively, using the *E. coli* enzyme. Alternatively, the two stereoisomers showed similar V_{max} values, at 7.9 μmol min^{-1} mg^{-1} for the L-isomer and 6.7 μmol min^{-1} mg^{-1} for the D. In contrast, Rosso and Adams[111] reported K_m values for the rat liver enzyme of 0.11 mM and 0.22 mM for L- and D-hydroxyketoglutarate, respectively, with a V_{max} ratio for the enantiomers of 8 : 5. Maitra and Dekker[112] reported K_m values of 1.0 mM for both isomers of hydroxyketoglutarate for the rat liver enzyme. Equilibrium constants ranging from 11.8 and 10.4 favoring cleavage to 0.8 and 1.3 favoring synthesis have been reported; the variability may be due to buffer dependence. Turner and co-workers[113] reported that the bacterial enzyme shows a narrower substrate specificity than does the mammalian enzyme. 2-Keto-4-hydroxybutyrate is accepted, although at only 8% of the rate of DL-hydroxyketoglutarate (Table 4). 2-Ketoglutarate, 2-keto-4-hydroxy-4-methylglutarate, 2-keto-valerate, 2-keto-4,5-dihydroxyvalerate, 2-keto-(L or D)-gluconate, 5-keto-D-gluconate, 2-keto-3-deoxy-galactonate, 5-keto-4-deoxyglucarate, acetoacetate, isocitrate, FDP, 2-deoxyribose, 2-deox-yribose-5-phosphate, and L-threonine are not accepted as substrates.

The stereospecificity of the reaction in the synthetic direction has also been explored. Using isotopically labeled pyruvate, Meloche[114] has shown that C-3 of pyruvate is deprotonated stereo-specifically (Scheme 21). These same studies showed a primary kinetic isotope effect of 6–7, suggesting that deprotonation of pyruvate C-3 is largely rate determining.

Mammalian hydroxyketoglutarate aldolase from several sources is tetrameric, with a subunit molecular weight near 36 kDa. The *E. coli* enzyme, however, is trimeric, with a subunit molecular weight of 22.2 kDa.[115] Various reports state that metal ions affect aldolase activity; no mechanistic role for such ions has been postulated.

Hydroxyketoglutarate aldolase from *E. coli* has been cloned, sequenced, and overexpressed. Dekker and co-workers[116–118] elucidated the primary amino acid sequence of both the aldolase active site and the entire protein through labeling and digestion studies. The enzyme consists of 213 amino acids and shows 65% homology with KDPG aldolase from *P. putida*. The *E. coli* aldolase has a pH

Table 4 Substrate specificity of hydroxyketoglutarate aldolases.

Substrate	Rat liver (%)	Bovine kidney (%)	E. coli (%)
DL-Hydroxyketoglutarate	100	100	100
L-Hydroxyketoglutarate	99	113	157
D-Hydroxyketoglutarate	101	113	13
2-Keto-3-deoxyglucarate	29	ND[a]	ND
2-Keto-4,5-dihydroxyvalerate	2	ND	0
Oxaloacetate	ND	12	ND
2-Keto-4-hydroxybutyrate	ND	ND	8

Source: Dekker *et al.*,[110] Rosso and Adams,[111] Maitra and Dekker,[112] and Floyd *et al.*[113] [a]ND: not determined.

Table 5 Substrate specificity of hydroxyketoglutarate aldolase from rat liver (reactions with glyoxylate).

Substrate	Rat liver (%)
Pyruvate	100
Oxaloacetate	3300
3-(4-Hydroxyphenyl)pyruvate	490
3-(4-Imidazole)pyruvate	290
3-Phenylpyruvate	450
3-Thiopyruvate	65
3-Hydroxypyruvate	100
3-Bromopyruvate	180
2-Ketobutyrate	260
2-Ketoglutarate	71
Pyruvaldehyde	290
Pyruvic acid, methyl ester	140
Pyruvic acid, ethyl ester	120
Acetaldehyde	94
Pyruvic acid	100

Source: Rosso and Adams[111] and Maitra and Dekker.[112]

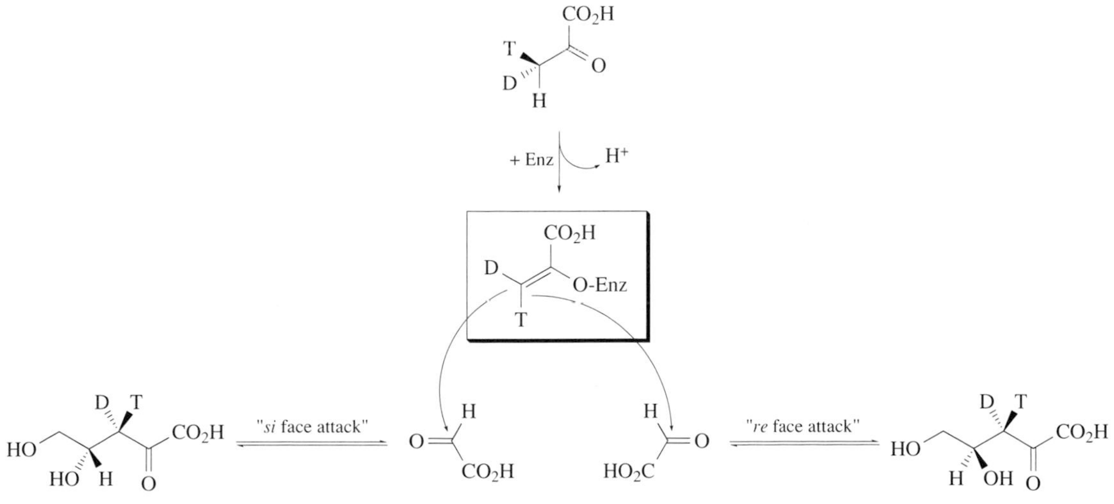

Scheme 21

optimum of 8.6 and does not require metal ion cofactors for activity. Labeling studies have shown that Lys133 is the essential Schiff base forming residue while Arg49 and Glu45 act as amphoteric proton donors/acceptors.[119,120] Competition studies show that pyruvate, glyoxylate, and hydroxyketoglutarate bind to the same active site; all three substrates inactivate an identical lysine on treatment with cyanoborohydride.

In 1992, Conway and co-workers[121] reported that hydroxyketoglutarate aldolase from *E. coli* was identical to the *E. coli* KDPG aldolase (EC 4.1.2.14, Section 3.13.8.4) based on cloning studies. Nonetheless, significantly different behaviors of the two enzymes have been reported.[113,122] Despite this observation, it remains difficult to ascertain the identities of the enzymes involved in many investigations. Significantly different, albeit overlapping, substrate specificities of those enzymes involved in both the synthesis and cleavage of hydroxyketoglutarate make unambiguous identification of these enzymes difficult.

3.13.2.7   4-Hydroxy-2-ketovalerate Aldolase

Aromatic hydrocarbons are important structures in fossil fuels as well as in both natural and anthropogenic products. The aromatic ring cleavage pathways of bacteria consist of two groups; the *ortho* cleavage pathway, which transforms the common intermediate 3-oxoadipate enol-lactone into succinate and acetyl CoA, and the *meta* cleavage pathway, which produces pyruvate and a short chain aldehyde. 4-Hydroxy-2-ketovalerate aldolase functions in aromatic *meta* cleavage pathways for the degradation of biphenyl, phenol, toluene, and cumene as well as *p*-cumate and naphthalene (Scheme 23).

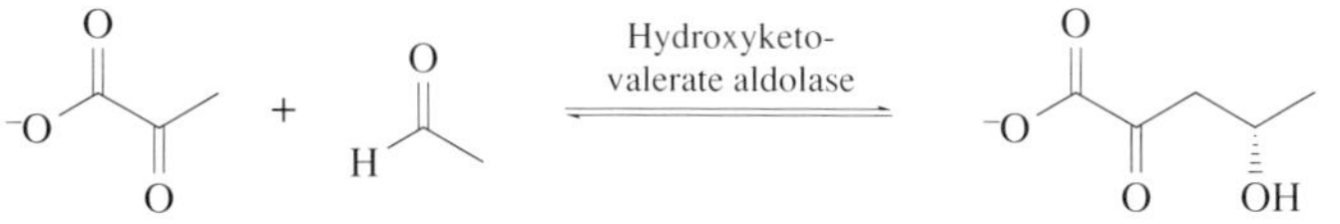

Scheme 22

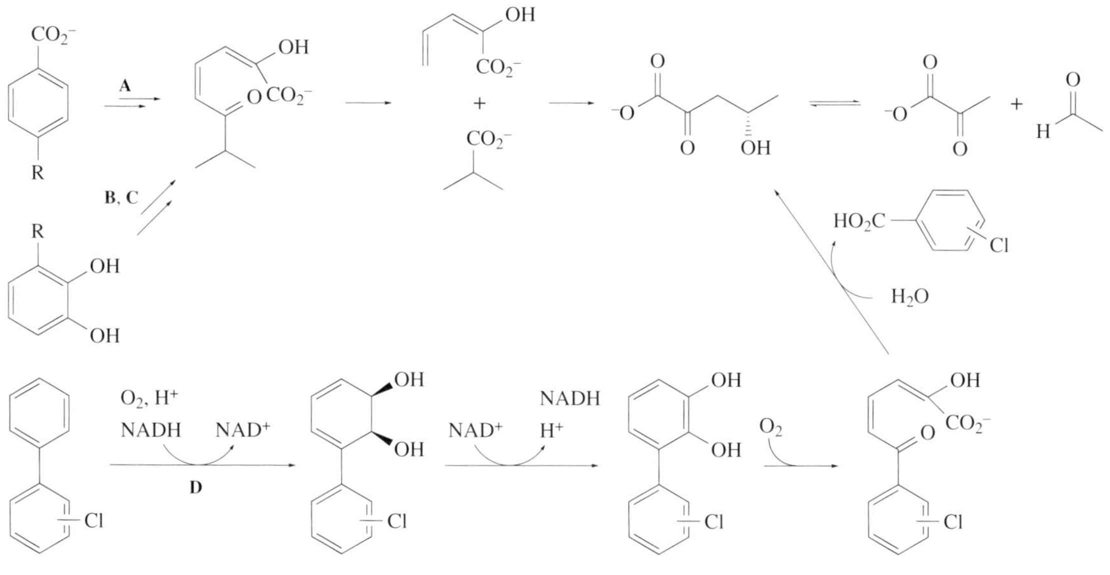

A: *p*-Cumate metabolism
B: *meta*-Fission path for substituted catechols
C: Toluene conversion to tricarboxylic acid cycle intermediates
D: Biphenyl degradation pathway

Scheme 23

Hydroxyketovalerate aldolase has been isolated from strains of *Pseudomonas* (*putida* and *fluorescens*), *E. coli*, and *Acinetobacter* sp. Powlowski and co-workers[123] purified the enzyme involved in phenol degradation from *Pseudomonas* sp. strain CF600. The enzyme shows a pH optimum of 8.8 and was stimulated by the addition of both Mn^{II} and NADH. The authors indicated that the stimulation by NADH results from either an allosteric site for the pyridine nucleotide or a pyridine nucleotide-induced conformational change in the closely associated dehydrogenase; both would lead to enzyme activation. The enzyme retains significant activity in the absence of divalent metals and is not inactivated by dialysis against EDTA. On this basis, the aldolase can tentatively be classified as a type I enzyme, although the definitive test—inactivation by pyruvate and sodium borohydride—has apparently not been attempted.

Burlingame and Chapman[124] investigated the stereospecificity of hydroxyketovalerate aldolase from three sources; *E. coli*, *P. putida*, and *Acinetobacter* sp. While the latter two enzymes show complete stereospecificity in the retroaldol cleavage, the *E. coli* aldolase shows no stereochemical preference. The *E. coli* enzyme may be either a single nonstereospecific aldolase or a mixture of two enzymes; distinction between these possibilities awaits further investigation.

Eaton[125] has cloned and sequenced the hydroxyketovalerate aldolase involved in *p*-cumate metabolism from *P. putida* F1. Lau and co-workers[126] have sequenced the hydroxyketovalerate aldolase involved in toluene degradation from this same strain. Hofer *et al.*[127] cloned and sequenced the hydroxyketovalerate aldolase from *Pseudomonas* sp. LB400 and identified this enzyme as a component of the biphenyl degradation pathway. Likewise, Kikuchi *et al.*[128] have sequenced the hydroxyketovalerate aldolase involved in biphenyl and polychlorinated biphenyl degradation from *Pseudomonas* sp. strain KKS102. This activity in hydroxyketovalerate aldolase cloning and sequencing should prove valuable for investigations of bacterial aromatic degradation pathways.

3.13.3 DIVALENT METAL-CATALYZED ALDOLASES

3.13.3.1 4-Hydroxy-2-keto-4-methylglutarate Aldolase

Aldolases that cleave 4-substituted-4-hydroxy-2-ketoglutarate have been isolated from both plants and bacteria. In 1962, Shannon and Marcus[129] reported the first such protein from peanut cotyledons. In 1972, another hydroxyketomethylglutarate aldolase was purified to a specific activity of 134 U mg^{-1} from *P. putida* grown on syringic acid as the sole carbon source.[130] Later, a similar protein showing a specific activity of 282 U mg^{-1} was isolated from *P. ochraceae* grown on phthalate. Both enzymes are hexameric with subunit molecular weights near 26 kDa. The isoelectric point of the *P. ochraceae* enzyme is 5.0; both enzymes exhibit high optimum pH values, near 9.0. Crude cell-free extracts of *P. testosteroni* have also been reported to cleave hydroxymethylketoglutarate, suggesting the protein may be widespread among pseudomonads.[131]

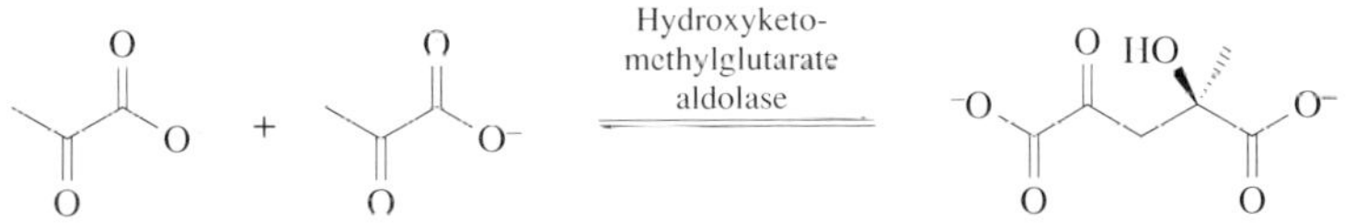

Scheme 24

The enzyme is apparently involved in gallic acid and protocatechuate degradation. In the former case, gallic acid is cleaved to 4-carboxy-4-hydroxy-2-ketoadipate which is subsequently cleaved by the aldolase to pyruvate and oxaloacetate.[132] In some pseudomonads, protocatechuate is converted to both 4-hydroxy-4-methyl-2-ketoglutarate and 4-carboxy-4-hydroxy-2-ketoadipate.[133,134] Both intermediates are cleaved by 4-hydroxy-2-keto-4-methylglutarate aldolase (EC 4.1.3.17; CAS 37290-65-6), in the former case to two equivalents of pyruvate and in the latter to pyruvate and oxaloacetate. The enzyme is induced in *P. ochraceae* grown on terephthalate and hydroxybenzoate, suggesting a role for the enzyme in the metabolism of these compounds as well.

The bacterial proteins shown a strict requirement for divalent metals, and in both cases Mg^{II} is most effective. The *putida* enzyme reportedly functions in the presence of Mn^{II} but not Ca^{II} or Zn^{II}, while the *ochraceae* enzyme is active in the presence of Mg^{II}, Mn^{II}, Co^{II}, Zn^{II}, and Cd^{II} but not Be^{II}, Ni^{II}, Ba^{II}, Hg^{II}, Sr^{II}, Cr^{III}, or Fe^{III} (Table 6). Similar Michaelis constants for Zn^{II} (170 μM) and hydroxymethylketoglutarate (290 μM) were reported for the *P. putida* enzyme. Unlike the peanut enzyme, neither of the bacteria enzymes requires reduced thiol.[130,131]

Both bacterial enzymes also catalyze aldol addition. Using the *P. ochraceae* enzyme, Maruyama[135–137] calculated equilibrium constants of 0.07 M for 4-hydroxy-4-methyl-2-ketoglutarate, and 0.03 M for 4-hydroxy-2-ketoglutarate. A value for 4-carboxy-4-hydroxy-2-ketoadipate could not be calculated since the ketoglutarate generated in the retroaldol reaction decarboxylates rapidly, driving the equilibrium towards cleavage. On the other hand, the peanut enzyme apparently cannot catalyze aldol addition, although it can exchange labeled pyruvate into hydroxymethylketoglutarate. On this basis, Maruyama[136] postulated that the enzymes operate by different mechanisms; the dependence of the peanut enzyme on reduced thiol bolsters this claim.

Table 6 Kinetic parameters for hydroxyketomethylglutarate aldolase from *P. ochraceae*.

Substrate	Metal	K_m (metal) (μM)	K_m (substrate) (μM)	V_{max} (U mg^{-1})
L-4-Carboxy-4-hydroxy-2-ketoadipate	Mg^{II}	32.3	8.8	262
	Mn^{II}	5.9	8.6	277
	Co^{II}	7.7	7.4	122
	Zn^{II}	1.8	6.5	67
	Cd^{II}	3.4	7.5	57
	Mg^{II}	154	97	69
($\pm$)-4-Hydroxy-4-methyl-2-ketoglutarate	Mg^{II}	1720	1090	185
($\pm$)-4-Hydroxy-2-ketoglutarate	Mg^{II}	2000	250	1.5

Source: Dagley[130] and Sparnins and Dagley.[131]

The enzymes show only moderately broad substrate specificity: 4-hydroxy-2-ketovalerate, citrate, 4-hydroxy-2-ketobutyrate, 2-ketoglutarate, and FDP are not cleaved by the *ochraceae* enzyme.[135] Reports of the stereospecificity of the enzyme vary. Tack *et al.*[138] reported that the *P. putida* enzyme was highly specific for the L-enantiomer of hydroxymethylketoglutarate. Alternatively, Maruyama[137] reports that although the enzyme shows a preference for the L-enantiomer of all three substrates, both enantiomers are ultimately utilized. No attempts were made to establish the level of asymmetric induction for the forward reaction.

3.13.4 ALDOLASES UTILIZING PHOSPHOENOLPYRUVATE

3.13.4.1 3-Deoxy-D-arabino-2-heptulosonic-7-phosphate Synthetase

In vivo, 3-deoxy-D-arabino-2-heptulosonic-7-phosphate (DAHP) synthetase (EC 4.1.2.15; CAS 9026-94-2) catalyzes the reaction of phosphoenolpyruvate and D-erythrose-4-phosphate to produce 7-phospho-2-keto-3-deoxy-D-arabinoheptanoate. DAHP synthetase is a key intermediate in the shikimate pathway of aromatic amino acid biosynthesis (Scheme 26).

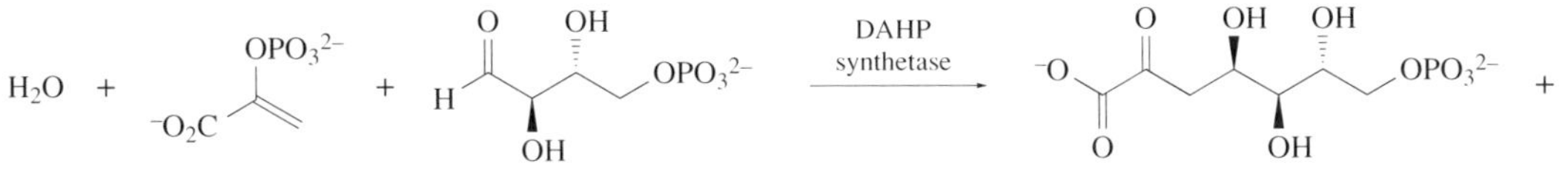

Scheme 25

Srinivasan and Sprinson[139,140] first detected DAHP synthetase in 1959 in *E. coli*. Since that time, DAHP synthetases have been isolated from a variety of plant, yeast, and bacterial cells including *Pisum sativum*,[141] *Vigna radiata*,[142] *Nicotiana silvestris*,[143] potatoes,[144,145] carrot cells,[146] tomato cells,[147] *Amycolaptosis methanolica*,[148] *Acinetobacter calcoaceticus*,[149,150] *Bacillus subtilis*,[151] *Saccharomyces cerevisiae*,[152] *Candida albicans*,[153] *Streptomyces*,[154,155] *Chlorella*,[156] and *Buchnera aphidicola*.[157] DAHP synthetases from a variety of sources have been cloned, including the *E. coli* enzyme.[158-160]

An interesting feature of DAHP synthetases is the existence of multiple isozymes in both plant and bacterial cells. A survey by Ganson *et al.*[161] of DAHP synthetases from a variety of plant sources indicates that a pair of isozymes may be universally present in plants. One of the two synthetases shows high substrate specificity and requires dithiothreitol for activity. This synthetase is stimulated fourfold by Mn^{II} and exhibits a pH optimum near 7.5. The second synthetase has greater substrate ambiguity, accepting glycolaldehyde and glyceraldehyde-3-phosphate at rates greater than that for erythrose-4-phosphate. The second isozyme also shows an absolute requirement for divalent metal cations and a high pH optimum near 9.0.

E. coli apparently expresses three isofunctional DAHP synthetases with varying activities.[162] The three enzymes appear to have evolved from a common precursor based on amino acid sequences. This range of isozymes allows the cell to modulate synthetic rates in response to the availability of various nutrients. All three synthetases are repressed by high levels of tyrosine, phenylalanine, and tryptophan. Michaelis constants for phosphoenolpyruvate have been determined and vary from 5.8 μM to 80 μM for the different isozymes. Because *E. coli* intracellular phosphoenolpyruvate

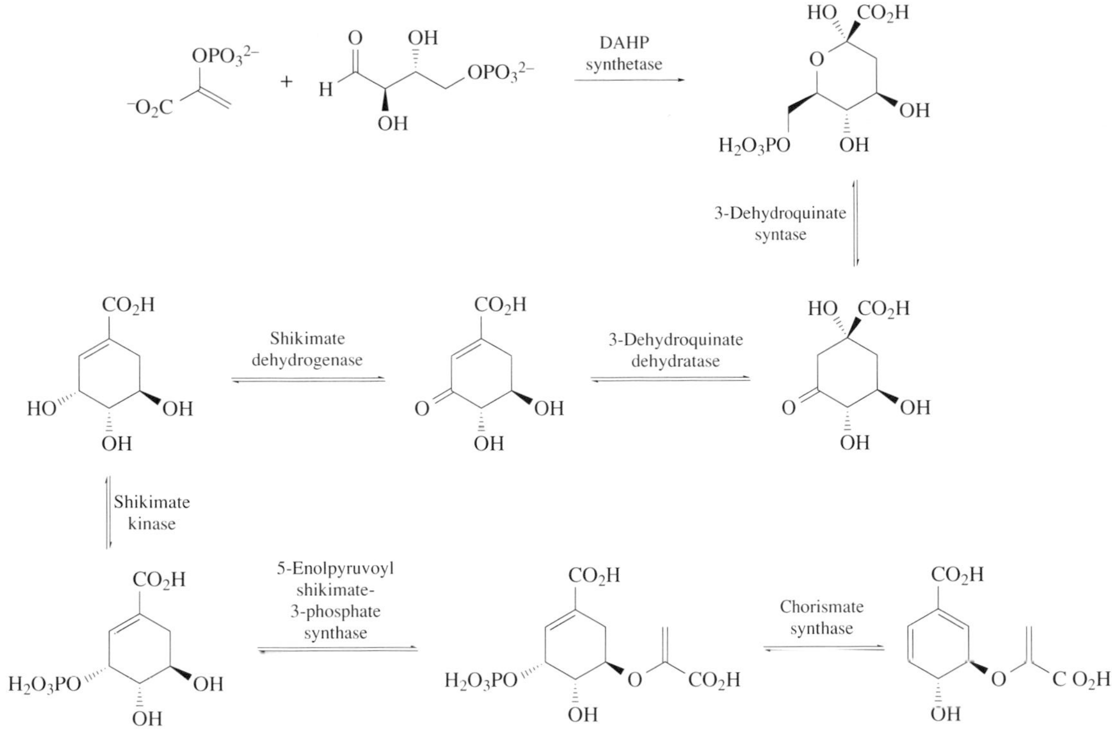

Scheme 26

concentrations are at least 88 µM, two isozymes of the synthetase exist entirely as the enzyme–phosphoenolpyruvate complex.

DAHP synthetase has been utilized to prepare DAHP (Scheme 27).[163] The range of cofactors and enzymes required for the synthesis render whole-cell synthesis more efficient than protocols utilizing soluble enzymes.

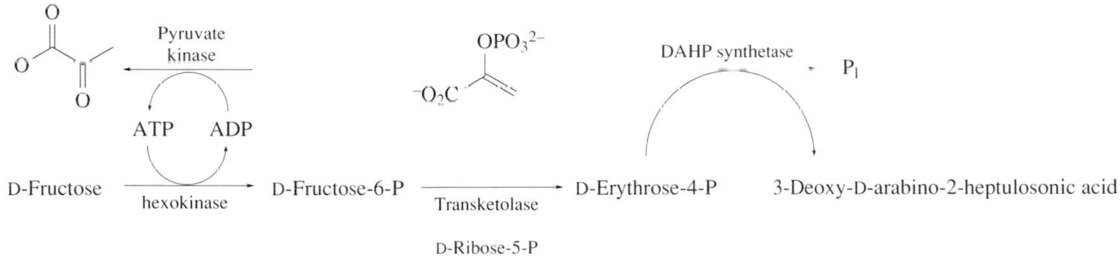

Scheme 27

3.13.4.2 3-Deoxy-D-manno-2-octulosonic Acid 8-Phosphate Synthase

3-Deoxy-D-manno-2-octulosonic acid 8-phosphate (KDO-8P) synthase (EC 4.1.2.16; CAS 9026-96-4) catalyzes the reaction between phosphoenolpyruvate and arabinose-5-phosphate to produce KDO-8P. The enzyme is a key component in the synthesis of bacterial lipopolysaccharide, in turn a major constituent of gram-negative bacterial cell wall.[164] The overall conversion involves isomerization of ribose-5-phosphate to arabinose-5-phosphate, reaction with phosphoenolpyruvate to form KDO-8P, dephosphorylation to KDO, activation of KDO as the cytidine monophosphate (CMP) derivative and transfer into lipopolysaccharide (Scheme 29). The enzyme was generally thought to be restricted to gram-negative prokaryotes, although some reports suggest that the enzyme is also present in several plants: KDO is known to occur in the rhamnogalacturonan-II pectin polysaccharide of many plant cell walls.[165] The study of plant KDO-8P synthase is complicated by the presence of DAHP synthetase (EC 4.1.2.15, Section 3.13.4.1); overlapping electrophile substrate specificities obscure the existence of distinct activities. Nonetheless, it seems clear on the basis of work by Jensen and co-workers[161] that the KDO-8P synthase does indeed represent a discrete and distinct enzyme in plants.

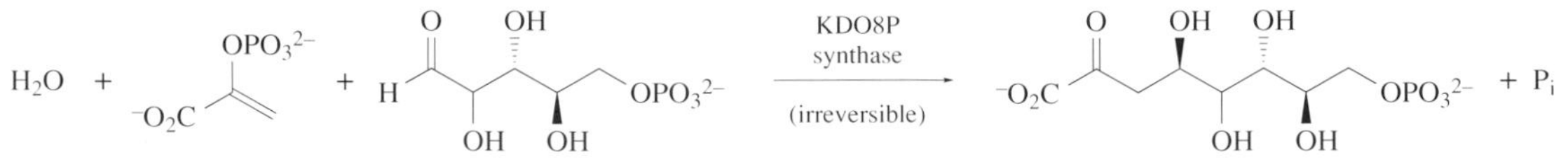

Scheme 28

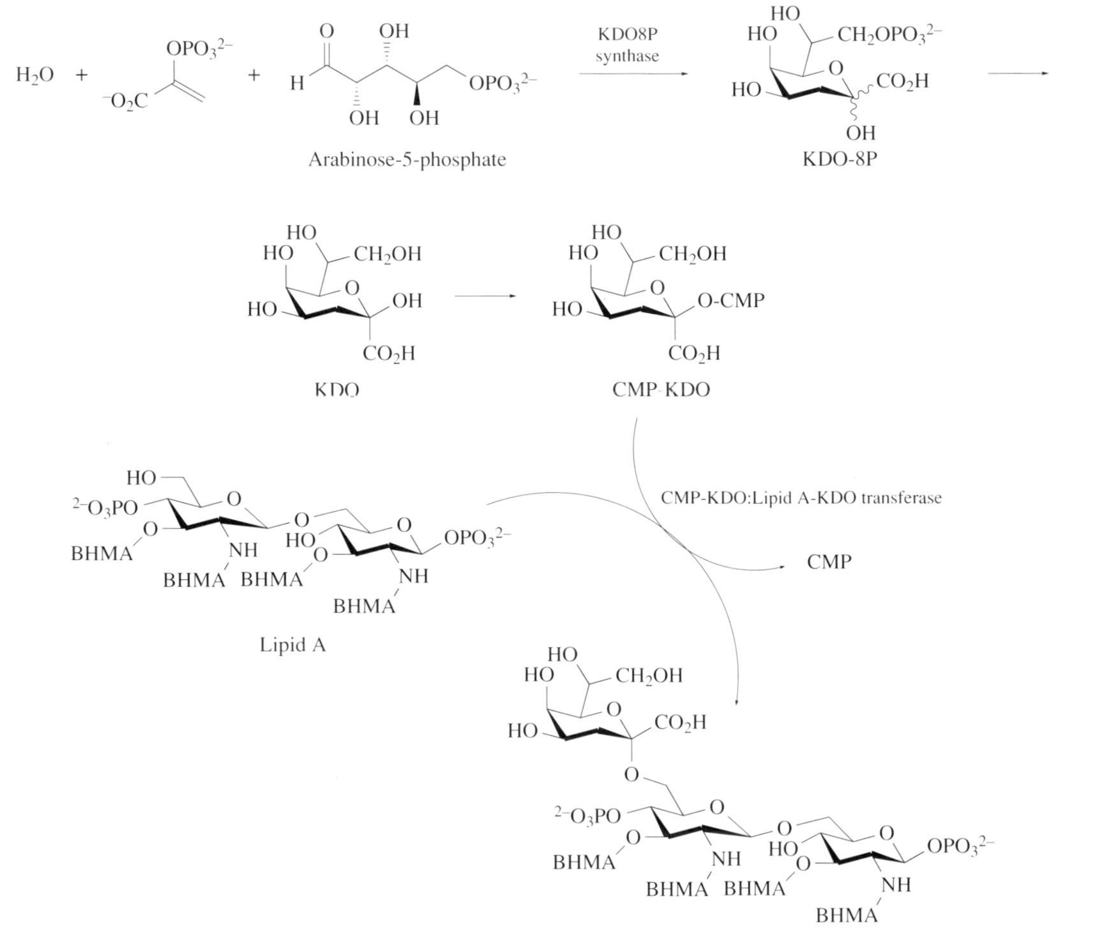

Scheme 29

KDO-8P synthase activity was first observed in 1959 by Levin and Racker[166] in *P. aeruginosa* extracts. The researchers purified the enzyme 30-fold to a specific activity of 3.80 U mg^{-1}. A comprehensive study of the purification of the *P. aeruginosa* enzyme has been reported by Ray.[167]

The *E. coli* enzyme has been purified 450-fold.[168,169] The enzyme exists as a trimer with a subunit molecular weight of 30.8 kDa. KDO-8P synthase displays behavior highly dependent on buffer; a pH optimum was observed at pH 4–6 in succinate buffer, while in glycine buffer the pH optimum was near 9.0. The optimum temperature for enzyme activity is 45 °C although substantial enzyme inactivation occurs at higher temperatures. Michaelis constants were determined for D-arabinose-5-phosphate (20 mM) and for phosphoenolpyruvate (6 mM). D-Ribose-5-phosphate is a competitive inhibitor of D-arabinose-5-phosphate with an apparent K_i of 1 mM. The synthase utilizes erythrose-4-phosphate at 28% of the rate of arabinose-5-phosphate at saturating substrate concentrations.

KDO-8P synthase from *E. coli* has been cloned by Woisetschläger *et al.*[170,171] and positive transformants show a threefold increase in specific activity of the enzyme over crude extracts of *E. coli* and *S. typhimurium* host cells. The enzyme has been sequenced and the aldolase gene localized; it belongs to an operon.[172] The synthase has been crystallized in two forms (α from polyethylene glycol and β from ammonium sulfate) by Tolbert *et al.*;[173] however, no structures have been reported to date. Interestingly, KDO-8P synthase and DAHP synthetase (EC 4.1.2.16, Section 3.13.4.2) from *E. coli* strain CB198 share a 17% sequence homology.

Rick *et al.*[174-177] reported a series of investigations on a mutant *S. typhimurium* deficient in KDO-8P synthase. The temperature sensitive mutant shows normal lipopolysaccharide synthesis below 30 °C, synthesis dependent on exogenous D-arabinose-5-phosphate at 30-42 °C, and an inability to conduct lipopolysaccharide synthesis above 42 °C. The mutant produced KDO-deficient lipid A increasingly with time, which ultimately led to growth inhibition. Later work by these authors reports a second mutation whose lethal expression is dependent on the inability of the mutant to synthesize a fully acylated and KDO-substituted lipid A portion of lipopolysaccharide at elevated temperatures; this failure results in cell death.

The mechanism of KDO-8P synthase has been investigated extensively by several groups, including Hedstrom and Abeles,[11] Baasov,[12,182] Dotson and co-workers,[181] and Rick and Young.[178] The mechanism is an ordered Bi–Bi scheme, with obligatory initial binding of phosphoenolpyruvate and initial release of inorganic phosphate.[178] Acceptance by the enzyme of 4-deoxy-arabinose-5-phosphate, which exists only in the open-chain form, demonstrates that the enzyme is specific for free aldehyde substrates. Because arabinose-5-phosphate exists in the open chain form to the extent of 1%, the "true" K_m for this substrate is 0.26 μM.

In 1988, Hedstrom and Abeles,[11] using ^{18}O labeled substrates, demonstrated that the enzyme catalyzed reaction proceeds with C—O, rather than the expected P—O, bond cleavage of phosphoenolpyruvate. Additionally, these researchers demonstrated that the C-2 oxygen of KDO-8P originates from water. A covalent enzyme/phosphoenolpyruvate intermediate was not detectable. Finally, no ^{32}P P_i exchange into phosphoenolpyruvate or scrambling of bridge ^{18}O to nonbridging positions in [^{18}O]phosphoenolpyruvate was observed, suggesting that phosphate cleavage is irreversible. Bromopyruvate was shown to inactivate the synthase; phosphoenolpyruvate protects against inactivation whereas arabinose-5-phosphate does not. Based on these observations, Abeles proposed a mechanism that invokes attack by water on phosphoenolpyruvate, which in turn facilitates attack on the *re* face of arabinose-5-phosphate. Loss of phosphate by C—O bond cleavage and cyclization complete the reaction (Scheme 30).

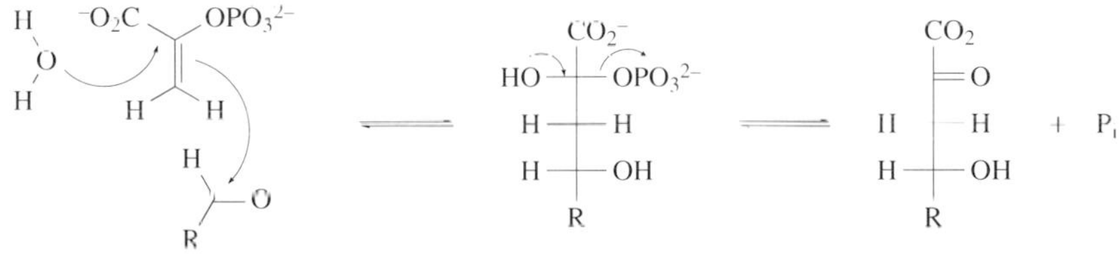

Scheme 30

Baasov and co-workers[12,179,180] have extended these mechanistic studies through the preparation of inhibitors. Several 2-deoxy analogues of KDO-8P were synthesized and probed as synthase inhibitors; the analogues bind to the enzyme and act as competitive inhibitors with respect to phosphoenolpyruvate, showing K_i values of 470 μM (α) and 303 μM (β); K_i of KDO-8P is 590 μM. Comparison of these values suggests that both anomers bind to the enzyme with a slight preference for β and that the C-2 hydroxy is not important for binding. This result suggests that the carboxylate binding site of the product is indistinct and the hydroxy and carboxylate binding sites may be interchangeable. Baasov *et al.*[12,179,180] proposed a mechanism in which phosphoenolpyruvate is attacked by the C-3 hydroxy of arabinose-5-phosphate, which in turn facilitates *concomitant* attack by phosphoenolpyruvate on the arabinose-5-phosphate aldehyde (Scheme 31). This concerted process initially yields a cyclic glycosyl phosphate, which in turn decomposes through an oxonium ion to the product hemiketal. The authors note that in addition to inhibition studies, this scheme is consistent with the observation that 4-deoxy, but not 3-deoxy, arabinose-5-phosphate is accepted as the electrophilic substrate by the synthase. The proposed cyclic intermediate can be mimicked by the isosteric phosphonate analogue 2,6-anhydro-3-deoxy-2β-phosphonylmethyl-8-phosphate-D-glycero-D-talooctonate which has the topological and electrostatic properties of the intermediate. Indeed, this mimic was found to be the most potent inhibitor of the enzyme known, with a K_i of 5 μM.

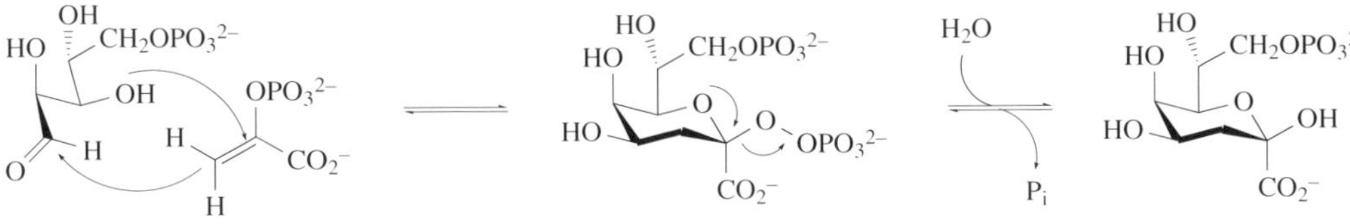

Scheme 31

In 1993 the groups of Dotson[181] and Baasov[182] independently reported the stereochemistry of the KDO-8P synthase addition. Both groups incubated *E* and *Z* deuterium labeled phosphoenolpyruvate with arabinose-5-phosphate and KDO-8P synthase from *E. coli* K12 pMW101. The *Z*-phosphoenolpyruvate produces the 3*S* configuration in the product KDO, while the *E* alkene produces predominantly the *R* product isomer. The synthase reaction therefore proceeds with stereospecific *si*-facial addition to C-3 of phosphoenolpyruvate.

The substrate specificity of the synthase requires further study; few unnatural substrates have been investigated. Pyruvate was not accepted in place of phosphoenolpyruvate, but 3-phosphoglycerate is accepted to some extent. Glyceraldehyde-3-phosphate, FDP, and sedoheptulose-1,7-diphosphate are all accepted as electrophiles at 10–25% the rate of ribose-5-phosphate.[169] KDO-8P has been synthesized preparatively from D-arabinose via the synthase in a coupled reaction with hexokinase, which phosphorylates arabinose.[183]

3.13.4.3   *N*-Acetylneuraminate Synthase

NeuAc synthase (EC 4.1.3.19; CAS 37290-66-7) catalyzes the reaction of *N*-acetyl-D-mannosamine and phosphoenolpyruvate to produce NeuAc; the enzyme is responsible for the biosynthesis of NeuAc in bacteria. The synthase has been detected in several bacterial strains including *Neisseria meningitidis*,[184] *E. coli*,[185] and several strains of *Corynebacteria*.[186]

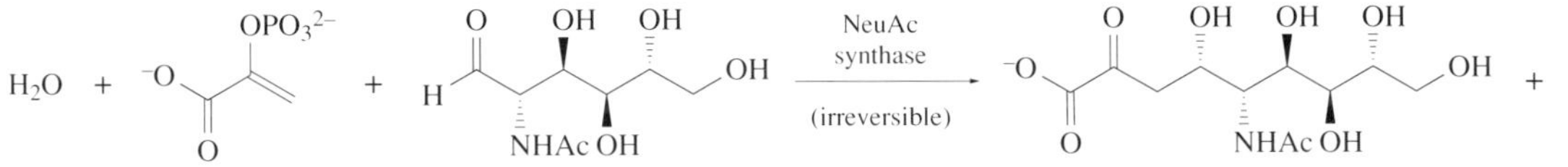

Scheme 32

In 1962, Blacklow and Warren[184] reported the first purification of the enzyme from *N. meningitidis*. Purification was achieved by acetone precipitation and alumina gel and DEAE cellulose chromatography. Manganese(II) is required for activity but can be replaced with 50% retention of activity by Mg^{II} and Co^{II}. EDTA completely inhibits activity. Reduced sulfhydryl groups are also required for optimal activity. The synthase has a specific activity of 0.061 U mg^{-1} with a pH optimum (Tris) of 8.0–8.4 at 37 °C. The K_m for *N*-acetylmannosamine is 6.25 mM while that for phosphoenolpyruvate is 0.042 mM.

Masson *et al.*[187] investigated a *N. meningitidis* isogenic mutant defective in polysaccharide production and showed that the production of surface sialic acid polysaccharide in serogroup B is directly related to the virulence of the *N. meningitidis* for mice. Isogenic mutants incapable of producing sialic acid were 20 000 times less virulent than wild-type organisms. Additionally, virulence and polysaccharide production are regained together in revertant strains. These researchers postulated a membrane-bound system for the biosynthesis of sialic acid based on the cellular localization of the enzymes.

Edwards *et al.*[188] located the genes encoding for the enzymes of capsular polysaccharide biosynthesis in *N. meningitidis* on a 5 kb DNA fragment within the chromosomal cps gene cluster. Sequencing revealed four open reading frames that comprise a transcriptional unit. The NeuAc synthase gene expressed for a protein of 38.3 kDa molecular weight.

N. meningitidis synthase accepts 3-phosphoglyceric acid as a phosphoenolpyruvate substitute; pyruvate, lactate, 2-phospholactate, and oxaloacetate are not accepted. *N*-acetyl-D-glucosamine, *N*-acetyl-D-galactosamine, *N*-acetylmannosamine-6-phosphate, glucosamine, galactosamine, mannosamine-6-phosphate, mannose, glucose, galactose, glucose-6-phosphate, ribose, ribose-5-phosphate, erythrose, rhamnose, arabinose, UDP-*N*-acetylglucosamine, UDP-glucose, and GDP-mannose were investigated as electrophilic substrates; none was accepted.[184] Brossmer and co-workers[189] utilized the *N. meningitidis* synthase to synthesize a 9-azido-9-deoxy derivative of NeuAc, indicating the possibility of a broader substrate spectrum than previously reported.

3.13.4.4 *N*-Acetylneuraminate-9-phosphate Synthase

N-Acetylneuraminate-9-phosphate (NeuAc9P) synthase (EC 4.1.3.20; CAS 9031-58-7) catalyzes the reaction of *N*-acetyl-D-mannosamine-6-phosphate and phosphoenolpyruvate to produce NeuAc9P. This enzyme, along with NeuAc9P phosphatase, participates in the final steps of the biosynthesis of NeuAc in mammals (Scheme 34). NeuAc in arterial cell walls may be involved in biological functions such as nonthrombogenicity of endothelial cells and the control of proliferation of smooth muscle cells.[190-192]

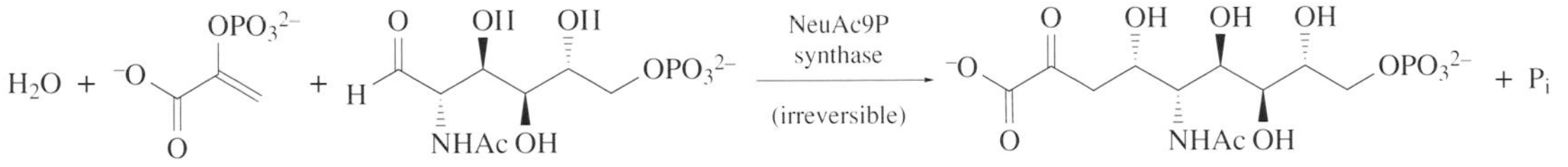

Scheme 33

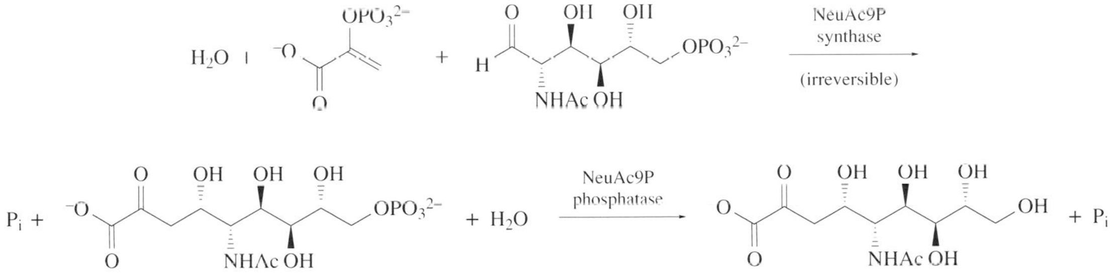

Scheme 34

NeuAc9P synthase has been isolated from a variety of mammalian sources including rat, pig, sheep, and human tissues. In 1961, Warren and Felsenfeld[193-195] first isolated this enzyme from bovine liver and salivary glands. Watson *et al.*[196,197] continued investigations with the enzyme from hog and sheep submaxillary glands, purifying the enzyme 800-fold by ammonium sulfate fractionation, DEAE cellulose and hydroxyapatite chromatography to a specific activity of 50 µmol mg^{-1}h^{-1}. The purified enzyme is extremely unstable and loses activity upon freezing. The enzyme also loses 50% of activity on storage at 4 °C over a period of 3 days. Investigation of the pH–activity relationship (Tris-HCl) reveals a bell-shaped curve with a pH optimum near 7.8. Magnesium(II) is essential but can be substituted, with some retention of activity, by Mn^{II}, Ni^{II}, Fe^{II}, Co^{II}, and Zn^{II}. Cu^{II}, Ca^{II}, and Al^{III} are inactive while EDTA inhibits the reaction completely. NeuAc9P synthase is also inhibited by 4-deoxy-*N*-acetylmannosamine.

Corfield *et al.*[198,199] investigated NeuAc9P synthase from mucosal cells of rat colon. Such cells are highly sialylated suggesting significant NeuAc9P synthase activity. In these cells, NeuAc biosynthesis occurs in the cytosol, cytidine monophosphate-NeuAc synthase is located in the nucleus, and sialyl transfer occurs in the Golgi membranes.

van Rinsum *et al.*[200] reported investigations of NeuAc9P synthase activity from various rat organs. Enzymes from the liver, kidney, spleen, brain, lung, muscle, erythrocytes, intestinal mucosa, salivary glands, pancreas, and thymus were investigated with the largest amount of activity detected in the salivary glands. Utilizing *N*-acetyl[^{14}C]mannosamine-6-phosphate, the researchers again determined that enzymatic activity is localized in the cytosolic fraction.

The substrate specificity for NeuAc9P synthase appears to be limited.[196,197] NeuAc9P synthase (hog and sheep) accepts erythrose-4-phosphate, arabinose-5-phosphate, ribose-5-phosphate, glucose-6-phosphate, mannose-6-phosphate, glucose-1-phosphate, fructose-6-phosphate, galactose-1-phosphate, galactose-6-phosphate, glucosamine-6-phosphate, *N*-acetylglycosamine-6-phosphate, *N*-acetylmannosamine, *N*-glycolylmannosamine, pyruvate, or oxaloacetate at <2% the rate of natural substrates.

3.13.5 THIAMINE-DEPENDENT ALDOLASES

3.13.5.1 D-Xylulose-5-phosphate Phosphoketolase

D-Xylulose-5-phosphate phosphoketolase (EC 4.1.2.9; CAS 9031-75-8) catalyzes the reaction of acetylphosphate and D-glyceraldehyde-3-phosphate to produce D-xylulose-5-phosphate. A thiamine cofactor and Mg^{II} are required for activity. *In vivo*, the xylulose phosphoketolase is involved in the metabolism of glucose and ribose in organisms lacking the Embden–Meyerhof–Parnas and Entner–Doudoroff pathway enzymes (Scheme 36). Xylulose phosphoketolase was initially identified by Heath *et al.*[201] in 1956 in heterofermentative *Lactobacilli* and subsequently by Sgorbati *et al.*[202] in *Bifidobacteria*. Low activities of the enzyme have also been detected in yeasts.[203]

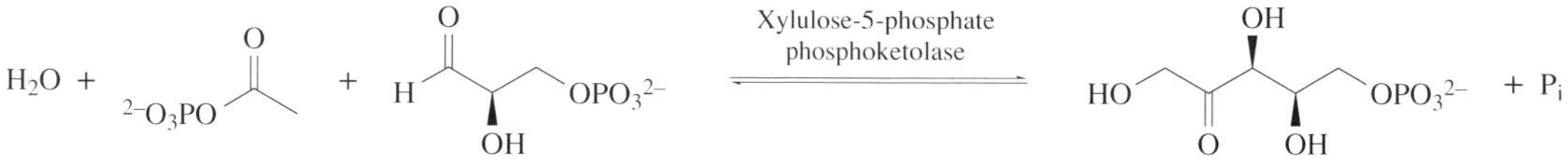

Scheme 35

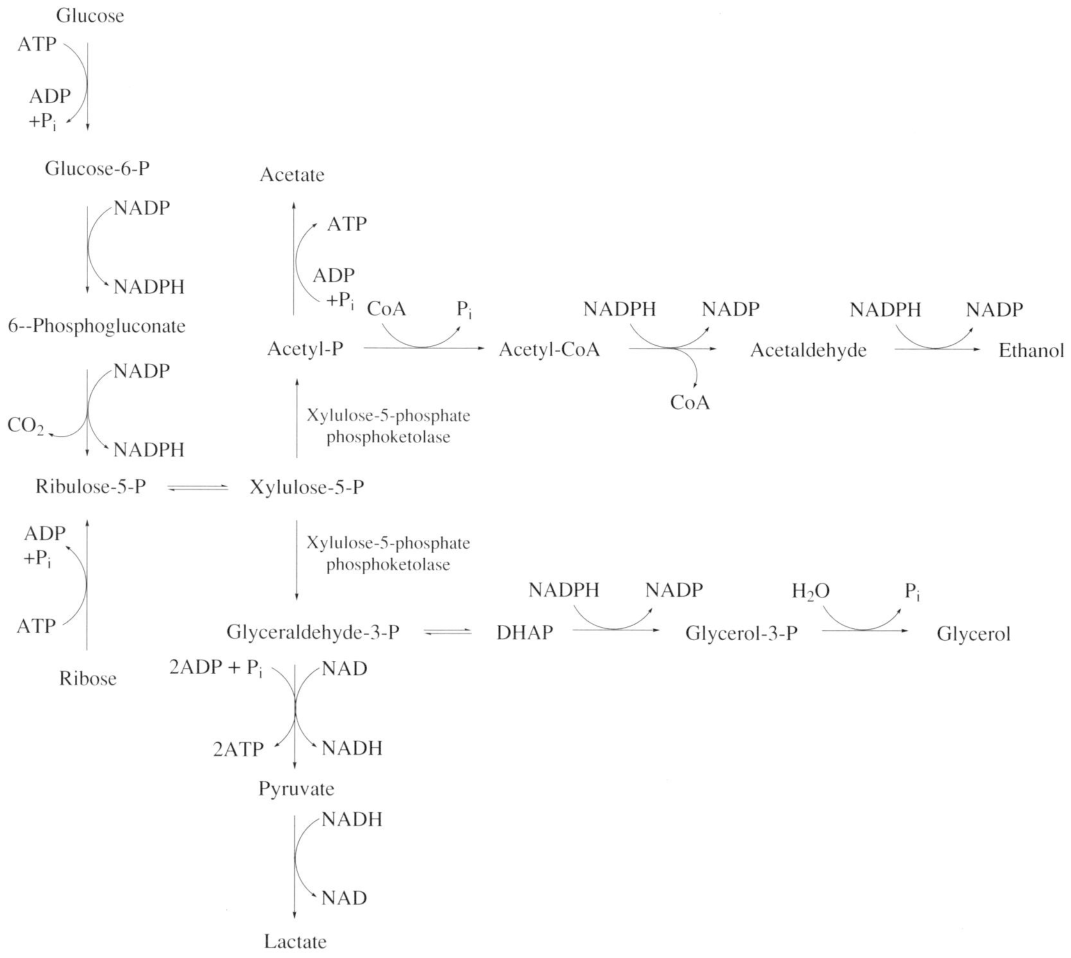

Scheme 36

In 1966, Goldberg *et al.*[204] isolated and purified the phosphoketolase from *Leuconostoc mesenteroides*. Purification was conducted by pH fractionation, streptomycin treatment, ammonium sulfate fractionation, and crystallization to a specific activity of $9.9\,U\,mg^{-1}$. Both xylulose-5-phosphate and fructose-6-phosphate are accepted as substrates by this enzyme. Sulfhydryl compounds including thioglycerol, thioethanol, cysteine, and glutathione accept the acetyl group to form thioesters at 10–20% of the rate of acetylphosphate formation. Xylulose phosphoketolase activity is stimulated 10–15% by EDTA and 15–20% by sodium borate. Inhibitors include *p*-hydroxymercuribenzoate, histidine, glyceraldehyde-3-phosphate, and erythrose-4-phosphate. The optimal pH range for activity is 5.8–7.6, and the purified enzyme is reportedly stable at $-55\,°C$ for several years. K_m for xylulose-5-phosphate is 4.7 mM while that for fructose-6-phosphate is 29 mM.

Singh *et al.*[205] investigated D-xylose fermentation in the yeast *Fusarium oxysporum* and detected xylulose phosphoketolase activity. Evans and Ratledge[206] likewise examined several yeasts and detected the phosphoketolase in 20 of the 25 strains assayed. No activity was observed in yeasts grown on glucose, indicating that this path is probably the major route of pentose dissimilation in such strains. Yeasts including *Candida* sp., *Kluyveromyces* sp., *Lipomyces staskeyi*, *Hansenula*, *Pichia media*, *Rhodosporium toruloides*, *Rhodotorola* sp., *S. cerevisiae*, *Trichosporum cutaneum*, *Wingea robertsi*, and *Yarrowia lipocytica* exhibit xylulose phosphoketolase activity. Lachke and Jeffries[207] also detected phosphoketolase activity in *Pachysolen tannophilus*. London and Chace[208] detected the pentitol metabolic route in *L. casei*, which also involves phosphoketolase. Additionally, Lees and Jago[209] detected xylulose phosphoketolase activity in *S. lactis*.

Ratledge and Botham[210] investigated the phosphoketolase in *Candida* 107, a lipid accumulating yeast. Xylulose phosphoketolase was not detected in normal cells but was present in high quantities in cells treated with toluene. The enzyme was highly unstable and not isolated, although some properties were investigated. As expected, thiamine and Mg^{II} were required for activity. The pH optimum was found to be 6.0 and the optimum temperature to be 30 °C. Inhibitors include NADH, NADPH, phosphoenolpyruvate, citrate, ATP, acetyl CoA, and dodecanoyl CoA. Greenly and Smith[211] also detected xylulose phosphoketolase in *Thiobacillus novellus* that does not accept fructose-6-phosphate as a substrate. The *novellus* phosphoketolase has a specific activity of 0.070 mmol $min^{-1}\,mg^{-1}$ and a K_m for xylulose-5-phosphate of 4.27 mM. The pH optimum is 6.0 while the optimum temperature is 43 °C.

3.13.5.2 Fructose-6-phosphate Phosphoketolase

Fructose-6-phosphate phosphoketolase (EC 4.1.2.22; CAS 37290-57-6) is an enzyme of the fructose-6-phosphate shunt (Scheme 38) characteristic of, but not restricted to, *Bifidobacteria*.[212] Fructose phosphoketolase catalyzes the cleavage of fructose-6-phosphate to acetylphosphate and D-erythrose-4-phosphate and is involved in the regulation of erythritol formation. This phosphoketolase was first investigated in partially purified form in 1958 in a mutant of *Acetobacter xylinum* by Schramm *et al.*[213] Other sources include *Candida* 107,[210] *Gardnerella vaginalis*,[214] saprophytic mycobacteria,[215] *Bifidobacteria* species[216,217] including *longum*, *animalis*, *globosum*, and *dentium*, and honeybee intestines.[202]

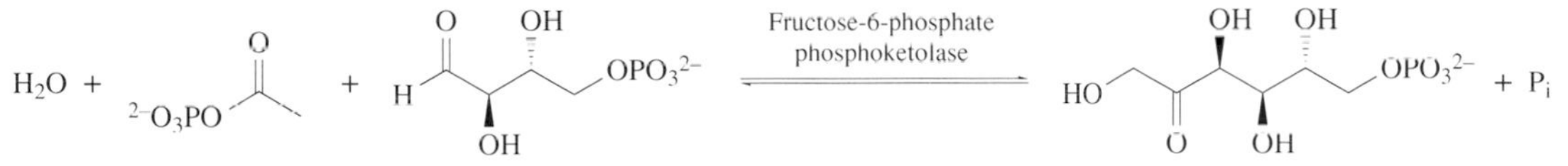

Scheme 37

The enzyme from *B. globosum* has been purified to a specific activity of 24 U mg^{-1} by Sgorbati *et al.*[202] The phosphoketolase is stable to temperatures of 57 °C, and shows a pH optimum of 5–6. The *globosum* enzyme requires Mg^{II} for stability and has a molecular weight of 290 kDa. The enzyme from *B. dentium* has been purified to a specific activity of 30 U mg^{-1}; this enzyme requires Mn^{II} for activity and exhibits a molecular weight of 160 kDa. The *dentium* enzyme retains 60% of maximum activity during 30 min at 57 °C and shows a pH optimum of 7. Kaster and Brown[217] investigated three different strains of anaerobic dextranase producing gram-positive rod-shaped bacteria from human dental plaque associated with root carious lesions. Each strain contained fructose phosphoketolase and was identified as *Bifidobacteria*, but not *B. dentium* as expected. Likewise, *B. longum* BB536 and *B. dentium* ATCC 27534 are human strains that have been investigated.[202] Copper(II) and mercuric acetate inhibit both enzymes. The molecular weight of all three enzymes was reported to be near 163 kDa.

Reports of substrate specificity indicate that this enzyme also accepts xylulose-5-phosphate.[212,213] Care must be taken when considering such studies since a distinct xylulose phosphoketolase (EC 4.1.2.9, Section 3.13.5.1) activity was not known at the time such studies were conducted.

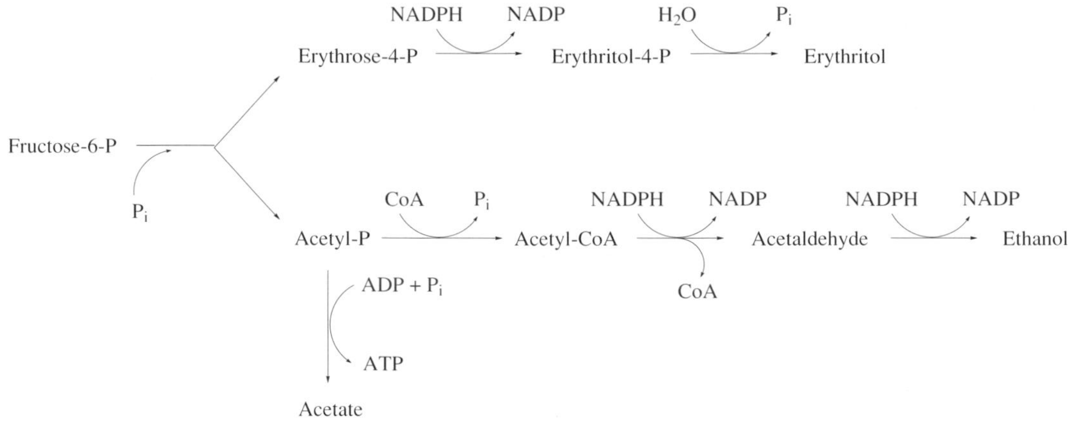

Scheme 38

3.13.5.3 Propion Synthase

Morimoto *et al.*[218,219] reported a thiamine-dependent enzyme in commercial baker's yeast that carries out an acyloin reaction with propanal. The enzyme was purified 270-fold to a final specific activity of $1 U mg^{-1}$. The authors note that the enzyme generates a variety of acyloin products, including propion, furoin, methylfuroin, acetoin, isobutyroin, and valeroin without coupling to a decarboxylation. The authors suggest that this property qualifies the activity as distinct from other enzymes known to catalyze the acyloin reaction with concomitant decarboxylation, such as pyruvate decarboxylase. No information regarding the stereochemical requirements of the enzyme was provided.

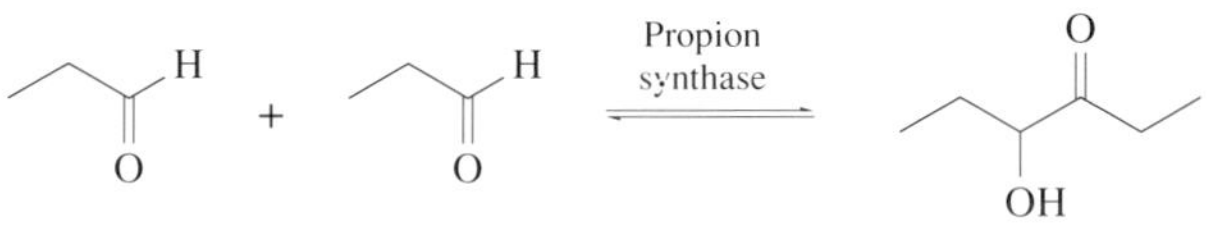

Scheme 39

The protein exhibits a molecular weight near 100 kDa (gel filtration chromatography and equilibrium ultracentrifugation). Enzyme activity is maximal near neutral pH (6.8–7.0) and is weakly enhanced by Fe^{II}, Mn^{II}, β-mercaptoethanol, and Mg^{II}. Propion synthase (EC 4.1.2.35) is markedly inhibited by Cu^{II}, Hg^{II}, Zn^{II}, Sn^{II}, and iodoacetic acid.

3.13.5.4 Benzoin Aldolase

Considerable research has been undertaken to establish the bacterial enzymes responsible for lignin degradation, both as isolated species and those organisms found in termite gut. In a single instance, a thiamine-dependent benzoin lyase (EC 4.1.2.38) has been isolated from *P. fluorescens* biovar I.[220] The 563 amino acid 58 kDa enzyme has been cloned into *P. putida* KT2440 and sequenced.[221] The protein shows moderate (20–28%) homology with other thiamine-dependent enzymes, especially acetolactate synthase and pyruvate decarboxylase. The enzyme is apparently inducible, and *P. fluorescens* biovar I grown on glucose does not express the protein.

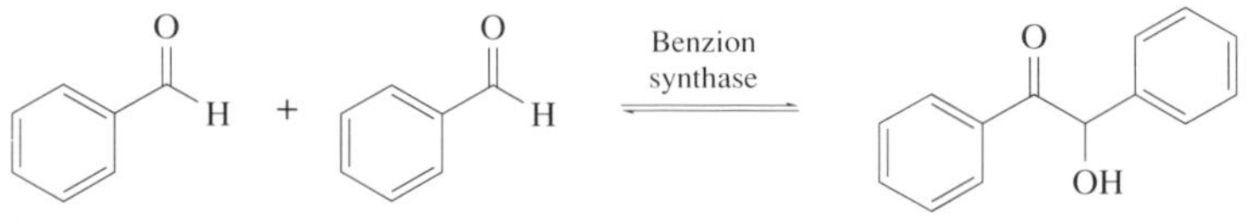

Scheme 40

A variety of substrates were examined (Figure 3): cleavage was detected only for benzoin and anisoin (4,4′-dimethoxybenzoin).[221] Kinetic analysis of these substrates yielded Michaelis constants of 9 mM and 32.5 mM, respectively. It is worthy of note, however, that all assays were performed at very low substrate concentrations (0.05–0.2 mM) and activity may be observable at higher substrate concentrations.

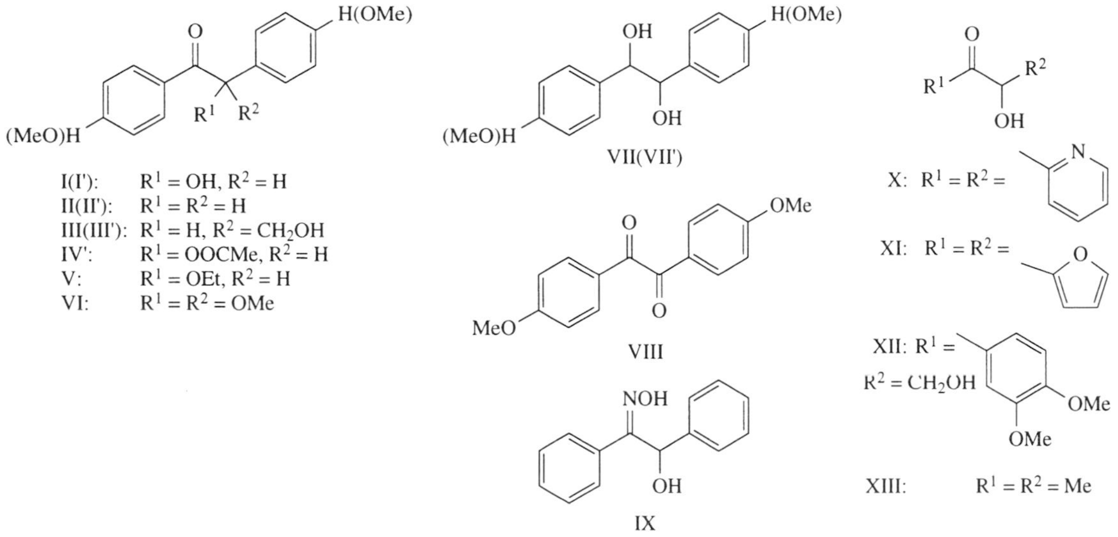

H(OMe)
O
(MeO)H
R1 R2
I(I′): R1 = OH, R2 = H
II(II′): R1 = R2 = H
III(III′): R1 = H, R2 = CH2OH
IV′: R1 = OOCMe, R2 = H
V: R1 = OEt, R2 = H
VI: R1 = R2 = OMe
OH
OH
VII(VII′)
OMe
MeO
VIII
NOH
IX
R1
R2
X: R1 = R2 =
N
XI: R1 = R2 =
XII: R1 =
R2 = CH2OH
XIII: R1 = R2 = Me

Figure 3 Substrate spectrum for benzoin aldolase.

The reaction proceeds to completion in the retro-acyloin direction, and attempted reaction of both benzaldehyde and *p*-methoxybenzaldehyde failed to yield acyloin products. No information was provided as to the stereochemical configuration of the substrates.

3.13.6 PYRIDOXAL-DEPENDENT ALDOLASES

3.13.6.1 L-Kynurenine Hydrolase (Kynureninase)

Kynureninase (EC 3.7.1.3; CAS 9024-78-6) is a pyridoxal-5′-phosphate dependent enzyme that catalyzes the hydrolytic cleavage of L-kynurenine to produce L-alanine and anthranilic acid; the latter species is subsequently metabolized to quinolinic acid. Quinolinic acid is neurotoxic and may be involved in the etiology of neurodegenerative diseases such as Huntington's chorea, epilepsy, and AIDS-related dementia.[222–224] *In vivo*, kynureninase plays a pivotal role in L-tryptophan catabolism in bacteria such as *Pseudomonas* (Scheme 42). Kynurenic acid itself is an endogenous antagonist acting at the glycine recognition site present on the *N*-methyl-D-aspartate receptor ion-channel.[225] Kynureninase also catalyzes the aldol reaction between benzaldehyde and L-alanine produced via the *in vivo* cleavage of L-kynurenine, yielding 2-amino-4-hydroxy-4-phenylbutanoic acid (Scheme 43). Kynureninase has been isolated from a variety of sources including mammalian liver,[226,227] *Suncus murinus*,[228] *Bombyx mori*,[229] *Neurospora crassa*,[230,231] *P. marginalis*,[232] *Rhizopus stolonifer*,[233] *Aspergillus niger*,[233] *Penicillum roqueforti*,[233] *P. fluorescens*,[233] *P. aureofaciens*,[234] and *S. parivulus*.[235] Aldol addition, however, has only been investigated with the *P. fluorescens* enzyme. The *P. marginalis* kynureninase has been crystallized.[232]

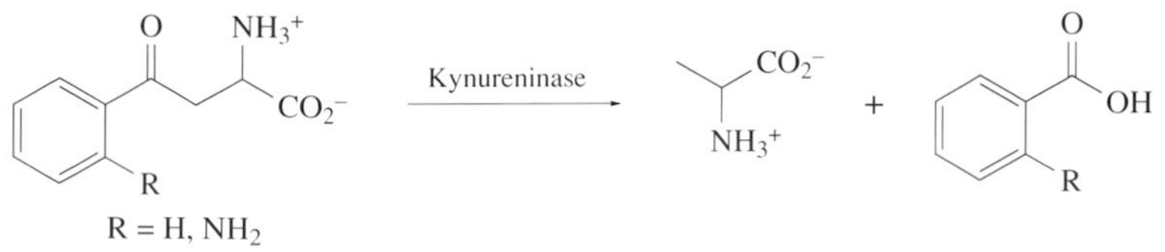

O
NH3+
CO2−
R
R = H, NH2
Kynureninase
CO2−
NH3+
+
O
OH
R

Scheme 41

The *P. marginalis* kynureninase was purified by a series of hydroxyapatite and diethylaminoethyl (DEAE) cellulose columns.[232] The enzyme can be stored in phosphate buffer at 4 °C for one week

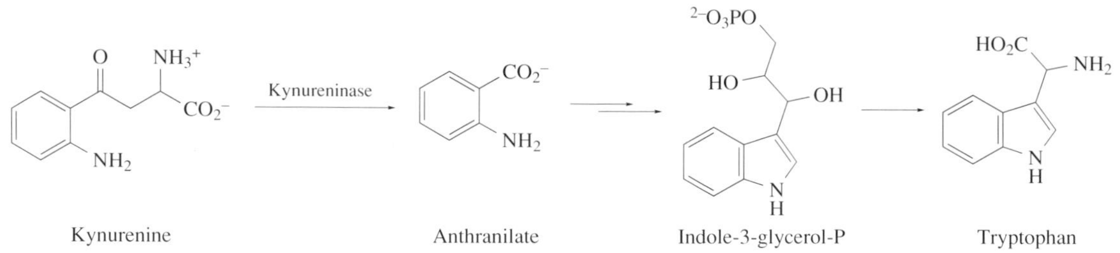

Scheme 42

R = H, NH₂

Scheme 43

with only 15% loss of activity and is stable for at least three weeks at $-20\,°C$. Kynureninase is active from pH 5.8–8.0, with maximum activity at pH 8.0. The molecular weight of the enzyme was reported to be $100\,kDa$.

Bild and Morris[236] first investigated the aldol-type reaction (see Scheme 43); however, they did not determine the stereochemistry at the new stereogenic center. Likewise, Tanizawa and Soda[237] reported the aldol cleavage of dihydro-L-kynurenine with no mention of substrate stereochemical requirements.

Phillips and Dua[238] reported the stereospecificity of both aldol and retro-aldol reactions. This work demonstrated that the aldol reaction with benzaldehyde provides a mixture of diastereomers with 4:1 in favor of the 2S,4R configuration to the 2S,4S configuration. The retro-aldol cleavage of dihydrokynurenine proved to be more selective with only the 2S,4R substrate converted.

Several investigations of the mechanism of kynureninase-catalyzed hydrolyses have been reported. The hydrolysis is assisted by a base which removes a proton from water to facilitate ketone hydration (Scheme 44) to produce anthranilic acid and the L-alanine enamine. Mechanistic investigations by Palcic *et al.*[239] detail scrambling of the α-proton between the α and β positions of L-alanine, indicating a polyprotic base, most likely a lysine ε-amino group. Kishore[240] investigated the base-catalyzed hydrolysis and reported the presence of a carboxylate group which can be modified by suicide substrate inhibitors. Based on these reports, there is debate in the literature concerning the number and identity of the bases involved in catalysis.

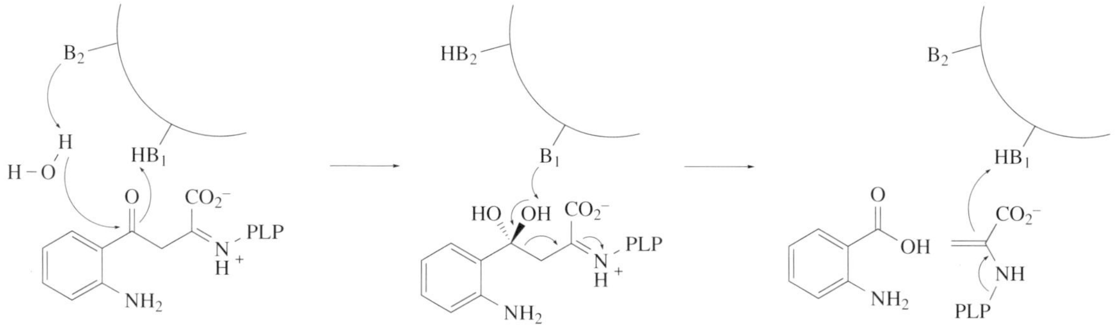

Scheme 44

Phillips and Dua[238] have proposed a mechanism for the retro-aldol reaction (Scheme 45). Such reactions are facilitated by the catalytic base involved in hydrolysis which assists catalysis by removing the proton from the hydroxy group.

Kynureninase inhibitors have been synthesized by several groups; Natalini *et al.*,[241] Phillips and co-workers,[242] and Kishore[240] investigated the inhibition of kynureninase with S-*m*-nitrobenzoyl-alanine, S-aryl-L-cysteine S,S-dioxides, and β-substituted amino acids, respectively.

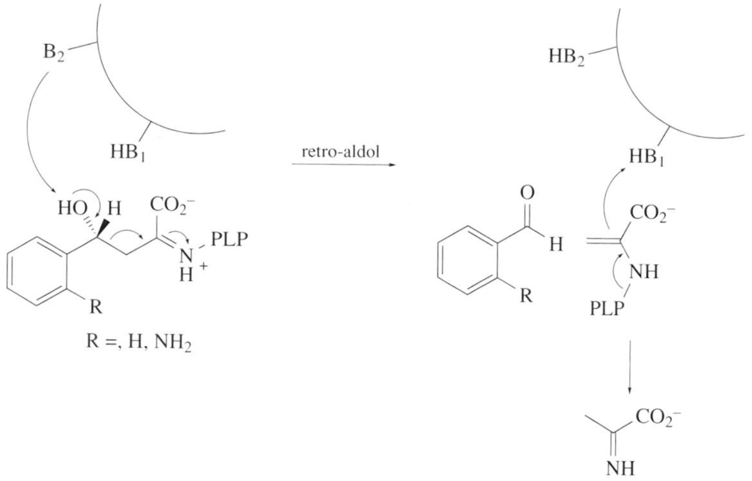

Scheme 45

A hydroxykynureninase distinct from the kynureninase in inducibility and kinetic properties has also been detected.[243] This enzyme may catalyze aldol-type reactions, providing access to products with *p*-substituted phenol products.

3.13.6.2 Threonine Aldolase, Allothreonine Aldolase, Serine Hydroxymethyltransferase, β-Phenylserine Aldolase, L-*threo*-(4-Hydroxyphenyl)serine Aldolase, and L *threo* (3,4-Dihydroxyphenyl)serine Aldolase

Aldol cleavage of threonine to glycine and acetaldehyde was first proposed by Knoop[244] in 1914. In 1949 Braunstein and Vilenkina[245] reported the existence of such an enzyme in various mammalian tissues. Since that time, a large class of enzymes has been found in animal, plant, bacterial, and fungal sources that are responsible for the retro-aldol cleavage of serine to glycine and formaldehyde and of threonine or allothreonine to glycine and acetaldehyde. In the latter case, formation of acetaldehyde is typically coupled to the formation of methylene-tetrahydrofolate: such activity is referred to as serine hydroxymethyltransferase (SHMT). In animals, the latter process is a major biosynthetic route to C_1-loaded folate, utilized in purine, thymidilate, and methionine biosynthesis.[246]

In bacteria and fungi, retro-aldol decomposition is one of several mechanisms of serine/threonine catabolism.[247] It has also been suggested that carbohydrate metabolism in some strains of *Clostridium* involves this enzyme.[248] In plants, the enzyme appears to function as part of the photorespiratory pathway, recovering carbon from the Calvin cycle lost by oxidation of ribulose-1,5-diphosphate.[249]

It is unclear whether the activities noted above can be ascribed to a single protein or to multiple enzymes, and various researchers have reported a contradictory body of evidence in this regard. In several cases, claims have been made that one or two of the activities can be separated; in other instances, the activities cannot be isolated independently of others. Some of these reports are outlined below. Further complicating the picture, a variety of other activities have been credited to this enzyme, including decarboxylation, transamination, and racemization. Additionally, Marcus and Dekker[250] reported that the 2-amino-3-ketobutyrate ligase from *E. coli* showed threonine aldolase activity. The Enzyme Commission have deleted allothreonine aldolase as a distinct enzyme.

Serine hydroxymethyltransferases/threonine aldolases have been isolated and examined from several sources. Complete amino acid sequences have been determined for the proteins from lamb,[251] rabbit (both cytosolic and mitochondrial),[252,253] *N. crassa*,[254] pea,[255] *E. coli*,[256] *Braja japonicum*,[257] *S. typhimurium*,[258] and *C. jejunji*.[259]

SHMT has been reviewed,[246-249] and our goal here is to draw attention to the important aspects of the enzymology rather than to be exhaustive. Below we outline some of those enzymes that have been isolated, highlighting the confusion regarding the identity of distinct enzymes.

3.13.6.2.1 Mammalian liver enzymes

Malkin and Greenberg[260] reported an isolation of a threonine/allothreonine aldolase from rat liver. Through a purification protocol including ammonium sulfate fractionation and hydroxyapatite

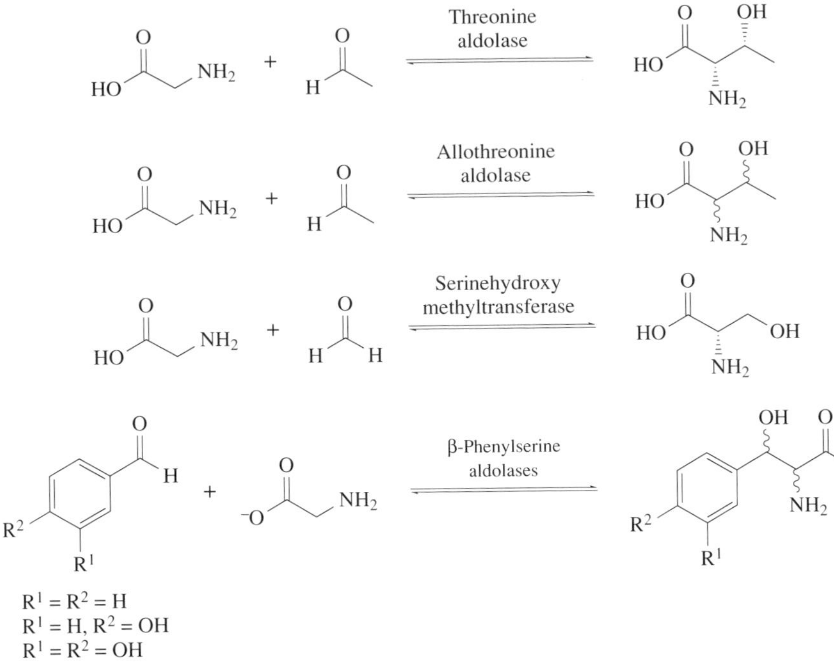

Scheme 46

gel adsorption the ratio of threonine/allothreonine aldolase activities remained constant, and the authors assigned both activities to a single enzyme. Michaelis constants of 20 mM and 0.2 mM were determined for threonine and allothreonine, respectively. In contrast, Palekar et al.[261] reported that the enzyme from rat liver does not possess allothreonine activity. Later, however, Reario-Sforza et al.[262] reported that careful ammonium sulfate fractionation allowed separation of the threonine and allothreonine aldolase activities. These authors also noted that the two activities showed different inhibition patterns with EDTA and HgCl$_2$. Finally, Thomas et al.[263] reported that the rat liver enzyme showed both allothreonine aldolase and SHMT activity, citing K_m values of 0.45 mM and 1 mM for serine and allothreonine, respectively.

Several researchers have examined threonine aldolase/SHMT from rabbit liver and concluded that the activities arise from the same or very closely related enzymes. Based on kinetic and stereochemical evidence, Schirch and Gross,[264] Fujioka[265] and Scrimgeour andHuennekens[266] suggested that both aldolase and SHMT activities were attributable to the same enzyme. Later, Akhtar and El-Obeid[267] reached the same conclusion based on chemical inactivation studies with chloroacetaldehyde, iodoacetamide, and bromopyruvate.

The study of these enzymes was further complicated by the findings of Akhtar and El-Obeid.[267] This group noted that while both the aldolase and SHMT activities arose from a single protein, cytoplasmic and mitochondrial isozymes exist and show significantly different properties. It is now clear that the existence of cytoplasmic and mitochondrial SHMT is a property of all eukaryotes. From rabbit, both enzymes show relatively low (2–5 U mg^{-1}) specific activities and effect reversible aldol cleavage of threonine and allothreonine. As had been noted by several researchers, the cleavage of serine to glycine and formaldehyde is accelerated roughly 100-fold by the presence of tetrahydrofolate. This acceleration appears to arise primarily from turnover: k_{cat} values for cleavage of serine in the presence and absence of tetrahydrofolate have been reported as $1.3 \, s^{-1}$ and 4.7×10^{-5} s^{-1}, respectively.

Ramesh and Rao[268] have reported an allosteric SHMT from monkey. The enzyme was detected in most tissue, including kidney, spleen, testes, lung, pancreas, heart, brain, and skeletal muscle, but the largest quantity and the highest specific activity of SHMT was detected in the tissues from liver cells. Subcellular localization of the enzyme showed that the protein is primarily cytosolic, although significant activity was detected in both mitochondrial and microsomal fractions. As is the case with other mammalian enzymes, the protein shows a low specific activity of 3.4 U mg^{-1}. Also in keeping with other mammalian enzymes, the protein is tetrameric with a subunit molecular weight of 52 kDa and requires both reduced thiol and EDTA for activity and stability. The enzyme exhibits a k_{cat} for serine cleavage in the presence of tetrahydrofolate of $1.7 \times 10^{-3} \, s^{-1}$. This protein shows threonine aldolase activity and cleaves the substrate in the absence of folate with a k_{cat} value of 0.76×10^{-3} s^{-1}. Allothreonine aldolase activity was not reported.

The protein from lamb liver has been investigated by several researchers. The enzyme is a homotetramer with a subunit molecular weight of 53 kDa. The cytosolic enzyme has been cloned and overexpressed in *E. coli*.[251] Schirch and co-workers[269] have investigated the kinetic properties of the enzyme with several natural and unnatural substrates. The protein shows both SHMT and threonine aldolase activity. In addition to accepting both threonine and allothreonine as substrates, diastereomers of β-phenylserine were accepted as substrates (Table 7).

Table 7   Kinetic properties of lamb liver SHMT.

Substrate	K_m (mM)	k_{cat} (s^{-1})
D/L-Allothreonine	1.3	0.56
L-Threonine	32	0.09
D/L-*erythro*-β-Phenylserine	9.5	21
D/L-*erythro*-β-Phenylserine methyl ester	70	29
D/L-*threo*-β-Phenylserine	84	7
D/L-*p*-Methoxy-*erythro*-β-phenylserine	20	31
D/L-*p*-Chloro-*erythro*-β-phenylserine	5	17
D/L-*m*-Chloro-*erythro*-β-phenylserine	1.9	14
D/L-*p*-Nitro-*erythro*-β-phenylserine	3	2.3

Source: Angelaccio *et al.*[269]

Ulevitch and Kallen and Ching and Kallen,[270–272] using substituted β-phenylserine derivatives, illustrated a linear free energy relationship between rate of cleavage and the Hammett σ-value for the phenyl substituent. Elaborating on this work, Matthews and Webb[273] noted that the large slope of the Hammett plot ($\rho = 0.93$) mandated that deprotonation and loss of the aldehyde to form the pyridoxal imine (see below) must be rate determining.

SHMT from porcine liver has also been extensively investigated.[274] This protein is also tetrameric and shows a subunit molecular weight of 53 kDa. Similar kinetic parameters were observed for natural and unnatural substrates as for other mammalian liver enzymes, notably lamb SHMT (Table 8). As is the case for other enzymes, tetrahydrofolate increases the rate of serine cleavage by a factor of 200 000 while the rate of cleavage of threonine and threonine derivatives is enhanced by less than a factor of 10.

Table 8   Kinetic properties of porcine liver SHMT.

Substrate	K_m (mM)	k_{cat} (s^{-1})
Serine	ND[a]	1.1×10^{-4}
Threonine	60	0.33
L-Allothreonine	1.4	0.37
4-Chloro-L-threonine	1.6	3.3×10^{-2}
D/L-*threo*-β-Phenylserine	ND	16

Source: Matthews *et al.*[274]   [a]ND: not determined.

Both the cytosolic and mitochondrial SHMT from humans have been cloned.[275] The cytosolic enzyme is a 483-residue protein of molecular weight 53 kDa while the mitochondrial protein contains 473 amino acids providing a molecular weight of 52.4 kDa. The proteins are highly homologous to other mammalian enzymes, showing 92% and 97% identity with the rabbit liver enzymes. The proteins also exhibit 43% sequence identity with the *E. coli* enzyme, suggesting a high sequence conservation. The mitochondrial and cytosolic proteins show 63% sequence identity to one another. Of particular interest concerning the human enzyme, abnormal SHMT activity has been detected in the temporal lobes of schizophrenics.[276]

In 1958, Bruns and Fiedler[277] reported a pyridoxal phosphate-dependent enzyme from human, rat, mice, guinea pig, pig, sheep, and cow livers and kidneys that converted both L-*threo*- and L-*erythro*-β-phenylserine to glycine and benzaldehyde. Their investigation focused on the incorporation of labeled phenylserine into the benzoyl moiety of urinary hippuric acid. Investigations of specificity indicated that neither diastereomers of the D-isomer was a substrate. Since that time, a variety of both substituted and unsubstituted phenylserine aldolases have been detected in mammalian and bacterial sources.[278–284] This activity has relevance in the treatment of Parkinson's disease since D/L-*threo*-dihydroxyphenylserine, a nonphysiological precursor of noradreniline, is cleaved to glycine and protocatechualdehyde by the enzyme.[285]

Again, it is unclear whether this activity is distinct from SHMT. In 1977, Ulevitch and Kallen[270] showed that a single protein from sheep liver cleaved both allothreonine and β-phenylserine. On the other hand, the same authors report that the rat liver SHMT does not cleave either *threo* or *erythro* isomers of β-phenylserine.[271] A β-dihydroxyphenylserine aldolase was investigated by Naoi *et al.*[278] from human brain. That enzyme showed a K_m value for L-*threo*-dihydroxyphenylserine of 10.6 mM and a V_{max} of 3.4 nmol min^{-1} mg protein^{-1}.

As with threonine and allothreonine aldolases, it is impossible at this point to determine whether retro-aldol reaction of β-phenylserine is attributable to SHMT or a separate activity. The two activities may be discrete in some sources while residing in a single protein with a broad substrate specificity in others. A 4-hydroxyphenylserine aldolase has been detected in *S. amakusaensis* that is stable, highly selective for the *threo* stereochemistry, and stereospecific for the 2*S*,3*R* configuration.[279] This activity is certainly distinct from SHMT, which shows selectivity for the L-configuration, but poor *erythro–threo* discrimination. Unambiguous determination of the identity of the proteins likely awaits investigations with recombinant proteins.

3.13.6.2.2 Other eukaryotic sources

Serine hydroxymethyltransferases from yeasts, including *S. cerevisiae*[286] and *C. humicola*[287] have been examined. Both mitochondrial and cytosolic *S. cerevisiae* SHMT have been cloned, and the sequences deduced.[288] The proteins show strong homology to one another and to other eukaryotic SHMT genes.

The *Candida* protein has been purified to homogeneity.[287,289] The crystalline protein shows an apparent molecular weight of 277 kDa and a subunit weight of 46 kDa, suggesting a hexameric structure. The same investigators observed threonine aldolase activity in *C. rugosa*, *C. guilliermondii*, and *C. utilis*.[289]

A SHMT has been isolated from *Euglena gracilis*.[290] The protein is apparently dimeric with a subunit molecular weight of 46 kDa. The protein shows a pH optimum near 7 and cleaves the L-isomers of threonine, allothreonine, and serine (Table 9). The D-isomers of the same substrates were not cleaved.

Table 9 Kinetic parameters of *Euglena* SHMT.

Substrate	K_m (mM)	V_{max} (mol min^{-1} mol enzyme^{-1})
L-Serine[a]	25	19.6
L-Threonine	13.9	45.0
L-Allothreonine	1.63	250

Source: Sakamoto *et al.*[290] [a]In the presence of folate.

3.13.6.2.3 Bacterial SHMT

SHMT/threonine aldolase activity has been observed in several bacterial strains. Dainty and Peel[291] first observed SHMT activity in *C. pasteurianum* and suggested that the enzyme was important in amino acid biosynthesis for *Clostridium* grown on carbohydrate as the sole carbon source (Scheme 47). Enzyme purified from this source exhibited a pH optimum near 7 and a K_m value for threonine of 0.4 mM. The enzyme cleaves allothreonine quantitatively at a rate much lower than threonine. Stöchlein and Schmidt[292] later reported that threonine and allothreonine were cleaved by different enzymes in *Clostridium*, presumably a threonine aldolase and a SHMT.

The *E. coli* SHMT has also been examined extensively. The protein has been cloned, sequenced, expressed, and subjected to mutagenesis.[256] The protein is dimeric with a subunit molecular weight of 46 kDa. As is true for other SHMT, oxidation of an active site cysteine abolishes activity, and reduced thiols are vital for full enzyme activity. The protein catalyzes retro-aldol cleavage of serine, threonine, and allothreonine (Table 10).

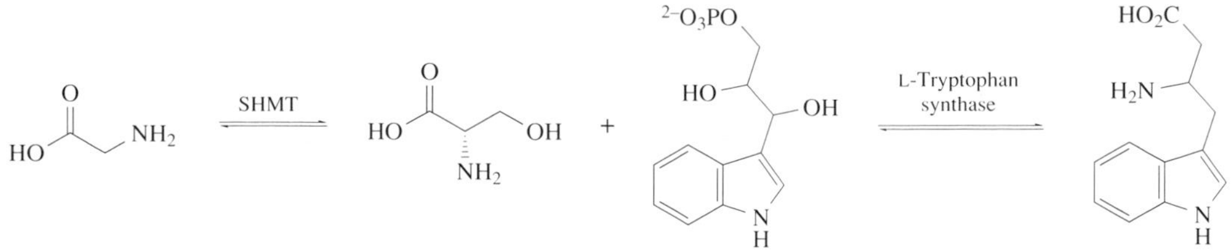

Scheme 47

Table 10 Kinetic properties of *E. coli* SHMT.

Substrate	K_m (mM)	k_{cat} (s^{-1})
Serine	0.3	640
L-Threonine	12	2.2
L-Allothreonine	1.5	30

Source: Plamann *et al.*[256]

SHMT activity has also been detected in a wide range of other microorganisms, including several strains of *Bacillus*,[293] *Pseudomonas*,[294-296] *Brevibacterium*,[297] and *Corynebacterium*;[298] at this point the enzyme should be considered ubiquitous in bacterial sources.

An unusual SHMT gene has been cloned from *Methylobacterium extorquens*.[299] This methylotroph is purported to express two SHMT proteins, one responsible for serine biosynthesis during growth on one-carbon sources, and a second responsible for the production of glycine and alkylated folate during growth on succinate.

3.13.6.2.4 SHMT from plants

As noted above, SHMT plays two important roles in plant metabolism—in one-carbon metabolism, and in photorespiration. A SHMT was isolated from maize seedlings by Masuda *et al.*,[300] in 1980. The enzyme is trimeric with a subunit molecular weight of 40 kDa. As with other enzymes, reduced thiol was required for full activity. The protein cleaves serine in the presence of tetrahydrofolate and L-allothreonine, but not L-threonine.

Neuburger and co-workers[249] reported mitochondrial and chloroplastic isoforms of a SHMT from spinach leaf. The proteins exist as homotetramers with subunit molecular weights near 50 kDa and show Michaelis constants for serine of roughly 1 mM; no additional specificity data were reported. A similar protein has been isolated from pea.[301]

The SHMT from *Solanum tuberosum* has been cloned, although data on the sequence were not reported.[302] The protein shows high sequence homology to other aldolases.

3.13.6.2.5 The mechanism of SHMT

Matthews and Drummond[303] have reviewed the mechanism of SHMT, and a detailed description of the mechanism will not be repeated here. The mechanism of retro-aldol cleavage of threonine/allothreonine is shown in Scheme 48. As noted above, Hammett plots using aryl-substituted substrates mandate that deprotonation be at least largely rate determining.

Cleavage of serine to glycine and formaldehyde is greatly enhanced by the presence of folate. Various labeling studies designed to probe the stereochemical fidelity of formaldehyde transfer yielded discordant results, and the precise role of folate in SHMT cleavage is unclear. Briefly, there remains debate regarding the possible role of folate as an actual enzyme prosthetic group or simply as a formaldehyde "sponge," designed to shuttle formaldehyde in and out of the enzyme active site. Matthews and Drummond[303] provide several mechanistic possibilities and discuss the relative merits of each.

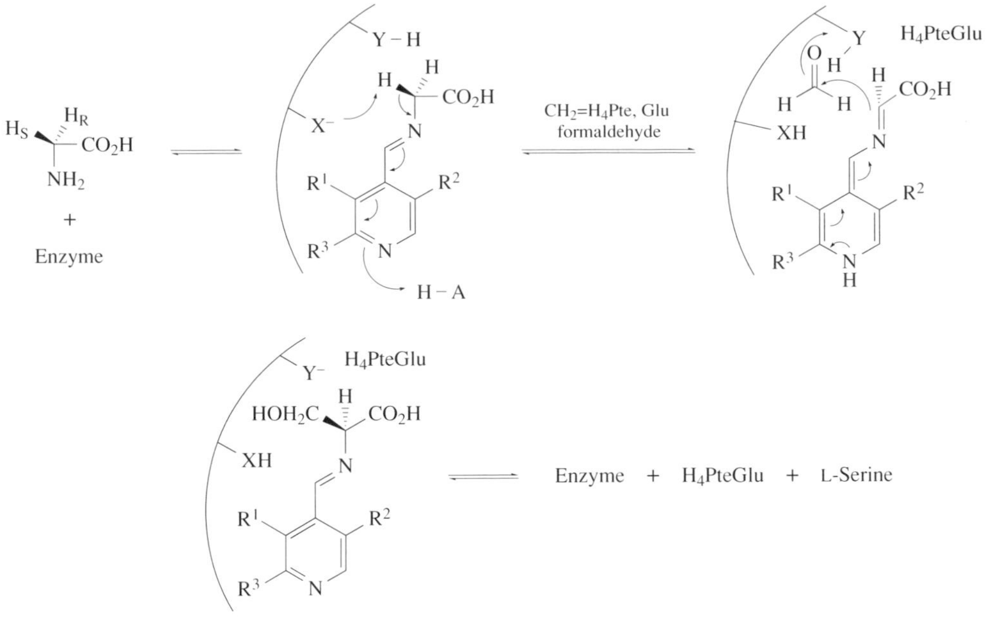

Scheme 48

3.13.6.3 2-Keto-3-deoxy-D-glucarate Aldolase

2-Keto-3-deoxy-D-glucarate aldolase (EC 4.1.2.20; CAS 37290-56-5) catalyzes aldol addition between pyruvate and tartronate semialdehyde to produce 2-keto-3-deoxy-D-glucarate. This enzyme is part of the D-glucarate catabolism pathway common to all glucarate-metabolizing members of *Enterobacter*.[304] Additionally, the enzyme has been isolated from *E. coli*,[305,306] *Pseudomonas* species,[307,308] *Methylophilus methanolovorus*,[309] and human serum.[310]

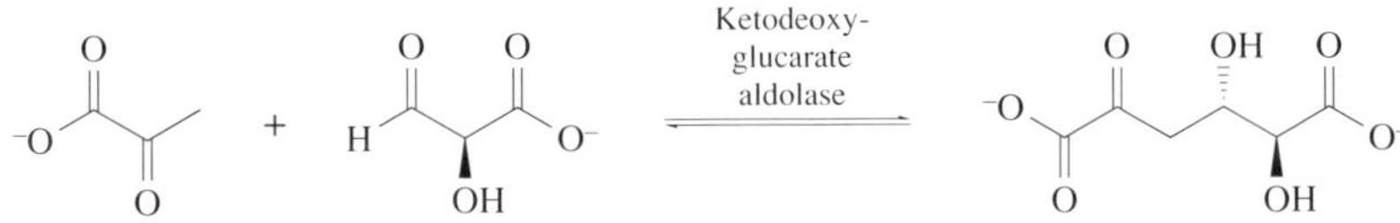

Scheme 49

The *Pseudomonas* aldolase was purified 45-fold by ammonium sulfate fractionation, DEAE cellulose and negative adsorption calcium phosphate gel chromatography.[307] The glucarate aldolase is a pyridoxal-dependent $\alpha_2\beta_2$ tetramer and requires Mg^{II} for activity. Fish and Blumenthal[305] reported that the aldolase from *E. coli* has a K_m value for 2-keto-3-deoxyglucarate of 3.2×10^{-4} M with the equilibrium lying in the cleavage direction. The glucarate aldolase does not cleave 2-keto-3-deoxy-gluconic, -galactonic, -heptonic, -octonic, or 6-phosphogluconic acids.

Meloche and Mehler[311] compared the mechanisms of liver hydroxyketoglutarate aldolase (EC 4.1.3.16, Section 3.13.6) and *E. coli* 2-keto-3-deoxyglucarate aldolase (Scheme 50); both enzymes catalyze the reaction of pyruvate and glyoxylate to produce hydroxyketoglutarate. Meloche notes that enolpyruvate formation is at least partially rate-limiting for both enzymes. *E. coli* glucarate aldolase produces racemic hydroxyketoglutarate apparently since glyoxylate is randomly oriented in the active site and the *re* or *si* faces are attacked with equal facility.

3.13.6.4 Sphinganine-1-phosphate Aldolase

Sphingolipids are metabolized *in vivo* to phosphorylethanolamine and a fatty aldehyde, generally palmitaldehyde. Both metabolites are ultimately converted to glycerophospholipids. The lipids are first phosphorylated by a kinase and then cleaved by the pyridoxal-dependent sphinganine-1-phosphate aldolase (EC 4.1.2.27; CAS 37290-61-2 (formerly 39391-27-0)).[312] The enzyme has been identified in a wide variety of mammalian organs, bacteria, and human fibroblast monolayers, and

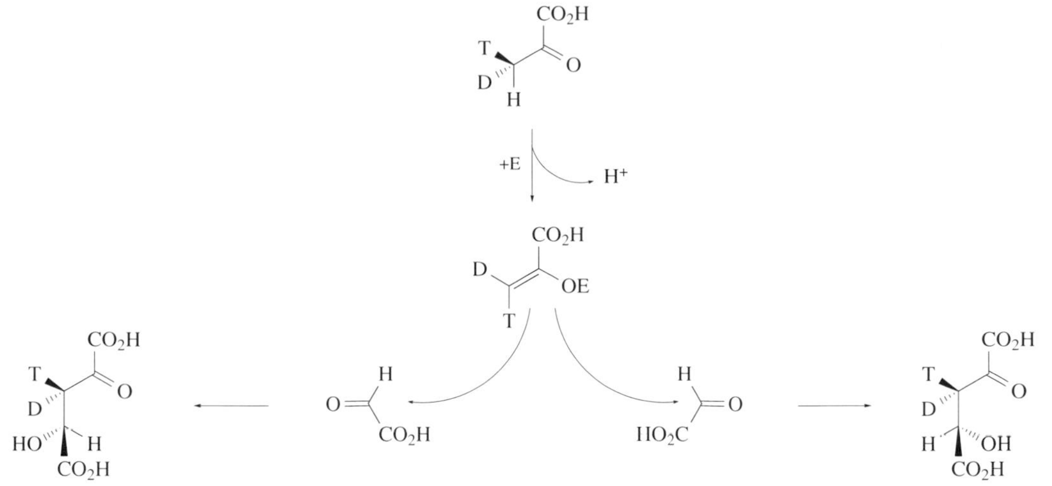

Scheme 50

is generally associated with the microsomal cell fraction.[313] Further fractionation has shown the enzyme to be associated with both the endoplasmic reticulum and mitochondria, although a later report disputes evidence of a mitochondrial enzyme. The enzyme is likely membrane bound with the catalytic unit facing the cytosol.[314–316]

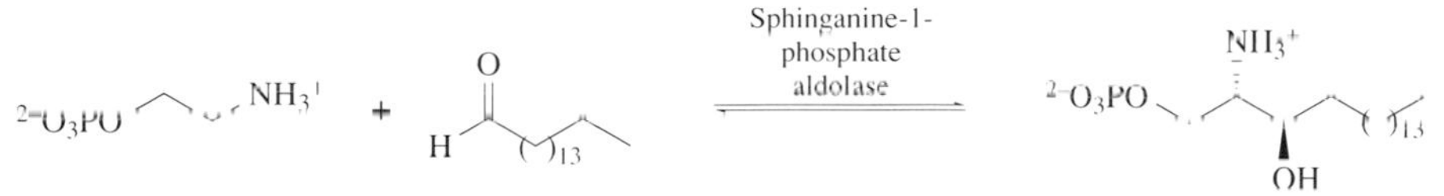

Scheme 51

The enzyme appears to be highly specific for the $2S,3R$ absolute configuration and the enantiomeric and diastereomeric versions of sphingolipid substrates are not converted or are converted only slowly.[317,318] The enzyme shows no more than weak specificity for the alkyl chain, and 4-D-hydroxysphinganine-1-phosphate and 1-desoxysphinganine-1-phosphonate are accepted.[319] The phosphonate analogue of the *in vivo* substrate is converted at 10% of the rate of the corresponding phosphate although with an essentially identical Michaelis constant. The phosphonate analogue is also a competitive inhibitor of the enzyme, with a K_i of 0.05 mM. *In vivo*, a variety of 2-amino-1,3-dihydroxyalkane or alkene substrates, including eicosadihydrosphingosine and short and medium chain-length substrates, are rapidly converted to ethanolamine and the corresponding aldehyde. Studies by Shimojo *et al.*[312,318] have shown that retro-aldol cleavage proceeds with stereospecific incorporation of a solvent hydrogen, generating the R-enantiomer of tritiated ethanolamine.

Enzymes from several sources exhibit a pH optimum near 7.5. A K_m value of 0.016 mM has been reported for sphinganine-1-phosphate for the enzyme from rat liver.[320] No information is available regarding molecular weight or subunit structure. In addition to a requirement for pyridoxal, the aldolase requires a reduced sulfhydryl group, and reagents designed for selective labeling of thiols (*p*-chloromercuribenzoate, iodoacetamine, *N*-ethylmaleimide) abolish activity. A mechanism has been proposed that utilizes the active site sulfhydryl residue for formation of a S,O-hemiacetal (Scheme 52).[321]

Other than sphingolipid catabolism, no *in vivo* function has been unambiguously assigned to the aldolase. However, it has been noted that, because sphingosine-phosphate has been proposed as a calcium-releasing messenger, the aldolase may have a regulatory function.[322]

3.13.6.5 3-Hydroxyaspartate Aldolase

Growth of *Micrococcus denitrificans* on glycollate as the sole carbon source induces the β-hydroxyaspartate pathway, the key enzymes of which are glyoxylate-L-aspartate aminotransferase, *erythro*-β-3-hydroxyaspartate aldolase (EC 4.1.3.14; CAS 37290-64-5), and *erythro*-β-hydroxy-

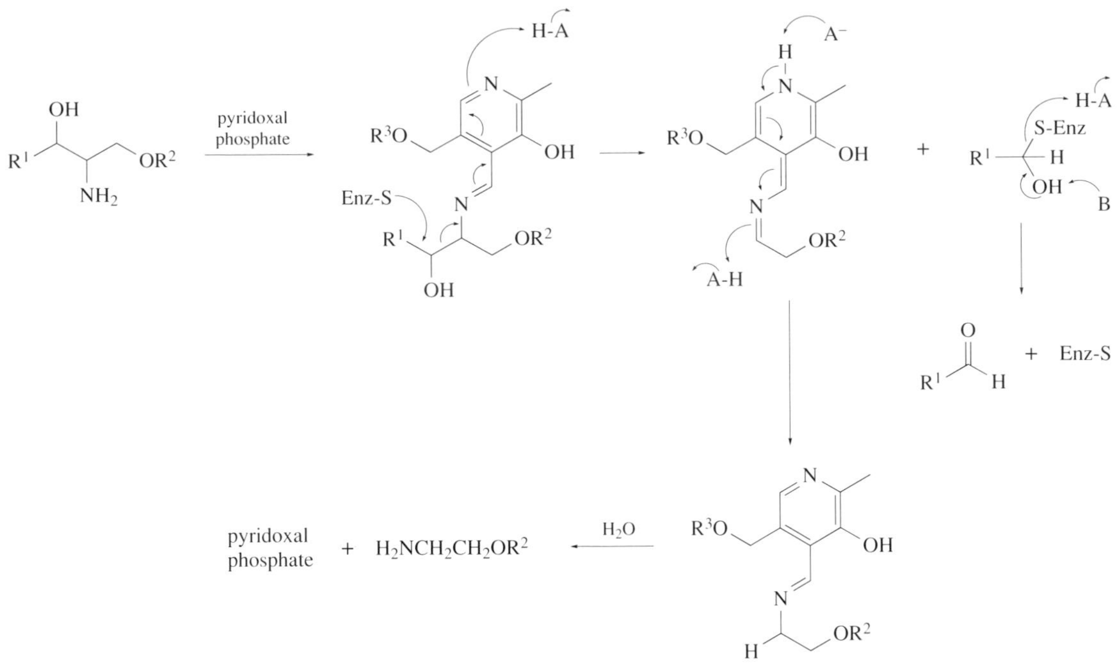

Aldolases

Scheme 52

aspartate dehydratase.[323–326] Together, these enzymes function to convert glyoxylate to oxaloacetate (Scheme 54).[327,328] In humans, protein C, a vitamin K-dependent regulator of blood coagulation, contains β-hydroxyaspartic acid in a domain homologous to the epidermal growth factor. β-Hydroxyaspartic acid has been proposed to play a direct role in calcium binding in protein C and related proteins.[329] Additionally, the acid has been proposed to chelate iron in factor IX, which participates in the intrinsic path of blood coagulation[330] and to act as a competitive antimetabolite of aspartic acid.[331]

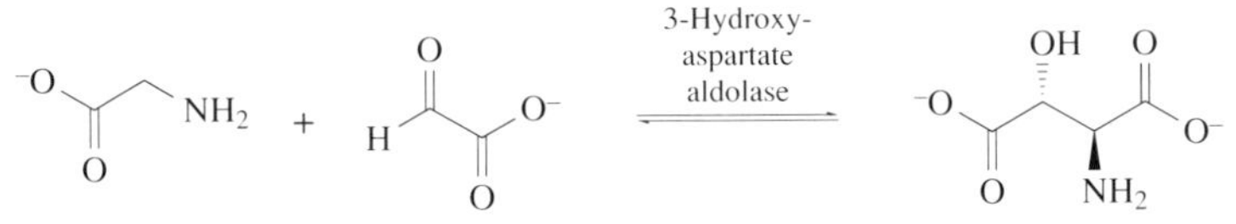

Scheme 53

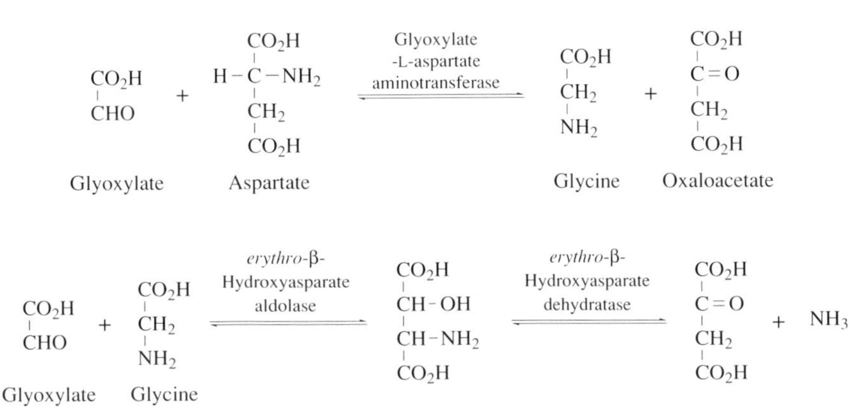

Scheme 54

An isolation protocol and initial report of the *M. denitrificans* 3-hydroxyaspartate aldolase specificity and properties has appeared in the literature.[323] The enzyme exhibits a broad pH optimum near 8 and a specific activity for *erythro*-3-hydroxy-L-aspartate of 0.2 U mg^{-1}. Substrates other than the natural substrate, in particular β-methyl-3-hydroxyaspartate (K_m of 2.8 mM) are accepted. Although the enzyme is specific for the *erythro* stereochemistry, racemic substrates were used throughout and no determination of absolute stereoselectivity was attempted.

threo-3-Hydroxyaspartic acid has been isolated from culture broths of *Arthrinium phaeosperumum* T-53, although the biosynthetic origin of this compound has not been elucidated.[327] Likewise, the presence of aspartate aminotransferase in *E. coli* indicates the potential for aldolase expression.[332]

3.13.6.6 Tyrosinephenol Lyase (β-Tyrosinase)

β-Tyrosinase (EC 4.1.99.2; CAS 9059-31-8) catalyzes the reaction of pyruvate, phenol, and ammonia to produce L-tyrosine. The lyase also catalyzes a variety of other reactions, including α,β-elimination, β-replacement, and racemization. *In vivo*, tyrosinase is responsible for the degradation of L-tyrosine; *in vitro*, this reaction is reversible. The enzyme has been isolated from a wide variety of bacterial sources including *Escherichia*,[333] *Proteus*,[334,335] *Aerobacter*,[336] *Clostridium*,[337] *Aeromonas*,[338] *Citrobacter*,[339] *Erwinia*,[340] and *Symbiobacterium*.[341] Nonbacterial sources include *Leptoglossus phyllopus* (hemipteran),[342] *Oxidus gracilis* (millipede),[343] and mouse albino melanocytes.[344]

Scheme 55

Tyrosinase is a pyridoxal-dependent enzyme and is activated by both a metal cofactor and NH_4^+. Well-known inhibitors of the enzyme include *o*- and *m*-substituted phenols, L-alanine, L-phenylalanine, and Na^+. The inhibition by Na^+ prompted an investigation of the potential inhibitory effects of other monovalent cations such as Li^+, K^+, Rb^+, Cs^+, and NH_4^+. In all cases, the rate of β-elimination is increased by the presence of cations. Interestingly, although most act as noncompetitive activators, Li^+ exerts no effect and Na^+ is inhibitory.[345]

The enzyme from *Symbiobacterium thermophilium*, an obligately symbiotic thermophile, was purified 300-fold via ammonium sulfate fractionation, anion exchange, hydroxyapatite, and hydrophobic chromatography by Suzuki *et al.*[346] The enzyme is stable to 80 °C with a K_m (L tyrosine) of 0.054 mM. Immobilized derivatives of pyridoxal were used by Ikeda and Fukui[347] for the purification of the lyase by affinity chromatography.

The crystal structure of tyrosinase from *Citrobacter intermedius* has been solved (Figure 4).[348] The lyase from *Erwinia herbicola* has also been crystallized, although a structure for this enzyme has yet to be reported.[349] The gene from *E. herbicola* has been cloned into *E. coli* and expressed at high levels and the transcriptional regulation of tyrosinase has been investigated.[350–352] The gene from *E. intermedia* has also been cloned and sequenced.[351]

Kiick and Phillips[353] investigated the mechanism of the tyrosinase-catalyzed reaction and reported that the retro-aldol reaction proceeds by the formation of an aldimine adduct between L-tyrosine and pyridoxal (Scheme 56). The α-proton of the substrate is abstracted by an enzyme-bound base to give a quinonoid structure. A second base subsequently abstracts the phenolic hydroxy proton while the first base protonates the aromatic C-4 to form the cyclohexadienone. Pyridoxal electron push, in cooperation with hydroxy electron pull, causes C—C bond rupture, releasing phenol.

Although turnover is not possible, L- and D-alanines are bound by tyrosinase. A common quinonoid intermediate in the reactions of L- and D-alanine was detected by Chen and Phillips.[354] Formation of this first quinonoid intermediate is the rate-determining step in racemization. A second quinonoid intermediate is formed in the reaction with L- but not D-alanine. The second intermediate occurs after a structural reorganization in the aromatic side chain binding site. Concurrently, Phillips and co-workers[353] and Faleeve *et al.*[355] determined, on the basis of presteady-state kinetics, that the second intermediate occurs only when the phenol undergoes a 90° reorientation. The phenol begins perpendicular to the pyridoxal-π system and reorients to be parallel following C—C bond rupture (Scheme 57).

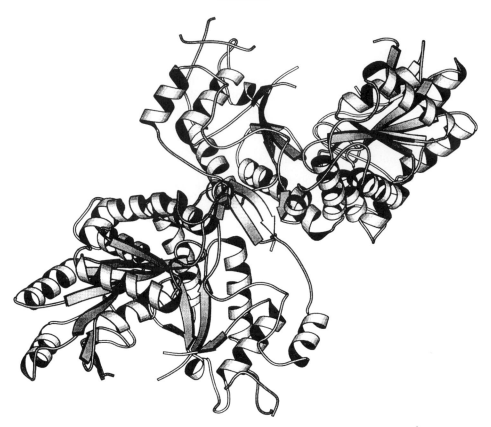

Figure 4   *Citrobacter intermedius* tyrosinase crystal structure at 2.3 Å.

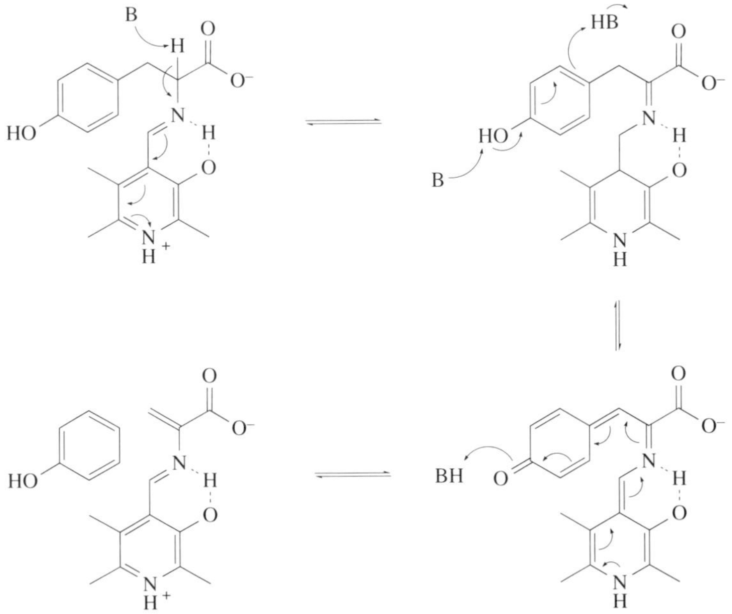

Scheme 56

   The nature of the active site has been elucidated through a variety of studies. Inactivation of an essential histidyl residue yields a holoenzyme which fails to catalyze the exchange of the α-hydrogen on L-alanine with tritiated water. This observation is consistent with a catalytic role for histidine at the active site of the enzyme, presumably removal of the α-hydrogen of the substrate. Mechanistic

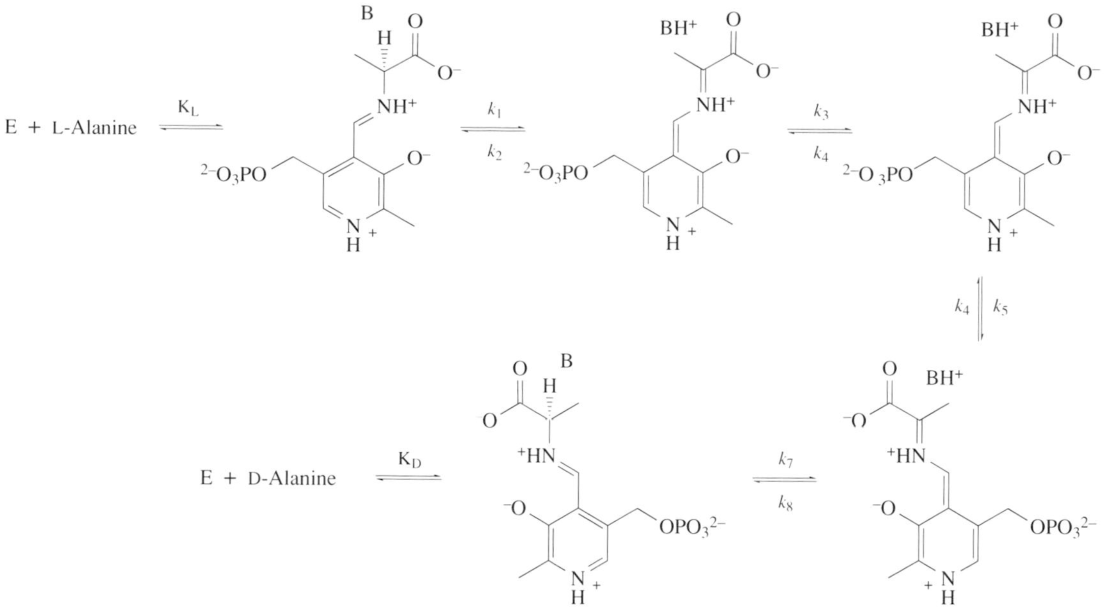

Scheme 57

investigations were also conducted through mutagenesis studies by Chen *et al.*[356] A Y71P mutant of *C. freundii* tyrosinase shows no turnover but does form stable quinonoid complexes in a biphasic fashion. These researchers report that the slow step in turnover is cofactor binding, while the fast step represents deprotonation/reprotonation. The mutation also results in an increase of 1700-fold in K_D for pyridoxal. Tyr70 thus apparently serves a dual role in catalysis: binding of cofactor in the absence of substrate and as a general acid catalyst in quinonoid leaving group elimination.

Double reciprocal plots of the extrapolated reaction velocity at infinite concentrations of one substrate and variable concentrations of the remaining two substrates (three substrate kinetics) and spectral studies suggest that the binding is ordered with ammonia binding first, followed by pyruvate and phenol.[353] The fact that proton exchange at C-3 of pyruvate occurs during incubation of the enzyme with ammonia and pyruvate supports the existence of an enzyme bound α-aminoacrylate intermediate. pH–activity studies further indicate that the formation of this intermediate is the rate-limiting step.

The factors responsible for both substrate affinity and specificity during catalysis were investigated by Faleev *et al.*[355] Broadly, a correlation exists between the free energy of inhibitor binding and side-chain hydrophobicity; binding efficiency increases with increasing side-chain hydrophobicity. Aspartic and glutamic acids are potent inhibitors although their side chains have low hydrophobicity, suggesting the presence of an electrophilic group in the active site that interacts with the terminal carboxylic group of inhibitors. Enzyme-catalyzed isotope exchange of α-protons with ^2H_2O was observed for L- but not D-amino acids. Thus, substrate specificity of tyrosinase is controlled during phenol elimination which requires a *p*-hydroxy group and is sensitive to the steric parameters of other ring substituents. When all the specificity requirements are met, α-proton abstraction is the rate-limiting step.

Although tyrosinase enzymes from *E. herbicola* and *C. freundii* catalyze the same reaction and have analogous mechanisms, substrate specificities differ. In order to investigate fully this aspect of catalysis, the pH dependence of kinetic parameters and the primary deuterium isotope effect were determined for both enzymes by Kiick and Phillips.[353] Primary deuterium isotope effects indicate that proton abstraction from the 2-position of the substrate is partially rate limiting for both enzymes. The *freundii* primary deuterium isotope effect is pH independent, indicating that tyrosine does not dissociate faster than it reacts. V_{max} for the *freundii* enzyme is also pH independent requiring that the substrate bind only the correctly protonated form of the enzyme. With the *herbicola* enzyme, both the primary deuterium isotope effect and V_{max} are pH dependent; thus, while the protonated or unprotonated enzyme can bind substrate, only the unprotonated Michaelis complex is catalytically competent.

Azido-substituted aromatic amino acids have been synthesized using tyrosinase by Kirk and co-workers.[357] For example, 2-azido-L-tyrosine was synthesized via a tyrosinase-catalyzed reaction between pyruvate and 3-azidophenol. Again, an unsubstituted position *para* to the phenolic hydroxy of the substrate is apparently the only significant structural requirement. Yamada and Kumagai[358,359] reported the synthesis of L-tyrosine analogues with tyrosinase from *E. intermedia*. Among the substrates capable of replacing phenol were pyrocatechol, resorcinol, *m*- and *o*-cresols, and *m*- and *o*-chlorophenols.

The *S. thermophilum* tyrosinase substrate specificity is also reportedly broad, with some substrates reacting at higher rates than that of L-tyrosine (Table 11).[360]

Table 11 Substrate specificity of tyrosinase from *Symbiobacterium thermophilum*.

Substrate[a]	Relative rate (%)
L-Tyrosine	100
L-Serine	15
L-Cysteine	10
D-Tyrosine	7
S-Methyl-L-cysteine	603
β-Chloro-L-alanine	1450
L-Alanine	0
L-Aspartate	0
L-Histidine	0
L-Homoserine	0
L-Methionine	0
L-Phenylserine	0
L-Threonine	0
L-Tryptophan	0
L-Valine	0
D-Serine	0
D-Alanine	0
D-Cysteine	0

Source: Suzuki *et al.*[360] [a]Reactions performed in 50 mM phosphate buffer, pH 8.0, with 0.016 U enzyme, 0.1 mM pyridoxal phosphate, and 2.5 mM substrate.

A variety of pyridoxal-dependent enzymes, including tyrosinase, have been immobilized on Sepharose by Ikeda and Ikeda.[347] This technique has been used synthetically to produce L-DOPA. In a similar fashion, the lyase has been enclosed in hollow-fiber reactors and used for the synthesis of L-tyrosine by Fuganti *et al.*[361] Cloned tyrosinase has also been used for the industrial scale production of L-DOPA from catechol.[362]

3.13.7 ALDOLASES OF UNDETERMINED MECHANISM

3.13.7.1 Ketotetrose Phosphate Aldolases (Erythulose-1-phosphate)

Ketotetrose phosphate aldolase (EC 4.1.2.2; CAS 9024-45-7) catalyzes the *in vivo* cleavage of erythulose-1-phosphate to dihydroxyacetone phosphate and formaldehyde. The enzyme plays a key role in carbohydrate metabolism, specifically erythritol metabolism which proceeds via a series of phosphorylated intermediates.[363]

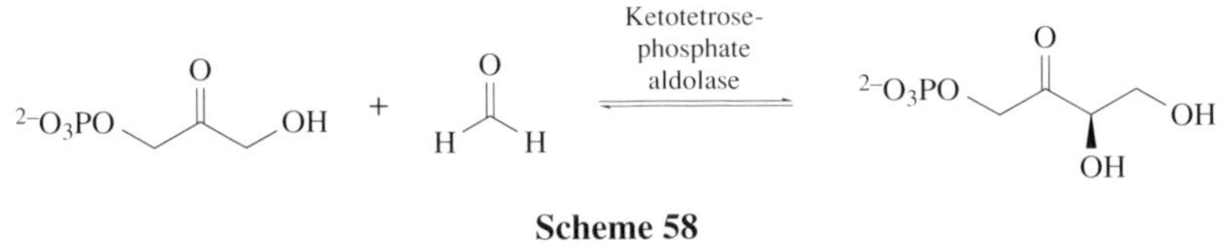

Scheme 58

Ketotetrose phosphate aldolase has been isolated from rat liver by Charlampous and co-workers[347,364] as well as from *Propionibacterium pentosaceum* by Wawszkiewicz.[365] The reaction is reversible although the equilibrium favors synthesis. The rat liver enzyme was purified to a specific activity of 9.4 U mg^{-1} by ammonium sulfate and citrate-ammonium sulfate fractionation, isoelectric precipitation, and adsorption on calcium phosphate gel. Divalent cations are required for activity and the pH optimum is 7.2 at 37 °C. The purified enzyme was reportedly stable for at least two weeks at −20 °C in buffer or in an ammonium sulfate suspension. A D-erythulose reductase has been detected in beef liver, possibly indicating the presence of a ketotetrose phosphate aldolase as well.

The substrate specificity of the aldolase from any source has not been investigated extensively. Acetaldehyde, glycoaldehyde, and glyceraldehyde were tested as unnatural aldehydes and were not accepted by rat liver aldolase.[347,364] The mechanism also requires investigation, and cloning of ketotetrose phosphate aldolase would facilitate such studies.

3.13.7.2 Ketopantoaldolase

Ketopantoaldolase (EC 4.1.2.12; CAS 9024-51-5) catalyzes the reaction of 2-keto-3-methyl-butanoate and formaldehyde to produce 2-dehydropantoate. The *in vivo* role of the enzyme is the synthesis of pantothenate during CoA biosynthesis. Ketopantoaldolase has been detected in several sources including *Kluyveromyces van der Walt*,[366] *E. coli*,[367] and *Aerobacter aerogenes*.[368]

Scheme 59

The enzyme is activated by divalent metal ions and activity is completely lost during dialysis against the chelator Versene, treatment with Dowex, or ammonium sulfate fractionation.[369] Activity is fully restored by the addition of divalent ions including Co^{II}, Mn^{II}, Ni^{II}, Mg^{II}, and Fe^{II}. Of the 11 species of *Kluyveromyces van der Walt* examined, five expressed the ketopantoate aldolase.[366] The enzymes were not purified nor were properties investigated.

The enzyme from *E. coli* has been isolated and purified 36-fold by heat treatment, protamine sulfate treatment, and ammonium sulfate fractionation.[367] The pH optimum is 7.8 and tris, gly-cylglycine, potassium phosphate, and glycerol buffers proved inhibitory, possibly due to metal-ion chelation.

Ketopantoaldolase appears to be specific for both substrates,[367,368] and pyruvate could not be substituted by 2-keto-3-methylbutanoate, nor could acetaldehyde be substituted by formaldehyde.

3.13.7.3 6-Phospho-5-keto-2-deoxygluconate Aldolase

Myoinositol is a commonly found cyclitol in a variety of legumes and woody plants, including pea and soybean nodules and redwood. A variety of microbial and fungal sources metabolize inositol and at least one report of a mammalian system of enzymes for cyclitol metabolism exists. Two distinct metabolic pathways have been proposed (Scheme 61): the first is initiated by a nicotinamide-dependent oxidoreductase (prokaryotes) while the second is initiated by an oxidase (eukaryotes).[370,371] The former pathway proceeds to 5-keto-2-deoxygluconophosphate which is subsequently cleaved to malonate semialdehyde and dihydroxyacetone phosphate by the 6-phospho-5-

keto-2-deoxygluconate aldolase (EC 4.1.2.29; CAS 62213-25-6). Malonate semialdehyde is subsequently decarboxylated to acetyl CoA while the latter metabolite enters normal glycolysis cycles.

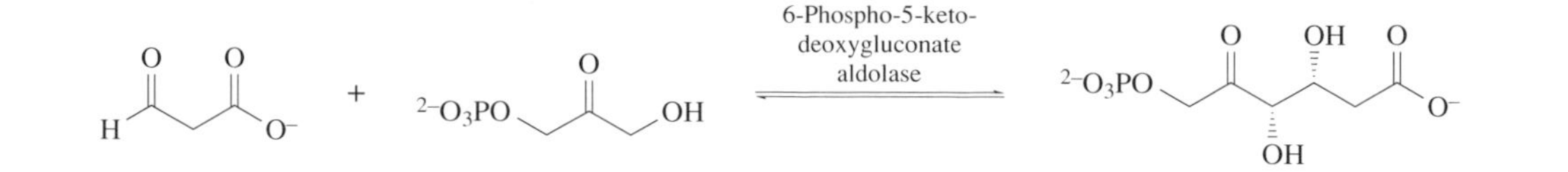

Scheme 60

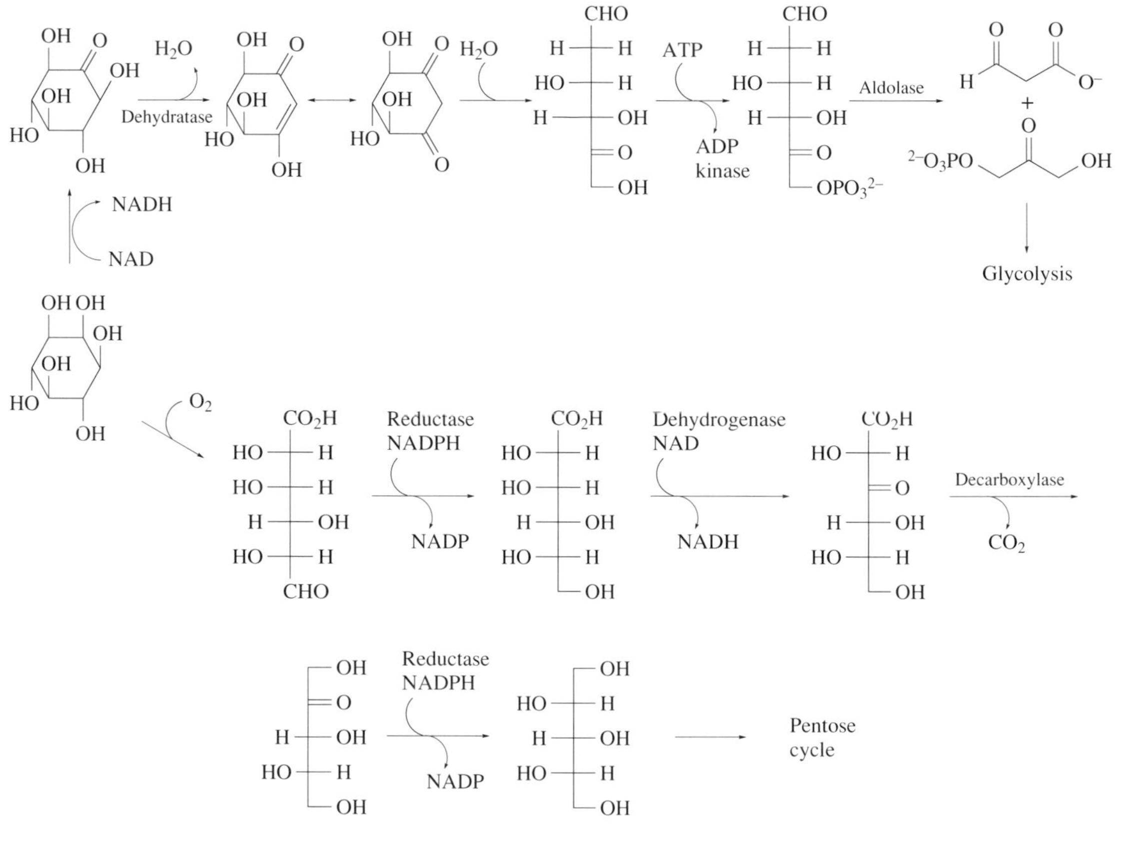

Scheme 61

The aldolase pathway was first elucidated and most extensively studied by Talbot and Seidler[372] and Magasanik[373,374] in *Aerobacter aerogenes* (later *Klebsiella*), although the aldolase from this source has not been isolated to date. The existence of the pathway has been inferred in *Rhizobium leguminosarum bv. viciae* by cloning studies.[375] The segment contains 12 open reading frames encoding a total of seven genes and apparently includes an aldolase-containing metabolic sequence of enzymes. By homology with other aldolases, the gene designated B65C encodes for the aldolase. Talbot and Seidler[372] detected the activity in a variety of organisms, including several strains of *Klebsiella*. Golubev[376] surveyed a wide variety of yeasts capable of growth on inositol and found in each case an oxidase, suggesting a glucuronate pathway of inositol metabolism. Fujita and co-workers[377] cloned a 15 kb section of the *B. subtilis* genome responsible for inositol utilization by this organism.

3.13.7.4 17-α-Hydroxyprogesterone Aldolase

17-α-Hydroxyprogesterone aldolase (EC 4.1.2.30; CAS 62213-24-5) is a key enzyme in the biosynthesis of testosterone from progesterone (Scheme 63), and is found in mammalian testes.[378–382] A purification of the rabbit enzyme has been reported.[382] The enzyme exhibits a pH optimum of 7.4

and a temperature optimum near 37 °C. No information was reported regarding the substrate or stereospecificity of the enzyme. The enzyme was apparently not assayed in the synthetic direction.

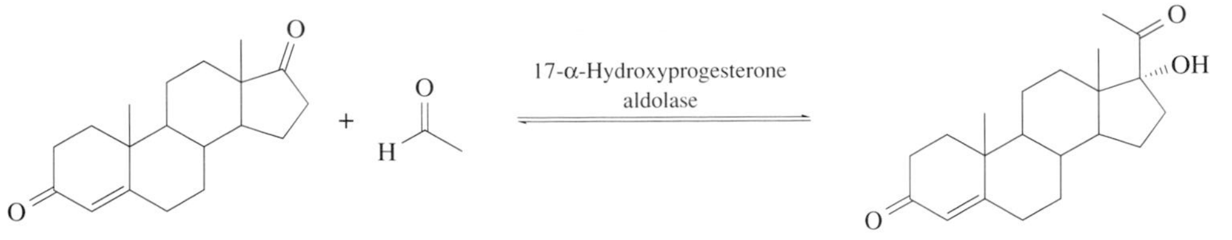

17-α-Hydroxyprogesterone aldolase

Scheme 62

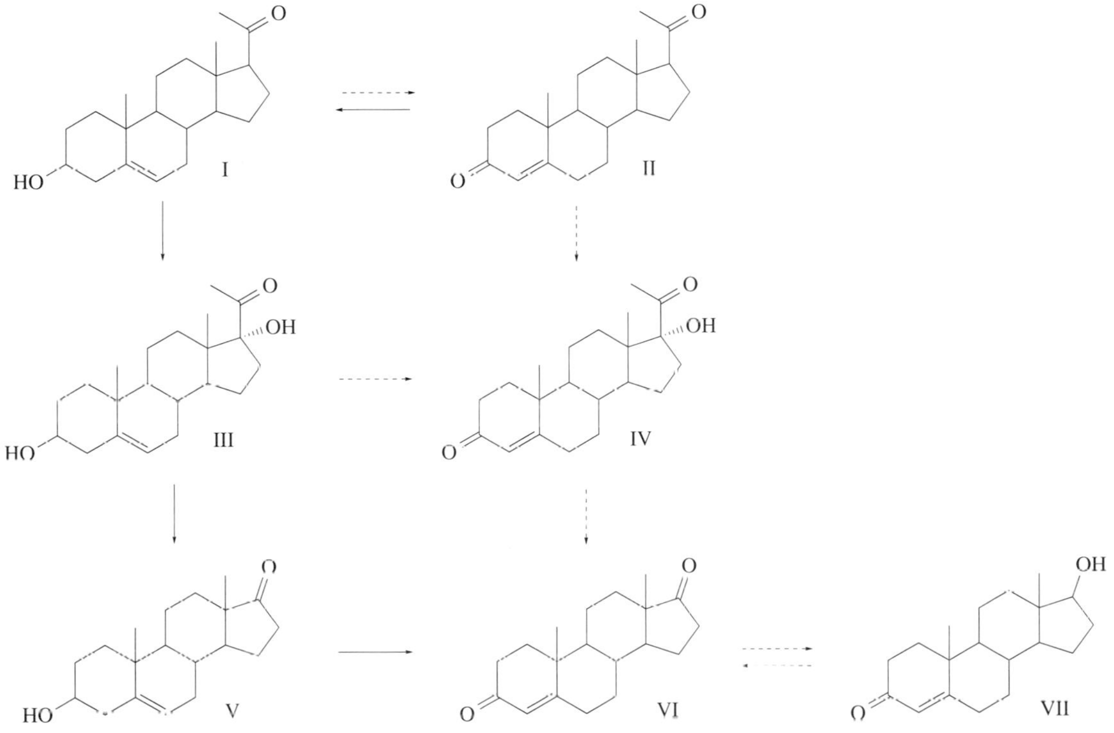

I, 3β-Hydroxy-5-pregen-20-one; II, Progesterone; III, 3β,17α-Dihydroxy-5-pregen-20-one; IV, 17α-Hydroxyprogesterone; V, 3β-Hydroxy-5-androsten-17-one; VI, 4-Androstene-3,17-dione; VII

Scheme 63

Numerous studies have been reported on the molecular origin of the hypogonadism frequently observed in alcoholic males. Despite a report that hydroxyprogesterone aldolase may be inhibited by acetaldehyde, the majority of published reports suggest that inhibition of 17-β-hydroxyprogesterone dehydrogenase arises from nicotinamide cofactor imbalances.[383–388]

3.13.7.5 Trimethylamine Oxide Aldolase and Dimethylaniline-*N*-oxide Aldolase

A variety of methylotrophic bacteria grow on methylamine as the sole carbon source. Virtually all such organisms convert trimethylamine to formaldehyde and ammonia (Scheme 65). It appears that obligate methylotrophs catalyze these conversions with a trimethylamine dehydrogenase, a flavoprotein that catalyzes direct oxidative deamination of trimethylamine to dimethylamine and formaldehyde. Alternatively, several facultative methylotrophs convert trimethylamine to trimethyl-amine oxide and in a subsequent step demethylate to dimethylamine and formaldehyde utilizing trimethylamine oxide aldolase (EC 4.1.2.32; CAS 72561-08-1). This pathway was first described for the facultative methylotroph *Bacillus* PM6,[389] and is apparently operative in several other organisms, including *P. aminovorans*,[390–393] bacterium 5B1,[394] and the pink pseudomonad *Pseudomonas* 3A2.[395]

Further oxidation of dimethylamine to methylamine and formaldehyde is catalyzed by a mixed function secondary amine oxidase system that may involve an aldolase.

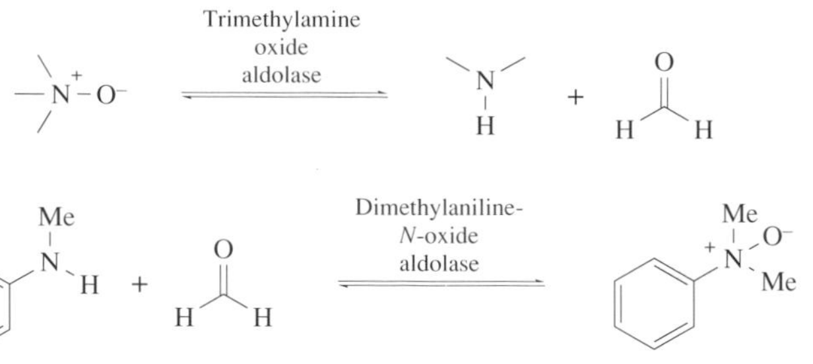

Scheme 64

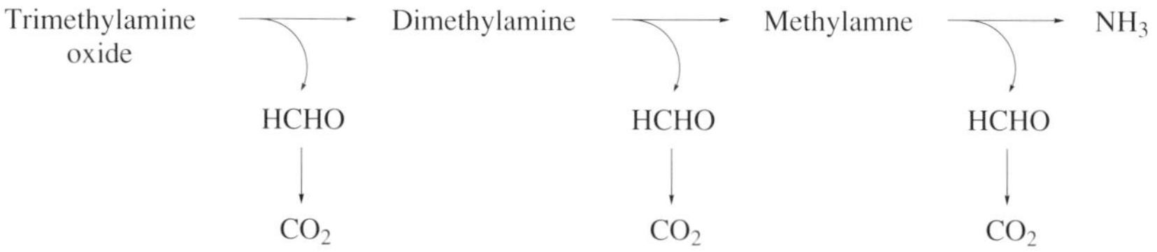

Scheme 65

Trimethylamine oxide aldolase has been purified from both *Bacillus* PM6[389] and *P. aminovorans*.[390–393] The former protein appears to be monomeric with a molecular weight near 45 kDa. The enzyme contains no known prosthetic groups but is strongly activated by the presence of Fe^{II}. The *Bacillus* enzyme also accepts benzyldimethylamine oxide and chloropromazine *N*-oxide as substrates. Both enzymes show K_m values for trimethylamine oxide near 2 mM and are active between pH 5.0 and 7.5. The reactions are postulated to involve an oxygen transfer reaction, followed by dealkylation. Craig *et al.*[396,397] have postulated a free radical mechanism.

A similar activity, dimethylaniline-*N*-oxide aldolase (EC 4.1.2.24; CAS 37290-58-7), was reported from pig liver microsomes by Machinist *et al.*[398] in 1966. The enzyme was reported to have a pH optimum near 7 and convert a variety of dialkylaryl amine-*N*-oxides (Table 12).

Table 12 Kinetic properties of porcine liver dimethylaniline-*N*-oxide aldolase.

Substrate	V_{rel}	K_m (mM)	V_{max}[a]
N,*N*-Dimethyl-1-naphthylamine-*N*-oxide	100	7	200
N,*N*-Dimethyl-*p*-toluidine-*N*-oxide	65	20	286
p-Chlorodimethylaniline-*N*-oxide	43	ND[b]	ND
N,*N*-Dimethylaniline-*N*-oxide	14	80	167
N-Ethyl-*N*-methylaniline-*N*-oxide	10	ND	ND

Source: Machinist *et al.*[398]   [a]mmol formaldehyde per mg protein per min.   [b]ND: not determined.

In 1982, Hlavica[399] argued that aldolase activity was not due to an enzyme activity, but rather the spontaneous dealkylation of aromatic amine *N*-oxides produced by cytochrome P-450. Indeed, compared to the nonenzymatic pathways for *N*-oxide dealkylation available to trialkylamine oxides, additional mechanistic possibilities are available to aromatic amine oxides. For example, Bamberger and Leyden observed formation of aminophenols from aniline *N*-oxides during thermolysis or treatment with either ferricytochrome C or ferrihemoglobin.[400] Similar observations using other aromatic amines were made by Terayama[401] (dimethylaminoazobenzene-*N*-oxide) and Coccia and Westerfeld[402] (chlorpromazine-*N*-oxide).

In 1989, Pandey *et al.*[403] reported the enzymatic demethylation of dimethylaniline by an isozyme of cytochrome P-450 from rabbit liver microsomes. Although some isozymes effectively dealkylate dimethylaniline, none produces measurable quantities of dimethylaniline-*N*-oxide. Based on these data, the authors concluded that dimethylaniline-*N*-oxide aldolase activity was in fact a distinct enzyme activity. Final resolution of this conflict will likely require cloning of the putative aldolase to ensure observation of pure enzyme activities.

3.13.7.6 Oxalomalonate Lyase

Acetobacter suboxydans apparently lacks a functional tricarboxylic acid pathway and alternative amino acid biosynthesis pathways are required.[404–408] In the 1960s, Cheldin and co-workers[409,410] proposed a biosynthetic route to glutamate that relies on enzymatic reaction of glyoxylate and oxaloacetate to yield α-hydroxy-γ-ketoglutarate. In 1966, this group reported the existence of such an aldolase (oxalomalonate lyase; EC 4.1.3.13; CAS 37290-63-4) in the vinegar producing *Acetobacter*. The enzyme exhibits a pH optimum near 6.0. α-Hydroxy-γ-ketoglutarate produced by enzymatic reaction was decarboxylated with hydrogen peroxide, yielding only D-malate, suggesting that the reaction is stereospecific. No reports of substrate specificity, retro-aldol reaction or mechanistic data are available in the literature to date.

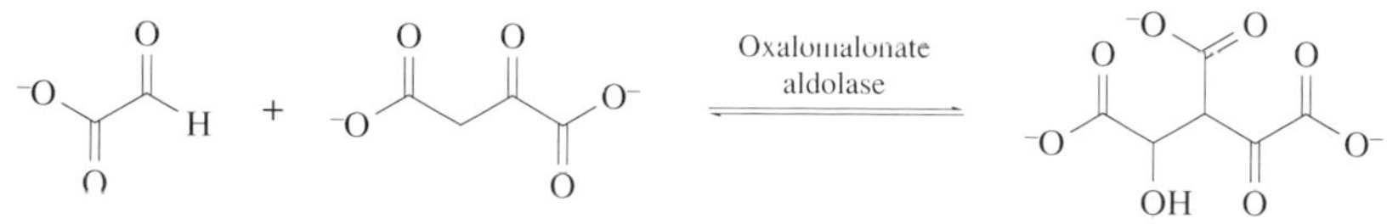

Scheme 66

3.13.7.7 4-Hydroxy-2-ketopimelate Aldolase

4-Hydroxy-2-ketopimelate aldolase is a widely distributed bacterial aldolase that catalyzes the cleavage of hydroxyketopimelate to pyruvate and succinic semialdehyde.[411] *In vivo*, the enzyme functions in aromatic catabolism; the enzyme is induced in bacteria grown on L-tyrosine or 4-hydroxyphenylacetate. The pathway has been observed in a variety of gram-positive and gram-negative strains of bacteria, including *Bacillus* sp.,[412] *Micrococcus lysodeikticus*,[411] *P. putida*,[413,414] *E. coli*,[415] and *Acinetobacter* sp.[416]

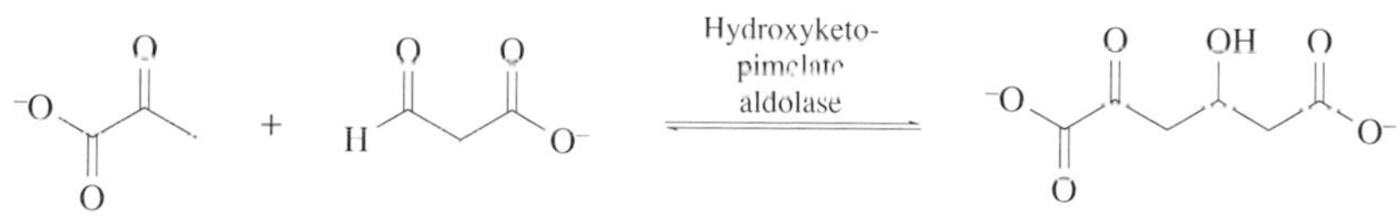

Scheme 67

The only isolation of hydroxyketopimelate aldolase reported to date was conducted in the laboratories of Dagley during the late 1970s and early 1980s.[417,418] Purification of the *Acinetobacter* enzyme consisted of heat treatment, ammonium sulfate fractionation, DEAE cellulose, and Sephadex chromatography. Pure enzyme provided a specific activity of $42.5\,U\,mg^{-1}$, a 181-fold purification of crude extracts. The maximum activity of hydroxyketopimelate aldolase is observed at pH 8.0. Metal ion cofactors are not required for activity. The enzyme exists as a hexamer with a total molecular weight of 158 kDa. 4-Hydroxy-2-ketovalerate is attacked at 3% of the rate of hydroxyketopimelate while 4-hydroxy-4-methyl-2-ketoglutarate and 4-carboxy-4-hydroxy-2-ketoadipate are not accepted as substrates.

3.13.7.8 3-Hexulose Phosphate Synthase

Kula and others[419–422] have reported a 3-hexulose phosphate synthase from *Methylomonas* M15 that catalyzes the synthesis of D-arabino-3-hexulose-6-phosphate. In addition to formaldehyde, several aliphatic and aromatic aldehydes are accepted, including propionaldehyde, which forms two diastereomers of 7,8-dideoxy-4-octulose-1-phosphate. The product has exclusively the S configuration at C-5 while the $6S$ and $6R$ diastereomers are generated in a ratio of $1:2.4$. No further investigations of this enzyme have appeared in the literature.

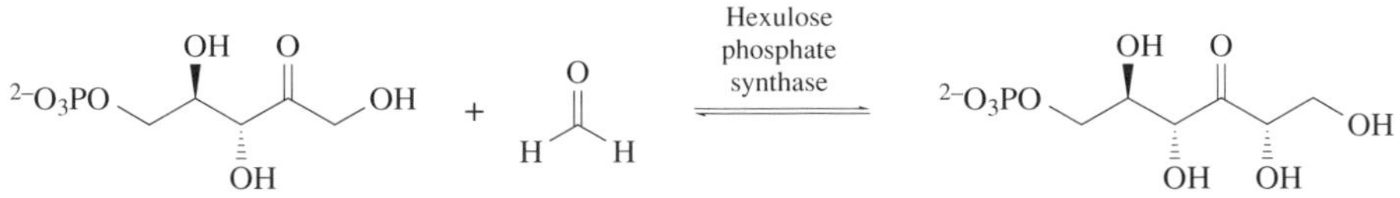

Scheme 68

3.13.8   ENZYMES OF GLYCOLYSIS

Both eukaryotes and prokaryotes convert glucose to pyruvate and lactate via glycolysis. In all species, the primary glycolytic pathway is the Embden–Meyerhof–Parnas route, converting glucose successively to glucose-6-phosphate, fructose-6-phosphate, and FDP. The latter metabolite is cleaved in a retro-aldol reaction to D-glyceraldehyde-3-phosphate and dihydroxyacetone phosphate by FDP aldolase. In some prokaryotes, parallel pathways exist for the metabolism of stereoisomers of fructose; thus, rhamnulose-1-phosphate aldolase, fuculose-1-phosphate aldolase and tagatose-1,6-diphosphate aldolases have been identified. FDP and tagatose-1,6-diphosphate aldolases each provide one equivalent of dihydroxyacetone phosphate and D-lactaldehyde, while rhamnulose-1-phosphate and fuculose-1-phosphate aldolases provide dihydroxyacetone phosphate and D-glyceraldehyde-3-phosphate.

In all species, a pentose shunt is operative as a secondary glycolytic route: this pathway is the major mechanism of glycolysis in red blood cells. In this route, glucose is successively oxidized, phosphorylated, and decarboxylated to the central metabolite ribulose-5-phosphate. The successive action of transketolase and transaldolase finally produces FDP and D-glyceraldehyde-3-phosphate, which are further metabolized through the Embden–Meyerhof pathway.

In prokaryotes a third metabolic route, the Entner–Doudoroff pathway, also exists. This pathway converts glucose to pyruvate and D-glyceraldehyde-3-phosphate through the successive intermediacy of gluconate, 6-phosphogluconate, and 2-keto-3-deoxy-6-phosphogluconate. A parallel pathway, the DeLey–Doudoroff path, exists in some microorganisms for the conversion of galactose to pyruvate and D-glyceraldehyde-3-phosphate.

All other sugars metabolized for energy by various organisms are converted to one of these feedstocks. Scheme 69 outlines the pathways of glycolysis that have been identified in prokaryotic and eukaryotic sources.

Below we describe the ten aldolases (including transketolase and transaldolase) that participate in glycolysis.

3.13.8.1   Transketolase

Transketolase (EC 2.2.1.1; CAS 9014-48-6) is a thiamine-dependent enzyme of the pentose phosphate pathway of glycolysis. This path forms a metabolic duct between the pentose/hexose phosphate shunt and the glycolytic pathway (Scheme 69). The pentose phosphate shunt serves two major roles: to generate pentose sugars for nucleic acid and amino acid biosynthesis and to recover carbon that enters the pentose phosphate shunt via the oxidative route. This latter function is associated with the conversion of glucose to ribulose-5-phosphate with the concomitant generation of two equivalents of NADPH, which in turn maintains the reducing environment of the cell and protects against oxidative stress. In bacteria, the pentose phosphate shunt serves the additional role of generating erythrose-4-phosphate for entry into the shikimate biosynthetic pathway.

Broadly, transketolase catalyzes cleavage of a carbon–carbon bond in keto sugars, transferring the glycolic aldehyde to aldose sugar acceptors. *In vivo*, transketolase catalyzes stereospecific carbon–carbon bond formation at two points in the pentose phosphate pathway: the transfer of a two-carbon ketol from D-xylulose-5-phosphate to D-ribose-5-phosphate, producing D-sedoheptulose-7-

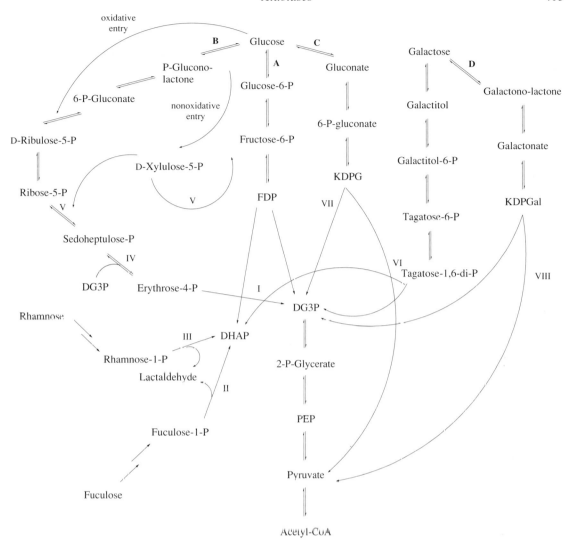

Scheme 69

I: Fructose-1,6-diphosphate aldolase
II: Fuculose-1-phosphate aldolase
III: Rhamnulose-1-phosphate aldolase
IV: Tagatose-1,6-diphosphate aldolase
V: Transketolase
VI: Transaldolase
VII: KDPG aldolase
VIII: KDPGal aldolase

A: Embden–Meyerhof–Parnas pathway
B: Pentose phosphate shunt
C: Entner–Doudoroff pathway
D: DeLey–Doudoroff pathway

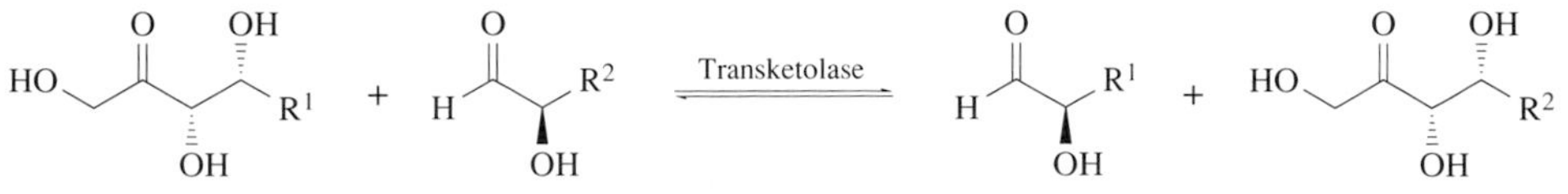

Scheme 70

phosphate, and the transfer of a two-carbon ketol from D-xylulose-5-phosphate to erythrose-4-phosphate, forming fructose-6-phosphate. *In vivo* and *in vitro*, these reactions are reversible. When hydroxypyruvate is the substrate *in vitro*, carbon dioxide is produced and the reaction is irreversible (Scheme 71).

Scheme 71

Transketolase is found in almost all animal and plant tissues as well as microorganisms.[423,424] The enzyme has been detected in *E. coli*,[425] *B. subtilis*,[426] *S. cerevisiae* (baker's yeast),[427] spinach,[428,429] tobacco,[430] rat and pig liver,[431] mouse cornea,[432] and a variety of human sources.[433] Purification schemes and properties of several such enzymes have been extensively investigated; several reviews of thiamine-dependent enzymes are available as well.[434,435] The *E. coli* transketolase, purified 30-fold to a specific activity of 23 U mg^{-1}, is stable when stored as crude cell extracts at 20 °C, pH 7.0, for at least one week.[425] This transketolase has also been immobilized for stability on Eupergit C acrylic beads with 40% retention of activity.[423] The baker's yeast enzyme was purified through a series of steps involving acetone, ethanol, and ammonium sulfate fractionation, heat treatment, DEAE cellulose chromatography, and crystallization to a specific activity of 20 U mg^{-1}.[427] Yeast transketolase is stable when stored in buffer at 4 °C, pH 7.6, or as an ammonium sulfate suspension. Rat and pig liver enzymes were purified by ammonium sulfate fractionation, DEAE cellulose, Sephadex, and hydroxyapatite chromatography to specific activities of 1.5 U mg^{-1} and 0.88 U mg^{-1}, respectively.[431] Both enzymes were stable when stored at 4 °C, pH 7.6–8.2. Additionally, the pig liver enzyme was stable at temperatures up to 40 °C.

Several transketolases have been sequenced including the enzymes from humans,[436] and *E. coli*.[437] The latter enzyme was cloned and expressed in a high copy plasmid. From this organism, 4 kg of enzyme with a specific activity of 30 U mg^{-1} was isolated from a 1500 L growth.

Limited structural information is available for the rat and pig liver enzymes. The rat liver enzyme is a dimer of identical subunits with a total molecular weight of 130 kDa. Pig liver transketolase has a molecular weight of 138 kDa and exists as an $\alpha_2\beta_2$ tetramer, with subunit molecular weights of 52 kDa and 29 kDa, respectively.[431]

Transketolase is a thiamine-dependent enzyme and the catalytic mechanism invokes a series of proton transfer steps (Scheme 72). Initially, proton abstraction at C-2 of the thiazolium ring of thiamine occurs via Glu418. Following release of the first product, the α-carbanion intermediate serves as a nucleophile in an attack on an electrophilic aldehyde, facilitated by protonation of the aldehydic oxygen of the substrate. His30 and His263 are postulated to form hydrogen bonds to the negatively charged oxygen atom generated at the transition state.

Structural information is available for both the yeast and *E. coli* enzymes. Baker's yeast transketolase is a homodimer with subunits of 680 amino acids and a molecular weight of 74 kDa (Figure 5).[438,439] Each subunit comprises three α/β-type domains and the dimer is formed via tight interactions between the amino terminal and central domains. The thiamine cofactor is bound at the subunit interface and is anchored to the protein via a divalent metal ion, coordinated to the oxygen of each phosphate group, Asp157, Asn187, the amide oxygen of residue 189, and a water molecule in an octahedral geometry. The phosphate groups form hydrogen bonds to His69, His263, and the amide nitrogen of Gly158. The cofactor thiazolium ring is located between domains and interacts hydrophobically with the protein. The thiamine pyrimidine ring is bound in a hydrophobic pocket via a hydrogen bond between the ring nitrogen and Glu418.

Wild-type and mutant structures of *E. coli* transketolase have been investigated by French and Ward.[440] As with the yeast enzyme, the dimer forms two active sites consisting of residues from both subunits. The active site is a funnel-shaped cleft with the thiamine cofactor located at the cleft base. Mutagenesis of Ile189, located at the base in a hydrophobic pocket above thiamine, to a smaller hydrophobic alanine residue leads to 80% reduction in activity compared to wild-type enzyme. Interestingly, the mutant enzyme has twofold greater affinity for thiamine than wild-type. Additionally, while mutant enzyme K_m for D-xylulose-5-phosphate is similar to that of wild-type, the K_m for D-ribose-5-phosphate is twofold lower than wild-type transketolase. Thus Ile189 has little effect on the affinity of transketolase for the donor substrate but does affect acceptor substrate and thiamine binding, implicating a possible role for Ile189 in the catalytic mechanism. In the yeast enzyme, an analogous Ile191 residue is postulated to play a role in the hydrophobic binding of thiamine.

Martin and co-workers[441] have investigated the role of conserved residues of transketolases from different species, focusing specifically on the human enzyme. These studies show that a conserved histidine at position 110 plays markedly different roles in catalysis in human and yeast enzymes. In

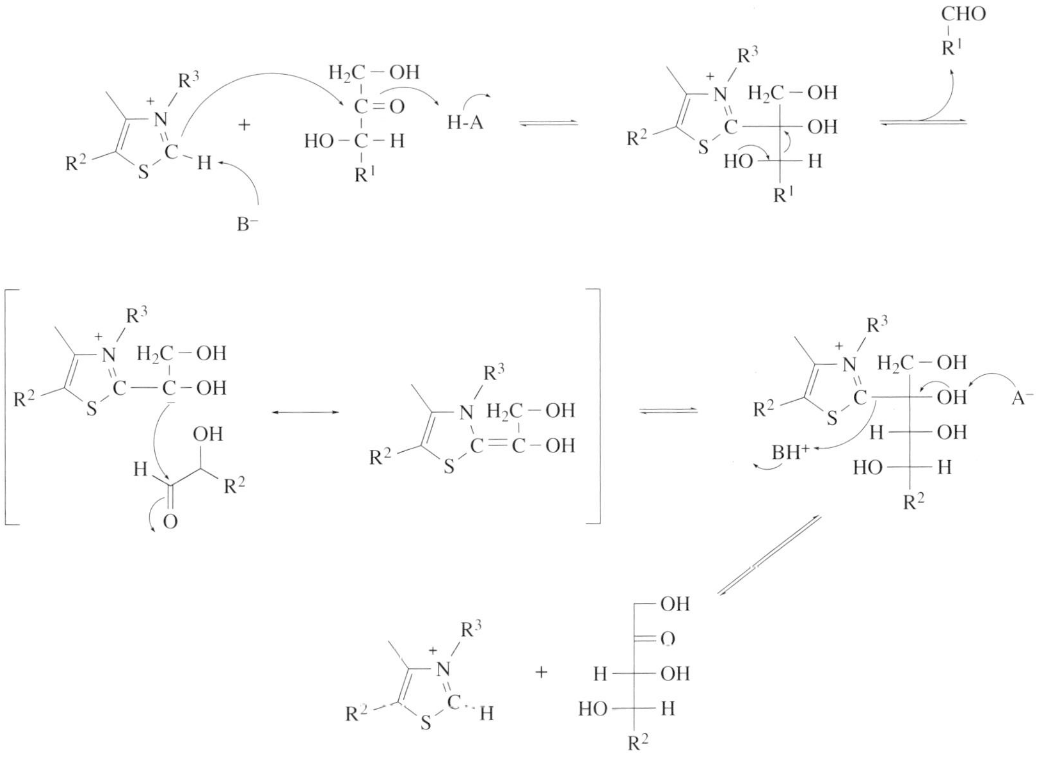

Scheme 72

the yeast enzyme, His110 aids in binding substrates, while in the human enzyme the same residue functions as a general base and abstracts a proton from protonated 4′-iminopyrimidine.

Transketolase exhibits broad acceptor substrate specificity. In general, compounds with an α-hydroxy group in the D-configuration are good substrates; however, α-unsubstituted and α-keto-aldehydes are also accepted. Isotopically labeled ketoses and the beetle pheromone (±)-*exo*-brevicomin have been prepared using transketolase-catalyzed reactions (Scheme 73).[442] Industrial scale production of the *E. coli* enzyme has been investigated as well.[425,443,444]

3.13.8.2 Transaldolase

Transaldolase (EC 2.2.1.2; CAS 9014-46-4) catalyzes the transfer of a three-carbon fragment from a phosphorylated ketose to an acceptor aldehyde. In glycolysis, the enzyme transfers dihydroxyacetone phosphate from sedoheptulose-7-phosphate to glyceraldehyde-3-phosphate, generating fructose-6-phosphate and erythrose-4-phosphate. This reaction is reversible and the equilibrium is near unity.

Transaldolase is a key enzyme in the metabolism of D-glucose and D-xylose (Scheme 69). The protein exists as three dimeric isozymes denoted I, II, and III. While isozymes I and III are homodimers, isozyme II is a hybrid formed by the exchange of the subunits of isozymes I and III. The formation of mixed isozyme is reversible and can be produced *in vitro* with pure I and III. Isozymes I and III display no kinetic or catalytic differences but do show structural variation at the amino acid level, perhaps indicative of unique genetic origins. Tsolas[445] has postulated that the two proteins may be activated in the presence of different sugars, one by hexoses and the other by pentoses.

Transaldolase has been isolated from *C. utilis*,[446] the methanogenic bacteria *Methanococcus voltae*,[447] *E. coli* K-12,[448] *S. cerevisiae*,[449] brewer's yeast,[450] human red blood and arterial wall cells,[451] potato,[452] rat tissues,[453] *Musca domestica*,[454] *Tetranychus telarius*,[455] spinach leaves,[456] *Euglena* and *Chlorella*,[457] *Chromatium* and *Chlorobium thiosulfatophilum*,[458] and a variety of tumors (Novikoff hepatoma, Krebs-ascites, and Walker).[459–462]

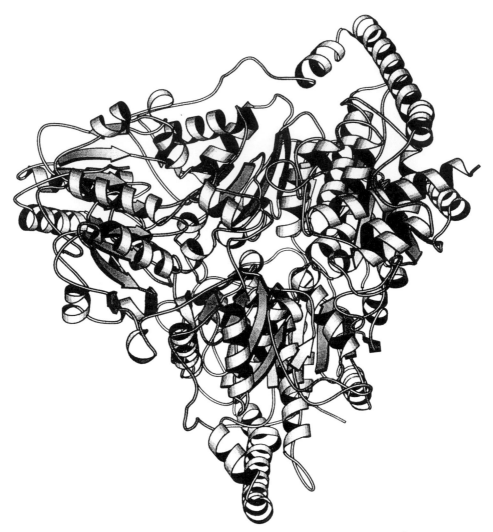

Figure 5 Baker's yeast transketolase crystal structure at 2.0 Å.

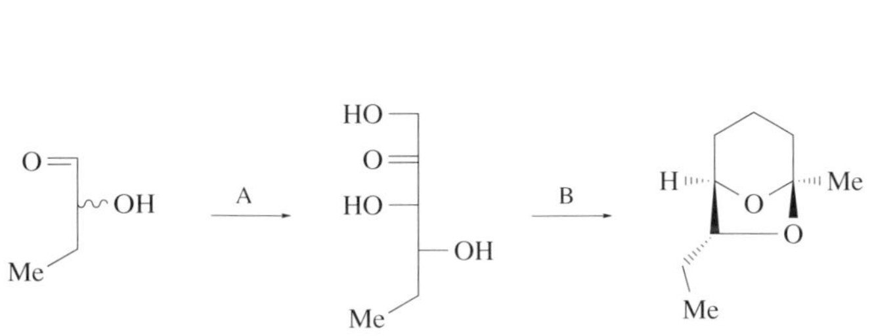

A: i, Transketolase, hydroxypyruvate, thiamine pyrophosphate, MgCl$_2$, pH 7.5;
B: Acetone, ZnI$_2$; ii, NaBH$_4$, EtOH; iii, NaIO$_4$, H$_2$O, acetone; iv, C; v, H$_2$, 5% Pd(OH)$_2$; vi, TsOH, CH$_2$Cl$_2$
C:

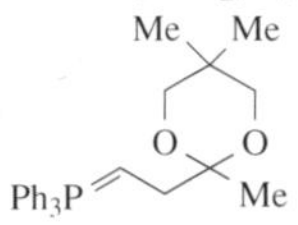

Scheme 73

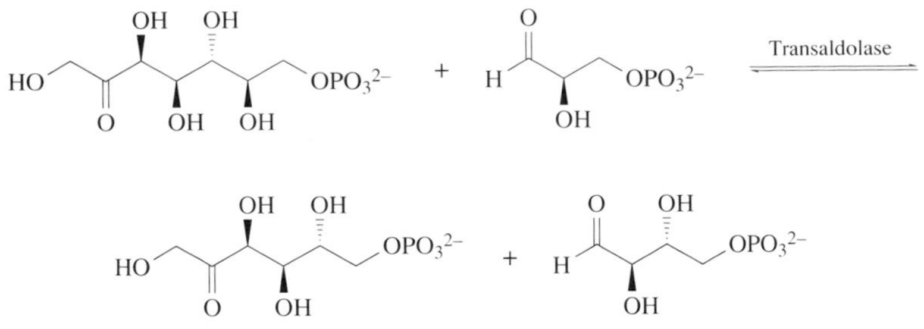

Scheme 74

The *E. coli* K-12 enzyme has been purified via ammonium sulfate fractionation and anion exchange chromatography to a specific activity of 60 U mg^{-1}.[448] The optimum pH was found to be 8.5 and the enzyme displayed activity over the temperature range 15–40 °C. The molecular weight of the protein is 70 kDa and the enzyme exists as a homodimer of 35 kDa subunits. Tris-HCl, phosphate, and sugars such as L-glyceraldehyde and D-arabinose-5-phosphate act as inhibitors. The enzyme from *E. coli* K-12 has been cloned into high copy vectors and overexpressed. The crystal structure for this enzyme has also been reported (Figure 6).[463,464]

Transaldolase isozymes I and III from *C. utilis* have been purified, crystallized, and sequenced.[446] Transaldolase from humans has also been cloned, sequenced, and expressed.[465] The human enzyme shows 58% sequence homology with the yeast protein; however, over several short blocks (11–15 amino acids) the sequence identity is 100%. These conserved regions may be important for enzyme structure and/or function. Comparative amino acid sequence analysis indicates that the yeast enzyme lacks several phosphorylation sites present in the human protein, suggesting that the two enzymes may be differentially regulated.

Transaldolase uses Schiff base formation for cleavage to the nucleophilic dihydroxyacetone. In 1960, Ricci and co-workers[466] isolated a transaldolase–dihydroxyacetone complex, representing the first direct demonstration of the existence of an enzyme–substrate complex. Ricci also demonstrated that dihydroxyacetone is linked to transaldolase via the C-2 rather than the C-3 hydroxymethylene and that lysine is the key reactive amino acid in the active site. An active site histidine also plays a mechanistic role. Following Schiff base formation and interconversion to the active ketamine, the histidine forms stabilizing hydrogen bonds to the carbanion. Ricci[467] also noted that transaldolase is a half-site enzyme and there is only one active site per dimer.

Mechanistic information has also been derived from mutagenesis studies.[468] The replacement of Lys142 by glutamine resulted in complete loss of enzymatic activity, identifying the Lys142 as the residue essential for catalysis. In the yeast *S. cerevisiae*, replacement of a Lys144 also resulted in complete loss of enzymatic activity.[469]

Transaldolase from *E. coli* catalyzes the transfer of dihydroxyacetone to a variety of acceptor aldehydes, including glyceraldehyde-3-phosphate, erythrose-4-phosphate, and nonphosphorylated trioses and tetraoses (Tables 13, 14).[459] In general, a *trans* configuration of the hydroxy groups at carbons 3 and 4 is required in the donor sugar.

3.13.8.3 Fructose-1,6-diphosphate Aldolase

The best studied aldolase is FDP aldolase (EC 4.1.2.13; CAS 9024-52-6). *In vivo*, this enzyme catalyzes the reversible reaction of D-glyceraldehyde-3-phosphate and dihydroxyacetone phosphate to generate FDP in the Embden–Meyerhof–Parnas metabolic pathway of glucose (Scheme 69). Two new stereogenic centers are formed stereospecifically; the stereochemistry of the vicinal diol produced is always D-*threo*. The groups of Whitesides and Wong[9,470,471] have thoroughly investigated FDP aldolase, particularly the rabbit muscle enzyme, and several reviews focus on this enzyme.

Mammalian muscle FDP aldolase is a Schiff base forming enzyme while enzyme from bacterial sources is metal-dependent. The aldolase is a ubiquitous and abundant enzyme, and has been isolated from a variety of mammalian[472–476] and plant sources.[477,478] The protein has also been

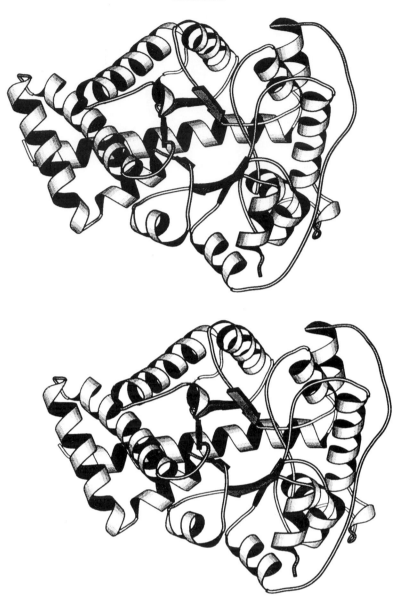

Figure 6 *E. coli* transaldolase crystal structure at 2.2 Å.

Table 13 Kinetic parameters of *E. coli* and yeast transaldolases.

Substrate	Yeast K_m (μM)	E. coli K_m (μM)	V_{rel} (%)
Fructose-6-phosphate	430	1200	ND[a]
Erythrose-4-phosphate	170	9	ND
DL-Glyceraldehyde-3-phosphate	220	38	8
Sedoheptulose-7-phosphate	180	280	5
Fructose	ND	>2 M	12

Source: Gumaa and McLean.[459] [a]ND: not determined.

isolated and cloned from several bacteria, including thermophilic eubacteria,[479,480] allowing reactions at higher temperatures. We refer the reader to several excellent reviews on the properties of FDP aldolases,[481,482] mechanistic investigations,[483] crystal structures, cloning,[484] and a plethora of synthetic applications.[485–492]

Table 14 Substrate specificity of *E. coli* transaldolase.

Substrate	V_{rel}
DL-Glyceraldehyde-3-phosphate	8
D-Erythrose-4-phosphate	100
D-Ribose-5-phosphate	0.8
D-Glyceraldehyde	7
D-Fructose-6-phosphate	100
D-Sedoheptulose-7-phosphate	5
D-Fructose	12

Source: Gumaa and McLean.[459]

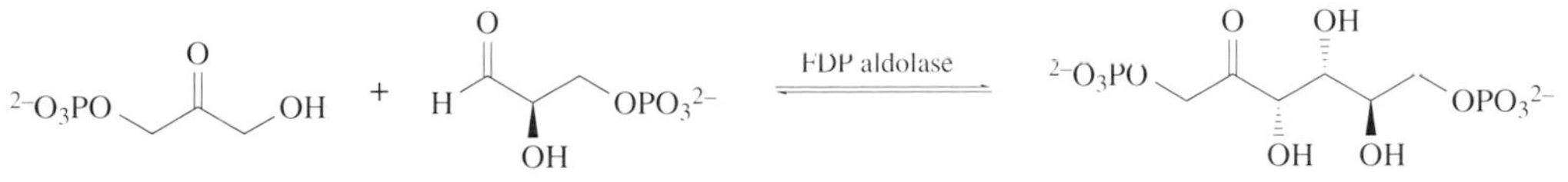

Scheme 75

3.13.8.4 2-Keto-3-deoxy-6-phosphogluconate Aldolase

In vivo, 2-keto-3-deoxy-6-phosphogluconate (KDPG) aldolase (EC 4.1.2.14; CAS 9024-53-7) catalyzes the reversible reaction of pyruvate with D-glyceraldehyde-3-phosphate to form KDPG. The equilibrium is roughly $1000\,M^{-1}$ in favor of synthesis. KDPG aldolase is a key component of the Entner–Doudoroff pathway of carbohydrate utilization. The enzyme has been reported in archaebacteria, eubacteria, and eukaryotes, but is especially prevalent in gram-negative bacteria, where it may operate in either catabolic or anabolic roles.[493] KDPG aldolase has been isolated from a variety of sources including *B. stearothermophilus*,[494] *Treponems saccharophilum*,[495] *Azospirillum amazonense*,[496] *Haloferax mediterranei* and *Haloarcula vallismortis*,[497] *Shewanella putrefaciens*,[498] *Helicobacter pylori*,[499] *Azotobacter vinelandii*,[500] *Zymomonas mobilis*,[501] *E. coli*,[502] *P. putida*,[503] *P. saccharophila*,[504] and *P. fluorescens*.[505]

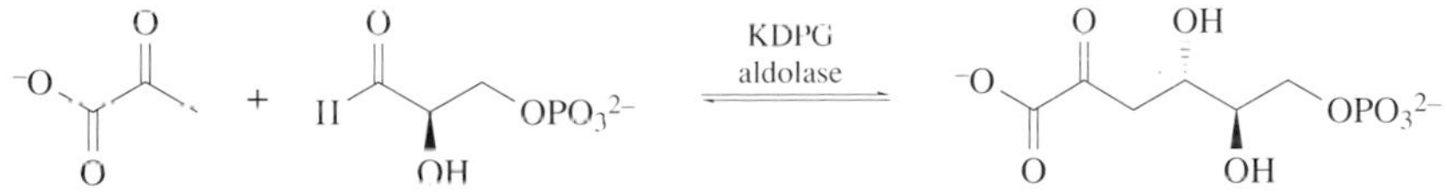

Scheme 76

In 1955, Kovachevich and Wood[505] first investigated carbohydrate metabolism by *P. fluorescens* and detected KDPG aldolase. They purified this enzyme via ammonium sulfate fractionation and calcium-phosphate gel adsorption to a specific activity of $112\,U\,mg^{-1}$, a 30-fold purification. The pH optimum of the enzyme is 7.5–8.5 with 50% retention of activity at pH 6.3 and 9.5. There is no coenzyme requirement and no increase in activity is observed on addition of metal ions. Other sources for KDPG aldolase examined by this group include *P. aeruginosa*, *E. coli*, *P. fragi*, *A. melanogenum*, and *Azotobacter vinelandii*.[505]

Taha and Deits[506] recently purified KDPG aldolase from *A. vinelandii*. The enzyme is a 70 kDa trimeric class I aldolase with a specific activity of $625\,\mu mol\,min^{-1}\,mg^{-1}$ and a K_m (KDPG) of 38 μM. This enzyme was purified via acid treatment and Sepharose and Sephadex chromatography to increase specific activity 500-fold. The enzyme also accepts hydroxyketoglutarate with a K_m of 39 μM and a V_{max} of $4.8\,mmol\,min^{-1}$, 140 times slower than KDPG. Hydroxyketoglutarate inhibits KDPG aldolase activity with a K_i of 0.17 mM.

Entner–Doudoroff pathways of carbohydrate metabolism have been observed in a variety of other organisms, and despite the fact that KDPG aldolases have not been isolated, they are presumed to be present. Martínez-Drets *et al.*[507] investigated the catabolism of carbohydrates in six strains of *Azospirillum amazonense*. None of the six strains showed Embden–Meyerhof–Parnas pathways during growth on sucrose, fructose, and glucose. *Azospirillae* are grouped into three subgroups:

those that metabolize hexoses via the Embden–Meyerhof–Parnas pathway, those that use both the Entner–Doudoroff and Embden–Meyerhof–Parnas pathways, and those that utilize the Entner–Doudoroff path exclusively. The specific activities of the latter aldolases range from 57 to 250 nmol $min^{-1} mg^{-1}$. Entner–Doudoroff pathway enzymes were also detected in *H. pylori* by Mendz *et al.*[508] *H. pylori* is the principal etiological agent of chronic gastritis, a contributing factor to peptic ulcer disease and the development of gastric cancer. The aldolase from *Treponema saccharophilum* sp. *nov.*, a large pectinolytic spirochete from bovine rumen, is involved in the anaerobic degradation of pectin and was found to be an inducible enzyme.[509]

Altekar and Rangaswamy[510] investigated Entner–Doudoroff pathways in halophilic archaebacteria. While many halophilic archaebacteria do not utilize carbohydrates, those that do utilize glucose do so via a modified Entner–Doudoroff path where oxidation precedes phosphorylation. These workers identified both *Haloferax mediterranei* and *Haloarcula vallismortis* as possessing constitutive KDPG aldolases.

The Entner–Doudoroff pathway was detected in *Shewanella putrefaciens*, a facultative methylotroph lacking an Embden–Meyerhof–Parnas path.[498] Strains were obtained from several different environments, many of which are suboxic or anoxic, and characterized as redox interface organisms on the basis of abundance at oxic/anoxic interfaces in the Baltic Sea. The carbon sources utilized include glucose, lactate, pyruvate, propionate, ethanol, acetate, formate, and a number of amino acids such as serine.

Knappmann and Kula[511] report that expression of KDPG aldolase in *Zymomonas mobilis* was enriched sixfold by treatment with ammonium sulfate and phosphoric acid. The *Z. mobilis* enzyme shows a K_m (KDPG) of 53 µM. The enzyme is stable for at least three weeks when purified and is activated by Mn^{II}, Ni^{II}, Mg^{II}, and Ca^{II}, which stabilize the active conformation of KDPG aldolase. Co^{II}, Ba^{II}, Zn^{II}, and Fe^{II} have no effect while Cu^{II} strongly inhibits activity.

Investigations of the KDPG aldolases from *E. coli*, *P. putida*, and *Z. mobilis* in our laboratories demonstrated that all three enzymes are remarkably stable to the addition of cosolvents, with >100% activity in the presence of either dimethylformamide or dimethyl sulfoxide.[122,512] pH–activity curves revealed that the three enzymes exhibit markedly different activities. The enzymes from both *Zymomonas* and *Pseudomonas* produce bell-shaped pH–activity curves with maxima near 7.5–8, indicating a requirement for two enzyme ionizable residues which presumably act as general acid and general base catalysts during turnover. In contrast, the enzyme from *Escherichia* exhibits a single transition pH–activity curve, apparently utilizing the solvent as the general acid during the aldol reaction.

The *Z. mobilis* enzyme has been crystallized, although a structure has not been reported. The enzyme is a 70 kDa trimer and shows a specific activity roughly double that of the *P. putida* enzyme (600 U mg^{-1} vs. 300 U mg^{-1}, respectively) although K_m for KDPG is elevated by a factor of 3.5 (0.25 mM vs. 0.07 mM, respectively).[474]

In 1982, Mavridis *et al.*[513] investigated the structure of KDPG aldolase from *P. putida* following crystallization. The protein folds to form an eight-stranded α/β-barrel structure similar to triosephosphate isomerase, the A-domain of pyruvate kinase, and Taka amylase. The interior of the enzyme folds regularly and forms an eight-stranded β-barrel of parallel chains similar to the fold of triosephosphate isomerase.[514] The exterior of the protein is composed primarily of α-helices; each subunit is an ellipsoid with dimensions of 25 Å × 40 Å × 40 Å. The barrel possesses a fairly severe twist around its cylindrical axis, and the strands are not strictly parallel. As a result, hydrogen bonding between the strands is poor or nonexistent except in the midsection of the twisted barrel. On the surface, there are several vacant cavities with hydrophobic residues extending into them. The active site Lys144 lies in a shallow depression near the end of the α/β-barrel cavity. The active site depression is 20 Å × 25 Å in width and 9 Å deep. The entrance of the cavity is lined with the apolar residues Leu145, Pro147, Phe169, and Pro171 from one subunit and Gly152′, Gly153′, and Ala155′ from another. There are some 14 residues in the immediate vicinity of Lys144 with side chains directed towards the putative active site, including His63, Arg142, Phe143, Pro147, Arg168, Phe169, Cys170, Pro171, and Trp196 of one subunit and Ala155′, Ala156′, Ile157′, Lys158′, and Phe160′ from a second. Together, the residues form an ellipsoid of 9 Å × 14 Å × 14 Å around Lys144. The hydrophobic environment of Lys144 renders the pK_a of the ε amino group unusually low and it is probably not protonated at physiological pH.

KDPG aldolase from *P. putida* is also trimeric with a molecular weight of 72–74 kDa.[515] The enzyme shows the highest specific activity of any aldolase and retains stability even in 0.1 M HCl.

Suzuki and Wood[516] determined the complete primary sequence of the homotrimer KDPG aldolase subunit of *P. putida*. The enzyme has 225 amino acid residues and a molecular weight near 24 kDa. Phibbs and co-workers[517] reported a tight clustering of the gene loci for glucose-6-phosphate dehydrogenase and KDPG aldolase in the 45–55 min chromosomal region. The same researchers have also detected a clustering of the genes specifying carbohydrate catabolism including KDPG aldolase in *P. aeruginosa*.[518,519]

The structure of the Entner–Doudoroff pathway operon and the molecular biology of gluconate metabolism have been studied extensively. The *eda* (aldolase) and *edd* (dehydratase) genes are closely linked and are 95% cotransducible.[520] Fradkin and Fraenkel[521] localized the *eda* gene to 35 min on the *E. coli* genetic map. Later, Conway *et al.*[522] identified loci for the *zwf* (glucose-6-phosphate dehydrogenase), *edd*, and *eda* genes. This latter group has now cloned, sequenced, and expressed both the *E. coli* and *Z. mobilis* KDPG aldolases.[523,524]

Meloche and co-workers[114,525–527] performed extensive investigations on the mechanism of KDPG aldolase of *P. putida* and *P. saccharophila*. In 1973, these workers first reported the inactivation of KDPG aldolase by the active site-directed reagent bromopyruvate. Further investigations revealed that the inactivation occurs when the γ-carboxy of protein-bound glutamate attacks C-3 of bromopyruvate displacing bromide and forming an ester linkage (Scheme 77). Glutamate must lie within the catalytic site although it is far removed in linear sequence from the active-site lysine. Likewise, 3*S*-bromoketobutyrate, but not 3*R*, inactivated the enzyme, indicating that the enzyme catalyzes protonation of the *re* face at C-3 of the enzyme-pyruvate enamine. The pyruvate-lysine ketimine can also be inactivated by hydride reduction. Such reduction can occur at either the *re* or *si* face of the ketimine carbon. The reduction stereochemistry of the Schiff base formed between pyruvate and catalytic lysine revealed stereoselective reduction with 56% *R* and 44% *S*. Thus, the reducing agent favors slightly the *si* face of the ketimine carbon.

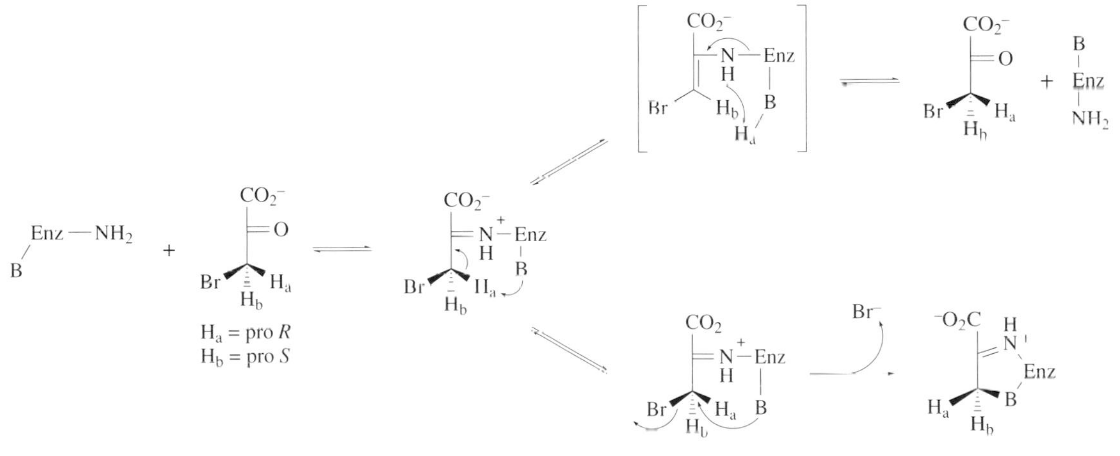

Scheme 77

Meloche *et al.*[114,527] also investigated the kinetics of the KDPG aldolase-catalyzed reaction and reported that C—C bond formation occurs considerably faster than C—^{3}H bond rupture. Studies with isotopically labeled pyruvate show a hydrogen isotope discrimination in deprotonation, suggesting that generation of enolpyruvate is at least partially rate limiting. 3*S*-(3 ^{3}H, ^{2}H, ^{1}H) and 3*R*-(3 ^{3}H, ^{2}H, ^{1}H) pyruvate were utilized in a KDPG aldolase-catalyzed reaction with D-glyceraldehyde-3-phosphate to yield 3*S* and 3*R* KDPG, respectively. Retention of configuration indicates attack of reagents (exchanging proton from water and D-glyceraldehyde-3-phosphate) from the same face of the enzyme-bound pyruvyl enamine.

KDPG aldolase has been investigated as a reagent for stereospecific aldol addition.[122] KDPG aldolases from *P. putida*, *Z. mobilis*, and *E. coli* all accept several unnatural aldehydes as electrophilic substrates at synthetically useful rates, providing access to highly and differentially functionalized α-keto acid products (Table 15). At least some unnatural nucleophiles are also accepted providing the aldehydic component is reactive enough. In contrast to other aldolases, simple aliphatic aldehydes are not accepted as electrophiles.

KDPG aldolase has been used in the synthesis of several carbohydrate-like compounds including 2-keto-3-deoxygluconate and D-*erythro*-3,6-dideoxy-2-hexulosonate from glyceraldehyde and D-lactaldehyde on a preparative scale.[122] KDPG aldolase from *E. coli* has also been used in preparative-scale syntheses of the nikkomycin amino acid (Scheme 78).[528]

Table 15 Substrate specificity of KDPG aldolases from *Pseudomonas putida*, *Escherichia coli* and *Zymomonas mobilis*.

Nucleophile	Electrophile	P. putida	E. coli	Z. mobilis
Pyruvate	D-glyceraldehyde	100	100	100
	L-glyceraldehyde	—[a]	—	+[b]
	DL-glyceraldehyde	+ +[c]	+ + +[d]	+ + +
	D-lactaldehyde	+ +	+ +	+
	L-lactaldehyde	—	—	+
	D-erythrose	+	+	+ +
	D-threose	—	+	+
	L-erythrose	—	—	+
	L-threose	+ +	+	+
	D-erythrose-4-phosphate	+ + + +[e]	+ + + +	+ + +
	D-ribose	—	+	+
	D-ribose-5-phosphate	+	+	+ +
	chloroacetaldehyde	+ + + +	+	+ + + +
	2,3-*O*-isopropylidine D-glyceraldehyde	—	—	—
	glycolaldehyde	+	+	+ +
	glyoxylate	+ + + +	+ + + +	+ + + +
	2-pyridine carboxaldehyde	+ + + +	+ + + +	+ + + +
	3-pyridine carboxaldehyde	—	+ +	+ + +
	4-pyridine carboxaldehyde	+	+ + + +	+ + + +
	2-thiophene carboxaldehyde	—	—	—
	3-thiophene carboxaldehyde	—	—	+
	2-furaldehyde	—	+	+ +
	3-furaldehyde	—	—	+
	2-chlorobenzaldehyde	—	—	—
	3-chlorobenzaldehyde	—	—	—
	4-chlorobenzaldehyde	—	—	—
	benzaldehyde	—	—	—
	valeraldehyde	—	—	—
	butyraldehyde	—	—	—
	acrolein	—	—	—
2-Ketobutyrate	D-glyceraldehyde-3-phosphate	+ +	+ + + +	+
3-Hydroxypyruvate	D-glyceraldehyde-3-phosphate	+	+	+
3-Fluoropyruvate	D-glyceraldehyde-3-phosphate	+	+	+ +

Source: Shelton *et al.*[122]   [a] —: nonsubstrate.   [b] +: 0.5–25%.   [c] + +: 25–50%.   [d] + + +: 50–100%.   [e] + + + +: >100%.

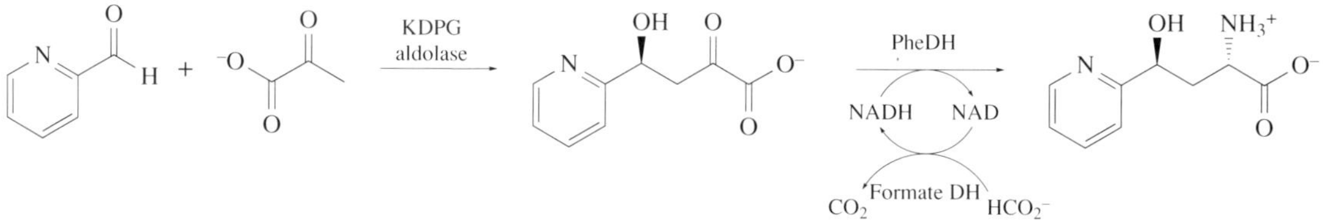

Scheme 78

3.13.8.5 Fuculose-1-phosphate Aldolase

In 1956, Green and Cohen[529] reported *E. coli* capable of growth on fucose as the sole carbon source. They further postulated that fucose was metabolized by conversion to the corresponding phosphorylated ketose fuculose-1-phosphate and cleaved to dihydroxyacetone phosphate and L-lactaldehyde. In 1961, Eagon[530] surveyed some 33 microorganisms for the ability to utilize fucose as the carbon source. Twelve of the 33, including *Aerobacter aerogenes*, *Agrobacterium tumefaciens*, *Corynebacterium fascians*, *Gaffkya tetragena*, *Rhizobium leguminosarum*, *Salmonella enterididis*, *Sarcina lutea*, *Serratia marcescens*, *Shigella sonnei*, *Sporocytophaga congregata*, and *Xanthomonas phaseoli* showed growth and were presumed to express a metabolic pathway that included a fuculose-1-phosphate aldolase (EC 4.1.2.17; CAS 9024-54-8). This same enzyme is apparently responsible for growth of some strains of *E. coli*[530–532] and *Klebsiella aerogenes*[533,534] on D-arabinose, which possesses the correct D-*erythro* stereochemistry for fuculose-1-phosphate aldolase cleavage.

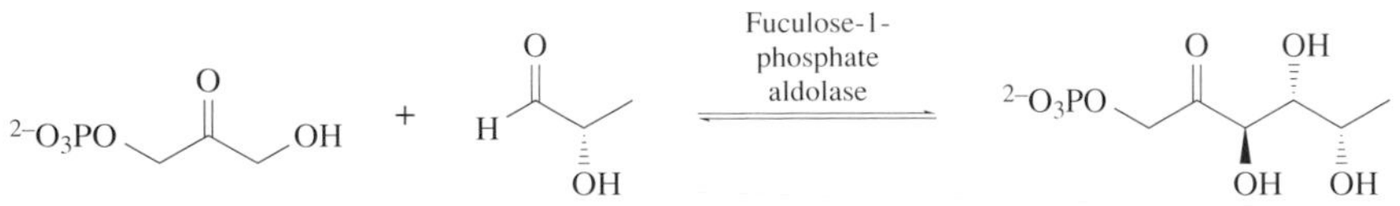

Scheme 79

In 1962, Heath and Ghalambor[535,536] reported an isolation of fuculose-1-phosphate aldolase from *E. coli*. The protein is a class II aldolase, and dialysis against EDTA abolished activity. Activity could be restored by addition of divalent manganese, magnesium, or calcium. The enzyme exhibits a rather sharp pH optimum near 7.4, and a Michaelis constant for fuculose-1-phosphate of 0.7 mM. The enzyme-catalyzed reaction is readily reversible. In the synthetic direction, fuculose-1-phosphate aldolase accepts a broad range of electrophilic substrates, although the enzyme is highly specific for dihydroxyacetone phosphate as the nucleophilic component.[537] The enzyme also accepts dihydroxyacetone in the presence of arsenate or vanadate, presumably through the formation of kinetically labile arsenate or vanadate esters. The enzyme shows high stereospecificity in reactions, producing only the D-*erythro* stereochemistry at the C-3/C-4 vicinal diol. Because of this broad substrate specificity and high stereospecificity, the enzyme has been used extensively as a synthetic catalyst.[538–546]

Lin and co-workers[547–549] reported the physical location of the L-fucose utilization genes in *E. coli* in 1984 and in 1989 the same group reported the cloning of the entire sequence of genes including *fucA*, which codes for the aldolase. In 1990, Sinskey, Whitesides, and co-workers[550] reported overexpression of this gene in *E. coli*, making available large amounts of the enzyme both for enzymatic studies and use in synthetic organic chemistry. The protein contains 215 amino acids with a molecular weight near 23.8 kDa. The active enzyme is tetrameric. In 1993, Dreyer and Schulz[551] reported a crystal structure of a zinc form of the fuculose-1-phosphate aldolase (Figure 7). Subsequently, Fessner *et al.*[552] reported the structural details of fuculose-1-phosphate aldolase. As with other aldolases, the active site of the enzyme is formed at the subunit interface. Each active site zinc is coordinated to three histidine residues (His92, His94, His155) and a single carboxylate (Glu73). In 1996, the same group reported the structure of the aldolase bound to the dihydroxyacetone phosphate mimic phosphoglycolohydroxamate, and on the basis of this structure proposed a mechanism for the reaction. The inhibitor binds the active site zinc through the primary hydroxy and enol hydroxy groups, displacing Glu73 as a zinc ligand. In a catalytic cycle, the latter residue in turn deprotonates the bound nucleophile, facilitating reaction with an electrophilic aldehyde (Scheme 80).

3.13.8.6 2-Keto-3-deoxy-L-arabonate Aldolase

Two alternative pathways have been identified for L-arabinose metabolism in *Pseudomonas* (Scheme 82). In both pathways, arabinose is oxidized to arabanoate and dehydrated to 2-keto-3-deoxy-L-arabonate. At this point, the pathways diverge. In the first pathway, operative in *P. saccharophila* and *P. fragi*, further dehydration and oxidation produces α-ketoglutarate which is subsequently cleaved. In the second route, 2-keto-3-deoxy-L-arabonate is cleaved by an aldolase to pyruvate and glycolaldehyde.[553,554]

In 1969, Anderson and Dahms[555] reported a 2-keto-3-deoxy-L-arabonate aldolase (EC 4.1.2.18; CAS 9076-49-7) in an unclassified pseudomonad designated MSU1. Later, the same authors reported that MSU1 utilizes the same enzyme in the metabolism of D-fucose, cleaving 2-keto-3-deoxyfuconate to pyruvate and D-lactaldehyde.[553] The enzyme shows a broad pH optimum near 8.2 and is a class II aldolase. Activity is abolished by dialysis against EDTA and can be restored with divalent manganese, cobalt, magnesium, and nickel salts. Substrate specificity tests showed that in the retro-aldol reaction only 2-keto-3-deoxy-L-arabonate or 2-keto-3-deoxy-D-fuconate were substrates; 2-keto-3-deoxygluconate and its 6-phosphate ester, 2-keto-3-deoxygalactonate and its 6-phosphate ester, 2-keto-4-hydroxyglutarate, and neuraminate were not accepted as substrates. Michaelis constants of 2.9 mM and 1.8 mM were reported for 2-keto-3-deoxyfuconate and 2-keto-3-deoxy-L-

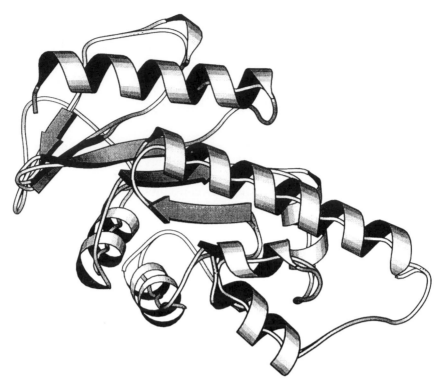

Figure 7   *E. coli* fuculose-1-phosphate aldolase crystal structure at 2.7 Å.

arabonate, respectively. The reaction is reversible, and catalyzes the reaction of pyruvate with both glycolaldehyde and D-lactaldehyde.

Subsequent to these studies, a 2-keto-3-deoxy-L-arabonate aldolase pathway for L-arabinose metabolism has been postulated for slow-growth *Rhizobium*, including *Rhizobium* strain BTAi1, *R. japonicum* and *Rhizobium* sp. strain 32H1.[556,557] In each case, enzyme assays were performed only with crude extracts and mechanistic and substrate specificity studies are unavailable.

3.13.8.7   Rhamnulose-1-phosphate Aldolase

The ability of *B. rhamnosifermentans* to grow on rhamnose as the sole carbon source was first noted by Kluyver and Schnellen[558] in 1937. Subsequently, rhamnulose-1-phosphate aldolase (EC 4.1.2.19; CAS 9054-58-4) was isolated and crystallized from *E. coli*.[559] The protein has also been isolated from *L. plantarum*.[560] Several other organisms are capable of growth on rhamnose and presumably express the aldolase,[561] including strains of *A. aerogenes*, *Agrobacterium tumerfaciens*, *B. megaterium*, *E. carotovora*, *Gaffkya tetragena*, *Proteus vulgaris*, *R. leguminosarum*, *S. dysenteriae*, and *Sporocytophaga congregata*. *Arthrobacter pyridinolis*,[562] *Pasteurella pestis*,[563] *S. typhimurium*,[564,565] *S. enteriditis*,[564] and *Sarcina lutea*[564] also express the aldolase.

The physical location of the *E. coli* rhamnulose-1-phosphate aldolase gene, designated *rhaD*, was determined in 1989 by Badía *et al.*[566] The gene was later sequenced, cloned, and expressed.[567] Cloning of the *rhaD* gene in *S. typhimurium* has also been reported.[568] The *E. coli* enzyme has been studied extensively.[569] The deduced amino acid sequence contains 274 amino acids corresponding to a molecular weight near 30.1 kDa. In active form, the protein is tetrameric. The enzyme exhibits a pH optimum near 7.5 and has an isoelectric point of 5.05. As a class II aldolase, the protein requires divalent zinc for activity.[570] In addition, there are various reports that monovalent ions, specifically Na^+, Cs^+, NH_4^+, Rb^+, and K^+, enhance both stability and activity. Michaelis constants for several substrates have been determined, including L-lactaldehyde (6.0 mM), dihydroxyacetone phosphate (3.0 mM), L-rhamnulose-1-phosphate (0.3 mM), D-sorbose-1-phosphate (1.8 mM), and L-xylulose-1-phosphate (0.2 mM).

Inhibitor studies have been conducted with the zinc chelator 1,10-phenanthroline.[571] This species acts as a competitive inhibitor of dihydroxyacetone phosphate but not L-lactaldehyde. The same inhibitor is competitive with rhamnulose-1-phosphate. Together, these findings indicate that both

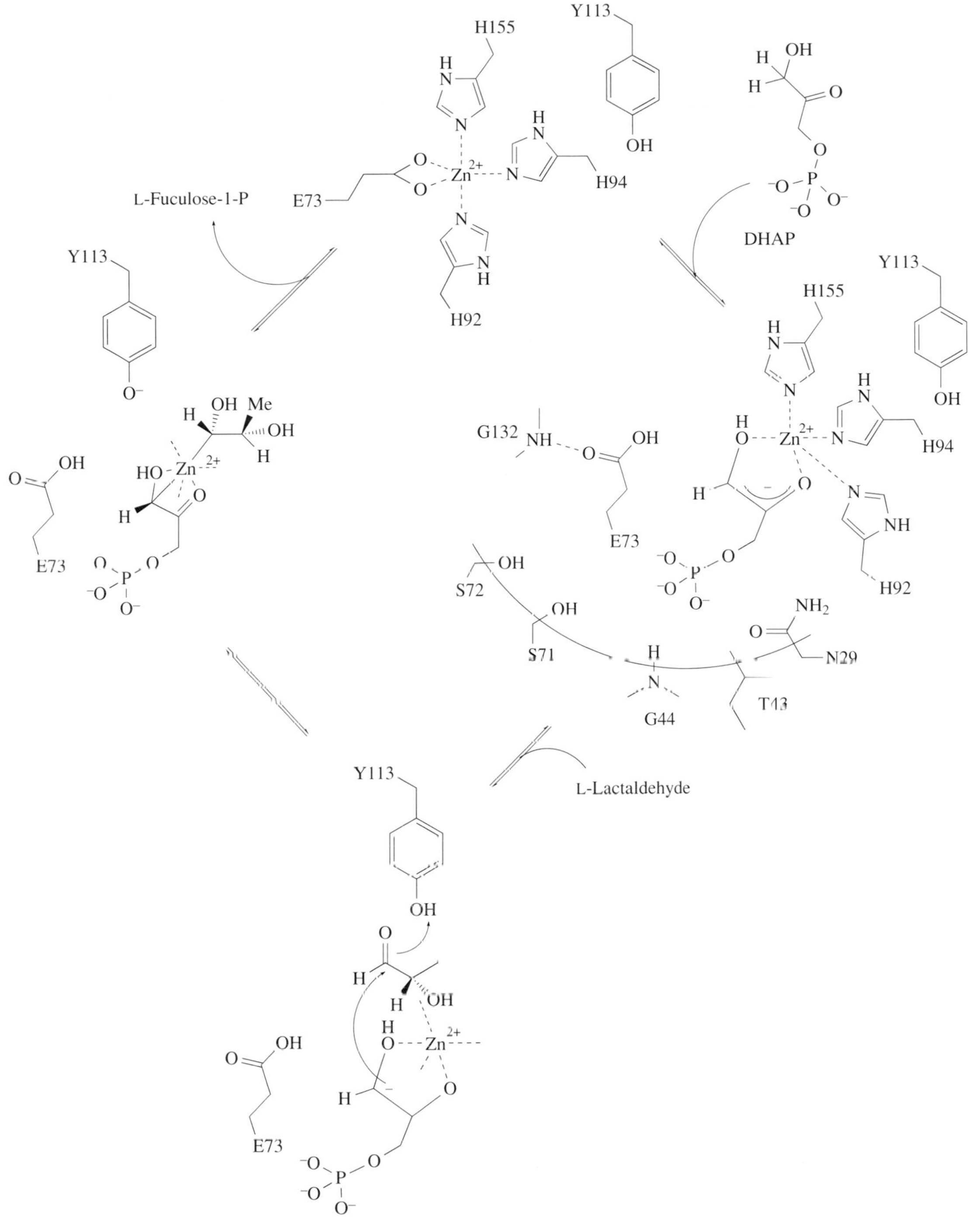

Scheme 80

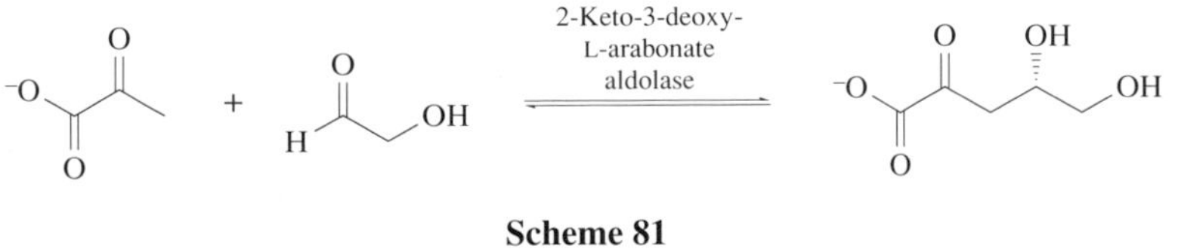

Scheme 81

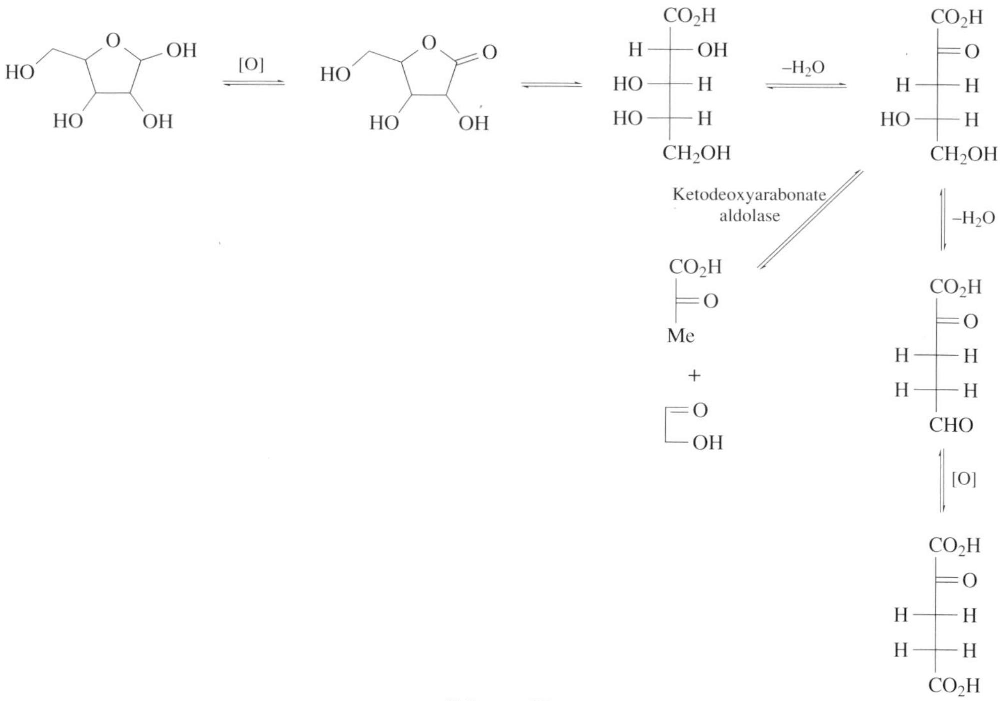

Scheme 82

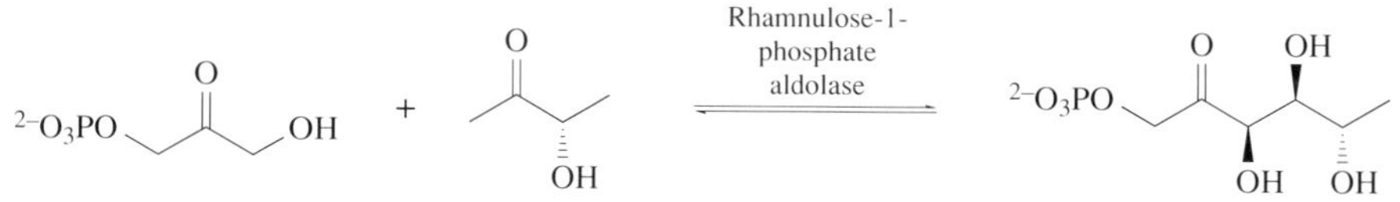

Scheme 83

dihydroxyacetone phosphate and rhamnulose-1-phosphate bind near the active site zinc and that the enzyme utilizes a ping-pong kinetic mechanism. Apparently, dihydroxyacetone phosphate binding is followed by release of a proton, binding of L-lactaldehyde and finally release of rhamnulose-1-phosphate.

The enzyme-catalyzed reaction is readily reversible, and a variety of aldehydes act as electrophilic substrates in aldol addition.[486] As is the case for all known dihydroxyacetone phosphate aldolases, the reaction is highly specific for the nucleophilic component and will not accept even minor modifications to the dihydroxyacetone phosphate skeleton. The enzyme is highly specific for the L-*threo* configuration at the C-3/C-4 vicinal diol in both the cleavage and reaction reactions. Because of the broad electrophile substrate spectrum and high stereospecificity, rhamnulose-1-phosphate aldolase has found utility in organic synthesis; the groups of Wong[486,572,573] and Fessner[34,575–576] in particular have exploited this activity.

3.13.8.8   2-Keto-3-deoxy-6-phosphogalactonate Aldolase

2-Keto-3-deoxy-6-phosphogalactonate (KDPGal) aldolase (EC 4.1.2.21; CAS 9030-99-3) is the enzyme complementary to KDPG aldolase (EC 4.1.2.14, Section 3.13.8.4) and is involved in the bacterial oxidative degradation of galactose, an important component of the outer cell wall of most gram-negative bacteria. The pathway of galactose metabolism involving KDPGal aldolase is known as the DeLey–Doudoroff pathway; ultimately D-galactose is metabolized to pyruvate and D-glyceraldehyde-3-phosphate (Scheme 69).[577] In the synthetic direction, KDPGal aldolase catalyzes the reaction of pyruvate with D-glyceraldehyde-3-phosphate to produce KDPGal, with the production of the R configuration at the newly formed stereogenic center. KDPGal aldolase has been detected in several sources including *Azotobacter vinelandii*,[578] *P. saccharophila*,[579] *P. fluorescens*,[580] *Rhizobium meliloti*,[581] *Mycobacterium* strains,[582] *Caulobacter crescentus*,[583] and *E. coli*.[584]

Scheme 84

Meloche and O'Connell[578] reported an isolation of KDPGal aldolase from *P. saccharophila*. The protein was separated from the inducible KDPG aldolase by a series of Sephadex and DEAE cellulose columns. The *P. saccharophila* enzyme is trimeric with a molecular weight of ~ 73 kDa. The enzyme is specific for the open-chain form of KDPGal, populated to 13% in solution. K_m for the open-chain sugar is 50 μM. Stating that the literature method of Meloche was not reproducible, Shuster[585] also reported an isolation of KDPGal aldolase from *P. saccharophila* grown on galactose. Acid precipitation, heat treatment, and ammonium sulfate precipitation, followed by DEAE cellulose chromatography yielded pure enzyme. The enzyme exhibits typical Michaelis–Menten kinetics with no substrate or product inhibition. K_m for KDPGal is 5×10^{-4} mM and the pH optimum is 7.8. The substrate specificity of the enzyme was investigated for retro-aldol cleavage; no cleavage was detectable for KDPG, 3-deoxy-6-phosphogalactose, glucometasaccharinic acid, or keto-3-deoxy-L-arabonate. None of these substrates acts as an inhibitor.

Deacon and Cooper[584] investigated *E. coli* KDPGal aldolase purified via ammonium sulfate fractionation and Sephadex chromatography. The enzyme is unstable following chromatography and all activity was lost after 4 days at 4 °C. The aldolase was present in high concentrations when grown on galactonate, but not detectable when grown on glycerol, gluconate, or galactose.

A mutant strain of *Caulobacter crescentus* has been reported in which KDPGal aldolase activity increased fivefold when grown on galactose.[583] In a second mutant, KDPGal aldolase was expressed constitutively. These studies indicate that the activities of the enzymes of galactose catabolism in *Caulobacter* are independently regulated despite the fact they may map a chromosomal cluster.

As with KDPG aldolase, the mechanism of the KDPGal aldolase catalyzed reaction involves a Schiff base intermediate with the protonated ketimine acting as the nucleophilic species. Meloche and Monti[586] also investigated the mechanism of this enzyme from *P. saccharophila*. As with KDPG aldolase, 3[³H₃]pyruvate and D-glyceraldehyde-3-phosphate react to form products 8–10 times faster than tritium is released to water. Pyruvate deprotonation is required for reaction, requiring a hydrogen isotope effect in enolpyruvate formation, which must then be at least partially rate-limiting for C—C synthesis. Like KDPG aldolase, KDPGal aldolase catalyzes C—C and C—H synthesis with retention of configuration at C-3, indicating an asymmetric active site with solutes approaching a single face of the bound pyruvyl enolate. KDPGal aldolase must have the opposite chirality of KDPG aldolase in terms of the respective aldehyde-specific portions, for correct orientation of the carbonyl face of the incoming D-glyceraldehyde-3-phosphate and generation of products with opposite C-4 configurations.

3.13.8.9 2-Keto-3-deoxy-D-xylonate Aldolase

In 1974, Dahms[587] reported a pathway of xylose metabolism in the unclassified pseudomonad MSU1 parallel to that previously demonstrated in the same organism for arabinose (EC 4.1.2.18, Section 3.13.6). Thus, D-xylose is oxidized to xylonate, dehydrated to 2-keto-3-deoxyxylonate and cleaved to pyruvate and glycolaldehyde. The enzyme is a class II aldolase (EC 4.1.2.28; CAS 55326-36-8), activated by Mn^{II}, Co^{II}, Mg^{II}, Zn^{II}, Ca^{II}, and Ni^{II}. The reaction is reversible, and the enzyme shows high substrate stereospecificity. 2-Keto-3-deoxy-L-arabonate, 2-keto-3-deoxy-D-fuconate, 2-keto-3-deoxygluconate and its 6-phosphate ester, and 2-keto-4-hydroxyglutarate were not cleaved at 5 mM concentrations. A Michaelis constant of 0.97 mM was reported for the natural substrate, 2-keto-3-deoxy-D-xylonate. No further reports on this aldolase from MSU1 or other organisms have been made.

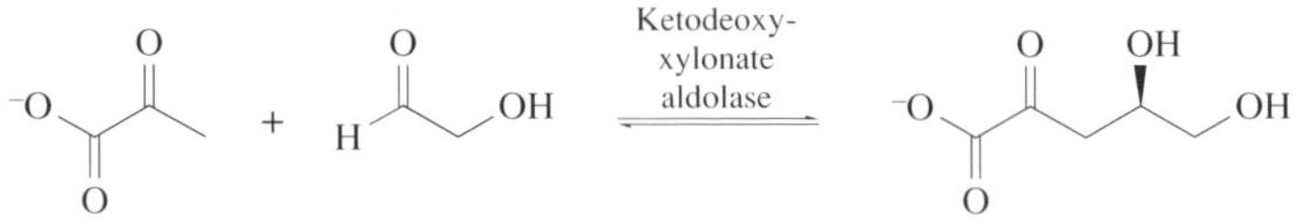

Scheme 85

3.13.8.10   Tagatose-1,6-diphosphate Aldolase

Most organisms metabolize galactose via Leloir conversion to glucose followed by entry into the Embden–Meyerhof–Parnas glycolysis pathway. In contrast, a number of organisms are capable of growth on galactitol as the sole carbon source, suggesting the existence of an alternate route. In 1976, Markwell *et al.*[588] reported that *K. pneumoniae* utilized galactitol by successive phosphorylation, oxidation to D-tagatose-6-phosphate, phosphorylation, and aldolase-catalyzed cleavage to D-glyceraldehyde-3-phosphate and dihydroxyacetone phosphate. This metabolic route is also utilized by some strains of *lactococci* for lactose metabolism. These organisms forego the Leloir conversion of galactose to glucose in favor of the more efficient lactose phosphotransferase system. In this pathway, lactose is phosphorylated prior to glycosidic cleavage to glucose and galactose-6-phosphate. The former monosaccharide is utilized in the familiar Embden–Meyerhof–Parnas pathway, while the latter is converted to tagatose-6-phosphate via an aldose–ketose isomerase, phosphorylated again to tagatose-1,6-diphosphate and finally cleaved enzymatically to dihydroxyacetone phosphate and D-glyceraldehyde-3-phosphate (Scheme 87).

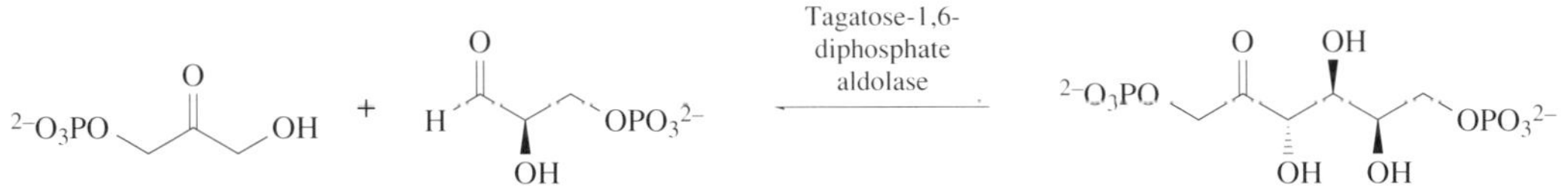

Scheme 86

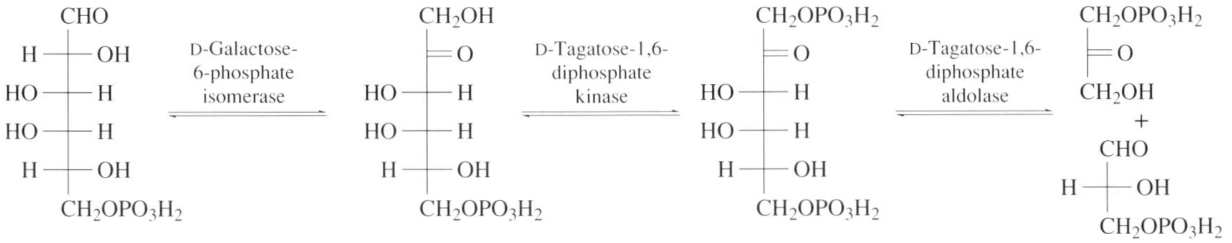

Scheme 87

In a series of papers dating to 1973, Bissett and Anderson[589–593] elucidated a glycolytic pathway for lactose and galactose in *S. aureus* containing a tagatose-1,6-diphosphate aldolase. These authors first identified the non-Leloir pathway for galactose metabolism in several strains of *Streptococcus*, including *S. lactis*, *S. cremoris*, and *S. diacetilactis*. Later, the same authors reported genetic evidence for the existence of the non-Leloir pathway in *S. aureus*, isolating organisms deficient in D-galactose-6-phosphate isomerase, D-tagatose-6-phosphate kinase, and tagatose-1,6-diphosphate aldolase. Finally, in 1980 these authors reported an isolation of the aldolase. Cloning and molecular biology of the tagatose-6-phosphate pathway operon from *S. aureus* was reported by Stewart and co-workers[594] in 1991.

In contrast to most bacterial aldolases, tagatose-1,6-diphosphate aldolase is a class I aldolase, inhibited by sodium borohydride and dihydroxyacetone phosphate but not EDTA.[588–593] The protein is apparently monomeric in its active form, with a molecular weight near 37 kDa. In pure form the enzyme shows a specific activity near 20 U mg^{-1}. The enzyme exhibits less than complete stereospecificity in retroaldol reaction, and both D-tagatose-1,6-diphosphate and D-fructose-1,6-diphosphate were cleaved by the enzyme. Fructose-1- or 6-phosphates were not substrates for the enzyme, nor were D-tagatose-6-phosphate, L-sorbose-1-phosphate, or a variety of aldose sugars. Michaelis constants of 1.5 mM and 2.5 mM were determined for D-tagatose-1,6-diphosphate and D-fructose-1,6-diphosphate, respectively. V_{max} for fructose-1,6-diphosphate is roughly half that of the tagatose diastereomer. The enzyme catalyzed reaction is reversible, although the *S. aureus* tagatose-1,6-diphosphate aldolase catalyzed reaction of dihydroxyacetone phosphate and D-glyceraldehyde-

3-phosphate produces a mixture of all four stereoisomeric sugars. L-glyceraldehyde-3-phosphate was not utilized as a substrate.

In 1982, Crow and Thomas[595] reported the isolation of tagatose-1,6-diphosphate aldolase from both *L. lactis* and *L. cremoris* (both formerly *Streptococcus*). The protein shows strong similarities to the *Staphylococcal* protein: it is monomeric with a molecular weight near 34.5 kDa and is a class I aldolase that shows a specific activity of 23 U mg^{-1}. The aldolase is only weakly specific for the L-*erythro* configuration of the C-3/C-4 vicinal diol. The enzyme also utilizes both tagatose and fructose diphosphates as substrates. Again, K_m is lower and V_{max} higher for tagatose (K_m 0.1 mM, V_{max} 44.4 U mg^{-1}) than for fructose (K_m 0.25 mM, V_{max} 22.3 U mg^{-1}). Although the *Streptococcal* enzyme shares an absolute specificity for the D-isomer of glyceraldehyde-3-phosphate in the synthetic reaction with the *Staphylococcal* enzyme, the *Streptococcal* tagatose aldolase produces a mixture of only fructose and tagatose diphosphates. No psicose or sorbose stereoisomers were detected in the mixture. The equilibrium between the isomeric ketoses was determined to be 7:1 in favor of the fructo product. Cloning and expression of the protein was reported by this group[596] as well as by van Rooijen *et al.*[597] A nonspecific coccal tagatose-1,6-diphosphate aldolase was cloned and overexpressed in *E. coli* by Wong and co-workers[598] in 1995.

In 1993, Fessner and co-workers[598] reported a tagatose-1,6-diphosphate aldolase from *E. coli* substantially different from those aldolases isolated from lactic acid bacteria. The *E. coli* protein is a homotetramer with a subunit molecular weight near 28 kDa. The protein is a type II aldolase, and shows no inhibition on treatment with sodium borohydride and dihydroxyacetone phosphate. Again, in contrast to the lactic acid bacterial aldolases, the protein is highly specific for the L-*erythro* stereochemistry and produces fructose-1,6-diphosphate from dihydroxyacetone phosphate and D-glyceraldehyde-3-phosphate at only 1% of the rate of production of tagatose-1,6-diphosphate, although the isomeric sugars show similar K_m values of 0.33 mM (tagatose-1,6-diphosphate) and 0.5 mM (fructose-1,6-diphosphate), respectively.

E. coli tagatose-1,6-diphosphate aldolase accepts a broad range of electrophilic substrates, although the enzyme is again highly specific for the nucleophile dihydroxyacetone phosphate. Fessner and co-workers[191,599,599,601] have exploited this broad substrate specificity, and utilized the *E. coli* enzyme for the preparation of a number of unnatural sugar derivatives. These investigators have noted, however, that the diastereoselectivity with unnatural electrophiles is frequently less than that observed for reaction with D-glyceraldehyde-3-phosphate.

ACKNOWLEDGMENTS

This research was supported by the National Institutes of Health (GM 48181). EJT is a Camille Dreyfus Teacher–Scholar and a fellow of the Alfred P. Sloan Foundation.

3.13.9 REFERENCES

1. H. J. M. Gijsen, L. Qiao, W. Fitz, and C.-H. Wong, *Chem. Rev.*, 1996, **96**, 443.
2. C.-H. Wong and G. M. Whitesides, "Enzymes in Synthetic Organic Chemistry," Pergamon, Oxford, 1994.
3. K. Drauz and H. Waldmann, "Enzyme Catalysis in Organic Synthesis: A Comprehensive Handbook," VCH Publishers, Weinheim, 1995.
4. C.-H. Wong, R. L. Halcomb, Y. Ichikawa, and T. Kajimoto, *Angew. Chem., Int. Ed. Engl.*, 1995, **34**, 412.
5. W.-D. Fessner, *Kontakte (Darmstadt)*, 1992, 3.
6. W.-D. Fessner, *Kontakte (Darmstadt)*, 1993, 23.
7. S. David, C. Augé, and C. M. Gautheron, *Adv. Carbohydr. Chem. Biochem.*, 1992, **49**, 175.
8. D. G. Drueckhammer, W. J. Hennen, R. L. Pederson, C. F. Barbas, III, C. M. Gautheron, T. Krach, and C.-H. Wong, *Synthesis*, 1991, 499.
9. E. J. Toone, E. S. Simon, M. D. Bednarski, and G. M. Whitesides, *Tetrahedron*, 1989, **45**, 5365.
10. G. M. Smith, A. S. Mildvan, and E. T. Harper, *Biochemistry*, 1980, **19**, 1248.
11. L. Hedstrom and R. Abeles, *Biochem. Biophys. Res. Commun.*, 1988, **157**, 816.
12. T. Baasov, S. Sheffer-Dee-Noor, A. Kohen, A. Jakob, and V. Belakhov, *Eur. J. Biochem.*, 1993, **217**, 991.
13. E. Racker, *J. Biol. Chem.*, 1951, **196**, 347.
14. D. P. Groth, *Methods Enzymol.*, 1966, **9**, 549.
15. J. D. Pollack, M. C. McElwain, D. DeSantis, J. T. Manolukas, J. G. Tully, C.-J. Chang, R. F. Whitcomb, K. J. Hackett, and M. V. Williams, Jr., *Int. J. Syst. Bacteriol.*, 1989, **39**, 406.
16. P. Hoffee, O. M. Rosen, and B. L. Horecker, *Methods Enzymol.*, 1966, **9**, 545.
17. S. Gonzalez, V. Morata DeAmbrosinbi, M. Manca DeNadra, A. Pesce De Ruiz Holgado, and G. Oliver, *J. Food Prot.*, 1994, **57**, 436.
18. P. Hoffee, *Methods Enzymol.*, 1975, **42**, 276.

19. H. H. Saxild, L. N. Andersen, and K. Hammer, *J. Bacteriol.*, 1996, **178**, 424.
20. P. L. Ipata, F. Sgarrella, and M. G. Tozzi, *Curr. Top. Cell. Regul.*, 1985, **26**, 419.
21. J. P. Petzel, M. C. McElwain, D. DeSantis, J. Manolukas, M. V. Williams, P. A. Hartman, M. J. Allison, and J. D. Pollack, *Arch. Microbiol.*, 1989, **152**, 309.
22. C. M. Fraser, J. D. Gocayne, O. White, M. D. Adams, R. A. Clayton, R. D. Fleischmann, C. J. Bult, A. R. Kerlavage, G. Sutton, J. M. Kelley, *et al.*, *Science*, 1995, **270**, 397.
23. S. Loechel, J. M. Inamine, and P.-C. Hu, *Nucleic Acids Res.*, 1989, **17**, 801.
24. D. DeSantis, V. V. Tryon, and J. D. Pollack, *J. Gen. Microbiol.*, 1989, **135**, 683.
25. R. D. Fleischmann, M. D. Adams, O. White, R. A. Clayton, E. F. Kirkness, A. R. Kerlavage, C. J. Bult, J.-F. Tomb, B. A. Dougherty, J. M. Merrick, *et al.*, *Science*, 1995, **269**, 496.
26. P. Valentin-Hansen, F. Boètius, K. Hammer-Jespersen, and I. Svendsen, *Eur. J. Biochem.*, 1982, **125**, 561.
27. V. Burland, G. Plunkett, III, H. J. Sofia, D. L. Daniels, and F. R. Blattner, *Nucleic Acids Res.*, 1995, **23**, 2105.
28. E. A. Stura, S. Ghosh, E. Garcia-Junceda, L. Chen, C.-H. Wong, and I. A. Wilson, *Proteins: Struct. Funct. Genet.*, 1995, **22**, 67.
29. L. Chen, D. P. Dumas, and C.-H. Wong, *J. Am. Chem. Soc.*, 1992, **114**, 741.
30. C. F. Barbas, III, Y.-F. Wang, and C.-H. Wong, *J. Am. Chem. Soc.*, 1990, **112**, 2013.
31. H. J. M. Gijsen and C.-H. Wong, *J. Am. Chem. Soc.*, 1994, **116**, 8422.
32. H. J. M. Gijsen and C.-H. Wong, *J. Am. Chem. Soc.*, 1995, **117**, 2947.
33. C.-H. Wong, E. Garcia-Junceda, L. Chen, O. Blanco, H. J. M. Gijsen, and D. H. Steensma, *J. Am. Chem. Soc.*, 1995, **117**, 3333.
34. W.-C. Chou, L. Chen, J.-M. Fang, and C.-H. Wong, *J. Am. Chem. Soc.*, 1994, **116**, 6191.
35. T. Kajimoto, L. Chen, K. K.-C. Liu, and C.-H. Wong, *J. Am. Chem. Soc.*, 1991, **113**, 6678.
36. T. Kajimoto, L. Chen, K. K.-C. Liu, and C.-H. Wong, *J. Am. Chem. Soc.*, 1991, **113**, 9009.
37. F. M. Unger, *Adv. Carbohydr. Chem. Biochem.*, 1981, **38**, 323.
38. M. A. Ghalambor, E. M. Levine, and E. C. Heath, *J. Biol. Chem.*, 1966, **241**, 3207.
39. M. A. Ghalambor and E. C. Heath, *J. Biol. Chem.*, 1966, **241**, 3216.
40. M. A. Ghalambor and E. C. Heath, *J. Biol. Chem.*, 1966, **241**, 3222.
41. B. R. Knappmann and M.-R. Kula, *Appl. Microbiol. Biotechnol.*, 1990, **33**, 324.
42. T. Sugai, G.-J. Shen, Y. Ichikawa, and C.-H. Wong, *J. Am. Chem. Soc.*, 1993, **115**, 413.
43. J. B. Mathis and G. M. Brown, *J. Biol. Chem.*, 1970, **245**, 3015.
44. P. Lopez and S. A. Lacks, *J. Bacteriol.*, 1993, **175**, 2214.
45. J. Slock, D. P. Stahly, C.-Y. Han, E. W. Six, and I. P. Crawford, *J. Bacteriol.*, 1990, **172**, 7211.
46. F. Volpe, S. P. Ballantine, and C. J. Delves, *Eur. J. Biochem.*, 1993, **216**, 449.
47. F. Volpe, S. P. Ballantine, and C. J. Delves, *Gene*, 1990, **160**, 41.
48. P. Kellam, W. S. Dallas, S. P. Ballantine, and C. J. Delves, *FEMS Microbiol. Lett.*, 1995, **134**, 165.
49. W. O. Taussen, "Basic Principles of Plant Bioenergetics," Academy of Sciences of the USSR, Moscow, 1950.
50. K.-M. Yen and C. M. Serdar, *CRC Crit. Rev. Microbiol.*, 1988, **15**, 247.
51. C. E. Cerniglia, *Adv. Appl. Microbiol.*, 1984, **30**, 31.
52. R. Lal, S. Lal, P. S. Dhanaraj, and D. M. Saxena, *Adv. Appl. Microbiol.*, 1995, **41**, 55.
53. M. C. Mahajan, P. S. Phale, and C. S. Vaidyanathan, *Arch. Microbiol.*, 1994, **161**, 425.
54. K.-M. Yen and I. C. Gunsalus, *Proc. Natl. Acad. Sci. USA*, 1982, **79**, 874.
55. R. A. Austen and N. W. Dunn, *J. Gen. Microbiol.*, 1980, **117**, 521.
56. E. A. Barnsley, *Appl. Environ. Microbiol.*, 1988, **54**, 428.
57. C. M. Morris and E. A. Barnsley, *Can. J. Microbiol.*, 1982, **28**, 73.
58. M. A. Connors and E. A. Barnsley, *J. Bacteriol.*, 1982, **149**, 1096.
59. T. R. Patel and E. A. Barnsley, *J. Bacteriol.*, 1980, **143**, 668.
60. E. A. Barnsley, *Biochem. Biophys. Res. Commun.*, 1976, **72**, 1116.
61. A. E. Kuhm, H.-J. Knackmuss, and A. Stolz, *J. Biol. Chem.*, 1993, **268**, 9484.
62. R. W. Eaton, *J. Bacteriol.*, 1994, **176**, 7757.
63. R. W. Eaton and P. J. Chapman, *J. Bacteriol.*, 1992, **174**, 7542.
64. J. Sanseverino, B. M. Applegate, J. M. H. King, and G. S. Sayler, *Appl. Environ. Microbiol.*, 1993, **59**, 1931.
65. H. Kiyohara, N. Takizawa, H. Date, S. Torigoe, and K. Yano, *J. Ferment. Bioeng.*, 1990, **69**, 54.
66. W. F. Guerin and G. E. Jones, *Appl. Environ. Microbiol.*, 1988, **54**, 937.
67. W. F. Guerin and G. E. Jones, *Appl. Environ. Microbiol.*, 1988, **54**, 929.
68. E. A. Barnsley, *J. Bacteriol.*, 1983, **154**, 113.
69. B. Nörtemann, A. E. Kuhm, H.-J. Knackmuss, and A. Stolz, *Arch. Microbiol.*, 1994, **161**, 320.
70. M. J. Larkin and M. J. Day, *J. Appl. Bacteriol.*, 1986, **60**, 233.
71. H. H. Higa, G. N. Roger, and J. C. Paulson, *Virology*, 1985, **144**, 279.
72. M. J. Phillips, E. Nudelman, F. C. A. Gaeta, M. Perez, A. K. Singhal, S. Hakomori, and J. C. Paulson, *Science*, 1990, **250**, 1130.
73. T. A. Springer and L. A. Lasky, *Nature (London)*, 1991, **349**, 196.
74. C. A. A. Van Boeckel, *Recl. Trav. Chim. Pays-Bas*, 1986, **105**, 35.
75. G. Yogees Waran and P. L. Salk, *Science*, 1981, **212**, 1514.
76. C. W. Dewitt and J. A. Rowe, *J. Bacteriol.*, 1961, **82**, 838.
77. K. A. Homer, S. Kelley, J. Hawkes, D. Beighton, and M. C. Grootveld, *Microbiology*, 1996, **142**, 1221.
78. G. H. De Vries and S. B. Binkley, *Arch. Biochem. Biophys.*, 1972, **151**, 234.
79. R. Schauer and M. Wember, *Hoppe-Seyler's Z. Physiol. Chem.*, 1971, **352**, 1517.
80. S. Nees, R. Schauer, and F. Mayer, *Hoppe-Seyler's Z. Physiol. Chem.*, 1976, **357**, 839.
81. S. B. Arden, W.-H. Chang, and L. Barksdale, *J. Bacteriol.*, 1972, **112**, 1206.
82. D. G. Comb and S. Roseman, *J. Biol. Chem.*, 1960, **235**, 2529.
83. R. Drzeniek, W. Scharmann, and E. Balke, *J. Gen. Microbiol.*, 1972, **72**, 357.
84. H. E. Müller and W. Mannheim, *Zentralbl. Bakteriol.*, 1995, **283**, 105.
85. R. Heimer and K. Meyer, *Proc. Natl. Acad. Sci. USA*, 1956, **42**, 728.

86. Y. Uchida, Y. Tsukada, and T. Sugimoro, *J. Biochem. (Tokyo)*, 1984, **96**, 507.
87. R. Schauer, *Adv. Carbohydr. Chem. Biochem.*, 1982, **40**, 131.
88. A. Varki, *Glycobiology*, 1992, **1**, 25.
89. R. Schauer, S. Kelm, G. Reuter, P. Roggentin, and L. Shaw, in "Biology of Sialic Acids," ed. R. Abraham, Plenum Press, New York, 1995, p. 7.
90. R. Schauer and M. Wember, *Biol. Chem. Hoppe-Seyler*, 1996, **377**, 293.
91. M. A. Ferrero, A. Reglero, M. Fernandez-Lopez, R. Ordas, and L. B. Rodriguez-Aparico, *Biochem. J.*, 1996, **317**, 157.
92. G. G. Lilley, M. von Itzstein, and N. Ivancic, *Protein Expr. Purif.*, 1992, **3**, 434.
93. T. Izard, M. C. Lawrence, R. L. Malby, G. G. Lilley, and P. M. Colman, *Structure*, 1994, **2**, 361.
94. C. M. Deijl and J. F. G. Vliegenthart, *Biochem. Biophys. Res. Commun.*, 1983, **111**, 668.
95. W. Baumann, J. Freidenreich, G. Weisshaar, R. Brossmer, and H. Friebolin, *Biol. Chem. Hoppe-Seyler*, 1989, **370**, 141.
96. S. David, A. Malleron, and B. Cavaye, *New J. Chem.*, 1992, **16**, 751.
97. U. Kragl, A. Gödde, C. Wandrey, N. Lubin, and C. Augé, *J. Chem. Soc., Perkin Trans. 1*, 1994, 119.
98. K. Aisaka, A. Igarashi, K. Yamaguchi, and T. Uwajima, *Biochem. J.*, 1991, **276**, 541.
99. T. Sugai, A. Kuboki, S. Hiramatsu, H. Okazaki, and H. Ohta, *Bull. Chem. Soc. Jpn.*, 1995, **68**, 3581.
100. A. Lubineau, C. Augé, C. Gautheron-Le Narvor, and J.-C. Ginet, *Bioorg. Med. Chem.*, 1994, **2**, 669.
101. P. Zhou, H. M. Salleh, and J. F. Honek, *J. Org. Chem.*, 1993, **58**, 264.
102. H. Nishihara and E. E. Dekker, *J. Biol. Chem.*, 1972, **247**, 5079.
103. T. S. M. Taha and T. L. Deits, *Biochem. Biophys. Res. Commun.*, 1994, **200**, 459.
104. E. E. Dekker and R. P. Kitson, *J. Biol. Chem.*, 1992, **267**, 10 507.
105. J. M. Scholtz and S. M. Schuster, *Bioorg. Chem.*, 1984, **12**, 229.
106. K. Kuratomi and K. Fukunaga, *Biochim. Biophys. Acta*, 1963, **78**, 617.
107. H. Hift and H. R. Mahler, *J. Biol. Chem.*, 1952, **198**, 901.
108. U. Maitra and E. E. Dekker, *J. Biol. Chem.*, 1963, **238**, 3660.
109. S. C. Gupta and E. E. Dekker, *J. Biol. Chem.*, 1984, **259**, 10 012.
110. E. E. Dekker, H. Nishihara, and S. R. Grady, *Methods Enzymol.*, 1975, **42**, 285.
111. R. G. Rosso and E. Adams, *J. Biol. Chem.*, 1967, **242**, 5524.
112. U. Maitra and E. E. Dekker, *J. Biol. Chem.*, 1964, **239**, 1485.
113. N. C. Floyd, M. H. Liebster, and N. J. Turner, *J. Chem. Soc., Perkin Trans. 1*, 1992, 1085.
114. H. P. Meloche, *Trends Biochem. Soc.*, 1981, **6**, 38.
115. J. K. Wang, E. E. Dekker, N. D. Lewinski, and H. C. Winter, *J. Biol. Chem.*, 1981, **256**, 1793.
116. R. V. Patil and E. E. Dekker, *J. Bacteriol.*, 1992, **174**, 102.
117. C. J. Vlahos and E. E. Dekker, *J. Biol. Chem.*, 1988, **263**, 11 683.
118. C. J. Vlahos and E. E. Dekker, *J. Biol. Chem.*, 1986, **261**, 11 049.
119. C. J. Vlahos and E. E. Dekker, *J. Biol. Chem.*, 1990, **265**, 20 384.
120. C. J. Vlahos, M. A. Ghalambor, and E. E. Dekker, *J. Biol. Chem.*, 1985, **260**, 5480.
121. S. F. Egan, R. Fliege, S. Tong, A. Shibata, R. E. Wolf, Jr., and T. Conway, *J. Bacteriol.*, 1992, **174**, 4638.
122. M. C. Shelton, I. C. Cotterill, S. T. A. Novak, R. M. Poonawala, S. Sudershan, and E. J. Toone, *J. Am. Chem. Soc.*, 1996, **118**, 2117.
123. J. Powlowski, L. Sahlman, and V. Shingler, *J. Bacteriol.*, 1993, **175**, 377.
124. R. Burlingame and P. J. Chapman, *J. Bacteriol.*, 1983, **155**, 424.
125. R. W. Eaton, *J. Bacteriol.*, 1996, **178**, 1351.
126. P. C. K. Lau, H. Bergeron, D. Lubbé, Y. Wang, R. Brousseau, and D. T. Gibson, *Gene*, 1994, **146**, 7.
127. B. Hofer, S. Backhaus, and K. N. Timmis, *Gene*, 1994, **144**, 9.
128. Y. Kikuchi, Y. Yasukochi, Y. Nagata, M. Fukuda, and M. Takagi, *J. Bacteriol.*, 1994, **176**, 4269.
129. L. M. Shannon and A. Marcus, *J. Biol. Chem.*, 1962, **237**, 3342.
130. S. Dagley, *Methods Enzymol.*, 1982, **90**, 272.
131. V. L. Sparnins and S. Dagley, *J. Bacteriol.*, 1975, **124**, 1374.
132. B. F. Tack, P. J. Chapman, and S. Dagley, *J. Biol. Chem.*, 1972, **247**, 6438.
133. S. Dagley, P. J. Geary, and J. M. Wood, *Biochem. J.*, 1968, **109**, 559.
134. P. J. Kersten, S. Dagley, J. W. Whittaker, D. M. Arciero, and J. D. Lipscomb, *J. Bacteriol.*, 1982, **152**, 1154.
135. K. Maruyama, *J. Biochem. (Tokyo)*, 1990, **108**, 327.
136. K. Maruyama, *J. Biochem. (Tokyo)*, 1991, **110**, 976.
137. K. Maruyama, *J. Biochem. (Tokyo)*, 1990, **108**, 334.
138. B. F. Tack, P. J. Chapman, and S. Dagley, *J. Biol. Chem.*, 1972, **247**, 6444.
139. P. R. Srinivasan and D. B. Sprinson, *J. Biol. Chem.*, 1959, **234**, 716.
140. D. B. Sprinson, P. R. Srinivasan, and M. Katagiri, *Methods Enzymol.*, 1962, **5**, 394.
141. M. Reinink and A. C. Borstlap, *Plant Sci. Lett.*, 1982, **26**, 167.
142. J. L. Rubin and R. A. Jensen, *Plant Physiol.*, 1985, **79**, 711.
143. C. A. Bonner and R. A. Jensen, *Plant Sci.*, 1991, **74**, 229.
144. J. Zhao and K. M. Herrmann, *Plant Physiol.*, 1992, **100**, 1075.
145. L. M. Weaver, J. E. B. P. Pinto, and K. M. Herrmann, *Bioorg. Med. Chem. Lett.*, 1993, **3**, 1421.
146. N. Suzuki, M. Sakuta, and S. Shimizu, *J. Plant Physiol.*, 1996, **149**, 19.
147. D. M. Tieman and A. K. Handa, *J. Am. Soc. Hortic. Sci.*, 1996, **121**, 52.
148. G. J. W. Euverink, G. I. Hessels, C. Franke, and L. Dijkhuizen, *Appl. Environ. Microbiol.*, 1995, **61**, 3796.
149. G. S. Byng, A. Berry, and R. A. Jensen, *Arch. Microbiol.*, 1985, **143**, 122.
150. S. Ahmad, B. Rightmire, and R. A. Jensen, *J. Bacteriol.*, 1986, **165**, 146.
151. O. Kurahashi, M. Noda-Watanabe, K. Sato, Y. Morinaga, and H. Enei, *Agric. Biol. Chem.*, 1987, **51**, 1785.
152. G. Paravicini, T. Schmidheini, and G. Braus, *Eur. J. Biochem.*, 1989, **186**, 361.
153. I. A. Shumilin, R. H. Kretsinger, and R. Bauerle, *Proteins: Struct. Funct. Genet.*, 1996, **24**, 404.
154. F. Stuart and I. S. Hunter, *Biochim. Biophys. Acta*, 1993, **1161**, 209.
155. G. E. Walker, B. Dunbar, I. S. Hunter, H. G. Nimmo, and J. R. Coggins, *Microbiology*, 1996, **142**, 1973.

156. C. A. Bonner, R. S. Fischer, R. R. Schmidt, P. W. Miller, and R. A. Jensen, *Plant Cell Physiol.*, 1995, **36**, 1013.
157. D. Kolibachuk, D. Rouhbakhsh, and P. Baumann, *Curr. Microbiol.*, 1995, **30**, 313.
158. T. Ogino, C. Garner, J. L. Markley, and K. M. Herrmann, *Proc. Natl. Acad. Sci. USA*, 1982, **79**, 5828.
159. J. Shultz, M. A. Hermodson, C. C. Garner, and K. M. Herrmann, *J. Biol. Chem.*, 1984, **259**, 9655.
160. W. D. Davies and B. E. Davidson, *Nucleic Acids Res.*, 1982, **10**, 4045.
161. R. J. Ganson, T. A. D'Amato, and R. A. Jensen, *Plant Physiol.*, 1986, **82**, 203.
162. A. J. Pittard, in "Escherichia coli and Salmonella," ed. F. C. Neidhardt, ASM Press, Washington, DC, 1996, p. 458.
163. L. M. Reimer, D. L. Conley, D. L. Pompliano, and J. W. Frost, *J. Am. Chem. Soc.*, 1986, **108**, 8010.
164. M. Inouye, "Bacterial Outer Membranes: Biogenesis and Functions," Wiley, New York, 1979.
165. R. L. Doong, S. Ahmad, and R. A. Jensen, *Plant Cell Environ.*, 1991, **14**, 113.
166. D. H. Levin and E. Racker, *J. Biol. Chem.*, 1959, **234**, 2532.
167. P. H. Ray, *Methods Enzymol.*, 1982, **83**, 525.
168. P. H. Ray, *J. Bacteriol.*, 1980, **141**, 635.
169. G. D. Dotson, R. K. Dua, J. C. Clemens, E. W. Wooten, and R. W. Woodard, *J. Biol. Chem.*, 1995, **270**, 13 698.
170. M. Woisetschläger, A. Hödl-Neuhofer, and G. Högenauer, *J. Bacteriol.*, 1988, **170**, 5382.
171. M. Woisetschläger and G. Högenauer, *J. Bacteriol.*, 1986, **168**, 437.
172. M. Woisetschläger and G. Högenauer, *Mol. Gen. Genet.*, 1987, **207**, 369.
173. W. D. Tolbert, J. R. Moll, R. Baurele, and R. H. Kretsinger, *Proteins: Struct. Funct. Genet.*, 1996, **24**, 407.
174. P. D. Rick and M. J. Osborn, *J. Biol. Chem.*, 1977, **252**, 4895.
175. P. D. Rick and D. A. Young, *J. Bacteriol.*, 1982, **150**, 456.
176. P. D. Rick and D. A. Young, *J. Bacteriol.*, 1982, **150**, 447.
177. P. D. Rick, B. A. Neumeyer, and D. A. Young, *J. Biol. Chem.*, 1983, **258**, 629.
178. A. Kohen, A. Jakob, and T. Baasov, *Eur. J. Biochem.*, 1992, **208**, 443.
179. S. Sheffer-Dee-Noor, V. Belakhov, and T. Baasov, *Bioorg. Med. Chem. Lett.*, 1993, **3**, 1583.
180. T. Baasov and A. Kohen, *J. Am. Chem. Soc.*, 1995, **117**, 6165.
181. G. D. Dotson, P. Nanjappan, M. D. Reily, and R. W. Woodard, *Biochemistry*, 1993, **32**, 12 392.
182. A. Kohen, R. Berkovich, V. Belakhov, and T. Baasov, *Bioorg. Med. Chem. Lett.*, 1993, **3**, 1577.
183. M. D. Bednarski, D. C. Crans, R. DiCosimo, E. S. Simon, P. D. Stein, and G. M. Whitesides, *Tetrahedron Lett.*, 1988, **29**, 427.
184. R. S. Blacklow and L. Warren, *J. Biol. Chem.*, 1962, **237**, 3520.
185. R. P. Silver, C. W. Finn, W. F. Vann, W. Aaronson, R. Schneerson, P. J. Kretchmer, and C. F. Garon, *Nature (London)*, 1981, **289**, 696.
186. H. E. Müller, *Zentralbl. Bakteriol. Parasitenk., Infektionskr. Hyg. Abt. 1: Orig. Reihe A*, 1973, **225**, 59.
187. L. Masson and B. E. Holbein, *J. Bacteriol.*, 1983, **154**, 728.
188. U. Edwards, A. Müller, S. Hammerschmidt, R. Gerardy-Schahn, and M. Frosch, *Mol. Microbiol.*, 1994, **14**, 141.
189. R. Brossmer, U. Rose, D. Kasper, T. L. Smith, H. Grasmuk, and F. M. Unger, *Biochem. Biophys. Res. Commun.*, 1980, **96**, 1282.
190. C. Hagemeier, P. Vischer, and E. Buddecke, *Biochem. Soc. Trans.*, 1986, **14**, 1173.
191. P. Görog, I. Schraufstätter, and G. V. R. Born, *Proc. R. Soc. London, B. Biol. Sci.*, 1982, **214**, 471.
192. J. Nilsson, T. Ksiazek, and J. Thyberg, *Exp. Cell. Res.*, 1982, **142**, 333.
193. L. Warren and H. Felsenfeld, *Biochem. Biophys. Res. Commun.*, 1961, **4**, 232.
194. L. Warren and H. Felsenfeld, *Biochem. Biophys. Res. Commun.*, 1961, **5**, 185.
195. L. Warren and H. Felsenfeld, *J. Biol. Chem.*, 1962, **237**, 1421.
196. D. R. Watson, G. W. Jourdian, and S. Roseman, *J. Biol. Chem.*, 1966, **241**, 5627.
197. D. Watson, G. W. Jourdian, and S. Roseman, *Methods Enzymol.*, 1966, **8**, 201.
198. A. P. Corfield, J. R. Clamp, and S. A. Wagner, *Biochem. Soc. Trans.*, 1983, **11**, 767.
199. A. P. Corfield, J. R. Clamp, and S. A. Wagner, *Biochem. J.*, 1985, **226**, 163.
200. J. van Rinsum, W. van Dijk, G. J. M. Hooghwinkel, and W. Ferwerda, *Biochem. J.*, 1984, **223**, 323.
201. E. C. Heath, J. Hurwitz, B. L. Horecker, and A. Ginsburg, *J. Biol. Chem.*, 1958, **231**, 1009.
202. B. Sgorbati, G. Lenaz, and F. Casalicchio, *Antonie van Leeuwenhoek*, 1976, **42**, 49.
203. D. A. Whitworth and C. Ratledge, *J. Gen. Microbiol.*, 1977, **102**, 397.
204. M. Goldberg, J. M. Fessenden, and E. Racker, *Methods Enzymol.*, 1966, **9**, 515.
205. A. Singh, P. K. Kumar, and K. Schuger, *Biochem. Int.*, 1992, **27**, 831.
206. C. T. Evans and C. Ratledge, *Arch. Microbiol.*, 1984, **139**, 48.
207. A. H. Lachke and T. W. Jeffries, *Enzyme Microb. Technol.*, 1986, **8**, 353.
208. J. London and N. M. Chace, *J. Bacteriol.*, 1979, **140**, 949.
209. G. J. Lees and G. R. Jago, *J. Dairy Res.*, 1976, **43**, 75.
210. C. Ratledge and P. A. Botham, *J. Gen. Microbiol.*, 1977, **102**, 391.
211. D. E. Greenley and D. W. Smith, *Arch. Microbiol.*, 1979, **122**, 257.
212. V. Scardovi and L. D. Trovatelli, *Ann. Microbiol. Enzymol.*, 1965, **15**, 19.
213. M. Schramm, V. Klybas, and E. Racker, *J. Biol. Chem.*, 1958, **233**, 1283.
214. F. Gavini, M. van Esbroeck, J. P. Touzel, A. Fourment, and H. Goossens, *Anaerobe*, 1996, **2**, 191.
215. E. L. Golovlev, N. V. Eroshina, and L. M. Baryshnikova, *Dokl. Akad. Nauk SSSR*, 1983, **272**, 490.
216. J.-P. Grill, J. Crociani, and J. Ballongue, *Curr. Microbiol.*, 1995, **31**, 49.
217. A. G. Kaster and L. R. Brown, *Infect. Immun.*, 1983, **42**, 716.
218. S. Morimoto, K. Azuma, T. Oshima, and M. Sakamoto, *J. Ferment. Technol.*, 1988, **66**, 7.
219. S. Morimoto, *Hakko Kogaku Zasshi*, 1972, **50**, 850.
220. B. González and R. Vicuña, *J. Bacteriol.*, 1989, **171**, 2401.
221. P. Hinrichsen, I. Gómez, and R. Vicuña, *Gene*, 1994, **144**, 137.
222. R. Schwarcz, E. Okuno, R. J. White, E. D. Bird, and W. O. Whetsell, Jr., *Proc. Natl. Acad. Sci. USA*, 1988, **85**, 4079.
223. S. Mazzari, C. Aldino, M. Beccaro, G. Toffano, and R. Schwarcz, *Brain Res.*, 1986, **380**, 309.
224. M. P. Heyes, B. J. Brew, A. Martin, R. W. Price, A. M. Salaazar, J. J. Sidtis, J. A. Yergey, M. M. Mouradian, A. E. Sadler, J. Keilp, D. Rubinow, and S. P. Markey, *Ann. Neurol.*, 1991, **29**, 202.

225. G. B. Watson, W. F. Hood, J. B. Monahan, and T. H. Lanthorn, *Neurosci. Res. Commun.*, 1988, **2**, 169.
226. S. M. El-Sewedy, G. A. Abdel-Tawab, M. H. Abdel-Daim, and M. F. El-Sawy, *Biochem. Pharmacol.*, 1972, **21**, 379.
227. G. Machill, A. Knapp, and W. Teichmann, *Clin. Chim. Acta*, 1972, **39**, 171.
228. T. Ishikawa, E. Okuno, J. Kawai, and R. Kido, *Comp. Biochem. Physiol. B*, 1989, **93**, 107.
229. H. Ogawa and K. Hasegawa, *Insect Biochem.*, 1980, **10**, 589.
230. K. Tanizawa, T. Yamamoto, and K. Soda, *FEBS Lett.*, 1976, **70**, 235.
231. K. Tanizawa and K. Soda, *J. Biochem. (Tokyo)*, 1979, **85**, 1367.
232. M. Moriguchi, T. Yamamoto, and K. Soda, *Biochemistry*, 1973, **12**, 2969.
233. A. S. Shetty and F. H. Gaertner, *J. Bacteriol.*, 1975, **122**, 235.
234. O. Salcher and F. Lingens, *J. Gen. Microbiol.*, 1980, **121**, 465.
235. T. Troost, M. J. M. Hitchcock, and E. Katz, *Biochim. Biophys. Acta*, 1980, **612**, 97.
236. G. S. Bild and J. C. Morris, *Arch. Biochem. Biophys.*, 1984, **235**, 41.
237. K. Tanizawa and K. Soda, *J. Biochem. (Tokyo)*, 1979, **86**, 1199.
238. R. S. Phillips and R. K. Dua, *J. Am. Chem. Soc.*, 1991, **113**, 7385.
239. M. M. Palcic, M. Antoun, K. Tanizawa, K. Soda, and H. G. Floss, *J. Biol. Chem.*, 1985, **260**, 5248.
240. G. M. Kishore, *J. Biol. Chem.*, 1984, **259**, 10 669.
241. B. Natalini, L. Mattoli, R. Pellicciari, R. Carpenedo, A. Chiarugi, and F. Moroni, *Bioorg. Med. Chem. Lett.*, 1995, **5**, 1451.
242. R. K. Dua, E. W. Taylor, and R. S. Phillips, *J. Am. Chem. Soc.*, 1993, **115**, 1264.
243. A. De Antoni, C. Costa, and G. Allergri, *Acta Vitaminol. Enzymol.*, 1975, **29**, 339.
244. F. Knoop, *Z. Physiol. Chem.*, 1914, **89**, 151.
245. A. E. Braunstein and G. Y. Vilenkina, *Dokl. Akad. Nauk. SSSR*, 1949, **66**, 243.
246. M. A. Karasek and D. M. Greenberg, *J. Biol. Chem.*, 1963, **227**, 191.
247. H. Yamada, H. Kumagai, T. Nagate, and H. Yoshida, *Agric. Biol. Chem.*, 1971, **35**, 1340.
248. R. H. Dainty, *Biochem. J.*, 1970, **117**, 585.
249. V. Besson, M. Neuberger, F. Rebeille, and R. Douce, *Plant Physiol. Biochem.*, 1995, **33**, 665.
250. J. P. Marcus and E. E. Dekker, *Biochim. Biophys. Acta*, 1993, **1164**, 299.
251. R. Usha, H. S. Savithri, and N. A. Rao, *Biochim. Biophys. Acta*, 1994, **1204**, 75.
252. F. Martini, B. Maras, P. Tanci, S. Angelaccio, S. Pascarella, D. Barra, F. Bossa, and V. Schirch, *J. Biol. Chem.*, 1989, **264**, 8509.
253. F. Martini, S. Angelaccio, S. Pascarella, D. Barra, F. Bossa, and V. Schirch, *J. Biol. Chem.*, 1987, **262**, 5499.
254. C. R. McClung, C. R. Davis, K. M. Page, and S. A. Denome, *Mol. Cell. Biol.*, 1992, **12**, 1412.
255. S. R. Turner, R. Ireland, C. Morgan, and S. Rawsthorne, *J. Biol. Chem.*, 1992, **267**, 13 528.
256. M. D. Plamann, L. T. Stauffer, M. L. Urbanowski, and G. L. Stauffer, *Nucleic Acids Res.*, 1983, **11**, 2065.
257. S. Rossbach and H. Hennecke, *Mol. Microbiol.*, 1991, **5**, 39.
258. M. L. Urbanowski, M. D. Plamann, L. T. Stauffer, and G. V. Stauffer, *Gene*, 1984, **27**, 47.
259. V. L. Chan and H. L. Bingham, *Gene*, 1991, **101**, 51.
260. L. I. Malkin and D. M. Greenberg, *Biochim. Biophys. Acta*, 1964, **85**, 117.
261. A. G. Palekar, S. S. Tate, and A. Meister, *J. Biol. Chem.*, 1973, **248**, 1158.
262. G. Reario Sforza, R. Pagani, and E. Marinello, *Eur. J. Biochem.*, 1969, **8**, 88.
263. N. R. Thomas, J. E. Rose, and D. Gani, *J. Chem. Soc., Perkin Trans. 1*, 1993, 2933.
264. N. Schirch and T. Gross, *J. Biol. Chem.*, 1968, **243**, 5651.
265. M. Fujioka, *Biochim. Biophys. Acta*, 1969, **185**, 338.
266. K. G. Scrimgeour and F. M. Huennekens, *Methods Enzymol.*, 1962, **5**, 838.
267. M. Akhtar and H. A. El-Obeid, *Biochim. Biophys. Acta*, 1972, **258**, 791.
268. K. S. Ramesh and N. A. Rao, *Biochem. J.*, 1980, **187**, 623.
269. S. Angelaccio, S. Pascarella, E. Fattori, F. Bossa, W. Strong, and V. Schirch, *Biochemistry*, 1992, **31**, 155.
270. R. J. Ulevitch and R. G. Kallen, *Biochemistry*, 1977, **16**, 5342.
271. R. J. Ulevitch and R. G. Kallen, *Biochemistry*, 1977, **16**, 5355.
272. W. M. Ching and R. G. Kallen, *Biochemistry*, 1979, **18**, 821.
273. H. K. Webb and R. G. Matthews, *J. Biol. Chem.*, 1995, **270**, 17 204.
274. R. G. Matthews, J. Ross, C. M. Baugh, J. D. Cook, and L. Davis, *Biochemistry*, 1982, **21**, 1230.
275. T. A. Garrow, A. A. Brenner, V. M. Whitehead, X.-N. Chen, R. G. Duncan, J. R. Korenberg, and B. Shane, *J. Biol. Chem.*, 1993, **268**, 11 910.
276. R. Waziri, S. Baruah, T. S. Hegwood, and A. D. Sherman, *Neurosci. Lett.*, 1990, **120**, 237.
277. F. H. Bruns and L. Fiedler, *Nature (London)*, 1958, **181**, 1533.
278. M. Naoi, T. Takahashi, N. Kuno, and T. Nagatsu, *Biochem. Biophys. Res. Commun.*, 1987, **143**, 482.
279. R. B. Herbert, B. Wilkinson, G. J. Ellames, and E. K. Kunec, *J. Chem. Soc., Chem. Commun.*, 1993, 205.
280. W. Maruyama, D. Nakahara, and M. Naoi, *J. Neural Transm. Park. Dis. Dement. Sect.*, 1994, **7**, 21.
281. A. Reches, V. Jackson-Lewis, and S. Fahn, *Naunyn-Schmiedeberg's Arch. Pharmacol.*, 1985, **331**, 202.
282. H. Misono, H. Okuda, K. Shin, S. Nagata, and S. Nagasaki, *Biosci. Biotechnol. Biochem.*, 1995, **59**, 339.
283. R. B. Herbert, B. Wilkinson, and G. J. Ellames, *Can. J. Chem.*, 1994, **72**, 114.
284. M. Bycroft, R. B. Herbert, and G. J. Ellames, *J. Chem. Soc., Perkin Trans. 1*, 1996, 2439.
285. A. Reches, *Clin. Neuropharmacol.*, 1985, **8**, 249.
286. J. B. McNeil, E. M. McIntosh, B. V. Taylor, F.-R. Zhang, S. Tang, and A. L. Bognar, *J. Biol. Chem.*, 1994, **269**, 9155.
287. H. Kumagai, T. Nagate, H. Yoshida, and H. Yamada, *Biochim. Biophys. Acta*, 1972, **258**, 779.
288. B. V. Taylor, J. B. McNeil, E. M. McIntosh, F. R. Zhang, and A. L. Bognar, in "Chemistry and Biology of Pteridines and Folates," eds. J. E. Ayling, M. G. Nair, and C. M. Baugh, Plenum Press, New York, 1993, p. 711.
289. H. Yamada, H. Kumagai, T. Nagate, and H. Yoshida, *Biochem. Biophys. Res. Commun.*, 1970, **39**, 53.
290. M. Sakamoto, T. Masuda, Y. Yanagimoto, Y. Nakano, and S. Kitaoka, *Agric. Biol. Chem.*, 1991, **55**, 2243.
291. R. H. Dainty and J. L. Peel, *Biochem. J.*, 1970, **117**, 573.
292. W. Stöchlein and H.-L. Schmidt, *Biochem. J.*, 1985, **232**, 621.
293. A. J. Willetts and J. M. Turner, *Biochim. Biophys. Acta*, 1971, **252**, 105.

294. S. C. Bell and J. M. Turner, *Biochem. J.*, 1977, **166**, 209.
295. M. Diaz-Diaz, O. P. Ward, J. Honek, and G. Lajoie, *Can. J. Microbiol.*, 1995, **41**, 438.
296. S. C. Bell and J. M. Turner, *Biochem. Soc. Trans.*, 1973, **1**, 678.
297. H. Ide, K. Hamaguchi, S. Kobata, A. Murakami, Y. Kimura, K. Makino, M. Kamada, S. Miyamoto, T. Nagaya, and K. Kamogawa, *J. Chromatogr.*, 1992, **596**, 203.
298. S. C. Bell and J. M. Turner, *Biochem. J.*, 1976, **156**, 449.
299. L. V. Chistoserdova and M. E. Lidstrom, *J. Bacteriol.*, 1994, **176**, 6759.
300. T. Masuda, M. Yoshino, I. Nishizaki, A. Tai, and H. Ozaki, *Agric. Biol. Chem.*, 1980, **44**, 2199.
301. F. Rebeille, M. Neuburger, and R. Douce, *Biochem. J.*, 1994, **302**, 223.
302. S. Kopriva and H. Bauwe, *Plant Physiol.*, 1995, **107**, 271.
303. R. G. Matthews and J. T. Drummond, *Chem. Rev.*, 1990, **90**, 1275.
304. P. W. Trudgill and R. Widdus, *Nature (London)*, 1966, **211**, 1097.
305. D. C. Fish and H. J. Blumenthal, *Methods Enzymol.*, 1966, **9**, 529.
306. D. C. Fish and H. J. Blumenthal, *Bacteriol. Proc.*, 1963, 110.
307. R. Jeffcoat, H. Hassall, and S. Dagley, *Biochem. J.*, 1969, **115**, 969.
308. R. Jeffcoat, *Biochem. J.*, 1974, **139**, 477.
309. N. V. Loginova, N. I. Govorukhina, and Y. A. Trotsenko, *Mikrobiologiya*, 1981, **50**, 305.
310. H. J. Blumenthal, V. L. Lucuta, and D. C. Blumenthal, *Anal. Biochem.*, 1990, **185**, 286.
311. H. P. Meloche and L. Mehler, *J. Biol. Chem.*, 1973, **248**, 6333.
312. T. Shimojo and G. J. Schroepfer, Jr., *Biochim. Biophys. Acta*, 1976, **431**, 433.
313. P. P. van Veldhoven and G. P. Mannaerts, *Biochem. J.*, 1994, **299**, 597.
314. W. Stoffel, D. LeKim, and G. Sticht, *Hoppe-Seyler's Z. Physiol. Chem.*, 1969, **350**, 1233.
315. P. P. van Veldhoven and G. P. Mannaerts, *J. Biol. Chem.*, 1991, **266**, 12 502.
316. W. Stoffel, E. Bauer, and J. Stahl, *Hoppe-Seyler's Z. Physiol. Chem.*, 1974, **355**, 61.
317. W. Stoffel and K. Bister, *Hoppe-Seyler's Z. Physiol. Chem.*, 1973, **354**, 169.
318. T. Shimojo, T. Akino, Y. Miura, and G. J. Schroepfer, Jr., *J. Biol. Chem.*, 1976, **251**, 4448.
319. W. Stoffel and M. Grol, *Chem. Phys. Lipids*, 1974, **13**, 372.
320. W. Stoffel and G. Assmann, *Hoppe Seyler's Z. Physiol. Chem.*, 1972, **353**, 965.
321. W. Stoffel, *Mol. Cell. Biochem.*, 1973, **1**, 147.
322. T. K. Ghosh, J. Bian, and D. L. Gill, *Science*, 1990, **248**, 1653.
323. H. L. Kornberg and J. G. Morris, *Nature (London)*, 1963, **197**, 456.
324. H. L. Kornberg and J. G. Morris, *Biochim. Biophys. Acta*, 1962, **65**, 537.
325. R. G. Gibbs and J. G. Morris, *Methods Enzymol.*, 1970, **17**, 981.
326. R. G. Gibbs and J. G. Morris, *Biochim. Biophys. Acta*, 1964, **85**, 501.
327. T. Ishiyama, T. Furuta, M. Takai, and Y. Okimoto, *J. Antibiot. (Tokyo)*, 1975, **28**, 821.
328. W. T. Jenkins, *Anal. Biochem.*, 1979, **93**, 134.
329. A.-K. Ohlin, G. Landes, P. Bourdon, C. Oppenheimer, R. Wydro, and J. Stenflo, *J. Biol. Chem.*, 1988, **263**, 19 240.
330. S. A. Fowler, D. Paulson, B. A. Owen, and W. G. Owen, *J. Biol. Chem.*, 1986, **261**, 4371.
331. M. Mokotoff, J. F. Bagaglio, and B. S. Parikh, *J. Med. Chem.*, 1975, **18**, 354.
332. H. Hayashi and H. Kagamiyama, *Biochemistry*, 1995, **34**, 9413.
333. S. Sawada, H. Kumagai, H. Yamada, and R. K. Hill, *J. Am. Chem. Soc.*, 1975, **97**, 4334.
334. H. Yoshida, T. Utagawa, H. Kumagai, and H. Yamada, *Agric. Biol. Chem.*, 1974, **38**, 2065.
335. H. Yoshida, H. Kumagai, H. Yamada, and H. Matsubara, *Biochim. Biophys. Acta*, 1975, **391**, 494.
336. H. Kumagai, S. Sejima, Y. Choi, H. Tanaka, and H. Yamada, *FEBS Lett.*, 1975, **52**, 304.
337. N. Brot, Z. Smit, and H. Weissbach, *Arch. Biochem. Biophys.*, 1965, **112**, 1.
338. G. M. Carman and R. E. Levin, *Appl. Environ. Microbiol.*, 1977, **33**, 192.
339. J. Polak and H. Brzeski, *Biotechnol. Lett.*, 1990, **12**, 805.
340. H. Kumagai, N. Kashima, H. Torii, H. Yamada, H. Enei, and S. Okumura, *Agric. Biol. Chem.*, 1972, **36**, 472.
341. S. Suzuki, T. Hirahara, J. K. Shim, S. Horinouchi, and T. Beppu, *Biosci. Biotechnol. Biochem.*, 1992, **56**, 84.
342. S. S. Duffey, J. R. Aldrich, and M. S. Blum, *Comp. Biochem. Physiol. B*, 1977, **56**, 101.
343. S. S. Duffey and M. S. Blum, *Insect Biochem.*, 1977, **7**, 57.
344. L. Larue and B. Mintz, *Somatic Cell Mol. Genet.*, 1990, **16**, 361.
345. T. V. Demidkina and I. V. Myagkikh, *Biochimie*, 1989, **71**, 565.
346. S. Suzuki, T. Hirahara, S. Horinouchi, and T. Beppu, *Agric. Biol. Chem.*, 1991, **55**, 3059.
347. S.-I. Ikeda and S. Fukui, *Methods Enzymol.*, 1979, **62**, 517.
348. A. A. Antson, B. V. Strokopytov, G. N. Murshudov, M. N. Isupov, E. H. Harutyunyan, T. V. Demidkina, D.-G. Vassylyev, Z. Dauter, H. Terry, and K. S. Wilson, *FEBS Lett.*, 1992, **302**, 256.
349. S. V. Pletnev, M. N. Isupov, Z. Dauter, K. S. Wilson, N. G. Faleev, E. G. Harutyunyan, and T. V. Demidkina, *Biochem. Mol. Biol. Int.*, 1996, **38**, 37.
350. F. Foor, N. Morin, and K. A. Bostian, *Appl. Environ. Microbiol.*, 1993, **59**, 3070.
351. Y. Kurusu, M. Fukushima, K. Kohama, M. Kobayashi, M. Terasawa, H. Kumagai, and H. Yukawa, *Biotechnol. Lett.*, 1991, **13**, 769.
352. H. Suzuki, T. Katayama, K. Yamamoto, and H. Kumagai, *Biosci. Biotech. Biochem.*, 1995, **59**, 2339.
353. D. M. Kiick and R. S. Phillips, *Biochemistry*, 1988, **27**, 7333.
354. H. Chen and R. S. Phillips, *Biochemistry*, 1993, **32**, 11 591.
355. N. G. Faleev, S. B. Ruvinov, T. V. Demidkina, I. V. Myagkikh, M. Y. Gololobov, V. I. Bakhmutov, and V. M. Belikov, *Eur. J. Biochem.*, 1988, **177**, 395.
356. H. Y. Chen, T. V. Demidkina, and R. S. Phillips, *Biochemistry*, 1995, **34**, 12 276.
357. D. Hebel, D. C. Furlano, R. S. Phillips, S. Koushik, C. R. Creveling, and K. L. Kirk, *Bioorg. Med. Chem. Lett.*, 1992, **2**, 41.
358. H. Yamada and H. Kumagai, *Pure Appl. Chem.*, 1978, **50**, 1117.
359. H. Kumagai, H. Yamada, S. Sawada, E. Schleicher, K. Mascaro, and H. G. Floss, *J. Chem. Soc., Chem. Commun.*, 1977, 85.

360. S. Suzuki, T. Hirahara, J.-K. Shim, S. Horinouchi, and T. Beppu, *Biosci. Biotech. Biochem.*, 1992, **56**, 84.
361. C. Fuganti, D. Ghiringhelli, D. Giangrasso, and P. Grasselli, *J. Chem. Soc., Chem. Commun.*, 1974, 726.
362. H. Yamada and H. Kumagai, *Adv. Appl. Microbiol.*, 1975, **19**, 249.
363. F. C. Charalampous, *Methods Enzymol.*, 1962, **5**, 283.
364. F. C. Charalampous and G. C. Mueller, *J. Biol. Chem.*, 1952, **201**, 161.
365. E. J. Wawszkiewicz, *Biochemistry*, 1968, **7**, 683.
366. D. G. Sidenberg and M.-A. Lachance, *Int. J. Syst. Bacteriol.*, 1983, **33**, 822.
367. W. K. Maas and H. J. Vogel, *J. Bacteriol.*, 1953, **65**, 388.
368. W. K. Maas and B. D. Davis, *J. Bacteriol.*, 1950, **60**, 733.
369. E. N. McIntosh, M. Purko, and W. A. Wood, *J. Biol. Chem.*, 1957, **228**, 499.
370. F. C. Charalampous, *J. Biol. Chem.*, 1960, **235**, 1286.
371. W. A. Anderson and B. Magasanik, *J. Biol. Chem.*, 1971, **246**, 5662.
372. H. W. Talbot, Jr. and R. J. Seidler, *Appl. Environ. Microbiol.*, 1979, **38**, 599.
373. B. Magasanik, *J. Am. Chem. Soc.*, 1951, **73**, 5919.
374. B. Magasanik, *Annu. Rev. Microbiol.*, 1994, **48**, 1.
375. P. S. Poole, A. Blyth, C. J. Reid, and K. Walters, *Microbiology*, 1994, **140**, 2787.
376. V. I. Golubev, *Microbiology*, 1989, **58**, 276.
377. K. Yoshida, H. Sano, Y. Miwa, N. Ogasawara, and Y. Fujita, *Microbiology*, 1994, **140**, 2289.
378. E. Nowotny, R. D. Sananez, G. Nattero, C. Yantorno, and M. G. Faillaci, *Hoppe-Seyler's Z. Physiol. Chem.*, 1974, **355**, 716.
379. V. V. Graef, Z. Zubrzycki, and K. Jarra, *Arzneim. Forsch.*, 1984, **34**, 1760.
380. D. Vega, M. DePiante-DePaoli, B. Pacheco-Rupil, and E. Nowotny, *Cell. Mol. Biol.*, 1983, **29**, 245.
381. S. Genti-Raimondi, E. Nowotny, M. Galmarini, and C. Yantorno, *Acta Physiol. Lat. Am.*, 1979, **29**, 107.
382. C. Norsten-Höög, T. Cronholm, S. H. G. Andersson, and J. Sjövall, *Biochem. J.*, 1992, **286**, 141.
383. H. Inano, K. Suzuki, M. Onoda, and K. Wakabayashi, *Carcinogenesis*, 1996, **17**, 355.
384. M. Rodamilans, M. J. Montez-Osaba, J. To-Figueras, F. Rivera-Fillat, M. Torra, P. Perez, and J. Corbella, *Toxicol. Lett.*, 1988, **42**, 285.
385. A. Akane, S. Fukushima, H. Shino, and Y. Fukui, *Alcohol Alcohol.*, 1988, **23**, 203.
386. T. V. Widenius, M. M. Orava, R. K. Vihko, R. H. Ylikahri, and C. J. P. Eriksson, *J. Steroid Biochem.*, 1987, **28**, 185.
387. P. M. D. Foster, L. V. Thomas, M. W. Cook, and D. G. Walters, *Toxicol. Lett.*, 1983, **15**, 265.
388. D. E. Johnston, Y.-B. Chiao, J. S. Gavaler, and D. H. van Thiel, *Biochem. Pharmacol.*, 1981, **30**, 1827.
389. P. A. Myers and L. J. Zatman, *Biochem. J.*, 1971, **121**, 10P.
390. C. A. Boulton and P. J. Large, *FEMS Microbiol. Lett.*, 1979, **5**, 159.
391. T. R. Jarman and P. J. Large, *J. Gen. Microbiol.*, 1972, **73**, 205.
392. C. A. Boulton, M. J. C. Crabbe, and P. J. Large, *Biochem. J.*, 1974, **240**, 253.
393. C. W. Bamforth and P. J. Large, *Biochem. J.*, 1977, **161**, 357.
394. C. G. VanGinkel, J. D. VanDijk, and A. G. M. Kroon, *Appl. Environ. Microbiol.*, 1992, **58**, 3083.
395. C. A. Boulton, G. W. Haywood, and P. J. Large, *J. Gen. Microbiol.*, 1980, **117**, 293.
396. J. C. Craig, F. P. Dwyer, A. N. Glazer, and E. C. Horning, *J. Am. Chem. Soc.*, 1961, **83**, 871.
397. J. C. Craig, N. Y. Mary, N. L. Goldman, and L. Wolf, *J. Am. Chem. Soc.*, 1964, **86**, 3866.
398. J. M. Machinist, W. H. Orme-Johnson, and D. M. Ziegler, *Biochemistry*, 1966, **5**, 2939.
399. P. Hlavica, *CRC Crit. Rev. Biochem.*, 1982, **12**, 39.
400. E. Bamberger and P. Leyden, *Ber. Dtsch. Chem. Ges.*, 1901, **34**, 12.
401. H. Terayama, *Gann*, 1963, **54**, 195.
402. P. F. Coccia and W. W. Westerfeld, *J. Pharmacol. Exp. Ther.*, 1967, **157**, 446.
403. R. N. Pandey, A. P. Armstrong, and P. F. Hollenberg, *Biochem. Pharmacol.*, 1989, **38**, 2181.
404. J. G. Hauge, T. E. King, and V. H. Cheldelin, *J. Biol. Chem.*, 1955, **214**, 11.
405. P. A. Kitos, C. H. Wang, B. A. Mohler, T. E. King, and V. H. Cheldelin, *J. Biol. Chem.*, 1958, **233**, 1295.
406. M. R. R. Rao, *Annu. Rev. Microbiol.*, 1957, **11**, 317.
407. J. A. Fewster, *Biochem. J.*, 1958, **69**, 582.
408. M. Röhr, *Naturwissenschaften*, 1961, **48**, 478.
409. Y. Sekizawa, M. E. Maragoudakis, S. S. Kerwar, M. Flikke, A. Baich, T. E. King, and V. H. Cheldin, *Biochem. Biophys. Res. Commun.*, 1962, **9**, 361.
410. Y. Sekizawa, M. E. Maragoudakis, T. E. King, and V. H. Cheldin, *Biochemistry*, 1966, **5**, 2392.
411. S. Dagley, *Bacteria*, 1978, **6**, 305.
412. V. L. Sparnins and P. J. Chapman, *J. Bacteriol.*, 1976, **127**, 362.
413. Y.-L. Lee and S. Dagley, *J. Bacteriol.*, 1977, **131**, 1016.
414. M. G. Barbour and R. C. Bayly, *J. Bacteriol.*, 1980, **142**, 480.
415. R. A. Cooper and M. A. Skinner, *J. Bacteriol.*, 1980, **143**, 302.
416. V. L. Sparnins, P. J. Chapman, and S. Dagley, *J. Bacteriol.*, 1974, **120**, 159.
417. P.-K. Leung, P. J. Chapman, and S. Dagley, *J. Bacteriol.*, 1974, **120**, 168.
418. S. Dagley, *Methods Enzymol.*, 1982, **90**, 277.
419. H. Sahm, H. Schütte, and M.-R. Kula, *Eur. J. Biochem.*, 1976, **66**, 591.
420. R. Beisswenger and M.-R. Kula, *Appl. Microbiol. Biotechnol.*, 1991, **34**, 604.
421. M. B. Kemp, *Biochem. J.*, 1974, **139**, 129.
422. R. Beisswenger, G. Snatzke, J. Thiem, and M.-R. Kula, *Tetrahedron Lett.*, 1991, **32**, 3159.
423. B. L. Horecker and P. Z. Smyrniotis, *J. Am. Chem. Soc.*, 1953, **75**, 1009.
424. E. Racker, G. de la Haba, and I. G. Leder, *J. Am. Chem. Soc.*, 1953, **75**, 1010.
425. K. G. Morris, M. E. B. Smith, N. J. Turner, M. D. Lilly, R. K. Mitra, and J. M. Woodley, *Tetrahedron: Asymm.*, 1996, **7**, 2185.
426. P. De Wulf, W. Soetaert, D. Schwengers, and E. J. Vandamme, *J. Ind. Microbiol.*, 1996, **17**, 104.
427. M. Kovina, M. Viryasov, L. Baratova, and G. Kochetov, *FEBS Lett.*, 1996, **392**, 293.
428. J. Bolte, C. Demuynck, and H. Samaki, *Tetrahedron Lett.*, 1987, **28**, 5525.

429. C. Demuynck, F. Fisson, I. Bennani-Baiti, H. Samaki, and J.-C. Mani, *Agric. Biol. Chem.*, 1990, **54**, 3073.
430. R.-M. Schmidt, M. Stitt, and U. Sonnewald (BASF A.-G.), *Eur. Pat.* 723017A2 (1996) (*Chem. Abstr.*, 1996, **125**, 161 128n).
431. S. F. Kim, B. Kim, J. Jeng, Y. Soh, C.-I. Bak, J.-W. Huh, and B. J. Song, *J. Biochem. Mol. Biol.*, 1996, **29**, 146.
432. C. M. Sax, C. Salamon, W. T. Kays, J. Guo, F. X. Yu, R. A. Cuthbertson, and J. Piatigorsky, *J. Biol. Chem.*, 1996, **271**, 33 568.
433. G. D. Schellenberg, N. M. Wilson, B. R. Copeland, and C. E. Furlong, *Methods Enzymol.*, 1982, **90**, 223.
434. L. O. Krampitz, *Annu. Rev. Biochem.*, 1969, **38**, 213.
435. R. Rej, *Ann. Clin. Lab. Sci.*, 1977, **7**, 455.
436. B. A. McCool, S. G. Plonk, P. R. Martin, and C. K. Singleton, *J. Biol. Chem.*, 1993, **268**, 1397.
437. G. A. Sprenger, *Biochim. Biophys. Acta*, 1993, **1216**, 307.
438. M. Nikkola, Y. Lindqvist, and G. Schneider, *J. Mol. Biol.*, 1994, **238**, 387.
439. U. Nilsson, L. Meshalkina, Y. Lindqvist, and G. Schneider, *J. Biol. Chem.*, 1997, **272**, 1864.
440. C. French and J. M. Ward, *Ann. N.Y. Acad. Sci.*, 1996, **799**, 11.
441. C. K. Singleton, J. J.-L. Wang, L. Shan, and P. R. Martin, *Biochemistry*, 1996, **35**, 15 865.
442. D. C. Myles, P. J. Andrulis, III, and G. M. Whitesides, *Tetrahedron Lett.*, 1991, **32**, 4835.
443. M. D. Lilly, R. Chauhan, C. French, M. Gyamerah, G. R. Hobbs, A. Humphrey, M. Isupov, J. A. Littlechild, R. K. Mitra, K. G. Morris, M. Rupprecht, N. J. Turner, J. M. Ward, A. J. Willetts, and J. M. Woodley, *Ann. N.Y. Acad. Sci.*, 1996, **782**, 513.
444. J. M. Woodley, R. K. Mitra, and M. D. Lilly, *Ann. N.Y. Acad. Sci.*, 1996, **799**, 434.
445. O. Tsolas, in "Isozymes, Int. Conf." 3rd edn., ed. C. Markert, New York, 1975, p. 767.
446. P. Srere, J. R. Cooper, M. Tabachnick, and E. Racker, *Arch. Biochem. Biophys.*, 1958, **74**, 295.
447. C. G. Choquet, J. C. Richards, G. B. Patel, and G. D. Sprott, *Arch. Microbiol.*, 1994, **161**, 481.
448. E. D. Bergmann, U. Z. Littauer, and B. E. Volcani, *Biochem. Biophys. Acta*, 1954, **13**, 288.
449. D. Couri and E. Racker, *Arch. Biochem. Biophys.*, 1959, **83**, 195.
450. B. L. Horecker and P. Z. Smyrniotis, *J. Am. Chem. Soc.*, 1953, **75**, 2021.
451. Z. Dische and E. Pollaczek, in "Abstracts of the 2nd International Congress on Biochemistry," Paris, 1952, p. 289.
452. C. P. Moehs, P. V. Allen, M. Friedman, and W. R. Belknap, *Plant Mol. Biol.*, 1996, **32**, 447.
453. K. A. Gumaa, F. Novello, and P. McLean, *Biochem. J.*, 1969, **114**, 253.
454. W. Chefurka, *Can. J. Biochem. Physiol.*, 1958, **36**, 83.
455. K. N. Mehrotra, *Comp. Biochem. Physiol.*, 1961, **3**, 184.
456. B. L. Horecker and P. Z. Smyrniotis, *J. Biol. Chem.*, 1955, **212**, 811.
457. E. Racker, in "The Enzymes," eds. P. D. Boyer, H. Lardy, and K. Myrback, Academic Press, New York, 1961, vol. 5, p. 407.
458. R. M. Smillie, N. Rigopoulos, and H. Kelly, *Biochim. Biophys. Acta*, 1962, **56**, 612.
459. K. A. Gumaa and P. McLean, *Biochem. J.*, 1969, **115**, 1009.
460. K. Brand, K. Deckner, and J. Musill, *Hoppe-Seyler's Z. Physiol. Chem.*, 1970, **351**, 213.
461. R. Venkataraman and E. Racker, *J. Biol. Chem.*, 1961, **236**, 1876.
462. B. L. Josephson and D. G. Fraenkel, *J. Bacteriol.*, 1969, **100**, 1289.
463. G. A. Sprenger, U. Schörken, G. Sprenger, and H. Sahm, *J. Bacteriol.*, 1995, **177**, 5930.
464. T. Yura, H. Mori, H. Nagai, T. Nagata, A. Ishihama, N. Fujita, K. Isono, K. Mizobuchi, and A. Nakata, *Nucleic Acids Res.*, 1992, **20**, 3305.
465. K. Banki, D. Halladay, and A. Perl, *J. Biol. Chem.*, 1994, **269**, 2847.
466. B. L. Horecker, S. Pontremoli, C. Ricci, and T. Cheng, *Proc. Natl. Acad. Sci. USA*, 1961, **47**, 1949.
467. B. L. Horecker, *Ital. J. Biochem.*, 1991, **40**, 1.
468. K. Banki and A. Perl, *FEBS Lett.*, 1996, **378**, 161.
469. T. Miosga, I. Schaaf-Gerstenschlager, E. Franken, and F. K. Zimmerman, *Yeast*, 1993, **9**, 1241.
470. C.-H. Wong and G. M. Whitesides, *J. Org. Chem.*, 1983, **48**, 3199.
471. E. J. Toone and G. M. Whitesides, in "Enzymes in Carbohydrate Synthesis, ACS Symposium Series 466," eds. M. D. Bednarksi and E. S. Simon, ACS Press, Washington, DC, 1991.
472. E. Grazi, T. Cheng, and B. L. Horecker, *Biochem. Biophys. Res. Commun.*, 1962, **7**, 250.
473. D. R. Tolan, A. B. Amsden, S. D. Putney, M. S. Urdea, and E. E. Penhoet, *J. Biol. Chem.*, 1984, **259**, 1127.
474. R. K. Scopes, *Biochem. J.*, 1977, **161**, 253.
475. A. J. Morris and D. R. Tolan, *J. Biol. Chem.*, 1993, **268**, 1095.
476. P. Izzo, P. Costanzo, A. Lupo, E. Rippa, G. Paolella, and F. Salvatore, *Eur. J. Biochem.*, 1988, **174**, 569.
477. M.-L. Valentin and J. Bolte, *Tetrahedron Lett.*, 1993, **34**, 8103.
478. I. Kruger and C. Schnarrenberger, *Eur. J. Biochem.*, 1983, **136**, 101.
479. H. A. O. Hill, R. R. Lobb, S. L. Sharp, A. M. Stokes, J. I. Harris, and R. S. Jack, *Biochem. J.*, 1976, **153**, 551.
480. C. De Montigny and J. Sygusch, *Eur. J. Biochem.*, 1996, **241**, 243.
481. J. P. Rasor, *Chim. Oggi*, 1995, **13**, 9.
482. T. Gefflaut, C. Blonski, J. Perie, and M. Wilson, *Prog. Biophys. Mol. Biol.*, 1995, **63**, 301.
483. S. Qamar, K. Marsh, and A. Berry, *Protein Sci.*, 1996, **5**, 154.
484. E. R. E. van den Bergh, S. C. Baker, R. J. Raggers, P. Terpstra, E. C. Woudstra, L. Dijkhuizen, and W. G. Meijer, *J. Bacteriol.*, 1996, **178**, 888.
485. G. C. Look, C. H. Fotsch, and C.-H. Wong, *Acc. Chem. Res.*, 1993, **26**, 182.
486. F. Moris-Varas, X.-H. Qian, and C.-H. Wong, *J. Am. Chem. Soc.*, 1996, **118**, 7647.
487. M.-J. Kim and I. T. Lim, *Synlett*, 1996, 138.
488. R. Duncan and D. G. Drueckhammer, *J. Org. Chem.*, 1996, **61**, 438.
489. R. Duncan and D. G. Drueckhammer, *J. Org. Chem.*, 1995, **60**, 7394.
490. H. J. M. Gijsen and C.-H. Wong, *Tetrahedron Lett.*, 1995, **36**, 7057.
491. O. Eyrisch and W.-D. Fessner, *Angew. Chem., Int. Ed. Engl.*, 1995, **34**, 1639.
492. H. J. M. Gijsen and C.-H. Wong, *J. Am. Chem. Soc.*, 1995, **117**, 7585.
493. T. Conway, *FEMS Microbiol. Rev.*, 1992, **103**, 1.

494. M. L. L. Martins and D. W. Tempest, *J. Gen. Microbiol.*, 1991, **137**, 1391.
495. B. J. Paster and E. Canale-Parola, *Appl. Environ. Microbiol.*, 1985, **50**, 212.
496. G. Martínez-Drets, E. Fabiano, and A. Cardona, *Appl. Environ. Microbiol.*, 1985, **50**, 183.
497. W. Altekar and V. Rangaswamy, *Arch. Microbiol.*, 1992, **158**, 356.
498. J. H. Scott and K. H. Nealson, *J. Bacteriol.*, 1994, **176**, 3408.
499. G. Mendz, S. L. Hazell, and B. P. Burns, *Arch. Biochem. Biophys.*, 1994, **312**, 349.
500. A. R. Lynn and J. R. Sokatch, *J. Bacteriol.*, 1984, **158**, 1161.
501. H. J. Rehm, "Industrielle Mikrobiologie," Springer, Berlin, Germany, 1971.
502. J. M. Pouyssegur and F. R. Stoeber, *Eur. J. Biochem.*, 1972, **30**, 479.
503. R. W. Hammerstedt, H. Möhler, K. A. Decker, and W. A. Wood, *J. Biol. Chem.*, 1971, **246**, 2069.
504. N. Entner and M. Doudoroff, *J. Biol. Chem.*, 1952, **196**, 853.
505. R. Kovachevich and W. A. Wood, *J. Biol. Chem.*, 1955, **213**, 757.
506. S. M. Taha and T. L. Deits, *Biochem. Biophys. Res. Commun.*, 1994, **200**, 459.
507. G. Martínez-Drets, E. Fabiano, and A. Cardona, *Appl. Environ. Microbiol.*, 1985, **50**, 183.
508. G. L. Mendz, S. L. Hazell, and B. P. Burns, *Arch. Biochem. Biophys.*, 1994, **312**, 349.
509. B. J. Paster and E. Canale-Parola, *Appl. Environ. Microbiol.*, 1985, **50**, 212.
510. W. Altekar and V. Rangaswamy, *Arch. Microbiol.*, 1992, **158**, 356.
511. B. R. Knappmann, A. Steigel, and M.-R. Kula, *Biotechnol. Appl. Biochem.*, 1995, **22**, 107.
512. S. T. Allen, G. R. Heintzelman, and E. J. Toone, *J. Org. Chem.*, 1992, **57**, 426.
513. I. M. Mavridis, M. H. Hatada, A. Tulinsky, and L. Lebioda, *J. Mol. Biol.*, 1982, **162**, 419.
514. J. S. Richardson, *Biochem. Biophys. Res. Commun.*, 1979, **90**, 285.
515. W. A. Wood, *Trends Biochem. Soc.*, 1977, 223.
516. N. Suzuki and W. A. Wood, *J. Biol. Chem.*, 1980, **255**, 3427.
517. R. A. Roehl, T. W. Feary, and P. V. Phibbs, Jr., *J. Bacteriol.*, 1983, **156**, 1123.
518. S. M. Cuskey and P. V. Phibbs, Jr., *J. Bacteriol.*, 1985, **162**, 872.
519. S. M. Cuskey, J. A. Wolff, P. V. Phibbs, Jr., and R. H. Olsen, *J. Bacteriol.*, 1985, **162**, 865.
520. P. Faik, H. L. Kornberg, and E. McEvoy-Bowe, *FEBS Lett.*, 1971, **19**, 225.
521. J. E. Fradkin and D. G. Fraenkel, *J. Bacteriol.*, 1971, **108**, 1277.
522. T. Conway, K. C. Yi, S. E. Egan, R. E. Wolf, Jr., and D. L. Rowley, *J. Bacteriol.*, 1991, **173**, 5247.
523. S. E. Egan, R. Fliege, S. Tong, A. Shibata, R. E. Wolf, Jr., and T. Conway, *J. Bacteriol.*, 1992, **174**, 4638.
524. T. Conway, R. Fliege, D. Jones-Kilpatrick, J. Liu, W. O. Barnell, and S. E. Egan, *Mol. Microbiol.*, 1991, **5**, 2901.
525. H. P. Meloche, G. R. Sparks, C. T. Monti, J. W. Waterbor, and T. H. Lademan, *Arch. Biochem. Biophys.*, 1980, **203**, 702.
526. C. T. Monti, J. W. Waterbor, and H. P. Meloche, *J. Biol. Chem.*, 1979, **254**, 5862.
527. H. P. Meloche, C. T. Monti, and R. A. Hogue-Angeletti, *Biochem. Biophys. Res. Commun.*, 1978, **84**, 589.
528. D. P. Henderson, M. A. Shelton, I. C. Cotterill, and E. J. Toone, *J. Org. Chem.*, 1997, **62**, 7910.
529. M. Green and S. S. Cohen, *J. Biol. Chem.*, 1956, **219**, 557.
530. R. G. Eagon, *J. Bacteriol.*, 1961, **82**, 548.
531. D. J. LeBlanc and R. P. Mortlock, *J. Bacteriol.*, 1971, **106**, 90.
532. F. A. Elsinghorst and R. P. Mortlock, *J. Bacteriol.*, 1988, **170**, 5423.
533. E. J. St. Martin and R. P. Mortlock, *J. Mol. Evol.*, 1977, **10**, 111.
534. E. J. St. Martin and R. P. Mortlock, *J. Bacteriol.*, 1980, **141**, 1157.
535. M. A. Ghalambor and E. C. Heath, *J. Biol. Chem.*, 1962, **237**, 2427.
536. M. A. Ghalambor and E. C. Heath, *Methods Enzymol.*, 1966, **9**, 538.
537. D. G. Drueckhammer, J. R. Durrwachter, R. L. Pederson, D. C. Crans, L. Daniels, and C.-H. Wong, *J. Org. Chem.*, 1989, **54**, 70.
538. W.-D. Fessner, G. Sinerius, A. Schneider, M. Dreyer, G. E. Schulz, J. Badia, and J. Aguilar, *Angew. Chem., Int. Ed. Engl.*, 1991, **30**, 555.
539. W.-C. Chou, L. Chen, J.-M. Fang, and C.-H. Wong, *J. Am. Chem. Soc.*, 1994, **116**, 6191.
540. W.-D. Fessner, A. Schneider, O. Eyrisch, G. Sinerius, and J. Badia, *Tetrahedron: Asymm.*, 1993, **4**, 1183.
541. K. K.-C. Liu, T. Kajimoto, L. Chen, Z. Zhong, Y. Ichikawa, and C.-H. Wong, *J. Org. Chem.*, 1991, **56**, 6280.
542. W. J. Lees and G. M. Whitesides, *Bioorg. Chem.*, 1992, **20**, 173.
543. W.-D. Fessner, J. Badia, O. Eyrisch, A. Schneider, and G. Sinerius, *Tetrahedron Lett.*, 1992, **33**, 5231.
544. Y.-F. Wang, Y. Takaoka, and C.-H. Wong, *Angew. Chem., Int. Ed. Engl.*, 1994, **33**, 1242.
545. C.-H. Wong, R. Alajarin, F. Moris-Varas, O. Blanco, and E. García-Junceda, *J. Org. Chem.*, 1995, **60**, 7360.
546. T. Kajimoto, L. Chen, K. K.-C. Liu, and C.-H. Wong, *J. Am. Chem. Soc.*, 1991, **113**, 6678.
547. Y.-M. Chen, Z. Lu, and E. C. Lin, *J. Bacteriol.*, 1989, **171**, 6097.
548. Y.-M. Chen, T. Chakrabarti, and E. C. Lin, *J. Bacteriol.*, 1984, **159**, 725.
549. Y.-M. Chen, Y. Zhu, and E. C. Lin, *Mol. Gen. Genet.*, 1987, **210**, 331.
550. A. Ozaki, E. J. Toone, C. von der Osten, A. J. Sinskey, and G. M. Whitesides, *J. Am. Chem. Soc.*, 1990, **112**, 4970.
551. M. K. Dreyer and G. E. Schulz, *J. Mol. Biol.*, 1996, **259**, 458.
552. W. D. Fessner, A. Schneider, H. Held, G. Sinerius, C. Walter, M. Hixon, and J. V. Schloss, *Angew. Chem.*, 1996, **35**, 2219.
553. A. S. Dahms and R. L. Anderson, *J. Biol. Chem.*, 1972, **247**, 2238.
554. W. A. Wood, in "Enzymes," 3rd edn., ed. P. Boyer, Academic Press, New York, 1972, vol. 5, p. 296.
555. A. S. Dahms and R. L. Anderson, *Biochem. Biophys. Res. Commun.*, 1969, **36**, 809.
556. F. O. Pedrosa and G. T. Zancan, *J. Bacteriol.*, 1974, **119**, 336.
557. M. D. Stowers and A. R. J. Eaglesham, *J. Gen. Microbiol.*, 1983, **129**, 3651.
558. A. J. Kluyver and C. Schnellen, *Enzymologia*, 1937, **4**, 7.
559. T.-H. Chiu and D. S. Feingold, *Biochemistry*, 1969, **8**, 98.
560. G. F. Domagk and R. Heinrich, *Biochem. Z.*, 1965, **341**, 420.
561. D. S. Feingold and P. A. Hoffee, in "Enzymes," 3rd edn., ed. P. Boyer, Academic Press, New York, 1972, vol. 7, p. 303.

562. S. L. Levinson and T. A. Krulwich, *J. Gen. Microbiol.*, 1976, **95**, 277.
563. E. Englesberg, *Arch. Biochem. Biophys.*, 1957, **71**, 179.
564. M. T. Akhy, C. M. Brown, and D. C. Old, *J. Appl. Bacteriol.*, 1984, **56**, 269.
565. S. Al-Zarban, L. Heffernan, J. Nishitani, L. Ransone, and G. Wilcox, *J. Bacteriol.*, 1984, **158**, 603.
566. J. Badía, L. Baldomà, J. Aguilar, and A. Boronat, *FEMS Microbiol. Lett.*, 1989, **53**, 253.
567. P. Moralejo, S. M. Egan, E. Hidalgo, and J. Aguilar, *J. Bacteriol.*, 1993, **175**, 5585.
568. J. Nishitani and G. Wilcox, *Gene*, 1991, **105**, 37.
569. H. Sawada and Y. Takagai, *Biochim. Biophys. Acta*, 1964, **92**, 26.
570. N. B. Schwartz, D. Abram, and D. S. Feingold, *Biochemistry*, 1974, **13**, 1726.
571. N. B. Schwartz and D. S. Feingold, *Bioinorg. Chem.*, 1973, **2**, 75.
572. R. Alajarín, E. García-Junceda, and C.-H. Wong, *J. Org. Chem.*, 1995, **60**, 4294.
573. I. Henderson, K. B. Sharpless, and C.-H. Wong, *J. Am. Chem. Soc.*, 1994, **116**, 558.
574. W.-D. Fessner and G. Sinerius, *Bioorg. Med. Chem.*, 1994, **2**, 639.
575. O. Eyrisch, M. Keller, and W.-D. Fessner, *Tetrahedron Lett.*, 1994, **35**, 9013.
576. W.-D. Fessner and G. Sinerius, *Angew. Chem., Int. Ed. Engl.*, 1994, **33**, 209.
577. W. A. Wood, in "Enzymes," 3rd edn., ed. P. Boyer, Academic Press, New York, 1972, vol. 5, p. 295.
578. T. Y. Wong and X.-T. Yao, *Appl. Environ. Microbiol.*, 1994, **60**, 2065.
579. H. P. Meloche and E. L. O'Connell, *Methods Enzymol.*, 1982, **90**, 263.
580. H. P. Meloche and C. T. Monti, *Biochemistry*, 1975, **14**, 3682.
581. A. Arias and C. Cerveñansky, *J. Bacteriol.*, 1986, **167**, 1092.
582. T. Szumilo, *J. Bacteriol.*, 1981, **148**, 368.
583. N. Kurn, I. Contreras, and L. Shapiro, *J. Bacteriol.*, 1978, **135**, 517.
584. J. Deacon and R. A. Cooper, *FEBS Lett.*, 1977, **77**, 201.
585. C. W. Shuster, *Methods Enzymol.*, 1966, **9**, 524.
586. H. P. Meloche and C. T. Monti, *J. Biol. Chem.*, 1975, **250**, 6875.
587. A. S. Dahms, *Biochem. Biophys. Res. Commun.*, 1974, **60**, 1433.
588. J. Markwell, G. T. Shimamoto, D. L. Bissett, and R. L. Anderson, *Biochem. Biophys. Res. Commun.*, 1976, **71**, 221.
589. D. L. Bissett and R. L. Anderson, *Biochem. Biophys. Res. Commun.*, 1973, **52**, 641.
590. D. L. Bissett and R. L. Anderson, *J. Bacteriol.*, 1974, **119**, 698.
591. D. L. Bissett and R. L. Anderson, *J. Bacteriol.*, 1974, **117**, 318.
592. D. L. Bissett and R. L. Anderson, *J. Biol. Chem.*, 1980, **255**, 8750.
593. R. L. Anderson and D. L. Bissett, *Methods Enzymol.*, 1982, **90**, 228.
594. E. L. Rosey, B. Oskouian, and G. C. Stewart, *J. Bacteriol.*, 1991, **173**, 5992.
595. V. L. Crow and T. D. Thomas, *J. Bacteriol.*, 1982, **151**, 600.
596. G. K. Y. Limsowtin, V. L. Crow, and L. E. Pearce, *FEMS Microbiol. Lett.*, 1986, **33**, 79.
597. R. J. van Rooijen, S. van Schalkwijk, and W. M. de Vos, *J. Biol. Chem.*, 1991, **266**, 7176.
598. E. García-Junceda, G.-J. Shen, T. Sugai, and C.-H. Wong, *Bioorg. Med. Chem.*, 1995, **3**, 945.
599. O. Eyrisch, G. Sinerius, and W.-D. Fessner, *Carbohydr. Res.*, 1993, **238**, 287.
600. W.-D. Fessner and C. Walter, *Angew. Chem., Int. Ed. Engl.*, 1992, **31**, 614.
601. W.-D. Fessner, J. Badía, O. Eyrisch, A. Schneider, and G. Sinerius, *Tetrahedron Lett.*, 1992, **33**, 5231.
602. W.-D. Fessner and O. Eyrisch, *Angew. Chem., Int. Ed. Engl.*, 1992, **31**, 56.

3.14
Starch and Glycogen Biosynthesis

JACK PREISS and MIRTA SIVAK
Michigan State University, East Lansing, MI, USA

3.14.1   INTRODUCTION

The biosynthesis of α-1,4-polyglucans is an important process by which living organisms accumulate energy reserves to be used when carbon skeletons and energy are not readily available from the environment. The main advantage of using polysaccharides as storage reserves is that, because of their high molecular weights and other physical properties, they have little effect on the internal osmotic pressure in the cell. The many linkages of amylopectin and glycogen (α-1,4-chains attached to the main molecule by α-1,6-glucosydic linkages) result in an organization of the molecule which facilitates the action of degradative enzymes, which work either on internal linkages or on the nonreducing ends. The degradative enzymes, whether amylases or phosphorylases, can act simultaneously at the many branches, speeding the conversion of the large polymers into glucose.

Starch is a mixture of amylose and amylopectin. Amylose is a relatively small (about 1000 residues) α-1,4-glucan which is largely linear, although it contains occasional α-1,6 branches. Amylopectin is a much larger molecule with a degree of polymerization of 10^4–10^5 residues and has frequent branch points. The chemical and physical aspects of the structure of the starch granule and its components, amylose and amylopectin, have been discussed in some excellent reviews;[1,2] for the structure of animal glycogen see the review by Manners.[3] Much less is known on the structures of bacterial and fungal glycogen,[4] but the data available indicate that, overall, the bacterial and fungal glycogen structures are quite similar to that of mammalian glycogen.[4] Muscle contains spherical β-particles 1500–4000 nm in diameter, while liver contains α-particles or rosettes which appear as aggregates of β-particles. A β-particle can contain as many as 6×10^4 glucose residues, joined in straight α-1,4 chains joined every 8–10 residues with α-1,6 linkages.

Even though the structures of bacterial and mammal glycogen are relatively similar, the similarity does not extend to the mechanism of their biosynthesis and, most important, the glucose donor for elongation of the primer glucan is different. This leads to differences in the mechanism of regulation: in mammals (but not in bacteria or in plants) covalent modification of an enzyme is involved in the control of the pathway.

The first synthesis *in vitro* of a polyglucan was obtained by Carl and Gerti Cori[5] using phosphorylase and glucose-1-P. For some time afterwards, this was the accepted route for the biosynthesis of polyglucans. However, starting in the 1950s, evidence emerged for a different biosynthetic pathway and, with the discovery of glycogen synthase in 1957,[6] it became clearer that biosynthesis and degradation of glycogen occur by different pathways, and that phosphorylase is, rather than synthetic, a degradative enzyme.

Although we sometimes use the term "animal" in this review, it is worth mentioning that the research discussed here deals mainly with mammal metabolism. Regulation of mammalian glycogen synthesis has been an object of intense study, not only by those seeking to understand glycogen synthesis itself, but also as a model for studying the effect of hormonal action on cellular regulation. A concise description of this important research area of cellular control is provided below, including *in vitro* experiments relevant to processes involved in physiological control.

Some reviews on bacterial glycogen synthesis[7-9] and on starch biosynthesis[10-19] discuss in more detail some of the areas presented in this chapter. For the regulation of mammalian glycogen synthesis see refs.[20-22]

3.14.2   THE ROLE OF GLYCOGEN AND STARCH

3.14.2.1   Glycogen in Bacteria

Glycogen is found in many bacteria, and usually accumulates in environmental conditions that limit growth and also offer excess carbon supply.[7,23-25] Glycogen accumulation has been shown to occur in the stationary phase of the growth cycle as a response to limitations in the supply of nitrogen, sulfur, or phosphate. Glycogen is not required for bacterial growth, and glycogen-deficient mutants grow as well as the wild-type strains. The biological functions of bacterial glycogen have been reviewed;[25] under nonfavorable conditions and when an alternative carbon source is not available, glycogen is probably utilized to preserve cell integrity. Bacteria require energy for maintenance under nongrowing conditions and this is defined as "energy of maintenance," the energy required for processes such as maintenance of motility and intracellular pH, chemotactic response, turnover of proteins and RNA, and osmotic regulation. In media lacking a carbon source, *Escherichia coli* and *Enterobacter aerogenes* containing glycogen do not degrade their RNA and protein components, while the glycogen-deficient bacteria release NH_3 for their nitrogen-containing components.[25] Glycogen-containing *E. aerogenes*, *E. coli*, and *Streptococcus mitis* also survive better than organisms having no glycogen. Another function for glycogen has been suggested in various *Clostridia* species; these organisms accumulate glycogen up to 60% of their dry weight before or during initiation of sporulation and, during spore formation, this glycogen is rapidly degraded.[26] Glycogen-deficient strains are poor spore formers, suggesting that glycogen serves as a source of carbon and energy for spore formation and maturation. Although these studies suggest that glycogen plays a role in bacterial survival, glycogen-rich *Sarcina lutea* cells die faster when starved in phosphate buffer than cells with no polysaccharide.[27]

3.14.2.2   Starch in Plants

Starch is present in almost all green plants and in many types of plant tissues and organs, e.g., leaves, roots, shoots, fruits, grains, and stems. Starch disappears from leaves exposed to low light or left in the dark for a long time (24–48 h), as documented as early as the nineteenth century.[28] Illumination of the leaf in bright light causes the reformation of starch granules in the chloroplast. This can be seen by staining the leaf with iodine[29] or by light or electron microscopy.[30] Carbon fixation during photosynthesis in the light leads to starch formation, and this starch is degraded in the dark. The products of starch degradation are used as an energy source and converted into sucrose, which is transported to other organs of the plant. Biosynthesis and degradation of starch in the leaf are therefore dynamic processes, and the starch content fluctuates during the day.

In fruit, storage organs, or seed, synthesis of starch occurs during the development and maturation of the tissue. Starch degradation in these tissues occurs at the time of sprouting or germination of the seed or tuber, or ripening of the fruit, where it is used as a source of both carbon and energy. Thus, degradative and biosynthetic processes in the storage of tissues may be temporally separated, although it is possible that some turnover of the starch molecule occurs during each phase of starch metabolism.

3.14.2.3   Glycogen in Animals

Glycogen is a major energy source in animal cells. In humans, for example, glycogen represents about 1% wet weight of the skeletal muscle and 5% of the liver; e.g., for a 70 kg man, glycogen reserves would total some 90 g in muscle and 350 g in the liver. The storage of glucose as glycogen is important for mammalian homeostasis, and the glycogen reserves in skeletal muscle and the liver have specific functions.

Hepatic glycogen is accumulated when excess glucose is available in the diet, and it is used to maintain a steady level of blood glucose; liver glycogen rarely supplies energy to the liver itself. A 70 kg man uses 180 g of glucose per day for those tissues that can only utilize carbohydrate as a source of energy, and about one-half of this is derived from hepatic glycogen. Therefore, hepatic glycogen synthesis and mobilization is dictated by blood glucose levels and is controlled by gluco-regulatory hormones, primarily glucagon, insulin, and glucocorticoids. In all other mammalian extrahepatic tissues, in contrast, glycogen stores are utilized for specific functions of the tissue.

Although the glycogen content of skeletal muscle is large, compared with that of liver, it is not directly available as a source of blood glucose. During exercise, lactate is formed from skeletal muscle by glycogenolysis and glycolysis. At rest, it is converted to glucose by the liver and kidney through gluconeogenesis, serving as a source of about 10–20% of the total blood glucose. Thus, muscle glycogen primarily serves as an energy source, broken down to provide energy for muscle contraction. Replenishment occurs when the diet is such that there is a concomitant increase in high blood glucose (hyperglycemia) and insulin.

3.14.3 SYNTHESIS OF BACTERIAL GLYCOGEN AND STARCH

After the discovery of UDP-Glc and other sugar nucleotides by Leloir and his collaborators, it was soon shown that particulate cell fractions from many sources, both prokaryotic and eukaryotic, could carry out glycosyl transfers from sugar nucleotides to suitable acceptors. Sugar nucleotides could be formed from sugar phosphates and nucleoside triphosphates, and were essential tools for interconversion and anabolism. The synthesis of polysaccharides may be divided into several steps: (a) the synthesis of the donor molecule from sugar 1-phosphates, (b) transfer of the glucosyl unit from the sugar nucleotide to the glucan primer, and (c) rearrangement of the polysaccharide.

3.14.3.1 Glycogen Synthesis in Bacteria

By 1964, it was clear that the glucose donor for glycogen synthesis in mammals was UDP-Glc, and for synthesis of starch in plants, ADP-Glc.[31] Sigal *et al.*[32] showed that several mutants of *E. coli* deficient in UDP-glucose pyrophosphorylase accumulated normal amounts of glycogen during growth in limiting nitrogen media. Thus, UDP-glucose could not be the glucosyl donor for glycogen synthesis. The synthesis of ADP-glucose (Equation (1)) is catalyzed by ADP-glucose (synthetase) pyrophosphorylase (2.7.7.27; ATP:α-D-glucose-1-phosphate adenylyltransferase). In 1964, it was also reported that extracts of several bacteria contained large activities of an ADP-glucose: α-1,4-D-glucan-4-α-glucosyl transferase, also known as the bacterial glycogen synthase (Equation (2)).[33,34] These same bacterial extracts also contained ADP-glucose pyrophosphorylase (Equation (1)), as previously found in plants.

$$ATP + \alpha\text{-glucose-1-P} \rightleftarrows ADP\text{-glucose} + PPi \tag{1}$$

$$ADP\text{-glucose} + glucan \rightarrow \alpha\text{-1,4-glucosyl-glucan} + ADP \tag{2}$$

$$\text{elongated } \alpha\text{-1,4-oligosaccharide chain-glucan} \rightarrow \alpha\text{-1,4-}\alpha\text{-1,6-branched-glucan} \tag{3}$$

About 10% of the linkages in bacterial glycogen are α-1,6. The formation of these linkages is catalyzed by an α-1,4-glucan branching enzyme [EC 2.4.1.18; 1,4-α-D-glucan 6-α-(1,4-α-glucano)-transferase] in Equation (3). Branching enzyme activity has been detected in *E. coli*,[35,36] *Arthrobacter globiformis*,[37] *Salmonella typhimurium*,[38] and *Streptococcus mitis*.[39] The branching enzyme genes from *E. coli*,[40] *Streptomyces aureofaciens*,[41] *Bacillus stearothermophilus*,[42,43] *Bacillus caldolyticus*,[44] and cyanobacteria[45,46] have been cloned.

There are other pathways leading to the formation of α-glucans in bacteria. For example, in certain bacteria, a glycogen-like glucan can be synthesized either directly from sucrose (Equation (4)) or from maltose (Equation (5)) or from glucose-1-phosphate via the phosphorylase reaction (reaction 6).[47]

$$\text{sucrose} + (\text{1,4-}\alpha\text{-D-glucosyl})_n \rightleftharpoons \text{D-fructose} + (\text{1,4-}\alpha\text{-D-glucosyl})_{n+1} \tag{4}$$

$$\text{maltose} + (\text{1,4-}\alpha\text{-D-glucosyl})_n \rightleftharpoons \text{D-glucose} + (\text{1,4-}\alpha\text{-D-glucosyl})_{n+1} \tag{5}$$

Amylosucrase, the enzyme catalyzing Equation (4), is found in *Neisseria* strains that, when grown on sucrose, accumulate large amounts of a glycogen-type polysaccharide.[48,49] Amylosucrase, however, is found only in a few bacterial species and is active only when there is sucrose in the media. *Neisseria* can metabolize exogenous sucrose but does not synthesize sucrose. Therefore, the observed accumulation of glycogen in *Neisseria* and in other microorganisms grown on carbon sources other than sucrose is not due to amylosucrase. Similarly, amylomaltase, the enzyme catalyzing Equation (5), is induced along with a number of other enzymes when several strains of *E. coli*, *Streptococcus mutans*, *Aerobacter aerogenes*,[50,51] *Streptococcus mitis*,[52] *Diplococcus pneumoniae*,[53] and *Pseudomonas stutzeri*[54] are grown on maltose or maltodextrins. The synthesis of amylomaltase, however, is repressed by glucose[55] and its activity therefore cannot account for the synthesis of glycogen in organisms grown on glucose as a carbon source.

Maltodextrin phosphorylase and glycogen phosphorylase (Equation (6)) occur in many bacteria and can catalyze the phosphorolysis and synthesis of α-4-glucosidic linkages present in α-1,4-glucans. But maltodextrin phosphorylase is induced only in the presence of maltodextrins, and for the microorganisms studied, glycogen phosphorylase activity is insufficient to account for their rate of glycogen accumulation.[56–58] Also, *E. coli* mutants deficient in maltodextrin phosphorylase accumulate maltodextrins, suggesting that the phosphorylase is involved in degradation (phosphorolysis), and not synthesis, of α-1,4-glucans.[59]

$$\alpha\text{-1,4-(glucosyl)}_n + \text{Pi} \rightleftharpoons \alpha\text{-glucose-1-P} + \alpha\text{-1,4-(glucosyl)}_{n-1} \tag{6}$$

Thus, all bacteria accumulating glycogen do so using the ADP-glucose pathway. Glycogen-deficient or glycogen-excess mutants of *E. coli* and *Salmonella typhimurium* (reviewed in references [7–9,24]) have been isolated and it has been shown that they are affected either in glycogen synthase activity or in ADP-glucose pyrophosphorylase activity or in both. Lists of bacteria containing glycogen and/or the glycogen biosynthetic enzymes have been compiled in past reviews;[25,60] for a report on the characterization of the ADP-Glc PPase of *Thermus caldophilus* Gk-24, see reference [61]. Glycogen accumulation is not restricted to any class of bacteria, being present in gram-negative or gram-positive types and even in archaebacteria.

3.14.3.2 Starch Synthesis in Plants and Algae

The reactions of starch synthesis, Equations (1)–(3), are essentially similar to those of glycogen synthesis in bacteria but the final product is different.

$$\text{ATP} + \alpha\text{-glucose-1-P} \rightleftharpoons \text{ADP-glucose} + \text{PPi} \tag{1}$$

$$\text{ADP-glucose} + \alpha\text{-1,4-glucan} \rightarrow \alpha\text{-1,4-glucosyl-}\alpha\text{-1,4-glucan} + \text{ADP} \tag{7}$$

$$\text{elongated } \alpha\text{-1,4-oligosaccharide chain} \rightarrow \alpha\text{-1,4-}\alpha\text{-1,6-branched-glucan} \tag{8}$$

Equation (7) is catalyzed by starch synthase (EC 2.4.1.21; ADP-glucose; 1,4-α-D-glucan 4-α-glucosyltransferase), the same reaction as the one catalyzed by the bacterial glycogen synthase (Equation (2)). Here we give it a different reaction number to stress that the final products, glycogen and starch, are different in structure. It has also been shown that there are isozymic forms of plant starch synthases (cited in references [11,12,62–66]) and branching enzymes (cited in references [11,12,62,67–73]). The starch synthase isozymes are different gene products and they seem to have different roles in the synthesis of the two polymers of starch, amylose and amylopectin. For example, in many different plants as well as in *Chlamydomonas reinhardtii*,[74–80] a granule-bound starch synthase is involved in the synthesis of amylose. Mutants defective in this enzyme, known as *waxy* mutants, make starch granules having only amylopectin.

Reaction (7) was first reported by Leloir *et al.*[81] with UDP-glucose as the glycosyl donor, but it was later shown that ADP-glucose was more efficient in terms of maximal velocity and K_m value.[31]

Leaf starch synthases and the soluble starch synthases of reserve tissues are specific for ADP-glucose. In contrast, the starch synthases bound to the starch granule in reserve tissues do have some activity with UDP-glucose, although much lower than that seen with ADP-glucose.

Equation (8) is catalyzed by branching enzyme, again a similar reaction to the one in bacteria (Equation (3)). However, the branch chains in amylopectin are longer (about 20–24 glucose units) and there is less branching in amylopectin ($\sim$5% of the glucosidic linkages are α-1,6) than in glycogen (10–13 glucose units long and 10% of linkages are α-1,6). Thus, the starch branching enzymes are likely to have different specificities with respect to chain transfer than those that branch glycogen, or the interaction of the starch branching enzymes with the starch synthases may be different from the interaction between the branching enzyme and glycogen synthase in bacteria.

At least one other enzyme, a debranching enzyme, is involved in synthesis of the starch granule and its polysaccharide components amylose and amylopectin.[82-84] This will be discussed in Section 3.14.6. Also to be discussed in Section 3.14.6 is the evidence suggesting that initiation of starch synthesis in plants may also involve a protein acceptor similar to glycogenin.

3.14.3.3   Glycogen Synthesis in Mammals

The glycogen synthetic pathway in mammals is shown below in Equations (9)–(12).

$$\text{UTP} + \alpha\text{-glucose-1-P} \rightleftarrows \text{UDP-glucose} + \text{PPi} \tag{9}$$

$$\text{UDP-glucose} + \text{apo-glycogenin} \rightarrow (\text{glucosyl})_{10\text{-}11}\text{-glycogenin} \tag{10}$$

$$\text{UDP-glucose} + \text{glucan} \rightarrow \alpha\text{-1,4-glucosyl-glucan} + \text{UDP} \tag{11}$$

$$\text{elongated } \alpha\text{-1,4-oligosaccharide chain-glucan} \rightarrow \alpha\text{-1,4-}\alpha\text{-1,6-branched-glucan} \tag{12}$$

The synthesis of the glucosyl-nucleotide donor, UDP-glucose (Equation (9)) is catalyzed by the enzyme UDP-glucose (synthetase) pyrophosphorylase (EC 2.7.7.9; UTP: α-D-glucose-1-phosphate uridylyltransferase). The equilibrium of the reaction towards UDP-glucose synthesis is close to 1 but the other product of the reaction, pyrophosphate, is immediately cleaved by the inorganic pyrophosphatase. Thus, formation of UDP-glucose is essentially irreversible in mammalian cells. In Equation (10), glucose is transferred from the sugar nucleotide to a protein, glycogenin, which is currently considered to participate in the initiation of glycogen synthesis.[85-87] The rabbit muscle glycogenin is a 37 kDa self-glucosylating protein of 332 amino acids,[88] that glycosylates the hydroxyl group of its tyrosine residue 194;[89-91] the reaction requires Mn^{2+} as a cofactor. Further self-glucosylation occurs and the initial glucose residue is glucosylated up to a chain containing 7 to 11 glucosyl units linked by α-1,4 glucosidic bonds. This glucosylated protein can then serve as a primer (glycogen and lower MW maltodextrins can also serve as primers) for Equation (11) which is catalyzed by glycogen synthase (EC 2.4.1.11; UDP-glucose-glycogen 4-α-glucosyl transferase). Chain elongation of the glucosylated glycogenin continues and branching enzyme (EC 2.4.1.18; 1,4-α-D-glucan 6-α-(1,4-α-glucano)-transferase) catalyzes the formation of α-1,6 branched points. The actual details of events between formation of the oligosaccharide-glycogenin primer and formation of the glycogen of a molecular size of 10^7 kDa are not known. In addition to the four reactions known for the initiation and synthesis of mammalian glycogen, other reactions converting glycogenin to glycogen have been postulated. A lower molecular weight (about 400 kDa) form of glycogen, proglycogen, that is insoluble in trichloroacetic acid (in contrast to glycogen) has been isolated from astrocytes of newborn rat brain and is believed to be an intermediate fraction between glycogenin and glycogen.[92,93] These authors postulated that there is a form of glycogen synthase named proglycogen synthase, which would glucosylate the already glucosylated glycogenin to synthesize the proglycogen, and another form of glycogen synthase which is the enzyme forming the 10^7 kDa macroglycogen from proglycogen,[92,93] but these hypothetical forms of glycogen and synthases remain to be characterized.

3.14.4 PROPERTIES OF THE BACTERIAL AND PLANT ENZYMES INVOLVED IN THE SYNTHESIS OF GLYCOGEN AND STARCH

3.14.4.1 ADP-glucose Pyrophosphorylases: Structure and Properties

A fundamental feature of the ADP-Glc PPases is that they are allosteric enzymes and their regulation, via activation by glycolytic intermediates and inhibition by AMP, ADP, or inorganic phosphate, is important for controlling synthesis of both bacterial glycogen and plant starch.

3.14.4.1.1 *Molecular weight and subunit structure*

The molecular weights of the homogeneous enzymes from *E. coli* B,[94,95] *R. rubrum*[96] and spinach leaf[97] have been estimated using sedimentation equilibrium ultracentrifugation, and the molecular weights of purified and partially purified enzymes from *Aeromnas hydrophila, Rhodospirillum molischianum, Rhodospirillum tenue, Rhodobacter spheroides, Rhodobacter gelatinosa, Rhodobacter viridis, Rhodospirillum fulvum, Rhodobacter acidophila, Rhodobacter globiformis, Salmonella typhimurium,* and *Serratia marcescens* have been estimated using ultracentrifugation on sucrose gradients.[98] A study of the subunit molecular weights of four bacterial enzymes showed that the native enzymes from bacterial sources are tetramers of similar subunits with subunits of molecular mass 45–51 kDa. In the case of *S. marcescens* enzyme, two molecular weight species are seen, one 96 kDa, the other 186 kDa.[99] If the subunit molecular weight of the *S. marcescens* enzyme is similar to the *E. coli* enzyme, it probably exists as homotetrameric and homodimeric forms in equilibrium

The ADP-Glc PPase gene from *E. coli* has been isolated, expressed,[100] and sequenced.[40] The calculated molecular weight of the protein as deduced from the nucleotide sequence is 48 762, which is in excellent agreement with the determined approximate molecular weight of 50 000.[94,95]

The plant enzyme studied in most detail with respect to kinetic properties and structure is that isolated from spinach leaf.[101–103] The kinetic properties of the ADP-glucose pyrophosphorylases from other leaf extracts, e.g., barley, butter lettuce, kidney bean, maize, peanut, rice, sorghum, sugar beet, tobacco, and tomato, are similar to those of the spinach leaf enzyme.[104]

The spinach leaf ADP-glucose pyrophosphorylase has been purified to homogeneity by preparative disk gel electrophoresis[97] and by hydrophobic chromatography.[103] The spinach leaf enzyme has a molecular mass of 206 000 and is composed of two different subunits, of molecular masses of 51 and 54 kDa.[104–107] These subunits can be distinguished not only by differences in their molecular mass but also with respect to amino acid composition, amino-terminal sequences, peptide patterns of the tryptic digests on HPLC, and antigenic properties. The two subunits are, therefore, quite different and most probably the products of two genes. In contrast, and as discussed above, bacterial ADP-Glc PPases including the cyanobacterial enzymes are homotetrameric, i.e., composed of four identical subunits of 50–55 kDa in mass depending on the species.[24]

Other plant ADP-Glc PPases have been studied in detail and they also have been shown to be composed of two dissimilar subunits. The maize endosperm ADP-Glc PPase, which has a molecular mass of 230 kDa, reacts with the antibody prepared against the native spinach leaf enzyme in immunoblot experiments.[108] The enzyme is composed of subunits of 55 and 60 kDa, which would correspond, respectively, to the spinach leaf 51 and 54 kDa.[108]

The studies of the maize endosperm mutants *shrunken* 2 (*sh* 2) and *brittle* 2 (*bt* 2) which are deficient in ADP-Glc PPase activity (reviewed by Preiss[11,12]) are also relevant. In immunoblotting experiments and using antibodies raised against the native (holoenzyme) and against each subunit of the spinach leaf enzyme, it was found that the mutant *bt* 2 endosperm lacks the 55 kDa subunit and the mutant *sh* 2 endosperm lacks the 60 kDa subunit. These results[108] strongly suggest that the maize endosperm ADP-Glc PPase is composed of two immunologically distinctive subunits and that the *Sh* 2 and *Bt* 2 mutations cause reduction in ADP-Glc PPase activity through the lack of one of the subunits; the *Sh* 2 gene would be the structural gene for the 60 kDa protein while the *Bt* 2 gene would be the structural gene for the 55 kDa protein. The isolation of an ADP-Glc PPase cDNA clone from a maize endosperm library[109] which hybridized with the small subunit cDNA clone from rice[110] is consistent with this hypothesis. This maize ADP-Glc PPase cDNA clone was found to hybridize to a transcript present in maize endosperm but absent in *bt* 2 endosperm. Thus, *bt* 2 would be a mutation of the structural gene of the 55 kDa subunit of the ADP-Glc PPase.

In short, data indicate that both seed and leaf ADP-glucose pyrophosphorylases are heterotetramers composed of two different subunits, and that, on the basis of immunoreactivity and sequence data,[111] there is corresponding homology between the subunits in the leaf enzyme and the subunits of reserve tissue enzyme.

3.14.4.1.2   *Reaction mechanism*

The substrate saturation curves of the *Rhodospirillum rubrum* ADP-glucose pyrophosphorylase are hyperbolic at low temperatures and in the presence of activator. The reaction mechanism was investigated using kinetic studies;[112] and intersecting reciprocal plots were obtained, indicating a sequential kinetic mechanism. Product initiation patterns eliminated all known sequential mechanisms except the ordered BiBi or Theorell–Chance mechanisms.[113] Small intercept effects suggested the existence of significant concentrations of central transitory complexes. Kinetic constants obtained in the study also favored the ordered BiBi mechanism. Moreover, studies using ATP-[^{32}P]pyrophosphate isotope exchange at equilibrium support a sequential-ordered mechanism, also indicating that ATP is the first substrate to bind and that ADP-glucose is the last product to dissociate from the enzyme.[112]

Binding of substrates and effectors of the *E. coli* B ADP-glucose pyrophosphorylase enzyme was studied by Haugen and Preiss.[114] Equilibrium dialysis showed that in the presence of 5 mM MgCl$_2$ and 1.5 mM fructose-1,6-bisphosphate, glucose-1-phosphate does not bind to the enzyme. However, ATP does bind, suggesting that the reaction mechanism of the *E. coli* B enzyme is similar to the *R. rubrum* enzyme in that there is ordered binding, with MgATP binding first and glucose-1-phosphate binding second. Chromium adenosine triphosphate (CrATP), a potent inhibitor of many enzymes which utilizes MgATP as a substrate,[115,116] is a potent competitive inhibitor of the *E. coli* ADP-glucose pyrophosphorylase.[114] When this inactive analogue is present in the glucose-1-phosphate binding equilibrium dialysis experiments, one mole of glucose-1-phosphate binds per mole of ADP-glucose pyrophosphorylase subunit.

Only two moles of MgATP or CrATP bind to the tetrameric protein in the absence of glucose-1-P. Thus, MgATP sites appear to exhibit half-site reactivity.[116,117] This is in contrast to equilibrium dialysis experiments using ADP-glucose, in which 4 moles of this substrate were found to bind to 1 mole of the tetrameric protein.[114] But when glucose-1-phosphate is present, 4 moles of CrATP bind to the tetrameric protein. Thus, it appears that in the pyrophosphorylase reaction mechanism in the synthesis direction, 2 moles of MgATP initially bind to the tetrameric protein; this permits the binding of glucose-1-phosphate to the four binding sites on the tetrameric protein. Further binding of the next 2 moles of MgATP may then follow, with concomitant catalysis occurring. Haugen and Preiss[114] proposed a mechanism for enzyme catalysis that explains some of the kinetic and binding properties in terms of an asymmetry in the distribution of the conformational states of the four identical subunits.[114]

3.14.4.1.3   *Location of substrate and effector sites*

Chemical modification can be used to obtain information on the catalytic mechanism and on the catalytic sites of the enzyme of interest. Chemical modification studies on the ADP-Glc PPases have involved the use of the following affinity labels, that can be radioactive, depending on the nature of the experiment:

   (i) pyridoxal-5-phosphate (PLP), an analogue of 3-PGA and of sugar phosphates;

   (ii) the photoaffinity substrate analogues, 8-azido-ATP and 8-azido-ADP-Glc (when 8-azido compounds are irradiated with UV light (257 nm), a nitrene radical is formed which can then react with electron-rich residues, inactivating the enzyme[118]);

   (iii) phenylglyoxal, for the identification of arginine residues, perhaps in binding of anionic substrates.

These kinds of studies have provided information on the catalytic and regulatory sites of the spinach ADP-Glc PPase and on the role of the large and small subunits. In ADP-Glc PPase from *E. coli*, Lys residue 195 has been identified as the binding site for the phosphate of glucose-1-P[119] and tyrosine residue 114 has been identified as involved in the binding of the adenosine portion of the other substrate, ATP.[120] Chemical modification and site-directed mutagenesis studies of the *E. coli* ADP-Glc PPase have provided evidence for the location of the activator binding site,[121,122] the inhibitor binding site,[123,124] and the substrate binding sites.[119,122,123] In these experiments pyridoxal-P was used as an analogue for the activator, fructose 1,6-bisP, and for the substrate, glucose-1-P. The photoaffinity reagent 8-azido-ATP ($8N_3ATP$), an ATP analogue was shown to be a substrate[120] and 8-azido AMP ($8N_3AMP$), an effective inhibitor analogue.[123,124] Since the amino acid sequence of the *E. coli* ADP-Glc PPase gene, *glgC*, is known, the identification of the amino acid sequence around the modified residue helped locate the modified residue within the primary structure of the enzyme. The amino-acid residue involved in binding the activator was Lys-39 and the amino-acid involved in binding the adenine portion of the substrates (ADP-Glc and ATP) was Tyr-114. Tyr-114 was also the major binding site for the adenine ring of the inhibitor, AMP. Because Lys-195 was protected from reductive phosphopyridoxylation when the substrate, ADP-Glc, was present, it was proposed that this Lys is also a part of the substrate binding site.

For the enzyme from *S. typhimurium*, the amino acid sequence has been deduced from the nucleotide sequence of the cloned gene.[125] Comparison of the nucleotide sequences of the *E. coli* and *S. typhimurium GlgC* genes shows a 80% identity, and 90% identity for the amino acid sequence. Most of the changes are conservative, and the amino acids known to be involved in the binding of substrates and allosteric effectors and those involved in maintaining allosteric function in the *E. coli* enzyme are all conserved in *S. typhimurium*.

The *E. coli* enzyme has been crystallized[126] but the crystals were of poor diffraction quality and were further damaged by exposure to X rays. For these reasons, it has not been possible to propose a three-dimensional structure of the enzyme. Crystallization, however, is an essential step in the determination of the structure of an enzyme and in the understanding of its mechanism of action and efforts in this direction continue in the authors' laboratory.

Because the plant native ADP-Glc PPases are tetrameric and composed of two different subunits, it is of interest to know why the two subunits are required for optimal catalytic activity. The enzyme must contain ligand-binding sites for the activator, 3PGA, and inhibitor, Pi, as well as catalytic sites for the two substrates, ATP and glucose-1-P, and it is possible that these sites could be located on different subunits. The overall amino acid sequence identity of the *E. coli* enzyme when aligned with the plant and cyanobacterial ADP-Glc PPases ranges from 30 to 33%.[111] In contrast, there is greater sequence identity when the *E. coli* ATP and glucose-1-P binding sites (Table 1) are compared with the corresponding sequences of the plant and cyanobacterial enzymes, suggesting that those sequences are still important in the plant enzyme, probably having the same function. Indeed, in a recent preliminary experiment with the potato tuber ADP-Glc PPase expressed in *E. coli*,[127] site-directed mutagenesis on the lysine residue K198 of the 50 kd subunit (equivalent to the *E. coli* ADP-Glc PPase K195) to a glutamate residue, increased the K_m for glucose-1-P from 57 μM to over 31 mM without any change in the K_m or K_a for the other substrates, Mg^{+2}, ATP or for activator, 3PGA. These results indicate an involvement of Lys residue 198 of the plant ADP-Glc PPase in the binding of glucose-1-P. In the case of the proposed ATP binding site, instead of tyrosine there is a phenylalanine residue in the corresponding sequences of the plant and cyanobacterial enzymes. Future site-directed mutagenesis and chemical modification studies should indicate if the WFQGTA-DAV region of the plant enzyme is a portion of the ATP binding region.

The binding site for pyridoxal phosphate in the small subunit (Table 2) was isolated, revealing a lysine residue close to the C-terminus which may be important for 3PGA activation.[130] When PLP is covalently bound, the plant ADP-Glc PPase is active and no longer requires 3PGA for activity. Moreover, the covalent binding of PLP is prevented by the allosteric effectors, 3PGA and Pi. These results show that the modified enzyme no longer requires an activator for maximal activity and that the covalent modification is prevented by the presence of the allosteric effectors and strongly indicate that the activator analogue, PLP, is binding at the activator site. Three lysine residues of the spinach leaf large subunit are also involved or are close to the binding site of pyridoxal-P and, presumably, of the activator, 3PGA.[131] The chemical modification of these Lys residues by pyridoxal-P was prevented by the presence of 3PGA during the reductive phosphopyridoxylation process and, in the case of the Lys residue of site 1 of the small subunit and site 2 of the large subunit, Pi also prevented them from being modified by reductive pyrixodylation.

Similar results were obtained with the *Anabaena* ADP-Glc PPase.[132] Chemical modification of the enzyme with PLP resulted in the cyanobacterial enzyme independent of activator for maximal activity; modification with PLP was prevented by 3PGA and Pi. The modified Lys residue was identified as Lys419 and the sequence adjacent to that residue is very similar to that observed for site 1 sequences of the higher plants (Table 2). Site-directed mutagenesis of Lys419 to either Arg, Ala, Gln, or Glu produced mutant enzymes having 25- to 150-fold lower apparent affinities for the activator than that of wild-type enzyme. No other kinetic constants such as K_m for substrates and the inhibitor, Pi, were affected, nor was the catalytic efficiency of the enzyme affected. These mutant

Table 1 Conservation in plant ADP-Glc PPases of the glucose-1-phosphate[119] of the ATP binding[7] sites present in *E. coli* ADP-Glc PPase.

Source	Glucose-1-P site	ATP site
Prokaryotes	195	114
E. coli	IIEFVEKP-AN	WYRGTADAV
S. typhimurium	**D*****_**	********
Agrobacterium tumefaciens	**D*I***-*D	**E******
Anabaena	V*D*S***KGE	*FQ******
Synechocystis	*TD*S***QGE	*FQ******
Plant small subunit		
A. thaliana (small subunit)	****A***KGE	*FQ******
Maize endosperm 54 kDa	****A***KGE	*FQ******
Potato tuber 50 kDa	****A***QGE	*FQ******
Rice seed (small subunit)	*V**A***KGE	*FQ******
Spinach leaf 51 kDa	****A***KGE	*FQ******
Tomato fruit	****A***QGQ	*FQ******
Wheat endosperm (small subunit)	****A***KGE	*FQ******
Vicia faba	****A***KGE	*FQ******
Plant large subunit		
Maize endosperm 60 kDa	VLQ*F***KGA	*FQ****SI
Potato tuber 51 kDa	VVQ*A***KGF	*FQ******
Spinach leaf 54 kDa	VLS*S***KGD	*FQ******
Wheat endosperm (large subunit)	VVQ*S*Q*KGD	*FR*****W

References to these sequences for the plant ADP-Glc PPases are in ref.[111] The sequences for the *Anabaena* enzyme is in ref.[128], for the *Synechocystis* enzyme in ref.[128] and for the wheat endosperm small subunit, in ref.[129] The numbers 195 and 114 correspond to Lys195 and Tyr114 of the *E. coli* enzyme and * signifies the same amino acid as in the *E. coli* enzyme.

Table 2 Plant and cyanobacterial ADP-glucose pyrophosphorylase activator binding sites.

	Activator site 1	Activator site 2
Cyanobacteria	419	382
Anabaena	SGIVVVLKNAVITDGTII	QRRAIIDKNAR
Synechocystis	NGIVVVIKNVTIADGTVI	IRRAIIDKNAR
Higher plants	Small subunit activator site 1	Large subunit activator site 2
Barley endosperm	SGIVTVIKDALLPSGTVI	ISNCIIDMNAR
Beta vulgaris tap root	SGIVTIIKDAMIPSGTVI	IKNCIIDINAR
Maize endosperm	GGIVTVIKDALLPSGTVI	IRNCIIDMNAR
Potato tuber	SGIVTVIKDALIPSGIII	IRKCIIDKNAK
Rice seed	SGIVTVIKDALLLAEQLY	
Spinach leaf	SGIVTVIKDALIPSGTVI	IKDAIIDKNAR
Tomato fruit	SGIVTVIKDALIPSGIVI	
Wheat leaf		IKRAIIDKNAR
Wheat seed	SGIVTVIKDALLPSGTVI	IQNCIIDKNAR

The sequences are listed in one letter code and were taken from ref.[111] and from references indicated in the text. The Lys residues that are in outline are those covalently modified by pyridoxal-P and the chemical modification prevented by 3PGA and Pi. In the case of the potato tuber enzyme the Lys residue was identified via site-directed mutagenesis experiments. The numbers 419 and 382 correspond to the Lys residues in the *Anabaena* ADP-Glc PPase subunit. Site 1 is present in the small subunit of the plant ADP-Glc PPase, while site 2 is present in the large subunit.

enzymes, however, were still activated to a great extent at higher concentrations of 3PGA, suggesting the presence of an additional binding site for the activator. The Lys419Arg mutant could be chemically modified with the activator analogue, PLP, causing a dramatic alteration in the allosteric properties of the enzyme which could be prevented by the presence of 3PGA or Pi during the chemical modification process. Lys382 was the residue modified and for this reason it was concluded that it is the additional site involved in the binding of the activator. As shown in Table 2, the sequence around Lys382 in the *Anabaena* enzyme is very similar to that seen for the higher plants' site 2 which is situated on the large subunit.

Site 1 corresponds to the lysyl residue near the C-terminus, Lys440, that is phosphopyridoxylated in the spinach leaf small subunit,[130] and corresponds to Lys441 in the potato tuber ADP-Glc PPase small subunit and to Lys468 in the rice seed small subunit. Site 2 is also situated close to the C-terminus, equivalent to Lys382 in the *Anabaena* ADP-Glc PPase and Lys404 of the potato tuber large subunit. Table 2 also shows that the amino acid sequence of the spinach leaf small subunit peptide containing the modified lysyl residue of site 1 is highly conserved in the rice seed, potato tuber, maize,[133] barley,[134] and wheat endosperm small subunits[129] and the *Anabaena*[128] and *Synechocystis*[135] ADP-Glc PPase subunits. Similarly, the amino acid sequence of site 2 of the spinach leaf large subunit is highly conserved in the large subunits of the potato tuber, maize[136] and barley endosperm,[137] wheat seed,[138] and wheat leaf ADP-Glc PPases.[138]

Phenylglyoxal inactivation of the enzyme can be prevented by 3PGA or by Pi, evidence that one or more arginine residues are present in the allosteric sites of the spinach leaf enzyme, and both subunits were labeled when [^{14}C]phenylglyoxal was used.[139] Where the Arg residue(s) is located in the sequence is unknown but there is a possibility it may be close to the Lys residue at activator site 2.

The cDNA clones that encode the putative mature forms of the large and small subunits of the potato tuber ADP-Glc PPase have been expressed together, using compatible vectors, in an *E. coli* mutant devoid of ADP-Glc PPase activity.[127,140] The ADP-GLC PPase activity expressed was high and the enzyme displayed catalytic and allosteric kinetic properties very similar to the ADP-Glc PPase purified from potato tuber.[140] Moreover, the enzyme activity was neutralized by antibody prepared against potato tuber and not by antibody prepared against the *E. coli* ADP-Glc PPase.[140] This expression system is a very useful tool to perform site-directed mutagenesis to further characterize the allosteric function of the lysyl residues identified via chemical modification with pyridoxal-P of the spinach enzyme. Indeed, in preliminary results, site-directed mutagenesis of Lys441 of the potato ADP-Glc PPase small subunit to Glu and Ala results in mutant enzymes with lower affinities for 3PGA, 30- to 83-fold, respectively.[141]

3.14.4.1.4   *Possible functions of the small and large subunits of the higher plants' ADP-Glc PPases*

The ability to express cDNA clones representing the potato tuber small and large subunits together in *E. coli*[127] to obtain a highly active enzyme has also enabled us to express the subunits separately to determine their specific functions. It was found that the potato tuber small subunit, when expressed alone, had high catalytic activity provided that the 3PGA concentration was increased to 20 mM. The 3PGA saturating concentration for the expressed transgenic or natural potato tuber heterotetrameric enzyme is 3 mM. It was found that the K_a of the transgenic enzyme in ADP-Glc synthesis is 0.16 mM, while that for the small subunit alone is 2.4 mM. Thus, the small subunit by itself has about 15-fold lower apparent affinity for the activator. Also, the small subunit is more sensitive to Pi inhibition that the transgenic heterotetrameric enzyme with an 8-fold lower K_i. The kinetics of 3PGA activation and the Pi inhibition were the main kinetic differences between the homotetrameric small subunit and the recombinant heterotetrameric ADP-Glc PPase. These results are consistent with those obtained for the *Arabidopsis thaliana* mutant ADP-Glc PPase lacking the large subunit where the enzyme had lower affinity for the activator and higher sensitivity towards Pi inhibition than the heterotetrameric normal enzyme.[142]

The potato tuber large subunit expressed by itself had negligible activity. Thus, it seems that the dominant function of the small subunit is catalysis while the dominant function of the large subunit is to modulate the sensitivity of the small subunit to allosteric activation and inhibition.

3.14.4.1.5   *Conservation and evolution of amino acid sequence of the ADP-Glc PPases from bacteria to higher plants*

A high degree of amino acid sequence identity is observed when comparing the sequences of corresponding ADP-Glc PPase subunits from different species, a result that could be predicted from the fact that the spinach leaf lower molecular weight subunit antibody reacts very well with the equivalent subunits of the enzymes from maize endosperm,[108,143] rice seed,[110,144] *A. thaliana*,[145] and potato tuber.[146] The antibody for the lower MW spinach leaf subunit does not react well with the larger subunits of the ADP-Glc PPase of other species and not much homology was expected between sequences of the small and large subunits. Indeed, the degree of identity between the large and small subunits (obtained by Edman degradation or deduced from nucleotide sequences of cDNAs or genomic DNA) is around 40 to 60%.[111] Sequence analyses indicate a greater identity between the 54 kDa subunit of the spinach leaf enzyme, the subunit coded by the *Sh* 2 gene from maize, and the subunit encoded by the cDNA insert, we7, from wheat endosperm[138] suggesting that the latter corresponds to the large molecular weight subunit of the ADP-Glc PPase.

Because of the relatively low, but certain, homology between the two subunits of the ADP-Glc PPase, it is speculated that they arose originally from the same gene. The bacterial ADP-Glc PPase has been shown to be a homotetramer composed of only one subunit.[24] The cyanobacterial ADP-Glc PPase has 3PGA as an allosteric activator and Pi as an inhibitor, similar to the enzyme from higher plants,[147] and unlike other bacterial ADP-Glc PPases. Both bacterial[7,24] and cyanobacterial[148] ADPglucose pyrophosphorylases are homotetrameric, unlike the higher plant enzymes, indicating that regulation by 3PGA and Pi is not related to the heterotetrameric nature of the higher plant enzyme. It is quite possible that during evolution there was duplication of the ADP-Glc PPase gene and then divergence of the genes produced two different genes coding for the two peptides, both required for optimal activity of the native higher plant enzyme.

As discussed above, the catalytic function can be assigned tentatively to the small subunit of the ADP-Glc PPase, and this is consistent with the identity and similarity in sequence between the small subunits isolated from different plants and tissues. In the case of the large subunit in which amino acid sequences have lower similarity than that observed for the small subunits, it is quite possible that the different large subunits lend different regulatory properties to the heterotetrameric ADP-Glc PPases of different species and/or tissues. This makes sense considering that the needs and amounts of starch required for each type of tissue are different. Thus, because sequences of the large subunits reflect their occurrences in different plant tissues, e.g., leaf, stem, guard cells, tuber, endosperm, root,[111] it is possible that these sequence differences give the enzyme from each tissue different regulatory properties.

3.14.4.2 Bacterial Glycogen Synthase

3.14.4.2.1 *Enzyme structure and properties*

The *E. coli* B[149] and *E. coli* K12[150] glycogen synthases have been purified to homogeneity. The enzyme from *E. coli* B is strongly absorbed to the hydrophobic resin 4-aminobutyl-Sepharose and is eluted only by solutions containing both 1 M KCl and 1 M maltose. The homogeneous enzyme has a specific activity of about 115 μmol of glucose transferred to glycogen per mg of protein per minute.[151,152] Subunit molecular weight, as determined by SDS–gel electrophoresis, is about 49 000, while sucrose density gradient centrifugation showed aggregated forms of 98 000, 135 000, and 185 000.[149] Thus, *E. coli* B glycogen synthase can exist as dimers, trimers, and tetramers. The bacterial glycogen synthase subunit is about one-half the size of the mammalian glycogen synthase subunit.

The bacterial glycogen synthase is different from the mammalian glycogen synthase (the latter is discussed in Section 3.14.8.2) in two major aspects. First, the bacterial enzyme exhibits no regulatory properties, and it does not exist in inactive or active forms as the mammalian glycogen synthase does. There is no evidence for either phosphorylation or dephosphorylation of the enzyme or for any other enzyme-catalyzed modification. The bacterial enzyme is not activated by glucose 6-phosphate or other glycolytic intermediates.[153–156] Second, the bacterial enzyme uses ADP-glucose as its physiological sugar nucleotide glucosyl donor[33] rather than UDP-glucose, the physiological glucosyl donor of the mammalian glycogen synthase reaction. In one report[157] the *E. coli* glycogen synthase has, with the sugar nucleotides UDP-glucose, CDP-glucose, GDP-glucose, and TDP-glucose, less than 1% of the activity measured for ADP-glucose. Deoxy-ADP glucose, a nonphysiological analogue of ADP-glucose, has been found to be active to the same extent as ADP-glucose with the *Arthobacter viscosus* glycogen synthase[153] as well as with the *E. coli* B enzyme.[154]

The K_m values reported for ADP-glucose range from 20 μM for the glycogen synthase from *M. smegmatis*[156] and 25–35 μM for the enzyme from *E. coli* B[149] to 0.4 mM for the *Pasteurella pseudotuberculosis* glycogen synthase.[155] The K_m for deoxy-ADP-glucose is 38 μM for *E. coli* B glycogen synthase and 27 μM for the *A. viscosus* enzyme. It has been shown that the bacterial glycogen synthase is very sensitive to inhibition by *p*-hydroxymercuribenzoate inhibition, and is also inhibited by ADP which competes with ADP-glucose, with a K_i (for the *E. coli* system) of 15 μM.[155]

Several α-glucans can serve as effective primers for the glycogen synthase, e.g., glycogen from animal and bacterial sources, amylose, and amylopectin. Lower molecular weight maltodextrins can also serve as primers; maltotriose is quite effective as a primer, and maltotetraose has been identified as the immediate product. Maltose is effective as a primer only at high concentrations, and maltotriose is the immediate product. Glucose is not active as an acceptor of glucose in the bacterial glycogen synthase reaction.

3.14.4.2.2 *Reversibility of the glycogen synthase reaction*

The glycogen synthase reaction has been shown to be reversible.[149] Formation of labeled ADP-glucose occurs from either [^{14}C]ADP or [^{14}C]glycogen. The ratio of ADP to ADP-glucose at equilibrium at 37 °C has been found to vary three-fold in the pH range 5.27–6.82, suggesting that in the formation of a new α-1,4-glucosidic bond, a proton is liberated. Since the pK_a of ADP^{2-} ionizing to $ADP^{3-} + H^+$ is about 6.4, a proton would be liberated in varying amounts in the range of pH 5.4 to 7.4, and in stoichiometric amounts above pH 7.4. If the pK_a of ADP^{2-} is assumed to be 6.4[158] then the ratio of ADP^{2-} to ADP-glucose can be calculated, and it is constant; in the range of pH 5.27–6.82 it was 45.8±4.5. The constancy of this equilibrium ratio suggests that ADP^{2-} is the reactive species in the reaction.

3.14.4.2.3 *Substrate and catalytic sites*

The structural gene for glycogen synthase, *glgA*, has been cloned from both *E. coli* and *S. typhimurium*[100,159] and the nucleotide sequence of the *E. coli glgA* gene has been determined.[160] It consists of 1431 bp specifying a protein of 477 amino acids with a MW of 52 412.

Some chemical modification studies[161] have been done that show two distinct sulfihydryl groups important for enzyme activity and protected by the primer, glycogen and the substrate, ADP-Glc, respectively. The reactive sulfihydryl residues are probably located at or near the binding sites for the substrates, glycogen and ADP-Glc.

An affinity analogue of ADP-Glc, adenosine diphosphopyridoxal (ADP-pyridoxal), was used to identify the ADP-Glc binding site.[162] Incubation of the enzyme with the analogue plus sodium borohydride led to an inactivated enzyme; the degree of inactivation correlated with the incorporation of about one mol of analogue per mol of enzyme subunit for 100% inactivation. After tryptic hydrolysis one labeled peptide was isolated and the modified Lys residue was identified as Lys15. The sequence, Lys–X–Gly–Gly, where lysine is the amino acid modified by ADP-pyridoxal, has been found to be conserved in the mammalian glycogen synthase[163,164] and in the plant starch synthases.

Furukawa *et al.*[165,166] performed site-directed mutagenesis experiments to determine structure–function relationships for a number of amino acids in the *E. coli* glycogen synthase. Substitution of other amino acids for Lys at residue 15 suggested that the Lys residue is mainly involved in binding the phosphate residue adjacent to the glycosidic linkage of the ADP-Glc and not in catalysis. The major effect of a mutation at residue 15 on the kinetics of the enzyme was an increased (30–50-fold) K_m for ADP-Glc when either Gln or Glu were the substituted amino acids. Substitution of Ala for Gly at residue 17 decreased the catalytic rate constant, k_{cat}, by about three orders of magnitude as compared with the wild-type enzyme. Substitution of Ala for Gly18 only decreased the rate constant by 3.2-fold. The K_m effect on the substrates, glycogen and ADP-Glc, were minimal. Furukawa *et al.*[165] postulated that the two glycyl residues in the conserved Lys–X–Gly–Gly sequence participated in the catalysis by assisting in maintaining the correct conformational change of the active site or by stabilizing the transition state.

Since the Lys15Gln mutant was still binding to the ADP-Glc and had appreciable catalytic activity, the ADP-pyridoxal modification was repeated and in this instance about 30 times higher concentration was needed for inactivation of the enzyme.[166] The enzyme was maximally inhibited by about 80% and tryptic analysis of the modified enzyme yielded one peptide containing the affinity analogue and with the sequence, Ala–Glu–Asn–modified Lys–Arg. The modified Lys was identified as Lys277. Site-directed mutagenesis of Lys277 to form a Gln mutant was done and the K_m for ADP-Glc was essentially unchanged but k_{cat} was decreased 140-fold. It was concluded that Lys residue 277 was more involved in the catalytic reaction than in substrate binding.

3.14.4.3 Starch Synthases

3.14.4.3.1 *Granule-bound and soluble forms*

Starch synthase activity can be measured as the transfer of [^{14}C]glucose from ADP-glucose into a prime such as rabbit glycogen or amylopectin.[167,168]

As first shown in our laboratory, citrate stimulates a reaction in the absence of added maltodextrin primer which is due to small amounts of endogenous primer strongly bound to the enzyme.[169–172] Citrate has been shown to increase the apparent affinity for the glucan primer.[171,172] Thus, only a minute amount of endogenous glucan (e.g., 6 nmol of anhydroglucose units) is needed per reaction mixture for the citrate-stimulated "unprimed reaction."

Many questions regarding starch biosynthesis remain to be answered, partly because of the difficulties inherent to the starch granule itself, which is insoluble in water and has a very complicated structure.[1] Synthesis of a starch granule has not been obtained *in vitro*. *In vivo*, it occurs by deposition on the granule surface by the concerted action of starch synthases and branching enzyme. On centrifugation of a crude extract, starch synthase activity is found associated with the starch granules or in the supernatant. It is assumed that this partition is a result of differences in the structure of the isozymes and/or differences in the role they play in the synthesis of the starch components, amylose and amylopectin. Also, for storage organs such as seeds and tubers, the starch granule is formed and grows for several weeks and it is likely that different isozymes vary in importance during this period. In maize endosperm there are at least four starch synthases, two soluble[169] and at least two granule-bound.[168] The number of isoforms may vary with the plant species and the developmental stage, but those that have been studied more carefully seem to have a similar number of isoforms. Indeed, as in the case of pea embryo, an isozyme of starch synthase, starch synthase II, can exist as soluble and starch-granule-bound.[173] The question remains whether the soluble and granule-bound forms are both functional. Indeed, Mu-Forster *et al.*[174] reported that in maize endosperm, more than 85% of the starch synthase I protein may be associated with the starch granule. This was determined by using antibody prepared against the starch synthase. However, no evidence was presented to indicate that the starch synthase I was active at that particulate stage. The cDNA clones that encode the two isozymes of granule-bound starch synthase of pea embryo are optimally expressed at different times during development;[79] while isozyme II is expressed in every organ, isozyme I is not expressed in roots, stipules, or flowers.[79]

Purification of the starch synthase in large amounts and to a high specific activity has proven to be difficult, and partly for this reason it has not been possible to find out how the enzymes interact to produce the two carbohydrates, amylose and amylopectin, that form the starch granule.

3.14.4.3.2 *Identification of the* **Waxy** *locus as the structural gene for the granule-bound starch synthase*

Genetic studies implicate one granule-bound starch synthase (GBSS) isoform in the synthesis of amylose. In *waxy* (*wx*) mutants there is virtually no amylose, GBSS activity is very low,[74,175,176] and the Wx protein is missing. The final product of the *Wx* locus is a protein of molecular weight 58 kDa associated with the starch granule.

Federoff *et al.* prepared cDNA clones homologous to *Wx* mRNA and, in subsequent experiments,[177] restriction endonuclease fragments containing part of the *Wx* locus were cloned from strains carrying the different *wx* alleles to further characterize the controlling insertion elements activator (ac) and dissociation (ds). Excision of the ds element from certain *wx* alleles produces two new alleles encoding for wx proteins with altered starch synthase activities.[178]

The DNA sequence of the *Wx* locus of *Zea mays* was determined by analysis of both a genomic and an almost full length cDNA clone;[179] the *Wx* locus from barley has been cloned and its DNA sequenced.[180] Amino-acid sequences are also available for rice,[181] potato,[77] cassava,[182] wheat,[129] and pea isozymes.[79]

Table 3 compares the deduced amino acid sequences in three regions from the potato, cassava, maize, barley, wheat, rice, and isozyme I of pea embryo wx clones, with the amino acid sequence for the *E. coli* glycogen synthase,[160] and the rice soluble starch synthase.[183] Region 1 starts with the first 27 amino acids of the *N*-terminus of the *E. coli* glycogen synthase. Thirteen amino acids are identical to the amino acid sequences deduced for the plant Wx proteins. Of particular significance is the sequence starting at residue Lys15 of the bacterial enzyme ...KTGGL.... The lysine in the bacterial glycogen synthase has been implicated in the binding of the substrate, ADP-glucose[162] on the basis of the chemical modification of that site by the substrate analogue ADP-pyridoxal (see Section 3.14.4.2.3). The similarity of sequences between the bacterial glycogen synthase, the soluble starch synthase and the Wx protein provides further evidence that the *Wx* gene is indeed the structural gene for the granule-bound starch synthase.

There are two other regions that are highly conserved in both the GBSSs and the *E. coli* glycogen synthase. In region II, only one or two amino acids of the 13 amino acids are different from the *E. coli* sequence and in region III, all the GBSSs are completely identical with respect to the amino acid sequence while the bacterial enzyme differs in only two of the nine amino acids, an Arg for a Ser and an Ala for a Val.

The genetic evidence points to the *Wx* locus as the structural gene for a starch synthase bound to the starch granule. However, direct biochemical evidence was lacking, mainly because of the difficulties involved in studying the proteins associated with starch. To solve this problem, starch was solubilized using amylases, and the starch proteins released into the supernatant were fractionated by chromatography on DEAE (diethyl amino ethyl).[168] The GBSSI was clearly associated with the Wx protein (recognized by its mobility on SDS polyacrylamide gels and its reaction with antibodies raised against the pure Wx protein) throughout purification. The molecular mass of the GBSSI, determined by gel filtration or by sucrose density gradients, was about 59 kDa.[184]

Because she was unable to detect starch synthase activity in the Waxy protein from pea endosperm, Smith[185] suggested that "the *waxy* protein of pea is not the major granule-bound starch synthase" and that a "re-examination, species by species, of the identity of the starch-granule-bound starch synthase ..." was required. Sivak *et al.*[184] however, found that starch extracted from developing embryos of pea contained starch synthase activity which was associated with the Waxy protein. The MW of the pea starch synthase is about 59 kDa, as determined by ultracentrifugation in sucrose density gradients. A pea granule-bound starch synthase preparation displayed a relatively high specific activity (over 10 μmol glucose incorporated per min per mg protein). This enzymatic fraction when subjected to SDS polyacrylamide gel electrophoresis migrated the same as the Waxy protein and gave a strong immunoblot with antibody prepared against the Waxy protein either from pea embryo or maize and only the Wx protein stain was visible.[184] Thus, the immunological data indicated that the activity assayed by Sivak *et al.*[184] was due to the granule-bound starch synthase (Waxy protein) and not due to the truncated soluble starch synthase of 60 kd detected by Edwards *et al.*[173] When the gene coding for the mature Waxy protein from maize kernel was expressed in *E.coli*, the recombinant protein had a MW similar to the maize protein as determined by SDS-PAGE, reacted with antibody raised against the plant protein, and had starch synthase activity.[186] Thus, the biochemical re-examination of starch synthase present in starch granules from two species, maize and pea, strengthens the genetic evidence supporting the role of the Wx protein as a granule-bound starch synthase with a major role in the determination of amylose content of starch.

It has been shown in many experiments involving anti-sense RNA in potato[78,187] and in rice,[188] that disappearance of amylose correlates very well with the loss of *Wx* gene expression. If the interior of the granule is devoid of branching enzyme or, if branching enzyme in the granule is not appreciably active, the presence of an active chain-elongating enzyme, i.e., starch synthase, without active branching enzyme present, could lead to amylose formation. However, this situation may be more complicated since more than one isozyme of the GBSS has been found for a number of plants. Also, it is quite possible that GBSS may also be involved in the initial formation of amylopectin near the exterior portion of the granule along with the soluble starch synthases. In *Chlamydomonas reinhardtii*, a *wx* mutant deficient in GBSS was isolated.[80] In this mutant, the isolated starch was not only deficient in amylose but also in one of the amylopectin fractions, amylopectin II. Amylopectin II has longer chains than the amylopectin I fraction, as indicated by the increase in λ_{max} of the glucan-I_2 complex. When GBSS is active, it would not be rate limiting and thus amylose and amylopectin would be normal components of the starch granule. When there is a loss of the major GBSS activity (e.g., in the *wx* mutant), the rate of formation of the amylose and initial amylopectin structures would be limiting and only the higher branched amylopectin I fraction would be present.

3.14.4.3.3 *Characterization of the soluble starch synthases*

Many plant systems have multiple forms of soluble starch synthases (SSS). In studies on spinach leaf, potato tuber, barley, maize and wheat endosperm, pea, rice, sorghum, and teosinte seeds,

Table 3 Regions within amino acid sequences of the *E. coli* glycogen synthase and the various granule-bound starch synthases.

	Region 1	Region 2	Region 3
E. coli glycogen synthase	1MQVLHVCSEMFPLLKTGGLADVIGALP	372VPSRFEPCGLTQL	397RTGGLADTV
Rice soluble starch synthase	20RSVVFVTGEASPYAKSGGLGDVCGSLP	372MPSRFEPCGLNQL	397GTGGLRDTV
Potato tuber wx protein	4MNLIFVGTEVGPWSKTGGLGDVLRGLP	397VPSRFEPCGLIQL	422STGGLVDTV
Cassava Wx protein	4MNLIFVGAEVGPWSKTGGLGDVLGGLP	398VPSRFEPCGLIQL	423STGGLVDTV
Maize Wx protein	5MNVVFVGAEMAPWSKTGGLGDVLGGLP	398VTSRFEPCGLIQL	423STGGLVDTV
Barley Wx protein	6MNLVFVGAEMAPWSKTGGLGDVLGGLP	396VTSRFEPCGLIQL	421STGGLVDTV
Wheat Wx protein	7MNLVFVGAEMAPWSKTGGLGDVLGGLP	410VTSRFEPCGLIQL	435STGGLVDTV
Rice Wx protein	6MNVVFVGAEMAPWSKTGGLGDVLGGLP	397VPSRFEPCGLIQL	422STGGLVDTV
Pea Wx protein I	1MSLVFVGAEVGPWSKTGGLGDVLGGLP	403IPSRFEPCGLIQL	428STGGLVDTV

The numbers preceding the sequence indicate the residue number from the *N*-terminus in the sequence. The sequence in outline form, KTGGL, has been shown for the *E. coli* glycogen synthase to be involved in binding of the sugar nucleotide substrate.[162] References to the other sequences may be obtained from ref.[17]

extracts have indicated the presence of at least two major forms of SSS (reviewed in refs.[11,12,17,189]) designated as types I and II because starch synthase I (SSSI) usually elutes from an anion exchange column at lower salt concentrations than starch synthase II (SSSII).

Although SSSI has been partly purified from maize kernels,[171,172] SSSII, a more unstable isoform, has been more difficult to purify. The properties observed for the isoforms of maize endosperm are representative of the properties of the corresponding enzyme forms in other plants (for a review see ref.[17]). The apparent affinity for ADP-glucose, measured by the K_m, is similar for the two forms. The maximal velocity of the type I enzyme is greater with rabbit liver glycogen than with amylopectin and the type II enzyme is less active with glycogen than with amylopectin. Citrate stimulation of the primed reaction is greater for type I than for type II. Both forms can use the oligosaccharides maltose and maltotriose as primers when present at high concentrations. SSSI seems to have more activity than SSSII with these acceptors.

The lower activity for SSSI with amylopectin as a primer as compared with glycogen, suggests that SSSI may prefer the short exterior chains (A-chains) that are more prevalent in glycogen than in amylopectin. The reverse may be true for SSSII where this isoform may prefer the longer chains (B-chains) seen in amylopectin. Differences in the apparent affinities with respect to primer were also noted. For example, the K_m of the type I enzyme for amylopectin is nine times lower than that of the type II enzyme. The type I enzyme is active without added primer in the presence of 0.5 M citrate, while the type II enzyme is inactive in these conditions. Citrate decreases the K_m of amylopectin for both types of enzymes: 160-fold for the type I enzyme and about 16-fold for the type II starch synthase with 0.5 M citrate.

The starch synthase isozymes in maize endosperm have different molecular masses. The GBSS isozyme I has a molecular mass of 60 kd, that of GBSSII is 95 kd, the SSSI has a molecular mass of 72 kDa, and that of SSSII is 95 kd (reviewed in ref.[17]). Mu *et al.*[190] have reported the molecular mass of maize endosperm SSSI as 76 kDa, which is near the value reported previously. These molecular mass values for the starch synthases are all higher than that of the *E. coli* glycogen synthase with a molecular weight of 52 kDa.[160] It appears that the maize endosperms SSSI and SSSII are immunologically distinct;[168] antibody prepared against maize endosperm SSSI showed very little reaction with SSSII in neutralization tests.

In summary, the maize SSSI and SSSII are different, and are distinguished on the basis of their physical, kinetic, and immunological properties; they are probably products of two different genes. Because of their different kinetic properties and different specificities with respect to primer activities, the two isoforms may have different functions in the formation of the starch granule. Purification of the isoforms to high specific activity and lack of interfering activities will facilitate the characterization of the isoforms with respect to primer specificity and interaction with isoforms of branching enzyme, supplying information about their role *in vivo*.

In rice, three isoforms of soluble starch synthase were separated by anion exchange chromatography which, in immunoblot, reacted with antibodies raised to the rice waxy protein.[183] After affinity chromatography of the active fractions, amino-terminal sequences were obtained for the protein bands of 55–57 kDa (separated by SDS-PAGE) that cross-reacted weakly with serum raised against the rice waxy protein. It is worth noting that this experimental approach does not exclude the possibility that other soluble starch isoforms were present which did not cross-react with the antiserum, and the authors indicate that other results suggest that another soluble starch synthase isoform, with a MW of 66 kDa, is also present in seed extracts.

Other forms of starch synthase may be present in plants. Marshall *et al.*[191] have reported the presence of a starch synthase, 140 kDa, in potato tubers which may account for 80% of the total soluble starch synthase activity. A cDNA representing the protein gene was isolated. Expression of an anti-sense mRNA caused a reduction of about 80% of the soluble starch synthase activity in the tuber extracts. The severe reduction in activity, however, had no effect on starch content or on the amylose/amylopectin ratio of the starch. There was a change in the morphology of the starch granules, suggesting an alteration in the starch structure. The specific alteration in structure causing the morphology change remains to be determined.

Baba *et al.*[183] isolated cDNA clones coding for the putative soluble starch synthase from maize from an immature rice seed library in λgt 11 using as probes synthetic oligonucleotides designed on the basis of the amino-terminal amino acid sequences available. The insert of about 2.5 kb was sequenced and shown to code for a 1878-nucleotide open reading frame. Comparison with the corresponding amino-terminal sequences led the authors to conclude that the protein is initially synthesized as a precursor, carrying a long transit peptide at the amino acid terminus and that the same gene would be expressed both in seeds and in leaves.

3.14.4.3.4 Chlamydomonas reinhardtii *mutants affected in starch synthesis*

Mutants of *Chlamydomonas* affected in starch synthesis have been induced and isolated: a mutant deficient in SSSII[192] and double mutants deficient both in GBSS and in SSSII.[193] These studies have provided important information on the function of these isozymes and their involvement in amylopectin biosynthesis. The mutant deficient in SSSII had 20–40% of the starch content of the wild-type organism, and the percent amylose of the starch increased from 25 to 55%. This mutant also contained a modified amylopectin which had an increased number of very short chains (2–7 DP) and a decrease of intermediate-size chains (8–60 DP). These results suggested that the SSSII is involved in the synthesis or maintenance of the intermediate-size chains present in amylopectin. The higher amylose content could be explained if a lesser branched amylose-like intermediate was a precursor for amylopectin synthesis, and the SSSII mutant could not effectively use this substrate. This amylose fraction may be more highly branched than the usual amylose, as the absorption spectrum of its I$_2$ complex has a lower maximal wavelength than the wild-type amylose fraction, suggesting that it contained more branching. The mutant amylose fraction may therefore have a greater amount of branched amylose intermediates on the route to amylopectin biosynthesis.

The double mutants defective in SSSII and a GBSS,[193] had an even lower starch content, 2–16% of the wild type, and the amount of starch present was inversely correlated with the severity of the GBSS defect of the double mutant. The authors suggest that the GBSS is required to form the basic structure of the amylopectin and that the effects of the absence of GBSS are exacerbated by the diminished SSSII activity. Of interest is that the SSSI may, in addition to a small amount of starch, synthesize a small water-soluble polysaccharide. Analysis of both fractions suggests that they may be intermediate in structure between amylopectin and glycogen with respect to extent of branching.

These studies of the *Chlamydomonas* mutants by Ball and co-workers[192,193] provide supporting evidence for involvement of the GBSS in amylopectin as well as in amylose synthesis, and suggest that an important function for SSSII would be its involvement in synthesis of the intermediate-size (B) branches in amylopectin.

3.14.4.4 Branching Enzymes from Plants and Bacteria

3.14.4.4.1 *Assay*

Branching enzyme can be assayed in a number of ways. The iodine assay is based on the decrease in absorbance of the glucan–iodine complex resulting from the branching of amylose or amylopectin. During incubation of the assay mixture containing amylose or amylopectin with enzyme, aliquots are taken at intervals and iodine reagent is added.[69,194] The decrease of absorbance is measured at 660 nm for amylose and, for amylopectin, at 530 nm. A unit of activity is defined as decrease in absorbance of 1.0 per min at 30 °C at the defined wavelength.

The phosphorylase-stimulation assay[194–196] is based on the stimulation of the "unprimed" (without added glucan) phosphorylase activity of the phosphorylase *a* from rabbit muscle as the branching enzyme present in the assay mixture increases the number of nonreducing ends available to the phosphorylase for elongation. One unit is defined as 1 µmol transferred from glucose-1-P per min at 30 °C.

The branch-linkage assay (BL assay[197]) is an assay that measures the number of branch chains formed by branching enzyme catalysis, rather than an indirect effect of its action as in the two assays described above. The enzyme fraction is incubated with the substrate, NaBH$_4$-reduced amylose. The reaction is then stopped by boiling and the product is incubated for debranching, with pure *Pseudomonas* isoamylase. Finally, the reducing power of the oligosaccharide chains transferred by the enzyme is measured by a modification of the Park–Johnson method. Amylose reduced with borohydride is used (rather than amylose itself) to lower the initial reducing power seen with amylose.

The branching-linkage assay is the most quantitative assay for branching enzyme, but the presence of impurities, like amylolytic activity, interferes the most with this assay. The phosphorylase-stimulation assay is the most sensitive. The I$_2$ assay is not very sensitive but allows the testing of branching enzyme specificity with various maltodextrins, providing information on the possible role of the different branching enzyme isoforms. It may be best to employ all three assays when studying the properties of the branching enzymes, but, above all, if reliable information is being sought, the branching enzymes must be purified to the extent that all degradative enzymes are eliminated before studying their properties.

3.14.4.4.2 The branching enzyme belongs to the α-amylase family

The structural gene of the *E. coli* branching enzyme (BE), *glgB*, has been cloned[198] and its complete nucleotide and deduced amino acid sequences determined. The information obtained was consistent with the amino acid analysis of the pure protein, the MW determined by SDS gel electrophoresis, and with the amino acid sequence analyses obtained of the amino terminal and of the various peptides obtained via chemical and proteolytic degradation. The gene consisted of 2181 bp, specifying a protein of 727 amino acids and with a MW of 84 231.

Romeo *et al.*[199] reported on the similarities observed between the amino acid sequences of the bacterial BE and those of amylolytic enzymes such as α-amylase, pullulanase, glucosyltransferase, and cyclodextrin glucanotransferase, particularly in the regions believed to be contacts between the substrate and the enzyme. Baba *et al.*[200] reported that there was marked conservation in the amino acid sequence of the four catalytic regions of amylolytic enzymes in maize endosperm BEI. As shown in Table 4, four regions that putatively constitute the catalytic regions of the amylolytic enzymes are conserved in the starch branching isoenzymes of maize endosperm, rice seed, and potato tuber and the glycogen branching enzymes of *E. coli*. Analysis of this high conservation in the α-amylase family has been considerably extended,[201,202] not only with respect to sequence homology but also in the prediction of the $(\beta/\alpha)_8$-barrel structural domains with a highly symmetrical fold of eight inner, parallel β-strands, surrounded by eight helices, in the various groups of enzymes in the family. The $(\beta/\alpha)_8$-barrel structural domain was determined from the crystal structure of some α-amylases and cyclodextrin glucanotransferases.

Table 4 Comparison of the primary structure of several branching enzymes with the four most conserved regions of the α-amylase family.

	Region 1		Region 2		Region 3		Region 4	
Maize endosperm BE I	277	DVVHSH	347	GFRFDGVTS	402	TVVAEDVS	470	CTAYAESHD
Maize endosperm BE II	315	DVVHSH	382	GFRFDGVTS	437	VTTGEDVS	501	CVTYAESHD
Potato tuber BE	355	DVVHSH	424	GFRFDGITS	453	VTMAEEST	545	CVTYAESHD
Rice seed BE 1	271	DVVHSH	341	GFRFDGVTS	396	TIVAEDVS	461	CVTYAESHD
Rice seed BE 3	337	DVVHSH	404	GFRFDGVTS	459	ITIGEDVS	524	CVTYAESHD
E. coli glycogen BE	335	DWVPGH	400	ALRVDAVAS	453	VTMAEEST	517	NVFLPLNHD
Barley 1 α-amylase	88	DIVINH	175	AWRLDFARG	201	LAVAEVWD	283	AATFVDNHD
Wheat 3 α-amylase	111	DIVINH	199	GWRFDFAKG	225	FVVGELYD	287	TVTFIDNHD
Porcine pancreas α-amylase	96	DIVINH	193	GFRLDASKII	229	FIFQEVID	292	ALVFVDNHD
B. subtilis α-amylase	100	DAVINH	171	GFRFDAAKH	204	FQYGEILQ	261	LVTWVEGHD
K. pneumonia pullulanase	620	DVVYNH	673	GFRFDLMGY	702	YFFGEGWD	826	VVNYVSKHD
B. sphaericus cyclodext.	238	DAVFNH	323	GSRLDVANE	350	IIVGEVWH	414	SFNLLCSHD

The sequences have been derived from references cited in the text. Only two examples of enzymes from the amylase family are shown. Svensson[201] compared over 40 enzymes comprising amylases, glucosidases, and various α-1,6-debranching enzymes, as well as four branching enzymes. The invariant amino acid residues are in bold letters; these are believed to be involved in catalysis.

The conservation of the putative catalytic sites of the α-amylase family in the glycogen and starch branching enzymes would be anticipated as the BE catalyzes two consecutive reactions in synthesizing α-1,6-glucosidic linkages: an α-1,4-glucosidic linkage in a 1,4-α-D-glucan is cleaved to form a nonreducing end oligosaccharide chain that is then transferred to a C-6 hydroxyl group of the same or another 1,4-α-D-glucan. It would be of interest to know whether the eight highly conserved amino acid residues of the α-amylase family are also functional in branching enzyme catalysis. Further experiments such as chemical modification and analysis of the three-dimensional structure of the BE are needed to determine the precise functions and nature of its catalytic residues and mechanism; some preliminary studies are discussed when the plant starch branching enzymes are reviewed in Section 3.14.4.4.4.

3.14.4.4.3 Plant branching enzyme isoforms

For some species, e.g., maize, the usual biochemical methodology has led to a clear separation and characterization of isoforms, which has been confirmed by genetic studies of mutants and isolation and sequencing of cDNA clones. For other species, e.g., potato and rice, the situation is still far from clear and awaits the use of different methodology.

In maize endosperm there are three branching enzyme isoforms.[69,194,203] Reports on other tissues are consistent with the presence of more than one isoform, as in the castor bean.[204] Purification of

BEI, IIa and IIB from maize kernels[69,199,203] no longer contained amylolytic activity.[69,197] Molecular weights were 82 000 for isoform I and 80 000 for isoforms IIa and IIb.[194,195]

There has been some progress[69,197] and some results using the different assays are summarized in Table 5. Takeda *et al.*[197] have analyzed the branched products made from amylose by each BE isoform. This was done by debranching the products of each reaction using isoamylase, followed by gel filtration. BEIIa and BEIIb are very similar in their affinity for amylose and the size of chain transferred. When presented with amyloses of different average chain length, the three BEs had higher activity with the longer chain amylose, but while BEI could still catalyze the branching of an amylose of average chain length (c.l.) of 197 with 89% of the activity shown with the c.l. of 405, the activity of BEII dropped sharply with chain length. The study of the reaction products showed that the action of BEIIa and BEIIb results in the transfer of shorter chains than those transferred by BEI. The action of the isoforms on amylopectin has been studied[69] and, of the three isoforms, BEI had the highest activity (using the iodine assay) in branching amylose (Table 5) and its rate of branching amylopectin was less than 10% of that with amylose. In contrast, the BEIIa and BEIIb isoforms branched amylopectin at twice the rate they branched amylose, and catalyzed branching of amylopectin at about 3 to 4 times the rate observed for BEI. These results are consistent with the results of Takeda *et al.*[197] in suggesting that BEI catalyzes the transfer of longer branched chains and that BEIIa and BEIIb catalyze transfer of shorter chains. Thus, it is quite possible that BEI may produce slightly branched polysaccharides which serve as substrates for enzyme complexes of BEII isoforms and starch synthases to synthesize amylopectin; BEII isoforms may play a major role in forming the short chains present in amylopectin. Also, BEI may be more involved in producing the more interior (B) chains of the amylopectin while BEIIa and BEIIb would be involved in forming the exterior (A) chains.

Table 5　Specific activities (units per mg protein) of maize endosperm branching enzyme isoforms as measured by different assay methods.

Assay method	Activity of the branching enzymes		
	BE I	BE IIa	BE IIb
Phosphorylase stimulation (a)	1332	795	927
Branching linkage assay (b)	2.4	0.32	0.33
Iodine stain assay (c)			
Amylose (c_1)	574	29.5	53
Amylopectin (c_2)	47	59	105
Ratio of activity			
a/b	555	2484	2809
a/c_1	2.3	27	18
a/c_2	49.8	13.5	8.8
c_2/c_1	0.03	2	2

For phosphorylase stimulation and branching linkage assays the units are μmol/min; for the iodine stain assay, a decrease of one absorbance unit per min.

Vos-Scheperkeuter *et al.*[176] purified a single form of branching activity of 79 kDa molecular mass in potato tuber. Antibodies were prepared to the native potato enzyme and they were found to react strongly with maize BEI and very weakly with maize BEIIb. In neutralization tests, the antiserum inhibited the activities of both the potato tuber BE and of maize BEI. It was concluded that the potato branching enzyme shows a high degree of similarity to the maize BEI and to a lesser extent with the other maize BE. However, it has not been determined whether potato tubers have two isoforms of branching enzyme such as BEI and BEII. Borovsky *et al.*[205] isolated from potato tubers a BE of molecular mass 85 kDa. This is close to the mass of 79 kDa found by Vos-Scheperkeuter *et al.*[176] It has been claimed that branching enzymes of molecular mass 97 and 103 kDa were isolated,[206,207] suggesting that the previous lower molecular mass values of 79 and 85 kDa were the result of proteolysis during purification of the 103 kDa BE. Khoshnoodi *et al.*[208] showed that limited proteolysis of the 103 kDa enzyme either with trypsin or chymotrypsin produced an enzyme, still fully active, of molecular size 80 kDa. Four cDNA clones have been isolated for BE, one for 91 to 99 kDa;[208–210] all of these allelic clones have sequences similar to the BEI type. It is still not resolved whether the 97 and 80 kDa proteins could be the products of different allelic forms of the BE gene or different BE genes. Also, the *sbeIc* allele codes for a mature enzyme of 830 amino acids and a molecular weight of 95 180; the *sbeIc* BE protein product, expressed in *E. coli*, migrates as a 103 kDa protein.[208]

When discussing the potato BE, it is worth noting that BEs isolated from other plants, and from bacteria and mammals, have molecular masses ranging from 75 to about 85 kDa. These molecular masses are consistent with the molecular weights obtained from deduced amino acid sequences obtained from isolated genes or cDNA clones.

Mizuno *et al.*[67] have reported four forms of branching enzyme from immature rice seeds that were separated by chromatography on DEAE-cellulose chromatography. Two of the forms, BE1 and BE2 (composed of BE2a and BE2b) were the major forms, while BE3 and BE4 were minor forms comprising less than 10% of the total branching enzyme activity. The MW of the branching enzymes were: BE1, 82 kDa; BE2a, 85 kDa, BE2b, 82 kDa; BE3, 87 kDa; BE4a, 93 kDa; and BE4b, 83 kDa. However, BE1, BE2a, and BE2b seem to be immunologically similar in their reaction to maize endosperm BEI antibody. Moreover, the rice seed BE1, BE2a, and BE2b had very similar *N*-terminal amino acid sequences. All three BEs had two *N*-terminal sequences, TMVXVVEEVDHLPIT and VXVVEEVDHLPITDL. The latter sequence is very similar to the first but lacking just the first two *N*-terminal amino acids. Thus, although these activities came out in separate fractions from the DEAE-cellulose column they seem to be the same protein on the basis of immunology and *N*-terminal sequences; BE2a, however, is 3 kDa larger. Antibody raised against BE3 reacted strongly against BE3 but not towards BE1 and BE2a,2b. Thus, rice endosperm, as noted for maize endosperm, may have just two different isoforms of BE. Because of the many isoforms existing for the rice seed branching enzymes, Yamanouchi and Nakamura[211] studied and compared the BEs from rice endosperm, leaf blade, leaf sheath, culm, and root. The BE activity could be resolved into two fractions, BE 1 and BE 2, and both fractions were found in all tissues studied in different ratios of activity. The specific activity of the endosperm activity either on the basis of fresh weight or protein was 100–1000-fold greater than other tissues studied. On native gel electrophoresis, rice endosperm BE 2 could be resolved into two fractions, BE 2a and BE 2b. Of interest was that upon electrophoresis of the other tissue BE 2 forms, only BE 2b was found. BE 2a was only detected in the endosperm tissue. It appears that in rice there may be tissue specific isoforms of BE.

Three forms of branching enzyme from developing hexaploid wheat (*Triticum aestivum*) endosperm have been partially purified and characterized.[212] Two forms are immunologically related to maize BE I and one form to maize BE II. The *N*-terminal sequences are consistent with these relationships. The wheat BE I_B gene is located on chromosome 7B while the wheat BE I_{AD} peptides genes are located on chromosomes 7A and 7D. The BE classes in wheat are differentially expressed during endosperm development in that BE II is constitutively expressed throughout the whole cycle, while BE I_B and BE I_{AD} are expressed in late endosperm development.

3.14.4.4.4 *Genetic studies on mutants deficient in branching enzyme*

There are some maize endosperm mutants which appear to increase the proportion of amylose of the starch granule. The starch granule contains about 25% of the polysaccharide as amylose with the rest as amylopectin. In contrast, *amylose extender* mutants may have as much as 55–70% of the polysaccharide as amylose and may have an amylopectin fraction with fewer branch points and with the branch chains longer in length compared with those of normal amylopectin. Results with the recessive maize endosperm mutant, amylose extender, *ae*, suggest that *Ae* is the structural gene for either branching enzyme IIa or IIb[195,213,214] as activity of BEI was not affected by the mutation. In gene dosage experiments, Hedman and Boyer[215] reported a near-linear relationship between increased dosage of the dominant *Ae* allele and BEIIb activity. Since the separation of form IIa from IIb was not very clear, it is possible that the *Ae* locus was also affecting the level of BEIIa.

MacDonald and Preiss[203] concluded that although some homology exists between the three starch branching enzymes, there are major differences in the structure of branching enzyme I when compared with IIa and IIb, as shown by its different reactivity with some monoclonal antibodies, and differences in amino acid composition and in proteolytic digest maps. It was also concluded[203] that branching enzymes IIa and IIB are very similar and perhaps the product of the same gene. However, studies by Fisher *et al.*[216,217] in analyzing 16 isogenic lines having independent alleles of the maize *ae* locus suggest that BEIIa and BEIIb are encoded by separate genes and the BEIIb enzyme is encoded by the *AE* gene. They isolated a cDNA clone labeled *Sbe* 2b, which had a cDNA predicted amino acid sequence at residues 58 to 65, exactly as the *N*-terminal sequence of the maize BEIIb that they had purified.[216] Moreover, they did not detect in *ae* endosperm extracts any mRNA with the *Sbe* 2b cDNA clone. Gao *et al.*[218] have reported the isolation of a cDNA clone encoding the putative BEIIa. The deduced amino acid sequence differed from BEIIb in having a 49-amino acid *N*-terminal extension and a region of substantial sequence divergence. However, the cDNA remains to be expressed to determine whether the encoded BE would have the enzymatic properties of BEIIa which are very similar to BEIIb. In the *ae* extracts some BE activity was observed that chromatographed as BEIIa.

The finding that the enzyme defect in the *ae* mutant is BEIIb is consistent with the finding that, *in vitro*, BEII is involved in transfer of small chains. Besides having an increase in amylose, the altered amylopectin structure in the *ae* mutant consists of fewer and longer chains and a smaller number of total chains. In other words, there are very few short chains.

The wrinkled pea has a reduced starch level of about 66–75% of that seen in the round seed, and, whereas the amylose content is about 33% in the round form, it is 60–70% in the wrinkled pea seed. Edwards *et al.*[219] measured the activities of several enzymes involved in starch metabolism in the wrinkled pea at four different developmental stages. In this variety it was found that branching enzyme activity was, at its highest, only 14% of that seen for the round seed. The other starch biosynthetic enzymes and phosphorylase had similar activities in the wrinkled and round seeds. These results were confirmed by Smith[220] who also showed that the *r* (*rugosus*) lesion (as found in the wrinkled pea of genotype *rr*) was associated with the absence of one isoform of branching enzyme. Edwards *et al.*[219] proposed that the reduction in starch content observed in the mutant seeds is caused indirectly by the reduction in BE activity through an effect on the starch synthase. The authors suggested that, in the absence of branching enzyme activity, the starch synthase forms an α-1-4-glucosyl elongated chain which is a poor glucosyl-acceptor (primer) for the starch synthase substrate, ADP-Glc, therefore decreasing the rate of α-1-4-glucan synthesis. Indeed, in a study of rabbit muscle glycogen synthase[221] it was found that continual elongation of the outer chains of glycogen caused it to become an ineffective primer, thus decreasing the apparent activity of the glycogen synthase. The observation that ADP-glucose in the wrinkled pea accumulates to higher concentrations than in the round or normal pea was considered evidence that activity of the starch synthase was restricted *in vivo*. Under optimal *in vitro* conditions, in which a suitable primer like amylopectin or glycogen is added, starch synthase activity in the wrinkled pea was equivalent to that found in the wild type.

Amylose extender mutants have been found in rice and studied;[222] the alteration of the starch structure is very similar to that reported for the maize endosperm *ae* mutants. The defect is BE3 isozyme and BE3 of rice is more similar in amino acid sequence to maize BEII than to BEI.[71,222] Thus, rice BE3 may catalyze the transfer of small chains rather than long chains.

The *r* locus of pea seed has been cloned by using an antibody towards one of the pea branching enzyme isoforms and screening a cDNA library.[223] It appears that the branching enzyme gene in the wrinkled pea contains an 800 bp insertion causing it to express an inactive branching enzyme. The authors indicated that the sequence of the 2.7 kb clone showed over a 50% homology to the glycogen-branching enzyme of *E. coli*[198] and proposed that the cDNA that they had cloned corresponded to the starch branching enzyme gene of pea seed. The *glg* B gene sequence has been determined for a cyanobacterium[46] and its deduced amino acid sequence has extensive similarity to the amino acid sequence (62% identical amino acids) in the middle area of the *E. coli* protein. It appears, therefore, that branching enzymes in nature have extensive homology irrespective of the degree of branching of their products, which is higher (about 10% α-1,6 linkages) in glycogen, the storage polysaccharide in enteric bacteria and in cyanobacteria, than in the amylopectin (about 5% α-1,6 linkages) present in higher plants.

The cDNA clones of genes representing different isoforms of branching enzyme have been isolated from potato tuber,[208–210] maize kernel (BE I[70,71,216,217]), cassava,[182] and rice seeds (branching enzyme I;[67,224] branching enzyme 3[222]). The cDNA clones of the maize BE I and BE II have been overexpressed in *Escherichia coli* and purified.[70,71,225] The transgenic enzymes had the same properties as seen with the natural maize endosperm BEs with respect to specific activity and specificity towards amylose and amylopectin.[69]

3.14.4.4.5 *Amino acid residues that are functional in branching enzyme catalysis*

As indicated in Table 4, four regions which constitute the catalytic regions of the amylolytic enzymes are conserved in the starch branching isoenzymes of maize endosperm, rice seed, potato tuber, and the glycogen branching enzymes of *E. coli*.[201,202] Of interest is that the eight highly conserved amino acid residues of the α-amylase family are indeed also functional in branching enzyme catalysis. Preliminary experiments,[226] where amino acid replacements were done by site-directed mutagenesis, suggest that the conserved Asp residues of regions two and four, Asp386 and Asp509, and the Glu residue 441 of region 3 (Table 4, in bold) are important for BE II catalysis. Their exact functions, however, are unknown and further experiments such as chemical modification and analysis of the three-dimensional structure of the BE would be needed to determine the precise functions and nature of its catalytic residues and mechanism. Arginine residues are also important, as suggested by chemical modification with phenylglyoxal,[227] as well as histidine residues, as suggested by chemical modification studies with diethyl pyrocarbonate.[228] Of interest would also be to determine the regions of the *C*- and *N*-termini which are dissimilar in sequence and in size in the various branching isoenzymes. It may be these areas that are important with respect to BE preference, with respect to substrate (amylose-like or amylopectin-like), as well as in size of chain transferred, or to the extent of branching.

3.14.4.5   Initiation of Starch Synthesis

Initiation of starch synthesis via a glucosyl-protein (as will be discussed later for glycogen synthesis in mammals) is a viable hypothesis. Tandecarz and Cardini[229] described a system which comprises at least two enzymatic reactions in which proplastid membranes from potato tuber glucosylate a membrane protein at a serine or threonine residue using UDP-glucose to form a glucoprotein. This product, a glucosylated 38 kDa protein, is used as an acceptor for a long chain of glucoses sequentially added in a α-1,4 bond using either ADP-glucose or UDP-glucose as donors. This system has been further characterized and one of the enzymes has been purified.[230,231] The potato enzyme catalyzes its own glycosylation;[232] the reaction requires Mn^{2+} (in a reaction similar to the self-glycosylation carried out by glycogenin in mammals). Although the enzymatic formation of the glucosyl-protein has been demonstrated in maize endosperm[233] not much information is available on the fate of the putative glucan protein.

3.14.5   THE PLASTIDS, SITE OF STARCH SYNTHESIS IN PLANTS

The site of starch synthesis in leaves and other photosynthetic tissues is the chloroplast. The starch formed during the day is degraded at night and the carbon is utilized to synthesize sucrose. The starch biosynthetic enzymes, i.e., ADP-Glc PPase, starch synthase, and branching enzyme, are present solely in the chloroplast (see refs.[12,234] for a review). For nonphotosynthetic tissues, there is ample information indicating that the starch biosynthetic enzymes are located within the amyloplast; for the ADP-Glc PPase see, for example, the studies on cultures of soybean cell,[235] wheat endosperm,[236] pea embryo,[220] and oilseed rape embryos.[237]

For nonphotosynthetic tissues, e.g., seeds and tubers, starch synthesis is also carried out in plastids. For example, the seed imports carbon and energy from the source (leaves) in the form of sucrose; in the seed, the site of starch synthesis is the amyloplast, a nonphotosynthetic organelle. Amyloplasts resemble chloroplasts in that they are enclosed by an envelope comprising two membranes[30] and in that they develop from proplastids. Sucrose has to be converted in another product before it can be taken up by the amyloplast, because the inner envelope is practically impermeable to sucrose. Although the events that lead to the flow of carbon into starch have been fairly well established for photosynthetic tissues, the situation is far less clear for storage tissues. This is because the amyloplast, the organelle in which starch is stored in sink tissues, is even more fragile than the chloroplast.

Although it is possible that many of the reactions occurring in chloroplasts also occur in the amyloplast, direct extrapolation is not possible. Indeed, the metabolism of the amyloplast, which is dependent on the cytosol for carbon and energy, is bound to differ, in many ways, from that of the chloroplast, which generates ATP and fixes CO_2. Information regarding amyloplast metabolism can be obtained in a number of ways, e.g., localization of the starch biosynthetic enzymes using immunocytochemical studies, measurement of enzyme activity in isolated amyloplasts, and measurement of uptake of labeled metabolites by isolated plastids.

To study the metabolism of a plastid, it is essential to isolate active plastids that are intact, free of cytosolic contamination and of other organelles, and in good yield. If the isolated plastids are good enough, they will provide reliable information on the enzymes present in them, what metabolites they can take up, at what rate, whether transport of a particular metabolite is passive or active, etc.

Keeling *et al.*[238] circumvented the problem by using a different approach: they supplied glucose or fructose labeled in [1-^{13}C] or in [6-^{13}C] to developing wheat endosperm and then examined the extent of redistribution of ^{13}C between carbons 1 and 6 in the starch glucosyl moieties. The redistribution was lower (12–20%) than would have been expected if carbon flow into starch had been by the C_3 pathway via triosephosphate isomerase. The authors then suggested that hexose monophosphates (rather than triose phosphates) were more likely to be the main source of energy and carbon for the amyloplast. Thus, the major carbon transport system for the wheat grain amyloplast would not involve triose-P but, most likely, hexose-P. If a triose-P/Pi transport system (like the one in the chloroplast envelope) were required for starch synthesis, the amyloplast should have fructose-1,6-biphosphatase, but Edwards and ap Rees[239] could not detect significant amounts of such an enzyme in amyloplasts from wheat endosperm. In search of a transport system capable of supplying carbon for starch synthesis, Tyson and ap Rees[240] incubated intact amyloplasts from wheat endosperm with different ^{14}C-labeled compounds, i.e., glucose, glucose-1-P, glucose-6-P, fructose-6-P, fructose-1,6-bisP, dihydroxyacetone-P, and glycerol-P. From these compounds, only glucose-1-P was incorporated into starch and this incorporation was dependent on the integrity of the amyloplast, a result consistent with the results of Keeling *et al.*[238] Direct import of six carbon compounds has also been reported for amyloplasts of potato, fava beans,[241] maize endosperm, and other tissues.[242] Hill and Smith[243] reported that glucose 6-phosphate was the preferred metabolite for starch synthesis by pea embryo amyloplasts and that ATP was also required. In pea roots[244,245] it was reported that a Pi translocator was active with dihydroxy-acetone-P, 3PGA, glucose-6-P, and P-enol-pyruvate.

Thus, there appears to be much diversity between the several translocators exchanging Pi with phosphorylated compounds (for review see refs.[242,246]) and it seems that the major transport system for most reserve nonphotosynthetic plant systems may be of hexose-P level and not triose-P; studies since the early 1990s are consistent with this view. A glucose-6-P translocator was found in intact plastids isolated from cauliflower buds[247–249] and from maize endosperm.[247] In amyloplasts isolated from a suspension of potato cultured cells, a glucose-1-P translocator was found.[250] In all these cases, the uptake of the hexose-P into the amyloplast was much higher than that of dihydroxy-acetone-P. Also, when cut spinach leaves were incubated in a 50 mM glucose solution for over 4 days, a glucose-6-P transporter was induced in the chloroplasts.[251] Figure 1 shows the interaction between the cytosol and amyloplast in reserve tissues by the hexose-P translocators.

Fruit chloroplasts assimilate CO_2 in much smaller quantities than those fixed in the leaf chloroplasts, and appear to import carbohydrates. The intact chloroplasts of green pepper fruits[252] and tomato fruit chloroplasts and chromoplasts[253] also have systems that translocate hexose-phosphates. Solubilized envelope proteins from tomato were reconstituted into liposomes. It was found that the liposomes containing leaf chloroplast proteins transported Pi, dihydroxy-acetone-P, and 3-PGA, and had low activity with glucose-6-P or glucose-1-P. However, the fruit chloroplast and chromoplast envelope proteins in addition to good translocation with the triose-phosphates also conferred activity with P-enol-pyruvate, glucose-6-P, and glucose-1-P.

The properties of the glucose-6-P translocator found in plastids of cauliflower bud and maize endosperm[247,248] and in chloroplasts from green pepper fruit[252] are interesting. For example, the translocator identified in the plastids from cauliflower bud has a molecular mass of 31.6 kDa;[247] transport of glucose-6-P was measured indirectly by incorporation of the label into the plastid starch, and was stimulated 6- to 40-fold by the presence of ATP and 3-PGA. The authors' interpretation was that ATP and 3PGA are needed for starch synthesis, and they postulated that in the cytosol these metabolites act as feedforward signals for starch synthesis. ATP and 3PGA would also be transported into the plastid and utilized for synthesis of ADP-glucose.

In the chloroplast, photosynthesis is the main source of ATP, but amyloplasts do not have that source of ATP. Glycolysis may then take on a more important function in the amyloplast than the one it has in the chloroplast, in that it would contribute to the production of amyloplastic ATP. Thus, the concentration of 3PGA could be an indicator of the ATP supply and of the availability of carbon in the amyloplast and, in this case, the regulatory effect of 3PGA on the ADP-Glc PPase from nonphotosynthetic tissues (i.e., stimulation and/or reversal of its inhibition by P_i) would have a physiological role similar to the one it has for the leaf enzyme.

As discussed above, authors using a variety of methods and plant systems reported that in nonphotosynthetic tissues the enzymes of starch biosynthesis appear to be restricted to the amyloplast. However, two reports have proposed that a significant portion of the ADP-Glc PPase activity may be present in the cytosol. Amyloplasts were isolated from wheat endosperm with intactness ranging from 41 to 89% by Thornbjørnsen *et al.*[254] The proportion of enzymatic activity recovered in the amyloplast fraction, in relation to total activity, was 13–17% for starch synthase and alkaline pyrophosphatase, and only 2.5% for the ADP-Glc PPase. On this basis, the authors calculated that

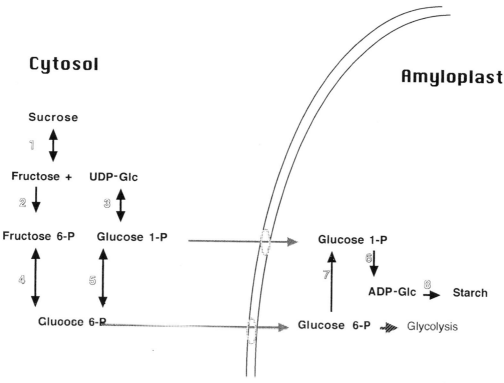

Figure 1 Interaction of cytosol and amyloplast with respect to sugar transport. The figure shows metabolism of sucrose to glucose-6-P and glucose-1-P in the cytosol and their transport into the amyloplast. The enzyme reactions shown are: 1, sucrose synthase; 2, fructokinase; 3, UDP-Glc PPase; 4, cytosolic P-hexoseisomerase; 5, cytosolic phosphoglucomutase; 6, ADP-Glc PPase; 7, plastidial phosphoglucomutase; and 8, starch synthase.

ADP-Glc PPase activity residing in the amyloplast was 15% of the total, and that the rest was in the cytosol. Immunological studies by the same authors detected two different isoforms of the ADP-Glc PPase, one mainly cytosolic and the other mainly plastidial. The authors indicated that there is an excess of ADP-Glc PPase activity in the amyloplast to account for the starch synthetic rate, and they were uncertain about the function of the putative cytosolic ADP-Glc PPase.

In another report,[255] preparations enriched in maize endosperm plastids contained 24 to 47% of the total activity of the plastid marker enzymes' starch synthase and alkaline pyrophosphatase, but they contained only 3% of the total ADP-Glc PPase activity. On this basis, the authors estimated that more than 95% of the ADP-Glc PPase activity was nonplastidial. Using antibodies prepared against the Bt2 subunit of the maize endosperm ADP-Glc PPase, they showed that most of the Bt2 protein was confined to the supernatant, and some was in the plastid. In *bt 2* mutant kernels, the cytosolic protein that reacted with the Bt2 antiserum was not detected but there was a plastidial form of ADP-Glc PPase. These data are somewhat different than those obtained by Miller and Chourey[256] and by Prisul[257] who, using immunogold labeling, detected the Bt2 protein in the amyloplast. If the data from Denyer *et al.*[255] are not artifactual; this would mean that there is more than one route for synthesis of ADP-Glc in maize endosperm. Most authors believe that carbon translocated into the plastid via a glucose-6-P translocator is converted to ADP-Glc by the action of the (plastidial) phosphoglucomutase and the ADP-Glc PPase.[247] Conversely, Denyer *et al.*[255] believe that some ADP-Glc synthesis goes on in the amyloplast, catalyzed via a plastidial ADP-Glc PPase, but, since in their model most of the ADP-Glc is synthesized in the cytosol, it must be translocated into the plastid for starch synthesis; thus, their model demands an ADP-Glc transporter. However, no protein with those properties has been identified. Although ADP-Glc uptake by the *Acer pseudoplatanus* amyloplasts has been reported,[258] Borchert *et al.*[245] and Batz *et al.*[259] showed that this ADP-Glc transport may be not physiologically relevant. *In vitro*, ADP-Glc may be translocated via the ATP/ADP translocator, but, since both ADP and ATP at concentrations lower than their physiological concentrations effectively inhibit ADP-Glc uptake (in pea root and cauliflower bud amyloplasts), *in vivo* transport of ADP-Glc by the ATP/ADP translocator is unlikely to be relevant. On the other hand, the hypothetical ADP-Glc transporter required by Denyer's model remains to be found.

It has been suggested that the *Bt* 1 gene product may be the ADP-Glc transporter. The *Bt* 1 gene encodes a plastidial membrane associated protein[260,261] whose deduced amino acid sequence shows similarity to known adenine nucleotide transporters.[262] The *bt* 1 mutant is starch deficient and shows a high level of ADP-Glc concentration in the endosperm compared with the normal endosperm.[263] However, this is highly speculative and the Bt 1 protein remains to be studied, characterized, and its function determined.

To conclude, although much has been learned about the characteristics of the plastidial barley and maize endosperm ADP-Glc PPases, more research is needed, particularly regarding hexose incorporation into starch. In the meantime, the interaction between cytosol and amyloplast can be summarized as shown in Figure 1, on the basis of reliable information.

3.14.6 SYNTHESIS OF AMYLOPECTIN *IN VIVO*: A HYPOTHESIS ASSIGNING SPECIFIC ROLES TO STARCH SYNTHASES, AND TO BRANCHING AND DEBRANCHING ENZYMES

The "basics" of starch biosynthesis, both in leaves and in storage tissues, have been described above. The regulation of the first enzyme in the pathway, the ADP-glucose pyrophosphorylase is well understood and will be discussed below; this knowledge has made it possible to increase the starch content of potato tuber and tomato fruit, the first time that an increase in the accumulation of a useful natural product has been achieved by genetic transformation.[225,264] Many details, however, are missing from the general picture so it is not possible to give a precise description of how the synthesis of the starch granule starts, how amylopectin and amylose are made, or why starch granules from different species differ in their size, number per cell, and composition.

The cluster model of amylopectin structure, as postulated by Hizukuri,[2,265] is the accepted one. A feature of the model is the clustering of the α-1,6 linkage branch points in certain regions of the amylopectin and the occurrence of B chains of varying sizes: B1, about 19 glucose units long; B2 with 41; B3 with 69; and B4, about 104–115 glucose units long. The number of B3 and B4 chains are few as compared to the number of B2 and B3 chains. The B1 chains only extend into one area of clusters while the B2, B3, and B4 chains extend into two, three, and four cluster areas of α-1,6 branch linkage areas, respectively, and these areas are separated by 39 to 44 glucose units.[265]

What roles do the starch synthases (SS) and branching enzyme isozymes play in the formation of the crystalline starch granule and amylopectin structures? How is amylose formed? Why are starch granules from different species different in size and in the number per cell? These differences most probably are related to the SS specificities in chain elongation and to the size transfer of glucose chain units by BE and where the α-1,6 bond is formed after transfer. As pointed out in the discussion of BEs,[69,72,197] BEI transfers long chains (40 to > 100 DP) while BEII only transfers shorter chains (6–14 DP). Amylose is the preferred substrate for BEI while amylopectin is the preferred substrate for BEII. Thus, BEI may be more involved in synthesis of the interior B chains while BEII is involved in the synthesis of exterior A and B1 chains. This mode of action was apparent when the isozymes were expressed in *E. coli*; maize BEI did transfer longer chains, and BEII transferred shorter chains.[197]

Ball and co-workers have isolated various mutants of *Chlamydomonas* deficient in starch synthase activities: a granule-bound starch synthase (GBSS) deficient mutant, a soluble starch synthase II (SSSII) deficient mutant,[192] and a double mutant deficient both in GBSS and in SSSII.[193] As indicated in the discussion of the starch synthases (Section 3.14.4.3), the mutant defective in SSSII had only 20–40% of the starch content found in the wild type and the amylose fraction increased from 25% to 55% of total starch. This mutant also had a different amylopectin, with an increased amount of short chains (2–7 DP), and a decrease of intermediate-size chains (8–60 DP). This suggests that the SSSII is involved in the synthesis or maintenance of the intermediate-size chains (mainly B chains) in amylopectin. The higher amylose content could be explained because of the failure of the SSSII-defective mutant to make extended chains.

Studies (prior to those on the GBSS of *Chlamydomonas*) had shown that a deficiency of GBSS activity in maize, rice, barley, and sorghum endosperms and in potato tuber resulted in a loss of the amylose fraction in the starch granule, with little effect on amylopectin. Thus, GBSS was considered to play a major role in amylose synthesis. This view was reinforced by transforming potato to produce antisense RNA (from a gene construct containing GBSS cDNA in reverse orientation); total suppression of GBSS activity led to tubers containing amylose-free starch,[78] and, once again, the amylopectin fraction did not appear to be affected.

The double mutants defective in SSSII and GBSS in *C. reinhardtii*,[193] however, had a starch content of only 2–16% of the wild-type. The severity of the defect of the double mutant led to an almost null mutant containing very little starch. For this reason, the authors suggested that GBSS is very important for synthesis not only of amylose, but also of the internal structure of the amylopectin; the effect of GBSS deficiency would be worsened by diminished SSSII activity. These studies using *Chlamydomonas* mutants provide evidence for the involvement of the GBSS not only in amylose but also in amylopectin synthesis, and suggest that one function for SSSII would be the synthesis of the intermediate-size B branch chains in amylopectin.

From all of these data, a possible route for the biosynthesis of amylopectin and amylose can be proposed, as shown in Figure 2. A reaction with the potential of being the one starting it all has been found in potato tuber,[231] and maize endosperm,[233] involving a transfer of glucose from UDP-Glc to serine or threonine residues. A glucosylated 38 kDa protein is formed that may serve as a primer for the synthesis of starch via the starch synthase reactions. After formation of this unbranched maltodextrin-protein, high rates of polysaccharide formation may occur at the surface of the developing starch granule, where GBSS, SSSII, and branching enzyme I interact with the glucosylated protein primer to form a branched α-glucan containing both long and intermediate-size chains; whether there is a reaction that transfers glucose from ADP-Glc to an acceptor protein has not been demonstrated.

PHASE

1.   UDP-Glc (ADP-Glc?) + acceptor protein — — —> glucosylated-protein + UDP

        BEI + SSSII
2.   ADP-Glc — — — — — — — —> Long and intermediate-size branched + ADP
        + GBSS   (B chains) glucan

                              BEII + SSSI
3.   Long and intermediate-size + ADP-Glc — — — — — —> addition of A and shorter
       branched (B chains) glucan   B chains for first
                                                         amylopectin cluster
                                                         structure

4.   repeat of phases 2 and 3 for synthesis of pro-amylopectin (phytoglycogen)

              debranching enzyme
5.   pro amylopectin — — — — — — — — — — — —> amylopectin + pro-amylose

                        GBSS
6.   pro-amylose + ADP-Glc — — — — — —> amylose + ADP

Figure 2   Proposed synthesis of amylose and amylopectin. Phase 1: initiation of α-glucan synthesis via synthesis of a glucoprotein primer for starch synthases. Phase 2: formation of the internal cluster structure of the ultimate amylopectin product by GBSS, SSSII AND BEI. Phase 3: formation of the external cluster structure (exterior A and B chains) by SSSI and BEII. Phase 4: continuous repeats of phase 2 and 3 reactions to form completion of a highly branched pro-amylopectin (photoglycogen). Phase 5: debranching of pro-amylopectin to form amylopectin which can now crystallize and the amylose primer, "pro-amylose." Phase 6: formation of amylose by elongation by GBSS.

Phase 2 in Figure 2 is postulated on the basis of the study of the polysaccharide structures in the *Chlamydomonas* SSSII and GBSS defective mutants, as well as the *ae* mutants of rice[222] and maize[213] which are defective in BEII. The BEII-deficient mutants have altered oligosaccharides with fewer branches and longer branched chains. In phase 3, SSSI and BEII are responsible for the synthesis of the A- and exterior B-chains to complete the first cluster region in the glucan. Continued synthesis in phase 4 is essentially a repeat of phases 2 and 3 to synthesize a highly branched α-glucan termed pro-amylopectin or phytoglycogen. This highly α-branched glucan is water soluble and noncrystalline. In phase 5, a debranching enzyme debranches the pre-amylopectin to form amylopectin which can now crystallize. In phase 6, the chains, liberated by the debranching action of the pro-amylopectin (phytoglycogen) are used as primers by GBSS to form amylose. Amylose synthesis may occur only inside the starch granule and only GBSS would be involved because it may be the only starch synthase present at the site of amylose synthesis. Inside the granule, branching enzyme activity is quite restricted and, therefore, the amylose would only be slightly branched. Possibly, the slight branching observed in the amylose fraction had occurred previously before the debranching phase, i.e., phase 5. Debranching of the pro-amylopectin may have liberated primer for GBSS that had some branch chains.

The reason for postulating a water-soluble pro-amylopectin is based on the existence of the *sugary* 1 mutation of maize endosperm which contains reduced amounts of amylopectin and starch granules. The mutant accumulates about 35% of its dry weight as phytoglycogen, a highly branched water-soluble polysaccharide.[82] The *sugary* 1 mutation was shown to be deficient in debranching enzyme activity. The evidence that the *sugary* 1 mutation affects the structural gene for a debranching enzyme is supported by the isolation of a cDNA of the *su* 1 gene. Its deduced amino acid sequence is most similar to a bacterial isoamylase.[83] It remains to be shown whether the *su* 1 gene product debranching enzyme activity is actually an isoamylase, or pullulanase or an R enzyme.[2] Moreover, the specificity of the reaction needs to be studied with respect to the factors that determine which α-1,6 linkages are cleaved and which remain resistant to debranching action. It is quite possible that the crowding of the α-1,6 linkages in the cluster region causes some steric difficulties for the debranching of the linkages in the cluster region, but this is only a hypothesis.

These reactions do not have to occur in perfect sequence and the phases may have some overlap, e.g., phases 2, 3, and 4 may overlap, and possibly even 5 and 6. However, the evidence, such as intermediate products formed by starch mutants of *Chlamydomonas* and of higher plants, supports the sequence of reactions shown in Figure 2 for amylopectin and amylose biosynthesis. Further experiments are certainly required to test the proposed scheme in Figure 2.

The data available on the localization of branching enzyme within the plastid have been obtained with potato[266] using antibodies raised against potato BE and immunogold electron microscopy. The enzyme (which would be the equivalent of the BEI isoform of maize, as discussed above) was found in the amyloplast, concentrated at the interface between stroma and starch granule, rather than throughout the stroma, as is the case with the ADP-glucose pyrophosphorylase.[267] This would explain how amylose synthesis is possible when the enzyme responsible for its formation, i.e., the Wx protein, one of the granule-bound starch synthases, is capable of elongating both linear and branched glucans. The spatial separation of branching enzyme and granule-bound starch synthase, even if only partial, would allow the formation of amylose without it being subsequently branched by the branching enzyme. However, even if spatial separation did not exist, starch crystallization would have the same effect, i.e., prevent further branching. Morell and Preiss[268] found about 5% of the total branching enzyme activity associated with the starch granule after amylase digestion. Whether the branching enzyme associated with the starch granule was similar to the soluble branching enzymes was not determined, but Preiss and Sivak,[17] using SDS polyacrylamide electrophoresis, found among the proteins present in maize and pea starch of maize starch, a polypeptide of about 80 kDa that reacted with antibodies raised against maize BEII. It is worth noting, however, that small amounts of BE are expected to sediment with the starch granule because of its affinity for the polysaccharide. These results have been confirmed by Mu-Forster *et al.*[174]

3.14.7 REGULATION OF THE SYNTHESIS OF BACTERIAL GLYCOGEN AND STARCH

3.14.7.1 Introduction

The main regulatory site for bacterial glycogen synthesis and for plant and algal starch synthesis is different from that of mammalian glycogen synthesis. The differences in the mode of regulation for the whole pathway are probably connected to the fact that the glucosyl donors are different. ADP-glucose is the glucosyl donor in the bacterial and plant α-glucan systems, and UDP-glucose, the glucosyl donor for mammalian glycogen synthesis, is also utilized for the synthesis of other sugar nucleotides, mainly UDP-galactose and UDP-glucuronate, precursors for the synthesis of several cellular constituents. Thus, the first unique reaction for mammalian glycogen synthesis is the glycogen synthase step, and it is here that both allosteric control and hormonal-mediated control are exerted. Conversely, in bacteria and in plants the only physiological function for ADP-glucose is to be a donor of glucose for α-1,4-glucosyl linkages, and for this reason, it is advantageous to conserve the ATP utilized for synthesis of the sugar nucleotide by regulating glucan synthesis at the level of ADP-glucose formation.

Over 50 ADP-Glc PPases (mainly bacterial but also plant) have been studied with respect to their regulatory properties and, in almost all cases, glycolytic intermediates activate ADP-Glc synthesis, while AMP, ADP, and/or P$_i$ are inhibitors. Glycolytic intermediates in the cell can be considered as indicators of carbon excess and therefore, under conditions of limited growth with excess carbon in the media, accumulation of glycolytic intermediates would be signals for the activation of ADP-Glc synthesis. Thus, the enzyme seems to be modulated by the availability of ATP in the cell and the presence of glycolytic intermediates.

3.14.7.2 Activators and Inhibitors of ADP-Glc PPase

Based on the differences observed in the metabolic intermediates that activate the ADP-glucose pyrophosphorylases studied so far, the enzymes may be classified into seven groups. These groups are listed in Table 6.

Table 6 ADP-glucose pyrophosphorylases from different plant, algal, and bacterial species, classified according to activator specificity.

Species	Activator(s)
Rhodospirillum spp. (*Rh. fulvum, Rh. molischianum, Rh photometricum, Rh. rubrum, Rh. tenue*), *Rhodocyclus purpureus*	Pyruvate
Aphanocapsa 6308, *Synechococcus* 6301, *Synechocystis* PCC6803, *Anabaena* PCC 7120, *Chlorella pyrenoidosa, Chlorella vulgaris, Chlamydomonas reinhardtii, Scenedesmus obliquus,* plant tissues (leaf or reserve tissue)	3-P-glycerate
Agrobacterium tumefaciens, Arthrobacter viscosus, Chlorobium limicola, Chromatium vinosum, Rhodobacter spp. (*R. gelatinosa, R. blastica, R. capsulata, R. palustris), Rhodomicrobium vannielii*	Pyruvate Fructose-6-P
Rhodobacter spp. (*R. gelatinosa, R. globiformis, R. spheroides*)	Pyruvate Fructose-6-P Fructose-1,6-bis-P
Aeromonas hydrophila, Micrococcus luteus, Mycobacterium smegmatis, Rhodopseudomonas viridis, Thermus caldophilus Gk-24	Fructose-6-P Fructose-1,6-bis-P
Citrobacter freundii, Edwardsiella tarda, Enterobacter aerogenes, Enterobacter cloacae, Escherichia aurescens, Escherichia coli, Klebsiella pneumoniae, Salmonella enteritidis, Salmonella typhimurium, Shigella dysenteriae	Fructose-1,6-bis-P Pyridoxal-5-P NADPH
Clostridium pasteurianum, Enterobacter hafniae, Serratia spp. (*Ser. liquifaceans, Ser. marcescens*)	None

A group of ADP-Glc PPases comprises the enteric bacteria (*Citrobacter freundii, Edwardsiella tarda, Escherichia coli, Escherichia aurescens, Enterobacter aerogenes, Enterobacter cloacae, Klebsiella pneumoniae, Salmonella enteritidis, Salmonella typhimurium, Shigella dysenteriae*) and they are activated by fructose-1,6-diphosphate, NADPH, and pyridoxal phosphate.[38,269-273]

Another group of ADP-Glc PPases with a different allosteric activator specificity comprises the phototrophic bacteria belonging to the genus previously called *Rhodopseudomonas*, now called *Rhodobacter*, and to *Rhodomicrobium* (*R. acidophila, R. capsulata, R. palustris, Rm. vanniellii*) or to the families Chromatiaceae and Chlorobiaceae. They have as activators both fructose-6-phosphate and pyruvate.[62,274] Included in this group are *Agrobacterium tumefaciens*[270] and *Arthrobacter viscosus*.[275-277]

The ADP-Glc PPase from the anaerobic photosynthetic *Rhodobacter viridis* is unusual in that pyruvate does not affect it, but it is activated by fructose-6-phosphate and fructose-1,6-diphosphate. ADP-Glc PPases with this activator specificity are another group, also found in nonphotosynthetic organisms such as Aeromonads, gram-negative facultative anaerobes, or the gram-positive aerobic organisms *Micrococcus luteus* and *Mycobacterium smegmatis*.[278,279]

The ADP-Glc PPases in another group are isolated from bacteria of the genus *Rhodospirillum* (*R. rubrum, R. molischianum, R. tenue, R. fulvum, R. photometricum*) and *Rhodocyclus purpureus*, and are activated solely by pyruvate.[62,96,280,281]

Another group of prokaryotes, this one capable of oxygenic photosynthesis, are the cyanobacteria, and their ADP-Glc PPases are activated by 3-phosphoglycerate (3PGA), the primary CO_2 fixation product of photosynthesis;[128,147,148] 3PGA is also the primary activator of the green algae[104,282] and higher plant ADP-Glc PPases.[11,17,283] Thus, the initial product of photosynthesis serves as an allosteric activator for the synthesis of a reserve product, glycogen, or starch. The specificity of the activation is the same irrespective of whether the enzyme is from a plant that uses the C3 pathway or the C4 pathway of photosynthesis. The ADP-glucose pyrophosphorylases of *Chlorella pyrenoidosa*, *Chlorella vulgaris*, *Scenedesmus obliquus*, and *Chlamydomonas reinhardtii*, and of the cyanobacteria are also activated by 3PGA and are inhibited by Pi. The presence of 3PGA also increases the apparent affinity of the spinach leaf enzyme for its substrates from 2- to 13-fold. Conversely, 22 μM Pi (in the absence of activator at pH 7.5) inhibits ADP-glucose synthesis by the spinach leave enzyme by 50%;[101] in the presence of 1 mM 3PGA, 50% inhibition required 1.3 mM phosphate. Thus, the activator decreased sensitivity to Pi inhibition about 450-fold; and Pi at 0.5 mM increased the concentration of 3PGA needed for activation.

The ADP-Glc PPases isolated from *R. sphaeroides*, *R. gelatinosa*, and *R. globiformis*[284–286] constitute a group and are activated by fructose-1,6-diphosphate in addition to fructose-6-phosphate and pyruvate.

There are some enteric organisms from the genus *Serratia* (*S. liquefaciens* and *S. marcescens*) and from *Enterobacter hafniae* that contain an enzyme that is not activated by any of the metabolites tested.[99] Another ADP-Glc PPase with no apparent activator is that isolated from *Clostridium pasteurianum*.[287] These enzymes may be classified as a seventh group.

3.14.7.3 Overlapping Specificities of the ADP-Glc Pyrophosphorylase Activator Binding Site

In the seven regulatory groups, the dominant activators are pyruvate, fructose-6-phosphate, 3PGA, and fructose-1,6-bisphosphate. All the ADP-Glc PPases isolated from the photosynthetic anaerobic bacteria (except for *R. viridis*) are activated by pyruvate. A number are also activated by fructose-6-phosphate, and a few are activated by a third metabolite, fructose-1,6-bisphosphate. Fructose-1,6-bisphosphate is an activator of the enzyme from enteric organisms, as well as those from the Aeromonads, *M. luteus* and *M. smegmatis*, but fructose-6-phosphate, an effective activator of the ADP-Glc PPases in these organisms, is not an activator for the enteric enzymes. Also, 3PGA, a highly effective activator for the enzyme from the cyanobacteria, green algae, and plant tissues, is a poor activator for the enteric enzymes. Conversely, fructose-1,6-bisphosphate activates the plant leaf enzymes, but less effectively than 3PGA. Whereas 28 μM fructose-1,6-bisphosphate is required for 50% of maximal stimulation of the spinach leaf enzyme at pH 8.5, only 10 μM 3PGA is required for the same effect.[101] Moreover, the maximum stimulation of V_{max} effected by fructose-1,6-bisphosphate (16-fold) is considerably less than that observed for 3PGA (58-fold) at pH 8.5. In contrast, fructose-1,6-bisphosphate stimulates (30-fold) the rate of ADP-glucose synthesis catalyzed by *E. coli* B ADP-Glc PPase, while the same concentration of 3PGA gives only a 1.5-fold stimulation.

This overlapping of specificity for the activators in the various ADP-Glc PPase classes suggests that the activator sites for the different groups are similar or related. Indeed, looking at the deduced amino acid sequences of many of the ADP-Glc PPases,[9,111] there is much similarity of amino acid sequences, particularly at the allosteric binding sites as well as the substrate binding sites. One may therefore speculate that, during evolution, mutation of the gene at the activator binding site of the ADP-Glc PPase occurred, thus modifying the activator specificity. The pressure for change may have come from a need of coordination (or compatibility) between the metabolite activator and the major carbon assimilation and dissimilation pathways of the organism.

Metabolites associated with energy metabolism, AMP, ADP, or inorganic phosphate, are inhibitors of the ADP-Glc PPases. The enteric ADP-Glc PPases,[97,272,288] including those found in the genus *Serratia*[272] and *Enterobacter hafniae*,[278] are very sensitive to AMP inhibition. The plant, algal, and cyanobacterial enzymes, however, are highly sensitive to inorganic phosphate.[17,62,101,104,128,148,289] Other ADP-Glc PPases are sensitive to either Pi, ADP, or AMP.[290] Thus, ADP-Glc and glycogen synthesis only proceeds when ATP availability is high (i.e., the energy state of the cell is high). Exceptions to this rule would be the enzymes from Aeromonads,[278] *M. smegmatis*,[279] *R. rubrum*,[96] and *R. molischianum*[281] in which Pi, ADP, and AMP, at concentrations of 5 mM or less, cause little or no inhibition. The peptide domains involved in the inhibitor binding sites remain to be identified. However, some data obtained with the *E. coli* enzyme[114,121,123,124,291] strongly suggest that the inhibitor binding site overlaps with both the activator and substrate ATP binding sites.

3.14.7.4 Effect of the Activators on Enzyme Kinetics

Kinetic studies on several ADP-Glc PPases show that the presence of the activator in reaction mixtures usually lowers the concentration of the substrates, ATP, glucose-1-phosphate, pyrophosphate ADP-glucose, and the cationic activator Mg^{2+} (or Mn^{2+}), required for 50% of maximal velocity (K_m or $S_{0.5}$), about 2- to 15-fold. The apparent affinity of the enzyme for the substrate is thus increased in the presence of activator. The activator also increases maximal velocity and k_{cat}, 2- to 60-fold depending on the pH and the particular ADP-Glc PPase studied. The prime function of the allosteric activator may be, however, to reverse the sensitivity of the enzyme to inhibition by Pi, AMP, or ADP, which are usually noncompetitive with the substrates. Indeed, for many ADP-Glc PPases, relatively high concentrations of the activator can completely reverse the inhibition caused by AMP, Pi, or ADP. A well-studied system is the *E. coli* B enzyme, where fructose-1,6-bisphosphate modulates the sensitivity to AMP inhibition.[288,292] The $S_{0.5}$ (concentration giving 50% inhibition) for 5'-AMP is about 70 μM at 1.7 mM fructose-1,6-bisphosphate. With lower concentrations of activator, however, the $S_{0.5}$ for AMP is lower and, e.g., at 60 μM fructose-1,6-bisphosphate, only 3.4 μM AMP is needed for 50% inhibition. For the *Rhodospirillum tenue* ADP-Glc PPase, the interaction between the activator, pyruvate, and the inhibitors inorganic phosphate, ADP and AMP, is illustrated in Table 7. Pyruvate increases the reaction rate threefold,[281] half-maximal stimulation ($A_{0.5}$) occurs at 28 μM. At low concentrations of activator (15 μM), 0.5 mM AMP inhibits by 90%, but this inhibition is almost completely negated by increasing concentrations of the activator, pyruvate. At 1 mM pyruvate, there is no inhibition. AMP, at 62 μM, inhibits the enzyme 50% but in the presence of pyruvate at 50 μM, a concentration giving 60% of the maximal velocity, the AMP concentration required for 50% inhibition, $I_{0.5}$, was 260 μM. AMP has two effects on the enzyme; it increases the $A_{0.5}$ value of pyruvate (from 28 μM to 140 μM in the presence of 0.5 mM AMP) and increases the sigmoidicity of the activation curve. The Hill coefficient increases from 1.0 in the absence of AMP to 1.8 in the presence of 0.5 mM AMP.

Table 7 Interaction of inhibitors, AMP, with the *Rhodospirillum tenue* ADP-Glc PPase with the activator, pyruvate.

Activator/Inhibitor	$I_{0.5}$[a] (mM)	ñ	$A_{0.5}$[b] (mM)	ñ
No activator	0.62	1.0		
Pyruvate, 0.015 mM	0.095	1.0		
Pyruvate, 0.05 mM	0.26	1.0		
No inhibitor			0.028	1.0
AMP, 0.2 mM			0.078	1.4
AMP, 0.5 mM			0.14	1.8

[a] $I_{0.5}$ is the concentration required for 50% inhibition. [b] $A_{0.5}$ is the concentration giving 50% of maximal activation.

Similar interactions are seen in the case of the potato tuber ADP-Glc PPase, as shown in Table 8, which also illustrates that relatively small changes in the concentrations of Pi and 3PGA lead to large effects on the rate of ADP-glucose synthesis, particularly at low concentrations of 3PGA where the activation is minimal in the presence of Pi. At 1.2 mM Pi and 0.2 mM 3PGA, ADP-Glc synthesis is inhibited over 95%. However, if the Pi concentration decreases by 33% to 0.8 mM and the 3PGA concentration increases by 50% to 0.3 mM there is an 8.5-fold increase in the rate of ADP-Glc synthesis. Conversely, at 0.4 mM 3PGA and 0.8 mM Pi, the rate of ADP-Glc synthesis is 7.5 nmol per 10 min; this is reduced to 2.2 nmol per 10 min (70% decrease) if only the 3PGA concentration decreases 50% to 0.2 mM. If the Pi concentration is also increased, to 1.2 mM (a 50% increase) then the synthetic rate falls to 0.65 nmol which is a reduction of ADP-Glc synthesis of 91%. The reason for the small changes in the effector concentrations giving such large effects in the synthetic rate is due to the sigmoidal nature of the curves particularly at low concentrations of

3PGA. Below, we will present strong evidence that the ratio of activator/inhibitor modulates the activity of ADP-Glc PPase not only *in vitro*, as in Table 8, but also *in vivo* in bacteria and in plants, thereby regulating the synthesis of α-1,4 glucans in these systems.

Table 8 Interaction of 3-P-glycerate and phosphate concentrations in modulating potato tuber ADP-Glc PPase activity.

[Orthophosphate] (mM)	0.4	0.8	1.2
[3PGA]	ADP-glucose formed		
(mM)	nmol per 10 min		
0.0	0.16	0.18	0.08
0.1	2.1	0.49	0.16
0.2	7.2	2.2	0.65
0.3	9.9	5.5	2.7
0.4	11.3	7.5	3.8
0.5	11.8	8.5	6.6
0.6	12.8	10.3	7.8
0.7	14.2	11.0	9.0
1.0	16.9	13	10.6

3.14.7.5 Experimental Evidence for the Role of ADP-Glc PPase in the Regulation of the Biosynthesis of α-1,4-Glucans

3.14.7.5.1 Plant systems

The leaf ADP-Glc PPase is highly sensitive to 3PGA, the primary product of CO_2 fixation by photosynthesis, and to inorganic phosphate (Pi). Thus, it has been suggested that these compounds play a significant role *in vivo* in regulating starch biosynthesis in higher plants and algae, and in regulating glycogen synthesis in cyanobacteria.[9,11–15,25,283,290] In the light, the concentrations of ATP and reduced pyridine nucleotides rise, leading to the formation of sugar phosphates from 3PGA. At the same time, concentrations of Pi decrease and that of 3PGA increases, thus increasing the activity of the ADP-Glc PPase and starch synthesis. Conversely, in the dark, phosphate concentration increases, while the concentrations of 3PGA, ATP, and reduced pyridine nucleotides decrease, leading to inhibition of ADP-glucose synthesis and therefore of starch synthesis.

Data on the correlation between changes in the cellular concentrations of Pi and/or 3PGA and starch content or rates of starch synthesis have been discussed in previous reviews,[11,12,189] and indicate that *in vivo*, 3PGA and Pi levels affect starch synthetic rates via modulation of ADP-Glc PPase activity. Since those reviews, experimental evidence of a different nature has been obtained indicating that the regulatory effects seen for the plant and algal ADP-Glc PPase are important in the determination of the rate of starch synthesis *in vivo*. Kacser–Burns control analysis methods[293,294] vary enzyme activity, either by using mutants deficient in that enzyme or by varying the physiological conditions and correlating the effect of these changes on the rate of a metabolic process (e.g., starch synthesis). If the enzyme activity is rate-limiting or important in controlling the metabolic process, then a large effect on that process should be seen. Conversely, if there is no (or little) effect, then the activity of that particular enzyme is not considered to be rate-limiting for the overall metabolic process being measured. The influence of an enzyme is quantified as a flux coefficient ratio. If the variation of enzyme activity determines a commensurate change in the rate of the process measured, a correlation ratio close to one should be observed.

The Kacser–Burns approach was used in the analysis of starch synthesis by mutants of *Arabidopsis thaliana*,[295,296] and showed that the leaf ADP-Glc PPase, and how it is regulated by 3PGA[296] are important determinants of the rate of starch synthesis *in vivo*.[295] *A. thaliana* mutant strains containing only 7% of the wild-type activity of ADP-Glc PPase, and a hybrid of mutant and wild type with 50% activity, had 90% and 39% of the wild type's starch synthetic rate, at high light intensity.[296] This is a fairly good correlation between the activity of the ADP-Glc PPase and the rate of starch synthesis; the flux control coefficient was determined to be 0.64.

Despite the fairly high value seen for the ADP-Glc PPase flux control coefficient, it is quite possible that it may be underestimated due to the allosteric properties of the enzyme. For flux control analysis, the value of the maximal enzyme activity measured is used in the calculations. In the case of an allosteric enzyme the potential maximal enzyme activity may not be as critical as the allosteric effector concentrations that determine the actual enzyme activity *in situ*. With ADP-Glc PPases, activation by 3PGA can be anywhere from 10- to 100-fold. Moreover, inhibition by the allosteric inhibitor, Pi and variations in the [3PGA]/[Pi] ratio could cause greater fluctuations in the potential maximal activity. Thus, flux coefficient control values based on only the enzyme's maximal activities of the *Arabidopsis thaliana* mutants and normal ADP-Glc PPases are likely to underestimate the regulatory potential of the ADP-Glc PPase reaction.

In experiments utilizing a *Clarkia xantiana* mutant, deficient in leaf cytosolic phosphoglucoseisomerase (with only 18% of the activity seen in the wild type), lower sucrose synthesis rates and increased starch synthesis rates were observed.[295] The chloroplast 3PGA concentration increased about 2-fold, suggesting that the increase of starch synthesis rate measured in the mutant deficient in cytosolic phosphoglucoseisomerase was due to activation of the ADP-Glc PPase by the increased 3PGA concentration and the 3PGA/Pi ratio.

Important evidence indicating that the *in vitro* activation of the ADP-Glc PPase is truly relevant *in vivo* comes from isolation of a class of mutants where the mutation directly affects the allosteric properties of the ADP-Glc PPase. Such mutants were found easily for the bacteria *E. coli* and *Salmonella typhimurium*.[8,14,24] Similar mutants have been found for *Chlamydomonas reinhardtii* and for maize endosperm. A significant finding was made by Ball *et al.*[297] who isolated a starch-deficient mutant of *C. reinhardtii* in which the defect was shown to be in the ADP-Glc PPase, which could not be effectively activated by 3PGA. The inhibition by Pi was similar to the wild type.[298] The starch deficiency was observed in the mutant whether the organism was grown photoautotrophically with CO_2 or in the dark with acetate as the carbon source. Thus, the allosteric mechanism seems to be operative for photosynthetic or nonphotosynthetic starch biosynthesis.

Another putative ADP-Glc PPase allosteric mutant from maize endosperm, which has 15% more dry weight (in addition to starch) than the normal endosperm, has been isolated and described by Giroux *et al.*[299] The mutant allosteric ADP-Glc PPase was less sensitive to Pi inhibition than the normal enzyme.

The *Chlamydomonas* starch-deficient mutant and higher dry weight maize endosperm mutant studies strongly suggest that the *in vitro* regulatory effects observed with the photosynthetic and nonphotosynthetic plant ADP-Glc PPases are highly functional *in vivo* and that ADP-Glc synthesis is rate-limiting for starch synthesis. Thus, data continue to accumulate showing the importance of the plant ADP-Glc PPase in regulating starch synthesis and that the allosteric effectors, 3PGA and Pi, are important *in vivo*, in photosynthetic as well as in nonphotosynthetic starch synthesis.

3.14.7.5.2 Bacterial systems

Some mutants of *E. coli* B,[81,99,269,290,300] *E. coli* K12,[270,301] and *Salmonella typhimurium* LT-2[38] affected in their ability to accumulate glycogen were isolated and it was found that their activities of glycogen biosynthetic enzymes were similar to the wild type. These mutants were shown to contain ADP-Glc PPases altered in their regulatory properties.

Table 9 shows the maximum amounts of glycogen accumulated by four mutants of *E. coli* B and two of *S. typhimurium* LT-2 as compared with the respective parent strains. In minimal media containing excess glucose, the rate of glycogen accumulated in SG5, CL1136 and 618 is 2-, 3.5- and 3.7-fold greater, respectively, than that found in *E. coli* B.[81] For the mutants SG5 and CL1136, glycogen synthesis was about 2-, 3.5- and 3.5-fold greater. The *S. typhimurium* LT-2 mutants JP51 and JP23 accumulate 67% and 25% more glycogen than the parent strain. The activities of the glycogen biosynthetic enzymes in the mutants and in the parent strains were similar and thus could not account for the increased rate of accumulation of glycogen present in the mutants. Furthermore, the *E. coli* B mutant ADP-Glc PPases have approximately the same apparent affinities as the parent enzyme for the substrates ATP and glucose-1-P and for the activator Mg^{2+}. Although the affinities of the *S. typhimurium* mutant enzymes for the substrates have not been studied, it seems likely that, just as in *E. coli* enzymes, the only properties changed are the apparent affinities for the allosteric effectors.

The concentration of fructose-1,6-bisphosphate required for 50% of maximal activation ($A_{0.5}$) is about threefold less for the SG5 ADP-Glc PPase, 4.5-fold less for the 618, and 12-fold less for the CL1136 enzymes (Table 9). The mutant ADP-Glc PPases also display greater apparent affinities for

Table 9 Glycogen content of *E. coli* and *S. typhimurium* LT2 expressing the wild-type or allosteric mutant ADP-Glc PPases.

Strain	Maximal glucan content (mg g^{-1} cell)^a	Fructose 1,6-bisP $A_{0.5}$ (µM)^b	AMP $I_{0.5}$ (µM)^c	Mutation
E. coli B	20	68	75	
Mutant SG14	8.4	820	500	Ala-44→Thr
Mutant SG5	35	22	170	Pro-295→Cys
Mutant 618	70	15	860	Gly-336→Asp
Mutant CL1136	74	5	68	Arg-67→Thr
S. typhimurium				
LT2	12	95	110	
Mutant JP23	15	not activated	250	
Mutant JP51	20	84	490	

^{a}The bacterial strains were grown in minimal media with 0.75% glucose and the data are expressed as maximal mg of anhydroglucose units per gram (wet wt) of cells in stationary phase. ^{b}$A_{0.5}$ is the fructose 1,6-bisP giving 50% of maximal activation. ^{c}$I_{0.5}$ is the AMP concentration required for 50% inhibition.

the activators NADPH and pyridoxal-5′ phosphate than the wild-type *E. coli* B enzyme. The mutant ADP-Glc PPases are also less sensitive to the allosteric inhibitor AMP (Table 9). The concentrations of AMP necessary to inhibit the mutant ADP-Glc PPases by 50% are much higher than those required for inhibition of the wild-type enzyme; also, at 1.5 mM fructose-1,6-bisphosphate, the mutant enzymes are less sensitive than the parent enzyme to inhibition.

The energy charge is defined as [ATP] + 1/2[ADP]/[ATP] + [ADP] + [AMP].[300] When assayed under equivalent conditions and energy charge values,[99,269] the CL1136 and SG5 ADP-Glc PPases have more activity than the *E. coli* B wild-type enzyme. At an energy charge of 0.7, the SG5 enzyme has about five times more activity than the *E. coli* B with 1.5 mM fructose-1,6-bisphosphate as the activator.[269] At a range of energy charge of 0.85–0.9 and with 0.75 mM fructose-1,6-bisP, in approximately physiological conditions, the SG5 enzyme shows about twice as much activity as the wild-type *E. coli* B enzyme. The CL1136 ADP-Glc PPase is almost fully active in the energy charge range of 0.75–1.0 with fructose-1,6-bisP concentrations of 0.75 mM and higher.[99] Significant activity (10- to 25-fold of maximal activity) is seen even at an energy charge of 0.4 with fructose-1,6-bisP concentrations of 0.75 mM or higher. In the absence of the activator and at an energy charge value of 0.75, the activity of the CL1136 enzyme is almost 30% of the maximal activity. In contrast, the *E. coli* B ADP-Glc PPase activity is less than 4% of the maximal at an energy charge of 0.65, even in the presence of 3 mM fructose-1,6-bisP. At an energy charge level of 0.75 and with 3.0 mM fructose-1,6-bisphosphate, the *E. coli* B enzyme exhibits only 11% of its maximal activity, while the CL1136 enzyme shows 98% of its maximal activity under these conditions. The energy charge range in various *E. coli* strains in physiological conditions is in the range 0.85–0.91.[301,302] The physiological concentrations of fructose-bisP are between 0.71 and 3 mM.[301,303] Under these conditions, the CL1136 enzyme shows maximal activity and is not sensitive either to change in energy charge or to fluctuations in fructose-1,6-bisP; conversely, the wild-type enzyme is very sensitive to these ranges of fructose-1,6-bisP concentration and energy charge.[99,269] These studies strongly suggest that the increased accumulation of glycogen in the mutants SG5 and CL1136 is due to alterations in the structure of the ADP-glucose pyrophosphorylases that lead to a greater affinity for the activators and a lower affinity for the inhibitor. The observed correlation between the relative sensitivities of *E. coli* B and SG5 and CL1136 ADP-Glc PPases to inhibitor and activator and the rates of glycogen accumulation is in agreement with the view that the cellular levels of the allosteric activators and inhibitors of ADP-Glc PPase modulate the rate of synthesis and accumulation of glycogen in the cell.

The *S. typhimurium* ADP-Glc PPase mutants are of some interest because their altered kinetic properties are different from those seen for the *E. coli* allosteric mutants. As indicated in Table 8, both activator and inhibitor constants ($A_{0.5}$ and $I_{0.5}$) are affected by the mutation in *E. coli* mutants SG5, 618 and CL1136. Conversely, the *S. typhimurium* LT-2 mutant JP51 enzyme has an $A_{0.5}$ value for fructose-1,6-bisphosphate similar to that of the parent strain ADP-Glc PPase. However, the JP51 enzyme has about a 5-fold higher $I_{0.5}$ value for AMP than the parent enzyme, thus suggesting that the lesser sensitivity to AMP inhibition is the reason for higher accumulation of glycogen in mutant JP51. The JP23 mutant enzyme, in contrast to the parent strain enzyme, is not activated by fructose bis-P, i.e., it is not dependent on fructose-1,6-bisphosphate for full activity.[38] The mutant enzyme is also less sensitive to AMP inhibition than the wild-type strain enzyme (Table 9) and, in addition, AMP cannot inhibit it by more than 60%, either in the presence or absence of fructose-1,6-bisP. If the concentration of AMP in cells is assumed to be 0.15 ± 0.05 mM, then the JP23 ADP-Glc PPase activity could vary between 50% and 60% of its maximal activity. Under similar conditions (1 mM fructose-1,6-bisphosphate and AMP concentration indicated above), the *S. typhimurium* LT-2 enzyme displays 5–50% of its total activity.[38] Since JP23 crude extracts have 50% of the activity of the extracts of LT-2, the JP23 enzyme activity would be 0.7–4.9 times as active as the LT-2 enzyme over this concentration range of AMP. The alteration of the sensitivity of the ADP-Glc PPase to AMP inhibition could, by itself, explain why mutant JP23 can accumulate more glycogen than LT-2, Dietzler *et al.*[301,303] showed that when the growth of *E. coli* W4597(K) ceased because of nitrogen limitation, glycogen accumulation rates increased about 3.3–4.2-fold. There was also a decrease in the concentration of fructose-1,6-bisphosphate of about 76% (from 3.1 mM to 0.71 mM) and an increase in the energy charge from 0.74 (in the exponential phase) to 0.87 (in the stationary phase). The total concentration of the adenylate pool (ATP+AMP+ADP) in exponential and stationary phase was 3 mM. Although the concentration of fructose-1,6-bisphosphate decreased by 76% from exponential to stationary phase, Dietzler *et al.*[301,303] concluded that the increase in energy charge would more than offset this decrease in the concentration of allosteric activator. The conclusion of Dietzler *et al.*[303] was based on the data of Govons *et al.*[269] which showed a 3-fold increase in *E. coli* B ADP-glucose pyrophosphorylase activity when the energy charge increased from 0.74 to 0.87, with a concomitant decrease of fructose-1,6-bisphosphate concentration from 1.5 mM to 0.5 mM. The total adenosine nucleotide pool was 2 mM in these *in vitro* experiments.

In various *E. coli* strains grown in nitrogen-limiting minimal media, with glucose as the carbon source, the total adenine nucleotide concentration is about 3 mM,[301,304,305] the fructose-1,6-bisP concentration ranges from 0.71 mM to 3.2 mM,[301,303–305] the mass action ratio of the adenylate kinase reaction ranges from 0.37 to 0.69, and energy charge values lie between 0.74 and 0.9. While glucose-1-phosphate concentrations in *E. coli* cells have not been determined, they have been calculated from the published glucose-6-phosphate values reported for *E. coli* HFr 139[303] and *E. coli* strains W4597(K) or G34(416), with the assumption that the phosphoglucomutase reaction is at equilibrium *in vivo* ($K_{equil} = 17$ at 37 °C). Determinations of the glucose-1-phosphate concentration range from 39 to 45 μM.

Using the above values of fructose-1,6-bisP, adenine pool concentrations, and an adenylate kinase mass action of 0.45, the response of the *E. coli* B ADP-Glc PPase to energy charge was calculated.[99,290] The *E. coli* B enzyme shows only 4% of its maximal velocity at an energy charge value of 0.75 with 0.75 mM fructose-1,6-bisphosphate. This increases to 37% of maximal velocity at energy charge 0.9. With 3 mM fructose-1,6-bisphosphate, the *E. coli* B enzyme activity is 11% of its maximal velocity at energy charge 0.75, and this increases to 78% maximal velocity at energy charge 0.9. The rate of ADP-Glc synthesis in extracts of exponential phase *E. coli* B, is sevenfold greater than the glycogen accumulation rate and about 10-fold greater in stationary phase cell extracts.[99,306] Thus only 10–14% of the maximal ADP-glucose synthetic activity is needed to account for the observed glycogen accumulation rates. Glycogen degradation rates in *E. coli* appear to be two orders of magnitude less than synthetic rates.[307] Thus, glycogen accumulation rates may be considered to be determined solely by glycogen synthetic activity. Taking into account that the calculated glucose-1-phosphate concentration is at subsaturating concentrations for the ADP-Glc PPase in *E. coli* B, the percentage of maximal activity can be calculated at various charge values and fructose-1,6-bisphosphate concentrations in the physiological range. Sufficient *E. coli* B enzyme activity to account for the glycogen accumulation rate in exponential phase would be seen at energy charges of 0.87 with 0.75 mM fructose-1,6-bisphosphate, or at 0.85 with 1.5–3.0 mM fructose-1,6-bisphosphate. In stationary phase, sufficient ADP-glucose synthesis rates are also observed at energy charge values of 0.8 with 2.5–3.0 mM fructose-1,6-bisP, 0.85 with 1.5 mM, or 0.86 with 0.75 mM.[99,290,306] Thus, the calculated ADP-glucose-synthesizing activity in *E. coli* B cells at the physiological range of energy charge and fructose-1,6-bisphosphate levels is sufficient to account for the observed glycogen accumulation rates. The fructose-1,6-bisP concentrations and energy charge values during growth of *E. coli* B are unknown. In contrast to *E. coli* W4597(K), Lowry *et al.*[305] have shown that the fructose-1,6-bisP concentrations increase slightly from 2.6 to 3.2 mM in another strain of *E. coli* K-12, Hfr 139, when it reaches stationary phase in the presence of excess glucose and limiting nitrogen; the energy charge remains essentially constant. The foregoing calculations also have assumed no compartmentation of the metabolites in relation to the possible compartmentation of glycogen biosynthetic enzymes. Also, the effects of other cations, anions, and other metabolites on ADP-Glc PPase activity have not been considered. Nevertheless, the correlation of ADP-Glc PPase activity with the known fructose-1,6-bisphosphate concentrations and energy charge values with the observed rates of glycogen accumulation appears to be quite good.

Dietzler et al.[304,308] designed some interesting experiments: mutants of *E. coli* W4597(K) and G34, were grown in different media (varying the carbon and/or the nitrogen source) that gave a 10-fold range in the rate of glycogen accumulation in the stationary phase. The different rates of glycogen accumulation found in the various nutrient conditions were linearly related to the square of the fructose-1,6-bisP concentration in the bacteria; ATP concentrations in the bacteria were the same and independent of the composition of the media. A relationship was found between the concentration of the cellular fructose-1,6-bisP and the rate of glycogen accumulation; the authors[308] fitted these data to the Hill equation and obtained an $A_{0.5}$ of 0.82 mM and a Hill slope value n_H of 2.08. These values are in agreement with the *in vitro* kinetic values obtained with the *E. coli* B ADP-Glc PPase at energy charge 0.85 ($A_{0.5} = 0.68$ mM; $n_H = 2.0$).[269] The work of Dietzler et al.[304,308] indicates that fructose-1,6-bisP is the physiological activator for the *E. coli* ADP-Glc PPase, and that this activation is relevant *in vivo*.

The fact that regulation of ADP-Glc PPase is relevant to accumulation of glycogen by bacteria is also supported by studies on the mutant SG14;[306,309] this mutant accumulates glycogen at 28% the rate of *E. coli* B (Table 9) and contains about 23-25% of the ADP-glycose-synthesizing activity of *E. coli* B.[306] The activity is still sixfold greater than that required for the observed rate of glycogen accumulation in SG14. The concentrations of ATP and Mg^{2+} required for 50% of maximal activity ($S_{0.5}$) are four- to fivefold higher for the SG14 enzyme than for the *E. coli* B enzyme. Whereas the $S_{0.5}$ values for ATP and Mg^{2+} are 0.38 and 2.3 mM, respectively, for the *E. coli* B enzyme in the presence of 1.5 mM fructose-1,6-bisphosphate, the $S_{0.5}$ values for ATP and Mg^{2+} are 1.6 and 10.8 mM, respectively, for the SG14 ADP-Glc PPase in presence of saturating fructose-1,6-bisP concentration (4.0 mM). Reports in the literature indicate that the ATP level in growing *E. coli* is approximately 2.4 mM[301,304,305] and the Mg^{2+} level is about 25–40 mM.[306] Therefore, the SG14 ADP-Glc PPase would be essentially saturated with respect to these substrates. The apparent affinities ($S_{0.5}$) of the *E. coli* B and SG14 enzymes for glucose-1-phosphate are about the same.[306] The major difference between the SG14 and *E. coli* B ADP-Glc PPases appears to lie in their sensitivities toward activation and inhibition. About 12 times as much fructose-1,6-bisphosphate is needed for 50% maximal stimulation of the SG14 ADP-Glc PPase ($A_{0.5} = 0.82$ mM) as for half-maximal stimulation of the *E. coli* B enzyme. The $A_{0.5}$ for pyridoxal phosphate for the SG14 enzyme (0.44 mM) is about 25 times higher than the $A_{0.5}$ observed for the *E. coli* B enzyme. Pyridoxal phosphate and fructose-1,6-bisphosphate stimulate ADP-glucose synthesis catalyzed by the *E. coli* B enzyme to about the same extent. However, the stimulation of the SG14 ADP-Glc PPase seen with pyridoxal phosphate is only half that elicited by fructose-1,6-bisphosphate.[306] Another notable difference is that NADPH does not stimulate the SG14 enzyme. Compounds structurally similar to NADPH, such as 1-pyrophosphorylribose-5-phosphate and 2'-PADPR, which are capable of activating the *E. coli* B ADP-Glc PPase, do not activate the SG14 enzyme. Since the apparent affinity of the SG14 enzyme for its activators is considerably lower than that observed for the *E. coli* ADP-Glc PPase, it was interesting to find that SG14 is capable of accumulating glycogen even at 28% the rate observed for the parent strain. This rate is accounted for by the relative insensitivity of the SG14 enzyme to inhibition by AMP.[306,309] The SG14 ADP-Glc PPase is much less sensitive than the parent strain enzyme to AMP inhibition in the concentration range of 0–0.2 mM. At a saturating concentration of fructose-1,6-bisP for the SG14 enzyme (4.0 mM), only 7% inhibition of the SG14 enzyme is observed at 0.2 mM AMP. The same concentration of AMP gives 40% inhibition of the *E. coli* B enzyme. At a concentration of fructose-1,6-bisP (1.5 mM) that gives 80% of maximal velocity for the SG14 enzyme, 0.2 mM 5'-AMP causes 80% and 33% inhibition of the *E. coli* B and SG14 enzymes, respectively. A decrease in fructose-1,6-bisphosphate to 1.0 or 0.5 mM further increases the sensitivity of the *E. coli* B ADP-Glc PPase activity to inhibition, while at these fructose-bisP concentrations the sensitivity of the SG14 ADP-Glc PPase to AMP remains the same or becomes less than that observed at 1.5 mM fructose-bisP. At concentrations of 0.5–1.0 mM of fructose-bisP, the *E. coli* B enzyme is inhibited 90% or more by 0.2 mM AMP and the inhibition of the SG14 enzyme ranges from 12% to 30%. Although the SG14 enzyme has a lower apparent affinity for its activators, it is also less sensitive to AMP inhibition. The two effects on activation and inhibition appear to compensate for each other and to allow SG14 to accumulate glycogen at about 28% the rate of the parent strain.[306] The data obtained from kinetic studies of the SG14 ADP-Glc PPase suggest that fructose-1,6-bisP is the most important physiological activator of the *E. coli* ADP-Glc PPase. This is based on the observation that NADPH is not an activator of the SG14 enzyme and that the concentration of pyridoxal phosphate needed for activation of the enzyme ($A_{0.5} = 0.44$ mM) is considerably higher than the concentration reported to be present in *E.*

coli B. The concentration or pyridoxal phosphate is 24–48 μM,[311] and in the cell most of this metabolite is probably protein bound, and unavailable for activation of the ADP-Glc PPase.[312] The concentration of fructose-1,6-bisphosphate in *E. coli* is about 0.71–3.2 mM,[301,303,304] and the $A_{0.5}$ of SG14 ADP-Glc PPase is 0.82 mM. The concentration of fructose-bisP is the *E. coli* cell is therefore sufficient for activation of ADP-glucose synthesis at the rates required for the observed glycogen accumulation rate in SG14.

3.14.7.6 Genetic Regulation of Bacterial Glycogen Synthesis

The activity of the glycogen biosynthetic enzymes in *E. coli* increases as cultures enter the stationary phase.[8,24,313] When cells are grown in an enriched medium containing yeast extract and 1% glucose, as cultures enter the stationary phase, the specific activities of ADP-Glc PPase and glycogen synthase increase 11- to 12-fold, and branching enzyme increases fivefold. In defined media, branching enzyme is fully induced in the exponential phase, with only about a twofold increase in specific activity of the ADP-Glc PPase and glycogen synthase when cells reach the stationary phase. The same phenomena are also seen with the glycogen biosynthetic enzymes in *S. typhimurium*.[38] Possibly the gene encoding the branching enzyme is regulated differently from the genes for ADP-Glc PPase and glycogen synthase. Cattaneo *et al.*[314] showed that the addition of inhibitors of RNA or protein synthesis to prestationary phase cultures prevented the enhancement of glycogen synthesis in the stationary phase, suggesting that the pathway is under transcriptional control.

The structural genes for glycogen biosynthesis are clustered in two adjacent operons, also containing genes for glycogen catabolism. The structural genes for glycogen synthesis are located at approximately 75 min on the *E. coli* K-12 chromosome, and the gene order at this location, as established by transduction, is *glgA–glgC–glgB–asd*.[315] These genes encode the enzymes glycogen synthase, ADP-Glc PPase, glycogen branching enzyme, respectively, and are close to *asd*, the structural gene for the enzyme aspartate semialdehyde dehydrogenase (EC 1.2.1.11).

The molecular cloning of the *E. coli glg* structural genes[100] greatly facilitated the study of the genetic regulation of bacterial glycogen biosynthesis; they were cloned into pBR322 via selection with the closely linked essential gene *asd*. Among the several *asd*$^+$ plasmid clones that were isolated, pOP12 was found to contain a 10.5 kb PstI fragment encoding the structural genes *glgC*, *glgA* and *glgB*. Romeo and Moore[316] developed a more general method for cloning α-1,4-glucan biosynthetic genes, based upon screening of clones with iodine vapor.

The arrangement of genes encoded by pOP12 has also been determined by deletion-mapping experiments,[100] and the nucleotide sequence of the entire *glg* gene cluster is known.[40,160,198,317,318] The genetic and physical map of the *E. coli* K-12 *glg* gene cluster is as follows:

Asd glgB GlgX GlgC GlgA GlgP GlpD

The continuous nucleotide sequence of over 15 kb of this region of the genome has been determined and includes the sequences of the flanking genes *asd*[319] and *glpD* (glycerol phosphate dehydrogenase).[320] This region of the *E. coli* K-12 chromosome is located at 4140 kb on the physical map of Kohara *et al.*[321] and is situated within the region 3584 to 3594 kb on version 6 of the physical map of Rudd *et al.* (original version, ref.[322]; version 6, ref.[323]).

Analysis of the nucleotide sequence showed that, in addition to the *glgC*, *glgA*, and *glgB* genes, pOP12 contains an open reading frame, *glgX*, situated between *glgB* and *glgC*, and a second ORF, *glgP*, originally referred to as *glgY*, located downstream from *glgA*.[324] The deduced amino acid sequence of *glgX* shows significant similarities to regions of the α-glucanases and transferases, including α-amylases, pullulanase, cyclodextrin glucanotransferase, glycogen branching enzyme, etc. The homologous regions include residues reported to be involved in substrate binding and cleavage by α-amylases and the amylase family.[201,202] A report has shown that the *glgX* gene when expressed has glycogen debranching enzyme activity,[325] with about 16 times more activity on a phosphorylase-limit dextrin product of glycogen than on native glycogen, confirming an earlier observation on the specificity of *E. coli* debranching enzyme.[326]

The *glgP* gene was identified through its homology with the phosphorylase from rabbit muscle glycogen,[199,318] and through its expression and characterization of its gene product[318] it was confirmed that it codes for a phosphorylase. Neither *glgX* nor *glgP* are needed for glycogen synthesis, suggesting that both may be more involved in glycogen catabolism.[199]

The organization of the gene cluster suggests that the *glg* genes may be transcribed as two randomly arranged operons, *glgBX* and *glgCAP*.[9] The *glgB* and *glgX* coding regions overlap by one base pair, *glgC* and *glgA* are separated by two base pairs, and genes *glgA* and *glgP* are separated by 18 base pairs. The close proximity of these genes would suggest translational coupling within the two proposed operons. However, a noncoding region of approximately 500 bp separates *glgB* and *glgC*. Transcriptions initiating upstream of *glgC* have been analyzed by S1 nuclease mapping[60] and studies of the regulation of the *glg* structural genes, using *lacZ* translational fusions and other approaches, are consistent with a two-operon arrangement for the *glg* gene cluster, in which the *glgCAP* and *glgBX* operons may be preceded by growth phase-regulated promoters.

Addition of exogenous cAMP to *E. coli* W4597(K) results in a modest enhancement in the rate of *in vivo* glycogen biosynthesis.[327,328] It was observed that the genes *cya*, encoding adenylate cyclase (EC 4.6.1.1), and *crp*, encoding cAMP-receptor protein (CRP), are required for optimal synthesis of glycogen, and that exogenous cAMP can restore glycogen synthesis in a *cya* strain but not in a *crp* mutant.[329]

Both cAMP and CRP are strong positive regulators of the expression of the *glgC* and *glgA* genes, but do not affect *glgB* expression.[60] Addition of cAMP and CRP to S-30 extracts with *in vitro* coupled transcription-translation reactions and pOP12 as the genetic template, resulted in about 25- and 10-fold increase in the expression of *glgC* and *glgA*, respectively, without affecting *glgB* expression. In reactions of completely defined composition, using the dipeptide synthesis assay, cAMP and CRP also enhanced the expression of *glgC* and *glgA* when encoded by either plasmids or restriction fragments.[330] The dipeptide synthesis assay measures protein expression by quantifying the formation of the first dipeptide of a specified gene product directed by a DNA template.[3,330] A restriction fragment containing *glgC* and 0.5 kb of DNA upstream noncoding region of *glgC*, was sufficient to permit cAMP-CRP regulated expression in the dipeptide synthesis assay, suggesting that the *glgC* gene contains its own cAMP-regulated promoter(s). Gel retardation analyses[60] demonstrated a CRP-binding site on a 243 bp restriction fragment from the upstream region of *glgC*. Potential consensus CRP-binding sequences within the *glgC* upstream region preceding both the *E. coli*[60] and the *S. typhimurium*[316] *glgC* genes have been identified.

More details on the genetic regulation of glycogen synthesis are described in some recent reviews.[8,9]

3.14.8 PROPERTIES OF THE GLYCOGEN BIOSYNTHETIC ENZYMES OF MAMMALS

3.14.8.1 UDP-glucose Pyrophosphorylase

The UDP-glucose pyrophosphorylase (UDP-Glc PPase), which catalyzes the synthesis of UDP-glucose, seems to be ubiquitous in nature. It was first demonstrated in yeast[331] and the enzyme has been isolated and characterized from bacteria, plants, and mammals.[332] UDP-Glc PPases have been highly purified from calf liver, human liver, lamb, goat, rabbit livers, and human erythrocytes (see review[332]). A molecular mass of 480 kDa was reported for the calf liver enzyme, with subunit of 60 kDa; the native enzyme is therefore an octamer of eight identical subunits. Other mammalian enzymes appear to have similar molecular masses. UDP-Glc PPases have an absolute requirement for a divalent cation, magnesium being the best cation for activity; Mn^{2+}, Ca^{2+}, and Ni^{2+} are active to some extent. The optimum pH is in the range of 7 to 9 and the equilibrium constant in the direction of UDP-glucose formation ranges from 0.15 to 0.34 for the animal, plant, and bacterial enzymes studied.[332] Although highly specific for UDP-glucose, the calf and human liver enzymes can also catalyze the pyrophosphorolysis of TDP glucose, CDP-glucose, GDP-glucose, UDP-galactose, UDP-xylose, and UDP-mannose to a small extent, with 0.1 to 2.2% of the rate shown with UDP-glucose.[333]

The reaction mechanism of UDP-Glc PPase has been studied with the enzyme from liver, erythrocytes, and *Acanthamoeba castellani*[332] and it appears to be an ordered BiBi mechanism. The nucleoside phosphate is both the first substrate to be added and the last product to be released in the mechanism.

$$UTP + E \rightleftarrows E\text{-}UTP + Glc\text{-}1\text{-}P \rightleftarrows E\text{-}UTP\text{-}Glc\text{-}1\text{-}P$$
$$E + UDP\text{-}Glc \rightleftarrows E\text{-}UDP\text{-}Glc + PPi \rightleftarrows E\text{-}UDP\text{-}Glc\text{-}PPi$$

UDP-glucose is the most potent inhibitor of the animal UDP-Glc PPase. Thus, its concentration possibly exerts some regulation of the enzyme. The inhibition appears to be competitive with UTP.

Roach *et al.*[334] suggested that the concentration ratio of UDP-glucose to UTP may be the most important determinant of UDP-Glc PPase activity. No other regulatory phenomenon has been associated with the mammalian enzyme and because UDP-glucose functions not only in the synthesis of glycogen but also in the synthesis of other sugar nucleotides (e.g., UDP-galactose and UDP-glucuronic acid), it would be expected that the dominant regulation of glycogen synthesis would occur at the level of glycogen synthase.

3.14.8.2 Glycogen Synthase

Mammalian glycogen synthase has been purified from several sources, e.g., skeletal and cardiac muscle, liver, adipose, and kidney.[335–340] Also, cDNA representing the structural gene of the enzyme has been isolated (human muscle,[341] rabbit muscle,[342] rat liver,[343] human liver,[344] yeast[345,346]) and the rabbit skeletal muscle cDNA has been exposed both in bacterial[347] and in COS cells.[348–350] The deduced amino acid sequences of the rabbit and human muscle glycogen synthases have 97% identity, while the rat and human liver enzymes are 92% similar. The two yeast glycogen synthases are 80% identical but they have only a 50% overall identity to the muscle glycogen synthases. The amino acid similarity between the enzymes from muscle and liver in humans is only 60%, with the lowest identity in the *N*- and *C*-terminal regions of the proteins. The human and rat liver enzymes are truncated by 32 to 34 amino acids compared with the rabbit and human muscle glycogen synthases. Glycogen synthases from all of these sources seem to be composed of identical (or very similar) subunits of MW 80 000–85 000. The native forms from liver or adipose tissue are aggregates of two identical subunits, whereas that from muscle contains four. Glycogen synthase exists in at least two forms: a phosphorylated form, arising from covalent modification of serine residues by ATP; and a dephosphorylated form, which can be obtained using phosphatase on the phosphorylated form (Figure 3). These two forms were originally named as the "a" (or I) (unphosphorylated) and "b" (or D) (phosphorylated) forms; the b form was dependent on glucose-6-P for activity, whereas the a form was active in the absence of glucose-6-P. The a and b forms can also be distinguished on the basis of K_m for the substrate, UDP-glucose; the b form usually has a higher K_m (lower apparent affinity) than the a form. It is apparent that the a form is the physiologically active form of the enzyme while the b form is an inactive form of the glycogen synthase.

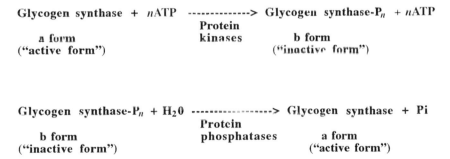

Figure 3 Covalent modification of glycogen synthase by phosphorylation by protein kinases and dephosphorylation via protein phosphatases.

3.14.8.3 Branching Enzyme

The branching enzymes from rat liver[351–353] and rabbit skeletal muscle[354–357] have been studied in some detail. Larner[352] showed that the enzyme catalyzed the formation of new (α-1-6)-a-D-glucosidic bonds with glycogen containing average chain lengths of 11 to 21 glucosyl residues. Using rat liver[353] and rabbit skeletal muscle[356] systems, it was found that the enzymes cleave linear chains of six or more glucosyl residues from the terminal portion of the outer chains of the α-1-4-α-D-glucan substrate, and then transfer and reattach the cleaved oligosaccharide portion in a 1-6-α-D-glucosidic linkage to the outer portions of the α-D glucan. The branching enzyme from rabbit skeletal muscle preferentially catalyzed the transfer of segments seven glucosyl units long when it acted on polysaccharides elongated by rabbit muscle glycogen synthase action *in vitro*.[355] The liver and muscle enzymes are also active on amylose and amylopectin molecules.[351,354] The rabbit muscle enzyme also catalyzed the formation of new 1-6-α-D-linkages in products formed by the phosphorylase from glucose-1-P, and also greatly stimulated the "unprimed" phosphorylase reaction.[357] Branching enzyme in a combined system with mammalian glycogen synthase stimulates the rate of transfer of glucose units in the presence of a minimal quantity of glycogen primer.[354] This stimulation is very similar to that observed for the bacterial branching enzyme on bacterial glycogen synthase activity, and for the stimulation of plant branching enzyme on starch synthase activity. There is some information about the minimal size of oligosaccharide chain that can be transferred by the action of branching enzyme; the relevant sections discuss this aspect of the action of the plant and bacterial enzymes. What is not known for any branching enzyme are the amino acid residues involved in determining the specificity of size of oligosaccharide cleaved and transferred, and the interbranch distance between formation of the new branch points; this important information remains to be determined.

Human glycogen storage disease ("type IV") is due to the absence of branching enzyme.[358,359] The glucan product isolated from the liver of these patients is an amylopectin-type polysaccharide. Thus, some branches have been formed, suggesting that some branching activity is present at low levels. Purified rabbit muscle branching enzyme can cause further branching of this α-D-glucan.[356] The rabbit skeletal muscle enzyme has been purified to near homogeneity.[356]

The apparent molecular weight of the mammal branching enzyme is about 92 kDa, as determined by sucrose density gradient centrifugation. Thus, it is similar to the molecular weights of plant and bacterial enzymes. The enzyme has a broad pH optimum in citrate buffer, 6.8–7.8, and is stimulated about twofold in 0.5 M sodium citrate at pH 7.0. The partially purified liver enzyme was also activated twofold by sodium citrate and ~ 1.7-fold by sodium borate.[351]

Yeast (*Saccharomyces cerevisiae*) is used in complementation studies. Branching enzyme genes have been isolated from yeast[360] and from a human hepatoma cell line.[361] The deduced amino acid sequence from the yeast *GLC3* gene was compared with the bacterial branching enzymes from *Escherichia coli*,[198] *Synechococcus* and *Bacillus stearothermophilus*[46] and there was only 8% identity and 42% similarity of the yeast BE with the prokaryotic sequences. The cDNA encoding the human branching enzyme could complement the yeast BE mutant, *glc3* and had a 67% identical amino acid sequence with the yeast BE.[361] The human gene was located on chromosome 3.

3.14.8.4 Glycogenin

Two reviews describing the discovery and characterization of glycogenin are available.[362,363] This protein, about 37 kDa in size, was first found to be covalently bound to glycogen[89] and also associated with glycogen synthase even after extensive purification of the rabbit muscle glycogen synthase.[364] Most of the glycogenin in rabbit muscle is considered to be covalently linked to glycogen. In liver, however, most of the glycogen and glycogenin are in free form and not associated with each other.[365] The first step in converting the apo-glycogenin into a primer for glycogen synthesis is an autoglucosylation of the glycogenin tyrosine residue 194 by UDP-glucose. The reaction absolutely requires either Mn^{2+} or Mg^{2+} for the autoglycosylation. Up to 7 to 11 glucosyl units can be attached to the glucosyl-tyrosine residue. The complete sequence of the rabbit muscle glycogenin has been elucidated.[88]

Glycogenin, when isolated and purified, contains a glycosylated tyrosine residue and so there was always a question whether the first glucosyl unit was due to autoglucosylation by glycogenin or whether there was another enzyme responsible for the first glycosylation. The rabbit muscle glycogenin was expressed in *Escherichia coli*[366,367] and purified. The Tyr194 residue was already glucosylated and contained from one to eight residues of glucose. The glycogenin could incorporate another 5 mol of glucose per mol of glycogenin if supplied with UDP-glucose; the K_m was 4.5 μM.

Isoamylase can remove the oligosaccharide chain from the tyrosine residue.[368] and pretreatment of the glucosylated glycogenin enhanced incorporation of labeled glucose from UDP-glucose. This suggested that glycogenin can self-glucosylate its tyrosine 194 residue. More direct evidence was obtained by expressing the glycogenin in an *E. coli* mutant deficient in UDP-Glc PPase activity,[369] resulting in production of a carbohydrate-free glycogenin, apoglycogenin. When UDP-xylose + Mn^{2+} was incubated with the glycogenin, one mole of xylose was incorporated per mole of glycogenin. With UDP-glucose, an average of eight glucose chains are added per glycogenin. However, upon release of the glucose chains by isoamylase, the size of the chains varied, with the predominant chains being in the 7 to 11 glucose units range. The production of a carbohydrate-free apo-glycogenin and its ability to self-glucosylate eliminates the need to invoke a separate enzyme for the addition of the first glucose residue to tyrosine 194.

Thus, glycogenin can catalyze two different glycosylation reactions. First, the glycosylation of a tyrosine hydroxyl group and then further glucosylations to form α-1,4 glucosidic linkages. UDP and UTP were found to be effective inhibitors.[370] Other pyrimidine base sugar nucleotides could substitute for UDP-glucose as glucosyl donors.[371] The rates of glycosylation using CDP-glucose and TDP-glucose were 71% and 33%, respectively, of the rate with UDP-Glc; both ADP-glucose and GDP-glucose were inactive. Meezan *et al.*[372] reported that UDP-xylose could also serve as a glycosyl donor with only one xylose molecule being transferred to glycogenin itself. No further chain growth could occur with either UDP-xylose or UDP-glucose. Other reactions were also catalyzed by glycogenin. The glycogenin could transfer glucose from UDP-glucose to exogenous substrates such as *p*-nitrophenyl-linked malto-oligosaccharides,[373] to tetradecyl-β-D-maltoside, octyl-β-D-maltoside, and dodecyl-β-D-maltoside.[374]

If the recombinant glycogenin is mutated from Tyr to Phe or Thr at residue 194, the enzyme loses its ability to self-glucosylate.[367,370,371] However, the Phe194 and Thr194 mutants are still able to glycosylate with UDP-glucose, dodecyl-β-D-maltoside[371] and *p*-nitrophenyl-linked malto-oligo-saccharides.[367] Also noted was the ability of the mutant and normal glycogenins to hydrolyze UDP-glucose at rates similar to self-glucosylation rates of the normal enzyme. This hydrolysis is competitive with the glucose transfer to *p*-nitrophenyl-linked maltoside.[367] The self-glucosylation, glucosylation of other acceptors, and hydrolysis all appear to be catalyzed by the same active center.

Glycogenin and the mutant proteins, Phe194 and Thr194, could also transfer glucose from UDP-glucose to maltose to form maltotriose.[260] However, no further conversion to a higher oligo-saccharide occurred. Analysis of the crystal structure by X-ray diffraction indicated that glycogenin existed as dimers.[260]

There have been some reports of a glycogenin in *E. coli*, similar to the mammalian glycogenin, involved in the initiation of glycogen synthesis. Barengo *et al.*[375] reported the formation of a labeled TCA insoluble fraction upon incubation of extracts of *E. coli* with UDP-glucose-[14]C. The radioactivity was solubilized by α-amylase suggesting that the label was an α-1,4 glucosyl oligosaccharide attached to a protein. Evidence was also presented to suggest that this labeled fraction was an intermediate in glycogen synthesis. Goldraij *et al.*[376] have also isolated a 31 kDa protein presumably bound to *E. coli* glycogen. Whether this 31 kDa protein is indeed the bacterial glycogenin remains to be established; at present the glycogenin of bacteria is not well characterized. Moreover, the sugar nucleotide donor for glycogen synthesis is ADP-glucose and not UDP-glucose. As will be shown later, mutants of *E. coli* defective in ADP-glucose PPase activity are deficient in glycogen. The involvement of UDP-Glc, if any, in glycogen synthesis would be restricted to synthesis of a glucosylated glycogenin but, as indicated before, deficient UDP-Glc PPase *E. coli* mutants have normal glycogen levels. It is possible, however, that the bacterial glycogenin may have a different sugar nucleotide glycosyl donor specificity.

3.14.8.4.1 *Genetic evidence indicating that glycogenin is required for glycogen synthesis*

In *Saccharomyces cerevisiae*, two genes, Glg1p and Glg2p, coding for self-glucosylating proteins encode proteins of 618 and 380 amino acids, respectively, and have 55% sequence identity over their *N*-terminal 258 amino acids. These two proteins, Glg1p and Glg2p, have amino acid sequences that are 33% and 34% identical, respectively, to that of rabbit muscle glycogenin in the *N*-terminal region of 258 amino acids.[377] Thus, they are larger than the muscle glycogenin which has 332 amino acids. The COOH termini of Glg1p and Glg2p are largely nonidentical in sequence except for two small segments of sequence similarity. Each contains a Tyr residue in correspondence with the rabbit muscle Tyr194. The residue in Glg1p and in Glg2p is Tyr232.

When the Glg1p and Glg2p genes were disrupted separately by homologous recombination, there was little effect on glycogen accumulation,[377] but loss of both genes caused the almost complete loss of glycogen. Glycogen synthase activity was normal in this double mutant so this was not the reason for the lack of glycogen synthesis. Glycogen synthesis was almost completely restored when the rabbit muscle glycogenin was expressed in the double Glg1p, Glg2p mutant, i.e., the mammalian glycogenin could complement the double mutant deficiency. These data indicate that the Glg1p and Glg2p genes were involved in the initiation of glycogen synthesis and this report[377] was the first presenting *in vivo* evidence of the requirement of a glycogenin in the biosynthesis of glycogen in eukaryotes.

3.14.9 REGULATION OF MAMMALIAN GLYCOGEN SYNTHESIS

3.14.9.1 General Considerations

As seen above, the site of regulation of glycogen synthesis in bacteria and starch synthesis in plants is at the ADP-Glc PPase step, and is different from the site of regulation in mammals. In mammalian systems, the regulatory enzyme is glycogen synthase. The difference in regulatory sites in the various systems may be linked to the difference in specificity for the glucosyl donor (ADP-glucose for the bacterial and plant α-D-glucan systems, UDP-glucose for mammals). As shown in Figure 4, UDP-glucose is utilized for the synthesis of other intermediaries required for the synthesis of many cellular constituents. The first unique reaction for mammalian glycogen synthesis, therefore, after synthesis of the glucosylated acceptor protein glycogenin, is the glycogen synthase step, where both allosteric control and covalent modification control are exerted. In contrast, in bacteria and plants, the only known function for ADP-glucose is the synthesis of α-1,4-D-glucosyl bonds in bacterial glycogen and starch. Thus, the prokaryote and plant cells regulate α-glucan synthesis at the level of ADP-glucose formation so as to conserve ATP utilized for synthesis of the sugar nucleotide.

$$\text{UTP} + \alpha\text{-glucose-1-P} \overset{1}{\rightleftarrows} \text{UDP-glucose} + \text{PPi}$$

$$\text{UDP-glucose} + \text{Dolichol-P} \overset{2}{\longrightarrow} \text{Dolichyl-P-glucose} + \text{UDP}$$

$$\text{UDP-glucose} + \alpha\text{-1,4-glucan} \overset{3}{\longrightarrow} \alpha\text{-1,4-glucosyl-glucan} + \text{UDP}$$

$$\text{UDP-glucose} + 2\text{NAD} \overset{4}{\longrightarrow} \text{UDP-glucuronate} + 2\text{NADH}$$

$$\text{UDP-glucose} \overset{5}{\rightleftarrows} \text{UDP-galactose}$$

Figure 4 Various reactions of UDP-glucose in mammalian cells: (1) UDP-glucose pyrophosphorylase; (2) UDP-glucose, Dolichyl-P glucosyl transferase; (3) glycogen synthase; (4) UDP-glucose dehydrogenase; and (5) UDP-glucose epimerase.

The findings that regulation occurs at the glycogen synthase step in mammalian systems and at the ADP-Glc PPase step in bacteria and plants is consistent with the concept that major regulation of a biosynthetic pathway does occur at the first unique step of the pathway. One can recognize also the need in mammalian systems for a type of regulation that involves an efficient, rapid on–off type of control of glycogen synthesis that permits synthesis of glycogen when carbohydrate or carbon is plentiful in the diet, but prevents synthesis and permits degradation of glycogen to occur during muscular contraction or during starvation. Such a mechanism involves covalent modification of the enzyme catalyzing the limiting reaction of the process in order to produce either inactive or active forms of the enzyme.

3.14.9.2 Regulation of Glycogen Synthase by Phosphorylation and Dephosphorylation

The ratio of activity of the glycogen synthase in absence of glucose-6-P/activity in presence of glucose-6-P has been used as a measure of the state of phosphorylation of the enzyme. The a form was recognized as the primary active species within the cell, especially as modulated by hormonal control. The cAMP-dependent protein kinase was shown to catalyze phosphorylation and, consequently, inactivation of this enzyme. Subsequent studies also indicated that on each subunit of the glycogen synthase there were multiple and unique phosphorylation sites. Originally, Smith et al.[378] reported that, if partially purified enzyme was incubated extensively with ATP and subsequently purified, the inactive glycogen synthase contained 6 mol of phosphate per subunit. Although this information did not initially gain general acceptance, Picton et al.[379,380] showed there are seven serine phosphorylation sites on the rabbit skeletal muscle glycogen synthase. Currently, the observed number of potential phosphorylation sites found *in vivo* is nine and more than ten sites can be phosphorylated *in vitro*.[381] Table 10 shows the various *in vivo* phosphorylation sites and the major protein kinases that are involved in the phosphorylation of those sites.

Table 10 *In vivo* phosphorylation sites in rabbit muscle glycogen synthase.

Phosphorylation site		
Residue	Name	Protein kinases
7	2	cAMP-dependent protein kinase
		calmodulin-dependent protein kinase II
		phosphorylase kinase
		protein kinase C
10		casein kinase I
640	3a	glycogen synthase kinase 3
		cAMP-dependent protein kinase
644	3b	glycogen synthase kinase 3
648	3c	glycogen synthase kinase 3
652	4	glycogen synthase kinase 3
		cAMP-dependent protein kinase
656	5	casein kinase II
697	1a	cAMP-dependent protein kinase
		protein kinase C
710	1b	cAMP-dependent protein kinase
		calmodulin-dependent protein kinase II

Modified from Roach,[22] who acknowledges the contributions by many investigators to the identification of the glycogen synthase phosphorylation sites and the glycogen synthase kinases. The sites listed in this table are only those that are known to be labeled *in vivo*. The protein kinases listed for the sites are those that phosphorylate the enzyme *in vitro*.

Phosphorylation of these sites is catalyzed by more than six kinases.[22,382,383] Phosphorylation at the different sites synergistically inactivates the enzyme; however, the effects observed may vary depending on the conditions used to assay the glycogen synthase. The effects of multisite phosphorylation would depend on the concentration of effectors used, such as glucose-6-P, or the substrate, UDP-glucose. For example, with preparations of enzyme containing 0.27–3.49 mol of alkali-labile phosphate per glycogen synthase subunit, Roach and Larner[382] reported that the $A_{0.5}$ (concentration of activator, glucose-6-P, needed for 50% of maximal activation) and $S_{0.5}$ (concentration of substrate, UDP-glucose, having 50% of maximal velocity) varied with phosphate content, from 3.3 µM to 2.7 mM and from 0.75 mM to at least 60 mM, respectively. Both parameters increased with phosphate content. The greatest absolute change occurred at values greater than 2 mol of phosphate bound per enzyme subunit. Plots of activity versus glucose-6-P became more sigmoidal with increasing enzyme phosphate content. Activation by glucose-6-P was related primarily to modulation of UDP-glucose affinity. Several inhibitors such as ATP, ADP, AMP, UDP, and Pi had increasing effects with enzyme of increasing alkali-labile phosphate content. These investigators presented a scheme in which glycogen synthase activity is sigmoidally and inversely dependent on the state of phosphorylation. The hormonal effects of insulin were counteractive to those of epinephrine, glucagon, and so on, respectively, decreasing or increasing the extent of phosphorylation. Inhibitors, such as UDP, ATP, and AMP, accentuate the inhibition by phosphorylation, whereas the activator, glucose-6-P and substrate, UDP-glucose diminished the extent of such inhibition.

As seen in Table 10, there are two phosphorylation sites at the *N*-terminal region of the rabbit skeletal muscle glycogen synthase. The remaining seven sites are situated at the *C*-terminal region. The cAMP-dependent protein kinase preferentially phosphorylates three sites, 1a, 1b, and 2.[384] The cAMP-dependent kinase can also phosphorylate sites 3a and 4 but at a much slower rate[385] and thus is not considered to be as important as glycogen synthase kinase for those sites. There are overlapping specificities among the different protein kinases for site 2 as phosphorylase kinase, calmodulin-dependent protein kinase, and protein kinase C also can phosphorylate this site.[21]

Of interest is the phosphorylation of site 5 catalyzed by casein kinase II. Picton *et al.*[380] showed that dephosphorylation at site 5 did not alter the regulatory kinetics of rabbit muscle glycogen synthase; nor did rephosphorylation of site 5 by casein kinase II (also called glycogen synthase kinase-5) affect the activity of the glycogen synthase. In other words, the site-5 phosphorylated glycogen synthase did not depend to any extent on glucose-6-P for maximal activity. However, the presence of phosphate at site 5 was necessary for the phosphorylation of site 3a, 3b, and 3c[386] by glycogen synthase kinase-3, which did increase the dependency of the glycogen synthase activity for glucose-6-P. The phosphate at site 5 appears to be highly stable as it is resistant to dephosphorylation by the rabbit muscle protein phosphatases but can be removed by potato acid phosphatase. Phosphorylation of sites 1a, 1b, or 2 did not require the presence of phosphate at site 5.

This observation was confirmed and extended by Fiol *et al.*[381,387] They synthesized a peptide corresponding to the rabbit muscle glycogen synthase amino acid sequence containing sites 3a, 3b, 3c, 4, and 5. Synergism was observed between casein kinase II phosphorylation of site 5 and phosphorylation of sites 3a, 3b, 3c, and site 4, by GSK-3 of the synthesized peptide. Indeed, phosphorylation of site 5 was obligatory for the phosphorylation of the four sites at 3 and 4 by glycogen synthase kinase-3. As seen in Table 11, the GSK-3 sites were regularly spaced every fourth residue in the motif SXXXS(P). It was also found that the phosphorylations by GSK-3 were ordered: first, site 4 was phosphorylated, followed by sites 3c, 3b, and finally 3a.[388] This was clearly shown by synthesizing a series of peptides where the sites 3a, 3b, 3c, and 4 were replaced, one at a time, with alanine. The alanine at site 4 peptide could not be phosphorylated at the site 3 residues even though it was phosphorylated at site 5 by GSK-3. Also it was observed that GSK-3 would not phosphorylate the serine residue at sites, amino-terminal of the site containing the alanine residue. With alanine replacing serine at site 3b, only sites 4 and 3c were phosphorylated. With alanine substituted at site 3c, only site 4 was phosphorylated. Thus, the multiple phosphorylation by GSK-3 was of an obligate order, first 4 then 3c, 3b, and then 3a. Most probably GSK-3 recognizes the motif SXXXS(P) and this may explain the need for GSK-5 to phosphorylate site 5. The sequential formation of new recognition sequences, SXXXS(P) at sites 5, 4, 3c, 3a, and 3b, would explain the ordered phosphorylation.

Table 11 Amino acid sequences of the regions corresponding to the phosphorylated sites of the glycogen synthase from rabbit muscle.

N-Terminal phosphorylation sites

```
        7    10
PLSRTLSVSSLPGL----
        (site 2)
```

C-Terminal phosphorylation sites

```
        640   644   648    652    656
RYPRPASVPPSPSLSRHSSPHCSEDEEEPRDGLPEEDGERYDEDEEAAKD
      (sites 3a    3b    3c     4      5)
                            697                    710
RRNIRAPQWPRRASCTSSSGGSKRSNSVDTSSLSTPSEP-------
            (sites 1a                    1b)
```

This interdependency of GSK-3 with GSK-5 has been defined as hierarchal phosphorylation.[22] Another example of the hierarchal phosphorylation is seen with cyclic-AMP-dependent protein kinase and casein kinase I. The cAMP-dependent protein kinase enhances phosphorylation of the glycogen synthase by casein kinase I.[389] The phosphorylation by casein kinase was serine residue 10.[342] Synthetic peptides based on the four phosphorylated regions in the muscle glycogen sequence (residues 694-707, 706-733, 1-14, and 636-662) were synthesized and phosphorylated. Casein kinase could not phosphorylate the unphosphorylated peptides but, if cAMP-dependent kinase phosphorylated peptides 694-707, 706-733, and 1-14, all three peptides were easily phosphorylated by the casein kinase I.[388] The greatest stimulation was seen with peptide 1-14. In the case of peptide 1-14 the phosphorylation site was at Ser10 and in the case of peptides 694-707 and 706-733, the phosphorylated residue was Thr713. However, the rate of phosphorylation was 20 to 40 times greater at Ser10 than Thr713. Moreover, whereas Ser10 was demonstrated to be phosphorylated *in vivo*, Thr713 was not.[22] Thus, the physiologically important site for phosphorylation by casein kinase I is considered to be Ser10 and this phosphorylation is considered to be dependent on an initial phosphorylation of Ser7 (site 2) by cAMP-dependent protein kinase.

The phosphorylation of peptide 636-662, the peptide encompassing glycogen synthase phosphorylation sites 3a, 3b, 3c, 4, and 5[388] is of great interest. Using the peptides where the alanine residue was substituted for the serine residue at the different phosphorylation sites, it was shown that serine residues 646 and 651 (Table 11) were phosphorylated by casein kinase I and this phosphorylation was significantly enhanced by prior phosphorylation of the sites 3a, 3b, 3c, 4, and 5.[389] However, it is still not clear whether these sites are phosphorylated *in vivo* and whether their phosphorylation by casein kinase I substantially affects the glycogen synthase activity. These studies, mainly by Roach's group, also indicate that the recognition of the serine phosphorylation site by casein kinase I is -S(P)-XXS- and for GSK-3 is SXXX-S(P).

3.14.9.3 Effect of Phosphorylation on Glycogen Synthase Activity and Relative Effects of Phosphorylation on Different Sites

3.14.9.3.1 *Studies* in vitro

Phosphorylation of the glycogen synthase activity leads to decreased activity but phosphorylations at different sites have different effects. Little or no inactivation is seen with phosphorylation at sites 5, 1a, and 1b, while site 2 phosphorylation gives moderate inactivation. A most potent inactivation is seen with phosphorylation by GSK-3 at sites 3a, 3b, and 3c.[21,22] Although site 5 phosphorylation does not cause any change in activity it is functional in regulation as site 3 cannot be phosphorylated unless there is an initial phosphorylation of site 5. Phosphorylation of sites 1a, 1b, and 2 by cAMP-dependent protein kinase leads only to a partial inactivation of the rabbit muscle glycogen synthase, and phosphorylation of site 2 did not decrease activity in rabbit[181] or rat liver glycogen synthase.[390] However, casein kinase I phosphorylation of Ser10, proceeding after phosphorylation of sites 1a, 1b, and 2 in rabbit muscle glycogen synthase and site 2 in liver glycogen synthase, causes a total inactivation of the muscle[181] and liver enzymes.[390] It should be pointed out that sites 1a and 1b are absent in rat liver[181,343,391] and human liver glycogen synthase.[344]

Thus, the secondary phosphorylations by casein kinase I of Ser residue 10, and by GSK-3 of sites 3a, 3b, 3c, and 4 have greater effects on the activity of glycogen synthase than the primary phosphorylations by cAMP-dependent kinase (sites 2, 1a, and 1b) and glycogen synthase kinase-5, and there may be two different routes to inactivating glycogen synthase, the GSK-3 sites and the casein kinase site.

It should also be mentioned that the above studies were done with a recombinant glycogen synthase expressed in *Escherichia coli*[347] and the same results were seen with phosphorylation site peptide analogues, namely, the dependency of phosphorylation by GSK-3 on the prior phosphorylation casein kinase II, the potent inactivation by GSK-3, the partial inactivation by phosphorylation by cAMP-dependent kinase, the stimulation of phosphorylation by casein kinase I by prior phosphorylation by phosphorylase kinase, and the greater inactivation of glycogen synthase after the combined phosphorylation by casein kinase I and cAMP-dependent protein kinase.

Further studies of the recombinant glycogen synthase expressed in *E. coli* allowed Wang and Roach[392] to generate mutant forms of the rabbit muscle glycogen synthase at GSK-3 phosphorylation sites, S640A, S644A, and S648A (sites 3a, 3b, and 3c, respectively). All three mutants had high $-/+$ glucose-6-P ratios of activity (0.8 to 0.9). Phosphorylation of the mutants was done with GSK-3 and casein kinase II. The mutants phosphorylated at sites 5 and 4 (mutant S648A) and at sites 3c, 4, and 5 (mutant S644A) had full activity. When sites 3b, 3c, 4, and 5 (mutant S640A) were phosphorylated, the activity ratio decreased modestly to about 0.6 to 0.7. When all sites were phosphorylated in the recombinant enzyme, the activity ratio decreased to 0.1. The results of this study demonstrated that phosphorylation site 3a, and to a lesser extent 3b, correlated with the inactivation of the glycogen synthase. The apparent affinity constant for the activator, glucose-6-P, was 2.4 μM and only increased appreciably, to 24 μM, when site 3a was phosphorylated with the other four sites.

3.14.9.3.2 *Studies* in vivo

The studies *in vitro* discussed above clearly showed that for mammal glycogen synthase (or at least for the rabbit muscle enzyme) the important phosphorylation sites for inactivation were sites 2 and Ser10 (site 2a), sites 3a and 3b. However, it is of interest to know which are the important phosphorylation sites under hormonal control *in vivo*. In rabbit muscle, enzyme intravenous insulin administration doubles the $-/+$ glucose-6-P activity ratio, with a decrease in phosphorylation of all the sites.[21,393,394] Epinephrine, which increases the phosphate content of the glycogen synthase, increases the phosphorylation at practically all the phosphorylation sites.[393,395–397]

With respect to the liver glycogen synthase, the hormones glucagon, vasopressin, and epinephrine all increased the phosphorylation of the peptide regions containing sites 2 and 3[21,22] and Akatsuka *et al.*[398] showed that glucagon promoted phosphorylation of the casein kinase I site now referred to as site 2a or equivalent liver Ser residue to muscle glycogen synthase residue, Ser10.

To study in detail the role of individual phosphorylation sites in the regulation of the rabbit muscle glycogen synthase, the enzyme was overexpressed in COS M9 cells.[348,349] The activity ratio of $-/+$ glucose-6-P was found to be very low, ~ 0.01, indicative of a high level of phosphorylation. Ser to Ala mutations were introduced singly, or in combinations, at the nine known phosphorylation sites and it was found that no single Ser to Ala mutation caused a substantial increase in the activity ratio.[349] It was shown that simultaneous mutations were needed at both regions of site 2, the *N*-terminal region, and site 3, the *C*-terminal region. The most effective combinations were mutations at site 3a (Ser640) or site 3b (Ser644) together with site 2 (Ser7). Double mutants, Ser640Ala-Ser7Ala, Ser644Ala-Ser7Ala and Ser10Ala (site 2a)-Ser640Ala, gave activity ratios of 0.59, 0.25, and 0.21, respectively. These results were consistent with site 2 phosphorylation being a prerequisite for phosphorylation of site 2a. In contrast with the results obtained *in vitro*, the mutation of site 5 (Ser656), although affecting phosphorylation at the sites 3 and 4, did not result in an increase in the activity ratio;[349] the authors proposed that in COS cells sites 3a and 3b may be phosphorylated by an alternative pathway independent of phosphorylation of site 5. Nevertheless, the COS cell data did show that the most important sites in regulation of glycogen synthase were sites 2, 2a, 3a, and 3b, a conclusion that was consistent with the *in vitro* data.

This system was studied further and it was shown that phosphorylation of sites 3a and 3b occurred even when mutations were made at sites 5, 4, and 3c;[399] thus, phosphorylation of sites 3a and 3b may occur via other protein kinases other than GSK-3. Evidence supporting this view has been obtained by Skurat and Roach[400] who mutagenized amino acid residues close to the phosphorylation sites, 3a and 3b, that may be important for a protein kinase to recognize and phosphorylate sites 3a and 3b; i.e., arginine residue 637 and proline residue 645. The mutants made were Arg637Gln and Pro645Ala, a double mutant, R637Q, S644A (site 3b), and two triple mutants, S7A (site 2), R637Q, S644A and S7A, S644A, P645A; in addition, the serine residue of sites 3c, 4, and 5 of these mutants were mutagenized to alanine to avoid possible phosphorylation of sites 3a and 3b by GSK-3. Mutation of Arg637 to Gln eliminated phosphorylation of site 3a, suggesting that Arg637 may be important for another protein kinase to recognize site 3a. The mutant Pro645Ala also eliminated phosphorylation of site 3b, suggesting a possible involvement of a "proline-directed protein kinase." Either mutation alone did not substantially increase the activation ratio, meaning that phosphorylation at either site plus phosphorylation at sites 2 and 2a produced a totally inactive enzyme; the triple mutant, S7A, R637Q, S644A, however, was active with an activity ratio of 0.62 while S7A, S644A, P645A had an activity ratio of 0.21.[400] The results also point out that in the COS cells, sites 2, 2a, 3a, and 3b are all important for regulation of glycogen synthase and, most significantly, suggest that sites 3a and 3b can be phosphorylated independently of one another by distinct protein kinases. Thus, the existence of three protein kinases is proposed for the phosphorylation of sites 3a and 3b,[400] as shown in Figure 5.

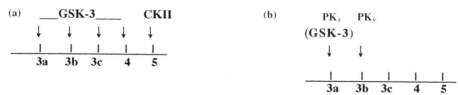

Figure 5   (a) Phosphorylation of sites 3a, 3b, 3c, and 4 by glucogen synthase kinase 3 and phosphorylation of site 5 by casein kinase II. Phosphorylations occur in sequence: first of site 5, then of sites 4, 3c, 3b, and last, of site 3a. (b) Phosphorylation of sites 3a and 3b by the putative protein kinases, PK$_x$ and PK$_y$, respectively. These phosphorylations are independent of each other and do not require prior phosphorylation of sites 3c, 4, and 5.

Thus, in the cell, multiple mechanisms involving at least three protein kinases in the regulation of sites 3 of glycogen synthase and two protein kinases for regulation of sites 2 may exist. The reason for redundant mechanisms of inactivation of glycogen synthase is unknown, but could be the way for integrating messages of a number of hormonal and signal transduction pathways. It should be pointed out, however, that the *in vivo* studies have been done in COS cells and whether the same phenomena, particularly the unidentified protein kinases, PK$_x$ and PK$_y$, are relevant to control of glycogen synthase in skeletal muscle, remains to be established.

Other interesting experiments done in the COS cell system indicated that overexpression of the glycogen biosynthetic enzymes, glycogen synthase, glycogenin, branching enzyme, and UDP-Glc

PPase, alone, did not lead to increased or overaccumulation of glycogen.[400] If, however, the glycogen synthase mutant S7A, S640A was overexpressed, then there was about a 2- to 2.5-fold increase in glycogen levels, suggesting that glycogen synthase activity was rate-limiting. However, the co-overexpression of the mutant glycogen synthase with the UDP-Glc PPase or with glycogenin led to even greater glycogen accumulation, i.e., another 34 to 70% increase. Thus, with overexpression of the hyperactive glycogen synthase, the synthesis of UDP-glucose or synthesis of the glycosylated acceptor protein may become the rate-limiting reactions.

3.14.9.4 Mechanism of Stimulation by Insulin of Glycogen Synthesis in Mammalian Skeletal Muscle

3.14.9.4.1 *Protein phosphatase 1*

Because the glycogen synthase is phosphorylated at multiple sites per subunit, it would be of interest to know whether one or several protein phosphatases catalyze dephosphorylation of these distinct sites. Four protein phosphatases are known; for background on the structures and biochemical properties of these phosphatases, see refs.[401–403] Of these phosphatases, protein phosphatase 1, which hydrolyzes mainly the phosphate of the β-subunit of phosphorylase kinase, has strong activity on the phosphate sites 1a, 2, 2a, 3a, 3b, 3c, and 4 of glycogen synthase and is the principal enzyme in dephosphorylating glycogen synthase, since both the phosphatase and glycogen synthase are usually bound to the glycogen particle. Protein phosphatase 2A, which has greater activity on the α-subunit phosphates of phosphorylase kinase than protein phosphatase 1, also has activity on the above glycogen synthase phosphate sites.

Of interest is that protein phosphatase 1, of 37 kDa, is associated with the glycogen particles *in vivo* when it is in a complex with a 160 kDa protein referred to as the G subunit.[404,405] When the protein phosphatase 1 is associated with the G subunit, it binds to the glycogen particle and is far more enzymatically active, under physiological conditions, in dephosphorylating glycogen synthase, glycogen phosphorylase, and phosphorylase kinase (enzymes that also bind to the glycogen particle) than when it is free.[406] The G subunit is phosphorylated by cAMP-dependent protein kinase *in vitro*[404,407] and in response to epinephrine *in vivo*.[408,409] Phosphorylation occurs at two serine sites, sites 1 and 2, separated by 18 residues and the phosphorylation of both sites causes a dissociation of the phosphatase from the G subunit.[407] The phosphatase is about five to eight times less active than the complex in dephosphorylating glycogen synthase and phosphorylase present in the glycogen particle and this lowering of the protein phosphatase activity is one way in which epinephrine stimulates glycogen breakdown and inhibits glycogen synthesis.

The dissociation of subunit G from the phosphatase correlates with the phosphorylation of site 2, and not with that of site 1;[407] and reassociation of the G subunit with the phosphatase occurs with dephosphorylation of site 2 by protein phosphatase 2A under conditions where site 1 still retains the phosphate residue as it is more resistant to dephosphorylation by protein phosphatase 2A. Thus, inactivation of the protein phosphatase is due to phosphorylation of site 2 and not site 1.

Parker et al.[393] showed that, in response to insulin administration, the phosphate released from glycogen synthase is mainly from sites 3a, 3b, and 3c, suggesting that insulin caused the inhibition of GSK-3 or activated protein phosphatase 1. As will be shown, both phenomena, activation of protein phosphatase 1 and inhibition of GSK-3, do occur due to insulin. Dent et al.[410] showed that the G protein phosphatase complex with phosphate mainly at site 1 had protein phosphatase activity which was now associated with the glycogen particle and about 2.5- to 3-fold higher activity than the dephosphorylated protein phosphatase. Moreover, a protein kinase was isolated from rabbit skeletal muscle that phosphorylated the G subunit at site 1 but not at site 2. This protein kinase also phosphorylates a ribosomal protein, S6, *in vitro*. The activity of this protein kinase was increased about 2-fold within 15 minutes after insulin administration, the same time frame for the increase in glycogen synthase activity.[393] The phosphorylation of the G subunit increased phosphatase activity about 2.8-fold on glycogen synthase and phosphorylase kinase. It was also determined that insulin administration stimulated *in vivo* phosphorylation of site 1 and not site 2. As shown in Figure 6, Dent et al.[410] proposed that the interaction of insulin with its membrane-bound receptor activated its tyrosine protein kinase, leading to an activation of a protein serine/threonine kinase-kinase which, in turn, phosphorylates and makes active a kinase labeled as the insulin-stimulated kinase

(ISPK). This ISPK then phosphorylates site 1 of the G subunit making the protein phosphatase 1-G complex more active. The active protein phosphatase then dephosphorylates the phosphate residues off sites 3a, 3b, and 3c, thus activating glycogen synthesis.

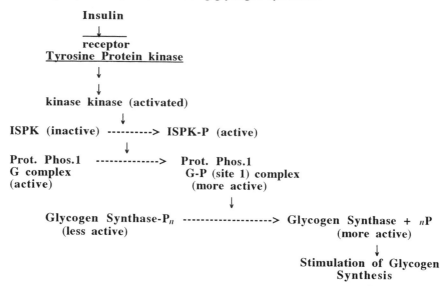

Figure 6 Activation of glycogen synthase by insulin, as proposed by Dent *et al.*[410] Binding of insulin to its receptor tyrosine protein kinase activates a series of protein kinases which finally activates a Ser/Thr protein kinase. This in turn phosphorylates the protein phosphatase 1-G protein complex, making the protein phosphatase activity more active. In turn this phosphatase dephosphorylates the glycogen synthase at sites 3a, 2, and 2a, increasing the glycogen synthase activity.

3.14.9.4.2 *Inactivation of glycogen synthase kinase-3*

Evidence has also been accumulated indicating that insulin can induce inactivation of glycogen synthase kinase-3 in many different cells.[411-413] The inactivation appears to be a phosphorylation catalyzed by protein kinase, also known as Akt/RAC,[414-416] which is regulated by an activated phosphatidylinositol (PI3) kinase.[417] Inactivation of the GSK-3 by protein kinase B is accompanied by phosphorylation of serine residue 9 of the GSK-3β and serine-21 of the GSK-3α isozymes *in vitro* or *in vivo*.[416] Although other protein kinases are known to phosphorylate and inactivate GSK-3,[416] it is believed that the activation by insulin of glycogen synthase via inactivation of GSK-3 is due to activation of protein kinase B. Insulin stimulation of L6 myotube cells caused a 10-fold increase in protein kinase B activity and decreased GSK-3 activity by 40–50%. The half-time for activation of protein kinase B was one minute, slightly faster than inhibition of GSK-3 which was 2 minutes; the inhibition of GSK-3 was reversed by incubation with protein phosphatase 2A.

Other studies, such as those done by Eldar-Finkelman *et al.*[418] on whole cells, also indicate the importance of GSK-3 in regulation glycogen synthase activity; mutants of GSK-3 were made at serine-9 (Ser9Ala and Ser9Glu). These mutants could not be phosphorylated or their GSK-3 activity made inactive by phosphorylation. The wild-type, normal, and mutant enzymes were expressed in 293 cells and their activity was determined. Cells expressing the S9A mutant, WT and S9E mutant GSK-3 had 2.6-, 1.8- and 2.0-fold higher GSK-3 activity, respectively, as compared with control cells. The higher activity of the S9A mutant suggested serine 9 as a key regulatory site for GSK-3 inactivation. However, substitution of glutamic acid for serine could not mimic the inactivation caused by the negative ion, phosphate, on the serine-9 residue. The effects of expressing the Wt and mutant GSK-3 mutants in the 293 cells on glycogen synthase $-$glucose-6-P$/+$glucose-6-P activity ratio was measured. A 50% reduction in the activity ratio was seen for the cells having the S9A mutant while a 20 to 30% decrease was observed in cells having the WT and S9E mutant. Thus, evidence was obtained that activation of GSK-3 is sufficient to inhibit glycogen synthase in intact cells and for supporting a physiological role for GSK-3 in regulating glycogen synthase.

In summary, insulin stimulates glycogen synthesis by activating protein phosphatase 1 and by inactivating glycogen synthase kinase-3. Thus, the inactivated glycogen synthase becomes active and at the same time phosphorylase a and phosphorylase kinase are inactivated due to their dephosphorylation by activated protein phosphatase 1, resulting in an overall increase in glycogen synthesis and decreased glycogen degradation.

Three lines of transgenic mice have been generated where the rabbit skeletal muscle glycogen synthase was overexpressed in mouse skeletal muscle.[419] The glycogen synthase expressed was the glycogen synthase-sites 2 and 3a mutant (Ser7Ala, Ser640Ala) so that the overexpression of the synthase could not be inactivated by phosphorylation. The glycogen synthase activity was expressed by as much as 10-fold with concomitant increases of up to 5-fold in glycogen content. The levels of UDP-glucose decreased markedly, consistent with the increase in glycogen synthase activity. Levels of the glycogen degradative enzyme, phosphorylase, increased up to 3-fold but the activity of the insulin-sensitive glucose transporter either remained unchanged or decreased. Thus, increasing active glycogen synthase caused an increase in glycogen content, supporting the conclusion that activation of glycogen synthase contributes to the increased accumulation of glycogen observed in response to insulin.

3.14.10 REFERENCES

1. W. R. Morrison and J. Karkalas, in "Methods in Plant Biochemistry," ed. P. M. Dey, Academic Press, London, 1990, vol. 2, p. 323.
2. S. Hizukuri, in "Carbohydrates in Food," ed. A.-C. Eliasson, Marcel Dekker, New York, 1995, p. 347.
3. D. J. Manners, *Carbohydr. Polym.*, 1991, **16**, 37.
4. D. J. Manners, in "The Yeasts," eds. A. H. Rose and J. S. Harrison, Academic Press, New York, 1971, vol. 2, p. 419.
5. C. F. Cori and G. R. Cori, *J. Biol. Chem.*, 1939, **131**, 397.
6. L. F. Leloir and C. E. Cardini, *J. Am. Chem. Soc.*, 1957, **79**, 6340.
7. J. Preiss and T. Romeo, *Adv. Microb. Physiol.*, 1989, **30**, 183.
8. J. Preiss and T. Romeo, *Prog. Nucleic Acids Res. Mol. Biol.*, 1994, **47**, 299.
9. J. Preiss, in "Escherichia coli and Salmonella typhimurium: Cellular and Molecular Biology," 2nd edn., ed. F. C. Neidhardt, Amer. Soc. Microbiol., Washington, DC, 1996, vol. 1, p. 1015.
10. C. Martin and A. M. Smith, *Plant Cell*, 1995, **7**, 971.
11. J. Preiss, in "The Biochemistry of Plants," ed. J. Preiss, Academic Press, New York, 1988, vol. 14, p. 184.
12. J. Preiss, in "Oxford Survey of Plant Molecular and Cellular Biology," ed. B. J. Miflin, Oxford University Press, Oxford, UK, 1991, vol. 7, p. 59.
13. J. Preiss, in "Frontiers in Carbohydrate Research," ed. R. Chandrasekaran, Elsevier Science Publishers, Essex, UK, 1992, vol. 2, p. 208.
14. J. Preiss, in "Biotechnology Annual Review," ed. R. El-Gewely, Elsevier Science, Amsterdam, The Netherlands, 1996, vol. 2, p. 259.
15. J. Preiss, *Carbohydrates in Europe*, 1996, **15**, 11.
16. J. Preiss, in "Engineering Improved Carbon and Nitrogen Resource Use Efficiency in Higher Plants," eds. C. Foyer and P. Quick, Taylor and Francis, London, 1997, p. 81.
17. J. Preiss and M. N. Sivak, in "Photoassimilate Distribution in Plants and Crops. Source Sink Relationships," eds. E. Zamski and A. A. Schaffer, Marcel Dekker, New York, 1996, p. 63.
18. M. N. Sivak and J. Preiss, in "Seed Development and Germination," eds. J. Kigel and G. Galili, Marcel Dekker, New York, 1995, p. 139.
19. A. M. Smith, K. Denyer, and C. R. Martin, *Plant Physiol.*, 1995, **107**, 673.
20. P. Cohen, in "The Enzymes," 3rd edn., eds. P. D. Boyer and E. G. Krebs, Academic Press, San Diego, CA, 1986, vol. 17, p. 461.
21. P. J. Roach, in "The Enzymes," 3rd edn., eds. P. D. Boyer and E. G. Krebs, Academic Press, San Diego, CA, 1986, vol. 17, p. 499.
22. P. J. Roach, *FASEB J.*, 1990, **4**, 2961.
23. E. A. Dawes and P. J. Senior, *Adv. Microb. Physiol.*, 1973, **10**, 135.
24. J. Preiss, *Annu. Rev. Microbiol.*, 1984, **38**, 419.
25. J. Preiss, in "Bacteria in Nature," eds. J. S. Poindexter and E. Leadbetter, Plenum, New York, 1989, vol. 3, p. 189.
26. B. M. Mackey and J. G. Morris, *J. Gen. Microbiol.*, 1971, **66**, 1.
27. I. G. Burleigh and E. A. Dawes, *Biochem. J.*, 1967, **102**, 236.
28. J. Sachs, in "Lectures of the Physiology of Plants," trans. H. M. Ward, Clarendon Press, Oxford, 1887, p. 304.
29. G. Edwards and D. A. Walker, in "C₃,C₄: Mechanisms and Cellular and Environmental Regulation of Photosynthesis," University of California Press, Berkeley, CA, 1983, p. 204.
30. N. P. Badenhuizen, in "The Biogenesis of Starch Granules in Higher Plants," Appleton-Century Crofts, New York, 1969.
31. E. Recondo and L. F. Leloir, *Biochem. Biophys. Res. Commun.*, 1961, **6**, 85.
32. N. Sigal, J. Cattaneo, and I. H. Segel, *Arch. Biochem. Biophys.*, 1964, **108**, 440.
33. E. Greenberg and J. Preiss, *J. Biol. Chem.*, 1964, **239**, 4314.
34. L. Shen, H. P. Ghosh, E. Greenberg, and J. Preiss, *Biochim. Biophys. Acta*, 1964, **89**, 370.
35. N. Sigal, J. Cattaneo, J. P. Chambost, and A. Favard, *Biochem. Biophys. Res. Commun.*, 1965, **20**, 616.
36. C. Boyer and J. Preiss, *Biochemistry*, 1977, **16**, 3693.
37. L. P. T. M. Zevenhuizen, *Biochim. Biophys. Acta*, 1964, **81**, 608.
38. K. E. Steiner and J. Preiss, *J. Bacteriol.*, 1977, **129**, 246.
39. G. J. Walker and J. E. Builder, *Eur. J. Biochem.*, 1971, **20**, 14.
40. P. A. Baecker, C. E. Furlong, and J. Preiss, *J. Biol. Chem.*, 1983, **258**, 5084.

41. D. Homerova and J. Kormanec, *Biochim. Biophys. Acta*, 1994, **1200**, 334.
42. J. A. K. W. Kiel, J. M. Boels, G. Beldman, and G. Venema, *Mol. Gen. Genet.*, 1991, **230**, 136.
43. H. Takata, T. Takaha, T. Kuriki, S. Okada, M. Takagi, and T. Imanaka, *Appl. Environ. Microbiol.*, 1994, **60**, 3096.
44. J. A. K. W. Kiel, J. M. Boels, G. Beldman, and G. Venema, *DNA Seq.*, 1992, 221.
45. J. A. K. W. Kiel, H. S. A. Elgersma, G. Beldman, J. P. M. J. Vossen, and G. Venema, *Gene*, 1989, **78**, 9.
46. J. A. K. W. Kiel, J. M. Boels, G. Beldman, and G. Venema, *Gene*, 1990, **89**, 77.
47. S. Hestrin, in "The Bacteria," eds. I. Gunsalas and R. Y. Stanier, Academic Press, New York, 1960, vol. 3, p. 373.
48. E. J. Hehre, *Adv. Enzymol.*, 1951, **11**, 297.
49. G. Okada and E. J. Hehre, *J. Biol. Chem.*, 1974, **249**, 126.
50. T. N. Palmer, G. Wöber, and W. J. Whelan, *Eur. J. Biochem.*, 1973, **39**, 601.
51. J. Monod and A. M. Torriani, *C.R. Acad. Sci. Paris*, 1948, **227**, 240.
52. G. J. Walker, *Biochem. J.*, 1966, **101**, 861.
53. S. Lacks, *Genetics*, 1968, **60**, 685.
54. G. Wober, *Hoppe-Seyler's Z. Physiol. Chem.*, 1973, **354**, 75.
55. J. Chao and C. J. Weathersbee, *J. Bacteriol.*, 1974, **117**, 181.
56. G. S. Chen and I. H. Segel, *Arch. Biochem. Biophys.*, 1968, **127**, 164.
57. G. S. Chen and I. H. Segel, *Arc. Biochem. Biophys.*, 1968, **127**, 175.
58. R. L. Khandelwal, T. N. Spearman, and I. R. Hamilton, *Arch. Biochem. Biophys.*, 1973, **154**, 295.
59. M. Schwartz, *J. Bacteriol.*, 1966, **92**, 1083.
60. T. Romeo and J. Preiss, *J. Bacteriol.*, 1989, **171**, 2773.
61. J. H. Ko, C. H. Kim, D.-S. Lee, and Y. S. Kim, *Biochem. J.*, 1996, **319**, 977.
62. J. Preiss and C. Levi, in "Proceedings of the 4th International Congress on Photosynthesis," eds. D. O. Hall, J. Coombs, and T. W. Goodwin, The Biochemistry Society, UK, 1978, p. 457.
63. K. Denyer and A. M. Smith, *Planta*, 1992, **186**, 609.
64. K. Denyer, C. Sidebottom, C. M. Hylton, and A. M. Smith, *Plant J.*, 1993, **4**, 191.
65. K. Denyer, C. M. Hylton, C. F. Jenner, and A. M. Smith, *Planta*, 1995, **196**, 256.
66. C. M. Hylton, K. Denyer, P. L. Keeling, M.-T. Chang, and A. M. Smith, *Planta*, 1996, **198**, 230.
67. K. Mizuno, K. Kimura, Y. Arai, T. Kawasaki, H. Shimada, and T. Baba, *J. Biochem. (Tokyo)*, 1992, **112**, 643.
68. M. Bhattacharyya, C. Martin, and A. M. Smith, *Plant Mol. Biol.*, 1993, **22**, 525.
69. H. P. Guan and J. Preiss, *Plant Physiol.*, 1993, **102**, 1269.
70. H. P. Guan, T. Baba, and J. Preiss, *Plant Physiol.*, 1994, **104**, 1449.
71. H. P. Guan, T. Baba, and J. Preiss, *Cell. Mol. Biol.*, 1994, **40**, 981.
72. H. Guan, T. Kuriki, M. Sivak, and J. Preiss, *Proc. Nat. Acad. Sci. USA*, 1995, **92**, 964.
73. R. A. Burton, J. D. Bewley, A. M. Smith, M. K. Bhattacharyya, H. Tatge, S. Ring, V. Bull, W. D. Hamilton, and C. Martin, *Plant J.*, 1995, **7**, 3.
74. O. E. Nelson, P. S. Chourey, and M. T. Chang, *Plant Physiol.*, 1978, **62**, 383.
75. M. Shure, S. Wessler, and N. Federoff, *Cell*, 1983, **35**, 225.
76. Y. Sano, *Theor. Appl. Genetics*, 1984, **68**, 467.
77. F. R. Van der Leij, R. G. Visser, A. S. Ponstein, E. Jacobsen, and W. J. Feenstra, *Mol. Gen. Genet.*, 1991, **228**, 240.
78. R. G. Visser, I. Somhorst, G. J. Kuipers, N. J. Ruys, W. J. Feenstra, and E. Jacobsen, *Mol. Gen. Genet.*, 1991, **225**, 289.
79. I. Dry, A. Smith, A. Edwards, M. Bhattacharyya, P. Dunn, and C. Martin, *Plant J.*, 1992, **2**, 193.
80. B. Delrue, T. Fontaine, F. Routier, A. Decq, J.-M. Wieruszeski, N. van den Koornhuyse, M.-L. Maddelein, B. Fournet, and S. Ball, *J. Bacteriol.*, 1992, **174**, 3612.
81. L. F. Leloir, M. A. Rongine deFekete, and C. E. Cardini, *J. Biol. Chem.*, 1961, **236**, 636.
82. D. Pan and O. E. Nelson, *Plant Physiol.*, 1984, **74**, 324.
83. M. G. James, D. S. Robertson, and A. M. Meyers, *Plant Cell*, 1995, **7**, 417.
84. Y. Nakamura, T. Umemoto, Y. Takahata, K. Komae, E. Amano, and H. Satoh, *Physiol. Plant.*, 1996, **97**, 491.
85. C. R. Krisman and R. Barengo, *Eur. J. Biochem.*, 1975, **52**, 117.
86. J. Lomako, W. M. Lomako, and W. J. Whelan, *FASEB J.*, 1988, **2**, 3097.
87. J. Pitcher, C. Smythe, D. G. Campbell, and P. Cohen, *Eur. J. Biochem.*, 1987, **169**, 497.
88. D. G. Campbell and P. Cohen, *Eur. J. Biochem.*, 1989, **185**, 119.
89. I. R. Rodriguez and W. J. Whelan, *Biochem. Biophys. Res. Commun.*, 1985, **132**, 829.
90. C. Smythe, F. B. Caudwell, M. Ferguson, and P. Cohen, *EMBO J.*, 1988, **7**, 2681.
91. J. Lomako, W. M. Lomako, W. J. Whelan, R. S. Dombro, J. T. Neary, and M. D. Norenberg, *FASEB J.*, 1993, **7**, 1386.
92. M. D. Alonso, J. Lomako, W. M. Lomako, and W. J. Whelan, *FASEB J.*, 1995, **9**, 1126.
93. J. Lomako, W. M. Lomako, and W. J. Whelan, *FEBS Lett.*, 1991, **279**, 223.
94. T. H. Haugen, A. Ishaque, A. K. Chatterjee, and J. Preiss, *FEBS Lett.*, 1974, **42**, 205.
95. T. H. Haugen, A. Ishaque, and J. Preiss, *J. Biol. Chem.*, 1976, **251**, 7880.
96. C. E. Furlong and J. Preiss, *J. Biol. Chem.*, 1969, **244**, 2539.
97. G. Ribereau-Gayon and J. Preiss, *Methods Enzymol.*, 1971, **23**, 618.
98. J. Preiss and D. A. Walsh, in "Biology of Carbohydrates," ed. V. Ginsburg, Wiley, New York, 1981, vol. 1, p. 199.
99. J. Preiss, C. Lammel, and E. Greenberg, *Arch. Biochem. Biophys.*, 1976, **174**, 105.
100. T. W. Okita, R. L. Rodriguez, and J. Preiss, *J. Biol. Chem.*, 1981, **256**, 6944.
101. H. P. Ghosh and J. Preiss, *J. Biol. Chem.*, 1966, **241**, 4491.
102. J. Preiss, H. P. Ghosh, and J. Wittkop, in "Biochemistry of Chloroplasts," ed. T. W. Goodwin, Academic Press, New York, 1965, vol. 2, p. 131.
103. L. Copeland and J. Preiss, *Plant Physiol.*, 1981, **68**, 996.
104. G. G. Sanwal and J. Preiss, *Arch. Biochem. Biophys.*, 1967, **9**, 454.
105. J. Preiss, M. Morell, B. Bloom, V. L. Knowles, and T. P. Lin, in "Progress in Photosynthesis Research, Proc. 7th Int. Congr. Photosynthesis, 1986," ed. J. Biggins, Nijhoff, Dordrecht, 1987, vol. 13, p. 693.
106. M. K. Morell, M. Bloom, V. Knowles, and J. Preiss, *Plant Physiol.*, 1987, **85**, 182.

107. M. Morell, M. Bloom, R. Larsen, T. W. Okita, and J. Preiss, in "Plant Gene Systems and Their Biology," eds. J. L. Key and L. McIntosh, Alan R. Liss, New York, 1987, vol. 62, p. 227.
108. J. Preiss, S. Danner, P. S. Summers, M. K. Morell, C. R. Barton, L. Yang, and M. Nieder, *Plant Physiol.*, 1990, **92**, 881.
109. C. Barton, L. Yang, M. Galvin, C. Sengupta-Gopalan, and T. Borelli, in "Regulation of Carbon and Nitrogen Reduction and Utilization in Maize," eds. J. C. Shannon, D. P. Knievel, and C. D. Boyer, American Society of Plant Physiologists, Rockville, MD, 1986, p. 363.
110. J. M. Anderson, J. Hnilo, R. Larson, T. W. Okita, M. Morell, and J. Preiss, *J. Biol. Chem.*, 1989, **264**, 12 238.
111. B. J. Smith-White and J. Preiss, *J. Mol. Evol.*, 1992, **34**, 449.
112. M. R. Paule and J. Preiss, *J. Biol. Chem.*, 1971, **246**, 4602.
113. W. W. Cleland, in "The Enzymes," 3rd edn., ed. P. D. Boyer, Academic Press, San Diego, CA, 1970, vol. 2, p. 1.
114. T. Haugen and J. Preiss, *J. Biol. Chem.*, 1979, **254**, 127.
115. M. L. DePamphilis and W. W. Cleland, *Biochemistry*, 1973, **122**, 3714.
116. C. A. Janson and W. W. Cleland, *J. Biol. Chem.*, 1974, **249**, 2572.
117. M. Lazdunski, *Curr. Top. Cell. Regul.*, 1972, **6**, 267.
118. J. Preiss, D. Cress, J. Hutny, M. K. Morell, M. Bloom, T. Okita, and J. Anderson, in "Biocatalysis in Agricultural Biotechnology," *ACS Symp. Ser.* eds. J. R. Whitaker and P. E. Sonnet, Amer. Chem. Soc., Washington, DC, 1989, vol. 389, p. 84.
119. M. A. Hill, K. Kaufmann, J. Otero, and J. Preiss, *J. Biol. Chem.*, 1991, **266**, 12 455.
120. Y. M. Lee and J. Preiss, *J. Biol. Chem.*, 1986, **261**, 1058.
121. T. F. Parsons and J. Preiss, *J. Biol. Chem.*, 1978, **253**, 6197.
122. T. F. Parsons and J. Preiss, *J. Biol. Chem.*, 1978, **253**, 7638.
123. C. E. Larsen and J. Preiss, *Biochemistry*, 1986, **25**, 4371.
124. C. E. Larsen, Y. M. Lee, and J. Preiss, *J. Biol. Chem.*, 1986, **261**, 15 402.
125. P. S. C. Leung and J. Preiss, *J. Bacteriol.*, 1987, **169**, 4355.
126. A. M. Mulichak, E. Skrzypczak-Jankun, T. J. Rydel, A. Tulinsky, and J. Preiss, *J. Biol. Chem.*, 1988, **263**, 17 237.
127. M. A. Ballicora, M. J. Laughlin, Y. Fu, T. W. Okita, G. F. Barry, and J. Preiss, *Plant Physiol.*, 1995, **109**, 245.
128. Y.-Y. Charng, G. Kakefuda, A. A. Iglesisas, W. J. Buikema, and J. Preiss, *Plant Mol. Biol.*, 1992, **20**, 37.
129. C. Ainsworth, J. Clark, and J. Balsdon, *Plant Mol. Biol.*, 1993, **22**, 67.
130. M. Morell, M. Bloom, and J. Preiss, *J. Biol. Chem.*, 1988, **263**, 633.
131. K. Ball and J. Preiss, *J. Biol. Chem.*, 1994, **269**, 24 706.
132. Y.-Y. Charng, A. A. Iglesias, and J. Preiss, *J. Biol. Chem.*, 1994, **269**, 24 107.
133. J. M. Bae, J. Giroux, and L. C. Hannah, *Maydica*, 1990, **35**, 317.
134. T. Thornbjørnsen, P. Villand, L. A. Kleczkowski, and O.-A. Olsen, *Biochem. J.*, 1996, **313**, 149.
135. G. Kakefuda, Y.-Y. Charng, A. A. Iglesias, L. McIntosh, and J. Preiss, *Plant Physiol.*, 1992, **99**, 359.
136. M. R. Bhave, S. Lawrence, C. Barton, and L. C. Hannah, *Plant Cell*, 1990, **2**, 581.
137. P. Villand, O.-A. Olsen, A. Killian, and L. A. Kleczkowski, *Plant Physiol.*, 1992, **100**, 1617.
138. M. R. Olive, R. J. Ellis, and W. W. Schuch, *Plant Mol. Biol.*, 1989, **12**, 525.
139. K. L. Ball and J. Preiss, *J. Protein Chem.*, 1992, **11**, 231.
140. A. A. Iglesias, G. F. Barry, C. Meyer, L. Bloksberg, P. A. Nakata, T. Greene, M. J. Laughlin, T. W. Okita, G. M. Kishore, and J. Preiss, *J. Biol. Chem.*, 1993, **268**, 1081.
141. J. Preiss, J. Sheng, Y. Fu, and M. A. Ballicora, in "Proc. 10th International Photosynthesis Congress," ed. P. Mathis, Kluwer Academic Publishers, Dordrecht, 1996, vol. 5, p. 47.
142. L. Li and J. Preiss, *Carbohydr. Res.*, 1992, **227**, 227.
143. W. C. Plaxton and J. Preiss, *Plant Physiol.*, 1987, **83**, 105.
144. H. B. Krishnan, C. D. Reeves, and T. W. Okita, *Plant Physiol.*, 1986, **81**, 642.
145. T.-P. Lin, T. Caspar, C. Somerville, and J. Preiss, *Plant Physiol.*, 1988, **86**, 1131.
146. T. W. Okita, P. A. Nakata, J. M. Anderson, J. Sowokinos, M. Morell, and J. Preiss, *Plant Physiol.*, 1990, **93**, 785.
147. C. Levi and J. Preiss, *Plant Physiol.*, 1976, **58**, 753.
148. A. A. Iglesias, G. Kakefuda, and J. Preiss, *Plant Physiol.*, 1991, **97**, 1187.
149. J. Fox, K. Kawaguchi, E. Greenberg, and J. Preiss, *Biochemistry*, 1976, **15**, 849.
150. J. Cattaneo, J. P. Chambost, and N. Creuzat-Sigal, *Arch. Biochem. Biophys.*, 1978, **190**, 85.
151. K. Kawaguchi, J. Fox, E. Holmes, C. Boyer, and J. Preiss, *Arch. Biochem. Biophys.*, 1978, **190**, 385.
152. J. Cattaneo, M. Magnan, and J. Bigliardi, *Arch. Biochem. Biophys.*, 1979, **196**, 449.
153. E. Greenberg and J. Preiss, *J. Biol. Chem.*, 1965, **240**, 2341.
154. J. Preiss and E. Greenberg, *Biochemistry*, 1965, **4**, 2328.
155. D. N. Dietzler and J. L. Strominger, *J. Bacteriol.*, 1973, **113**, 946.
156. A. D. Elbein and M. Mitchell, *J. Bacteriol.*, 1973, **113**, 863.
157. E. Holmes and J. Preiss, *Arch. Biochem. Biophys.*, 1979, **196**, 436.
158. R. M. Bock, in "Enzymes," 2nd edn., eds. P. D. Boyer, H. Lardy, and K. Myrback, Academic Press, New York, 1960, vol. 2, p. 3.
159. P. S. C. Leung and J. Preiss, *J. Bacteriol.*, 1987, **169**, 4349.
160. A. Kumar, C. E. Larsen, and J. Preiss, *J. Biol. Chem.*, 1986, **261**, 16 256.
161. E. Holmes and J. Preiss, *Arch. Biochem. Biophys.*, 1982, **216**, 736.
162. K. Furukawa, M. Tagaya, M. Inouye, J. Preiss, and T. Fukui, *J. Biol. Chem.*, 1990, **265**, 2086.
163. M. Tagaya, K. Nakano, and T. Fukui, *J. Biol. Chem.*, 1985, **260**, 6670.
164. A. M. Mahrenholz, Y. Wang, and P. J. Roach, *J. Biol. Chem.*, 1988, **263**, 10 561.
165. K. Furukawa, M. Tagaya, K. Tanaziwa, and T. Fukui, *J. Biol. Chem.*, 1993, **268**, 23 837.
166. K. Furukawa, M. Tagaya, K. Tanaziwa, and T. Fukui, *J. Biol. Chem.*, 1994, **269**, 868.
167. F. D. Macdonald and J. Preiss, *Plant Physiol.*, 1983, **73**, 175.
168. F. D. Macdonald and J. Preiss, *Plant Physiol.*, 1985, **78**, 849.
169. J. L. Ozbun, J. S. Hawker, and J. Preiss, *Biochem. Biophys. Res. Commun.*, 1971, **43**, 631.
170. J. L. Ozbun, J. S. Hawker, and J. Preiss, *Biochem. J.*, 1972, **126**, 953.

171. C. Boyer and J. Preiss, *Plant Physiol.*, 1979, **64**, 1039.
172. C. Pollock and J. Preiss, *Arch. Biochem. Biophys.*, 1980, **204**, 578.
173. A. Edwards, J. Marshall, K. Denyer, C. Sidebottom, R. G. Visser, C. Martin, and A. M. Smith, *Plant Physiol.*, 1996, **112**, 89.
174. C. Mu-Forster, R. Huang, J. R. Powers, R. W. Harriman, M. Knight, G. W. Singletary, P. L. Keeling, and B. P. Wasserman, *Plant Physiol.*, 1996, **111**, 821.
175. O. E. Nelson and H. W. Rines, *Biochem. Biophys. Res. Commun.*, 1962, **9**, 297.
176. G. H. Vos-Scheperkeuter, J. G. de Wit, A. S. Ponstein, W. J. Feenstra, and B. Witholt, *Plant Physiol.*, 1989, **90**, 75.
177. N. Fedoroff, S. Wessler, and M. Shure, *Cell*, 1983, **35**, 235.
178. S. R. Wessler, G. Baran, M. Varagona, and S. L. Dellaporta, *EMBO J.*, 1986, **5**, 2427.
179. R. B. Klösgen, A. Gierl, Z. Schwartz-Sommer, and H. Saedler, *Mol. Gen. Genet.*, 1986, **203**, 237.
180. W. Rhode, D. Becker, and F. Salamini, *Nucleic Acids Res.*, 1988, **16**, 7185.
181. Y. Wang, A. W. Bell, M. A. Hermodson, and P. J. Roach, *J. Biol. Chem.*, 1986, **261**, 16 909.
182. S. N. I. M. Salehuzzaman, E. Jacobsen, and R. G. F. Visser, *Plant Mol. Biol.*, 1992, **20**, 809.
183. T. Baba, M. Nishihara, K. Mizuno, T. Kawasaki, H. Shimada, E. Kobayashi, S. Ohnishi, K. Tanaka, and Y. Arai, *Plant Physiol.*, 1993, **103**, 565.
184. M. N. Sivak, M. Wagner, and J. Preiss, *Plant Physiol.*, 1993, **103**, 1355.
185. A. M. Smith, *Planta*, 1990, **182**, 599.
186. M. N. Sivak, J. P. Guan, and J. Preiss, unpublished results.
187. A. G. J. Kuipers, E. Jacobsen, and R. G. F. Visser, *Plant Cell*, 1994, **6**, 43.
188. H. Shimada, Y. Tada, T. Kawasaki, and T. Fujimura, *Theor. Appl. Genet.*, 1993, **86**, 665.
189. J. Preiss and C. Levi, in "The Biochemistry of Plants: Carbohydrates, Structure and Function," ed. J. Preiss, Academic Press, New York, 1980, vol. 3, p. 371.
190. C. Mu, C. Harn, Y.-T. Ko, G. W. Singletary, P. L. Keeling, and B. P. Wasserman, *Plant J.*, 1994, **6**, 151.
191. J. Marshall, C. Sidebottom, M. Debet, C. Martin, A. M. Smith, and A. Edwards, *Plant Cell*, 1996, **8**, 1121.
192. T. Fontaine, C. D'Hulst, M.-L. Maddelein, F. Routier, T. M. Pepin, A. Decq, J.-M. Wieruszeski, B. Delrue, N. Van Den Koornhuyse, J.-P. Bossu, B. Fournet, and S. Ball, *J. Biol. Chem.*, 1993, **268**, 16 223.
193. M.-L. Maddelein, N. Libessart, F. Bellanger, B. Delrue, C. D'Hulst, N. Van Den Koornhuyse, T. Fontaine, J.-M. Wieruszeski, A. Decq, and S. Ball, *J. Biol. Chem.*, 1994, **269**, 25 150.
194. C. D. Boyer and J. Preiss, *Carbohydr. Res.*, 1978, **61**, 321.
195. C. D. Boyer and J. Preiss, *Biochem. Biophys. Res. Commun.*, 1978, **80**, 169.
196. J. S. Hawker, J. L. Ozbun, H. Ozaki, E. Greenberg, and J. Preiss, *Arch. Biochem. Biophys.*, 1974, **160**, 530.
197. Y. Takeda, H. P. Guan, and J. Preiss, *Carbohydr. Res.*, 1993, **240**, 253.
198. P. A. Baecker, E. Greenberg, and J. Preiss, *J. Biol. Chem.*, 1986, **26**, 8738.
199. T. Romeo, A. Kumar, and J. Preiss, *Gene*, 1988, **70**, 363.
200. T. Baba, K. Kimura, K. Mizuno, H. Etoh, Y. Ishida, O. Shida, and Y. Arai, *Biochem. Biophys. Res. Commun.*, 1991, **181**, 87.
201. B. Svensson, *Plant Mol. Biol.*, 1994, **25**, 141.
202. H. M. Jesperson, E. A. MacGregor, B. Henrissat, M. R. Sierks, and B. Svensson, *J. Protein Chem.*, 1993, **12**, 791.
203. F. D. MacDonald and J. Preiss, *Plant Physiol.*, 1985, **78**, 849.
204. W. Goldner and H. Beevers, *Phytochem.*, 1989, **28**, 1809.
205. D. Borovsky, E. E. Smith, and W. J. Whelan, *Eur. J. Biochem.*, 1975, **59**, 615.
206. A. Blennow and G. Johansson, *Phytochem.*, 1991, **30**, 437.
207. J. Khoshnoodi, B. Ek, L. Rask, and H. Larsson, *FEBS Lett.*, 1993, **332**, 132.
208. J. Khoshnoodi, A. Blennow, B. Ek, L. Rask, and H. Larsson, *Eur. J. Biochem.*, 1996, **242**, 148.
209. J. Kossmann, R. G. F. Visser, B. Müller-Röber, L. Willmitzer, and U. Sonnewald, *Mol. Gen. Genet.*, 1991, **230**, 39.
210. P. Poulsen and J. D. Kreiberg, *Plant Physiol.*, 1993, **102**, 1053.
211. H. Yamanouchi and Y. Nakamura, *Plant Cell Physiol.*, 1992, **33**, 985.
212. M. K. Morell, A. Blennow, B. Kosar-Hashemi, and M. S. Samuel, *Plant Physiol.*, 1997, **113**, 201.
213. C. D. Boyer and J. Preiss, *Plant Physiol.*, 1981, **67**, 1141.
214. J. Preiss and C. D. Boyer, in "Mechanisms of Saccharide Polymerization and Depolymerization," ed. J. J. Marshall, Academic Press, New York, 1980, p. 161.
215. K. D. Hedman and C. D. Boyer, *Biochem. Genet.*, 1982, **20**, 483.
216. D. K. Fisher, C. D. Boyer, and L. C. Hannah, *Plant Physiol.*, 1993, **102**, 1045.
217. D. K. Fisher, M. Gao, K.-N. Kim, C. D. Boyer, and M. J. Guiltinan, *Plant Physiol.*, 1996, **110**, 611.
218. M. Gao, D. K. Fisher, K.-N. Kim, J. C. Shannon, and M. J. Guiltinan, *Plant Physiol.*, 1997, **114**, 69.
219. J. Edwards, J. H. Green, and T. ap Rees, *Phytochem.*, 1988, **27**, 1615.
220. A. M. Smith, *Planta*, 1988, **175**, 270.
221. J. Carter and E. E. Smith, *Carbohydr. Res.*, 1978, **61**, 395.
222. K. Mizuno, T. Kawasaki, H. Shimada, H. Satoh, E. Kobayashi, S. Okamura, Y. Arai, and T. Baba, *J. Biol. Chem.*, 1993, **268**, 19 084.
223. M. K. Bhattacharyya, A. M. Smith, T. H. Noel-Ellis, C. Hedley, and C. Martin, *Cell*, 1990, **60**, 115.
224. Y. Nakamura and H. Yamanouchi, *Plant Physiol.*, 1992, **99**, 1265.
225. J. Preiss, D. Stark, G. F. Barry, H. P. Guan, Y. Libal-Weksler, M. N. Sivak, T. W. Okita, and G. M. Kishore, in "Proceedings of Symposium, Improvement of Cereal Quality by Genetic Engineering," eds. R. J. Henry and J. A. Ronalds, Plenum, New York, 1994, p. 115.
226. T. Kuriki, H. Guan, M. Sivak, and J. Preiss, *J. Protein Chem.*, 1996, **15**, 305.
227. H. Cao and J. Preiss, *J. Protein Chem.*, 1996, **15**, 291.
228. K. Funane, N. Libessart, D. Stewart, T. Michishita, and J. Preiss, *J. Protein Chem.*, 1998, in press.
229. J. S. Tandecarz and C. E. Cardini, *Biochim. Biophys. Acta*, 1978, **543**, 423.
230. S. Moreno, C. E. Cardini, and J. S. Tandecarz, *Eur. J. Biochem.*, 1986, **157**, 539.
231. S. Moreno, C. E. Cardini, and J. S. Tandecarz, *Eur. J. Biochem.*, 1987, **162**, 609.
232. F. J. Ardila and J. S. Tandecarz, *Plant Physiol.*, 1992, **99**, 1342.

233. A. Rothschild and J. S. Tandecarz, *Plant Sci.*, 1994, **97**, 119.
234. T. W. Okita, *Plant Physiol.*, 1992, **100**, 560.
235. F. D. Macdonald and T. ap Rees, *Biochim. Biophys. Acta*, 1983, **755**, 81.
236. G. Entwistle and T. ap Rees, *Biochem. J.*, 1988, **255**, 391.
237. F. Kang and S. Rawsthorne, *Plant J.*, 1994, **6**, 795.
238. P. L. Keeling, J. R. Wood, R. H. Tyson, and I. G. Bridges, *Plant Physiol.*, 1988, **87**, 311.
239. J. Edwards and T. ap Rees, *Phytochem.*, 1988, **25**, 2033.
240. R. H. Tyson and T. ap Rees, *Planta*, 1988, **175**, 33.
241. R. Viola, H. V. Davies, and A. R. Chudeck, *Planta*, 1991, **183**, 202.
242. H. W. Heldt, U.-I. Flügge, and S. Borchert, *Plant Physiol.*, 1991, **95**, 341.
243. L. M. Hill and A. M. Smith, *Planta*, 1991, **185**, 91.
244. S. Borchert, H. Grosse, and H. W. Heldt, *FEBS Lett.*, 1989, **253**, 183.
245. S. Borchert, J. Harborth, D. Schünemann, P. Hoferichter, and H. W. Heldt, *Plant Physiol.*, 1993, **101**, 303.
246. U. I. Flügge and H. W. Heldt, *Ann. Rev. Plant Physiol. Plant Mol. Biol.*, 1991, **42**, 129.
247. H. E. Neuhaus, O. Batz, E. Thom, and R. Scheibe, *Biochem. J.*, 1993, **296**, 395.
248. H. E. Neuhaus, G. Henrichs, and R. Scheibe, *Plant Physiol.*, 1993, **101**, 573.
249. O. Batz, R. Scheibe, and H. E. Neuhaus, *Biochem. J.*, 1993, **294**, 15.
250. H. Kosegarten and K. Mengel, *Physiol. Plant*, 1994, **91**, 111.
251. W. P. Quick, R. Scheibe, and H. E. Neuhaus, *Plant Physiol.*, 1995, **109**, 113.
252. O. Batz, R. Scheibe, and H. E. Neuhaus, *Planta*, 1995, **196**, 50.
253. D. Schünemann and S. Borchert, *Bot. Acta*, 1994, **107**, 461.
254. T. Thornbjørnsen, P. Villand, K. Denyer, O.-A. Olsen, and A. M. Smith, *Plant J.*, 1996, **10**, 243.
255. K. Denyer, F. Dunlap, T. Thornbjørnsen, P. Keeling, and A. M. Smith, *Plant Physiol.*, 1996, **112**, 779.
256. M. E. Miller and P. S. Chourey, *Planta*, 1995, **197**, 522.
257. J. Brangeou, A. Reyss, and J.-L. Prisul, *Plant Physiol. Biochem.*, **35**, 847
258. J. Pozueta-Romero, M. Frehner, A. M. Viale, and T. Akazawa, *Proc. Natl. Acad. Sci. USA*, 1991, **88**, 5769.
259. O. Batz, U. Maass, G. Heinrichs, R. Scheibe, and H. E. Neuhaus, *Biochim. Biophys. Acta*, 1994, **1200**, 148.
260. Y. Cao, L. K. Steinrauf, and P. J. Roach, *Arch. Biochem. Biophys.*, 1995, **319**, 293.
261. T. D. Sullivan and Y. Kaneko, *Planta*, 1995, **196**, 477.
262. T. D. Sullivan, L. I. Strelow, C. A. Illingworth, R. L. Phillips, and O. E. Nelson, Jr., *Plant Cell*, 1991, **3**, 1337.
263. J. C. Shannon, F.-M. Pien, and K.-C. Lui, *Plant Physiol.*, 1996, **110**, 835.
264. D. M. Stark, K. P. Timmerman, G. F. Barry, J. Preiss, and G. M. Kishore, *Science*, 1992, **258**, 287.
265. S. Hizukuri, *Carbohydr. Res.*, 1986, **147**, 342.
266. A. M. Kram, G. T. Oostergetel, and E. F. J. Van Bruggen, *Plant Physiol.*, 1993, **101**, 237.
267. W. T. Kim, V. R. Franceschi, T. W. Okita, N. L. Robinson, M. Morell, and J. Preiss, *Plant Physiol.*, 1989, **91**, 217.
268. M. Morell and J. Preiss, unpublished experiments.
269. S. Govons, N. Gentner, E. Greenberg, and J. Preiss, *J. Biol. Chem.*, 1973, **248**, 1731.
270. N. Creuzat-Sigal, M. Latil-Damotte, J. Cattaneo, J. Puig, A. Favard, and C. Frixon, in "Biochemistry of the Glycosidic Linkage," Proc. Symp., 1971, eds. R. Piras and H. G. Pontis, Academic Press, New York, 1972, p. 647.
271. G. Ribereau-Gayon, A. Sabraw, C. Lammel, and J. Preiss, *Arch. Biochem. Biophys.*, 1971, **142**, 675.
272. J. Preiss, L. Shen, E. Greenberg, and N. Gentner, *Biochemistry*, 1996, **5**, 1833.
273. N. Gentner, E. Greenberg and J. Preiss, *Biochem. Biophys. Res. Commun.*, 1969, **36**, 373.
274. L. Eidels, P. L. Edelman, and J. Preiss, *Arch. Biochem. Biophys.*, 1970, **140**, 60.
275. L. Shen and J. Preiss, *Biochem. Biophys. Res. Commun.*, 1964, **17**, 424.
276. L. Shen and J. Preiss, *J. Biol. Chem.*, 1965, **240**, 2334.
277. L. Shen and J. Preiss, *Arch. Biochem. Biophys.*, 1966, **116**, 375.
278. S. G. Yung, M. Paule, R. Beggs, E. Greenberg, and J. Preiss, *Arch. Microbiol.*, 1984, **138**, 1.
279. D. Lapp and A. D. Elbein, *J. Bacteriol.*, 1972, **112**, 327.
280. S. G. Yung and J. Preiss, *J. Bacteriol.*, 1981, **147**, 101.
281. J. Preiss and E. Greenberg, *J. Bacteriol.*, 1981, **147**, 711.
282. A. A. Iglesias, Y.-Y. Charng, S. Ball, and J. Preiss, *Plant Physiol.*, 1994, **104**, 1287.
283. J. Preiss, *Ann. Rev. Plant Physiol.*, 1982, **33**, 431.
284. J. Preiss, E. Greenberg, T. F. Parsons, and J. Downey, *Arch. Microbiol.*, 1980, **126**, 21.
285. E. Greenberg, J. E. Preiss, M. Van Boldrick, and J. Preiss, *Arch. Biochem. Biophys.*, 1983, **220**, 594.
286. S.-G. Yung and J. Preiss, *J. Bacteriol.*, 1982, **151**, 742.
287. R. L. Robson, R. M. Robson, and J. G. Morris, *Biochem. J.*, 1974, **144**, 503.
288. N. Gentner and J. Preiss, *J. Biol. Chem.*, 1968, **243**, 5882.
289. G. Sanwal, E. Greenberg, J. Hardie, E. C. Cameron, and J. Preiss, *Plant Physiol.*, 1968, **43**, 417.
290. J. Preiss, in "Advances in Enzymology and Related Areas of Molecular Biology," ed. A. Meister, Wiley, New York, 1978, vol. 46, p. 317.
291. A. Kumar, T. Tanaka, Y. M. Lee, and J. Preiss, *J. Biol. Chem.*, 1988, **263**, 14634.
292. N. Gentner and J. Preiss, *Biochem. Biophys. Res. Commun.*, 1967, **27**, 417.
293. H. Kacser and J. A. Burns, *Symp. Soc. Exp. Biol.*, 1973, **27**, 65.
294. H. Kacser, in "The Biochemistry of Plants," ed. D. D. Davies, Academic Press, New York, 1987, vol. 11, p. 39.
295. H. E. Neuhaus, A. L. Kruckeberg, R. Feil, and M. Stitt, *Planta*, 1989, **178**, 110.
296. H. E. Neuhaus and M. Stitt, *Planta*, 1990, **182**, 445.
297. S. Ball, T. Marianne, L. Dirick, M. Fresnoy, B. Delrue, and A. Decq, *Planta*, 1991, **185**, 17.
298. Iglesias and J. Preiss, unpublished results.
299. M. J. Giroux, J. Shaw, G. Barry, B. G. Cobb, T. Greene, T. Okita, and L. C. Hannah, *Proc. Nat. Acad. Sci. USA*, 1996, **93**, 5824.
300. D. E. Atkinson, in "The Enzymes," 3rd edn., ed. P. D. Boyer, Academic Press, New York, 1970, vol. 1, p. 461.
301. D. N. Dietzler, C. J. Lais, and M. P. Leckie, *Arch. Biochem. Biophys.*, 1974, **160**, 14.
302. J. S. Swedes, R. J. Sedo, and D. E. Atkinson, *J. Biol. Chem.*, 1975, **250**, 6930.

303. D. N. Dietzler, M. P. Leckie, and C. J. Lais, *Arch. Biochem. Biophys.*, 1973, **156**, 684.
304. D. N. Dietzler, M. P. Leckie, C. J. Lais, and J. L. Magnani, *Arch. Biochem. Biophys.*, 1974, **162**, 602.
305. O. H. Lowry, J. Carter, J. B. Ward, and L. Glaser, *J. Biol. Chem.*, 1971, **246**, 6511.
306. J. Preiss, E. Greenberg, and A. Sabraw, *J. Biol. Chem.*, 1975, **250**, 7631.
307. T. Holme and H. Palmstierna, *Acta Chemica Scand.*, 1956, **10**, 578.
308. D. N. Dietzler, M. P. Leckie, C. J. Lais, and J. L. Magnani, *J. Biol. Chem.*, 1975, **250**, 2383.
309. J. Preiss, A. Sabraw, and E. Greenberg, *Biochem. Biophys. Res. Commun.*, 1971, **42**, 180.
310. S. Silver, *Proc. Nat. Acad. Sci. USA*, 1969, **62**, 764.
311. W. B. Dempsey, *Biochim. Biophys. Acta*, 1972, **264**, 344.
312. W. B. Dempsey and L. J. Arcement, *J. Bacteriol.*, 1971, **107**, 580.
313. E. G. Krebs and J. Preiss, in "International Review of Science, Biochemistry of Carbohydrates, Ser. One," ed. W. J. Whelan, University Park Press, Baltimore, MD, 1975, vol. 5, p. 337.
314. J. Cattaneo, M. Damotte, N. Sigal, F. Sanchez-Medina, and J. Puig, *Biochem. Biophys. Res. Commun.*, 1969, **34**, 694.
315. M. Latil-Damotte and C. Lares, *Mol. Gen. Genet.*, 1977, **150**, 325.
316. T. Romeo and J. Moore, *Nucleic Acids Res.*, 1991, **19**, 3452.
317. A. Kumar, P. Ghosh, Y. M. Lee, M. A. Hill, and J. Preiss, *J. Biol. Chem.*, 1989, **264**, 10 464.
318. F. Yu, Y. Jen, E. Takeuchi, M. Inouye, H. Nakayama, M. Tagaya, and T. Fukui, *J. Biol. Chem.*, 1988, **263**, 13 706.
319. C. Haziza, P. Stragier, and J.-C. Patte, *EMBO J.*, 1982, **1**, 379.
320. D. Austin and T. J. Larson, *J. Bacteriol.*, 1991, **173**, 101.
321. Y. Kohara, K. Akiyama, and K. Isono, *Cell*, 1987, **50**, 495.
322. K. E. Rudd, W. Miller, J. Ostell, and D. A. Benson, *Nucleic Acids Res.*, 1990, **18**, 313.
323. K. Rudd, personal communication.
324. T. Romeo, J. Black, and J. Preiss, *Current Microbiol.*, 1990, **21**, 131.
325. H. Yang, M. Y. Liu, and T. Romeo, *J. Bacteriol.*, 1996, **178**, 1012.
326. R. Jeanningros, N. Creuzat-Sigal, C. Frixon, and J. Cattaneo, *Biochim. Biophys. Acta*, 1975, **438**, 186.
327. D. N. Dietzler, M. P. Leckie, J. L. Magnani, M. J. Sughrue, P. E. Bergstein, and W. L. Steinheim, *J. Biol. Chem.*, 1979, **254**, 8308.
328. D. N. Dietzler, M. P. Leckie, W. L. Sternheim, T. L. Taxman, J. M. Ungar, and S. E. Porter, *Biochem. Biophys. Res. Commun.*, 1977, **77**, 1468.
329. M. P. Leckie, R. H. Ng, S. E. Porter, D. R. Compton, and D. N. Dietzler, *J. Biol. Chem.*, 1983, **258**, 3813.
330. J. Urbanowski, P. Leung, H. Weissbach, and J. Preiss, *J. Biol. Chem.*, 1983, **258**, 2782.
331. A. Munch-Petersen, H. M. Kalckar, E. Cutolo, and E. E. B. Smith, *Nature*, 1953, **172**, 1036.
332. R. L. Turnquist and R. G. Hansen, in "The Enzymes," 3rd edn., ed. P. D. Boyer, Academic Press, New York, 1973, vol. 8, p. 51.
333. J. K. Knop and R. G. Hansen, *J. Biol. Chem.*, 1970, **245**, 2499.
334. P. J. Roach, K. R. Warren, and D. E. Atkinson, *Biochemistry*, 1975, **14**, 5445.
335. T. R. Soderling, J. P. Hickenbottom, E. M. Reimann, F. L. Hunkeler, D. A. Walsh, and E. G. Krebs, *J. Biol. Chem.*, 1970, **245**, 6317.
336. N. E. Brown and J. Larner, *Biochim. Biophys. Acta*, 1971, **242**, 69.
337. C. Nakai and J. A. Thomas, *J. Biol. Chem.*, 1975, **250**, 4081.
338. D. C. Lin and H. L. Segal, *J. Biol. Chem.*, 1973, **248**, 7007.
339. S. D. Killilea and W. J. Whelan, *Biochemistry*, 1976, **15**, 1349.
340. R. E. Miller, E. A. Miller, B. Fredholm, J. B. Yellin, R. D. Eichner, S. E. Mayer, and D. Steinberg, *Biochemistry*, 1975, **14**, 2481.
341. M. F. Browner, K. Nakano, A. G. Bang, and R. J. Fletterick, *Proc. Natl. Acad. Sci. USA*, 1989, **86**, 1443.
342. W. Zhang, M. F. Browner, R. G. Fletterick, A. A. DePaoli-Roach, and P. J. Roach, *FASEB J.*, 1989, **3**, 2532.
343. G. Bai, Z. Zhang, R. Werner, F. Q. Nuttall, A. W. H. Tan, and E. Y. C. Lee, *J. Biol. Chem.*, 1990, **265**, 7843.
344. F. Q. Nuttall, M. C. Gannon, G. Bai, and E. Y. C. Lee, *Arch. Biochem. Biophys.*, 1994, **311**, 443.
345. I. Farkas, T. A. Hardy, A. A. DePaoli-Roach, and P. J. Roach, *J. Biol. Chem.*, 1990, **265**, 20 879.
346. I. Farkas, T. A. Hardy, M. G. Goebl, and P. J. Roach, *J. Biol. Chem.*, 1991, **266**, 15 602.
347. W. Zhang, A. A. DePaoli-Roach, and P. J. Roach, *Arch. Biochem. Biophys.*, 1993, **304**, 219.
348. A. V. Skurat, Y. Cao, and P. J. Roach, *J. Biol. Chem.*, 1993, **268**, 14 701.
349. A. V. Skurat, Y. Wang, and P. J. Roach, *J. Biol. Chem.*, 1994, **269**, 25 534.
350. A. V. Skurat, H.-L. Peng, H.-Y. Chang, J. F. Cannon, and P. J. Roach, *Arch. Biochem. Biophys.*, 1996, **328**, 283.
351. C. R. Krisman, *Biochim. Biophys. Acta*, 1962, **65**, 307.
352. J. Larner, *J. Biol. Chem.*, 1953, **202**, 491.
353. W. Verhue and H. G. Hers, *Biochem. J.*, 1966, **99**, 222.
354. B. I. Brown and D. H. Brown, in "Methods in Enzymology," eds. E. F. Neufeld and V. Ginsburg, Academic Press, New York, 1966, vol. 8, p. 395.
355. D. H. Brown and B. I. Brown, *Biochim. Biophys. Acta*, 1966, **130**, 263.
356. W. B. Gibson, B. I. Brown and D. H. Brown, *Biochemistry*, 1971, **10**, 4253.
357. B. Illingworth, D. H. Brown, and C. F. Cori, *Proc. Natl. Acad. Sci. USA*, 1961, **47**, 469.
358. B. I. Brown and D. H. Brown, *Proc. Natl. Acad. Sci. USA*, 1966, **56**, 725.
359. B. I. Brown and D. H. Brown, in "Carbohydrate Metabolism and Its Disorders," eds. F. Dickens, P. J. Randle, and W. J. Whelan, Academic Press, London, 1968, vol. 2, p. 123.
360. V. J. Thon, C. Vigneron-Lesens, T. Marianne-Pepin, J. Montreuil, A. Decq, C. Rachez, S. G. Ball, and J. F. Cannon, *J. Biol. Chem.*, 1992, **267**, 15 224.
361. V. J. Thon, M. Khalil, and J. F. Cannon, *J. Biol. Chem.*, 1993, **268**, 7509.
362. W. J. Whelan, *Bioessays*, 1986, **5**, 136.
363. C. Smythe and P. Cohen, *Eur. J. Biochem.*, 1991, **200**, 625.
364. H. G. Nimmo, C. G. Proud, and P. Cohen, *Eur. J. Biochem.*, 1976, **68**, 21.
365. M. C. Gannon and F. Q. Nuttall, *Trends in Glycosci. Glycotechnol.*, 1996, **8**, 183.
366. E. Viskupic, Y. Cao, W. Zhang, C. Cheng, A. A. Depaoli-Roach, and P. J. Roach, *J. Biol. Chem.*, 1992, **267**, 25 759.

367. M. D. Alonzo, J. Lomako, W. M. Lomako, and W. J. Whelan, *J. Biol. Chem.*, 1995, **270**, 15 315.
368. J. Lomako, W. M. Lomako, and W. J. Whelan, *Carbohydr. Res.*, 1992, **227**, 331.
369. M. D. Alonzo, J. Lomako, W. M. Lomako, W. J. Whelan, and J. Preiss, *FEBS Lett.*, 1994, **352**, 222.
370. Y. Cao, A. M. Mahrenholz, A. A. Depaoli-Roach, and P. Roach, *J. Biol. Chem.*, 1993, **268**, 14 687.
371. M. D. Alonso, J. Lomako, W. M. Lomako, and W. J. Whelan, *FEBS Lett.*, 1994, **342**, 38.
372. E. Meezan, S. Ananth, S. Manzella, P. Campbell, S. Siegal, D. J. Pillion, and L. Roden, *J. Biol. Chem.*, 1994, **269**, 11 503.
373. J. Lomako, W. M. Lomako, and W. J. Whelan, *FEBS Lett.*, 1990, **264**, 13.
374. S. M. Manzella, L. Rodén, and E. Meezan, *Glycobiology*, 1995, **5**, 263.
375. R. Barengo, M. Flavia, and C. R. Krisman, *FEBS Lett.*, 1975, **53**, 274.
376. A. Goldraij, M. C. Miozzo, and J. A. Curtino, *Biochem. Mol. Biol. Int.*, 1993, **30**, 453.
377. C. Cheng, J. Mu, I. Farkas, D. Huang, M. G. Goebl, and P. J. Roach, *Mol. Cell. Biol.*, 1995, **15**, 6632.
378. C. H. Smith, N. E. Brown, and J. Larner, *Biochim. Biophys. Acta*, 1971, **242**, 81.
379. C. Picton, A. Aitken, T. Bilham, and P. Cohen, *Eur. J. Biochem.*, 1982, **124**, 37.
380. C. Picton, J. Woodgett, B. Hemmings, and P. Cohen, *FEBS Lett.*, 1982, **150**, 191.
382. P. J. Roach and J. Larner, *J. Biol. Chem.*, 1976, **251**, 1920.
383. P. Cohen, D. Yellowlees, A. Aitken, A. Donella-Deana, B. A. Hemmings, and P. J. Parker, *Eur. J. Biochem.*, 1982, **124**, 21.
384. N. Embi, P. J. Parker, and P. Cohen, *Eur. J. Biochem.*, 1981, **115**, 405.
385. V. S. Sheorain, J. D. Corbin, and T. R. Soderling, *J. Biol. Chem.*, 1985, **260**, 1567.
386. D. B. Rylatt, A. Aitken, T. Bilham, G. D. Condon, N. Embi, and P. Cohen, *Eur. J. Biochem.*, 1980, **107**, 529.
387. C. Fiol, A. M. Mahrenholz, A. Wang, R. W. Roeske, and P. J. Roach, *J. Biol. Chem.*, 1987, **262**, 14 042.
381. C. J. Fiol, A. Wang, R. W. Roeske, and P. J. Roach, *J. Biol. Chem.*, 1990, **265**, 6061.
388. H. Flotow and P. J. Roach, *J. Biol. Chem.*, 1989, **264**, 9126.
389. H. Flotow, P. R. Graves, A. Q. Wang, C. J. Fiol, R. W. Roeske, and P. J. Roach, *J. Biol. Chem.*, 1990, **265**, 14 264.
390. K. P. Haung, A. Akatsuka, T. J. Singh, and K. R. Blake, *J. Biol. Chem.*, 1983, **258**, 7094.
391. Y. Wang, M. Camici, F. T. Lee, Z. Ahmad, A. A. Depaoli-Roach, and P. J. Roach, *Biochim. Biophys. Acta*, 1986, **888**, 225.
392. Y. Wang and P. J. Roach, *J. Biol. Chem.*, 1993, **268**, 23 876.
393. P. J. Parker, F. B. Caudwell, and P. Cohen, *Eur. J. Biochem.*, 1983, **130**, 227.
394. J. C. Lawrence, Jr., and J. N. Zhang, *J. Biol. Chem.*, 1994, **269**, 11 595.
395. P. J. Parker, N. Embi, F. B. Caudwell, and P. Cohen, *Eur. J. Biochem.*, 1982, **124**, 47.
396. L. Poulter, S. G. Ang, B. W. Gibson, D. H. Williams, C. F. B. Holmes, F. B. Caudwell, and P. Cohen, *Eur. J. Biochem.*, 1988, **175**, 497.
397. V. S. Sheorain, H. Juhl, M. Bass, and T. R. Soderling, *J. Biol. Chem.*, 1984, **259**, 7024.
398. A. Akatsuka, T. J. Singh, H. Nakabayashi, M. C. Lin, and K. P. Huang, *J. Biol. Chem.*, 1985, **260**, 3239.
399. A. V. Skurat and P. J. Roach, *J. Biol. Chem.*, 1995, **270**, 12 491.
400. A. V. Skurat and P. J. Roach, *Biochem. J.*, 1996, **313**, 45.
401. L. M. Ballou and E. H. Fischer, in "The Enzymes," 3rd edn., eds. P. D. Boyer and E. G. Krebs, Academic Press, San Diego, CA, 1986, vol. 17, p. 311.
402. P. Cohen, *Ann. Rev. Biochem.*, 1989, **58**, 453.
403. P. Cohen and P. T. Cohen, *J. Biol. Chem.*, 1989, **264**, 21 435.
404. P. Strålfors, A. Hiraga, and P. Cohen, *Eur. J. Biochem.*, 1985, **149**, 295.
405. M. J. Hubbard and P. Cohen, *Eur. J. Biochem.*, 1989, **180**, 457.
406. M. J. Hubbard and P. Cohen, *Eur. J. Biochem.*, 1989, **186**, 711.
407. M. J. Hubbard and P. Cohen, *Eur. J. Biochem.*, 1989, **186**, 701.
408. C. MacKintosh, D. G. Campbell, A. Hiraga, and P. Cohen, *FEBS Lett.*, 1988, **234**, 189.
409. P. Dent, D. G. Campbell, F. B. Caudwell, and P. Cohen, *FEBS Lett.*, 1990, **259**, 281.
410. P. Dent, A. Lavoinne, S. Nakielny, F. B. Caudwell, P. Watt, and P. Cohen, *Nature*, 1990, **348**, 302.
411. G. I. Welsh and C. G. Proud, *Biochem. J.*, 1993, **294**, 625.
412. D. A. Cross, D. R. Alessi, J. R. Vandenheede, H. E. McDowell, H. S. Hundal, and P. Cohen, *Biochem. J.*, 1994, **303**, 21.
413. G. I. Welsh, E. J. Foulstone, S. W. Young, J. M. Tavar'e, and C. G. Proud, *Biochem. J.*, 1994, **303**, 15.
414. A. Bellacosa, J. R. Testa, S. P. Staal, and P. N. Tsichlis, *Science*, 1991, **254**, 274.
415. P. J. Coffer and J. R. Woodgett, *Eur. J. Biochem.*, 1991, **201**, 475.
416. D. A. Cross, D. R. Alessi, P. Cohen, M. Andjelkovich, and B. A. Hemmings, *Nature*, 1995, **378**, 785.
417. T. F. Franke, S. I. Yang, T. O. Chan, K. Datta, A. Kazlauskas, D. K. Morrison, D. R. Kaplan, and P. N. Tsichlis, *Cell*, 1995, **81**, 727.
418. H. Eldar-Finkelman, G. M. Argast, O. Foord, E. H. Fischer, and E. G. Krebs, *Proc. Natl. Acad. Sci. USA*, 1996, **93**, 10 228.
419. J. Manchester, A. V. Skurat, P. Roach, S. D. Hauschka, and J. C. Lawrence, Jr., *Proc. Natl. Acad. Sci. USA*, 1996, **93**, 10 707.

3.15
Biosynthesis of Pectins and Galactomannans

DEBRA MOHNEN

University of Georgia, Athens, GA, USA

3.15.1 INTRODUCTION

All plant cells are surrounded by an extracellular matrix, composed of polysaccharides and proteins, which is typically referred to as the cell wall. Two types of wall are distinguished based on the developmental state of the cells: the primary wall and the secondary wall. The primary wall surrounds growing plant cells, meristematic cells such as cambium cells, and cells in succulent tissues such as leaf or fruit parenchyma and collenchyma cells.[1,2] Primary wall is also found in the junction between cells known as the middle lamellae and at the outer edges of secondary walls, since new wall is laid down continually at the plasma membrane and the older wall is pushed outward.[3] Cells which differentiate in order to perform specialized functions often have walls with altered polysaccharide composition and morphology that may be lignified. Such walls are known as secondary walls.[1,2]

The primary wall is composed of ~90% carbohydrate and 10% protein.[4] The carbohydrate originates from the polysaccharides cellulose, hemicellulose, and pectin. A comparison of the types of hemicellulosic polysaccharides and the relative amounts of pectin polysaccharides in the cell walls from diverse plants has led to the proposal that two general types of primary wall exist.[5] Type I primary walls, which contain ~22–35% pectin, are found in all *Dicotyledonae* and in some *Monocotyledonae*.[5,6] Type II walls, which contain ~10% pectin, are found in the grass family (*Poaceae*) of the *Monocotyledonae* which include agricultural crops such as corn, rye, oats, and wheat, and in closely related monocot families.[5–9] Pectin biosynthesis is generally studied in plants with type I walls.[10–21]

The polysaccharides in the primary wall are either homopolymers such as cellulose (i.e., a polymer of β-1,4-linked glucose) and homogalacturonan (i.e., a polymer of α-1,4-linked galacturonic acid) or heteropolymers such as the hemicellulose xyloglucan or the pectins rhamnogalacturonan I and rhamnogalacturonan II.[5] Although the heteropolymers contain more than one type of monosaccharide, most of them have a homopolymer backbone. For example, the structurally complex pectin polysaccharide rhamnogalacturonan II has a backbone of α-1,4-linked galacturonic acid. The pectic polysaccharide rhamnogalacturonan I, on the other hand, is a heteropolymer that does not have a homopolymer backbone but rather has a backbone composed of a disaccharide repeat [→4)-α-D-GalpA-(1→2)-α-L-Rhap-(1→] (Gal = galactose, Rha = rhamnose). Since many cell wall polysaccharides are homopolymers or have a homopolymer backbone, it is believed that they are synthesized by the sequential synthesis of at least a part of the homopolymer backbone, followed by the sequential addition of side chain sugar residues.

Pectin is abundant in primary walls and is greatly reduced in quantity in secondary walls.[2] Therefore, this review will concentrate on the biosynthesis of pectin in cells rich in primary walls. However, one class of secondary wall structures that will be discussed are the galactomannans. Galactomannans are food reserve polysaccharides found in the endosperm cell wall of leguminous seeds.[22–24] Such abundant and water soluble polysaccharides are referred to as gums or mucilages and have been extensively studied because of their commercial importance in the food industry.[22] The galactomannans represent a class of cell wall polysaccharides for which biochemical studies of the synthases in plant extracts have yielded particularly meaningful information about how the coordinated activity of multiple glycosyltransferases results in the synthesis of species-specific complex polysaccharide structures. As such, the approach used to study galactomannan biosynthesis offers a paradigm that is useful for the study of pectin biosynthesis.

In this review our current level of understanding about the biochemistry of the biosynthesis of pectins and galactomannans will be summarized. The glycosyltransferases that catalyze the formation of each of the specific types of glycosyl linkages of the pectic polysaccharides and the enzymes that modify the polymer during biosynthesis will be discussed. Progress in cloning the genes for the enzymes that synthesize the nucleotide–sugar substrates for cell wall polysaccharide biosynthesis, and efforts to clone the genes for the glycosyltransferases will be outlined. Primary literature and previous reviews on pectin biosynthesis,[25,26] nucleotide sugar transformation,[7,27–31] plant cell wall biosynthesis,[9,25,32–44] and cell wall structure[3,5–7,45] have been drawn upon.

3.15.2 STRUCTURE OF PECTIN

Pectins are plant cell wall polysaccharides that contain D-galacturonic acid.[6] The structure of D-galacturonic acid and the ten other types of glycosyl residues found in pectin are given in Figure 1. Structural analysis of the pectic polysaccharides isolated from dicotyledonous and mono-cotyledonous plants, as well as gymnosperms and some lower plants, has led to the identification of three polysaccharides found in all pectin: homogalacturonan (HGA) (Figure 2 and (1)), rhamno-galacturonan I (RG-I) ((2)–(13)), and rhamnogalacturonan II (RG-II) (14).[1,6] The pectic xylo-galacturonans[51–53] and apiogalacturonans[22,54,55] are only present in some plants, and thus are not classified as typical pectic polysaccharides.[1] Tissues, such as bark and seeds[22] and roots[56] from some plants produce water- or mild alkali-extractable pectic-like polysaccharides referred to as gums. The designation of these polysaccharides as gums[22] or root mucilage[56] is due to their ease of extraction compared with pectin in the wall and because of their relative abundance.[22] Since these gum "pectins" are not strictly from the cell wall, and because their overall structure resembles that of pectin from the wall, they are not dealt with separately in this review.

$$\rightarrow4)\text{-}\alpha\text{-D-Gal}p\text{A-}(1\rightarrow4)\text{-}\alpha\text{- D-Gal}p\text{A-}(1\rightarrow4)\text{-}\alpha\text{-D-Gal}p\text{A-}(1\rightarrow4)\text{-}\alpha\text{- D-Gal}p\text{A-}(1\rightarrow$$

(1)

3.15.2.1 Homogalacturonan

Homogalacturonan (Figure 2 and structure (1)) is a linear homopolymer of 1,4-linked α-D galacturonic acid which is partially methyl-esterified at the C-6 carboxyl[6,46] and may also contain other unidentified esters.[57,58] The length of the HGA in pectin remains a matter of debate although values ranging from a degree of polymerization (DP) of 30[22] to 200[5] have been reported. The distribution of methylesters in HGA is not known[6] although evidence has been obtained that the distribution of nonesterified galacturonosyl residues is not random.[46] Homogalacturonan isolated from some plants (e.g., sugar beet and potato) is O-acetylated at C-3.[47–50] Portions of the HGA from some plants such as apple,[49,52] cotton and watermelon,[60] carrot,[53] and pea[61] contain β-D-xylose linked to C-3 of GalA.[45,53,59,60] Such regions of xylosylated HGA are referred to as xylogalacturonan.

3.15.2.2 Rhamnogalacturonan I

RG-I is a branched pectic polysaccharide that accounts for 7–14% of the primary wall. RG-I contains a backbone of up to 100 repeats of the disaccharide [→4)-α-D-GalpA-(1→2)-α-L-Rhap-(1→] (2).[1,6,62–65] The galacturonic acid may be O-acetylated at C-3 or C-2.[3,6,50,66,67] The average molecular weight of RG-I from sycamore has been estimated to be between 10^5–10^6 Da.[6] Between 20–80% of the rhamnosyl residues are substituted at C-4, and occasionally at C-3, with oligo-saccharide side chains composed mostly of arabinosyl and/or galactosyl residues.[1,6,68] These side chains, which are referred to as galactans (3)–(7),[6,68,69] arabinans (8)–(9),[5,67,68] and arabinogalactans (10)–(13),[1,5,6,22,68] range in size from one to 50 or more glycosyl residues.[1,6,70] The number and type of different side chains in RG-I have not been determined and it is not known how much the structure of RG-I varies in different species, in different cell types, or during different stages of development.[5,71] Nonetheless, several side chains (3)–(13) have been identified in RG-I and these will be used as representative structures in this review. The representative side chains (3)–(4) and (8)–(13) or portions of side chains (5)–(7) of RG-I are shown in the line formulas (3),[6,68] (4),[6,68] (5),[69] (6),[69] (7),[69] (8),[5,68] (9),[5] (10),[6,68] (11),[5] (12),[22,45,72–74] and (13).[5] The oligosaccharides with rhamnitol at the reducing end (3)–(4), (8), and (10) represent fragments released by selective cleavage of galactosyluronic acid residues in the backbone of RG-I by treatment of RG-I with lithium in ethylenediamine[75,76] followed by reduction of products to yield a mixture of oligoglycosyl alditols.[68] Thus, the rhamnitol at the reducing end represents the rhamnose from the backbone of RG-I. Structures (5)–(7) represent acidic oligosaccharides released from RG-I following partial acid hydrolysis and are proposed to be terminal oligosaccharides from galactosyl-containing side chains of RG-I.[69]

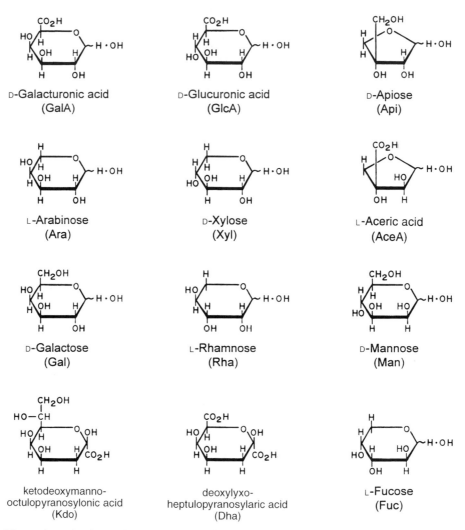

Figure 1 Haworth projections of the sugars found in pectin. The pyranose form of arabinose shown is the cyclic form in the biosynthetic substrate UDP-L-Ara; however, the arabinosyl residues in pectin are in the furanose form. Mannose is not a typical component of pectin; however, it is included owing to its importance for the synthesis of rhamnose and fucose and because it will be discussed with regard to galactomannan biosynthesis. The abbreviations in parentheses are used in this review.

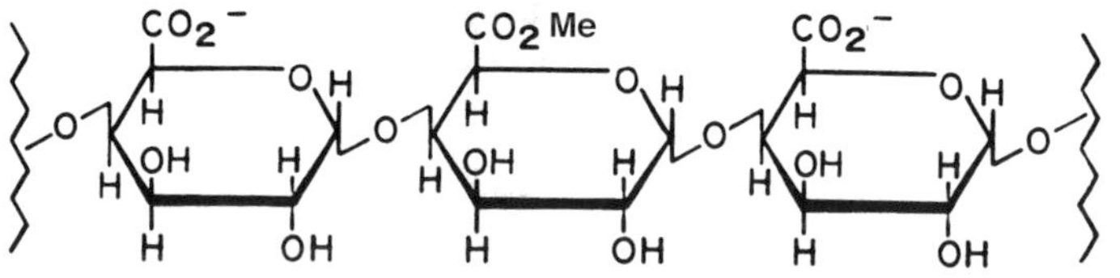

Figure 2 Structure of homogalacturonan (HGA). HGA in the wall is partially methyl-esterified at C-6[6,46] and may contain acetyl esters at C-2 or C-3.[47–50]

→4)-α-D-GalpA-(1→2)-α-L-Rhap-(1→

(2)

β-D-Galp-(1→6)-β-D-Galp-(1→4))-β-D-Galp-(1→4)-Rhamnitol

(3)

α-L-Fucp-(1→2)-β-D-Galp-(1→4)-β-D-Galp-(1→4)-Rhamnitol   β-D-GlcpA-(1→6)-D-Galp

(4)   **(5)**

4-*O*-Me-β-D-GlcpA-(1→6)-D-Galp   β-D-GlcpA-(1→4)-D-Galp   Ara-(1→4)-Rhamnitol

(6)   **(7)**   **(8)**

α-L-Araf
1
↓
2
→5)-α-L-Araf-(1→5)-α-L-Araf-(1→5)-α-L-Araf-(1→5)-α-L-Araf-(1→
3   3
↑   ↑
1   1
α-L-Araf   α-L-Araf
3
↑
1
α-L-Araf

(9)

L-Araf-(1→5)-α-L-Araf-(1→2)-α-L-Araf-(1→3)-β-D-Galp-(1→4)-Rhamnitol

(10)

→4)-β-D-Galp-(1→4)-β-D-Galp-(1→4)-β-D-Galp-(1→4)-β-D-Galp-(1→4)-β-D-Galp-(1→
3   3
↑   ↑
1   1
α-L-Araf   α-L-Araf

(11)

→4)-β-D-Galp-(1→4)-β-D-Galp-(1→4)-β-D-Galp-(1→4)-β-D-Galp-(1→
3
↑
1
L-Araf-(1→5)-L-Araf

(12)

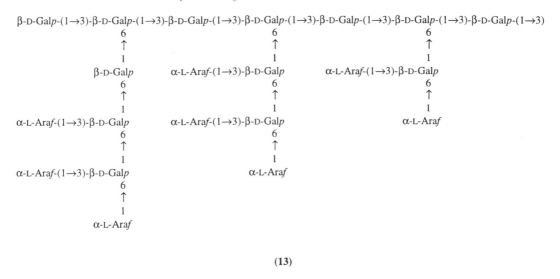

(13)

The galactans in RG-I may contain only galactosyl residues (**3**) or may also contain other neutral glycosyl residues (**4**).[6] Some galactans may also contain GalA[6,22] and/or GlcA (**5**) and (**7**)[22,69] or 4-*O*-methyl-GlcA (**6**)[69] residues and some have β-1,6-branching.[22] The size and specific linkages in the galactose-containing side branches of RG-I vary depending upon the species.[6]

Arabinans in RG-I are individual or linear chains of L-arabinofuranosyl (Ara*f*, where *f* stands for furanosyl) residues (**8**) or chains of 1,5-linked-α-L-Ara*f* that are substituted at O-3 and occasionally at O-2 with additional Ara*f* residues (**9**).[22,70,77]

The arabinogalactans have been divided into type I and type II arabinogalactans. The type I arabinogalactans (**10**)–(**12**) have a 1→4-linked β-D-galactan backbone while the type II arabinogalactans (**13**) have a 1→3-linked β-D-galactan backbone and are generally highly branched.[1,5,6,22] It has been reported that type II arabinogalactans may be associated with glucuronomannoglycans[22] and there are reports suggesting that mannose may be a component in some pectins, probably as a side branch of RG-I.[22] However, the precise structural role of mannose in pectin has not yet been clearly demonstrated and thus, mannose is not included as a component in pectin in this review. Most of the arabinan and galactans in the wall and some of the arabinogalactans are covalently attached to RG-I.[1] However, at least some of the type II arabinogalactan is associated with the family of proteoglycans known as arabinogalactan proteins (AGPs).[78–80] AGPs are hydroxyproline-rich proteins that are located at the plasma membrane, the cell wall, or in the media surrounding suspension-cultured cells.[79–82] Since the exact structures of the arabinogalactan II-type side chains of RG-I are not known, it will be assumed for the purposes of this review that the structures shown in (**10**)–(**13**) are representative of those found in RG-I. The cross-reactivity of the antibody CCRC-M7 with both RG-I and arabinogalactan proteins,[83] and the fact that the antibody reacts against arabinosylated (1→6)-β-D-galactans which are likely to contain arabinose linked to O-3 of the galactose,[84] add additional evidence that the arabinogalactan II structure is found in RG-I. Pectic polysaccharides from dicots in the *Chenopodiaceae* family such as spinach (*Spinacia oleracea*) and sugar beet (*Beta vulgaris*) are esterified with phenolics such as ferulic acid.[6,85,86] The feruloylation occurs on galactose and arabinose residues which are likely to be substituents in the side branches of RG-I.[6,86–88] RG-I has been shown to have the same general structure in different plants, although RG-I in the *Poaceae* lack fucose.[6,71] RG-I-like polysaccharides are present in gums of some plants including *Combretum leonense*,[89] *Sterculia* spp., *Khaya* spp., *Astragalus* spp., *Cochlospermum gossypium*, and *Rhizophora mangle*.[22]

3.15.2.3 Rhamnogalacturonan II

RG-II is a complex polysaccharide that accounts for ∼4% of the wall in dicots and less than 1% of the wall in monocots.[6] RG-II contains 11 different types of glycosyl residues including both methyletherified (e.g., 2-*O*-Me-xylose and 2-*O*-Me-fucose) and *O*-acetylated glycosyl residues (e.g., 3-*O*- or 4-*O*-Ac-fucose) in 19 different linkages (**14**).[90,91] RG-II also contains the unusual sugars aceric acid (3-*C*-carboxy-5-deoxy-L-xylose),[92] Kdo (2-keto-3-deoxy-D-manno-octulopyranosylonic acid,[93] and Dha (3-deoxy-D-lyxo-2-heptulopyranosylaric acid)[94] (see Figure 1). RG-II has a backbone of nine α-1,4-linked D-galactosyluronic acid residues with structurally complex side chains attached to C-2 and/or C-3.[1,6,90,91,93–99] A representative structure of RG-II[90,91] is shown in (**14**). The order of the four side chains is not known and is arbitrarily assigned. The structure shown is likely to be only a partial structure (O'Neill, Darvill and Albersheim, unpublished results).[100] RG-II is present in the walls of all plants and its structure is conserved. RG-II in the wall is now known to be complexed with borate and exists as a dimer that is cross-linked by a borate diester.[90,101–105]

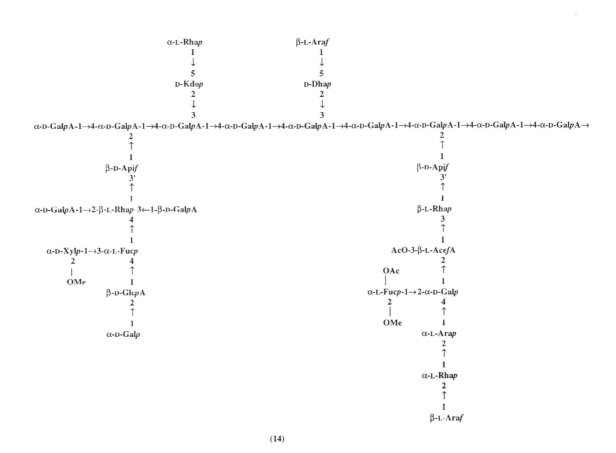

(**14**)

3.15.2.4 Association of HGA, RG-I, and RG-II in the Wall

It is not known whether HGA, RG-I, and RG-II exist as one polysaccharide linked together, and thus may be synthesized sequentially, or whether they are synthesized independently and either are not covalently cross-linked in the wall or are linked postbiosynthetically. Since RG-II has an HGA backbone and is extended by HGA on both the nonreducing and the reducing end,[91] it is usually assumed that RG-II is synthesized by substitution onto a HGA backbone via the action of glycosyltransferases, methyltransferases, and acetyltransferases. The question of whether RG-I is linked to HGA remains open. Many authors have suggested that RG-I contains HGA[106] since RG-I is released from walls treated with endopolygalacturonase, an enzyme that cleaves HGA stretches.[64,107] The fact that endopolygalacturonase treatment of cell walls solubilizes both HGA and substantial amounts of RG-I and RG-II also provides evidence that these three polysaccharides are covalently linked.[107]

3.15.3   SUBCELLULAR LOCALIZATION OF PECTIN BIOSYNTHETIC ENZYMES

Plant cell wall polysaccharides are synthesized at one of two different locations in the cell. Cellulose is synthesized by macromolecular complexes at the plasma membrane.[39,108] In contrast, pectin and the hemicelluloses are synthesized in the Golgi apparatus and then transported via vesicles to the plasma membrane and inserted into the wall.[109] However, little is known about the initiation of cell wall polysaccharide biosynthesis and thus, it is possible that pectin or hemicellulose synthesis may begin in the endoplasmic reticulum.[32,110]

The intracellular location of plant cell wall polysaccharide biosynthesis has been studied using various techniques. For example, electron microscopy[111,112] has been used to localize cellulose synthase complexes at the plasma membrane,[113,114] but has not been useful for identifying cell wall polysaccharide biosynthetic complexes in the Golgi system. Some of the earliest evidence for the localization of pectin and hemicellulose biosynthesis within the intracellular membranes was the autoradiographic identification of radiolabeled polysaccharides in Golgi cisternae and their chase from the Golgi to the cell wall by nonradiolabeled precursors.[115,116] These studies were extended using cells that were grown in the presence of radiolabeled glucose. Golgi-enriched fractions from these cells were collected and the radioactive glycosyl residues in the Golgi fraction were identified as Gal, Ara, and GalA, the major glycosyl residues found in the pectic polysaccharides.[117] Further studies involved the *in vivo* labeling of cells with radiolabeled sugar, or the *in vitro* labeling of cellular membranes with radiolabeled nucleotide sugars, followed by fractionation of the membranes and the localization of either radiolabeled polysaccharides, or specific glycosyltransferase activity, respectively, in the Golgi membranes[21,32,110] (Sterling, Wolff, Norambuena, Orellano, and Mohnen, unpublished results).[118]

More recently, immunocytochemistry of thin cell sections using antibodies directed against cell wall carbohydrate epitopes[109,119–121] has been used to show where, within the Golgi apparatus, specific polysaccharide epitopes are synthesized. This approach has confirmed that pectins are synthesized in the Golgi and *trans*-Golgi network[109] and also indicated that the synthesis of specific pectic carbohydrate epitopes is sublocalized within the Golgi. For example, the HGA- and the RG-I-like epitopes are present in both the *cis*- and medial-Golgi and the synthesis of these epitopes may, at least in some cell types, begin as early as the *cis*-Golgi,[122,123] and continue into the medial Golgi.[109,119,123] The esterification of HGA appears to occur in the medial and *trans*-Golgi,[109,123–126] while more extensive branching of pectin occurs in the *trans*-Golgi cisternae.[109,123] It is important to note, however, that the precise Golgi compartment in which specific pectic epitopes localize differs in different cell types (e.g., epidermal, vs. cortical, vs. peripheral root cap cells)[109,122,127] and has been referred to as "tissue-specific retailoring."[109] Thus, the localization and/or regulation of the pectin biosynthetic enzymes is likely to be complex and cell-type specific.[128] Moreover, the absence of a specific carbohydrate epitope may be due to masking of the epitope, rather than to a lack of the synthesis of the epitope; thus, immunolocalization of the biosynthetic enzymes is required to confirm the proposed location of pectin biosynthesis in any given cell type.[129]

It is generally believed, based on *in vivo* labeling and biochemical fractionation studies,[130] and on immunolocalization studies with antibodies reactive against relatively unesterified HGA (JIM 5;[127,131] PGA/RG-I;[122,132,133] 2F4[134]) and relatively esterified HGA (JIM 7[127]), that HGA is synthesized in the *cis* and medial Golgi,[109,135] becomes methylesterified in the medial and *trans* Golgi[123,124] and is transported via vesicles to the plasma membrane and inserted into the wall as a highly methyl-esterified polymer.[5,136,137] The demethylesterification of HGA by pectin methylesterase is believed to occur in the wall or cell plate[137] and results in the generation of more acidic HGA.[126,130,138–140] This is supported by the charge and chemical nature of polysaccharides produced following *in vivo* labeling of cells with radiolabeled glucose,[130] by the localization of esterified HGA throughout the cell wall,[124–127,136–138] and by the frequently observed localization of relatively unesterified HGA to the middle lamella and/or the outer surface of the wall.[122,124,126,127,130–133,136,138,141–143] In addition, the absence of unesterified HGA epitopes in the *trans*-Golgi vesicles, frequently observed in immunological studies,[123,124] supports the hypothesis that HGA is inserted into the wall in a highly esterified form. However, some cell types show a different localization of unesterified HGA as an apparent result of "tissue specific retailoring." For example, clover root epidermal cells, which secrete slime to the root cap,[120] and melon callus cells[125] contain unesterified HGA in the *trans*-Golgi. Also, some immunocytochemical studies have identified unesterified HGA epitopes at the plasma membrane–cell wall interface, suggesting that HGA can also be inserted into the wall in a relatively unesterified form and that HGA is not necessarily synthesized as highly esterified HGA.[127,129] Thus, the extent to which HGA is esterified during synthesis remains an open question.

In summary, it appears that pectin biosynthesis occurs in a *cis*- to *trans*-direction in the Golgi and it is likely that this spatial organization is due to a subcompartmentalization of the glycosyltransferases and modifying enzymes within the Golgi complex.[109,123]

3.15.4 NUCLEOTIDE SUGAR SUBSTRATES FOR GLYCOSYLTRANSFERASES

Two types of enzyme are required for the synthesis of plant cell wall polysaccharides. The first type catalyzes the production of the energetically activated glycosyl residue substrates for polysaccharide synthesis, while the second group transfers the glycosyl residues from the activated donors onto a growing polysaccharide.[28,33] To date, all evidence suggests that nucleoside diphosphate sugars (NDP-sugars) are the activated glycosyl residue substrates used for the synthesis of cell wall polysaccharides.[2,9,27–30,43] Table 1 lists the major NDP-sugars known to be used as substrates for the synthesis of pectins.[27,28,30] The known NDP-sugars contain either uracil or guanine, although, based on analogy to bacteria,[144,145] it can be hypothesized that Kdo[144,146] and Dha may be synthesized via CMP-linked sugars. Two different pathways are known for the synthesis of NDP-sugars: the nucleotide interconversion pathway (see Figure 3) and the so called salvage pathway (for reviews of these pathways see Carpita,[7] Hassid,[28] Feingold and co-workers[27,29,30]).

Table 1 Nucleotide sugars required for the synthesis of pectin.[a]

Nucleotide sugar	Immediate precursor	Immediate biosynthetic enzyme	EC number	Gene-cloned from plants, Reference
UDP-D-GlcA	UDP-D-Glc	UDP-glucose 6-dehydrogenase	1.1.1.22	155
UDP-D-GalA	UDP-D-GlcA	UDP-glucuronate 4-epimerase	5.1.3.6	
UDP-D-Xyl	UDP-D-GlcA	UDP-glucuronate decarboxylase	4.1.1.35	
UDP-L-Ara	UDP-D-Xyl	UDP-arabinose 4-epimerase	5.1.3.5	
UDP-Apiose	UDP-GlcA	UDP-Apiose/UDP-Xyl synthase		
UDP-D-Gal	UDP-D-Glc	UDP-glucose 4-epimerase	5.1.3.2	184
UDP-L-Rha	UDP-D-Glc (three steps)	[1]UDP-glucose 4,6-dehydratase [2]UDP-4-keto-L-rhamnose 3,5-epimerase [3]UDP-4-keto-L-rhamnose reductase	4.2.1.76	
GDP-L-Fuc	GDP-D-Man (three steps)	[1]GDP-D-Man-4,6-dehydratase [2]GDP-4-keto-6-deoxy-D-Man-3,5-epimerase [3]GDP-4-keto-L-fucose reductase	4.2.1.47	186 (for step one enzyme)
?CMP–Kdo	D-ribulose 5-phosphate (four steps)	[1]Ara-5-phosphate isomerase [2]Kdo-8-phosphate synthetase [3]Kdo-8-phosphate phosphatase [4]CMP-Kdo synthetase	5.3.1.13 4.1.2.16 2.7.7.38	
?CMP–Dha	?	?		
?XXX–aceric acid	?	?		

[a]Those nucleotides marked with a ? are only proposed to be the authentic nucleotide sugar substrate; however, specific proof is still wanting for their presence in plants. The order corresponds to the order of "secondary" nucleotide *de novo* synthesis via interconversion pathways starting with the "primary" nucleotide sugar UDP-Glc or GDP-Man.[30] Note that enzymes in the salvage pathway are not shown (see text). Enzyme names are as in Webb.[147] See previous reviews by Feingold and co-workers[29,30] and Shea *et al.*[140] for a summary of early biochemical studies on these enzymes.

The predominant pathway for the synthesis of NDP-sugars in growing tissues is their *de novo* synthesis from UDP-Glc (or GDP-Man) via the nucleotide sugar interconversion pathway (see Figure 3).[28,30] Nucleotide biosynthesis in this pathway is most easily described by considering the NDP-sugars as either primary NDP-sugars or secondary nucleotide sugars.[30] The primary NDP-sugars are ADP-Glc, TDP-Glc, UDP-Glc, and GDP-Man, with only the latter two being involved in pectin biosynthesis. UDP-Glc and GDP-Man originate from D-fructose-6-phosphate, D-glucose-1-phosphate (D-Glc-1-P) or sucrose produced by photosynthesis or starch breakdown or supplied *in vitro* as carbohydrate sources.[30] The secondary NDP-sugars are formed by the enzymatic modification of the glycosyl residue of the primary NDP-sugars.[30]

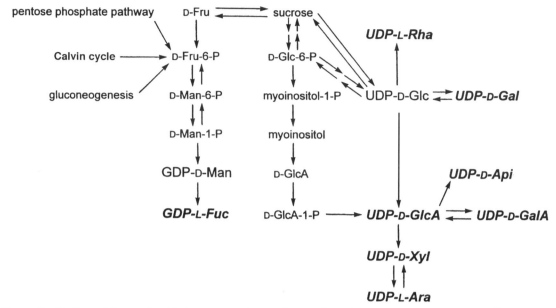

Figure 3 Outline of the nucleotide interconversion pathway adapted from Feingold and Barber.[30] The nucleotide sugars involved in pectin biosynthesis are italicized (Fru = fructose).

The salvage pathway is an alternative pathway for the synthesis of the NDP-sugars. The glycosyl residues L-Ara, D-Gal, D-Man, D-GalA, D-GlcA, L-Rha, L-Fuc, D-Glc, and D-Xyl can be recycled from the wall via transformation into sugar-1-phosphates by C-1-pyrophosphorylases.[27–30] The sugar-1-phosphates are converted into nucleotide sugars by pyrophosphorylases that catalyze the transfer of the nucleoside monophosphate from the nucleoside triphosphate (NTP) onto the phosphate of the sugar-1-phosphate (sugar-P) with the release of pyrophosphate (PPi): sugar-P + NTP → NDP-sugar + PPi.[27] This pathway of synthesis of UDP-sugars is referred to as the salvage pathway since it facilitates the reutilization of glycosyl residues following turnover of the wall. An alternative pathway for the synthesis of UDP-GlcA in plants is the myoinositol pathway in which myoinositol, produced from D-Glc-6-P, is oxidatively cleaved to produce D-glucuronate.[28,30] The D-GlcA is subsequently phosphorylated and uridylated to produce UDP-GlcA.

Figure 3 shows the nucleotide interconversion pathway for the synthesis of the NDP-sugars required for pectin biosynthesis.[30] Since GalA is quantitatively the most abundant glycosyl residue found in pectin, discussion of the nucleotide interconversion pathway will begin with UDP-GlcA, the substrate for UDP-GalA synthesis.

3.15.4.1 UDP-D-Glucuronic Acid

UDP-GlcA is produced either by the uridylation of GlcA-1-P via the myoinositol pathway or by the oxidation of UDP-Glc catalyzed by UDP-Glc-6-dehydrogenase[148] (see also Feingold and Avigad[29] for review of early literature on this topic). The oxidation of C-6 of UDP-Glc by UDP-Glc-6-dehydrogenase (UDP-GlcDH) (EC 1.1.1.22) occurs via a four-electron oxidation at C-6 and is accompanied by the reduction of two moles of NAD^+.[29] The reaction is believed to be an ordered reaction in which UDP-Glc binds to the enzyme, followed by the binding of NAD^+.[29] After reduction of the first bound NAD^+, NADH is released and the second NAD^+ is bound, reduced, and released.[29,149] Bovine UDP-Glc-6-dehydrogenase has been particularly well studied.[29,150–154] The bovine enzyme consists of six 52 kDa subunits[150,152,153] with 1 mol of substrate bound per 2 mol of enzyme, thus exhibiting "half-of-the-site" behavior.[151] A UDP-GlcDH has been partially purified 1000-fold from pea,[155] 12-fold from germinating lily (*Lilium longiflorum*) pollen,[156] and 341-fold from elicitor-treated French bean (*Phaseolus vulgaris* L) cell suspensions.[157] The K_m (Michaelis constant) for UDP-Glc for the enzymes from the three plant sources were 70 μM, 300 μM, and 5.5 mM, respectively, while the K_m for NAD^+ was 115 μM, 400 μM, and 20 μM. All eukaryotic UDP-GlcDHs are cooperatively inhibited by UDP-Xyl, suggesting that UDP-Xyl is a feedback inhibitor of the enzyme.[29,149] A cDNA clone for UDP-GlcDH has been cloned from soybean.[158] The soybean enzyme has a predicted molecular weight of 52.9 kDa (480 amino acids) and appears to be encoded by a single copy gene that is highly homologous to the cloned bovine UDP-GlcDH gene.[153] The soybean gene has a conserved NAD-binding site motif and contains the catalytic Cys residue.[153,158] The transcript for the gene is highly expressed in young growing tissues and is expressed at greatly reduced levels in more mature tissues.[158] This suggests that the expression of the enzyme in tissues actively synthesizing wall is controlled, at least partially, at the level of the transcript production or stability. Previous work by Bolwell and co-workers led to the identification and purification of a putative UDP-GlcDH from French bean.[157] The characteristics of the cloned UDP-GlcDH from soybean are not consistent with the purified UDP-GlcDH from French bean[157] which has a molecular weight of 40 kDa, copurifies with alcohol dehydrogenase activity, and is preferentially located in cells that make secondary walls. It remains to be determined whether the 40 kDa protein from French bean represents a bona fide multifunctional UDP-GlcDH preferentially expressed during secondary wall synthesis,[157] or whether it represents an alcohol dehydrogenase that plays little or no role in the formation of UDP-GlcA *in planta*. Stewart and Copeland[159] purified and characterized UDP-GlcDH from soybean (*Glycine max* L.) nodules. UDP-GlcDH was a cytosolic enzyme with a K_m for UDP-Glc and NAD$^+$ of 50' μM and 120' μM, respectively, values comparable to other plant UDP-GlcDHs. Soybean nodule UDP-GlcDH has a pH optimum of 8.4, a monomeric mass of 47' kDa and a native mass of 272' kDa, suggesting the enzyme functions as a hexamer. UDP-GlcDHs from soybean seeds and shoots, and from wheat and canola seedlings have similar kinetic properties as the enzyme in soybean nodules and were active in rapidly growing tissues.

3.15.4.2 UDP-D-Galacturonic Acid

The formation of UDP-GalA by the four-epimerization of UDP-GlcA is catalyzed by UDP-glucuronate-4-epimerase (EC 5.1.3.6).[29,148,160–163] Crude plant extracts containing UDP-GlcA-4-epimerase have been used to make radiolabeled UDP-GalA via the four-epimerization of radiolabeled UDP-GlcA.[148,161–163] The enzyme activity has been recovered as both soluble and membrane-bound enzyme.[148] A UDP-glucuronate-4-epimerase has been partially purified from the blue green alga *Anabaena flos-aquae*.[165] The *Anabaena* enzyme has a K_m for UDP-GlcA of 37 μM, a pH optimum of 8.5, and an equilibrium constant of 2.6 for UDP-GalA formation. The epimerase has not been purified to homogeneity nor has its gene been cloned.

3.15.4.3 UDP-D-Xylose

The formation of UDP-Xyl by the decarboxylation of UDP-GlcA is catalyzed by UDP-GlcA carboxylase (EC 4.1.1.35).[29,148,166–167] UDP-GlcA carboxylase contains a tightly bound NAD$^+$ and the reaction proceeds via a UDP-4-ketohexose intermediate.[29] UDP-GlcA carboxylase has been partially purified from wheat germ[168] into two isoenzymes of 21 000 kDa that do not require exogenous NAD$^+$ and have pH optimums of 7.0. The activity of both isozymes was activated at low (< 100 μM) concentrations of UDP-GlcA, indicating cooperative allosteric regulation of the isozymes by UDP-GlcA.[168] The apparent K_m for UDP-GlcA of the fully activated isozymes were 0.18 mM and 0.53 mM, respectively. Both isozymes are allosterically inhibited by UDP-Xyl,[168] indicating a likely level of regulation of the activity of UDP-GlcA carboxylase *in vivo*. UDP-GlcA carboxylase is found in both a soluble and membrane bound form.[166] The membrane bound UDP-GlcA carboxylase from soybean has a pH optimum of 6.0–7.5 and an apparent K_m of 240 μM for UDP-GlcA[166] while the soluble UDP-GlcA carboxylase has a K_m of 700 μM.[166] There is evidence supporting the localization of at least some of the enzyme within the lumen of the Golgi.[166]

3.15.4.4 UDP-L-Arabinose

The formation of UDP-L-Ara by the four-epimerization of UDP-Xyl is catalyzed by UDP-arabinose-4-epimerase (EC 5.1.3.5).[29,148,169,170] The enzyme has been purified at least 20-fold from wheat germ.[169] The partially purified enzyme has a pH optimum of 8.0, a K_m of 1.5 mM for UDP-Xyl and 0.5 mM for UDP-L-Ara.[169] Interestingly, Ara is in the pyranose form in UDP-Ara, yet arabinofuranose is the predominant form of this sugar in the wall polysaccharides, proteoglycans, arabinans, and arabinogalactan proteins.[7,27,29] It is currently believed that during polysaccharide biosynthesis arabinosyltransferase(s) must catalyze ring rearrangement before formation of the glycosidic bond.[7]

3.15.4.5 UDP-D-Apiose

UDP-apiose is formed from UDP-GlcA by a decarboxylation and rearrangement catalyzed by a NAD$^+$-dependent UDP-apiose/UDP-Xyl synthase.[30,171–174] As the name indicates, the enzyme also produces UDP-Xyl from UDP-GlcA,[172] complicating the purification of UDP-apiose following *in vitro* synthesis. For example, the ratio of UDP-apiose to UDP-Xyl formed by the enzyme was 1.4 in Tris–HCl (trishydroxymethylaminomethane–HCl) buffer (pH 8.2) and 2.7 in potassium phosphate buffer (pH 8.2).[172] The inability to separate the UDP-apiose synthase activity from the UDP-Xyl synthase activity in a 1400-fold purified protein preparation, led Matern and Grisebach[172] to conclude that the protein is multifunctional. The mechanism of UDP-apiose formation is believed to include the formation of an L-*threo*-4-pentosulose intermediate common to both UDP-apiose and UDP-Xyl formation, followed by ring contraction and epimerization.[172] UDP-apiose/UDP-Xyl synthase has been partially purified from *Lemna minor*[175] and from parsley.[172] The enzyme from *Lemna minor* had optimum activity at ~1 mM NAD$^+$ and pH 8.0–8.3.[175] The enzyme from parsley is composed of two proteins: a 65 kDa protein composed of two identical 34 kDa subunits and an 86 kDa protein containing two identical 44 kDa subunits.[172] The 65 kDa protein could be separated from the 86 kDa protein by fractionation of the synthase over diethylaminoethyl cellulose (DEAE cellulose) in the presence of urea.[172] The 86 kDa protein contained all the enzyme activity. The 86 kDa protein bound 0.5 mol of UDP-GlcA per mol of protein and, in the presence of UDP-GlcA, bound 0.5 mol NAD$^+$ in a ratio of 0.5 mol mol^{-1} of catalytic protein.[172] The 65 kDa protein was enzymatically inactive, although it was required for stability of the 86 kDa protein.[172]

3.15.4.6 UDP-L-Rhamnose

Tobacco,[176] mung bean,[177] and *Silene dioica*[178] leaves and cultures of *Chlorella pyrenoidasa*[179] have been used as a source of an NADH-dependent enzyme preparation that is able to convert UDP-D-Glc to UDP-L-rhamnose (UDP-L-Rha) (reviewed by Feingold and co-workers[29,30]). UDP-4-keto-6-deoxy-D-Glc was shown to be an intermediate in the conversion.[176,178,179] Based on analogy to the biosynthetic pathway for the synthesis of deoxythymidine diphosphate-rhamnose (dTDP-rhamnose) from dTDP-glucose in bacteria, a pathway catalyzed by enzymes encoded by rfb genes,[180] the following biosynthetic scheme is proposed for plants.[29,178] UDP-L-Rha synthesis is catalyzed by the conversion of UDP-D-Glc to UDP-4-keto-6-deoxy-Glc by UDP-glucose-4,6-dehydratase (EC 4.2.1.76). The UDP-4-keto-6-deoxy-Glc is then epimerized to UDP-4-keto-6-deoxy-L-Man by UDP-4-keto-L-rhamnose-3,5-epimerase. Finally, UDP-L-Rha is formed by the reduction of UDP-4-keto-6-deoxy-L-Man catalyzed by UDP-4-ketorhamnose reductase. None of these enzymes has been purified to homogeneity or cloned in plants.

3.15.4.7 UDP-D-Galactose

The formation of UDP-Gal by the epimerization of UDP-Glc is catalyzed by UDP-Glc-4-epimerase (EC 5.1.3.2).[29,181] UDP-4-ketohexose is an intermediate in the reaction.[182,183] The enzyme has been extensively studied in bacteria and its structure has been determined by X-ray crystallography at 2.5 Å resolution.[184] The bacterial enzyme is composed of two identical 39.5 kDa proteins,[184,185] each of which contains NAD$^+$ as a cofactor.[184] The size of the enzyme and the tightness of the binding of NAD$^+$ to the enzyme varies in different organisms. For example, the UDP-D-Glc-4-epimerase from *Candida pseudotropicalis* consists of two identical 60 kDa subunits with one NAD$^+$ tightly bound per active enzyme molecule, while the bovine enzyme is a monomer of 40 kDa that requires exogenous NAD$^+$ for activity,[182,186] suggesting that NAD$^+$ is not tightly bound to the enzyme in mammals (reviewed by Feingold and Avigad[29]). The UDP-Glc-4-epimerase from leaves of *Vicia faba* was shown to be a soluble cytoplasmic protein with a pH optimum of 8.8 and a K_m for UDP-Gal of 95 μM.[187] A gene for UDP-Glc-4-epimerase in plants has been cloned in *Arabidopsis* and shown to encode the epimerase via complementation of the gal10 mutant of

Saccharomyces cerevisiae and by expression of the recombinant protein in *E. coli*.[188] The *Arabidopsis* gene encodes a protein of 39 kDa with a broad pH optimum from 7.0 to 9.55 and a K_m for UDP-Glc of 110 μM.[188] Dörmann and Benning[189] produced transgenic *Arabidopsis* plants that express the cDNA for UDP-Glc-4-epimerase in a sense and antisense orientation. Although these plants expressed an increase or decrease, respectively, in both the mRNA and the enzyme levels, there was no detectable effect on plant growth and morphology, or in the levels of UDP-Glc, UDP-Gal, galactolipids and cell wall galactose content in leaves and stems of soil grown plants. These results, and the fact that UDP-Glc-4-epimerase activity could not be reduced to less than 10% of wild types levels in any antisense line tested, raises the possibility that at least one other UDP-Glc-4-epimerase gene exists in *Arabidopsis*.

3.15.4.8 GDP-L-Fucose

The NDP-sugar precursor for GDP-L-Fuc is GDP-D-Man.[190] Soluble enzyme preparations from different plant species, including *Phaseolus vulgaris*, have been shown to convert GDP-D-Man to GDP-L-Fuc.[190] In *Phaseolus* the reaction requires NADPH or NADH, has a pH optimum from 6.9 to 7.8, and an apparent K_m for GDP-D-Man of 160 μM.[190] The reaction requires three enzyme activities and it has been shown that GDP-Man is oxidized at C-4 and reduced at C-6 by GDP-D-Man-4,6-dehydratase (EC 4.2.1.47) to form GDP-4-keto-6-deoxy-D-mannose.[29,190] It is proposed that the GDP-4-keto-6-deoxy-D-Man is bound tightly to the GDP-4-keto-6-deoxy-D-Man-3,5-epimerase to produce a GDP-4-keto-6-deoxy-L-galactose intermediate.[29,191] The GDP-4-keto-6-deoxy-L-galactose is proposed to stay bound to the epimerase–intermediate complex and act as a substrate for the GDP-4-keto-L-fucose reductase,[29] producing GDP-L-fucose.[29] The gene for the GDP-D-mannose-4,6-dehydratase has been cloned in *Arabidopsis*[191] and shown to complement the *Arabidopsis MUR1* mutant, confirming that the *MUR1* mutant is indeed mutated in the GDP-D-mannose-4,6-dehydratase gene. The cloned gene encodes a protein of 41.9 kDa.

3.15.4.9 Activated Precursors for Kdo, Dha, and Aceric Acid

The pathways for the biosynthesis in plants of the activated glycosyl donors of Kdo, Dha, and aceric acid are not known. In contrast, the pathway for the synthesis of the activated donor of Kdo, CMP–Kdo, has been extensively studied in bacteria.[144,145,192–195] By analogy to the bacterial pathway, it is proposed that the synthesis of CMP–Kdo in plants begins with D-ribulose-5-phosphate, a metabolite in the Calvin carbon fixation cycle (i.e., the reductive pentose phosphate pathway) and the oxidative pentose phosphate pathway.[196] D-Ribulose-5-phosphate is isomerized into D-arabinose-5-phosphate by D-arabinose-5-phosphate isomerase.[144] The D-arabinose-5-phosphate is condensed with phosphoenolpyruvate to form Kdo-8-phosphate (2-dehydro-3-deoxy-D-octonate-8-phosphate) in a reaction catalyzed by Kdo-8-phosphate synthetase (2-dehydro-3-deoxyphosphooctonate aldolase).[144,197] Kdo-8-phosphate synthetase has been identified in eight different plant species and has been partially purified,[197] thus lending some credence to this proposed pathway. The enzyme has a pH optimum of 6.2, a K_m for arabinose-5-phosphate of 0.27 mM and a K_m for phosphoenolpyruvate of 35 μM.[197] The 8-phosphate is likely to be removed by Kdo-8-phosphate phosphatase to produce Kdo and pyrophosphate.[144] In a reaction catalyzed by CMP–Kdo synthetase,[144] the Kdo reacts with cytidine 5'-triphosphate (CTP) to form CMP–Kdo, the proposed substrate for the Kdo transferase that catalyzes the addition of Kdo to RG-II. There is no direct information available regarding the identity or the biosynthetic pathway for the activated glycosyl donor for Dha (3-deoxy-D-lyxo-2-heptulosaric acid). However, a cytosolic form of 3-deoxy-D-arabinoheptulosonate-7-phosphate synthase that has a wide substrate specificity has been identified in plants.[198] It has been proposed that this enzyme could catalyze the condensation of phosphoenolpyruvate with threose to generate a precursor of Dha.[198] Alternatively, the activated donor of Dha could be synthesized via interconversion of CMP–Kdo to CMP–Dha through oxidation and decarboxylation reactions.

3.15.4.10 Location of the Enzymes that Catalyze NDP Sugar Interconversion

The enzymes required for the interconversion of NDP-sugars are reported to be located in the cytosol or bound to the cytosolic side of the Golgi membranes.[29,166] However, some evidence for a

Golgi intralumenal location of UDP-glucuronate carboxylase and UDP-arabinose-4-epimerase has been reported.[166] It has also been proposed that in animal cells UDP-Xyl, a substrate for proteoglycan synthesis, has a lumenal location in both the Golgi and the endoplasmic reticulum (ER).[199,200] Thus, it is possible that a subset of the NDP sugar interconversion enzymes are located in the lumen of the Golgi and act in concert with the glycosyltransferases to synthesize polysaccharides.

3.15.4.11   Translocation and Metabolism of NDP Sugars

It has been suggested that, as in animals, nucleotide sugar translocators transfer NDP-sugars to the lumen of the ER and Golgi where the glycosyltransferases catalyze polymerization.[33] Orellana and co-workers have provided evidence for the existence of a Golgi-localized UDP-Glc:uridine nucleotide antiport in pea and the possible existence of UDP-Gal and UDP-Xyl transporters.[201] Hayashi *et al.*[166] also provided evidence for the existence of a UDP-Xyl transporter in soybean Golgi membranes. Thus, NDP-sugars located in the cytosol may gain access to the lumen of the Golgi by UDP sugar:UMP/UDP antiporters[201] in a manner analogous to such UDP-sugar antiporters in animal systems.[33,202] In plants, as in animals, it is believed that the NDP that is released from the NDP-sugar upon transfer of the glycosyl residue to a carbohydrate acceptor, is hydrolyzed into UMP and phosphate by a Golgi localized nucleoside diphosphatase (NDPase).[203,204] An NDPase has been solubilized from rice cells with the detergent Triton X-100.[203] The enzyme hydrolyzes UDP, GDP, and inosine diphosphate with K_m values of 0.5. 0.67, and 0.48 mM, respectively,[203] but has greatly reduced activity on ADP, CDP, and thymidine 5′-diphosphate. Thus, the rice NDPase is able to hydrolyze NDP from those NDP-sugars involved in pectin biosynthesis, a finding consistent with the existence of NDP-sugar:UMP antiporters. A latent UDPase (uridine diphosphatase) from *Pisum sativum* has been localized to the Golgi and shown to be an integral membrane protein with its active site facing the lumen of the Golgi.[204] As noted above, while it is believed that most NDP-sugars are synthesized in the cytosol and thus would require a mechanism for transport into the Golgi, other NDP sugars may actually be synthesized from their precursor NDP sugar within the lumen of the Golgi.[166]

3.15.5   PECTIN GLYCOSYLTRANSFERASES

Glycosyltransferases catalyze the addition of glycosyl residues to a growing oligo/polysaccharide. Multiple glycosyltransferases may be required for the initiation, elongation, and termination steps of polymer synthesis.[33,39] For example, individual enzymes that form the primer for polysaccharide synthesis[205–207] may be required. The existence of protein, lipid, or polysaccharide primers for chain initiation has been suggested,[208,209] but no detailed mechanism has yet been determined for chain initiation of plant cell wall polysaccharides.[33,39] Little is known about the termination of plant cell wall polysaccharide biosynthesis; however, it has been hypothesized that the rate of membrane vesicle movement and fusion with the plasma membrane may play a role in determining chain length.[33]

A reasonable assumption, when considering how many distinct glycosyltransferases are required for the synthesis of a polysaccharide, is that a distinct enzyme will be required for each distinct glycosyl linkage formed between each two unique glycosyl residues. Furthermore, if it is assumed that the addition of the glycosyl residues occurs at the nonreducing end of the polysaccharide, as has been shown for β-glucan synthesis[210] and for homogalacturonan synthesis,[18] then the glycosyl residues located at the nonreducing end of the oligosaccharide/polysaccharide acceptor will be specifically recognized by the glycosyltransferase. Such a specificity of glycosyltransferases occurs during the synthesis of glycoproteins. For example, the core region of *N*-linked oligosaccharides of yeast glycoproteins contains two *N*-acetylglucosaminyl residues and 8–13 mannosyl residues and may be further mannosylated to contain up to 200 mannoses.[211] Although only four types of mannosyl residue linkages (β-1,4; α-1,2-; α-1,3-; and α-1,6-) are present, it is believed that at least seven distinct mannosyltransferases are required for synthesis of the glycoproteins due to the different nonreducing glycosyl residues that must be recognized to become mannosylated during synthesis.[211]

There are at least 46 glycosyltransferases (see Table 2) required for the synthesis of pectin, based on the one linkage–one enzyme assumption and on the structure of pectin shown in (**1**)–(**14**). The list of glycosyltransferases includes five galacturonosyltransferases, five rhamnosyltransferases, nine galactosyltransferases, fifteen arabinosyltransferases, three fucosyltransferases, two xylosyl-transferases, three glucuronosyltransferases, and single apiosyl-, Kdo-, Dha-, and aceronosyl-transferases. The different enzymes in each class have been numbered to facilitate their identification in this review. Two of the enzymes, galacturonosyltransferase-V and rhamnosyltransferase-V, are noted with a "?" since these enzymes would be required to synthesize the glycosyl residues at the hypothesized "junctions" of HGA and RG-I. However, since no oligosaccharide structures con-

Table 2 List of glycosyltransferases "required" for pectin biosynthesis.

Type of glycosyl-transferase	Working[a] number	Parent polymer[b]	Enzyme[c,d]		Ref. for structure	Line formula
			Acceptor substrate	*Enzyme activity*		
D-GalAT	I	HGA	*GalA α-1,4-GalA	α-1,4-GalAT	6	(**1**)
D-GalAT	II	RG-I	L-Rha α-1,4-GalA	α-1,2-GalAT	6, 62, 63	(**2**)
D-GalAT	III	RG-II	L-Rha β-1,3-Apif	α-1,2-GalAT	5, 91	(**14**)
D-GalAT	IV	RG-II	L-Rha β-1,3-Apif	β-1,3-GalAT	5, 91	(**14**)
D-GalAT	V?	RG-I/HGA	GalA α-1,2 L-Rha	α-1,4-GalAT		
L-RhaT	I	RG-I	GalA α-1,2-L-Rha	α-1,4-L-RhaT	6, 62, 63	(**2**)
L-RhaT	II	RG-II	Apif β-1,2-GalA	β-1,3-L-RhaT	5, 91	(**14**)
L-RhaT	III	RG-II	Kdo 2,3GalA	α-1,5-L-RhaT	5, 91	(**14**)
L-RhaT	IV	RG-II	L-Ara α-1,4-Gal	α-1,2-L-RhaT	5, 91	(**14**)
L-RhaT	V?	HGA/RG-I	GalA α-1,4-GalA	α-1,4-L-RhaT		
D-GalT	I	RG-I	L-Rha α-1,4-GalA	β-1,4-GalT	6, 68	(**3**)
D-GalT	II	RG-I	Gal β-1,4-Rha	β-1,4-GalT	6, 68	(**3**)
D-GalT	III	RG-I	Gal β-1,4-Gal	β-1,4-GalT	6, 22, 45, 68, 72–74	(**11**) (**12**)
D-GalT	IV	RG-I	Gal β-1,4-Gal	β-1,6-GalT	6, 68	(**3**)
D-GalT	V	RG-I/AGP	Gal β-1,3-Gal	β-1,3-GalT	5	(**13**)
D-GalT	VI	RG-I/AGP	Gal β-1,3-Gal	β-1,6-GalT	5	(**13**)
D-GalT	VII	RG-I/AGP	Gal β-1,6-Gal β-1,3-Gal	β-1,6-GalT	5	(**13**)
D-GalT	VIII	RG-II	GlcA β-1,2-Fuc	α-1,2-GalT	5, 91	(**14**)
D-GalT	IX	RG-II	L-AcefA β-1,3-Rha	α-1,2-GalT	5, 91	(**14**)
L-AraT	I	RG-I	Gal β-1,4-Rha	α-1,3-L-AraT	6, 68	(**10**)
L-AraT	II	RG-I	L-Ara α-1,3-Gal	α-1,2-L-AraT	6, 68	(**10**)
L-AraT	III	RG-I	L-Ara α-1,2 Ara	1,5-L-AraT	6, 68	(**10**)
L-AraT	IV	RG-I	L-Rha α-1,4-GalA	1,4-AraT	68	(**8**)
L-AraT	V	RG-I	L-Ara α-1,5-Ara	α-1,5-L-AraT	5	(**9**)
L-AraT	VI	RG-I	L-Ara α-1,5-Ara	α-1,2-L-AraT	5	(**9**)
L-AraT	VII	RG-I	L-Ara α-1,5-Ara	α-1,3 L AraT	5	(**9**)
L-AraT	VIII	RG-I	L-Ara γ-1,3-Ara	α-1,3-L-AraT	5	(**9**)
L-AraT	IX	RG-I	Gal β-1,4-Gal	α-1,3-L-AraI	5, 6, 22, 45, 72–74	(**11**) (**12**)
L-AraT	X	RG-I	L-Ara-1,3-Gal	1,5-L-AraT	22, 45, 72–74	(**12**)
L-AraT	XI	RG-I/AGP	Gal β-1,6-Gal	α-1,3-L-AraT	5	(**13**)
L-AraT	XII	RG-I/AGP	Gal β-1,6-Gal	α-1,6-L-AraT	5	(**13**)
L-AraT	XIII	RG-II	Dha 2,3-GalA	β-1,5-L-AraT	5, 91	(**14**)
L-AraT	XIV	RG-II	Gal α-1,2-L-AcefA	α-1,4-L-AraT	5, 91	(**14**)
L-AraT	XV	RG-II	L-Rha α-1,2-L-Ara	β-1,2-L-AraT	5, 91	(**14**)
L-FucT	I	RG-I	Gal β-1,4-Gal	α-1,2-L-FucT	6, 68	(**4**)
L-FucT	II	RG-II	L-Rha β-1,3-Apif	α-1,4-L-FucT	5, 91	(**14**)
L-FucT	III	RG-II	Gal α-1,2-L-AcefA	α-1,2-L-FucT	5, 91	(**14**)
D-ApifT	I	RG-II	GalA α-1,4-GalA	β-1,2-ApifT	5, 91	(**14**)
D-XylT	I	RG-II	L-Fuc α-1,4-L-Rha	α-1,3-XylT	5, 91	(**14**)
D-XylT	II	HGA	GalA α-1,4-GalA	β-1,3-XylT	45, 53, 59, 60	
D-GlcAT	I	RG-I	Gal …	β-1,6-GlcAT	69	(**5**)
D-GlcAT	II	RG-I	Gal …	β-1,4-GlcAT	69	(**7**)
D-GlcAT	III	RG-II	L-Fuc α-1,4-L-Rha	β-1,4-GlcAT	5, 91	(**14**)
D-KdoT	I	RG-II	GalA α-1,4-GalA	2,3-KdoT	5, 91	(**14**)
D-DhaT	I	RG-II	GalA α-1,4-GalA	2,3-DhaT	5, 91	(**14**)
L-AcefA	I	RG-II	L-Rha β-1,3-Apif	β-1,3-AcefAT	5, 91	(**14**)

[a]The order of the Roman numbers for different members of the same groups has been given based on the structure of pectin and on the assumption that HGA is synthesized first, followed by RG-I backbone and RG-II side branches of HGA. The numbers are given so as to facilitate comparison of the enzymes. [b]HGA: homogalacturonan; RG-I: rhamnogalacturonan I; RG-II: rhamnogalacturonan II. [c]All sugars are D sugars and have pyranose rings unless otherwise indicated. [d]Unless noted: enzyme adds to the glycosyl residue on the left *.

taining these "junctions" have been reported, the presence of these enzymes remains hypothetical. It should be noted that there are examples in which a single enzyme can transfer a glycosyl residue to more than one type of nonreducing end "disaccharide acceptor." The polysialic acid synthase ST8Sia II transfers sialic acid in an α-2,8 linkage onto the α-2,3-linked sialic acid on the *N*-glycans of neural cell adhesion molecule (NCAM), and also catalyzes the polysialylation of the subsequent α-2,8-linked oligomeric chain.[212] Also, a single enzyme, Kdo transferase (3-deoxy-D-manno-octulosonic acid transferase), has been shown to catalyze the addition of two[213] or three[214] uniquely linked Kdo residues during the synthesis of the lipopolysaccharide core structure in *E. coli*[213] and *Chlamydia trachomatis*,[214] respectively. Thus, the number of distinct enzymes listed in Table 2 may be an overestimation since it is possible that a single enzyme may catalyze more than one of the reactions shown. Nonetheless, the glycosyltransferase activities listed in Table 2 represent the range of activities required for pectin synthesis. The challenge is to determine how many unique enzymes are required to cover all of the listed activities.

The goal of pectin biosynthesis studies is to describe biochemically the location, structure, and function of the individual glycosyltransferases and modifying enzymes that synthesize the polysaccharides. Such studies are required in order to understand how multiple enzymes function together to produce the complex pectic polymers found in the wall. Ultimately, to understand the function of these enzymes, both in wall synthesis and in plant growth and development, the genes for the enzymes need to be cloned and the effect of their modified expression in plants elucidated. Although, in theory, one could attempt to identify pectin biosynthetic enzymes by searching for genes that have sequence similarity to homologous enzymes in other eukaryotes,[215] this general approach for the identification of novel glycosyltransferases has been only moderately successful.[216] There are a limited number of cases where novel mammalian glycosyltransferases have been identified by sequence similarity to related glycosyltransferases.[217–219] These successes, however, have been restricted to glycosyltransferases within "rather small catalytically related families."[220] Several plant putative cellulose synthase genes have been identified based on regions of amino acid sequence similarity to microbial cellulose synthase genes.[221,222] This use of computer-facilitated comparisons of known genes with the predicted protein sequences in gene databases, so-called "*in silico* hybridization,"[222] could be useful for the identification of pectin biosynthetic genes for which homologous or related genes have already been identified in other organisms. Once identified, however, definitive proof of the identity of such putative glycosyltransferases would still require the expression of the cloned enzyme(s) in an active form, an exceedingly difficult task for polysaccharide synthesizing glycosyltransferases.[223] In addition, the absence of any cloned galacturonosyltransferase from any organism limits this approach for an entire class of pectin biosynthetic enzymes (i.e., the galacturonosyltransferases). Thus, most investigators studying pectin cell wall biosynthesis are still faced with the daunting task of purifying the desired protein in sufficient quantities to obtain amino acid sequence information for use in constructing polymerase chain reaction (PCR) primers to clone the gene.

Many different glycosyltransferase activities involved in higher plant wall biosynthesis have been identified in cell free membrane fractions, but in only a few cases has glycosyltransferase activity been retained in detergent-solubilized preparations, and in even fewer cases have any purified polypeptides been identified as plant cell wall glycosyltransferases.[38,39] A summary of the progress in identifying and characterizing glycosyltransferases that synthesize pectin follows (Table 3).

3.15.5.1 Galacturonosyltransferases

The substrate for homogalacturonan (polygalacturonate) synthesis is UDP-galacturonic acid.[10,234] In the 1960s, Hassid and co-workers synthesized homogalacturonan using cell-free extracts.[10,234,235] The addition of radiolabeled UDP-GalA to particulate fractions from mung bean, tomato, or turnip, resulted in the synthesis of a trichloroacetic acid (TCA) or ethanol insoluble and hot water soluble product.[234,235] This product, when deesterified, was completely hydrolyzed to D-galacturonic acid by a polygalacturonase-containing crude enzyme preparation from *Penicillium*. The partial degradation of the product with expolygalacturonic acid transeliminase yielded unsaturated digalacturonic acid.[235] Also, treatment of the biosynthesized material with fungal pectinase released mono-, di-, and trigalacturonic acid.[234] The particulate HGA-synthesizing enzyme preparation from mung bean (*Phaseolus aureus*) was subsequently shown to have an apparent Michaelis constant of 1.7 μM for UDP-GalA, required 1.7 mM MnCl$_2$ for maximum activity and, at a substrate concentration of 35 μM, catalyzed the polymerization of GalA residues at a rate of 4.7 nmol min^{-1}

Table 3 Comparison of Michaelis constant and pH optimum of glycosyltransferases and methyltransferases involved in, or potentially involved in, pectin biosynthesis.[a]

GlycosylT[b]	Plant	K_m[c] (μM)	pH optimum	Ref.
GalAT-I	mung bean	1.7	6.0	10
GalAT-I	pea	n.d.[d]	6.0	14
GalAT-I	sycamore	770	n.d.	11
GalAT-1	tobacco	8.9	7.8	17
GalAT-I (sol)[e]	tobacco	37	6.3–7.8	19
GalT-?[f]	mung bean	5.7	7.0–7.2	225
GalT-?[f]	flax	38	6.5	226
AraT-?[f]	mung bean	n.d.	6–6.5	227
AraT-?[f]	bean	178	n.d.	12, 228
FucT[g]	pea	80	6.0–7.0	229, 230, 231
FucT[h]	radish	170	6.8	248
ApiT[i]	*Lemna minor*	4.9	5.7	250
HGA-MT[j]	mung bean	60	6.6–7.0	13, 224, 233
PMT[k]	flax	10–30[l]	n.d.	15, 232
PMT (sol)	flax	0.5[l]	7.1	21
HGA-MT	tobacco	38[l]	7.8	20

[a]Unless indicated, all enzymes are measured in particulate preparations. [b]The abbreviations for the glycosyltransferases are as in Table 1. [c]K_m for the appropriate nucleotide sugar unless specified. [d]n.d.: not determined. [e](sol): detergent-solubilized enzyme. [f]?: The product of the enzyme has not been sufficiently characterized to state definitely that the enzyme synthesizes pectin. [g]This FucT is involved in xyloglucan synthesis. Its potential role in RG-I synthesis has not been tested. [h]This FucT is thought to synthesize arabinogalactan proteins. Its potential role in RG-I synthesis has not been tested. [i]This ApiT synthesizes apiogalacturonan. Its potential role in RG-II synthesis has not been tested. [j]HGA-MT: homogalacturonan-methyltransferase. [k]PMT: pectin methyltransferase. [l]K_m is for SAM (S-adenosyl-L-methionine).

mg^{-1} protein.[10] The enzyme responsible for this activity was named polygalacturonate α-4-galacturonosyltransferase (EC 2.4.1.43) (PGA-GalAT). Attempts to solubilize the enzyme(s) by digitonin treatment or saponification resulted in total loss of activity.[10]

Cumming and Brett subsequently reported limited solubilization of PGA-GalAT from pea (*Pisum sativum*) using the detergent lauryldimethylamine oxide (LDAO).[14] The pea enzyme had a pH optimum of 6.0 and synthesized a product of > 100 000 Da based on exclusion from Sepharose CL6B.[14] The product produced by the enzyme was identified as HGA since it was degraded into low molecular weight material upon treatment with polygalacturonase and released galacturonic acid upon acid hydrolysis.[14] No subsequent reports on the characterization or purification of this protein have been published.

Bolwell *et al.*[11] studied PGA-GalAT in particulate preparations from sycamore (*Acer pseudoplatanus*). The K_m for UDP-GalA was 770 μM, significantly higher than that reported in other systems. The enzyme was judged to be HGA since 88% of the 50% ethanol-insoluble product was digested into galacturonic acid upon treatment with pectinase and 79% of the product was recovered as galacturonic acid following acid hydrolysis.[11]

In vitro studies of homogalacturonan biosynthesis have been hampered in the past due to the time and expense required to synthesize radiolabeled UDP-galacturonic acid. A reproducible and facile method to prepare UDP-[14C]GalA[163] has been developed and used to identify and characterize PGA-GalAT in cell free membrane preparations of tobacco (*Nicotiana tabacum* L. cv Samsun) cell suspension cultures.[17] The incubation of UDP-[14C]GalA with tobacco microsomal membranes resulted in a time-dependent incorporation of [14C]galacturonic acid into a chloroform–methanol-precipitable and 65% ethanol-insoluble product. The enzyme in particulate preparations has a pH optimum of 7.8, a temperature optimum of 25–30 °C, an apparent K_m for UDP-GalA of 8.9 μM and a V_{max} (maximum velocity) of 150 pmol min^{-1} mg^{-1} protein. The intact product produced by the membrane-bound enzyme had a molecular mass of ~105 000 Da based on gel filtration chromatography and comparison with dextran standards. Treatment of the product with base to hydrolyze ester linkages (e.g., methyl esters) followed by digestion with a homogeneous endo-polygalacturonase (EPGase) and separation by TLC and high performance anion exchange chromatography (HPAEC) showed that up to 89% of [14C]-labeled product cochromatographed with mono-, di-, and trigalacturonic acid, indicating that at least 89% of the product contained contiguous 1,4-linked α-D-galactosyluroic acid residues. This proportion of galacturonic acid matches very closely the proportion of galacturonic acid calculated to be present solely in homogalacturonan based on the structure of homogalacturonans, RG-I and RG-II, and their relative amounts in the primary cell wall.[8] Optimal EPGase-fragmentation of the HGA product required prior base treatment, suggesting that 45–67% of the galacturonic acid residues in the newly synthesized HGA are esterified. At least 40% of the base sensitive linkages were shown to be methyl esters by comparing the sensitivity of base-treated and pectin methylesterase-treated products to fragmentation by EPGase. Taken together these results show that the product synthesized in microsomal membranes of tobacco cell suspensions by PGA-GalAT in the presence of UDP-[^{14}C]GalA and endogenous acceptors is primarily partially esterified HGA.

The PGA-GalAT from tobacco has recently been solubilized in an active form.[19] Solubilization of the enzyme requires a detergent (e.g., 3-[(3-cholamidopropyl)dimethylammonio]-1-propane sulfonate (CHAPS)) and a metal chelator (e.g., EDTA). The solubilized PGA-GalAT requires exogenous HGA acceptors of DP > 9, is active over a broad pH range from 6.3–7.8, has a K_m for UDP-GalA of 37 μM and a V_{max} of 290 pmol min^{-1} mg^{-1} protein. Incubation of the solubilized enzyme with a HGA acceptor of DP 15 resulted in the synthesis of an HGA of DP 16. Thus, the solubilized galacturonosyltransferase did not display the processivity expected for a polymer synthase. The release of radiolabeled galacturonic acid upon treatment of the product with purified exopolygalacturonase confirmed that the enzyme catalyzed the addition of galacturonic acid to HGA in an α-1,4-linkage and is thus a PGA-GalAT. The solubilized PGA-GalAT was shown to add galacturonic acid onto the nonreducing end of the HGA acceptor,[18] based on the fact that reducing-end biotinylated homogeneous oligomers of homogalacturonan[236] function as exogenous acceptors, while nonreducing end 4,5-unsaturated homogalacturonan oligomers, generated by cleavage of polygalacturonic acid with pectate lyase, do not.[18] The reason for the apparent loss of processivity of the solubilized galacturonosyltransferase (putative PGA-GalAT) is not known. The solubilization of the enzyme may disrupt a polymer synthase complex required for processivity, or the exogenous HGA acceptor may not contain sufficient structural information to convert PGA-GalAT into a processive mode of action.[19]

The cell specific and developmental regulation of PGA-GalAT was measured in membrane preparations from cambial cells and both differentiating and differentiated xylem cells of the sycamore tree (*Acer pseudoplatanus*).[11] The specific activity of PGA-GalAT was highest in the cambium cells, being two- to three-fold higher in cambium than in developing xylem cells and six- to ten-fold higher in cambium than in differentiated xylem.[11] This pattern of PGA-GalAT activity is consistent with the presence of pectin as a major polysaccharide component in cells with primary walls (e.g., cambium cells), and with the greatly reduced levels of pectin in cells with secondary walls (e.g., xylem cells). The results are also consistent with the hypothesis that a primary point of regulation of cell wall synthesis is at the level of the glycosyltransferases. PGA-GalAT activity was also measured along the length of pea (*Pisum sativum*) epicotyls.[14] The greatest activity was found in those top portions of the epicotyl that were undergoing active elongation.[14]

Based on the early studies of pectin biosynthesis[10,224,237] and on the author's work on homo-galacturonan synthesis,[17,238] the following model for HGA synthesis is suggested. The model states that PGA-GalAT is localized on the lumenal side of the Golgi and that it catalyzes the transfer of galacturonic acid from UDP-GalA onto a growing HGA chain that becomes partially esterified. It is hypothesized that the UDP-GalA enters the Golgi lumen by a UDP-GalA:UMP antiport in a manner analogous to other UDP sugar antiporters in animals[33,202] and similarly to the UDP-Glc:UMP antiport in plants.[239] The model also predicts that a UDP-GlcA-4-epimerase is either free in the cytoplasm or bound to the cytoplasmic side of the Golgi. UDP-GlcA-4-epimerase catalyzes the production of UDP-GalA which is then transported into the Golgi by the putative UDP-GalA:UMP antiport. The UDP produced following transfer of GalA to HGA by PGA-GalAT is hydrolyzed by a NDPase present in the Golgi.[203] The UMP produced by NDPase is transported out of the Golgi by the UDP-GalA:UMP antiport.

The available evidence further supports a HGA biosynthesis model in which UDP-galacturonic acid is a substrate for synthesis of HGA in the ER/Golgi, and HGA is subsequently methylesterified in the ER/Golgi, perhaps by an enzyme complex.[224] The particulate fractions from mung bean[13] and tobacco[17] that contained the PGA-GalAT activity also contained an enzyme that transferred the [^{14}C]-labeled methyl groups from S-adenosyl-L-methionine to the carboxyl groups of poly-galacturonic acid.[13,20] The model postulating that HGA is first synthesized and subsequently methyl-ated is supported by data demonstrating that the formation of HGA in both mung bean and tobacco increases that rate of methylesterification of HGA.[17,237] The model is further supported by the observation that the rate of HGA formation from UDP-GalA is not dependent on exogenous S-adenosyl-L-methionine[17,237] and by the inability of plant extracts to incorporate UDP-methyl-D-galacturonic acid into HGA.[10]

No studies have been reported on the other galacturonosyltransferases that synthesize the back-bone of RG-I or the side branches of RG-II.

3.15.5.2 Rhamnosyltransferases

There have been no reported studies of the rhamnosyltransferases that synthesize either the backbone of RG-I or the side chains of RG-II. A complementary DNA (cDNA) clone for a rhamnosyltransferase (UDP rhamnose: anthocyanidin-3-glucoside-L-rhamnosyltransferase) from *Petunia hydrida* that catalyzes the rhamnosylation of flavonol-3-*O*-D-glucosides has been isolated.[240] Whether or not this gene will be useful in identifying rhamnosyltransferases involved in the synthesis of RG-I or RG-II remains to be determined. A 1-2-rhamnosyltransferase (UDP-rhamnose: flavonone-7-*O*-glucoside-2″-*O*-rhamnosyltransferase) from citrus has been purified to homogeneity[241] and its developmental regulation studied.[242] Although this rhamnosyltransferase would not be expected to function in pectin biosynthesis, the eventual cloning of the gene would offer a second plant UDP-Rha:rhamnosyltransferase that might allow the identification of conserved motifs useful for "fishing out" rhamnosyltransferases involved in pectin biosynthesis.

3.15.5.3 Galactosyltransferases

A cell-free particulate preparation from mung bean (*Phaseolus aureus*) shoots incubated with UDP-[^{14}C]Gal was shown to catalyze the synthesis of a water-soluble radiolabeled product with a size of at least 4600 Da.[225] Mild acid hydrolysis of the product yielded dimers and trimers of galactose, suggesting that the product was a galactan. However, neither the linkage of the galactose within the polymer nor the identity of the acceptor (i.e., RG-I, AGPs, etc.) was determined. The K_m for UDP-Gal was 5.8 µM, the pH optimum was 7.0–7.2, and the enzyme was stimulated by $MgCl_2$. Galactosyltransferase activity was highest in 4–5 day old seedlings.[225]

Galactosyltransferases in mung bean were further studied by Panayotatos and Villemez.[243] Particulate preparations from mung bean hypocotyls catalyzed the incorporation of [^{14}C]Gal from UDP-[^{14}C]Gal into at least two different Gal-containing and alkali-insoluble radiolabeled products. Chemical and enzymatic hydrolysis showed that 20% of the product was β-1,4-linked galactan and a smaller percentage of the product was β 1,3-linked galactan. It is possible that the galactan synthesized was a component of RG-I, although this was not definitively shown.

A galactosyltransferase that catalyzes the addition of galactose in an α-1,6-linkage onto a β-1,4-linked mannan has been characterized by Reid and co-workers.[244,245] Since there are no known α-1,6-galactose linkages in pectin, and little if any mannose in pectin, this galactosyltransferase will not be discussed. Rather, it will be described in the section on galactomannan synthesis.

Galactosyltransferases have been studied in cell suspensions of flax (*Linum usitatissimum* L.).[16,226] Incubation of particulate preparations with UDP-[^{14}C]Gal yielded a 70% ethanol precipitated product with a size range from 5000–50 000 Da.[226] The K_m for production of product using microsomal membranes was 38 µM for UDP-Gal and the pH optimum was 6.5. Galactosyltransferase in suspension-cultured flax cells showed two peaks of activity, one at 2–3 days after transfer to fresh medium in the lag before active growth and a second peak at 6–7 days after transfer during active growth of the cells.[16] Based on methylation analysis of the total polysaccharide present (i.e., radiolabeled + nonradiolabeled) in the absence of a radioactive detector, the authors suggested that a β-1,4-galactan component of pectin (e.g., RG-I) was the main component in the water-solubilized product while β-1,3;β-1,6-galactan was the main component of the alkali-solubilized product.[226] These results would support the conclusion that the galactosyltransferases studied in the particulate preparations in flax include at least one β-1,4-galactosyltransferase involved in the synthesis of RG-I and a β-1,3-galactosyltransferase and β-1,6-galactosyltransferase that may be involved in the synthesis of galactans in RG-I and/or AGPs.[16] However, this conclusion awaits methylation analysis specifically of the radiolabeled product in order to confirm that the radioactive galactose was incorporated into the suggested polymers in the suggested linkages. Galactosyltransferase activity was recovered following solubilization of the enzyme from membranes using the detergents CHAPS and Brij-35.[246] Solubilized galactosyltransferase activity did not require exogenous polysaccharide acceptors.[246] The purification of the galactosyltransferase(s) has not been reported.

3.15.5.4 Arabinosyltransferases

A particulate preparation from dark grown mung bean (*Phaseolus aureus*) shoots, resuspended in 0.1% of the detergent Triton-X100 in order to inhibit xylosyltransferase activity, produced a radiolabeled polysaccharide that contained 99% of the radioactivity as arabinose.[227] The linkage of the arabinose in the arabinan product was not determined, although the authors suggested that ~24-40% of the product may have been associated with pectin based on its acidic nature. The pH optimum of the enzyme(s) was between 6 and 6.5, the enzyme required 7 mM Mn^{2+} for activity, and was inhibited by EDTA, Hg^{2+}, and Cu^{2+}.

Bolwell and Northcote[12] reported an arabinosyltransferase in bean (*Phaseolus vulgaris*) hypocotyl and callus which produced a product that was suggested to be covalently attached to pectin based on its degradation into arabinose and arabinobiose by a commercially available pectinase. However, since purified enzymes were not used, the possibility that the pectinase also contained arabinase cannot be ruled out. Thus, the identity of the precise linkage(s) and identity of the polysaccharide in which the arabinose is located (e.g., in RG-I vs. hemicellulose) requires further study. The arabinosyltransferase activity was not stimulated by UDP-galacturonic acid, lending no direct evidence for association with pectin. The incorporation of arabinose into product required Mn^{2+} and had a K_m for UDP-Ara of 178 μM.[12] The activity was inhibited by 80% by 0.1% Triton,[110] complicating the solubilization of the enzyme. However, the arabinosyltransferase was partially purified from *Phaseolus vulgaris* by solubilizing the enzyme from membranes with 1% reduced Triton X-100 and separating the enzyme by anion exchange chromatography prior to assaying for enzyme activity.[228] The latter step was necessary to recover activity since it facilitated the removal of pyrophosphorylases and phosphatases that hydrolyzed the UDP-[1-^{3}H]arabinose.[228] The arabinosyltransferase was purified 73-fold and eluted as a 70 000 Da protein based on size exclusion chromatography.[228] A monoclonal antibody (hybridoma 2C3), raised against Golgi and endoplasmic reticulum membranes and that reacts with a 70 000 Da protein, is able to inhibit arabinosyltransferase activity,[247] further supporting the identity of the 70 kDa protein as an arabinosyltransferase.

The arabinosyltransferase activity in hypocotyls was greatest during the stage of rapid extension while in callus cultures the greatest activity was associated with the most active stages of cell division occurring immediately after the lag period following transfer of cells to fresh medium.[12] The arabinosyltransferase activity in bean hypocotyl and callus was shown by membrane fractionation studies to be primarily in the Golgi, although some activity was also associated with the endoplasmic reticulum.[110] The timing of the expression of arabinosyltransferase activity is consistent with the greatest activity at the times of greatest pectin biosynthesis and a reduction in enzyme activity at the onset of secondary wall formation.[227] This correlation was further substantiated by the finding that arabinosyltransferase activity was high in bean cell suspensions grown in a media that stimulated cell division and cell growth without differentiation and secondary wall formation, while conversely, bean cell suspensions grown on media that induced differentiation of xylem cells (i.e., cells with secondary wall formation) showed comparatively less arabinosyltransferase activity.[248] The induction of arabinosyltransferase activity in cell suspensions during active growth was inhibited by the transcription inhibitor actinomycin D and by the translation inhibitor D-2-(4-methyl-2,6-dinitro-anilino)-*N*-methylpropionamide, suggesting that induction of arabinosyltransferase requires both transcription and translation.[248] The plant hormone auxin was reported to activate the incorporation of arabinose into pectin,[249] suggesting that auxin may have a role in regulating the levels of arabinosyltransferase.

3.15.5.5   Fucosyltransferases

No reports specifically targeting the α-1,2-L-fucosyltransferase(s) or the α-1,4-L-fucosyltransferase required for the synthesis of RG-I (**4**) and RG-II (**14**) have been published. However, several fucosyltransferases involved in the synthesis of xyloglucan or polysaccharide gums from root cap cells have been identified.[215,229,230,250–252] Maize (*Zea mays*) root cap cells secrete a mucilage that appears to be a modified form of pectin that contains relatively high levels of fucose.[253] The fucosyltransferase activity in these cells requires GDP-fucose as a substrate and colocalizes with endoplasmic reticulum and Golgi membranes upon subcellular membrane fractionation.[251] The linkage and structure of the endogenous acceptor for these fucosyltransferase(s) was not determined, and thus it is not known whether one or more of the fucosyltransferase(s) studied in root cap cells catalyzes the addition of fucose to RG-I or RG-II.

The fucosyltransferase that catalyzes the addition of fucose in an α-1,2-linkage onto the galactose of xyloglucan has been identified in membranes from etiolated pea (*Pisum sativum*) stems.[229] The solubilized enzyme has a K_m for GDP-Fuc of 80 μM, a pH optimum of 6–7, and an apparent molecular weight of 150 kDa.[230,231] Since the fucosyltransferase transfers fucose as an α-1,2-linkage onto a β-linked galactose, it is conceivable that the xyloglucan fucosyltransferase could also catalyze the transfer of fucose onto RG-I (**4**). To date, RG-I acceptors have not been used to test this possibility directly. Once the gene for the xyloglucan fucosyltransferase is cloned, its sequence may be useful in the identification of the pectin fucosyltransferases.

A second unique α-1,2-fucosyltransferase, believed to be involved in the synthesis of arabinogalactan proteins from radish (*Raphanus sativus* L.), has been identified.[254] This fucosyltransferase catalyzes the transfer of Fuc in an α-1,2-linkage from GDP-L-Fuc onto the nonreducing terminal Ara of a pyridylaminated (PA) trisaccharide acceptor: L-Ara*f*-α-(1-3)-D-Gal-β-(1-6)-D-Gal-PA.[254] The enzyme in membrane preparations has a pH optimum of 6.8, a K_m for GDP-Fuc of 170 μM, a K_m for the AraGalGalPA acceptor of 3.7 mM, and requires 0.1% detergent and 5 mM Mn^{2+} for maximal activity.[254] The "L-Ara*f*-specific" fucosyltransferase was localized in the Golgi and shown to have a developmental regulation in seedlings, roots, hypocotyls, and leaves distinct from the xyloglucan fucosyltransferase. It is not known whether the fucosyltransferase will transfer Fuc in an α-1,2-linkage onto RG-I or RG-II acceptors.

3.15.5.6 Apiosyltransferases

No studies of the β-(1-2)-apiosyltransferase that transfers apiose from UDP-D-apiose specifically onto the HGA backbone of RG-II (**14**) have been reported. However, some aquatic monocotyledonous plants produce a modified pectic polysaccharide known as apiogalacturonan[54,174] in which apiose or apiobiose (D-Api*f*-β-(1-3)-D-apiose) is attached to O-2 or O-3 of HGA. There is some evidence that the glycosidic linkage of apiose to HGA is in the β configuration, although this needs to be confirmed.[174] The *in vivo* synthesis of apiogalacturonan has been studied in vegetative fronds of *Spirodela polyrrhiza*.[255] Pan and Kindel[256] identified and characterized D-apiosyltransferase in cell-free particulate preparations from duckweed (*Lemna minor*). The enzyme catalyzed the transfer of [^{14}C]apiose from UDP-[^{14}C]apiose to endogenous acceptor in the particulate membrane preparations. The pH optimum of the enzyme was 5.7 and the K_m for UDP-apiose was 4.9 μM.[256] The rate of apiosyltransferase activity could be increased two-fold by the addition of UDP-GalA,[256] and the product synthesized in the presence of UDP-GalA bound more tightly to the anion exchanger, DEAE-Sephadex,[257] suggesting that the apiosyltransferase was transferring apiose onto a growing HGA chain. The product was soluble in 1% ammonium oxalate, a solubility comparable to apiogalacturonans located from the wall,[257] and the size of the solubilized product was reduced by treatment with a fungal pectinase, as expected if the apiose is added to HGA. Acid hydrolysis (pH 4) of [^{14}C]-labeled solubilized product resulted in the release of 21% of the incorporated [^{14}C]apiose as monomeric [^{14}C]apiose and 25% of the incorporated [^{14}C]apiose as [^{14}C]apiobiose,[257] the expected side chains of apiogalacturonan.[257] The instability of the enzyme precluded its purification by Mascaro and Kindel.[257]

3.15.5.7 Xylosyltransferases

There have been no reports of the identification of the α-(1-3)-xylosyltransferase that transfers xylose from UDP-Xyl onto the L-Fuc-α-(1-4)-L-rhamnosyl portion of the side branch of RG-II (**14**).

A xylosyltransferase was identified by Kindel and co-workers during their study of apiogalacturonan synthesis.[256,257] Although the product produced by this enzyme was not extensively characterized, at least some of the radioactive xylose appeared to be incorporated into apiogalacturonan and/or HGA. Such an enzyme could be a candidate for the xylosyltransferase that synthesizes xylogalacturonan.

Several xylosyltransferases involved in the synthesis of nonpectic wall polysaccharides have been identified. The α-(1-6)-xylosyltransferase that catalyzes the addition of xylose onto the β-1,4-linked glucan backbone of xyloglucan has been identified in particulate preparations from soybean,[166,258] pea,[259,260] sycamore maple,[261] and French bean.[208] Xylosyltransferases that catalyze the synthesis of xylans of undetermined linkage have been identified in particulate preparations from mung bean shoots[227] and from bean (*Phaseolus vulgaris*) hypocotyls.[12,110] Two of these xylosyltransferases have been partially purified from bean[228] and both appear to be involved in secondary wall synthesis. A β-(1-4)-xylosyltransferase involved in the synthesis of the secondary wall hemicellulose glucuronoxylan in pea has been studied in particulate[262–265] and detergent-solubilized[266] preparations. None of the genes for any of the nonpectic xylosyltransferases have been cloned and thus are not available for sequence search strategies for identifying pectin biosynthesis xylosyltransferases.

3.15.5.8 Glucuronosyltransferases

There have been no reported studies of the β-(1-4)-glucuronosyltransferase that transfers GlcA from UDP-GlcA to the L-Fuc-α-(1-4)-L-Rha portion of the side branch of RG-II (**14**). There have also been no reports of β-(1-6)-glucuronosyltransferase that transfers GlcA onto galactose in a side branch of RG-I, (**5**) and (**6**). An α-(1-2)-glucuronosyltransferase that synthesizes the hemicellulose glucuronoxylan has been studied in particulate[262–264,267] and solubilized fractions[266] from pea; however, no gene for this enzyme has yet been cloned.

3.15.5.9 Kdo-transferase, Dha-transferase, and Acerosyltransferases

There have been no reported studies of the Kdo-transferase, Dha-transferase, and aceronosyltransferases involved in RG-II synthesis. The genes for Kdo-transferases involved in lipopolysaccharide biosynthesis have been identified in bacteria (*Escherichia coli*)[268] and *Chlamydia trachomatis*.[214] These Kdo transferases, which use CMP–Kdo as a nucleotide sugar substrate, may be useful for a sequence similarity approach to identify plant Kdo-transferases involved in RG-II synthesis.

3.15.6 NONGLYCOSYLTRANSFERASE-PECTIN BIOSYNTHETIC ENZYMES

The enzyme activities required for the synthesis of pectin include, in addition to glycosyltransferases, several methyltransferases and acetyltransferases that modify the glycosyl residues in the polysaccharide (Table 4). The best studied of these modifying enzymes is the methyltransferase that transfers a methyl group from *S*-adenosyl-L-methionine (SAM) to the C-6 of galacturonic acid in homogalacturonan. This enzyme has been referred to in the literature as pectin methyltransferase. However, since pectin contains several different methylated glycosyl residues such as methyletherified glycosyl residues in RG-II (see Table 4), we will refer to the enzyme that methylesterifies homogalacturonan as HGA methyltransferase (HGA-MT) in those cases where it has been shown that the methylated product is HGA. In other cases, where the product characterization has been less rigorous, we will refer to the enzyme activity as pectin methyltransferase (PMT).

Table 4 List of nonglycosyltransferases "required" for the synthesis of pectin.

Type of transferase	Parent polymer[a]	Enzyme activity	Enzyme acceptor[b] substrate	Ref.[c]	Line formula
methylT	HGA	HGA-methyltransferase	GalA α-1,4-GalA$_{(n)}$	15, 20, 233	
acetylT	HGA	HGA: GalA-3-*O*-acetyltransferase	GalA α-1,4-GalA$_{(n)}$	47–50	
acetylT	RG-I	RG-I: GalA-3-*O*/2-*O*-acetyltransferase	GalA α-1,2-L-Rha α-1,4$_{(n)}$	3, 6, 50, 66, 67	
methylT	RG-I	RG-I: GlcA-4-*O*-methyltransferase	GlcA β-1,6-Gal	69	(**6**)
methylT	RG-II	RG-II: xylose-2-*O*-methyltransferase	D-Xyl α-1,3-L-Fuc	5, 91	(**14**)
methylT	RG-II	RG-II: fucose-2-*O*-methyltransferase	L-Fuc α-1,2-D-Gal	5, 91	(**14**)
acetylT	RG-II	RG-II: fucose-acetyltransferase	L-Fuc α-1,2-D-Gal	5, 91	(**14**)
acetylT	RG-II	RG-II: aceric acid-3-*O*-acetyltransferase	L-Ace*f*A β-1,3-L-Rha	5, 91	(**14**)

[a]HGA: homogalacturonan; RG-I: rhamnogalacturonan I; RG-II: rhamnogalacturonan II. [b]All sugars are D sugars and have pyranose rings unless otherwise indicated. [c]Reference is for the enzyme activity, when available.

The degree of methylesterification of HGA varies during cell culture[232,269] and during development.[270] For example, in the walls of young cells, HGA is highly methylesterified while in the walls of older cells HGA has a lower degree of esterification.[232] The differences in the degree of methylesterification of pectins are believed to be controlled by the activities of PMT in the Golgi apparatus[15] and pectin methylesterase (PME) in the cell wall.[271] It is not known whether the methylation of HGA during pectin synthesis occurs randomly along the HGA chain or in a defined pattern such as, for example, the methylesterification of blocks of HGA residues along the chain.

HGA-MT activity was first identified in particulate preparations from mung bean seedlings.[224,233] The HGA-MT from mung bean catalyzed the transfer of [14]CH$_3$ from SAM to give a product that released [^{14}C]methanol when treated with PME or base,[233] thus providing evidence that the enzyme methylated HGA. Furthermore, since the rate of HGA-MT activity in mung bean membranes increased in the presence of UDP-GalA, it appeared that HGA synthesized in the membranes was the methyl acceptor.[237] No sensitivity of the methylated product to cleavage by *endo*-polygalacturonase (EPGase) or pectin lyase, however, was reported and the HGA-MT from mung bean has not been solubilized or purified. The HGA-MT in membranes from mung bean had a pH optimum of 6.6–7.0[233] and a K_m for SAM of 60 μM.[13]

Putative HGA-MT activity has also been detected in particulate preparations from flax (*Linum usitatissiumum* L.) hypocotyls[15] and suspension-cultured flax cells.[232] The PMT from flax catalyzed the synthesis of a product from which [^{14}C]methanol was released by treatment with 1 M sodium hydroxide.[15] Studies to establish that the enzyme transfers [14]CH$_3$ from SAM specifically to HGA have not yet been reported. The PMT activity in membranes fractionates at a density consistent with a localization in the Golgi.[21] The apparent K_m for SAM by PMT in flax membranes was 10–30 μM.[21] The PMT in flax freeze-thawed microsomes is stimulated by exogenous pectins of both low (0.1) and high (0.5) degrees of esterification.[272]

The PMT from flax was solubilized from membranes, partially purified, and characterized.[21] The apparent K_m of the solubilized enzyme for SAM was 0.5 μM and the pH optimum was 7.1.[21] Solubilized HGA-MT was stimulated by exogenous polygalacturonic acid (PGA). The apparent K_m for PGA was 0.5–0.7 mg mL^{-1} (equivalent to 57–79 μM, assuming an average DP of 50 for PGA). The solubilized PMT from flax was separated into several peaks of activity, ranging from apparent masses of roughly 5000–150 000 Da.[21] Both basic and neutral isoforms of solubilized PMT were detected.[21] The specificity of the PMT activities for the type of pectin substrate methylated was not ascertained.[21] However, the fact that the solubilized flax enzyme is activated by exogenous PGA supports its identity as HGA-MT.[21] The activity of solubilized PMT on PGA-, RG-I- and RG-II-acceptors has been reported.[273] Contrary to an earlier report,[21] the pH optimum of solubilized PMT for the methylation of PGA was reported to be acidic (5.5), while the pH optimum with RG-II as exogenous acceptor was 7.0 (with high activity from pH 6.5–8.0), and the pH optimum with RG-I was broad ranging from 6.0–8.0.[273] The amount of homogalacturonan covalently attached to the RG-I and RG-II was not given. Therefore, it is not known whether the incorporation of methyl groups reflected the methylesterification of HGA "tails" of the RG-I and RG-II, or rather, the methyletherification of nongalacturonic acid residues in the RG-I and RG-II.

The observation that solubilized PMT from flax incorporates methyl groups into different types of pectins (e.g., HGA, RG-I, and RG-II) at different pH optima[273,274] suggests that unique PMTs exist that specifically methylate the different pectic polysaccharides to form methylesters or methylethers. Methyltransferases that catalyze the methyletherification of RG-II to produce 2-*O*-methylxylose (14) and 2-*O*-methylfucose[90,275] (14) and methyltransferases that catalyze the production of 4-*O*-methylglucuronic acid in a side branch of RG-I[276] (6) may be required to synthesize pectin. Alternatively, the glycosyl residues could be methylated at the level of the nucleotide sugar and subsequently transferred onto RG-II in a premethylated form.

A PMT that methylates homogalacturonan (HGA-MT) has been identified and characterized in membrane preparations from tobacco (*Nicotiana tabacum* L. cv Samsun).[20] The tobacco HGA-MT transfers [^{14}C]methyl from SAM to the C-6 carboxyl group of homogalacturonan, and does not use 5-methyltetrahydrofolate as an alternative methyl donor. The pH optimum for HGA-MT in tobacco membranes is 7.8, the apparent K_m for SAM was 38 μM and the V_{max} is 0.81 pmol s^{-1} mg^{-1} protein. The PMT was shown to be a HGA-MT since at least 59% of the radioactivity in the product was released by mild base treatment and by enzymatic hydrolysis using purified pectin methylesterase. The released radioactivity was identified as methanol by fractionation over a Rezex ROA-organic acid column.[20] The product produced by HGA-MT in membranes using endogenous acceptor could be cleaved by a purified *endo*-polygalacturonase into fragments that migrated on TLC similarly to small oligomers of HGA.[20] The membrane-bound enzyme was not stimulated by exogenous PGA or pectin, however, HGA-MT activity in membranes was stimulated by UDP-galacturonic acid, a substrate for HGA synthesis. Tobacco HGA-MT has been solubilized from membranes (Goubet and Mohnen, unpublished results) although the characteristics of the solubilized enzyme have not yet been reported.

There have been no reported studies of the other methyltransferases or acetyltransferases that modify HGA (HGA-acetyltransferase), RG-I (RG-I GlcA acetyltransferase) (6) or RG-II (RG-II xylose-methyltransferase, RG-II fucose-methyltransferase, RG-II fucose-acetyltransferase, RG-II aceric acid acetyltransferase) (14) (see Table 4).

3.15.7   DIRECTION OF PECTIN POLYSACCHARIDE BIOSYNTHESIS

Cell wall polymers are assumed to be synthesized by a glycosyltransferase-catalyzed addition of monosaccharides from a nucleotide sugar to the nonreducing end of a growing polymer.[38] Direct evidence for this assumption, however, is limited. Henry and Stone[210] used pulse-chase experiments to show that β-glucans are synthesized from the nonreducing end of the primer. Scheller *et al.*[18] used reducing end and nonreducing end modified exogenous acceptors to show that solubilized PGA-GalAT catalyzes the transfer of GalA to the nonreducing end of homogalacturonan acceptors. It is not known whether the oligosaccharide side chains of RG-II are attached to be homogalacturonan backbone by the sequential addition of individual glycosyl residues, or whether the side chains are synthesized on a lipid intermediate and subsequently transferred to the HGA backbone. Similarly, it is not known how the side branches of RG-I are attached to Rha of the alternating GalA-Rha backbone.

3.15.8   REGULATION OF PECTIN BIOSYNTHESIS

Pectin biosynthesis must be a highly regulated process. A coordinated regulation of multiple enzymes must occur when cells make the transition from primary to secondary wall synthesis since the level of total pectin in most secondary walls is greatly reduced compared with primary walls.[2] The reduction in galacturonosyltransferase activity in sycamore cambium cells as they differentiate into xylem cells,[11] and the reduction in arabinosyltransferase activity in bean suspension cultures during xylem differentiation,[248] is an example of the downregulation of pectin biosynthetic enzymes during the transition to secondary wall formation. There must also be a mechanism for the regulation of specific glycosyltransferases and modifying enzymes since immunocytochemistry with anti-carbohydrate antibodies demonstrates that specific pectic carbohydrate epitopes are regulated in a cell type and development-specific manner.[83,122,127,128] The composition and overall structure of pectin is also modified in response to environmental stresses such as osmotic stress.[277–279] Also, changes in the neutral side chains of RG-I have been associated with the ability of carrot (*Daucus carota* L.) cells to adhere to each other,[280] a property which is believed to be important for the embryogenic competence of cultured cells. The amount of xylosylation of HGA has also been associated with the ability of suspension-cultured cells to adhere to one another.[48] The dissociation (friability) of sugar beet (*Beta vulgaris* L.) callus cells has also been associated with increased levels of acetylation of pectin in the walls of these cells.[136] Thus, it is likely that a fine-tuned regulation of specific pectin biosynthetic enzymes during development contributes to such fine scale changes in pectin structure.

There is biochemical evidence, based on the *in vivo* labeling of tissues and cell cultures with radiolabeled glucose and inositol, in the presence and absence of plant hormones, that the plant hormone auxin may regulate cell wall synthesis.[249,281–284] For example, Ray and co-workers have shown that auxin (indoleacetic acid) promotes the synthesis of pectin and hemicellulose synthesis in oat (*Avena sativa*) coleoptiles.[285,286] Furthermore, the activity of pectin biosynthetic enzymes is increased during auxin-induced cell division and cell elongation.[248,249,285,286] The induction of arabinosyltransferase, which occurs in actively growing bean cell suspensions,[248] is inhibited by transcription and translation inhibitors.[248] This suggests that the induction of arabinosyltransferase is regulated at both the level of transcription and translation.

A precise understanding of the molecular basis of the regulation of pectin biosynthesis will require the generation and use of antibodies against specific pectin biosynthetic glycosyltransferases and modifying enzymes and the identification of the genes for the specific enzymes. With these tools we will be able to ask whether pectin biosynthesis is regulated at the level of enzyme transcript or the enzyme,[12,170] by the level and/or availability of the nucleotide sugars,[158,170,255,287,288] via allosteric regulation of the enzymes, via competition of the enzymes for common substrates, or by other mechanisms.

3.15.9 GALACTOMANNANS

Galactomannans are storage polysaccharides found in endosperm cell walls and cell lumen in seeds from legumes.[24,244] The galactomannans function as a reserve carbohydrate source during seed germination and protect the seed from desiccation, and are used as thickeners and stabilizers in the food industry.[289] The galactomannans have a linear backbone of 1,4-linked β-D-mannosyl residues with single α-D-galactosyl residues attached at species-specific frequencies to C-6 of the backbone (**15**).[22,24,244] The degree of galactose substitution of the mannan backbone ranges from between 25% and 97% with a higher degree of galactose substitution occurring in the more phylogenically advanced legumes.[244,245,290]

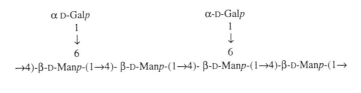

(**15**)

A galactomannan mannosyltransferase, whose activity and timing correlate with the amount of galactomannan biosynthesized in the seed, was first identified in particulate extracts from endosperm of fenugreek seed (*Trigonella*).[291] The mannosyltransferase transferred [^{14}C]mannose from GDP-[^{14}C]mannose to produce a soluble galactomannan product.[291] UDP-Gal was also included in the reactions in order to facilitate galactomannan biosynthesis. The product was shown to be galactomannan by selective precipitation with borate ions and with the transition ions Cr^{3+}, Mn^{2+}, and Fe^+, and by its ability to bind to the galactosyl-binding *Ricinus communis* lectin (RCA-II: castor bean hemagglutinin).[291] Subsequent studies using particulate preparations from both fenugreek (*Trigonella foenum graecum* L.) and guar (*Cyamopsis tetragonoloba* [L.] Tabu) endosperm led to the identification of a GDP-Man-dependent mannosyltransferase and a UDP-Gal-dependent galactosyltransferase which catalyze the synthesis of galactomannans.[244] The mannosyltransferase, in the presence of a divalent cation such as Mg^{2+}, Mn^{2+}, or Ca^{2+} and GDP-[^{14}C]-D-Man, synthesized a radiolabeled linear 1,4-linked β-D mannan, as demonstrated by the recovery of mannose, mannobiose, mannotriose, and oligomers of mannan following the hydrolysis of the product by a purified endo-β-D-mannanase.[244] When particulate enzyme preparations were incubated with UDP-D-Gal and GDP-Man in the presence of Mn^{2+}, a largely water-soluble product was formed that was shown, using purified endo-β-D-mannanase and α-D-galactosidase, to contain galactose connected by an α-1,6-linkage to a linear 1,4-linked β-D-mannan backbone. The transfer of galactose from UDP-Gal to the galactomannan chain by galactosyltransferase required the simultaneous transfer of mannose from GDP-Man. This was demonstrated by the lack of galactosyltransferase activity when only UDP-Gal was used as a substrate[244] and by the inability of galactosyltransferase to use galactomannan that had been previously produced by incubation of particulate preparations with GDP-Man as substrate. The results showed that the galactosyltransferase required the presence of a growing mannan chain and the associated transfer of a mannosyl residue catalyzed by mannosyltransferase.[244]

The degree of substitution of the mannan with galactose can be regulated *in vitro* by adjusting the relative concentrations of GDP-Man and UDP-Gal,[244] suggesting that the ratio of galactose to mannose in galactomannan produced *in vivo* may be regulated by the levels of the nucleotide sugars. However, it was subsequently shown that the regulation of the glycosyltransferases themselves was also important for determining the species-specific ratio of mannose to galactose in galactomannans.[245,289,292] Three leguminous species, fenugreek (*Trigonella foenum-graecum*), guar (*Cyamopsis tetragonoloba*), and *Senna* (*Senna occidentalis*), that form, respectively, galactomannan with high (96%), medium (65%), and low (30%) galactan substitution,[245,289,292] were used to show that the maximum degree of substitution of the mannan backbone with galactose is regulated by the activity and/or specificity of the glycosyltransferases.[245] Furthermore, in some species such as *Senna*, the degree of substitution of the galactomannan in the wall is regulated by the postbiosynthetic removal of galactose from the galactomannan by α-galactosidase.[292] Based on a statistical analysis of the oligosaccharide fragmentation pattern produced by enzymatic hydrolysis of the galactomannan products produced by fenugreek, guar, and *Senna* with endo-(1→4)-β-D-mannanase, it is proposed that galactosyltransferase recognizes at least three consecutive mannose residues in the mannan backbone.[245] The mannose residues recognized include the mannose to which the galactose is added and the two neighboring mannoses toward the reducing end.[245] It is further proposed that the relative rates of the galactosyltransferase activity in the different species *in vivo* lead to the observed species-specific difference in the degree of galactose substitution of the mannose backbone.[245]

The importance of the studies of the galactomannan galactosyltransferase for studies of pectin biosynthesis are two-fold. First, the work of Reid and co-workers[245] clearly indicates that a wealth of information can be obtained by a detailed and careful biochemical analysis of glycosyltransferases in particulate preparations. Specifically, these researchers have generated detailed models regarding the mechanism of how multiple glycosyltransferases coordinate their activity to produce complex and species-specific polysaccharides. Such models are critically important for understanding cell wall polysaccharide biosynthesis and clearly indicate the importance of studying the biosynthesis reaction in cell extracts. It is particularly noteworthy that these researchers have produced these models without the prior cloning of the genes or the purification of the enzymes. In recent times there is a tendency to believe that the "gene must be cloned" before progress in understanding the function of enzymes *in vivo* can be made. While it is clear that all available molecular tools, including the genes, should be gathered to address any biological problem, a narrow focus on the gene, particularly when studying polysaccharides which are secondary rather than primary gene products, is unlikely to be the most fruitful approach.

A second aspect of the galactomannan research that should be useful for studies of pectin biosynthesis is the fact that the identification of some glycosyltransferases may require the active synthesis of a growing polysaccharide chain that is dependent upon a different nucleotide sugar (or other activated sugar) than that used by the specific glycosyltransferase of interest. Such coordinated synthesis of more than one type of glycosyl residue has also been demonstrated for some of the glycosyltransferases involved in xyloglucan biosynthesis[208,258–260,293] and in glucuronoxylan synthesis,[264] and will be likely to be the case for the synthesis of the RG-I backbone and possibly for the synthesis of some of the oligosaccharide side chain residues in RG-II. It is also possible that a coordination of the synthesis of the HGA backbone with the methylesterification and/or acetylation of some of the galacturonosyl residues in the backbone may be required in order to achieve HGA polymer synthesis *in vitro*.

3.15.10   FUTURE PROSPECTS

The structure of the pectic polysaccharides is now known in sufficient detail for studies of the biosynthetic enzymes to be approachable. Biochemical studies by a limited number of research groups have led to the identification, solubilization, and partial purification of a small number of pectin biosynthetic enzymes. The importance of the cell wall in plant growth and development warrants more research in the study of the glycosyltransferases and modifying enzymes that synthesize pectin, with the targeted goal of manipulating pectin synthesis in order to elucidate the function of pectin in the plant. Most certainly, the genes for the biosynthetic enzymes must be cloned for this to occur. A combined biochemical and molecular strategy should allow the cloning of the first pectin biosynthetic genes within the next few years. The precise strategies that will be successful remain to be determined, but it is likely that it will be necessary to take a multidisciplinary approach that includes the use of cell wall mutants,[294] plant material enriched in pectin,[295–299] and DNA sequence similarities between closely related enzyme families,[221,222] which are all backed by a firm understanding of the biochemistry of biosynthetic reactions themselves.

ACKNOWLEDGMENTS

I thank Malcolm O'Neill for a critical reading of this review and for many helpful discussions, Carol L. Gubbins Hahn for drawing some of the figures, Karen Howard for editorial help, and members of my laboratory and colleagues at the Complex Carbohydrate Research Center for their helpful discussions. I particularly thank Stefan, Tiffany, and Samantha Eberhard for their patience. The NRI Competitive Grants Program (USDA Award No. 94-37304-1103), NSF Division of International Programs (NSF Award No. INT-9722509), and Hercules Incorporated are gratefully acknowledged for financial support of the work of my laboratory on pectin biosynthesis.

3.15.11 REFERENCES

1. P. Albersheim, A. G. Darvill, M. A. O'Neill, H. A. Schols, and A. G. J. Voragen, in "Pectins and Pectinases," eds. J. Visser and A. G. J. Voragen, Elsevier, Amsterdam, 1996, p. 47.
2. G. B. Fincher and B. A. Stone, in "Encyclopedia of Plant Physiology New Series. Plant Carbohydrates II," eds. W. Tanner and F. A. Loewus, Springer-Verlag, Berlin, 1981, vol. 13B, p. 68.
3. A. Bacic, P. J. Harris, and B. A. Stone, in "The Biochemistry of Plants," ed. J. Preiss, Academic Press, New York, 1988, vol. 14, p. 297.
4. M. McNeil, A. G. Darvill, S. C. Fry, and P. Albersheim, *Annu. Rev. Biochem.*, 1984, **53**, 625.
5. N. C. Carpita and D. M. Gibeaut, *Plant J.*, 1993, **3**, 1.
6. M. O'Neill, P. Albersheim, and A. Darvill, in "Methods in Plant Biochemistry," ed. P. M. Dey, Academic Press, London, 1990, vol. 2, p. 415.
7. N. C. Carpita, *Annu. Rev. Plant Physiol. Plant Mol. Biol.*, 1996, **47**, 445.
8. D. Mohnen, R. L. Doong, K. Liljebjelke, G. Fralish, and J. Chan, in "Pectins and Pectinases," eds. J. Visser and A. G. J. Voragen, Elsevier, Amsterdam, 1996, p. 109.
9. S. C. Fry, in "Oxford Surveys of Plant Molecular and Cell Biology," ed. B. J. Miflin, Oxford University Press, Oxford, 1985, vol. 2, p. 1.
10. C. L. Villemez, A. L. Swanson, and W. Z. Hassid, *Arch. Biochem. Biophys.*, 1966, **116**, 446.
11. G. P. Bolwell, G. Dalessandro, and D. H. Northcote, *Phytochemistry*, 1985, **24**, 699.
12. G. P. Bolwell and D. H. Northcote, *Planta*, 1981, **152**, 225.
13. H. Kauss and W. Z. Hassid, *J. Biol. Chem.*, 1967, **242**, 3449.
14. C. M. Cumming and C. T. Brett, in "Cell Walls '86. Proceedings of the Fourth Cell Wall Meeting, Paris, September 10–12, 1986," eds. B. Vian, D. Reis, and R. Goldberg, Université Pierre et Marie Curie—Ecole Normale Supérieure, Paris, 1986, p. 360.
15. M. P. Vannier, B. Thoiron, C. Morvan, and M. Demarty, *Biochem. J.*, 1992, **286**, 863.
16. F. Goubet and C. Morvan, *Plant Cell Physiol.*, 1994, **35**, 719.
17. R. L. Doong, K. Liljebjelke, G. Fralish, A. Kumar, and D. Mohnen, *Plant Physiol.*, 1995, **109**, 141.
18. H. V. Scheller, R. L. Doong, B. L. Ridley, and D. Mohnen, submitted for publication.
19. R. L. Doong and D. Mohnen, *Plant J.*, 1998, **13**, 363.
20. F. Goubet, L. N. Council, and D. Mohnen, *Plant Physiol.*, 1998, **116**, 337.
21. M.-P. Bruyant-Vannier, A. Gaudinet-Schaumann, T. Bourlard, and C. Morvan, *Plant Physiol. Biochem.*, 1996, **34**, 489.
22. A. M. Stephen, in "The Polysaccharides," ed. G. O. Aspinall, Academic Press, New York, 1983, vol. 2, p. 97.
23. G. A. Towle and R. L. Whistler, in "Phytochemistry," ed. L. P. Miller, Van Nostrand-Reinhold, New York, 1973, vol. I, p. 198.
24. J. S. G. Reid, in "Biochemistry of Plant Cell Walls," eds. C. T. Brett and J. R. Hillman, Cambridge University Press, Cambridge, 1985, p. 259.
25. H. Kauss, in "Annual Proceedings of the Phytochemical Society, Plant Carbohydrate Biochemistry," ed. J. B. Pridham, Academic Press, London, 1974, vol. 10, p. 191.
26. D. H. Northcote, in "ACS Symposium Series, Chemistry and Function of Pectins," eds. M. L. Fishman and J. J. Jen, American Chemical Society, Washington, DC, 1986, vol. 310, p. 134.
27. W. Z. Hassid, E. F. Neufeld, and D. S. Feingold, *Proc. Natl. Acad. Sci. USA*, 1959, **45**, 905.
28. W. Z. Hassid, *Annu. Rev. Plant Physiol.*, 1967, **18**, 253.
29. D. S. Feingold and G. Avigad, in "The Biochemistry of Plants," ed. J. Preiss, Academic Press, New York, 1980, vol. 3, p. 101.
30. D. S. Feingold and G. A. Barber, in "Methods in Plant Biochemistry," ed. P. M. Dey, Academic Press, London, 1990, vol. 2, p. 39.
31. O. Gabriel and L. Van Lenten, in "International Review of Biochemistry: Biochemistry of Carbohydrates II," ed. D. J. Manners, International Publishers in Science and Medicine, Baltimore, MD, 1978, vol. 16, p. 1.
32. D. H. Northcote, in "Biochemistry of Plant Cell Walls," eds. C. T. Brett and J. R. Hillman, Cambridge University Press, Cambridge, 1985, p. 177.
33. D. P. Delmer and B. A. Stone, in "The Biochemistry of Plants," ed. J. Preiss, Academic Press, Orlando, FL, 1988, vol. 14, p. 373.
34. G. P. Bolwell, *Phytochemistry*, 1988, **27**, 1235.
35. W. Z. Hassid, *Science*, 1969, **165**, 137.
36. M. C. Ericson and A. D. Elbein, in "The Biochemistry of Plants," ed. J. Preiss, Academic Press, New York, 1980, vol. 3, p. 589.
37. S. Levy and L. A. Staehelin, *Curr. Opin. Cell Biol.*, 1992, **4**, 856.
38. D. M. Gilbert and N. C. Carpita, *FASEB J.*, 1994, **8**, 904.
39. K. Iiyama, T. B. T. Lam, P. J. Meikle, K. Ng, D. I. Rhodes, and B. A. Stone, in "Forage Cell Wall Structure and Digestibility: International Symposium," eds. H. G. Jung, D. R. Buxton, R. D. Hatfield, and J. Ralph, American Chemical Society, Washington, DC, 1993, p. 621.
40. K. W. Waldron and C. T. Brett, in "Biochemistry of Plant Cell Walls," eds. C. T. Brett and J. R. Hillman, Cambridge University Press, Cambridge, 1985, p. 79.
41. G. P. Bolwell, *Int. Rev. Cytol.*, 1993, **146**, 261.
42. C. Brett and K. Waldron, 'Physiology and Biochemistry of Plant Cell Walls," Unwin Hyman, London, 1990.
43. D. W. James, Jr., J. Preiss, and A. D. Elbein, in "The Polysaccharides," ed. G. O. Aspinall, Academic Press, New York, 1985, vol. 3, p. 107.
44. D. H. Northcote, in "ACS Symposium Series: Plant Cell Wall Polymers," eds. N. G. Lewis and M. G. Paice, American Chemical Society, Washington, DC, 1989, vol. 399, p. 1.
45. G. O. Aspinall, in "The Biochemistry of Plants," ed. J. Preiss, Academic Press, New York, 1980, vol. 3, p. 473.
46. A. J. Mort, F. Qiu, and N. O. Maness, *Carbohydr. Res.*, 1993, **247**, 21.

47. F. M. Rombouts and J. F. Thibault, in "ACS Symposium Series: Chemistry and Function of Pectins," eds. M. L. Fishman and J. J. Jen, American Chemical Society, Washington, DC, 1986, vol. 310, p. 49.
48. T. Ishii, *Mokuzai Gakkaishi*, 1995, **41**, 669.
49. J. A. De Vries, A. G. J. Voragen, F. M. Rombouts, and W. Pilnik, in "ACS Symposium Series: Chemistry and Function of Pectins," eds. M. L. Fishman and J. J. Jen, American Chemical Society, Washington, DC, 1986, vol. 310, p. 38.
50. T. Ishii, *Plant Physiol.*, 1997, **113**, 1265.
51. H. O. Bouveng, *Acta Chem. Scand.*, 1965, **19**, 953.
52. H. A. Schols, E. Vierhuis, E. J. Bakx, and A. G. J. Voragen, *Carbohydr. Res.*, 1995, **275**, 343.
53. A. Kikuchi, Y. Edashige, T. Ishii, and S. Satoh, *Planta*, 1996, **200**, 369.
54. D. A. Hart and P. K. Kindel, *Biochem. J.*, 1970, **116**, 569.
55. L. Cheng and P. K. Kindel, *Carbohydr. Res.*, 1997, **301**, 205.
56. M. Tomoda, Y. Suzuki, and N. Satoh, *Chem. Pharm. Bull.*, 1979, **27**, 1651.
57. J. A. Brown and S. C. Fry, *Plant Physiol.*, 1993, **103**, 993.
58. M. C. McCann, J. Shi, K. Roberts, and N. C. Carpita, *Plant J.*, 1994, **5**, 773.
59. H. A. Schols, E. J. Bakx, D. Schipper, and A. G. J. Voragen, *Carbohydr. Res.*, 1995, **279**, 265.
60. L. Yu and A. J. Mort, in "Pectins and Pectinases," eds. J. Visser and A. G. J. Voragen, Elsevier, Amsterdam, 1996, p. 79.
61. C. M. G. C. Renard, R. M. Weightman, and J.-F. Thibault, *Int. J. Biol. Macromol.*, 1997, **21**, 155.
62. J. M. Lau, M. McNeil, A. G. Darvill, and P. Albersheim, *Carbohydr. Res.*, 1985, **137**, 111.
63. S. Eda, K. Miyabe, Y. Akiyama, A. Ohnishi, and K. Kato, *Carbohydr. Res.*, 1986, **158**, 205.
64. M. McNeil, A. G. Darvill, and P. Albersheim, *Plant Physiol.*, 1980, **66**, 1128.
65. J. An, L. Zhang, M. A. O'Neill, P. Albersheim, and A. G. Darvill, *Carbohydr. Res.*, 1994, **264**, 83.
66. P. Komalavilas and A. J. Mort, *Carbohydr. Res.*, 1989, **189**, 261.
67. P. Lerouge, M. A. O'Neill, A. G. Darvill, and P. Albersheim, *Carbohydr. Res.*, 1993, **243**, 359.
68. J. M. Lau, M. McNeil, A. G. Darvill, and P. Albersheim, *Carbohydr. Res.*, 1987, **168**, 245.
69. J. An, M. A. O'Neill, P. Albersheim, and A. G. Darvill, *Carbohydr. Res.*, 1994, **252**, 235.
70. P. Lerouge, M. A. O'Neill, A. G. Darvill, and P. Albersheim, *Carbohydr. Res.*, 1993, **243**, 359.
71. J. R. Thomas, A. G. Darvill, and P. Albersheim, *Carbohydr. Res.*, 1989, **185**, 279.
72. G. O. Aspinall, R. Begbie, A. Hamilton, and J. N. C. Whyte, *J. Chem. Soc., Perkin Trans. 1*, 1967, 1065.
73. M. Morita, *Agric. Biol. Chem.*, 1965, **29**, 564.
74. M. Morita, *Agric. Biol. Chem.*, 1965, **29**, 626.
75. A. J. Mort and W. D. Bauer, *J. Biol. Chem.*, 1982, **257**, 1870.
76. J. M. Lau, M. McNeil, A. G. Darvill, and P. Albersheim, *Carbohydr. Res.*, 1987, **168**, 219.
77. S. S. Bhattacharjee and T. E. Timell, *Can. J. Chem.*, 1965, **43**, 758.
78. A. E. Clarke, R. L. Anderson, and B. A. Stone, *Phytochemistry*, 1979, **18**, 521.
79. G. B. Fincher, B. A. Stone, and A. E. Clarke, *Annu. Rev. Plant Physiol.*, 1983, **34**, 47.
80. A. Bacic, H. Du, B. A. Stone, and A. E. Clarke, *Essays Biochem.*, 1996, **31**, 91.
81. M. D. Serpe and E. A. Nothnagel, *Plant Physiol.*, 1995, **109**, 1007.
82. A. M. Gane, D. Craik, S. L. A. Munro, G. J. Howlett, A. E. Clarke, and A. Bacic, *Carbohydr. Res.*, 1995, **277**, 67.
83. G. Freshour, R. P. Clay, M. S. Fuller, P. Albersheim, A. G. Darvill, and M. G. Hahn, *Plant Physiol.*, 1996, **110**, 1413.
84. W. Steffan, P. Kovac, P. Albersheim, A. G. Darvill, and M. G. Hahn, *Carbohydr. Res.*, 1995, **275**, 295.
85. S. C. Fry, *Annu. Rev. Plant Physiol.*, 1986, **37**, 165.
86. T. Ishii, *Plant Sci.*, 1997, **127**, 111.
87. S. C. Fry, *Biochem. J.*, 1982, **203**, 493.
88. S. C. Fry, *Planta*, 1983, **157**, 111.
89. G. O. Aspinall and V. P. Bhavanandan, *J. Chem. Soc.*, 1965, **488**, 2693.
90. M. A. O'Neill, D. Warrenfeltz, K. Kates, P. Pellerin, T. Doco, A. G. Darvill, and P. Albersheim, *J. Biol. Chem.*, 1996, **271**, 22 923.
91. M. A. O'Neill, D. Warrenfeltz, K. Kates, P. Pellerin, T. Doco, A. G. Darvill, and P. Albersheim, *J. Biol. Chem.*, 1997, **272**, 3869.
92. M. W. Spellman, M. McNeil, A. G. Darvill, P. Albersheim, and K. Henrick, *Carbohydr. Res.*, 1983, **122**, 115.
93. W. S. York, A. G. Darvill, M. McNeil, and P. Albersheim, *Carbohydr. Res.*, 1985, **138**, 109.
94. T. T. Stevenson, A. G. Darvill, and P. Albersheim, *Carbohydr. Res.*, 1988, **179**, 269.
95. A. J. Whitcombe, M. A. O'Neill, W. Steffan, P. Albersheim, and A. G. Darvill, *Carbohydr. Res.*, 1995, **271**, 15.
96. V. Puvanesarajah, A. G. Darvill, and P. Albersheim, *Carbohydr. Res.*, 1991, **218**, 211.
97. M. W. Spellman, M. McNeil, A. G. Darvill, P. Albersheim, and A. Dell, *Carbohydr. Res.*, 1983, **122**, 131.
98. L. D. Melton, M. McNeil, A. G. Darvill, P. Albersheim, and A. Dell, *Carbohydr. Res.*, 1986, **146**, 279.
99. T. T. Stevenson, A. G. Darvill, and P. Albersheim, *Carbohydr. Res.*, 1988, **182**, 207.
100. M. A. O'Niell, A. G. Darvill, and P. Albersheim, 1998, unpublished results.
101. T. Matoh, S. Kawaguchi, and M. Kobayashi, *Plant Cell Physiol.*, 1996, **37**, 636.
102. M. Kobayashi, T. Matoh, and J.-I. Azuma, *Plant Physiol.*, 1996, **110**, 1017.
103. T. Ishii and W. Matsunaga, *Carbohydr. Res.*, 1996, **284**, 1.
104. S. Kaneko, T. Ishii, and T. Matsunaga, *Phytochemistry*, 1997, **44**, 243.
105. M. Takasaki, S. Kawaguchi, M. Kobayashi, K. Takabe, and T. Matoh, in "Boron in Soils and Plants," eds. R. W. Bell and B. Rerkasem, Kluwer, Netherlands, 1997, 243.
106. M. C. Jarvis, *Plant Cell Environ.*, 1984, **7**, 153.
107. C. M. S. Carrington, L. C. Greve, and J. M. Labavitch, *Plant Physiol.*, 1993, **103**, 429.
108. D. P. Delmer and Y. Amor, *Plant Cell*, 1995, **7**, 987.
109. L. A. Staehelin and I. Moore, *Annu. Rev. Plant Physiol. Plant Mol. Biol.*, 1995, **46**, 261.
110. G. P. Bolwell and D. H. Northcote, *Biochem. J.*, 1983, **210**, 497.
111. R. M. Brown, Jr., J. H. M. Willison, and C. L. Richardson, *Proc. Natl. Acad. Sci. USA*, 1976, **73**, 4565.
112. R. M. Brown, Jr., *J. Cell Sci. Suppl.*, 1985, **2**, 13.
113. S. C. Mueller, R. M. Brown, Jr., and T. K. Scott, *Science*, 1976, **194**, 949.

114. S. C. Mueller and R. M. Brown, Jr., *Planta*, 1982, **154**, 489.
115. D. H. Northcote, *Endeavor*, 1971, **30**, 26.
116. D. H. Northcote and J. D. Pickett-Heaps, *Biochem. J.*, 1966, **98**, 159.
117. P. J. Harris and D. H. Northcote, *Biochim. Biophys. Acta*, 1971, **237**, 56.
118. J. Sterling, C. Wolff, L. Norambuena, A. Orellano, and D. Mohnen, 1997, unpublished results.
119. P. J. Moore, K. M. M. Swords, M. A. Lynch, and L. A. Staehelin, *J. Cell Biol.*, 1991, **112**, 589.
120. T. Hoson, *Int. Rev. Cytol.*, 1991, **130**, 233.
121. J. P. Knox, *Protoplasma*, 1992, **167**, 1.
122. M. A. Lynch and L. A. Staehelin, *J. Cell Biol.*, 1992, **118**, 467.
123. G. F. Zhang and L. A. Staehelin, *Plant Physiol.*, 1992, **99**, 1070.
124. D. J. Sherrier and K. A. Vanden Bosch, *Plant J.*, 1994, **5**, 185.
125. B. Vian and J.-C. Roland, *Biol. Cell*, 1991, **71**, 43.
126. F. Liners and P. Van Cutsem, *Protoplasma*, 1992, **170**, 10.
127. J. P. Knox, P. J. Linstead, J. King, C. Cooper, and K. Roberts, *Planta*, 1990, **181**, 512.
128. N. J. Stacey, K. Roberts, N. C. Carpita, B. Wells, and M. C. McCann, *Plant J.*, 1995, **8**, 891.
129. P. J. Casero and J. P. Knox, *Protoplasma*, 1995, **188**, 133.
130. R. W. Stoddart and D. H. Northcote, *Biochem. J.*, 1967, **105**, 45.
131. K. A. VandenBosch, D. J. Bradley, J. P. Knox, S. Perotto, G. W. Butcher, and N. J. Brewin, *EMBO J.*, 1989, **8**, 335.
132. P. J. Moore and L. A. Staehelin, *Planta*, 1988, **174**, 433.
133. P. J. Moore, A. G. Darvill, P. Albersheim, and L. A. Staehelin, *Plant Physiol.*, 1986, **82**, 787.
134. F. Liners, J.-J. Letesson, C. Didembourg, and P. Van Cutsem, *Plant Physiol.*, 1989, **91**, 1419.
135. R. Cacan, C. Villers, M. Bélard, A. Kaiden, S. S. Krag, and A. Verbert, *Glycobiology*, 1992, **2**, 127.
136. F. Liners, T. Gaspar, and P. Van Cutsem, *Planta*, 1994, **192**, 545.
137. L. Dolan, P. Linstead, and K. Roberts, *J. Exp. Bot.*, 1997, **48**, 713.
138. P. Marty, R. Goldberg, M. Liberman, B. Vian, Y. Bertheau, and B. Jouan, *Plant Physiol. Biochem.*, 1995, **33**, 409.
139. Y. Q. Li, F. Chen, H. F. Linskens, and M. Cresti, *Sex Plant Reprod.*, 1994, **7**, 145.
140. E. M. Shea, D. M. Gibeaut, and N. C. Carpita, *Planta*, 1989, **179**, 293.
141. F. Liners, J.-F. Thibault, and P. Van Cutsem, *Plant Physiol.*, 1992, **99**, 1099.
142. Q. Liu and A. M. Berry, *Protoplasma*, 1991, **163**, 93.
143. F. Bonfante-Fasolo, B. Vian, S. Perotto, A. Faccio, and J. P. Knox, *Planta*, 1990, **180**, 537.
144. F. M. Unger, *Adv. Carbohydr. Chem. Biochem.*, 1981, **38**, 323.
145. C. R. H. Raetz, *Annu. Rev. Biochem.*, 1990, **59**, 129.
146. C. Whitfield and M. A. Valvano, *Adv. Microb. Physiol.*, 1993, **35**, 135.
147. E. C. Webb (ed.), "Enzyme Nomenclature," Recommendations (1992) of the Nomenclature Committee of the International Union of Biochemistry and Molecular Biology, San Diego, Academic Press, 1992.
148. D. S. Feingold, E. F. Neufeld, and W. Z. Hassid, *J. Biol. Chem.*, 1960, **235**, 910.
149. R. E. Campbell, R. F. Sala, I. Van De Rijn, and M. E. Tanner, *J. Biol. Chem.*, 1997, **272**, 3416.
150. P. A. Gainey, T. C. Pestell, and C. F. Phelps, *Biochem. J.*, 1972, **129**, 821.
151. J. S. Franzen, P. Marchetti, R. Ishman, J. Ashcom, and D. S. Feingold, *Biochem. J.*, 1978, **173**, 701.
152. R. Jaenicke, R. Rudolph, and D. S. Feingold, *Biochemistry*, 1986, **25**, 7283.
153. J. Hempel, J. Perozich, H. Romovacek, A. Hinich, I. Kuo, and D. S. Feingold, *Protein Sci.*, 1994, **3**, 1074.
154. J. Zalitis and D. S. Feingold, *Arch. Biochem. Biophys.*, 1969, **132**, 457.
155. J. L. Strominger and L. W. Mapson, *Biochem. J.*, 1957, **66**, 567.
156. M. D. Davies and D. B. Dickinson, *Arch. Biochem. Biophys.*, 1972, **152**, 53.
157. D. Robertson, C. Smith, and G. P. Bolwell, *Biochem. J.*, 1996, **313**, 311.
158. R. Tenhaken and O. Thulke, *Plant Physiol.*, 1996, **112**, 1127.
159. D. C. Stewart and L. Copeland, *Plant Physiol.*, 1998, **116**, 349.
160. H. Ankel and R. F. Tischer, *Biochim. Biophys. Acta*, 1969, **178**, 415.
161. E. F. Neufeld, D. S. Feingold, and W. Z. Hassid, *J. Am. Chem. Soc.*, 1958, **80**, 4430.
162. E. J. Mitcham, K. C. Gross, and B. P. Wasserman, *Phytochem. Anal.*, 1991, **2**, 112.
163. K. Liljebjelki, R. Adolphson, K. Baker, R. L. Doong, and D. Mohnen, *Anal. Biochem.*, 1995, **225**, 296.
164. B. T. Eidson, J. Chan, A. S. Vandersall, L. E. Elvebak II, J. J. Smith, R. L. Doong, and D. Mohnen, *Plant Physiol.*, 1996, **111S**, 101.
165. M. A. Gaunt, U. S. Maitra, and H. Ankel, *J. Biol. Chem.*, 1974, **249**, 2366.
166. T. Hayashi, T. Koyama, and K. Matsuda, *Plant Physiol.*, 1988, **87**, 341.
167. D. J. Hannapel, *Am. Potato J.*, 1991, **68**, 179.
168. K. V. John, J. S. Schutzbach, and H. Ankel, *J. Biol. Chem.*, 1977, **252**, 8013.
169. D.-F. Fan and D. S. Feingold, *Plant Physiol.*, 1970, **46**, 592.
170. D. Robertson, B. A. McCormack, and G. P. Bolwell, *Biochem. J.*, 1995, **306**, 745.
171. D. Baron, U. Streitberger, and H. Grisebach, *Biochim. Biophys. Acta*, 1973, **293**, 526.
172. U. Matern and H. Grisebach, *Eur. J. Biochem.*, 1977, **74**, 303.
173. P. K. Kindel and R. R. Watson, *Biochem. J.*, 1973, **133**, 227.
174. R. R. Watson and N. S. Orenstein, *Adv. Carbohydr. Chem. Biochem.*, 1975, **31**, 135.
175. P. K. Kindel, D. L. Gustine, and R. R. Watson, *Fed. Proc. Amer. Soc. Exp. Biol.*, 1971, **30**, 1117.
176. G. A. Barber, *Arch. Biochem. Biophys.*, 1963, **103**, 276.
177. G. A. Barber, *Biochem. Biophys. Res. Commun.*, 1962, **8**, 204.
178. J. Kamsteeg, J. Van Brederode, and G. Van Nigtevecht, *FEBS Lett.*, 1978, **91**, 281.
179. G. A. Barber and M. T. Y. Chang, *Arch. Biochem. Biophys.*, 1967, **118**, 659.
180. G. Stevenson, B. Neal, D. Liu, M. Hobbs, N. H. Packer, M. Batley, J. W. Redmond, L. Lindquist, and P. Reeves, *J. Bacteriol.*, 1994, **176**, 4144.
181. D.-F. Fan and D. S. Feingold, *Plant Physiol.*, 1969, **44**, 599.
182. E. S. Maxwell, *J. Biol. Chem.*, 1957, **229**, 139.
183. U. S. Maitra and H. Ankel, *Proc. Natl. Acad. Sci. USA*, 1971, **68**, 2660.

184. A. J. Bauer, I. Rayment, P. A. Frey, and H. M. Holden, *Proteins*, 1992, **12**, 372.
185. D. B. Wilson and D. S. Hogness, *J. Biol. Chem.*, 1969, **244**, 2132.
186. C. R. Geren and K. E. Ebner, *J. Biol. Chem.*, 1977, **252**, 2082.
187. B. Königs and E. Heinz, *Planta*, 1974, **118**, 159.
188. P. Dormann and C. Benning, *Arch. Biochem. Biophys.*, 1996, **327**, 27.
189. P. Dörman and C. Benning, *Plant J.*, 1998, **13**, 641.
190. T. H. Liao and G. A. Barber, *Biochim. Biophys. Acta*, 1971, **230**, 64.
191. C. P. Bonin, I. Potter, G. F. Vanzin, and W.-D. Reiter, *Proc. Natl. Acad. Sci. USA*, 1997, **94**, 2085.
192. C. Rosenow, I. S. Roberts, and K. Jann, *FEMS Microbiol. Lett.*, 1995, **125**, 159.
193. T. Baasov and A. Kohen, *J. Am. Chem. Soc.*, 1995, **117**, 6165.
194. S. Jelakovic, K. Jann, and G. E. Schulz, *FEBS Lett.*, 1996, **391**, 157.
195. C. Pazzani, C. Rosenow, G. J. Boulnois, D. Bronner, K. Jann, and I. S. Roberts, *J. Bacteriol.*, 1993, **175**, 5978.
196. H. Mohr and P. Schopfer, "Plant Physiology," Springer-Verlag, Berlin, 1995, p. 1.
197. R. L. Doong, S. Ahmad, and R. A. Jensen, *Plant Cell Environ.*, 1991, **14**, 113.
198. R. L. Doong, J. E. Gander, R. J. Ganson, and R. A. Jensen, *Physiol. Plant*, 1992, **84**, 351.
199. A. E. Kearns, B. M. Vertel, and N. B. Schwartz, *J. Biol. Chem.*, 1993, **268**, 11 097.
200. B. M. Vertel, L. M. Walters, N. Flay, A. E. Kearns, and N. B. Schwartz, *J. Biol. Chem.*, 1993, **268**, 11 105.
201. P. Munoz, L. Norambuena, and A. Orellana, *Plant Physiol.*, 1996, **112**, 1585.
202. C. B. Hirschberg and M. D. Snider, *Annu. Rev. Biochem.*, 1987, **56**, 63.
203. T. Mitsui, M. Honma, T. Kondo, N. Hashimoto, S. Kimura, and I. Igaue, *Plant Physiol.*, 1994, **106**, 119.
204. A. Orellana, G. Neckelmann, and L. Norambuena, *Plant Physiol.*, 1997, **114**, 99.
205. C. Smythe, F. B. Caudwell, M. Ferguson, and P. Cohen, *EMBO J.*, 1988, **7**, 2681.
206. J. Lomako, W. M. Lomako, and W. J. Whelan, *FASEB J.*, 1990, **4**, A711.
207. J. Lomako, W. M. Lomako, and W. J. Whelan, *FEBS Lett.*, 1990, **264**, 13.
208. R. E. Campbell, C. T. Brett, and J. R. Hillman, *Biochem. J.*, 1988, **253**, 795.
209. T. Hayashi and G. Maclachlan, *Phytochemistry*, 1984, **23**, 487.
210. R. J. Henry and B. A. Stone, *Carbohydr. Polym.*, 1985, **5**, 1.
211. A. Herscovics and P. Orlean, *FASEB J.*, 1993, **7**, 540.
212. N. Kojima, Y. Tachida, Y. Yoshida, and S. Tsuji, *J. Biol. Chem.*, 1996, **271**, 19 457.
213. C. J. Belunis and C. R. H. Raetz, *J. Biol. Chem.*, 1992, **267**, 9988.
214. C. J. Belunis, K. E. Mdluli, C. R. H. Raetz, and F. E. Nano, *J. Biol. Chem.*, 1992, **267**, 18 702.
215. A. E. DeRocher and K. Keegstra, *Plant Physiol.*, 1994, **105S**, 59.
216. I. M. Saxena, R. M. Brown, Jr., M. Fevre, R. A. Geremia, and B. Henrissat, *J. Bacteriol.*, 1995, **177**, 1419.
217. B. W. Weston, P. L. Smith, R. J. Kelly, and J. B. Lowe, *J. Biol. Chem.*, 1992, **267**, 24 575.
218. B. W. Weston, R. P. Nair, R. D. Larsen, and J. B. Lowe, *J. Biol. Chem.*, 1992, **267**, 4152.
219. S. Natsuka, K. M. Gersten, K. Zenita, R. Kannagi, and J. B. Lowe, *J. Biol. Chem.*, 1994, **269**, 16 789.
220. S. Natsuka and J. B. Lowe, *Curr. Opin. Struct. Biol.*, 1994, **4**, 683.
221. J. R. Pear, Y. Kawagoe, W. E. Schreckengost, D. P. Delmer, and D. M. Stalker, *Proc. Natl. Acad. Sci. USA*, 1996, **93**, 12 637.
222. S. Cutler and C. Somerville, *Curr. Biol.*, 1997, **7**, 108.
223. M. M. Palcic, *Methods Enzymol.*, 1994, **230**, 300.
224. H. Kauss, A. L. Swanson, R. Arnold, and W. Odzuck, *Biochim. Biophys. Acta*, 1969, **192**, 55.
225. J. M. McNab, C. L. Villemez, and P. Albersheim, *Biochem. J.*, 1968, **106**, 355.
226. F. Goubet and C. Morvan, *Plant Cell Physiol.*, 1993, **34**, 1297.
227. W. Odzuck and H. Kauss, *Phytochemistry*, 1972, **11**, 2489.
228. M. W. Rodgers and G. P. Bolwell, *Biochem. J.*, 1992, **288**, 817.
229. A. Camirand, D. Brummell, and G. Maclachlan, *Plant Physiol.*, 1987, **84**, 753.
230. R. Hanna, D. A. Brummell, A. Camirand, A. Hensel, E. F. Russell, and G. A. Maclachlan, *Arch. Biochem. Biophys.*, 1991, **290**, 7.
231. A. Faïk, C. Chileshe, J. Sterling, and G. Maclachlan, *Plant Physiol.*, 1997, **114**, 245.
232. A. Schaumann, M.-P. Bruyant-Vannier, F. Goubet, and C. Morvan, *Plant Cell Physiol.*, 1993, **34**, 891.
233. H. Kauss, A. L. Swanson, and W. Z. Hassid, *Biochem. Biophys. Res. Commun.*, 1967, **26**, 234.
234. T.-Y. Lin, A. D. Elbein, and J. C. Su, *Biochem. Biophys. Res. Commun.*, 1966, **22**, 650.
235. C. L. Villemez, T.-Y. Lin, and W. Z. Hassid, *Proc. Natl. Acad. Sci. USA*, 1965, **54**, 1626.
236. B. L. Ridley, M. D. Spiro, J. Glushka, P. Albersheim, A. Darvill, and D. Mohnen, *Anal. Biochem.*, 1997, **249**, 10.
237. H. Kauss and A. L. Swanson, *Z. Naturforsch.*, 1969, **24**, 28.
238. R. L. Doong, J. J. Smith, and D. Mohnen, *Plant Physiol.*, 1996, **111S**, 101.
239. C. Penel and H. Greppin, *Plant Physiol. Biochem.*, 1996, **34**, 479.
240. F. Brugliera, T. A. Holton, T. W. Stevenson, E. Farcy, C.-Y. Lu, and E. C. Cornish, *Plant J.*, 1994, **5**, 81.
241. M. Bar-Peled, E. Lewinsohn, R. Fluhr, and J. Gressel, *J. Biol. Chem.*, 1991, **266**, 20 953.
242. M. Bar-Peled, R. Fluhr, and J. Gressel, *Plant Physiol.*, 1993, **103**, 1377.
243. N. Panayotatos and C. L. Villemez, *Biochem. J.*, 1973, **133**, 263.
244. M. Edwards, P. V. Bulpin, I. C. M. Dea, and J. S. G. Reid, *Planta*, 1989, **178**, 41.
245. J. S. G. Reid, M. Edwards, M. J. Gidley, and A. H. Clark, *Planta*, 1995, **195**, 489.
246. F. Goubet, Ph.D. Dissertation, Université de Rouen, 1994, p. 1.
247. G. P. Bolwell and D. H. Northcote, *Planta*, 1984, **162**, 139.
248. G. P. Bolwell and D. H. Northcote, *Biochem. J.*, 1983, **210**, 509.
249. P. H. Rubery and D. H. Northcote, *Biochim. Biophys. Acta*, 1970, **222**, 95.
250. D. W. James, Jr. and R. L. Jones, *Plant Physiol.*, 1979, **64**, 909.
251. D. W. James, Jr. and R. L. Jones, *Plant Physiol.*, 1979, **64**, 914.
252. G. Maclachlan, B. Levy, and V. Farkas, *Arch. Biochem. Biophys.*, 1992, **294**, 200.
253. C. Zurzolo and E. Rodriguez-Boulan, *Science*, 1993, **260**, 550.
254. H. Misawa, Y. Tsumuraya, Y. Kaneko, and Y. Hashimoto, *Plant Physiol.*, 1996, **110**, 665.

255. J. M. Longland, S. C. Fry, and A. J. Trewavas, *Plant Physiol.*, 1989, **90**, 972.
256. Y.-T. Pan and P. K. Kindel, *Arch. Biochem. Biophys.*, 1977, **183**, 131.
257. L. J. Mascaro, Jr. and P. K. Kindel, *Arch. Biochem. Biophys.*, 1977, **183**, 139.
258. T. Hayashi and K. Matsuda, *J. Biol. Chem.*, 1981, **256**, 11 117.
259. P. M. Ray, *Biochim. Biophys. Acta*, 1980, **629**, 431.
260. A. R. White, Y. Xin, and V. Pezeshk, *Biochem. J.*, 1993, **294**, 231.
261. A. R. White, Y. Xin, and V. Pezeshk, *Physiol. Plant*, 1993, **87**, 31.
262. M. C. Hobbs, M. H. P. Delarge, E. A.-H. Baydoun, and C. T. Brett, *Biochem. J.*, 1991, **277**, 653.
263. M. C. Hobbs, E. A.-H. Baydoun, M. H. P. Delarge, and C. T. Brett, *Biochem. Soc. Trans.*, 1991, **19**, 245S.
264. E. A.-H. Baydoun, K. W. Waldron, and C. T. Brett, *Biochem. J.*, 1989, **257**, 853.
265. E. A.-H. Baydoun and C. T. Brett, *J. Exp. Bot.*, 1997, **48**, 1209.
266. K. W. Waldron, E. A. H. Baydoun, and C. T. Brett, *Biochem. J.*, 1989, **264**, 643.
267. E. A.-H. Baydoun, M. C. Hobbs, M. H. P. Delarge, M. J. Farmer, K. W. Waldron, and C. T. Brett, *Biochem. Soc. Trans.*, 1991, **19**, 250S.
268. T. Clementz and C. R. H. Raetz, *J. Biol. Chem.*, 1991, **266**, 9687.
269. M. C. Jarvis, W. Forsyth, and H. J. Duncan, *Plant Physiol.*, 1988, **88**, 309.
270. V. Vreeland, S. R. Morse, R. H. Robichaux, K. L. Miller, S.-S. T. Hua, and W. M. Laetsch, *Planta*, 1989, **177**, 435.
271. J. Gaffe, C. Morvan, A. Jauneau, and M. Demarty, *Phytochemistry*, 1992, **31**, 761.
272. T. Bourlard, A. Schaumann-Gaudinet, M.-P. Bruyant-Vannier, and C. Morvan, *Plant Cell Physiol.*, 1997, **38**, 259.
273. T. Bourlard, P. Pellerin, and C. Morvan, *Plant Physiol. Biochem.*, 1997, **35**, 623.
274. T. Bourlard, M. P. Bruyant-Vannier, A. Gaudinet-Schaumann, B. Thoiron, and C. Morvan, in "CW95 7th Cell Wall Meeting—Abstracts and Programme," eds. I. Zarra and G. Revilla, Universidad de Santiago de Compostela, Santiago, Spain, 1995, p. 181.
275. A. Darvill, M. McNeil, and P. Albersheim, *Plant Physiol.*, 1978, **62**, 418.
276. J. An, M. A. O'Neill, P. Albersheim, and A. G. Darvill, *Carbohydr. Res.*, 1994, **252**, 235.
277. N. M. Iraki, R. A. Bressan, P. M. Hasegawa, and N. C. Carpita, *Plant Physiol.*, 1989, **91**, 39.
278. N. M. Iraki, N. Singh, R. A. Bressan, and N. C. Carpita, *Plant Physiol.*, 1989, **91**, 48.
279. N. M. Iraki, R. A. Bressan, and N. C. Carpita, *Plant Physiol.*, 1989, **91**, 54.
280. A. Kikuchi, Y. Edashige, T. Ishii, T. Fujii, and S. Satoh, *Planta*, 1996, **198**, 634.
281. P. M. Ray and D. B. Baker, *Nature*, 1962, **195**, 1322.
282. D. A. Brummell and J. L. Hall, *Physiol. Plant*, 1985, **63**, 406.
283. U. Kutschera and W. R. Briggs, *Proc. Natl. Acad. Sci. USA*, 1987, **84**, 2747.
284. A. A. Abdul-Baki and P. M. Ray, *Plant Physiol.*, 1971, **47**, 537.
285. D. B. Baker and P. M. Ray, *Plant Physiol.*, 1965, **40**, 345.
286. P. M. Ray and D. B. Baker, *Plant Physiol.*, 1965, **40**, 353.
287. S.-i. Amino, Y. Takeuchi, and A. Komamine, *Physiol. Plant*, 1985, **64**, 111.
288. D. Robertson, I. Beech, and G. P. Bolwell, *Phytochemistry*, 1995, **39**, 21.
289. J. S. G. Reid, M. E. Edwards, M. J. Gidley, and A. H. Clark, *Biochem. Soc. Trans.*, 1992, **20**, 23.
290. J. S. G. Reid and H. Meier, *Z. Pflanzenphysiol.*, 1970, **62**, 89.
291. J. M. Campbell and J. S. G. Reid, *Planta*, 1982, **155**, 105.
292. M. Edwards, C. Scott, M. J. Gidley, and J. S. G. Reid, *Planta*, 1992, **187**, 67.
293. R. Gordon and G. Maclachlan, *Plant Physiol.*, 1989, **91**, 373.
294. W.-D. Reiter, C. Chapple, and C. R. Somerville, *Plant J.*, 1997, **12**, 335.
295. B. Wells, M. C. McCann, E. Shedletzky, D. Delmer, and K. Roberts, *J. Microsc.*, 1994, **173**, 155.
296. E. Shedletzky, M. Shmuel, D. P. Delmer, and D. T. A. Lamport, *Plant Physiol.*, 1990, **94**, 980.
297. E. Shedletzky, M. Shmuel, T. Trainin, S. Kalman, and D. Delmer, *Plant Physiol.*, 1992, **100**, 120.
298. K. C. Vaughn, J. C. Hoffman, M. G. Hahn, and L. A. Staehelin, *Protoplasma*, 1996, **194**, 117.
299. K. C. Vaughn and N. A. Durso, *Suppl. Plant Physiol.*, 1997, **114**, 87.

3.16
Celluloses

RAJAI H. ATALLA

USDA Forest Service and University of Wisconsin, Madison, WI, USA

3.16.1 INTRODUCTION

Cellulose is, in many respects, among the most challenging of the polysaccharides. Although it is a dominant component in the vast majority of plant forms and has a number of vital biological functions, and although cellulose based materials have been part of daily life for many millennia, our understanding of its nature remains incomplete. This circumstance is not the consequence of lack of interest in investigations of cellulose, for the effort to develop a scientific characterization of cellulosic matter is well over 150 years old. The constraints on our understanding of cellulose and its phenomenology are rather rooted in the complexity of its behavior and the inadequacy of our conceptual frameworks and methodologies for the characterization of so unusual a substance; perhaps more than any other common chemical species, its chemistry and its entry into biological processes are as much a function of its state of aggregation as of its primary structure.

In the modern era, the beginning of the search for scientific understanding can be associated with the work of the French agriculturalist Anselm Payen, who introduced the term cellulose in 1842 to denote the neutral structural polysaccharides of plant tissue; he viewed the isolated matter, which had an elemental composition similar to that of the saccharides, as an aggregated form of "dextrose." By the turn of the century, when methods for discrimination between the monosaccharides were established, cellulose was recognized to be an aggregated, oligomeric form of anhydroglucose that remained the subject of controversy. Within the context of the polymer hypothesis, developed in the early decades of the twentieth century, cellulose has been recognized as the linear β-(1-4)-linked homopolymer of anhydroglucose (**1**). More recently, it has been noted that it is more accurate to define it as the homopolymer of anhydrocellobiose, since such definition explicitly incorporates the identity of the linkage. Given this definition, it might be anticipated that the nature and the chemical behavior of cellulose could be understood in terms of the chemistry of its monomeric constituents together with considerations arising from its polymeric nature. The reality, however, is that the chemical or physical transformations of particular samples of cellulose are as much reflections of their history and their states of aggregation as they are of the chemistry of their monomeric constituents. Indeed, this is the basis for the title *Celluloses* given to this chapter in reflection of the great diversity of the forms of cellulose. The unusual nature of cellulose in its aggregated states is perhaps best represented by its relationship to water; a primary structure with three hydroxyl groups per pyranose ring would, in the normal course of events, be expected to be quite soluble in water and aqueous media. But it is in fact essentially insoluble, and this characteristic is but one of the many unusual patterns of behavior of the aggregated states of cellulose.

(1)

In this chapter, the author is concerned with delineating the frontiers of our understanding of cellulose, particularly with respect to its native forms and the biological processes into which they enter. While the presentation is relevant to the industrial utilization of cellulose, because it addresses the nature of the native forms of many of the feedstocks used as well as the effects of processes of isolation on structure, particular attention will be given to questions confronting the research community undertaking studies of cellulose in biological contexts, including biogenesis, biological function, and biodegradation.

Progress in the characterization of cellulose has been intimately intertwined with advances in methods of chemical analysis and the development of the molecular hypothesis, with the accumulation of additional observations on celluloses from an ever-expanding variety of sources, and with the development of new conceptual frameworks for interpreting the behavior of macromolecules. The evolution in understanding the nature of cellulose has continued during recent decades and will likely be ongoing for some time to come. One objective of this chapter is to capture the central features of this evolution in order to place recent studies of cellulose in a historical perspective and to suggest directions for future work.

One of the major challenges in any experimental program involving the use of cellulose as a substrate is the development of an adequate characterization of the tertiary structure of cellulosic samples. Careful specification of the history of particular samples of cellulose or precise definition of their states of aggregation are usually essential to reproducibility of experimental observations. Traditionally, this reality has been dealt with in experimental programs by reliance on the use of readily available types of pure celluloses such as cotton, or filter paper made from cotton linters, or, frequently in later years, microcrystalline celluloses prepared by acid hydrolysis of different cellulosic substrates. While this provides a point of reference for other investigators who are likely to have access to similar celluloses, it begs the question of an adequate definition of the state of aggregation of the cellulose; the morphology of these celluloses is complex and quite variable at supramolecular levels in ways that are not easily detected or characterized. Another important objective of this chapter is therefore to address the question of aggregation and tertiary structure and the methodologies for their characterization. In addition to seeking to clarify issues of structure in the context of experimental studies of cellulosic substrates, the methodologies will also be discussed in relation to characterization of the wide variety of native forms of cellulose that are as diverse as the species of organisms that produce them.

We begin with a historical perspective, with some reflection on the different stages in the progress towards better understanding of the nature of cellulose. This is not intended to provide a duplication of some of the very comprehensive discussions of the history of ideas concerning the identity and structure of cellulose. Rather it is intended to illustrate how often the inadequacy of the paradigms available from the chemistry of a particular period has resulted in constraints that have limited progress and often resulted in major detours along the path to clearer understanding of the phenomenology. The discussion is intended to alert us to the reality that the unusual character of cellulose requires innovation in concept as well as in methodology, and to establish a basis for analyzing the conceptual barriers that remain to be dealt with in the next cycle of studies of the nature of celluloses.

The discussion of structure will focus on the results of investigations carried out during the past three decades, though some of the important contributions from earlier periods will not be neglected. Among the important developments has been the introduction of spectroscopic methods to complement information derived from traditional crystallographic methods for characterization of the aggregated states of cellulose. In this arena, the use of solid-state ^{13}C NMR spectroscopy and Raman spectroscopy have resulted in significant new insights concerning differences in the states of aggregation of different celluloses. The application of lattice imaging methods in electron microscopic studies of the algal celluloses has also been an important development during the last decade. Nevertheless, a number of questions concerning the structure of cellulose remain outstanding.

The challenge of defining structure emerges most clearly in an examination of the wide range of native celluloses that occur in nature. They occur in the cell walls of the vast majority of plant forms and are highly ordered structures that are elaborately integrated into biological structural tissues. They are also produced by a number of other organisms, including bacteria and some classes of marine life. Yet the types of order that prevail do not lend themselves to description in terms of the traditional concepts that have been developed for the description of organization in the solid states of inorganic or inanimate organic substances. Although cellulose is often regarded as crystalline, review of observations of its many different forms will reveal that the classical definitions of crystalline order for both molecular and polymeric materials are not well suited for the characterization of cellulose, although they may provide a basis for some useful measures of its organization.

The biogenesis of cellulose has also presented a major challenge to investigators. Early studies focused on the biosynthesis of bacterial celluloses as these systems lent themselves more readily to examination by the available methodologies. A significant number of electron microscopic examinations of the process was carried out. Considerable effort has also been invested in studies of biogenesis in selected higher plants, cotton being the primary one among these. The pathways for biosynthesis were established by administration of radiolabeled precursors and following their entry into the structure of cellulose. A significant effort has been invested in pursuing genetic encoding of the biosynthetic enzymes. But much remains unknown in this arena as well.

The diversity of celluloses and the difficulty in characterizing their states of aggregation have been complicating factors in studies directed at understanding the action of different agents on cellulosic substrates whether the agents be chemical or biological. Because of the heterogeneity of the substrates, issues of accessibility arise, and these, in turn, are intimately related to states of aggregation. The literature on biodegradation of cellulose and on the action of cellulolytic enzymes has many examples of the effects of sample history and structure on its response. The literature on chemical modification in many ways parallels that on the action of cellulolytic enzymes. Indeed, it goes further in the sense that differences in responses to the action of chemical modifying reagents have been used to characterize the differences between states of aggregation of cellulosic samples with different histories.

A wide range of methodologies of modern polymer science rely on the measurement of properties of the polymers in solution. The considerable difficulty in solubilizing cellulose has been a major hindrance to developing adequate methodologies. Almost all of the solvent systems for cellulose are multicomponent solvents, many of them acting by forming metastable derivatives. The majority of them are aqueous solutions of reagents that enter into strong associative interactions with cellulose.

Some are used in some of the established analytical procedures for characterizing celluloses. Others are used in industrial processes for regeneration of celluloses to form a variety of products. A number of organic solvent-based systems have also been developed. Some are used in analytical procedures and as media for some chemical derivatization reactions. One is also used for the regeneration of cellulose in fiber form. It is also of interest that a number of solutions, some aqueous and others nonaqueous, are capable of swelling cellulose without resulting in dissolution. They are of particular interest because many of them can penetrate the highly ordered semicrystalline domains sufficiently to facilitate a solid-state transformation that can result in a different type of lattice order in the semicrystalline domains. However, the solvent systems and their behavior are beyond the scope of this chapter.

The utilization of cellulosic materials is as old as civilization and predates by millennia any effort to understand its nature at a constitutional level. A number of uses established in past millennia remain at the heart of major industries. Those most commonly cited are those components of the textile industry that are based on cotton, and the pulp and paper industry. Of course the utilization of wood remains central also in the daily life of most cultures, and cellulose is the key structural component in wood. Here again, the technology is beyond the scope of a treatise on natural products.

In the years ahead it is likely that advances in methodologies for the characterization of biological order will be important. New instrumental methods will no doubt be developed allowing more detailed characterization of the diversity of celluloses. Advances in molecular biology related to both biogenesis and biodegradation are also likely to play key roles in advancing understanding of the nature of celluloses and their many entries into the biosphere. The author looks ahead and attempts to anticipate the pre-occupations of cellulose science that are likely to be dominant in the decades ahead.

3.16.2   HISTORICAL PERSPECTIVE

Progress in understanding the nature of cellulose can be viewed as occurring in three distinct though overlapping stages. The first can be defined as the period between the early work of Payen, recognizing the carbohydrate-like character of the key constituents of plant cell walls, and establishment of the polymer hypothesis in the second and third decades of the twentieth century. Efforts during this period were focused on establishing the primary structure of cellulose, that is, the pattern of covalent bonds in the cellulose molecule. Though modern day cellulose scientists take for granted knowledge of this structure, the limitations of both methodologies and conceptual frameworks resulted in extension of this period from the time of Payen to the 1920s, when the polymer hypothesis received confirmation from a number of sources, and methods for distinguishing the different monosaccharides were well developed.

The second period can be regarded as the interval between the general acceptance of the polymer hypothesis, in the mid-1920s, and the 1970s when a number of new methodologies for characterizing cellulose were introduced and provided a basis for rationalizing the great diversity of native celluloses. During this second period the general thrust of most structural studies was towards establishing the secondary and tertiary structures of cellulose and was based, in large measure, on application of the methods of crystallography. Much of the effort was implicitly based on the premise that a single crystal structure, identified as cellulose I, was basic to all native celluloses. However, there was very little consensus as to the fine points of this structure. A second crystallographic structure, identified as cellulose II, was regarded as the basic constituent of both regenerated and mercerized celluloses; this latter structure did not receive as much attention as that of cellulose I, in part because it was of limited significance to biological questions and in part because specimens of cellulose II of relatively high crystallinity could not be prepared.

The second period also witnessed the broad application of the principles of polymer science to the study of cellulose and its derivatives. Many aspects of the phenomenology of cellulose were interpreted by analogy with the phenomenology of its derivatives and that of some of the relatively more simple synthetic polymers that had become of considerable interest, both commercially and scientifically, early in this period.

In addition to advances in understanding the nature of cellulose as a substance, the second period saw the beginning of serious inquiries and many important contributions to new understanding of the role of cellulose in the biological systems within which it occurs. There were many investigations of the nature of its organization at the microscopic level within the cell walls of many plant forms. Important studies of the processes of biosynthesis and biodegradation of cellulose were also undertaken. These will not be included in this historical perspective which is primarily concerned with issues of structure and organization in cellulose in its many aggregated states, and with the conceptual frameworks available for their description.

The third and current period in studies of the structure of native celluloses can be viewed as beginning with the discovery that all plant celluloses are composites of two closely related crystalline lattice forms, I_α and I_β, that occur in distinctive proportions characteristic of the species producing the particular cellulose under observation. The uniqueness of the blend to each particular native form provides a basis for resolving some of the earlier conflicts in interpretations of crystallographic data which had been acquired from celluloses derived from different organisms. In addition to the advances in understanding issues of structure, the current period has witnessed progress in understanding both the biosynthesis and biodegradation of cellulose, as well as in characterizing some of the most interesting of cellulose derivatives. This third stage in investigations of the many aspects of the nature of cellulose and its entry into biological and industrial processes is likely to continue for many decades.

In parallel with the progress in understanding the nature of cellulose, there were, of course, many advances in the industrial technologies for the utilization of cellulose, the majority of which predate knowledge of its molecular structure. The two primary industries based on the utilization of cellulose in its native forms are those concerned with the production of textiles from different plant fibers, and with the production of paper from cellulosic fibers isolated primarily by pulping of woody species. In addition, the latter half of the nineteenth century and the first half of the twentieth century witnessed significant industrial utilization of cellulose derivatives in the manufacture of a wide array of products; although a number of the products fabricated from cellulose derivatives have been displaced by ones based on synthetic polymers, many cellulose derivatives remain important products with a wide range of applications. These three major industrial sectors have provided, over the years, considerable motivation for many of the research efforts that have resulted in advancing the understanding of cellulose. The technologies based on the utilization of cellulose and their histories are covered by broad categories of literature dedicated to each of them.

3.16.2.1 The Early Period (1840–1920)

An excellent account of the development of the early ideas concerning the nature of cellulose is presented by Purves.[1] It is very well complimented by the historical introduction in the treatise by Hermans,[2] and, to a more limited degree, at least with respect to cellulose, by the historical introduction in the classical treatise on polymer science by Flory.[3] The analysis presented here is based on these sources as well as some of the early texts on cellulose chemistry.[4,5]

Payen's seminal contribution arose from recognition that plant tissues from a number of different sources, when adequately "purified," resulted in substances that, when submitted to elemental analyses, appeared to have proportions of carbon, oxygen, and hydrogen that were similar to those of the sugars and starch. This led him to the view that another polysaccharide related to starch was one of the key constituents of plant cell walls. He was the first to argue that the different elemental compositions found in other plant tissues, particularly lignified ones, were the consequence of inadequate removal of other substances that encrusted the primary cell wall constituent. His view was further reinforced by the observation that the purified cell wall substances, when subjected to acid hydrolysis, produced dextrorotatory reducing sugars not unlike those obtained upon hydrolysis of starch. He proposed that all plant cell walls contain a uniform chemical constituent made up of glucose residues; he viewed it as isomeric with starch. Since starch stained readily with iodine, while the cell wall residue gave a similar color only upon swelling with sulfuric acid, he regarded the cell wall substance as a more highly aggregated isomer and named it cellulose. Payen is likely to have based his conclusions on studies of cellulose from sources such as cotton or flax, wherein the cellulose is not heavily encrusted with constituents that would be more difficult to remove. Though he clearly was not aware of the possibility that other polysaccharides could co-exist with cellulose in the cell wall, Payen's fundamental insight concerning the existence of a primary structural component in all plant cell walls has proved to be an accurate one.

The next major advance in understanding the constitution of cell walls came in the decade between 1880 and 1890, when methods for discriminating between the monosaccharides were developed, and many celluloses that were comprehended within Payen's definition were shown to contain other sugars. Soon thereafter Schulze, who observed that the majority of the polysaccharides that contain residues other than glucose are more readily hydrolyzed, suggested that the more resistant component of the cell wall be identified as the cellulose. Thus, the approach adopted by Schulze followed that of Payen; both were influenced in this direction by the adoption of cotton as the standard for cellulose, a standard that continues in many laboratories to this day. The other polysaccharides in the cell walls of most plants were then categorized as the hemicelluloses. In due course, the noncarbohydrate components of the cell wall were recognized to be chemically distinct and to include lignin in addition to a variety of minor constituents that were extractable with organic solvents.

It should be noted that the advances made during this period occurred in an environment wherein a number of investigators in plant science resisted the categorizations of the constituents of plant cell walls outlined above and gave preference to the view that the lignocelluloses derived from different species represented different chemical individuals. This view was reinforced by the observations that cell wall preparations from different sources responded differently in staining reactions that have long been used to distinguish between chemical constituents of biological tissues. This controversy is in many respects similar to others that have arisen over the years and that are rooted in the differences between patterns of aggregation of cell wall constituents at the supramolecular level. Resolution of the issue of the distinctness of cellulose as a constituent was finally accomplished when X-ray diffractometery demonstrated that diffraction patterns derived from samples of wood and other plant tissues before and after delignification were essentially the same.

Other major influences during the early period were the ongoing developments in organic chemistry and the many investigations associated with the industrial utilization of cellulose and its derivatives. The situation is best described in an excerpt from Purves:[1]

> Concurrently, structural organic chemistry in general was being built up with brilliant success by arguments based on the analyses of substances which were chemical individuals. But cellulose and its derivatives were generally fibrous, insoluble materials which had no sharp melting or boiling points and which gave negative or indefinite results in molecular-weight determinations. Deprived of the usual criteria of homogeneity and purity, investigators in the cellulose field very frequently made the mistake of assuming these qualities in their preparations on insufficient evidence and accordingly attached molecular meanings to their analytical data.

In these circumstances it was not surprising that different investigators, using preparative procedures that may have differed in detail, would ascribe different formulas to what would have been similar chemical entities.

The other group of scientific activities that had an important influence on the interpretations of the phenomenology of cellulosic preparations were those associated with the development of colloid chemistry, which was one of the key areas of physical chemical investigation. The similarities between the behavior of cellulose and its solutions and solutions of its derivatives, on the one hand, and those of colloidal systems, on the other, led many to conclude that the cellulosics were also colloidal. Among the many similarities were observations of colligative properties including osmotic pressures, the abnormal viscosity behavior, and the manner in which quantitative data can depend on the previous history of a particular sample. This led to the proposal of many different structures for cellulose that were based on the premise that it is an aggregate of one or another of a number of oligomeric structures that were consistent with both the elemental analysis and the results of hydrolytic action by acids.

The matter of the primary structure of cellulose was brought to closure in the early part of the twentieth century when crystallographic studies suggested an extended linear molecule and the methodologies of carbohydrate chemistry had advanced to the point that permethylation analyses supported the view that the primary linkage between the anhydroglucose units is of the β-1,4 type. In the closure of this brief overview of the early period it is well to include the closing paragraph from the essay by Purves.[1] He noted:

> It may seem at first sight surprising that the association theory retained its grip over the intellects of so many workers for so long a period. But it must be remembered that their intuition, although wrong in one dimension, was certainly right in two. Since the time when the thread-like nature of the cellulose molecule was placed beyond question, many researchers have been concerned with the association and interplay of the secondary valence forces that radiate laterally from the cellulose chain.

As will be clear from the following discussions, the question of the lateral associations of the cellulose chains remains one of the key issues in studies of the secondary and tertiary structures of cellulose to this day.

3.16.2.2 The Middle Period (1920–1975)

The middle period in the effort to understand the nature of cellulose was focused on exploration of the lateral organization of the long chain molecules that were now almost universally accepted. It also included the beginning of important investigations of the role of cellulose in the biological contexts in which it occurs. The exploration of structure proceeded at two levels. The first was the solution of the crystallographic problem, the second was the effort to apply some of the ideas of colloid science to understanding the lateral organization of native celluloses at levels that were an order of magnitude greater than those addressed by the solution of the crystallographic problem. The work of this period is covered in a number of important reviews. Among the most comprehensive reviews of the crystallographic work are those by Jones[6] and by Tonessen and Ellefsen.[7] The matter of lateral organization at the fibrillar level has been discussed most extensively in the treatise by Hermans[2] and in the classics on cell wall structure by Preston[8] and Frey-Wyssling,[9] although all three have also presented extensive discussions of the crystallographic problem. Atalla[10] has also presented an overview of the same structural questions in light of the early developments of the applications of spectroscopic methods. The reader is referred to these sources for much more detailed discussions of the many issues addressed in the effort to reconcile the wide range of diverse observations during this period. The classic treatises by Preston[8] and Frey-Wyssling[9] also include extensive discussions of the biological dimensions of cellulose science. In the following, the author first focuses on the crystallographic problem. He then examines issues of lateral aggregation at the next higher level of organization.

3.16.2.2.1 Crystallographic studies

From the present perspective, it would appear that exploration of structure during the middle period was dominated by the newly developed methods of crystallography. Though it was rarely stated explicitly, analyses of the diffractometeric data implicitly sought definition of both the secondary and tertiary structures of the molecules. For clarity it is well to note here that the secondary structure represents definition of the relative organization of the repeat units in an individual chain, that is, the conformations of the chain, while the tertiary structure describes the arrangement of the molecules relative to each other in a particular state of aggregation. It is in the nature of crystallographic methods that definition of the structure of a unit cell in a crystal lattice of a polymeric material implies knowledge of the conformations of the individual molecules as well as their arrangement relative to each other. The distinction between these two levels of structure was of little note until spectroscopic methods became important.

The procedures for structural studies on cellulose have much in common with investigations of structure in polymers in general. In most instances diffractometric data are not sufficient for a solution of the structure in a manner analogous to that possible for lower molecular weight compounds which can be made to form single crystals. It becomes necessary, therefore, to complement diffractometric data with structural information derived from studies carried out on monomers or oligomers.

Kakudo and Kasai[11] have summarized the central problem well:

> There are generally less than 100 independently observable diffractions for all layer lines in the x-ray diagram of a fibrous polymer. This clearly imposes limitations on the precision which can be achieved in polymer structure analysis, especially in comparison with the 2000 or more diffractions observable for ordinary single crystals. However, the molecular chains of the high polymer usually possess some symmetry of their own, and it is often possible to devise a structural model of the molecular chain to interpret the fiber period in terms of the chemical composition by comparison with similar or homologous substances of known structure. Structural information from methods other than x-ray diffraction (e.g., infrared and NMR spectroscopy) are also sometimes helpful in devising a structural model of the molecular chain. The majority of the structural analyses which have so far been performed are based on models derived in this way. This is, of course, a trial and error method.

Similar perspectives have been presented by Arnott,[12] Atkins,[13] and Tadokoro.[14,15]

An acceptable fit to the diffractometric data is not the ultimate objective, however. Rather it is the development of a model that possesses a significant measure of validity as the basis for organization, explanation, and prediction of experimental observations. With respect to this criterion, the models of cellulose which were developed in the middle period leave much to be desired, for their capacity to integrate and unify the vast array of information concerning celluloses was limited indeed.

Quite early in the X-ray diffractometric studies of cellulose, it was recognized that its crystallinity is polymorphic. It was established that native cellulose, on the one hand, and both regenerated and mercerized celluloses, on the other, represent two distinct crystallographic allomorphs.[16] Little has transpired since the early studies to change these perceptions. There has been, however, little agreement regarding the structures of the two forms. For example, Petitpas *et al.*[17] have suggested, on the basis of extensive analyses of electron-density distributions from X-ray diffractometric measurements, that chain conformations are different in celluloses I and II. In contrast, Norman[18] has interpreted the results of his equally comprehensive X-ray diffractometric studies in terms of similar conformations for the two allomorphs.

At a more basic level than comparison of celluloses I and II, the stucture of the native form itself has remained in question. Among studies in the 1970s, for example, Gardner and Blackwell,[19] in their analysis of the structure of cellulose from *Valonia ventricosa*, assumed a lattice belonging to the $P2_1$ space group, with the twofold screw axis coincident with the molecular chain axis. Hebert and Muller,[20] on the other hand, in an electron diffractometric study of a number of celluloses including *Valonia*, confirmed the findings of earlier investigators who found no systematic absences of the odd order reflections forbidden by the selection rules of $P2_1$, and concluded that the cellulose unit cells do not belong to that space group.

Even when $P2_1$ is taken to be the appropriate space group, the question of chain polarity remains. As noted by Jones,[21] and by Howsmon and Sisson,[22] the structure initially proposed by Meyer and Mark[23] assumed that the chains were parallel in polarity. The structure later proposed by Meyer and Misch[24] was based on the reasoning that the rapidity of mercerization and its occurrence without dissolution required that the polarity of the chains be the same in both celluloses I and II. It was reasoned further that regeneration of cellulose from solution is most likely to result in precipitation in an antiparallel form, and that the similarity between X-ray diffraction patterns of mercerized and regenerated cellulose required that they have the same polarity. It was thus inferred that native cellulose must also have an antiparallel structure.

The premise that regeneration from solution in the antiparallel mode of crystallization is more probable than the parallel mode was shown to be false within a decade of its first presentation.[25] In spite of this finding, the antiparallel organization of molecules suggested by Meyer and Misch remained the point of departure for most subsequent investigators.

When the models incorporating antiparallel arrangement of the chains are extended to native cellulose, they pose serious questions concerning proposed mechanisms for the biosynthesis of cellulose. Plausible mechanisms for simultaneous synthesis and aggregation of antiparallel chains are more difficult to envision. It is perhaps for this reason that proposals of parallel structures for native cellulose developed during the 1970s were embraced by investigators of the mechanism of biosynthesis.

As the last cycle of crystallographic studies during the middle period remains the basis of much discussion of the structure of cellulose in the wider arena of cellulose science, it is well to review the issues arising in greater detail. This will provide a clearer basis for assessing the implications of the new information concerning structure that has been developed since that time.

As noted by Kakudo and Kasai,[11] the primary difficulty in structural studies on polymeric fibers is that the number of reflections usually observed in diffractometric studies are quite limited. In the case of cellulose it is generally difficult to obtain more than 50 reflections. Consequently, it becomes necessary to minimize the number of structural coordinates to be determined from the data by adopting plausible assumptions concerning the structure of the monomeric entity. The limited scattering data are then used to determine the orientation of the monomer units with respect to each other. In the majority of diffractometric studies of cellulose published so far, the monomeric entity has been chosen as the anhydroglucose unit. Thus, structural information from single crystals of glucose is implicitly incorporated in analyses of the structure of cellulose. The coordinates which are adjusted in search of a fit to the diffractometric data include those of the primary alcohol group at C-6, those of the glycosidic linkage, and those defining the positions of the chains relative to each other.

In addition to selection of the structure of the monomer as the basis for defining the internal coordinates of the repeat unit, the possible structures are usually further constrained by taking advantage of any symmetry possessed by the unit cell. The symmetry is derived from the systematic absence of reflections which are forbidden by the selection rules for a particular space group. In the case of cellulose, the simplification usually introduced is the application of the symmetry of space group $P2_1$, which includes a twofold screw axis parallel to the direction of the chains. The validity of this simplification remained the subject of controversy, however, because the reflections which are disallowed under the selection rules of the space group are in fact frequently observed. In most of the studies, these reflections, which are usually weak relative to the other main reflections, were assumed to be negligible. The controversy continued in part because the relative intensities can be influenced by experimental conditions such as the periods of exposure of the diffractometric plates. Furthermore, the disallowed reflections tend to be more intense in electron diffractometric measurements than in X-ray diffraction measurements. Thus, more often than not, investigators using electron diffraction challenged the validity of the assumption of twofold screw axis symmetry.

The key assumption with respect to symmetry, however, is not the existence of the twofold screw axis as an element of the symmetry of the unit cell, but rather the additional assumption that this axis coincides with the axis of the molecular chains of cellulose. This latter assumption has implicit in it a number of additional constraints on the possible structures which can be derived from the data. It requires that adjacent anhydroglucose units are related to each other by a rotation of 180 degrees about the axis, accompanied by a translation equivalent to half the length of the unit cell in that direction; it is implicit, therefore, that adjacent anhydroglucose units are symmetrically equivalent and, correspondingly, that alternating glycosidic linkages along the chain are symmetrically equivalent.

If the assumption concerning coincidence of the twofold screw axis and the molecular chain axis were relaxed, the diffractometric patterns would admit nonequivalence of alternate glycosidic linkages along the molecular chain, as well as the nonequivalence of adjacent anhydroglucose units. This possibility has been ignored, however, in large part because it requires expansion of the number of internal coordinates which have to be determined from the diffractometric data.

The assumptions concerning the symmetry of the unit cell noted above have been the basis of the last cycle of refinements of the structure of cellulose I. In one such refinement,[19] the forbidden reflections were simply assumed negligible, and the intensity data from *Valonia* cellulose were used to arrive at a final structure. In another study, the inadequate informational content of the diffractometric data was complemented with analyses of lattice packing energies;[26] the final structures were constrained to minimize the packing energy as well as optimizing the fit to the diffractometric data. Here the assumptions implicit in the weighting of the potential functions that are used in the energy calculations further complicate the interpretations; they can result in unacceptable hard sphere overlap of the two protons on the two carbons that anchor the glycosidic linkage. Furthermore, as has been noted by French,[27] the structures derived in these two studies, though both based on parallel chain arrangements, are nevertheless very different crystal structures. When the same convention is applied to defining the axes of the crystal lattice, the structure most favored in one analysis is strongly rejected in the other. Furthermore, neither of these is strongly favored over yet a third, antiparallel structure.[28]

It is fair to say that at the conclusion of the middle period, many of the complexities of the crystallographic problem had been explored but there was little consensus concerning the details of the structure, particularly for cellulose I. The publication during the 1970s of two crystallographic models with parallel chains was embraced by the community of research workers concerned with the biosynthesis of cellulose. However, there remained significant elements of the crystallographic data sets that could not be rationalized in terms of any of the models that were current.

In addition to the crystallographic studies focused on the structures of celluloses I and II, the application of X-ray diffractometry in investigations of cellulose revealed the occurrence of two other polymorphic forms of cellulose that are generally recognized, namely celluloses III and IV.[29] Cellulose III is most commonly prepared by treating cellulose with anhydrous ammonia, while cellulose IV is prepared by treating cellulose in glycerol at temperatures well above 200 °C. These received less attention than forms I and II, by and large because they were not encountered in any of the major commercial applications of cellulose.

3.16.2.2.2 *Lateral aggregation*

As noted earlier, in concluding his essay, Purves wrote that "the secondary valence forces that radiate laterally from the cellulose chain" had become the subject of investigation by many researchers. These secondary forces are now understood to be the hydrogen bonding associated with the hydroxyl groups and the van der Waals interactions involving the hydrophobic faces of the pyranose rings. Together with the distinctive conformational characteristics of the cellulose molecule itself, the secondary forces are responsible for the diversity of patterns that occur in the aggregated forms of cellulose, particularly in the native state. The diversity of these patterns had emerged from microscopic examination of the native forms of cellulose, first, at the resolution levels of optical microscopes and, later at levels made accessible by electron microscopes.

The application of electron microscopy, in particular, revealed that the aggregation of celluloses seemed to be subject to additional organizing influences at a scale that is an order of magnitude higher than that of the unit-cells of the crystallographic structure. It was recognized that this next higher level of organization, the supermolecular level, was the one at which the distinctiveness of the different native forms manifests itself. Aggregation at this level had become the subject of intensive study in two areas not unrelated to the study of cellulose; these were colloid science and the physical chemistry of macromolecules, both of which provided helpful concepts and methodologies that were welcomed by students of the nature of celluloses.

Studies of the organization of cellulose at the supermolecular level during the middle period have been the subject of a number of excellent comprehensive discussions to which the interested reader is referred. Those by Hermans[2] and by Ellefsen and Tonessen[30] focus on the physical, chemical and colloidal aspects of organization, while those by Frey-Wyssling[9] and Preston[8] address the issues from the perspective of plant biology.

A review of the research literature addressing questions of lateral organization in native cellulose suggests a continuing search for a suitable conceptual framework. Though it has not been explicitly stated, the challenge was to reconcile the fundamental premises of crystallographic models, that is, an infinite linear repetition of the unit cell in the direction of the chain axis, with the reality that at the level of 5–10 nm in lateral extension, most native celluloses were revealed to posses a fibrillar organization that is rarely linear. Rather, it most often had curvature in three dimensions usually reflecting the native morphology of the cell walls from which the cellulose was isolated. It was to bridge this gap that cellulose scientists turned their attention to the structural levels addressed within colloid science and the physical–chemical aspects of macromolecular science.

Within the context of colloid science, the concept of micellar organization had been developed to address the phenomenology of a number of classes of systems wherein substances were dispersed at a scale of 5–500 nm. Some of these systems were dispersions of particulate substances in liquid media, while others were dispersions of amphiphilic organic substances in aqueous media. They had in common the occurrence of large interfacial areas which were such that a significant fraction of the matter in the dispersed phase was at or near the interface. The term micelle was introduced to describe the individual units or domains of the dispersed phase. Since the dimensions of the micelles were usually dependent on the nature of the dispersed phase and on the microenvironment within a colloidal dispersion, the conceptual framework that was developed for the analysis of colloidal phenomena was thought to be applicable to characterization of the aggregation of cellulose.

With respect to the organization of native celluloses, particularly those from higher plant forms, the concept of micellar organization was adapted to describe organization in the two directions perpendicular to the chain axis. The organization of the fibrils seemed to be consistently the same for samples of cellulose isolated from a particular species. Thus, it appeared that the manner of assembly of the cellulose chains resulted in a pattern of aggregation that is determined by the microenvironment of the cell wall during biogenesis. A number of speculative models of the organization of cellulose chains within the fibrils were put forth; the majority of them were directed at devising patterns that were consistent with both the organization of chains in the unit cell and with the less linear organization at the next higher scale of organization. Most of the models included a component of disordered cellulose, in part to account for the difference between the geometry of the unit cells and that of the fibrils.

The models can be regarded as in two categories in that some of them focus on micellar organization in the lateral dimension while the others also include the longitudinal dimension. The first attempt to replicate a cross-section of the fibrils did not address the third dimension, which is parallel to the chain axis. The micellar organization is envisioned in two dimensions. Many studies were devoted to deriving estimates of the lateral dimensions of the crystalline micellar domains in native celluloses on the basis of the broadening of the primary reflections in X-ray diffraction patterns. In most instances the dimensions derived from analysis of the line shapes in the diffraction patterns were of the same magnitude as those derived from electron microscopic observations of the most finely dispersed forms of the native celluloses obtained through sonication of the aggregates.

The second category of models was based on a longitudinal view of the fibrils. The manner in which these models reconciled the geometry of the unit cell with that of the fibrils was based on introduction of the concept of micellization in the longitudinal dimension. Here, the micellar dimensions adopted in the direction of the chains were assumed to be much larger than in the lateral dimensions. In this context, the fibrils were viewed as consisting of domains of extended order connected by disordered domains. The disordered domains were often characterized as amorphous. As such, they do not impose a geometric constraint at the next higher level of organization, thus allowing for the differences in geometry between the fibrillar level and the unit cell level. The term "fringed micelle" was introduced to describe the individual ordered domains surrounded and connected by disordered domains.

The concept of fringed micelle was also introduced in descriptions of the structures of cellulose derivatives and synthetic polymers capable of partial crystallization. Such polymers, usually described as semicrystalline, aggregate as microcrystalline domains embedded within domains in which the molecules are less well ordered. In these systems the dimensions of the crystalline domains are dependent on the conditions prevailing during the process of aggregation and in this respect are not unlike colloidal systems. The crystalline domains in these polymers were thus also viewed as micellar in nature. These crystalline domains have many similarities to the crystalline domains in both regenerated and mercerized cellulose. Because of the relationship between native and mercerized cellulose, many of these similarities were also ascribed to native cellulose; the degree to which this is appropriate remains in serious question.

3.16.2.2.3 *Infrared spectroscopic studies*

The contribution of infrared spectroscopy to studies of structure in the middle period was complementary to the crystallographic observations; it has been reviewed by Blackwell and Marchessault.[31] The information derived was in two areas. In studies where the dichroism of infrared absorption of oriented specimens was measured, proposals of particular hydrogen-bonding schemes within the crystalline domains were made. The differences between the spectra of celluloses I and II were explained in terms of differences in the packing of molecular chains and associated variations in the hydrogen-bonding patterns.

The second and perhaps more significant arena in which infrared spectroscopy provided important new insights was in the finding that, based on their absorption in the OH stretching region, native celluloses could be classified into two categories.[32] The first, consisting of algal and bacterial celluloses, was identified as group I_A, while the second, including celluloses from such standards as cotton and ramie, which are representative of higher plant celluloses, were identified as I_B. Though this finding pointed to a higher degree of complexity in the structure of native celluloses, the crystallographic studies continued to seek a unique unit cell assumed characteristic of all native celluloses.

Infrared absorption measurements were also used to explore the partition between crystalline and noncrystalline domains through the use of exposure to D_2O vapor to determine the accessible portion of the cellulose which was deemed noncrystalline.[33] IR spectroscopy was used in yet another more practical application as the basis of a crystallinity index by Nelson and O'Connor.[34,35]

Since closure of the middle period, the development of new information from analyses of both the Raman and solid-state ^{13}C NMR spectra have resulted in reassessments of the crystallographic problem; these will be considered in a subsequent section (3.16.3.2.1).

3.16.2.3 The current period (1975–Present)

Studies during the current period will be discussed in greater detail in the following sections. It is useful, however, to place them in perspective in relation to the work reported in the middle period. Much of the work in the middle period incorporated a criterion long honored in scientific studies, namely, William of Ockham's principle of economy, which requires that the most simple hypothesis consistent with observations should always be adopted. This is perhaps best represented by the crystallographic studies where the most simple structures that account for the majority of the diffractometric data were the ones adopted. In biological systems, where individuality may be reflected in very subtle differences, such as those noted in the infrared spectra, this approach can divert attention from important observations. Such has been the case in the study of native celluloses.

The beginning of the current period can be associated with the first application of some new spectroscopic methods to the characterization of cellulose, and the findings that the observed spectra could not be reconciled with the crystal structures derived from the diffractometric data. These led to reassessment of the degree of precision that can be ascribed to the structures based on the limited amount of diffractometric data, and the questions outlined above concerning the degree to which the reported structures are constrained by and in fact reflect assumptions with respect to symmetry introduced at the outset.

The first questions arose on the basis of Raman spectral observations that pointed to distinctive differences between the skeletal conformations of celluloses I and II. The differences could not be accounted for on the basis of the conformations prescribed by the twofold screw axes of symmetry coincident with the chains in both celluloses I and II. The first efforts to rationalize the Raman spectra suggested that the true conformations may be represented by small departures from the twofold screw axis, with the departures from the twofold screw axis symmetry being a right-handed departure in the instance of cellulose I and a left-handed departure in the case of cellulose II. More careful consideration of this proposal led to the conclusion that it would not be consistent with the rapidity of the transformation from cellulose I to cellulose II in the course of mercerization of native celluloses.

Further examination of the Raman spectra and comparisons with the spectra of dimeric structures led to an alternative interpretation of the departures from twofold screw axis symmetry. It was proposed that the departures were small alternating right-handed and left-handed departures centered at dihedral angles of alternating glycosidic linkages along the chains. Such small departures from the twofold screw axis symmetry would result in diffraction patterns that are predominantly consistent with the $P2_1$ the selection rules, with disallowed reflections appearing quite weak under most conditions of observation. An important implication of this proposal, however, is that the basic repeat unit in the crystal structures must be viewed as the anhydrocellobiose unit rather than the anhydroglucose unit; all earlier studies had been based on the latter premise.

The proposal that anhydrocellobiose is the basic repeat unit of physical structure and the implicit corollary that alternating glycosidic linkages are not symmetrically equivalent were one of the motivations for turning to the method of solid-state ^{13}C NMR, which was then becoming available for application to polymeric systems. Although the solid-state ^{13}C NMR spectra could not be interpreted as conclusive evidence of nonequivalence of alternating glycosidic linkages, they were consistent with such an interpretation. But introduction of the use of solid-state ^{13}C NMR had even more profound implications with respect to a deeper understanding of the diversity of native celluloses.

Based on studies of a wide range of native celluloses from plant sources, it was proposed that the crystalline domains of all of these forms were composites of two different forms, I_α and I_β, that occur in proportions that are distinctive to the particular species producing the cellulose. Though the proposal of alternating glycosidic linkages remains controversial, the finding that native celluloses, are composites of the I_α and I_β forms has been widely accepted and has become the basis for many investigations directed at more detailed characterization of native celluloses.

This discovery of the I_α and I_β forms provided a deeper understanding of the source of the categorization developed earlier on the basis of infrared spectral observations in the OH stretching region. More fundamental perhaps, it also provided a basis for resolution of a paradox first posed by Cross and Bevan[4] almost a century ago when the crystalline aspects of the character of cellulose were first recognized on the basis of its birefringence. They had noted, "The root idea of crystallography is identical invariability, (while) the root idea of the world of living matter is essential individual variation," clearly recognizing that the uniformity of structure characteristic of crystalline order may be inconsistent with the diversity of celluloses from different biological sources. The uniqueness of the blend or integration of the I_α and I_β forms in the cellulose of a particular organism thus provides the basis for the "essential individual variation" alluded to by Cross and Bevan.[4]

The discovery of the occurrence of the two forms of cellulose, I_α and I_β, became the point of departure for studies that led to other important advances during this period. Some involved further refinements of the solid-state ^{13}C NMR spectral analysis. Others were in the area of electron microscopic characterization of cellulose. The lattice imaging technique provided important evidence of the homogeneity of order within crystalline domains that were composites of the I_α and I_β forms of cellulose. Special methods for staining the reducing end groups of cellulosic chains addressed the question of parallel vs. antiparallael alignment of the chains in the crystalline domains. Greater sophistication in application of the methods of electron diffraction has opened up the possibility of a better understanding of the nature of the two forms of native cellulose, and Raman spectral evidence pointed to the similarities and differences between them.

The progress in understanding the nature of structure and aggregation in celluloses during the current period will be described and discussed in greater detail in the following section.

3.16.3 STRUCTURES

As noted earlier, the beginning of the current period of studies on the structure of cellulose was marked by the reintroduction of unit cell models based on parallel alignment of the cellulose molecular chains,[19,26] not unlike those abandoned by Meyer and Misch[24] in the 1930s, but also incorporating bending of the glycosidic linkage to allow the intramolecular hydrogen bond, as suggested by Hermanns.[2] The new models were not consistent with each other, however, apart from the fact that both were based on parallel alignment of the cellulose chains. As French[27] pointed out, they were also not strongly preferred over an antiparallel structure. In the analysis by French,[27] it was recognized that the source of the inconsistency was not so much that the different laboratories were using different computational approaches as it was that the different diffractometric data sets were gathered from different samples and represented different intensities for the same reflections. All of these studies were undertaken before the variability of the crystalline forms of native celluloses was revealed through high-resolution solid-state ^{13}C NMR investigations.

The new crystallographic models also remained in question because the analyses on which they were based incorporated a level of symmetry in the unit cell that was inconsistent with some of the diffractometeric data. Some of the reflections that are consistently observed in electron diffraction patterns and are disallowed by the selection rules for the space group $P2_1$[20] were ignored in these crystallographic analyses. In addition to the disallowed reflections in the electron diffraction patterns that placed the crystallographic models in question, new spectral evidence was developed pointing to the need for further refinement of the structural models, particularly for native celluloses. The models derived from the crystallographic studies could not rationalize many features of the spectral data known to be quite sensitive to structural variations.

On the other hand, electron microscopic studies based on new staining techniques, specific to the reducing end groups of the polysaccharides, confirmed the parallel alignment of molecular chains within the microfibrils in native celluloses. These findings were confirmed further by the manifestation, at the electron microscopic level, of the action of cellulases specific to the nonreducing end group; they were clearly active at only one end of each microfibril. These observations were regarded as confirmation of the most recent crystallographic models. The remaining questions at the time, therefore, were concerned with the degree to which the symmetry of space group $P2_1$ is consistent with the other structure-sensitive observations.

It is well to revisit the issue of levels of structure at this point and clarify the levels at which the different investigative methods are most sensitive. The crystallographic models, which represent coordinates of the atoms in the unit cell, represent the most complete possible specification of structure because they include primary, secondary, and tertiary structures. Indeed, crystallographic studies of the monosaccharides and related structures provide the basis for much information concerning bond lengths and bond angles, as well as conformations in saccharide structures. As noted in the previous section, however, for polymeric systems the diffractometric data is far more limited than for a single crystal of a low molecular weight compound, so that diffraction data from a polymer must be complemented by information from other structure-sensitive methods. An acceptable model must rationalize not only the diffractometric data, which for cellulose is quite limited in comparison to the number of coordinates that must be specified in a definition of the unit cell, but it must also be such that it can be reconciled with information derived from measurements known to be sensitive to other levels of structure.

The new spectral evidence that must be rationalized by any acceptable structure came from two methodologies that are most sensitive to structure at the secondary and tertiary levels. These are Raman spectroscopy and solid-state ^{13}C NMR spectroscopy, both of which were applied to cellulosic samples for the first time during the 1970s. The exploration of spectra measurable by these two methods can provide significant information concerning both secondary and tertiary structures in the solid state. Because the spectral features observed are also sensitive to the molecular environment, they are influenced by the degree of symmetry of the aggregated state. Hence, they provide another avenue for exploration of the applicability of the symmetry of space group $P2_1$ to the structures of the solid state.

Though Raman spectroscopy was first developed in the 1930s, it could not be applied to optically heterogeneous samples like cellulose until lasers could be used as sources for excitation and until new generations of monochromators and detectors became available in the early 1970s. High-resolution solid-state ^{13}C NMR spectroscopy also first became available in the 1970s, and one of its most important areas of application has been in investigations of polymeric systems. Though these two methodologies were, in the first instances, applied separately to cellulose, they were eventually found to be complementary and facilitated development of the foundation for continuing investigations of the nature of native celluloses. The following sections will reflect this evolution. They begin with the early applications of Raman spectroscopy that focused on issues of molecular conformation, and with the high-resolution solid-state ^{13}C NMR investigations that led to discovery of the I_α and I_β forms of native celluloses. They will be followed by a discussion of more recent applications of these methods together with infrared spectroscopy to further explore the nature of this duality in native celluloses.

In addition to the spectroscopic investigations, the use of electron microscopy to explore the structures of native cellulose benefited from the development and application of new techniques that were adapted to investigations of cellulose. The application of stains specific to the reducing end groups of cellulose noted above allowed the issues of parallel and antiparallel chain orientation to be explored in new ways. The method of lattice imaging was also applied to assessment of the homogeneity of microfibrils of native cellulose. Finally, electron diffraction from microdomains in individual microfibrils became an important tool in exploring the nature of I_α and I_β duality. A discussion of these developments will follow discussion of the spectral studies.

Yet another class of investigations that can shed light on the nature of cellulose are based on computational analysis of molecular models using the programs that have become increasingly available in recent decades. Such studies are frequently limited with respect to development of conclusive results because of assumptions that are implicitly or explicitly incorporated in the computational programs in order to make them more tractable, or because of constraints of symmetry that are frequently imposed in the course of the formulation of the models of cellulose that are chosen for analysis. However, they remain of value because they can provide insights concerning the classes of molecular interactions that are likely to be dominant in determining the structures of cellulose and the nature of its interactions with other constituents that occur with it in its native state. A brief overview of some work in this area will be presented.

To complement these studies of cellulose, some analyses of the structures of the oligomers have also been considered. Though these pertain primarily to the structure of cellulose II, they provide some insights with respect to the influence of packing in a lattice on a sequence of anhydroglucose units. As Kakudo and Kasai[11] have noted, the structures of the oligomers can provide important insights concerning the manner of organization of the monomers in the polymeric chains.

3.16.3.1 Spectroscopy

Spectroscopic observations are important in structural investigations because they can provide information which is complementary to that derived from diffractometric data. Such information is particularly important in instances such as that of cellulose, where the diffractometric data alone cannot provide sufficient information to serve as the basis for a structure determination. While the information derived from spectra is not directly related to the absolute values of the coordinates of atoms within the unit cell, it is particularly sensitive to the values of the internal coordinates, which in turn define molecular structures and conformations. In addition, in Raman spectroscopy and infrared spectroscopy, distinctive features associated with the crystal structure can provide critical information. In the case of solid-state ^{13}C NMR, the spectral information is also sensitive to nonequivalences in the environments of chemically equivalent atoms within adjacent monomeric repeat units of the primary structure. Thus, the spectral information provides a basis for testing the degrees of equivalence of structures. Very often also, specific spectral features can be identified with particular functional groups defined by distinctive sets of internal coordinates.

While spectral analyses cannot provide direct information concerning the structures, they establish criteria that any structure must meet to be regarded as an adequate model. The information from spectroscopic studies represents one of the major portions of the phenomenology that any acceptable structural model must rationalize. Reconciliation of the spectral observations with the model derived from crystallographic investigations provides tests of consistency of the proposed structures in the sense set forth by Kakudo and Kasai.[11]

3.16.3.1.1 *Raman spectroscopic studies and conformational questions*

Raman spectroscopy is the common alternative to infrared spectroscopy for investigating molecular vibrational states and vibrational spectra. It has enjoyed a significant revival since the development of laser sources for excitation of the spectra. In laser-excited Raman spectroscopy, a sample is exposed to the monochromatic laser beam and the scattered light is analyzed. While most of the scattering consists of Rayleigh scattered photons of the same frequency as that of the laser, a small fraction of the scattered photons undergo an energy exchange with the scattering molecules and are shifted in frequency relative to that of the exciting source; these are the Raman scattered photons. The magnitudes of the shifts in the frequencies of the Raman scattered photons correspond to the vibrational frequencies of the molecules in the sample under investigation. Thus, the information contained within a Raman spectrum, with respect to the vibrational frequencies of the scattering molecules, is similar to that contained in an infrared spectrum.

Both Raman and infrared spectroscopy provide information about chemical functionality, molecular conformation, and hydrogen bonding. Raman spectroscopy, however, has some important advantages in the study of biological materials. The key advantage arises from the different bases for activity of molecular vibrations in Raman and infrared spectra. That is, whereas activity in the infrared region requires finite transition moments involving the permanent dipoles of the bonds undergoing vibrations, activity in the Raman spectrum requires finite transition moments involving the polarizabilities of the bonds. Thus, in infrared spectroscopy the exchange of energy between the molecules and the exciting field is dependent on the presence of an oscillating permanent dipole. In Raman spectroscopy, in contrast, the exciting field induces a dipole moment in the molecule and the induced moment then becomes the basis for exchange of energy with the exciting field. It is useful in this context to view bonds in terms of Pauling's classification along a scale between the two extremes of polar and covalent.[36] Bonds that are highly polar and possess relatively high dipole moments tend, when they undergo vibrational transitions, to result in bands that are intense in the infrared and relatively weak in Raman spectra. Conversely, bonds that are primarily covalent in character and have a relatively high polarizability generally result in bands that are intense in the Raman spectra but are relatively weak in the infrared. This is perhaps best illustrated by the fact that O_2 and N_2, which are homonuclear and without permanent dipoles, have very intense Raman spectra though they are inactive in infrared absorption, while H_2O, with a high permanent dipole moment, is a very strong absorber in the infrared but a very weak Raman scatterer. With respect to cellulose, the OH groups of cellulose and those of adsorbed water are dominant in many of the spectral features in infrared spectra. In contrast, the skeletal C—C bonds and the C—H bonds dominate the Raman spectra. A further simplification in the Raman spectra results from the circumstance that the selection rules forbidding activity of overtone and combination bands are more rigidly adhered to than is the case in infrared spectra so that the bands observed in Raman spectra are usually confined to the fundamental modes of the molecules under investigation.[37]

Raman spectroscopy possesses another important advantage in comparison to infrared spectroscopy in the study of biological systems; optical heterogeneities do not present serious problems. In infrared spectroscopy, the key observable is the degree of attenuation of an incident infrared beam in comparison to a reference beam. Thus, if any process other than absorption can result in attenuation of the infrared beam, interpretation of the spectra can be complicated. Since the refractive index of a substance can undergo large excursions in the neighborhood of strong absorption bands, the Rayleigh scattering losses will vary with frequency in the infrared absorption region of the spectrum, and they can cause anomalous features in the infrared spectra. Furthermore, in the case of cellulosic materials, the optical heterogeneities generally have dimensions in the same range as the wavelengths of the infrared absorption in the region of the fundamental vibrations. In consequence, Rayleigh scattering of the incident infrared beam can be significant and frequency dependent. In Raman spectroscopy, in contrast, variations in the refractive index do not present a difficulty since the excitation frequency and the frequencies of the Raman scattered photons are far removed from any absorption bands. It is therefore easier to record meaningful Raman spectra from samples such as cellulose, even though they may cause a high level of Rayleigh scattering.

In the context of studies on the structure of cellulose, the key advantage of Raman spectroscopy is the degree of its sensitivity to the skeletal vibrations of the cellulose molecule, with the mode of packing in the lattice having only secondary effects. This sensitivity is a consequence of the reality that most of the skeletal bonds are C—C and C—O bonds, both of which have relatively high polarizabilities and, hence, high Raman scattering coefficients. The minimal contribution of packing effects arises from the low scattering coefficients of the highly polar OH groups, which are the functionalities that are most directly involved in intermolecular associations. The result is that intramolecular variations such as changes in internal coordinates have a significantly greater influence on the Raman spectra than variations in intermolecular associations.

These considerations were paramount in the first detailed examination and comparison of the Raman spectra of celluloses I and II;[38] the spectra are shown in Figure 1. It was concluded that the differences between the spectra, particularly in the low frequency region, could not be accounted for in terms of chains possessing the same conformation but packed in different ways in the different lattices. As noted earlier, that had become the accepted rationalization of the differences between celluloses I and II developed from diffractometric studies of these two most common allomorphs. The analyses of the Raman spectra led to the proposal that two different stable conformations of the cellulose chains occur in the different allomorphs.

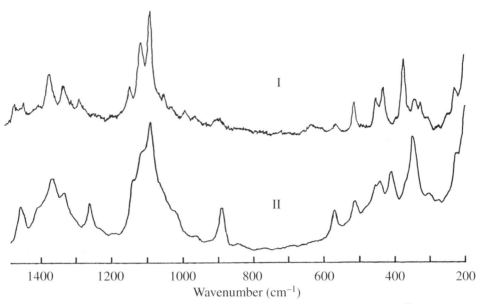

Figure 1 Raman spectra of high crystallinity celluloses I and II.[38]

Raman spectra of native cellulose had previously been reported and discussed by analogy with the spectra of glucose,[39] but the significant differences between celluloses I and II had not been addressed, nor did the assignments consider the complexity of coupling between the different fundamental modes. To establish a basis for assessing the differences between celluloses I and II, Atalla and co-workers undertook an extensive series of studies of model compounds of increasing complexity.[40–47] The studies included comprehensive normal coordinate analyses of the molecular vibrations of each of the groups of model compounds based on complementary infrared and Raman spectra. The objective of these analyses was to establish the degree to which the different classes of vibrational motions contribute to the spectral features in the different regions of the spectrum. Such a comprehensive approach was necessary because the skeletal bond systems that occur in the structures of carbohydrates are made up predominantly of C—C and C—O bonds, which possess similar reduced masses and vibrational force constants and, hence, have very similar vibrational frequencies. In consequence a high degree of coupling occurs between the vibrations, with the result that very few of the vibrational modes are localized within specific bonds or functional groups. Thus, the traditional group frequency approach common in the assignment of infrared and Raman spectra is of very limited use except in the case of vibrations localized in the bonds of hydrogen atoms bonded to much heavier atoms such as oxygen or carbon. On the other hand, the normal coordinate analyses allow identification of the degree to which the vibrations of each of the internal coordinates contributes to each of the observed bands. Since the coupling of the vibrations is very sensitive to changes in the bond angles and the dihedral angles associated with the bonds whose vibrations are coupled, the normal coordinate analyses allow detailed and systematic exploration of the effects of differences in skeletal conformations on the bands associated with particular vibrations.

The compounds chosen to establish the foundations for analysis of the spectra of cellulose began with the 1,5-anhydropentitols[40,41] and continued with the pentitols and erythritol,[42] the pentoses,[43] the inositols,[44,45] and three hexoses including glucose, mannose and galactose.[46,47] For each group of compounds, the force fields were refined against the observed frequencies of a subset of the group until a satisfactory fit was obtained. The force fields were validated by successful prediction of the spectra of compounds not included in the refinements. Furthermore, the calculated potential energy distributions were reasonable in comparison with the group frequency literature on carbohydrates. The analyses were thus successful in developing a physically meaningful force field specifically tailored to the carbohydrates. The hexose field was then used to extend the normal coordinate method to the cello-oligodextrins,[48] the vibrational spectra of which more closely resemble the spectra of cellulose. Because the number of vibrational degrees of freedom greatly exceeds the number of bands that can be resolved in the spectra of the cello-oligodextrins, it was neither possible nor meaningful to refine the force constants further in this context. The distribution of calculated frequencies was in qualitative agreement with the observed spectra. The force field derived from the hexoses thus appears to provide a good basis for interpreting the spectra of the cello-oligodextrins and cellulose.

With respect to the question concerning the conformations of celluloses I and II, it is useful to consider first some of the pertinent information developed from the normal coordinate analyses, particularly with respect to the classes of molecular motions associated with the different spectral features. The region below 1500 cm^{-1} was the primary focus of the early exploration because the intense bands clustered at about 2900 cm^{-1} can be identified with the C—H stretching vibrations and the region beyond 3000 cm^{-1} is clearly associated with the O—H stretching vibrations. In addition to the C—H and O—H stretching vibrations, the internal deformation of the methylene group on C-6 is the only vibration which closely approximates a group or local mode in the usual sense implicit in discussions of assignments of vibrational spectra; the HCH bending vibration usually occurs above 1450 cm^{-1}. In all other bands at frequencies below 1450 cm^{-1}, the normal coordinate analysis indicated that the vibrations are so highly coupled that, in most instances, no single internal coordinate contributes more than 20% of the potential energy change associated with any particular frequency, though in a few instances contributions were as high as 40%. Thus, the traditional group frequency approach to assignment of vibrational spectra, which is based on the concept of local modes, is generally not applicable in this region in the spectra of saccharides. It is necessary instead to focus on the classes of internal motions that are associated with the different frequency ranges and to interpret the spectra in terms of the influence that variations in the internal coordinates can have on the coupling between different types of vibrational deformations.

For analysis of the spectra of celluloses, it is possible to classify the groups of features in the different spectral regions in terms of the types of internal deformations that make their maximum contributions to bands in those regions. The bands between 1200 and 1450 cm^{-1} are due to modes involving considerable coupling between methine bending, methylene rocking and wagging, and COH in-plane bending motions; these are angle bending coordinates involving one bond to a hydrogen atom and the other to a heavy atom. Significant contributions from ring stretching begin below 1200 cm^{-1} and these modes, together with C—O stretching motions, dominate between 950 and 1150 cm^{-1}. Below 950 cm^{-1}, angle bending coordinates involving heavy atoms only (i.e., CCC, COC, OCC, OCO) begin to contribute, though ring and C—O stretches and the external bending modes of the methylene group may be major components as well. The region between 400 and 700 cm^{-1} is dominated by the heavy atom bending, both C—O and ring modes, although some ring stretching coordinates still make minor contributions. In some instances, O—H out-of-plane bending motions may make minor contributions in this region as well. Between 300 and 400 cm^{-1}, the ring torsions make some contributions, and below 300 cm^{-1} they generally dominate.

In addition to the above generalized categorization concerning modes that occur in one or another of the model compound systems used in the normal coordinate analyses, the spectrum of cellulose will have contributions due to modes centered at the glycosidic linkage. The computations based on the cello-oligodextrins indicate that these modes are strongly coupled with modes involving similar coordinates in the adjacent anhydroglucose rings. The contributions of the different classes of internal coordinates to the different bands are presented in greater detail elsewhere.[49]

As noted above and shown in Figure 1, the differences between the Raman spectra of celluloses I and II are quite significant, particularly in the region of the skeletal bending modes of vibration. In the region above 800 cm^{-1} the differences are most obvious with respect to the relative intensities of the bands and the broadening of some of the bands upon conversion from cellulose I to cellulose II. In the region below 700 cm^{-1}, in contrast, the main features are quite different in the two spectra; these differences are even more evident in the spectra of single fibers which will be presented later.

In the analyses of the spectra of model compounds, changes of the magnitude indicated in Figure 1 were associated exclusively with the occurrence of differences in conformations. It seemed very probable therefore that the differences between the spectra of celluloses I and II reflect a change in molecular conformation accompanying the transition from one form to the other. Since the basic ring structure is not expected to change,[50] it would appear that variations in the dihedral angles at the glycosidic linkages provide the only opportunity for conformational differences. Because of the controversy surrounding similar conclusions based on crystallographic studies carried out in the early 1960s,[17,18] a number of experimental and theoretical avenues for validating this interpretation were pursued.

The first consideration was whether a multiplicity of stable conformations is consistent with the results of conformational energy calculations that were available at the time.[50,51] In both studies, the potential energy surfaces were found to possess multiple minima. When the additional constraint of a repeat length of approximately 0.515 nm per anhydroglucose unit was added, two minima representing both left and right-handed departures from the twofold helix appeared to be likely loci of the stable conformations. It was noted in this context that these two minima were close to the positions of the dihedral angles of the glycosidic linkages in cellobiose and methyl-β-cellobioside, respectively, as these were determined from crystallographic studies.[52,53]

Next, inquiry was made into the degree to which changes in the dihedral angles about the bonds in the glycosidic linkage could influence the modes of vibration responsible for the spectral features in the different regions of the spectra. Two approaches were adopted for this purpose. The first was based on examining the Raman spectra of polysaccharide polymers and oligomers that were known to occur in different conformations. The second was a theoretical one based on an adaptation of the matrix perturbation treatment used by Wilson *et al.*[54] to discuss the effects of isotopic substitution on infrared and Raman spectra.

The polysaccharide systems chosen for investigation were among those most closely related to cellulose in the sense that they are the α-(1-4)-linked polymers and oligomers of anhydroglucose. They included amylose and two of its cyclic oligomers, with primary emphasis on the latter, the α- and β-Schardinger dextrins, often also known as cyclohexa- and cyclohepta-amylose. The structures of the two oligomers differ in that the values of the dihedral angles about the bonds of the glycosidic linkages have to change to accommodate the different number of monomer units. Comparison of the Raman spectra of the cyclic dextrins showed that the differences between them were quite minor in the regions above 800 cm^{-1}, but they were quite significant in the lower frequency region dominated by the skeletal bending and torsional modes. The differences were similar in kind and distribution to the differences between celluloses I and II. It was also noted that in earlier studies of the Raman spectra of amylose,[55] it had been observed that forms V_a and V_h, which are very similar in conformation but had different levels of hydration, had almost identical spectra. In contrast, form B, which is known to have a distinctly different helix period, was found to have a spectrum that differs from those of forms V_a and V_h in a manner approximating the differences between the two cyclic oligodextrins. Taken together, the observations of the Raman spectra of the amyloses support the interpretation of the differences between the Raman spectra of celluloses I and II as pointing to differences in the chain conformations localized at the glycosidic linkages.

In the theoretical analysis, the method of Wilson *et al.*[54] was adapted to explore the consequences of variations in the dihedral angles about the bonds in the glycosidic linkage. Changes in the dihedral angles were found to influence skeletal stretching and bending modes primarily through changes in some of the corresponding off-diagonal terms in the kinetic energy matrix **G**. Examination of the general expressions for these terms[56] reveals that only one of the four classes of terms that influence stretching is sensitive to the value of the dihedral angle, and it is a class representing stretch–bend interactions. The interactions influencing the bending modes, in contrast, are more sensitive to the dihedral angles. Among these, three of the four classes of bend–bend interactions change with the dihedral angles; these are in addition to the stretch–bend interaction cited earlier, which would also influence the bending modes. Finally, the majority of the terms involving torsional coordinates are sensitive to variation in the dihedral angles. These considerations led to the conclusion that the skeletal bending and torsional modes are altered to a greater degree than the skeletal stretching modes when the dihedral angles associated with the glycosidic linkage undergo variations. When translated to spectral features in the Raman spectra, these observations point to major differences in the low-frequency region below 700 cm^{-1}, and minor ones in the fingerprint region between 900 and 1500 cm^{-1}. These are indeed precisely the types of differences observed in comparisons of the spectra of celluloses I and II.

One final consideration that was addressed is the possibility that rotations of the primary alcohol group at C-6 could account for the spectral differences seen in the spectra of celluloses I and II and in the spectra of the amyloses. The normal coordinate analyses of the hexoses showed that rotations about the C-5—C-6 bond can result in minor variations in the region below 600 cm^{-1} but that the major impact of such rotations is expected in the spectral region above 700 cm^{-1}.[46,47] With all of the above considerations in mind, it became clear that the only plausible rationalization of the differences between the Raman spectra of celluloses I and II had to be based on the possibility that differences between the skeletal conformations were the key.

The first effort to rationalize differences in conformation was based on the results of the conformational energy mappings that were available at the time.[50,51] The key points derived from those analyses, which have been confirmed by additional studies,[57,58] were that the two energy minima associated with variations in the dihedral angles of the glycosidic linkage correspond to relatively small left- and right-handed departures from glycosidic linkage conformations that are consistent with twofold helical symmetry. The minima also represented values of the dihedral angles that were very similar to those reported for cellobiose and methyl β-cellobioside on the basis of crystallographic analyses.[52,53] The relationship between the different conformations is represented in Figure 2, which was adapted by Atalla[59] from a diagram first presented by Reese and Skerret.[50] It is a ψ/ϕ map presenting different categories of information concerning the conformation of the anhydrocellobiose unit as a function of the values of the two dihedral angles about the bonds in the glycosidic linkage. ψ is defined as the dihedral angle about the bond between C-4 and the glycosidic linkage oxygen and ϕ as the dihedral angle about the bond between C-1 and the glycosidic linkage oxygen. The parallel lines indicated by $n = 3(L)$, 2, and 3(R) represent values of the dihedral angles that are consistent with a left-handed threefold helical conformation, a twofold helical conformation, and a right-handed threefold helical conformation, respectively; a twofold helical conformation inherently does not have a handedness to it. The dashed contours represent conformations that have the indicated repeat period per anhydroglucose unit; the innermost represents a period of 5.25 Å corresponding to 10.5 Å per anhydrocellobiose unit. The two dotted lines indicate conformations corresponding to values of 2.5 and 2.8 Å for the distance between the two oxygen atoms anchoring the intramolecular hydrogen bond between the C-3 hydroxyl group of one anhydroglucose unit and the ring oxygen of the adjacent unit; the values bracket the range wherein hydrogen bonds are regarded as strong. The two domains defined by solid lines on either side of the twofold helix line ($n = 2$) represent the potential energy minima calculated by Rees and Skerret for the different conformations of cellobiose. Finally, the points marked by J and W represent the structure of cellobiose determined by Chu and Jeffry[52] and the structure of methyl-β-cellobioside determined by Ham and Williams.[53] The key point to be kept in mind in relation to this diagram is that structures along the twofold helix line and with a repeat period of 10.3 Å per anhydrocellobiose unit possess an unacceptable degree of overlap between the van der Waals radii of the protons on either side of the glycosidic linkage.

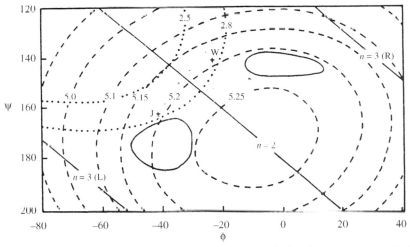

Figure 2 ψ/ϕ map. (— — —) Loci of structures with constant anhydroglucose repeat periods; ($\cdots$) loci of structures of constant intramolecular hydrogen bond O—O distances; (———) contours of potential energy minima based on nonbonded interactions in cellobiose. J, cellobiose; W, β-methylcellobioside. $n = 2$, the twofold helix line; $n = 3$ the threefold helix lines, (R) right-handed, (L) left-handed. The Meyer–Misch structure is at $\psi = 180$, $\phi = 0$.[59]

Taking into consideration that it had earlier been speculated that the diffractometric data were consistent with a helix of much longer period that may be degenerate with the twofold helix[7] and that it was commonly assumed in polymer crystallography that chemically equivalent units were symmetrically equivalent,[3] Atalla proposed that the structure of cellulose I may be represented as a right-handed helix of long period with dihedral values similar to those in methyl-β-cellobioside and that the structure of cellulose II may be represented as a left-handed helix of long period with dihedral values similar to those of cellobiose.[60] Further observations and considerations led to revision of this proposal both with respect to the equivalence of successive units in the polymeric chains and with respect to the nature of the departures from twofold helix symmetry.

The first revision followed from observations and comparisons of the Raman spectra of cellulose II and of cellobiose in the O—H stretching region.[59] The latter showed a single sharp band superimposed on a broader background, and the band was identified with the O—H stretching vibration of the isolated intramolecular hydrogen bond revealed in the crystal structure;[52] it occurs between the hydroxyl group on C-3 of the reducing anhydroglucose unit and the ring oxygen of the nonreducing unit. The spectrum of cellulose II revealed two such sharp bands in the same region; similar bands were observed in the spectra of the cello-oligodextrins.[48] Since the frequency at which such bands occur is very sensitive to the distance between the oxygen atoms that anchor the hydrogen bond, it appeared that the structure of cellulose II must incorporate intramolecular hydrogen bonds with two distinct values of the O—O distance. This led to the proposal that successive units in the structure are not equivalent, and that, as a consequence, alternating glycosidic linkages have different sets of dihedral angles defining their coordinates.[59] Thus, the dimeric anhydrocellobiose was regarded as the repeat unit of physical structure rather than the anhydroglucose unit. These conclusions, based on the Raman spectra in the O—H region, were confirmed when the solid-state ^{13}C NMR spectra became available[61] as splittings were observed in the resonances associated with C-1 and C-4, which anchor the glycosidic linkage. The occurrence of these splittings is indicative of the presence of nonequivalent glycosidic linkages within the structure; the NMR spectra will be considered in greater detail in the following section.

The second revision, which flowed from the first one, was rooted in the recognition that when alternating glycosidic linkages are admitted as an option, and when anhydrocellobiose is viewed as the repeat unit of the structure, the alternating glycosidic linkages no longer needed to have the same sense of departure from the twofold helix. That is, it was now possible to consider structures wherein the nonequivalent glycosidic linkages are alternating left- and right-handed departures from the twofold helix. Such structures would be ribbon like and could appear to approximate the twofold helix. The possibility of such structures would resolve one objection that was raised with respect to the earlier proposal,[62] namely, that the transition from cellulose I to cellulose II would require a remarkable degree of rotation of the molecular chains about their axes as they were converted from cellulose I to cellulose II during mercerization. The proposal incorporating the second revision also has the advantage that it can be reconciled with much of the diffractometeric data. That the departures from twofold helical symmetry are relatively small may explain the weakness of the reflections that are disallowed by the selection rules of space group $P2_1$.

Based on the considerations outlined, the model that was adopted as a basis for continuing explorations of the spectra of cellulose was based on the proposal that the glycosidic linkages alternated between small left- and right-handed departures from the twofold helical conformation. Thus, the differences between the conformations of celluloses I and II now had to be understood in terms of differences in the internal organization of the anhydrocellobiose units that were the basic units of structure.[63,64]

In search of a rationalization of the changes in the internal organization of the cellobiose unit associated with the transition from cellulose I to cellulose II, Atalla drew on the analogy with the structures of cellobiose and methyl-β-cellobioside, which are represented in Figure 3. The methyl-β-cellobioside, which has values of dihedral angles corresponding to a right-handed departure from the twofold helix, also has a bifurcated intramolecular hydrogen bond in which the proton from the C-3 hydroxyl group appears to be located between the ring oxygen and the primary alcohol oxygen at C-6 of the adjacent unit. This bifurcation is in part responsible for the absence of a sharp OH band in the OH region of the spectrum of the methyl-β-cellobioside. Atalla suggested that such bifurcated intramolecular hydrogen bonds may occur in connection with every other glycosidic linkage in a molecule of native cellulose; these bifurcated hydrogen bonds would be associated with those glycosidic linkages that have values of dihedral angles representing right-handed departures from the twofold helix in a manner not unlike those in methyl-β-cellobioside. The action of mercerizing agents was seen as resulting in the disruption of the bifurcated OH bonds, thus allowing the glycosidic linkages to relax to slightly greater departure from the twofold helix.[63,64] Such an explanation would also be consistent with the observation that the two HCH bending bands in the Raman spectra of native celluloses collapse into a single band upon mercerization, suggesting a nonequivalence of the two primary alcohol groups in native cellulose and a shift closer to equivalence upon mercerization. It is also consistent with the greater splitting of the resonances associated with C-1 and C-4 seen in the solid-state ^{13}C NMR spectra of cellulose II to be discussed in the following section. While the evidence supporting this proposal is strong, it is not conclusive and thus awaits further confirmation. Atalla also introduced the terms k_I and k_{II} to designate the conformations corresponding to celluloses I and II; the term k_o was introduced to describe cellulose in a disordered state.[65]

More detailed characterizations of the Raman spectra of a number of celluloses were undertaken at a later point. These are best considered, however, after presentation of some of the important conclusions derived from the CP/MAS solid-state ^{13}C NMR studies that were undertaken soon after some of the early studies of the Raman spectra.

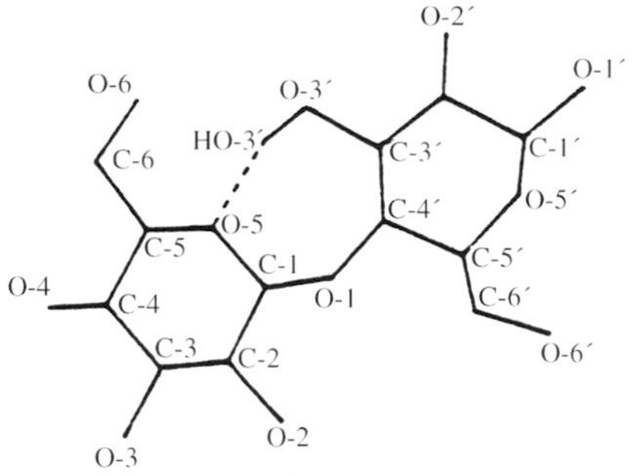

Figure 3 Structures of β-cellobiose and β-methylcellobioside.

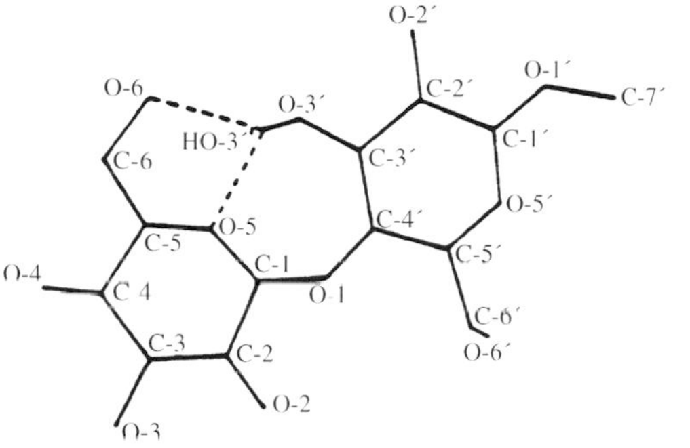

3.16.3.1.2 Solid-state ^{13}C NMR spectroscopic studies and the two forms of native cellulose I_α and I_β

Though applied to cellulose later than Raman spectroscopy, high-resolution solid-state ^{13}C NMR has provided perhaps the most significant new insights regarding the structures of cellulose, particularly in its native state. Earlier applications of NMR to the study of solids had been based on exploration of dipolar interactions between pairs of magnetic nuclei and did not afford the possibility of acquisition of spectral information characteristic of molecular architecture in a manner that was parallel to that possible with liquid-state NMR spectra. The development of high-resolution solid-state NMR spectroscopy and its application to polymeric materials grew from complementary application of a number of procedures that had been developed in NMR spectroscopy. The first is proton carbon cross-polarization (CP) that is used to enhance sensitivity to the low-abundance ^{13}C nucleus. This was combined with high-power proton decoupling to eliminate the strong dipolar interaction between the ^{13}C nuclei and neighboring protons. Finally, the angular dependence of the chemical shift, or chemical shift anisotropy, is overcome by spinning the sample about an axis at a special angle to the direction of the magnetic field, commonly referred to as the magic angle, the procedure denoted by (MAS). The combined application of these procedures, usually designated by (CP/MAS), results in the acquisition of spectra that contain isotropic chemical shift information analogous to that obtained from liquid-state ^{13}C NMR with proton decoupling. One key difference between the two is that molecular motion in the liquid state is sufficiently rapid that chemically equivalent carbons result in single resonances, whereas in the solid-state spectra, because the molecules are immobile, the isotropic chemical shift is sensitive to the molecular environment. As a result, chemically equivalent carbons that occur in sites that are not magnetically equivalent, within an aggregated solid state, may have differences between their chemical shifts.

In summary, the most important characteristic of the spectra acquired using the (CP/MAS) ^{13}C NMR technique is that, if they are acquired under optimal conditions, they can have sufficient resolution so that chemically equivalent carbons that occur in magnetically nonequivalent sites can be distinguished. In the present context, the corresponding carbons in different anhydroglucose units would be regarded as chemically equivalent. If they are not also symmetrically equivalent, that is, if they occur in different environments or if the anhydroglucose rings possess different conformations, within the rings, at the glycosidic linkage, or at the primary alcohol group, the carbons will not have magnetically equivalent environments and will therefore result in distinctive resonances in the NMR spectrum. The fundamental challenge in the application of this method is to achieve a level of resolution sufficient to distinguish nonequivalences between chemically equivalent carbons, because the magnetic nonequivalence can result in variations in the chemical shift that are small relative to the shifts determined by the primary chemical bonding pattern.

Another important feature of the (CP/MAS) ^{13}C NMR technique is that for a system such as cellulose, which consists of rather rigid hydrogen-bonded molecules and in which all carbons have directly bonded protons, the relative intensities of the resonances are expected to correspond to the proportion of the particular carbons giving rise to them. Thus, the intensities arising from each of the six carbons in the anhydroglucose ring are expected to be equal. This is an important characteristic that is central to the analysis and interpretation of the information contained within the spectra.

The first applications of the new technique to cellulose[61,66] demonstrated resolution of multiple resonances for some of the chemically equivalent carbons in the anhydroglucose units. It became clear that rationalization of the spectra that were observed would provide valuable additional information concerning the structure of the celluloses investigated. The first step in such a rationalization was the assignment of the resonances which appear in the spectra. The assignments, which have been discussed in a number of reports,[61,66-71] were based on comparisons with solution spectra of cello-oligosaccharides and of a low DP cellulose.[72] They are indicated in Figure 4, which shows a spectrum of cotton linters.[73] Beginning at the upfield part of the spectrum, the region between 60 and 70 ppm is assigned to C-6 of the primary alcohol group. The next cluster of resonances, between 70 and 81 ppm, is attributed to C-2, C-3, and C-5, the ring carbons other than those anchoring the glycosidic linkage. The region between 81 and 93 ppm is associated with C-4 and that between 102 and 108 ppm with C-1, the anomeric carbon.

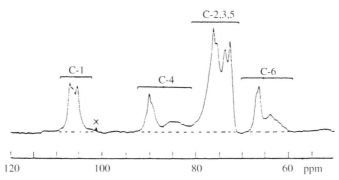

Figure 4 ^{13}C CP-MAS spectrum of cotton linters. The horizontal bars indicate the spectral ranges of the corresponding carbon sites in the anhydroglucose monomer unit of cellulose. The "X" marks the position of the small first spinning sideband of linear polyethylene, which was added as a chemical shift reference. The polyethylene center band (not shown) occurs at 33.63 ppm; the zero of reference for chemical shifts is liquid tetramethylsilane. Note the existence of both broad and narrow resonance features.[73]

In one of the first reports on application of the technique to studies of different celluloses, the splittings of the resonances of C-4 and C-1 in the spectrum of cellulose II (Figure 5) were regarded as confirmation of the occurrence of nonequivalent glycosidic linkages that had earlier been proposed on the basis of comparison of the Raman spectra of cellulose II and cellobiose in the O—H stretching region.[61] These splittings were also observed in the CP/MAS spectra of the cello-oligodextrins, which crystallize in a lattice very similar to that of cellulose II. In that context the splittings were attributed to the occurrence of nonequivalent cellulose molecules in the same unit cell.[69] However, such an interpretation leaves open the question as to why the resonances for carbons 2, 3, and 5 do not display similar splittings. If the splittings were indeed due to nonequivalent molecules it would be anticipated that those carbons nearest to the boundaries of the molecule would be the most affected. The carbons anchoring the glycosidic linkage, that is, C-1 and C-4, are the ones most removed from adjacent molecules, yet they also display the greatest splittings.

Interpretation of the spectra of native celluloses presented an even more challenging task. In the spectrum of cotton linters (Figure 4), the two resonance regions associated with C-6 and C-4 include sharper resonances overlapping broader upfield wings. After excluding the possibility that the broader wings could arise entirely from molecular mobility,[66,67] the wings were attributed to cellulose chains in two categories of environment. The first includes all chains located at the surfaces of cellulose microfibrils, which, because of their occurrence at the boundary, are less constrained with respect to the conformations they can adopt. The surfaces are regarded as regions of limited two-dimensional order. The importance of this category of order had earlier been demonstrated in a

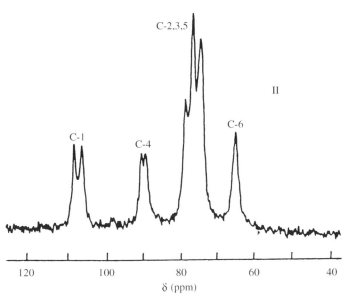

Figure 5 The CP/MAS ^{13}C spectrum of high crystallinity cellulose II recorded at relatively low resolution. Chemical shifts are shown in parts per million relative to Me$_4$Si. Assignment of the C-1, C-4, and C-6 resonances are based on pertinent liquid-state spectra.[61]

study of different native celluloses undertaken by Earl and VanderHart.[67] The celluloses had natural fibril diameters varying between 3.5 and 20 nm and it was shown that the areas of the upfield wings of C-4 and C-6 declined as the surface to volume ratio declined. The second category of environments contributing to the upfield wings is that of chains in regions within which the incoherence of order is not limited to two dimensions. Here, the dispersion of the frequencies at which resonances occur may arise from conformational differences, variations in bond geometries, changes in hydrogen bonding patterns, and nonuniformities in neighboring chain environments. These possibilities arise because in such regions the molecular chains are free to adopt a wider range of conformations than the ordering in a crystal lattice or its boundaries would allow.

Although the obvious upfield wings of the C-4 and C-6 resonances are the most direct evidence for the cellulose chains in less ordered environments, it is expected that the chains in these environments make similar contributions to the resonance regions associated with the other carbons. In the region of C-1, the contribution appears to be primarily underneath the sharper resonances, though a small component appears to extend towards 104 ppm. Similarly, it is expected that the contribution from chains in the less ordered environments underlies the sharper resonances of the C-2, C-3, and C-5 cluster.

The relative contributions of the two categories of environment to the intensity of the upfield wings was assessed in a careful analysis of the C-4 wing.[73] It was demonstrated that part of the wing could be correlated with the range of the C-4 resonance in amorphous cellulose prepared by ball milling. It was therefore assigned to cellulose chains occurring in the second type of environment, that is, domains wherein the incoherence of order is extended in all three dimensions. The other part of the wing was attributed to chains at the surfaces of the fibrils and, on the basis of these comparisons, it was concluded that approximately 50% of the wing is contributed by cellulose chains in each of the two types of less ordered environments described in the preceding paragraph. Though the upfield wing of C-4 is the basis of this allocation of intensities, it can be assumed that the relative contributions are similar for the upfield wing of C-6 and for the component that appears to underlie the sharper resonances at C-1. It is also expected that these domains contribute to the total intensity of the C-2, C-3, and C-5 cluster between 70 and 81 ppm.

The sharper resonances in the C-6 and C-4 regions, centered at 66 and 90 ppm, respectively, each appear to consist of more than one resonance line even though the resolution is not sufficient to distinguish the components well. The C-6 resonance seems to include at least two components while the C-4 resonance appears to include three closely spaced component lines. These multiplicities were interpreted as arising from carbons in cellulose molecules within the interior of crystalline domains and therefore taken as evidence of the occurrence of chemically equivalent carbons in different magnetic environments within the crystalline domains.

The region between 102 and 108 ppm, attributed to C-1, also reveals multiplicity and sharp resonance features. Here, however, the shoulder is very limited. It appears that the resonances associated with the two categories of disordered domains described above lie underneath the sharp resonances associated with the interior of the crystalline domains. It can be concluded that, in most instances, the dispersion of frequencies associated with the disorder is small relative to the shift associated with the character of the anomeric carbon C-1, while that is not the case for the shifts associated with C-4 and C-6. Alternatively, it may be evidence that, because of the anomeric effect, the internal coordinates surrounding C-1 are much less flexible within the range of possible conformational variations than the other internal coordinates.

In search of a rationalization of the splittings observed in the sharp resonances, (CP/MAS) ^{13}C NMR spectra of a wide variety of samples of cellulose I were recorded. Some of these are shown in Figure 6. They include (a) ramie fibers (b), cotton linters (c), hydrocellulose prepared from cotton linters by acid hydrolysis (d), a low-DP regenerated cellulose I (e), cellulose from *Acetobacter xylinum* (f), and cellulose from the cell wall of *Valonia ventricosa*, an alga. While similar observations were reported in a number of studies,[61,63,67–71] their implications with respect to structure were more fully developed in the work of VanderHart and Atalla,[73,74] which provides the basis for the following discussion.

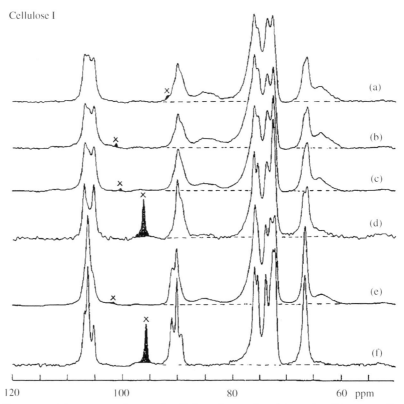

Figure 6 The ^{13}C CP/MAS spectra of several cellulose I samples: (a) ramie; (b) cotton linters; (c) hydrocellulose from cotton linters; (d) a low DP regenerated cellulose I: (e) *Acetobacter xylinum* cellulose; (f) *Valonia ventricosa* cellulose. Note the varied fine structure particularly at C-1 and C-4. Signal-to-noise variation due to limited amount of some samples. In that instance more polyethylene was added so the side band intensity increased. No line broadening or resolution enhancement techniques were applied in the acquisition of the spectra (after VanderHart and Atalla[73]).

All of the spectra shown in Figure 6 (a)–(f) are of celluloses that occur in relatively pure form in their native states and require relatively mild isolation procedures. The most striking feature in these spectra, when viewed together, is the variation in the patterns of the multiplets at C-1, C-4, and C-6. These resonances, which are viewed as arising from chains in the interior of crystalline domains, appear to be unique to the particular celluloses; among the native forms they appear to be distinctive of the source species. The first attempt to rationalize the spectra was in terms of information that they might provide concerning the unit cell of the structure of cellulose I. However, it soon became obvious that such a rationalization was not possible because the relative intensities within the multiplets were not constant nor were they in ratios of small whole numbers as would be the case if the same unit cell prevailed throughout the crystalline domains. The conclusion was that the multiplicities were evidence of site heterogeneity within the crystalline domains and that, therefore, native celluloses must be composites of more than one crystalline form.

Further rationalization of the spectra required a careful analysis of the mutliplets at C-1, C-4, and C-6 and the variations of the relative intensities of the lines within each multiplet among the spectra of the different celluloses. In addition to excluding a single crystal form on the basis of the considerations noted above, it was also possible to exclude the possibility of three different forms with each contrbuting a line to the more complex multiplets. Thus, a decomposition of the spectra on the basis of two distinct crystalline forms was pursued. The results of the decomposition are shown as spectra (b) and (c) in Figure 7, and were designated as the I_α and I_β forms of native cellulose; this designation was chosen in order to avoid the possibility of confusion with the I_A and I_B forms that had earlicr been defined in terms of differences in the appearance of the O—H bands in different types of native celluloses.[22,31] Spectrum (a) was acquired from a high-crystallinity sample of cellulose II and is included so as to distinguish the heterogeneity of crystalline forms occurring in the different forms of cellulose I from the long-known polymorphic variation of the crystallinity of cellulose.

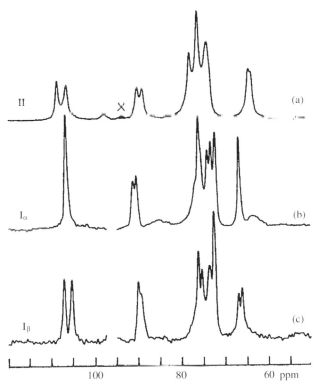

Figure 7 Comparison of the ^{13}C CP-MAS spectrum (a) of a low-DP cellulose II sample and the spectra (b) and (c) corresponding, respectively, to the two proposed crystalline forms of cellulose I, namely I_α and I_β. Spectra (b) and (c) were obtained by taking linear combinations of the low-DP and *Acetobacter* cellulose spectra. Discontinuities in spectra (b) and (c) occur where the polyethylene sidebands would have appeared. The I_α spectrum still contains a significant amount of non-I_α resonances as shown by the visible C-4 and C-6 upfield wings. Multiplicities of the C-1, C-4, and C-6 narrower resonances ought to indicate unit cell inequivalences.[73]

Spectra (b) and (c) in Figure 7 were in fact derived from appropriate linear combinations of the spectra of the low-DP cellulose I (d) and of the *Acetobacter xylinum* cellulose (e) in Figure 6. Though they represent the best approximations to the two forms of cellulose postulated, they cannot be regarded as representative of the pure forms as they do not adequately reflect the component of the cellulose at the surfaces of the crystalline domains. Spectrum 7(b) does have some intensity in the upfield wings of C-4 and C-6, but spectrum 7(c) has very little evidence of such wings. There is very little question, however, that the sharp components of spectra 7(b) and 7(c) include the key features in the spectra of the I_α and I_β forms. It is of interest to note here that among the distinct resonances of the I_α form at C-1, C-4, and C-6, only the one at C-4 appears to be split, while for the I_β form all three resonances associated with these carbons show splitting, with the one at C-1 the most pronounced.

In an effort to further validate the proposal that the I_α and I_β forms were the primary constituents of native celluloses, VanderHart and Atalla undertook another extensive study to exclude the possibility that experimental artifacts contributed to the key spectral features assigned to the two forms.[75] A number of possible sources of distinctive spectral features were explored. The first was the question whether surface layers associated with crystalline domains within particular morphological features in the native celluloses could give rise to features other than those of the core crystalline domains. The second was whether variations in the anisotropic bulk magnetic susceptibility associated with different morphologies could contribute distinctive spectral features. Exploration of the spectra of higher plant celluloses with different native morphologies revealed very little difference in the essential features of the spectra, even after the samples had been subjected to acid hydrolysis. Furthermore, it was concluded that the I_α component of higher plant celluloses was sufficiently low that some question was raised as to whether it occurs at all in these higher plant celluloses. In this context, it was also concluded that, in higher plant celluloses, the lineshapes of the I_β form at C-4 could only be reconciled with a unit cell possessing more than four anhydroglucose residues per unit cell.

Attention was then directed to analysis of the spectra of algal celluloses wherein the I_α component is the dominant one. Relaxation experiments confirmed that the essential spectral features identified with the two crystalline forms of cellulose were characteristic of the core crystalline domains; when measurements were conducted such that magnetization of the surface domains was first allowed to undergo relaxation, very little change in the spectral features was observed. The relaxation experiments suggested that domains consisting of both the I_α and I_β forms have equal average proximity to the surface. One possible interpretation of these observations, that the two forms are very intimately mixed, was ruled out at that time on the basis of hydrolysis experiments, the results of which are now in question.

Two groups of modifying experiments were carried out with the algal celluloses. In the first, the algal celluloses were subjected to severe mechanical action in a Waring blender. In the second, the algal celluloses were subjected to acid hydrolysis, in 4 N HCl for 44 h at 100 °C. While the mechanical action resulted in some reduction in the proportion of the I_α form, acid hydrolysis resulted in a dramatic reduction, sufficient indeed to make the spectra seem like those of the higher plants, except that resolution of the spectral lines was much enhanced relative to that observed in the spectra of even the purest higher plant celluloses. The samples subjected to hydrolysis wherein the recovery varied between 12 and 22%, were examined by electron microscopy and shown to have lateral dimensions not unlike those of the original samples. These observations were interpreted to imply that the I_α form is more susceptible to hydrolysis than the I_β form. An earlier study of the effect of hydrolysis, under similar conditions but for only 4 h, had been carried out with cellulose from *Rhizoclonium heiroglyphicum* with no discernible effect on the spectra.[76] The difference in duration of the hydrolysis may well have been the key factor. Both of these observations and their interpretations had been presented, however, before it was recognized that exposure of celluloses with relatively high contents of the I_α form to elevated temperatures can result in its conversion to the I_β form.[77] When the possibility that the I_α content of the algal cellulose had been converted to the I_β form is taken into account, the results of the relaxation experiments of VanderHart and Atalla cited above can be reinterpreted as indicating intimate mixing of the I_α and I_β forms within the crystalline domains of the algal celluloses.

VanderHart and Atalla also took advantage of the spectra derived from the acid hydrolyzed samples of the algal cellulose to generate more highly resolved representative spectra of the I_α and I_β forms. These are shown in Figure 8 where it is clear that even in the spectrum representative of the I_α form the upfield wings of the C-4 and C-6 resonances are reduced to a minimum. With the completion of this study by VanderHart and Atalla, most of the questions about the possibility that the spectral features were the results of artifacts were put to rest, and the hypothesis that all native celluloses belong to one or to a combination of these forms was generally accepted.

With the above resolution of the questions concerning the nature of native celluloses in mind, it was possible to classify these celluloses with respect to the relative amounts of the I_α and I_β forms occurring in the celluloses produced by particular species. It emerged in these early studies that the celluloses from more primitive organisms such as *Valonia ventricosa* and *Acetobacter xylinum* are predominantly of the I_α form, while those from higher plants such as cotton and ramie are predominantly of the I_β form. As noted earlier, the nomenclature chosen was intended to avoid confusion with the I_A and I_B forms previously used to classify the celluloses on the basis of their infrared spectra in the OH stretching region. In relation to that classification, the NMR spectra

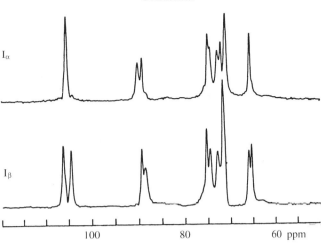

Figure 8 Alternative candidates for the spectra of celluloses I_α (top) and I_β (bottom) derived from linear combinations of the spectra of I_α rich *Cladophera glomerata*, before and after acid hydrolysis, which resulted in a I_β rich cellulose.[75]

suggest that the I_A group has the I_α form as its dominant component, while the I_B group is predominantly of the I_β form.

With the recognition that all plant celluloses are composites of the I_α and I_β forms, it was possible to rationalize many of the earlier difficulties in developing suitable structural models. It became clear that the efforts to reconcile the diffraction patterns in terms of a unique unit cell for native celluloses were frustrated by the reality that the celluloses were composites of two crystalline forms that were blended in different proportions in celluloses produced by different organisms. As noted earlier, the finding that celluloses from different biological sources represented unique blends of the two different forms also resolved the paradox posed early in this century by Cross and Bevan[4] when they contrasted the observation of individuality in biological systems with the invariance of crystalline forms.[4]

3.16.3.2 Further Studies of Structures in Cellulose

With the wide acceptance of the proposal of the two crystalline forms (I_α and I_β) came the challenge of understanding the differences between them and their relationship to each other within the morphology of native cellulosic tissues. A number of complementary approaches were pursued by different investigators in the search for answers to these questions. Some were based on further application of solid-state ^{13}C NMR to the study of different celluloses as well as to celluloses that had been subjected to different modifying treatments. Others were based on application of Raman and IR spectroscopy to new classes of cellulosic samples. Others still were based on refinement of electron microscopic and electron diffractometric methods. The results of these investigations will be presented in summary.

3.16.3.2.1 Raman and infrared spectroscopic studies

The categorization of native celluloses into the I_A and I_B group by Howsmon and Sisson[22] and Blackwell and Marchessault[31] on the basis of the appearance of the OH stretching region of their infrared spectra suggested that the hydrogen bonding patterns within the crystalline domains may be part of the key to the differences between the two forms of native cellulose. This was, in fact, confirmed in the course of more detailed investigations of the Raman spectra carried out on single oriented fibers of native celluloses[78] and in a comprehensive study of the infrared spectra of a number of celluloses of the two forms.[79] The Raman spectral investigations were part of a broader study directed primarily at assigning the bands associated with the skeletal vibrational motions and at exploring the differences between celluloses I and II.[78] They differed from earlier Raman spectral studies in that the spectra were recorded with a Raman microprobe on which individual fibers could be mounted for spectral investigation. With this system it was also possible to explore the variation of intensity of the bands as the polarization of the exciting laser beam was rotated relative to the axis of the fibers.

The observed spectra are shown in Figures 9 and 10, each of which includes six spectra. Figure 9 shows the region between 250 and 1500 cm^{-1}, while Figure 10 shows the region above 2600 cm^{-1}; the region between 1500 and 2600 cm^{-1} does not contain any spectral features. The spectra in Figures 9 and 10 are of native and mercerized ramie fibers and native *Valonia ventricosa*, and they are recorded with both parallel and perpendicular polarization of the exciting laser beam. Those identified as 0° spectra were recorded with polarization of the electric vector of the exciting laser beam parallel to the direction of the fiber axes, while those identified as 90° spectra were recorded with polarization of the electric vector of the laser perpendicular to the fiber axes. The ramie fibers are known to have the molecular chains parallel to the fiber axes; the *Valonia ventricosa* fibers were prepared by drawing the cell wall in order to align the microfibrils within it.

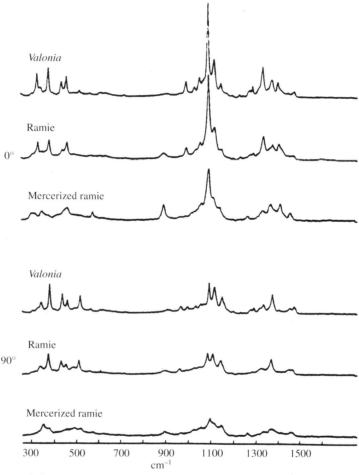

Figure 9 Comparison of the Raman spectra from *Valonia*, ramie, and mercerized ramie (low-frequency region). Spectra were recorded with the electric vector at both 0° and 90°.[78]

A number of features in the spectra are noteworthy with respect to earlier discussions. The first is a comparison of the spectra of *Valonia ventricosa* and ramie. It is clear that, apart from a broadening of the bands in the ramie spectra, because of the smaller lateral dimensions of the crystalline domains, the spectra are very similar except in the OH stretching region. This was interpreted as evidence that the chain conformations in both the I$_\alpha$ and I$_\beta$ forms are the same but that the hydrogen bonding patterns between the chains are different within the two forms. This interpretation is more clearly demonstrated in a comparison of the spectra of *Valonia ventricosa* and *Halocynthia* presented below.

The second feature worthy of note is the dramatic difference between the spectra of native (cellulose I) and mercerized (cellulose II) ramie fibers, particularly in the low-frequency region. This was taken as further confirmation that the conformations of cellulose I and cellulose II must differ sufficiently to result in significant alteration of the coupling patterns between the internal vibrational

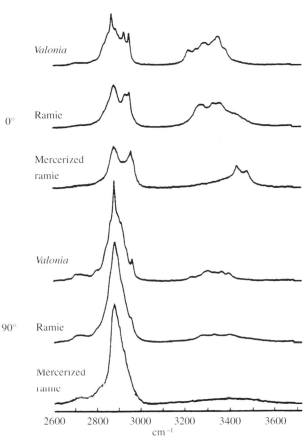

Figure 10 Comparison of the Raman spectra from *Valonia*, ramie, and mercerized ramie (high-frequency region). Spectra were recorded with the electric vector at both 0° and 90°.[78]

modes of the pyranose rings in the molecular chains. It is interesting in this connection to also compare the intensities of the band at 1098 cm^{-1} in the spectra of the two forms of ramie. The band is clearly less intense in the spectrum of the mercerized sample, suggesting that a conformational change which reduces coupling of the skeletal motions has occurred. The 1098 cm^{-1} band is the strongest skeletal band and it is the most intense feature in the spectrum when polarization of the exciting radiation is parallel to the chain direction. The sensitivity of the Raman spectra to the orientation of an intramolecular vibrational motion is also illustrated in the intensity of the methine CH stretching band at about 2889 cm^{-1}. It is most intense with the electric vector of the exciting radiation at 90° to the chain axis, an orientation which is parallel to that of the methine CH bonds of the pyranose rings.

Finally, in light of the discussion of the nonequivalence of adjacent anhydroglucose units and the corresponding nonequivalence of alternating glycosidic linkages, the OH region in the 0° spectrum of mercerized ramie is of particular interest. It shows the two distinct sharp bands that provide evidence of the presence of nonequivalent intramolecular hydrogen bonds in concert with the alternating glycosidic linkages along the chain; the hydrogen bonds are oriented parallel to the chain direction. This alternation clearly stands out most distinctly in cellulose II. These distinct bands cannot be attributed to nonequivalent chains as the difference in frequency implies a difference in the O$\cdots$O distances between the oxygen atoms anchoring the hydrogen bond as well as a difference in the dihedral angles ψ and ϕ of the associated glycosidic linkages. Nonequivalent chains would have different periods in the chain direction if they possessed twofold helical symmetry.

Infrared spectral studies of the I$_\alpha$ and I$_\beta$ forms were carried out by Sugiyama *et al.*[79] on a number of different native celluloses of both forms. Furthermore, it included examination of a number of I$_\alpha$ rich celluloses that were converted to the I$_\beta$ form through the annealing process first reported by Hori and co-workers.[80] In order to complement the infrared spectra, Sugiyama *et al.*[81] recorded electron diffraction patterns for the samples, which allowed classification of the celluloses through comparison with the diffraction patterns acquired in an earlier electron diffractometeric study which will be discussed in greater detail in a subsequent section (3.16.3.2.3).

The key finding emerging from examination of the infrared spectra of the different forms was that the only differences noted were in bands clearly associated with the OH group. This was also true of the changes observed upon conversion of the I_α form to the I_β form through annealing. The bands associated with both the differences in native forms and with the effects of transformation were observed in both the O—H stretching region above 3000 cm^{-1} and the O—H out-of-plane bending region between 650 and 800 cm^{-1}. It was reported that spectra of the I_α form had distinctive bands at 3240 and 750 cm^{-1}, while the spectra of the I_β form had distinctive bands at 3270 and 710 cm^{-1}. Furthermore, it was observed that the band at 3240 cm^{-1} appears to be polarized parallel to the direction of the fibril orientation, while the band at 3270 cm^{-1} is not polarized. Among the low-frequency bands, the one at 710 cm^{-1} appears to be polarized perpendicular to the fibril direction, while the one at 750 cm^{-1} is not polarized. It was also observed that upon transformation of the I_α rich celluloses to the I_β form through annealing, the corresponding bands changed accordingly. The authors concurred with the interpretation of the differences between the two forms suggested by Wiley and Atalla and concluded that I_α to I_β transformation corresponded primarily to a rearrangement of the hydrogen bond system within the structures and that the two structures appeared to have very similar conformations. The infrared spectral studies by Sugiyama *et al.* are particularly interesting because they included the spectra of both *Valonia* and *Halocynthia*, the Raman spectra of which have been investigated at high resolution.[82]

The Raman spectra of *Valonia macrophysa* and *Halocynthia* (tunicate) celluloses acquired by Atalla *et al.*[82] are shown in Figure 11. These particular spectra are of interest because *V. macrophysa* is known to be predominantly the I_α form, while *Halocynthia* is predominantly of the I_β form. Comparison of their spectra can be more rigorous than was possible in the earlier work of Wiley and Atalla[78] because the lateral dimensions of the fibrils of both forms are of the order of 20 nm, with the result that their spectra show equal resolution of the bands in all regions of the spectrum. It is to be noted that their spectra are essentially identical in all of the regions associated with skeletal vibrations of all types as well as regions associated with most of the vibrations involving C—H bonds, whether in the bending or stretching regions. Indeed, the primary differences between the two spectra are in the broad complex bands that occur in the O—H stretching region, and these differences are not unlike those noted in the earlier Raman spectral studies described above. In addition, the weak band at about 840 cm^{-1} in the spectrum of *V. macrophysa* has no corresponding band in the spectrum of *Halocynthia*; this is the band attributed to the out-of-plane bending vibrations of hydrogen bonded OH groups. There are also minor differences in the relative intensities of the methylene C—H stretches and H—C—H bending vibrations, but these are the natural consequences of different hydrogen bonding patterns for the hydroxyl group at C-6. Comparison of the two spectra reinforces the interpretation presented earlier, on the basis of spectra in Figures 9 and 10, leading to the conclusion that the only difference between the I_α and I_β forms is in the pattern of hydrogen bonding. Thus, Raman spectral comparison of the two forms is entirely consistent with that reported for the infrared spectra of these highly crystalline celluloses. It must be kept in mind, of course, that the bands associated with OH group vibrations are not expected to coincide in the Raman and infrared spectra; because of the different bases for activity in the two different spectral approaches to measurement of vibrational frequencies, Raman-active vibrational modes are frequently silent in the infrared and vice versa. This, of course, is also true for the skeletal bands.

In view of the considerable variation observed in the Raman spectra of celluloses as a result of changes in molecular conformations, there can be little question that the spectra in Figure 11 indicate that the conformations of the cellulose molecules in *Valonia* and *Halocynthia* are essentially identical. It is also important to note that the Raman spectra of the celluloses from *V. macrophysa* and *Valonia ventricosa*, both of which have been used in different studies as representative of the I_α form, are effectively indistinguishable in all regions of the spectra. This is also true of the Raman spectra of celluloses from the algae *Cladophera glomerata* and *Rhizoclonium heirglyphicum*, which have also been used in many studies as representative of celluloses that are predominantly of the I_α form.

In summary then, the Raman and infrared spectral studies undertaken after discovery of the composite nature of native celluloses point to the conclusion that the only difference between the two forms is in the pattern of hydrogen bonding between chains that possess identical conformations. Yet electron microscopic and electron diffractometric studies to be described in greater detail in a following section (3.16.3.2.3) have led to conclusions that the two forms represent two crystalline phases with different crystal habits.[81] It is therefore important to consider what information may be developed from the vibrational spectra with regard to this question.

The key conclusion drawn from the electron diffractometeric data was that the I_α form represents

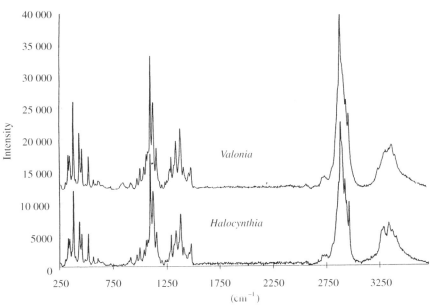

Figure 11 Raman spectra of I_α rich *Valonia* and I_β rich *Halcynthia* celluloses. Because both have fibrils of large lateral dimensions, the spectra of both are well resolved and provide a better basis for comparison of the spectra of I_α rich and I_β rich celluloses [82]

a triclinic phase with one chain per unit cell, while the I_β form represents a monoclinic phase with two chains per unit cell. Furthermore, the symmetry of the monoclinic phase appeared to be that of space group $P2_1$. It has been recognized recently[82] that such a proposal is not consistent with the vibrational spectra. While it was not possible to have full confidence in this conclusion on the basis of the earlier spectral data because of the differences in the level of resolution between the spectra of ramie and *Valonia* celluloses, the spectra shown in Figure 11 are of sufficiently high resolution and sufficiently similar that the comparisons can indeed be made with confidence.

The key issue is that when crystal structures possess more than one molecule per unit cell, and the molecules have the same vibrational frequencies, the vibrational modes of the unit cell become degenerate. Under these circumstances, couplings will arise between equivalent modes in the different molecules, and it is generally observed that such couplings result in splittings of the bands associated with key vibrational modes. The type of coupling that is relevant in the case of cellulose is that described as correlation field splitting.[83] This effect arises because, as a result of the coupling, the vibrations of a particular mode in the two molecules will now occur at two frequencies that are different from those of the isolated molecule; one of the two new frequencies will have the modes in the two different molecules in phase with each other, while the other will have the modes out of phase with each other. Such correlation field effects result in doublets with a splitting of 10–15 cm^{-1} in some modes of crystalline polyethylenes having two chains per unit cell. Since no evidence of such splittings occurs in the *Halocynthia* spectrum shown in Figure 11, it must be concluded that the I_β form cannot have more than one molecule per unit cell. Nor can it be suggested that the two molecules in a monoclinic unit cell are nonequivalent and may have modes that are at different frequencies because the skeletal bands in the *Halocynthia* spectrum are essentially identical to those in the *Valonia* spectrum. Furthermore, this similarity was also reported in the infrared spectra observed by Sugiyama *et al.*[81] Thus, it is clear that the vibrational spectra, both Raman and infrared, point to the conclusion that both the I_α and I_β forms have only one molecule per unit cell. This conclusion of course raises the question as to why the crystallographic data has been viewed for so long as pointing to a two-chain unit cell with the symmetry of space group $P2_1$. This is an issue that is best addressed after the results of electron diffractometric studies have been described in greater detail.

3.16.3.2.2   *Solid-state* ^{13}C *NMR spectroscopic studies*

It is not surprising that the methodology which first provided the basis for understanding the composite nature of native celluloses in terms of the I_α and I_β duality has continued to be the one most often used for seeking deeper understanding of the differences between native celluloses derived from different biological sources. This has been facilitated by the greater availability of solid-state ^{13}C NMR spectrometer systems and by the relative simplicity of the procedures for acquiring the spectra from cellulosic samples. The studies undertaken on the basis of further examination of the solid-state ^{13}C NMR spectra of celluloses are in a number of categories. The first group are focused on further examination of the spectra of different native celluloses, in part aided by mathematical procedures for deconvolution of the spectra or for resolution enhancement. Another group rely on exploring the spectral manifestations of native celluloses that have been modified in different ways. Yet a third approach is based on investigation of celluloses subjected to different but well-known procedures for inducing structural transformations in the solid aggregated state of cellulose. Since the approaches adopted by some groups of investigators in their acquisition and analysis of the spectra have been different, the following discussions will be structured to reflect these differences. The work of the different groups with regard to native cellulose and its response to a variety of treatments will be explored first to the extent that it illuminates questions concerning the nature of native celluloses. This will be followed by an examination of the manifestations of a broader category of structural changes induced by different treatments known to alter the states of aggregation of cellulose. In selecting the investigations to be emphasized in this discussion, the author focuses on studies that provide insight into the variations of the states of aggregation with the history of particular celluloses, both with respect to source and processes of isolation and transformation.

The group at the Kyoto University Institute for Chemical Research carried out important studies that were complementary to those undertaken by VanderHart and Atalla.[73-75] A number of other groups have made valuable contributions. Since a number of questions concerning the nature of the I_{α} and I_{β} forms remain outstanding, it is useful to begin with an overview of the findings of different groups. These will then make it possible to view results of studies using other methods in a clearer perspective.

The early studies by the Kyoto University group have been well summarized in a report that addresses the key points that were the focus of their investigation.[84] In a careful analysis of the chemical shifts of the C-1, C-4, and C-6 carbons in the CP/MAS spectra of monosaccharides and disaccharides for which crystallographic structures were available, Horii *et al.*[84] recognized a correlation between the chemical shifts and the dihedral angles defined by the bonds associated with these particular carbons. In particular, with respect to C-6, they demonstrated a correlation between the chemical shift of the C-6 resonance and the value of the dihedral angle χ defining the orientation of the OH group at C-6 relative to the C-4—C-5 bond in the pyranose ring. This correlation is of value in interpretation of the solid-state ^{13}C NMR spectra with respect to structure as well as discussion of the implications of splittings of the C-6 resonances observed in some of the spectra.

Of even greater interest, in light of the discussions of deviations from twofold screw axis symmetry in some of the structures, it was observed that the chemical shifts of C-1 and C-4 are correlated with the dihedral angles about the glycosidic linkage. In particular, there was a correlation between the shift of C-1 and the dihedral angle ϕ about the C-1—O bond and a correlation between the shift of C-4 and the dihedral angle ψ about the O—C-4 bond. As the spectra published in the earliest studies did not have sufficient resolution to reveal the splittings of the resonances of C-1 and C-4, the possibility of occurrence of nonequivalent glycosidic linkages was not addressed at that time.

In addition to analysis of the correlation between the chemical shifts and the dihedral angles, the Kyoto group investigated the distribution of cellulosic matter between crystalline and noncrystalline domains on the basis of measurements of the relaxation of magnetization associated with the different features of the spectra. By measurement of the values of the spin–lattice relaxation times $T_1(C)$ associated with the different spectral features, they developed a quantification of the degree of crystallinity in the different celluloses. They also undertook analysis of the lineshapes of the different resonances, particularly that of the C-4 resonance. The lineshape analysis was based on deconvolution of the spectral features into combinations of Lorentzian functions centered at the assigned shifts for the particular resonances. It is to be noted that the use of Lorentzian functions, which can be justified at a fundamental level in the case of spectra from molecules in solution, has no basis in any fundamental understanding of the phenomenology of acquisition of the solid-state ^{13}C NMR spectra. However, since deconvolution into Lorentzians has been found to be a useful tool in assessing the spectral features in the spectra of cellulose, its use has continued. The qualifications that must be kept in mind when it is used have been addressed by VanderHart and Campbell.[85]

In the early studies by Horii *et al.*,[84] all of the upfield wing of the C-4 resonance was attributed to molecules in noncrystalline domains. On this basis, lineshape analysis of the C-4 resonance of different native celluloses did not seem consistent with the model proposed by VanderHart and

Atalla[73] with respect to the composite nature of native celluloses. In later studies, when Horii *et al.* took note of the fact that, in the study by VanderHart and Atalla, approximately half of the upfield wing of C-4 in the spectra of higher plant celluloses was attributed to the surface molecules of crystalline domains, Yamamoto and Horii.[86] indicated that their results confirm the proposal of VanderHart and Atalla. It is to be noted that in their early reports in this area, Hori *et al.* used the designations I_b and I_a to designate the different groups of celluloses in which the I_α and I_β forms were dominant. However, in their later studies they have adopted the I_α and I_β designations that are designed to avoid confusion with the categories first introduced by Howsmon and Sisson[16] discussed earlier.

In pursuit of further understanding of the I_α and I_β duality, Horii et al. explored the effects of transformative treatments on the solid-state ^{13}C NMR spectra. The first group of studies were directed at the effects of annealing, first in saturated steam[77] and later in aqueous alkaline solutions (0.1 N NaOH) selected to avoid hydrolytic decomposition of the cellulose.[87,88] In summary, the key findings were that the celluloses wherein the I_α form is dominant are substantially transformed into the I_β form when conditions are established so as to allow the transformation to be complete. The cellulose representative of the I_α form that was used for these studies was *Valonia macrophysa*. The effects of the annealing treatment are demonstrated in Figure 12 which shows the progression in the degree of conversion as the treatment temperature is increased. Each of the treatments were for 30 min in aqueous alkaline solution. These results, of course, point to the susceptibility of the I_α form to conversion to the I_β form, suggesting that the latter is the more stable form. To test this hypothesis, a sample of tunicate cellulose, which had earlier been shown to be of the I_β form by Belton *et al.*,[89] was also annealed in an aqueous alkaline solution at 260 °C; it showed little change as a result of the annealing.[88]

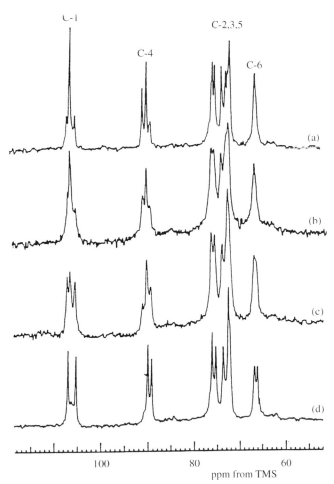

Figure 12 50 MHz CP/MAS ^{13}C NMR spectra of *Valonia* cellulose annealed at different temperatures in 0.1 N NaOH solution: (a) original; (b) 220 °C; (c) 240 °C; (d) 260 °C.[87]

Additional studies by the Kyoto group relied on solid-state ^{13}C NMR to explore the effects of different variables on the structure of cellulose.[90] They will be considered in a later section. It is in order, in the present context, to note briefly the results of one study in which celluloses from *Acetobacter xylinum* cultures were investigated. One of the variables explored was the temperature of the culture; it was observed that lower temperatures favored formation of the I_α form at the expense of the I_β form. This finding raises a fundamental question regarding the possibility that the variation of the balance between the two forms is, in part, an adaptive response to changes in the environment.

In 1990, Newman and Hemmingson[91] began to combine some additional methods of processing the ^{13}C NMR spectral data with those that had been used previously such as monitoring the value of T_1 (C) associated with the different spectral features. While these procedures incorporate a significant degree of empiricism, they have facilitated rationalization of the spectral features of a number of native celluloses and are therefore valuable contributions to the repertoire of methods available for interpreting the ^{13}C NMR spectra of native celluloses. It must be noted, however, that the application of these methods has been complemented in the work of Newman and Hemmingson by a considerable degree of awareness of the complexity of the structures of both native and processed celluloses, so that their application by others needs to be approached with this awareness in mind.

Two innovations were introduced into studies of cellulose in this work. The first was the adoption of an alternative resolution enhancement protocol that was justified primarily on the basis of its success in isolating certain features of the spectra. The second and perhaps more helpful innovation was the application of a procedure that relies on the variation in the proton relaxation times to edit the spectra in order to separate contributions from domains wherein the molecules had different degrees of molecular mobility. In lignocellulosic materials that have not been chemically fractionated through the application of appropriate isolation procedures, the possibility of separation of the subspectra from domains with different degrees of molecular mobility provides a distinct advantage. Thus, the procedure affords the opportunity to examine native tissues that include cellulose without the need to isolate the cellulose, and thereby to develop some useful information concerning the nature of the cellulose under conditions that more closely approximate its native state.

Resolution enhancement was accomplished by convoluting the free induction decay with a function having the form:

$$\mathbf{f}(t) = \exp\{\mathbf{a}t^2 - \mathbf{b}t^3\}$$

This function differs from that used by other investigators in earlier studies in that both exponents have been incremented by 1. The rationalization was that the function used in earlier studies was the one usually used for resolution enhancement in liquids and that there is no fundamental basis for its application to solids. Furthermore, the function adopted was found to be better suited to the studies of cellulose. Parameters $\mathbf{a}$ and $\mathbf{b}$ were selected to enhance resolution "without decreasing the signal-to-noise ratio to unacceptable levels." Here again, it should be noted that this procedure needs to be applied with caution as it is in the nature of the function adopted for the convolution that it can create artifacts in the spectra. Applied without supporting information from complementary methods, it can be misleading.

The procedure based on variability in rates of proton spin relaxation[92] was used to generate spectra that were described as "proton spin-relaxation edited" or PSRE spectra. Relaxation of the protons was allowed to occur for a particular period prior to the cross-polarization step in the CP-MAS protocol; these were referred to as "delayed-contact" experiments. Linear combinations of spectra acquired with and without the delay were then generated to describe the spectra of the domains within which the different populations of protons occurred. This procedure requires adjustment of parameters used in generating the linear combinations of the spectra. The guideline used for optimization was "to maximize mutual discrimination between signals at 89 ppm (C-4 in crystal interior cellulose) and 80 ppm (C-4 in noncrystalline cellulose) without allowing any signal to become inverted."

The results of this process are well illustrated in application of this approach to acquisition of spectra from Avicel microcrystalline cellulose illustrated in Figures 13 and 14. Figure 13 shows the normal CP-MAS spectrum together with one acquired with a proton relaxation time of 10 ms. Figure 14 shows the PSRE subspectra of the crystalline and noncrystalline domains generated through linear combination of the two spectra shown in Figure 13, with the linear combination optimized through trial and error to meet the criterion quoted above. It is clear that application of this procedure reveals that a significant fraction of the upfield wings of C-4 and C-6 is associated with domains wherein the molecular mobility is lower than in the more disordered regions.

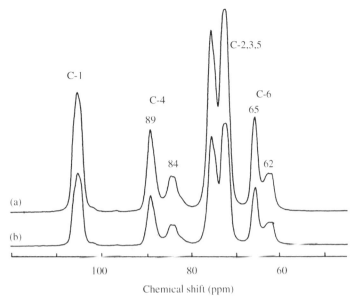

Figure 13 ^{13}C NMR spectra of microcrystalline cellulose I: (a) normal CP/MAS NMR spectrum, (b) delayed contact spectrum with $T = 10$ ms.[97]

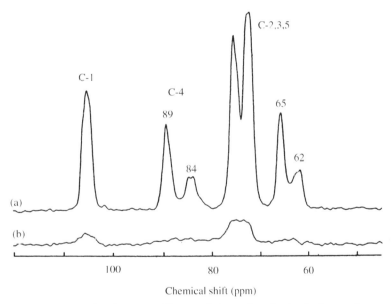

Figure 14 PSRE subspectra of microcrystalline cellulose I, obtained from the spectra shown in Figure 13, selected for domains with (a) $T_{1\rho}(H) = 20$ ms and (b) $T_{1\rho}(H) = 13$ ms.[92]

Application of the resolution enhancement procedure to the data for the more crystalline fraction in Figure 14 results in the spectra shown in Figure 15. Here, both the benefits and the risks of resolution enhancement procedures are demonstrated. The possibility of artifacts is suggested by the low-level peaks that occur in regions wherein no resonances occur in the unenhanced spectra. The magnitude of these artifacts is such that it raises uncertainty about the relative intensities of features that are recognized as real because they coincide with features known to occur in unedited and unenhanced spectra. On the other hand, resolution enhancement clearly demonstrates that the contribution of the less mobile component of the upfield wing of C-4 is a doublet. This doublet has been attributed to molecules on crystalline surfaces representing different faces of the crystalline domains. The resolution enhanced spectrum also gives credibility to the suggestion of a doublet in the upfield wing of C-6.

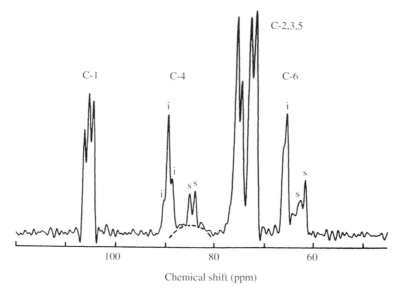

Figure 15 Resolution-enhanced PSRE subspectrum of crystalline domains in Avicel microcrystalline cellulose I. Signals assigned to crystal interior and crystal surface cellulose are labeled i and s, respectively. A broken line outlines a broad signal assigned to poorly-ordered surfaces.[92]

Newman *et al.*[93–96] have used resolution enhanced spectra in attempts to quantify the relative amounts of I_α and I_β forms in a number of native celluloses.[93–96] In such instances the results must be viewed with caution precisely because of the uncertainties concerning the manner in which the enhancement procedure can distort the relative intensities of the spectral features associated with each of the forms and because in the spectral regions characteristic of both C-1 and C-4, there is significant overlap between the spectral features of the I_α and I_β forms. Nevertheless, when applied with care and on the basis of comparisons with spectra of well-characterized species, this approach can be useful for indicating trends.

A different approach to mathematical analysis of the solid-state ^{13}C NMR spectra of celluloses was introduced by the group at the Swedish Forest Products Laboratory (STFI).[97] They took advantage of statistical multivariate data analysis techniques that had been adapted for use with spectroscopic methods. Principal component analysis (PCA) was used to derive a suitable set of subspectra from the CP/MAS spectra of a set of well-characterized cellulosic samples. The relative amounts of I_α and I_β forms and the crystallinity index for these well-characterized samples were defined in terms of the integrals of specific features in the spectra. These were then used to derive subspectra of the principal components, which in turn were used as the basis for a partial least squares analysis of the experimental spectra. Once the subspectra of the principal components are validated by relating their features to the known measures of variability, they become the basis for analysis of the spectra of other cellulosic samples that were not included in the initial analysis.

Here again, it must be recognized that although the methodology has proved to be quite useful in characterizing both native and processed celluloses, it incorporates some arbitrarily defined measures of some of the variables investigated. Thus, quantification of the variables in the context of this approach must be viewed as an aid in comparison of different samples rather than as an absolute measure of these variables. For example, the crystallinity index was defined in terms of the ratio of the integral between 86 and 92 ppm to the integral between 80 and 92 ppm. This, of course, ignores the fact that some component of the integral between 80 and 86 ppm must be attributed to surface crystalline domains. In the same way, the relative amounts of the I_α and I_β forms are defined in terms of arbitrarily chosen partitions of the integral between 103.3 and 106.4 ppm. The part between 104.4 and 105.4 is taken as a measure of the I_α component and the sum of the parts between 103.3 and 104.4 ppm and between 105.4 and 106.4 ppm as a measure of the I_β component; this measure of the partition between the I_α and I_β forms clearly ignores the contribution of the noncrystalline component underlying all of the resonances associated with C-1.

The procedures developed were applied to the analysis of a variety of wood pulps.[98] While they indicate a higher level of the I_α form than had been suggested in previous studies of wood pulps, the relative proportions of the I_α and I_β forms seemed to be the same in most of the pulps, with the major differences between them appearing to reside in the variation in the amorphous fraction.

In later studies[99,100] that are more significant to fundamental understanding of the nature of native celluloses, the spectra of some highly crystalline celluloses were analyzed by nonlinear least-squares fitting with a combination of Lorentzian and Gaussian functions. These analyses led to the conclusion that the *Halocynthia* celluloses do in fact contain a limited amount of the I_α form, contrary to earlier reports that it is exclusively the I_β form. Furthermore, the analyses detected within the envelope of the C-4 resonances give some indication of the occurrence of a "paracrystalline" component. On the basis of spin relaxation measurements, this component was judged to be somewhat more mobile than the I_α and I_β components but clearly less mobile than the amorphous fraction or the fractions of crystalline cellulose at the surface of microfibrils. On this basis it was described as a paracrystalline core structure.

The application of nonlinear least-squares decomposition into a combination of Lorentzian and Gaussian functions incorporates a significant element of empiricism as it requires judgments to be made with respect to which parameters are to be adjusted to achieve the best fit. In addition as noted earlier, the use of Lorentzian functions is not based on any fundamental understanding of the phenomenology of spectral acquisition in solid-state ^{13}C NMR. Yet, in the comparison of the native celluloses, it represents a useful tool that can provide insights into the differences between the types of order and organization that prevail in the different native forms. This is well illustrated in Table 1, taken from Larsson *et al.*,[100] wherein they report the variation of the types of order among a variety of the naturally occurring forms of cellulose.

Table 1 Quantitations made by nonlinear least-squares fitting of the C-4 region in the ^{13}C NMR spectra. All values are relative intensities in percent, and values given in parentheses are standard errors.[100]

Cellulose source	Cellulose I_α	Cellulose I_β	Paracrystalline cellulose	Surface cellulose	Amorphous cellulose	Hemicellulose and cellulose oligomers	Unassigned signal intensity
Wood	9.1 (0.7)		31.1 (0.7)	6.0 (0.8)	53.7 (0.9)		
Cotton	4.2 (0.5)	27.6 (1.8)	33.1 (1.3)	7.8 (1.2)	24.1 (2.0)	3.1 (1.6)	
Halocynthia	9.6 (1.8)	61.2 (11.6)	12.5 (3.5)		16.7 (1.6)		
Cladophora	51.7 (1.1)	29.4 (1.6)	4.5 (0.8)		11.4 (1.0)		3.1 (0.4)
Valonia	55.9	25.1	7.0		7.0		5.0

The types of native celluloses represented cover the wide range of states of aggregation that occur naturally. The most highly crystalline are those of *Halocynthia*, *Cladophera*, and *Valonia*, all of which have microfibrils that have lateral dimensions of 20 nm or more. Among these it appears that *Valonia* and *Cladophera*, both algal celluloses, are the most crystalline in that they have the lowest content of paracrystalline and amorphous components, and for both of them the I_α form is the dominant component although the content of the I_β form ranges from 45 to 60% of that of the I_α form. The *Halocynthia* cellulose appears to have a somewhat higher level of the paracrystalline and amorphous fractions, and the partition between the I_α and I_β forms is clearly shifted much more toward the I_β form. For all three of the highly crystalline celluloses, the value of the surface cellulose content is not reported. It is clear that the resonances at 83.2 and 84.1 ppm, which are associated with the different ordered celluloses at the surfaces of the microfibrils, are within the limits of error.

The results for cotton are particularly interesting because almost half of the intensity of the C-4 resonance that had earlier been regarded as the crystalline component[73,75,84] is reported by Larsson *et al.* to be in the category of the paracrystalline form.[100] The surface celluloses are clearly measurable although a small fraction, and the amorphous fraction is almost as large as the paracrystalline fraction.

The results for wood perhaps show most clearly that this different approach to analysis of the spectra provides a different perspective. Here the I_α and I_β forms together represent a fraction that is one-third that of the paracrystalline form, which is the largest fraction of cellulose. The surface celluloses also are clearly measurable. No value is reported for the amorphous fraction, which is presumably included in the fraction labeled "hemicellulose and cellulose oligomers."

Though the partitioning of the cellulose content in Table 1 cannot be regarded as indicating absolute values of the different fractions, it represents an important step forward in recognizing the different categories of order with respect to the level of molecular mobility associated with them. Furthermore, it highlights the fact that the celluloses which are produced in different functional contexts in different organisms can have a wide range of variation in their states of aggregation and the levels of mobility associated with these states. Thus, it is clear that the mechanisms for formation of cellulose can result in different types of organization at the nanoscale level and, as a consequence, a wide range of properties.

3.16.3.2.3 *Electron microscopic studies*

The use of electron microscopy in the study of celluloses, particularly in their native state, has resulted in important advances beginning with investigations that were undertaken at the time of introduction of the earliest electron microscopes. The early work has been ably reviewed by a number of authors.[101,102] Of particular note among these is the coverage of the subject in the treatise by Preston.[8] Here the author focuses on studies that have been important to advancing understanding of the structure of cellulose at the submicroscopic level.

The earliest and most significant observations, from a structural perspective, were those by Hieta et al.[103] where they applied a staining method incorporating a chemistry that requires the presence of reducing end groups. They observed that when whole microfibrils of *Valonia* were viewed, only one end of each microfibril was stained. This clearly indicated that the molecular chains were parallel as the reducing ends of the cellulose chains occurred together at one end of the fibrils. Had the structure been one with an antiparallel arrangement of cellulose chains, it would have been expected that the reducing end groups would occur with equal frequency at both ends of the microfibril with the result that both ends would be equally stained.

The conclusions of Hieta et al.[104] were independently confirmed by another method introduced by Chanzy and Henrissat wherein the microfibrils were subjected to the action of a cellobiohydrolase that is specific in its action on the nonreducing ends of the cellulose chains. They observed a clear narrowing of the tips of the microfibrils to a triangular form at only one end of each microfibril. In this instance the action was at the nonreducing ends but the observations were equally convincing evidence that the chains are aligned in a parallel arrangement in these microfibrils.

These early studies were focused on microfibrils from algal celluloses that, because of their larger lateral dimensions, could be more easily visualized in detail. More recently, the technique of specific staining of reducing end groups was adapted for application to cotton microfibrils by Maurer and Fengel.[105] In addition to application of the technique to examination of native cellulose, Maurer and Fengel applied the method to examination of microfibrils of mercerized cellulose (cellulose II) for which they also observed staining at only one end of the microfibrils. This last observation, which indicates a parallel chain structure in cellulose II, is very much in contrast to the crystallographic models that point to an antiparallel structure for this form of cellulose. It reinforces the view that the structure of cellulose II still has many uncertainties associated with it, in spite of the many theoretical analyses that have attempted to rationalize the antiparallel form.

In yet another important set of investigations by Sugiyama et al.[106–108] reported at approximately the same time, it was demonstrated that lattice images could be recorded from the microfibrils of *V. macrophysia*. The first images captured were based on lateral observation of the microfibrils;[106,107] an example is shown in Figure 16. Later, the techniques were refined to allow the acquisition of lattice images of cross-sections of microfibrils[108] as shown in Figure 17. The significance of these observations is that it is now possible to demonstrate conclusively that the microfibrils are uniform in formation and that there is no evidence that they are constituted of smaller subunits which aggregate together to form the individual microfibrils that are observed in the electron micrographs. Thus, the observations resolved some of the questions that had arisen earlier concerning the interpretation of electron micrographs of native celluloses;[8,9] the findings of Sugiyama et al. were the first direct evidence that the approximately 20×20 nm cross-sections were not composed of distinguishable smaller subunits. It should be noted, however, that the electron diffraction processes responsible for formation of the lattice images are dominated by organization of the heavy atoms in the molecular chains and would be insensitive to any nonuniformity in the hydrogen bonding patterns within the interior of the 20×20 nm fibrils. The homogeneity of the microfibrils revealed in the lattice images is an issue that needs to be revisited in the context of discussions of biogenesis, for in each instance the homogeneous crystalline domains clearly include a much larger number of cellulose chains than could possibly arise from the individual membrane complexes associated with the biogenesis of cellulose.

Later studies by Sugiyama et al. were based on electron diffraction and were directed at addressing questions concerning the nature of the differences between the I_α and I_β forms of cellulose. In a landmark study,[109] electron diffraction patterns were recorded from *Valonia macrophysia* in both its

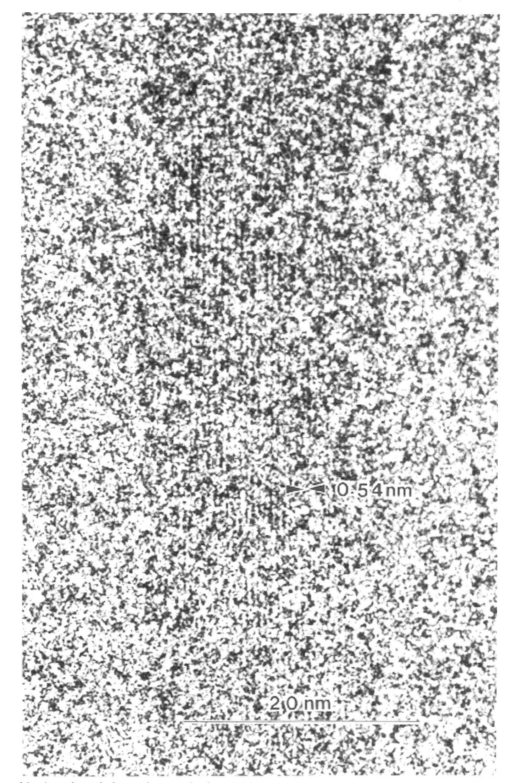

Figure 16 An enlarged photomicrograph of a cellulose microfibril of *Valonia macrophysa* with 0.54 nm lattice lines clearly visible (after Sugiyama[107]).

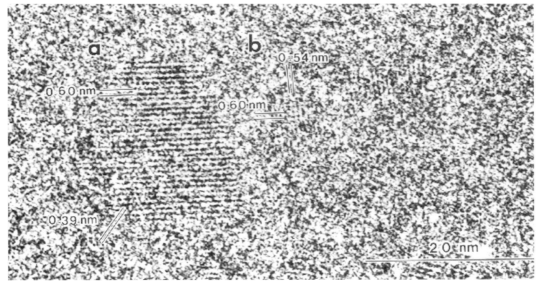

Figure 17 Lattice images od the lateral faces of microfibrils of *Valonia macrophysa* showing more than two sets of lattice fringes indicated (after Sugiyama[108]).

native state, wherein the I_α and I_β forms occur in their natural relative proportions, and after annealing using the process first reported by Horii and co-workers,[80] which converts the I_α form into the I_β form. The native material, which is predominantly the I_α form, was shown to produce a complex electron diffraction pattern similar to that which had earlier led Honjo and Watanabe to propose an eight-chain unit cell. In sharp contrast, the annealed sample, which is essentially all of the I_β form, produced a more simple and symmetric pattern that could be indexed in terms of a two-chain monoclinic unit cell. The observed patterns are shown in Figure 18. Figure 18(a) shows the diffraction pattern of the native forms, while Figure 18(b) shows how the diffraction pattern is transformed upon annealing. It is the latter that is identified with the I_β form and which has been interpreted to indicate a monoclinic unit cell. Figures 18(c) and 18(d) are schematic representations of the spots in the diffraction diagrams (a) and (b) and show more clearly how the diffraction pattern is transformed by annealing; the spots marked with arrows are those that disappear upon annealing. Upon separating the diffraction pattern of the I_β form from the original pattern, it was possible to identify the components of the original pattern that could be attributed to the I_α form, and it was found to correspond to a triclinic unit cell.

In this first report concerning the differences between the diffraction patterns of the I_α and I_β forms, the positioning of the chains within the monoclinic unit cell associated with the I_β form was left open. Two possibilities were regarded as consistent with the diffraction patterns, the first with the twofold screw axes coincident with the molecular chains, the second with the twofold screw axes between the chains. Both possibilities were consistent with the occurrence of nonequivalent anhydroglucose units. The triclinic unit cell associated with the I_α form was also viewed as consistent with two possibilities, the first a two-chain unit cell and the second an eight-chain unit cell similar to the one first proposed by Honjo and Watanabe.[110]

In a later study by Sugiyama et al.,[111] the possibilities were narrowed. It was stated that the monoclinic unit cell corresponding to the I_β form was viewed as one wherein the chains were coincident with the twofold screw axes. It was also indicated that the pattern of the triclinic unit cell corresponding to the I_α form appeared consistent with a unit cell with only one chain per unit cell. In both instances the rationale for these determinations was not presented.

Another interesting group of observations reported in the second electron diffraction study by Sugiyama et al.[111] was interpreted as evidence of the occurrence of the two forms of cellulose in separate domains within the same microfibrils. It was reported that the subsets of reflections associated with the two different forms could be observed separately or in combination along the length of an individual microfibril within domains 50 μm from each other. This set of observations was interpreted as indicating an alternation between the I_α and I_β forms along the length of an individual microfibril. Such an interpretation of course raises questions concerning the processes of biogenesis particularly since the relative proportions of the two forms of cellulose have been found to be invariant for a particular species as long as the procedures used for isolation of the cellulose

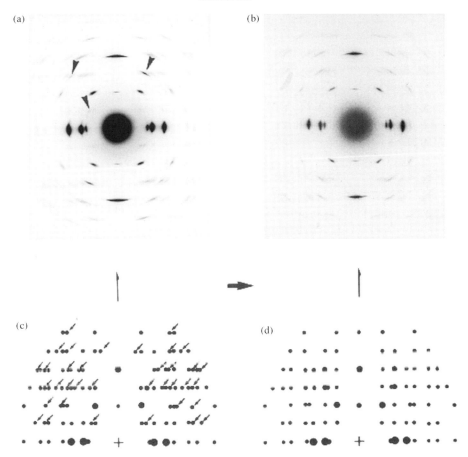

Figure 18 Typical electron diffraction patterns of *Valonia macrophysa* before (a) and after (b) annealing. In the schematic representations of the patterns the spots marked with arrows correspond to the reflections that disappear during the annealing treatment (after Sugiyama[109]).

do not incorporate exposure to conditions that can result in transformation of the I_α form into the I_β form.

The observation of different domains producing different diffraction patterns along the same microfibril can be envisioned as arising in two ways. The first is the possibility that the microfibril which was used to acquire the diffraction patterns had a limited amount of curvature or twist to it so that the angle between the electron beam and the unit cell axes was not constant. This could result in differences of relative intensities of diffraction spots from different planes and, given the short duration of the exposures, result in an unintended editing of the diffraction patterns. Thus, only those diffraction spots that are intense enough to be observed at a particular angle will be detected, while weaker ones go unseen. For example, if the lattice structure first suggested by Honjo and Watanabe[110] is the true one characteristic of the algal celluloses, diffraction patterns observed at different angles would result in different degrees of enhancement of the different subsets of the total diffraction pattern. This would also be true if the three-dimensional organization of the chains is more appropriately viewed as a superlattice. Indeed it is possible that the lattice structure first proposed by Honjo and Watanabe[110] represents the unit cell of such a superlattice.

Such an interpretation of these observations is consistent with earlier observations by Chanzy *et al.*,[112] wherein an electron microscopic image of microfibrils of algal celluloses was formed by use of a technique based on diffraction contrast. It resulted in images of the algal microfibrils that had alternating dark and bright domains which appeared to be of the order of 50 nm in length. This suggests that the Bragg angle associated with a particular set of reflections is not likely to be coherently ordered relative to the electron beam in domains that are more than 50 nm in length. Given that the coherence of orientation relative to the electron beam was not found to extend beyond 50 nm, it would appear unlikely that it would remain invariant over a distance of 50 μm.

An alternative interpretation of the observation is that alternation of the I_α and I_β forms is real, as proposed by Sugiyama *et al.*,[111] and it reflects an assembly process that is not yet sufficiently well understood. It has been suggested that mechanical stress can facilitate transformation of the I_α form into the I_β form and that formation of the I_β form may arise from mechanical deformations of the fibrils in the course of deposition; as they emerge from the plasma membrane they are required, in most instances, to be bent to be parallel with the plane of the cell wall. If this is indeed the source of the reported alternation of the I_α and I_β forms along the microfibril, it would raise questions concerning the uniqueness of the balance between the I_α and I_β forms that seems to be characteristic of particular species.

3.16.3.3 Computational Modeling

Computational modeling has become an important aid in advancing understanding of complex molecular systems. Molecular modeling methods have allowed exploration of many different factors and interactions at the molecular level and the degree to which they may contribute to the phenomenology of different molecular systems. Computational modeling has found particular favor in the analyses of large molecules of biological origin, and of course, cellulose and its oligomers have attracted some attention in this arena. It is valuable to briefly review some of the efforts directed at advancing the understanding of cellulose because, in addition to providing insights regarding the contributions of different classes of interactions, they illustrate the reality that the results of analyses can often be the consequences of assumptions and premises introduced at the outset, rather than conclusions that can provide definitive answers to questions under exploration. It has been the author's experience that conclusions concerning the structures that are most favored for different celluloses change when new sets of potential functions are introduced into the computational programs and as different approaches to finding the most favored structures are adopted.

As alluded to earlier, the analysis by Rees and Skerret[50] was one of the first computational efforts to explore the constraints on the freedom of variation inherent in the structure of cellobiose. It relied on a potential function that is focused on van der Waals interactions in order to establish the degree to which domains within ψ/ϕ space may be excluded by hard sphere overlap. The key finding was that approximately 95% of ψ/ϕ space was indeed excluded from accessibility on the basis of hard sphere overlaps that were unacceptable in the sense that they required particular atoms associated with the region of the glycosidic linkage to be significantly closer to each other than the sum of the van der Waals radii. Upon mapping the energy associated with allowable conformations, they found that the two regions indicated by the solid line contours in Figure 2 represented energy minima close to the conformations defined by the twofold helical constraint. The boundaries of the acceptable region are not very far removed from the domains within the contours; the region between the two domains along the twofold helix line ($n = 2$) was not excluded by hard sphere overlap but it did represent a saddle point in the potential energy surface.

The next group of computational studies did incorporate hydrogen bonding energies as well as the van der Waals interactions. Whether they exhibited the double minima, and the degree to which the double minima were pronounced, depended in large measure on the relative weighting given to the different types of nonbonded interactions. In many, particularly those relying on the potential energy functions incorporated in the linked atom least-squares (LALS) programs, the weighting was based on fitting the potential functions to optimize the match between computed structures for small molecules for which the crystal structures were known from crystallographic studies. The results were that some showed very shallow minima off the twofold helix line;[51] the twofold helical structures were then rationalized on the basis that the departure represented a small difference in the energies that were regarded as within the error of the computation. When the criterion for quality of fit is chosen as a global minimum of the potential energy, without attention to the fact that it may incorporate unacceptable hard sphere overlaps, the results of the computational analysis can be misleading.

Later studies did not incorporate disproportionate weighting of the different types of nonbonded interactions,[57,58] and the result is perhaps best illustrated in the mapping of the potential energy for cellobiose shown in Figure 19, taken from the study of cello-oligomers by Henrissat *et al.*[58] In this instance, for the purposes of visualization, the ψ/ϕ map presents the mirror image of the potential energy surface computed for cellobiose. While well more than two minima are shown in this mapping, it should be noted that only the two corresponding to the crystal structures of cellobiose and methyl-β-cellobioside, marked by arrows, are within the boundaries established in the analysis by Rees and Skerret[50] described earlier. The other minima correspond to conformations that are more favorable to hydrogen bonding, but with relatively high energies associated with the van der Waals interactions pointing to severe hard sphere overlap.

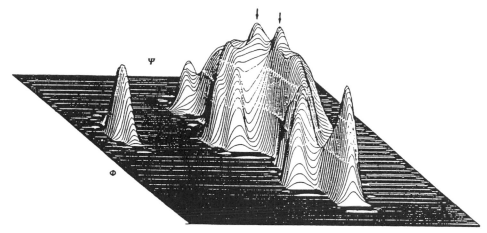

Figure 19 Perspective drawing of the three-dimensional shape of the mirror image of the conformational energy well for the full angular range of Φ and Ψ. The volume was constructed using the following scheme: $V(\Phi,\Psi) > 15\ \mathrm{kcal^{-1}\ mol^{-1}}$: $V_p(\Phi,\Psi) = 0$; $V(\Phi,\Psi) < 15\ \mathrm{kcal^{-1}\ mol^{-1}}$: $V_p(\Phi,\Psi) = -(V(\Phi,\Psi)-15)$, V being the energy expressed relative to the minimum. Proceeding from top to bottom of the three-dimensional shape, note the very low energy region (the arrows point towards the conformations observed for crystalline cellobiose and methyl-β-D-cellobioside). The 5–10 $\mathrm{kcal^{-1}\ mol^{-1}}$ energy contours correspond to the light gray region of the volume.[58]

There is little question that the results of modeling the cellobiose system reflect constraints on the freedom of rotation about the bonds of the glycosidic linkage. This is confirmed by the observation that the most favored structures indeed coincide with the crystal structures of cellobiose and methyl-β-cellobioside, both of which have been established on the basis of single crystal studies that are not in question. It is at the next higher level of structure in oligomers that the complexity of the potential functions and the multiplicity of adjustable and variable parameters begin to introduce considerable uncertainty about the results and the degree to which they reflect the real system. In the work of Henrissat *et al.*,[58] for example, when the analyses of the cellotriose and cellotetraose are carried out, the conclusions drawn are based on additional support from other sources of information on structure, and even with the additional information from other sources, the conclusions can only be presented as the most plausible, given the information available.

It is beyond the scope of this chapter to discuss the many computational modeling studies that have been undertaken wherein the subjects of the modeling exercise were chains of cellulose consisting of four or more anhydroglucose units. This is particularly so since a number of groups have entered the field of inquiry and are, in most instances, still refining their computational methods to factor in the eccentricities of cellulose as a molecule. Some observations on the different approaches are in order, however. The modeling programs are by and large of two types, although some investigators use combinations of the two approaches.

The first and the earlier of the two approaches adopted is the one identified as molecular mechanics, although it is in fact not a dynamic model but rather is based on a search for coordinates corresponding to potential energy minima. Its weakness is that it combines in one potential energy function interactions corresponding to all of the levels of structure alluded to earlier. Thus, it combines searches for equilibrium values of bond lengths together with the search for optimum packing of the molecules on the basis of different possible secondary and tertiary structures. In such computations it is to be expected that energy contributions of the interatomic bonds of the primary structure will be the dominant component of the potential energy. As a result, it is not unusual for the variations in energy associated with the hard sphere overlaps at the glycosidic linkage to be within the uncertainties of the calculations. In such cases, the search for global minima can point to structures with unrealistic local conformations.

The second approach, which is usually identified as molecular dynamics simulation, is based on integration of the equations of motion for all of the atoms in the system. Thus, they differ from the earlier studies in that the search is not for an equilibrium structure that represents a global minimum of potential energy but rather an exploration of the range of movement of the atoms that make up the structure under the influence of the prevailing forces of intramolecular and intermolecular interactions, which are represented by complex potential functions that include terms appropriate for each of the interactions. The coordinates of the atoms are recorded at regular intervals and are used to determine the range of variation of the different internal coordinates within a molecule and the degree to which the molecules in an aggregate move relative to each other. Here again, the assumptions introduced at the outset can be primary determinants of the results. In some studies using this approach, for example, the force field used relies on the united-atom approximation wherein all aliphatic hydrogen atoms attached to a carbon atom are not included except as variations on the terms representing the carbon atoms in the potential function. While this is a useful and adequate approximation in the vast majority of instances where it is applied, its use with cellulose ignores the constraints of hard sphere overlaps on the glycosidic linkage. On the other hand, the programs designed to carry out the dynamic simulation have been modified to allow introduction of the constraint that the bond lengths are fixed in value, so that the energy associated with the primary structure no longer dominates the potential energy as is the case in the molecular mechanics calculations. It is therefore likely to be more successful in advancing understanding of the variability of the secondary and tertiary structures of cellulose.

Since many of the programs that have been used in the computational modeling studies of cellulose were, in the first instance, developed for a far wider range of molecular modeling studies, it is not surprising that they are not particularly well adapted to cellulose at the present time. If progress in computational modeling of complex molecules continues to advance as it has in the past decade, it is anticipated that within the next decade or two, some of these programs will be sufficiently flexible that allowances can be made for the special features in the structure of cellulose associated with the glycosidic linkage and that it will be possible to explore more widely the variability of secondary and tertiary structures associated with the primary structure that is fairly well established. It should then be possible to rationalize the wide variation in the patterns of aggregation of cellulose. The molecular dynamics simulation approach also has the advantage that it allows the exploration of interactions between cellulosic surfaces and associated liquids such as water.

The reader interested in exploring the different approaches to computational modeling of cellulose in greater depth can find them in the publications of Hardy and Sarko *et al.*,[113] Kroon-Battenberg *et al.*,[114,115] Heiner *et al.*,[116,117] O'Sullivan and co-workers[118] and references cited therein.

3.16.3.4 Polymorphy in Cellulose

As was noted earlier, one of the discoveries from the diffractometric studies of cellulose undertaken during the middle period was that it can occur in a number of allomorphic forms in the solid state, each producing distinctive X-ray diffractometric patterns.[30] In addition to the cellulose II form, which has been discussed extensively, two other forms are well recognized; these are cellulose III and cellulose IV. It is of interest to consider them briefly because they reflect the capacity of cellulose to aggregate in a wide variety of secondary and tertiary structures, and because some of the higher plant celluloses produce diffraction patterns that are not unlike those of cellulose IV. Furthermore, they reflect the tendency for some of the celluloses to retain some memory of their earlier states of aggregation in a manner not yet understood.

Cellulose III is of little interest from a biological perspective except to the extent that its behavior may reveal some of the interesting characteristics of the native celluloses from which it can be prepared. It can be prepared from either native cellulose or from cellulose II by treatment with anhydrous liquid ammonia at temperatures near $-30\ °C$. It produces distinctive X-ray patterns, Raman spectra, and solid-state ^{13}C NMR spectra. The most interesting characteristic it has is that it can be restored to the original form by treatment in boiling water. Because of this characteristic it is common to designate samples of cellulose III as either III_I or III_{II} to indicate both the source material and the form that will be recovered if the cellulose is boiled in water. In the case of native celluloses, transformation to the III_I form and back to the I form also has the unusual effect of converting those which have the I_α form dominant, such as those from algal sources, into forms in which the I_β form is dominant. This effect, first reported by Chanzy *et al.*,[119] is accompanied by partitioning of the algal microfibrils into smaller ones that are closer in lateral dimensions to those characteristic of higher plants. The solid-state ^{13}C NMR spectra also then appear more like those of the higher plants. No such changes have been reported for native celluloses in which the I_β form is dominant. These behaviors by cellulose III point to a memory effect with respect to the secondary and tertiary structures of cellulose that remains very much a mystery at the present time.

Cellulose IV is most often described as the high-temperature cellulose because it can be prepared by exposing the source cellulose to temperatures in the vicinity of $260\ °C$ while it is immersed in glycerol. In this preparation, its structure depends on whether it is prepared from cellulose I or cellulose II; hence the frequent designation as IV_I or IV_{II}. When prepared from cellulose I, it is first converted to the III_I form prior to treatment at high temperature in glycerol. When prepared from cellulose II, it can be produced directly from the II form or via the III_{II} form as an intermediate. However, in the instance of cellulose IV, there are no known procedures that allow restoration to the original form; the use of the different designations reflects some differences in the diffraction patterns observed from the two different forms. Furthermore, most of its reported preparations from native forms of cellulose have been from higher plant celluloses wherein the I_β form is dominant and the lateral dimensions of the native microfibrils are quite small; it is not at all clear that treatment of microfibrils of larger lateral dimensions such as those of *Valonia* or *Halocynthia* will result in such changes.

In addition to its preparation by heating at 260 °C in glycerol, cellulose IV has been recovered when cellulose is regenerated from solution at elevated temperatures. This has been observed with solutions in phosphoric acid regenerated in boiling water or in ethylene glycol or glycerol at temperatures above 100 °C.[120] It has also been observed upon regeneration from the dimethyl-sulfoxide–paraformaldehyde solvent system at elevated temperatures.[120] In yet another exploration of high-temperature effects on the aggregation of cellulose, it was found that when amorphous celluloses are prepared under anhydrous conditions and then induced to crystallize by exposure to water, exposure at elevated temperatures resulted in the formation of cellulose IV rather than cellulose II, which is the form usually obtained upon crystallization at room temperature.[121]

The samples of cellulose IV obtained through regeneration from solution were shown to have Raman spectra that could be represented as linear combinations of the spectra of celluloses I and II, suggesting that it may be a mixed lattice in which molecules with two different secondary structures coexist. This possibility is consistent with the earlier conclusion that both cellulose I and cellulose II have ribbon like structures that depart to a limited degree from a twofold helix but in different ways. It is not at all implausible that molecules so similar in shape could coexist in the same lattice.

One of the complications in interpreting observations of the occurrence of cellulose IV is that its X-ray diffraction powder pattern is very similar to that of cellulose I. The 020 reflection is nearly identical to that of cellulose I and the 110 and the $1\bar{1}0$ reflections collapse into a single reflection approximately midway between those of cellulose I. As a result, many of the less well-ordered native celluloses produce X-ray patterns that could equally well be interpreted as indicating cellulose I, but with inadequate resolution of the 110 and $1\bar{1}0$ reflections, or cellulose IV. They are usually characterized as indicating cellulose I because they represent celluloses derived from native sources. Indeed, when cellulose IV was first observed, it was thought to be a less ordered form of cellulose I.

The close relationship between cellulose IV and the native state is also reflected in reports of its observation in the native state of primary cell wall celluloses. These were observations based on electron diffraction studies of isolated primary cell wall celluloses.[122]

3.16.3.5 Oligomers and Structure at the Nanoscale Level

The oligomers of cellulose, which are essentially insoluble at the octamer level, have been the subject of a number of investigations because their secondary and tertiary structures seem to converge to structures similar to those of cellulose II. The X-ray powder patterns are quite similar to those of cellulose II,[123] and the vibrational spectra, both Raman and infrared, converge to that of cellulose at the tetramer.[48] The high-resolution solid-state ^{13}C NMR also converge to that of cellulose II, although at the tetramer the resonance of the anomeric carbon on the reducing end is still distinct.[58,69]

In one of the studies cited,[58] a multidisciplinary approach was used to explore the structures and led to the conclusion that the molecules of the celotetraose were arranged in an antiparallel manner, while another, based on X-ray diffractometry, concluded that a parallel structure is equally probable.[123] These studies also indicated that the dihedral angles of the glycosidic linkage departed systematically from those of a twofold helix. As noted earlier, the recent results of Maurer and Fengel[105] point to a parallel alignment of the molecular chains in the structure of cellulose II. Thus, in spite of many studies, the structures of both cellulose II and the oligomers remain the subject of many unanswered questions.

In relation to the structure of the polymer, the questions arising are the degree to which the observations on the oligomers can inform the interpretation of observations on the polymer. The key question is the point at which convergence of the properties to those of the polymer can be anticipated. While it has been demonstrated for cellulose II only, it is anticipated that convergence of the vibrational spectra to those of the polymer at the level of the tetramer would also be true for celluloses I_α and I_β if it were possible to isolate tetramers in those forms. Thus, in terms of the basic structural units, the vibrational coupling patterns of four anhydroglucose units are sufficient to define the majority of bands in the polymer. These observations, when taken together with the observations of the X-ray powder patterns, suggest that the secondary and tertiary structures are well defined by the organization of eight chains over a length equivalent to that of the tetramer. That is, a volume defined by dimensions equal to twice the dimensions of the most simple unit cells in all three dimensions. The dimensions are of the order of 1.6–2.1 nm, and this level of structure will be defined as organization at the nanoscale level.

3.16.3.6   From Nanoscale to Microscale: Supramolecular Organization in Celluloses

Structure at the level next beyond that defined as nanoscale is the one at which most of the distinctive organization of native celluloses is most clearly expressed. It is also the level most readily accessible through electron microscopy. Here we define it as the microscale level and associate with it molecular organization at the scale ranging between 2 and 50 nm. We have already discussed some aspects of structure at this level in distinguishing the different classes of samples of native cellulose that have been the subject of the studies presented above in overview. We will discuss structure at the microscale level and its implications in greater detail in the next section.

Before summarizing our discussions of structure and considering those of native celluloses, it is important to examine a number of questions that arise at the transition point from nanoscale to microscale for any system, but which have not been adequately addressed in the case of cellulose. To the author's knowledge, these questions have not been addressed in prior reviews of the structure of cellulose. They are of two classes though not unrelated to each other. The first class deals with the degrees of departure from the approximation of the infinite lattice linearly extended in three dimensions implicit in the structural models at the nanoscale level. The second class is concerned with the validity of regarding the aggregated states of cellulose, particularly in its native state, as phases in the traditional sense.

In most native celluloses the departures from the infinite linearly extended lattice are evidently very close to the nanoscale level. It has already been noted and it is well established that lateral order rarely exceeds 6 nm except in the case of some algal celluloses and those of the tunicates, which can have lateral dimensions of up to 25 nm. In most instances curvature of the elementary fibrils begins to manifest itself at these levels. Both of these departures from the infinitely extended linear lattice are usually inherent in the nature of organization of the biological tissue of which the native cellulose is a constituent and an integral part. They are manifestations of the biological order of the particular native cellulose. Yet the effects of these departures from a linear lattice structure on both diffractometric and spectroscopic measurements are indistinguishable from the effects of random disorder as it might occur in systems that are not of a biological origin. In the context of cellulose science, this reality has most often been dealt with by defining that which is not ordered in a linear lattice as disordered or amorphous. Such an approach ignores the fact that in most states of aggregation, amorphous systems are usually assumed to be homogeneously disordered. It has been quite common in the literature on cellulose to attribute to the nonordered components of native cellulosic substances the properties of such amorphous substances. This is an issue that must be kept in mind as individual native celluloses are considered below.

The second class of questions is a logical consequence of the limitation of lateral dimensions. They are associated with the fact that when the lateral dimensions are as small as they are in most native celluloses, a significant fraction of the substance lies at the surface, so that the criterion for describing the aggregates as a separate phase is no longer adhered to. This criterion is that the amount of substance at the surface be vanishingly small in comparison with the amount of substance within the interior of the phase.[124] Given this circumstance, the application of stability criteria derived from traditional thermodynamic analyses must be approached with caution.

It is also appropriate at this point to address a fallacy commonly recurring in the literature with respect to the relative stability of celluloses I and II. Cellulose I is frequently regarded as the "metastable" form of cellulose because, upon regeneration from solution, cellulose II is usually the form recovered, and because upon swelling in strong caustic solution followed by deswelling, which is the process of mercerization, cellulose II is also recovered. Cellulose I, which has been prepared by regeneration from cellulose solutions at elevated temperatures,[125,126] though the procedure is not easy to reproduce and there remain a number of factors involved in the regeneration that are not well understood.

3.16.3.7   Chemical Implications of Structure

It was noted earlier that an acceptable fit to the diffractometric data is not the ultimate objective of structural studies. Rather it is the development of a model that possesses a significant measure of validity and usefulness as the basis for organizing, explaining, and predicting the results of experimental observations. In the sections above, the new and evolving conceptual framework for describing the structures of cellulose was described in relation to spectral observations. It is also important to consider the degree to which the structural information that has been developed above may be useful as the basis for advancing understanding of the response of celluloses to chemical reagents and enzyme systems. It is useful first to review briefly past work directed at rationalizing the responses to such agents.

The vast majority of studies of the chemistry of cellulose has been directed at the preparation of cellulose derivatives with varying degrees of substitution depending on the desired product. Sometimes the goal is to prepare a cellulose derivative that possesses properties which differ significantly from those of the native form, some derivatives are water-soluble, others are thermoplastic, and others still are used as intermediates in processes for the regeneration of cellulose in the form of films or fibers. At other times the objective is to introduce relatively small amounts of substitution to modify the properties of the cellulosic substrate without it losing its macroscopic identity or form such as fiber or microcrystalline powder or regenerated filament or film. All such modification processes begin with a heterogeneous reaction system which may or may not eventually evolve into a homogeneous system as the reaction progresses. Thus, in all chemical investigations that begin with cellulose as one of the ingredients, issues associated with heterogeneous reaction systems arise. Understandably, the one that has been dominant in most investigations is the question of accessibility of the cellulose.

A variety of methods has been developed to relate accessibility to microstructure. Almost all of them begin with the premise that the cellulose can be regarded as having a crystalline fraction and a disordered or amorphous fraction. It is then assumed that the amorphous or disordered fraction is accessible while the crystalline fraction is not. In some instances, the portion of crystalline domains at the surface is regarded as accessible and it is therefore included as part of the disordered fraction. In other instances, the particular chemistry is thought to occur only in the disordered fraction and the surfaces of crystalline domains are not included. The different approaches have been reviewed by Bertoniere and Zeronian,[127] who regard the different approaches as alternative methods for measuring the degree of crystallinity or the crystalline fraction in the particular celluloses.

A number of different chemical and physical approaches is described by Bertoniere and Zeronian. The first is based on acid hydrolysis followed by quantification of the weight loss due to dissolution of glucose, cellobiose, and the soluble oligomers.[128] This method is thought to incorporate some error in quantification of the crystalline domains because chain cleavage upon hydrolysis can facilitate crystallization of chain molecules that had been kept in disorder due to entanglement with other molecules. Another method is based on monitoring the degree of formylation of cellulose when reacted with formic acid to form the ester.[129] In this method, progress of the reaction with cellulose is compared with a similar reaction with starch, which provides a measure of the possibility of formylation in a homogeneous system wherein the issue of accessibility does not arise.

In another method developed by Rowland and co-workers,[130–133] accessible hydroxyl groups are tagged through reaction of the particular cellulose with N,N-diethylaziridinium chloride to produce a DEAE-cellulose. This is then hydrolyzed, subjected to enzyme action to remove the untagged glucose, silylated, and subjected to chromatographic analysis. This method has the added advantage that it can be used to explore the relative reactivity of the different hydroxyl groups. It is usually observed that the secondary hydroxyl group on C-2 is the most reactive and the one at C-3, the least reactive, with the primary hydroxyl at C-6 having a reactivity approaching that of the group on C-2 under some conditions. Here, of course, steric effects are also factors in these substitution reactions.

Among the physical methods discussed by Bertoniere and Zeronian[127] are those based on sorption and solvent exclusion. One of the earliest studies relying on the use of sorption as a measure of accessibility was the classical study by Mann and Marinan[33] wherein deuterium exchange with protons was monitored. The cellulose was exposed to D_2O vapor for a period sufficient to attain equilibrium and then the degree of exchange was measured by observation of the infrared spectra. Comparison of the bands associated with the OD stretching vibration to those associated with the OH stretching vibration provided a measure of the relative amounts of accessible and inaccessible hydroxyl groups. Another approach to monitoring availability to adsorbed molecules is measurement of moisture regain upon conditioning under well-defined conditions as described by Zeronian and co-workers.[134]

The method of solvent exclusion has been used to explore issues of accessibility on a somewhat larger scale. An approach pioneered by Stone and Scallan[135] and Stone *et al.*[136] relied on static measurement using a series of oligomeric sugars and dextrans of increasing size to establish the distribution of pore sizes in different preparations of a variety of native celluloses. These, however, provide measures of accessibility beyond the nanoscale and were indeed designed to explore pores closer to the lower range of what we have defined as the microscale.

While methods for characterizing celluloses on the basis of their accessibility have been useful, they do not provide a basis for understanding the level of structure at which the response of a particular cellulose is determined. This follows from the rather simple categorization of the substrate cellulose into ordered and disordered fractions corresponding to the fractions thought to be crystalline and those that are not. This classification does not allow discrimination between effects that have their origin at the level of secondary structure and those that arise from the nature of the tertiary structure. Thus, in terms of chemical reactions, this approach does not facilitate separation of steric effects that follow from the conformation of the molecule as it is approached by a reacting species, from effects of accessibility which are inherently a consequence of the tertiary structure.

The possibility of advancing understanding of the chemical implications of structure is best illustrated in the context of hydrolytic reactions. Among the patterns that emerge fairly early in any examination of the published literature on acid hydrolysis and enzymatic degradation of cellulose are the many similarities in the response to the two classes of hydrolytic agents. In both instances a rapid initial conversion to glucose and cellodextrins is followed by a period of relatively slower conversion, the rate of conversion in the second period depending on the prior history of the cellulosic substrate. In general, the nonnative polymorphic forms are degraded more rapidly during this second phase. In addition, it is found that the most crystalline or highly ordered of the native celluloses are particularly resistant to attack, with the most highly crystalline regions converted much more slowly than any of the other forms of cellulose.

The relationship of the patterns of hydrolytic susceptibility to the range of conformational variation discussed above can be interpreted in terms of contrast between the states of the glycosidic linkage in cellobiose and β-methylcellobioside. The differences between the states that are likely to contribute to the differences in observed reactivity are of two types. The first is differences in the steric environment of the glycosidic linkage, particularly with respect to activity of the C-6 group as a steric hindrance to, or as a potential promoter of, proton transfer reactions, depending on its orientation relative to the adjacent glycosidic linkage. The second type of difference is electronic in nature and involves readjustment of hybridization of the bonding orbitals at the oxygen in the linkage. It is worthwhile examining the potential contribution of each of these effects.

The effect of steric environment emerges most simply from examination of scale models of the cello-oligodextrins. They reveal that when C-6 is positioned in a manner approximating the structure in β-methylcellobioside, the methylene hydrogens are so disposed that they contribute significantly to creation of a hydrophobic protective environment for the adjacent glycosidic linkage. If, however, rotation about the C-5—C-6 bond is allowed, the primary hydroxyl group can come into proximity with the linkage and provide a potential path for more rapid proton transfer.

If, as suggested earlier on the basis of spectral data, the orientation of some C-6 groups in native cellulose is locked in by their participation in a bifurcated hydrogen bond to the hydroxyl group on C-3, they may contribute to the higher degree of resistance to hydrolytic action. Access to the linkage oxygen would be through a relatively narrow solid angle, barely large enough to permit entry of the hydronium ions that are the primary carriers of protons in acidic media.[137] If, on the other hand, the C-6 group has greater freedom to rotate, as is likely to be the case in cellulose II, the hindrance due to the methylene hydrogens can be reduced and, in some orientations, the oxygen of the primary hydroxyl group may provide a tunneling path for transfer of protons from hydronium ions to the glycosidic linkage. This would result in greater susceptibility of nonnative celluloses to hydrolytic attack.

The hypothesis concerning steric effects in acid hydrolysis has as its corollary the proposal that the role of the C_1 component in cellulase enzyme system complexes is to disrupt engagement of the C-6 oxygen in the bifurcated intramolecular hydrogen bond and thus permit rotation of the C-6 group into a position more favorable to hydrolytic attack.

The key role of C-6 in stabilizing the native cellulose structures is supported by findings concerning the mechanism of action of the dimethylsulfoxide–paraformaldehyde solvent system for cellulose, which is quite effective for solubilizing even the most crystalline of celluloses. The crucial step in the mechanism that has been established for this system is substitution of a methylol group on the primary hydroxyl at the C-6 carbon.[138,139]

The effect of conformation on the electronic structure of the linkage is also likely to be a factor with respect to its susceptibility to hydrolytic attack. Though there is no basis for anticipating the directions of this effect at this time, it is well to consider it from a qualitative perspective. First it is clear that the hybrids of oxygen orbitals involved in the bonds to carbon must be nonequivalent because the bond distances differ to a significant degree.[52,53] The angle of approximately 116° imposed on the linkage is likely to result in greater differences between the bonding orbitals and the lone pair orbitals than might be expected in a typical glycosidic linkage that is free from strain. Among themselves, the lone pair orbitals are likely to be different because of their different disposition with respect to the ring oxygen adjacent to C-1 in the linkage; the differences may be small and subtle, but they are no less real. Given these many influences on the nature of the hybridization at the oxygen in the linkage, it seems most unlikely that they would remain unaltered by changes in the dihedral angles of the magnitude of the difference between cellobiose and β-methylcellobioside. Hence a difference in electronic character must be expected.

At present it is not possible to estimate the magnitude of the effects discussed, nor to speculate about the direction of the change in relative reactivity of the glycosidic linkage in the two different conformations. Yet it is clear that differences can be anticipated and they may be viewed, within limits of course, as altering the chemical identity of the glycosidic linkage as its conformation changes. It remains for future studies to define the differences more precisely.

The points raised with regard to the influence of conformation on factors that determine the pathways for chemical reaction have not been specific subjects of investigation because methods for characterizing secondary structure as apart from the tertiary structure have not been available. It has also been true that suitable conceptual frameworks have not been available for developing the questions beyond the levels of the order–disorder duality. With the development of the approaches outlined above for exploring and distinguishing between matters of secondary and tertiary structure, it is quite likely that in the years ahead it will be possible to achieve a higher level of organization of information concerning the chemistry of cellulose.

With respect to questions of tertiary structure, the key issue introduced by the new structural information, and one that has not been explored at all to date, is whether the different hydrogen bonding patterns associated with the I_α/I_β duality have associated with them differences between the reactivity of the hydroxyl groups involved. It is not clear at this time how experiments exploring such effects might be carried out so as to separate issues associated with the differences between the hydrogen bonding patterns from issues associated with differences in accessibility.

3.16.3.8 Cellulose Structures in Summary

Before turning to the discussion of biological aspects in the next section, it is helpful to summarize where studies at the nanoscale level stand at the present time, and to assess the degree of confidence with which one can use their conclusions as the basis for further discussion.

From crystallographic studies, based on both X-ray and electron diffraction measurements, it can be concluded that the secondary structures of native celluloses are ribbon-like conformations approximating twofold helical structures. Their organization into crystallographic unit cells remains uncertain, however. The monoclinic space group $P2_1$, with two chains per unit cell, has been proposed for both the earlier studies prior to discovery of the I_α/I_β duality in native forms, and also for the I_β form. The I_α form is thought to possess a triclinic unit cell structure. Some important questions remain regarding the degree to which these are adequate representations of the organization of the crystalline domains in native celluloses. The majority of crystallographic studies also point to parallel alignment of the cellulose chains in the native celluloses, and this conclusion has been confirmed by electron micrographic observations. Also for cellulose II, the structures derived from X-ray diffraction data suggest a ribbon-like secondary structure approximating twofold helical organization and, in this instance, antiparallel alignment of the chains. For cellulose II though, the antiparallel proposal has been contradicted by recent electron micrographic observations. The unit cell organization of space group $P2_1$, with two chains per unit cell, has also been suggested for cellulose II, though the degree of confidence is even less than that with respect to the structures of cellulose I.

The early Raman spectroscopic studies clearly could not be reconciled with the premise that both cellulose I and cellulose II possess twofold helical conformations as the crystallographic studies had suggested. The Raman spectra, together with some corresponding infrared spectra, also pointed to the probability that the repeat unit of the structure of crystalline celluloses is anhydrocellobiose, so that alternating nonequivalent glycosidic linkages occur within each chain. To preserve the ribbon-like structural approximation, the different conformations of celluloses I and II were rationalized as incorporating glucosidic linkages that represent alternate left- and right-handed departures from the twofold helical structure, with those in cellulose II representing somewhat larger departures from the twofold helical conformation than those in cellulose I.

The introduction of high-resolution solid-state ^{13}C NMR spectral analyses into the study of celluloses resulted in resolution of one of the fundamental mysteries in the variability of native celluloses by establishing that all native celluloses are composites of two forms. These were identified as the I_α and I_β forms to distinguish them from the I_A and I_B categories that had been introduced more than three decades earlier to distinguish the celluloses produced by algae and bacteria from those produced by higher plants. The correspondence between the two classifications is that those in the I_A category have the I_α form as the dominant component, while those in the I_B category are predominantly of the I_β form. The nature of the difference between the I_α and I_β forms remains the subject of serious inquiry. Recognition of the I_α/I_β duality has facilitated a significant amount of additional research seeking to establish the balance between the two forms in a wide range of higher plant celluloses.

In later studies, the Raman spectra and corresponding infrared spectra indicated that the primary differences between the I_α and I_β forms of native cellulose were in the pattern of hydrogen bonding. Furthermore, the Raman spectra of the two forms raise questions as to whether the structures can possess more than one molecule per unit cell since there is no evidence of any correlation field splittings of any of the bands in the spectra of the two forms.

Electron microscopic studies relying on agents that act selectively either at the reducing or the nonreducing end groups of the cellulose chains have provided convincing evidence that cellulose chains are aligned parallel to one another in native cellulose. More recently, similar evidence has been presented supporting the view that alignment of the chains is also parallel in cellulose II. Other electron microscopic studies using the methods of lattice imaging have been used to demonstrate that the highly ordered microfibrils derived from algal celluloses represent homogeneous lattice structures with respect to the diffraction planes defined by the organization of their heavy atoms.

Electron diffraction studies carried out on algal celluloses after discovery of the I_α/I_β duality have been interpreted to indicate that the two forms may alternate along the length of individual microfibrils. These observations can also be interpreted as manifestations of the slow twisting about the long axis that has been observed in other studies of similar algal celluloses.

The possibility of the coexistence of the I_α and I_β forms within a superlattice structure has been suggested in the context of studies intended to mimic the conditions of biogenesis. These will be examined in greater detail in relation to the discussions of native celluloses and their biogenesis.

This discussion of structure has so far focused on issues of structure at the nanoscale level, identified as corresponding to domains that are of the order of 2 nm in dimension. Organization at the microscale level, defined as the range between 2 and 50 nm, requires consideration of a number of issues that have not been dealt with adequately in the literature on structures of cellulose. These include the well-recognized departures from a linear lattice, which have been generally regarded as measures of disorder when in fact they are more appropriately regarded as indicators of the nonlinear organization in a biological structure. Another issue arising at the microscale level is associated with the occurrence of significant fractions of cellulose molecules at the surface of the microfibrils of most native celluloses, particularly in the case of higher plants. It is a pertinent question as to whether the microfibrillar structure can be viewed as a separate phase in the traditional sense and whether criteria developed for the stability of homogeneous phases in the context of classical thermodynamics can have meaning when applied to native cellulosic structures. These issues arise in relation to discussions of native celluloses and their biogenesis.

Finally, the new developments with respect to structure, which facilitate separate though complementary focus on secondary and tertiary structures, provide a basis for exploring the relationships between structure and reactivity in new ways. The framework for more detailed characterization of the states of aggregation should permit more systematic exploration of the influence of source and history of cellulosic samples on their entry into chemical and biological processes.

3.16.4 BIOLOGICAL ASPECTS

The biological aspects of cellulose have usually been incorporated briefly in most prior reviews of the chemistry of cellulose. In view of the advances with respect to the characterization of celluloses, particularly in their native forms, and the significant progress in understanding processes of biogenesis and biodegradation, it is particularly appropriate to devote a section to the biological aspects of cellulose in a treatise on natural products chemistry. It is well at the outset to present the perspective that informs our discussion. When viewed in the context of biology, it is increasingly obvious that cellulose can no longer be regarded as another semicrystalline polymer, the phenomenology of which needs to fit within the traditional paradigms of polymer science. Even though investigations of its nature and its derivatives contributed to foundations of the polymer hypothesis and were the basis for developing many of the principles of polymer science, it is now important to recognize its distinctive biological functions and acknowledge that it is a remarkable and unusual biological molecule with characteristics that allow the formation of unique cellulosic structures within particular species and often within different tissues of the same organism. Rather than viewing cellulose as a linear homopolymer capable of crystallization under different sets of conditions, the biological celluloses must be regarded as important constituents of living matter formed through self-assembly of this unusual molecule in a wide variety of forms that are unique to the organisms within which it occurs. In a wider range of biological contexts, cellulose is also capable of co-aggregating with other constituents of the tissues within which it is present. It is therefore important to adopt an approach that is not entirely reductive with respect to the role of cellulose in living matter. It is with this in mind that this section of the chapter begins with a discussion of the processes of biogenesis of cellulose, beginning with the formation of the primary structure through the biosynthetic pathways and continuing with the formation of microstructure which is at the heart of the differences between native celluloses from different biological sources and tissues.

The discussion of biogenesis is followed by a discussion of a few of the native forms of cellulose that have been the subject of extensive investigation. Some of these forms have been studied primarily because they facilitate fundamental understanding of the nature of native celluloses, while others have been characterized exhaustively because of their commercial importance. New among the former group are the celluloses formed by the tunicates, which are marine organisms that have simply been regarded as an interesting curiosity of little interest. The finding that these celluloses represent the only naturally occurring highly ordered forms that are almost pure I_β types has dictated that they receive greater attention than has been the case in the past.

Within the biosphere, the biogenesis of cellulose is balanced by a wide variety of processes that result in its conversion to other forms of biomass. To reflect this, Section 3.16.4.3 discusses the biological processes of disassembly or biodegradation of cellulose as these processes are part of the key to understanding the role of cellulose within the carbon cycle. The biodegradation or bioconversion of cellulose occurs in two major contexts. The first is the natural disassembly of plant matter that has come to the end of its natural life cycle; the disassembly is carried out by microorganisms that are particularly well adapted to this purpose. The second is in the context of the function of cellulose as an important nutrient in the diet of ruminants and herbivores. A different class of microorganisms that are symbiotic with the animal hosts appear to be adapted to this function. But entirely apart from the primary processes of recycling of the celluloses, it is also now increasingly recognized that the disassembly and reassembly of cellulose may be an important part of the morphogenetic processes of plant tissues during their formation and growth, thus adding yet another dimension to studies of the disassembly of cellulose.

Finally, from a systems perspective, it is in order to compare the levels of complexity that are recognized in discussions of the different aspects of the biological processes in which cellulose plays a central role. The questions arise, for example, whether the complexities of the systems that have evolved for the disassembly of cellulose are adequately reflected in our understanding of the processes for its assembly? To the extent that the processes of disassembly reflect diversity in the secondary and tertiary structures of the native celluloses, should consideration of these levels of structure be an integral part of the analyses of structure formation? Are investigations focused on the formation of primary structure adequate to the task of unraveling the mysteries of the formation of native celluloses, or are the mechanisms of formation of secondary and tertiary structures as central to the answers as the processes of polymerization? These questions are more appropriately addressed at the conclusion of this section, but their statement at the outset will help place the discussions in perspective.

3.16.4.1 Biogenesis

The processes involved in the biogenesis of native celluloses are complex and highly coupled as formation of the primary structure at the nanoscale level is coordinated with the development of organization of secondary and tertiary structures at the microscale level. The primary structure is, by definition, the same for all cellulose-forming organisms, and it may well be that the processes of formation in different groups of organisms have much in common. It is not plausible, however, at the present time to speculate regarding the degree of commonality between the formation of primary structure in the celluloses that occur in the tunics of marine organisms and the processes of the plant kingdom. In contrast with primary structure, secondary and tertiary structures appear distinctive for particular species and, for all but unicellular organisms, can vary with tissue type. In higher plant cell walls, for example, the tertiary structures of the celluloses may differ among the different layers of the wall. In all instances wherein a secondary wall occurs, differences are observed between the primary and secondary walls, and in many plants the secondary wall is layered with significant variation in the organization of the celluloses within the different layers. For these reasons, the study of biogenesis has occurred at two different levels. The first is concerned with biosynthesis or the chemical pathways that lead to formation of the primary structure through the polymerization of glucose. At this level the issues are identification of the precursors and the enzyme systems involved in the processes of polymerization; investigations in this arena are primarily biochemical. At the next level, the focus is on ultrastructure, that is, the aggregated forms of native cellulose and the fibrillar organization that manifests itself at different levels within the morphological features characteristic of the particular cell wall or organism; studies addressing issues at this level rely more heavily on electron microscopy and seek to advance understanding of the secondary and tertiary structures and related phenomena at the microscale level. It is at this level also that many of the methods discussed in the previous section are applied. Though much of the investigation of celluloses at this level is biologically oriented and beyond the scope of this chapter, the self-assembly of cellulose during biogenesis and its modulation by particular chemical reagents are key to understanding the nature of native celluloses. Thus, the ultrastructural level is central to the chemistry of cellulose.

3.16.4.1.1 Biosynthesis

This discussion of biosynthesis will be confined to celluloses of the plant kingdom and related microorganisms that have been regarded as appropriate model systems. Studies of biosynthetic pathways have focused on investigation of model systems, with the expectation that much of what is learned of the biosynthesis of cellulose in the model systems will be common to the broader group of organisms that can produce cellulose. The system studied most often is the cellulose-producing bacterium *Acetobacter xylinum*. This system has two degrees of simplification relative to plant models. The first is that the processes of aggregation of the cellulose are more simple. The fibrils are not part of a cell wall but rather associate to form a pellicle that does not possess a higher level of order beyond the constraints associated with formation parallel to the surface of the culture medium. In most cellulose-producing organisms, in contrast, the cellulose is an integral part of a cell wall and its aggregation into fibrils is subject to additional constraints as it is coordinated with the development of higher levels of organization. The second advantage of *Acetobacter xylinum* over plant models is that it does not produce the β-(1-3)-linked glucan callose, formation of which is part of the wound response process of plants. Since the process of isolation of synthase enzymes obviously requires dramatic disruption of the integrity of the cells, plant systems usually shift to the formation of callose whenever the cells are subjected to enzyme isolation procedures. As a consequence, it has not been possible so far to isolate fully functional cellulose synthase enzyme systems from higher plants. The isolates instead produce glucans that are either the β-(1-3)-linked callose or include mixed linkages, most commonly the β-(1-3)- and β-(1-4)-linked polymer. These polymers also tend to aggregate in insoluble forms and have often been mistaken for cellulose on the basis of their insolubility and hydrolysis to produce glucose.

The biochemistry of cellulose synthesis by *Acetobacter xylinum* has been investigated in considerable detail by Benziman, Delmer, and their associates.[140-145] Important advances have been facilitated by isolation of the cellulose synthase from this organism in a highly active form.[143-145] This has allowed examination of the different factors involved in the synthesis and has established an important point of departure for additional studies exploring cellulose synthesis by other more complex systems. Availability of the synthase has provided access to the genetic encoding for the enzyme system and its constituents, allowing exploration of the degree to which related structures may occur in appropriate protein extracts from other organisms. This, in turn, has led to identification of such genes in a number of important plant systems.

It has been recognized for some time that the precursor in cellulose synthesis is a sugar nucleotide, although there had been some uncertainty regarding the specific one involved in cellulose.[146] Following considerable accumulation of evidence in support of this view, it is now generally accepted that, while other nucleotides may be involved in the synthesis of other polysaccharides or oligosaccharides, UDP glucose (uridine 5'-(α-D-glucopyranosyl pyrophosphate)) is the carrier of glucose for the formation of cellulose.[147] Assembly of the polymer from the glucose portion of UDP glucose is accomplished through a sequence of coordinated steps that seem to be orchestrated by the synthase complex. Northcote[148] has suggested that, in addition to the primary linkage formation by a glucosyltransferase enzyme system, the functions of the cellulose synthase include transportation of UDP glucose from the cytoplasm to the outer surface of the membrane and binding of both donor and acceptor molecules in proper orientation for the action of transglucosylase. In addition, he has suggested the occurrence of transmembrane proteins that associate the synthase complex with cytoskeletal elements which facilitate directed movement of the complex and orientation of the microfibrils and other subsidiary proteins that may serve in the assembly of the complex and its location on the membrane. Though Northcote was discussing cellulose synthesis in higher plants, most of the functions he describes are also expected to be present in the synthase of *Acetobacter xylinum*.

After isolation of the synthase in a highly active form, Benziman and associates pursued an understanding of the key factors in conservation of the activity of the synthase upon isolation.[149-152] This led them to discover the occurrence of a unique activator, cyclic diguanylic acid (c-di-GMP), and enzyme systems involved in regulating its level. They identified one enzyme, diguanilate cyclase, responsible for the formation of c-di-GMP and two phosphodiesterases, which, acting in sequence, degrade it. The balance between the two processes regulates the level of c-di-GMP, which in turn regulates the activity of cellulose synthase. Another important observation flowing from investigation of the isolated synthase was the demonstration that no covalently bonded intermediates were required for the process of assembly of the cellulose chains. It was concluded that "catalysis from UDP-glucose occurs via a direct substitution mechanism, in which the phosphoester-activating group at the anomeric carbon of one glucose residue is displaced by the C-4 hydroxyl group of another glucosyl residue, inverting the α configuration to form a β-glucosidic bond." This is in contrast to other proposed mechanisms that had implicated lipid-linked oligosaccharide intermediates in the synthesis of cellulose.[146] The relative simplicity of the mechanism proposed by Benziman and associates makes more plausible the possibility of a multienzyme complex wherein a number of synthase systems act in concert to produce cellulose chains that can then enter into the self-assembly process so characteristic of native celluloses.

As noted, another important advance was identification of the genes responsible for encoding the structures of the synthase components. This provided a basis for exploration of the degree to which similar genes may occur in higher plants. While such genes have not been found, ones with considerable similarity with respect to the domains thought to code for the binding of UDP-glucose and for catalysis of the β-(1-4)-linkage formation have been identified for cotton (*Gossypium hirsutum*) and rice (*Oryza sative*).[153] Though this represents an important advance, the findings cannot yet be regarded as conclusive since higher plants are likely to contain many other enzyme systems capable of forming β-(1-4)-linkages during the biosynthesis of the majority of the hemicelluloses, and it is not clear that the genes identified are not related to this group of enzymes which are thought to function within the Golgi apparatus rather than at the cell membrane.

Similar genes have been associated with cellulose formation in *Arabidopsis thaliana*.[154] In this instance, a mutant defective in a gene that is similar to those identified for cotton was found to be associated with the production of lower levels of cellulose when *Arabidopsis* is grown at 31 °C when compared to controls grown at 23 °C. Instead, under the conditions of elevated temperature, the mutant produced a glucan, identified as a soluble β-(1-4)-glucan, in amounts approximately equivalent to the deficiency in cellulose production. Thus, there is little question that the gene encodes a function that is essential to formation of aggregated microfibrillar cellulose, but since a β-(1-4)-

glucan is in fact formed, the encoded function is clearly associated with some more subtle trans-formation that is at an intermediate stage in the aggregation of cellulose in the unique native form. It is generally recognized that unbranched β-1,4-glucans are inherently insoluble at the octamer,[155] so the anomalous glucan is likely to include limited amounts of branching at a level sufficient to make it soluble.[156] More detailed characterization of this anomalous glucan must be accomplished before the full role of the gene can be defined. It is clear that the observations represent important advances towards better understanding of the biogenesis of cellulose at the molecular level, but much remains to be explored before the processes of cellulose formation in higher plants are fully understood.

With increased interest in the tunicate celluloses, it will be of considerable interest to compare the pathways of cellulose biosynthesis when there has been sufficient progress in understanding these systems.

3.16.4.1.2 *Ultrastructural organization*

Ultrastructural studies of native celluloses have focused most often on organization of the microfibrils, which are the basic units of structure at the level next above that of the unit cells in the lateral direction. At this level also, *Acetobacter xylinum* has been the model system of choice in explorations of the processes of biogenesis. Many other native celluloses from algal and plant sources, and, more recently, from the tunicates, have been investigated, but while they are important from a biological perspective, they do not lend themselves as readily to addressing questions of formation at the level of molecular aggregation. Rather, the patterns of formation are complicated by microscale processes associated with cell wall development; these include the influence of other cell wall constituents on the aggregation of cellulose as well as the constraints of cell wall geometry and the geometry of the assembly complexes, which can impose particular patterns of deformation during the process of aggregation. They also include poorly understood relationships between organization of the microtubules of the cytoskeleton and orientation of the microfibrils in plant cell walls. With *A. xylinum*, in contrast, in the absence of the conditions and the constraints of the cell wall environment, the patterns of aggregation reflect more than anything else the self-assembly characteristic inherent in the nature of cellulose, and to an uncertain degree the influence of the assembly complexes at the cell wall membrane that are thought to play a key role in the aggregation of cellulose molecules in the characteristic cellulose I form.

It is useful here to reiterate the issues that are associated with the distinction between the self-assembly of biological macromolecules within the context of native tissues and their crystallization if they are in an isolated form, because the distinctions are subtle and not immediately obvious and can result in arrival at misleading conclusions in some otherwise important contributions. The key point is that a structure does not have to be crystalline in order to result in diffraction patterns, whether with X-rays or electron beams. The prerequisite rather is a periodicity of structure over domains wherein the coherence of order is sufficient in extension to result in diffraction. With modern instruments it is now possible to acquire such patterns from very small domains, and it is usually implicit in most discussions of such observations of diffraction that the substrate is crys-talline. The concept of crystallization in the traditional sense, in contrast, though also implying a process of self-assembly, carries with it the notion of phase separation. As noted in an earlier section, the nanoscale and microscale structures encountered in native celluloses rarely if ever meet the criteria for separate phases established on the basis of statistical physics. It is therefore important to avoid ascribing to the highly ordered native celluloses the characteristics that devolve from the definition of a crystalline phase in the classical sense. For this reason, and to avoid confusion, an effort will be made to describe the ordering of cellulose as self-assembly even in reference to prior work wherein the authors cited spoke of crystallization.[161]

Though the processes of formation of cellulose by *A. xylinum* had been investigated quite extensively in earlier studies,[157–160] Brown and his associates[161,162] were the first to focus attention on the relationship of cellulose fibrils to structures in the walls of the bacterial cells in the course of biogenesis. They first established that linear arrays of terminal complexes on the cell surface were associated with the formation of cellulose ribbons produced by the individual cells. In the course of this work they discovered that certain dyes and fluorescent brightening agents resulted in dramatic changes in the patterns of aggregation.[163–167] They also studied in greater detail an effect that had been reported earlier by Ben-Hayim and Ohad[160] that certain cellulose derivatives of limited substitution also modified the processes of aggregation. The investigations by Brown and associates were directed at elucidating the relationship between the process of polymerization and self-assembly. These results allow clarification of the levels of structure at which the native organization of the cellulose first manifests itself.

Prior to their exploration of the patterns of aggregation of bacterial celluloses, Brown and co-workers[168] had demonstrated a clear association between the points of deposition of cellulose microfibrils and the cellulose assembly complexes at the plasma membrane for a number of organisms. It was generally observed that *A. xylinum* and a variety of cellulose-forming algae possessed linear arrays of terminal complexes,[169] while higher plant organisms were more frequently found to have the terminal complexes organized into rosettes consisting of six globules each.[170] Though the exact relationship between the terminal complexes and the processes of synthesis have not been established, it is generally accepted that the complexes play a role in the ordering or self-assembly of the cellulose deposited. In most other organisms, observation of the complexes by electron microscopy requires freeze fracture of the cell wall in the plane of the membrane, so that the role of the assembly complexes in biogenesis has to be inferred from adjacency of the ends of the cellulose microfibrils. In the case of *A. xylinum*, the association of cellulose synthesis with the assembly complexes is obvious from electron micrographs that can be prepared from systems wherein the bacteria are cultured directly on microscope grids allowing observation without any disruption of the association between the microfibrils and the terminal complexes embedded in the cell wall. In the same manner, it is also possible to explore the effects of modifying agents without the complications of other effects that may be associated with cell wall geometry or other cell wall components.

When grown under normal conditions, that is, in the absence of any agents that can interact with the cellulose at the molecular level, the cellulose appears to be produced through a linear array of complexes aligned parallel to the cell axis. Under such conditions the cellulose appears in the form of a ribbon that is 40–60 nm wide as shown in Figure 20; the ribbon appears to have a twist with a periodicity of 0.6–0.9 μm. The ribbons appear to consist of smaller fibrillar subunits which are thought to be 6–7 nm in lateral dimension; these are smaller than the microfibrils of many of the algae, but larger than the subunits in higher plants.[171] X-ray diffractometry reveals that these cellulosic aggregates possess a higher degree of order coherence than the pure celluloses from higher plants. As noted earlier, the bacterial celluloses in their native form are shown by solid-state ^{13}C NMR to contain both I_α and I_β forms of cellulose with the I_α form being dominant.

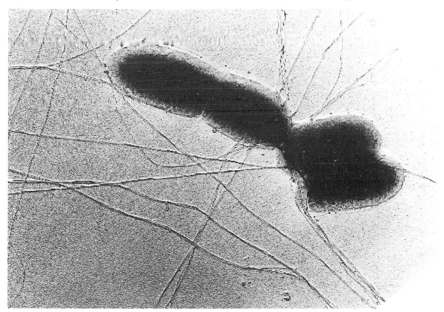

Figure 20   Several normal ribbons of cellulose synthesized by *Acetobacter xylinum*. Twists occur at several points, and in some regions smaller fibrillar subunits are visible (after Haigler[171]).

When the cellulose is formed in the presence of modifying agents, the fibrillar organization is altered to varying degrees. Among the most informative studies are those carried out with fluorescent brightening agents (FBAs) which have conjugated multiple aromatic rings in a planar configuration that facilitates their association with the cellulose molecules in nascent fibrils. The brightening agent studied most extensively is FBA 28, usually identified as Calcofluor White ST or Tinopal LPW. Its chemical formula is 4,4'-bis[4-anilino-6-bis(2-hydroxyethyl)amino-S-triazin-2-ylamino]-2,2'-stil-benedisulfonic acid; CI40622. It has been used in two classes of experiments. When added in small amounts (1–4 μM), the FBA alters the pattern of formation so that the cellulose aggregates as very fine fibrils, approximately 1.5 nm in diameter and they are extruded perpendicular to the cell axis and the plane of the cell wall. It is thought that these very fine fibrils represent aggregation of the cellulose molecules emerging from a single synthase complex system at the cell wall. When the concentration of FBA is raised to 8–250 μM, the fibrillar structure is no longer observed and the cellulose appears to form into many small sheets. Studies of these sheets in the wet state show them to be noncrystalline. One of the most interesting observations from a structural perspective is the transformation that is observed if the noncrystalline sheets are washed in water at a pH of 3.0. The result is that the sheets are transformed into helices of microfibrils about 1.5 nm in diameter as shown in Figure 21.

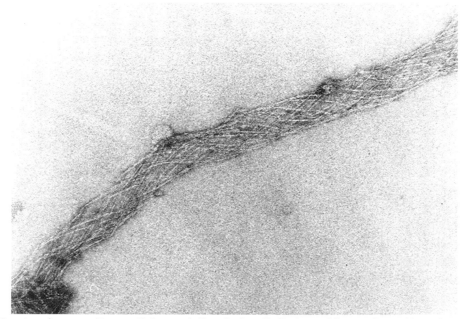

Figure 21   *Acetobacter xylinum* cellulose: a helix of microfibrils formed by gentle washing of the dye from extended sheets formed in its presence. The microfibrils contain cellulose I crystallites with reduced size compared to controls. All of these microfibrils would have been part of one normal ribbon (after Haigler[171]).

   The cellulose derivative that was found to have the greatest influence on aggregation of the cellulose from *A. xylinum* is carboxymethyl cellulose (CMC) with a degree of substitution (DS) of 0.7 relative to a maximum possible value of 3.0, and a molecular weight of 90 000 Da. Its effect on the pattern of aggregation was not unlike the lower levels of addition of the FBA in that it resulted in limitations on the aggregation of the ribbon. However, the fibrillar dimensions were somewhat larger than with low levels of FBA and of the order of the lateral dimensions inferred from X-ray diffractometry, in the range of 6–12 nm. The crystallinity of the cellulose formed under these conditions appeared to be similar to that of the controls, suggesting that the action of CMC was at the level of aggregates larger than the 1.5 nm fibrils.

   A number of the observations described above have implications with respect to the processes of structure formation at the secondary and tertiary levels. The finding that in the presence of higher levels of FBA the aggregation was not crystalline, but that upon washing, the cellulose I form is recovered, suggests that the cellulose molecules are extruded from the assembly complex in a stable secondary structure typical of the cellulose I form and that it can persist even in the absence of crystallization. Upon washing away the FBA, the cellulose molecules organized in this conformation can then aggregate to form the tertiary structure characteristic of cellulose I. The implication is that the secondary structure is inherently stable or that it is stabilized by association with the FBA molecules that are interleaved between the cellulose chains. At lower concentrations of FBA, where the 1.5 nm fibrils are formed, the implication is that the FBA can be associated with the surfaces of these small fibrils but is not sufficient in amount to penetrate between the individual chains. Both of these observations point to a self-assembly characteristic of cellulose molecules in this conformation. This self-assembly seems to occur at the level of association of the chains into the

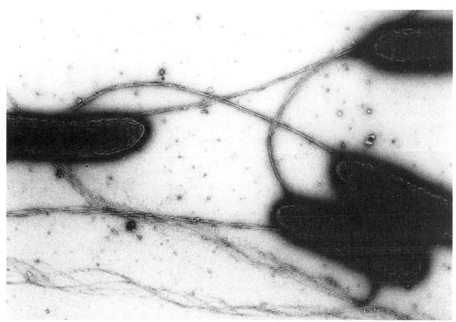

Figure 22 *Acetobacter xylinum* cellulose: the subunits of the normal ribbon remain separate when synthesized in the presence of corboxymethyl cellulose. They may lie closer together, thereby resembling normal ribbons, but close inspection shows separated or sometimes highly splayed subunits (after Haigler[171]).

1.5 nm fibrils when the FBA is not sufficient to associate with all of the chains, but it also seems to function at the next higher level where, in the absence of the FBA, the fibrils associate to form coherently ordered microfibrils at the 7 nm and higher levels, and beyond that in assembly of the fibrils into the ribbons usually produced in control cultures.

Thus, it can be said that the driving force in the process of aggregation is the tendency for self-assembly inherent in the nature of cellulose as it is formed in its native state. A corollary that can be inferred from this observation is that the assembly complexes are the determinants of the secondary structure of the cellulose and that the secondary structure in turn determines the tertiary structure. The process of aggregation of native cellulose can be envisioned as occurring at a number of hierarchic levels, and agents that can modify the processes of aggregation can act at one or more of these levels. In this context, the FBA can be regarded as capable of acting at the most basic level if it is present in sufficient concentration, but at lower concentrations it acts at the next level, that of the 1.5 nm fibrils. CMC can then be viewed as acting at one level above that of the FBA, resulting in the formation of the 8–12 nm fibrils.

In a manner not unlike that of the modifying agents, it can be expected that in the cell walls of higher plants, if constituents that associate strongly with cellulose are present, they may well modify the aggregation of cellulose as it is formed and extruded from the assembly complexes at the cell wall membrane. Though such an effect cannot be demonstrated directly in plants, the addition of selected cell wall constituents to cultures of *A. xylinum* does provide a basis for exploring such effects. In a number of studies wherein hemicelluloses have been added to cultures of *A. xylinum*, precisely such an effect is in fact observed. Studies have been carried out for xyloglucans,[172-174] xylans,[172,173] mannans,[172,173] and glucomannans.[175] In all instances the character of the aggregates of bacterial celluloses was shifted towards greater similarity to the aggregates of cellulose found to occur in higher plants. It was also demonstrated that some of these hemicelluloses, particularly the xyloglucan and the glucomannan, co-aggregate with the cellulose so intimately that they cannot be fully extracted even when quite strong caustic solutions are used for the extractions. Finally, it has been observed that the balance between the I_α and the I_β forms, as revealed in the solid-state ^{13}C NMR spectra, is shifted by the hemicelluloses so that the hemicellulose-modified bacterial celluloses have spectra very similar to the celluloses from higher plants. This has led to the proposal that one of the functions of the hemicelluloses is regulation of the aggregation of celluloses in higher plants.

The view of the processes of association presented here, elaborate as they may be, clearly do not yet represent the full complexity of the system for the assembly of native cellulose. This is perhaps best illustrated by observations by Hirai *et al.*[176] on the effects of temperature. They observed that by allowing *A. xylinum* to grow at 4 °C they were able to induce effects not unlike those that Haigler *et al.*[164–167] observed as a result of the addition of FBAs. The effect was equally reversible; that is, when grown at 4 °C the cultures produced noncrystalline bands of cellulose, but when the temperature was increased to 28 °C the production of cellulose reverted to the normal fibrillar form. The full implications of these observations have not yet been developed, but when taken together with observations of the effects of temperature on the biogenesis of cellulose in the *Arabidopsis thaliana* mutant, they clearly suggest that the stability of the synthases or their capacity to generate celluloses in native conformations is more sensitive to temperature than had heretofore been recognized.

The genetic regulation of aggregation at the microscale level is also illustrated in observations by Sugiyama *et al.*[177,178] Both in studies of algal celluloses[177] and in a systematic examination of approximately 40 tunicates,[178] they observed clear correlations between the I_α/I_β ratios of the celluloses and the species from which they were derived. Among the algal celluloses, there was also an indication that the I_α and I_β forms were spatially organized into nanodomains that varied in the relative distribution of the two forms and that the distributions were species specific. Here again, the full implications of these observations have yet to be developed, but they leave little question that the secondary and tertiary structures of native celluloses are also genetically regulated and that this regulation is at least as sensitive to environmental perturbations as is the formation of the primary structure.

It is worth noting at this point that the formation of the relatively large fibrils of celluloses from the algae and the tunicates are perhaps one of the clearest examples of the tendency of cellulose to self-assemble. When the lateral dimensions are of the order of 20 nm and it is known that the assembly complexes are of smaller dimension, the formation of coherently ordered extended structures points more clearly to self-assembly in a biological sense than to phase separation in the classical sense.

3.16.4.2 Native Celluloses

It was noted earlier that the chemistry of cellulose and its entry into biological processes are as much a function of its state of aggregation as of its primary structure. This reality is manifested as a variability of the response of different celluloses to the action of chemical reagents or cellulolytic enzyme systems. While the origins of this variability have generally been cast in terms of differences in accessibility, this has been primarily because quantitative measures of the organization of the celluloses at the microscale level have been elusive. In search of reproducibility in experimental studies, the vast majority of chemical and biological investigations of cellulose have relied on a handful of readily available substrates that occur in relatively pure form. The substrates most commonly used are cotton, cotton linters, and commercially available microcrystalline celluloses; the cotton linters are to be found in most chemical laboratories in the form of filter paper. Explorations of the effects of aggregation on chemical or biochemical reaction pathways have usually focused on changes that can be induced in the cellulosic substrates through mechanical action to induce varying degrees of decrystallization and disorder or by one or another of the chemical treatments that are known to swell or alter the ordered domains of the native forms. Quantitative measures of the organization of the different aggregated states have usually been cast in terms of accessibility to chemical reagents or in terms of variations in the degree of ordering as quantified by one or another of a number of empirical methods. With the development of some of the new methods for characterizing secondary and tertiary structures described in the preceding section, it is now likely that more studies can be carried out on a wider variety of native forms, and that the relationships between native states of aggregation and the response of the celluloses to the action of chemical and biological agents will be more readily characterized. It is useful, therefore, to develop an overview of what is known of organization in some of the commonly investigated native celluloses.

When considering the structures of native celluloses, it is important to distinguish between two categories of native forms. The first category includes the celluloses that occur in relatively pure form in their native state and which can be isolated using procedures that are relatively mild and that do not perturb or alter the state of aggregation to any significant degree. These are the celluloses that have been the subject of most structural studies. The second category, which includes the vast majority of naturally occurring celluloses, consists of celluloses that are an integral part of the complex architecture of cell walls of higher plants and are intimately blended with the other cell wall constitutents. They are most intimately blended with other cell wall polysaccharides, which are primarily other β-(1-4)-linked pentosans or hexosans with varying degrees of limited branching, collectively identified as the hemicelluloses. The celluloses and other cell wall polysaccharides are together also quite intimately integrated with cell wall lignins. Celluloses in this second category undergo varying degrees of change in their state of aggregation during isolation. However, these changes have not been well recognized or characterized in most instances. One of the major challenges yet to be met in future studies is the definition and characterization of the native state of such celluloses prior to application of isolating procedures that are disruptive to cell wall structure. The usual practice in most prior work on higher plant celluloses has been to assume that the structures of the pure celluloses are adequate models of these more complex forms of cellulose. While for some applications such an approximation may be useful, it has resulted in models of cell wall architecture that ignore the intimacy of the blending of cellulose with the hemicelluloses and lignin. These models will need revision to incorporate findings that the long-accepted two-phase model of plant cell walls is no longer adequate for rationalizing the organization of constituents of plant cell walls. In the following discussion we will first address the issues that complicate characterization of the first category of celluloses, that is, those that naturally occur in a relatively pure state. We will then examine the more complex states of aggregation of celluloses that are so intimately blended with other cell wall constitutents that more severe conditions are necessary for their isolation.

3.16.4.2.1 *Pure native celluloses*

Much of the discussion in the preceding section has focused on secondary and tertiary structure at the level of the unit cell, with characteristic dimensions of the order of 1 nm, and on the degree of equivalence or nonequivalence of different monomeric or dimeric units within the unit cell. Yet, the unit cell is the foundational concept in the description of crystalline order, which, as noted earlier, is usually predicated on the notion of infinite linear extension of the lattices in all three directions. It is necessary therefore to consider the implications of organization in cellulose at the next level beyond the unit cell; for most native celluloses it is the level at which departures from linear extension of the lattice of the aggregated celluloses is first manifested.

The most common approach, and the one that must be transcended to achieve clearer definition of order in celluloses, is based on categorization of that which is not crystalline as disordered. A number of methods for measurement of the "degree of crystallinity" of celluloses have been established on the basis of this classification. While this approach may allow easy methods of measurements that may be useful for some applications, it is fundamentally misleading because the "nonordered" fraction is usually regarded as disordered and identified as amorphous. The approach is particularly misleading with respect to native celluloses because the methodologies are frequently predicated on the assumption that the nonordered or amorphous fraction is homogeneously disordered, and that is clearly not the case in native celluloses. The reality is that most native celluloses are highly organized, often in a hierarchy of morphological features that are defined at the different scales of observation of the native tissue within which the cellulose occurs.

The difficulty and a measure of the paradoxical situations that arise are best illustrated through comparison of order in algal and cotton cell walls. By all common measures of crystallinity, it is very easy to ascribe to algal celluloses, such as those from *Valonia*, a higher degree of order than one would ascribe to cotton cellulose. Yet at the microscopic level it is clear that the fibrils of cellulose in cotton are more highly organized than those of *Valonia*, and into much more complex morphological features. In cotton, the fibrils are organized into lamella within which the microfibrils are helically wound at $45°$ to the fiber axis and wherein the direction of the helix is reversed at regular intervals. Furthermore, the lamella occur in concentric diurnal rings with no known correlation between organization of the microfibrils in the different lamella. In *Valonia*, in contrast, the microfibrils are linear and are of much more limited curvature because of the much larger diameters of the *Valonia* cells. The fundamental difference is that to accommodate the much more complex and hierarchic organization of the cellulose in cotton fibers, the microfibrils need to have smaller diameters and are subject to higher degrees of curvature than the microfibrils of *Valonia*. These two factors have a significant influence on the results of measurements used as the basis for most of the common methods for quantification of the crystallinity of cellulose. A number of consequences follow from each of the two factors and it is well to consider their implication for different methods of measurement.

The smaller diameter of cotton microfibrils has been reported as in the range of 3.5–4.5 nm. If, for the purposes of discussion, one were to assume a unit cell of the dimensions generally accepted for two-chain unit cells, one would find that two-thirds of the unit cells are at the surface of the microfibril. For *Valonia*, in contrast, only one in six would be at the surface. The placement of a unit cell at the surface reduces the packing constraints on the range of values that the internal coordinates defining the secondary structure or conformation can assume. Thus, the freedom of the secondary structure to depart from that in an ideal lattice conformation is far greater for cellulose molecules in a cotton microfibril than for a *Valonia* microfibril. This effect of microfibril diameter is clearly manifested in Figures 9 and 10 where the Raman spectra of ramie and *Valonia* can be contrasted. Ramie has a fibril diameter of about 5–6 nm, which is slightly larger than that of cotton. Furthermore, in the case of ramie the microfibrils are more like *Valonia* with respect to their degree of curvature so that their departure from the ideal lattice is primarily due to the smaller lateral dimensions. Yet it is clear from the broadening of the Raman spectra of ramie fibers relative to those of *Valonia* that the values of the internal coordinates defining the conformation are more broadly distributed than is the case for *Valonia*. This broadening of the distribution of the values of the internal coordinates results in a broadening of the distribution of the frequencies of the normal modes of vibration relative to their values in an ideal lattice.

The effects of microfibril diameter and the presence of significant fractions of cellulose chains near the surface are also manifested in the solid-state ^{13}C NMR spectra. As noted earlier, it was shown by Earl and VanderHart[67] that the areas of the upfield wings of the resonances of C-4 and C-6 increase as the microfibril diameter decreases, indicating that surface chains were subject to different conformational constraints. This variation is seen in Figure 6 where the upfield wing for C-4 is essentially at the level of noise for the spectrum of *Valonia* while its area increases in the celluloses known to have microfibrils of smaller diameters. In the studies by VanderHart and Atalla[73-75] and in those carried out by Newman and Hemmingson,[92] the contribution of the crystalline surfaces to the upfield wings of C-4 was clearly recognized. Newman's interpretation of a doublet resolved in this region as reflecting different crystalline surfaces of the microfibrils suggests that the microfibrils must be viewed as having polyhedral cross-sections; this is in contrast to the premise that the microfibrils may be regarded as cylindrical which is implicit in many discussions of the microfibrils of higher plants. Figure 6 also reveals another manifestation of the effects of microfibril diameter in that the same type of broadening as described above for the Raman spectra in Figures 9 and 10 is obvious in the lineshapes of the NMR resonances. Those that are narrowest and most clearly resolved are those associated with *Valonia*.

In X-ray diffractometry, the method used most often to characterize order in celluloses, the most common measurement method is the acquisition of a powder pattern. A number of different measures of coherence of order are derivable from such patterns. One is based on the width at half-height for the 020 peak at approximately 22.6°, which is inversely related to the lateral dimensions of the crystalline domain by the Debye–Scherer equation. A second is based on the intensity of the 020 peak relative to the intensity at $2\theta = 18°$, which is taken as representative of the disordered fraction. Yet another, used less often on its own, but sometimes measured to test the consistency of the other two measurements, is the position of the peak of the 020 reflection. It is usually highest for the most highly ordered celluloses, which are expected to have the fewest defects and the highest densities in the crystalline domains.

For all of the above approaches to characterizing the states of aggregation of native cellulose, the curvature of the microfibrils has an effect not unlike that of the smaller diameters. To the extent that the curvature is a measure of departure from the infinitely extended linear lattice, it induces in the internal coordinates departures from their values in an ideal lattice. It also results in a reduction of the degree of coherence in the relationships between unit cells as they are further apart from each other within an individual microfibril. Though curvature clearly contributes to broadening of the lines in the Raman spectra and the solid-state ^{13}C NMR resonances, its most serious misinterpretation occurs in the context of efforts to quantify the crystallinity of cellulose on the basis of X-ray powder patterns, where its contribution to the broader background is taken as an indication of homogeneous disorder.

The effects of microfibril diameter and curvature are noted here, not to discredit the methods that are commonly used to quantify order in different celluloses, but rather to suggest that the interpretation of results derived from such measurements must be approached with caution. It is clear that the types of order that occur in native celluloses at higher levels of organization are not readily described in mathematical terms and therefore it is not easy to relate the experimental measurements to the degrees of departure from the linearly extended infinite lattice. Yet it is also clear that the categorization of that which is not linearly ordered as disordered or amorphous can be quite misleading, particularly in the context of native celluloses. One of the major challenges for the years ahead is the development of a conceptual framework that is adequate for description of the more complex forms of order that occur in native celluloses.

3.16.4.2.2 *Complex native celluloses*

The term complex native celluloses is used here to indicate the class of native celluloses that occur naturally in intimate combination with other plant cell wall constituents, most often in higher plants. It has been generally assumed that the cell wall structure is adequately represented as a two-phase system, one consisting of the microfibrils of pure cellulose, the other a blend of all of the other constituents. Furthermore, it was implicit in this view that the cellulose phase is not unlike the pure native celluloses and that it is possible to isolate the cellulose from such cell walls by removing the other components, leaving behind the cellulose in a condition that approximates its native state. While these assumptions remain the basis of most discussions of plant cell wall structure, there is now clear evidence that the assumptions must be reassessed. The reassessments need to be at two levels. The first, which is perhaps more readily resolved, addresses the question whether it is possible to remove the other cell wall constituents without disrupting the native organization of the cellulose in its native form. The second raises the more fundamental question as to whether the two-phase model is at all valid in the context of the native state.

The generally accepted model of the cell wall in plants is well represented by Preston's chapter on "General principles of wall architecture",[8] where it is stated that:

> All plant cell walls are multiphase systems which for simplicity we may regard as two-phase. All walls known contain a crystalline polysaccharide of specific composition embedded in a matrix consisting usually of the wide variety of polysaccharides and other compounds already dealt with. As already mentioned, the crystalline polysaccharide of most plants is cellulose.

For primary cell wall structure, a model by Carpita and Gibaut[179] has received wide acceptance, but it also implicitly incorporates the two-phase view of the cell wall matrix. In both contexts, the two-phase view is based on the knowledge that cellulose occurs in the cell wall and that some form of insoluble polysaccharide can be isolated, that can be shown to include significant amounts of β-(1-4)-linked anhydroglucose units, and that in some instances it produces X-ray diffraction patterns typical of cellulose. It is therefore important to consider whether the steps intermediate between the native state and the isolated cellulose alter the state of aggregation, and to examine whether detection of the diffraction pattern of cellulose is an adequate criterion for the definition of cellulose.

The first explorations of this issue were undertaken in the course of assessment of industrial processes for the isolation of cellulose from woody tissue through commercial pulping operations. It had earlier been demonstrated that with mercerized celluloses, the degree of ordering could be enhanced by annealing at temperatures below those used in commercial pulping.[180] The question arose as to whether exposure to the elevated temperatures used in commercial pulping operations (170–190 °C) can have a similar influence on the celluloses isolated from woody tissue wherein the cellulose occurs in intimate association with the cell wall hemicelluloses, particularly in the secondary wall. Samples of wood were subjected to delignification using the acid-chlorite process at 70 °C and this was followed by treatment with 4% NaOH to remove the extractable hemicelluloses. A control sample was then subjected to an X-ray diffractometric measurement, while another was sealed in a buffer solution and exposed to a temperature cycle simulating the temperature cycle applied to the wood chips and the pulping liquors during commercial pulping. The X-ray diffractometric pattern reflected a significant increase in the degree of ordering of the cellulose as a result of exposure to the elevated temperature,[181] suggesting that removal of the hemicelluloses had facilitated a transformation in the state of aggregation of the cellulose. While this observation has many implications, the key one in the present context is that the state of aggregation of most commercially available celluloses derived from complex celluloses does not reflect the native state.

Though the observation concerning commercially available celluloses may seem relevant only to the uses of cellulose in commercial applications, it is of considerable importance in cellulose research because it has become quite common for research workers to rely on microcrystalline celluloses that are commercially available from suppliers of chemical reagents. The microcrystalline celluloses are, in the vast majority of instances, prepared from commercial dissolving pulps through acid hydrolysis followed by mechanical disruption and dispersion. The dissolving pulps, in turn, are derived from the complex celluloses that occur in woody species so that in their native state they are intimately blended with hemicelluloses. The majority of hemicelluloses are degraded and extracted in the course of the pulping cycles, but it is rare to find a microcrystalline cellulose that is entirely free of hemicellulose residues.

There are of course more fundamental reasons for exploring the nature of complex celluloses. In a number of studies, the association between hemicelluloses and cellulose was explored in contexts that simulated the processes of biogenesis. Here again, the model system used was *A. xylinum*. The work represented an extension of the studies by Haigler *et al.*[164–167] concerning the effects of cellulose derivatives on the aggregation of celluloses produced by *A. xylinum*. In these studies the polymers added were not cellulose derivatives, but rather hemicelluloses known to occur in higher plants.[172–174] The clear indication in all of these studies was that the hemicelluloses associated with the cellulose during its biogenesis. It was clear that the association is at the most elementary level in that it resulted in co-aggregation of the hemicelluloses with cellulose resulting in what has been defined as complex celluloses. Furthermore, it appeared that each of the hemicelluloses modified the process of aggregation in a different manner so that the resulting complex celluloses were different from each other when characterized by methods that are sensitive to variations in secondary and tertiary structure.

With respect to understanding the nature of complex celluloses, two features which emerged in the studies of these model systems are important and suggest the possibility of errors in interpretations of measurements in studies of plant celluloses. The first is related to the degree to which the crystallization habit of cellulose is dominant even when significant amounts of hemicelluloses are co-aggregated with it. For example, it was observed that X-ray powder patterns recorded for the complex celluloses appeared very similar to those usually recorded for less ordered celluloses.[172–174] This phenomenon is not unusual in the context of the X-ray diffractometry of polymers where it is often observed that in polymer blends or in block copolymer systems, the diffraction pattern is often that of the dominant component though it may appear somewhat modified.[11] This effect is also noted in the solid-state ^{13}C NMR spectra of the complex celluloses; the major features of the spectra are very similar to those of the pure celluloses when the latter are less ordered.

The second feature explored in the studies of complex celluloses generated with the *A. xylinum* system is that of the degree to which the hemicelluloses are integrated into the structure of cellulose. One of the observations in these studies was that although the hemicelluloses were added at the level of 0.5% in the culture medium, uptake into the structure of the cellulose was at the level of 30% for the xyloglucan and 24% in the case of the glucomannan. It was also observed that extraction with 17.5% NaOH resulted in removal of less than half of the amount of hemicellulose incorporated into the complex cellulose. In order to carry out comparisons with the traditional α-cellulose content test usually applied to commercial celluloses derived from wood pulps, the complex celluloses prepared from *A. xylinum* would have had to be subjected to an alkaline pulping process first; this was not done. The difficulty of extracting the hemicelluloses is clearly indicative of incorporation of the hemicelluloses at an early stage of aggregation. It would appear that the hemicelluloses, both with β-(1-4)-linked backbones, are able to enter into the self-assembly process of the cellulose, quite probably at the earliest stage of aggregation intermediate between the 1.5 nm fibrils reported by Haigler[171] and the level above. The intimacy of incorporation of the hemicelluloses and the difficulty of extracting them may well have been the basis for the opinion expressed by many senior researchers in the field of cellulose science late into the first half of this century,[182] which held that some of the hemicellulose derived sugars are covalently bonded and an integral part of the cellulose backbone.

The effects of polymeric additives on cultures of *Acetobacter xylinum* were also studied by Yamamoto and Horii[86] who also detected shifts in the balance between the I_α and I_β forms that were consistent with earlier reports based on other observational methods,[171–174] as well as on the basis of ^{13}C NMR.[175] The earlier studies had shown that polymers with a β-1,4 glucan backbone tend to shift the structures to ones more similar to higher plant celluloses. Yamamoto and Hori found that addition of xyloglucan and the sodium salt of carboxymethylcellulose at 2% and 2.5%, respectively, reduced the I_α content of the bacterial cellulose to a significant degree, while the addition of polyvinyl alcohol or polyethylene glycol had no effect. As noted in an earlier section (3.16.3.2.2), they also observed an effect of temperature that was quite interesting in that lower temperatures favored a higher level of the I_α form.

In another investigation of celluloses in plants, Hackney, Atalla, and VanderHart examined a wide range of cell wall constituents.[183,184] The general pattern which emerged was that in many primitive algae, where the other cell wall polysaccharides tended to be primarily acidic or pectic, there was very little evidence of co-aggregation of the polysaccharides with cellulose, and the celluloses isolated were sufficiently free of other matter that they could be classified among the pure celluloses. At higher levels of evolution, there emerged a pattern of occurrence of neutral polysaccharides with β-(1-4)-linked backbones, together with the cellulose. In some instances the other polysaccharides were the dominant components. However, in the vast majority of instances, the cellulose was the dominant component. It appeared that at a threshold level of 65% cellulose content, the aggregation habit of cellulose became the one that was detected by both X-ray powder patterns and solid-state ^{13}C NMR. In all instances, it was clear that while the X-ray patterns and the NMR spectra were the distinctive ones characteristic of native cellulose, they reflected a much lower level of ordering than would be observed from similar observations of pure celluloses.

The observations on complex celluloses are particularly important in relation to the common practice of using diffraction patterns to distinguish celluloses isolated from plant cell walls using mild extraction procedures. It needs to be noted in future studies that although patterns characteristic of cellulose are observed, the isolated insoluble substance may contain significant amounts of hemicelluloses as minor components and yet retain the characteristic aggregative properties of cellulose as characterized by X-ray diffraction, solid-state ^{13}C NMR spectroscopy, or on the basis of insolubility in acid media, the latter of course in the absence of strong hydrolytic conditions.

In relation to the discussion of complex celluloses, it is also helpful to revisit some of the relatively recent solid-state ^{13}C NMR studies of these celluloses. Newman *et al.*[185] applied the procedures they developed to investigations of a number of native celluloses of particular interest from a biological perspective, all of which are in the class of complex celluloses as they have been defined here. These include primary wall celluloses from apple cell walls[185] and from the leaves of *Arabidopsis thaliana*[186] and, in 1997, tissue from the silver tree fern *Cyathea dealbata*.[187] In the studies of primary wall celluloses from both apple cells and *A. thaliana* leaves, it was possible to resolve two resonances associated with the surfaces of fibrils and to suggest that the I_α and I_β forms occur in similar proportions. The intensities of the resonances associated with the surfaces of fibrils, when compared with the intensities of resonances associated with crystalline interiors, suggested lateral dimensions of the fibrils of 2.5–3.0 nm. These were then interpreted in terms of fibrils consisting of 23 chains of cellulose molecules, 14 of which were at the surface and nine in the interior; of the interior molecules, eight were only one layer removed from the surface. While the observations and their interpretation are not questioned here within the context of the accepted models of order in native celluloses, they clearly point to the need to ask again whether the traditional concepts associated with crystalline order are meaningful when so much of the substance under examination is at the surface. It would appear that some of the questions arising in the earlier periods of research on cellulose have yet to be answered.

Newman's study of the cellulose from fibrous material from the silver tree fern *C. dealbata* did not address the dimensions of the fibrils, but suggested the occurrence of similar proportions of the I_α and I_β forms. This, taken together with the earlier observations on primary wall celluloses, may raise questions concerning the earlier generalizations that higher plant celluloses have the I_β form as the dominant component. The question, however, must remain open until the possibility of quantitation on the basis of resolution-enhanced spectra is more firmly established. The question concerning quantitation of the relative amounts of the I_α and I_β forms on the basis of resolution-enhanced spectra also arises with respect to the comparison of celluloses from softwoods and hardwoods carried out by Newman.[188]

It is also useful to review the results of Larsson *et al.* concerning the solid-state ^{13}C NMR spectra of the celluloses they investigated within the present perspective.[100] It emerges that the wood celluloses, which are the only complex celluloses examined by them, are those which appear to have as their largest fractions the category characterized as paracrystalline and the category of hemicelluloses and cellulose oligomers. The I_α and I_β forms together represent only 9.1% of the total. Cotton cellulose, which is a pure cellulose but possesses a very complex morphology, has approximately 32% of the I_α and I_β forms, with the remaining large fractions belonging almost equally to the categories of paracrystalline and amorphous and surface fractions. The *Halocynthis*, *Cladophera*, and *Valonisa* celluloses are clearly among the pure celluloses, but the fact that the microfibrils need not enter into a more complex morphological architecture allows the largest fraction of the cellulose to retain the coherence of order of the molecular chains within the fibrils.

With the development of new methods to probe the states of aggregation, it is quite likely that new approaches will be developed to establish the relationships between microscale order and the properties of the celluloses as substrates. The key will be movement from characterization of the level of order in terms of a partitioning between fully ordered and disordered towards a more detailed account of the distribution among the different categories of order that prevail within the context of a particular state of aggregation.

New approaches to characterization of celluloses may well be based on the variability in secondary and tertiary structures. Variations in the secondary structure are most likely to manifest themselves as steric effects, which are issues associated with reaction pathways at the molecular level. Differences in tertiary structure, in turn, are expected to have their primary influence on accessibility to reagents, which is an issue arising from heterogeneity at the microscale level. Because of the difficulty in separating these effects, efforts to examine their separate influences have been quite limited. New methods for categorizing order in celluloses are likely to be developed on the basis of spectroscopic measurements.

3.16.4.3 Biodegradation

Next to biogenesis, the biodegradation of cellulose is the most significant life process connecting cellulose with the biosphere, for it is an essential component in the pathway to the re-entry of dead plant and algal biomass into the processes of life. The biodegradation is, in most instances, the result of action of microorganisms adapted to the degradation of algal and plant biomass. Through the action of hydrolytic enzymes, the cellulose and other cell wall polysaccharides are converted to glucose and other pyranoses or oligosaccharides which then become available as nutrients for the host organisms producing the enzymes. The biodegradation occurs in two broad categories of environments. The first is in the digestive systems of a wide range of ruminant and herbivorous organisms where the by-products of hydrolysis become available to both the microorganisms producing the cellulolytic enzymes and the host organisms. The second and broader category is the wide range of environments in which biomass is degraded in open fields and aquatic environments as part of the cyclic death and renewal of annual or perennial plants, and of shrubs, dead trees, and foliage on the forest floor. This second broader category also includes many of the plant pathogenic microorganisms that secrete hydrolytic enzymes as an integral part of their process of penetrating and metabolizing the constituents of the plants that they colonize.

The processes of biodegradation of cellulose and related cell wall polysaccharides have been investigated quite extensively both from a fundamental perspective in search of deeper understanding of the mechanisms of their action and within the context of efforts toward practical application of this understanding to protect celluloses from decay or, when specific enzymes of known and controllable action can be isolated, to use them to modify cellulosic fibers in a number of commercial applications. For example, cellulases have been used to modify the properties of cellulosic fibers in textile applications, and they have been used to facilitate removal of inks from recycled cellulosic pulp fibers in the manufacture of paper. The xylanases from some of the same organisms have been used to enhance the bleachability of virgin kraft pulp fibers in the manufacture of paper.

Here the author discusses briefly some of the fundamental aspects of cellulolytic action by organisms that have developed the capacity to disassemble cellulose through evolutionary adaptation to environments wherein lignocellulosic biomass represents the primary source of nutrients for the organisms. The emphasis will be on the functional aspects of cellulolytic systems. Much more comprehensive discussions of the structures and protein chemistry of the enzyme systems are available to the interested reader.[189–193] The enzyme systems that complement the cellulolytic systems by disassembling other constituents of plant biomass are beyond the scope of this chapter.

While the occurrence of cellulolytic systems was recognized late in the nineteenth century, mechanistic explorations of the processes began in the middle of the twentieth century. The pioneering work of Rees and Mandels[194] established that in the instance of native celluloses used in textile applications such as cotton or linen, which have highly ordered crystalline domains as a major fraction, a complex enzyme system was necessary for breaking down the cellulose. They were able to separate two active fractions that appeared to act in synergy with each other to disassemble the celluloses. One fraction, identified as C_x, was found to be capable of hydrolyzing celluloses that had been modified by the application of swelling treatments such as, for example, Walseth cellulose,[195] which is prepared by swelling with phosphoric acid. But acting alone, it was not effective for hydrolyzing cotton. When the other fraction identified as C_1 was added, it appeared to act in synergy with the C_x fraction to bring about hydrolysis. It was proposed that the C_1 fraction activates the crystalline cellulose so that it becomes susceptible to attack by the C_x fraction.

While the C_1–C_x hypothesis represented an important advance in recognizing that the action of cellulolytic organisms proceeded through the action of multienzyme systems, it was soon recognized that most cellulolytic organisms produce a wider array of enzymes that act in synergy to disassemble the cellulose. Eriksson and co-workers[196,197] suggested that the C_1 enzyme is in reality an "endwise acting" exoglucanase enzyme that can detach cellobiose units from the nonreducing ends of the cellulose chain. The C_x enzyme was viewed as an endoglucanase that can attack and hydrolyze the β-(1-4)-linkages at any point in an exposed chain. Thus, the new nonreducing chain ends would become available for the action of the exoglucanase and therein lay the basis for the synergy. In addition it was recognized that a (1-4)-β-glucosidase was also present in the cellulolytic system to convert the cellobiose and other soluble cellodextrins to glucose. Soon thereafter exoglucanases were isolated from a number of cellulose-degrading fungi.

With the work of this period it became clear that the enzyme systems of most cellulolytic organisms included a wider variety of enzymes with each optimized for a different aspect of the disassembly of cellulose molecules and their ordered aggregates. In addition to the participation of a number of different hydrolases, each with a specific function in relation to the overall process, some oxidative enzymes were also identified although their function is not as well understood. The review by Eriksson and Wood provides a good overview of cellulolytic enzyme systems as they were understood in terms of function, as well as the synergies between them.[198] Since that time advances in protein chemistry have added a different dimension to studies of cellulolytic systems, for in addition to defining them by function it has become possible to classify them by structure. The additional structural information has made possible explorations of relationships between structure and function. One of the most comprehensive reviews is that by Tomme *et al.*;[191] the perspective presented therein forms the basis of the following brief overview and additional discussions.

Beyond the realization that most cellulolytic microorganisms rely on a multiplicity of enzymes that act in synergy, one of the central results of structural studies is the recognition that the enzymes are modular in organization. While there are some exceptions, the majority consist of three key domains. The first is a cellulose binding domain (CBD), the primary function of which is to adsorb the enzyme on the substrate. In addition, in some instances the CBD appears to condition the surface to enhance its susceptibility to attack by the catalytic domain which is responsible for the hydrolysis. These two modules are connected by a third domain referred to as the linker region. The structural information has also allowed the classification of cellulases from different organisms into a number of structurally related families on the basis of amino acid sequence similarities. The tabulation by Tomme *et al.*[191] includes 17 such families.

Another important finding is related to the manner in which the cellulases of a particular cellulolytic system function to complement each other. Among those described by Tomme *et al.* as "Noncomplexed systems," the organism secretes into the environment a mix of endoglucanases and exoglucases or cellobiohydrolases which act in synergy upon the substrate. Such systems are generally characteristic of aerobic fungi and bacteria. The most extensively studied system is *Trichoderma reesei*, which secretes two major cellobiohydrolases, CBHI and CBHII, two major endoglucanases, EGI and EGII, and at least two low molecular weight endoglucanases EGIII and EGV. In addition, some β-glucosidases are secreted to relieve product inhibition by hydrolyzing cellobiose and other soluble celloligosaccharides. Another extensively studied system is that used by the fungus *Phanaerochete chrysosporium* which is of particular interest because it possesses multiple CBHI genes.[199] The expression of the different genes in response to differences in the cellulosic substrates[200,201] reflects a higher order in the mechanism of induction than had heretofore been recognized.

The second major group of cellulolytic systems are the complexed systems wherein the cellulases are organized in relation to each other so that their action can be collective. They are typically produced by anaerobic organisms including bacteria and fungi that colonize anaerobic environments such as the rumen and hind-gut of herbivores, composting biomass, and sewage. Among the most interesting of the anaerobic bacteria are those such as the *Clostridium* spp. where the complexed enzymes occur in distinct high molecular weight protein complexes called cellulosomes. The cellulolytic systems of *Clostridium thermocellum* have been studied extensively by Lamed, Bayer and co-workers.[202-205] In these systems the cellullases appear spatially organized through their association with a special protein referred to as a scafoldin. Its function appears to be to localize the different cellulases so that they act in concert. Furthermore, it has been demonstrated that this protein includes a CBD so that it can position the associated cellulases in relation to the cellulosic substrate. The cellulosomes appear to be exocellular organelles protruding from the surfaces of the cells, but they also seem to detach and attach themselves to the cellulosic substrates. It is of particular interest that the specific activity on crystalline celluloses appears to be 50-fold higher than the extracellular system produced by *Trichoderma reesei*. There has been significant progress in understanding the structure and functional organization of cellulosomes and these have been discussed in several recent reviews.[205-207]

Two other arenas of research on the natural disassembly of celluloses are worthy of note here. Recent studies of the action of selected cellulases on relatively pure celluloses have demonstrated a greater susceptibility of the I_α form to attack although the I_β form was also susceptible to attack but in different patterns.[208,209] There is little question that further pursuit of these patterns will add to understanding of both the native structures and the nature of cellulolytic action.

The second area that is of equal importance to the processes of biodegradation discussed above but is beyond the scope of this chapter is the role of endoglucanases in life processes of plants. Cellulase activity has been detected in both growing and senescing plants.[210] To the extent that native celluloses are the primary structural components of plant tissue, it is evident that changes in form cannot take place without both assembly and disassembly on a continuing basis.

3.16.5 FUTURE DIRECTIONS

In Section 3.16.2, it was noted that at the beginning of the twentieth century studies of cellulose played an important role in formulating the polymer hypothesis. In the main, cellulose has been investigated as one of the natural homopolymers within the framework of polymer science; much of the work has been based on the premise that a single ideal structure was the key to understanding cellulose. The discovery that all native celluloses are composites of two native forms and that the mixture of these two forms is species specific will open the possibility for future focus on the individuality of native celluloses. The findings that microorganisms can detect subtle differences between celluloses resulting in the expression of different isozymes for their disassembly[200,201] suggests that the interfaces between cellulose science and biology and microbiology will be fruitful ones for research in the twenty-first century.

It is anticipated that the physics and chemistry of nanodomains will advance considerably during the early part of the twenty-first century, in part to accommodate the drive towards miniaturization and in part due to the development of instrumental methods capable of greater resolution of structure than is possible at the present time. It is likely that these advances in scientific methods in general will have parallels in the arena of cellulose science. It is likely that as a consequence there will be greater recognition of the individuality of native celluloses as unique composites at the nanoscale level, and that this in large measure is a reflection of the capacity of the cellulose molecule for self-assembly in different patterns depending on the microenvironment within which the assembly takes place.

Finally, from a systems perspective, it will be in order to compare the levels of complexity that are recognized in the discussions of the different aspects of the biological processes in which cellulose plays a central role. The question arises, for example, whether the complexities of the systems that have evolved for the disassembly of cellulose, as in the case of cellulosomes, need be more complex than the systems that assemble cellulose? Or conversely, is the fact that systems as complex as the cellulosomes have evolved for disassembly a measure of the complexity of assembly? To the extent that the processes of disassembly frequently reflect diversity in the secondary and tertiary structures of the native celluloses, it would appear that a complete picture of the biogenesis will require a better understanding of the processes of formation of secondary and tertiary structure, and that investigations focused on the formation of primary structure are not adequate for the task of unraveling the mysteries of formation of native celluloses as they occur in higher plants. It is inviting to speculate that the assembly complexes usually visualized through electron microscopy are organelles at least as complex as the cellulosomes. Though Northcote did not explicitly suggest this,[148] his description of the many functions that must be coordinated to produce the primary structure suggests the orchestrated action of multiple enzyme systems, and his focus was on primary structure formation. If one then considers the need for regulation of secondary and tertiary structure formation, it seems plausible that the assembly complexes are more than synthases that are simply of a higher order of complexity than those observed in *Acetobacter xylinum*. Rather, it would appear that they are more highly regulated, complex assemblies of subsystems analogous to the bacterial synthase. The fact that they appear to be stable only in the context of living cells poses serious challenges in the search for deeper understanding of the biogenesis of celluloses of higher organisms. It may well be that investigations of model systems in tissue culture will prove to be the key to understanding these systems.

ACKNOWLEDGMENTS

The author's work in this area has been supported over the years by the USDA Forest Service, the Division of Energy Biosciences of the Office of Basic Energy Sciences of the US Department of Energy and by the Institute of Paper Chemistry, Appleton, Wisconsin (now the Institute of Paper Science and Technology, Atlanta, Georgia); all are gratefully acknowledged. The author's many collaborators in the area of Raman spectroscopy, whose work has been cited, have contributed immeasurably during many discussions of the different aspects of the work. The long standing collaboration with Dr. David VanderHart, of the National Institute of Standards and Technology, has been particularly important to much of the progress in the arena of solid state ^{13}C NMR studies of cellulose. Dr. VanderHart's expertise in this arena and his wisdom in addressing complex issues have been of great aid to the author. Discussions with colleagues over the years have also contributed to the author's perspective. Of particular value have been discussions with Dr. A. Isogai, Dr. C. Haigler, Dr. H. Chanzy, Dr. J. Sugiyama, Dr. R. A. J. Warren, Dr. T. Teeri, and Dr. L. Viikari; all are gratefully acknowledged.

3.16.6 REFERENCES

1. C. B. Purves, in "Cellulose and Cellulose Derivatives, Pt. I," eds. E. Ott, H. M. Spurlin, and M. W. Graffline, Wiley, New York, 1954, p. 29.
2. P. H. Hermans, "Physics and Chemistry of Cellulose Fibers," Elsevier, New York, 1949.
3. P. J. Flory, "Principles of Polymer Chemistry," Cornell University Press, Ithaca, NY, 1953.
4. C. F. Cross and E. J. Bevan, Researches on Cellulose (III), Longmans, Green & Co., London, 1912.
5. E. Heuser, The Chemistry of Cellulose, Wiley, New York, 1944.
6. D. W. Jones, in "Cellulose and Cellulose Derivatives, Pt. IV," eds. N. M. Bikales and L. Segal, Wiley, New York, 1971, p. 117.
7. B. A. Tonessen and O. Ellefsen, in "Cellulose and Cellulose Derivatives, Pt. IV," eds. N. M. Bikales and L. Segal, Wiley, New York, 1971, p. 265.
8. R. D. Preston, "The Physical Biology of Plant Cell Walls," Chapman and Hall, London, 1974.
9. A. Frey-Wyssling, "The Plant Cell Wall," Gebruder Borntrager, Berlin, 1976.
10. R. H. Atalla, in "The Structures of Celluloses," ed. R. H. Atalla, ACS Symposium Series No. 340, American Chemical Society, Washington, DC, 1987 p. 1.
11. M. Kakudo and N. Kasai, "X-Ray Diffraction by Polymers," Elsevier, New York, 1972, p. 285.
12. S. Arnott, in "Fiber Diffraction Methods," ACS Symposium Series No. 141, American Chemical Society, Washington, DC, 1980, p. 1.
13. E. D. T. Atkins, in "Fiber Diffraction Methods," ACS Symposium Series No. 141, American Chemical Society, Washington, DC, 1980, p. 31.
14. H. Tadokoro, in "Fiber Diffraction Methods," ACS Symposium Series No. 141, American Chemical Society, Washington, DC, 1980, p. 43.
15. H. Tadokoro, "Structure of Crystalline Polymers," Wiley, New York, 1979, p. 6.
16. J. A. Howsmon and W. A. Sisson, in "Cellulose and Cellulose Derivatives, Pt. I," eds E. Ott, H. M. Spurlin, and M. W. Graffline, Wiley, New York, 1954, p. 231.
17. T. Petipas, M. Oberlin, and J. Mering, *J. Polym. Sci. C*, 1963, **2**, 423.
18. M. Norman, *Text. Res. J.*, 1963, **33**, 711.
19. K. H. Gardner, and J. Blackwell, *Biopolymers*, 1974, **13**, 1975.
20. J. J. Hebert and L. L. Muller, *J. Appl. Polym. Sci.*, 1974, **18**, 3373.
21. D. W. Jones, in "Cellulose and Cellulose Derivatives, Pt. IV," eds. N. M. Bikales and L. Segal, Wiley, New York, 1971, p. 138.
22. J. A. Howsmon and W. A. Sisson, in "Cellulose and Cellulose Derivatives, Pt. I," eds. E. Ott, H. M. Spurlin, and M. W. Graffline, Wiley, New York, 1954, p. 237.
23. K. H. Meyer and H. Z. Mark, *Physik. Chem.*, 1929, **B2**, 115.
24. K. H. Meyer and L. Misch, *Helv. Chim. Acta*, 1937, **20**, 232.
25. F. T. Pierce, *Trans. Faraday Soc.*, 1946, **42**, 545.
26. A. Sarko and R. Muggli, *Macromolecules*, 1974, **7**, 486.
27. A. D. French, W. A. Roughead, and D. P. Miller, in "The Structures of Celluloses," ed. R. H. Atalla, ACS Symposium Series No. 340, American Chemical Society, Washington, DC, 1987, p. 15.
28. A. D. French, *Carbohydrate Res.*, 1978, **61**, 67.
29. R. H. Marchessault and P. R. Sundararajan, in "The Polysaccharides," ed. G. O. Aspinall, Academic Press, New York, 1983, vol. 2, p. 11.
30. O. Ellefsen and B. A. Tonessen, in "Cellulose and Cellulose Derivatives, Pt. IV," eds. N. M. Bikales and L. Segal, Wiley, New York, 1971, p. 151.
31. J. Blackwell and R. H. Marchessault, in "Cellulose and Cellulose Derivatives, Pt. IV," eds. N. M. Bikales and L. Segal, Wiley, New York, 1971, p. 1.
32. H. J. Marinan and J. Mann, *J. Polym. Sci.*, 1956, **21**, 301.
33. J. Mann and H. J. Marinan, *J. Polym. Sci.*, 1958, **32**, 357.
34. M. L. Nelson and R. T. O'Connor, *J. Appl. Polym. Sci.*, 1964, **8**, 1311.
35. M. L. Nelson and R. T. O'Connor, *J. Appl. Poly. Sci.*, 1964, **8**, 1325.
36. L. Pauling, "The Nature of the Chemical Bond," 3rd edn., Cornell University Press, Ithaca, NY, 1960, p. 65.
37. L. A. Woodward, "Introduction to the Theory of Molecular Vibrations and Vibrational Spectroscopy," University Press, Oxford, 1972, p. 344.
38. R. H. Atalla, *Appl. Polym. Symp.*, 1976, **28**, 659.
39. J. Blackwell, P. D. Vasko, and J. L. Konig, *J. Appl. Phys.*, 1970, **41**, 4375.
40. L. J. Pitzner, Ph.D. Dissertation, The Institute of Paper Chemistry, Appleton, WI, 1973.
41. L. J. Pitzner and R. H. Atalla, *Spectrochim. Acta Part A*, 1975, **31A**, 911.
42. G. M. Watson, Ph.D. Dissertation, The Institute of Paper Chemistry, Appleton, WI, 1974.
43. S. L. Edwards, Ph.D. Dissertation, The Institute of Paper Chemistry, Appleton, WI, 1976.

44. R. M. Williams, Ph.D. Dissertation, The Institute of Paper Chemistry, Appleton, WI, 1977.
45. R. M. Williams and R. H. Atalla, *J. Phys. Chem.*, 1984, **88**, 508.
46. H. A. Wells, Ph.D. Dissertation, The Institute of Paper Chemistry, Appleton, WI, 1977.
47. H. A. Wells and R. H. Atalla, *J. Mol Struct.*, 1990, **224**, 385.
48. K. P. Carlson, Ph.D. Dissertation, The Institute of Paper Chemistry, Appleton, WI, 1978.
49. H. Wiley and R. H. Atalla, *Carbohyd. Res.*, 1987, **160**, 113.
50. D. A. Rees and R. J. Skerret, *Carbohyd. Res.*, 1968, **7**, 334.
51. A. Sarko and R. Muggli, *Macromolecules* 1974, **7**, 486.
52. S. S. C. Chu and G. A. Jeffrey, *Acta Cryst.*, 1968, **B24**, 830.
53. J. T. Ham and D. G. Williams, *Acta Cryst.*, 1970, **B29**, 1373.
54. E. B. Wilson, Jr., J. C. Decius, and P. C. Cross, "Molecular Vibrations: The Theory of Infrared and Raman Vibrational Spectra," McGraw-Hill, New York, 1955, p. 188.
55. J. J. Cael, J. L. Koenig, and J. Blackwell, *Carbohydr. Res.*, 1973, **29**, 123.
56. J. C. Decius, *J. Chem. Phys.*, 1948, **16**, 1025.
57. S. Melberg and K. Rasmussen, *Carbohydr. Res.*, 1979, **71**, 25.
58. B. Henrissat, S. Perez, I. Tvaroska, and W. T. Winters, in "The Structures of Celluloses," ed. H. Atalla, ACS Symposium Series No. 340, American Chemical Society, Washington, DC, 1987 p. 38.
59. R. H. Atalla, *Adv. Chem. Ser.*, 1979, **181**, 55.
60. R. H. Atalla, *Appl. Polym. Symp.*, 1976, **28**, 659.
61. R. H. Atalla, J. C. Gast, D. W. Sindorf, V. J. Bartuska, and G. E. Maciel, *J. Am. Chem. Soc.*, 1980, **102**, 3249.
62. D. Page, personal communication.
63. R. H. Atalla, in "Proceedings of the International Symposium on Wood and Pulping Chemistry," SPCI Rept. No. 38, Stockholm, 1981, vol. 1, p. 57.
64. R. H. Atalla, in "Structure, Function and Biosynthesis of Plant Cell Walls," eds. W. M. Dugger and S. Bartinicki-Garcia, American Society of Plant Physiologists, Rockville, MD, 1984, p. 381.
65. R. H. Atalla, *J. Appl. Pol. Symp.*, 1983, **37**, 295.
66. W. L. Earl and D. L. VanderHart, *J. Am. Chem. Soc.*, 1980, **102**, 3251.
67. W. L. Earl and D. L. VanderHart, *Macromolecules* 1981, **14**, 570.
68. D. L. VanderHart, Deductions about the morphology of wet and wet beaten cellulose from solid state ^{13}C NMR, NBSIR 82–2534, National Bureau of Standards, 1982.
69. R. L. Dudley, C. A. Fyfe, P. J. Stephenson, Y. Deslandes, G. K. Hamer, and R. H. Marchessault, *J. Am. Chem. Soc.*, 1983, **105**, 2469.
70. F. Horii, A. Hirai, and R. Kitamaru, *Polym. Bull.*, 1982, **8**, 163.
71. G. E. Maciel, W. L. Kolodziejski, M. S. Bertran, and B. E. Dale, *Macromolecules*, 1982, **15**, 686.
72. J. C. Gast, R. H. Atalla, and R. D. McKelvey, *Carbohydr. Res.*, 1980, **84**, 137.
73. D. L. VanderHart and R. H. Atalla, *Macromolecules*, 1984, **17**, 1465.
74. R. H. Atalla and D. L. VanderHart, *Science* 1984, **223**, 283.
75. D. L. VanderHart and R. H. Atalla, in "The Structures of Celluloses," ed. R. H. Atalla, ACS Symposium Series No. 340, American Chemical Society, Washington, DC, 1987 p. 88.
76. R. H. Atalla, R. E. Whitmore, and D. L. VanderHart, *Biopolymers*, 1985, **24**, 421.
77. F. Horii, H. Yamamoto, R. Kitamaru, M. Tanahashi, and T. Higuchi, *Macromolecules*, 1987, **20**, 2946.
78. J. H. Wiley and R. H. Atalla, in "The Structures of Celluloses," ed. R. H. Atalla, ACS Symposium Series No. 340, American Chemical Society, Washington, DC, 1987 p. 151.
79. J. Sugiyama, R. Yuong, and H. Chanzy, *Macromolecules*, 1991, **24**, 4168.
80. H. Yamamoto, F. Horii, and H. Odani, 1989, **22**, 4130.
81. J. Sugiyama, T. Okano, H. Yamamoto, and F. Horii, *Macromolecules*, 1990, **23**, 3196.
82. R. H. Atalla, J. Hackney, U. P. Agarwal, and A. Isogai, to be published.
83. D. I. Bower and W. F. Maddams, "The Vibrational Spectroscopy of Polymers," Cambridge University Press, Cambridge, 1989.
84. F. Horii, A. Hirai, and R. Kitamaru, in "The Structures of Celluloses," ed. R. H. Atalla, ACS Symposium Series No. 340, American Chemical Society, Washington, DC, 1987, p. 119.
85. D. L. VanderHart and G. C. Campbell, *J. Magn. Reson.*, 1998, **134**, 88.
86. H. Yamamoto and F. Horii, *Macromolecules*, 1993, **26**, 1313.
87. H. Yamamoto, F. Horii, and H. Odani, *Macromolecules*, 1989, **22**, 4130.
88. H. Yamamoto and F. Horii, *Macromolecules*, 1993, **26**, 1313.
89. P. S. Belton, S. F. Tanner, N. Cartier, and H. Chanzy, *Macromolecules*, 1989, **22**, 1615.
90. H. Yamamoto and F. Horii, *Cellulose*, 1994, **1**, 57.
91. R. H. Newman and J. A. Hemmingson, *Holzforschung*, 1990, **44**, 351.
92. R. H. Newman and J. A. Hemmingson, *Cellulose*, 1995, **2**, 95.
93. R. H. Newman, *J. Wood Chem. Tech.*, 1994, **14**, 451.
94. R. H. Newman, M. A. Ha, and L. D. Melton, *J. Agric. Food. Chem.*, 1994, **42**, 1402
95. R. H. Newman, L. M. Davies, and P. J. Harris, *Plant Physiol.*, 1996, **111**, 475.
96. R. H. Newman, *Cellulose*, 1997, **4**, 269.
97. H. Lenholm, T. Larsson, and T. Iversen, *Carbohydr. Res.*, 1994, **261**, 119.
98. H. Lennholm, T. Larsson, and T. Iversen, *Carbohydr. Res.*, 1994, **261**, 119.
99. T. Larsson, U. Westermark, and T. Iversen, *Carbohydr. Res.*, 1995, **278**, 339.
100. T. Larsson, K. Wickholm, and T. Iversen, *Carbohydr. Res.*, 1997, **302**, 19.
101. C.W. Hock, in "Cellulose and Cellulose Derivatives, Pt. 1," eds. E. Ott, H. M. Spurlin, and M. W. Grafflin, Wiley, New York, 1954, p. 347.
102. F. F. Morehead, in "Cellulose and Cellulose Derivatives, Pt. IV," eds. N. M. Bikales and L. Segal, Wiley, New York, 1971, 213.
103. K. Hieta, S. Kuga, and M. Usuda, *Biopolymers*, 1984, **23**, 1807.
104. H. Chanzy and B. Henrissat, *FEBS Lett.*, 1985, **184**, 285.

105. A. Maurer and D. Fengel, *Holz als Roh- und Werkstaff*, 1992, **50**, 493.
106. J. Sugiyama, H. Harada, Y. Fujiyoshi, and N. Uyeda, *Mokuzai Gakkaishi*, 1984, **30**, 98.
107. J. Sugiyama, H. Harada, Y. Fujiyoshi, and N. Uyeda, *Mokuzai Gakkaishi*, 1985, **31**, 61.
108. J. Sugiyama, H. Harada, Y. Fujiyoshi, and N. Uyeda, *Planta*, 1985, **166**, 161.
109. J. Sugiyama, T. Okano, H. Yamamoto, and F. Horii, *Macromolecules*, 1990, **23**, 3196.
110. G. Honjo and M. Watanabe, *Nature*, 1958, **181**, 326.
111. J. Sugiyma, R. Vuong, and H. Chanzy, *Macromolecules*, 1991, **24**, 4168.
112. E. Roche and H. Chanzy, *J. Biol. Macromol.*, 1981, **3**, 201.
113. B. J. Hardy and A. Sarko, *Polym. Prepr.*, 1995, **36**, 640.
114. L. M. J. Kroon-Batenburg, B. Bouma, and J. Kroon, *Macromolecules*, 1996, **29**, 5695.
115. L. M. J. Kroon-Batenburg and J. Kroon, *Glycoconjugate J.*, 1997, **14**, 677.
116. A. P. Heiner, J. Sugiyama, and O. Teleman, *Carbohyd. Res.*, 1995, **273**, 207.
117. A. P. Heiner and O. Teleman, *Langmuir*, 1997, **13**, 511.
118. M. S. Baird, A. C. O'Sullivan, and W. B. Banks, *Cellulose*, 1998, **5**, 89.
119. H. Chanzy, B. Henrissat, M. Vincendon, S. Tanner, and P. S. Belton, *Carbohydr. Res.*, 1987, **160**, 1.
120. R. H. Atalla, B. E. Dimick, and S. C. Nagel, in "Cellulose Chemistry and Technology," ed. J. C. Arthur, Jr., ACS Symp. Series No. 40, 1977, p. 30.
121. R. H. Atalla, J. D. Ellis, and L. R. Schroeder, *J. Wood Chem. Technol.*, 1984, **4**, 465.
122. H. Chanzy, K. Imada, A. Mollard, R. Vuong, and F. Barnoud, *Protoplasma*, 1979, **100**, 303.
123. A. Sakthivel, A. D. French, B. Eckhardt, and R. A. Young, in "The Structures of Celluloses," ed. R. H. Atalla, ACS Symposium Series No. 340, American Chemical Society, Washington, DC, 1987, p. 68.
124. L. D. Landau and E. M. Lifshitz, "Statistical Physics", Addison-Wesley, Reading, MA, 1958.
125. R. H. Atalla and S. C. Nagel, *Science*, 1974, **185**, 522.
126. R. E. Whitmore and R. H. Atalla, *Int. J. Biol. Macromol.*, 1985, **7**, 182.
127. N. R. Bertoniere and S. H. Zeronian, in "The Structures of Celluloses," ed. R. H. Atalla, ACS Symposium Series, No. 340, American Chemical Society, Washington, DC, 1987, p. 255.
128. S. P. Rowland and E. J. Robserts, *J. Polym. Sci., A-1*, 1972, **10**, 2447.
129. R. F. Nickerson, *Text. Res. J.*, 1951, **21**, 195.
130. S. P. Rowland, E. J. Roberts, and C. P. Wade, *Text. Res. J.*, 1969, **39**, 530.
131. S. P. Rowland, in "Modified Cellulosics," eds. R. M. Rowell and R. A. Young, Academic Press, New York, 1978, p. 147.
132. S. P. Rowland, E. J. Roberts, and J. L. Bose, *J. Polym. Sci., A-1*, 1971, **9**, 1431.
133. S. P. Rowland, E. J. Roberts, J. L. Bose, and C. P. Wade, *J. Polym. Sci., A-1*, 1971, **9**, 1623.
134. S. H. Zeronian, M. L. Coole, K. W. Alger, and J. M. Chandler, *J. Appl. Polym. Sci., Appl. Pol. Symp.*, 1983, **37**, 1053.
135. J. E. Stone and A. M. Scallan, *Pulp. Pap. Mag. Can.*, 1968, **69**, 69.
136. J. E. Stone, E. Treiber, and E. Abrahamson, *TAPPI*, 1969, **52**, 108.
137. R. P. Bell, "The Proton in Chemistry," Chapman & Hall, London, 2nd edn., 1973.
138. D. C. Johnson, M. D. Nicholson, and F. C Haigh, *Appl. Polym. Symp.*, 1976, **28**, 931.
139. D. C. Johnson and M. D. Nicholoson, *Cellul. Chem. Technol.*, 1977, **11**, 349.
140. D. P. Delmer, M. Benziman, and E. Padan, *Proc. Natl. Acad. Sci. USA*, 1982, **79**, 5282.
141. M. Benziman, C. H. Haigler, R. M. Brown, Jr., A. R. White, and K. M. Cooper, *Proc. Natl. Acad. Sci. USA*, 1980, **77**, 6678.
142. Y. Aloni and M. Benziman, in "Cellulose and Other Natural Polymer Systems," ed. R. M. Brown, Plenum, New York, 1982, p. 341.
143. Y. Aloni, D. P. Delmer, and M. Benziman, *Proc. Natl. Acad. Sci. USA*, 1982, **79**, 6448.
144. M. Benziman, Y. Aloni, und D. P. Delmer, *J. Appl. Polym. Sci., Appl. Pol. Symp.*, 1983, **37**, 131.
145. Y. Aloni, R. Cohen, M. Benziman, and D. P. Delmer, *J. Biol. Chem.*, 1983, **258**, 4419.
146. H. Hori and A. D. Elbein, in "Biosynthesis and Biodegradation of Wood Components," ed. T. Higuchi, Academic Press, New York, 1985, p. 109.
147. Y. Kawagoe and D. P. Delmer, in "Genetic Engineering," ed. J. K. Setlow, Plenum, New York, 1997, vol. 19, p. 63.
148. D. H. Northcote, in "Biosynthesis and Biodegradation of Cellulose," eds. C. H. Haigler and P. J. Weimer, Marcel Dekker, New York, 1991, p. 165.
149. P. Ross, Y. Aloni, C. Weinhouse, D. Michaeli, P. Weinberger-Ohana, R. Mayer, and M. Benziman, *FEBS Lett.*, 1985, **186**, 191.
150. P. Ross, Y. Aloni, C. Weinhouse, D. Michaeli, P. Weinberger-Ohana, R. Mayer, and M. Benziman, *Carbohyd. Res.*, 1986, **149**, 101.
151. P. Ross, C. Weinhouse, Y. Aloni, D. Michaeli, P. Weinberger-Ohana, R. Mayer, S. Braun, E. de Vroom, G. A. van der Marel, J. H. van Boom, and M. Benziman, *Nature*, 1987, **325**, 279.
152. P. Ross, R. Mayer, and M. Benziman, in "Biosynthesis and Biodegradation of Cellulose," eds. C. H. Haigler and P. J. Weimer, Marcel Dekker, New York, 1991, p. 219.
153. J. R. Pear, Y. Kawagoe, W. E. Schreckengost, D. P. Delmer, and D. M. Stalker, *Proc. Natl. Acad. Sci. USA*, 1996, **93**, 12637.
154. T. Arioli, L. Peng, A. S. Betzner, J. Burn, W. Wittke, W. Herth, C. Camilleri, H. Hofte, J. Plazinski, R. Birch, A. Cork, J. Glover, J. Redmond, and R. E. Williamson, *Science*, 1998, **279**, 717.
155. B. A. Tonnesen and O. Ellefsen, in "Cellulose and Cellulose Derivatives," Pt. IV, eds. N.M. Bikales and L. Segal, Wiley, New York, 1971, vol. 5, p. 265.
156. R. H. Atalla, *Science*, 1998, **282**, 591.
157. J. R. Colvin, in "The Fromation of Wood in Forest Trees," ed. M. H. Zimmerman, Academic Press, New York, 1964, p. 189.
158. J. R. Colvin, in "Cellulose and Cellulose Derivatives," Pt. IV, eds. N.M. Bikales and L. Segal, Wiley, New York, 1971, p. 695.
159. I. Ohad, D. Danon, and S. Hestrin, *J. Cell Biol.*, 1962, **12**, 31.
160. G. Ben-Hayim and I. Ohad, *J. Cell Biol.*, 1965, **25**, 191.

161. R. M. Brown, Jr., J. H. M. Willison, and C. L. Richardson, *Proc. Natl. Acad. Sci. USA*, 1976, **73**, 4565.
162. T. E. Bureau and R. M. Brown, Jr., *Proc. Natl. Acad. Sci. USA*, 1987, **84**, 6985.
163. C. H. Haigler and R. M. Brown, Jr., *J. Cell Biol.*, 1979, **83**, 70a.
164. C. H. Haigler, R. M. Brown, Jr., and M. Benziman, *Science*, 1980, **210**, 903.
165. M. Benziman, C. H. Haigler, R. M. Brown, Jr., A. R. White, and K. M. Cooper, *Proc. Natl. Acad. Sci. USA*, 1980, **77**, 6678.
166. R. M. Brown, Jr., C. H. Haigler, and K. Cooper, *Science*, 1982, **218**, 1141.
167. R. M. Brown, Jr., C. H. Haigler, J. Suttie, A. R. White, E. M. Roberts, C. A. Smith, T. Ito, and K. H. Cooper, *J. Appl. Polym. Sci., Appl. Polym. Symp.*, 1983, **37**, 33.
168. R. M. Brown and D. Montezinos, *Proc. Natl. Acad. Sci. USA*, 1976, **73**, 143.
169. H. Quader, in "Biosynthesis and Biodegradation of Cellulose," eds. C. H. Haigler and P. J. Weimer, Marcel Dekker, New York, 1991, p. 51.
170. A. M. C. Emons, in "Biosynthesis and Biodegradation of Cellulose," eds. C. H. Haigler and P. J. Weimer, Marcel Dekker, New York, 1991, p. 71.
171. C. H. Haigler, in "Biosynthesis and Biodegradation of Cellulose," eds. C. H. Haigler and P. J. Weimer, Marcel Dekker, New York, 1991, p. 99.
172. K. I. Uhlin, Doctoral Dissertation, The Institute of Paper Science and Technology, Atlanta, GA, 1990.
173. K. I. Uhlin, R. H. Atalla, and N. S. Thompson, *Cellulose*, 1995, **2**, 129.
174. R. H. Atalla, J. M. Hackney, I. Uhlin, and N. S. Thompson, *Int. J. Biol. Macromol.*, 1993, **15**, 109.
175. J. M. Hackney, R. H. Atalla and D. L. VanderHart, *Int. J. Biol. Macromol.*, 1994, **16**, 215
176. A. Hirai, M. Tsuji, and F. Horii, *Cellulose*, 1997, **4**, 239.
177. T. Imai and J. Sugiyama, *Macromolecules*, 1998, **31**, 6275.
178. T. Okamoto, J. Sugiyama, and T. Itoh, *Wood Res.*, 1996, **83**, p. 27.
179. N. C. Carpita and D. M. Gibeaut, *Plant J.*, 1993, **3**, 1.
180. R. H. Atalla and S. C. Nagel, *J. Polym. Sci., Polym. Lett.*, 1974, **12**, 565.
181. R. H. Atalla and R. Whitmore, *J. Polym. Sci., Polym. Lett.*, 1978, **16**, 601.
182. E. Heuser, in "Nature of the Chemical Components of Wood," ed. C. J. West, TAPPI Monograph Series No. 6, Technical Association of the Pulp and Paper Industry, New York, 1948, p. 8.
183. R. H. Atalla and J. M. Hackney, in "Hierarchically Structured Materials," eds. I. A. Aksay, E. Baer, M. Sarikaya, and D. A. Tirrell, Materials Research Society Symposium Proceedings, 1992, vol. 255, p. 387.
184. R. H. Atalla, J. M. Hackney, and D. L. VanderHart, unpublished.
185. R. H. Newman, M. A. Ha, and L. D. Melton, *J. Agric. Food Chem.*, 1994, **42**, 1402.
186. R. H. Newman, L. M. Davies, and P. J. Harris, *Plant Physiol.*, 1996, **111**, 475.
187. R. H. Newman, *Cellulose*, 1997, **4**, 269.
188. R. H. Newman, *J. Wood Chem. Technol.*, 1994, **14**, 451.
189. T. M. Wood, in "Biosynthesis and Biodegradation of Cellulose," eds. C. H. Haigler and P. J. Weimer, Marcel Dekker, New York, 1991, p. 491.
190. P. Rapp and A. Beermann, in "Biosynthesis and Biodegradation of Cellulose," eds. C. H. Haigler and P. J. Weimer, Marcel Dekker, New York, 1991, p. 535.
191. P. Tomme, R. A. J. Warren, and N. R. Gilkes, *Adv. Microb. Physiol.*, 1995, **37**, 1.
192. T. T. Teeri, *Trends Biotechnol.*, 1997, **15**, 160.
193. E. A. Bayer, E. Morag, R. Lamed, S. Yaron, and Y. Shoham, in "Carbohydrases from *Trichoderma reesei* and other Organisms," eds. M. Clayssens, W. Nerinckx, and K. Piens, The Royal Society of Chemistry, London, 1998, p. 39.
194. E. T. Reese and M. Mandels, in "Cellulose and Cellulose Derivatives," Pt. V, eds. N.M. Bikales and L. Segal, Wiley, New York, 1971, p. 1079.
195. C. S. Walseth, *TAPPI*, 1952, **35**, 228.
196. K.-E. Eriksson, *Adv. Chem. Series*, 1969, **95**, 83.
197. K.-E. Eriksson and B. Petterssen, *Eur. J. Biochem.*, 1975, **51**, 213.
198. K.-E. Eriksson and T. M. Wood, in "Biosynthesis and Biodegradation of Wood Components," ed. T. Higuchi, Academic Press, New York, 1985, p. 469.
199. P. F. G. Sims, M. S. Soares-Felipe, Q. Wang, M. E. Gent, C. Tempelaars, and P. Broda, *Mol. Microbiol.*, 1994, **12**, 209.
200. M. A. Vallim, B. J. H. Janse, J. Gaskell, A. A. Pizzirani-Kleiner, and D. Cullen, *Appl. Environ. Microbiol.*, 1998, **64**, 1924.
201. D. Cullen and R. H. Atalla, unpublished.
202. R. Lamed, J. Naimark, E. Morgenstern, and E. A. Bayer, *J. Bacteriol.*, 1987, **169**, 3792.
203. R. Lamed and E. A. Bayer, in "Biochemistry and Genetics of Cellulose Degradation," eds. J.-P. Aubert, P. Beguin, and J. Millet, Academic Press, London, 1988, p. 101
204. R. Lamed and E. A. Bayer, in "Biosynthesis and Biodegradation of Cellulose," eds. C. H. Haigler and P. J. Weimer, Marcel Dekker, New York, 1991, p. 377.
205. E. A. Bayer, E. Morag, and R. Lamed, *Trends Biotechnol.*, 1994, **12**, 379.
206. P. Beguin and J.-P. Aubert, *FEMS Microbiol. Rev.*, 1994, **13**, 25.
207. C. R. Felix and L. G. Ljungdahl, *Ann. Rev. Microbiol.*, 1993, **47**, 791.
208. N. Hayashi, J. Sugiyama, T. Okano, and M. Ishihara, *Carbohydr. Res.*, 1998, **305**, 109.
209. N. Hayashi, J. Sugiyama, T. Okano, and M. Ishihara, *Carbohydr. Res.*, 1998, **305**, 261.
210. G. Maclachlan and S. Carrington, in "Biosynthesis and Biodegradation of Cellulose," eds. C. H. Haigler and P. J. Weimer, Marcel Dekker, New York, 1991, p. 599.

3.17
Hemicelluloses

ABIGAIL GREGORY and G. PAUL BOLWELL
Royal Holloway and Bedford New College, Egham, UK

3.17.1 INTRODUCTION

Hemicelluloses are a significant component of the plant cell wall. In contrast to other plant-derived polysaccharides such as cellulose, starch, pectins, and gums, hemicelluloses do not have a comparable commercial value. However, they are a subject of importance since they can affect the extraction of cellulose and they also make significant contributions to wood and fibre quality. Some soluble hemicelluloses are important dietary constituents. Hemicelluloses are typically defined as components that can be precipitated by ethanol after extraction from the cell wall by dilute alkali. In such procedures they are extracted after depletion of the pectin content of the walls by aqueous solvents and calcium chelators. Use of alkali, however, makes it difficult to maintain the structural integrity of many polysaccharides during their extraction from the cell wall matrix. As an alternative to alkali extraction, enzymes can be used to dissociate hemicellulose fragments from the cell wall matrix. For example, xylanohydrolases and glucuronoxylanases can be used as probes to resolve glucuronoxylan and glucuronoarabinoxylan structures.[1] As a consequence of such a chemical definition, the family of hemicelluloses include noncellulosic polysaccharides other than starch or fructans that are abundant in the ariel and normally lignified tissues of higher land plants. Such a definition has been extended to the soluble polymers of the endosperm and materials from the roots. Ultimately, however, they are best defined by their structures that make them distinct from the pectins. Here there is a fairly clear and widespread understanding of what constitutes a hemicellulose, and the contents of this chapter reflect this agreement. This chapter is primarily about the biosynthesis of hemicelluloses. However, some discussion of their structure and role is necessary to fully understand the considerable problems in elucidating their synthesis and assembly. Thus, it is important to know the full range of linkages present and their frequency in the nascent polysaccharide and any modifications that take place after the primary synthesis in order to understand the complexity of the biosynthetic processes.

3.17.2   THE HEMICELLULOSE FAMILY

3.17.2.1   Primary Wall Hemicelluloses

While whole plants have been extensively studied for the range of polysaccharides present in their walls the best in-depth studies have used suspension-cultured cells as source material. Some of the extracellular polysaccharides and glycoproteins which accumulate in the medium of cell suspension cultures are structurally similar to components of primary plant cell walls, and are, therefore, useful sources of material for structural characterization of wall polysaccharides.[2] Xyloglucan is probably the best-studied matrix cell wall polysaccharide. The basic alternating heptasaccharide/ nonasaccharide repeat sequence has been recognized for some time. Chemical analysis has extended the understanding of the repeat structures within the polymer. Xyloglucans are highly branched polysaccharides of the primary wall that are associated with cellulose and are structurally related in that they consist of β-1,4-linked Glc$p$ residues, about 75% of which are substituted at C-6 with α-D-xyl$p$ residues.[3] Some of the α-D-xyl$p$ residues are substituted at C-2 with β-D-Gal$p$ or an α-L-Fuc$p$-1,2-β-D-Gal$p$ substituted with O-acetyl groups.[4] A whole series of oligosaccharides that was generated enzymatically have been characterized and reveal rarer structures. About 2% of the residues are L-Ara$p$ which have been located at the C-2 of 2,4,6-linked Glc$p$ residues and at the C-3 of 3-linked β-Xyl$p$ residues.[5,6] There are, therefore, at least eight linkages that have to be accounted for in the enzymology of xyloglucan synthesis.

The equivalent hemicellulosic polymer of similar abundance in the walls of monocotyledon walls is glucuronoarabinoxylan, which is a minor constituent of dicot primary walls. Glucurono-arabinoxylans consist of backbones of 1,4-linked-β-D-xyl$p$ residues, of which C-2 or C-3 are substituted with arabinosyl-, galactosyl-, and glucuronosyl-rich side chains. A highly substituted glucuronoarabinoxylan (i.e., six out of seven xylosyl units) is associated with the maximum growth rate of coleoptiles.[7] Nishitani and Nevins[8] found a sequence-dependent xylanase which could be used as a probe to further resolve the structures of glucuronoxylan and glucuronoarabinoxylan. In this case, there are greater than eight linkages that probably have to be accounted for in understanding the biosynthesis. Additionally, some monocots have been shown to contain small levels of fucosylated xyloglucan, as in the case of cell suspension cultures of *Festuca arundinacea*.[9] Other hemicelluloses of monocot walls are the mixed 1,3-1,4-linked-β-Glc$p$ glucans. Their general structure was also deduced enzymatically using a sequence-dependent endoglycanase, a β-D-glucanohydrolase from *Bacillus subtilis* that catalyzes the hydrolysis of a 1,4-β-D-glucosyl linkage only if preceded by a 1,3-β-D-linked glucosyl unit on the nonreducing side.[10] Some galactans have also been considered hemicellulosic on the basis of extractability.[11]

3.17.2.2 Hemicelluloses of the Secondary Wall

Xylans are heterogeneous polymers of the secondary wall. They broadly consist of 1,4-linked β-D-xyl$p$ residues, substituted with 4-O-methyl-D-glucuronic acid, D-glucuronic acid, arabinose, and acetate. There are at least six types of linkages involved in their biosynthesis. In dicots, the 4-O-methylglucuronic acid side chain is the main substituent attached to C-2 positions, and hardwood xylans are acetylated. The acetyl content is variable and is about 10% for beechwood and birchwood, for example,[12] but may be as high as 70%. L-ara$f$ residues are less abundant and attached to some C-3 positions. Gymnosperm xylans contain less of these arabinosyl units and are not acetylated. In graminaceous xylans, arabinosyl units predominate. Glucomannans are also a feature of gymnosperm walls, while they are of minor occurrence in angiosperms. These polysaccharides consist of 1,4-linked β-D-mannose residues in which a significant proportion (20–50%) are replaced by D-glucosyl units. Only mannose is found contiguously though. The distribution of the sugars is also not regular. Small amounts of D-galactose have been detected, and some residues are acetylated.

3.17.3 LOCALIZATION AND ASSOCIATION WITH OTHER CELL WALL COMPONENTS

Although it is not the purpose of this chapter to deal with detailed structural aspects of hemicelluloses it is important that aspects of subcellular localization and interactions with other cell wall components are considered in relation to biosynthesis since assembly may involve regulatory aspects. The definitions with regard to whether hemicelluloses are characteristic of primary or secondary wall, based on extractability, also need to be modified in relation to tissue and subcellular localization. Our understanding of these aspects has been greatly improved by the development of immunolocalization techniques, particularly using immunogold. These studies have revealed subtleties, and it is apparent that xyloglucan is not confined to the primary wall, and is a substantial component of the vessel walls of some species, while xylans may also be found in small amounts in primary walls. Furthermore, when studied in relationship to other wall components particularly pectins an additional complexity to deposition and assembly of the wall has been revealed. One of the earliest studies[13] revealed that in primary walls of suspension-cultured sycamore cells, xyloglucan was localized throughout the entire wall whereas the pectin rhamnogalactan-I was restricted to the middle lamella and was especially evident at the junctions between cells. While such differences may reflect the accessibility of the reacting epitope, it is more likely to reflect a hidden complexity to wall biosynthesis and deposition. The xyloglucan could be further localized to the cellulose microfibril region of growing walls.[14] A further study showed that xylan as distinct from xyloglucan could be detected in primary walls but not cell plates, but was present in large amounts in secondary thickening.[15] Freshour *et al.*[16] have confirmed that xyloglucan and rhamnogalacturonan 1 and/or arabinogalactan proteins are among the first components laid down in the newly synthesized wall. Immunocytochemical studies using antibodies to secreted polysaccharides with specific sugar epitopes have demonstrated that different polysaccharides are synthesized in different types of Golgi. For example, in sycamore cell suspension cultures which were high pressure frozen/freeze substituted, it was seen that assembly of pectic polysaccharides involved cis, medial and trans types of Golgi cisternae, whereas the synthesis of xyloglucan was confined to trans-Golgi cisternae and the trans-Golgi network.[17,18]

Much of the early work in cell wall analysis was carried out by deconstructing the cell wall and subsequently analyzing the components chemically and microscopically. An alternative approach is to construct networks *in vitro* using cell wall polymers. In the cell wall, xyloglucan is believed to associate with cellulose at or near the point of cellulose synthesis. The cellulose-producing bacteria *Acetobacter aceti* ssp. *xylinum* and tamarind xyloglucan have been used to provide microscopic evidence for the generation of cross-bridges between cellulose ribbons produced in the presence of xyloglucan.[19] These cross-links can also be produced abiotically, but incorporation of xyloglucan was found to be higher in an actively growing system. From 80% to 85% of the xyloglucan adopted a rigid conformation similar to that of cellulose, and the molecules probably align with the cellulose chains themselves. The remaining xyloglucan was found to be more mobile, and appeared to be assigned to cross-bridges with a twisted backbone conformation. Surprisingly, fucose residues were not found to be essential for network formation. Many grasses have enriched levels of aromatic substances such as esters of hydroxycinnamates, ferulate, and *p*-coumarate in nonlignified walls. Glucuronarabinoxylans are cross-linked by esterified and etherified phenolics. Ferulate and *p*-coumarate esters are attached to O-5 of arabinosyl units of glucuronoarabinoxylans, and it has been demonstrated that feruloylated chains are cross-linked by formation of 5,5-diferulate.[20] Ferulic acid forms ether–ester bridges, while coumarate, cinnamate, and ferulic acid all form single esters. Work by Wende and Fry[21,22] has demonstrated that the *O*-feruloyl group at position 5 frequently has an additional Xyl*p* group attached. This Xyl residue significant changed the molecular environment of the feruloyl group, affecting susceptibility to alkaline hydrolysis and to enzyme-catalyzed reactions and also to oxidation by endogenous plant peroxidases involved in cell wall cross-linking. Dimers can also be formed oxidatively and photochemically between phenolic acids, thus forming cross-links between polysaccharides. Evidence for lignin–arabinoxylan linkages has also been obtained in barley straw and perennial rye-grass extractions.[23] The site of this modification is probably in the wall, but it may take place in the endomembrane system during the secretory mechanism. Protein–xyloglucan associations have also been postulated, though the majority of evidence so far generated concerns pectin–protein linkages via galacturonyl bonds involving the carboxyl group of pectins.[24,25]

Other associations have been studied in secondary wall. There is evidence that during the assembly, the synthesis and deposition of xylan is intimately linked with that of cellulose.[26] The only examples of specific secondary wall proteins localized directly, other than inferred from cDNA sequence, have been shown in Loblolly pine,[27] the hypocotyl of the French bean,[28] and in differentiating *Zinnia* cells.[29] The French bean protein proved to be hydroxyproline-poor and glycosylated, being recognized by wheat germ agglutinin and localized to tracheary elements, xylary, and phloem fibres, and can be localized in secondary walls induced in bean cultures. Alkaline extracts of hemicellulose showed that this protein could still be detected immunologically, indicating a strong association with xylan. Other lysine-rich proteins have been recognized from their cognate cDNAs. The epitope conferring recognition by wheat germ agglutinin is a particular feature of secondary thickenings.[30] Another epitope found in secondary thickenings of *Zinnia* cells was recognized by a monoclonal antibody (JIM 13) which is specific for an arabinogalactan protein. There are probably arabinogalactan proteins specifically expressed in xylem.[31] As yet there is no defined function for these proteins, although they may also form structural associations with secondary wall xylan. The deposition of lignin in these walls involves generation of mesomeric phenoxy radicals from the hydroxycinnamyl alcohol precursors. These will rapidly form linkages with the polysaccharides of the wall, which may take place randomly. Other factors govern whether the lignin encrusting some of the xylan will be of the *p*-hydroxyphenylpropane, guaiacyl, or syringyl type.

3.17.4 SITES OF BIOSYNTHESIS

Early radiolabeling studies using autoradiography served to localize wall matrix polysaccharide biosynthesis with the Golgi. Since this early work, subcellular membrane fractionation followed by characterization of enzyme distribution has confirmed this, and has been further supported more recently by immunogold localization of nascent polysaccharides but not as yet the enzymes themselves.

3.17.4.1 Subcellular Fractionation

The distribution of enzymes involved in matrix polysaccharide biosynthesis has long been studied in membrane preparations. Using marker enzymes, peaks of activity have mainly been localized to Golgi fractions. More recently, improvements in membrane fractionation have allowed study of the distribution throughout the Golgi stack.[32,33] Golgi secretory vesicles can be separated from dictyosomes by rate–zonal centrifugation, and this has been used to study the location of the biosynthesis of xyloglucan. In the case of pea microsomal membranes it was found that the secretory vesicles possessed high levels of xyloglucan fucosyltransferase activity, but lacked significant xyloglucan xylosyltransferase activity. This contrasted with the total dictyosomal membranes, which possessed both fucosyl and xylosyltransferase activities. Further centrifugation, using a shallower gradient for separation, showed that the lighter dictyosomal membranes exhibited primarily xylosyltransferase activity, and the denser membranes exhibited both xylosyl and fucosyltransferase activities. This differential localization of function indicates that the lighter dictyosomal membranes, which perhaps correspond to the cis-cisternae, are where the xyloglucan glucose–xylose backbone is initiated. The denser membranes, which perhaps correspond to the medial and trans-cisternae, are where the backbone is completed and fucosylation begins, and the fucosylation is completed within the secretory vesicles during transport to the cell wall.[32]

3.17.4.2 Immunolocalization

As mentioned earlier, sites of synthesis have also been deduced from the localization of nascent polysaccharides to the Golgi. Moore *et al.*[17] have demonstrated that individual Golgi stacks simultaneously process glycoproteins and complex polysaccharides. However, antibodies to xyloglucan and rhamnogalacturonan 1 were found to bind to mutually exclusive subsets of Golgi cisternae. The labeling pattern for xyloglucan was consistent with its assembly in the trans-Golgi network and departure from the monensin-sensitive trans-Golgi network. This contrasted with the localization of the rhamnogalacturonan 1-type polysaccharides to the cis- and medial cisternae and their departure from a monensin-sensitive, medial cisternal compartment. This segregation of polysaccharide synthases within the Golgi complex ensures that hybrid polysaccharides are not synthesized, and provides a means by which the individual polysaccharides can be separated into types of secretory vesicles and targeted to different cell wall domains.[17]

3.17.5 ENZYMOLOGY

Plant polysaccharide synthases are still poorly characterized. Considering the amount of cell wall synthesized, this is an enormous gap in our knowledge of plant metabolism. These have proven difficult enzymes to work with, but some advances continue to be made slowly.

3.17.5.1 General Features

Nucleotide diphosphate sugars (mostly UDP derivatives) are used as the sugar donors for polysaccharide synthases. The availability of nucleotide diphosphate sugars does appear to be under some fine control[34] (see also Section 3.17.7). Although the synthesis of matrix polysaccharides occurs in the endoplasmic reticulum, Golgi bodies and vesicles, the synthesis of the nucleotide diphosphate sugars occurs in the cytosol. Since membranes are generally impermeable to sugar nucleotides, there must be some transport mechanism by which they are able to cross. Interestingly, sucrose synthase, which is believed to have a role in the production of UDP-glucose for glucan synthesis, has been shown to be associated with the plasma membrane,[35] and, thus, may channel carbon directly from sucrose to cellulose and or callose synthases in the plasma membrane. Unlike protein synthesis, which takes place along the residues of mRNA, there is no similar template for polysaccharide synthesis. Thus, there are many intriguing questions about the regulation of polysaccharide synthesis and the coordination of synthesis of heteropolymers which involve the action of more than one synthase.[36] In the listed examples of glycosyltransferases involved in polysaccharide synthesis there is no evidence for proteinaceous primers or lipid intermediates, so their involvement is traditionally considered unlikely. Most of the available evidence suggests that sugar residues are added directly from the sugar nucleotide to a growing chain rather than via a lipid intermediate. However, in the case of maize root mucilage there is clear evidence that the polysaccharide is made as a glycoprotein,[37] and here the protein may be acting as a primer, with the involvement of glycolipid intermediates.[38] However, the mucilage is not a cell wall polysaccharide and would not necessarily be synthesized in the same way.[39] Pulse chase studies have indicated that the transit time from endoplasmic reticulum to cell wall is somewhere in the region of 20–30 min. The initiating reaction for complex polysaccharides is therefore not yet elucidated. In the case of homopolymers, these can be synthesized by membranes where a complete system including donors would be present, but the fact that there are examples of the same product being synthesized by solubilized membranes which would disrupt the action of lipid donors is also some evidence that these are not required. The few examples of synthesis by relatively pure enzymes also suggests that protein primers are not required and that plant polysaccharide synthases are unique in their synthetic mechanisms. Even the classic storage polysaccharide glycogen requires a protein primer, glycogenin, whereas the starch polymers appear to be made by a variety of mechanisms.[40] With the initiation of the core oligosaccharides these can then be decorated by other glycosyltransferases. Whether each linkage found in polysaccharides (and, additionally, glycoproteins) requires a separate glycosyltransferase is of some debate. This would require an enormous number of genes. We are very hampered at the moment from lack of identification of glycosyltransferase genes, but some of the unknown expressed sequence tags must code for them. However, the existence of hundreds of glycosyltransferase genes in a single genome seems unlikely, so there must be some redundancy in the linkage made by each glycosyltransferase so that the glycosidic bond made is governed in part by the orientation of the hydroxyl groups on the receiving oligosaccharide. The β-1,3-1,4 mixed glucan synthase is a case in point. We have shown that the environment which the glucan synthase is in can also be important. Bean callose synthase makes β-1,3-glucans in a totally aqueous environment, while increasing the hydrophobicity with organic solvents leads to production of a mixed-linkage glucan.[41] The number of glycosyltransferases required to make a complex hemicellulose therefore remains unknown, and although a number have been characterized, we are still relatively ignorant when compared with our knowledge of the biosynthesis of other plant products.

3.17.5.2 Characterization and Purification of Glycosyltransferases

3.17.5.2.1 *Xyloglucan synthases*

Xyloglucan is one of the most prominent matrix polysaccharides in nongraminaceous plant cell walls, making up 20% of total primary cell wall polysaccharides. As stated earlier, it is composed of a linear glucan backbone with regular side chain additions of xylose and galactose and sometimes fucose and arabinose. A tentative mechanism by which the elaboration of the xylose–glucose backbone in the pea is initiated in lighter dictyosomal membranes, backbone synthesis is concluded, fucosylation begins in denser dictyosomal membranes, and fucosylation is completed in Golgi secretory vesicles is described in more detail in Section 3.17.4.1.[32]

(i) *Glucosyltransferase*

Several different 1,4-β-glucan 4-β-glucosyltransferases are present in plant cells. Glucan synthase I may make glucan chains similar to those in cellulose. Glucan synthase II is a 1,3-β-glucan 3-β-glucosyltransferase that makes the wound polymer callose. Xyloglucan glucosyltransferase may produce the glucan backbone for xyloglucan. In cell membranes isolated from etiolated pea tissue,[42,43] incorporation of glucose from UDP-D-[¹⁴C]glucose into polysaccharides with linkages consistent with synthesis of xyloglucan was demonstrated.

(ii) *Xylosyltransferase*

Xylosyltransferase involved in xyloglucan synthesis uses UDP-xylose as a substrate to add xylosyl side chain residues to the C-6 position of backbone glucosyl residues. This has been examined in a suspension-cultured dwarf French bean.[44] A particulate enzyme preparation was shown to incorporate xylose from UDP-D-[¹⁴C]xylose into xyloglucan, but the xylosyltransferase was found to be almost inactive unless in the presence of UDP-glucose. The reaction was stimulated by divalent cations, Mn^{2+} being the most effective. However, glucose could be incorporated into xyloglucan in the absence of UDP-D-xylose, and the backbone of the xyloglucan could be extended without the simultaneous addition of side chains. Also, attempted proteinase digestions indicated that the nascent xyloglucan was closely associated with protein, perhaps suggesting the involvement of a protein primer in this case.

(iii) *Galactosyltransferase*

Little work has been carried out on galactosyltransferases with respect to xyloglucan biosynthesis, though there is evidence for several galactan synthases in flax cell suspension cultures, and their role in the synthesis of various galactans has been investigated.[45] During cell growth, pectic β-1,4-galactan was synthesized mainly during the lag phase whereas the greatest amount of hemicellulose

β-1,3-β-1,6-galactan was detected during the growth phase.[46] Two peaks of galactan synthase activity were detected that catalyzed the synthesis of β-1,4-galactan at pH 8 during the lag phase and the synthesis of β-1,3-β-1,4-galactan at pH 5 during the growth phase. At the end of the growth phase both activities were negligible. The galactan synthase responsible for the biosynthesis of a pectic β-1,4-galactan has also been characterized from the mung bean,[47] and had an apparent pH optimum of 6.5 in the presence of Mg^{2+}.

(iv) Fucosyltransferase

In dicotyledonous plants, fucosylated xyloglucan is the major hemicellulose of primary cell walls, where it is bound firmly to and between cellulose microfibrils. Hanna *et al.*[48] first described the solubilization and characterization of a plant 1,2-α-fucosyltransferase, a xyloglucan 1,2-α-fucosyltransferase from pea microsomes. The K_m for fucosyl transfer to tamarind xyloglucan by the membrane-bound or solubilized (0.3% w/v Chaps) enzyme was about 80 mM, which was 10-fold greater than the K_m for transfer to endogenous pea nascent xyloglucan. Optimum activity was between pH 6 and 7, and the solubilized enzyme showed no requirement for, or stimulation by, added cations or phospholipids. The fucosyltransferase was shown to have an M_r of 150 000 by gel entrapment.

A novel α-L-fucosyltransferase has also been isolated from the microsomal fraction of primary roots from 6 d old radish seedlings.[49] This enzyme transfers L-fucose from GDP-L-Fuc to O-2 of an α-L-arabinofuranosyl residue and undergoes development and organ-specific expression in root tissue, whereas the L-Fuc transfer to tamarind xyloglucan can be detected in microsomal fractions from various organs in developing radish plants. Two distinct α-L-fucosyltransferases with different acceptor specificities are associated from Golgi membranes from primary roots, but hypocotyl Golgi membranes completely lack the enzyme specific for the L-Araf residue.

(v) Arabinosyltransferases

Arabinose is a rare substitution of xyloglucan, contributing about 2%, and 1,2 linked to Glc*p* or 1,3 linked to xyl*p* residues. However, most arabinosyltransferases that have been characterized are involved in pectic arabinan synthesis (1,2, 1,3, and 1,5 linkages), or O-glycosylation of hydroxy-proline-rich proteins and arabinogalactan proteins.[50]

3.17.5.2.2 *Xylan synthases*

Xylosyltransferases involved in secondary wall xylan synthesis have been described from the French bean. Membrane fractions from bean hypocotyl and callus were shown to incorporate xylose from UDP-α-D-xylose into xylan, and the control of xylan synthase was studied during xylogenesis in stele and in xylogenesis induced in callus tissue.[51] Xylan synthase was shown to be induced during the period of secondary thickening of the cell wall and was also shown to be correlated with the induction of phenylalanine ammonia lyase and with lignin synthesis. No lipid or proteinaceous intermediates were found, and glycosylations were neither stimulated by added dolichyl phosphate nor inhibited by compounds that usually prevent transfers involving polyprenylphosphate intermediates.[52,53] The French bean xylan synthases could be solubilized, and two isoforms were purified. One, an M_r 40 000 form, could be purified to apparent homogeneity. The relative recovery of the two peaks of xylosyltransferases was dependent on the age at which the bean hypocotyls were harvested. At 8 d, the relative activity of the peak designated XS1 was greater than that of XS2, whereas between 10–12 d the relative activities were reversed. It is likely that both these forms were associated with xylan synthesis since the glucan backbone necessary for xyloglucan biosynthesis was not present in the reaction mixture. Polysaccharide analysis showed that xylan was present as about 15% of the total hemicellulose at the peak of XS1 activity whereas xylans constituted about 60% of total hemicellulose at the peak of XS2 activity.[54]

Xylan synthase activity has also been examined in differentiated xylem cells of sycamore trees.[55] The K_m of the synthetase for UDP-D-xylose was 0.4 mM. Enzyme activity was not enhanced in the presence of detergent or EDTA, but was stimulated by Mg^{2+} and Mn^{2+}. Increased xylan production during differentiation of xylem cells was exerted by a sixfold increase in xylan synthase activity during the period of maturation of the cells. In sycamore and poplar, the activities of the decarboxylase and dehydrogenase also increased during differentiation, and the decarboxylase activity was always higher than that of the dehydrogenase, which is probably rate-limiting.[56] However, the dehydrogenase step can be bypassed using a direct route from glucose via myoinositol to UDP-D-glucuronic acid. There is evidence for the coexistence of both pathways with changing importance of either route during plant development.[57] Thus, the decarboxylase also becomes an important control step for the production of UDP-xylose as a substrate for xylan synthesis.

3.17.5.2.3 Glucuronoarabinoxylan synthases

The primary wall of dividing grass cells is markedly different from that of most other plants, being composed of cellulose microfibrils interlaced predominantly by glucuronoarabinoxylans. These polysaccharides are made up of linear chains of 1,4-α-D-xylan substituted with *t*-Ara units at *O*-3 and *t*-GlcA units at *O*-2 of the xylosyl backbone units. The degrees of substitution and alkali extractability (i.e., strength of binding to the cellulose fibrils) can be used to differentiate between the different glucuronoarabinoxylans. Arabinose has been shown to be hydrolyzed from the glucuronoarabinoxylans, leaving behind relatively unsubstituted xylans which are capable of tighter hydrogen bonding to the other matrix polymers.[58] There have been few reports of the arabinosyl- or glucuronosyltransferases.

3.17.5.2.4 β-1,3-1,4-Glucan synthases

Mixed-linkage glucans are hemicellulosic components of the cell walls of a number of grasses. Since the first demonstration of plant β-glucan synthesis *in vitro*, characterization and purification of the relevant enzymes has been slow, as the polysaccharide synthases are tightly membrane bound and hard to purify. Also, the process of cell extraction causes the induction of callose synthesis and almost entirely wipes out cellulose synthesis. Attempts to purify mixed linkage glucans have been made from a number of monocots. In barley, a mixed-linkage glucan is present as a major component of the cell walls of the aleurone and starchy endosperm and also of the coleoptile cell walls. The properties of the membrane-bound glucan synthases in barley have been studied with regard to temperature, pH, cofactors, and substrate concentration.[59] The synthesis of both β-glucans was optimal at 20 °C. In tris HCl buffer, the pH optima for 1,3-1,4-β-glucan synthesis and 1,3-β-glucan synthesis were 8.5 and 7, respectively. Both synthases required Mg^{2+} ions: 1,3-β-glucan synthase at 2 mM and 1,3-1,4-β-glucan synthase at 200 mM. 1,3-β-Glucan synthesis also required calcium. The K_m with respect to UDP-glucose was 1.5 mM for 1,3-β-glucan synthase and 1 mM for 1,3-1,4-β-glucan synthase.

3.17.5.2.5 Glucomannan synthases

Gluco- and galactomannans are major polysaccharides found in the hemicellulose fractions of gymnosperms. They also occur in much smaller amounts in the cell wall matrix of angiosperms. Glucomannan synthases have been described from the pea and pine. Particulate membrane preparations isolated from cambial cells and differentiating and differentiated xylem cells of pine trees were shown to synthesize [^{14}C]glucans using either GDP-D-[U-^{14}C]glucose or UDP-D-[U-^{14}C]glucose as glycosyl donors.[60] Although these glucans had β-1,3 and β-1,4 linkages in the approximate ratio of 1:1, the distribution of linkages varied according to the substrate used. For example, the synthesis of the mixed glucan from GDP-D-glucose was changed to that of the β-1,4 linkages in the presence of increased levels of GDP-D-mannose. The glucan formed from the UDP-D-glucose was not affected by any concentration of GDP-D-mannose. The apparent K_m and V_{max} of the glucan synthase for GDP-D-glucose were 6.8 mM and 5.08 mM, respectively. No lipid intermediates were detected during the synthesis of either glucan or glucomannan.

Membrane fractions and digitonin-solubilized enzymes prepared from stem segments of pea seedlings were found to catalyze the synthesis of a β-1,4-mannan from GDP-D-mannose, a mixed β-1,3- and β-1,4-glucan from GDP-D-glucose, and a β-1,4-glucomannan from both GDP-D-mannose and GDP-D-glucose. The ratio of glucose to mannose found in the product was shown to be dependent on the ratio of the amounts of the two different substrates supplied to the enzyme, which indicated that *in vivo* the availability of the substrates at the catalytic site(s) of the transferase was an important factor in the final composition of the compound.[61] In gymnosperm membranes, there is an insoluble GDP-D-mannose 2-epimerase present, which was not found in the enzyme preparations made from pea internode tissue. Thus, the mechanisms of control of glucomannan synthase are different in the pea than those observed in the pine. The control of the ratio of the two substrates is presumably either dependent on the rate of synthesis of the respective nucleotide sugars in the cytoplasm or on the transport mechanisms by which the nucleotide sugars are taken up into the endomembrane system. Piro and Leucci[62] have also shown that glucomannan synthesis is inhibited by light essentially during the biogenesis of the primary cell wall, and have suggested that light may regulate the synthesis of glucomannan by reducing the expression of mRNA encoding for the glucomannan synthase. The galactomannans present in the hemicellulosic fraction of endosperm cell walls of leguminous seeds all have a common structural pattern, that is, a 1,4-β-linked D-mannan chain substituted laterally by 1,6-α-linked D-galactose residues.[63] Variations in galactose content and/or distribution within the legume–seed galactomannans clearly arise from differences either in the pathway of biosynthesis or in the postsynthetic processing of the galactomannan molecules. Galactomannan biosynthesis has been studied in the developing endosperms of fenugreek, guar, and senna[64] (see also Chapter 3.15). Labeled galactomannans with a range of mannose/galactose residues were synthesized *in vitro* using membrane-bound preparations from all three plants, and the substitution patterns compared.

3.17.6 REGULATION OF HEMICELLULOSE BIOSYNTHESIS

3.17.6.1 Regulation of Flux

Since xyloglucan is the major hemicellulose of primary wall, and xylan and its variants are the major hemicelluloses of the secondary wall, there are gross shifts in hemicellulose biosynthesis during differentiation. Additionally, during growth and expansion of the primary wall there are changes in the flux into xyloglucan, and it is modified actually in the wall. Fluxes in polysaccharide biosynthesis and metabolism can be measured by the flow and accumulation of radioactivity from labeled sugar substrates into specific polysaccharides and cellular compartments. Cell cultures, and in particular suspension cultures, can be manipulated in such a way that they offer an excellent model to study polysaccharide biosynthesis. Alternatively, another approach to gauging the rates of cell wall biosynthesis can be through assaying component enzymes and correlating their specific activities with pathway fluxes in relation to developmental events. In its most sophisticated form, these data can be combined into control analysis which can be used to identify key regulatory reactions. This has not yet been applied to wall polysaccharide biosynthesis. Even without this rigorous analysis, clues exist as to the type of control mechanisms involved. Cell culture systems that have been particularly well studied for expansion growth are carrot and spinach, while bean, tobacco, and, most spectacularly, the differentiation of mesophyll cells of *Zinnia elegans* into tracheids have been used to study xylogenesis. These systems can be manipulated experimentally to study control mechanisms involved in synthesis and modification of hemicelluloses. A more descriptive method has been to take successive scrapings through the cambial and xylem initial and differentiating layers during the spring growth period of a number of tree species such as sycamore and, most productively, Loblolly pine. The latter source has been used to clone the cognate genes of the enzymes of the whole lignification pathway and a number of vascular-specific proteins.[65]

The actual synthesis of cell wall polysaccharides is a Golgi-based process with the exception of the glucans, cellulose, and callose, which are synthesized at the plasmalemma. All these glycosyltransferases remain rather poorly described but progress is being made in their characterization.[66] When cells of the French bean are induced to form secondary walls, arabinosyltransferase activity catalyzed by an M_r 70 000 Golgi-localized enzyme is reduced, indicating a cessation of pectin synthesis.[52–54] Similarly, there was a loss of polygalacturonate acid synthase in sycamore cells on differentiation.[50] On the other hand, xylosyltransferase activity involved in xylan synthesis, which is probably catalyzed by an M_r 40 000 protein,[54] is seen to rapidly increase several-fold in cells grown in induction medium.[52,53] This increase in xylan synthase activity reflects that seen in differentiating sycamore[50] and for enzymes responsible for the synthesis of other hemicelluloses such as glucomannan.[60,61] In cultures, the kinetics of appearance of these examples of glycosyltransferase activities correlates to the changes observed between the induced and noninduced cell walls. The cell walls found in the French bean system which have undergone xylogenic-like changes in composition are also reminiscent of those found in differentiated *Zinnia* cells.[67] Similarly, in the *Zinnia* system, xylosyltransferase activities have also been correlated to the increase in xylan synthesis.[68] Furthermore, these observations made in French bean cells have also been made in developing bean hypocotyls, which would tend to confirm that cell culture systems provide a valid model for xylogenic differentiation.[53] Other systems such as flax, which accumulates high levels of xylan, may be more amenable to study of the biosynthesis of these hemicelluloses.[69] Cellulose synthesis has not been measured directly due to the difficulty in measuring this enzyme activity, but there is known to be increased cellulose deposition in *Zinnia*.[70,71]

Although it is generally assumed that the major controlling factor in the qualitative production of cell wall polysaccharides resides in the complement of the membrane-bound synthases,[72] the underlying enzymes involved in the supply of UDP-sugars, as substrates for the synthases, may affect the overall balance of cell wall polysaccharide biosynthesis. An intimate association has been demonstrated between sucrose synthase, thought to be the major enzyme in the provision of UDP-glucose from sucrose, with plasmalemma and with cellulose and callose synthesis.[35] Immunolocalization demonstrates an up-regulation of sucrose synthase in differentiating tracheary elements of *Zinnia*.[73] In French bean cells, however, sucrose synthase declined in activity, and it seems more likely that in these cells starch is a major source for wall polysaccharides.[34] This work also showed that UDP-glucose dehydrogenase increased in activity, which correlated with the xylogenic-like changes found in cell walls of induced French bean cells. The total enzyme activity may reflect the participation of two enzyme systems. The 50 kDa form, which shows high sequence similarity to mammalian UDP-glucose dehydrogenase[57] and has a K_m of 0.2 mM, is most highly expressed in root tissues and epicotyls of soybean. A less specific 40 kDa form, which is an isoform of alcohol dehydrogenase, has a K_m of 5.5 mM and has been isolated from the French bean.[74] There may also be contributions through the inositol pathway. The less specific UDP-glucose dehydrogenase form was immunolocalized to developing xylem and phloem of bean hypocotyl. The combination of these enzymes is probably a key regulatory step since the reaction is thought to be committal as it is irreversible. The product, UDP-glucuronic acid, can readily be converted to UDP-xylose by the action of a decarboxylase. This enzyme also appeared to be under developmental control but to a lesser extent than the dehydrogenase.[34] UDP-xylose can exert negative-feedback control on the activity of the purified dehydrogenase.[74] In comparison with the plant, manipulating cell cultures to become xylogenic makes it possible to envisage more easily how different aspects of polysaccharide biosynthesis interact. Not only are there gross changes in the type of cell wall carbohydrate polymers synthesized but these are tightly regulated with respect to the metabolism of the necessary UDP-sugars required. Moreover, this is timed to coincide with the cytoskeletal changes involving microtubules which are required to deposit the newly formed polymers into the forming secondary wall. This is especially so in cells which display highly architectured arrays of secondary thickenings characteristic of xylem vessels and tracheids.

3.17.6.2 Extracellular Signals

This biochemical regulation is potentially modulated by an array of extracellular signals. The most common method of inducing differentiation-related changes is to increase the ratio of cytokinin to auxin and to raise the level of sucrose in the medium. In fact, manipulation of many of these factors can induce morphological changes in cell tissue cultures including shape, such as elongation and adhesion and clump size as in cytokinin constitutive cultures of tobacco. Tracheids can also be found spontaneously in these cultures. There is still an absolute requirement for auxin in all systems studied.[75] This requirement for auxin may be sufficient to initiate tracheary formation in the absence of cytokinin, and in such systems the level of endogenous cytokinin production does not appear to be a limiting factor. Both auxin and cytokinin administration to cell cultures stimulate synthesis *de novo* of a large number of proteins.[72] In cells of sycamore and *Haplopapus gracilis*, low levels of auxin fed to cultures in pulses are more effective in promoting differentiation than a single addition. This differentiation in *Haplopapus* was also dependent on mitochondrial protein synthesis.[76]

Sucrose has a known influence on the expression of a number of genes.[77] These include proteinase inhibitors,[78,79] chalcone synthase,[80] aminopeptidases, and pathogenesis-related proteins.[81] In some cases, a sugar response element in the promoters has been identified, and a common mechanism of sugar-sensing resulting in the repression of photosynthetic genes and the activation of stress-related and pathogenesis-related genes has been proposed.[77] The interrelatedness of the response to various stimuli is compounded when the effect of constitutive cytokinin expression in inducing stress-response genes is considered,[82] and suspension cultured calli derived from these lines develop tracheids. Perturbation of sugar metabolism by viruses or in transgenic plants is accompanied by the induction of stress-response genes.[81] In the study by Herbers *et al.*[78] on the effect of the accumulation of soluble sugars in cells, 1-aminocyclopropane-1-carboxylic acid (ACC) oxidase was also induced. Carbohydrates have been reported to induce ethylene production,[83] and this has been investigated with respect to sucrose concentration in the lettuce pith system.[84] Ethylene has been implicated in vascular differentiation in a number of studies.[75] The ethylene pathway of signal transduction is now one of the best understood, especially in relation to stress.[85] It is striking how those genes classically turned on by wounding, β-1,3-glucanase, proline-rich proteins, chitinase, and hydroxyproline-rich glycoproteins are also induced by the expression of the Tcyt gene in transgenic tobacco, leading to high endogenous levels of cytokinin.[82] Lignification genes are also stimulated by stress[72] and, for example, a major antifungal hybrid chitin-binding protein of the French bean is expressed in developing xylem at the plasmalemma wall interface.[86] The analogy between the wounding response and xylogenesis has often been raised. It may well be that some of the transduction components first identified in the ethylene response have pleiotropic effects and also participate in xylogenesis. Although not measured directly yet, the synthesis of hemicelluloses will also be subjected to these signals during xylogenesis.

Other low molecular mass compounds have also been implicated as signals working in the developmental pathway. Experiments with the *Zinnia* system using conditioned medium on the cessation of cell expansion before the differentiation phase suggest an involvement of oligosaccharins.[87] Oligosaccharins have been implicated in modulating growth patterns in a large number of systems.[88] Most have been characterized as regulators of expansion growth while others affect development, including floral initiation. However, oligogalacturonides induce stress-related lignification[89] while xylanase treatment induces ethylene production.[90] Transformed tobacco cells undergoing tracheid formation have an extractable xylanase present in their cell walls which is absent from control cultures and could be involved in generating such morphogenic signals.[91] Experiments involving the use of conditioned medium augurs well for the identification of modulatory signals in vascular differentiation. In comparison, characterization of these would be extremely difficult in whole plants. Experiments using conditioned medium have been particularly successful in studies of somatic embryogenesis, and have indicated an involvement for wall-derived signals in the initiation of formation of other cell types related to their relative position. Nevertheless, some of these signals will influence vascular differentiation, and, therefore, hemicellulose biosynthesis.[92] There is some evidence that nod-like factors may be involved. This was based on the observation that addition of an M_r 32 000 chitinase could release globular embryos arrested in development and that this could also be mimicked by the addition of rhizobial nod factors.[93,94] More recent work suggests that putative endogenous nod factors may affect the balance between the action of auxin and cytokinin[95] and may thus influence vascular differentiation also. An enhanced production of chitinase is required,[96] and this is a feature of transformed cells expressing high endogenous cytokinin levels.[82,97] Certainly, the occurrence of potential substrates for chitinases is highest in secondary walls.[30,93]

3.17.6.3 Polysaccharide Turnover and Wall Assembly

Turnover and modification of wall polysaccharides is a feature of developmental change, and this is true also in tissue cultures.[98] Use of these systems has shown that xyloglucans continue to be modified after synthesis and deposition. The expansion phase of cambial development is an event that probably involves polymer turnover and rearrangement of existing structures in addition to a massive increase in biosynthesis and deposition of new material. A number of extracellular enzymes and proteins have been implicated in these processes. Glucanases,[99] xyloglucan endotransglycosylases and endoxyloglucan transferases,[100–105] and expansins[106–109] have all been championed as the major influence. All these could potentially act upon xyloglucan. The relative contribution of each of these to the modification of hemicellulose has not been absolutely determined, although a large number of hydrolases have been demonstrated in many tissue cultures.[100]

An extensive study was ongoing in 1998 to explore the complex changes in the glycan structures of the wall in *Zinnia*.[29] Changes in the secretion and turnover of pectins, xyloglucan, and the arabinogalactan epitopes of arabinogalactan proteins, show that a rhamnogalacturonan appears around the time of determination, while a specific arabinogalactan protein appears later and accumulates in secondary thickenings. The fucose-containing epitope on xyloglucan disappears just before the onset of secondary thickening. Such studies may reveal which polysaccharides are targeted for the generation of possible modulatory signals. A fucosidase appears to be one hydrolase that is activated. There may be some involvement in the production of oligosaccharin signals as feedback or monitoring signals for the cells as to the stage on the differentiation pathway that has been reached. Indeed, induction of other hydrolases such as chitinase appears to be necessary for somatic embryogenesis in carrot cells where cell expansion is a process that appears inversely related to this phenomenon.[92] In *Zinnia*, expansion also appears to need completion before differentiation.[110] Elongation has also been modeled in a carrot system[111] and in *Catharanthus*.[112] Differences were seen in the arabinose- and galactose-containing components, and while these may reflect changes in neutral pectins, arabinogalactan proteins have also been implicated. Arabinogalactan proteins have been characterized from carrot lines and have been shown to influence somatic embryogenesis.[113,114] Mapping of these and their influence on the differentiation of *Zinnia* mesophyll cells is underway and may lead to a deeper understanding of the role of these proteins in morphogenesis.[29]

The polymers which are exocytosed into the extracellular compartment have to be deposited at defined points along the plasmalemma to give rise to the characteristic architecture of the secondary cell wall during xylogenesis. Use of the *Zinnia* system in the presence of inhibitors of cellulose biosynthesis suggests that the secreted polymers assemble in a self-perpetuating cascade.[26] Normal secondary cell wall thickenings contain cellulose, xylan and lignin as well as specific proteins. When cells were treated with either 2,6-dichlorobenzonitrile or isoxaben at concentrations that inhibit cellulose biosynthesis at the sites of secondary thickening then xylan- and glycine-rich proteins could not be detected immunologically. At lower inhibitor concentrations where some cellulose synthesis occurred, xylan- and glycine-rich proteins could be detected between thickenings but were not assembled, indicating that a whole population of components was required to allow self-assembly.[26] Once the thickenings are established it is only then that lignin deposition occurs.

3.17.7 IDENTIFICATION OF GENES INVOLVED IN HEMICELLULOSE BIOSYNTHESIS

Full understanding of the regulation of hemicellulose biosynthesis requires molecular probes for analysis of transcript levels. Acquisition of cDNAs also allows the potential for genetic modification of important crop plants. It also allows the cloning of the cognate genes and analysis of the promoter structure and regulation. Compared with many other pathways the identification of genes involved in hemicellulose biosynthesis has been slow. Very few have been cloned. However, the enzyme controlling the biosynthesis of UDP-glucuronic acid, UDP-glucose dehydrogenase has been cloned from soybean by an antibody screening procedure.[57] The sequence was found to be highly homologous to that from bovine liver, being identical to 61% and homologous to more than 77%. The sequence contains a cofactor binding site for NAD, and the various motifs of the enzymes are totally conserved, so a similar three-dimensional structure of the plant dehydrogenase would be expected to be similar to that of the bovine enzyme. Northern blot analysis was used to show that the gene was expressed highly in root tips and lateral roots and to a lesser extent int he epicotyl and expanding leaves. Expression in the main root, hypocotyl, and in mature leaves was much lower. From this, Tenhaken and Thulke[57] concluded that UDP-glucose dehydrogenase plays an important role in providing hemicellulose precursors in roots and expanding leaves. The low expression of the gene in other parts of the plant can be explained either by a low demand for UDP-glucuronic acid-derived sugars in these relatively well-differentiated cells or by the possibility that the inositol oxidation pathway is utilized in these tissues. Obviously, the expression of the genes involved in the alternative pathway needs to be examined to make a direct comparison, and this has been made possible by the cloning of inositol 1-phosphate from the tomato.[115]

With respect to genes and their products which could have direct influence on hemicelluloses, a large number of xyloglucan transglycosylases have been cloned. Highly conserved cDNAs encoding xyloglucan-endo transglycosylases (XETs) have been isolated from vegetative tissues of the azuki bean, the soybean, the tomato, *Arabidopsis*, and wheat seedlings.[116] However, two divergent XETs have been isolated from *Nasturtium*, which exhibited mutually exclusive patterns of expression,[117] suggesting different roles *in vivo*. The *Arabidopsis* TCH4 gene has been shown to encode an XET which acts on the major hemicellulose of the plant cell wall. In addition to TCH4, an extensive XET-related gene family has been reported in *Arabidopsis*, with all eight cDNAs sharing between 46% and 79% sequence identity and the corresponding, predicted proteins sharing from 37% to 84% identity.[118] All eight proteins included potential N-terminal signal sequences, and most had a conserved motif that is also found in *Bacillus* β-glucanase. This gene family was found to be differentially regulated, thus demonstrating the flexibility of the cell wall hemicellulose content to respond to differing environmental cues.

What is lacking conspicuously, however, is the cloning of any plant polysaccharide synthases, with the exception of the bacterial cellulose synthase homologues from cotton and rice.[119] However, there are a number of options in attempting to access genes involved in hemicellulose biosynthesis.

3.17.7.1 Differential Screening and Differential Display

The most promising developments using this technology have used cDNA libraries from differentiating *Zinnia* cells and from Loblolly pine cambium. As a result, there is, however, a myriad of information regarding the hydroxyproline-rich glycoproteins, proline-rich proteins, glycine-rich proteins, lipid transfer proteins, and cysteine proteases which has been based upon studies using tissue printing, *in situ* hybridizations, and differential screening of cDNA libraries.[26,27,120–127] Many of these have been related to secondary wall formation but more recently, novel induced genes have been identified. Programmes such as differential screening of libraries[121,124,125,128] and, more recently, differential display[129] have identified additional types of cDNAs coding for genes with no known function or with sequence similarity to regulatory proteins in other organisms. Some of these may be identified as glycosyltransferases once the cognate protein sequences become available.

3.17.7.2 EST Database

The EST database is a depository of several thousand expressed sequence tags—or ESTs—which have been produced and analyzed. As such, it can be used to screen for genes which should be present in *Arabidopsis* by comparison with known nonplant genes. ESTs were used to assemble the amino acid sequence of UDP-glucose dehydrogenase in *Arabidopsis*,[57] when a computer search of several unknown *Arabidopsis* EST clones pulled up almost identical sequences to the soybean protein. Using this computer approach, more than 90% of the *Arabidopsis* sequence could be predicted. This illustrates the power of EST screening. A search of the EST database also reveals the presence of several putative glycosyltransferases and regulatory proteins, including β-mannosyltransferase, galactosyltransferase-associated protein kinase, UDP-glucose 6 dehydrogenase, UDP-glucose 4-epimerase, sucrose synthase, cellulose synthase, glycosyltransferase, and galactokinase. Potentially, the availability of protein amino acid sequences from glycosyltransferases involved in hemicellulose biosynthesis will facilitate the cloning of the cognate genes through EST database searching. Functionality of the proteins coded for by these ESTs can also be further explored through analysis of transformants using antisense or partial sense suppression.

3.17.7.3 Mutational Analysis

Mutated *Arabidopsis* plants have been screened for alterations in their polysaccharide composition to elucidate the roles of individual polysaccharides in cell wall structure and function.[130] Of the 5200 mutagenized plants screened for alterations in polysaccharide composition, 38 mutant lines were found which had heritable changes. Of these, five lines were almost completely devoid of L-fucose in shoot-derived cell wall material, whereas in the wild-type plants, fucose accounted for 0.5% of the dry weight of cell wall material. Mutant plants were distinguished from the wild type by dwarfing, characterized by shorter petioles, shorter internodes, less height, and reduced apical dominance, unless they were grown with a supplement of L-fucose, in which case they were identical to wild-type plants. It was suggested that the absence of fucose from the side chains of xyloglucan prevented its stable binding to the cellulose fibrils, thus weakening the cell walls. However, Whitney *et al.*[19] found that fucose was not necessary for network formation between xyloglucan and hemicellulose in studies involving the *in vitro* assembly of these polysaccharides. Further investigation[131] demonstrated that the replacement of L-fucose by L-galactose had no significant effect on the biological activity of oligosaccharides derived from xyloglucan. However, the mutant *Arabidopsis* plants were slightly smaller and displayed more fragile stems. This approach could be adopted to screen *Arabidopsis* mutants for reduced xylose content.

3.17.7.4   Reverse Genetics

Acquiring protein sequence not only allows searching the *Arabidopsis* ESTs but also the possibility of cloning homologous genes from other species by a polymerase chain reaction approach. Work is ongoing by the authors to obtain sequence information from purified xylan synthase in order to adopt this approach. This approach proved successful for the cloning of xyloglucan-modifying enzymes. Endoxyloglucan transferase was cloned, and the cDNA sequenced[116] from several plants following its purification from *Vigna angularis* or bean.[132] In the five plant species, the amino acid sequence of the mature proteins was conserved in the range 71–90% throughout their length. The consensus sequence for *N*-linked glycosylation and four cysteine residues was conserved in all five species.

3.17.8   HEMICELLULOSES, A TARGET FOR MANIPULATION?

The biosynthesis of the plant cell wall is an important target for plant biotechnology as it is the basic resource for all areas that involve dietary fibre and structural and commercial use of wood fibre. The properties of the cell wall are important determinants of fibre quality in paper making. Modern printing technology requires speciality papers with a smooth finish and appropriate tear strength. The xylan and xyloglucan components of the cell wall influence these properties. To date, this problem has been overcome by modifying processing procedures to provide some uniformity of product. However, this is expensive in terms of waste of unsuitable material, use of energy and reagents, and inefficiency in general. Approximately 1.2 Mha of forest is grown for pulp in the EU. Taking the UK as an example, pulp imports for use in paper are in excess of 400 000 tonnes at a value of great than £250 tonne^{-1}.[133] Therefore, even very modest improvements in efficiency would have clear benefit. An opportunity exists for understanding the relatively short pathway to matrix polysaccharides and its manipulation. Manipulating cell wall composition has been previously limited to engineering extracellular enzymes. There is a future opportunity to test the feasibility of engineering walls through modification of biosynthetic processes. This would have important general implications in addition to the production of a new valuable resource.

The relatively short pathways leading to the pool of UDP-sugars for hemicellulose biosynthesis is an obvious target. However, the formation of UDP-glucuronate from UDP-glucose is a complex step with at least three possible enzyme systems: a high specificity and specific UDP-glucose dehydrogenase,[57] a low-specificity, high-K_m but vascular-specific dehydrogenase,[74] and the inositol pathway.[134] The decarboxylase and UDP-xylose epimerase steps are likely to be less complex enzymatically. Down-regulating these steps by antisense may lead to less hemicellulose in the walls without being detrimental to the plant. This is an approach that may lead to improved cellulose extraction. A more direct way to manipulate hemicellulose biosynthesis would be to target the polysaccharide synthases themselves. The chances of manipulating the actual polysaccharide synthases entirely rests on the cloning of the cognate genes, which has not been achieved as yet.

Nevertheless, the technology has been applied with success to engineering lignin for improved cellulose extraction. In comparison with this work carried out on reducing lignification, tobacco is the model plant of choice for the development of the technology and studying of the effects of manipulation on fibre quality in the first instance due to faster production of experimental material. The likely tree species to be manipulated first would be eucalyptus as a tropical species, and poplar as a temperate species. In these, xylan would be the most desirable target. In gymnosperms, the equivalent target would be glucomannan. The feasibility of manipulating glucomannan biosynthesis in Loblolly pine is at present unknown. GDP-mannose is produced by a separate pathway, and the extent of the characterization of the biosynthetic system would appear to be more limited than xyloglucan or xylan biosynthesis. Both partial sense and antisense could be used to generate xylose-deficient wall polysaccharide plants. Manipulating the dehydrogenase and decarboxylase step should lower the pool of UDP-xylose, and affect xyloglucan and xylan levels. Expression of the dehydrogenase is a likely factor in overall control of flux, so antisense may down-regulate matrix polysaccharides in general while relatively increasing cellulose. Antisense xylan synthase should affect xylan content only. If the cDNAs are derived from sources other than tobacco initially, they can still be tested since heterologous antisense manipulation has been carried out with success on the phenylpropanoid pathway in lignification by a number of groups. The availability of various antibody or cDNA probes would allow verification of down-regulation. Transformants with reduced lignification have also been tested with success with paper companies. Incidentally, one study involving down-regulation of a particular step in the lignification pathway also affected hemicellulose levels. The effect of *O*-methyltransferase (OMT) cDNA modulation on cell wall composition and ultrastructure has been analyzed using antisense technology.[135] Antisense cDNA expression inhibited OMT activity by 92%, whereas sense constructs led to either 98% inhibition or overexpression of OMT activity. OMT-depleted stems showed decreased hemicellulose content, but unchanged lignin content. This suggests that altering the hemicellulose content transgenically is a feasible and worthwhile goal to pursue. A number of companies are conducting field trials with transgenic poplar, eucalyptus and Loblolly pine. The feasibility of all this still requires absolute identification of the cognate genes of hemicellulose biosynthesis.

3.17.9 REFERENCES

1. K. Nishitani and D. J. Nevins, *Food Hydrocoll.*, 1991, **5**, 197.
2. I. M. Sims and A. Bacic, *Phytochem.*, 1995, **38**, 1397.
3. W. S. York, H. van Halbeek, A. G. Darvill, and P. Albersheim, *Carbohydr. Res.*, 1990, **200**, 9.
4. W. S. York, J. E. Oates, H. van Halbeek, A. G. Darvill, P. Albersheim, P. R. Tiller, and A. Dell, *Carbohydr. Res.*, 1988, **173**, 113.
5. L. L. Kiefer, W. S. York, P. Albersheim, and A. G. Darvill, *Carbohydr. Res.*, 1990, **197**, 139.
6. M. Hisamatsu, G. Impallomeri, W. S. York, P. Albersheim, and A. G. Darvill, *Carbohydr. Res.*, 1991, **211**, 117.
7. N. C. Carpita and D. Whittern, *Carbohydr. Res.*, 1986, **146**, 129.
8. K. Nishitani and D. J. Nevins, *J. Biol. Chem.*, 1991, **266**, 6539.
9. G. J. McDougall and S. C. Fry, *J. Plant Physiol.*, 1994, **143**, 591.
10. R. G. Staudte, J. R. Woodward, G. B. Fincher, and B. A. Stone, *Carbohydr. Polym.*, 1983, **3**, 299.
11. F. Goubet, I. Bourlard, R. Giraut, C. Alexandre, M. C. Vandevelde, and C. Morvan, *Carbohydr. Polymers*, 1995, **27**, 221.
12. P. Biely, J. Puls, and H. Schneider, *FEBS Lett.*, 1985, **186**, 80.
13. P. J. Moore, A. G. Darvill, P. Albersheim, and L. A. Staehelin, *Plant Physiol.*, 1986, **82**, 787.
14. P. J. Moore and L. A. Staehelin, *Planta*, 1988, **174**, 433.
15. D. H. Northcote, R. Davey, and J. Lay, *Planta*, 1989, **178**, 353.
16. G. Freshour, R. P. Clay, M. S. Fuller, P. Albersheim, A. G. Darvill, and M. G. Hahn, *Plant Physiol.*, 1996, **110**, 1413.
17. P. J. Moore, K. M. M. Swords, M. A. Lynch, and L. A. Staehelin, *J. Cell Biol.*, 1991, **112**, 589.
18. G. F. Zhang and A. L. Staehelin, *Plant Physiol.*, 1992, **99**, 1070.
19. S. E. C. Whitney, J. E. Brigham, A. H. Darke, J. S. G. Reid, and M. J. Gidley, *Plant J.*, 1995, **8**, 491.
20. N. C. Carpita, *Annu. Rev. Plant Physiol. Plant Mol. Biol.*, 1996, **47**, 445.
21. G. Wende and S. C. Fry, *Phytochem.*, 1997, **44**, 1011.
22. G. Wende and S. C. Fry, *Phytochem.*, 1997, **44**, 1019.
23. G. Wallace, W. R. Russell, J. A. Lomax, M. C. Jarvis, C. Lapierre, and A. Chesson, *Carbohydr. Res.*, 1995, **272**, 41.
24. J. A. Brown and S. C. Fry, *Plant Physiol.*, 1993, **103**, 993.
25. J.-B. Kim and N. C. Carpita, *Plant Physiol.*, 1992, **98**, 646.
26. J. G. Taylor and C. H. Haigler, *Acta Bot. Neerland.*, 1993, **42**, 153.
27. W. Bao, D. M. O'Malley, and R. R. Sederoff, *Proc. Natl. Acad. Sci. USA*, 1992, **89**, 6604.
28. P. Wojtaszek and G. P. Bolwell, *Plant Physiol.*, 1995, **108**, 1001.
29. N. J. Stacey, K. Roberts, N. C. Carpita, B. Wells, and M. C. McCann, *Plant J.*, 1995, **8**, 891.
30. N. Benhamou and A. Asselin, *Biol. Cell*, 1989, **67**, 341.
31. C. A. Loopstra and R. R. Sederoff, *Plant Mol. Biol.*, 1995, **27**, 277.
32. D. A. Brummell, A. Camirand, and G. A. Maclachlan, *J. Cell Science*, 1990, **96**, 705.
33. M. C. Hobbs, M. H. P. Delarge, E. A. H. Baydoun, and C. T. Brett, *Biochem. J.*, 1991, **277**, 653.
34. D. Robertson, I. Beech, and G. P. Bolwell, *Phytochemistry*, 1995, **39**, 21.
35. Y. Amor, C. H. Haigler, S. Johnson, M. Wainscott, and D. P. Delmer, *Proc. Natl. Acad. Sci. USA*, 1995, **92**, 9353.
36. C. J. Smith, in "Plant Biochemistry and Molecular Biology," eds. P. J. Lea and R. C. Leegood, Wiley, Chichester, 1993, ch. 4, p. 73.
37. J. R. Green and D. H. Northcote, *Biochem. J.*, 1978, **170**, 599.
38. J. R. Green and D. H. Northcote, *Biochem. J.*, 1979, **178**, 661.
39. K. W. Waldron and C. T. Brett, in "Biochemistry of Plant Cell Walls," eds. C. T. Brett and J. R. Hillman, Cambridge University Press, London, 1985, p. 79.
40. O. Nelson and D. Pan, *Annu. Rev. Plant Physiol. Plant Mol. Biol.*, 1995, **46**, 475.
41. A. C. Gregory, M. Kerry, and G. P. Bolwell, unpublished data.
42. A. R. White, Y. Xin, and V. Pezeshk, *Biochem. J.*, 1993, **294**, 231.
43. A. R. White, Y. Xin, and V. Pezeshk, *Physiol. Plant*, 1993, **87**, 31.
44. R. E. Campbell, C. T. Brett, and J. R. Hillman, *Biochem. J.*, 1988, **253**, 795.
45. F. Goubet and C. Morvan, *Plant Cell Physiol.*, 1993, **34**, 1297.
46. F. Goubet and C. Morvan, *Plant Cell Physiol.*, 1994, **35**, 719.
47. L. S. Brickell and J. S. Grant Reid, in "Pectins and Pectinases. Progress in Biotechnology," eds. J. Visser and A. G. J. Voragan, Elsevier, Amsterdam, 1995, vol. 14, p. 127.
48. R. Hanna, D. A. Brummell, A. Camirand, A. Hensel, E. F. Russell, and G. A. Maclachlan, *Arch. Biochem. Biophys.*, 1991, **290**, 7.
49. H. Misawa, Y. Tsumuraya, Y. Kaneko, and Y. Hashimoto, *Plant Physiol.*, 1996, **110**, 665.
50. G. P. Bolwell, G. Dalessandro, and D. H. Northcote, *Phytochem.*, 1985, **24**, 699.
51. G. P. Bolwell and D. H. Northcote, *Planta*, 1981, **152**, 225.
52. G. P. Bolwell and D. H. Northcote, *Biochem. J.*, 1983, **210**, 497.

53. G. P. Bolwell and D. H. Northcote, *Biochem. J.*, 1983, **210**, 509.
54. M. W. Rodgers and G. P. Bolwell, *Biochem. J.*, 1992, **288**, 817.
55. G. Dalessandro and D. H. Northcote, *Planta*, 1981, **151**, 53.
56. G. Dalessandro and D. H. Northcote, *Planta*, 1981, **151**, 61.
57. R. Tenhaken and O. Thulke, *Plant Physiol.*, 1996, **112**, 1127.
58. D. M. Gibeaut and N. C. Carpita, *Plant Physiol.*, 1991, **97**, 551.
59. M. Becker, C. Vincent, and J. S. G. Reid, *Planta*, 1995, **195**, 331.
60. G. Dalessandro, G. Piro, and D. H. Northcote, *Planta*, 1986, **169**, 564.
61. G. Piro, A. Zuppa, G. Dalessandro, and D. H. Northcote, *Planta*, 1993, **190**, 206.
62. G. Piro and M. R. Leucci, *Giorn. Bot. Ital.*, 1993, **127**, 1211.
63. J. S. G. Reid, *Adv. Bot. Res.*, 1985, **11**, 125.
64. J. D. Bewley and J. S. Grant Reid, in "Biochemistry of Storage Carbohydrates in Green Plants," eds. P. M. Dey and R. A. Dixon, Academic Press, London, 1985, vol. 265, p. 289.
65. Whetton and Sederoff, *Plant Cell*, 1995, **7**, 1001.
66. D. M. Gibeaut and N. C. Carpita, *FASEB J.*, 1994, **8**, 904.
67. E. Ingold, M. Sugiyama, and A. Komamine, *Plant Cell Physiol.*, 1988, **29**, 295.
68. K. Suzuki, E. Ingold, M. Sugiyama, and A. Komamine, *Plant Cell Physiol.*, 1991, **32**, 303.
69. N. Carpita, personal communication.
70. J. G. Taylor, T. P. Owen, Jr., L. T. Koonce, and C. H. Haigler, *Plant J.*, 1992, **2**, 959.
71. K. Suzuki, E. Ingold, N. Sugiyama, H. Fukuda, and A. Komamine, *Physiol. Plant*, 1992, **86**, 43.
72. G. P. Bolwell, *Int. Rev. Cytol.*, 1993, **146**, 261.
73. C. H. Haigler, personal communication.
74. D. Robertson, C. G. Smith, and G. P. Bolwell, *Biochem. J.*, 1996, **313**, 311.
75. H. Fukuda, *Int. Rev. Cytol.*, 1992, **136**, 289.
76. W. Kuternozinska and M. Pilipowicz, *Biol. Plant.*, 1993, **35**, 307.
77. J.-C. Jang and J. Sheen, *Plant Cell*, 1994, **6**, 1665.
78. R. Johnson and C. A. Ryan, *Plant Mol. Biol.*, 1990, **14**, 527.
79. S. R. Kim, M. A. Costa, and G. An, *Plant Mol. Biol.*, 1991, **17**, 973.
80. H. Tsukaya, T. Ohshima, S. Naito, M. Chino, and Y. Komeda, *Plant Physiol.*, 1991, **97**, 1414.
81. K. Herbers, G. Monke, R. Badur, and U. Sonnewald, *Plant Mol. Biol.*, 1995, **29**, 1027.
82. J. Memelink, J. H. C. Hoge, and R. A. Schilperoort, *EMBO J.*, 1989, **6**, 3579.
83. S. Philosoph-Hadas, S. Meir, and N. Aharoni, *Plant Physiol.*, 1985, **78**, 139.
84. J. Warren-Wilson, L. W. Roberts, P. M. Warren-Wilson, and P. M. Gresshoff, *Ann. Bot.*, 1994, **73**, 65.
85. J. Ecker, *Science*, 1995, **268**, 667.
86. D. J. Millar, A. R. Slabas, C. Sidebottom, C. G. Smith, A. K. Allen, and G. P. Bolwell, *Planta*, 1992, **187**, 176.
87. A. W. Roberts and C. H. Haigler, personal communication.
88. S. Aldington and S. C. Fry, *Adv. Bot. Res.*, 1993, **19**, 1.
89. R. J. Bruce and C. A. West, *Plant Physiol.*, 1989, **91**, 889.
90. Y. Fuchs, A. Saxena, H. R. Gamble, and J. D. Anderson, *Plant Physiol.*, 1989, **89**, 138.
91. G. P. Mitchell and G. P. Bolwell, unpublished data.
92. S. C. De Vries, A. K. J. de Jong, and F. A. van Engelen, *Biochem. Soc. Symp.*, 1994, **60**, 43.
93. A. J. De Jong, L. Cordewener, F. Lo Schiavo, M. Terzi, J. Vanderkerkhove, A. Van Kammen, and S. C. De Vries, *Plant Cell*, 1992, **4**, 425.
94. A. J. De Jong, R. Heidstra, H. P. Spaink, M. V. Hartog, E. A. Meijer, T. Hendricks, F. Lo Schiavo, M. Terzi, T. Bisseling, A. Van Kammen, and S. C. De Vries, *Plant Cell*, 1993, **5**, 615.
95. E. D. L. Schmidt, A. J. de Jong, and S. C. de Vries, *Plant Mol. Biol.*, 1994, **26**, 1305.
96. A. J. De Jong, T. Hendriks, E. A. Meijer, M. Penning, F. Lo Schiavo, M. Terzi, A. B. van Kammen, and S. C. de Vries, *Develop. Genet.*, 1995, **16**, 332.
97. G. P. Mitchell and G. P. Bolwell, unpublished data.
98. N. C. Carpita and D. M. Gibeaut, *Plant J.*, 1993, **3**, 1.
99. G. Maclachlan and S. Carrington, in "Biosynthesis and Biodegradation of Cellulose," eds. C. H. Haigler and P. J. Weimer, Dekker, New York, 1991, p. 599.
100. S. C. Fry, *Annu. Rev. Plant Physiol. Plant Mol. Biol.*, 1995, **46**, 497.
101. S. C. Fry, R. C. Smith, K. F. Renwick, D. J. Martin, S. K. Hodge, and K. J. Matthews, *Biochem. J.*, 1992, **282**, 821.
102. P. R. Hetherington and S. C. Fry, *Plant Physiol.*, 1993, **103**, 987.
103. I. Potter and S. C. Fry, *J. Exp. Bot.*, 1994, **45**, 1703.
104. W. Xu, M. M. Purugganan, D. H. Polisensky, D. M. Antosiewicz, S. C. Fry, and J. Braam, *Plant Cell*, 1995, **7**, 1555.
105. K. Nishitani, *J. Plant Res.*, 1995, **108**, 137.
106. D. J. Cosgrove and Z. C. Li, *Plant Physiol.*, 1993, **103**, 1321.
107. Z. C. Li, D. J. Durachko, and D. J. Cosgrove, *Planta*, 1993, **191**, 349.
108. S. McQueen-Mason, D. M. Durachko, and D. J. Cosgrove, *Plant Cell*, 1992, **4**, 1425.
109. S. J. McQueen-Mason, S. C. Fry, D. M. Durachko, and D. J. Cosgrove, *Planta*, 1993, **190**, 327.
110. A. W. Roberts and C. H. Haigler, *Plant Physiol.*, 1994, **105**, 699.
111. H. Masuda, Y. Ozeki, S.-i. Amino, and A. Komamine, *Physiol. Plant*, 1984, **62**, 65.
112. K. Suzuki, S.-i. Amino, Y. Takeuchi, and A. Komamine, *Plant Cell Physiol.*, 1990, **31**, 7.
113. T. C. Baldwin, M. C. McCann, and K. Roberts, *Plant Physiol.*, 1993, **103**, 115.
114. M. Kreuger and G. J. van Holst, *Planta*, 1993, **189**, 243.
115. G. E. Gillaspy, J. S. Keddie, K. Oda, and W. Gruissem, *Plant Cell*, 1995, **7**, 2175.
116. K. Okazawa, Y. Sato, T. Nakagawa, K. Asada, I. Kato, E. Tomita, and K. Nishitani, *J. Biol. Chem.*, 1993, **268**, 25364.
117. J. K. C. Rose, D. A. Brummell, and A. B. Bennett, *Plant Physiol.*, 1996, **110**, 493.
118. W. Xu, P. Campbell, A. K. Vargheese, and J. Braam, *Plant J.*, 1996, **9**, 879.
119. J. R. Pear, Y. Kawagoe, W. E. Schreckengost, D. P. Delmer, and D. M. Stalker, *Proc. Natl. Acad. Sci. USA*, 1996, **93**, 12637.

120. Z.-H. Ye, R. E. Kneusel, K. Matern, and J. E. Varner, *Plant Cell*, 1991, **6**, 1427.
121. Z.-H. Ye and J. E. Varner, *Plant Physiol.*, 1993, **103**, 805.
122. Z.-H. Ye and J. E. Varner, *Plant Mol. Biol.*, 1996, **30**, 1233.
123. B. Keller, *Plant Physiol.*, 1993, **101**, 1127.
124. T. Demura and H. Fukuda, *Plant Physiol.*, 1993, **103**, 815.
125. T. Demura and H. Fukuda, *Plant Cell*, 1994, **6**, 967.
126. A. Minami and H. Fukuda, *Plant Cell Physiol.*, 1995, **36**, 1599.
127. A. M. Showalter, *Plant Cell*, 1993, **5**, 9.
128. Z.-H. Ye and J. E. Varner, *Proc. Natl. Acad. Sci. USA*, 1994, **91**, 6539.
129. L. T. Koonce and C. H. Haigler, personal communication.
130. W.-D. Reiter, C. C. S. Chapple, and C. R. Somerville, *Science*, 1993, **261**, 1032.
131. E. Zablackis, W. S. York, M. Pauly, S. Hantus, W.-D. Reiter, C. C. S. Chapple, P. Albersheim, and A. Darvill, *Science*, 1996, **272**, 1808.
132. K. Nishitani and R. Tominaga, *J. Biol. Chem.*, 1992, **267**, 21 058.
133. F. C. Hummell, in "Biomass Forestry in Europe: A Strategy for the Future," eds. F. C. Hummel, W. Palz, and G. Grassi, Elsevier Applied Science, London, 1988, p. 256.
134. F. Loewus, M. S. Chen, and M. W. Loewus, in "Biogenesis of Plant Wall Polysaccharides," eds. F. A. Loewus, Academic Press, New York, 1973, p. 1027.
135. V. Mab, C. Migne, A. Cornu, M. P. Maillot, E. Grenet, J. M. Besle, R. Atanassova, F. Matz, and M. Legrand, *J. Sci. Food Agric.*, 1996, **72**, 385.

3.18
The Nature and Function of Lignins

NORMAN G. LEWIS and LAURENCE B. DAVIN
Washington State University, Pullman, WA, USA

and

SIMO SARKANEN
University of Minnesota, St Paul, MN, USA

3.18.1   INTRODUCTION

The land masses on earth began to be colonized by plants between 400 and 450 million years ago. That they were able to do so was in no small measure dependent upon the evolution of the biochemical pathway leading to the lignin biopolymers.[1] These ostensibly three-dimensional cross-linked aromatic macromolecules are laid down within and between the extracellular matrix cell wall components of vascular plants which are composed of polysaccharides and proteins. In as far as they can facilitate redistribution of stresses, lignins are thought to contribute to the compressive strengths of plant cell walls without influencing the tensile properties of the composite structures.

Lignin biosynthesis was originally supposed by Karl Freudenberg to occur within plant cell walls through the random free-radical coupling of the monomeric precursors, the monolignols (viz. *p*-coumaryl (**1**), coniferyl (**2**), and sinapyl (**3**) alcohols (Figure 1)). According to Freudenberg, the only enzymatic requirement, as far as lignin biopolymer assembly *in vivo* is concerned, was provision of one-electron oxidative capacity through peroxidase/H₂O₂ and/or laccase/O₂.[2,3] Thus, the view at the time was that lignification relied upon a unique biochemical mechanism to the extent that it was not reliant upon any proteinaceous control over its ultimate configuration. Indeed, in 1967, John Harkin, who had previously worked in Freudenberg's laboratories, described lignification in the following terms.

> There are many similarities between lignin and playing cards. In lignin there are certain fixed amounts of definite structures just as there are aces, deuces, etc., in a pack of cards, which can be shuffled into any order inside the pack ... It is as if thousands of packs of cards had been put together and mixed and then separated into random bundles, some with more cards and some with fewer than a normal pack ... nothing can be said of the fortuitous order of the cards in a single pack.[4]

However, a point was reached some three decades later where it was no longer possible to ignore clear indications that lignification, in harmony with the formation of all other biopolymers, rests upon a carefully controlled nonrandom mechanism of macromolecular assembly.

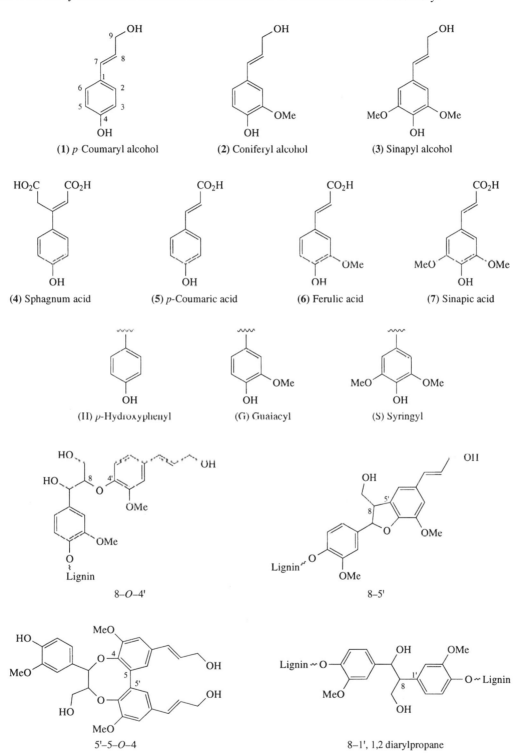

Figure 1 Aromatic constituents (**1**) to (**7**) present in certain plant types, the common denotation of aromatic rings, **H**, **G** and **S** and various lignin interunit linkages, exemplified in dimers and trimers.

However they may be formed, lignins endow vascular plants with three selective advantages for their growth and development on land. First, they consolidate the strengths of crystalline cellulose and, in so doing, enable plants to grow to far greater heights. Second, their hydrophobic character provides the means by which xylem tissues can conduct water over large distances without significant loss through absorption or evaporation. Finally, the deposition of lignins confers upon plant cell walls greater resistance to degradation, and thus creates a barrier against opportunistic pathogens, microorganisms and herbivores alike.

As the second most abundant group of biopolymers after cellulose, lignins (Latin: *lignum* = wood) are integral cell wall components occurring in all vascular tissues/bundles. For illustrative purposes, Figure 2 juxtaposes a typical cross-section of a lignified woody plant with an example of the sort of heights and longevity that can be attained with lignification. It should be noted, however, that lignins are presumed to be absent from primitive algae, fungi, and mosses, even though the latter contain a number of somewhat related phenolic constituents, e.g., sphagnum acid (**4**) and other substances in mosses, which have occasionally been confused with lignins.[5-7]

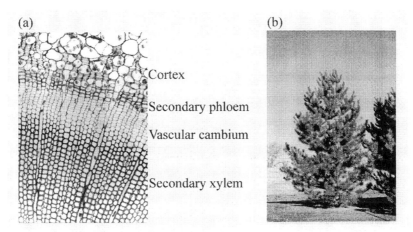

Figure 2 Plant cell wall reinforcement through lignification. (a) Cross-section of the woody gymnosperm, Douglas fir (*Pseudotsuga menziesii*), showing the lignified secondary xylem elements essential for water conduction, and in helping offset compressive forces acting on the stem, and (b) Conifer tree. Douglas fir which can grow up to 100 m in height and live for several hundred years.

Depending upon the plant species (pteridophytes, gymnosperms, or angiosperms), lignin contents can vary enormously. For example, in woody gymnosperms (conifers), they typically represent about 20–35% of the secondary xylem (wood) cells, the remainder being composed of cellulose (~40%), hemicelluloses (~30%), and small amounts of proteins. Woody angiosperm dicotyledons, on the other hand, generally contain about 20% lignin, while herbaceous plants have even lower lignin contents (1–20%).[8]

Gymnosperm lignins are mainly formed from coniferyl alcohol (**2**), together with small proportions of *p*-coumaryl alcohol (**1**), whereas woody angiosperms generally result from coniferyl (**2**) and sinapyl (**3**) alcohols in roughly equal amounts, as well as small quantities of (**1**) as before. Monocotyledons harbor the additional complexity of *p*-hydroxycinnamic acids, e.g., *p*-coumaric (**5**), ferulic (**6**), and sinapic (**7**) acids, being associated with their lignin biopolymers. Believed to be the last *structural* components to be deployed in vascular tissues, lignins are presumably not only in intimate physical contact with the hemicelluloses but are also thought to be covalently linked to them.[9] Covalent linkages of lignins to cell-wall proteins could (conceivably) occur, and it was proposed in the 1970s that an early step in lignification might be cross-linking with proteins in the primary wall.[10]

Figure 3 shows a representation proposed in 1974 by Nimz[11] of what a lignin biopolymer macromolecular fragment was envisaged to look like in angiosperm dicotyledons. This structural depiction was based on degradative analytical studies and ^{13}C NMR spectroscopy, even though the constituent products were often isolated in very low yield. (Note that in this and other lignin preparations, the terms *p*-hydroxyphenyl (**H**), guaiacyl (**G**), and syringyl (**S**) are respectively used to denote the three types of aromatic rings in the monomer residues (see Figure 1).) As enlarged upon in Section 3.18.7, the monomer residues themselves are linked to one another in as many as 10 different ways through C–O bonds (primarily alkyl aryl and diaryl ether structures) and C–C bonds (variously involving alkyl and aryl moieties). Remarkably, one type of interunit linkage, the 8–*O*–4′ alkyl aryl ether, is much more prevalent than any of the others, constituting one half or more of the structures present between adjacent monomer residues.[12] Otherwise, 8–5′ phenyl-coumarans, 5′–5–*O*–4 dibenzodioxocins and 8–1′ 1,2-diarylpropanes (or a progenitor) together constitute the second most frequent kinds of interunit linkages (see Figure 1). These characteristics are certainly unusual and serve to underscore the uniqueness of lignins among biopolymers.

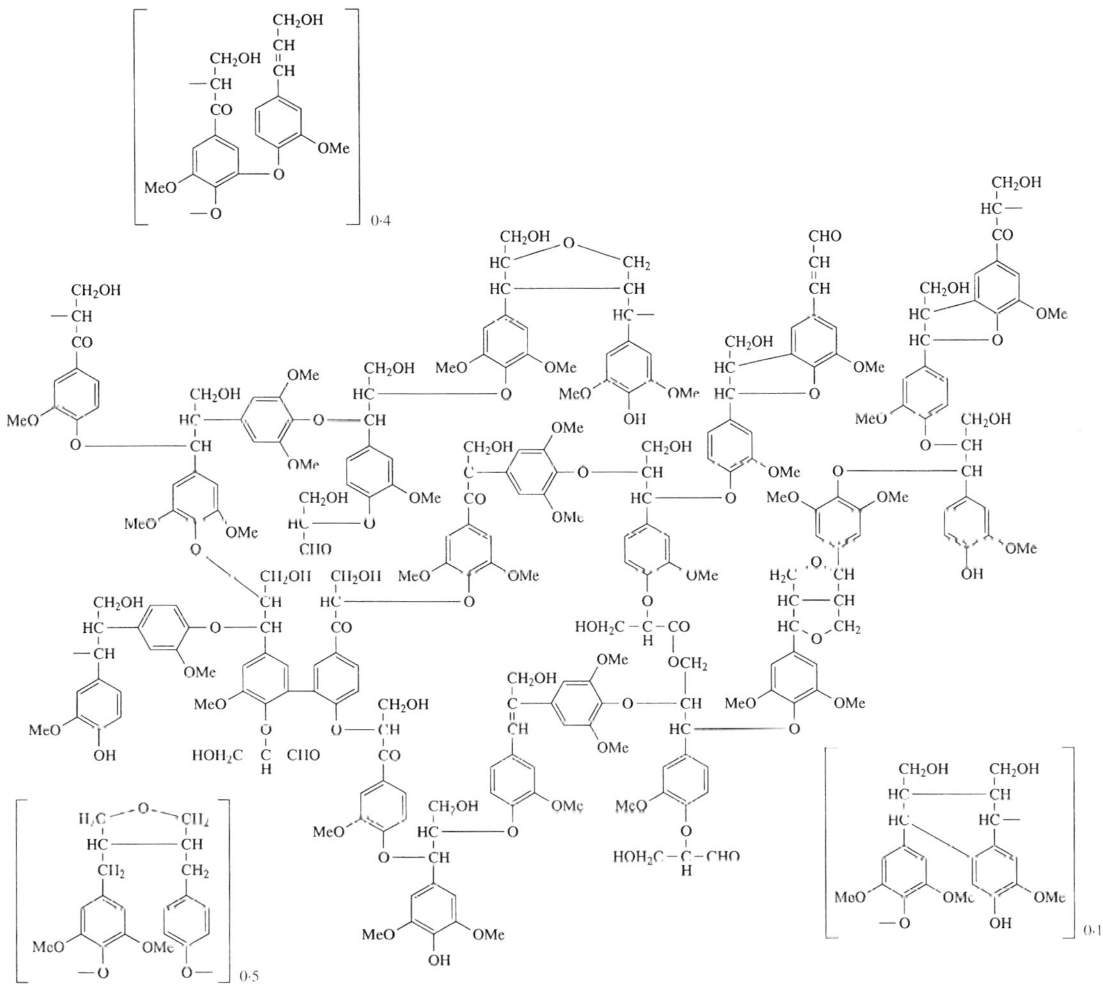

Figure 3   Beechwood lignin constitution (after Nimz, 1974[11]).

This chapter addresses various aspects of lignification, some of which are only partly understood or resolved at the time of writing. These include how the phenylpropanoid pathway to the lignins and related natural products might have evolved, how the lignin monomers themselves are formed, and how the lignin biopolymers are spatially and temporally related to other cell wall constituents. Particular attention is given to the rapidly evolving view about how lignin assembly *in situ* is controlled during vascular plant cell wall formation, as well as a discussion of the attempts to modify lignin content and composition in transgenic plants. In this regard, much of the past and current research in the lignin field has focused upon three practical goals: first, the identification of factors that influence monolignol composition in lignin; second, the determination of what affects the monolignol content in lignins, and third, the clarification of how lignin configuration is established *in vivo*. The first two subjects are predicated upon considerations that a lignin higher in sinapyl alcohol (**3**)-derived monomer residues will be more readily amenable to being removed from wood by existing pulping processes, and that an overall reduction in lignin content will be more cost-effective for pulp/paper manufacture. Analogous modifications in grasses are expected to render animal feedstocks more readily digestible. Lastly, although reviewed elsewhere, it is important to remember that lignin biopolymers in the cell walls of rotting wood are only slowly degraded in the natural environment, and thus they occupy a pivotal position in the carbon cycle.[13]

3.18.2 EVOLUTION OF THE PHENYLPROPANOID PATHWAY AND LIGNIN BIOSYNTHETIC CAPACITY

In relation to their aquatic counterparts, vascular plants required additional traits for their growth and survival which were gained largely through elaboration of the phenylpropanoid-(acetate) pathway(s) leading to the lignins, lignans, hydroxycinnamic acids, flavonoids, suberins and other phenolic components.[1] As shown in Figure 4, the phenylpropanoid pathway is essentially initiated via deamination of the aromatic amino acid, phenylalanine (8), and, in some instances, tyrosine (9), through catalysis by phenylalanine ammonia lyase (PAL) and a presumed tyrosine ammonia lyase (TAL), respectively.

The new traits furnished by the different branches of phenylpropanoid metabolism gave vascular plants a number of advantages for land adaptation: these include a high water potential facilitating active metabolism in desiccating environments and effective mechanisms for obtaining and transporting water and nutrients. Additionally, they provide capabilities of minimizing the effects of (sudden) variations in temperature, humidity, and light (particularly UV radiation), and an ability to withstand and modulate forces of compression acting upon the plant structure. They also further provide the capacity of altering growth characteristics (e.g., stem realignment) in response to changes in perceived gravitational vectors, as well as affording defense mechanisms conferring resistance to pathogens and herbivores.

The notion that vascular plants originated from algal progenitors has largely stemmed from comparisons with the green algal group, Charophyceae, which include the *Coleochaete*, *Nitella*, and *Spirogyra* species (reviewed by Lewis and Davin[1]). Such comparisons revealed not only ultra-structural similarities[14] for both the algae and vascular plants, but also a relatively high homology between group II introns in the tRNA^{Ala} and tRNA^{Ile} genes of chloroplast DNAs.[15]

In order to account for the emergence of vascular plants, it has been proposed that ancestral green algal precursors separately evolved to give the primitive bryophytes and the primitive tracheophytes, respectively.[16] These, in turn, afforded the modern fungi, mosses, liverworts, and hornworts, and the extant pteridophytes, gymnosperms, and angiosperms. Such processes were envisaged to have occurred by the development of specialized elongated cells (so-called hemitracheids), through which water conduction occurred, eventually giving the present day hydroids (water conducting cells in mosses) and the tracheids.[17] However, the fossil record established thus far does not clearly demonstrate how this evolutionary progression from algal to terrestrial vascular plants occurred.[18]

From a chemotaxonomical perspective, the earliest indication of plant lignins came from the analysis of Silurian plant fossils (i.e., *Eohostimella*) which yielded *traces* of presumed lignin derived products released during alkaline nitrobenzene oxidation, e.g., to give *p*-hydroxybenzaldehyde (21) and vanillin (22).[17,19] From these results, it was concluded that Silurian plant fossils were forerunners of the early gymnosperms, and that lignin biosynthesis was coincident with the appearance of vascular plant tissues in the late Silurian/early Devonian period.[17,19] However, it should be cautioned that nitrobenzene oxidation is not specific for lignins, but will also release *p*-hydroxybenzaldehyde (21) and vanillin (22) from other moieties such as *p*-coumarate (5) and ferulate (6). Whatever the case, despite the appeal of this being a possible evolutionary starting point for lignin biosynthesis, the fossil record is still too incomplete for any rigorous conclusions to be drawn.

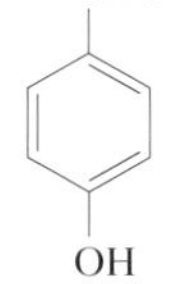

(21) *p*-Hydroxybenzaldehyde 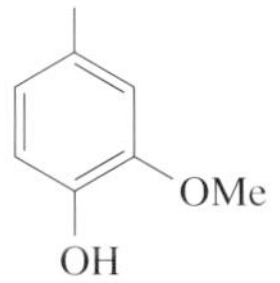   (22) Vanillin

Nevertheless, the evolutionary progression of the phenylpropanoid pathway, leading in certain cases to lignins, is still best understood by comparing the related chemical constituents within extant algae, bryophytes (mosses, liverworts, hornworts, and takakiophytes), fungi, pteridophytes (e.g., ferns), gymnosperms and angiosperms, respectively.

3.18.2.1 Algae

There is no evidence for lignins in algae.[20] The overall phenylpropanoid pathway (Figure 4) is generally considered to be absent, although phenylalanine ammonia-lyase is thought to be present

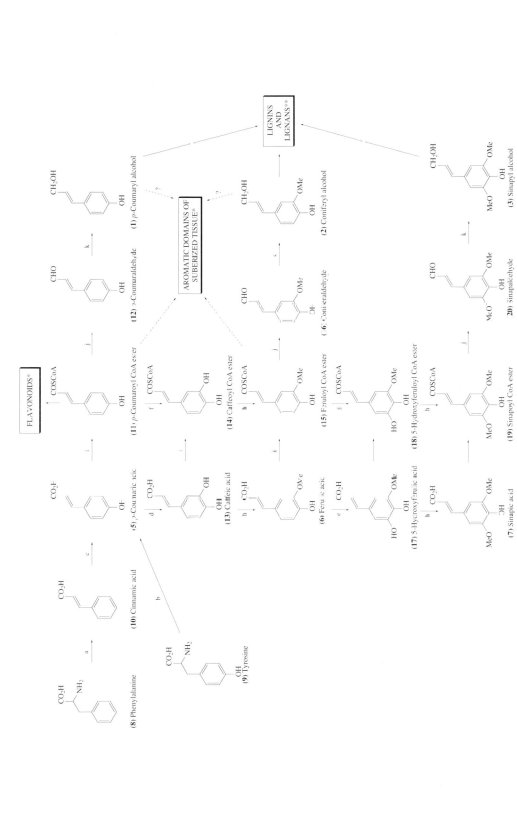

Figure 4 Simplified denotation of phenylpropanoid metabolism in vascular plants. (a) Phenylalanine ammonia-lyase, (b) presumed tyrosine ammonia-lyase, (c) 4-cinnamate hydroxylase and its associated reductase, (d) *p*-coumarate 3-hydroxylase, (e) ferulic 5-hydroxylase and associated reductase, (f) *p*-coumaroyl CoA 3-hydroxylase, (g) feruloyl CoA 5-hydroxylase, (h) hydroxycinnamoyl *O*-methyltransferases, (i) hydroxycinnamoyl CoA ligases, (j) cinnamcyl-CoA oxidoreductase, and (k) cinnamyl alcohol dehydrogenase. *: plus acetate pathway for both flavonoids and suberins, **: main source of lignan skeleta.

in *Dunaliella*.[21] Yet, despite the lack in algae of the essential biochemical capabilities for monolignol formation, various reports have claimed that lignins are present in brown algae, such as *Fucus vesiculosus*[22] and *Cystoseira barbata*.[23,24] These reports were later shown to be incorrect, since the "lignins" were not actually derived from monolignols. Instead they were "1,3,5-trihydroxy phenolic-like" substances (**23**),[25] presumably of acetate origin. The occurrence of lignins has also been claimed in *Staurastrum*[26] and *Coleochaete*[27] sp., but this has never been substantiated.

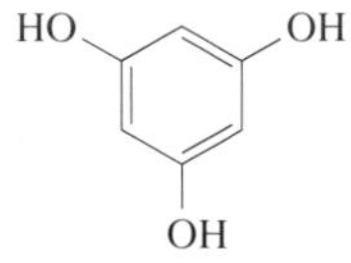

(**23**) Phloroglucinol

3.18.2.2 Bryophytes

The Bryophyte category currently includes the Bryophyta (mosses), Hepatophyta (liverworts), Anthocerotophyta (hornworts), and Takakiophyta (takakiophytes),[28] none of which have been shown to contain lignin. Lignins were, however, originally claimed to be present in mosses,[5–7] but a more exhaustive examination of both mosses and giant mosses established that the claims were incorrect. Mosses instead contain phenolic substances such as sphagnum acid (**4**) (Figure 1),[29] whereas giant mosses have "1,3,5-trihydroxy phenolic-like" (**23**) polymers of unknown constitution,[30] which could well be acetate-derived. Some mosses do, however, have a partially developed phenylpropanoid pathway, and can biosynthesize flavone *C*- and *O*-glycosides, biflavonols, dihydroflavones, etc.[31] These are presumed to be effective in helping to limit the deleterious effects of ultraviolet radiation. Related flavonoids are also present in liverworts and hornworts.[31]

3.18.2.3 Fungi

Lignins are absent from fungi. However, certain fungi have a partially developed phenyl-propanoid-acetate pathway leading to cinnamic (**10**) and *p*-coumaric (**5**) acids. Depending on the organism, these can "condense" with acetate-derived moieties to yield styrylpyrones, such as hispidin (**24**), *bis*-noryangonin (**25**), hymenoquinone (**26**), and its leuco derivative (**27**), e.g., in the genera *Gymnopilus*, *Polyporus*, *Pholiota*, *Phellinus*, and *Hymenochaete* (Basidiomycetes).[32] Interestingly, cell-free extracts from the *Polyporus hispidus* fruiting body can convert hispidin (**24**) into presumed cell wall polymers, which have occasionally been confused with lignins.[33,34]

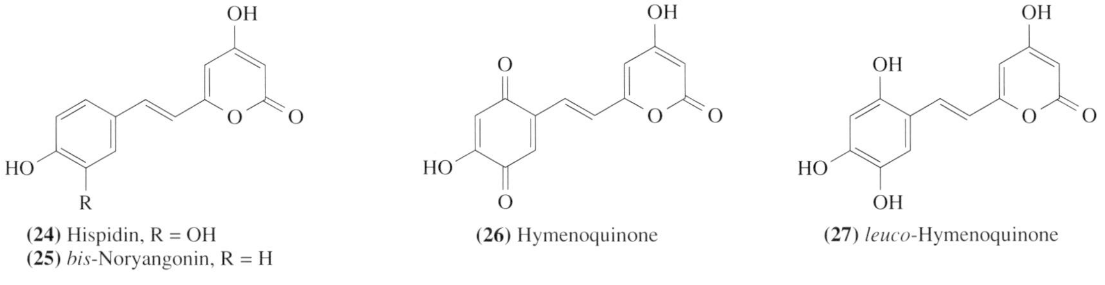

(**24**) Hispidin, R = OH
(**25**) *bis*-Noryangonin, R = H

(**26**) Hymenoquinone

(**27**) *leuco*-Hymenoquinone

Certain fungi, however, have the ability to *O*-methylate phenols giving, e.g., methyl *p*-methoxy-cinnamate (**28**)[32,35,36] and isoferulic acid (**29**)[32] in *Lentinus lepideus*, and veratryl alcohol (**30**)[37] in *Phanerochaete chrysosporium*. This regiospecificity in *O*-methylation, however, reveals an important evolutionary distinction between fungi and the higher plants. Although both can methylate caffeic acid (**13**), only the higher plants seem to be able to synthesize ferulic acid (**6**) via *O*-methylation of the 3-OH group. Accordingly, the regiospecificity of methylation may have been crucial in elaborating that part of the phenylpropanoid pathway which leads to the lignins in higher plants.

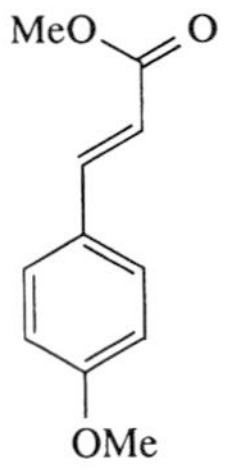

(28) Methyl *p*-methoxycinnamate 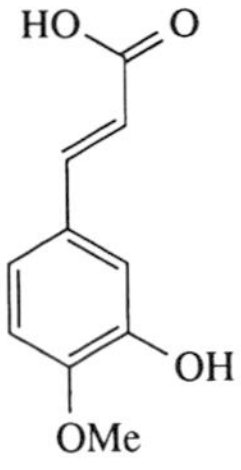   (29) Isoferulic acid 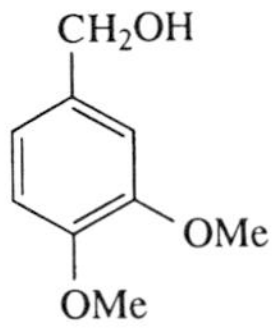   (30) Veratryl alcohol

3.18.2.4 Pteridophyta

Lignins may have first emerged with the appearance of the pteridophytes. The Pteridophyta can be traced back to the Devonian period, with members of the Rhacophytales, for example, sharing common features, such as secondary xylem, with the extant ferns and gymnosperms (see Lewis and Davin[1]).

Chemotaxonomically, still very little is known about extant pteridophytes (i.e., ferns), clubmosses, and horsetails, although it is reported, on the basis of results from nitrobenzene oxidations, that they are lignified (discussed in Lewis and Yamamoto[20]). For example, it is thought that the Cyatheaceae (e.g., *Dicksonia squarrosa*) and Adiantaceae (e.g., *Pteris podophylla*) families contain lignins derived mainly from *p*-coumaryl (1) and coniferyl (2) alcohols, whereas the Dennstaedtiaceae (i.e., *Dennstaedtia bipinnata*) contain syringyl (S) lignins.[38] Lignins may also be present in the Lycopodiaceae, Selaginellaceae, and Psilotopsida.[20] Perhaps the most compelling evidence for a fully functional phenylpropanoid pathway, at least to *p*-coumaryl (1) and coniferyl (2) alcohols, is the presence of the presumed coniferyl alcohol (2) derived lignans, dihydrodehydrodiconiferyl alcohol 9-*O*-β-D-glucoside (31) and lariciresinol 9-*O*-β-D-glucoside (32) in *Pteris vittata*.[39]

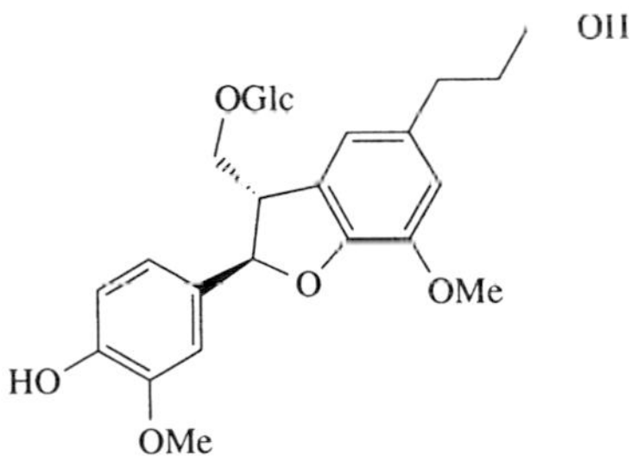

(31) *trans*-Dihydrodehydrodiconiferyl alcohol 9-*O*-β-D-glucoside   (32) Lariciresinol-9-*O*-β-D-glucoside

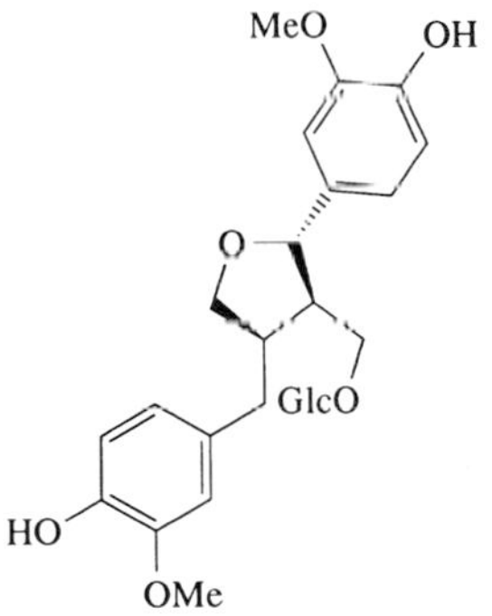

Hence, with the emergence of the pteridophytes, three new biochemical features had appeared, namely position-specific methylation of caffeic acid (13) (or its CoA-derivative (14)) to give ferulic acid (6) (or feruloyl-CoA (15)), synthesis of sinapic acid (7) (or its CoA-derivative (19)) and the stepwise reduction of the corresponding CoA-derivatives (11), (15), and (19) to afford the monolignols (1), (2), and (3). Accordingly, the stage was now set for other plant forms to exploit these biochemical developments.

3.18.2.5 Gymnosperms and Angiosperms

Most current knowledge about lignins is derived from analyses of the gymnosperm and angiosperm lignins. Yet, the precise origin (ancestors) and sequential progression (evolution) of both the present-day gymnosperms and angiosperms remain quite obscure, even though the latter are the most prevalent forms of plant life (>200 000 species). Available evidence, however, indicates that both the angiosperm dicotyledons and monocotyledons were well-entrenched by the Cretaceous period.[28] Nevertheless, modern gymnosperms and angiosperms have fully functional phenylpropanoid pathways to the monolignols (1)–(3), which apparently differ only in whether or not sinapyl alcohol (3) is involved in lignin formation: most gymnosperm lignins are derived from *p*-coumaryl (1) and coniferyl (2) alcohols, whereas the angiosperm lignins incorporate all three monolignols (1)–(3).

In summary, while the existing fossil record does not yet provide any effective means for determining how the evolution of vascular plants progressively occurred, the chemotaxonomical data obtained from extant plants have revealed important insights. The biochemical pathways to the monolignols and lignins are apparently absent in algae, bryophytes, fungi, and mosses. On the other hand, the first evidence for a fully functional monolignol forming pathway, and hence the capacity to biosynthesize lignins, appears to be associated with the pteridophytes. This adaptation was then fully exploited by the various plant families within the gymnosperms and angiosperms (dicotyledons and monocotyledons).

3.18.3 MONOLIGNOL BIOSYNTHESIS

3.18.3.1 Enzymological and Molecular Aspects of Monolignol Formation

The biochemical pathway to the monolignols (Figure 4) is generally considered as being initiated from phenylalanine (8)/tyrosine (9). Depending upon the species, tissue, or cell type, however, this pathway can afford a wide range of phenylpropanoid and phenylpropanoid-acetate metabolites (e.g., flavonoids, suberins, coumarins, and so forth), in addition to the monolignols, lignins, and lignans (see Davin and Lewis[40]). It must be recognized, though, that many of these "overlapping" branchpoint pathways (Figure 4) in phenylpropanoid metabolism (e.g., to flavonoids) are actually both temporally and spatially distinct from those affording the lignins. Nevertheless, the presence of overlapping biochemical pathways, even if in different cell types, has complicated the study of phenylpropanoid metabolism leading to the lignins.

Numerous investigations have been directed towards selectively altering the traits of particular plants (e.g., in regard to lignin composition and/or content) by attempting to alter the expression of certain presumed lignin specific genes. Meaningful progress can only be made in this area when explicit account has been taken care of among the enzymatic steps beyond phenylalanine (8)/tyrosine (9). As can be seen in Figure 4, the enzymatic conversions involved in monolignol biosynthesis include deamination, aromatic ring hydroxylations, CoA ligations, *O*-methylation(s), and consecutive reduction of the cinnamoyl CoA esters and the resulting aldehydes.[20,40] Interestingly, formation of the monolignols, and hence the lignins/lignans, represent some of the metabolically most expensive end-products in the plant kingdom.[41]

3.18.3.2 Nitrogen Availability, Metabolic Pool Sizes and Control Points in Monolignol Biosynthesis

Prior to describing the enzymes and genes involved in monolignol formation, some detailed consideration of two central questions about plant cells undergoing active monolignol biosynthesis is necessary. The first concerns nitrogen availability during phenylalanine/tyrosine turnover, and the second involves the metabolic pool sizes and control points during active phenylpropanoid (monolignol) biosynthesis. Both are important, since they influence the type and amounts of monolignols (1) to (3) and, hence, lignins that are formed.

As discussed in Chapter 1.22, Volume 1, both phenylalanine (8) and tyrosine (9) are derived from the shikimate/chorismate (33/34) pathway. During their formation (Scheme 1), nitrogen is only introduced at the point where the action of prephenate aminotransferase gives rise to arogenate (36), i.e., just prior to aromatic amino acid (8) and (9) formation.[42] Given that lignins can constitute up to 30% of the dry weight of lignified plant tissues,[43] however, an equivalent amount of intermediate phenylalanine (8) and/or tyrosine (9) must be formed during the course of their biosynthesis. It is, therefore, quite striking that, when the aromatic amino acids are subsequently conscripted into phenylpropanoid metabolism, the nitrogen is immediately removed during the transformations to cinnamic (**10**) and/or *p*-coumaric (**5**) acids, where an equimolar amount of ammonium ion is released.

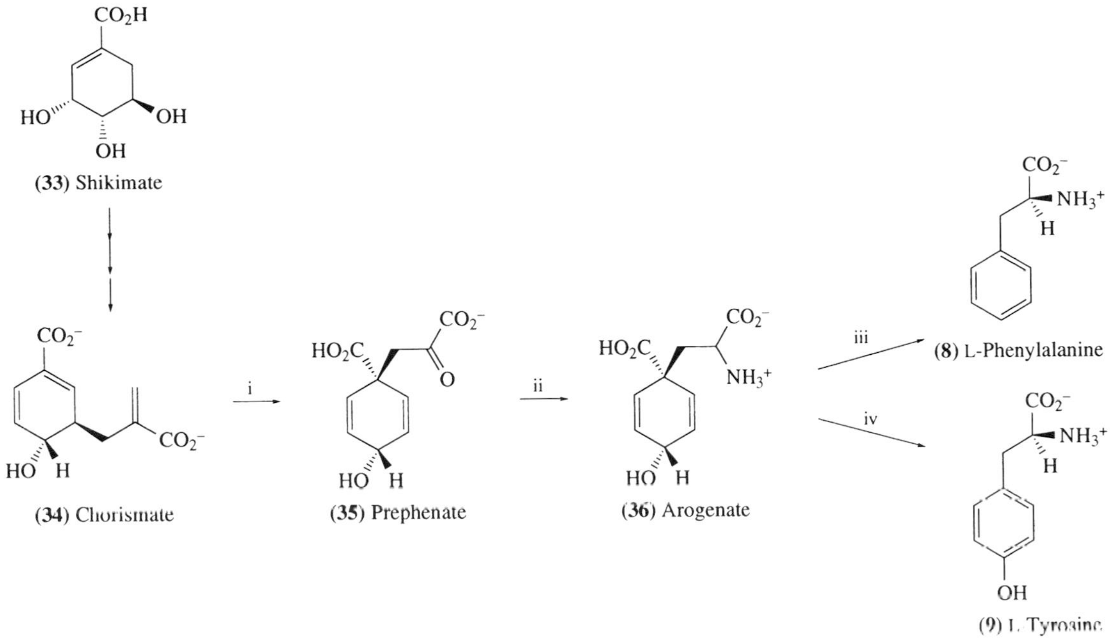

i, chorismate mutase; ii, prephenate aminotransferase; iii, arogenate dehydratase; iv, arogenate dehydrogenase

Scheme 1

The biochemical fate of the released ammonium ion was only established in the mid-1990s,[43–47] and this has revealed that a nitrogen recycling mechanism is required to sustain phenylpropanoid metabolism. This was determined largely by a study of loblolly pine (*Pinus taeda*) cell suspension cultures,[43] which can be induced, in a cell-bathing medium containing 8% glucose/20 mM KI, both to form and secrete *p*-coumaryl (**1**) and coniferyl (**2**) alcohols. Moreover, this occurs without any major interference by metabolites from related phenylpropanoid (acetate) pathways, such as that to the flavonoids. During the induction process, PAL activity is amplified from the limit of detectability to ca. 10 pkat mg^{-1} protein within 12 h, a level that was essentially maintained over 96 h. This provided the first system, therefore, for studying nitrogen metabolism in a monolignol-forming system as well as the factor(s) influencing monolignol formation and composition.

Thus, using a judicious combination of various ^{15}N-labeled amino acid precursors (^{15}N-L-Phe, ^{15}N-L-Glu, ^{15}NH$_4^+$, ^{15}N-L-Gln) and specific enzyme inhibitors for PAL, glutamine synthase (GS) and glutamine 2-oxoketoglutarate amino transferase (GOGAT), the biochemical fate of the released ammonium ion was established (Figure 5). During active phenylpropanoid metabolism, the NH$_4^+$ liberated from ^{15}N-Phe was rapidly assimilated via the GS/GOGAT system to regenerate glutamate. This, in turn, served as nitrogen donor for arogenate (**36**) formation, and hence regeneration of phenylalanine (**8**) (Figure 6(a)). The effects of the specific inhibitors of GS, GOGAT, and PAL confirmed that this nitrogen recycling was indeed operative (see Figures 6(b)–6(d)): when the PAL inhibitor L-2-aminooxy-3-phenylpropanoic acid (AOPP) was added to the culture medium of *P. taeda* cells, [^{15}N]Phe (**8**) was no longer metabolized (Figure 6(b)). On the other hand, in the presence of L-methionine-*S*-sulfoximine (MSO), the ammonium ion released was no longer assimilated, and thus ^{15}NH$_4^+$ was the only other ^{15}N signal detected, revealing that inhibition of GS had occurred (Figure 6(c)). In contrast, when the GOGAT inhibitor, azaserine (AZA), was employed, the [^{15}N] species now released during active phenylpropanoid metabolism was [^{15}N]-glutamine (δ-amide), revealing that the ammonium ion had been assimilated via the action of GS (Figure 6(d)).[43] These findings have also been confirmed in potato tuber slices, yams, and fungi suggesting a general mechanism to be operative.[44–47]

Significantly, the metabolic pool sizes of arogenate (**36**) and phenylalanine (**8**) were always very low, regardless of whether the *P. taeda* cells were induced or not.[43] Indeed, [^{15}N]phenylalanine (**8**) formed *de novo* was only detected when [^{15}N]-Gln and [^{15}N]-L-Glu were used as substrates, and only

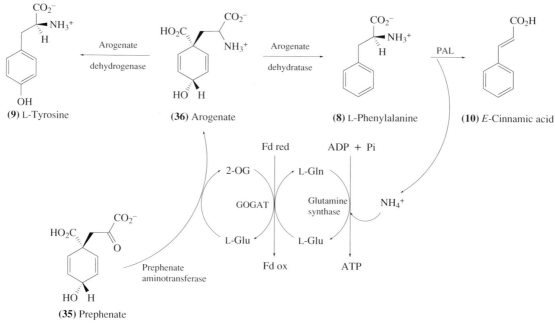

Figure 5   Nitrogen recycling during active phenylpropanoid metabolism (after Lewis and co-workers[43–47]).

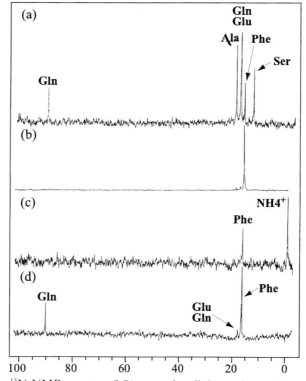

Figure 6   Representative ^{15}N-NMR spectra of *Pinus taeda* cellular amino acid extracts after administration of (a) 10 mM ^{15}N-Phe (**8**) for 96 h, (b) 10 mM ^{15}N-Phe and 0.1 mM L-AOPP for 96 h, (c) 10 mM ^{15}N-Phe and 5 mM MSO for 72 h, and (d) 10 mM ^{15}N-Phe and 5 mM AZA for 72 h (after Lewis *et al.*[43]).

under conditions when PAL activity was inhibited by AOPP. Taken together, these findings explain how active phenylpropanoid metabolism affording the monolignols can be perpetuated, via this nitrogen recycling mechanism, without any additional requirement for nitrogen on the part of the lignifying cell(s). These observations, however, suggest that phenylalanine (**8**) biosynthesis and hence its *availability* might be rate-limiting in this segment of the overall pathway.

The inducible *P. taeda* monolignol-forming system also provided the *first* opportunity to determine whether monolignol biosynthesis was under control of a single rate-limiting step or was under multipoint modulation. The strategy adopted for identifying control points in the pathway took two approaches: the first entailed measurement of metabolic pool sizes of various intermediates during active phenylpropanoid metabolism, and the second involved examination of the fates of exogenously provided pathway intermediates in terms of their accumulation within the cells, and their effect on subsequent (downstream) metabolic processes.[48] Prior to this point, virtually every step in the monolignol-forming pathway had been arbitrarily described as being rate-limiting.

Table 1 shows average metabolic pool sizes of intracellular intermediates beyond phenylalanine (**8**) to the monolignols (**1**) and (**2**), and Figure 7 shows the accumulation over time of *p*-coumaryl (**1**) and coniferyl (**2**) alcohols secreted into the cell "bathing" solution. Within the cells, however, only cinnamic (**10**), *p*-coumaric (**5**), and caffeic acids (**13**), *p*-coumaryl (**1**) and coniferyl (**2**) alcohols were detectable, but not the corresponding CoA esters (**11**) and (**15**), ferulic acid (**6**), *p*-coumaraldehyde (**12**), or coniferaldehyde (**16**).

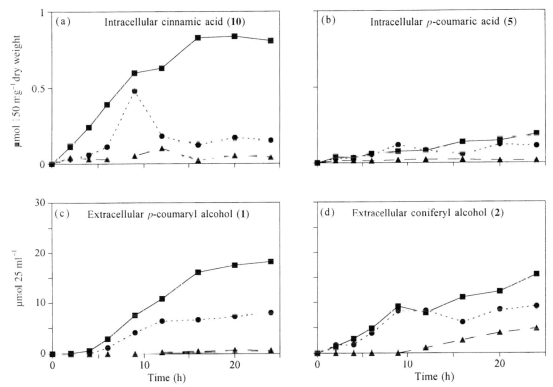

Figure 7 Effect of exogenously provided Phe (**8**) on: intracellular levels of (a) cinnamic (**10**) and (b) *p*-coumaric (**5**) acids in *Pinus taeda* cells, and corresponding extracellular accumulation of (c) *p*-coumaryl (**1**) and (d) coniferyl (**2**) alcohols. ▲ 0 μmol, ● 5 μmol, and ■ 40 μmol Phe (**8**), respectively. Note that higher amounts of Phe (**8**) (>40 μmol) had no additional effect on levels of metabolites examined.

The effect of exogenously providing various phenylpropanoid pathway intermediates to the induced *P. taeda* cells was also very instructive. As can be seen in Figures 7(c) and 7(d), increasing phenylalanine (**8**) availability (0 to 40 μmol) resulted in an approximately 26- and 3-fold increase in extracellular monolignol (**1**) and (**2**) release, after 24 h, as well as an ~16 and 10-fold increase in intracellular cinnamate (**10**) and *p*-coumarate (**5**) levels (Figures 7(a) and 7(b)), respectively. Furthermore, as the amounts of exogenously provided phenylalanine (**8**) were increased (0 to 40 μmol), there was an accompanying increase in the relative amounts of *p*-coumaryl (**1**) and coniferyl alcohols (**2**) secreted into the extracellular "bathing" medium. Of particular interest was the observation that, at low levels of Phe (**8**) availability, the *p*-coumaryl alcohol (**1**)/coniferyl alcohol (**2**) ratio approximated 1:8, whereas at higher concentrations (up to 40 μmol) the relative amount of *p*-coumaryl alcohol (**1**) increased ultimately to give an approximately 1:1 ratio with respect to coniferyl alcohol (**2**). On the other hand, none of the intracellular pool sizes for caffeic acid (**13**), *p*-coumaryl (**1**) or coniferyl (**2**) alcohols were significantly affected; nor did other intermediates (viz. the CoA esters (**11, 15**), ferulic acid (**6**), *p*-coumaraldehyde (**12**), or coniferaldehyde (**16**)) build up to detectable levels. Additionally, administration individually of cinnamic (**10**), *p*-coumaric (**5**),

The Nature and Function of Lignins

Table 1 Typical pool sizes of intracellular phenyl-propanoid metabolites in *Pinus taeda* cell suspension cultures[a].

Metabolite	Amount nmol 100 mg^{-1} cells[b] $(DW)^c$
Cinnamic acid (**10**)	66.7
p-Coumaric acid (**5**)	20.0
Caffeic acid (**13**)	16.7
Ferulic acid (**6**)	4.7
p-Coumaroyl CoA (**11**)	n.d.[d]
Caffeoyl CoA (**14**)	n.d.
Feruloyl CoA (**15**)	n.d.
p-Coumaryl aldehyde (**12**)	n.d.
Coniferyl aldehyde (**16**)	n.d.
p-Coumaryl alcohol (**1**)	20.0
Coniferyl alcohol (**2**)	20.0

[a]*P. taeda* cells were incubated in 8% sucrose 20 mM KI for various time intervals. [b]Data show average values. [c]DW = dry weight. [d]n.d.: not detected.

caffeic (**13**), and ferulic (**6**) acids had essentially no significant effect on monolignol biosynthesis either intracellularly or extracellularly. Instead, they were primarily converted into other derivatives, such as glucose esters. It would, therefore, appear that these acids, although taken up by the induced *P. taeda* cells, were either not transported to the requisite subcellular locations for further processing and/or were derivatized by way of a detoxification or a storage mechanism.

However, administration individually of *p*-coumaraldehyde (**12**) and coniferaldehyde (**16**) to the induced *P. taeda* cells, up to levels > 100 000 times above their detection limits, gave very different results (see Figures 8(a) and 8(b)). In this case, a massive and rapid conversion into the corresponding monolignols (**1**) and (**2**), respectively, was observed (Figures 8(c) and 8(d)). Moreover, no intracellular accumulation of the corresponding aldehydes (**12**) or (**16**) was noted until nonphysiological levels (~1.2 mM) were reached. This, in turn, revealed that *P. taeda* cells have sufficient capacity to reduce very high levels of *p*-hydroxycinnamyl aldehydes (**12**) or (**16**) and, thus, this reductive step was not rate-limiting. Moreover, it was also significant that both cinnamaldehydes (**12**) and (**16**) were reduced at essentially the same rate, indicating that the reductive process utilized both substrates equally effectively.

Taken together, these data suggest that the monolignol-forming pathway is controlled by multisite modulation rather than a single "rate-limiting" step. There are three reasons for this: firstly, all plant tissues examined to date, including the *P. taeda* cell cultures,[43] have low or undetectable phenylalanine (**8**) and arogenate (**36**) pool sizes, thus indirectly suggesting that phenylalanine (**8**) availability may be important. Secondly, elevating the level of available phenylalanine (**8**) not only results in a significant increase in the extracellular amounts of the monolignols, *p*-coumaryl (**1**) and coniferyl (**2**) alcohols, but it also has a profound effect on their relative ratios. However, while the levels of both intracellular cinnamate (**10**) and *p*-coumarate (**5**) increased with the availability of phenylalanine (**8**), the pool sizes of the other intracellular metabolites were essentially unaffected. Thirdly, subsequent reductive transformations, particularly those brought about by cinnamyl alcohol dehydrogenase, were not rate-limiting, given the rapid and complete conversion of the cinnamyl aldehydes (**12**) and (**16**) to the corresponding monolignols (**1**) and (**2**)—even when the precursor aldehydes (**12**) and (**16**) were supplied at very high (nonphysiological) levels.

It is, therefore, reasonably evident that phenylalanine (**8**) availability for phenylpropanoid metabolism must be rate-limiting as a consequence, presumably, of the rate of nitrogen recycling that gives phenylalanine (**8**) via regeneration of arogenate (**36**). However, when phenylalanine (**8**) is exogenously provided in increasing amounts, elevated intracellular levels result of both cinnamate (**10**) and *p*-coumarate (**5**), respectively. Moreover, as these intracellular levels rise, increased extracellular *p*-coumaryl (**1**) and coniferyl (**2**) alcohol secretion also occurs, the former predominating at higher Phe (**8**) concentrations. Thus, as cinnamate (**10**) and *p*-coumarate (**5**) levels are elevated intracellularly, the conversion of cinnamate (**10**) into *p*-coumarate (**5**) is initially rate-limiting, until the rate of *p*-coumarate (**5**) formation exceeds that of caffeate (**13**) biosynthesis. Under these conditions, more *p*-coumarate (**5**) is then reduced to the corresponding monolignol (**1**), thus altering the ratio of *p*-coumaryl (**1**) to coniferyl (**2**) alcohol. This is shown in Scheme 2.

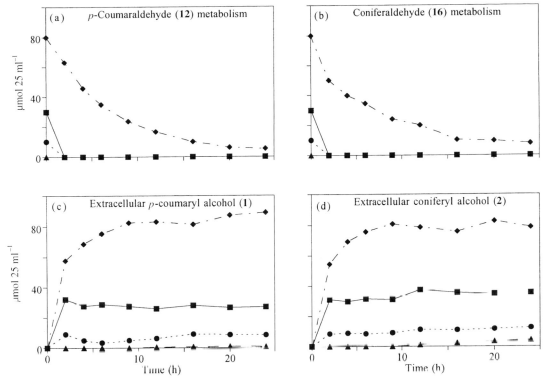

Figure 8 Time course of metabolism of exogenously provided (a) *p*-coumaraldehyde (**12**) and (b) coniferaldehyde (**16**) by *Pinus taeda* cells, and subsequent extracellular secretion of (c) *p*-coumaryl (**1**) and (d) coniferyl (**2**) alcohols. ▲, 0 μmol; ●, 10 μmol; ■, 30 μmol; and ◆, 80 μmol of aldehydes (**12**) and (**16**), respectively.

Phenylalanine ammonia-lyase is largely a cytosolic enzyme, as are the *O*-methyltransferases, CoA ligases, cinnamoyl CoA reductases, and cinnamyl alcohol dehydrogenases. Cinnamic acid 4-hydroxylase and presumably *p*-coumaric acid 3-hydroxylase are, on the other hand, membrane bound (see Davin and Lewis[40]). It therefore seems evident, on the basis of metabolite accumulation and turnover, that the two hydroxylation steps largely determine the ratio of *p*-coumarate (**5**) and caffeate (**13**) released into the cytosol. Clearly, precisely how cinnamate (**10**) is apportioned into *p*-coumaric (**5**) and caffeic (**13**) acids represents the single most important determinant of how monolignol composition is controlled.

Now that these metabolic findings have elucidated how the overall pathway is modulated, it is next fitting to describe the enzymes and genes involved in monolignol formation.

3.18.3.3 Enzymes and Genes in the Monolignol Forming Pathway: A Historical Perspective

Each of the biochemical transformations involved in monolignol (**1**)–(**3**) formation is described below in terms of a historical account about the discovery and purification of the pertinent enzyme, the proposed enzymatic mechanism, gene cloning, and recombinant protein studies. The (presumed) roles of particular isoenzymes in lignin formation in different plant species also receives emphasis to the extent that current information allows (discussed in Section 3.18.4).

3.18.3.3.1 *Phenylalanine and tyrosine ammonia-lyases*

The deamination of phenylalanine (**8**) and tyrosine (**9**) to afford *E*-cinnamic (**10**) and *E*-*p*-coumaric (**5**) acids, respectively, together with stoichiometric amounts of ammonium ion, represents by far the most extensively studied enzymatic conversion in secondary metabolism. Table 2 summarizes various milestone achievements in the study of phenylalanine ammonia-lyase (E.C. 4.3.1.5).

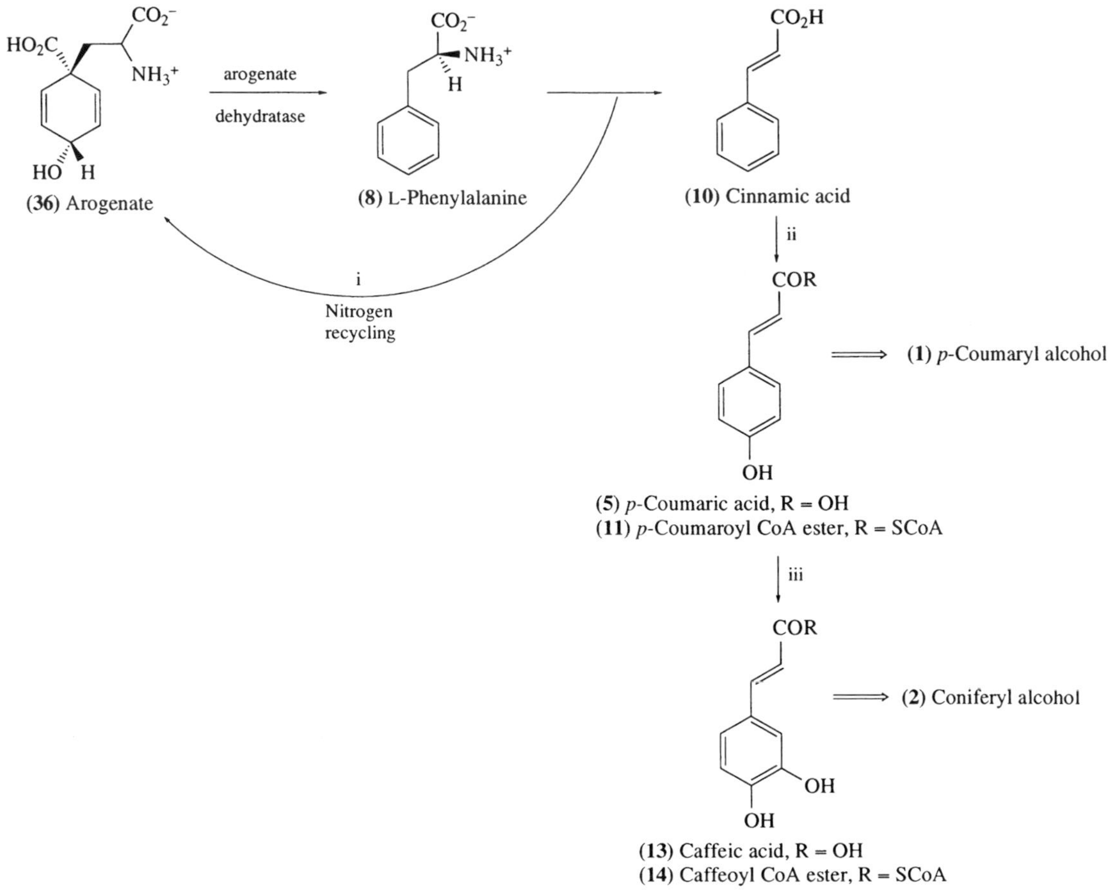

i, regeneration of phenylalanine; ii, cinnamate 4-hydroxylase; iii, *p*-coumarate 3-hydroxylase

Scheme 2

Initially, radiochemical tracer studies by Neish, Brown and co-workers in the latter half of the 1950s established that lignins were derived from phenylalanine (**8**) and cinnamic acid (**10**) in both monocotyledons and dicotyledons, and also partly from tyrosine (**9**) in the monocots.[49-54] Their studies thus led to the suggestion that the conversion of Phe (**8**) and Tyr (**9**) respectively into cinnamic (**10**) and *p*-coumaric (**5**) acids might occur in one of two ways (Scheme 3). Although consecutive transamination, reduction of the corresponding α-keto acid and dehydration of the resulting product was initially favored as the more likely sequence (pathway A),[53] it was recognized that direct deamination (pathway B) could not be excluded.[49] Indeed, the second of the two alternatives had seen precedence in the deamination of aspartate, β-methyl aspartate and histidine.[55]

Koukol and Conn subsequently clarified the matter in 1960 by describing an enzymatic process in sweet clover, which catalyzed the direct conversion of phenylalanine (**8**) into cinnamic acid (**10**) *in vitro*.[56] The enzyme deemed responsible was first described as phenylalanase, then phenylalanine deaminase, and finally PAL. This conversion was further confirmed with a 28-fold partially purified PAL preparation from barley (*Hordeum vulgare*),[57] where the product *trans*-cinnamic acid (**10**) was identified on the basis of its UV-spectrum and the equimolar co-product, ammonium ion, was determined by nesslerization.[58] An analogous study was carried out by Neish, while working in Conn's laboratory, on a 40-fold purified TAL (then called tyrase) preparation from barley,[59] which converted L-tyrosine (**9**) into *p*-coumaric acid (**5**) and ammonium ion. Neither PAL nor the TAL had any metal ion or other cofactor requirements and, from the perceived losses in relative enzymatic activities during prolonged storage, it was considered at that time that both were distinct enzymes.

PAL ($M_r \sim 330\,000$) was subsequently purified 310-fold to near homogeneity from sliced potato tubers (*Solanum tuberosum*).[60] Interestingly, the tuber tissues had no measurable PAL activity until exposed to fluorescent light,[61] the maximum activity being reached within 20–30 h of exposure. PAL induction in potato tubers is now known to be accompanied by a number of downstream metabolic

Table 2 Phenylalanine (tyrosine) ammonia-lyase: milestones.

Milestone	Year	Investigator(s)	Achievement	Ref.
Phenylalanine–cinnamic acid relationship	1960	Neish	Phenylalanine (**8**) conversion to cinnamic acid (**10**); two mechanistic possibilities (Scheme 3) proposed	49
PAL discovery	1960 1961	Koukol and Conn	Partial purification of PAL (TAL) from sweet clover and barley	56, 57
TAL discovery	1961	Neish	Partial purification of TAL from barley (in the Conn laboratory)	59
PAL induced by light	1965	Zucker	Demonstration of induction of PAL activity in potato tubers when exposed to light	61
Protein purified	1968	Havir and Hanson	PAL purified to (near) homogeneity from light-treated potato tuber slices	60
Active site determination	1968	Havir and Hanson	Dehydroalanine residue[a] proposed as active site	64
Biochemical mode of catalysis				
Loss of pro-*S* hydrogen from substrate	1971	Hanson *et al.* Ife and Haslam	Demonstration of pro-*S* hydrogen, based on comparable studies with histidine ammonia-lyase	77, 78
Putative anionic mechanism	1970	Hanson and Havir	Mechanism proposed based on presumed existence of a dehydroalanine residue in active site	80
Putative electrophilic mechanism	1995	Schuster and Rétey	Mechanism proposed based on site-directed mutagenesis	81
Common catalytic site for Phe/Tyr in monocots	1971	Havir *et al.*	PAL/TAL catalyzing enzyme purified from maize	68
	1997	Rösler *et al.*	Recombinant maize PAL protein with demonstrable PAL/TAL activity	69
Gene cloning[b]	1985	Edwards *et al.*	Cloned first PAL cDNA	70
Inhibition	1976–1977	Amrhein *et al.*	AOPP and AOP shown to be competitive inhibitors	82, 83

[a] Inactivity based on comparative results from histidine ammonia-lyase.[79] [b] This was the first bona fide example of PAL cloning. Previous claims of PAL cloning were not substantiated.[71,72]

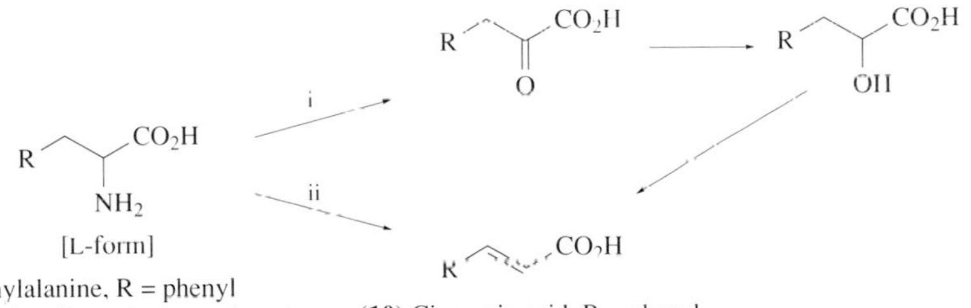

i, transamination, reduction and dehydration; ii, direct deamination

(**8**) Phenylalanine, R = phenyl
(**9**) Tyrosine, R = *p*-hydroxyphenyl

(**10**) Cinnamic acid, R = phenyl
(**5**) *p*-Coumaric acid, R = *p*-hydroxyphenyl

Scheme 3

processes, involving chlorogenic acid, alkyl ferulate, and suberin biosynthesis, and perhaps even a very low level of lignification.[62,63] Furthermore, two forms of phenylalanine ammonia-lyase were purified to >80% purity from potato tuber.[60] The purified PAL forms ostensibly exhibited apparent K_m (38–260 µM) and V_{max} values which doubled with phenylalanine (**8**) concentration,[64] although it is unclear as to what the authors believed these observations to signify. Their investigations did, however, demonstrate the reversibility of the PAL-catalyzed reaction in transforming cinnamic acid (**10**) into L-phenylalanine (**8**) to the extent of ⩾6% overall conversion.[64] The reversibility of the enzyme-catalyzed transformation has also been demonstrated with PAL from fungi (*Ustilago hordei*)[65] and yeast (*Rhodotorula glutinis*).[66]

One of the central questions that subsequently emerged was whether PAL and TAL activities in monocots depend upon a single protein, or whether two distinct enzymes are involved, as had been suggested earlier (reviewed by Camm and Towers[67]). Indeed, different plant systems appeared to have distinctive PAL and TAL catalytic properties, although neither activity was ever fully resolvable from the other chromatographically. That the enzyme could utilize both substrates was shown with a L-phenylalanine ammonia-lyase preparation purified from maize (*Zea mays*), which converted phenylalanine (**8**) and tyrosine (**9**) into cinnamic (**10**) and *p*-coumaric (**5**) acids: the kinetic data gave K_m values of 270 μM (Phe) and 29 μM (Tyr), with the V_{max} value being eight times higher for L-Phe (**8**).[68] It was, therefore, concluded that the maize PAL possessed the dual capacity of converting L-Phe (**8**) and L-Tyr (**9**) to the corresponding (hydroxy)cinnamic acids (**10**) and (**5**). This conclusion was further confirmed by Rösler *et al.* using a recombinant maize PAL protein (Zm PAL1). At pH 8.7, this converted L-Phe (**8**) and L-Tyr (**9**) into the respective products (**10**) and (**5**) (pH 8.7, K_m 270 ± 20 μM (Phe) (**8**), 19 ± 1 μM (Tyr) (**9**), respectively), with the turnover number being about 10 times higher for Phe (**8**).[69] These data were thus in very good agreement with those of Havir and Hanson,[68] although there were some discrepancies between the pH optima (7.7 vs. 8.7). The latter could presumably arise from differences between the kinetic behavior of a purified isoform[69] and that of an isoenzyme mixture.[68]

It is quite noteworthy that phenylalanine ammonia lyase was first cloned from elicitor-treated cells of bean (*Phaseolus vulgaris*) by Edwards *et al.* in 1985.[70] Its cDNA encoded a polypeptide which was immunoprecipitated with monospecific antibodies raised against the bean PAL. Previous claims that a PAL cDNA clone had been obtained from parsley (*Petroselinum crispum*)[71,72] were later shown to be incorrect, since it had no significant sequence homology with the bean PAL cDNA,[70] or other parsley PAL cDNAs subsequently obtained.[73]

The phenylalanine/tyrosine ammonia-lyases obtained from various sources exist as tetramers with molecular weights which, depending upon the species, range from 240–330 kDa, these being composed of 55–85 kDa subunits.[74] With no metal ion or co-factor requirement, the catalytic mode of action was of considerable interest. In agreement with findings previously obtained from histidine ammonia-lyase,[75,76] it was established that the pro-3S hydrogen, from either phenylalanine (**8**) or tyrosine (**9**), was abstracted during the transformation to cinnamic (**10**) or *p*-coumaric (**5**) acids, respectively, and ammonium ion (Scheme 4), via a *trans* elimination.[77,78]

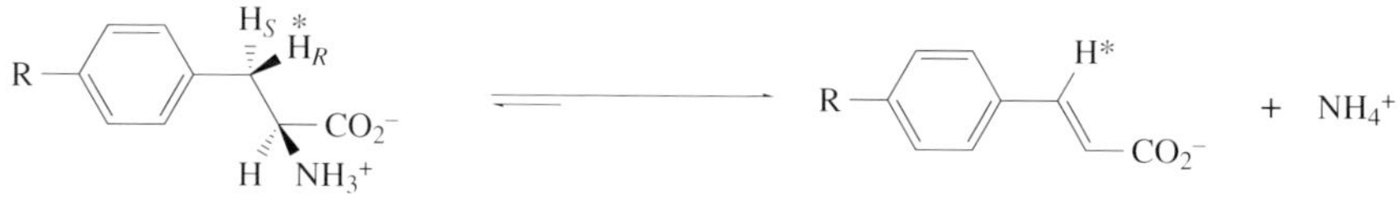

(**8**) Phe, R = H
(**9**) Tyr, R = OH

(**10**) Cinnamic acid, R = H
(**5**) *p*-Coumaric acid, R = OH

Scheme 4

Studies of that other deaminase, histidine ammonia-lyase,[75,76,79] revealed that it contained a dehydroalanine active site residue which, when reduced with NaB^3H_4, gave inactive enzyme and released alanine-t_4 upon acid hydrolysis. A comparable situation held with respect to the active site residue in PAL,[80] and an enzymatic mechanism (Scheme 5; after Schuster and Rétey[81]) was proposed involving nucleophilic attack by the amino group of phenylalanine (**8**) on the dehydroalanine. In this instance, however, abstraction of the H_S proton was envisaged to generate the putative anionic intermediate depicted.

The mechanism has been challenged. The essential dehydroalanine residue is formed post-translationally at the Ser-202 amino acid residue, and site-directed mutagenesis—where Ser-202 was replaced by alanine and threonine—has brought a new perspective to the catalytic mechanism involved.[81] Although these mutations resulted in loss of PAL activity, the alanine-202 and threonine-202 containing PAL-derived proteins were, nevertheless, capable of catalyzing the deamination of *m*-tyrosine and L-4-nitrophenylalanine, respectively. Accordingly, these findings were said to infer that, in wild-type PAL, the initial enzyme-substrate complex is formed via electrophilic attack on the dehydroalanine active site residue (Scheme 6; after Schuster and Rétey[81]), with subsequent hydride abstraction and deamination taking place to give cinnamate (**10**) and ammonium ion, as shown.

Lastly, PAL is inhibited by the phenylalanine analogues, 2-aminooxy-3-phenyl-propionic acid (AOPP) and 3-aminooxy propionic acid (AOP), where *in vitro* experiments gave K_i values of 1.4 nM and 0.10 mM, respectively, for the inhibition of PAL activities.[82,83] Both are competitive inhibitors. To date no X-ray structures for PAL/TAL have been reported. This would be useful, particularly if the crystal structure contained protein complexed with either inhibitor. An X-ray structure of aspartase (L-aspartate ammonia-lyase) has, however, been obtained from *Escherichia coli*.[84] This was crystallized in its tetrameric form, but unfortunately without substrate or inhibitor. Accordingly, the precise active site residues involved in deamination and orientation and binding of the substrate in aspartase remain elusive.

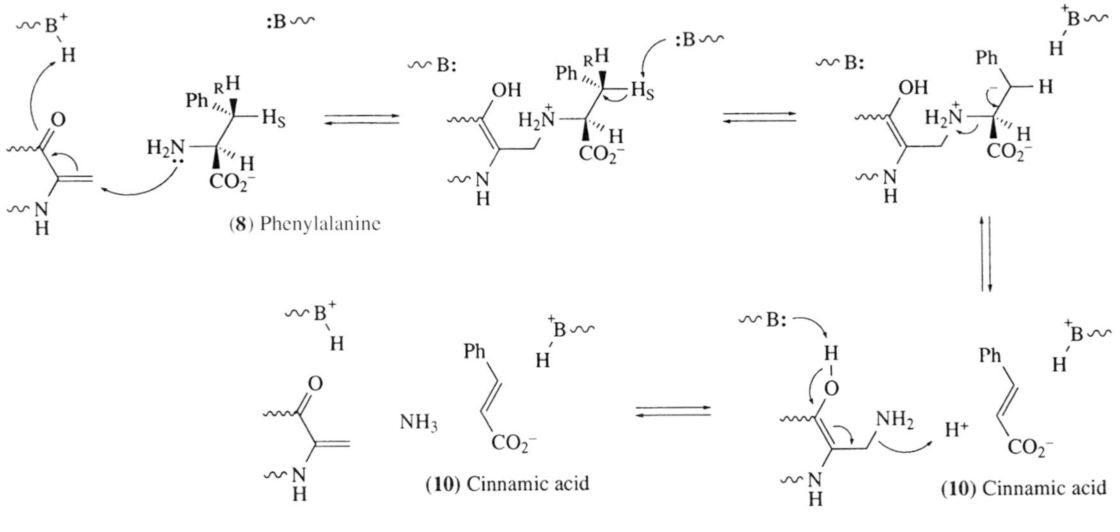

Scheme 5

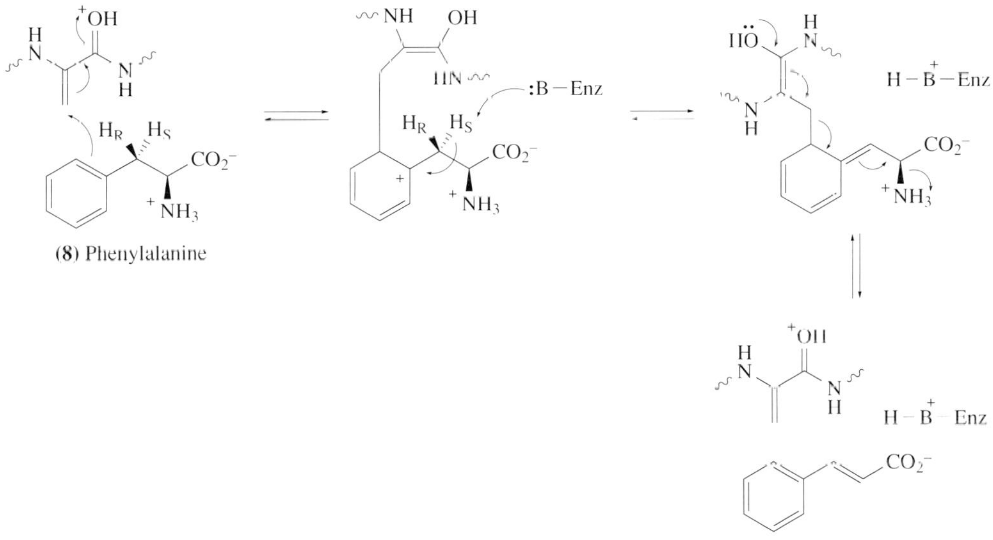

Scheme 6

3.18.3.3.2 *(Hydroxy)cinnamate hydroxylases and NADPH-dependent reductase(s)*

(i) Cinnamate 4-hydroxylase

Cinnamate 4-hydroxylase (E.C. 1.14.13.11) is a cytochrome P450 hydroxylase, which acts *in vivo* with an associated NADPH-dependent reductase to regiospecifically introduce a hydroxyl group *para* to the propenoic side-chain of cinnamate (**10**). The overall catalytic cycle adapted from Segall *et al.*[85] is shown in Figure 9 and, as before, milestone achievements in the study of this enzyme are summarized (Table 3).

The discovery of cinnamate 4-hydroxylase (C4H) can be traced to investigations by David Russell and Eric Conn, who reported that a presumed microsomal fraction, isolated at $104\,000\,g$ from pea seedlings, catalyzed the NADPH-dependent conversion of [8-^{14}C] cinnamic acid (**10**) into [8-^{14}C]*p*-coumaric acid (**5**).[86] On the other hand, previous reports of an acetone powder from spinach leaves catalyzing this same conversion[87,88] were subsequently disproved.[89]

The C4H from pea seedlings ($K_m = 17\,\mu M$ with respect to cinnamate) was demonstrated to be an

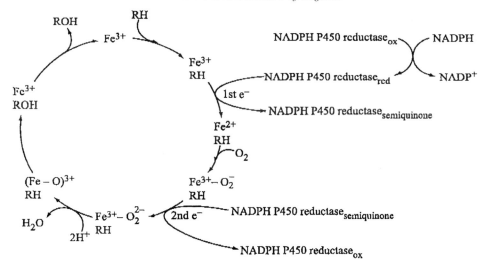

R = Cinnamic (**10**) or ferulic (**6**) acids.

Figure 9 Proposed catalytic cycle for cytochrome P450 oxidase(s) and associated NADPH-dependent reductase(s) in formation of *p*-coumaric acid (**5**) and 5-hydroxyferulic acid (**17**) from cinnamic acid (**10**) and ferulic acid (**6**), respectively (after Segall *et al.*[85]).

Table 3 Cinnamate-4-hydroxylase and related oxidases: milestones.

Milestone	Year	Investigator(s)	Achievement	Ref.
O_2-requiring enzymes for hydroxylation	1955	Mason *et al.*	Demonstration that phenolases were O_2-requiring	98
		Hayaishi *et al.*	Demonstration that pyrocatechase was O_2-requiring	99
(Heme) pigment/CO-binding	1958	Klingenberg	(Heme) pigment and CO-binding in rat liver microsomal fraction	91
Heme protein discovery	1962 1964	Omura and Sato	(Heme) protein containing CO-binding pigments from rabbit liver	92–94
	1962	Hashimoto *et al.*	(Heme) protein containing CO-binding pigments from rabbit liver	95
Demonstration of a cytochrome P450 catalytic function	1963	Estabrook *et al.*	First physiological function of cytochrome P450 demonstrated in the hydroxylation of 17-hydroxy-progesterone	96
NIH shift	1967	Guroff *et al.*	Tritium migration observed during hydroxylation of the aromatic ring to the *ortho*-position	101
Discovery of cinnamate-4-hydroxylase (C4H)	1967	Russell and Conn	Pea seedling microsomes catalyzed NADPH-dependent hydroxylation of cinnamic acid (**10**)	86
C4H catalyzed NIH shift	1968	Russell *et al.*	Tritium migration observed during C4H hydroxylation of cinnamic acid (**10**)	100
Cytochrome P450 nature of C4H	1971	Russell	O_2/NADPH-dependence/CO and azide inhibition of C4H (while in Conn laboratory)	89
	1974	Potts *et al.*	Cytochrome P450 detected spectrophotometrically	90
C4H purification	1991	Gabriac *et al.*	C4H purified to apparent homogeneity from Jerusalem artichoke tubers	113
C4H cloning	1993	Mizutani *et al.*	Mung bean cytochrome P450 gene thought to encode C4H (but no recombinant protein authentication)	115
		Teutsch *et al.*	Jerusalem artichoke cytochrome P450 with recombinant C4H activity in yeast expression system enzyme assay (data not shown until a year later)	116
		Fahrendorf and Dixon	Alfalfa C4H authentication, in terms of sequenced clone and recombinant protein activity	119

O_2-requiring monooxygenase and/or mixed function oxidase.[89] This was based on the observation that it catalyzed the conversion of an O_2/NADPH-dependent hydroxylation reaction which, on account of its effective inhibition ($\sim 70\%$) by CO ($1:1$, $CO:O_2$) and azide (78% at 10 mM), respectively, exhibited a requirement for a cytochrome P450 prosthetic group. The cytochrome P450 nature of C4H was later established spectrophotometrically using a microsomal preparation from sorghum (*Sorghum vulgare*) seed.[90]

Earlier mammalian studies had previously established the nature of cytochrome P450 hydroxylation reactions, beginning with the discovery of a pigment in rat liver microsomes that absorbs light at 450 nm upon CO binding when reduced with dithionite.[91] Evidence for the existence of a hemoprotein came later during 1962–64 through studies by Omura and Sato[92–94] and Hashimoto *et al.*[95] that used a CO-binding pigment from rabbit liver. At about the same time, Ron Estabrook and others reported the first physiological function of a cytochrome P450 enzyme with the hydroxylation of 17-hydroxyprogesterone into cortexolone.[96] Moreover, by 1969, the purification and reconstitution of a cytochrome P450 monooxygenase system had been fully achieved using liver microsomes.[97] However, the discovery of a molecular oxygen-dependent hydroxylation by oxygenases in general can actually be traced back to the pioneering studies of Mason *et al.*[98] and Hayaishi *et al.*[99] in 1955 using phenolase and pyrocatechase, respectively.

Russell *et al.*[100] also demonstrated, using [4-^{3}H]cinnamic acid (10) as a substrate, that hydroxylation catalyzed by C4H occurred in a manner whereby $\sim 90\%$ of the tritium was retained, it having migrated to the *ortho* position relative to the hydroxyl group. This observation was in keeping with the presumption that hydroxylation induced migrations occur through an epoxide intermediate as shown (Scheme 7). This migration, trivially called the NIH shift, had been previously reported in a study of the hydroxylation of various aromatic phenols by oxygenases.[101] There is, however, some question as to whether this mechanism is indeed operative, since evidence for a radical cation mechanism and/or direct addition of ferryl oxygen to the aromatic ring has emerged from other studies.[102–104]

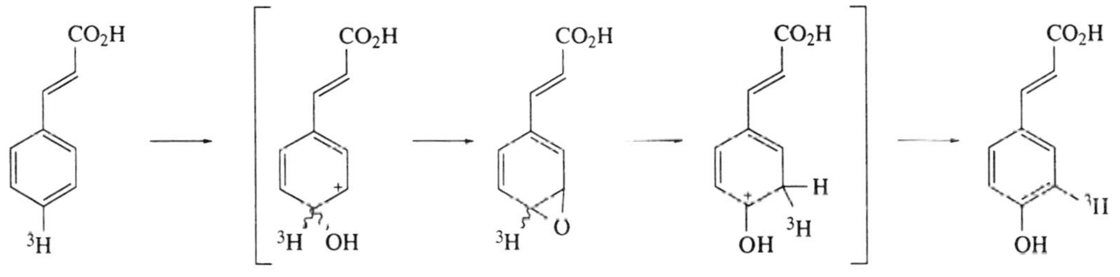

(10) Cinnamic acid   (5) *p*-Coumaric acid

Scheme 7

The purification of C4H and its cloning, as indeed for all the plant cytochrome P450 enzymes, lagged significantly behind the corresponding mammalian studies. For example, the first two plant cytochrome P450s that were purified did not even have any known catalytic function. These were obtained from tulips in 1985[105] and from avocado in 1989,[106] with subsequent cloning of one from avocado being reported in 1990.[107] To place this into a more general context, of the 154 cytochrome P450 clones known by 1991, only the avocado cDNA was from a plant source.[108]

On the other hand, the first example of a C4H to be purified to apparent homogeneity was also reported in 1991, following a series of attempts that spanned nearly two decades.[109–112] It was obtained from Jerusalem artichoke (*Helianthus tuberosus*) tuber tissue, which was wounded in order to induce its C4H activity.[113] The purified C4H so obtained had an M_r of $\sim 57\,000$ and $K_m = 1.5$ µM with respect to cinnamate (10), its catalytic activity being effectively inhibited ($\sim 98\%$) by anti C4H immunoglobulin G (IgG) antibodies. This study was then shortly followed by the purification of C4H from etiolated mung bean (*Vigna radiata*) seedlings.[114]

Reports about the cloning of cDNAs encoding C4H appeared in three articles that had been almost contemporaneously submitted for publication. The first report by Mizutani *et al.* (submitted 25 December 1992) dealing with mung bean lacked any recombinant protein activity data to authenticate the clone's function.[115] The second by Teutsch *et al.* (submitted three days later) described a Jerusalem artichoke C4H whose recombinant protein, when expressed in yeast, apparently converted cinnamic acid (10) into *p*-coumarate (5), although no supporting data were provided.[116] These enzymatic data were, however, published later.[117,118] The third submission (received 12 February 1993), by Fahrendorf and Dixon, established a cDNA sequence of one clone from a small family of C4H genes in alfalfa (*Medicago sativa*), and more importantly provided evidence of recombinant protein activity authenticating that the clone did indeed encode C4H.[119] In all three cases, the gene sequences encoded proteins of $M_r \sim 57\,000$–$58\,000$.

At the time of writing, there was no X-ray structure for C4H, and thus what was known about the binding of prosthetic groups and substrate came essentially from the analysis of other nonplant cytochrome P450s[120,121] which had been crystallized.[122–127] Based on such comparisons, Teutsch *et al.* proposed that, in the Jerusalem artichoke C4H, Thr310 has an essential function in stabilizing the dioxygen binding groove in helix 1, which is located several residues upstream from the oxygen binding site. Additionally, the highly conserved heme-binding domain, which has the general motif of Phe (Gly/Ser) X GlyX (His/Arg) X Cys X Gly X (Ile/Leu/Phe/Ala),[121] is centered around Cys[447]. It certainly will be of interest to obtain an X-ray structure for C4H and determine how it compares with other cytochrome P450s.

Figure 9 depicts the general cytochrome P450 catalytic cycle according to Segall *et al.*[85] As can be seen, the steps involved include binding of substrate, e.g., cinnamic acid (**10**), and reduction of the heme Fe^{3+} ion to Fe^{2+} through electron transfer from the corresponding NADPH-dependent P450 reductase. Oxygen binding then occurs to give the $Fe^{2+}(O_2)$ complex which is reductively transformed to the peroxoiron (III) complex, i.e., $Fe^{3+}(O_2^{2-})$. The O—O bond is then protonated and cleaved, resulting in the formation of a reactive iron–oxo species (believed to be $Fe^{4+}=O$-porphyrin$^{+\bullet}$) and incorporation of the distal oxygen into H_2O. The iron–oxo complex transfers an oxygen atom into the substrate (e.g., cinnamic acid (**10**)), the process being followed by dissociation of the product (e.g., *p*-coumaric acid (**5**)) from the enzyme and completion of the catalytic cycle.[85,128]

(ii) NADPH-dependent reductase

As shown in Figure 9, the NADPH-dependent reductase is an integral part of the catalytic machinery involved in aromatic ring hydroxylation to give *p*-coumaric acid (**5**). Milestone achievements in the study of this enzyme are summarized in Table 4.

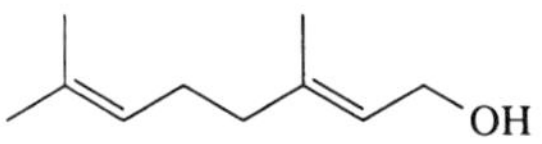

(**37**) Geraniol

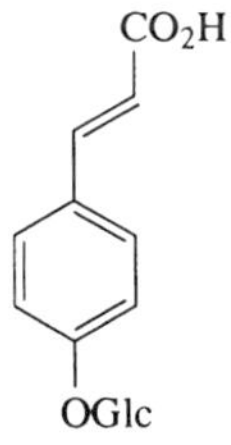

(**38**) *p*-Coumaric acid D-glucose

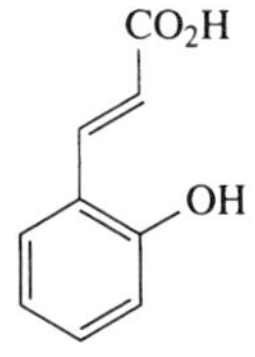

(**39**) *o*-Coumaric acid

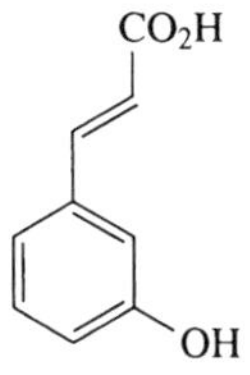

(**40**) *m*-Coumaric acid

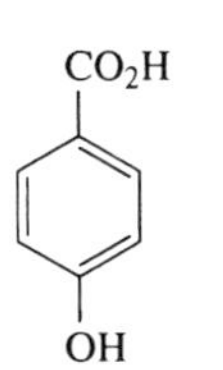

(**41**) *p*-Hydroxybenzoic acid

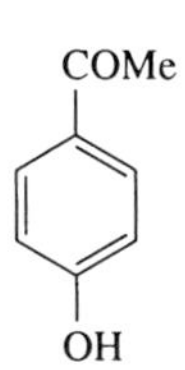

(**42**) *p*-Hydroxyacetophenone

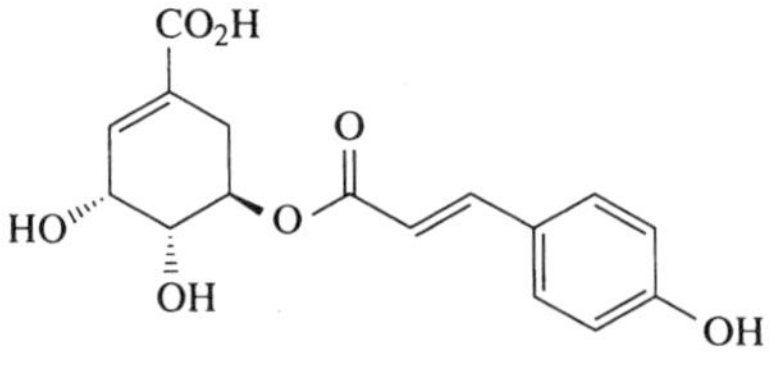

(**43**) *E*-5–*O*–(4-coumaroyl)shikimate

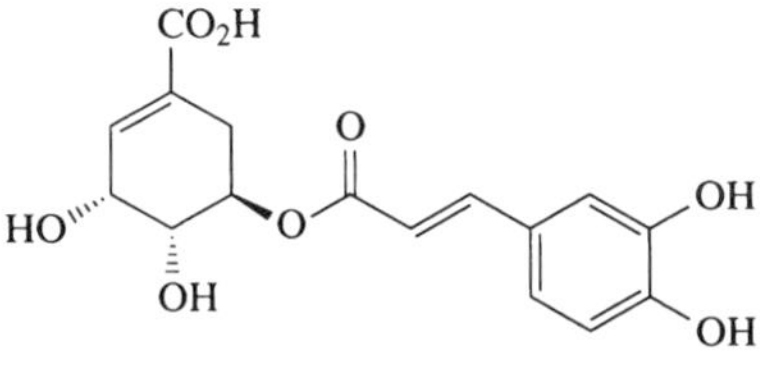

(**44**) *E*-5–*O*–(4-caffeoyl)shikimate

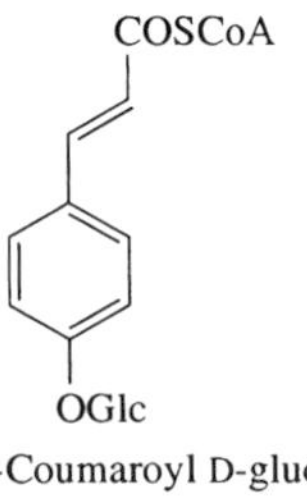

(**45**) *p*-Coumaroyl D-glucose

As in the case of C4H, much of the knowledge about the associated NADPH-dependent reductase is drawn from mammalian and yeast studies. In this regard, the first reports of NADPH cytochrome

Table 4 NADPH-dependent cytochrome P450 reductases: milestones.

Milestone	Year	Investigator(s)	Achievement	Ref.
Discovery of cytochrome c reductase				
in yeast	1940	Haas *et al.*	Detection of cytochrome c reductase activity in	129
in pig liver	1950	Horecker	yeast and subsequently in liver extracts	130
Flavoprotein in microsomal fraction	1962	Williams and Kamin	Discovery of NADPH-dependent cytochrome c reductase (flavoprotein) in microsomes	131
		Phillips and Langdon		132
FAD and FMN prosthetic groups	1973	Iyanagi and Mason	Established FAD and FMN prosthetic groups in cytochrome c reductase by fluorimetric assay, paper chromatography and apoflavodoxin reconstitution	133
Purification to homogeneity	1975	Digman and Strobel	Purification to homogeneity (SDS–PAGE, single band)	134
Mechanism of electron transfer during catalysis	1978	Vermilion and Coon	Analysis of results obtained from spectrophotometric study	135
Identification of the membrane binding domain	1979	Gum and Strobel	Partial digestion of protein in reconstituted vesicles and identification of membrane binding	137
		Black *et al.*	domain	136
Purification of first plant cytochrome P450 reductase	1985	Madyastha and Coscia	Purification of presumed reductase involved in geraniol (**37**) hydroxylation	143
Determination of nucleotide sequence of cytochrome P450 reductase from rat liver and determination of FAD and FMN binding domains	1985	Porter and Kasper	Deduced amino acid sequence in agreement with sequences obtained from purified rat liver cytochrome P450 reductase, and identification of FAD/FMN binding domains	138
Determination of cytochrome c binding domain	1986	Nisimoto	Cross-linking study using 1 ethyl-3-(3-dimethylamino-propyl)carbodiimide established cytochrome c binding domain	139
Purification of C4H P450 reductase	1986	Benveniste *et al.*	Reconstitution of cinnamate 4-hydroxylase activity	144
X-ray structure of cytochrome P450 reductase	1995	Djordjevic *et al.*	Crystal structure of cytochrome P450 reductase and its analysis	141
	1997	Wang *et al.*		142

c reductases were recorded in 1940 and 1950 from yeast[129] and pig liver,[130] respectively. The flavoprotein nature of this microsomal reductase was established later,[131,132] and by 1973 these flavin cofactors had been identified as FAD and FMN[133] prior to purification of the protein itself.[134] In 1978, Vermilion and Coon established that the flow of electrons occurs from NADPH to FAD to FMN to cytochrome P450.[135] The membrane-binding domain of the NADPH cytochrome P450 reductase was next identified by analysis of the peptide remaining in the membrane following proteolytic cleavage of the catalytic portion of the vesicle-reconstituted enzyme.[136,137] By 1985, the nucleotide sequence had been obtained[138] and the FMN and FAD binding domains had also been identified. Since then the cytochrome P450 binding region has been deduced[139,140] and a three-dimensional structure for a NADPH cytochrome P450 reductase was reported during 1994–97.[141,142]

Purification and cloning of the cinnamate 4-hydroxylase P450 reductase again lagged behind the corresponding mammalian and yeast studies. Indeed, it was only in 1979 that a plant NADPH-dependent cytochrome P450 reductase was first purified, the enzyme having been obtained from *Catharanthus roseus* in a study associated with delineating geraniol (**37**) hydroxylation.[143] Seven years later, a cytochrome P450 reductase, presumed to be associated with C4H, was purified from Jerusalem artichoke (*H. tuberosus*).[144] From 1993 to 1997, various clones of plant Cyt P450 reductases were obtained from *V. radiata*,[145] *C. roseus*,[146] *Arabidopsis thaliana*,[147] and *P. crispum*,[148] among which the one from the latter was associated with C4H.

(iii) 4-Coumarate 3-hydroxylase

In as far as both C4H and ferulate 5-hydroxylase (*vide infra*) are cytochrome P450 oxidases, 4-coumarate 3-hydroxylase might be expected to be of the same catalytic type. However, unambiguous efforts to identify and characterize the protein(s) involved in either caffeic acid (**13**) or caffeoyl CoA

(**14**) formation have not been completely successful. Accordingly, how caffeoyl moieties are actually formed remains an open question at the time of writing (Figure 10).

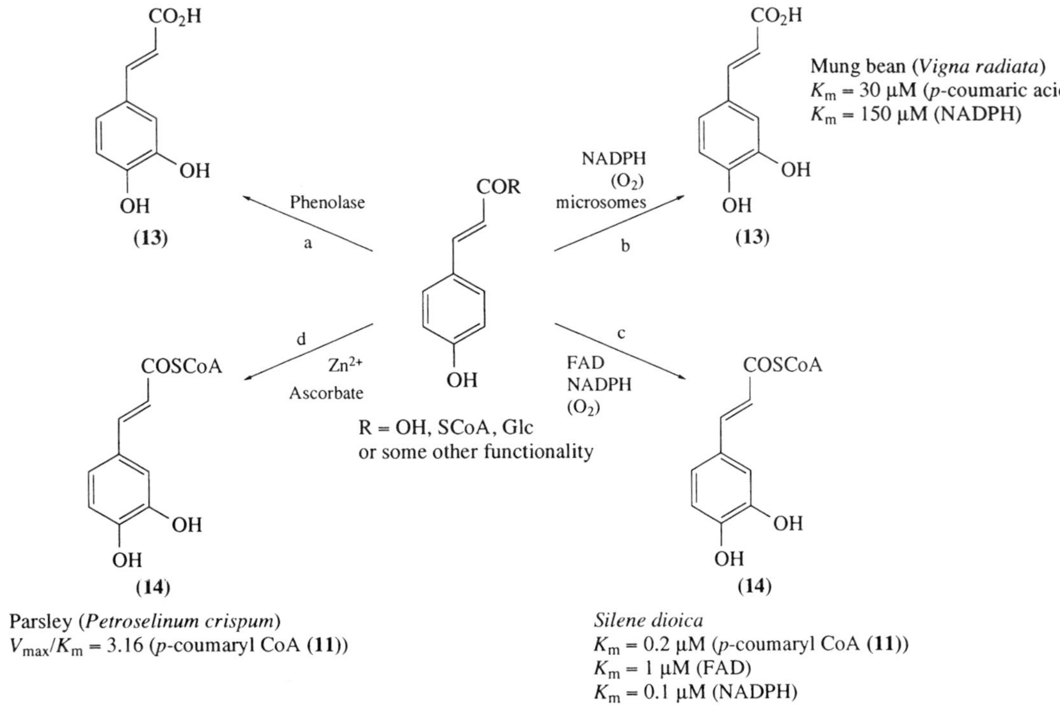

Figure 10 Representation of various enzyme preparations reportedly catalyzing 3-hydroxylation reactions in different plants. (a) Phenolase/nonspecific hydroxylase, (b) specific hydroxylation of *p*-coumaric acid (**5**) using NADPH-dependent microsomal preparations from mung bean (*Vigna radiata*), (c) soluble FAD dependent *p*-coumaroyl CoA (**11**) hydroxylase from *Silene dioica*, and (d) Zn^{2+}/ascorbate dependent oxidation of *p*-coumaroyl CoA (**11**) using soluble enzyme preparation from parsley (*Petroselinum crispum*).

Patil and Zucker initially reported that a soluble potato "phenolase" preparation catalyzed the conversion of *p*-coumaric acid (**5**) into caffeic acid (**13**),[149] and Sato next suggested that this enzyme was present in chloroplasts.[150] Vaughan and Butt purified (~ 1000 fold) a presumed cytosolic 4-coumarate 3-hydroxylase from spinach beet (*Beta vulgaris*) leaves,[151] but subsequent characterization of these proteins revealed that they were nonspecific polyphenol oxidases (PPOs), which gave cause for doubt that they were the actual proteins involved. Indeed, this suspicion grew further from subsequent studies using mung bean (*V. radiata*) seedlings, the polyphenol oxidase activity of which was inhibited by tentoxin but nevertheless still formed caffeic acid (**13**) derivatives.[152,153]

In a continuation of this work, Kojima and Takeuchi[154] reported that a mung bean preparation, obtained from cell organelles sedimenting in a sucrose gradient between the mitochondria and endoplasmic reticulum, was capable of catalyzing a NADPH-dependent conversion of *p*-coumaric acid (**5**) into caffeic acid (**13**) ($K_m = 30$ μM and 150 μM with respect to *p*-coumaric acid (**5**) and NADPH, respectively). On the other hand, *E*-cinnamic acid (**10**), its glucose derivative *p*-coumaric acid D-glucose (**38**), *o*- and *m*-coumaric acids (**39**) and (**40**), L-Tyr (**9**), ferulic acid (**6**), *p*-hydroxy-benzoic acid (**41**), and *p*-hydroxyacetophenone (**42**) did not serve as substrates, indicating that this protein exhibited a high degree of specificity for *p*-coumaric acid (**5**). Accordingly, these data might be interpreted to suggest that a specific cytochrome P450 oxidase was involved in caffeic acid (**13**) formation.

However, a number of other protein preparations have been claimed to be capable of regio-specifically introducing a 3-hydroxyl group into *p*-coumaric acid (**5**) or its derivatives. For example, a soluble FAD-dependent enzyme preparation from *Silene dioica* petals has been reported which converts *p*-coumaroyl CoA (**11**) into caffeoyl CoA (**14**),[155] but this has not been shown to have any general significance in the plant kingdom. Additionally, Heller and Kühnl[156] have described a cytochrome P450 dependent conversion of *E*-5-*O*-(4-coumaroyl)shikimate (**43**) into its caffeic acid derivative (**44**), but again the same reservations hold about the general significance of this conversion. The enzymological basis of caffeic acid (**13**) biosynthesis also assumed a new level of complexity with the report by Kneusel *et al.*[157] about a crude soluble protein preparation from cultured parsley cells which preferentially converted *p*-coumaroyl CoA (**11**) into caffeoyl CoA (**14**), in the presence of Zn^{2+} and ascorbate. Nevertheless, the differences in V_{max}/K_m between both the CoA ester (**11**) and the free acid (**5**) were very small (3.16 vs. 2.36), thus raising some question as to the general applicability of their findings.

Finally, a soluble *p*-coumaroyl-D-glucose hydroxylase was purified to apparent homogeneity ($M_r \sim 33\,000$) from sweet potato (*Ipomea batatas*) roots.[158] However, this protein was again nonspecific towards the substrate and indeed exhibited quite high K_m values (K_m = 1.5, 49, and 24 mM with respect to *p*-coumaroyl-D-glucose (**45**), D,L-DOPA, and caffeic acid (**13**), respectively). Since then Wang *et al.*,[159] in a related study of a *p*-coumaroyl CoA hydroxylase from *Lithospermum erythrorhizon* cell cultures, have concluded that both this and the sweet potato enzyme are nonspecific polyphenol oxidases that differ from the protein responsible for hydroxylation of *p*-coumaric acid (**5**) (derivatives) *in vivo*.

Thus, in summary, while it is uncertain how caffeoyl derivatives are formed during active phenylpropanoid metabolism, the overall pattern of evidence still points to the involvement of a specific cytochrome P450 enzyme in their formation.

(iv) Ferulate 5-hydroxylase

In early studies, it was unclear whether caffeoyl intermediates (**13**) or (**14**) were hydroxylated to give the 3,4,5-trihydroxycinnamoyl derivatives (**46**)/(**47**) which would then be methylated to give the syringyl components of lignins in angiosperms, or whether the sinapoyl intermediate was instead formed via hydroxylation of feruloyl moieties (**6**)/(**15**) (Figures 4 and 11). This situation was complicated further by the paucity of any data revealing whether either 3,4,5-trihydroxycinnamoyl (**46**)/(**47**) or 5-hydroxyferuloyl (**17**/**18**) derivatives accumulate *in planta*. These questions were resolved in two mutually reinforcing studies as discussed below, and milestone achievements in these investigations are described in Table 5.

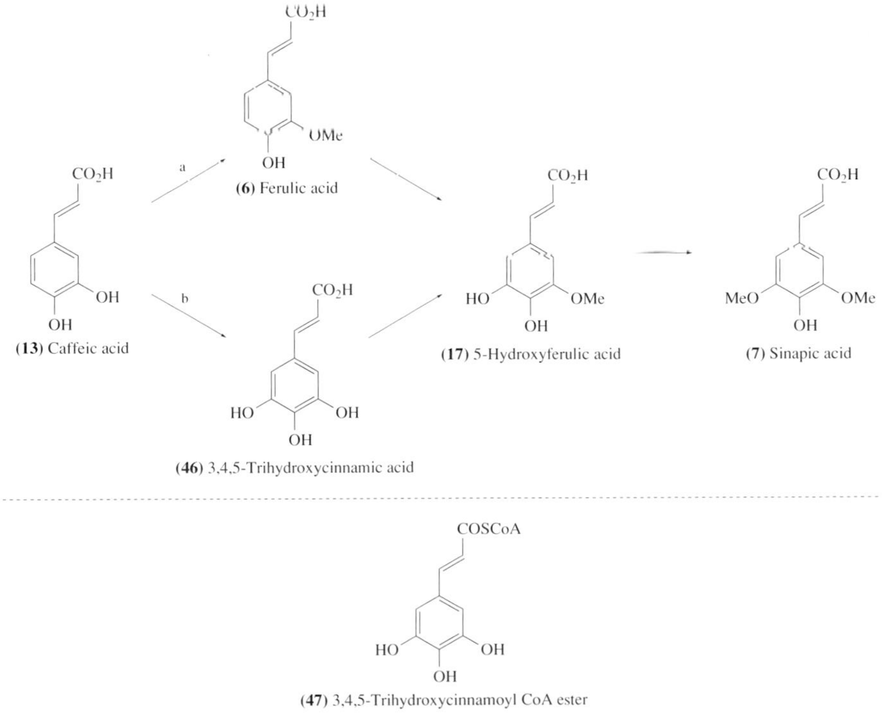

Figure 11 Early postulates for sinapic acid (**7**) formation. (a) Methylation, hydroxylation, and methylation and (b) hydroxylation and consecutive methylation.

Table 5   Ferulate 5-hydroxylase: milestones (see Table 3 for background to cytochrome P450).

Milestone	Year	Investigator(s)	Achievement	Ref.
Discovery of ferulate-5-hydroxylase	1985	Grand	Detection of F5H activity in poplar tissue, and deduction that C4H and F5H were distinct enzymes	160
Discovery of 5-hydroxyferulic acid (**17**)	1987	Ohashi *et al.*	Determination that 5HFA is a cell wall bound natural product in barley and maize	161
Discovery of *Arabidopsis* mutant lacking F5H	1992	Chapple *et al.*	Identified *Arabidopsis thaliana* mutant lacking sinapoyl malate (**48**)	162
T-DNA tagging to obtain putative F5H clone	1996	Meyer *et al.*	Complementation of *A. thaliana* with F5H under control of CaMV 35S promoter restoring sinapoyl malate (**48**) biosynthesizing ability	163

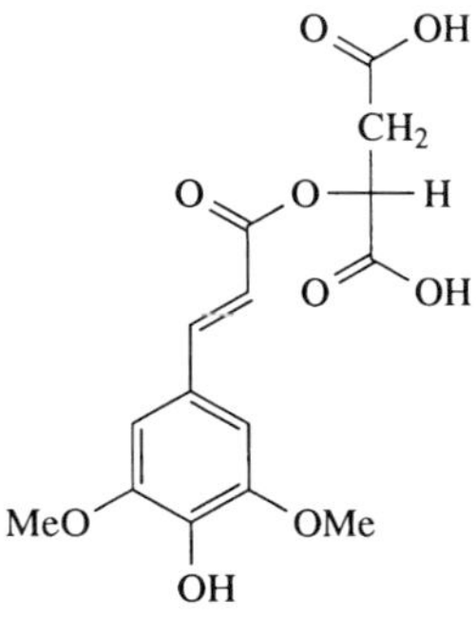

(**48**) Sinapoyl malate

The first breakthrough entailed the detection of ferulate 5-hydroxylase (F5H) in xylem- and sclerenchyma-enriched tissue from poplar (*Populus × euramericana*), which revealed an O_2/NADPH-dependent activity that, in its partial inhibition by CO, suggested a cytochrome P450 enzyme.[160] In agreement with this conclusion, the crude enzyme was a microsomal (100 000*g* precipitate) preparation which gave apparent K_m values of 6.3 and 1.9 μM with respect to ferulic acid (**6**) and cinnamic acid (**10**), respectively. Interestingly, the F5H activity was highest in the sclerenchyma tissue which is considered to possess a lignin rich in syringyl moieties.[160] Furthermore, although both C4H and F5H activities were detected in both xylem and sclerenchyma preparations, each activity was differentially inhibited by addition of *p*-coumaric acid (**5**) and 5-hydroxyferulic acid (**17**), thereby suggesting the presence of two distinct enzymes.

The next advance came with the demonstration that 5-hydroxyferulic acid (**17**) was in fact a natural product, this being discovered in maize (*Z. mays*) and barley (*H. vulgare*) cell wall preparations.[161]

Attempts to isolate ferulate 5-hydroxylase through conventional purification approaches proved unsuccessful in various laboratories. However, in a quite unrelated study using *A. thaliana*, a *fah1* mutant was discovered which appeared to lack ferulate 5-hydroxylase activity[162] on the basis of an absence of detectable sinapoyl malate (**48**) accumulation in the mutant plant.

Subsequent T-DNA tagging was then used to attempt to clone the putative F5H gene[163] which ultimately gave a 1560 bp open reading frame (ORF) corresponding to an enzyme of $M_r \sim 58\,728$, that incorporated an eight amino acid motif between Pro450 and Gly460 which was assumed to be involved in heme binding. Evidence for the clone's authenticity as an F5H was not obtained by direct expression of the recombinant protein and assay of its activity, but instead via complementation of the *fah1* mutant with the F5H gene under control of the CaMV 35S promoter. This manipulation was considered to restore the plant's ability to biosynthesize sinapoyl malate (**48**), but the evidence was only through TLC comparison of UV absorbing substances from the corresponding plant extract relative to an authentic standard. Accordingly, nothing is yet known about whether either ferulic acid (**6**) or feruloyl CoA (**15**) functions as the preferred substrate for this hydroxylation (cf.

p-coumarate hydroxylase), or indeed about any other aspect of substrate specificity. Nor has there, at the time of writing, been any study devoted to identifying the associated NADPH-dependent reductase.

3.18.3.3.3 *Hydroxycinnamyl and Hydroxycinnamoyl O-methyltransferases*

Distinct plant proteins catalyze the *meta*-directing, regiospecific methylation of particular hydroxycinnamic acids (EC. 2.1.1.68) and hydroxycinnamoyl CoA esters (EC. 2.1.1.104), respectively (Figure 12). Milestone achievements in the field of (plant) hydroxycinnamoyl (CoA) OMTs are summarized in Table 6.

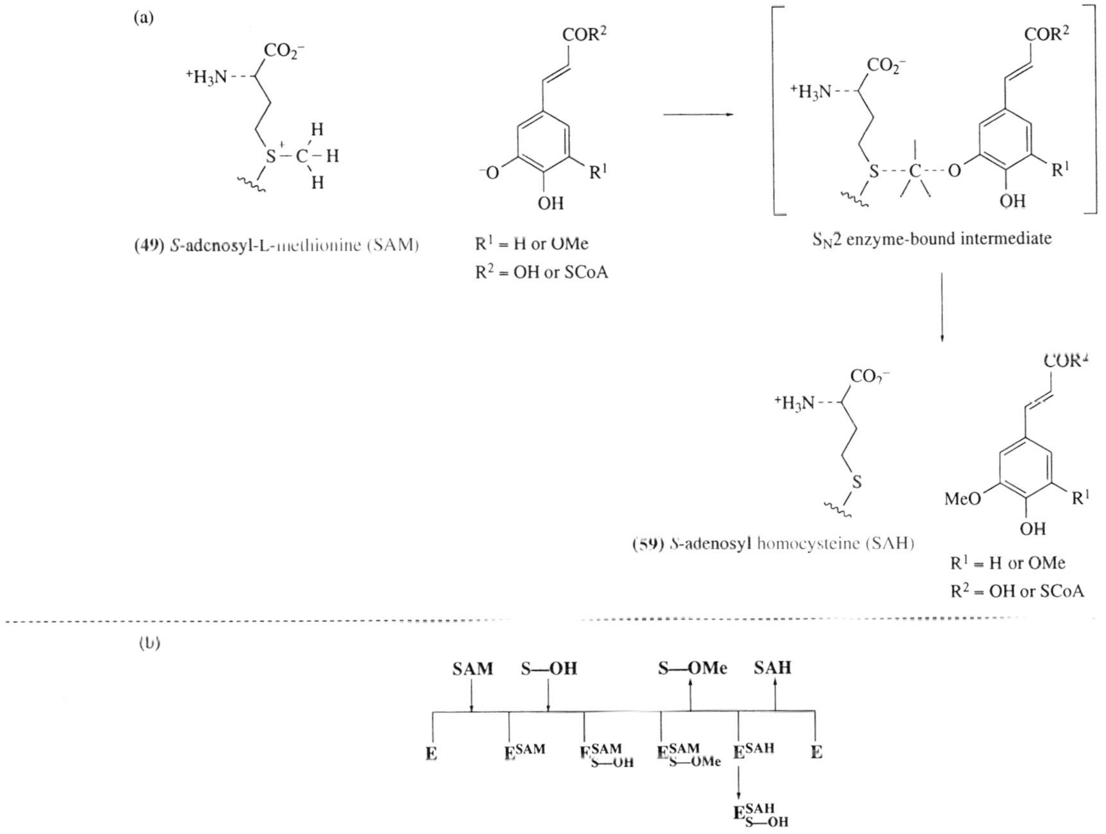

Figure 12 Proposed (a) mechanism for *O*-methylation of hydroxycinnamic acids (CoA esters) via an enzyme-bound S$_N$2 intermediate[208] and (b) ordered bi-bi mechanism according to Sharma and Brown.[202]

As before, the discovery of *O*-methyltransferase (OMT) catalyzed reactions (so-called *trans*-methylation), and recognition that *S*-adenosyl methionine (SAM) (**49**) serve as a cofactor can be traced to the mammalian literature. For example, in experiments with rabbits administered deutero-methionine, the methyl group of anserine (**50**) was established to be methionine (**51**) derived,[164] as was the methyl group of epinephrine (**52**).[165] Thereafter, Cantoni determined that the actual biological methylation agent was SAM (**49**).[166]

Recognition that the methyl group in lignins was derived from (methyl)methionine was first made in 1952 by Richard Byerrum and John Flokstra.[167,168] This determination essentially followed related studies on the origins of the methyl group in the alkaloid, nicotine (**53**), in *Nicotiana rustica*,[169,170] as well as that of the methoxyl group of ricinine (**54**) in castor beans (*Ricinus communis*),[171] and the methylenedioxy group(s) of protopine (**55**) in *Dicentra* hybrids.[172]

Thus, Byerrum and Flokstra demonstrated that D,L-[^{14}C]methionine (**51**) served as a precursor of the methoxyl groups in the lignin of barley (*H. vulgare*) plants. This was proven by cleaving the methoxyl groups with HI to afford ^{14}CH$_3$I, which was then trapped as its triethyl (^{14}CH$_3$)-methyl ammonium iodide derivative, as well as by alkaline nitrobenzene oxidative degradation to give [^{14}C]syringaldehyde (**56**).[167,168]

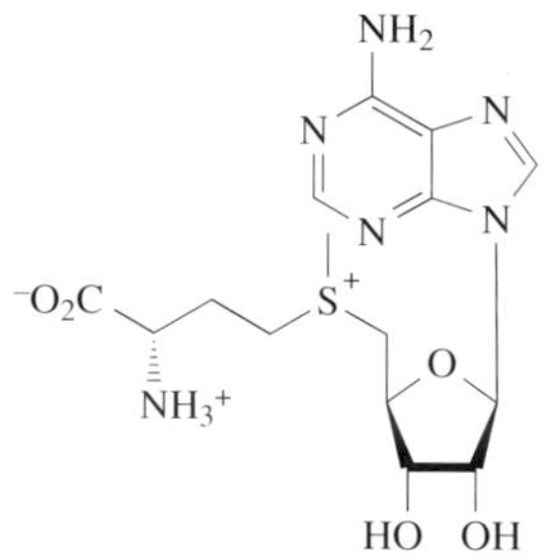

(49) *S*-Adenosylmethionine (SAM)

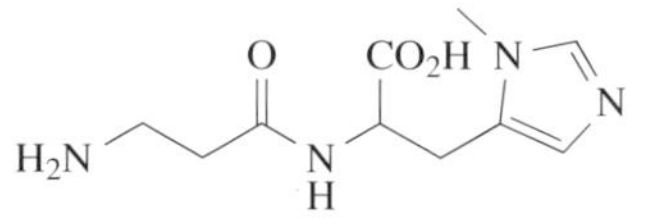

(50) Anserine

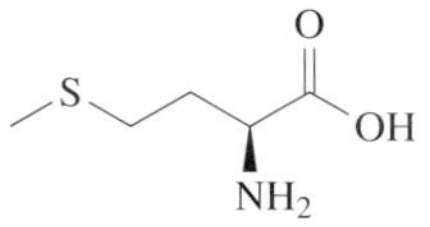

(51) Methionine

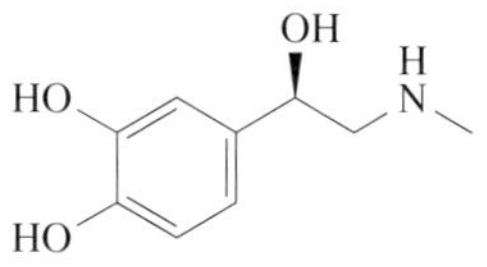

(52) Epinephrine

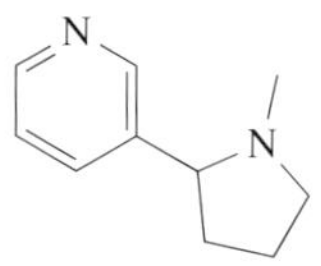

(53) Nicotine

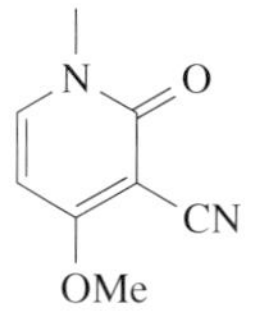

(54) Ricinine

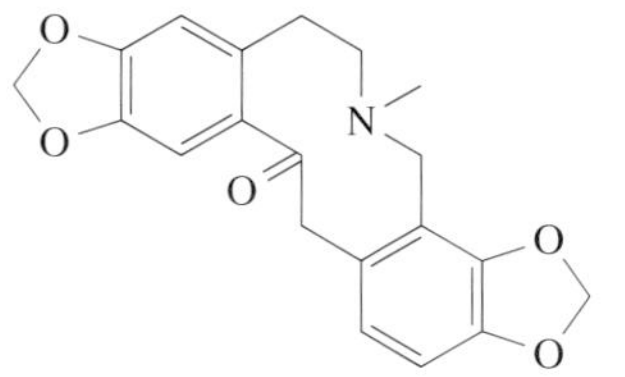

(55) Protopine

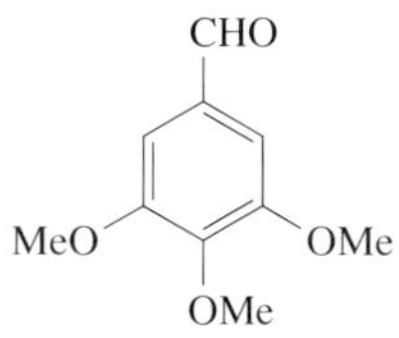

(56) Syringaldehyde

It was another ten years or so in 1954 before hydroxycinnamyl OMT activity in cell-free plant extracts was described. The studies in plants again lagged well behind mammalian studies, which had documented in 1958 the enzymatic *O*-methylation of caffeic acid (**13**) by rat liver homogenates to afford ferulic acid (**6**),[173] as well as the corresponding transformations of related substances.[174] In the subsequent plant studies, it was found that a SAM-dependent *meta*-specific conversion of caffeic acid (**13**) into ferulic acid (**6**) could be achieved with apple tree cambial sap crude cell-free extracts,[175] as well as with cell-free preparations from *Pittosporum crassifolia*[175] and pampas grass (*Cortaderia selloana*).[176] Subsequently, Dieter Hess, using acetone powders from the angiosperms, *Rosa* sp., *Petunia hybrida*, *Streptocarpas hybrida*, *Gaillarda picta sanguinea*, and *Avena sativa*, documented the differential conversion of both caffeic (**13**) and 5-hydroxyferulic (**17**) acids into ferulic (**6**) and sinapic (**7**) acids, respectively.[177] Interestingly, based on the product distribution, the relative efficacy of the *O*-methylations slightly favored formation of sinapic acid (**7**).

Approximately twenty years elapsed before a plant OMT was purified to apparent homogeneity,[178] this period having been punctuated by a series of unsuccessful attempts to purify completely OMTs from various plant sources.[179–186] On the other hand, considerable progress had been made in the purification of rat liver catechol OMT (EC. 2.1.1.6) through the expedient of employing either dithiothreitol (DTT) or thioethanol to stabilize the enzyme preparations.[187,188] Indeed, catechol OMT was readily purified to apparent homogeneity, and kinetic studies suggested, on the basis of K_m values, that a bi-bi-random mechanism was operative for the SAM-dependent methylation of epinephrine (**52**).[189]

The studies with partially purified (60-fold) OMT preparations from the plant source, soybean (*Glycine max*) cell suspension cultures, nevertheless, yielded interesting substrate specificity information (Table 7),[183,184] as well as the observation that distinct OMTs were involved in caffeic acid (**13**)/5-hydroxyferulic acid (**17**) methylation and flavonoid biosynthesis (converting luteolin (**57**) into chrysoeriol (**58**)), respectively.[183] This, therefore, extended earlier findings by Hess that angiosperm OMTs preferentially methylated 5-hydroxyferulic acid (**17**), albeit only slightly, compared to caffeic acid (**13**).[177] Analogous findings with bamboo[180] had resulted in similar conclusions, although they disconcertingly conflicted with earlier claims made by the same authors.[179]

Table 6 Hydroxycinnamyl *O*-methyltransferases (and related OMTs): milestones.

Milestone	Year	Investigator(s)	Achievement	Ref.
Existence of *trans*-methylation reactions	1941	du Vigneaud *et al.*	Discovery that anserine (**50**) methyl group in rabbits was derived from methionine (**51**)	164
Determination that methyl group of lignins was (methyl) methionine derived	1952	Byerrum and Flokstra	Radiolabeling/degradation studies showed methyl group of lignins to be methionine derived	167
Discovery of SAM as methyl donor	1953	Cantoni	Established enzymatic conversion of L-methionine (**51**), in presence of ATP, to SAM (**49**)	166
Enzymatic *O*-methylation of caffeic acid (**13**) by catechol OMT in rat liver to give ferulic acid (**6**)	1958	Pellerin and D'Iorio	Substrate specificity study with various catechols gave conversion of caffeic acid (**13**) into ferulic acid (**6**) by rat liver preparation	173
Cell-free conversion of caffeic acid (**13**) into ferulic acid (**6**) with plant extracts	1963	Finkle and Nelson	Cell-free preparations from apple cambial tissue convert caffeic acid (**13**) into ferulic acid (**6**)	175
Distinct *O*-methylation patterns of caffeic (**13**) and 5-hydroxyferulic (**17**) acids by plant cell-free preparations	1965	Dieter Hess	Recognition of dual methylation by crude OMTs, and slight substrate preference for 5-hydroxyferulic acid (**17**) compared to caffeic acid (**13**)	177
Evidence for two distinct OMTs in a plant system involved in caffeic (**13**) and 5-hydroxyferulic (**17**) acid methylation and that in flavonoid formation	1976	Poulton *et al.*	Soybean cell suspension cultures harbor a caffeic acid (**13**)/5HFA (**17**) OMT, and a distinct OMT converting luteolin (**57**) into chrysoeriol (**58**)	183, 184
Purification of first hydroxycinnamic acid OMT, and separation into three isoforms	1987	Hermann *et al.*	Purification and characterization of three tobacco OMTs	178
Discovery of caffeoyl CoA OMT, and that it utilizes only caffeoyl CoA (**14**) and not caffeic acid (**13**)	1989	Pakusch *et al.*	Parsley cell suspension cultures harbor a specific caffeoyl CoA OMT, which is also present in *Zinnia* and tobacco. Caffeic acid (**13**) is not used as substrate	190
Cloning of caffeoyl CoA OMT	1991	Schmitt *et al.*	First hydroxycinnamoyl CoA OMT cloned	193
Cloning of hydroxycinnamic acid OMT (using 5-hydroxyferulic acid (**17**) preferentially)	1991	Gowri *et al.* Bugos *et al.*	Hydroxycinnamic acid OMT clones obtained from alfalfa and aspen. Each clone was authenticated by recombinant protein assay	192 195
Ordered bi-bi mechanism	1979	Sharma and Brown	Ordered bi-bi mechanism proposed based on affinity chromatographic behavior of OMTs	202
	1989	De Carolis and Ibrahim	Characterization of caffeic acid *O*-methyltransferase	203
	1991	Pakusch and Matern	Characterization of caffeoyl CoA *O*-methyltransferase	205
X-ray crystal structure and identification of general SAM-binding domain and active site	1994	Vidgren *et al.*	Crystallization and analysis of catechol OMT	206

Table 7 Kinetic parameters for different substrates of partially purified soybean OMT.[184]

Substrate	K_m (µM)	V (pkat mg^{-1} protein)	V/K_m	Rel. V/K_m
Caffeic acid (**13**)	133	72	0.54	1.0
5-Hydroxyferulic acid (**17**)	55	142	2.60	4.8
S-Adenosyl methionine (**49**)	15	64	4.27	7.9

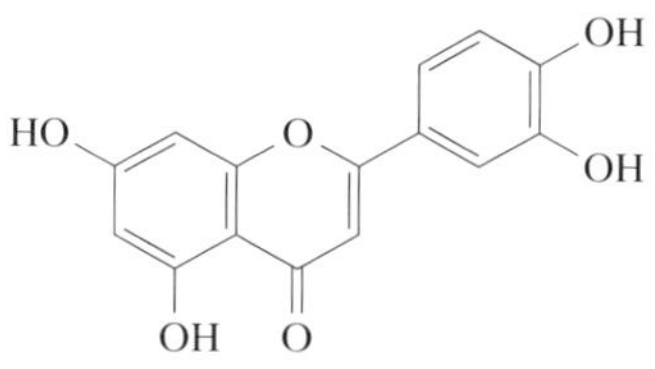

(**57**) Luteolin

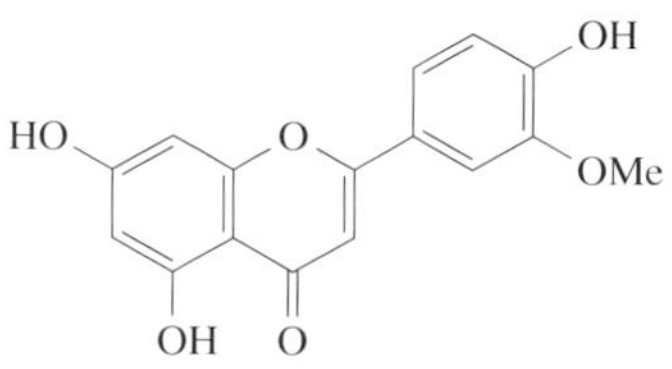

(**58**) Chrysoeriol

Next, two partially purified (4–90-fold) OMT preparations from the gymnosperms, Japanese black pine (*Pinus thunbergii*)[181] and cedar (*Thuja orientalis*),[186] were reported to utilize caffeic acid

(13) more effectively as a methylation substrate compared to 5-hydroxyferulic acid (17) (Table 8). These findings were then conscripted by the authors in an attempt to explain the differences in monomer residue composition between the gymnosperm (predominantly guaiacyl) and angiosperm (primarily guaiacyl–syringyl) lignins. These arguments were put forward even though the differences in relative V/K_m were less than an order of magnitude (and hence insignificant) and before any insight had been developed about how the preceding hydroxylation reactions occurred to give caffeic (13) and 5-hydroxyferulic (17) acids.

Table 8 Kinetic parameters for different substrates of partially purified *Thuja* OMT preparations.[186]

Substrate	K_m (µM)	V (pkat mg^{-1} protein)	V/K_m	Rel. V/K_m
Caffeic acid (13)	130	130	1.0	4.35
5-Hydroxyferulic acid (17)	330	77	0.23	1.00

Substantial progress in the hydroxycinnamyl OMT field, which now appears to have yielded some clarification about the circumstances operative *in vivo* in both angiosperms and gymnosperms, had to await purification and gene cloning of the enzymes themselves. In this regard, the first bona fide demonstration of purification to apparent homogeneity for a plant hydroxycinnamic acid OMT was achieved with the isolation of three distinct forms of the enzyme ($M_r \sim 29$, 42, and 43 kDa) from tobacco.[178] Shortly thereafter, Pakusch *et al.*[190,191] made an important discovery to the effect that a dimeric 48 kDa OMT, isolated from pathogen elicited parsley (*Petroselinum crispum*) cell suspension cultures, utilized caffeoyl CoA (14) as a substrate. This implied the existence of two quite distinct OMT types, one preferentially methylating caffeoyl CoA (14) and the other 5-hydroxyferulic acid (17).

That two different OMTs did indeed exist was subsequently established in the more or less contemporaneously submitted reports about the cloning of hydroxycinnamate OMT genes from alfalfa (*Medicago sativa*),[192] parsley (*P. crispum*),[193,194] aspen (*Populus tremuloides*),[195] and poplar (*Populus trichocarpa × Populus deltoides*).[196,197] Authenticity of the hydroxycinnamyl OMT genes from alfalfa and aspen were demonstrated through expression of their recombinant proteins in *Escherichia coli*. As expected from the earlier work of Hess,[177] the recombinant hydroxycinnamic acid OMTs demonstrated a slight substrate preference for 5-hydroxyferulic acid (17) compared to caffeic acid (13) ($\sim 2:1$). Significantly, in addition to the OMT utilizing caffeoyl CoA (14) in parsley, comparable genes were also found in *Zinnia elegans*[198,199] and tobacco.[200] *Zinnia*, however, also possessed a hydroxycinnamic acid OMT,[199] which preferentially utilized 5-hydroxyferulic acid (17), in agreement again with the idea that two quite distinct OMTs are operative *in planta*. Interestingly, the genes encoding the hydroxycinnamic acid OMT and hydroxycinnamoyl CoA OMT display little sequence identity and homology with respect to each other.

These data, therefore, could be interpreted as suggesting that methylation of caffeoyl CoA (14) affords feruloyl CoA (15), which is then metabolized into coniferyl alcohol (2), whereas formation of sinapic acid (7) occurs via methylation of 5-hydroxyferulic acid (17). Indeed, additional evidence for this suggestion has been obtained from an analysis of lignin-modified brown-rib maize (*Z. mays*) mutants and results from other transgenic plant forms.

More recently, a third class of hydroxycinnamate OMT has been proposed to exist in Monterey pine (GenBank accession number U70873) and loblolly pine (*P. taeda*),[201] respectively. The latter purportedly effectively converts both caffeic (13) and 5-hydroxyferulic (17) acids and their corresponding CoA esters (14) and (18) into the resulting methylated derivatives (6), (7), (15), and (19), and thus would differ from all OMTs described before. However, evidence for this claim relied solely upon thin-layer chromatographic analysis of the putative methylation products that were formed during incubation with either crude loblolly pine cell-free extracts or with crude recombinant protein expressed in *Pichia pastoris*.[201] It was moreover quite disconcerting that titration of presumed antibodies raised against this protein only partially inhibited (16–61%) the purported hydroxycinnamate OMT, the extent of which varied with the substrate used. This contrasts markedly with typical antibody titrations where essentially complete inhibition is obtained, e.g., the >98% inhibition for cinnamate-4-hydroxylase, as described in Section 3.18.3.3.2. These observations, therefore, raise questions about the ultimate significance of the OMT findings. Furthermore, it was asserted that this gymnosperm OMT must differ from that described by Shimada *et al.* in 1972,[181] since the ratio of products derived from methylation of caffeic acid (13) and 5-hydroxyferulic (17) acid were 1.4:1 and 3.3:1, respectively. While such differences are too small for any meaningful operational distinctions to be made, the authors had even confused the work with Japanese black pine (*P. thunbergii*)[181] as being the same as that with loblolly pine (*P. taeda*).[201] It has already been pointed out that different gymnosperm species differentially methylate caffeic acid (**13**) and 5-hydroxyferulic acid (**17**) at rates giving ratios ranging from 1.12:1 to 33.3:1![40]

In an earlier study by Flohe and Schwabe (1970), it was concluded that a random bi-bi mechanism was operative as far as the action of catechol OMTs is concerned.[189] This working hypothesis was challenged, however, in 1979 by Sharma and Brown[202] in a study of OMTs present in *Ruta graveolens*. That study employed various affinity chromatography matrices containing AH-Sepharose-4B linked to *S*-adenosyl-L-homocysteine (SAH) (**59**), 5-(3-carboxypropanamide)xanthotoxin (**60**), and ferulic acid (**6**) ligands, respectively. The elution profiles of three distinct OMTs, namely furanocoumarin 5- and 8-OMTs as well as caffeic acid OMT, revealed that binding of SAM (**49**) induces binding of the specific phenolic substrates. Accordingly, Sharma and Brown provisionally suggested that an ordered bi-bi mechanism must be operative as shown in Figure 12 while emphasizing that "although we believe the order of product released to be the more probable one in the light of available evidence, we do not maintain that it has yet been rigorously established."

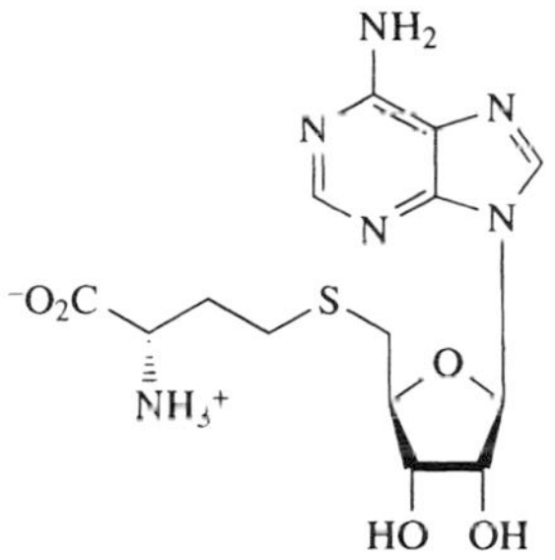

(**59**) *S*-Adenosylhomocysteine (SAH) 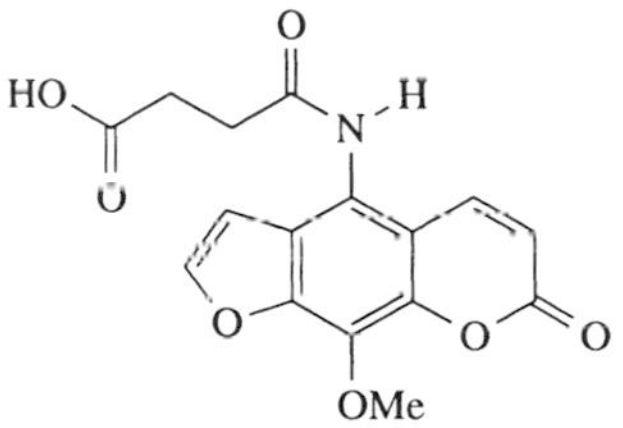 (**60**) 5-(3-carboxypropanamide)xanthoxin

That an ordered bi-bi mechanism involving initial SAM (**49**) binding was, indeed, operative was subsequently confirmed by the work of De Carolis and Ibrahim[203] in a detailed study of 5-hydroxyferulic acid OMT (K_m – 20 μM and 55 μM for 5-hydroxyferulic acid (**17**) and SAM (**49**), respectively) from cabbage (*Brassica oleracea*). Similar conclusions were reached later by Meng and Campbell.[204] Kinetic analysis of the corresponding caffeoyl CoA OMT from parsley cell suspension cultures also demonstrated that an ordered bi-bi mechanism is in effect for caffeoyl CoA methylation, thereby suggesting a consistent overall mechanistic pattern for these enzymes.[205]

Furthermore, although no X-ray structure for a plant OMT had yet become available at the time of writing, the Mg^{2+} and SAM-binding domains and active site had been identified from both the X-ray structure of rat liver catechol OMT[206] and comparisons of consensus sequences of rat, human and pig catechol OMTs. Actually, the X-ray crystal structure of a DNA methyl transferase exhibits a similar SAM (**49**) binding domain as well.[207]

Catechol OMT crystallization was carried out in the presence of a catechol OMT competitive inhibitor (3,5-dinitrocatechol (**61**)), Mg^{2+}, and SAM (**49**) at 2.0 Å resolution, and Figure 13 shows a stereo view of its active site when complexed with 3,5-dinitrocatechol (**61**). As can be seen, Mg^{2+} complexation occurs via coordination with the side-chain oxygens of Asp141, Asp169, and Asp170. Although not shown, the adenine group of SAM (**49**) (which binds after Mg^{2+}) has van der Waals contacts/interactions with Trp143, His142, and Met91, with the ribose ring lying parallel to Trp143. Hydrogen bonding of the methionine carboxyl oxygens occurs with the backbone Val42 nitrogen and H_2O, whereas the methionine amino group is also hydrogen bonded to Asp141, Gly66 (main chain oxygen), and Ser72 (side chain oxygen). Other interactions include orientation by Met40 of the sulfur containing the electrophilic methyl group of SAM (**49**) near to the nucleophilic phenol group of the catechol. Both phenolic hydroxyl groups are then complexed with Mg^{2+}, bringing one phenolic group juxtapositioned to the donor methyl group of SAM (**49**) (Figure 13). This phenolic group is, therefore, surrounded by three positively charged groups (Mg^{2+}, S—$Ch_3^{[+]}$, and Lys144), thereby promoting phenoxide ion formation, whereas the proton of the other phenolic hydroxyl group is stabilized by Glu199 to prevent anion formation. Mechanistically, methyl transfer from the sulfur atom of SAM (**49**) to the phenoxide ion involves an S_N2-like transition state,[208] and methylation occurs with inversion of configuration as demonstrated using chirally labeled

[C¹H²H³H]SAM.[209] That 56 plant SAM-dependent OMTs harbor similar SAM-binding domains has been established by standard computational sequence comparisons.[210]

(a)

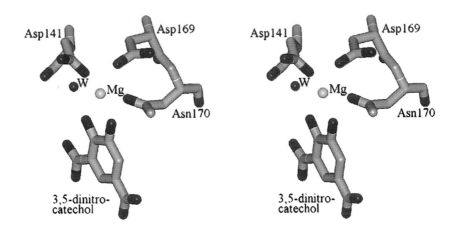

(b)

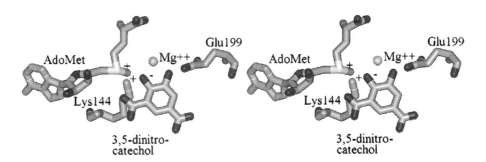

Figure 13 Stereo view of the active site of catechol *O*-methyltransferase. The enzyme was crystallized in complex with Mg²⁺ and the inhibitor 3,5-dinitrocatechol (**61**), which were clearly defined in the electron density maps. (a) The binding site for Mg²⁺. Here the substrate position is occupied by 3,5-dinitrocatechol; the sphere marked "W" represents a water molecule. (b) Binding of 3,5-dinitrocatechol in the active site of catechol *O*-methyltransferase. The most important residues that interact with the bound inhibitor are shown (reproduced with permission from *Nature*, 1994, **368**, 354).[206]

(**61**) 3,5-Dinitrocatechol

3.18.3.3.4 *Hydroxycinnamoyl CoA ligases*

This type of enzyme (EC. 6.2.1.12) catalyzes the stepwise conversion of hydroxycinnamic acids *in planta* into the corresponding CoA esters via the intermediate AMP derivatives. This kind of reaction was demonstrated to occur in mammalian systems in 1953,[211] with the report that a pig kidney cortex soluble preparation, in the presence of ATP/CoASH, converted benzoic acid (62) into benzoyl CoA (63). The initial discovery of the activation of organic acids as their CoA esters was, in turn, traceable to the study of Fritz Lipmann (see Lipmann[212] for review), as was the determination of the structure of CoASH. A mechanism of CoA ligation was proposed by Berg (1956) in the formation of acetyl coenzyme A.[213] It was suggested that ATP first binds to the enzyme followed by substrate (acetate), with pyrophosphate being released with the adenosyl 5′-acylphosphate remaining enzyme bound. Subsequent binding of CoASH thereupon resulted in release of acetyl CoA and AMP.

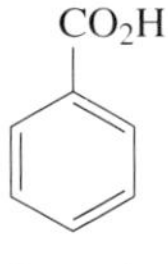

CO$_2$H

COSCoA

(62) Benzoic acid    (63) Benzoyl CoA ester

Figure 14 shows the overall biochemical mechanism of 4-coumaroyl CoA ligation involving the three substrates and forming three products, while the expository developments in the study of this enzyme are summarized in Table 9. 4-Coumarate:CoA ligase belongs to the carboxylic acid:CoA ligase subfamily including acetyl CoA synthetase; O-succinyl:CoA ligase; long-chain fatty acid acyl-CoA synthetase, bile acid CoA ligase; carnitine:CoA ligase; acyl CoA synthetase and 4-chlorobenzoate:CoA ligase.[214]

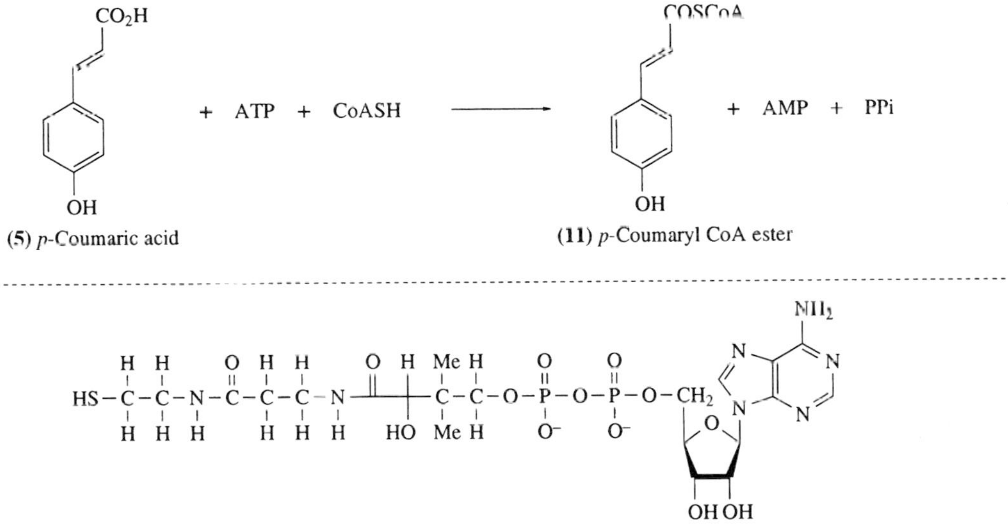

CO$_2$H

+ ATP + CoASH   ⟶   + AMP + PPi

COSCoA

OH   OH

(5) p-Coumaric acid   (11) p-Coumaryl CoA ester

Coenzyme A

Figure 14   Representation of the overall CoA ligation reaction.

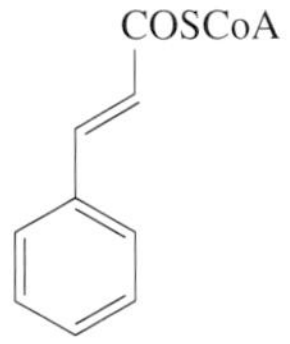

COSCoA

(64) Cinnamoyl CoA ester

Neish and co-workers[53] originally suggested that hydroxycinnamoyl CoA esters (11), (15), and (19) might be intermediates in the reduction of the hydroxycinnamic acids (5), (6), and (7) to the corresponding monolignols (1), (2), and (3). This suggestion was not, however, substantiated by either of two initial experimental reports.[215,216] Higuchi and Brown,[215] for example, thought that acetone powders from cambial cell-free extracts of *Pinus strobus* and wheat (*Triticum aestivum*) were able to catalyze the reduction of ferulic acid (6) directly into coniferyl alcohol (2) via coniferyl aldehyde (16). This result, however, could not be duplicated.[217,218]

Table 9 Hydroxycinnamoyl CoA ligases (and related enzymes): milestones.

Milestone	Year	Investigator(s)	Achievement	Ref.
Determination of CoASH structure and function	1954	Lipmann	Established CoASH structure	212
Intermediacy of AMP intermediate during CoA ester formation	1956	Berg	Identification of steps involved in CoA ligase reaction affording acetyl CoA	213
Proposed involvement of hydroxycinnamoyl CoA esters	1959	Brown *et al.*	Adapted from analogous observations in other biochemical pathways	53
	1966	Gross and Zenk	Indirect evidence for involvement of hydroxycinnamoyl CoA esters	219
Hydroxycinnamate CoA ligation reaction demonstrated *in plants*	1970	Walton and Butt	Conversion of cinnamic acid (**10**) into cinnamoyl CoA (**64**)	220
	1970	Hahlbrock and Grisebach	CoA ligation of various hydroxycinnamic acids with partially purified enzyme preparations	221
Cloning of near full-length *p*-coumaryl CoA ligase	1987	Douglas *et al.*	Cloning of near full length cDNAs for two 4-CL isoenzymes	232
	1988	Lozoya *et al.*		233
Definition of ATP/AMP binding domains	1991	Becker-André *et al.*	Sequence homology comparison	238

That the CoA esters were, in fact, intermediates on the way to the monolignols was established in a series of papers beginning with the synthesis of hydroxycinnamic acid CoA esters (**11**), (**15**), and (**19**), which provided indirect evidence for their involvement as intermediates in monolignol formation.[219] A very preliminary study[220] next suggested that cinnamic acid (**10**), in the presence of ATP and CoASH, was converted into cinnamoyl CoA (**64**) using cell-free extracts from spinach beet (*Beta vulgaris*). Subsequent investigations with cell-free preparations from parsley[221] demonstrated the conversion of cinnamic (**10**), *p*-coumaric (**5**), and ferulic (**6**) acids into their CoA esters (**64**), (**11**), and (**15**). During 1972–73 the reduction of ferulic acid (**6**) to coniferyl alcohol (**2**) was further confirmed to occur through the intermediacy of feruloyl CoA (**15**) using cell-free extracts from *Salix alba* and *Forsythia* tissues, respectively.[218,222] These studies were followed by the work of Rhodes *et al.* who obtained an initial set of kinetic data for hydroxycinnamoyl CoA (**11**) and (**15**) formation using cell-free extracts from *Brassica napo-brassica* root tissue (K_m = 14, 31, 19, 21, and 2.6 μM with respect to *p*-coumaric acid (**5**), ferulic acid (**6**), ATP, Mg^{2+}, and CoA, respectively).[223]

The next decade or so witnessed several attempts to purify a hydroxycinnamoyl CoA ligase to apparent homogeneity, but purification was only really beginning to be achieved in 1981 and 1983 using parsley (*P. hortense*)[224] and spruce (*Picea abies*)[225] preparations, respectively. Nevertheless, the earlier studies with partially purified enzyme samples afforded several interesting findings. These included the observation that soybean cell suspension cultures[226] and petunia (*Petunia hybrida*)[227] both harbored two distinct hydroxycinnamoyl CoA ligase isoenzymes; that the fungus *Polyporus hispidus*[228] contained light-inducible hydroxycinnamate CoA ligase activity; and that CoA ligase activity was apparently absent in *Sphagnum* (bryophyte) and *Equisitum*, *Pteris wimsetti*, and *Selaginella kraussiana* (pteridophyte) tissues.[229] This latter finding was quite unexpected given that pteridophytes were assumed to contain lignins and lignans. Interestingly, sinapic acid (**7**) did not serve as a substrate for CoA ligation using partially purified CoA ligase preparations from *Acer saccharinum*, *Forsythia suspensa*, and *Taxus baccata*,[229] an observation that extended to the CoA ligase from parsley.[230] The latter, however, was also assayed with 5-hydroxyferulic acid (**17**) which was effectively converted into 5-hydroxyferuloyl CoA (**18**). These findings would, therefore, seem to suggest that 5-hydroxyferuloyl CoA (**18**) also serves as a methylation substrate during the formation of syringyl residues in lignins.

The plant 4-coumarate CoA ligases have $M_r \sim$ 55 000–67 000 depending upon the origin of the enzyme. Although partial clones were obtained at first,[231] near full length cDNAs for 4CL were secured in 1988 for the two 4CL isoenzymes of parsley.[232,233] By comparison, the primary structure of succinyl CoA synthetase in *E. coli* was obtained in 1985,[234] with its X-ray structure also established.[235,236] Unfortunately, to date no X-ray crystal structure for any 4-coumarate CoA ligase has yet been obtained, and thus no explicit understanding of how its catalytic mechanism is specifically achieved is available. Nevertheless, from comparisons of numerous ATP-binding enzymes in both prokaryotes and eukaryotes, putative AMP- and ATP-binding domains have been deduced,[237] i.e., Gly Glu Ile Cys Ile Arg Gly and Tyr (Phe) Ser Ser Gly Thr Thr Gly Leu Pro Lys Gly, the cysteine residue of the former being assumed to participate directly in ATP binding, although this has not been proven.[238]

Surprisingly little has also been done rigorously to define the precise biochemical conversions associated with 4-coumarate ligase catalysis. On the other hand, a study with 4-chlorobenzoate:CoA ligase established the intermediacy of the acyl adenylate, identified by characterization of the 4-chlorobenzoic acid AMP adduct obtained via ligase-catalyzed reaction of MgATP and 4-chlorobenzoic acid (**65**).[239] In the presence of CoASH, this is then converted into the corresponding CoA ester (**66**) (Scheme 8). An analogous situation may be occurring for *p*-coumarate CoA ligase, but this needs to be confirmed.

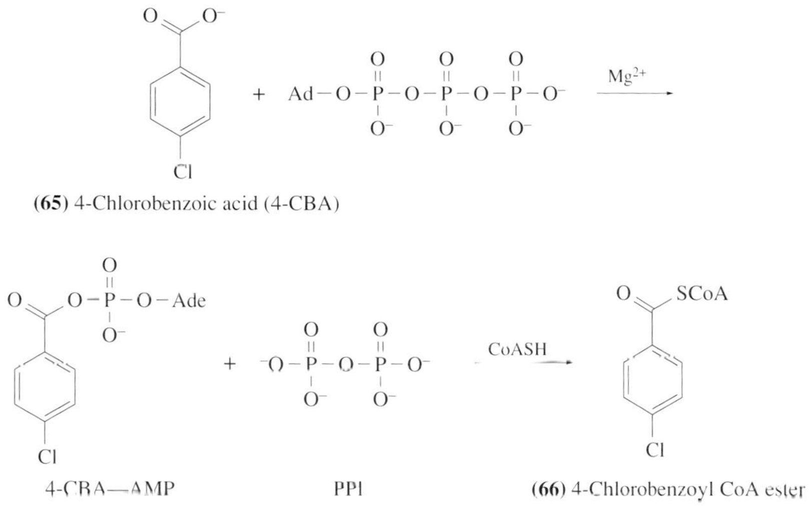

(**65**) 4-Chlorobenzoic acid (4-CBA)

4-CBA—AMP   PPi   (**66**) 4-Chlorobenzoyl CoA ester

Scheme 8

Lastly, in recognizing that different plant species harbor more than one 4CL isoenzyme, various isoforms from soybean,[240] pea (*Pisum sativum*),[241] petunia,[242] parsley,[233] and poplar[243] were examined for their substrate specificities with respect to *p*-coumaric (**5**), caffeic (**13**), and ferulic (**6**) acids. These studies revealed that a particular isoenzyme in soybean,[240] petunia,[242] and poplar[243] was capable of ligating *p*-coumaric (**5**) and caffeic (**13**) acids but not sinapic acid (**7**). On the other hand, other isoenzymes in these species effectively ligated sinapic acid (**7**), implying that different isoenzymes fulfill distinct physiological functions.

3.18.3.3.5 *Cinnamoyl CoA:NADP-dependent oxidoreductases*

Cinnamoyl CoA:NADP oxidoreductase (EC. 1.2.1.44) catalyzes the four-electron, NADPH-dependent reduction of *p*-hydroxycinnamoyl CoA derivatives (**11**), (**15**), and (**19**) to afford the corresponding *p*-hydroxycinnamaldehydes (**12**), (**16**), and (**20**). Scheme 9 shows the overall mechanism of hydride transfer, and Table 10 highlights the more important stages through which the study of this class of enzymes has developed. As for all enzymes described so far in the monolignol forming pathway, much of the knowledge about this class of oxidoreductases is derived from extensive studies on mammalian, yeast, and bacterial systems. For example, a hamster gene encoding 3-hydroxy-3-methylglutaryl coenzyme A reductase (HMGCoA reductase) had become available by 1984.[244]

Following the original suggestion that hydroxycinnamoyl CoA esters (**11**), (**15**), and (**19**) might serve as intermediates in the reduction of the carboxylic acids to the monolignols (**1**)–(**3**),[53] Mansell *et al.* subsequently demonstrated the ATP/CoASH/Mg^{2+}/NADPH-dependent conversion of ferulic acid (**6**) into coniferyl alcohol (**2**) using *Salix alba* cell-free extracts.[218] Next, Gross *et al.*[222] discovered the existence of cinnamoyl CoA:NADP oxidoreductase, initially named feruloyl CoA reductase, which was first detected in *Salix alba* cell-free extracts and which catalyzed the NADPH-dependent conversion of feruloyl CoA (**15**) into coniferyl aldehyde (**16**). Using a *Forsythia* cell-free preparation, Gross and Kreiten[245] then demonstrated that cinnamoyl CoA:NADP oxidoreductase was a type B reductase, since it abstracted [^{3}H] only from [4B-^{3}H]NADPH and not the [4A-^{3}H]NADPH (Scheme 9). Designation of type A and type B reductases stemmed from elegant work in the early 1950s on the stereospecificity of the NADPH-dependent reductive processes catalyzed by alcohol dehydrogenase[246,247] where H$_A$ = proR and H$_B$ = proS hydrogens, respectively.

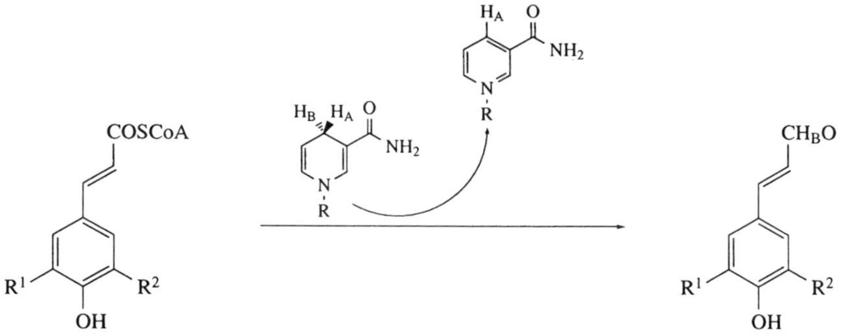

(11) *p*-Coumaroyl CoA ester, $R^1 = R^2 = H$
(15) Feruloyl CoA ester, $R^1 = H, R^2 = OMe$
(19) Sinapoyl CoA ester, $R^1 = R^2 = OMe$

(12) *p*-Coumaryl aldehyde, $R^1 = R^2 = H$
(16) Coniferyl aldehyde, $R^1 = H, R^2 = OMe$
(20) Sinapyl aldehyde, $R^1 = R^2 = OMe$

Scheme 9

Table 10   Cinnamoyl CoA:NADP oxidoreductase: milestones.

Milestone	Year	Investigator(s)	Achievement	Ref.
Postulate that hydroxycinnamoyl CoA esters are intermediates leading to monolignols	1959	Brown *et al.*	Recognition of analogy to other aliphatic CoA derivatives	53
Demonstrated intermediacy of ferulic acid (**6**) in coniferyl alcohol (**2**) biosynthesis	1972	Mansell *et al.*	Conversion of ferulic acid (**6**) to coniferyl alcohol (**2**) by *Salix* cell-free extracts	218
Discovery of cinnamoyl CoA:NADP oxidoreductase	1973	Gross *et al.*	Conversion of feruloyl CoA (**15**) into coniferyl aldehyde (**16**)	222
Type B reductase classification	1975	Gross and Kreiten	NADPH-dependent reductase abstraction of [4B-³H]NADPH during feruloyl CoA (**15**) reductio in	245
Purification of cinnamoyl CoA:NADP oxidoreductase	1976	Wengenmayer *et al.*	Cinnamoyl CoA reductase partially purified (1600-fold) from soybean (*Glycine max*) cell suspension culture	248
	1981	Lüderitz and Grisebach	*Picea abies* cambial sap cinnamoyl CoA reductase purified 4084-fold; single band in SDS–PAGE	249
	1984	Sarni *et al.*	Poplar stems purified 595-fold; single band in SDS–PAGE	250
Existence of cinnamoyl CoA:NADP oxidoreductase isoforms and differences in substrate specificity	1981	Lüderitz and Grisebach	Soybean and spruce contain different isoforms; soybean and spruce enzymes have very different specificities for sinapoyl CoA (**19**)	249
Cloning of first CoA ester reductase gene (3-hydroxy-3-methylglutaryl CoA reductase, HMG CoA reductase)	1984	Reynolds *et al.*	Bacteriophage λ genomic library from hamster UT-1 cells gave first HMG CoA reductase	244
X-ray crystal structure of HMG CoA reductase	1995	Lawrence *et al.*	*Pseudomonas mevalonii* HMG CoA reductase crystal structure to 3.0 Å resolution	253
Cloning of first cinnamoyl CoA:NADP oxidoreductase, and identification of putative NADPH and substrate binding domains	1997	Lacombe *et al.*	First plant cinnamoyl CoA:NADP oxidoreductase, and was isolated from *Eucalyptus gunnii*	252

Over the next decade, three cinnamoyl CoA:NADP oxidoreductases, possessing $M_r \sim 36\,000$–$38\,000$ and lacking discrete subunits, were purified to apparent homogeneity from soybean (*G. max*) cell suspension cultures,[248] spruce (*P. abies*) cambial sap,[249] and poplar (*Populus euramericana*) stems,[250] respectively. In all three examples, both gel permeation chromatography data and SDS–PAGE analysis suggested that the CoA reductases were monomers. The relative substrate specificities of spruce (gymnosperm) and soybean (angiosperm) cinnamoyl CoA:NADP oxidoreductase[249] were also compared. As shown in Table 11, the oxidoreductases from both sources displayed the following order of substrate preference: feruloyl CoA (**15**) > sinapoyl CoA (**19**) > *p*-coumaroyl

CoA (**11**), where the largest difference lay in the relative rate of sinapoyl CoA (**19**) conversion. That is to say, the gymnosperm and angiosperm cinnamoyl CoA reductases, displayed differences that were greater than two orders of magnitude from one another when sinapoyl CoA (**19**) was used as a substrate. Interestingly, *p*-coumaryl CoA (**11**) was not a particularly effective substrate in comparison with either of the other two CoA esters (**15**) or (**19**).

Table 11 Kinetic parameters for cinnamoyl CoA:NADP oxidoreductase.[249]

Source of enzyme	*Substrate*	K_m (µM)	V (nkat mg^{-1} protein)	V/K_m
Spruce cambial sap	Feruloyl CoA (**15**)	5.2	404	78.0
	NADPH	2.3		176.0
	Sinapoyl CoA (**19**)	28.6	44	1.5
	NADPH	4.0		11.0
	p-Coumaroyl CoA (**11**)	43.5	27	0.6
	NADPH	3.0		9.0
Soybean cells	Feruloyl CoA (**15**)	3.3	1690	512.0
	NADPH	2.3		735.0
	Sinapoyl CoA (**19**)	2.1	464	221.0
	NADPH	2.2		211.0
	p-Coumaroyl CoA (**11**)	15.5	54	3.5
	NADPH	3.4		16.0

Some years later, cinnamoyl CoA:NADP oxidoreductase was again isolated, in this case from *Eucalyptus gunnii*,[251] and cloned for the first time.[252] Its gene has an open reading frame (ORF) of 1008 nucleotides, which encodes a 336 amino acid polypeptide chain corresponding to $M_r \sim 36\,500$; this sequence apparently displayed 88.5% and 85.9% overall homology to the corresponding proteins from alfalfa and maize, although no sequence data were provided. Functional expression of the *Eucalyptus* reductase was also demonstrated in *E. coli* BL 21 (DE 3) cells,[252] hence verifying its gene sequence, with K_m values of 35, 28, 20, and 19 µM being reported with respect to feruloyl (**15**), sinapoyl (**19**), and *p*-coumaroyl (**11**) CoA esters, and NADPH, respectively. Comparison of its amino acid sequence with other reductases revealed a 35 amino acid domain, close to the N-terminal, that is putatively involved in NADPH binding. A substrate binding domain at Lys158 and/or Lys165 and Cys162 was also proposed, but this will require proof through, for example, site-directed mutagenesis.

At the present time, there is no X-ray structure for any cinnamoyl CoA:NADP oxidoreductase to facilitate further elucidation of how the enzyme actually catalyzes this reductive process. Nevertheless, consideration of the X-ray structure (3 Å resolution) for *Pseudomonas mevalonii* 3-hydroxy-3-methylglutaryl coenzyme A (HMG-CoA) reductase, which was obtained in 1995, may offer some important insights.[253] The HMG-CoA reductase was found to be a dimer possessing two open cavities at the dimer interface, these embodying two active sites. (According to the authors, the dimeric configuration was viewed as being essential for catalytic activity, whereas cinnamoyl CoA:NADP oxidoreductase is believed to be monomeric.) Five important domains were identified in the HMG-CoA reductase, specifically Glu X X X Gly X X X X Pro (for HMG-CoA binding), a Glu83 loop (for catalysis involving Glu83 and HMG-CoA binding), Asp Ala Met Gly X Asn and Gly X X Gly Gly X Thr (for NAD(H) binding), and a His381 residue (for catalysis). The X-ray structure revealed that, when HMG-CoA binds, the scissile bond of the thioester is also within reach of the glutamyl side-chain of Glu83, and that binding of the NADH places the nicotinamide ring adjacent to the thioester's scissile bond. Indeed, at the active site, both the HMG-CoA and C-4 of the nicotinamide ring are within 3.5 Å of each other, with both the Glu83 and His381 residues being implicated in catalysis.[254–257] Two other noteworthy observations were made: first, the HMG-CoA reductase lacked the usual dinucleotide binding fold, which was replaced with a four-stranded antiparallel β-sheet with two helical cross-overs lying across the side of the structure: the second was that HMG-CoA reductases from higher eukaryotic organisms are activated by phosphorylation of a single serine moiety located only six residues from the catalytic histidine.[258–260] Accordingly, from these findings some interesting developments can be anticipated; certainly it will be instructive to compare any forthcoming X-ray structure for cinnamoyl CoA:NADP oxidoreductase with HMG-CoA reductase and any others that become available.

3.18.3.3.6 Cinnamyl alcohol dehydrogenases

The cinnamyl alcohol dehydrogenases (CAD, EC. 1.1.1.195) convert *p*-hydroxycinnamaldehydes (**12**), (**16**), and (**20**) into *p*-coumaryl (**1**), coniferyl (**2**), and sinapyl (**3**) alcohols, thus catalyzing the final step in monolignol biosynthesis. Scheme 10 shows the overall two electron hydride transfer mechanism operative for cinnamyl alcohol dehydrogenase. It is a Type A reductase, abstracting the pro-*R* hydride from NADPH (H_A = pro-*R*, H_B = pro-*S*). Table 12 highlights some of the more important stages in the study of this class of enzymes. Since CAD belongs to the ubiquitous family of alcohol dehydrogenases, a very substantial body of literature about this enzyme class existed prior to the discovery of CAD during 1972–73.[218,261]

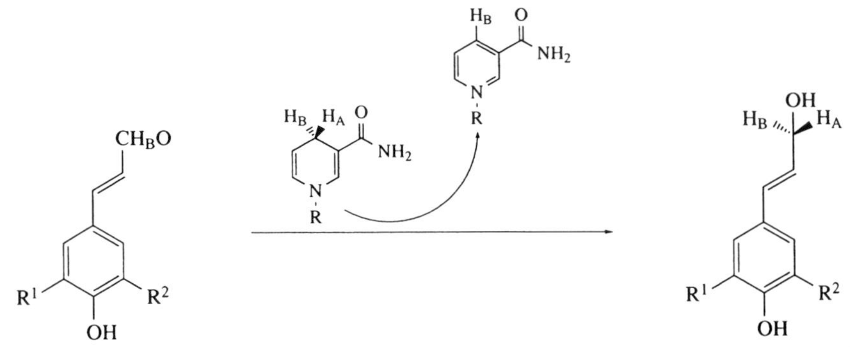

(12) *p*-Coumaryl aldehyde, $R^1 = R^2 = H$
(16) Coniferyl aldehyde, $R^1 = H, R^2 = OMe$
(20) Sinapyl aldehyde, $R^1 = R^2 = OMe$

(1) *p*-Coumaryl alcohol, $R^1 = R^2 = H$
(2) Coniferyl alcohol, $R^1 = H, R^2 = OMe$
(3) Sinapyl alcohol, $R^1 = R^2 = OMe$

Scheme 10

Table 12   Cinnamyl alcohol dehydrogenase: milestones.

Milestone	Year	Investigator(s)	Achievements	Ref.
Postulate that hydroxycinnamaldehydes are monolignol intermediates	1959	Brown *et al.*	Demonstration based on radiolabeling experiments	53
Intermediacy of ferulic acid (6) in coniferyl alcohol (2) biosynthesis	1972	Mansell *et al.*	Conversion of ferulic acid (6) to coniferyl alcohol (2) by *Salix alba* cell-free extracts	218
Discovery of cinnamyl alcohol dehydrogenase (CAD)	1973	Gross *et al.*	Conversion of coniferyl aldehyde (16) into coniferyl alcohol (2)	222
Purification to apparent homogeneity	1974	Mansell *et al.*	CAD purified (604-fold) from *Forsythia suspensa*	262
X-ray crystal structure of horse liver alcohol dehydrogenase (HADH)	1970	Brändén *et al.*	Low resolution model of HADH	286 287
	1973	Brändén *et al.*	High resolution model of HADH	
Established Type A reductase nature of CAD	1974	Mansell *et al.*	³H from [4A-³H]NADPH was abstracted during reduction of coniferyl aldehyde (16) to coniferyl alcohol (2)	262
Reverse reaction of CAD removes pro-*R* hydrogen of coniferyl alcohol (2)	1978	Klischies *et al.*	pro-*R* hydrogen removed from coniferyl alcohol (2) in agreement with horse liver and yeast alcohol dehydrogenase	263, 264
Profound substrate differences between CAD of (woody) gymnosperms and angiosperms	1981	Lüderitz and Grisebach	Spruce CAD utilizes sinapyl aldehyde (20) quite poorly relative to soybean which uses both coniferyl (16) and sinapyl (20) aldehydes almost equally well	249
First authentic clone reported for CAD[a]	1992	Knight *et al.*	Purified CAD from tobacco, and cloned a putative CAD. Gene sequence comparison with yeast and *Aspergillus* ADH1	279
Verification of catalytic activity of recombinant CAD2 (from *Eucalyptus*)	1993	Grima-Pettenati *et al.*	Use of tobacco CAD cDNA probe to clone CAD in *Eucalyptus*. Functional expression was established in *E. coli*	282
Molecular modeling of CAD 3D-structure and definition of its active site/binding regions	1993	McKie *et al.*	Molecular modeling comparison with HADH, with postulation of amino acid residue substitution in active site facilitating monolignol binding	285

[a]Original claims that CAD had been cloned were not substantiated in print: in the first case a presumed CAD encoding gene actually encoded malic enzyme,[270,272,273] and in the second only very small partial sequences were given[276] with the full sequence appearing 3 years later (EMBL accession number Z37991).

As described earlier (Section 3.18.3.3.4), Mansell *et al.* had convincingly demonstrated that ferulic acid (**6**) could be converted into coniferyl alcohol (**2**) using *Salix alba* cell-free extracts supplemented with ATP, CoASH, Mg^{2+}, and NADPH.[218] At the same time, preliminary evidence was obtained for the concomitant formation of coniferyl aldehyde (**16**). Stöckigt *et al.* extended this finding by demonstrating that *Forsythia* cell-free extracts also catalyzed the corresponding conversion of *p*-coumaric acid (**5**) into *p*-coumaryl alcohol (**1**).[261]

The actual detection of cinnamyl alcohol dehydrogenase, however, came with the subsequent demonstration that coniferyl aldehyde (**16**) could be converted into coniferyl alcohol (**2**) using a NADPH-requiring *Salix alba* cell-free preparation.[222] Mansell *et al.*[262] then purified (604-fold) the CAD enzyme to apparent homogeneity, as far as disk gel electrophoresis was concerned, to give a preparation which only acted upon cinnamyl alcohols. Preliminary characterization gave Michaelis' constants of 52, 16, 35, and 70 µM with respect to NADPH, ADP, coniferyl aldehyde (**16**), and *p*-coumaryl aldehyde (**12**), respectively. Unlike cinnamoyl CoA:NADP oxidoreductase, this enzyme was a type A reductase in that [^{3}H] was abstracted from [4A-^{3}H]NADPH, but not [4B-^{3}H]NADPH (Scheme 10).[262] It was later established that, in the reverse reaction, only the 9-pro*R* hydrogen of coniferyl alcohol (**2**) was removed during its conversion into coniferyl aldehyde (**16**),[263] in harmony with the mechanism of action determined for horse liver and yeast alcohol dehydrogenases.[264]

Activity staining of slab gel electrophoretic patterns representing cell-free extracts from 89 plant species was also undertaken and this suggested that cinnamyl alcohol dehydrogenases were ubiquitous throughout the bryophytes, pteridophytes, gymnosperms, and angiosperms, although the activity within the bryophytes examined was notably lower. However, since it has not been demonstrated that bryophytes possess monolignol synthesizing capacity, this may indicate that the assay conditions did not always reliably discriminate between CAD and other dehydrogenase systems *in planta*.[265] The studies of Mansell *et al*[265] also suggested that CAD isozymes might be present in certain organisms, in agreement with earlier findings by Wyrambik and Grisebach in 1975, who reported two putative CAD isozymes in soybean cell-suspension cultures.[266]

The next decade (1977–88) witnessed relatively few contributions to the subject of CAD.[249,250,267–269] Of those that appeared, Grisebach[249,267] suggested, on the basis of preliminary inhibition studies (discussed later), that an ordered bi-bi mechanism might be operative for the CAD-catalyzed reaction. These studies also led to recognition of substantial differences in the substrate specificities of CAD preparations from spruce (*Picea abies*) cambial sap and soybean cell suspension cultures (Table 13).[249] In the case of the spruce (gymnosperm) CAD purified 1600-fold, both coniferyl (**16**) and *p*-coumaryl (**12**) aldehydes were effective substrates, whereas sinapyl aldehyde (**20**) was not. On the other hand, soybean CAD exercised no significant differential effect on any of the three substrates (**12**), (**16**), and (**20**). A similar trend in regard to substrate specificity was noted later by Kutsuki *et al.*[268] using a 20-fold purified extract containing Japanese black pine (*P. thunbergii*) CAD.

Table 13 Substrate specificity of partially purified spruce (*Picea abies*) and soybean (*Glycine max*) cinnamyl alcohol dehydrogenase.[249]

CAD Preparation	Substrate	K_m (µM)	V (nkat mg^{-1} protein)	V/K_m
Spruce cambial sap	Coniferyl aldehyde (**16**)	3.6	1720	479
	NADPH	1.7		1010
	Sinapyl aldehyde (**20**)	83.0	167	2
	NADPH	1.9		88
	p-Coumaryl aldehyde (**12**)	12.5	2860	229
	NADPH	11.9		240
Soybean cell suspension (partially purified)	Coniferyl aldehyde (**16**)	1.7	25^a	14
	NADPH	7.2		3
	Sinapyl aldehyde (**20**)	4.3	21	5
	p-Coumaryl aldehyde (**16**)	9.1	26	3

^{a}For the pure enzyme, V_{max} was reported to be about 1500 nkat mg^{-1} protein.

By the mid-1980s, attention had been directed towards the cloning of CAD, and this resulted in several unexpected developments. Initially encouraging reports about the cloning of the CAD gene from bean (*Phaseolus vulgaris*)[270] were tempered by the lack of any functional expression data for recombinant CAD, and hence verification of the clone's authenticity. Nevertheless, this cDNA was then used, for example, to study induction of the monolignol-forming pathway during laticifer-specific gene expression in *Hevea brasiliensis* (rubber tree).[271] Subsequent gene sequence comparisons, however, led to a suspicion that the clone harbored, instead of CAD, a malic enzyme;[272] the suspicion was unambiguously confirmed later in 1994.[273]

In 1991 sundry contemporaneous reports were submitted on the purification of CAD including from tobacco (*Nicotiana tabacum*) stems, on 15 May,[274] wheat (*Triticum aestivum*), on 3 June,[275] loblolly pine (*Pinus taeda*), on 9 July,[276] *Aralia cordata*, on 18 November,[277] and *Eucalyptus gunnii* Hook xylem, on 5 December.[278] Of these contributions, the loblolly pine paper claimed in its title that the CAD had been cloned. However, only a partial N-terminal sequence (47 amino acid residues) and one short (13 amino acid) internal sequence were provided, this being somewhat comparable in length to the partial sequence data disclosed by Hibino *et al.*[277] A loblolly pine CAD cDNA sequence was actually never formally reported until more than three years later (29 September 1994, EMBL accession number Z37991).

The first verifiable cDNA sequence encoding CAD was obtained from tobacco[279] whose ORF contained 1419 nucleotides encoding a protein of 357 amino acids ($M_r \sim 38\,760$). Analysis of sequence comparisons for alcohol dehydrogenases (e.g., yeast and *Aspergillus* ADH1) revealed putative NADPH-binding (GlyXGlyXXGly) and Zn-binding/ligation domains. Other cDNA gene sequences for CAD followed about $1\frac{1}{2}$ years later beginning with disclosures on *Aralia cordata*[280] and Norway spruce (*Picea abies*),[281] and then *Eucalyptus gunnii*.[282,283] The *Eucalyptus* CAD (CAD 2) was obtained using the tobacco CAD cDNA clone as a probe, and functional expression of the CAD2 recombinant protein in *E. coli* finally verified that the enzyme had, indeed, truly been cloned.

With recombinant CAD available, the first real opportunity was provided to study its enzymology, a matter of importance given that there are very few aromatic alcohol dehydrogenases present in nature. Aliphatic alcohol dehydrogenases, on the other hand, from mammalian, yeast, and other systems have long been studied, and numerous X-ray structures were available for comparative evaluation. (For an excellent historical account see the article by Brändén and Eklund[284] which discusses the various alcohol dehydrogenase X-ray structures obtained since 1970.)

Molecular modeling was thus next undertaken in order to compare the eucalyptus CAD2 with other alcohol dehydrogenases,[285] particularly horse-liver alcohol dehydrogenase (HADH).[286–289] Indeed, based on its homology to HADH, a putative 3D structure for CAD2 was obtained and analyzed. Table 14 summarizes some of the important domains identified with respect to catalysis and cofactor binding in CAD2 as a result of this comparison. It was found that the three Zn1 catalytic residues are fully conserved between CAD2 and HADH, being located at Cys 47, His69, and Cys163, as well as at Cys46, His67, and Cys174, respectively; Zn1 itself provided the fourth coordination bond. Zn2 (noncatalytic zinc), on the other hand, is coordinated to four cysteine residues at Cys100, Cys103, Cys106, and Cys114, this being a well conserved motif for HADH and other dehydrogenases. Additionally, the NADPH-binding domain, the so-called Rossmann fold domain, begins just before the highly conserved Gly Leu Gly Gly Phe Gly motif at Gly188, and the serine (Ser212) residue in CAD2 was also deemed to be involved in recognition of the NADPH cofactor; the aspartyl residue (ASP223) fulfills this function for NADH binding to HADH by H-bonding with the adenosine ribosyl 2'-OH group. The arginine (Arg217) was also considered to be instrumental in stabilizing the NADPH phosphate group, and the residues neighboring the nicotinamide ring were well conserved in CAD2. On the other hand, the residues (305 to 320) at the putative dimer interface of HADH showed only limited homology when compared with CAD.

Table 14 Important domains/amino acid residues in *Eucalyptus gunnii* cinnamyl alcohol dehydrogenase, deduced from sequence homology comparison.

Feature	Region/domain
Zn1-binding (catalytic zinc)residues	Cys[47], His[69], Cys[163]
Zn2-binding (structural zinc) lobe	[88]Gly Asp Arg Val Gly Thr Gly Ile Val Val Gly Cys
Four cysteine residues (bold) are involved in Zn binding	**Cys** Arg Ser **Cys** Ser Pro **Cys** Asn Ser Asp Gln Glu Gln Tyr **Cys**[114]
NADP(H) nucleotide-binding domain or Rossman motif	Gly Leu Gly Gly Phe Gly
Residues neighboring nicotinamide ring	Gly[268] Gly[274]

Taken together, the three-dimensional molecular modeling suggested a structure for CAD2 similar to that of HADH and some of the salient features of the presumed active site in CAD2 are summarized in Table 15 and Figure 15. From other studies of alcohol dehydrogenases, it is well known that the "floor" of the substrate binding pocket contains the catalytic zinc and its corresponding binding residues, as well as the NADPH nicotinamide ring which occupies one face within the pocket. From the molecular modeling, it was concluded that substitution of various HADH residues (Phe93, Phe110, Leu116, and Ile318) in CAD2 by Ile95, Tyr113, Trp119, and Phe298 could enhance monolignol substrate binding. This was based on a number of considerations.

Replacement of Leu116 (HADH) with Trp119 (CAD2) gave additional space both in the pocket and a contact face (at the benzene ring of tryptophan) for interaction with the phenolic ring of the monolignol. In a completely analogous manner, replacement of Ile318 (HADH) by Phe298 provides a second aromatic ring, and hence the ability for the monolignol substrate to form a "molecular" sandwich between the two. Replacement of Phe110 (HADH) with Tyr113 (CAD2) enables the phenolic group of the monolignol substrate to H-bond with Tyr113 (side-chain) and Trp119 (carbonyl group). Docking was then considered to be further facilitated by substitution of Phe93 (HADH) by Ile95 (CAD2), thereby enhancing further interaction with Trp113, with the substrate methyl groups being accommodated on the surface of the active site. Indeed, these constraints, enforced by the molecular modeling parameters, provide an explanation for the observed Class-A stereospecificity of cinnamyl alcohol dehydrogenase (Figure 15) as shown in Scheme 10. Thus, in the absence of any definitive X-ray structure for CAD2, a reasonable working model has been created to account for the mechanism of CAD catalysis.

Table 15 Corresponding ligand-binding sites of HADH and CAD2 (after McKie *et al.*[285]).

HADH	CAD2	Comments on role
Phe93	Ile95	Hydrophobic substrate-binding pocket
Thr94	Val96	
Phe319	Ile299	
Phe110	Tyr113	Potential H-bonded interaction with substrate from CAD2
Ser177	Asn120	
Leu116	Trp119	The aromatic rings of Trp119 and Phe298 could sandwich the phenyl moiety of the CAD2 substrate
Ile318	Phe298	
Ser48	Ser49	Proton shuttle
His51	His52	
Asp223	Ser212	Residues associated with cofactor specificity
Lys228	Arg217	

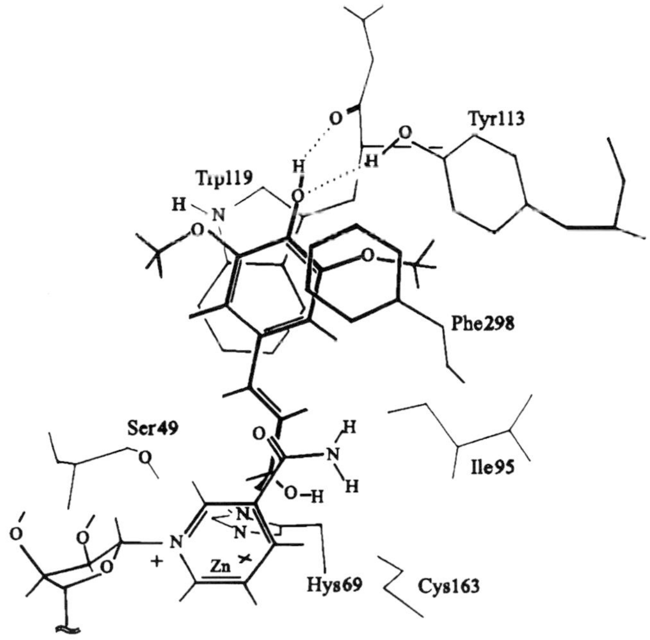

Figure 15 View of the substrate binding region of CAD2 from *Eucalyptus* showing the phenolic ring of sinapyl alcohol (**3**) sandwiched between the essentially coplanar layers of Trp119 (behind) and Phe298 (above). The phenolic OH group of the substrate can make hydrogen-bonds to the side-chain of Tyr113 and the backbone carbonyl group of Trp119. The pyridinium ring of NADP$^+$ is located above the scissile C—H bond of the alcohol substrate, explaining the observed[263,267] class-A stereospecificity of the reaction. The catalytic Zn is also shown, close to its ligands His69 and Cys163. The structure shown was modeled on the known three-dimensional coordinates of horse liver alcohol dehydrogenase[288] (reproduced by permission from *Biochim. Biophys. Acta*, 1993, **1202**, 61).

Availability of recombinant CAD2 protein has also facilitated detailed comparison of its steady-state kinetic behavior with a site-directed mutation replacing serine (Ser212) with an aspartate residue.[290] This, in turn, furnished an opportunity for altering cofactor binding since Ser212 was considered to be involved in recognition of the NADPH cofactor.[290] Steady-state kinetic data suggested rapid equilibrium birandom kinetics, for both wild-type CAD2 recombinant protein and the mutagenized Ser212Asp CAD2. This is illustrated in Figure 16, and contrasts with Wyrambik and Grisebach's earlier suggestion that CAD catalysis involved an ordered bi-bi mechanism.

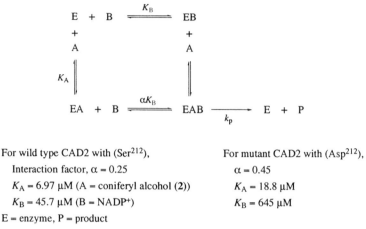

For wild type CAD2 with (Ser[212]),

    Interaction factor, $\alpha = 0.25$

    $K_A = 6.97\ \mu M$ (A = coniferyl alcohol (**2**))

    $K_B = 45.7\ \mu M$ (B = NADP$^+$)

E = enzyme, P = product

For mutant CAD2 with (Asp[212]),

    $\alpha = 0.45$

    $K_A = 18.8\ \mu M$

    $K_B = 645\ \mu M$

Figure 16 Illustration of rapid equilibrium birandom kinetic scheme for cinnamyl alcohol dehydrogenase (CAD2) (after Lauvergeat *et al.*[290]).

3.18.4 TISSUE AND SUBCELLULAR LOCALIZATION OF MONOLIGNOL PATHWAY ENZYMES AND ASSOCIATED GENE EXPRESSION

Section 3.18.3.3 summarized the pathway to the monolignols from phenylalanine (**8**), and the advances hitherto made in comprehensively delineating its associated enzymology. Much still remains to be done in terms of fully elucidating the mechanism of the biochemical transformations involved (e.g., through X-ray crystallographic studies) for each individual step. It was quite interesting, however, that none of the Michaelis parameters (K_m values) characteristic of the enzymes from phenylalanine ammonia-lyase to cinnamyl alcohol dehydrogenase varied by more than just over an order of magnitude (see Table 16). (Note, though, that *p*-coumarate 3-hydroxylase has not yet been unambiguously identified, and hence its K_m was unavailable at the time of writing.) Nevertheless, although the kinetic parameters for the various enzymes were typically obtained with preparations from different organisms rather than from any one plant species, the data might seem to argue against any single kinetic barrier being operative between phenylalanine (**8**) and the monolignols (**1**)–(**3**). On the other hand, studies of phenylalanine (**8**) metabolism with *P. taeda* cell cultures (Section 3.18.3.2) suggested that not only Phe (**8**) availability, but also metabolite through-put via cinnamate 4-hydroxylase and *p*-coumarate 3-hydroxylase, were rate-limiting *in situ*.

This section is focused on evaluating the current understanding of enzyme localization at the subcellular level from Phe (**8**) to the monolignols (**1**)–(**3**), and what has so far emerged about their associated gene expression. Initially, localization of the enzymes was tentatively established from studies of both relative enzyme solubilities and organelle fractionation, although these have now been largely replaced by immunolocalization approaches. A further complication is that many of the enzymatic steps (depending upon the organism) are controlled by (multi) gene families. That is, yet another level of complexity confounds the evaluation of the spatial and temporal organization of the pathway(s) leading to the lignin(s). However, such difficulties are beginning to be resolved by using gene specific probes for studying gene expression associated with lignin deposition.

In deference to the reader with little botanical experience, Figure 17 illustrates both the names and locations of common plant components together with the highlighting of lignified and presumed lignin-containing tissues. These are intended to serve as an aid in understanding the potential complexities associated with the protein localization and gene expression involved in lignification. Moreover, it should be remembered that different vascular plant tissues and/or cell-types can also biosynthesize distinct phenylpropanoid-derived substances (e.g., hydroxycinnamic acids, flavonoids) which need not be associated with lignin formation, either temporally or spatially. Additionally, no

Table 16 Apparent Michaelis constants for monolignol pathway enzymes from phenylalanine (**8**).

Enzyme	K_m (µM)	Plant species	Ref.
Phenylalanine ammonia-lyase	38 (Phe (**8**))	*Solanum tuberosum*	64
Cinnamate 4-hydroxylase	1.5 (cinnamic acid (**10**))	*Helianthus tuberosus*	113
Ferulate 5-hydroxylase	6.3 (ferulic acid (**6**))	*Populus × euramericana*	160
Hydroxycinnamoyl CoA *O*-methyltransferase	32 (caffeoyl CoA (**14**))	*Petroselinum crispum*	190
Hydroxycinnamic acid *O*-methyltransferase	40 (caffeic acid (**13**))	*Populus tremuloides*	204
	19 (5-hydroxyferulic acid (**17**))		
Hydroxycinnamyl-CoA ligase	11 (*p*-coumaric acid (**5**))	*Picea abies*	225
	10 (ferulic acid (**6**))		
Cinnamoyl CoA reductase	43.5 (*p*-coumaroyl CoA (**11**))	*Picea abies*	249
	5.2 (feruloyl CoA (**15**))		
	28.6 (sinapoyl CoA (**19**))		
Cinnamyl alcohol dehydrogenase	12.5 (*p*-coumaryl aldehyde (**12**))	*Picea abies*	249
	3.6 (coniferyl aldehyde (**16**))		
	83.0 (sinapyl aldehyde (**20**))		

single plant species has yet been comprehensively characterized in regard to monolignol pathway enzyme localization and gene expression in all phases of its life cycle. While a short life-span organism such as *Arabidopsis* may have some utility in this regard, it nevertheless will be quite limited in the types of lignified tissues that it forms, as well as in the variety of related metabolites that are generated.

3.18.4.1 Phenylalanine Ammonia-lyase

Depending upon the species involved, PAL has been reported to be the product of either a single gene or a multigene family. A multigene origin would intuitively seem to be more likely, however, because PAL contributes to different metabolic roles in development and defense. For example, in loblolly pine (*P. taeda*) developing xylem, PAL was described by Whetten and Sederoff as being derived from a single copy in the genome,[291] whereas another investigation by Ellis and co-workers identified five distinct classes of PAL sequences in the closely related species jack pine (*Pinus banksiana*).[292] This latter finding would seem to suggest that the report of a single PAL gene in loblolly pine needs to be further confirmed, as the result could easily be an artifact arising from high stringency washing. Indeed, the contrasting claims emphasize the need for circumspection in regard to reports that PAL exists as a single form and hence originates from one gene, e.g., in species such as bamboo (*Bambusa oldhami* Munro) shoots,[293] sunflower (*Helianthus annuus*) hypocotyls,[294] and strawberry (*Fragaria ananassa* Duch) fruit.[295] In contrast, southern blot analyses have shown that multiple PAL gene families are present in, for example, fungal elicitor-treated alfalfa (*M. sativa*),[296,297] bean (*P. vulgaris*),[298–300] and parsley (*P. crispum*),[75,301] as well as in rice (*Oryza sativa*),[302] *A. thaliana*,[303] pea (*Pisum sativum*),[304] potato (*S. tuberosum*),[305] tobacco (*N. tabacum*),[306] grape (*Vitis vinifera*),[307] aspen (*Populus kitakamiensis*),[308] poplar (*Populus trichocarpa* × *P. deltoides*),[309] and jack pine (*P. banksiana*).[292]

In the 1970s, Czichi and Kindl[310] postulated that distinct phenylalanine (**8**) pools were involved either in specific branches of phenylpropanoid metabolism and/or in protein biosynthesis. Although the technology applied at the time was unable to prove this hypothesis unambiguously, the existence of multigene PAL families and presumably distinct sites of PAL subcellular localization (e.g., cytosol, plastid, mitochondria, and microbodies[74]) seemed to be in accord with this suggestion. It is, therefore, very important that both the localization and physiological roles of each form of PAL, encoded by each PAL gene, be fully determined and identified. The progress made so far in this direction is described below; it began with organelle fractionation studies before progressing to protein immunolocalization and, most recently, to the analysis of the tissue-specificity of PAL promoter expression. Although no direct criticism is intended of these multifarious studies, their significance would have been substantially furthered had they been precisely and definitively correlated with identification of the downstream metabolites so formed.

3.18.4.1.1 Organelle fractionation

Early organelle fractionation studies first suggested that PAL was located in peroxisome fractions and glyoxysomes, as well as in the cytosol and particulate (microsomal) fractions including the

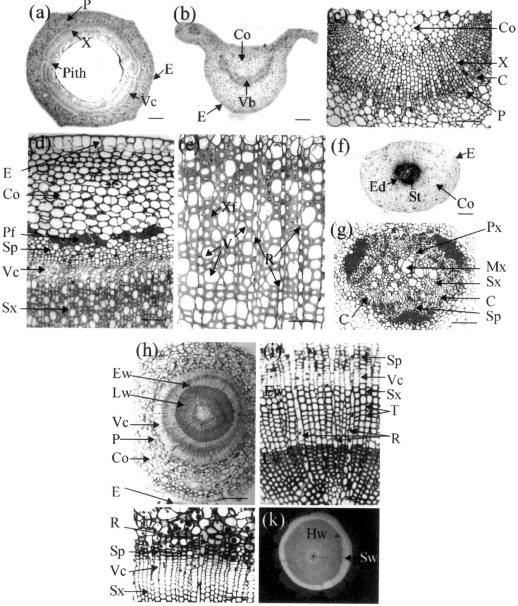

Figure 17 Miscellaneous plant components and highlighted lignin and presumed lignin containing tissues. (a) cross section of *Forsythia intermedia* mature stem (bar = 50 μm); (b) cross section of *Forsythia intermedia* petiole (bar = 100 μm); (c) enlargement of petiole vascular bundle (bar = 10 μm); (d) enlargement of mature stem of *F. intermedia* (bar = 5 μm); (e) secondary xylem of *F. intermedia* (bar = 2.5 μm); (f) pea (*Pisum sativum*) root cross section (bar = 50 μm); (g) enlargement of pea root (bar = 10 μm); (h) cross section of 1 year old Douglas fir seedling (bar = 70 μm); (i) enlargement of Douglas fir stem secondary xylem (bar = 20 μm); (j) enlargement of Douglas fir secondary phloem and vascular combium (bar = 20 μm); and (k) cross section of a tamarack larch (*Larix laricina*). C, cambium; Co, cortex; E, epidermis; Ed, endodermis; Hw, heartwood; Mx, metaxylem; P, phloem; Pf, phloem fiber; Px, protoxylem; R, ray cell; Sp, secondary phloem, Sw, sapwood; St, stele: Sx, secondary xylem; V, vessel; Vc, vascular cambium; Vb, vascular bundle; T, tracheid; X, xylem; Xf, xylem fiber. Lignified regions are associated with secondary xylem (tracheids, vessels, fibers, certain ray cells), phloem fibers, and vascular bundles.

thylakoid membranes. The claims of peroxisome and glyoxysome localization for PAL resulted from organelle fractionation using spinach (*Spinacia oleracea*) leaves[311] and germinating castor bean (*R. communis*) endosperm, respectively,[312] whereas the occurrence in cytosolic (~90–97%) and particulate microsomal (3–10%) preparations was detected through studies with buckwheat (*Fagopyrum esculentum* Moench) hypocotyls[313] and potato (*S. tuberosum*) tubers.[314,315] In the latter case, the potato tuber microsomal fractions apparently were supposedly capable of converting L-[4-

[³H]Phe (**8**) into both *o*- and *p*-coumaric acids (**39**) and (**5**), respectively.[315] PAL activity (~13% of total) was also thought to be associated with microsomes from 5-day old castor bean (*R. communis*) endosperm;[316] moreover, PAL was deemed to be present in chloroplasts of *Cucumis melo*[317] and thylakoid membranes of the alga, *Dunaliella marina*,[318] as well as watercress (*Nasturtium officinale*), *Astilbe chinensis*, and *Hydrangea macrophylla*.[319]

3.18.4.1.2   *PAL protein and mRNA localization* in situ

The next phase of establishing PAL localization more explicitly came with the application of anti-PAL polyclonal antibodies to loblolly pine cell suspension cultures,[320] French bean,[321] *Z. elegans*,[322] aspen (*P. kitakamiensis*),[323] and Norway spruce (*Picea abies*),[324] as well as the examination of PAL mRNA expression via *in situ* hybridization using parsley.[325] In the loblolly pine cultures,[320] immunolocalization (immunogold labeling with silver enhancement) revealed that PAL was mainly cytosolic, with some limited binding being detected to the endomembranes, e.g., the endoplasmic reticulum, but *not* in the chloroplasts and thylakoid membranes as previously suggested (Figure 18).[317–319] Additionally, certain cells displayed a tendency for PALs being localized adjacent to the plasma membrane, an observation that, as discussed later, was also made with PAL mRNA during *in situ* hybridization using parsley (*P. crispum*).[325] These findings seem, therefore, consistent with

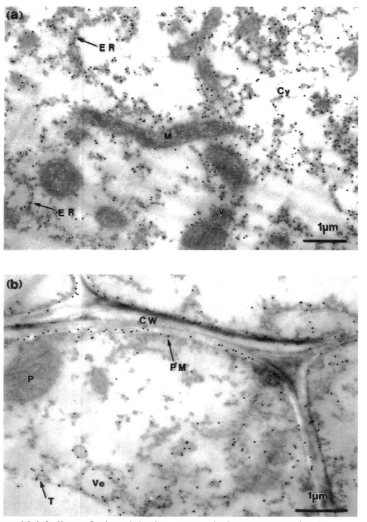

Figure 18 Immunogold labeling of phenylalanine ammonia-lyase present in *Pinus taeda* cell suspension cultures grown on medium containing 2,4-D. The antibodies employed were provided by Whetten and Sederoff (North Carolina State University, NC, USA), which had been raised against loblolly pine PAL isolated from cambium tissue. P, plastid; M, mitochondria; T, tonoplast; Ve, vesicle; Cy, cytosol; ER, endoplasmic reticulum.

the deployment of specific pools of phenylalanine (**8**) for different metabolic purposes, although this needs to be more definitively demonstrated at both the subcellular and tissue-specific levels.

In agreement with the foregoing observations, PAL was located in both the cytosol and endo-membrane fractions in cells adjacent to the differentiating and mature xylem of 8-day old French bean hypocotyls.[321] (Note, however, that in this case the antibodies used for visualization recognized a number of PAL isoforms (based on relative molecular mass), as well as their degradation products.[326]) On the other hand, in differentiating *Z. elegans* cells which form tracheary elements from their mesophyll cells,[322] PAL was reported to be located on either secondary cell wall thick-enings, Golgi-derived bodies, or in the cytoplasm. However, in that particular study, pre-immune serum controls displayed significant nonspecific labeling. Indeed, the described cell-wall binding may have simply resulted from cell leakage or fragmentation giving rise to artifacts.

The *in situ* hybridization of parsley PAL mRNA also gave somewhat analogous results to that in loblolly pine although these varied depending upon the stage of xylem cell development.[325] In young developing xylem of the stele of a hypocotyl, for example, the PAL mRNAs were observed throughout the entire cytosol while, as tracheid formation proceeded to maturity, they were detect-able only in areas adjacent to the developing cell wall, i.e., suggesting a more ordered grouping of PAL during lignin deposition.

Next, although lacking any precise correlation with metabolite formation, PAL localization studies were extended further in bean hypocotyls,[321] young aspen (*P. kitakamiensis*) stems,[323] Norway spruce (*Picea abies*) bark,[324] and parsley (*P. crispum*) seedlings.[327] In the case of the 8-day old bean hypocotyls,[321] PAL localization was examined at various developmental stages, i.e., sites 1, 2, and 3, corresponding to young (1), more mature (2), and mature (3) xylem tissues, respectively (see Figure 19). Using this approach, the young tissue was only lightly immunolabeled, as would be expected since these tissues had only begun to lignify. At the progressively more mature sites, however, labeling became heavier in both the metaxylem and adjacent xylem parenchyma cells as lignification proceeded. Furthermore, in the most mature tissue, labeling was also noted in the phloem in addition to the differentiating xylem and mature xylem, whereas it was absent from the nonlignified pith, cambium, and cortex tissues.

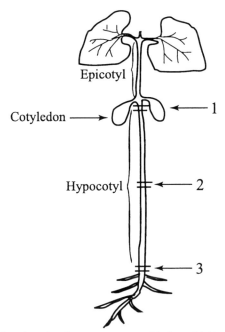

Figure 19 Schematic drawing showing the three segments of 8-day old French bean seedlings used to study PAL localization. (After Smith *et al.*[321])

In an analogous manner, PAL antibodies were employed to determine the gross cell-specificity of the enzyme in a 1-year old aspen seedling, via examination of regions between the various internodes (Figures 20(a) to 20(f)).[323] As can be seen, PAL was barely detectable in the protoxylem (Figures 20(a) and 20(b)), but was quite pronounced in the differentiating xylem zone undergoing lignification (Figures 20(d) and 20(e)). Indeed, the highest intensity was in the xylem tracheary elements whereas ray parenchyma cells were much less intense at this developmental stage. On the other hand, with mature Norway spruce (Figures 21(a) to 21(e)), PAL was mainly located in the ray cells (Figures 21(c) and 21(d)), these being involved in formation of unidentified phenols, presumed to be tannins, accumulating in axial parenchyma cell vacuoles (Figures 21(a) and 21(b)).[324]

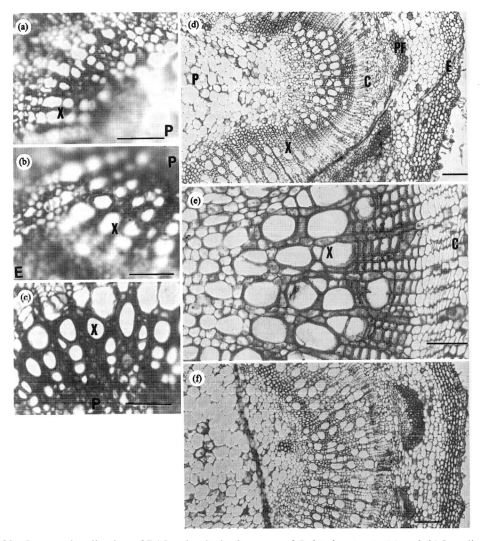

Figure 20 Immunolocalization of PAL subunits in the stems of *P. kitakamiensis.* (a) and (b) Localization of PAL in young stems; × 400, bar = 30 µm. (c) Control reaction in a young stem using the antiserum against total proteins from *E. coli* cells harboring recombinant PAL fragment; × 400, bar = 30 µm. (d) Localization of PAL in an old stem; × 100, bar = 100 µm. (e) Higher magnification of part of (d); × 400, bar = 5 µm. (f) Control reaction in an old stem; × 100, bar = 100 µm. The positions of pith (P), phloem fibers (PF), xylem (X), cambium (C), and epidermis (E) are indicated (reproduced by permission from *Planta*, 1996, **200**, 13).

Lastly, PAL antibodies were also employed in order to localize sites of induction of PAL activity when young parsley seedlings were challenged with zoospores of the soybean-pathogenic fungus, *Phytophthora megasperma* f.sp. *glycinea*.[327] Areas adjacent to the fungal invasion, as well as the nonvascular epidermis of primary leaf petioles, represented the main site(s) of PAL induction. These areas were essentially coincident with the deposition of various induced UV/blue fluorescent substances which were thought, but not proven, to represent furanocoumarins and other wall-bound phenolic components.

3.18.4.1.3 PAL gene promoter expression

The most recent attempts to unravel both the temporal and spatial localization, and associated metabolic functions, of specific PAL genes has involved examining the tissue-specific expression of

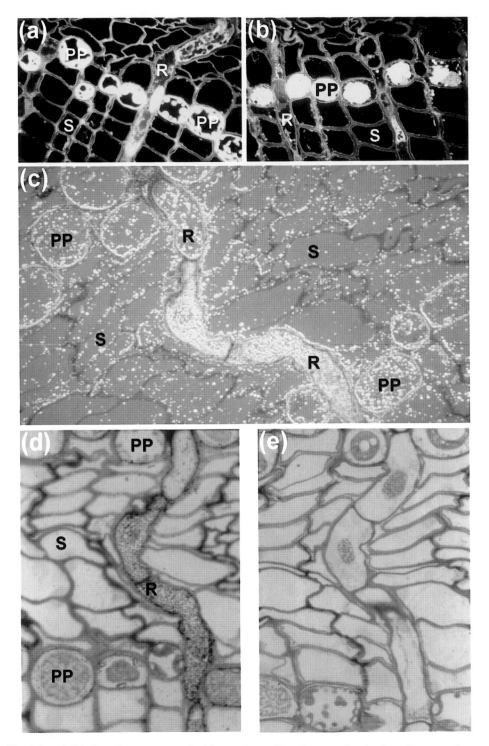

Figure 21 (a) and (b) Autofluorescence of phloem tissue (fourth row of polyphenolic parenchyma, PP) in Norway spruce clones 409 and 589. (a) Clone 589. The globules within the vacuole of PP cells show very intense fluorescence. (b) Clone 409. The large phenolic bodies in the vacuoles of PP cells fluoresce intensely. (c) to (e) Immunocytochemical localization of phenylalanine ammonia-lyase (PAL) in Norway spruce secondary phloem studied with anti-PAL immune serum ((c) and (d)), and pre-immune serum negative control (e). In the confocal microscopic image of phloem cross-section (c), silver grains (label) have been imaged by reflected light and are orange whereas on bright field microscopy image the same grains are black (d). PAL specific labeling is seen throughout the cytoplasm of the polyphenolic parenchyma (PP) cell layers and ray parenchyma (R) of clone 589 (c) and (d). PP, polyphenolic parenchyma; S, sieve cell; R, ray parenchyma (reproduced by permission from *American Journal of Botany*, 1998, **85**, 601).

various PAL-promoter elements in tobacco, potato, *Arabidopsis*, and parsley, respectively. This has met with only limited success so far, since the studies were all essentially phenomenological in that they lacked any definitive correlation with associated metabolic processes. These studies were, in turn, derived from an earlier investigation which identified three PAL genes in French bean, of which the two cloned, gPAL 2 and gPAL 3, had significant differences in their 5' and 3' flanking regions, suggesting differential regulation.[299] Indeed, gPAL 2 was induced by fungal elicitor treatment and mechanical wounding in *P. vulgaris*, whereas gPAL 3 activity was only increased by the latter.[328]

Nevertheless, expression of the bean gPAL 2 promoter (encoding ~1000 bp of the 5' flanking region) fused to the β-glucuronidase (GUS) reporter gene was investigated using transgenic potato and tobacco plants.[328] In the case of tobacco, its stems were harvested at maturity, with tissues arbitrarily classified as either young (Y), obtained from transverse sections just below the apex and containing primary xylem with discrete vascular bundles, more mature (MM) stems containing complete vascular bundles, or mature xylem (MX), which had undergone secondary thickening (Figure 22). With young (Y) tissue, GUS staining was only observed in the differentiating lignifying xylem elements of primary xylem (Figure 22(a)), whereas with the more mature (MM) stems this occurred in xylem which had completed vascular cylinder development, as well as in lignifying fibers associated with the internal/external phloem (Figure 22(b)). This GUS staining was also positively correlated with the presumed progressive phases of lignin deposition, as suggested by phloroglucinol (23)-treatment (Figure 22(c)), which is sometimes used histochemically to indicate the presence of coniferyl aldehyde "end-groups" in lignins. (Note, however, that this reagent can also give comparable color reactions with other phenolic components, including lignans.[20]) The mature fully lignified xylem (MX) cells, while staining positive with phloroglucinol (23), did not, however, display GUS activity (Figure 22(d)). On the other hand, there was very significant GUS staining noted in radial ray files which the authors were unable to explain, since these are not typically considered to be involved in lignification. Additionally, strong GUS staining was also detected in epidermal regions (epidermis and trichomes), which may be lignified and/or suberized, as well as in the root hairs and root vascular system (data not shown). A somewhat related observation apparently occurred with potato, where positive GUS staining was noted in the tuber tissue xylem elements, even though no visual evidence was furnished in support of this contention.[328] Wound-healing (suberizing) tuber tissue also displayed significant GUS activity in specific cells near or adjacent to the healing layer (discussed in Section 3.18.9). However, some of these tissues are not obviously involved in lignin formation and, for example, wound-healing potato tubers also biosynthesize other phenylpropanoid derived metabolites, e.g., alkyl ferulates, feruloyl tyramines, etc.[63]

Other attempts to document PAL-promoter expression in *Arabidopsis*[303] and parsley[329] seedlings were very limited in scope, since they lacked any correlation with the actual metabolic processes involved. Accordingly, future PAL promoter expression studies must be correlated with actual metabolite formation, particularly when much of the expression observed is not exclusively lignin-related. Yet, surprisingly, these preliminary results have dampened expectations that specific PAL genes may have specific roles (e.g., in lignin formation in vascular tissue) during plant growth/development and in plant defense. On the other hand, it is premature to dismiss this possibility arbitrarily until appropriate comprehensive and definitive correlations have been established, in terms of the actual tissue and cell-specific deposition of the various metabolites formed in a particular plant species.

3.18.4.2 (Hydroxy)cinnamate Hydroxylases and NADPH-dependent Reductase(s)

3.18.4.2.1 Cinnamate 4-hydroxylase

Cinnamate 4-hydroxylase is considered to be the product of a small gene family in *Z. elegans*,[330] alfalfa (*M. sativa*),[119,331] and poplar (*P. kitakamiensis*),[332] whereas in Jerusalem artichoke (*H. tuberosus*),[116] pea (*P. sativum*),[333] and *A. thaliana*[334,335] only a single gene copy has so far been reported.

C4H was first demonstrated to be associated with microsomal preparations obtained from pea.[86,89,100] Further confirmation that it was present in microsomes was subsequently obtained from studies using potato[314] and Jerusalem artichoke tubers,[109] as well as cucumber (*Cucumis sativus*) cotyledons[310] and French bean cell suspension cultures.[336]

As in the study of PAL, the subcellular location and tissue-specificity of C4H was investigated in

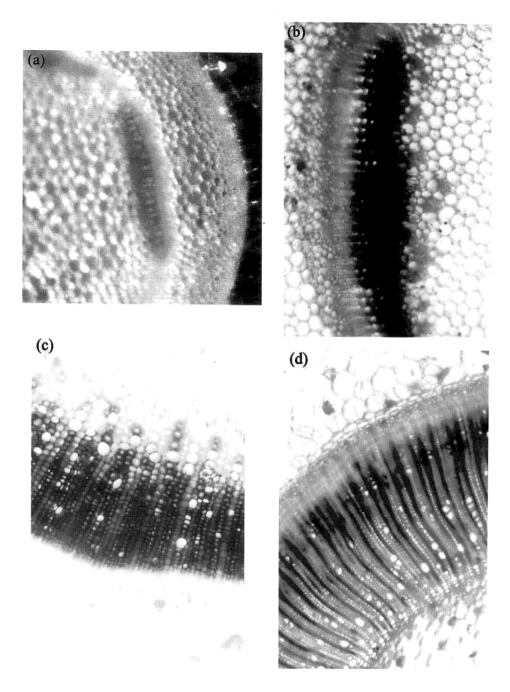

Figure 22 Histochemical localization of PAL, as evidenced by fused GUS activity in tobacco xylem tissue. (a) Transverse section of the apical region of a stem of a young (Y) tobacco plant stained with X-glucuronide: staining can be seen over the developing xylem cells while the vascular ring is not yet fully developed (× 50). (b) Transverse section of a stem of older tobacco plant, stained with X-glucuronide, showing more mature (MM) xylem that has formed a complete vascular cylinder: staining is most intense over the older cells of the xylem, with lighter staining over fiber cells associated with the external phloem also being observed (× 75). (c) Transverse section of a stem showing more mature (MM) xylem, stained with phloroglucinol to reveal lignin; some staining can also be seen in the fiber cells of the internal and external phloem (× 75). (d) Transverse section of stem of mature xylem (MX) in a tobacco transformant, stained with X-glucuronide; the xylem has undergone substantial secondary thickening, and staining can be seen in distinct radial files of cells (× 75) (reproduced by permission from the *EMBO J.*, 1989, **8**, 1899).

French bean hypocotyls, *Z. elegans* and *A. thaliana* using immunocytochemical localization, Northern blotting, tissue printing, and promoter expression. Thus, in 8-day old French bean hypocotyls,[321] immunogold labeling with silver enhancement using polyclonal antibodies established that C4H was, as expected, membrane-bound, being localized in both the ER and throughout the Golgi stacks. Next, tissue-specific immunolocalization with the different developmental sites 1–3 in the hypocotyl (Figure 19), as before, revealed that site 1 contained C4H in the young xylem, whereas at sites 2 and 3 it was present in the maturing/mature xylem as well as in cells adjacent to the metaxylem.[321] Additionally, C4H was detected in epidermal cells following wounding, thus once more demonstrating, as for PAL, that its expression/subcellular localization was not restricted to domains of lignin deposition.

Various tissue-specific localization patterns have been noted for C4H in *Z. elegans* and *A. thaliana*. With *Z. elegans*, Northern (mRNA) blots established that its C4H mRNA was expressed in stems, roots, and flower buds, and, to a lesser extent, in leaves. Moreover, with 4-week old *Zinnia* plantlet stems (Figure 23), tissue print hybridization[330] revealed that first internode tissues contained C4H mRNA in differentiating lignifying xylem cells but not in developing phloem fibers (Figures 23(a) and 23(b)) whereas, at the second internode, C4H mRNA hybridization occurred in both (Figures 23(c) and 23(d)). On the other hand, the most mature (third internode) stage gave intense hybridization only with the differentiating xylem in vascular xylem bundles (Figures 23(e) and 23(f)).

Northern blots of C4H mRNA in *A. thaliana*, however, displayed expression in root, leaf, inflorescent stem, flower, and silique tissues, the ages of which ranged from 3 weeks to 5 weeks old. An ~ 3 kb upstream promoter sequence from C4H was also examined in order to identify any tissue-specific promoter expression through the use of the GUS reporter gene strategy.[335] In this way, it was observed that cotyledons displayed GUS staining throughout their epidermal and mesophyll cells, as well as in the ray parenchyma and vascular tissues. Intense but diffuse GUS staining was observed in developing primary leaves even though this was not restricted to the vascular system, while root tissue was intensely stained from the hypocotyl/root interface to just below the root tip. Staining was also noted in both the lignifying stem xylem and sclerified parenchyma, whereas in flower tissues it was present in the sepal vasculature, particularly just below the stigmatic surface as well as in mature siliques. How much of this tissue expression is lignin specific remains to be determined.

Lastly, Table 17 summarizes what was known at the time of writing about targeting sequences for each of the enzymatic processes involved in the monolignol pathway from phenylalanine (**8**) onwards.[115,163,334,335,337–340] As can be seen, C4H has an N-terminal 30 amino acid hydrophobic stretch, which, from studies with numerous cytochrome P450s, is known to help position the protein at the cytoplasmic membrane surface (see Nelson and Strobel[341]).

Nothing had yet been described at the time of writing about the corresponding NADPH-dependent reductase(s), and their subcellular and tissue locations. However, two cDNAs encoding two isoforms of cyt P450 reductase in *A. thaliana* were heterologously expressed in both *Saccharomyces cerevisiae* (ATR1 and ATR2)[147] and *Spodoptera frugiperda* (AR1 and AR2),[342] with AR1 and AR2 being nearly identical to ATR1 and ATR2, respectively. Both purified recombinant NADPH-dependent reductases were fully competent when incubated with cinnamate 4-hydroxylase and cinnamic acid (**10**) as the substrate. Indeed, they gave comparable levels of enzymatic activity, with both being presumed to originate from single copy genes. Subsequently, mRNA blot analyses revealed that AR1 was constitutively expressed, whereas AR2 was induced by wounding and incident light, with its induction preceding those of PAL and cinnamate 4-hydroxylase. A(T)R2 also has an N-terminal poly(Ser/Thr) stretch which is not observed in any other cyt P450 reductase, and which is reminiscent of a chloroplastic targeting sequence.[147] Additionally, both A(T)R1 and A(T)R2 contain hydrophobic segments within the N-terminal portion that are presumed also to participate in the binding of the reductase to the microsomal membrane. It will ultimately be of interest to determine how both C4H and its reductase are able to be closely juxtaposed in order that the catalytic cycle can be achieved.

3.18.4.2.2 p-Coumarate 3-hydroxylase

As described earlier (Section 3.18.3.3.2), the actual enzyme involved in caffeic acid (**13**)/caffeoyl CoA (**14**) formation had not yet been unambiguously identified at the time of writing. Accordingly, it is not yet possible to elucidate the tissue-specific expression of this enzyme. However, it would be anticipated that this would also be a membrane-associated cytochrome P450.

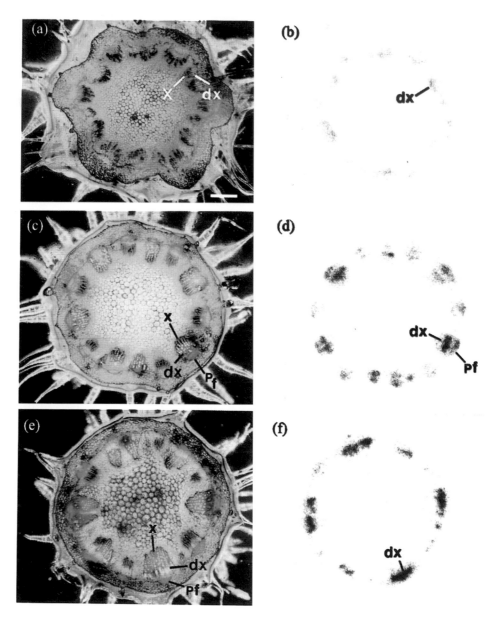

Figure 23 Tissue print hybridization of C4H mRNA in *Zinnia* stems. Internodes from 4-week old *Zinnia* plants were sectioned with a double-edge razor blade and sections were printed onto a nylon membrane for transfer of RNAs. 35S-labeled C4H antisense RNA probe was used to hybridize the membrane for detection of C4H mRNA. At left are toluidine blue-stained sections from the first (a), second (c), and third (e) internodes; at right are the corresponding C4H mRNA localizations for the first (b), second (d), and third (f) internodes. dx Denotes differentiating xylem, pf phloem fibers, and x xylem. Scale bar equals 0.5 mm for all panels (reproduced by permission from *Plant Science*, 1996, **121**, 133).

3.18.4.2.3 Ferulate 5-hydroxylase

The ferulate 5-hydroxylase (F5H) gene was obtained by T-DNA tagging (see Section 3.18.3.3.2), the corresponding protein's physiological and catalytic function being inferred through complementation of an F5H-lacking mutant with this same gene.[343] As is the case for C4H, F5H also contains a putative ER targeting sequence (see Table 17). Nothing, however, has yet been established about either its tissue-specific expression or subcellular location. It will be revealing to establish whether in *Arabidopsis*, for example, this only occurs in sclerified parenchyma cells between the vascular bundles of the rachis, since these are considered to be involved in the biosynthesis of syringyl lignins (derived from sinapyl alcohol (**3**)).[162]

Table 17 Targeting and membrane domain localization for cinnamate 4-hydroxylase and ferulate 5-hydroxylase.

N terminal amino acid sequence			Plant species	Ref.
Cinnamate 4-hydroxylase				
MDLLLIEKTL	VALFAAIIGA	ILISKLRGKK	*H. tuberosus*	116
MDLLLLEKTL	LGLFLAAVVA	IVVSKLRGKR	*V. radiata*	115
MMDFVLLEKA	LLGLFIATIV	AITISKLRGK	*P. crispum*	337
MDLLLLEKSL	IAVFVAVILA	TVISKLRGKK	*A. thaliana*	334, 335
MDLLLLEKTL	LGLFAAIIVA	SIVSKLRGKK	*C. roseus*	338
FTVYGEHWRT	MRTIMNLPFF	FKKGVHNYST	*P. vulgaris*	339
MTKLLHSYFS	IPFSPFYVSI	PIATVLFVLI	*P. vulgaris*	340
Ferulate 5-hydroxylase				
MESSISQTLS	KLSDPTTSLV	IVVSLFIFIS	*A. thaliana*	163

3.18.4.3 Hydroxycinnamoyl *O*-Methyltransferases

As described in Section 3.18.3.3.3, there are at least two distinct types of hydroxycinnamoyl OMTs. One apparently preferentially methylates caffeoyl CoA (**14**), and hence may be directed towards coniferyl alcohol (**2**) biosynthesis, whereas the other seems to utilize 5 hydroxyferulic acid (**17**) more effectively, suggesting a role in sinapyl alcohol (**3**) formation.

Genes encoding caffeoyl CoA OMTs have been obtained from parsley (*P. crispum*),[193,194] *Stellaria longipes*,[344] tobacco (*N. tabacum*),[200,345] *Z. elegans*,[198] *A. thaliana*,[346] *Vitis vinifera*,[347] *Eucalyptus gunnii*,[348] *Mesembryanthemum crystallinum*,[349] poplar (*Populus balsamifera* subsp. *trichocarpa*),[350] as well as aspen (*P. tremuloides*[351] and *P. kitakamiensis*[352]). As established for PAL and C4H, the OMTs in these species are currently thought to be the products of either small multigene families[193,198,344] or single genes.[193,351] Additionally, genes encoding 5-hydroxyferulic acid/caffeic acid OMT were identified in aspen,[195,351,353] poplar,[196,197] alfalfa,[192] *Z. elegans*,[199] tobacco,[354] ryegrass,[355] *Capsicum chinense*,[356] *Clarkia breweri*,[357] *Chrysosplenium americanum*,[358] *Prunus dulcis*,[359] barley,[360] and maize[361,362] and are also considered to be derived from either multigene families[192,195,197,353,354] or single genes.[197,361,362] However, profound caution has to be exercised in ensuring that the single gene models are indeed correct, as previously explained for PAL in pine species (see Section 3.18.4.1).

At present, perhaps the most pressing issue, in regard to each class of OMT, involves the determination of their precise functions *in planta* both physiologically and as far as substrate utilization is concerned. Equally important, of course, are their temporal and spatial (subcellular) localizations. Currently, these questions are only partly resolved, primarily because some of the most definitive experiments have yet to be carried out (at least to the point where the results have been reported). Nevertheless, the progress made to the time of writing is summarized below.

3.18.4.3.1 *Hydroxycinnamoyl CoA OMT (caffeoyl CoA OMT)*

The full extent to which hydroxycinnamoyl CoA OMT is present in the plant kingdom has yet to be fully determined. Their genes have, however, been identified in over ten species in the angiosperms[193,198,199,344,351,363] and perhaps, based on a partial sequence, in the gymnosperms as well (*Pinus pinaster*).[364] Enzymatic activities have also been detected in carnation (*Dianthus caryophyllus*), safflower (*Cartharmus tinctorius*), carrot (*Daucus carota*), and *Ammi majus*.[193]

Nothing is yet known about the subcellular localization of any hydroxycinnamoyl CoA OMT, and studies have instead focused only upon mRNA localization (Northern blots and tissue printing). For example, the temporal and spatial appearance of hydroxycinnamoyl CoA mRNAs have been examined in parsley,[193] *Zinnia*,[198,199] *Stellaria*,[344] and aspen.[351] In parsley, Northern blot analyses of roots and leaves suggested an equivalent level of expression in both tissues, although, curiously, the corresponding levels of expression in stems were not individually reported. Hydroxycinnamoyl CoA OMT mRNA expression was also subsequently noted within differentiating xylem elements of parsley petioles,[193] whereas among the stem, root, and leaf tissues of *S. longipes* (ages not provided), highest levels were observed in stems and to a lesser extent in the leaves, but none in roots.[344]

Perhaps the hitherto most significant finding in regard to hydroxycinnamoyl CoA (HCoA) OMT mRNA localization was obtained from *Z. elegans*. Initially, its mRNA was observed in all lignifying cell types,[199] including stems, roots and flower buds, although it was nearly absent in the leaf tissue.[198] Subsequent tissue prints of 4–6-week old *Z. elegans* plant sections, however, established that HCoA

OMT mRNA expression was fully coordinated with the development of lignified tracheids (Figure 24). With 6-week old tissues (first and second internodes), it was essentially located only in the differentiating xylem but not in the developing phloem fibers (Figure 24(a), (b) and (d), (e)), whereas at the third internode (harvested after 4 weeks), extensive transcription was observed in both differentiating xylem and phloem fibers (Figure 24(g), (h)). No expression was detected in the cambium region. Hydroxycinnamoyl CoA OMT mRNA expression was also observed in the stems of 3–4-year old aspen (*P. tremuloides*), but this manifestation was not examined further in much detail.[351]

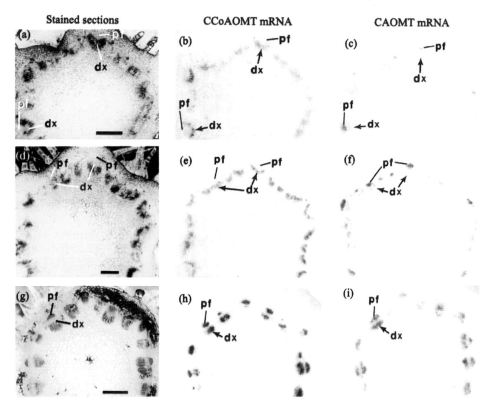

Figure 24 Tissue-print hybridization of the CCoAOMT mRNA and the CAOMT mRNA in *Zinnia* stems. Six-week old ((a) and (d)) and 4-week old (g) *Zinnia* plants were used for tissue prints. RNAs from the mirror sides of a cut were transferred onto two separate membranes and probed with the 35*S*-labeled CCoAOMT and CAOMT antisense RNA probes, respectively. The sections were stained with toluidine blue so that lignified walls stain blue. Anatomy (a) and the CCoAOMT and CAOMT mRNA localization ((b) and (c), respectively) of a section from the first internode. Anatomy (d) and the CCoAOMT and CAOMT mRNA localization ((e) and (f), respectively) of a section from the second internode. Anatomy (g) and the CCoAOMT and CAOMT mRNA localization ((h) and (i), respectively) of a section from the third internode. dx Denotes differentiating xylem and pf phloem fibers. Bars = 0.5 mm (reproduced by permission from *Plant Physiology*, 1995, **108**, 459).

Although the overall significance is as yet unknown, it has also been discovered that the genomic sequence of the unicellular cyanobacterium, *Synechocystis* sp. strain PCC 6803, embodies only a single CoA OMT and no other OMT type.[365] This suggests that the original appearance of caffeoyl CoA OMT may have preceded the evolution of vascular plants, in contrast to that of the molecular pathway to sinapic acid (**7**) and sinapyl alcohol (**3**) which apparently evolved later.

3.18.4.3.2 *Hydroxycinnamic acid OMT (5-hydroxyferulic acid/caffeic acid OMT)*

As in the case of hydroxycinnamoyl CoA OMT, the delineation of the temporal and spatial relationship of 5-hydroxyferulic acid/caffeic acid OMT gene expression and protein localization is only partly complete. The techniques employed have also included, to varying extents, Northern blots, tissue-printing, *in situ* hybridization, and promoter tissue specific expression using aspen,[195] poplar,[197] maize,[362,366] and *Z. elegans*,[199] respectively. As described below, the results hitherto obtained seem to vary between species and even between different *Populus* (poplar/aspen) types. As before, no correlation was definitively established with the actual metabolic end products.

With aspen (*P. tremuloides*), gross Northern blot analyses of 3–4-year old plants revealed that hydroxycinnamic acid OMT was mainly present in developing xylem, and to a lesser extent in so-called bark tissue, but was not detectable in leaves.[195] Additionally, both hydroxycinnamic acid OMT and hydroxycinnamoyl CoA OMT activity and the expression of their corresponding mRNAs followed similar patterns during the growing season. On the other hand, Northern blots from poplar (*Populus trichocarpa × P. deltoides*) indicated that hydroxycinnamoyl OMT was present in both xylem and leaves, although to a much lesser extent in the latter (~50 times less); surprisingly, no comparative examination of hydroxycinnamoyl CoA OMT expression was attempted in this case.[197]

Alfalfa transcript levels, however, as analyzed by Northern blots, differed significantly from those observed in poplar.[192] The alfalfa plants were harvested at weekly intervals over a period of 9 weeks: while the highest level of mRNA expression was again noted in stems, a build-up of hydroxycinnamyl OMT mRNA levels was also clearly observed in root, bud, flower, nodule, leaf, and petal tissues. However, while no metabolic explanation for this observation has been forthcoming, this should be relatively straightforward to resolve since most of the alfalfa phenolics (lignins, isoflavonoids, etc.) are quite well characterized.

A most significant finding was also obtained in *Zinnia*: expression of the hydroxycinnamic acid OMT did not show an absolute correlation with the induction of lignification during tracheid formation (Figure 24).[199] Instead, tissue-print hybridization analyses of 6-week old first and second *Zinnia* internodes showed that the mRNA for the enzyme was abundant in developing phloem fibers rather than in differentiating (lignifying) xylem (Figures 24(a), 24(c), 24(d), and 24(f)) whereas in the third internode (4 weeks old) it was present in both cell types (Figures 24(f) and 24(i)). Immunogold labeling of hydroxycinnamyl OMT in *Zinnia* stems also revealed that the enzyme was mainly present in the phloem fibers of xylem fiber vessels, and to a lesser extent in differentiating vessels.[199] Thus, the expression of hydroxycinnamic acid OMT mRNAs was both temporally and spatially separate from that associated with the hydroxycinnamoyl CoA OMT, a situation that is indicative of quite distinct metabolic roles.

Moreover, there are also at least two distinct hydroxycinnamic acid OMTs present in maize (*Z. mays*), of which one has ~78% similarity to the aspen (*P. tremuloides*) isoenzyme, whereas the other exhibits much lower similarity (~51%).[361,362] With respect to the former,[361] Northern blot analysis initially revealed its mRNA expression in the root elongation zone, and in the mesocotyl of differentiated roots, although lower levels of transcripts were observed in nodes, coleoptiles, adult roots, root meristematic tips, flowers, seeds, embryos, and leaves. Its promoter expression (1955 bp upstream) was also investigated using the GUS reporter strategy,[366] which localized it to various areas, of which only some are involved in lignification, i.e., the meta-xylem of the stem and exodermis of roots, as well as the vascular bundles extending from the node into the leaf. Somewhat comparable observations were made when the same promoter region was introduced into tobacco.

On the other hand, the second hydroxycinnamic acid OMT displayed a very different pattern of expression.[362] Northern blots revealed that its mRNA was present in the endodermal cells which had not yet synthesized the (suberized) Casparian strip, whereas *in situ* hybridization revealed the involvement of both the developing endodermis (Casparian strip) and the exodermis. These observations thus suggest the involvement of this specific OMT in suberization.

Lastly, the *Zinnia* findings are beginning to establish quite clearly that the hydroxycinnamic acid OMTs and hydroxycinnamoyl CoA OMTs present in plants have quite distinct gene expression and enzyme localization patterns. Provisionally, it would appear that the hydroxycinnamoyl CoA OMT is involved in the formation of the so-called guaiacyl lignins in the developing xylem, which is known to precede deposition of the syringyl units (see Section 3.18.6.4). On the other hand, the activation of hydroxycinnamic acid OMT occurs initially only in the phloem fibers, which are considered to be syringyl rich. Accordingly, provided that only lignification is being observed in both developing xylem and phloem fibers, it will be of considerable importance to ascertain why each is initially activated differentially for each cell type, but then eventually both are expressed (at the third internode phase) during the later stages.

3.18.4.4 Hydroxycinnamoyl CoA Ligase

As described in Section 3.18.3.3.4, earlier studies had established the presence of 4-coumarate CoA ligase (4CL) isoenzymes in soybean,[240] pea,[241] petunia,[242] parsley,[233] and poplar,[243] of which several displayed different substrate specificities, thereby suggesting distinct metabolic roles. Since that time, a number of genes that encode (or that are presumed to encode) 4CLs have been reported from, e.g., parsley,[325,329,367-369] tobacco,[370] soybean,[371] *Lithospermum erythrorhizon*,[372] *A. thaliana*,[373] loblolly pine,[374] aspen,[375] rice,[376] potato,[238] and maize.[377] As before, these were classified into either small multigene[232,233,238,370-372,375] or single-gene families.[374]

Surprisingly, in several of these studies, 4CL was occasionally described as the "last step in general phenylpropanoid metabolism."[325,367] This view fails to consider that all vascular plants contain, at the very least, lignins derived from *p*-coumaryl (1) and coniferyl (2) alcohols. Hence, *p*-coumarate CoA ligase (4CL) is not a "last step," given that there are subsequently two other general reductive steps which are necessary for monolignol biosynthesis (see Figure 4). Indeed, this "final step" ascription may originate from those authors' fixation with furanocoumarin/flavonoid formation in parsley cell suspension cultures, which utilizes *p*-coumaroyl CoA (11) to initiate a branchpoint in the pathway.

In terms of subcellular localization of the 4CL isoenzymes, nothing has yet been described. Much has been done, though, in regard to various plant 4CL mRNAs and the associated promoter expression in species such as parsley,[325,329,367-369] potato,[238] tobacco,[369,370] *Arabidopsis*,[373] and aspen.[375] Once again, however, no definitive correlation with any of the presumed metabolic products has been attempted, other than through (occasional) histochemical staining of presumed lignified zones. Initially, expression studies suggested that the distinct 4CL isoenzymes in a particular plant species did not substantially differ in their mRNA expression profiles. For example, in potato,[238] Northern blots indicated that both 4CL-1 and 4CL-2 isoenzymes were concurrently present in stems, leaves, tubers, roots, anthers, petals, and the corolla. Moreover, with 18-day old parsley seedlings, using a probe which did not distinguish between 4CL-1 and 4CL-2 isoforms, *in situ* localization established the presence of 4CL mRNA in epithelial cells, oil ducts, vascular bundles, and, to a lesser extent, in phloem and apex, including apical meristem, ground meristem and emerging leaf primordium. As would be expected, with increasing maturation of the hypocotyls, the 4CL mRNA signals in both vascular bundles and oil ducts decreased greatly as the biosynthetic processes neared completion.

Subsequent studies suggested that both parsley 4CLs might, in fact, be differentially expressed, with the observation that 4CL-1 was strongly activated in root tissue, whereas 4CL-2 predominated in flowering stem regions. In a somewhat analogous manner, distinct soybean mRNA transcripts, presumed to encode various 4CLs, also displayed distinct expression patterns in roots when infected with incompatible and compatible races of the fungus *Phytophthora megasperma* f.sp. *glycinea*.[371]

On the other hand, previous 4CL promoter expression studies using transgenic tobacco and parsley, respectively,[367,368] had been interpreted as leading to the conclusion that there was little difference between the 4CL-1 and 4CL-2 promoter regions, since both were activated by UV and fungal elicitation. Accordingly, only the GUS-fusion expression pattern of the 4CL-1 promoter 597 bp fragment was examined in parsley and tobacco tissues that ranged from 3–4-week old seedlings to 4-month old plants. Significantly, no expression was noted in epidermal cells of either transgenic plant species,[367,368] while in contrast PAL was expressed in these cells.[328] A number of other tissues did, however, display 4CL promoter expression, which predominated in xylem tissue containing differentiating tracheids while also occurring in the primary xylem of axillary buds and developing leaf veins. As the plant tissue matured, no staining was observed in the fully differentiated xylem or mature leaf petioles, and active expression again became restricted to files of ray parenchyma cells as previously noted for PAL in tobacco (Figure 22).[328] GUS staining was also noted in the subapical cells of emerging lateral roots, in floral tissues, developing seeds (in a single epidermal cell layer), and in epidermal cells of mature stigmas and pigmented petal regions.[367] In regard to the foregoing observations, however, it would have been even more informative to have examined the expression of the 4CL-2 promoter, given the lack of 4CL-1 promoter expression in the epidermal cells.

Subsequently, two distinct 4CL-1 genes were obtained from aspen,[375] one form apparently being primarily expressed in the epidermal region and the other localized in vascular (xylem) tissues of both stems and petioles (incorrectly described as leaf tissue in the original paper) (Figure 25). While this may explain differences in tissue specificity between distinct 4CL isoenzymes, it needs to be established if such an observation holds true for all stages of development. It would also be instructive to identify the actual metabolic processes in which the two isoenzymes participate.

Interestingly, both aspen 4CL isoforms were obtained in recombinant form.[375] However, the specific activities for the putatively pure proteins (no details of purification having been provided) were very low and did not differ by even so much as an order of magnitude from that of an ∼72-fold purified preparation from petunia.[242] Moreover, since no attempts were made to measure such fundamental kinetic parameters as V_{max}, it remains unclear what the putative substrate preference might signify.

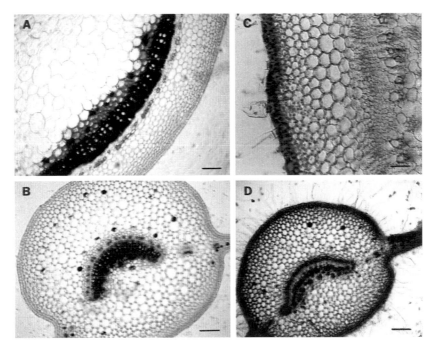

Figure 25 Histochemical analysis of 4-coumarate CoA ligase from aspen fused to the GUS reporter gene in transgenic tobacco. Transverse sections of (a) stem (bar = 100 μm) and (b) petiole (bar = 100 μm) excised from Pt4CL1p-GUS transgenic tobacco, and transverse sections of (c) stem (bar = 250 μm) and (d) petiole (bar = 100 μm) from Pt4CL2p-GUS transgenic tobacco were stained for GUS activity (reproduced by permission from the *Proc. Natl. Acad. Sci. USA*, 1998, **95**, 5407).

Lastly, in contrast to the distinct 4CL-isoforms in parsley, soybean, and aspen, *Arabidopsis* is currently viewed as only having a single 4CL gene.[373] Accordingly, this species was also investigated in order to ascertain the sites of both its mRNA and its promoter expression. mRNA expression was observed about 2 days after germination, being localized mainly in the roots with very low levels in the cotyledons, while the 4CL promoter was expressed within 3 days in the developing xylem of the cotyledons, stems, and roots, although it was not restricted to vascular tissue. As plants reached maturity, however, 4CL mRNA expression was noted in "bolting" stems, mature leaves (low levels), and floral buds, whereas promoter expression was highest in the xylem of bolting stems, as well as in anthers, flower buds (prior to anthesis), and in developing siliques (data not shown).[373] Given the difficulties noted previously, however, in identifying specific tissue and/or metabolic roles for hydroxycinnamate CoA ligase, it seems premature to conclude that *Arabidopsis* hydroxycinnamate CoA ligase is encoded by a single gene. On the other hand, it is even more important, fully to define and delineate the metabolic processes and tissue dependence(s) for the various isoenzymes obtained thus far.

3.18.4.5 Cinnamoyl CoA:NADP(H)-dependent Oxidoreductases

Cinnamoyl CoA:NADP(H)-dependent oxidoreductases (CCR) are also occasionally described as being lignin specific,[252] even though they are involved in the formation of other biochemically distinct metabolites, such as the lignans (see Chapter 1.25, Volume 1). They are presumed to be part of a small gene family, being apparently encoded by a single gene in eucalyptus (*E. gunnii*),[252] and by two in maize (*Z. mays*).[378] As is the case for the "soluble" hydroxycinnamyl OMTs and hydroxycinnamoyl CoA ligases, there has been no reported attempt to determine its subcellular location in any cell type. Northern blot analyses, however, with 5-year old eucalyptus (*E. gunnii*) showed that CCR mRNA transcripts were equally abundant in young stems, roots, and leaves, although attempts to define this further through *in situ* hybridization were unsuccessful due to high background levels.[252] Accordingly, efforts were next directed to employing this technique to study

3-month old poplar (*Populus tremula* × *P. alba*), where signals were only observed in differentiating xylem but not the secondary phloem, flower, or young periderm tissues.[252] As in the case of 4CL isoforms, these studies should have been extended to embrace difference stages of plant development as well as to identify the precise metabolic significance of the enzymatic activities in question.

With the two distinct maize CCRs (designated as Zm CCR1 and Zm CCR2),[378] gene specific probes were designed for each, with Northern analyses then being performed on the (maize) leaves, roots, tassels, and different stem internodes harvested at the flowering stage. Accordingly, Zm CCR1 was present in seminal and adventitious roots, leaves, and stalk (stem), whereas Zm CCR2 was expressed at low levels in root tissue. Again, while no correlation was established with metabolite formation, the authors proposed that each may have a different function: Zm CCR2 was contemplated as being involved in suberin or lignan deposition, whereas Zm CCR1 might participate in lignin biosynthesis.

3.18.4.6 Cinnamyl Alcohol Dehydrogenase

As in the case of CCR, cinnamyl alcohol dehydrogenase (CAD) has occasionally been claimed to be "lignin-specific."[379] This characterization even continues at the time of writing though various plant species, such as western red cedar (*Thuja plicata*), can form ~20% of its dry weight as coniferyl alcohol (**2**)-derived lignans, e.g., plicatic acid (**67**), in addition to the lignin polymers themselves.[380] Moreover, other lignans such as dehydrodiconiferyl alcohol (**68**) (glucosides) are ubiquitous throughout the plant kingdom, and these are also (of course) monolignol-derived.[381] As described in Chapter 1.25, Volume 1, the lignans constitute a biochemical class of plant phenolic metabolites that is quite distinct from the lignins.

CAD has been cloned from a variety of species, including tobacco,[279] loblolly pine,[382] eucalyptus,[282] Norway spruce,[281] alfalfa,[383] poplar,[383] *Arabidopsis thaliana*,[384] *Zinnia elegans*,[385] and *Aralia cordata*,[280] and is thought to exist in single-gene[281,382] and multigene families.[279,280,282,386] As in the case of all the previous enzymatic steps, documenting both the temporal and spatial (sub)cellular organization of the CAD protein(s) and the associated gene expression has been centrally important. Yet attempts to identify the subcellular localization of CAD, *Zinnia* and poplar (*Populus tremula* × *P. alba*) gave different results. For example, in differentiating *Zinnia* tracheids,[322] it was concluded, using polyclonal antibodies raised against the *Aralia* CAD, that the enzyme was mainly present in secondary wall thickenings and, to a lesser extent, in the cytosol, particularly in vesicles. (A similar conclusion by these authors had also been reached for PAL, even though controls showed nonspecific labeling.) On the other hand, polyclonal antibodies raised against eucalyptus CAD (the CAD-2 isoform), when applied to 4-month old poplar plants, yielded results that suggested its presence in the cambium, developing xylem cells, ray cells, and phloem fibers.[387] Interestingly, in the young xylem cells, antibody localization also indicated that CAD accumulated in zones adjacent to primary cell walls which were undergoing secondary thickening and hence presumably lignification. Such an observation appears to be consistent with the localization of PAL near the plasma membrane during putative lignification of parsley cells,[325] perhaps reflecting a legacy of metabolite partitioning.

The tissue and cell specificity of CAD gene expression has also been investigated, but with mixed results.[388–390] The first report involved tobacco, where ~2.6 kb of its promoter region was fused to the GUS reporter gene.[388] Although no data were disclosed, it was claimed that CAD promoter expression, as visualized by GUS-staining, was highest in the roots but lower in stems and leaves. In stem sections, GUS activity was highest in the region encompassing the second to the fourth internodes but it decreased overall with maturation of the lignifying cells until ultimately it was only detectable in the ray parenchyma of the most mature cells, as had previously been noted for PAL (Figure 22).[328] GUS-activity was also present in stem trichomes, axillary buds and leaf veins and trichomes, as well as in the base of flowers (cf. PAL), but it was absent in the cortex, epidermis, and pith which are not lignified.

The same methodology, using a 2.5 kb promoter region of the eucalyptus CAD (CAD-2 isoform), was applied to the study of 3-month old transformed poplar (*Populus tremuloides* × *P. alba*) plants but gave different results depending upon the actual protocol employed.[389] In that study, poplar tissues were first fixed and then assayed for β-glucuronidase activity, whereupon staining was essentially only noted in the ray (parenchyma) cells located between xylem vessels. A comparable observation was made in petiole tissue. Accordingly, these observations prompted the authors initially to conclude that lignin precursors were being biosynthesized and transported into lignifying cells via parenchyma ray cells through cell-to-cell cooperation.

On the other hand, when the sequence of fixation and β-glucuronidase staining was reversed, different observations were made.[390] These are highlighted in Figure 26 and reveal the differences encountered when fixation preceded GUS-staining and vice versa. With the original sequence in protocol (Figure 26(a) and (c)), GUS-staining was again predominantly observed in the ray parenchyma cells and, to a lesser extent, in the phloem fiber cells, sclerids, and periderm. When the sequence was reversed, however, staining was extremely prominent throughout the vascular cambium/differentiating xylem zone and primary xylem/pith zone (Figure 26(b) and (d)), as well as in (results not shown) the periderm layer. These findings are, however, interpreted by the authors of the present chapter as providing additional evidence to the effect that cells undergoing lignification are themselves fully competent to produce lignin. It will be instructive to fully determine the role of the ray parenchyma cells, and to what extent they are involved in the formation of other metabolites, e.g., the lignans.

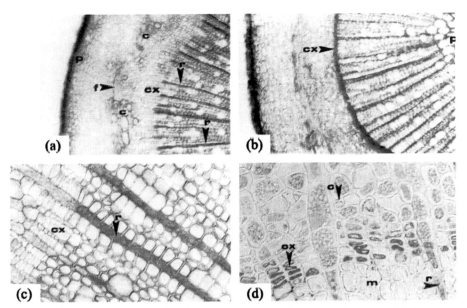

Figure 26 CAD 2 promoter-GUS biochemistry, transverse section of poplar stem. (a) Sample preparation technique 1, hand section, × 100: GUS activity (blue coloration) is present in xylem rays [r], parenchyma cells [c], surrounding groups of phloem fibers [f], and cells interior to the periderm [p]; no coloration can be seen in the vascular cambium/differentiating xylem zone [cx]. (b) Same as in (a), but using sample preparation technique 2, vibratome section, × 100: GUS expression pattern similar to (a) above plus strong activity in vascular cambium/differentiating xylem zone [cx] and primary xylem/pith zone [p]. (c) Same as in (a), vibratome section, × 400: GUS activity restricted to rays [r], no activity in vascular cambium/differentiating xylem [cx]. (d) Same as in (b), paraffin embedded semithin (10 μm) section counterstained with Safranin O, × 400: GUS activity present in cells of vascular cambium/differentiating xylem [cx] and xylem ray cells (r); faint activity can also be detected in remains of cytoplasm in more mature xylem cells [m] undergoing autolysis and lignification, as well as in phloem cells [c] (reproduced by permission from *Plant Physiology*, 1997, **113**, 321).

3.18.4.7 Summary

At the time of writing, the preceding empirical account on enzyme localization and gene expression in the monolignol forming pathway from Phe (**8**) has revealed four noteworthy features. First, care must be exercised in describing any enzymatic step as being lignin specific, since related, yet distinct, metabolic processes may also be active even within the so-called lignifying tissues, e.g., to the lignans. This matter is discussed in detail in Volume 1, Chapter 25.

Second, as secondary xylem maturation and lignin deposition proceeds, cells forming tracheids, vessels and fibers are clearly fully competent to form the required monolignols, and hence the lignin biopolymers. Thus, there is no obvious need for cell-to-cell cooperation involving monolignol (glucoside) transport.

Third, cinnamyl alcohol dehydrogenase is apparently expressed in many different tissues and cell types, e.g., the trichomes and flower vasculature, as well as in gametophytes.[276] While this needs to be clarified in terms of whether lignification is involved or not, particular attention must now be given to other metabolic processes, e.g., to the lignans and dihydrocinnamyl alcohols, as well as to their physiological roles, e.g., in plant defenses.

Fourth, as expected, there is now a growing recognition that different isoenzymes have distinct metabolic functions. This is perhaps best illustrated with the various hydroxycinnamic acid OMTs, and hydroxycinnamoyl CoA OMTs, respectively, which have presumed roles in both lignin and suberized tissue formation. Additionally, the differences between the localization of the distinct hydroxycinnamoyl CoA OMTs and the hydroxycinnamic acid OMTs is striking (see Figure 24), and while unique metabolic functions for each in guaiacyl and syringyl lignin formation respectively, this too remains to be proven. It does, however, suggest an intricacy in monolignol formation that hitherto had not been envisaged for lignin biopolymer formation. In a somewhat analogous manner, the distinct CoA ligase isoenzymes and associated genes appear to be both differentially localized in different tissues, e.g., the epidermis versus secondary xylem forming tissues, and the metabolic significance of this needs also to be established.

Routine techniques are now in place to help establish how these variable metabolic processes are differentially initiated, regulated, and controlled. One major difficulty in this regard, however, remains, namely that no single plant species has yet been examined comprehensively, at all stages of development, in order to ascertain how the phenylpropanoid pathway is expressed and controlled.

3.18.5 MONOLIGNOL PARTITIONING, INTRACELLULAR TRANSPORT, GLUCOSYLTRANSFERASES, AND β-GLUCOSIDASES

Originally Freudenberg[2,391] proposed that in the cambial tissues the monolignols (**1**)–(**3**) were converted into the corresponding glucosides, *p*-coumaryl alcohol glucoside (**71**), coniferin (**72**), and syringin (**73**). These were envisaged then to diffuse into cells undergoing lignification where they were supposed to migrate into the secondary cell walls. According to that view, the action of a β-glucosidase would regenerate the monolignols which would then be polymerized into the lignin biopolymers. This perspective about lignification, however, never took into account the formation of other metabolites, such as the lignans, even though spruce (*Picea abies*) contained them in the same "cambial" tissues being examined.[392] Nor did these authors recognize that lignifying cells were fully competent to biosynthesize lignin precursors directly. Indeed, Holgar Erdtman stated as early as 1957 that "already the uneven distribution of lignin in the cell walls may serve as an indication of the latter view."[393] Later, Marcinowski and Grisebach[394] concluded that turnover of coniferin (**72**) in *P. abies* could only account for part of lignin biopolymer deposition.

It is worthwhile briefly to consider the reasons why it was concluded that the monolignols accumulate in cambial cells prior to subsequent diffusion into those cells undergoing lignification. Indeed, a contribution entitled "the glucosides of cambial sap of spruce (*Picea abies*)" from Freudenberg and Harkin sheds some light on the matter,[392] the following excerpt from which gives the lively flavor of plant biochemical research during a bygone era.

The peeled trunk (3 feet long) was stood on its end on a galvanized iron grating over a large polyethylene dish and washed down with boiling water containing 2% formalin, in order to inactivate the enzymes. This arrangement must be used to prevent the sap from being sucked up by capillarity by the vessels at the end of the trunk. The cambial tissue around the surface of the trunk was then scraped off and the sap and tissues washed into the dish with more boiling formaldehyde solution. The inside of the strips of bark were worked up separately as the sap from this contained greater quantities of hydroxymatairesinol... The sticky solution of cambial sap was separated from tissue material by filtration through cloth, and the residue extracted with warm dilute formaldehyde solution... The benzene extract was shown to contain a small amount of coniferyl alcohol and the main phenolic substances which are formed on dehydrogenation of the latter compound with laccase *in vitro*. The butanol extract (1.2 g) contained larger amounts of the same materials, some of which were identified by paper chromatography using the solvent mixtures described by Freudenberg and Lehmann. These included coniferyl alcohol, coniferyl aldehyde, dehydrodiconiferyl alcohol, pinoresinol, guaiacylglycerol-β-coniferyl ether, and surprisingly, relatively large amounts of guaiacylglycerol-β-pino-resinol ether, a substance only recently recognized as a trimeric intermediate of lignification.

Undoubtedly these substances must be regarded as occurring naturally... Matairesinol was also present, probably extracted from the resin ducts or from knots or the stumps of the branches in the wood.

The important point to note is that both monolignols and dimeric lignans co-occurred in the same tissue(s) under investigation, although to what extent artifacts were formed was beyond the documentation provided. The authors also concluded later in the paper that *p*-coumaryl alcohol glucoside (**71**), coniferin (**72**), and syringin (**73**) were present in the cambial sap, although it is worth noting that syringin (**73**) has never since been reported to occur in the Pinaceae.

The present section thus intends to evaluate the still incomplete body of knowledge as to how monolignol partitioning occurs, as well as to the possible participation of both glucosyltransferases and β-glucosidases in the pathways involved. The main issue to be borne in mind in clarifying the matter is that cells forming lignified walls (i.e., tracheids, vessels, fibers) are quite capable of synthesizing monolignols (**1**)–(**3**) from phenylalanine (**8**); however, the complication is that other metabolic processes also utilize monolignols, but whether they co-occur in cells that become tracheids, vessels, and fibers remains to be determined.

3.18.5.1 Monolignol Partitioning

Monolignols (**1**)–(**3**) can be differentially partitioned (see Figure 27 and Schemes 11–13) into several distinct biochemical pathways affording the lignins, the (oligomeric) lignans (e.g., (**67**), (**68**), (**74**)–(**77**), (**82**)–(**100**), (**108**)–(**112**)), the monolignol glucosides ((**71**)–(**73**), (**78**), (**80**), (**81**)), the dihydrocinnamyl alcohols ((**69**), (**70**), (**107**)), and, probably, the alkyl phenols ((**101**)–(**106**)) such as (iso)eugenol (**102**) or (**105**).

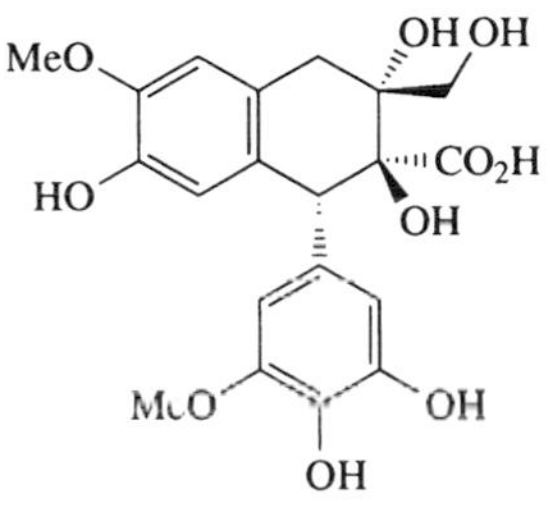

(**67**) Plicatic acid

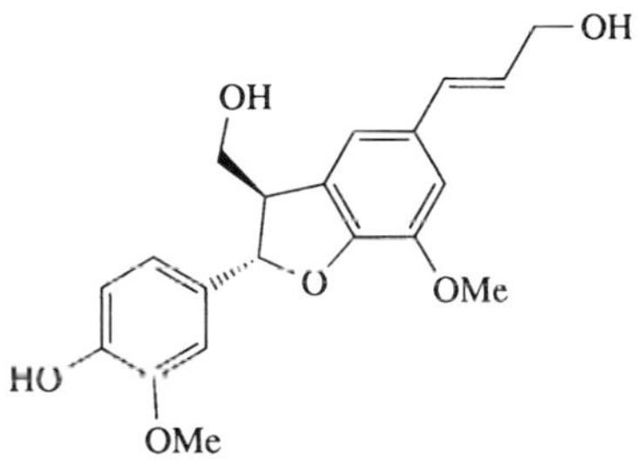

(**68**) Dehydrodiconiferyl alcohol

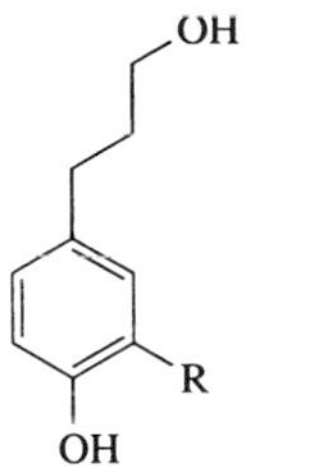

(**69**) R = H, Dihydro *p*-coumaryl alcohol
(**70**) R = OMe, Dihydroconiferyl alcohol

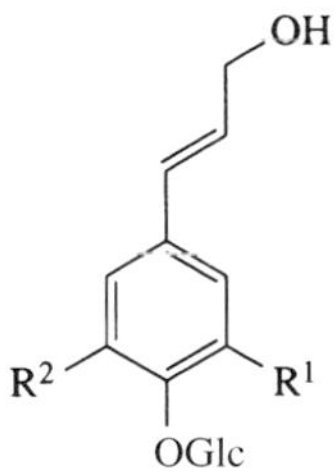

(**71**) R^1 = R^2 = H, *p*-Coumaryl alcohol glucoside
(**72**) R^1 = OMe, R^2 = H, Coniferin
(**73**) R^1 = R^2 = OMe, Syringin

Monolignol partitioning is a phenomenon worthy of emphasis because of a tendency in the field to arbitrarily to correlate all biochemical processes involving monolignols (e.g., glucoside formation) with lignin deposition. Lewis and co-workers have pioneered several studies devoted to elucidating the biochemical fate of the monolignols (**1**)–(**3**) and the corresponding 8—8′, 8—5′ and 8—*O*—4′ lignans in various plant species. From these studies, more than 10 new enzymatic steps were characterized in monolignol/lignan metabolism. They are, however, biochemically distinct from the corresponding transformations affording the lignins, even although ostensibly lignifying tissues are sometimes implicated. The reader is referred to Chapter 1.25, Volume 1 for a summary of the progress hitherto made in identifying and characterizing the relevant proteins, enzymes, and genes.

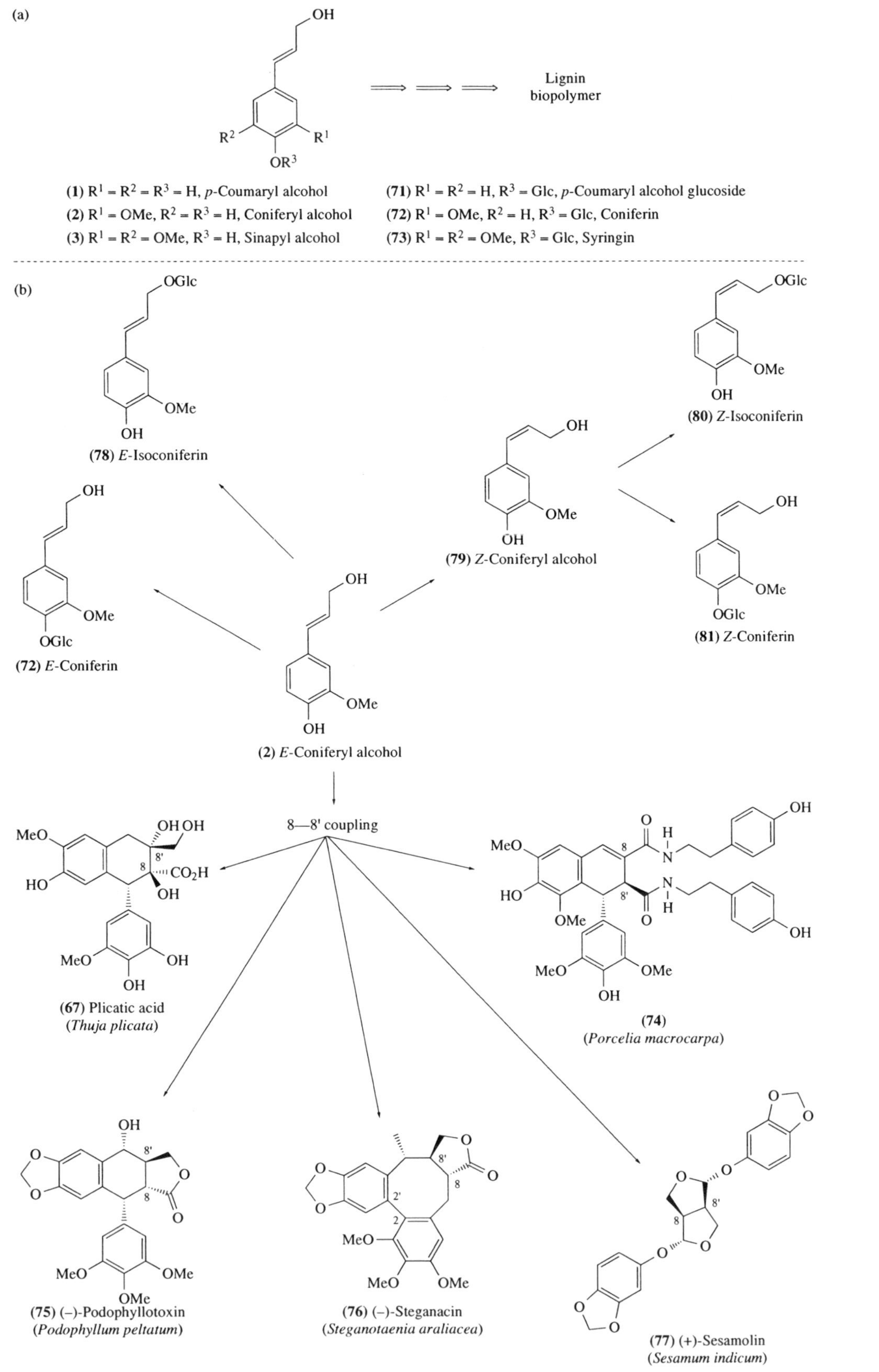
(a)
OH
R^2
R^1
OR^3
Lignin biopolymer
(1) $R^1 = R^2 = R^3 = H$, *p*-Coumaryl alcohol
(2) $R^1 = OMe$, $R^2 = R^3 = H$, Coniferyl alcohol
(3) $R^1 = R^2 = OMe$, $R^3 = H$, Sinapyl alcohol
(71) $R^1 = R^2 = H$, $R^3 = Glc$, *p*-Coumaryl alcohol glucoside
(72) $R^1 = OMe$, $R^2 = H$, $R^3 = Glc$, Coniferin
(73) $R^1 = R^2 = OMe$, $R^3 = Glc$, Syringin
(b)
(78) *E*-Isoconiferin
(80) *Z*-Isoconiferin
(79) *Z*-Coniferyl alcohol
(72) *E*-Coniferin
(81) *Z*-Coniferin
(2) *E*-Coniferyl alcohol
8—8' coupling
(67) Plicatic acid
(*Thuja plicata*)
(74)
(*Porcelia macrocarpa*)
(75) (–)-Podophyllotoxin
(*Podophyllum peltatum*)
(76) (–)-Steganacin
(*Steganotaenia araliacea*)
(77) (+)-Sesamolin
(*Sesamum indicum*)

Figure 27 Biochemical pathways involving monolignols to: (a) lignins and (b) 8–8′ linked lignans and *E*- and *Z*-monolignol derivatives.

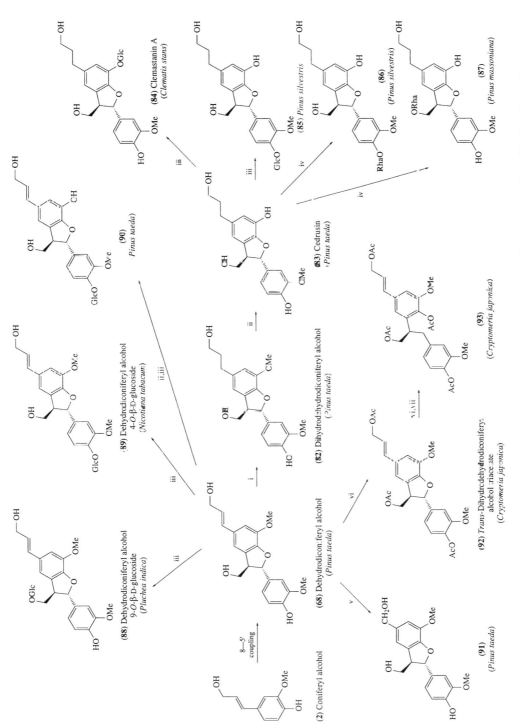

Scheme 11

i, allylic reduction; ii, regiospecific demethylation; iii, regiospecific glucosylation; iv, regiospecific rhamnosation; v, C-7–C-8 cleavage; vi, acetylation; vii, regiospecific reduction

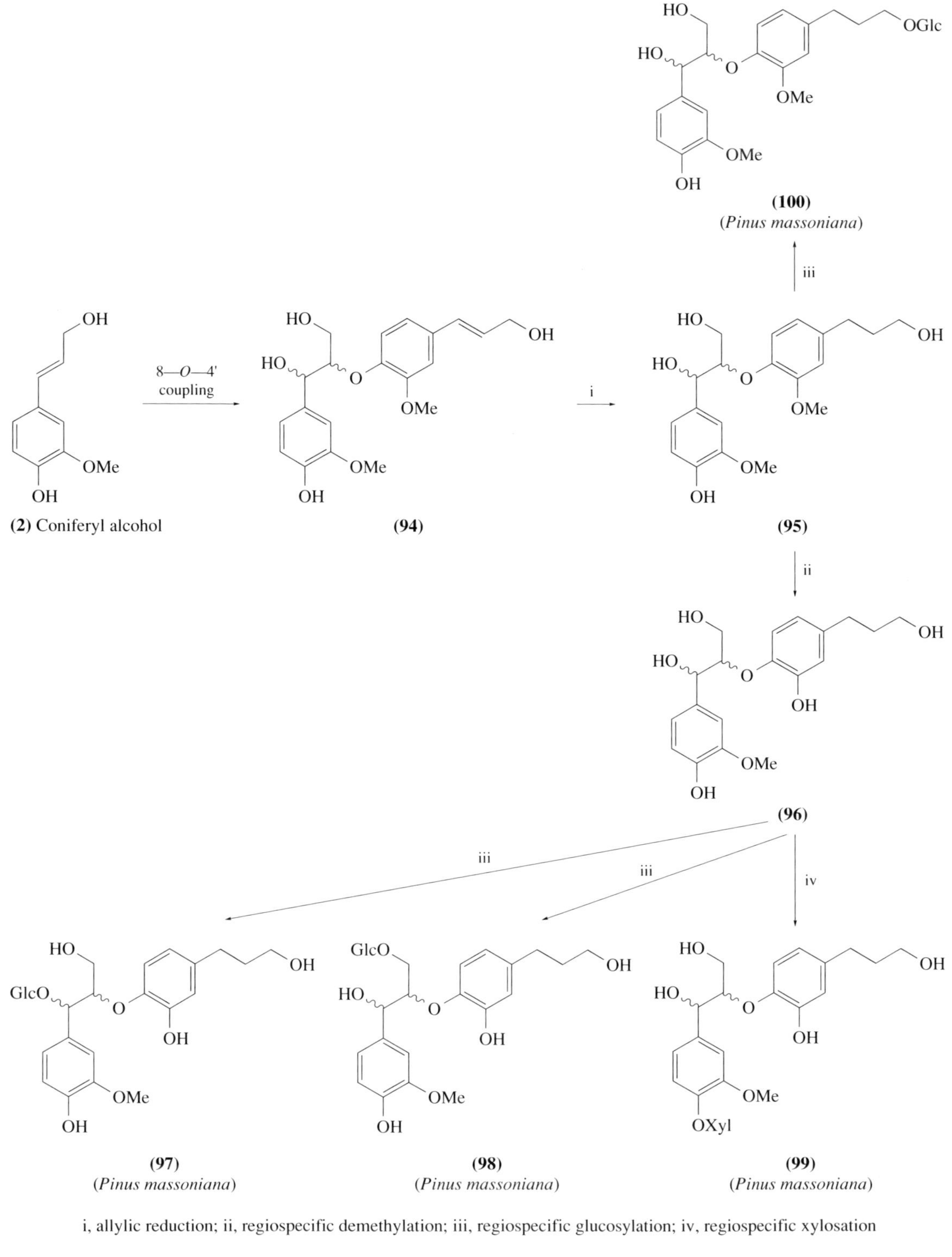

(100)
(*Pinus massoniana*)

(2) Coniferyl alcohol   **(94)**   **(95)**

(96)

(97)   **(98)**   **(99)**
(*Pinus massoniana*)   (*Pinus massoniana*)   (*Pinus massoniana*)

i, allylic reduction; ii, regiospecific demethylation; iii, regiospecific glucosylation; iv, regiospecific xylosation

Scheme 12

3.18.5.2 Intracellular Monolignol Transport Leading to Lignification

How intracellular monolignol transport occurs, and the biochemical nature of the monomers being transported into the cell wall, was still not unambiguously resolved at the time of writing. There appear to be four different possible alternatives (Figure 28): (i) monolignols **(1)**–**(3)** are transported directly to the plasma membrane and are then released into the cell wall; (ii) the

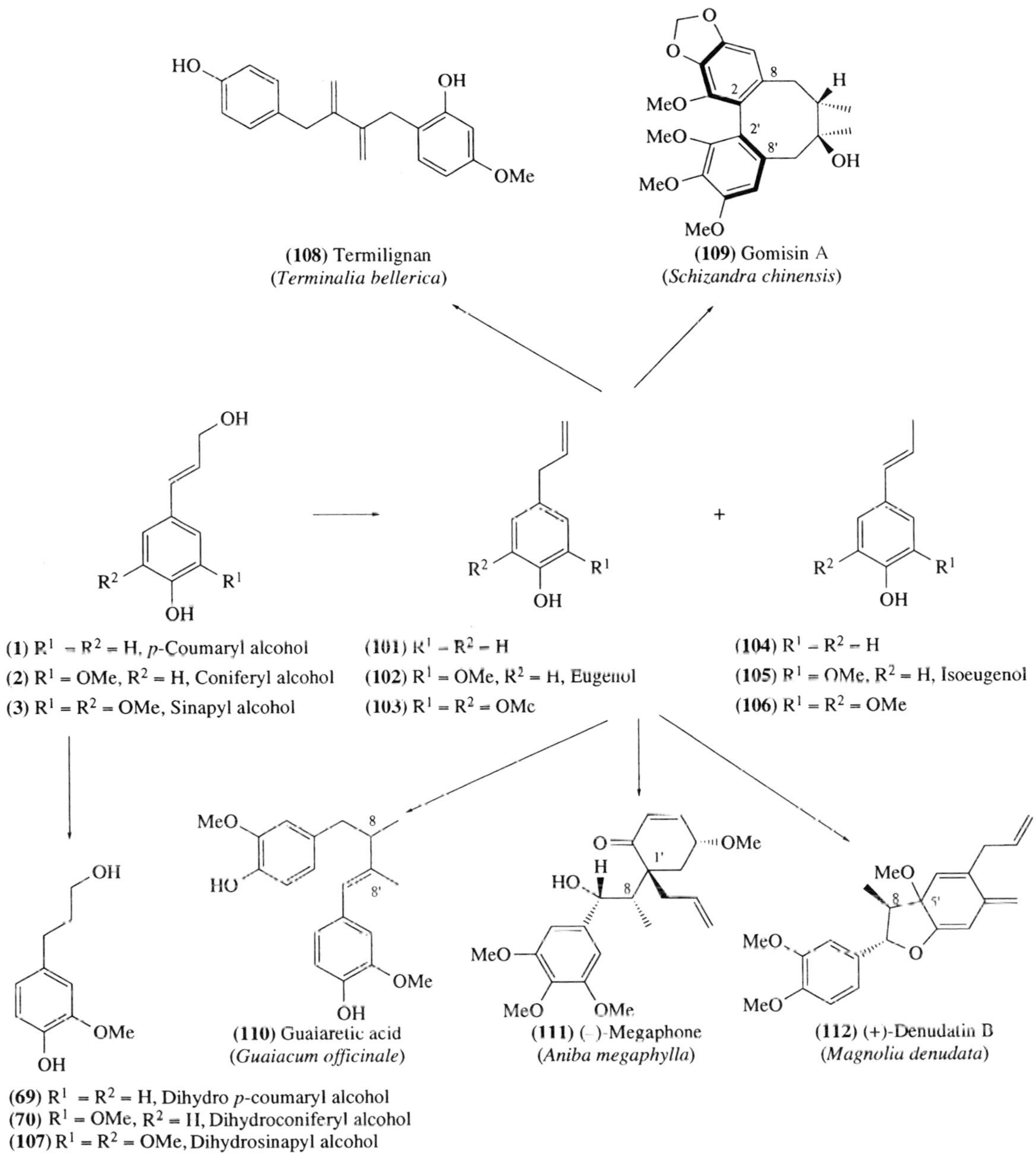

(108) Termilignan
(*Terminalia bellerica*)

(109) Gomisin A
(*Schizandra chinensis*)

(1) R¹ = R² = H, *p*-Coumaryl alcohol
(2) R¹ = OMe, R² = H, Coniferyl alcohol
(3) R¹ = R² = OMe, Sinapyl alcohol

(101) R¹ = R² = H
(102) R¹ = OMe, R² = H, Eugenol
(103) R¹ = R² = OMe

(104) R¹ = R² = H
(105) R¹ = OMe, R² = H, Isoeugenol
(106) R¹ = R² = OMe

(110) Guaiaretic acid
(*Guaiacum officinale*)

(111) (−)-Megaphone
(*Aniba megaphylla*)

(112) (+)-Denudatin B
(*Magnolia denudata*)

(69) R¹ = R² = H, Dihydro *p*-coumaryl alcohol
(70) R¹ = OMe, R² = H, Dihydroconiferyl alcohol
(107) R¹ = R² = OMe, Dihydrosinapyl alcohol

Scheme 13

monolignols **(1)**–**(3)** are correspondingly transported as their glucosides **(71)**–**(73)**; (iii) both the monolignols **(1)**–**(3)** and the monolignol glucosides **(71)**–**(73)** can be involved, but to different extents; or (iv) some unknown transport mechanism is operative of a nature broadly comparable to the one responsible for targeting flavonoids into the vacuole.[395] Of these, (i) currently appears to be the most likely in view of the following observations.

There have been a limited number of studies directed towards identifying the subcellular organelles where the monolignols **(1)**–**(3)** are formed and from which they are transported to the cell wall. For example, Pickett-Heaps[396] described an ultrastructural study of xylem cell wall deposition in very young wheat coleoptiles (0.2–0.5 cm shoot length) where the uptake and metabolism of both radiolabeled phenylalanine **(8)** and [³H]cinnamic acid **(10)** were followed to the point of incorporation into the presumed lignin. The results, which are still occasionally cited,[397] suggested that vesicles containing lignin monomers appeared to be fusing with the plasma membranes of lignifying tracheids (or more precisely tracheary elements). These interpretations, however, need to be con-

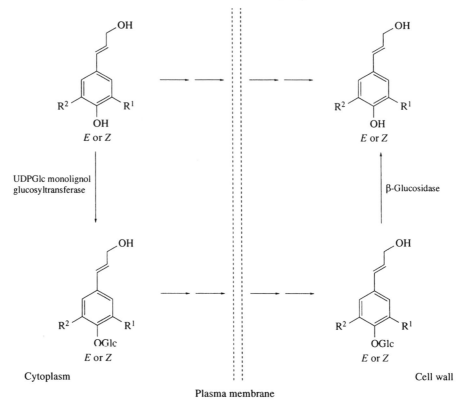

Figure 28 Postulated monomer transport into lignifying cell walls. (R^1, R^2 = H or OMe).

sidered carefully, since nitrobenzene oxidation of the putatively lignified material released only a single radiolabeled constituent with a paper chromatographic migration profile similar to that of *p*-hydroxybenzaldehyde (**21**). Significantly, neither radiolabeled vanillin (**22**) nor syringaldehyde (**56**) were detected. Given the young age of the plant material, these data seem more consistent with the metabolism of [³H]cinnamic acid (**10**) into *p*-coumarate (**5**), a common cell wall constituent of wheat.[398] To the extent that wheat lignins contain all three monolignols (**1**)–(**3**), nitrobenzene oxidation should have liberated all three aldehydes ((**21**), (**22**), and (**56**)) had the tissue reached any degree of maturity in lignification. Thus, it is not certain that lignin biosynthesis was effectively examined by the work of Pickett-Heaps.

Deposition of the cell-wall components (cellulose, hemicellulose, and lignins) in the differentiating xylem of *Cryptomeria japonica* was examined by Takabe *et al.*[399,400] These workers concluded, from autoradiography of tissues previously administered the L-[2,4,6-³H]amino acid (**8**), that the Phe-derived radioactivity was primarily distributed intracellularly throughout the Golgi-bodies, vesicles derived from same, as well as on both the rough and smooth endoplasmic reticulum (rER and sER). Of these, only the vesicles and sER were recognized as fusing with the plasma membrane, and thus they were thought to represent the conduit for lignin monomer transport to the cell wall interface; the rER, on the other hand, was not considered further since it is involved in protein biosynthesis.

The progression of lignin biosynthesis in tracheids, vessels, and fibers is accompanied by the ultimate destruction of the cytoplasm during the final stages of wall maturation.[401] This last phase is often viewed as occurring through a mechanism marked by increasing vacuolarization, which pushes the cytoplasmic components towards the plasma membrane (PM). Often, so-called "lipid" or "oil-bodies" appear to migrate toward the PM at this stage, and these could also be associated with lignin monomer transport. Consequently, there appear to be at least three alternative possibilities that could describe how monolignols are transported; hence it is premature to make any categorical conclusions about how transport of the lignin monomers to the cell wall actually occurs.

Moreover, the chemical identities of the monomers being transported has not been unambiguously established. That these are the monolignols (**1**)–(**3**) is currently favored by the present authors on the basis of the following considerations. First, monolignol glucosides that have been partitioned into the cell wall, and subsequently hydrolyzed by a β-glucosidase, would generate an equimolar amount of glucose relative to the monolignols and lignin monomer residues. There is, however, no evidence for anything like it. Second, in the study of *p*-coumaryl (**1**) and coniferyl (**2**) alcohol biosynthesis in *P. taeda* cell suspension cultures, only the monolignols (**1**)–(**3**), but not their glucosides (**71**)–(**73**), were detected intracellularly (see Section 3.18.3.2). The monolignols (**1**)–(**3**) were observed in the extracellular medium which, incidently, also possessed measurable β-glucosidase activity.[402] Third, in a study of the developing xylem of jack pine (*P. banksiana*) and *Pinus strobus*,[403] the coniferin (**72**) detected was mainly present in the vacuoles of protoplasts presumed to be primarily derived from ray cells. These cells are not, however, proven to be exclusively involved in lignification (see Section 3.18.4.6). Fourth, in a study of the metabolism and turnover of coniferin (**72**) in *Picea abies*, Marcinowski and Grisebach[394] deduced that just a part of the lignin might be derived from coniferin (**72**), which itself accumulated only to low levels in *P. abies* seedlings (50–70 nanomoles per seedling) and was apparently turned over very slowly (60–120 h). Fifth, although β-glucosidases have been detected in all gymnosperms examined, they have only been found in a limited number of angiosperm plant species and not in others such as poplar (B. E. Ellis, personal communication).

Thus, the question of intracellular monolignol transport to the cell walls still awaits resolution, which will require the following steps: establishing the chemical identities of the lignin monomers present in vesicles, sER, and/or oil-lipid bodies; determining the cell types involved in monolignol glucoside formation, and establishing whether only specialized cells are involved such as the ray parenchyma; documenting the subcellular localization of monolignol glucosyltransferase(s), elucidating whether or not monolignol glucosides contribute to lignin deposition in tracheids, vessels, and fibers; and mapping the overall distribution of β-glucosidases in the plant kingdom, and their particular cell specificities and subcellular locations.

3.18.5.3 Monolignol UDPGlc-Glucosyltransferase(s)

The most frequently encountered monolignol glucoside is coniferin (**72**), which had been purified from *Abies excelsa*, *A. pectinata*, *Pinus strobus*, *P. cembra*, and *Larix europaea* as early as 1874. Its structure was determined at that time by Tiemann and Haarman,[404] and treatment with emulsin (β-glucosidase) converted it to coniferyl alcohol (**2**) and glucose.[405] Since then it has been established that both the *E*- and *Z*-forms of the monolignol glucosides, e.g., the *E*-coniferyl (**2**) and *Z*-coniferyl (**79**) alcohol derivatives, exist in plants, the latter having been reported in American beech (*Fagus grandifolia* Ehrh) bark.[406,407] Interestingly, glucose can be linked at either the phenolic oxygen atom (4-position) or at the C-9 terminal hydroxyl group[407,408] to give, for example, isoconiferin (**78**).

Monolignol uridine diphosphate glucose (UDPGlc) glucosyltransferases (EC 2.4.1.111) catalyze the conversion of monolignols (**1**)–(**3**) into the corresponding glucosides (**71**)–(**73**) (Figure 29), although none has yet been cloned or sequenced. The study of the monolignol UDPGlc-glucosyltransferases has lagged far behind that of other glucosyltransferases, such as those catalyzing DNA glucosylation in the T4 phage.[409] Indeed, the first report of a monolignol UDP-glucosyltransferase appeared only in 1976 and described a partially purified preparation from Paul's Scarlet rose.[410] This activity was later detected in numerous angiosperms and pteridophytes, and possibly the bryophytes,[411] although the latter do not appear to biosynthesize monolignols (as discussed earlier in Section 3.18.2.2 and reference [1]).

The first monolignol glucosyltransferase purified to apparent homogeneity (∼1700-fold) was obtained from *Picea abies* cambial sap in 1982.[412] It had a M_r ∼ 50 000 and was believed to exist as a monomer. It displayed a preference for coniferyl alcohol (**2**) (K_m = 250 μM, V/K_m = 3400 nkat l^{-1} mol^{-1} mg^{-1}, with K_m = 250 μM for UDPGlc), coniferin (**72**) being the only product formed without any isoconiferin (**78**) being detected.[412] Mechanistically, a mono-iso-ordered bi-bi catalytic transformation was proposed as shown in Figure 29. On the other hand, sinapyl alcohol (**3**) (K_m = 580 μM, V/K_m = 230 nkat l^{-1} mol^{-1} mg^{-1}) was the least preferred substrate and *p*-coumaryl alcohol (**1**) was not effectively utilized (∼7% of the activity observed with respect to coniferyl alcohol (**2**)).

Since then there have only been a few reports about monolignol glucosyltransferases involving crude cell-free extracts from *F. grandifolia*[413] and *P. taeda*,[40] and an account of partially purified enzymes from jack pine (*P. banksiana*) and *P. strobus*.[414] In the study of *F. grandifolia*, it was demonstrated that *Z*-coniferyl alcohol (**79**) was the preferred substrate relative to its *E*-form (**2**)

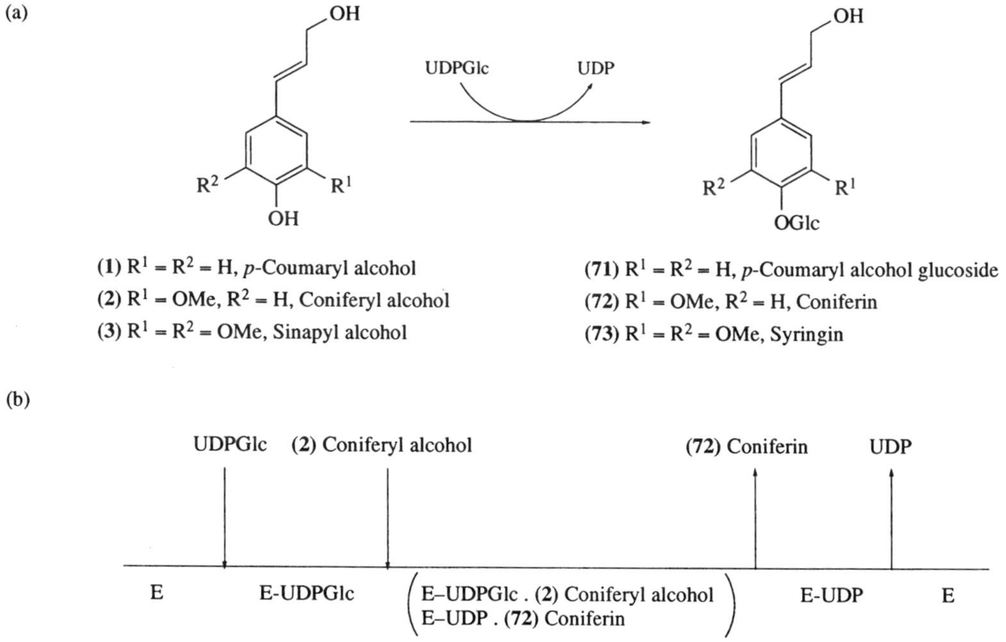

(a)

(1) $R^1 = R^2 = H$, *p*-Coumaryl alcohol
(2) $R^1 = OMe$, $R^2 = H$, Coniferyl alcohol
(3) $R^1 = R^2 = OMe$, Sinapyl alcohol

(71) $R^1 = R^2 = H$, *p*-Coumaryl alcohol glucoside
(72) $R^1 = OMe$, $R^2 = H$, Coniferin
(73) $R^1 = R^2 = OMe$, Syringin

(b)

Figure 29   (a) Monolignol UDPGlc glucosyltransferase reaction. (b) Proposed ordered bi-bi mechanism of monolignol glucosyltransferase. E, enzyme. Redrawn from ref. 412.

(5.7% vs. 0.2% conversion during the same period)[413] whereas with *P. taeda* *E*-coniferyl alcohol (2) was only slightly preferred (57% vs. 27% conversion).[40]

No X-ray crystal structure was available at the time of writing for any monolignol glucosyltransferase, although others, such as the bacteriophage T7 β-glucosyltransferase (EC 2.4.1.27), had been described and may give some useful mechanistic insights. The latter catalyzes the transfer of glucose from UDPGlc to a 5-hydroxymethylcytosine residue in DNA.[415] As shown in Figure 30, the UDPGlc is considered to reside within the two characteristic domains of similar topology on the glucosyltransferase, where the uracyl moiety is hydrogen bonded through the hydroxyl group at C-4 and N-3 of the ring to Ile[238] (main chain nitrogen and oxygen) (Figure 31). Hydrogen bonding of the ribose ring occurs through its C-2 and C-3 hydroxyl groups with the Glu[272] carboxyl side chain, and the three phosphate oxygens are also hydrogen bonded with the guanidinium groups of the arginine residues (at 191, 195 and 269, respectively). The attached glucose residue is presumed to rest in a pocket near the O-5P of the terminal phosphate group. Mechanistically, it is envisaged that nucleophilic attack takes place at C-1 of the glucose via attack by the 5-hydroxymethyl group of the modified cytosine substrate, which in turn is thought to be activated by various neighboring residues. Thus, an X-ray crystal structure of the monolignol glucosyltransferase(s) will be of considerable interest from a comparative point of view.

Lastly, a preliminary study of the subcellular localization of the presumed spruce (*P. abies*) monolignol UDP-glucosyltransferases was carried out in 10 day old seedlings, where it was noted that the glucosyltransferase activity increased rapidly after 6 days, reached a maximum level by the 10th day, and then decreased back to the base value by day 16.[416] Nevertheless, immunofluorescent localization of the corresponding antibodies revealed that the enzyme was mainly present in the epidermal/subepidermal layers, as well as to a lesser extent in the vascular bundles.

It should also be recognized, however, that plants also contain a wide array of lignan glucosides, e.g., (84)–(89), (97)–(100), which are clearly generated following monolignol coupling and/or subsequent enzymatic modification. For example, in a single species such as pine (*Pinus massoniana*), glycosylation of guaiacylglycerol 8–*O*–4′ coniferyl alcohol ether (94) residues can occur differentially even at C-4, C-7, C-9, and C-9′ (see Scheme 12), in addition to the customary glucosylation resulting in coniferin (72) formation. This diversity in the derivatization patterns again underscores how poorly the roles of glucosylation *in vivo* are understood, and emphasizes the need for circumspection in assigning any physiological function as far as lignification is concerned.

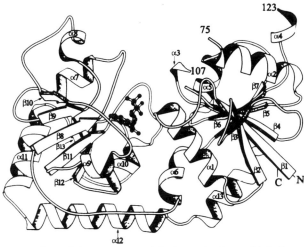

Figure 30 A ribbon representation of the structure of the bacteriophage T4 β-glucosyltransferase. The atoms for the UDP portion of the substrate are shown in ball and stick representation[415] (reproduced by permission from *EMBO Journal*, 1994, **13**, 3413).

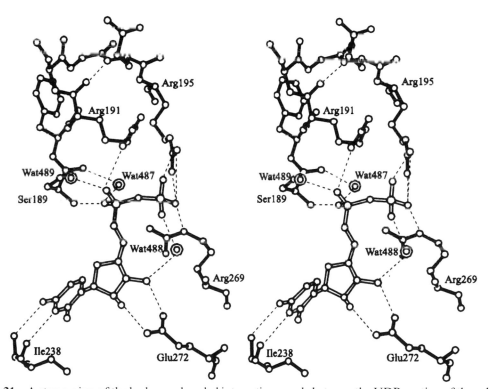

Figure 31 A stereo-view of the hydrogen bonded interactions made between the UDP portion of the substrate and the surrounding bacteriophage T4 protein and water molecules. The UDP portion of the substrate is shown in open bonds and the protein is shown in closed bonds. Three water molecules are shown by double circles. Hydrogen bonded interactions are shown by dotted lines (reproduced by permission from *EMBO Journal*, 1994, **13**, 3413).

3.18.5.4 β-Glucosidase

β-Glucosidases are important members of the large class of glycosyl hydrolases, and have been found in five of the 57 families known.[417] They are defined as enzymes capable of hydrolyzing β-glucosidic linkages present in either disaccharides, oligosaccharides, or so-called conjugated glucosides, such as coniferin (**72**) and syringin (**73**). The earliest report of a β-glucosidase was one about emulsin in almonds which acts on the cyanogenic glucosides;[418] the overall modes of action for this enzyme and the monolignol glucoside β-glucosidases are shown in Figure 32.

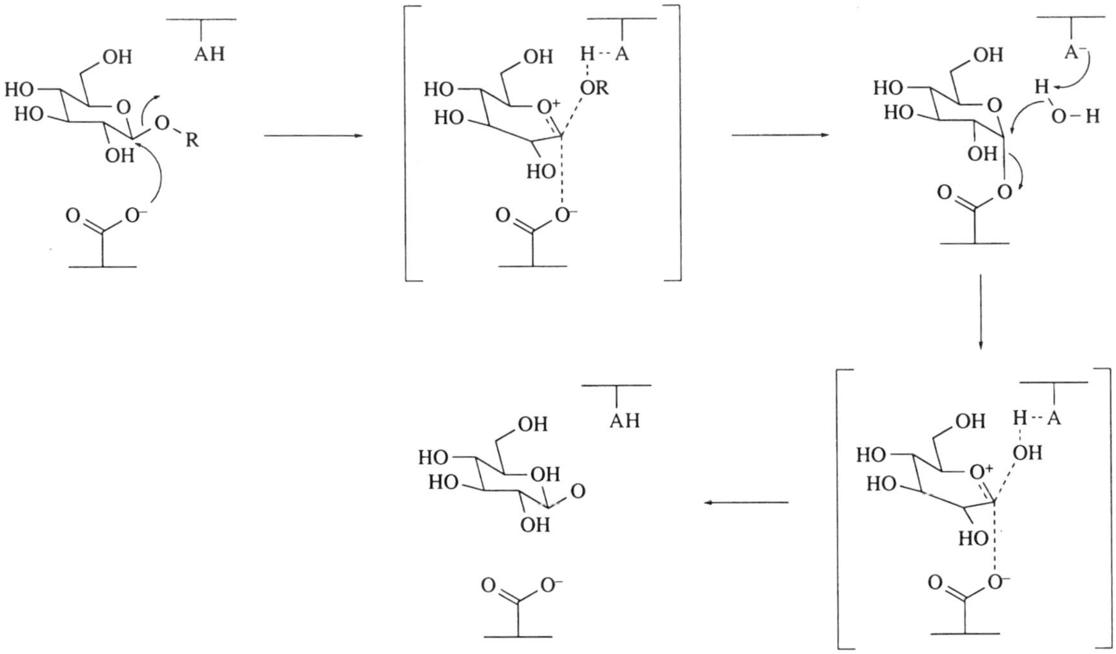

Figure 32 Proposed mechanism of action of β-glucosidase, using the mechanism described for *Agrobacterium faecalis* β-glucosidase for illustrative purposes.[426] R = glucoside conjugate.

According to the view of Freudenberg, the β-glucosidases are essential for lignin formation; they were thought to be colorimetrically detectable in spruce (*P. abies*) cambial sap through staining with indican.[391,419] Later β-glucosidase preparations, that were also presumed to be connected with lignification, were obtained from chick pea (*Cicer arietinum*)[420] and spruce (*P. abies*)[421] seedlings. Each exhibited interesting differences in substrate specificity: in chick pea, *E*-coniferin (**72**) was the preferred substrate with respect to *E*-syringin (**73**) (V/K_m = 120 and 6, respectively), whereas the spruce β-glucosidase utilized both substrates (**72**) and (**73**) equally well (V/K_m = 110 and 67, respectively). Subsequent immunofluorescent localization of the spruce β-glucosidase in *P. abies* seedlings was also quite instructive; it was detected in all cell types from the epidermal layer to the pith, although it predominated slightly in the epidermal and vascular bundles. On the other hand, phloroglucinol-HCl staining gave a color reaction only with the epidermis, endodermis (Casparian strip), and vascular bundles of the immature seedlings,[422] suggesting that localization of the components reacting with phloroglucinol-HCl and the β-glucosidase might only be partly coincident.

A third member of the monolignol glucoside β-glucosidase group of enzymes has been purified from the differentiating xylem of lodgepole pine (*Pinus contorta*), together with two others which were detected through their ability to hydrolyze 4-methyl umbelliferyl β-D-glucoside.[397,423] The lodgepole pine β-glucosidase hydrolyzed both coniferin (**72**) and syringin (**73**) effectively, although the former substrate was appreciably preferred, in a manner analogous to that noted for spruce β-glucosidase. Table 18 shows the range of substrate versatility reported for the enzyme.

The *P. contorta* β-glucosidase has been purified to apparent homogeneity as a protein with $M_r \sim 60$ kDa,[423] the gene of which was subsequently cloned and the corresponding recombinant protein obtained.[424] Apparently it possesses a 23 N-terminal amino acid sequence reminiscent of a secretory signal peptide. However, whether the protein is targeted to an extracellular location in the cell wall, or attached to an intracellular membrane, remains to be determined. There is only one other gymnosperm β-glucosidase that has been reported: this enzyme is found in jack pine (*P. banksiana*),[425] but it apparently has a larger molecular mass and also an N-terminal sequence having no apparent homology to any other known glucosidase.[397]

No detailed enzymatic analysis of the lodgepole pine coniferin β-glucosidase has yet been

Table 18 Substrate specificity of purified coniferin (**72**) β-glucosidase from *Pinus contorta*.[397]

Substrate	Relative activity (coniferin (**72**) = 100)	K_m (mM)
Coniferin (**72**)	100	0.18
Syringin (**73**)	51	0.29
4-Methyl umbelliferyl-β-glucoside	16	2.30
2-Nitrophenyl β-glucoside	57	—
4-Nitrophenyl β-glucoside	26	1.90
5[4-(β-D-glucopyranosyloxy)-3-methoxyphenyl-methylene]-2-thioxothiazolidine-4-one-3-ethanoic acid	45	—
Salicin	10	—
4-Methyl umbelliferyl-α-glucoside	1	—
2-Nitrophenyl β-galactoside	12	—
4-Nitrophenyl β-cellobioside	not detected	—

described. Nevertheless, much is known about β-glucosidases in terms of their general catalytic mechanisms. This is illustrated in the scheme proposed by Withers (Figure 32),[426,427] which closely resembles a version advanced much earlier by D. E. Koshland in 1953.[428] The hydrolytic cleavage is considered to occur through a double displacement mechanism that is based upon a detailed mechanistic study of the β-glucosidase from *Agrobacterium faecalis*[426] coupled with results from site-directed mutagenesis.[427] Thus, initial binding of the substrate to the enzyme occurs followed by nucleophilic attack of a functional group in the active site onto the anomeric carbon forming the glucosyl-enzyme intermediate. Subsequent hydrolysis through general base catalysis then takes place, with attack on the resulting anomeric center by water to give β-glucopyranose as the enzyme is returned to its resting state.

This mechanism was substantiated further through both X-ray diffraction analyses and site-directed mutagenesis. In this connection, X-ray crystal structures have been reported for four family I glycosyl hydrolyases: a phospho-β-galactosidase from *Lactococcus lactis*,[429] a cyanogenic β-glucosidase from *Trifolium repens*,[430] a myrosinase from *Sinapis alba*,[431] and a β-glucosidase A from *Bacillus polymyxa*.[417] The latter was isolated as a crystalline octamer (130 × 130 × 130 Å), for which the X-ray diffraction analysis was performed at 2.4 Å resolution. Figure 33 depicts the *B. polymyxa* β-glucosidase subunit, the complexation of which with the inhibitor D-glucono-1,5-lactone (**113**) (which can also exist in its D gluconic acid and D-glucono-1,4-lactone forms) provided further support for the catalytic mechanism proposed earlier (Figure 32). As can be seen, both Glu166 and Glu352 are positioned (Figure 33) close to the anomeric carbon on the monosaccharide (or, in this case, its analogue), with the Glu352 pointing in such a direction as to facilitate nucleophilic attack. Hydrogen bonding of D-glucono-1,5-lactone (**113**) also occurs through its C-2 hydroxyl group with both Glu352 and the His121 side-chain; on the other hand, Trp296 is hydrogen bonded to the C-5 hydroxyl group, where it is thought to stabilize the putative oxocarbenium ion. (Indeed in the *A. faecalis* β-glucosidase, site-directed mutagenesis of the Glu170 and Trp298 residues gave a stable glucosyl-enzyme intermediate with $t_{1/2} > 7$ h.) Lastly, Trp406 is hydrogen bonded to the C-3 hydroxyl group, whereas Gln20 and Glu405 are hydrogen bonded to the hydroxyl groups at C-3 and C-4, and at C-4 and C-6, respectively. Analogous amino acid residues required for catalytic function have also been identified for the *P. contorta* β-glucosidase, but these are Glu and Asp rather than Trp.

3.18.6 LIGNINS IN PLANT TISSUES: RELATIONSHIP WITH OTHER CELL WALL COMPONENTS

A central, but unanswered, question in plant biology is how the various types of cell walls that characterize the distinguishable structures and domains in plants are formed, whether lignified or not. This includes the primary walls and secondary xylem encompassing, for example, the ray parenchyma, tracheids, vessels, fibers, and suberized layers. From a biochemical perspective, such processes are poorly understood not only in terms of the biosynthesis of individual cell wall components (cellulose, hemicellulose, lignins, and associated proteins) but also how the overall assembly of the various extracellular matrices is achieved. Indeed, the full delineation of these processes, which are directly or indirectly controlled by the living organism, represents *the* major challenge facing plant biology. Accordingly, this section intends briefly to summarize the most

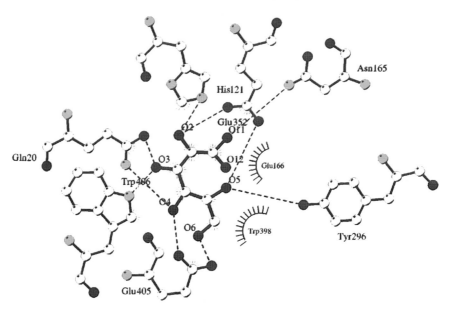

Figure 33 Interaction of *Bacillus polymyxa* β-glucosidase A with inhibitor D-glucono-1,5-lactone (**113**), shown in black. Schematic diagram of the possible enzyme-ligand hydrogen bond network involved in the recognition of the substrate (reproduced by permission from *J. Mol. Biol.*, 1998, **275**, 491).[417]

common variations in lignin configuration during the formation and maturation of particular cell (wall) types and tissues.

The so-called "normal" woody secondary xylem tissue (Figures 17(i) and 17(j)) is formed when a tree stem grows vertically (Figure 34(a)), parallel to the earth's gravitational field. This tissue differs in appearance and texture from that of the vascular bundles (Figure 17(b)) in herbaceous plants and grasses. Nevertheless, the formation of all of these vascular tissues follows a particular outline that exhibits little variation: after cell division, the periphery of plant cells is first reinforced with a pliable "extracellular matrix" called the primary wall. The cell then undergoes enlargement, whereupon secondary wall polysaccharide biosynthesis ensues followed by lignin formation. As can be seen in Figure 34(b), the secondary xylem cell walls can be separated into three layers designated (rather unimaginatively) S_1, S_2, and S_3 respectively.[432] The nature of these layers has long held the interest of plant biochemists, who have mounted intensive efforts to elucidate the structural and chemical characteristics of the polysaccharide, lignin, and (minute levels of) proteinaceous components present, and establish their functional interrelationships. Furthermore, it has been reasonably well confirmed that the basic cell wall architecture is first achieved primarily through polysaccharide and protein deposition, and that lignification predominantly occurs after the various cell wall layers have been largely constructed.[433]

3.18.6.1 Cell Wall Polysaccharide Heterogeneity

(Sections 3.18.6.1 and 3.18.6.2 have in large part been excerpted by S. Sarkanen with permission from reference 400. Copyright 1989 American Chemical Society.)

It has been clearly established that particular plant polysaccharides are differentially laid down in distinct layer-like domains during cell wall development, which in turn significantly influence the fundamental architecture of the intact assembly; their composition, however, can vary both within and between the tissues of different species as briefly discussed below.

The heterogeneous nature of the polysaccharides deposited during normal (wood) cell wall assembly was determined in pioneering work by Meier and Wilkie[434] and two years later by Meier.[435] Those studies led to the realization that the various plant cell wall layers are not homogenous in terms of their polysaccharide constituents. This was established by isolating radial sections from the differentiating xylem of *Pinus sylvestris*, *Picea abies*, and *Betula verrucosa*, and individually separating, with a micromanipulator, subcellular fractions which were subsequently analyzed. Paper chromatographic evaluation of softwood tracheid monosaccharide compositions released from the different polysaccharide fractions of the various wall layers, revealed that the outer part of the S_2

(a)

(c)

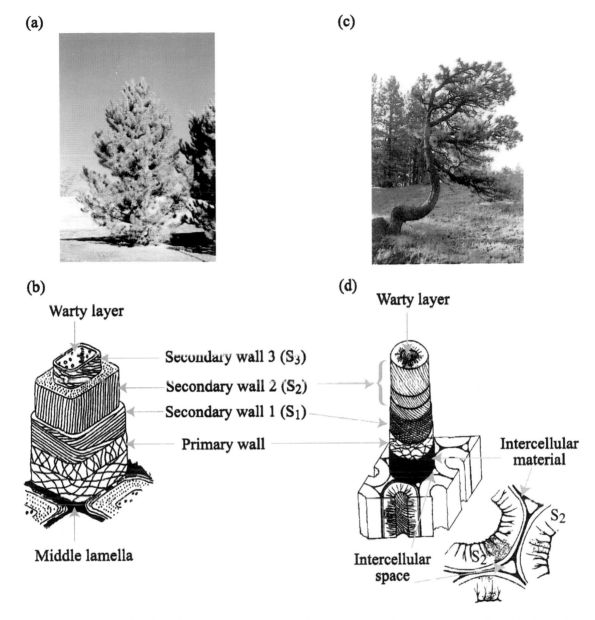

(b)

Warty layer

Secondary wall 3 (S$_3$)

Secondary wall 2 (S$_2$)

Secondary wall 1 (S$_1$)

Primary wall

Middle lamella

(d)

Warty layer

Intercellular material

Intercellular space

S$_2$

S$_2$

Figure 34 (a) Example of conifer (gymnosperm) growing parallel to the earth's gravitational field and predominantly forming "normal" wood; (b) schematic model of the cell-wall structure of softwood tracheids and hardwood libriform fibers;[432] (c) conifer (gymnosperm) with bent stem section. Plant has partially regained realignment via formation of reinforcing reaction wood. In conifer, this occurs by reinforcing the underside of the stem with compression wood; and (d) schematic model of the cell-wall structure of typical compression wood tracheids.[466]

layer was richest in cellulose, whereas the S$_3$ layer was highest in arabinoglucuronoxylan with the glucomannan content gradually increasing towards the lumen. On the other hand, in hardwood fibers, the inner parts of the S$_2$ and S$_3$ layers were richest in cellulose, while the S$_1$ and the outer part of the S$_2$ layers had a high glucuronoxylan content. Côté *et al.*[436] also examined the polysaccharide distribution in *Abies balsamea* tracheids according to the method of Meier and reached similar conclusions.

Further insight into subcellular polysaccharide heterogeneity came from the studies of Larson[437,438] who administered $^{14}CO_2$ photosynthetically to *Pinus resinosa*, divided the differentiating xylem into several fractions, and measured the radioactivity of each cell wall component. From these studies it was concluded that, as tracheid maturation progressed, deposition of xylose-containing biopolymers increased, whereas those containing mannose remained relatively constant while those composed primarily of arabinose and/or galactose decreased considerably.

Takabe *et al.*[439] also examined polysaccharide deposition during cell wall formation using gas chromatographic analysis of fractions separated by microfractionation so as to be enriched in cells at different developmental stages. Accordingly, the differentiating xylem of *C. japonica* was separated into twelve fractions, for each of which the neutral monosaccharide content and type was determined. Again cellulose was mainly deposited in the middle part of the S_2 layer, whereas the hemicelluloses (mannans and xylan) principally accumulated in the S_1 and the outer part of the S_2 and S_3 layers. These results were further confirmed by transmission electron microscopy (TEM) of differentiating xylem stained with PATAg,[400] a technique believed to be specific for hemicelluloses. Under these conditions, heavy staining of the S_1, outer S_2, and S_3 layers, and the so-called warty layer was noted.

Additionally, Takabe *et al.* investigated the incorporation of [U-^{14}C]glucose into the poly-saccharides during cell wall formation.[440] After the labeled sugar had been incorporated, the differentiating xylem tissue of *C. japonica* was separated into eight fractions representing different cell wall developmental stages. Each of these was subjected to mild acid hydrolysis. The monosaccharides thus released from the polysaccharides were separated by thin-layer chromatography to visualize the individual radioactivity content for each sugar by autoradiography and its measurement by scintillation counting. The distribution of radioactivity among the individual sugars was in good agreement with previous analyses of the biopolymers, i.e., cellulose deposition mainly occurred in the middle part of the S_2 to S_3 developmental stage, whereas xylan deposition appeared in the S_1 to early S_2 and again in the S_3 developmental layers. Mannan deposition, on the other hand, occurred by-and-large throughout secondary cell wall biosynthesis and not during formation of the primary wall.

3.18.6.2 Temporal and Spatial Deposition of Lignins in the Cell Walls of Gymnosperms and Angiosperms

Lignins are biosynthesized in the various plant cell wall domains primarily after polysaccharide deposition is complete, as was first determined *in situ* by monitoring the onset and spatial deposition of lignins during secondary xylem formation. Eventually it was unexpectedly found that the various monolignols were differentially incorporated into the various cell wall domains and tissues during lignification.

Early investigations of the lignification of differentiating *Pinus radiata* tracheids by Wardrop[441] used ultraviolet microscopy, where it was observed that lignin formation was initiated at the cell corners of the primary wall; it was then extended into the middle lamella and secondary cell wall regions. These findings were subsequently confirmed and elaborated by Imagawa *et al.*[442] from UV-photomicrographic analyses of the differentiating xylem of *Larix leptolepis*. Lignin accumulation began in the intercellular layer at the cell corners and pit borders, and was then extended to the radial/tangential middle lamellae, and finally progressed toward the lumen.

Although this UV-microscopic approach provided substantial insight about lignification and lignin distribution through the cell walls, it was of limited utility because of its low resolving power compared to electron microscopy. Accordingly, Wardrop[443,444] subsequently carried out TEM studies with ultrathin sections of *Eucalyptus elaeophora*, specimens of which were fixed with potassium permanganate; it was concluded that lignin formation was initiated in the middle lamella at the cell corners and, subsequently, in the outer part of the S_1 layer. It then proceeded along the middle lamella contour, through the primary wall and into the secondary wall. Kutscha and Schwarzmann[445] examined *Abies balsamea* tracheids also by TEM and showed that lignification was initiated in the middle lamella between the pit borders of adjacent tracheids, before progressing into the pit borders themselves and the cell corners. At the cell corners, it was manifested in either the outer portion of the primary wall or the middle lamella, and then moved into the cell corner region of the S_1 layer, leaving the primary wall unlignified before beginning to reach towards the lumen. This technique, which relies on permanganate fixation, can however cause swelling of both the cell and the cell wall, and result in the extraction of many cell components. Moreover, Kishi *et al.*[446] reported that the staining intensity produced by permanganate did not reflect true lignin content, thus casting some doubt on the foregoing results.

In an attempt to overcome these difficulties, Saka and Thomas[447] examined *Pinus taeda* tracheids by the SEM-EDXA technique, and concluded that lignification was indeed initiated in the cell corner middle lamella and compound middle lamella regions during S_1 layer formation. Subsequently, rapid lignin deposition occurred in both regions, while secondary wall lignification began when the middle lamella lignin concentration was approaching 50% of its maximum, and then proceeded toward the lumen.

Lignification was also investigated in *C. japonica* tracheids by Takabe *et al.*,[399,400,448] as well as in Japanese black pine (*Pinus thunbergii*).[449] This study involved administration of tritiated phenylalanine (**8**) to the tissues as a lignin precursor, followed by analysis through a combination of UV-microscopy, light microscopy/autoradiography, and TEM coupled with appropriate chemical treatments of the ultrathin sections. Using UV-absorption microscopic measurements, lignification was first detected in the compound middle lamella of the tracheid at the S_1 developmental stage; it then increased during secondary wall thickening, becoming essentially constant after the S_3 stage. On the other hand, UV-absorption in the secondary wall first appeared in the other portion of the tracheid during the S_2 stage, and then spread slowly toward the lumen during subsequent stages. The tracheid, during the final part of cell wall formation, ultimately showed uniform absorption throughout the secondary wall. It was also evident that the labeled lignin precursors derived from phenylalanine (**8**) were rapidly incorporated into the compound middle lamella lignin, whereas they entered more slowly into the secondary wall lignin.[400] Additionally, incorporation into compound middle lamella lignin took place during the S_1 and S_2 developmental stages, while incorporation into secondary wall lignin occurred mainly after the S_3 stage. In the latter case, the radioactivity was distributed throughout the secondary wall, suggesting that monolignols were being continuously supplied to all regions of the secondary wall. Thus, the lignin content in the secondary wall appeared gradually to increase by ostensibly uniform coupling of monolignol radicals.

These findings were further supported by analyses of the lignin skeleton from differentiating *C. japonica* xylem remaining after treatment of ultrathin sections with hydrofluoric acid, following resin extraction,[449] so as to remove polysaccharides effectively without undue cell wall swelling. Thus, it was concluded that lignification was initiated at the outer surface of the primary wall in the cell corners, just before S_1 formation, and then proceeded to the intercellular layer and intercellular regions between the cell corners. Remarkably, when a tracheid was adjacent to a ray parenchyma, the cell corner region on the ray parenchyma side seemed to be lignified earlier than that on the opposite side. It was also interesting that secondary wall lignification was initiated at the S_1 cell corner region during the S_1 developmental stage before proceeding into the unlignified S_1 layers.[449] Moreover, when a tracheid was adjacent to a ray parenchyma, lignification of the S_1 layer was also initiated earlier on the ray side. After that, it gradually spread towards the lumen, lagging behind cell wall thickening. Lignin deposition then predominated after the S_3 stage, while being less active during the S_1, S_2, and S_3 developmental stages. The lignin content of the secondary wall became fairly constant in the final stage of cell wall formation, though the warty layer was more highly lignified.[449]

3.18.6.3 Lignin Heterogeneity in Gymnosperm Cell Walls

In a study of black spruce (*Picea mariana*) tracheid lignification *in situ* using UV-microscopy, Goring and co-workers[450,451] demonstrated that the lignin concentration was highest in the middle lamella regions between adjacent tracheids compared to that in secondary walls (85–100% vs. 22%), although the overall total lignin content was higher ($\sim$72–80%) in the secondary wall due to its greater volume (Figure 35). A subsequent study on tissue fractions thought to represent the middle lamella and secondary wall, respectively, of *P. mariana* suggested that the methoxyl group content of the secondary wall lignin was 1.7 times higher than that of the middle lamella, implying that the latter embodies more *p*-coumaryl (**1**) alcohol derived units in its lignin.[452] Although this tissue fractionation method is open to question,[453] Terashima and co-workers,[453,454] using radiolabeled monolignol precursors and autoradiography in Japanese black pine (*Pinus thunbergii*), also concluded that a similar pattern of monolignol deposition occurred. The microautoradiograms obtained from their examination of differentiating xylem tissues indicated that monolignol residues were incorporated into consecutive stages of cell wall formation in the order of *p*-coumaryl followed by guaiacyl units.[454] The distribution of silver grains in the autoradiograms suggested that *p*-coumaryl alcohol (**1**) derived lignin is formed mainly in the compound middle lamella of differentiating pine tracheids[454] during the early stages of lignification. Analogous results were obtained with gingko (*Ginkgo biloba*).[455]

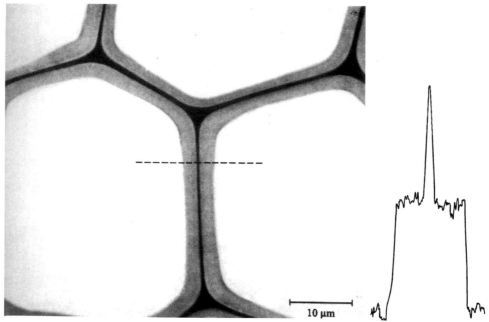

Figure 35 Cross-section of Epon embedded tracheids of black spruce earlywood photographed in ultraviolet light of wavelength 240 nm. The densitometer tracing was taken across the wall on the dotted line.[451] Lignin contents estimated in the secondary wall and middle lamella (cell corner) are 22 and 85%, respectively (reproduced with permission from *Wood Science and Technology*, 1969, **3**, 117).

3.18.6.4 Lignin Heterogeneity in Cell Walls of Woody Angiosperm Dicotyledons

The determination of lignin distributions in woody angiosperm (hardwood) cell walls is complicated by the fact that the biopolymers contain both guaiacyl and syringyl residues. For example, as far as ultraviolet microscopy is concerned, the absorptivity of the guaiacyl residue at the customary monitoring wavelength of 280 nm is about four times higher than that of the syringyl unit, and the local maximum in the absorbance of the latter in the 270–280 nm range lies at lower wavelength and is less distinct. Nevertheless, on the basis of these spectral differences, it was suggested by Goring and co-workers in a study of white birch (*Betula papyrifera*) that the lignin located in the fiber wall was mainly of the syringyl type, while vessel wall lignin contained primarily guaiacyl residues.[456,457] Similarly, in an extensive study of 14 hardwood species, it was found that syringyl residues became more predominant in the walls of fiber and ray cells as the methoxyl content increased, while the walls and cell corner regions of vessels contained mostly guaiacyl residues, except when the methoxyl contents were at their highest, whereupon the syringyl character of the lignin became more evident.[458] These findings immediately prompted Fergus and Goring in 1970 to speculate that the lignin precursors originate within the differentiating cell and not from the cambial zone by diffusion.[456]

These seminal observations were satisfactorily confirmed by other techniques. For example, when the three ^{3}H-labeled monolignol glucosides (**71**)–(**73**) were administered to magnolia (*Magnolia kobus* KC), lilac (*Syringa vulgaris*), and poplar, the radioactivities of each appeared to be incorporated into the newly formed xylem in the order of *p*-coumaryl, guaiacyl, and syringyl units at consecutive stages of wall formation and, hence, lignification.[459,460] Again, the *p*-coumaryl alcohol (**1**) unit was incorporated only at the earliest stages of lignification involving the cell corner and middle lamella, but was not evidently found in the secondary wall. Moreover, because the vessel cell walls apparently lignify earlier than fiber walls, the vessel wall lignin contains larger amounts of *p*-hydroxyphenyl (H) and guaiacyl (G) units than the fiber wall, though deposition of guaiacyl lignin continued throughout the early to late stages in both kinds of walls. Syringyl (S) lignin, on the other hand, was deposited mainly during the middle and late stages of lignification, i.e., mostly in the secondary walls of fibers, though it also may have participated to a very small extent in the formation of middle lamella lignin. These observations are in general accord with those furnished by ultraviolet microscopic spectrophotometry,[456–458] as well as by chemical characterization of the cell wall fractions[461–463] including scanning electron microscopy analysis, and the differential levels of expression of *O*-methyltransferases in vessels and fibers.[198,199] Similar patterns of sequential monolignol incorporation and the morphological localization of the resulting lignins have also been observed in rice plants.[464] A general scheme for the deposition of lignin residues is juxtaposed in Figure 36, with assembly of other cell wall components.

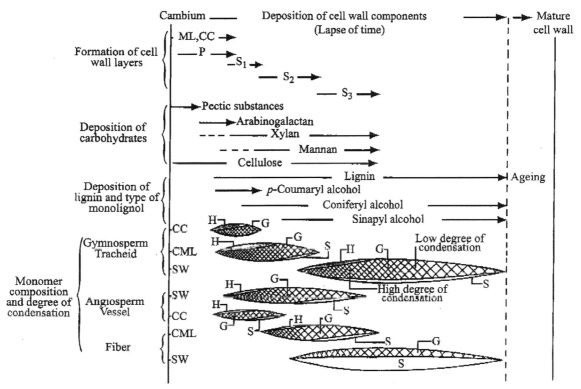

Figure 36 A schematic representation of the processes of deposition of cell-wall components and the heterogenous formation of protolignin macromolecule. ML, middle lamella; CC, cell corner; P, primary wall; CML, compound middle lamella; S_1, S_2, and S_3, outer, middle, and inner layer of secondary wall; H, G, and S, p-hydroxyphenyl, guaiacyl-, and syringylpropane units. The term condensation refers to the extent of C–C interunit linkages in lignin, versus C–O–C. (After Fukushima and Terashima.[460])

Thus, microautoradiography combined with selective radiolabeling of specific structural units in lignins has been employed to monitor the growth of protolignin macromolecules in specific morphological regions of differentiating xylem.[454,455,459,465] From these studies, lignification was found to proceed in three distinct stages, each preceded by deposition of carbohydrates. The first stage occurred at the cell corners and middle lamella after the cellulose microfibrils, mannan, and xylan are deposited into the S_2 layer. The major onset of lignification occurs in the third stage after deposition of cellulose microfibrils in the S_3 layer has started. Thus, at least phenomenologically, lignin is never formed in the absence of carbohydrates. This basic pattern for the deposition of carbohydrates and lignin seems to hold generally among the gymnospermous and angiospermous trees.[460]

3.18.6.5 Lignins in Reaction Wood

It is generally believed that when the stems of grasses and herbaceous (non-woody) plants are perturbed from their upright "equilibrium" positions, they react by increasing or promoting longitudinal growth on the lower side. On the other hand, woody plants generate tissues known as "reaction wood" under such conditions (Figure 34(c)). These tissues differ greatly depending upon whether the woody plant is a gymnosperm or an angiosperm, with the former generally giving rise to compression wood and the latter to tension wood.[466] Interestingly, the formation of these tissues is inducible: for example, they are formed when woody plant (tree) stems are displaced away from alignment with the gravitational field, or when branches are growing or otherwise placed under an increased load (e.g., through accumulation of ice, snow, etc.).

In gymnosperms, compression wood is formed on the undersides of branches, being generated through increased radial growth in these domains. This growth profile follows along the grain of the branch so as to exert an axial pressure and thus help establish its designated orientation. Similarly, when a woody stem is displaced from vertical alignment, particular cells are induced to generate compression wood tissue on the underside. Slowly, until recovery of stem alignment occurs, this reinforcement tends to bend the leaning stem back until vertical orientation of the photosynthetic canopy is again achieved. Compression wood formation is quite nicely illustrated (Figure 37) with the example of a redwood (*Sequoia sempervirens*) which valiantly attempted to regain a vertical alignment from a leaning orientation over a period of several hundred years. Clearly, the reaction wood tissue growth at the underside is extensive.

Figure 37 Cross-section of *Sequoia sempervirens* stem. Picture shows pith and opposite wood, as well as extensive compression wood formation.

As shown in Figure 34(d), the tracheid cells that are conscripted to the formation of compression wood have been "reprogrammed" where their walls now differ substantially from those of "normal" wood. Taking loblolly pine (*Pinus taeda*) as an example, the tracheids are rounder and shorter (~ 10–40%) than "normal" xylem cells, with thicker cell walls and consequently almost twice the specific gravity. As shown in Figure 34(d), the S_2 layer is now deeply fissured, the S_3 layer is absent and intercellular spaces (previously the region of the compound middle lamella) are now abundant. In terms of their biopolymeric constituents, the cellulose content is lower (see Table 19),[466,467] its degree of polymerization is smaller and it is apparently less crystalline. Correspondingly, the overall lignin content is higher as shown, for example, for compression wood from *P. radiata*,[468] Douglas fir (*Pseudotsuga menziesii*),[466,469] and *Pinus pinaster*.[470,471] Interestingly, the frequency of units derived from *p*-coumaryl alcohol (**1**) is thought to increase from 5% to 30%. Indeed, in the latter regard, results from thioacidolysis have implied that $\sim 90\%$ of the *p*-coumaryl alcohol (**1**) derived units were ether linked through C-8, thereby leaving the phenolic hydroxyl groups unmodified.[470,471]

Table 19 Chemical composition of normal and compression woods of *Pinus taeda*. All values are in per cent of unextracted wood.[466,467]

Component	Normal wood	Compression wood
Extractives	2.7	2.5
Lignin	28.3	35.2
Cellulose	45.7	34.6
Pentosans	12.4	12.2
Solubility in:		
Hot water	1.8	2.0
1% Sodium hydroxide	9.9	12.6

It may be worthwhile speculating about how certain cells are conscripted to forming compression wood, in terms of the attendant signal transduction mechanisms within the structure of the living organism itself. Presumably, this can only result from local stress gradients which, upon exceeding a particular threshold, engender macromolecular conformational changes that ultimately influence the biosynthetic pathways in such a way as to produce the reinforcing tissue. Indeed, Lewis *et al.* embarked upon an investigation of whether Douglas fir and loblolly pine seedlings, while in a microgravity environment on the Space Shuttle *Columbia* (STS-78, July 1996), would engender compression wood formation via mechanical stresses. This straightforward experiment was devised to establish that compression wood formation was indeed not restricted to the consequences of a change in the gravitational vector experienced by the plant. As expected, it was observed (Figure 38) that compression wood formation did occur when mechanical bending was introduced in microgravity,[472] confirming that plants are capable of reacting to local stress gradients. Figure 38 shows some Douglas fir compression wood that was formed in the microgravity environment on the Space Shuttle as a result.

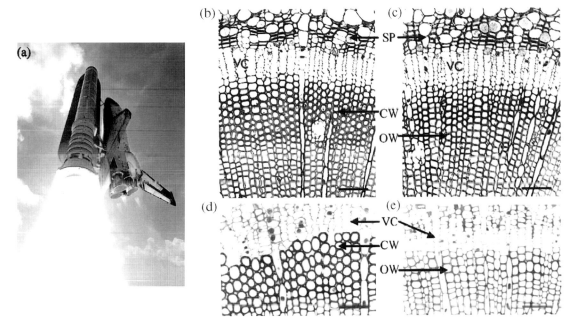

Figure 38 (a) Space Shuttle Columbia during take off of STS78 mission (June 1996); (b)–(e) One-year old *Douglas fir* seedling stems were bent at 45° in microgravity, and held in this orientation for 15 days. Analysis of stem cross-sections by light microscopy showing formation of (b) compression wood, CW (bar = 5 μm); (c) opposite wood, OW (bar = 5 μm); and (d) enlargement of compression wood (bar = 2 μm), and (e) opposite wood. VC, vascular cambium and SP, secondary phloem.

On the other hand, woody dicotyledonous angiosperms generate a different kind of reaction wood tissue known as tension wood. In this case, radial growth generally, but not always, occurs on the upper side of the branch or leaning stem. Thus a tensile stress is exerted in an attempt to regain a particular alignment or orientation. Here the fibers, rather than the vessels (which are typically fewer, smaller, and mechanically weaker), undergo changes in configuration. From an ultrastructural perspective (Figure 39), the tension wood fibers display a thick innermost gelatinous (G) layer at the lumen surface, possessing axially oriented, crystalline cellulose, which replaces the S_3 layer. The so-called G-layer is thought to be unlignified, although in some instances the lignin content of the other layers is purportedly reduced or increased.[466] The S_2 layer is also much smaller, and may sometimes be smaller than the S_1 layer.

3.18.6.6 Lignins in Grasses and Herbaceous Plants

Interest in grass lignins has, to a large extent, been fueled by the idea that lignins and covalently bound phenolic compounds limit forage cell-wall digestibility.[473] However, the substances described as grass lignins exhibit high solubilities in aqueous alkaline solution[474] and in this respect are quite unlike lignins encountered in woody tissues.

This subsection attempts to summarize our rather incomplete knowledge of lignins, and presumed

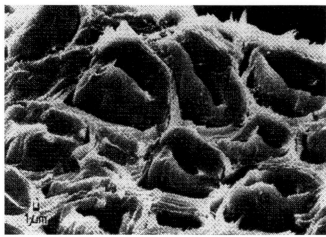

Figure 39 Transverse plane of tension wood in poplar (*Populus* sp.). The thick G-layers (GL) of the tension wood fibres are partly separated from the residual cell-walls, and are innermost layers between the adjacent walls. SEM micrograph (reproduced by permission of Walter de Gruyter from Wood Chemistry, Ultrastructure, Reactions, 1984).

lignins, in grasses and herbaceous species. Initial emphasis was placed upon examining monomer variations in different species, this being subsequently extended to the study of lignin compositional differences at particular stages of plant maturity. Immunocytochcmical approachcs have also revealed quite distinct differences in the nature of inter-unit linkages between lignins in various subcellular layers. While these are discussed below, it should be recognized that there remains a great paucity of experimental data on both the enzymology and molecular biology of lignification in grasses and herbaceous plants, as well as in our understanding of the actual lignin-forming processes in the individual cell types.

3.18.6.6.1 *Monomer compositions and ester/ether linkages*

Initially, Creighton and Hibbert[475] in 1944 reported finding *p*-hydroxybenzaldehyde (**21**), vanillin (**22**) and syringaldehyde (**56**) in ratios of approximately 3:9:5 among the alkaline nitrobenzene oxidation products of maize stalk meal (*Z. mays*). Since this represented the first isolation of free *p*-hydroxybenzaldehyde (**21**) purportedly derived from oxidative cleavage of a lignin, its formation was proposed as a means for distinguishing between mono- and dicotyledons. This suggestion thus extended the basis that had already been established by Hibbert *et al.*[476] for characterizing lignins in gymnosperms and angiosperms as being guaiacyl (**G**) and guaiacyl-syringyl (**G-S**) rich, respectively: in grasses, on the other hand, *p*-hydroxyphenyl (**H**), guaiacyl (**G**) and syringyl (**S**) units apparently contributed to roughly similar extents to the compositions of their constituent lignins (if indeed they were real lignins).

In 1955, *p*-coumaric acid (**5**) was then isolated from sugarcane (*Saccharum officinarum*) lignin by treatment with aqueous 1 M hydroxide,[477] and rudimentary ultraviolet and infrared spectral analyses suggested that its liberation was due to hydrolysis of ester groups. Small amounts of *p*-hydroxybenzoic (**114**), vanillic (**115**), syringic (**116**) and ferulic (**6**) acids were apparently also detected. Some 35 years later, the cell walls of sugarcane parenchyma and vascular bundles were found to contain sinapic acid (**7**) in addition to *p*-coumaric (**5**) and ferulic (**6**) acids.[478] The last of the three was deposited at an early stage in the lignification process, whereas *p*-coumaric (**5**) and sinapic (**7**) acids accumulated steadily as lignification progressed.

CO_2H

R^2 R^1

OH

(**114**) $R^1 = R^2 = H$, *p*-Hydroxybenzoic acid
(**115**) $R^1 = OMe$, $R^2 = H$, Vanillic acid
(**116**) $R^1 = R^2 = OMe$, Syringic acid

Wheat straw (*T. aestivum*) lignin similarly yielded *p*-coumaric (**5**) and ferulic (**6**) acids accompanied by traces of vanillic (**115**) and syringic (**116**) acids.[477] Moreover, at least 93% of the *p*-coumaric acid (**5**) in a milled wheat straw lignin preparation was shown to be alkali-labile, and thus was linked to the macromolecular structure by ester bonds; the remainder could only be liberated through acid-catalyzed hydrolysis.[479] On the other hand, 35% of the ferulic acid (**6**) residues were resistant to alkaline hydrolysis, but these too were capable of being liberated by acidolysis.[479] The existence of ether bonds to ferulic acid (**6**) was confirmed by ^{13}C NMR spectroscopic examination of a lignin sample produced by alkaline treatment of the material extracted in aqueous 90% dioxane from the milled straw after hydrolysis with cellulases. Thus it was suggested that etherified ferulate esters might act as crosslinks between the lignin macromolecules and arabinose residues in arabinoglucuronoxylans.[479] In this regard it was pointed out that, *a priori*, ether bonds to ferulic acid (**6**) moieties could be formed through either direct radical coupling or nucleophilic addition to intermediate quinone methides during monolignol dehydropolymerization.[479] Such an effect could contribute to cell wall rigidity in much the same way as the dehydrodimerization of ferulate esters that produces crosslinks between hemicellulose chains through diferulate (**117**) bridges.[480]

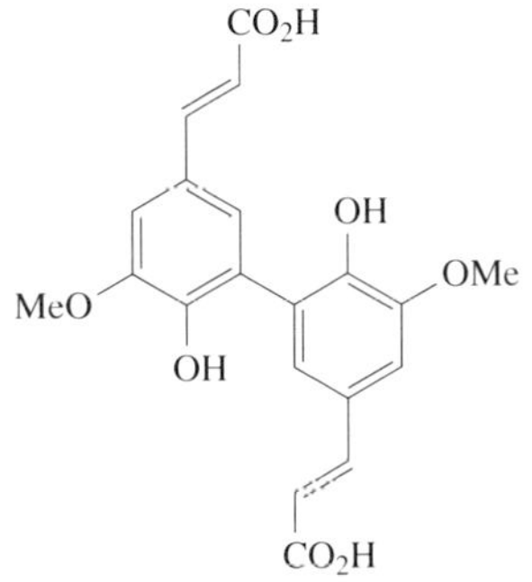

(**117**) Diferulic acid

That indeed the hydroxycinnamic acids can be bound to the grass hemicelluloses has been firmly established by a number of studies. For example, when wheat bran cell walls were treated with a mixture of polysaccharide hydrolases (from *Oxyporus* spp.), 2-*O*-[5-*O*-(*trans*-feruloyl)-β-L-arabinofuranosyl]-D-xylopyranose (**118**) was released as the major feruloyl compound.[481] The same feruloylated disaccharide (**118**) was similarly isolated from other graminaceous cell walls also, in particular barley (*Hordeum vulgare*) straw and the leaves of wheat and Italian ryegrass (*Lolium multiflorum*).

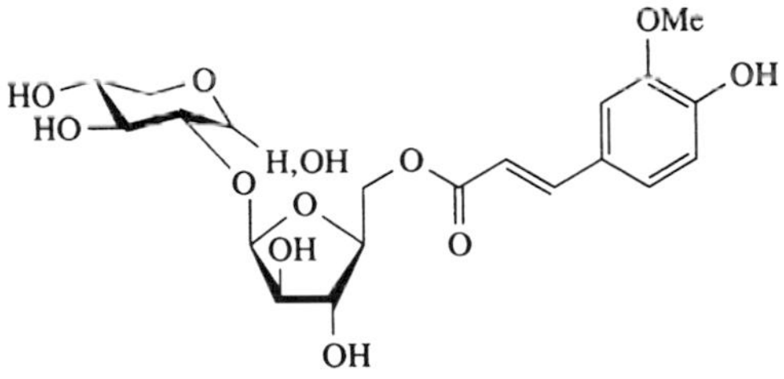

(**118**) 2-*O*-[5-*O*-(*trans*-feruloyl)-β-L-arabinofuranosyl]-D-xylopyranose

Shortly thereafter, a feruloylated trisaccharide identified as *O*-(5-*O*-feruloyl-α-L-arabinofuranosyl)-(1 → 3)-*O*-β-D-xylopyranosyl-(1 → 4)-D-xylopyranose (**119**) was isolated from bagasse by treating a lignin–carbohydrate complex with cellulases.[482] The same feruloylated trisaccharide (**119**) was subsequently obtained from growing *Z. mays* shoot cell walls, showing that the feruloyl residues occur at the 3-*O*-arabinofuranosyl branchpoints of the (1 → 4)-linked β-D-xylopyranose backbone in arabinoglucuronoxylan.[483] Before this, however, two feruloylated disaccharides, 4-*O*-(6-*O*-feruloyl-β-D-galactopyranosyl)-D-galactose (**120**) and 3-*O*-(3-*O*-feruloyl)-α-L-arabinofuranosyl)-L-arabinose (**121**), had been isolated from cell walls of suspension-cultured spinach (*Spinacia oleracea*) cells;[484] structures not shown. It had been suggested that such disaccharidic fragments were derived from pectin.[485] There could, therefore, be a significant difference between monocots and dicots in terms of the polysaccharides to which feruloyl residues are ester-linked: in the former, it is the arabinoglucuronoxylan, whereas in the latter the arabinan and galactan portions of the pectin seem to be involved.

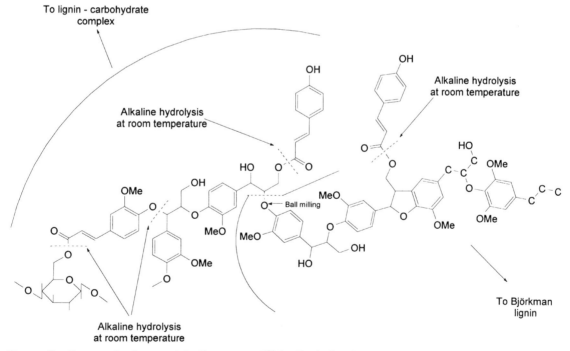

(119) *O*-(5-*O*-*trans*-feruloyl-α-L-arabinofuranosyl)-(1→3)-*O*-β-D-xylopyranosyl-(1→4)-D-xylopyranose

Using aqueous hydroxide under conditions engendering the selective release of ester- and ether-linked hydroxycinnamic acids, respectively, studies were mounted to determine whether there are any differences in their bonding patterns with each stage of cell wall development.[479,486] Thus, wheat internodes were sectioned prior to examination into four segments representing consecutive stages of maturity. At all stages of development, most of the *p*-coumaric acid (**5**) was esterified and increased in quantity with internode maturity. Ferulic acid (**6**), on the other hand, participated as frequently in ether as in ester linkages. The relative amount of etherified ferulic acid (**6**) increased from the youngest to the next of the four internode segments, at which point it had attained a more or less constant value; etherification of ferulic acid (**6**) was thus held to be an early event during maturation.[486] In contrast, the relative amount of esterified ferulic acid (**6**) decreased as the internodes matured; as a result, the lignin extracted from mature milled wheat internodes contained ferulic acid (**6**) that was mostly etherified.

It was suggested, then, that these ferulic acid (**6**) residues were involved in crosslinking the polysaccharides and lignin.[479,486] Consequently, alkaline saponification of a lignin-ether-ferulate-ester-polysaccharide bridge would have been expected, if this model was correct, to release large quantities of lignin bearing etherified ferulic acid (**6**) molecules (Figure 40). Indeed the amount of lignin soluble in aqueous dioxane was empirically observed to increase by an order of magnitude as a result.[486]

Figure 40   Structural scheme originally proposed[486] for lignin in wheat internode cell walls. The benzyl aryl ether linking a lignin monomer residue to a ferulate ester moiety should instead have been an 8–8′ C—C bond.[490]

In Figure 40 some of the 9-positions of the lignin monomer residues are also *p*-coumaroylated. That these are the positions which are currently believed to be exclusively involved in ester bonds to the *p*-coumaroyl moieties (**5**) was confirmed by a careful ^{13}C-^{1}H correlative NMR spectroscopic comparison of a lignin fraction from maize stem internode rind tissue with a synthetic dehydro-polymerisate produced from coniferyl alcohol (**2**) and coniferyl *p*-coumarate (**122**).[487] However, it was subsequently established that the *p*-coumarate (**5**) residues were actually primarily linked to sinapyl alcohol (**3**) derived moieties, i.e., it appeared as though sinapyl *p*-coumarate (**123**) esters had been incorporated into the lignin.[488] Nevertheless, these results were generally in good agreement with those of Shimada *et al.*,[489] who had previously demonstrated that *p*-coumaric acid (**5**) was mainly linked to the terminal CH_2OH (C-9) position of the lignins in the Graminaceae, including bamboo.

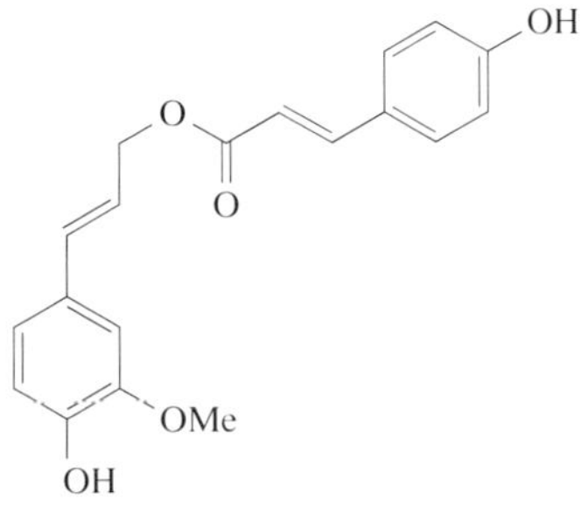

(122) Coniferyl *p*-coumarate 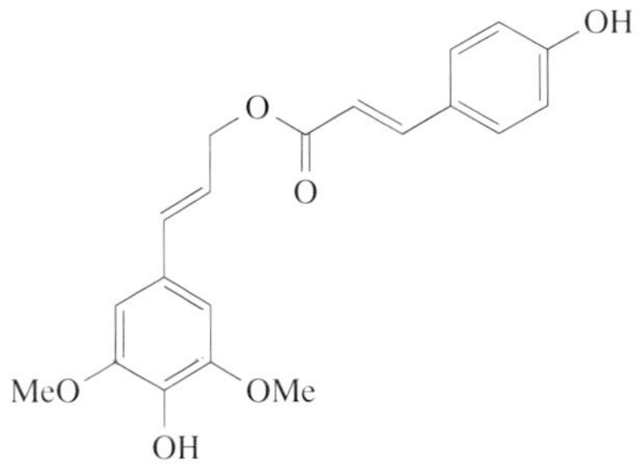   **(123)** Sinapyl *p*-coumarate

On the other hand, the feruloyl moiety in Figure 40 (extreme left) that is ester-linked to the polysaccharide chain participates through its 4-*O* atom in a benzyl aryl ether bond to the 7' position of a guaiacyl residue present in the lignin macromolecule. This would have been formed by nucleophilic addition of the feruloyl 4-*O* to the corresponding quinone methide intermediate produced during monolignol dehydropolymerization.[4,90,494] However, some quite compelling evidence has been secured for the occurrence of radical coupling between the ferulate (**6**) and lignin monomer residues *in vivo*. Long-range C–H correlations in the HMBC spectrum of uniformly ^{13}C-labeled ryegrass (*Lolium multiflorum* Lam.) lignin have diagnosed the existence of 8—8' bonds from the feruloyl to the guaiacyl (**G**) and syringyl (**S**) moieties, respectively,[490] whereas coupling with the 5- and 4-*O*-positions in the lignin monomer residues was conspicuously absent in the preparations examined. Thus it has been suggested that polysaccharide-bound ferulates (**6**) may act as initiation sites for lignification: the lignin units that have participated in such coupling reactions must originally have possessed intact 4-hydroxycinnamyl structures.[490]

Clearly, phenolic carboxylic acid residues are ubiquitous in grass lignins and contribute in an important way to their macromolecular configurations. They have been detected and examined to various extents on innumerable occasions. For example, ^{13}C-NMR spectroscopy of the dioxane lignin from one reed internode (*Arundo donax*) revealed a remarkably high incidence (0.28 per aromatic unit) of esterified *p*-coumaric acid (**5**) residues.[491] The existence of both ester and ether-linked *p*-coumaric acid (**5**) (0.83% and 0.11%) and ferulic acid (**6**) (0.24% and 0.20%) in the cell walls of rice (*Oriza sativa*) stems was reported in 1986 for the first time.[492] Indeed, phenolic carboxylic acids were found to be deposited continuously in the cell walls of rice plants as lignification progressed.[493] The *p*-coumaric (**5**) and ferulic (**6**) acids appeared to be esterified mainly to the lignin and polysaccharides, respectively, and the deposition of ferulic acid (**6**) predominated over *p*-coumaric acid (**5**) in the early stages of cell wall formation.[493] In this respect, the rice plant appears not to be very different from wheat.[486] However, sinapic acid (**7**) was found to occur at levels comparable to ferulic acid (**6**) in the cell walls of rice plants but it seemed to be ester-linked to the lignin components present.[493]

The presence of the hydroxycinnamic acids, and the relatively facile procedures required for lignin removal from grasses, perhaps imply that the lignins being laid down are quite distinct from that of the lignins in woody tissues. This will, however, only be resolved by establishing the enzymology and gene expression involved in lignification for each cell type in such plant species, and determining how the lignins in vessels, fibers, etc., differ or compare to those present in woody tissues.

3.18.6.6.2 *Immunocytochemical probes of lignin heterogeneity*

In addition to the gross variations in lignin monomeric compositions (H:G:S ratios), another approach has employed immunocytochemistry to probe whether there are any differences in the nature of interunit linkages of the lignin(s) in distinct cell-wall layers. This possibility was taken into consideration given that it has been known for some time that various lignin preparations are also heterogenous with respect to the nature of their interunit linkages,[494,495] i.e., whether they are C–O–C (so-called uncondensed) or C–C (condensed) bonded. In fact, Table 20 shows an example of the different frequencies of C–O–C and C–C interunit linkages in a (gymnosperm) spruce (*P. abies*) lignin-derived preparation.[496] Moreover, it has also been known since 1956 that interunit frequencies in *synthetic* macromolecular "lignin" chains can vary according to whether the mono-lignol is introduced altogether at once ("Zulaufverfahren") or added gradually ("Zutropfverfahren") to an oxidative enzyme bringing about dehydrogenative polymerization *in vitro*.[494,497] In this way, the Zulaufverfahren preparation typically leads to the formation of a larger proportion of "condensed" linkages in the resulting dehydropolymerisate, while the Zutropfverfahren method engenders more "uncondensed" macromolecules because it restricts dehydrogenative coupling of the monomers to occurring primarily with the growing ends of the lignin chains.[495]

These synthetic "lignin" preparations, "Zulauf" and "Zutropf" have been used to obtain poly-clonal antibodies in rabbits, e.g., against "Zutropf" synthetic *p*-hydroxyphenyl (**H**), guaiacyl (**G**) and guaiacyl-syringyl (**GS**) lignin-like preparations,[497,498] respectively. These in turn have been employed to examine lignin interunit heterogeneity at the ultrastructural level with both maize (*Z. mays*)[498,499] and wheat (*T. aestivum*).[500] Thus, the apparent concentration of guaiacyl-syringyl (**GS**) lignin was observed to increase in maize parenchyma cell walls during maturation of the internode tissue, whereas very little labeling was noted with the anti-(**G**) probe. On the other hand, the strongest labeling was observed throughout the lengths of the fibers within the internode with the anti-(**GS**) probe, suggesting that the "Zutropf" guaiacyl-syringyl (**GS**) lignin is biosynthesized continuously in these cells.

In the maize fibers, apparent labeling of *p*-hydroxyphenyl (**H**) lignin was faint in the early stages of lignification but intensified steadily with secondary thickening of the cell wall. However, the only consistent labeling observed was obtained in the cell corners of both the youngest and oldest tissues in the maize internode with the anti-(**H**) probe.[499] No significant labeling with the anti-(**G**) probe was detected in the younger fibers at the base of the maize internode, but heterogeneous distributions of guaiacyl (**G**) lignins were evident in the polylamellar cell walls of mature fibers.[499] Labeling by the anti-(**G**) probe was uniformly weak in most cell types, which consequently were assumed to embody a fairly high degree of "Zulauf" characteristic in the guaiacyl (**G**) lignin. The sole notable exceptions were the metaxylem vessels where the apparent concentration of "Zutropf" guaiacyl (**G**) lignin remained approximately constant all along the maturing maize internode.[499]

Immunocytochemical labeling of wheat (*T. aestivum*) straw tissue sections with polyclonal anti-bodies raised against "Zulauf" and "Zutropf" synthetic lignin-like preparations revealed patterns,[500] that were in part complementary to those observed in maize. The "Zulauf" guaiacyl (**G**) lignin was noted to be present at fairly uniform concentrations in the cell corners, middle lamellae and secondary walls of the fibers, whereas "Zulauf" guaiacyl-syringyl (**GS**) lignin was observed at somewhat lower levels in the fiber cell corners and middle lamellae only. On the other hand, "Zutropf" guaiacyl-syringyl (**GS**) lignin was observed only in the secondary walls of fiber and other cell types e.g., protoxylem, vessels, etc.). A rather striking example of ultrastructural lignin heterogeneity was also revealed within the metaxylem vessel secondary cell walls: the four constituent layers visible after potassium permanganate staining exhibited a tendency toward a mutually exclusive alternating preference for labeling by the "Zulauf" and "Zutropf" anti-(**GS**) probes, respectively (Figure 41).[500]

Although the epitopes recognized by the polyclonal antibodies have not been explicitly identified, the differential labeling of maize and wheat cell wall domains clearly reveals the existence of heterogeneous distributions of lignins among the various constituent zones. Accordingly, these results show that lignification is far from uniform within any cross-section of a maize internode, this being attained in a well-defined spatially and temporally controlled manner. Much, however, obviously remains to be learned about macromolecular lignin assembly processes before the basis for these findings can be fully understood. The results do, nevertheless, underscore the levels of control that are exercised during lignification within the different cell types at their various stages of maturity.

3.18.6.7   Lignins in Mistletoe

The mistletoe are dicotyledonous photosynthetic hemi-parasites, which belong to the Santales, and are members of the Rosidae encompassing both the Lorantheacae and Viscaceae.[501] Of these,

Table 20 Early estimates of relative proportion of various inter-unit C–O–C and C–C linkages in aqueous 90% dioxane soluble lignin from spruce (*Picea abies*), estimated by analytical degradation.[a]

Carbon skeleton	Inter-unit linkage	Proportion (%)
	Arylglycerol-8-aryl ether 8—*O*—4'	48
	Glyceraldehyde-2-aryl ether from 8—*O*—4' after 8"—1 coupling	2
	Acyclic benzyl aryl ether 7—*O*—4'	6–8
	Phenylcoumaran 8—5'	9–12
	8—6' (or 8—2') Structures	2.5–3
	Biphenyl 5—5'	9.5–11
	Diaryl ether 4—*O*—5'	3.5 4
	1,2-diarylpropane 8—1'	7
	Pinoresinol or divanillyl tetrahydrofuran 8—8'	2
	Quinone ketal 1—*O*—4'	Trace

[a] Deduced from acid-catalyzed hydrolysis and permanganate oxidation.[496]

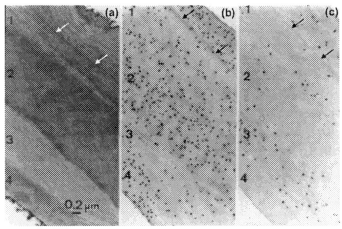

Figure 41 Lignin distribution within wheat straw cell-wall vessel metaxylem. (a) $KMnO_4$ staining reveals four zones of unequal intensity; (b) anti-GS (Zutropf or "uncondensed") antibody gives strong positive labeling with layers 2 and 4; (c) anti-GS (Zulauf or "condensed") antibody labels layers 1 and 3 which were only weakly reactive to the Zutropf antiserum. 1, 2, 3, and 4 refer to concentric secondary wall layers of metaxylem from the middle lamella (arrows) inward to the lumen. Visualization with protein-A-5 nm gold with silver enhancement for both immunolabeling studies.

the decorative mistletoes, used primarily for the festive season, are obtained from *Phoradendron* (N. America) and *Viscum* (Europe and West Asia) species. It was, therefore, of some interest to establish the nature of their lignins. In this pursuit, Freudenberg summarized very unusual properties of mistletoe lignin in a *Science* article entitled "Lignin: its Constitution and Formation from *p*-Hydroxycinnamyl Alcohols".[502] That account described an unpublished investigation by J. M. Harkin, working in Freudenberg's group at the time, which claimed that the lignin extracted from mistletoe (*Viscum album*) was of a coniferous (guaiacyl (**G**)) type when the plant was grown on pine (*Pinus sylvestris*), but was of a guaiacyl-syringyl (**GS**) nature when grown on deciduous trees, such as hawthorn (*Crataegus oxyacantha*)! This quite remarkable account prompted Freudenberg to declare that "these are examples of roles that lignin plays in taxonomy."

Such claims would imply that the hemi-parasitic mistletoe was able to suck up the monolignol precursors from either the cambial or lignifying regions of the host, translocate them to the appropriate lignifying cells in the mistletoe tissues, and then form the corresponding lignin type mimicking that of the host. Freudenberg[502] extended these claims further by indicating that it was unclear as to whether either the monolignols (**1**)–(**3**) or monolignol glucosides (**71**)–(**73**) were taken up by the mistletoe, given that no β-glucosidase activity could be detected in the hemi-parasite as revealed by indigo staining with indican. The significance of that observation became subsequently unclear. In later studies by Grisebach, it was observed that the detection of β-glucosidase activity, as evidenced by immunofluorescence, in various cross-sections of plants, was not spatially coincident with Freudenberg's indigo staining technique test.[422]

Indeed, none of the Freudenberg mistletoe lignin claims, of such an exceptional nature, were ever corroborated with experimental data of any kind. Moreover, experimental validation to the Freudenberg mistletoe lignin host-dependent postulate was not to be forthcoming in studies by others.[503-505] European mistletoe (*V. album*) was again collected from both angiosperms (apple, poplar) and gymnosperms (pine, spruce), and subjected to a comprehensive series of analyses, including histochemistry, nitrobenzene oxidation, thermofractography and NMR spectroscopy (of Björkman lignins). This established that for all of the host-hemi-parasite combinations examined, the mistletoe lignins isolated were of a guaiacyl-syringyl (**GS**) nature, in accordance with its angiosperm character.

That this was indeed the case was further validated nearly three years later in a contribution from the Higuchi research group, which conducted a comparable investigation on the lignin of European mistletoe (*V. album*) and obtained similar evidence that it was a guaiacyl-syringyl lignin.[506] It is perhaps also worth reflecting that although the Higuchi group never identified any of the actual enzymatic steps in either monolignol biosynthesis or subsequent monolignol coupling, they continually served the quite useful function of validating and extending many of the pioneering contributions by others.

Thus, in conclusion, mistletoe lignin unexceptionally belongs to the guaiacyl-syringyl class of lignins, the nature of whose host-hemi-parasitic interrelationship continues to be investigated even to this day.[507] This unusual saga does, however, reveal Freudenberg's uncritical acceptance of preliminary results from his own research group, whenever they appeared to support his notion about the random assembly of lignins.

3.18.7 FORMATION OF LIGNIN MACROMOLECULES *IN VIVO* AND *IN VITRO*

The preceding sections have summarized our current understanding of monolignol biosynthesis, as well as having provided some insight into the documented differences in lignin content and composition among various plant forms and tissues. However, when ideas about lignification were first developed, it was not known that the differences encountered in lignin monomer composition and content were a function of, for example, variations in the cell type and cell-wall layer (e.g., of tracheids, vessels and phloem fibers). Thus the early attempts to rationalize how lignin biosynthesis might be occurring suffered from three major drawbacks: first, lignin analyses primarily used woody tissues containing more than one cell-wall type and frequently even contained both sapwood and heartwood. Hence, the results obtained were, at best, only average determinations; second, these studies did not take into account the consequences of competition between biochemical pathways utilizing monolignols and related phenylpropanoids, e.g., hydroxycinnamic acids; third, the available methods suffered from being capable of providing only limited information about the nature of interunit linkages between the respective monomer residues in the biopolymer, a difficulty which remains even at the time of writing. Nevertheless, in spite of these limitations, Holgar Erdtman had developed a theory in the 1930s that lignin formation results from dehydrogenative polymerization of monolignol-derived intermediates.[508]

3.18.7.1 Early Monolignol Dehydrogenation Studies *in vitro*

The involvement of monolignols in lignification can first be traced back to studies by Peter Klason,[509,510] who concluded by 1908 that lignin was a condensation product of coniferyl alcohol (**2**) and oxyconiferyl alcohol.[511] Kürschner later speculated in 1925 that lignin was derived from the polymerization of coniferin (**73**).[512,513]

Holgar Erdtman, however, obtained a result in 1932 that was to prove of central importance to lignin biochemistry.[508,514] By that time, oxidants such as FeCl$_3$ were already known formally to remove a hydrogen atom from a phenol to form a free radical, which could undergo radical coupling to produce a biphenyl compound or diphenyl ether. Indeed, isoeugenol (**105**) had been oxidized earlier with either FeCl$_3$ or a mushroom oxidase, and Cousin and Hérissey[515] had obtained a product, to which the biphenyl structure (**124**) had been ascribed (Figure 42).[516] On the other hand, a phenol with a side chain in the *o*- or *p*-position, bearing a double bond conjugated with the aromatic ring, was envisaged by Erdtman to undergo coupling at the 8-position to give the 8–8′ linked dimer dehydroguaiaretic acid (**125**). Erdtman's experiments subsequently refuted both structures, in place of which the racemic (±)-phenylcoumaran, dehydrodiisoeugenol (**126**), was obtained as an entity that arises from 8–5′ radical coupling as shown.[508,514] This led Erdtman[508,514] to propose that lignins are formed by dehydrogenation of phenolic precursors carrying conjugated 7,8-double bonds, like coniferyl alcohol (**2**). The structure of dehydrodiisoeugenol (**126**) was confirmed almost a decade later by Freudenberg and Richtzenhain.[517] It is also noteworthy that Freudenberg and Dürr[518] had (incorrectly) speculated that the phenylcoumaran structures might be derived from rearrangement of putative primary lignin chains linked in a linear fashion through 9–*O*–4′ bonds.

During the latter part of the 1930s, the oxidation of coniferyl alcohol (**2**) with FeCl$_3$ was found to result in an amorphous product[519] which gave, *inter alia*, veratric acid and isohemipinic acid (**127**) upon treatment with hot alkali followed by methylation and permanganate oxidation. Since the same products were obtained from dehydrodiisoeugenol (**126**) and from lignins, these results tended to support the idea that lignins might be formed by the dehydrogenation of coniferyl alcohol (**2**) and related compounds, as Erdtman had suggested.

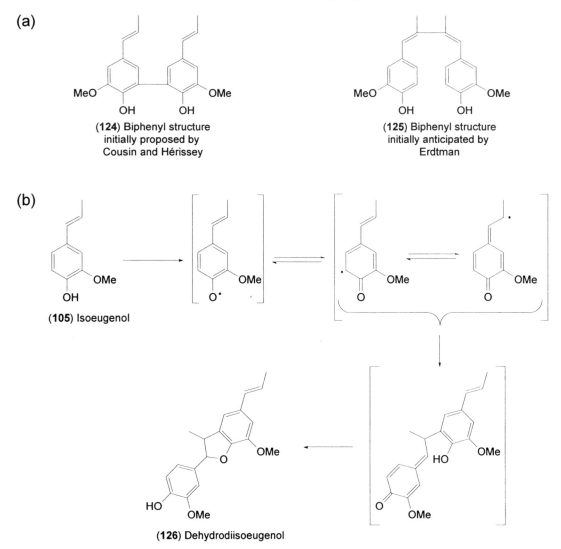

Figure 42 Dehydrogenative coupling of isoeugenol (**105**). (a) Products initially envisaged by Cousin and Hérissey (compound (**124**)) and by Erdtman (compound (**125**)). (b) Erdtman's discovery that dehydro-diisoeugenol (**126**) was formed led to the dehydrogenation theory for monolignol polymerization.

(**127**) Isohemipinic acid

In the 1940s and 1950s, a substantial effort was mounted to examine the outcome of monolignol coupling reactions *in vitro* using various crude oxidative enzyme preparations obtained from organisms, some of which did not even biosynthesize lignins. In any event, the dehydrogenation of coniferyl alcohol (**2**) with, for example, O_2 catalyzed by a mushroom oxidase preparation[520] or the laccase present in the crude mushroom enzyme,[521,522] or with H_2O_2 in the presence of a peroxidase,[523] gave a dehydrogenative polymerisate with some similarities to the lignin extracted by aqueous 90% dioxane from softwood meal (see Table 20). Among the initial dimeric, trimeric and oligomeric products formed, it proved possible to suggest structures for about 30 different compounds. Apart from the uncharacterized (supposedly) polymeric material (45%), the major products were (±)-dehydrodiconiferyl alcohols (**68**) (26%), (±)-pinoresinols (**128**) (13%) and (±)-guaiacylglycerol 8–

O–4′ coniferyl alcohol ethers (**94**) (9%).[524] Small amounts of the trimer (**129**), tetramer (**130**), and pentamer (**131**) were also subsequently reported;[525,526] Figure 43 illustrates the radical–radical coupling processes involved, and subsequent reactions affording the higher oligomers. As can be seen, compounds (**129**)–(**131**) incorporate a trifunctional branchpoint that has arisen from nucleophilic addition of the phenolic hydroxyl group onto either a monolignol (e.g., (**2**) or an 8 5′ linked dilignol (**94**) to the C-7 of the intermediate methylene quinone formed from 8–*O*–4′ coupling between two coniferyl alcohol radicals) (Figure 43(e)). It should be noted, however, that it has not been possible in isolated lignin samples to detect in such amounts the presence of 7–*O*–4′ acyclic benzyl aryl ether linkages which contribute so prominently to the skeleta of these oligolignols. This could be due to the relatively harsh conditions required for lignin solubilization, which might result in the cleavage of such labile bonds.

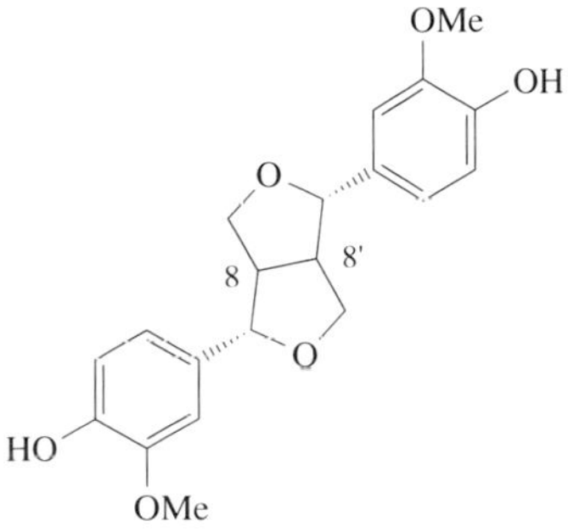

(**128**) Pinoresinol

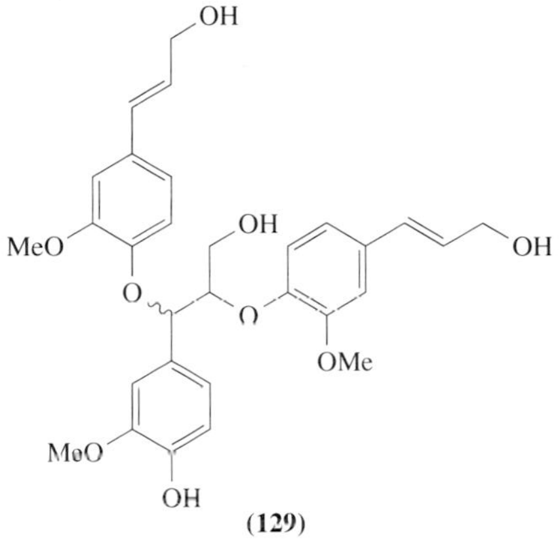

(**129**)

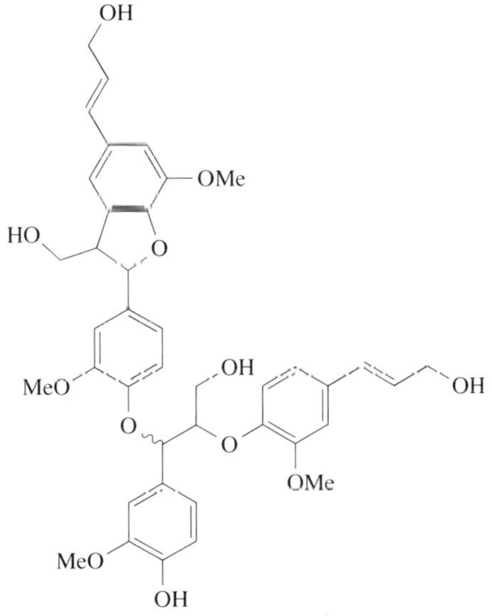

(**130**)

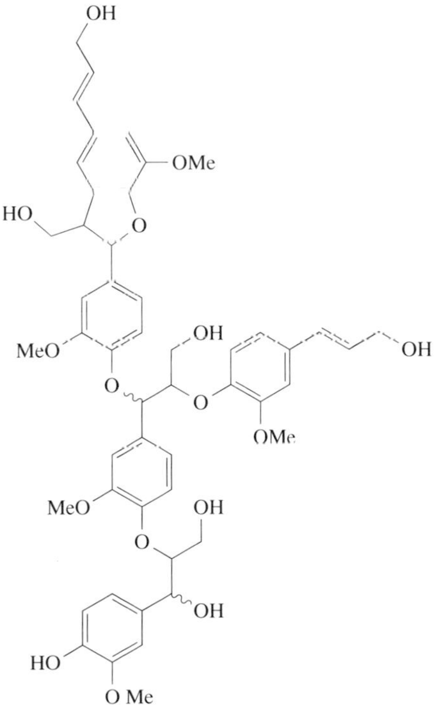

(**131**)

There emerged over time a quite distressing variability in the yields that were reported by Freudenberg for the principal dilignols (**68**), (**94**), and (**128**) formed from the *in vitro* dehydrogenation of coniferyl alcohol (**2**).[527,528] Indeed, the results were conspicuously revised[495] without identifiable experimental corroboration[528] in such a way as to favor the proportion of the 8–*O*–4′ linked dimers (**94**) formed relative to the other two dilignols (**68**) and (**128**). For example, in 1957, the 8–*O*–4′

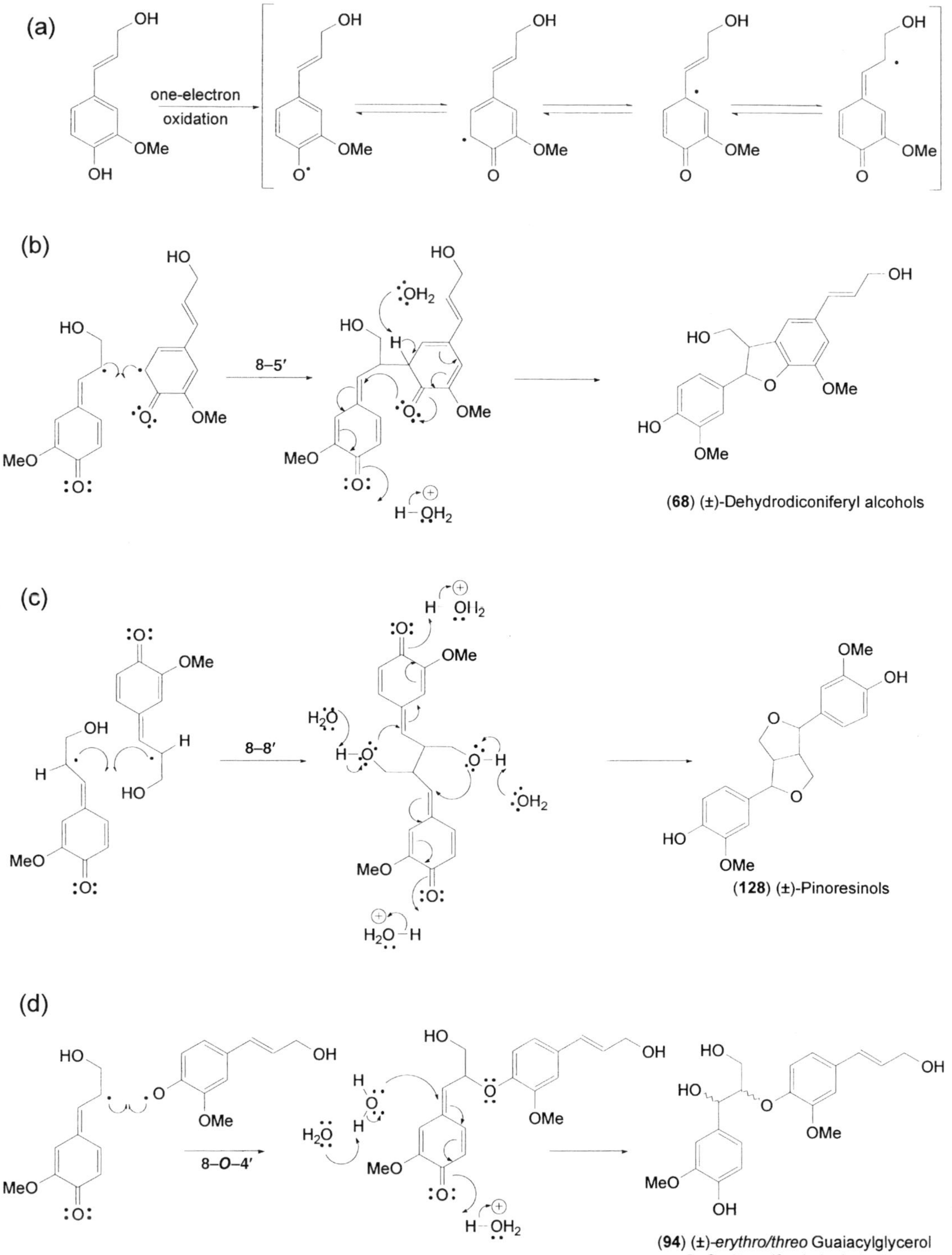

Figure 43 One-electron oxidation of coniferyl alcohol (**2**) *in vitro*. (a) Hybrid resonance forms (free radical) resonance; (b)–(d): 8–5′, 8–8′ and 8–*O*–4′ coupling modes, with either intramolecular cyclization or nucleophilic addition of H₂O at C-7; (e) nucleophilic attack of 8–*O*–4′ quinone methide intermediate with various phenoxide anions to give oligolignols (**129**)–(**131**).

linked dilignols were cited[529] on the basis of Freudenberg's reviews[494,530,531] as being the predominant products (~ 70%) resulting from the dehydrogenative dimerization of coniferyl alcohol (**2**) *in vitro*, whereas all previous experimental evidence had affirmed that the 8–5′ linked dilignol (**68**) actually predominated.[524]

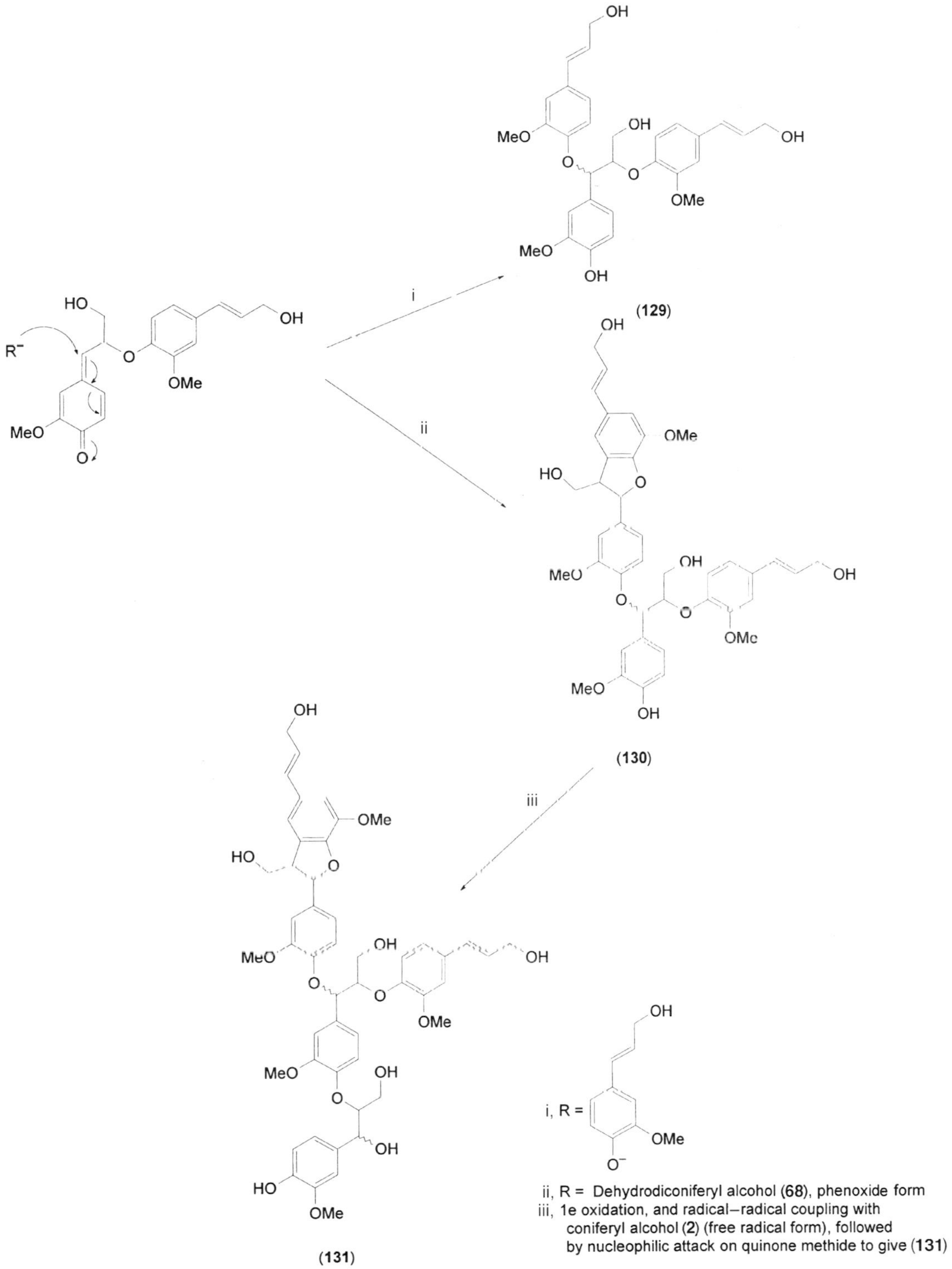

Figure 43 (continued)

The changes in the product distribution claimed to be formed *in vitro* by Freudenberg seemed closely to follow or mirror the results obtained by other workers in the analysis of actual lignified tissues. Thus analytical degradative studies had begun to indicate that 8–O–4′ linkages represented about 50% of the total in softwood lignins,[529] and the dilignol frequencies reported by Freudenberg in various reviews began to change in concert with these findings without support from any new data.

However there is no inherent contradiction between these findings and the observation that 8–O–4′ linkages are present in less than 20% of dilignols dehydrogenatively formed from coniferyl alcohol (**2**) *in vitro*. Macromolecular lignin assembly in plant cell-walls is thought primarily to involve dehydrogenative coupling between monolignols and monomer residues on growing biopolymer chains,[495] where at least one (usually the C-8) position is already covalently linked to another unit. What occurs under such circumstances is reflected in the outcome of dehydrogenative coupling between coniferyl alcohol (**2**) and the dimers dehydrodiconiferyl alcohol (**68**),[532] *epi*-pinoresinol (**132**)[533] and guaiacylglycerol 8–O–4′ coniferyl alcohol ether (**94**),[532] respectively, where 8–O–4′ linkages were found, by Freudenberg and coworkers during the early 1960s, to predominate in all three cases. Thus the insistence, by Freudenberg at the end of the 1950s,[2,391,533] upon the indistinguishability of lignin biopolymers and monolignol dehydropolymerisates produced *in vitro* no longer dictated that 8–O–4′ linked dilignols had to be the most frequent among the dehydrodimers formed from coniferyl alcohol (**2**). Indeed, this may have accounted for the somewhat more moderate assertion made by Freudenberg in 1968[528] that guaiacylglycerol 8–O–4′ coniferyl alcohol ether (**94**) *probably* surpasses in yield that of pinoresinol (**128**) and dehydrodiconiferyl alcohol (**68**).

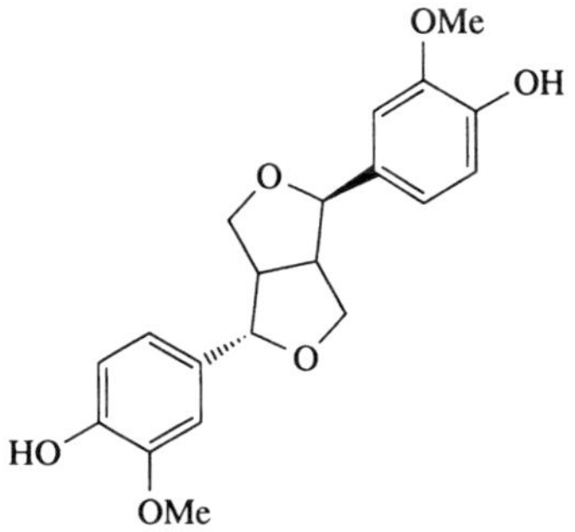

(**132**) *Epi*-pinoresinol

3.18.7.2   Enzymes and Proteins Implicated in the Dehydrogenative Coupling of Monolignols

From the early 1950s considerable attention has been given to the nature of the oxidative enzymes responsible for generating the free-radical intermediates involved in lignin biosynthesis; this section thus assesses the current evidence for the possible involvement of various enzyme and protein candidates for the final dehydropolymerization step.

It is quite remarkable that no fewer than five enzymes have been variously implicated in the dehydrogenative coupling of monolignols to form the lignins, i.e. peroxidase,[2,391,528,534–538] laccase,[2,391,522,528,534,539–541] peroxidase *and* laccase,[2,391,528,542,543] coniferyl alcohol oxidase,[544–546] (poly) phenol oxidase,[547] and even cytochrome oxidase.[548] The reason for this is that ostensibly each can oxidize the monolignols (**1**)–(**3**) to generate the corresponding resonance stabilized free-radical intermediates via one-electron oxidation. In all cases, the resulting free-radical intermediates are then envisaged to diffuse away from the site of catalysis and to undergo random free-radical coupling (Figure 43(a)–(d)). However, the resulting products obtained *in vitro* never duplicated lignin structures and, in some cases, there was not even any evidence for the formation of polymeric components.

There is no other instance where the assembly of a biopolymer has been purported to be capable of appropriating the activity of any one of five distinct enzymes in an identical capacity during the same biochemical step. Perhaps this is a consequence of the fact that the individual coupling processes, through which lignin macromolecules are assembled from monolignol radicals, have not generally been considered to be under direct enzymatic control. A further complication is that there are no *in vitro* assay conditions which satisfactorily assess whether actual lignin biopolymer formation has occurred or not. However, in *Zinnia elegans* tissues[549] and *Pinus taeda* suspension cultures,[536] H_2O_2 generation (and thus, presumably, peroxidase activity) exhibited a strong temporal correlation with presumed lignin formation. Similarly, incubation with H_2O_2 markedly increased, within only 1 h, the epitope concentration to antilignin antibodies in *Z. mays* coleoptile cell-walls.[550]

The subsections which follow evaluate in turn each of the claims for the involvement of particular enzymatic transformations in lignification, from the perspective of the following four criteria:

(i) the enzyme must be able to convert mono-, oligo-, and polylignols into their free-radical derivatives;

(ii) the enzyme must be both temporally and spatially correlated with sites of lignin biosynthesis;

(iii) the enzyme, in the presence of the requisite co-factors and/or other proteins, must be demonstrably capable of converting the monolignols into macromolecular lignin chains; and

(iv) the enzyme must unequivocally be demonstrated as essential for lignin biosynthesis, e.g., through loss of function.

3.18.7.2.1   Copper-containing oxidases

Various copper-containing oxidases have been proposed as being involved in lignin biosynthesis, namely (poly)phenol oxidases [including coniferyl alcohol oxidase] and plant laccases.

(i) Polyphenol oxidases

Polyphenol oxidases (PPOs) are ubiquitous in higher plants, and are also present in bacteria, fungi, and animals. They are bi-copper, oxygen-dependent, metalloenzymes which catalyze either one or two electron oxidations of phenols.[551]

In general terms, the polyphenol oxidases can be subdivided into two broad groups, the so-called cresolase class (EC. 1.14.8.1) which is able to o-hydroxylate monophenols ultimately affording the corresponding o-diquinones, and catecholase (EC. 1.10.3.2) which converts diphenols into o-diquinones.

Plant polyphenol oxidases are typically proteins with M_rs of ~ 40–72 kDa,[551] the physiological functions of which *in planta* remain poorly understood. Several have been found to be associated with root plastids, potato amyloplasts, leucoplasts, etioplasts, chloroplasts, and plastid-like particles; they are generally present in tubers, leaves, storage roots, fruits, and floral parts. However, there is no convincing data linking rigorously purified and characterized PPOs to any specific role in lignification.

Yet, by 1953, Freudenberg had suggested, from the use of crude enzyme preparations,[419,520,552–555] that monolignol oxidation leading to lignin formation was under the influence of polyphenol oxidase and oxygen. Interestingly, this preparation, variously referred to as phenol dehydrogenase, catechol oxidase, and mushroom dehydrogenase and reductase, respectively, was from the press-sap extract of a mushroom (e.g., *Agaricus campestris*), which does not even biosynthesize the lignins.

Mason and Cronyn[547] examined further the putative role of purified mushroom oxidase in lignin formation, no details of purification having been provided other than that it possessed 150 units/mg catechol oxidase activity. However, they concluded that this enzyme preparation was not responsible for monolignol oxidation because of the negligible oxygen consumption observed. Rather, they suggested that the oxidase present in Freudenberg's crude mushroom extracts was some other enzymatic species.

Higuchi later suggested, based on studies with crude enzyme preparations from bamboo-shoots (*Phyllostachys heterocycla, P. nigra* var Heronis, and *P. rediculata*),[535] as well as with other crude homogenates from Japanese lacquer mushroom (*Psalliota campestris*),[522] that the oxidase in Freudenberg's mushroom extracts might instead have been a laccase.

(ii) Coniferyl alcohol oxidase

Polyphenol oxidases again re-emerged as possible contenders for involvement in lignification by 1992. In 1990, the research group of Savidge,[544,545,556–559] and later that of McDougall,[546,560,561] reported the existence of a "coniferyl alcohol oxidase" in plant tissues undergoing secondary wall thickening and lignification. The activity was first detected in jack pine (*P. strobus*),[544] on the basis of histochemical staining using p-anisidine (**133**), which suggested its presence in cells undergoing S_1, S_2, and S_3 thickening and lignification. It was not, however, detected, either prior to secondary thickening and lignin biosynthesis or after cell maturation and autolysis had reached completion. Somewhat comparable results were obtained with histochemical staining of cross-sections of sitka spruce (*Picea sitchensis*) using ABTS (2,2′ azinobis-[3-ethylbenzthiazoline-6-sulfonate]) for visualization purposes.[560]

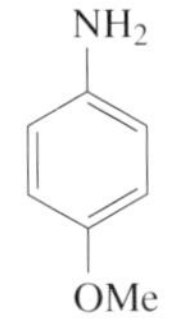

NH$_2$

OMe

(133) *p*-anisidine

The reported properties of the *P. strobus* coniferyl alcohol oxidase are summarized in Table 21.[558] As can be seen, it is thought to be a glycoprotein existing in its native state as a tetramer, with a deglycosylated subunit of M$_r$ ~67 000. Based on graphite furnace atomic absorption spectroscopic analysis, it is thought to contain one copper atom per protein molecule. However, the copper atom was apparently not detectable by EPR spectroscopy, suggesting it to be Type 3 (discussed below in subsection iii). Like so many other oxidases present in plants, it can convert coniferyl alcohol (**2**) into a range of products, two of which were demonstrated to be the lignans, dehydrodiconiferyl alcohol (**68**) and pinoresinol (**128**). Little is yet known about this enzyme in terms of its basic kinetic parameters (K_m, V_{max}), although preliminary characterization suggests its overall activity to be quite low (oxidation rate of 300 nmol h^{-1} mg^{-1} protein at 70 μM coniferyl alcohol (**2**) concentration); Savidge classifies it as a catechol oxidase (E.C. 1.10.3.1).

Table 21 Summary of findings for CAO in conifers.

	Characteristic	*Result*	*Reference*
1.	Amino acid composition:	Asx (8.5), Glx (10.3), Ser (6.3), Gly (9.8), His (2.2), Arg (4.7), Thr (5.2), Ala (9.7), Pro (5.6), Tyr (2.7), Val (7.6), Met (2.8), Ile (4.7), Leu (9.4), Phe (4.9), Lys (5.5)	559
2.	N-terminal sequence:	H$_2$N-X E L A Y S P P Y X P S	559
3.	Molecular weights:		
	Native CAO:	107 500, by SDS PAGE	545
	Native CAO:	426 000, by native PAGE	unpublished
	Deglycosylated CAO:	67 000, by SDS PAGE	545
4.	Deduced glycan content:	38%	545,559
5.	Copper content:	1 atom per protein molecule	545
6.	Copper type:	Type 3 (EPR silent)	559
7.	Activity studies:		
	Temperature optimum:	30 C	545
	pH optimum:	6.3	545
	Oxidation rate:	0.31 μmol h^{-1} mg^{-1} (70 μM coniferyl alcohol (**3**))	559
	Oxidation products:	pinoresinol (**128**)	544,556,557
		dehydrodiconiferyl alcohol (**68**)	544,556,557
	Requirement for O$_2$:	N$_2$ inactivation	544,556,557
		CO inactivation	unpublished
		No H$_2$O$_2$ enhancement	544,556,557

On the other hand, the sitka spruce "coniferyl alcohol oxidase" has not been purified much beyond about 56-fold purification, this preparation exhibiting (partially) both phenoloxidase and laccase properties.[560] It was reported to convert coniferyl alcohol (**2**), over extended periods of time (~48 h), into a precipitate (no yield disclosed) which was thought to be possibly lignin-like. EPR analysis (data not given) of this partially purified "coniferyl alcohol oxidase"[561] suggested the presence of both Type 2 and Type 1 copper atoms (see subsection iii below), thus suggesting a quite different protein from that described by Savidge. However, there were at least eight major bands in the oxidase preparation, as revealed by SDS-PAGE and silver staining, which precludes meaningful interpretation of the nature of this oxidase at present.

Clearly, the characterization of the putative coniferyl alcohol oxidase has not progressed past a preliminary stage, and the following needs to be carried out before a role in lignification can be accepted as plausible:

(i) establishing whether pinoresinol (**128**) and dehydrodiconiferyl alcohol (**68**) are terminal products of the enzymatic reaction, or are further oxidized to macromolecular lignin chains;

(ii) identifying its subcellular location;

(iii) determining its fundamental catalytic characteristics (such as K_m, V_{max}, substrate specificity, etc.);

(iv) elucidating whether the products obtained from catalysis are racemic or not;

(v) delineating whether the enzyme is truly involved in lignification, or in some other biochemical process such as in lignan biosynthesis; and

(vi) ascertaining the actual identity of the enzyme through gene sequence comparisons.

(iii) Laccase

Whether or not the copper-containing glycoproteins, the laccases (EC. 1.10.3.2), have a role in lignification has been an ongoing debate since the 1950s,[391,520,522,534,535,537,540,543,562–565] and it is still not resolved. The first report of the occurrence of laccase was made in 1883 by Hikorokuro Yoshida, who examined a milky secretion, know as urishi, from the Japanese lacquer tree,[566] which is still used as a base for fine furniture lacquer finishes. This secretion could be separated into an alcohol soluble portion ("urishic acid") and a nitrogenous component (diastase = enzyme) which, when combined together in the presence of O_2 and water, resulted in lacquer formation.

More than a decade later, in 1894, Bertrand made a similar discovery in a study of Indo-Chinese lacquer tree (*Rhus succedanea*) secretions,[567] and named the diastase as laccase and its substrate as "laccol"; the components of laccol were subsequently identified as the catechol derivatives (**134**)–(**137**) shown.[568] The discovery of laccase was thus not correlated with lignification in any manner; indeed, its lacquer hardening function still remains as the only truly verifiable physiological function of laccase.

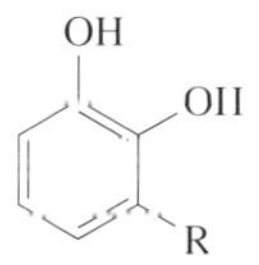

(**134**) R = (CH$_2$)$_{14}$–CH$_3$
(**135**) R = (CH$_2$)$_7$–CH=CH–(CH$_2$)$_5$–CH$_3$
(**136**) R = (CH$_2$)$_7$–CH=CH–CH$_2$–CH=CH–(CH$_2$)$_2$–CH$_3$
(**137**) R = (CH$_2$)$_7$–CH=CH–CH=CH–CH$_2$–CH=CH–CH$_3$

Later, Keilin and Mann, in the 1930s and 1940s, carried out a series of pioneering studies on plant laccases. These were purified first from the Indo-Chinese lacquer tree (*R. succedanea*),[569] as well as from Japanese (*R. vernicifera*) and Burmese (*Malanorrhea usitata*) lacquer trees, respectively.[570] The Indo-Chinese laccase was a copper-containing metalloenzyme consisting of approximately 45% protein, with the remainder being of a polysaccharide nature. Interestingly, the various laccase preparations obtained displayed broad substrate versatilities, in terms of being able to oxidize various *o*- and *p*-phenols, diamines, and aminophenols. Although the laccases are now known to belong to the so-called "blue enzymes" due to the presence of copper,[571] Keilin and Mann had originally incorrectly attributed this color to a pigmented impurity.

The involvement of copper for enzymatic activity was established with the Indo-Chinese lacquer laccase; loss of catalytic activity occurred following treatment with cyanide ion and subsequent dialysis, but was restored by addition of small amounts of cupric sulfate.[572] Takao Nakamura later demonstrated that the *R. vernicifera* laccase ($M_r \sim 120\,000$) contained four copper atoms.[573] Various laccases are found not only in plants and fungi,[574] but they may also possibly be present in insects[575] and bacteria (i.e., *Bacillus sphoericus*).[576]

The possible role of plant oxidases (laccases, peroxidases and other polyphenol oxidases) in lignification was proposed in the late 1940s by the Russian researchers.[577,579] Later, Legrand[579] suggested that the mushroom (*Psalliota campestris*) press-sap extract, which had been demonstrated earlier to convert coniferyl alcohol **2** into dehydrodiconiferyl alcohol (**68**),[520] likely contained laccase. In 1957, following the papers of Legrand[579] and Mason and Cronyn,[547] Higuchi was the next to suggest that Freudenberg's "phenol dehydrogenase" might also be a laccase.[535] To explore this possibility, laccase preparations (actually crude homogenates)[522] were obtained from the Japanese lacquer tree and wild mushroom (*Lactarius piperatus*), respectively, which oxidized *p*-phenyl-enediamine (**138**), hydroquinone (**139**), catechol (**140**) and *p*-cresol (**141**). These preparations were also considered to be able to oxidize coniferyl alcohol (**2**); firstly, on the basis of oxygen consumption,[535] and secondly via coniferyl alcohol (**2**) conversion into an ill-defined precipitate, a so-called dehydrogenative polymerisate (DHP), for which no yield was reported.[562] However, in that same paper, Higuchi and Ito concluded, based on relative peroxidase and laccase activities in bamboo shoots, that "the enzyme responsible for the biosynthesis of lignin from coniferyl alcohol (2) might probably be peroxidase". At around the same time, Freudenberg et al.[534] reported that a "highly active laccase" (crude homogenate from either mushroom or filtered spruce cambial sap) was able to convert coniferyl alcohol (2) into various colored products which were not investigated further. By 1959, Freudenberg had concluded[391] that lignin was derived from the monolignols (1)–(3) via the action of O_2/laccases and H_2O_2-dependent peroxidases.

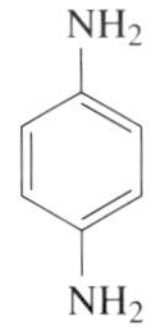

| (138) Phenylenediamine | (139) Hydroquinone | (140) Catechol | (141) p-Cresol |

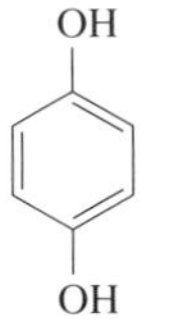

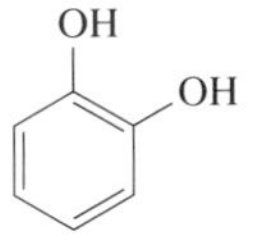

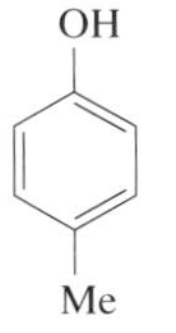

The matter then became further confused when Wataru Nakamura purified the blue laccase from the Japanese lacquer tree,[580] but was unable to observe formation of a lignin-like substance from coniferyl alcohol (2), based on turbidity measurements; this result was interpreted to signify a lack of DHP formation and thus to constitute proof that laccase was not involved in lignin biosynthesis. Nakamura did not, however, establish whether other oxidative coupling occurred, e.g., to give lignan dimers, such as dehydrodiconiferyl alcohol (68). Later, Harkin and Obst in a *Science* article entitled "Lignification in trees: indication of exclusive peroxidase participation"[563] appeared to provide further evidence in support of the lack of involvement of laccase via use of syringaldazine (142) as a histochemical staining reagent; this was employed since, when incubated in the presence of either O_2/laccase or H_2O_2-peroxidase, syringaldazine (142) can be converted into the purple tetramethoxyazo-p-methylene quinone (143) (Equation 1).

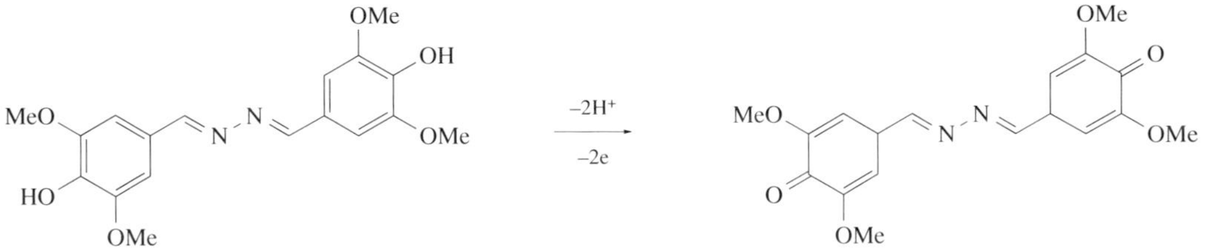

When using this reagent to stain cross-sections of freshly microtomed surfaces of branch and sapling tissues of seven angiosperms, including sugar maple (*Acer saccharum* Marsh), quaking aspen (*P. tremuloides*) and eastern cottonwood (*P. deltoides*), no chromophoric reaction was apparently observed with syringaldazine (142) until H_2O_2 was added. The researchers thus concluded that laccase was not present in these species, and hence that peroxidase was exclusively involved in lignification. Indeed, according to these researchers "We have applied syringaldazine (142) to finally resolve the lignification oxidase question."

The assertions of Harkin and Obst were not be substantiated in further studies: firstly, a laccase purified from *Schinus molle*,[581,582] could not be detected using the syringaldazine (142) reagent, casting some doubt on the original methods[563] employed. Secondly, an extracellular laccase was purified to apparent homogeneity ($M_r \sim 97\,000$) from sycamore (*Acer*) cell suspension cultures, this finding being in contrast to the Harkin and Obst claim which had concluded that laccases were not present in *Acer* species.[583] On the other hand, since the purified *Acer* laccase was capable of oxidizing a wide range of phenols, it was again proposed that it was probably involved in lignin biosynthesis;[583] however, monolignols were not examined as substrates, and the sycamore cells in suspension culture do not biosynthesize either monolignols or lignins in detectable amounts.

Polyclonal antibodies, raised against the sycamore deglycosylated laccase ($M_r \sim 66\,000$), were next employed in order to determine the subcellular location of the corresponding enzyme in sycamore (*A. pseudoplatanus*).[539] However, the antibodies actually cross-reacted with a smaller ~ 56 kDa polypeptide (assumed to be a laccase), and attempts to titrate these against various sycamore laccase preparations failed to inhibit enzymatic activity. Nevertheless, using these polyclonal antibodies, immunogold localization suggested that the cross-reacting protein was present in epidermal tissues of sycamore stems, as well as in the lignified walls of protoxylem and metaxylem of the petioles (but not the vascular parenchyma). Immunolabeling was also noted in the periderm phloem fibers, and in the stems of "wood vessels and wood fibers". In order to understand the significance of these findings, however, it would be useful to establish whether the laccase was indeed the entity being immunodetected in that study.

Other investigations have also proposed a role for laccase in lignification, but again unequivocal proof is lacking. Sterjiades *et al.*[540,542] purified the sycamore laccase ($M_r \sim 115$ kDa) and examined its capacity to oxidize monolignols (**1**)–(**3**) via a spectrophotometric assay. Some preliminary evidence was also provided that a DHP was obtained with each monolignol in the presence of laccase; however, the products were not characterized in terms of molecular size, i.e., as to whether they were polymeric. On the other hand, ^{13}C-NMR spectroscopic analyses of these same preparations, showed that the 8–*O*–4′ linkage was essentially absent, in contrast to its dominance ($\sim 60\%$) in angiosperm lignins. Since then an acidic laccase from sycamore maple has been cloned and its gene sequence reported;[584] the cloned gene did indeed encode a laccase with 4 highly conserved copper binding sites.

In another preliminary study, a presumed laccase ($M_r \sim 90.5$ kDa) putatively associated with lignification was reported to have been isolated from loblolly pine (*P. taeda*) cambial tissue,[541] with the evidence given for its identity being the blue color, an absorbance at 610 nm (so-called Type 1 copper) and the EPR spectrum (data not provided). To date no gene encoding a gymnosperm laccase has yet been described, although three *P. taeda* cDNAs were reported by the same researchers as putatively encoding three blue copper oxidases; however, these reportedly had 51.8% to 55.35% identity to that of cell-wall bound ascorbate oxidase although the sequence data was not given.[565] The staining reagents, diaminofluorene (DAF) (**144**) and syringaldazine (**142**), were then also employed[541] in an effort to demonstrate that the *P. taeda* oxidase activity of interest was in "differentiating xylem near the cambium"; however, whether the staining reaction detected resulted from the same oxidase as purified, and/or whether it had a role in lignification, was not established. Moreover, the substrate specificity of this 90.5 kDa oxidase revealed a much higher preference for the non-physiological substrate phenylhydrazine (1920 nkat mg^{-1} protein), than for either *p*-coumaryl (**1**), coniferyl (**2**), or sinapyl (**3**) alcohols (5, 72 and 47 nkat mg^{-1} protein, respectively). Additionally, the K_m values for two of the monolignols, coniferyl (**2**) and sinapyl (**3**) alcohols, were high, namely 12 and 25 mM, respectively, whereas with *p*-coumaryl alcohol (**1**) no K_m could be obtained due to the low rate of catalysis encountered. Indeed, this observation prompted Barcelo to state "With these high K_m values, it is difficult to imagine what concentration of cinnamyl alcohols it would be necessary to reach its lignifying cell walls to saturate laccase during the oxidation of cinnamyl alcohols to lignin like compounds."[537] Furthermore, neither this nor related studies have reported kinetic data for any of the resulting lignan dimers or oligomers; these are expected to be even more slowly oxidized.

(**144**) Diaminofluorene

It has also been proposed that laccases might participate in the early stages of lignification; however, the kinetic data described above does not appear to agree with this interpretation: if *p*-coumaryl alcohol (**1**) is indeed initially incorporated into the lignin of the cell corners/middle lamella,[453,455,456] it seems to be a contradiction that no K_m value for *p*-coumaryl alcohol (**1**) with laccase can even be obtained. Accordingly, it seems premature to insist upon an involvement of laccase in lignification on the basis of such findings.

Nor is there any rigorous evidence for polymer formation when monolignols, such as coniferyl alcohol (**2**), are incubated with laccase: for example, the dehydrogenative polymerisate formed by the loblolly pine oxidase, actually only afforded various unidentified substances, attributed as being dimers and perhaps oligomers. Even ignoring linkage patterns,[541] these do not constitute lignins. On the other hand, Chen examined the product distribution obtained following coniferyl alcohol (**2**) oxidation with a *Rhus* laccase over a period of 145 h: this afforded the well-known lignan dimers, dehydrodiconiferyl alcohol (**68**) (21%), pinoresinol (**128**) (26%), guaiacylglycerol 8–*O*–4′ coniferyl alcohol ether (**94**) (2%), and 40% of unidentified oligomers;[585] and the results obtained are consistent with the known catalytic properties of laccase.[586] Chen also reported that no DHP was obtained using the plant laccase,[585] thus returning its tentative role in lignification to the same level of uncertainty as it had in the 1950s.

Furthermore, no laccase (or laccase gene) has yet been reported as present in monocotyledons (see Lewis and Davin[1] and EMBL/GenBank/DDBJ databases) which, if correct, also argues against any general role in lignification. Indeed, Ros Barcelo[537] has suggested that laccase involvement in lignification might simply be an evolutionary vestige, whose function was subsequently displaced by the peroxidases (see Section 3.18.7.2.2). Another alternative, however, is that laccases fulfill functions quite distinct from that of lignification.[537]

(a) *General catalytic mechanism.* The broad substrate specificity of laccases is one of the major complicating factors in attempting to define their physiological functions. With monolignols (**1**)–(**3**), the laccases catalyze the one-electron oxidation of the substrate to afford the corresponding free-radical forms. Using coniferyl alcohol (**2**) as an example, the free-radical intermediates are then presumed to diffuse away from the enzyme active site where radical–radical coupling can occur to afford, at least initially, the racemic lignans, dehydrodiconiferyl alcohol (**68**), pinoresinol (**128**) and guaiacylglycerol 8–*O*–4′ coniferyl alcohol ether (**94**), respectively. Equation 2 shows the laccase catalytic cycle with monolignols, which involves four consecutive one-electron oxidations, with concomitant reduction of molecular oxygen to give H_2O.

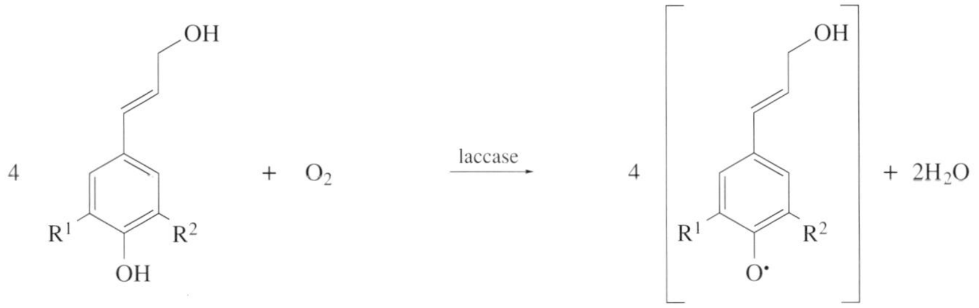

$$R^1, R^2 = H \text{ or } OMe \qquad\qquad\qquad (2)$$

The catalytic cycle of both plant and fungal laccases has been investigated extensively.[586–589] However, even today the precise mechanism of electron transfer within the enzyme remains to be fully delineated. Nevertheless, the laccases contain four copper-containing sites which are involved in the catalytic cycle (see Reinhammar[587]), where the copper atoms exist as both a mononuclear (Type 1) species and a trinuclear cluster (Type 2 and Type 3), respectively.[591,592] In *R. vernicifera* laccase, for example, the mononuclear Cu^{2+} (so-called Type 1 copper and EPR-detectable) is considered to be coordinated to two histidines, one cysteine and one methionine;[589] with the characteristic blue color at 610 nm[593] being due to a cysteine $S \rightarrow Cu^{2+}$ charge transfer complex. The Type 1 copper is also the site of binding of the first organic substrate molecule.[594,595] The trinuclear copper cluster, on the other hand, contains a Type 2 Cu^{2+} (EPR-detectable), coordinated to two histidyl imidazole ligands, and an antiferromagnetically paired Type 3 $[Cu^{2+}–Cu^{2+}]$ copper (EPR silent), with each presumed to be linked to three histidyl imidazole ligands;[589] the two Type 3 copper atoms are linked via a bridging hydroxyl ligand. Using the ~100 kDa laccase isolated from *F. intermedia* as an example, which contains ~40% polysaccharide, its corresponding gene was cloned, and Figure 44 shows its full length sequence; underlined are the four highly conserved copper-binding domains.

It is considered that the Type 2 Cu^{2+} site is the presumed site of dioxygen binding,[594,595] where dioxygen (molecular oxygen) is converted first into a peroxide-like intermediate, and then into an oxygenated bridge between the Type 2 and Type 3 coppers prior to reduction to H_2O.[596,597]

Although there is no crystal structure yet available for any plant laccase, X-ray structures of the related ascorbate oxidase[589,598] and a fungal (*Coprinus cinereus*) laccase[599] have been reported; the latter, however, lacks its Type 2 copper atom which results in a distorted coordination for the Type 3 copper atoms. Nevertheless, the fungal laccase (*C. cinereus*) crystals were obtained in monomeric form and resolved at 2.2 Å.[599] In that case, the β-barrel molecular architecture is similar to that of the ascorbate oxidase monomer: the Type 1 copper of the fungal laccase is located in domain 3, whereas the two Type 3 copper atoms are located at the interface between domains 1 and 3. The Type 1 copper center of the fungal laccase is trigonally coordinated to two histidines (His396 and His457) and Cys452. That is, in contrast to the *Rhus* laccase, there is no direct coordination to a methionine, this being replaced by a leucine (Leu462). This in turn suggests that the methionine is not essential for catalysis. The Type 1 copper atom is also approximately 6 Å away from the

```
-140   GCGTCTATATATTACATATATTTATGCATATACTCGAAACCATTCTGCAAATAAACAAGTCCTACTAGACAGCCT   -66

 -65   AAATACAGACAAAAAACTCGGAGGGGTTCATATGAACATATCTTTATTGTGAACATTGAGTTAAATATGAAGTTT    9
   1                                                                      M  K  F     3

  10   TCTCTCTTACATTTAATAGGTTTTCTTCTGCTAGGAGGAGTGCTAGTGCTTCCTCTTCATGCTGCTCTTATCACT   84
   4   S  L  L  H  L  I  G  F  L  L  L  G  G  V  L  V  L  P  L  H  A  A  L  I  T      28

  85   CGCCATAGGTTTGTGTTGACAGATACTCCATTCACAAGGCTGTGCAGCAACAAGAGCATATTTGTAGTAAATGGA  159
  29   R  H  R  F  V  L  T  D  T  P  F  T  R  L  C  S  N  K  S  I  F  V  V  N  G      53

 160   CAATTTCCAGGGCCAACTATATATGCCACTGAAGGAGATACAATAATCGTTGATGTCATTAACCAACCAAGCGAA  234
  54   Q  F  P  G  P  T  I  Y  A  T  E  G  D  T  I  I  V  D  V  I  N  Q  P  S  E      78

 235   AATGTAACCATTCACTGGCATGGAGTGAAGCAACCTAGATATCCATGGTCAGATGGTCCTAATTATATTACTCAG  309
  79   N  V  T  I  H  W  H  G  V  K  Q  P  R  Y  P  W  S  D  G  P  N  Y  I  T  Q     103

 310   TGCCCCATACAACCCGGAGCAAACTTTAGCCAAAAGATTATACTTTCGGACGAGATAGGAACTTTGTGGTGGCAC  384
 104   C  P  I  Q  P  G  A  N  F  S  Q  K  I  I  L  S  D  E  I  G  T  L  W  W  H     128

 385   GCCCACAGCGACTGGTCACGTGCCACCGTGCACGGTGCAATCGTTATTCGTCCCAAGAATAACAGTAATTATCCT  459
 129   A  H  S  D  W  S  R  A  T  V  H  G  A  I  V  I  R  P  K  N  N  S  N  Y  P     153

 460   TTTCGTACGCCTGATGCGGAAGCGACAATCATATTAGGGGAATGGTGGAAAAGTGATATACGGGCAGTTCAAAAT  534
 154   F  R  T  P  D  A  E  A  T  I  I  L  G  E  W  W  K  S  D  I  R  A  V  Q  N     178

 535   GAGTTTCTTGGTAACGGAGGGGACGCAAATGTTTCTGATGCTTTCCTCATAAATGGTCAACCTGGTGATCTCTAT  609
 179   E  F  L  G  N  G  G  D  A  N  V  S  D  A  F  L  I  N  G  Q  P  G  D  L  Y     203

 610   CCATGCTCAAGATCAGATACATATAATCTGACGGTGGAATCTGGTAAAACTTATTTGATTAGAATGATTAATGCT  684
 204   P  C  S  R  S  D  T  Y  N  L  T  V  E  S  G  K  T  Y  L  I  R  M  I  N  A     228

 685   GTGATGAACACCATTATGTTTTTTTCCATTGCAAACCACAGTGTGACCGTTGTTGGATCAGATGCTGCCTACACC  759
 229   V  M  N  T  I  M  F  F  S  I  A  N  H  S  V  T  V  V  G  S  D  A  A  Y  T     253

 760   AAGCCCCTCAAAAGTGATTATATTACCATCTCTCCTGGACAAACCATTGATTTCTTGTTGCAAGCTAACCAAACA  834
 254   K  P  L  K  S  D  Y  I  T  I  S  P  G  Q  T  I  D  F  L  L  Q  A  N  Q  T     278

 835   CCTAGTCACTACTACATGGCTGCCCGTGCTTATGCTGTTGCAGGAAACTTCGATAACACGACCACAACAGCTATT  909
 279   P  S  H  Y  Y  M  A  A  R  A  Y  A  V  A  G  N  F  D  N  T  T  T  T  A  I     303

 910   ATCCCCTACAAAGGAAATTATACAGCTCCATCGTCGCCATCGTTTCCAAATCTTCCAGGGTTCAATGATACAAAT  984
 304   I  R  Y  K  G  N  Y  T  A  P  S  S  P  S  F  P  N  L  P  G  F  N  D  T  N     328

 985   GCTTCAGTTAATTTCACATACAGACTAAGAAGTTTAGGAAATAAAAATTATCCAGTCGACGTTCCTAAAAATGTG 1059
 329   A  S  V  N  F  T  Y  R  L  R  S  L  G  N  K  N  Y  P  V  D  V  P  K  N  V     353

1060   ACTGATAAACTCTTATTTACATTTTCCATCAATTTAACCCCTTGTCCTAATAATTCATGTGCTGGCCCTTTCAAT 1134
 354   T  D  K  L  L  F  T  F  S  I  N  L  T  P  C  P  N  N  S  C  A  G  P  F  N     378

1135   GAGCGCTTTCGAGCAAGTGTGAATAACATTACTTTTGTTCCGCCTACCATTGCCATTCTTCAAGCTTATTACCAA 1209
 379   E  R  F  R  A  S  V  N  N  I  T  F  V  P  P  T  I  A  I  L  Q  A  Y  Y  Q     403

1210   CGCATCAGAAATGTCTACAGTAATAACTTTCCTAGTAATCCACCATTTACTTTCAATTATACTTCAGATATTATC 1284
 404   R  I  R  N  V  Y  S  N  N  F  P  S  N  P  P  F  T  F  N  Y  T  S  D  I  I     428

1285   CCTCGAGACCTATGGAGGCCTCAGAATGGTACAGAAGTAAAAGTTCTTAAATATAATTCTACAGTGGAGATTGTT 1359
 429   P  R  D  L  W  R  P  Q  N  G  T  E  V  K  V  L  K  Y  N  S  T  V  E  I  V     453

1360   TTCCAGGGAACAAATATACTTGCAGGAATAGATCATCCTATACATTTGCATGGACAGAGCTTCTATGTAGTTGGG 1434
 454   F  Q  G  T  N  I  L  A  G  I  D  H  P  I  H  L  H  G  Q  S  F  Y  V  V  G     478

1435   TGGGGACTTCCAAACTTTAACAATGCCACAGATCCATTGAATTATAATCTTGTCGACCCTCCTCTAATGAACACG 1509
 479   W  G  L  G  N  F  N  N  A  T  D  P  L  N  Y  N  L  V  D  P  P  L  M  N  T     503

1510   ATCGCGGTTCCTGTAAGTGGTTGGACAGCAGTCAGATTCAAAGCAAGTAATCCTGGAGTTTGGTTGTTGCATTCT 1584
 504   I  A  V  P  V  S  G  W  T  A  V  R  F  K  A  S  N  P  G  V  W  L  L  H  C     528

1585   CATTTAGAACGCCATTTAAGTTGGGGAATGGACATGGTATTCATTACCCAAAATGGTGAGGGAAAAAATGAAAGA 1659
 529   H  L  E  R  H  L  S  W  G  M  D  M  V  F  I  T  Q  N  G  E  G  K  N  E  R     553

1660   ATCTTGCCTCCACCTCCAGATATGCCTCCATGTTGATGCTTGCCGGCACATTGCTAGTTTCACATTCTACCGACA 1734
 554   I  L  P  P  P  P  D  M  P  P  C  *                                           565

1735   AGCTTGTACTACATGTTCAAGACGTAACACTTATATTTGTAATTTGTTTGTAATACTAGTATACGATTTGTATGG 1809

1810   ATCAATTGCTATTAGTATCATTAGAAAAATAAATTTATGTAGCTCAAAAAAAAAAAAAAAAAAA 1872
```

Figure 44 Full length cDNA sequence of a *Forsythia intermedia* laccase isoenzyme; copper atom binding domains are underlined.

molecular surface at the "floor" of a shallow depression (~6 Å deep and 12 Å wide), the size of which is presumed to account for the broad substrate specificity of the fungal laccase. Additionally, X-ray studies of both ascorbate oxidase[589,598] and azurin[600] revealed that the mononuclear Type 1 copper is circa 12–14 Å away from the Type 2/Type 3 trinuclear copper cluster, where electron transfer from the Type 1 copper site to the trinuclear (Type 2 and Type 3) copper cluster is considered to occur via a conserved Cys-His pathway.[598] Figure 45 illustrates the binuclear Type 3 copper cluster as obtained for the fungal laccase (*C. cinereus*).[599]

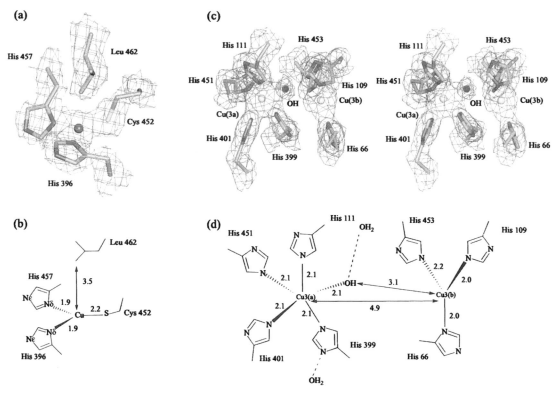

Figure 45 (a) The Type 1 copper center of the *Coprinus cinereus* laccase. The electron density shown is a 2.2 Å resolution maximum-likehood weighted $2F_o - F_c$ synthesis, in divergent stereo, and is contoured at a level of approximately 0.4 e Å^{-3}. (b) A schematic diagram showing the Type 1 site in laccase. The bond lengths shown were not restrained to any target value during the refinement. (c) The binuclear Type 3 copper site. The two Type 3 atoms and the water/hydroxide molecule are shown as yellow and red spheres respectively. The electron density shown is at 2.2 Å resolution with maximum-likehood weighted $2F_o - F_c$ synthesis and is contoured at a level of -0.4 e Å^{-3} (red) and a weighted $F_o - F_c$ synthesis, calculated prior to the incorporation of the "bridging" ligand, and contoured at 4σ (blue). (d) A schematic diagram of the Type 3 copper coordination in the *C. cinereus* laccase.[599]

3.18.7.2.2   *(Per)oxidases and H_2O_2 generation*

Two classes of iron-containing oxidases have been proposed as involved in monolignol polymerization leading to the lignins, namely cytochrome oxidases and peroxidases. Of these, the evidence for the former was very weak. On the other hand, the peroxidases remain the most likely candidates of all of the oxidases in providing the oxidative capacity required for monolignol oxidation during lignin biopolymer formation.

(i) Cytochrome oxidase

A preliminary report in 1964 implicated the involvement of cytochrome oxidase in lignification. This was based upon a histochemical test, known as the G-Nadi reaction (employing *N,N*-dimethyl *p*-phenylenediamine and α-naphthol), which gave a blue coloration with spruce cambial tissue.[548] It is quite surprising that a role for cytochrome oxidase was ever envisaged in cell-wall lignification, given that the enzyme resided in the mitochondria and not in the cell-wall. The claim was disputed shortly thereafter, as it was found that horseradish peroxidase, in the presence of H_2O_2, also afforded this blue color reaction with the same staining reagents. That is, the histochemical "detection" of cytochrome oxidase was instead interpreted simply to be a consequence of interfering peroxidase activity.[601]

(ii) Peroxidases

The peroxidases are a ubiquitous group of metalloenzymes (glycoproteins) found in plants, with a heme prosthetic group. They display broad substrate versatilities and are frequently present in various isoforms; they can oxidize monolignols into their corresponding free-radical forms, with H_2O_2 being used as a co-substrate.

The detection of peroxidases can be traced back to Louis Antoine Planche in 1810, who observed that an alcohol tincture of guaiac resin (from the "guayaco" plant) generated a distinct blue coloration when exposed to various plant extracts, including that of horseradish (see Fruton[602]). From 1848–1855, Schönbein extended these studies by showing that this blue coloration could result from the mixing of guaiac resin with both animal and plant tissues, as well as with H_2O_2.[603,604] The name, peroxidase, was finally introduced near the turn of the century by Linossier[605] initially to describe an oxidase-free preparation from leucocysts requiring H_2O_2.

Later, in the 1930s, Kuhn, Hand and Florkin had obtained clues as to the nature of the protoporphyrin prosthetic group of peroxidase by noting a direct proportional relationship between its activity and the characteristic Soret band absorption.[606] By 1937, Keilin and Mann had not only isolated horseradish peroxidase, but they also identified the protohemin nature of its prosthetic group, as well as demonstrating the formation of two peroxidase-peroxide compounds.[607] Shortly thereafter in 1942, Theorell[608] obtained horseradish peroxidase in crystalline form. It was not until 1976, however, when the first amino acid sequence of a plant peroxidase, in this case horseradish, was described,[609] with the subsequent cloning of various horseradish peroxidase genes being reported in 1988.[610]

The earliest suggestion of the involvement of peroxidase(s) in lignification came from studies of Manskaya.[577,578,611–613] Indeed, it was she who first recognized that the so-called cambial sap and young woody tissue of vascular plants possessed both peroxidase and oxidase activities,[577,578] which were somehow correlated with lignification. Indeed, during 1947[578] Manskaya had already measured changes in gross peroxidase and polyphenol oxidase activities of pine "sap" taken at different time points throughout the year. These were compared with coniferin (72) concentrations, and she concluded that these enzymes were involved in lignification. By 1948, the effects of individually treating both coniferyl alcohol (2) and coniferin (72) with peroxidase and polyphenol oxidases *in vitro* were also initiated.[611,612] By 1952 "lignin-like" precipitates were next obtained from coniferyl alcohol (2), using H_2O_2-dependent crude peroxidase preparations from pine and a polyphenol oxidase from sugar beet root "sternchen" tissue, respectively.[613]

In the USA, Siegel[614] also thought that peroxidases were involved in the lignification process, this first being considered following the observation that excised axes from red kidney bean seeds converted eugenol (102) into lignin-like substances, a process being greatly enhanced by H_2O_2 addition. The peroxidase activity was also localized to the vascular cylinder, the site of lignification. Related studies by Siegel[615,616] added to the idea that peroxidases were responsible for monolignol oxidation, and hence formation of lignin-like substances *in vitro*.

One of Freudenberg's major accomplishments was in the extension of these preliminary findings of the other researchers. This was achieved by demonstrating, for example, that coniferyl (2) and sinapyl (3) alcohols, when incubated with peroxidase in the presence of H_2O_2, were converted into a range of products, including the racemic lignans (68), (96), (128), and ($\pm$) syringaresinols (145), respectively.[391,527] Higuchi also provided additional confirmatory evidence for the role of peroxidase in lignin formation; this was initially based on histochemical staining of lignifying bamboo shoots (*P. heterocycla*, *P. nigra*, and *P. reticulata*) with benzidine, pyrogallol, α-naphthol, and guaiacol (i.e. nonmonolignol substrates) in the presence of H_2O_2,[535] and then by formation of a so-called DHP lignin-like precipitate formed when coniferyl alcohol (2) was incubated with horseradish peroxidase and H_2O_2.[562]

For a number of decades that followed, however, it seemed that any inclination to probe the biochemical mechanisms associated with lignin biopolymer assembly *in vivo* were more or less halted. The reasons for this seem to be multifarious: the first resulted from the dogma that the so-called synthetic DHP-lignins were identical to native lignins, whereas in fact they were not (see Lewis and Davin[12]). The second was that the actual oxidative enzymes responsible seemed to be almost incidental, given that O_2/laccases, H_2O_2-peroxidases and other phenol oxidases could ostensibly all be utilized. A third factor was that many of the studies that followed focused only upon formation of the synthetic dehydropolymerisates and not lignin proper, and the fourth was that, in studies of actual lignification, various non-physiological chromophoric substrates were used rather than monolignols (see Lewis and Yamamoto[20]). These facts thus did not lend themselves to establishing, in any fruitful manner, the precise nature of the oxidative enzymes involved in monolignol oxidation and in lignin biopolymer assembly.

However, following the histochemical staining study by Harkin and Obst,[563] Mäder *et al.*[617,618] established the existence of various peroxidase forms in *N. tabacum* tissue cultures. These were conveniently separated into four main groups based on relative electrophoretic mobilities, namely the so-called anodic (G I and G II) and cathodic (G III and G IV) forms, respectively. Of these, the G I group more readily oxidized *p*-coumaryl (**1**) and coniferyl (**2**) alcohols than the G II group, although both were considered as being cell-wall associated "acidic" peroxidases. On the other hand, the G IV group was virtually inactive with either substrate, being designated as a vacuolar ["neutral or basic"] peroxidase. Comparable studies, with lupin (*Lupinus albus*) hypocotyls, that were presumed to be lignifying, also established the occurrence of two acidic A_1 and A_2, and two basic, B_1 and B_2, peroxidase isoforms. The acidic forms were considered to be extracellular, based on relative binding abilities to membranaceous and cell-wall fractions.[619] Moreover, a study of their kinetic properties, and the failure to detect any laccase activity in this species using 3,3′-diamino benzidine, prompted Ros Barcelo to propose that only peroxidase was responsible for lignification during the secondary thickening of *Lupinus* xylem vessels: the K_m and V_{max} values for both coniferyl alcohol (**2**) and H_2O_2 were obtained for the acidic peroxidase, and the values seemed to be in agreement with this contention. That is, coniferyl alcohol (**2**) and H_2O_2 had K_m values of 10 μM and 16 μM, whereas the V_{max}'s obtained were 12 nmol and 53 nmol of coniferyl alcohol (**2**) oxidized per minute, respectively. Although no precise correlation of the acidic peroxidase(s) with lignification was made, these kinetic data were in accord with previous K_m values for the other enzymes in the monolignol (**1**)–(**3**) forming pathway (see Table 16). This contrasts markedly to that of the *P. taeda* oxidase ($K_m \sim 12\,000$ μM for coniferyl alcohol (**2**)) described earlier.[541]

Other peroxidase isoforms have also been considered to have roles in lignification, although unequivocal proof is lacking. For example, an anionic peroxidase (~ 42 kDa) was obtained from sycamore maple (*A. pseudoplatanus*) cell suspension cultures, which also afford the laccase previously described.[620] Its kinetic data varied substantially from that of the lupin peroxidase, with K_m values for *p*-coumaryl (**1**), coniferyl (**2**), and sinapyl (**3**) alcohols being 1300, 700, and 290 μM, respectively. Lastly, five distinct isoenzymes (P1–P5) were purified[621] from *Zinnia* cells undergoing tracheary element formation. Of these, P1, P3, and two forms of P5 (A and B) appeared to be involved in tracheid differentiation. Although no formal kinetic parameters were measured, the coniferyl alcohol (**2**) used as a substrate was preferred in every case over either *p*-coumaryl (**1**) or sinapyl (**3**) alcohols. Moreover, the P3 form was the only one able to oxidize sinapyl alcohol (**3**), albeit poorly. These data suggest that the different isoenzymes may, therefore, have specific roles in monolignol oxidation.

In summary, the proposed involvement of peroxidase(s) in lignification rests mainly on several lines of indirect evidence:

(i) the peroxidases are ubiquitous cell-wall components found in all vascular plants; this has not been demonstrated for laccases;

(ii) the kinetic data (K_m and V_{max}) demonstrate peroxidases will outcompete laccases for substrate, provided the co-factor H_2O_2 is supplied;

(iii) the peroxidases can act on *p*-coumaryl alcohol (**1**) whereas this is a poor substrate for all known plant laccases;

(iv) *P. taeda* cell suspension cultures can be induced to form an extracellular precipitate and a lignin (or lignin-like) material in its cell-wall. Addition of KI to the cells, however prevents this polymerization from occurring, and instead the monolignols (**1**) and (**2**) are secreted into the medium.[536] This occurs because the KI is an effective H_2O_2 scavenger, which in turn suggests involvement of a H_2O_2 requiring peroxidase system in monolignol coupling, at least with the *P. taeda* cell cultures;

(v) there are numerous examples in the literature of a temporal correlation with H_2O_2 generation and presumed lignification e.g., in *Z. elegans*[549] and;

(vi) the epitope concentration to lignin antibodies in *Z. mays* coleoptile cell-walls was increased when exogenous H_2O_2 was provided.[550] Together, these observations, seem to suggest that the peroxidase(s) have a crucial oxidative role in the lignin-forming process.

(a) *Catalytic mechanism.* Given the ubiquity of peroxidases in all living systems, there has been much interest in defining both the catalytic cycle and the three-dimensional structures of various peroxidases. In this context, the plant peroxidase superfamily can be conveniently subdivided into three broad classes (I, II, and III).[622,623] Of these, the Class I peroxidases are intracellular, with representatives including the pea cytosolic ascorbate oxidase[624] and yeast cytochrome c peroxidase,[625] respectively; indeed, the latter was the first peroxidase X-ray structure to be obtained. The Class II and Class III peroxidases are, on the other hand, generally extracellular: examples of Class III peroxidases include horseradish peroxidase,[626] peanut peroxidase,[627] and barley grain peroxidase,[622] each of whose X-ray structures have been obtained. Horseradish peroxidase is of particular note since it is the most frequently employed to engender formation of synthetic dehydropolymerisates from the monolignols (1)–(3)—even though the tissue from which it is obtained is hardly even lignified. The Class II peroxidases[622] are represented by examples such as the extracellular manganese-peroxidase from the fungus, *Phanaerochaete chrysosporium.*

The general scheme for peroxidase catalysis is shown in Figure 46. Initial binding of hydrogen peroxide first occurs, followed by O—O bond cleavage, with loss of H_2O to afford the radical cation (${}^+_{\bullet}$) intermediate, the so-called compound I $(Fe^{4+})^{\bullet}$. Compound II formation then next occurs via donation of an electron from the substrate (e.g., a monolignol) to afford compound II, which receives a second electron (e.g., from another substrate monolignol) thereby returning the enzyme to its resting ferric state (Fe^{3+}) with concomitant loss of H_2O.

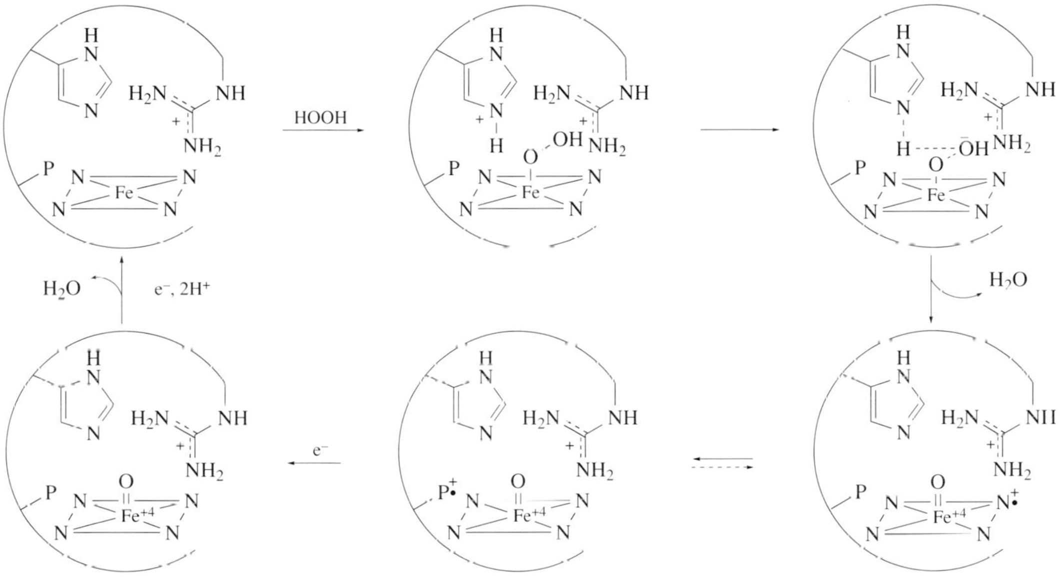

Figure 46 A general catalytic mechanism for classical peroxidases. The heme group is represented by the four nitrogen iron ligands. P stands for a protein residue. With some exceptions, only one of the two compound I structures shown in the scheme is detected for a given peroxidase. (Redrawn from reference [628]).

(b) *Crystal structure of class II peroxidases.* With the structures of three Class III extracellular glycoprotein peroxidases established, i.e., from peanut,[627] horseradish,[626] and barley,[622] respectively, it is quite useful to consider their overall structure in terms of the mode of catalysis. Using the horseradish peroxidase c crystal structure, obtained at 2.1 Å resolution, as an example,[626] two domains are clearly defined: domain I (housing the N-terminal domain related to heme, and consisting of helices A–D), and domain II (the C-terminal proximal domain with helices F–J); helix E interconnects both.

Binding of the heme occurs in a crevice between both domains, being sandwiched between helix B and C-terminus of helix F. A proximal histidine residue (His170), within helix F, is ligated to Fe^{3+} through N-2 of the imidazole ring, and indirectly through ligand Thr171 to the distal Ca^{2+} site. Additionally, the N-δ1 of His170 is hydrogen-bonded to Asp247, an interaction which is considered to increase the basicity of the histidine group. Figure 47 shows both the distal heme cavity, and the hydrogen bonding network, where a Ca^{2+} (required to maintain the structural environment of the heme) is hepta-coordinated to carbonyl and side-chain oxygens, as well as to Asp43.

Arg38, Phe41, and His42 make up the distal peroxide binding pocket, with both the Arg and His residues considered to function in a concerted manner in the acid-base catalyzed cleavage of the peroxide (O—O) bond. As shown in Figure 47, the incoming peroxide donates one proton to the N-ε2 of His42, with the negative charge of the peroxide (OH^{-}) being stabilized in the Arg38 residue. The proton from His42 is then transferred to the peroxide terminal oxygen with cleavage of the O—

(a)

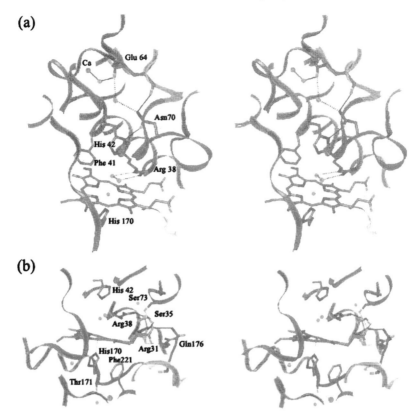

(b)

Figure 47   (a) Distal heme cavity and associated hydrogen-bonding network for horseradish peroxidase c. The protein backbone is shown in a ribbon representation. Side chains of distal pocket residues (above heme in this view) from helix A, Phe41, Arg38, and His42, and proximal histidine His170 are also shown along with the distal waters. The extended hydrogen-bonding network from His42 through Asn70, water, Glu64, and coordinating water to distal Cα (green) is also shown. An only partially occupied distal water molecule, also hydrogen bonding to the distal Pro139 O is not shown. (b) Anchoring of the heme group, linkage from distal pocket to proximal pocket and linkage from proximal pocket to proximal calcium site. This figure shows the hydrogen-bonding network from the distal Arg, through water, to the propionate of the heme. From the helix A, in the background, Arg31 and Ser35 make hydrogen bonds to the propionate deepest in the cavity. In the foreground, Ser73 and Gln176 make hydrogen bonds to the other propionate. The backbone oxygen of Gln176, making the link to the hydrogen-bonding network of the proximal pocket is not shown. The two water molecules of the proximal pocket are shown. Phe221 is seen to "stack" with the proximal His170. Thr171, which is a ligand to the proximal calcium, is also shown.[626]

O bond affording the [Fe^{4+}] species and H$_2$O; the dioxygen bond cleavage requires two electrons, these being abstracted from both Fe and the porphyrin, respectively.

   The aromatic monolignol substrate, required for the two consecutive one-electron reductions of the enzyme, approaches via a substrate access channel at the exposed δ-meso heme edge of the protein (Figure 48). As shown, there is a distinctive hydrophobic area due to the presence of Phe68, Phe142, and Phe179. Indeed, a Phe residue has long been speculated as the site of a pre-electron transfer complex between the substrate and a charged transient intermediate,[629,630] and the X-ray data strongly suggests that Phe179 fulfills that role.

   (c) *H$_2$O$_2$ generation.* In addition to the uncertainty as to how monolignol oxidation is actually achieved, numerous biochemical processes have also been proposed for the formation of the required co-substrate for peroxidase catalysis, i.e., extracellular H$_2$O$_2$. Indeed, how it is formed is still not resolved. Nevertheless, two criteria need to be met for H$_2$O$_2$ biosynthesis. The first is the generation of H$_2$O$_2$ in amounts sufficient for the peroxidase-catalyzed oxidation of monolignols (**1**)–(**3**) and the second is whether its formation is temporally and spatially associated with the peroxidase activity and lignin deposition. In this context, various researchers have described both temporal and spatial correlative relationships linking H$_2$O$_2$ detection to lignifying tissues, in agreement with a role for peroxidase in lignification.[549,631,632]

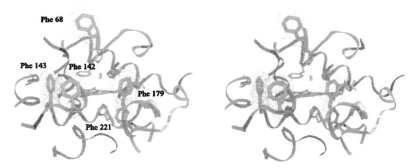

Figure 48 Critical Phe residues guarding the substrate access channel with their accessible surfaces contoured. The entrance to the reactive heme edge is viewed edge on showing the presumed substrate interaction site. Phe68 is situated above the distal pocket, Phe142 to the left and Phe179 to the right of the substrate access channel.[626]

There are a number of possible mechanisms for H_2O_2 biosynthesis in the various, but distinct, compartments of vascular plants. These include processes in the chloroplast via superoxide (O_2^-) quenching by superoxide dismutase (SOD), in the peroxisomes during glycollate oxidation, and in the mitochondria via putative NADH-dependent transformations.[633] There are also at least three known means of H_2O_2 biosynthesis within the plant cell-walls: the first and most extensively described has only been reported thus far in the *Gramineae* and *Leguminosae* families. This involves (poly)amine oxidase(s) which, in the presence of oxygen, convert(s) spermidine/spermine/putrescine into their corresponding aldehydes, or cyclization products thereof, with concomitant formation of H_2O_2 (Scheme 14).[634–637] These amine oxidases are located in vascular elements, e.g., the apoplasts of sclerenchymatous fibers, xylem, and vascular parenchyma in species such as *Z. mays*, *Cicer arietinum*,[634] oat (*Avena sativa*),[635] and lentil (*Lens culinaris*).[636] However, they still have no known physiological function and their limited distribution may not be relevant to a role in lignin biosynthesis.

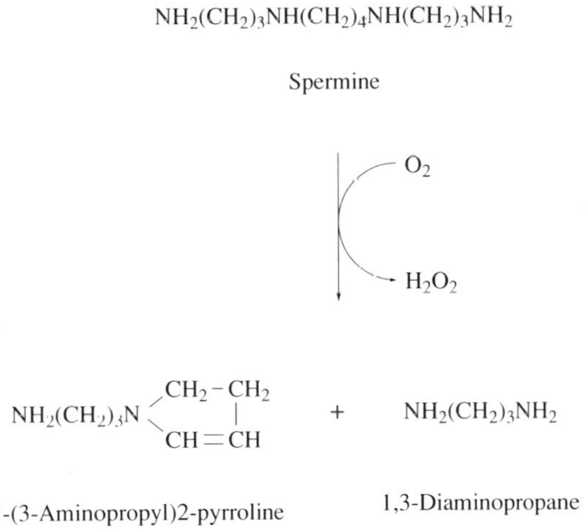

$$NH_2(CH_2)_3NH(CH_2)_4NH(CH_2)_3NH_2$$

Spermine

O_2

H_2O_2

$$NH_2(CH_2)_3N \underset{CH=CH}{\overset{CH_2-CH_2}{<}} \quad + \quad NH_2(CH_2)_3NH_2$$

1-(3-Aminopropyl)2-pyrroline   1,3-Diaminopropane

Scheme 14

A second enzymatic process proposed to engender cell-wall H_2O_2 biosynthesis has two variants, these being based on findings reported with horseradish (*Armoracia lapathifolia*, Gilib.)[638,639] and *F. suspensa*,[640] respectively. Thus, a malate-oxaloacetate shuttle was proposed to be operative between the cytoplasm and plasmalemma (Figure 49), this putatively being required for a cell-wall bound malate dehydrogenase to complete its catalytic cycle. Oxygen involvement was then invoked to complete the conversion of NADH to NAD, with the superoxide anion (O_2^-) so generated undergoing disproportionation to give H_2O_2. Indeed, it was reported that in spinach (*S. oleracea*) hypocotyl seedlings, a superoxide generating NAPDH oxidase was detected in vascular tissues via the use of nitroblue tetrazolium (NBT) as a histochemical staining reagent; the latter could be converted into the colored formozan, but only if SOD activity was inhibited. Additionally, CuZn-SOD was also apparently immunohistochemically detected, this being generally located where lignin was deposited, as evidenced by phloroglucinol staining.[641] Bolwell *et al.* have also provided evidence for a membrane bound NAD(P)H oxidase for biosynthesis of H_2O_2.[642] The H_2O_2 so formed is then putatively supplied as a co-factor for peroxidase-catalyzed conversions giving the polymeric lignins.

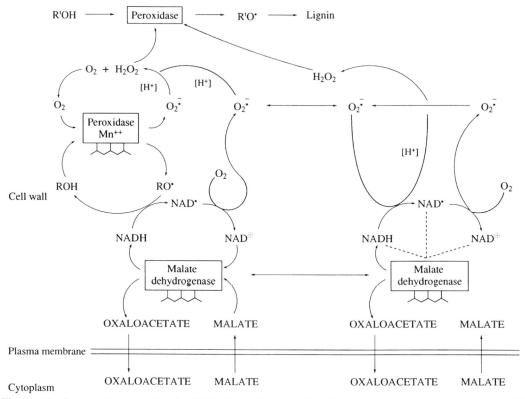

Figure 49   Proposed mechanism for H_2O_2 biosynthesis in the cell-wall: Malate-oxaloacetate shuttle.[639]

Efforts by the original investigators, however, to solubilize the putative cell-wall bound malate dehydrogenase failed.[638–640] In a later histochemical study, Goldberg *et al.*[643] reported that NADH oxidation by *N. tabacum* apoplasts occurred rapidly if malate and oxaloacetate were added, and that NADH oxidation was required for H_2O_2 synthesis to occur.

A variation of this mechanism is based on the observation that NADH oxidation can apparently be increased by addition of Mn^{2+} and phenolics.[644] Gross *et al.* also noted a Mn^{2+} and phenol dependence and proposed that malate dehydrogenase was associated with not one, but two, distinct H_2O_2-generating species (Figure 49).[639] As a further complication, other investigators proposed that H_2O_2 generation is a *direct* consequence of an unusual NADH-requiring peroxidase which, during its catalytic cycle, generates H_2O_2.[634,645] There is precedence for this suggestion, since a H_2O_2-synthesizing, nonheme, nonmetal peroxidase has been found in *Streptococcus faecilis*,[646] although not yet in plants.

Other proposed mechanisms for H_2O_2-synthesis include that of phenol oxidase catalysis,[646] auto-oxidation[646] as well as peroxidase-catalyzed IAA oxidation, where H_2O_2 is generated for monolignol polymerization. With all of these distinct processes being described for H_2O_2 biosynthesis for lignin formation, more definitive data is required in order to demonstrate whether or not a general process is in effect *in planta* for lignin formation.

3.18.7.2.3   *Dirigent proteins*

The preceding subsections have described the various lines of evidence for each of the distinct oxidases, which have been assumed, in some manner, to be associated with lignifying tissues. However, all of these oxidase classes have one major limitation in common: none have the ability to designate the outcome of inter-unit coupling between the respective monolignol, oligolignol, or polylignol-derived radical intermediates during lignin biosynthesis. Indeed, this could not be expected to occur if the molecule has more than one potential coupling site, and if the free-radical intermediate is no longer bound to the enzyme. Accordingly, none of the oxidative enzymes described can thus readily account for the preponderance of either 8–*O*–4' linkages in native lignins, or for the stereoselective coupling modes noted, for example, in lignan biosynthesis leading to the 8–8' linked (+)-pinoresinol (**128**)[12,586] (see Chapter 1.25, Volume 1). Instead, the coupling products that the oxidases engender can only result from "random" coupling.

Yet in spite of the limitations noted for the sundry oxidases, nature extensively deploys free-radical biochemistry, whether to form lignins, lignans, suberins, insect cuticles, melanins, fungal fruiting bodies, and so forth.[586] In Chapter 1.25, Volume 1 it was described how the outcome of phenoxy free-radical coupling could be stipulated, using the example of (+)-pinoresinol (**128**) formation. Indeed, through the involvement of a dirigent protein (Latin: *dirigere*, to guide or to align), stereoselective 8–8' coupling occurred even though the free-radical intermediates themselves were generated by either a non-specific oxidase (laccase) or an inorganic oxidant.[586] This is illustrated in Equation 3. Thus, in the *absence* of the dirigent protein random coupling occurs where the initial coupling modes with *E*-coniferyl alcohol (**2**) are 8–5' ≫ 8–8' ≫ 8–*O*–4', with all resulting products being racemic (Figures 43(b)–43(d)). However, in the presence of the dirigent protein, only stereoselective 8–8' coupling occurred to give the optically pure (+)-pinoresinol (**128**). Moreover, based on existing kinetic data, a mechanism of free-radical capture by the dirigent protein was envisaged to account for the stereoselectivity observed.[586] The gene encoding the glycosylated protein was subsequently cloned and found to encode a protein of ~18 kDa, and the functional recombinant dirigent protein was expressed in a *Spodoptera*/baculovirus system thereby verifying the authenticity of the gene sequence.[647,648] Significantly, its gene had no sequence homology with any other encoding a protein or enzyme of known function, i.e., it is unique. The major ramification of this discovery is that nature has indeed developed mechanisms for precisely stipulating the outcome of monolignol coupling, even if the oxidative capacity itself is non-specific.

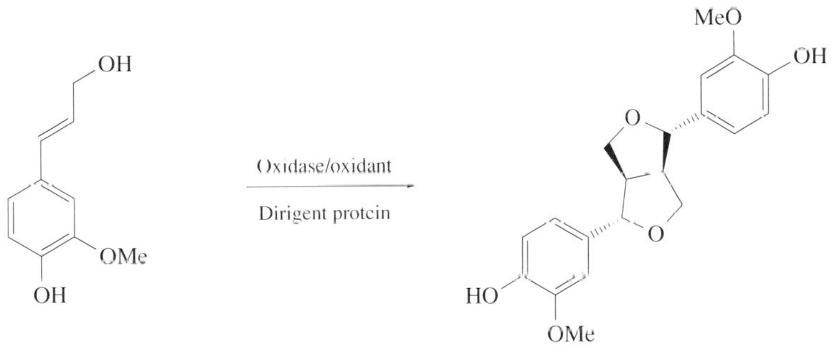

(3)

(**2**) Coniferyl alcohol   (**128**) (+) Pinoresinol

Thus, the existence of this new class of protein active sites is important, not only for lignan biosynthesis but, it is assumed, for lignin formation as well. Accordingly, the remaining subsections build on this discovery in the following ways: description of the primary lignin structures known thus far, including identification of recently established lignin inter-unit substructures; the preliminary evidence for a template polymerization mechanism, and the hypothesis that dirigent protein array(s) stipulate the precise formation of the initial macromolecular chain(s), to be biosynthesized in any lignin domain.

3.18.7.3 Primary Lignin Structures *in vivo*

The determination of the macromolecular structures of lignins *in vivo* has been traditionally confounded by the fact that analytical degradation has always destroyed some part or another of the biopolymer. It has consequently been difficult to ascertain which chemical features are artifacts and what structural *motifs* might not have survived degradation at all. Nevertheless, classical estimates of the frequencies in softwood lignins of different inter-unit linkages, based on acid-catalyzed hydrolysis and oxidation with permanganate, are listed in Table 20.[496] Indeed, the original quantitative estimates for the contributions of structural features discerned in 1-D (inverse-gated proton decoupled) ^{13}C NMR spectra of lignin samples were typically compared to the corresponding values from analytical degradation as a basis for confirming their reliability. These frequencies were considered to be reasonably accurate for more than two decades until the early 1990s, when 2-D

NMR spectroscopic results began to challenge the established views about the frequencies and configurations of certain fairly prominent linkages (*vide infra*).

It is evident from the data compiled in Table 20 that 8–*O*–4′ alkyl aryl ethers would represent about 50% of the linkages between monomer residues in softwood lignin macromolecules, while the others are much less frequent. Broadly speaking, the same kinds of inter-unit linkages were reported to occur in hardwood lignins (Figure 3), which are composed primarily of guaiacyl (**G**) and syringyl (**S**) units, derived from coniferyl (**2**) and sinapyl (**3**) alcohol, respectively (Figure 1), in roughly equal proportions. However here linkages do not occur to the 5-positions in syringyl units, which are already occupied by methoxyl groups; such a restriction does not apply to softwood lignins where syringyl units are usually absent. Consequently, the predominant 8–*O*–4′ alkyl aryl ethers constitute an even higher proportion of all the inter-unit linkages in hardwood (>60%) than in softwood (~50%) lignins.[496]

Certain structural features included in the ostensible hardwood lignin fragment of Figure 3 may be the result of artifacts from analytical degradation, such as the 7–8′ C—C bond for example. In general, the attribution of a particular kind of inter-unit linkage between adjacent monomer residues in lignin macromolecules is considered to be more reliable if the basic carbon skeleton appears to remain intact under a range of analytical degradative conditions.

The arrangements of the (approximately) ten different inter-unit linkages (Table 20) may be expected to have a substantial impact upon the heterogeneity of macromolecular lignin structures. Although roughly half are of the same 8–*O*–4′ alkyl aryl ether type, they nevertheless span a much larger range of structures than the monomer residues themselves—which differ only in aromatic methoxyl substitution patterns. Thus it is the *sequence of inter-unit linkages*, rather than monomer residues, that is more likely to delineate the primary structure of a macromolecular lignin chain.

3.18.7.4 Mechanism of Formation of Predominant Inter-Unit Linkages

The monolignol radicals form C—C and C—O bonds between one another, but the factors governing the distributions of the different bond types in lignin macromolecules have not been fully elucidated. Moreover, no reports now claim that the same proportions of inter-unit linkages as observed in native lignins can be produced through the dehydrogenative polymerization of monolignols *in vitro*. It has not even been possible accurately to account for the ratios of inter-unit linkages among the dehydrodimers formed *in vitro* on the basis of the unpaired electron densities on the atomic centers of the interacting monolignol radicals. In ESR spectra, the hyperfine coupling constants to adjacent protons suggest that the unpaired electron spin densities at the C-5 and C-8 in coniferyl alcohol (**2**) derived radicals are very similar to one another.[649] Furthermore, molecular orbital calculations tend to indicate that the unpaired electron density on the (ESR-silent) O-4 is appreciably lower than on C-5 and C-8 (T. J. Elder, Auburn University, personal communication, 1997). These findings are in general accord with the observation that 8–5′ and 8–8′ linked dilignols (**68**) and (**128**) predominate over 8–*O*–4′ linked dimers (**94**) in product mixtures resulting from the enzyme-catalyzed dehydrodimerization of coniferyl alcohol (**2**) *in vitro*.[524,650,651]

The mechanisms for the formation of the 8–5′, 8–8′, and 8–*O*–4′ linked dehydrodimers are illustrated in Figure 43. It has been suggested[652] that these processes involve the formation of an initial π-complex (omitted from Figure 43) which is reversibly transformed to one of the three (or more) possible σ-complexes (viz. the intermediates shown in Figure 43), where the principal C—C or C—O bonds characteristic of the respective inter-unit linkages have been formed.[653] The reversibility of σ-complex formation is not altogether unreasonable in as far as it had been held to be very likely that the coupling of two *p*-methylphenoxy radicals to give a dimeric cyclohexadienone is a reversible process.[654] However, the heats of formation deduced from molecular orbital calculations have indicated that the stabilities of the σ-complexes directly formed from 8–5′, 8–8′ and 8–*O*–4′ coupling between coniferyl alcohol (**2**) derived radicals may all be very similar,[655] and thus the regioselectivity of these processes would seem to be under kinetic control.

Freudenberg reported in the early 1960s that 7–*O*–4′ acyclic benzyl aryl ether linkages (such as those in oligolignols (**129**)–(**131**)) are readily formed *in vitro* through nucleophilic addition by a phenolic hydroxyl group on a mono- or oligolignol to the C-7 of an intermediate methylene quinone (see Figure 43(e)).[525,526] Even though it has subsequently been possible to confirm[656] Freudenberg's original findings in this regard, acyclic benzyl aryl ether structures have been difficult to detect in lignin-like preparations, spectroscopically.[657] Minute quantities of 7–*O*–4′ linkages have apparently been detected in a ^{13}C-enriched poplar lignin,[658] but no evidence for the presence of acyclic benzyl aryl ether structures has been observed in softwood lignins. Of course it is possible that these labile benzyl ethers were all cleaved even under the relatively mild conditions employed to isolate the lignin samples investigated.

It has long been claimed that there are covalent bonds between lignins and hemicelluloses in woody plant cell-walls. The evidence for this has arisen from studies of lignin–carbohydrate complexes isolated from milled wood by variations of the methods developed by Björkman.[659,660] Enzyme catalyzed hydrolytic cleavage of the polysaccharides in preparations of such lignin–carbohydrate complexes has been used to enrich those monosaccharide units in the material thought to be directly bound to the lignin.[661] The lignin–carbohydrate bonds in lignin–carbohydrate complexes from Norway spruce (*Picea abies*) were thought to have been formed by either of two plausible mechanisms.[661] One involved the (traditional) nucleophilic addition of a monosaccharide hydroxyl group to an intermediate methylene quinone arising, during lignin biosynthesis, from the coupling of a lignol radical through the C-8[662] leading, ultimately, to a nonphenolic benzyl ether structure at the C-7.[661] The other invoked mechanically initiated coupling reactions that could occur during the ball-milling stage[663] of lignin–carbohydrate complex isolation.

It has been pointed out[656] that carbohydrate addition to intermediate methylene quinones is not necessarily so likely to occur in lignifying plant cell-walls because such a process has not been observed in aqueous solutions without nonaqueous cosolvent. Whatever the case, in ryegrass ferulate moieties that are bound through ester groups to α-L-arabinofuranose units of arabinoxylan are also 8 8′ linked to some of the lignin monomer residues.[490] The virtual absence of 8–O–4′ and 8–5′ linkages between the ferulate and lignin moieties has been taken to indicate that the former act as initiation sites for lignification in the primary cell-walls of grasses:[664] coupling at the O–4′ and C–5′ would only occur with lignin residues that are already linked through the C–8′.

3.18.7.4.1   5′–5–O–4 Dibenzodioxocin structures

Biphenyl 5–5′ structures[665] occur more frequently as inter-unit linkages in softwood lignins than had originally been thought (Table 20); they are now estimated to comprise about 25% of the bonds between adjacent monomer residues.[666] The oxidation of *o,o*-dihydroxybiphenyl model compound (**145**) with equimolar quantities of coniferyl alcohol (**2**) by H$_2$O$_2$ in the presence of horseradish peroxidase produces 8-membered ring dibenzodioxocins (**146**) in the reaction mixtures (Scheme 15).[667,668] The correlation peaks at 4.84/84.20 and 4.15/82.5 ppm in the HMQC spectra of these structures coincide exactly with otherwise unassignable features in the corresponding ^{1}H-detected multiple quantum ^{1}H-^{13}C correlation spectra of softwood Björkman[659] lignin preparations.

A conspicuous aspect of the delignification of wood to produce pulp for making paper is that, under alkaline conditions, the process occurs in (three) distinct phases. An initial rapid phase precedes the slower phase that involves the bulk of the lignin.[669] It has been popularly assumed that the rapid initial delignification phase is governed by readily hydrolyzable inter-unit linkages, of which acyclic benzyl aryl ethers attached to phenolic lignin residues would be the most labile among those listed in Table 20. However 2-D NMR spectroscopic studies have failed to reveal more than traces of such linkages in isolated lignin preparations, but the 5–5′–O–4 dibenzodioxocin structure embodies both a benzyl (7″–O–4) and an alkyl (8″–O–4) aryl ether bond within the 8-membered ring. The latter is substantially more reactive than a normal 8–O–4′ linkage under alkaline pulping conditions.[670]

Thus a model compound dibenzodioxocin (**146**) undergoes 50% cleavage to dehydro-dipropylguaiacol (**147**) in aqueous 1 M NaOH at 140 °C after 3 h.[670] In contrast, under the same conditions guaiacylglycerol-β-guaiacyl ether (**148**), the prototypical phenolic 8–O–4′ alkyl aryl ether model dimer, yields about 15% cleavage products; compounds not shown.[671] Unfortunately it had not been reported at the time of writing how the dibenzodioxocin model compound behaves when the aqueous alkaline solution contains bisulfide at concentrations comparable to those used during kraft pulping.

Acyclic 7–O–4′ benzyl aryl ethers are very rapidly cleaved under acidic conditions whether the benzyl moiety is phenolic or not.[672] Accordingly the dibenzodioxocin structure readily reacts with dilute acid to form a new 7-membered ring, releasing a phenolic hydroxyl group in the process. Hereby (**149**) is transformed into (**150**) (Equation 4).[673]

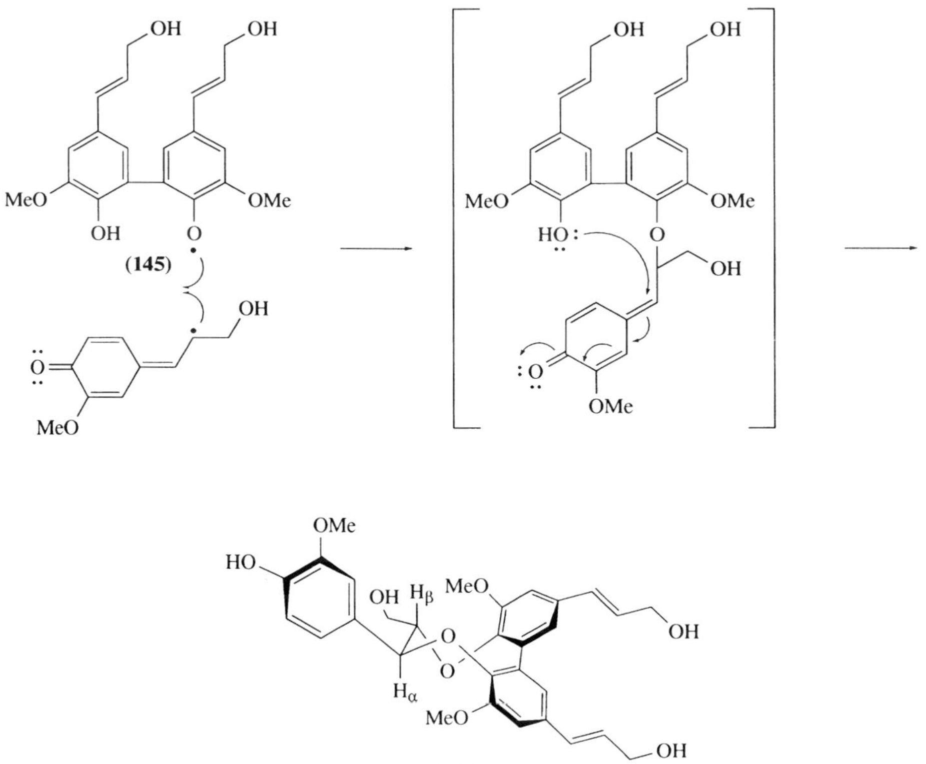

Scheme 15

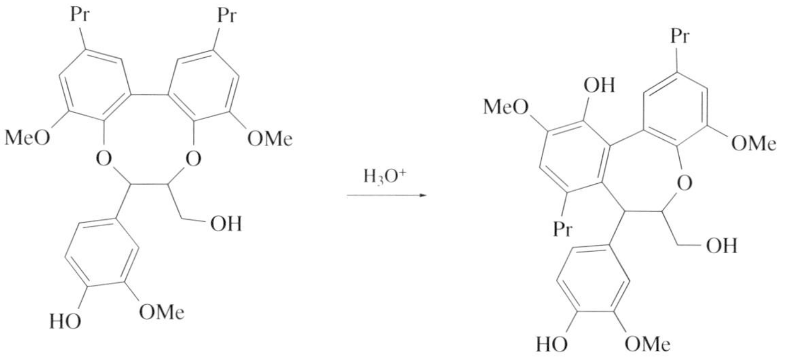

(4)

The foregoing findings, then are provisionally in accord with the possibility that, in lignins, the reactivity under acidic and alkaline conditions of the most labile inter-unit linkages, which had been previously identified as acyclic 7–*O*–4′ benzyl aryl ether bonds, could possibly reside in the behavior of 5′–5–*O*–4 benzodioxocin structures in the biopolymer.

3.18.7.4.2   8–1′ Structures

Evidence in lignins for the existence of 1,2-diarylpropane structures embodying 8–1′ inter-unit linkages appeared in 1965.[674,675] Compounds derived from 1,2-diarylpropane skeleta have contributed prominently to the product mixtures formed during the analytical degradation of many lignin preparations (Table 20). The mechanism of their formation entails the loss of the 3-carbon side chain originally occupying the 1′-position of the unit to which the other radical has coupled at its C-8. However, estimates obtained from NMR spectroscopy for the frequencies of such structures have been very low in both softwood and hardwood lignins, and thus the significance of 8–1′ linkages in the native biopolymers has been uncertain.

In the 1,2-diarylpropane-1,3-diols liberated by acidic hydrolysis from methylated (CH_2N_2) spruce wood shavings, the O–4′ predominantly remained underivatized while the O–4 was almost always methylated.[676] This was taken to mean that the 8–1′ structures are present in lignins as terminal units attached through the O–4′ to the macromolecular chains by linkages that are sensitive to acid. On the other hand, a later interpretation suggested that the biopolymer might instead embody tetrahydrofuran-3-spiro-4′-cyclohexadienone (**152**) structures as progenitors of 8–1′ linkages. However, in a trimeric fraction resulting from the consecutive treatment of milled pine (*Pinus taeda*) wood with acetyl bromide and Zn dust, two aryl isochroman (**153**) derivatives were found that had presumably arisen from migration of the side chain originally at the 1′-position to the 6′-position (see Scheme 16 for an example).[677]

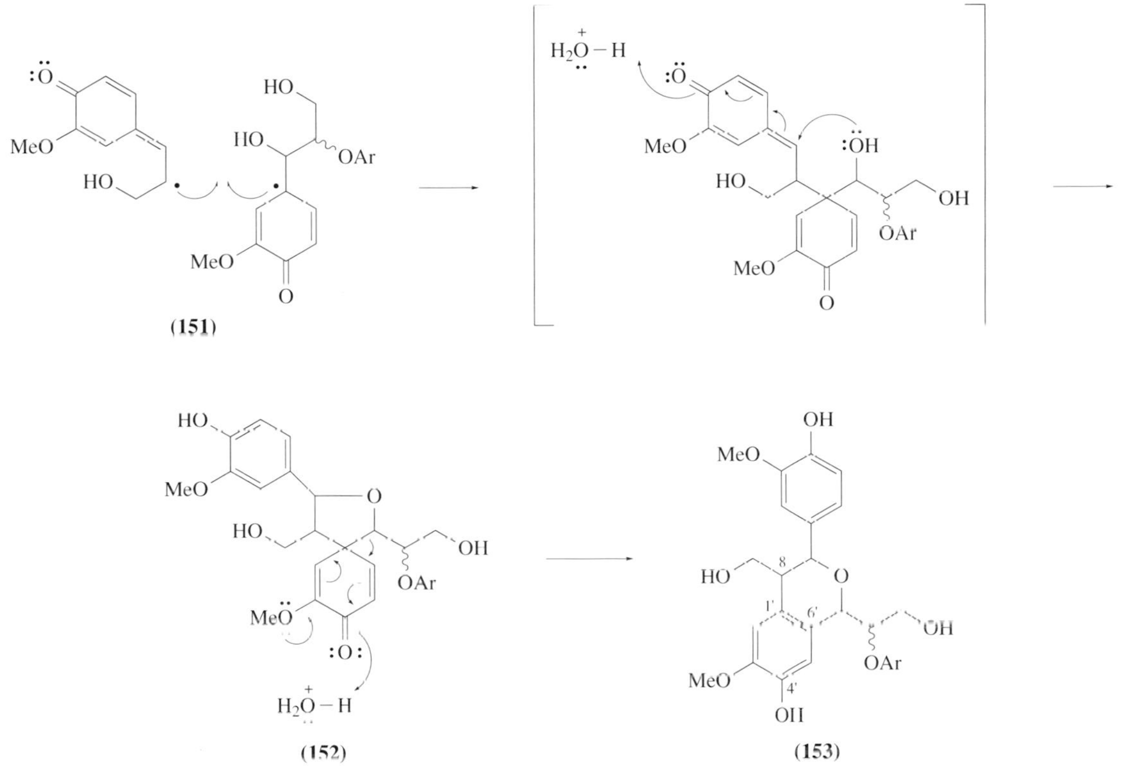

Scheme 16

Although TOCSY correlations appeared to confirm their existence in isolated *P. taeda* Björkman lignins, the aryl isochroman (**153**) structures were present at levels too low to account for the quantities of 8–1′ derived products obtained by analytical degradative methods.[677] Thus the possibility remains that the tetrahydrofuran-3-spiro-4′-cyclohexadienone (**152**) may be the progenitor *in situ* of aryl isocroman derivatives.

3.18.7.5 Structural Scheme for Macromolecular Softwood Lignin Chains

The recent developments in apprehending lignin configurations have been incorporated into a scheme that was compiled[656] to describe representative fragments of the constituent macromolecular chains (Scheme 17). It is worth emphasizing the more important aspects of the comparison between the structural features in this scheme and those previously thought to reflect macromolecular lignin configurations adequately (Table 20). However, this is still undeniably incomplete in that it fails to account for the distributions of products obtained from the degradation of lignin preparations under both acidic and alkaline conditions.

As shown, the predominant inter-unit linkage in both cases is the 8–*O*–4′ alkyl aryl ether, but the acyclic 7–*O*–4′ benzyl aryl ether is missing entirely from the latter depiction. The 8–5′ phenylcoumaran and 4–*O*–5′ diaryl ether structures (the latter serving as branchpoints) are incorporated into both schemes, but the more recent version includes 5–5′ biphenyl bonds only as part of the 5′–5–*O*–4 dibenzodioxocin structures (which also act as prominent branchpoints). The 8–1′ linkage is portrayed in Scheme 17 both as a 1,2-diarylpropane with the corresponding glyceraldehyde-2-aryl ether endgroup, and parenthetically also as the proposed progenitorial tetrahydrofuran-3-spiro-4′-cyclohexadienone structure.

The 8–8′ linkage in Scheme 17 has been incorporated (without any particular justification) as a pinoresinol rather than a divanillyl tetrahydrofuran structure, while the 8–6′ bond in Table 20 has been omitted altogether because of the (unconfirmed) possibility that it may have been confused with the result of C–7′ side chain migration to the 6′-position after 8–1′ coupling. Otherwise the two macromolecular lignin fragments, with a total of 25 monomer residues altogether, are too small to depict the frequencies of minor structural features (such as the 1–*O*–4′ quinone ketal in Table 20) accurately.

3.18.7.6 Secondary Structure of Lignins

It would be surprising if reasonable facsimiles of native macromolecular lignin chains could be produced *de novo* from monolignols in homogeneous solution, regardless of how ingeniously the dehydropolymerizing conditions are manipulated and adjusted. Indeed, the formation of no other biopolymer has been understood to occur in this way, and the preceding sections have identified only some of the shortcomings inherent in such a working model.

The first question to arise is what the minimum set of additional instructions must be for generating native lignin configurations, and how this information is recognized at the molecular level. Since no lignin primary structures have yet been determined explicitly, it is impossible *a priori* to advance provisional working hypotheses about how such a fundamental characteristic of macromolecular lignin chains could be realized and replicated if it were nonrandom.

Actually a great deal more has emerged about the secondary structures of lignins, which are probably created in a well-defined way[678] as a direct consequence of the macromolecular assembly processes themselves.[679] Analyses of Raman spectra acquired using suitably polarized incident laser beams have led to the suggestion that the aromatic rings of the lignin matrices exposed in *Picea marina* tracheid cross-sections tend, in many regions, to be orientated in directions parallel to the cell-wall lumen surface.[678] This may have some bearing on the finding that the lignin domains in *Pinus radiata* cell walls maintain a more or less uniform density (with respect to potassium permanganate staining) as they expand.[679] This result suggests that there are pronounced noncovalent interactions between the polymeric lignin chains: such an effect cannot arise from a crosslink density of only 0.052[680] involving tetrafunctional branch points[681] in the macromolecular structure.

The manner in which the lignin domains expand is strongly affected by the configurations of the carbohydrate matrices undergoing lignification. In the middle lamella region, the peripheries of discrete lignin "particles" expand uniformly but the material enclosed within the boundaries remains at constant density; only when neighboring domains meet are the intervening spaces filled in.[679] The polysaccharides in the primary cell-wall region are mainly randomly oriented cellulose microfibrils that are not arranged into the kinds of lamellae characteristic of the secondary wall.[682] On the other hand, the expansion of lignin domains within the secondary wall occurs much more rapidly in directions that are aligned with, rather than perpendicular to, the axes of the cellulose microfibrils.[679]

It has been suggested that lignification patterns are determined by well-defined distributions of "initiation sites", which became an integral part of the cell wall structure as it was formed; in this respect the cell protoplast exerts control over their placement.[679] Since lignin biosynthesis begins in the middle lamella before spreading into the secondary wall, premature dehydrogenative polymerization of the monolignols, as they diffuse from the protoplast into the cell-wall, must be precluded by some feature of the overall process until they are recognized by complementary initiation sites.[679] Hereby the protoplast could prescribe the topochemical characteristics of the lignin being biosynthesized in different locations within the cell wall:[679] the initiation sites could have been deployed in specific non-uniform distributions that discriminate between the different monolignols secreted during consecutive stages of lignification.[460] If so, the precise nature of these "initiation sites" will be central to an understanding of macromolecular lignin assembly in mechanistic terms.

Whatever the case, lignin polymer chains *in situ* are though to be organized into compact domains where the aromatic rings tend to be parallel to one another. Furthermore, the empirical molecular

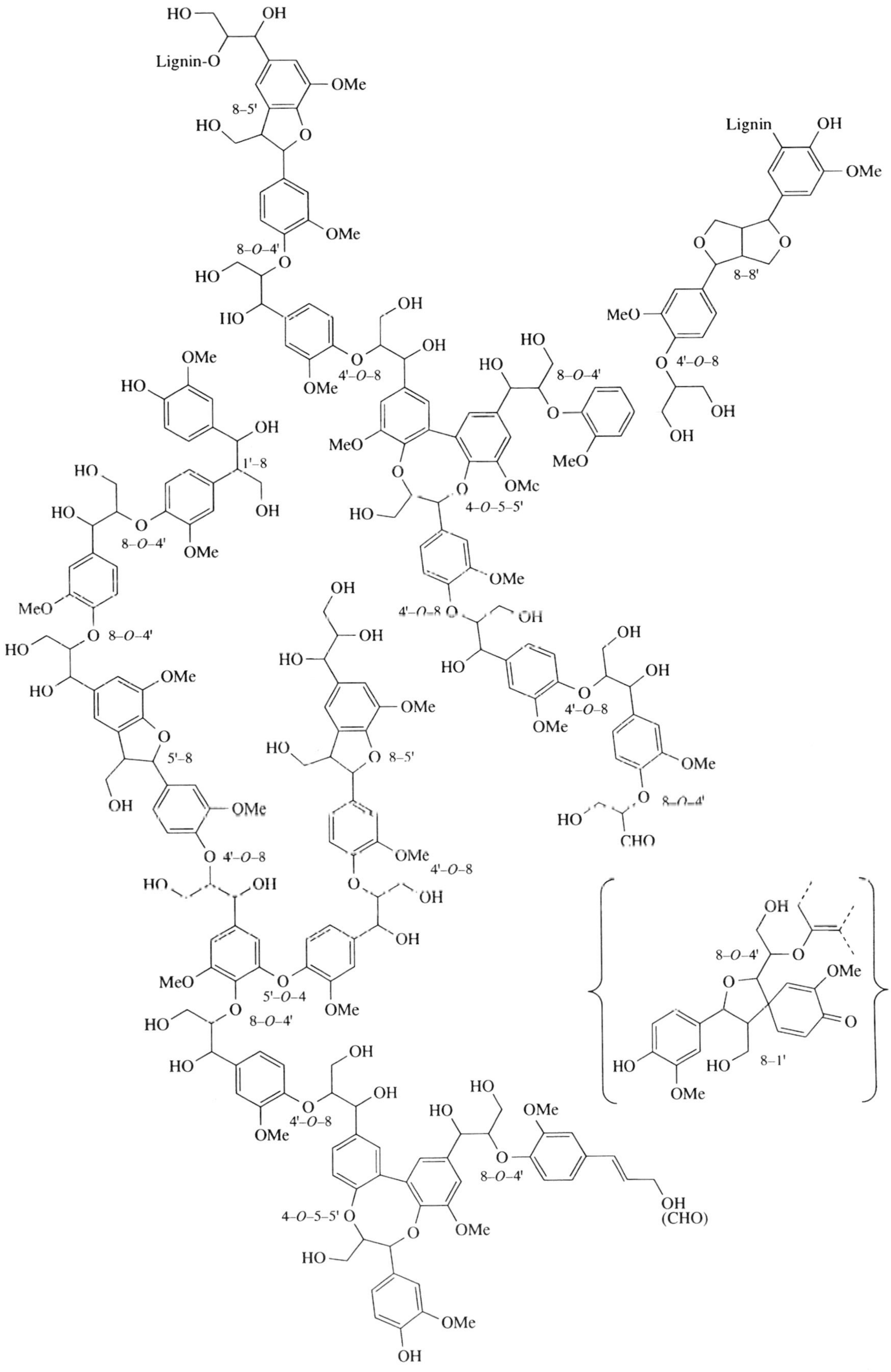

Scheme 17

weight independence of mono-molecular film thickness[683] and the molecular dimensions visible in electron micrographs[684] are together consistent with the working hypothesis that lignin derivatives are comprised of disk-like macromolecular fragments cleaved from lamellar parent structures which are themselves ~2 nm thick. The relative placement of the monomer residues could either be a direct consequence of specific orientational requirements for dehydrogenative coupling between monolignols and the growing polymer chains, or arise from subsequent conformational changes in the lignin macromolecules themselves. Preliminary work has suggested that the secondary structures of native lignins are governed by a direct template polymerization mechanism through which the monolignols are dehydrogenatively linked into macromolecular chains.[685]

3.18.7.7 Template Polymerization in Lignin Biosynthesis

An extended Zutropfverfahren was devised to examine whether high molecular weight lignin components are capable of influencing noncovalently the molecular weight distributions of dehydro-polymerisates formed from a monolignol *in vitro*.[685] Thus coniferyl alcohol (**2**) and H_2O_2 were very gradually introduced into a homogeneous solution containing low levels of horseradish peroxidase activity in the presence and absence of dissolved softwood-derived lignin macromolecules ($M_w = 206\,000$, 1.0×10^{-8} M initial concentration) for periods ranging between 20 h and 80 h. During this time interval the (coupled and free) monolignol units attained molar concentrations in solution which correspondingly varied from values 3100 to 8500 times greater than that of the pre-existing macromolecular lignin components; the lignin macromolecules that were originally present were presumed to possess configurations very similar to native lignin biopolymer chains.[685]

The molecular weight distributions of the resulting dehydropolymerisates affirmed that, at low concentrations in homogeneous solution, the macromolecular lignin components promoted formation of high molecular weight entities during the peroxidase-catalyzed dehydrogenative coupling of coniferyl alcohol (**2**).[685] Indeed, the high molecular weight dehydropolymerisate components themselves seemed to exhibit an autocatalytic effect upon the rate of their formation even though their molar concentrations remained far below those of the smaller species.

It is not considered possible for the formation of the high molecular weight dehydropolymerisate species to have arisen from preferential radical coupling of the mono- (and oligo-) lignols with the growing ends of the lignin macromolecules that were originally present. Indeed, an almost identical result was obtained when the coniferyl alcohol and H_2O_2 were gradually introduced into a homogeneous solution, where the same horseradish peroxidase activity as before was being maintained in the presence of *methylated* macromolecular lignin components ($M_w = 15\,400$, 2.7×10^{-8} M initial concentration). The configurations of these macromolecular species were again considered to be very similar to those of native lignin chains, except that all phenolic hydroxyl groups had been methylated. In the time interval during which the process was monitored, the coupled and free monolignol units together reached molar concentrations in solution which varied between values 1100 and 3100 times greater, respectively, than that of the methylated lignin components.

It is clearly evident from the molecular weight distributions of the dehydropolymerisates (Figure 50) that, at comparably low levels, the methylated lignin macromolecules are about as effective as the unmethylated macromolecular lignin components in eliciting the formation of high molecular weight species during the dehydrogenative coupling of coniferyl alcohol (**2**). In this case, however, there is no obvious concern about covalent participation in the process on the part of the lignin macromolecules since they had been methylated. As before, the high molecular weight dehydropolymerisate components being produced seemed to autocatalyze the rate of their own formation (Figure 50) even though their molar concentrations remained far below those of the smaller species.

The coupling between a coniferyl alcohol radical and another (mono- or oligolignol) radical occupying adjacent sites on a macromolecular lignin component must, of course, compete with the corresponding process in open solution. In this respect results of the kind depicted in Figure 50 were sustained only under limiting conditions where there was little opportunity for coupling to occur away from the surfaces of pre-existing lignin macromolecules or macromolecular assemblies. However, when the radicals become more numerous as the peroxidase concentration was increased, coupling in open solution did become more frequent.[685]

The mechanism through which pre-existing macromolecular lignin components act as templates[686] in promoting the formation of high molecular weight dehydropolymerisate species is likely to be governed by strong nonbonded orbital interactions with the (mono- and oligolignol) radicals: the unmethylated and methylated lignin macromolecules managed to elicit remarkably similar effects

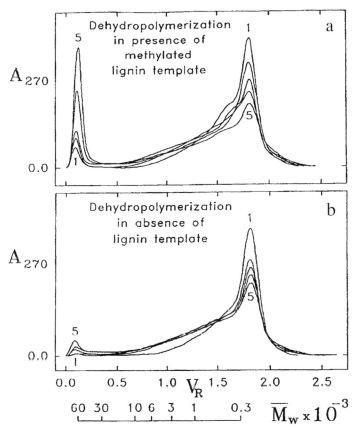

Figure 50 Molecular weight distributions of dehydropolymerisates successively formed under limiting Zutropfverfahren conditions from monolignol in (a) presence and (b) absence of methylated macromolecular lignin template after (1) 20 h, (2) 50 h, (3) 70 h, (4) 75 h, and (5) 80 h. (Sephadex G100/aqueous 0.10 M NaOH.) (Reproduced by permission from *Phytochemistry*, 1997, **45**, 911.)

despite vast differences in their molecular weights (M_w = 206 000 and 15 400, respectively) and solubility characteristics.[685] The low monolignol radical concentration prevailing in lignifying plant cell walls would tend to ensure that template polymerization predominates *in vivo* because there would be essentially no opportunity for coupling to occur away from the surfaces of macromolecular lignin chains. The nonbonded orbital interactions between the aromatic template moieties and the radicals about to undergo coupling also would engender generally parallel orientations of the benzene rings in the resulting biopolymer, much as observed in the lignin of *P. marina* tracheids.[678]

Moreover, the biosynthesis of new polymer chains could entail direct replication of macromolecular lignin primary structures[685] if the relative positions of the aromatic rings of monomer residues are uniquely determined by the intervening linkage. Postcoupling reductive processes, however, such as those leading to divanillyl tetrahydrofuran (Table 20), dihydrodehydrodiconiferyl alcohol (**68**), shonanin, and isoshonanin moeties,[536] cannot be direct consequences of template polymerization, and it is indeed not clear that such structures actually become an integral part of the native biopolymer. Certainly it would be very unlikely for them to be distributed in an incidental manner along macromolecular lignin chains.

3.18.7.8 Origin of Lignin Primary Structures

Before describing how the primary structures of lignins could be propagated, it should be borne in mind that there is considerable ultrastructural heterogeneity of lignins within plant cell walls or, more specifically, there are substantial variations among the sequences of inter-unit linkages along polymeric lignin chains (*vide supra*: Section 3.18.6.6.2). The question arises as to whether there could be any uniformity of macromolecular configuration within a given ultrastructurally distinct lignin domain. It has traditionally been thought that the biosynthetic process itself cannot exert any direct control over lignin primary structure. However this view has been called into question by two possibilities that macromolecular lignin assembly could entail the immediate replication of polymer chain configuration through a template polymerization mechanism,[685] and that monolignol coupling can be stipulated through the involvement of a dirigent protein.

Accordingly, a fixed lignin primary structure would certainly be preserved within each individual locus or "particle" that has developed from a single "initiation site".[679] The question arises as to how far this uniformity extends among all such loci occupying any particular ultrastructurally distinct lignin domain. The answer may lie in the nature of each initiation site that gives rise to a discrete lignin locus. Do these sites simply harbor an oxidative enzyme ready to polymerize monolignols dehydrogenatively, or are they more complicated in character?

The configurations of the initiation sites for lignification in plant cell walls are almost certainly related to the active sites of dirigent proteins through which regio- and stereoselectivity is prescribed for the critical phenoxy-radical coupling steps during lignan biosynthesis.[586,650] These dirigent proteins, while lacking their own catalytically active centers, apparently capture specific monolignol-derived free radicals in such a way as to juxtapose them into the correct relative orientations for regio- and stereoselective coupling.[586] Thus the active sites of dirigent proteins interact in a highly distinctive way with the radicals formed by single-electron oxidation of specific monolignols; it would be very surprising if they were relegated exclusively to the lignan biosynthetic pathway and not involved in macromolecular lignin assembly at all.

However, it should be emphasized that the same dirigent proteins, which govern the regio- and stereospecific phenoxy-radical coupling events leading to lignans, cannot themselves participate in the dehydropolymerization of monolignols to form lignins. As the first step in a sequence of dehydropolymerizing events, the dimerization of monolignols would be incompatible with the ultimate formation of the same frequencies of inter-unit linkages as found in native lignins. Rather, dehydrogenative coupling of the monomers must be able to occur preferentially with the growing ends of macromolecular lignin chains[495] if the outcome is to encompass what is presently known about the full range of dehydropolymerisate structures occurring *in vivo*.

Thus any protein(s) controlling macromolecular lignin configuration must proffer arrays of adjacent lignol–radical binding sites of the same types as those on the dirigent proteins that play a central role in lignan biosynthesis.[586] These dirigent arrays would facilitate the assembly of progenitorial lignin macromolecules with specific primary structures formed through preferential dehydrogenative coupling between the individual monolignols and the growing ends of the polymer chains.[687] Once formed, the progenitorial lignin macromolecules will not be displaced from the contiguous arrays of dirigent sites on the proteins as the next dehydropolymerisate chains begin to be assembled; their primary structures could instead be replicated through a direct template polymerization mechanism.[685]

Since dirigent proteins are capable of prescribing both the regio- and stereospecificity inherent in the dehydrodimerization of monolignols to lignans,[586] it might be expected that a contiguous dirigent array of lignol–radical coupling sites would produce some net optical activity in macromolecular lignin chains. It has long been claimed, however, that there is no sign of this in the native biopolymer. Further comment about the matter should await a more complete characterization of the template polymerization process because there are a number of possibilities. For example, individual macromolecular lignin chains might be largely internally compensated as far as net optical activity is concerned, or their replication could partly engender enantiomeric configurations of inter-unit linkages that alternate from one strand to the next within any particular locus under the control of a single initiation site.

The first question that arises is, of course, whether any dirigent arrays of lignol–radical coupling sites have yet been discovered. In this connection, it is of interest that three kinds of structural proteins in plant cell walls have been characterized according to their amino acid compositions and repeating sequences.[688] They are the extensions of hydroxyproline-rich glycoproteins (HRGPs), proline-rich proteins (PRPs) and glycine-rich proteins (GRPs). Some compelling observations have been recorded through the use of polyclonal antibodies raised against a 33 000 molecular weight PRP from *Glycine max*[689] that is comprised almost completely of a decameric amino acid repeat sequence, ProHypValTyrLysProHypValGluLys.[690] By these means the co-localization of PRP epitopes and lignins has been established both spatially and temporally in developing cell walls of the *Z. mays* coleoptile[550] and in secondary walls of differentiating protoxylem elements in the *G. max* hypocotyl.[691]

It has been proposed that PRPs could act as scaffolds for initiating lignification at their tyrosine residues,[691] but it is well known that polyphenols—including monomers such as propyl gallate—associate noncovalently with PRPs at their proline residues,[692] the first in a ProPro sequence being a much preferred binding site. Therefore, there are likely to be favorable noncovalent interactions between PRPs and macromolecular lignin chains in plant cell-walls. At the time of writing it was not clear how proline residues in polypeptide chains interact with free radicals derived from mono-, oligo-, and polylignols; yet the question is quite relevant to whether the primary structures of macromolecular lignin chains could be encoded in PRPs or other similar cell-wall proteins. Here it may be more than coincidental that the dirigent protein isolated from *Forsythia suspensa*, which is responsible for coupling two coniferyl-alcohol-derived free radicals to form (+)-pinoresinol,[586] possesses a single ProProValGlyArg sequence near the middle of the polypeptide chain.[690] This is reminiscent of the decameric repeat ProHypValTyrLysProHypValGluLys that almost completely comprises the PRP[690] the localization of which has been both spatially and temporally correlated with lignin in the *Z. mays* coleoptile[550] and *G. max* hypocotyl.[691]

Moreover, whether lignin primary structure is encoded in PRPs or other structural proteins, dirigent proteins have been detected in the secondary xylem cell-wall regions of *F. intermedia* stems (Figure 51)[647,687] that witness the initial stages of lignin deposition.[679] They have been discovered through immunochemical localization studies employing polyclonal antibodies raised against the *F. intermedia* [recombinant] dirigent protein that brings about 8–8′ coupling between coniferyl alcohol (2) derived radicals; these polyclonal antibodies primarily label the outer part [S₁ layer] of the *F. intermedia* secondary xylem cell walls.[687] Since these regions have no obvious connection with lignan biosynthesis, such findings implicate the involvement of arrays of dirigent protein sites in the lignin biosynthesis in a most striking manner. Evidently, the polyclonal antibodies here recognize their targets whether the latter are isolated sites on individual proteins, or parts of dirigent arrays. Thus it seems that a beginning has been made in apprehending the fundamental mechanism through which macromolecular lignin configurations are created *in vivo*.

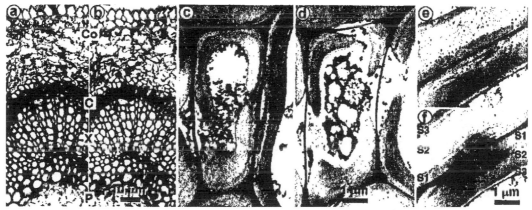

Figure 51 Light microscopy (a) and (b) and transmission microscopy (c) to (f) immuno-gold labeling of *F. intermedia* young stem using polyclonal antiserum against the recombinant dirigent protein. (a), (c), and (e), labeling with the immune-serum. (b), (d), and (f), Control with the pre-immune serum on same zones of tissues on serial sections. Co, cortex; C, cambium; X, xylem; P, pith cavity; S₁, S₂, S₃, sub-layers of the secondary wall from the outer part (near the middle lamella) to the inner part (near the lumen).[647,687]

3.18.8 CONCLUDING REMARKS

With the discovery of dirigent proteins and the detection of dirigent protein epitopes in lignifying tissues, it has become evident that macromolecular lignin chains depend on such protein sites for their configurations to be established. The journey to elucidate the details of the biochemical mechanisms involved now lies ahead.

It has been (repeatedly) emphasized that monolignols are the biosynthetic precursors of both the (structural) lignins and (non-structural) lignans, the latter being abundant components in various plant tissues including certain heartwoods, seeds, etc. Chapter 1.25 (Volume 1) summarizes their quite distinct biochemical pathways which, nonetheless, can also be associated with lignified tissues.

There is an important difference between the ways in which the stereo- and regiospecificities of the interunit linkages are determined in lignins and lignans. While the dirigent proteins alone control the basic configurations of the linkages in lignans, the lignins are envisaged to be formed largely by a template polymerization mechanism through which the first macromolecule biosynthesized in each lignifying domanin is replicated. The primary configuration of their progenitorial lignin chain, on the other hand, is dictated by an array of protein sites.

Lastly, transgenic plants have been steadily becoming more important in terms of the potential for modifying lignin *in situ*. A compelling review of recent developments in the field by Baucher *et al.*[379] appeared only a few months before the present chapter was completed.

ACKNOWLEDGMENTS

The authors thank the United States Department of Energy (DE-FG03-97ER20259), the National Science Foundation (MCB09631980), the National Aeronautics and Space Administration (NAG100164), the United States Department of Agriculture (9603622), McIntire-Stennis, the Arthur M. and Kate Eisig Tode Foundation and the Lewis B. and Dorothy Cullman and G. Thomas Hargrove Center for Land Plant Adaptation Studies for generous support. Acknowledgement for support of this work is also made to the Vincent Johnson Lignin Research Fund at the University of Minnesota, and the Minnesota Agricultural Experiment Station. [Paper No. 984436802 of the Scientific Journal Series of the Minnesota Agricultural Experiment Station, funded through Minnesota Agricultural Experiment Station project No. 43-68, supported by Hatch Funds.] Sections 3.18.6.1 and 3.18.6.2 have in large part been reproduced by Simo Sarkanen with permission from reference 400.

3.18.9   REFERENCES

1. N. G. Lewis and L. B. Davin, in "Isopentenoids and Other Natural Products: Evolution and Function", ACS Symposium Series, vol. 562, ed. W. D. Nes, American Chemical Society, Washington, DC, 1994, p. 202.
2. K. Freudenberg, *Bull. Soc. Chim. Fr.*, 1959, 1748.
3. K. Freudenberg, *J. Polym. Sci.*, 1960, **48**, 371.
4. J. M. Harkin, in "Oxidative Coupling of Phenols", eds. W. I. Taylor and A. R. Battersby, Marcel Dekker, New York, 1967, vol. 1, p. 243.
5. D. E. Bland, A. Logan, M. Menshun, and S. Sternhell, *Phytochemistry*, 1968, **7**, 1373.
6. V. C. Farmer and R. I. Morrison, *Geochim. Cosmochim. Acta*, 1964, **28**, 1537.
7. S. M. Siegel, *Am. J. Bot.*, 1969, **56**, 175.
8. K. V. Sarkanen and H. L. Hergert, in "Lignins—Occurrence, Formation, Structure and Reactions", eds. K. V. Sarkanen and C. H. Ludwig, Wiley Interscience, New York, 1971, p. 43.
9. Ö. Eriksson, D. A. I. Goring, and B. O. Lindgren, *Wood Sci. Technol.*, 1980, **14**, 267.
10. F. W. Whitmore, *Plant Sci. Lett.*, 1978, **13**, 241.
11. H. Nimz, *Angew. Chem., Int. Ed. Engl.*, 1974, **13**, 313.
12. N. G. Lewis and L. B. Davin, in "Lignin and Lignan Biosynthesis", ACS Symposium Series, vol. 697, eds. N. G. Lewis and S. Sarkanen, American Chemical Society, Washington, DC, 1998, p. 334.
13. T. K. Kirk and R. L. Farrell, *Annu. Rev. Microbiol.*, 1987, **41**, 465.
14. G. L. Stebbins and G. J. C. Hill, *Am. Nat.*, 1980, **115**, 342.
15. J. R. Manhart and J. D. Palmer, *Nature*, 1990, **345**, 268.
16. L. E. Graham, "Origin of Land Plants", Wiley, New York, 1993.
17. K. J. Niklas, in "Biochemical Aspects of Evolutionary Biology", ed. M. H. Nitecki, The University of Chicago Press, Chicago, IL, 1982, p. 29.
18. T. N. Taylor, *Taxon*, 1988, **37**, 805.
19. K. J. Niklas and L. M. Pratt, *Science*, 1980, **209**, 396.
20. N. G. Lewis and E. Yamamoto, *Annu. Rev. Plant Physiol. Plant Mol. Biol.*, 1990, **41**, 455.
21. W. Löffelhardt, B. Ludwig, and H. Kindl, *Hoppe-Seyler's Z. Physiol. Chem.*, 1973, **354**, 1006.
22. V. M. Reznikov, M. F. Mikhaseva, and M. A. Zil'bergleit, *Chem. Nat. Compd.*, 1978, **14**, 554.
23. I. V. Dovgan and E. I. Medvedeva, *Chem. Nat. Compd.*, 1983, **19**, 81.
24. I. V. Dovgan and E. I. Medvedeva, *Chem. Nat. Compd.*, 1983, **19**, 84.
25. M. A. Ragan, *Phytochemistry*, 1984, **23**, 2029.
26. D. Gunnison and M. Alexander, *Appl. Microbiol.*, 1975, **29**, 729.
27. C. F. Delwiche, L. E. Graham, and N. Thomson, *Science*, 1989, **245**, 399.
28. T. N. Taylor and E. L. Taylor, "The Biology and Evolution of Fossil Plants", Prentice Hall, Englewood Cliffs, NJ, 1993.
29. R. Tutschek, *Z. Pflanzenphysiol.*, 1975, **76**, 353.
30. M. A. Wilson, J. Sawyer, P. G. Hatcher, and H. E. Lerch III, *Phytochemistry*, 1989, **28**, 1395.
31. K. R. Markham, in "Bryophytes. Their Chemistry and Chemical Taxonomy", *Proc. Phytochem. Soc. Eur.*, eds. H. D. Zinsmeister and R. Mues, Oxford University Press, New York, 1990, vol. 29, p. 143.
32. C.-K. Wat and G. H. N. Towers, in "Recent Advances in Phytochemistry", vol. 12, eds. T. Swain, J. B. Harborne and C. F. Van Suínare, Plenum, New York, 1979, p. 371.
33. J. D. Bu'Lock and H. G. Smith, *Experientia*, 1961, **17**, 553.
34. J. D. Bu'Lock, P. R. Leeming, and H. G. Smith, *J. Chem. Soc.*, 1962, 2085.
35. H. Shimazono, W. J. Schubert, and F. F. Nord, *J. Am. Chem. Soc.*, 1958, **80**, 1992.

36. M. Shimada, A. Ohta, H. Kurosaka, T. Hattori, T. Higuchi, and M. Takahashi, in "Plant Cell Wall Polymers. Biogenesis and Biodegradation", ACS Symposium Series, vol. 399, eds. N. G. Lewis and M. G. Paice, American Chemical Society, Washington, DC, 1989, p. 412.
37. M. Shimada, F. Nakatsubo, T. K. Kirk, and T. Higuchi, *Arch. Microbiol.*, 1981, **129**, 321.
38. K. J. Logan and B. A. Thomas, *New Phytol.*, 1985, **99**, 571.
39. T. Satake, T. Murakami, Y. Saiki, and C.-M. Chen, *Chem. Pharm. Bull.*, 1978, **26**, 1619.
40. L. B. Davin and N. G. Lewis, in "Recent Advances in Phytochemistry", vol. 26, eds. H. A. Stafford and R. K. Ibrahim, Plenum, New York (and references therein), 1992, p. 325.
41. N. G. Lewis and E. Yamamoto, in "Chemistry and Significance of Condensed Tannins", eds. R. W. Hemingway and J. J. Karchesy, Plenum, New York, 1989, p. 23.
42. R. A. Jensen, in "Recent Advances in Phytochemistry", vol. 20, ed. E. E. Conn, Plenum, New York, 1986, p. 57.
43. P. S. van Heerden, G. H. N. Towers, and N. G. Lewis, *J. Biol. Chem.*, 1996, **271**, 12 350.
44. R. A. Razal, S. Ellis, S. Singh, N. G. Lewis, and G. H. N. Towers, *Phytochemistry*, 1996, **41**, 31.
45. S. Singh, N. G. Lewis, and G. H. N. Towers, *Polyphénols Actual.*, 1997, **16**, 16.
46. G. H. N. Towers, S. Singh, P. S. van Heerden, J. Zuiches, and N. G. Lewis, in "Lignin and Lignan Biosynthesis", ACS Symposium Series, vol. 697, eds. N. G. Lewis and S. Sarkanen, American Chemical Society, Washington, DC, 1998, p. 42.
47. S. Singh, N. G. Lewis, and G. H. N. Towers, *J. Plant Physiol.*, 1998, **153**, 316.
48. A. M. Anterola, H. van Rensburg, P. van Heerden, L. B. Davin, and N. G. Lewis, 1998 (submitted).
49. A. C. Neish, *Annu. Rev. Plant Physiol.*, 1960, **11**, 55.
50. S. A. Brown and A. C. Neish, *Can. J. Biochem. Physiol.*, 1955, **33**, 948.
51. S. A. Brown and A. C. Neish, *Can. J. Biochem. Physiol.*, 1956, **34**, 769.
52. D. Wright, S. A. Brown, and A. C. Neish, *Can. J. Biochem. Physiol.*, 1958, **36**, 1037.
53. S. A. Brown, D. Wright, and A. C. Neish, *Can. J. Biochem. Physiol.*, 1959, **37**, 25.
54. D. R. McCalla and A. C. Neish, *Can. J. Biochem. Physiol.*, 1959, **37**, 537.
55. H. Tabor, A. H. Mehler, O. Hoyaishi, and J. White, *J. Biol. Chem.*, 1952, **196**, 121.
56. J. Koukol and E. E. Conn, in "Abstracts of the Pacific Slope Biochemical Conference, Davis, CA, 1960."
57. J. Koukol and E. E. Conn, *J. Biol. Chem.*, 1961, **236**, 2962.
58. J. C. Bock and S. R. Benedict, *J. Biol. Chem.*, 1915, **20**, 47.
59. A. C. Neish, *Phytochemistry*, 1961, **1**, 1.
60. E. A. Havir and K. R. Hanson, *Biochemistry*, 1968, **7**, 1896.
61. M. Zucker, *Plant Physiol.*, 1965, **40**, 779.
62. M. Bernards, M. L. Lopez, J. Zajicek, and N. G. Lewis, *J. Biol. Chem.*, 1995, **270**, 7382.
63. M. A. Bernards and N. G. Lewis, *Phytochemistry*, 1998, **47**, 915.
64. E. A. Havir and K. R. Hanson, *Biochemistry*, 1968, **7**, 1904.
65. P. V. Subba Rao, K. Moore, and G. H. N. Towers, *Can. J. Biochem.*, 1967, **45**, 1863.
66. S. Takaç, B. Akay, and T. H. Özdamar, *Enzyme Microb. Technol.*, 1995, **17**, 445.
67. E. L. Camm and G. H. N. Towers, *Phytochemistry*, 1973, **12**, 961.
68. E. A. Havir, P. D. Reid, and H. V. Marsh, Jr., *Plant Physiol.*, 1971, **48**, 130.
69. J. Rösler, F. Krekel, N. Amrhein, and J. Schmid, *Plant Physiol.*, 1997, **113**, 175.
70. K. Edwards, C. L. Cramer, G. P. Bolwell, R. A. Dixon, W. Schuch, and C. J. Lamb, *Proc. Natl. Acad. Sci. USA*, 1985, **82**, 6731.
71. D. N. Kuhn, J. Chappell, and K. Hahlbrock, in "Structure and Function of Plant Genomes", eds. O. Ciferri and L. Dure III, Plenum Press, London, 1983, p. 329.
72. D. N. Kuhn, J. Chappell, A. Boudet, and K. Hahlbrock, *Proc. Natl. Acad. Sci. USA*, 1984, **81**, 1102.
73. R. Lois, A. Dietrich, K. Hahlbrock, and W. Schulz, *EMBO J.*, 1989, **8**, 1641.
74. K. R. Hanson and E. A. Havir, in "The Biochemistry of Plants: Secondary Plant Products", ed. E. E. Conn, Academic Press, New York, 1981, vol. 7, p. 577.
75. I. L. Givot, T. A. Smith, and R. H. Abeles, *J. Biol. Chem.*, 1969, **244**, 6341.
76. R. B. Wickner, *J. Biol. Chem.*, 1969, **244**, 6550.
77. K. R. Hanson, R. H. Wightman, J. Staunton, and A. R. Battersby, *J. Chem. Soc., Chem. Commun.*, 1971, 185.
78. R. Ife and E. Haslam, *J. Chem. Soc. (C)*, 1971, 2818.
79. T. A. Smith, F. H. Cordelle, and R. H. Abeles, *Arch. Biochem. Biophys.*, 1967, **120**, 724.
80. K. R. Hanson and E. A. Havir, *Arch. Biochem. Biophys.*, 1970, **141**, 1.
81. B. Schuster and J. Retey, *Proc. Natl. Acad. Sci. USA*, 1995, **92**, 8433.
82. N. Amrhein, K.-H. Gödeke, and V. I. Kefeli, *Ber. Dtsch. Bot. Ges.*, 1976, **89**, 247.
83. N. Amrhein and K.-H. Gödeke, *Plant Sci. Lett.*, 1977, **8**, 313.
84. W. Shi, J. Dunbar, M. M. K. Jayasekera, R. E. Viola, and G. K. Farber, *Biochemistry*, 1997, **36**, 9136.
85. M. D. Segall, M. C. Payne, W. Ellis, G. T. Tucker, and N. Boyes, *Chem. Res. Toxicol.*, 1998, **11**, 962.
86. D. W. Russell and E. E. Conn, *Arch. Biochem. Biophys.*, 1967, **122**, 256.
87. P. M. Nair and L. C. Vining, *Phytochemistry*, 1965, **4**, 161.
88. P. M. Nair and L. C. Vining, *Phytochemistry*, 1965, **4**, 401.
89. D. W. Russell, *J. Biol. Chem.*, 1971, **246**, 3870.
90. J. R. M. Potts, R. Weklych, E. E. Conn, and J. Rowell, *J. Biol. Chem.*, 1974, **249**, 5019.
91. M. Klingenberg, *Arch. Biochem. Biophys.*, 1958, **75**, 376.
92. T. Omura and R. Sato, *J. Biol. Chem.*, 1962, **237**, 1375.
93. T. Omura and R. Sato, *J. Biol. Chem.*, 1964, **239**, 2370.
94. T. Omura and R. Sato, *J. Biol. Chem.*, 1964, **239**, 2379.
95. Y. Hashimoto, T. Yamano, and H. S. Mason, *J. Biol. Chem.*, 1962, **237**, 3843.
96. R. W. Estabrook, D. Y. Cooper, and O. Rosenthal, *Biochem. Z.*, 1963, **338**, 741.
97. A. Y. H. Lu, K. W. Junk, and M. J. Coon, *J. Biol. Chem.*, 1969, **244**, 3714.
98. H. S. Mason, W. L. Fowlks, and E. Peterson, *J. Am. Chem. Soc.*, 1955, **77**, 2914.
99. O. Hayaishi, M. Katagiri, and S. Rothberg, *J. Am. Chem. Soc.*, 1955, **77**, 5450.

100. D. W. Russell, E. E. Conn, A. Sutter, and H. Grisebach, *Biochim. Biophys. Acta*, 1968, **170**, 210.
101. G. Guroff, J. W. Daly, D. M. Jerina, J. Renson, B. Witkop, and S. Udenfriend, *Science*, 1967, **157**, 1524.
102. L. Castle, J. R. L. Smith, and G. V. Buxton, *J. Mol. Catal.*, 1980, **7**, 235.
103. L. T. Burka, T. M. Plucinski, and T. L. MacDonald, *Proc. Natl. Acad. Sci. USA*, 1983, **80**, 6680.
104. K. Korzekwa, W. Trager, M. Gouterman, D. Spangler, and G. H. Loew, *J. Am. Chem. Soc.*, 1985, **107**, 4273.
105. K. Higashi, K. Ikeuchi, M. Obara, Y. Karasaki, H. Hirano, S. Gotoh, and Y. Koga, *Agric. Biol. Chem.*, 1985, **49**, 2399.
106. D. P. O'Keefe and K. J. Leto, *Plant Physiol.*, 1989, **89**, 1141.
107. K. R. Bozak, H. Yu, R. Sirevåg, and R. E. Christoffersen, *Proc. Natl. Acad. Sci. USA*, 1990, **87**, 3904.
108. D. W. Nebert, D. R. Nelson, M. J. Coon, R. W. Estabrook, R. Feyereisen, Y. Fujii-Kuriyama, F. J. Gonzalez, F. P. Guengerich, I. C. Gunsalus, E. F. Johnson, J. C. Loper, R. Sato, M. R. Waterman, and D. J. Waxman, *DNA Cell Biol.*, 1991, **10**, 1.
109. I. Benveniste and F. Durst, *C. R. Acad. Sci. Paris Série D*, 1974, **278**, 1487.
110. I. Benveniste, J.-P. Salaun, and F. Durst, *Phytochemistry*, 1977, **16**, 69.
111. D. Reichhart, J.-P. Salaün, I. Benveniste, and F. Durst, *Plant Physiol.*, 1980, **66**, 600.
112. B. Gabriac, I. Benveniste, and F. Durst, *C. R. Acad. Sci. Paris, Ser. 3*, 1985, **301**, 753.
113. B. Gabriac, D. Werck-Reichhart, H. Teutsch, and F. Durst, *Arch. Biochem. Biophys.*, 1991, **288**, 302.
114. M. Mizutani, D. Ohta, and R. Sato, *Plant Cell Physiol.*, 1993, **34**, 481.
115. M. Mizutani, E. Ward, J. DiMaio, D. Ohta, J. Ryals, and R. Sato, *Biochem. Biophys. Res. Commun.*, 1993, **190**, 875.
116. H. G. Teutsch, M. P. Hasenfratz, A. Lesot, C. Stoltz, J.-M. Garnier, J.-M. Jeltsch, F. Durst, and D. Werck-Reichhart, *Proc. Natl. Acad. Sci. USA*, 1993, **90**, 4102.
117. M. A. Pierrel, Y. Batard, M. Kazmaier, C. Mignotte-Vieux, F. Durst, and D. Werck-Reichhart, *Eur. J. Biochem.*, 1994, **224**, 835.
118. P. Urban, D. Werck-Reichhart, H. G. Teutsch, F. Durst, S. Regnier, M. Kazmaier, and D. Pompon, *Eur. J. Biochem.*, 1994, **222**, 843.
119. T. Fahrendorf and R. A. Dixon, *Arch. Biochem. Biophys.*, 1993, **305**, 509.
120. T. L. Poulos, *Curr. Opin. Struct. Biol.*, 1995, **5**, 767.
121. S. Graham-Lorence and J. A. Peterson, *FASEB J.*, 1996, **10**, 206.
122. T. L. Poulos, B. C. Finzel, and A. J. Howard, *J. Mol. Biol.*, 1987, **195**, 687.
123. R. Raag, S. A. Martinis, S. G. Sligar, and T. L. Poulos, *Biochemistry*, 1991, **30**, 11 420.
124. K. G. Ravichandran, S. S. Boddupalli, C. A. Hasermann, J. A. Peterson, and J. Deisenhofer, *Science*, 1993, **261**, 731.
125. C. A. Hasemann, K. G. Ravichandran, J. A. Peterson, and J. Deisenhofer, *J. Mol. Biol.*, 1994, **236**, 1169.
126. J. R. Cupp-Vickery and T. L. Poulos, *Nat. Struct. Biol.*, 1995, **2**, 144.
127. S.-Y. Park, H. Shimizu, S.-i. Adachi, A. Nakagawa, I. Tanaka, K. Nakahara, H. Shoun, E. Obayashi, H. Nakamura, T. Iizuka, and Y. Shiro, *Nat. Struct. Biol.*, 1997, **4**, 827.
128. J. T. Groves, and Y.-Z. Han, in "Cytochrome P450. Structure, Mechanism and Biochemistry", ed. P. R. Ortizde Montellano, Plenum, New York, 1995, p. 3.
129. E. Haas, B. L. Horecker, and T. R. Hogness, *J. Biol. Chem.*, 1940, **136**, 747.
130. B. L. Horecker, *J. Biol. Chem.*, 1950, **183**, 593.
131. C. H. Williams, Jr. and H. Kamin, *J. Biol. Chem.*, 1962, **237**, 587.
132. A. H. Phillips and R. G. Langdon, *J. Biol. Chem.*, 1962, **237**, 2652.
133. T. Iyanagi and H. S. Mason, *Biochemistry*, 1973, **12**, 2297.
134. J. D. Dignam and H. W. Strobel, *Biochem. Biophys. Res. Commun.*, 1975, **63**, 845.
135. J. L. Vermilion and M. J. Coon, *J. Biol. Chem.*, 1978, **253**, 8812.
136. S. D. Black, J. S. French, C. H. Williams, Jr., and M. J. Coon, *Biochem. Biophys. Res. Commun.*, 1979, **91**, 1528.
137. J. R. Gum and H. W. Strobel, *J. Biol. Chem.*, 1979, **254**, 4177.
138. T. D. Porter and C. B. Kasper, *Proc. Natl. Acad. Sci. USA*, 1985, **82**, 973.
139. Y. Nisimoto, *J. Biol. Chem.*, 1986, **261**, 14 232.
140. S. G. Nadler and H. W. Strobel, *Arch. Biochem. Biophys.*, 1991, **290**, 277.
141. S. Djordjevic, D. L. Roberts, M. Wang, T. Shea, G. W. Camitta, B. S. S. Masters, and J. J. P. Kim, *Proc. Natl. Acad. Sci. USA*, 1995, **92**, 3214.
142. M. Wang, D. L. Roberts, R. Paschke, T. M. Shea, B. S. S. Masters, and J. J. P. Kim, *Proc. Natl. Acad. Sci. USA*, 1997, **94**, 8411.
143. K. M. Madyastha and C. J. Coscia, *J. Biol. Chem.*, 1979, **254**, 2419.
144. I. Benveniste, B. Gabriac, and F. Durst, *Biochem. J.*, 1986, **235**, 365.
145. M. S. Shet, K. Sathasivan, M. A. Arlotto, M. C. Mehdy, and R. W. Estabrook, *Proc. Natl. Acad. Sci. USA*, 1993, **90**, 2890.
146. A. H. Meijer, M. I. Lopes Cardoso, J. T. Voskuilen, A. de Waal, R. Verpoorte, and J. H. C. Hoge, *Plant J.*, 1993, **4**, 47.
147. P. Urban, C. Mignotte, M. Kazmaier, F. Delorme, and D. Pompom, *J. Biol. Chem.*, 1997, **272**, 19 176.
148. E. Koopmann and K. Hahlbrock, *Proc. Natl. Acad. Sci. USA*, 1997, **94**, 14 954.
149. S. S. Patil and M. Zucker, *J. Biol. Chem.*, 1965, **240**, 3938.
150. M. Sato, *Phytochemistry*, 1966, **5**, 385.
151. P. F. T. Vaughan and V. S. Butt, *Biochem. J.*, 1969, **113**, 109.
152. K. C. Vaughn and S. O. Duke, *Physiol. Plant.*, 1981, **53**, 421.
153. S. O. Duke and K. C. Vaughn, *Physiol. Plant.*, 1982, **54**, 381.
154. M. Kojima and W. Takeuchi, *J. Biochem. (Tokyo)*, 1989, **105**, 265.
155. J. Kamsteeg, J. van Brederode, P. M. Verschuren, and G. van Nigtevecht, *Z. Pflanzenphysiol.*, 1981, **102**, 435.
156. W. Heller and T. Kühnl, *Arch. Biochem. Biophys.*, 1985, **241**, 453.
157. R. E. Kneusel, U. Matern, and K. Nicolay, *Arch. Biochem. Biophys.*, 1989, **269**, 455.
158. M. Tanaka and M. Kojima, *Arch. Biochem. Biophys.*, 1991, **284**, 151.
159. Z.-X. Wang, S.-M. Li, R. Löscher, and L. Heide, *Arch. Biochem. Biophys.*, 1997, **347**, 249.
160. C. Grand, *FEBS Lett.*, 1984, **169**, 7.

161. H. Ohashi, E. Yamamoto, N. G. Lewis, and G. H. N. Towers, *Phytochemistry*, 1987, **26**, 1915.
162. C. S. Chapple, T. E. Vogt, B. E. Ellis, and C. R. Somerville, *Plant Cell*, 1992, **4**, 1413.
163. K. Meyer, J. C. Cusumano, C. Somerville, and C. C. S. Chapple, *Proc. Natl. Acad. Sci. USA*, 1996, **93**, 6869.
164. V. du Vigneaud, M. Cohn, J. P. Chandler, J. R. Schenck, and S. Simmonds, *J. Biol. Chem.*, 1941, **140**, 625.
165. E. B. Keller, R. A. Boissonnas, and V. du Vigneaud, *J. Biol. Chem.*, 1950, **183**, 627.
166. G. L. Cantoni, *J. Biol. Chem.*, 1953, **204**, 403.
167. R. U. Byerrum and J. H. Flokstra, *Fed. Proc.*, 1952, **11**, 131.
168. R. U. Byerrum, J. H. Flokstra, L. J. Dewey, and C. D. Ball, *J. Biol. Chem.*, 1954, **210**, 633.
169. S. A. Brown and R. U. Byerrum, *J. Am. Chem. Soc.*, 1952, **74**, 1523.
170. R. U. Byerrum and R. E. Wing, *J. Biol. Chem.*, 1953, **205**, 637.
171. M. Dubeck and S. Kirkwood, *J. Biol. Chem.*, 1952, **199**, 307.
172. M. Sribney and S. Kirkwood, *Nature*, 1953, **171**, 931.
173. J. Pellerin and A. D'Iorio, *Can. J. Biochem. Physiol.*, 1958, **36**, 491.
174. J. Axelrod and R. Tomchick, *J. Biol. Chem.*, 1958, **233**, 702.
175. B. J. Finkle and R. F. Nelson, *Biochim. Biophys. Acta*, 1963, **78**, 747.
176. B. J. Finkle and M. S. Masri, *Biochim. Biophys. Acta*, 1964, **85**, 167.
177. D. Hess, *Z. Pflanzenphysiol. Bd.*, 1965, **53**, 460.
178. C. Hermann, M. Legrand, P. Geoffroy, and B. Fritig, *Arch. Biochem. Biophys.*, 1987, **253**, 367.
179. T. Higuchi, M. Shimada, and H. Ohashi, *Agric. Biol. Chem.*, 1967, **31**, 1459.
180. M. Shimada, H. Ohashi, and T. Higuchi, *Phytochemistry*, 1970, **9**, 2463.
181. M. Shimada, H. Fushiki, and T. Higuchi, *Phytochemistry*, 1972, **11**, 2657.
182. H. Kuroda, M. Shimada, and T. Higuchi, *Phytochemistry*, 1975, **14**, 1759.
183. J. Poulton, H. Grisebach, J. Ebel, B. Schaller-Hekeler, and K. Hahlbrock, *Arch. Biochem. Biophys.*, 1976, **173**, 301.
184. J. Poulton, K. Hahlbrock, and H. Grisebach, *Arch. Biochem. Biophys.*, 1976, **176**, 449.
185. H. Kuroda, M. Shimada, and T. Higuchi, *Phytochemistry*, 1981, **20**, 2635.
186. H. Kutsuki, M. Shimada, and T. Higuchi, *Mokuzai Gakkaishi*, 1981, **27**, 39.
187. K.-P. Schwabe, *Biochem. Doplomarbeit, Physiol. Chem. Inst.*, 1969, 19.
188. M. Assicot and C. Bohuon, *Eur. J. Biochem.*, 1970, **12**, 490.
189. L. Flohe and K.-P. Schwabe, *Biochim. Biophys. Acta*, 1970, **220**, 469.
190. A. E. Pakusch, R. Kneusel, and U. Matern, *Arch. Biochem. Biophys.*, 1989, **271**, 488.
191. A. E. Pakusch, U. Matern, and E. Schiltz, *Plant Physiol.*, 1991, **95**, 137.
192. G. Gowri, R. C. Bugos, W. H. Campbell, C. A. Maxwell, and R. A. Dixon, *Plant Physiol.*, 1991, **97**, 7.
193. D. Schmitt, A.-E. Pakusch, and U. Matern, *J. Biol. Chem.*, 1991, **266**, 17416.
194. B. Grimmig and U. Matern, *Plant Mol. Biol.*, 1997, **33**, 323.
195. R. C. Bugos, V. L. C. Chiang, and W. H. Campbell, *Plant Mol. Biol.*, 1991, **17**, 1203.
196. B. Dumas, J. van Doorsselaere, J. Gielen, M. Legrand, B. Fritig, M. van Montagu, and D. Inze, *Plant Physiol.*, 1992, **98**, 796.
197. J. van Doorsselaere, B. Dumas, M. Baucher, B. Fritig, M. Legrand, M. van Montagu, and D. Inze, *Gene*, 1993, **133**, 213.
198. Z.-H. Ye, R. E. Kneusel, U. Matern, and J. E. Varner, *Plant Cell*, 1994, **6**, 1427.
199. Z.-H. Ye and J. E. Varner, *Plant Physiol.*, 1995, **108**, 459.
200. F. Martz, S. Maury, G. Pinçon, and M. Legrand, *Plant Mol. Biol.*, 1998, **36**, 427.
201. L. Li, J. L. Popko, X.-H. Zhang, K. Osakabe, C.-J. Tsai, C. P. Joshi, and V. L. Chiang, *Proc. Natl. Acad. Sci. USA*, 1997, **94**, 5461.
202. S. K. Sharma and S. A. Brown, *Can. J. Biochem.*, 1979, **57**, 986.
203. E. De Carolis and R. K. Ibrahim, *Biochem. Cell Biol.*, 1989, **67**, 763.
204. H. Meng and W. H. Campbell, *Arch. Biochem. Biophys.*, 1996, **330**, 329.
205. A. E. Pakusch and U. Matern, *Plant Physiol.*, 1991, **96**, 327.
206. J. Vidgren, L. A. Svensson, and A. Liljas, *Nature*, 1994, **368**, 354.
207. X. Cheng, S. Kumar, J. Posfai, J. Pflugrath, and R. Roberts, *Cell*, 1993, **74**, 299.
208. M. F. Hegazi, R. T. Borchardt, and R. L. Schowen, *J. Am. Chem. Soc.*, 1979, **101**, 4359.
209. R. W. Woodard, M. O. Tsai, H. G. Floss, P. A. Crooks, and J. K. Coward, *J. Biol. Chem.*, 1980, **255**, 9124.
210. C. P. Joshi and V. L. Chiang, *Plant Mol. Biol.*, 1998, **37**, 663.
211. D. Schachter and J. V. Taggart, *J. Biol. Chem.*, 1953, **203**, 925.
212. F. Lipmann, *Science*, 1954, **120**, 855.
213. P. Berg, *J. Biol. Chem.*, 1956, **222**, 991.
214. D. Dunaway-Mariano and P. C. Babbitt, *Biodegradation*, 1994, **5**, 259.
215. T. Higuchi and S. A. Brown, *Can. J. Biochem. Physiol.*, 1963, **41**, 613.
216. H. Grisebach, W. Barz, K. Hahlbrock, S. Kellner, and L. Patschke, in "Biosynthesis of Aromatic Compounds", Proc. Meet. Fed. Eur. Biochem. Soc., ed. G. Billek, Pergamon, Oxford, 1966, 2nd edn., vol. 3, p. 25.
217. S. A. Brown, *BioScience*, 1969, **19**, 115.
218. R. L. Mansell, J. Stöckigt, and M. H. Zenk, *Z. Pflanzenphysiol.*, 1972, **68**, 286.
219. G. G. Gross and M. H. Zenk, *Z. Naturforsch.*, 1966, **21b**, 683.
220. E. Walton and V. S. Butt, *J. Exp. Bot.*, 1970, **21**, 887.
221. K. Hahlbrock and H. Grisebach, *FEBS Lett.*, 1970, **11**, 62.
222. G. G. Gross, J. Stöckigt, R. L. Mansell, and M. H. Zenk, *FEBS Lett.*, 1973, **31**, 283.
223. M. J. C. Rhodes and L. S. C. Wooltorton, *Phytochemistry*, 1973, **12**, 2381.
224. H. Ragg, D. N. Kuhn, and K. Hahlbrock, *J. Biol. Chem.*, 1981, **256**, 10061.
225. T. Lüderitz, G. Schatz, and H. Grisebach, *Eur. J. Biochem.*, 1982, **123**, 583.
226. K. H. Knobloch and K. Hahlbrock, *Plant Med. Sup.*, 1975, 102.
227. R. Ranjeva, R. Faggion, and A. M. Boudet, *Physiol. Veg.*, 1975, **13**, 725.
228. C. P. Vance, A. M. D. Nambudiri, C.-K. Wat, and G. H. N. Towers, *Phytochemistry*, 1975, **14**, 967.
229. G. G. Gross, R. L. Mansell, and M. H. Zenk, *Biochem. Physiol. Pflanzen.*, 1975, **168**, 41.

230. K. H. Knobloch and K. Hahlbrock, *Arch. Biochem. Biophys.*, 1977, **184**, 237.
231. K. Fritzemeier, C. Cretin, E. Kombrink, F. Rohwer, J. Taylor, D. Scheel, and K. Hahlbrock, *Plant Physiol.*, 1987, **85**, 34.
232. C. Douglas, H. Hoffmann, W. Schulz, and K. Hahlbrock, *EMBO J.*, 1987, **6**, 1189.
233. E. Lozoya, H. Hoffmann, C. Douglas, W. Schultz, D. Scheel, and K. Hahlbrock, *Eur. J. Biochem.*, 1988, **176**, 661.
234. D. Buck, M. E. Spencer, and J. R. Guest, *Biochemistry*, 1985, **24**, 6245.
235. W. T. Wolodko, M. N. G. James, and W. A. Bridger, *J. Biol. Chem.*, 1984, **259**, 5316.
236. W. T. Wolodko, M. E. Fraser, M. N. G. James, and W. A. Bridger, *J. Biol. Chem.*, 1994, **269**, 10883.
237. A. Bairoch, PROSITE: A Dictionary of Sites and Patterns in Proteins, Release 8.0, 1991, Medical Biochemistry Department, University of Geneva, Switzerland.
238. M. Becker-André, P. Schulze-Lefert, and K. Hahlbrock, *J. Biol. Chem.*, 1991, **266**, 8551.
239. K.-H. Chang and D. Dunaway-Mariano, *Biochemistry*, 1996, **35**, 13478.
240. K. H. Knoblock and K. Hahlbrock, *Eur. J. Biochem.*, 1975, **52**, 311.
241. P. J. Wallis and M. J. C. Rhodes, *Phytochemistry*, 1977, **16**, 1891.
242. R. Ranjeva, A. M. Boudet, and R. Faggion, *Biochimie*, 1976, **58**, 1255.
243. C. Grand, A. Boudet, and A. M. Boudet, *Planta*, 1983, **158**, 225.
244. G. A. Reynolds, S. K. Basu, T. F. Osborne, D. J. Chin, G. Gil, M. S. Brown, J. L. Goldstein, and K. L. Luskey, *Cell*, 1984, **38**, 275.
245. G. G. Gross and W. Kreiten, *FEBS Lett.*, 1975, **54**, 259.
246. H. F. Fisher, E. E. Conn, B. Vennesland, and F. H. Westheimer, *J. Biol. Chem.*, 1953, **202**, 687.
247. F. A. Loewus, F. H. Westheimer, and B. Vennesland, *J. Am. Chem. Soc.*, 1953, **75**, 5018.
248. H. Wengenmayer, J. Ebel, and H. Grisebach, *Eur. J. Biochem.*, 1976, **65**, 529.
249. T. Lüderitz and H. Grisebach, *Eur. J. Biochem.*, 1981, **119**, 115.
250. F. Sarni, C. Grand, and A. M. Boudet, *Eur. J. Biochem.*, 1984, **139**, 259.
251. D. Goffner, M. M. Campbell, C. Campargue, M. Clastre, G. Borderies, A. Boudet, and A. M. Boudet, *Plant Physiol.*, 1994, **106**, 625.
252. E. Lacombe, S. Hawkins, J. Van Doorsselaere, J. Piquemal, D. Goffner, O. Poeydomenge, A. M. Boudet, and J. Grima-Pettenati, *Plant J.*, 1997, **11**, 429.
253. C. M. Lawrence, V. W. Rodwell, and C. V. Stauffacher, *Science*, 1995, **268**, 1758.
254. T. C. Jordan-Starck and V. W. Rodwell, *J. Biol. Chem.*, 1989, **264**, 17919.
255. Y. Wang, B. G. Darnay, and V. W. Rodwell, *J. Biol. Chem.*, 1990, **265**, 21634.
256. B. G. Darney, Y. Wang, and V. W. Rodwell, *J. Biol. Chem.*, 1992, **267**, 15064.
257. B. G. Darney and V. W. Rodwell, *J. Biol. Chem.*, 1993, **268**, 8429.
258. P. R. Clarke and D. G. Hardie, *EMBO J.*, 1990, **9**, 2439.
259. P. R. Clarke and D. G. Hardie, *FEBS Lett.*, 1990, **269**, 213.
260. G. Gillespie and D. G. Hardie, *FEBS Lett.*, 1992, **306**, 59.
261. J. Stöckigt, R. L. Mansell, G. G. Gross, and M. H. Zenk, *Z. Pflanzenphysiol.*, 1973, **70 S**, 305.
262. R. L. Mansell, G. G. Gross, J. Stöckigt, H. Franke, and M. H. Zenk, *Phytochemistry*, 1974, **13**, 2427.
263. M. Klischies, J. Stockigt, and M. H. Zenk, *Phytochemistry*, 1978, **17**, 1523.
264. A. R. Battersby, J. Staunton, and H. R. Wiltshire, *J. Chem. Soc., Perkin Trans. 1*, 1975, 1156.
265. R. L. Mansell, G. R. Babbel, and M. H. Zenk, *Phytochemistry*, 1976, **15**, 1849.
266. D. Wyrambik and H. Grisebach, *Eur. J. Biochem.*, 1975, **59**, 9.
267. D. Wyrambik and H. Grisebach, *Eur. J. Biochem.*, 1979, **97**, 503.
268. H. Kutsuki, M. Shimada, and T. Higuchi, *Phytochemistry*, 1982, **21**, 19.
269. C. Grand, F. Sarni, and A. M. Boudet, *Planta*, 1985, **163**, 232.
270. M. H. Walter, J. Grima-Pettenati, C. Grand, A. M. Boudet, and C. J. Lamb, *Proc. Natl. Acad. Sci. USA*, 1988, **85**, 5546.
271. A. Kush, E. Goyvaerts, M. L. Chye, and N. H. Chua, *Proc. Natl. Acad. Sci. USA*, 1990, **87**, 1787.
272. M. H. Walter, J. Grima-Pettenati, C. Grand, A. M. Boudet, and C. J. Lamb, *Plant Mol. Biol.*, 1990, **15**, 525.
273. M. H. Walter, J. Grima-Pettenati, and C. Feuillet, *Eur. J. Biochem.*, 1994, **224**, 999.
274. C. Halpin, M. E. Knight, J. Grima-Pettenati, D. Goffner, A. Boudet, and W. Schuch, *Plant Physiol.*, 1992, **98**, 12.
275. C. Pillonel, P. Hunziker, and A. Binder, *J. Exp. Bot.*, 1992, **43**, 299.
276. D. M. O'Malley, S. Porter, and R. R. Sederoff, *Plant Physiol.*, 1992, **98**, 1364.
277. T. Hibino, D. Shibata, T. Umezawa, and T. Higuchi, *Phytochemistry*, 1993, **32**, 565.
278. D. Goffner, I. Joffroy, J. Grima-Pettenati, C. Halpin, M. E. Knight, W. Schuch, and A. M. Boudet, *Planta*, 1992, **188**, 48.
279. M. E. Knight, C. Halpin, and W. Schuch, *Plant Mol. Biol.*, 1992, **19**, 793.
280. T. Hibino, D. Shibata, J. Q. Chen, and T. Higuchi, *Plant Cell Physiol.*, 1993, **34**, 659.
281. H. Galliano, M. Cabane, C. Eckerskorn, F. Lottspeich, H. Sandermann, Jr., and D. Ernst, *Plant Mol. Biol.*, 1993, **23**, 145.
282. J. Grima-Pettenati, C. Feuillet, D. Goffner, G. Borderies, and A. M. Boudet, *Plant Mol. Biol.*, 1993, **21**, 1085.
283. C. Feuillet, A. M. Boudet, and J. Grima-Pettenati, *Plant Physiol.*, 1993, **103**, 1447.
284. C.-I.Brändén and H. Eklund, *Experientia*, 1980, **S36**, 40.
285. J. H. McKie, R. Jaouhari, K. T. Douglas, D. Goffner, C. Feuillet, B. J. Grima-Pettenati, A. M. M. Baltas, and L. Gorrichon, *Biochim. Biophys. Acta*, 1993, **1202**, 61.
286. C.-I. Brändén, E. Zeppezauer, T. Boiwe, G. Söderlund, B. O. Söderburg, and B. Nordström, in "Pyridine Nucleotide Dependent Dehydrogenases", ed. H. Sund, Springer-Verlag, Berlin, 1970, 129.
287. C.-I. Brändén, H. Eklund, B. Nordström, T. Boiwe, G. Söderlund, E. Zeppezauer, I. Ohlsson, and Å. Åkeson, *Proc. Natl. Acad. Sci. USA*, 1973, **70**, 2439.
288. H. Eklund, B. Nordström, E. Zeppezauer, G. Söderlund, I. Ohlsson, T. Boiwe, B.-O. Söderberg, O. Tapia, C.-I. Brändén, and Å. Åkeson, *J. Mol. Biol.*, 1976, **102**, 27.
289. H. Eklund, J.-P. Samama, L. Wallén, C.-I. Brändén, Å. Åkeson, and T. A. Jones, *J. Mol. Biol.*, 1981, **146**, 561.
290. V. Lauvergeat, K. Kennedy, C. Feuillet, J. H. McKie, L. Gorrichon, M. Baltas, A. M. Boudet, J. Grima-Pettenati, and K. T. Douglas, *Biochemistry*, 1995, **34**, 12426.

291. R. W. Whetten and R. R. Sederoff, *Plant Physiol.*, 1992, **98**, 380.
292. S. L. Butland, M. L. Chow, and B. E. Ellis, *Plant Mol. Biol.*, 1998, **37**, 15.
293. R.-Y. Chen, T.-C. Chang, and M.-S. Liu, *Agric. Biol. Chem.*, 1988, **52**, 2137.
294. J. Jorrin, R. López-Valbuena, and M. Tena, *Biochim. Biophys. Acta*, 1988, **964**, 73.
295. N. K. Given, M. A. Venis, and D. Grierson, *J. Plant Physiol.*, 1988, **133**, 31.
296. J. Jorrin and R. A. Dixon, *Plant Physiol.*, 1990, **92**, 447.
297. G. Gowri, N. L. Paiva, and R. A. Dixon, *Plant Mol. Biol.*, 1991, **17**, 415.
298. G. P. Bolwell, J. N. Bell, C. L. Cramer, W. Schuch, C. J. Lamb, and R. A. Dixon, *Eur. J. Biochem.*, 1985, **149**, 411.
299. C. L. Cramer, K. Edwards, M. Dron, X. Liang, S. L. Dildine, G. P. Bolwell, R. A. Dixon, C. J. Lamb, and W. Schuch, *Plant Mol. Biol.*, 1989, **12**, 367.
300. X. Liang, M. Dron, C. L. Cramer, R. A. Dixon, and C. J. Lamb, *J. Biol. Chem.*, 1989, **264**, 14 486.
301. W. Schulz, H.-G. Eiben, and K. Hahlbrock, *FEBS Lett.*, 1989, **258**, 335.
302. E.-i. Minami, Y. Ozeki, M. Matsuoka, N. Koizuka, and Y. Tanaka, *Eur. J. Biochem.*, 1989, **185**, 19.
303. S. Ohl, S. A. Hedrick, J. Chory, and C. J. Lamb, *Plant Cell*, 1990, **2**, 837.
304. S. Kawamata, T. Yamada, Y. Tanaka, P. Sriprasertsak, H. Kato, Y. Ichinose, H. Kato, T. Shiraishi, and H. Oku, *Plant Mol. Biol.*, 1992, **20**, 167.
305. H. J. Joos and K. Hahlbrock, *Eur. J. Biochem.*, 1992, **204**, 621.
306. T. Fukasawa-Akada, S.-d. Kung, and J. C. Watson, *Plant Mol. Biol.*, 1996, **30**, 711.
307. F. Sparvoli, C. Martin, A. Scienza, G. Gavazzi, and C. Tonelli, *Plant Mol. Biol.*, 1994, **24**, 743.
308. Y. Osakabe, K. Osakabe, S. Kawai, Y. Katayama, and N. Morohoshi, *Plant Mol. Biol.*, 1995, **28**, 1133.
309. R. Subramaniam, S. Reinold, E. K. Molitor, and C. J. Douglas, *Plant Physiol.*, 1993, **102**, 71.
310. U. Czichi and H. Kindl, *Planta*, 1977, **134**, 133.
311. H. Ruis and H. Kindl, *Phytochemistry*, 1971, **10**, 2627.
312. H. Ruis and H. Kindl, *Hoppe-Seyler's Z. Physiol. Chem.*, 1970, **351 S**, 1425.
313. N. Amrhein and M. H. Zenk, *Z. Pflanzenphysiol.*, 1971, **64**, 145.
314. E. L. Camm and G. H. N. Towers, *Phytochemistry*, 1973, **12**, 1575.
315. U. Czichi and H. Kindl, *Planta*, 1975, **125**, 115.
316. H. D. Gregor, *Z. Pflanzenphysiol.*, 1976, **77**, 454.
317. B. Monties, *C. R. Acad. Sci. Paris, Série C*, 1974, **278**, 1465.
318. U. Czichi and H. Kindl, *Hoppe-Seyler's Z. Physiol. Chem.*, 1975, **356 S**, 475.
319. W. Löffelhardt and H. Kindl, *Hoppe-Seyler's Z. Physiol. Chem.*, 1975, **356 S**, 487.
320. H. van Rensburg, A. M. Anterola, L. H. Levine, L. B. Davin, and N. G. Lewis, eds. W. G. Glasser, R. Northey, and T. P. Schultz, ACS Symp. Ser., Washington D.C., 1999, vol. XX, p. (in press).
321. C. G. Smith, M. W. Rodgers, A. Zimmerlin, D. Ferdinando, and G. P. Bolwell, *Planta*, 1994, **192**, 155.
322. J. Nakashima, T. Awano, K. Takabe, M. Fujita, and H. Saiki, *Plant Cell Physiol.*, 1997, **38**, 113.
323. Y. Osakabe, K. Nanto, H. Kitamura, S. Kawai, Y. Kondo, T. Fujii, K. Takabe, Y. Katayama, and N. Morohoshi, *Planta*, 1996, **200**, 13.
324. V. R. Franceschi, T. Krekling, A. A. Berryman, and E. Christiansen, *Am. J. Bot.*, 1998, **85**, 601.
325. S.-C. Wu and K. Hahlbrock, *Z. Naturforsch.*, 1992, **47c**, 591.
326. G. P. Bolwell, J. Sap, C. L. Cramer, C. J. Lamb, W. Schuch, and R. A. Dixon, *Biochim. Biophys. Acta*, 1986, **881**, 210.
327. W. Jahnen and K. Hahlbrock, *Planta*, 1988, **173**, 197.
328. M. Bevan, D. Shufflebottom, K. Edwards, R. Jefferson, and W. Schuch, *EMBO J.*, 1989, **8**, 1899.
329. R. Lois and K. Hahlbrock, *Z. Naturforsch.*, 1992, **47c**, 90.
330. Z.-H. Ye, *Plant Sci.*, 1996, **121**, 133.
331. R. A. Dixon, M. J. Harrison, and N. L. Paiva, *Physiol. Plant.*, 1995, **93**, 385.
332. S. Kawai, A. Mori, T. Shiokawa, S. Kajita, Y. Katayama, and N. Morohoshi, *Biosci. Biotech. Biochem.*, 1996, **60**, 1586.
333. M. R. Frank, J. M. Deyneka, and M. A. Schuler, *Plant Physiol.*, 1996, **110**, 1035.
334. M. Mizutani, D. Ohta, and R. Sato, *Plant Physiol.*, 1997, **113**, 755.
335. D. A. Bell-Lelong, J. C. Cusumano, K. Meyer, and C. Chapple, *Plant Physiol.*, 1997, **113**, 729.
336. M. W. Rodgers, A. Zimmerlin, D. Werck-Reichhart, and G. P. Bolwell, *Arch. Biochem. Biophys.*, 1993, **304**, 74.
337. E. Logemann, M. Parniske, and K. Hahlbrock, *Proc. Natl. Acad. Sci. USA*, 1995, **92**, 5905.
338. M. Hotze, G. Schröder, and J. Schroder, *FEBS Lett.*, 1995, **374**, 345.
339. S. C. Jupe, D. Werck-Reichhart, and G. P. Bolwell, EMBL accession number Y09449, submitted 14 November 1996.
340. S. C. Jupe, D. Werck-Reichhart, and G. P. Bolwell, EMBL accession number Y09447, submitted 14 November 1996.
341. D. R. Nelson and H. W. Strobel, *J. Biol. Chem.*, 1988, **263**, 6038.
342. M. Mizutani and D. Ohta, *Plant Physiol.*, 1998, **116**, 357.
343. K. Meyer, A. M. Shirley, J. C. Cusumano, D. A. Bell-Lelong, and C. C. S. Chapple, *Proc. Natl. Acad. Sci. USA*, 1998, **95**, 6619.
344. X.-H. Zhang and C. C. Chinnappa, *J. Biosci.*, 1997, **22**, 161.
345. G. Busam, B. Grimmig, R. E. Kneusel, and U. Matern, *Plant Physiol.*, 1997, **113**, 1003.
346. M. Bevan, M. Weichselgartner, B. Fartmann, K. Granderath, D. Dauner, A. Herzl, S. Neumann, J. Hoheisel, T. Jesse, L. Heijnen, P. Vos, H. W. Mewes, K. F. X. Mayer, and C. Schueller, EMBL accession number AL021961, submitted 15 October 1998, 1998.
347. G. Busam, H. Kassemeyer, and U. Matern, Swiss-Prot accession number Q43237, submitted September 1995, 1995.
348. P. Rech, J. Grima-Pettenati, and A. M. Boudet, Swiss-Prot accession number O04854, submitted April 1997, 1997.
349. C. B. Michalowski and H. J. Bohnert, Genbank accession number AF053553, submitted 13 March 1998, 1998.
350. H. Meyermans, W. Ardiles-Diaz, M. Van Montagu, and W. Boerjan, EMBL accession number AJ224896, submitted 10 March 1998, 1998.
351. H. Meng and W. H. Campbell, *Plant Mol. Biol.*, 1998, **38**, 513.
352. S. Kawai and M. Maruyama, DDBJ accession number AB000408, submitted 10 January 1997, 1997.
353. T. Hayakawa, K. Nanto, S. Kawai, Y. Katayama, and N. Morohoshi, *Plant Sci.*, 1996, **113**, 157.
354. L. Pellegrini, P. Geoffroy, B. Fritig, and M. Legrand, *Plant Physiol.*, 1993, **103**, 509.

355. F. M. McAlister, C. L. D. Jenkins, and J. M. Watson, *Aust. J. Plant Physiol.*, 1998, **25**, 225.
356. J. Curry, M. Mendoza, and M. A. O'Connell, Genbank accession number AF081214, submitted 28 July 1998, 1998.
357. J. Wang and E. Pichersky, *Plant Physiol.*, 1997, **114**, 1567.
358. A. Gauthier, P. J. Gulick, and R. K. Ibrahim, Genbank accession number U16793, submitted 1 November 1994, 1994.
359. J. Garcia-Mas, R. Messeguer, P. Arus, and P. Puigdomenech, EMBL accession number X83217, submitted 5 December 1994, 1994.
360. J. E. Lee, A. Kleinhofs, A. Graner, S. Wegener, B. Parthier, and M. Löbler, *DNA Seq.*, 1997, **7**, 357.
361. P. Collazo, L. Montoliu, P. Puigdomenech, and J. Rigau, *Plant Mol. Biol.*, 1992, **20**, 857.
362. B. M. Held, H. Wang, I. John, E. S. Wurtele, and J. T. Colbert, *Plant Physiol.*, 1993, **102**, 1001.
363. H. Meng and W. H. Campbell, *Plant Physiol.*, 1995, **108**, 1749.
364. P. Costa, N. Bahrman, J. Frigerio, A. Kremer, and C. Plomion, Swiss-Prot accession number P81081, submitted October 1997, 1997.
365. T. Kaneko, S. Sato, H. Kotani, A. Tanaka, E. Asamizu, Y. Nakamura, N. Miyajima, M. Hirosawa, M. Sugiura, S. Sasamoto, T. Kimura, T. Hosouchi, A. Matsuno, A. Muraki, N. Nakazaki, K. Naruo, S. Okumura, S. Shimpo, C. Takeuchi, T. Wada, A. Watanabe, M. Yamada, M. Yasuda, and S. Tabata, *DNA Res.*, 1996, **3**, 109.
366. M. Capellades, M. A. Torres, I. Bastisch, V. Stiefel, F. Vignols, W. B. Bruce, D. Peterson, P. Puigdomenech, and J. Rigau, *Plant Mol. Biol.*, 1996, **31**, 307.
367. K. D. Hauffe, U. Paszkowski, P. Schulze-Lefert, K. Hahlbrock, J. L. Dangl, and C. J. Douglas, *Plant Cell*, 1991, **3**, 435.
368. C. J. Douglas, K. D. Hauffe, M.-E. Ites-Morales, M. Ellard, U. Paszkowski, K. Hahlbrock, and J. L. Dangl, *EMBO J.*, 1991, **10**, 1767.
369. S. Reinold, K. D. Hauffe, and C. J. Douglas, *Plant Physiol.*, 1993, **101**, 373.
370. D. Lee and C. J. Douglas, *Plant Physiol.*, 1996, **112**, 193.
371. A. Uhlmann, and J. Ebel, *Plant Physiol.*, 1993, **102**, 1147.
372. K. Yazaki, A. Ogawa, and M. Tabata, *Plant Cell Physiol.*, 1995, **36**, 1319.
373. D. Lee, M. Ellard, L. A. Wanner, K. R. Davis, and C. J. Douglas, *Plant Mol. Biol.*, 1995, **28**, 871.
374. K. S. Voo, R. W. Whetten, D. M. O'Malley, and R. R. Sederoff, *Plant Physiol.*, 1995, **108**, 85.
375. W.-J. Hu, A. Kawaoka, C.-J. Tsai, J. Lung, K. Osakabe, H. Ebinuma, and V. L. Chiang, *Proc. Natl. Acad. Sci. USA*, 1998, **95**, 5407.
376. Y. Zhao, S. D. Kung, and S. K. Dube, *Nucleic Acids Res.*, 1990, **18**, 6144.
377. J. R. Vincent and R. L. Nicholson, *Physiol. Mol. Plant Pathol.*, 1987, **30**, 121.
378. M. Pichon, I. Courbou, M. Beckert, A. M. Boudet, and J. Grima-Pettenati, *Plant Mol. Biol.*, 1998, **38**, 671.
379. M. Baucher, B. Monties, M. Van Montagu, and W. Boerjan, *Crit. Rev. Plant Sci.*, 1998, **17**, 125.
380. M. Fujita, D. R. Gang, L. B. Davin, and N. G. Lewis, *J. Biol. Chem.*, 1999, **274**, 618.
381. D. R. Gang, M. Fujita, L. B. Davin, and N. G. Lewis, in "Lignin and Lignan Biosynthesis", ACS Symposium Series, vol. 697, eds. N. G. Lewis and S. Sarkanen, Washington, DC, 1998, p. 389.
382. J. J. MacKay, W. Liu, R. Whetten, R. R. Sederoff, and D. M. O'Malley, *Mol. Gen. Genet.*, 1995, **247**, 537.
383. J. van Doorsselaere, M. Baucher, C. Feuillet, A. M. Boudet, M. Van Montagu, and D. Inze, *Plant Physiol. Biochem.*, 1995, **33**, 105.
384. M. Baucher, J. Van Doorsselaere, J. Gielen, M. Van Montagu, D. Inze, and W. Boerjan, *Plant Physiol.*, 1995, **107**, 285.
385. Y. Sato, T. Watanabe, A. Komamine, T. Hibino, D. Shibata, M. Sugiyama, and H. Fukuda, *Plant Physiol.*, 1997, **113**, 425.
386. A. M. Boudet, C. LaPierre, and J. Grima-Pettenati, *New Phytol.*, 1995, **129**, 203.
387. J. Samaj, S. Hawkins, V. Lauvergeat, J. Grima-Pettenati, and A. Boudet, *Planta*, 1998, **204**, 437.
388. M. H. Walter, J. Schaaf, and D. Hess, *Acta Hortic.*, 1994, **381**, 162.
389. C. Feuillet, V. Lauvergeat, C. Deswarte, G. Pilate, A. Boudet, and J. Grima-Pettenati, *Plant Mol. Biol.*, 1995, **27**, 651.
390. S. Hawkins, J. Samaj, V. Lauvergeat, A. Boudet, and J. Grima-Pettenati, *Plant Physiol.*, 1997, **113**, 321.
391. K. Freudenberg, *Nature*, 1959, **183**, 1152.
392. K. Freudenberg and J. M. Harkin, *Phytochemistry*, 1963, **2**, 189.
393. H. Erdtman, *Ind. Eng. Chem.*, 1957, **49**, 1385.
394. S. Marcinowski and H. Grisebach, *Phytochemistry*, 1977, **16**, 1665.
395. Z.-S. Li, M. Alfenito, P. A. Rea, V. Walbot, and R. A. Dixon, *Phytochemistry*, 1997, **45**, 689.
396. J. D. Pickett-Heaps, *Protoplasma*, 1968, **65**, 181.
397. D. P. Dharmawardhana and B. E. Ellis, in "Lignin and Lignan Biosynthesis," eds. N. G. Lewis and S. Sarkanen, ACS Symposium Series, Washington, DC, 1998, vol. 697, p. 76.
398. E. Yamamoto, G. H. Bokelman, and N. G. Lewis, in "Plant Cell Wall Polymers: Biogenesis and Biodegradation," eds. N. G. Lewis and M. G. Paice, ACS Symposium Series, Washington, DC, 1989, vol. 399, p. 68.
399. K. Takabe, M. Fujita, H. Harada, and H. Saiki, *Mokuzai Gakkaishi*, 1985, **31**, 613.
400. K. Takabe, K. Fukazawa, and H. Harada, in "Plant Cell Wall Polymers: Biogenesis and Biodegradation," eds. N. G. Lewis and M. G. Paice, ACS Symposium Series, Washington, DC, 1989, vol. 399, p. 47.
401. Unpublished observations, 1998.
402. T. L. Eberhardt, Ph.D. Dissertation, Virginia Polytechnic Institute and State University, 1992.
403. V. Leinhos and R. A. Savidge, *Can. J. For. Res.*, 1993, **23**, 343.
404. F. Tiemann and W. Haarmann, *Ber. Dtsch. Chem. Ges.*, 1874, **7**, 608.
405. F. Tiemann, *Ber. Dtsch. Chem. Ges.*, 1875, **8**, 1127.
406. N. G. Lewis, M. E. J. Inciong, H. Ohashi, G. H. N. Towers, and E. Yamamoto, *Phytochemistry*, 1988, **27**, 2119.
407. E. Morelli, R. N. Rej, N. G. Lewis, G. Just, and G. H. N. Towers, *Phytochemistry*, 1986, **25**, 1701.
408. J. Harmatha, H. Lübke, I. Rybarik, and M. Mähdalik, *Collect. Czech. Chem. Commun.*, 1978, **43**, 774.
409. S. R. Kornberg, S. B. Zimmerman, and A. Kornberg, *J. Biol. Chem.*, 1961, **236**, 1487.
410. R. K. Ibrahim and H. Grisebach, *Arch. Biochem. Biophys.*, 1976, **176**, 700.
411. R. K. Ibrahim, *Z. Pflanzenphysiol.*, 1977, **85**, 253.
412. G. Schmid and H. Grisebach, *Eur. J. Biochem.*, 1982, **123**, 363.

413. E. Yamamoto, M. E. J. Inciong, L. B. Davin, and N. G. Lewis, *Plant Physiol.*, 1990, **94**, 209.
414. H. Förster, and R. Savidge, "Proceedings of the 22nd Annual Meeting of the Plant Growth Regulation Society of America, 1995," p. 402.
415. A. Vrielink, W. Rüger, H. P. C. Driessen, and P. S. Freemont, *EMBO J.*, 1994, **13**, 3413.
416. G. Schmid, D. K. Hammer, A. Ritterbusch, and H. Grisebach, *Planta*, 1982, **156**, 207.
417. J. Sanz-Aparicio, J. A. Hermoso, M. Martinez-Ripoll, J. L. Lequerica, and J. Polaina, *J. Mol. Biol.*, 1998, 491.
418. J. Liebig, and F. Wöhler, *Annalen*, 1837, **22**, 11.
419. K. Freudenberg, H. Reznik, H. Boesenberg, and D. Rasenack, *Chem. Ber.*, 1952, **85**, 641.
420. W. Hösel, E. Surholt, and E. Borgmann, *Eur. J. Biochem.*, 1984, **84**, 487.
421. S. Marcinowski and H. Grisebach, *Eur. J. Biochem.*, 1978, **87**, 37.
422. S. Marcinowski, H. Falk, D. K. Hammer, B. Hoyer, and H. Grisebach, 1979, **144**, 161.
423. D. P. Dharmawardhana, B. E. Ellis, and J. E. Carlson, *Plant Physiol.*, 1995, **107**, 331.
424. D. P. Dharmawardhana, Ph. D. Dissertation, University of British Columbia, 1996.
425. V. Leinhos, P. Udagama-Randeniya, and R. A. Savidge, *Phytochemistry*, 1994, **37**, 311.
426. J. B. Kempton and S. G. Withers, *Biochemistry*, 1992, **31**, 9961.
427. Q. Wang, D. Trimbur, R. Graham, R. A. J. Warren, and S. G. Withers, *Biochemistry*, 1995, **34**, 14554.
428. D. E. Koshland, *Biol. Rev.*, 1953, **28**, 416.
429. C. Wiesmann, G. Beste, W. Hengstenberg, and G. E. Schulz, *Structure*, 1995, **3**, 961.
430. T. Barrett, C. G. Suresh, S. P. Tolley, E. J. Dodson, and M. A. Hughes, *Structure*, 1995, **3**, 951.
431. W. P. Burmeister, S. Cottaz, H. Driguez, R. Iori, S. Palmeieri, and B. Henrissat, *Structure*, 1997, **5**, 663.
432. D. Fengel and G. Wegener, "Wood Chemistry, Ultrastructure, Reactions," Walter de Gruyter, Berlin, 1984.
433. E. Ingold, M. Sugiyama, and A. Komamine, *Physiol. Plant.*, 1990, **78**, 67.
434. H. Meier and K. C. B. Wilkie, *Holzforschung*, 1959, **13**, 177.
435. H. Meier, *J. Polym. Sci.*, 1961, **51**, 11.
436. E. A. Côté, Jr, N. P. Kutscha, B. W. Simson, and T. E. Timell, *Tappi*, 1968, **51**, 33.
437. P. R. Larson, *Holzforschung*, 1969, **23**, 17.
438. P. R. Larson, *Tappi*, 1969, **52**, 2170.
439. K. Takabe, M. Fujita, H. Harada, and H. Saiki, *Mokuzai Gakkaishi*, 1983, **29**, 183.
440. K. Takabe, M. Fujita, H. Harada, and H. Saiki, *Mokuzai Gakkaishi*, 1984, **30**, 103.
441. A. B. Wardrop, *Tappi*, 1957, **40**, 225.
442. H. Imagawa, K. Fukazawa, and S. Ishida, *Res. Bull. Coll. Expt. Forests, Hokkaido Univ.*, 1976, **33**, 127.
443. A. B. Wardrop, in "Lignins. Occurrence, Formation, Structure and Reactions," eds. K. V. Sarkanen and C. H. Ludwig, Wiley Interscience, New York, NY, 1971, p. 19.
444. A. B. Wardrop, *Appl. Poly. Symp.*, 1976, **28**, 1041.
445. N. P. Kutscha and J. M. Schwarzmann, *Holzforschung*, 1975, **29**, 79.
446. K. Kishi, H. Harada, and H. Saiki, *Bull. Kyoto Univ. Forests*, 1982, **54**, 209.
447. S. Saka and R. J. Thomas, *Wood Sci. Technol.*, 1982, **16**, 167.
448. K. Takabe, M. Fujita, H. Harada, and H. Saiki, *Res. Bull. Coll. Expt. Forests, Hokkaido Univ.*, 1986, **43**, 783.
449. K. Takabe, M. Fujita, H. Harada, and H. Saiki, *Mokuzai Gakkaishi*, 1981, **27**, 249.
450. J. A. N. Scott, A. R. Procter, B. J. Fergus, and D. A. I. Goring, *Wood Sci. Technol.*, 1969, **3**, 73.
451. B. J. Fergus, A. R. Procter, J. A. N. Scott, and D. A. I. Goring, *Wood Sci. Technol.*, 1969, **3**, 117.
452. P. Whiting, and D. A. I. Goring, *Wood Sci. Technol.*, 1982, **16**, 261.
453. H. L. Hardell, G. J. Leary, M. Stoll, and U. Westermark, *Sven. Papperstidn.*, 1980, **83**, 44.
454. N. Terashima, and K. Fukushima, *Wood Sci. Technol.*, 1988, **22**, 259.
455. K. Fukushima and N. Terashima, *Holzforschung*, 1991, **45**, 87.
456. B. J. Fergus, and D. A. I. Goring, *Holzforschung*, 1970, **24**, 113.
457. B. J. Fergus, and D. A. I. Goring, *Holzforschung*, 1970, **24**, 118.
458. Y. Musha, and D. A. I. Goring, *Wood Sci. Technol.*, 1975, **9**, 45.
459. N. Terashima, K. Fukushima, and K. Takabe, *Holzforschung*, 1986, **40**, 101.
460. N. Terashima, and K. Fukushima, in "Plant Cell-wall Polymers: Biogenesis and Biodegradation," eds. N. G. Lewis and M. G. Paice, ACS Symposium Series, Washington DC, 1989, vol. 399, p. 160.
461. T. J. Eom, G. Meshitsuka, and J. Nakano, *Mokuzai Gakkaishi*, 1987, **33**, 576.
462. H.-L. Hardell, G. J. Leary, M. Stoll, and U. Westermark, *Svensk Papperstidn.*, 1980, **83**, 71.
463. N. S. Cho, J. Y. Lee, G. Meshitsuka, and J. Nakano, *Mokuzai Gakkaishi*, 1980, **26**, 527.
464. L. He and N. Terashima, *Mokuzai Gakkaishi*, 1989, **35**, 116.
465. N. Terashima, K. Fukushima, Y. Sano, and K. Takabe, *Holzforschung*, 1988, **42**, 347.
466. T. E. Timell (ed.) "Compression Wood in Gymnosperms," Springer-Verlag, Berlin, 1986.
467. M. Y. Pillow and M. W. Bray, *Pap. Trade J.*, 1935, **101**, 361.
468. D. E. Bland, *Holzforshung*, 1961, **15**, 103.
469. M. A. Latif, Ph.D. Thesis., University of Seattle, 1968.
470. C. LaPierre, B. Monties, and C. Rolando, *Holzforschung*, 1988, **42**, 409.
471. C. LaPierre and C. Rolando, *Holzforshung*, 1988, **42**, 1.
472. M. Kwon, D. Bedgar, W. Piastuch, L. B. Davin, and N. G. Lewis, submitted for publication.
473. R. D. Hartley and C. W. Ford, in "Plant Cell-wall Polymers—Biogenesis and Biodegradation," eds. N. G. Lewis and M. G. Paice, ACS Symposium Series, Washington DC, 1989, vol. 399, p. 137.
474. E. Beckmann, O. Liesche, and F. Lehmann, *Biochem. Z.*, 1923, **139**, 491.
475. R. H. J. Creighton and H. Hibbert, *J. Amer. Chem. Soc.*, 1944, **66**, 37.
476. R. H. J. Creighton, R. D. Gibbs, and H. Hibbert, *J. Amer. Chem. Soc.*, 1944, **66**, 32.
477. D. C. C. Smith, *Nature*, 1955, **176**, 267.
478. L. He and N. Terashima, *J. Wood. Chem. Technol.*, 1990, **10**, 435.
479. A. Scalbert, B. Monties, J.-Y. Lallemand, E. Guittet, and C. Rolando, *Phytochemistry*, 1985, **24**, 1359.
480. H. U. Markwalder and H. Neukom, *Phytochemistry*, 1976, **15**, 836.
481. M. M. Smith and R. D. Hartley, *Carbohydrate Res.*, 1983, **118**, 65.

482. A. Kato, J.-i. Azuma, and T. Koshijima, *Chem. Lett.*, 1983, 137.
483. Y. Kato and D. J. Nevins, *Carbohydrate Res.*, 1985, **137**, 139.
484. S. C. Fry, *Biochem. J.*, 1982, **203**, 493.
485. S. C. Fry, *Planta*, 1983, **157**, 111.
486. K. Iiyama, T. B. T. Lam, and B. A. Stone, *Phytochemistry*, 1990, **29**, 733.
487. J. Ralph, R. D. Hatfield, S. Quideau, R. F. Helm, J. H. Grabber, and H.-J. G. Jung, *J. Amer. Chem. Soc.*, 1994, **116**, 9448.
488. J. H. Grabber, S. Quideau and J. Ralph, *Phytochemistry*, 1996, **43**, 1189.
489. M. Shimada, T. Fukuzuka, and T. Higuchi, *Tappi*, 1971, **54**, 72.
490. J. Ralph, J. H. Grabber, and R. D. Hatfield, *Carbohydrate Res.*, 1995, **275**, 167.
491. C. P. Neto, A. Seca, A. M. Nunes, M. A. Coimbra, F. Domingues, D. Evtuguin, A. Silvestre, and J. A. S. Cavaleiro, *Ind. Crops Prod.*, 1997, **6**, 51.
492. U. Sharma, J. M. Brillouet, A. Scalbert, and B. Monties, *Agronomie*, 1986, **6**, 265.
493. L. He and N. Terashima, *Mokuzai Gakkaishi*, 1989, **35**, 123.
494. K. Freudenberg, *Angew. Chem.*, 1956, **68**, 508.
495. K. V. Sarkanen, in "Lignins—Occurrence, Formation, Structure and Reactions", eds. K. V. Sarkanen and C. H. Ludwig, Wiley Interscience, New York, NY, 1971, vol. p. 95.
496. E. Adler, *Wood. Sci. Technol.*, 1977, **11**, 169.
497. O. Faix, *Holzforschung*, 1986, **40**, 273.
498. K. Ruel, O. Faix, and J. P. Joseleau, *J. Trace Microprobe Techniques*, 1994, **12**, 247.
499. J.-P. Joseleau and K. Ruel, *Plant Physiol.*, 1997, **114**, 1123.
500. V. Burlat, K. Ambert, K. Ruel, and J.-P. Joseleau, *Plant Physiol. Biochem.*, 1997, **35**, 645.
501. D. J. Mabberley (ed.) "The Plant-Book. A Portable Dictionary of the Higher Plants," Cambridge University Press, Cambridge, 1993.
502. K. Freudenberg, *Science*, 1965, **148**, 595.
503. H. Becker and H. Nimz, *Z. Pflanzenphysiol*, 1974, **72**, 52.
504. H. Nimz, H.-D. Luedemann, and H. Becker, *Z. Pflanzenphysiol*, 1974, **73**, 226.
505. E. Stahl, F. Karig, U. Brögmann, H. Nimz, and H. Becker, *Holzforschung*, 1973, **27**, 89.
506. H. Kuroda and T. Higuchi, *Phytochemistry*, 1976, **15**, 1511.
507. B. A. Fineran, *Protoplasma*, 1997, **198**, 186.
508. H. Erdtman, *Liebigs Ann. Chem.*, 1933, **503**, 283.
509. P. Klason, *Svensk Kem. Tidskr*, 1897, **9**, 133.
510. P. Klason, *Bericht über die Hauptversammlung des Vereins der Zellstoff- und Papier Chemiker*, 1908, 52.
511. P. Klason, *Arkiv. Kemi., Mineral. Geol.*, 1908, **3**, 1.
512. K. Kürschner, in "Chemischer und Chemisch-technischer," ed. F. B. Ahrens, Verlag von Ferdinand Enke., Stuttgart, 1926, vol. XXVIII, p. 71.
513. K. Kürschner and W. Schramek, *Tech. Chem. Papier-Zellstoff-Fabr.*, 1932, **29**, 35.
514. H. Erdtman, *Biochem. Z.*, 1933, **258**, 172.
515. H. Cousin and H. Hérissey, *C. R. Acad. Sci.*, 1908, **146**, 1413.
516. H. Hérissey and G. Doby, *J. Pharm. Chim.*, Paris (6), 1090, **30**, 289.
517. K. Freudenberg and H. Richtzenhain, *Liebigs Annalen*, 1942, **552**, 126.
518. K. Freudenberg, *J. Chem. Educ.*, 1932, **9**, 1171.
519. K. Freudenberg, *Fortschritte der Chem. Org. Naturstoffe*, 1939, **2**, 1.
520. K. Freudenberg and H. Richtzenhain, *Chem. Ber.*, 1943, **76**, 997.
521. K. Freudenberg, G. Grion, and J. M. Harkin, *Angew. Chem.*, 1958, **70**, 743.
522. T. Higuchi, *J. Biochem.*, 1958, **45**, 515.
523. K. Freudenberg, K. Jones, and H. Renner, *Chem. Ber.*, 1963, **96**, 1844.
524. K. Freudenberg and H. Schlüter, *Chem. Ber.*, 1955, **88**, 617.
525. K. Freudenberg and M. Friedmann, *Chem. Ber.*, 1960, **93**, 2138.
526. K. Freudenberg and H. Tausend, *Chem. Ber.*, 1964, **97**, 3418.
527. K. Freudenberg, *Adv. Chem. Ser.*, 1966, **59**, 1.
528. K. Freudenberg, in "Constitution and Biosynthesis of Lignin," eds. K. Freudenberg and A. C. Neish, Springer-Verlag, New York, NY, 1968, p. 47.
529. E. Adler, *Ind. Eng. Chem.*, 1957, **49**, 1377.
530. K. Freudenberg, *Angew. Chem.*, 1956, **68**, 84.
531. K. Freudenberg, in "Fortschritte d. Chemie Org. Naturstoffe," ed. L. Zechmeister, Springer Verlag, Vienna, 1954, vol. 11, p. 43.
532. K. Freudenberg and H. Tausend, *Chem. Ber.*, 1963, **96**, 2081.
533. K. Freudenberg and H. Nimz, *Chem. Ber.*, 1962, **95**, 2057.
534. K. Freudenberg, J. M. Harkin, M. Reichert, and T. Fukuzumi, *Chem. Ber.*, 1958, **91**, 581.
535. T. Higuchi, *Physiol. Plant.*, 1957, **10**, 356.
536. M. Nose, M. A. Bernards, M. Furlan, J. Zajicek, T. L. Eberhardt, and N. G. Lewis, *Phytochemistry*, 1995, **39**, 71.
537. A Ros Barcelo, *Int. Rev. Cyt.*, 1997, **176**, 87.
538. A Ros Barcelo, *Protoplasma*, 1995, **186**, 41.
539. A Driouich, A.-C. Lainé, B. Vian, and L. Faye, *Plant J.*, 1992, **2**, 13.
540. R. Sterjiades, J. F. D. Dean, and K.-E. L. Eriksson, *Plant Physiol.*, 1992, **99**, 1162.
541. W. Bao, D. M. O'Malley, R. Whetten, and R. R. Sederoff, *Science*, 1993, **260**, 672.
542. K. Freudenberg, *Chem. Ber.*, 1959, **92**, 89.
543. R. Sterjiades, J. F. D. Dean, G. Gamble, D. S. Himmelsbach, and K.-E. L. Eriksson, *Planta*, 1993, **190**, 75.
544. R. Savidge and P. Udagama-Randeniya, *Phytochemistry*, 1992, **31**, 2959.
545. P. V. Udagama-Randeniya and R. Savidge, *Electrophoresis*, 1994, **15**, 1072.
546. G. J. McDougall, D. Stewart, and I. M. Morrison, *Planta*, 1994, **194**, 9.
547. H. S. Mason and M. Cronyn, *J. Amer. Chem. Soc.*, 1955, **77**, 491.

548. H. Koblitz and D. Koblitz, *Nature*, 1964, **204**, 199.
549. P. D. Olson and J. E. Varner, *Plant J.*, 1993, **4**, 887.
550. G. Müsel, T. Schindler, R. Bergfeld, K. Ruel, G. Jacquet, C. Lapierre, V. Speth, and P. Schopfer, *Planta*, 1997, **201**, 146.
551. J. C. Steffens, E. Harel, and M. D. Hunt, in "Recent Advances in Phytochemistry," eds. B. E. Ellis, G. W. Kuroki, and H. A. Stafford, Plenum Press, New York, 1994, vol. 28 p. 275.
552. K. Freudenberg and W. Heimberger, *Chem. Ber.*, 1950, **83**, 519.
553. K. Freudenberg and H. Dietrich, *Chem. Ber.*, 1953, **86**, 1157.
554. K. Freudenberg and F. Bittner, *Chem. Ber.*, 1953, **86**, 155.
555. K. Freudenberg, R. Kraft, and W. Heimberger, *Chem. Ber.*, 1951, **84**, 472.
556. R. A. Savidge and P. V. Randeniya, *Biochemical Society Transactions*, 1992, **20**, 229S.
557. R. A. Savidge, *Polyphénols Actualités*, 1992, **8**, 12.
558. R. A. Savidge, P. V. Udagama-Randeniya, Y. Xu, V. Leinhos, and H. Förster, in "Lignin and Lignan Biosynthesis," eds. N. G. Lewis and S. Sarkanen, ACS Symposium Series, Washington, DC, 1998, vol. 697, p. 109.
559. P. V. Udagama-Randeniya and R. A. Savidge, *Trees*, 1995, **10**, 102.
560. A. Richardson, D. Stewart, and G. J. McDougall, *Planta*, 1997, **203**, 35.
561. N. Deighton and G. J. McDougall, *Phytochemistry*, 1998, **48**, 601.
562. T. Higuchi and Y. Ito, *J. Biochem.*, 1958, **45**, 575.
563. J. M. Harkin and J. R. Obst, *Science*, 1973, **180**, 296.
564. J. F. D. Dean, P. R. LaFayette, C. Rugh, A. H. Tristram, J. T. Hoopes, K.-E. L. Eriksson and S. A. Merkle, in "Lignin and Lignan Biosynthesis," eds. N. G. Lewis and S. Sarkanen, ACS Symposium Series, Washington, DC, 1998, vol. 697, p. 96.
565. D. M. O'Malley, R. Whetten, W. Bao, C. L. Chen, and R. R. Sederoff, *Plant J.*, 1993, **4**, 751.
566. H. Yoshida, *J. Chem. Soc.*, 1883, **XLIII**, 472.
567. M. G. Bertrand, *Bulletin de la Société Chimique de Paris*, 3ᶜ série, 1894, **XI**, 717.
568. S. V. Sunthankar and C. R. Dawson, *J. Amer. Chem. Soc.*, 1954, **76**, 5070.
569. D. Keilin and T. Mann, *Nature*, 1939, **143**, 23.
570. D. Keilin and T. Mann, *Nature*, 1940, **145**, 304.
571. W. G. Levine, in "The Biochemistry of Copper," eds. J. Peisach, P. Aisen, and W. E. Blumberg, Academic Press, New York, NY, 1966, vol. p. 371.
572. A Tissières, *Nature*, 1948, **162**, 340.
573. T. Nakamura, *Biochim. Biophys. Acta*, 1958, **30**, 44.
574. J. F. D. Dean and K.-E. L. Eriksson, *Holzforschung*, 1994, **48**, 21.
575. S. O. Anderson, in "Comprehensive Insect Physiology, Biochemistry and Pharmacology," eds. G. A. Kerkut and L. I. Gilbert, Pergamon, New York, NY, 1985, vol. 3, p. 59.
576. H. Claus and Z. Filip, *Microbiol. Res.*, 1997, **152**, 209.
577. S. M. Manskaya, *Biokhim. Vinodeliya Akad. Nauk SSSR, Shornik*, 1947, **1**, 32.
578. S.M. Manskaya, *Uspekhi Sovremennoi Biol.*, 1947, **23**, 203.
579. G. Legrand, *Compt. Rend.*, 1955, **240**, 249.
580. W. Nakamura, *J. Biochem.*, 1967, **62**, 54.
581. N. Bar-Nun, A. M. Mayer, and N. Sharon, *Phytochemistry*, 1981, **20**, 407.
582. D. M. Joel, I. Marbach, and A. M. Mayer, *Phytochemistry*, 1978, **17**, 796.
583. R. Bligny and R. Douce, *Biochem. J.*, 1983, **209**, 489.
584. P. R. LaFayette, K.-E. L. Eriksson, and J. F. D. Dean, *Plant Physiol.*, 1995, **107**, 667.
585. C.-L. Chen, in "Lignin and Lignan Biosynthesis," eds. N. G. Lewis and S. Sarkanen, ACS Symposium Series, Washington, DC, 1998, vol. 697, p. 255.
586. L. B. Davin, H.-B. Wang, A. L. Crowell, D. L. Bedgar, D. M. Martin, S. Sarkanen and N. G. Lewis, *Science*, 1997, **275**, 362.
587. B. Reinhammar, in "Copper Proteins and Copper Enzymes," ed. R. Lontie, 1984, vol. 3, p. 1.
588. A. Messerschmidt, A. Rossi, R. Ladenstein, R. Huber, M. Bolognesi, G. Gatti, A. Marchesini, R. Petruzzelli, and A. Finazzi-Agro, *J. Mol. Biol.*, 1989, **206**, 513.
589. A. Messerschmidt and R. Huber, *Eur. J. Biochem.*, 1990, **187**, 341.
590. W. B. Mims, J. L. Davis, and J. Peisach, *Biophys. J.*, 1984, **45**, 755.
591. M. D. Allendorf, D. J. Spira, and E. I. Solomon, *Proc. Natl. Acad. Sci., USA*, 1985, **82**, 3063.
592. D. J. Spira-Solomon, M. D. Allendorf and E. I. Solomon, *J. Am. Chem. Soc.*, 1986, **108**, 5318.
593. J. L. Cole, L. Avigliano, L. Morpurgo, and E. I. Solomon, *J. Am. Chem. Soc.*, 1991, **113**, 9080.
594. L.-E. Andréasson and B. Reinhammar, *Biochim. Biophys. Acta*, 1976, **445**, 579.
595. L.-E. Andréasson and B. Reinhammar, *Biochim. Biophys. Acta*, 1979, **568**, 145.
596. J. L. Cole, G. O. Tan, E. K. Yang, K. O. Hodgson, and E. I. Solomon, *J. Am. Chem. Soc.*, 1990, **112**, 2243.
597. J. L. Cole, D. P. Ballou, and E. I. Solomon, *J. Am. Chem. Soc.*, 1991, **113**, 8544.
598. A. Messerschmidt, R. Ladenstein, R. Huber, M. Bolognesi, L. Avigliano, R. Petruzzelli, A. Rossi, and A. Finazzi-Agró, *J. Mol. Biol.*, 1992, **224**, 179.
599. V. Ducros, A. M. Brzozowski, K. S. Wilson, S. H. Brown, P. Østergaard, P. Schneider, D. S. Yaver, A. H. Pedersen, and G. J. Davies, *Nature Structural Biology*, 1998, **5**, 310.
600. A. G. Sykes, *Adv. Inorg. Chem.*, 1991, **36**, 377.
601. J. Lipetz, *J. Histochem. Cytochem.*, 1965, **13**, 300.
602. J. S. Fruton (ed.) "Molecules and Life. Historical Essays on the Interplay of Chemistry and Biology," Wiley-Interscience, New York, NY, 1972.
603. C. F. Schönbein, *Ann. Phys.*, 1848, **75**, 357.
604. C. F. Schönbein, *J. Prakt. Chem.*, 1863, **89**, 323.
605. G. Linossier, *Compt. Rend. Soc. Biol*, 1898, **50**, 373.
606. R. Kuhn, D. B. Hand, and M. Florkin, *Z. Physiol. Chem.*, 1931, **201**, 255.
607. D. Keilin and T. Mann, *Proc. Royal Soc.*, 1937, **122B**, 119.

608. H. Theorell, *Arkiv. Kemi, Mineral. Geol.*, 1942, **16A**, 1.
609. K. G. Welinder, *FEBS Lett.*, 1976, **72**, 19.
610. K. Fujiyama, H. Takemura, S. Shibayama, K. Kobayashi, J.-K. Choi, A. Shinmyo, M. Takano, Y. Yamada, and H. Okada, *Eur. J. Biochem.*, 1988, **173**, 681.
611. S. M. Manskaya, *Doklady Akad. Nauk SSSR*, 1948, **62**, 369.
612. S. M. Manskaya, *Trudy Konferents. Vysokomolekulyar Soedineniyam* (*High-Mol. Compounds*) 4-*oi Konf. Moskow*, 1948, 158.
613. S. M. Manskaya and M. S. Bardinskaya. *Biokhimiya*, 1952, **17**, 711.
614. S. M. Siegel, *Physiol. Plant.*, 1953, **6**, 134.
615. S. M. Siegel, *Physiol. Plant.*, 1954, **7**, 41.
616. S. M. Siegel, *Physiol. Plant.*, 1955, **8**, 20.
617. M. Mäder, Y. Meyer, and M. Bopp, *Planta*, 1975, **122**, 259.
618. P. Schloss, C. Walter, and M. Mäder, *Planta*, 1987, **170**, 225.
619. A. Ros Barcelo, R. Munoz, and F. Sabater, *Plant Sci.*, 1989, **63**, 31.
620. J. F. D. Dean, R. Sterjiades and K.-E. L. Eriksson, *Physiol. Plant.*, 1994, **92**, 233.
621. Y. Sato, M. Sugiyama, A. Komamine, and H. Fukuda, *Planta*, 1995, **196**, 141.
622. A. Henriksen, K. G. Welinder, and M. Gajhede, *J. Biol. Chem.*, 1998, **273**, 2241.
623. K. G. Welinder, *Curr. Opin. Struct. Biol.*, 1992, **2**, 388.
624. W. R. Patterson and T. L. Poulos, *Biochemistry*, 1995, **34**, 4331.
625. T. L. Poulos, S. T. Freer, R. A. Alden, S. L. Edwards, U. Skogland, K. Takio, B. Eriksson, N.-h. Xuong, T. Yonetani, and J. Kraut, *J. Biol. Chem.*, 1980, **255**, 575.
626. M. Gajhede, D. J. Schuller, A. Henriksen, A. T. Smith, and T. L. Poulos, *Nat. Struct. Biol.*, 1997, **4**, 1032.
627. D. J. Schuller, N. Ban, R. B. van Huystee, A. McPherson, and T. L. Poulos, *Structure*, 1996, **4**, 311.
628. P. R. Ortiz de Montellano, *Annu. Rev. Pharmacol. Toxicol.*, 1992, **32**, 89.
629. N. C. Veitch and R. J. P. Williams, *Eur. J. Biochem.*, 1990, **189**, 351.
630. N. C. Veitch and R. J. P. Williams, in "Biochemical, Molecular and Physiological Aspects of Plant Peroxidases," eds. J. Lobarzewski, H. Greppin, C. Penel, and T. Gaspar, University of Geneva, Geneva, 1991, p. 99.
631. Y. Czaninski, R. M. Sachot, and A. M. Catesson, *Ann. Bot.*, 1993, **72**, 547.
632. A. Catesson, Y. Czaninski, and B. Monties, *C. R. Acad. Sci. Series D*, 1978, **286**, 1787.
633. E. F. Elstner, *Ann. Rev. Plant Physiol.*, 1982, **33**, 73.
634. R. Angelini and R. Federico. *J. Plant Physiol.*, 1989, **135**, 212.
635. R. Kaur-Sawhney, H. E. Flores, and A. W. Galston, *Plant Physiol*, 1981, **68**, 494.
636. A. Rossi, R. Petruzzelli, and A. F. Agró, *FEBS*, 1992, **301**, 253.
637. T. A. Smith, *Biochem. Biophys. Res. Commun.*, 1970, **41**, 1452.
638. G. G. Gross, *Phytochemistry*, 1977, **16**, 319.
639. G. G. Gross, C. Janse, and E. F. Elstner, *Planta*, 1977, **136**, 271.
640. G. G. Gross and C. Janse, *Z. Pflanzenphysiol Bd.*, 1977, **84**, 447.
641. K. Ogawa, S. Kanematsu, and K. Asada, *Plant Cell Physiol.*, 1997, **38**, 1118.
642. G. P. Bolwell, V. S. Butt, D. R. Davies, and A. Zimmerlin, *Free Rad. Res.*, 1995, **23**, 517.
643. R. Goldberg, T. Le, and A. M. Catesson, *J. Exp. Bot.*, 1985, **36**, 503.
644. T. Akazawa and E. E. Conn, *J. Biol. Chem.*, 1958, **232**, 403.
645. G. J. McDougall, *Phytochemistry*, 1992, **31**, 3385.
646. R. P. Ross and A. Claiborne, *J. Mol. Biol.*, 1991, **221**, 857.
647. D. R. Gang, M. A. Costa, M. Fujita, A. T. Dinkova-Kostova, H. B. Wang, V. Burlat, W. Martin, S. Sarkanen, L. B. Davin, and N. G. Lewis, 1999, (submitted).
648. N. G. Lewis, L. B. Davin, A. T. Dinkova-Kostova, D. R. Gang, M. Fujita, and S. Sarkanen, *US Pat.* PCT/US97/20391 (Recombinant Pinoresinol/Laccaresinol Reductase, Recombinant Dirigent Protein and Methods of Use, p. 147).
649. W. R. Russell, A. R. Forrester, A. Chesson, and M. J. Burkitt, *Arch. Biochem. Biophys.*, 1996, **332**, 357.
650. L. B. Davin and N. G. Lewis, *An. Acad. Bras. Cienc*, 1995, **67 (Sup. 3)**, 363.
651. K. Okusa, T. Miyakoshi, and C.-L. Chen, *Holzforschung*, 1996, **50**, 15.
652. D. R. Armstrong, C. Cameron, D. C. Nonhebel, and P. G. Perkins, *J. Chem. Soc., Perkin Trans. II*, 1983, 563
653. F. Chioccara, S. Poli, B. Rindone, T. Pilati, G. Brunow, P. Pietikäinen, and H. Setälä, *Acta Chem. Scand.*, 1993, **47**, 610.
654. R. A. Anderson, D. T. Dalgleish, D. C. Nonhebel, and P. L. Pauson, *J. Chem. Res.*, 1977, **(S)** 12–13, **(M)** 0201.
655. T. J. Elder and R. M. Ede, *Proc. 8th Internat. Symp. Wood Pulp. Chem.*, 1995, **I**, 115.
656. G. Brunow, I. Kilpeläinen, J. Sipilä, K. Syrjänen, P. Karhunen, H. Setälä, and P. Rummakko, in "Lignin and Lignan Biosynthesis," eds. N. G. Lewis and S. Sarkanen, ACS Symposium Series, Washington, DC, 1998, vol. 697, p. 131.
657. R. M. Ede and G. Brunow, *J. Org. Chem.*, 1992, **57**, 1477.
658. I. Kilpelainen, E. Ammalahti, G. Brunow, and D. Robert, *Tetrahedron Lett.*, 1994, **35**, 9267.
659. A. Björkman, *Svensk Papperstidn.*, 1956, **59**, 477.
660. A. Björkman, *Svensk Papperstidn.*, 1957, **60**, 243.
661. T. Iversen, *Wood. Sci. Technol.*, 1985, **19**, 243.
662. K. Freudenberg and J. M. Harkin, *Chem. Ber.*, 1960, **93**, 2814.
663. T. Kleinert, *Papier*, 1970, **24**, 623.
664. J. H. Grabber, J. Ralph, and R. D. Hatfield, in "Lignin and Lignan Biosynthesis," eds. N. G. Lewis and S. Sarkanen, ACS Symposium Series, Washington, DC, 1998, vol. 697, p. 163.
665. J. C. Pew, *J. Org. Chem.*, 1963, **28**, 1048.
666. M. Drumond, M. Aoyama, C.-L. Chen, and D. Robert, *J. Wood. Chem. Technol.*, 1989, **9**, 421.
667. P. Karhunen, P. Rummakko, J. Sipilä, G. Brunow, and I. Kilpäläinen, *Tetrahedron Lett.*, 1995, **36**, 169.
668. P. Karhunen, P. Rummakko, J. Sipilä, G. Brunow, and I. Kilpäläinen, *Tetrahedron Lett.*, 1995, **36**, 4501.
669. J. Gierer and I. Norén, *Holzforschung*, 1980, **34**, 197.
670. P. Karhunen, J. Mikkola, and G. Brunow, *Proc. 9th Internat. Symp. Wood Pulp. Chem.*, 1997, **R1**, 1.
671. G. E. Miksche, *Acta Chem. Scand.*, 1973, **27**, 1355.

672. J. Gierer, *Svensk Papperstidn.*, 1970, **73**, 571.
673. P. Karhunen, P. Rummakko, A. Pajunen, and G. Brunow, *J. Chem. Soc., Perkin Trans. I*, 1996, 2303.
674. K. Lundquist and G. E. Miksche, *Tetrahedron Lett.*, 1965, 2131.
675. H. Nimz, *Chem. Ber.*, 1965, **98**, 3153.
676. G. Gellerstedt and L. Zhang, *Nordic Pulp Pap. Res. J.*, 1991, **6**, 136.
677. J. Ralph, J. Peng, and F. Lu, *Tetrahedron Lett.*, 1998, **39**, 4963.
678. R. H. Atalla and U. P. Agarwal, *Science*, 1985, **227**, 636.
679. L. A. Donaldson, *Wood Sci. Technol.*, 1994, **28**, 111.
680. J. F. Yan, F. Pla, R. Kondo, M. Dolk, and J. L. McCarthy, *Macromolecules*, 1984, **17**, 2137.
681. F. Pla and J. F. Yan, *J. Wood. Chem. Technol.*, 1984, **4**, 285.
682. A. B. Wardrop, in "The Formation of Wood in Forest Trees," ed. M. H. Zimmermann, Academic Press, New York, NY, 1964, vol. p. 87.
683. D. A. I. Goring, in "Cellulose Chemistry and Technology," ed. J. C. Arthur, Jr., ACS Symposium Series, Washington DC, 1977, vol. 48, 273.
684. D. A. I. Goring, R. Vuong, C. Gancet, and H. Chanzy, *J. Appl. Polym. Sci.*, 1979, **24**, 931.
685. S.-y. Guan, J. Mlynár, and S. Sarkanen, *Phytochemistry*, 1997, **45**, 911.
686. R. A. Volpe, and H. L. Frisch, *Macromolecules*, 1987, **20**, 1747.
687. V. Burlat, M. Kwon, L. B. Davin, and N. G. Lewis, *Plant J.*, 1999, (submitted).
688. Z.-H. Ye, Y.-R. Song, A. Marcus, and J. E. Varner, *Plant J.*, 1991, **1**, 175.
689. A. Marcus, J. Greenberg, and V. Averyhart-Fullard, *Physiol. Plant.*, 1991, **81**, 273.
690. K. Datta, A. Schmidt, and A. Marcus, *Plant Cell*, 1989, **1**, 945.
691. U. Ryser, M. Schorderet, G.-F. Zhao, D. Studer, K. Ruel, G. Hauf, and B. Keller, *Plant J.*, 1997, **12**, 97.
692. N. J. Baxter, T. H. Lilley, E. Haslam, and M. P. Williamson, *Biochemistry*, 1997, **36**, 5566.

3.19
Condensed Tannins

DANEEL FERREIRA, REINIER J. J. NEL, and RIAAN BEKKER
University of the Orange Free State, Bloemfontein, South Africa

3.19.1   INTRODUCTION

The oligomeric proanthocyanidins (condensed tannins) constitute one of the most ubiquitous groups of all plant phenolics.[1-11] Leucoanthocyanidins are defined[2,4,5] as monomeric proanthocyanidins which produce anthocyanidins (1) by cleavage of a C—O bond on heating with mineral acid. The oligomeric proanthocyanidins are flavan-3-ol oligomers which produce anthocyanidins by cleavage of a C—C bond under strongly acidic conditions. The exceptional concentrations of these compounds in the barks and heartwoods of a variety of tree species have resulted in their commercial extraction with the initial objective of applying the extracts in leather manufacture. Together with the biflavonoids they represent the two major classes of complex C_6–C_3–C_6 secondary metabolites. The bi- and tri-flavonoids[12] are products of oxidative coupling of flavones, flavonols, dihydroflavonols, flavanones, isoflavones, aurones, and chalcones and thus consistently possess a carbonyl function at C-4 or its equivalent in every constituent flavonoid unit. The oligomeric proanthocyanidins, on the contrary, usually originate by coupling at C-4 (C-ring) of an electrophilic flavanyl unit, generated from a flavan-4-ol[10] or a flavan-3,4-diol,[6] to a nucleophilic flavanyl moiety, often a flavan-3-ol. However, the limits between the biflavonoids and the oligomeric proanthocyanidins have become somewhat arbitrary since an increasing number of "mixed" dimers, for example, flavan-3-ol → dihydroflavonol, and "nonproanthocyanidins" comprising oxidatively coupled flavan-3-ols have been identified.[10,11] Compounds possessing at least one flavan or flavan-3-ol constituent unit are discussed in this chapter.

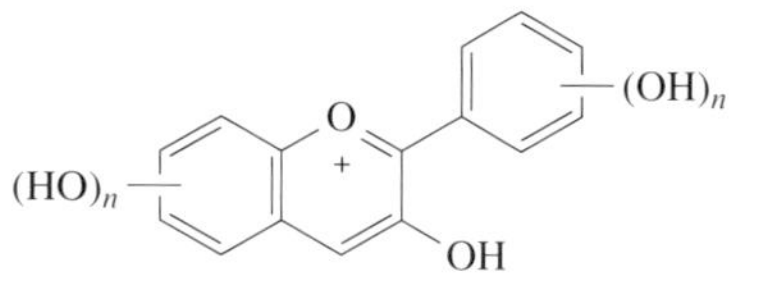

(1)

The biological significance (e.g., the protection of plants from insects, diseases, and herbivores) and most of the current uses (e.g., leather manufacture) and promising new uses (e.g., as pharmaceuticals and wood preservatives) of the oligomeric proanthocyanidins rest on their complexation with other biopolymers (e.g., proteins and carbohydrates, or metal ions). When taken in conjunction with the growing realization of the importance of these compounds as antioxidants in the human diet (e.g., the "French paradox," an apparent compatibility of a high fat diet with a low incidence of coronary atherosclerosis), this has led to a sharp increase in research effort. The review that follows demonstrates that, although considerable progress has been made in definition of proanthocyanidin structure, understanding of the properties of these complex compounds is still rather limited.

3.19.2   NOMENCLATURE

The system of nomenclature proposed by Hemingway *et al.*[13] and extended by Porter[10] is applied consistently and is briefly summarized as follows:

(i) The names of the basic flavan units are given in Table 1. All flavan-3-ols in this list possess a $2R,3S$ configuration, for example, catechin (2). Those with a $2R,3R$ configuration are prefixed with "epi," for example, epicatechin (3). Units possessing a $2S$ configuration are differentiated by the enantio (*ent*) prefix.

(ii) The flavanoid skeleton is drawn and numbered in the way as illustrated for the catechins (2) and (3).

(iii) The location of the interflavanyl bond in dimers and oligomers is denoted within parentheses as in the carbohydrates. The configuration of the interflavanoid bond at C-4 is denoted as α or β as in the IUPAC (International Union of Pure and Applied Chemistry) rules. Thus, the familiar procyanidin B-7 (4) is named epicatechin-(4β→6)-catechin, the analogous prodelphinidin (5) is

Table 1 Proanthocyanidin nomenclature: proanthocyanidin type and names of monomer units.

Proanthocyanidin	Monomer	Hydroxylation pattern						
		3	5	7	8	3'	4'	5'
Procassinidin	Cassiaflavan	H	H	OH	H	H	OH	H
Proapigeninidin	Apigeniflavan	H	OH	OH	H	H	OH	H
Proluteolinidin	Luteoliflavan	H	OH	OH	H	OH	OH	H
Protricetinidin	Tricetiflavan	H	OH	OH	H	OH	OH	OH
Prodistenidin	Distenin	OH	OH	OH	H	H	H	H
Propelargonidin	Afzelechin	OH	OH	OH	H	H	OH	H
Procyanidin	Catechin	OH	OH	OH	H	OH	OH	H
Prodelphinidin	Gallocatechin	OH	OH	OH	H	OH	OH	OH
Proguibourtinidin	Guibourtinidol	OH	H	OH	H	H	OH	H
Profisetinidin	Fisetinidol	OH	H	OH	H	OH	OII	H
Prorobinetinidin	Robinetinidol	OH	H	OH	H	OH	OH	OH
Proteracacinidin	Oritin	OH	H	OH	OH	H	OH	H
Promelacacinidin	Mesquitol	OH	H	OH	OH	OH	OH	H
Propeltogynidin	Peltogynane (**8**)	OCH_2—	H	OH	H	H	OH	OH
Promopanidin	Mopanane (**9**)	OCH_2—	H	OH	H	OH	OH	H

After: Porter.[10]

named epigallocatechin-($4\beta\rightarrow6$)-gallocatechin, and the corresponding 2*S* enantiomer (**6**) is named *ent*-epicatechin-($4\alpha\rightarrow8$)-*ent*-catechin.

(iv) A-type proanthocyanidins are often wrongly named due to the fact that the DEF-unit in, for example, dimeric analogues, is rotated through 180°. The proposed system[10,14] cognizant of this aspect will thus be used. Proanthocyanidin A-2 (**7**) is hence named epicatechin-($2\beta\rightarrow7$, $4\beta\rightarrow8$)-epicatechin.

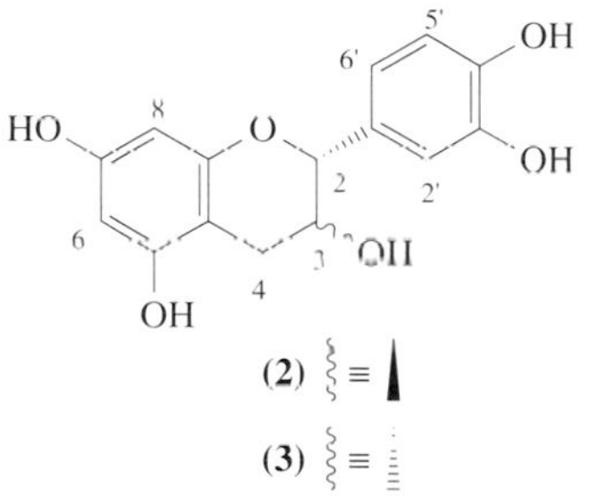

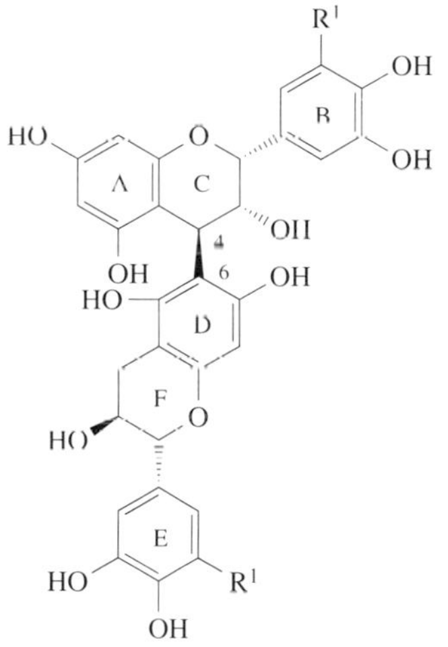

(**2**) ⧧ ≡ ▮

(**3**) ⧧ ≡ ⋮

(**4**) R^1 = H

(**5**) R^1 = OH

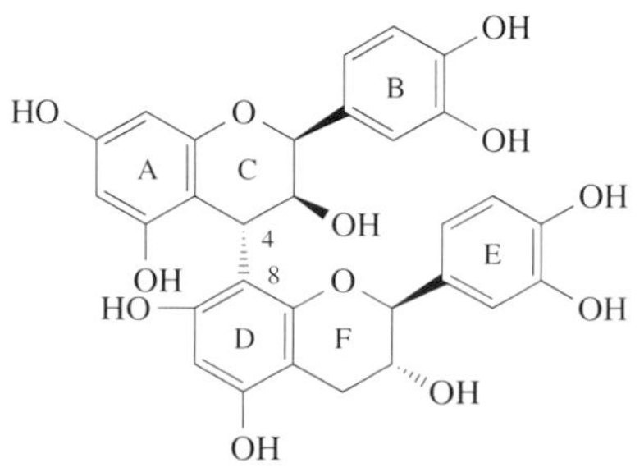

(**6**)

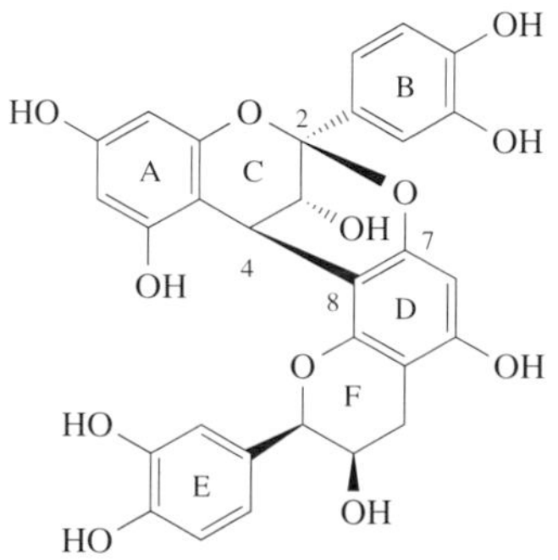

(**7**)

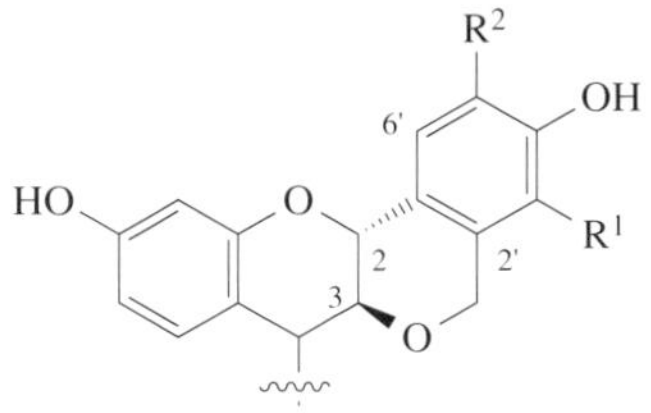

(8) $R^1 = H$, $R^2 = OH$
(9) $R^1 = OH$, $R^2 = H$

3.19.3 FLAVAN-3-OLS, FLAVAN-3,4-DIOLS, FLAVAN-4-OLS, AND FLAVANS AS BUILDING BLOCKS FOR OLIGOMERIC PROANTHOCYANIDINS

Owing to the purported role of the flavan-3-ols and flavans as nucleophilic chain-terminating units and of the flavan-3,4-diols and flavan-4-ols (leucoanthocyanidins) as electrophilic chain-extender units in the biosynthesis of the oligomeric proanthocyanidins,[5,10,15] these four classes of compounds are included in this discussion. In addition, knowledge of the chemistry of the constituent flavanyl moieties is often essential to progress in understanding of the oligomeric proanthocyanidins.

3.19.3.1 Biosynthesis

The biosynthesis of flavonoids from the malonate/shikimate level to dihydroflavonols (e.g., dihydroquercetin) is now firmly established (Scheme 1: biosynthesis of monomeric precursors to oligomeric proanthocyanidin). Many of the enzyme systems effecting the transformations have been isolated and the reactions demonstrated in cell-free media.[16] The sequence of steps leading from the dihydroflavonols to the direct biogenetic precursors, the flavan-3,4-diols and flavan-3-ols, has been also largely established, as is indicated in Scheme 1.

There were originally three major objections to a general acceptance of the dihydroflavonol → flavan-3,4-diol → flavan-3-ol route to oligomeric proanthocyanidin formation. First, flavan-3,4-diols with a phloroglucinol pattern A-ring have not yet been identified in natural sources.[17] The 5-deoxy analogues on the contrary are well known and widespread natural products.[5,10] Second, nearly all natural dihydroflavonols are of a 2,3-*trans*-2R,3R configuration[18] leading to the controversy regarding the natural occurrence of 2,3-*cis*-dihydroflavonols on thermodynamic considerations.[19] Third, it is observed that the constituent flavanyl units of oligomeric proanthocyanidins often display contradictory stereochemistry and/or oxygenation patterns to the bottom terminal flavan-3-ol moiety.

These objections have led to the proposal that cyclization of naturally occurring α-hydroxy-chalcones may lead to the simultaneous formation of 2,3-*cis*- and 2,3-*trans*-dihydroflavonols.[15] The intermediacy of a symmetrical flav-3-en-3-ol intermediate in the α-hydroxychalcone pathway *en route* to the (3S)- or (3R)-flavan-3-ols has also been postulated.[17]

Establishment of the sequence in Scheme 1 as the favored biosynthetic route to proanthocyanidins was triggered by the synthesis of catechin-4α-ol via metal hydride reduction of (2R,3R)-dihydro-quercetin in ethanol.[20,21] This synthesis was subsequently used to establish the intermediacy of leucocyandin (catechin-4α-ol) in catechin[22–24] and anthocyanidin[22] formation. In particular, it was established[24] that cell suspension cultures of *Pseudotsuga menziesii* callus contained an NADPH-dependent reductase capable of converting (2R,3R)-dihydroquercetin into catechin-4β-ol.

Evidence that an equivalent sequence also leads to 2,3-*cis*-proanthocyanidins has been obtained by the isolation of several examples of 2,3-*cis*-dihydroflavonols. (2R,3S)-7,8,3′,4′-Tetrahydroxy-dihydroflavonol coexists in *Acacia melanoxylon* with flavan-3,4-diols exhibiting identical oxygenation pattern and C-ring stereochemistry,[25] while (2R,3S)-2,3-*cis*-dihydroquercetin-3-O-β-D-glucopyranoside was obtained from *Taxillus kaempferi*,[26] and the 3-O-β-D-xylosides of (2R,3S)- and (2S,3R)-*cis*-dihydroquercetin from *Thujopsis dolabrata*.[27] The most significant observation of the latter investigation was the stability of the 2,3-*cis*-aglycons resulting from hydrolysis with hesperidinase in aqueous medium.

The flavan-4-ols are presumably derived biosynthetically by a single reduction step from a flavanone. A second reduction step would then lead to the flavans. Such a hypothesis is supported

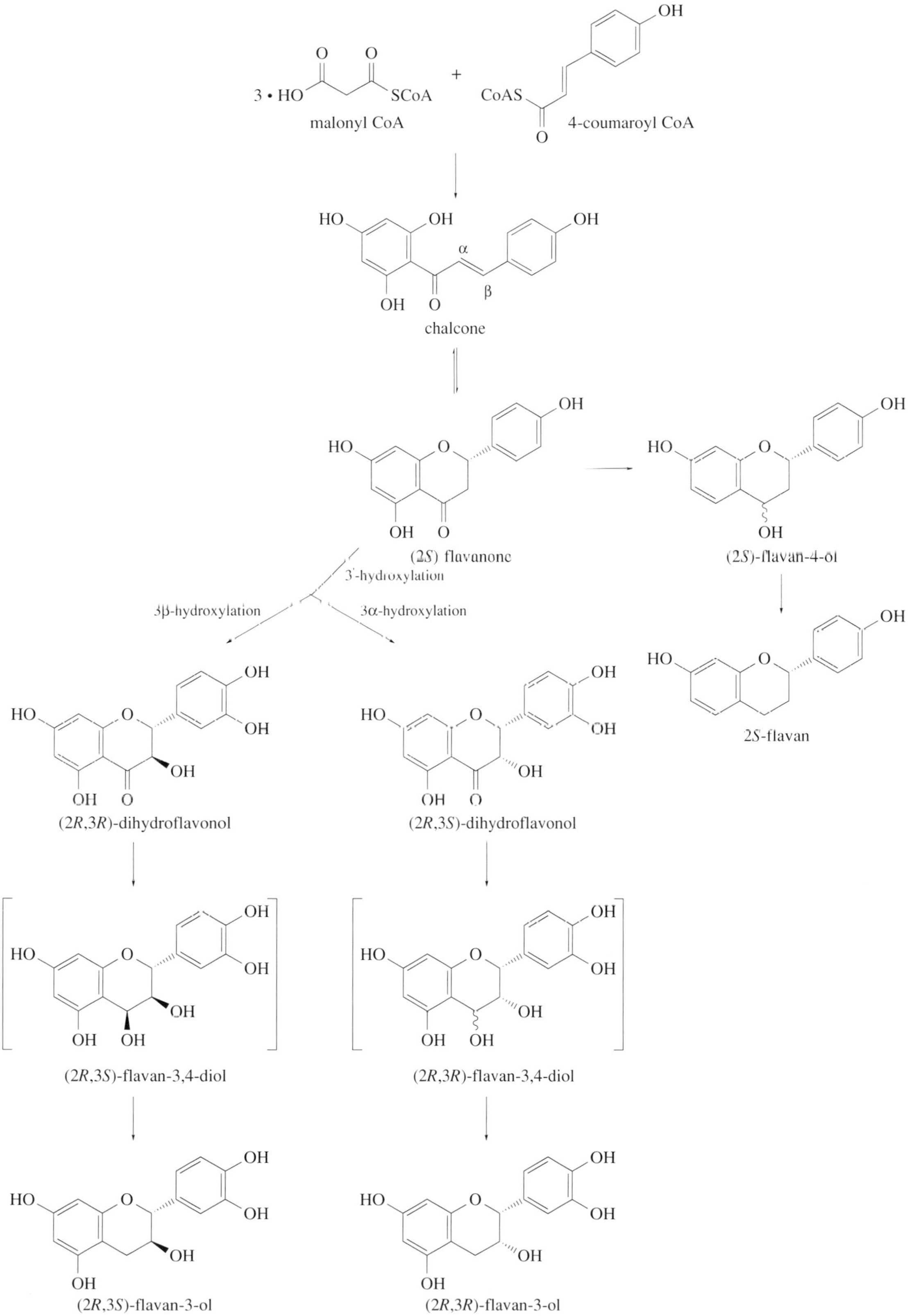
3 • HO
SCoA
malonyl CoA
+
CoAS
4-coumaroyl CoA
chalcone
(2S) flavanone
3'-hydroxylation
3β-hydroxylation
3α-hydroxylation
(2S)-flavan-4-ol
2S-flavan
(2R,3R)-dihydroflavonol
(2R,3S)-dihydroflavonol
(2R,3S)-flavan-3,4-diol
(2R,3R)-flavan-3,4-diol
(2R,3S)-flavan-3-ol
(2R,3R)-flavan-3-ol

Scheme 1

by the common cooccurrence of flavans and flavanones of identical hydroxylation pattern.[5,10] All the flavans whose ORD or c.d. spectra have been studied possess the 2*S* absolute configuration, as would be expected from the flavanone origin.

3.19.3.2 Flavan-3-ols

The chemistry of the flavan-3-ols is intimately linked to progress in the understanding of the oligomeric proanthocyanidins. Catechin (**2**) and epicatechin (**3**) constitute the predominant chain-terminating units of oligomeric proanthocyanidins and their structures must be traced back to the pioneering work by Freudenberg and co-workers[1,28,29] over many years, and to the final confirmation of their absolute configuration.[30,31,32] The stereochemistry of gallocatechin (**10**) was subsequently related to that of catechin (**2**).[33] Known naturally occurring flavan-3-ols and their derivatives (e.g., simple esters and *O*-glycosides) as well as their general properties and chemistry have been reviewed by Freudenberg and Weinges,[1] Weinges *et al.*,[2] Haslam,[4] Porter,[5,10] Hemingway,[6] and Ferreira and Bekker.[11]

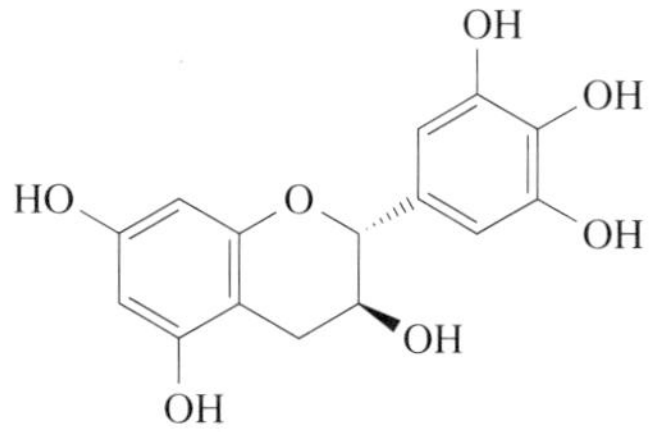

(**10**)

The most important features of the flavan-3-ols pertaining to the chemistry of the oligomeric proanthocyanidins are the nucleophilicity of their A-rings, the aptitude of their heterocyclic rings to cleavage and subsequent rearrangement, the susceptibility of analogues with pyrocatechol- or pyrogallol-type B-rings to phenol oxidative coupling, and the conformational mobility of their pyran rings. These aspects will be fully dealt with in the appropriate sections that follow.

3.19.3.3 Flavans

In contrast to the ubiquitous distribution of flavonoids substituted at C-3 and/or C-4 of their heterocyclic rings, the unsubstituted analogues (2-phenylchromans) are more rarely found, presumably due to their instability in solution. The naturally occurring flavans,[5,10,34] for example, compound (**11**) obtained by methylation of a natural product,[35,36] consistently possess A-ring hydroxylation patterns typical of the nucleophilic precursors of oligomeric proanthocyanidins. These compounds accordingly feature as the chain-terminating units in a variety of non-proanthocyanidins (Section 3.19.5).

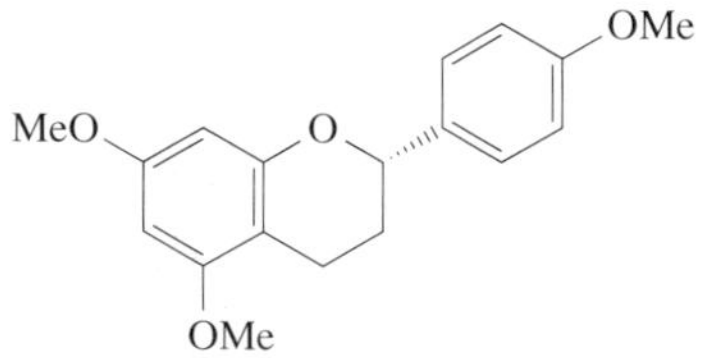

(**11**)

The flavans co-occur with chalcones,[35,37–39] flavanones,[37,39] flavan-3,4-diol,[39] flavonols,[40] and 1,3-diphenylpropanes.[35,40] Many natural flavans are lipid soluble and appear to be leaf-surface constituents. A number are phytoalexins, while the flavan from *Lycoris radiata* bulbs was found to be an antifeedant for the larvae of the yellow butterfly.[41] The so-called "dragons blood" resins afforded a variety of mono- and dimeric flavans.[5] The sensitivity of these flavans to oxidation to form stable quinone methides, for example, dracorubin (**12**), largely accounts for the intense red colors of these resins.

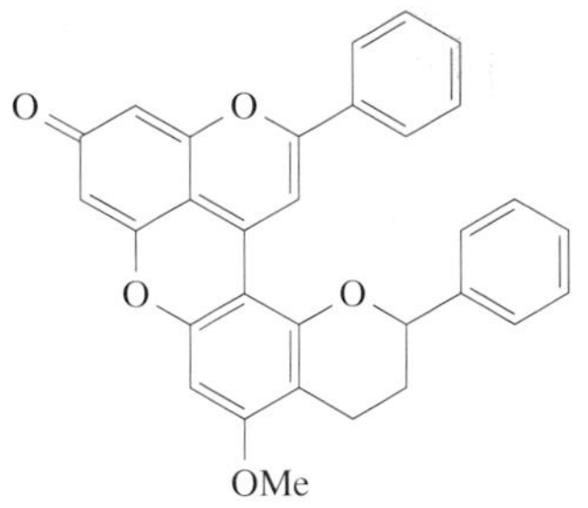

(12)

3.19.3.4 Flavan-3,4-diols

Other important units involved in oligomeric proanthocyanidin structure are the flavan-3,4-diols. The first of these to be isolated, melacacidin (**13**) (epimesquitol-4α-ol), was recognized by King and Bottomley,[42,43] while the first semisynthesis from dihydroflavonols was a leucorobinetinidin (**16**) (robinetinidol-4α-ol) by Freudenberg and Roux.[44,45] This stimulated rapid progress in the chemistry of leucoanthocyanidins particularly through the work of Clark-Lewis and co-workers[46–50] and Roux and co-workers.[51–55]

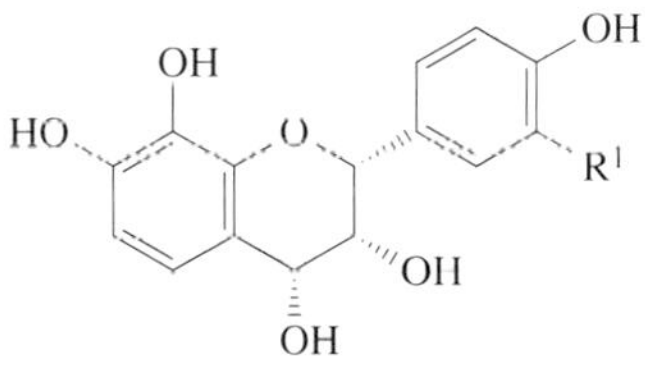

(13) R^1 = OH
(14) R^1 = H

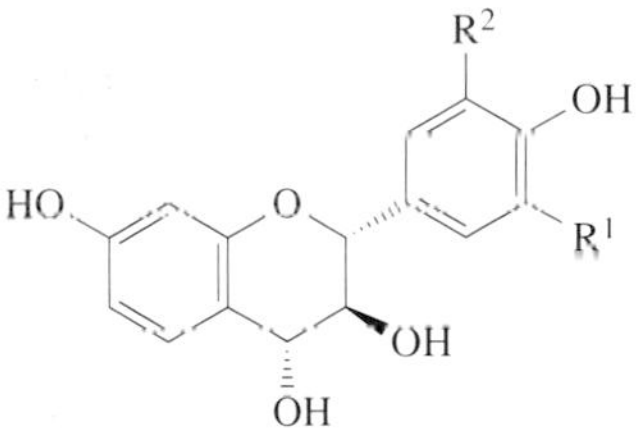

(15) R^1 = H, R^2 = OH
(16) R^1 – R^2 – OH
(17) R^1 = R^2 = H

The predominant feature of the flavan-3,4-diols relating to the chemistry of oligomeric proanthocyanidins is their role as precursors of flavan-4-carbocations or A-ring quinone methide electrophiles. The stability of the carbocations is dependent on the degree of delocalization of the positive charge over the A-ring. From simple chemical concepts it may be predicted that such delocalization will be most effective for C-4 carbocations (see structure (**21**) in Scheme 2) derived from flavan-3,4-diols with phloroglucinol-type A-rings (**18**)–(**20**), intermediate in efficiency for resorcinol-type leuco compounds (**15**)–(**17**), and still less effective for pyrogallol-type melacacidins (**13**) and teracacidins (**14**). These concepts provide a simple rationale for the striking instability of leucocyanidins (**18**), leucodelphinidins (**19**), and leucopelargonidins (**20**), and hence their absence from natural sources containing oligomers derived from them. This contrasts with the stability and wide distribution of the natural 5-deoxy analogues (**13**)–(**17**).

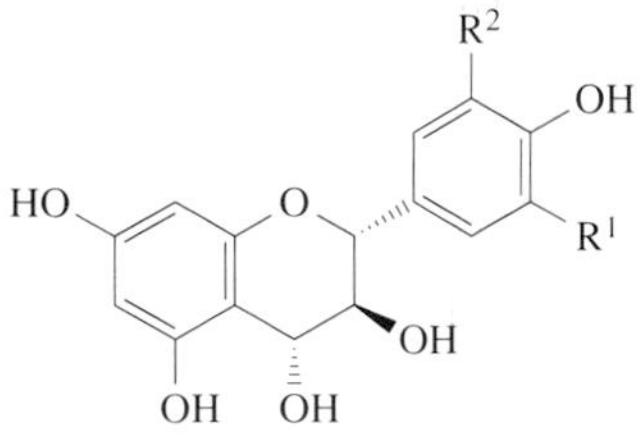

(18) R^1 = H, R^2 = OH
(19) R^1 = R^2 = OH
(20) R^1 = R^2 = H

The potential of the B-ring to contribute towards stabilizing C-4 carbocations of type (**21**) via an A-conformation (**24**) has been overlooked for a long time. First proposed by Brown and Shaw,[56] recognized by the authors on several occassions,[57–62] and formally designated A-conformer by Porter *et al.*,[63] this represents a half-chair/sofa conformation for the pyran ring in which the 2-aryl group occupies an axial (24) as opposed to the "customary" equatorial orientation in the E-conformer (21). The profound effect of the B-ring in additionally stabilizing C-4 carbocations via an A-conformation was strikingly demonstrated by the different rates of condensation observed for leucorobinetinidin (16),[64] mollisacacidin (15),[64] and guibourtacidin (17).[65] Owing to the conformational mobility of the pyran heterocycle, benzylic carbocations of type (21) may hence be additionally stabilized by charge donation from the B-ring (Scheme 2). The more electron-rich pyrogallol function in the leucorobinetinidin carbocations ((21)⇌(24)) is more effective than the pyrocatechol functionality in mollisacacidin analogues ((22)⇌(25)) and the monooxygenated moiety in the leucoguibourtinidin ions ((23)⇌(26)) hence leading to condensation rates decreasing in the order (16) > (15) > (17). A similar donation of B-ring charge may also contribute towards stabilization of the electron deficiency at C-4 of a quinone methide (27), favored by some[66–68] to be the electrophile of choice in oligomeric proanthocyanidin formation.

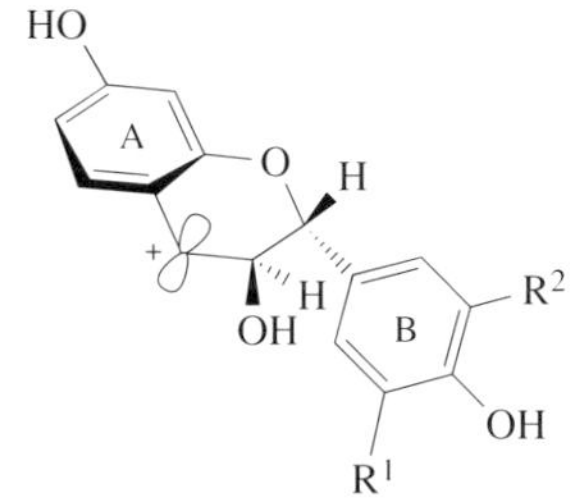

(21) R¹ = R² = OH
(22) R¹ = H, R² = OH
(23) R¹ = R² = H

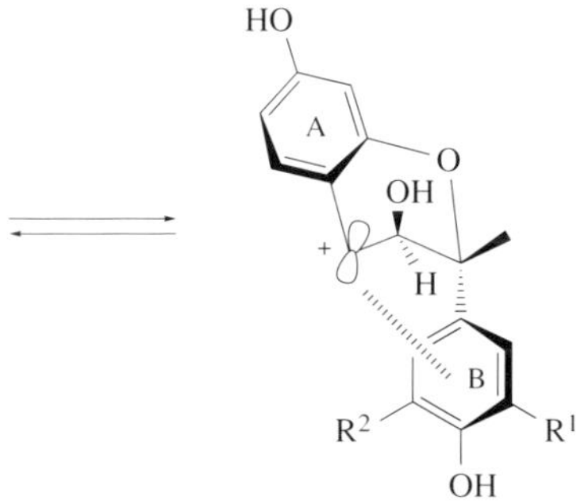

(24) R¹ = R² = OH
(25) R¹ = H, R² = OH
(26) R¹ = R² = H

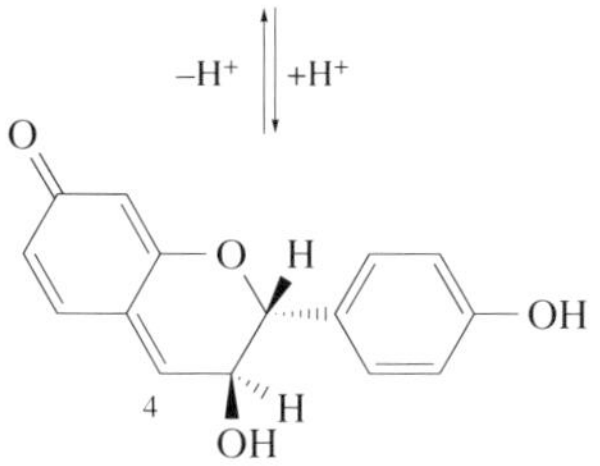

−H⁺ ⇅ +H⁺

(27)

Scheme 2

Peltogynol (28) and mopanol (29), which also possess the potential for C-4 carbocation or A-ring quinone methide formation, are nonreactive under conditions which readily promote aromatic substitutions with flavan-3,4-diols as electrophiles.[69] Forcing conditions are required for promoting condensations of (28) and (29) with nucleophilic phenols and the reactions are characterized by low yields.[69,70] The increased energy requirements for these condensation reactions similarly result from the C-rings of these compounds being restricted to an (E) C-3 sofa conformation of type (21) by the D-ring, hence eliminating contributions by an A-conformer of type (24) towards a decrease in the activation energy.[70]

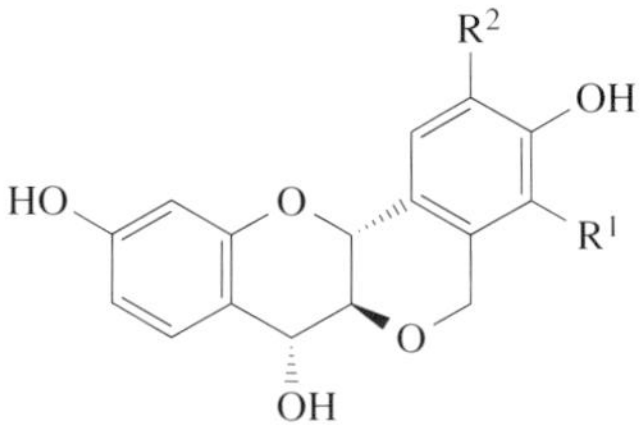

(28) R¹ = H, R² = OH
(29) R¹ = OH, R² = H

The stereochemistry at C-3 and C-4 also influences the reactivity of flavan-3,4-diols as incipient electrophiles. Analogues possessing 4-axial hydroxy groups are susceptible to facile ethanolysis

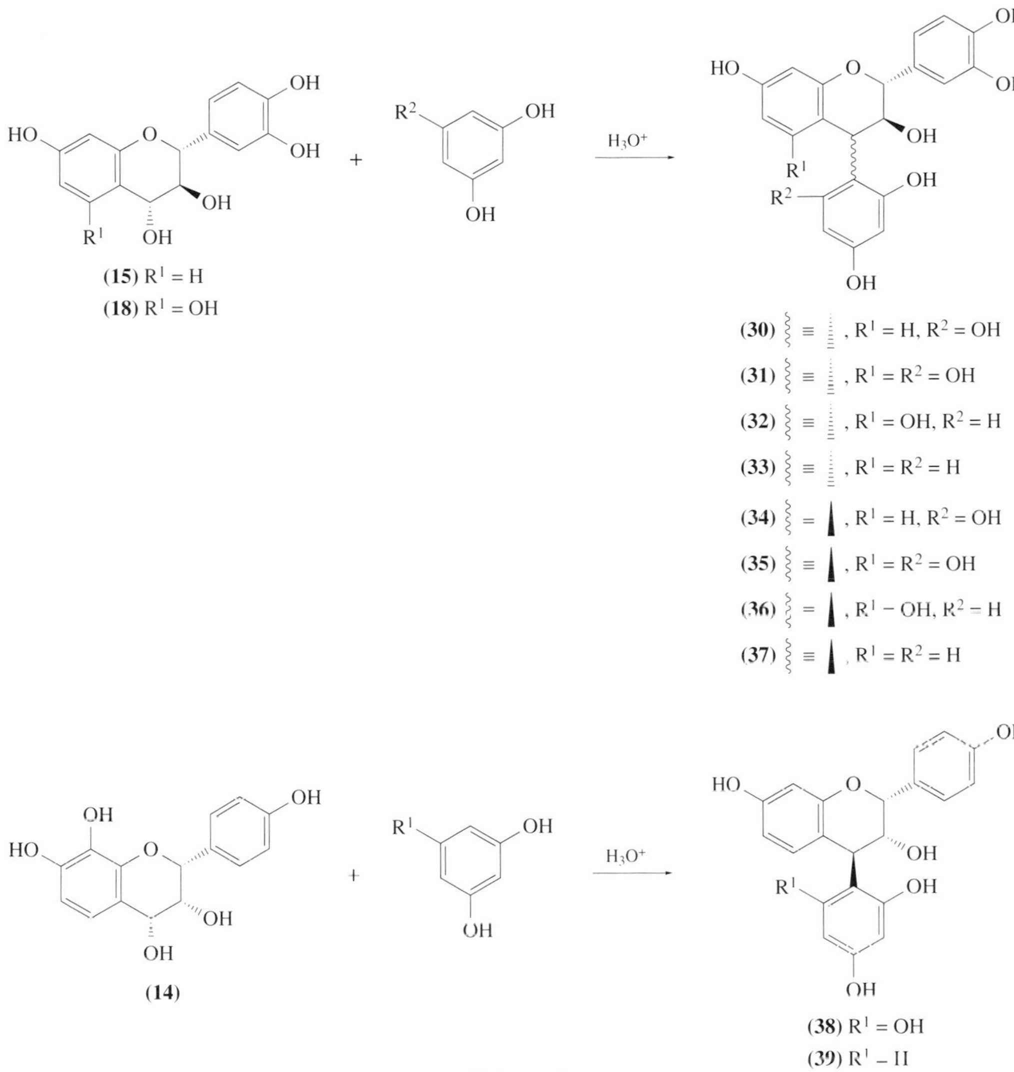

(15) R¹ = H
(18) R¹ = OH

(30) ≡, R¹ = H, R² = OH
(31) ≡, R¹ = R² = OH
(32) ≡, R¹ = OH, R² = H
(33) ≡, R¹ = R² = H
(34) =, R¹ = H, R² = OH
(35) ≡, R¹ = R² = OH
(36) ≡, R¹ = OH, R² = H
(37) ≡, R¹ = R² = H

(14)

(38) R¹ = OH
(39) R¹ = H

Scheme 3

under mild acidic conditions, while those with 4-equatorial hydroxy functions are less prone to solvolytic reactions.[71] Such differences are explicable in terms of the enhanced leaving group ability of the C-4 hydroxy group due to overlapping of the developing *p*-orbital with the π-system of the A-ring.[4,71,72] Axial C-3 hydroxy groups may further stabilize C-4 carbocations by formation of a protonated epoxide intermediate.[4]

The stereochemical course of condensation of flavan-3,4-diols with nucleophilic phenols (phloroglucinol and resorcinol) (Scheme 3) is largely controlled by the configuration of the C-3 hydroxy group and to a lesser extent the C-5 hydroxy group. Thus, substitution at C-4 of flavan-3,4-diols with 2,3-*trans* configuration (e.g., mollisacacidin (15) and its 5-oxy analogue, leucocyanidin (18) both with 2R,3S absolute configuration, proceeds stereoselectively to afford 3,4-*trans*((30)–(33)) and 3,4-*cis*-4-arylflavan-3-ols (34)–(37) in the proportions of ~1.5–2:1.[73,74] By contrast both phloroglucinol and resorcinol are captured with complete stereoselectivity by the carbocation generated from (2R,3R)-2,3-*cis*-3,4-*cis*-teracacidin (14) to give the 2,3-*cis*-3,4-*trans*-4-arylflavan-3-ols (38) and (39) with inversion of configuration at C-4.

Assuming that the carbocationic intermediates possess sofa conformations, nucleophilic attack on the ion with 2R,3R-2,3-*cis* configuration (40) proceeds from the less hindered "upper" side, presumably with neighboring group participation of the 3-axial hydroxy in an E-conformation and by the 2-axial B-ring in an A-conformation. The reaction with a 2,3-*trans* carbocation (41) is directed preferentially from the less hindered "lower" side, that is reaction proceeds with a moderate degree of stereoselectivity. It should be emphasized that the controversy[5] regarding the intermediacy of a

C-4 carbocation, for example (**40**), or an A-ring quinone methide, for example (**27**), in the acid-mediated condensation of flavan-3,4-diols with capture nucleophiles is actually irrelevant to the stereochemical course of the coupling step, since C-4 is in either species sp^2 hybridized with similar heterocyclic ring geometry. The formation of A-ring quinone methide intermediates nevertheless constitutes a viable mechanism for the condensation of 4-substituted flavans over a wide range of pH values.[66–68]

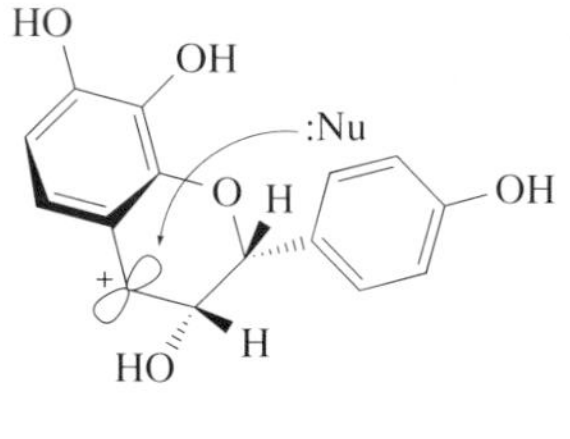

(40) Complete stereoselectivity 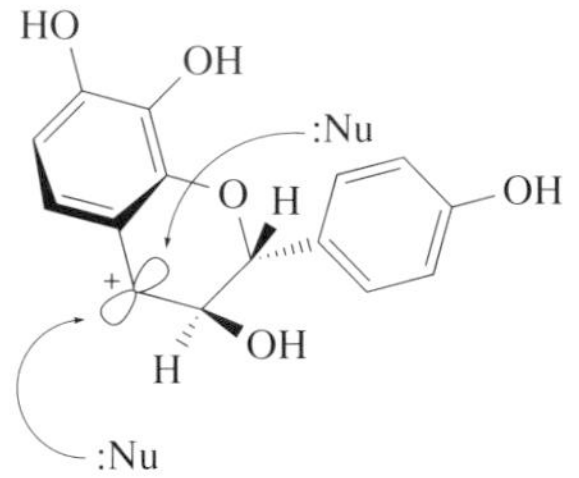   **(41)** Moderate stereoselectivity

The reactions of leucocyanidin (**18**) are more highly directed to the 3,4-*trans*-4-arylflavan-3-ols (**31**) and (**32**)[20,75,76] owing to increased steric constraint by the C-5 hydroxy group. Coupling of leucocyanidin (**18**) and catechin (**2**) afforded the 3,4-*trans*-4-linked dimers with no evidence for any 3,4-*cis* analogues.[77] This result was consistent with the common occurrence of (4α→8)-bis-catechin (procyanidin B-3), catechin-(4α→8)-epicatechin (procyanidin B-4), and their (4α→6)-isomers in lower yields from plants containing procyanidins. However, the synthesis and natural occurrence in low proportions of procyanidins exhibiting 2,3-*trans*-3,4-*cis* linkages have been demonstrated[78–82] (see also Section 3.19.4.3). Principles similar to those advanced here apparently also govern the stereochemistry of reactions of flavan-3,4-diols with sulfur[56,83] and oxygen[84] nucleophiles, and of the solvolysis of (2R,3R)-2,3-*cis*-procyanidin oligomers in the presence of thiols, phloroglucinol, or flavan-3-ols.[85–87] The known naturally occurring flavan-3,4-diols and their derivatives, as well as additional aspects of their properties and chemistry, have been reviewed by Haslam,[4] Porter,[5,10] Hemingway,[6] and Ferreira and co-workers.[8,11]

3.19.3.5   Flavan-4-ols

The flavan-4-ols may be chemically, and presumably also biosynthetically, derived by a single reduction step from a flavanone (Scheme 1). Following the first isolation of 4′,5,7-trimethoxy-2,4-*trans*-flavan-4-ol (**42**) by Lam and Wrang,[88] only a limited number of these metabolites and their glycosides have since been identified and are listed by Porter.[5,10] All the aglycones were found to possess the 2,4-*trans* stereochemistry using ^{1}H NMR analysis[50] for heterocyclic ring proton couplings. Since the majority of the flavan-4-ols coexist with the corresponding flavanones they presumably all are of 2S,4R absolute configuration. The flavan-4-ol glycosides (e.g., erubin A (**43**) and B (**44**)[89] and triphyllin A (**45**) and B (**46**)[90]) in contrast, possess the 2R,4S stereochemistry Gluc = β-D-glucopyranosyl. In both these cases the normal (2S)-flavanone occurred together with the flavan-4-ols. The flavan-4-ols apparently also serve as incipient electrophiles for a limited but growing series of unique oligomers possessing flavan chain-extender units.[36,91–94] These will be dealt with in Section 3.19.4.11.

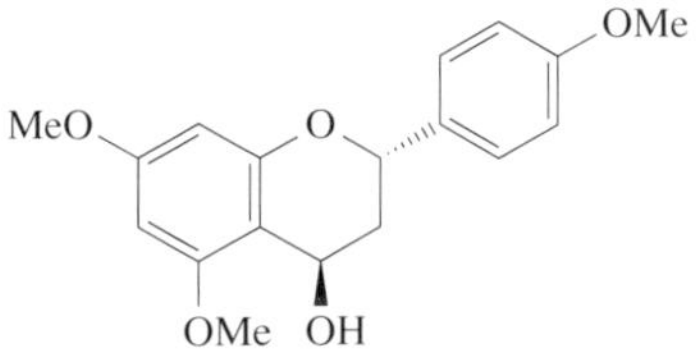

(42) 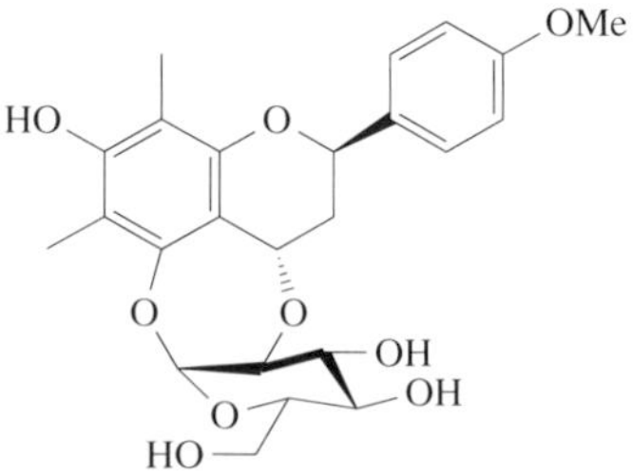   (43)

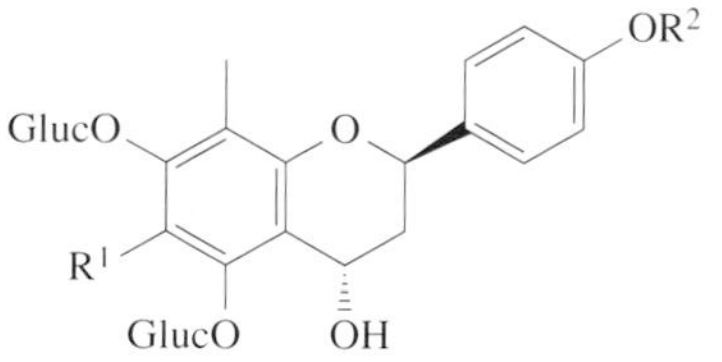

(**44**) R^1 = R^2 = Me
(**45**) R^1 = CH$_2$OH, R^2 = Me
(**46**) R^1 = CH$_2$OH, R^2 = H

3.19.4 OLIGOMERIC PROANTHOCYANIDINS

Following relatively closely upon the initial flavan-3,4-diol chemistry was the recognition of natural oligomeric leucocyanidins by Forsyth and Roberts[95] from the cocoa bean (*Theobroma cacao*) and by Freudenberg and Weinges[96,97] from *Gleditschia triancanthos* and *Crataegus oxyacantha* during the early 1960s. Although the initial structures, involving C—O—C interflavanyl linkages, were eventually disproved, it nevertheless paved the way for the announcement of the first C—C linked leucocyanidin biflavanoids from avocado seed[98] (*Persea gratissima*), cola nuts[99] (*Cola acuminata*), and the strawberry[100] (*Fagaria vesca*).

However, the precise structural details of these biflavanoids were not described. This conformed to the trend at the time when the more significant work on oligoflavanoids was inevitably limited to an analytical approach involving biflavanoids.[101-105] Inhibiting factors have been the complexity of oligoflavanoid extract composition and the consequent problem of their isolation and purification, the lack of a universal method of both synthesis and of assessing the absolute stereochemistry at the point of the interflavan linkage, the need to contend with the phenomenon of dynamic rotational isomerism[101] about interflavanoid bonds during NMR spectral investigations, and the lack of knowledge regarding the points of bonding at nucleophilic centers of the flavan-3-ol chain extender units. Notable exceptions to the above approach were the earlier attempts by Geissman and Yoshimura[106] and by Weinges *et al.*[107,108] to synthesize procyanidin biflavanoid derivatives. However, by their very nature, these synthetic methods would not permit their extension to higher oligomers.

The aforementioned impediments to progress in the chemistry of oligomeric proanthocyanidins necessitated reappraisal of the fundamental principles that control the chemical behavior of this complex group of natural products. The two main developments that contributed most significantly towards understanding the intricate chemistry of these compounds are based upon the acid-catalyzed thiolytic cleavage of the interflavanyl bond(s) in the 5-oxygenated (A-ring) proanthocyanidins, and on the premise that flavan-3,4-diols as potential electrophiles and nucleophilic flavan-3-ols are involved in initiating the condensation that would lead to the formation of oligomers.

3.19.4.1 B-type Proanthocyanidins

Proanthocyanidins of the B-type are characterized by singly linked flavanyl units, usually between C-4 of the flavan-3-ol chain extender unit and C-6 or C-8 of the chain-terminating moiety. The number of oligomeric proanthocyanidins (dimers to pentamers) has steadily increased to more than 300. A compilation of known analogues may be found in Porter,[5,10] Hemingway and co-workers,[6,9] and Ferreira and Becker.[11] In view of the major influence of the semisynthetic approach towards the development of the chemistry of these secondary metabolites this will be addressed in some detail.

3.19.4.2 Semisynthesis of Oligoflavanoids

3.19.4.2.1 Selection of precursors

In those flavanoid metabolic pools which possess the potential for condensed tannin formation, flavan-3,4-diols, (e.g., (**15**)) when considered as *p*-hydroxybenzyl alcohols, represent structural units capable of generating C-4 carbocations, (e.g., (**22**))[100] under mild acidic conditions. These may subsequently be trapped via interaction with the potent nucleophilic centers of the ubiquitous flavan-3-ols, (e.g., (**2**))[100] which usually exhibit *meta*-substituted A-rings. This approach is demonstrated in Scheme 4 for the semisynthesis of oligomeric profisetinidins, where the initial condensation step affords predominantly the fisetinidol-(4α→8)- and -(4β→8)-catechin biflavanoids (**47**) and (**48**), and to a lesser extent also the (4,6)-regiomers (**49**) and (**50**).[75,76] Substitution at the remaining and more potent nucleophilic site of the D-ring compared to that of the A-ring of these biflavanoids by carbocation (**22**) would then lead to the "angular" triflavanoids (**51**)–(**54**).[109,110]

3.19.4.2.2   *Flavan-3,4-diols and flavan-4-thioethers as incipient electrophiles*

The principles governing the role of the flavan-3,4-diols, via their C-4 carbocations (e.g., (**21**)), as a source of the chain extender units in the semisynthetic approach to oligoflavanoids were discussed comprehensively in Section 3.19.3.4.

The C-4 thioethers of flavan-3-ols (e.g., (**62**)) were equally important and fulfilled the decisive role in development of the chemistry of the 5-oxy (A-ring) analogues (procyanidins, prodelphinidins, and propelargonidins) as the incipient electrophiles. In studies initiated by Brown and co-workers[111,112] into the reactions of thioacetic acid with flavan-4-ol model compounds and its possible use as a selective splitting agent for the oligomeric flavanoids of common heather (*Calluna vulgaris*) and applied by others to the same effect,[84,113–117] Haslam and co-workers[85] eventually used phenyl-methanethiol (toluene-α-thiol) in ethanol–acetic acid finally to establish the structures of the pro-cyanidin dimers B-1 (**55**)–B-4 (**58**). The philosophy of such a protocol is summarized for procyanidin B-1 in Scheme 5.

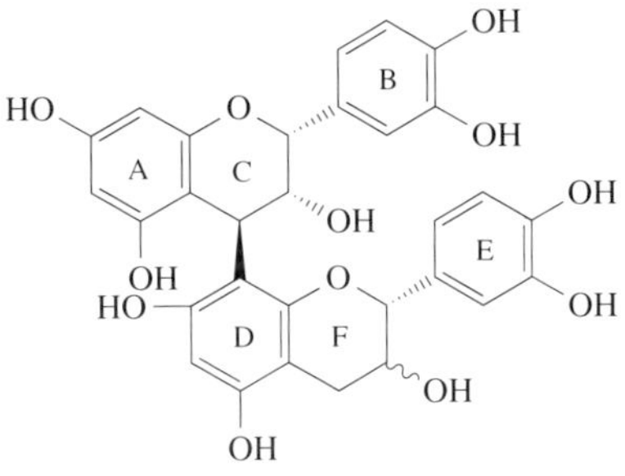

(**55**) ⁅ ≡ ▮
(**56**) ⁅ ≡ ⋮

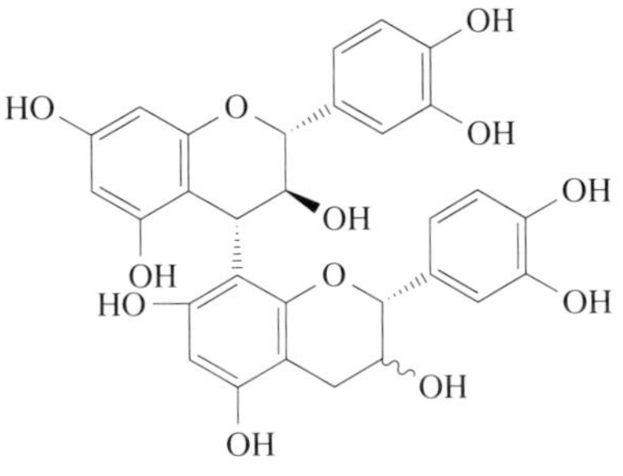

(**57**) ⁅ ≡ ▮
(**58**) ⁅ ≡ ⋮

Thus, protonation of the phloroglucinol D-ring[98] leads to intermediate (**59**) which is susceptible to cleavage of the interflavanyl linkage under the influence of the powerful electron donating A-ring to give the chain-terminating DEF unit, catechin (**2**), and the chain-extender ABC moiety as the C-4 carbocation (**60**)/A-ring quinone methide (**61**). Trapping with the capture nucleophile, phenylmethanethiol, then gives the 4β-benzyl thioether (**62**) in a highly stereoselective manner which may be desulfurized to the epicatechin DEF unit (**3**). Procyanidins with a catechin ABC unit, for example, (**57**) and (**58**), afford a mixture of the 4α- and 4β-benzyl thioethers (**63**) and (**64**).

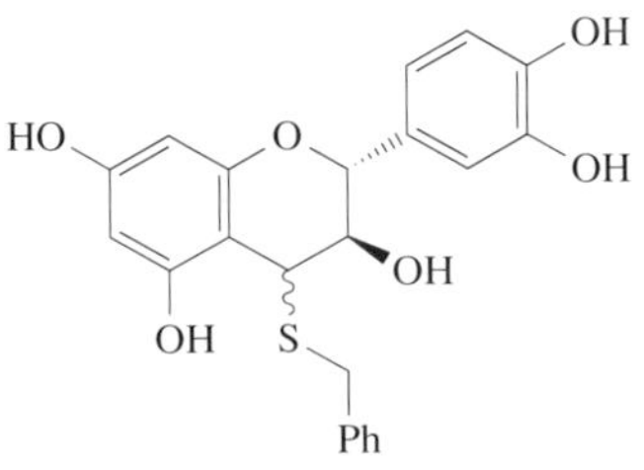

(**63**) ⁅ ≡ ⋮
(**64**) ⁅ ≡ ▮

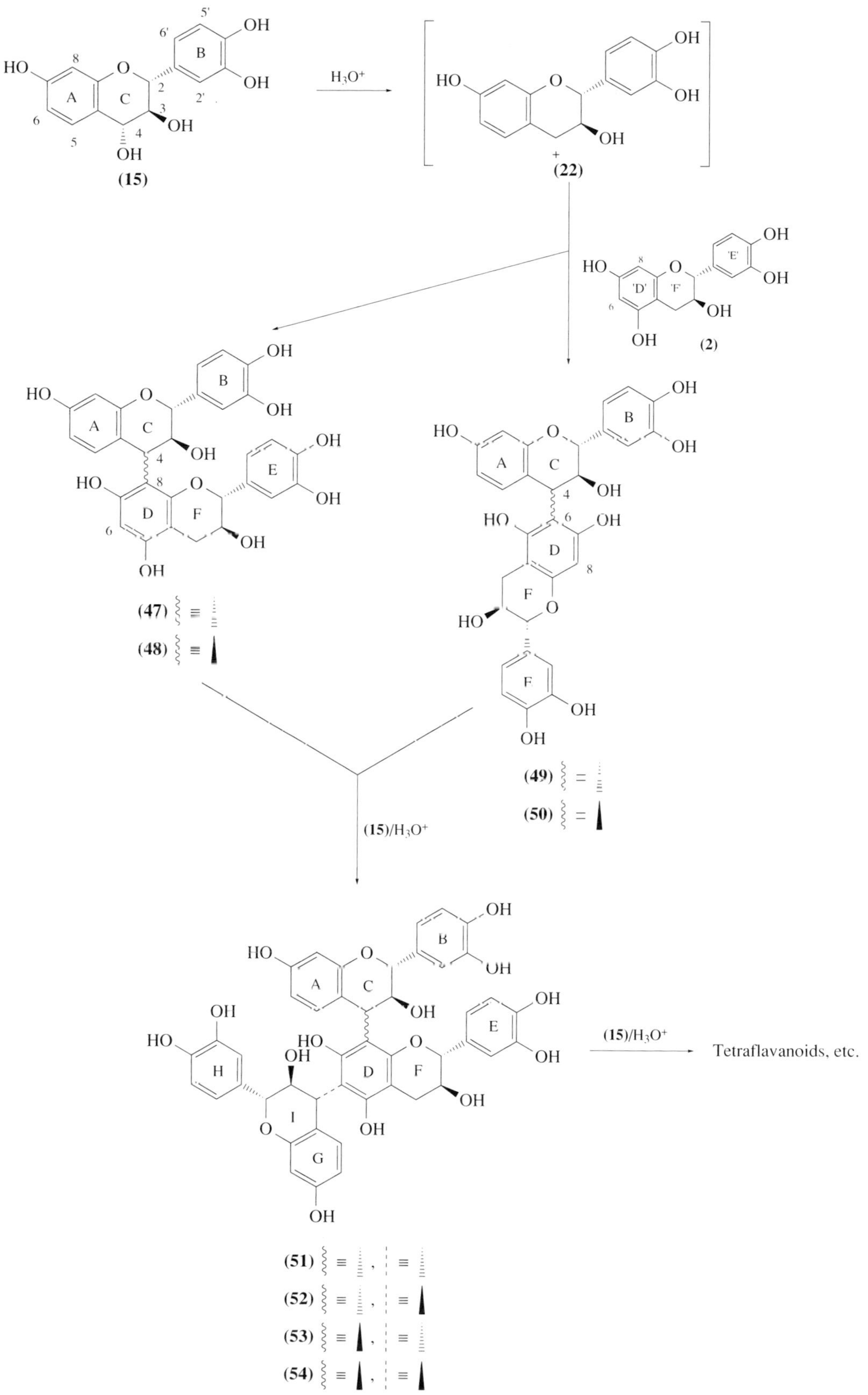

Scheme 4

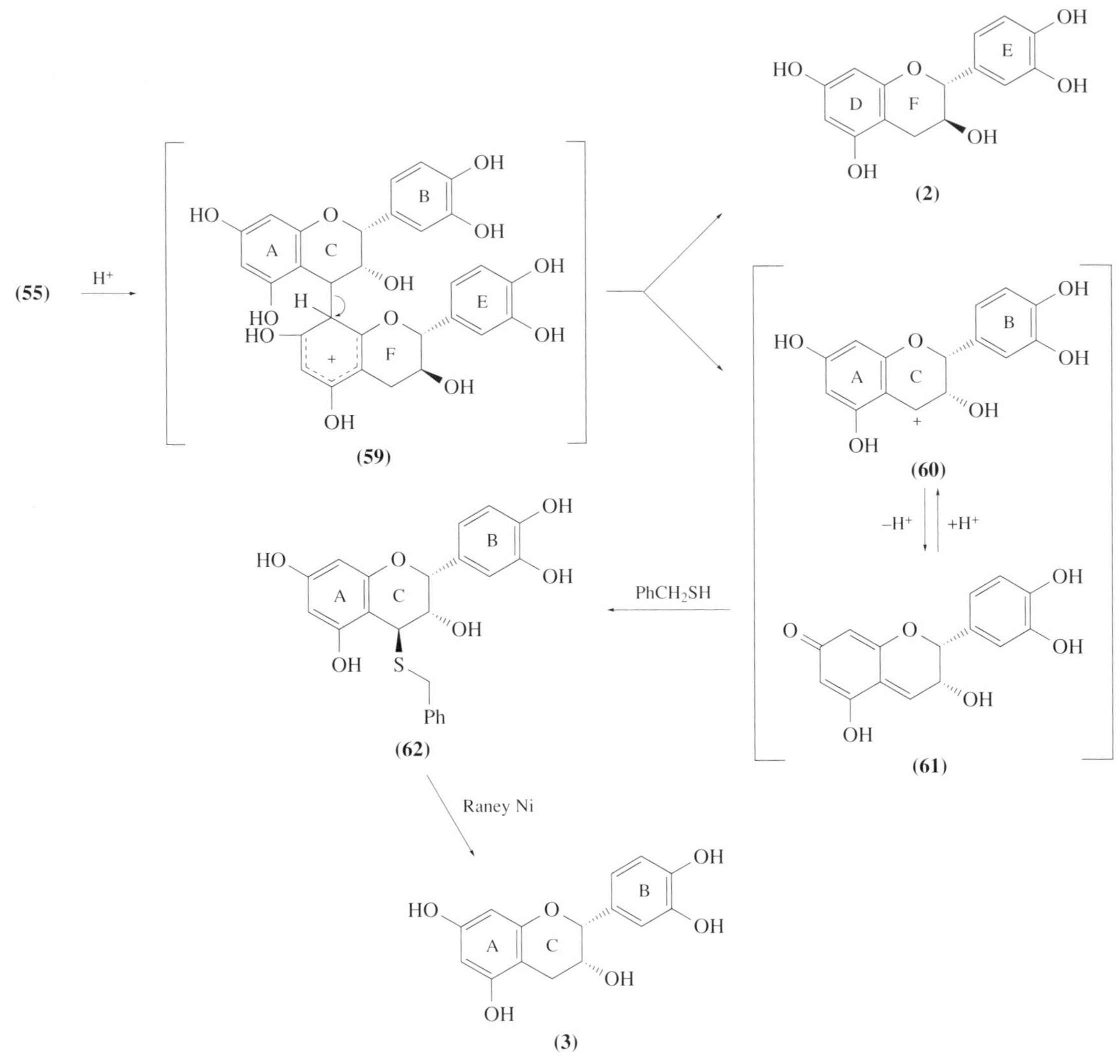

Scheme 5

These flavan-3-ol C-4 benzyl thioethers were used as incipient electrophiles in the synthesis of procyanidin B-1 (**55**) for example, by condensation of (**62**) and catechin under acidic conditions.[118] The thioethers may also be used in the same sense under basic conditions (see Section 3.19.4.3). Haslam and co-workers[85,86] have elegantly used the combined results of the degradative, synthetic, and analytical (especially ^{13}C NMR) techniques to establish unequivocally the structures and absolute configuration of these important procyanidin-type biflavanoids. Earlier, Weinges *et al.*[104] successfully assigned the structures and some of the stereochemical details of procyanidins B-1–B-4 on the basis principally of an examination of the ^{1}H NMR spectra of their decaacetates.

Leucocyanidin (**18**), available via reduction of dihydroquercetin,[20,21,75–77] was eventually also used as the incipient electrophile in the semisynthesis of procyanidin oligomers.

3.19.4.2.3 *Nucleophilic flavanoids*

The biomimetic pool from which oligoflavanoids with C-4(sp^3) — C-6/-8(sp^2) interflavanyl linkages originate presumably contains a variety of potential nucleophilic units. Despite this, the majority of the oligomers contain a chain-terminating unit comprising a nucleophilic C-4 deoxyflavan-3-ol with phloroglucinol A-ring. Most prominent amongst these are catechin (**2**) and epicatechin (**3**), which feature almost ubiquitously, while 5-deoxy analogues with their reduced nucleophilicity such as fisetinidol (**65**), robinetinidol (**66**), and mesquitol (**67**) represent less important "terminal" lower units (see Porter,[5,10] Hemingway,[6] and Ferreira and Bekker[11]).

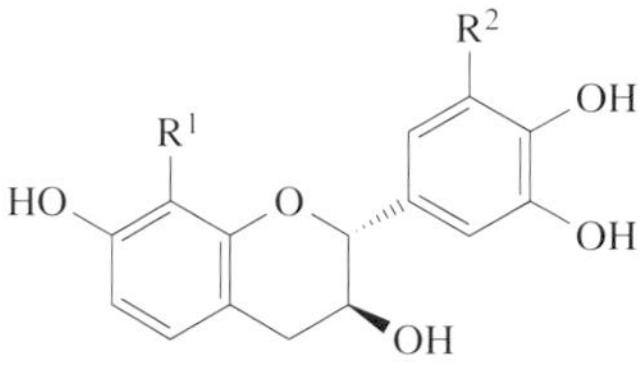

(**65**) R[1] = R[2] = H

(**66**) R[1] = H, R[2] = OH

(**67**) R[1] = OH, R[2] = H

Flavonoids possessing C-4 carbonyl functions exhibit reduced nucleophilicities of their aromatic A-rings. By the same token, the inductive effect of the 4-hydroxy function of flavan-3,4-diols or of the C-4 carbocation resulting from its protonation reduces their innate tendency for self-condensation.[69,119] Examples where the heterocycles of the terminal "lower" units are oxygenated are nevertheless increasing and include dimeric and a trimeric profisetinidins with terminal flavan-3,4-diol function,[67,101,102,119] (e.g., compounds (**68**) and (**69**)), a promelacacinidin (**70**) with an epi-mesquitol 4β-ol terminal unit,[120] profisetinidins with constituent dihydroflavonol units (e.g., (**71**) and (**72**))[121] and a series of nine prorobinetinidins based on flavan-3,4-diol, dihydroflavonol, flavonol (e.g., (**73**)), and a flavone chain-terminating unit from the heartwood of the locust tree (*Robinia pseudacacia*),[122] In this natural source, the flavan-3,4-diol, leucorobinetinidin (**16**), as the incipient electrophile for prorobinetinidin biosynthesis coexists with a variety of monomeric flavonoids invariably possessing C 4 oxygenation. The locust tree therefore represents a rare metabolic pool where oligomer formation has to occur via the action of the very potent electrophile (**16**)[119] on chain-terminating units apparently lacking the nucleophilicity that is associated with natural sources in which oligomeric proanthocyanidin formation is paramount. Such reduced nucleophilicity of the A-ring functionality of flavan-3,4-diols presumably also explains the genesis of the unique series of proanthocyanidins possessing C—O—C interflavanyl linkages[123–125] (e.g., compounds (**74**)–(**77**)). Here, the heterocyclic hydroxy groups apparently serve as nucleophiles to trap the C-4 carbocationic or equivalent intermediates.

3.19.4.2.4 *Bonding positions at nucleophilic centers*

One of the more important problems which hampered progress of the chemistry of oligomeric proanthocyanidins was differentiation between the alternatives of C-4 → C-8 and C-4 → C-6 (cf. structures (**47**) and (**49**)) interflavanoid linkages in those instances where the lower DEF-unit possesses a phloroglucinol-type D-ring. This has led to the development of a solvent shift method for methoxy function by Pelter and co-workers,[126,127] a method that is based on the absolute chemical shift of the "residual" proton on the D-ring by Roux *et al.*,[128–131] and the differentiation of (4→6)- and (4→8)-regiomers of permethylaryl ethers on the basis of nuclear Overhauser effect (NOE) difference spectroscopy.[132] However, the sophisticated pulse sequences of modern NMR spectrometers routinely permit the definition of the bonding position via long-range COSY (homonuclear correlation spectroscopy) and HETCOR (heteronuclear correlation spectroscopy) experiments.

No special attention has been given to referencing the multitude of physical methods that are applicable to condensed tannin structure since these are well documented in the appropriate references, for example, reference to the *Carbon-13 NMR of Flavonoids* by Agrawal may be found in the review by Porter.[10] The same applies for chromatographic methods.

By contrast, bonding to flavan-3-ols with resorcinol-type A-rings (e.g., fisetinidol (**65**)) occurs preferentially at the C-6 position.[75,76] Despite this preference, analogues where C-8 (A-ring) and C-6 (B-ring) of fisetinidol (**65**) and *ent*-epifisetinidol served as nucleophilic centers have been identified. Amongst these are the (4α→8)-linked profisetinidins (**78**) and (**79**),[133] the guibourtinidol-(4α→8)-fisetinidol (**80**),[133] the C → E-ring profisetinidins (**81**) and (**82**),[134] and the guibourtinidol-(4α→6′)-fisetinidol (**83**).[133]

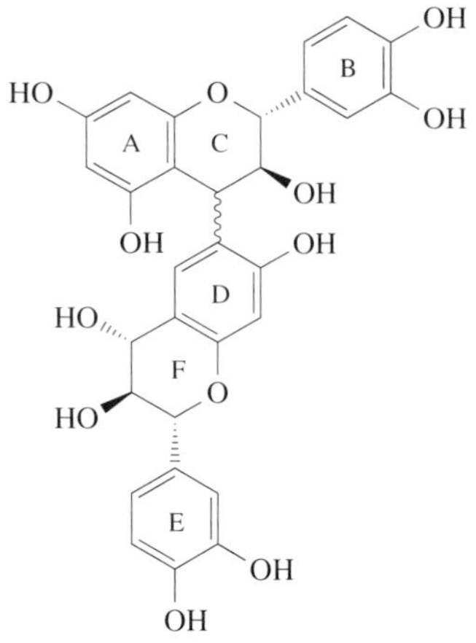

Condensed Tannins

(68) $\}$ ≡ ┊┊┊

(69) $\}$ ≡ ▌

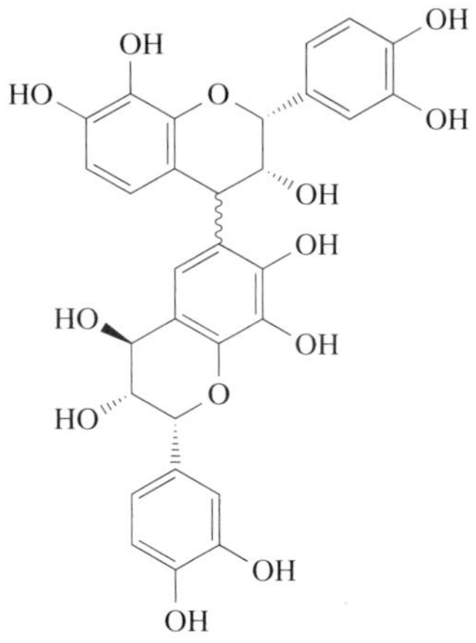

(70)

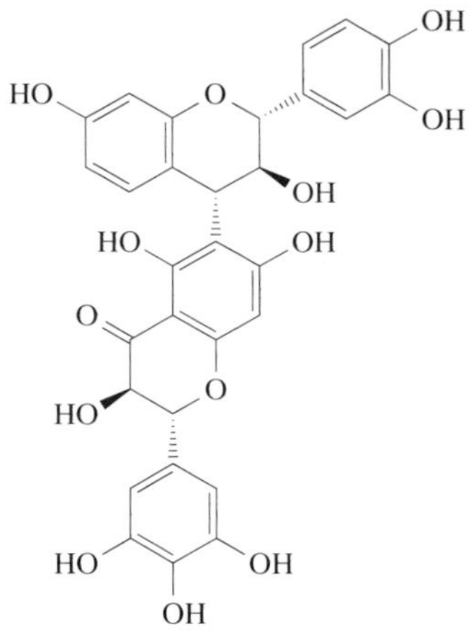

(71) $\}$ ≡ ┊┊┊

(72) $\}$ ≡ ▌

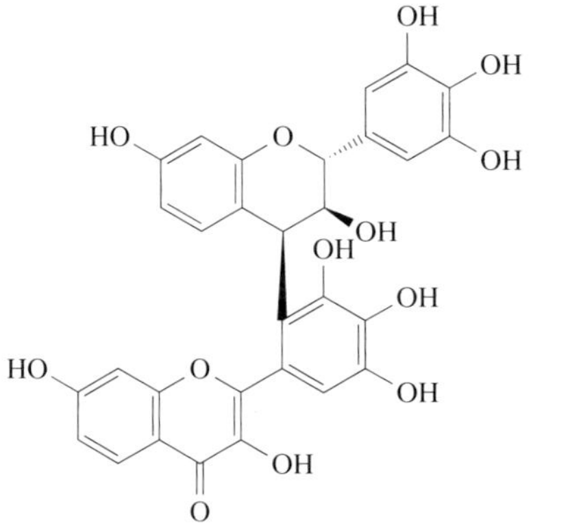

(73)

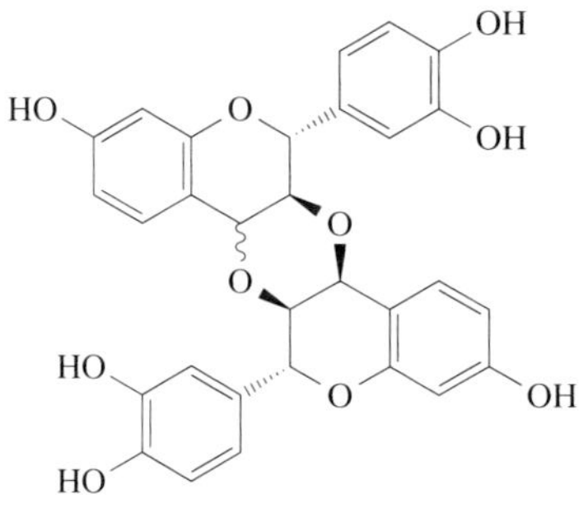

(74) $\}$ ≡ ▌

(75) $\}$ ≡ ┊┊┊

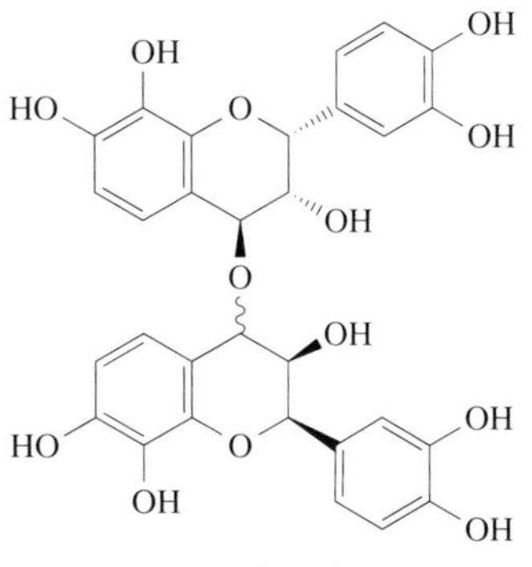

(76) $\}$ ≡ ┊┊┊

(77) $\}$ ≡ ▌

It should be emphasized that the oligoflavanoids exhibiting "abnormal" coupling patterns, for example, (78)–(83), are usually encountered in natural sources which do not possess significant concentrations of flavan-3-ols with phloroglucinol-type A-rings, or which are devoid of flavan-3-ols. The latter situation prevails in the heartwood of *Guibourtia coleosperma* where stilbenes replaced flavan-3-ols as nucleophiles in the biosynthetic sequence leading to a series of guibourtinidol–stilbene oligomers, for example, (84) and (85).[135,136]

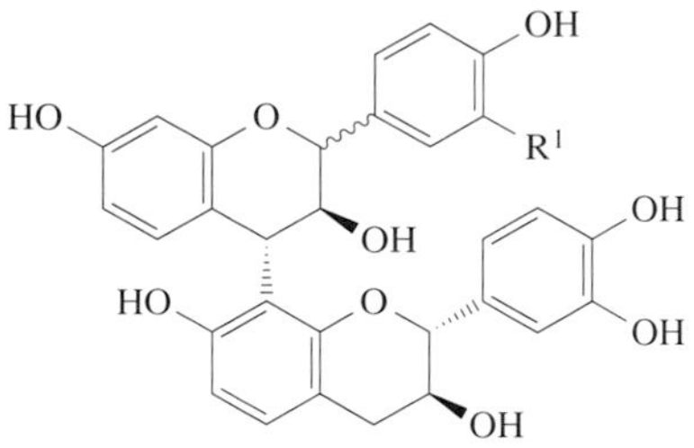

(78) $\zeta \equiv \vdots$, R^1 = OH

(79) $\zeta \equiv$ ▮ , R^1 = OH

(80) $\zeta \equiv \vdots$, R^1 = H

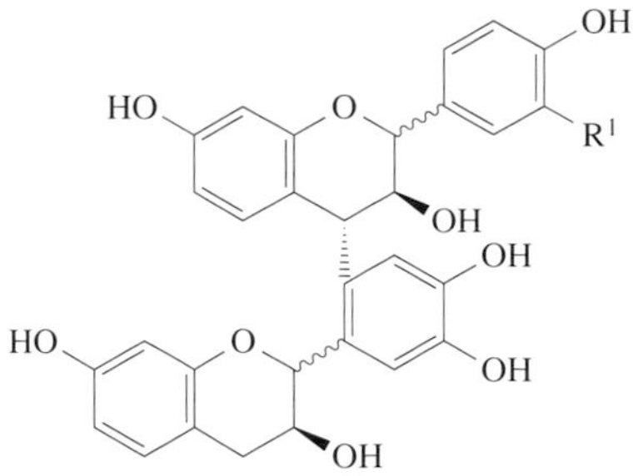

(81) $\zeta \equiv \vdots$, R^1 = OH

(82) $\zeta \equiv$ ▮ , R^1 = OH

(83) $\zeta \equiv \vdots$, R^1 = H

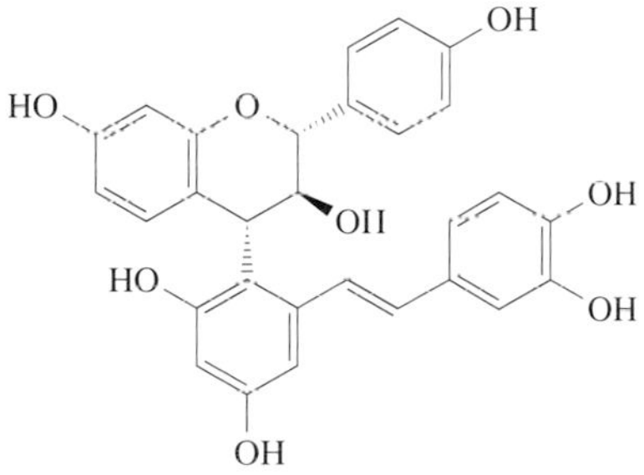

(84)

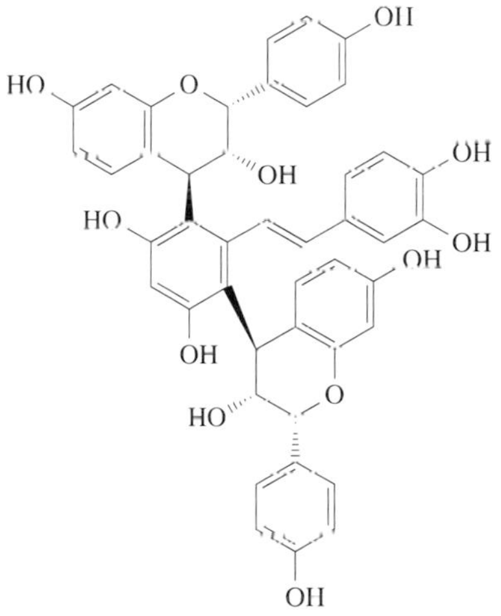

(85)

3.19.4.2.5 *Absolute configuration at C-4 of oligoflavanoids*

A direct method of establishing the absolute configuration at C-4 of oligomeric proanthocyanidins seriously impeded development in this field. Apart from defining the ideal conditions for acid-catalyzed condensation of flavan-3,4-diols and nucleophilic phenolic nuclei, the derivatives of the optically pure 4-arylflavan-3-ols in Scheme 3 offered the opportunity of formulating a chiroptical rule which defines the absolute configuration at C-4 of flavanoid units of this type and hence in biflavanoids and higher oligomers. The c.d. bands of the 4-arylflavan-3-ols and other proanthocyanidins are much more intense than those of their constituent flavan units because of the close proximity of the A- and D-ring chromophores (Snatzke's second chiral sphere)[137] in contrast to the more remote locality of the A- and B-ring chromophores in monomeric flavans (Snatzke's third chiral sphere). Thus, the absolute configuration of the interflavanyl bond could be correlated with the sign of the c.d. band in the 220–240 nm region (probably a ^1L_a transition), a positive sign being correlated with a 4β (86) and negative with a 4α (87) configuration, regardless of the configuration of the rest of the molecule.[73–76,138] The c.d. method supplements a previous indirect method based on ^{13}C NMR chemical-shift differences.[86]

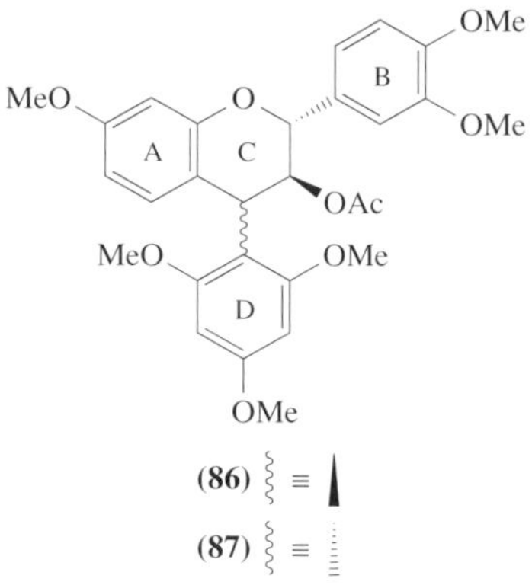

However, 4-arylflavan-3-ol derivatives with a 2,3-*cis*-3,4-*cis* configuration (e.g., (**88**)[139] and its enantiomer)[60] and also some with all-*trans* configurations do not obey the aromatic quadrant rule,[140] hence leading to exceptions to the aforementioned observations. Analogues which do not conform to this otherwise simple rule usually exhibit "abnormal" ¹H NMR coupling constants for the protons of the heterocyclic ring which have been ascribed to boat conformations for these rings. Owing to the high energy requirements,[63] involvement of a boat conformation must, however, be rejected. Such deviations in coupling constants and also the exceptions to the aromatic quadrant rule are more accurately explained in terms of an equilibrium between E- and A-conformers[63] of the heterocycle of 4-arylflavan-3-ols and related compounds (see also Section 3.19.6).

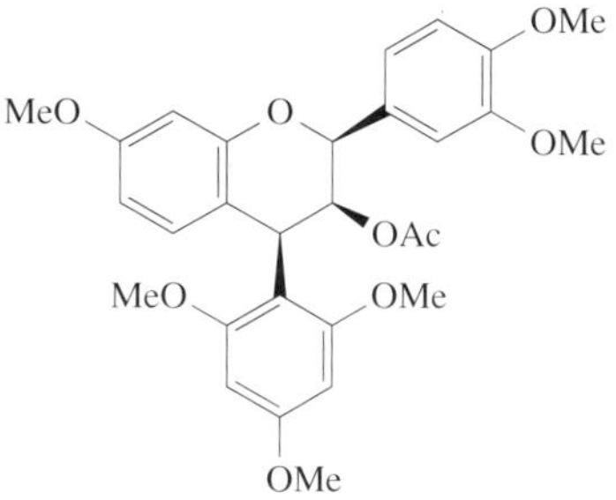

3.19.4.3   Procyanidins (3,5,7,3′,4′-Pentahydroxylation)

The procyanidins, representing one of the most important group of oligomeric proanthocyanidins, are broadly distributed in the leaves, fruit, bark, and less commonly the wood of a wide spectrum of plants.[5–7,10] Analogues with epicatechin chain-extender units [2R,3R-(2,3-*cis*)], for example, procyanidin B-1 (**55**), occur most frequently and invariably possess (4β→8)- and/or (4β→6)-interflavanyl bonds. With a single exception,[82] the procyanidins with catechin (2R,3S-(2,3-*trans*)) chain-extender units (e.g., procyanidin B-3 (**57**)) display 4α-bonds. A growing number of the rare series of procyanidins with *ent*-epicatechin chain-extender units, and hence with (4α→8)- and/or (4α→6)-interflavanoid bonds (e.g., compound (**89**)) have been identified.[5,6,10,11,141] The vast majority of the naturally occurring procyanidins have either catechin or epicatechin as terminal units.[5,6,10,11] A limited number, however, are based on *ent*-epicatechin (e.g., *ent*-epicatechin-(4α→8)-*ent*-epicatechin (**89**))[141] epiafzelechin (e.g., *ent*-epicatechin-(4α→8)-epiafzelechin (**90**)),[142] and *ent*-epiafzelechin (e.g., epicatechin-(4β→8)-*ent*-epiafzelechin (**91**)).[142] Such a preference for catechin and epicatechin constituent units is also reflected at the tri-, tetra-, and pentameric levels where only one exception, that is, the 3-O-galloylepicatechin-(4β→8)-3-O-galloyl-*ent*-epicatechin-(4α→8)-*ent*-epicatechin (**92**), has thus far been documented.[141] A significant number of new entries[5,10,11] are derivatized via, for example, 3,4,5-trihydroxybenzoylation (galloylation) and glycosylation, hence stressing the relevance of the flavan-3-ols exhibiting similar derivatization (see Section 3.19.3.2).

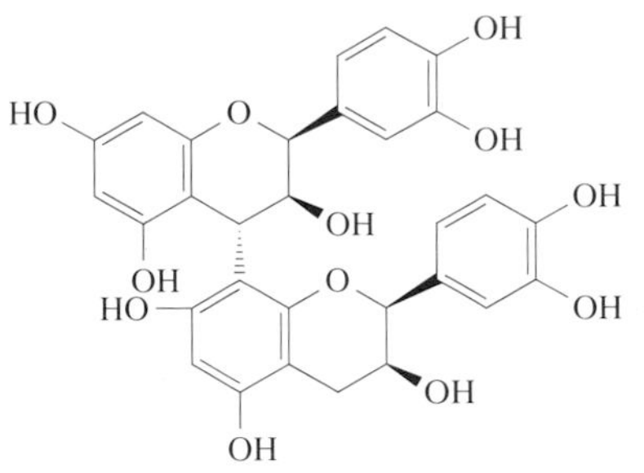

(89)

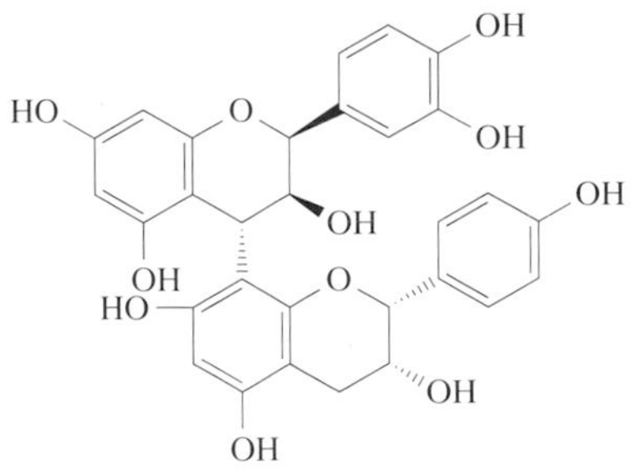

(90)

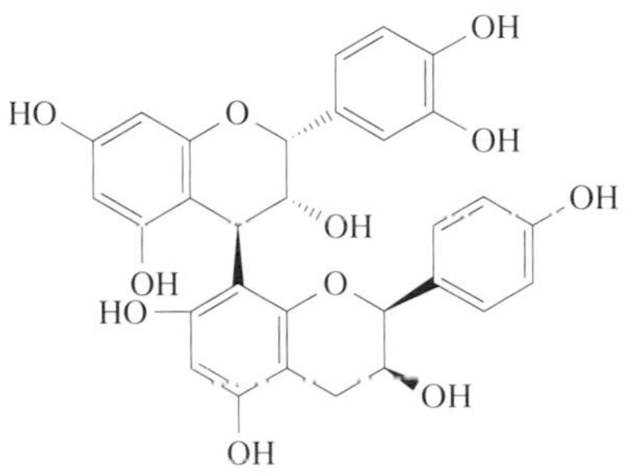

(91)

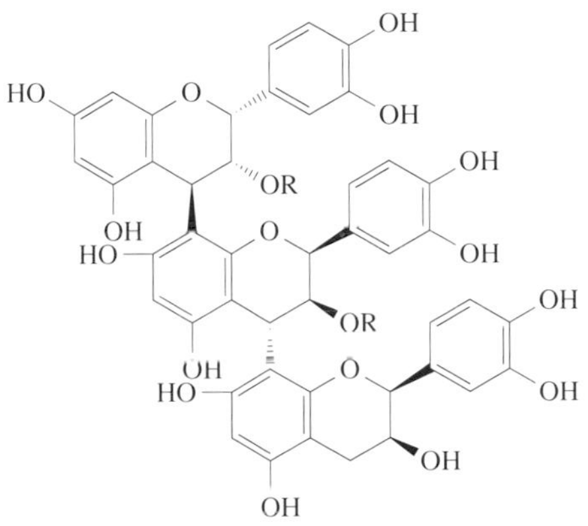

(92) R = galloyl

Whereas leucocyanidin (18) readily serves as the source of (2*R*,3*S*)-2,3-*trans* chain-extender units in the semisynthesis of oligomeric procyanidins,[8,20,21,75,76,77] the absence of a flavan-3,4-diol or dihydroflavonol analogue of epicatechin in natural sources has impeded the semisynthetic approach towards procyanidins with (2*R*,3*R*)-2,3-*cis* constituent units. This problem has been circumvented via acid-catalyzed thiolysis (phenylmethanethiol or benzenethiol)[13,87,143] of polymeric procyanidins with epicatechin chain-extender units, especially those from *Pinus* species, and the subsequent utilization of the C-4β thioethers (e.g., (62)) as electrophiles in condensation with appropriate phenolic nucleophiles.[77,68,86] Conversion of the thioether (62) into the A-ring quinone methide (61) (cf. Scheme 5) under mild basic conditions and subsequent trapping by epicatechin-(4β→6)-catechin (93) enabled Foo and Hemingway[68] to synthesize the first "branched" procyanidin trimer (94) in higher yield than the "linear" analogue (95) (Scheme 6), suggesting that naturally occurring procyanidin polymers may be highly branched,[68] despite the fact that such analogues have not yet been encountered in natural sources. The synthesis of the first procyanidin with a 3,4-*cis* configuration[78,79] has, however, similarly preceded the first recognition[82] of the (4β→8)-bis-catechin in nature. Phloroglucinol offers significant advantages over the use of thiols in the above solvolysis reactions and the resultant epicatechin-(4β→2)-phloroglucinol-type adducts have contributed considerably towards progress in this field.[144–146]

The utility of the aforementioned approaches towards the synthesis of oligomeric procyanidins is, however, limited by the lability of the interflavanyl bond under both acidic[98,147–149] and basic conditions.[66] The ability to carry out condensations in mild alkaline conditions, that is, at pH levels below 9, where the interflavanyl bond is relatively stable, nevertheless presents considerable advantages for the synthesis of procyanidins, for example, "branched" trimer (94).[68]

3.19.4.3.1 *Base-catalyzed rearrangement of flavan-3-ols and of procyanidins in the presence of external nucleophiles*

The facile epimerization of epicatechin (3) to *ent*-catechin (99) and of *ent*-epicatechin (98) to catechin (2) in basic or neutral solution is well established.[150–152] The mechanism proposed by Mehta and Whalley[153] (Scheme 7) proceeding through ionization of the 4'-hydroxy group and B-ring

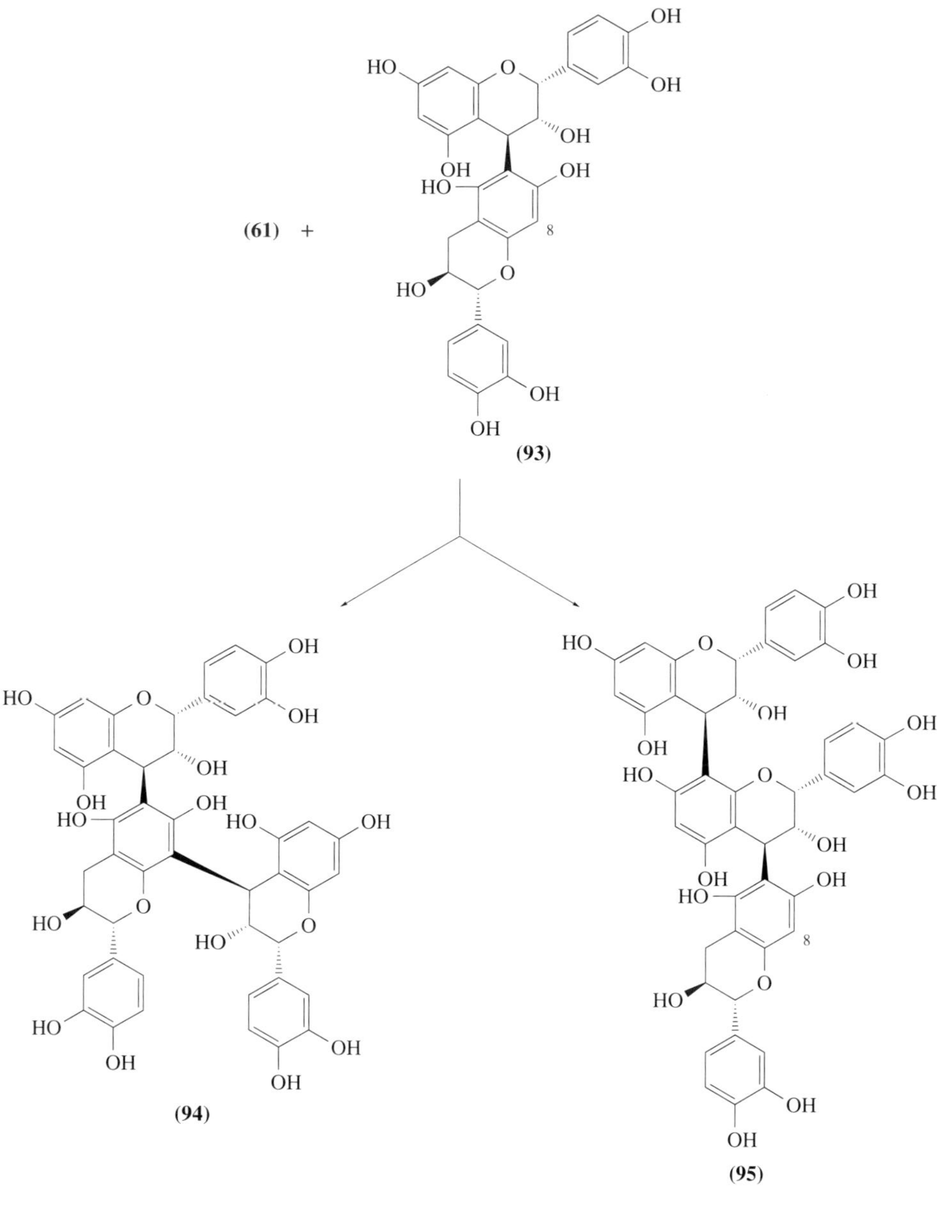

(61)   +

(93)

(94)

(95)

Scheme 6

quinone methide intermediates (**96**) and (**97**) via a reversible Michael addition is supported by the fact that catechin tetra-*O*-methyl ether remains unchanged after prolonged heating in alkaline solution. The quinone methide (**96**) presumably also serves as precursor to the formation of (+)-catechinic acid (**100**)[154,155] via an ionic (two-electron) mechanism through interaction of the equivalent of C-8 (A-ring) and the *si*-face at C-2.

However, Powell and co-workers[156] established that opening of the pyran ring of catechin (**2**) for epimerization or nucleophilic addition is greatly retarded by the total exclusion of oxygen. On these premises it was suggested that the formation of epimerization and rearrangement products at alkaline pH may proceed through a one-electron (radical) mechanism, initiated by the formation of a 4′-oxy radical via autoxidation of catechin.[157,158]

Ent-epicatechin (**98**) and *ent*-catechin (**99**) were synthesized in gram quantities from catechin (**2**) and epicatechin (**3**), respectively, by either brief treatment with a strongly basic solution and rapid quenching of the reaction, or by prolonged heating in a neutral solution.[159] First-order kinetics were observed by Wellons and co-workers[160] for the rates of epimerization of catechin (**2**) and epicatechin (**3**), and for the rate of conversion of catechin (**2**) to catechinic acid (**100**) over the pH range 5.4–

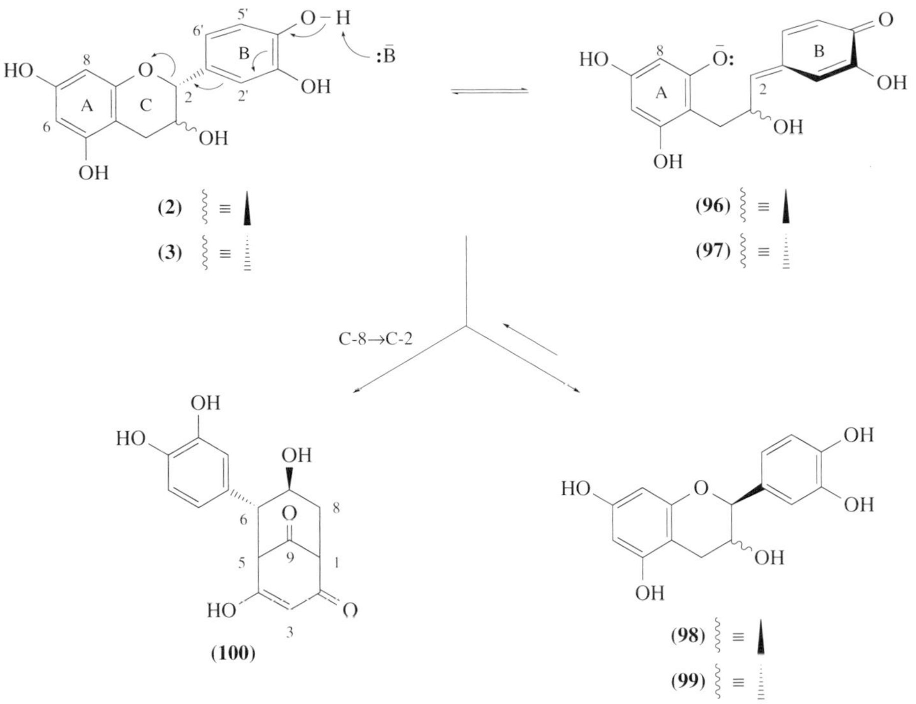

Scheme 7

11.0 and the temperature range 34–100 °C. At low pH, rate coeffient k (epimerization) $\gg k$ (rearrangement), and epimerization approaches an equilibrium in which catechin (**2**) predominates over *ent*-epicatechin (**98**). Near pH 11 and at elevated temperatures, k (epimerization) is only slightly greater than k (rearrangement), and the rapid, irreversible formation of catechinic acid under these conditions determines the product composition.

Owing to the fact that many of the industrial applications of polymeric proanthocyanidins involve their dissolution and/or reaction at alkaline pH,[161,162] their base-catalyzed transformations have received considerable attention, especially by Hemingway and co-workers.[163,164] The interflavanoid bond of the proanthocyanidins with phloroglucinol A-rings is extremely susceptible to cleavage under mild alkaline conditions.[66,67] At pH 12 and ambient temperature, in the presence of an excess of phenylmethanethiol as capture nucleophile, procyanidins, (e.g., B-1 (**55**)) are subject to rapid base-catalyzed cleavage of both the interflavanyl and ether linkage of the pyran ring to form mono- or dibenzyl sulfide derivatives (**103**) and (**109**) via the intermediates indicated in Scheme 8.[163] Base-catalyzed loss of phenylmethanethiol and tautomeric rearrangement led to the formation of the propan-2-one derivatives (**104**) and (**110**), the former compound eventually serving as the precursor of the indan derivative (**105**). The predominant formation of the 1,3-bisbenzylthio-propan-2-ol (**109**) implies that interflavanoid bond cleavage precedes opening of the pyran ring, genesis of the propanone derivative (**110**) being explicable in terms of stereoselective reactions at C-4 and C-2 of the quinone methides (**106**) and (**108**) formed sequentially from the upper 2,3-*cis*-flavan-3-ol unit of procyanidin B-1 (**55**).

The transformation of the mono- and dibenzyl sulfide derivatives (**103**) and (**109**) to the propan-2-one derivatives (**104**) and (**110**) via the appropriate enolic intermediates[163,164] represents a chemical analogy for Haslam's proposed flav-3-en-3-ol intermediate[17,165,166] in the biogenesis of 2,3-*cis*-procyanidins. Such an intermediate would permit inversion of the absolute configuration at C-3 in the reduction products of (2R,3R)-2,3-*trans*-dihydroquercetin.

The A- and B-ring quinone methides (**106**) and (**102**) also feature prominently in the base-catalyzed reactions of polymeric procyanidins (e.g., (epicatechin)$_n$-(4β→8)-catechin) with phloroglucinol as external capture nucleophile at ambient temperatures to initiate the formation of "complex" catechinic acid derivatives.[13,167,168] These results assist in explaining the low aldehyde reactivity and the acidity of polymeric procyanidins that have been extracted from plant tissue or reacted at alkaline pH. Although these reactions are detrimental to the reactivity of the polymeric proanthocyanidins extractable from conifer tree barks in applications such as their use in wood adhesives, alkaline

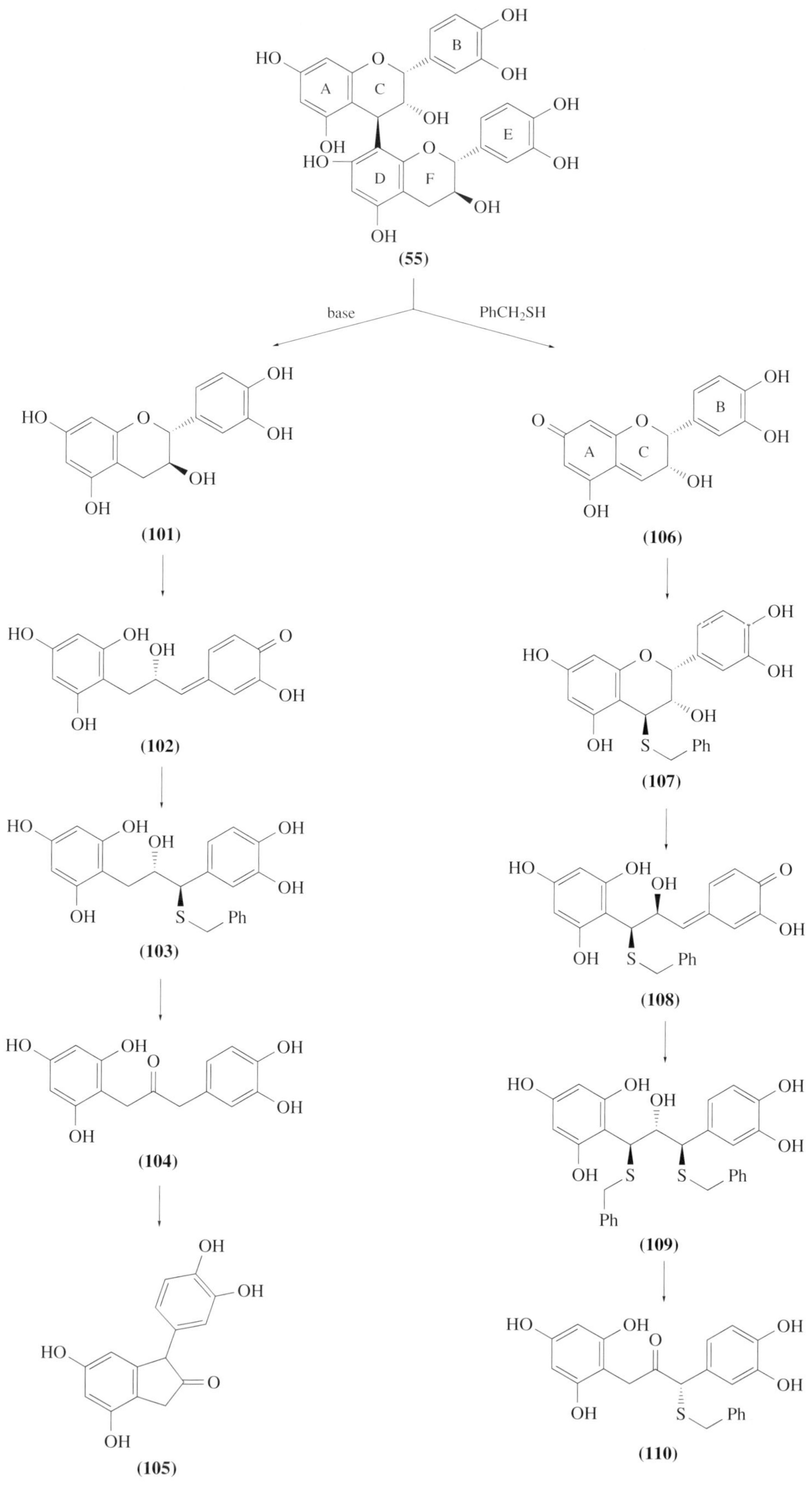

Condensed Tannins

Scheme 8

solutions can be used, even in cold-setting phenolic adhesive systems, provided that these rearrangement reactions are properly controlled.[161,162]

In the absence of external capture nucleophiles, the course of the base-catalyzed transformation of dimeric procyanidins differs substantially from that described in Scheme 8. The principles controlling the chemistry of the ensuing unique pyran ring rearrangements will be dealt with in Section 3.19.4.6.2 and were demonstrated for procyanidin B-2 (**56**)[169] and B-3 (**57**).[170]

3.19.4.4 Prodelphinidins (3,5,7,3′,4′,5′-Hexahydroxylation)

Even though the prodelphinidins are represented in the polymeric proanthocyanidins of a broad spectrum of plants,[7] their numbers are limited compared with those of the procyanidins. The known naturally occurring prodelphinidins and their derivatives, for example, 3-*O*-gallates, have been reviewed by Porter[5,10], Hemingway,[6] and Ferreira and Bekker.[11] A notable feature of these compounds is that they often occur as "mixed" oligomers, for example, as the "mixed" procyanidin–prodelphinidin trimer (**111**).[171] Prodelphinidin gallates also constitute the principal proanthocyanidins in green tea.[172]

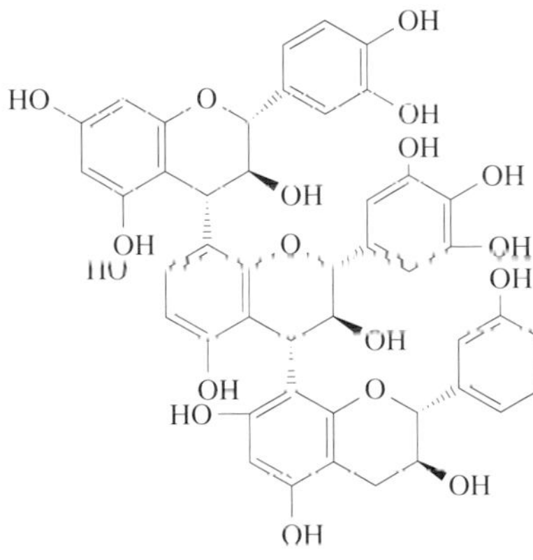

(**111**)

Application[173] of the acid-catalyzed thiolytic cleavage to gain insight into the structure of the polymeric proanthocyanidins from pecan nut pith, known to be comprised of epigallocatechin, gallocatechin, and epicatechin chain-extender units in the approximate ratios of 5:2:1 with either catechin or gallocatechin as terminal units, consistently afforded significant amounts of phloro glucinol and a mixture of 1,3-dithiobenzyl-2,4,5,6-tetrahydroxyindane diastereomers (**118**). Such a conversion is demonstrated in Scheme 9 for a typical prodelphinidin (**112**) with 2,3-*cis* configuration of the chain-extender units. Thiolytic cleavage of the prodelphinidin (**112**) gives the 4β-thiobenzyl ether (**113**) which is protonated at the electron-rich phloroglucinol A-ring to afford intermediate (**114**) with a labile C-4—C-10 bond which then ruptures under the influence of the electron-donating thiobenzyl group. This process is unique and represents the equivalent of the cleaving of the interflavanyl bond under acidic conditions, but under the influence of an external sulfur nucleophile. Rearrangement of the intermediate sulfonium ion (**115**) leads to the formation of the indane diastereomeric mixture (**116**) with its labile benzylic ether linkage which is cleaved, with the release of phloroglucinol, to carbocation (**117**). Reaction of the latter with the capture nucleophile phenylmethanethiol then affords the mixture of 1,3-dithiobenzyl-2,4,5,6-tetrahydroxyindane diastereomers (**118**). These results invalidate the use of extended thiolysis to provide meaningful estimates of the molecular weight of polymeric proanthocyanidins. It also calls into question the use of thiolysis as a means of obtaining quantitative information on the composition of mixed proanthocyanidin polymers.

3.19.4.5 Propelargonidins (3,5,7,4′-Tetrahydroxylation)

The propelargonidins[7] constitute a smaller group of 5-oxygenated (A-ring) proanthocyanidins‧ and are well represented in the fruit and leaves of *Cassia fistula*,[80,126] for example, *ent*-epiafzelechin-(4α→8)-epiafzelechin (**119**). They also feature in a few mixed procyanidin–propelargonidin oligo-

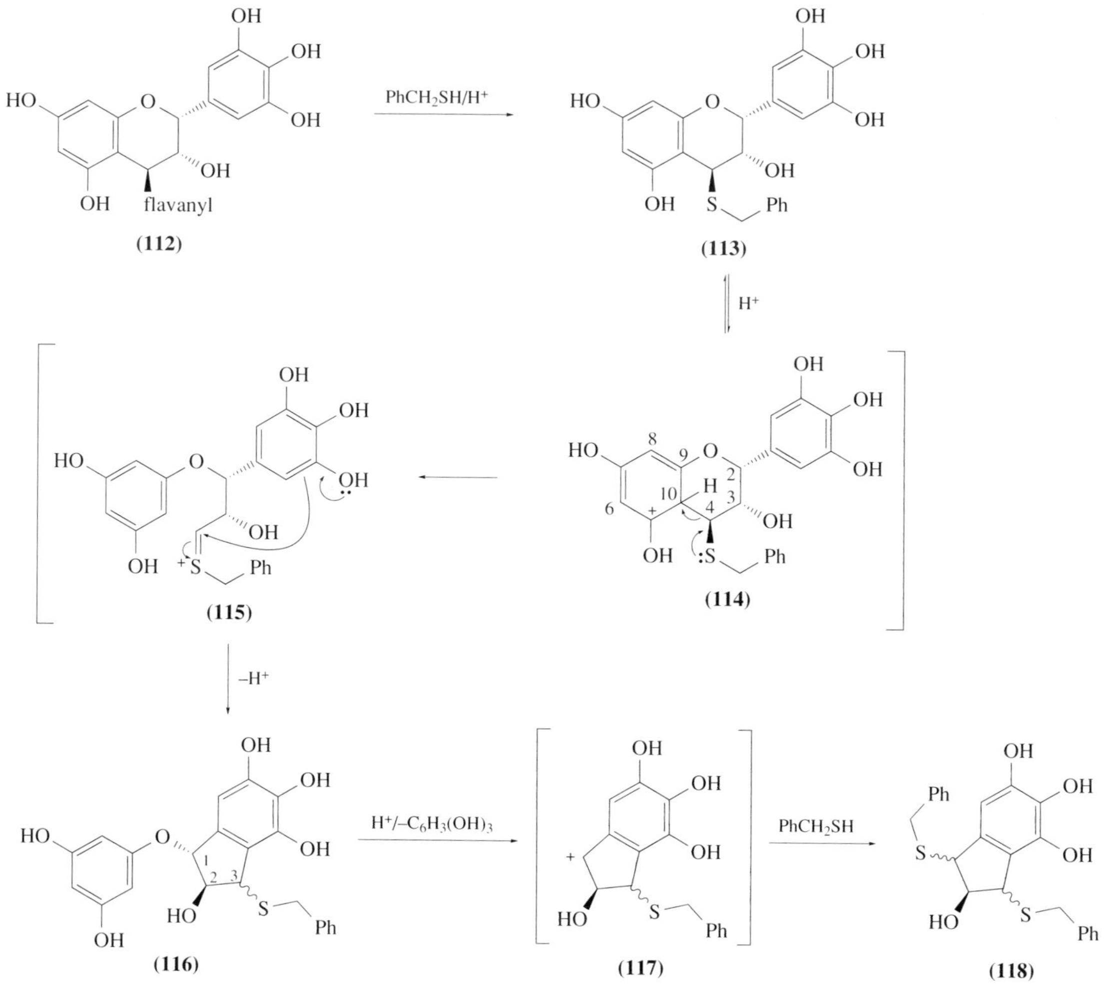

Scheme 9

mers,[174,175] and quite prominently as constituent units in A-type proanthocyanidins (see Section 3.19.4.12). The naturally occurring analogues have been listed by Porter,[5,10] Hemingway,[6] and Ferreira and Bekker.[11]

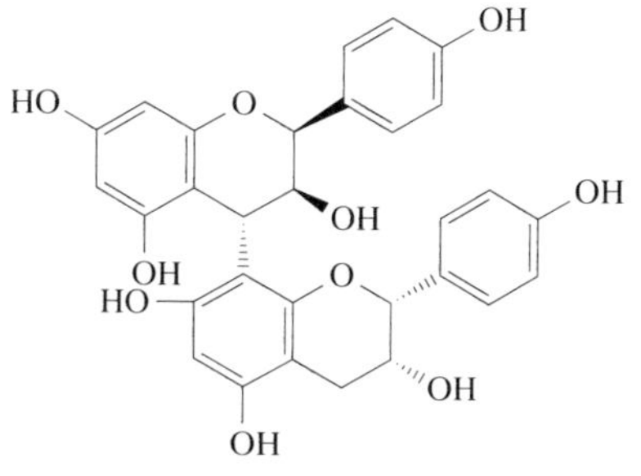

(119)

3.19.4.6   Profisetinidins (3,7,3′,4′-Tetrahydroxylation)

The profisetinidins are the most important polymeric proanthocyanidins of commerce, representing the major constituents of wattle (*Acacia mearnsii*) and quebracho (*Schinopsis* species). They occur widely among the *Acacia* species[176] and are arguably the most comprehensively studied group of oligomeric proanthocyanidins.[3-11] The profisetinidins from the heartwoods of the *Acacia* species usually contain analogues with (2*R*,3*S*)-2,3-*trans*-fisetinidol chain-extender units (e.g., (**47**)), while those from *Schinopsis* and *Rhus* contain the (2*S*,3*R*)-2,3-*trans-ent*-fisetinidol constituent units. The structures of the vast majority of these secondary metabolites have been rigorously established via spectroscopic and semisynthetic methods[179-183] according to the protocol depicted in Scheme 4.

Catechin represents the predominant chain-terminating moiety of the oligomeric profisetinidins. A considerable number that are based on fisetinidol,[133] *ent*-epifisetinidol,[184] guibourtinidol,[133,134] afzelechin,[134] epicatechin,[182,183] dihydroflavonols,[121] and robinetinidol,[121] for example, fisetinidol-(4β→6)-robinetinidol (**120**), have been identified. The synthesis[121] of compound (**120**) represented an interesting variation on the general theme since it offered the first opportunity of establishing the course of coupling between a flavan-3,4-diol and a flavan-3-ol in which the nucleophilicity of the B-ring is comparable to that of the resorcinol-type A-ring. Acid-mediated coupling of mollisacacidin (**15**) and robinetinidol afforded the fisetinidol-(4β→6)- and -(4α→6)-robinetinidols (**120**) and (**121**), and indeed also the unique fisetinidol-(4α→2')-robinetinidol (**122**). The latter compound represents the first *in vitro* example where the B-ring of the flavan-3-ol competes as nucleophile with the resorcinol A-ring in coupling with the flavan-3,4-diol-derived C-4 carbocationic intermediate. All efforts to induce coupling at C-8 (A-ring) or C-6 (B-ring), that is, to synthesize compounds like (**78**)–(**80**) and (**81**)–(**83**), respectively, have hitherto failed.[133,134]

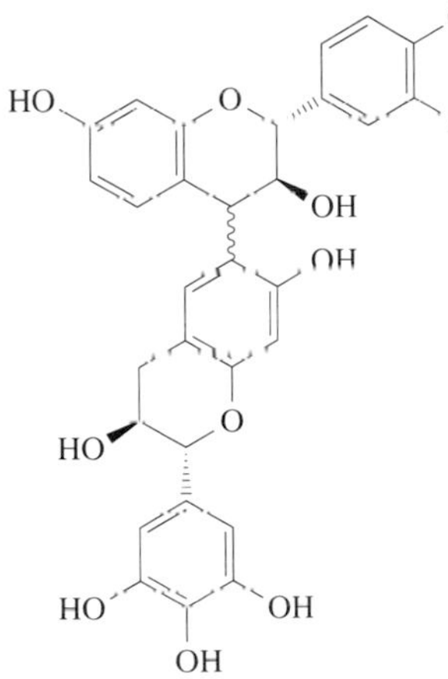

(**120**)   ⟨ ≡ |

(**121**)   ⟨ ≡ ⋮

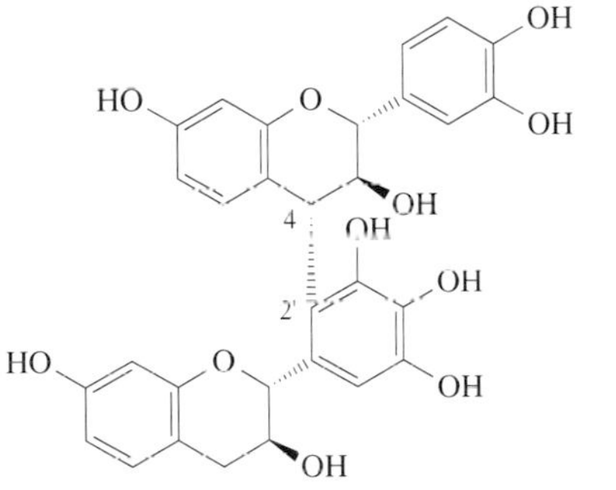

(**122**)

3.19.4.6.1 *Reductive cleavage of the interflavanyl bond in profisetinidins*

In contrast to the readily occurring cleavage of the interflavanyl bond in proanthocyanidins which exhibit C-5 oxygenation of the A-ring of their chain-extender units, this $C(sp^3)$—$C(sp^2)$ bond in the 5-deoxy series of compounds (e.g., the fisetinidol-(4→8)- and -(4→6)-catechin profisetinidins (**47**)–(**50**)) is remarkably stable and has hitherto resisted all efforts at cleavage in a controlled fashion. Such a stable interflavanyl bond has adversely affected both the structural investigation of the polymeric proanthocyanidins in black wattle bark and of those from other commercial sources, as well as the establishment of the absolute configuration of the chain-terminating flavan-3-ol moiety in the 5-deoxyoligoflavanoids. It has been demonstrated that the interflavanyl bond in both the procyanidins and profisetinidins and their permethylaryl ethers are readily subject to reductive cleavage with sodium cyanoborohydride, Na(CN)BH$_3$, in trifluoroacetic acid at 0 °C.[185,186]

Separate reduction of profisetinidin biflavanoids (**47**), (**48**), and (**49**), with varying interflavanyl bond strengths, afforded the chain-terminating flavan-3-ol unit, catechin (**2**), and the 1,3-diaryl-propan-2-ol (**127**) from reductive cleavage of the C-ring of the chain-extender unit (Scheme 10). Protonation of the electron-rich phloroglucinol D-ring in profisetinidins (**47**) and (**48**) and con-comitant delivery of the equivalent of a hydride ion at C-2 of the C-ring of intermediate (**123**) effect the concurrent rupture of the pyran C-ring and of the C-4—C-8 bond to give catechin (**2**) and the

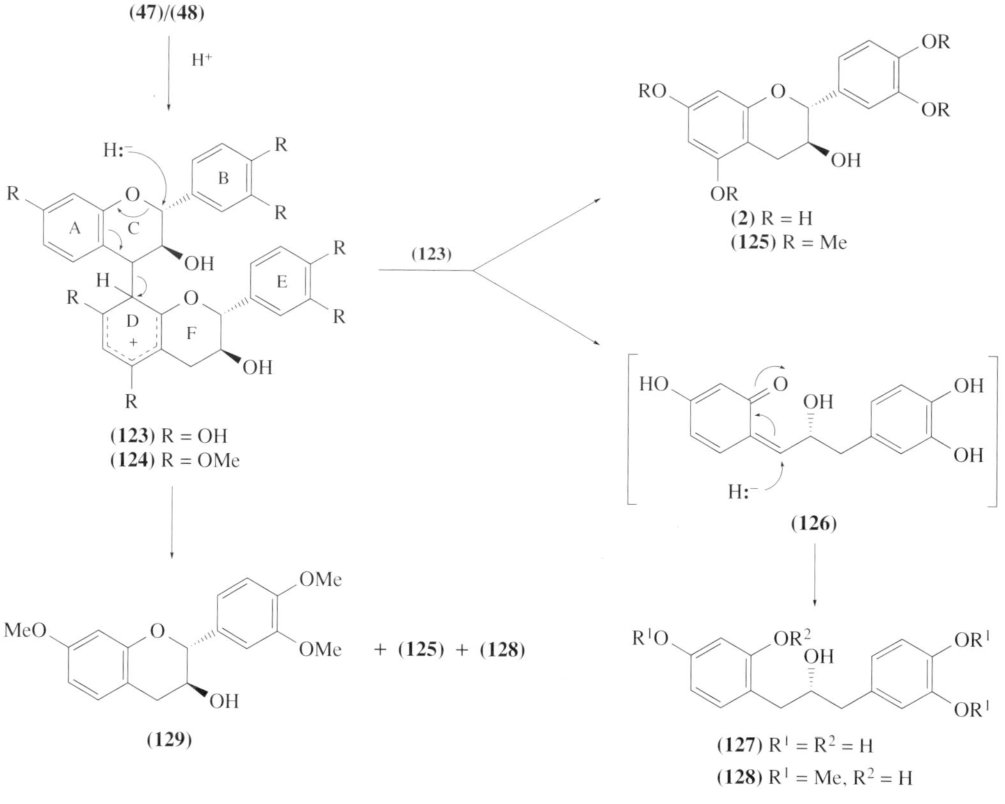

(47)/(48)

Scheme 10

o-quinone methide intermediate (126) which is subsequently reduced to the 1,3-diarylpropan-2-ol (127). Similar reduction of the permethylaryl ether afforded tetra-*O*-methylcatechin (125), the 1,3-diarylpropan-2-ol (128), and tri-*O*-methylfisetinidol (129) via intermediate (124). The mechanism for bond cleavage was rigorously corroborated using Na(CN)BD$_3$ and a variety of other profisetinidin biflavanoids.[185,186] The mild conditions effecting simple cleavage of the strong interflavanyl bonds in the profisetinidins (47)–(49) also cause rupture of the same bonds in procyanidins B-1 and B-3 (55) and (57) and of their permethylaryl ethers, without the concomitant "opening" of the C-ring.

3.19.4.6.2　Base-catalyzed pyran ring rearrangement of oligomeric profisetinidins

The natural occurrence and synthesis of a novel class of C-ring isomerized oligomeric flavanoids, termed phlobatannins, was demonstrated in the mid-1980s.[187,188] These 3,4,9,10-tetrahydro-2*H*,8*H*-pyrano[2,3-*f*]chromenes (e.g., (134)) are characterized by the "liberated" resorcinol moieties from the A/C-ring arrangement of the parent biflavanoid (e.g., (47)) and by the conspicuous absence of the effects of dynamic rotational isomerism in the ¹H NMR spectra of their permethylaryl ether diacetates at ambient temperatures.

Initial identification of the pyran-ring rearranged profisetinidins was followed by recognition of additional members of this class of oligoflavanoids from the heartwoods of *Colophospermum mopane*,[133,189–195] *G. coleosperma*,[57,58,190,191] *Baikiaea plurijuga*,[57,58,191,196] *Julbernardia globiflora*,[196] and the commercially important extract of the bark of *A. mearnsii*.[197] Since the usual methods of differentiating regioisomeric bi- and triflavanoids and the establishment of absolute configuration are less reliable for the phlobatannins, a concise synthetic protocol was developed to establish the complex structures of these natural products. The principles of the proposed route for the formation of tetrahydropyrano[2,3-*f*]chromenes from fisetinidol-(4→8)-catechin profisetinidins are summarized in Scheme 11.

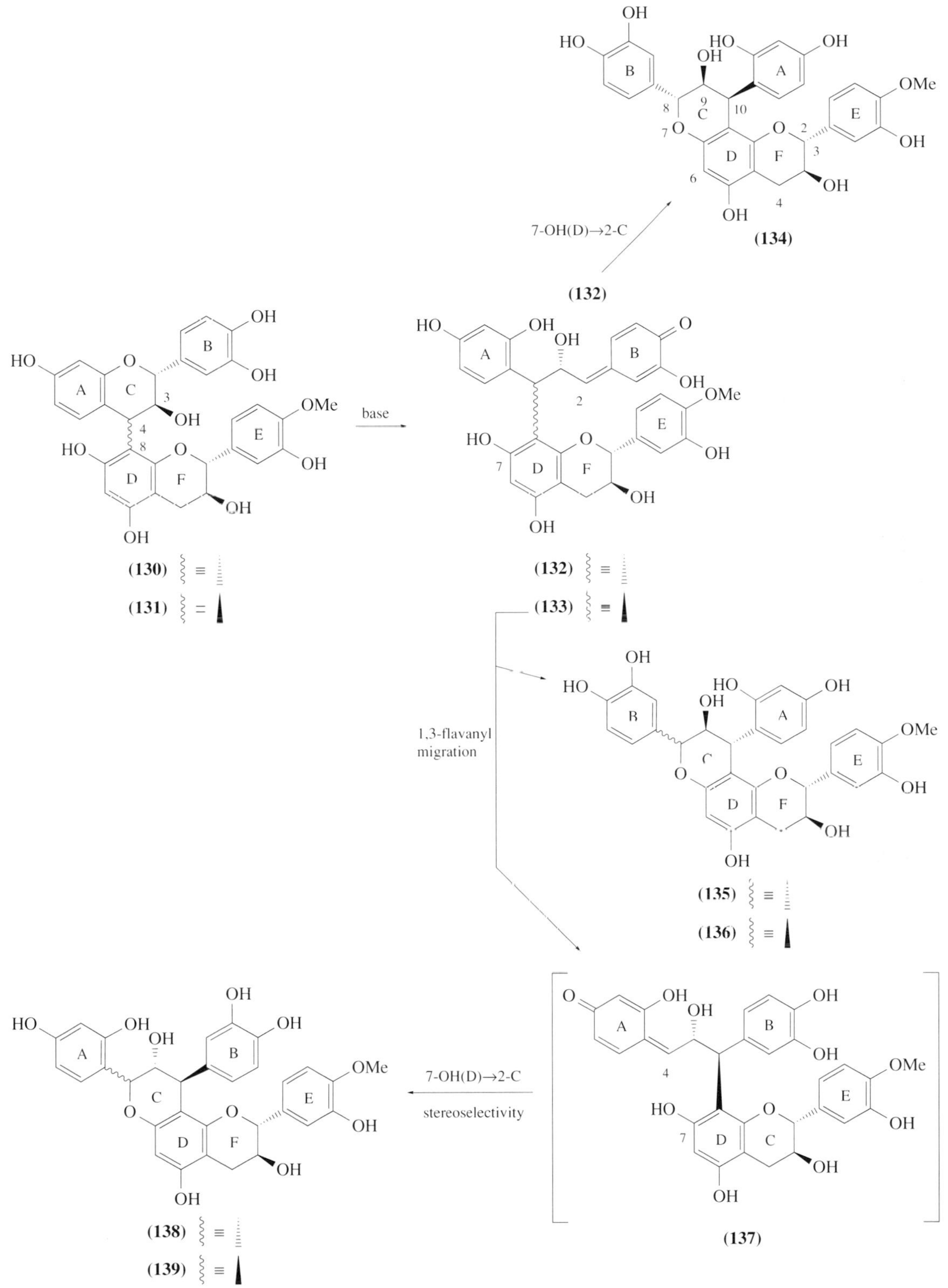

Scheme 11

Biflavanoids, protected at 4-OH(E) in order to prevent the unwanted side reactions that are associated with the formation of an E-ring quinone methide[57,58] (e.g., the fisetinidol-(4→8)-catechin profisetinidins (**130**) and (**131**)), are susceptible to base-catalyzed cleavage of the C-ring with the formation of the B-ring quinone methides (**132**) and (**133**). Quinone methide (**132**) which is derived from the dimer with a 3,4-*trans* (C-ring) configuration undergoes a highly stereoselective recyclization involving 7-OH(D) and the *re*-face at C-2 to give the tetrahydropyrano[2,3-*f*]chromene (**134**). This process thus invariably leads from the 3,4-*trans* configuration in the parent biflavanoid (**130**) to the 9,10-*cis* arrangement in the phlobatannin (**134**). Besides its stereoselective recyclization involving 7-OH(D) and both the *re*- and *si*-faces at C-2 to give the tetrahydropyrano[2,3-*f*]chromenes (**135**) and (**136**), the quinone methide (**133**) is also susceptible to an unusual 1,3-migration of the catechin DEF-unit to give the A-ring quinone methide (**137**).[57,58] Stereoselective recyclization involving 7-OH(D) and 4-C then gives the tetrahydropyrano[2,3-*f*]chromenes (**138**) and (**139**) with interchanged resorcinol A- and pyrocatechol B-rings, and with inverted absolute configuration at C-9(C), compared with the arrangement prevailing in the "normal" analogues (**135**) and (**136**).[52,58,60] Quinone methides with phloroglucinol-type A-rings, that is, those derived from procyanidin B-2 (**56**)[152] and B-3 (**57**),[153] additionally undergo 1,3-migration of this phloroglucinol moiety under the influence of the electron-releasing D-ring, hence initiating the formation of a complex series of 2-flavanyl-4-aryl-3,4-dihydro-2*H*-benzopyrans.[169,170] Profisetinidins with (4→6)-interflavanyl linkages (e.g., (**140**)) are transformed by base into the regioisomeric tetrahydropyrano[2,3-*h*]- and -[3,2-*g*]chromenes (**141**) and (**142**) (Scheme 12).[57,58]

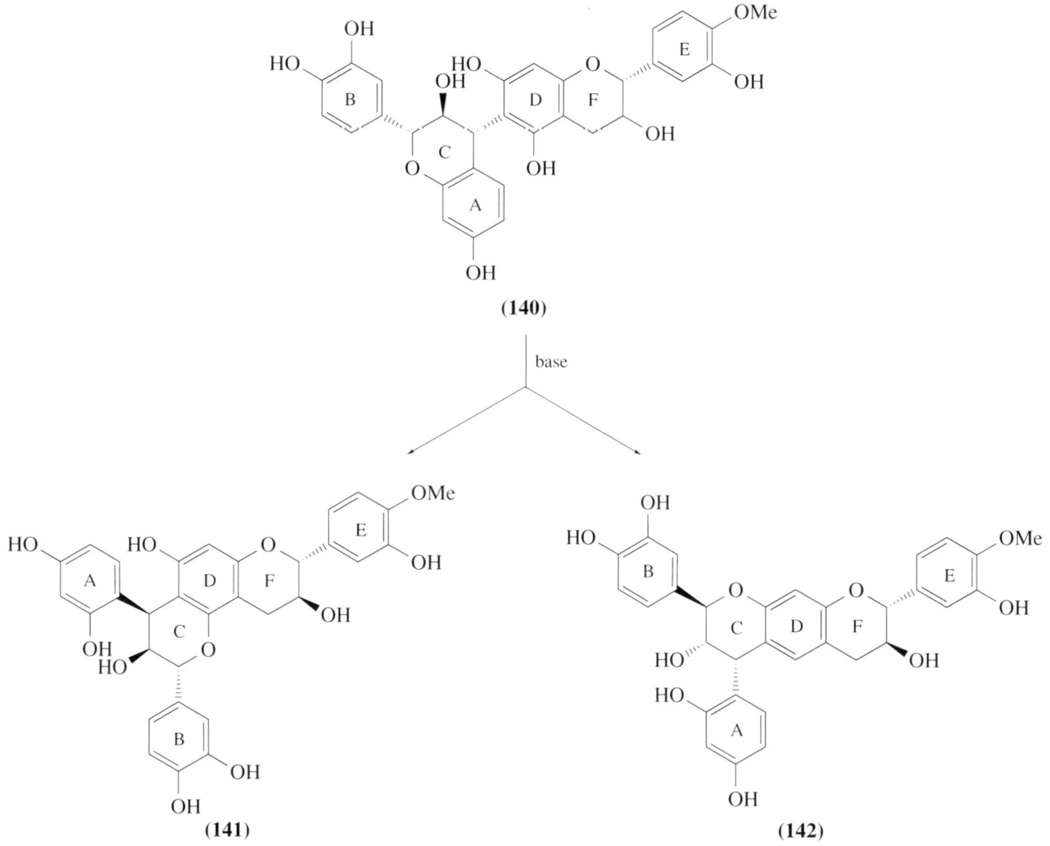

Scheme 12

The aforementioned principles also govern the base-catalyzed C-ring isomerization of trimeric profisetinidins,[190,192–195] (e.g., fisetinidol-(4α→6)-catechin-(8→4β)-fisetinidol (**143**)). Analogues possessing constituent chain-extender units with 3,4-*cis* stereochemistry (ABC unit in structure (**143**)) are similarly subject to extensive 1,3-migrations and thus to the formation of exceptionally complex reaction mixtures.[192] This has led to the development of a more controlled synthesis that is based upon the repetitive formation of the interflavanyl bond and pyran ring rearrangement of the chain-extender unit under mild basic conditions.[198]

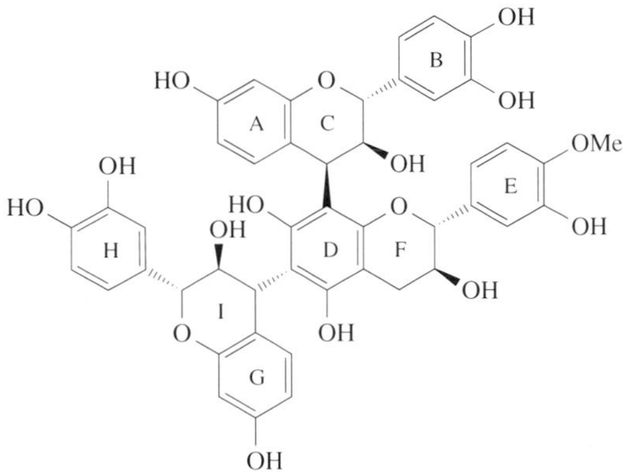

(143)

Collectively, the work described in this section and in Section 3.19.4.3.1 is of fundamental importance to an understanding of the chemistry of oligomeric proanthocyanidins in basic solution. It provides a basis for the commercial utilization of proanthocyanidins, and also an understanding of the *post mortem* processes involved in the aging of these biopolymers in wood and bark. The recognition of the phlobatannins also contributes to a rational explanation of the much reduced solubility of "aged" proanthocyanidins in aqueous solvents. The phlobatannins all exhibit the characteristic structural features that are essential for the use of "Mimosa" extract in cold-setting adhesives and leather-tanning applications;[199] thus, their abundant presence in the bark extract[197] may well explain the industrial utility of this important renewable resource.

3.19.4.7 Prorobinetinidins (3,7,3′,4′,5′-Pentahydroxylation)

The prorobinetinidins also feature prominently in wattle bark extract and thus complement the profisetinidins in this industrial resource.[5,6,11] A number of analogues based on epigallocatechin and its 3-*O*-gallate have been identified in the bark of *Stryphnodendron adstringens*.[200] Their semisynthesis from leucorobinetinidin (16) as incipient electrophile and the appropriate nucleophilic flavanoids,[75,76,122] and their behavior under basic conditions[57,58,188] are controlled by the same principles that were discussed for the profisetinidins.

3.19.4.8 Proguibourtinidins (3,7,4′-Trihydroxylation)

The pro- and leucoguibourtinidins with their 7,4′-dihydroxyphenolic functionality represent a relatively rare group of oligomeric proanthocyanidins[5,6,10] which, while occurring as minor components in Australian *Acacia* species,[176] predominate in the Southern African species *G. coleosperma*,[55,135,136] *J. globiflora*,[127] *Acacia luederitzil*,[201,202] and *C. mopane*.[65,133,134] Notable amongst these compounds are those analogues possessing a 3,4,3′,5′-tetrahydroxystilbene terminating unit (e.g., proguibourtinidins (84) and (85) from *G. coleosperma*).[135,136] The heartwood of *C. mopane* contains analogues that are based upon the 5-deoxyflavan-3-ols, fisetinidol and *ent*-epifisetinidol as chain-terminating units (e.g., proguibourtinidins (144)–(147)).[65] These compounds were difficult to synthesize by the usual protocol presumably as a result of the low nucleophilicity of the 5-deoxy-flavan-3-ols and, more importantly, of reduced reactivity of the flavan-3,4-diol due to the poor ability of the monooxygenated B-ring to stabilize an intermediate carbocation via an A-conformation (26).

A guibourtinidol–epiafzelechin dimer was obtained from *C. fistula* sapwood for which a (4α→8)-interflavanyl linkage was assumed but not proven,[203] while an epiguibourtinidol-(4β→8)-epicatechin was identified in *Dalbergia monetaria* bark.[183] Two novel analogues based upon afzelechin and epiafzelechin were obtained from *Cassia abbreviata*.[204]

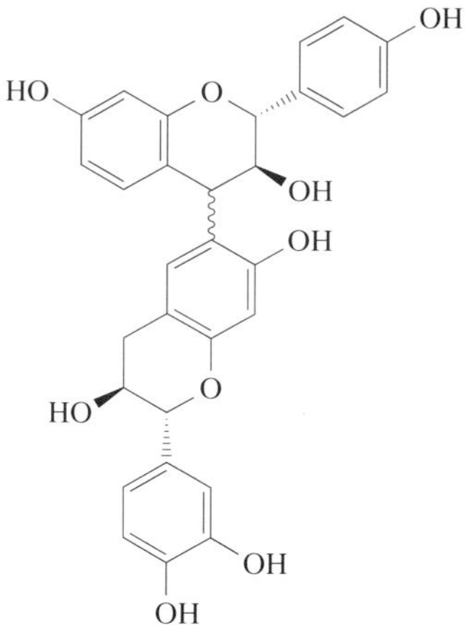

(144) $\}$ ≡ ▮   (146) $\}$ ≡ ▮

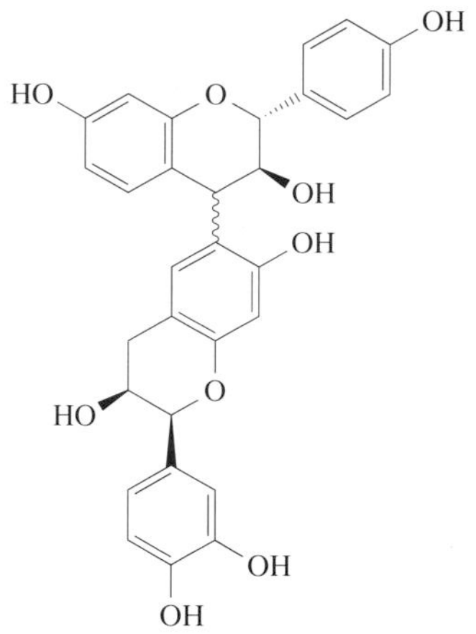

(145) $\}$ ≡ ┊   (147) $\}$ ≡ ┊

3.19.4.9 Proteracacinidins (3,7,8,4′-Tetrahydroxylation) and Promelacacinidins (3,7,8,3′,4′-Pentahydroxylation)

Whilst considerable progress has been made in the chemistry and structures of proanthocyanidins based on the phloroglucinol and resorcinol A-ring flavanoids, those oligomers with pyrogallol-type A-rings remain largely unexplored. Although the flavan-3,4-diol melacacidin (13), its C-4 epimer, isomelacacidin, and teracacidin (14) are present in a large number of *Acacia* species,[176,205] their corresponding proanthocyanidin oligomers are more sparsely populated. The additional hydroxy function at C-8 presumably counteracts electron release from the 7-hydroxy group, thus reducing the tendency of flavan-3,4-diols (13) and (14) to form C-4 carbocations or A-ring quinone methides which are essential for initiating condensation. These considerations led to suggestions[114,206] that oligomers composed of pyrogallol-type A-ring moieties are unlikely to exist and that the polymers that co-occur with these flavan-3,4-diols are probably oxidation products.

Several studies[73,74,207] have since demonstrated that the flavan-3,4-diols (13) and (14) are susceptible to facile condensation with phenolic nuclei under mild acidic conditions to give 4-arylflavan-3-ols of types (38) and (39) (cf. Scheme 3), such a phenomenon suggesting that the formation of natural proanthocyanidins of the 7,8-dihydroxyflavanoid pattern is not chemically prohibited. These observations have subsequently led to identification of several promelacacinidins from *Prosopis glandulosa*[132] and *A. melanoxylon*[120,125] (e.g., compounds (76) and (77)).

The first four dimeric proteracacinidins were recently isolated from the heartwoods of *Acacia galpinii* and *Acacia caffra*. The compounds were epioritin-(4β→6)-epioritin-4α-ol (148) and its C-4(F) epimer (149),[208] the doubly linked *ent*-oritin-(4β→7, 5→6)-epioritin-4α-ol (150),[209] and *ent*-oritin-(4β→5)-epioritin-4β-ol (151).[210] The structures of dimers (148) and (149) were confirmed by the acid-catalyzed self-condensation of their apparent biogenetic flavan-3,4-diol precursor, epioritin-4α-ol, which co-occurs in the heartwood of *A. galpinii*. Besides samarangenins A and B[211] the doubly linked proteracacinidin (150) represents only the third proanthocyanidin where the interflavanyl linkages are presumably established by a combined one-electron (5→6 bond) and two-electron (7→0→4 bond) process.

The natural occurrence of these promelacacinidins and proteracacinidins clearly demonstrates that the pyrogallol A-ring function is sufficiently reactive for nucleophilic condensation to take place and also to facilitate C-4 carbocation formation from an associated flavan-3,4-diol, hence initiating the formation of proanthocyanidin oligomers, the (4→0→4)-coupled analogues (76) and (77) further extending the phenomenon of heterogeneity of the interflavanyl linkage among natural proanthocyanidins.

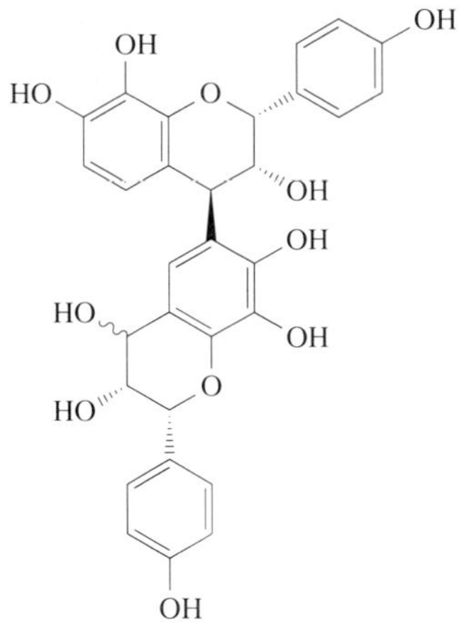

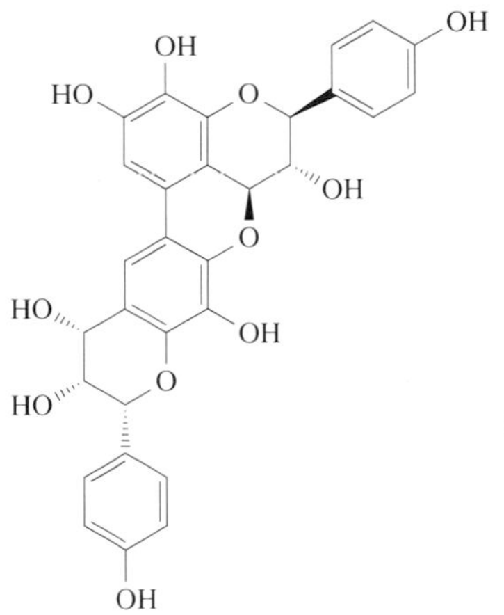

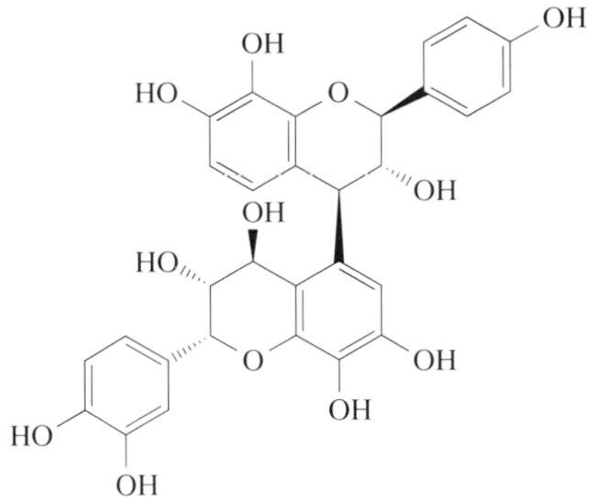

(148) 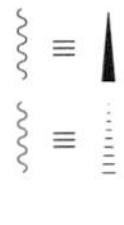 ⟨ ≡ ▮

(149) ⟨ ≡ ⋮

(150)

(151)

3.19.4.10 Propeltogynidins and Promopanidins

Peltogynol (peltogynan-4α-ol) (**28**) and mopanol (mopanan-4α-ol) (**29**) feature as incipient electrophiles in the formation of a novel series of propeltogynidins (e.g., peltogynane-(4β→6)-fisetinidol (**152**)) and promopanidins (e.g., mopanane-(4β→6)-fisetinidol (**153**)) in the heartwood of *C. mopane*.[70] Peltogynol also serves as the nucleophile via coupling at the B ring in the formation of the profisetinidin, fisetinidol-(4α→6′)-peltogynan-4α-ol, while *ent*-epimopanone is apparently involved in a radical coupling process with the hitherto unknown guibourtinidol to give a unique guibourtinidol-3′-*O*-4′-*ent*-epimopanone.[70]

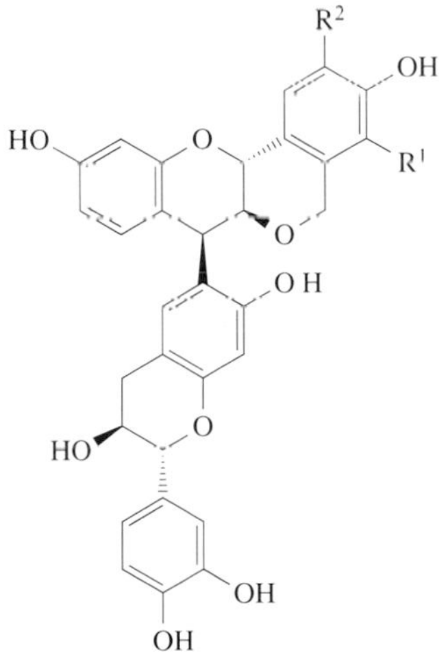

(152) R^1 = H, R^2 = OH
(153) R^1 = OH, R^2 = H

3.19.4.11 Procassinidins (7,4′-Dihydroxylation) and Prodistenidins (3,5,7-Trihydroxylation)

Like the propeltogynidins and promopanidins, the procassinidins and prodistenidins represent new classes of oligomeric proanthocyanidins based respectively on a (2*S*)-7,4′-dihydroxyflavan and a (2*R*,3*S*)-5,7-dihydroxyflavan-3-ol chain-extender unit. The procassinidins[91] (e.g., cassiaflavan-(4α→8)- and -(4β→8)-epiafzelechins (**154**) and (**155**)) are restricted to the four possible (4α,β)-diastereomeric pairs of cassiaflavan linked to epiafzelechin, and a mixed trimer, cassiaflavan-(4β→

8)-epiafzelechin-(4β→8)-epiafzelechin. The dimers were conveniently prepared by reduction of the (2S)-flavanone, liquiritigenin, to the flavan-4-ol diastereomeric mixture and subsequent coupling to epiafzelechin in acidic medium.[91] The procassinidins complement the series of 5-deoxy-(A)-proanthocyanidins. The prodistenidins are limited to a single dimer, epidistenin-(4β→8)-epidistenin (156), and the related trimer (epidistenin-(4β→8))₂-epidistenin.[212]

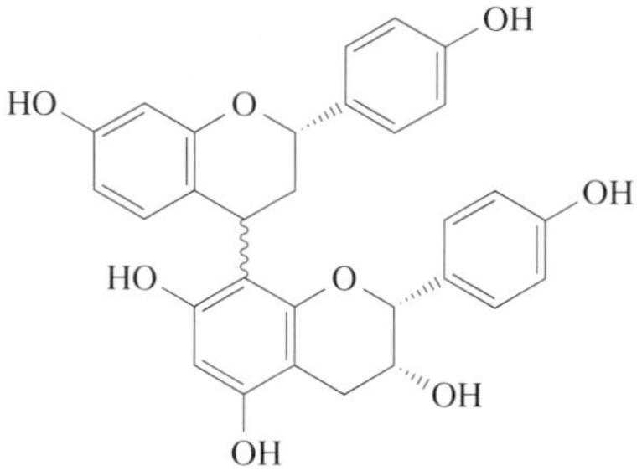

(154) ≡ |

(155) ≡ ⋮

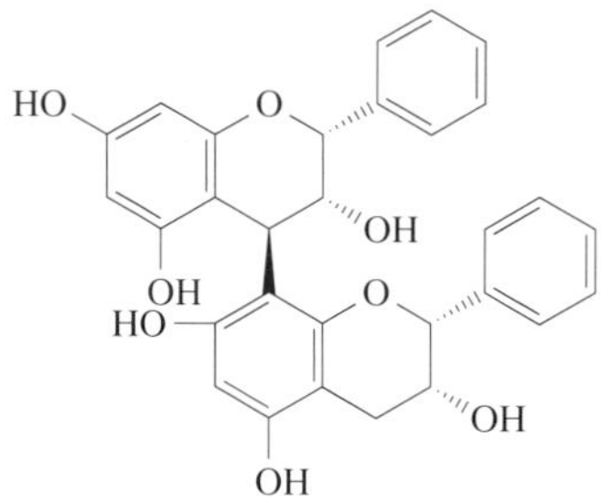

(156)

3.19.4.12 A-type Proanthocyanidins

In contrast to proanthocyanidins of the B-type where the constituent flavanyl units are linked via only one bond, analogues of the A-class possess an unusual second ether linkage to C-2 of the T-unit. This feature introduces a high degree of conformational stability which culminates in high quality and unequivocal NMR spectra, conspicuously free of the effects of dynamic rotational isomerism. Compounds of this class are readily recognizable from the characteristic AB-doublet ($^3J_{3,4}$ = 3–4 Hz) of C-ring protons in the heterocyclic region of their ^{1}H NMR spectra irrespective of the 3,4-relative configuration,[213,214] and may possess either (2α,4α)- or (2β,4β)-double interflavanyl linkages, which are reliably assessed by means of chiroptical data.[215]

Since the first isolation[216] and structure elucidation of proanthocyanidin A-2 (7) (epicatechin-(4β→8, 2β→7)-epicatechin),[213,217] a considerable number of these unique secondary metabolites

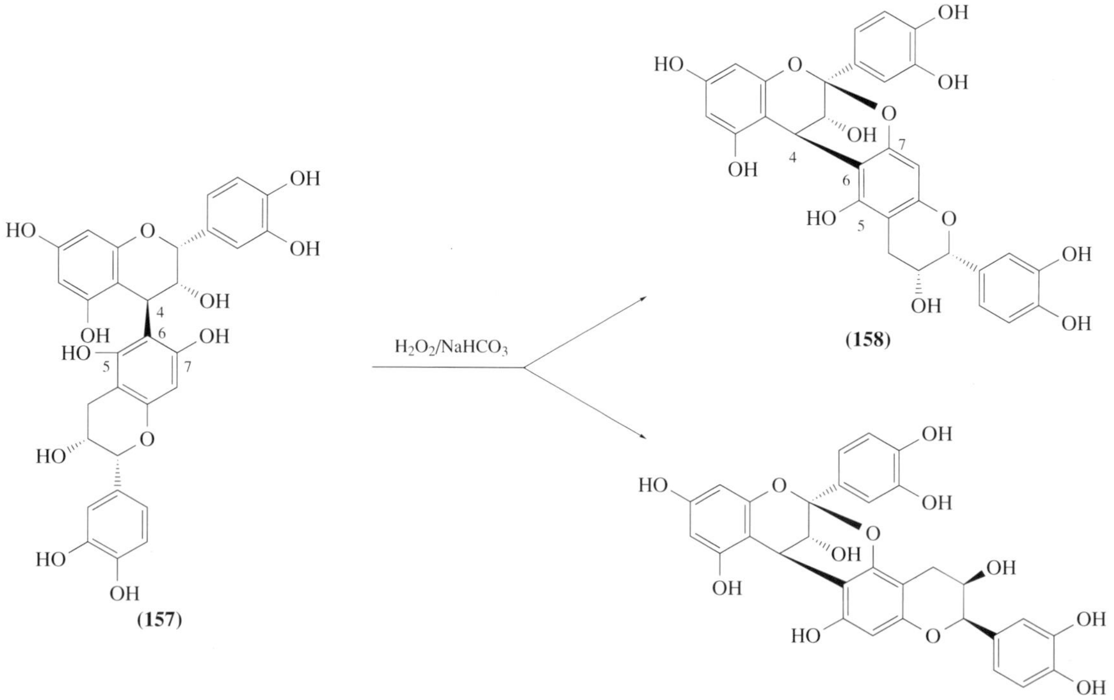

$H_2O_2/NaHCO_3$

(157)

(158)

(159)

Scheme 13

with profound biological activity[218] have been reported. Constituent units include the catechins, afzelechins, epigallocatechin, *ent*-apigeniflavan, and kaempferol, which may possess *O*-methyl, *O*-glycoside, and *O*-gallate substituents (see Porter,[5,10] and Ferreira and Becker[11] for a comprehensive summary of known analogues). The regiochemistry of these compounds is limited to three possible modes, that is, $(2 \rightarrow 7, 4 \rightarrow 8)$, $(2 \rightarrow 7, 4 \rightarrow 6)$, and $(2 \rightarrow 5, 4 \rightarrow 6)$, as is demonstrated for the double β-linkages in proanthocyanidin A-2 (**7**), A-6 (**158**), and A-7 (**159**), respectively. Proanthocyanidins A-6 (**158**) and A-7 (**159**) were the first A-type compounds possessing $(4 \rightarrow 6)$-interflavanyl linkages and were independently obtained from *Aesculus hippocastanum*[219] and *T. cacao*.[220] The structures of compounds (**158**) and (**159**) were elegantly elucidated by their semisynthesis via oxidative conversion of procyanidin B-5 (**157**) (Scheme 13).[219]

A considerable number of oligomers containing both A- and B-type linkages have been isolated from a variety of sources.[5,10,11] Elucidation of these complex structures, up to the pentameric level, initially[219,221] rested largely on the products of hydrolysis with phenylmethanethiol, as is demonstrated in Scheme 14 for aesculitannin G (**160**). Acid-catalyzed thiolytic cleavage of the central B-type interflavanyl bond gave proanthocyanidin A-2 (**7**) and its 4′β-benzylthioether (**161**) which was transformed into the former by desulfurization with Raney Ni.

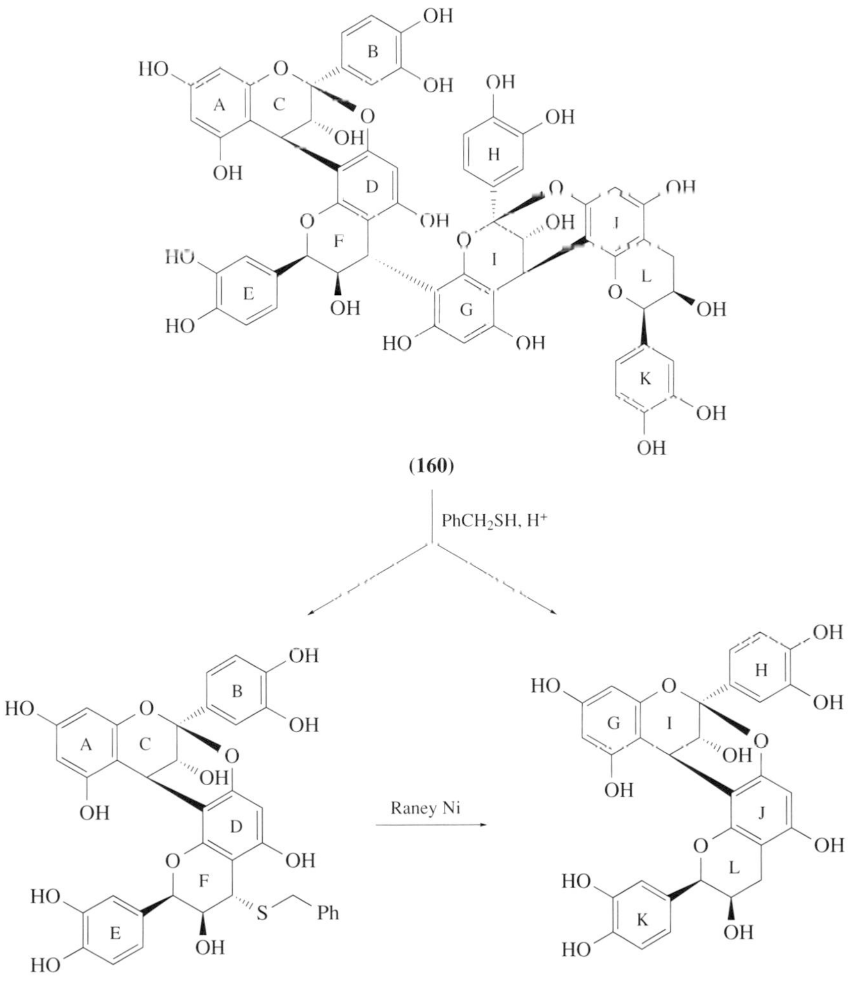

(**160**)

PhCH$_2$SH, H$^+$

Raney Ni

(**161**)   (**7**)

Scheme 14

The stem bark of *Pavetta owariensis* has proven to be an extremely productive source of A-type proanthocyanidins which were dubbed the pavetannins.[222–226] The structures were elucidated by using physical techniques, especially ^{1}H and ^{13}C NMR spectroscopy, and where sample quantities were sufficient the method of thiolytic cleavage (Scheme 14). The ^{1}H NMR spectra of most of the analogues also containing a β-type interflavanyl bond were complicated due to the adverse effects of dynamic rotational isomerism. These[222–226] and other papers[227–229] contain comprehensive ^{1}H and ^{13}C NMR data and should thus feature as key references for future work in this field.

The conspicuous identity of $^3J_{HH}$ values for 3-H and 4-H (C-ring) in analogues possessing 3,4-*trans* and 3,4-*cis* relative configurations,[197,213,214] is explicable in terms of the conformational rigidity of the bicyclic ring system which culminates in very similar dihedral angles between these protons in the stereoisomers (**162**) and (**163**). The establishment of the absolute configuration at C-3 of the F-ring is greatly simplified by application of the modified Mosher's method[230,231] that was developed for a series of flavan-3-ols and 4-arylflavan-3-ols as models for representative classes of oligomeric proanthocyanidins.[232,233]

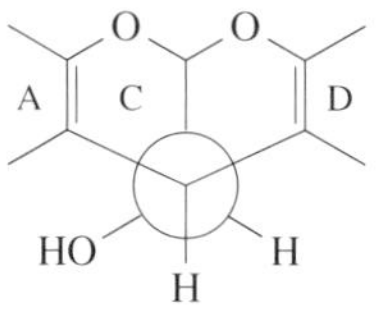

(**162**) (3,4-*trans*) 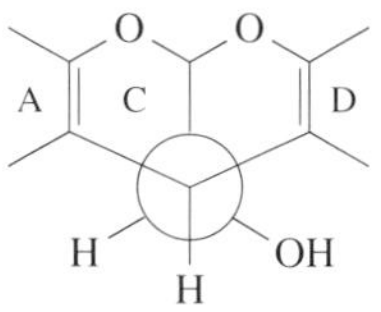   (**163**) (3,4-*cis*)

Owing to the close structural relationship between proanthocyanidin A-2 (**7**) and procyanidin B-2 (**56**), Porter[5] has proposed a biosynthetic pathway for the conversion of B- to A-type proanthocyanidins which involves an enzyme-mediated hydroxylation at C-2 (C-ring of (**56**)). Such a proposal is presumably supported by the ease of *in vitro* transformation, albeit in low yield, of B- to A-type compounds using H_2O_2–$NaHCO_3$[198,234–236] or molecular oxygen[169,214] as oxidants.

3.19.5 NONPROANTHOCYANIDINS WITH FLAVAN OR FLAVAN-3-OL CONSTITUENT UNITS

3.19.5.1 Daphnodorins, Genkwanols, and Wikstrols

In addition to the extensive range of di- and trimeric oligoflavanoids with rearranged C-rings (Section 3.19.4.6.2), larixinol (**164**),[237] genkwanols A–C (e.g., genkwanol B (**165**)), and daphnodorins A–C, D-1, D-2, and E–I (e.g., daphnodorins A (**166**) and B (**167**)) represent rearranged biflavanoid metabolites comprising either flavan or flavan-3-ol and 4,2′,4′,6′-tetrahydroxychalcone (chalconaringenin) constituent units. Progress in the knowledge of these compounds comes entirely from the work of Taniguchi and Baba[238] and the late Kozawa[239] (see also Porter[5,10] and Ferreira and Becker[11] for a full list of their contributions). The structures and absolute configurations, where applicable, were established by the collective utilization of ^{1}H- and ^{13}C-NMR data, X-ray analysis, and the modified Mosher's method. The basic carbon framework of an afzelechin or an apigeniflavan moiety coupled at C-8 to the α-carbon of a chalconaringenin unit is evident in the structures of some of the daphnodorins and genkwanols. This is best illustrated (Scheme 15) by the acid-catalyzed rearrangement of daphnodorin A (**166**) into the atropisomeric 5,7,4′-trihydroxyflavone-(3→8)-(2*S*)-5,7,4′-trihydroxyflavans, daphnodorins D-1 (**170**) and D-2, and of daphnodorin B (**167**) into the naturally occurring (3→8)-coupled atropisomeric wikstrols A (**171**) and B, via the intermediate ketals (**168**) and (**169**).[239]

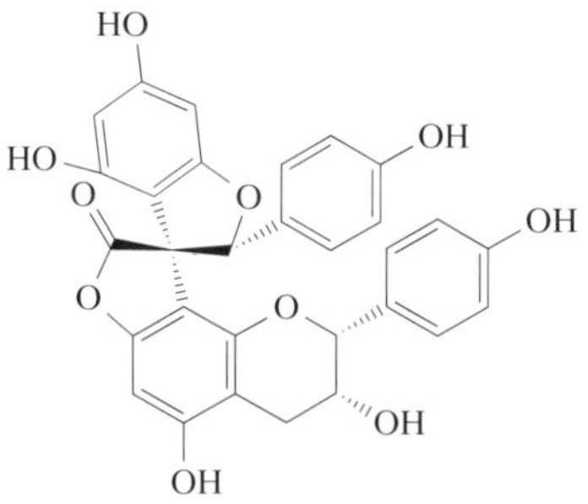

(**164**) 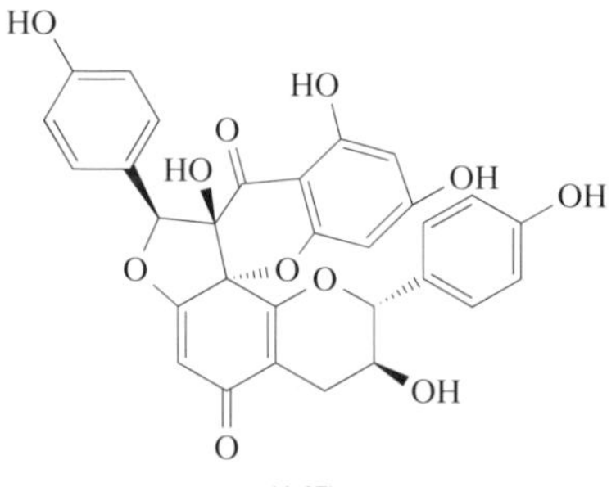   (**165**)

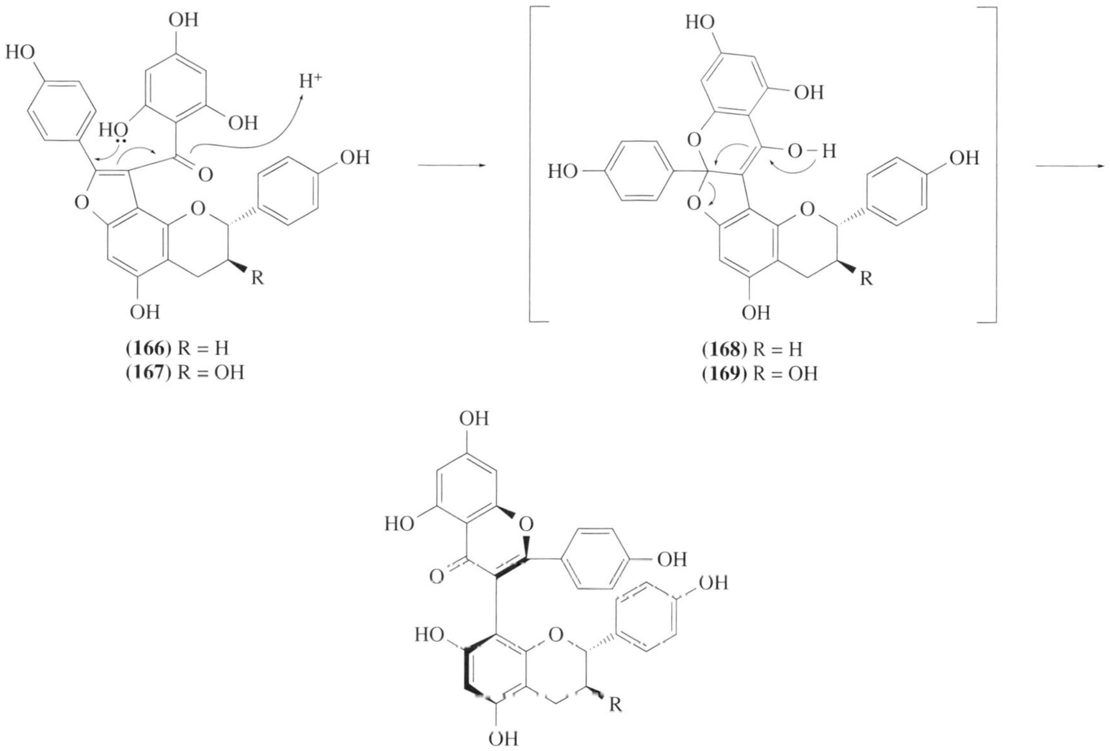

(166) R = H
(167) R = OH

(168) R = H
(169) R = OH

(170) R = H
(171) R = OH

Scheme 15

3.19.5.2 Gambiriins

A series of catechin or *ent*-epicatechin metabolites, the gambiriins, have been isolated from *Uncaria gambir* leaves and twigs.[240,241] These (e.g., gambiriin A-1 **(172)**) represent the products of condensation between the flavan-3-ols and C-3 of the 1,3-diarylpropan-2-ol analogues[242] of catechin. The gambiriins are in fact identical with the products of acid self-condensation of catechin (see Weinges *et al.*[2]).

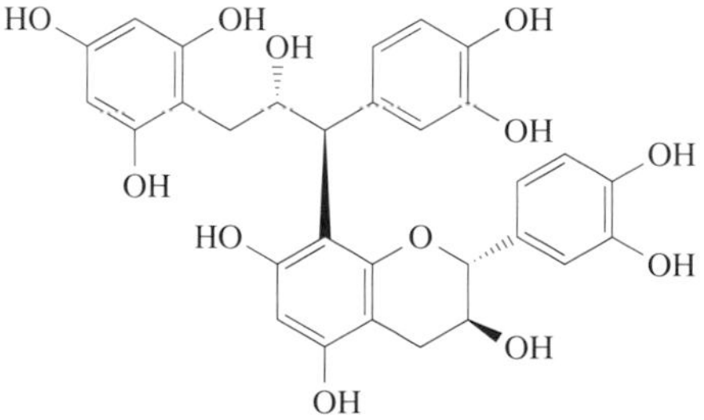

(172)

3.19.5.3 Deoxotetrahydrochalcone Flavans

The dragon's blood tree, *Dracaena cinnabari*, contains two unique "nonproanthocyanidins," the biflavanoid cinnabarone **(173)**,[243] comprising a *retro*-dihydrochalcone and a deoxotetrahydrochalcone constituent unit, and the triflavanoid damalachawin **(174)**,[244] comprising a flavan and two deoxotetrahydrochalcone moieties. Definition of the absolute configuration at the stereocenters of both (173) and (174) was, however, not attempted. These unique products were accompanied by the novel (2S)-7,3'-dihydroxy-4'-methoxyflavan.[245]

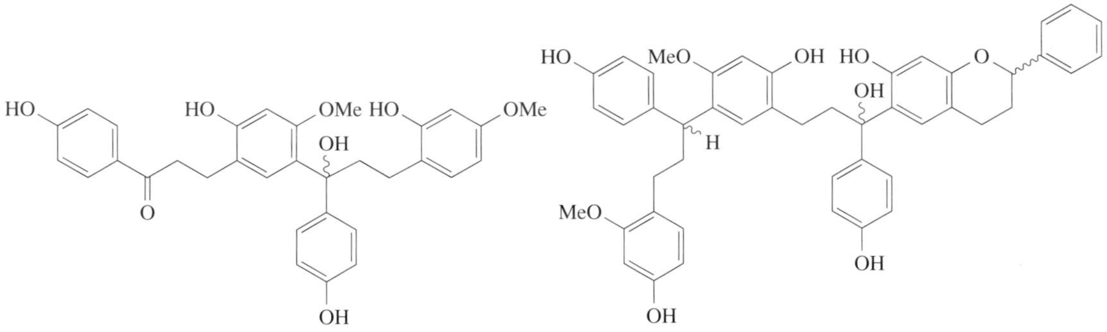

(173)   (174)

3.19.5.4 Flavan-3-ol Oligomers via Phenol Oxidative Coupling

Although oxidative coupling of flavonoids is an established natural phenomenon affecting mainly flavones and flavanones,[6,12] participation by flavan-3-ols in this mode of condensation is less common. Examples of the latter involve (2'→8)-coupling of catechin via the respective B- and A-rings to give "dehydrodicatechins" and higher oligomers.[2,246–248] Several studies have since dealt with the oxidation of catechin with polyphenoloxidase.[249]

Of special interest amongst this class of flavan-3-ol oligomers are the theaflavins and thearubigins which play such a key role in the quality of black tea.[250] These compounds do not occur in the green leaf, but are formed during the black tea manufacturing process by enzymatic oxidative transformation of some of the flavan-3-ols present in the fresh green leaf. The principal components of the green tea leaf are epicatechin (3) and epigallocatechin, and their respective 3-O-gallate esters. These precursors are then transformed into the theaflavin- and thearubigin-type pigments of black tea during "fermentation" which involves a random enzyme-catalyzed oxidative coupling process.

The polymeric thearubigins have been recognized as being in part polymeric proanthocyanidins.[147,251] The theaflavin fraction, on the contrary, comprises a considerable number of compounds including theaflavin (175), its 3-O- and 3'-O-monogallates, and corresponding O,O-digallate, and theaflavic and isotheaflavic acids.[252] Theaflavin (175) may be considered to be formed by oxidative coupling of epicatechin (3) and epigallocatechin in a normal type of benztropolone synthesis.[253,254] The isolation[255,256] of theasinensins A–G, a series of (2'→2')-biphenyl linked combinations of epicatechin and epigallocatechin (e.g., theasinensins A and B (177) and (178)), gives credence to the proposals regarding the genesis of theaflavin (175) and isotheoflavin (176). Nonaka, Nishioka and co-workers have since also identified other fermentation products, for example, oolongtheanin,[256]

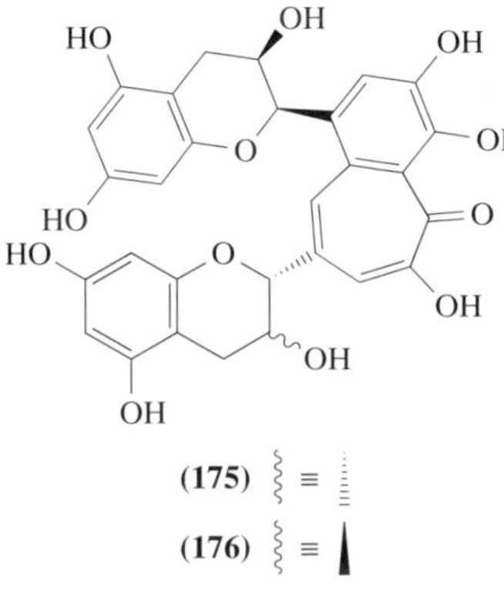

(175) $\}$ ≡ ⋮

(176) $\}$ ≡ ▮

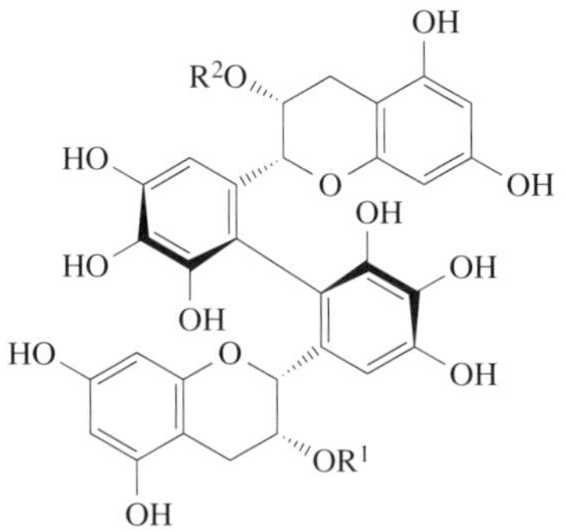

(177) R^1 = R^2 = gallate
(178) R^1 = gallate, R^2 = H

theogallinin, theaflavonin, and desgalloyltheaflavonin.[257] The *R* absolute configuration of the atropisomeric biphenyl linkages of the latter three compounds was established by comparison of their c.d. data with those of theasinensins C and E having *R* and *S* chirality, respectively. The same authors also proposed an informative schematic representation of the enzymatic conversion of the polyphenols in green tea.

In the heartwood of *P. glandulosa* the promelacacinidins, mesquitol-(4α→8)-catechin and (4α→6)-bis-mesquitol, are accompanied by a series of oxidatively coupled analogues with mesquitol (**67**) serving as the key precursor.[132] Amongst these are the (5→6)-bismesquitol, the two atropisomeric mesquitol-(5→8)-catechins, and the four atropisomeric (5→6, 5→8)-*m*-terphenyl-type triflavan-3-ols of general structure (**179**). Oligomeric structures were confirmed by biomimetic oxidative coupling[132,258] involving mesquitol and catechin. NOE difference spectroscopy elegantly permitted definition of the absolute configuration at the point of the interflavanyl linkage in the permethylaryl ether acetates of the (5→8)-biphenyl-type flavan-3-ols,[132] and of the four *m*-terphenyl analogues of type (**179**).[259] These compounds possess the characteristic intense c.d. bands anticipated for biaryl compounds, hence facilitating definition of their absolute configurations.

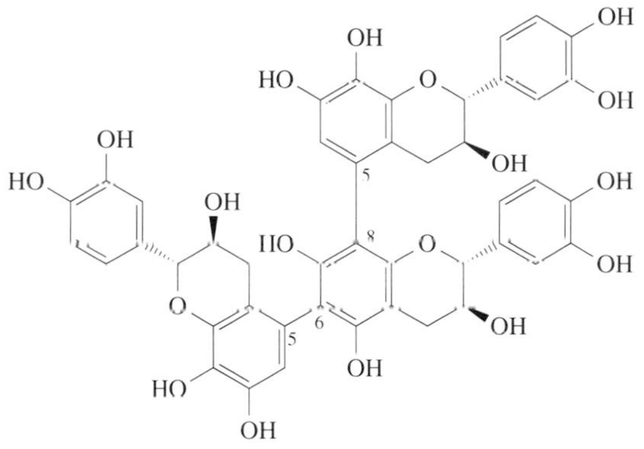

(179)

3.19.5.5 "Complex Tannins"

The term "complex tannin" appears to be established as a descriptor for the class of polyphenols in which a flavan-3-ol unit, representing a constituent unit of the condensed tannins, is connected to a hydrolyzable (gallo- or ellagi-) tannin through a carbon–carbon linkage. Since the first demonstration of their natural occurrence in 1985,[260] a considerable number of these unique secondary metabolites have been identified.[5,10,11] They are all of similar structure, where a *C*-glycoside is formed between C-6 or C-8 of the flavan-3-ol (catechin, epicatechin, gallocatechin) and C-1 of the glucose portion of the ellagitannin. This is often difficult to recognize in their sometimes highly modified structures. Additions to this series of compounds together with progress in the elucidation of their chemistry came exclusively from the groups of Nonaka and Nishioka[260,262] and of Okuda and Yoshida[261,263,264] in Japan. A few key examples are selected to demonstrate some of the more fundamental issues.

Malabathrins A (**180**), E (**181**), and F (**182**) are composed of a *C*-glucosidic ellagitannin and a C—C coupled epicatechin moiety.[261] The *S* chirality for both the hexahydroxydiphenoyl (HHDP) (hexahydroxybiphenyldicarbonyloxy) groups in compound (**180**) was deduced from its c.d. curve which exhibited positive and negative Cotton effects at 233 and 262 nm, respectively, and the structure was unequivocally confirmed by acid-catalyzed condensation of the ellagitannin, casuarinin, and 3-*O*-galloepicatechin. Structure elucidation of malabathrin E (**181**) and of its regioisomer, malabathrin F (**182**), was performed by comparison of ¹H- and ¹³C-NMR data with those of mongolicain A, which also possesses a cyclopentenone moiety linked to glucose C-1. The *S* absolute configuration of the HHDP group in malabathrin E (**181**) was established by methanolysis of its permethylaryl ether, which gave dimethyl hexamethoxydiphenate with an [α]_D value of −27°. Comparison of the c.d. data of malabathrin F (**182**) with those of (**181**) confirmed the same *S* chirality for the HHDP moiety in the former compound. The cyclopentenone moiety in, for example, compound (**181**) is regarded as the product of oxidative conversion of the HHDP group at O-2—O-3 of the glucosyl unit in, for example, compound (**180**).[261]

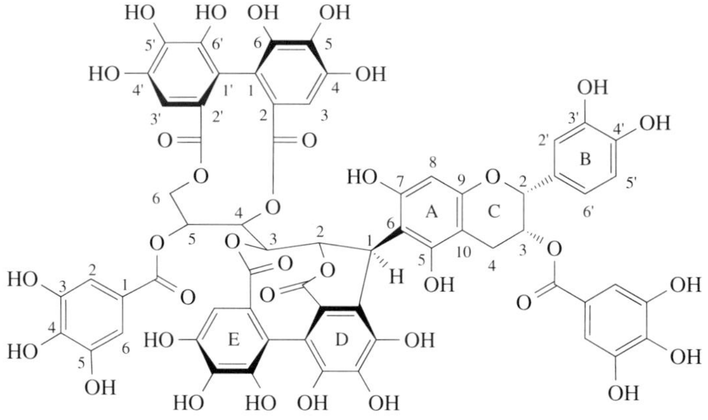

(**180**)

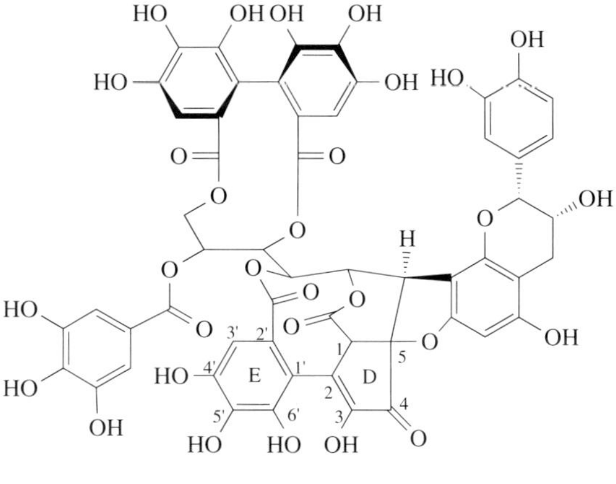

(**181**)

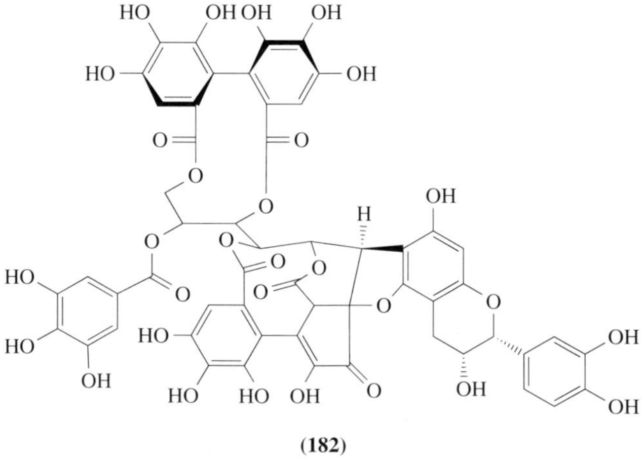

(**182**)

The novel psiguavin (**183**),[262] in which the B-ring of the gallocatechin unit is extensively rearranged, is considered to be derived biosynthetically from eugenigrandin A (partial structure (**184**)) by successive oxidation of the pyrogallol B-ring, benzylic acid-type rearrangement, and decarboxylation, followed by oxidative coupling, as is indicated in Scheme 16.[262]

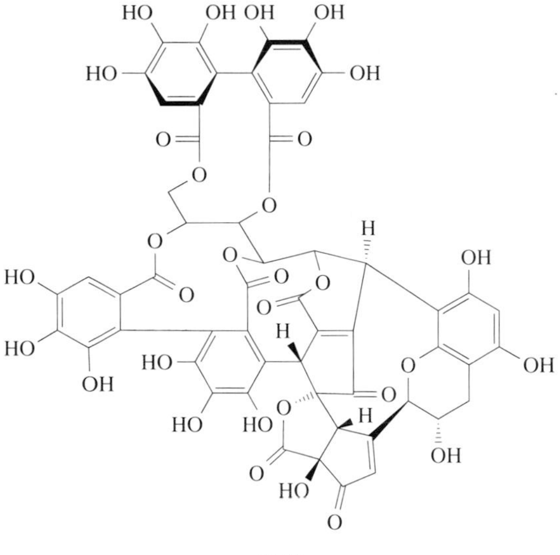

(183)

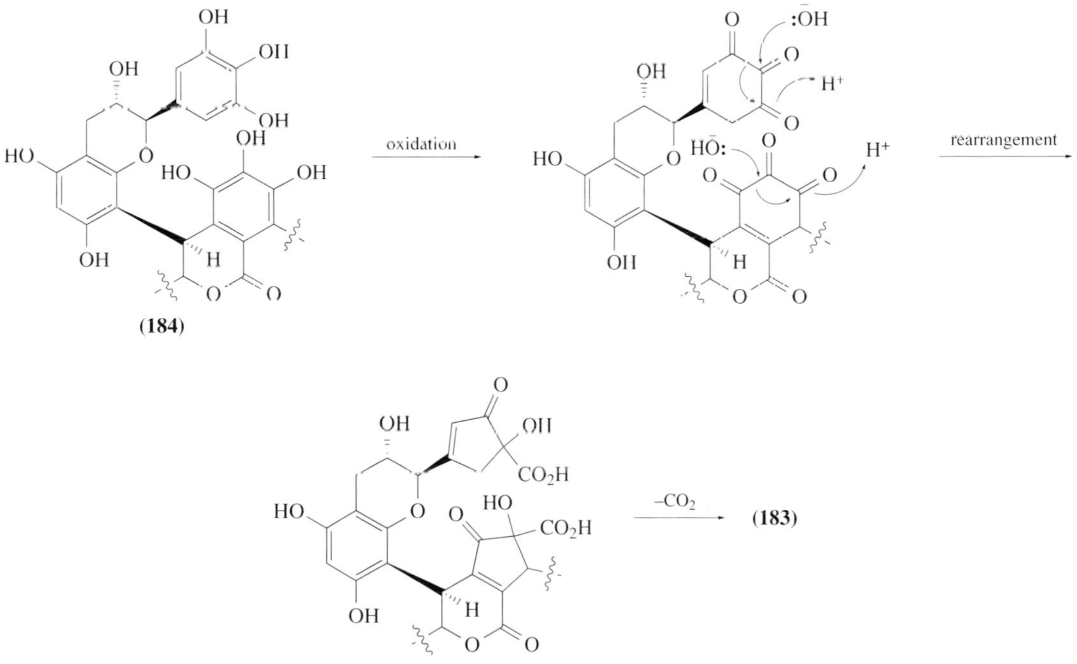

(184)

(183)

Scheme 16

Camelliatannins C (**185**),[263] D,[264] and E (**186**)[263] possess structural features that are unique among the complex tannin group of natural products. Compounds (**185**) and (**186**) with their C-6 and C-8 substituted epicatechin moieties, respectively, represent the first examples lacking a C—C bond between C-1 of glucose and the HHDP group at O-2 of the glucose unit. These bonds could, however, be readily formed by treatment of analogues (**185**) and (**186**) with polyphosphoric acid, hence transforming them into camelliatannins B and A, respectively.[263] Camelliatannins C (**185**) and E (**186**) may thus be considered as precursors to the "normal" type of complex tannins and may be anticipated to co-occur in the plant sources containing the latter class of metabolites. Camelliatannin D,[264] a new inhibitor of bone resorption, represents the first example of a complex tannin composed of dimeric hydrolyzable tannin and flavan-3-ol constituent units. Further details regarding the natural occurrence and chemistry of this class of compounds are available in Porter,[5,10] and Ferreira and Bekker.[11]

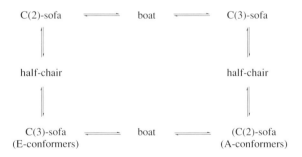

(185) R = 6-epicatechin
(186) R = 8-epicatechin

3.19.6 CONFORMATION OF PROANTHOCYANIDINS

Conformational analysis of proanthocyanidin oligomers is in principle concerned with the conformation of the pyran heterocycle and with the phenomenon of conformational isomerism due to restricted rotation about the interflavanyl linkage(s). The advent of ^{1}H NMR spectroscopy enabled Clark-Lewis *et al.*[265] to propose C-ring conformations approximating a half-chair, with the B-ring in an equatorial position, for a series of flavans with phenolic groups protected by methylation and with various heterocyclic ring substituents. Numerous ^{1}H NMR investigations have since borne out these findings. X-ray crystallographic studies of epicatechin (**3**),[266] the 8-bromo-5,7,3′,4′-tetramethyl ether derivatives of catechin[130] and epicatechin,[4] and leucocyanidin (**18**)[21] generally supported the NMR conclusions. Realization of the fact that the conformational itinerary of the heterocyclic rings involves a dynamic equilibrium between E- and A-conformers[63] had a substantial impact in this field[8] and has led to an increased utilization of relevant molecular modeling calculations in an effort to address some of the many unexplained phenomena that still exist.

The conformational equilibration of the C-ring of flavan-3-ols may be described by the equilibrium shown in Scheme 17.

```
C(2)-sofa    ⇌    boat    ⇌    C(3)-sofa

    ↕                              ↕

half-chair                     half-chair

    ↕                              ↕

C(3)-sofa    ⇌    boat    ⇌    (C(2)-sofa
(E-conformers)                 (A-conformers)
```

Scheme 17

E- and A-conformers are those with the B-ring equatorial or axial, respectively.[267,268] Figure 1[63] depicts the ground-state energy conformations which may be adopted by the flavan heterocycle with the hatched line indicating the projection of the A-ring. Figure 2 gives the relative stereochemistry of groups at C-2 and C-3 for the E- and A-conformations of catechin (**2**) and epicatechin (**3**). The conformations are viewed in the sense indicated by the arrow in structures (**2**) and (**3**) and the solid line in structures (**187**)–(**190**) represents the A-ring plane.

The boat conformation represents the high-energy transition state for the interconversion of E- and A-conformers. An unequal conformational energy for these conformers is manifested by an unequal population of the two states, the one with the lower energy being populated to a greater extent. ^{1}H NMR measurements in conjunction with theoretical calculations[63] demonstrated that the

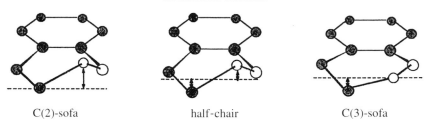

C(2)-sofa   half-chair   C(3)-sofa

Figure 1 Ground-state energy conformations of the flavan heterocycle.

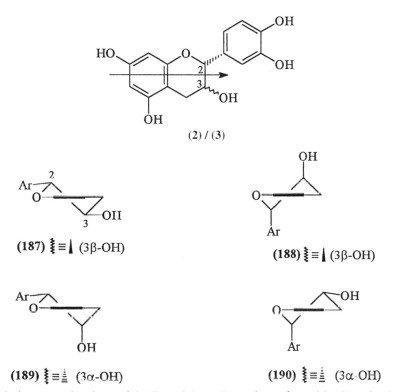

Figure 2 Relative stereochemistry of the E- and A-conformations of catechin (**2**) and epicatechin (**3**).

E:A ratio for catechin (**2**) and epicatechin (**3**) were 62:38 and 86:14, respectively. Acetylation of the 3-hydroxy group stabilized the A-conformation and altered the ratio to 48:52.

It was suggested that substitution at C-4 of catechin or epicatechin by a hydroxy or aryl substituent would strongly favour the E-conformation due to the tendency to minimize 1,3-diaxial interactions and the pseudoallylic or A (1,3)-strain effect.[63] In derivatives of 4-resorcyl-5-oxyflavan-3-ols, however, coupling constants of heterocyclic protons for analogues with a 2,4-*cis* arrangement of B- and D-rings (e.g., the 2,3-*trans*-3,4-*trans*- and 2,3-*cis*-3,4-*cis*- compounds (**191**) and (**192**)), are not reconcilable with dihedral angles.[60] Such conformational behavior also results in reversal of the Cotton effects in the 220–250 nm region of their c.d. spectra predicted by the aromatic quadrant rule.[140] The effect of A-strain on the 2,4-*cis* E-conformers[61,269] is reflected in a tendency of the pyran ring towards a C-2 sofa conformation, hence decreasing the C-3—C-4—C-10—C-9 torsion angle and the out-of-plane distance of C-3. The latter represents an effective increase in the torsion angle between 5-OMe(A) and the 4-resorcyl group and, therefore, alleviation of the A-strain. This effect is absent for E-conformers of 4-resorcyl-5-oxyflavan-3-ols with 2,4-*trans* B- and D-rings culminating in a tendency towards a C-3 sofa conformation for the C-ring and the absence of irregularities regarding their ¹H-NMR and c.d. data.

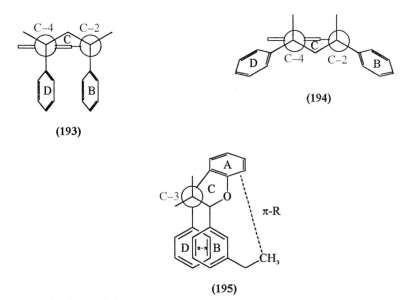

(191)   (192)

Maximum relief from A-strain, however, is achieved for the aforementioned 2,4-*cis*-isomers by inversion of the pyran ring to an A-conformer.[61,269] While 1,3-diaxal arrangements are commonly avoided on energetic grounds in terms of a classical stereochemical approach, A-conformers (**193**) as opposed to E-conformers (**194**) for these isomers appear to be an exception by virtue of the aromaticity and associated geometry of the 1,3-diaxal 2,4-biphenyl substituents, which are stacked

(193)

(194)

(195)

Figure 3 Newman projections of the 2,4-*cis*-resorcyl-5-oxyflavan-3-ols, illustrating an E-conformer (**194**) viewed along the C-2—O-1/C-4—C-10 bonds with the A-ring depicted horizontally; the analogous A-conformer (**193**); and the A-conformer viewed along the C-3—C-4 bond, showing the offset face-to-face arrangement of the B- and D-rings and the π–π and π–alkyl interactions. Substituents have been omitted for clarity.

parallel to the O-1—C-2 bond at an interplanar distance of $\sim 3.5^{\circ}$ by MMXP (Figure 3). Even though this geometry is achieved by MMXP, principally via van der Waals and electrostatic interactions, it also conforms with an offset face-to-face arrangement (**195**) required for π-stacking (stabilizing π–σ attraction).[270] This is probably further reinforced by a π–CH interaction[271] between 3-OMe(B) and the π-system of the A-ring (see structure (**195**)).

(4→8)-Linked procyanidin dimers possess detectable conformational isomers resulting from steric interactions in the vicinity of the interflavanoid bond[86,272] which exhibit two sets of ¹H NMR signals[159] as well as heterogeneous fluorescence decay.[273] Conformations (**196**) and (**197**) of the 2,3-*cis*- and -*trans*-dimers, respectively, correspond to that in which the C-4 proton eclipses the aromatic A-ring of the lower flavanyl unit, with the bulky C-2 and C-4(axial) substituents occupying positions of least steric interaction with the lower flavan unit. In CDCl₃, conformation (**199**) predominates in 2,3-*trans* dimer peracetates whereas conformers (**196**) and (**198**) occur in nearly equal proportions for 2,3-*cis* dimer decaacetates.[86,189] These models have also led to the proposal[274] of a transition polarization diagram to account for the sign of the short-wavelength c.d. couplet of dimeric procyanidins.

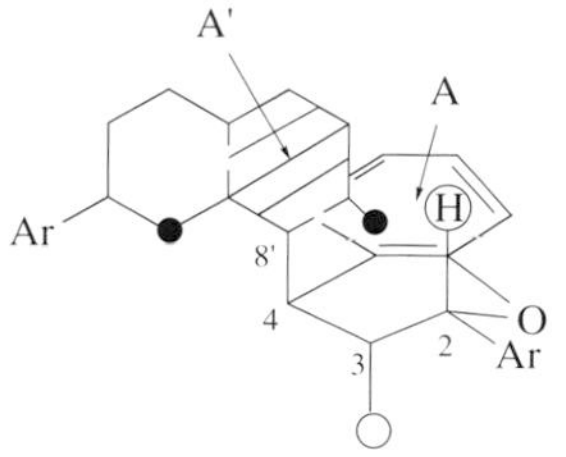

(196)
2,3-*cis*

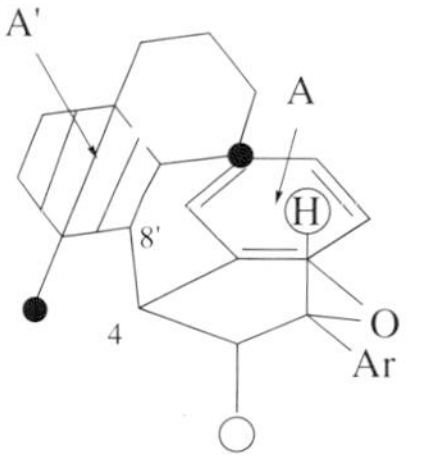

(198)
2,3-*cis*

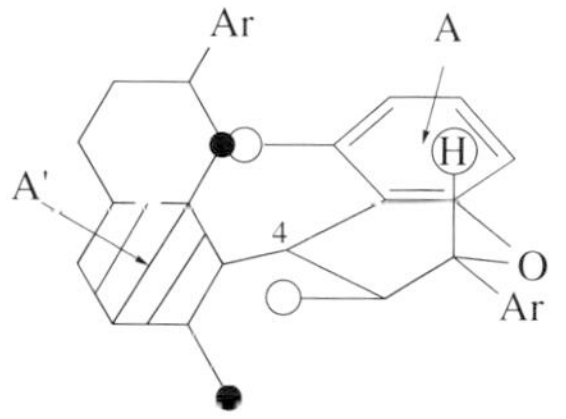

(197)
2,3-*trans*

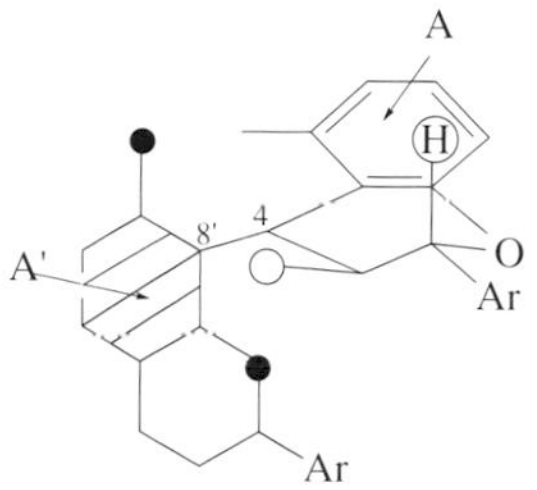

(199)
2,3-*trans*

It was demonstrated[275] that crystalline tetra-*O*-methylcatechin exists in two different conformations. The substituents of the pyran ring attain equatorial orientations of the C-3—O-H and C-4′—O—Me bonds, which result from optimization of intermolecular hydrogen bonding. In solution, however, the A- and E-conformations are present in approximately 40:60 relative populations, hence mutating the heterocyclic proton dihedral angles, which roughly explains the observed $^3J_{2,3}$ ^{1}H NMR value of 8.1 Hz. A GMMX conformational search routine, in contrast, gives an ensemble of conformations that reflect the Boltzmann-averaged heterocyclic ring conformations of tetra-*O*-methylcatechin.[276] This approach led to the prediction of all three coupling constants of heterocyclic protons of this compound with a remarkable degree of accuracy (observed: $J_{2,3} = 8.1$, $J_{3,4eq} = 5.5$, $J_{3,4ax} = 9.0$ Hz; calculated: $J_{2,3} = 8.15$, $J_{3,4eq} = 5.25$, $J_{3,4ax} = 9.84$ Hz). This method thus possesses obvious advantages compared with the approach that assumes a distribution of time spent between A- and E-conformer idealized states.

The presence or absence of a C-5 hydroxy group at the A-ring has a profound influence on the reactivity of the C-6 and C-8 positions of flavan-3-ols with electrophiles as well as the stability of the interflavanyl bond in oligomeric proanthocyanidins. In an investigation by Tobiason *et al.*,[277] the crystal structure, conformational analyses, and charge density distributions of *ent*-epifisetinidol **(200)** were studied as a model for the profisetinidin class of oligomeric proanthocyanidins and the results compared with those obtained for epicatechin **(3)**. Molecular modeling and molecular orbital analyses of compound **(200)** gave good predictions of the approximate "reverse half-chair" conformation found for the crystal structure. MNDO and AM1 analyses of HOMO electron densities permitted the same authors[277] to explain for the first time the remarkable degree of regioselectivity at C-6 (A-ring) which is observed when 5-deoxyflavan-3-ols are treated with electrophilic reagents.[75,76]

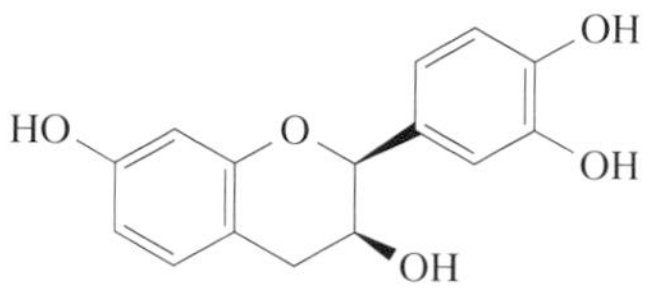

(200)

The C-5 hydroxy group of the A-ring also influences the fluorescence properties of the procyanidin monomers and dimers compared with those of analogous compounds in the profisetinidin series.[278] There is a measurable heterogeneity in the fluorescence of fisetinidol (5-deoxy) in contrast to the simpler fluorescence of the procyanidin monomers, catechin **(2)** and epicatechin **(3)**. This heterogeneity is attributed to differences in the photophysical properties of the aromatic A- and B-rings in fisetinidol which are larger in this compound than in catechin and epicatechin. In the absence of a conformational constraint that forces the occupation of a single rotational isomer at the inter-flavanoid bond, dimeric procyanidins and profisetinidins exhibit heterogeneous decay of fluor-escence, which can be used to assign the populations of the two rotamers at the interflavanyl bond in the procyanidins but not in the profisetinidin series of compounds.

The conformation of the heterocyclic rings in the upper and lower flavan-3-ol units, and the conformations of major and minor rotational isomers about the interflavanyl linkages, were recently assessed for a series of methyl ether acetate derivatives of diastereomeric dimeric profisetinidins by application of COSY and NOE experiments.[62] These results indicated that the eight fisetinidol- and *ent*-fisetinidol-(4→6)- and -(4→8)-catechins were present in two rotameric forms. They are depicted by formulations (**201**) and (**202**) for the fisetinidol-(4α→8)-catechins, and by (**203**) and (**204**) for the fisetinidol-(4α→6)-catechins, conformations (**201**) and (**204**) being the more crowded or more com-pact conformations. A conspicuous preference of all compounds for these compressed con-

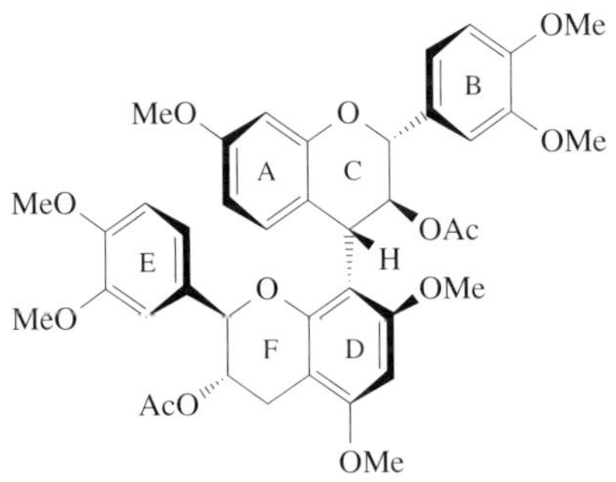

(**201**)

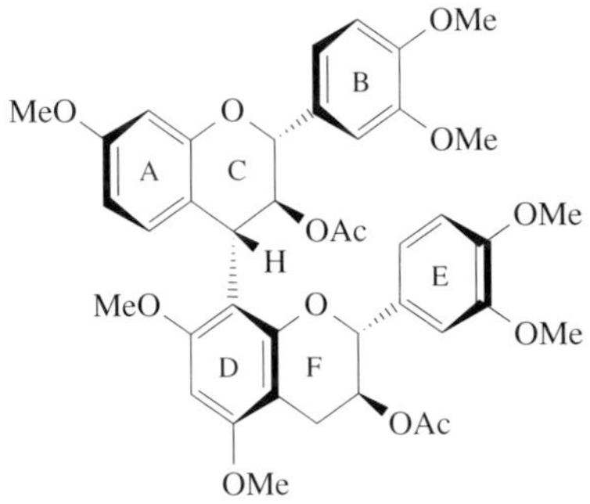

(**202**)

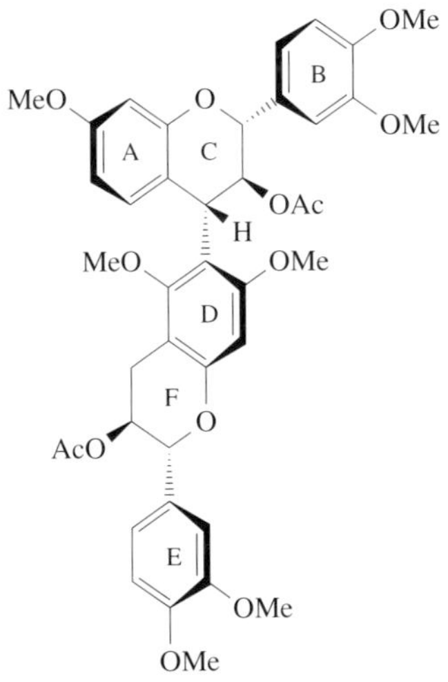

(**203**)

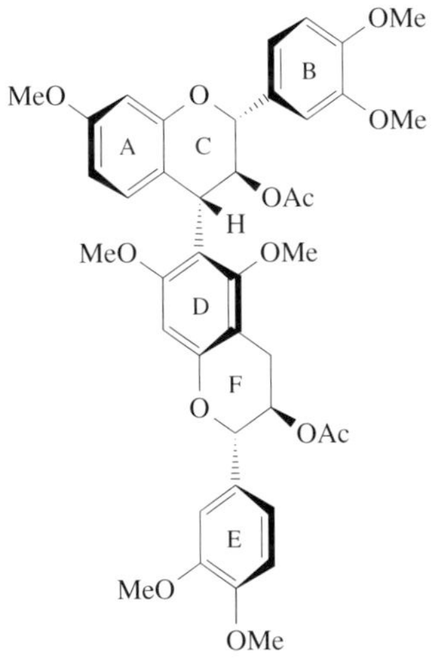

(**204**)

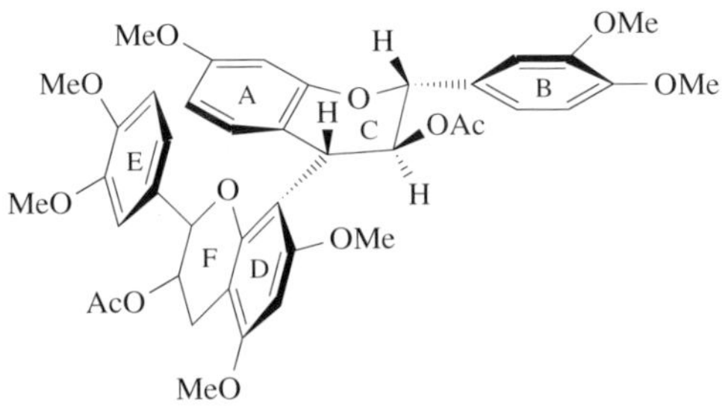

(**205**)

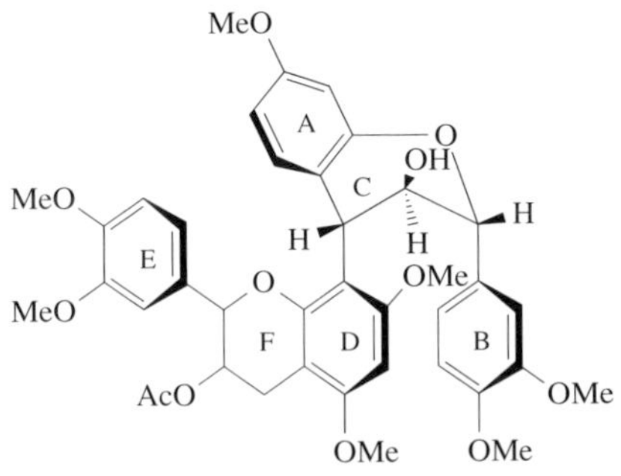

(**206**)

formations was observed, presumably to minimize the surface area of the molecule, and hence solute–solvent contact.[159] The heterocyclic ring in the "upper" chain-extender unit was predominantly in an E-conformation (**205**) rather than an A-conformation (**206**), that is, half-chair in the 2R,3S isomers and a "reverse" half-chair in the 2S,3R isomers, while the pyran ring conformation of the terminal catechin unit, although mostly in an E-conformation, was often represented by significant proportions of A-conformers.

Proanthocyanidin polymers are of considerable commercial importance due to their potential as a renewable source of useful chemicals,[6] their probable use by plants as a defense mechanism,[279] and their formation of complexes with a variety of naturally occurring and synthetic polymers.[280–283] Owing to the purported importance of the conformation of oligoflavanoids towards these phenomena, continued study in this regard would provide a much needed foundation to understanding the intricate physicochemical nature of these interactions.

3.19.7 ASTRINGENCY

Polyphenolic compounds, including the oligomeric proanthocyanidins, have a harsh astringent taste and produce in the palate a feeling of roughness, dryness, and constriction.[284] These substances thus contribute significantly if not uniquely to the astringency of wines, fruits and fruit juices, teas, and other beverages. The primary reaction whereby astringency develops is via precipitation of proteins and mucopolysaccharides in the mucous secretions.[285] Mammalian herbivores produce unique proline-rich salivary proteins (PRPs) which have a high affinity for polyphenols.[286] In humans these PRPs appear to be essential and are present in amounts which reflect the approximate level of polyphenols and related phenolics in the normal diet. It has thus been suggested that the PRPs constitute the first line of defense against polyphenols in the digestive tract.[286] This has led to intensive investigations of the action of polysaccharides and proline-rich peptides in the moderation of astringent response. The current status of knowledge in this regard is summarized by Luck *et al.*[285] A paper by Helsper *et al.*[287] discussing the trypsin inhibitor activity of the proanthocyanidins from *Vicia faba* L. (faba beans) is also informative.

The fruits of astringent Japanese persimmon cultivars are edible after artificial removal of the astringency by treatment with ethanol vapor, carbon dioxide gas, or warm water.[288] During these anaerobic treatments, acetaldehyde accumulates and concomitantly the water-soluble oligoflavanoids are gradually changed chemically via the formation of 1,1-ethylidene bridges into insoluble forms to decrease the astringency. This phenomenon[288] was confirmed by "removal" of the astringency from persimmon fruit using ethanol and the subsequent thiol-promoted ($HSCH_2CH_2OH/H^+$) degradation of the insolubilized proanthocyanidin polymers to give 4α-(2-hydroxymethylsulfanyl)-6- and -8-[1-(2-hydroxyethylsulfanyl)ethyl]flavan-3-ols (e.g., the 8-substituted epigallocatechin derivative (**207**).

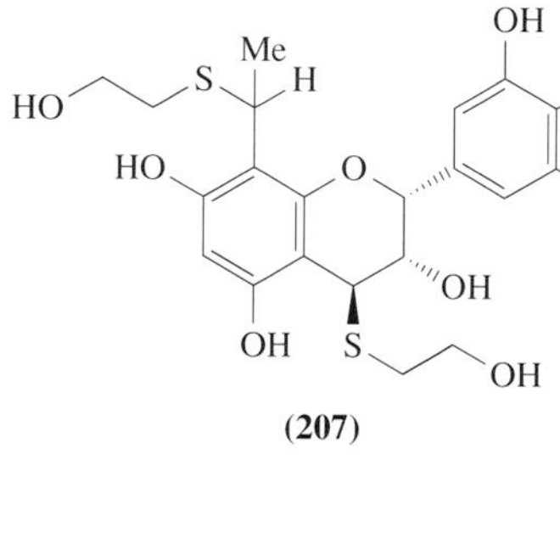

(**207**)

3.19.8 ROLE OF POLYPHENOLS AS CHEMOPREVENTERS

It has been well established that dietary factors play a major role in the development of human chronic diseases, such as cardiovascular disorders and cancer.[289,290] Human diet, in addition to the essential nutrients, contains a number of natural nonnutritional components, some of which may provide protection against these chronic diseases.[291,292] These compounds, described either as functional foods, nutraceuticals, chemopreventers, or phytochemicals, can be found in many types of foods, with polyphenols, including the condensed and hydrolyzable tannins, from vegetables, fruits, tea, and red wine providing the most apparent beneficial effects for human health.[291,293]

The major beneficial component in green tea, epigallocatechin-3-gallate (**208**), possesses strong antioxidant properties.[294–296] The anticarcinogenic activity of compound (**208**) may be related to several factors, that is, its effect on the tumor promotion stage of cancer processes,[297] its effect on DNA-adduct formation and on the scavenging of free radicals,[298] or the increase of antioxidant activities.[299] It should, however, be emphasized that the concentrations of potent antioxidants like the epigallocatechin derivative (**208**) are significantly reduced in the manufacturing of "black tea" via "fermentation" of green tea, by oxidative conversion into the theaflavin fraction (see Section 3.19.5.4).

(**208**)

Following several independent reports about the "French paradox", an apparent compatibility of a high-fat diet with a low incidence of coronary atherosclerosis, the beneficial effect of regular drinking of red wine has been the subject of investigation by a number of researchers.[300–303] Besides the beneficial effects of the moderate intake of alcohol (5–10 g day^{-1}) for adult consumers, components other than alcohol contribute significantly to the beneficial effects of red wine on coronary heart disease.[304]

Several phenolic compounds in red wine exert potent antioxidant effects,[301] and as such may act as chemopreventive components. In addition to a number of complementary biological mechanisms, the phenolic substances in red wine such as the catechins, epicatechins, quercetin (5,7,3′,4′-tetrahydroxyflavonol),[305] resveratrol (3,4,3′,5′-tetrahydroxystilbene), anthocyanins, and procyanidins,[302,306,307] were presumed to be responsible for the beneficial effects of red wine on coronary heart disease.[308] A red wine extract from which the ethanol had been removed inhibited the oxidation of low-density lipoprotein and thus presumably thrombotic phenomena.[301] Additional information may be extracted from a number of good review articles,[292,309–312] and also from some original articles.[313–315]

3.19.9   REFERENCES

1. K. Freudenberg and K. Weinges, in "The Chemistry of Flavonoid Compounds," ed. T. A. Geissman, Macmillan, New York, 1962, p. 197.
2. K. Weinges, W. Bähr, W. Ebert, K. Göritz, and H.-D. Marx, *Fortschr. Chem. Org. Naturst.*, 1969, **27**, 158.
3. E. Haslam, in "The Flavonoids," eds. J. B. Harborne, T. J. Mabry, and H. Mabry, Academic Press, New York, 1975, p. 505.
4. E. Haslam, in "The Flavonoids—Advances in Research," eds. J. B. Harborne and T. J. Mabry, Chapman and Hall, London, 1982, p. 417.
5. L. J. Porter, in "The Flavonoids—Advances in Research since 1980," ed. J. B. Harborne, Chapman and Hall, London, 1988, p. 21.
6. R. W. Hemingway, in "Natural Products of Woody Plants 1," ed. J. W. Rowe, Springer-Verlag, New York, 1989, p. 571.
7. L. J. Porter, in "Natural Products of Woody Plants 1," ed. J. W. Rowe, Springer-Verlag, New York, 1989, p. 651.
8. D. Ferreira, J. P. Steynberg, D. G. Roux, and E. V. Brandt, *Tetrahedron*, 1992, **48**, 1743.
9. R. W. Hemingway and P. E. Laks (eds.), "Plant Polyphenols. Synthesis Properties, Significance," Plenum Press, New York, 1992.
10. L. J. Porter, in "The Flavonoids—Advances in Research since 1986," ed. J. B. Harborne, Chapman and Hall, London, 1994, p. 23.
11. D. Ferreira and R. Bekker, *Nat. Prod. Rep.*, 1996, **13**, 411.
12. H. Geiger, in "The Flavonoids—Advances in Research since 1986," ed. J. B. Harborne, Chapman and Hall, London, 1994, p. 95.
13. R. W. Hemingway, L. Y. Foo, and L. J. Porter, *J. Chem. Soc., Perkin Trans. 1*, 1982, 1209.
14. H. Kolodziej, D. Ferreira, G. Lemière, T. De Bruyne, L. Pieters, and A. J. Vlietinck, *J. Nat. Prod.*, 1993, **56**, 1199.
15. D. G. Roux and D. Ferreira, *Phytochemistry*, 1974, **13**, 2039.

16. J. Ebel and K. Hahlbrock, in "The Flavonoids—Advances in Research," eds. J. B. Harborne and T. J. Mabry, Chapman and Hall, London, 1982, p. 641.
17. D. Jacques, C. T. Opie, L. J. Porter, and E. Haslam, *J. Chem. Soc., Perkin Trans. 1*, 1977, 1637.
18. B. A. Bohm, in "The Flavonoids—Advances in Research Since 1980," ed. J. B. Harborne, Chapman and Hall, London, 1988, p. 329.
19. J. W. Clark-Lewis and W. Korytnyk, *J. Chem. Soc.*, 1958, 2367.
20. L. J. Porter and L. Y. Foo, *Phytochemistry*, 1982, **21**, 2947.
21. L. J. Porter, R. Y. Wong, and B. G. Chan, *J. Chem. Soc., Perkin Trans. 1*, 1985, 1413.
22. W. Heller, G. Forkmann, L. Britsch, and H. Grisebach, *Planta*, 1985, **165**, 284.
23. K. N. Kristiansen, *Carlsberg Res. Commun.*, 1986, **51**, 51.
24. H. A. Stafford, H. H. Lester, and L. J. Porter, *Phytochemistry*, 1985, **24**, 333.
25. L. Y. Foo, *Phytochemistry*, 1987, **26**, 813.
26. A. Sakurai, K. Okada, and Y. Okamura, *Bull. Chem. Soc. Jpn.*, 1982, **55**, 3051.
27. G. Nonaka, Y. Goto, J. Kinjo, T. Nohara, and I. Nishioka, *Chem. Pharm. Bull.*, 1987, **35**, 1105.
28. K. Freudenberg, H. Fikentscher, M. Harder, and O. Schmidt, *Justus Liebigs Ann. Chem.*, 1925, **444**, 135.
29. K. Freudenberg, *Sci. Proc. R. Dublin Soc.*, 1956, **27**, 153.
30. A. J. Birch, J. W. Clark-Lewis, and A. V. Robertson, *J. Chem. Soc.*, 1957, 3586.
31. E. Hardegger, H. Gempeler, and A. Zuest, *Helv. Chim. Acta*, 1957, **40**, 1819.
32. A. Zuest, F. Lohse, and E. Hardegger, *Helv. Chim. Acta*, 1960, **43**, 1274.
33. W. Mayer and G. Bauni, *Justus Liebigs Ann. Chem.*, 1958, **611**, 264.
34. K. S. Saini and S. Ghosal, *Phytochemistry*, 1984, **23**, 2415.
35. A. J. Birch and M. Salahuddin, *Tetrahedron Lett.*, 1964, 2211.
36. A. J. Birch, C. J. Dahl, and A. Pelter, *Tetrahedron Lett.*, 1967, 481.
37. G. Cardillo, L. Merlini, and G. Nasini, *J. Chem. Soc. C*, 1971, 3967.
38. R. Braz Filho, M. S. da Silva, and O. R. Gottlieb, *Phytochemistry*, 1980, **19**, 1195.
39. R. Sahai, S. K. Agarwal, and R. P. Rastogi, *Phytochemistry*, 1980, **19**, 1560.
40. R. Braz Filho, P. P. Diaz, and O. R. Gottlieb, *Phytochemistry*, 1980, **19**, 455.
41. A. Numata, T. Takemura, H. Ohbayashi, T. Katsuno, K. Yamamoto, K. Sato, and S. Kobayashi, *Chem. Pharm. Bull.*, 1983, **31**, 2146.
42. F. E. King and W. Bottomley, *Chem. Ind. (London)*, 1953, 1368.
43. F. E. King and W. Bottomley, *J. Chem. Soc.*, 1954, 1399.
44. K. Freudenberg and D. G. Roux, *Naturwissenschaften*, 1954, **41**, 450.
45. D. G. Roux and K. Freudenberg, *Justus Liebigs Ann. Chem.*, 1958, **613**, 56.
46. J. W. Clark-Lewis and G. F. Katekar, *Proc. Chem. Soc.*, 1960, 345.
47. J. W. Clark-Lewis, G. F. Katekar, and P. I. Mortimer, *J. Chem. Soc.*, 1961, 499.
48. J. W. Clark-Lewis and I. Dainis, *Aust. J. Chem.*, 1964, **17**, 1170.
49. J. W. Clark-Lewis and I. Dainis, *Aust. J. Chem.*, 1967, **20**, 2191.
50. J. W. Clark-Lewis, *Aust. J. Chem.*, 1968, **21**, 2059.
51. S. E. Drewes and D. G. Roux, *Biochem. J.*, 1964, **90**, 343.
52. S. E. Drewes and D. G. Roux, *Biochem. J.*, 1965, **94**, 482.
53. S. E. Drewes and D. G. Roux, *Biochem. J.*, 1965, **96**, 681.
54. S. E. Drewes and D. G. Roux, *Biochem. J.*, 1966, **98**, 493.
55. H. M. Saayman and D. G. Roux, *Biochem. J.*, 1965, **96**, 36.
56. B. R. Brown and M. R. Shaw, *J. Chem. Soc., Perkin Trans. 1*, 1974, 2036.
57. J. P. Steynberg, J. F. W. Burger, D. A. Young, E. V. Brandt, J. A. Steenkamp, and D. Ferreira, *J. Chem. Soc., Chem. Commun.*, 1988, 1055.
58. J. P. Steynberg, J. F. W. Burger, D. A. Young, E. V. Brandt, J. A. Steenkamp, and D. Ferreira, *J. Chem. Soc., Perkin Trans. 1*, 1988, 3323, 3331.
59. J. A. Steenkamp, J. C. S. Malan, and D. Ferreira, *J. Chem. Soc., Perkin Trans. 1*, 1988, 2179.
60. J. P. Steynberg, J. F. W. Burger, D. A. Young, E. V. Brandt, and D. Ferreira, *Heterocycles*, 1989, **28**, 923.
61. J. P. Steynberg, E. V. Brandt, and D. Ferreira, *J. Chem. Soc., Perkin Trans. 2*, 1991, 1569.
62. J. P. Steynberg, E. V. Brandt, D. Ferreira, C. A. Helfer, W. L. Mattice, D. Gornik, and R. W. Hemingway, *Magn. Reson. Chem.*, 1995, **33**, 611.
63. L. J. Porter, R. Y. Wong, M. Benson, B. G. Chan, V. N. Vishwanadhan, R. D. Gandour, and W. L. Mattice, *J. Chem. Res.*, 1986, (S), 86; (M), 830.
64. P. M. Viviers, J. J. Botha, D. Ferreira, D. G. Roux, and H. M. Saayman, *J. Chem. Soc., Perkin Trans. 1*, 1983, 17.
65. J. C. S. Malan, P. J. Steynberg, J. P. Steynberg, D. A. Young, B. C. B. Bezuidenhoudt, and D. Ferreira, *Tetrahedron*, 1990, **46**, 2883.
66. R. W. Hemingway and L. Y. Foo, *J. Chem. Soc., Chem. Commun.*, 1983, 1035.
67. M. R. Attwood, B. R. Brown, S. G. Lisseter, C. L. Torrero, and P. M. Weaver, *J. Chem. Soc., Chem. Commun.*, 1984, 177.
68. L. Y. Foo and R. W. Hemingway, *J. Chem. Soc., Chem. Commun.*, 1984, 85.
69. F. R. van Heerden, E. V. Brandt, D. Ferreira, and D. G. Roux, *J. Chem. Soc., Perkin Trans. 1*, 1981, 2483.
70. J. C. S. Malan, D. A. Young, J. P. Steynberg, and D. Ferreira, *J. Chem. Soc., Perkin Trans. 1*, 1990, 219.
71. J. W. Clark-Lewis and P. I. Mortimer, *J. Chem. Soc.*, 1960, 4106.
72. L. Y. Foo and H. Wong, *Phytochemistry*, 1986, **25**, 1961.
73. J. J. Botha, D. Ferreira, and D. G. Roux, *J. Chem. Soc., Chem. Commun.*, 1978, 698.
74. J. J. Botha, D. A. Young, D. Ferreira, and D. G. Roux, *J. Chem. Soc., Perkin Trans. 1*, 1981, 1213.
75. J. J. Botha, D. Ferreira, and D. G. Roux, *J. Chem. Soc., Chem. Commun.*, 1978, 700.
76. J. J. Botha, D. Ferreira, and D. G. Roux, *J. Chem. Soc., Perkin Trans. 1*, 1981, 1235.
77. J. A. Delcour, D. Ferreira, and D. G. Roux, *J. Chem. Soc., Perkin Trans. 1*, 1983, 1711.
78. J. A. Delcour, E. J. Serneels, D. Ferreira, and D. G. Roux, *J. Chem. Soc., Perkin Trans. 1*, 1985, 669.
79. H. Kolodziej, *Phytochemistry*, 1985, **24**, 2460.

80. H. Kolodziej, *Phytochemistry*, 1986, **25**, 1209.
81. H. Kolodziej, *Phytochemistry*, 1990, **29**, 1671.
82. S. Schleep, H. Friedrich, and H. Kolodziej, *J. Chem. Soc., Chem. Commun.*, 1986, 392.
83. B. R. Brown and J. A. H. MacBride, *J. Chem. Soc.*, 1964, 3822.
84. I. C. du Preez, D. Ferreira, and D. G. Roux, *J. Chem. Soc. C*, 1971, 336.
85. R. S. Thompson, D. Jacques, E. Haslam, and R. J. N. Tanner, *J. Chem. Soc., Perkin Trans. 1*, 1972, 1387.
86. A. C. Fletcher, L. J. Porter, E. Haslam, and R. K. Gupta, *J. Chem. Soc., Perkin Trans. 1*, 1977, 1628.
87. R. W. Hemingway, J. J. Karchesy, G. W. McGraw, and R. A. Wielesek, *Phytochemistry*, 1983, **22**, 275.
88. J. Lam and P. Wrang, *Phytochemistry*, 1975, **14**, 1621.
89. N. Tanaka, T. Sada, T. Murakami, Y. Saiki, and C. M. Chen, *Chem. Pharm. Bull.*, 1984, **32**, 490.
90. N. Tanaka, T. Murakami, H. Wada, A. B. Gutierrez, Y. Saiki, and C. M. Chen, *Chem. Pharm. Bull.*, 1985, **33**, 5231.
91. S. Morimoto, G. Nonaka, R. Chen, and I. Nishioka, *Chem. Pharm. Bull.*, 1988, **36**, 39.
92. Y. Kasahara and H. Hikino, *Heterocycles*, 1983, **20**, 1953.
93. L. Camarda, L. Merlini, and G. Nasini, "*Studies in Organic Chemistry: Proceedings International Bioflavonoid Symposium*," Munich, 1981, p. 311.
94. E. Malan, A. Sireeparsad, E. Swinny, and D. Ferreira, *Phytochemistry*, 1997, **44**, 529.
95. W. G. C. Forsyth and J. B. Roberts, *Biochem. J.*, 1960, **74**, 374.
96. K. Freudenberg and K. Weinges, *Tetrahedron Lett.*, 1961, 267.
97. K. Freudenberg and K. Weinges, *Angew. Chem.*, 1962, **74**, 182.
98. T. A. Geissman and H. F. Dittmar, *Phytochemistry*, 1965, **4**, 359.
99. K. Weinges and K. Freudenberg, *J. Chem. Soc., Chem. Commun.*, 1965, 220.
100. L. L. Creasy and T. Swain, *Nature (London)*, 1965, **208**, 151.
101. S. E. Drewes, D. G. Roux, J. Feeney and S. H. Eggers *J. Chem. Soc., Chem. Commun.*, 1966, 368.
102. S. E. Drewes, D. G. Roux, S. H. Eggers, and J. Feeney, *J. Chem. Soc.* 1967, 1217.
103. S. E. Drewes, D. G. Roux, H. M. Saayman, S. H. Eggers, and J. Feeney, *J. Chem. Soc. C*, 1967, 1302.
104. K. Weinges, K. Kaltenhäuser, H. D. Marx, E. Nader, F. Nader, J. Perner, and D. Seiler, *Justus Liebigs Ann. Chem.*, 1968, **711**, 184.
105. K. Weinges, K. Göritz, and F. Nader, *Justus Liebigs Ann. Chem.*, 1968, **715**, 164.
106. T. A. Geissman and N. N. Yoshimura, *Tetrahedron Lett.*, 1966, 2669.
107. K. Weinges and J. Perner, *J. Chem. Soc., Chem. Commun.* 1967, 351.
108. K. Weinges, J. Perner, and H. D. Marx, *Chem. Ber.*, 1970, **103**, 2344.
109. J. J. Botha, D. Ferreira, and D. G. Roux, and W. E. Hull *J. Chem. Soc., Chem. Commun.*, 1979, 510.
110. J. J. Botha, P. M. Viviers, D. A. Young, I. C. du Preez, D. Ferreira, D. G. Roux, and W. E. Hull, *J. Chem. Soc., Perkin Trans. 1*, 1982, 527.
111. M. J. Betts, B. R. Brown, P. E. Brown, and W. T. Pike, *J. Chem. Soc., Chem. Commun.*, 1967, 1110.
112. M. J. Betts, B. R. Brown, and M. R. Shaw, *J. Chem. Soc. C*, 1969, 1178.
113. I. C. du Preez, T. G. Fourie, and D. G. Roux, *J. Chem. Soc., Chem. Commun.*, 1971, 333.
114. T. G. Fourie, I. C. du Preez, and D. G. Roux, *Phytochemistry*, 1972, **11**, 1763.
115. F. Delle-Monache, F. Ferrari, and G. B. Marini-Bettolo, *Gazz. Chim. Ital.*, 1971, **101**, 387.
116. K. D. Sears and R. L. Casebier, *J. Chem. Soc., Chem. Commun.*, 1968, 1437.
117. K. D. Sears and R. L. Casebier, *Phytochemistry*, 1970, **9**, 1589.
118. E. Haslam, *J. Chem. Soc., Chem. Commun.*, 1974, 594.
119. P. M. Viviers, D. A. Young, J. J. Botha, D. Ferreira, D. G. Roux, and W. E. Hull, *J. Chem. Soc., Perkin Trans. 1*, 1982, 535.
120. L. Y. Foo, *J. Chem. Soc., Chem. Commun.*, 1986, 236.
121. J. C. S. Malan, D. A. Young, J. A. Steenkamp, and D. Ferreira, *J. Chem. Soc., Perkin Trans. 1*, 1988, 2567.
122. J. Coetzee, J. P. Steynberg, P. J. Steynberg, E. V. Brandt, and D. Ferreira, *Tetrahedron*, 1995, **51**, 2339.
123. S. E. Drewes and A. H. Ilsley, *J. Chem. Soc. C.*, 1969, 897.
124. D. A. Young, D. Ferreira, and D. G. Roux, *J. Chem. Soc., Perkin Trans. 1*, 1983, 2031.
125. L. Y. Foo, *J. Chem. Soc., Chem. Commun.*, 1989, 1505.
126. A. Pelter and P. I. Amenechi, *J. Chem. Soc. C* 1969, 887.
127. A. Pelter, P. I. Amenechi, R. Warren, and S. H. Harper, *J. Chem. Soc. C*, 1969, 2572.
128. H. K. L. Hundt and D. G. Roux, *J. Chem. Soc., Chem. Commun.*, 1978, 696.
129. H. K. L. Hundt and D. G. Roux *J. Chem. Soc., Perkin Trans. 1*, 1981, 1227.
130. D. W. Engel, M. Hattingh, H. K. L. Hundt, and D. G. Roux, *J. Chem. Soc., Chem. Commun.*, 1978, 695.
131. K. Kolodziej, D. Ferreira, and D. G. Roux, *J. Chem. Soc., Perkin Trans. 1*, 1984, 343.
132. E. Young, E. V. Brandt, D. A. Young, D. Ferreira, and D. G. Roux, *J. Chem. Soc., Perkin Trans. 1*, 1986, 1737.
133. J. C. S. Malan, J. A. Steenkamp, J. P. Steynberg, D. A. Young, E. V. Brandt, and D. Ferreira, *J. Chem. Soc., Perkin Trans. 1*, 1990, 209.
134. J. A. Steenkamp, J. C. S. Malan, D. G. Roux, and D. Ferreira, *J. Chem. Soc., Perkin Trans. 1*, 1988, 1325.
135. J. P. Steynberg, D. Ferreira, and D. G. Roux, *Tetrahedron Lett.*, 1983, **24**, 4147.
136. J. P. Steynberg, D. Ferreira, and D. G. Roux *J. Chem. Soc., Perkin Trans. 1*, 1987, 1705.
137. G. Snatzke, M. Katjar, and F. Snatzke, in "Fundamental Aspects and Recent Developments in Optical Rotatory Dispersion and Circular Dichroism," eds. F. Ciardelli and P. Salvadori, Heyden, London, 1973, p. 148.
138. M. W. Barrett, W. Klyne, P. M. Scopes, A. C. Fletcher, L. J. Porter, and E. Haslam, *J. Chem. Soc., Perkin Trans. 1*, 1979, 2375.
139. J. H. van der Westhuizen, D. Ferreira, and D. G. Roux, *J. Chem. Soc., Perkin Trans. 1*, 1981, 1220.
140. G. G. DeAngelis and W. C. Wildman, *Tetrahedron*, 1969, **25**, 5099.
141. F. Geiss, M. Heinrich, D. Hunkler, and H. Rimpler, *Phytochemistry*, 1995, **39**, 635.
142. Y. Kashiwada, H. Iizuka, K. Yoshioka, R.-F. Chen, G.-i Nonaka, and I. Nishioka, *Chem. Pharm. Bull*, 1990, **38**, 888.
143. J. J. Karchesy and R. W. Hemingway, *J. Agric. Food Chem.*, 1980, **28**, 222.
144. Z. Czochanska, L. Y. Foo, R. H. Newman, and L. J. Porter, *J. Chem. Soc., Perkin Trans. 1*, 1980, 2278.
145. L. Y. Foo and L. J. Porter, *J. Chem. Soc., Perkin Trans. 1*, 1978, 1186.

146. L. J. Porter, R. H. Newman, L. Y. Foo, H. Wong, and R. W. Hemingway, *J. Chem. Soc., Perkin Trans. 1*, 1982, 1217.
147. A. G. Brown, W. B. Eyton, A. Holmes, and W. D. Ollis, *Phytochemistry*, 1969, **8**, 2333.
148. R. W. Hemingway, G. W. McGraw, J. J. Karchesy, L. Y. Foo, and L. J. Porter, *J. Appl. Polym. Sci.: Appl. Polym. Symp.*, 1983, **37**, 967.
149. J. F. Beart, T. H. Lilley, and E. Haslam, *J. Chem. Soc., Perkin Trans. 2*, 1985, 1439.
150. K. Freudenberg, O. Bohme, and L. Purrmann, *Chem. Ber.*, 1922, **55**, 1734.
151. K. Freudenberg and L. Purrmann, *Chem. Ber.*, 1923, **56**, 1185.
152. K. Freudenberg and L. Purrmann, *Justus Liebigs Ann. Chem.*, 1924, **437**, 274.
153. P. P. Mehta, and W. B. Whalley *J. Chem. Soc.*, 1963, 5327.
154. K. D. Sears, R. L. Casebier, H. L. Hergert, G. H. Stout, and L. E. McCandlish, *J. Org. Chem.*, 1974, **39**, 3244.
155. P. Courbat, A. Weith, A. Albert, and A. Pelter, *Helv. Chim. Acta*, 1977, **60**, 1665.
156. J. A. Kennedy, M. H. G. Munro, H. K. J. Powell, L. J. Porter, and L. Y. Foo, *Aust. J. Chem.*, 1984, **37**, 885.
157. J. A. Kuhnle, J. J. Windle, and A. C. Waiss, *J. Chem. Soc. B*, 1969, 613.
158. O. N. Jensen and J. A. Pedersen, *Tetrahedron*, 1983, **39**, 1609.
159. L. Y. Foo and L. J. Porter, *J. Chem. Soc., Perkin Trans. 1*, 1983, 1535
160. P. Kiatgrajai, J. D. Wellons, L. Gollob, and J. D. White, *J. Org. Chem.*, 1982, **47**, 2910.
161. A. Pizzi, "Wood Adhesives: Chemistry and Technology," Marcel Dekker, New York, 1983.
162. R. E. Kreibich and R. W. Hemingway, in "Proceedings of IUFRONTRI Symposium on Wood Adhesives," CSIR, Pretoria, 1985, vol. 17, p. 1.
163. P. E. Laks and R. W. Hemingway, *J. Chem. Soc., Perkin Trans. 1*, 1987, 465.
164. R. W. Hemingway and P. E. Laks, *J. Chem. Soc., Chem. Commun.*, 1985, 746.
165. E. Haslam, *Phytochemistry*, 1977, **16**, 1625.
166. R. V. Platt, C. T. Opie, and E. Haslam, *Phytochemistry*, 1984, **23**, 2211.
167. P. E. Laks, R. W. Hemingway, and A. H. Conner, *J. Chem. Soc., Perkin Trans. 1*, 1987, 1875.
168. P. E. Laks and R. W. Hemingway, *Holzforschung*, 1987, **41**, 287.
169. J. F. W. Burger, H. Kolodziej, R. W. Hemingway, J. P. Steynberg, D. A. Young, and D. Ferreira, *Tetrahedron*, 1990, **46**, 5733.
170. J. P. Steynberg, B. C. B. Bezuidenhoudt, J. F. W. Burger, D. A. Young, and D. Ferreira, *J. Chem. Soc., Perkin Trans. 1*, 1990, 203.
171. Y. Cai, F. J. Evans, M. F. Roberts, J. D. Phillipson, M. H. Zenk, and Y. Y. Gleba, *Phytochemistry*, 1991, **30**, 2033.
172. G. i Nonaka, R. Sakai, and I. Nishioka, *Phytochemistry*, 1984, **23**, 1753.
173. G. W. McGraw, J. P. Steynberg, and R. W. Hemingway, *Tetrahedron Lett.*, 1993, **34**, 987.
174. C.-B. Cui, Y. Tezuka, T. Kikuchi, H. Nakano, T. Tamaoki, and J.-H. Park, *Chem. Pharm. Bull.*, 1991, **39**, 2179.
175. C.-B. Cui, Y. Tezuka, H. Yamashita, T. Kikuchi, H. Nakano, T. Tamaoki, and J. H. Park, *Chem. Pharm. Bull.*, 1993, **41**, 1491.
176. M. D. Tindale and D. G. Roux, *Phytochemistry*, 1969, **8**, 1713.
177. E. V. Brandt, D. A. Young, H. Kolodziej, D. Ferreira, and D. G. Roux, *J. Chem. Soc., Chem. Commun.*, 1986, 913
178. E. V. Brandt, D. A. Young, D. Ferreira, and D. G. Roux, *J. Chem. Soc., Perkin Trans. 1*, 1987, 2353.
179. J. A. Steenkamp, D. Ferreira, D. G. Roux, and W. E. Hull, *J. Chem. Soc., Perkin Trans. 1*, 1983, 23.
180. P. M. Viviers, H. Kolodziej, D. A. Young, D. Ferreira, and D. G. Roux, *J. Chem. Soc., Perkin Trans. 1*, 1983, 2555.
181. D. A. Young, D. Ferreira, and D. G. Roux, *J. Polym. Sci., Part A: Polym. Chem.*, 1986, **24**, 835.
182. J. P. Steynberg, J. F. W. Burger, J. C. S. Malan, A. Cronjé, D. A. Young, and D. Ferreira, *Phytochemistry*, 1990, **29**, 275
183. D. S. Nunes, A. Haag, and H.-J. Bestmann, *Phytochemistry*, 1989, **28**, 2183.
184. J. C. S. Malan, J. A. Steenkamp, D. A. Young, and D. Ferreira, *Tetrahedron*, 1989, **45**, 7859.
185. P. J. Steynberg, J. P. Steynberg, B. C. B. Bezuidenhoudt, and D. Ferreira, *J. Chem. Soc., Chem. Commun.*, 1994, 31.
186. P. J. Steynberg, J. P. Steynberg, B. C. B. Bezuidenhoudt, and D. Ferreira, *J. Chem. Soc., Perkin Trans. 1*, 1995, 3005.
187. J. A. Steenkamp, J. P. Steynberg, E. V. Brandt, D. Ferreira, and D. G. Roux, *J. Chem. Soc., Chem. Commun.*, 1985, 1678.
188. J. P. Steynberg, D. A. Young, J. F. W. Burger, D. Ferreira, and D. G. Roux, *J. Chem. Soc., Chem. Commun.*, 1986, 1013.
189. J. C. S. Malan, D. A. Young, J. P. Steynberg, and D. Ferreira, *J. Chem. Soc., Perkin Trans. 1*, 1990, 227.
190. J. P. Steynberg, J. A. Steenkamp, J. F. W. Burger, D. A. Young, and D. Ferreira, *J. Chem. Soc., Perkin Trans. 1*, 1990, 235.
191. J. P. Steynberg, J. F. W. Burger, A. Cronjé, S. L. Bonnet, J. C. S. Malan, D. A. Young, and D. Ferreira, *Phytochemistry*, 1990, **29**, 2979.
192. S. L. Bonnet, J. P. Steynberg, B. C. B. Bezuidenhoudt, C. M. Saunders, and D. Ferreira, *Phytochemistry*, 1996, **43**, 215.
193. S. L. Bonnet, J. P. Steynberg, B. C. B. Bezuidenhoudt, C. M. Saunders, and D. Ferreira, *Phytochemistry*, 1996, **43**, 229.
194. S. L. Bonnet, J. P. Steynberg, B. C. B. Bezuidenhoudt, C. M. Saunders, and D. Ferreira, *Phytochemistry*, 1996, **43**, 241.
195. S. L. Bonnet, J. P. Steynberg, B. C. B. Bezuidenhoudt, C. M. Saunders, and D. Ferreira, *Phytochemistry*, 1996, **43**, 253.
196. P. J. Steynberg, J. F. W. Burger, B. C. B. Bezuidenhoudt, J. P. Steynberg, M. S. van Dyk, and D. Ferreira, *Tetrahedron Lett.*, 1990, **31**, 2059.
197. A. Cronjé, J. P. Steynberg, E. V. Brandt, D. A. Young, and D. Ferreira, *J. Chem. Soc., Perkin Trans. 1*, 1993, 2467.
198. C. M. Saunders, S. L. Bonnet, J. P. Steynberg, and D. Ferreira, *Tetrahedron*, 1996, **52**, 6003.
199. A. Pizzi, E. Orovan, and F. A. Cameron, *Holzforschung*, 1988, **46**, 67.
200. J. Palazzo de Mello, F. Petereit, and A. Nahrstedt, *Phytochemistry*, 1996, **42**, 857.
201. I. C. du Preez, A. C. Rowan, and D. G. Roux, *J. Chem. Soc., Chem. Commun.*, 1970, 492.
202. D. Ferreira, I. C. du Preez, J. C. Wijnmaalen, and D. G. Roux, *Phytochemistry*, 1985, **24**, 2415.
203. A. D. Patil and V. H. Deshpande, *Indian J. Chem.*, 1982, **21B**, 626.

204. E. Malan, E. Swinny, D. Ferreira, and P. J. Steynberg, *Phytochemistry*, 1996, **41**, 1209.
205. J. W. Clark-Lewis and L. J. Porter, *Aust. J. Chem.*, 1972, **25**, 1943.
206. E. Malan and D. G. Roux, *Phytochemistry*, 1975, **14**, 1835.
207. L. Y. Foo, *J. Chem. Soc., Chem. Commun.*, 1985, 1273.
208. E. Malan and A. Sireeparsad, *Phytochemistry*, 1995, **38**, 237.
209. E. Malan, A. Sireeparsad, J. F. W. Burger, and D. Ferreira, *Tetrahedron Lett.*, 1994, **35**, 7415.
210. E. Malan, *Phytochemistry*, 1995, **40**, 1519.
211. G.-i. Nonaka, Y. Aiko, K. Aritake, and I. Nishioka, *Chem. Pharm. Bull.*, 1992, **40**, 2671.
212. K. Hori, T. Satake, Y. Saiki, T. Murakami, and C. M. Chen, *Chem. Pharm. Bull.*, 1988, **36**, 4301.
213. D. Jacques, E. Haslam, G. R. Bedford, and D. Greatbanks, *J. Chem. Soc., Perkin Trans. 1*, 1974, 2663.
214. A. Cronjé, J. F. W. Burger, E. V. Brandt, H. Kolodziej, and D. Ferreira, *Tetrahedron Lett.*, 1990, **31**, 3789.
215. H. Hikino, N. Shimoyama, Y. Kasahara, M. Takahashi, and C. Konno, *Heterocycles*, 1982, **19**, 1381.
216. W. Mayer, L. Goll, E. M. von Arndt, and A. Mannschreck, *Tetrahedron Lett.*, 1966, 429.
217. P. H. van Rooyen and H. J. P. Redelinghuys, *S. Afr. J. Chem.*, 1983, **36**, 49.
218. A. M. Balde, L. M. van Hoof, L. A. C. Pieters, D. A. R. Van den Berghe, and A. J. Vlietinck, *Phytother. Res.*, 1990, **4**, 182.
219. S. Morimoto, G. Nonaka, and I. Nishioka, *Chem. Pharm. Bull.*, 1987, **35**, 4717.
220. L. J. Porter, Z. Ma, and B. G. Chan, *Phytochemistry*, 1991, **30**, 1657.
221. G.-i. Nonaka, S. Morimoto, and I. Nishioka, *J. Chem. Soc., Perkin Trans. 1*, 1983, 2139.
222. A. M. Balde, L. A. Pieters, A. Gergely, H. Kolodziej, M. Claeys, and A. J. Vlietinck, *Phytochemistry*, 1991, **30**, 337.
223. A. M. Balde, L. A. Pieters, V. Wray, H. Kolodziej, D. A. Vanden Berghe, M. Claeys, and A. J. Vlietinck, *Phytochemistry*, 1991, **30**, 4129.
224. A. M. Balde, T. De Bruyne, L. Pieters, M. Claeys, D. Vanden Berghe, A. Vlietinck, V. Wray, and H. Kolodziej, *J. Nat. Prod.*, 1993, **56**, 1078.
225. A. M. Balde, T. De Bruyne, L. Pieters, H. Kolodziej, D. Vanden Berghe, M. Claeys, and A. Vlietinck, *Phytochemistry*, 1995, **38**, 719.
226. A. M. Balde, T. De Bruyne, L. Pieters, H. Kolodziej, D. Vanden Berghe, M. Claeys, and A. Vlietinck, *Phytochemistry*, 1995, **40**, 933.
227. A. C. Irizar, M. F. Fernandez, A. G. Conzales, and A. G. Ravelo, *J. Nat. Prod.*, 1992, **55**, 450.
228. A. G. Conzales, A. C. Irizar, A. G. Ravelo, and M. F. Fernandez, *Phytochemistry*, 1992, **31**, 1432.
229. C. Santos-Buelga, H. Kolodzieij, and D. Treutter, *Phytochemistry*, 1995, **38**, 499.
230. J. A. Dale and H. S. Mosher, *J. Am. Chem. Soc.*, 1973, **95**, 512.
231. G. R. Sullivan, J. A. Dale, and H. S. Mosher, *J. Org. Chem.*, 1973, **38**, 2143.
232. A. F. Hundt, J. F. W. Burger, J. P. Steynberg, J. A. Steenkamp, and D. Ferreira, *Tetrahedron Lett.*, 1990, **31**, 5073.
233. W. Rossouw, A. F. Hundt, J. A. Steenkamp, and D. Ferreira, *Tetrahedron*, 1994, **50**, 12477.
234. G.-i. Nonaka, S. Morimoto, J. Kinjo, T. Nohara, and I. Nishioka, *Chem. Pharm. Bull.*, 1987, **35**, 149.
235. S. Morimoto, G.-i. Nonaka, and I. Nishioka, *Chem. Pharm. Bull.*, 1985, **33**, 4338; 1988, **36**, 33.
236. F. Hashimoto, G.-i. Nonaka, and I. Nishioka, *Chem. Pharm. Bull.*, 1989, **37**, 3255.
237. Z. Shen, C. P. Falshaw, E. Haslam, and M. J. Begley, *J. Chem. Soc., Chem. Commun.*, 1985, 1135.
238. M. Taniguchi and K. Baba, *Phytochemistry*, 1996, **42**, 1447.
239. K. Baba, M. Taniguchi, and M. Kozawa, *Phytochemistry*, 1994, **37**, 879.
240. G.-i. Nonaka and I. Nishioka, *Chem. Pharm. Bull.*, 1980, **28**, 3145.
241. T. Tanaka, G.-i. Nonaka, and I. Nishioka, *Phytochemistry*, 1983, **22**, 2575.
242. A. Kijjoa, A. M. Giesbrecht, O. R. Gottlieb, and H. E. Gottlieb, *Phytochemistry*, 1981, **20**, 1385.
243. M. Masaoud, H. Ripperger, U. Himmelreich, and G. Adam, *Phytochemistry*, 1995, **38**, 751.
244. U. Himmelreich, M. Masaoud, G. Adam, and H. Ripperger, *Phytochemistry*, 1995, **39**, 949.
245. M. Masaoud, H. Ripperger, A. Porzel, and G. Adam, *Phytochemistry*, 1995, **38**, 745.
246. K. Weinges, and W. Ebert, *Phytochemistry*, 1968, **7**, 153.
247. K. Weinges, W. Ebert, D. Huthwelker, H. Mattauch, and K. Perner, *Justus Liebigs Ann. Chem.*, 1969, **726**, 114.
248. K. Weinges, H. Mattauch, C. Wilkins, and D. Frost, *Justus Liebigs. Ann. Chem.*, 1971, **754**, 124.
249. S. Guyot, J. Vercauteren, and V. Cheynier, *Phytochemistry*, 1996, **42**, 1279.
250. E. A. H. Roberts and R. F. Smith, *J. Sci. Food Agric.*, 1963, **14**, 689.
251. A. G. Brown, W. B. Eyton, A. Holmes and W. D. Ollis, *Nature (London)*, 1969, **221**, 742.
252. D. T. Coxon, A. Holmes, and W. D. Ollis, *Tetrahedron Lett.*, 1970, 5247.
253. A. Critchlow, E. Haslam, R. D. Haworth, P. B. Tinker, and N. M. Waldron, *Tetrahedron*, 1967, **23**, 2829.
254. L. Horner, K. H. Weber, and W. Durckheimer, *Chem. Ber.*, 1961, **94**, 2881.
255. G.-i. Nonaka, O. Kawahara, and I. Nishioka, *Chem. Pharm. Bull.*, 1983, **31**, 3906.
256. F. Hashimoto, G.-i. Nonaka, and I. Nishioka, *Chem. Pharm. Bull.*, 1988, **36**, 1676.
257. F. Hashimoto, G.-i. Nonaka, and I. Nishioka, *Chem. Pharm. Bull*, 1992, **40**, 1383.
258. D. A. Young, E. Young, D. G. Roux, E. V. Brandt, and D. Ferreira, *J. Chem. Soc., Perkin Trans. 1*, 1987, 2345.
259. E. V. Brandt, D. A. Young, E. Young, and D. Ferreira, *J. Chem. Soc., Perkin Trans. 2*, 1987, 1365.
260. G.-i. Nonaka, H. Nishimura, and I. Nishioka, *J. Chem. Soc., Perkin Trans. 1*, 1985, 163.
261. T. Yoshida, F. Nakata, K. Hosotani, A. Nitta, and T. Okuda, *Chem. Pharm. Bull.*, 1992, **40**, 1727.
262. T. Tanaka, N. Ishida, M. Ishimatsu, G.-i. Nonaka, and I. Nishioka, *Chem. Pharm. Bull.*, 1992, **40**, 2092.
263. T. Hatano, L. Han, S. Taniguchi, T. Shingu, T. Okuda, and T. Yoshida, *Chem. Pharm. Bull.*, 1995, **43**, 1629.
264. T. Hatano, L. Han, S. Taniguchi, T. Okuda, Y. Kiso, T. Tanaka, and T. Yoshida, *Chem. Pharm. Bull.*, 1995, **43**, 2033.
265. J. W. Clark-Lewis, L. M. Jackman, and T. M. Spotswood, *Aust. J. Chem.*, 1964, **17**, 632.
266. F. R. Fronczek, G. Gannuch, W. L. Mattice, F. L. Tobiason, J. L. Broeker, and R. W. Hemingway, *J. Chem. Soc., Perkin Trans. 2*, 1984, 1611.
267. F. Baert, R. Fouret, M. Sliwa, and H. Sliwa, *Tetrahedron*, 1980, **36**, 2765.
268. F. Baert, R. Fouret, M. Sliwa, and H. Sliwa *Acta Crystallogr., Sect. B*, 1983, **39**, 444.
269. J. P. Steynberg, E. V. Brandt, M: J. H. Hoffman, R. W. Hemingway, and D. Ferreira, in "Plant Polyphenols: Synthesis, Properties, Significance," eds. R. W. Hemingway, P. E. Laks, and S. J. Branham, Plenum Press, New York, 1992, p. 501.

270. C. A. Hunter and J. K. M. Sanders, *J. Am. Chem. Soc.*, 1990, **112**, 5525.
271. M. Nishio and M. Hirota, *Tetrahedron*, 1989, **45**, 7201.
272. A. C. Fletcher, L. J. Porter, and E. Haslam, *J. Chem. Soc., Chem. Commun.*, 1976, 627.
273. W. R. Bergmann, M. D. Barkley, R. W. Hemingway, and W. L. Mattice, *J. Am. Chem. Soc.*, 1987, **109**, 6614.
274. W. Gaffield, L. Y. Foo, and L. J. Porter, *J. Chem. Res. (S)*, 1989, 144.
275. F. R. Fronczec, R. W. Hemingway, G. W. McGraw, J. P. Steynberg, C. A. Helfer, and W. L. Mattice, *Biopolymers*, 1993, **33**, 275.
276. F. L. Tobiason and R. W. Hemingway, *Tetrahedron Lett.*, 1994, **35**, 2137.
277. F. L. Tobiason, F. R. Fronczec, J. P. Steynberg, E. C. Steynberg, and R. W. Hemingway, *Tetrahedron*, 1993, **49**, 5927.
278. C. A. Helfer, J.-S. Sun, M. A. Matties, W. L. Mattice, R. W. Hemingway, J. P. Steynberg, and L. A. Kelly, *Polym. Bull.*, 1995, **34**, 79.
279. E. Haslam, *Biochem. J.*, 1974, **139**, 285.
280. H. I. Oh, J. E. Hoff, G. S. Armstrong, and L. A. Haff, *J. Agric. Food Chem.*, 1980, **28**, 394.
281. A. Hagerman and L. G. Butler, *J. Biol. Chem.*, 1981, **256**, 4494.
282. W. R. Bergmann and W. L. Mattice, *ACS Symp. Ser.*, 1987, **358**, 162.
283. L. F. Tilstra, H. Maeda, and W. L. Mattice, *J. Chem. Soc., Perkin Trans. 2*, 1988, 1613.
284. E. Haslam and T. H. Lilley, *Crit. Rev. Food Sci. Nutr.*, 1988, **27**, 1.
285. G. Luck, H. Liao, N. J. Murray, H. R. Grimmer, E. E. Warminski, M. P. Williamson, T. H. Lilley, and E. Haslam, *Phytochemistry*, 1994, **37**, 357.
286. H. Mechanso, L. G. Butler, and D. M. Carlson, *Annu. Rev. Nutr.*, 1987, **7**, 423.
287. J. P. F. G. Helsper, H. Kolodziej, J. M. Hoogendijk, and A. van Norel, *Phytochemistry*, 1993, **34**, 1255.
288. T. Tanaka, R. Takahashi, I. Kouno, and G.-i. Nonaka, *J. Chem. Soc., Perkin Trans. 1*, 1994, 3013.
289. "*Diet, Nutrition and Cancer*," National Academy Press, Washington, DC, 1982.
290. M. G. L. Hertog, E. J. M. Feskens, P. C. H. Hollman, M. B. Katan, and D. Kromhout, *Lancet*, 1993, **342**, 1007.
291. L. W. Wattenberg, *Cancer Res.*, 1992, **52** (Suppl.), 2085S.
292. B. Stavric, *Clin. Biochem.*, 1994, **27**, 319.
293. H. N. Graham, *Prev. Med.*, 1992, **21**, 334.
294. H. Fujiki, S. Yoshizawa, T. Horiuchi, M. Suganama, J. Yatsunami, S. Nishiwaki, S. Okabe, R. Nishiwaki-Matsushina, T. Okuda, and T. Sugimura, *Prev. Med.*, 1992, **21**, 503.
295. T. Kada, K. Kaneko, S. Matsuzaki, T. Matsuzaki, and Y. Hara, *Mutat. Res.*, 1985, **150**, 127.
296. S. K. Katiyar, R. Agarwal, Z. Y. Wang, A. K. Bathia, and H. Mukhtar, *Nutr. Cancer*, 1992, **18**, 73.
297. J. E. Klaunig, *Prev. Med.*, 1992, **21**, 510.
298. Z. Y. Wang, S. J. Cheng, Z. C. Zhou, M. Athar, W. A. Khan, D. R. Bickers, and H. Mukhtar, *Mutat. Res.*, 1989, **223**, 273.
299. S. G. Kahn, S. K. Katiayar, R. Agarwal, and H. Mukhtar, *Cancer Res.*, 1992, **52**, 4050.
300. G. Ford, "The French Paradox and Drinking for Health," Wine Appreciation Guild, San Francisco, 1993.
301. E. N. Frankel, J. Kanner, J. B. German, E. Parks, and J. E. Kinsella, *Lancet*, 1993, **341**, 454.
302. E. N. Frankel, A. L. Waterhouse, and J. E. Kinsella, *Lancet*, 1993, **341**, 1103.
303. S. Renaud and M. de Lorgeril, *Lancet*, 1992, **339**, 1523.
304. J. E. Kinsella, J. Kanner, E. N. Frankel, and B. German, in "Proceedings, Potential Health Effects of Components of Plant Foods and Beverages in the Diet," University of California, Davis, 1992, p. 107.
305. B. Stavric, *Clin. Biochem.*, 1994, **27**, 245.
306. M. Bourzeix, D. Weyland, and N. Heredia, *Bull. O.I.V.*, 1986, **59**, 1171.
307. T. Fuleki, *Grower*, 1993, **43**, 7.
308. J. E. Kinsella, E. N. Frankel, B. German, and J. Kanner, *Food Technol.*, 1993, **47**, 85.
309. M. Namiki, *Crit. Rev. Food Sci. Nutr.*, 1990, **29**, 273.
310. B. Stavric, *Food Chem. Toxicol.*, 1994, **32**, 79.
311. M. T. Huang, T. Ferraro, and C.-T. Ho, "Cancer Chemoprevention by Phytochemicals in Fruits and Vegetables," ACS Symposium Series 546, American Chemical Society, Washington, DC, 1994, p. 2.
312. M.-T. Huang and T. Ferraro, "Phenolic Compounds in Food and Cancer Prevention," ACS Symposium Series 507, American Chemical Society, Washington, DC, 1992, p. 8.
313. J. M. R. da Silva, N. Darmon, Y. J. Fernandez, and S. E. Mitjavila, *J. Agric. Food Chem.*, 1991, **39**, 1549.
314. X. M. Gao, E. M. Perchellet, H. U. Gali, L. Rodriquez, R. W. Hemingway, and J.-P. Perchellet, *Int. J. Oncol.*, 1994, **5**, 285.
315. A. Scalbert, *Phytochemistry*, 1991, **30**, 3875.

3.20
Biosynthesis of Hydrolyzable Tannins

GEORG G. GROSS
Universität Ulm, Germany

3.20.1 INTRODUCTION

Plants have always been an indispensable factor in human life, not only as nutrients but also as a rich source of chemicals that are required for a wide array of medical, artistic, culinary, or technological purposes. In contrast to the fully recognized importance of alkaloids or essential oils as curatives, the role of plant phenolics—which perhaps constitute the most abundant class of natural plant products—has been widely neglected. This applies particularly to plant tannins, for example plant polyphenols which have been used by humankind over several thousand years for the conversion of raw animal hides to leather,[1] and also as chemicals for the production of inks or dyes. They have also been used as curatives, for instance oak galls preserved by the eruption of Mount Vesuvius in a shop in Herculaneum, Italy were likely sold for use in medicines.[2] It was only at the beginning of the twentieth century that intensive chemical studies on the nature of such substances were conducted, particularly by E. Fischer, K. Freudenberg, M. Bergmann, or P. Karrer, to mention some of the most prominent names. Though some principal structural features of Chinese and Turkish gallotannins were elaborated in these investigations it became apparent that the then available analytical armament was insufficient to tackle the evidently tremendous complexity of these plant constituents, with the consequence that it was impossible to isolate individual compounds from the substance mixtures as a prerequisite for the identification of their structures. It is not surprising that interest in this field waned considerably, and only in the 1950s, in response to the beginning of development of advanced separation techniques and sensitive analytical procedures, a remarkable renaissance started with the outstanding investigations of O. Th. Schmidt and W. Mayer[3] on ellagitannins, and which was continued by several highly active research groups, mainly in England and Japan, including, among many others, the laboratories of E. C. Bate-Smith, T. Swain, E. Haslam, T. Okuda, and I. Nishioka.

The results of these and many other efforts have provided pictures of the exact structures of an innumerable host of hydrolyzable tannins and related compounds, and also of their evolution and distribution in the plant kingdom.[4-6] It is understandable that, emerging from that solid basis, emphasis in this field also began to be directed towards related, far-reaching practical challenges like the potential role of tannins in traditional and modern medicine[7-11] or in ecological systems,[12-15] or was focused on more academic considerations like the biosynthetic routes involved in the biosynthesis of these complex molecules.[16] Concerning this latter question, it must be emphasized that it has become common practice to tackle such questions by enzyme studies. This technique has the advantage that it not only allows the unequivocal identification of metabolic intermediates, but also provides otherwise inaccessible information about "activated" intermediates as an indispensable tool for the elucidation of biochemical reaction mechanisms; consequently, the results described in this chapter on the biosynthesis of hydrolyzable tannins have almost exclusively been obtained by this method. One major problem of such *in vitro* investigations is the pronounced tanning potential of the enzyme substrates and products (cf. Section 3.20.2), that is their tendency to bind to proteins which results in the risk of denaturation, precipitation and hence, inactivation of enzymes. Fortunately, as documented in this chapter, the enzymes of the pathways examined were remarkably resistant to their unfavorable substrates and products, while major problems were encountered only in the preparation and characterization of these compounds.

3.20.2 CLASSIFICATION AND STRUCTURAL PRINCIPLES

Before discussing the biochemical events involved in the formation of hydrolyzable tannins it is appropriate to describe briefly their structural principles as an indispensable prerequisite for the understanding of the questions, and also of their solutions, related to the biogenesis of these natural products. According to an already classical definition that was formulated by Freudenberg in 1920,[17] plant tannins are usually divided into the flavonoid derived condensed tannins (nowadays often referred to as proanthocyanidins owing to the liberation of colored anthocyanidins upon treatment with alcoholic mineral acid),[4] and into hydrolyzable tannins which are the subject of this chapter. They are characterized by a central polyol moiety (usually β-D-glucopyranose, but also hamamelose, shikimic acid, quinic acid, or cyclitols have been identified)[18] whose hydroxy groups are typically esterified with gallic acid (3,4,5-trihydroxybenzoic acid) (1). Stepwise substitution, beginning with the monogalloylglucose, β-glucogallin (1-*O*-galloyl-β-D-glucopyranose) (2), leads via a series of so-called "simple esters" to 1,2,3,4,6-penta-*O*-galloyl-β-D-glucopyranose (3). This ester is regarded as the immediate precursor of both subclasses of hydrolyzable tannins, that is gallotannins and ellagitannins. The former are characterized by the introduction of additional galloyl residues, linked to the pentagalloylglucose core (3) via so-called *meta*-depside bonds (4) and reaching total substitution degrees of at least 10–12 galloyl units as has been shown for the tannins from *Rhus semialata* (Chinese gallotannin),[19] *Quercus infectoria* (Turkish gallotannin),[20] or *Paeonia albiflora* (synonym *P. lactiflora*).[21,22] It should be noted that evidence, based on NMR spectroscopy, arose from these studies that natural gallotannins may also contain *para*-depsides, in addition to the *meta*-bonds traditionally known from the literature; this view, however, still awaits supporting experiments.

Ellagitannins, in contrast, are the result of oxidative processes that lead to the introduction of secondary C—C linkages between spatially adjacent galloyl groups of (3). In the case of the preferred $^4C_1$ conformation of the D-glucose core, this event usually takes place between the galloyl groups at C-2/C-3 and C-4/C-6, yielding tellimagrandin II (5) and casuarictin (6) (Scheme 1); however, 1,2 and 1,6 coupling have also been observed.[18] In other plant families (3) may also adopt the energetically less favorable $^1C_4$ conformation that remains fixed after 2,4 and 3,6 galloyl coupling;

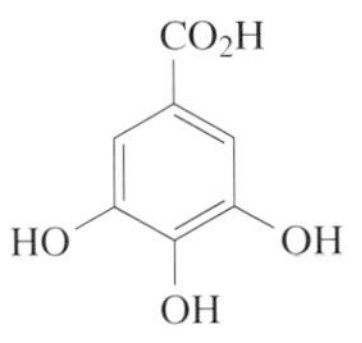

(1)

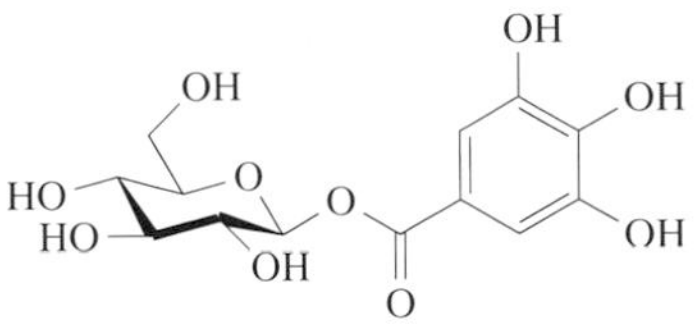

(2)

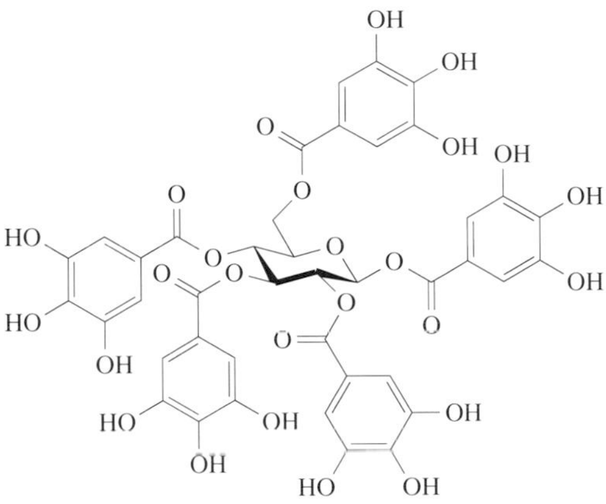

(3)

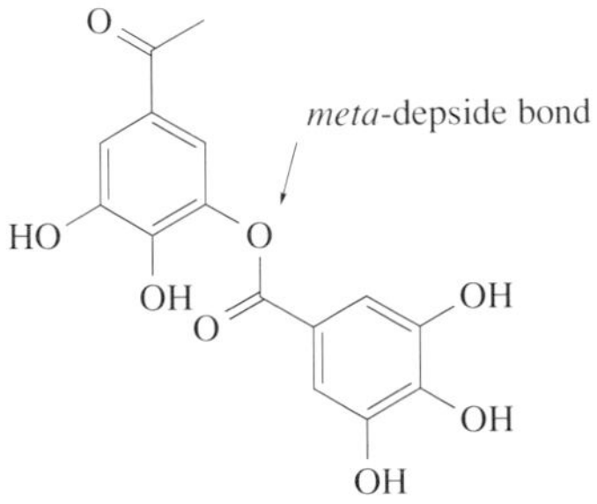

meta-depside bond

(4)

alternatively 1,6 and 2,4 coupling may occur.[16,23] In all these reactions, characteristic (*R*) or (*S*)-3,4,5,3',4',5'-hexahydroxydiphenoyl (HHDP) residues (**8**) are formed. After their eventual hydrolytic release from the tannin core, the resulting free diphenic acid cyclizes spontaneously to the stable, extremely insoluble dilactone, ellagic acid (**9**), and this typical degradation product gave its name to this whole class of natural products.

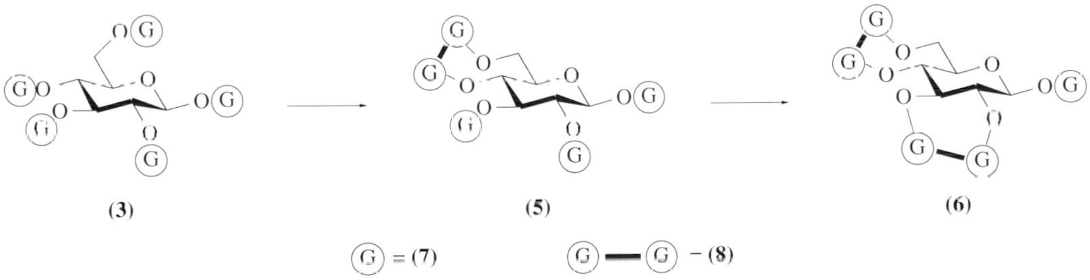

(3)   (5)   (6)

(G) = (7)   (G)—(G) – (8)

Scheme 1

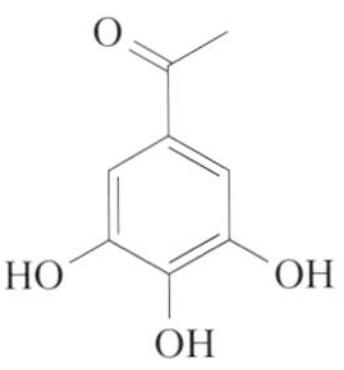

(7) 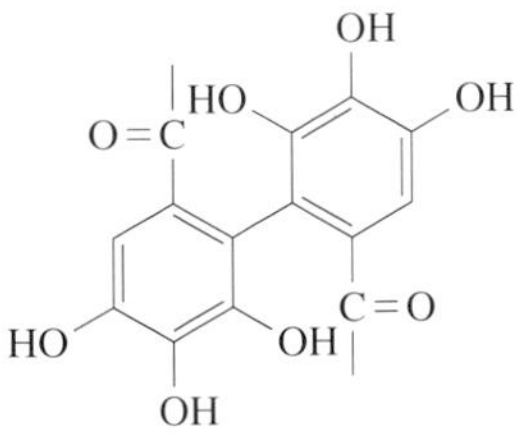   (8) 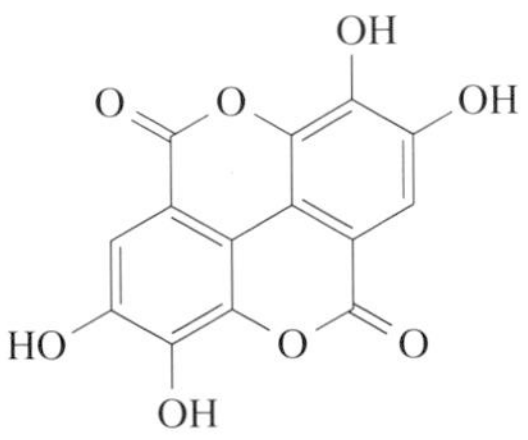   (9)

In contrast to gallotannins, ellagitannins have a strong tendency to combine to higher aggregates. The intermolecular continuation of the above reported intramolecular oxidation processes will thus lead to ellagitannin dimers, and subsequently oligomers, that are interconnected via nonahydroxy-triphenic acid (**10**). More common, however, is the participation of a phenolic OH-group in such oxidation reactions which lead, depending on the nature of reactants, to a linkage via dehydrodigallic acid (**11**) by the coupling of two galloyl residues, or to valoneic acid (**12**) and its isomer, sanguisorbic acid (**13**), when a galloyl group combines with a diphenoyl moiety.[4,16,24] All these latter compounds are thus characterized by the occurrence of aryl-*O*-aryl bridges.

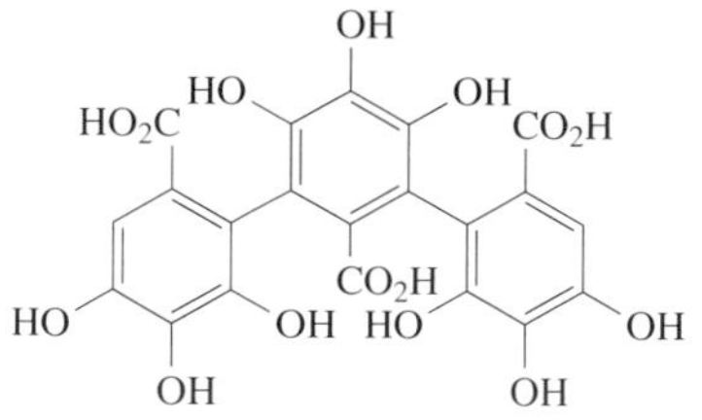

(10)

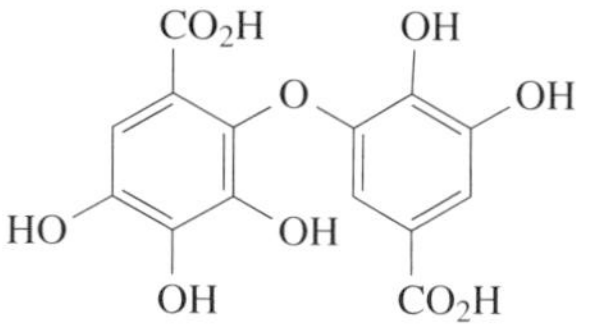

(11)

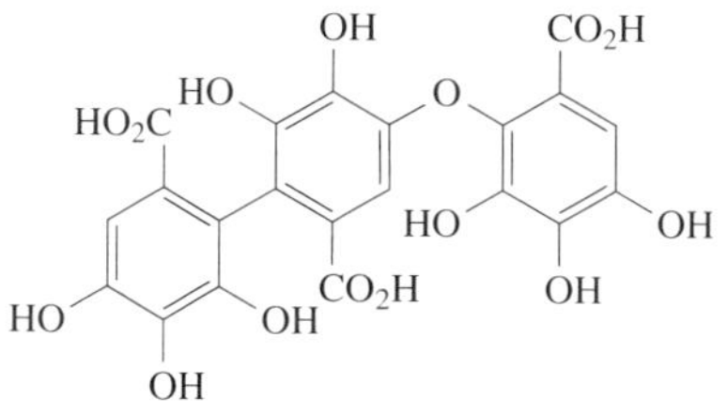

(12)

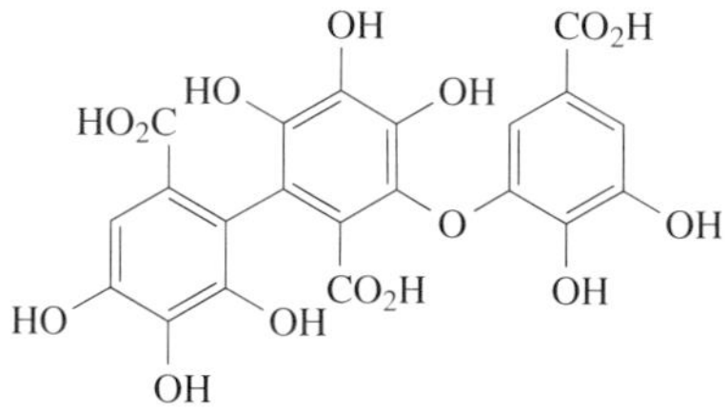

(13)

Innumerable variations of these fundamental structural principles have been discovered in higher plants, but their discussion is beyond the scope of this chapter. There is only one aspect that should be discussed briefly, namely the relation between the structure of hydrolyzable tannins and their reactivity with proteins, as this property can be a decisive factor for the success of *in vitro* enzyme studies. It has been recognized that hydrogen bonding is less important in the interaction between proteins and polyphenols compared with hydrophobic bonding.[25] The intensity of such protein–tannin complexations was found to be largely governed by the substitution degrees of galloyl-glucoses, that is of both their molecular weights and the total number of phenolic hydroxys. While mono- and digalloylglucoses had practically no tanning capacity, significantly increasing inhibition of the enzyme β-glucosidase, paralleled by an increasing relative astringency, was observed for the series tri- < tetra- < pentagalloylglucose (3) with molecular weights of 636, 788, and 940, respectively. Extremely pronounced effects were observed in the presence of the dimeric ellagitannins rugosin-D (1874 M_r) and sanguin H-6 (1870 M_r).[1,4,26] Consequently, natural polyphenols with a molecular weight < $\sim$ 600 are ineffective for the tannage of hides, and there is also an upper limit ($\sim$ 3000 M_r) when molecular diffusion into the collagen fibers becomes difficult.[1] However, the sheer amount of available phenolic residues cannot be the sole controlling factor because it was found that the inhibitory effect of the monomeric ellagitannin casuarictin (6) was appreciably lower than that of pentagalloylglucose (3) although both compounds possess identical numbers of phenolic hydroxy groups and similar molecular weights of 936 and 940, respectively.[1,4] It was concluded that the free spatial orientation of the reactive groups must be important, and this may be one reason for the reported affinity differences of structural isomers of galloylglucoses with the same substitution degree.[27]

Generally, such negative influences on the activity of enzymes in cell-free assays were initially a major concern in our studies on the biosynthesis of hydrolyzable tannins. It became evident, however, that the enzymes related to this biogenetic sequence must possess some resistance against the tanning properties of their polyphenolic substrates and products. It appears, for instance, that the catalytic activities of the enzymes catalyzing the synthesis of tetra- and pentagalloylglucose are maintained to a certain extent even after their association with these tannous compounds, as indicated by the observation that the supernatants of enzyme assay mixtures were devoid of both substrates and reaction products which, however, could be recovered after extracting the protein that had been precipitated during the incubation period.

It is obvious that the above outlined knowledge of the chemical configuration and natural distribution of hydrolyzable tannins was also suitable to stimulate considerations of their biogenetic relationships. Plausible pathways were proposed on this basis,[16] which, as depicted in Scheme 2, are conveniently subdivided into several major sections that all comprise specific challenging questions, namely: (i) the still only partially understood biosynthetic route(s) to gallic acid (1) as the principal phenolic unit; (ii) the origin of β-glucogallin (2) as the first specific intermediate in the pathway to hydrolyzable tannins; (iii) the conversion of this monoester to pentagalloylglucose (3), comprising manifold questions including the nature of galloyl donating agents or the exact structures of the participating intermediates; and finally the secondary transformations of this pivotal intermediate to yield either (iv) gallotannins by depsidical attachment of additional galloyl groups, or to form (v) ellagitannins by oxidative C—C and C—O coupling of adjacent galloyl residues.

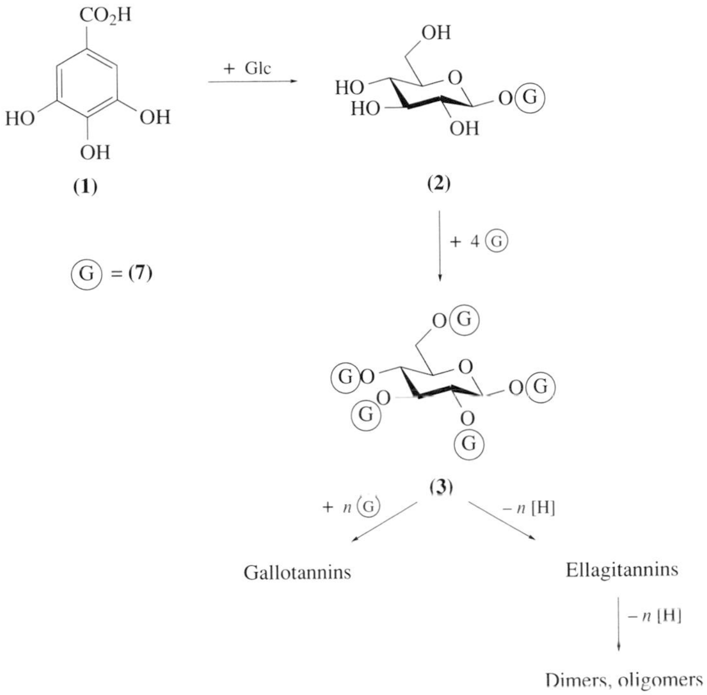

Scheme 2

Little experimental evidence has been available on the natural intermediates and biochemical events of the entire pathway, as documented, for instance, by a review article published in 1985 on the biosynthesis of tannins.[28] It must be attributed to subsequent intensive enzymatic investigations that this unsatisfying situation has changed considerably. As reported in later sections, detailed insights into many of the above questions have been provided by this laborious but also highly evidential technique.

3.20.3 ORIGIN OF GALLIC ACID

It is now generally accepted that benzoic acids (phenylcarboxylic acids) are produced in higher plants by degradative processes, either by true catabolism of complex natural products (e.g., flavonoids) or, in more anabolically orientated metabolic sequences, by side-chain degradation of cinnamic acids (phenylacrylic acids); the substitution pattern of the benzoic acid is thus determined by that of the cinnamate precursor. The exact mechanism of this conversion has long been a matter of dispute, and in 1996 it was proven that the 30 year old proposal of Zenk[29] of a cinnamoyl-CoA dependent β-oxidation sequence is indeed realized in higher plants.[30] However, particular problems have always been encountered with respect to gallic acid (**1**). In spite of numerous investigations, the biosynthesis of this widespread plant constituent is still highly enigmatical. Some essentials of the conflicting proposals, which usually were the result of feeding experiments with putative precursors, are depicted in Scheme 3. A rather conventional pathway (route a) assuming CoA-dependent β-oxidation of 3,4,5-trihydroxycinnamic acid (**18**) to yield (**1**) was formulated by Zenk.[31] The major objection to his proposal was the fact that this precursor, thought to be produced by hydroxylation of caffeic acid (**17**), has never been identified as a natural product and was thus occasionally regarded as the "missing cinnamic acid."[32] A variation of this proposal (route b in Scheme 3) avoided this problem by putting the side-chain degradation one step forward, resulting in the putative sequence, caffeic acid (**17**) → protocatechuic acid (**19**) →gallic acid (**1**).[33] A quite different pathway (Scheme 3, route c) was proposed by other authors after tracer experiments with the fungus *Phycomyces* and various higher plants; their results led to the postulate of a direct aromatization of shikimic acid (**15**) or a biogenetically closely related compound, most likely 3-dehydroshikimic acid (**14**).[34,35]

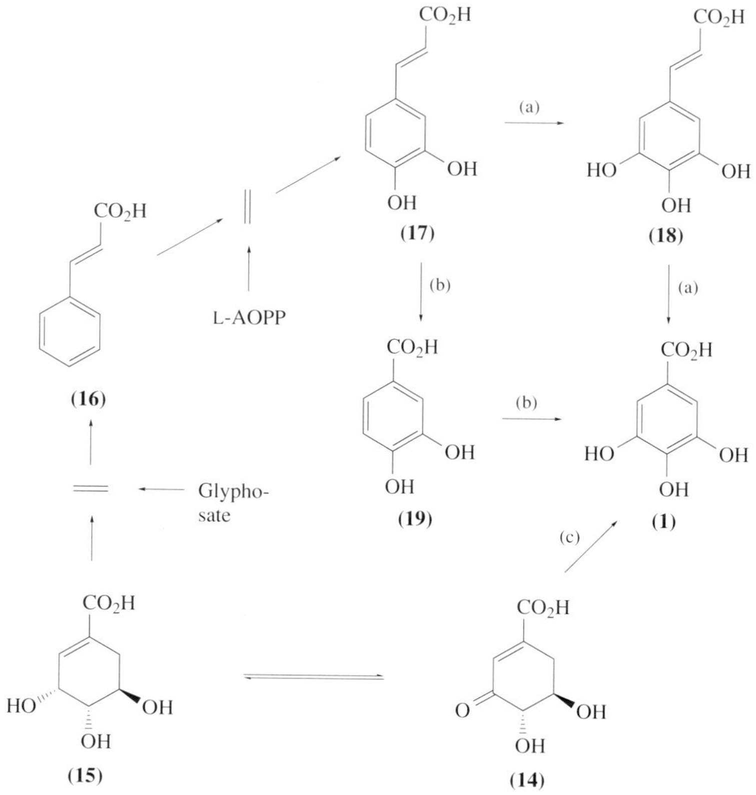

CO_2H (17)

CO_2H (18)

CO_2H (16)

L-AOPP

CO_2H (19)

CO_2H (1)

CO_2H (15)

Glyphosate

CO_2H (14)

(a) (b) (c)

Scheme 3

Such contradictory results are only one example of the usual ambiguity arising from tracer experiments. However, even enzyme studies—which are regarded as more definitive—were not very helpful in this instance. Two short communications on work with cell-free systems from mung bean seedlings[36] and leaves from *Pelargonium*[37] suggested the route, 3-dehydroshikimic acid (**14**) → protocatechuic acid (**19**) → gallic acid (**1**). Unfortunately, these preliminary results were never corroborated for higher plants, but the bacterium *Escherichia coli* was found capable of converting 3-dehydroshikimic acid (**14**) via protocatechuic acid (**19**) to aliphatic degradation products.[38] Also the participation of chemical transformations in such processes cannot be totally excluded; for example, it has been noted that dehydroshikimic acid (**14**) was converted to protocatechuic acid (**19**) by heating with HCl[39] or to gallic acid (**1**) by mild oxidants such as Cu^{2+}.[32,40]

No clear-cut decision was obtained after feeding experiments with specifically carboxyl-labeled shikimic acid (**15**); by this approach, at least in theory, a simple discrimination between the two competing alternatives should be possible: when the biosynthetic route via cinnamic acid (**16**) is operative, gallic acid (**1**) must have lost the radioactive carboxyl group, while this label is retained in the direct aromatization of the alicyclic C_6—C_1 precursor to (**1**). Such studies have shown that 4-hydroxybenzoic acid was exclusively formed via C_6—C_3 precursors in cell cultures from *Lithospermum*, as indicated by the complete loss of the carboxyl of the administered precursor, [1,7-^{13}C]shikimic acid.[41] On the other hand, young tea shoots converted carboxyl-labeled shikimic (**15**), 5-dehydroshikimic (**14**), or dehydroquinic acid directly to gallic acid (**1**), but also used C_6—C_3 precursors for the formation of this compound.[42] The simplest explanation for these contradictory phenomena was the assumption of at least two different pathways, a conclusion that has also been drawn by other authors after feeding experiments with leaves from *Acer* and *Rhus* and which again suggested two routes to gallic acid (**1**), with the preferential route depending upon leaf age and plant species investigated.[43,44]

Other experiments were performed with herbicides that are known effectively to block the synthesis of shikimic acid derived phenolics. L-AOPP (L-2-aminooxy-3-phenylpropionic acid) specifically inhibits the deamination of L-phenylalanine (**16**) to cinnamic acid, while glyphosate [*N*-(phosphonomethyl)glycine], a phosphoenolpyruvate analogue, blocks the activity of 5-enolpyruvyl-shikimate dehydrogenase, a key enzyme of the shikimate pathway. In the presence of these inhibitors, significantly elevated levels of hydroxybenzoic acids were observed, while derivatives that are known to be formed exclusively via phenylpropanoids (e.g., methoxybenzoic acids) were found to be unaffected.[45-47] Summarizing the conflicting evidence discussed above, it appears most plausible to regard the direct aromatization of 5-dehydroshikimic acid (**14**) at least as a significant, if not the predominant, route to gallic acid.

Strong evidence supporting this conclusion has been obtained by feeding [¹³C]glucose to cultures of the fungus *Phycomyces blakesleeanus* and to leaves of the dicotyledonous tree *Rhus typhina*, followed by determination of isotope distributions of isolated gallic acid and aromatic amino acids and interpretation of the resulting isotopomer patterns by a retrobiosynthetic approach.[48] The data showed that gallic acid was derived in both species from an early intermediate of the shikimate pathway, most probably 5-dehydroshikimate (**14**). Notably, the carboxyl group of gallic acid was found to originate from a C_6—C_1 intermediate of the shikimate pathway and not from the side-chain of a C_6—C_3 metabolite, that is phenylalanine or hydroxylated cinnamic acids, thus ruling out routes a and b in Scheme 3 as major pathways. It was concluded that dehydrogenation of 5-dehydroshikimate (**14**) in both the fungus and the plant was the predominant pathway to gallic acid. However, the available data could not exclude an alternative route to gallic acid by dehydration of (**14**) to protocatechuic acid (**19**) and subsequent introduction of a third phenolic OH-group by a monooxygenase.

3.20.4 BIOSYNTHESIS OF β-GLUCOGALLIN

The naturally occurring phenolic ester β-glucogallin (1-*O*-galloyl-β-D-glucopyranose) (**2**) was first isolated from Chinese rhubarb (*Rheum officinale*) in 1903[4] and is regarded as the primary metabolite in the biosynthesis of hydrolyzable tannins.[16] Consideration of the formation of this ester has to take into account that, for thermodynamic reasons, the participation of an "activated" intermediate (i.e., a compound with a high group-transfer potential) has to be postulated for such a reaction (reverse esterase reactions work nicely under laboratory conditions, but not in nature). This requirement can be met in two ways, either by reaction of an energy-rich galloyl derivative with free glucose, or by the participation of an activated glucose derivative (most likely the ubiquitous UDP-glucose) that combines with the free acid. When the studies discussed in this paper were begun it was already known that esterification of phenolic acids with different hydroxylated compounds proceeded via acyl-coenzyme A (CoA) intermediates. Since the very first report in this field by Stöckigt and Zenk in 1974,[49] dealing with the caffeoyl-CoA dependent synthesis of the ubiquitous plant depside chlorogenic acid (3-*O*-caffeoylquinate), a wide variety of related esters have been recognized to be formed analogously.[50,51]

It thus appeared conceivable that galloyl-CoA (**20**) might represent the energy-rich metabolite required for the biosynthesis of β-glucogallin (**2**). To test this hypothesis, this then unknown thioester was synthesized chemically.[52] As summarized in Scheme 4, gallic acid (**1**) was converted to 4-*O*-β-D-glucosidogallic acid (**21**) to block the reactive phenolic hydroxy groups, followed by transformation to *N*-succinimidyl-4-*O*-β-D-glucosidogallate (**22**) in the presence of DCC. Subsequent transacylation of (**22**) with CoA yielded 4-*O*-β-D-glucosidogalloyl-CoA (**12**) from which galloyl-CoA (**20**) was liberated by treatment with the enzyme β-glucosidase. In enzymatic studies with cell-free extracts from higher plants, however, no evidence has been found, to date, that galloyl-CoA (**20**) was involved in the biosynthesis of β-glucogallin (**2**) or its higher galloylated derivatives. Instead, the second of the above alternatives was found to be realized in nature, i.e., the (reversible) reaction of free gallic acid (**1**) with UDP-glucose to yield β-glucogallin (**2**) and UDP (Equation (1)). The enzyme catalyzing this reaction was detected in leaves of *Quercus robur*[53] and partially purified from cell-free extracts of *Q. rubra*.[54] The glucosyltransferase had a molecular weight of 68 kDa, a pH optimum at 6.5–7.0, and an optimum temperature of 40 °C. UDP-glucose was found to act as the exclusive sugar donor while numerous benzoic and, at significantly lower rates, cinnamic acids could serve as acceptor molecules. According to the best substrate, vanillic acid, the systematic name UDP-glucose: vanillate 1-*O*-glucosyltransferase (EC 2.4.1.-) was proposed; however, it was concluded that the physiological role of the enzyme is the formation of β-glucogallin.

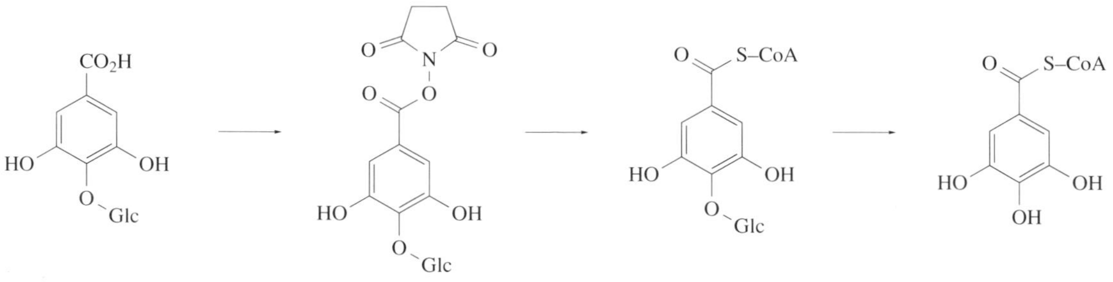

(21)   **(22)**   **(23)**   **(20)**

Scheme 4

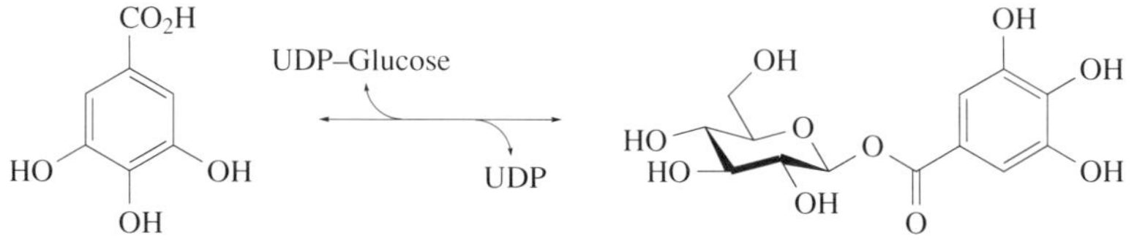

(1)   **(2)**   (1)

In light of the available evidence, the existence of this glucosyltransferase is not surprising; numerous enzymes have meanwhile been identified from various plant sources that all catalyze the formation of phenolic 1-*O*-acylglucoses according to the mechanism of Equation (1) (cf.[50,51]), and it appears that UDP-glucose must be regarded as the general activated donor required for the esterification of glucose with phenolic acids.

3.20.5   "SIMPLE" GALLOYLGLUCOSE ESTERS—FROM β-GLUCOGALLIN TO PENTAGALLOYLGLUCOSE

3.20.5.1   The Main Pathway

It was discovered very early in the initial phases of the investigations of the biosynthesis of gallotannins that cell-free extracts from young oak leaves were able to form digalloylglucose and trigalloylglucose in reaction mixtures which contained β-glucogallin (**2**) as the sole substrate. This result suggested that (**2**) must have played an unusual dual role, according to which it acted not only as an acceptor substrate, as expected, but apparently also as the energy-rich acyl donor required for such a reaction. Again, no evidence for the participation of galloyl-CoA (**20**) was obtained in this conversion.[55] Subsequent investigations led to the isolation of a partially purified enzyme from oak leaves that specifically transferred a galloyl residue from (**2**) to the glucose-6-OH position of the acceptor molecule, to yield 1,6-di-*O*-galloyl-β-D-glucose (**24**) under the concomitant liberation of 1 mol glucose as deacylated by-product, as shown in Equation (2). The enzyme had a molecular weight of about 400 kDa, temperature and pH optima of 30 °C and 6.5, respectively, and was stable between pH 4.5 and pH 6.0.[56] Substrate specificity studies also revealed that the closely related esters 1-*O*-protocatechuoyl-β-D-glucose (**28**) and 1-*O*-hydroxybenzoyl-β-D-glucose (**27**) could serve as substrates, with relative activities of 58 and 8% compared with (**2**).[57] This new enzyme catalyzing

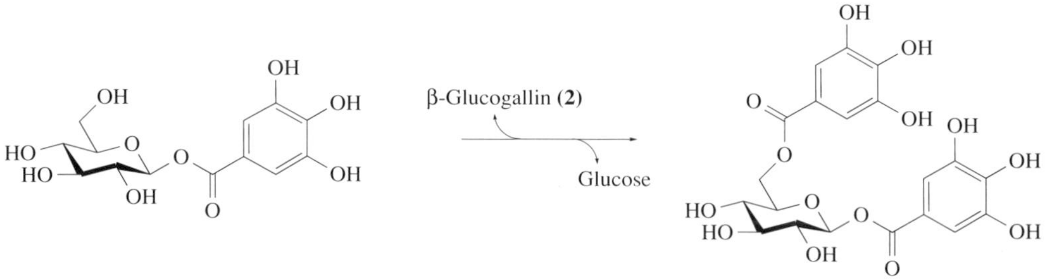

(2)   **(24)**   (2)

an unusual "disproportionation" reaction was thus named β-glucogallin: β-glucogallin 6-O-galloyltransferase (EC 2.3.1.90).

With respect to thermodynamic considerations, the identification of such an enzyme activity was fairly surprising because the group-transfer potential of 1-O-acylglucose esters was then considered to be comparatively low. It was known that the hemiacetal phosphate of glucose-1-phosphate has a ΔG^0 of about -21 kJ mol^{-1}, whereas the rather inert ester linkage of isomeric glucose-6-phosphate has a ΔG^0 of only about -10.5 to -12.5 kJ mol^{-1}.[58] This question has been solved, meanwhile, by the finding[59] that the related ester 1-O-sinapoyl-β-D-glucose (**30**) has an unexpectedly high group-transfer potential of -35.7 kJ mol^{-1}, a value that is comparable to the well-known data for acyl-CoA thioesters (ca. -36 kJ mol^{-1}), and it is reasonable to assume that the ΔG^0 of β-glucogallin (**2**) is in the same order of magnitude. Consequently, glucose esters lacking the energy-rich 1-O-acyl group should be unable to serve as acyl donors; the results from substrate specificity studies with cell-free enzyme preparations have demonstrated this conclusion on several occasions (see Section 3.20.5.2).

The above observations on the role of β-glucogallin in transacylation reactions coincided with similar results from other laboratories on the intermediacy of 1-O-acylglucoses in the enzymatic esterification of numerous phenolic acids (Table 1). Interestingly enough, an enzyme from *Ipomoea*[61] was found among those that produced chlorogenic acid and related depsides from hydroxycinnamoylglucoses and free quinic acid, thus presenting evidence of a novel pathway as an alternative to the long-established acyl-CoA dependent biosynthesis of these compounds.[49] It is evident from this data that the widely neglected or underestimated phenolic 1-O-acylglucose esters, often previously regarded as metabolically inert compounds, occupy a prominent position in the secondary metabolism of higher plants which is at least comparable to that of the generally acknowledged role of acyl-CoA esters.

Table 1 Biosynthesis of phenolic esters by acylglucose-dependent acyltransferases and related enzymes.[a]

Donor substrate	Acceptor	Product	Ref.
1-O-p-Coumaroylglucose	D-Quinate	p-Coumaroylquinate	60
1-O-p-Coumaroylglucose	*meso*-Tartarate	p-Coumaroyl-*meso*-tartarate	61
1-O-Caffeoylglucose	D-Quinate	Chlorogenate	62
1-O-Sinapoylglucose	L-Malate	Sinapoyl-L-malate	63
1-O-Sinapoylglucose	Choline	Sinapoylcholine (sinapine)	64
1-O-Sinapoylglucose	1-O-Sinapoylglucose	1,2-Disinapoylglucose	65
3-O-Caffeoyl-D-quinate (chlorogenate)	Glucarate	Caffeoylglucarate	66
3-O-Caffeoyl-D-quinate (chlorogenate)	Chlorogenate	3,5-Dicaffeoylquinate (isochlorogenate)	67
1-O-Indolylacetylglucose	*myo*-Inositol	Indolylacetyl-*myo*-inositol	68
1-O-Indolylacetylglucose	Glycerol	Indolylacetylglycerol	69
1-O-Hydroxycinnamoylglucose	Betanidinglycosides	Acylated betacyanins	70
1-O-Hydroxycinnamoylglucose	Anthocynanins	Acylated anthocyanins	71

[a]Not included are the enzymes related to the biosynthesis of hydrolyzable tannins; their properties are specified in the text.

Initial enzyme studies had already led to the assumption that digalloylglucose might be transformed to trigalloylglucose by a continuation of the mechanism described, that is by galloylation with β-glucogallin (**2**) serving as the donor substrate.[55] This proposal was substantiated with enzyme extracts from staghorn sumac (*R. typhina*) leaves that catalyzed the highly position-specific galloylation of the 2-hydroxy of the substrate, 1,6-digalloylglucose (**24**), to yield 1,2,6-trigalloylglucose (**25**) (Scheme 5).[72] The enzyme, which was also detected in oak leaves, was partially purified and named β-glucogallin: 1,6-digalloyl-β-D-glucose 2-O-galloyltransferase (EC 2.3.1.-).[73] It was found to have a molecular weight of about 450 kDa,[74] while in earlier experiments a molecular weight of about 750 kDa had been estimated;[73] the discrepancy is probably due to initial solubility problems of this protein. The enzyme had an optimum temperature of 50 °C, an optimal pH of 5.0–5.5, and maximal stability was observed between pH 3.4 and pH 5.8. Substrate specificity studies revealed that, besides several unphysiological substrates, the β-glucogallin (**2**) analogues 1-O-p-hydroxybenzoylglucose (**27**) and 1-O-protocatechuoylglucose (**28**) could also act as potent acyl donors, exhibiting relative activities of 54% and 93%, respectively, compared with (**2**). It was further found that tri-O-protocatechuoylglucose was efficiently formed upon incubation of (**28**) as donor together with (**29**) as acceptor (46% relative activity).[73]

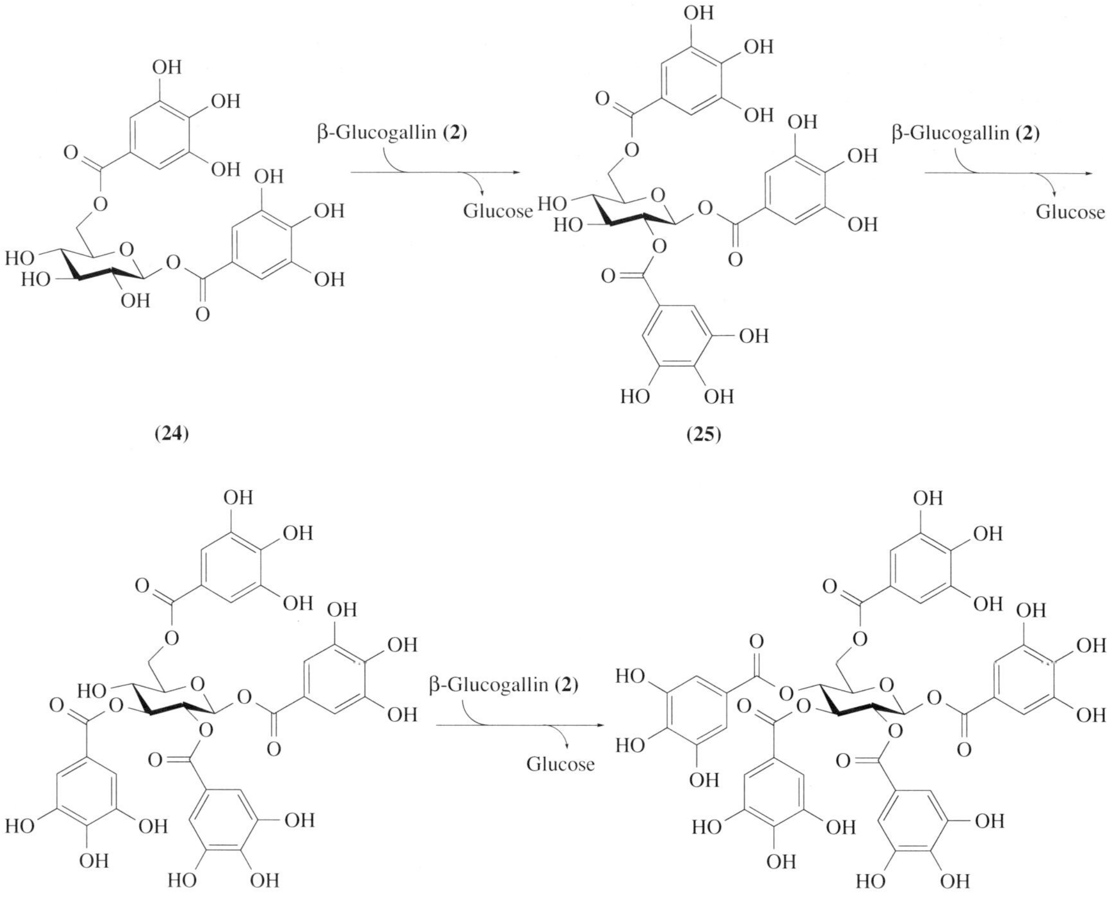

(24) **(25)**

(26) **(3)**

Scheme 5

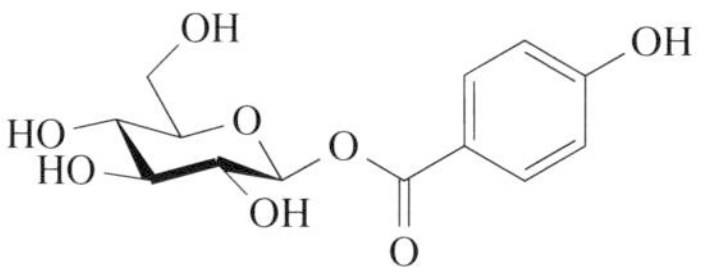

(27)

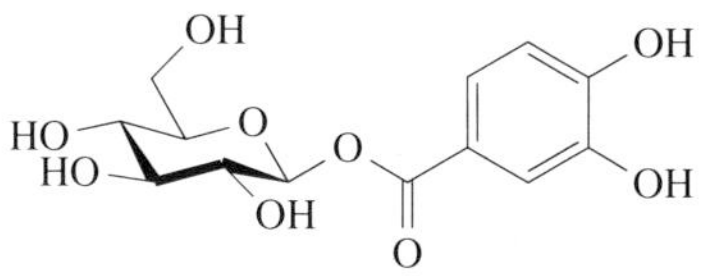

(28)

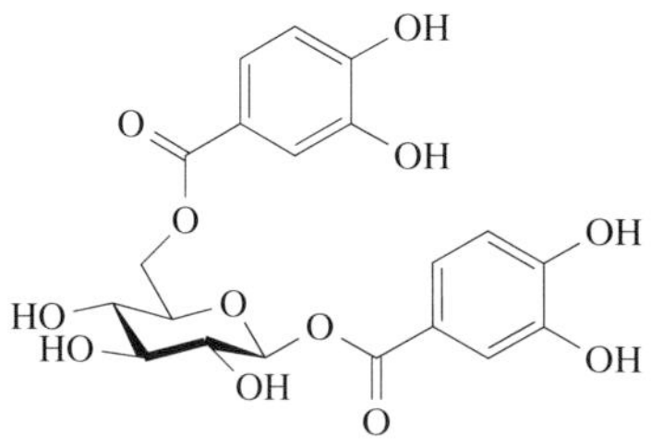

(29)

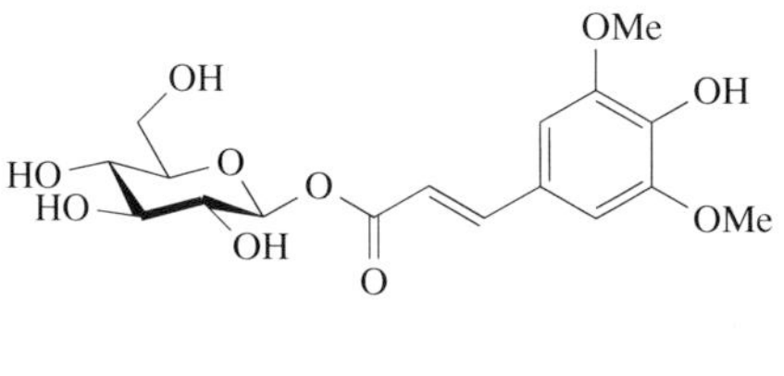

(30)

On the basis of these results, it appears almost trivial to report that *β*-glucogallin (**2**) was also found to function as galloyl donor in the acylation of 1,2,6-trigalloylglucose (**25**) to 1,2,3,6-tetragalloylglucose (**26**), followed by the analogous conversion of this intermediate to the final metabolite of the entire sequence, 1,2,3,4,6-pentagalloyl-*β*-D-glucose (**3**) (Scheme 5). The tetragalloylglucose forming enzyme (β-glucogallin: 1,2,6-tri-*O*-galloyl-β-D-glucose 3-*O*-galloyl-transferase; EC 2.3.1.-) was initially detected in sumac leaves, but was partially purified from green acorns of pedunculate oak (*Q. robur*, synonym *Q. pedunculata*). It had a molecular weight of ca. 380 kDa, pH and temperature optima of 6.0 and 55 °C, respectively, and was most stable between pH 4.0 and pH 6.5. In addition to the natural substrates β-glucogallin (2) (donor) and 1,2,6-trigalloylglucose (25) (acceptor), the isomer 1,3,6-trigalloylglucose—which is not an intermediate in the biosynthesis of hydrolyzable tannins in oak or sumac—was an extremely efficient acceptor molecule; in both cases, however, 1,2,3,6-tetragalloylglucose (26) was the sole reaction product.[75]

The subsequent enzyme, β-glucogallin: 1,2,3,6-tetra-*O*-galloyl-β-D-glucose 4-*O*-galloyltransferase (EC 2.3.1.-), which catalyzed the formation of 1,2,3,4,6-pentagalloylglucose (3) was partially purified from young oak leaves; it depended strictly on (26) as acceptor, whereas the 1,2,4,6-isomer was inactive. This transferase (molecular weight ca. 206 kDa) was stable between pH 5.0 and pH 6.5, exhibiting highest activities at pH 6.3 and 40 °C.[76] The enzyme has been purified more than 1000-fold to apparent homogeneity, as shown by polyacrylamide-gel electrophoresis in the presence of sodium dodecylsulfate. Only one single protein band corresponding to a molecular weight of 65 kDa was found in these experiments, which suggested that the native protein was a homotetramer composed of four identical subunits.[77]

3.20.5.2 Side Reactions

The results of the enzyme studies described above provide the impression of a clear and logically constructed pathway which is characterized by the fact that only one common acyl donor, β-glucogallin (2), is required in the individual steps. However, this attractive picture was blurred by the discovery of additional and apparently β-glucogallin-independent side reactions. Ambiguous and partially contradictory results were encountered in the purification and closer characterization of the 2-*O*-galloyltransferase that catalyzed the conversion of 1,6-digalloylglucose (24) to 1,2,6-trigalloylglucose (25) (Equation (2)).[72] The initially unexplainable problems were clarified only after recognizing the unexpected existence of an interfering enzyme that obviously effected the formation of the same product (25) but evidently without any requirement for the established acyl donor, β-glucogallin (2).[78] According to Equation (3), this novel enzyme represented another example of a "disproportionation" reaction, in this instance, however, with the participation of two molecules of 1,6-digalloylglucose (24) which are converted to 1,2,6-trigalloylglucose (25) and anomeric 6-galloylglucose as a partially deacylated by-product. The enzyme could be purified from sumac leaves almost 1700-fold; it had a molecular weight of 56 kDa, was stable between pH 4.5 and pH 6.5, and most active at pH 5.9 and 40 °C; it was named 1,6-digalloylglucose: 1,6-digalloylglucose 2-*O*-galloyltransferase (EC 2.3.1.-).[78] While this transferase also accepted the structurally closely related 1,6-diprotocatechuoylglucose (29) as an efficient substrate (relative activity 37%), no reaction occurred in the presence of 3,6-di-*O*-galloylglucose, a fact that underlined the above discussed essential role of the energy-rich 1-*O*-acyl bond and that agreed with the results from earlier experiments with 6-*O*-galloylglucose.[56,79] Similar observations were made with an enzyme from oak, where β-glucogallin could be replaced by 1,6-digalloylglucose (24) as acyl donor in the esterification of trigalloylglucoses to 1,2,3,6-tetragalloylglucose (26) according to Equations (4) and (5).[75]

$$\text{1,6-Digalloylglucose} + \text{1,6-digalloylglucose} \longrightarrow \text{1,2,6-trigalloylglucose} + \text{6-galloylglucose} \quad (3)$$
$$\text{(24)} \qquad\qquad \text{(24)} \qquad\qquad\qquad \text{(25)}$$

$$\text{1,6-Digalloylglucose} + \text{1,2,6-trigalloylglucose} \longrightarrow \text{1,2,3,6-tetragalloylglucose} + \text{6-galloylglucose} \quad (4)$$
$$\text{(24)} \qquad\qquad \text{(25)} \qquad\qquad\qquad \text{(26)}$$

$$\text{1,6-Digalloylglucose} + \text{1,3,6-trigalloylglucose} \longrightarrow \text{1,2,3,6-tetragalloylglucose} + \text{6-galloylglucose} \quad (5)$$
$$\text{(24)} \qquad\qquad \text{(25)} \qquad\qquad\qquad \text{(26)}$$

These findings provoked assumptions that, besides β-glucogallin, higher substituted galloylglucoses could generally act as acyl donors, provided that they were bearing the energetically indispensable 1-*O*-galloyl group. This view was corroborated by a series of substrate specificity studies with the β-glucogallin-independent 2-*O*-galloyltransferase from *Rhus* (see above) which led to the following order of reactivities: 1,6-di- (100% relative activity) > 1,3,6-tri > 1,2,6-tri > 1,2,3,6-tetra- > 1,2,4,6-tetra- > 1,2,3,4,6-pentagalloylglucose (0%). In response to increasing bulkiness of the higher sub-stituted compounds, tetragalloylglucoses were only very poor donors, and pentagalloylglucose was completely inactive.[78] Moreover, it was evident that the relative activities of substrates bearing a 3-*O*-galloyl group were significantly higher. This applied particularly to 1,3,6-trigalloylglucose; it should be emphasized, however, that this compound,, as well as 1,2,4,6-tetragalloylglucose, does not represent a natural gallotannin precursor in *Rhus* or *Quercus* (cf. Section 3.20.5.3).

These findings of the donor potential of multiply substituted 1-*O*-acylglucoses are consistent with the discussed possibility that the frequently naturally occurring galloylglucoses with an unacylated anomeric hydroxy group might originate from such processes.[4] However, interference of these events with simple hydrolysis of the comparatively labile 1-*O*-acyl bond of galloylglucoses cannot be excluded, and it is conceivable that enzymatic degradation also occurs, for example by tannase-like plant enzymes (cf. Section 3.20.8). It should finally be mentioned that galloylation of acylglucoses bearing a free anomeric OH-group has never been observed *in vitro*, a fact indicating that the existence of an acyl residue at this specific position might be indispensable, not only for the thermodynamic requirements emphasized above, but also to ensure the correct identification of these esters as galloyltransferase substrates.

Consequently, in addition to the earlier recognized β-glucogallin-dependent pathway, β-gluco-gallin-independent side reactions must also now be taken into account as processes that eventually contribute significantly to the formation of various hydrolyzable tannins. However, owing to the observed decreasing reactivity of galloylglucoses in response to increasing substitution, the 1-mono- and 1,6-diesters still represent the predominating acyl donors, while the reactions of the higher analogues remain more or less negligible.[89]

Finally, some characteristics of a completely different galloyltransferase should be briefly dis-cussed here. Cell-free extracts from oak leaves were found to catalyze a very unusual exchange reaction between β-glucogallin (**2**) (or related 1-*O*-esters) and free D-glucose.[55] This reaction, which was detectable only by the use of labeled substrates, as shown in Equation (6), proceeded optimally with 1-*O*-benzoylglucose and had a specific requirement for D-glucose as acceptor. The partially purified enzyme was thus named 1-*O*-benzoylglucose: D-glucose 1-*O*-benzoyltransferase (EC 2.3.1.-); it had a molecular weight of 380 kDa, a pH optimum of 6.6, and a temperature optimum of 30 °C.[79,80] The physiological significance of this acyltransferase reaction is still obscure; however, it was successfully employed for the convenient and economic preparation of radiolabeled β-glucogallin (cf. Section 3.20.9).

$$1\text{-}O\text{-Galloyl-}\beta\text{-}D\text{-glucose} \;+\; D\text{-}[^{14}C]\text{glucose} \;\rightleftharpoons\; 1\text{-}O\text{-Galloyl-}\beta\text{-}D\text{-}[^{14}C]\text{glucose} \;+\; D\text{-glucose} \qquad (6)$$

$$\quad\quad (\mathbf{2}) \qquad\qquad\qquad\qquad\qquad\qquad\qquad\qquad\qquad\qquad\qquad\qquad (\mathbf{2})$$

3.20.5.3   General Characteristics of the Pathway

On the basis of the structures and of the natural distribution of numerous galloylglucose esters, a metabolic pathway had been postulated by Haslam and co-workers in 1982,[16] which comprised the then hypothetical sequence β-glucogallin (**2**) → 1,6-digalloylglucose (**24**) → 1,2,6-trigalloylglucose (**25**) → 1,2,3,6- (**26**) or 1,2,4,6-tetragalloylglucose → 1,2,3,4,6-pentagalloylglucose (**3**). The above reported enzyme studies were not only suitable to clarify the nature of the activated intermediates required for these conversions, they could also demonstrate that, at least in oak and sumac, the route to pentagalloylglucose did not involve ramifications because the putative intermediate 1,2,4,6-tetragalloylglucose was found to play no role in this biosynthetic sequence. Instead, a strictly linear pathway, as depicted in Scheme 6, has been found to be realized in nature. One of the most striking findings in this connection was certainly the discovery of the surprisingly pronounced position-specificity of each of the individual galloylation steps in this sequence, that is the series 1-OH > 6-OH > 2-OH > 3-OH > 4-OH for the enzymatic substitution of the glucose core. Interestingly, an identical sequence of reactivities has been determined for the chemical esterification of the hydroxy groups of β-glucose in studies with 1-benzyl- or 1-methyl-β-D-glucopyranose. As plausible reasons for the apparent reactivity differences, it has been discussed that, after the preferred hemiacetal-OH at C-1, the primary 6-OH is more reactive than the residual secondary hydroxys; the 2-OH among these is the most reactive one due to an activating effect of the neighboring anomeric center. Discrimination between the theoretically equivalent hydroxys at C-3 and C-4 occurs by the fact that access to the 4-OH group adjacent to the already substituted bulky 6-position is sterically hindered, resulting in a higher relative activity of the 3-OH group.[81–83]

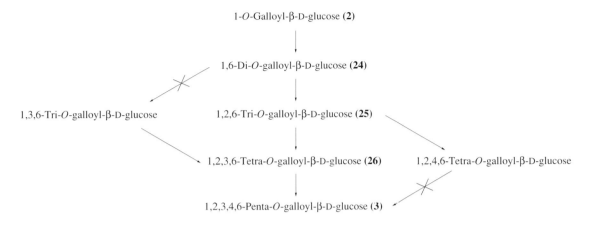

Scheme 6

Concerning the abortive side reactions included in Scheme 6, it has to be emphasized that enzymatic galloylation of 1,6-digalloylglucose (**24**) by enzyme preparations from *Rhus* yielded trace amounts (about 1%) of 1,3,6-trigalloylglucose, in addition to the 1,2,6-substituted main product (**25**).[74] Though enzyme assays had shown that this compound was transformed efficiently to 1,2,3,6-tetragalloylglucose (**26**) in the subsequent step,[75] it is impossible to assign any importance to this alternative *in vivo* due to the negligible supply of precursor. Cell-free extracts from sumac leaves were also found to produce a certain amount (ca. 4%) of 1,2,4,6-tetragalloylglucose as by-product, besides the main metabolite, 1,2,3,6-tetragalloylglucose (**26**);[74] however, the enzyme catalyzing the next step to the pentagalloylglucose level exhibited no affinity towards this compound.[75]

Some final remarks on the enzymes involved in the pathway from gallic acid (**1**) to pentagalloylglucose (**3**) appear appropriate at this stage. Though some of the above described enzyme activities have been isolated from oak, while others were from sumac, it should be emphasized that supplementary experiments were carried out to secure that the principles of the pathway from gallic acid to pentagalloylglucose were identical in both plant species.[74] Generally, a pronounced uniformity of their basic properties was apparent: their pH-optima lay around pH 6.5, they were most stable in slightly acidic media (pH 4–6), their temperature dependencies showed optima at 40–50 °C, Q_{10} values were around 1.8–2.0 and activation energies 30–50 kJ mol^{-1}, and they commonly displayed an unusual cold tolerance, as expressed by residual reaction rates of 10–25% at 0 °C. Moreover, a pronounced trend to unusually high molecular weights of about 260–450 kDa became evident. Exceptions were the glucogallin-synthesizing glucosyltransferase (Equation (1))[53,54,84] and the "β-glucogallin-independent" 1,6-digalloylglucose-"disproportionating" galloyltransferase (Equation (3))[78] with molecular weights of 68 kDa and 56 kDa, respectively. In the case of the latter enzyme, also a quite different Q_{10} value of 3.0, equivalent to an activation energy of 74.5 kJ mol^{-1}, was determined, a finding that contributed to the interpretation that this enzyme did not belong to the central "β-glucogallin-dependent" metabolic route to pentagalloylglucose. It must be stressed in this connection, however, that the terms "β-glucogallin-dependent" or "β-glucogallin-independent" are only simplifications for practical purposes; in the last analysis, all aromatic residues of galloylglucoses are directly or indirectly derived from β-glucogallin. Thus, the entire class of hydrolyzable tannins must be considered a very homogeneous group of natural plant products that emerges from only one specific constituent surrounding the central polyol moiety.

3.20.6 BIOSYNTHESIS OF GALLOTANNINS

Fortunately, the negative statement of a review article, published in 1989, that "virtually nothing is known about the formation of the characteristic *meta*-depside bond"[50] of gallotannins has since been superseded. Only speculations were then possible about the nature of the activated galloyl derivative required in such reactions, that is whether, by analogy to the preceding steps, β-glucogallin (**2**) also served in these subsequent conversions as the energy-rich intermediate or whether, owing to the markedly differing nature of the phenolic OH-group to be substituted instead of the aliphatic hydroxys of glucose, other compounds that were thought to have a much higher group-transfer potential, for instance galloyl-CoA (**20**), were utilized.

Shortly after the formulation of the above statement, it was discovered that partially purified enzyme extracts from sumac leaves catalyzed the efficient galloylation of 1,2,3,4,6-pentagalloyl-glucose (**3**), affording a mixture of numerous higher substituted products.[85] These compounds were interpreted as hexa-, hepta-, octa-, and nonagalloylglucoses, together with traces of decagalloyl-glucose, and eventually even of undecagalloylglucose, by graphical analysis of the results obtained by normal-phase HPLC[21] of the reaction products (Figure 1). According to previous observations with "simple" galloyl esters[86] or oligomeric hydrolyzable tannins,[87] plots of the presumptive degree of galloylation of the reaction product versus the logarithm of their retention time resulted in a characteristic straight line, and the data achieved coincided perfectly with those obtained with the available authentic references. By analogy to the preceding pathway to pentagalloylglucose, β-glucogallin (**2**) was again found to serve as the specific galloyl donor in these conversions, thus demonstrating the predominating significance of this ester for the entire biogenetic route to gallotannins. Time-course experiments demonstrated the expected sequential synthesis of hexa-, hepta-, and octagalloylglucoses (Figure 2), at the expense of the substrates β-glucogallin (**2**) and pentagalloylglucose (**3**).[85]

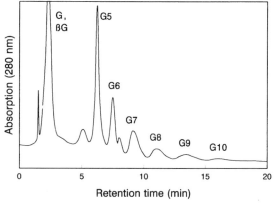

Figure 1 Normal-phase HPLC of enzymatically formed gallotannins. Reaction products were analyzed on Si-60 silica-gel columns with the solvent system *n*-hexane–methanol–tetrahydrofuran–formic acid (55:33:11:1; oxalic acid 400 mg l⁻¹);[19,21] under these conditions the components were eluted according to their galloylation degrees. G, Gallic acid (**1**); βG, β-glucogallin (**2**); G5–G10, tetra- to decagalloylglucoses.

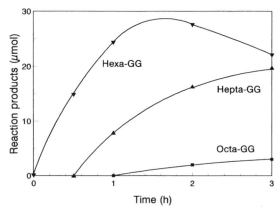

Figure 2 Time course of the *in vitro* synthesis of gallotannins. Analyses were performed according to Figure 1. GG, galloylglucose.

Additional evidence for the presumed gallotannin nature of the reaction products arose from degradation studies with purified material. Treatment with fungal tannase resulted in complete hydrolysis, leaving gallic acid (**1**) as the sole phenolic component. Upon methanolysis, that is, under conditions known to cleave exclusively the *meta*-depside bonds of gallotannins, the hexa-, hepta-, and octagalloylglucose fractions were degraded to 1,2,3,4,6-pentagalloylglucose (**3**) and methyl gallate in molar ratios of 1:1, 1:2, and 1:3, respectively. Hence, these three fractions must have had a 1,2,3,4,6-pentagalloylglucose core to which one, two, or three galloyl moieties had been attached by depside linkages.[85] (It is worth mentioning in this connection that the occurrence of methyl or ethyl gallate in tannin preparations is regarded as being indicative of the formation of artifacts due to exposure to alcoholic solvents in the course of the isolation protocol.)

Conclusive proof of the proposed reactions was obtained when the products of scaled-up enzyme assays were isolated by chromatography on Sephadex LH-20 and subsequent semipreparative reversed-phase HPLC on octadecyl-substituted silica gel.[21] The *in vitro* formation of three hexa-galloylglucoses, four heptagalloylglucoses, and numerous higher substituted derivatives was demonstrated by this means.[85,88] The hexagalloylglucoses and one of the heptagalloylglucoses among these could be obtained in sufficiently high quantity and purity to allow detailed ^{1}H and ^{13}C NMR spectroscopy studies which resulted in the unequivocal identification of the hexagalloylglucoses as 1,2,4,6-tetra-*O*-galloyl-3-*O*-digalloyl-β-D-glucose (**31**), 1,3,4,6-tetra-*O*-galloyl-2-*O*-digalloyl-β-D-glucose (**32**), and 1,2,3,6-tetra-*O*-galloyl-4-*O*-digalloyl-β-D-glucose (**33**), and of the heptagalloyl-glucose as 1,2,4,6-tetra-*O*-galloyl-3-*O*-trigalloyl-β-D-glucose (**34**). The structures of two other heptagalloylglucoses, 1,4,6-tri-*O*-galloyl-2,3-di-*O*-digalloyl-β-D-glucose (**35**) and 1,3,6-tri-*O*-galloyl-2,4-di-*O*-digalloyl-β-D-glucose (**36**) are to date not yet totally proven owing to interference with severe acyl migration during the purification sequence (this phenomenon is perhaps one of the major reasons for the often conflicting results observed in this field); the fourth heptagalloylglucose with the putative structure 1,2,6-tri-*O*-galloyl-3,4-di-*O*-digalloyl-β-D-glucose (**37**) was formed only in trace amounts in the enzyme reaction mixtures.[88] The most prominent signals of the ^{1}H NMR spectra of these compounds were the sharp 2s-peaks in the 7.0–7.1 δ ppm region of the aryl-hydrogen atoms (H-2,6) of pentagalloylglucose (**3**).[4] After substitution with further galloyl groups via *meta*-depside linkages, these signals were both split into two characteristic doublets owing to the resulting asymmetry of the proximal galloyl residue, and shifted to significantly higher δ ppm values of 7.25–7.55 (Table 2). The aryl-2,6 hydrogens of the newly introduced distal residues, in contrast, generally displayed such a typically sharp 2s-peak at 7.1 ppm that this signal could be used like an internal marker in the NMR spectra.[51,88] Considering the minimal quantities of material required for ^{1}H NMR spectroscopy, knowledge of these characteristics should contribute to the more facile and unequivocal identification of gallotannins.

Table 2 ^{1}H NMR chemical shifts of galloyl residues of pentagalloylglucose (**3**) and enzymatically derived gallotannins (**31**)–(**34**).[a]

| Compound | Aryl H-2,6 of galloyl group located at glucose carbon | | | | | |
	1	2	3	4	6	X^b
(**3**)	7.10	7.00	6.96	7.04	7.17	—
(**32**)	7.16	*7.34, 7.28*	6.96	7.03	7.22	7.10
(**31**)	7.15	7.00	*7.30, 7.23*	7.05	7.22	7.10
(**33**)	7.16	6.99	6.96	*7.35, 7.33*	7.25	7.09
(**34**)	7.15	7.00	*7.29, 7.26*[c] *7.56, 7.49*[d]	7.05	7.28	7.09

[a] δ_H values in ppm TMS, measured in d$_6$-acetone.
[b] X denotes the distal galloyl residue of *meta*-depside side chains.
[c] Proximal galloyl residue of the trigalloyl side chain.
[d] Central galloyl residue of the trigalloyl side chain.

It is important to note that the structures and, within certain limits, the relative amounts of the reaction products obtained *in vitro* with cell-free extracts from *R. typhina* were identical to those of the *in vivo* formed gallotannins that had earlier been isolated from the related species *R. semialata*.[19] It was a particular feature of this "Chinese gallotannin" that the C-1 and C-6 positions remained free of depsidic residues throughout, in contrast to the similar gallotannins from *Q. infectoria*[20]

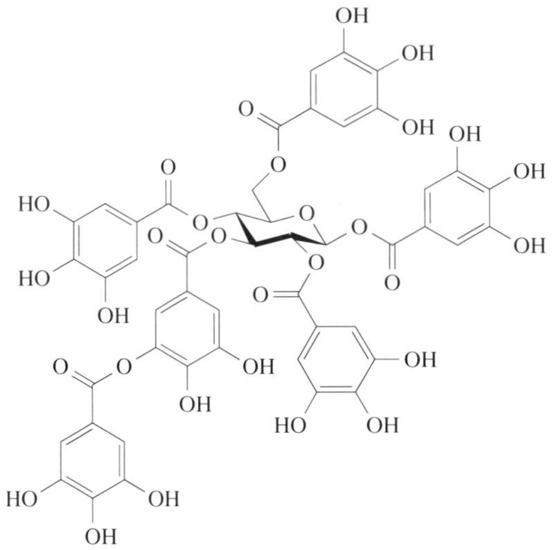

(31)

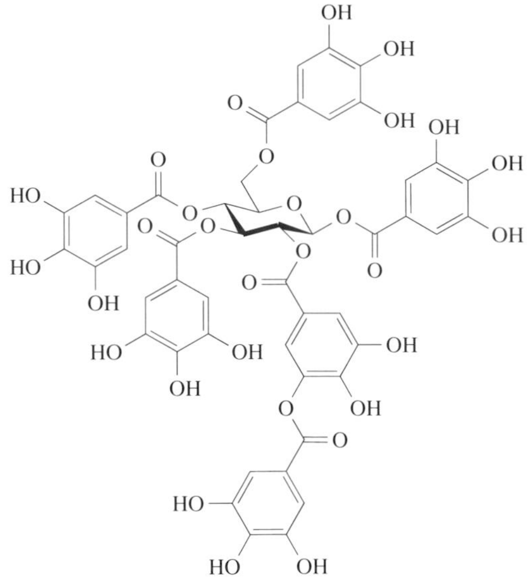

(32)

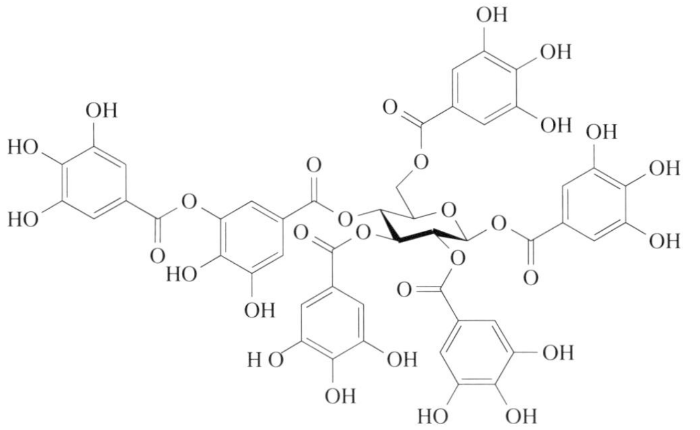

(33)

("Turkish gallotannin") or *P. lactiflora*[21,22] where depside bonds also occurred at C-6. Only the C-1 position thus appears to be generally devoid of depside substituents. As a result of these observations, it was postulated that the biosynthesis of gallotannins in the genus *Rhus* proceeds generally as summarized in Equation (7).[85]

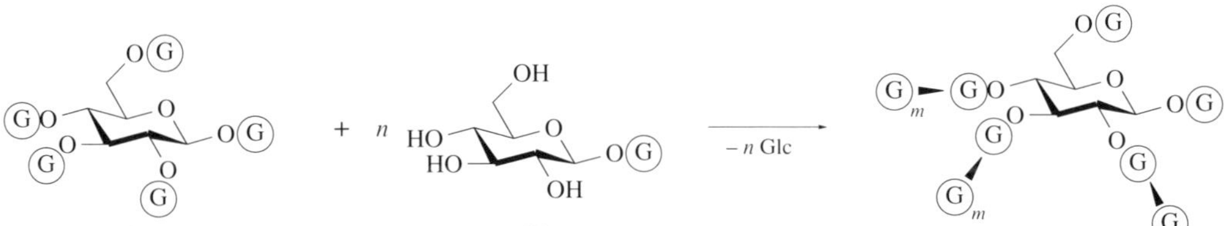

$$(3) \quad + \quad n \ (2) \quad \xrightarrow{- n \ \text{Glc}} \quad \qquad (7)$$

Ⓖ = (7); Ⓖ►Ⓖ = (4); Glc = glucose; m = 0, 1, 2, ...

It is the objective of current research to purify the numerous reaction products as a prerequisite to the identification of their exact position in the metabolism of gallotannins. Preliminary enzyme studies devoted to this question,[88,89] whose results are summarized in Scheme 7, have shown that pentagalloylglucose (3) was preferentially converted to the hexagalloylglucoses (31) and (32), in contrast to galloylation in the 4-position to give (33). Further experiments with pure hexa-galloylglucoses as acceptor substrates revealed that galloylation of (31) in the 3-position to (34) was about three times more efficient than 2-substitution yielding (35). In contrast, *in vitro* acylation of

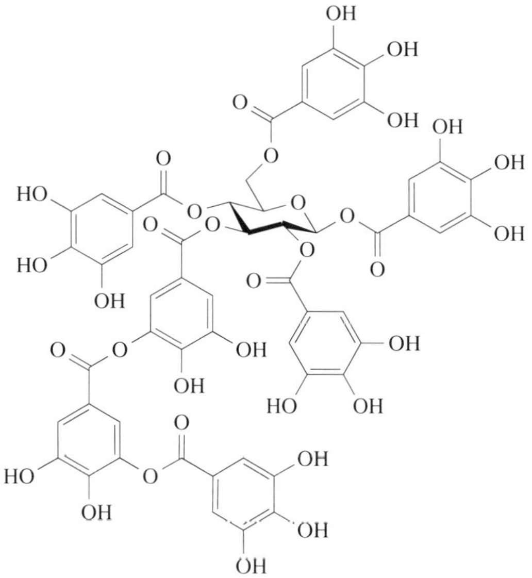

(34)

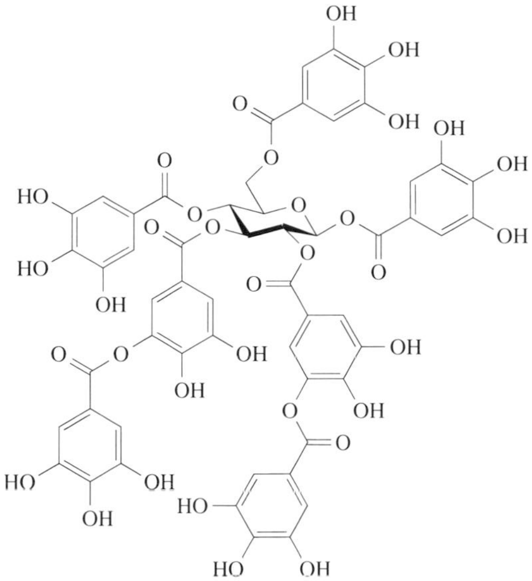

(35)

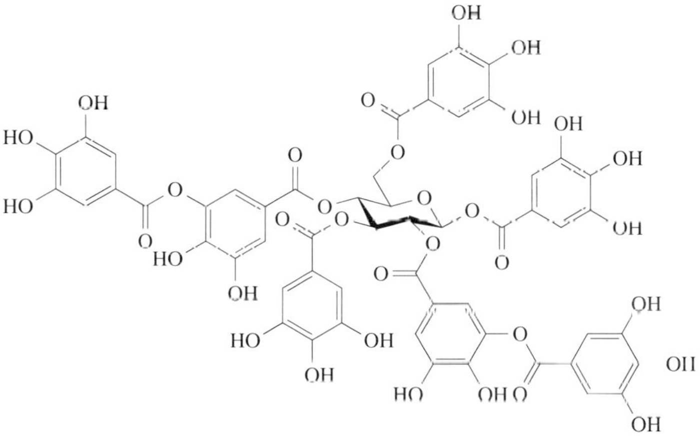

(36)

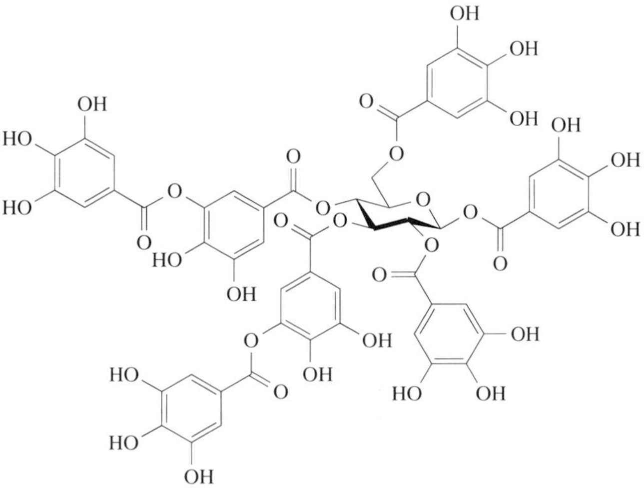

(37)

(32) was equally effective in the formation of (35) and (36). Finally, (33) was more or less exclusively converted to (36); the already mentioned negligible role of (37) was thus confirmed in these experiments. It remained completely unclear, however, whether these individual reactions were catalyzed by specific enzymes, or whether one, or a few, of these β-glucogallin-utilizing galloyltransferases were involved in the pathway.

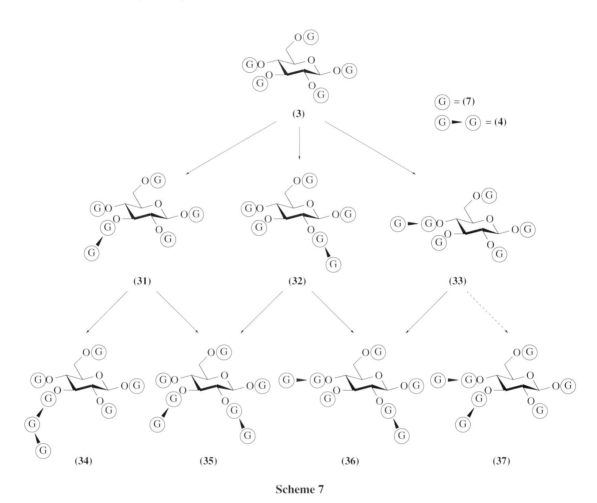

Scheme 7

Enzyme studies devoted to this challenging question with cell-free extracts from sumac (*R. typhina*) leaves revealed the existence of several isoenzymes that catalyzed the *in vitro* acylation of pentagalloylglucose (3). Among these, three galloyltransferases were isolated and separated according to their different molecular weights of ca. 360 000, 290 000, and 170 000, respectively (transferases A, B, and C). Galloyltransferase C has been purified to apparent homogeneity and was found to consist of four identical subunits of M_r 42 000.[90] Substrate specificity studies showed that pentagalloylglucose (3) was the preferred acceptor which was converted to hexa-, hepta-, and octagalloylglucoses in a ratio of 30 : 10 : 1. Closer analysis showed that 3-*O*-digalloyl-1,2,4,6-tetra-*O*-galloyl-β-D-glucose (31) was the predominantly formed hexagalloylglucose. Further experiments in which the standard substrate, pentagalloylglucose, had been replaced by the hexagalloylglucoses (31)–(33) revealed that (31), the main product formed from pentagalloylglucose in the preceding step, was almost exclusively acylated to the heptagalloylglucose, 3-*O*-trigalloyl-1,2,4,6-tetra-*O*-galloyl-β-D-glucose (34). Galloyltransferases A and B have, to date, been partially purified and characterized. While their principal properties were roughly similar to those of enzyme C, they were found to preferentially acylate the 4-position of pentagalloylglucose to hexagalloylglucose (33), followed by substitution of the 2-position of (33) to heptagalloylglucose (36). A minor pathway with an inverse substitution sequence led to the series (3) → (32) → (36), plus a trace activity towards (35). In summary, it is evident that galloyltransferases A and B promoted galloylation at positions 2 and 4 of the galloylglucose core, while transferase C was specific in acylating the 3-position.

3.20.7 OXIDATION OF PENTAGALLOYLGLUCOSE TO ELLAGITANNINS

In contrast to the rather limited occurrence of gallotannins, ellagitannins have been found to occur in many plant families,[16] and they also display a much more pronounced structural variability, owing to the chance of bridging a wide variety of different positions of the basic pentagalloylglucose (3) core by characteristic hexahydroxydiphenoyl (HHDP) (8) residues (cf. Section 3.20.2). Additionally, their strong tendency to form dimeric and oligomeric derivatives contributes significantly to the vast number of compounds in this class of natural plant products; it is estimated that over 500 ellagitannins have been identified thus far.[91] This pronounced structural diversity has traditionally attracted the interest of organic chemists for decades; extreme attention, however, is being paid to these compounds because of evidence which arose from ethnopharmacological studies which indicated that ellagitannins display manifold striking biological activities; accordingly, one has to ascribe to them a new role as extremely promising chemotherapeutic agents, for example as antimutagens, anticarcinogens, or virus inhibitors.[87,92] This intriguing property stimulated not only increasing numbers of reports on the isolation and characterization of such compounds, but also on the chemical synthesis of ellagitannin components, for example the (S)-HHDP (8) unit[92,93] or its dehydrodigalloyl (11) equivalent,[94] of complete ellagitannins like sanguin H-5,[91] permethyltellimagrandin II,[95] tellimagrandin I (38),[96] and pedunculagin (39),[97] or of the chemical transformation of the ellagitannin geraniin to chebulagic acid.[98]

Unfortunately, no similar progress can be recorded for the biosynthesis of ellagitannins. In the 1930s it was postulated[99] that the HHDP (8) residues of ellagitannins originated from the dehydrogenation of gallic acid esters, a view that was again explicitly formulated in 1956 by one of the leading scientists in this field, O. Th. Schmidt.[3] Combining these ideas with the known structures of common ellagitannins, a metabolic sequence was proposed in which 1,2,3,4,6-pentagalloylglucose (3) was sequentially oxidized to tellimagrandin II (5) and casuarictin (6) as primary metabolites, with side reactions yielding the tetraaryl derivatives tellimagrandin I (38) and pedunculagin (39) (Scheme 8).[4,100] Numerous attempts to unravel the mechanism of these conversions have been published in the past; these were carried out either by chemical oxidants (e.g., O_2, Fe^{3+})[101] or by *in vitro* studies with the fungal enzyme system lactase/O_2[102,103] or the plant enzyme system peroxidase/H_2O_2,[101,104,105] utilizing gallic acid (1), methyl gallate, β-glucogallin (2), or 3,6-digalloylglucose as substrates. In all cases, ellagic acid (9) was formed as a typical product, indicating that oxidative aryl coupling via hexahydroxydiphenic acid had occurred, even in experiments with pentagalloylglucose (3);[101] however, the formation of true ellagitannins, bearing a glucose-bound HHDP (8) group, was never observed.

Considering the negative results obtained with these ubiquitous and rather nonspecific enzymes, it was attempted to search for more specific enzymes that catalyzed the oxidative transformation of pentagalloylglucose to ellagitannins. Leaves, green acorn cups, or cotyledons of pedunculate oak were chosen as easily available enzyme sources in these investigations, as this plant is known to synthesize the above simple ellagitannins (cf. Scheme 8) as precursors of complex subsequent derivatives.[4] Pentagalloylglucose was found to be partially hydrolyzed in these studies, but was also converted to manifold unidentified higher and lower molecular weight derivatives, as shown by analysis of the reaction products by reversed-phase HPLC. Neither with soluble enzyme fractions nor with microsomal preparations, however, could the oxidation to ellagitannins be proven, regardless of whether the enzyme assays had been supplemented with NAD, NADP, FAD, FMN, or ubiquinone as possible electron acceptors, or whether the reactions were conducted under aerobic or anaerobic conditions.[88]

Considering the obviously minimal chances for the direct identification (e.g., by HPLC) of ellagitannins among numerous reaction products and by-products, formed by the oxidation of pentagalloylglucose with crude enzyme preparations, it became apparent that an analytical system was required that combined high sensitivity with general applicability to a multitude of different reaction conditions. These requirements were met by a strategy that utilized ellagic acid (9) as a general indicator for the formation of ellagitannins. Reaction mixtures thought to contain oxidatively formed ellagitannins were depleted of free ellagic acid that originated from nonspecific side reactions, followed by hydrolysis of the ellagitannin reaction products. After spontaneous lactonization of liberated HHDP residues (which hence originated exclusively from true ellagitannins), the amount of this second ellagic acid fraction represented a measure of the total reaction rate. When galloyl-labeled pentagalloylglucose was used in such experiments, it was thus possible to discriminate between endogenous and newly synthesized ellagitannins.

For this purpose, labeled [U-^{14}C]pentagalloylglucose of high specific radioactivity was prepared by photoassimilation of $^{14}CO_2$ in leaves of *R. typhina*.[106] In broad screening programs with this

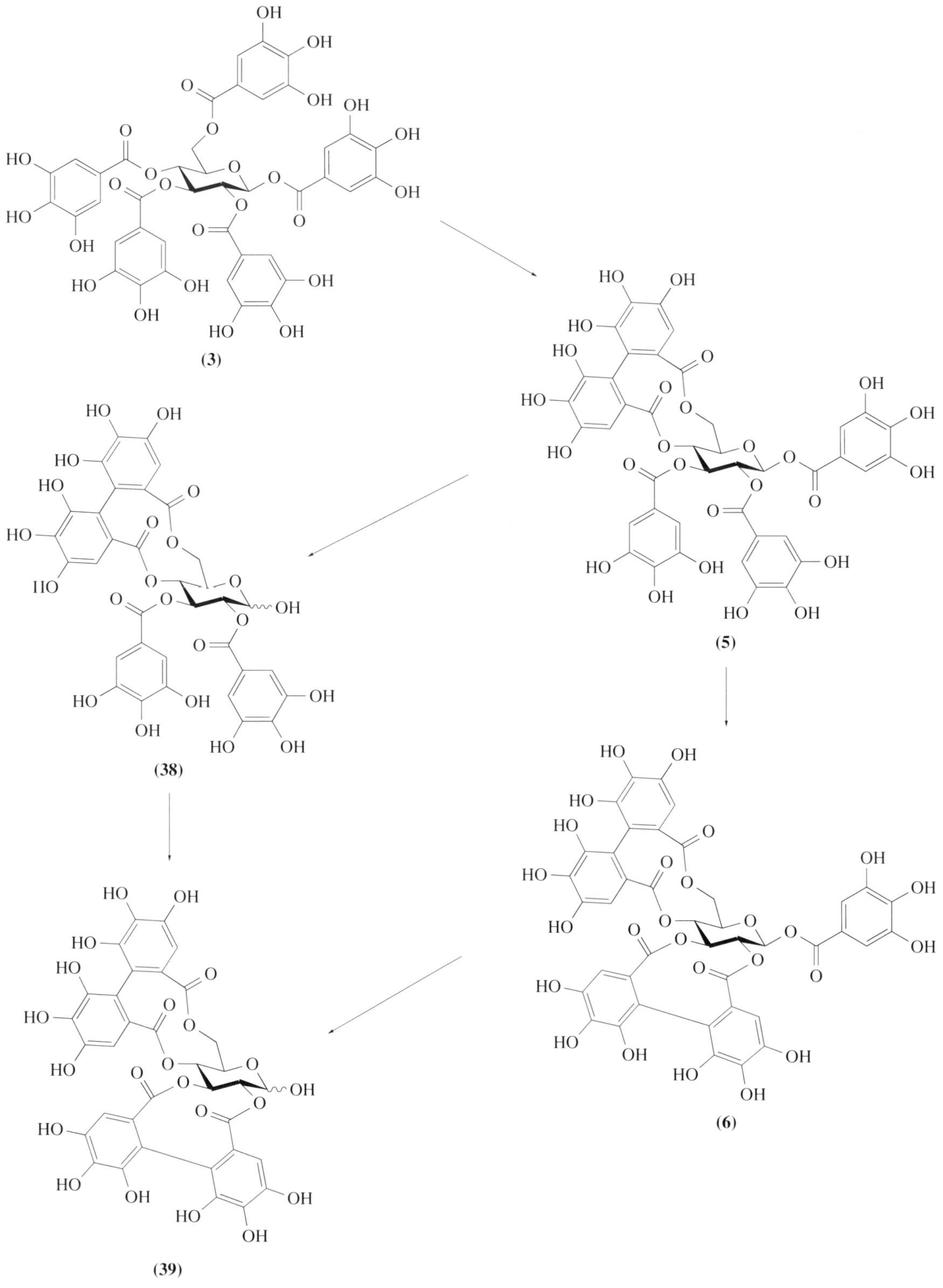

Scheme 8

radioactive substrate, cell-free extracts from leaves of *Q. robur*, and later of *Tellima grandiflora* (saxifragaceae) which characteristically contain significant amounts of tellimagrandin I (**38**) and tellimagrandin II (**5**), were examined for their ability to synthesize ellagitannins. Finally, a soluble protein was isolated and partially purified from *T. grandiflora* that formed, according to the above strategy, [^{14}C]ellagic acid in the presence of FMN as hydrogen acceptor; this result was indicative of the preceding *in vitro* formation of ellagitannins. Subsequent HPLC analysis of large-scale reaction mixtures with this enzyme fraction, containing unlabeled pentagalloylglucose and FMN as substrates, revealed the presence of several unknown products, among which the major component was purified and identified as tellimagrandin II (**5**) by cochromatography with reference compounds, degradation experiments and particularly by ^{1}H and ^{13}C NMR spectroscopy.[89,107] It was thus concluded that a soluble enzyme from *T. grandiflora* had catalyzed the FMN-dependent oxidation of pentagalloylglucose (**3**) to tellimagrandin II (**5**) as the primary intermediate in the pathways to more complex ellagitannins.[89]

Severe inconsistencies were encountered upon detailed analyses, however, which finally led to a reinvestigation of the whole system with [^{14}C]-labeled pentagalloylglucose. These studies could confirm the above reported formation of [^{14}C]ellagic acid, but also produced evidence that tellimagrandin II (**5**) was completely devoid of radioactivity. This highly surprising discrepancy could be explained by the observation that hydrolytically released [^{14}C]gallic acid was prone to undergo chemical oxidation in the presence of FMN in a phosphate buffered milieu, while no such effect occurred with tris(hydroxymethyl)aminomethane as buffer. On the other hand, the isolated unlabeled tellimagrandin II (**5**) was found to originate from the native plant material; due to its tanning, that is protein binding potential (cf. Section 3.20.2), this compound was carried through the entire "enzyme" purification sequence in close and obviously selective association with a limited number of individual proteins. Hence, it also contaminated the enzyme assays, where it was finally released by exchange with excess pentagalloylglucose present as substrate, thus falsely indicating the *in vitro* formation of an ellagitannin.[107] This finding is in accord with considerations that tannins exhibit some specificities in tannin–protein interactions, which eventually reflect physiologically significant functions.[108]

It thus has to be stated that the old and challenging question regarding the mechanism of ellagitannin biosynthesis is still highly elusive, and it is evident that sophisticated techniques and unconventional new strategies will be required for its eventual clarification. It is interesting to note in this connection that this unsatisfying situation finds its exact counterpart in the unknown formation of dimeric or oligomeric proanthocyanidins;[109] the reported structures of dimers resulting from (+)-catechin coupling with polyphenoloxidase[110] were not consistent with those of the natural products.

3.20.8 ENZYMATIC DEGRADATION OF HYDROLYZABLE TANNINS

It is a common tradition that investigations of the metabolism of secondary plant products are usually focussed on the events involved in their biosynthesis while the reverse catabolic reactions are mostly neglected. This situation certainly does not apply to plant tannins; many decades ago it was already recognized that polyphenolic substances were easily degraded by various micro-organisms. As reviewed,[111] this property is particularly common among ascomycetous and basidiomycetous fungi, for example *Aspergillus*, *Penicillium*, *Fomes*, *Polyporus*, *Poria*, and *Trametes*; it must be emphasized, however, that this ability to metabolize tannins is mostly directed towards the catabolism of hydrolyzable tannins, that is gallotannins and ellagitannins. Tannin degradation has also been observed by yeasts and bacteria,[111] including an isolated anaerobic ruminal bacterium that transforms tannins via gallic acid into pyrogallol.[112]

The enzyme catalyzing these degradative processes is known as tannase (tannin acyl hydrolase; EC 3.1.1.20). It typically cleaves the ester bonds between the aryl moieties of gallotannins and the aliphatic hydroxys of the central polyol with the liberation of gallic acid (**1**), but also of the hexahydroxydiphenoyl (**8**), dehydrodigalloyl (**11**), or valoneoyl (**12**) residues of ellagitannins ("esterase" activity in Scheme 9).[111] In addition, the depsidic ester linkages (**4**) characteristic of gallotannins are hydrolyzed by tannase ("depsidase" activity). It should be stressed that these two reactions must not be confused with the widely occurring activity of glucosidases that cleave the acetal bond of phenolic glucosides, for instance the artificial chromogenic enzyme substrate *p*-nitrophenyl β-D-glucose pyranoside (**40**). Galloyl residues are usually more easily hydrolyzed than other groups, and it is also evident that the position of the substituents on the glucose core significantly affects their reactivity.[111] It has been a matter of dispute for decades whether the esterase and depsidase activities of tannase are due to two separate enzymes or to only one enzyme catalyzing both reactions. Attempts to clarify this question with enzyme preparations from *Aspergillus niger* revealed the existence of several tannase isoenzymes in this organism that all displayed both esterase and tannase activities in different ratios.[113] The problem can be regarded as solved by reports of the purification of tannases to apparent homogeneity, isolated from the ascomycetes *A. niger*[114] and *Cryphonectria* (*Endothia*) *parasitica*,[115] which were found to exhibit esterase as well as depsidase activity, indicating that the simultaneous expression of both enzyme activities is an inherent property of tannase.

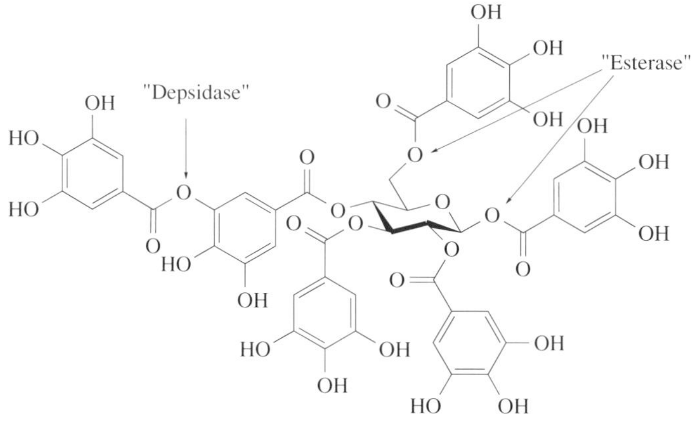

Scheme 9

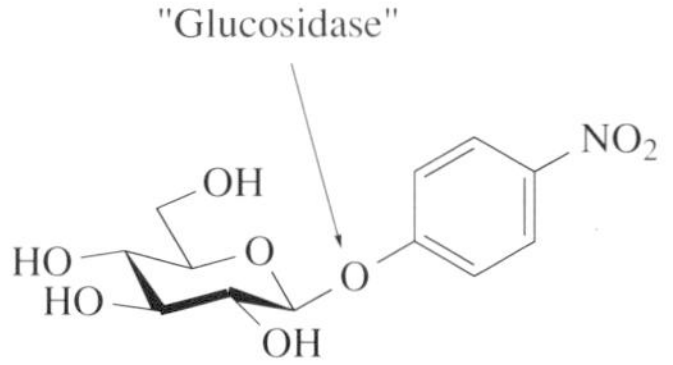

Currently available data show that the basic properties of fungal tannases are rather uniform. The enzymes from *A. niger*,[114] *Aspergillus flavus*,[116] *Aspergillus oryzae*,[117] *Penicillium chrysogenum*,[118] *C. parasitica*,[115] and yeast[119] were all characterized by rather high molecular weights, ranging from ca. 190 kDa to 240 kDa. For the enzymes from *A. niger* and *C. parasitica*, subunits of 53 kDa and 58 kDa, respectively, were determined. Their pH optima were usually around 5.0–5.5, and they were also most stable in this pH range; optimal temperatures for the enzyme reactions lay at about 40 °C. The preferred substrates were differently substituted galloylglucoses, tannic acid, Chinese gallotannin, but also smaller molecules like methylgallate.

It should not be overlooked that tannase may also be useful for industrial applications. Some examples are suggestions for the production of gallic acid (**1**) by fermentation of tara tannin with *A. niger*,[120] the use of a tannase preparation from *Trametes versicolor*[121] for the reduction of tannin contents in meal from double zero (low glucosinolate/low erucic acid) rapeseed varieties such as canola to allow its use as a source of food-grade proteins in animal feed,[122] or the synthesis of the widely used food antioxidant, propylgallate, by transesterification of tannic acid with propanol.[123]

Concerning the occurrence of tannases in higher plants, only one report has been published in the past on such an enzyme from divi-divi fruit pods (*Caesalpinia coriaria*, fabaceae).[124] It was discovered that cell-free extracts from leaves of *Q. robur* contained an esterase that actively hydrolyzed β-glucogallin (**2**) and related galloyl esters.[57] This enzyme could be purified more than 1900-fold to apparent homogeneity. Its molecular weight was determined by various gel-filtration methods as 150 kDa and 300 kDa, respectively. Two protein bands were also detected after native polyacrylamide-gel electrophoresis, while denaturing electrophoresis in the presence of sodium dodecylsulfate revealed the existence of only one protein with an M_r of 75 kDa. It was concluded that the native enzyme existed as both a dimeric and a tetrameric protein derived from identical subunits.[125]

Detailed studies were carried out on the substrate-specificity of this enzyme with a wide array of pure compounds, in contrast to the above investigations on fungal tannases in which only few and rather ill-defined substrates like tannic acid had been used in most cases. It was found that the hydrolase was absolutely inactive with methyl and 1-*O*-glucose esters of variously substituted cinnamic acids, substrates with nitro-substitution of the aromatic acid moiety, and also with phenolic glucosides like (**40**). Hydrolysis occurred with simple galloyl esters (methyl, ethyl, propyl gallate), naphthyl acetate (but not with naphthyl propionate or butyrate), mono- to hexasubstituted galloylglucoses, variously ring-substituted 1-*O*-phenylcarboxyl-β-D-glucoses (which were the most active substrates), and depsides like *meta*-digallic acid or chlorogenic acid. By analogy to previous observations with fungal tannases,[111] a clear sequence of reactivities according to the position of the substituents on the glucose core was observed, with 1-*O*-derivatives being the preferentially hydrolyzed residues. Methyl, ethyl and propyl gallate, β-glucogallin, and pentagalloylglucose displayed sigmoidal substrate saturation curves, while the other active substrates followed normal Michaelis–Menten kinetics. Summarizing these properties, it is evident that the enzyme from oak leaves exhibited pronounced esterase and depsidase activity towards galloylglucoses and related compounds and that its properties closely resembled those of the above reported fungal enzymes; it thus appeared right to regard it as a new tannase of plant origin.[125]

Finally, it appears appropriate to discuss briefly the ecological significance of such a plant tannase. Though polymerization of tannins to insoluble derivatives or association with pectins has been discussed as major deastringency mechanisms in fruit ripening,[126] it is reasonable to assume that this enzyme could contribute to these processes by loss of astringency via degradation of tannins. In green leaves, however, the role of tannase is much less apparent. It has been recognized, for instance, that condensed tannins of *Acacia nigrescens* acted as antidefoliate agents against browsing by giraffe,[127] and after studies with *Epilobium*, *Cornus*, or *Alnus* it was concluded that the soluble galloylglucoses and ellagitannins present in these plants were important in the defense against ruminants.[128] The existence of tannin-degrading enzymes in leaves would thus not make much sense. In contrast to herbivorous mammals, however, the situation with insects could be quite different. The feeding deterrent role traditionally ascribed to tannins due to their astringency, causing reduced palatability of plant parts, has come under criticism; evidence has been presented for the hypothesis that the ellagitannin geraniin preferentially acted as protoxin that released insect growth inhibitors, particularly ellagic acid, upon hydrolytic cleavage.[129]

Accordingly, hydrolyzable tannins would play a dual protective role in plant–herbivore interactions, being active not only by direct protection against herbivorous animals, but also indirectly in the form of their degradation products. The occurrence of tannase in green leaves would thus significantly contribute to the latter process. It is easy to visualize that loss of cellular compartmentalization under the attack of an insect predator brings this enzyme into contact with its tannin substrates, thus causing the release of harmful products. Analogous defense strategies are well-documented for many plants that utilize secondary metabolites like *o*-coumaroyl glucosides, cyanogenic glucosides, or glucosinolates as precursors of hydrolytically released toxins.[128] Eventually, the system tannin–tannase has to be added to the list of chemical defense mechanisms in higher plants.

3.20.9 PREPARATION OF GALLOYLGLUCOSES

In most *in vitro* investigations of biochemical pathways, problems are encountered not only regarding the activity and stability of the enzymes catalyzing the metabolic steps, but also with respect to the supply of substrates required in the studies, one serious obstacle being commercial unavailability that enforces laborious chemical syntheses or isolation from natural sources. As this situation applied explicitly to the studies discussed above on the biosynthesis of hydrolyzable tannins, it appeared appropriate to end this chapter with a short survey of procedures allowing the preparation of these rare chemicals. The occurrence, isolation and characterization of a host of tannins and tannin precursors is extensively documented in the relevant literature; here in this section only those methods will be mentioned that have been recognized as practical for routine work on the basis of several years of laboratory experience.

β-Glucogallin (**2**) is of particular importance because of its dual role as both primary intermediate and predominant acyl donor in the entire biogenetic sequence to producing gallotannins. Consequently, this ester is required for enzyme studies not only in considerable amounts, but also in various labeled forms for special analytical applications. While the isolation of *β*-glucogallin from natural sources, e.g., rhizomes of rhubarb (*Rheum*) or leaves of sumac (*Rhus*) where it occurs as a minor constituent, is not advisable, both chemical and enzymatic syntheses have been developed to meet the above requirements. For the chemical synthesis of *β*-glucogallin, *α*-acetobromoglucose was dehalogenated with silver carbonate; the resulting 2,3,4,6-tetra-*O*-acetyl-D-glucose was esterified with the acid chloride of triacetylgallic acid. The protecting groups of the resulting heptaacetylated 1-*O*-galloylglucose were finally removed by treatment with sodium methylate, yielding analytically pure *β*-glucogallin.[53] This ester and related 1-*O*-benzoyglucoses have also been synthesized enzymatically;[84] a major advantage of this technique, which is normally applicable only for small-scale preparations, is the avoidance of all problems related to the introduction and elimination of protecting groups which makes this method particularly interesting for the synthesis of radioactively labeled compounds. [^{14}C-glucosyl]*β*-Glucogallin has thus been prepared from UDP-D-[^{14}C]glucose with glucosyltransferase from oak leaves (cf. Section 3.20.4).[55] Considering the enormous costs of the labeled substrate, an economic and convenient alternative employing drastically cheaper D-[^{14}C]glucose was developed later for the synthesis of glucosyl-labeled *β*-glucogallin, utilizing the *β*-glucogallin–glucose exchange reaction described in Section 3.20.5.2.[79] For other applications,

β-glucogallin labeled in the galloyl moiety was required. Again, this ester was accessible by enzymatic esterification of UDP-glucose and free [^{14}C]gallic acid,[55] the commercially unavailable labeled substrate being isolated from sumac leaves after photoassimilation of $^{14}CO_2$.[130]

Due to difficulties in obtaining suitable partially protected glucose units, chemical syntheses play no role in the preparation of higher substituted galloylglucoses, except for 1,2,3,4,6-penta-galloylglucose (3) which is easily obtained from free glucose (see below). Instead, isolation from natural tannin sources is recommended, but enzymatic syntheses have also been developed. The rare ester, 1,6-di-O-galloyl-β-D-glucose (24) was extracted from rhubarb roots (*Rhizoma rhei*), purified by chromatography on Sephadex LH-20 and reversed-phase HPLC, and finally crystallized from water.[131] Yields up to 180 mg of pure (24) were thus obtained from 3 kg of plant material.[72] This laborious and time-consuming procedure was replaced by an efficient enzymatic method using β-glucogallin: β-glucogallin 6-O-galloyltransferase (see Section 3.20.5.1) which had been immobilized on a Phenyl-Sepharose column. By simply cycling a buffered solution of the substrate, β-glucogallin, through this "enzyme reactor", followed by adsorption of the reaction product on a subsequent column with RP-18 silica gel while unreacted substrate passed through the column and was recycled, 60 mg of pure 1,6-digalloylglucose (24) could be conveniently synthesized within 5 days, including the time required for the preparation of the enzyme.[57]

When sufficient amounts of (24) were easily accessible by this method, this diester was also used for the enzymatic transformation to 1,2,6-tri-O-galloyl-β-D-glucose (25) with an acid-precipitated insoluble pellet of β-glucogallin-dependent 2-O-galloyltransferase from sumac leaves (cf. Section 3.20.5.1). This pellet was simply suspended in buffer, supplemented with substrates and incubated until maximal conversion was reached (ca. 5 h). The enzyme was easily collected by centrifugation and resuspended in fresh substrate solution; its stability was usually sufficient for at least four incubation cycles. The reaction product (25) was purified on small columns of RP-18 silica gel in excellent yields.[75]

Commercially available tannin was found to be a good source of 1,2,3,6-tetra-O-galloyl-β-D-glucose (26). Starting with only 30 g of crude product, 240 mg of pure ester were obtained after chromatography on Sephadex LH-20 in ethanol[21] and isocratic reversed-phase HPLC.[76]

1,2,3,4,6-Penta-O-galloyl-β-D-glucose (3) is known to be an abundant constituent of many plant tannins from which it can easily be isolated by the above mentioned procedures. However, chemical methods provide an interesting alternative because unprotected glucose can be used. By this strategy, gram quantities of pentagalloylglucose could be prepared by reacting triacetylgalloylchloride with D-glucose and subsequent hydrolysis of the protecting acetyl groups from penta(triacetylgalloyl)-β-D-glucose.[55] The limiting step in this procedure was the final purification of the reaction product by chromatography on Sephadex LH-20. As already mentioned (see Section 3.20.7), (3) was also prepared in radioactively labeled form as [U-^{14}C]pentagalloylglucose. This was achieved by sub-jecting sumac leaves to photoassimilation with $^{14}CO_2$ in the presence of the herbicide glyphosate. From the extracted tri- to decagalloylglucoses, pure labeled (3) was isolated by Sephadex LH-20 chromatography and subsequent semipreparative reversed-phase HPLC.[106]

Detailed procedures have been published for the purification of gallotannins, including the hexagalloylglucoses (31)–(33) and the heptagalloylglucoses (34)–(37) mentioned in Section 3.20.6. The methods comprise acetone extraction of dried plant material from *R. semialata*,[19] *Q. infectoria*,[20] or *Paeonia trichocarpa*,[22] partitioning against ethyl acetate, prepurification on Sephadex LH-20,[21] and final purification by preparative reversed-phase HPLC with acetonitrile–water mixtures being supplemented with oxalic acid (2–3 g l^{-1}) to give sharper peaks in the chromatograms.

3.20.10 CONCLUSIONS AND PERSPECTIVES

As discussed in this chapter, the principles of the reactions catalyzing the metabolic route from gallic acid to pentagalloylglucose, the pivotal intermediate in this biosynthesis of hydrolyzable tannins, have been unraveled by enzyme studies, and this applies also to the subsequent stepwise transformation of this precursor along the gallotannin branch of this class of polyphenolic natural products. As discussed above in detail, earlier assumptions on the mechanism(s) involved in the biosynthesis of ellagitannins, the naturally predominating second class of hydrolyzable tannins, proved to be no longer tenable in light of evidence arising from investigations devoted to this challenging question. Consequently, it will be necessary to put greater emphasis in the future on this important facet in the area of phenolic plant constituents, and it can only be hoped that new ideas and sensitive techniques will emerge that are suitable to tackle this old enigma.

However, many other important questions also deserve interest, for example elucidation of the details of the structure and biosynthesis of complex gallotannins, of the regulatory aspects related to the formation and eventual degradation of hydrolyzable tannins, or of the cellular localization of the tannin-synthesizing enzymes and their products. Other challenges include the physiological or eventual ecological role of tannase, or the molecular structure of the enzymes related to the formation of hydrolyzable tannins. The latter aspect is particularly fascinating if one considers the high tanning (i.e., protein precipitating and enzyme denaturing) potential of most of the galloyl-glucoses which are involved in these pathways as enzyme substrates and products, and it appears feasible to ascribe particular structural features to these enzymes that evidently are so remarkably resistant against the deleterious properties of these phenolic plant constituents.[131]

The final paragraph of this chapter is devoted to the presentation of an important analogue to the biosynthetic route from gallic acid (**1**) to pentagalloylglucose (**3**) discussed in Section 3.20.5. Glandular trichomes of wild tomato (*Lycopersicon penellii*) leaves secrete high concentrations of unusual epicuticular acylated sugars that possess pronounced activity as insect repellents. Their structure is characterized by a glucose moiety which is esterified at positions 2, 3, and 4 with straight and branched fatty acids of short and medium chain length. The biosynthesis of these acylsugars was found to be independent of acyl-CoA esters as activated components but to proceed via 1-*O*-acyl-β-D-glucose.[132] Enzyme assays with isobutyric acid (**41**) catalyzed the formation of 1-*O*-isobutyroyl-β-D-glucopyranose (**42**) in the presence of UDP-glucose.[133,134] This primary metabolite was found to serve as the energy-rich donor for a series of subsequent transacylation reactions (cf. Scheme 10) yielding 1,3 di (**43**), 1,2,3 tri (**44**), and 1,2,3,4-tetra-*O*-isobutyroylglucose (**45**).[134] The transferase catalyzing this sequence has been characterized as a serine carboxypeptidase-like enzyme that, however, possessed a synthetic function.[135] A final acyl exchange reaction of the tetra-substituted derivative with free glucose afforded the characteristic anomeric 2,3,4-tri-*O*-acylglucose (**46**) under the liberation of (**42**).[134] The striking similarity of this pathway to the pattern of

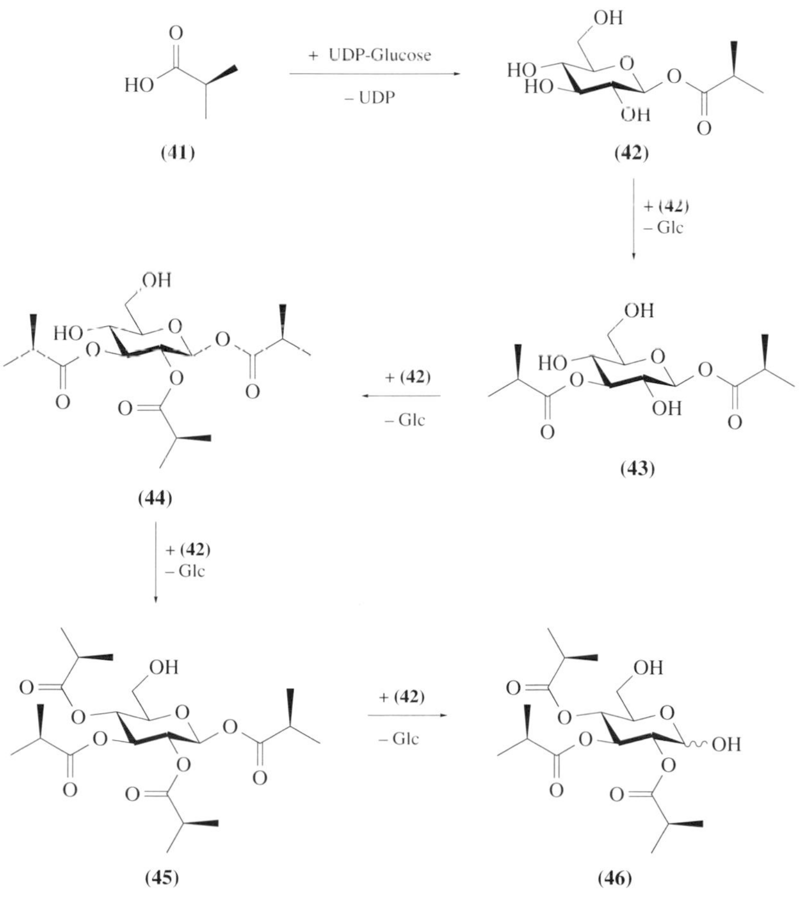

Scheme 10

galloylglucose biosynthesis also extended to other facets; e.g., parallels to the exchange reaction between β-glucogallin (**2**) and free D-glucose (cf. Section 3.20.5.2, Equation (6)) were identified in acyl transfer reactions of (**42**) with D-glucose, yielding 1- or 3-*O*-acylglucoses under the liberation of glucose.[132] In the light of these results it can be concluded that not only phenolic 1-*O*-acylglucoses occupy a central role as activated intermediates in secondary metabolism, as already discussed in Section 3.20.5.1, but that this property must be ascribed now also to their aliphatic analogues, thus corroborating the presumed general importance of these ester acetals.

ACKNOWLEDGMENTS

I am indebted to many colleagues and co-workers who contributed to the investigations of my laboratory discussed above in the form of active participation, helpful suggestions, or supply of indispensable chemicals. Generous financial support by the Deutsche Forschungsgemeinschaft, the Fonds der Chemischen Industrie, and research grants from the University of Ulm is gratefully acknowledged. This chapter is dedicated to Professor Meinhart H. Zenk on the occasion of his 65th birthday.

3.20.11   REFERENCES

1. E. Haslam, *J. Soc. Leather Technol. Chem.*, 1988, **72**, 45.
2. H. G. Larew, *Economic Botany*, 1987, **41**, 33.
3. O. T. Schmidt and W. Mayer, *Angew. Chem.*, 1956, **68**, 103.
4. E. Haslam, "Plant Polyphenols. Vegetable Tannins Revisited," Cambridge University Press, Cambridge, 1989.
5. E. C. Bate-Smith, *Phytochemistry*, 1984, **23**, 945.
6. T. Okuda, T. Yoshida, and T. Hatano, *Phytochemistry*, 1993, **32**, 507.
7. E. Haslam, T. H. Lilley, Y. Cai, R. Martin, and D. Magnolato, *Planta Medica*, 1989, **55**, 1.
8. T. Okuda, T. Yoshida, and T. Hatano, in "Phenolic Compounds in Food and Their Effects on Health II. Antioxidants and Cancer Prevention," eds. M.-T. Huang, C.-T. Ho, and C. Y. Lee, *ACS Symp. Ser.*, Washington, DC, 1992, Vol. 507, p. 160.
9. H. Goto, Y. Shimada, Y. Akechi, K. Kohta, M. Hattori, and K. Terasawa, *Planta Medica*, 1996, **62**, 436.
10. E. Haslam, *J. Nat. Prod.*, 1996, **59**, 205.
11. H. Nakashima, K. Ichiyama, F. Hirayama, K. Uchino, M. Ito, T. Saitoh, M. Ueki, N. Yamamoto, and H. Ogawara, *Antiviral Res.*, 1996, **30**, 95.
12. A. Scalbert and E. Haslam, *Phytochemistry*, 1987, **26**, 3191.
13. T. Ozawa, T. H. Lilly, and E. Haslam, *Phytochemistry*, 1987, **26**, 2937.
14. A. Scalbert, *Phytochemistry*, 1991, **30**, 3875.
15. M. D. Hunter and J. C. Schultz, *Oecologia*, 1993, **94**, 195.
16. E. A. Haddock, R. K. Gupta, M. K. Al-Shafi, K. Layden, E. Haslam, and D. Magnolato, *Phytochemistry*, 1982, **21**, 1049.
17. K. Freudenberg, "Die Chemie der natürlichen Gerbstoffe," Springer, Berlin, 1920, p. 4.
18. L. J. Porter, in "Methods in Plant Biochemistry," ed. J. B. Harborne, Academic Press, London, 1989, Vol. 1, p. 389.
19. M. Nishizawa, T. Yamagishi, G.-I. Nonaka, and I. Nishioka, *J. Chem. Soc., Perkin Trans. 1*, 1982, 2963.
20. M. Nishizawa, T. Yamagishi, G.-I. Nonaka, and I. Nishioka, *J. Chem. Soc., Perkin Trans. 1*, 1983, 961.
21. M. Nishizawa, T. Yamagishi, G.-I. Nonaka, and I. Nishioka, *Chem. Pharm. Bull.*, 1980, **28**, 2850.
22. M. Nishizawa, T. Yamagishi, G.-I. Nonaka, I. Nishioka, T. Nagasawa, and H. Oura, *Chem. Pharm. Bull.*, 1983, **31**, 2593.
23. E. A. Haddock, R. K. Gupta, and E. Haslam, *J. Chem. Soc., Perkin Trans. 1*, 1982, 2535.
24. R. K. Gupta, S. M. K. Al-Shafi, K. Layden, and E. Haslam, *J. Chem. Soc., Perkin Trans. 1*, 1982, 2525.
25. E. Haslam, *J. Nat. Prod.*, 1996, **59**, 205.
26. H. Kawamoto, F. Nakatsubo, and K. Murakami, *Phytochemistry*, 1996, **41**, 1427.
27. H. Kawamoto, F. Nakatsubo, and K. Murakami, *Phytochemistry*, 1995, **40**, 1503.
28. W. E. Hillis, in "Biosynthesis and Biodegradation of Wood Components," ed. T. Higuchi, Academic Press, Orlando, FL, 1985, p. 325.
29. M. H. Zenk, in "Biosynthesis of Aromatic Compounds," ed. G. Billek, Pergamon Press, Oxford, 1966, p. 45.
30. R. Löscher and L. Heide, *Plant Physiol.*, 1994, **106**, 271.
31. M. H. Zenk, *Z. Naturforsch.*, 1964, **19b**, 83.
32. E. Haslam, *Fortschr. Chem. Org. Naturst.*, 1982, **41**, 1.
33. S. Z. El-Basyouni, D. Chen, R. K. Ibrahim, A. C. Neish, and G. H. N. Towers, *Phytochemistry*, 1964, **3**, 485.
34. E. E. Conn and T. Swain, *Chem. Ind.*, 1961, 592.
35. D. Cornthwaite and E. Haslam, *J. Chem. Soc.*, 1965, 3008.
36. T. N. Tateoka, *Bot. Mag. Tokyo*, 1968, **81**, 103.
37. N. Kato, M. Shiroya, S. Yoshida, and M. Hasegawa, *Bot. Mag. Tokyo*, 1968, **81**, 506.
38. K. D. Draths and J. W. Frost, *J. Am. Chem. Soc.*, 1991, **113**, 9361.
39. K. H. Scharf, M. H. Zenk, D. K. Onderka, M. Carroll, and H. G. Floss, *J. Chem. Soc., Chem. Commun.*, 1971, 765.
40. P. F. Knowles, R. D. Haworth, and E. Haslam, *J. Chem. Soc.*, 1961, 1854.
41. L. Heide, H. G. Floss, and M. Tabata, *Phytochemistry*, 1989, **28**, 2643.

42. R. Saijo, *Agric. Biol. Chem.*, 1983, **47**, 455.
43. N. Ishikura, *Experientia*, 1975, **31**, 1407.
44. N. Ishikura, S. Hayashida, and K. Tazaki, *Bot. Mag. Tokyo*, 1984, **97**, 355.
45. N. Amrhein, H. Topp, and O. Joop, *Plant Physiol.*, 1984, **75s**, 18.
46. J. Lydon and S. O. Duke, *J. Agric. Food Chem.*, 1988, **36**, 813.
47. J. M. Becerill, S. O. Duke, and J. Lydon, *Phytochemistry*, 1989, **28**, 695.
48. I. Werner, A. Bacher, and W. Eisenreich, *J. Biol. Chem.*, 1997, **272**, 25 474.
49. J. Stöckigt and M. H. Zenk, *FEBS Lett.*, 1974, **42**, 131.
50. G. G. Gross, in "Plant Cell Wall Polymers: Biogenesis and Biodegradation," eds. N. G. Lewis and M. G. Paice, *ACS Symp. Ser.*, Washington, DC, 1989, Vol. 399, p. 108.
51. G. G. Gross, in "Phenolic Metabolism in Plants," eds. H. A. Stafford and R. K. Ibrahim, *Rec. Adv. Phytochem.* Plenum Press, New York, 1992, Vol. 26, p. 297.
52. G. G. Gross, *Z. Naturforsch.*, 1982, **37c**, 778.
53. G. G. Gross, *FEBS Lett.*, 1982, **148**, 67.
54. G. G. Gross, *Phytochemistry*, 1983, **22**, 2179.
55. G. G. Gross, *Z. Naturforsch.*, 1983, **38c**, 519.
56. S. W. Schmidt, K. Denzel, G. Schilling, and G. G. Gross, *Z. Naturforsch.*, 1987, **42c**, 87.
57. G. G. Gross, K. Denzel, and G. Schilling, *Z. Naturforsch.*, 1990, **45c**, 37.
58. M. R. Atkinson and R. K. Morton, in "Comparative Biochemistry. Free Energy and Biological Function," eds. M. Florkin and H. S. Mason, Academic Press, New York, 1960, Vol. II, p. 1.
59. H. P. Mock and D. Strack, *Phytochemistry*, 1993, **32**, 575.
60. M. Kojima and R. J. A. Villegas, *Agric. Biol. Chem.*, 1984, **48**, 2397.
61. D. Strack, J. Heilemann, B. Boehnert, L. Grotjahn, and V. Wray, *Phytochemistry*, 1987, **26**, 107.
62. R. J. A. Villegas and M. Kojima, *J. Biol. Chem.*, 1986, **261**, 8729.
63. W. Gräwe, P. Bachhuber, H. P. Mock, and D. Strack, *Planta*, 1992, **187**, 236.
64. W. Gräwe and D. Strack, *Z. Naturforsch.*, 1986, **41c**, 28.
65. D. Strack, B. Dahlbender, L. Grotjahn, and V. Wray, *Phytochemistry*, 1984, **23**, 657.
66. B. Dahlbender and D. Strack, *Phytochemistry*, 1986, **25**, 1043.
67. R. J. A. Villegas, T. Shimokawa, H. Okuyama, and M. Kojima, *Phytochemistry*, 1987, **26**, 1577.
68. J. M. Kesy and R. S. Bandurski, *Plant Physiol.*, 1990, **94**, 1598.
69. S. Kowalczyk and R. S. Bandurski, *Plant Physiol.*, 1990, **94**, 4.
70. M. Bokern, S. Heuer, and D. Strack, *Bot. Acta*, 1992, **105**, 146.
71. W. E. Gläßgen and H. U. Seitz, *Planta*, 1992, **186**, 582.
72. K. Denzel, G. Schilling, and G. G. Gross, *Planta*, 1988, **176**, 135.
73. G. G. Gross and K. Denzel, *Z. Naturforsch.*, 1991, **46c**, 389.
74. K. Denzel, Ph.D. Thesis, University of Ulm, 1995.
75. S. Hagenah and G. G. Gross, *Phytochemistry*, 1993, **32**, 637.
76. J. Cammann, K. Denzel, G. Schilling, and G. G. Gross, *Arch. Biochem. Biophys.*, 1989, **273**, 58.
77. P. Grundhöfer and G. G. Gross, manuscript in preparation.
78. K. Denzel and G. G. Gross, *Planta*, 1991, **184**, 285.
79. K. Denzel, S. Weisemann, and G. G. Gross, *J. Plant Physiol.*, 1988, **133**, 113.
80. G. G. Gross, S. W. Schmidt, and K. Denzel, *J. Plant Physiol.*, 1986, **126**, 173.
81. J. M. Sugihara, *Adv. Carbohydr. Chem.*, 1953, **8**, 1.
82. E. Reinefeld and D. Ahrens, *Justus Liebigs Ann. Chem.*, 1971, **747**, 39.
83. J. M. Williams and A. C. Richardson, *Tetrahedron*, 1967, **23**, 1369.
84. S. Weisemann, K. Denzel, G. Schilling, and G. G. Gross, *Bioorg. Chem.*, 1988, **16**, 29.
85. A. S. Hofmann and G. G. Gross, *Arch. Biochem. Biophys.*, 1990, **283**, 530.
86. I. Krajci and G. G. Gross, *Phytochemistry*, 1987, **26**, 141.
87. T. Okuda, T. Yoshida, and T. Hatano, *J. Nat. Prod.*, 1989, **52**, 1.
88. A. S. Hofmann, Ph.D. Thesis, University of Ulm, 1996.
89. G. G. Gross, *Acta Horticulturae*, 1994, **381**, 74.
90. R. Niemetz and G. G. Gross, *Phytochemistry*, 1998, in press.
91. K. S. Feldman and A. Sambandam, *J. Org. Chem.*, 1995, **60**, 8171.
92. K. S. Feldman and S. M. Ensel, *J. Am. Chem. Soc.*, 1994, **116**, 3357.
93. T. D. Nelson and A. I. Meyers, *J. Org. Chem.*, 1994, **59**, 2577.
94. K. S. Feldman, S. Quideau, and H. M. Appel, *J. Org. Chem.*, 1996, **61**, 6656.
95. B. H. Lipshutz, Z.-P. Liu, and F. Kayser, *Tetrahedron Lett.*, 1994, **35**, 5567.
96. K. S. Feldman, S. M. Ensel, and R. D. Minard, *J. Am. Chem. Soc.*, 1994, **116**, 1742.
97. K. S. Feldman and R. S. Smith, *J. Org. Chem.*, 1996, **61**, 2606.
98. T. Tanaka, I. Kouno, and G.-I. Nonaka, *Chem. Pharm. Bull.*, 1996, **44**, 34.
99. H. Erdtman, *Svensk. Kem. Tidskr.*, 1935, **47**, 223.
100. T. Hatano, R. Kira, M. Yoshizaki, and T. Okuda, *Phytochemistry*, 1986, **25**, 2786.
101. W. Mayer, E. H. Hoffmann, N. Lösch, H. Wolf, B. Wolter, and G. Schilling, *Justus Liebigs Ann. Chem.*, 1984, 929.
102. D. E. Hathway, *Biochem. J.*, 1957, **67**, 445.
103. W. Flaig and K. Haider, *Planta Medica*, 1961, **9**, 123.
104. M. Y. Kamel, N. S. Saleh, and A. M. Ghazy, *Phytochemistry*, 1977, **16**, 521.
105. F. Pospíšil, M. Cvikrová, and M. Hrubcová, *Biol. Plantar. (Praha)*, 1983, **25**, 373.
106. H. Rausch and G. G. Gross, *Z. Naturforsch.*, 1996, **51c**, 473.
107. H. Rausch and G. G. Gross, unpublished results.
108. T. N. Asquith and L. G. Butler, *Phytochemistry*, 1986, **25**, 1591.
109. S. Singh, J. McCallum, M. R. Koupai-Abyazani, G. H. N. Towers, B. A. Bohm, A. D. Muir, and M. Y. Gruber, *Polyphénols Actualités*, 1994, **11**, 18.
110. S. Guyot, J. Vercauteren, and V. Cheynier, *Phytochemistry*, 1996, **42**, 1279.

111. A. Scalbert, *Phytochemistry*, 1991, **30**, 3875.
112. K. E. Nelson, A. N. Pell, P. Schofield, and S. Zinder, *Appl. Environ. Microbiol.*, 1995, **61**, 3293.
113. E. Haslam and J. E. Stangroom, *Biochem. J.*, 1966, **99**, 28.
114. C. Barthomeuf, F. Regerat, and H. Pourrat, *J. Ferment. Bioeng.*, 1994, **77**, 320.
115. G. M. Farias, C. Gorbea, J. R. Elkins, and G. J. Griffin, *Physiol. Mol. Plant Pathol.*, 1994, **44**, 51.
116. H. Yamada, O. Adachi, M. Watanabe, and N. Sato, *Agric. Biol. Chem.*, 1968, **32**, 1070.
117. S. Ibuchi, Y. Minoda, and K. Yamada, *Agric. Biol. Chem.*, 1972, **36**, 1553.
118. G. S. Rajakumar and S. C. Nandy, *Appl. Environ. Microbiol.*, 1983, **46**, 525.
119. K. Aoki, R. Shinke, and A. Nishira, *Agric. Biol. Chem.*, 1976, **40**, 79.
120. H. Pourrat, F. Regerat, A. Pourrat, and D. Jean, *J. Ferment. Technol.*, 1985, **63**, 401.
121. J. Archambault, K. Lacki, and Z. Duvnjak, *Biotechnol. Lett.*, 1996, **18**, 771.
122. L. Lorusso, K. Lacki, and Z. Duvnjak, *Biotechnol. Lett.*, 1996, **18**, 309.
123. A. Gaathon, Z. Gross, and M. Rozhanski, *Enzyme Mikrob. Technol.*, 1989, **11**, 604.
124. W. Madhavakrishna and S. M. Bose, *Bull. Cent. Leather Res. Inst. Madras*, 1961, **8**, 153 [CA 56: 11759f (1962)].
125. J. U. Niehaus and G. G. Gross, *Phytochemistry*, 1997, **45**, 1555.
126. T. Ozawa, T. H. Lilley, and E. Haslam, *Phytochemistry*, 1987, **26**, 2937.
127. D. Furstenburg and W. Van Hoven, *Comp. Biochem. Physiol.*, 1994, **107A**, 425.
128. C. T. Robbins, T. A. Hanley, A. E. Hagerman, O. Hjeljord, D. L. Baker, C. C. Schwartz, and W. W. Mautz, *Ecology*, 1987, **68**, 98.
129. J. A. Klocke, B. Van Wagenen, and M. F. Balandrin, *Phytochemistry*, 1986, **25**, 85.
130. P. Matile, *Naturwissenschaften*, 1984, **71**, 18.
131. G. G. Gross, in "Basic Life Science: Plant Polyphenols," eds. R. W. Hemingway and P. E. Laks, Plenum Press, New York, 1992, Vol. 59, p. 43.
132. G. S. Ghangas and J. C. Steffens, *Proc. Natl. Acad. Sci. USA*, 1993, **90**, 9911.
133. G. S. Ghangas and J. C. Steffens, *Arch. Biochem. Biophys.*, 1995, **316**, 370.
134. J.-P. Kŭai, G. S. Ghangas, and J. C. Steffens, *Plant Physiol.*, 1997, **115**, 1581.
135. J. C. Steffens, unpublished results.

Author Index

This Author Index comprises an alphabetical listing of the names of the authors cited in the text and the references listed at the end of each chapter in this volume.

Each entry consists of the author's name, followed by a list of numbers, for example

Templeton, J. L., 366, 385[233] (350, 366), 387[370] (363)

For each name, the page numbers for the citation in the reference list are given, followed by the reference number in superscript and the page number(s) in parentheses of where that reference is cited in the text. Where a name is referred to in text only, the page number of the citation appears with no superscript number. References cited in both the text and in the tables are included.

Although much effort has gone into eliminating inaccuracies resulting from the use of different combinations of initials by the same author, the use by some journals of only one initial, and different spellings of the same name as a result of the transliteration processes, the accuracy of some entries may have been affected by these factors.

Aaronson, W., 434[185] (390)
Aas-Eng, D. A., 61[97] (46, 47), 62[98] (46, 47)
Aasheim, H. C., 61[97] (46, 47), 62[98] (46, 47)
Abbas, S. A., 83[32] (76, 77, 78)
Abbott, S. D., 292[252] (277)
Abdel-Daim, M. H., 435[226] (395)
Abdel-Tawab, G. A., 435[226] (395)
Abdul-Baki, A. A., 527[284] (520)
Abe, F., 357[28] (313, 314, 315), 357[29] (314, 315), 358[80] (313, 315), 359[117] (316, 321)
Abe, H., 157[51] (134)
Abe, M., 83[86] (76), 125[62] (109, 123)
Abeijon, C., 293[319] (286, 288)
Abeles, R. H., 290[78] (255), 735[75] (634), 735[79] (633, 634)
Abeles, R., 370, 389, 431[11] (370, 389)
Abernethy, J. L., 83[52] (75)
Abo-Ghalia, M., 290[120] (259)
Abraham, E. P., 292[234] (275)
Abraham, R., 433[89] (378)
Abrahamson, E., 597[136] (576)
Abram, D., 440[570] (426)
Abubakirov, N. K., 358[76] (313, 315), 361[293] (328)
Acevedo, E., 106[131] (102)
Achenbach, H., 357[20] (314, 317), 359[175] (317, 322)
Achtman, M., 239[564] (226), 239[565] (226)
Acton, S. L., 365[511] (356)
Actor, P., 289[3] (242, 274), 289[18] (245)
Adachi, H., 291[186] (267), 291[191] (268)
Adachi, K., 157[30] (132, 134)
Adachi, S.-i., 736[127] (638)
Adams, D. E., 83[32] (73)

Adams, E., 290[61] (254), 290[63] (254), 382, 433[111] (382, 383)
Adams, M. D., 237[453] (220, 221), 363[426] (343, 344), 364[467] (345), 432[22] (373), 432[25] (373)
Adams, R. J., 361[237] (328)
Adermann, K., 83[61] (75)
Adhya, S., 234[230] (203), 234[231] (203)
Adir, J., 158[99] (142)
Adler, B., 62[146] (48), 63[184] (51, 55), 65[270] (55), 235[313] (208), 235[316] (208, 217), 364[135] (344, 346)
Adler, E., 742[496] (700, 723, 724), 742[529] (706, 707)
Adolphson, R., 525[163] (507, 513)
Aebi, M., 309[55] (303)
Agarwal, U. P., 596[82] (558, 559), 745[678] (728, 731)
Agro, A. F., 744[636] (721)
Agterberg, M., 61[55] (44, 58, 59)
Aguilar, J., 439[538] (425), 440[566] (426), 440[567] (426)
Aharoni, N., 614[83] (609)
Ahgren, A., 362[328] (330)
Ahlert, J., 364[455] (345, 354)
Ahmad, S., 433[150] (386), 434[165] (388), 526[197] (509)
Ahmad, Z., 495[391] (485)
Ahmed, K., 361[276] (328), 361[277] (328)
Ahmed, S. P., 158[106] (143)
Ahrens, F. B., 742[512] (703)
Aiello, L. P., 84[91] (76)
Aignesberger, A., 106[134]
Ainscough, R., 63[174] (56)
Ainsworth, C., 491[129] (450, 451, 454)
Aisaka, K., 381, 433[98] (381)
Aisen, P., 743[571] (711)

Aitken, A., 495[379] (482), 495[383] (483)
Aizawa, S., 128[219] (123)
Ajito, K., 361[267] (327), 361[268] (327)
Ajl, S. J., 232[106] (189)
Akamatsu, S., 65[292] (56, 57)
Akamatsu, Y., 230[20] (182)
Akane, A., 437[385] (411)
Akatsuka, A., 485, 495[390] (485), 495[398] (486)
Akay, B., 735[66] (633)
Akazawa, T., 493[258] (465), 744[644] (722)
Åkeson, Å., 738[787] (656), 738[288] (654, 656)
Akhtar, M., 398, 435[267] (398)
Akhy, M. T., 440[564] (426)
Akino, T., 436[318] (403)
Akita, E., 357[27] (314)
Akiyama, A., 33[88] (21)
Akiyama, K., 494[321] (477)
Akiyama, Y., 524[63] (499, 511)
Aksay, I. A., 598[183] (590)
Al-Hendy, A., 234[239] (203, 227), 235[314] (208, 213), 236[383] (215, 227), 364[438] (344, 346, 347)
Al-Qahtani, A., 309[56] (303)
Al-Zarban, S., 440[565] (426)
Alajarin, R., 439[545] (425), 440[572] (428)
Alavi, A., 128[221] (124), 357[6] (312)
Albers, J. J., 233[215] (202)
Albers-Schönberg, G., 358[81] (315, 355)
Albersheim, P., 503, 523[1] (498, 499, 502, 503), 523[4] (498), 524[62] (499, 511), 524[64] (499), 525[133] (504), 526[225] (513, 515), 526[236] (514), 527[275] (519), 527[276] (519), 613[3] (600), 613[4] (600), 615[131] (611)
Albert, M. J., 359[115] (316), 359[116] (316)

Assmann, G., 436[320] (403)
Astrin, K., 33[58] (19)
Aszalos, A., 360[221] (323)
Atalla, R. H., 535, 547, 548, 552, 554, 558, 560, 561, 588, 590, 595[10] (535), 595[27] (537, 541), 596[45] (544), 596[47] (544, 546), 597[120] (573), 597[121] (573), 598[173] (585, 590), 598[174] (585, 590), 745[678] (728, 731)
Atanassova, R., 615[135] (613)
Athanassiadis, A., 34[116] (23, 26), 34[119] (23)
Atkin, C. L., 362[316] (329)
Atkins, E. D. T., 535, 595[13] (535)
Atkinson, D. E., 493[300] (473, 474), 493[302] (474), 494[334] (479)
Atkinson, P. H., 32[19] (18, 21), 33[87] (21), 34[100] (22)
Ator, M., 362[352] (332)
Atta, M., 362[349] (332), 364[463] (343)
Attridge, S., 235[284] (207)
Aubert, J.-P., 598[203] (593), 598[206] (593)
Aucken, H. M., 232[122] (190), 232[123] (190)
Augé, C., 431[7] (368), 433[97] (380, 381), 433[100] (381)
Auger, M., 64[254] (54)
Augeron, C., 84[145] (80)
Auling, G., 361[300] (328, 345)
Austen, K. F., 176[76] (170)
Austen, R. A., 432[ff] (375)
Austin, D., 494[320] (477)
Austin, E. A., 239[547] (224, 225)
Averyhart-Fullard, V., 745[689] (732)
Avidor, I., 65[318] (55, 59)
Avigad, G., 506, 508, 523[29] (498, 505, 506, 507, 508, 509)
Avigliano, L., 743[595] (714), 743[598] (708, 714, 715)
Avramovic, O., 363[374] (334), 363[402] (338)
Awano, T., 739[322] (661, 662, 674)
Axamawaty, M. T. H., 157[58] (134)
Axelrod, J., 737[174] (644)
Axford, J. S., 357[6] (312)
Ayala, J. A., 290[105] (258), 291[153] (264, 267)
Ayling, J. E., 435[288] (400)
Aziz, A., 66[331] (57), 66[346] (57)
Azuma, J.-I., 524[102] (503), 742[482] (697)
Azuma, K., 434[218] (394)
Azuma, N., 106[143] (102)

Baasov, T., 370, 389, 390, 431[12] (370, 389), 434[178] (389), 434[179] (389), 526[193] (509)
Baba, T., 457, 459, 490[67] (445, 461, 462), 490[70] (445, 462), 492[183] (454, 457), 492[200] (459), 780
Babbel, G. R., 738[265] (655)
Babbitt, G. E., 357[41] (314, 315, 317)
Babbitt, P. C., 737[214] (649)
Babczinski, P., 82[9] (70)
Babia, T., 106[107] (100)
Bacher, A., 363[411] (340)
Bachmann, B. J., 231[80] (187)
Bacic, A., 523[3] (498, 499, 518), 524[80] (502), 524[82] (502), 613[2] (600)
Back, U., 230[3] (181)
Backhaus, S., 433[127] (385)
Backstrom, G., 177[105] (173)
Baczynskyj, L., 360[202] (317, 322)

Badía, J., 426, 439[538] (425), 439[540] (425), 440[566] (426), 440[601] (431)
Baddiley, J., 289[12] (243)
Badenhuizen, N. P., 489[30] (443, 463)
Badet, B., 234[272] (206), 289[35] (251), 289[36] (251), 290[66] (254), 290[68] (254), 292[235] (275), 292[236] (275)
Badet-Denisot, M. A., 289[37] (251)
Badur, R., 614[81] (608, 609)
Bae, J. M., 491[133] (451)
Baecker, P. A., 489[40] (444, 447, 477), 492[198] (459, 462, 477, 480)
Baenziger, J. U., 105[47] (94), 126[72] (109, 123, 124)
Baer, E., 598[183] (590)
Bagaglio, J. F., 436[331] (404)
Baguley, B. C., 360[232] (323)
Bahrman, N., 740[364] (669)
Bai, G., 494[343] (479, 485), 494[344] (479, 485)
Baich, A., 437[409] (413)
Bailey, M. J. A., 239[556] (226), 239[567] (226)
Bailly, P., 84[106] (78)
Bainbridge, B., 238[304] (222)
Baird, M. S., 597[118] (572)
Bairoch, A., 738[237] (650)
Bak, C.-I., 438[431] (416)
Baker, D. B., 527[281] (520), 527[285] (520)
Baker, K., 525[163] (507, 513)
Baker, M. A., 32[4] (14), 64[241] (54), 83[76] (76, 77), 83[78] (76, 77, 79), 84[131] (79), 84[133] (79), 85[150] (80)
Baker, M., 158[100] (142)
Baker, P. J., 230[37] (182)
Baker, S. C., 438[484] (420)
Bakhaeva, G. P., 358[68] (313, 315, 317, 322)
Bakhmutov, V. I., 436[355] (405, 407)
Bakker, H., 61[55] (44, 58, 59)
Baky, F. I., 524[52] (499), 524[59] (499, 511)
Balaji, P. V., 61[77] (45)
Baldenius, K. U., 361[260] (326)
Baldomà, L., 440[566] (426)
Baldwin, T., 308[26] (298)
Baldwin, T. C., 614[113] (610)
Balke, E., 432[83] (378)
Ball, C. D., 737[168] (643)
Ball, K. L., 491[131] (449), 491[139] (451)
Ball, S., 458, 466, 473, 490[80] (445, 455), 492[192] (458, 466), 492[193] (458, 466, 467), 493[287] (470), 493[297] (473)
Ball, S. G., 494[360] (480)
Ballantine, S. P., 432[46] (375), 432[47] (375)
Ballicora, M. A., 491[127] (449, 451), 491[141] (451)
Ballongue, J., 434[216] (393)
Ballou, D. P., 363[398] (337, 339, 340), 363[409] (339, 340), 743[597] (708, 714)
Ballou, L. M., 495[401] (487)
Balsdon, J., 491[129] (450, 451, 454)
Baltas, A. M. M., 738[285] (654, 656)
Baltas, M., 738[290] (658)
Baltz, R. H., 364[455] (345, 354), 365[499] (351), 365[503] (352)
Balys, M. M., 83[50] (75)
Bamberger, E., 437[400] (412)
Bamforth, C. W., 437[393] (411, 412)
Ban, N., 744[627] (718, 719)
Bandarian, V., 363[398] (337, 339, 340)
Banerjee, A., 290[95] (256)
Banerjee, D. K., 293[279] (281)

Banerjee, P., 126[121] (111, 112, 113, 121), 127[134] (112), 128[202] (120)
Bang, A. G., 494[341] (479)
Banki, K., 438[465] (419), 438[468] (419)
Banks, W. B., 597[118] (572)
Bannister, B., 359[132] (316, 321)
Banville, D. L., 360[234] (323), 360[235] (323)
Bao, W., 613[27] (602, 611), 742[541] (708, 713, 718), 743[565] (711, 713)
Baptiste, J., 158[110] (142)
Bar-Nun, N., 743[581] (712)
Bar-Peled, M., 526[241] (515), 526[242] (515)
Bar-Shavit, Z., 160[240] (156)
Baran, G., 492[178] (454)
Baratova, L., 437[427] (416)
Barbas III, C. F., 431[8] (368), 432[30] (373)
Barbé, J., 363[420] (343, 344), 363[424] (343, 344), 364[461] (343), 364[463] (343)
Barber, G. A., 523[30] (498, 505, 506, 508), 525[176] (508), 525[177] (508), 526[190] (509)
Barbieri, G., 359[174] (317, 322)
Barbour, M. G., 437[414] (413)
Barcza, S., 358[103] (316)
Bardinskaya., M. S., 744[613] (717)
Bardsley, B., 293[295] (283)
Barengo, R., 481, 490[85] (446), 495[375] (481)
Barker, B. J., 357[41] (314, 315, 317)
Barker, H. A., 364[464] (343)
Barker, H. M., 158[108] (143)
Barker, R., 126[104] (109)
Barksdale, L., 236[391] (216), 432[81] (378)
Barlow, G. J., 293[265] (280)
Barlow, J. J., 83[72] (76, 77, 78)
Barna, J. C. J., 292[213] (272, 281)
Barnell, W. O., 439[524] (423)
Barner, M., 83[78] (76, 77, 79)
Barnes, S., 234[212] (202)
Barnoud, F., 597[122] (573)
Barnsley, E. A., 375, 432[56] (375, 378), 432[57] (375)
Baron, D., 525[171] (508)
Baron, L. S., 234[737] (203), 236[404] (216)
Barondes, S. H., 127[171] (117), 158[170] (143)
Barr, K., 236[368] (214), 236[370] (214)
Barra, D., 435[252] (397), 435[253] (397)
Barranger, J. A., 106[143] (102)
Barras, C., 230[8] (181)
Barrett, T., 741[430] (687)
Barry, G. F., 491[127] (449, 451), 491[140] (451), 492[225] (462, 466), 493[264] (466)
Barry, G., 493[299] (473)
Bartinicki-Garcia, S., 596[64] (548)
Bartolo, D. C., 83[69] (76)
Barton, C., 491[108] (447, 451), 491[109] (447), 491[136] (451)
Bartuska, V. J., 596[61] (548, 550, 552)
Bartz, Q. R., 359[118] (316, 321)
Baruah, S., 435[276] (399)
Baryshnikova, L. M., 434[215] (393)
Barz, W., 737[216] (649)
Basbaum, C., 82[19] (71)
Bass, M., 495[397] (485)
Bassi, R., 106[127] (101)
Bast, B. J. E. G., 61[95] (46, 47)
Basten, J. E. M., 11[16] (2)
Bastin, D. A., 235[289] (207, 208, 216),

47, 59), 62[102] (46, 47), 65[318] (55, 59), 65[319] (55, 59), 66[321] (55, 59), 66[338] (57, 59), 82[2] (70), 84[117] (79), 84[118] (79), 125[46] (109, 112, 119, 121), 127[148] (113, 119, 121), 127[162] (114, 119), 128[198] (120), 128[204] (121, 124), 432[71] (378), 432[72] (378)

Paulsson, M., 175[4] (161, 163, 171, 172, 174)
Paulus, T. J., 365[497] (351)
Paulus, W., 175[26] (165)
Pauly, M., 615[131] (611)
Pauson, P. L., 744[654] (724)
Pavanaram, S. K., 358[59] (313, 315)
Pavelka, Jr., M. J., 236[364] (213)
Payne, M. C., 735[85] (635, 638)
Pazur, J. H., 235[308] (208)
Pazzani, C., 237[477] (221), 526[195] (509)
Peach, M. J., 156[24] (131, 134, 135)
Peak, I. R. A., 239[581] (227, 228)
Pear, J. R., 526[221] (512, 522), 597[153] (581), 614[119] (611)
Pearce, L. E., 440[596] (431)
Pearce, L., 156[25] (131, 135)
Pearse, A., 106[112] (100)
Pearson, M. J., 293[302] (283, 285)
Pearson, N. B., 238[546] (224)
Pearson, W. H., 158[84] (139, 140), 158[85] (139)
Pedersen, A. H., 743[599] (714, 715)
Pedersen, S. S., 232[146] (193, 194)
Pederson, R. L., 431[8] (368), 439[537] (425)
Pedrosa, F. O., 439[556] (426)
Peel, J. L., 400, 435[291] (400)
Pegg, L. E., 160[240] (156)
Pegg, W., 63[188] (50)
Pegues, J. C., 238[523] (223)
Peisach, J., 743[571] (711), 743[590]
Pelczar, H., 62[136] (46), 84[135] (79)
Pellegrini, L., 739[354] (669)
Pellerin, J., 737[173] (644, 645)
Pellerin, P., 524[90] (503, 519), 524[91] (503, 511, 518), 527[273] (519)
Pelletier, S. W., 156[3] (130, 138), 156[4] (130, 134, 141), 157[33] (132)
Pellicciari, R., 435[241] (396)
Peloso, J. S., 357[41] (314, 315, 317)
Penco, S., 359[174] (317, 322)
Penel, C., 526[239] (514), 744[630] (720)
Peng, H.-L., 234[233] (203), 234[250] (204), 494[350] (479)
Peng, J., 745[677] (727)
Peng, K. C., 64[232] (54)
Peng, L., 597[154] (581)
Penhoet, E. E., 438[473] (419)
Penn, C. W., 239[572] (227)
Penning, M., 614[96] (609)
Penttilä, L., 84[101] (78)
Pepin, T. M., 492[192] (458, 466)
Percheron, F., 357[10] (312)
Percival, E., 357[25] (313, 314), 357[26] (313, 314)
Percy, C., 63[174] (56)
Pereira, A. C. de S., 156[25] (131, 135)
Perez, M., 128[198] (120), 432[72] (378), 66[338] (57, 59)
Perez, P., 437[384] (411)
Perez, S., 596[58] (546, 570, 571, 573)
Perez-Vilar, J., 83[37] (73)
Perie, J., 438[482] (420)
Perillo, N. L., 83[89] (76)
Perini, J. M., 85[151] (80)

Perkins, H. R., 289[1] (242, 243), 289[33] (249), 361[272] (327)
Perkins, P. G., 744[652] (724)
Perl, A., 438[465] (419), 438[468] (419)
Perng, G. S., 63[184] (51, 55), 66[368] (59)
Perotto, S., 525[131] (504), 525[143] (504)
Perozich, J., 525[153] (506, 507)
Perry, M. B., 231[55] (184), 231[56] (184), 232[105] (187, 189, 217), 232[113] (189, 218), 232[114] (189), 239[606] (229)
Persson, K., 233[173] (195), 238[488] (221)
Pestell, T. C., 525[150] (506)
Petcher, T. J., 359[124] (316)
Peter, H. H., 359[146] (317)
Peters, S., 82[30] (73, 75), 82[31] (73, 74), 159[143] (146)
Peters, T., 11[8] (1)
Petersen, F., 230[43] (183)
Peterson, D., 740[366] (670, 671)
Peterson, E., 735[98] (636, 637)
Peterson, J. A., 736[121] (638), 736[124] (638)
Peterson, J. R., 35[210] (31)
Peterson, P. A., 159[158] (147), 35[211] (31)
Petersson, K., 232[132] (191)
Petersson, L., 362[318] (329)
Petipas, T., 595[17] (536, 545)
Petit, J. M., 65[294] (56)
Petrescu, S. M., 35[217] (31)
Pctricoin III, E. F., 238[579] (223), 238[530] (223)
Petroni, E. A., 235[337] (210)
Petruzzelli, R., 743[588] (714), 743[598] (708, 714, 715), 744[636] (721)
Petryniak, B., 65[299] (56), 65[307] (56, 58), 66[350] (58)
Petsko, G. A., 290[69] (254), 290[74] (254, 255)
Petterssen, B., 598[197] (592)
Pettitt, J. M., 61[71] (45, 50), 61[72] (45, 50)
Petursson, S., 35[203] (31), 158[107] (143)
Petzel, J. P., 432[21] (373)
Pew, J. C., 744[665] (725)
Pezeshk, V., 527[260] (517, 522), 527[261] (517), 613[42] (604), 613[43] (604)
Pezzanite, J. O., 360[196] (317, 322)
Pfister, J. A., 158[87] (140)
Pfister, M., 230[18] (181)
Pflugrath, J., 737[207] (647)
Phale, P. S., 432[53] (375, 378)
Phelps, C. F., 525[150] (506)
Phibbs, Jr., P. V., 422, 439[517] (422), 439[518] (423)
Philippon, A., 293[306] (285)
Phillips, A. H., 736[132] (639)
Phillips, L. M., 66[321] (55, 59)
Phillips, M. J., 432[72] (378)
Phillips, M. L., 65[318] (55, 59), 65[320] (55, 59), 66[338] (57, 59), 66[339] (57), 128[198] (120)
Phillips, R. L., 493[262] (466)
Phillips, R. S., 396, 405, 407, 435[238] (396), 435[242] (396), 436[353] (405, 407), 436[354] (405)
Philosoph-Hadas, S., 614[83] (609)
Piastuch, W., 741[472] (695)
Piatigorsky, J., 438[432] (416)
Piau, J. P., 65[309] (56)
Pichersky, E., 740[357] (669)
Pichon, M., 740[378] (673, 674)
Pickard, M. A., 292[249] (276), 292[250] (276)
Pickcring, N. J., 82[15] (71, 77)

Pickett-Heaps, J. D., 525[116] (504), 681, 740[396] (681)
Picton, C., 482, 483, 495[379] (482), 495[380] (482, 483)
Pien, F.-M., 493[263] (466)
Piens, K., 598[193] (592)
Pieper, P. A., 363[401] (338, 339), 363[410] (339)
Piepersberg, W., 11[27] (7), 363[394] (337, 344, 351, 353), 364[455] (345, 354), 365[504] (353), 365[505] (354)
Pier, G. B., 234[268] (206)
Pierce, F. T., 595[25] (536)
Pierce, M., 35[190] (29), 63[184] (51, 55), 64[232] (54), 64[233] (54), 65[270] (55), 66[323] (55), 66[368] (59), 84[90] (76), 126[112] (111)
Pierrel, M. A., 736[117] (637)
Pierson, D. E., 234[240] (204)
Pietikäinen, P., 744[653] (724)
Piez, K. A., 175[4] (161, 163, 171, 172, 174)
Pigman, W., 357[9] (312), 357[10] (312)
Pilarski, L. M., 82[25] (72)
Pilate, G., 740[389] (674)
Pilati, T., 744[653] (724)
Pilipcinec, E., 239[551] (225)
Pilipowicz, M., 614[76] (608)
Pilkiewicz, F., 156[16] (131, 138)
Pillarisetti, J., 176[35] (166)
Piller, F., 83[74] (76), 83[77] (76), 84[100] (78), 84[106] (78), 106[105] (99), 126[77] (109)
Piller, V., 83[74] (76)
Pillion, D. J., 495[372] (481)
Pillonel, C., 738[275] (656)
Pillow, M. Y., 741[467] (694)
Pilnik, W., 524[49] (499, 518)
Pilotti, Å., 357[48] (314)
Pinçon, G., 737[200] (646, 669)
Pinhal, M. A., 176[57] (169, 174)
Pinto, B. M., 11[8] (1)
Pinto, J. E. B. P., 433[145] (386)
Piqué, N., 238[490] (221)
Piquemal, J., 738[252] (652, 653, 673, 674)
Piras, R., 493[270] (469, 473)
Piriou, F., 357[21] (314, 317)
Piro, G., 607, 614[60] (606, 607), 614[61] (607)
Pissowotzki, K., 363[394] (337, 344, 351, 353)
Pitcher, J., 490[87] (446)
Pitt, T. L., 232[123] (190), 232[125] (190)
Pittard, A. J., 434[162] (386)
Pittelkow, M. R., 126[125]
Pitto, M., 125[48] (109, 112, 119)
Pitts, B. J. R., 361[286] (328)
Pitzner, L. J., 595[40] (544), 595[41] (544)
Pizza, C., 358[79] (313)
Pizzirani-Kleiner, A. A., 598[200] (593, 594)
Pla, F., 745[680] (728), 745[681] (728)
Pla, J., 290[105] (258)
Plaas, A. H., 176[55] (168)
Plamann, M. D., 435[256] (397, 400), 435[258] (397)
Planche, L. A., 717
Plancke, Y., 82[18] (71)
Plapp, R., 289[22] (246)
Platt, F. M., 33[68] (20), 35[204] (31)
Plaxton, W. C., 491[143] (451)
Plazinski, J., 597[154] (581)
Pletnev, S. V., 436[349] (405)

Subject Index

PHILIP AND LESLEY ASLETT
Marlborough, Wiltshire, UK

Every effort has been made to index as comprehensively as possible, and to standardize the terms used in the index in line with the IUPAC Recommendations. In view of the diverse nature of the terminology employed by the different authors, the reader is advised to search for related entries under the appropriate headings.

The index entries are presented in letter by letter alphabetical sequence. Compounds are normally indexed under the parent compound name, with the substituent component separated by a comma of inversion. An entry with a prefix/locant is filed after the same entry without any attachments, and in alphanumerical sequence. For example, 'diazepines', '1,4-diazepines', and '2,3-dihydro-1,4-diazepines' will be filed as:-

    diazepines
    1,4-diazepines
    1,4-diazepines, 2,3-dihydro-

The Index is arranged in set-out style, with a maximum of three levels of heading. Location references refer to volume number (in bold) and page number (separated by a comma); major coverage of a subject is indicated by bold, elided page numbers; for example;

    triterpene cyclases, **299–320**
        amino acids, 315

See cross-references direct the user to the preferred term; for example,

    olefins *see* alkenes

See also cross-references provide the user with guideposts to terms of related interest, from the broader term to the narrower term, and appear at the end of the main heading to which they refer, for example,

    thiones
        see also thioketones

profisetinidins, 772
proguibourtinidins, 775
guibourtinidol, as profisetinidin precursor, 771
guibourtinidol-3′-*O*-4′-*ent*-epimopanone, biosynthesis, 777
guibourtinidol-(4α→6′)-fisetinidol, structure, 761
Gymnospermae
  lignins, 620
    cell wall deposition, 690
    evolution, 625
    heterogeneity, 691

HA *see* hyaluronan (HA)
HADH *see* horse-liver alcohol dehydrogenase (HADH)
Haemophilus influenzae
  choline uptake, 229
  invasion capacity correlation, 229
  lipopolysaccharides, 5, 180
    biosynthesis, 228
  nrd genes, 345
HAF *see* human amniotic fluid (HAF)
Halocynthia spp.
  celluloses
    NMR spectra, 565
    Raman spectra, 558
HAS *see* hyaluronan synthase (HAS)
HAS genes
  *HAS*1, expression, 174
  *HAS*2, expression, 174
  *HAS*3, expression, 174
  occurrence, 174
heart disease, red wine benefits, 792
heart muscle, glycogen synthase (mammalian), 479
Helianthus annuus, phenylalanine ammonia-lyase, 659
Helianthus tuberosus
  trans-cinnamate 4-monooxygenase, 637
    localization, 669
Helicobacter pylori, 2-dehydro-3-deoxyphosphogluconate aldolase, 421
hemicelluloses
  biochemistry, **599–615**
  biosynthesis, 2, 8
    extracellular signals, 608
    gene identification, 610
    polysaccharide turnover, 609
    regulatory mechanisms, 607
    sites, 602
  definitions, 599
  differential screening, 611
  enzymology, 603
  expressed sequence tag databases, 611
  families, 600
  immunolocalization, 603
  importance of, 599
  localization, 601
  manipulation, 612
  mutational analysis, 611
  primary wall, 600
  reverse genetics, 612
  secondary wall, 601
  structure, 9
  subcellular fractionation, 602
hemolysins, biosynthesis, 227
HEMPAS *see* hereditary erythroblastic multinuclearity with a positive acidified serum (HEMPAS)
heparan sulfate (HS)
  binding, 4
  biosynthesis, 2, 172
  substitution, 161
heparan sulfate proteoglycans (HSPGs), fate of, 172
heparin
  binding, 4
  biosynthesis, 2, 172

substitution, 161
Hepaticae, and lignin evolution, 624
hepatocytes, lipopolysaccharide biosynthesis, 181
hepatomas
  fucosyltransferases, 116
  and β-1,4-mannosyl-glycoprotein β-1,4-*N*-acetylglucosaminyltransferase levels, 53
heptitol, 2,6-diamino-2,6-imino-7-*O*-(β-D-glucopyranosyl)-D-glycero-L-gulo- (MDL), as glucosidase inhibitor, 153
DL-*glycero*-D-*manno*-heptitol, 2,5-dideoxy-2,5-imino-(homoDMDP)
  structure, 130
  toxicity, 140
L-*Glycero*-D-*manno*-heptose, structure, 187
heptosyltransferase I, catalysis, 224
heptosyltransferase II, encoding, 224
hereditary erythroblastic multinuclearity with a positive acidified serum (HEMPAS)
  etiology, 27, 150
  lysis test, 59
Hevea brasiliensis, cinnamyl-alcohol dehydrogenases, 655
hexahydroxybiphenyldicarbonyloxy groups, chirality, 783
hexahydroxydiphenoyl groups, chirality, 783
3,4,5,3′,4′,5′-hexahydroxydiphenoyl residues, 800
hexaploid wheat *see Triticum aestivum*
hexosaminoglycans, in *Escherichia coli*, 185
L-*arabino*-hexose, 4,*O*-isobutyryl-2,6-dideoxy-3-*C*-methyl- *see* L-olivomycose E
L-*arabino*-hexose, 2,3,6-trideoxy-3-*C*-methyl-4-*O*-methyl-3-nitro- *see* L-evernitrose
hexose chains, lipopolysaccharides, 226
hexose monophosphates, as energy source, 463
hexose-1-phosphate nucleotidyltransferases, roles, in 6-deoxyhexose biosynthesis, 333
hexose phosphates
  translocation, 464
  translocator, 464
hexoses
  in cellulose studies, 544
  in *Serratia marcescens*, 190
D-*arabino*-3-hexulose-6-phosphate, biosynthesis, 413
3-hexulose phosphate synthase, biochemistry, 413
HGA *see* homogalacturonan (HGA)
HGA methyltransferase (HGA-MT)
  activity, 519
  occurrence, 519
  roles, in pectin biosynthesis, 518
HGA-MT *see* HGA methyltransferase (HGA-MT)
H glycolipid, synthesis, 115
high-density lipoprotein (HDL), lipopolysaccharide attenuation, 181
Hinc-repeats, in *rfb* gene clusters, 218
hispidin, biosynthesis, 624
histidase *see* histidine ammonia-lyase
histidinase *see* histidine ammonia-lyase
histidine ammonia-lyase, active sites, 634
histidine α-deaminase *see* histidine ammonia-lyase
histidine(s), deamination, 632
HIV *see* human immunodeficiency virus (HIV)
HMG-CoA reductase(s) *see* hydroxymethylglutaryl-CoA reductase(s)
HNK-1 epitope, roles, 122
Hogness box *see* TATA box
homoDMDP *see* dl-*glycero*-D-*manno*-heptitol, 2,5-dideoxy-2,5-imino- (homoDMDP)
homogalacturonan (HGA)
  O-acetylation, 499
  biosynthesis, 8, 504, 510, 522
    models, 514
  in cell wall, 498, 503
  methylester distribution, 499

WITHDRAWAL